BIOLOGY | THE UNITY & DIVERSITY OF LIFE

THIRTEENTH EDITION

CECIE STARR   RALPH TAGGART   CHRISTINE EVERS   LISA STARR

BROOKS/COLE
CENGAGE Learning

Australia • Brazil • Japan • Korea • Mexico • Singapore • Spain • United Kingdom • United States

BROOKS/COLE
CENGAGE Learning·

**Biology: The Unity and Diversity of Life,
Thirteenth Edition, International Edition**
Cecie Starr, Ralph Taggart, Christine Evers,
Lisa Starr

Publisher: Yolanda Cossio

Senior Acquisitions Editor, Life Sceinces:
    Peggy Williams

Assistant Editor: Shannon Holt

Editorial Assistant: Sean Cronin

Media Editor: Lauren Oliveira

Marketing Manager: Tom Ziolkowski

Marketing Coordinator: Jing Hu

Marketing Communications Manager: Linda Yip

Content Project Manager: Hal Humphrey

Design Director: Rob Hugel

Art Director: John Walker

Print Buyer: Karen Hunt

Rights Acquisitions Specialist: Dean Dauphinais

Production Service: Grace Davidson &
    Associates

Text Designer: John Walker

Photo Researcher: Myrna Engler Photo
    Research Inc.

Text Researcher: Pablo D'Stair

Copy Editor: Anita Wagner

Illustrators: Gary Head, ScEYEnce Studios,
    Lisa Starr

Cover Image: Red-eyed tree frog, Costa Rica;
    © Paul Souders | WorldFoto | Getty Images.

Compositor: Lachina Publishing Services

International Edition:
ISBN-13: 978-1-111-58098-8
ISBN-10: 1-111-58098-7

Cengage Learning International Offices

Asia
www.cengageasia.com
tel: (65) 6410 1200

Australia/New Zealand
www.cengage.com.au
tel: (61) 3 9685 4111

Brazil
www.cengage.com.br
tel:(55) 11 3665 9900

India
www.cengage.co.in
tel: (91) 11 4364 1111

Latin America
www.cengage.com.mx
tel: (52) 55 1500 6000

UK/Europe/Middle East/Africa
www.cengage.co.uk
tel: (44) 0 1264 332 424

**Represented in Canada by
Nelson Education, Ltd.**
www.nelson.com
tel: (416) 752 9100 / (800) 668 0671

Cengage Learning is a leading provider of customized learning solutions with office locations around the globe, including Singapore, the United Kingdom, Australia, Mexico, Brazil, and Japan. Locate your local office at: **www.cengage.com/global**.

For product information and free companion resources: **www.cengage.com/ international**

Visit your local office: **www.cengage.com/global**

Visit our corporate website: **www.cengage.com**

Purchase any of our products at your local college store or at our preferred online store **www.CengageBrain.com**.

Printed in China
1 2 3 4 5 6 7 16 15 14 13 12

CONTENTS IN BRIEF

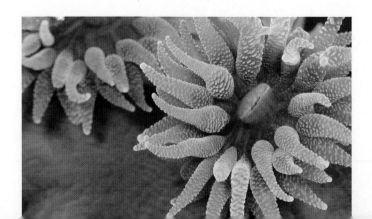

DETAILED CONTENTS

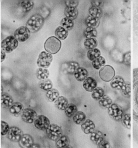
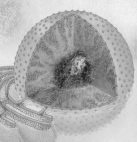

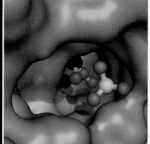

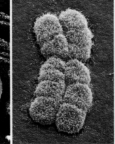

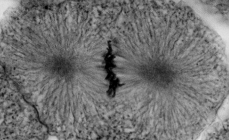
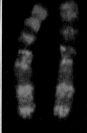

UNIT III   PRINCIPLES OF EVOLUTION

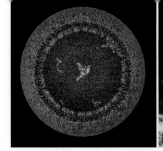

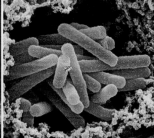

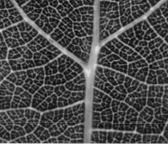
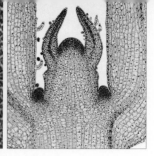

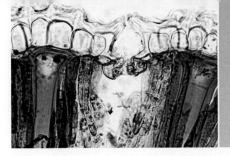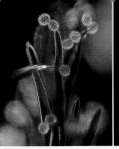

UNIT VI   HOW ANIMALS WORK

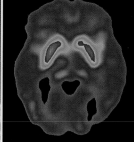

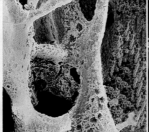

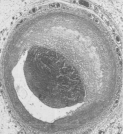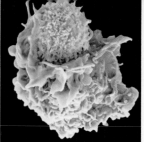

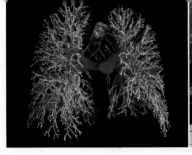

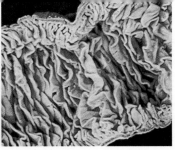

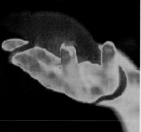

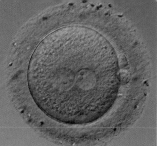

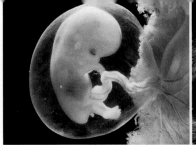

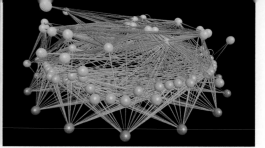

48 Human Impact on the Biosphere

Preface

This edition of *Biology: The Unity and Diversity of Life* includes a wealth of new information about recent discoveries in biology (details can be found in the *Power Bibliography*, which lists journal articles and other references we used in the revision process). Descriptions of current research, along with photos and videos of scientists who carry it out, underscore the concept that science is an ongoing endeavor carried out by a diverse community of people. We discuss not only what was discovered, but also how the discoveries were made, how our understanding has changed over time, and what remains undiscovered.

At the same time, we provide a thorough, accessible introduction to well-established concepts and principles that underpin modern biology. We examine every topic from an evolutionary perspective and emphasize the connections between all forms of life. Throughout the book, we revised text and diagrams to make difficult concepts easier for students to grasp.

We also continue to focus on real world applications of the topics we discuss. This edition provides expanded coverage of human health and diseases. It also covers in more detail the many ways in which human activities can alter the environment and threaten both human health and the biodiversity of our planet.

CHANGES TO THIS EDITION

Learning Road Map Each chapter opens with a *Learning Road Map* that places the chapter in a larger context. It notes which previously discussed material the chapter draws upon, highlights key concepts discussed within the chapter, and looks ahead to later sections that refer back to the chapter's concepts. For added impact, each *Learning Road Map* overlays an intriguing full-page opening photo. The photo reappears within the chapter, together with a caption that explains its content.

Opening Essay/Essay Revisited We now devote the first section of each chapter to an essay about an environmental or health application of that chapter's theme. The final section revisits this topic as a reminder of the chapter's relevance to the real world.

Section-Based Links to Earlier Concepts *Links to Earlier Concepts* appear at the start of most sections. These links emphasize connections between biological topics and facilitate review of material that has not been fully mastered. For example, the heading of the section about membrane potential refers back to earlier coverage of potential energy and of transport proteins.

Section-Based Glossaries In addition to a full glossary of terms at the end of the book, each section now has a *Section-Based Glossary* that includes all boldface terms for that section. A student can simply glance at this glossary for a quick check of a term's meaning.

Step-by-Step Graphics Integrated With Text We have long explained biological processes using figures accompanied by step-by-step captions. In this edition, we've also integrated stepwise descriptions of many graphics into the text, using colored, numbered balls to make the connections.

Visual Chapter Summary To further enhance visual learning, we've added graphics to each section of the chapter summary. As before, each chapter summary includes all key terms from that chapter in boldface.

Chapter-Specific Changes Every chapter was extensively revised for clarity; this edition has more than 200 new photos and over 250 new or updated figures. A section-by-section guide to content and figures is available upon request, but we summarize the highlights here.

• *Chapter 1, Introduction to Biology* Revised and expanded coverage of critical thinking, scientific process, and philosophy of science. Material on classification systems moved to this chapter.

• *Chapter 2, Chemical Basis of Life* New opening essay discusses mercury toxicity and prevalence; revisions emphasize electron behavior in atoms as it relates to ions and bonding.

• *Chapter 3, Molecules* Importance of protein structure now exemplified with prions.

• *Chapter 4, Cell Structure and Function* Expanded discussion and new photos of archeans.

• *Chapter 5, Metabolism* Chapter now includes material on cell membrane function; new section on cofactors, including ATP.

• *Chapter 6, Photosynthesis* Biofuels essay revised to include food supply debate; revisited section on climate change and CO_2 now includes atmospheric change after photosynthesis evolution.

• *Chapter 7, Cells Release Energy* Introductory section now discusses the relationship between the evolution of oxygenic photosynthesis and aerobic respiration. Third stage art revised to connect with summary illustration.

• *Chapter 8, DNA Structure and Function* This chapter was moved forward in book to allow a more complete explanation of genes and inheritance; now includes material on chromosome structure and mutations. New opener essay about cloned 9/11 rescue dog Trackr.

• *Chapter 9, DNA to Protein* Introductory essay revised to include RIPs other than ricin; material on sickle cell anemia and other effects of beta globin mutations added.

• *Chapter 10, Control Over Genes* Theme of evolutionary connections strengthened throughout; male sex determination and riboswitches added to existing sections; new epigenetics section added.

• *Chapter 11, Cells Reproduce* New fluorescence micrograph of HeLa cells shows how defects in cell cycle controls lead to cancer; section on telomeres added.

• *Chapter 12, Meiosis* Artwork updated; new fluorescence micrograph illustrates checkpoint proteins in meiosis.

• *Chapter 13, Patterns of Inherited Traits* Opening essay is now on inheritance of CF mutation. Environmental effects section rewritten to include epigenetics and pelage cycle.

• *Chapter 14, Human Inheritance* Opening essay on human skin color moved to this chapter. Chapter reorganized to strengthen introduction to human genetics; genetic screening section updated; many photos replaced with newer versions.

• *Chapter 15, Genomes* New opening section on personal genetic testing; updated genomics section includes expanded material on DNA profiling. New brainbow photo.

• *Chapter 16, Evidence of Evolution* New graphics include a page from Darwin's evolution journal and updated paleogeography illustrations.

• *Chapter 17, Processes of Evolution* New illustrated examples include allopatric speciation in snapping shrimp, sympatric speciation in Lake Victoria cichlids, ring species; coevolution of an ant and a butterfly.

• *Chapter 18, Information About Species* Expanded and updated phylogeny/cladistics section; *Hox* gene comparison material updated with new examples and photos.

• *Chapter 19, Early Evolution* Opening essay focuses on astrobiology. Added micrographs and discussion of bacteria with internal membranes. Improved figure illustrates step on the road to life.

• *Chapter 20, Viruses and Prokaryotes* Opening essay discusses SIV and evolution of HIV. More detailed depiction of HIV replication. Coverage of prions moved to Chapter 3. New figure shows viral recombination.

• *Chapter 21, Protists* New opening essay about malaria. New step-by-step figure illustrating *Plasmodium* life-cycle. Clearer, simpler illustration of *Chlamydamonas* life cycle. Choanoflagellates now introduced here.

• *Chapter 22, Plants* Essay about Nobel Prize winner W. Mathai's work. Life cycle figures revised.

• *Chapter 23, Fungi* New opening essay about the threat wheat stem rust poses to food supplies. Life cycle figures revised for clarity. Additional photos of fungal diversity.

• *Chapter 24, Invertebrates* Improved or new graphics of animal evolutionary tree, types of body cavities, crab life cycle, grasshopper anatomy.

• *Chapter 25, Chordates* New evolutionary tree diagram, graphic of tunicate. Discussion of fish evolution revised. New graphic illustrates major mammal orders. Material on human evolution moved to separate chapter.

• *Chapter 26, Human Evolution* Opening essay describes recent evidence that Neanderthals mated with *Homo sapiens*. The chapter covers primate groups and the latest discoveries regarding human ancestry.

• The chapter *Plants and Animals—Common Challenges* has been deleted and material previously covered therein is now integrated into other chapters.

• *Chapter 27, Plant Tissues* New introductory essay on carbon sequestration by plants; reorganized tissue section now includes thumbnail photos; many revised illustrations and new micrographs.

• *Chapter 28, Plant Nutrition* Plant nutrient table revised to include functions; soil water uptake artwork revised; new artwork and micrographs of xylem tubes; expanded discussion of stomata function.

• *Chapter 29, Plant Reproduction* Material on early plant development, senescence, and dormancy moved to this chapter and expanded. Many new photos.

• *Chapter 30, Plant Development* Extensively revised and expanded to reflect paradigm shifts driven by recent research. New introductory essay on health benefits of cocoa; new section introducing plant hormones added; material on major hormones expanded and reorganized by section; new section on plant stress responses added.

• *Chapter 31, Animal Organ Systems* New graphics depicting levels of organization, endocrine versus exocrine glands. Added discussion of carcinomas, vitiligo, loss of body hair in human evolution.

• *Chapter 32, Neural Control* Added latest research on effects of Ecstasy use. Additional info about the role of neuroglia in disease. Split-brain research deleted.

• *Chapter 33, Sensory System* Updated discussion of olfaction with new figure. Increased coverage of diversity of visual systems. More about effects of noise pollution.

• *Chapter 34, Endocrine System* New graphic of human endocrine system. Added section about thymus. Added information about pthalates as endocrine disruptors.

• *Chapter 35, Support and Movement* New essay about effects of myostatin. Improved figures depicting muscle contraction.

• *Chapter 36, Circulation* Updated art throughout. Added illustration of a stent.

• *Chapter 37, Immune System* Innate responses material expanded to two sections that include updated illustrations and photos; also added neutrophil nets. Expanded sections on immune dysfunction, vaccines.

• *Chapter 38, Respiration* New essay about carbon monoxide poisoning. New graphic of CO_2 transport.

• *Chapter 39, Nutrition and Digestion* New essay about the importance of gut bacteria. Discussion of stomach and small intestine revised to improve flow.

• *Chapter 40, Internal Environment* New figures for urine formation and renin–angiotensin–aldosterone system. New coverage of hibernation.

• *Chapter 41, Reproductive Systems* New graphics depicting sperm formation, female genitals. Section about FSH and twins deleted.

• *Chapter 42, Development* Updated essay with disucssion of "octomom" and early opposition to IVF.

• *Chapter 43, Behavior* Added discussions of kinesis and taxis, epigenetic effects, behavioral plasticity.

• *Chapter 44, Ecology* New essay about Canada goose population explosion. Human population material updated.

• *Chapter 45, Biodiversity* New studies of competition cited. Revised discussion of disturbance and exotic species.

• *Chapter 46, Ecosystems* New essay on phosphate pollution. Section on carbon cycle revised and updated.

• *Chapter 47, Biosphere* Added information about latest deep ocean life. New biome photos.

• *Chapter 48, Human Impact on the Biosphere* Biological magnification, ozone depletion and pollution now covered in this section

Student and Instructor resources

Test Bank   Nearly 4,000 test items, ranked according to difficulty and consisting of multiple-choice (organized by section heading), selecting the exception, matching, labeling, and short answer exercises. Includes selected images from the text. Available on the instructor's companion website, login.cengage.com.

ExamView®   Create, deliver, and customize tests (both print and online) in minutes with this easy-to-use assessment and tutorial system. Each chapter's end-of-chapter material is also included.

Instructor's Resource Manual   Includes chapter outlines, objectives, key terms, lecture outlines, suggestions for presenting the material, classroom and lab enrichment ideas, discussion topics, paper topics, possible answers to critical thinking questions, answers to data analysis activities, and more. Available on the instructor's companion website, login.cengage.com.

The Brooks/Cole Biology Video Library 2010 featuring BBC Motion Gallery   Looking for an engaging way to launch your lectures? The Brooks/Cole series features short high-interest segments: Pesticides: Will More Restrictions Help or Hinder?; A Reduction in Biodiversity; Are Biofuels as Green as They Claim?; Bone Marrow as a New Source for the Creation of Sperm; Repairing Damaged Hearts with Patients' Own Stem Cells; Genetically Modified Virus Used to Fight Cancer; Seed Banks Helping to Save Our Fragile Ecosystem; The Vanishing Honeybee's Impact on Our Food Supply.

Biology CourseMate   Cengage Learning's Biology CourseMate brings course concepts to life with interactive learning, study, and exam preparation tools that support the printed textbook, or the included ebook. With CourseMate, professors can use the included Engagement Tracker to assess student preparation and engagement. Use the tracking tools to see progress for the class as a whole or for individual students. Visit www.cengagebrain.com for more information.

Premium eBook   This complete online version of the text is integrated with multimedia resources and special study features, providing the motivation that so many students need to study and the interactivity they need to learn. New to this edition are pop-up tutors, which help explain key topics with short video explanations.

Acknowledgments

Writing, revising, and illustrating a biology textbook is a major undertaking for two full-time authors, but our efforts constitute only a small part of what is required to produce and sell this one. We are truly fortunate to be part of a huge team of very talented people who are as committed as we are to creating and disseminating an exceptional science education product.

Biology is not dogma; paradigm shifts are a common outcome of the fantastic amount of research in the field. Thus, ideas about what material should be taught and how best to present that material to students changes even from one year to the next. It is only with the ongoing input of our many academic reviewers and advisors (see *opposite page*) that we can continue to tailor this book to the needs of instructors and students while integrating new information and models. We continue to learn from and be inspired by these dedicated educators. A special thank-you goes to Michael Plotkin for his thoughtful, thorough input on current paradigms in evolutionary biology.

On the production side of our team, the indispensable Grace Davidson orchestrated the flow of countless files, photos, and illustrations while managing schedules, budgets, and whatever else happened to be on fire at the time. Grace, thank you for your continued patience and dedication. Photoresearch for this book presents a special challenge because we often request images from unusual or difficult-to-reach sources. Paul Forkner, thank you for your persistence and determination to track down many of the exceptional photos that make this book unique. Copyeditor Anita Wagner and proofreader Diane Miller, your valuable suggestions kept our text clear and concise.

At Cengage Learning, the brilliant John Walker created this book's highly appealing new design. Yolanda Cossio, thank you for continuing to support us and for encouraging our efforts to innovate and improve. Peggy Williams, we are as always grateful for your enthusiastic, thoughtful guidance, and for your many travels (and travails) on behalf of our books.

Thanks to Hal Humphrey our Cengage Production Manager, Tom Ziolkowski our Marketing Manager, Lauren Oliveira who creates our exciting technology package, Shannon Holt who managed all the print supplements for this edition, and Editorial Assistant Sean Cronin.

LISA STARR AND CHRISTINE EVERS
August 20011

1 Introduction to Biology

LEARNING ROADMAP

Where you have been Whether or not you have studied biology, you already have an intuitive understanding of life on Earth because you are part of it. Every one of your experiences with the natural world—from the warmth of the sun on your skin to the love of your pet—contributes to that understanding.

Where you are now

The Science of Nature
We can understand life by studying it at many levels, starting with its component atoms and molecules and extending to interactions of organisms with their environment.

Life's Unity
All living things must have ongoing inputs of energy and raw materials; all sense and respond to change; and all have DNA that guides their development and functioning.

Life's Diversity
Observable characteristics vary tremendously among organisms. Various classification systems help us keep track of the differences.

How We Study Life
Carefully designed experiments help researchers unravel cause-and-effect relationships in complex natural systems.

What Science Is (and Is Not)
Science addresses only testable ideas about observable events and processes. It does not address the untestable, including beliefs and opinions.

Where you are going This book parallels nature's levels of organization, from atoms to the biosphere. Learning about the structure and function of atoms and molecules will prime you to understand how living cells work. Learning about processes that keep a single cell alive can help you understand how multicelled organisms survive. Knowing what it takes for organisms to survive can help you see why and how they interact with one another and their environment.

1.1 The Secret Life of Earth

In this era of detailed satellite imagery and cell phone global positioning systems, could there possibly be any places left on Earth that humans have not yet explored? Actually, there are plenty of them. As recently as 2005, for example, helicopters dropped a team of scientists into the middle of a vast and otherwise inaccessible cloud forest covering the top of New Guinea's Foja Mountains. Within a few minutes, the explorers realized that their landing site, a dripping, moss-covered swamp, had been untouched by humans. Team member Bruce Beehler remarked, "Everywhere we looked, we saw amazing things we had never seen before. I was shouting. This trip was a once-in-a-lifetime series of shouting experiences."

How did the explorers know they had landed in uncharted territory? For one thing, the forest was filled with plants and animals previously unknown even to native peoples that have long inhabited other parts of the region. During the next month, the team members discovered many new species, including a rhododendron plant with flowers the size of a plate and a frog the size of a pea. They also came across hundreds of species that are on the brink of extinction in other parts of the world, and some that supposedly had been extinct for decades. The animals had never learned to be afraid of humans, so they could easily be approached. A few were discovered as they casually wandered through campsites (**Figure 1.1**).

New species are discovered all the time, often in places much more mundane than Indonesian cloud forests. How do we know what species a particular organism belongs to? What is a species, anyway, and why should discovering a new one matter to anyone other than a scientist? You will find the answers to such questions in this book. They are part of the scientific study of life, biology, which is one of many ways we humans try to make sense of the world around us.

Trying to understand the immense scope of life on Earth gives us some perspective on where we fit into it. For example, hundreds of new species are discovered every year, but about 20 species become extinct every minute in rain forests alone—and those are only the ones we know about. The current rate of extinctions is about 1,000 times faster than normal, and human activities are responsible for the acceleration. At this rate, we will never know about most of the species that are alive on Earth today. Does that matter? Biologists think so. Whether or not we are aware of it, humans are intimately connected with the world around us. Our activities are profoundly changing the entire fabric of life on Earth. These changes are, in turn, affecting us in ways we are only beginning to understand.

Ironically, the more we learn about the natural world, the more we realize we have yet to learn. But don't take our word for it. Find out what biologists know, and what they do not, and you will have a solid foundation upon which to base your own opinions about how humans fit into this world. By reading this book, you are choosing to learn about the human connection—your connection—with all life on Earth.

biology The scientific study of life.

Figure 1.1 Explorers found hundreds of rare species and dozens of new ones during recent survey expeditions to the Foja Mountain cloud forest (*top*).

Bottom, Paul Oliver discovered this tree frog (*Litoria*) perched on a sack of rice during a particularly rainy campsite lunch. The explorers dubbed the new species "Pinocchio frog" after the Disney character because the male frog's long nose inflates and points upward during times of excitement.

1.2 Life Is More Than the Sum of Its Parts

- Biologists study life by thinking about it at different levels of organization.
- The quality of life emerges at the level of the cell.

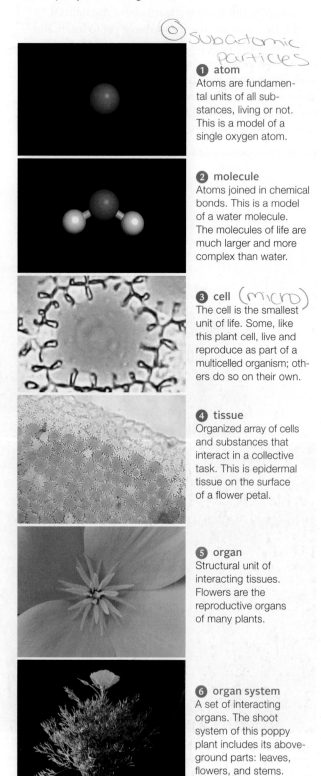

Subatomic particles

1 atom
Atoms are fundamental units of all substances, living or not. This is a model of a single oxygen atom.

2 molecule
Atoms joined in chemical bonds. This is a model of a water molecule. The molecules of life are much larger and more complex than water.

3 cell (micro)
The cell is the smallest unit of life. Some, like this plant cell, live and reproduce as part of a multicelled organism; others do so on their own.

4 tissue
Organized array of cells and substances that interact in a collective task. This is epidermal tissue on the surface of a flower petal.

5 organ
Structural unit of interacting tissues. Flowers are the reproductive organs of many plants.

6 organ system
A set of interacting organs. The shoot system of this poppy plant includes its aboveground parts: leaves, flowers, and stems.

Figure 1.2 Animated Levels of life's organization.

Biologists study all aspects of life, past and present. What, exactly, is the property we call "life"? We may never actually come up with a good definition, because living things are too diverse, and they consist of the same basic components as nonliving things. When we try to define life, we end up only identifying properties that differentiate living from nonliving things.

Complex properties, including life, often emerge from the interactions of much simpler parts. For an example, take a look at these drawings:

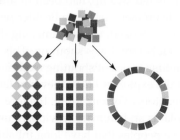

The property of "roundness" emerges when the parts are organized one way, but not other ways. Characteristics of a system that do not appear in any of the system's components are called emergent properties. The idea that structures with emergent properties can be assembled from the same basic building blocks is a recurring theme in our world, and also in biology.

Through the work of biologists, we are beginning to understand a pattern in life's organization. That organization occurs in successive levels, with new emergent properties appearing at each level (**Figure 1.2**).

Life's organization starts with interactions among atoms. **Atoms** are fundamental units of matter, building blocks of all substances **1**. Atoms join as **molecules** **2**. There are no atoms unique to living things, but there are unique molecules. In today's world, only living things make the "molecules of life," which are lipids, proteins, DNA, RNA, and complex carbohydrates. The emergent property of "life" appears at the next level, when many molecules of life become organized as a cell **3**. A **cell** is the smallest unit of life. Cells survive and replicate themselves using energy, raw materials, and information in their DNA. Some cells live and reproduce independently. Others do so as part of a multicelled organism. An **organism** is an individual that consists of one or more cells. A poppy plant is an example of a multicelled organism **7**.

In most multicelled organisms, cells make up tissues **4**. The cells of a **tissue** are typically specialized, and they are organized in a particular pattern. The arrangement allows the cells to collectively perform a special function such as protection from injury (dermal tissue), movement (muscle tissue), and so on. An **organ** is an organized array of tissues that collectively carry out a particular task or set of tasks ❺. For example, a flower is an organ of reproduction in plants; a heart, an organ that pumps blood in animals. An **organ system** is a set of organs and tissues that interact to keep the individual's body working properly ❻. Examples of organ systems include the aboveground parts of a plant (the shoot system), and the heart and blood vessels of an animal (the circulatory system).

A **population** is a group of individuals of the same type, or species, living in a given area ❽. An example would be all of the California poppies in California's Antelope Valley Poppy Reserve. At the next level, a **community** consists of all populations of all species in a given area. The Antelope Valley Reserve community includes California poppies and all other plants, animals, microorganisms, and so on ❾. Communities may be large or small, depending on the area defined.

The next level of organization is the **ecosystem**, or a community interacting with its physical and chemical environment ❿. The most inclusive level, the **biosphere**, encompasses all regions of Earth's crust, waters, and atmosphere in which organisms live ⓫.

atom Fundamental building block of all matter.
biosphere All regions of Earth where organisms live.
cell Smallest unit of life.
community All populations of all species in a given area.
ecosystem A community interacting with its environment.
~~emergent property~~ A characteristic of a system that does not appear in any of the system's component parts.
molecule An association of two or more atoms.
organ In multicelled organisms, a grouping of tissues engaged in a collective task.
organ system In multicelled organisms, set of organs engaged in a collective task that keeps the body functioning properly.
organism Individual that consists of one or more cells.
population Group of interbreeding individuals of the same species that live in a given area.
tissue In multicelled organisms, specialized cells organized in a pattern that allows them to perform a collective function.

Take-Home Message

How do living things differ from nonliving things?

» All things, living or not, consist of the same building blocks: atoms. Atoms join as molecules.

» The unique properties of life emerge as certain kinds of molecules become organized into cells.

» Higher levels of life's organization include multicelled organisms, populations, communities, ecosystems, and the biosphere.

» Emergent properties occur at each successive level of life's organization.

❼ **multicelled organism**
Individual that consists of one or more cells. Cells of this California poppy plant are part of its two organ systems: aboveground shoots and belowground roots.

❽ **population**
Group of single-celled or multicelled individuals of a species in a given area. This population of California poppy plants is in California's Antelope Valley Poppy Reserve.

organisms

❾ **community**
All populations of all species in a specified area. These plants are part of the Antelope Valley Poppy Reserve community.

organism & plants

❿ **ecosystem**
A community interacting with its physical environment through the transfer of energy and materials. Sunlight and water sustain the natural community in the Antelope Valley.

Biomes

⓫ **biosphere**
The sum of all ecosystems: every region of Earth's waters, crust, and atmosphere in which organisms live. The biosphere is a finite system, so no ecosystem in it can be truly isolated from any other.

⑫ *Gaia*

1.3 How Living Things Are Alike

- Continual inputs of energy and the cycling of materials maintain life's complex organization.
- Organisms sense and respond to change.
- All organisms use information in the DNA they inherited from their parent or parents to develop and function.

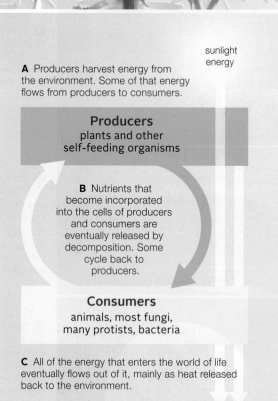

A Producers harvest energy from the environment. Some of that energy flows from producers to consumers.

sunlight energy

Producers
plants and other self-feeding organisms

B Nutrients that become incorporated into the cells of producers and consumers are eventually released by decomposition. Some cycle back to producers.

Consumers
animals, most fungi, many protists, bacteria

C All of the energy that enters the world of life eventually flows out of it, mainly as heat released back to the environment.

Figure 1.3 Animated The one-way flow of energy and cycling of materials in the world of life. The photo (*top*) shows a producer acquiring energy and nutrients from the environment, and consumers acquiring energy and nutrients by eating the producer.

Even though we cannot define "life," we can intuitively understand what it means because all living things share a set of key features. All require ongoing inputs of energy and raw materials; all sense and respond to change; and all have DNA that guides their functioning.

Organisms Require Energy and Nutrients

Not all living things eat, but all require energy and nutrients on an ongoing basis. Both are essential to maintain life's organization and functioning. **Energy** is the capacity to do work. A **nutrient** is a substance that an organism needs for growth and survival but cannot make for itself.

Organisms spend a lot of time acquiring energy and nutrients. However, what type of energy and nutrients are acquired varies considerably depending on the type of organism. The differences allow us to classify all living things into two categories: producers and consumers. **Producers** make their own food using energy and simple raw materials they get directly from their environment. Plants are producers that use the energy of sunlight to make sugars from water and carbon dioxide (a gas in air), a process called **photosynthesis**. By contrast, **consumers** cannot make their own food. They get energy and nutrients by feeding on other organisms. Animals are consumers. So are decomposers, which feed on the wastes or remains of other organisms. The leftovers of consumers' meals end up in the environment, where they serve as nutrients for producers. Said another way, nutrients cycle between producers and consumers.

Energy, however, is not cycled. It flows through the world of life in one direction: from the environment, through organisms, and back to the environment. The flow of energy maintains the organization of each individual, and it is also the basis of how organisms interact with one another and their environment. It is a one-way flow because with each transfer, some energy escapes as heat. Cells do not use heat to do work. Thus, all of the energy that enters the world of life eventually leaves it, permanently (**Figure 1.3**).

Organisms Sense and Respond to Change

An organism cannot survive for very long in a changing environment unless it adapts to the changes. Thus, every living thing has the ability to sense and respond to conditions both inside and outside of itself (**Figure 1.4**). For example, after you eat, the sugars from your meal enter your bloodstream. The added sugars set in

Figure 1.4 Organisms sense and respond to stimulation. This baby orangutan is laughing in response to being tickled. Apes and humans make different sounds when being tickled, but the airflow patterns are so similar that we can say apes really do laugh.

motion a series of events that causes cells throughout the body to take up sugar faster, so the sugar level in your blood quickly falls. This response keeps your blood sugar level within a certain range, which in turn helps keep your cells alive and your body functioning.

The fluid in your blood is part of your body's internal environment, which consists of all body fluids outside of cells. Unless that internal environment is kept within certain ranges of composition, temperature, and other conditions, your body cells will die. By sensing and adjusting to change, you and all other organisms keep conditions in the internal environment within a range that favors cell survival. **Homeostasis** is the name for this process, and it is one of the defining features of life.

Organisms Use DNA

With little variation, the same types of molecules perform the same basic functions in every organism. For example, information encoded in an organism's **DNA** (deoxyribonucleic acid) guides the ongoing metabolic activities that sustain the individual through its lifetime. Such activities include **development**: the process by which the first cell of a new individual becomes a multicelled adult; **growth**: increases in cell number, size, and volume; and **reproduction**: processes by which individuals produce offspring.

Individuals of every natural population are alike in certain aspects of their body form and behavior, an outcome of shared information encoded in DNA. Orangutans look like orangutans and not like caterpillars because they inherited orangutan DNA, which differs from caterpillar DNA in the information it carries. **Inheritance** refers to the transmission of DNA to offspring. All organisms receive their DNA from one or more parents.

DNA is the basis of similarities in form and function among organisms. However, the details of DNA molecules differ, and herein lies the source of life's diversity. Small variations in the details of DNA's structure give rise to differences among individuals, and also among types of organisms. As you will see in later chapters, these differences are the raw material of evolutionary processes.

consumer Organism that gets energy and nutrients by feeding on tissues, wastes, or remains of other organisms.

development Multistep process by which the first cell of a new individual becomes a multicelled adult.

DNA Deoxyribonucleic acid; carries hereditary information that guides development and functioning.

energy The capacity to do work.

growth In multicelled species, an increase in the number, size, and volume of cells.

homeostasis Set of processes by which an organism keeps its internal conditions within tolerable ranges.

inheritance Transmission of DNA to offspring.

nutrient Substance that an organism needs for growth and survival, but cannot make for itself.

photosynthesis Process by which producers use light energy to make sugars from carbon dioxide and water.

producer Organism that makes its own food using energy and simple raw materials from the environment.

reproduction Processes by which individuals produce offspring.

Take-Home Message

How are all living things alike?

» A one-way flow of energy and a cycling of nutrients sustain life's organization.

» Organisms sense and respond to conditions inside and outside themselves. They make adjustments that keep conditions in their internal environment within a range that favors cell survival, a process called homeostasis.

» Organisms develop and function based on information encoded in their DNA, which they inherit from their parents. DNA is the basis of similarities and differences in form and function.

1.4 How Living Things Differ

■ There is great variation in the details of appearance and other observable characteristics of living things.

Living things differ tremendously in their observable characteristics. Various classification schemes help us organize what we understand about the scope of this variation, which we call Earth's **biodiversity**.

For example, organisms can be classified into broad groups depending on whether they have a **nucleus**, which is a sac with two membranes that encloses and protects a cell's DNA. **Bacteria** (singular, bacterium) and **archaea** (singular, archaeon) are two types of organisms whose DNA is not contained within a nucleus. All bacteria and archaea are single-celled, which means each organism consists of one cell (**Figure 1.5A,B**). As a group, they are also the most diverse organisms: Different kinds are producers or consumers in nearly all regions of the biosphere. Some inhabit such extreme environments as frozen desert rocks, boiling sulfurous lakes, and nuclear reactor waste. The first cells on Earth may have faced similarly hostile challenges to survival.

Traditionally, organisms without a nucleus have been called prokaryotes, but this designation is an informal one. Despite their similar appearance, bacteria and archaea are less related to one another than we had once thought. Archaea are actually more closely related to **eukaryotes**, organisms whose DNA is contained within a nucleus. Some eukaryotes live as individual cells; others are multicelled (**Figure 1.5C**). Eukaryotic cells are typically larger and more complex than bacteria or archaea.

Structurally, **protists** are the simplest eukaryotes. As a group they vary a great deal, from single-celled consumers to giant, multicelled producers.

Most **fungi** (singular, fungus), such as the types that form mushrooms, are multicelled eukaryotes. Many are decomposers. All are consumers that secrete substances that break down food outside of the body. Their cells then absorb the released nutrients.

animal Multicelled consumer that develops through a series of stages and moves about during part or all of its life cycle.
archaeon Member of a group of single-celled organisms that lack a nucleus but are more closely related to eukaryotes than to bacteria.
bacterium Member of the most diverse and well-known group of single-celled organisms that lack a nucleus.
biodiversity Scope of variation among living organisms.
eukaryote Organism whose cells characteristically have a nucleus.
fungus Single-celled or multicelled eukaryotic consumer that digests material outside its body, then absorbs released nutrients.
nucleus Double-membraned sac that encloses a cell's DNA.
plant A multicelled, typically photosynthetic producer.
protist Member of a diverse group of simple eukaryotes.

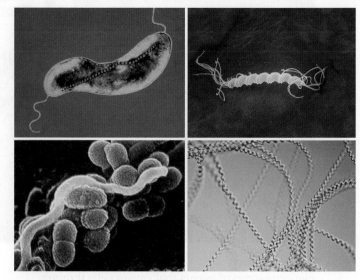

A Bacteria are the most numerous organisms on Earth. *Clockwise from upper left*, a bacterium with a row of iron crystals that acts like a tiny compass; a common resident of cat and dog stomachs; spiral cyanobacteria; types found in dental plaque.

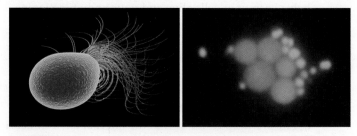

B Archaea resemble bacteria, but are more closely related to eukaryotes. *Left*, an archaeon from volcanic ocean sediments. *Right*, two types of archaea from a seafloor hydrothermal vent.

Figure 1.5 Animated Representatives of life's diversity.

Plants are multicelled eukaryotes that live on land or in freshwater environments. Nearly all are photosynthetic producers. Besides feeding themselves, plants and other photosynthesizers also serve as food for most of the other organisms in the biosphere.

Animals are multicelled consumers that ingest tissues or juices of other organisms. Herbivores graze, carnivores eat meat, scavengers eat remains of other organisms, parasites withdraw nutrients from the tissues of a host, and so on. Animals develop through a series of stages that lead to the adult form. All kinds actively move about during at least part of their lives.

Take-Home Message

How do organisms differ from one another?

» Organisms differ in their details; they show tremendous variation in observable characteristics, or traits.

» We divide Earth's biodiversity into broad groups based on traits such as having a nucleus or being multicellular.

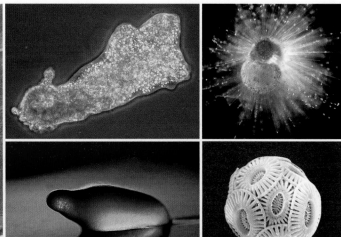

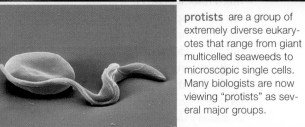

protists are a group of extremely diverse eukaryotes that range from giant multicelled seaweeds to microscopic single cells. Many biologists are now viewing "protists" as several major groups.

plants are multicelled eukaryotes, most of which are photosynthetic. Nearly all have roots, stems, and leaves. Plants are the primary producers in land ecosystems.

fungi are eukaryotes. Most are multicelled. Different kinds are parasites, pathogens, or decomposers. Without decomposers such as fungi, communities would be buried in their own wastes.

animals are multicelled eukaryotes that ingest tissues or juices of other organisms. All actively move about during at least part of their life.

C Eukaryotes are single-celled or multicelled organisms whose DNA is contained within a nucleus.

1.5 Organizing Information About Species

- Each type of organism, or species, is given a unique name.
- We define and group species based on shared traits.

Each time we discover a new **species**, or unique kind of organism, we name it. **Taxonomy**, a system of naming and classifying species, began thousands of years ago. However, naming species in a consistent way did not become a priority until the eighteenth century. At the time, European explorers and naturalists who were just discovering the scope of life's diversity started having more and more trouble communicating with one another because species often had multiple names.

For example, the dog rose (a plant native to Europe, Africa, and Asia) was alternately known as briar rose, witch's briar, herb patience, sweet briar, wild briar, dog briar, dog berry, briar hip, eglantine gall, hep tree, hip fruit, hip rose, hip tree, hop fruit, and hogseed—and those are only the English names! Species often had multiple scientific names too, in Latin that was descriptive but often cumbersome. The scientific name of the dog rose was *Rosa sylvestris inodora seu canina* (odorless woodland dog rose), and also *Rosa sylvestris alba cum rubore, folio glabro* (pinkish white woodland rose with smooth leaves).

An eighteenth-century naturalist, Carolus Linnaeus, standardized a two-part naming system that we still use today. By the Linnaean system, every species is given a unique two-part scientific name. The first part is the name of the **genus** (plural, genera), a group of species that share a unique set of features. The second part is the **specific epithet**. Together, the genus name plus the specific epithet designate one species. Thus, the dog rose now has one official name, *Rosa canina*, that is recognized worldwide.

Genus and species names are always italicized. For example, *Panthera* is a genus of big cats. Lions belong to the species *Panthera leo*. Tigers belong to a different species in the same genus (*Panthera tigris*), and so do leopards (*P. pardus*). Note how the genus name may be abbreviated after it has been spelled out once.

Today we rank species into ever more inclusive categories. Each rank, or **taxon** (plural, taxa), is a group of organisms that share a unique set of features. The categories above species—genus, family, order, class, phylum, kingdom, and domain—are the higher taxa (**Figure 1.6**). Each higher taxon consists of a group of the next lower taxon. Using this system, we can sort all life into a few categories (**Figure 1.7**).

A Rose by Any Other Name . . .

The individuals of a species share a unique set of observable characteristics, or **traits**. For example, giraffes normally have very long necks, brown spots

domain	Eukarya	Eukarya	Eukarya	Eukarya	Eukarya
kingdom	Plantae	Plantae	Plantae	Plantae	Plantae
phylum	Magnoliophyta	Magnoliophyta	Magnoliophyta	Magnoliophyta	Magnoliophyta
class	Magnoliopsida	Magnoliopsida	Magnoliopsida	Magnoliopsida	Magnoliopsida
order	Apiales	Rosales	Rosales	Rosales	Rosales
family	Apiaceae	Cannabaceae	Rosaceae	Rosaceae	Rosaceae
genus	*Daucus*	*Cannabis*	*Malus*	*Rosa*	*Rosa*
species	*carota*	*sativa*	*domestica*	*acicularis*	*canina*
common name	wild carrot	marijuana	apple	prickly rose	dog rose

Figure 1.6 Taxonomic classification of five species that are related at different levels. Each species has been assigned to ever more inclusive groups, or taxa: in this case, from genus to domain.
Figure It Out: Which of the plants shown here are in the same order?

Answer: Marijuana, apple, prickly rose, and dog rose are all in the order Rosales.

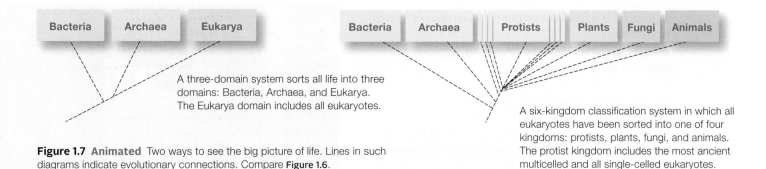

A three-domain system sorts all life into three domains: Bacteria, Archaea, and Eukarya. The Eukarya domain includes all eukaryotes.

A six-kingdom classification system in which all eukaryotes have been sorted into one of four kingdoms: protists, plants, fungi, and animals. The protist kingdom includes the most ancient multicelled and all single-celled eukaryotes.

Figure 1.7 Animated Two ways to see the big picture of life. Lines in such diagrams indicate evolutionary connections. Compare **Figure 1.6**.

on white coats, and so on. These are morphological traits (*morpho*– means form). Individuals of a species also share physiological traits, and they respond the same way to certain stimuli, as when hungry giraffes feed on tree leaves. These are behavioral traits.

A species is assigned to higher taxa based on some subset of heritable traits it shares with other species. That assignment may change as we discover more about the species and the traits involved. For example, Linnaeus grouped plants by the number and arrangement of reproductive parts, a scheme that resulted in odd pairings such as castor-oil plants with pine trees. Having more information today, we place these plants in separate phyla.

Traits vary a bit within a species, such as eye color does in people. However, there are often tremendous differences between species. Think of petunias and whales, beetles and emus, and so on. Such species look very different, so it is easy to tell them apart. Species that share a more recent ancestor may be much more difficult to distinguish (**Figure 1.8**).

How do we know whether similar-looking organisms belong to different species? The short answer is that we rely on whatever information we have. For example, early naturalists studied anatomy and distribution—essentially the only methods available at the time. Thus, species were named and classified according to what they looked like and where they lived. Today's biologists have at their disposal an array of techniques much more sophisticated than those of the eighteenth century. They are able to study and compare traits that the early naturalists did not even know about, including biochemical ones such as DNA sequences. As you will see in Section 18.4, such comparisons are deepening our understanding of the shared evolutionary history among species.

Evolutionary biologist Ernst Mayr defined a species as one or more groups of individuals that potentially can interbreed, produce fertile offspring, and do not interbreed with other groups. This "biological species concept" is useful in many cases, but it is not universally applicable. For example, we may never know whether separate populations could interbreed even if they did get together. As another example, populations often continue to interbreed even as they diverge, so the exact moment at which two populations become two species is often impossible to pinpoint. We return to speciation and how it occurs in Chapter 17, but for now it is important to remember that a "species" is a convenient but artificial construct of the human mind.

genus A group of species that share a unique set of traits; also the first part of a species name.
species Unique type of organism.
specific epithet Second part of a species name.
taxon A group of organisms that share a unique set of features.
taxonomy The science of naming and classifying species.
trait An observable characteristic of an organism or species.

Figure 1.8 Four butterflies, two species: Which are which?

The *top* row shows two forms of the species *Heliconius melpomene*; the *bottom* row, two forms of *Heliconius erato*.

H. melpomene and *H. erato* never cross-breed. Their alternate but similar patterns of coloration evolved as a shared warning signal to predatory birds that these butterflies taste terrible.

Take-Home Message

How do we keep track of all the species we know about?

» Each species has a unique, two-part scientific name.

» Classification systems group species on the basis of shared, inherited traits.

1.6 The Science of Nature

■ Judging the quality of information before accepting it is called critical thinking.

■ Scientists practice critical thinking by testing predictions about how the natural world works.

Thinking About Thinking

Most of us assume that we do our own thinking, but do we, really? You might be surprised to find out how often we let others think for us. For instance, a school's job, which is to impart as much information as possible to students, meshes perfectly with a student's job, which is to acquire as much knowledge as possible. In this rapid-fire exchange of information, it is sometimes easy to forget about the quality of what is being exchanged. Anytime you accept information without questioning it, you let someone else think for you.

Critical thinking is the deliberate process of judging the quality of information before accepting it. "Critical" comes from the Greek *kriticos* (discerning judgment). When you use critical thinking, you move beyond the content of new information to consider supporting evidence, bias, and alternative interpretations. How does the busy student manage this? Critical thinking does not necessarily require extra time, just a bit of extra awareness. There are many ways to do it. For example, you might ask yourself some of the following questions while you are learning something new:

What message am I being asked to accept?
Is the message based on facts or opinion?
Is there a different way to interpret the facts?
What biases might the presenter have?
How do my own biases affect what I'm learning?

Table 1.1 The Scientific Method

1. Observe some aspect of nature.

2. Think of an explanation for your observation (in other words, form a hypothesis).

3. Test the hypothesis.

   a. Make a prediction based on the hypothesis.

   b. Test the prediction using experiments or surveys.

   c. Analyze the results of the tests (data).

4. Decide whether the results of the tests support your hypothesis or not (form a conclusion).

5. Report your results to the scientific community.

A Sequencing the human genome.

Figure 1.9 Examples of research in the field of biology.

Such questions are a way of being conscious about learning. They can help you decide whether to allow new information to guide your beliefs and actions.

How Science Works

Critical thinking is a big part of **science**, the systematic study of the observable world and how it works (**Figure 1.9**). A scientific line of inquiry usually begins with curiosity about something observable, such as, say, a decrease in the number of birds in a particular area. Typically, a scientist will read about what others have discovered before making a **hypothesis**, a testable explanation for a natural phenomenon. An example of a hypothesis would be, "The number of birds is decreasing because the number of cats is increasing." Making a hypothesis this way is an example of **inductive reasoning**, which means arriving at a conclusion based on one's observations. Inductive reasoning is the way we come up with new ideas about groups of objects or events.

A **prediction**, or statement of some condition that should exist if the hypothesis is correct, comes next. Making predictions is called the if–then process, in which the "if" part is the hypothesis, and the "then" part is the prediction. Using a hypothesis to make a prediction is a form of **deductive reasoning**, or logical process of using a general premise to draw a conclusion about a specific case.

Next, a scientist will devise ways to test a prediction. Tests may be performed on a **model**, or analogous system, if working with an object or event directly is not possible. For example, animal diseases are often used as models of similar human diseases. Careful

B Looking for fungi in atmospheric dust.

C Improving the efficiency of biofuel production from agricultural wastes.

D Studying the ecological benefits of weedy buffer zones on farms.

observations are one way to test predictions that flow from a hypothesis. So are **experiments**: tests designed to support or falsify a prediction. A typical experiment explores a cause-and-effect relationship.

Researchers investigate causal relationships by changing or observing **variables**, which are characteristics or events that can differ among individuals or over time. An **independent variable** is defined or controlled by the person doing the experiment. A **dependent variable** is an observed result that is supposed to be influenced by the independent variable. For example, an independent variable in an investigation of hangover preventions may be the administration of artichoke extract before alcohol consumption. The dependent variable in this experiment would be the severity of the forthcoming hangover.

Biological systems are complex, with many interacting variables. It can be difficult to study one variable separately from the rest. Thus, biology researchers often test two groups of individuals simultaneously. An **experimental group** is a set of individuals that have a certain characteristic or receive a certain treatment. This group is tested side by side with a **control group**, which is identical to the experimental group except for one independent variable: the characteristic or the treatment being tested. Any differences in experimental results between the two groups should be an effect of changing the variable.

Test results—**data**—that are consistent with the prediction are evidence in support of the hypothesis. Data inconsistent with the prediction are evidence that the hypothesis is flawed and should be revised.

A necessary part of science is reporting one's results and conclusions in a standard way, such as in a peer-reviewed journal article. The communication gives other scientists an opportunity to evaluate the information for themselves, both by checking the conclusions drawn and by repeating the experiments.

Forming a hypothesis based on observation, and then systematically testing and evaluating the hypothesis are collectively called the scientific method (Table 1.1).

control group In an experiment, a group of individuals who are not exposed to the independent variable being tested.
critical thinking Judging information before accepting it.
data Experimental results.
deductive reasoning Using a general idea to make a conclusion about a specific case.
dependent variable In an experiment, a variable that is presumably affected by the independent variable being tested.
experiment A test designed to support or falsify a prediction.
experimental group In an experiment, a group of individuals who are exposed to an independent variable.
hypothesis Testable explanation of a natural phenomenon.
independent variable Variable that is controlled by an experimenter in order to explore its relationship to a dependent variable.
inductive reasoning Drawing a conclusion based on observation.
model Analogous system used for testing hypotheses.
prediction Statement, based on a hypothesis, about a condition that should exist if the hypothesis is correct.
science Systematic study of the observable world.
scientific method Making, testing, and evaluating hypotheses.
variable In an experiment, a characteristic or event that differs among individuals or over time.

Take-Home Message

What is science?

» The scientific method consists of making, testing, and evaluating hypotheses. It is a way of critical thinking, or systematically judging the quality of information before allowing it to guide one's beliefs and actions.

» Experiments measure how changing an independent variable affects a dependent variable.

1.7 Examples of Experiments in Biology

■ Researchers unravel cause-and-effect relationships in complex natural processes by changing one variable at a time.

There are many different ways to do research, particularly in biology. Some biologists do surveys; they observe without making hypotheses. Some make hypotheses and leave the experimentation to others. However, despite a broad range of subject matter, scientific experiments are typically designed in a consistent way. Experimenters try to change one independent variable at a time, and see what happens to a dependent variable.

To give you a sense of how biology experiments work, we summarize two published studies here.

Potato Chips and Stomachaches

In 1996 the U.S. Food and Drug Administration (the FDA) approved Olestra®, a fat replacement manufactured from sugar and vegetable oil, as a food additive. Potato chips were the first Olestra-containing food product on the market in the United States. Controversy about the food additive soon raged.

Many people complained of intestinal problems after eating the chips, and thought that the Olestra was at fault. Two years later, researchers at the Johns Hopkins University School of Medicine designed an experiment to test whether Olestra causes cramps.

The researchers predicted that if Olestra indeed causes cramps, then people who eat Olestra will be more likely to get cramps than people who do not. To test the prediction, they used a Chicago theater as a "laboratory." They asked 1,100 people between the ages of thirteen and thirty-eight to watch a movie and eat their fill of potato chips. Each person got an unmarked bag that contained 13 ounces of chips.

In this experiment, the individuals who got Olestra-containing potato chips constituted the experimental group, and individuals who got regular chips were the control group. The independent variable was the presence or absence of Olestra in the chips.

A few days after the experiment was finished, the researchers contacted all of the people and collected any reports of post-movie gastrointestinal problems. Of 563 people making up the experimental group, 89 (15.8 percent) complained about cramps. However, so did 93 of the 529 people (17.6 percent) making up the control group—who had eaten the regular chips.

People were about as likely to get cramps whether or not they ate chips made with Olestra. These results did not support the prediction, so the researchers concluded that eating Olestra does not cause cramps (Figure 1.10).

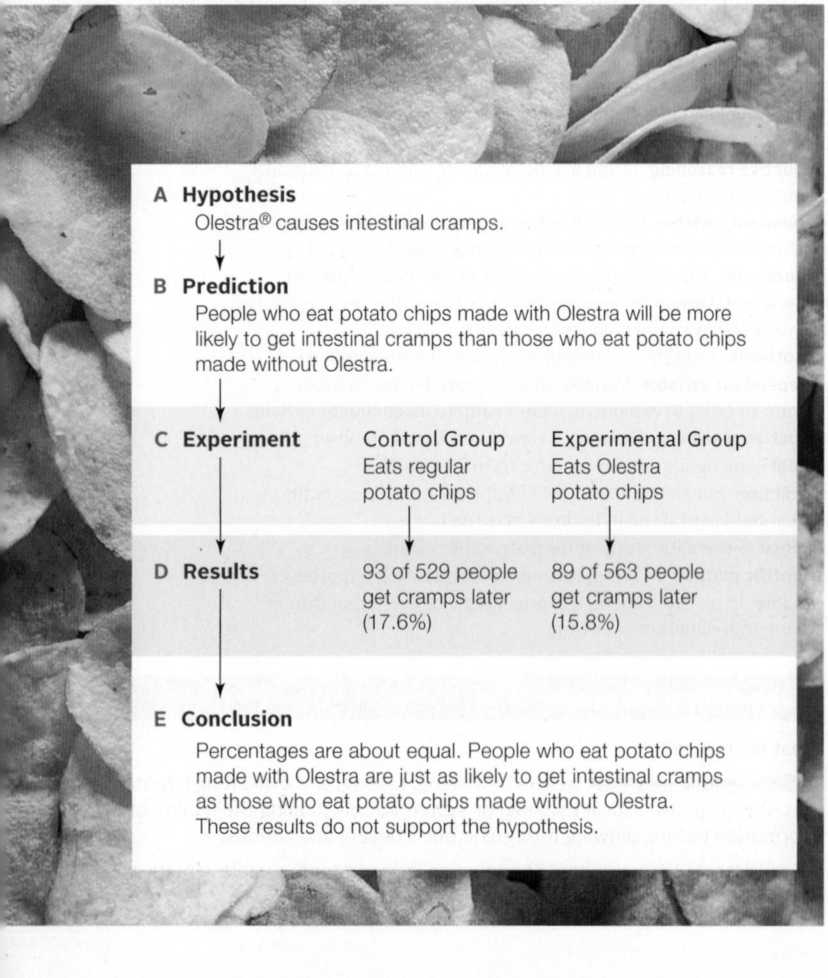

Butterflies and Birds

The peacock butterfly is a winged insect named for the large, colorful spots on its wings. In 2005, researchers reported the results of experiments investigating whether certain behaviors help peacock butterflies defend themselves against insect-eating birds. The researchers began with two observations. First, when a peacock butterfly rests, it folds its wings, so only the dark underside shows (Figure 1.11A). Second, when a butterfly sees a predator approaching, it repeatedly flicks its wings open and closed, while also moving the hindwings in a way that produces a hissing sound and a series of clicks.

A Hypothesis
Olestra® causes intestinal cramps.

B Prediction
People who eat potato chips made with Olestra will be more likely to get intestinal cramps than those who eat potato chips made without Olestra.

C Experiment	Control Group Eats regular potato chips	Experimental Group Eats Olestra potato chips
D Results	93 of 529 people get cramps later (17.6%)	89 of 563 people get cramps later (15.8%)

E Conclusion
Percentages are about equal. People who eat potato chips made with Olestra are just as likely to get intestinal cramps as those who eat potato chips made without Olestra. These results do not support the hypothesis.

Figure 1.10 The steps in a scientific experiment to determine if Olestra causes cramps. A report of this study was published in the *Journal of the American Medical Association* in January 1998. **Figure It Out:** What was the dependent variable in this experiment?

Answer: Whether or not a person got cramps

Figure 1.11 Peacock butterfly defenses against predatory birds. (**A**) With wings folded, a resting peacock butterfly looks a bit like a dead leaf. (**B**) When a bird approaches, the butterfly repeatedly flicks its wings open and closed, a behavior that exposes brilliant spots and produces hissing and clicking sounds.

Researchers tested whether the butterfly's behavior deters blue tits (**C**). They painted over the spots of some butterflies, cut the sound-making part of the wings on other butterflies, and did both to a third group; then the biologists exposed each butterfly to a hungry bird.

The results, listed in **Table 1.2**, support the hypotheses that peacock butterfly spots and sounds can deter predatory birds.

Figure It Out: What percentage of butterflies with no spots and no sound survived the test?   Answer: 20 percent

Table 1.2   Results of Peacock Butterfly Experiment*

Wing Spots	Wing Sound	Total Number of Butterflies	Number Eaten	Number Survived	
Spots	Sound	9	0	9	(100%)
No spots	Sound	10	5	5	(50%)
Spots	No sound	8	0	8	(100%)
No spots	No sound	10	8	2	(20%)

* Proceedings of the Royal Society of London, Series B (2005) 272: 1203–1207.

The researchers were curious about why the peacock butterfly flicks its wings. After they reviewed earlier studies, they came up with two hypotheses that might explain the wing-flicking behavior:

1. Although wing-flicking probably attracts predatory birds, it also exposes brilliant spots that resemble owl eyes (**Figure 1.11B**). Anything that looks like owl eyes is known to startle small, butterfly-eating birds, so exposing the wing spots might scare off predators.
2. The hissing and clicking sounds produced when the peacock butterfly moves its hindwings may be an additional defense that deters predatory birds.

The researchers used their hypotheses to make the following predictions:

1. If peacock butterflies startle predatory birds by exposing their brilliant wing spots, then individuals with wing spots will be less likely to get eaten by predatory birds than those without wing spots.
2. If peacock butterfly sounds deter predatory birds, then sound-producing individuals will be less likely to get eaten by predatory birds than silent individuals.

The next step was the experiment. The researchers used a marker to paint the wing spots of some but-terflies black, and scissors to cut off the sound-making part of the hindwings of others. A third group had their wing spots painted and their hindwings cut. The researchers then put each butterfly into a large cage with a hungry blue tit (**Figure 1.11C**) and watched the pair for thirty minutes.

Table 1.2 lists the results of the experiment. All of the butterflies with unmodified wing spots survived, regardless of whether they made sounds. By contrast, only half of the butterflies that had spots painted out but could make sounds survived. Most of the silenced butterflies with painted-out spots were eaten quickly. The test results confirmed both predictions, so they support the hypotheses. Predatory birds are indeed deterred by peacock butterfly sounds, and even more so by wing spots.

Take-Home Message

Why do biologists perform experiments?

» Natural processes are often very complex and influenced by many interacting variables.

» Experiments help researchers unravel causes of complex natural processes by focusing on the effects of changing a single variable.

1.8 Analyzing Experimental Results

■ Biology researchers often experiment on subsets of a group. Results from such an experiment may differ from results of the same experiment performed on the whole group.

■ Science is, ideally, a self-correcting process because scientists check one another's work.

Sampling Error

Researchers can rarely observe all individuals of a group. For example, the explorers you read about in Section 1.1 did not—and could not—survey every uninhabited part of the Foja Mountains. The cloud forest itself cloaks more than 2 million acres, so surveying all of it would take unrealistic amounts of time and effort. Besides, tromping about even in a small area can damage delicate forest ecosystems.

Given such limitations, researchers often look at subsets of an area, a population, or an event. They test or survey the subset, then use the results to make generalizations. However, generalizing from a subset is risky because a subset may not be representative of the whole.

For example, the golden-mantled tree kangaroo was first discovered in 1993 on a single forested mountaintop in New Guinea. For more than a decade, the species was never seen outside of that habitat, which is getting smaller every year because of human activities. Thus, the golden-mantled tree kangaroo was considered to be one of the most endangered animals on the planet. Then, in 2005, the New Guinea explorers discovered that this kangaroo species is fairly common in the Foja Mountain cloud forest (**Figure 1.12**). As a result,

Figure 1.12 Biologist Kris Helgen holds a golden-mantled tree kangaroo he found during the 2005 expedition to the Foja Mountains cloud forest in New Guinea.

This kangaroo species is extremely rare in other regions of the world, so it was thought to be critically endangered prior to the 2005 survey expedition.

biologists now believe its future is secure, at least for the moment.

Making generalizations from testing or surveying a subset is risky because of sampling error. Sampling error is a difference between results obtained from testing a subset of a group, and results from testing the whole group. Sampling error may be unavoidable, as illustrated by the example of the golden-mantled tree kangaroo. However, knowing how it can occur helps researchers design their experiments to minimize it. For example, sampling error can be a substantial problem with a small subset (**Figure 1.13**), so experimenters try to start with a relatively large sample, and they typically repeat their experiments.

A Natalie, blindfolded, randomly plucks a jelly bean from a jar. The jar contains 120 green and 280 black jelly beans, so 30 percent of the jelly beans in the jar are green, and 70 percent are black.

B The jar is hidden from Natalie's view before she removes her blindfold. She sees one green jelly bean in her hand and assumes that the jar must hold only green jelly beans.

C Still blindfolded, Natalie randomly picks out 50 jelly beans from the jar. She ends up picking out 10 green and 40 black ones.

D The larger sample leads Natalie to assume that one-fifth of the jar's jelly beans are green (20 percent) and four-fifths are black (80 percent). This sample more closely approximates the jar's actual green-to-black ratio of 30 percent to 70 percent. The more times Natalie repeats the sampling, the greater the chance she has of guessing the actual ratio.

Figure 1.13 How sample size affects sampling error.

To understand why such practices reduce the risk of sampling error, think about what happens each time you flip a coin. There are two possible outcomes: The coin lands heads up, or it lands tails up. Thus, the chance that the coin will land heads up is one in two (1/2), which is a proportion of 50 percent. However, when you flip a coin repeatedly, it often lands heads up, or tails up, several times in a row. With just 3 flips, the proportion of times that heads actually land up may not even be close to 50 percent. With 1,000 flips, the proportion of times that the coin lands heads up is likely to be close to 50 percent.

In cases like flipping a coin, it is possible to calculate **probability**: the measure, expressed as a percentage, of the chance that a particular outcome will occur. That chance depends on the total number of possible outcomes. For instance, imagine that 10 million people enter a random drawing to win a car. There are 10 million possible outcomes, so each person has the same probability of winning the item: 1 in 10 million, or (an extremely improbable) 0.00001 percent.

Analysis of experimental data often includes calculations of probability. If a result is very unlikely to have occurred by chance alone, it is said to be **statistically significant**. In this context, the word "significant" does not refer to the result's importance. It means that a complicated statistical analysis shows the result has a very low probability (typically, less than a 5 percent chance) of being skewed by sampling error. In science, every result—even a statistically significant one—has a probability of being incorrect.

Variation in a set of data is often shown as error bars on a graph (**Figure 1.14**). Depending on the graph, error bars may indicate variation around an average for one sample set, or the difference between two sample sets.

Bias in Interpreting Results

Experimenting with a single variable apart from all others is not often possible, particularly when studying humans. For example, remember that the people who participated in the Olestra experiment were chosen randomly, which means the study was not controlled for gender, age, weight, medications taken, and so on. Such variables may well have influenced the results.

probability The chance that a particular outcome of an event will occur; depends on the total number of outcomes possible.
sampling error Difference between results derived from testing an entire group of events or individuals, and results derived from testing a subset of the group.
statistically significant Refers to a result that is statistically unlikely to have occurred by chance.

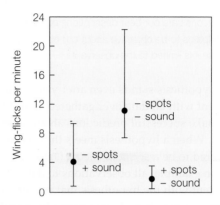

Figure 1.14 Example of error bars in a graph. This particular graph was adapted from the peacock butterfly research described in Section 1.7.

The researchers recorded the number of times each butterfly flicked its wings in response to an attack by a bird. The dots represent average frequency of wing flicking for each sample set of butterflies. The error bars that extend above and below the dots indicate the range of values—the sampling error.

Figure It Out: What was the fastest rate at which a butterfly with no spots or sound flicked its wings?

Answer: 22 times per minute

Humans are by nature subjective, and scientists are no exception. Experimenters risk interpreting their results in terms of what they want to find out. That is why they often design experiments to yield quantitative results, which are counts or some other data that can be measured or gathered objectively. Such results minimize the potential for bias, and also give other scientists an opportunity to repeat the experiments and check the conclusions drawn from them.

This last point gets us back to the role of critical thinking in science. Scientists expect one another to recognize and put aside bias in order to test their hypotheses in ways that may prove them wrong. If a scientist does not, then others will, because exposing errors is just as useful as applauding insights. The scientific community consists of critically thinking people trying to poke holes in one another's ideas. Their collective efforts make science a self-correcting endeavor.

Take-Home Message

How do scientists avoid potential pitfalls of sampling error and bias when doing research?

» Researchers minimize sampling error by using large sample sizes and by repeating their experiments.

» A statistical analysis can show the probability that a result has occurred by chance alone.

» Science is a self-correcting process because it is carried out by an aggregate community of people systematically checking one another's ideas.

1.9 The Nature of Science

■ Scientific theories are our best descriptions of reality.
■ Science helps us to be objective about our observations, in part because it is limited to the observable.

Suppose a hypothesis stands even after years of tests. It is consistent with all data ever gathered, and it has helped us make successful predictions about other phenomena. When a hypothesis meets these criteria, it is considered to be a scientific theory (**Table 1.3**).

To give an example, all observations to date have been consistent with the hypothesis that matter consists of atoms. Scientists no longer spend time testing this hypothesis for the compelling reason that, since we started looking 200 years ago, no one has discovered matter that doesn't consist of atoms. Thus, scientists use the hypothesis, now called atomic theory, to make other hypotheses about matter and the way it behaves.

Scientific theories are our best descriptions of reality. However, they can never be proven absolutely, because to do so would necessitate testing under every possible circumstance. For example, in order to prove atomic theory, the atomic composition of all matter in the universe would have to be checked—an impossible task even if someone wanted to try.

Like all hypotheses, a scientific theory can be disproven by a single observation or result that is inconsistent with it. For example, if someone discovers a form of matter that does not consist of atoms, atomic theory would have to be revised. The potentially falsifiable nature of scientific theories means that science has a built-in system of checks and balances. A theory is revised until no one can prove it to be incorrect. For example, the theory of evolution, which states that

change occurs in a line of descent over time, still holds after a century of observations and testing. As with all other scientific theories, no one can be absolutely sure that it will hold under all possible conditions, but it has a very high probability of not being wrong. Few other theories have withstood as much scrutiny.

You may hear people apply the word "theory" to a speculative idea, as in the phrase "It's just a theory." This everyday usage of the word differs from the way it is used in science. Speculation is an opinion, belief, or personal conviction that is not necessarily supported by evidence. A scientific theory is different. By definition, it is supported by a large body of evidence, and it is consistent with all known facts.

A scientific theory also differs from a law of nature, which describes a phenomenon that has been observed to occur in every circumstance without fail, but for which we do not have a complete scientific explanation. The laws of thermodynamics, which describe energy, are examples. We know how energy behaves, but not exactly why it behaves the way it does.

The Limits of Science

Science helps us be objective about our observations in part because of its limitations. For example, science does not address many questions, such as "Why do I exist?" Answers to such questions can only come from within as an integration of the personal experiences and mental connections that shape our consciousness. This is not to say subjective answers have no value, because no human society can function for long unless its individuals share standards for making judgments,

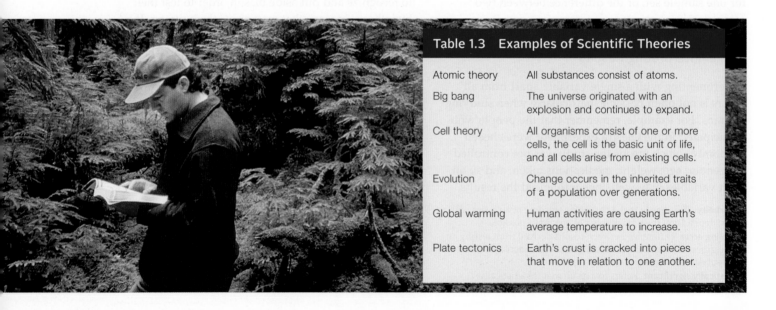

Table 1.3	Examples of Scientific Theories
Atomic theory	All substances consist of atoms.
Big bang	The universe originated with an explosion and continues to expand.
Cell theory	All organisms consist of one or more cells, the cell is the basic unit of life, and all cells arise from existing cells.
Evolution	Change occurs in the inherited traits of a population over generations.
Global warming	Human activities are causing Earth's average temperature to increase.
Plate tectonics	Earth's crust is cracked into pieces that move in relation to one another.

The Secret Life of Earth (revisited)

Of an estimated 100 billion species that have ever lived, at least 100 million are still with us. That number is only an estimate because we are still discovering them. For example, a mouse-sized opossum and a cat-sized rat turned up on a return trip to the Foja Mountains. Other recent surveys have revealed a new species of leopard in Borneo; lemurs and sucker-footed bats in Madagascar; birds in the Philippines; a jackal in Egypt; monkeys in Tanzania, Brazil, and India;

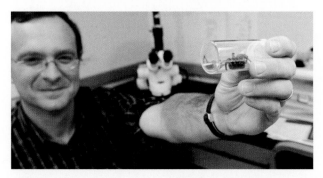

Figure 1.15 The discoverer of a new species typically has the honor of naming it. Dr. Jason Bond holds a new species of spider he discovered in California in 2008. Bond named the spider *Aptostichus stephencolberti*, after TV personality Stephen Colbert.

insects and spiders in California; a giant crayfish in Tennessee; a rat-eating plant in the Philippines; carnivorous sponges near Antarctica; whales, sharks, giant jellylike animals, fishes, and other aquatic wildlife; and scores of plants and single-celled organisms. Most were discovered by biologists simply trying to find out what lives where.

Biologists discover thousands of new species per year (**Figure 1.15**). Each is a reminder that we do not yet know all of the organisms living on our own planet. We don't even know how many to look for. The vast information about the 1.8 million species we do know about changes so quickly that collating it has been impossible—until recently. A web site titled the Encyclopedia of Life is intended to be an online reference source and database of species information maintained by collaborative effort. See its progress at www.eol.org.

How would you vote? Substantial populations of some species currently listed as endangered may exist in unexplored areas. Should we wait to protect endangered species until all of Earth has been surveyed?

---

even if they are subjective. Moral, aesthetic, and philosophical standards vary from one society to the next, but all help people decide what is important and good. All give meaning to our lives.

Neither does science address the supernatural, or anything that is "beyond nature." Science neither assumes nor denies that supernatural phenomena occur, but scientists often cause controversy when they discover a natural explanation for something that was thought to have none. Such controversy often arises when a society's moral standards are interwoven with its understanding of nature.

For example, Nicolaus Copernicus concluded in 1540 that Earth orbits the sun. Today that idea is generally accepted, but during Copernicus's time the prevailing belief system had Earth as the immovable center of the universe. In 1610, astronomer Galileo Galilei published evidence for the Copernican model of the solar system, an act that resulted in his imprisonment. He was publicly forced to recant his work, spent the rest of his life under house arrest, and was never allowed to publish again.

law of nature Generalization that describes a consistent natural phenomenon for which there is incomplete scientific explanation.
scientific theory Hypothesis that has not been disproven after many years of rigorous testing.

As Galileo's story illustrates, exploring a traditional view of the natural world from a scientific perspective is often misinterpreted as a violation of morality. As a group, scientists are no less moral than anyone else. However, they follow a particular set of rules that do not necessarily apply to others: Their work concerns only the natural world, and their ideas must be testable in ways others can repeat.

Science helps us communicate our experiences without bias. As such, it may be as close as we can get to a universal language. We are fairly sure, for example, that the laws of gravity apply everywhere in the universe. Intelligent beings on a distant planet would likely understand the concept of gravity. We might well use gravity or another scientific concept to communicate with them, or anyone, anywhere. The point of science, however, is not to communicate with aliens. It is to find common ground here on Earth.

Take-Home Message

Why does science work?

» Checks and balances inherent in the scientific process help researchers to be objective about their observations.

» Because a scientific theory is revised until no one can prove it wrong, it is our best way of describing reality.

I PRINCIPLES OF CELLULAR LIFE

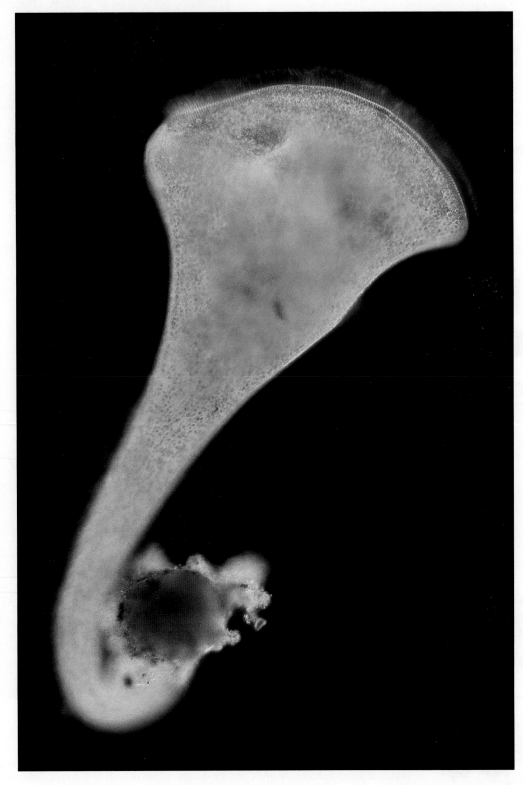

Staying alive means securing energy and raw materials from the environment. Shown here, a living cell of the genus *Stentor*. This protist has hairlike projections around an opening to a cavity in its body, which is about 2 millimeters long. Its "hairs" of fused-together cilia beat the surrounding water. They create a current that wafts food into the cavity.

2 Chemical Basis of Life

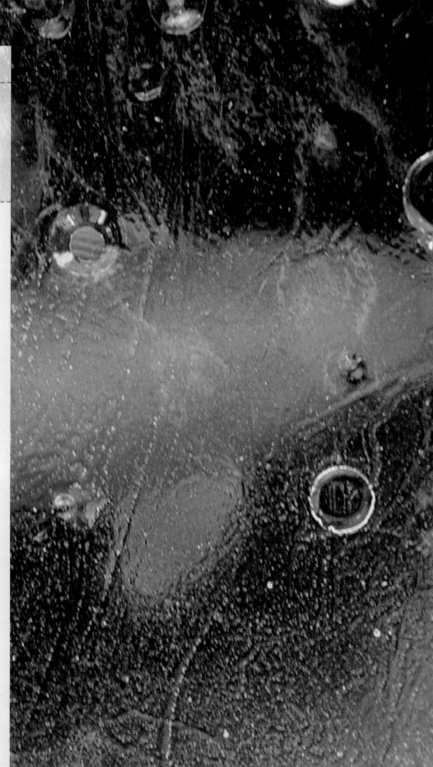

LEARNING ROADMAP

Where you have been In this chapter, you will explore the first level of life's organization—atoms—as you encounter the first example of how the same building blocks, arranged different ways, form different products (Section 1.2). You will also see one aspect of homeostasis, the process by which organisms keep themselves in a state that favors cell survival (Section 1.3).

Where you are now

Atoms and Elements
Atoms, the building blocks of all matter, differ in their numbers of protons, neutrons, and electrons. Atoms of an element have the same number of protons.

Why Electrons Matter
Whether and how an atom interacts with other atoms depends on the number of electrons it carries. An atom with an unequal number of electrons and protons is an ion.

Atoms Bond
Atoms of many elements interact by acquiring, sharing, and giving up electrons. Interacting atoms may form ionic, covalent, or hydrogen bonds.

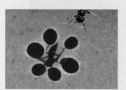

Water
Hydrogen bonding among individual molecules gives water properties that make life possible: temperature stabilization, cohesion, and the ability to dissolve many other substances.

Hydrogen Power
Most of the chemistry of life occurs in a narrow range of pH, so most fluids inside organisms are buffered to stay within that range.

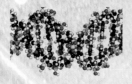

Where you are going Electrons will come up again as you learn how energy drives metabolism, especially in photosynthesis (Chapter 6) and respiration (Chapter 7). Hydrogen bonding is critical for the molecules of life (3.4, 3.6–3.8); the properties of water, for membranes (Section 4.2 and Chapter 5), plant nutrition and transport (28.4), and temperature regulation (40.9). You will also see how radioisotopes are used to date rocks and fossils (16.6), and the dangers of free radicals (8.6) and acid rain (48.5).

2.1 Mercury Rising

Actor Jeremy Piven, best known for his Emmy-winning role on the television series *Entourage*, began starring in a Broadway play in 2008. He quit suddenly after two shows, citing medical problems. Piven explained that he was suffering from mercury poisoning caused by eating too much sushi. The play's producers and his co-actors were skeptical. The playwright ridiculed Piven, saying he was leaving to pursue a career as a thermometer. But mercury poisoning is no laughing matter.

Mercury is a naturally occurring metal. Most of it is safely locked away in rocky minerals, but volcanic activity and other geologic processes release it into the atmosphere. So do human activities, especially burning coal (**Figure 2.1**). Once airborne, mercury can drift long distances before settling to Earth's surface. There, microbes combine it with carbon to form a substance called methylmercury.

Mercury does not cross skin or mucous membranes very well, but methylmercury does so easily. In water, it ends up in the tissues of aquatic organisms. All fish and shellfish contain methylmercury. Humans contain it too, mainly as a result of eating seafood.

Once mercury enters the body's internal environment, it damages the nervous system, brain, kidneys, and other organs. A dose as low as 3 micrograms per kilogram of body weight (about 200 micrograms for an average-sized adult) can cause tremors, itching or burning sensations, and loss of coordination. Exposure to larger amounts can result in thought and memory impairment, coma, and death. The developing brain is particularly sensitive to mercury because the metal interferes with nerve formation. Thus, mercury is acutely toxic in infants, and it causes long-term neurological effects in children. Mercury in a pregnant woman's blood passes to her unborn child, along with a legacy of permanent developmental problems.

The U.S. Food and Drug Administration requires that foods contain less than 1 part per million of mercury, and for the most part they do. However, it takes months or even years for mercury to be cleared from the body, so it can build up to high levels if even small amounts are ingested on a regular basis. That is why large predatory fish have a lot of mercury in their tissues. It is also why the U.S. Environmental Protection Agency recommends that adults ingest less than 0.1 microgram of mercury per kilogram of body weight per day. For an average-sized person, that limit works out to be about 7 micrograms per day, which is not a big amount if you eat seafood. A two-ounce piece of sushi tuna typically contains about 40 micrograms of mercury, and the occasional piece has many times that amount. It doesn't matter if the fish is raw, grilled, or canned, because mercury is unaffected by cooking. Eat a medium-sized tuna steak, and you could be getting more than 700 micrograms of mercury along with it.

With this chapter, we turn to the first of life's levels of organization: atoms. Interactions between atoms make the molecules that sustain life, and also some that destroy it.

Figure 2.1 Atmospheric fallout from coal-fired power plant emissions is now the biggest cause of mercury pollution. Mercury can accumulate to toxic levels in the tissues of tuna and other large predatory fish.

2.2 Start With Atoms

■ The behavior of elements, which make up all living things, starts with the structure of individual atoms.

■ The number of protons in the atomic nucleus defines the element, and the number of neutrons defines the isotope.

■ Link to atoms 1.2

Life's unique characteristics start with the properties of different atoms, tiny particles that are building blocks of all substances. Even though atoms are about 20 million times smaller than a grain of sand, they consist of even smaller subatomic particles. Positively charged protons (p^+) and uncharged neutrons occur in an atom's core, or nucleus. Negatively charged electrons (e^-) move around the nucleus (**Figure 2.2**). Charge is an electrical property: opposite charges attract, and like charges repel.

Atoms differ in their number of subatomic particles. The number of protons in an atom's nucleus is called the atomic number, and it determines the type of atom, or element. Elements are pure substances, each consisting only of atoms that have the same number of protons in their nucleus. For example, the atomic number of carbon is 6, so all atoms with six protons in their nucleus are carbon atoms, no matter how many electrons or neutrons they have. A chunk of carbon consists only of carbon atoms, and all of those atoms have six protons.

The same elements that make up a living body also occur in nonliving things, but their proportions differ. For example, a human body contains a much larger

Figure 2.2 Atoms consist of electrons moving around a core, or nucleus, of protons and neutrons.

Models such as this diagram cannot show what atoms really look like. Electrons zoom around in fuzzy, three-dimensional spaces about 10,000 times bigger than the nucleus.

● proton
● neutron
○ electron

proportion of carbon atoms than sand or seawater. Why? Unlike sand or seawater, a body consists of a very high proportion of the molecules of life, which in turn consist of a high proportion of carbon atoms.

Knowing the numbers of electrons, protons, and neutrons in atoms helps us predict how elements will behave. In 1869, chemist Dmitry Mendeleyev arranged the elements known at the time by their chemical properties (**Figure 2.3**). The arrangement, which he called the periodic table, turned out to be by atomic number, even though subatomic particles would not be discovered until the early 1900s. Gaps in the table allowed Mendeleyev to predict the existence of elements that had not yet been discovered.

In the periodic table, each element is represented by a symbol that is typically an abbreviation of the element's Latin or Greek name. For instance, Pb (lead) is short for *plumbum;* the word "plumbing" is related—ancient Romans made their water pipes with lead. Carbon's symbol, C, is from *carbo,* the Latin word for coal (which is mostly carbon).

atom Particle that is a fundamental building block of all matter.

atomic number Number of protons in the atomic nucleus; determines the element.

charge Electrical property. Opposite charges attract, and like charges repel.

electron Negatively charged subatomic particle that occupies orbitals around an atomic nucleus.

element A pure substance that consists only of atoms with the same number of protons.

isotopes Forms of an element that differ in the number of neutrons their atoms carry.

mass number Total number of protons and neutrons in the nucleus of an element's atoms.

neutron Uncharged subatomic particle in the atomic nucleus.

nucleus Core of an atom; occupied by protons and neutrons.

periodic table Tabular arrangement of the elements by atomic number.

proton Positively charged subatomic particle that occurs in the nucleus of all atoms.

radioactive decay Process by which atoms of a radioisotope emit energy and/or subatomic particles when their nucleus spontaneously disintegrates.

radioisotope Isotope with an unstable nucleus.

tracer Substance with a detectable component, such as a molecule labeled with a radioisotope.

Figure 2.3 Animated The periodic table and its creator, Dmitry Mendeleyev. Until he came up with the table, Mendeleyev was known mainly for his extravagant hair (he cut it only once per year).

Atomic numbers are shown above the element symbols. Some of the symbols are abbreviations for their Latin names. For instance, Pb (lead) is short for *plumbum;* the word "plumbing" is related—ancient Romans made their water pipes with lead. Appendix IV shows the table in more detail.

1 H																	2 He
3 Li	4 Be											5 B	6 C	7 N	8 O	9 F	10 Ne
11 Na	12 Mg											13 Al	14 Si	15 P	16 S	17 Cl	18 Ar
19 K	20 Ca	21 Sc	22 Ti	23 V	24 Cr	25 Mn	26 Fe	27 Co	28 Ni	29 Cu	30 Zn	31 Ga	32 Ge	33 As	34 Se	35 Br	36 Kr
37 Rb	38 Sr	39 Y	40 Zr	41 Nb	42 Mo	43 Tc	44 Ru	45 Rh	46 Pd	47 Ag	48 Cd	49 In	50 Sn	51 Sb	52 Te	53 I	54 Xe
55 Cs	56 Ba	71 Lu	72 Hf	73 Ta	74 W	75 Re	76 Os	77 Ir	78 Pt	79 Au	80 Hg	81 Tl	82 Pb	83 Bi	84 Po	85 At	86 Rn
87 Fr	88 Ra	103 Lr	104 Rf	105 Db	106 Sg	107 Bh	108 Hs	109 Mt	110 Ds	111 Rg	112 Uub	113 Uut	114 Uuq	115 Uup	116 Uuh		118 Uuo

		57 La	58 Ce	59 Pr	60 Nd	61 Pm	62 Sm	63 Eu	64 Gd	65 Tb	66 Dy	67 Ho	68 Er	69 Tm	70 Yb		
		89 Ac	90 Th	91 Pa	92 U	93 Np	94 Pu	95 Am	96 Cm	97 Bk	98 Cf	99 Es	100 Fm	101 Md	102 No		

Isotopes and Radioisotopes

Atoms of an element do not differ in the number of protons, but they can differ in the number of other subatomic particles. Those that differ in the number of neutrons are called isotopes. We define isotopes by their mass number, which is the total number of protons and neutrons in their nucleus. Mass number is written as a superscript to the left of an element's symbol. For example, the most common isotope of carbon has six protons and six neutrons, so it is designated ^{12}C, or carbon 12. The other naturally occurring isotopes of carbon are ^{13}C (six protons, seven neutrons), and ^{14}C (six protons, eight neutrons).

Carbon 14 is a radioisotope, or radioactive isotope. Atoms of a radioisotope have an unstable nucleus that breaks down spontaneously. As a nucleus breaks down, it emits radiation—subatomic particles, energy, or both—a process called radioactive decay. The atomic nucleus cannot be altered by ordinary means, so radioactive decay is unaffected by external factors such as temperature, pressure, or whether the atoms are part of molecules.

Each radioisotope decays at a predictable rate into predictable products. For example, when carbon 14 decays, one of its neutrons splits into a proton and an electron. The nucleus emits the electron as radiation. Thus, an atom with eight neutrons and six protons (^{14}C) becomes an atom with seven neutrons and seven protons, which is nitrogen (^{14}N):

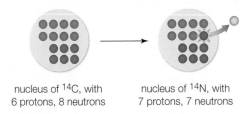

nucleus of ^{14}C, with   nucleus of ^{14}N, with
6 protons, 8 neutrons   7 protons, 7 neutrons

This process is so predictable that we can say with certainty that about half of the atoms in any sample of ^{14}C will be ^{14}N atoms after 5,730 years. Researchers use the predictability of radioactive decay when they estimate the age of a rock or fossil by measuring its isotope content (we return to this topic in Section 16.6).

All isotopes of an element generally have the same chemical properties regardless of the number of neutrons in their atoms. The consistent chemical behavior means that organisms use atoms of one isotope the same way that they use atoms of another. Thus, radioisotopes can be used to study biological processes. A tracer is any substance with a detectable component, such as a molecule in which an atom (such as ^{12}C) has been replaced with a radioisotope (such as ^{14}C).

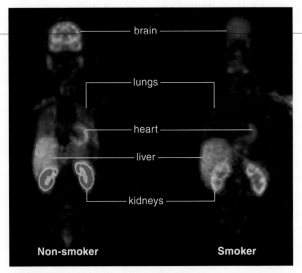

Figure 2.4 Animated PET scans. The result of a PET scan is a digital image of a process in the body's interior. These two PET scans show the activity of a molecule called MAO-B in the body of a non-smoker (*left*) and a smoker (*right*). The activity is color-coded from *red* (highest activity) to *purple* (lowest). Low MAO-B activity is associated with violence, impulsiveness, and other behavioral problems.

A radioactive tracer can be delivered into a biological system such as a cell, body, or ecosystem, and then followed as it moves through the system with instruments that detect radiation.

Radioactive tracers are widely used in research. A famous example is a series of experiments carried out by Melvin Calvin and Andrew Benson. These researchers synthesized carbon dioxide with ^{14}C, then let green algae take up the radioactive gas. Using instruments that detect electrons emitted by the radioactive decay of ^{14}C, they tracked carbon through steps by which the algae—and all plants—make sugars.

Radioisotopes have medical applications as well. For example, PET (short for positron-emission tomography) helps us "see" a functional process inside the body. By this procedure, a radioactive sugar or other tracer is injected into a patient, who is then moved into a scanner. Inside the patient's body, cells with differing rates of activity take up the tracer at different rates. The scanner detects radioactive decay wherever the tracer is, then translates that data into an image (**Figure 2.4**).

Take-Home Message

What are the basic building blocks of all matter?

» All matter consists of atoms, tiny particles that in turn consist of electrons moving around a nucleus of protons and neutrons.

» An element is a pure substance that consists only of atoms with the same number of protons. Isotopes are forms of an element that have different numbers of neutrons.

» Unstable nuclei of radioisotopes disintegrate spontaneously (decay) at a predictable rate to form predictable products.

2.3 Why Electrons Matter

■ Whether an atom will interact with other atoms depends on how many electrons it has.
■ Link to building blocks of life 1.2

Electrons are really, really small. How small are they? If they were as big as apples, you would be about 3.5 times taller than our solar system is wide. Simple physics explains the motion of, say, an apple falling from a tree, but electrons are so tiny that such everyday physics does not explain their behavior.

For example, electrons carry energy, but only in incremental amounts. An electron can gain energy only by absorbing the exact amount needed to boost it to the next energy level. Likewise, it can lose energy only by emitting the exact difference between two energy levels. This concept will be especially important to remember in later chapters, when you learn about how cells harvest and release energy.

A typical atom has approximately the same number of electrons as protons. The atoms of many elements have a lot of electrons zipping around the same nucleus. However, despite moving at nearly the speed of light (300,000 kilometers per second, or 670 million miles per hour), electrons that are part of an atom never collide. Why not? Electrons avoid one another

because they occupy different orbitals, which are defined volumes of space around the nucleus.

To understand how orbitals work, imagine that an atom is a multilevel apartment building with a nucleus in the basement. Each "floor" of the building corresponds to a certain energy level, and each has a certain number of "rooms" (orbitals) available for rent.

Two electrons can occupy each room. Pairs of electrons populate rooms from the ground floor up; in other words, they fill orbitals from lower to higher energy levels. The farther an electron is from the nucleus in the basement, the greater its energy. An electron can move to a room on a higher floor if an energy input gives it a boost, but it immediately emits the extra energy and moves back down.

Shell models help us visualize how electrons populate atoms (**Figure 2.5**). In this model, nested "shells" correspond to successively higher energy levels. Thus, each shell includes all of the rooms on one floor of our atomic apartment building.

We draw a shell model of an atom by filling shells with electrons (represented as balls or dots), from the innermost shell out, until there are as many electrons as the atom has protons. There is only one room on the first floor, one orbital at the lowest energy level

---

A The first shell corresponds to the first energy level, and it can hold up to 2 electrons. Hydrogen has one proton, so it has 1 electron and 1 vacancy. A helium atom has 2 protons, 2 electrons, and no vacancies. The number of protons in each model is shown.

first shell

1 proton ⟶ 1
1 electron ⟶
hydrogen (H)

2
helium (He)

B The second shell corresponds to the second energy level, and it can hold up to 8 electrons. Carbon has 6 protons, so its first shell is full. Its second shell has 4 electrons, and four vacancies. Oxygen has 8 protons and two vacancies. Neon has 10 protons and no vacancies.

second shell

6
carbon (C)

8
oxygen (O)

10
neon (Ne)

C The third shell, which corresponds to the third energy level, can hold up to 8 electrons. A sodium atom has 11 protons, so its first two shells are full; the third shell has one electron. Thus, sodium has seven vacancies. Chlorine has 17 protons and one vacancy. Argon has 18 protons and no vacancies.

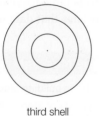
third shell

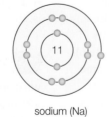
11
sodium (Na)

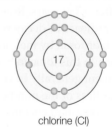
17
chlorine (Cl)

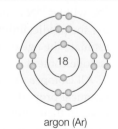
18
argon (Ar)

Figure 2.5 Animated Shell models. Each circle (shell) represents all orbitals at one energy level. A model is filled with electrons from the innermost shell out, until there are as many electrons as protons. Atoms with vacancies (room for additional electrons) in their outermost shell tend to get rid of them.
Figure It Out: Which of these models have unpaired electrons in their outer shell?

Answer: Hydrogen, carbon, oxygen, sodium, and chlorine

(**Figure 2.5A**). It fills up first. In hydrogen, the simplest atom, one electron occupies that room. Helium, with two protons, has two electrons that fill the room—and the first shell. In larger atoms, more electrons rent the second-floor rooms (**Figure 2.5B**). When the second floor fills, more electrons rent third-floor rooms (**Figure 2.5C**), and so on.

About Vacancies

When an atom's outermost shell is filled with electrons, we say that it has no vacancies. Any atom is in its most stable state when it has no vacancies. Helium, neon, and argon are examples of elements with no vacancies. Atoms of these elements are chemically stable, which means they have no tendency to interact with other atoms. Thus, these elements occur most frequently in nature as solitary atoms.

By contrast, when an atom's outermost shell has room for another electron, it has a vacancy. Atoms with vacancies tend to get rid of them by interacting with other atoms; in other words, they are chemically active. For example, the sodium atom (Na) depicted in **Figure 2.5C** has one electron in its outer (third) shell, which can hold eight. With seven vacancies, we can predict that this atom is chemically active.

In fact, this particular sodium atom is not just active, it is extremely so. Why? The shell model shows that a sodium atom has an unpaired electron, but electrons like to occupy orbitals in pairs. Thus, with some exceptions, solitary atoms that have unpaired electrons are not very stable. These atoms are called **free radicals**. Most free radicals have such a strong tendency to interact with other atoms that they exist only briefly in that state. As you will see in later chapters, such interactions make free radicals dangerous to life.

A sodium atom can easily lose its one unpaired electron. After that happens, the atom's second shell —which is full of electrons—is its outermost. Thus, no vacancies remain. This is a very stable configuration for sodium, so it is not surprising that the vast majority of sodium atoms on Earth have 11 protons and 10 electrons. Atoms like this, with an unequal number of protons and electrons, are called **ions**. Ions carry a net (or overall) charge. The negative charge of an elec-

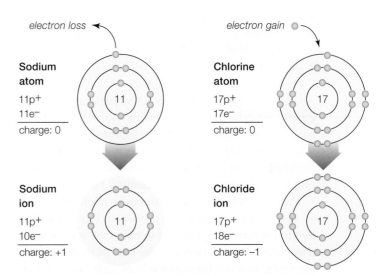

A A sodium atom (Na) becomes a positively charged sodium ion (Na$^+$) when it loses the electron in its third shell. The atom's full second shell is now its outermost, so it has no vacancies.

B A chlorine atom (Cl) becomes a negatively charged chloride ion (Cl$^-$) when it gains an electron and fills the vacancy in its third, outermost shell.

Figure 2.6 Animated Ion formation. **Figure It Out:** Does a chloride ion have an unpaired electron?

Answer: No

tron is the same magnitude as the positive charge of a proton, so the two charges cancel one another. Thus, an atom with the same number of electrons and protons carries no net charge. Sodium ions (Na$^+$) offer an example of how atoms gain a positive charge by losing an electron (**Figure 2.6A**).

Other atoms gain a negative charge by accepting an electron. For example, an uncharged chlorine atom has 17 protons and 17 electrons. Its outer shell can hold eight electrons, but it has only seven. This atom has one vacancy and one unpaired electron, so we can predict—correctly—that it is chemically very active. A chlorine atom easily fills its third shell by accepting an electron. When that happens, the atom becomes a chloride ion (Cl$^-$) with 17 protons, 18 electrons, and a net negative charge (**Figure 2.6B**).

free radical Atom with an unpaired electron.
ion Atom that carries a charge because it has an unequal number of protons and electrons.
shell model Model of electron distribution in an atom.

Take-Home Message

Why do atoms interact?

» An atom's electrons are the basis of its chemical behavior.

» Shells represent all electron orbitals at one energy level in an atom. When the outermost shell is not full of electrons, the atom has a vacancy.

» Atoms with vacancies tend to interact with other atoms.

2.4 Chemical Bonds: From Atoms to Molecules

- Chemical bonds link atoms into molecules.
- The characteristics of a chemical bond arise from the properties of the atoms taking part in it.
- Link to building blocks of life 1.2

An atom can get rid of vacancies by participating in a chemical bond, which is an attractive force that arises between two atoms when their electrons interact. When atoms interact, they often form molecules. A molecule consists of atoms held together in a particular number and arrangement by chemical bonds.

Water is an example of a substance made of molecules. Each water molecule consists of three atoms: two hydrogen atoms bonded to the same oxygen atom (**Figure 2.7**). A water molecule consists of two or more elements, so it is a compound. Other molecules, includ-

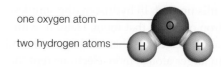

one oxygen atom
two hydrogen atoms

Figure 2.7 The water molecule. Each water molecule has two hydrogen atoms bonded to the same oxygen atom.

ing molecular oxygen (a gas in air), have atoms of one element only.

The term "bond" applies to a continuous range of atomic interactions. However, we can categorize most bonds into distinct types based on their different properties. Which type forms depends on the atoms taking part in the molecule.

Ionic Bonds Two ions may stay together by the mutual attraction of their opposite charges, an association called an ionic bond. Ionic bonds can be quite strong. Ionically bonded sodium and chloride ions make sodium chloride (NaCl), which we know as common table salt. A crystal of this substance consists of a lattice of sodium and chloride ions interacting in ionic bonds (**Figure 2.8A**).

Ions retain their respective charges when participating in an ionic bond (**Figure 2.8B**). Thus, one "end" of an ionic bond has a positive charge, and the other "end" has a negative charge. Any such separation of charge into distinct positive and negative regions is called polarity (**Figure 2.8C**).

A sodium chloride molecule is polar because the chloride ion keeps a very strong hold on its extra electron. In other words, it is strongly electronegative. Electronegativity is a measure of an atom's ability to pull electrons away from another atom. Electronegativity is not the same as charge. Rather, an atom's electronegativity depends on its size, how many vacancies it has, and what other atoms it is interacting with.

An ionic bond is completely polar because the atoms participating in it have a very large difference in electronegativity. When atoms with a lower difference

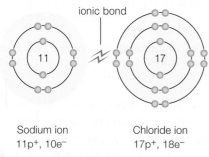

ionic bond

Sodium ion
11p⁺, 10e⁻

Chloride ion
17p⁺, 18e⁻

A The mutual attraction of opposite charges holds a sodium ion and a chloride ion together in an ionic bond.

Cl⁻ Na⁺

B Each crystal of table salt consists of many sodium and chloride ions locked together in a cubic lattice by ionic bonds.

C Ions taking part in an ionic bond retain their charge, so the molecule is polar. One side is positively charged (here represented by a *blue* overlay); the other side is negatively charged (*red* overlay).

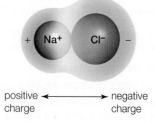

+ Na⁺ Cl⁻ −

positive ←——→ negative
charge   charge

Figure 2.8 Animated An example of ionic bonding: table salt, or NaCl.

chemical bond An attractive force that arises between two atoms when their electrons interact.
compound Molecule that has atoms of more than one element.
covalent bond Chemical bond in which two atoms share a pair of electrons.
electronegativity Measure of the ability of an atom to pull electrons away from other atoms.
ionic bond Chemical bond that consists of a strong mutual attraction between ions of opposite charge.
molecule Group of two or more atoms joined by chemical bonds.
polarity Separation of charge into positive and negative regions.

Molecular hydrogen (H—H)

Two hydrogen atoms, each with one proton, share two electrons in a nonpolar covalent bond.

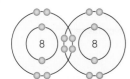

Molecular oxygen (O═O)

Two oxygen atoms, each with eight protons, share four electrons in a double covalent bond.

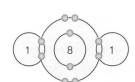

Water molecule (H—O—H)

Two hydrogen atoms share electrons with an oxygen atom in two polar covalent bonds. The oxygen exerts a greater pull on the shared electrons, so it has a slight negative charge. Each hydrogen has a slight positive charge.

Figure 2.9 Animated Covalent bonds, in which atoms fill vacancies by sharing electrons. Two electrons are shared in each covalent bond. When sharing is equal, the bond is nonpolar. When one atom exerts a greater pull on the electrons, the bond is polar.

Table 2.1 Representing Covalent Bonds in Molecules		
Common name	Water	Familiar term.
Chemical name	Dihydrogen monoxide	Describes elemental composition.
Chemical formula	H_2O	Indicates unvarying proportions of elements. Subscripts show number of atoms of an element per molecule. The absence of a subscript means one atom.
Structural formula	H—O—H 	Represents each covalent bond as a single line between atoms. The bond angles may also be represented.
Structural model		Shows relative sizes and positions of atoms in three dimensions.
Shell model		Shows how pairs of electrons are shared in covalent bonds.

in electronegativity interact, they tend to form chemical bonds that are less polar than ionic bonds.

Covalent Bonds In a covalent bond, two atoms share a  pair of electrons, so that each atom's vacancy becomes partially filled (**Figure 2.9**). Sharing electrons links two atoms just as sharing earphones links two friends (*inset*). Covalent bonds can be stronger than ionic bonds, but they are not always so.

Table 2.1 shows different ways of representing covalent bonds. In structural formulas, a line between two atoms represents a single covalent bond, in which two atoms share one pair of electrons. For example, molecular hydrogen (H_2) has one covalent bond between hydrogen atoms (H—H). Two, three, or even four covalent bonds may form between atoms when they share multiple pairs of electrons. For example, two atoms sharing two pairs of electrons are connected by two covalent bonds. Such double bonds are represented by a double line between the atoms. A double bond links the two oxygen atoms in molecular oxygen (O═O). Three lines indicate a triple bond, in which two atoms share three pairs of electrons. A triple covalent bond links the two nitrogen atoms in molecular nitrogen (N≡N).

Double and triple bonds are not represented in structural models, which show positions and relative sizes of the atoms in three dimensions. Covalent bonds are shown as one stick connecting two balls, which represent atoms. Elements are usually coded by color:

carbon   hydrogen   oxygen   nitrogen   phosphorus

Atoms share electrons unequally in a polar covalent bond. A bond between an oxygen atom and a hydrogen atom in a water molecule is an example. The more electronegative atom (the oxygen, in this case) pulls the electrons a little more toward its side of the bond, so that atom bears a slight negative charge. The less electronegative atom (the hydrogen) at the other end of the bond bears a slight positive charge. Covalent bonds in compounds are usually polar. By contrast, atoms participating in a nonpolar covalent bond share electrons equally. There is no difference in charge between the two ends of such bonds. Molecular hydrogen (H_2), oxygen (O_2), and nitrogen (N_2) are examples.

Take-Home Message

How do atoms interact in chemical bonds?

» A chemical bond forms between atoms when their electrons interact. A chemical bond may be ionic or covalent depending on the atoms taking part in it.

» An ionic bond is a strong mutual attraction between two ions of opposite charge.

» Atoms share a pair of electrons in a covalent bond. When the atoms share electrons unequally, the bond is polar.

2.5 Hydrogen Bonds and Water

- The unique properties of liquid water arise because of the water molecule's polarity.
- Extensive hydrogen bonds form among water molecules.

Hydrogen Bonding

Life evolved in water. All living organisms are mostly water, many of them still live in it, and all of life's chemistry is carried out in water. What is so special about water? Water has unique properties that arise from the two polar covalent bonds in each water molecule. Overall, the molecule has no charge, but the oxygen atom carries a slight negative charge, and the hydrogen atoms carry a slightly positive charge. Thus, the molecule itself is polar (**Figure 2.10**).

The polarity of individual water molecules attracts them to one another. The slight positive charge of a hydrogen atom in one water molecule is drawn to the slight negative charge of an oxygen atom in another. This type of interaction is called a hydrogen bond. A hydrogen bond is an attraction between a covalently bonded hydrogen atom and an electronegative atom taking part in a separate polar covalent bond (**Figure 2.11A**). Like ionic bonds, hydrogen bonds form by the mutual attraction of opposite charges. However, unlike ionic bonds, hydrogen bonds do not make molecules out of atoms, so they are not chemical bonds.

Being on the weaker end of the spectrum of atomic interactions, hydrogen bonds form and break much more easily than covalent or ionic bonds. Even so, many of them form, and collectively they are quite strong. As you will see in Chapter 3, hydrogen bonds stabilize the characteristic structures of biological molecules such as DNA and proteins. They also form in tremendous numbers among water molecules (**Figure 2.11B**). The extensive hydrogen bonding among water molecules gives special properties to liquid water.

Water's Special Properties

Water Has Cohesion Molecules of some substances resist separating from one another, and the resistance gives rise to a property called cohesion. Water has cohesion because hydrogen bonds collectively exert a continuous pull on its individual molecules. You can see cohesion in water as surface tension, which means that the surface of liquid water behaves a bit like a sheet of elastic (*inset*).

Figure 2.10 Polarity of the water molecule. Each of the hydrogen atoms in a water molecule bears a slight positive charge (represented by a *blue* overlay). The electronegative oxygen atom carries a slight negative charge (*red* overlay).

Cohesion is a part of many processes that sustain multicelled bodies. As one example, water molecules constantly escape from the surface of liquid water as vapor, a process called evaporation. Evaporation is resisted by hydrogen bonding among water molecules. In other words, overcoming water's cohesion takes energy. Thus, evaporation sucks energy in the form of heat from liquid water, lowering the water's surface temperature. Evaporative water loss can help you and some other mammals cool off when you sweat in hot, dry weather. Sweat, which is about 99 percent water, cools the skin as it evaporates.

Cohesion works inside organisms, too. For instance, plants continually absorb water from soil as they grow. Water molecules evaporate from leaves, and replacements are pulled upward from roots. Cohesion makes it possible for columns of liquid water to rise from roots to leaves inside narrow pipelines of vascular tissue. In some trees, these pipelines extend upward for hundreds of feet. Section 28.4 returns to this topic.

Water Stabilizes Temperature Temperature is a way to measure the energy of molecular motion. All molecules jiggle nonstop, and they jiggle faster as they absorb heat. Hydrogen bonding keeps water molecules from jiggling as much as they would otherwise, so

cohesion Property of a substance that arises from the tendency of its molecules to resist separating from one another.
evaporation Transition of a liquid to a gas.
hydrogen bond Attraction between a covalently bonded hydrogen atom and another atom taking part in a separate covalent bond.
hydrophilic Describes a substance that dissolves easily in water.
hydrophobic Describes a substance that resists dissolving in water.
mixture An intermingling of two or more types of molecules.
salt Ionic compound that releases ions other than H^+ and OH^- when it dissolves in water.
solute A dissolved substance.
solution Homogeneous mixture.
solvent Substance that can dissolve other substances.
temperature Measure of molecular motion.

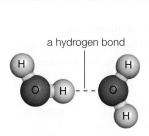

a hydrogen bond

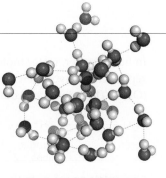

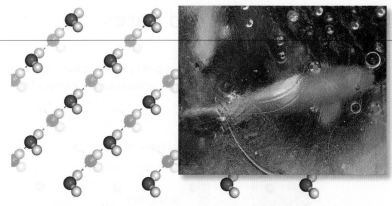

A A hydrogen bond is an attraction between a hydrogen atom and an electronegative atom taking part in a separate polar covalent bond.

B The many hydrogen bonds that form among water molecules impart special properties to liquid water.

C Hydrogen bonds lock water molecules in a rigid lattice in ice. The molecules in this lattice pack less densely than in liquid water, which is why ice floats on water. A covering of ice can insulate water underneath it, thus keeping aquatic organisms from freezing during harsh winters.

Figure 2.11 Animated Hydrogen bonds and water.

it takes more heat to raise the temperature of water compared with other liquids. Temperature stability is an important part of homeostasis, because most of the molecules of life function properly only within a certain range of temperature.

Below 0°C (32°F), water molecules do not jiggle enough to break hydrogen bonds, and they become locked in the rigid, lattice-like bonding pattern of ice (**Figure 2.11C**). Individual water molecules pack less densely in ice than they do in water, so ice floats on water. Sheets of ice that form on the surface of ponds, lakes, and streams can insulate the water under them from subfreezing air temperatures. Such "ice blankets" protect aquatic organisms during cold winters.

Water Is an Excellent Solvent The polarity of the water molecule and its ability to form hydrogen bonds make water an excellent solvent, which means that many other substances easily dissolve in it. Substances that dissolve easily in water are hydrophilic (water-loving). Ionic solids such as sodium chloride (NaCl) dissolve in water because the slight positive charge on each hydrogen atom in a water molecule attracts negatively charged ions (Cl⁻), and the slight negative charge on the oxygen atom attracts positively charged ions (Na⁺). Hydrogen bonds among water molecules are collectively stronger than ionic bonds, so the solid dissolves as water molecules tug apart and surround the ions:

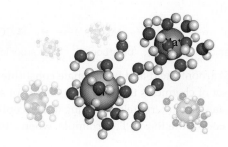

Sodium chloride is called a salt because it releases ions other than H⁺ and OH⁻ when it dissolves in water. When a substance such as NaCl dissolves, its component ions disperse among the molecules of liquid, and it becomes a solute. A homogeneous mixture such as salt dissolved in water is a solution. Chemical bonds do not form between molecules of solute and solvent, so the proportions of the two substances in a solution can vary. A mixture is an intermingling of substances.

Nonionic solids such as sugars dissolve easily in water because their molecules can form hydrogen bonds with water molecules. Hydrogen bonding dissolves a solid substance by pulling its molecules away from one another and keeping them apart.

Water does not interact with hydrophobic (water-dreading) substances such as oils. Oils consist of nonpolar molecules, and hydrogen bonds do not form between nonpolar molecules and water. When you mix oil and water, the water breaks into small droplets, but quickly begins to cluster into larger drops as new hydrogen bonds form among its molecules. The bonding excludes molecules of oil and pushes them together into drops that rise to the surface of the water. The same interactions occur at the thin, oily membrane that separates the watery fluid inside cells from the watery fluid outside of them. As you will see in Chapter 4, the organization of membranes—and of life—starts with such interactions.

Take-Home Message

What gives water the special properties that make life possible?

» Extensive hydrogen bonding among water molecules arises from the polarity of the individual molecules.

» Hydrogen bonding among water molecules imparts cohesion to liquid water, and the ability to stabilize temperature and dissolve many substances.

2.6 Acids and Bases

- pH is a measure of the number of hydrogen ions in a fluid.
- Most biological processes occur within a narrow range of pH, typically around pH 7.
- Link to Homeostasis 1.3

more acidic

0	battery acid
1	gastric fluid
2	acid rain lemon juice cola vinegar
3	orange juice tomatoes, wine bananas
4	
5	beer bread black coffee urine, tea, typical rain
6	corn butter milk
7	pure water
8	blood, tears egg white seawater
9	baking soda detergents Tums
10	toothpaste hand soap milk of magnesia
11	household ammonia
12	hair remover
13	bleach oven cleaner
14	drain cleaner

more basic

Figure 2.12 A pH scale. Here, *red* dots signify hydrogen ions (H^+) and *blue* dots signify hydroxyl ions (OH^-). Also shown are approximate pH values for some common solutions.

This pH scale ranges from 0 (most acidic) to 14 (most basic). A change of one unit on the scale corresponds to a tenfold change in the amount of H^+ ions.

Figure It Out: What is the approximate pH of cola?

Answer: 2.5

In liquid water, water molecules spontaneously separate into hydrogen ions (H^+) and hydroxide ions (OH^-). These ions can combine again to form water:

$$H_2O \longrightarrow H^+ + OH^- \longrightarrow H_2O$$

water   hydrogen   hydroxide   water
         ions   ions

Concentration refers to the amount of a particular solute that is dissolved in a given volume of fluid. Hydrogen ion concentration is a special case. We measure the number of hydrogen ions in a solution using a value called pH. When the number of H^+ ions is the same as the number of OH^- ions, the pH of the solution is 7, or neutral. The pH of pure water (but not rainwater or seawater) is 7. The higher the number of hydrogen ions, the lower the pH. A one-unit decrease in pH corresponds to a tenfold increase in the number of H^+ ions, and a one-unit increase corresponds to a tenfold decrease in the number of H^+ ions (**Figure 2.12**).

One way to get a sense of pH is to taste dissolved baking soda (pH 9), distilled water (pH 7), and lemon juice (pH 2). Nearly all of life's chemistry occurs near pH 7. Most of your body's internal environment (tissue fluids and blood) stays between pH 7.3 and 7.5.

Substances called acids give up hydrogen ions when they dissolve in water, so these substances can lower the pH of a solution and make it acidic (below pH 7). Bases accept hydrogen ions when they dissolve in water, so they can raise the pH of a solution and make it basic, or alkaline (above pH 7).

Strong acids ionize completely in water to give up all of their H^+ ions; weak acids give up only some of them. Strong bases can remove all of the H^+ ions from a weak acid. Hydrochloric acid (HCl) is an example of a strong acid: its H^+ and Cl^- ions stay separated in water. Inside your stomach, the H^+ from HCl makes gastric fluid acidic (pH 1–2). Carbonic acid is an example of a weak acid. It forms when carbon dioxide gas dissolves in the fluid portion of human blood:

$$H_2O + CO_2 \longrightarrow H_2CO_3$$

        carbon dioxide   carbonic acid

Carbonic acid dissociates into a hydrogen ion and a bicarbonate ion, which in turn recombine to form carbonic acid:

$$H_2CO_3 \longrightarrow H^+ + HCO_3^- \longrightarrow H_2CO_3$$

carbonic acid   bicarbonate   carbonic acid

Together, carbonic acid and bicarbonate constitute a buffer, a set of chemicals that can keep the pH of a solution stable by alternately donating and accepting ions that contribute to pH. For example, when a base

Mercury Rising (revisited)

All ecosystems now have detectable effects of air pollution, but many of those effects are not as well understood as acid rain. We do know that the amount of mercury in Earth's waters is rising. The concentration of mercury in the Pacific Ocean is predicted to double within forty years. We also know that this rise is occurring as a consequence of human activities that release mercury into the atmosphere. In 2009 alone, we released more than 2,500 tons of mercury worldwide.

All human bodies, too, now have detectable amounts of mercury. Some of it comes from dental fillings, imported skin-bleaching cosmetics, and broken fluorescent lamps. However, most comes from eating contaminated fish and shellfish. Tuna harvested from the Pacific Ocean accounts for almost half of the mercury in the average human body.

How would you vote? An advisory from the U.S. Food and Drug Administration warns children, and women who are pregnant, nursing, or planning to conceive, to avoid eating tuna because of high mercury levels. According to the FDA, a healthy, average-sized adult may safely consume up to 14 micrograms per day of mercury. A typical 6-ounce can of albacore tuna contains about 60 micrograms. Canned tuna companies say that most of that mercury comes from natural sources, and consider the federal advisory to be sufficient. Do you think cans of tuna should carry labels warning of mercury content?

is added to an unbuffered fluid, the number of OH⁻ ions increases, so the pH rises. However, if the fluid is buffered, the addition of base causes the buffer to release H^+ ions. These combine with OH^- ions to form water, which has no effect on pH. Excess hydrogen ions combine with the buffer, so they do not contribute to pH. Thus, the pH of a buffered fluid stays the same when base or acid is added.

Under normal circumstances, the fluids inside cells (as well as those inside bodies) stay within a consistent range of pH because they are buffered. For example, excess OH^- in blood combines with the H^+ from carbonic acid to form water, which does not contribute to pH. Excess H^+ in blood combines with bicarbonate, so it does not affect pH. This exchange of ions keeps the blood pH stable, but only up to a certain point. A buffer can neutralize only so many ions. Even slightly more than that limit and the pH of the fluid will change dramatically.

Most biological molecules can function properly only within a narrow range of pH. Even a slight deviation from that range can halt cellular processes, so buffer failure can be catastrophic in a biological system. For instance, when breathing is impaired suddenly, carbon dioxide gas accumulates in tissues, and too much carbonic acid forms in blood. The resulting decline in blood pH may cause the person to enter a coma (a dangerous level of unconsciousness). By contrast, hyperventilation (sustained rapid breathing) causes the body to lose too much CO_2. The loss results in a rise in blood pH. If blood pH rises too much, prolonged muscle spasm (tetany) or coma may occur.

Burning fossil fuels such as coal releases sulfur and nitrogen compounds that affect the pH of rain and other forms of precipitation. Water is not buffered, so the addition of acids or bases has a dramatic effect. In places with a lot of fossil fuel emissions, the rain and fog can be more acidic than vinegar. The corrosive effects of this acid rain is visible in urban areas (*right*). Acid rain also drastically changes the pH of water in soil, lakes, and streams. Such changes can overwhelm the buffering capacity of fluids inside organisms, with lethal effects. We return to acid rain in Section 48.5.

acid Substance that releases hydrogen ions in water.
base Substance that accepts hydrogen ions in water.
buffer system Set of chemicals that can keep the pH of a solution stable by alternately donating and accepting ions that contribute to pH.
concentration The number of molecules or ions per unit volume of a solution.
pH A measure of the number of hydrogen ions in a fluid.

Take-Home Message

Why are hydrogen ions important in biological systems?

» pH reflects the number of hydrogen ions in a fluid. Most biological systems function properly only within a narrow range of pH.

» Acids release hydrogen ions in water; bases accept them. Salts release ions other than H^+ and OH^-.

» Buffers help keep pH stable. Inside organisms, they are part of homeostasis.

3 Molecules

LEARNING ROADMAP

Where you have been Having learned about atomic interactions (Section 2.3), you are now in a position to understand the structure of the molecules of life. Keep the big picture in mind by reviewing Section 1.2. You will be building on your knowledge of covalent bonding (2.4), acids and bases (2.6), and the effects of hydrogen bonds (2.5).

Where you are now

Structure Dictates Function
Cells build complex carbohydrates and lipids, proteins, and nucleic acids. These organic compounds have functional groups attached to a backbone of carbon atoms.

Carbohydrates
Cells use carbohydrates as structural materials, for fuel, and to store and transport energy. Cells build different kinds from the same sugar monomers, bonded in different patterns.

Lipids
Lipids are the main structural component of all cell membranes. Cells use them to make other compounds, to store energy, and as waterproofing or lubricating substances.

Proteins
Proteins are the most diverse molecules of life. They include enzymes and structural materials. A protein's function arises from and depends on its structure.

Nucleotides and Nucleic Acids
Nucleotides are building blocks of nucleic acids; some have additional roles in metabolism. DNA and RNA are part of a cell's system of storing and retrieving heritable information.

Where you are going In Chapter 4, you will read more about how lipids and proteins are the structural and functional parts of cell membranes. Chapter 5 explores enzymes and metabolic reactions, including phosphate-group transfers by nucleotides. Chapter 7 details how cells break down carbohydrates for energy. Chapter 8 revisits DNA structure and function, and Chapter 9 returns to protein synthesis. You will see in Chapter 10 how gene expression can be influenced by functional groups and RNA.

3.1 Fear of Frying

The human body requires only about a tablespoon of fat each day to stay healthy, but most people in developed countries eat far more than that. The average American eats about 70 pounds of fat per year, which may be part of the reason why the average American is overweight. Being overweight increases one's risk for many chronic illnesses. However, the total quantity of fat in the diet may be less important than the types of fats. Fats are more than inert molecules that accumulate in strategic areas of our bodies. They are major constituents of cell membranes, and as such they have powerful effects on cell function.

The typical fat molecule has three fatty acids, each with a long chain of carbon atoms that can vary a bit in structure. Fats with a certain arrangement of hydrogen atoms around those carbon chains are called *trans* fats (**Figure 3.1**). Small amounts of *trans* fats occur naturally in red meat and dairy products. However, most of the *trans* fats that humans eat come from partially hydrogenated vegetable oil, an artificial food product.

Hydrogenation is a manufacturing process that adds hydrogen atoms to vegetable oils in order to change them into solid fats. Procter & Gamble Co. developed partially hydrogenated soybean oil in 1908 as a substitute for the more expensive solid animal fats they had been using to make candles. However, the demand for candles began to wane as more households in the United States became wired for electricity, and P & G began to look for another way to sell its proprietary fat. Partially hydrogenated vegetable oil looks a lot like lard, so in 1911 the company began aggressively marketing it as a revolutionary new food: a solid cooking fat with a long shelf life, mild flavor, and lower cost than lard or butter.

By the mid-1950s, hydrogenated vegetable oil had become a major part of the American diet. It was (and still is) found in many manufactured and fast foods: french fries, butter substitutes, cookies, crackers, cakes and pancakes, peanut butter, pies, doughnuts, muffins, chips, granola bars, breakfast bars, chocolate, microwave popcorn, pizzas, burritos, chicken nuggets, fish sticks, and so on.

For decades, hydrogenated vegetable oil was considered more healthy than animal fats because it was made from plants, but we now know otherwise. The *trans* fats in hydrogenated vegetable oils raise the level of cholesterol in our blood more than any other fat, and they directly alter the function of our arteries and veins. The effects of such changes are quite serious. Eating as little as 2 grams a day of hydrogenated vegetable oils increases a person's risk of atherosclerosis

(hardening of the arteries), heart attack, and diabetes. A small serving of french fries made with hydrogenated vegetable oil contains about 5 grams of *trans* fat.

All organisms consist of the same kinds of molecules, but small differences in the way those molecules are put together can have big effects in a living organism. With this concept, we introduce you to the chemistry of life. This is your chemistry. It makes you far more than the sum of your body's molecules.

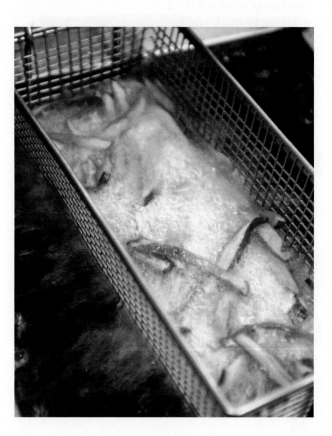

A Oleic acid, a *cis* fatty acid.

B Elaidic acid, a *trans* fatty acid.

Figure 3.1 *Trans* fats, an unhealthy food. A fat has three fatty acid tails, each a long carbon chain. Hydrogenation changes the arrangement of hydrogen atoms around double bonds in those chains from (**A**) a *cis* configuration to (**B**) a *trans* configuration. This small difference in structure makes a big difference in our bodies.

3.2 Organic Molecules

- All of the molecules of life are built with carbon atoms.
- We use different models to highlight different aspects of the same molecule.
- Links to Elements 2.2, Representing Molecules 2.4

Carbon—The Stuff of Life

Living things are mainly oxygen, hydrogen, and carbon. Most of the oxygen and hydrogen are in the form of water. Put water aside, and carbon makes up more than half of what is left.

The carbon in living organisms is part of the molecules of life—complex carbohydrates, lipids, proteins, and nucleic acids. These molecules consist primarily of hydrogen and carbon atoms, so they are said to be organic. The term is a holdover from a time when such molecules were thought to be made only by living things, as opposed to the "inorganic" molecules that formed by nonliving processes. The term persists, even though we now know that organic compounds were present on Earth long before organisms were, and we can also make them synthetically in laboratories.

Carbon's importance to life starts with its versatile bonding behavior. Each carbon atom can form covalent bonds with one, two, three, or four other atoms. Depending on the other elements in the resulting molecule, such bonds may be polar or nonpolar. Many organic compounds have a backbone—a chain of carbon atoms—to which other atoms attach. The ends of a backbone may join so that the carbon chain forms one or more ring structures (**Figure 3.2**). Such versatility means that carbon atoms can be assembled and remodeled into a variety of organic compounds.

Representing Structures of Organic Molecules

As you will see in the next few sections, the function of an organic molecule depends on its structure.

A Carbon's versatile bonding behavior allows it to form a variety of structures, including rings.

B Carbon rings form the framework of many sugars, starches, and fats, such as those found in doughnuts.

Figure 3.2 Carbon rings.

Researchers routinely make models of organic molecules such as proteins to study structure–function relationships, surface properties, changes during synthesis or biochemical processes, and molecular recognition. A molecule's structure can be modeled in various ways. The different models allow us to visualize different characteristics of the same molecule.

Structural formulas of organic molecules can be quite complex, even when the molecules are relatively small (**Figure 3.3A**). Thus, formulas of organic molecules are typically simplified. Some of the bonds may not be shown, but they are implied. Hydrogen atoms bonded to a carbon backbone may also be omitted, and other atoms as well. Carbon ring structures such as the ones that occur in glucose and other sugars are often represented as polygons. If no atom is shown at a corner or at the end of a bond, a carbon atom is implied there (**Figure 3.3B**).

Ball-and-stick models show the positions of individual atoms in three dimensions (**Figure 3.3C**). Single, double, and triple covalent bonds are all shown as one stick connecting two balls, which represent atoms. Ball size reflects relative sizes of the atoms, and ball color indicates the element according to a standard code. The most common color scheme shows carbon

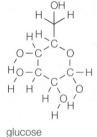 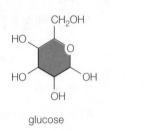  

glucose · glucose · glucose · glucose · glucose

A A structural formula for an organic molecule can be very complicated.

B Structural formulas of organic molecules are often simplified by using polygons as symbols for rings, and omitting the labels for some atoms.

C A ball-and-stick model shows the arrangement of atoms in three dimensions.

D A space-filling model shows a molecule's overall shape.

Figure 3.3 Modeling an organic molecule.

in black, oxygen in red, hydrogen in white, nitrogen in blue, and sulfur in yellow.

Space-filling models represent atomic volume most accurately (**Figure 3.3D**). This type of model shows the overall shape of an organic molecule. Atoms in space-filling models may be color-coded by element using the same scheme as ball-and-stick models.

Many organic molecules are so large that ball-and-stick or space-filling models of them may be incomprehensible. **Figure 3.4** shows three different ways to represent the same molecule, hemoglobin, a very large protein that colors your blood red. Hemoglobin transports oxygen to tissues throughout the body of all vertebrates (animals that have a backbone). A ball-and-stick or space-filling model of hemoglobin is so complicated that many interesting features of this molecule are obscured (**Figure 3.4A**).

To reduce visual complexity, other types of molecular models do not depict individual atoms. Surface models of large molecules can reveal large-scale features, such as folds or pockets, that can be difficult to see when individual atoms are shown. For example, in the surface model of hemoglobin in **Figure 3.4B**, you can see folds of the molecule that cradle hemes. Hemes are carbon ring structures that often have an iron atom at their center. They are part of many important proteins that you will encounter in this book.

Very large molecules such as hemoglobin are often shown as ribbon models. Such models highlight different features of the structure, such as coils or sheets. In a ribbon model of hemoglobin (**Figure 3.4C**), you can see that the protein consists of four coiled components, each of which folds around a heme.

Such structural details are clues to how a molecule functions. For example, hemoglobin, which is the main oxygen-carrier in vertebrate blood, has four hemes. Oxygen binds at the hemes, so each hemoglobin molecule can carry up to four molecules of oxygen.

organic Describes a compound that consists primarily of carbon and hydrogen atoms.

A A space-filling model of hemoglobin.

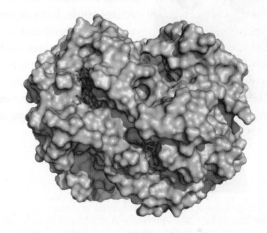

B A surface model of the same molecule reveals crevices and folds that are important for its function. Heme groups, in *red*, are cradled in pockets of the molecule.

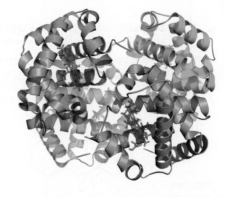

C A ribbon model of hemoglobin shows all four heme groups, also in *red*, held in place by the molecule's coils.

Figure 3.4 Visualizing the structure of hemoglobin, the oxygen-transporting molecule in red blood cells. Models that show individual atoms usually depict them color-coded by element. Other models may be shown in various colors, depending on which features are being studied.

Take-Home Message

How are all of the molecules of life alike?

» The molecules of life (carbohydrates, lipids, proteins, and nucleic acids) are organic, which means they consist mainly of carbon and hydrogen atoms.

» The structure of an organic molecule starts with its carbon backbone, a chain of carbon atoms that may form a ring.

» We use different models to represent different characteristics of a molecule's structure. Considering a molecule's structural features gives us insight into how it functions.

3.3 Molecules of Life—From Structure to Function

■ The function of organic molecules in biological systems begins with their structure.

■ Links to Ions 2.3, Polarity 2.4, Acids and bases 2.6

All biological systems are based on the same organic molecules, a similarity that is one of many legacies of life's common origin. However, the details of those molecules differ among organisms. Just as atoms bonded in different numbers and arrangements form different molecules, simple organic building blocks bonded in different numbers and arrangements form different versions of the molecules of life. Cells assemble complex carbohydrates, lipids, proteins, and nucleic acids from small organic molecules. These small organic molecules—simple sugars, fatty acids, amino acids, and nucleotides—are called monomers when they are used as subunits of larger molecules. Molecules that consist of multiple monomers are called polymers. The remainder of this chapter introduces the different types of biological molecules and the monomers from which they are built.

What Cells Do to Organic Compounds

Cells build polymers from monomers, and break down polymers to release monomers. Any process in which a molecule changes is called a reaction. Reactions that run constantly inside cells help them stay alive, grow, and reproduce. Metabolism refers to reactions and all other activities by which cells acquire and use energy as they make and break apart organic compounds (Figure 3.5A). Metabolism requires enzymes, which are organic molecules (usually proteins) that speed up reactions without being changed by them.

Many metabolic reactions build large organic molecules from smaller ones. With condensation, an enzyme covalently bonds two molecules together. Water (H—O—H) usually forms as a product of condensation when a hydroxyl group (—OH) from one of the molecules combines with a hydrogen atom (—H) from the other molecule (Figure 3.5B). Hydrolysis, which is the reverse of condensation, breaks apart large organic molecules into smaller ones (Figure 3.5C). Hydrolysis enzymes break a bond by attaching a hydroxyl group to one atom and a hydrogen atom to the other. The —OH and —H come from a water molecule, so this reaction requires water. We will revisit enzymes and metabolic reactions in Chapter 5.

Functional Groups

An organic molecule that consists only of hydrogen and carbon atoms is called a hydrocarbon. Methane, the simplest kind, is one carbon atom bonded to four hydrogen atoms. We use larger kinds for fuel. Hydrocarbons are generally nonpolar. Functional groups, which are clusters of atoms covalently bonded to a carbon atom of an organic molecule, impart additional chemical properties such as polarity or acidity.

All of the molecules of life have at least one functional group. The chemical behavior of these molecules arises mainly from the number, kind, and arrangement of their functional groups. Table 3.1 lists a few functional groups that are common in biological molecules.

The hydroxyl group mentioned earlier is an example of a functional group. Hydroxyl groups add polar character to organic compounds such as alcohols and

A Metabolism refers to processes by which cells acquire and use energy as they make and break down molecules. Humans and other consumers break down the molecules in food. They use energy and raw materials from the breakdown to maintain themselves and to build new components.

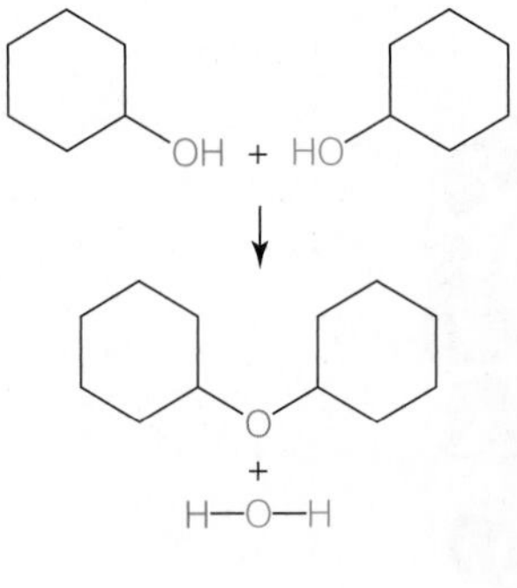

B Condensation. Cells build a large molecule from smaller ones by this reaction.

An enzyme removes a hydroxyl group from one molecule and a hydrogen atom from another. A covalent bond forms between the two molecules, and water also forms.

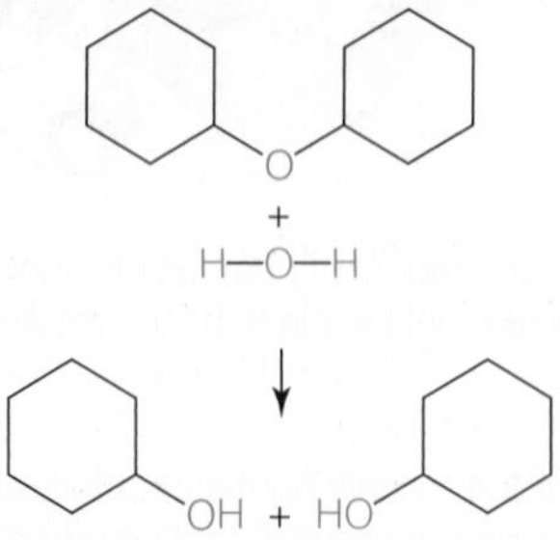

C Hydrolysis. Cells split a large molecule into smaller ones by this water-requiring reaction.

An enzyme attaches a hydroxyl group and a hydrogen atom (both from water) at the cleavage site.

Figure 3.5 **Animated** Metabolism. Two common reactions by which cells build and break down organic molecules are shown.

simple sugars. Methyl groups add nonpolar character. Adding a methyl group to a molecule reduces the solubility of a compound and may dampen the character of another functional group. Methyl groups added to DNA act like an "off" switch for this molecule; acetyl groups act like an "on" switch. Acetyl groups also carry two carbons from one molecule to another in many metabolic reactions.

Aldehyde and ketone groups are part of simple sugars. Some sugars convert to a ring form when the highly reactive aldehyde group on their terminal carbon reacts with a hydroxyl group on another carbon (**Figure 3.6**). Carboxyl groups make amino acids and fatty acids acidic; amine and amide groups make nucleotide bases basic.

Reactions that transfer phosphate groups from one molecule to another also transfer energy. Bonds between sulfhydryl groups stabilize the structure of many proteins, including those that make up human hair. Heat and some kinds of chemicals can temporarily break sulfhydryl bonds, which is why we can curl straight hair and straighten curly hair.

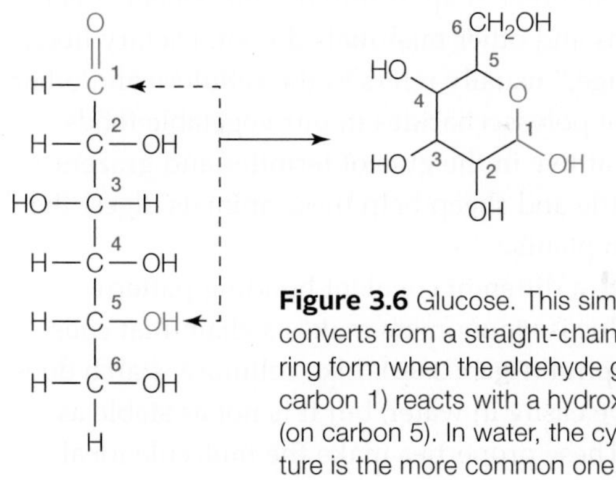

Figure 3.6 Glucose. This simple sugar converts from a straight-chain into a ring form when the aldehyde group (on carbon 1) reacts with a hydroxyl group (on carbon 5). In water, the cyclic structure is the more common one.

condensation Process by which enzymes build large molecules from smaller subunits; water also forms.
enzyme Compound (usually a protein) that speeds up a reaction without being changed by it.
functional group A group of atoms bonded to a carbon of an organic compound; imparts a specific chemical property to the molecule.
hydrocarbon Compound that consists only of carbon and hydrogen atoms.
hydrolysis Process by which an enzyme breaks a molecule into smaller subunits by attaching a hydroxyl group to one part and a hydrogen atom to the other.
metabolism All the enzyme-mediated chemical reactions by which cells acquire and use energy as they build and break down organic molecules.
monomers Molecules that are subunits of polymers.
polymer Molecule that consists of multiple monomers.
reaction Process of molecular change.

Table 3.1 Some Functional Groups in Biological Molecules

Group	Structure	Character	Found in:
acetyl		polar, acidic	some proteins, coenzymes
aldehyde		polar, reactive	simple sugars
amide		weakly basic, stable, rigid	proteins nucleotide bases
amine	—NH₂	very basic	nucleotide bases amino acids
carboxyl		very acidic	fatty acids amino acids
hydroxyl	—OH	polar	alcohols sugars
ketone		polar, acidic	simple sugars nucleotide bases
methyl	—CH₃	nonpolar	fatty acids some amino acids
phosphate		polar, reactive	nucleotides DNA RNA phospholipids proteins
sulfhydryl	—SH	reacts with metals, forms rigid disulfide bonds	cysteine many cofactors

Take-Home Message

How do organic molecules work in living systems?

» All life is based on the same organic compounds: complex carbohydrates, lipids, proteins, and nucleic acids.

» By processes of metabolism, cells assemble these molecules of life from monomers. They also break apart polymers into component monomers.

» Functional groups impart chemical characteristics to organic molecules. Such groups contribute to the function of biological molecules.

» An organic molecule's structure dictates its function in biological systems.

3.4 Carbohydrates

- Carbohydrates are the most plentiful biological molecules.
- Cells use some carbohydrates as structural materials; they use others for fuel, or to store or transport energy.
- Link to Hydrogen bonds 2.5

Carbohydrates are organic compounds that consist of carbon, hydrogen, and oxygen in a 1:2:1 ratio. Cells use different kinds as structural materials, for fuel, and for storing and transporting energy.

Carbohydrates in Biological Systems

Simple Sugars "Saccharide" is from *sacchar*, a Greek word that means sugar. Monosaccharides (one sugar unit) are the simplest type of carbohydrate, but they have extremely important roles as components of larger molecules. Common monosaccharides have a backbone of five or six carbon atoms, one carbonyl group, and two or more hydroxyl groups. Enzymes can easily break the bonds of monosaccharides to release energy (we will return to carbohydrate metabolism in Chapter 7). Monosaccharides are very soluble in water, so they move easily throughout the water-based internal environments of all organisms.

Monosaccharides that are components of the nucleic acids DNA and RNA have five carbon atoms. Glucose has six. Glucose can be used as a fuel to drive cellular processes, or as a structural material to build larger molecules. It can also be used as a precursor, or starting material, that is remodeled into other molecules. For example, cells of plants and many animals make vitamin C from glucose. Human cells are unable to make vitamin C, so we need to get it from our food.

Short-Chain Carbohydrates An oligosaccharide is a short chain of covalently bonded monosaccharides (*oligo*– means a few). Disaccharides consist of two sugar monomers. The lactose in milk is a disaccharide, with one glucose and one galactose unit. Sucrose, the most plentiful sugar in nature, has a glucose and a fructose

unit (**Figure 3.7**). Sucrose extracted from sugarcane or sugar beets is our table sugar. Oligosaccharides with three or more sugar units are often attached to lipids or proteins that have important functions in immunity.

Complex Carbohydrates The "complex" carbohydrates, or polysaccharides, are straight or branched chains of many sugar monomers—often hundreds or thousands of them. There may be one type or many types of monomers in a polysaccharide.

The most common polysaccharides are cellulose, glycogen, and starch. All consist of glucose monomers, but their chemical properties differ substantially. Why? The answer begins with differences in patterns of covalent bonding that link their glucose monomers.

Cellulose, the major structural material of plants, is the most abundant biological molecule in the biosphere. It consists of long, straight chains of glucose monomers. Hydrogen bonds lock the chains into tight, sturdy bundles (**Figure 3.8A**). In plants, these tough cellulose fibers act like reinforcing rods that help stems resist wind and other forms of mechanical stress.

Cellulose does not dissolve in water, and it is not easily broken down. Some bacteria and fungi make enzymes that break it apart into its component sugars, but humans and other mammals do not. Dietary fiber, or "roughage," usually refers to the cellulose and other indigestible polysaccharides in our vegetable foods. Bacteria that live in the guts of termites and grazers such as cattle and sheep help these animals digest the cellulose in plants.

In starch, a different covalent bonding pattern between glucose monomers makes a chain that coils up into a spiral (**Figure 3.8B**). Like cellulose, starch does not dissolve easily in water, but it is not as stable as cellulose. These properties make the molecule ideal for storing chemical energy in the watery, enzyme-filled interior of plant cells. Most plants make much more glucose than they can use. The excess is stored as starch inside cells that make up roots, stems, and

glucose   +   fructose   ⟶   sucrose   +   water

Figure 3.7 The synthesis of a sucrose molecule is an example of a condensation reaction. You are already familiar with sucrose—it is common table sugar.

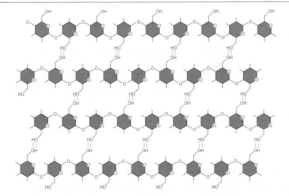

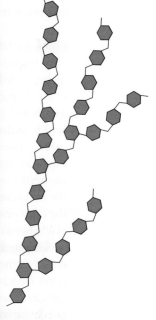

A Cellulose, a structural component of plants. Chains of glucose units stretch side by side and hydrogen-bond at many —OH groups. The hydrogen bonds stabilize the chains in tight bundles that form long fibers. Very few types of organisms can digest this tough, insoluble material.

B In amylose, one type of starch, a series of glucose units form a chain that coils. Starch is the main energy reserve in plants, which store it in their roots, stems, leaves, fruits, and seeds (such as coconuts).

C Glycogen. In humans and other animals, this polysaccharide functions as an energy reservoir. It is stored in muscles and in the liver.

Figure 3.8 Structure of (**A**) cellulose, (**B**) starch, and (**C**) glycogen, and their typical locations in a few organisms. All three carbohydrates consist only of glucose units, but the different bonding patterns that link the subunits result in substances with very different properties.

leaves. However, because it is insoluble, starch cannot be transported out of the cells and distributed to other parts of the plant. When sugars are in short supply, hydrolysis enzymes break the bonds between starch's monomers to release glucose subunits. Humans also have enzymes that hydrolyze starch, so this carbohydrate is an important component of our food.

The covalent bonding pattern in glycogen forms highly branched chains of glucose monomers (**Figure 3.8C**). In animals, glycogen is the sugar-storage equivalent of starch in plants. Muscle and liver cells store it to meet a sudden need for glucose. These cells break down glycogen to release its glucose subunits.

carbohydrate Molecule that consists primarily of carbon, hydrogen, and oxygen atoms in a 1:2:1 ratio.
cellulose Polysaccharide; major structural material in plants.
disaccharide Polymer of two sugar subunits.
glycogen Polysaccharide; energy reservoir in animal cells.
monosaccharide Simple sugar; monomer of polysaccharides.
polysaccharide Polymer of many monosaccharides.
starch Polysaccharide; energy reservoir in plant cells.

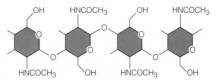

Figure 3.9 Chitin. This polysaccharide strengthens the hard parts of many small animals, such as crabs.

Chitin is a polysaccharide similar to cellulose. Its monomers are glucose with a nitrogen-containing carbonyl group (**Figure 3.9**). Long, unbranching chains of these monomers are linked by hydrogen bonds. As a structural material, chitin is durable, translucent, and flexible. It strengthens hard parts of many animals, including the outer cuticle of crabs, beetles, and ticks, and it reinforces the cell wall of many fungi.

Take-Home Message

What are carbohydrates?

» Simple carbohydrates (sugars), bonded together in different ways, form various types of complex carbohydrates.

» Cells use carbohydrates for energy or as structural materials.

3.5 Greasy, Oily—Must Be Lipids

■ Triglycerides, phospholipids, waxes, and steroids are lipids common in biological systems.

Lipids are fatty, oily, or waxy organic compounds. They vary in structure, but all are hydrophobic. Many lipids incorporate fatty acids, which are small organic molecules that consist of a hydrocarbon "tail" topped with a carboxyl group "head" (Figure 3.10). The tail of a fatty acid is hydrophobic (hence the name "fatty"), but the carboxyl group (the "acid" part of the name) makes the head hydrophilic. You are already familiar with the properties of fatty acids because these molecules are the main component of soap. The hydrophobic tails of the fatty acids in soap attract oily dirt, and the hydrophilic heads dissolve the dirt in water.

Saturated fatty acids have only single bonds in their tails. In other words, their carbon chains are fully saturated with hydrogen atoms (Figure 3.10A). Saturated fatty acid tails are flexible and they wiggle freely. The tails of unsaturated fatty acids have one or more double bonds that limit their flexibility (Figure 3.10B,C). Figure 3.1 shows how these bonds are *cis* or *trans*, depending on the way the hydrogens are arranged around them.

Lipids in Biological Systems

Fats The carboxyl group of a fatty acid easily forms bonds with other molecules. Fats are lipids with one, two, or three fatty acids bonded to glycerol, which is an alcohol. A fatty acid attaches to a glycerol via its

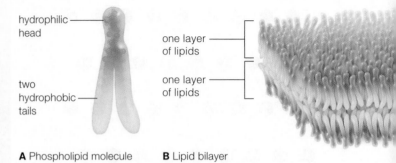

A Phospholipid molecule   **B** Lipid bilayer

Figure 3.12 Phospholipids as components of cell membranes. A double layer of phospholipids—the lipid bilayer—is the structural foundation of all cell membranes.

carboxyl group head. When it does, the fatty acid loses its hydrophilic character. When three fatty acids attach to a glycerol, the resulting molecule, which is called a triglyceride, is entirely hydrophobic (Figure 3.11A).

Because they are hydrophobic, triglycerides do not dissolve easily in water. Most "neutral" fats, such as butter and vegetable oils, are examples. Triglycerides are the most abundant and richest energy source in vertebrate bodies. They are concentrated in adipose tissue that insulates and cushions body parts.

Animal fats are saturated, which means they consist mainly of triglycerides with three saturated fatty acid tails. Saturated fats tend to remain solid at room temperature because their floppy saturated tails can pack tightly together. Most vegetable oils are unsaturated, which means these fats consist mainly of triglycerides with one or more unsaturated fatty acid tails. Each

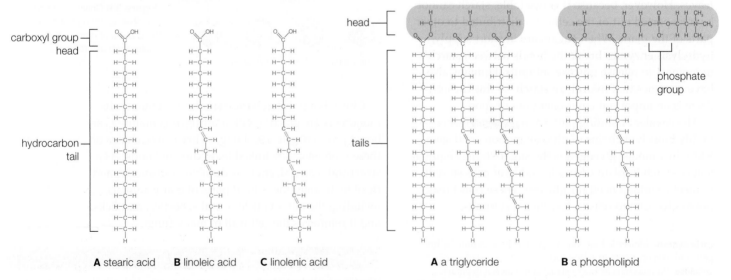

A stearic acid   **B** linoleic acid   **C** linolenic acid

Figure 3.10 Fatty acids. (**A**) The tail of stearic acid is fully saturated with hydrogen atoms. (**B**) Linoleic acid, with two double bonds, is unsaturated. The first double bond occurs at the sixth carbon from the end of the tail, so linoleic acid is called an omega-6 fatty acid. Omega-6 and (**C**) omega-3 fatty acids are "essential fatty acids": Your body does not make them, so they must come from food.

A a triglyceride   **B** a phospholipid

Figure 3.11 Animated Lipids with fatty acid tails. (**A**) Fatty acid tails of a triglyceride are attached to a glycerol head. (**B**) Fatty acid tails of a phospholipid are attached to a glycerol head with a phosphate group. **Figure It Out: Is the triglyceride saturated or unsaturated?** Answer: Unsaturated

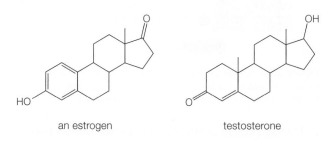

an estrogen

testosterone

female
wood duck

male
wood duck

Figure 3.13 Estrogen and testosterone, steroid hormones that cause different traits to arise in males and females of many species such as wood ducks (*Aix sponsa*), pictured at *right*.

double bond in a fatty acid tail makes a rigid kink. Kinky tails do not pack tightly, so unsaturated fats are typically liquid at room temperature. The partially hydrogenated vegetable oils that you learned about in Section 3.1 are an exception. They are solid at room temperature. The special *trans* double bond keeps fatty acid tails straight, allowing them to pack tightly just like saturated fats do.

Phospholipids A phospholipid has two fatty acid tails and a head that contains a phosphate group (**Figure 3.11B**). The tails are hydrophobic, but the highly polar phosphate group makes the head very hydrophilic. The opposing properties of a phospholipid molecule give rise to cell membrane structure. Phospholipids are the most abundant lipids in cell membranes, which have two layers of lipids (**Figure 3.12**). The heads of one layer are dissolved in the cell's watery interior, and the heads of the other layer are dissolved in the cell's fluid surroundings. In such lipid bilayers, all of the hydrophobic tails are sandwiched between the hydrophilic heads. You will read more about the structure of cell membranes in Chapters 4 and 5.

Waxes A wax is a complex, varying mixture of lipids with long fatty acid tails bonded to long-chain alcohols or carbon rings. The molecules pack tightly, so the resulting substance is firm and water-repellent. Plant leaves secrete a layer of waxes that helps restrict water loss and keeps out parasites and other pests. Secreted waxes also protect, lubricate, and soften our skin and hair. Waxes, together with fats and fatty acids, make feathers waterproof. Bees store honey and raise new generations of bees inside honeycomb made from wax that they secrete.

Steroids All eukaryotic cell membranes contain steroids, lipids with a rigid backbone of four carbon rings and no fatty acid tails. Cholesterol, the most common steroid in animal cell membranes, also serves as a precursor for bile salts (which help digest fats), vitamin D (required to keep teeth and bones strong), and steroid hormones. Estrogens and testosterone, hormones that govern reproduction and the development of sexual traits, are steroid hormones (**Figure 3.13**).

fat Lipid that consists of a glycerol molecule with one, two, or three fatty acid tails.
fatty acid Organic compound that consists of a chain of carbon atoms with an acidic carboxyl group at one end. Carbon chain of saturated types has single bonds only; that of unsaturated types has one or more double bonds.
lipid Fatty, oily, or waxy organic compound.
lipid bilayer Double layer of lipids arranged tail-to-tail; structural foundation of all cell membranes.
phospholipid A lipid with a phosphate group in its hydrophilic head, and two nonpolar fatty acid tails; main constituent of eukaryotic cell membranes.
saturated fatty acid Fatty acid that contains no carbon–carbon double bonds.
steroid Type of lipid with four carbon rings and no fatty acid tails.
triglyceride A fat with three fatty acid tails.
unsaturated fatty acid Fatty acid that has one or more carbon–carbon double bonds in its tail.
wax Water-repellent mixture of lipids with long fatty acid tails bonded to long-chain alcohols or carbon rings.

Take-Home Message

What are lipids?

» Lipids are fatty, waxy, or oily organic compounds. Common types include fats, phospholipids, waxes, and steroids.

» Triglycerides are lipids that serve as energy reservoirs in vertebrate animals.

» Phospholipids are the main lipid component of cell membranes.

» Waxes are lipid components of water-repelling and lubricating secretions.

» Steroids are lipids that occur in cell membranes. Some are remodeled into other molecules.

3.6 Proteins—Diversity in Structure and Function

- All cellular processes involve proteins, the most diverse biological molecules.
- Cells build thousands of different proteins by stringing together amino acids in different orders.
- Link to Covalent bonding 2.4

Of all biological molecules, proteins are the most diverse in both structure and function. Structural proteins support cell parts and, as part of tissues, multicelled bodies. Feathers, hooves, and hair, as well as bones and other body parts, consist mainly of structural proteins. A tremendous number of different proteins, including some structural types, are active participants in all processes that sustain life. Most enzymes that drive metabolic reactions are proteins. Proteins move substances, help cells communicate, and defend the body.

From Structure to Function

Amino Acids Cells can make all of the thousands of different kinds of proteins they need from only twenty kinds of monomers called amino acids. Proteins are polymers of amino acids. An amino acid is a small organic compound with an amine group, a carboxyl group (the acid), and one or more atoms called an "R group." In most amino acids, all three groups are attached to the same carbon atom (**Figure 3.14**).

Peptide Bonds Protein synthesis involves covalently bonding amino acids into a chain. For each type of protein, instructions coded in DNA specify the order of the component amino acids.

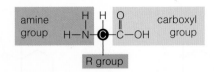

Figure 3.14 Generalized structure of amino acids. The complete structures of the twenty most common amino acids found in eukaryotic proteins are shown in Appendix I.

During protein synthesis, a condensation reaction links the amine group of one amino acid to the carboxyl group of the next. The bond that forms between the two amino acids is called a peptide bond (**Figure 3.15**). Enzymes repeat this bonding process hundreds or thousands of times, so a long chain of amino acids (a polypeptide) forms (**Figure 3.16**). The linear sequence of amino acids in the polypeptide is called the protein's primary structure ❶. You will learn more about protein synthesis in Chapter 9.

The Structure–Function Relationship One of the fundamental ideas in biology is that structure dictates function. This idea is particularly appropriate as applied to proteins, because a protein's biological activity arises from and depends on its shape.

There are several levels of protein structure beyond amino acid sequence. Even before a polypeptide is finished being synthesized, it begins to twist and fold as hydrogen bonds form among the amino acids of the chain. This hydrogen bonding may cause parts of the polypeptide to form flat sheets or coils (helices), patterns that constitute a protein's secondary structure ❷. The primary structure of each type of protein is unique, but most proteins have sheets and coils.

A Two amino acids (here, methionine and serine) are joined by condensation. A peptide bond forms between the carboxyl group of the methionine and the amine group of the serine.

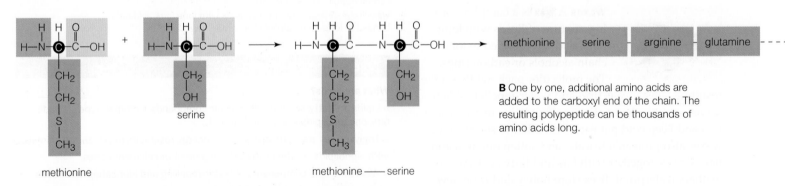

B One by one, additional amino acids are added to the carboxyl end of the chain. The resulting polypeptide can be thousands of amino acids long.

Figure 3.15 Animated Polypeptide formation. Chapter 9 offers a closer look at protein synthesis.
Figure It Out: What functional group forms with a peptide bond?   Answer: Amide

lysine | glycine | glycine | arginine

1 A protein's primary structure consists of a linear sequence of amino acids (a polypeptide chain). Each type of protein has a unique primary structure.

Figure 3.16 Animated Protein structure.

2 Secondary structure arises as a polypeptide chain twists into a coil (helix) or sheet held in place by hydrogen bonds between different parts of the molecule. The same patterns of secondary structure occur in many different proteins.

3 Tertiary structure occurs when a chain's coils and sheets fold up into a functional domain such as a barrel or pocket. In this example, the coils of a globin chain form a pocket.

4 Some proteins have quaternary structure, in which two or more polypeptide chains associate as one molecule. Hemoglobin, shown here, consists of four globin chains (*green* and *blue*). Each globin pocket now holds a heme group (*red*).

Much as an overly twisted rubber band coils back upon itself, hydrogen bonding between different parts of a protein make it fold up into even more compact domains. A domain is a part of a protein that is organized as a structurally stable unit. Such units are part of a protein's overall three-dimensional shape, or tertiary structure **3**. Tertiary structure is what makes a protein a working molecule. For example, sheets or coils of some proteins curl up into a barrel shape (*inset*). A barrel domain often functions as a tunnel for small molecules, allowing them to pass, for example, through a cell membrane. Globular domains of enzymes form chemically active pockets that can make or break bonds of other molecules.

Many proteins also have quaternary structure, which means they consist of two or more polypeptide chains that are in close association or covalently bonded together. Most enzymes and many other proteins consist of two or more polypeptide chains that collectively form a roughly spherical shape **4**.

Some proteins aggregate by many thousands into much larger structures, with their polypeptide chains organized into strands or sheets. The keratin in your hair is an example **5**. Some fibrous proteins contribute to the structure and organization of cells and tissues. Others, such as the actin and myosin filaments in muscle cells, are part of the mechanisms that help cells, cell parts, and multicelled bodies move.

Sugars or lipids bonded to proteins make glycoproteins and lipoproteins, respectively. The molecules that allow a tissue or a body to recognize its own cells are glycoproteins, as are other molecules that help cells interact in immune responses. Some lipoproteins form when enzymes covalently bond lipids to a protein.

5 Many proteins aggregate by the thousands into much larger structures, such as the keratin filaments that make up hair.

Other lipoproteins are aggregate structures that consist of variable amounts and types of proteins and lipids. These particles carry fats and cholesterol through the bloodstream. Low-density lipoprotein, or LDL, transports cholesterol out of the liver and into cells. High-density lipoprotein, or HDL (*right*), ferries the cholesterol that is released from dead cells back to the liver.

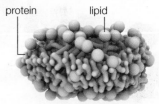

protein   lipid

an HDL particle

amino acid Small organic compound that is a subunit of proteins. Consists of a carboxyl group, an amine group, and a characteristic side group (R), all typically bonded to the same carbon atom.
peptide bond A bond between the amine group of one amino acid and the carboxyl group of another. Joins amino acids in proteins.
polypeptide Chain of amino acids linked by peptide bonds.
protein Organic compound that consists of one or more chains of amino acids (polypeptides).

Take-Home Message

What are proteins?

» Proteins are chains of amino acids. The order of amino acids in a polypeptide chain dictates the type of protein.

» Polypeptide chains twist and fold into coils, sheets, and loops, which fold and pack further into functional domains.

» A protein's shape is the source of its function.

3.7 Why Is Protein Structure So Important?

■ Changes in a protein's shape may have drastic health consequences.

Protein shape depends on hydrogen bonds and other interactions that heat, some salts, shifts in pH, or detergents can disrupt. Such disruption causes proteins to denature, which means they lose their shape. Once a protein's shape unravels, so does its function.

You can see denaturation in action when you cook an egg. A protein called albumin is a major component of egg white. Cooking does not disrupt the covalent bonds of albumin's primary structure, but it does destroy the hydrogen bonds that maintain the protein's shape. When a translucent egg white turns opaque, the albumin has been denatured. For a very few proteins, denaturation is reversible if normal conditions return, but albumin is not one of them. There is no way to uncook an egg.

Prion diseases such as mad cow disease (bovine spongiform encephalitis, or BSE) in cattle, Creutzfeldt–Jakob disease in humans, and scrapie in sheep, are the dire aftermath of a protein that changes shape. These infectious diseases may be inherited, but more often they arise spontaneously. All are characterized by relentless deterioration of mental and physical abilities that eventually causes death (Figure 3.17A).

All prion diseases begin with a protein that occurs normally in mammals. One such protein, PrPC, is found in cell membranes throughout the body. This copper-binding protein is especially abundant in brain cells, but we still know very little about what it does. Very rarely, a PrPC protein spontaneously misfolds so that it loses some of its coils. In itself, a single misfolded protein molecule would not pose much of a threat. However, when this particular protein misfolds, it becomes a prion, or infectious protein. The altered shape of a misfolded PrPC protein somehow causes normally folded PrPC proteins to misfold too. Because each protein that misfolds becomes infectious, the number of prions increases exponentially.

The shape of misfolded PrPC proteins allows them to align tightly into long fibers. Fibers of aggregated prion proteins begin to accumulate in the brain (Figure 3.17B). The water-repellent patches grow as more prions form, and they begin to disrupt brain cell function, causing symptoms such as confusion, memory loss, and lack of coordination. Tiny holes form in the brain as its cells die. Eventually, the brain becomes so riddled with holes that it looks like a sponge.

In the mid-1980s, an epidemic of mad cow disease in Britain was followed by an outbreak of a new variant of Creutzfeldt–Jakob disease (vCJD) in humans. Researchers isolated a prion similar to the one in scrapie-infected sheep from cows with BSE, and also from humans affected by the new type of Creutzfeldt–Jakob disease. How did the prion get from sheep to cattle to people? Prions are not denatured by cooking or typical treatments that inactivate other types of infectious agents. The cattle became infected by the prion after eating feed prepared from the remains of scrapie-infected sheep, and people became infected by eating beef from the infected cattle.

Two hundred people have died from vCJD since 1990. The use of animal parts in livestock feed is now banned in many countries, and the number of cases of BSE and vCJD has since declined. Cattle with BSE still turn up, but so rarely that they pose little threat to human populations.

denature To unravel the shape of a biological molecule.
prion Infectious protein.

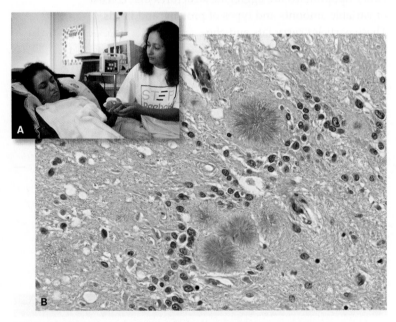

Figure 3.17 Variant Creutzfeldt–Jakob disease (vCJD).

(**A**) Charlene Singh, here being cared for by her mother, was one of the three people who developed symptoms of vCJD disease while living in the United States. Like the others, Singh most likely contracted the disease elsewhere; she spent her childhood in Britain. She was diagnosed in 2001, and she died in 2004.

(**B**) Slice of brain tissue from a person with vCJD. Characteristic holes and prion protein fibers radiating from several deposits are visible.

Take-Home Message

Why is protein structure important?

» A protein's function depends on its structure.

» Conditions that alter a protein's structure may also alter its function.

» Protein shape unravels if hydrogen bonds are disrupted.

3.8 Nucleic Acids

- Nucleotides are subunits of DNA and RNA. Some have roles in metabolism.
- Links to Inheritance 1.3, Diversity 1.4, Hydrogen bonds 2.5

Nucleotides are small organic molecules that function as energy carriers, enzyme helpers, chemical messengers, and subunits of DNA and RNA. Each consists of a sugar with a five-carbon ring, bonded to a nitrogen-containing base and one or more phosphate groups. The nucleotide ATP (adenosine triphosphate) has a row of three phosphate groups attached to its ribose sugar (Figure 3.18A). When the outer phosphate group of an ATP is transferred to another molecule, energy is transferred along with it. You will read about such phosphate-group transfers and their important metabolic role in Chapter 5.

Nucleic acids are polymers, chains of nucleotides in which the sugar of one nucleotide is joined to the phosphate group of the next (Figure 3.18B). An example

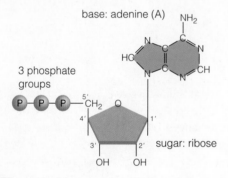

base: adenine (A)

3 phosphate groups

sugar: ribose

A ATP, a nucleotide monomer of RNA, and also an essential participant in many metabolic reactions.

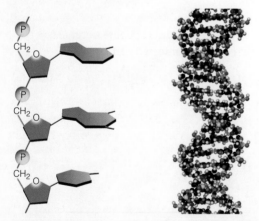

B A chain of nucleotides is a nucleic acid. The sugar of one nucleotide is covalently bonded to the phosphate group of the next, forming a sugar–phosphate backbone.

C DNA consists of two chains of nucleotides, twisted into a double helix. Hydrogen bonding maintains the three-dimensional structure of this nucleic acid.

Figure 3.18 Animated Nucleic acid structure.

Fear of Frying (revisited)

Trans fatty acids are relatively rare in unprocessed foods, so it makes sense from an evolutionary standpoint that our bodies may not have enzymes to deal with them efficiently. We do have enzymes that hydrolyze *cis* fatty acids, but these enzymes have difficulty breaking down *trans* fatty acids—a problem that may be a factor in the ill effects of *trans* fats on our health.

How would you vote? All prepackaged foods in the United States are now required to list *trans* fat content, but may be marked "zero grams of trans fats" even when a single serving contains up to half a gram. Should hydrogenated oils be banned from all food?

is RNA, or ribonucleic acid, named after the ribose sugar of its component nucleotides. RNA consists of four kinds of nucleotide monomers, one of which is ATP. RNA molecules carry out protein synthesis, which we discuss in detail in Chapter 9.

DNA, or deoxyribonucleic acid, is a nucleic acid named after the deoxyribose sugar of its component nucleotides. A DNA molecule consists of two chains of nucleotides twisted into a double helix (Figure 3.17C). Hydrogen bonds between the nucleotides hold the two chains together.

Each cell starts life with DNA inherited from a parent cell. That DNA contains all information necessary to build a new cell and, in the case of multicelled organisms, an entire individual. The cell uses the order of nucleotide bases in DNA—the DNA sequence—to guide production of RNA and proteins. Parts of the sequence are identical or nearly so in all organisms, but most is unique to a species or an individual (Chapter 8 returns to DNA structure and function).

ATP Adenosine triphosphate. Nucleotide that consists of an adenine base, a ribose sugar, and three phosphate groups.
DNA Deoxyribonucleic acid. Nucleic acid that carries hereditary information about traits; consists of two nucleotide chains twisted in a double helix.
nucleic acid Single- or double-stranded chain of nucleotides joined by sugar–phosphate bonds; for example, DNA, RNA.
nucleotide Monomer of nucleic acids; has a five-carbon sugar, a nitrogen-containing base, and phosphate groups.
RNA Ribonucleic acid. Some types have roles in protein synthesis.

Take-Home Message

What are nucleotides and nucleic acids?

» Nucleotides are monomers of the nucleic acids DNA and RNA. Some have additional roles.

» DNA's nucleotide sequence encodes heritable information.

» RNA plays several important roles in the process by which a cell uses the instructions written in its DNA to build proteins.

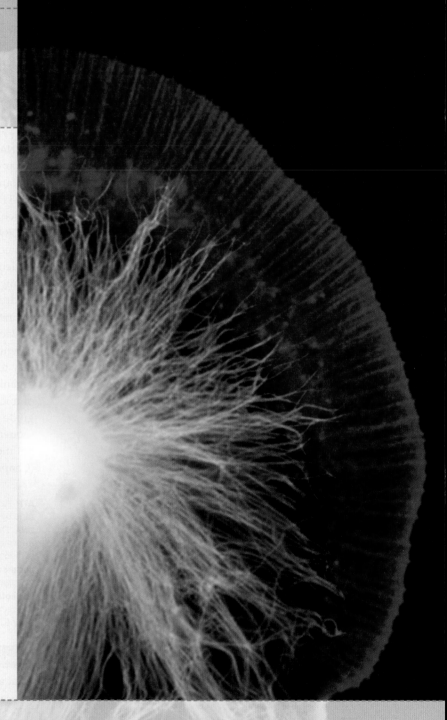

LEARNING ROADMAP

Where you have been Reflect on the Section 1.2 overview of life's levels of organization. You will now begin to see how the nonliving molecules of life—carbohydrates (3.4), lipids (3.5), proteins (3.6, 3.7), and nucleic acids (3.8)—form cells (1.2) and carry out functions that define life (1.3). You will also see an application of tracers (2.2), and revisit the philosophy of science (1.9).

Where you are now

What All Cells Have In Common
Every cell has a plasma membrane that separates its interior from the exterior environment. The interior contains cytoplasm, DNA, and other structures.

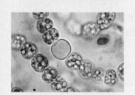

The "Prokaryotes"
Archaea and bacteria have no nucleus. In general, they are smaller and structurally simpler than eukaryotic cells, but they are by far the most numerous.

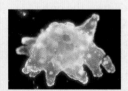

Eukaryotic Cells
Protists, plants, fungi, and animals are eukaryotes. Cells of these organisms differ in internal parts and surface specializations, but all start out life with a nucleus.

Organelles of Eukaryotes
Membranes around eukaryotic organelles maintain internal environments that allow these structures to carry out specialized functions within a cell.

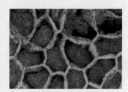

Other Cell Components
Cytoskeletal elements organize and move cells and cell components. Cells secrete protective and structural materials. Cell junctions connect cells to one another.

Where you are going Chapter 5 explores cell function; Chapter 6, photosynthesis; and Chapter 7, aerobic respiration. We revisit protein synthesis in Chapter 9, and control over it in Chapter 10. Some cellular structures are required for cell division (Chapter 11). Human genetic disorders return in Chapter 14. Chapter 19 details the structures, evolution, and metabolism of bacteria and archaea; Chapter 21, protists. Cell structures introduced in this chapter return in context of the physiology of animals (Chapters 31–42) and plants (Chapters 27–30). You will see some medical consequences of biofilms in Section 37.3.

4.1 Food for Thought

We find bacteria at the bottom of the ocean, high up in the atmosphere, miles underground—essentially anywhere we look. Mammalian intestines typically harbor fantastic numbers of them, but bacteria are not just stowaways there. Intestinal bacteria make vitamins that mammals cannot, and they crowd out more dangerous germs. Cell for cell, bacteria that live in and on a human body outnumber the person's own cells by about ten to one.

Escherichia coli is one of the most common intestinal bacteria of warm-blooded animals. Only a few of the hundreds of types, or strains, of *E. coli*, are harmful. One that is called O157:H7 (**Figure 4.1**) makes a potent toxin that can severely damage the lining of the human intestine. After ingesting as few as ten O157:H7 cells, a person may become ill with severe cramps and bloody diarrhea that lasts up to ten days. In some people, complications of O157:H7 infection result in kidney failure, blindness, paralysis, and death. Each year, about 73,000 people in the United States become infected with *E. coli* O157:H7, and more than 60 of them die.

E. coli O157:H7 can live in the intestines of other animals—mainly cattle, deer, goats, and sheep—apparently without sickening them. Humans are exposed to the bacteria when they come into contact with feces of animals that harbor it, for example, by eating contaminated ground beef. During slaughter, meat can come into contact with feces. Bacteria in the feces stick to the meat, then get thoroughly mixed into it during the grinding process. Unless contaminated meat is cooked to at least 71°C (160°F), live bacteria will enter the digestive tract of whoever eats it.

People also become infected by eating fresh fruits and vegetables that have come into contact with animal feces. Washing contaminated produce with water does not remove *E. coli* O157:H7, because the bacteria are sticky. In 2006, more than 200 people became ill and 3 died after eating fresh spinach grown in a field close to a cattle pasture. Water contaminated with manure may have been used to irrigate the field.

The economic impact of such outbreaks, which occur with disturbing regularity, extends beyond the casualties. Growers lost $50–100 million recalling fresh spinach after the 2006 outbreak. In 2007, about 5.7 million pounds of ground beef were recalled after 14 people were sickened. More than 2.8 million pounds of meat products were recalled between November 2009 and January 2010. Food growers and processors are now using procedures that they hope will reduce *E. coli* O157:H7 outbreaks. Some meats and produce are now tested for pathogens before sale, and improved documentation should allow a source of contamination to be pinpointed more quickly.

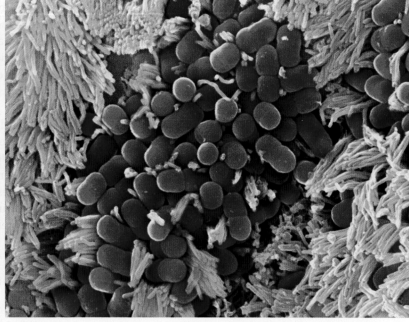

Figure 4.1 *E. coli* O157:H7 bacteria (*above*, *red*) on intestinal cells (*tan*) of a small child. This type of bacteria can cause a serious intestinal illness in people who eat foods contaminated with it, such as ground beef or fresh produce (*left*).

4.2 Cell Structure

- All cells have a plasma membrane and cytoplasm, and all start out life with DNA.
- Links to Life's organization 1.2, Prokaryotes and eukaryotes 1.4, Lipid bilayer 3.5, DNA 3.8

You learned in Section 1.2 that the cell is the smallest unit that shows the properties of life. What, exactly, is a cell? Cells vary in shape and in what they do, but all share certain organizational and functional features: a plasma membrane, cytoplasm, and DNA (**Figure 4.2**).

Components of All Cells

Plasma Membrane Every cell has a plasma membrane, an outer membrane that separates the cell's contents from its environment. A plasma membrane is selectively permeable, which means it allows only certain materials to cross. Thus, it controls exchanges between the cell and its environment. Phospholipids (Section 3.5) are the most abundant type of lipid in cell membranes. Many different proteins embedded in a bilayer or attached to one of its surfaces carry out membrane functions (**Figure 4.3**). For example, some proteins form channels through a bilayer; others pump substances across it. You will see

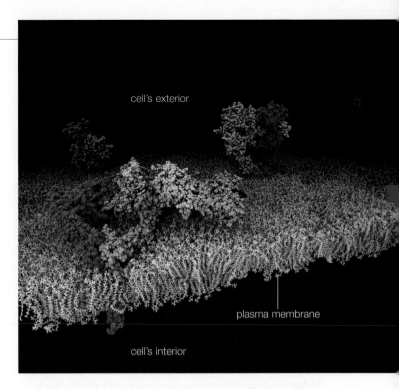

Figure 4.3 Preview of cell membranes. A plasma membrane separates a cell from its external environment. Proteins (in color) embedded in the lipid bilayer (*gray*) carry out special membrane functions. Chapter 5 returns to membrane structure and function.

in Chapter 5 how a membrane's function arises from its mosaic structure of lipids and proteins.

Cytoplasm The plasma membrane encloses a fluid or jellylike mixture of water, sugars, ions, and proteins called cytoplasm. Some or all of a cell's metabolism occurs in the cytoplasm, and the cell's internal components, including organelles, are suspended in it. Organelles are structures that carry out special metabolic functions inside a cell. Membrane-enclosed organelles compartmentalize tasks such as building, modifying, and storing substances.

DNA All cells start life with DNA, though a few types of cells lose it as they mature. Only eukaryotic cells have a nucleus (plural, nuclei), an organelle with a double membrane that contains the cell's DNA. A few prokaryotes have lipid bilayers enclosing their DNA, but the structure of this membrane differs from a nuclear membrane. In most prokaryotes, the DNA is suspended directly in cytoplasm.

Constraints on Cell Size

Almost all cells are too small to see with the naked eye. Why? The answer begins with the processes that keep a cell alive. A living cell must exchange substances with its environment at a rate that keeps pace

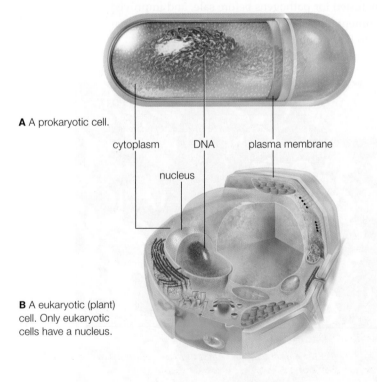

A A prokaryotic cell.

cytoplasm   DNA   plasma membrane

nucleus

B A eukaryotic (plant) cell. Only eukaryotic cells have a nucleus.

Figure 4.2 Animated Overview of the general organization of a cell.

Archaea are similar to bacteria in overall structure; both are generally much smaller than eukaryotic cells. If the two cells depicted here had been drawn to the same scale, the prokaryotic cell would be about this big:

with its metabolism. These exchanges occur across the plasma membrane, which can handle only so many exchanges at a time. The rate of exchange across a plasma membrane depends on its surface area: the bigger it is, the more substances can cross the membrane during a given interval. Cell size is limited by a physical relationship called the **surface-to-volume ratio**. By this ratio, an object's volume increases with the cube of its diameter, but its surface area increases only with the square. Apply the surface-to-volume ratio to a round cell. As **Figure 4.4** shows, when a cell expands in diameter, its volume increases faster than its surface area does. Imagine that a round cell expands until it is four times its original diameter. The volume of the cell has increased 64 times (4^3), but its surface area has increased only 16 times (4^2). Each unit of plasma membrane must now handle exchanges with four times as much cytoplasm ($64 \div 16 = 4$). If the cell gets too big, the inward flow of nutrients and the outward flow of wastes across that membrane will not be fast enough to keep the cell alive.

Surface-to-volume limits also affect the body plans of multicelled species. For example, small cells attach end to end to form strandlike algae, so that each can interact directly with its surroundings. Muscle cells in your thighs are as long as the muscle in which they occur, but each is thin, so it exchanges substances efficiently with fluids in the tissue surrounding it.

Cell Theory

Nearly all cells are so small that no one even knew they existed until after the first microscopes were invented. In 1665, Antoni van Leeuwenhoek, a Dutch draper, wrote about the tiny moving organisms he spied in rainwater, insects, fabric, sperm, feces, and other samples. In scrapings of tartar from his teeth, Leeuwenhoek saw "many very small animalcules, the motions of which were very pleasing to behold." He (incorrectly) assumed that movement defines life, but (correctly) concluded that the moving animalcules he saw were alive. The term "cell" was coined when Robert Hooke, a contemporary of Leeuwenhoek, magnified a piece of thinly sliced cork. Hooke named the tiny compartments he observed "cellae," a Latin word for the small chambers that monks lived in.

In the 1820s, botanist Robert Brown was the first to identify a cell nucleus. Matthias Schleiden, another botanist, hypothesized that a plant cell is an independent living unit even when it is part of a plant. Schleiden compared notes with the zoologist Theodor Schwann, and both concluded that the tissues of ani-

Diameter (cm)	2	3	6
Surface area (cm^2)	12.6	28.2	113
Volume (cm^3)	4.2	14.1	113
Surface-to-volume ratio	3:1	2:1	1:1

Figure 4.4 Three examples of the surface-to-volume ratio. This physical relationship between increases in volume and surface area constrains cell size and shape.

mals as well as plants are composed of cells and their products. Together, the two scientists recognized that cells have a life of their own even when they are part of a multicelled body.

Later, physiologist Rudolf Virchow realized that all cells he studied descended from another living cell. These and many other observations yielded four generalizations that today constitute the **cell theory**:

1. Every living organism consists of one or more cells.
2. The cell is the structural and functional unit of all organisms. A cell is the smallest unit of life, individually alive even as part of a multicelled organism.
3. All living cells arise by division of preexisting cells.
4. Cells contain hereditary material, which they pass to their offspring during division.

The cell theory, first articulated in 1839 by Schwann and Schleiden and later revised, remains an important foundation of modern biology.

cell theory Theory that all organisms consist of one or more cells, which are the basic unit of life; all cells come from division of preexisting cells; and all cells pass hereditary material to offspring.
cytoplasm Semifluid substance enclosed by a cell's plasma membrane.
nucleus Organelle with two membranes that holds a eukaryotic cell's DNA.
organelle Structure that carries out a specialized metabolic function inside a cell.
plasma membrane A cell's outermost membrane.
surface-to-volume ratio A relationship in which the volume of an object increases with the cube of the diameter, and the surface area increases with the square.

Take-Home Message

How are all cells alike?

» All cells start life with a plasma membrane, cytoplasm, and a region of DNA, which, in eukaryotic cells only, is enclosed by a nucleus.

» The surface-to-volume ratio limits cell size and influences cell shape.

» Observations of cells led to the cell theory: All organisms consist of one or more cells; the cell is the smallest unit of life; each new cell arises from another cell; and a cell passes hereditary material to its offspring.

4.3 How Do We See Cells?

■ We use different types of microscopes to study different aspects of organisms, from the smallest to the largest.

■ Link to Electrons and Tracers 2.2

Most cells are 10–20 micrometers in diameter, about fifty times smaller than the unaided human eye can perceive. One micrometer (μm) is one-thousandth of a millimeter, which is one-thousandth of a meter (**Table 4.1**). We use microscopes to observe cells and other objects in the micrometer range of size.

Light microscopes use visible light to illuminate samples (**Figure 4.5A**). As you will learn in Chapter 6, all light travels in waves. This property makes light bend when it passes through curved glass lenses.

Inside a light microscope, curved lenses focus light that passes through a specimen, or bounces off of one, into a magnified image. Photographs of images enlarged with a microscope are called micrographs, and you will find many of them in this book.

Phase-contrast microscopes shine light through specimens. Most cells are nearly transparent, so their internal details may not be visible unless they are first stained, or exposed to dyes that only some cell parts soak up. Parts that absorb the most dye appear darkest. Staining results in an increase in contrast (the difference between light and dark) that allows us to see a greater range of detail (**Figure 4.6A**). Surface details can be revealed by reflected light (**Figure 4.6B**).

With a fluorescence microscope, a cell or a molecule serves as the light source; it fluoresces, or emits energy in the form of light, when a laser beam is focused on it. Some molecules fluoresce naturally (**Figure 4.6C**). More typically, researchers attach a light-emitting tracer (Section 2.2) to the cell or molecule of interest.

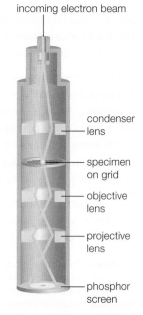

path of light rays (bottom to top) to eye

prism that directs rays to ocular lens

ocular lens

objective lenses

specimen stage

condenser lens

illuminator

focusing knob

light source (in base)

A A compound light microscope.

incoming electron beam

condenser lens

specimen on grid

objective lens

projective lens

phosphor screen

B A transmission electron microscope (TEM).

Figure 4.5 Animated Examples of microscopes.

Table 4.1	Common Units of Length		
Unit		**Equivalent**	
		Meter	**Inch**
centimeter	cm	1/100	0.4
millimeter	mm	1/1000	0.04
micrometer	μm	1/1,000,000	0.00004
nanometer	nm	1/1,000,000,000	0.00000004
meter	m	100 cm 1,000 mm 1,000,000 μm 1,000,000,000 nm	

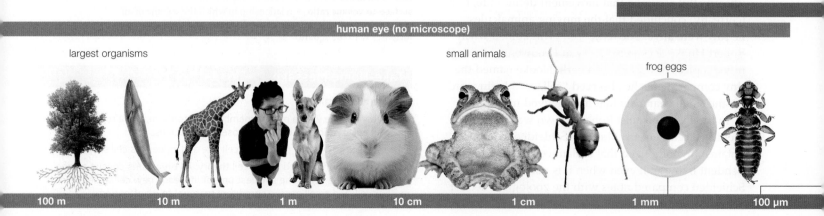

human eye (no microscope)

largest organisms

small animals

frog eggs

100 m   10 m   1 m   10 cm   1 cm   1 mm   100 μm

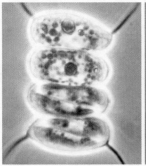

A Phase-contrast light microscopes yield high-contrast images of transparent specimens. Dark areas have taken up dye.

B Reflected light microscopes capture light reflected from the surface of specimens.

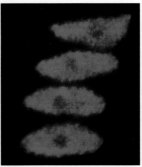

C This fluorescence micrograph shows fluorescent light emitted by chlorophyll molecules in the cells.

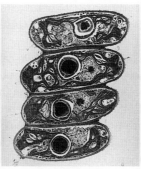

D Transmission electron micrographs reveal fantastically detailed images of internal structures.

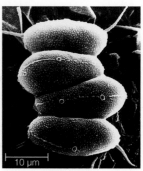

E Scanning electron micrographs show surface details. SEMs may be artificially colored to highlight specific details.

Figure 4.6 Different microscopes can reveal different characteristics of the same aquatic organism—a green alga (*Scenedesmus*). Try estimating the size of one of these algal cells by using the scale bar.

Structures smaller than about 200 nanometers across appear blurry under light microscopes because the wavelength of light—the distance from the peak of one wave to the peak behind it—limits the power of even the best light microscope. To observe objects of this size range clearly, we would have to switch to an electron microscope. Electron microscopes use magnetic fields to focus a beam of electrons onto a sample (**Figure 4.5B**). Electrons travel in wavelengths much shorter than those of visible light, so these microscopes resolve details thousands of times smaller than light microscopes do. Transmission electron microscopes direct electrons through a thin specimen, and the specimen's internal details appear as shadows in the resulting image (**Figure 4.6D**). Scanning electron microscopes direct a beam of electrons back and forth across the surface of a specimen that has been coated with a thin layer of metal such as gold. The irradiated metal emits electrons and x-rays, which are converted into an image of the surface (**Figure 4.6E**). Figure 4.7 compares the resolving power of light microscopes and electron microscopes with that of the human eye.

Take-Home Message

How do we see cells?

» Most cells are visible only with the help of microscopes.

» Different types of microscopes reveal different aspects of cell structure.

Figure 4.7 Relative sizes. *Below*, the diameter of most cells is in the range of 1 to 100 micrometers. **Table 4.1** shows conversions among units of length; also see Units of Measure, Appendix VIII. **Figure It Out:** Which one is smallest: a protein, a lipid, or a water molecule?

Answer: A water molecule

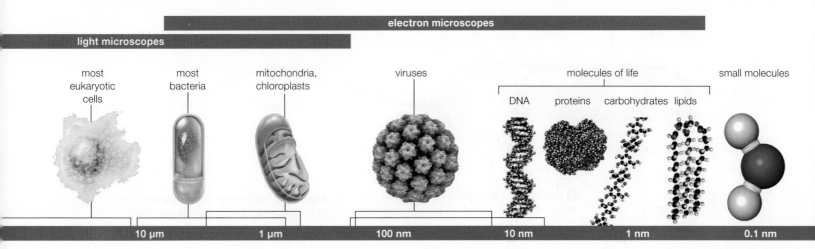

4.4 Introducing "Prokaryotes"

■ Bacteria and archaea are the prokaryotes.

■ Links to Prokaryotes 1.4, Polysaccharides 3.4, ATP 3.8

All bacteria and archaea are single-celled, and none have a nucleus. Outwardly, they appear so similar that archaea were once thought to be an unusual group of bacteria. Both were classified as prokaryotes, a word that means "before the nucleus." By 1977, it had become clear that archaea are more closely related to eukaryotes than to bacteria, so they were given their own separate domain. The term "prokaryote" is now considered an informal designation. (You will read more about the discovery of archaea in Chapter 20.)

As a group, bacteria and archaea are the smallest and most metabolically diverse forms of life that we know about. They inhabit nearly all of Earth's environments, including some very hostile places. The two kinds of cells differ in structure and metabolism. Chapter 20 revisits them in more detail; here we present an overview of their structure (**Figures 4.8** and **4.9**).

Most bacteria and archaea are not much bigger than a few micrometers. None has a complex internal framework, but protein filaments under the plasma membrane reinforce the cell's shape. Such filaments also act as scaffolding for internal structures. The cytoplasm of these cells contains many **ribosomes** ❶ (organelles upon which polypeptides are assembled), and in some species, additional organelles. **Plasmids** also occur in cytoplasm. These small circles of DNA carry a few genes (units of inheritance) that can provide advantages, such as resistance to antibiotics. The cell's remaining genes typically occur on one large circular molecule of DNA located in an irregularly shaped region of cytoplasm called the **nucleoid** ❷. Some nucleoids are enclosed by a membrane.

The plasma membrane ❸ of all bacteria and archaea selectively controls which substances move into and

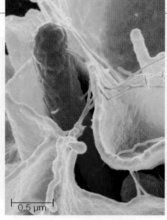
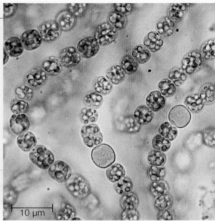

A *Salmonella typhimurium* cells (*red*) use their pili to invade human cells (*yellow*).

B Ball-shaped *Nostoc* are freshwater photosynthetic bacteria. The cells in each strand stick together in a sheath of their own jellylike secretions.

Figure 4.9 A sampling of bacteria (*this page*) and archaea (*facing page*).

out of the cell, as it does for eukaryotic cells. The plasma membrane bristles with proteins that carry out important metabolic processes. For example, part of the plasma membrane of cyanobacteria (**Figure 4.9B**) folds into the cytoplasm. Molecules that carry out photosynthesis are embedded in this membrane, just as they are in the inner membrane of chloroplasts, which are organelles specialized for photosynthesis in eukaryotic cells (we return to chloroplasts in Section 4.9).

A durable **cell wall** surrounds the plasma membrane of nearly all bacteria and archaea ❹. Dissolved substances easily cross this permeable layer on the way to and from the plasma membrane. Most archaean cell walls consist of proteins; most bacterial cell walls consist of a polymer of peptides and polysaccharides.

Sticky polysaccharides form a slime layer or capsule around the wall of many types of bacteria ❺. The sticky capsule helps these cells adhere to many types of

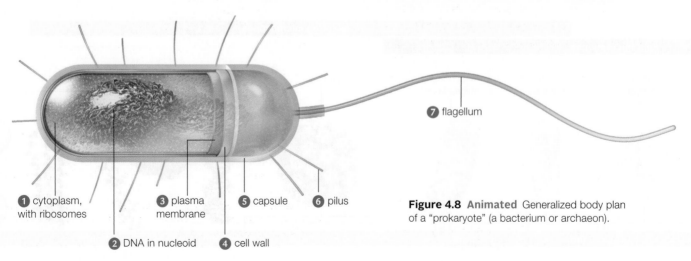

❶ cytoplasm, with ribosomes

❸ plasma membrane

❺ capsule

❻ pilus

❼ flagellum

❷ DNA in nucleoid

❹ cell wall

Figure 4.8 Animated Generalized body plan of a "prokaryote" (a bacterium or archaeon).

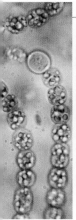

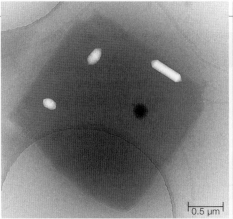

C The square archaea *Haloquadra walsbyi* thrives in brine pools saltier than soy sauce. Gas-filled organelles (*white* structures) buoy these highly motile cells, which can aggregate into flat sheets reminiscent of tile floors.

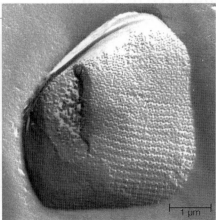

D *Ferroglobus placidus* prefers superheated water spewing from the ocean floor. The durable composition of archaeal lipid bilayers and cell walls (note the gridlike texture) keeps their membranes intact at extreme heat and pH.

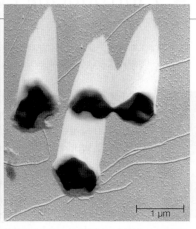

E *Metallosphaera prunae*, a species of archaea discovered in a smoking pile of ore at a uranium mine, prefers high temperatures and low pH. (*White* shadows are an artifact of electron microscopy.)

surfaces (such as spinach leaves and meat), and it also offers some protection against predators and toxins.

Many bacteria and archaea have one or more flagella projecting from their surface. Flagella (singular, flagellum) are long, slender cellular structures used for motion ❼. A bacterial flagellum rotates like a propeller that drives the cell through fluid habitats, such as an animal's body fluids. Some prokaryotes also have protein filaments called pili (singular, pilus) projecting from their surface ❻. Pili help cells cling to or move across surfaces (**Figure 4.9A**). One kind, a "sex" pilus, attaches to another bacterium and then shortens. The attached cell is reeled in, and a plasmid is transferred from one cell to the other through the pilus.

Biofilms

Bacteria often live so close together that an entire community shares a layer of secreted polysaccharides and proteins. A communal living arrangement in which single-celled organisms live in a shared mass of slime is called a biofilm (**Figure 4.10**). In nature, a biofilm usually consists of multiple species, all entangled in their own mingled secretions. It may include bacteria, algae, fungi, protists, and archaea. Participating in a biofilm

---

biofilm Community of microorganisms living within a shared mass of slime.
cell wall Semirigid but permeable structure that surrounds the plasma membrane of some cells.
flagellum Long, slender cellular structure used for motility.
nucleoid Region of cytoplasm where the DNA is concentrated inside a bacterium or archaeon.
pilus Protein filament that projects from the surface of some bacteria and archaea.
plasmid Small circle of DNA in some bacteria and archaea.
ribosome Organelle of protein synthesis.

---

allows the cells to linger in a favorable spot rather than be swept away by fluid currents, and to reap the benefits of living communally. For example, rigid or netlike secretions of some species serve as permanent scaffolding for others; species that break down toxic chemicals allow more sensitive ones to thrive in polluted habitats that they could not withstand on their own; and waste products of some serve as raw materials for others. Later chapters discuss medical implications of biofilms, including dental plaque.

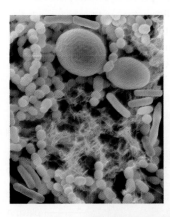

Figure 4.10 Oral bacteria in dental plaque, a biofilm. Three species of bacteria (*tan, green*) and a yeast (*red*) stick to one another and to teeth via a gluelike mass of shared, secreted polysaccharides (*pink*). Other secretions of these organisms cause cavities and periodontal disease.

Take-Home Message

How are bacteria and archaea alike?

» Bacteria and archaea do not have a nucleus. Most kinds have a cell wall around their plasma membrane. The permeable wall reinforces and imparts shape to the cell body.

» The structure of bacteria and archaea is relatively simple, but as a group these organisms are the most diverse forms of life. They inhabit nearly all regions of the biosphere.

» Some metabolic processes occur at the plasma membrane of bacteria and archaea. They are similar to complex processes that occur at certain internal membranes of eukaryotic cells.

4.5 Introducing Eukaryotic Cells

■ Eukaryotic cells carry out much of their metabolism inside organelles enclosed by membranes.

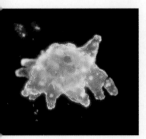

A single-celled eukaryote (a protist)

All protists, fungi, plants, and animals are eukaryotes. Some of these organisms are independent, free-living cells (*left*); others consist of many cells working together as a body. All of these cells start life with a nucleus (*eu*– means true; *karyon* means nut, or kernel) and other organelles (**Table 4.2**). Typical components of a eukaryotic cell include the endomembrane system, mitochondria, and a cytoskeleton, which is a dynamic "skeleton" of proteins (*cyto*– means cell). Certain cell types have additional components.

Membranes around eukaryotic organelles control the types and amounts of substances that enter and exit them. Such control maintains a special internal environment that allows the organelle to carry out its particular function. That function may be isolating toxic or sensitive substances from the rest of the cell, moving substances through cytoplasm, maintaining fluid balance, or providing a favorable environment for a special process.

Much as interactions among organ systems keep an animal body running, interactions among organelles keep a cell running. Substances shuttle from one kind of organelle to another, and to and from the plasma membrane. Some metabolic pathways take place in a series of different organelles.

Typical animal and plant cells are shown in **Figure 4.11**. We explore components of these cells (**Figure 4.12**) in the remaining sections of this chapter.

Table 4.2 Components of Eukaryotic Cells

Name	Main Function
Organelles with membranes	
Nucleus	Protecting, controlling access to DNA
Endoplasmic reticulum (ER)	Routing, modifying new polypeptide chains; synthesizing lipids
Golgi body	Modifying new polypeptide chains; sorting, shipping proteins and lipids
Vesicles	Transporting, storing, or digesting substances in a cell
Mitochondrion	Making ATP by glucose breakdown
Chloroplast	Photosynthesis in plants, some protists
Lysosome	Intracellular digestion
Peroxisome	Inactivating toxins
Vacuole	Storage
Organelles without membranes	
Ribosomes	Assembling polypeptide chains
Centriole	Anchor for cytoskeleton
Other components	
Cytoskeleton	Contributes to cell shape, internal organization, movement

Take-Home Message

What do all eukaryotic cells have in common?

» All eukaryotic cells start life with a nucleus and other membrane-enclosed organelles.

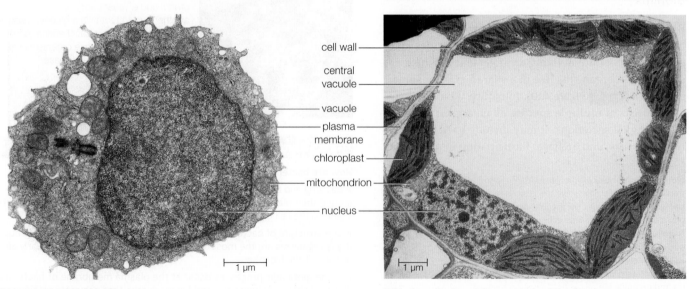

cell wall
central vacuole
vacuole
plasma membrane
chloroplast
mitochondrion
nucleus

1 µm

1 µm

A Human white blood cell.

B Photosynthetic cell from a blade of timothy grass.

Figure 4.11 Transmission electron micrographs of two eukaryotic cells.

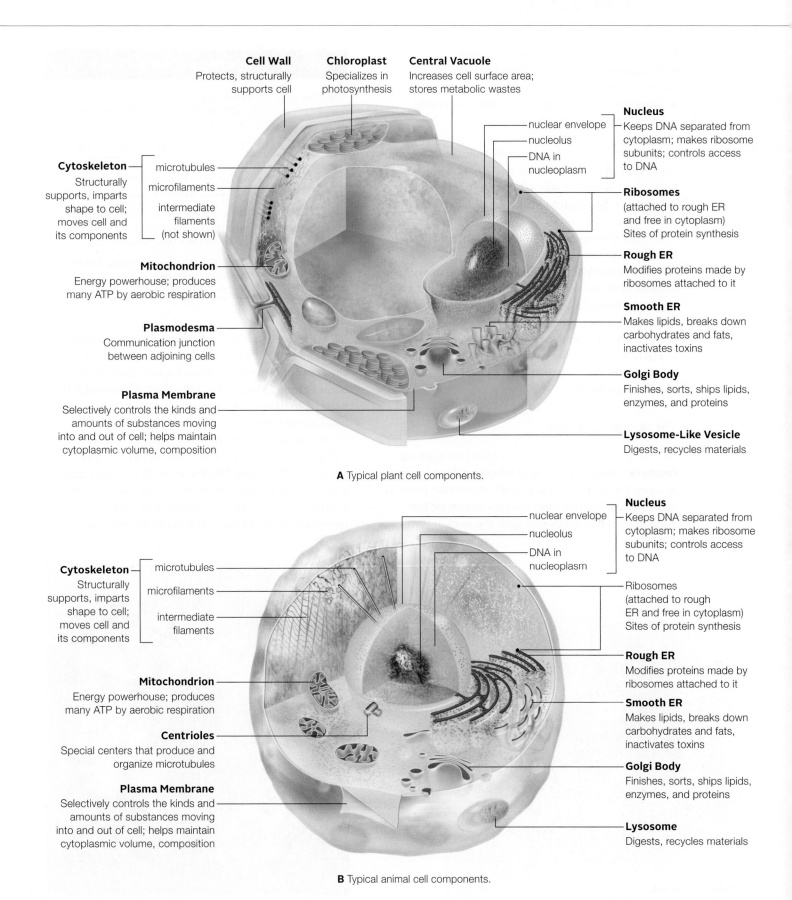

Cell Wall
Protects, structurally supports cell

Chloroplast
Specializes in photosynthesis

Central Vacuole
Increases cell surface area; stores metabolic wastes

nuclear envelope
nucleolus
DNA in nucleoplasm

Nucleus
Keeps DNA separated from cytoplasm; makes ribosome subunits; controls access to DNA

Cytoskeleton
Structurally supports, imparts shape to cell; moves cell and its components

microtubules
microfilaments
intermediate filaments (not shown)

Ribosomes
(attached to rough ER and free in cytoplasm) Sites of protein synthesis

Mitochondrion
Energy powerhouse; produces many ATP by aerobic respiration

Rough ER
Modifies proteins made by ribosomes attached to it

Plasmodesma
Communication junction between adjoining cells

Smooth ER
Makes lipids, breaks down carbohydrates and fats, inactivates toxins

Golgi Body
Finishes, sorts, ships lipids, enzymes, and proteins

Plasma Membrane
Selectively controls the kinds and amounts of substances moving into and out of cell; helps maintain cytoplasmic volume, composition

Lysosome-Like Vesicle
Digests, recycles materials

A Typical plant cell components.

nuclear envelope
nucleolus
DNA in nucleoplasm

Nucleus
Keeps DNA separated from cytoplasm; makes ribosome subunits; controls access to DNA

Cytoskeleton
Structurally supports, imparts shape to cell; moves cell and its components

microtubules
microfilaments
intermediate filaments

Ribosomes
(attached to rough ER and free in cytoplasm) Sites of protein synthesis

Mitochondrion
Energy powerhouse; produces many ATP by aerobic respiration

Rough ER
Modifies proteins made by ribosomes attached to it

Centrioles
Special centers that produce and organize microtubules

Smooth ER
Makes lipids, breaks down carbohydrates and fats, inactivates toxins

Golgi Body
Finishes, sorts, ships lipids, enzymes, and proteins

Plasma Membrane
Selectively controls the kinds and amounts of substances moving into and out of cell; helps maintain cytoplasmic volume, composition

Lysosome
Digests, recycles materials

B Typical animal cell components.

Figure 4.12 Animated Organelles and structures typical of (**A**) plant cells and (**B**) animal cells.

4.6 The Nucleus

- All of a eukaryotic cell's DNA is in its nucleus.
- The nucleus keeps eukaryotic DNA away from potentially damaging reactions in the cytoplasm.
- The nuclear envelope controls when DNA is accessed.
- Link to Fibrous proteins 3.6

As molecules go, DNA is gigantic. Unraveled and stretched out, the DNA in the nucleus of a single human cell would be about 2 meters (6–1/2 feet) long. All of that DNA fits into a nucleus about six microns in diameter.

A cell's nucleus serves two important functions. First, it keeps the cell's genetic material—its DNA—safe from metabolic processes that might damage it. Isolated in its own compartment, the cell's DNA stays separated from the bustling activity of the cytoplasm. Second, a nucleus controls the passage of certain molecules across its membrane.

Figure 4.13 shows the components of the nucleus. **Table 4.3** lists their functions. Let's zoom in on the individual components.

The Nuclear Envelope

The nuclear membrane, which is called the nuclear envelope, consists of two lipid bilayers folded together as a single membrane (**Figure 4.14**). Membrane proteins embedded in the two lipid bilayers aggregate into thousands of tiny pores that span the nuclear enve-

Table 4.3	Components of the Nucleus
Nuclear envelope	Pore-riddled double membrane that controls which substances enter and leave the nucleus
Nucleoplasm	Semifluid interior portion of the nucleus
Nucleolus	Rounded mass of proteins and copies of genes for ribosomal RNA used to construct ribosomal subunits
Chromatin	Total collection of all DNA molecules and associated proteins in the nucleus; all of the cell's chromosomes
Chromosome	One DNA molecule and many proteins associated with it

lope. The pores are anchored by the nuclear lamina, a dense mesh of fibrous proteins that supports the inner surface of the envelope. Some bacteria have membranes around their DNA, but we do not consider the bacteria to have nuclei because there are no pores in these membranes.

As you will see in Chapter 5, large molecules, including RNA and proteins, cannot cross a lipid bilayer on their own. Nuclear pores function as gateways for these molecules to enter and exit a nucleus. Protein synthesis offers an example of why this movement is important. Protein synthesis occurs in

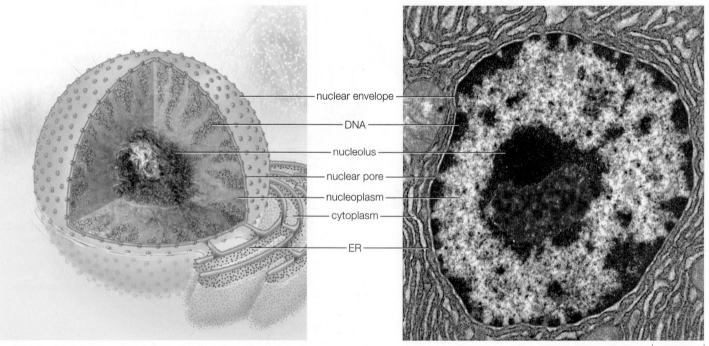

nuclear envelope
DNA
nucleolus
nuclear pore
nucleoplasm
cytoplasm
ER

1 μm

Figure 4.13 Animated The cell nucleus. TEM at *right*, nucleus of a mouse pancreas cell.

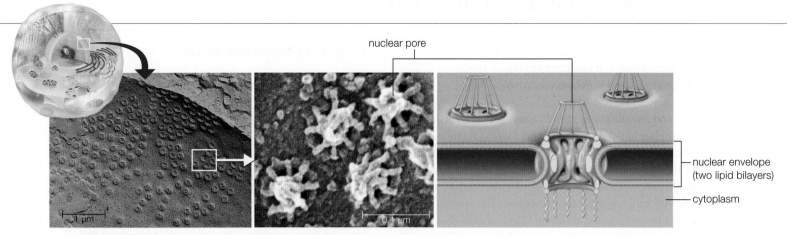

nuclear pore

nuclear envelope (two lipid bilayers)

cytoplasm

A The outer surface of this nuclear envelope was split apart to reveal the pores that span the two lipid bilayers.

B Each nuclear pore is an organized cluster of membrane proteins that selectively allows certain substances to cross it on their way into and out of the nucleus.

Figure 4.14 **Animated** Structure of the nuclear envelope.

cytoplasm, and it requires the participation of many molecules of RNA. RNA is produced in the nucleus. Thus, RNA molecules must move from nucleus to cytoplasm, and they do so through nuclear pores. Proteins must move through the pores in the other direction, because RNA synthesis, which occurs in the nucleus, requires proteins produced in cytoplasm.

A cell can regulate the amounts and types of proteins it makes at a given time by selectively restricting the passage of certain molecules through nuclear pores. Later chapters return to details of protein synthesis and controls over it, as well as how molecules cross membranes.

The Nucleolus

The nucleus contains at least one nucleolus (plural, nucleoli), a dense, irregularly shaped region of proteins and nucleic acid where subunits of ribosomes are produced. The subunits pass through nuclear pores into the cytoplasm, where they join and become active in protein synthesis.

The DNA

The nuclear envelope encloses nucleoplasm, a viscous fluid, similar to cytoplasm, in which the cell's DNA is suspended. Most eukaryotic cells have a number of DNA molecules. That number is characteristic of the type of organism and the type of cell, but it varies widely among species. For instance, the nucleus of a normal oak tree cell contains 12 DNA molecules; a human body cell, 46; and a king crab cell, 208. Each molecule of DNA, together with its many attached

proteins, is called a chromosome. DNA molecules, together with their associated proteins, are collectively called chromatin.

Chromosomes are anchored to and organized by the nuclear lamina. They change in appearance over the lifetime of a cell. When a cell is not dividing, its chromatin is invisible in light micrographs. Just before a cell divides, the DNA in each chromosome is cop-

ied, or duplicated. As cell division begins, the duplicated chromosomes condense. As they do, they become visible in micrographs in their characteristic "X" forms (*inset*). In later chapters, we will look in more detail at the dynamic structure and function of eukaryotic chromosomes.

chromatin Collective term for DNA molecules together with their associated proteins.
chromosome A structure that consists of DNA and associated proteins; carries part or all of a cell's genetic information.
nuclear envelope A double membrane that constitutes the outer boundary of the nucleus. Pores in the membrane control which substances can cross.
nucleolus In a cell nucleus, a dense, irregularly shaped region where ribosomal subunits are assembled.
nucleoplasm Viscous fluid enclosed by the nuclear envelope.

Take-Home Message

What is the function of the cell nucleus?

» A nucleus protects and controls access to a eukaryotic cell's DNA.

» The nuclear envelope is a double lipid bilayer. Proteins embedded in it control the passage of molecules between the nucleus and the cytoplasm.

4.7 The Endomembrane System

■ The endomembrane system is a set of organelles that makes, modifies, and transports proteins and lipids.

■ Links to Lipids 3.5, Polypeptide formation 3.6

The **endomembrane system** is a series of interacting organelles between nucleus and plasma membrane. Its main function is to make lipids, enzymes, and other proteins destined for secretion, or for insertion into cell membranes. It also destroys toxins, recycles wastes, and has other specialized functions. The system's components vary among different types of cells, but here we present the most common ones (Figure 4.15).

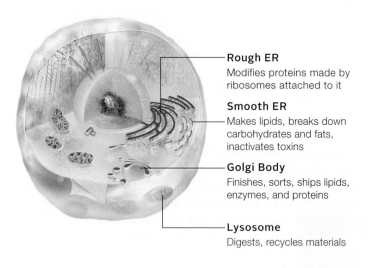

Rough ER
Modifies proteins made by ribosomes attached to it

Smooth ER
Makes lipids, breaks down carbohydrates and fats, inactivates toxins

Golgi Body
Finishes, sorts, ships lipids, enzymes, and proteins

Lysosome
Digests, recycles materials

Figure 4.15 Animated Endomembrane system, where many proteins are modified and lipids are built. These molecules are sorted and shipped to cellular destinations or to the plasma membrane for export.

A Series of Interacting Vessels

Endoplasmic Reticulum Part of the endomembrane system is an extension of the nuclear envelope called **endoplasmic reticulum**, or **ER.** ER forms a continuous compartment that folds into flattened sacs and tubes. Two kinds of ER, rough and smooth, are named for their appearance in electron micrographs. Thousands of ribosomes transiently attached to the outer surface of rough ER give this organelle its "rough" appearance. These ribosomes make polypeptides that thread into the interior of the ER as they are assembled ❶. Inside the ER, the polypeptides fold and take on their tertiary structure. Some of them become part of the ER membrane itself.

Cells that make, store, and secrete proteins have a lot of rough ER. For example, ER-rich gland cells in the pancreas make and secrete enzymes that help digest meals in the small intestine.

Smooth ER has no ribosomes on its surface ❷, so it does not make protein. Some of the polypeptides made in the rough ER end up as enzymes in the smooth ER. These enzymes assemble most of the lipids that form the cell's membranes. They also break down carbohydrates, fatty acids, and some drugs and poisons. In skeletal muscle cells, one type of smooth ER stores calcium ions and has a role in muscle contraction.

Vesicles Small, membrane-enclosed, saclike **vesicles** form in great numbers, in a variety of types, either on their own or by budding ❸. Many vesicles transport

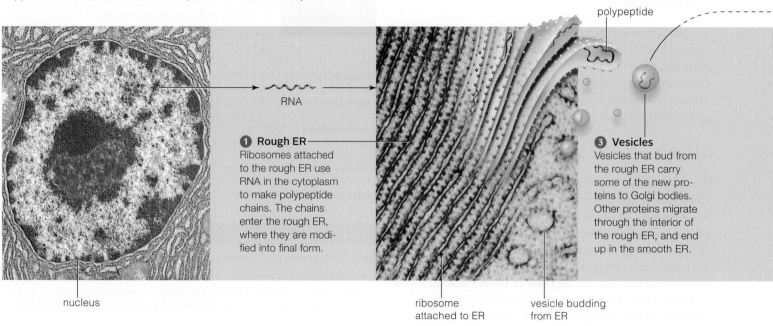

polypeptide

RNA

❶ Rough ER
Ribosomes attached to the rough ER use RNA in the cytoplasm to make polypeptide chains. The chains enter the rough ER, where they are modified into final form.

❸ Vesicles
Vesicles that bud from the rough ER carry some of the new proteins to Golgi bodies. Other proteins migrate through the interior of the rough ER, and end up in the smooth ER.

nucleus

ribosome attached to ER

vesicle budding from ER

substances from one organelle to another, or to and from the plasma membrane. Some form as a patch of plasma membrane sinks into the cytoplasm. Others bud from the ER or Golgi membranes and transport substances to the plasma membrane for export.

Vesicles that are a bit like trash cans collect and dispose of waste, debris, or toxins. Enzymes in ER-derived peroxisomes break down fatty acids, amino acids, and toxins such as alcohol. **Lysosomes** that bud from Golgi bodies take part in intracellular digestion. They contain powerful enzymes that can break down carbohydrates, proteins, nucleic acids, and lipids. Vesicles in white blood cells or amoebas deliver ingested bacteria, cell parts, and other debris to lysosomes for destruction.

central vacuole Fluid-filled vesicle in many plant cells.
endomembrane system Series of interacting organelles (endoplasmic reticulum, Golgi bodies, vesicles) between nucleus and plasma membrane; produces lipids, proteins.
endoplasmic reticulum (ER) Organelle that is a continuous system of sacs and tubes; extension of the nuclear envelope. Smooth ER makes lipids and breaks down carbohydrates and fatty acids; rough ER modifies polypeptides made by ribosomes on its surface.
Golgi body Organelle that modifies polypeptides and lipids; also sorts and packages the finished products into vesicles.
lysosome Enzyme-filled vesicle that functions in intracellular digestion.
peroxisome Enzyme-filled vesicle that breaks down amino acids, fatty acids, and toxic substances.
vacuole A fluid-filled organelle that isolates or disposes of waste, debris, or toxic materials.
vesicle Small, membrane-enclosed, saclike organelle; different kinds store, transport, or degrade their contents.

Vacuoles are vesicles that appear empty under a microscope. They form by the fusion of multiple vesicles, and have various functions depending on cell type. Many isolate or dispose of waste, debris, and toxins. Amino acids, sugars, ions, wastes, and toxins accumulate in the water-filled interior of a plant cell's large central vacuole. Fluid pressure in a central vacuole keeps plant cells plump, so stems, leaves, and other structures stay firm. As you can see in **Figure 4.11B**, a central vacuole typically takes up 50 to 90 percent of a plant cell's interior, with cytoplasm confined to a narrow zone between it and the plasma membrane.

Golgi Bodies Some vesicles fuse with and empty their contents into a Golgi body. This organelle has a folded membrane that typically looks like a stack of pancakes ❹. Enzymes in a Golgi body put finishing touches on proteins and lipids that have been delivered from the ER. They attach phosphate groups or oligosaccharides, and cleave certain polypeptides. The finished products are sorted and packaged into new vesicles that carry them to lysosomes or to the plasma membrane ❺.

Take-Home Message

What is the endomembrane system?

» The endomembrane system includes rough and smooth endoplasmic reticulum, vesicles, and Golgi bodies.

» This series of organelles works together mainly to synthesize and modify cell membrane proteins and lipids.

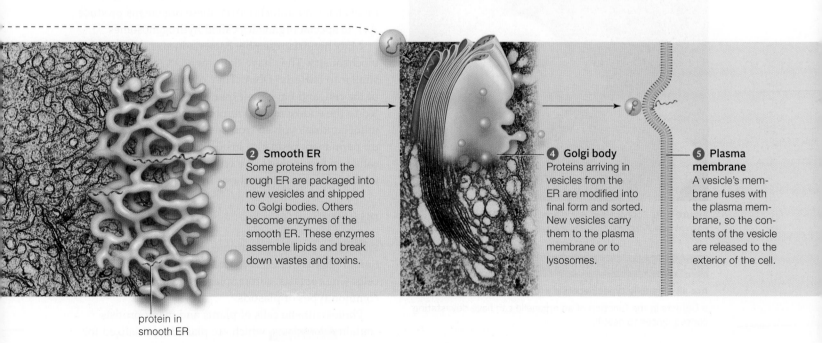

protein in smooth ER

❷ **Smooth ER**
Some proteins from the rough ER are packaged into new vesicles and shipped to Golgi bodies. Others become enzymes of the smooth ER. These enzymes assemble lipids and break down wastes and toxins.

❹ **Golgi body**
Proteins arriving in vesicles from the ER are modified into final form and sorted. New vesicles carry them to the plasma membrane or to lysosomes.

❺ **Plasma membrane**
A vesicle's membrane fuses with the plasma membrane, so the contents of the vesicle are released to the exterior of the cell.

4.8 Lysosome Malfunction

■ When lysosomes do not work properly, some cellular materials are not properly recycled, with devastating results.

Lysosomes serve as vessels for waste disposal and recycling. Enzymes inside them break large molecules into smaller subunits that the cell can use as building material or eliminate. A deficiency or malfunction in one of these enzymes can cause molecules that would normally get broken down to accumulate instead.

Cells continually make and break down gangliosides, a kind of lipid. The turnover is especially brisk during early development. In Tay–Sachs disease, the lysosomal enzyme responsible for ganglioside breakdown misfolds and is destroyed. Affected infants usually seem normal for the first few months, but symptoms appear as gangliosides accumulate to higher and higher levels inside their nerve cells. Within three to six months the child becomes irritable, listless, and may have seizures. Blindness, deafness, and paralysis follow. Affected children usually die by age five (Figure 4.16).

Tay–Sachs is a genetic disease, so it is heritable. It occurs in all populations, but it is most common in Jews of eastern European descent. Researchers continue to explore potential therapies such as blocking ganglioside synthesis, using gene therapy to deliver a normal version of the missing enzyme to the brain, or infusing normal blood cells from umbilical cords. All treatments are still considered experimental, and Tay–Sachs is still incurable.

Figure 4.16 Conner Hopf was diagnosed with Tay–Sachs disease at age 7–1/2 months. He died at 22 months.

Take-Home Message

Are all organelle types necessary for life?

» Defects in the function of an organelle can have devastating consequences to health.

4.9 Other Organelles

■ Eukaryotic cells make most of their ATP in mitochondria.
■ Plastids function in storage and photosynthesis in plants and some types of algae.
■ Links to Metabolism 3.3, ATP 3.8

Mitochondria

A mitochondrion (plural, mitochondria) is a type of eukaryotic organelle that specializes in making ATP (Figure 4.17). Aerobic respiration, an oxygen-requiring series of reactions that proceeds in mitochondria, extracts more energy from organic compounds than any other metabolic pathway. With each breath, you are taking in oxygen mainly for the mitochondria in your trillions of aerobically respiring cells.

Typical mitochondria are between 1 and 4 micrometers in length. Some are branched. These organelles can change shape, split in two, and fuse together. Each has two membranes, one highly folded inside the other. This arrangement creates two compartments. During aerobic respiration, hydrogen ions accumulate between the two membranes. The buildup causes the ions to flow across the inner mitochondrial membrane, and this flow drives ATP formation (Chapter 7 returns to the details of aerobic respiration).

Nearly all eukaryotic cells have mitochondria, but the number varies by the type of cell and by the organism. For example, a yeast cell might have only one mitochondrion, but a human skeletal muscle cell may have a thousand or more. Cells that have a very high demand for energy tend to have many mitochondria. Some eukaryotes that live in oxygen-free environments have no mitochondria; these organisms produce ATP in special organelles called hydrogenosomes.

Mitochondria resemble bacteria in size, form, and biochemistry. They have their own DNA, which is similar to bacterial DNA. They divide independently of the cell, and have their own ribosomes. Such clues led to the endosymbiont hypothesis, which states that mitochondria evolved from aerobic bacteria that took up permanent residence inside a host cell, a process called endosymbiosis. We will explore evidence for the endosymbiont hypothesis in Section 19.6.

Plastids

Plastids are double-membraned organelles that function in photosynthesis or storage in plant and algal cells. Chloroplasts, chromoplasts, and amyloplasts are common types of plastids.

Photosynthetic cells of plants and many protists contain chloroplasts, which are plastids specialized for photosynthesis. Most chloroplasts are oval or disk-shaped (**Figure 4.18A**). Each has two outer membranes enclosing a semifluid interior, the stroma, that contains enzymes and the chloroplast's own DNA. In the stroma, a third, highly folded membrane forms a single, continuous compartment. The folded membrane, which resembles stacks of flattened disks, is called the thylakoid membrane (**Figure 4.18B**). Photosynthesis occurs at this membrane.

The thylakoid membrane incorporates many pigments, the most abundant of which is chlorophyll. The abundance of green chlorophylls in plant cell chloroplasts is the reason most plants are green. During photosynthesis, chlorophylls and other molecules in the thylakoid membrane harness the energy in sunlight to drive the synthesis of ATP. The ATP is then used inside the stroma to build carbohydrates from carbon dioxide and water. (Chapter 6 describes the process of photosynthesis in more detail.) In many ways, chloroplasts resemble photosynthetic bacteria, and like mitochondria they may have evolved by endosymbiosis.

Chromoplasts make and store pigments other than chlorophylls. They have an abundance of carotenoids,

a pigment that colors many flowers, leaves, fruits, and roots red or orange. For example, as a tomato ripens, its green chloroplasts are converted to red chromoplasts, and the color of the fruit changes. The *inset* shows carotenoid-containing chromoplasts in cells of a red bell pepper.

Amyloplasts are unpigmented plastids. Typical amyloplasts store starch grains, and are notably abundant in starch-storing cells of stems, tubers (underground stems), and seeds. Starch-packed amyloplasts are dense and heavy compared to cytoplasm. In some plant cells, they function as gravity-sensing organelles.

chloroplast Organelle of photosynthesis in the cells of plants and many protists.
mitochondrion Organelle that produces ATP by aerobic respiration in eukaryotes.
plastid Category of double-membraned organelle in plants and algal cells. Different types specialize in storage or photosynthesis; e.g., chloroplast, amyloplast.

Take-Home Message

What eukaryotic organelles are specialized for producing ATP?

» Mitochondria are eukaryotic organelles that produce ATP from organic compounds in reactions that require oxygen.

» Chloroplasts are plastids that carry out photosynthesis in cells of plants and many protists.

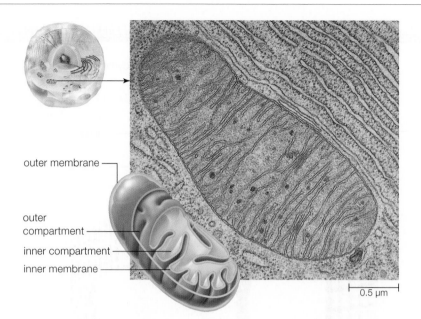

outer membrane

outer compartment

inner compartment

inner membrane

0.5 µm

Figure 4.17 Animated Sketch and transmission electron micrograph of a mitochondrion. This organelle specializes in producing large quantities of ATP.

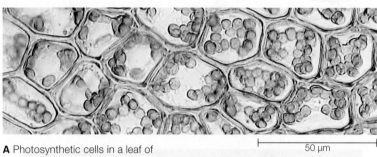

A Photosynthetic cells in a leaf of *Plagiomnium ellipticum*, a moss.

50 µm

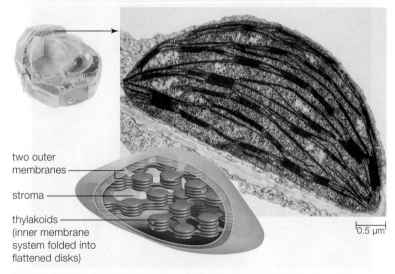

two outer membranes

stroma

thylakoids (inner membrane system folded into flattened disks)

0.5 µm

B Each chloroplast is enclosed by two outer membranes. Photosynthesis occurs at a much-folded inner membrane. The transmission electron micrograph shows a chloroplast from a tobacco leaf (*Nicotiana tabacum*). The lighter patches are nucleoids where DNA is stored.

Figure 4.18 The chloroplast.

4.10 The Dynamic Cytoskeleton

- Eukaryotic cells have an extensive and dynamic internal framework called a cytoskeleton.
- Links to Protein structure and function 3.6, 3.7

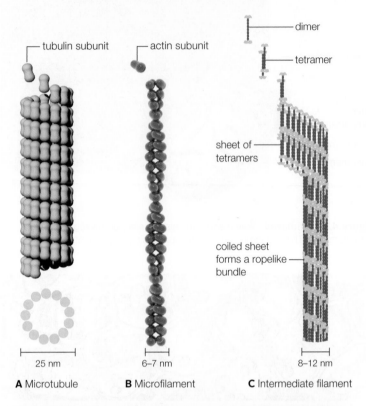

A Microtubule — tubulin subunit — 25 nm

B Microfilament — actin subunit — 6–7 nm

C Intermediate filament — dimer, tetramer, sheet of tetramers, coiled sheet forms a ropelike bundle — 8–12 nm

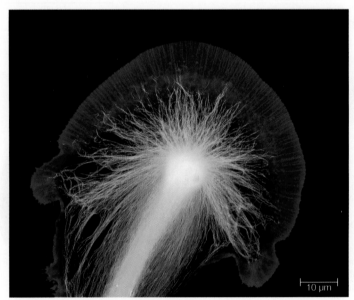

D A fluorescence micrograph shows microtubules (*yellow*) and actin microfilaments (*blue*) in the growing end of a nerve cell. These cytoskeletal elements support and guide the cell's lengthening.

10 μm

Figure 4.19 Animated Cytoskeletal elements.

Between the nucleus and plasma membrane of all eukaryotic cells is a system of interconnected protein filaments collectively called the cytoskeleton. Elements of the cytoskeleton reinforce, organize, and move cell structures, and often the whole cell. Some are permanent; others form only at certain times.

Microtubules are long, hollow cylinders that consist of subunits of the protein tubulin (**Figure 4.19A**). They form a dynamic scaffolding for many cellular processes, rapidly assembling when they are needed, disassembling when they are not. For example, before a eukaryotic cell divides, microtubules assemble, separate the cell's duplicated chromosomes, then disassemble. As another example, microtubules that form in the growing end of a young nerve cell support and guide its lengthening in a particular direction.

Microfilaments are fibers that consist primarily of subunits of the globular protein actin (**Figure 4.19B**). They strengthen or change the shape of eukaryotic cells. Crosslinked, bundled, or gel-like arrays of them make up the cell cortex, a reinforcing mesh under the plasma membrane. Actin microfilaments that form at the edge of a cell drag or extend it in a certain direction (**Figure 4.19D**). Myosin and actin microfilaments interact to bring about muscle cell contraction.

Intermediate filaments that support cells and tissues are the most stable elements of the cytoskeleton (**Figure 4.19C**). These filaments form a framework that lends structure and resilience to cells and tissues. Some kinds underlie and reinforce membranes. The nuclear lamina, for example, consists of intermediate filaments of fibrous proteins called lamins.

Among many accessory molecules associated with cytoskeletal elements are motor proteins, which move cell parts when energized by a phosphate-group transfer from ATP. A cell is like a bustling train station, with molecules and structures being moved continuously throughout its interior. Motor proteins are like freight trains, dragging their cellular cargo along tracks of dynamically assembled microtubules and microfilaments (**Figure 4.20**). Motor proteins also move external structures such as flagella and cilia. Eukaryotic flagella are structures that whip back and forth to propel cells such as sperm (*right*) through fluid. They have a different structure and type of motion than flagella of bacteria and archaea. Cilia (singular, cilium) are short, hairlike structures that project from the

flagellum

head

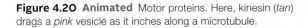

Figure 4.20 Animated Motor proteins. Here, kinesin (*tan*) drags a *pink* vesicle as it inches along a microtubule.

surface of some cells. Cilia are usually more profuse than flagella. The coordinated waving of many cilia propels cells through fluid, and stirs fluid around stationary cells. A motor protein called dynein interacts with organized arrays of microtubules to bring about movement of eukaryotic flagella and cilia. A special 9+2 array of microtubules extends lengthwise through these motile structures (**Figure 4.21**). The array consists of nine pairs of microtubules ringing another pair in the center. The microtubules grow from a barrel-shaped organelle called the **centriole,** which remains below the finished array as a **basal body.**

Amoebas and other types of eukaryotic cells form **pseudopods**, or "false feet." As these temporary, irregular lobes bulge outward, they move the cell and engulf a target such as prey. Elongating microfilaments force the lobe to advance in a steady direction. Motor proteins that are attached to the microfilaments drag the plasma membrane along with them.

basal body Organelle that develops from a centriole.
cell cortex Reinforcing mesh of cytoskeletal elements under a plasma membrane.
centriole Barrel-shaped organelle from which microtubules grow.
cilium Short, movable structure that projects from the plasma membrane of some eukaryotic cells.
cytoskeleton Dynamic framework of protein filaments that support, organize, and move eukaryotic cells and their internal structures.
intermediate filament Stable cytoskeletal element that structurally supports cells and tissues.
microfilament Reinforcing cytoskeletal element; a fiber of actin subunits.
microtubule Cytoskeletal element involved in cellular movement; hollow filament of tubulin subunits.
motor protein Type of energy-using protein that interacts with cytoskeletal elements to move the cell's parts or the whole cell.
pseudopod A temporary protrusion that helps some eukaryotic cells move and engulf prey.

Take-Home Message

What is a cytoskeleton?

» A cytoskeleton of protein filaments is the basis of eukaryotic cell shape, internal structure, and movement.

» Microtubules organize eukaryotic cells and help move their parts. Networks of microfilaments reinforce their surfaces. Intermediate filaments strengthen and maintain the shape of animal cells and tissues.

» When energized by ATP, motor proteins move along tracks of microtubules and microfilaments. As part of cilia, flagella, and pseudopods, they can move the whole cell.

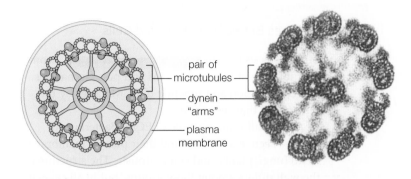

A Sketch and micrograph of a eukaryotic flagellum, cross-section. Like a cilium, it contains a 9+2 array: a ring of nine pairs of microtubules plus one pair at its core. Stabilizing spokes and linking elements that connect to the microtubules keep them aligned in a radial pattern. (Plasma membrane not visible in the micrograph.)

pair of microtubules
dynein "arms"
plasma membrane

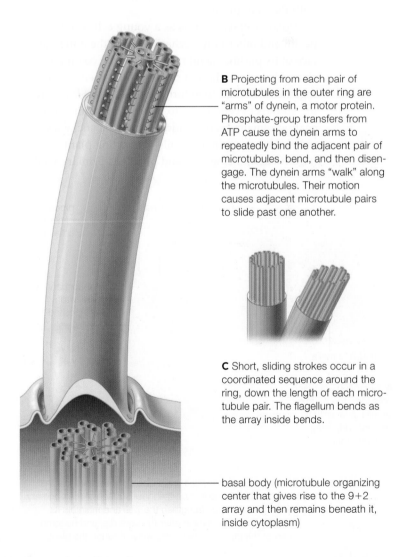

B Projecting from each pair of microtubules in the outer ring are "arms" of dynein, a motor protein. Phosphate-group transfers from ATP cause the dynein arms to repeatedly bind the adjacent pair of microtubules, bend, and then disengage. The dynein arms "walk" along the microtubules. Their motion causes adjacent microtubule pairs to slide past one another.

C Short, sliding strokes occur in a coordinated sequence around the ring, down the length of each microtubule pair. The flagellum bends as the array inside bends.

basal body (microtubule organizing center that gives rise to the 9+2 array and then remains beneath it, inside cytoplasm)

Figure 4.21 Animated Mechanism of movement of eukaryotic flagella and cilia.

4.11 Cell Surface Specializations

■ Many cells secrete materials that form a covering or matrix outside their plasma membrane.

■ Links to Tissue 1.2, Chitin 3.4

Matrixes Between and Around Cells

Many cells secrete an extracellular matrix (ECM), a complex mixture of fibrous proteins and polysaccharides. The composition and function of ECM varies by the type of cell that secretes it.

A cell wall is an example of ECM. Bacteria and archaea secrete a wall around their plasma membrane, as do fungi, plants, and some protists. The structure of the wall differs among these groups, but in all cases it protects, supports, and imparts shape to the cell. The cell wall is also porous: Water and solutes easily cross it on the way to and from the plasma membrane. Cells could not live without exchanging these substances with their environment.

A plant cell wall forms as a young cell secretes pectin and other polysaccharides onto the outer surface of its plasma membrane. The sticky coating is shared between adjacent cells, and it cements them together. Each cell then forms a primary wall by secreting strands of cellulose into the coating. Some of the coating remains as the middle lamella, a sticky layer in between the primary walls of abutting plant cells (**Figure 4.22A**). Being thin and pliable, a primary wall

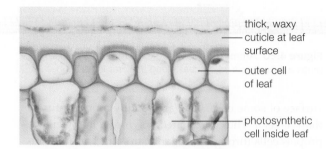

Figure 4.23 A plant ECM. Section through a plant leaf showing cuticle, a protective covering of deposits secreted by living cells.

- thick, waxy cuticle at leaf surface
- outer cell of leaf
- photosynthetic cell inside leaf

allows a growing plant cell to enlarge and change shape. At maturity, cells in some plant tissues stop enlarging and begin to secrete material onto the primary wall's inner surface. These deposits form a firm secondary wall (Figure 4.22B). One of the materials deposited is lignin, an organic compound that makes up as much as 25 percent of the secondary wall of cells in older stems and roots. Lignified plant parts are stronger, more waterproof, and less susceptible to plant-attacking organisms than younger tissues.

A cuticle is a type of ECM secreted by cells at a body surface. The cuticle secreted by cells at the surface of a leaf or stem consists of waxes and proteins (**Figure 4.23**). Plant cuticle helps a plant retain water and fend off insects. The cuticle of crabs, spiders, and other arthropods is mainly chitin, a tough polysaccharide.

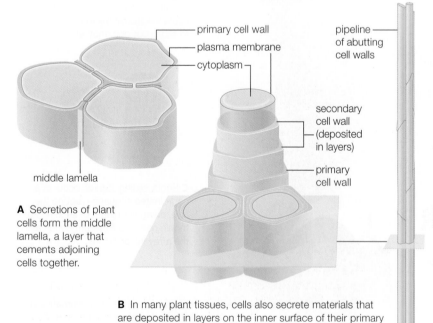

- primary cell wall
- plasma membrane
- cytoplasm
- secondary cell wall (deposited in layers)
- primary cell wall
- pipeline of abutting cell walls

middle lamella

A Secretions of plant cells form the middle lamella, a layer that cements adjoining cells together.

B In many plant tissues, cells also secrete materials that are deposited in layers on the inner surface of their primary wall. These layers strengthen the wall and maintain its shape. The walls remain after the cells die, and become part of the pipelines that carry water through the plant.

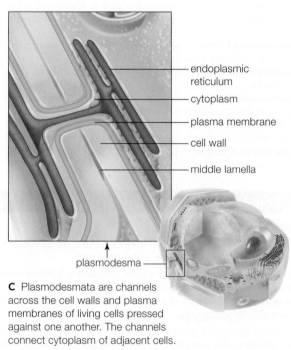

- endoplasmic reticulum
- cytoplasm
- plasma membrane
- cell wall
- middle lamella

plasmodesma

C Plasmodesmata are channels across the cell walls and plasma membranes of living cells pressed against one another. The channels connect cytoplasm of adjacent cells.

Figure 4.22 Animated Some characteristics of plant cell walls.

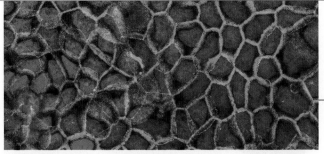

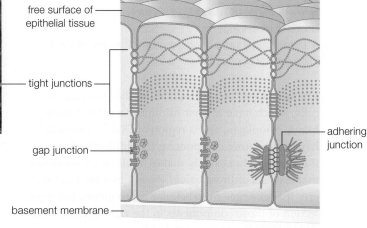

Figure 4.24 Animated Three types of cell junctions in animal tissues: tight junctions, gap junctions, and adhering junctions.

In the micrograph *above*, a profusion of tight junctions (*green*) seals abutting surfaces of kidney cell membranes and forms a waterproof tissue. The DNA in each cell nucleus appears *red*.

Inside animal bodies, other types of ECM support and organize tissues, and some have roles in cell signaling. For example, bone is an ECM composed mostly of the fibrous protein collagen, and hardened by deposits of calcium and phosphorus.

Cell Junctions

ECM does not prevent a cell from interacting with other cells or the surroundings. In multicelled species, such interaction occurs by way of cell junctions, which are structures that connect a cell to other cells and to the environment. Cells send and receive ions, molecules, or signals through some junctions. Other kinds help cells recognize and stick to each other and to extracellular matrix.

In plants, open channels called plasmodesmata (singular, plasmodesma) extend across cell walls, connecting the cytoplasm of adjoining cells (**Figure 4.22C**). Substances such as water, nutrients, and signaling molecules can flow quickly from cell to cell through these cell junctions.

Three types of cell junctions are common in animal bodies (**Figure 4.24**). In tissues that line body surfaces and internal cavities, rows of proteins that form tight junctions between plasma membranes prevent body fluids from seeping between adjacent cells. To cross these tissues, fluid must pass directly through the cells. For example, an abundance of tight junctions in the lining of the stomach normally keeps acidic fluid from leaking out. If a bacterial infection damages this lining, acid and enzymes can erode the underlying layers. The result is a painful peptic ulcer.

Strong adhering junctions composed of adhesion proteins snap cells to one another. They also connect microfilaments and intermediate filaments inside the cell to ECM outside the cell. Skin and other tissues

that are subject to abrasion or stretching have a lot of adhering junctions. These cell junctions also strengthen contractile tissues such as heart muscle.

Gap junctions form channels that connect the cytoplasm of adjoining cells, thus permitting ions and small molecules to pass directly from the cytoplasm of one cell to another. By opening or closing, these channels allow entire regions of cells to respond to a single stimulus. Heart muscle and other tissues in which the cells perform a coordinated action have many of these communication channels.

adhering junction Cell junction composed of adhesion proteins; anchors cells to each other and extracellular matrix.
cell junction Structure that connects a cell to another cell or to extracellular matrix.
cuticle Secreted covering at a body surface.
extracellular matrix (**ECM**) Complex mixture of cell secretions; supports cells and tissues; has roles in cell signaling.
gap junction Cell junction that forms a channel across the plasma membranes of adjoining animal cells.
lignin Material that stiffens cell walls of vascular plants.
plasmodesmata Cell junctions that connect the cytoplasm of adjacent plant cells.
primary wall The first cell wall of young plant cells.
secondary wall Lignin-reinforced wall that forms inside the primary wall of a plant cell.
tight junctions Arrays of fibrous proteins; join epithelial cells and collectively prevent fluids from leaking between them.

Take-Home Message

What structures form on the outside of eukaryotic cells?

» Cells of many protists, nearly all fungi, and all plants have a porous wall around the plasma membrane. Animal cells do not have walls.

» Plant cell secretions form a waxy cuticle that helps protect the exposed surfaces of soft plant parts.

» Cell secretions form extracellular matrixes between cells in many tissues.

» Cells make structural and functional connections with one another and with extracellular matrix in tissues.

4.12 The Nature of Life

■ We define life by describing the set of properties that is unique to living things.
■ Links to Life's levels of organization 1.2, Homeostasis 1.3, Philosophy of science 1.9

In this chapter, you learned about the structure of cells, which have at their minimum a plasma membrane, cytoplasm, and a region of DNA. Most cells have many other components in addition to these things. But what exactly makes a cell, or an organism that consists of them, alive? We say that life is a property that emerges from cellular components, but a collection of those components in the right amounts and proportions is not necessarily alive.

We know intuitively what "life" is, but defining it unambiguously is challenging, if not impossible. We can more easily describe what sets the living apart from the nonliving, but even that can be tricky. For example, living things have a high proportion of the organic molecules of life, but so do the remains of dead organisms in seams of coal. Living things use energy to reproduce themselves, but computer viruses, which are arguably not alive, can do that too. So how do biologists, who study life as a profession, describe life? The short answer is that their best description is very long. It consists of a list of properties associated with things we know to be alive. You already know about two of these properties of organisms:

1. They make and use the organic molecules of life.
2. They consist of one or more cells.

The remainder of this book details the other properties of living things:

3. They engage in self-sustaining biological processes such as metabolism and homeostasis.
4. They change over their lifetime, for example by growing, maturing, and aging.
5. They use DNA as their hereditary material.
6. They have the collective capacity to change over successive generations, for example by adapting to environmental pressures.

Collectively, these properties characterize living things as different from nonliving things.

Carbon, hydrogen, oxygen, and other atoms of organic molecules are the stuff of you, and us, and all of life. Yet it takes far more than organic molecules to complete the picture. Life continues only as long as a continuous flow of energy sustains its organization, because it takes energy to assemble molecules into cells. Life is no more and no less than a marvelously complex system for prolonging order. Sustained by energy inputs, it continues by a defining capacity for self-reproduction. With energy and the hereditary codes of DNA, matter becomes organized, generation after generation. Even with the death of individuals, life elsewhere is prolonged. With each death, molecules are released and may be cycled as raw materials for new generations.

Take-Home Message

What is life?

» We describe the characteristic of "life" in terms of a set of properties unique to living things.

» In living things, the molecules of life are organized as one or more cells that engage in self-sustaining biological processes.

» Organisms use DNA as their hereditary material.

» Living things change over lifetimes, and over generations.

Food for Thought (revisited)

One food safety measure involves sterilization, which kills *E. coli* O157:H7 and other bacteria. Recalled, contaminated ground beef is typically cooked or otherwise sterilized, then processed into ready-to-eat products. Raw beef trimmings are effectively sterilized when sprayed with ammonia and ground to a paste. The resulting meat product is routinely added as a filler to hamburger patties, fresh ground beef, hot dogs, lunch meats, sausages, frozen entrees, canned foods, and other items sold to quick service restaurants, hotel and restaurant chains, institutions, and school lunch programs.

Meat, poultry, milk, and fruits sterilized by exposure to radiation are available in supermarkets. By law, irradiated foods must be marked with the symbol on the *right*, but foods sterilized with chemicals are not currently required to carry any disclosure. Some worry that sterilization may alter food or leave harmful chemicals in it. Whether any health risks are associated with consuming sterilized foods is unknown.

How would you vote? Some think the safest way to protect consumers from food poisoning is by sterilizing food with chemicals or radiation to kill any bacteria that may be in it. Others think we should prevent food from getting contaminated in the first place by enacting stricter laws governing farming practices and food preparation. Both approaches add cost to our food. Which approach would you choose?

Table 4.4 Summary of Typical Components of Prokaryotic and Eukaryotic Cells

Cell Component	Main Functions	Bacteria, Archaea	Eukaryotes			
			Protists	Fungi	Plants	Animals
Cell wall	Protection, structural support	✔*	✔*	✔	✔	–
Plasma membrane	Control of substances moving into and out of cell	✔	✔	✔	✔	✔
Nucleus	Physical separation of DNA from cytoplasm	–**	✔	✔	✔	✔
DNA	Encoding of hereditary information	✔	✔	✔	✔	✔
Nucleolus	Assembly of ribosome subunits		✔	✔	✔	✔
Ribosome	Protein synthesis	✔	✔	✔	✔	✔
Endoplasmic reticulum (ER)	Initial modification of polypeptide chains; lipid synthesis	–	✔	✔	✔	✔
Golgi body	Final modification of proteins, lipid assembly, and packaging of both for use inside cell or export	–	✔	✔	✔	✔
Lysosome	Intracellular digestion	–	✔	✔*	✔*	✔
Mitochondrion	Aerobic production of ATP	–	✔	✔	✔	✔
Hydrogenosome	Anaerobic production of ATP	–	✔*	✔*	–	✔*
Photosynthetic pigments	Light-to-energy conversion	✔*	✔*	–	✔	–
Chloroplast	Photosynthesis; some starch storage	–	✔*	–	✔	–
Central vacuole	Increasing cell surface area; storage	–	–	✔*	✔	–
Bacterial flagellum	Locomotion through fluid surroundings	✔*	–	–	–	–
Eukaryotic flagellum or cilium	Locomotion through or motion within fluid surroundings	–	✔*	✔*	✔*	✔
Cytoskeleton	Cell shape; internal organization; basis of cell movement and, in many cells, locomotion	✔	✔*	✔*	✔*	✔

*Known to be present in some species.

** One or two lipid bilayers surround the DNA of some species

5 Metabolism

LEARNING ROADMAP

Where you have been In this chapter, you will gain insight into the one-way flow of energy (Section 1.3) through the world of life (1.2). You will learn more about specific types of energy (2.5) and laws that describe energy in general (1.9) as you revisit the structure and function of atoms (2.3), molecules (2.4, 3.2, 3.3–3.6), and cells (4.6, 4.7, 4.10, 4.11).

Where you are now

Energy Flow
Each time energy is transferred, some of it disperses. Organisms maintain their organization only by continually harvesting energy from the environment.

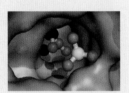

How Enzymes Work
Enzymes increase the rate of reactions. Cofactors assist enzymes, and environmental factors such as temperature, salinity, and pH can influence enzyme function.

The Nature of Metabolism
Energy-driven sequences of enzyme-mediated reactions build, convert, and dispose of materials in cells. Controls that govern steps in these pathways quickly shift cell activities.

Movement of Fluids
Gradients drive the directional movements of substances across membranes. Water tends to diffuse across cell membranes to regions where solute concentration is higher.

Membrane Transport
Transport proteins that work with or against gradients adjust or maintain solute concentrations. Particles and substances in bulk are moved across membranes inside vesicles.

Where you are going The next two chapters detail two major metabolic pathways by which cells store and retrieve energy in chemical bonds: photosynthesis (Chapter 6) and aerobic respiration (Chapter 7). You will reencounter metabolic reactions in the context of cancer (Chapter 10) and inherited disease (Chapter 14); and membrane proteins in processes of immunity (Chapter 37). Later chapters return to the topic of membrane transport and calcium ions in cell signaling (Chapters 30, 32, and 35), and how enzymes participate in digestion (Chapter 39).

5.1 A Toast to Alcohol Dehydrogenase

Although most college students are under the legal drinking age, alcohol use and abuse continues to be the most serious drug problem on college campuses in the United States. Before you drink, consider what you are consuming. A bottle of beer, a glass of wine, or a shot of vodka all contain the same amount of alcohol or, more precisely, ethanol.

Ethanol molecules move quickly from the stomach and small intestine into the bloodstream. Almost all of the ethanol ends up in the liver, which is a large organ in the abdomen. Liver cells have impressive numbers of enzymes (Section 3.3). One of them, alcohol dehydrogenase, helps rid the body of ethanol and other toxic compounds (**Figure 5.1**).

Ethanol harms the liver. Breaking down ethanol produces molecules that directly damage liver cells, so the more a person drinks, the fewer liver cells are left to do the breaking down. Ethanol also interferes with normal processes of metabolism. For example, oxygen that would ordinarily take part in breaking down fatty acids is diverted to breaking down ethanol. As a result, fats tend to accumulate as large globules in the tissues of heavy drinkers.

Long-term, heavy drinking can cause alcoholic hepatitis, a disease characterized by inflammation and destruction of liver tissue. It can also lead to cirrhosis, or scarring of the liver. (The term cirrhosis is from the Greek *kirros*, meaning orange-colored, after the abnor-mal skin color of people with the disease.) Eventually, the liver of a heavy drinker may just quit working, with dire health consequences. The liver is the largest gland in the human body, and it has many important functions. In addition to breaking down fats and tox-ins, it helps regulate the body's blood sugar level, and it makes proteins that are essential for blood clotting, immune function, and maintaining the solute balance of body fluids. Loss of these functions can be deadly.

Heavy drinking is dangerous in the short term too. Tens of thousands of undergraduate students have been polled about their drinking habits in recent sur-veys. More than half of them reported that they regu-larly drink five or more alcoholic beverages within a two-hour period—a self-destructive behavior called binge drinking. Consuming large amounts of alcohol in a brief period of time does far more than damage one's liver. Aside from the related 500,000 injuries from accidents, the 600,000 assaults by intoxicated students, 100,000 cases of date rape, and 400,000 incidences of unprotected sex among students, binge drinking is responsible for killing or causing the death of more than 1,700 college students every year. Ethanol is toxic: If you put more of it into your body than your enzymes can deal with, then you will die.

With this sobering example, we invite you to learn about how and why your cells break down organic compounds, including toxic molecules such as ethanol.

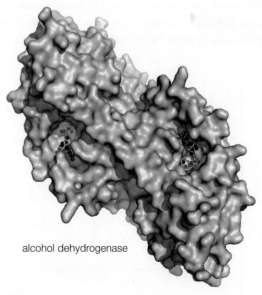

alcohol dehydrogenase

Figure 5.1 Alcohol metabolism. Alcohol dehydrogenase helps the body break down toxic alcohols such as ethanol. This enzyme makes it possible for humans to drink beer, wine, and other alcoholic beverages.

5.2 Energy in the World of Life

■ Sustaining life's organization requires ongoing energy inputs.
■ Links to Life's organization 1.2, Energy 1.3, Laws of nature 1.9, Chemical bonding 2.4, ATP 3.8

Energy Disperses

We define energy as the capacity to do work, but this definition is not very satisfying. Even brilliant physicists who study it cannot say what energy is, exactly. However, even without a perfect definition, we have an intuitive understanding of energy just by thinking about familiar forms of it, such as light, heat, electricity, and motion. We also understand intuitively that one form of energy can be converted to another. Think about how a lightbulb changes electricity into light, or how an automobile changes gasoline into the energy of motion, which is also called kinetic energy.

The formal study of heat and other forms of energy is called thermodynamics (*therm* is a Greek word for heat; *dynam* means energy). Thermodynamics researchers discovered that the total amount of energy before and after every conversion is always the same. In other words, energy cannot be created or destroyed—a phenomenon that we call the first law of thermodynamics. Remember, a law of nature describes something that occurs without fail, but our explanation of why it occurs is incomplete (Section 1.9).

Energy also tends to spread out, or disperse, until no part of a system holds more than another part. In a kitchen, for example, heat always flows from a hot pan to cool air until the temperature of both is the same. We never see cool air raising the temperature of a hot pan. Entropy is a measure of how much the energy of a particular system has become dispersed. Let's use the

Figure 5.3 It takes more than 10,000 pounds of soybeans and corn to raise a 1,000-pound steer. Where do the other 9,000 pounds go?

About half of the steer's food is indigestible. The animal's body breaks down molecules in the remaining half to access energy stored in chemical bonds. Only about 15% of that energy goes toward building body mass. The rest is lost during energy conversions, as metabolic heat.

hot pan in a cool kitchen as an example of a system. As heat flows from the pan into the air, the entropy of the system increases (**Figure 5.2**). Entropy continues to increase until the heat is evenly distributed throughout the kitchen, and there is no longer a net (or overall) flow of heat from one area to another. Our system has now reached its maximum entropy with respect to heat. The tendency of entropy to increase is the second law of thermodynamics. This is the formal way of saying that energy tends to spread out spontaneously.

Biologists use the concept of entropy as it applies to chemical bonding, because energy flow in living systems occurs mainly by the making and breaking of chemical bonds. How is entropy related to chemical bonding? Think about it just in terms of motion. Two unbound atoms can vibrate, spin, and rotate in every direction, so they are at high entropy with respect to motion. A covalent bond between the atoms restricts their movement, so they are able to move in fewer ways than they did before bonding. Thus, the entropy of two atoms decreases when a bond forms between them. Such entropy changes are part of the reason why some reactions occur spontaneously and others require an energy input, as you will see in Section 5.4.

Energy's One-Way Flow

Work occurs as a result of energy transfers. As an example, a plant cell makes glucose (work) by transferring light energy to molecules that use the energy to convert carbon dioxide and water to glucose. This particular energy transfer involves the conversion of light energy to chemical energy. Most other types of cellular work occur by the transfer of chemical energy from one molecule to another.

As you learn about such processes, remember that every time energy is transferred, a bit of it disperses. The energy lost from the transfer is usually in the form

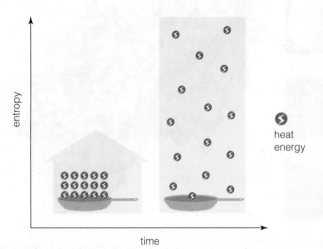

heat energy

Figure 5.2 Entropy. Entropy tends to increase, but the total amount of energy in any system always stays the same.

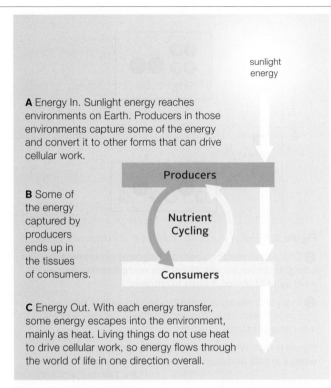

A **Energy In.** Sunlight energy reaches environments on Earth. Producers in those environments capture some of the energy and convert it to other forms that can drive cellular work.

Producers

Nutrient Cycling

Consumers

B Some of the energy captured by producers ends up in the tissues of consumers.

C **Energy Out.** With each energy transfer, some energy escapes into the environment, mainly as heat. Living things do not use heat to drive cellular work, so energy flows through the world of life in one direction overall.

Figure 5.4 Animated Energy flows from the environment into living organisms, then back to the environment. The flow drives a cycling of materials among producers and consumers.

Figure 5.5 Potential energy. By opposing the downward pull of gravity, the rope attached to the rock prevents the man from falling. Similarly, a chemical bond keeps two atoms together by opposing entropy.

of heat. As a simple example, a typical incandescent lightbulb converts about 5 percent of the energy of electricity into light. The remaining 95 percent of the energy ends up as heat that radiates from the bulb.

Dispersed heat is not very useful for doing work, and it is not easily converted to a more useful form of energy (such as electricity). Because some of the energy in every transfer disperses as heat, and heat is not useful for doing work, we can say that the total amount of energy available for doing work in the universe is always decreasing.

Is life an exception to this inevitable flow? An organized body is hardly dispersed. Energy becomes concentrated in each new organism as the molecules of life organize into cells. Even so, living things constantly use energy—to grow, to move, to acquire nutrients, to reproduce, and so on—and some energy is lost in every one of these processes (**Figure 5.3**). Unless those losses are replenished with energy from another source, the complex organization of life will end.

Most of the energy that fuels life on Earth comes from the sun. In our world, energy flows from the sun, through producers, then consumers (**Figure 5.4**). During this journey, energy is transferred many times. With each transfer, some energy escapes as heat until, eventually, all of it is permanently dispersed. However, the second law of thermodynamics does not say how

quickly the dispersal has to happen. Energy's spontaneous dispersal is resisted by chemical bonds. The energy in chemical bonds is a type of potential energy, because it can be stored (**Figure 5.5**). Think of all the bonds in the countless molecules that make up your skin, heart, liver, fluids, and other body parts. Those bonds hold the molecules, and you, together—at least for the time being.

energy The capacity to do work.
entropy Measure of how much the energy of a system is dispersed.
first law of thermodynamics Energy cannot be created or destroyed.
kinetic energy The energy of motion.
potential energy Stored energy.
second law of thermodynamics Energy tends to disperse spontaneously.

Take-Home Message

What is energy?

» Energy is the capacity to do work. It can be transferred between systems or converted from one form to another, but it cannot be created or destroyed.

» Energy disperses spontaneously.

» Some energy is lost during every transfer or conversion.

» Organisms can maintain their complex organization only as long as they replenish themselves with energy they harvest from someplace else.

5.3 Energy in the Molecules of Life

■ All cells store and retrieve energy in chemical bonds of the molecules of life.

■ Links to Chemical bonds 2.4, Chemical reactions 3.3, Carbohydrates 3.4, Nucleotides 3.8

Chemical Bond Energy

Chemical reactions change molecules into other molecules. During a chemical reaction, one or more **reactants** (molecules that enter a reaction) become one or more **products** (molecules that remain at the reaction's end). Intermediate molecules may form between reactants and products.

We show a chemical reaction as an equation in which an arrow points from reactants to products:

$$2H_2 \quad + \quad O_2 \quad \longrightarrow \quad 2H_2O$$
$$\text{(hydrogen)} \quad \text{(oxygen)} \quad \quad \text{(water)}$$

A number before a chemical formula in such equations indicates the number of molecules; a subscript indicates the number of atoms of that element per molecule. Note that atoms shuffle around in a reaction, but they never disappear: The same number of atoms that enter a reaction remain at the reaction's end (**Figure 5.6**).

Every chemical bond holds energy. The amount of energy held by a particular bond depends on which elements are taking part in it. For example, the covalent bond between an oxygen and hydrogen atom in any water molecule always holds the same amount of energy. That is the amount of energy required to break the bond, and it is also the amount of energy released when the bond forms.

Bond energy and entropy both contribute to a molecule's free energy, which is the amount of energy that is available ("free") to do work.

In most reactions, the free energy of the reactants differs from the free energy of the products (**Figure 5.7**). If the reactants have less free energy than the products, the reaction will not proceed without a net energy input. Such reactions are **endergonic**, which means "energy in" ❶. If the reactants have more free energy than the products, the reaction will end with a

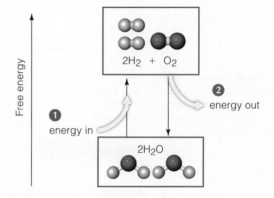

Figure 5.7 Energy inputs and outputs in chemical reactions.

❶ Endergonic reactions convert molecules with lower free energy to molecules with higher free energy, so they require a net input of energy in order to proceed.

❷ Exergonic reactions convert molecules with higher free energy to molecules with lower free energy, so they end with a net energy output.

Figure It Out: Which law of thermodynamics explains energy inputs and outputs in chemical reactions?

Answer: The first law

net release of free energy. Such reactions are **exergonic**, which means "energy out" ❷.

Why Earth Does Not Go Up in Flames The molecules of life release energy when they combine with oxygen. Think of how a spark ignites tinder-dry wood in a campfire. Wood is mostly cellulose, which consists of long chains of repeating glucose monomers (Section 3.4). The spark initiates a reaction that converts cellulose in the wood and oxygen in the air to water and carbon dioxide. The reaction is highly exergonic, which means it releases a lot of energy—enough to initiate the same reaction with other cellulose and oxygen molecules. That is why a campfire continues to burn after it has been lit.

Earth is rich in oxygen—and in potential exergonic reactions. Why doesn't it burst into flames? Luckily, chemical bonds do not break without at least a small input of energy, even in an exergonic reaction. **Activation energy** is the minimum amount of energy

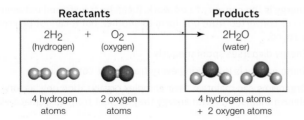

Figure 5.6 Animated Chemical bookkeeping. In equations that represent chemical reactions, reactants are written to the left of an arrow that points to the products.

A number before a formula indicates the number of molecules. Atoms may shuffle around in a reaction, but the same number of atoms that enter the reaction remain at the reaction's end.

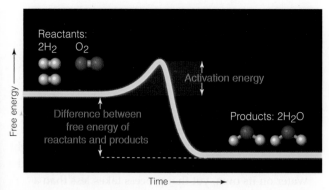

Figure 5.8 **Animated** Activation energy. Most reactions, including exergonic ones such as burning wood cellulose, will not begin without an input of activation energy, which is shown in the graph *above* as a bump in a free energy hill. Reactants in this example have more energy than the products. Activation energy keeps this and other exergonic reactions from starting spontaneously.

required to get a chemical reaction started. Activation energy is a bit like a hill that reactants must climb before they can coast down the other side to products (**Figure 5.8**).

Both endergonic and exergonic reactions have activation energy, but the amount varies with the reaction. Consider guncotton (nitrocellulose), a highly explosive derivative of cellulose. Christian Schönbein accidentally discovered a way to manufacture it when he used his wife's cotton apron to wipe up a nitric acid spill on his kitchen table, then hung it up to dry next to the oven. The apron exploded. Being a chemist in the 1800s, Schönbein immediately tried marketing guncotton as a firearm explosive, but it proved to be too unstable to manufacture. So little activation energy is needed to make guncotton react with oxygen that it tends to explode unexpectedly. Several manufacturing plants burned to the ground before guncotton was abandoned for use as a firearm explosive. The substitute? Gunpowder, which has a higher activation energy for a reaction with oxygen.

Energy In, Energy Out Cells store free energy by running endergonic reactions that build organic compounds (**Figure 5.9A**). Plant cells do this when they

make sugars during photosynthesis. Cells harvest free energy by running exergonic reactions that break the bonds of organic compounds (**Figure 5.9B**). Most cells do this when they carry out the overall reactions of aerobic respiration, which convert glucose and oxygen to carbon dioxide and water for a net output of energy. You will see in the next three sections how cells use free energy released from exergonic reactions to drive endergonic ones.

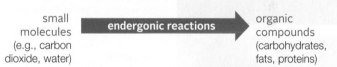

A Cells store free energy in the bonds of organic compounds.

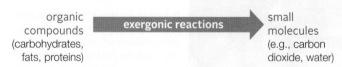

B Cells retrieve free energy stored in the bonds of organic molecules.

Figure 5.9 How cells store and retrieve free energy.

activation energy Minimum amount of energy required to start a reaction.
endergonic Describes a reaction that requires a net input of free energy to proceed.
exergonic Describes a reaction that ends with a net release of free energy.
product A molecule that remains at the end of a reaction.
reactant A molecule that enters a reaction.

Take-Home Message

How do cells use energy?

» Activation energy is the minimum amount of energy required to start a chemical reaction.

» Endergonic reactions cannot run without a net input of energy. Exergonic reactions end with a net release of energy.

» Cells store energy in chemical bonds by running endergonic reactions that build organic compounds. They harvest energy by breaking the bonds.

5.4 How Enzymes Work

■ Enzymes make specific reactions occur much faster than they would on their own.
■ Links to Temperature 2.5, pH 2.6, Enzymes 3.3, Protein structure 3.6, Denaturation 3.7

The Need for Speed

In cells, the making and breaking of chemical bonds requires enzymes. Why? Consider that centuries might pass before a sugar molecule would become CO_2 and water on its own, yet that process takes less than a millisecond inside your cells. Enzymes make the difference. In a process called catalysis, an enzyme makes a reaction run much faster than it would on its own. Enzymes are unchanged by participating in a reaction, so they can work again and again.

Some enzymes are RNAs, but most are proteins. Each kind recognizes specific reactants, or substrates, and alters them in a specific way. For instance, the enzyme hexokinase adds a phosphate group to the

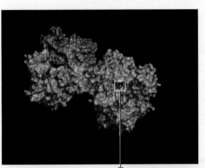

active site

enzyme

A Like other enzymes, hexokinase has active sites that bind and alter specific substrates. A model of the whole enzyme is shown to the *left*.

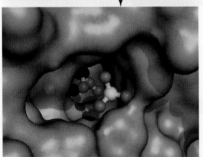

reactant(s)

B A close-up shows glucose and phosphate meeting in the active site. The microenvironment of the site favors a reaction between the two molecules.

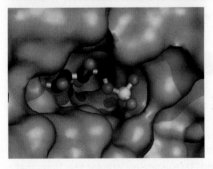

product(s)

C Here, the glucose has bonded with the phosphate. The product of this reaction, glucose-6-phosphate, is shown leaving the active site.

Figure 5.10 An example of an active site. This one is in a hexokinase, an enzyme that adds a phosphate group to glucose and other six-carbon sugars.

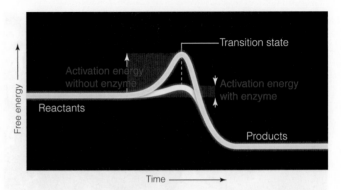

Transition state

Activation energy without enzyme

Activation energy with enzyme

Reactants

Products

Time

Figure 5.11 Animated An enzyme enhances the rate of a reaction by lowering its activation energy. **Figure It Out: Is this reaction endergonic or exergonic?**   Answer: Exergonic

hydroxyl group on the sixth carbon of glucose. Such specificity occurs because an enzyme's polypeptide chains fold up into one or more active sites, which are pockets where substrates bind and where reactions proceed (**Figure 5.10**). An active site is complementary in shape, size, polarity, and charge to the enzyme's substrate. This fit is the reason why each enzyme acts in a specific way on specific substrates.

The Transition State

When we talk about activation energy, we are really talking about the energy required to break the bonds of the reactants. Depending on the reaction, that energy may force substrates close together, redistribute their charge, or cause some other change. The change brings on the transition state, when substrate bonds reach their breaking point and the reaction will run spontaneously to product. Enzymes can help bring on the transition state by lowering activation energy (**Figure 5.11**). They do this by the following four mechanisms, which work alone or in combination.

Helping Substrates Get Together Binding at an active site brings two or more substrates close together. The closer the substrates are to one another, the more likely they are to react with one another.

Orienting Substrates in Positions That Favor Reaction On their own, substrates collide from random directions. By contrast, binding at an active site positions substrates so they align appropriately for a reaction.

Inducing a Fit Between Enzyme and Substrate By the induced-fit model, an enzyme's active site is not quite complementary to its substrate. Interacting with a substrate molecule causes the enzyme to change shape so that the fit between them improves. The improved fit may result in a stronger bond between enzyme and substrate, or it may better bring on the substrate's transition state.

Shutting Out Water Molecules Metabolism occurs in water-based fluids, but water molecules can interfere with certain reactions. The active sites of some enzymes repel water, and keep it away from the reactions.

Effects of Temperature, pH, and Salinity

Adding energy in the form of heat boosts free energy, which is one reason why molecular motion increases with temperature. The greater the free energy of reactants, the closer a reaction is to its activation energy. Thus, the rate of an enzymatic reaction typically increases with temperature. However, that rule of thumb holds only up to a point, because an enzyme denatures above a characteristic temperature. Then, the reaction rate falls sharply as the shape of the enzyme changes and it stops functioning (**Figure 5.12**).

The pH tolerance of enzymes varies. Most enzymes in the human body have an optimal pH between 6 and 8, which is perhaps not surprising as most body fluids have a pH in this range. Other enzymes work far outside of this range. Consider the enzyme pepsin, which works best between pH 1 and 2 (**Figure 5.13**). Pepsin in stomach fluid cleaves peptide bonds that hold amino acids together in proteins. Stomach fluid is very acidic, with a pH of 2. During digestion, the contents of the stomach pass into the small intestine, where the pH rises to about 9. Pepsin denatures at pH 5.5, so it becomes inactivated in the small intestine. There, other enzymes that tolerate the higher pH, including hexokinase and trypsin, take over digestion.

The activity of many enzymes is also influenced by the amount of salt in the surrounding fluid. Too little salt, and ionic functional groups in different parts of the enzyme attract one another so strongly that the enzyme's shape changes. Too much salt interferes with the hydrogen bonds that hold the enzyme in its characteristic shape, and the enzyme denatures.

active site Of an enzyme, pocket in which substrates bind and a reaction occurs.
catalysis The acceleration of a reaction by a molecule that is unchanged by participating in the reaction.
induced-fit model The concept that substrate binding to an active site of an enzyme improves the fit between the two molecules.
substrate A molecule that is specifically acted upon by an enzyme.
transition state Point during a reaction at which substrate bonds reach their breaking point and the reaction will run spontaneously.

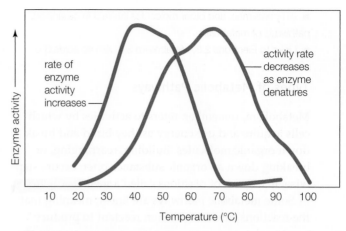

■ DNA polymerase from *E. coli*
■ DNA polymerase from *Thermus aquaticus*

Figure 5.12 Enzyme activity varies with temperature. The rate of an enzymatic reaction typically increases with temperature, up to a point. After that point, the rate falls sharply as the enzyme denatures. *E. coli* inhabits the gut (normally 37°C); *T. aquaticus* lives in hot springs, preferably at 70°C.

Graphs like the one shown here are called activity profiles. Each enzyme has a characteristic temperature/activity profile curve.

Figure It Out: At what temperature does the *E. coli* DNA polymerase work fastest?
Answer: About 37°C

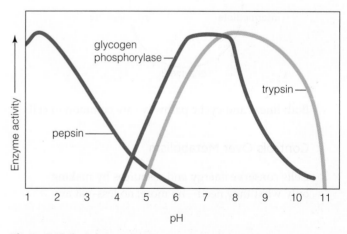

Figure 5.13 Each enzyme functions best within a characteristic range of pH. On either side of that range, the reaction rate falls sharply and the enzyme denatures.

Figure It Out: What is the optimal pH for glycogen phosphorylase?
Answer: This enzyme functions best within a range of pH between 6 and 8.

Take-Home Message

How do enzymes work?

» Enzymes greatly enhance the rate of specific reactions.

» Binding at an enzyme's active site causes a substrate to reach its transition state. In this state, the substrate's bonds are at the breaking point.

» Each enzyme works best at certain temperatures, pH, and salt concentration.

5.5 Metabolism—Organized, Enzyme-Mediated Reactions

■ ATP, enzymes, and other molecules interact in organized pathways of metabolism.

■ Links to Electrons 2.3, Metabolism 3.3, Amino acids 3.6

Types of Metabolic Pathways

Metabolism, remember, refers to activities by which cells acquire and use energy as they build and break down organic molecules. Building, rearranging, or breaking down an organic substance often occurs step-wise, in a series of reactions called a metabolic pathway.

Some metabolic pathways are linear, meaning that the reactions run straight from reactant to product:

Other metabolic pathways are cyclic. In a cyclic pathway, the last step regenerates a reactant for the first step:

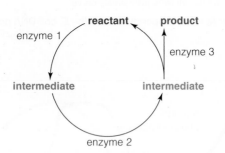

Both linear and cyclic pathways are common in cells.

Controls Over Metabolism

Cells conserve energy and resources by making only what they need—no more, no less—at any

Figure 5.14 Feedback inhibition. In this example, three kinds of enzymes act in sequence to convert a substrate to a product, which inhibits the activity of the first enzyme.

Figure It Out: Is this metabolic pathway cyclic or linear?

Answer: This is a linear pathway.

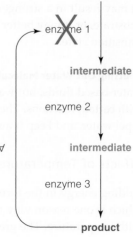

given moment. How does a cell adjust the types and amounts of molecules it produces? Several mechanisms help a cell maintain, raise, or lower its production of thousands of different substances. For example, reactions do not only run from reactants to products. Many also run in reverse at the same time, with some of the products being converted back to reactants. The rates of the forward and reverse reactions often depend on the concentrations of reactants and products: A high concentration of reactants pushes the reaction in the forward direction, and a high concentration of products pushes it in the reverse direction.

Other mechanisms more actively regulate enzymes. Certain molecules in a cell govern how fast enzyme molecules are made, or influence the activity of enzymes that have already been built. For example, the end product of a series of enzymatic reactions may inhibit the activity of one of the enzymes in the series, an effect called feedback inhibition (Figure 5.14).

Some regulatory molecules or ions activate or inhibit an enzyme by binding directly to its active site.

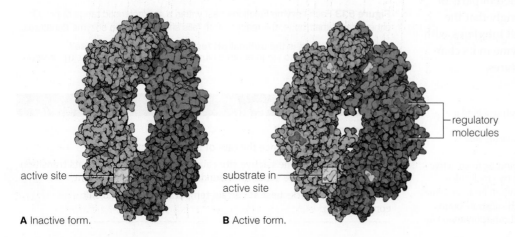

A Inactive form.　**B** Active form.

Figure 5.15 Animated An example of an allosteric effect.

(A) Pyruvate kinase is an enzyme that consists of four identical polypeptide chains. Each chain has an active site and a binding site for a regulatory molecule.

(B) When regulatory molecules (*red*) bind to the allosteric sites, the overall shape of the enzyme changes. The change makes the enzyme functional. Here, substrate molecules (*yellow*) are bound to the active sites. Small *green* balls are magnesium ions.

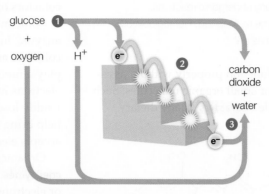

B In cells, the same overall reaction occurs in a stepwise fashion that involves an electron transfer chain. Energy is released in amounts that cells are able to harness for cellular work.

1 An input of activation energy splits glucose into carbon dioxide, electrons, and hydrogen ions (H^+).

2 Electrons lose energy as they move through an electron transfer chain. Energy released by electrons is harnessed for cellular work.

3 Electrons, hydrogen ions, and oxygen combine to form water.

A *Left*, glucose and oxygen react (burn) when exposed to a spark. Energy is released all at once as light and heat when CO_2 and water form.

Figure 5.16 Animated Comparing uncontrolled and controlled energy release.

Others bind to **allosteric** sites, which are regions of an enzyme (other than the active site) where regulatory molecules bind. *Allo–* means other, and *steric* means structure. Binding of an allosteric regulator alters the shape of the enzyme in a way that enhances or inhibits its function (**Figure 5.15**).

Redox Reactions

The bonds of organic molecules hold a lot of energy that can be released in a reaction with oxygen. One such reaction—burning—releases the energy of organic molecules all at once, explosively (**Figure 5.16A**). Cells use oxygen to break the bonds of organic molecules. However, they have no way to harvest the explosive burst of energy that occurs during burning. Instead, they break the molecules apart in several steps that release the energy in small, manageable increments. Most of these steps are oxidation–reduction reactions, or redox reactions for short. In a **redox reaction**, one molecule accepts electrons (it becomes reduced) from another molecule (which becomes oxidized). To remember what reduced means, think of how the negative charge of an electron reduces the charge of a recipient molecule. Redox reactions are also called electron transfers.

In the next two chapters, you will learn about the importance of redox reactions in electron transfer chains. An **electron transfer chain** is an organized series of reaction steps in which membrane-bound arrays of enzymes and other molecules give up and accept

electrons in turn. Electrons are at a higher energy level when they enter a chain than when they leave. Electron transfer chains can harvest the energy given off by an electron as it drops to a lower energy level (**Figure 5.16B**).

Many coenzymes deliver electrons to electron transfer chains in photosynthesis and aerobic respiration. Energy released at certain steps in those chains helps drive the synthesis of ATP.

allosteric Describes a region of an enzyme that can bind a regulatory molecule and is not the active site.
electron transfer chain Array of enzymes and other molecules that accept and give up electrons in sequence, thus releasing the energy of the electrons in usable increments.
feedback inhibition Mechanism in which a change that results from some activity decreases or stops the activity.
metabolic pathway Series of enzyme-mediated reactions by which cells build, remodel, or break down an organic molecule.
redox reaction Oxidation–reduction reaction, in which one molecule accepts electrons (it becomes reduced) from another molecule (which becomes oxidized).

Take-Home Message

What are metabolic pathways?

» Metabolic pathways are sequences of enzyme-mediated reactions. Some are biosynthetic; others are degradative.

» Control mechanisms enhance or inhibit the activity of many enzymes. The adjustments help cells produce only what they require in any given interval.

» Many metabolic pathways involve electron transfers, or redox reactions.

» Redox reactions occur in electron transfer chains. The chains are important sites of energy exchange in photosynthesis and aerobic respiration.

5.6 Cofactors in Metabolic Pathways

- Most enzymes require cofactors.
- Energy in ATP drives many endergonic reactions.
- Links to Electrons and free radicals 2.3, Hemoglobin 3.2, Nucleotides 3.8, Peroxisomes 4.7

Most enzymes do not function properly without assistance from metal ions or small organic molecules. Such enzyme helpers are called cofactors. Many dietary

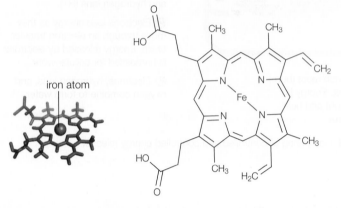

iron atom

Figure 5.17 Heme, modeled two ways. This small organic molecule participates in catalysis of many enzymatic reactions. In other contexts, it carries oxygen (such as in hemoglobin molecules), or electrons (such as in cytochrome enzymes). The iron atom at its center (Fe) can accept and donate electrons.

Figure It Out: Is heme a cofactor or a coenzyme?   Answer: It is both.

Table 5.1 Some Cofactors in Biological Systems

Cofactor	Example of Function
Metal Ions	
Iron	Redox reactions in hemes
Magnesium	Stabilizes reaction intermediate (hexokinase)
Manganese	Oxidizes water in photosynthesis
Zinc	Structural and catalytic roles in alcohol dehydrogenase
Selenium	Redox reaction in peroxide breakdown (in cytoplasm)
Copper	Reduces oxygen in mitochondrial electron transfer chain
Coenzymes	
NADH, NAD^+	Electron carrier in glycolysis
NADP, NADPH	Electron, hydrogen atom carrier in redox reactions of phosynthesis
FAD, FADH, $FADH_2$	Electron carrier in photoreceptors, redox reactions of aerobic respiration
CoA	Acetyl group ($COCH_3$) carrier in glycolysis
Heme	Electron carrier in hydrogen peroxide breakdown
Ascorbic acid	Electron carrier in peroxide breakdown (in lysosomes)
Biotin	CO_2 carrier in fatty acid synthesis

vitamins and minerals are essential because they are cofactors or are precursors for them.

Some metal ions stabilize the structure of an enzyme, in which case the enzyme denatures if the cofactors are removed. In other cases, metal cofactors play a functional role in a reaction by interacting with electrons in nearby atoms. Atoms of metal elements readily lose or gain electrons, so a metal cofactor can help bring on the transition state when it donates or accepts electrons.

Organic molecules that are cofactors are called coenzymes. Coenzymes carry chemical groups, atoms, or electrons from one reaction to another in metabolic pathways. In some reactions, they are tightly bound to the enzyme. In others, they participate as separate molecules. Unlike enzymes, many coenzymes are modified by taking part in a reaction, then become regenerated in other reactions. A few examples of cofactors are listed in **Table 5.1**. Consider NAD^+ (nicotinamide adenine dinucleotide), a coenzyme derived from niacin (vitamin B_3). NAD^+ is converted to its reduced form, NADH, by accepting electrons and hydrogen atoms. NAD^+, the oxidized form, is regenerated when the electrons and hydrogen atoms are removed from NADH in a different reaction.

We can use catalase, a peroxisome enzyme, as an example of how cofactors work. Catalase, like hemoglobin (Section 3.2), has four hemes. A heme is a small organic compound with an iron atom at its center (**Figure 5.17**). Catalase's active site holds its substrate, hydrogen peroxide (H_2O_2), close to a heme. Two H_2O_2 molecules alternately oxidize and then reduce the heme's iron atom, an interaction that causes the molecules to break down.

Hydrogen peroxide forms during normal metabolic reactions that use oxygen. It is highly reactive, easily oxidizing organic molecules or forming dangerous oxygen radicals that do. By neutralizing H_2O_2, cata-lase prevents it from reacting with (and destroying) the molecules of life. Thus, we call it an antioxidant, which means it interferes with the oxidation of other molecules. Antioxidants reduce the amount of

antioxidant Substance that prevents oxidation of other molecules.
ATP The nucleotide adenosine triphosphate.
ATP/ADP cycle Process by which cells regenerate ATP. ADP forms when ATP loses a phosphate group, then ATP forms again as ADP gains a phosphate group.
coenzyme An organic molecule that is a cofactor.
cofactor A metal ion or a coenzyme that associates with an enzyme and is necessary for its function.
phosphorylation Transfer of a phosphate group from one molecule to another.

oxidative damage that cells sustain. Oxidative damage is associated with many diseases, including cancer, diabetes, atherosclerosis, stroke, and neurodegenerative problems such as Alzheimer's disease.

ATP—A Special Coenzyme

A reaction in which a phosphate group is transferred from one molecule to another is a phosphorylation. In cells, the nucleotide coenzyme ATP (adenosine triphosphate) often couples reactions that release energy with reactions that require it. ATP has three phosphate groups (**Figure 5.18A**), and the bonds between these groups hold a lot of energy. When a phosphate group is transferred to or from a nucleotide, energy is transferred along with it. Thus, the nucleotide can receive energy from an exergonic reaction, and it can also donate energy that contributes to the "energy in" part of an endergonic reaction.

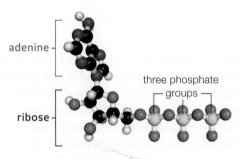

A Structure of ATP.

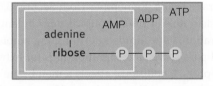

B After ATP loses one phosphate group, the nucleotide is ADP (adenosine diphosphate); after losing two phosphate groups, it is AMP (adenosine monophosphate).

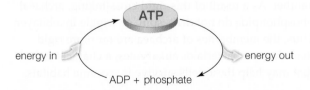

C ATP forms by endergonic reactions that phosphorylate ADP. ADP forms again when a phosphate group from ATP is used to phosphorylate another molecule. Energy from such transfers drives endergonic reactions that are the stuff of cellular work.

Figure 5.18 ATP, an important energy currency in cells.

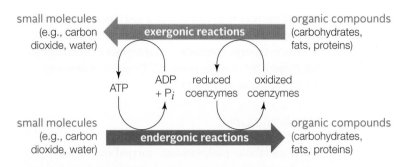

Figure 5.19 ATP and coenzymes couple energy-requiring reactions with energy-releasing ones. P_i stands for inorganic phosphate group.

ATP forms and enzymes are reduced in exergonic reactions. Energy stored in ATP and in the reduced coenzymes helps drive endergonic reactions. Compare **Figure 5.18C**.

ADP (adenosine diphosphate) forms when an enzyme transfers a phosphate group from ATP to another molecule (**Figure 5.18B**). Cells constantly run this reaction in order to drive a variety of endergonic reactions. Thus, they constantly replenish their stockpile of ATP—by running exergonic reactions that phosphorylate ADP. The cycle of using and replenishing ATP is called the ATP/ADP cycle (**Figure 5.18C**). Because ATP is such an important currency in a cell's energy economy, we use a cartoon coin to symbolize it.

Figure 5.19 shows how the ATP/ADP cycle couples endergonic reactions with exergonic ones. Cells harvest energy from organic compounds by running metabolic pathways that break them down. Energy that cells harvest in these pathways is not "released" to the environment, but rather stored in the high-energy phosphate bonds of ATP molecules and in electrons carried by reduced coenzymes. Both the ATP and the reduced cofactors that form in these pathways can be used to drive many different kinds of endergonic reactions, including pathways that build carbohydrates, fats, and proteins from smaller molecules.

Take-Home Message

How do cofactors work?

» Cofactors associate with enzymes and assist their function.

» Metal ions stabilize the structure of many enzymes. They also participate in some enzymatic reactions by donating or accepting electrons.

» Many coenzymes carry chemical groups, atoms, or electrons from one reaction to another.

» The formation of ATP from ADP is an endergonic reaction. ADP forms again when a phosphate group is transferred from ATP to another molecule. Energy from such transfers drives cellular work.

5.7 A Closer Look at Cell Membranes

■ A cell membrane is organized as a lipid bilayer with many proteins embedded in it and attached to its surfaces.

■ Links to Emergent properties 1.2, Enzymes 3.3, Lipid bilayer 3.5, Proteins 3.6, Nuclear pores 4.6, Cytoskeletal elements 4.10, Extracellular matrix 4.11

Membrane Lipids

Many metabolic pathways are carried out in cell membranes, the structural and functional foundation of which is the lipid bilayer (**Figure 5.20A**). Lipid bilayers consist mainly of phospholipids (Section 3.5). The polar head of a phospholipid interacts with water molecules; the nonpolar fatty acid tails do not. As a result of these influences, phospholipids swirled into water spontaneously organize themselves into a lipid bilayer sheet or bubble. A cell is essentially a lipid bilayer bubble filled with fluid (*left*).

Other molecules, including cholesterol and proteins, are embedded in or associated with the lipid bilayer of every cell membrane. Most of these molecules move around the membrane more or less freely. A cell membrane behaves like a two-dimensional liquid of mixed composition, so we describe it as a **fluid mosaic**. The "mosaic" part of the name comes from a cell membrane's mixed composition of lipids and pro-

A In a watery fluid, phospholipids spontaneously line up into two layers, tails to tails. This lipid bilayer spontaneously shapes itself into a sheet or a bubble. It is the basic structural and functional framework of all cell membranes. Many types of proteins and other lipids intermingle among the phospholipids.

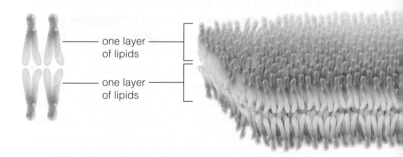

one layer of lipids

one layer of lipids

Figure 5.20 Animated Cell membrane structure. (**A**) Organization of phospholipids in cell membranes. *Opposite*, (**B–E**) Examples of common membrane proteins.

teins. The fluidity occurs because the phospholipids in a typical cell membrane are not bonded to one another. They stay organized as a bilayer as a result of collective hydrophobic and hydrophilic attractions, which, on an individual basis, are relatively weak. Thus, phospholipids in a bilayer drift sideways and spin around their long axis, and their tails wiggle.

Different kinds of cells may have different kinds of membrane lipids. Remember from Section 3.5 that the fatty acid tails of phospholipids vary in length and saturation. Glycolipids intermingling among the phospholipids serve as recognition tags for extracellular receptors. Cholesterol affects a membrane's fluidity and permeability. Archaea do not even build their phospholipids with fatty acids. Instead, they use molecules that have reactive side chains, so the tails of archaeal phospholipids form covalent bonds with one another. As a result of this rigid crosslinking, archaeal phospholipids do not drift, spin, or wiggle in a bilayer. Thus, the membranes of archaea are far more rigid than those of bacteria or eukaryotes, a characteristic that may help these cells survive in extreme habitats.

Membrane Proteins

Many types of proteins are associated with a cell membrane (**Table 5.2**). These proteins can be assigned to one of two categories, depending on the way they

Table 5.2	Common Types of Membrane Proteins	
Category	**Function**	**Examples**
Passive transporters	Allow ions or small molecules to cross a membrane to the side where they are less concentrated. Open or gated channels.	Porins; glucose transporter
Active transporters	Pump ions or molecules through membranes to the side where they are more concentrated. Require energy input, as from ATP.	Calcium pump; serotonin transporter
Receptors	Initiate change in a cell activity by responding to an outside signal (e.g., by binding a signaling molecule or absorbing light energy).	Insulin receptor; B cell receptor
Adhesion proteins	Help cells stick to one another, to cell junctions, and to extracellular matrix.	Integrins; cadherins
Recognition proteins	Identify cells as self (belonging to one's own body or tissue) as opposed to nonself (foreign to the body).	MHC molecules
Enzymes	Speed a specific reaction. Membranes provide a relatively stable reaction site for enzymes that work in series with other molecules.	Cytochrome c oxidase of mitochondria

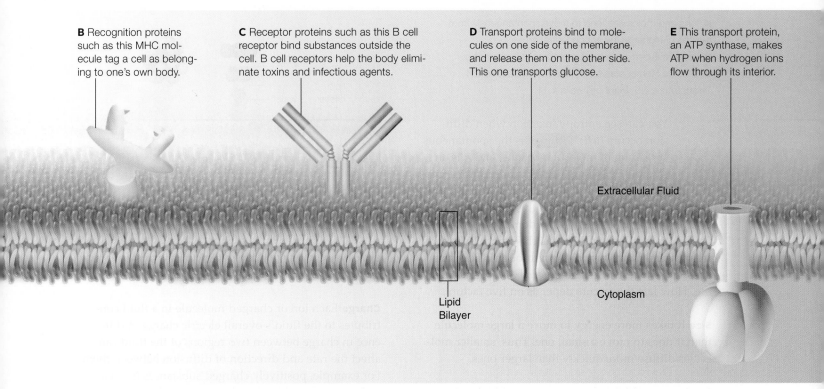

B Recognition proteins such as this MHC molecule tag a cell as belonging to one's own body.

C Receptor proteins such as this B cell receptor bind substances outside the cell. B cell receptors help the body eliminate toxins and infectious agents.

D Transport proteins bind to molecules on one side of the membrane, and release them on the other side. This one transports glucose.

E This transport protein, an ATP synthase, makes ATP when hydrogen ions flow through its interior.

Extracellular Fluid

Cytoplasm

Lipid Bilayer

are attached to the lipid bilayer. Peripheral membrane proteins temporarily attach to one of the lipid bilayer's surfaces by way of interactions with lipids or other proteins. Integral membrane proteins permanently attach to a bilayer. Some have hydrophobic regions that span the entire bilayer. Such domains anchor the protein in the membrane, and some form channels through it.

A cell membrane physically separates an external environment from an internal one, but that is not its only task. Each type of protein in a membrane imparts a specific function to it. Thus, a cell membrane can have different characteristics depending on the proteins in it. For example, the plasma membrane has certain proteins that no internal cell membrane has. Adhesion proteins in the plasma membrane fasten cells together in animal tissues. Recognition proteins function as identity tags for a cell type or individual (**Figure 5.20B**). Being able to recognize "self" imparts the potential ability to distinguish nonself (foreign) cells or particles. Receptor proteins bind to specific extracellular substances such as hormones or toxins, or to molecules on another cell's plasma membrane (**Figure 5.20C**). Binding triggers a change in the cell's activities that may involve metabolism, movement, division, or even cell death.

Additional proteins occur on all cell membranes. Many peripheral membrane proteins are enzymes.

Transport proteins move specific substances across a membrane, typically by forming a channel through it (**Figure 5.20D,E**). We return to the topic of transport proteins in Section 5.9.

Some proteins stay put, including those that cluster as nuclear pores. Cytoskeletal elements lock these and other proteins in place. For example, plasma membrane adhesion proteins connect intermediate filaments of adjacent cells in animal tissues. This arrangement strengthens a tissue, and it can constrain certain membrane proteins to an upper or lower surface of the cells.

adhesion protein Membrane protein that helps cells stick together in animal tissues.
fluid mosaic Model of a cell membrane as a two-dimensional fluid of mixed composition.
receptor protein Plasma membrane protein that binds to a particular substance outside of the cell.
recognition protein Plasma membrane protein that identifies a cell as belonging to self (one's own body).
transport protein Protein that passively or actively assists specific ions or molecules across a membrane.

Take-Home Message

What is a cell membrane?

» The structural foundation of all cell membranes is the lipid bilayer.

» Adhesion proteins, recognition proteins, transport proteins, receptors, and enzymes embedded in or associated with the lipid bilayer impart functionality to a cell membrane.

5.8 Diffusion and Membranes

■ Ions and molecules tend to move spontaneously from regions of higher to lower concentration.
■ Water diffuses across cell membranes by osmosis.
■ Links to Homeostasis 1.3, Ions 2.3, Molecules 2.4, Temperature 2.5, Plant cell walls 4.11

You can see entropy in action when diffusion occurs (*left*). **Diffusion** is the spontaneous spreading of molecules or ions, and it is an essential way in which substances move into, through, and out of cells. Molecules and ions are always jiggling, and this internal movement causes them to randomly collide with one another millions of times each second. Rebounds from the collisions propel them through a liquid or gas, with the result being a gradual and complete mixing. How fast this occurs depends on five factors:

Size It takes more energy to move a large molecule than it does to move a small one. Thus, smaller molecules diffuse more quickly than larger ones.

Temperature Molecules move faster at higher temperature, so they collide more often. Thus, the higher the temperature, the faster the rate of diffusion.

Concentration How much of a solute is dissolved in a given amount of fluid is the solute's concentration. A difference in solute concentration between adjacent regions of solution is a concentration gradient. Solutes tend to diffuse "down" their concentration gradient, from a region of higher concentration to one of lower concentration. Why? When molecules are more crowded, they collide more often. Thus, during a given interval, more molecules get bumped out of a region of higher concentration than get bumped into it.

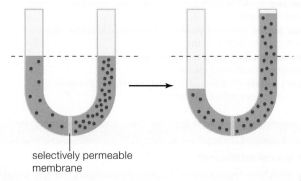

selectively permeable membrane

Figure 5.21 Animated Osmosis. Water moves across a selectively permeable membrane that separates two fluids of differing solute concentration. The fluid volume changes in the two compartments as water follows its gradient and diffuses across the membrane.

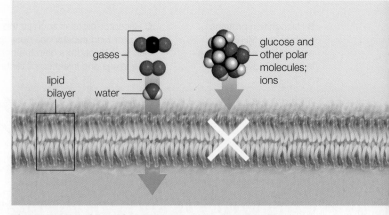

Figure 5.22 Animated Selective permeability of lipid bilayers. Hydrophobic molecules, gases, and water molecules can cross a lipid bilayer on their own. Ions in particular and most polar molecules such as glucose cannot.

Charge Each ion or charged molecule in a fluid contributes to the fluid's overall electric charge. A difference in charge between two regions of the fluid can affect the rate and direction of diffusion between them. For example, positively charged substances (such as sodium ions) will tend to diffuse toward a region with an overall negative charge.

Pressure Diffusion may be affected by a difference in pressure between two adjoining regions. Pressure squeezes molecules together, and molecules that are more crowded collide and rebound more frequently. Thus, diffusion occurs faster at higher pressures.

Semipermeable Membranes

Tonicity refers to the total concentration of solutes in fluids separated by a selectively permeable membrane (such membranes allow some substances, but not others, to cross). When the overall solute concentrations of the two fluids differ, the fluid with the lower concentration of solutes is said to be **hypotonic** (*hypo–*, under). The other one, with the higher solute concentration, is **hypertonic** (*hyper–*, over). Fluids that are **isotonic** have the same overall solute concentration.

A typical selectively permeable membrane allows water to cross it, but not solutes. When this type of membrane separates two fluids that are not isotonic, water will diffuse across the membrane, and move from the hypotonic fluid into the hypertonic one (**Figure 5.21**). The diffusion will continue until the two fluids are isotonic, or until some pressure against the hypertonic fluid counters it. The diffusion of water across a membrane is so important in biology that it is given a special name: **osmosis**.

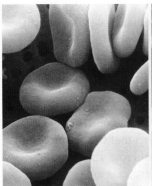 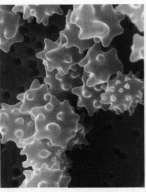 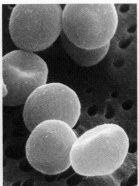 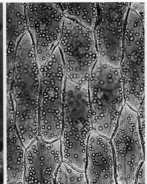 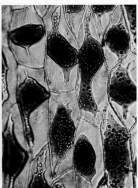

A Red blood cells immersed in an isotonic solution do not change in volume. The fluid portion of blood is typically isotonic with cytoplasm.

B Red blood cells immersed in a hypertonic solution shrivel up because more water diffuses out of the cells than into them.

C Red blood cells immersed in a hypotonic solution swell up because more water diffuses into the cells than out of them.

D Cells in an iris petal, plump with water. The cytoplasm extends to the cell wall.

E Cells from a wilted iris petal. The cytoplasm shrank, and the plasma membrane moved away from the wall.

Figure 5.23 Effects of tonicity in human red blood cells (**A–C**), and iris petal cells (**D,E**).

A lipid bilayer is one type of selectively permeable membrane. Water can cross it, but ions and most polar molecules cannot (**Figure 5.22**). If a cell's cytoplasm is hypertonic with respect to the fluid outside of its plasma membrane, water diffuses into it. If the cytoplasm is hypotonic with respect to the fluid on the outside, water diffuses out. In either case, the solute concentration of the cytoplasm may change. If it changes enough, the cell's enzymes will stop working, with lethal results. Most free-living cells, and some that are part of multicelled organisms, have built-in mechanisms that compensate for differences in tonicity between cytoplasm and external fluid. In cells with no such mechanism, the volume—and solute concentration—of cytoplasm will change as water diffuses into or out of the cell (**Figure 5.23A–C**).

Turgor

Cell walls of plants and many protists, fungi, and bacteria can resist an increase in the volume of cytoplasm even in hypotonic environments. In the case of plant cells, cytoplasm usually contains more solutes than soil water does. Thus, water usually diffuses from soil into a plant—but only up to a point. Rigid walls keep plant cells from expanding very much. Osmosis can cause pressure to build up inside these cells. Pressure that a fluid exerts against a structure that contains it is called turgor. When enough pressure builds up inside a plant cell, water stops diffusing into its cytoplasm. The amount of turgor that stops osmosis is called osmotic pressure.

Osmotic pressure keeps walled cells plump, just as high air pressure inside a tire keeps it inflated.

A young land plant can resist gravity to stay erect because its cells are plump with cytoplasm. When soil dries out, it loses water but not solutes, so the concentration of solutes increases in whatever water remains in it. If soil water becomes hypertonic with respect to cytoplasm, water will start diffusing out of the plant's cells, so their cytoplasm shrinks (**Figure 5.23D,E**). As turgor inside the cells decreases, the plant wilts.

concentration Number of molecules or ions per unit volume.
concentration gradient Difference in concentration between adjoining regions of fluid.
diffusion Spontaneous spreading of molecules or ions in a liquid or gas.
hypertonic Describes a fluid that has a high overall solute concentration relative to another fluid.
hypotonic Describes a fluid that has a low overall solute concentration relative to another fluid.
isotonic Describes two fluids with identical solute concentrations.
osmosis The diffusion of water across a selectively permeable membrane in response to a concentration gradient.
osmotic pressure Amount of turgor that prevents osmosis into cytoplasm or other hypertonic fluid.
turgor Pressure that a fluid exerts against a wall, membrane, or other structure that contains it.

Take-Home Message

What influences the movement of ions and molecules?

» Molecules or ions tend to diffuse into an adjoining region of fluid in which they are not as concentrated.

» The steepness of a concentration gradient as well as temperature, molecular size, charge, and pressure affect the rate of diffusion.

» Osmosis is a net diffusion of water between two fluids that differ in solute concentration and are separated by a selectively permeable membrane.

» Fluid pressure that a solution exerts against a membrane or wall influences the osmotic movement of water.

5.9 Membrane Transport Mechanisms

■ Many types of molecules and ions can cross a lipid bilayer only with the help of transport proteins.

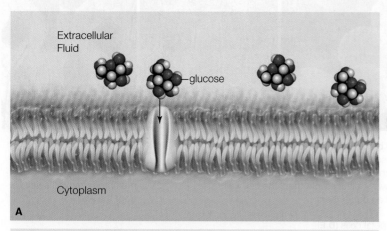

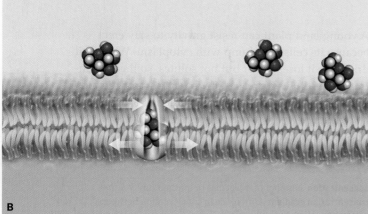

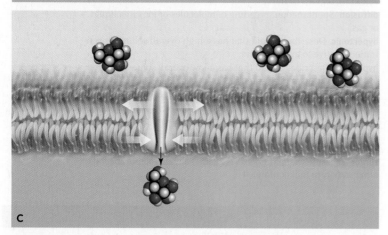

Figure 5.24 Animated Passive transport of glucose.

(**A**) A glucose molecule (here, in extracellular fluid) binds to a glucose transporter in the plasma membrane.

(**B**) Binding causes the transport protein to change shape.

(**C**) The glucose molecule detaches from the transport protein on the other side of the membrane (here, in cytoplasm), and the protein resumes its original shape.

Figure It Out: In this example, which fluid is hypotonic: extracellular fluid or the cytoplasm?
Answer: Cytoplasm

Gases, water, and small nonpolar molecules can diffuse directly across a lipid bilayer. Most other molecules, and ions in particular, cross only with the help of membrane transport proteins. Each type of transport protein can move a specific ion or molecule. Glucose transporters only transport glucose; calcium pumps only pump calcium; and so on. The specificity of transport proteins means that the amounts and types of many substances that cross a membrane depend on which transport proteins are embedded in it. Transport protein specificity also allows cells to control the volume and composition of their fluid interior by moving particular solutes one way or the other across their membranes.

For example, a glucose transporter in a plasma membrane can bind to a molecule of glucose, but not to a molecule of phosphorylated glucose. Enzymes in cytoplasm phosphorylate glucose as soon as it enters the cell. Phosphorylation prevents the molecule from moving back through the glucose transporter and leaving the cell.

Passive Transport

In **passive transport**, the movement of a solute (and the direction of the movement) through a transport protein is driven entirely by the solute's concentration gradient. For this reason, passive transport is also called facilitated diffusion. The solute simply binds to the passive transport protein, and the protein releases it to the other side of the membrane (**Figure 5.24**).

A glucose transporter is an example of a passive transport protein. This protein changes shape when it binds to a molecule of glucose. The shape change moves the solute to the opposite side of the membrane, where it detaches from the transport protein. Then, the transporter reverts to its original shape. Some passive transporters do not change shape; they form permanently open channels through a membrane. Others are gated, which means they open and close in response to a stimulus such as a shift in electric charge or binding to a signaling molecule.

Active Transport

Solute concentrations may shift in extracellular fluid or in cytoplasm. Maintaining a particular solute's concentration at a certain level often means transporting the solute against its gradient, to the side of a membrane where it is more concentrated. Pumping a solute against its gradient takes energy. In **active transport**, a transport protein uses energy to pump a solute against

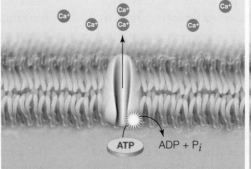

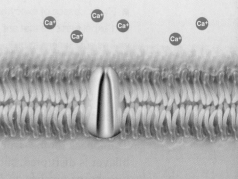

A Two calcium ions bind to the transport protein.

B Energy in the form of a phosphate group is transferred from ATP to the protein. The transfer causes the protein to change shape so that it ejects the calcium ions to the opposite side of the membrane.

C After it loses the calcium ions, the transport protein resumes its original shape.

Figure 5.25 Active transport. This is a model of a plasma membrane calcium pump.

its gradient across a cell membrane. After a solute binds to an active transporter, an energy input (often in the form of a phosphate-group transfer from ATP) changes the shape of the protein. The change causes the transporter to release the solute to the other side of the membrane.

A **calcium pump** is an example of an active transporter. This protein moves calcium ions across cell membranes (**Figure 5.25**). Calcium ions act as potent messengers inside cells, and many enzymes have allosteric sites that bind these ions. Thus, their presence in cytoplasm is tightly regulated. Calcium pumps in the plasma membrane of all eukaryotic cells can keep the concentration of calcium in cytoplasm thousands of times lower than it is in extracellular fluid.

Cotransporters are active transport proteins that move two substances at the same time, in the same or opposite directions across a membrane. Nearly all of the cells in your body have cotransporters called sodium–potassium pumps (**Figure 5.26**). Sodium ions (Na⁺) in the cytoplasm diffuse into the pump's open channel and bind to its interior. A phosphate-group transfer from ATP causes the pump to change shape. Its channel opens to extracellular fluid, where it releases the Na⁺. Then, potassium ions (K⁺) from extracellular fluid diffuse into the channel and bind to its interior. The transporter releases the phosphate

group and reverts to its original shape. The channel opens to the cytoplasm, where it releases the K⁺.

Bear in mind, the membranes of all cells, not just those of animals, have active transporters. For example, active transporters in plant leaf cells pump sugars into tubes that distribute them throughout the plant body.

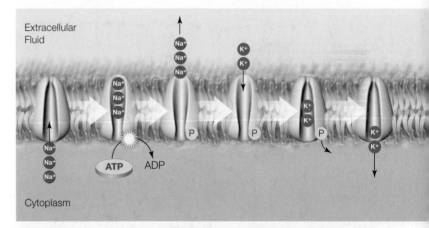

Figure 5.26 Cotransport. This model shows how a sodium–potassium pump transports sodium ions (Na⁺, *purple*) from cytoplasm to extracellular fluid, and potassium ions (K⁺, *green*) in the other direction across the plasma membrane. A phosphate-group transfer from ATP provides energy for the transport.

Take-Home Message

How do molecules or ions that cannot diffuse through a lipid bilayer cross a cell membrane?

» Transport proteins help specific molecules or ions to cross cell membranes.

» In passive transport, a solute binds to a protein that releases it on the opposite side of the membrane. The movement is driven by the solute's concentration gradient.

» In active transport, a transport protein pumps a solute across a membrane against its concentration gradient. The movement is driven by an energy input, as from ATP.

active transport Energy-requiring mechanism in which a transport protein pumps a solute across a cell membrane against its concentration gradient.
calcium pump Active transport protein; pumps calcium ions across a cell membrane against their concentration gradient.
passive transport Mechanism by which a concentration gradient drives the movement of a solute across a cell membrane through a transport protein. Requires no energy input.

5.10 Membrane Trafficking

■ By processes of exocytosis and endocytosis, cells take in and expel particles that are too big for transport proteins, as well as substances in bulk.

■ Links to Lipoproteins 3.6, Endomembrane system 4.7, Cytoskeleton 4.10

Endocytosis and Exocytosis

Think back on the structure of a lipid bilayer. When a bilayer is disrupted, such as when part of the plasma membrane pinches off as a vesicle, it seals itself. Why? The disruption exposes the nonpolar fatty acid tails of the phospholipids to their watery surroundings.

Remember, in water, phospholipids spontaneously rearrange themselves so that their tails stay together. When a patch of membrane buds, its phospholipid tails are repelled by water on both sides. The water "pushes" the phospholipid tails together, which helps round off the bud as a vesicle, and also seals the rupture in the membrane.

Patches of membrane constantly move to and from the cell surface as components of vesicles (**Figure 5.27**). The formation and movement of vesicles involves motor proteins and requires ATP.

By exocytosis, a vesicle moves to the cell's surface, and the protein-studded lipid bilayer of its membrane fuses with the plasma membrane. As the exocytic vesicle loses its identity, its contents are released to the surroundings.

There are three pathways of endocytosis, but they all take up substances near the cell's surface. A small patch of plasma membrane balloons inward, and then it pinches off after sinking farther into the cytoplasm. The membrane patch becomes the outer boundary of an endocytic vesicle, which delivers its contents to an organelle or stores them in a cytoplasmic region.

With receptor-mediated endocytosis, molecules of a hormone, vitamin, mineral, or another substance bind to receptors on the plasma membrane. A shallow pit forms in the membrane patch under the receptors. The pit sinks into the cytoplasm and closes back on itself, and in this way it becomes a vesicle (**Figure 5.28**).

The term "receptor-mediated endocytosis" is a bit misleading, because receptors also function in phagocytosis. Phagocytosis ("cell eating") is an endocytic

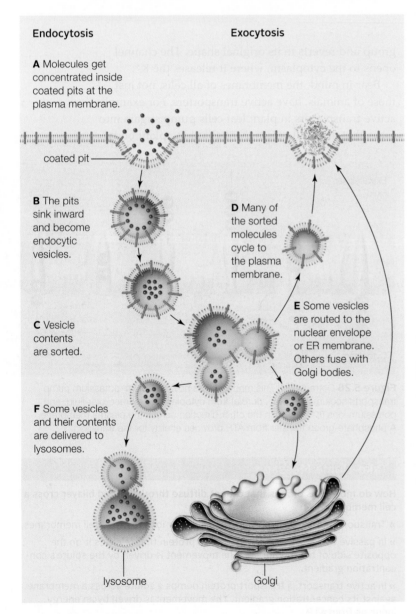

Endocytosis

A Molecules get concentrated inside coated pits at the plasma membrane.

coated pit

B The pits sink inward and become endocytic vesicles.

C Vesicle contents are sorted.

F Some vesicles and their contents are delivered to lysosomes.

lysosome

Exocytosis

D Many of the sorted molecules cycle to the plasma membrane.

E Some vesicles are routed to the nuclear envelope or ER membrane. Others fuse with Golgi bodies.

Golgi

Figure 5.27 Animated Membrane trafficking: endocytosis and exocytosis.

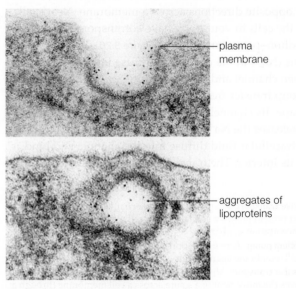

plasma membrane

aggregates of lipoproteins

Figure 5.28 Endocytosis of lipoproteins.

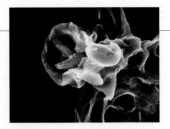

A Pseudopods of a white blood cell surround *Tuberculosis* bacteria (*red*).

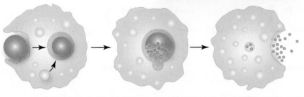

B Endocytic vesicle forms.

C Lysosome fuses with vesicle; enzymes digest pathogen.

D Cell uses the digested material or expels it.

Figure 5.29 Animated Phagocytosis. A phagocytic cell's pseudopods (extending lobes of cytoplasm) surround bacteria. The plasma membrane above the bulging lobes fuses and forms an endocytic vesicle. Once inside the cytoplasm, the vesicle fuses with a lysosome, which digests its contents.

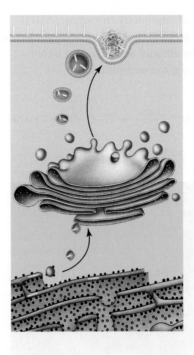

Figure 5.30 How membrane proteins become oriented to the inside or the outside of the cell.

Proteins of the plasma membrane are assembled in the ER, and finished inside Golgi bodies. The proteins (shown in *white*) become part of vesicle membranes that bud from the Golgi. The membrane proteins automatically become oriented in the proper direction when the vesicles fuse with the plasma membrane.

Figure It Out: What process does the upper arrow represent?

Answer: Exocytosis

pathway in which phagocytic cells such as amoebas engulf microorganisms, cellular debris, or other particles. In animals, macrophages and other white blood cells engulf and digest pathogenic viruses and bacteria, cancerous body cells, and other threats (**Figure 5.29A**). Phagocytosis begins when receptors bind to a particular target. The binding causes microfilaments to assemble in a mesh under the plasma membrane. The microfilaments contract, forcing some cytoplasm and plasma membrane above it to bulge outward as a lobe, or pseudopod. Pseudopods engulf a target and merge as a vesicle, which sinks into the cytoplasm and fuses with a lysosome (**Figure 5.29B,C**). Enzymes in the lysosome break down the vesicle's contents. The resulting molecular bits may be recycled by the cell, or expelled by exocytosis (**Figure 5.29D**).

Pinocytosis is an endocytic pathway that brings materials in bulk into the cell. It is not as selective as receptor-mediated endocytosis. An endocytic vesicle forms around a small volume of the extracellular fluid regardless of the kinds of substances dissolved in it.

Membrane Cycling

The composition of a plasma membrane begins in the ER. There, membrane proteins and lipids are made and modified, and both become part of vesicles that transport them to Golgi bodies for final modification. The finished proteins and lipids are repackaged as new vesicles that travel to the plasma membrane and fuse with it. The lipids and proteins of the vesicle membrane become part of the plasma membrane. This is the process by which new plasma membrane forms.

Figure 5.30 shows what happens when an exocytic vesicle fuses with the plasma membrane. Golgi bodies package membrane proteins facing the inside of a vesicle, so after the vesicle fuses with the plasma membrane, the proteins face the extracellular environment.

As long as a cell is alive, exocytosis and endocytosis continually replace and withdraw patches of its plasma membrane. If the cell is not enlarging, the total area of the plasma membrane remains more or less constant. Membrane lost as a result of endocytosis is replaced by membrane arriving as exocytic vesicles.

endocytosis Process by which a cell takes in a small amount of extracellular fluid by the ballooning inward of its plasma membrane.
exocytosis Process by which a cell expels a vesicle's contents to extracellular fluid.
phagocytosis "Cell eating"; an endocytic pathway by which a cell engulfs particles such as microbes or cellular debris.
pinocytosis Endocytosis of bulk materials.

Take-Home Message

How do large particles and bulk substances move into and out of cells?

» In exocytosis, a cytoplasmic vesicle fuses with the plasma membrane and releases its contents to the outside of the cell.

» In endocytosis, a patch of plasma membrane sinks inward and forms a vesicle in the cytoplasm.

» Phagocytosis is an endocytic pathway by which cells engulf particles such as microorganisms.

A Toast to Alcohol Dehydrogenase (revisited)

In the human body, alcohol dehydrogenase (ADH) converts ethanol to acetaldehyde, an organic molecule even more toxic than ethanol and the most likely source of various hangover symptoms:

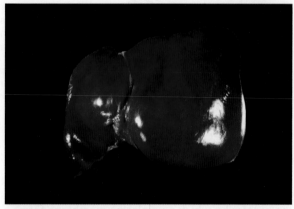

A Normal, healthy human liver.

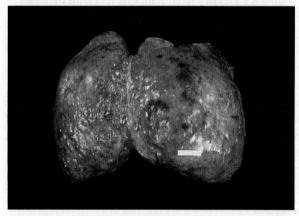

B Cirrhotic liver.

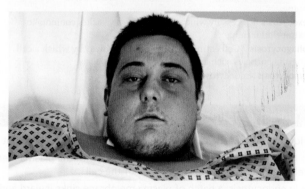

C Gary Reinbach, who died in 2009 from alcoholic liver disease shortly after this photograph was taken. His orange skin is a symptom of cirrhosis. He was 22 years old.

Figure 5.31 Alcoholic liver disease.

A different enzyme, aldehyde dehydrogenase (ALDH), very quickly converts the toxic acetaldehyde to non-toxic acetate:

Thus, the overall pathway of ethanol metabolism in humans is:

In the average adult human body, this metabolic pathway can detoxify between 7 and 14 grams of ethanol per hour. The average alcoholic beverage contains between 10 and 20 grams of ethanol, which is why having more than one drink in any two-hour interval may result in a hangover.

In most organisms, the main function of alcohol dehydrogenase is to detoxify the tiny quantities of alcohols that form in some metabolic pathways. In animals, the enzyme also detoxifies alcohols made by gut-inhabiting bacteria, and those in foods such as ripe fruit. Despite the small amounts of alcohol that humans encounter naturally, our bodies make at least nine different kinds of alcohol dehydrogenase. It is interesting to speculate about why so many of them have evolved.

We do understand how defects in the ADH enzyme affect our alcohol metabolism. For example, if a person's ADH is overactive, acetaldehyde accumulates faster than ALDH can detoxify it:

People with an overactive form of ADH become flushed and feel ill after drinking even a small amount of alcohol. The unpleasant experience may be part of the reason that these people are less likely to become alcoholic than other people.

Having underactive ALDH also causes acetaldehyde to accumulate:

Underactive ALDH is associated with the same effect—and the same protection from alcoholism—as overactive ADH. Both types of variant enzymes are common in people of Asian descent. For this reason, the alcohol flushing reaction is informally called "Asian flush."

Having an underactive ADH enzyme has the opposite effect. It slows alcohol metabolism, so people with low ADH activity may not feel the ill effects of drinking alcoholic beverages as much as other people. When these people drink alcohol, they have a tendency to become alcoholics. Compulsive, uncontrolled drinking damages an alcoholic's health and social relationships. The study mentioned in Section 5.1 showed that one-quarter of the undergraduate students who binged also had other signs of alcoholism.

Alcoholics will continue to drink despite the knowledge that doing so has tremendous negative consequences. In the United States, alcohol abuse is the leading cause of cirrhosis of the liver. The liver becomes so scarred, hardened, and filled with fat that it loses its function (**Figure 5.31A,B**). It stops making the protein albumin, so the solute balance of body fluids is disrupted, and the legs and abdomen swell with watery fluid. It cannot remove drugs and other toxins from the blood, so they accumulate in the brain—which impairs mental functioning and alters personality. Restricted blood flow through the liver causes veins to enlarge and rupture, so internal bleeding is a risk. The damage to the body results in a heightened susceptibility to diabetes and liver cancer. Once cirrhosis has been diagnosed, a person has about a 50 percent chance of dying within 10 years (**Figure 5.31C**).

How would you vote? Transplantation is a last-resort treatment for a failed liver, but even so there are not nearly enough donors for everyone who needs a liver transplant. Liver failure can be a result of factors that are generally  beyond an individual's control, such as inherited disease, cancer, illness, or infection. The damage to an alcoholic's liver is self-inflicted. Should lifestyle be a factor in deciding who gets a donated liver for transplant?

6 Photosynthesis

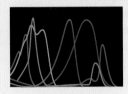

6.1 Biofuels

Today, the expression "food is fuel" is not just about eating. With fossil fuel prices soaring, there is an increasing demand for biofuels, which are oils, gases, or alcohols made from organic matter that is not fossilized. Much of the material currently used for biofuel production in the United States consists of food crops—mainly corn, soybeans, and sugarcane. Growing these crops in large quantities is typically expensive and damaging to the environment, and using them to make biofuel competes with our food supply. The diversion of food crops to biofuel production may be contributing to the worldwide increases in food prices we are seeing now.

How did we end up competing with our vehicles for food? We both run on the same fuel: energy that plants have stored in chemical bonds. Fossil fuels such as petroleum, coal, and natural gas formed from the remains of ancient swamp forests that decayed and compacted over millions of years. These fuels consist mainly of molecules originally assembled by ancient plants. Biofuels—and foods—consist mainly of molecules originally assembled by modern plants.

Autotrophs harvest energy directly from the environment, and obtain carbon from inorganic molecules (*auto*– means self; –*troph* refers to nourishment). Plants and most other autotrophs make their own food by photosynthesis, a process in which they use the energy of sunlight to assemble carbohydrates—sugars—from carbon dioxide and water. Directly or indirectly, photosynthesis also feeds most other life on Earth. Animals and other heterotrophs get energy and carbon by breaking down organic molecules assembled by other organisms (*hetero*– means other). We and almost all other organisms sustain ourselves by extracting energy from organic molecules that photosynthesizers make.

A lot of energy is locked up in the chemical bonds of molecules made by plants. That energy can fuel heterotrophs, as when an animal cell powers ATP synthesis by breaking the bonds of sugars. It can also fuel our cars, which run on energy released by burning biofuels or fossil fuels. Both processes are fundamentally the same: They release energy by breaking the bonds of organic molecules. Both use oxygen to break those bonds, and both produce carbon dioxide.

Corn and other food crops are rich in oils, starches, or sugars that can be easily converted to biofuels. The starch in corn kernels, for example, can be enzymatically broken down to glucose, which is converted to ethanol by heterotrophic bacteria or yeast. Making biofuels from other types of plant matter requires additional steps, because these materials contain a higher proportion of cellulose. Breaking down this tough, insoluble carbohydrate to its glucose monomers adds a lot of cost to the biofuel product. Researchers are currently trying to find cost-effective ways to break down the abundant cellulose in fast-growing weeds such as switchgrass (**Figure 6.1**), and agricultural wastes such as wood chips, wheat straw, cotton stalks, and rice hulls.

autotroph Organism that makes its own food using carbon from inorganic molecules such as CO_2, and energy from the environment.
heterotroph Organism that obtains energy and carbon from organic compounds assembled by other organisms.
photosynthesis Metabolic pathway by which most autotrophs capture light energy and use it to make sugars from CO_2 and water.

Figure 6.1 Biofuels. *Above*, Ratna Sharma and Mari Chinn are researching ways to reduce the cost of producing biofuel from renewable sources such as wild grasses and agricultural wastes. *Right*, switchgrass (*Panicum virgatum*) grows wild in North American prairies.

6.2 Sunlight as an Energy Source

■ Photosynthetic organisms use pigments to capture the energy of sunlight.

■ Links to Electrons 2.3, Bonding 2.4, Carbohydrates 3.4, Chromoplasts 4.9, Energy 5.2, Metabolism 5.5

Energy flow through nearly all ecosystems on Earth begins when photosynthesizers intercept energy from the sun. Harnessing the energy of sunlight for work is a complicated business, or we would have been able to do it in an economically sustainable way by now. Plants do it by converting light energy to chemical energy, which they and most other organisms use to drive cellular work. The first step involves capturing light. In order to understand how that happens, you have to understand a little about the nature of light.

Most of the energy that reaches Earth's surface is in the form of visible light, so it should not be surprising that visible light is the energy that drives photosynthesis. Visible light is a very small part of a large spectrum of electromagnetic energy radiating from the sun. Like all other forms of electromagnetic energy, light travels in waves, moving through space a bit like waves moving across an ocean. The distance between the crests of two successive waves is a **wavelength**. Light wavelength is measured in nanometers (nm).

Visible light travels in wavelengths between 380 and 750 nm (**Figure 6.2A**). Our eyes perceive all of these wavelengths combined as white light, and particular wavelengths in this range as different colors. White light separates into its component colors when it passes through a prism, or raindrops that act as tiny prisms. A prism bends longer wavelengths more than it bends shorter ones, so a rainbow of colors forms.

Light travels in waves, but it is also organized in packets of energy called photons. A photon's energy and its wavelength are related, so all photons traveling at the same wavelength carry the same amount of energy. Photons that carry the least amount of energy travel in longer wavelengths; those that carry the most energy travel in shorter wavelengths (**Figure 6.2B**). Photons of wavelengths shorter than about 380 nanometers carry enough energy to alter or break the chemical bonds of DNA and other biological molecules. That is why UV (ultraviolet) light, x-rays, and gamma rays are a threat to life.

Pigments: The Rainbow Catchers

Photosynthesizers use pigments to capture light. A **pigment** is an organic molecule that selectively absorbs light of specific wavelengths. Wavelengths of light that are not absorbed are reflected, and that reflected light gives each pigment its characteristic color.

Chlorophyll a is the most common photosynthetic pigment in plants, and also in photosynthetic protists and bacteria. Chlorophyll *a* absorbs violet and red

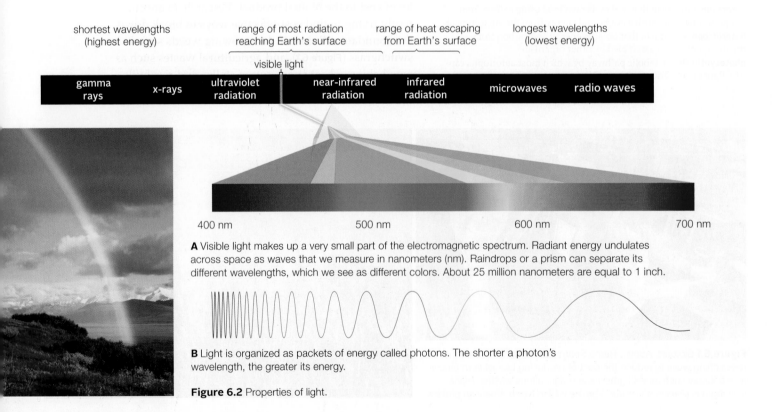

A Visible light makes up a very small part of the electromagnetic spectrum. Radiant energy undulates across space as waves that we measure in nanometers (nm). Raindrops or a prism can separate its different wavelengths, which we see as different colors. About 25 million nanometers are equal to 1 inch.

B Light is organized as packets of energy called photons. The shorter a photon's wavelength, the greater its energy.

Figure 6.2 Properties of light.

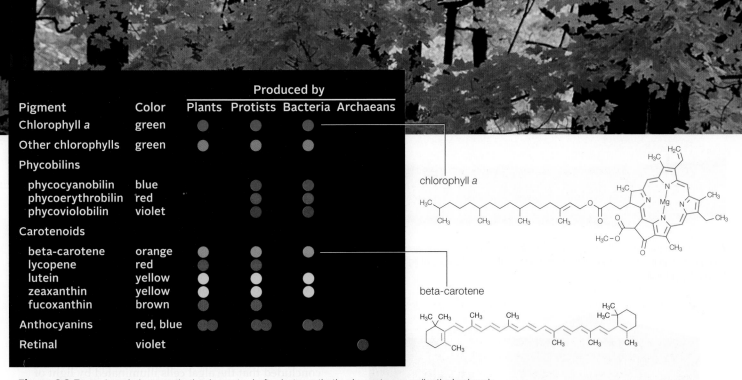

Pigment	Color	Produced by			
		Plants	Protists	Bacteria	Archaeans
Chlorophyll *a*	green	●	●	●	
Other chlorophylls	green	●	●	●	
Phycobilins					
phycocyanobilin	blue		●	●	
phycoerythrobilin	red		●	●	
phycoviolobilin	violet		●	●	
Carotenoids					
beta-carotene	orange	●	●	●	
lycopene	red	●	●		
lutein	yellow	●	●	●	
zeaxanthin	yellow	●	●	●	
fucoxanthin	brown	●	●		
Anthocyanins	red, blue	●●	●●	●●	
Retinal	violet				●

chlorophyll *a*

beta-carotene

Figure 6.3 Examples of photosynthetic pigments. *Left*, photosynthetic pigments can collectively absorb almost all visible light wavelengths.

Right, the light-catching part of a pigment (shown in color) is the region in which single bonds alternate with double bonds. These and many other pigments are derived from evolutionary remodeling of the same compound, as is heme (Section 5.6). Heme is a red pigment.

light, so it appears green to us. Accessory pigments, including other chlorophylls, work together with chlorophyll *a* to harvest a wide range of light wavelengths for photosynthesis. The colors of a few of the 600 or so known accessory pigments are shown in **Figure 6.3**.

Accessory pigments are multipurpose molecules. Antioxidant properties help protect plants and other organisms from the damaging effects of UV light in the sun's rays; appealing colors attract animals to ripening fruit or pollinators to flowers. You may already be familiar with some of these molecules. For example, carrots are orange because they contain beta-carotene (β-carotene). A tomato's color changes from green to red as it ripens because its chlorophyll-containing chloroplasts develop into lycopene-containing chromoplasts (Section 4.9). Roses are red and violets are blue because of their anthocyanin content.

Most photosynthetic organisms use a combination of pigments for photosynthesis. In plants, chlorophylls are usually so abundant that they mask the colors of other pigments, so leaves typically appear green. The green leaves of many plants change color during autumn because they stop making pigments in preparation for a period of dormancy. Chlorophyll breaks down faster than the other pigments, so the leaves turn red, orange, yellow, or violet as their chlorophyll content declines and their accessory pigments become visible.

The light-trapping part of a pigment is an array of atoms in which single bonds alternate with double bonds. Electrons in such arrays easily absorb photons, so pigment molecules function a bit like antennas that are specialized for receiving light energy of only certain wavelengths.

Absorbing a photon excites electrons. Remember, an energy input can boost an electron to a higher energy level (Section 2.3). The excited electron returns quickly to a lower energy level by emitting the extra energy. As you will see, photosynthetic cells can capture energy emitted from an electron returning to a lower energy level. Arrays of chlorophylls and other photosynthetic pigments in these cells hold on to the energy by passing it back and forth. When the energy reaches a special pair of chlorophylls, the reactions of photosynthesis begin.

chlorophyll a Main photosynthetic pigment in plants.
pigment An organic molecule that can absorb light of certain wavelengths.
wavelength Distance between the crests of two successive waves.

Take-Home Message

How do photosynthesizers absorb light?

» Energy radiating from the sun travels through space in waves and is organized as packets called photons.

» The spectrum of radiant energy from the sun includes visible light. Humans perceive different wavelengths of visible light as different colors. The shorter the wavelength, the greater the energy.

» Pigments absorb light at specific wavelengths. Photosynthetic species use pigments such as chlorophyll *a* to harvest the energy of light for photosynthesis.

6.3 Exploring the Rainbow

■ Photosynthetic pigments work together to harvest light of different wavelengths.

■ Link to Experiments 1.6

A Light micrograph of cells in a strand of *Chladophora*. Theodor Engelmann used this green alga in an early experiment to show that certain colors of light are best for photosynthesis.

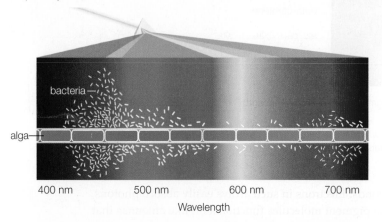

400 nm   500 nm   600 nm   700 nm

Wavelength

B Engelmann directed light through a prism so that bands of colors crossed a water droplet on a microscope slide. The water held a strand of *Chladophora* and oxygen-requiring bacteria. The bacteria clustered around the algal cells that were releasing the most oxygen—the ones that were most actively engaged in photosynthesis. Those cells were under red and violet light. These results constituted one of the first absorption spectra.

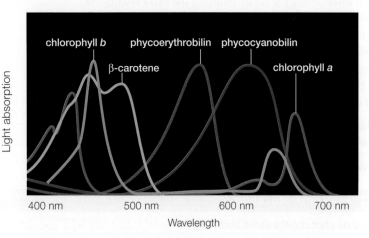

400 nm   500 nm   600 nm   700 nm

Wavelength

C Absorption spectra of chlorophylls *a* and *b*, β-carotene, and two phycobilins reveal the efficiency with which these pigments absorb different wavelengths of visible light. Line color indicates the characteristic color of each pigment.

Figure 6.4 Animated Discovery that photosynthesis is driven by particular wavelengths of light.

Figure It Out: Of the five pigments represented in C, which three are the main photosynthetic pigments in *Chladophora*?

Answer: Chlorophyll *a*, chlorophyll *b*, and β-carotene

In 1882, botanist Theodor Engelmann designed an experiment to test his hypothesis that the color of light affects the rate of photosynthesis. It had long been known that photosynthesis releases oxygen, so Engelmann used the amount of oxygen released by photosynthetic cells as a measure of how much photosynthesis was occurring in them. He used a prism to divide a ray of light into its component colors, then directed the resulting spectrum across a single strand of *Chladophora*, a photosynthetic alga (**Figure 6.4A**), suspended in a drop of water.

Oxygen-sensing equipment had not yet been invented, so Engelmann used oxygen-requiring bacteria to show him where the oxygen concentration in the water was highest. The bacteria moved through the water and gathered mainly where violet or red light fell across the strand of algae (**Figure 6.4B**). Engelmann concluded that the algal cells illuminated by light of these colors were releasing the most oxygen—a sign that violet and red light are the best for driving photosynthesis in these cells.

Engelmann's experiment allowed him to correctly identify the colors of light (red and violet) that are most efficient at driving photosynthesis in *Chladophora*. His results constituted an absorption spectrum, a graph that shows how efficiently a substance absorbs the different wavelengths of light. Peaks in the graph indicate wavelengths of light that the substance absorbs best (**Figure 6.4C**). Engelmann's results represent the combined absorption spectra of all the photosynthetic pigments in *Chladophora*.

Most photosynthetic organisms use a combination of pigments to drive photosynthesis, and the combination differs by species. Why? Different proportions of wavelengths in sunlight reach different parts of Earth. The particular set of pigments a species makes is an adaptation to the particular wavelengths of light available in its native habitat. For example, water absorbs light between wavelengths of 500 and 600 nm less efficiently than other wavelengths. Pigments in algae that live deep underwater absorb light in the range of 500–600 nm, which is the range that is most abundant in deep water. Phycobilins are the most common pigments in these algae.

Take-Home Message

Why do cells use more than one photosynthetic pigment?

» A combination of pigments allows a photosynthetic organism to most efficiently capture the particular range of light wavelengths that reaches the habitat in which it evolved.

6.4 Overview of Photosynthesis

- In plants and other photosynthetic eukaryotes, photosynthesis occurs in chloroplasts.
- Photosynthesis occurs in two stages.
- Links to Plastids 4.9, Metabolic Pathways 5.5

The **chloroplast** is an organelle that specializes in photosynthesis in plants and many protists (**Figure 6.5**). Plant chloroplasts have two outer membranes, and are filled with a semifluid matrix called **stroma.** Stroma contains the chloroplast's own DNA, some ribosomes, and an inner, much-folded **thylakoid membrane.** The folds of a thylakoid membrane typically form stacks of disks (thylakoids) that are connected by channels. The space inside all of the disks and channels is one continuous compartment.

Photosynthesis is often summarized by the following equation:

$$6\,CO_2 + 6H_2O \xrightarrow{\text{light energy}} C_6H_{12}O_6 + 6O_2$$

carbon dioxide   water   glucose   oxygen

However, photosynthesis is not a single reaction. It is a metabolic pathway, a series of many reactions that occur in two stages. The first stage is carried out by molecules embedded in the thylakoid membrane. It is driven by light, so the collective reactions of this stage are called the **light-dependent reactions.**

Two alternative sets of light-dependent reactions, a noncyclic and a cyclic pathway, both convert light energy to chemical bond energy of ATP. The noncyclic pathway, which is the main one in chloroplasts, yields NADPH and O_2 in addition to ATP:

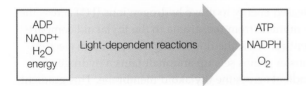

ADP
NADP+
H_2O
energy
— Light-dependent reactions →
ATP
NADPH
O_2

The reactions of the second stage of photosynthesis, which run in the stroma, break apart molecules of carbon dioxide and water, and reassemble their atoms as glucose. Light energy does not power these synthesis reactions, so they are collectively called the **light-independent reactions.** They run on energy delivered by coenzymes produced in the first stage:

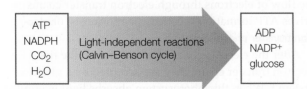

ATP
NADPH
CO_2
H_2O
— Light-independent reactions (Calvin–Benson cycle) →
ADP
NADP+
glucose

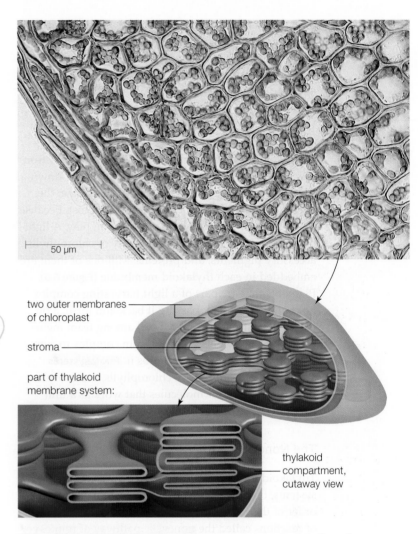

two outer membranes of chloroplast

stroma

part of thylakoid membrane system:

thylakoid compartment, cutaway view

Figure 6.5 Animated The chloroplast: site of photosynthesis in the cells of typical leafy plants. The micrograph shows chloroplast-stuffed cells in a leaf of *Plagiomnium ellipticum*, a type of moss.

chloroplast Organelle specialized for photosynthesis in plants and some protists.
light-dependent reactions First stage of photosynthesis; convert light energy to chemical energy of ATP and NADPH.
light-independent reactions Second stage of photosynthesis; use ATP and NADPH to assemble sugars from water and CO_2.
stroma Semifluid matrix between the thylakoid membrane and the two outer membranes of a chloroplast.
thylakoid membrane A chloroplast's highly folded inner membrane system; forms a continuous compartment in the stroma.

Take-Home Message

Where in a eukaryotic cell do the reactions of photosynthesis take place?

» In the first stage of photosynthesis, light energy drives the formation of ATP and NADPH, and oxygen is released. In eukaryotic cells, these light-dependent reactions occur at the thylakoid membrane of chloroplasts.

» The second stage of photosynthesis, the light-independent reactions, occur in the stroma of chloroplasts. ATP and NADPH drive the synthesis of carbohydrates from water and carbon dioxide.

6.5 Light-Dependent Reactions

■ The reactions of the first stage of photosynthesis convert the energy of light to the energy of chemical bonds.

■ Links to Electrons and energy levels 2.3, Chloroplasts 4.9, Energy 5.2, Electron transfer chains 5.5, Membrane properties 5.7, Gradients 5.8

When a pigment in a thylakoid membrane absorbs a photon, the photon's energy boosts one of the pigment's electrons to a higher energy level. The electron quickly emits the extra energy and drops back down to its unexcited state. In a thylakoid membrane, the energy emitted by excited electrons is not lost, because light-harvesting complexes can keep it in play. A light-harvesting complex is a circular array of chlorophylls, accessory pigments, and proteins. Millions of them are embedded in each thylakoid membrane (**Figure 6.6**). Pigments that are part of a light-harvesting complex hold on to energy by passing it back and forth, a bit like volleyball players pass a ball among team members. The energy gets volleyed from complex to complex until a photosystem absorbs it. Photosystems are groups of hundreds of chlorophylls, accessory pigments, and other molecules that work as a unit to begin the chemical reactions of photosynthesis.

The Noncyclic Pathway

Thylakoid membranes contain two kinds of photosystems, type I and type II, which were named in the order of their discovery. They work together in a set of reactions called the noncyclic pathway of photosynthesis (**Figure 6.7A**). These reactions begin when energy being passed among light-harvesting complexes reaches a photosystem II (**Figure 6.8**). At the center of each photosystem are two very closely associated chlorophyll *a* molecules (a "special pair"). When a

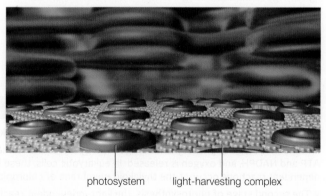

photosystem   light-harvesting complex

Figure 6.6 Artist's view of some of the components of the thylakoid membrane as seen from the stroma. Molecules of electron transfer chains and ATP synthases are also present, but not shown for clarity.

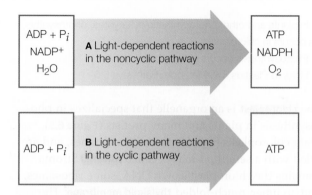

Figure 6.7 Summary of the inputs and outputs of photosynthesis.

photosystem absorbs energy, electrons are ejected from its special pair ❶. These electrons immediately enter an electron transfer chain in the thylakoid membrane.

A photosystem can lose only a few electrons before it must be restocked with more. Where do replacements come from? Photosystem II gets more electrons by removing them from water molecules in the thylakoid compartment. This reaction causes the water molecules to dissociate into hydrogen ions and oxygen ❷. The released oxygen diffuses out of the cell as O_2 gas. This and any other process by which a molecule is broken apart by light energy is called photolysis.

The actual conversion of light energy to chemical energy occurs when a photosystem donates electrons to an electron transfer chain ❸. Light does not take part in chemical reactions, but electrons do. The electrons pass from one molecule of the chain to the next in a series of redox reactions. With each reaction, the electrons release a bit of their extra energy.

Molecules of the electron transfer chain use the released energy to move hydrogen ions (H^+) across the membrane, from the stroma to the thylakoid compartment ❹. Thus, the flow of electrons through electron transfer chains sets up and maintains a hydrogen ion gradient across the thylakoid membrane. This gradient motivates hydrogen ions in the thylakoid compartment to move back into the stroma. However, ions cannot diffuse through lipid bilayers (Section 5.8). H^+ leaves the thylakoid compartment only by flowing through membrane transport proteins called ATP synthases ❼.

Hydrogen ion flow through an ATP synthase causes this protein to attach a phosphate group to ADP, so ATP forms in the stroma ❽. The process by which the flow of electrons through electron transfer chains drives ATP formation is called chemiosmosis, or **electron transfer phosphorylation**.

After the electrons have moved through the first electron transfer chain, they are accepted by a photosystem I. When this photosystem absorbs light energy,

In the figure:

A Light-dependent reactions in the noncyclic pathway

ADP + P_i, NADP$^+$, H_2O → ATP, NADPH, O_2

B Light-dependent reactions in the cyclic pathway

ADP + P_i → ATP

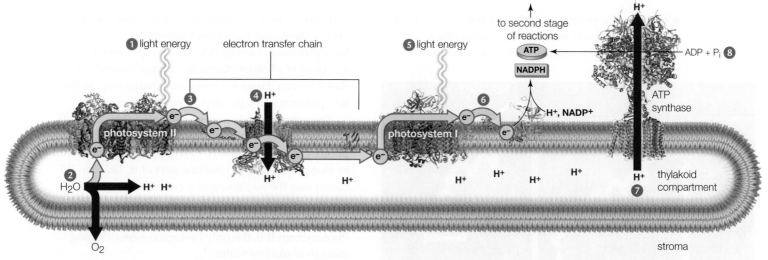

 Figure 6.8 Animated Noncyclic pathway of photosynthesis. Electrons that travel through two different electron transfer chains end up in NADPH, which delivers them to sugar-building reactions in the stroma. The cyclic pathway (not shown) uses a third type of electron transfer chain.

1 Light energy ejects electrons from a photosystem II.

2 The photosystem pulls replacement electrons from water molecules, which break apart into oxygen and hydrogen ions. The oxygen leaves the cell as O_2.

3 The electrons enter an electron transfer chain in the thylakoid membrane.

4 Energy lost by the electrons as they move through the electron transfer chain is used to pump hydrogen ions from the stroma into the thylakoid compartment. A hydrogen ion gradient forms across the thylakoid membrane.

5 Light energy ejects electrons from a photosystem I. Replacement electrons come from an electron transfer chain in the thylakoid membrane.

6 The electrons move through a second electron transfer chain, then combine with NADP+ and H+, so NADPH forms.

7 Hydrogen ions in the thylakoid compartment are propelled through the interior of ATP synthases by their gradient across the thylakoid membrane.

8 Hydrogen ion flow causes ATP synthases to attach phosphate to ADP, so ATP forms in the stroma.

electrons are ejected from its special pair of chlorophylls **5**. These electrons enter a different electron transfer chain. At the end of this second chain, the coenzyme NADP+ accepts the electrons along with H+, so NADPH forms **6**:

$$NADP^+ + 2e^- + H^+ \longrightarrow NADPH$$

ATP and NADPH continue to form as long as electrons continue to flow through transfer chains in the thylakoid membrane, and electrons flow through the chains as long as water, NADP+, and light are plentiful. The flow of electrons slows—and so does ATP and NADPH production—at night, or when water or NADP+ is scarce.

The Cyclic Pathway

At high oxygen levels, or when NADPH accumulates in the stroma, the noncyclic pathway backs up and stalls. Even when the noncyclic pathway is not running, cells continue producing ATP by photosynthesis with the cyclic pathway (**Figure 6.7B**). This pathway involves photosystem I and an electron transfer chain that cycles electrons back to it. The chain that acts in the cyclic pathway uses electron energy to move hydrogen

ions into the thylakoid compartment. The resulting hydrogen ion gradient drives ATP formation, just as it does in the noncyclic pathway. However, NADPH does not form, because electrons at the end of this chain are accepted by a photosystem I, not NADP+. Oxygen (O_2) does not form either, because photosystem I does not rely on photolysis to resupply itself with electrons.

electron transfer phosphorylation Process in which electron flow through electron transfer chains sets up a hydrogen ion gradient that drives ATP formation. Also called chemiosmosis.
photolysis Process by which light energy breaks down a molecule.
photosystem Cluster of pigments and proteins that converts light energy to chemical energy in photosynthesis.

Take-Home Message

What happens during the light-dependent reactions of photosynthesis?

» In the light-dependent reactions, chlorophylls and other pigments in the thylakoid membrane transfer the energy of light to photosystems.

» Absorbing energy causes photosystems to eject electrons that enter electron transfer chains in the membrane. The flow of electrons through the transfer chains sets up hydrogen ion gradients that drive ATP formation.

» In the noncyclic pathway, oxygen is released and electrons end up in NADPH.

» A cyclic pathway involving only photosystem I allows the cell to continue making ATP even when the noncyclic pathway is not running. NADPH does not form, and oxygen is not released.

6.6 Energy Flow in Photosynthesis

■ Energy flow in the light-dependent reactions is an example of how organisms harvest energy from their environment.

■ Links to Energy in metabolism 5.3, Redox reactions 5.5

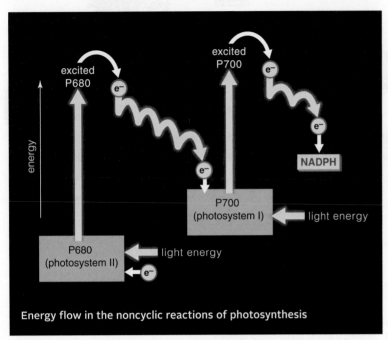

Energy flow in the noncyclic reactions of photosynthesis

A The noncyclic pathway involves a one-way flow of electrons from water, to photosystem II, to photosystem I, to NADPH. As long as electrons continue to flow through the two electron transfer chains, H⁺ continues to be carried across the thylakoid membrane, and ATP and NADPH keep forming. Light provides the energy boosts that keep the pathway going.

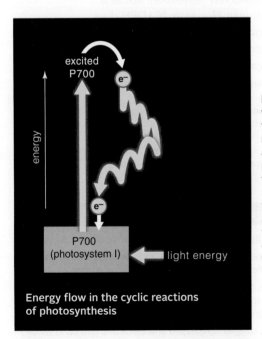

Energy flow in the cyclic reactions of photosynthesis

B In the cyclic pathway, electrons ejected from photosystem I are returned to it. As long as electrons continue to pass through its electron transfer chain, H⁺ continues to be carried across the thylakoid membrane, and ATP continues to form. Light provides the energy boost that keeps the cycle going.

Figure 6.9 Energy flow in the light-dependent reactions of photosynthesis. The P700 in photosystem I absorbs photons of a 700-nanometer wavelength. The P680 of photosystem II absorbs photons at 680 nanometers. Energy inputs boost P700 and P680 to an excited state in which they lose electrons.

A recurring theme in biology is that organisms use energy harvested from the environment to drive cellular processes. Energy flow in the light-dependent reactions of photosynthesis is a classic example of how that happens. **Figure 6.9** compares energy flow in the two pathways of light-dependent reactions.

The simpler cyclic pathway evolved first, and still operates in nearly all photosynthesizers. Later, the photosynthetic machinery in some organisms evolved so that photosystem II became part of it. That modification was the beginning of a combined sequence of reactions that pulls electrons away from water molecules, with the release of hydrogen ions and oxygen. Photosystem II is the only biological system strong enough to oxidize water.

In the noncyclic reactions, electrons that leave photosystem II do not return to it. Instead, they end up in NADPH, a powerful reducing agent (electron donor). In the cyclic reactions, electrons lost from photosystem I are cycled back to it. No NADPH forms, and no oxygen is released. In both the cyclic and noncyclic reactions, molecules in electron transfer chains use electron energy to shuttle H⁺ across the thylakoid membrane. Hydrogen ions accumulate in the thylakoid compartment, forming a gradient that powers ATP synthesis.

Today, the plasma membrane of different species of photosynthetic bacteria incorporates either type I or type II photosystems. Cyanobacteria use both types, as do plants and all photosynthetic protists. Which of the two pathways predominates at any given time depends on the organism's immediate metabolic demands for ATP and NADPH.

Having the alternate pathways is efficient, because cells can direct energy to producing NADPH and ATP or to producing ATP alone. NADPH accumulates when it is not being used, such as when sugar production declines during cold winters. The excess NADPH backs up the noncyclic pathway, so the cyclic pathway predominates. The cell still makes ATP, but not NADPH. When sugar production is in high gear, NADPH is being used quickly. It does not accumulate, and the noncyclic pathway is the predominant one.

Take-Home Message

How does energy flow during the light-dependent reactions of photosynthesis?

» Light provides energy inputs that keep electrons moving through electron transfer chains.

» Energy lost by electrons as they move through the chains sets up a hydrogen ion gradient that drives the synthesis of ATP alone, or ATP and NADPH.

6.7 Light-Independent Reactions

■ The chloroplast is a sugar factory operated by enzymes of the Calvin–Benson cycle. The cyclic, light-independent reactions are the "synthesis" part of photosynthesis.

■ Links to Carbohydrates 3.4, Phosphorylation 5.6

The enzyme-mediated reactions of the Calvin–Benson cycle build sugars in the stroma of chloroplasts (Figure 6.10). These reactions are light-independent because light energy does not power them. Instead, they run on ATP and NADPH that formed in the light-dependent reactions.

Light-independent reactions use carbon atoms from CO_2 to make sugars. Extracting carbon atoms from an inorganic source and incorporating them into an organic molecule is a process called carbon fixation. In most plants, photosynthetic protists, and some bacteria, the enzyme rubisco fixes carbon by attaching CO_2 to five-carbon RuBP (ribulose bisphosphate) ❶.

The six-carbon intermediate that forms by this reaction is unstable, so it splits right away into two three-carbon molecules of PGA (phosphoglycerate). Each PGA receives a phosphate group from ATP, and hydrogen and electrons from NADPH. Thus, ATP energy and the reducing power of NADPH convert each molecule of PGA into a molecule of PGAL (phosphoglyceraldehyde), a phosphorylated sugar ❷.

In later reactions, two or more of the three-carbon PGAL molecules can be combined and rearranged to form larger carbohydrates. Glucose, remember, has six carbon atoms (*left*). To make one glucose molecule, six CO_2 must be attached to six RuBP molecules, so twelve PGAL form. Two PGAL combine to form one glucose molecule ❸. The ten remaining PGAL regenerate the starting compound of the cycle, RuBP ❹.

Plants can use the glucose they make in the light-independent reactions as building blocks for other organic molecules, or they can break it down to access the energy held in its bonds. However, most of the glucose is converted at once to sucrose or starch by other pathways that conclude the light-independent reactions. Excess glucose is stored in the form of starch grains inside the stroma of chloroplasts. When sugars

Calvin–Benson cycle Light-independent reactions of photosynthesis; cyclic carbon-fixing pathway that forms sugars from CO_2.
carbon fixation Process by which carbon from an inorganic source such as carbon dioxide gets incorporated into an organic molecule.
rubisco Ribulose bisphosphate carboxylase. Carbon-fixing enzyme of the Calvin–Benson cycle.

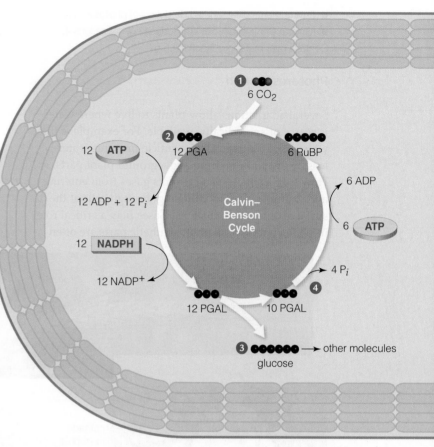

Figure 6.10 Animated Light-independent reactions of photosynthesis. The sketch shows a cross-section of a chloroplast with the light-independent reactions cycling in the stroma.

The steps shown are a summary of six cycles of the Calvin–Benson reactions. Black balls signify carbon atoms. Appendix V details the reaction steps.

❶ Six CO_2 diffuse into a photosynthetic cell, and then into a chloroplast. Rubisco attaches each to a RuBP molecule. The resulting intermediates split, so twelve molecules of PGA form.

❷ Each PGA molecule gets a phosphate group from ATP, plus hydrogen and electrons from NADPH. Twelve PGAL form.

❸ Two PGAL combine to form one glucose molecule.

❹ The remaining ten PGAL receive phosphate groups from ATP. The transfer primes them for endergonic reactions that regenerate the 6 RuBP.

are needed in other parts of the plant, the starch is broken down to sugar monomers and exported from the cell.

Take-Home Message

What happens during the light-independent reactions of photosynthesis?

» The light-independent reactions of photosynthesis run on the bond energy of ATP and the energy of electrons donated by NADPH. Both molecules formed in the light-dependent reactions.

» Collectively called the Calvin–Benson cycle, these carbon-fixing reaction use hydrogen (from NADPH), and carbon and oxygen (from CO_2) to build sugars.

6.8 Adaptations: Different Carbon-Fixing Pathways

■ Environments differ, and so do details of photosynthesis.
■ Links to Central vacuole 4.7, Surface specializations 4.11,
Controls over metabolic reactions 5.5

Photorespiration

Several adaptations allow plants to live where water
is scarce or sporadically available. For example, a thin,
waterproof coating called a cuticle prevents water
loss by evaporation from aboveground plant parts.
However, a cuticle also prevents gases from entering
and exiting a plant by diffusing through cells at the
surfaces of leaves and stems. Gases play a critical role
in photosynthesis, so photosynthetic parts are often
studded with tiny, closable gaps called **stomata** (singular, stoma). When stomata are open, carbon dioxide
for the light-independent reactions can diffuse from
air into the plant's photosynthetic tissues, and oxygen
produced by the light-dependent reactions can diffuse
from photosynthetic cells into the air.

Plants that use only the Calvin–Benson cycle to fix
carbon are called **C3 plants**, because three-carbon PGA
is the first stable intermediate to form in the light-
independent reactions. C3 plants typically conserve
water on dry days by closing their stomata. However,
when stomata are closed, oxygen produced by the
light-dependent reactions cannot escape from the plant.
Oxygen that accumulates in photosynthetic tissues lim-

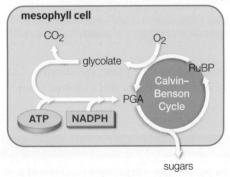

A *Hordeum vulgare* (barley) and other C3 plants have
two kinds of mesophyll cells: palisade and spongy.
The light-dependent and light-independent reactions
occur in both cell types.

A In *Eleusine coracana* (a type of millet) and other C4 plants,
carbon is fixed the first time in mesophyll cells, which are near
air spaces in the leaf. Carbon fixation occurs for the second
time in bundle-sheath cells, which ring the leaf veins.

B C4 plants.
Oxygen also
builds up inside
leaves when sto-
mata close during
photosynthesis.

An additional
pathway in these
plants keeps the
CO₂ concentra-
tion high enough
in bundle-sheath
cells to prevent
photorespiration.

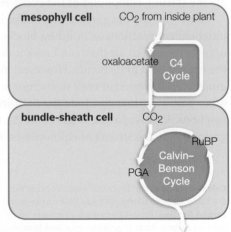

B On dry days, stomata close and oxygen accumulates
inside leaves. The excess causes rubisco to attach oxygen instead of carbon to RuBP. This is photorespiration,
and it makes sugar production inefficient in C3 plants.

Figure 6.11 Animated Light-independent reactions
in C3 plants.

Figure 6.12 Animated Light-independent reactions
in C4 plants.

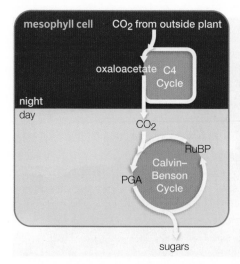

A A CAM plant: *Crassula argentea*, or jade plant.

B CAM plants open stomata and fix carbon using a C4 pathway at night. When the plant's stomata are closed during the day, the organic compounds made during the night are converted to CO_2 that enters the Calvin–Benson cycle.

Figure 6.13 Animated Light-independent reactions in CAM plants.

its sugar production. Why? Oxygen and CO_2 are both substrates of rubisco, so they compete for its active site. In a pathway called **photorespiration**, rubisco attaches oxygen (instead of carbon) to RuBP. Carbon dioxide is produced, so the plant loses carbon instead of fixing it (**Figure 6.11**). In addition, ATP and NADPH are used to convert the pathway's intermediates to a molecule that can enter the Calvin–Benson cycle, so extra energy is required to make sugars on dry days. C3 plants compensate for rubisco's inefficiency by making a lot of it: Rubisco is the most abundant protein on Earth.

C4 Plants

Over the past 50 to 60 million years, an additional set of reactions that compensates for rubisco's inefficiency evolved independently in many plant lineages. Plants that use the additional reactions also close stomata on dry days, but their sugar production does not decline. Examples are corn, switchgrass, and bamboo. We call these plants **C4 plants** because four-carbon oxaloacetate is the first stable intermediate to form in their carbon-fixation reactions (**Figure 6.12**).

C4 plants fix carbon twice, in two kinds of cells. The first set of light-independent reactions occurs in mesophyll cells, where carbon is fixed by an enzyme that does not use oxygen even when the oxygen level

is high. The resulting intermediate is transported to bundle-sheath cells, where an ATP-requiring reaction converts it to carbon dioxide. Rubisco fixes carbon for a second time as the CO_2 enters the Calvin–Benson cycle in the bundle-sheath cells. The C4 cycle keeps the CO_2 level near rubisco high, so it minimizes photorespiration. C4 plants use more ATP than C3 plants do, but on dry days they can make more sugar.

CAM Plants

Succulents, cacti, and other **CAM plants** have a carbon-fixing pathway that allows them to conserve water even in desert regions where the daytime temperatures can be extremely high. CAM stands for crassulacean acid metabolism, after the Crassulaceae family of plants in which this pathway was first studied (**Figure 6.13**). Like C4 plants, CAM plants use a C4 cycle in addition to the Calvin–Benson cycle, but the reactions occur at different times rather than in different cells. Stomata on a CAM plant open at night, when the C4 cycle fixes carbon from CO_2 in the air. The product of the cycle, a four-carbon acid, is stored in the cell's central vacuole. When the stomata close the next day, the acid moves out of the vacuole and becomes broken down to CO_2, which enters the Calvin–Benson cycle.

C3 plant Type of plant that uses only the Calvin–Benson cycle to fix carbon.
C4 plant Type of plant that minimizes photorespiration by fixing carbon twice, in two cell types.
CAM plant Type of C4 plant that conserves water by fixing carbon twice, at different times of day.
photorespiration Reaction in which rubisco attaches oxygen instead of carbon dioxide to ribulose bisphosphate.
stomata Gaps that open on plant surfaces; allow water vapor and gases to diffuse across the epidermis.

Take-Home Message

How do carbon-fixing reactions vary?

» When stomata are closed, oxygen builds up inside leaves of C3 plants. Rubisco then can attach oxygen (instead of carbon dioxide) to RuBP. This reaction, photorespiration, reduces the efficiency of sugar production so it can limit the plant's growth.

» Plants adapted to dry conditions limit photorespiration by fixing carbon twice. C4 plants separate the two sets of reactions in space; CAM plants separate them in time.

Biofuels (revisited)

The first cells on Earth did not tap into sunlight. They were **chemoautotrophs** that extracted energy and carbon from simple molecules in the environment, such as hydrogen sulfide and methane. Both gases were plentiful in the nasty brew that was Earth's early atmosphere (**Figure 6.14A**).

About 1.7 billion years ago, cyclic photophosphorylation evolved in the first **photoautotrophs** (photosynthetic autotrophs), and sunlight offered these organisms an essentially unlimited supply of energy. Not long afterward, the cyclic pathway became modified in some organisms. The new pathway, noncyclic photophosphorylation, split water molecules into hydrogen and oxygen.

Organisms that used the cyclic pathway were tremendously successful. Molecular oxygen, which had previously been very rare in the atmosphere, began accumulating. From that time on, Earth's atmosphere—and the world of life—would never be the same. As you will see in Chapter 7, the change in the composition of the atmosphere put tremendous selection pressure on early life, effectively spurring the evolution of aerobic respiration.

Earth's atmosphere is changing again, and this time, the level of carbon dioxide is increasing. To understand why, think about your own body. A human body is about 9.5 percent carbon by weight, which means that you contain an enormous number of carbon atoms. Where did all of those carbon atoms come from?

Heterotrophs, remember, ingest tissues of other organisms to get raw materials and energy. Thus, your body is built from organic compounds obtained from other organisms. The carbon atoms in those compounds may have passed through other heterotrophs before you ate them, but at some point they were part of photosynthetic organisms. Photosynthesizers strip carbon from carbon dioxide, then use the atoms to build organic compounds. Your carbon atoms—and those of most other organisms—were recently part of Earth's atmosphere, in molecules of CO_2.

Plants in particular are the starting point for nearly all of the carbon-based compounds that humans eat (**Figure 6.14B**). Each year, they collectively remove enough CO_2 from the atmosphere to produce about 220 billion tons of sugar, enough to make about 300

chemoautotroph Organism that makes its own food using carbon from inorganic sources such as carbon dioxide, and energy from chemical reactions.
photoautotroph Photosynthetic autotroph.

A An artist's view of Earth's early atmosphere, which was abundant in gases such as methane, sulfur, ammonia, and chlorine.

B Today, photosynthesis is now the main pathway by which energy and carbon enter the web of life. The plants in this orchard are producing oxygen and carbon-rich parts (apples) at the Jerzy Boyz farm in Chelan, Washington.

Figure 6.14 Then and now—a view of how our atmosphere was irrevocably altered by photosynthesis.

quadrillion sugar cubes. That is a lot of sugar—and a lot of carbon atoms!

Photosynthesis removes carbon dioxide from the atmosphere, and locks its carbon atoms inside organic compounds. When photoautotrophs and other aerobic organisms break down organic compounds for energy, carbon atoms are released in the form of CO_2, which then reenters the atmosphere. Since photosynthesis evolved, these two processes have constituted a more or less balanced cycle of the biosphere. You will learn more about the carbon cycle in Section 46.7. For now,

Figure 6.15 Visible evidence of fossil fuel emissions in the atmosphere: the sky over Los Angeles a sunny day.

know that the amount of carbon dioxide that photosynthesis removes from the atmosphere is roughly the same amount that organisms release back into it. At least it was, until humans came along.

As early as 8,000 years ago, humans began burning forests to clear land for agriculture. When trees and other plants burn, most of the carbon locked in their tissues is released into the atmosphere as carbon dioxide. Fires that occur naturally release carbon dioxide the same way. Today, we are burning a lot more than our ancestors ever did. In addition to wood, we are burning fossil fuels—coal, petroleum, and natural gas—to satisfy our greater and greater demands for energy. Fossil fuels are the organic remains of ancient organisms. When we burn these fuels, the carbon that has been locked in them for hundreds of millions of years is released back into the atmosphere, mainly as carbon dioxide.

Our activities have put Earth's atmospheric cycle of carbon dioxide out of balance. We are adding far more CO_2 to the atmosphere than photosynthetic organisms

are removing from it. Today, we release about 28 billion tons of carbon dioxide into the atmosphere each year, more than ten times the amount we released in the year 1900. Most of it comes from burning fossil fuels (**Figure 6.15**). How do we know? Researchers can determine how long ago the carbon atoms in a sample of CO_2 were part of a living organism by measuring the ratio of different carbon isotopes in it (you will read more about radioisotope dating techniques in Section 16.6). These results are correlated with fossil fuel extraction, refining, and trade statistics.

Tiny pockets of Earth's ancient atmosphere remain in Antarctica, preserved in snow and ice that have been accumulating in layers, year after year, for the last 15 million years. Air and dust trapped in each layer reveal the composition of the atmosphere that prevailed when the layer formed. These layers tell us that the atmospheric CO_2 level was relatively stable for about 10,000 years before the industrial revolution began in the mid-1800s. Since then, the CO_2 level has been steadily rising. In 2008, it was higher than it had been in at least *15 million years*.

Atmospheric carbon dioxide affects Earth's climate, so this increase in CO_2 is contributing to global climate change. We are seeing a warming trend that mirrors the increase in CO_2 levels: Earth is now the warmest it has been for 12,000 years. The trend is affecting biological systems everywhere. Life cycles are changing: birds are laying eggs earlier; plants are flowering earlier than usual; mammals are hibernating for shorter periods. Migration patterns and habitats are also changing. These changes may be too fast for many species, and the rate of extinctions is rising.

Under normal circumstances, extra carbon dioxide stimulates photosynthesis, which means extra carbon dioxide uptake. However, changes in temperature and moisture patterns as a result of global warming are offsetting this benefit because they are proving harmful to plants and other photosynthesizers.

Making biofuel production economically feasible is a high priority for today's energy researchers. Biofuels are a renewable source of energy: We can always make more of them simply by growing more biomass. Unlike fossil fuels, biofuels do not contribute to global climate change: growing plant matter for fuel recycles carbon that is already in the atmosphere.

How would you vote? Biofuels manufactured from crops currently cost more than gasoline, but they are renewable energy sources and have fewer emissions. Would you pay a premium to drive a vehicle that runs on biofuels?

LEARNING ROADMAP

Where you have been This chapter focuses on energy flow (Section 5.2) in metabolic reactions and pathways (3.3, 5.5) that degrade carbohydrates (3.4). Some of the reactions occur in mitochondria (4.9). You will revisit free radicals (2.3), lipids (3.5), proteins (3.6), electron transfer chains (5.5, 6.5), coenzymes (5.6, 6.6), membrane transport (5.8, 5.9), and photosynthesis (6.4).

Where you are now

Energy From Carbohydrates
Various pathways convert the chemical energy of carbohydrates to the chemical energy of ATP. Aerobic respiration yields the most ATP from the breakdown of a glucose molecule.

Glycolysis
Aerobic respiration and anaerobic fermentation start in the cytoplasm with glycolysis, a pathway that converts glucose to two pyruvate molecules and yields two ATP.

How Aerobic Respiration Ends
Eukaryotes break down pyruvate to CO_2 in mitochondria. Many coenzymes are reduced; these deliver electrons and hydrogen ions to ATP-producing electron transfer chains.

How Fermentation Pathways End
Fermentation ends in the cytoplasm, where organic molecules accept electrons from pyruvate. The net yield of ATP is small compared with that from aerobic respiration.

Other Metabolic Pathways
Many different pathways can convert dietary lipids and proteins to molecules that may enter glycolysis or the Krebs cycle.

Where you are going Chapter 14 explains why disorders such as Friedreich's ataxia can be inherited. In Section 11.6, you will see examples of how metabolic pathways are disrupted in cancer cells. We return to plant adaptations for photosynthesis in Chapter 27, muscle function in Chapter 35, how the body acquires oxygen for respiration Chapter 38, and digestion and nutrition in Chapter 39.

7.1 Mighty Mitochondria

Leah Chalcraft (**Figure 7.1**) started to lose her sense of balance and coordination when she was five years old; six years later she was in a wheelchair. Her little brother Joshua could not walk by the time he was eleven, and became blind soon afterward. Both had heart problems; both had spinal fusion surgery.

Leah and Josh's symptoms are those of a genetic disorder called Friedreich's ataxia, the origins of which began billions of years ago. Before the noncyclic pathway of photosynthesis evolved, molecular oxygen had been a very small component of Earth's early atmosphere. Afterward, the new abundance of atmospheric oxygen exerted tremendous selection pressure on organisms that existed at the time. Oxygen reacts with metals, including enzyme cofactors, and free radicals form during those reactions (Section 2.3). The ancient cells had no way to detoxify the radicals, so most of them quickly died out. Only a few types persisted in deep water, muddy sediments, and other anaerobic (oxygen-free) habitats. New metabolic pathways that evolved in the survivors detoxified oxygen radicals. Cells with such pathways were the first aerobic organisms—they could live in the presence of oxygen.

One of the new pathways, aerobic respiration, put the reactive properties of oxygen to use. In modern eukaryotic cells, most of the aerobic respiration pathway takes place inside mitochondria (Section 4.9). Like chloroplasts, mitochondria have an internal folded membrane system that allows them to make ATP very efficiently. Electron transfer chains in this membrane set up hydrogen ion gradients that power ATP synthesis. At the end of these chains, electrons are transferred to oxygen molecules.

ATP participates in almost all cellular reactions, so a cell benefits from making a lot of it. However, aerobic respiration is a risky business. When an oxygen molecule (O_2) accepts electrons from an electron transfer chain, it dissociates into oxygen atoms. Most of the atoms immediately combine with hydrogen ions and end up in water molecules. Occasionally, however, an oxygen atom escapes this final reaction. The atom has an unpaired electron, so it is a free radical.

Mitochondria cannot detoxify free radicals, so they rely on antioxidant enzymes and vitamins in the cell's cytoplasm to do it for them. The system works well, at least most of the time. However, a genetic disorder or an unfortunate encounter with a toxin or pathogen can result in a missing antioxidant, or a defective component of the mitochondrial electron transfer chain. In either case, free radicals accumulate and destroy first the function of mitochondria, then the cell.

Hundreds of disorders are associated with free radical damage in mitochondria, and more are being discovered all the time. Nerve and brain cells, which require a lot of ATP, are particularly affected. Symptoms can range from mild to major neurological deficits, blindness, strokes, seizures, and disabling muscle weakness. In Friedreich's ataxia, a protein called frataxin is not properly transported into mitochondria. This protein regulates the synthesis of iron-containing proteins in mitochondrial electron transfer chains. Iron atoms that were supposed to be incorporated into the proteins accumulate. When too much iron accumulates in mitochondria, too many free radicals form, and these destroy the molecules of life faster than they can be repaired or replaced. Eventually, the mitochondria stop working, and the cell dies. The resulting symptoms cause many of those affected, including Josh Chalcraft, to die as young adults.

aerobic Involving or occurring in the presence of oxygen.
anaerobic Occurring in the absence of oxygen.

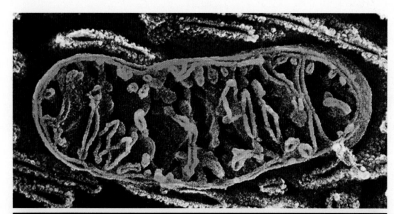

Figure 7.1 Sister, brother, and mitochondria. *Top*, a mitochondrion's folded internal membrane is the source of its function. *Bottom*, Leah and Joshua Chalcraft. Both developed Friedreich's ataxia, a genetic disorder in which excessive free radical formation in mitochondria causes a progressive loss of motor and sensory function. Assistive equipment has allowed them to go to school and to work in productive jobs. Josh died in 2009, at age 24.

7.2 Overview of Carbohydrate Breakdown Pathways

■ Photoautotrophs use the ATP they produce by photosynthesis to make sugars.

■ Most organisms, including photoautotrophs, make ATP by breaking down sugars and other organic compounds.

■ Links to Energy flow 5.2, Metabolic pathways and redox reactions 5.5, Coenzymes 5.6, Photosynthesis 6.4, Electron transfer phosphorylation 6.5, Reducing agents 6.6

In Chapter 6, you learned how photoautotrophs capture energy from the sun, and store it in the form of carbohydrates. Photoautotrophs and most other organisms use energy stored in carbohydrates to run the diverse reactions that sustain life. However, carbohydrates rarely participate in such reactions, so how do cells harness their energy? In order to use the energy stored in carbohydrates, cells must first transfer it to molecules such as ATP, which does participate in many of the energy-requiring reactions that a cell runs. The energy transfer occurs when cells break the bonds that hold a carbohydrate together. Free energy released as those bonds are broken drives ATP synthesis.

Aerobic Respiration

A few different pathways yield ATP by breaking down carbohydrates, but a typical eukaryotic cell uses aerobic respiration most of the time. This equation summarizes the overall pathway of aerobic respiration:

$$C_6H_{12}O_6 \ + \ 6O_2 \longrightarrow 6CO_2 \ + \ 6H_2O$$

glucose · · · · oxygen · · · · carbon dioxide · · · · water

Note that aerobic respiration requires oxygen (a by-product of photosynthesis), and it produces carbon dioxide and water (the same raw materials that photosynthesizers use to make sugars). With this connection, the cycling of carbon, hydrogen, and oxygen through the biosphere comes full circle (**Figure 7.2**).

Aerobic respiration begins with a set of reactions in the cytoplasm (**Figure 7.3A**). These reactions, which are collectively called glycolysis, convert one six-carbon molecule of glucose into two molecules of pyruvate, an organic compound with a three-carbon backbone.

Aerobic respiration continues with two more stages that occur inside mitochondria. During the second stage, the pyruvate is converted to acetyl–CoA, which enters the Krebs cycle. Carbon dioxide that forms in the second-stage reactions leaves the cell (**Figure 7.3B**).

Electrons and hydrogen ions released by the reactions of the first two stages are picked up by two coenzymes, NAD+ (nicotinamide adenine dinucleotide) and FAD (flavin adenine dinucleotide). When these

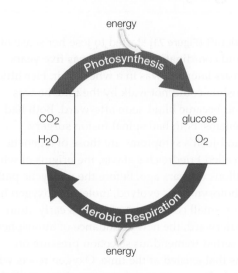

Figure 7.2 The global connection between photosynthesis and aerobic respiration. Note the cycling of materials, and the one-way flow of energy (compare **Figure 5.5**).

two coenzymes are carrying electrons and hydrogen, they are reduced (Section 5.6), and we refer to them as NADH and FADH$_2$.

Few ATP form during the first two stages of aerobic respiration. The big payoff occurs in the third stage, after coenzymes give up electrons and hydrogen to electron transfer chains. These chains power electron transfer phosphorylation, the process in which hydrogen ion gradients drive ATP formation (**Figure 7.3C**). Oxygen in mitochondria accepts electrons and H+ at the end of the transfer chains, so water forms.

Anaerobic Fermentation Pathways

Most types of eukaryotic cells use aerobic respiration exclusively, or they use it most of the time. Many bacteria, archaea, and protists harvest energy from carbohydrates by anaerobic pathways of fermentation. Fermentation and aerobic respiration begin with the same reactions—glycolysis—in the cytoplasm. After glycolysis, the pathways diverge (**Figure 7.4**). Reactions that conclude fermentation pathways occur in cytoplasm, and do not include electron transfer chains. Electrons are accepted by organic molecules (not

aerobic respiration Oxygen-requiring metabolic pathway that breaks down carbohydrates to produce ATP.
fermentation Metabolic pathway that breaks down carbohydrates to produce ATP; does not require oxygen.
glycolysis Set of reactions in which glucose or another sugar is broken down to two pyruvate for a net yield of two ATP.
pyruvate Three-carbon end product of glycolysis.

Aerobic Respiration

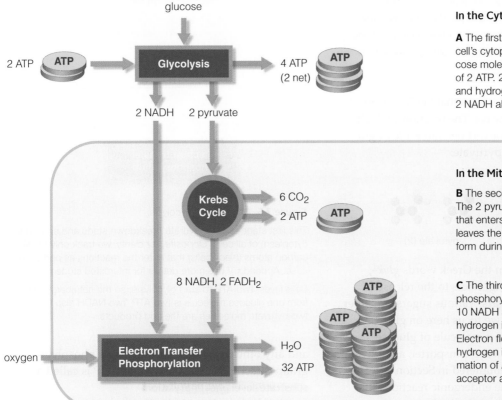

In the Cytoplasm

A The first stage, glycolysis, occurs in the cell's cytoplasm. Enzymes convert a glucose molecule to 2 pyruvate for a net yield of 2 ATP. 2 NAD^+ combine with electrons and hydrogen ions during the reactions, so 2 NADH also form.

In the Mitochondrion

B The second stage occurs in mitochondria. The 2 pyruvate are converted to a molecule that enters the Krebs cycle. CO_2 forms and leaves the cell. 2 ATP, 8 NADH, and 2 $FADH_2$ form during the reactions.

C The third and final stage, electron transfer phosphorylation, occurs inside mitochondria. 10 NADH and 2 $FADH_2$ donate electrons and hydrogen ions to electron transfer chains. Electron flow through the chains sets up hydrogen ion gradients that drive the formation of ATP. Oxygen is the final electron acceptor at the end of the chains.

Figure 7.3 Animated Overview of aerobic respiration. The reactions start in the cytoplasm and end in mitochondria.

Figure It Out: What is aerobic respiration's typical net yield of ATP?

Answer: 38 − 2 = 36 ATP per glucose

oxygen) at the end of the reactions, so fermentation pathways do not require oxygen to proceed.

Fermentation provides enough energy to sustain many anaerobic species, including bacteria, fungi, and single-celled protists that inhabit sea sediments, ani-

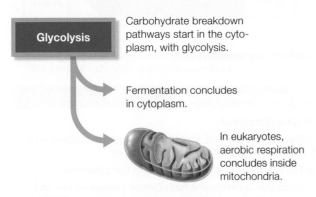

Figure 7.4 Animated Where the different pathways of carbohydrate breakdown start and end.

Carbohydrate breakdown pathways start in the cytoplasm, with glycolysis.

Fermentation concludes in cytoplasm.

In eukaryotes, aerobic respiration concludes inside mitochondria.

mal guts, improperly canned food, sewage treatment ponds, or deep mud. Some of these organisms, including the bacteria that cause botulism, cannot tolerate aerobic conditions, and will die when exposed to oxygen. Fermentation also helps cells of aerobic species produce ATP under anaerobic conditions, but aerobic respiration is a much more efficient way of harvesting energy from carbohydrates. You and other multicelled organisms could not live without its higher yield.

Take-Home Message

How do cells access the chemical energy in carbohydrates?

» Most cells convert the chemical energy of carbohydrates to the chemical energy of ATP by aerobic respiration or fermentation. Aerobic respiration and fermentation pathways start in cytoplasm, with glycolysis.

» Fermentation is anaerobic and ends in the cytoplasm.

» Aerobic respiration requires oxygen. In eukaryotes, it ends in mitochondria.

7.3 Glycolysis—Glucose Breakdown Starts

- The reactions of glycolysis convert one molecule of glucose to two molecules of pyruvate for a net yield of two ATP.
- An energy investment of ATP is required to start glycolysis.
- Links to Hydrolysis 3.3, Glucose 3.4, Endergonic reactions and phosphorylation 5.3, Gradients 5.8, Glucose transporter 5.9, Calvin–Benson cycle 6.7

Glycolysis is a series of reactions that begins carbohydrate breakdown pathways. The reactions, which occur in the cytoplasm, convert one molecule of glucose to two molecules of pyruvate:

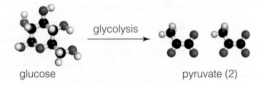

glucose   glycolysis   pyruvate (2)

The word glycolysis (from the Greek words *glyk–*, sweet; and *–lysis*, loosening) refers to the release of chemical energy from sugars. Various sugars can enter glycolysis, but for clarity we focus here on glucose.

Glycolysis begins when a molecule of glucose enters a cell through a glucose transporter, a passive transport protein you encountered in Section 5.9. The cell invests two ATP in the endergonic reactions that begin the pathway (**Figure 7.5**). In the first reaction, a phosphate group is transferred from ATP to the glucose, thus forming glucose-6-phosphate ❶. A model of hexokinase, the enzyme that catalyzes this reaction, is pictured in Section 5.4.

Unlike glucose, glucose-6-phosphate does not pass through glucose transporters in the plasma membrane, so it is trapped inside the cell. Almost all of the glucose that enters a cell is immediately converted to glucose-6-phosphate. This phosphorylation keeps the glucose concentration in the cytoplasm lower than it is in the fluid outside of the cell. By maintaining this concentration gradient across the plasma membrane, the cell favors uptake of even more glucose.

Glycolysis continues as glucose-6-phosphate accepts a phosphate group from another ATP, then splits in two ❷, forming two PGAL (phosphoglyceraldehyde, a phosphorylated sugar that also forms during the Calvin–Benson cycle). A second phosphate group is attached to each PGAL, so two PGA (phosphoglycerate) form ❸. During the reaction, two electrons and a hydrogen ion are transferred from each PGAL to NAD⁺, so two NADH form.

Next, a phosphate group is transferred from each PGA to ADP, so two ATP form ❹. Two more ATP form when a phosphate group is transferred from another pair of intermediates to two ADP ❺. This

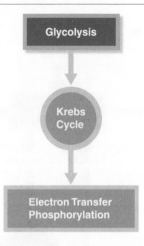

Figure 7.5 Animated Glycolysis.

This first stage of carbohydrate breakdown starts and ends in the cytoplasm of all cells. *Opposite*, for clarity, we track only the six carbon atoms (*black* balls) that enter the reactions as part of glucose. Appendix V has more details for interested students.

Cells invest two ATP to start glycolysis, so the net energy yield from one glucose molecule is two ATP. Two NADH also form, and two pyruvate molecules are the end products.

and any other reaction that transfers a phosphate group directly from a substrate to ADP is called a substrate-level phosphorylation.

Glycolysis ends with the formation of two three-carbon pyruvate molecules. These products may now enter the second-stage reactions of either aerobic respiration or fermentation. Remember, two ATP were invested to initiate the reactions of glycolysis. A total of four ATP form, so the net yield is two ATP per molecule of glucose that enters glycolysis ❻. Two NAD⁺ also pick up hydrogen ions and electrons, thereby becoming reduced to NADH:

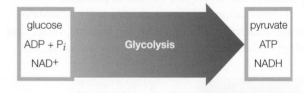

glucose   ADP + P$_i$ NAD+   **Glycolysis**   pyruvate   ATP   NADH

substrate-level phosphorylation A reaction that transfers a phosphate group from a substrate directly to ADP, thus forming ATP.

Take-Home Message

What is glycolysis?

» Glycolysis is the first stage of carbohydrate breakdown in both aerobic respiration and fermentation.

» The reactions of glycolysis occur in the cytoplasm.

» Glycolysis converts one molecule of glucose to two molecules of pyruvate, with a net energy yield of two ATP. Two NADH also form.

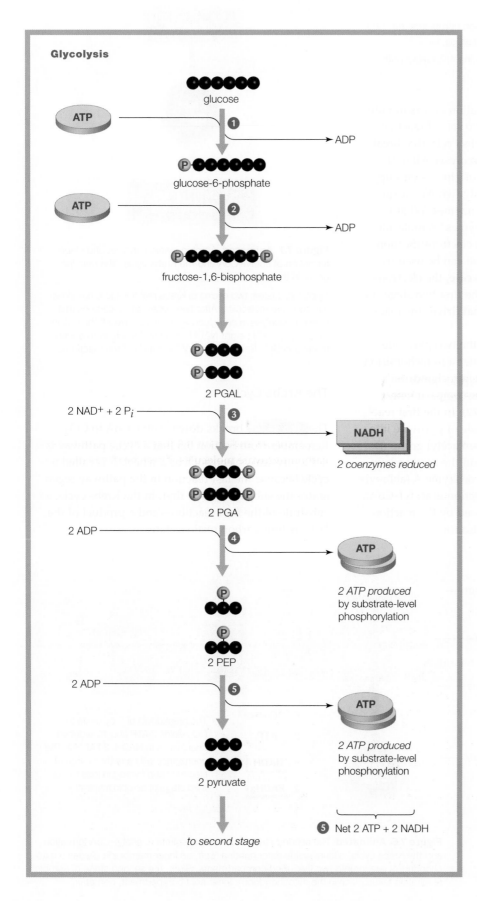

Glycolysis

glucose

ATP → ① → ADP

glucose-6-phosphate

ATP → ② → ADP

fructose-1,6-bisphosphate

2 PGAL

2 NAD⁺ + 2 P$_i$ → ③

NADH

2 coenzymes reduced

2 PGA

2 ADP → ④

ATP

2 ATP produced by substrate-level phosphorylation

2 PEP

2 ADP → ⑤

ATP

2 ATP produced by substrate-level phosphorylation

2 pyruvate

to second stage

⑤ Net 2 ATP + 2 NADH

ATP-Requiring Steps

① An enzyme (hexokinase) transfers a phosphate group from ATP to glucose, forming glucose-6-phosphate.

② A phosphate group from a second ATP is transferred to the glucose-6-phosphate. The resulting molecule is unstable, and it splits into two three-carbon molecules. The molecules are interconvertible, so we will call them both PGAL (phosphoglyceraldehyde).

Two ATP have now been invested in the reactions.

ATP-Generating Steps

③ Enzymes attach a phosphate to the two PGAL, and transfer two electrons and a hydrogen ion from each PGAL to NAD⁺. Two PGA (phosphoglycerate) and two NADH are the result.

④ Enzymes transfer a phosphate group from each PGA to ADP. Thus, two ATP have formed by substrate-level phosphorylation.

The original energy investment of two ATP has now been recovered.

⑤ Enzymes transfer a phosphate group from each of two intermediates to ADP. Two more ATP have formed by substrate-level phosphorylation.

Two molecules of pyruvate form at this last reaction step.

⑥ Summing up, glycolysis yields two NADH, two ATP (net), and two pyruvate for each glucose molecule.

Depending on the type of cell and environmental conditions, the pyruvate may enter the second stage of aerobic respiration or it may be used in other ways, such as in fermentation.

7.4 Second Stage of Aerobic Respiration

■ The second stage of aerobic respiration completes the breakdown of glucose that began in glycolysis.
■ Links to Acetyl groups 3.3, Mitochondria 4.9, Cyclic pathways 5.5

The second stage of aerobic respiration occurs in mitochondria (**Figure 7.6**). It includes two sets of reactions, acetyl–CoA formation and the Krebs cycle, that break down pyruvate, the product of glycolysis. All of the carbon atoms that were once part of glucose end up in CO_2, which departs the cell. Only two ATP form, but the reactions reduce many coenzymes. What is so important about reduced coenzymes? A molecule becomes reduced when it receives electrons (Section 5.5), and electrons carry energy that can be used to drive endergonic reactions. In this case, the electrons picked up by coenzymes during the first two stages of aerobic respiration carry energy that drives the reactions of the third stage.

The second stage begins when the two pyruvate molecules formed by glycolysis enter a mitochondrion. Pyruvate is transported across the mitochondrion's inner membrane and into the inner compartment, which is called the matrix (**Figure 7.7**). In the first reaction, an enzyme splits each molecule of pyruvate into a molecule of CO_2 and a two-carbon acetyl group ❶. The CO_2 diffuses out of the cell, and the acetyl group combines with a molecule called coenzyme A (abbreviated CoA). The product of this reaction is acetyl–CoA. Electrons and hydrogen ions released by the reaction combine with NAD^+, so NADH also forms.

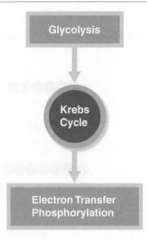

Figure 7.7 Animated Aerobic respiration's second stage: formation of acetyl–CoA and the Krebs cycle. The reactions occur in the mitochondrion's matrix.

Opposite, it takes two cycles of Krebs reactions to break down two pyruvate molecules. After two cycles, all six carbons that entered glycolysis in one glucose molecule have left the cell, in six CO_2. Two ATP, eight NADH, and two $FADH_2$ form during the two cycles. See Appendix V for details of the reactions.

The Krebs Cycle

The **Krebs cycle** breaks down acetyl–CoA to CO_2. Remember from Section 5.5 that a cyclic pathway is not a physical object, such as a wheel. It is called a cycle because the last reaction in the pathway regenerates the substrate of the first. In the Krebs cycle, a substrate of the first reaction—and a product of the last—is four-carbon oxaloacetate.

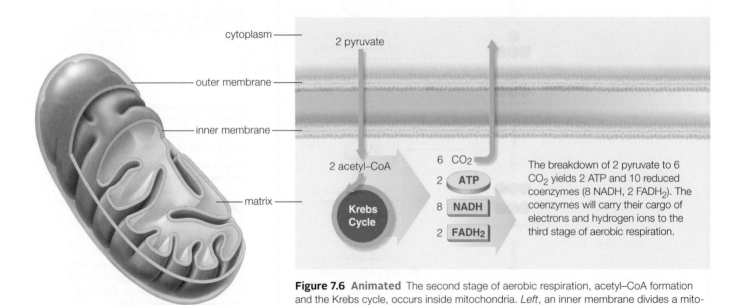

Figure 7.6 Animated The second stage of aerobic respiration, acetyl–CoA formation and the Krebs cycle, occurs inside mitochondria. *Left*, an inner membrane divides a mitochondrion's interior into two fluid-filled compartments. *Right*, the second stage of aerobic respiration takes place in the mitochondrion's innermost compartment, or matrix.

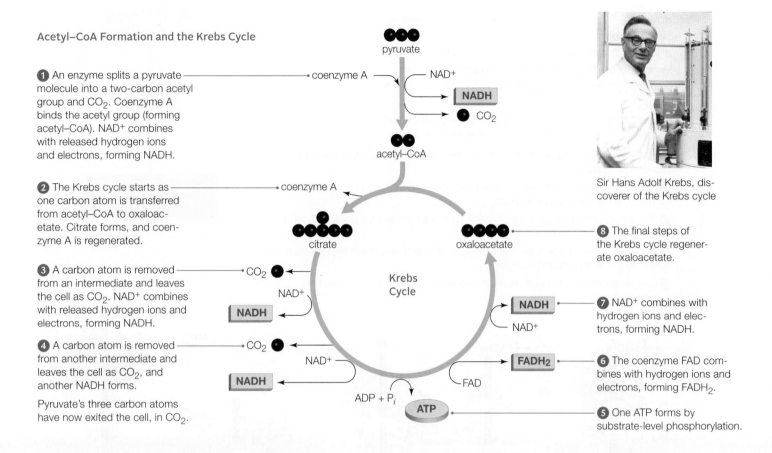

Acetyl–CoA Formation and the Krebs Cycle

1 An enzyme splits a pyruvate molecule into a two-carbon acetyl group and CO_2. Coenzyme A binds the acetyl group (forming acetyl–CoA). NAD^+ combines with released hydrogen ions and electrons, forming NADH.

2 The Krebs cycle starts as one carbon atom is transferred from acetyl–CoA to oxaloacetate. Citrate forms, and coenzyme A is regenerated.

3 A carbon atom is removed from an intermediate and leaves the cell as CO_2. NAD^+ combines with released hydrogen ions and electrons, forming NADH.

4 A carbon atom is removed from another intermediate and leaves the cell as CO_2, and another NADH forms.

Pyruvate's three carbon atoms have now exited the cell, in CO_2.

pyruvate

coenzyme A — NAD^+

NADH

CO_2

acetyl–CoA

coenzyme A

citrate

oxaloacetate

Krebs Cycle

CO_2

NAD^+

NADH

CO_2

NAD^+

NADH

NAD^+

NADH

NAD^+

FADH2

FAD

$ADP + P_i$

ATP

Sir Hans Adolf Krebs, discoverer of the Krebs cycle

8 The final steps of the Krebs cycle regenerate oxaloacetate.

7 NAD^+ combines with hydrogen ions and electrons, forming NADH.

6 The coenzyme FAD combines with hydrogen ions and electrons, forming FADH2.

5 One ATP forms by substrate-level phosphorylation.

During each cycle of Krebs reactions, two carbon atoms of acetyl–CoA are transferred to four-carbon oxaloacetate, forming citrate, the ionized form of citric acid **2**. The Krebs cycle is also called the citric acid cycle after this first intermediate. In later reactions, two CO_2 form and depart the cell. Two NAD^+ are reduced when they accept hydrogen ions and electrons, so two NADH form **3** and **4**. ATP forms by substrate-level phosphorylation **5**, and FAD **6** and another NAD^+ **7** are reduced. The final steps of the pathway regenerate oxaloacetate **8**.

Remember, glycolysis converted one glucose molecule to two pyruvate, and these were converted to two acetyl–CoA when they entered the matrix of a mitochondrion. There, the Krebs cycle reactions convert the two molecules of acetyl–CoA to six CO_2. At this point in aerobic respiration, the carbon backbone of one glucose molecule has been broken down completely: Six carbon atoms have left the cell, in six CO_2. Two ATP formed, which adds to the small net yield of glycolysis. However, six NAD^+ were reduced to six NADH, and two FAD were reduced to two FADH2.

In total, two ATP form and ten coenzymes (eight NAD^+ and two FAD) are reduced during aerobic respiration's second stage. Add the two NAD^+ reduced in glycolysis, and the breakdown of a glucose molecule has a big potential payoff. Twelve reduced coenzymes will deliver electrons (and the energy they carry) to the third stage reactions of aerobic respiration:

pyruvate		CO_2
$ADP + P_i$	**Acetyl–CoA formation and the Krebs cycle**	ATP
NAD^+		NADH
FAD		FADH2

Krebs cycle Cyclic pathway that, along with acetyl–CoA formation, breaks down pyruvate to carbon dioxide during aerobic respiration.

Take-Home Message

What happens during the second stage of aerobic respiration?

» The second stage of aerobic respiration, acetyl–CoA formation and the Krebs cycle, occurs in the inner compartment (matrix) of mitochondria.

» The pyruvate that formed in glycolysis is converted to acetyl–CoA and CO_2. Krebs cycle reactions break down the acetyl–CoA to CO_2.

» For two pyruvate molecules broken down in the second-stage reactions, two ATP form, and ten coenzymes (eight NAD^+; two FAD) are reduced.

7.5 Aerobic Respiration's Big Energy Payoff

■ Many ATP are formed during the third and final stage of aerobic respiration.

■ Links to Thermodynamics 5.2, Electron transfer chains 5.5, Membrane proteins 5.7, Selective permeability of cell membranes 5.8, Electron transfer phosphorylation 6.5

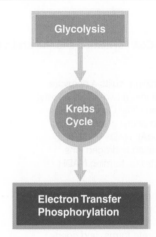

Figure 7.8 The third and final stage of aerobic respiration, electron transfer phosphorylation, occurs at the inner mitochondrial membrane.

① NADH and FADH$_2$ deliver electrons to electron transfer chains in the inner mitochondrial membrane.

② Electron flow through the chains causes hydrogen ions (H$^+$) to be pumped from the matrix to the intermembrane space.

③ The activity of the electron transfer chains causes a hydrogen ion gradient to form across the inner mitochondrial membrane.

④ Hydrogen ion flow back to the matrix through ATP synthases drives the formation of ATP from ADP and phosphate (P$_i$).

⑤ Oxygen combines with electrons and hydrogen ions at the end of the electron transfer chains, so water forms.

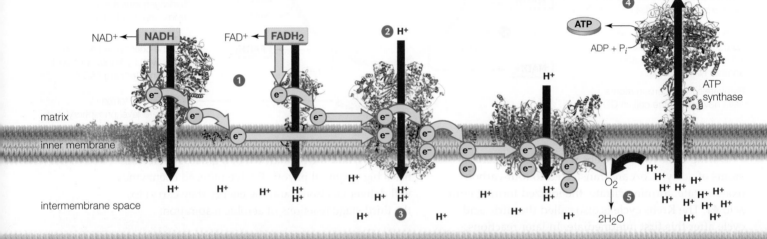

The third stage of aerobic respiration, electron transfer phosphorylation, occurs at the inner mitochondrial membrane (**Figure 7.8**). The reactions begin with the coenzymes NADH and FADH$_2$, which became reduced in the first two stages of aerobic respiration. These coenzymes donate their cargo of electrons and hydrogen ions to electron transfer chains embedded in the inner mitochondrial membrane **①**. As the electrons pass through the chains, they give up energy little by little (Section 5.5). Some molecules of the transfer chains harness that energy to actively transport hydrogen ions across the inner membrane, from the matrix to the intermembrane space **②**. The ions accumulate in the intermembrane space, so a hydrogen ion gradient forms across the inner mitochondrial membrane **③**.

This gradient attracts hydrogen ions back toward the matrix. However, the ions cannot diffuse through a lipid bilayer on their own (Section 5.8). Hydrogen ions cross the inner mitochondrial membrane only by flowing through ATP synthases embedded in the membrane. The flow of hydrogen ions through ATP synthases causes these proteins to attach phosphate groups to ADP, so ATP forms **④**.

Oxygen accepts electrons at the end of the mitochondrial electron transfer chains **⑤**. Aerobic respiration, which literally means "taking a breath of air," refers to oxygen as the final electron acceptor in this pathway. When oxygen accepts electrons, it combines with H$^+$ to form water, which is one product of the third stage.

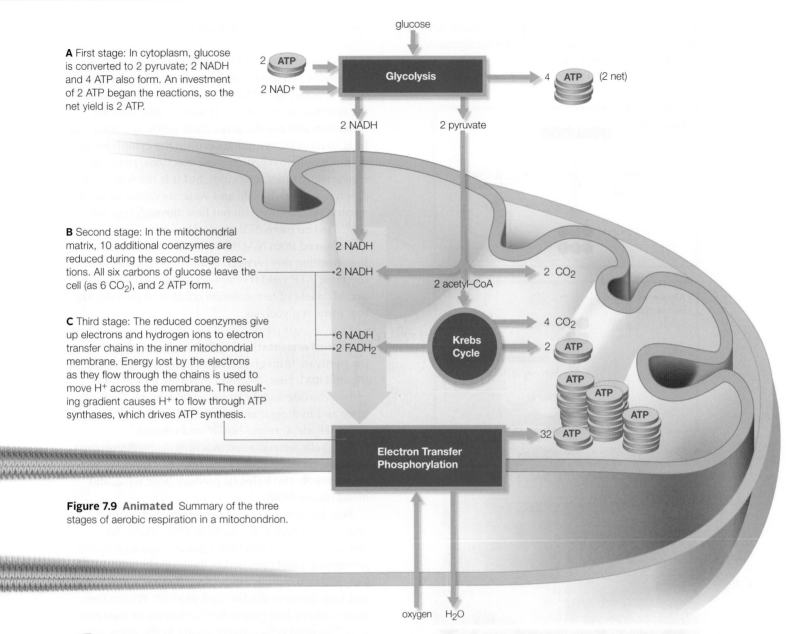

A First stage: In cytoplasm, glucose is converted to 2 pyruvate; 2 NADH and 4 ATP also form. An investment of 2 ATP began the reactions, so the net yield is 2 ATP.

glucose

2 ATP

2 NAD+

Glycolysis

4 ATP (2 net)

2 NADH   2 pyruvate

B Second stage: In the mitochondrial matrix, 10 additional coenzymes are reduced during the second-stage reactions. All six carbons of glucose leave the cell (as 6 CO_2), and 2 ATP form.

2 NADH

2 NADH   2 CO_2

2 acetyl–CoA

C Third stage: The reduced coenzymes give up electrons and hydrogen ions to electron transfer chains in the inner mitochondrial membrane. Energy lost by the electrons as they flow through the chains is used to move H+ across the membrane. The resulting gradient causes H+ to flow through ATP synthases, which drives ATP synthesis.

6 NADH   4 CO_2
2 FADH₂

Krebs Cycle

2 ATP

ATP
ATP
ATP
ATP

32 ATP

Electron Transfer Phosphorylation

oxygen H_2O

Figure 7.9 Animated Summary of the three stages of aerobic respiration in a mitochondrion.

For each glucose molecule that enters aerobic respiration, four ATP form in the first- and second-stage reactions. The twelve coenzymes reduced in these two stages deliver enough H+ and electrons to fuel the synthesis of about thirty-two ATP in the third stage. Thus, the breakdown of one glucose molecule typically yields thirty-six ATP (**Figure 7.9**). However, the overall yield varies. Factors such as cell type affect it. For example, the typical yield of aerobic respiration in brain and skeletal muscle cells is thirty-eight ATP, not thirty-six. Remember that some energy dissipates with every transfer (Section 5.2). Even though aerobic respiration is a very efficient way of retrieving energy from carbohydrates, about 60 percent of the energy harvested in this pathway disperses as metabolic heat.

Take-Home Message

What happens during the third stage of aerobic respiration?

» In aerobic respiration's third stage, electron transfer phosphorylation, energy released by electrons flowing through electron transfer chains is ultimately captured in the attachment of phosphate to ADP.

» The third stage reactions begin when coenzymes that were reduced in the first and second stages deliver electrons and hydrogen ions to electron transfer chains in the inner mitochondrial membrane.

» Energy released by electrons as they pass through electron transfer chains is used to pump H+ from the mitochondrial matrix to the intermembrane space. The H+ gradient that forms across the inner mitochondrial membrane drives the flow of hydrogen ions through ATP synthases, which results in ATP formation.

» About thirty-two ATP form during the third stage reactions, so a typical net yield of all three stages of aerobic respiration is thirty-six ATP per glucose.

7.6 Fermentation

■ Fermentation pathways break down carbohydrates without using oxygen. The final steps in these pathways regenerate NAD+ but do not produce ATP.

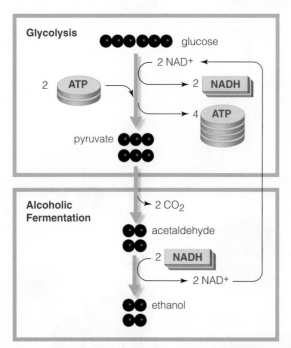

A Alcoholic fermentation begins with glycolysis, and the final steps regenerate NAD+. The net yield of these reactions is two ATP per molecule of glucose (from glycolysis).

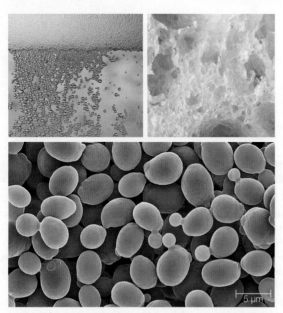

B One product of alcoholic fermentation in *Saccharomyces* cells (ethanol) makes beer alcoholic; another (CO_2) makes it bubbly. Holes in bread are pockets where CO_2 released by fermenting *Saccharomyces* cells accumulated in the dough. The micrograph shows budding *Saccharomyces* cells.

Figure 7.10 Animated Alcoholic fermentation.

Two Fermentation Pathways

Aerobic respiration and fermentation begin with the same set of glycolysis reactions in cytoplasm. Again, two pyruvate, two NADH, and two ATP form during the reactions of glycolysis. However, after that, fermentation and aerobic respiration pathways differ. The final steps of fermentation occur in the cytoplasm and do not require oxygen. In these reactions, pyruvate is converted to other molecules, but it is not fully broken down to carbon dioxide and water as occurs in aerobic respiration. Electrons do not flow through transfer chains, so no more ATP forms. However, electrons are removed from NADH, so NAD+ is regenerated. Regenerating this coenzyme allows glycolysis—and the small ATP yield it offers—to continue. Thus, the net ATP yield of fermentation consists of the two ATP that form in glycolysis.

Alcoholic Fermentation In alcoholic fermentation, the pyruvate from glycolysis is converted to ethanol (**Figure 7.10A**). First, 3-carbon pyruvate is split into carbon dioxide and 2-carbon acetaldehyde. Then, electrons and hydrogen are transferred from NADH to the acetaldehyde, forming NAD+ and ethanol.

Alcoholic fermentation in a fungus, *Saccharomyces cerevisiae*, sustains these yeast cells as they grow and reproduce. It also helps us produce beer, wine, and bread (**Figure 7.10B**).

Beer brewers typically use germinated, roasted, and crushed barley as a carbohydrate source for *Saccharomyces* fermentation. Ethanol produced by the fermenting yeast cells makes the beer alcoholic, and CO_2 makes it bubbly. Hops plant flowers add flavor and help preserve the finished product. Winemakers start with crushed grapes for *Saccharomyces* fermentation. The yeast cells convert sugars in the grape juice to ethanol.

Bakers take advantage of alcoholic fermentation by *Saccharomyces* cells to make bread from flour, which contains starches and a protein called gluten. When flour is kneaded with water, the gluten polymerizes in long, interconnected strands that make the resulting dough stretchy and resilient. Yeast cells in the dough produce CO_2 as they ferment the starches. The gas accumulates in bubbles that are trapped by the gluten mesh. As the bubbles expand, they cause the dough to rise. The ethanol produced by the fermentation reactions evaporates during baking.

Lactate Fermentation In lactate fermentation, the electrons and hydrogen ions carried by NADH are transferred directly to pyruvate. This reaction converts pyruvate to 3-carbon lactate (the ionized form of lactic acid), and also converts NADH to NAD⁺ (**Figure 7.11A**).

Some lactate fermenters spoil food, but we use others to preserve it. For instance, *Lactobacillus* bacteria break down lactose in milk by fermentation. We use this bacteria to produce dairy products such as buttermilk, cheese, and yogurt. Yeast species ferment and preserve pickles, corned beef, sauerkraut, and kimchi.

Animal skeletal muscles, which move bones, consist of cells fused as long fibers. The fibers differ in how they make ATP. Red fibers have many mitochondria and produce ATP by aerobic respiration. These fibers sustain prolonged activity such as marathon runs. They are red because they have an abundance of myoglobin, a protein that stores oxygen for aerobic respiration in these fibers (**Figure 7.11B**).

White muscle fibers contain few mitochondria and no myoglobin, so they do not carry out a lot of aerobic respiration. Instead, they make most of their ATP by lactate fermentation. This pathway makes ATP quickly but not for long, so it is useful for quick, strenuous activities such as sprinting (**Figure 7.11C**) or weight lifting. The low ATP yield does not support prolonged activity. That is one reason why chickens cannot fly very far: Their flight muscles consist mostly of white fibers (thus, the "white" breast meat). Chickens fly only in short bursts. More often, a chicken walks or runs. Its leg muscles consist mostly of red muscle fibers, the "dark meat." Most human muscles consist of a mixture of white and red fibers, but the proportions vary among muscles and among individuals. Great sprinters tend to have more white fibers. Great marathon runners tend to have more red fibers. Section 31.5 offers a closer look at skeletal muscle fibers and how they work.

alcoholic fermentation Anaerobic carbohydrate breakdown pathway that produces ATP and ethanol. Begins with glycolysis; end reactions regenerate NAD⁺ so glycolysis can continue.
lactate fermentation Anaerobic carbohydrate breakdown pathway that produces ATP and lactate.

Take-Home Message

What is fermentation?

» ATP can form by carbohydrate breakdown in fermentation pathways, which are anaerobic.

» The end product of lactate fermentation is lactate. The end product of alcoholic fermentation is ethanol. Both pathways have a net yield of two ATP per glucose molecule. The ATP forms during glycolysis.

» Fermentation reactions regenerate the coenzyme NAD⁺, without which glycolysis (and ATP production) would stop.

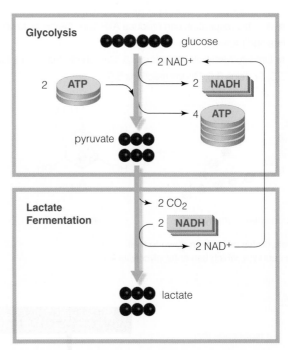

A Lactate fermentation begins with glycolysis, and the final steps regenerate NAD⁺. The net yield of these reactions is two ATP per molecule of glucose (from glycolysis).

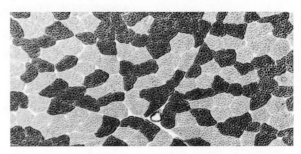

B Lactate fermentation occurs in white muscle fibers, visible in this micrograph of human thigh muscle. The red fibers, which make ATP by aerobic respiration, sustain endurance activities.

C Intense activity such as sprinting quickly depletes oxygen in muscles. Under anaerobic conditions, ATP is produced mainly by lactate fermentation in white muscle fibers. Fermentation does not make enough ATP to sustain this type of activity for long.

Figure 7.11 Animated Lactate fermentation.

7.7 Alternative Energy Sources in Food

- Aerobic respiration can produce ATP from the breakdown of fats and proteins.
- Links to Metabolic reactions 3.3, Carbohydrates 3.4, Lipids 3.5, Proteins 3.6, Redox reactions 5.5

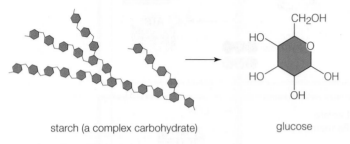

starch (a complex carbohydrate)   glucose

A Complex carbohydrates are broken down to their monosaccharide subunits, which can enter glycolysis ❶.

Energy From Dietary Molecules

Glycolysis converts glucose to pyruvate, and the Krebs cycle reactions transfer electrons from pyruvate to coenzymes. In other words, glucose becomes oxidized (it gives up electrons) and coenzymes become reduced (they accept electrons). Oxidizing an organic molecule can break the covalent bonds of its carbon backbone. Aerobic respiration fully oxidizes glucose, completely dismantling it carbon by carbon.

Cells also dismantle other organic molecules by oxidizing them. Complex carbohydrates, fats, and proteins in food can be converted to molecules that enter glycolysis or the Krebs cycle (**Figure 7.12**). As in glucose metabolism, many coenzymes are reduced, and the

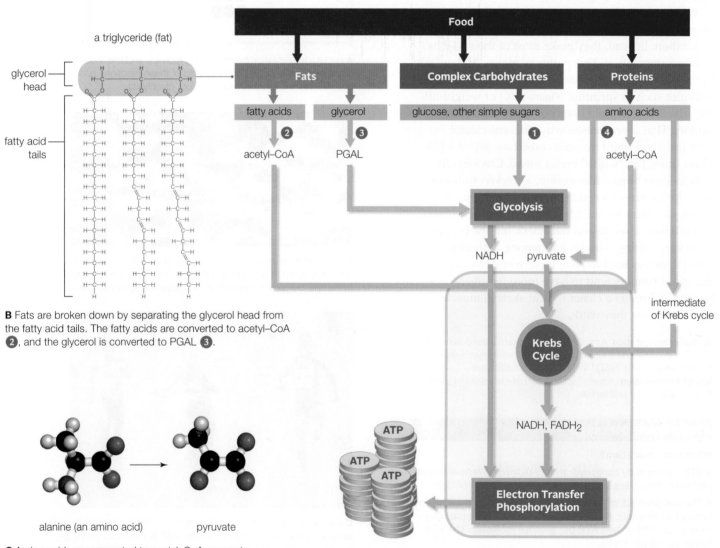

B Fats are broken down by separating the glycerol head from the fatty acid tails. The fatty acids are converted to acetyl–CoA ❷, and the glycerol is converted to PGAL ❸.

alanine (an amino acid)   pyruvate

C Amino acids are converted to acetyl–CoA, pyruvate, or an intermediate of the Krebs cycle ❹.

Figure 7.12 Animated A variety of organic compounds from food can enter the reactions of aerobic respiration.

Mighty Mitochondria (revisited)

At least 83 proteins are directly involved in the electron transfer chains of electron transfer phosphorylation in mitochondria. A defect in any one of them—or in any of the thousands of other proteins used by mitochondria, such as frataxin (*right*)—can wreak havoc in the body. About one in 5,000 people suffer from a known mitochondrial disorder. New research is showing that mitochondrial defects may be involved in many other illnesses such as diabetes, hypertension, Alzheimer's and Parkinson's disease, and even aging.

How would you vote? Developing new drugs is costly, which is why pharmaceutical companies tend to ignore Friedreich's ataxia and other disorders that affect relatively few people. Should governments fund private companies that research potential treatments for rare disorders?

---

energy of the electrons they carry ultimately drives the synthesis of ATP in electron transfer phosphorylation.

Complex Carbohydrates The digestive system breaks down starch and other complex carbohydrates to monosaccharide subunits (**Figure 7.12A**). These sugars are quickly taken up by cells and converted to glucose-6-phosphate, which continues in glycolysis ❶.

Unless ATP is being used quickly, its concentration rises in the cytoplasm. A high concentration of ATP causes glucose-6-phosphate to be diverted away from glycolysis and into a pathway that forms glycogen. Liver and muscle cells especially favor the conversion of glucose to glycogen, and these cells contain the body's largest stores of it. Between meals, the liver maintains the glucose level in blood by converting the stored glycogen to glucose.

Fats A fat molecule has a glycerol head and one, two, or three fatty acid tails (Section 3.5). Cells dismantle fat molecules by first breaking the bonds that connect the glycerol with the fatty acids (**Figure 7.12B**). Nearly all cells in the body oxidize free fatty acids by splitting their long backbones into two-carbon fragments. These fragments are converted to acetyl–CoA, which can enter the Krebs cycle ❷. The glycerol is converted to PGAL by enzymes in liver cells. PGAL, remember, is an intermediate of glycolysis ❸.

On a per carbon basis, fats are a richer source of energy than carbohydrates. Carbohydrate backbones have many oxygen atoms, so they are partially oxidized. A fat's long fatty acid tails are hydrocarbon chains that typically have no oxygen atoms bonded to them, so they have a longer way to go to become oxidized—more reactions are required to fully break them down. Coenzymes accept electrons in these oxidation reactions. The more reduced coenzymes that form, the more electrons can be delivered to the ATP-forming machinery of electron transfer phosphorylation.

What happens if you eat too many carbohydrates? When the blood level of glucose gets too high, acetyl–CoA is diverted away from the Krebs cycle and into a pathway that makes fatty acids. That is why excess dietary carbohydrate ends up as fat.

Proteins Some enzymes in your digestive system split dietary proteins into their amino acid subunits, which are absorbed into the bloodstream. Cells use the amino acids to build proteins or other molecules. Even so, when you eat more protein than your body needs, the amino acids become broken down further. The amino (NH_3^+) group is removed, and it becomes ammonia (NH_3), a waste product that the body eliminates in urine. The carbon backbone is split, and acetyl–CoA, pyruvate, or an intermediate of the Krebs cycle forms, depending on the amino acid (**Figure 7.12C**). Cells can divert these organic molecules into the Krebs cycle ❹.

Take-Home Message

Can the body use organic molecules other than glucose for energy?

» Complex carbohydrates, fats, and proteins can be oxidized in aerobic respiration to yield ATP.

» First the digestive system and then individual cells convert molecules in food into intermediates of glycolysis or the Krebs cycle.

II PRINCIPLES OF INHERITANCE

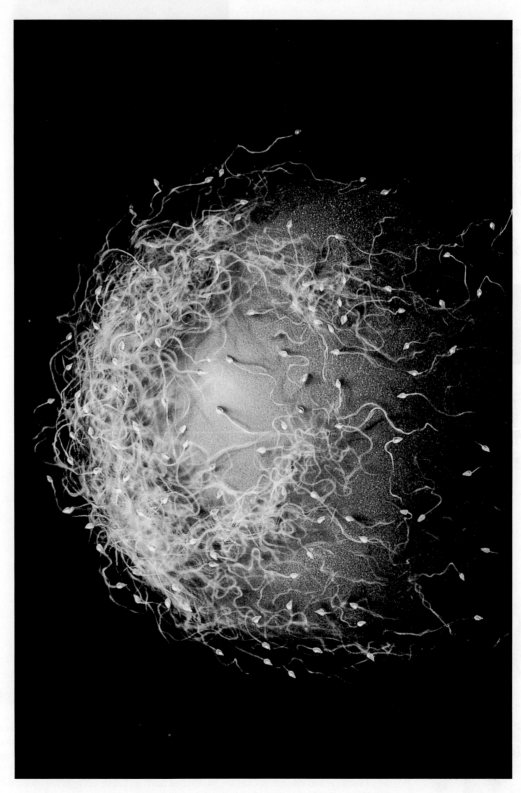

Human sperm, one of which will penetrate this mature egg and so set the stage for the development of a new individual in the image of its parents. This exquisite art is based on a scanning electron micrograph.

LEARNING ROADMAP

Where you have been Radioisotope tracers (Section 2.2) were used in research that led to the discovery that DNA (3.8), not protein (3.6), is the hereditary material of all organisms. This chapter revisits free radicals (2.3), the cell nucleus (4.6) and metabolism (5.4–5.6). Your knowledge of carbohydrate ring numbering (3.3) will help you understand DNA replication.

Where you are now

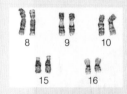

Chromosomes
The DNA of a eukaryotic cell is divided among a characteristic number of chromosomes that differ in length and shape. Sex chromosomes determine an individual's gender.

Discovery of DNA's Function
The work of many scientists over more than a century led to the discovery that DNA is the molecule that stores hereditary information.

Structure of DNA
A DNA molecule consists of two long chains of nucleotides coiled into a double helix. The order of nucleotide bases in the chains differs among individuals and among species.

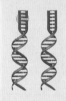

DNA Replication
Before a cell divides, it copies its DNA. Newly forming DNA is monitored for errors, most of which are corrected quickly. Uncorrected errors are mutations.

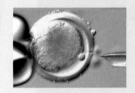

Cloning Animals
Clones of adult animals are commonly produced for research and agriculture. The techniques are far from perfect, and the practice continues to raise serious ethical questions.

Where you are going You will revisit concepts of DNA structure and function presented in this chapter many times, particularly when you learn about how genetic information is converted into structural and functional parts of a cell (Chapter 9), and how cells control that process (Chapter 10). Chromosome structure will turn up again in the context of cell division in Chapters 11 and 12, and in human inheritance and disease in Chapter 14. Viruses such as bacteriophage will be explained in more detail in Chapter 20, and stem cells in Chapter 31.

8.1 A Hero Dog's Golden Clones

On September 11, 2001, an off-duty Canadian police officer, Constable James Symington, drove his search dog Trakr from Nova Scotia to Manhattan. Within hours of arriving, the dog led rescuers to the spot where the fifth and final survivor of the World Trade Center attacks was buried. She had been clinging to life, pinned under rubble from the building where she had worked. Symington and Trakr helped with the search and rescue efforts for three days nonstop, until Trakr collapsed from smoke and chemical inhalation, burns, and exhaustion (**Figure 8.1**).

Trakr survived the ordeal, but later lost the use of his limbs from a degenerative neurological disease probably linked to toxic smoke exposure at Ground Zero. The hero dog died in April 2009, but his DNA lives on in his genetic copies—his **clones**. Symington's essay about Trakr's superior nature and abilities as a search and rescue dog won the Golden Clone Giveaway, a contest to find the world's most clone-worthy dog. Trakr's DNA was shipped to Korea, where it was inserted into donor dog eggs, which were then implanted into surrogate mother dogs. Five puppies, all clones of Trakr, were delivered to Symington in July 2009.

Many adult animals have been cloned besides Trakr, but cloning mammals is still unpredictable and far from routine. Typically, less than 2 percent of the implanted embryos result in a live birth. Of the clones that survive, many have serious health problems.

Why the difficulty? Even though all cells of an individual inherit the same DNA, an adult cell uses only a fraction of it compared with an embryonic cell. To make a clone from an adult cell, researchers must reprogram its DNA to function like the DNA of an egg. Even though we are getting better at doing that, we still have a lot to learn.

So why do we keep trying? The potential benefits are enormous. Already, cells of cloned human embryos are helping researchers unravel the molecular mechanisms of human genetic diseases. Such cells may one day be induced to form replacement tissues or organs for people with incurable diseases. Endangered animals might be saved from extinction; extinct animals may be brought back. Livestock and pets are already being cloned commercially.

Perfecting the methods for cloning animals brings us closer to the possibility of cloning humans, both technically and ethically. For example, if cloning a lost cat for a grieving pet owner is acceptable, why would it not be acceptable to clone a lost child for a grieving parent? Different people have very different answers to such questions, so controversy over cloning continues to rage even as the techniques improve. Understanding the basis of heredity—what DNA is and how it works—will help inform your own opinions about the issues surrounding cloning.

clone Genetically identical copy of an organism.

Figure 8.1
James Symington and his dog Trakr at Ground Zero, September 2001.

8.2 Eukaryotic Chromosomes

2 µm

- The DNA in a eukaryotic cell nucleus is organized as one or more chromosomes.
- Links to DNA 3.8, Cell nucleus 4.6

Stretched out end to end, the DNA molecules in a human cell would be about 2 meters (6.5 feet) long. That is a lot of DNA to fit into a nucleus that is less than 10 micrometers in diameter! Proteins structurally organize the DNA and help it pack tightly into a small nucleus.

Each DNA molecule is organized together with proteins as a structure called a **chromosome** (**Figure 8.2**). The DNA of a typical eukaryotic cell is divided among several chromosomes that differ in length and shape ❶. During most of the cell's life, each chromosome consists of one DNA strand. When the cell prepares to divide, it duplicates all of its chromosomes, so that both of its offspring will get a full set (we return to details of DNA replication in Section 8.5). At that point, each (duplicated) chromosome has two DNA strands—**sister chromatids**—attached to one another at a constricted region called the **centromere**;

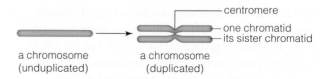

a chromosome (unduplicated) → a chromosome (duplicated) / centromere / one chromatid / its sister chromatid

A duplicated chromosome in its most condensed form consists of two long, tangled filaments (the sister chromatids) bunched into a characteristic X shape ❷. A closer look reveals that each filament is actually a hollow tube formed by coils, like an old-style telephone cord ❸. The coils consist of a twisted fiber ❹ of DNA wrapped twice at regular intervals around "spools" of proteins called **histones** ❺. In micrographs, these DNA–histone spools look like beads on a string. Each "bead" is a **nucleosome**, the smallest unit of chromosomal organization in eukaryotes. As you will see in Section 8.4, the DNA molecule consists of two strands twisted into a double helix ❻.

Chromosome Number

The total number of chromosomes in a eukaryotic cell—the **chromosome number**—is a characteristic of the species. For example, the chromosome number of oak trees is 12, so the nucleus of a cell from an oak tree contains 12 chromosomes. The chromosome number of king crab cells is 208, so they have 208 chromosomes. That of human body cells is 46, so human body cells have 46 chromosomes.

Actually, human body cells have two of each type of chromosome, which means that their chromosome number is **diploid** (2n). A technique called karyotyping reveals an individual's diploid complement of chromosomes. With this technique, cells taken from the individual are treated to make the chromosomes condense, and then stained so the individual chromosomes can be seen under a microscope. A micrograph of a single cell is digitally rearranged so the images of the chromosomes are lined up by centromere location,

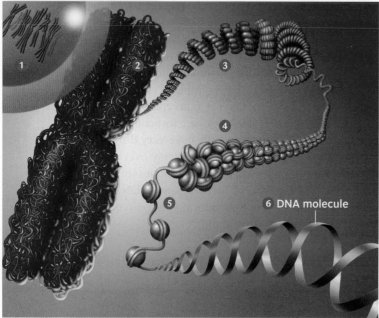

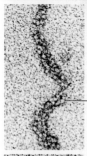

❶ The DNA in the nucleus of a eukaryotic cell is typically divided up into a number of chromosomes.

❷ At its most condensed, a duplicated chromosome packs tightly into an X shape.

❸ A chromosome unravels as a single fiber, a hollow cylinder formed by coiled coils.

❹ The coiled coils consist of a long molecule of DNA and proteins associated with it.

❺ At regular intervals, the DNA molecule is wrapped twice around a core of histone proteins. In this "beads-on-a-string" structure, the "string" is the DNA, and each "bead" is called a nucleosome.

❻ The DNA molecule itself has two strands twisted into a double helix.

6 DNA molecule

Figure 8.2 Animated Zooming in on chromosome structure. Tight packing allows a lot of DNA to fit into a very small nucleus.

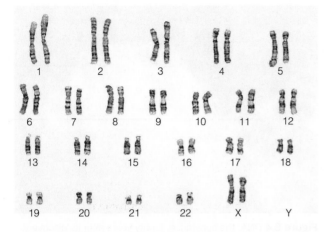

A A karyotype. This one shows 22 pairs of autosomes and a pair of X chromosomes.

Figure 8.3 Human chromosomes. **Figure It Out:** Is the karyotype in (A) from the cell of a female or a male?

Answer: A female

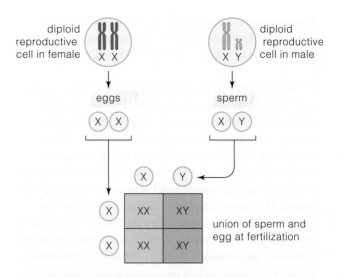

B Pattern of sex determination in humans. The grid shows how the different combinations of sex chromosomes result in offspring that are female (*pink*) or male (*blue*).

and arranged according to size, shape, and length. The finished array constitutes the individual's **karyotype** (**Figure 8.3A**). A karyotype shows how many chromosomes are in the individual's cells, and can also reveal major structural abnormalities.

Autosomes and Sex Chromosomes

All except one pair of a diploid cell's chromosomes are **autosomes**, which are the same in both females and males. The two autosomes of each pair have the same length, shape, and centromere location. They also hold information about the same traits. Think of them as two sets of books on how to build a house. Your father gave you one set. Your mother had her own ideas

about wiring, plumbing, and so on. She gave you an alternate set that says slightly different things about many of those tasks.

Members of a pair of sex chromosomes differ between females and males. The differences determine an individual's sex. The sex chromosomes of humans are called X and Y. Inheriting two X chromosomes makes an individual female, and inheriting one X and one Y makes an individual male (**Figure 8.3B**). Thus, the body cells of human females contain two X chromosomes (XX); those of human males contain one X and one Y chromosome (XY).

XX females and XY males are the rule among fruit flies, mammals, and many other animals, but there are other patterns. In butterflies, moths, birds, and certain fishes, males have two identical sex chromosomes, and females do not. Environmental factors (not sex chromosomes) determine sex in some species of invertebrates, turtles, and frogs. As an example, the temperature of the sand in which sea turtle eggs are buried determines the sex of the hatchlings.

autosome Any chromosome other than a sex chromosome.
centromere Constricted region in a eukaryotic chromosome where sister chromatids are attached.
chromosome A structure that consists of DNA and associated proteins; carries part or all of a cell's genetic information.
chromosome number The sum of all chromosomes in a cell of a given type.
diploid Having two of each type of chromosome characteristic of the species (2*n*).
histone Type of protein that structurally organizes eukaryotic chromosomes.
karyotype Image of an individual's complement of chromosomes arranged by size, length, shape, and centromere location.
nucleosome A length of DNA wound twice around a spool of histone proteins.
sex chromosome Member of a pair of chromosomes that differs between males and females.
sister chromatid One of two attached DNA molecules of a duplicated eukaryotic chromosome.

Take-Home Message

What are chromosomes?

» A chromosome consists of a molecule of DNA that is structurally organized by proteins. The organization allows the DNA to pack tightly.

» A eukaryotic cell's DNA is divided among some characteristic number of chromosomes, which differ in length and shape.

» Members of a pair of sex chromosomes differ between males and females. Chromosomes that are the same in males and females are called autosomes.

8.3 The Discovery of DNA's Function

■ Investigations that led to our understanding that DNA is the molecule of inheritance reveal how science advances.

■ Links to Radioisotopes 2.2, Proteins 3.6, DNA 3.8

In 1869, chemist Johannes Miescher isolated an acidic substance from nuclei (**Figure 8.4**). That substance, which was composed mostly of nitrogen and phosphorus, was deoxyribonucleic acid, or DNA. Sixty years later, Frederick Griffith was trying to make a vaccine for pneumonia. He isolated two strains (types) of *Streptococcus pneumoniae*, a bacteria that causes pneumonia. He named one strain R, because it grows in Rough colonies. He named the other strain S, because it grows in Smooth colonies. Griffith used both strains in a series of experiments that did not lead to the development of a vaccine, but did reveal a clue about inheritance (**Figure 8.5**).

First, he injected mice with live R cells ❶. The mice remained healthy despite having live R cells in their blood, so the R strain was harmless.

Second, he injected other mice with live S cells ❷. The mice died. The S strain caused pneumonia.

Third, he exposed S cells to high temperature. Mice injected with the heat-killed S cells did not die ❸.

Fourth, he mixed live R cells with heat-killed S cells. Mice injected with the mixture died ❹. Blood samples drawn from them teemed with live S cells.

What happened in the fourth experiment? If heat-killed S cells in the mix were not really dead, then mice injected with them in the third experiment would have died. If the harmless R cells had changed into killer cells on their own, then mice injected with R cells in experiment 1 would have died.

The simplest explanation was that heat had killed the S cells, but had not destroyed their hereditary material, including whatever specified "kill mice." That material had been transferred from the dead S cells to the live R cells, which put it to use. The trans-

Figure 8.4 DNA, the substance. Eighty years after its discovery, DNA was determined to be the material of heredity. The photo shows DNA extracted from human cells.

formation was permanent and heritable. Even after hundreds of generations, the descendants of transformed R cells were infectious.

What substance had caused the transformation? In 1940, Oswald Avery and Maclyn McCarty set out to identify that substance, which they termed the "transforming principle," by a process of elimination. The researchers made an extract of S cells that contained only lipid, protein, and nucleic acids. The S cell extract could still transform R cells after it had been treated with lipid- and protein-destroying enzymes. Thus, the transforming principle could not be lipid or protein, so Avery and McCarty realized that the substance they were seeking must be nucleic acid—RNA or DNA. The S cell extract could still transform R cells after treatment with RNA-degrading enzymes, but not after treatment with DNA-degrading enzymes. DNA had to be the transforming principle.

The result surprised Avery and McCarty, who, along with most other scientists, had assumed that proteins were the substance of heredity. After all,

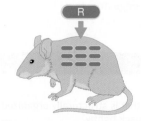

❶ Mice injected with live cells of harmless strain R do not die. Live R cells in their blood.

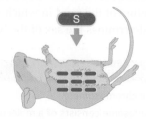

❷ Mice injected with live cells of killer strain S die. Live S cells in their blood.

❸ Mice injected with heat-killed S cells do not die. No live S cells in their blood.

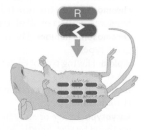

❹ Mice injected with live R cells plus heat-killed S cells die. Live S cells in their blood.

Figure 8.5 Animated Summary of results from Fred Griffith's experiments. The hereditary material of harmful *Streptococcus pneumoniae* cells transformed harmless cells into killers.

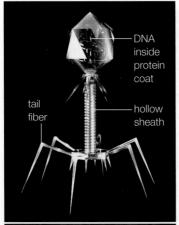

A *Top*, model of a bacteriophage. *Bottom*, micrograph of three viruses injecting DNA into an *E. coli* cell.

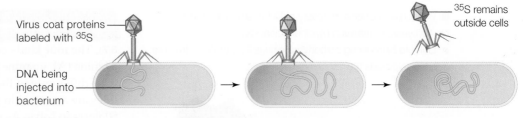

B In one experiment, bacteria were infected with virus particles that had been labeled with a radio-isotope of sulfur (^{35}S). The sulfur had labeled only viral proteins. The viruses were dislodged from the bacteria by whirling the mixture in a kitchen blender. Most of the radioactive sulfur was detected in the viruses, not in the bacterial cells. The viruses had not injected protein into the bacteria.

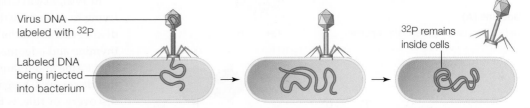

C In another experiment, bacteria were infected with virus particles that had been labeled with a radioisotope of phosphorus (^{32}P). The phosphorus had labeled only viral DNA. When the viruses were dislodged from the bacteria, the radioactive phosphorus was detected mainly inside the bacterial cells. The viruses had injected DNA into the cells—evidence that DNA is the genetic material of this virus.

Figure 8.6 Animated The Hershey–Chase experiments. Alfred Hershey and Martha Chase tested whether the genetic material injected by bacteriophage into bacteria is DNA, protein, or both. The experiments were based on the knowledge that proteins contain more sulfur (S) than phosphorus (P), and DNA contains more phosphorus than sulfur.

traits are diverse, and proteins were thought to be the most diverse biological molecules. Other molecules seemed too uniform. The two scientists were so skeptical that they published their results only after they had convinced themselves, by years of painstaking experiments, that DNA was hereditary material. They were also careful to point out that they had not proven DNA was the *only* hereditary material.

Confirmation of DNA's Function

By 1950, researchers had discovered **bacteriophage**, a type of virus that infects bacteria (**Figure 8.6A**). Like all viruses, these infectious particles carry hereditary information about how to make new viruses. After a virus infects a cell, the cell starts making new virus particles. Bacteriophages inject genetic material into bacteria, but was that material DNA, protein, or both?

Alfred Hershey and Martha Chase found the answer to that question by exploiting the long-known properties of protein (high sulfur content) and DNA (high phosphorus content). They cultured bacteria in growth medium containing a sulfur isotope, ^{35}S. In this medium, the protein (but not the DNA) of bacteriophage that infected the bacteria became labeled

with the ^{35}S tracer. Hershey and Chase allowed the labeled viruses to infect a fresh culture of unlabeled bacteria. They knew from electron micrographs that bacteriophages attach to bacteria by their slender tails. They reasoned it would be easy to break this precarious attachment, so they whirled the virus–bacteria mixture in a kitchen blender. The researchers then separated the bacteria from the virus-containing fluid, and measured the ^{35}S content of each separately. The fluid contained most of the ^{35}S. Thus, the viruses had not injected protein into the bacteria (**Figure 8.6B**).

Hershey and Chase repeated the experiment using an isotope of phosphorus, ^{32}P, which labeled the DNA (but not the proteins) of the bacteriophage. This time, infected bacteria contained most of the isotope. The viruses had injected DNA into the bacteria (**Figure 8.6C**). This experiment confirmed that DNA, not protein, carries hereditary information.

bacteriophage Virus that infects bacteria.

Take-Home Message

What is the molecular basis of inheritance?

» DNA is the material of heredity common to all life on Earth.

8.4 The Discovery of DNA's Structure

■ Watson and Crick's discovery of DNA's structure was based on 150 years of research by other scientists.

■ Links to Numbering carbohydrate rings 3.3, Protein structure 3.6, Nucleic acids 3.8

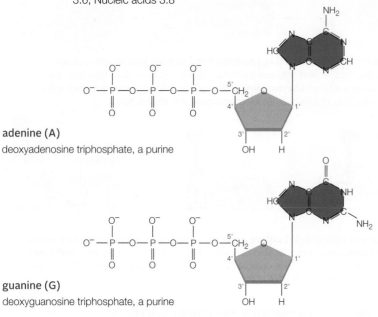

adenine (A)
deoxyadenosine triphosphate, a purine

guanine (G)
deoxyguanosine triphosphate, a purine

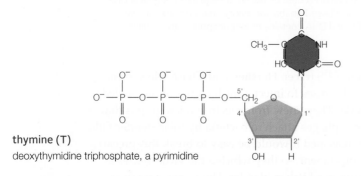

thymine (T)
deoxythymidine triphosphate, a pyrimidine

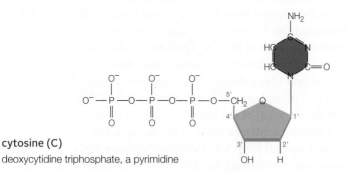

cytosine (C)
deoxycytidine triphosphate, a pyrimidine

Figure 8.7 Animated The four nucleotides in DNA. Adenine and guanine are called purines; thymine and cytosine, pyrimidines. All have three phosphate groups, a deoxyribose sugar (*orange*), and a nitrogen-containing base (*blue*) after which it is named. Biochemist Phoebus Levene identified the structure of the bases and how they are connected in nucleotides in the early 1900s. Levene worked with DNA for almost 40 years.

Numbering the carbons in the sugar rings (Section 3.3) allows us to keep track of the orientation of nucleotide chains (see **Figure 8.8**). The orientation is important in processes such as DNA replication.

A DNA nucleotide has a five-carbon sugar, three phosphate groups, and a nitrogen-containing base (**Figure 8.7**). The four kinds of nucleotides found in DNA—adenine (A), guanine (G), thymine (T), and cytosine (C)—differ only in their base. Just how these nucleotides are arranged in DNA was a puzzle that took over 50 years to solve. As molecules go, DNA is gigantic, and chromosomal DNA has a complex structural organization. Both factors made the molecule difficult to work with, given the laboratory methods of the time.

In 1950, Erwin Chargaff, one of many researchers trying to determine the structure of DNA, made two discoveries about the molecule. First, the amounts of thymine and adenine are identical, as are the amounts of cytosine and guanine (A = T and G = C). We call this discovery Chargaff's first rule. Chargaff's second discovery, or rule, is that the DNA of different species differs in its proportions of adenine and guanine.

Meanwhile, American biologist James Watson and British biophysicist Francis Crick, both at Cambridge University, had been sharing ideas about the structure of DNA. Linus Pauling, Robert Corey, and Herman Branson had just discovered the helical secondary structure that occurs in many proteins, and Watson and Crick suspected that the DNA molecule was also a helix. The two spent many hours arguing about the size, shape, and bonding requirements of the four kinds of nucleotides that make up DNA. They pestered chemists to help them identify bonds they might have overlooked, fiddled with cardboard cutouts, and made models from scraps of metal connected by suitably angled "bonds" of wire.

At King's College in London, biochemist Rosalind Franklin had also been working on the structure of DNA. Like Crick, Franklin specialized in x-ray crystallography, a technique in which x-rays are directed through a purified and crystallized substance. Atoms in the substance's molecules scatter the x-rays in a pattern that can be captured as an image. Researchers can use the pattern to calculate the size, shape, and spacing between any repeating elements of the molecules—all of which are details of molecular structure.

Franklin discovered that "wet" and "dry" DNA samples have different shapes. She made the first clear x-ray diffraction image of "wet" DNA, the form that occurs in cells. From the information in that image, she calculated that DNA is very long compared to its 2-nanometer diameter. She also identified a repeating pattern every 0.34 nanometer along its length, and another every 3.4 nanometers.

Franklin's image and data came to the attention of Watson and Crick, who now had all the information they needed to build a model of the DNA helix—one with two nucleotide chains running in opposite directions, and paired bases inside (**Figure 8.8**). Bonds between the sugar of one nucleotide and the phosphate of the next form the backbone of each chain. Hydrogen bonds between the internally positioned bases hold the two strands together. Only two kinds of base pairings form: A to T, and G to C, which explains the first of Chargaff's rules. Most scientists had assumed (incorrectly) that the bases had to be on the outside of the helix, because they would be more accessible to DNA-copying enzymes that way. You will see in the next section how DNA replication enzymes access the bases on the inside of a double helix.

DNA's Base-Pair Sequence

Just two kinds of base pairings give rise to the incredible diversity of traits we see among living things. How? Even though DNA is composed of only four bases, the order in which one base pair follows the next—the **DNA sequence**—varies tremendously among species (which explains Chargaff's second rule). For example, a piece of DNA from any organism might be:

one base pair

Note how the two strands match. They are complementary, which means each base on one strand pairs with a suitable partner base on the other. This bonding pattern (A to T, G to C) is the same in all molecules of DNA. However, the DNA sequence differs among species, and even among individuals of the same species. The information encoded by that sequence is the basis of traits that define species and distinguish individuals. Thus DNA, the molecule of inheritance in every cell, is the basis of life's unity. Variations in its base sequence are the foundation of life's diversity.

DNA sequence Order of nucleotide bases in a strand of DNA.

Take-Home Message

What is the structure of DNA?

» A DNA molecule consists of two nucleotide chains (strands) running in opposite directions and coiled into a double helix. Internally positioned nucleotide bases hydrogen-bond between the two strands. A pairs with T, and G with C.

» The sequence of bases along a DNA strand is genetic information. DNA sequences vary among species and among individuals. This variation is the basis of life's diversity.

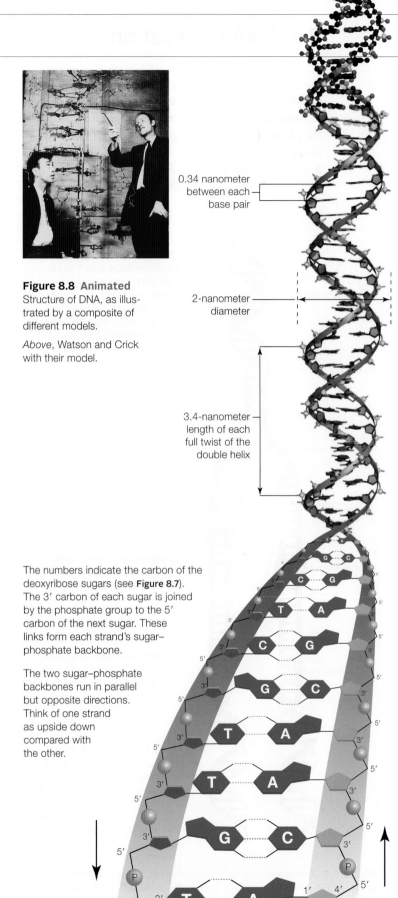

Figure 8.8 Animated
Structure of DNA, as illustrated by a composite of different models.

Above, Watson and Crick with their model.

0.34 nanometer between each base pair

2-nanometer diameter

3.4-nanometer length of each full twist of the double helix

The numbers indicate the carbon of the deoxyribose sugars (see **Figure 8.7**). The 3' carbon of each sugar is joined by the phosphate group to the 5' carbon of the next sugar. These links form each strand's sugar–phosphate backbone.

The two sugar–phosphate backbones run in parallel but opposite directions. Think of one strand as upside down compared with the other.

8.5 DNA Replication

- A cell copies its DNA before it reproduces. Each of the two strands of DNA in the double helix is replicated.
- DNA replication is an energy-intensive process that requires many enzymes, including DNA polymerase, and other molecules.
- Links to Functional groups 3.3, Enzymes 5.4, Metabolic pathways 5.5, Phosphate groups and energy transfers 5.6

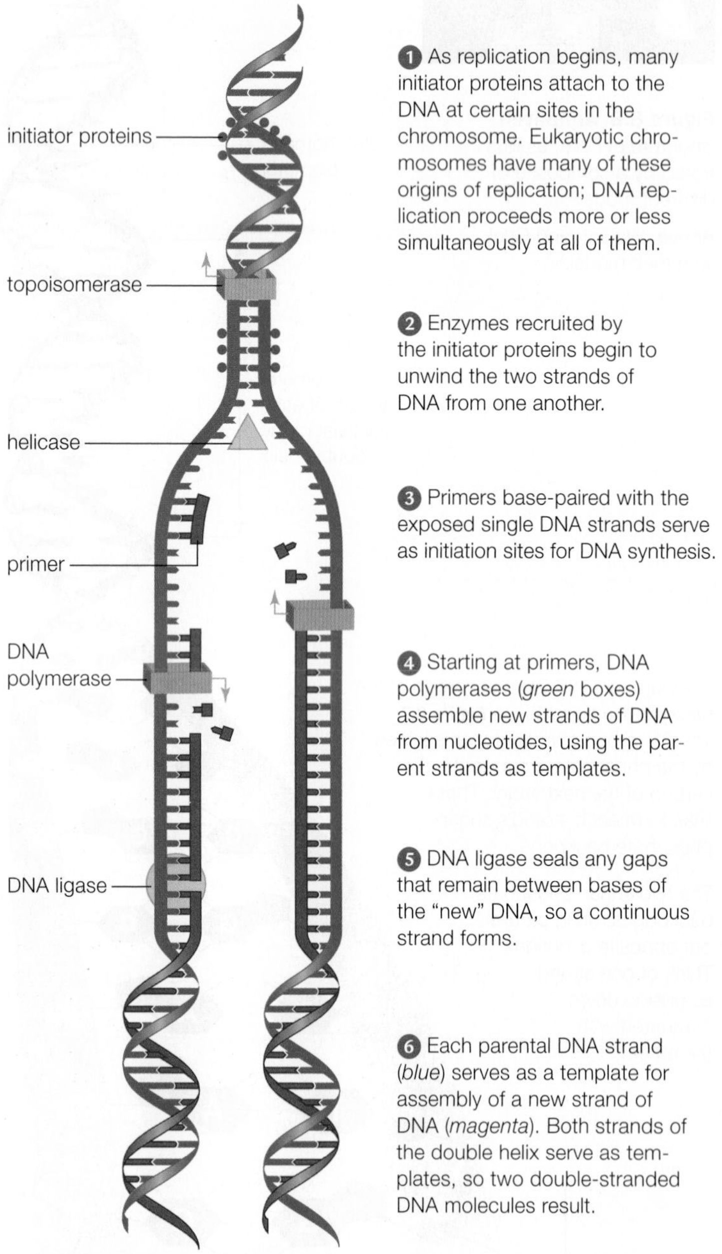

initiator proteins

topoisomerase

helicase

primer

DNA polymerase

DNA ligase

1 As replication begins, many initiator proteins attach to the DNA at certain sites in the chromosome. Eukaryotic chromosomes have many of these origins of replication; DNA replication proceeds more or less simultaneously at all of them.

2 Enzymes recruited by the initiator proteins begin to unwind the two strands of DNA from one another.

3 Primers base-paired with the exposed single DNA strands serve as initiation sites for DNA synthesis.

4 Starting at primers, DNA polymerases (*green* boxes) assemble new strands of DNA from nucleotides, using the parent strands as templates.

5 DNA ligase seals any gaps that remain between bases of the "new" DNA, so a continuous strand forms.

6 Each parental DNA strand (*blue*) serves as a template for assembly of a new strand of DNA (*magenta*). Both strands of the double helix serve as templates, so two double-stranded DNA molecules result.

Figure 8.9 Animated DNA replication, in which a double-stranded molecule of DNA is copied in entirety. Two double-stranded DNA molecules form; one strand of each is parental (old), and the other is new, so DNA replication is said to be semiconservative. *Green* arrows show the direction of synthesis for each strand. The Y-shaped structure of a DNA molecule undergoing replication is called a replication fork.

During most of its life, a cell contains only one set of chromosomes. However, when the cell reproduces, it must contain two sets of chromosomes: one for each of its future offspring. **DNA replication** is the energy-intensive process by which a cell copies its DNA. The process requires the coordinated action of a host of enzymes and other molecules that first open the double helix to expose the internally positioned bases, and then string together free nucleotides into new strands of DNA based on the sequence of those bases.

A cell's genetic information consists of the order of nucleotide bases—the DNA sequence—of its chromosomes. Descendant cells must get an exact copy of that information, or inheritance will go awry. Thus, each chromosome is copied in entirety, and the two chromosomes that result from replication are duplicates of the parent molecule.

A Host of Participants

In eukaryotes, DNA replication necessarily takes place in the nucleus. An enzyme called **DNA polymerase** is central to the process. There are several types of DNA polymerases, but all can assemble a new strand of DNA from free nucleotides. Each type requires one strand of DNA that serves as a template, or guide, for the synthesis of another. Each also requires a primer in order to initiate DNA synthesis. A **primer** is a short, single strand of DNA or RNA that is complementary to a targeted DNA sequence.

Before DNA replication, a chromosome consists of one molecule of DNA—one double helix (**Figure 8.9**). As replication begins, a large number of proteins called initiators bind to certain sequences of nucleotide bases in the DNA **1**. Initiator proteins recruit molecules that pry apart the two strands of the DNA. An enzyme called topoisomerase untwists the double helix, and another enzyme, helicase, breaks the hydrogen bonds holding the double helix together. The two DNA strands begin to unwind from one another as these enzymes move along the double helix **2**.

Another enzyme constructs short RNA primers that base-pair with the opened, single-stranded areas of the DNA **3**. The primers serve as attachment points for DNA polymerases, which begin assembling new strands of DNA on each of the parent strands. As a DNA polymerase moves along a strand of DNA, it uses the sequence of bases as a template, or guide, to assemble a new strand of DNA from free nucleotides **4**. The polymerase follows base-pairing rules: It adds a T to the end of the new DNA strand when it reaches an A in the template DNA strand; it adds a G when it reaches a C; and so on. Thus, the base sequence of each new strand of DNA is complementary to its template (parental) strand. The enzyme DNA ligase seals any gaps, so the new DNA strands are continuous.

Numbering the carbons in nucleotides allows us to keep track of the DNA strands in a double helix, because each strand has an unbonded 5′ carbon at one end and an unbonded 3′ carbon at the other:

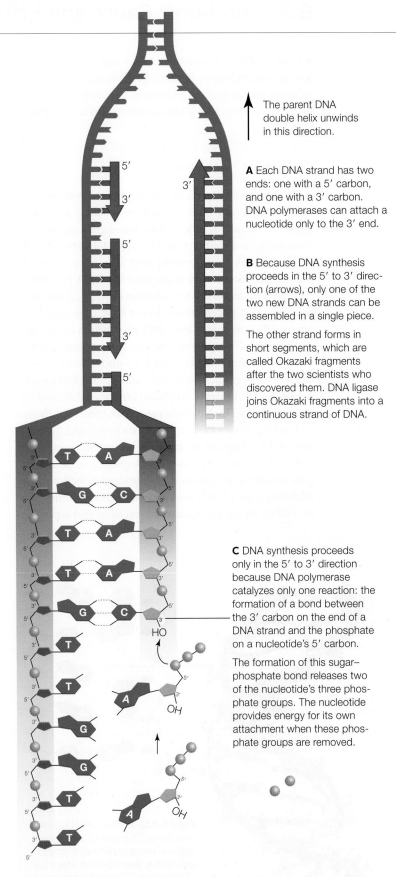

DNA polymerases can attach a free nucleotide only to the 3′ end of a DNA strand. Thus, only one of the two new strands of DNA can be synthesized continuously during DNA replication. Synthesis of the other strand occurs in segments, in the direction opposite that of unwinding (**Figure 8.10**). DNA ligase joins those segments into a continuous strand of DNA ❺.

Each nucleotide provides energy for its own attachment to the end of a growing strand of DNA. Remember from Section 5.6 that the bonds between phosphate groups hold a lot of free energy. Two of a nucleotide's three phosphate groups are removed when it is added to a DNA strand. Breaking those bonds releases enough energy to drive the attachment.

As each new DNA strand lengthens, it winds with its template strand into a double helix. So, after replication, two double-stranded molecules of DNA have formed ❻. One strand of each molecule is parental (old), and the other is new; hence the name of the process, **semiconservative replication.** Each new strand of DNA is complementary in sequence to one of the two parent strands, so both double-stranded molecules produced by DNA replication are duplicates of the parent molecule.

DNA ligase Enzyme that seals gaps in double-stranded DNA.
DNA polymerase DNA replication enzyme. Uses a DNA template to assemble a complementary strand of DNA.
DNA replication Process by which a cell duplicates its DNA before it divides.
primer Short, single strand of DNA that base-pairs with a targeted DNA sequence.
semiconservative replication Describes the process of DNA replication, which produces two copies of a DNA molecule: one strand of each copy is new, and the other is a strand of the original DNA.

Take-Home Message

How is DNA copied?

» DNA replication is an energy-intensive process by which a cell copies its chromosomes.

» Each strand of the double helix serves as a template for synthesis of a new, complementary strand of DNA.

The parent DNA double helix unwinds in this direction.

A Each DNA strand has two ends: one with a 5′ carbon, and one with a 3′ carbon. DNA polymerases can attach a nucleotide only to the 3′ end.

B Because DNA synthesis proceeds in the 5′ to 3′ direction (arrows), only one of the two new DNA strands can be assembled in a single piece.

The other strand forms in short segments, which are called Okazaki fragments after the two scientists who discovered them. DNA ligase joins Okazaki fragments into a continuous strand of DNA.

C DNA synthesis proceeds only in the 5′ to 3′ direction because DNA polymerase catalyzes only one reaction: the formation of a bond between the 3′ carbon on the end of a DNA strand and the phosphate on a nucleotide's 5′ carbon.

The formation of this sugar–phosphate bond releases two of the nucleotide's three phosphate groups. The nucleotide provides energy for its own attachment when these phosphate groups are removed.

Figure 8.10 Discontinuous synthesis of DNA. This close-up of a replication fork shows that only one new DNA strand is assembled continuously.

Figure It Out: What do the yellow balls represent? *Answer: Phosphate groups*

8.6 Mutations: Cause and Effect

■ DNA repair mechanisms correct most replication errors.

■ In science, as in other professions, public recognition for a discovery does not always include all contributors.

■ Links to Electron energy levels and free radicals 2.3, Radiant energy 6.2

DNA Repair Mechanisms

A DNA molecule is not always replicated with perfect fidelity. Sometimes the wrong base is added to a growing DNA strand; at other times, bases get lost, or extra ones are added. Either way, the new DNA strand will no longer match up perfectly with its parent. Some of these replication errors occur after the DNA becomes damaged by exposure to radiation or toxic chemicals, because DNA polymerases do not copy damaged DNA very well. However, most replication errors occur simply because DNA polymerases catalyze a tremendous number of reactions very quickly—about 50 nucleotides per second in eukaryotes, and up to 1,000 bases per second in bacteria. Mistakes are inevitable, and some types of DNA polymerases make many of them.

Luckily, most DNA polymerases proofread their own work. They correct mismatches by reversing the synthesis reaction to remove a mispaired nucleotide, then resuming synthesis in the forward direction. In addition, a set of enzymes and other proteins function as a DNA repair mechanism by excising and replacing

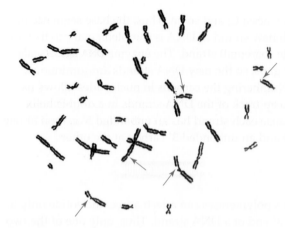

Figure 8.11 DNA damage. *Red* arrows show major breaks in chromosomes of a human white blood cell after exposure to ionizing radiation. Pieces of broken chromatids are often lost during DNA replication.

damaged or mismatched bases in DNA before replication begins.

When proofreading and repair mechanisms fail, an error becomes a **mutation**, a permanent change in the DNA sequence. Repair enzymes cannot recognize a mutation after it has been replicated, because it will base-pair properly with the new strand of DNA. Thus, the mutation becomes perpetuated with every subsequent cell division. A mutation in a body cell may cause that cell to become cancerous; mutations in cells that form eggs or sperm can be passed to offspring. However, not all mutations are dangerous. As you will see in later chapters, mutations that give rise to variations in traits are the raw material of evolution.

Environmental Causes of Mutations

Many environmental agents cause mutations. For example, electromagnetic energy with a wavelength shorter than 320 nanometers—gamma rays, x-rays, and most ultraviolet (UV) light—can knock electrons out of atoms. Such ionizing radiation breaks chromosomes into pieces that get lost during DNA replication (**Figure 8.11**). It also damages DNA indirectly when it penetrates living tissue, because it leaves a trail of destructive free radicals in its wake (Section 2.3).

UV light in the range of 320–400 nanometers can boost electrons to a higher energy level, but not enough to knock them out of atoms. UV light in this range is still dangerous, because it has enough energy to open up the double bond in the ring of a pyrimidine base (thymine or cytidine). The open ring can form a covalent bond with the ring of an adjacent pyrimidine base. The resulting pyrimidine dimer makes a kink in the DNA strand (**Figure 8.12**). DNA polymerase tends to

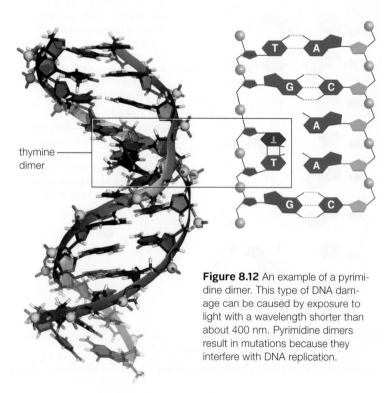

thymine dimer

Figure 8.12 An example of a pyrimidine dimer. This type of DNA damage can be caused by exposure to light with a wavelength shorter than about 400 nm. Pyrimidine dimers result in mutations because they interfere with DNA replication.

copy the kinked part incorrectly during replication, so a mutation becomes introduced into the DNA.

Mutations that arise as a result of pyrimidine dimers cause skin cancer. Exposing unprotected skin to sunlight increases the risk of cancer because its UV wavelengths cause pyrimidine dimers to form. For every second a skin cell spends in the sun, 50–100 pyrimidine dimers form in its DNA.

Some natural or synthetic chemicals also cause mutations. For instance, some of the fifty-five or more carcinogenic (cancer-causing) chemicals in tobacco smoke transfer small hydrocarbon groups to the nucleotide bases in DNA. The altered bases mispair during replication, or prevent replication entirely. Both events increase the chance of mutation. Many environmental pollutants, including those found in tobacco smoke, are converted by the body to other compounds that are easier to excrete. Some of the intermediates in these pathways bind irreversibly to DNA, thus causing replication errors that lead to mutation.

The Short Story of Rosalind Franklin

By the time Rosalind Franklin arrived at King's College, she was an expert x-ray crystallographer. She had solved the structure of coal, which is complex and unorganized (as is DNA). Like Pauling, she had built three-dimensional molecular models. Her assignment was to investigate DNA's structure. No one had told her Maurice Wilkins was already doing the same thing just down the hall. No one had told Wilkins about Franklin's assignment; he assumed she was a technician hired to do his x-ray crystallography work. And so a clash began. Wilkins thought Franklin displayed an appalling lack of deference that technicians of the era usually accorded researchers. To Franklin, Wilkins seemed oddly overinterested in her work and inexplicably prickly.

Wilkins and Franklin had been given identical samples of DNA carefully prepared by Rudolf Signer. Franklin's meticulous work with hers yielded the first clear x-ray diffraction image of DNA as it occurs in cells (**Figure 8.13**). She gave a presentation on this work in 1952. DNA, she said, had two chains twisted into a double helix, with a backbone of phosphate groups on the outside, and bases arranged in an unknown way on the inside. She had calculated DNA's diameter, the distance between its chains and between its bases, the angle of the helix, and the number of bases in each coil. Crick, with his crystallography background, would have recognized the significance of the work—if he had been there. Watson was in the audience but he

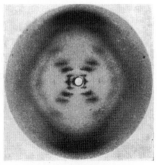

Figure 8.13 Rosalind Franklin and her x-ray diffraction image of DNA. This image was the final link in a long chain of clues that led to the discovery of DNA's structure. Making the image likely contributed to the cancer that took Franklin's life.

was not a crystallographer, and he did not understand the implications of Franklin's x-ray diffraction image or her calculations.

Franklin started to write a research paper on her findings. Meanwhile, and perhaps without her knowledge, Watson reviewed Franklin's x-ray diffraction image with Wilkins, and Watson and Crick read a report detailing Franklin's unpublished data. Crick, who had more experience with molecular modeling than Franklin, immediately understood what the image and the data meant. Watson and Crick used that information to build their model of DNA.

On April 25, 1953, Franklin's paper appeared third in a series of articles about the structure of DNA in the journal *Nature*. It supported with solid experimental evidence Watson and Crick's theoretical model, which appeared in the first article of the series.

Rosalind Franklin died in 1958, at the age of 37, of ovarian cancer probably caused by extensive exposure to x-rays during her work. At the time, the link between x-rays, mutations, and cancer was not fully understood. Because the Nobel Prize is not given posthumously, Franklin did not share in the 1962 honor that went to Watson, Crick, and Wilkins for the discovery of the structure of DNA.

mutation Permanent change in the nucleotide sequence of DNA.

Take-Home Message

What are mutations?

» Permanent changes in a DNA sequence are mutations.

» DNA damage by environmental agents such as UV light and chemicals can result in mutations, because damaged DNA is not replicated very well.

» Proofreading and repair mechanisms usually maintain the integrity of a cell's genetic information by fixing damaged DNA or correcting mispaired bases.

8.7 Animal Cloning

■ Various reproductive interventions produce genetically identical individuals.

The word "cloning," which means making an identical copy of something, can refer to deliberate interventions in reproduction that produce an exact genetic copy of an organism. Genetically identical organisms occur all the time in nature, arising most often by the process of asexual reproduction (which we discuss in Chapter 11). Embryo splitting, another natural process, results in identical twins. The first few divisions of a fertilized egg form a ball of cells that sometimes splits spontaneously. If both halves continue to develop independently, identical twins result. Artificial embryo splitting has been routine in research and animal husbandry for decades. With this technique, a ball of cells is grown from a fertilized egg in a laboratory. The ball is teased apart into two halves, each of which goes on to develop as a separate embryo. The embryos are implanted in surrogate mothers, who give birth to identical twins. Artificial twinning and any other technology that yields genetically identical individuals is called **reproductive cloning**.

Twins get their DNA from two parents that typically differ in their DNA sequence. Thus, although twins produced by embryo splitting are identical to one another, they are not identical to either parent. When animal breeders want an exact copy of a specific individual, they may turn to a cloning method that starts with a single cell taken from an adult organism. Such procedures present more of a technical challenge than embryo splitting. Unlike a fertilized egg, a body cell from an adult will not automatically start dividing. It must first be tricked into rewinding its developmental clock.

All cells descended from a fertilized egg inherit the same DNA. Thus, the DNA in each living cell of an individual is like a master blueprint that contains enough information to build an entirely new individual. As different cells in a developing embryo start using different subsets of their DNA, they differentiate (become different in form and function). Differentiation is usually a one-way path in animal cells. Once a cell specializes, all of its descendant cells will be specialized the same way. By the time a liver cell, muscle cell, or other specialized cell forms, most of its DNA has been turned off, and is no longer used.

To clone an adult, scientists must first transform one of its differentiated cells into an undifferentiated cell by turning its unused DNA back on. In **somatic cell nuclear transfer (SCNT)**, a researcher removes the nucleus from an unfertilized egg, then inserts into the egg a nucleus from an adult animal cell (**Figure 8.14**). A somatic cell is a body cell, as opposed to a reproductive cell (*soma* is a Greek word for body). If all goes well, the egg's cytoplasm reprograms the transplanted DNA to direct the development of an embryo, which is then implanted into a surrogate mother. The animal

A A cow egg is held in place by suction through a hollow glass tube called a micropipette. DNA is identified by a *purple* stain.

B Another micropipette punctures the egg and sucks out the DNA. All that remains inside the egg's plasma membrane is cytoplasm.

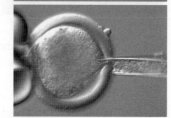

C A new micropipette prepares to enter the egg at the puncture site. The pipette contains a cell grown from the skin of a donor animal.

D The micropipette enters the egg and delivers the skin cell to a region between the cytoplasm and the plasma membrane.

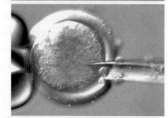

E After the pipette is withdrawn, the donor's skin cell is visible next to the cytoplasm of the egg. The transfer is now complete.

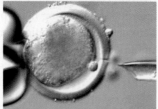

F An electric current causes the foreign cell to fuse with and empty its nucleus into the cytoplasm of the egg. The egg begins to divide, and an embryo forms that may be transplanted into a surrogate mother.

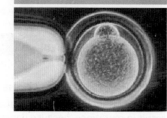

Figure 8.14 Animated Somatic cell nuclear transfer, using cattle cells. This series of micrographs was taken by scientists at Cyagra, a company that specializes in cloning livestock.

A Hero Dog's Golden Clones (revisited)

Trakr's clones were produced using SCNT. The ability to clone dogs is a recent development, but the technique is not. Scottish geneticist Ian Wilmut made headlines in 1997 when his team cloned an adult sheep using SCNT. The cloned lamb, Dolly, was genetically identical to the sheep that had donated an udder cell.

At first, Dolly looked and acted like a normal sheep, but she died early. Dolly may have had health problems because she was a clone.

SCNT has also been used to clone mice, rats, rabbits, pigs, cattle, goats, sheep, horses, mules, deer, cats, a camel, a ferret, a monkey, and a wolf. Many of the clones are unusually overweight or have enlarged organs. Cloned mice develop lung and liver problems, and almost all die prematurely. Cloned pigs tend to limp and have heart problems. One never did develop a tail or, even worse, an anus.

How would you vote? Some view sickly or deformed clones as unfortunate but acceptable casualties of animal cloning research that yields medical advances for humans. Should animal cloning be banned?

that is born to the surrogate is genetically identical with the donor of the nucleus.

SCNT is now a common practice among people who breed prized livestock. Among other benefits, many more offspring can be produced in a given time frame by cloning than by traditional breeding methods. Cloned animals have the same championship features as their DNA donors (**Figure 8.15**). Offspring can also be produced from a donor animal that is castrated or even dead.

As the techniques become routine, cloning a human is no longer only within the realm of science fiction. SCNT is already being used to produce human embryos for research, a practice called **therapeutic cloning**. Researchers harvest undifferentiated (stem) cells from the cloned human embryos. Cells from

these embryos are being used to study, among many other things, how fatal diseases progress. For example, embryos created using cells from people with genetic heart defects will allow researchers to study how the defect causes developing heart cells to malfunction. Such research may ultimately lead to treatments for people who suffer from fatal diseases. (We return to the topic of stem cells and their potential medical benefits in Chapter 31.) Reproductive cloning of humans is not the intent of such research, but if it were, SCNT would indeed be the first step toward that end.

Human eggs are difficult to come by. They also come with a hefty set of ethical dilemmas. Thus, researchers have started trying to make hybrid embryos using adult human cells and eggs from other animals, a technique called interspecies nuclear transfer, or iSCNT.

reproductive cloning Technology that produces genetically identical individuals.

somatic cell nuclear transfer (SCNT) Method of reproductive cloning in which genetic material is transferred from an adult somatic cell into an unfertilized, enucleated egg.

therapeutic cloning The use of SCNT to produce human embryos for research purposes.

Take-Home Message

What is cloning?

» Reproductive cloning technologies produce clones: genetically identical individuals.

» The DNA inside a living cell contains all the information necessary to build a new individual.

» Somatic cell nuclear transfer (SCNT) is a reproductive cloning technology in which nuclear DNA of an adult donor is transferred to an egg with no nucleus. The hybrid cell develops into an embryo that is genetically identical to the adult donor.

» Therapeutic cloning uses SCNT to produce human embryos for research.

Figure 8.15 Champion Holstein dairy cow Nelson's Estimate Liz (*right*) and her clone, Nelson's Estimate Liz II (*left*), who was produced by somatic cell nuclear transfer in 2003. Liz II had already begun to win championships by the time she was one year old.

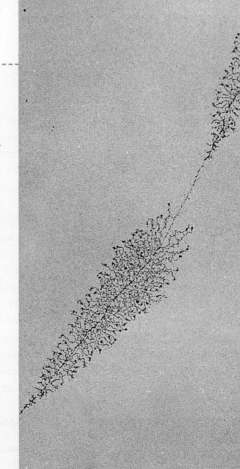

LEARNING ROADMAP

Where you have been Your knowledge of chromosomes (Section 8.2) and base pairing (8.4) will help you understand how cells use nucleic acids (3.8) to build proteins (3.6). You will revisit hydrophobicity (2.5), hemoglobin (3.2), pathogenic bacteria (4.1), organelles (4.4, 4.7), radicals and cofactors (5.6), enzyme function (3.3, 5.4), DNA replication (8.5), and mutations (8.6).

Where you are now

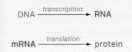

Gene Expression
The sequence of amino acids in a polypeptide is determined by a gene. The conversion of information in DNA to protein occurs in two steps: transcription and translation.

DNA to RNA: Transcription
During transcription, one strand of DNA serves as a template for assembling a single, complementary strand of RNA (a transcript). Each transcript is an RNA copy of a gene.

RNA
Messenger RNA carries DNA's protein-building instructions. Sixty-four mRNA codons represent the genetic code. Two other types of RNA translate that code.

RNA to Protein: Translation
Translation converts a sequence of codons in mRNA to a sequence of amino acids in a polypeptide. Transfer RNAs and ribosomes carry out the process.

Altered Proteins
Some mutations change a gene's DNA sequence so that an altered protein forms, an outcome that can have drastic consequences.

Where you are going What you learn in this chapter about genes will be the foundation for concepts of gene expression (Chapter 10), inheritance (Chapters 13 and 14), and genetic engineering (Chapter 15). Chapters 16 and 17 will show you how mutations are the raw material of natural selection and other processes of evolution. You will also revisit hemoglobin, sickle-cell anemia, and the circulatory system in Chapter 36, and immunity in Chapter 37.

9.1 The Aptly Acronymed RIPs

Ricin is a naturally occurring protein that is highly toxic: A dose as small as a few grains of salt can kill an adult human, and there is no antidote. Ricin effectively deters beetles, birds, mammals, and other animals from eating seeds of the castor-oil plant (*Ricinus communis*), which grows wild in tropical regions worldwide and is widely cultivated. Castor-oil seeds are the source of castor oil, an ingredient in plastics, cosmetics, paints, soaps, polishes, and many other items. After the oil is extracted from the seeds, the ricin is typically discarded along with the leftover seed pulp.

The lethal effects of ricin were known as long ago as 1888, but using ricin as a weapon is now banned by most countries under the Geneva Protocol. However, controlling its production is impossible, because it takes no special skills or equipment to manufacture the toxin from easily obtained raw materials. Thus, ricin appears periodically in the news. For example, at the height of the Cold War, the Bulgarian writer Georgi Markov had defected to England and was working as a journalist for the BBC. As he made his way to a bus stop on a London street, an assassin used the tip of a modified umbrella to jam a small, ricin-laced ball into Markov's leg. Markov died in agony three days later. Plotted or thwarted terrorist activities involving ricin have made the news almost every year since then.

Ricin is called a ribosome-inactivating protein (RIP) because it inactivates ribosomes, the organelles that assemble amino acids into proteins. RIPs occur in some bacteria, mushrooms, algae, and many plants (including food crops such as tomatoes, barley, and spinach). Many of these proteins are not particularly toxic because they do not cross intact cell membranes very well. Those that do, including ricin, have a domain that binds tightly to carbohydrates on plasma membranes (**Figure 9.1**). Binding causes the cell to take up the RIP by endocytosis. Once inside the cell, the second domain of the RIP—an enzyme—begins to inactivate ribosomes. One molecule of ricin can inactivate more than 1,000 ribosomes per minute. If enough ribosomes are affected, protein synthesis grinds to a halt. Proteins are critical to all life processes, so cells that cannot make them die very quickly.

Someone who inhales ricin can die from low blood pressure and respiratory failure within a few days of exposure, but very few people actually encounter this toxin. Other RIPs are more prevalent. Shiga toxin, made by *Shigella dysenteriae* bacteria, causes dysentery. *E. coli* bacteria, including the strain O157:H7 (Section 4.1), make enterotoxin, an RIP that is the source of symptoms associated with food poisoning.

 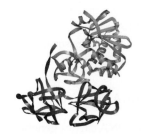

A Ricin and castor-oil seeds.

B Abrin and seeds of the rosary pea plant.

 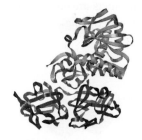

C Cinnamomin and camphor tree seeds.

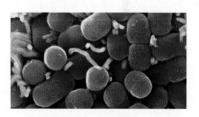 

D Enterotoxin and *Escherichia coli*.

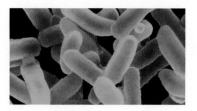 

E Shiga toxin and *Shigella dysenteriae*.

Figure 9.1 A few ribosome-inactivating proteins. The structure of RIPs are strikingly similar. One of their polypeptide chains (*red*) helps the molecule cross a cell's plasma membrane. The other chain (*orange*) destroys the cell's capacity for protein synthesis.

9.2 DNA, RNA, and Gene Expression

■ Transcription converts information in a gene to RNA; translation converts information in an mRNA to protein.
■ Links to Proteins 3.6, Nucleic acids 3.8, Ribosomes 4.4, Enzymes 5.4, DNA sequence and replication 8.5

Converting a Gene to an RNA

You learned in Chapter 8 that DNA carries genetic information. How does a cell convert that information into structural and functional components? Let's start with the nature of the information itself.

DNA is like a book, an encyclopedia that contains all of the instructions for building a new individual. You already know the alphabet used to write the book: the four letters A, T, G, and C, for the four nucleotide bases adenine, thymine, guanine, and cytosine. Each strand of DNA consists of a chain of those four kinds of nucleotides. The linear order, or sequence, of the four bases in the strand is the genetic information. All of a cell's RNA and protein products are encoded by subsets of that DNA sequence called genes.

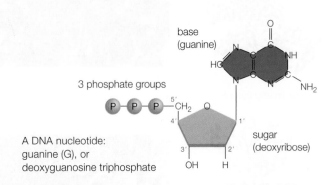

A DNA nucleotide:
guanine (G), or
deoxyguanosine triphosphate

A Guanine, one of the four nucleotides in DNA. The others (adenine, uracil, and cytosine) differ only in their component bases (*blue*). Three of the four bases in RNA nucleotides are identical to the bases in DNA nucleotides.

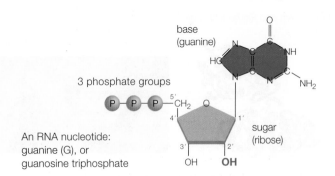

An RNA nucleotide:
guanine (G), or
guanosine triphosphate

B Compare the RNA nucleotide guanine. The only difference between the DNA and RNA versions of guanine (or adenine, or cytosine) is that RNA has a hydroxyl group at the 2' carbon of the sugar (shown in *red*).

Figure 9.2 Comparing nucleotides of DNA and RNA.

Converting the information encoded by a gene into a product starts with RNA synthesis, or transcription. By this process, enzymes use the nucleotide sequence of a gene as a template to synthesize a strand of RNA (ribonucleic acid):

$$\text{DNA} \xrightarrow{\text{transcription}} \text{RNA}$$

RNA usually occurs in a single-stranded form that is structurally similar to a single strand of DNA. For example, both are chains of four kinds of nucleotides. Like a DNA nucleotide, an RNA nucleotide has three phosphate groups, a sugar, and one of four bases. However, DNA and RNA nucleotides are slightly different (**Figure 9.2**). The two nucleic acids are named after their component sugars, ribose and deoxyribose, which differ in one functional group. Three of the bases (adenine, cytosine, and guanine) are the same in RNA and DNA nucleotides, but the fourth base in RNA is uracil, not thymine as it is in DNA (**Figure 9.3**).

Despite these small differences in structure, DNA and RNA have very different functions. DNA's only role is to store a cell's heritable information. By contrast, a cell transcribes several kinds of RNAs, and each kind has a different function. MicroRNAs are important in gene control, which is the subject of the next chapter. Three types of RNA have roles in protein synthesis. Ribosomal RNA (rRNA) is the main component of ribosomes, structures upon which polypeptide chains are built (Section 4.4). Transfer RNA (tRNA) delivers amino acids to ribosomes, one by one, in the order specified by a messenger RNA (mRNA).

Converting mRNA to Protein

Messenger RNA was named for its function as the "genetic messenger" between DNA and protein. A protein-building message is encoded within an mRNA's sequence by sets of three nucleotide bases, "genetic words" that follow one another along the length of the nucleotide strand. Like the words of a sentence, a series of these genetic words can form a meaningful parcel of information—in this case, the sequence of amino acids of a protein.

By the process of translation, the protein-building information in an mRNA is decoded (translated) into a sequence of amino acids. The result is a polypeptide chain that twists and folds into a protein:

$$\text{mRNA} \xrightarrow{\text{translation}} \text{protein}$$

Sections 9.4 and 9.5 describe how rRNA and tRNA interact to translate the sequence of base triplets in an mRNA into the sequence of amino acids in a

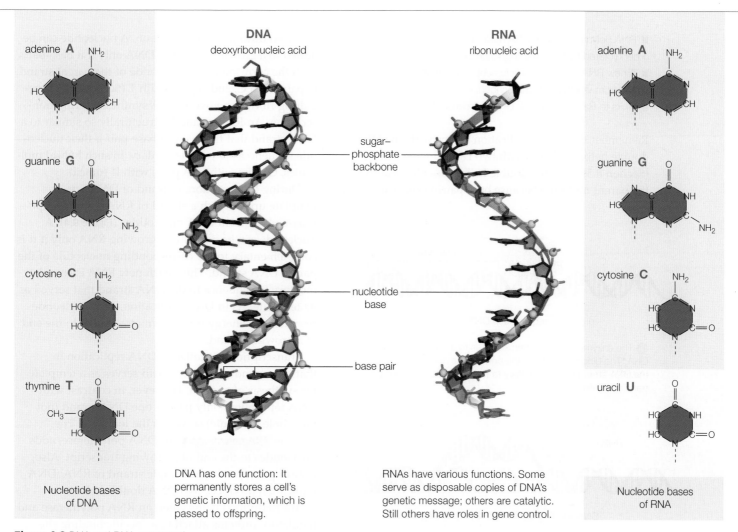

DNA
deoxyribonucleic acid

RNA
ribonucleic acid

adenine **A**

guanine **G**

cytosine **C**

thymine **T**

Nucleotide bases
of DNA

sugar–
phosphate
backbone

nucleotide
base

base pair

adenine **A**

guanine **G**

cytosine **C**

uracil **U**

Nucleotide bases
of RNA

DNA has one function: It
permanently stores a cell's
genetic information, which is
passed to offspring.

RNAs have various functions. Some
serve as disposable copies of DNA's
genetic message; others are catalytic.
Still others have roles in gene control.

Figure 9.3 DNA and RNA compared.

protein. Transcription and translation are steps in gene expression, the process by which genetic information encoded by a gene is converted into a structural or functional part of a cell or a body:

$$DNA \xrightarrow{\text{transcription}} mRNA \xrightarrow{\text{translation}} protein$$

gene DNA sequence that encodes an RNA or protein product.
gene expression Process by which the information in a gene becomes converted to an RNA or protein product.
messenger RNA (mRNA) A type of RNA that carries a protein-building message.
ribosomal RNA (rRNA) A type of RNA that becomes part of ribosomes.
transcription Process by which an RNA is assembled from nucleotides using the base sequence of a gene as a template.
transfer RNA (tRNA) A type of RNA that delivers amino acids to a ribosome during translation.
translation Process by which a polypeptide chain is assembled from amino acids in the order specified by an mRNA.

A cell's DNA sequence contains all of the information it needs to make the molecules of life. Each gene encodes an RNA, and different types of RNAs interact to assemble proteins from amino acids (Section 3.6). Some of those proteins are enzymes that assemble lipids and complex carbohydrates from simple building blocks (Section 3.3), replicate DNA (Section 8.5), and make RNA (as you will see in the next section).

Take-Home Message

What is the nature of genetic information carried by DNA?

» Genetic information occurs in genes, which are DNA sequences that encode instructions for building RNA or protein products.

» A cell transcribes the nucleotide sequence of a gene into RNA.

» Although RNA is structurally similar to a single strand of DNA, the two types of molecules differ functionally.

» A messenger RNA (mRNA) carries a protein-building code in its nucleotide sequence. rRNAs and tRNAs interact to translate that code into a protein.

9.3 Transcription: DNA to RNA

■ RNA polymerase links RNA nucleotides into a chain, in the order dictated by the base sequence of a gene.

■ A new RNA strand is complementary in sequence to the DNA strand from which it was transcribed.

■ Links to Base pairing 8.4, DNA replication 8.5

Remember that DNA replication begins with one DNA double helix and ends with two DNA double helices (Section 8.5). The two double helices are identical to the parent molecule because base-pairing rules are

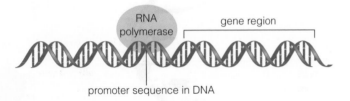

1 The enzyme RNA polymerase binds to a promoter in the DNA. The binding positions the polymerase near a gene. Only the DNA strand complementary to the gene sequence will be translated into RNA.

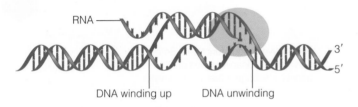

2 The polymerase begins to move along the gene and unwind the DNA. As it does, it links RNA nucleotides in the order specified by the base sequence of the complementary (noncoding) DNA strand. The DNA winds up again after the polymerase passes. The structure of the "opened" DNA at the transcription site is called a transcription bubble, after its appearance.

followed during DNA replication. A nucleotide can be added to a growing strand of DNA only if it base-pairs with the corresponding nucleotide of the parent strand: G pairs with C, and A pairs with T (Section 8.4). Base-pairing rules also govern RNA synthesis during transcription. An RNA strand is structurally so similar to a DNA strand that the two can base-pair if their nucleotide sequences are complementary. In such hybrid molecules, G pairs with C; A pairs with U (uracil).

During transcription, a strand of DNA acts as a template upon which a strand of RNA—a transcript—is assembled from RNA nucleotides (**Figure 9.4**). A nucleotide can be added to a growing RNA only if it is complementary to the corresponding nucleotide of the parent DNA strand. Thus, each new RNA is complementary in sequence to the DNA strand that served as its template. As in DNA replication, each nucleotide provides the energy for its own attachment to the end of a growing strand.

Transcription is similar to DNA replication in that one strand of a nucleic acid serves as a template for synthesis of another. However, in contrast with DNA replication, only part of one DNA strand (not the whole molecule) serves as the template. The enzyme **RNA polymerase**, not DNA polymerase, adds nucleotides to the end of a growing transcript. Also, transcription produces a single strand of RNA; DNA replication produces two DNA double helices.

Transcription begins when an RNA polymerase and regulatory proteins attach to a specific binding site in the DNA called a **promoter** **1**. Binding positions the polymerase at a transcription start site close to the gene. Typically, only one of the two DNA strands in a double helix encodes a gene. The strand that is complementary to the gene sequence (the noncod-

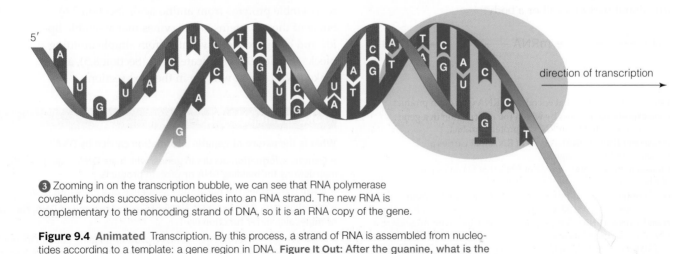

3 Zooming in on the transcription bubble, we can see that RNA polymerase covalently bonds successive nucleotides into an RNA strand. The new RNA is complementary to the noncoding strand of DNA, so it is an RNA copy of the gene.

Figure 9.4 Animated Transcription. By this process, a strand of RNA is assembled from nucleotides according to a template: a gene region in DNA. **Figure It Out: After the guanine, what is the next nucleotide that will be added to this growing strand of RNA?** Answer: Another guanine (G)

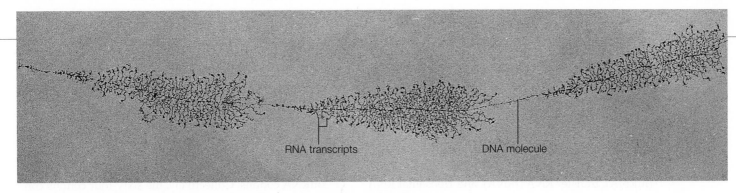

Figure 9.5 Typically, many RNA polymerases simultaneously transcribe the same gene, producing a structure often called a "Christmas tree" after its shape. Here, three genes next to one another on the same chromosome are being transcribed. **Figure It Out:** Are the polymerases transcribing this DNA molecule moving from left to right or from right to left? Answer: Left to right

ing strand) is the one that serves as the template for transcription. The polymerase starts moving along the DNA, in the 3' to 5' direction over the gene region ❷. As it moves, the polymerase unwinds the double helix just a bit so it can "read" the base sequence of the noncoding DNA strand. The polymerase joins free RNA nucleotides in the order dictated by that DNA sequence. As in DNA replication, the synthesis is directional: An RNA polymerase adds nucleotides only to the 3' end of the growing strand of RNA.

When the polymerase reaches the end of the gene region, the DNA and the new RNA are released. RNA polymerase follows base-pairing rules, so the new RNA strand is complementary in base sequence to the DNA strand from which it was transcribed ❸. It is an RNA copy of a gene, the same way that a paper transcript of a conversation carries the same information in a different format. Typically, many polymerases transcribe a particular gene region at the same time, so many new RNA strands can be produced very quickly (**Figure 9.5**).

Post-Transcriptional Modifications

In eukaryotes, transcription takes place in the nucleus. Eukaryotes also modify their RNA inside the nucleus, then ship it to the cytoplasm. Just as a dressmaker may snip off loose threads or add bows to a dress before it leaves the shop, so do eukaryotic cells tailor their RNA before it leaves the nucleus.

For example, most eukaryotic genes contain intervening sequences called **introns.** Introns are removed

in chunks from a newly transcribed RNA before it leaves the nucleus. Sequences that stay in the RNA are **exons** (**Figure 9.6**). Exons can be rearranged and spliced together in different combinations. By such **alternative splicing**, one gene can encode different proteins.

New transcripts that will become mRNAs are further tailored after splicing. A modified guanine "cap" gets attached to the 5' end of each. Later, the cap will help the mRNA bind to a ribosome. A tail of 50 to 300 adenines is also added to the 3' end of a new mRNA; hence the name, poly-A tail. The tail is a signal that allows an mRNA to be exported from the nucleus. As you will see in Chapter 10, an mRNA's poly-A tail also helps regulate the timing and speed of its translation.

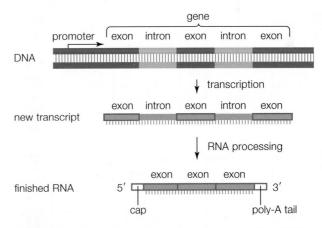

Figure 9.6 Animated Post-transcriptional modification of RNA. Introns are removed and exons spliced together. Messenger RNAs also get a poly-A tail and modified guanine "cap."

alternative splicing RNA processing event in which some exons are removed or joined in alternate combinations.
exon Nucleotide sequence that is not spliced out of RNA during processing.
intron Nucleotide sequence that intervenes between exons and is excised during RNA processing.
promoter In DNA, a sequence to which RNA polymerase binds.
RNA polymerase Enzyme that carries out transcription.

Take-Home Message

How is RNA assembled?

» In transcription, RNA polymerase uses the nucleotide sequence of a gene region in a chromosome as a template to assemble a strand of RNA.

» The new strand of RNA is a copy of the gene from which it was transcribed. Post-transcriptional modification of RNA occurs in the nucleus of eukaryotes.

9.4 RNA and the Genetic Code

- Base triplets in an mRNA encode a protein-building message. Ribosomal RNA and transfer RNA translate that message into a polypeptide chain.
- Links to Polypeptides 3.6, Ribosomes 4.4, Catalysis 5.4

DNA stores heritable information about proteins, but making those proteins requires messenger RNA (mRNA), transfer RNA (tRNA), and ribosomal RNA (rRNA). The three types of RNA interact to translate DNA's information into a protein.

mRNA—The Messenger

An mRNA is essentially a disposable copy of a gene. Its job is to carry DNA's protein-building information to the other two types of RNA for translation.

That protein-building information consists of a linear sequence of genetic "words" spelled with an alphabet of the four bases A, C, G, and U. Each of the genetic "words" carried by an mRNA is three bases long, and each is a code—a **codon**—for a particular amino acid. There are four possible bases in each of the three positions of a codon, so there are a total of sixty-four (4^3) mRNA codons. Collectively, the sixty-four codons constitute the **genetic code** (**Figure 9.7**). The sequence of the three nucleotides in a base triplet determines which amino acid the codon specifies. For instance, the codon AUG codes for the amino acid methionine (met), and UGG codes for tryptophan (trp).

One codon follows the next along the length of an mRNA, so the order of codons in an mRNA determines the order of amino acids in the polypeptide that will be translated from it. Thus, the base sequence of a gene is transcribed into the base sequence of an mRNA, which is in turn translated into an amino acid sequence (**Figure 9.8**).

Twenty naturally occurring amino acids are encoded by the sixty-four codons in the genetic code. Some amino acids are specified by more than one codon. For instance, GAA and GAG both code for glutamic acid. Other codons signal the beginning and end of a protein-coding sequence. For example, the first AUG in an mRNA is the signal to start translation in most species. AUG also happens to be the codon for methionine, so methionine is always the first amino acid in new polypeptides of such organisms. UAA, UAG, and UGA do not specify an amino acid. They

first base	second base U	second base C	second base A	second base G	third base
U	UUU phe	UCU ser	UAU tyr	UGU cys	U
U	UUC phe	UCC ser	UAC tyr	UGC cys	C
U	UUA leu	UCA ser	UAA stop	UGA stop	A
U	UUG leu	UCG ser	UAG stop	UGG trp	G
C	CUU leu	CCU pro	CAU his	CGU arg	U
C	CUC leu	CCC pro	CAC his	CGC arg	C
C	CUA leu	CCA pro	CAA gln	CGA arg	A
C	CUG leu	CCG pro	CAG gln	CGG arg	G
A	AUU ile	ACU thr	AAU asn	AGU ser	U
A	AUC ile	ACC thr	AAC asn	AGC ser	C
A	AUA ile	ACA thr	AAA lys	AGA arg	A
A	AUG met	ACG thr	AAG lys	AGG arg	G
G	GUU val	GCU ala	GAU asp	GGU gly	U
G	GUC val	GCC ala	GAC asp	GGC gly	C
G	GUA val	GCA ala	GAA glu	GGA gly	A
G	GUG val	GCG ala	GAG glu	GGG gly	G

ala	alanine (A)	leu	leucine (L)
arg	arginine (R)	lys	lysine (K)
asn	asparagine (N)	met	methionine (M)
asp	aspartic acid (D)	phe	phenylalanine (F)
cys	cysteine (C)	pro	proline (P)
glu	glutamic acid (E)	ser	serine (S)
gln	glutamine (Q)	thr	threonine (T)
gly	glycine (G)	trp	tryptophan (W)
his	histidine (H)	tyr	tyrosine (Y)
ile	isoleucine (I)	val	valine (V)

Figure 9.7 The genetic code. Each codon in mRNA is a set of three nucleotide bases. In the large chart, the *left* column lists a codon's first base, the *top* row lists the second, and the *right* column lists the third. Sixty-one of the triplets encode amino acids; the remaining three are signals that stop translation. The amino acid names that correspond to abbreviations in the chart are listed *above*. **Figure It Out: Which codons specify the amino acid lysine (lys)?**

Answer: AAA and AAG

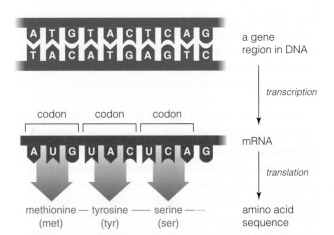

a gene
region in DNA

transcription

mRNA

translation

methionine — tyrosine —— serine ·····   amino acid
(met)   (tyr)   (ser)   sequence

Figure 9.8 Example of the correspondence between DNA, RNA, and proteins. A DNA strand is transcribed into mRNA, and the codons of the mRNA specify a chain of amino acids.

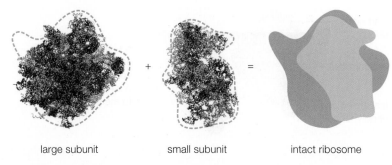

large subunit   small subunit   intact ribosome

Figure 9.9 Animated Ribosome structure. An intact ribosome consists of a large and a small subunit. Protein components are shown in *green*.

are signals that stop translation, so they are called stop codons. A stop codon marks the end of the protein-coding sequence in an mRNA.

The genetic code is highly conserved, which means that most organisms use the same code and probably always have. Bacteria, archaea, and some protists have a few codons that differ from the typical code, as do mitochondria and chloroplasts—a clue that led to a theory of how these two organelles evolved (we return to this topic in Section 19.5).

rRNA and tRNA—The Translators

Ribosomes and tRNAs interact to translate the genetic code carried by an mRNA into a polypeptide. Both are abundant in cytoplasm.

A ribosome has one large and one small subunit. Each subunit consists mainly of rRNA, with some associated structural proteins (**Figure 9.9**). During translation, a large and a small ribosomal subunit converge as an intact ribosome on an mRNA.

A tRNA has two attachment sites. The first is an anticodon, a triplet of nucleotides that base-pairs with an mRNA codon (**Figure 9.10**). The other attachment site binds to an amino acid—the one specified by the codon. Transfer RNAs with different anticodons carry different amino acids. You will see in the next section how those tRNAs deliver amino acids, one after the next, to a ribosome during translation of an mRNA.

anticodon Set of three nucleotides in a tRNA; base-pairs with mRNA codon.
codon In mRNA, a nucleotide base triplet that codes for an amino acid or stop signal during translation.
genetic code Complete set of sixty-four mRNA codons.

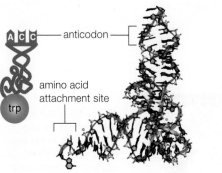

anticodon

amino acid
attachment site

trp

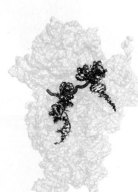

A Icon and model of the tRNA that carries the amino acid tryptophan. Each tRNA's antico-don is complementary to an mRNA codon. Each also carries the amino acid specified by that codon. *Pink* balls are magnesium ions.

B During translation, tRNAs dock at an intact ribosome (only the small subunit is shown, in *tan*). The anticodons of two tRNAs have lined up with complementary codons of an mRNA (*red*).

Figure 9.10 tRNA structure.

Ribosomal RNA is one example of RNA with enzymatic activity: The rRNA of a ribosome, not the protein, catalyzes the formation of a peptide bond between amino acids. As the amino acids are delivered, the ribosome joins them via peptide bonds into a new polypeptide (Section 3.6). Thus, the order of codons in an mRNA—DNA's protein-building message—becomes translated into a new protein.

Take-Home Message

What roles do mRNA, tRNA, and rRNA play during translation?

» mRNA carries protein-building information. The bases in mRNA are "read" in sets of three during protein synthesis. Most of these base triplets (codons) code for amino acids. The genetic code consists of all sixty-four codons.

» Ribosomes, which consist of two subunits of rRNA and proteins, assemble amino acids into polypeptide chains.

» A tRNA has an anticodon complementary to an mRNA codon, and it has a binding site for the amino acid specified by that codon. Transfer RNAs deliver amino acids to ribosomes.

9.5 Translation: RNA to Protein

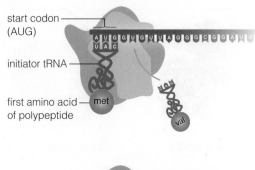

start codon (AUG)

initiator tRNA

first amino acid of polypeptide

1 Ribosome subunits and an initiator tRNA converge on an mRNA. A second tRNA binds to the second codon.

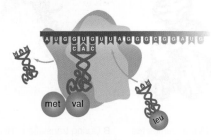

peptide bond

2 A peptide bond forms between the first two amino acids.

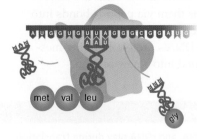

3 The first tRNA is released and the ribosome moves to the next codon. A third tRNA binds to the third codon.

4 A peptide bond forms between the second and third amino acids.

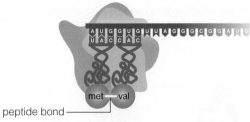

5 The second tRNA is released and the ribosome moves to the next codon. A fourth tRNA binds the fourth codon.

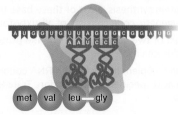

6 A peptide bond forms between the third and fourth amino acids.

The process repeats until the ribosome encounters a stop codon in the mRNA.

■ Translation converts the information carried by an mRNA into a new polypeptide chain.

■ The order of codons in an mRNA determines the order of amino acids in the polypeptide chain translated from it.

■ Links to Peptide bonds 3.6, Energy in metabolism 5.6

Translation, the second part of protein synthesis, proceeds in three stages: initiation, elongation, and termination. It occurs in the cytoplasm, which has many free amino acids, tRNAs, and ribosomal subunits. In eukaryotes, the initiation stage begins when an mRNA leaves the nucleus and a small ribosomal subunit binds to it (**Figures 9.11** and **9.12A**). Next, the anticodon of a special tRNA called an initiator base-pairs with the first AUG codon of the mRNA. Then, a large ribosomal subunit joins the small subunit **1**.

In the elongation stage, the ribosome assembles a polypeptide chain as it moves along the mRNA. The initiator tRNA carries the amino acid methionine, so the first amino acid of the new polypeptide chain is methionine. Another tRNA brings the second amino acid to the complex as its anticodon base-pairs with the second codon in the mRNA. The ribosome catalyzes formation of a peptide bond between the first two amino acids **2**.

The first tRNA is released and the ribosome moves to the next codon. Another tRNA brings the third amino acid to the complex as its anticodon base-pairs with the third codon of the mRNA **3**. A peptide bond forms between the second and third amino acids **4**.

The second tRNA is released and the ribosome moves to the next codon. Another tRNA brings the fourth amino acid to the complex as its anticodon base-pairs with the fourth codon of the mRNA **5**. A peptide bond forms between the third and fourth amino acids **6**. The new polypeptide chain continues to elongate as the ribosome catalyzes peptide bonds between amino acids delivered by successive tRNAs.

Termination occurs when the ribosome reaches a stop codon in the mRNA. The mRNA and the polypeptide detach from the ribosome, and the ribosomal subunits separate from each other. Translation is now complete. The new polypeptide will either join the pool of proteins in the cytoplasm, or enter rough ER of the endomembrane system (Section 4.7).

Figure 9.11 Animated Translation. Translation initiates when ribosomal subunits and an initiator tRNA converge on an mRNA. tRNAs deliver amino acids in the order dictated by successive codons in the mRNA. The ribosome links the amino acids together as it moves along the mRNA, so a polypeptide forms and elongates. Translation terminates when the ribosome reaches a stop codon.

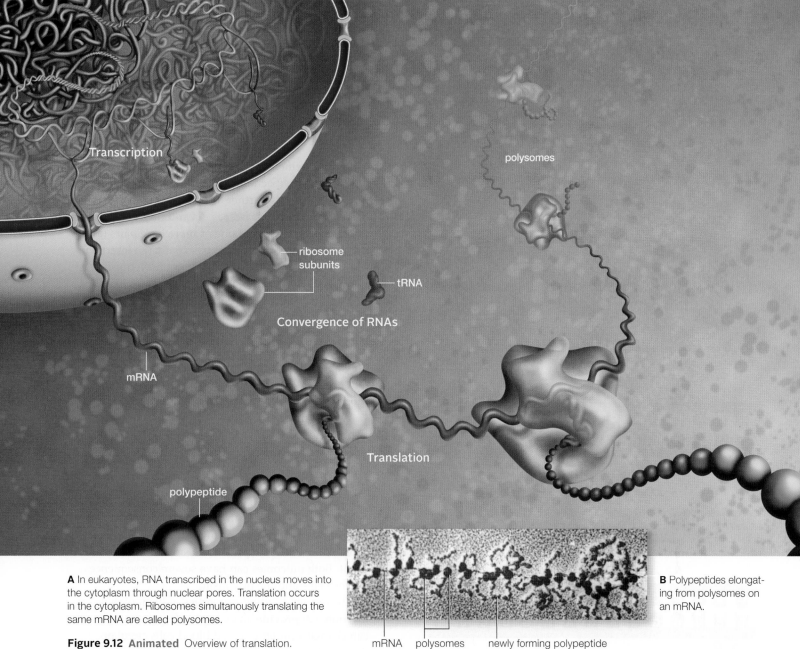

A In eukaryotes, RNA transcribed in the nucleus moves into the cytoplasm through nuclear pores. Translation occurs in the cytoplasm. Ribosomes simultanously translating the same mRNA are called polysomes.

Figure 9.12 **Animated** Overview of translation.

B Polypeptides elongating from polysomes on an mRNA.

mRNA   polysomes   newly forming polypeptide

In cells that are making a lot of protein, many ribosomes may simultaneously translate the same mRNA, in which case they are called polysomes (**Figure 9.12B**). In bacteria and archaea, transcription and translation both occur in the cytoplasm, and these processes are closely linked in time and space. Translation begins before transcription ends, so a transcription "Christmas tree" is often decorated with polysome "balls."

Given that many polypeptides can be translated from one mRNA, why would any cell also make many copies of an mRNA? Compared with DNA, RNA is not very stable. An mRNA may last only a few minutes before it is disassembled by enzymes in cytoplasm. The fast turnover allows cells to adjust their protein synthesis quickly in response to changing needs.

Translation is energy intensive. That energy is provided mainly in the form of phosphate-group transfers from the RNA nucleotide GTP (shown in **Figure 9.2B**) to molecules involved in the process.

Take-Home Message

How is mRNA translated into protein?

» Translation is an energy-requiring process that converts protein-building information carried by an mRNA into a polypeptide.

» During initiation, an mRNA, an initiator tRNA, and two ribosome subunits join.

» During elongation, amino acids are delivered to the complex by tRNAs in the order dictated by successive mRNA codons. As they arrive, the ribosome joins each to the end of the polypeptide chain.

» Termination occurs when the ribosome reaches a stop codon in the mRNA. The mRNA and the polypeptide are released, and the ribosome disassembles.

9.6 Mutated Genes and Their Protein Products

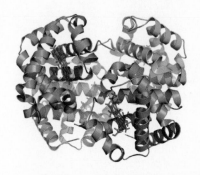

A Hemoglobin, an oxygen-binding protein in red blood cells. This protein consists of four polypeptides: two alpha globins (*blue*) and two beta globins (*green*). Each globin has a pocket that cradles a heme (*red*). Oxygen molecules bind to the iron atom at the center of each heme.

16 17 18 19 20 21 22 23 24 25 26 27 28 29 30

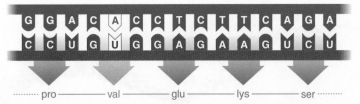

.... pro ———— glu ———— glu ———— lys ———— ser ....

B Part of the DNA (*blue*), mRNA (*brown*), and amino acid sequence (*green*) of human beta globin. Numbers indicate the position of the base pair in the coding sequence of the mRNA.

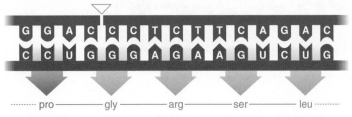

.... pro ———— val ———— glu ———— lys ———— ser ....

C A base-pair substitution replaces a thymine with an adenine. When the altered mRNA is translated, valine replaces glutamic acid as the sixth amino acid of the polypeptide. Hemoglobin with this form of beta globin is called HbS, or sickle hemoglobin.

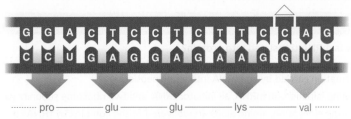

.... pro ———— gly ———— arg ———— ser ———— leu ....

D A deletion of one base pair causes the reading frame for the rest of the mRNA to shift, so a different protein product forms. This frameshift results in a defective beta globin chain. The outcome is beta thalassemia, a genetic disorder in which a person has an abnormally low amount of hemoglobin.

.... pro ———— glu ———— glu ———— lys ———— val ....

E An insertion of one base pair causes the reading frame for the rest of the mRNA to shift, so a different protein product forms. This frameshift results in a defective beta globin chain. The outcome is beta thalassemia.

Figure 9.13 Animated Examples of mutations in the human beta globin gene.

■ If the nucleotide sequence of a gene changes, it may result in an altered gene product, with harmful effects.

■ Links to Hydrophobicity 2.5, Hemoglobin 3.2, Protein structure 3.6, Free radicals and heme cofactors 5.6, Chromosomes 8.2, DNA replication 8.5, Mutations 8.6

Mutations, remember, are permanent changes in a DNA sequence (Section 8.6). A mutation in which one nucleotide and its partner are replaced by a different base pair is called a **base-pair substitution**. Other mutations involve the loss of one or more base pairs (a **deletion**) or the addition of extra base pairs (an **insertion**).

Mutations are relatively uncommon events in a normal cell. For example, the chromosomes in a diploid human cell collectively consist of about 7×10^9 base pairs, any of which may become mutated each time that cell divides. However, only about 175 bases actually do change after a division. On top of that, only about 3 percent of the cell's DNA encodes protein products, so there is a low probability that any of those mutations will be in a protein coding region.

When a mutation does occur in a protein coding region, the redundancy of the genetic code offers the cell a margin of safety. For example, a mutation that changes a UCU codon to UCC in an mRNA may not have further effects, because both codons specify serine. Mutations in a gene that have no effect on the gene's product are said to be silent. Other mutations are not silent: They may change an amino acid in a protein, or result in a premature stop codon that shortens it. Both outcomes can have severe consequences for the cell—and the organism.

The oxygen-binding properties of hemoglobin (Section 3.2) provide an example of how a mutation can change a protein. As red blood cells circulate through the lungs, the hemoglobin proteins inside of them bind to oxygen molecules. The cells then travel to other regions of the body, and the hemoglobin releases its oxygen cargo wherever the oxygen level is low. When the red blood cells return to the lungs, the hemoglobin binds to more oxygen.

Hemoglobin's structure allows it to bind and release oxygen. The protein consists of four polypeptides called globins (**Figure 9.13A**). Each globin folds around a heme, a cofactor with an iron atom at its center (Section 5.6). Oxygen molecules bind to hemoglobin at those iron atoms.

In adult humans, two alpha globin chains and two beta globin chains make up each hemoglobin molecule. Defects in either of the polypeptide chains can cause a condition called anemia, in which a person's blood is deficient in red blood cells or in hemoglobin.

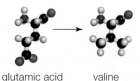

glutamic acid   valine

A A base-pair substitution results in the abnormal beta globin chain of sickle hemoglobin (HbS). The sixth amino acid in such chains is valine, not glutamic acid. The difference causes HbS molecules to form rod-shaped clumps that distort normally round blood cells into sickle shapes.

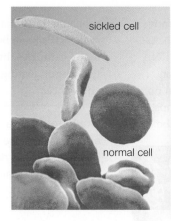

sickled cell

normal cell

B The sickled cells clog small blood vessels, causing circulatory problems that result in damage to many organs. Destruction of the cells by the body's immune system results in anemia.

Tionne "T-Boz" Watkins of the music group TLC is a celebrity spokesperson for the Sickle Cell Disease Association of America. She was diagnosed with sickle-cell anemia as a child.

Figure 9.14 Animated Sickle-cell anemia.

Both outcomes limit the blood's ability to carry oxygen, and the resulting symptoms range from mild to life-threatening.

Sickle-cell anemia, a type of anemia that is most common in people of African ancestry, arises because of a base-pair substitution in the beta globin gene. The substitution causes the body to produce a version of beta globin in which the sixth amino acid is valine instead of glutamic acid (**Figure 9.13B,C**). Hemoglobin with this mutation in its beta globin is called sickle hemoglobin, or HbS.

Unlike glutamic acid, which carries a negative charge, valine carries no charge. As a result of that one substitution, a tiny patch of the beta globin polypeptide changes from hydrophilic to hydrophobic, which in turn causes the hemoglobin's behavior to change slightly. HbS molecules stick together and form large, rodlike clumps under certain conditions. Red blood cells that contain the clumps become distorted into a crescent (sickle) shape (**Figure 9.14**). Sickled cells clog tiny blood vessels, thus disrupting blood circulation throughout the body. Over time, repeated episodes of sickling can damage organs and cause death.

A different type of anemia, beta thalassemia, is caused by the deletion of the twentieth base pair in the coding region of the beta globin gene (**Figure 9.13D**). Like many other deletions, this one causes a **frameshift,** in which the reading frame of the mRNA codons shifts. A frameshift usually has drastic consequences because it garbles the genetic message, just as incorrectly grouping a series of letters garbles the meaning of a sentence:

The cat ate the rat.
T hec ata tet her at.
Th eca tat eth era t.

The frameshift caused by the beta globin deletion results in a polypeptide that differs drastically from normal beta globin. Hemoglobin molecules do not assemble correctly with the altered polypeptides, which are only 18 amino acids long (the normal beta globin chain has 147 amino acids). This outcome is the source of anemia in beta thalassemia.

Beta thalassemia can also be caused by insertion mutations, which, like deletions, often cause frameshifts (**Figure 9.13E**). Insertion mutations are often caused by the activity of transposable elements, which are segments of DNA that can move spontaneously within or between chromosomes. Transposable elements can be hundreds or thousands of base pairs long, so when one interrupts a gene it becomes a major insertion that changes the gene's product. Transposable elements are common in the DNA of all species; about 45 percent of human DNA consists of transposable elements or their remnants.

base-pair substitution Mutation in which a single base pair changes.
deletion Mutation in which one or more base pairs are lost.
frameshift Mutation that causes the reading frame of mRNA codons to shift.
insertion Mutation in which one or more base pairs become inserted into DNA.
transposable element Segment of chromosomal DNA that can spontaneously move to a new location.

Take-Home Message

What happens after a gene becomes mutated?

» Mutations that result in an altered protein can have drastic consequences.

» A base-pair substitution may change an amino acid in a protein, or shorten it by introducing a premature stop codon.

» Frameshifts that occur after an insertion or deletion change an mRNA's codon reading frame, so they garble its protein-building instructions.

The Aptly Acronymed RIPs (revisited)

The enzyme in ricin inactivates ribosomes by removing a particular adenine base from one of the rRNAs in the heavy subunit. That adenine is part of an RNA binding site for proteins that help with elongation. After the base has been removed, a ribosome can no longer bind to those proteins, and elongation stops.

However, the main function of RIPs may not be destroying ribosomes. Many function in immune signaling, but it is their antiviral and anti-

cancer activity that has researchers abuzz. Plants that make RIPs have been used as traditional medicines for many centuries, but only recently have Western scientists begun testing these compounds to combat HIV and cancer.

How would you vote? Exposure to ricin is unlikely, but extremists continue to attempt using it for terrorist activities. Researchers have developed a vaccine against ricin. Do you want to be vaccinated?

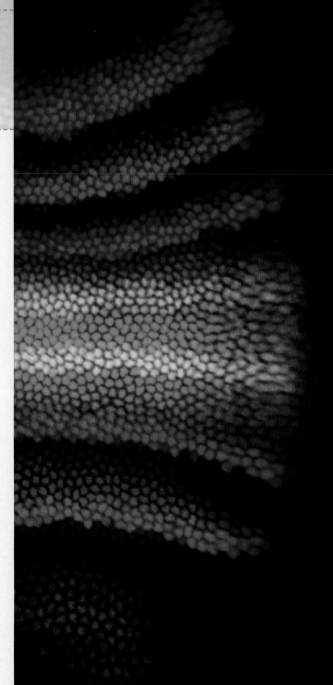

LEARNING ROADMAP

Where you have been This chapter explores metabolism (Sections 5.5, 5.6) in context of gene expression (9.2). You will be applying what you know about DNA: chromosomes (8.2), structure (8.4), replication (8.5), mutations and how they occur (9.6), and genetic reprogramming (8.7), as well as transcription (9.3) and translation (9.5). You will also revisit functional groups (3.3), carbohydrates (3.4), glycolysis (7.3), and fermentation (7.6).

Where you are now

Mechanisms of Gene Control
Gene expression changes in response to changing conditions both inside and outside the cell. Selective gene expression during development results in differentiation.

Master Genes
The orderly, localized expression of homeotic and other master genes gives rise to the body plan of complex multicelled organisms.

Examples in Eukaryotes
Genes that govern X chromosome inactivation and male sex determination in mammals, and flower formation in plants, offer examples of gene control in eukaryotes.

Examples in Prokaryotes
Most prokaryotic gene controls bring about fast adjustments in the rate of transcription in response to changes in external conditions.

Epigenetics
New research is revealing how gene expression changes that arise in response to environmental pressures can be passed to an individual's future offspring.

Where you are going Chapter 11 discusses the cell cycle and how controls over it go awry in cancer; Chapter 12 returns to the genetic basis of reproduction. Chapter 13 explores how inheritance works, and Chapter 14, inheritance patterns in humans. Genomics, the study of genomes, is explained in Section 15.5, and evolutionary processes involving changes in DNA sequence will be discussed more fully in Chapter 17. Section 19.3 discusses the hypothesis that RNA, not DNA, was genetic material in the distant past. Plant gene controls are discussed in more detail in Chapter 30.

10.1 Between You and Eternity

You are in college, your whole life ahead of you. Your risk of developing cancer is as remote as old age, an abstract statistic that is easy to forget. "There is a moment when everything changes—when the width of two fingers can suddenly be the total distance between you and eternity." Robin Shoulla wrote those words after being diagnosed with breast cancer. She was seventeen. At an age when most young women are thinking about school, friends, parties, and potential careers, Robin was dealing with radical mastectomy: the removal of a breast, all lymph nodes under the arm, and skeletal muscles in the chest wall under the breast. She was pleading with her oncologist not to use her jugular vein for chemotherapy and wondering if she would survive to see the next year (**Figure 10.1**).

Robin's ordeal became part of a statistic, one of more than 200,000 new cases of breast cancer diagnosed in the United States each year. About 5,700 of those cases occur in women and men under thirty-four years of age.

Every second, millions of cells in your skin, bone marrow, gut lining, liver, and elsewhere are dividing and replacing their worn-out, dead, and dying predecessors. They do not divide at random; in normal cells, growth and division is tightly regulated. When this control fails, cancer is the outcome.

Cancer is a multistep process in which abnormally growing and dividing cells disrupt body tissues. Mechanisms that normally keep cells from getting overcrowded in tissues are lost, so cancer cell populations may reach extremely high densities. Unless chemotherapy, surgery, or another procedure eradicates them, cancer cells can put an individual on a painful road to death. Each year, cancers cause 15 to 20 percent of all human deaths in developed countries alone.

Cancer typically begins with a mutation in a gene whose product is part of a system of stringent controls over cell growth and division. Such controls govern when and how fast specific genes are transcribed and translated. A cancer-causing mutation may be inherited, or it may be a new one, as when DNA becomes damaged by environmental agents. If the mutation alters the gene's protein product so that it no longer works properly, one level of control over the cell's growth and division has been lost. You will be considering the impact of gene controls in chapters throughout this book, and also in some chapters of your life.

Robin Shoulla survived. Radical mastectomy is rarely performed today, but it was her only option. Now, seventeen years later, she has what she calls a normal life: career, husband, children. Her goal as a cancer survivor: "To grow very old with gray hair and spreading hips, smiling."

normal cells in organized clusters

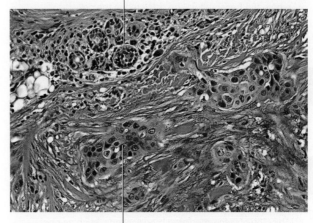

disorganized clusters of malignant cells

Figure 10.1 A case of breast cancer. *Top*, Robin Shoulla. *Bottom*, this light micrograph shows irregular clusters of cancer cells that have infiltrated milk ducts in human breast tissue. Diagnostic tests revealed abnormal cells such as these in Robin's body when she was just seventeen years old.

cancer Disease that occurs when the uncontrolled growth of body cells that can invade other tissues physically and metabolically disrupts normal function.

10.2 Switching Genes On and Off

- Gene controls govern the kinds and amounts of substances that are present in a cell at any given time.
- Links to Functional groups 3.3, Chromatin 4.6, Phosphorylation 5.6, Glycolysis 7.3, Histones 8.2, Gene expression 9.2, Transcription 9.3, Translation 9.5, Globin 9.6

All of the cells in your body are descended from the same fertilized egg, so they all contain the same DNA with the same genes. However, each cell rarely uses more than 10 percent of its genes at one time. Some of those genes affect structural features and metabolic pathways common to all cells. Others are expressed only by certain subsets of your cells. For example,

enhancer

Figure 10.3 Hypothetical part of a chromosome that contains a gene. Molecules that affect the rate of transcription of the gene bind at promoter (*yellow*) or enhancer (*green*) sequences.

most of your body cells express genes that encode the enzymes of glycolysis, but only your immature red blood cells express genes that encode globin.

Differentiation, the process by which cells of a multicelled organism become specialized, occurs as different cell lineages begin to express different subsets of their genes during development. Which genes a cell uses determines the molecules it will produce, which in turn determines what kind of cell it will be.

Gene Controls

Control over which genes are expressed at a particular time is crucial for proper development of complex, multicelled bodies. It also allows individual cells to respond appropriately to changes in their environment. The "switches" that turn gene expression on or off are called gene controls—molecules or structures that start, enhance, slow, or stop the individual steps of gene expression (**Figure 10.2**).

Transcription Many gene controls affect whether and how fast a gene is transcribed into RNA ❶. Proteins called transcription factors affect the rate of transcription by binding directly to DNA at promoters or other special nucleotide sequences. Whether and how fast a gene is transcribed depends on which transcription factors are bound to the DNA. A repressor slows or stops transcription by interfering with RNA polymerase binding to a promoter. An activator speeds up transcription when it binds. Many activators work by helping RNA polymerase attach to the promoter. Some speed transcription by binding to DNA sequences called enhancers (**Figure 10.3**). An enhancer is not necessarily close to the gene it affects, and may even be on a different chromosome.

Chromatin structure also affects transcription: RNA polymerase can only attach to DNA that has been unwound from histones (Section 8.2). Certain modifications to histone proteins change the way they interact with the DNA that wraps around them. Some modifications make them release their grip on the DNA; others make them tighten it. For example, adding an acetyl group to a histone loosens the DNA wrapped around it, so enzymes that acetylate histones encourage transcription. Conversely, adding a methyl

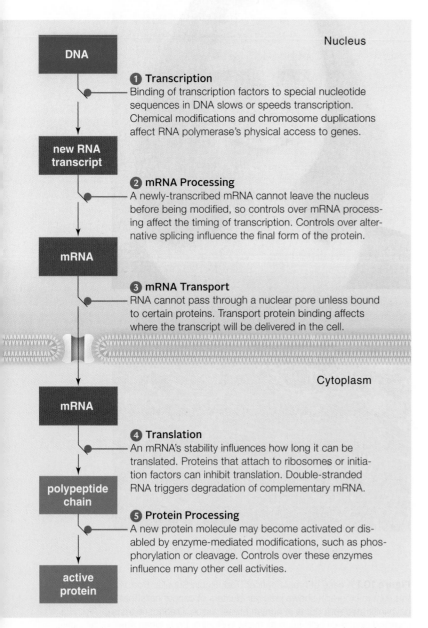

Nucleus

DNA

❶ **Transcription**
Binding of transcription factors to special nucleotide sequences in DNA slows or speeds transcription. Chemical modifications and chromosome duplications affect RNA polymerase's physical access to genes.

new RNA transcript

❷ **mRNA Processing**
A newly-transcribed mRNA cannot leave the nucleus before being modified, so controls over mRNA processing affect the timing of transcription. Controls over alternative splicing influence the final form of the protein.

mRNA

❸ **mRNA Transport**
RNA cannot pass through a nuclear pore unless bound to certain proteins. Transport protein binding affects where the transcript will be delivered in the cell.

Cytoplasm

mRNA

❹ **Translation**
An mRNA's stability influences how long it can be translated. Proteins that attach to ribosomes or initiation factors can inhibit translation. Double-stranded RNA triggers degradation of complementary mRNA.

polypeptide chain

❺ **Protein Processing**
A new protein molecule may become activated or disabled by enzyme-mediated modifications, such as phosphorylation or cleavage. Controls over these enzymes influence many other cell activities.

active protein

Figure 10.2 Animated Points of control over eukaryotic gene expression.

↳ transcription start site   ↲ transcription end

group to a histone can make the DNA wind more tightly around it, so enzymes that methylate histones discourage transcription.

In a few types of cells, the number of copies of a gene affects how fast its product is made. DNA in these cells is copied repeatedly. The result is a set of

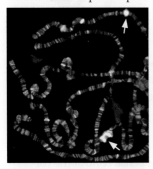

polytene chromosomes, each of which consists of hundreds or thousands of side-by-side copies of the same DNA molecule. Transcription of one gene occurs simultaneously on all of the identical DNA strands, and is visible as a puff on the chromosome (*above*). Transcription produces a lot of mRNA that can be translated quickly into a lot of protein. Polytene chromosomes are common in immature amphibian eggs, storage tissues of some plants, and saliva gland cells of some insect larvae.

mRNA Processing As you know, before eukaryotic mRNAs leave the nucleus, they are modified—spliced, capped, and finished with a poly-A tail (Section 9.3). Controls over these modifications can affect the form of a protein product and when it will appear in the cell ❷. For example, controls that determine which exons are spliced out of an mRNA affect which form of a protein will be translated from it.

mRNA Transport In eukaryotes, transcription occurs in the nucleus, and translation in the cytoplasm. Thus, mRNA transport out of the nucleus is another potential point of control ❸. A new RNA can pass through pores of the nuclear envelope only after it has been processed appropriately. Mechanisms that delay processing of an mRNA also delay its appearance in the cytoplasm, and thereby delay its translation.

Transport of mRNA outside the nucleus also influences gene expression. A short base sequence near an mRNA's poly-A tail is like a zip code. Proteins that attach to the zip code drag the mRNA along cytoskeletal elements to the organelle or area of the cytoplasm specified by the code. Other proteins that attach to the zip code region prevent translation of the mRNA before it reaches its destination. Localization of mRNA to a particular area of a cell allows growth or movement in a specific direction. In an egg, it is crucial for proper development of a forthcoming embryo.

Translational Control Most controls over eukaryotic gene expression affect translation ❹. Many of these controls govern the production or function of the various molecules that carry out the process. Others affect mRNA stability: The longer an mRNA lasts, the more protein can be made from it. Enzymes begin to disassemble a new mRNA as soon as it arrives in the cytoplasm. The fast turnover allows cells to adjust their protein synthesis quickly in response to changing needs. How long an mRNA persists depends on its base sequence, the length of its poly-A tail, and which proteins are attached to it.

MicroRNAs inhibit translation. Part of a microRNA folds back on itself and forms a small double-stranded region. By a process called RNA interference, any double-stranded RNA (including a microRNA) is cut up into small bits that are taken up by special enzyme complexes. These complexes destroy every mRNA in a cell that can base-pair with the bits. So, expression of a microRNA complementary in sequence to a gene inhibits expression of that gene.

Post-Translational Modification Many newly synthesized polypeptide chains must be modified before they become functional ❺. For example, some enzymes become active only after they have been phosphorylated (Section 5.6). Such post-translational modifications inhibit, activate, or stabilize many molecules, including the enzymes that participate in transcription and translation.

activator Regulatory protein that increases the rate of transcription when it binds to a promoter or enhancer.
differentiation The process by which cells become specialized.
enhancer Binding site in DNA for proteins that enhance the rate of transcription.
repressor Regulatory protein that blocks transcription.
transcription factor Regulatory protein that influences transcription; e.g., an activator or repressor.

Take-Home Message

What is gene control?

» Gene controls consist of molecules and structures that can start, enhance, slow, or stop individual steps of gene expression.

» Most cells of multicelled organisms differentiate as they start expressing a unique subset of their genes. Which genes a cell expresses depends on the type of organism, its stage of development, and environmental conditions.

10.3 Master Genes

■ Cascades of gene expression govern the development of a complex, multicelled body.

As an animal embryo develops, its differentiating cells form tissues, organs, and body parts. Some cells that alternately migrate and stick to other cells develop into nerves, blood vessels, and other structures that weave through the tissues. Events like these fill in the body's details, and all are driven by cascades of master gene expression. **Master genes** encode products that affect the expression of many other genes. Expression of a master gene causes other genes to be expressed, with the final outcome being the completion of an intricate task such as the formation of an eye.

Pattern formation is the process by which a complex body forms from local processes in an embryo. Pattern formation begins when maternal mRNAs are delivered to opposite ends of an unfertilized egg as it forms. These mRNAs are translated only after the egg is fertilized, and then their protein products diffuse away in gradients that span the entire developing embryo. Depending on where a nucleus falls within the gradients, one or another set of master genes will be transcribed inside of it. The products of those master genes also form in gradients that span the embryo. Still other master genes are transcribed depending on where a nucleus falls within these gradients, and so on.

Such regional gene expression during development results in a three-dimensional map that consists of overlapping concentration gradients of master gene products. Which master genes are transcribed at any given time changes, so the map changes too.

Figure 10.5
A homeodomain. The protein product of a homeotic gene called *antennapedia* (*gold*) is shown attached to a promoter sequence in a fragment of DNA. The region of the protein that recognizes and binds directly to regulatory sequences in DNA is the homeodomain.

Eventually, the products of some master genes cause undifferentiated cells to differentiate, and specialized tissues and structures are the outcome (**Figure 10.4**).

Homeotic Genes

Master genes called **homeotic genes** control the formation of specific body parts (eyes, legs, and so on) during development. All homeotic genes encode transcription factors with a homeodomain, which is a region of about sixty amino acids that can bind to a promoter or some other sequence of nucleotides in a chromosome (**Figure 10.5**).

Long before body parts develop, certain master genes trigger the transcription of homeotic genes in particular parts of the embryo. Products of the different homeotic genes cause cells in those parts to differentiate into tissues that form specific structures such as a wing, a leg, or an eye.

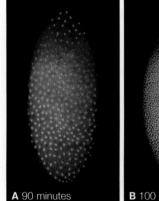 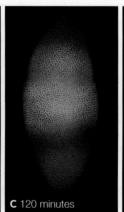 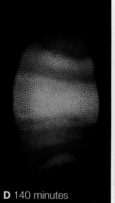 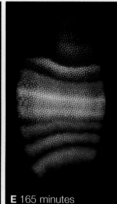 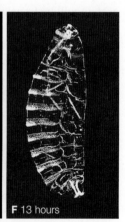

A 90 minutes B 100 minutes C 120 minutes D 140 minutes E 165 minutes F 13 hours

A, B The master gene *even-skipped* is expressed (in *red*) only where two maternal gene products (*blue* and *green*) overlap.

C–E The products of several master genes, including the two shown here in *green* and *blue*, confine the expression of *even-skipped* (*red*) to seven stripes.

F Seven segments that develop later correspond to the position of the *even-skipped* stripes.

Figure 10.4 How gene expression control makes a fly, as illuminated by segmentation. Expression of different master genes is shown by different colors in fluorescence microscopy images of whole *Drosophila* embryos at successive stages of development. Bright dots are individual nuclei.

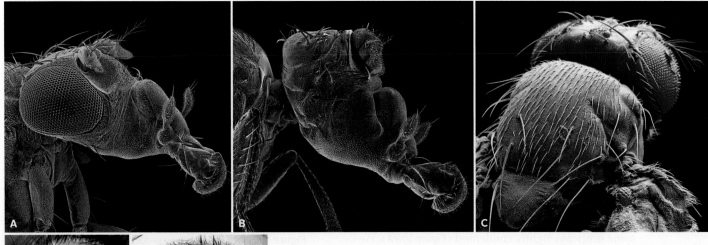

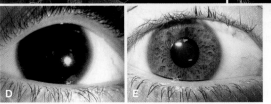

Figure 10.6 *Eyeless*: the eyes have it. (**A**) A normal fruit fly has large, round eyes. (**B**) A fruit fly with a mutation in its *eyeless* gene develops with no eyes. (**C**) Eyes form wherever the *eyeless* gene is expressed in fly embryos—here, on the head and wing.

Humans, mice, squids, and other animals have a gene called *PAX6*. In humans, *PAX6* mutations result in missing irises, a condition called aniridia (**D**). Compare a normal iris (**E**). *PAX6* is so similar to *eyeless* that it triggers eye development when expressed in fly embryos.

The function of many homeotic genes has been discovered by manipulating their expression, one at a time. Researchers inactivate a homeotic gene by introducing a mutation or deleting it entirely, an experiment called a knockout. An organism that carries the knocked-out gene may differ from normal individuals, and the differences are clues to the function of the missing gene product.

normal fly head

groucho mutation

Researchers often name homeotic genes based on what happens in their absence. For instance, fruit flies with a mutated *eyeless* gene develop with no eyes (**Figure 10.6A,B**). *Dunce* is required for learning and memory. *Wingless*, *wrinkled*, and *minibrain* are self-explanatory. *Tinman* is necessary for development of a heart. Flies with a mutated *groucho* gene have extra bristles above their eyes (*inset*). One gene was named *toll*, after what its German discoverer exclaimed upon seeing the disastrous effects of the mutation (*toll* means "cool!" in German slang).

Homeotic genes control development by the same mechanisms in all multicelled eukaryotes, and many are interchangeable among different species. Thus, we can infer that they evolved in the most ancient eukaryotic cells. Homeodomains often differ among species only in conservative substitutions (one amino acid has replaced another with similar chemical properties).

Consider the *eyeless* gene. Eyes form in embryonic fruit flies wherever this gene is expressed, which is typically in tissues of the head. If the *eyeless* gene is expressed in another part of the developing embryo, eyes form there too (**Figure 10.6C**). Humans, squids, mice, fish, and many other animals have a gene called *PAX6*, which is very similar to the *eyeless* gene. In humans, mutations in *PAX6* cause eye disorders such as aniridia, in which a person's irises are underdeveloped or missing (**Figure 10.6D,E**). *PAX6* works the same way in animals of different phyla. For example, if the *PAX6* gene from a human or mouse is inserted into an *eyeless* mutant fly, it has the same effect as the *eyeless* gene: An eye forms wherever it is expressed. Such studies are evidence of shared ancestry among these evolutionarily distant animals.

homeotic gene Type of master gene; its expression controls formation of specific body parts during development.
knockout An experiment in which a gene is deliberately inactivated in a living organism.
master gene Gene encoding a product that affects the expression of many other genes.
pattern formation Process by which a complex body forms from local processes during embryonic development.

Take-Home Message

How do genes control development?

» Development is orchestrated by cascades of master gene expression in embryos.

» The expression of homeotic genes during development governs the formation of specific body parts. Homeotic genes that function in similar ways across taxa are evidence of shared ancestry.

10.4 Examples of Gene Control in Eukaryotes

■ Selective gene expression gives rise to many traits.
■ Links to Chromosomes and sex determination 8.2, Mutations 9.6

Gene controls influence many traits that are characteristic of humans and other eukaryotic organisms, as the following examples illustrate.

X Chromosome Inactivation

In humans and other mammals, a female's cells each contain two X chromosomes, one inherited from her mother, the other one from her father. One X chromosome is always tightly condensed (**Figure 10.7A**). We call the condensed X chromosomes "Barr bodies," after Murray Barr, who discovered them. Condensation prevents transcription, so most of the genes on a Barr body are not expressed. This **X chromosome inactivation** ensures that only one of the two X chromosomes in a female's cells is active. According to a theory called **dosage compensation**, X chromosome inactivation equalizes expression of X chromosome genes between the sexes. The body cells of male mammals (XY) have one set of X chromosome genes. Body cells of female mammals have two sets, but female embryos do not develop properly when both sets are expressed.

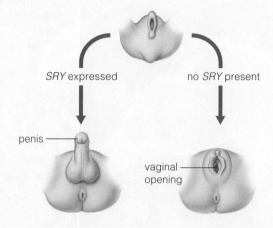

Figure 10.8 Development of reproductive organs in human embryos. An early embryo appears neither male nor female. *SRY* gene expression determines what reproductive organs form.

X chromosome inactivation occurs when an embryo is a ball of about 200 cells. In humans and many other mammals, it occurs independently in every cell of a female embryo. The maternal X chromosome may get inactivated in one cell, and the paternal or maternal X chromosome may get inactivated in a cell next to it. Once the selection is made in a cell, all of that cell's descendants make the same selection as they continue dividing and forming tissues. As a result of the inactivation, a female mammal is a "mosaic" for the expression of genes on the X chromosome. She has patches of tissue in which genes on the maternal X chromosome are expressed, and patches in which genes on the paternal X chromosome are expressed (**Figure 10.7B**).

How does just one of two X chromosomes get inactivated? A gene called *XIST* is transcribed on only one of the two X chromosomes. The gene's product, a long, noncoding RNA, sticks to the chromosome that expresses the gene. The RNA coats the chromosome, and by an unknown mechanism causes it to condense into a Barr body. Thus, transcription of the *XIST* gene keeps a chromosome from transcribing other genes. The other chromosome does not express *XIST*, so it does not get coated with RNA; its genes remain available for transcription. How a cell "chooses" which X chromosome will express *XIST* remains unknown.

Male Sex Determination in Humans

The human X chromosome carries 1,336 genes. Some of those genes are associated with sexual traits, such as the distribution of body fat and hair. However, most of the genes on the X chromosome govern nonsexual traits such as blood clotting and color perception. Such

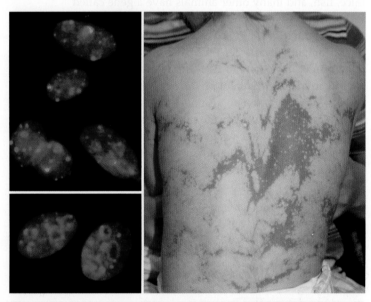

A Inactivated X chromosomes—Barr bodies—show up as *red* spots in the nucleus of these XX cells (*top*). Compare the nucleus of two XY cells (*bottom*).

B Visible evidence of mosaic tissues in a human female. One of this girl's X chromosomes carries a mutation that causes incontinentia pigmenti, a disorder that affects the skin, hair, nails, and teeth. This chromosome is active in darker patches of skin. Her other X chromosome, which does not carry the mutation, is active in the lighter patches of skin.

Figure 10.7 **Animated** X chromosome inactivation.

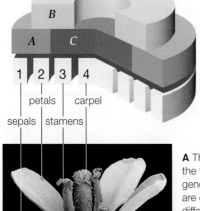

B Mutations in *ABC* genes result in malformed flowers.

Top left, right: Flowers of plants with *A* gene mutations have no petals, and no structures in place of the missing petals.

Bottom left: Flowers of plants with *B* gene mutations have sepals instead of petals.

Bottom right: Flowers of plants with *C* gene mutations have petals instead of sepals and carpels. Compare the normal flower in **A**.

A The pattern in which the floral identity genes *A*, *B*, and *C* are expressed affects differentiation of cells growing in whorls in the plant's tips. Their gene products guide expression of other genes in cells of each whorl; a flower results.

Figure 10.9 Animated Control of flower formation, as revealed by mutations in *Arabidopsis thaliana*.

genes are expressed in both males and females. Males, remember, also inherit one X chromosome.

The human Y chromosome carries only 307 genes, but one of them is *SRY*—the master gene for male sex determination in mammals. Its expression in XY embryos triggers the formation of testes, which are male gonads (**Figure 10.8**). Some of the cells in these primary male reproductive organs make testosterone, a sex hormone that controls the emergence of male secondary sexual traits such as facial hair, increased musculature, and a deep voice. We know that *SRY* is the master gene that controls emergence of male sexual traits because mutations in this gene cause XY individuals to develop external genitalia that appear female. An XX embryo has no Y chromosome, no *SRY* gene, and much less testosterone, so primary female reproductive organs (ovaries) form instead of testes. Ovaries make estrogens and other sex hormones that will govern the development of female secondary sexual traits, such as enlarged, functional breasts, and fat deposits around the hips and thighs.

Flower Formation

When it is time for a plant to flower, populations of cells that would otherwise give rise to leaves instead differentiate into floral parts—sepals, petals, stamens, and carpels. How does the switch happen? Three sets of master genes called *A*, *B*, and *C* guide the development of specialized parts of a flower. These genes are switched on by environmental cues such as seasonal changes in the length of night, as you will see in Section 30.9.

At the tip of a shoot, cells form whorls of tissue, one over the other like layers of an onion. Cells in each whorl give rise to different tissues depending on which of their *ABC* genes are activated (**Figure 10.9**). Studies of the phenotypic effects of mutations elucidated the function of these genes. In the outer whorl, only the *A* genes are switched on, and their products trigger events that cause sepals to form. Cells in the next whorl express both *A* and *B* genes; they give rise to petals. Cells farther in express *B* and *C* genes; they give rise to stamens, the structures that produce male reproductive cells. The cells of the innermost whorl express only the *C* genes; they give rise to carpels, the structures that produce female reproductive cells.

dosage compensation Theory that X chromosome inactivation equalizes gene expression between males and females.
X chromosome inactivation Shutdown of one of the two X chromosomes in the cells of female mammals.

Take-Home Message

What are some examples of gene control in eukaryotes?

» X chromosome inactivation balances expression of X chromosome genes between female (XX) and male (XY) mammals.

» *SRY* gene expression triggers the development of male traits in mammals.

» In plants, expression of *ABC* master genes governs development of the specialized parts of a flower.

10.5 Examples of Gene Control in Prokaryotes

■ Bacteria control gene expression mainly by adjusting the rate of transcription.

■ Links to Carbohydrates 3.4, Controls over metabolism 5.5, Lactate fermentation 7.6

Bacteria and archaea do not undergo development, so these cells have no need of master genes. However, they do respond to environmental fluctuations by adjusting gene expression. For example, when a preferred nutrient becomes available, a bacterium begins transcribing genes whose products allow the cell to use that nutrient. When the nutrient is no longer available, transcription of those genes stops. Thus, the cell does not waste energy and resources producing gene products that are not needed at a particular moment.

Bacteria control gene expression mainly by adjusting the rate of transcription. Genes that are used together often occur together on the chromosome, one after the other. A single promoter precedes the genes, so all are transcribed together into a single RNA strand. Thus, their transcription is controllable in a single step. Transcription control may occur at an operator, a region of DNA that serves as a binding site for a repressor. (Repressors, remember, stop transcription.) A promoter and one or more operators that together control transcription of multiple genes are collectively called an operon. Operons occur in bacteria, archaea, and eukaryotes.

The Lac Operon

Escherichia coli lives in the gut of mammals, where it dines on dissolved nutrients traveling past. Its carbohydrate of choice is glucose, but it can also use other sugars such as the lactose in milk. The operon that controls lactose metabolism in *E. coli* is called the lac operon (**Figure 10.10**). This operon includes three genes and a promoter flanked by two operators ❶.

Lactose consists of two monosaccharide monomers: glucose and galactose. The three genes in the

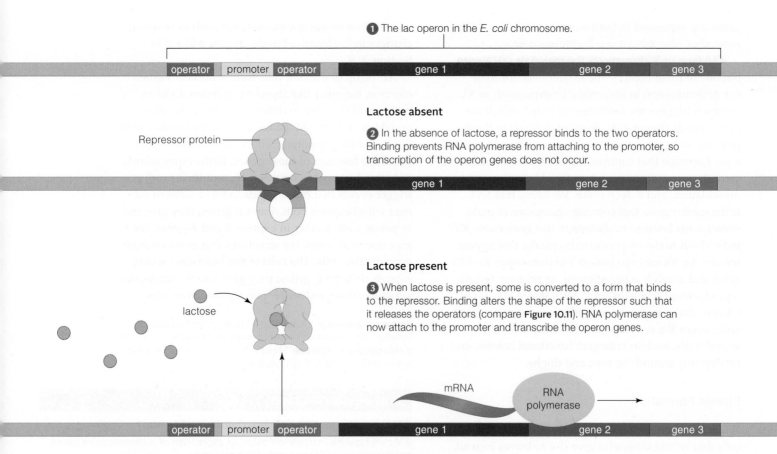

❶ The lac operon in the *E. coli* chromosome.

| operator | promoter | operator | gene 1 | gene 2 | gene 3 |

Lactose absent

❷ In the absence of lactose, a repressor binds to the two operators. Binding prevents RNA polymerase from attaching to the promoter, so transcription of the operon genes does not occur.

Repressor protein

| gene 1 | gene 2 | gene 3 |

Lactose present

❸ When lactose is present, some is converted to a form that binds to the repressor. Binding alters the shape of the repressor such that it releases the operators (compare **Figure 10.11**). RNA polymerase can now attach to the promoter and transcribe the operon genes.

lactose

mRNA

RNA polymerase

| operator | promoter | operator | gene 1 | gene 2 | gene 3 |

Figure 10.10 Animated Example of gene control in bacteria: the lactose operon on a bacterial chromosome. The operon consists of a promoter flanked by two operators, and three genes for lactose-metabolizing enzymes. **Figure It Out:** What portion of the operon binds RNA polymerase when lactose is present?

Answer: The promoter

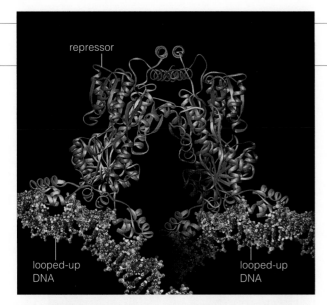

repressor

looped-up DNA   looped-up DNA

Figure 10.11 Model of the lactose operon repressor, shown here bound to operators. Binding twists the bacterial chromosome into a loop, which in turn prevents RNA polymerase from binding to the lac operon promoter.

lac operon encode proteins that allow *E. coli* to harvest glucose monomers from lactose. Bacteria conserve resources by making these proteins only when lactose is present. When lactose is not present, a repressor binds to the two operators, and lactose-metabolizing genes stay switched off. A repressor molecule that binds to the operators twists the region of DNA with the promoter into a loop (**Figure 10.11**). RNA polymerase cannot bind to the twisted-up promoter, so it cannot transcribe the operon genes ❷.

When lactose is present in the gut, some of it is converted to another sugar that binds to the repressor and changes its shape. The altered repressor can no longer stay attached to the DNA; it releases the operators and the looped DNA unwinds. The promoter is now accessible to RNA polymerase, and transcription begins ❸.

In bacteria, glucose metabolism requires fewer enzymes than lactose metabolism does, so it requires less energy and fewer resources. Accordingly, when both lactose and glucose are present, the cells will use up all of the available glucose before switching to lactose metabolism. How does a cell ignore the presence of one sugar while it uses another? Another level of gene control over the lac operon regulates this metabolic switch, but how it works is still being debated even after decades of research.

Lactose Intolerance

Like *E. coli*, humans and other mammals break down lactose into monosaccharide subunits, but most do so only when they are young. An individual's ability to digest lactose declines at a species-specific age. In most humans, the switch occurs at about age five, with

a programmed shutdown of the gene encoding lactase. The resulting decline in production of this lactose-metabolizing enzyme results in a common condition known as lactose intolerance.

Cells in the intestinal lining secrete lactase into the small intestine, where the enzyme cleaves lactose into its glucose and galactose monomers. Monosaccharides are absorbed directly by the small intestine, but lactose and other disaccharides are not. Thus, when lactase secretion slows, lactose passes undigested through the small intestine. Lactose ends up in the large intestine, which hosts huge numbers of *E. coli* and a variety of other bacteria. These resident organisms respond to the abundant sugar supply by switching on their lac operons. Carbon dioxide, methane, hydrogen, and other gaseous products of their various fermentation reactions accumulate quickly in the large intestine, distending its wall and causing pain. The other products of their metabolism (undigested carbohydrates) disrupt the solute–water balance inside the large intestine, and diarrhea results.

Not everybody is lactose intolerant. Many people of northern and central European ancestry carry a mutation in one of the genes responsible for programmed lactase shutdown. These people make enough lactase to continue drinking milk without problems into adulthood.

Riboswitches

Some bacterial mRNAs regulate their own translation with riboswitches, which are small sequences of RNA nucleotides that bind to a target molecule. The binding affects translation of the mRNA. Consider what happens when bacteria make vitamin B_{12}. The enzymes involved in synthesis of this vitamin are produced from mRNAs that have riboswitches. These particular riboswitches bind to vitamin B_{12}. Binding changes the shape of the mRNA so that ribosomes can no longer attach to it, and translation stops. This example also illustrates feedback inhibition, in which an end product inhibits its own production.

operator Part of an operon; a DNA binding site for a repressor.
operon Group of genes together with a promoter–operator DNA sequence that controls their transcription.

Take-Home Message

Do bacteria control gene expression?

» In bacteria, the main gene expression controls regulate gene expression in response to shifts in nutrient availability and other environmental conditions.

» Prokaryotes can regulate gene expression using operons and riboswitches.

10.6 Epigenetics

- Methylations and other modifications that accumulate in DNA during an individual's lifetime can be passed to offspring.
- Links to DNA structure 8.4, DNA replication 8.5, Environmental causes of mutations 8.6, Reprogramming DNA 8.7, Transposable elements 9.6

Heritable Methylations

You learned in Section 10.2 that methylation of histone proteins silences DNA transcription. Direct methylation of DNA also suppresses gene expression, but in an often more permanent manner than histone modification. The *XIST* gene offers an example. The one active X chromosome in cells of female mammals does not express this gene because its promoter is heavily methylated. Methylation also suppresses the activity of transposable elements, so it is important for the stability of a cell's chromosomes. Cancer is often associated with the loss of methylation.

Between 3 and 6 percent of the DNA in a normal, differentiated body cell is methylated. Methyl groups are most often attached to a cytosine that is followed by a guanine (**Figure 10.12A**), but which of these cytosines are methylated and which are not varies by the individual. The pattern starts early. For instance, individuals conceived during a famine end up with an unusually low number of methyl groups attached to certain genes. One of those genes encodes a hormone that fosters prenatal growth and development. The resulting increase in expression of this gene may offer a survival advantage in a poor nutritional environment.

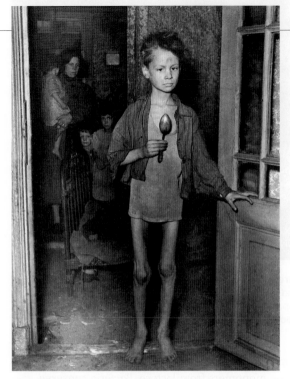

Figure 10.13 A cause of epigenetic changes. In 1944, a supply blockade followed by an unusually harsh winter caused a severe famine in the Nazi-occupied Netherlands. Grandsons of boys who endured the famine (such as the one pictured in this photo) lived about 32 years longer than grandsons of boys who ate well during the same winter.

An individual's DNA also acquires methyl groups on an ongoing basis. Once a particular base has become methylated in a cell's DNA, it will usually stay methylated in all of the cell's descendants (**Figure 10.12B**). Genes actively expressed in early development become silenced as their promoters get methylated during differentiation. Environmental factors, including the chemicals in cigarette smoke, add more methyl groups. Methylation also occurs by chance during DNA replication, so cells that divide a lot tend to have more methyl groups in their DNA than inactive cells. Thus, environmental factors that result in a gene being expressed can have long-term effects.

When an organism reproduces, it passes its chromosomes to offspring. The methylation of parental chromosomes is normally "reset" in the first cell of the new individual, with new methyl groups being added and old ones being removed. This reprogramming does not remove all of the parental methyl groups, however, so methylations acquired during an individual's lifetime can be passed to future offspring.

Any heritable changes in gene expression that are not due to changes in the underlying DNA sequence are said to be **epigenetic**. Epigenetic inheritance can adapt offspring to an environmental stressor much more quickly than evolutionary processes. Epigenetic

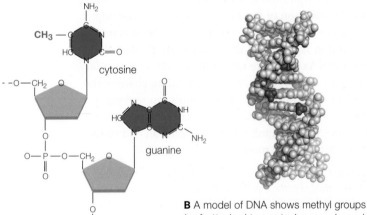

A In the DNA of differentiated cells, a methyl group (*red*) is most often attached to a cytosine that is followed by a guanine.

B A model of DNA shows methyl groups (*red*) attached to a cytosine–guanine pair on complementary DNA strands. When the cytosine on one strand is methylated, enzymes methylate the cytosine on the other strand. This is why a methylation tends to persist in a cell's descendants.

Figure 10.12 DNA methylation.

An effective cancer treatment must eliminate cancerous cells from a person's body. However, cancer cells are body cells, so drugs and other therapies that kill them also kill normal body cells. There is often a fine line between eliminating cancer from a patient's body, and killing the patient.

A normal cell has layers upon layers of gene expression controls and fail-safe mechanisms that determine when the division occurs and when it does not. This finely tuned system becomes unbalanced in a cancer cell, so that some genes are expressed at a higher level than they should be, and some are expressed at a lower level. For three decades, researchers have been studying the system and how it goes awry in cancer, because understanding how a cancer cell differs from a normal cell at a molecular level offers our best chance at finding a cure.

In the early 1980s, researchers discovered that mutations in some genes predispose individuals to develop certain kinds of cancer. Some of the genes are tumor suppressors, so named because tumors are more likely to occur when these genes mutate. Two examples are *BRCA1* and *BRCA2*: A mutated version of one or both of these genes is often found in breast and ovarian cancer cells. Because mutations in genes such as *BRCA* can be inherited, cancer is not only a disease of the elderly, as Robin Shoulla's story illustrates. Robin is one of the unlucky people who carry mutations in both *BRCA1* and *BRCA2*.

If a *BRCA* gene mutates in one of three especially dangerous ways, a woman has an 80 percent chance of developing breast cancer before the age of seventy. *BRCA* genes are master genes whose protein products help maintain the structure and number of chromosomes in a dividing cell. The multiple functions of these proteins are still being unraveled. We do know they participate directly in DNA repair (Section 8.6),

so any mutations that alter this function also alter the cell's capacity to repair damaged DNA. Other mutations are likely to accumulate, and that sets the stage for cancer.

The products of *BRCA* genes also bind to receptors for the hormones estrogen and progesterone, which are abundant on cells of breast and ovarian tissues. Binding suppresses transcription of growth factor genes in these cells. Among other things, growth factors stimulate cells to divide during normal, cyclic renewals of breast and ovarian tissues. When a mutation alters a *BRCA* gene so that its product cannot bind to hormone receptors, the cells overproduce growth factors. Cell division goes out of control, and tissue growth becomes disorganized. In other words, cancer develops.

Researchers recently found that the RNA product of the *XIST* gene localizes abnormally in breast cancer cells. In those cells, both X chromosomes are active. It makes sense that two active X chromosomes would have something to do with abnormal gene expression, but why the RNA product of an unmutated *XIST* gene does not localize properly in cancer cells remains a mystery.

Mutations in the *BRCA1* gene may be part of the answer. The researchers found that the protein product of the *BRCA1* gene physically associates with the RNA product of the *XIST* gene. They were able to restore proper *XIST* RNA localization—and proper X chromosome inactivation—by restoring the function of the *BRCA1* gene product in breast cancer cells.

How would you vote? Some women at high risk of developing breast cancer opt for preventive breast removal. Many of them never would have developed cancer. Should the surgery be restricted to cancer treatment?

marks such as methylation patterns are not evolutionary because the underlying DNA sequence does not change, but these marks may persist for generations after an environmental stressor has faded. For example, even when corrected for social factors, grandsons of boys who endured a winter of famine when they were 6 years old (**Figure 10.13**) lived about 32 years longer than the grandsons of boys who overate at the same age. Nine-year-old boys whose fathers smoked cigarettes before age 11 are very overweight compared with boys whose fathers did not smoke in childhood. In these and other studies, the effect was sex-limited:

boys were affected by lifestyle of individuals in the paternal line; girls, by individuals in the maternal line. Animal studies have confirmed the epigenetic effects seen in human historical data.

epigenetic Refers to heritable changes in gene expression that are not the result of changes in DNA sequence.

Take-Home Message

Can gene expression be inherited?

» Epigenetic marks in chromosomal DNA, including DNA methylations acquired during an individual's lifetime, can be passed to offspring.

11 Cells Reproduce

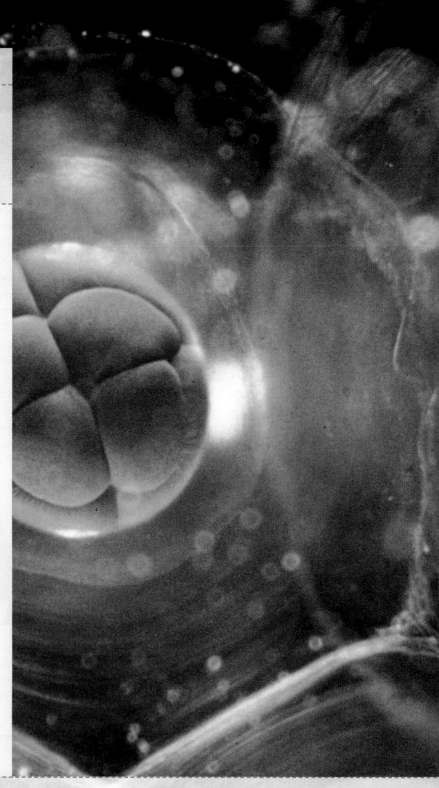

LEARNING ROADMAP

Where you have been Be sure you understand cell structure (Sections 4.2, 4.6, 4.10, 4.11), chromosomes (8.2), and DNA replication (8.5) before beginning this chapter. What you know about phosphorylation (5.6), membrane proteins (5.7), fermentation (7.6), mutations (8.6), and gene control (10.2) will help you understand how cancer (10.1) develops. You will also revisit animal cloning (8.7) and knockout experiments (10.3).

Where you are now

The Cell Cycle
A cell cycle starts when a new cell forms by division of a parent cell, and ends when the cell completes its own division. Built-in checkpoints control the timing and rate of the cycle.

Mitosis
Mitosis divides the nucleus and maintains the chromosome number. Four sequential stages parcel the cell's duplicated chromosomes into two new nuclei.

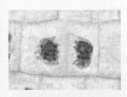

Cytoplasmic Division
After nuclear division, the cytoplasm may divide, so one nucleus ends up in each of two new cells. The division proceeds by a different mechanism in animal and plant cells.

Mitotic Clocks
Built into eukaryotic chromosomes are DNA sequences that protect the cell's genetic information. Degradation of these sequences is associated with cell death and aging.

The Cell Cycle Gone Awry
On rare occasions, checkpoint mechanisms fail, and cell division becomes uncontrollable. Tumor formation and cancer are outcomes.

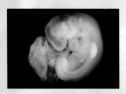

Where you are going Mitosis is the basis of reproduction of single-celled eukaryotes (Chapters 21 and 23). In multicelled eukaryotes, it has a role in reproduction and development (Chapters 29 and 42) as well as growth and tissue repair (Chapter 31). You will be comparing mitosis with meiosis in Chapter 12. The HPV virus will come up again in context of sexually transmitted diseases (Section 41.10) and as a cause of cervical cancer (Section 37.1). Other cancers are discussed in Chapters 31, 32, and 34.

Each human starts out as a fertilized egg. By the time of birth, that cell has given rise to about a trillion cells, all organized as a human body. Even in an adult, billions of cells divide every day as new cells replace worn-out ones. However, despite the ability of human cells to continue dividing as part of a body, they tend to divide a limited number of times and die within weeks when grown in the laboratory.

Researchers started trying to coax human cells to keep dividing outside of the body in the mid-1800s. Why? Many human diseases occur only in human cells. Immortal cell lineages—cell lines—would allow researchers to study human diseases (and potential cures for them) without experimenting on people.

At Johns Hopkins University, George and Margaret Gey were among the researchers trying to culture human cells. They had been working on the problem for almost thirty years when, in 1951, their assistant Mary Kubicek prepared a sample of human cancer cells. Mary named the cells HeLa, after the first and last names of the patient from whom the cells had been taken.

The HeLa cells began to divide, again and again. The cells were astonishingly vigorous, quickly coating the inside of their test tube and consuming their nutrient broth. Four days later, there were so many cells that the researchers had to transfer them to more tubes. The cell populations increased at a phenomenal rate. The cells were dividing every twenty-four hours and coating the inside of the tubes within days.

Sadly, cancer cells in the patient were dividing just as fast. Only six months after she had been diagnosed with cervical cancer, malignant cells had invaded tissues throughout her body. Two months after that, Henrietta Lacks, a young African American woman from Baltimore, was dead.

Although Henrietta passed away, her cells lived on in the Geys' laboratory (**Figure 11.1**). The Geys were able to grow poliovirus in HeLa cells, a practice that enabled them to find out which strains of the virus cause polio. That work was a critical step in the development of polio vaccines, which have since saved millions of lives.

Henrietta Lacks's immortal cells, frozen away in tiny tubes and packed in Styrofoam boxes, continue to be shipped among laboratories all over the world. They are still widely used to investigate cancer, viral growth, protein synthesis, the effects of radiation, and countless other processes important in medicine and research. HeLa cells helped several researchers win Nobel Prizes, and some even traveled into space for experiments on the Discoverer XVII satellite.

Henrietta Lacks was just thirty-one, a wife and mother of five, when runaway cell divisions killed her. Her legacy continues to help people, through her cells that are still dividing, again and again, more than fifty years after she died. Understanding why cancer cells are immortal—and why we are not—begins with understanding the structures and mechanisms that cells use to divide.

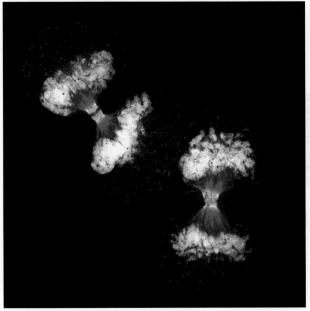

Figure 11.1 HeLa cells, an immortal legacy of cancer victim Henrietta Lacks (*left*).

The fluorescence micrograph (*far left*) shows some of the internal components of two HeLa cells in the process of dividing. *Blue* and *green* indicate the location of two proteins that help microtubules (*red*) attach to chromosomes (*white*).

Defects in these and other proteins that orchestrate cell division result in descendant cells with too many or too few chromosomes, an outcome that is a hallmark of cancer.

11.2 Multiplication by Division

- Division of a eukaryotic cell typically occurs in two steps: nuclear division followed by cytoplasmic division.
- The sequence of stages through which a cell passes during its lifetime is called the cell cycle.
- Links to Cell structure 4.2, Nucleus 4.6, Chromosomes 8.2, DNA replication 8.5, Gene controls 10.2

The Life of a Cell

Just as the series of events in an animal's life is called a life cycle, the events that occur from the time a cell forms until the time it divides is called a **cell cycle** (**Figure 11.2**). A typical cell spends most of its life in **interphase ❶**. During this phase, the cell increases its mass, roughly doubles the number of its cytoplasmic components, and replicates its DNA in preparation for division. Interphase consists of three stages:

❷ G1, the first interval (or gap) of cell growth, before DNA replication

❸ S, the time of synthesis (DNA replication)

❹ G2, the second interval (or gap), when the cell prepares to divide

Gap intervals were named because outwardly they seem to be periods of inactivity. Actually, most cells going about their metabolic business are in G1. Cells preparing to divide enter S, when they copy their DNA. During G2, they make the proteins that will drive cell division. Once the S phase begins, DNA replication usually proceeds at a predictable rate and ends before mitosis begins.

The remainder of the cell cycle consists of the division process itself. When a cell divides, both of its two

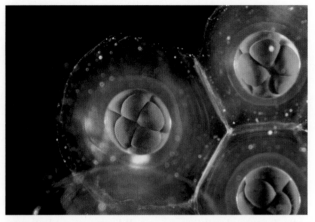

Figure 11.3 A multicelled eukaryote develops by repeated cell divisions. This photo shows early frog embryos, each a product of three mitotic divisions of the nucleus of one fertilized egg.

Figure It Out: Each of these embryos has how many nuclei?

Answer: Eight

cellular offspring end up with a blob of cytoplasm and some DNA. Each of the offspring of a eukaryotic cell inherits its DNA packaged inside a nucleus. Thus, a eukaryotic cell's nucleus has to divide before its cytoplasm does.

Mitosis is a nuclear division mechanism that maintains the chromosome number ❺. In multicelled species, mitosis and cytoplasmic division increase cell number during development (**Figure 11.3**), and replace damaged or dead cells later in life. In many eukaryotes, a single individual can reproduce by mitosis and cytoplasmic division, a process called **asexual reproduction**. Bacteria and archaea also reproduce asexually, but they do it by binary fission, a separate mechanism that we will consider in Section 20.6.

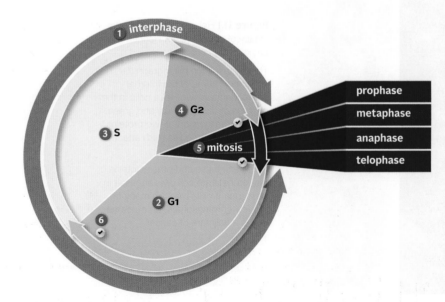

❶ A cell spends most of its life in interphase, which includes three stages: G1, S, and G2.

❷ G1 is the interval of growth before DNA replication. The cell's chromosomes are unduplicated.

❸ S is the time of synthesis, during which the cell copies its DNA.

❹ G2 is the interval after DNA replication and before mitosis. The cell prepares to divide during this stage.

❺ The nucleus divides during mitosis, the four stages of which are detailed in the next section. After mitosis, the cytoplasm may divide. The cycle begins anew, in interphase, for each descendant cell.

❻ Built-in checkpoints ✓ stop the cycle from proceeding until certain conditions are met.

Figure 11.2 Animated The eukaryotic cell cycle. The relative length of the intervals differs among cells.

Chromosomes During Division

Remember from Section 8.2 that diploid cells have two sets of chromosomes. For example, human body cells have 46 chromosomes, two of each type. Except for a pairing of sex chromosomes (XY) in males, the chromosomes of each pair are homologous. **Homologous chromosomes** have the same length, shape, and genes (*hom*– means alike). Typically, each member of a pair was inherited from one of two parents.

With mitosis followed by cytoplasmic division, a diploid cell produces two diploid offspring. Both offspring have the same chromosome number as the parent. However, if only the total number of chromosomes mattered, then one of the cell's offspring might get, say, two pairs of chromosome 22 and no chromosome 9. A cell cannot function properly without a full complement of DNA, which means it needs to have *a copy of each* chromosome.

With mitosis, each of the two new nuclei that form receives a complete set of chromosomes. When a cell is in G1, each of its chromosomes consists of one double-stranded DNA molecule (**Figure 11.4A**). The cell replicates its DNA in S, so by G2, each of its chromosomes consists of two double-stranded DNA molecules (**Figure 11.4B**). These molecules stay attached to one another at the centromere (as sister chromatids) until mitosis is almost over. The next section shows how the four stages of mitosis—prophase, metaphase, anaphase, and telophase—parcel sister chromatids into separate nuclei. When the cytoplasm divides, these new nuclei are packaged into separate cells (**Figure 11.4C**). Each new cell has a full complement of unduplicated chromosomes, and each starts the cell cycle over again in G1 of interphase.

Controls Over the Cell Cycle

Most differentiated cells in a human body are in G1, and some types never progress past that stage. Most of your nerve cells, skeletal muscle cells, heart muscle cells, and fat-storing cells have been in G1 since birth. When a cell divides—and when it does not—is determined by a host of gene expression controls (Section 10.2). Like the accelerator of a car, some of these controls cause the cell cycle to advance. Like brakes, others stop the cycle from proceeding. Controls over those controls function as built-in checkpoints that allow problems to be corrected before the cycle proceeds ❻. Protein products of "checkpoint genes" monitor whether a cell's DNA has been copied completely, whether it is damaged, and even whether

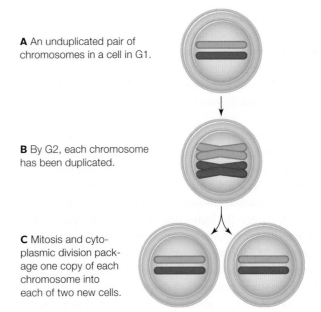

A An unduplicated pair of chromosomes in a cell in G1.

B By G2, each chromosome has been duplicated.

C Mitosis and cytoplasmic division package one copy of each chromosome into each of two new cells.

Figure 11.4 How mitosis maintains the chromosome number.

enough nutrients to support division are available. These checkpoint proteins interact with other cell cycle controls to delay or stop the cell cycle while simultaneously enhancing transcription of genes involved in fixing a problem. After the problem is corrected, the brakes are lifted and the cell cycle proceeds. If the problem stays uncorrected, other checkpoint proteins cause the cell to self-destruct. (You will read more about cell suicide, or apoptosis, in Section 42.4.).

asexual reproduction Reproductive mode by which offspring arise from a single parent only.
cell cycle A series of events from the time a cell forms until its cytoplasm divides.
homologous chromosomes Chromosomes with the same length, shape, and set of genes.
interphase In a eukaryotic cell cycle, the interval between mitotic divisions when a cell enlarges, roughly doubles the number of its cytoplasmic components, and replicates its DNA.
mitosis Nuclear division mechanism that maintains the chromosome number. Basis of body growth and tissue repair in multi-celled eukaryotes; also asexual reproduction in some plants, animals, fungi, and protists.

Take-Home Message

What is cell division and why does it occur?

» The sequence of stages through which a cell passes during its lifetime (interphase, mitosis, and cytoplasmic division) is called the cell cycle.

» A eukaryotic cell reproduces by division: nucleus first, then cytoplasm. Each descendant cell receives a set of chromosomes and some cytoplasm.

» The nuclear division process of mitosis is the basis of body size increases and tissue repair in multicelled eukaryotes. It is also the basis of asexual reproduction in single-celled and some multicelled eukaryotes.

» Gene expression controls advance, delay, or block the cell cycle in response to internal and external conditions.

11.3 A Closer Look at Mitosis

■ When a nucleus divides by mitosis, each new nucleus has the same chromosome number as the parent cell.

■ The four main stages of mitosis are prophase, metaphase, anaphase, and telophase.

■ Links to Cytoskeletal elements 4.10, Chromosome condensation 8.2, Transcriptional control 10.2

During interphase, a cell's chromosomes are loosened to allow transcription and DNA replication. Loosened chromosomes are spread out, so they are not easily visible under a light microscope (**Figure 11.5**). In preparation for nuclear division, the chromosomes begin to pack tightly ❶. Transcription and DNA replication stop as the chromosomes condense into their most compact "X" forms (Section 8.2). Tight condensation keeps the chromosomes from getting tangled and breaking during nuclear division.

A cell reaches **prophase**, the first stage of mitosis, when its chromosomes have condensed so much that they are visible under a light microscope ❷. "Mitosis" is from *mitos*, the Greek word for thread, after the threadlike appearance of the chromosomes during nuclear division.

In most animal cells, a centrosome organizes microtubules as they form. A typical centrosome consists of two centrioles, which are barrel-shaped organelles that help microtubules assemble (Section 4.10), surrounded by a region of dense cytoplasm. The centrosome gets duplicated just before prophase. Then, during prophase, one of the two centrosomes moves to the opposite side of the cell. Microtubules begin to extend from both centrosomes and form a **spindle**, a dynamic network of microtubules that moves chromosomes. Motor proteins traveling along the microtubules help the spindle extend in appropriate directions. Plant cells have spindles and structures that organize them, but no centrosomes.

During prophase, the spindle penetrates the nuclear region as the nuclear envelope breaks up. Some microtubules of the spindle stop lengthening when they reach the middle of the cell. Others lengthen until they reach a chromosome and attach to it at the centromere. By the end of prophase, one sister chromatid of each chromosome has become attached to microtubules extending from one spindle pole, and the other sister has become attached to microtubules extending from the other spindle pole ❸.

The opposing sets of microtubules then begin a tug-of-war by adding and losing tubulin subunits. As the microtubules lengthen and shorten, they push and pull the chromosomes. When all the microtubules are the same length, the chromosomes are aligned midway

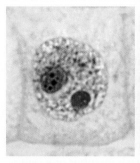
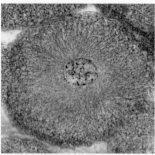

Onion root cell   **Whitefish embryo cell**

Figure 11.5 Animated Mitosis. Micrographs *here* and *opposite* show cells from a plant (onion root, *left*), and an animal (whitefish embryo, *right*). *This page*, interphase cells are shown for comparison, but interphase is not part of mitosis.

Opposite page, the stages of mitosis. The drawings show a diploid (2n) animal cell. For clarity, only two pairs of chromosomes are illustrated, but nearly all eukaryotic cells have more than two. The two chromosomes of the pair inherited from one parent are *pink*; the two chromosomes from the other parent are *blue*.

between the spindle poles ❹. The alignment marks **metaphase** (from *meta*, ancient Greek for between).

During **anaphase**, microtubules of the spindle separate the sister chromatids of each duplicated chromosome, and move them toward opposite spindle poles ❺. Each DNA molecule is now a separate chromosome.

Telophase begins when the two clusters of chromosomes reach the spindle poles ❻. Each cluster has the same number and kinds of chromosomes as the parent cell nucleus had: two of each type of chromosome, if the parent cell was diploid. A new nucleus forms around each cluster as the chromosomes loosen up again, after which telophase (and mitosis) is over.

anaphase Stage of mitosis during which sister chromatids separate and move to opposite spindle poles.
metaphase Stage of mitosis at which the cell's chromosomes are aligned midway between poles of the spindle.
prophase Stage of mitosis during which chromosomes condense and become attached to a newly forming spindle.
spindle Dynamically assembled and disassembled network of microtubules that moves chromosomes during nuclear division.
telophase Stage of mitosis during which chromosomes arrive at the spindle poles and decondense, and new nuclei form.

Take-Home Message

What is the sequence of events in mitosis?

» Each chromosome in a cell's nucleus was duplicated before mitosis begins, so each consists of two DNA molecules.

» Mitosis proceeds in four stages: prophase, metaphase, anaphase, and telophase.

» During these four stages, sister chromatids of the duplicated chromosomes are separated and packaged into two new nuclei. Each new nucleus contains the same number and types of chromosomes as the parent cell.

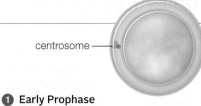

centrosome

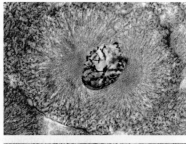

1 Early Prophase
Mitosis begins. In the nucleus, the DNA begins to appear grainy as it starts to condense. The centrosome gets duplicated.

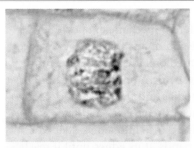

2 Prophase
The duplicated chromosomes become visible as they condense. One of the two centrosomes moves to the opposite side of the nucleus. The nuclear envelope breaks up.

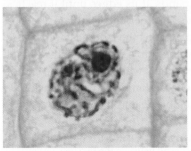

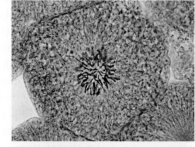

3 Transition to Metaphase
The nuclear envelope is gone, and the chromosomes are at their most condensed. Spindle microtubules assemble and bind to chromosomes at the centromere. Sister chromatids are attached to opposite spindle poles.

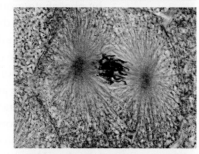

microtubule of spindle

4 Metaphase
All of the chromosomes are aligned midway between the spindle poles.

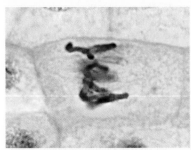

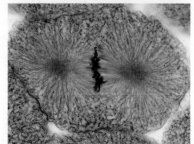

5 Anaphase
Spindle microtubules separate the sister chromatids and move them toward opposite spindle poles. Each sister chromatid has now become an individual, unduplicated chromosome.

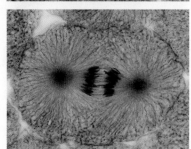

6 Telophase
The chromosomes reach the spindle poles and decondense. A nuclear envelope forms around each cluster, and mitosis ends.

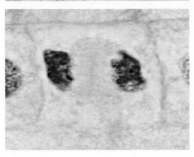

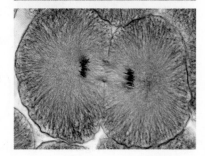

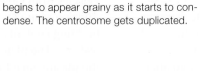

11.4 Cytokinesis: Division of Cytoplasm

■ In most eukaryotes, the cell cytoplasm divides between late anaphase and the end of telophase. The mechanism of division differs between plants and animals.

■ Links to Cytoskeleton 4.10, Primary wall of plant cells 4.11

A cell's cytoplasm usually divides after mitosis, so two cells form. The process of cytoplasmic division, which is called **cytokinesis**, differs among eukaryotes.

Typical animal cells pinch themselves in two after nuclear division (**Figure 11.6**). How? The spindle begins to disassemble during telophase ❶. The cell cortex, which is the mesh of cytoskeletal elements just under the plasma membrane, includes a band of actin and myosin filaments that wraps around the cell's midsection. The band is called a contractile ring because it contracts when its component proteins are energized by phosphate-group transfers from ATP. When the ring contracts, it drags the attached plasma membrane inward ❷. The sinking plasma membrane is visible on the outside of the cell as an indentation between the former spindle poles ❸. The indentation, a **cleavage furrow**, advances around the cell and deepens until the cytoplasm (and the cell) is pinched in two ❹. Each of the two cells formed by this division has its own nucleus and some of the parent cell's cytoplasm; each is enclosed by a plasma membrane.

Dividing plant cells face a particular challenge because they have stiff cell walls on the outside of their plasma membrane (Section 4.11). Accordingly, plant cells have a different mechanism of cytokinesis (**Figure 11.7**). By the end of anaphase, a set of short microtubules has formed on either side of the future plane of division. These microtubules now guide vesicles from Golgi bodies and the cell surface to the division plane ❺. There, the vesicles and their wall-building contents start to fuse into a disk-shaped **cell plate** ❻. The plate expands at its edges until it reaches the plasma membrane ❼. When the cell plate attaches to the membrane, it partitions the cytoplasm. In time, the cell plate will develop into a primary cell wall that merges with the parent cell's wall. Thus, by the end of division, each of the descendant cells will be enclosed by its own plasma membrane and its own cell wall ❽.

cell plate After nuclear division in a plant cell, a disk-shaped structure that forms a cross-wall between the two new nuclei.
cleavage furrow In a dividing animal cell, the indentation where cytoplasmic division will occur.
cytokinesis Cytoplasmic division.

Take-Home Message

How do eukaryotic cells divide?

» After mitosis, the cytoplasm of the parent cell typically is partitioned into two descendant cells, each with its own nucleus.

» In animal cells, a contractile ring pinches the cytoplasm in two. In plant cells, a cell plate that forms midway between the spindle poles partitions the cytoplasm when it reaches and connects to the parent cell wall.

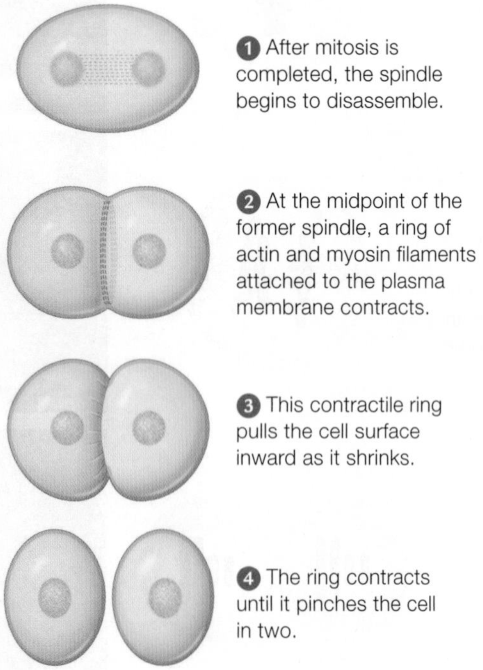

---

❶ After mitosis is completed, the spindle begins to disassemble.

❷ At the midpoint of the former spindle, a ring of actin and myosin filaments attached to the plasma membrane contracts.

❸ This contractile ring pulls the cell surface inward as it shrinks.

❹ The ring contracts until it pinches the cell in two.

Figure 11.6 Animated Cytoplasmic division of an animal cell.

❺ The future plane of division was established before mitosis began. Vesicles cluster here when mitosis ends.

❻ As the vesicles fuse with each other, they form a cell plate along the plane of division.

❼ The cell plate expands outward along the plane of division. When it reaches the plasma membrane, it attaches to the membrane and partitions the cytoplasm.

❽ The cell plate matures as two new cell walls. These walls join with the parent cell wall, so each descendant cell becomes enclosed by its own cell wall.

Figure 11.7 Animated Cytoplasmic division of a plant cell.

11.5 Marking Time With Telomeres

- Telomeres protect eukaryotic chromosomes from losing genetic information at their ends.
- Shortening telomeres are associated with aging.
- Links to DNA replication 8.5, Animal cloning 8.7, Knockout experiments 10.3

The very ends of the DNA strands composing eukaryotic chromosomes consist of noncoding sequences called **telomeres** (**Figure 11.8**). Vertebrate telomeres have a short DNA sequence, 5′-TTAGGG-3′, repeated perhaps thousands of times—"junk" sequences that provide a buffer against the loss of more valuable DNA internal to the chromosomes. This is particularly important because the chromosomes of a normal eukaryotic body cell shorten by about 100 bases with each replication. The shortening occurs in part because DNA polymerase can synthesize DNA only in the 3′ to 5′ direction, and it can start synthesis only at the 3′ hydroxyl group of a primer. Thus, a polymerase cannot copy the last hundred bases or so of the 3′ end of a chromosome. Each new DNA strand ends up a bit shorter than its parent.

When telomeres get too short, checkpoint gene products come into play to halt the cell cycle. The cell stops dividing, and it dies shortly thereafter. Thus, body cells can divide only a certain number of times. Scientists speculate that this limitation is a fail-safe mechanism because cells sometimes lose control over the cell cycle and begin to divide over and over. A limit on the number of divisions keeps such cells from overrunning the body, an outcome that, as you will see in the next section, has dangerous consequences.

In an adult, a few normal cells—stem cells—retain the ability to divide indefinitely. Descendants of these cells replace cell lineages that eventually die out. Stem cells are immortal because they continue to make telomerase, a molecule that consists of a noncoding RNA and a replication enzyme. The enzyme uses the RNA as a template to extend telomere repeat sequences at the 3′ end of a chromosome. Telomerase reverses telomere shortening that normally occurs after DNA replication.

Mice that have had their telomerase enzyme knocked out age prematurely. Their tissues degenerate much more quickly than those of normal mice, and their life expectancy declines to about half that of a normal mouse. When one of these knockout mice is close to the end of its shortened life span, rescuing the function of its telomerase enzyme results in lengthened telomeres. The rescued mouse also regains vitality: Decrepit tissue in its brain and other organs repairs itself and begins to function normally, and the

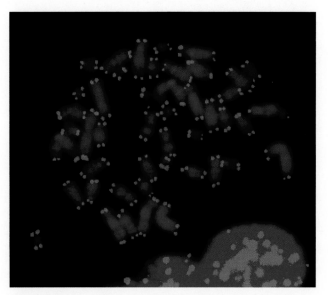

Figure 11.8 Telomeres. The *pink* dots at the end of each DNA strand in these duplicated chromosomes are telomere sequences.

once-geriatric individual even begins to reproduce again. This and other research in animals suggests that the built-in limit on cell divisions may be part of the mechanism that sets an organism's life span.

The very first cloned animal, Dolly the sheep, had telomeres that were abnormally short. When Dolly was only two years old, her telomeres were as short as those of a six-year-old sheep—the exact age of the adult animal that had been her genetic donor. By the time Dolly was five, she was as fat and arthritic as a twelve-year-old sheep. The following year, she contracted a lung disease that is typical of geriatric sheep, and had to be euthanized.

Dolly's early demise may well have been the result of short telomeres, but scientists are careful to point out that short telomeres could be an effect of aging rather than a cause. They also caution that while telomerase holds therapeutic promise for rejuvenation of aged tissues, it can also be dangerous: Cancer cells characteristically express high levels of the molecule.

telomere Noncoding, repetitive DNA sequence at the end of chromosomes; protects the coding sequences from degradation.

Take-Home Message

What is the function of telomeres?

» Telomeres at the ends of chromosomes provide a buffer against loss of genetic information.

» Telomeres shorten with every cell division in normal body cells. When they get too short, the cell stops dividing and dies.

11.6 When Mitosis Becomes Pathological

- On rare occasions, controls over cell division are lost and a neoplasm forms.
- Cancer develops as cells of a neoplasm become malignant.
- Links to Phosphorylations and energy 5.6, Membrane proteins 5.7, Environmental causes of mutation 8.6, Fermentation 7.6, Cancer 10.1

Sometimes a checkpoint gene mutates so that its protein product no longer works properly. In other cases, the controls that regulate its production fail, and a cell makes too much or too little of its product. When enough checkpoint mechanisms fail, a cell loses control over its cell cycle. The cell may skip interphase, so division occurs over and over with no resting period. Signaling mechanisms that make an abnormal cell die may stop working. The problem is compounded

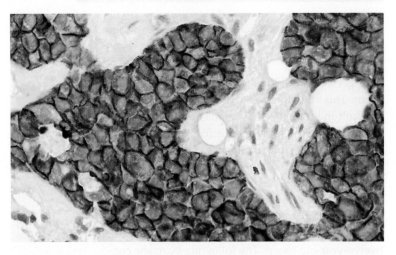

Figure 11.9 An oncogene. In this section of human breast tissue, phosphorylated EGF receptor is stained *brown*. Normal cells are the ones with lighter staining. The heavily stained cells have formed a neoplasm; the abnormal overabundance of the phosphorylated EGF receptor means that mitosis is being continually stimulated in these cells. The EGF receptor is overproduced or overactive in most neoplasms.

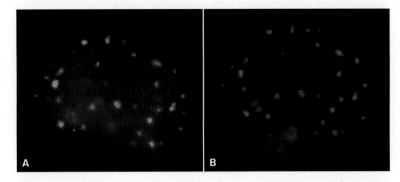

Figure 11.10 Checkpoint genes in action. Radiation damaged the DNA inside this nucleus. (**A**) *Green* dots pinpoint the location of the product of a gene called *53BP1*; (**B**) *red* dots show the location of the *BRCA1* gene product. Both proteins have clustered around the same chromosome breaks in the same nucleus; both function to recruit DNA repair enzymes. The integrated action of these and other checkpoint gene products blocks mitosis until the DNA breaks are fixed.

because checkpoint malfunctions are passed along to the cell's descendants, which form a **neoplasm**, an accumulation of cells that lost control over how they grow and divide.

A neoplasm that forms a lump in the body is called a **tumor**, but the two terms are sometimes used interchangeably. Once a tumor-causing mutation has occurred, the gene it affects is called an oncogene. An **oncogene** is any gene that helps transform a normal cell into a tumor cell (Greek *onkos*, or bulging mass). Some oncogene mutations can be passed to offspring, which is a reason that some types of tumors tend to run in families.

Genes encoding proteins that promote mitosis are called **proto-oncogenes** because mutations can turn them into oncogenes. A gene that encodes the epidermal growth factor (EGF) receptor is an example of a proto-oncogene. **Growth factors** are molecules that stimulate a cell to divide and differentiate. The EGF receptor is a plasma membrane protein that phosphorylates itself upon binding EGF. When phosphorylated, the receptor initiates a series of intracellular events that moves the cell cycle out of interphase and into mitosis. Mutations can result in an EGF receptor that always phosphorylates itself, so it stimulates mitosis even when EGF is not present. Most neoplasms carry mutations resulting in an overactivity or overabundance of this particular receptor (**Figure 11.9**).

Checkpoint gene products that inhibit mitosis are called tumor suppressors because tumors form when they are missing. The products of the *BRCA1* and *BRCA2* genes (Chapter 10) are examples of tumor suppressors. These proteins regulate, among other things, the expression of DNA repair enzymes (**Figure 11.10**). Mutations in *BRCA* genes are often found in cells of neoplasms. As another example, viruses such as HPV (human papillomavirus) cause a cell to make proteins that interfere with its own tumor suppressors. Infection with HPV causes skin growths called warts, and some kinds are associated with neoplasms that form on the cervix.

Cancer

Benign neoplasms such as ordinary skin moles are not dangerous (**Figure 11.11**). They grow very slowly, and their cells retain the plasma membrane adhesion proteins that keep them properly anchored to the other cells in their home tissue ❶.

A malignant neoplasm is one that gets progressively worse, and is dangerous to health. The disease called cancer occurs when the abnormally dividing cells of a malignant neoplasm disrupt body tissues, both physically and metabolically. Malignant cells typically display the following three characteristics:

First, like cells of all neoplasms, malignant cells grow and divide abnormally. Controls that usually keep cells from getting overcrowded in tissues are lost, so their populations may reach extremely high densities with cell division occurring very rapidly. The number of small blood vessels that transport blood to the growing cell mass also increases abnormally.

Second, the cytoplasm and plasma membrane of malignant cells are altered. The cytoskeleton may be shrunken, disorganized, or both. Malignant cells typically have an abnormal chromosome number, with some chromosomes present in multiple copies, and others missing or damaged. The balance of metabolism is often shifted, as in an amplified reliance on ATP formation by fermentation rather than aerobic respiration.

Altered or missing proteins impair the function of the plasma membrane of malignant cells. For example, these cells do not stay anchored properly in tissues because their plasma membrane adhesion proteins are defective or missing ❷. Malignant cells can slip easily into and out of vessels of the circulatory and lymphatic systems ❸. By migrating through these vessels, the cells establish neoplasms elsewhere in the body ❹. The process in which malignant cells break loose from their home tissue and invade other parts of the body is called **metastasis**. Metastasis is the third hallmark of malignant cells.

Unless chemotherapy, surgery, or another procedure eliminates malignant cells from the body, they can put an individual on a painful road to death. Each year, cancer causes 15 to 20 percent of all human deaths in developed countries. The good news is that mutations in multiple checkpoint genes are required to transform a normal cell into a malignant one, and these mutations may take a lifetime to accumulate. Lifestyle choices such as not smoking and avoiding exposure of unprotected skin to sunlight can reduce one's risk of acquiring mutations in the first place. Some neoplasms can be detected with periodic screening procedures such as Pap tests or dermatology exams (**Figure 11.12**). If detected early enough, many types of malignant neoplasms can be removed before metastasis occurs.

growth factor Molecule that stimulates mitosis and differentiation.
metastasis The process in which cancer cells spread from one part of the body to another.
neoplasm An accumulation of abnormally dividing cells.
oncogene Gene that helps transform a normal cell into a tumor cell.
proto-oncogene Gene that, by mutation, can become an oncogene.
tumor A neoplasm that forms a lump.

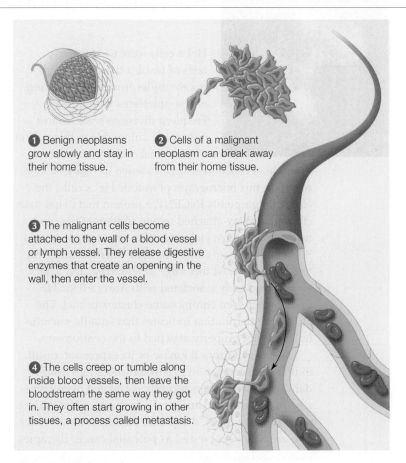

❶ Benign neoplasms grow slowly and stay in their home tissue.

❷ Cells of a malignant neoplasm can break away from their home tissue.

❸ The malignant cells become attached to the wall of a blood vessel or lymph vessel. They release digestive enzymes that create an opening in the wall, then enter the vessel.

❹ The cells creep or tumble along inside blood vessels, then leave the bloodstream the same way they got in. They often start growing in other tissues, a process called metastasis.

Figure 11.11 Animated Benign and malignant neoplasms.

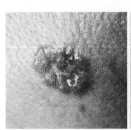 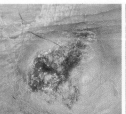 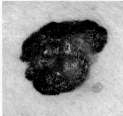

A Basal cell carcinoma is the most common type of skin cancer. This slow-growing, raised lump may be uncolored, reddish-brown, or black.

B The second most common form of skin cancer is a squamous cell carcinoma. This pink growth, firm to the touch, grows under the surface of skin.

C Melanoma, a malignant neoplasm of skin cells, spreads fastest. Cells form dark, encrusted lumps that may itch or bleed easily.

Figure 11.12 Skin cancer can be detected early with periodic screening.

Take-Home Message

What is cancer?

» Cancer is a disease that occurs when the abnormally dividing cells of a neoplasm physically and metabolically disrupt body tissues.

» A malignant neoplasm results from mutations in multiple checkpoint genes.

» Although some mutations are inherited, lifestyle choices and early intervention can reduce one's risk of cancer.

Henrietta's Immortal Cells (revisited)

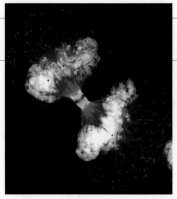

HeLa cells were used in early tests of taxol, a drug that keeps microtubules from disassembling and so interferes with mitosis. Frequent divisions make cancer cells more vulnerable to this poison than normal cells. A more recent example of cancer research is shown in **Figure 11.1** (and *above*). In this micrograph of mitotic HeLa cells, the *blue* stain pinpoints INCENP, a protein that helps sister chromatids stay attached to one another at the centromere. The *green* stain identifies an enzyme, Aurora B kinase, that helps attach spindle microtubules (*red*) to centromeres. At this stage of telophase, INCENP should be closely associated with Aurora B kinase midway between chromosome clusters (*white*). The abnormal distribution indicates that spindle microtubules are not properly attached to the centromeres.

Defects in Aurora B kinase or its expression result in unequal segregation of chromosomes into descendant cells. Researchers recently correlated overexpression of Aurora B in cancer cells with shortened patient survival rates. Thus, drugs that inhibit Aurora B function are now being tested as potential cancer therapies.

Despite the invaluable cellular legacy of Henrietta Lacks, her body rests in an unmarked grave in an unmarked cemetery. These days, physicians and researchers are required to obtain a signed consent form before they take tissue samples from a patient. No such requirement existed in the 1950s, when it was common for doctors to experiment on patients without their knowledge or consent. Thus, the young resident who was treating Henrietta Lacks's cancerous cervix probably never even thought about asking permission before he took a sample of it. That sample was the one that the Geys used to establish the HeLa cell line. No one in Henrietta's family knew about the cells until 25 years after she died. HeLa cells are still being sold worldwide, but her family has not received any compensation to date.

How would you vote? You can legally donate—but not sell—your own organs and tissues. However, companies can profit from research on donated organs or tissues, and also from cell lines derived from these materials. Companies that do so are not obligated to share their profits with the donors. Should profits derived from donated tissues or cells be shared with donors or their families?

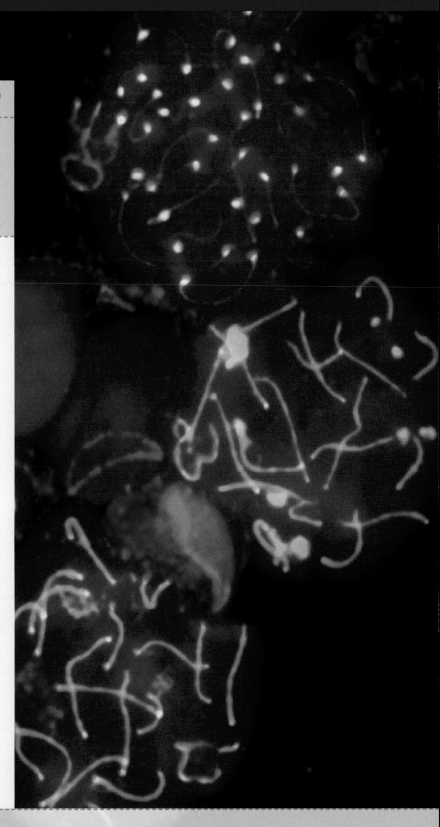

LEARNING ROADMAP

Where you have been Be sure you understand how eukaryotic chromosomes are organized (Sections 8.2 and 11.2) and how genes work (9.2). You will draw on your knowledge of DNA replication (8.5), cytoplasmic division (11.4), and cell cycle controls (11.6) as we compare meiosis with mitosis (11.3). This chapter also revisits clones (8.1, 8.7), the effects of mutation (9.6), and telomeres (11.5).

Where you are now

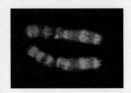

Sexual or Asexual Reproduction
In asexual reproduction, one parent transmits its genes to offspring. In sexual reproduction, offspring inherit genes from two parents who usually differ in some number of alleles.

Stages of Meiosis
Meiosis is a nuclear division process that occurs only in cells set aside for sexual reproduction. Meiosis reduces the chromosome number by sorting chromosomes into four new nuclei.

Recombinations and Shufflings
During meiosis, homologous chromosomes swap segments, then are randomly sorted into separate nuclei. Both processes lead to novel combinations of alleles among offspring.

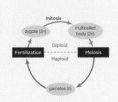

Sexual Reproduction in Life Cycles
Gametes form by different mechanisms in males and females, but meiosis is part of both processes. In most plants, spores occur between meiosis and gamete formation.

Mitosis and Meiosis Compared
Similarities suggest that meiosis originated by evolutionary remodeling of mechanisms that already existed for mitosis and, before that, for repairing damaged DNA.

Where you are going Meiosis is the basis of sexual reproduction, a topic covered in detail in Chapters 29 (plants) and 41 (animals). The variation in traits among individuals of sexually reproducing species arises from differences in alleles, a concept we revisit in context of inheritance in Chapters 13 and 14, and natural selection in Chapter 17.

12.1 Why Sex?

If the function of reproduction is the perpetuation of one's genes, then an asexual reproducer would seem to win the evolutionary race. In asexual reproduction, a single individual gives rise to offspring. All of the individual's genes are passed to all of its offspring. **Sexual reproduction** mixes up the genes of two parents (**Figure 12.1**), so only about half of each parent's genetic information is passed to offspring.

So why sex? Consider that all offspring of asexual reproducers are clones of their parent, so all are adapted the same way to an environment—and all are equally vulnerable to changes in it. By contrast, the offspring of sexual reproducers have different combinations of many traits. As a group, their diversity offers them flexibility: a better chance of surviving environmental change than clones. Some of them may have a particular combination of traits that suits them perfectly to a change. The variation among offspring of sexual reproducers is part of the reason that beneficial mutations tend to spread through sexually reproducing populations more quickly than asexual ones.

Other organisms are part of the environment, and they, too, can change. Think of predator and prey, say, foxes and rabbits. If one rabbit is better than others at outrunning the foxes, it has a better chance of escaping, surviving, and passing to its offspring the genes that help it evade foxes. Thus, over many generations, rabbits may get faster. If one fox is better than others at outrunning faster rabbits, it has a better chance of eating, surviving, and passing to its offspring genes that help it catch faster rabbits. Thus, over many generations, the foxes may tend to get faster. As one species changes, so does the other—an idea called the Red Queen hypothesis, after Lewis Carroll's book *Through the Looking Glass*. In the book, the Queen of Hearts tells Alice, "It takes all the running you can do, to keep in the same place."

Perhaps the most important advantage of sexual reproduction involves harmful mutations that inevitably occur. Collectively, the offspring of sexual reproducers have a better chance of surviving the effects of a harmful mutation that arises in the population. This is because, with asexual reproduction, individuals bearing a harmful mutation necessarily pass it to all of their offspring. By contrast, sexual reproduction mixes up the genetic material of two individuals. If an individual bearing a harmful mutation reproduces, it is very unlikely that all of the individual's offspring will inherit the mutation. Thus, all else being equal, harmful mutations accumulate in an asexually reproducing population much more quickly than in a sexually reproducing one.

Sexual reproduction generates new combinations of traits in fewer generations than does asexual reproduction. The process inherent to sexual reproduction that gives rise to this variation is **meiosis**, a nuclear division mechanism that halves the chromosome number.

meiosis Nuclear division process that halves the chromosome number. Basis of sexual reproduction.
sexual reproduction Reproductive mode by which offspring arise from two parents and inherit genes from both.

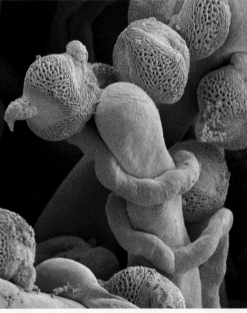

Figure 12.1 Moments in the stages of sexual reproduction of humans (*left*) and plants (*right*). Sexual reproduction mixes up the genetic material of two organisms.

In flowering plants, pollen grains (*orange*) germinate on flower carpels (*yellow*). Pollen tubes with male gametes inside grow from the grains down into tissues of the ovary, which house the flower's female gametes.

12.2 Meiosis Halves the Chromosome Number

■ Sexual reproduction mixes up alleles from two parents.

■ Meiosis, the basis of sexual reproduction, is a nuclear division mechanism that occurs in immature reproductive cells of eukaryotes.

■ Links to Clones 8.1 and 8.7, Eukaryotic chromosomes 8.2, DNA replication 8.5, Genes 9.2, Mutations 9.6, Homologous chromosomes and asexual reproduction and mitosis 11.2

Introducing Alleles

The **somatic** (body) cells of humans and many other sexually reproducing organisms are diploid—they contain pairs of chromosomes (Section 8.2). One chromosome of each pair is typically maternal, and the other is paternal. Except for a pairing of nonidentical sex chromosomes, homologous chromosomes carry the same set of genes (**Figure 12.2A**).

If the DNA sequences of homologous chromosomes were identical, then sexual reproduction would produce clones, just like asexual reproduction does. But the two chromosomes of a homologous pair are typically not identical. Why not? Mutations that inevitably occur in chromosomes change their DNA sequence. Over time, unique mutations accumulate in separate lines of descent, and some of those mutations occur in gene regions. Thus, any gene may differ a bit in sequence from the corresponding gene on the homologous chromosome (**Figure 12.2B**). Different forms of the same gene are called **alleles**.

Alleles may encode slightly different forms of the gene's product. Such differences influence thousands of traits. For example, the beta globin gene you encountered in Section 9.6 has more than 700 alleles: one that causes sickle-cell anemia, one that causes beta thalassemia, and so on. The beta globin gene is one of

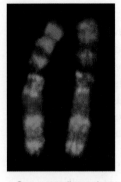

Genes occur in pairs on homologous chromosomes.

The members of each pair of genes may be identical, or they may differ slightly, as alleles.

A Corresponding colored patches in this fluorescence micrograph indicate corresponding DNA sequences in a homologous chromosome pair. These chromosomes carry the same set of genes.

B Homologous chromosomes carry the same series of genes, but the DNA sequence of any one of those genes might differ just a bit from that of its partner on the homologous chromosome.

Figure 12.2 Animated Genes on chromosomes. Different forms of a gene are called alleles.

about 20,000 human genes, and most genes have multiple alleles. Allele differences among individuals are one reason that the members of a sexually reproducing species do not all look identical. The offspring of sexual reproducers inherit new combinations of alleles, which is the basis of new combinations of traits.

What Meiosis Does

Sexual reproduction involves the fusion of haploid reproductive cells—**gametes**—from two parents. In both plants and animals, gametes form inside special reproductive structures or organs (**Figure 12.3**). Division

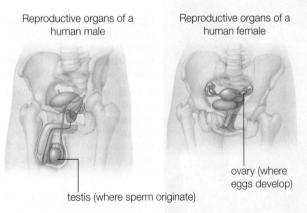

Reproductive organs of a human male

Reproductive organs of a human female

testis (where sperm originate)

ovary (where eggs develop)

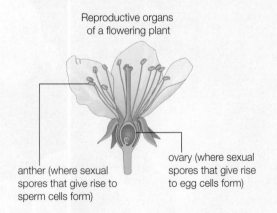

Reproductive organs of a flowering plant

anther (where sexual spores that give rise to sperm cells form)

ovary (where sexual spores that give rise to egg cells form)

Figure 12.3 Animated Examples of reproductive organs. In human ovaries or testes, meiosis in germ cells produces gametes called eggs and sperm. In the anthers or ovaries of flowering plants, meiosis produces haploid spores that give rise to gametes by mitosis.

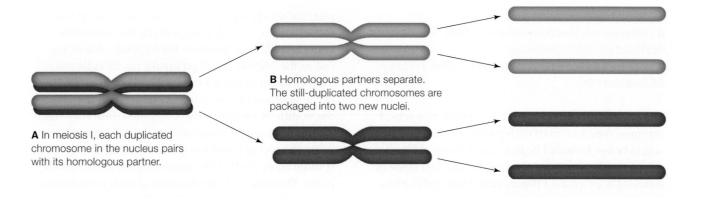

A In meiosis I, each duplicated chromosome in the nucleus pairs with its homologous partner.

B Homologous partners separate. The still-duplicated chromosomes are packaged into two new nuclei.

C Sister chromatids separate in meiosis II. The now unduplicated chromosomes are packaged into four new nuclei.

Figure 12.4 How meiosis halves the chromosome number.

of immature reproductive cells called **germ cells** gives rise to gametes. Meiosis in animal germ cells gives rise to eggs (female gametes) or sperm (male gametes). In plants, haploid germ cells form by meiosis. Gametes form when these cells divide by mitosis.

Gametes have a single set of chromosomes, so they are **haploid** (*n*): Their chromosome number is half of the diploid (*2n*) number (Section 8.2). Human body cells are diploid, with 23 pairs of homologous chromosomes. Meiosis of a human germ cell (*2n*) normally produces gametes with 23 chromosomes: one of each pair (*n*). The diploid chromosome number is restored at **fertilization**, when two haploid gametes (one egg and one sperm, for example) fuse to form a **zygote**, the first cell of a new individual.

The first part of meiosis is similar to mitosis. A cell duplicates its DNA before either nuclear division process starts. As in mitosis, the microtubules of a spindle move the duplicated chromosomes to opposite spindle poles. However, meiosis sorts the chromosomes into new nuclei not once, but twice, so it results in the formation of four haploid nuclei. The two consecutive nuclear divisions are called meiosis I and meiosis II:

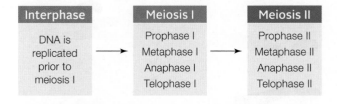

Interphase	Meiosis I	Meiosis II
DNA is replicated prior to meiosis I	Prophase I Metaphase I Anaphase I Telophase I	Prophase II Metaphase II Anaphase II Telophase II

In some cells meiosis II occurs immediately after meiosis I. In others, a period of protein synthesis—but no DNA synthesis—intervenes between the divisions.

During meiosis I, every duplicated chromosome aligns with its homologous partner (**Figure 12.4A**). Then the homologous chromosomes are pulled away from

one another (**Figure 12.4B**). After homologous chromosomes separate, each ends up in one of two new nuclei. At this stage, the chromosomes are still duplicated (the sister chromatids are still attached to one another).

During meiosis II, the sister chromatids of each chromosome are pulled apart, so each becomes an individual, unduplicated chromosome (**Figure 12.4C**). The chromosomes are sorted into four new nuclei. With one unduplicated version of each chromosome, the new nuclei are all haploid (*n*).

Thus, meiosis partitions the chromosomes of one diploid nucleus (*2n*) into four haploid (*n*) nuclei. The next section zooms in on the details of this process.

alleles Forms of a gene with slightly different DNA sequences; may encode slightly different versions of the gene's product.
fertilization Fusion of two gametes to form a zygote.
gamete Mature, haploid reproductive cell; e.g., an egg or a sperm.
germ cell Immature reproductive cell that gives rise to haploid gametes when it divides.
haploid Having one of each type of chromosome characteristic of the species.
somatic Relating to the body.
zygote Diploid cell formed by fusion of two gametes; the first cell of a new individual.

Take-Home Message

Why do populations that reproduce sexually tend to have the most variation in heritable traits?

» The nuclear division process of meiosis is the basis of sexual reproduction in eukaryotes. It precedes the formation of gametes or spores.

» Meiosis halves the diploid (*2n*) chromosome number, to the haploid number (*n*). When two gametes fuse at fertilization, the chromosome number is restored. The resulting zygote has one set of chromosomes from each parent.

» Corresponding genes on homologous chromosomes may vary in sequence as alleles.

» Alleles are the basis of traits. Sexual reproduction mixes up alleles from two parents.

12.3 Visual Tour of Meiosis

- Meiosis halves the chromosome number.
- During meiosis, chromosomes of a diploid nucleus become distributed into four haploid nuclei.
- Links to Diploid chromosome number 8.2, DNA replication 8.5, Mitosis 11.2

The nucleus of a diploid (2n) cell contains two sets of chromosomes, one from each parent. DNA replication occurs before meiosis I begins, so each one of the chromosomes has two sister chromatids. The first stage of meiosis I is prophase I (**Figure 12.5**). During this phase, the chromosomes condense, and homologous chromosomes align tightly and swap segments (more about

segment-swapping in the next section). The centrosome gets duplicated along with its two centrioles. One centriole pair moves to the opposite side of the cell as the nuclear envelope breaks up. Spindle microtubules begin to extend from the centrosomes ❶. Some of the microtubules stop lengthening when they reach the middle of the cell. Others lengthen until they reach a chromosome and attach to it at the centromere.

By the end of prophase I, microtubules of the spindle connect all of the chromosomes to the spindle poles. The maternal chromosome of each homologous pair is attached to one spindle pole, and the paternal chromosome to the other. The opposing sets of microtubules then begin a tug-of-war by adding and losing tubulin subunits. As the microtubules extend and shrink, they push and pull the chromosomes. At metaphase I, the microtubules are the same length, and the chromosomes are aligned midway between the spindle

Figure 12.5 Animated Meiosis. Two pairs of chromosomes are illustrated in a diploid (2n) animal cell. Homologous chromosomes are indicated in *blue* and *pink*. Micrographs show meiosis in a lily plant cell (*Lilium regale*).

Figure It Out: During which phase of meiosis does the chromosome number become reduced?

Answer: Anaphase I

Meiosis I | One diploid nucleus to two haploid nuclei

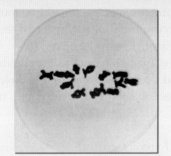
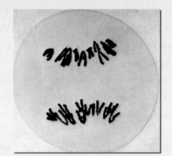
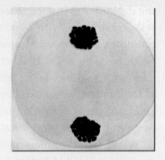

❶ Prophase I. Homologous chromosomes condense, pair up, and swap segments. Spindle microtubules attach to them as the nuclear envelope breaks up.

❷ Metaphase I. The homologous chromosome pairs are aligned midway between spindle poles. The two chromosomes of each pair are attached to opposite spindle poles.

❸ Anaphase I. Homologous chromosomes separate and begin heading toward the spindle poles.

❹ Telophase I. A complete set of chromosomes reaches each spindle pole. A nuclear envelope forms around each set, so two haploid (n) nuclei form.

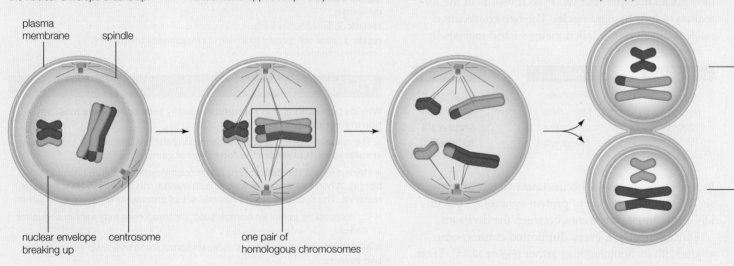

plasma membrane — spindle

nuclear envelope breaking up — centrosome

one pair of homologous chromosomes

poles ❷. In anaphase I, the spindle microtubules pull the homologous chromosomes of each pair apart and toward opposite spindle poles ❸. The two sets of chromosomes reach the spindle poles during telophase I ❹, and a new nuclear envelope forms around each cluster of chromosomes as the DNA loosens up. The two new nuclei are haploid (*n*); each contains one set of (duplicated) chromosomes. The cytoplasm may divide at this point, producing two haploid cells.

DNA replication does not occur before meiosis II begins. During prophase II, the chromosomes condense as a new spindle forms ❺. One centriole moves to the opposite side of each new nucleus, and the nuclear envelopes break up. By the end of prophase II, microtubules connect the chromosomes to the spindle poles. Each chromatid is now attached to one spindle pole; its sister is attached to the other. The microtubules lengthen and shorten, pushing and pulling the

chromosomes as they do. At metaphase II, the microtubules are the same length, and the chromosomes are aligned midway between the spindle poles ❻. In anaphase II, the spindle microtubules pull the sister chromatids apart ❼. Each chromosome now consists of one molecule of DNA. During telophase II, the chromosomes (now unduplicated) reach the spindle poles ❽. New nuclear envelopes form around the four clusters of chromosomes as the DNA loosens up. Each of the four haploid (*n*) nuclei that form contains one set of unduplicated chromosomes. The cytoplasm may divide at this point, producing four haploid cells.

Take-Home Message

What happens to a cell during meiosis?

» During meiosis, the nucleus of a diploid (2*n*) cell divides twice. Four haploid (*n*) nuclei form, each with a full set of chromosomes—one of each type.

Meiosis II Two haploid nuclei to four haploid nuclei

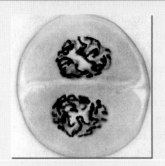 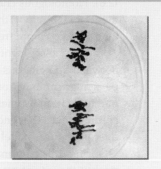  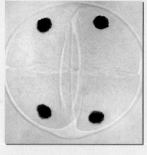

❺ Prophase II. The chromosomes condense. Spindle microtubules attach to each sister chromatid as the nuclear envelope breaks up.

❻ Metaphase II. The (still duplicated) chromosomes are aligned midway between poles of the spindle.

❼ Anaphase II. All sister chromatids separate. The now unduplicated chromosomes head to the spindle poles.

❽ Telophase II. A complete set of chromosomes reaches each spindle pole. A new nuclear envelope encloses each set, so four haploid (*n*) nuclei form.

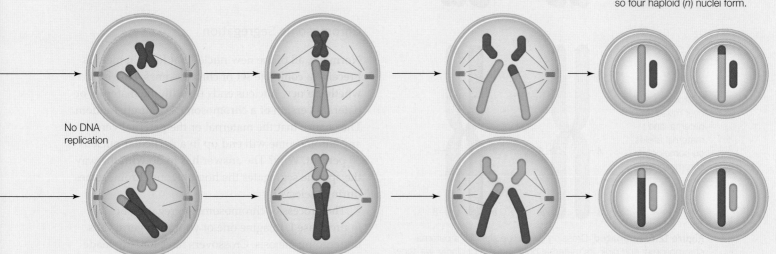

No DNA replication

12.4 How Meiosis Introduces Variations in Traits

■ Crossovers and the random sorting of chromosomes into gametes result in new combinations of traits among offspring of sexual reproducers.

■ Link to Chromosome structure 8.2

The previous section mentioned briefly that duplicated chromosomes swap segments with their homologous partners during prophase I. It also showed how spindle microtubules align and then separate homologous chromosomes during anaphase I. Both events introduce novel combinations of alleles into gametes.

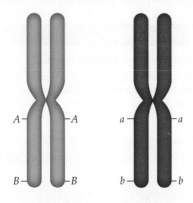

A Here, we focus on only two of the many genes on a chromosome. In this example, one gene has alleles *A* and *a*; the other has alleles *B* and *b*.

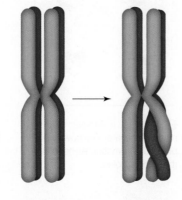

B Close contact between homologous chromosomes promotes crossing over between nonsister chromatids. Paternal and maternal chromatids exchange corresponding pieces.

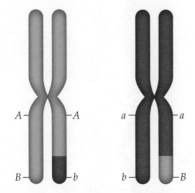

C Crossing over mixes up paternal and maternal alleles on homologous chromosomes.

Figure 12.6 Animated Crossing over. *Blue* signifies a paternal chromosome, and *pink*, its maternal homologue. For clarity, we show only one pair of homologous chromosomes and one crossover, but more than one crossover may occur in each chromosome pair.

Along with fertilization, these events contribute to the variation in combinations of traits among the offspring of sexually reproducing species.

Crossing Over in Prophase I

Early in prophase I of meiosis, all chromosomes in a germ cell condense. When they do, each is drawn close to its homologue. The chromatids of one homologous chromosome become tightly aligned with the chromatids of the other along their length:

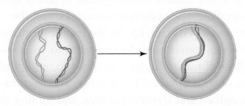

This tight, parallel orientation favors **crossing over**, a process in which a chromosome and its homologous partner exchange corresponding pieces of DNA (**Figure 12.6**). Homologous chromosomes may swap any segment of DNA along their length, although crossovers tend to occur more frequently in certain regions.

Swapping segments of DNA shuffles alleles between homologous chromosomes. It breaks up the particular combinations of alleles that occurred on the parental chromosomes, and makes new ones on the chromosomes that end up in gametes. Thus, crossing over introduces novel combinations of traits among offspring. It is a normal and frequent process in meiosis, but the rate of crossing over varies among species and among chromosomes. In humans, between 46 and 95 crossovers occur per meiosis, so on average each chromosome crosses over at least once.

Chromosome Segregation

Normally, all of the new nuclei that form in meiosis I receive a complete set of chromosomes. However, whether a new nucleus ends up with the maternal or paternal version of a chromosome is entirely random. The chance that the maternal or the paternal version of any chromosome will end up in a particular nucleus is 50 percent. Why? The answer has to do with the way the spindle segregates the homologous chromosomes during meiosis I.

The process of chromosome segregation begins in prophase I. Imagine one of your own germ cells undergoing meiosis. Crossovers have already made genetic mosaics of its chromosomes, but for simplicity let's put crossing over aside for a moment. Just call the

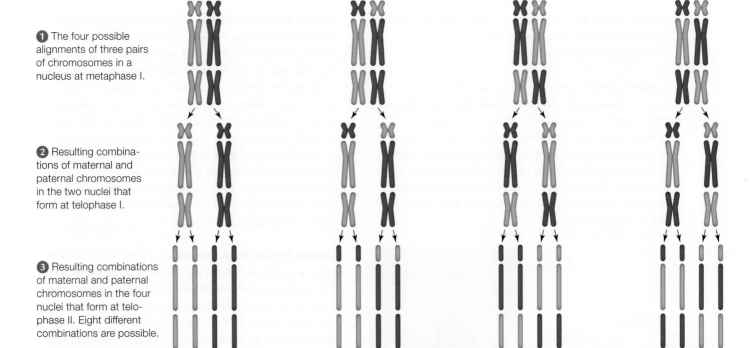

1 The four possible alignments of three pairs of chromosomes in a nucleus at metaphase I.

2 Resulting combinations of maternal and paternal chromosomes in the two nuclei that form at telophase I.

3 Resulting combinations of maternal and paternal chromosomes in the four nuclei that form at telophase II. Eight different combinations are possible.

Figure 12.7 Animated Hypothetical segregation of three pairs of chromosomes in meiosis I. Maternal chromosomes are *pink*; paternal, *blue*. Which chromosome of each pair gets packaged into which of the two new nuclei that form at telophase I is random. For simplicity, no crossing over occurs in this example, so all sister chromatids are identical.

twenty-three chromosomes you inherited from your mother the maternal ones, and the twenty-three from your father the paternal ones.

During prophase I, microtubules fasten your cell's chromosomes to the spindle poles. Chances are fairly slim that all of the maternal chromosomes get attached to one pole and all of the paternal chromosomes get attached to the other. Microtubules extending from a spindle pole bind to the centromere of the first chromosome they contact, regardless of whether it is maternal or paternal. Each homologous partner gets attached to the opposite spindle pole. Thus, there is no pattern to the attachment of the maternal or paternal chromosomes to a particular pole.

Now imagine that your germ cell has just three pairs of chromosomes (**Figure 12.7**). By metaphase I, those three pairs of maternal and paternal chromosomes are divvied up between the two spindle poles in one of four ways **1**. In anaphase I, homologous chromosomes separate and are pulled toward opposite spindle poles. In telophase I, a new nucleus forms around the chromosomes that cluster at each spindle pole. Each nucleus contains one of eight possible combinations of maternal and paternal chromosomes **2**.

In telophase II, each of the two nuclei divides and gives rise to two new haploid nuclei. The two new nuclei are identical because no crossing over occurred

in our hypothetical example, so all of the sister chromatids were identical. Thus, at the end of meiosis in this cell, two (2) spindle poles have divvied up three (3) chromosome pairs. The resulting four nuclei have one of eight (2^3) possible combinations of maternal and paternal chromosomes **3**.

Cells that give rise to human gametes have twenty-three pairs of homologous chromosomes, not three. Each time a human germ cell undergoes meiosis, the four gametes that form end up with one of 8,388,608 (or 2^{23}) possible combinations of homologous chromosomes. That number does not even take into account crossing over, which mixes up the alleles on maternal and paternal chromosomes, or fusion with another gamete at fertilization. Are you getting an idea of why such fascinating combinations of traits show up among the generations of your own family tree?

crossing over Process in which homologous chromosomes exchange corresponding segments during prophase I of meiosis.

Take-Home Message

How does meiosis introduce variation in combinations of traits?

» Crossing over is recombination between nonsister chromatids of homologous chromosomes during prophase I. It makes new combinations of parental alleles.

» Homologous chromosomes can be attached to either spindle pole in prophase I, so each homologue can end up in either one of the two new nuclei.

12.5 From Gametes to Offspring

■ Aside from meiosis, the details of gamete formation and fertilization differ among plants and animals.

■ Link to Cytoplasmic division 11.4

Gametes are the specialized reproductive cells of sexual reproducers. All are haploid, but they differ in other details. For example, human male gametes—sperm—have one flagellum (Section 4.10). Opossum sperm have two, and roundworm sperm have none. A flowering plant's male gamete consists simply of a nucleus. We leave details of reproduction for later chapters, but you will need to know a few concepts before you get there.

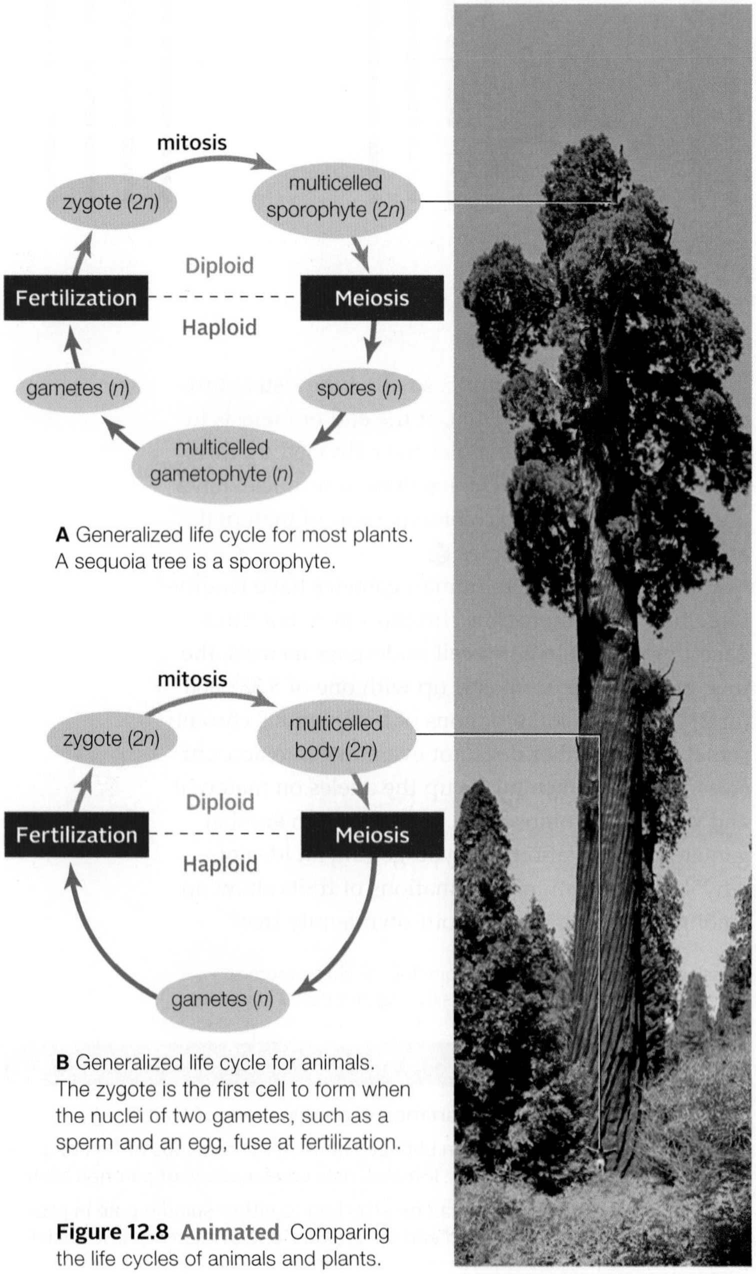

A Generalized life cycle for most plants. A sequoia tree is a sporophyte.

B Generalized life cycle for animals. The zygote is the first cell to form when the nuclei of two gametes, such as a sperm and an egg, fuse at fertilization.

Figure 12.8 Animated Comparing the life cycles of animals and plants.

Gamete Formation in Plants Two kinds of multicelled bodies form during the life cycle of a plant: sporophytes and gametophytes. A **sporophyte** is typically diploid, and spores form by meiosis in its specialized parts. Spores consist of one or a few haploid cells. These cells undergo mitosis and give rise to **gametophytes**, multicelled haploid bodies inside which one or more gametes form. A sequoia tree is an example of a sporophyte (**Figure 12.8A**). Male and female gametophytes develop inside different types of cones that form on each tree. In flowering plants, gametophytes form in flowers, which are specialized reproductive shoots of the sporophyte body.

Gamete Formation in Animals In a typical animal life cycle, a zygote develops into a multicelled individual that produces gametes (**Figure 12.8B**). Animal gametes arise by meiosis of diploid germ cells. The germ cell of a male animal (**Figure 12.9**) develops into a primary spermatocyte ❶. Meiosis I in a primary spermatocyte results in two haploid secondary spermatocytes ❷, which undergo meiosis II and become four spermatids ❸. Each spermatid then matures as a **sperm** ❹.

The germ cell of a female animal (**Figure 12.10**) develops into a primary oocyte, which is an immature egg ❺. This cell undergoes meiosis and division, as occurs with a primary spermatocyte. Two haploid cells form when the primary oocyte divides after meiosis I. However, the cytoplasm of a primary oocyte divides unequally, so these haploid cells differ in size and function. One of the cells is called a first polar body. The other cell, the secondary oocyte, is much larger because it gets nearly all of the parent cell's cytoplasm ❻. The larger cell undergoes meiosis II and cytoplasmic division, which again is unequal ❼. One of the two cells that forms is a second polar body. The other cell gets most of the cytoplasm and matures into a female gamete, which is called an ovum (plural, ova), or **egg**.

Polar bodies are not nutrient-rich or plump with cytoplasm, and generally do not function as gametes. In time they degenerate. Their formation ensures that the egg will have a haploid chromosome number, and also will get enough metabolic machinery to support early divisions of the new individual.

Fertilization Meiosis produces haploid gametes. When two gametes fuse at fertilization, the resulting zygote is diploid. Thus, meiosis halves the chromosome number, and fertilization restores it. If meiosis did not precede fertilization, the chromosome number would double with every generation. An individual's

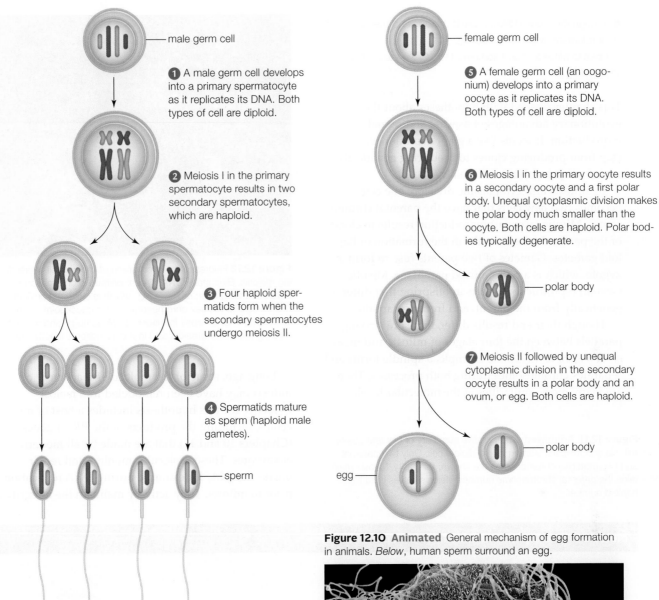

① A male germ cell develops into a primary spermatocyte as it replicates its DNA. Both types of cell are diploid.

male germ cell

② Meiosis I in the primary spermatocyte results in two secondary spermatocytes, which are haploid.

③ Four haploid spermatids form when the secondary spermatocytes undergo meiosis II.

④ Spermatids mature as sperm (haploid male gametes).

sperm

Figure 12.9 Animated General mechanism of sperm formation in animals.

female germ cell

⑤ A female germ cell (an oogonium) develops into a primary oocyte as it replicates its DNA. Both types of cell are diploid.

⑥ Meiosis I in the primary oocyte results in a secondary oocyte and a first polar body. Unequal cytoplasmic division makes the polar body much smaller than the oocyte. Both cells are haploid. Polar bodies typically degenerate.

polar body

⑦ Meiosis II followed by unequal cytoplasmic division in the secondary oocyte results in a polar body and an ovum, or egg. Both cells are haploid.

polar body

egg

Figure 12.10 Animated General mechanism of egg formation in animals. *Below,* human sperm surround an egg.

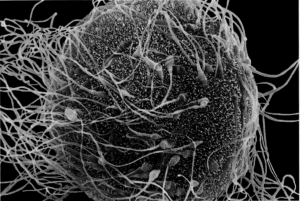

set of chromosomes is like a fine-tuned blueprint that must be followed exactly, page by page, in order to build a body that functions normally. As you will see in Chapter 14, chromosome number changes can have drastic consequences, particularly in animals.

egg Mature female gamete, or ovum.
gametophyte A haploid, multicelled body in which gametes form during the life cycle of land plants and some algae.
sperm Mature male gamete.
sporophyte Diploid, spore-producing stage of a plant life cycle.

Take-Home Message

How does meiosis fit into the life cycle of plants and animals?

» Meiosis and cytoplasmic division precede the development of haploid gametes in animals and spores in plants.

» The union of two haploid gametes at fertilization results in a diploid zygote.

12.6 Mitosis and Meiosis—An Ancestral Connection?

■ Though they have different results, mitosis and meiosis are fundamentally similar processes.

■ Links to Mitosis 11.3, Telomeres 11.5, Cell cycle controls 11.6

This chapter opened with hypotheses about the evolutionary advantages of asexual and sexual reproduction. It seems like a giant evolutionary step from producing clones to producing genetically varied offspring, but was it really?

By mitosis and cytoplasmic division, one cell becomes two new cells that have the parental chromosomes. Mitotic (asexual) reproduction results in clones of the parent. Meiosis results in the formation of haploid gametes. Gametes of two parents fuse to form a zygote, which is a cell of mixed parentage. Meiotic (sexual) reproduction results in offspring that differ genetically from the parent, and from one another.

Though their end results differ, there are striking parallels between the four stages of mitosis and meiosis II (Figure 12.11). As one example, a spindle forms and separates chromosomes during both processes. There are many more similarities at the molecular level.

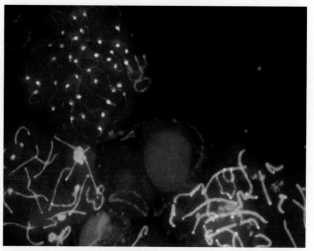

Figure 12.12 Fluorescence micrograph of mouse cell nuclei during meiosis. *Blue* stain: DNA; *green:* paired-up chromosomes before crossing over; *red:* BRCA1, which is clustered around telomeres and the sex chromosomes. This checkpoint protein is involved in DNA repair in mitosis (DNA damage) and meiosis (sealing crossover breaks), and in X chromosome inactivation.

Long ago, some of the molecular machinery of mitosis may have been remodeled into meiosis. Evidence for this hypothesis includes a host of molecules, including the products of the *BRCA* genes (Chapters 10 and 11) that are made by all modern eukaryotes. These molecules monitor and repair breaks in DNA, for example during DNA replication prior to mitosis. They actively maintain the integrity of

Figure 12.11 **Animated** Comparative summary of key features of mitosis and meiosis, starting with a diploid cell. Only two paternal and two maternal chromosomes are shown, for clarity. Mitosis maintains the parental chromosome number. Meiosis halves it, to the haploid number.

Meiosis I One diploid nucleus to two haploid nuclei

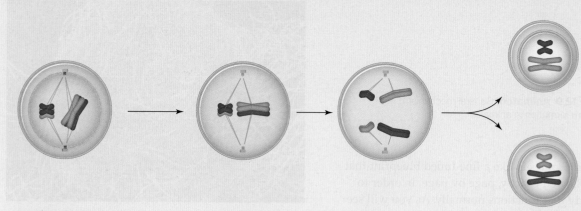

Prophase I
- Chromosomes condense.
- Homologous chromosomes pair.
- Crossovers occur (not shown).
- Spindle forms and attaches chromosomes to spindle poles.
- Nuclear envelope breaks up.

Metaphase I
- Chromosomes align midway between spindle poles.

Anaphase I
- Homologous chromosomes separate and move toward opposite spindle poles.

Telophase I
- Chromosome clusters arrive at spindle poles.
- New nuclear envelopes form.
- Chromosomes decondense.

a cell's chromosomes. The very same set of molecules monitor and fix the breaks in homologous chromosomes during crossing over in prophase I of meiosis (**Figure 12.12**). Some of them function as checkpoint proteins in both mitosis and meiosis, so mutations that affect them or the rate at which they are made can affect the outcomes of both nuclear division processes.

In anaphase of mitosis, sister chromatids are pulled apart. What would happen if the connections between the sisters did not break? Each duplicated chromosome would be pulled to one or the other spindle pole—which is exactly what happens in anaphase I of meiosis. Sexual reproduction probably originated by mutations that affected mitosis. As you will see in later chapters, the remodeling of existing processes into new ones is a common evolutionary theme.

Take-Home Message

Are the processes of mitosis and meiosis related?

» Meiosis may have evolved by the remodeling of existing mechanisms of mitosis.

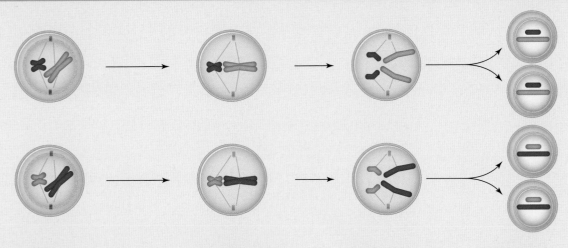

Mitosis One diploid nucleus to two diploid nuclei

Prophase
- Chromosomes condense.
- Spindle forms and attaches chromosomes to spindle poles.
- Nuclear envelope breaks up.

Metaphase
- Chromosomes align midway between spindle poles.

Anaphase
- Sister chromatids separate and move toward opposite spindle poles.

Telophase
- Chromosome clusters arrive at spindle poles.
- New nuclear envelopes form.
- Chromosomes decondense.

Meiosis II Two haploid nuclei to four haploid nuclei

Prophase II
- Chromosomes condense.
- Spindle forms and attaches chromosomes to spindle poles.
- Nuclear envelope breaks up.

Metaphase II
- Chromosomes align midway between spindle poles.

Anaphase II
- Sister chromatids separate and move toward opposite spindle poles.

Telophase II
- Chromosome clusters arrive at spindle poles.
- New nuclear envelopes form.
- Chromosomes decondense.

Why Sex? (revisited)

There are a few all-female species of fishes, reptiles, and birds in nature, but not mammals. In 2004, researchers fused two mouse eggs in a test tube and made an embryo using no DNA from a male. The embryo developed into Kaguya, the world's first fatherless mammal (*right*). The mouse grew up healthy,

engaged in sex with a male mouse, and gave birth to offspring. The researchers wanted to find out if sperm was required for normal development.

How would you vote? Researchers made a "fatherless" mouse by fusing two eggs. Should they be prevented from trying the process with human eggs?

LEARNING ROADMAP

Where you have been You may want to review what you know about traits (Chapter 1), chromosomes (Section 8.2), genes (9.2), sexual reproduction (12.1), alleles (12.2), and meiosis (12.3, 12.4). You will revisit probability and sampling error (1.8), laws of nature (1.9), protein structure (3.6), pigments (6.2), clones (8.7), gene control (10.2, 11.6), and epigenetics (10.6).

Where you are now

Where Modern Genetics Started
Gregor Mendel discovered that inherited traits are specified in units. The units, which are distributed into gametes in predictable patterns, were later identified as genes.

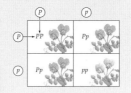

Insights from Monohybrid Crosses
Tracing inheritance patterns of single traits led to the discovery that during meiosis, pairs of genes on homologous chromosomes separate and end up in different gametes.

Insights from Dihybrid Crosses
Tracing inheritance patterns of two unrelated traits led to the discovery that pairs of genes often segregate into gametes independently of how other gene pairs are distributed.

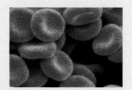

Variations on Mendel's Theme
An allele may be partly dominant over a nonidentical partner, or codominant with it. Multiple genes may influence a trait; some genes influence many traits.

Complex Variations in Traits
Environmental factors can alter the expression of genes that influence a trait. Many traits appear in a continuous range of forms.

Where you are going We return to melanin and human skin color in Section 14.1, and human genetic disorders in the rest of Chapter 14. The complex interplay between genes and the environment is further explored in an evolutionary context in Chapter 17. Signaling pathways in the body that lead to changes in gene expression comprise stimulus, perception, and response (Sections 31.9 and 33.2). Chapter 29 details reproduction in flowering plants. Neurons and how they work are the topic of Chapter 32, with neurological disorders in Section 32.6.

13.1 Menacing Mucus

In 1988, researchers discovered a gene that, when mutated, causes cystic fibrosis (CF). Cystic fibrosis is the most common fatal genetic disorder in the United States. The gene in question, *CFTR*, encodes a protein that moves chloride ions out of epithelial cells. Sheets of these cells line the passageways and ducts of the lungs, liver, pancreas, intestines, reproductive system, and skin. When the CFTR protein pumps chloride ions out of these cells, water follows the ions by osmosis. The two-step process maintains a thin film of water on the surface of the epithelial sheets. Mucus slides easily over the wet sheets of cells.

The most common mutation in CF is a deletion of three base pairs—one codon that specifies phenylalanine as the 508th amino acid of the CFTR protein. This deletion, which is called *ΔF508*, prevents proper membrane trafficking of CFTR so that newly assembled polypeptides are stranded in the endoplasmic reticulum. The altered protein still functions properly, but it never reaches the cell surface to do its job.

One outcome is that the transport of chloride ions out of epithelial cells is disrupted. If not enough chloride ions leave the cells, not enough water leaves them either, so the surfaces of epithelial cell sheets are not as wet as they should be. Mucus that normally slips and slides through the body's tubes sticks to the walls of the tubes instead. Thick globs of mucus accumulate and clog passageways and ducts throughout the body. Breathing becomes difficult as the mucus obstructs the smaller airways of the lungs.

The CFTR protein also functions as a receptor that alerts the body to the presence of bacteria. Bacteria bind to CFTR. The binding triggers endocytosis, which speeds the immune system's defensive responses. Without the CFTR protein on the surface of epithelial cell linings, disease-causing bacteria that enter the ducts and passageways of the body can persist there. Thus, chronic bacterial infections of the intestine and lungs are hallmarks of cystic fibrosis. Daily routines of posture changes and thumps on the chest and back help clear the lungs of some of the thick mucus, and antibiotics help control infections, but there is no cure. Even with a lung transplant, most cystic fibrosis patients live no longer than thirty years, at which time their tormented lungs usually fail (**Figure 13.1**).

About 1 in 25 people carry the *ΔF508* mutation in one of their two copies of the *CFTR* gene, but most of them do not realize it because they do not have the symptoms of cystic fibrosis. The disease occurs only when a person inherits two mutated genes, one from each parent. This unlucky event occurs in about 1 of 3,300 births worldwide.

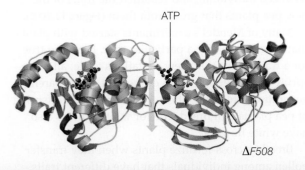

ATP

ΔF508

Figure 13.1 Cystic fibrosis. *Left*, model of the CFTR protein. The parts shown here are a pair of ATP-driven motors that widen or narrow a channel (*gray* arrow) across the plasma membrane. The tiny part of the protein that is deleted in most people with cystic fibrosis is shown in *green*.

Below, a few of the many young victims of cystic fibrosis, which occurs most often in people of northern European ancestry. At least one young person dies every day in the United States from complications of this disease.

Lindsay, 22   Savannah, 19   Cody, 23   Ben, 23   Jeff, 21   Brandon, 18

13.2 Mendel, Pea Plants, and Inheritance Patterns

■ Recurring patterns of inheritance offer observable evidence of how heredity works.

■ Links to Traits Chapter 1, Diploid 8.2, Genes 9.2, Alleles and haploid 12.2

In the nineteenth century, people thought that hereditary material must be some type of fluid, with fluids from both parents blending at fertilization like milk into coffee. However, the idea of "blending inheritance" failed to explain what people could see with their own eyes. Children sometimes have traits such as freckles that do not appear in either parent. A cross between a black horse and a white one does not produce gray offspring.

The naturalist Charles Darwin did not accept the idea of blending inheritance, but he could not come up with an alternative even though inheritance was central to his theory of natural selection (we return to natural selection in Chapter 16). Neither Darwin nor anyone else at the time knew that hereditary information (DNA) is divided into discrete units (genes), an insight that is critical to understanding how heredity really works. However, even before Darwin presented his theory, someone had been gathering evidence that would support it. Gregor Mendel, an Austrian monk, had been carefully breeding thousands of pea plants (Figure 13.2A). By meticulously documenting the passage of traits from one generation to the next, Mendel had been collecting evidence of how inheritance works.

Mendel's Experimental Approach

Mendel studied variation in the traits of the garden pea plant, *Pisum sativum*. This species is naturally self-fertilizing, which means that its flowers produce male gametes (in pollen) and female gametes (in carpels) that form viable embryos when they meet up.

In order to study inheritance, Mendel had to breed particular individuals together, then observe and document the traits of their offspring. Control over the reproduction of an individual pea plant begins with preventing it from self-fertilizing. Mendel did this by removing a flower's pollen-bearing anthers, then brushing its carpel with pollen from another plant (Figure 13.2B,C). He collected the seeds from the cross-fertilized individual, and recorded the traits of the new pea plants that grew from them (Figure 13.2D,E).

Many of Mendel's experiments started with plants that "breed true" for a particular trait. Breeding true for a trait means that, new mutations aside, all offspring have the same form of the trait as the parent(s), generation after generation. For example, all offspring of pea plants that breed true for white flowers also have white flowers.

Breeders cross-fertilize plants when they transfer pollen among individuals that have different traits. As you will see in the next section, Mendel discovered that the traits of the offspring of cross-fertilized pea plants often appear in predictable patterns. Mendel's meticulous work tracking pea plant traits led him to conclude (correctly) that hereditary information passes from one generation to the next in discrete units.

Inheritance in Modern Terms

DNA was not proven to be hereditary material until the 1950s (Section 8.3), but Mendel discovered its units, which we now call genes, almost a century before then. Today, we know that individuals of a

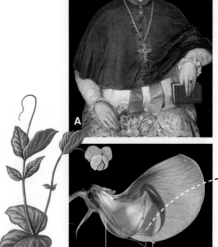
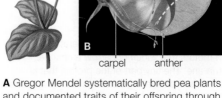

carpel   anther

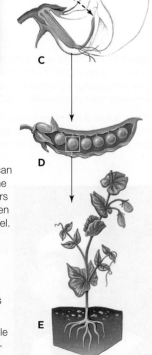

A Gregor Mendel systematically bred pea plants and documented traits of their offspring through many generations.

B Garden pea flower, cut in half. Male gametes form in pollen grains produced by anthers, and female gametes form in carpels. Experimenters can control the transfer of hereditary material from one flower to another by snipping off a flower's anthers (to prevent the flower from self-fertilizing), and then brushing pollen from another flower onto its carpel.

C In this example, pollen from a plant that has purple flowers is brushed onto the carpel of a white-flowered plant.

D Later, seeds develop inside pods of the cross-fertilized plant. An embryo in each seed develops into a mature pea plant.

E Every plant that arises from the cross has purple flowers. Predictable patterns such as this are evidence of how inheritance works.

Figure 13.2 Animated Breeding garden pea plants (*Pisum sativum*), which can self-fertilize or cross-fertilize.

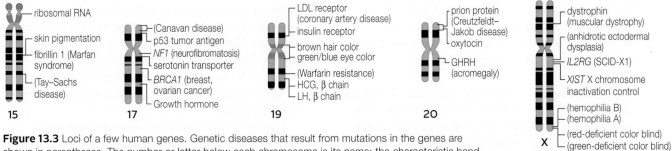

Figure 13.3 Loci of a few human genes. Genetic diseases that result from mutations in the genes are shown in parentheses. The number or letter below each chromosome is its name; the characteristic banding patterns appear after staining. Appendix VI has a similar map of all 23 human chromosomes.

species share certain traits because their chromosomes carry the same genes. Offspring tend to look like their parents because they inherited their parents' genes.

Each gene occurs at a specific location, or **locus** (plural, loci), on a particular chromosome (**Figure 13.3**). Somatic cells of humans and other animals are diploid: They have pairs of genes, on pairs of homologous chromosomes. In most cases, both genes of a pair are expressed (Section 9.2).

The two genes of a pair may be identical, or they may vary as alleles (Section 12.2). Organisms that breed true for a specific trait probably have identical alleles of the gene governing that trait. An individual with identical alleles of a gene is said to be **homozygous** for the allele. The particular set of alleles that an individual carries is called the individual's **genotype**.

Alleles are the major source of variation in a trait. New alleles arise by mutation (Section 8.6). A mutation may cause a trait to change, as when a gene that causes flowers to be purple mutates so the resulting flowers are white. "White-flowered" is an example of a **phenotype**, which refers to one or more aspects of an individual's observable traits. Any mutated gene is an allele, whether or not it affects phenotype.

Hybrids are offspring of a cross, or mating, between individuals that breed true for different forms of a trait. A hybrid has two different alleles of a gene, so it is said to be **heterozygous** for the alleles (*hetero–*, mixed). In many cases, the effect of one allele influences the effect of the other, and the outcome of this interaction is visible in the hybrid phenotype. An allele is **dominant** when its effect masks that of a **recessive** allele paired with it. Usually, a dominant allele is represented by an italic capital letter such as *A*; a recessive allele, with a lowercase italic letter such as *a*.

Mendel crossed plants that breed true for purple flowers with plants that breed true for white flowers. All offspring of these crosses have purple flowers, but Mendel did not know why. We now understand that one gene governs purple and white flower color in pea plants. The allele that specifies purple (let's call it *P*) is dominant over the allele that specifies white (*p*). Thus, a pea plant homozygous for two dominant alleles (*PP*) has purple flowers; one homozygous for two recessive alleles (*pp*) has white flowers. A plant heterozygous at this gene locus (*Pp*) has purple flowers (**Figure 13.4**).

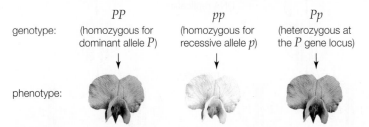

	PP	*pp*	*Pp*
genotype:	(homozygous for dominant allele *P*)	(homozygous for recessive allele *p*)	(heterozygous at the *P* gene locus)
phenotype:			

Figure 13.4 Genotype gives rise to phenotype. In this example, the dominant allele *P* specifies purple flowers; the recessive allele *p*, white flowers.

Figure It Out Which individual is a hybrid?   Answer: The heterozygous one

dominant Refers to an allele that masks the effect of a recessive allele paired with it.
genotype The particular set of alleles carried by an individual.
heterozygous Having two different alleles of a gene.
homozygous Having identical alleles of a gene.
hybrid The offspring of a cross between two individuals that breed true for different forms of a trait; a heterozygous individual.
locus Location of a gene on a chromosome.
phenotype An individual's observable traits.
recessive Refers to an allele with an effect that is masked by a dominant allele on the homologous chromosome.

Take-Home Message

How do alleles contribute to traits?

» Gregor Mendel indirectly discovered the role of alleles in inheritance by carefully breeding pea plants and tracking traits of their offspring.

» Genotype refers to the particular set of alleles carried by an individual's somatic cells. Phenotype refers to the individual's observable traits. Genotype is the basis of phenotype.

» A homozygous individual has two identical alleles at a particular locus. A heterozygous individual has nonidentical alleles at the locus.

» Dominant alleles mask the effects of recessive ones in heterozygous individuals.

13.3 Mendel's Law of Segregation

■ Pairs of genes on homologous chromosomes separate during meiosis, so they end up in different gametes.

■ Links to Probability and sampling error 1.8, Laws of nature 1.9, Meiosis 12.3, Chromosome segregation 12.4

When homologous chromosomes separate during meiosis, the gene pairs on those chromosomes separate too. Each gamete that forms carries only one of the two genes of a pair (**Figure 13.5**). Thus, plants homozygous for the dominant allele (*PP*) can only make gametes that carry the allele *P* ❶. Plants homozygous for the recessive allele (*pp*) can only make gametes that carry the allele *p* ❷. If these homozygous plants are crossed (*PP* × *pp*), only one outcome is possible: A gamete carrying a *P* allele meets up with a gamete carrying a *p* allele ❸. All of the offspring of this cross have one of each allele, so their genotype is *Pp*. A grid called a **Punnett square** is helpful for predicting the genetic and phenotypic outcomes of crosses ❹. In this example, all offspring of the cross carry the dominant allele *P*, so all have purple flowers.

This pattern is so predictable that it can be used as evidence of a dominance relationship between alleles.

Breeding experiments use such patterns to reveal genotype. In a **testcross**, an individual that has a dominant trait (but an unknown genotype) is crossed with an individual known to be homozygous recessive. The pattern of traits among the offspring of the cross can reveal whether the tested individual is heterozygous or homozygous.

For example, we may do a testcross between a purple-flowered pea plant (which could have a genotype of either *PP* or *Pp*) and a white-flowered pea plant (*pp*). If all of the offspring of this cross had purple flowers, we could be reasonably certain that the genotype of the purple-flowered parent was *PP*.

Dominance relationships between alleles determine the phenotypic outcome of a **monohybrid cross**, in which individuals identically heterozygous for one gene—*Pp*, for example—are bred together or self-fertilized. The frequency at which the two traits appear among the offspring of this cross depends on whether one of the alleles is dominant over the other.

To produce identically heterozygous individuals for a monohybrid cross, we would start with two individuals that breed true for two different forms of a trait. In pea plants, flower color (purple and white) is one example of a trait with two distinct forms, but there are many others. Mendel investigated seven of them: stem length (tall and short), seed color (yellow and green), pod texture (smooth and wrinkled), and so on (**Table 13.1**). A cross between the two true-breeding individuals yields offspring that are identically heterozygous for the alleles that govern the trait. When these F₁ (first generation) hybrids are crossed, the frequency at which the two traits appear in the F₂ (second generation) offspring offers information about a dominance relationship between the two alleles. F is an abbreviation for filial, which means offspring.

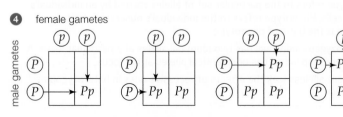

gametes (P)   gametes (p)

zygote (Pp)

female gametes

male gametes

Figure 13.5 Gene segregation. Homologous chromosomes separate during meiosis, so the pairs of genes they carry separate too. Each of the resulting gametes carries one of the two members of each gene pair. For clarity, only one set of chromosomes is shown.

❶ All gametes made by a parent homozygous for a dominant allele carry that allele.

❷ All gametes made by a parent homozygous for a recessive allele carry that allele.

❸ If these two parents are crossed, the union of any of their gametes at fertilization produces a zygote with both alleles. All offspring of this cross will be heterozygous.

❹ This outcome is easy to see with a Punnett square. Parental gametes are listed in circles on the top and left sides of a grid. Each square is filled with the combination of alleles that would result if the gametes in the corresponding row and column met up.

Table 13.1 Mendel's Seven Pea Plant Traits

Trait:	Recessive Form	Dominant Form
Seed Shape:	Round	Wrinkled
Seed Color:	Yellow	Green
Pod Texture:	Smooth	Wrinkled
Pod Color:	Green	Yellow
Flower Color:	Purple	White
Flower Position:	Along Stem	At Tip
Stem Length:	Tall	Short

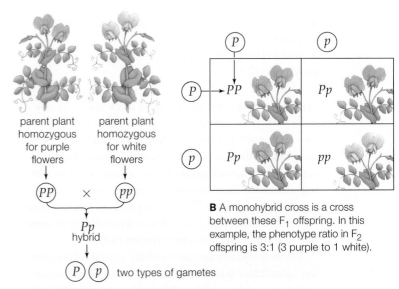

parent plant homozygous for purple flowers

PP

×

parent plant homozygous for white flowers

pp

Pp hybrid

P p two types of gametes

A All of the F$_1$ offspring of a cross between two plants that breed true for different forms of a trait are identically heterozygous (Pp). These offspring make two types of gametes: P and p.

B A monohybrid cross is a cross between these F$_1$ offspring. In this example, the phenotype ratio in F$_2$ offspring is 3:1 (3 purple to 1 white).

Figure 13.6 Animated A monohybrid cross. **Figure It Out:** In this example, how many possible genotypes are there in the F$_2$ generation?

Answer: Three: PP, Pp, and pp

A cross between purple-flowered heterozygous plants (Pp) offers an example. Each plant can make two types of gametes: ones that carry a P allele, and ones that carry a p allele (**Figure 13.6A**). So, in a monohybrid cross between Pp plants ($Pp \times Pp$), the two types of gametes can meet up in four possible ways at fertilization:

Possible Event		Probable Outcome
Sperm P meets egg P	→	zygote genotype is PP
Sperm P meets egg p	→	zygote genotype is Pp
Sperm p meets egg P	→	zygote genotype is Pp
Sperm p meets egg p	→	zygote genotype is pp

Three of four possible outcomes of this cross include at least one copy of the dominant allele P. Each time fertilization occurs, there are 3 chances in 4 that the resulting offspring will inherit a P allele, and have purple flowers. There is 1 chance in 4 that it will

inherit two recessive p alleles, and have white flowers. Thus, the probability that a particular offspring of this cross will have purple or white flowers is 3 purple to 1 white, represented as a ratio of 3:1 (**Figure 13.6B**).

If the probability of one individual inheriting a particular genotype is difficult to imagine, think about probability in terms of the phenotypes of many offspring. In this example, there will be roughly three purple-flowered plants for every white-flowered one. The 3:1 pattern is an indication that the purple and white flower colors are specified by alleles with a clear dominant–recessive relationship: Purple is dominant, and white is recessive.

The phenotype ratios in F$_2$ offspring of Mendel's monohybrid crosses were all close to 3:1. These results became the basis of his **law of segregation**, which we state here in modern terms: Diploid cells carry pairs of genes, on pairs of homologous chromosomes. The two genes of each pair separate from one another during meiosis, so they end up in different gametes.

law of segregation The members of each pair of genes on homologous chromosomes end up in different gametes during meiosis.
monohybrid cross Cross between two individuals identically heterozygous for one gene; for example $Aa \times Aa$.
Punnett square Diagram used to predict the genetic and phenotypic outcome of a cross.
testcross Method of determining genotype by tracking a trait in the offspring of a cross between an individual of unknown genotype and an individual known to be homozygous recessive.

Take-Home Message

What is Mendel's law of segregation?

» Diploid cells carry pairs of genes, on pairs of homologous chromosomes. The two genes of each pair are separated from each other during meiosis, so they end up in different gametes.

13.4 Mendel's Law of Independent Assortment

- Many gene pairs tend to sort into gametes independently of one another.
- Links to Meiosis 12.3, Crossing over and chromosome segregation 12.4

A monohybrid cross allows us to track alleles of one gene pair. What about alleles of two gene pairs? How two gene pairs get sorted into gametes depends partly on whether the two genes are on the same chromosome. When homologous chromosomes separate during meiosis, either member of the pair can end up in a particular nucleus. Thus, gene pairs on one chromosome tend to sort into gametes independently of gene pairs on other chromosomes (Figure 13.7).

Punnett squares are particularly useful for predicting inheritance patterns of two or more genes simultaneously, such as with a dihybrid cross. In a **dihybrid cross**, individuals identically heterozygous for alleles of two genes (dihybrids) are crossed. As with a monohybrid cross, the pattern of traits seen in the offspring of the cross depends on the dominance relationships between alleles of the genes.

To make a dihybrid cross, we would start with individuals that breed true for two different traits. Let's use a gene for flower color (*P*, purple; *p*, white) and one for plant height (*T*, tall; *t*, short). Assume that

these two genes are on separate chromosomes. **Figure 13.8** shows a dihybrid cross starting with a parent plant that breeds true for purple flowers and tall stems (*PPTT*), and one that breeds true for white flowers and short stems (*pptt*). Each homozygous plant makes only one type of gamete ❶. So, all offspring from a cross between these parent plants (*PPTT* × *pptt*) will be dihybrids (*PpTt*) with purple flowers and tall stems ❷.

Four combinations of *P* and *T* alleles are possible in the gametes of *PpTt* dihybrids ❸. If two *PpTt* plants are crossed (a dihybrid cross, *PpTt* × *PpTt*), the four types of gametes can combine in sixteen possible ways at fertilization ❹. Those sixteen genotypes would result in four different phenotypes. Nine would be tall with purple flowers, three would be short with purple flowers, three would be tall with white flowers, and one would be short with white flowers. Thus, the ratio of these phenotypes is 9:3:3:1.

Mendel discovered the 9:3:3:1 ratio of phenotypes among the offspring of his dihybrid crosses, but he had no idea what it meant. He could only say that "units" specifying one trait (such as flower color) are inherited independently of "units" specifying other traits (such as plant height). In time, Mendel's hypothesis became known as the **law of independent assortment**, which we state here in modern terms: During meiosis, the two

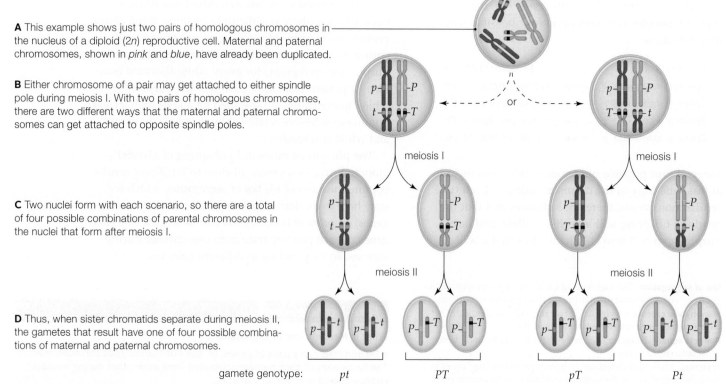

A This example shows just two pairs of homologous chromosomes in the nucleus of a diploid (2n) reproductive cell. Maternal and paternal chromosomes, shown in *pink* and *blue*, have already been duplicated.

B Either chromosome of a pair may get attached to either spindle pole during meiosis I. With two pairs of homologous chromosomes, there are two different ways that the maternal and paternal chromosomes can get attached to opposite spindle poles.

C Two nuclei form with each scenario, so there are a total of four possible combinations of parental chromosomes in the nuclei that form after meiosis I.

D Thus, when sister chromatids separate during meiosis II, the gametes that result have one of four possible combinations of maternal and paternal chromosomes.

gamete genotype: *pt*   *PT*   *pT*   *Pt*

Figure 13.7 Animated Independent assortment.

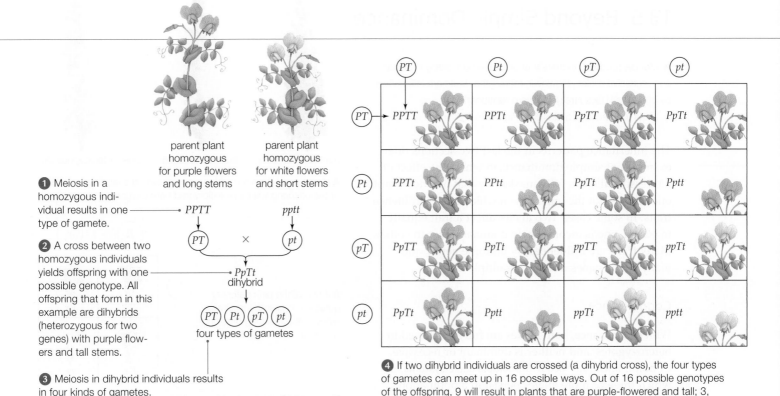

① Meiosis in a homozygous individual results in one type of gamete.

parent plant homozygous for purple flowers and long stems
parent plant homozygous for white flowers and short stems

PPTT   *pptt*

(PT) × (pt)

② A cross between two homozygous individuals yields offspring with one possible genotype. All offspring that form in this example are dihybrids (heterozygous for two genes) with purple flowers and tall stems.

PpTt dihybrid

(PT) (Pt) (pT) (pt)
four types of gametes

③ Meiosis in dihybrid individuals results in four kinds of gametes.

④ If two dihybrid individuals are crossed (a dihybrid cross), the four types of gametes can meet up in 16 possible ways. Out of 16 possible genotypes of the offspring, 9 will result in plants that are purple-flowered and tall; 3, purple-flowered and short; 3, white-flowered and tall; and 1, white-flowered and short. Thus, the ratio of phenotypes in a dihybrid cross is 9:3:3:1.

Figure 13.8 Animated A dihybrid cross between plants that differ in flower color and plant height. *P* and *p* stand for dominant and recessive alleles for flower color; *T* and *t*, dominant and recessive alleles for height.
Figure It Out: What do the flowers inside the boxes represent?
Answer: Phenotypes of F_2 offspring

genes of a pair tend to be sorted into gametes independently of how other gene pairs are sorted into gametes.

Mendel published his results in 1866, but apparently his work was read by few and understood by no one at the time. In 1871 he became abbot of his monastery, and his pioneering experiments ended. He died in 1884, never to know that they would be the starting point for modern genetics.

The Contribution of Crossovers

It makes sense that gene pairs on different chromosomes would assort independently into gametes, but what about gene pairs on the same chromosome? Mendel studied seven genes in pea plants, which have seven chromosomes. Was he lucky enough to choose one gene on each of those chromosomes? As it turns out, some of the genes Mendel studied *are* on the same chromosome. The genes are far enough apart that crossing over occurs between them very frequently—so frequently that they tend to assort into gametes independently, just as if they were on different chromosomes.

By contrast, genes that are very close together on a chromosome do not assort independently, because crossing over does not happen very often between

them. Such genes are said to be linked. Alleles of some linked genes stay together during meiosis more frequently than others, an effect due to the relative distance between the genes. Genes that are closer together get separated less frequently by crossovers. Thus, the closer together any two genes are on a chromosome, the more likely gametes will be to receive parental combinations of alleles of those genes. Genes are said to be tightly linked if the distance between them is relatively small.

All of the genes on a single chromosome are called a **linkage group**. Peas have 7 different chromosomes, so they have 7 linkage groups. Humans have 23 different chromosomes, so they have 23 linkage groups.

dihybrid cross Cross between two individuals identically heterozygous for two genes; for example *AaBb* × *AaBb*.
law of independent assortment During meiosis, members of a pair of genes on homologous chromosomes get distributed into gametes independently of other gene pairs.
linkage group All genes on a chromosome.

Take-Home Message

What is Mendel's law of independent assortment?

» Each member of a pair of genes on homologous chromosomes tends to be distributed into gametes independently of how other genes are distributed during meiosis.

13.5 Beyond Simple Dominance

■ Mendel focused on traits that are based on clearly dominant and recessive alleles. However, many other traits are influenced by alleles with less straightforward relationships.

■ Links to Fibrous proteins 3.6, Pigments 6.2

The inheritance patterns in the last two sections offer examples of simple dominance, in which the effect of a recessive allele is fully masked by that of a dominant one. This is not the only way in which alleles influence traits. In some cases, two alleles affect a trait equally; in others, one is incompletely dominant over the other. Many traits are influenced by multiple genes, and many single genes influence multiple traits.

Codominance

With **codominance**, both alleles are fully expressed in heterozygotes, and neither is dominant or recessive. Codominance may occur in **multiple allele systems**, in which three or more alleles persist in a population. The three alleles of the *ABO* gene are an example. An enzyme encoded by this gene modifies a carbohydrate on the surface of human red blood cell membranes. The *A* and *B* alleles encode slightly different versions of the enzyme, which in turn modify the carbohydrate differently. The *O* allele has a mutation that prevents its enzyme product from becoming active at all.

The two alleles you carry for the *ABO* gene determine the form of the carbohydrate on your blood cells, and that carbohydrate is the basis of your blood type (**Figure 13.9**). The *A* and the *B* alleles are codominant when paired. If your genotype is *AB*, then you have both versions of the enzyme, and your blood is type AB. The *O* allele is recessive when paired with either the *A* or *B* allele. If your genotype is *AA* or *AO*, your blood is type A. If your genotype is *BB* or *BO*, it is type B. If you are *OO*, it is type O.

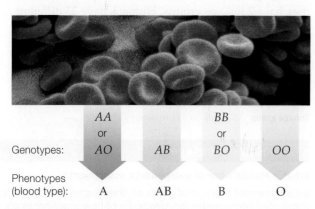

Genotypes:	*AA* or *AO*	*AB*	*BB* or *BO*	*OO*
Phenotypes (blood type):	A	AB	B	O

Figure 13.9 Combinations of alleles that are the basis of human blood type.

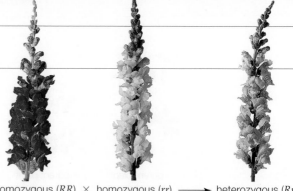

homozygous (*RR*) × homozygous (*rr*) ⟶ heterozygous (*Rr*)

A Cross a red-flowered with a white-flowered snapdragon, and all of the offspring will be pink-flowered heterozygotes.

B If two of the pink heterozygotes are crossed, the phenotypes of the resulting offspring will occur in a 1:2:1 ratio.

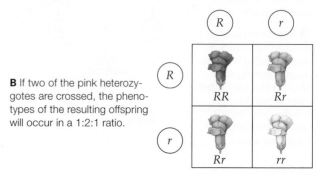

	R	*r*
R	*RR*	*Rr*
r	*Rr*	*rr*

Figure 13.10 **Animated** Incomplete dominance in heterozygous (*pink*) snapdragons. An allele that affects red pigment is paired with a "white" allele. **Figure It Out: Is the experiment in (B) a monohybrid or dihybrid cross?**   Answer: A monohybrid cross

Receiving incompatible blood cells in a transfusion is very dangerous, because the immune system usually attacks red blood cells bearing molecules that do not occur in one's own body. The attack can cause the blood cells to clump or burst, a transfusion reaction with potentially fatal consequences. People with type O blood can donate blood to anyone else, so they are called universal donors. However, they can receive transfusions of type O blood only. People with type AB blood can receive a transfusion of any blood type, so they are called universal recipients.

Incomplete Dominance

With **incomplete dominance**, one allele is not fully dominant over the other, so the heterozygous phenotype is somewhere between the two homozygous phenotypes. A gene that influences flower color in snapdragon plants is an example. There is one copy of this gene on each homologous chromosome; both copies are expressed. One allele (*R*) encodes an enzyme that makes a red pigment. The enzyme encoded by a mutated allele (*r*) cannot make any pigment. Plants homozygous for the *R* allele (*RR*) make a lot of red pigment, so they have red flowers. Plants homozygous for the *r* allele (*rr*) do not make any pigment at all, so their flowers are white. Heterozygous plants (*Rr*) make only enough red pigment to color their flowers pink (**Figure 13.10A**). A cross between two pink-flowered heterozygous plants yields red-, pink-, and white-flowered offspring in a 1:2:1 ratio (**Figure 13.10B**).

Epistasis and Pleiotropy

Some traits are affected by multiple gene products, an effect called polygenic inheritance or **epistasis**. For example, several gene products affect the coat color of a Labrador retriever, which can be black, yellow, or brown (**Figure 13.11**). One gene is involved in the synthesis of the pigment melanin. A dominant allele of the gene specifies black fur, and its recessive partner specifies brown fur. A dominant allele of a different gene causes melanin to be deposited in fur, and its recessive partner reduces melanin deposition.

A **pleiotropic** gene is one that influences multiple traits. Mutations in such genes are associated with complex genetic disorders such as sickle-cell anemia (Section 9.6) and cystic fibrosis. For example, thickened mucus in cystic fibrosis patients affects the entire body, not just the respiratory tract. The mucus clogs ducts that lead to the gut, which results in digestive problems. Male CF patients are typically infertile because their sperm flow is hampered by the thickened secretions.

Marfan syndrome is another example of a genetic disorder caused by mutation in a pleiotropic gene.

	EB	*Eb*	*eB*	*eb*
EB	*EEBB*	*EEBb*	*EeBB*	*EeBb*
Eb	*EEBb*	*EEbb*	*EeBb*	*Eebb*
eB	*EeBB*	*EeBb*	*eeBB*	*eeBb*
eb	*EeBb*	*Eebb*	*eeBb*	*eebb*

Figure 13.11 Epistasis in dogs. Epistatic interactions among products of two gene pairs affect coat color in Laborador retrievers. All dogs with an *E* and *B* allele have black fur. Those with an *E* and two recessive *b* alleles have brown fur. All dogs homozygous for the recessive *e* allele have yellow fur.

In this case, the gene encodes fibrillin. Long fibers of this protein impart elasticity to tissues of the heart, skin, blood vessels, tendons, and other body parts. Mutations in the fibrillin gene result in tissues that form with defective fibrillin or none at all. The largest blood vessel leading from the heart, the aorta, is particularly affected. Muscle cells in the aorta's thick wall do not work very well, and the wall itself is not as elastic as it should be. The aorta expands under pressure, so the lack of elasticity eventually makes it thin and leaky. Calcium deposits accumulate inside. Inflamed, thinned, and weakened, the aorta can rupture abruptly during exercise.

Marfan syndrome is particularly difficult to diagnose. Affected people are often tall, thin, and loose-jointed, but there are plenty of tall, thin, loose-jointed people that do not have the syndrome. Symptoms may not be apparent, so many people die suddenly and early without ever knowing they had the disorder (Figure 13.12).

Figure 13.12 Marfan syndrome. Basketball star Haris Charalambous died suddenly in 2006 when his aorta burst during warm-up exercises. He was 21.

Charalambous was very tall and lanky, with long arms and legs—traits that are valued in professional athletes such as basketball players. These traits are also associated with Marfan syndrome.

About 1 in 5,000 people are affected by Marfan syndrome worldwide. Like many of them, Charalambous did not realize he had the syndrome.

codominant Refers to two alleles that are both fully expressed in heterozygotes and neither is dominant over the other.
epistasis Effect in which a trait is influenced by the products of multiple genes.
incomplete dominance Effect in which one allele is not fully dominant over another, so the heterozygous phenotype is between the two homozygous phenotypes.
multiple allele system Gene for which three or more alleles persist in a population.
pleiotropic Refers to a gene that influences multiple traits.

Take-Home Message

Are all alleles dominant or recessive?

» An allele may be fully dominant, incompletely dominant, or codominant with its partner on a homologous chromosome.

» In epistasis, two or more gene products influence a trait.

» The product of a pleiotropic gene influences two or more traits.

13.6 Nature and Nurture

■ Variations in traits are not always the result of differences in alleles. Many traits are also influenced by environmental factors.

■ Links to Gene control 10.2, Epigenetics 10.6, Tumor suppressors and HPV 11.6, Advantages of sexual over asexual reproduction 12.1

The phrase "nature vs. nurture" refers to a centuries-old debate about whether human behavioral traits arise from one's genetics (nature) or from environmental factors (nurture). It turns out that both play a role. The environment affects the expression of many genes, which in turn affects phenotype—including behavioral traits. We can summarize this thinking with the following equation:

$$\text{genotype} + \text{environment} \longrightarrow \text{phenotype}$$

The science of epigenetics (Section 10.6) is revealing that the environment has an even greater contribution to this equation than most biologists had suspected. Environmentally driven changes in gene expression patterns can be permanent and heritable. Such changes are implemented by gene controls such as chromatin modifications and RNA interference that act on the DNA itself (Section 10.2).

As an example of how the environment influences gene controls, consider DNA methylation.

Figure 13.13 Example of environmental effects on animal phenotype. The color of the snowshoe hare's fur varies by season. In summer, the fur is brown (*left*); in winter, white (*right*).

Environmental cues can initiate cell-signaling pathways (you will learn more about such pathways in later chapters). Some cell-signaling pathways end with methyl groups being removed from or added to particular regions of DNA. The change in methylation enhances or suppresses gene expression in those regions. Diet, stress, exercise, drugs, and exposure to toxins such as tobacco, alcohol, arsenic, and asbestos affect DNA methylation patterns.

Examples of Environmental Effects on Phenotype

Most of the research into environmental effects on phenotype concerns diseases and disorders. However, mechanisms that adjust phenotype in response to external cues are part of an individual's normal ability to adapt to environmental change, as the following examples illustrate.

Seasonal Changes in Coat Color Seasonal changes in temperature and the length of day affect the production of melanin and other pigments that color the skin and fur of many animals. These animals have different color phases in different seasons (**Figure 13.13**). Hormonal signals triggered by changes in daylength cause the fur to be shed, and different types and amounts of pigments to be deposited in fur that grows back. The resulting change in phenotype offers these animals seasonally appropriate camouflage from predators.

Effect of Altitude on Yarrow Yarrow is a type of plant useful for genetics experiments because it grows easily from cuttings. All cuttings of a plant have the same genotype, so experimenters know that genes are not

A Mature cutting at high elevation (3,060 meters above sea level)

B Mature cutting at mid-elevation (1,400 meters above sea level)

C Mature cutting at low elevation (30 meters above sea level)

Figure 13.14 Experiment showing environmental effects on phenotype in yarrow (*Achillea millefolium*). Cuttings from the same parent plant were grown in the same kind of soil at three different elevations.

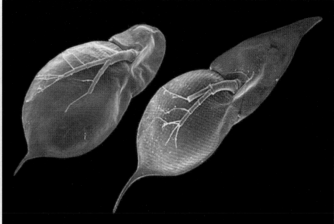

A Light micrograph of a living water flea.

Figure 13.15 Environmental effect on phenotype of the water flea (*Daphnia pulex*).

B Electron micrographs comparing *Daphnia* body form that develops in the presence of few predators (*left*) with the form that develops in the presence of many predators (*right*). Note the difference in the length of the tail spine and the pointiness of the head. Chemicals emitted by the water flea's insect predators provoke the change.

the basis for any phenotypic differences among them. Genetically identical yarrow plants grow to different heights at different altitudes (**Figure 13.14**). More challenging temperature, soil, and water conditions are typically encountered at higher altitudes. Differences in altitude are also correlated with changes in the reproductive mode of yarrow: Plants at higher altitude tend to reproduce asexually, and plants at lower altitude reproduce sexually. Plasticity of phenotype gives these plants an ability to thrive in diverse habitats.

Alternative Phenotypes in Water Fleas The water flea is a microscopic freshwater relative of shrimp (**Figure 13.15A**). Water fleas have different phenotypes depending on whether the aquatic insects that prey on them are present (**Figure 13.15B**). Individual water fleas can also switch between asexual and sexual reproduction depending on environmental conditions. During the early spring, competition is scarce in their freshwater pond habitats. At that time, the fleas reproduce rapidly by asexual means, giving birth to large numbers of female offpsring that quickly fill the ponds. Later in the season, pond water becomes warmer, saltier, and more crowded; competition for resources is also more intense. Under these conditions, some of the water fleas start giving birth to males, and then reproducing sexually. The increased genetic diversity of sexually produced offspring may offer the population an advantage in an environment that presents a greater challenge to survival.

Mood Disorders in Humans We have known for a long time that environment is a factor in schizophrenia,

bipolar disorder, depression, and probably other mood disorders as well. Certain mutations are associated with these disorders, but not all people with the mutations end up with a mental illness; environment also plays a part. Recent discoveries in animal models are beginning to reveal mechanisms by which the environment influences our mental state, and the extent to which it does.

For example, learning and memory are correlated with dynamic and rapid chromatin modifications in brain cells. Mood is, too. Stress-induced depression causes methylation-based silencing of a particular nerve growth factor. Some antidepressants work by reversing this methylation. As another example, rats whose mothers are not very nurturing end up anxious and having a reduced resilience for stress as adults. The difference between these rats and ones who had nurturing maternal care is traceable to chromatin modifications that result in a lower than normal level of another nerve growth factor. Drugs can reverse these chromatin modifications—and their effects. We do not know which human genes are correlated with mental state, but the implication of such research is that future treatments for many disorders will involve deliberate modification of epigenetic marks in one's DNA.

Take-Home Message

Is genotype the only factor that gives rise to phenotype?

» The environment influences gene expression, and therefore can alter phenotype.

» Cell-signaling pathways link environmental cues with epigenetic marks such as methylation and other chromatin modifications.

13.7 Complex Variation in Traits

■ Individuals of most species vary in some of their shared traits. Many traits show a continuous range of variation.

■ Link to Genetically identical organisms 8.7

You know by now that individuals of a species typically vary in many of their shared traits. The pea plant phenotypes that Mendel studied appeared in two or three forms, which made them easy to track through generations. All are single-gene traits (one gene determines the trait). However, many other traits do not appear in distinct forms with predictable inheritance patterns. Such traits are often the result of complex interactions between several genes, with environmental influences on top of those interactions (we return to this topic in Chapter 17, as we consider some evolutionary consequences of variation in phenotype). Tracking traits with complex variation presents a special challenge, which is why the genetic basis of many of them has not yet been completely unraveled.

Continuous Variation

Some traits occur in a range of small differences that is called **continuous variation**. Continuous variation can be an outcome of polygenic inheritance (epistasis), in which multiple genes affect a single trait. The more genes and environmental factors that influence a trait, the more continuous is its variation.

How do we determine whether a trait varies continuously? First, we divide the total range of phenotypes into measurable categories, such as inches of height (**Figure 13.16A,B**). Next, we count how many individuals of a group fall into each category; this count gives the relative frequencies of phenotypes across our range of measurable values. Finally, we plot the data as a bar chart (**Figure 13.16C**). A graph line around the top of the bars shows the distribution of values for the trait. If the line is a bell-shaped curve, or **bell curve**, the trait varies continuously.

59 60 61 62 63 64 65 66 67 68 69 70 71

A Height (in inches) of female biology students

Figure 13.16 Example of continuous variation. Biology students at the University of Florida were divided into categories of one-inch increments in height and counted.

63 64 65 66 67 68 69 70 71 72 73 74 75 76 77

B Height (in inches) of male biology students

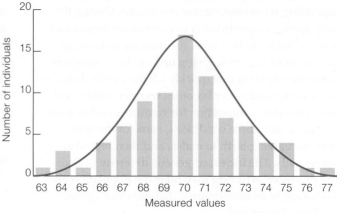

C Graphing the resulting data produces a bell-shaped curve, an indication that height varies continuously. This graph represents the data collected from male biology students, shown in (**B**).

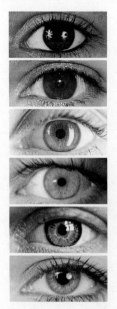

Human skin color varies continuously (a topic we return to in Chapter 14), as does human eye color. The colored part of the eye is the iris, a doughnut-shaped, pigmented structure. Iris color, like skin color, is the result of epistasis among gene products that make and distribute melanins. The more melanin deposited in the iris, the less light is reflected from it. Dark irises have dense melanin deposits that absorb almost all light, and reflect almost none. Melanin deposits are not as extensive in brown eyes, which reflect some light. Green and blue eyes have the least amount of melanin, so they reflect the most light.

Regarding the Unexpected Phenotype

Most organic molecules are made in metabolic pathways involving many enzymes. Genes encoding those enzymes can mutate in any number of ways, so their products may function within a spectrum of activity that ranges from excessive to not at all. Genes encoding products that regulate expression mutate as well, so the same gene product can be expressed at different levels in different individuals. Thus, the end product of a pathway can be produced within a range of concentration and activity. Such variations are associated

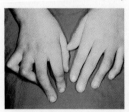

with unpredictable traits such as camptodactyly, in which finger shape and movement are abnormal. Any or all fingers on either or both hands may be affected, to varying degrees (*left*). Camptodactyly is heritable, but its manifestation can differ even within the same family tree.

bell curve Bell-shaped curve; typically results from graphing frequency versus distribution for a trait that varies continuously.
continuous variation Range of small differences in a shared trait.

Take-Home Message

Do all traits occur in distinct forms?

» The more genes and other factors that influence a trait, the more continuous is its range of variation.

» Unpredictable phenotypes arise from genes with a range of expression among individuals.

Menacing Mucus (revisited)

The allele most commonly associated with cystic fibrosis, ΔF508, is eventually lethal in homozygous individuals, but not in those who are heterozygous. This allele is codominant with the normal one, so both copies of the gene are expressed in heterozygous individuals. Such individuals make enough of the normal CFTR protein to have normal chloride ion transport.

The ΔF508 allele is at least 50,000 years old and very common: 1 in 25 people carry it in some populations. Why has this allele persisted for so long and at such high frequency if it is dangerous? The ΔF508 allele may be the lesser of two evils because it offers heterozygous individuals a survival advantage against certain deadly infectious diseases. The unmutated CFTR protein triggers endocytosis when it binds to bacteria. This process is an essential part of the body's immune response to bacteria in the respiratory tract.

However, the same function of CFTR allows bacteria to enter cells of the gastrointestinal tract, where they can be deadly. For example, endocytosis of *Salmonella typhi* (shown at *left*) into epithelial cells lining the gut results in a dangerous infection called typhoid fever. The ΔF508 mutation alters the CFTR protein so that bacteria can no longer be taken up by intestinal cells. People that carry it may have a decreased susceptibility to typhoid fever and other bacterial diseases that begin in the intestinal tract.

How would you vote? Tests for predisposition to genetic disorders are now available. Do you support legislation preventing discrimination based on the results of such tests?

LEARNING ROADMAP

Where you have been This chapter revisits dominance (Sections 13.2, 13.5, and 13.7), gene expression (9.2, 9.3), mutations (9.6), meiosis (12.3), gametes (12.5), and chromosome replication and repair (8.2, 8.5, 8.6). Sampling error (1.8), proteins (3.6), cell components (4.6, 4.8, 4.10, 4.11), metabolism (5.5), membrane receptors (5.7), pigments (6.2), gene control (10.4, 10.6), telomeres (11.5), and oncogenes (11.6) also turn up again.

Where you are now

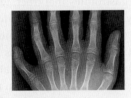

Tracking Traits in Humans
Inheritance patterns in humans are revealed by traits that crop up in family trees. The traits are typically genetic abnormalities or syndromes associated with a genetic disorder.

Autosomal Inheritance
Traits associated with dominant alleles on human autosomes appear in every generation. Traits associated with recessive alleles on human autosomes can skip generations.

Sex-Linked Inheritance
Traits associated with alleles on the X chromosome tend to affect more men than women. Men cannot pass such alleles to a son; carrier mothers bridge affected generations.

Changes in Chromosome Structure and Number
In humans, large-scale changes in the structure or number of autosomes or sex chromosomes usually result in a genetic disorder.

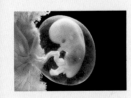

Genetic Testing
Genetic testing provides information about the risk of passing a harmful allele to offspring. Prenatal testing can reveal a genetic abnormality or disorder in a developing fetus.

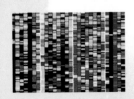

Where you are going Individual genetic disorders such as muscular dystrophy are discussed in later chapters, in the context of the systems that they affect. Chapter 15 returns to the topic of human chromosomes as part of genomics and genetic engineering. Chapter 17 explores evolutionary adaptations and factors that influence the frequency of alleles in a population. The cells and other structural components of human skin are covered in detail in Section 31.8. Chapters 41 and 42 return to processes of human reproduction and development.

14.1 Shades of Skin

The color of human skin begins with skin cell organelles called melanosomes. Melanosomes make two types of melanin: one brownish-black; the other, reddish. Most people have about the same number of melanosomes in their skin cells. Variations in skin color occur because the kinds and amounts of melanins vary among people, as does the formation, transport, and distribution of the melanosomes in the skin.

Variations in skin color may have evolved as a balance between vitamin production and protection against harmful UV radiation. Dark skin would have been beneficial under the intense sunlight of the African savannas where humans first evolved. Melanin acts as a natural sunscreen because it prevents UV radiation in sunlight from breaking down folate, a vitamin essential for normal sperm formation and embryonic development. Children born to light-skinned women exposed to high levels of sunlight have a heightened risk of birth defects.

Early human groups that migrated to regions with colder climates were exposed to less sunlight. In these regions, lighter skin color would have been beneficial. Why? UV radiation stimulates skin cells to make a molecule the body converts to vitamin D. Where sunlight exposure is minimal, UV radiation damage is less of a risk than vitamin D deficiency, which has serious health consequences for developing fetuses and children. People with dark, UV-shielding skin have a high risk of this deficiency in regions where little sunlight reaches Earth's surface.

Skin color, like most other human traits, has a genetic basis. More than 100 gene products are involved in the synthesis of melanin, and the formation and deposition of melanosomes. Mutations in at least some of these genes may have contributed to regional variations in human skin color. Consider a gene on chromosome 15, *SLC24A5*, that encodes a transport protein in melanosome membranes. Nearly all people of African, Native American, or east Asian descent carry the same allele of this gene. Between 6,000 and 10,000 years ago, a mutation gave rise to a different allele. The mutation, a single base-pair substitution, changed the 111th amino acid of the transport protein from alanine to threonine. The change results in less melanin—and lighter skin color—than the original African allele does. Today, nearly all people of European descent carry this mutated allele.

A person of mixed ethnicity may make gametes that contain different combinations of alleles for dark and light skin. It is fairly rare that one of those gametes contains all of the alleles for dark skin, or all of the alleles for light skin, but it happens (**Figure 14.1**).

Skin color is only one of many human traits that vary as a result of single nucleotide mutations. The small scale of such changes offers a reminder that all of us share the genetic legacy of common ancestry.

Figure 14.1 Variation in human skin color (*left*) begins with differences in alleles inherited from parents. *Above*, fraternal twin girls Kian and Remee, born in 2006. Both of the children's grandmothers are of European descent, and have pale skin. Both of their grandfathers are of African descent, and have dark skin. The twins inherited different alleles of some of the genes that affect skin color from their mixed-race parents, who, given the appearance of their children, must be heterozygous for those alleles.

14.2 Human Chromosomes

- Geneticists study inheritance patterns in humans by tracking genetic disorders and abnormalities through families.
- Charting genetic connections with pedigrees reveals inheritance patterns of certain traits.
- Links to Sampling error 1.8, Chromosomes 8.2, Dominance 13.2, Complex inheritance patterns 13.7

Some organisms, including pea plants and fruit flies, are ideal for genetic analysis. They have relatively few chromosomes, they reproduce quickly under controlled conditions, and breeding them poses few ethical problems. It does not take long to track a trait through many generations. Humans, however, are a different story. Unlike flies grown in laboratories, we humans live under variable conditions, in different places, and we live as long as the geneticists who study us. Most of us select our own mates and reproduce if and when we want to. Our families tend to be on the small side, so sampling error (Section 1.8) is a major factor in human genetics studies. Thus, geneticists often use historical records to track genetic disorders and abnormalities that run in families. These researchers make and study standardized charts of genetic connections called **pedigrees** (Figure 14.2). Pedigree analyses can reveal whether a trait is associated with a dominant or recessive allele, and whether the allele is on an autosome or a sex chromosome. Pedigree analysis also allows geneticists to determine the probability that a trait will recur in future generations of a family or a population.

Types of Genetic Variation

Some easily observed human traits follow Mendelian inheritance patterns. Like the flower color of pea plants, these traits are controlled by a single gene with two alleles, one dominant and the other recessive. For example, some people have earlobes that attach at their base, and others have earlobes that dangle free. The allele for unattached earlobes is dominant; the allele for attached earlobes is recessive. Similarly, an allele that specifies a cleft chin is dominant over the allele for a smooth chin, and the allele for dimples is dominant over that for no dimples. Someone who is homozygous for two recessive alleles of the *MC1R* gene makes the reddish kind of melanin but not the brownish-black kind, so this person has red hair.

Single genes on autosomes or sex chromosomes also govern more than 6,000 genetic abnormalities

pedigree Chart of family connections that shows the appearance of a trait through generations.

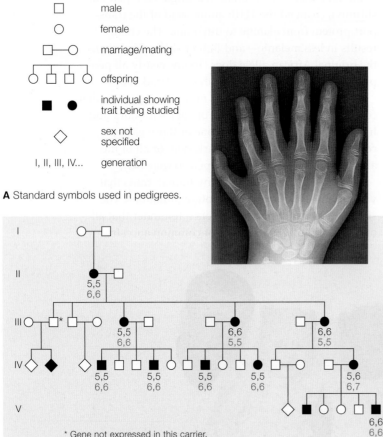

A Standard symbols used in pedigrees.

□	male
○	female
□—○	marriage/mating
○□□○	offspring
■ ●	individual showing trait being studied
◇	sex not specified
I, II, III, IV...	generation

* Gene not expressed in this carrier.

B A pedigree for polydactyly, which is characterized by extra fingers, toes, or both. The *black* numbers signify the number of fingers on each hand; the *red* numbers signify the number of toes on each foot. Though it occurs on its own, polydactyly is also one of several symptoms of Ellis–van Creveld syndrome.

C *Right*, pedigree for Huntington's disease, a progressive degeneration of the nervous system. Researcher Nancy Wexler and her team constructed this extended family tree for nearly 10,000 Venezuelans. Their analysis of unaffected and affected individuals revealed that a dominant allele on human chromosome 4 is the culprit. Wexler has a special interest in the disorder: It runs in her family.

Figure 14.2 Animated Pedigrees.

and disorders. **Table 14.1** lists a few examples. A genetic abnormality is a rare or uncommon version of a trait, such as having six fingers on a hand or having a web between two toes. Genetic abnormalities are not inherently life-threatening, and how you view them is a matter of opinion. By contrast, a genetic disorder sooner or later causes medical problems that may be severe. A genetic disorder is often characterized by a specific set of symptoms (a syndrome). In general, much more research focuses on genetic disorders than on other human traits, because what we learn helps us develop treatments for affected people.

The next two sections of this chapter focus on inheritance patterns of human single-gene disorders, which affect about 1 in 200 people. Keep in mind that these inheritance patterns are the least common kind. Most human traits, including skin color, are polygenic (influenced by multiple genes) and some have epigenetic contributions or causes. Many genetic disorders are like this, including diabetes, asthma, obesity, cancers, heart disease, and multiple sclerosis. The inheritance patterns of these disorders are complex, and despite intense research our understanding of the genetics behind them remains incomplete. For example, mutations on almost every chromosome have been found in people with autism, a developmental disorder, but not all people who carry these mutations have autism. Appendix VI shows a map of human chromosomes with the locations of some alleles known to play a role in genetic disorders and other human traits.

Alleles that give rise to severe genetic disorders are generally rare in populations, because they compromise the health and reproductive ability of their bearers. Why do they persist? Mutations periodically reintroduce them. In some cases, a codominant allele offers a survival advantage in a particular environment. You learned about one example, the $\Delta F508$ allele that causes cystic fibrosis, in Chapter 13: People heterozygous for this codominant allele are protected from infection by bacteria that cause typhoid fever. You will see additional examples in later chapters.

Take-Home Message

How do we study inheritance patterns in humans?

» Human inheritance patterns are often studied by tracking genetic abnormalities or disorders through family trees.

» A genetic abnormality is a rare version of an inherited trait. A genetic disorder is an inherited condition that causes medical problems.

» Some human genetic traits are governed by single genes and are inherited in a Mendelian fashion. Many others are influenced by multiple genes and epigenetics.

Table 14.1 Patterns of Inheritance for Some Genetic Abnormalities and Disorders

Disorder or Abnormality	Main Symptoms
Autosomal dominant inheritance pattern	
Achondroplasia	One form of dwarfism
Aniridia	Defects of the eyes
Camptodactyly	Rigid, bent fingers
Familial hypercholesterolemia	High cholesterol level; clogged arteries
Huntington's disease	Degeneration of the nervous system
Marfan syndrome	Abnormal or missing connective tissue
Polydactyly	Extra fingers, toes, or both
Progeria	Drastic premature aging
Neurofibromatosis	Tumors of nervous system, skin
Autosomal recessive inheritance pattern	
Albinism	Absence of pigmentation
Hereditary methemoglobinemia	Blue skin coloration
Cystic fibrosis	Abnormal glandular secretions leading to tissue and organ damage
Ellis–van Creveld syndrome	Dwarfism, heart defects, polydactyly
Fanconi anemia	Physical abnormalities, marrow failure
Friedreich's ataxia	Progressive loss of motor and sensory function
Galactosemia	Brain, liver, eye damage
Hereditary hemochromatosis	Iron overload damages joints, organs
Phenylketonuria (PKU)	Mental impairment
Sickle-cell anemia	Anemia, adverse pleiotropic effects
Tay–Sachs disease	Deterioration of mental and physical abilities; early death
X-linked recessive inheritance pattern	
Androgen insensitivity syndrome	XY individual but having some female traits; sterility
Red–green color blindness	Inability to distinguish red from green
Hemophilia	Impaired blood clotting ability
Muscular dystrophies	Progressive loss of muscle function
X-linked anhidrotic dysplasia	Mosaic skin (patches with or without sweat glands); other ill effects
X-linked dominant inheritance pattern	
Fragile X syndrome	Intellectual, emotional disability
Incontinentia pigmenti	Abnormalities of skin, hair, teeth, nails, eyes; neurological problems
Changes in chromosome number	
Down syndrome	Mental impairment; heart defects
Turner syndrome (XO)	Sterility; abnormal ovaries, sexual traits
Klinefelter syndrome	Sterility; mild mental impairment
XXX syndrome	Minimal abnormalities
XYY condition	Mild mental impairment or no effect
Changes in chromosome structure	
Chronic myelogenous leukemia (CML)	Overproduction of white blood cells; organ malfunctions
Cri-du-chat syndrome	Mental impairment; abnormal larynx

14.3 Examples of Autosomal Inheritance Patterns

■ An allele is inherited in an autosomal dominant pattern if the trait it specifies appears in heterozygous people.

■ An allele is inherited in an autosomal recessive pattern if the trait it specifies appears only in homozygous people.

■ Links to Protein structure 3.6, Nuclear envelope 4.6, Lysosomes in Tay–Sachs disease 4.8, Cytoskeletal elements 4.10, Cell membrane receptors 5.7, Autosomes 8.2, DNA replication 8.5, DNA repair 8.6, Gene expression 9.2, RNA processing 9.3, Mutations 9.6, Growth factors 11.6, Inheritance 13.2, Codominance and pleiotropy 13.5

The Autosomal Dominant Pattern

A dominant allele on an autosome is expressed in people who are heterozygous for it as well as those who are homozygous. Traits governed by such alleles tend to appear in every generation, and they affect both sexes equally. When one parent is heterozygous, and the other is homozygous for the recessive allele, each of their children has a 50 percent chance of inheriting the dominant allele and displaying the trait associated with it (Figure 14.3A).

Achondroplasia A form of hereditary dwarfism, achondroplasia, offers an example of an autosomal dominant disorder (a disorder caused by a dominant allele on an autosome). Mutations associated with achondroplasia

occur in a gene for a growth hormone receptor. By an unknown mechanism, the mutations cause the receptor, a negative regulator of bone development, to be overly active. About 1 out of 10,000 people are heterozygous for one of these mutations. As adults, these people are, on average, about four feet, four inches (1.3 meters) tall, and have abnormally short arms and legs relative to other body parts (Figure 14.3B).

An allele that causes achondroplasia can be passed to children because its expression does not interfere with reproduction, at least in heterozygous people. The homozygous condition results in severe skeletal malformations that cause early death.

Huntington's Disease Another autosomal dominant allele causes Huntington's disease, in which involuntary muscle movements increase as the nervous system slowly deteriorates. Mutations associated with this disorder alter a gene for a cytoplasmic protein whose function is still unknown. The mutations are insertions in which the same three nucleotides become repeated many, many times in the gene's sequence. The mutated gene encodes an oversized protein product that is chopped into pieces inside nerve cells of the brain. The pieces clump together, and large aggregates that accumulate in the cytoplasm eventually prevent the cells from functioning properly. Brain

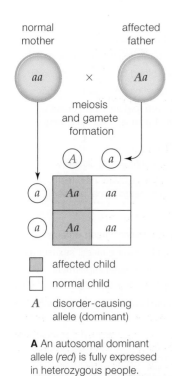

A An autosomal dominant allele (*red*) is fully expressed in heterozygous people.

B Achondroplasia affects Ivy Broadhead (*left*), as well as her brother, father, and grandfather.

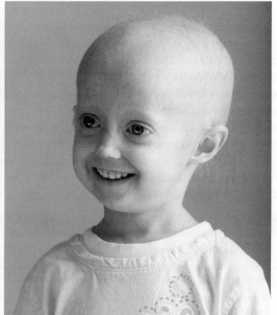

C Symptoms of Hutchinson–Gilford progeria are already evident in Megan Nighbor at age 5.

Figure 14.3 Animated Autosomal dominant inheritance.

cells involved in movement, thinking, and emotion are particularly affected. In the most common form of Huntington's, symptoms do not start until after the age of thirty, and affected people die during their forties or fifties. With this and other late-onset disorders, people tend to reproduce before symptoms appear, so the allele is often passed unknowingly to children.

Hutchinson–Gilford Progeria Hutchinson–Gilford progeria is an autosomal dominant disorder characterized by drastically accelerated aging. It is usually caused by a mutation in the gene for lamin A, a protein subunit of intermediate filaments that support the nuclear membrane. Lamins also have roles inside the nucleus, for example in mitosis, DNA synthesis and repair, and transcriptional regulation. The mutation is a base-pair substitution that adds a signal for a splice site. The resulting lamin A protein is too short and cannot be processed correctly after translation. Pleiotropic effects of the allele include a grossly abnormal architecture of the nuclear envelope. Nuclear pore complexes do not assemble properly, and membrane proteins localize to the wrong side of the envelope. The function of the nucleus as protector of chromosomes and gateway for transcription is severely impaired.

DNA damage accumulates quickly in the cells of affected people, and outward symptoms of the disorder begin to appear before age two. Skin that should be plump and resilient starts to thin, muscles weaken, and bones that should lengthen and grow stronger soften. Premature baldness is inevitable (**Figure 14.3C**). Most people with the disorder die in their early teens as a result of a stroke or heart attack brought on by hardened arteries, a condition typical of advanced age. Progeria does not run in families because affected people do not usually live long enough to reproduce.

The Autosomal Recessive Pattern

A recessive allele on an autosome is expressed only in homozygous people, so traits associated with the allele tend to skip generations. Both males and females are equally affected. People heterozygous for the allele are carriers, which means that they have the allele but not the trait. Any child of two carriers has a 25 percent chance of inheriting the allele from both parents (**Figure 14.4A**). Being homozygous for the allele, such children would have the trait.

Albinism Albinism, a genetic abnormality characterized by an abnormally low level of melanin, is inherited in an autosomal recessive pattern. Mutations

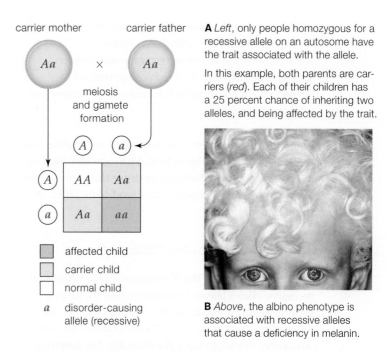

A *Left*, only people homozygous for a recessive allele on an autosome have the trait associated with the allele.

In this example, both parents are carriers (*red*). Each of their children has a 25 percent chance of inheriting two alleles, and being affected by the trait.

B *Above*, the albino phenotype is associated with recessive alleles that cause a deficiency in melanin.

Figure 14.4 Animated Autosomal recessive inheritance.

associated with the albino phenotype occur in genes involved in the production of melanin. Depending on which gene is mutated, skin, hair, or eye pigmentation may be reduced or missing. The most dramatic form of the phenotype is caused by mutations that disable the enzyme tyrosinase. The skin is typically very white and does not tan, and the hair is white. The irises lack pigment, so they appear red due to the visibility of the underlying blood vessels (**Figure 14.4B**). Melanin in the retina plays a role in vision, so people with this phenotype tend to have reduced visual acuity and other problems with vision.

Tay–Sachs Disease Tay–Sachs disease is another example of an autosomal recessive disorder. Mutations in the gene for an enzyme of lysosomes are responsible for the disease (Section 4.8). In the general population, about 1 in 300 people is a carrier for a Tay–Sachs allele, but the incidence is ten times higher in some groups, such as Jews of eastern European descent.

Take-Home Message

How do we know when a trait is associated with an allele on an autosome?

» Persons heterozygous for an allele inherited in an autosomal dominant pattern have the associated trait. The trait tends to appear in every generation.

» With an autosomal recessive inheritance pattern, only persons who are homozygous for an allele have the associated trait. The trait tends to skip generations.

14.4 Examples of X-Linked Inheritance Patterns

■ Traits associated with recessive alleles on the X chromosome appear more frequently in men than in women.

■ A man cannot pass an X chromosome allele to a son.

■ Links to Cytoskeleton 4.10, Extracellular matrix 4.11, Pigments 6.2, X chromosome inactivation 10.4, Homozygous and heterozygous 13.2

Many genetic disorders are associated with alleles on the X chromosome (**Figure 14.5**). Most are inherited in a recessive pattern, probably because those caused by dominant X chromosome alleles tend to be lethal in male embryos.

The X-Linked Recessive Pattern

A recessive allele on the X chromosome (an X-linked recessive allele) leaves two clues when it causes a genetic disorder. First, the disorder appears in males more often than in females. This is because all males who carry the allele have the disorder, but heterozygous females do not (**Figure 14.6A**). Due to random X chromosome inactivation, only about half of a heterozygous female's cells express the recessive allele. The other half express the dominant, normal allele that she carries on her other X chromosome, and this expression can mask the effects of the recessive allele. Males,

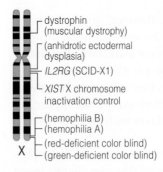

Figure 14.5 The human X chromosome. This chromosome carries about 2,000 genes—almost 10 percent of the total. Most X chromosome alleles that cause genetic disorders are inherited in a recessive pattern. A few disorders are listed (in parentheses).

dystrophin (muscular dystrophy)
(anhidrotic ectodermal dysplasia)
IL2RG (SCID-X1)
XIST X chromosome inactivation control
(hemophilia B)
(hemophilia A)
(red-deficient color blind)
(green-deficient color blind)

having only one X chromosome, do not carry a normal allele that can mask the effects of a recessive allele. Second, an affected father does not pass an X-linked recessive allele to a son, because all children who inherit their father's X chromosome are female. Thus, a heterozygous female is always the bridge between an affected male and his affected grandson.

Red–Green Color Blindness The pattern of X-linked recessive inheritance shows up among individuals who have some degree of color blindness (**Figure 14.6B,C**). The term refers to a range of conditions in which an individual cannot distinguish among some or all colors in the spectrum of visible light. Color vision depends on the proper function of pigment-containing receptors in the eyes. Most of the genes

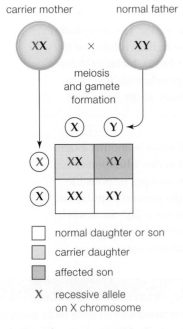

carrier mother × normal father

meiosis and gamete formation

normal daughter or son
carrier daughter
affected son
X recessive allele on X chromosome

A In this example of X-linked inheritance, the mother carries a recessive allele on one of her two X chromosomes (*red*).

B A view of color blindness. The image on the *left* shows how a person with red–green color blindness sees the image on the *right*. The perception of blues and yellows is normal; red and green appear similar.

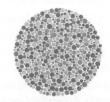

You may have one form of red–green color blindness if you see a 7 in this circle instead of a 29.

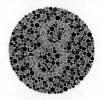

You may have another form of red–green color blindness if you see a 3 instead of an 8 in this circle.

C Part of a standardized test for color blindness. A set of 38 of these circles is commonly used to diagnose deficiencies in color perception.

Figure 14.6 Animated X-linked recessive inheritance.

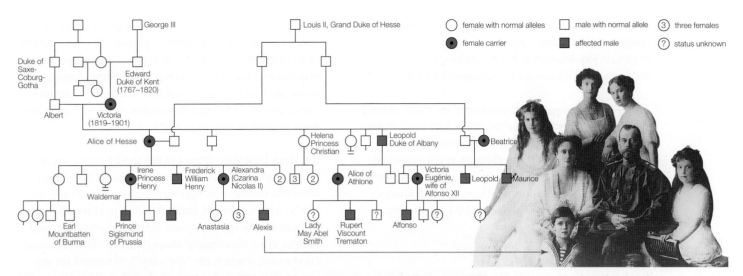

Figure 14.7 A classic case of X-linked recessive inheritance: a partial pedigree of the descendants of Queen Victoria of England. At one time, the recessive X-linked allele that resulted in hemophilia was present in eighteen of Victoria's sixty-nine descendants, who sometimes intermarried. Of the Russian royal family members shown, the mother (Alexandra Czarina Nicolas II) was a carrier.

Figure It Out: How many of Alexis's siblings were affected by hemophilia A? Answer: None

involved in color vision are on the X chromosome, and mutations in those genes often result in altered or missing receptors. Normally, humans can sense the differences among 150 colors. People who have red–green color blindness see fewer than 25 colors because receptors that respond to red and green wavelengths are weakened or absent. Some confuse red and green; others see green as gray.

Duchenne Muscular Dystrophy Duchenne muscular dystrophy (DMD) is an X-linked recessive disorder characterized by muscle degeneration. It is caused by mutations in the X chromosome gene for dystrophin, a cystoskeletal protein that links actin microfilaments in cytoplasm to a complex of proteins in the plasma membrane. This complex structurally and functionally links the cell to extracellular matrix. When dystrophin is absent, the entire complex is unstable. Muscle cells, which are subject to stretching, are particularly affected. Their plasma membrane is easily damaged, and they become flooded with calcium ions. Eventually, the muscle cells die and become replaced by fat cells and connective tissue.

DMD affects about 1 in 3,500 people, almost all of them boys. Symptoms begin between ages three and seven. Anti-inflammatory drugs can slow the progression of DMD, but there is no cure. When an affected boy is about twelve, he will begin to use a wheelchair and his heart muscle will start to fail. Even with the best care, he will probably die before age thirty, from a heart disorder or respiratory failure (suffocation).

Hemophilia A Hemophilia A is an X-linked recessive disorder that interferes with blood clotting. Most of us have a blood clotting mechanism that quickly stops bleeding from minor injuries. That mechanism involves factor VIII, a protein product of a gene on the X chromosome. Bleeding can be prolonged in males who carry a mutation in this gene, or in females who are homozygous for one (heterozygous females make enough factor VIII to have a clotting time that is close to normal). Affected people tend to bruise very easily, but internal bleeding is their most serious problem. Repeated bleeding inside the joints disfigures them and causes chronic arthritis.

In the nineteenth century, the incidence of hemophilia A was relatively high in royal families of Europe and Russia, probably because the common practice of inbreeding kept the allele in their family trees (**Figure 14.7**). Today, about 1 in 7,500 people is affected, but that number may be rising because the disorder is now a treatable one. More affected people are living long enough to transmit the mutated allele to children.

Take-Home Message

How do we know when a trait is associated with an allele on an X chromosome?

» Men who have an X-linked recessive allele have the trait associated with the allele. Heterozygous women do not, because they have a dominant allele on their second X chromosome that can mask the affects of the recessive allele. Thus, the trait appears more often in men.

» Men transmit an X-linked allele to their daughters, but not to their sons.

14.5 Heritable Changes in Chromosome Structure

■ Chromosome structure rarely changes, but when it does, the outcome can be severe or lethal.

■ Links to Protein structure 3.6, Karyotyping 8.2, Deletions 9.6, *SRY* gene 10.4, Telomeres 11.5, Oncogenes 11.6, Meiosis 12.3

Large-scale changes in chromosome structure usually have drastic effects on health; about half of all miscarriages are due to chromosome abnormalities of the developing embryo. These changes occur spontaneously in nature, but can also be induced by exposure to chemicals or radiation. The scale of such changes often allows them to be detected by karyotyping.

Types of Chromosome Changes

Duplication Even normal chromosomes have DNA sequences that are repeated two or more times. These repetitions are called **duplications** (**Figure 14.8A**). Duplications are an outcome of unequal crossing over between homologous chromosomes during prophase I of meiosis. When homologous chromosomes align side by side, their DNA sequences may misalign at some point along their length. In this case, the crossover

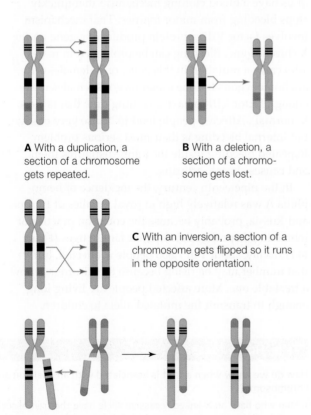

A With a duplication, a section of a chromosome gets repeated.

B With a deletion, a section of a chromosome gets lost.

C With an inversion, a section of a chromosome gets flipped so it runs in the opposite orientation.

D With a translocation, a broken piece of a chromosome gets reattached in the wrong place. This example shows a reciprocal translocation, in which two chromosomes exchange chunks.

Figure 14.8 Large-scale changes in chromosome structure.

deletes a stretch of DNA from one chromosome and splices it into the homologous partner. The probability of misalignment is greater in regions where the same sequence of nucleotides is repeated. Some duplications, such as the mutations that cause Huntington's, cause genetic abnormalities or disorders. Others have been evolutionarily important.

Deletion Deletions (**Figure 14.8B**) tend to have severe consequences on health. Most mutations associated with Duchenne muscular dystrophy are deletions in the X chromosome. A small deletion in chromosome 5 shortens life span, impairs mental functioning, and results in an abnormally shaped larynx. This disorder, cri-du-chat (French for "cat's cry"), is named for the sound that affected infants make when they cry.

Inversion With an **inversion**, part of the sequence of DNA within the chromosome becomes oriented in the reverse direction, with no molecular loss (**Figure 14.8C**). An inversion may not affect a carrier's health if it does not interrupt a gene or gene control region, because the individual's cells still have their full complement of genetic material. However, fertility may be affected because inverted chromosomes mispair during meiosis. Crossovers between mispaired chromosomes can produce large deletions or duplications that reduce the viability of forthcoming embryos. People who carry an inversion may not know about it until they are diagnosed with infertility and their karyotype is checked.

Translocation If a chromosome breaks, the broken part may get attached to a different chromosome, or to a different part of the same one. This structural change is called a **translocation**. Most translocations are reciprocal, or balanced, which means that two nonhomologous chromosomes exchange broken parts (**Figure 14.8D**). A reciprocal translocation between chromosomes 8 and 14 is the usual cause of Burkitt's lymphoma, an aggressive cancer of the immune system. This translocation moves a proto-oncogene to a region that is vigorously transcribed in immune cells, with disastrous results. Many other reciprocal translocations have no adverse effects on health, but, like inversions, they can affect fertility. During meiosis, translocated chromosomes pair abnormally and segregate improperly; about half of the resulting gametes carry major duplications or deletions. If one of these gametes unites with a normal gamete at fertilization, the resulting embryo almost always dies. As with inversions, people who carry a translocation may not know about it until they have difficulty with fertility.

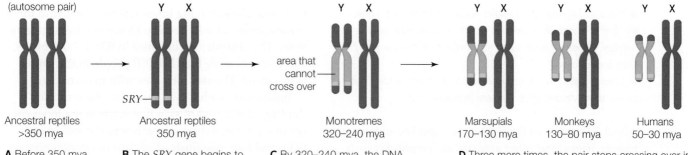

A Before 350 mya, sex was determined by temperature, not by chromosome differences.

B The *SRY* gene begins to evolve 350 mya. The DNA sequences of the chromosomes diverge as other mutations accumulate.

C By 320–240 mya, the DNA sequences of the chromosomes are so different that the pair can no longer cross over in one region. The Y chromosome begins to get shorter.

D Three more times, the pair stops crossing over in yet another region. Each time, the DNA sequences of the chromosomes diverge, and the Y chromosome shortens. Today, the pair crosses over only at a small region near the ends.

Figure 14.9 Evolution of the Y chromosome. Today, the *SRY* gene determines male sex. Homologous regions of the chromosomes are shown in *pink*; mya, million years ago. Monotremes are egg-laying mammals; marsupials are pouched mammals.

Chromosome Changes in Evolution

As you can see, large-scale alterations in chromosome structure may reduce an individual's fertility. Individuals who are heterozygous for such changes may not be able to produce offspring at all. However, individuals homozygous for an inversion sometimes become the founders of new species. It may seem as if this outcome would be exceedingly rare, but it is not. Speciation can and does occur by large-scale changes in chromosomes. Karyotyping and DNA sequence comparisons show that the chromosomes of all species contain evidence of major structural alterations. For example, duplications have often allowed a copy of a gene to mutate while the original carried out its unaltered function. The multiple and strikingly similar globin chain genes of humans and other primates apparently evolved by this process. Four globin chains associate in each hemoglobin molecule (Section 9.6). Different alleles specify different versions of the chains. Which versions of the chains get assembled into a hemoglobin molecule determines the oxygen-binding characteristics of the resulting protein.

As another example, X and Y chromosomes were once homologous autosomes in reptilelike ancestors of mammals (**Figure 14.9**). Ambient temperature probably determined the gender of those organisms, as it still does in turtles and some other modern reptiles. About 350 million years ago, a gene on one of the two homologous chromosomes mutated. The change, which was the beginning of the male sex determination gene *SRY*, interfered with crossing over during meiosis. A reduced frequency of crossing over allowed the chromosomes to diverge around the changed region. Mutations began to accumulate separately in the two chromosomes. Over evolutionary time, the chromosomes became so different that they no longer crossed over at all in the changed region, so they diverged even more. Today, the Y chromosome is much smaller than the X, and only retains about 5 percent homology with it. The Y crosses over mainly with itself—by translocating duplicated regions of its own DNA.

Some chromosome structure changes contributed to differences among closely related organisms, such as apes and humans. Human somatic cells have twenty-three pairs of chromosomes, but those of chimpanzees, gorillas, and orangutans have twenty-four. Thirteen human chromosomes are almost identical with chimpanzee chromosomes. Nine more are similar, except for some inversions. One human chromosome matches up with two in chimpanzees and the other great apes (**Figure 14.10**). During human evolution, two chromosomes evidently fused end to end and formed our chromosome 2. How do we know? The region where the fusion occurred contains the remnants of a telomere.

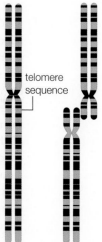

Figure 14.10
Human chromosome 2 compared with chimpanzee chromosomes 2A and 2B.

duplication Repeated section of a chromosome.
inversion Structural rearrangement of a chromosome in which part of it becomes oriented in the reverse direction.
translocation Structural change of a chromosome in which a broken piece gets reattached in the wrong location.

Take-Home Message

Does chromosome structure change?

» A segment of a chromosome may be duplicated, deleted, inverted, or translocated. Such a change is usually harmful or lethal, but may be conserved in the rare circumstance that it has a neutral or beneficial effect.

14.6 Heritable Changes in Chromosome Number

■ Occasionally, abnormal events occur before or during meiosis, and new individuals end up with the wrong chromosome number. Consequences range from minor to lethal changes in form and function.

■ Links to Bias 1.8, Chromosomes 8.2, X chromosome inactivation 10.4, Meiosis 12.3, Gamete formation 12.5

About 70 percent of flowering plant species, and some insects, fishes, and other animals, are **polyploid**, which means that they have three or more complete sets of chromosomes. Cells in some adult human tissues are normally polyploid, but inheriting more than two full sets of chromosomes is invariably fatal in humans.

Less than 1 percent of children are born with a diploid chromosome number that differs from the normal 46. Chromosome number changes often arise through **nondisjunction**, in which chromosomes do not separate properly during nuclear division. Nondisjunction during meiosis (**Figure 14.11**) can affect chromosome number at fertilization. For example, if a normal gamete (n) fuses with one that has an extra chromosome ($n+1$), the resulting zygote will have three copies of one type of chromosome and two of every other type ($2n+1$), a condition called trisomy. If an $n-1$ gamete fuses with a normal n gamete, the zygote will be $2n-1$, a condition called monosomy. Trisomy and monosomy are examples of **aneuploidy**, in which an individual's cells have too many or too few copies of a chromosome.

Autosomal Change and Down Syndrome

Autosomal aneuploidy is usually fatal in humans, but there are exceptions. A few trisomic humans are born alive, but only those with trisomy 21 have a high probability of surviving infancy. A newborn with three chromosomes 21 will develop Down syndrome (**Figure 14.12**). This disorder occurs once in 800 to 1,000 births, and it affects more than 350,000 people in the United States alone. The risk increases with maternal age.

Individuals with Down syndrome have upward-slanting eyes, a fold of skin that starts at the inner corner of each eye, a deep crease across the sole of each palm and foot, one (instead of two) horizontal furrows on their fifth fingers, and other outward symptoms. Not all of these outward symptoms develop in every individual. That said, people with the disorder tend to have moderate to severe mental impairment and heart problems. Their skeleton grows and develops abnormally, so older children have short body parts, loose joints, and misaligned bones of the fingers, toes, and hips. The muscles and reflexes are weak, and motor skills such as speech develop slowly. With medical care, affected individuals live about fifty-five years. Early training can help them learn to care for themselves and to take part in normal activities.

Change in the Sex Chromosome Number

Nondisjunction also causes alterations in the number of X and Y chromosomes, with a frequency of about 1 in 400 live births. Most often, such alterations lead to mild difficulties in learning and impaired motor skills such as a speech delay. These problems may be so subtle that the condition is never diagnosed.

Female Sex Chromosome Abnormalities Individuals with Turner syndrome have an X chromosome and no corresponding X or Y chromosome (XO). The

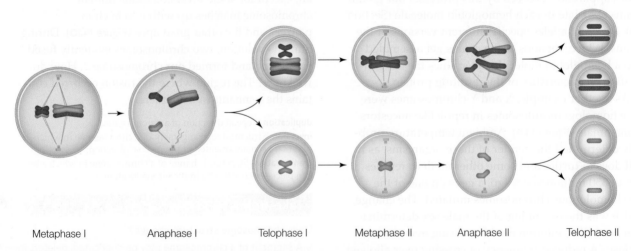

| Metaphase I | Anaphase I | Telophase I | Metaphase II | Anaphase II | Telophase II |

Figure 14.11 Nondisjunction, which can occur during anaphase I or II of meiosis. Of the two pairs of homologous chromosomes shown here, one fails to separate during anaphase I. A zygote that forms from one of the resulting gametes will have an abnormal chromosome number.

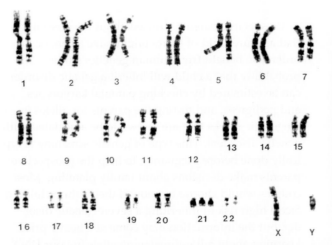

Figure 14.12 Down syndrome, genotype and phenotype. **Figure It Out: Is** the karyotype from an individual who is male or female? Answer: Male (XY)

syndrome is thought to arise most frequently as an outcome of inheriting an unstable Y chromosome from the father. The zygote starts out being genetically male, with an X and a Y chromosome. The Y chromosome breaks up and is lost during early development, so the embryo continues to develop as a female.

There are fewer people affected by Turner syndrome than other chromosome abnormalities: Only about 1 in 2,500 newborn girls has it. The risk does not increase with maternal age. XO individuals grow up well proportioned but short, with an average height of four feet, eight inches (1.4 meters). Their ovaries do not develop properly, so they do not make enough sex hormones to become sexually mature. Development of secondary sexual traits such as breasts is also inhibited.

A female may inherit multiple X chromosomes, a condition called XXX syndrome. XXX syndrome occurs in about 1 of 1,000 births; as with Down syndrome the risk increases with maternal age. Only one X chromosome is typically active in female cells, so having extra X chromosomes usually does not cause physical or medical problems, but the syndrome is associated with mild mental impairment.

Male Sex Chromosome Abnormalities About 1 out of every 500 males has an extra X chromosome (XXY). The resulting disorder, Klinefelter syndrome, develops at puberty. XXY males tend to be overweight, tall, and have mild mental impairment. They make more estrogen and less testosterone than normal males. This hormone imbalance causes affected men to have small testes and prostate glands, low sperm counts, sparse facial and body hair, high-pitched voices, and enlarged breasts. Testosterone injections during puberty can reverse these feminized traits.

About 1 in 1,000 males is born with an extra Y chromosome (XYY), a result of nondisjunction of the Y chromosome during sperm formation. Adults tend to be taller than average and have mild mental impairment, but most are otherwise normal.

XYY men were once thought to be predisposed to a life of crime. This misguided view was based on sampling error (too few cases in narrowly chosen groups such as prison inmates) and bias (the researchers who gathered the karyotypes also took the personal histories of the participants). That view was disproven in 1976, when a geneticist reported results from his study of 4,139 tall males, all twenty-six years old, who had registered for military service. The military had given the men physical examinations and intelligence tests, and had also recorded social and economic status, education, and any criminal convictions. Only twelve of the males studied were XYY, which meant that the "control group" had more than 4,000 males. The only findings? Mentally impaired, tall males who engage in criminal deeds are more likely to get caught, irrespective of karyotype.

aneuploidy A chromosome abnormality in which an individual's cells carry too many or too few copies of a particular chromosome.
nondisjunction Failure of sister chromatids or homologous chromosomes to separate during nuclear division.
polyploid Having three or more of each type of chromosome characteristic of the species.

Take-Home Message

What are the effects of chromosome number changes?

» Nondisjunction can change the number of autosomes or sex chromosomes in gametes. Such changes usually cause genetic disorders in offspring.

» Sex chromosome abnormalities are usually associated with some degree of learning difficulty and motor skill impairment.

14.7 Genetic Screening

■ Our understanding of human inheritance can provide prospective parents with information about the health of their future children.

■ Links to Probability 1.8, Amino acids 3.6, Metabolic pathways 5.5, Epigenetic marks 10.6

Studying human inheritance patterns has given us many insights into how genetic disorders arise and progress, and how to treat them. Surgery, prescription drugs, hormone replacement therapy, and dietary controls can minimize and in some cases eliminate the symptoms of a genetic disorder. Some disorders can be detected early enough to start countermeasures before symptoms develop. For example, most hospitals in the United States now screen newborns for mutations in the gene for phenylalanine hydroxylase, an enzyme that catalyzes the conversion of the amino acid phenylalanine to tyrosine. Without a functional form of this enzyme, the body becomes deficient in tyrosine, and phenylalanine accumulates to high levels. The imbalance inhibits protein synthesis in the brain, which in turn results in the severe neurological symptoms characteristic of phenylketonuria, or PKU. Restricting all intake of phenylalanine can slow the progression of PKU, so routine early screening has resulted in fewer individuals suffering from the symptoms of the disorder.

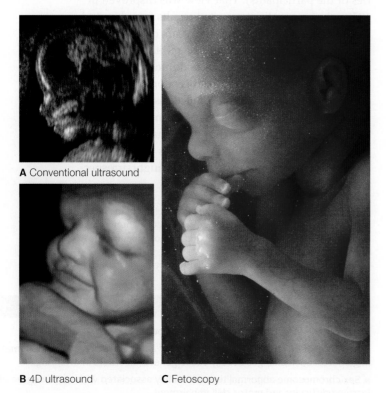

A Conventional ultrasound

B 4D ultrasound   **C** Fetoscopy

Figure 14.13 Three ways of imaging a developing human fetus.

Prospective parents worried about the possibility that a future child of theirs might have a genetic disorder also benefit from human genetics studies. The probability that a child will inherit a genetic disorder can be estimated by checking parental karyotypes and pedigrees, and testing the parents for alleles or epigenetic marks that are known to be associated with genetic disorders. This type of genetic screening is typically done before pregnancy, to help the prospective parents make decisions about family planning. Most couples would choose to know if their future children face a high risk of inheriting a severe genetic disorder, but the information may come at a heavy price. Learning about a life-threatening allele in your DNA can be devastating.

Prenatal Diagnosis

Genetic screening can also be done post-conception, in which case it is called prenatal diagnosis. Prenatal means before birth. The term 'embryo' applies until eight weeks after fertilization, after which 'fetus' is appropriate. Prenatal diagnosis checks for physical and genetic abnormalities in an embryo or fetus. It can often reveal the presence of a genetic disorder in an unborn child. More than 30 conditions, including aneuploidy, hemophilia, Tay–Sachs disease, sickle-cell anemia, muscular dystrophy, and cystic fibrosis, are detectable prenatally. If the disorder is treatable, early detection allows the newborn to receive prompt and appropriate treatment. A few defects are even surgically correctable before birth. Prenatal diagnosis also gives parents time to prepare for the birth of an affected child, and an opportunity to decide whether to continue with the pregnancy or terminate it.

As an example of how prenatal diagnosis works, consider a thirty-five-year-old woman who becomes pregnant. Her doctor will probably use a noninvasive procedure, obstetric sonography, in which ultrasound waves directed across the woman's abdomen form images of the fetus's developing limbs and internal organs (**Figure 14.13A,B**). There is no detectable risk to the pregnancy with ultrasound, and the images may reveal physical defects associated with a genetic disorder, in which case a more invasive technique would be recommended for further diagnosis. Fetoscopy yields images of the fetus that are much higher in resolution than ultrasound images (**Figure 14.14C**). With this procedure, sound waves are pulsed from inside the mother's uterus. Samples of tissue or blood are often be taken at the same time, and some corrective surgeries can be performed as well.

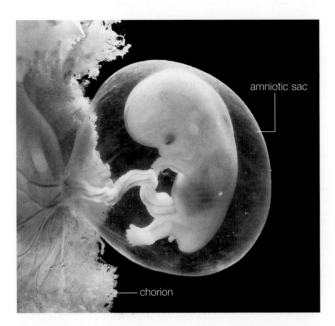

Figure 14.14 An 8-week-old fetus. With amniocentesis, fetal cells shed into the fluid inside the amniotic sac are tested for genetic disorders. Chorionic villus sampling tests cells of the chorion, which is part of the placenta.

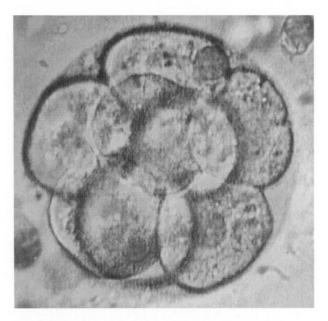

Figure 14.15 Clump of cells formed by three mitotic divisions after *in vitro* fertilization. All eight of the cells are identical and one can be removed for genetic analysis to determine whether the embryo carries any genetic defects.

Human genetics studies show that our thirty-five-year-old woman has about a 1 in 80 chance that her baby will be born with a chromosomal abnormality, a risk more than six times higher than when she was twenty years old. Thus, even if no abnormalities were detected by ultrasound, she probably will be offered a more thorough diagnostic procedure, amniocentesis, in which a small sample of fluid is drawn from the amniotic sac enclosing the fetus (**Figure 14.14**). The fluid contains cells shed by the fetus, and those cells can be tested for genetic disorders. Chorionic villus sampling (CVS) can be done earlier than amniocentesis. With this technique, a few cells from the chorion are removed and tested for genetic disorders. The chorion is a membrane that surrounds the amniotic sac and helps form the placenta, an organ that allows substances to be exchanged between mother and embryo.

An invasive procedure often carries a risk to the fetus. For example, if a punctured amniotic sac does not reseal itself quickly, too much fluid may leak out of it, resulting in miscarriage. The risks vary by the procedure. Amniocentesis has improved so much that, in the hands of a skilled physician, the procedure no longer increases the risk of miscarriage. CVS occasionally disrupts the placenta's development and thus causes underdeveloped or missing fingers and toes in 0.3 percent of newborns. Fetoscopy raises the miscarriage risk by a whopping 2 to 10 percent, so it is rarely performed unless surgery or another medical procedure is required before the baby is born.

Preimplantation Diagnosis

Couples who discover they are at high risk of having a child with a genetic disorder may opt for reproductive interventions such as *in vitro* fertilization. With this procedure, sperm and eggs taken from prospective parents are mixed in a test tube. If an egg becomes fertilized, the resulting zygote will begin to divide. In about forty-eight hours, it will have become an embryo that consists of a ball of eight cells (**Figure 14.15**). All of the cells in this ball have the same genes, but none has yet committed to being specialized one way or another. Doctors can remove one of these undifferentiated cells and analyze its genes, a procedure called preimplantation diagnosis. The withdrawn cell will not be missed. If the embryo has no detectable genetic defects, it is inserted into the woman's uterus to continue developing. Many of the resulting "test-tube babies" are born in good health.

Take-Home Message

How do we use what we know about human inheritance?

» Genetic testing can provide prospective parents with information about the health of their future children.

Shades of Skin (revisited)

Chinese and Europeans do not share any skin pigmentation allele that does not also occur in other populations. However, most people of Chinese descent carry a particular allele of the *DCT* gene, the product of which helps convert tyrosine to melanin. Few people of European or African descent have this allele. Taken together, the distribution of the *SLC24A5* and *DCT* genes suggests that (1) an African population was ancestral to both the Chinese and Europeans, and (2) Chinese and European populations separated before their pigmentation genes mutated and their skin color changed.

How would you vote? Physical attributes with a genetic basis, including skin color, are often used to define "race." Are twins such as Kian and Remee of different races?

LEARNING ROADMAP

Where you have been This chapter builds on your understanding of DNA's structure (Sections 8.3, 8.4,13.2, 14.7) and replication (8.5). Clones (8.1), gene expression (9.2, 9.3), and knockouts (10.3) are important in genetic engineering, particularly in research on human traits (13.7) and genetic disorders (Chapter 14). You will revisit tracers (2.2), triglycerides (3.5), denaturation (3.7), β-carotene (6.2), the lac operon (10.5), and cancer (11.6).

Where you are now

DNA Cloning
Researchers routinely make recombinant DNA by cutting and pasting together DNA from different species. Plasmids and other vectors can carry foreign DNA into host cells.

Finding Needles in Haystacks
Genetic engineering research relies on techniques such as hybridization and PCR for isolating and identifying particular fragments of DNA.

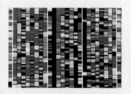

DNA Sequencing
Sequencing reveals the linear order of nucleotides in DNA. Comparing genomes offers insights into human genes and evolution. DNA sequence can be used to identify Individuals.

Genetic Engineering
Genetic engineering, the directed modification of an organism's genes, is a routine part of research and industry. Genetically modified organisms are now quite common.

Gene Therapy
Genetic engineering continues to be tested in medical applications. It also continues to raise ethical questions about modifying the human genome.

Where you are going DNA sequence comparisons help researchers study evolutionary relationships (Section 18.4). The escape of transgenic genes into the environment is an example of gene flow, an evolutionary process described in Section 17.8. Thermophilic bacteria, including *Thermus aquaticus*, are discussed in Section 20.7. Genetically engineered plants are useful in phytoremediation (Section 28.1), and many vaccines (37.12) are genetically engineered. Research using specific genetically engineered animals is explained in relevant chapters.

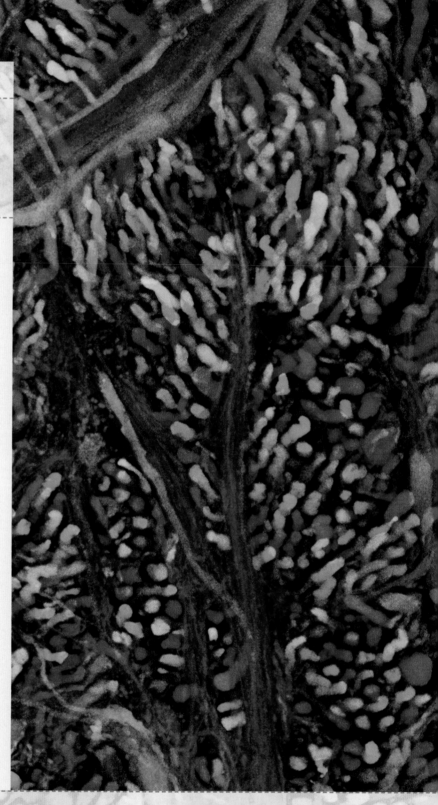

15.1 Personal DNA Testing

About 99 percent of your DNA is exactly the same as everyone else's. If you compared your DNA with your neighbor's, about 2.97 billion nucleotides of the two sequences would be identical; the remaining 30 million or so are sprinkled throughout your chromosomes, mainly as single nucleotide differences. The sprinkling is not entirely random because some regions of DNA vary less than others. Such conserved regions are of particular interest to researchers because they are the ones most likely to have an essential function. When a conserved sequence does vary among people, the variation tends to be in particular nucleotides. A nucleotide difference carried by a measurable percentage of a population, usually above 1 percent, is called a **single-nucleotide polymorphism**, or SNP (pronounced "snip").

Alleles of most genes differ by single nucleotides, and differences in alleles are the basis of the variation in human traits that makes each individual unique (Section 12.2). Thus, SNPs account for many of the differences in the way humans look, and they also have a lot to do with differences in the way our bodies work—how we age, respond to drugs, weather assaults by pathogens and toxins, and so on.

Consider a gene, *APOE*, that specifies apolipoprotein E, a protein component of lipoprotein particles (Section 3.5). One allele of this gene, *ε4*, is carried by about 25 percent of people. Nucleotide 4,874 of this allele is a cytosine instead of the normal thymine, a SNP that results in a single amino acid change in the protein product of the gene. How this change affects the function of apolipoprotein E is unclear, but we do know that having the *ε4* allele increases one's risk of developing Alzheimer's disease later in life, particularly in people homozygous for it.

About 4.5 million SNPs in human DNA have been identified, and that number is growing every day. A few companies are now offering to determine some of the SNPs you carry (**Figure 15.1**). The companies extract your DNA from the cells in a few drops of spit, then analyze it for SNPs.

Personal genetic testing may soon revolutionize medicine by allowing physicians to customize treatments on the basis of an individual's genetic makeup. For example, an allele associated with a heightened risk of a particular medical condition could be identified long before symptoms actually appear. People with that allele could then be encouraged to make lifestyle changes known to delay the onset of the condition. For some conditions, treatment that begins early enough may prevent symptoms from developing at all. Physicians could design treatments to fit the way a condition is likely to progress in the individual, and also to prescribe only those drugs that will work in the person's body.

single-nucleotide polymorphism (SNP) One-base DNA sequence variation carried by a measurable percentage of a population.

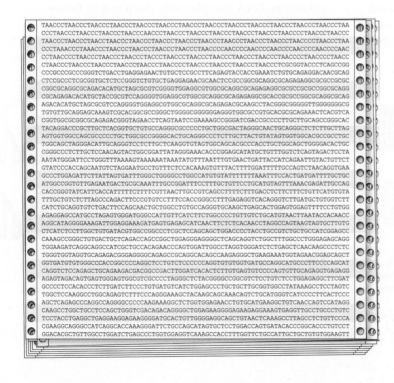

Figure 15.1 Personal genetic testing. *Above*, a SNP-chip. Personal DNA testing companies use chips like this one to analyze their customers' chromosomes for SNPs. This chip, shown actual size, reveals which versions of 1,140,419 SNPs occur in the DNA of four individuals at a time. Only about 1 percent of the 3 billion bases in a person's DNA (*left*) are unique to the individual.

15.2 Cloning DNA

- Researchers cut up DNA from different sources, then paste the resulting fragments together.
- Cloning vectors can carry foreign DNA into host cells.
- Links to Plasmids 4.4, Clones 8.1, Discovery of DNA structure 8.3, Base pairing and directionality of DNA strands 8.4, DNA ligase 8.5, mRNA 9.2, Introns 9.3, The lac operon 10.5

Cutting and Pasting DNA

In the 1950s, excitement over the discovery of DNA's structure gave way to frustration: No one could determine the order of nucleotides in a molecule of DNA. Identifying a single base among thousands or millions of others turned out to be a huge technical challenge. A seemingly unrelated discovery offered a solution. Werner Arber, Hamilton Smith, and their coworkers discovered why some bacteria are resistant to infection by bacteriophage (Section 8.3). Special enzymes inside these bacteria chop up any injected viral DNA before it has a chance to integrate into the bacterial chromosome. The enzymes restrict viral growth; hence their name, **restriction enzymes**. A restriction enzyme cuts DNA wherever a specific nucleotide sequence occurs (**Figure 15.2**). For example, the enzyme *Eco* RI (named after *E. coli*, the bacteria from which it was isolated) cuts DNA at the sequence GAATTC ❶. Other restriction enzymes cut different sequences.

The discovery of restriction enzymes allowed researchers to cut chromosomal DNA into manageable chunks. It also allowed them to combine DNA fragments from different organisms. How? Many restriction enzymes, including *Eco* RI, leave single-stranded tails on DNA fragments ❷. Researchers realized that complementary tails will base-pair, regardless of the source of the DNA ❸. The tails are called "sticky

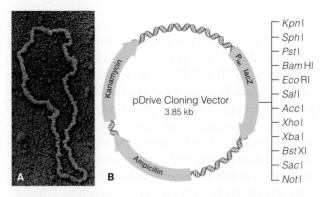

Figure 15.3 Plasmid cloning vectors. (**A**) Micrograph of a plasmid. (**B**) A commercial plasmid cloning vector. Restriction enzyme recognition sequences are indicated on the *right* by the name of the enzyme that cuts them. Foreign DNA can be inserted into the vector at these sites. Bacterial genes (*gold*) help researchers identify host cells that take up a vector with inserted DNA. This vector carries two antibiotic resistance genes and the lac operon.

ends," because two DNA fragments stick together when their matching tails base-pair. The enzyme DNA ligase (Section 8.5) can be used to seal the gaps between base-paired sticky ends, so continuous DNA strands form ❹.

Thus, using appropriate restriction enzymes and DNA ligase, researchers can cut and paste DNA from different sources. The result, a hybrid molecule of DNA from two or more organisms, is called **recombinant DNA**. Making recombinant DNA is the first step in **DNA cloning**, a set of laboratory methods that uses living cells to mass-produce specific DNA fragments. For example, researchers often insert fragments of eukaryotic DNA into plasmids (Section 4.4). Before a bacterium divides, it copies any plasmids it carries along with its chromosome, so both descendant cells get one of each. If a plasmid carries a fragment of for-

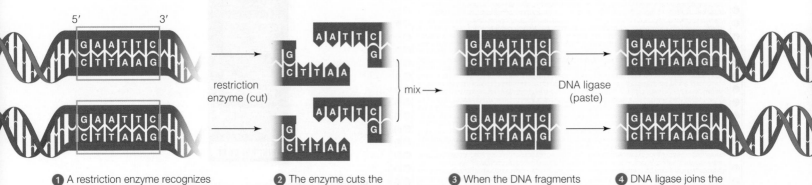

❶ A restriction enzyme recognizes a specific base sequence (*orange* boxes) in DNA from two sources.

❷ The enzyme cuts the DNA into fragments. This enzyme leaves sticky ends.

❸ When the DNA fragments from the two sources are mixed together, matching sticky ends base-pair with each other.

❹ DNA ligase joins the base-paired DNA fragments. Molecules of recombinant DNA are the result.

Figure 15.2 Animated Using restriction enzymes to make recombinant DNA.
Figure It Out: Why did the enzyme cut both strands of DNA?
Answer: Because the recognition sequence occurs on both strands.

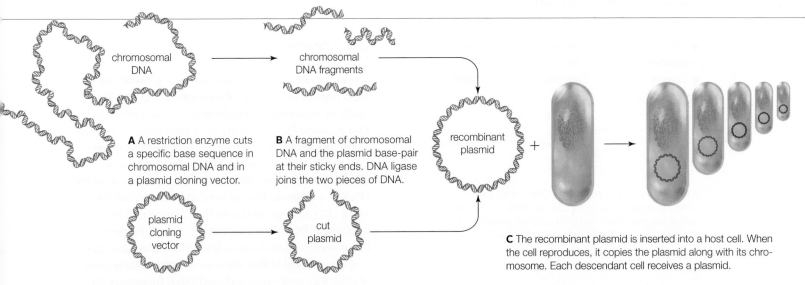

A A restriction enzyme cuts a specific base sequence in chromosomal DNA and in a plasmid cloning vector.

B A fragment of chromosomal DNA and the plasmid base-pair at their sticky ends. DNA ligase joins the two pieces of DNA.

chromosomal DNA

chromosomal DNA fragments

recombinant plasmid

plasmid cloning vector

cut plasmid

C The recombinant plasmid is inserted into a host cell. When the cell reproduces, it copies the plasmid along with its chromosome. Each descendant cell receives a plasmid.

Figure 15.4 Animated An example of cloning. Here, a fragment of chromosomal DNA is inserted into a plasmid.

eign DNA, that fragment gets copied and distributed to descendant cells along with the plasmid. Thus, plasmids can be used as **cloning vectors**, which are molecules that carry foreign DNA into host cells (**Figure 15.3**). A host cell into which a cloning vector has been inserted can be grown in the laboratory (cultured) to yield a huge population of genetically identical cells, or clones (Section 8.1). Each clone contains a copy of the vector (**Figure 15.4**). A DNA fragment carried by the vector can be collected in large quantities by harvesting it from the many clones.

cDNA Cloning

Remember from Section 9.3 that eukaryotic DNA contains introns. Unless you are a eukaryotic cell, it is not very easy to determine which parts of eukaryotic DNA encode gene products. Researchers who study gene expression in eukaryotes often start with mature mRNA, because intron sequences are excised during post-transcriptional processing.

An mRNA cannot be cut with restriction enzymes or pasted with DNA ligase, because these enzymes

cDNA DNA synthesized from an RNA template by the enzyme reverse transcriptase.
cloning vector A DNA molecule that can accept foreign DNA and get replicated inside a host cell.
DNA cloning Set of procedures that uses living cells to make many identical copies of a DNA fragment.
recombinant DNA A DNA molecule that contains genetic material from more than one organism.
restriction enzyme Type of enzyme that cuts specific nucleotide sequences in DNA.
reverse transcriptase An enzyme that uses mRNA as a template to make a strand of cDNA.

work only on double-stranded DNA. Thus, research with mRNA often involves transcribing the mRNA back into DNA. This process requires **reverse transcriptase**, a replication enzyme that uses an RNA template to assemble a strand of complementary DNA, or **cDNA**:

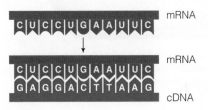

mRNA

mRNA

cDNA

DNA polymerase added to the mixture strips the RNA from the hybrid molecule as it copies the cDNA into a second strand of DNA. The outcome is a double-stranded DNA version of the original mRNA:

DNA

cDNA

*Eco*RI recognition site

Like any other double-stranded DNA, this one may be cut with restriction enzymes and pasted into a cloning vector using DNA ligase.

Take-Home Message

What is DNA cloning?

» DNA cloning uses living cells to mass-produce particular DNA fragments. Restriction enzymes cut DNA into fragments, then DNA ligase seals the fragments into cloning vectors. Recombinant DNA molecules result.

» A cloning vector that holds foreign DNA can be introduced into a living cell. When the host cell divides, it gives rise to huge populations of genetically identical cells (clones), each of which contains a copy of the foreign DNA.

15.3 Isolating Genes

- DNA libraries and the polymerase chain reaction (PCR) help researchers isolate particular DNA fragments.
- Links to Tracers 2.2, Denaturation 3.7, Base pairing 8.4, DNA replication 8.5

A Individual bacterial cells from a DNA library are spread over the surface of a solid growth medium. The cells divide repeatedly and form colonies—clusters of millions of genetically identical descendant cells.

B A piece of special paper pressed onto the surface of the growth medium will bind some cells from each colony.

C The paper is soaked in a solution that ruptures the cells and releases their DNA. The DNA clings to the paper in spots mirroring the distribution of colonies.

D A radioactive probe is added to the liquid bathing the paper. The probe hybridizes (base-pairs) with any spot of DNA that contains a complementary sequence.

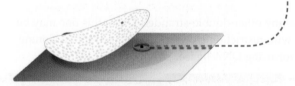

E The paper is pressed against x-ray film. The radioactive probe darkens the film in a spot where it has hybridized. The spot's position is compared to the positions of the original bacterial colonies. Cells from the colony that corresponds to the spot are cultured, and their DNA is harvested.

Figure 15.5 Animated Nucleic acid hybridization. In this example, a radioactive probe helps identify a colony of bacteria that host a targeted sequence of DNA.

DNA Libraries

The entire set of genetic material—the **genome**—of most organisms consists of thousands of genes. To study or manipulate a single gene, researchers must first separate the gene from all of the others. They often begin by cutting an organism's DNA into pieces, and then cloning all the pieces. The result is a genomic library, a set of clones that collectively contain all of the DNA in a genome. Researchers also harvest mRNA, make cDNA copies of it, and then clone the cDNA. The resulting cDNA library represents only those genes being expressed at the time the mRNA was harvested.

Genomic and cDNA libraries are **DNA libraries**, sets of cells that host various cloned DNA fragments. In such libraries, a cell that contains a particular DNA fragment of interest is mixed up with thousands or millions of others that do not. One way to find that one clone among the others—a needle in a haystack—involves **probes**, which are fragments of DNA or RNA labeled with a tracer (Section 2.2). For example, researchers may synthesize a short chain of nucleotides based on the known DNA sequence of a gene, then attach a radioactive phosphate group to it. The nucleotide sequences of the probe and the gene are complementary, so the two can base-pair. Base pairing between DNA (or DNA and RNA) from more than one source is called **nucleic acid hybridization**. When the probe is mixed with DNA from a library, it will base-pair with (hybridize to) the gene, but not to other DNA (**Figure 15.5**). Researchers pinpoint a clone that hosts the gene by detecting the label on the probe. That clone is isolated and cultured, and its DNA can be extracted in bulk.

PCR

The **polymerase chain reaction (PCR)** is a technique used to mass-produce copies of a particular section of DNA without having to clone it in living cells (**Figure 15.6**). The reaction can transform a needle in a haystack—that one-in-a-million fragment of DNA—into a huge

DNA library Collection of cells that host different fragments of foreign DNA, often representing an organism's entire genome.
genome An organism's complete set of genetic material.
nucleic acid hybridization Base-pairing between DNA or RNA from different sources.
polymerase chain reaction (PCR) Method that rapidly generates many copies of a specific section of DNA.
primer Short, single strand of DNA designed to hybridize with a DNA fragment.
probe Short fragment of DNA labeled with a tracer; designed to hybridize with a nucleotide sequence of interest.

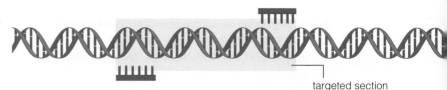

Figure 15.6 Animated Two rounds of PCR. Each cycle of this reaction can double the number of copies of a targeted sequence of DNA. Thirty cycles can make a billion copies.

targeted section

stack of needles with a little hay in it. The starting material for PCR is any sample of DNA with at least one molecule of a targeted sequence. It might be DNA from, say, a mixture of 10 million different clones, a sperm, a hair left at a crime scene, or a mummy—essentially any sample that has DNA in it.

The PCR reaction is based on DNA replication (Section 8.5). First, the starting material is mixed with DNA polymerase, nucleotides, and primers. **Primers** are short single strands of DNA that base-pair with a certain DNA sequence. In PCR, two primers are made. Each base-pairs with one end of the section of DNA to be amplified, or mass-produced ❶. Researchers expose the reaction mixture to repeated cycles of high and low temperatures. High temperatures disrupt the hydrogen bonds that hold the two strands of a DNA double helix together (Section 8.4), so every molecule of DNA unwinds and becomes single-stranded ❷. As the temperature of the reaction mixture is lowered, the single DNA strands hybridize with complementary partner strands, and double-stranded DNA forms again.

The DNA polymerases of most organisms denature at the high temperatures required to separate DNA strands. The kind that is used in PCR reactions, *Taq* polymerase, is from *Thermus aquaticus*. This bacterial species lives in hot springs and hydrothermal vents, so its DNA polymerase is necessarily heat-tolerant. *Taq* polymerase recognizes hybridized primers as places to start DNA synthesis ❸. Synthesis proceeds along the template strand until the temperature rises and the DNA separates into single strands ❹. The newly synthesized DNA is a copy of the targeted section.

When the mixture cools, the primers rehybridize, and DNA synthesis begins again. The number of copies of the targeted section of DNA can double with each cycle of heating and cooling ❺. Thirty PCR cycles may amplify that number a billionfold.

Take-Home Message

How do researchers study one gene in the context of many?

» Researchers isolate one gene from the many other genes in a genome by making DNA libraries or with PCR.

» Probes are used to identify one clone that hosts a DNA fragment of interest among many other clones in a DNA library.

» PCR quickly mass-produces copies of a particular section of DNA.

❶ DNA (*blue*) with a targeted sequence is mixed with primers (*pink*), nucleotides, and heat-tolerant *Taq* DNA polymerase.

❷ When the mixture is heated, the double-stranded DNA separates into single strands. When the mixture is cooled, some of the primers base-pair with the DNA at opposite ends of the targeted sequence.

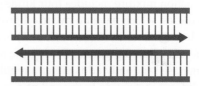

❸ *Taq* polymerase begins DNA synthesis at the primers, so it produces complementary strands of the targeted DNA sequence.

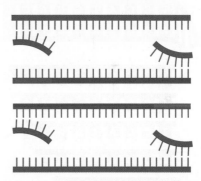

❹ The mixture is heated again, so all double-stranded DNA separates into single strands. When it is cooled, primers base-pair with the targeted sequence in the original template DNA and in the new DNA strands.

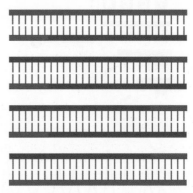

❺ Each round of PCR reactions can double the number of copies of the targeted DNA section.

15.4 DNA Sequencing

- DNA sequencing reveals the order of nucleotide bases in a section of DNA.
- Links to Tracers 2.2, Nucleotides 8.4, DNA replication 8.5

Technique

Researchers determine the order of bases in DNA with **DNA sequencing** (**Figure 15.7**). The most common method is similar to DNA replication. The DNA to be sequenced (the template) is mixed with nucleotides, a primer, and DNA polymerase. Starting at the primer, the polymerase joins free nucleotides into a new complementary strand of DNA, in the order dictated by the sequence of the template. Remember that DNA polymerase can add a nucleotide only to the hydroxyl group on the 3′ carbon of a DNA strand. The sequencing reaction mixture includes four kinds of dideoxynucleotides, which have no hydroxyl group on their 3′ carbon ❶. Each kind (A, C, G, or T) is labeled with a different colored pigment. During the reaction, the polymerase randomly adds either a regular nucleotide or a dideoxynucleotide to the end of a growing DNA strand. If it adds a dideoxynucleotide, the 3′ carbon of the strand will not have a hydroxyl group, so synthesis of the strand ends there ❷.

The reaction produces millions of DNA fragments of different lengths—incomplete copies of the starting DNA ❸. Each fragment ends with a dideoxynucleotide that is complementary to the template sequence. For example, if the tenth base in the template DNA was thymine, then any newly synthesized fragment that is 10 bases long ends with a dideoxyadenine.

The fragments are then separated by **electrophoresis**. With this technique, an electric field pulls the DNA fragments through a semisolid gel. DNA fragments of different sizes move through the gel at different rates. The shorter the fragment, the faster it moves, because shorter fragments slip through the tangled molecules of the gel faster than longer fragments do. All fragments of the same length move through the gel at the same speed, so they gather into bands. All fragments in a given band have the same dideoxynucleotide at their ends, and the pigment labels now impart distinct colors to the bands ❹. Each color designates one of the four dideoxynucleotides, so the order of colored bands in the gel represents the DNA sequence ❺.

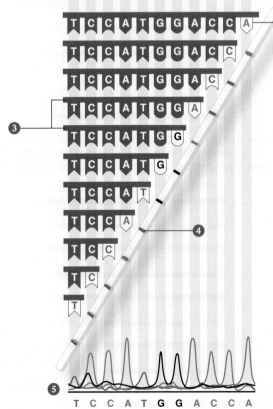

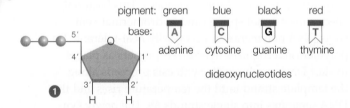

Figure 15.7 Animated DNA sequencing method in which DNA polymerase is used to incompletely replicate a section of DNA.

❶ Sequencing depends on dideoxynucleotides (compare **Figure 9.2**) to terminate DNA replication. Each of the four dideoxynucleotides (A, C, G, or T) is labeled with a different colored pigment.

❷ DNA polymerase uses a section of DNA as a template to synthesize new strands of DNA. Synthesis of each new strand stops when a dideoxynucleotide is added.

❸ Millions of DNA fragments of different lengths form. All are copies of the template DNA, but most are incomplete.

❹ Electrophoresis separates the fragments according to length. Fragments of identical length gather into bands. All of the DNA strands in each band end with the same dideoxynucleotide; thus, each band is the color of that dideoxynucleotide's tracer pigment.

❺ A computer detects and records the color of successive bands on the gel (see **Figure 15.8** for an example). The order of colors of the bands represents the sequence of the template DNA.

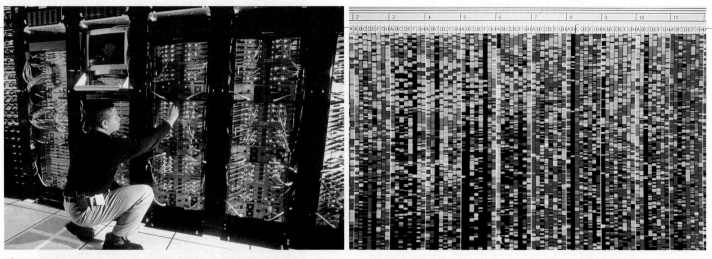

Figure 15.8 Human genome sequencing. *Left*, some of the supercomputers used to assemble the sequence of the human genome at Venter's Celera Genomics in Maryland. Information in Celera's SNP database is the basis of many new genetic tests. *Right*, a human DNA sequence, raw data.

The Human Genome Project

The sequencing method described above was invented in 1975. Ten years later, it had become so routine that people were debating about sequencing the entire human genome. Knowing the sequence would have huge potential payoffs for medicine and research, but sequencing 3 billion bases was a daunting proposition. It would require at least 6 million sequencing reactions, a task that, given techniques available at the time, would have taken 50 years to complete. Opponents said the effort would divert too much attention and funding from more urgent research. However, sequencing techniques kept getting better, and with each improvement more bases could be sequenced in less time. Automated (robotic) DNA sequencing and PCR had just been invented. Both were still too cumbersome and expensive to be useful in routine applications, but they would not be so for long. Waiting for faster technologies seemed the most efficient way to sequence the genome, but just how fast did they need to be before the project should begin?

A few privately owned companies decided not to wait, and started sequencing. One of them intended to patent the sequence after it was determined. This development provoked widespread outrage, but it also spurred commitments in the public sector. In 1988, the National Institutes of Health (NIH) effectively annexed the project by hiring James Watson (of DNA structure fame) to head an official Human Genome Project, and providing $200 million per year to fund it. A consortium formed between the NIH and international institutions that were sequencing different parts of the genome. Watson set aside 3 percent of the funding for studies of ethical and social issues arising from the

research. He later resigned over a patent disagreement, and geneticist Francis Collins took his place.

Amid ongoing squabbles over patent issues, Celera Genomics formed in 1998. With biologist Craig Venter at its helm, the company intended to commercialize genetic information. Celera invented faster techniques for sequencing genomic DNA (**Figure 15.8**), because the first to have the complete sequence had a legal basis for patenting it. The competition motivated the public consortium to move its efforts into high gear.

Then, in 2000, U.S. President Bill Clinton and British Prime Minister Tony Blair jointly declared that the sequence of the human genome could not be patented. Celera kept sequencing anyway. Celera and the public consortium separately published about 90 percent of the sequence in 2001. By 2003, fifty years after the discovery of the structure of DNA, the sequence of the human genome was officially completed. Researchers have not discovered what all of the genes encode, only where they are in the genome. What do we do with this vast amount of data? The next step is to find out what the sequence means.

DNA sequencing Method of determining the order of nucleotides in DNA.
electrophoresis Technique that separates DNA fragments by size.

Take-Home Message

How is the order of nucleotides in DNA determined?

» With DNA sequencing, a strand of DNA is partially replicated. Electrophoresis is used to separate the resulting fragments by length.

» Improved sequencing techniques and worldwide efforts allowed the human genome sequence to be determined.

15.5 Genomics

■ Comparing the human genome sequence with that of other species is helping us understand how the human body works.
■ Unique sequences of genomic DNA can be used to distinguish an individual from all others.
■ Links to Lipoproteins 3.6, DNA replication 8.5, Knockouts 10.3, Locus 13.2, Complex variation in traits 13.7, Chromosome structural changes 14.5

It took 15 years to sequence the human genome for the first time, but the techniques have improved so much that sequencing an entire genome now takes a few weeks. Anyone can now pay to have their genome sequenced. However, even though we are able to determine the sequence of an individual's genome, it will be a long time before we understand all the information coded within that sequence.

The human genome contains a massive amount of seemingly cryptic data. One way to decipher it is by comparing it to genomes of other organisms, the premise being that all organisms are descended from shared ancestors, so all genomes are related to some extent. We see evidence of such genetic relationships simply by comparing the raw sequence data, which, in some regions, is extremely similar across many species (**Figure 15.9**).

The study of genomes is called **genomics**, a broad field that encompasses whole-genome comparisons, structural analysis of gene products, and surveys of small-scale variations in sequence. Genomics is providing powerful insights into evolution. For example, comparing primate genomes revealed how speciation can occur by structural changes in chromosomes (Section 14.5). Comparing genomes also showed that changes in chromosome structure do not occur randomly. Rather, if a chromosome breaks, it tends to do so in a particular spot. Human, mouse, rat, cow, pig, dog, cat, and horse chromosomes have undergone several translocations at these breakage hot spots during evolution. In humans, chromosome abnormalities that contribute to the progression of cancer also occur at the very same hot spots.

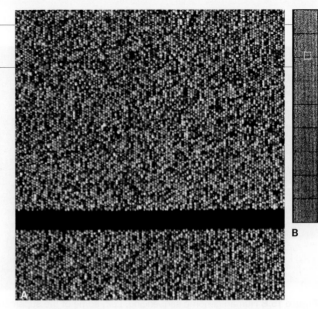

Figure 15.10 SNP-chip analysis. (**A**) Each spot is a region where the individual's genomic DNA has hybridized with one SNP. A *red* or *green* dot means that the individual is homozygous for a SNP; a combined signal (*yellow* dot) indicates heterozygosity. (**B**) The entire chip tests for 550,000 SNPs. The small *white* box indicates the magnified portion shown in **A**.

Comparing the coding regions of genomes offers medical benefits. We have learned the function of many human genes by studying their counterpart genes in other species. For instance, researchers comparing human and mouse genomes discovered a human version of a mouse gene, *APOA5*, that encodes a lipoprotein (Section 3.6). Mice with an *APOA5* knockout have four times the normal level of triglycerides in their blood. The researchers then looked for—and found—a correlation between *APOA5* mutations and high triglyceride levels in humans. High triglycerides are a risk factor for coronary artery disease.

DNA Profiling

As you learned in Section 15.1, only about 1 percent of your DNA is unique. The shared part is what makes you human; the differences make you a unique member of the species. In fact, those differences are so unique that they can be used to identify you. Identifying an individual by his or her DNA is called **DNA profiling**.

```
758 GATAATCCTGTTTTGAACAAAAGGTCAAATTGCTGAATAGAAA-GTCTTGATTAACTAAAAGATGTACAAAGTGGAATTA 836  Human
752 GATAATCCTGTTTTGAACAAAAGGTCAAATTGCTGAATAGAAA-GTCTTGATTAACTAAAAGATGTACAAAGTGGAATTA 830  Mouse
751 GATAATCCTGTTTTGAACAAAAGGTCAAATTGCTGAATAGAAA-GTCTTGATTAACTAAAAGATGTACAAAGTGGAATTA 829  Rat
754 GATAATCCTGTTTTGAACAAAAGGTCAAATTGCTGAATAGAAA-GTCTTGATTAACTAAAAGATGTACAAAGTGGAATTA 832  Dog
782 GATAATCCTGTTTTGAACAAAAGGTCAAATTGCTGAATAGAAA-GTCTTGATTAACTAAAAGATGTACAAAGTGGAATTA 860  Chicken
758 GATAATCCTGTTTTGAACAAAAGGTCAAATTGCTGAATAGAAA-GTCTTGATTAAGTAAAAGATGTACAAAGTGGAATTA 836  Frog
823 GATAATCCTGTTTTGAACAAAAGGTCAGATTGCTGAATAGAAAAGGCTTGATTAAAGCAGAGATGTACAAAGTGGACGCA 902  Zebrafish
763 GATAATCCTGTTTTGAACAAAAGGTCAAATTGTTGAATAGAGACGCTTTGATAAAGCGGAGGAGGTACAAAGTGGGACC- 841  Pufferfish
```

Figure 15.9 Genomic DNA alignment. This is a region of the gene for a DNA polymerase. Differences are highlighted. The chance that any two of these sequences would randomly match is about 1 in 10^{46}.

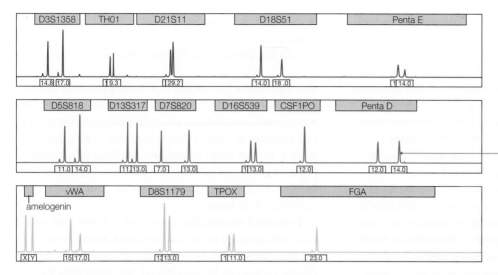

Figure 15.11 An individual's short tandem repeat profile. Tested regions of the individual's DNA are indicated in gray boxes (D3S1358, TH01, and so on).

A peak's location on the x-axis corresponds to the length of the DNA fragment amplified (a measure of the number of repeats). Peak height reflects the amount of DNA.

Remember, human body cells are diploid. Two peaks appear on a profile when the two members of a chromosome pair carry a different number of repeats. For example, this individual has 12 repeats at the Penta D region on one chromosome, and 14 repeats on the other.

One type of DNA profiling involves SNP-chips (one is shown in **Figure 15.1**). A SNP-chip contains a tiny glass plate with microscopic spots of DNA stamped on it. The DNA sample in each spot is a short, synthetic single strand with a unique SNP sequence. When an individual's genomic DNA is washed over a SNP-chip, it hybridizes only with DNA spots that have a matching SNP sequence. Probes reveal where the genomic DNA has hybridized—and which SNPs are carried by the individual (**Figure 15.10**).

Another method of DNA profiling involves analysis of **short tandem repeats**, sections of DNA in which a series of 4 or 5 nucleotides is repeated several times in a row. Short tandem repeats tend to occur in predictable spots, but the number of repeats in each spot differs among individuals. For example, one person's DNA may have fifteen repeats of the bases TTTTC at a certain locus. Another person's DNA may have this sequence repeated only twice in the same locus. Such repeats slip spontaneously into DNA during replication, and their numbers grow or shrink over generations. Unless two people are identical twins, the chance that they have identical short tandem repeats in even three regions of DNA is 1 in a quintillion (10^{18}), which is far more than the number of people on Earth. Thus, an individual's array of short tandem repeats is, for all practical purposes, unique.

Analyzing a person's short tandem repeats begins with PCR, which is used to copy ten to thirteen particular regions of chromosomal DNA known to have repeats. The lengths of the copied DNA fragments differ among most individuals, because the number of tandem repeats in those regions also differs. Thus, electrophoresis can be used to reveal an individual's unique array of short tandem repeats (**Figure 15.11**).

Short tandem repeat analysis will soon be replaced by full genome sequencing, but for now it continues to be a common DNA profiling method. Geneticists compare short tandem repeats on Y chromosomes to determine relationships among male relatives, and to trace an individual's ethnic heritage. They also track mutations that accumulate in populations over time by comparing DNA profiles of living humans with those of ancient ones. Such studies are allowing us to reconstruct population dispersals that happened long ago.

Short tandem repeat profiles are routinely used to resolve kinship disputes, and as evidence in criminal cases. Within the context of a criminal or forensic investigation, DNA profiling is called DNA fingerprinting. As of May 2011, the database of DNA fingerprints maintained by the Federal Bureau of Investigation contained the short tandem repeat profiles of 9.5 million offenders, and had been used in over 100,000 criminal investigations. DNA fingerprints have also been used to identify the remains of almost 300,000 people, including the individuals who died in the World Trade Center on September 11, 2001.

DNA profiling Identifying an individual by analyzing the unique parts of his or her DNA.
genomics The study of genomes.
short tandem repeats In chromosomal DNA, sequences of 4 or 5 bases repeated multiple times in a row.

Take-Home Message

How do we use what we know about human gene sequences?

» Analysis of the human genome sequence is yielding new information about human genes and how they work.

» DNA profiling identifies individuals by the unique parts of their DNA.

15.6 Genetic Engineering

- Bacteria and yeast are the most common genetically engineered organisms.
- Link to Gene expression 9.2

Traditional cross-breeding methods can alter genomes, but only if individuals with the desired traits will interbreed. Genetic engineering takes gene-swapping to an entirely different level. **Genetic engineering** is a laboratory process by which an individual's genome is deliberately modified. A gene may be altered and reinserted into an individual of the same species, or a gene from one species may be transferred to another to produce an organism that is **transgenic**. Both methods result in a **genetically modified organism**, or **GMO**.

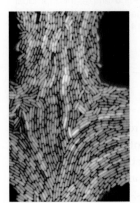

Genetically modified bacteria expressing a jellyfish gene emit green light.

The most common GMOs are bacteria and yeast. These cells have the metabolic machinery to make complex organic molecules, and they are easily modified. For example, the *E. coli* on the *left* have been modified to produce a fluorescent protein from jellyfish. The cells are genetically identical, so the visible variation in fluorescence among them reveals differences in gene expression. Such differences may help us discover why some bacteria of a population become dangerously resistant to antibiotics, and others do not.

Bacteria and yeast have been modified to produce medically important proteins. People with diabetes were among the first beneficiaries of such organisms. Insulin for their injections was once extracted from animals, but it provoked an allergic reaction in some people. Human insulin, which does not provoke allergic reactions, has been produced by transgenic *E. coli* since 1982. Slight modifications of the gene have also yielded fast-acting and slow-release forms of human insulin.

Engineered microorganisms also produce proteins used in food manufacturing. For example, cheese is traditionally made with an extract of calf stomachs, which contain the enzyme chymotrypsin. Most cheese manufacturers now use chymotrypsin produced by genetically engineered bacteria. Other GMO-produced enzymes improve the taste and clarity of beer and fruit juice, slow bread staling, or modify fats.

Take-Home Message

What is genetic engineering?

» Genetic engineering is the deliberate alteration of an individual's genome, and it results in a genetically modified organism (GMO).

» A transgenic organism carries a gene from a different species. Transgenic bacteria and yeast are used in research, medicine, and industry.

15.7 Designer Plants

- Genetically engineered crop plants are widespread in the United States.
- Links to β-carotene 6.2, Promoters 9.3

As crop production expands to keep pace with human population growth, it places unavoidable pressure on ecosystems everywhere. Irrigation leaves mineral and salt residues in soils. Tilled soil erodes, taking topsoil with it. Runoff clogs rivers, and fertilizer in it causes algae to grow so fast that fish suffocate. Pesticides can be harmful to humans, other animals, and beneficial insects such as bees.

Pressured to produce more food at lower cost and with less damage to the environment, many farmers have begun to rely on genetically modified crop plants. Genes can be introduced into plant cells by way of electric or chemical shocks, or by blasting them with DNA-coated micropellets. Another way involves *Agrobacterium tumefaciens* bacteria. *A. tumefaciens* carries a plasmid with genes that cause tumors to form on infected plants; hence the name Ti plasmid (for Tumor-inducing). Researchers replace the tumor-inducing genes with foreign or modified genes, then use the plasmid as a vector to deliver the desired genes into plant cells. Whole plants can be grown from plant cells that integrate a recombinant plasmid into their chromosomes (**Figure 15.12**).

Some genetically modified crop plants carry genes that impart resistance to devastating plant diseases. Others offer improved yields, such as a strain of transgenic wheat that has twice the yield of unmodified wheat. GMO crops such as Bt corn and soy help farmers use smaller amounts of toxic pesticides. Organic farmers often spray their crops with spores of Bt (*Bacillus thuringiensis*), a bacterial species that makes a protein toxic only to some insect larvae. Researchers transferred the gene encoding the Bt protein into plants. The engineered plants produce the Bt protein, but otherwise they are essentially identical to unmodified plants. Larvae die shortly after eating their first and only GMO meal. Farmers can use much less pesticide on crops that make their own (**Figure 15.13**).

Transgenic crop plants are also being developed for Africa and other impoverished regions of the world. Genes that confer drought tolerance and insect resistance are being introduced into plants such as corn,

genetic engineering Process by which deliberate changes are introduced into an individual's genome.
genetically modified organism (GMO) Organism whose genome has been modified by genetic engineering.
transgenic Refers to a genetically modified organism that carries a gene from a different species.

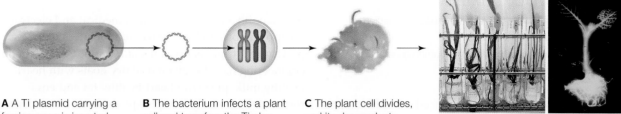

A A Ti plasmid carrying a foreign gene is inserted into an *Agrobacterium tumefaciens* bacterium.

B The bacterium infects a plant cell and transfers the Ti plasmid into it. The plasmid DNA becomes integrated into one of the cell's chromosomes.

C The plant cell divides, and its descendants form an embryo. Several embryos are sprouting from this mass of cells.

D Each embryo develops into a transgenic plant that expresses the foreign gene. The glowing tobacco plant is expressing a gene from fireflies.

Figure 15.12 Animated Using the Ti plasmid to make a transgenic plant.

beans, sugarcane, cassava, cowpeas, banana, and wheat. The resulting GMO crops may help people who rely on agriculture for food and income.

Genetic modifications can make food plants more nutritious. For example, rice plants have been engineered to make β-carotene, an orange photosynthetic pigment that is remodeled by cells of the small intestine into vitamin A. These rice plants carry two genes in the β-carotene synthesis pathway: one from corn, the other from bacteria. One cup of the engineered rice seeds—grains of Golden Rice—has enough β-carotene to satisfy a child's daily need for vitamin A.

The USDA Animal and Plant Health Inspection Service (APHIS) regulates the introduction of GMOs into the environment. At this writing, APHIS has deregulated seventy-eight crop plants, which means the plants are approved for unregulated use in the United States. Worldwide, more than 330 million acres are currently planted in GMO crops, the majority of which are corn, sorghum, cotton, soy, canola, and alfalfa engineered for resistance to the herbicide glyphosate. Rather than tilling the soil to control weeds, farmers can spray their fields with glyphosate, which kills the weeds but not the engineered crops.

After long-term, widespread use of glyphosate, weeds resistant to the herbicide are becoming more common. The engineered gene is also appearing in wild plants and in nonengineered crops, which means that recombinant genes can (and do) escape into the environment. The genes are probably being transferred from transgenic plants to nontransgenic ones via pollen carried by wind or insects.

Many people are opposed to any GMO. Some worry that our ability to tinker with genetics has surpassed our ability to understand the impact of the tinkering. Controversy raised by GMO use invites you to read the research and form your own opinions. The alternative is to be swayed by media hype (the term "Frankenfood," for instance), or by reports from possibly biased sources (such as herbicide manufacturers).

Figure 15.13 Genetically modified crops can help farmers use less pesticide. *Top*, the Bt gene conferred insect resistance to the genetically modified plants that produced this corn. *Bottom*, corn produced by unmodified plants is more vulnerable to insect pests.

Take-Home Message

Are genetically modified plants used as commercial crops?

» Genetically modified crop plants can help farmers be more productive while reducing overall costs.

» The widespread use of GMO crops has had unintended environmental effects. Herbicide resistant weeds are now common, and recombinant genes have spread to wild plants and non-GMO crops.

15.8 Biotech Barnyards

■ Genetically engineered animals are invaluable in medical research and in other applications.
■ Links to Knockout experiments 10.3, Human genetic disorders Chapter 14

Traditional cross-breeding has produced animals so unusual that transgenic animals may seem a bit mundane by comparison (**Figure 15.14A**). Cross-breeding is also a form of genetic manipulation, but many transgenic animals would probably never have occurred without laboratory intervention (**Figure 15.14B,C**).

The first genetically modified animals were mice. Today, such mice are commonplace, and they are invaluable in research (**Figure 15.15**). For example, we have discovered the function of human genes (including the *APOA5* gene discussed in Section 15.5) by inactivating their counterparts in mice. Genetically modified mice are also used as models of human diseases. For example, researchers inactivated the molecules involved in the control of glucose metabolism, one by one, in mice. Studying the effects of the knockouts has resulted in much of our current understanding of how diabetes works in humans.

Genetically modified animals also make proteins that have medical and industrial applications. Various transgenic goats produce proteins used to treat cystic fibrosis, heart attacks, blood clotting disorders, and even nerve gas exposure. Milk from goats transgenic for lysozyme, an antibacterial protein in human milk, may protect infants and children in developing countries from acute diarrheal disease. Goats transgenic for a spider silk gene produce the silk protein in their milk; researchers can spin this protein

into nanofibers that are useful in medical and electronics applications. Rabbits make human interleukin-2, a protein that triggers divisions of immune cells. Genetic engineering has also given us dairy goats with heart-healthy milk, pigs with heart-healthy fat and environmentally friendly low-phosphate feces, extra-large sheep, and cows that are resistant to mad cow disease.

Many people think that genetically engineering livestock is unconscionable. Others see it as an extension of thousands of years of acceptable animal husbandry practices. The techniques have changed, but not the intent: We humans continue to have a vested interest in improving our livestock.

Knockouts and Organ Factories

Millions of people suffer with organs or tissues that are damaged beyond repair. In any given year, more than 80,000 of them are on waiting lists for an organ transplant in the United States alone. Human donors are in such short supply that illegal organ trafficking is now a common problem.

Pigs are a potential source of organs for transplantation, because pig and human organs are about the same in both size and function. However, the human immune system battles anything it recognizes as nonself. It rejects a pig organ at once, because it recognizes proteins and carbohydrates on the plasma membrane of pig cells. Within a few hours, blood coagulates inside the organ's vessels and dooms the transplant. Drugs can suppress the immune response, but they also render organ recipients particularly vulnerable to infection. Researchers have produced genetically

A The genes of this chicken were modified using a traditional method: cross-breeding. Featherless chickens survive in deserts where cooling systems are not an option.

B This genetically engineered goat is transgenic for a human gene. Its milk contains human antithrombin III (a protein that inhibits blood clotting).

C The genetically engineered pig on the *left* is transgenic for a bacterial gene. The gene's product, a yellow fluorescent protein, is visible in all of its tissues. A nontransgenic littermate is on the *right*.

Figure 15.14 Examples of genetically modified animals.

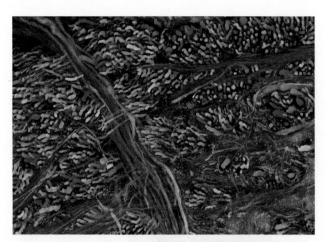

Figure 15.15 Example of how genetically engineered animals are useful in research. Mice transgenic for multiple pigments ("brainbow mice") are allowing researchers to map the complex neural circuitry of the brain. Individual nerve cells in the brain stem of a brainbow mouse are visible in this fluorescence micrograph.

modified pigs that lack the offending molecules on their cells. The human immune system may not reject tissues or organs transplanted from these pigs.

Transferring an organ from one species into another is called **xenotransplantation**. Critics of xenotransplantation are concerned that, among other things, pig-to-human transplants would invite pig viruses to cross the species barrier and infect humans, perhaps with catastrophic results. Their concerns are not unfounded. Evidence suggests that some of the worst pandemics arose when animal viruses adapted to replicate in human hosts.

Tinkering with the genes of animals raises a host of ethical dilemmas. For example, mice, monkeys, and other animals that have been genetically modified to carry mutations associated with certain human diseases often suffer the same terrible symptoms of these conditions as humans do. However, these animals are allowing researchers to study—and test treatments for—conditions such as multiple sclerosis, cystic fibrosis, diabetes, cancer, and Huntington's disease without experimenting on humans.

xenotransplantation Transplantation of an organ from one species into another.

Take-Home Message

Why do we genetically engineer animals?

» Animals that would be impossible to produce by traditional breeding methods are being created by genetic engineering.

» Most engineered animals are used for medical applications and medical research.

» The technique also offers a way to improve livestock.

15.9 Safety Issues

■ The first transfer of foreign DNA into bacteria ignited an ongoing debate about potential dangers of transgenic organisms that enter the environment.

When James Watson and Francis Crick presented their model of DNA in 1953, they ignited a global blaze of optimism. The very book of life seemed to be open for scrutiny. In reality, no one could read it. New techniques would have to be invented before that book would become readable.

Twenty years later, Paul Berg and his coworkers discovered how to make recombinant organisms by fusing DNA from two species of bacteria. Researchers now had the tools to be able to study its sequence in detail. They began to clone DNA from many different organisms. The technique of genetic engineering was born, and suddenly everyone was worried about it. Researchers knew that DNA itself was not toxic, but they could not predict with certainty what would happen each time they fused genetic material from different organisms. Would they accidentally make a superpathogen? Could they make a new, dangerous form of life by fusing DNA of two normally harmless organisms? What if an engineered organism escaped from the laboratory and transformed other organisms?

In a remarkably quick and responsible display of self-regulation, scientists reached a consensus on new safety guidelines for DNA research. Adopted at once by the NIH, these guidelines included precautions for laboratory procedures. They covered the design and use of host organisms that could survive only under the narrow range of conditions inside the laboratory. Researchers stopped using DNA from pathogenic or toxic organisms for recombinant DNA experiments until proper containment facilities were developed.

Now, all genetic engineering should be done under these laboratory guidelines, but the rules are not a guarantee of safety. We are still learning about escaped GMOs and their effects, and enforcement is a problem. For example, the expense of deregulating a GMO is prohibitive for endeavors in the public sector. Thus, most commercial GMOs were produced by large, private companies—the same ones that typically wield tremendous political influence over the very government agencies charged with regulating them.

Take-Home Message

Is genetic engineering safe?

» Guidelines for DNA research have been in place for decades in the United States and other countries. Researchers are expected to comply, but the guidelines are not a guarantee of safety.

15.10 Genetically Modified Humans

■ We as a society continue to work our way through the ethical implications of applying new DNA technologies.

■ The manipulation of individual genomes continues even as we are weighing the risks and benefits of this research.

■ Links to Proto-oncogenes and cancer 11.6, Locus 13.2, Human genetic disorders 14.2

Getting Better

We know of more than 15,000 serious genetic disorders. Collectively, they cause 20 to 30 percent of infant deaths each year, and account for half of all mentally impaired patients and a fourth of all hospital admissions. They also contribute to many age-related disorders, including cancer, Parkinson's disease, and diabetes. Drugs and other treatments can minimize the symptoms of some genetic disorders, but gene therapy is the only cure. **Gene therapy** is the transfer of recombinant DNA into an individual's body cells, with the intent to correct a genetic defect or treat a disease. The transfer, which occurs by way of lipid clusters or genetically engineered viruses, inserts an unmutated gene into an individual's chromosomes.

Human gene therapy is a compelling reason to embrace genetic engineering research. It is now being tested as a treatment for heart attack, sickle-cell anemia, cystic fibrosis, hemophilia A, Parkinson's disease, Alzheimer's disease, several types of cancer, and inherited diseases of the eye, the ear, and the immune system. The results are encouraging. For example, little Rhys Evans (**Figure 15.16**) was born with SCID-X1, a severe X-linked genetic disorder that stems from a mutated allele of the *IL2RG* gene. The gene encodes a receptor for an immune signaling molecule. Children affected by this disorder can survive only in germ-free isolation tents, because they cannot fight infections. In the late 1990s, researchers used a genetically engineered virus to insert unmutated copies of *IL2RG* into cells taken from the bone marrow of twenty boys with SCID-X1. Each child's modified cells were infused back into his bone marrow. Within months of their treatment, eighteen of the boys left their isolation tents for good. Rhys was one of them. Gene therapy had permanently repaired their immune systems.

Getting Worse

Manipulating a gene within the context of a living individual is unpredictable even when we know its sequence and locus. No one, for example, can predict where a virus-injected gene will become integrated into a chromosome. Its insertion might disrupt other

Figure 15.16
Rhys Evans, who was born with SCID-X1. His immune system has been permanently repaired by gene therapy.

genes. If it interrupts a gene that is part of the controls over cell division, then cancer might be the outcome. Five of the twenty boys treated with gene therapy for SCID-X1 have since developed a type of bone marrow cancer called leukemia, and one of them has died. The researchers had wrongly predicted that cancer related to the gene therapy would be rare. Research now implicates the very gene targeted for repair, especially when combined with the virus that delivered it. Apparently, integration of the modified viral DNA activated nearby proto-oncogenes (Section 11.6) in the children's chromosomes.

Getting Perfect

The idea of selecting the most desirable human traits, **eugenics**, is an old one. It has been used as a justification for some of the most horrific episodes in human history, including the genocide of 6 million Jews during World War II. Thus, it continues to be a hotly debated social issue. For example, using gene therapy to cure human genetic disorders seems like a socially acceptable goal to most people. However, imagine taking this idea a bit further. Would it also be acceptable to engineer the genome of an individual who is within a normal range of phenotype in order to modify a particular trait? Researchers have already produced mice that have improved memory, enhanced learning ability, bigger muscles, and longer lives. Why not people?

Given the pace of genetics research, the eugenics debate is no longer about how we would engineer desirable traits, but how we would choose the traits that are desirable. Realistically, cures for many severe but rare genetic disorders will not be found, because the financial return will not even cover the cost of the research. Eugenics, however, might just turn a profit.

Personal DNA Testing (revisited)

The results of SNP analysis by a personal DNA testing company also include estimated risks of developing conditions associated with your particular set of SNPs. For example, the test will probably determine whether you are homozygous for one allele of the *MC1R* gene. If you are, the company's report will tell you that you have red hair. Very few SNPs have such a clear cause-and-effect relationship as the *MC1R* allele for red hair, however. Most human traits are polygenic, and many are also influenced by environmental factors such as lifestyle (Section 13.6). Thus, although a DNA test can reliably determine the SNPs in an individual's genome, it cannot reliably predict the effect of those SNPs on the individual.

For example, if you carry one ε4 allele of the *APOE* gene, a DNA testing company cannot tell you whether you will develop Alzheimer's disease later in life. Instead, the company will report your lifetime risk of developing the disease, which is about 29 percent, as compared with about 9 percent for someone who has no ε4 allele.

What does a 29 percent lifetime risk of developing Alzheimer's disease mean? The number is a probability statistic; it means that, on average, 29 of every 100 people who have the ε4 allele eventually get the disease. Having a high risk does not mean you are certain to end up with Alzheimer's, however. Not everyone who develops the disease has the ε4 allele, and not everyone with the ε4 allele develops Alzheimer's disease. Other factors, including epigenetic methylation of DNA, contribute to the disease.

How would you vote? The plunging cost of genetic testing has spurred an explosion of companies offering personal DNA sequencing and SNP profiling. The results of such testing may in some cases be of clinical use, for example in diagnosis of early-onset genetic disorders, or in predicting how an individual will respond to certain medications. However, we are still at an extremely early stage in our understanding of how genes contribute to most conditions, particularly age-related disorders such as Alzheimer's disease. Geneticists believe that it will be five to ten more years before we can use genotype to accurately predict an individual's risk of these conditions. Until then, should genetic testing companies be prohibited from informing clients of their estimated risk of developing such disorders based on SNPs?

How much would potential parents pay to be sure that their child will be tall or blue-eyed? Would it be okay to engineer "superhumans" with breathtaking strength or intelligence? How about a treatment that can help you lose that extra weight, and keep it off permanently? The gray area between interesting and abhorrent can be very different depending on who is asked. In a survey conducted in the United States, more than 40 percent of those interviewed said it would be fine to use gene therapy to make smarter and cuter babies. In one poll of British parents, 18 percent would be willing to use it to keep a child from being aggressive, and 10 percent would use it to keep a child from growing up to be homosexual.

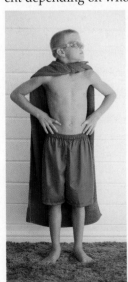

Getting There

Some people are adamant that we must never alter the DNA of anything. The concern is that gene therapy puts us on a slippery slope that may result in irreversible damage to ourselves and to the biosphere. We as a society may not have the wisdom to know how to stop once we set foot on that slope. One is reminded of our peculiar human tendency to leap before we look. And yet, something about the human experience allows us to dream of such things as wings of our own making, a capacity that carried us into space.

In this brave new world, the questions before you are these: What do we stand to lose if serious risks are not taken? And, do we have the right to impose the consequences of taking such risks on those who would choose not to take them?

eugenics Idea of deliberately improving the genetic qualities of the human race.
gene therapy Treating a genetic defect or disorder by transferring a normal or modified gene into the affected individual.

Take-Home Message

Can people be genetically modified?

» Genes can be transferred into a person's cells to correct a genetic defect or treat a disease. However, the outcome of altering a person's genome remains unpredictable given our current understanding of how the genome works.

Two male frigate birds (*Fregata minor*) in the Galápagos Islands, far from the coast of Ecuador. Each male inflates a gular sac, a balloon of red skin at his throat, in a display that may catch the eye of a female. The males lurk together in the bushes, sacs inflated, until a female flies by. Then they wag their head back and forth and call out to her. Like other structures that males use only in courtship, the gular sac probably is an outcome of sexual selection—one of the topics you will read about in this unit.

LEARNING ROADMAP

Where you have been You may wish to review critical thinking (Section 1.6) before reading this chapter, which explores a clash between belief and science (1.9). What you know about alleles (12.2) will help you understand natural selection. Dating rocks and fossils depends on the properties of radioisotopes (2.2). You will see an example of DNA sequence comparisons (15.5).

Where you are now

Emergence of Evolutionary Ideas
Nineteenth-century naturalists investigating the global distribution of species discovered patterns that could not be explained within the framework of traditional belief systems.

A Theory Takes Form
Evidence of evolution, or change in lines of descent, led Charles Darwin and Alfred Wallace to develop a theory of how traits that define each species change over time.

Evidence From Fossils
The fossil record provides physical evidence of past changes in many lines of descent. We use the property of radioisotope decay to determine the age of rocks and fossils.

Influential Geologic Forces
Over millions of years, slow movements of Earth's outer crust have affected land, oceans, and atmosphere. The changes have profoundly influenced life's evolution.

The Geologic Time Scale
By studying rock layers and fossils in them, we can correlate geologic and evolutionary events. The correlation helps explain the distribution of species, past and present.

Where you are going The next two chapters continue the theme of evolutionary processes, including how natural selection works (17.4–17.7), how tectonic plate movements can affect evolution (Section 17.10), and what comparative morphology (18.3) and DNA sequence comparisons (18.4) can tell us about shared evolutionary history. A continent's climate is affected by its position on the globe, as you will see in Chapter 47.

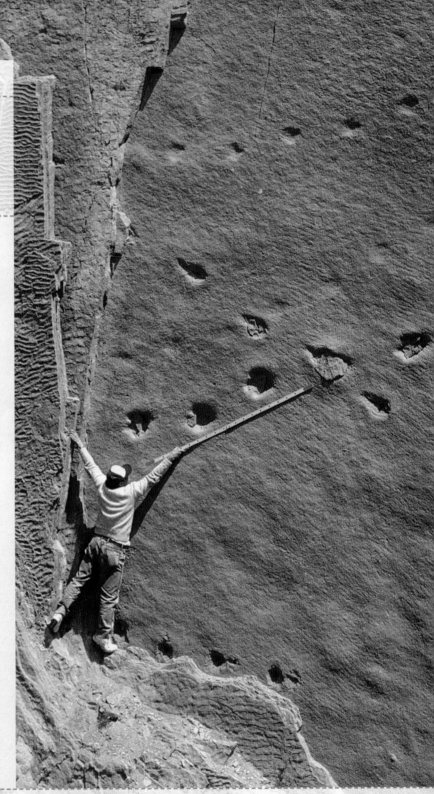

16.1 Reflections of a Distant Past

How do you think about time? Perhaps you can conceive of a few hundred years of human events, maybe a few thousand, but how about a few million? Envisioning the very distant past requires an intellectual leap from the familiar to the unknown.

One way to make that leap involves, surprisingly, asteroids. Asteroids are small planets hurtling through space. They range in size from 1 to 1,500 kilometers (roughly 0.5 to 1,000 miles) across. Millions of them orbit the sun between Mars and Jupiter—cold, stony leftovers from the formation of our solar system. Asteroids are difficult to see even with the best telescopes, because they do not emit light. Many cross Earth's orbit, but most of those pass us by before we know about them. Some have not passed us at all.

The mile-wide Barringer Crater in Arizona is difficult to miss (**Figure 16.1**). An asteroid 45 meters (150 feet) wide made this impressive pockmark in the desert sandstone when it slammed into Earth 50,000 years ago. The impact was 150 times more powerful than the bomb that leveled Hiroshima.

No humans were in North America at the time of the impact. If there were no witnesses, how can we know anything about what happened? We often reconstruct history by studying physical evidence of events that took place long ago. Geologists were able to infer the most probable cause of the Barringer Crater by analyzing tons of meteorites, melted sand, and other rocky clues at the site.

Similar evidence points to even larger asteroid impacts in the more distant past. For example, fossil hunters have long known about a **mass extinction**, or permanent loss of major groups of organisms, that occurred 65.5 million years ago. The event is marked by an unusual, worldwide layer of rock called the K–T boundary layer. There are plenty of dinosaur fossils below this layer. Above it, in rock layers that were deposited more recently, there are no dinosaur fossils, anywhere. A gigantic impact crater—274 kilometers (170 miles) across and 1 kilometer (3,000 feet) deep— off the coast of what is now the Yucatán Peninsula dates to about 65.5 million years ago. Coincidence? Many scientists say no. The asteroid that made the Yucatán crater had to be at least 20 kilometers (12 miles) wide when it hit. The impact of an asteroid that size would have been *40 million times* more powerful than the one at Barringer. The scientists are inferring that the impact caused a global catastrophe of sufficient scale to wipe out the dinosaurs.

mass extinction Simultaneous loss of many lineages from Earth.

You are about to make an intellectual leap through time, to places that were not even known a few centuries ago. We invite you to launch yourself from this premise: Natural phenomena that occurred in the past can be explained by the very same physical, chemical, and biological processes that operate today. That premise is the foundation for scientific research into the history of life. The research represents a shift from experience to inference—from the known to what can only be surmised—and it gives us astonishing glimpses into the distant past.

Figure 16.1 Evidence to inference: What made the Barringer Crater (*top*)? Rocky evidence points to a 300,000-ton asteroid that collided with Earth 50,000 years ago. *Bottom*, the K–T boundary layer, a unique layer of rock that formed 65.5 million years ago, worldwide. The layer marks an abrupt transition in the fossil record that implies a mass extinction. The *red* pocketknife gives an idea of scale.

16.2 Early Beliefs, Confounding Discoveries

■ Belief systems are influenced by the extent of our understanding of the natural world. Those that are inconsistent with systematic observations tend to change over time.

■ Link to Beliefs and science 1.9

The seeds of biological inquiry were taking hold in the Western world more than 2,000 years ago, as the Greek philospher Aristotle made connections between observations in an attempt to explain the order of the natural world. Like few others of his time, Aristotle viewed nature as a continuum of organization, from lifeless matter through complex plants and animals. Aristotle was one of the first **naturalists**, people who observe life from a scientific perspective.

By the fourteenth century, Aristotle's earlier ideas about nature had been transformed into a rigid view of life, in which a "great chain of being" extended from the lowest form (snakes), through humans, to spiritual beings. Each link in the chain was a species, and each was said to have been forged at the same time, in one place and in a perfect state. The chain itself was complete and continuous. Because everything that needed to exist already did, there was no room for change. Once every species had been discovered, the meaning of life would be revealed.

European naturalists that embarked on globe-spanning survey expeditions brought back tens of thousands of plants and animals from Asia, Africa, North and South America, and the Pacific Islands. Each newly discovered species was carefully catalogued as another link in the chain of being. By the late 1800s, naturalists such as Alfred Wallace were

A African milk barrel (*Euphorbia horrida*), native to the Great Karoo Desert of South Africa

B Saguaro cactus (*Carnegiea gigantea*), native to the Sonoran Desert of Arizona

Figure 16.3 Similar-looking but unrelated species that are native to distant geographic realms.

seeing patterns in where species live and how they might be related, and had started to think about the natural forces that shape life. These naturalists were pioneers in **biogeography**, the study of patterns in the geographic distribution of species. Some of the patterns raised questions that could not be answered within the framework of prevailing belief systems. For example, globe-trotting explorers had discovered plants and animals living in extremely isolated places. The isolated species looked suspiciously similar to species living across vast expanses of open ocean, or on the other side of impassable mountain ranges. Could different species be related? If so, how could the related species end up geographically isolated?

A Emu, native to Australia   **B** Rhea, native to South America   **C** Ostrich, native to Africa

Figure 16.2 Similar-looking, related species that are native to distant geographic realms. The three types of ratite birds are unlike most other birds in several traits, including their long, muscular legs and their inability to fly.

For example, the three birds in **Figure 16.2** live on different continents, but they share a set of unusual features. These flightless birds sprint about on long, muscular legs in flat, open grasslands about the same distance from the equator. All raise their long necks to watch for predators. Wallace thought that the shared set of unusual traits might mean that these three birds descended from a common ancestor (and he was right), but he had no idea how they could have ended up on different continents.

Naturalists of the time also had trouble classifying organisms that are very similar in some features, but different in others. For example, the plants in **Figure 16.3** are native to different continents. Both live in hot deserts where water is seasonally scarce. Both have rows of sharp spines that deter herbivores, and both store water in their thick, fleshy stems. However, their reproductive parts are very different, so these plants cannot be as closely related as their outward appearance might suggest.

Observations such as these are part of **comparative morphology**, the study of body plans and structures among groups of organisms. Organisms that are outwardly very similar may be quite different internally; think, for example, of fishes and porpoises. Organisms that differ greatly in outward appearance may be similar in underlying structure. For example, a human arm, a porpoise flipper, and a bat wing have comparable internal bones, as Section 18.3 explains.

Comparative morphology in the nineteenth century revealed body parts that have no apparent func-

tion, an idea that added to the naturalists' confusion. According to prevailing beliefs, every species had been created in a perfect state. If that were so, then why were there useless parts such as leg bones in snakes (which do not walk), or the vestiges of a tail in humans (**Figure 16.4**)?

Fossils were puzzling too. A **fossil** is the remains or traces of an organism that lived in the ancient past—physical evidence of ancient life. Geologists mapping rock formations exposed by erosion or quarrying had discovered identical sequences of rock layers in dif-

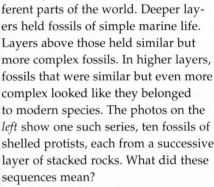

ferent parts of the world. Deeper layers held fossils of simple marine life. Layers above those held similar but more complex fossils. In higher layers, fossils that were similar but even more complex looked like they belonged to modern species. The photos on the *left* show one such series, ten fossils of shelled protists, each from a successive layer of stacked rocks. What did these sequences mean?

Fossils of many animals that had no living representatives were also being unearthed. If the animals had been perfect at the time of creation, then why had they become extinct?

Taken as a whole, the accumulating findings from biogeography, comparative morphology, and geology did not fit with prevailing beliefs of the nineteenth century. If species had not been created in a perfect state (and extinct species, fossil sequences, and "useless" body parts implied that they had not), then perhaps species had indeed changed over time.

biogeography Study of patterns in the geographic distribution of species and communities.
comparative morphology Study of body plans and structures among groups of organisms.
fossil Physical evidence of an organism that lived in the ancient past.
naturalist Person who observes life from a scientific perspective.

Take-Home Message

How did observations of the natural world change our thinking in the nineteenth century?

» Increasingly extensive observations of nature in the nineteenth century did not fit with prevailing belief systems.

» The cumulative findings from biogeography, comparative morphology, and geology led to new ways of thinking about the natural world.

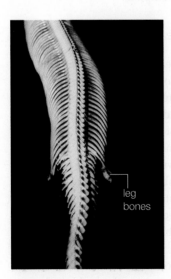
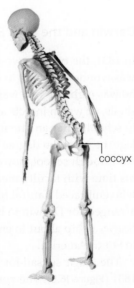
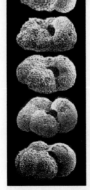

coccyx

leg bones

A Pythons and boa constrictors have tiny leg bones, but snakes do not walk.

B We humans use our legs, but not our coccyx (tail bones).

Figure 16.4 Animated Vestigial body parts.

16.3 A Flurry of New Theories

- In the 1800s, many scholars realized that life on Earth had changed over time, and began to think about what could have caused the changes.
- Link to Critical thinking and how science works 1.6

Squeezing New Evidence Into Old Beliefs

In the nineteenth century, naturalists were faced with increasing evidence that life on Earth, and even Earth itself, had changed over time. Around 1800, Georges Cuvier, an expert in zoology and paleontology, was trying to make sense of the new information. He had observed abrupt changes in the fossil record, and knew that many fossil species seemed to have no living counterparts. Given this evidence, he proposed an idea startling for the time: Many species that had once existed were now extinct. Cuvier also knew about evidence that Earth's surface had changed. For example, he had seen fossilized seashells on mountainsides far

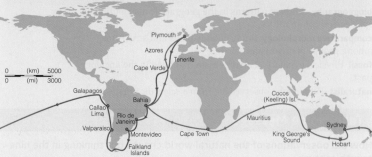

Figure 16.5 Voyage of the HMS *Beagle*. With Darwin aboard as ship's naturalist, the vessel (*top*) originally set sail to map the coast of South America, but ended up circumnavigating the globe (*bottom*). The path of the voyage is shown from *red* to *blue*. Darwin's detailed observations of the geology, fossils, plants, and animals he encountered on this expedition changed the way he thought about evolution.

from modern seas. Like most others of his time, he assumed Earth's age to be in the thousands, not billions, of years. He reasoned that geologic forces unlike those known today would have been necessary to raise seafloors to mountaintops in this short time span. Catastrophic geological events would have caused extinctions, after which surviving species repopulated the planet. Cuvier's idea came to be known as **catastrophism**. We now know it is incorrect; geologic processes have not changed over time.

Another scholar, Jean-Baptiste Lamarck, was thinking about processes that drive **evolution**, or change in a line of descent. A line of descent is also called a **lineage**. Lamarck thought that a species gradually improved over generations because of an inherent drive toward perfection, up the chain of being. The drive directed an unknown "fluida" into body parts needing change. By Lamarck's hypothesis, environmental pressures cause an internal need for change in an individual's body, and the resulting change is inherited by offspring.

Try using Lamarck's hypothesis to explain why a giraffe's neck is very long. We might predict that some short-necked ancestor of the modern giraffe stretched its neck to browse on leaves beyond the reach of other animals. The stretches may have even made its neck a bit longer. By Lamarck's hypothesis, that animal's offspring would inherit a longer neck. The modern giraffe would have been the result of many generations that strained to reach ever loftier leaves. Lamarck was correct in thinking that environmental factors affect a species' traits, but his understanding of how inheritance works was incomplete.

Darwin and the HMS *Beagle*

In 1831, the twenty-two-year-old Charles Darwin was wondering what to do with his life. Ever since he was eight, he had wanted to hunt, fish, collect shells, or watch insects and birds—anything but sit in school. After an attempt to study medicine in college, he earned a degree in theology from Cambridge. All through school, however, Darwin spent most of his time with faculty members and other students who embraced natural history. Botanist John Henslow arranged for Darwin to become a naturalist aboard the *Beagle*, a ship about to embark on a survey expedition to South America.

The *Beagle* set sail for South America in December 1831 (**Figure 16.5**). The young man who had hated school and had no formal training in science quickly became an enthusiastic naturalist. During the *Beagle*'s five-year voyage, Darwin found many unusual fos-

The text of this journal page reads as follows:

"Let a pair be introduced and increase slowly, from many enemies, so as often to intermarry who will dare say what result
—
According to this view animals on separate islands ought to become different if kept long enough apart with slightly differing circumstances. — Now Galapagos Tortoises, Mocking birds, Falkland Fox, Chiloe fox, — Inglish and Irish Hare"

Figure 16.6 Charles Darwin and a page from his 1836 notes on the "Transmutation of Species."

sils, and saw diverse species living in environments that ranged from the sandy shores of remote islands to plains high in the Andes. Along the way, he read the first volume of a new and popular book, Charles Lyell's *Principles of Geology*. Lyell was a proponent of what became known as the **theory of uniformity**, the idea that gradual, repetitive change had shaped Earth. For many years, geologists had been chipping away at the sandstones, limestones, and other types of rocks that form from accumulated sediments in lakebeds, river bottoms, and ocean floors. These rocks held evidence that gradual processes of geologic change operating in the present were the same ones that operated in the distant past.

The theory of uniformity held that strange catastrophes were not necessary to explain Earth's surface. Gradual, everyday geologic processes such as erosion could have sculpted Earth's current landscape over great spans of time. The theory challenged the prevailing belief that Earth was 6,000 years old. According to traditional scholars, people had recorded everything that happened in those 6,000 years—and in all that

time, no one had mentioned seeing a species evolve. However, by Lyell's calculations, it must have taken millions of years to sculpt Earth's surface. Darwin's exposure to Lyell's ideas gave him insights into the geologic history of the regions he would encounter on his journey. Was millions of years enough time for species to evolve? Darwin thought that it was (**Figure 16.6**).

catastrophism Now-abandoned hypothesis that catastrophic geologic forces unlike those of the present day shaped Earth's surface.
evolution Change in a line of descent.
lineage Line of descent.
theory of uniformity Idea that gradual repetitive processes occurring over long time spans shaped Earth's surface.

Take-Home Message

How did new evidence change the way people in the nineteenth century thought about the history of life?

» In the 1800s, fossils and other evidence led some naturalists to propose that Earth and the species on it had changed over time. The naturalists also began to reconsider the age of Earth.

» Darwin's detailed observations of nature during a five-year voyage around the world changed his ideas about how evolution occurs.

16.4 Darwin, Wallace, and Natural Selection

■ Darwin's observations of species in different parts of the world helped him understand a driving force of evolution.
■ Link to Alleles and traits 12.2

Darwin sent to England the thousands of specimens he had collected on his voyage. Among them were fossil glyptodons from Argentina. These armored mammals are extinct, but they have many traits in common with modern armadillos (**Figure 16.7**). For example, armadillos live only in places where glyptodons once lived. Like glyptodons, armadillos have helmets and protective shells that consist of unusual bony plates. Could the odd shared traits mean that glyptodons were ancient relatives of armadillos? If so, perhaps traits of their common ancestor had changed in the line of descent that led to armadillos. But why would such changes occur?

A Key Insight—Variation in Traits

Back in England, Darwin pondered his notes and fossils. He read an essay by one of his contemporaries, economist Thomas Malthus. Malthus had correlated increases in the size of human populations with episodes of famine, disease, and war. He proposed the idea that humans run out of food, living space, and other resources because they tend to reproduce beyond the capacity of their environment to sustain them. When that happens, the individuals of a population must either compete with one another for the limited resources, or develop technology to increase their productivity. Darwin realized that Malthus's ideas had wider application: All populations, not just human ones, must have the capacity to produce more individuals than their environment can support.

Reflecting on his journey, Darwin started thinking about how individuals of a species are not always identical; they often vary a bit in the details of shared traits such as size, coloration, and so on. He saw such variation among many of the finch species that live on isolated islands of the Galápagos archipelago. This island chain is separated from South America by 900 kilometers (550 miles) of open ocean, so most species living on the islands did not have the opportunity for interbreeding with mainland populations. The Galápagos island finches resembled finch species in South America, but many of them had unique traits that suited them to their particular island habitat.

Darwin was familiar with dramatic variations in traits that breeders of dogs and horses had produced through hundreds of years of selective breeding, or **artificial selection**. He recognized that an environment could similarly select traits that make individuals of a population suited to it. It dawned on Darwin that having a particular version of a variable trait might give an individual an advantage over competing members of its species. The trait might enhance the individual's ability to survive and reproduce in its particular environment. Darwin realized that in any population, some individuals have traits that make them better suited to their environment than others. In other words, individuals of a natural population vary in fitness. We define **fitness** as the degree of adaptation to a specific environment, and measure it as relative genetic contribution to future generations. A trait that enhances an individual's fitness is called an evolutionary **adaptation**, or **adaptive trait**.

Over many generations, individuals with the most adaptive traits tend to survive longer and reproduce more than their less fit rivals. Darwin understood that

A A modern armadillo, about a foot long.

B Fossil of a glyptodon, an automobile-sized mammal that existed from 2 million to 15,000 years ago.

Figure 16.7 Ancient relatives: the armadillo and the glyptodon. Though widely separated in time, these animals share a restricted distribution and unusual traits, including a shell and helmet of keratin-covered bony plates—a material similar to crocodile and lizard skin. (The fossil in (**B**) is missing its helmet.) Their unique shared traits were a clue that helped Darwin develop a theory of evolution by natural selection.

Figure 16.8 Alfred Wallace, codiscoverer of natural selection.

this process, which he called **natural selection**, could be a process by which evolution occurs. If an individual has a form of a trait that makes it better suited to an environment, then it is better able to survive. If an individual is better able to survive, then it has a better chance of living long enough to produce offspring. If individuals that bear an adaptive, heritable trait produce more offspring than those that do not, then the frequency of that trait will tend to increase in the population over successive generations. **Table 16.1** summarizes this reasoning in modern terms.

Great Minds Think Alike

Darwin wrote out his ideas about natural selection, but let ten years pass without publishing them. In the meantime, Alfred Wallace, who had been studying wildlife in the Amazon basin and the Malay Archipelago, wrote an essay and sent it to Darwin for advice. Wallace's essay outlined evolution by natural selection—the very same theory as Darwin's. Wallace had written earlier letters to Darwin and Lyell about patterns in the geographic distribution of species; he too had connected the dots. Wallace is now called the father of biogeography (**Figure 16.8**).

In 1858, just weeks after Darwin received Wallace's essay, the theory of evolution by natural selection was presented at a scientific meeting. Both Darwin and Wallace were credited as authors. Wallace was still in the field and knew nothing about the meeting, which Darwin did not attend. The next year, Darwin published *On the Origin of Species*, which laid out detailed evidence in support of the theory. Many people had

already accepted the idea of descent with modification, or evolution. However, there was a fierce debate over the idea that evolution occurs by natural selection. Decades would pass before experimental evidence from the field of genetics led to its widespread acceptance in the scientific community.

As you will see in the remainder of this chapter, the theory of evolution by natural selection is supported by and helps explain the fossil record as well as similarities in the form, function, and biochemistry of living things.

adaptation (**adaptive trait**) A heritable trait that enhances an individual's fitness in a particular environment.
artificial selection Process whereby humans choose traits that they favor in a domestic species by selective breeding.
fitness Degree of adaptation to an environment, as measured by an individual's relative genetic contribution to future generations.
natural selection A process in which environmental pressures result in the differential survival and reproduction of individuals of a population who vary in the details of shared, heritable traits.

Take-Home Message

What is natural selection?

» Natural selection is a process in which individuals of a population who vary in the details of shared, heritable traits survive and reproduce with differing success as a result of environmental pressures.

» Traits favored by natural selection are said to be adaptive. An adaptive trait increases the chances that an individual bearing it will survive and reproduce.

16.5 Fossils: Evidence of Ancient Life

■ Fossils are remnants or traces of organisms that lived in the ancient past. They give us clues about evolutionary relationships.

■ The fossil record holds clues to life's evolution, but it will always be incomplete.

A Fossil skeleton of an ichthyosaur that lived about 200 million years ago. These marine reptiles were about the same size as modern porpoises, breathed air like them, and probably swam as fast, but the two groups are not closely related.

B Extinct wasp (*Leptofoenus pittfieldae*) encased in amber. This 9-mm-long insect lived about 20 million years ago.

C Fossilized leaf of a 260-million-year-old seed fern (*Glossopteris*).

D Fossilized footprint of a theropod, a name that means "beast foot." This group of carnivorous dinosaurs, which includes the familiar *Tyrannosaurus rex*, arose about 250 million years ago. Modern birds are descended from them.

E Coprolite (fossilized feces). Fossilized food remains and parasitic worms inside coprolites offer clues about the diet and health of extinct species. A foxlike animal excreted this one.

Figure 16.9 Examples of fossils.

Even before Darwin's time, fossils were recognized as stone-hard evidence of earlier forms of life. Most fossils are mineralized bones, teeth, shells, seeds, spores, or other body parts (**Figure 16.9A–C**). Trace fossils such as footprints and other impressions, nests, burrows, trails, eggshells, or feces are evidence of an organism's activities (**Figure 16.9D,E**).

How Do Fossils Form?

The process of fossilization typically begins when an organism or its traces become covered by sediments or volcanic ash. Mineral-rich groundwater seeps into the remains, infiltrating spaces around and inside of them. Metal ions and other inorganic compounds dissolved in the water gradually replace minerals in bones and other hard tissues. Mineral particles crystallize and settle out of the groundwater inside cavities and impressions; detailed imprints of internal and external structures can form this way. Sediments that slowly accumulate on top of the site exert increasing pressure. After a very long time, extreme pressure and mineralization transform the remains into rock.

Most fossils are found in layers of sedimentary rock such as sandstone and shale (**Figure 16.10A**). Sedimentary rocks form as rivers wash silt, sand, volcanic ash, and other materials from land to sea. Mineral particles in the materials settle on seafloors in horizontal layers that vary in thickness and composition. After hundreds of millions of years, the layers of sediments become compacted into layers of rock.

Biologists study layers of sedimentary rock in order to understand the historical context of fossils (**Figure 16.10B**). Usually, the deeper layers in a stack were the first to form, and those closest to the surface formed most recently. Thus, in general, the deeper the layer of sedimentary rock, the older the fossils it contains. A layer's composition relative to other layers holds information about local and global events that were occurring as it formed; the K–T boundary layer discussed in Section 16.1 is one example. Minerals in the layers allow us to estimate the age of fossils in them (more about dating fossils in the next section). The relative thicknesses of the different layers can provide other clues. For instance, layers of sedimentary rock deposited during ice ages tend to be thinner than other layers. Why? During the ice ages, tremendous volumes of water froze and became locked in glaciers. Sedimentation slowed as rivers and seas dried up. When the climate warmed, the glaciers melted, and rivers began to flow again. As sedimentation resumed, the layers became thicker again.

The Fossil Record

We have fossils for more than 250,000 known species. Considering the current range of biodiversity, there must have been many millions more, but we will never know all of them. Why not? The odds are against finding evidence of an extinct species, because fossils are relatively rare. When an organism dies, its remains are often quickly obliterated by scavengers. Organic materials decompose in the presence of moisture and oxygen, so remains that escape scavenging can endure only if they dry out, freeze, or become encased in an air-excluding material such as sap, tar, or mud. Remains that do become fossilized are often deformed, crushed, or scattered by erosion and other geologic assaults.

In order for us to know about an extinct species that existed long ago, we have to find a fossil of it. At least one specimen had to be buried before it decomposed or something ate it. The burial site had to escape destructive geologic events, and it had to be a place accessible enough for us to investigate (**Figure 16.10C**).

Most ancient species had no hard parts to fossilize, so we do not find much evidence of them. Unlike bony fishes or the hard-shelled mollusks, for example, jellyfishes and soft worms do not show up much in the fossil record, although they were probably much more common. Also think about relative numbers of organisms. Fungal spores and pollen grains are typically released by the billions. By contrast, the earliest humans lived in small bands and few of their offspring survived. The odds of finding even one fossilized human bone are much smaller than the odds of finding a fossilized fungal spore. Finally, imagine two species, one that existed only briefly and the other for billions of years. Which is more likely to be represented in the fossil record?

Despite these challenges, the fossil record is substantial enough to help us reconstruct large-scale patterns in the history of life.

Take-Home Message

What are fossils?

» Fossils are evidence of organisms that lived in the remote past, a stone-hard historical record of life. The oldest are usually in the deepest sedimentary rocks.

» The fossil record will never be complete. Geologic events obliterated much of it. The rest of the record is slanted toward species that had hard parts, lived in dense populations with wide distribution, and persisted for a long time.

» Even so, the fossil record is substantial enough to help us reconstruct patterns and trends in the history of life.

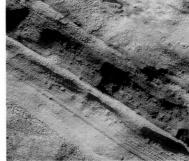

A Two types of sedimentary rock. Sandstones (*left*) consist of compacted grains of sand or minerals; shales (*right*) consist of ancient compacted clay or mud.

B Cindy Looy and Mark Sephton climb the walls of Butterloch Canyon, Italy, to look for fossilized fungal spores in a 251-million-year-old layer of rock.

C High in the Andes, a scientist measures the stride of a carnivorous dinosaur.

Figure 16.10 Fossils are most often found in layered sedimentary rock. This type of rock forms over hundreds of millions of years, often at the bottom of a sea. Geologic processes can tilt the rock and lift it far above sea level, where the layers become exposed by the erosive forces of water and wind.

16.6 Filling In Pieces of the Puzzle

■ Radiometric dating reveals the age of rocks and fossils.
■ New fossil discoveries are continually filling the gaps in our understanding of the ancient history of many lineages.
■ Links to Radioisotopes 2.2, Compounds 2.4, DNA sequence comparisons 15.5

Radiometric Dating

Remember from Section 2.2 that a radioisotope is a form of an element with an unstable nucleus. Atoms of a radioisotope become atoms of other elements as their nucleus disintegrates. The predictable products of radioactive decay are called daughter elements.

Radioactive decay is not influenced by temperature, pressure, chemical bonding state, or moisture; it is influenced only by time. Thus, like the ticking of a perfect clock, each type of radioisotope decays at a constant rate. The time it takes for half of a radioisotope's atoms to decay is called half-life (Figure 16.11A). Half-life is a characteristic of each radioisotope. For example, radioactive uranium 238 decays into thorium 234, which decays into something else, and so on until it becomes lead 206. The half-life of the decay of uranium 238 to lead 206 is 4.5 billion years.

The predictability of radioactive decay can be used to find the age of a volcanic rock (the date it solidified). Rock deep inside Earth is hot and molten, so atoms

swirl and mix in it. Rock that reaches the surface cools and hardens. As the rock cools, minerals crystallize in it. Each kind of mineral has a characteristic structure and composition. For example, the mineral zircon (*right*) consists primarily of ordered arrays of zircon silicate molecules ($ZrSiO_4$). Some of the molecules in a zircon crystal have uranium atoms substituted for zirconium atoms, but never lead atoms. Thus, new zircon crystals that form as molten rock cools contain no lead. However, uranium decays into lead at a predictable rate. Thus, over time, uranium atoms disappear from a zircon crystal, and lead atoms accumulate in it. The ratio of uranium atoms to lead atoms in a zircon crystal can be measured precisely. That ratio can be used to calculate how long ago the crystal formed (its age).

We have just described **radiometric dating**, a method that can reveal the age of a material by measuring its content of a radioisotope and daughter elements. The oldest known terrestrial rock, a tiny zircon crystal from Australia, is 4.404 billion years old.

Recent fossils that still contain carbon can be dated by measuring their carbon 14 content (**Figure 16.11B–D**). Most of the ^{14}C in a fossil will have decayed after about 60,000 years. The age of fossils older than that can be estimated by dating volcanic rock formations above and below the fossil-containing layer.

zircon

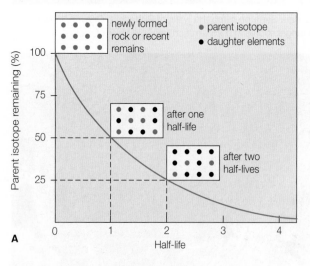

B Long ago, trace amounts of ^{14}C and a lot more ^{12}C were incorporated into the tissues of a nautilus. The carbon atoms were part of organic molecules in the nautilus's food. ^{12}C is stable and ^{14}C decays, but the proportion of the two isotopes in the nautilus's tissues remained the same. Why? As long as it was alive, the nautilus continued to gain both types of carbon atoms in the same proportions from its food.

C When the nautilus died, it stopped eating, so its body stopped gaining carbon. The ^{12}C atoms already in its tissues were stable, but the ^{14}C atoms (represented as *red* dots) were decaying into nitrogen atoms. Thus, over time, the amount of ^{14}C decreased relative to the amount of ^{12}C. After 5,370 years, half of the ^{14}C had decayed; after another 5,370 years, half of what was left had decayed, and so on.

D Fossil hunters discover the fossil and measure its ^{14}C and ^{12}C content—the number of atoms of each isotope. The ratio of those numbers can be used to calculate how many half-lives passed since the organism died. For example, if the ^{14}C to ^{12}C ratio is one-eighth of the ratio in living organisms, then three half-lives must have passed since the nautilus died. Three half-lives of ^{14}C is 16,110 years.

Figure 16.11 Animated Radiometric dating.

(**A**) *Above*, half-life.

(**B–D**) *Right*, an example of carbon 14 dating. ^{14}C that forms continually in the atmosphere offsets its radioactive decay into ^{14}N, so the ratio of ^{14}C to ^{12}C atoms in the atmosphere remains constant. Both isotopes combine with oxygen to become CO_2, which enters food chains by way of photosynthesis.

Figure It Out: How much of any radioisotope remains after two half-lives have passed?   Answer: 25 percent.

Finding a Missing Link

The discovery of intermediate forms of cetaceans (an order of animals that includes whales, dolphins, and porpoises) provides an example of how scientists use fossil finds to piece together evolutionary history. For some time, evolutionary biologists predicted that the ancestors of modern cetaceans walked on land, then took up life in the water again. Evidence in support of this line of thinking includes a set of distinctive features of the skull and lower jaw that cetaceans share with some kinds of ancient carnivorous land animals. DNA sequence comparisons also indicate that the ancient animals were probably artiodactyls, hooved mammals with an even number of toes (two or four) on each foot (**Figure 16.12A**). Modern representatives of the artiodactyl lineage include hippopotamuses, camels, pigs, deer, sheep, and cows.

Until recently, we had no fossils demonstrating gradual changes in skeletal features that accompanied a transition of whale lineages from terrestrial to aquatic life. Researchers knew there were intermediate forms because they had found a representative fossil of an ancient whale-like skull, but without a complete skeleton the rest of the story remained speculative.

Then, in 2000, Philip Gingerich and his colleagues found two of the missing links when they unearthed complete skeletons of ancient whales: a fossil *Rodhocetus kasrani* excavated from a 47-million-year-old rock formation in Pakistan (**Figure 16.12B**); and a fossil *Dorudon atrox*, from 37 million-year-old sandstone in Egypt (**Figure 16.12C**). Whalelike skull bones and intact ankle bones were present in the same skeletons. The ankle bones have distinctive features in common with those of extinct and modern artiodactyls. Modern cetaceans do not have even a remnant of an ankle bone (**Figure 16.12D**). With their in-between ankle bones, *Rodhocetus* and *Dorudon* were probably offshoots of the ancient artiodactyl-to-modern whale lineage as it transitioned back to life in water. The proportions of *Rodhocetus*'s limbs, skull, neck, and thorax indicate it swam with its feet, not its tail. Like modern whales, the 5-meter (16-foot) *Dorudon* was clearly a fully aquatic tail-swimmer: The entire hindlimb was only about 12 centimeters (5 inches) long, much too small to have supported the animal out of water.

half-life Characteristic time it takes for half of a quantity of a radioisotope to decay.
radiometric dating Method of estimating the age of a rock or fossil by measuring the content and proportions of a radioisotope and its daughter elements.

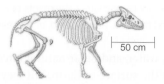

A 30-million-year-old *Elomeryx*. This small terrestrial mammal was a member of the same artiodactyl group that gave rise to hippopotamuses, pigs, deer, sheep, cows, and whales.

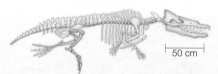

B *Rodhocetus kasrani*, an ancient whale, lived about 47 million years ago. Its distinctive ankle bones point to a close evolutionary connection to artiodactyls.

Compare the ankle bones of a modern artiodactyl, a pronghorn antelope (*top*), with those of a *Rodhocetus* (*bottom*). Artiodactyls are defined by the unique "double-pulley" shape of the bone that forms the lower part of their ankle joint.

C *Dorudon atrox*, an ancient whale that lived about 37 million years ago. Its tiny, artiodactyl-like ankle bones were much too small to have supported the weight of its body on land, so this mammal had to be fully aquatic.

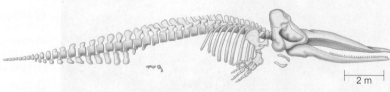

D Modern cetaceans such as the sperm whale have remnants of a pelvis and leg, but no ankle bones.

Figure 16.12 Links in the ancient lineage of whales. Ancestors of whales and other cetaceans probably walked on land. The drawings highlight comparable hindlimb bones in *blue*.

Take-Home Message

Why do biologists care about the age of rocks and fossils?

» The predictability of radioisotope decay can be used to estimate the age of rock layers and fossils in them.

» Radiometric dating helps evolutionary biologists retrace changes in ancient lineages over time.

16.7 Drifting Continents, Changing Seas

■ Over billions of years, movements of Earth's outer layer have changed the land, atmosphere, and oceans.

Wind, water, and other natural forces continuously sculpt the surface of Earth, but they are only part of a bigger picture of geologic change. Earth itself also changes dramatically. For instance, the Atlantic coasts of South America and Africa seem to "fit" like jigsaw puzzle pieces. By an early (1912) theory, all continents that exist today were once part of a bigger supercontinent—**Pangea**—that had split into fragments and drifted apart. The idea explained why the same types of fossils occur in identical rock formations on both sides of the Atlantic Ocean.

At first, most scientists did not accept this theory, which was called continental drift. However, evidence that supported the model kept turning up. For instance, molten rock deep inside Earth wells up and solidifies on the surface. Some iron-rich minerals become magnetic as they solidify, and their magnetic poles align with Earth's poles when they do. If continents never moved, then all of these ancient rocky magnets would be aligned north-to-south, like compass needles. Indeed, the magnetic poles of the rock formations are aligned—but not north-to-south. The poles of rock formations on different continents point in all different directions. Either Earth's magnetic poles veer dramatically from their north–south axis, or the continents wander.

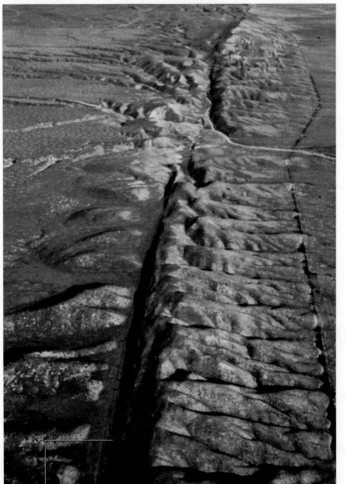

Figure 16.13 Animated Plate tectonics. Huge pieces of Earth's outer layer of rock slowly drift apart and collide. As these plates move, they convey continents around the globe. The current configuration of the plates is shown in Appendix VII.

❶ At oceanic ridges, plumes of molten rock welling up from Earth's interior drive the movement of tectonic plates. New crust spreads outward as it forms on the surface, forcing adjacent tectonic plates away from the ridge and into trenches elsewhere.

❷ At trenches, the advancing edge of one plate plows under an adjacent plate and buckles it.

❸ Faults are ruptures in Earth's crust where plates meet. The diagram shows a rift fault, in which plates move apart. The aerial photo (*left*) shows about 4.2 kilometers (2.6 miles) of the San Andreas Fault, which extends 1,300 km (800 miles) through California. The San Andreas fault is a boundary between two tectonic plates slipping against one another in opposite directions.

❹ Plumes of molten rock rupture a tectonic plate at what are called "hot spots." The Hawaiian Islands have been forming from molten rock that continues to erupt from a hot spot under the Pacific (tectonic) Plate.

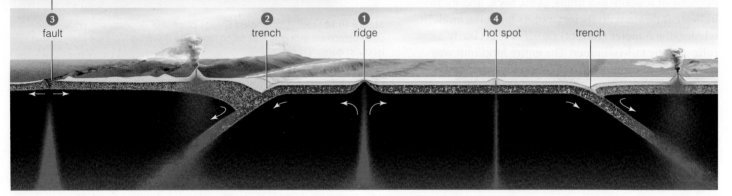

Deep-sea explorers also discovered that ocean floors are not as static and featureless as had been assumed. Immense ridges stretch thousands of kilometers across the seafloor (**Figure 16.13**). Molten rock spewing from the ridges pushes old seafloor outward in both directions ❶, then hardens into new seafloor as it cools. Elsewhere, old seafloor plunges into deep trenches ❷.

By the 1960s, such discoveries had swayed the skeptics, and the theory of continental drift—which by then had been renamed **plate tectonics**—was generally accepted. By this theory, Earth's relatively thin outer layer of rock is cracked into immense plates, like a huge cracked eggshell. Molten rock emanating from an undersea ridge or continental rift at one edge of a plate pushes old rock at the opposite edge into a trench. The movement is like that of a colossal conveyor belt that transports continents on top of it to new locations (**Figure 16.14**). The plates move no more than 10 centimeters (4 inches) a year—about half as fast as your toenails grow—but it is enough to carry a continent all the way around the world after 40 million years or so.

Evidence of tectonic movement is all around us, in faults ❸ and various other geologic features of our landscapes. For example, volcanic island chains (archipelagos) form as a plate moves across an undersea hot spot. Hot spots are places where a narrow plume of molten rock wells up from deep inside Earth and ruptures a plate ❹.

The fossil record also provides evidence in support of plate tectonics. Consider an unusual geologic formation that occurs in a belt across Africa. The sequence of rock layers in this formation is so complex that it is quite unlikely to have formed more than once, but identical sequences also occur in huge belts that span India, South America, Africa, Madagascar, Australia, and Antarctica. The most likely explanation for the wide distribution is that the formations were deposited together on a single continent that later broke up. This explanation is supported by fossils in the rock layers: the remains of the seed fern *Glossopteris* (pictured in **Figure 16.9C**), whose seeds were too heavy to float or to be windblown over an ocean; and an early reptile (*Lystrosaurus*), whose body was not built for swimming between continents.

Glossopteris disappeared in a mass extinction event 251 million years ago, and *Lystrosaurus* disappeared

6 million years after that. Both organisms were extinct millions of years before Pangea formed. This and other evidence suggests that both organisms evolved together on a different supercontinent, one that predated Pangea. The older supercontinent, which we now call **Gondwana**, included most of the landmasses that currently exist in the Southern Hemisphere as well as India and Arabia. Many modern species, including the ratite birds pictured in **Figure 16.2**, live only in places that were once part of Gondwana.

After Gondwana formed about 500 million years ago, it wandered across the South Pole, then drifted north until it merged with other continents to form Pangea about 270 million years ago.

We now know that at least five times since Earth's outer layer of rock solidified 4.55 billion years ago, a single supercontinent formed and then split up again. As you will see in later chapters, the resulting changes have had a profound impact on the course of life's evolution. A continent's climate changes—often dramatically so—along with its position on Earth. Colliding continents physically separate organisms living in oceans, and bring together those that had been living apart on land. As continents break up, they separate organisms living on land, and bring together ones that had been living in separate oceans. Such changes are a major driving force of evolution, as you will see in Chapter 17.

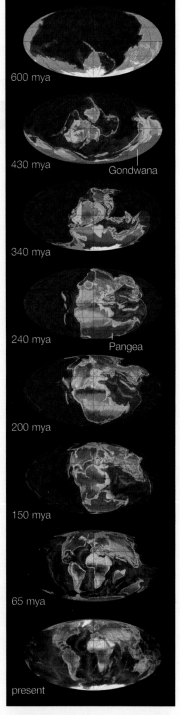

600 mya

430 mya   Gondwana

340 mya

240 mya   Pangea

200 mya

150 mya

65 mya

present

Figure 16.14 A series of reconstructions of the drifting continents.

Gondwana Supercontinent that existed before Pangea, more than 500 million years ago.
Pangea Supercontinent that formed about 250 million years ago.
plate tectonics Theory that Earth's outer layer of rock is cracked into plates, the slow movement of which rafts continents to new locations over geologic time.

Take-Home Message

How has Earth changed over geologic time spans?

» Over geologic time, movements of Earth's crust have caused dramatic changes in continents and oceans. These changes profoundly influenced the course of life's evolution.

■ Transitions in the fossil record are boundaries for great intervals of the geologic time scale.

Radiometric dating and fossils allow us to recognize similar sequences of sedimentary rock layers around the world (**Figure 16.15**). Transitions between layers mark boundaries between great intervals of time in the **geologic time scale**, which is a chronology of Earth's history. Each layer's composition offers clues about conditions on Earth during the time the layer was deposited. Fossils in the layers are a record of life during that period of time.

Eon	Era	Period	Epoch	mya	Major Geologic and Biological Events
Phanerozoic	Cenozoic	Quaternary	Recent	0.01	Modern humans evolve. Major extinction event is now under way.
			Pleistocene	1.8	
		Tertiary	Pliocene	5.3	Tropics, subtropics extend poleward. Climate cools; dry woodlands and grasslands emerge. Adaptive radiations of mammals, insects, birds.
			Miocene	23.0	
			Oligocene	33.9	
			Eocene	55.8	
			Paleocene		
	Mesozoic	Cretaceous	Late	65.5 ◄	Major extinction event, perhaps precipitated by asteroid impact. Mass extinction of all dinosaurs and many marine organisms.
				99.6	
			Early		Climate very warm. Dinosaurs continue to dominate. Important modern insect groups appear (bees, butterflies, termites, ants, and herbivorous insects including aphids and grasshoppers). Flowering plants originate and become dominant land plants.
				145.5	
		Jurassic			Age of dinosaurs. Lush vegetation; abundant gymnosperms and ferns. Birds appear. Pangea breaks up.
				199.6 ◄	Major extinction event
		Triassic			Recovery from the major extinction at end of Permian. Many new groups appear, including turtles, dinosaurs, pterosaurs, and mammals.
				251 ◄	Major extinction event
	Paleozoic	Permian			Supercontinent Pangea and world ocean form. Adaptive radiation of conifers. Cycads and ginkgos appear. Relatively dry climate leads to drought-adapted gymnosperms and insects such as beetles and flies.
				299	
		Carboniferous			High atmospheric oxygen level fosters giant arthropods. Spore-releasing plants dominate. Age of great lycophyte trees; vast coal forests form. Ears evolve in amphibians; penises evolve in early reptiles (vaginas evolve later, in mammals only).
				359 ◄	Major extinction event
		Devonian			Land tetrapods appear. Explosion of plant diversity leads to tree forms, forests, and many new plant groups including lycophytes, ferns with complex leaves, seed plants.
				416	
		Silurian			Radiations of marine invertebrates. First appearances of land fungi, vascular plants, bony fishes, and perhaps terrestrial animals (millipedes, spiders).
				443 ◄	Major extinction event
		Ordovician			Major period for first appearances. The first land plants, fishes, and reef-forming corals appear. Gondwana moves toward the South Pole and becomes frigid.
				488	
		Cambrian			Earth thaws. Explosion of animal diversity. Most major groups of animals appear (in the oceans). Trilobites and shelled organisms evolve.
				542	
Proterozoic					Oxygen accumulates in atmosphere. Origin of aerobic metabolism. Origin of eukaryotic cells, then protists, fungi, plants, animals. Evidence that Earth mostly freezes over in a series of global ice ages between 750 and 600 mya.
				2,500	
Archaean and earlier					3,800–2,500 mya. Origin of bacteria and archaea.
					4,600–3,800 mya. Origin of Earth's crust, first atmosphere, first seas. Chemical, molecular evolution leads to origin of life (from protocells to anaerobic single cells).

Figure 16.15 Animated The geologic time scale correlated with sedimentary rock exposed by erosion in the Grand Canyon. Transitions between layers of sedimentary rock mark great time spans in Earth's history (not to the same scale). mya: millions of years ago. Dates are from the International Commission on Stratigraphy, 2007.

A *Above*, we can reconstruct some of the events in the history of life by studying rocky clues in the layers. *Red* triangles mark times of great mass extinctions. "First appearance" refers to appearance in the fossil record, not necessarily the first appearance on Earth; we often discover fossils that are significantly older than previously discovered specimens.

Kaibab Limestone

Toroweap Formation

Permian

Coconino Sandstone

Hermit Shale

Esplanade Sandstone

Wescogame Formation

Carboniferous

Manakacha Formation

Watahomigi Formation

Redwall Limestone

Temple Butte Formation

Muav Limestone

Cambrian

Bright Angel Shale

Tapeats Sandstone

Proterozoic

Chuar Group*

Nankoweap Formation*

Unkar Group*

Vishnu Basement Rocks

*Layers not visible in this
view of the Grand Canyon

B Each rock layer has a composition and set of fossils that reflect events during its deposition. For example, Coconino Sandstone, which stretches from California to Montana, is mainly weathered sand. Ripple marks and reptile tracks are the only fossils in it. Many think it is the remains of a vast sand desert, like the Sahara is today.

geologic time scale Chronology of Earth's history.

Take-Home Message

What is the geologic time scale?

» The geologic time scale is a chronology of Earth's history that correlates geologic and evolutionary events of the ancient past.

Reflections of a Distant Past (revisited)

The K–T boundary was named after the periods it separates: the Cretaceous (K) and Tertiary (T). K stands for *Kreide*, which is German for chalk (the Cretaceous layer is especially abundant in this sedimentary rock). The K–T boundary layer is unusually rich in iridium, an element rare on Earth's surface but common in asteroids. After finding the iridium, researchers looked for evidence of an asteroid big enough to cover the entire Earth with its debris. They looked for—and found—the Yucatán Peninsula impact crater. It is so big that no one had realized it was a crater before. The K–T boundary layer also contains small glass spheres called tektites, and shocked quartz—rocks that form

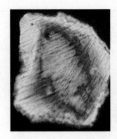

when sand or quartz (respectively) undergoes a sudden, violent application of extreme pressure. As far as we know, the only processes on Earth that produce shocked quartz (*inset*) and tektites are atomic bomb explosions and meteorite impacts.

How would you vote? Many theories and hypotheses about events in the ancient past are necessarily based on traces left by those events, not on data collected by direct observations. Is indirect evidence ever enough to prove a theory about a past event?

LEARNING ROADMAP

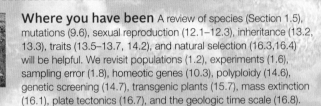

Where you have been A review of species (Section 1.5), mutations (9.6), sexual reproduction (12.1–12.3), inheritance (13.2, 13.3), traits (13.5–13.7, 14.2), and natural selection (16.3, 16.4) will be helpful. We revisit populations (1.2), experiments (1.6), sampling error (1.8), homeotic genes (10.3), polyploidy (14.6), genetic screening (14.7), transgenic plants (15.7), mass extinction (16.1), plate tectonics (16.7), and the geologic time scale (16.8).

Where you are now

Microevolution
Individuals of a population inherit different alleles, the basis of differences in phenotype. An allele may increase or decrease in frequency in a population, a change called microevolution.

Patterns of Natural Selection
Natural selection drives microevolution. Depending on the population and its environment, natural selection can shift or maintain the range of variation in heritable traits.

Other Processes of Microevolution
With genetic drift, change can occur in a line of descent by chance alone. Gene flow counters the evolutionary effects of mutation, natural selection, and genetic drift.

How Species Arise
Speciation typically starts after gene flow ends. Microevolutionary events lead to genetic divergences, which are reinforced as reproductive isolation mechanisms evolve.

Macroevolution
Patterns of genetic change that involve more than one species include the origin of major groups, one species giving rise to many, and mass extinctions.

Where you are going Chapter 18 explores the techniques we use to keep track of species and evolutionary patterns. Later chapters return to polyploid plants (Section 29.9), the genetic basis of behavior (Section 43.2), mating behaviors (Section 43.7), examples of natural selection at work in populations (Section 44.6), coevolved species (45.6), and competition and other interactions between species (Chapter 45).

Slipping in and out of the pages of human history are rats—*Rattus*—the most notorious of mammalian pests. Rats thrive in urban centers, where garbage is plentiful and natural predators are not. The average city in the United States sustains about one rat for every ten people. Part of their success stems from an ability to reproduce very quickly: Rat populations can expand within weeks to match the amount of garbage available for them to eat.

Unfortunately for human populations, rats can carry pathogens and parasites associated with infectious diseases such as bubonic plague and typhus. They chew through walls and wires, and eat or foul 20 to 30 percent of our total food production (**Figure 17.1**). In total, rats cost us about $19 billion annually.

For decades, people have been fighting back with dogs, traps, ratproof storage facilities, and poisons that include arsenic and cyanide. Baits laced with warfarin, an organic compound that interferes with blood clotting, were popular in the 1950s. Rats that ate the poisoned baits died within days after bleeding internally or losing blood through cuts or scrapes. Warfarin was extremely effective, and compared to other rat poisons, it had much less impact on harmless species. It quickly became the rodenticide of choice.

In 1958, however, a Scottish researcher reported that warfarin was not working against some rats. Similar reports from other European countries followed. About twenty years later, about 10 percent of rats caught in urban areas of the United States were resistant to warfarin. What happened?

To find out, researchers compared rats that were resistant to warfarin with those who were not. They traced the difference to a gene called *VKORC1*. A particular mutation in this gene was very common among warfarin-resistant rat populations, but very rare among vulnerable ones. Warfarin inhibits the gene's product, an enzyme that recycles vitamin K after it has been used to activate blood clotting factors. The mutations made the enzyme less active, but also insensitive to warfarin.

"What happened" was evolution by natural selection. As warfarin exerted pressure on rat populations, the warfarin-resistance allele became adaptive, and the rat populations changed. Rats that had an unmutated *VKORC1* gene died after eating warfarin. The lucky ones that had a warfarin-resistance allele survived and passed it to their offspring. The rat populations recovered quickly, and a higher proportion of rats in the next generation carried the allele. With each onslaught of warfarin, the frequency of the allele in the rat populations increased.

Selection pressures can and often do change. When warfarin resistance increased in rat populations, people stopped using warfarin. The frequency of the warfarin-resistance allele in rat populations declined, probably because rats that carry the allele are not as healthy as ones that do not. Now, savvy exterminators in urban areas know that the best way to control a rat infestation is to exert another kind of selection pressure: Remove their source of food, which is usually garbage. Then the rats will eat each other.

Figure 17.1 Rats as pests. *Left*, rats infesting rice fields in the Philippine Islands ruin more than 20 percent of the crop. *Right*, rats thrive wherever people do. Spreading poisons around buildings and soil does not usually exterminate rat populations, which recover quickly. Rather, the practice selects for rats that are resistant to the poisons.

17.2 Individuals Don't Evolve, Populations Do

- Mutations in individuals are the source of new alleles in a population's gene pool.
- A change in an allele's frequency in a population is called microevolution.
- Links to Population 1.2, Mutations in genes 9.6, Red Queen hypothesis 12.1, Alleles 12.2, Mendelian inheritance 13.3, Complex variation in traits 13.5–13.7, Evolution 16.3

Alleles in Populations

Section 1.2 introduced a **population** as a group of interbreeding individuals of the same species in some specified area. The individuals of a population (and a species) share certain features. For example, giraffes normally have long necks, brown spots on white coats, and so on. These are examples of morphological traits (*morpho–* means form). Individuals of a species also share physiological traits, such as metabolic activities. They also respond the same way to certain stimuli, as when hungry giraffes feed on tree leaves. These are behavioral traits.

Individuals of a population have the same traits because they have the same genes. However, almost every shared trait varies a bit among individuals of a population (**Figure 17.2**). Alleles of the shared genes are the basis of this variation. Many traits have two or more distinct forms, or morphs. A trait with only two forms is dimorphic (*di–* means two). Purple and white flower color in the pea plants that Gregor Mendel studied is an example of a dimorphic trait (Section 13.3). Dimorphic flower color occurs in this case because the interaction of two alleles with a clear dominance relationship gives rise to the trait. Traits with more than two distinct forms are polymorphic (*poly–*, many). Human blood type, which is deter-

mined by the codominant *ABO* alleles, is an example (Section 13.5). The genetic basis of traits that vary continuously among the individuals of a population is typically quite complex (Sections 13.5–13.7). Any or all of the genes that influence such traits may have multiple alleles.

In earlier chapters, you learned about genetic events that contribute to the variation in shared traits we see among individuals of a population (**Table 17.1**). Mutation is the original source of new alleles. Other events shuffle alleles into different combinations, and what a shuffle that is! There are $10^{116,446,000}$ possible combinations of human alleles. Not even 10^{10} people are living today. Unless you have an identical twin, it is extremely unlikely that another person with your precise genetic makeup has ever lived, or ever will.

An Evolutionary View of Mutations

Being the original source of new alleles, mutations are worth another look, this time in context of their impact on populations. We cannot predict when or in which individual a particular gene will mutate. We can, however, predict the average mutation rate of a species, which is the probability that a mutation will occur in a given interval. In the human species, that rate is about 2.2×10^{-9} mutations per base pair per year. In other words, about 70 nucleotides in the human genome sequence change every decade.

Many mutations give rise to structural, functional, or behavioral alterations that reduce an individual's chances of surviving and reproducing. Even one biochemical change may be devastating. For instance, the skin, bones, tendons, lungs, blood vessels, and other vertebrate organs incorporate the protein collagen. If one of the genes for collagen mutates in a way that changes the protein's function, the entire body may

Table 17.1	Sources of Variation in Traits Among Individuals of a Species
Genetic Event	**Effect**
Mutation	Source of new alleles
Crossing over at meiosis I	Introduces new combinations of alleles into chromosomes
Independent assortment at meiosis I	Mixes maternal and paternal chromosomes
Fertilization	Combines alleles from two parents
Changes in chromosome number or structure	Transposition, duplication, or loss of chromosomes

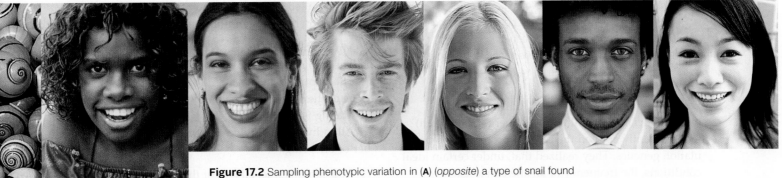

Figure 17.2 Sampling phenotypic variation in (**A**) (*opposite*) a type of snail found on islands in the Caribbean, and (**B**) humans. The variation in shared traits among individuals is mainly an outcome of variations in alleles that influence those traits.

be affected. A mutation such as this can change phenotype so drastically that it results in death, in which case it is a **lethal mutation**.

A **neutral mutation** changes the base sequence in DNA, but the alteration has no effect on survival or reproduction. It neither helps nor hurts the individual. For instance, if you carry a mutation that keeps your earlobes attached to your head instead of swinging freely, attached earlobes should not in itself stop you from surviving and reproducing as well as anybody else. So, natural selection does not affect the frequency of this particular mutation in a population.

Occasionally, a change in the environment favors a mutation that had previously been neutral or even somewhat harmful. The warfarin-resistance gene in rats is an example. Even if a beneficial mutation bestows only a slight advantage, its frequency tends to increase in a population over time. This is because natural selection operates on traits with a genetic basis. With natural selection, remember, environmental pressures result in an increase in the frequency of a beneficial trait in a population over generations (Section 16.4). Mutations have been altering genomes for billions of years, and they are still at it. Cumulatively, they have given rise to Earth's staggering biodiversity. Think about it: The reason you do not look like an avocado or an earthworm or even your next-door neighbor began with mutations that occurred in different lines of descent.

Allele Frequencies

Together, all the alleles of all the genes of a population comprise a pool of genetic resources—a **gene pool**. Members of a population breed with one another more often than they breed with members of other populations, so their gene pool is more or less isolated. **Allele frequency** refers to the abundance of a particular allele among the individuals of a population. Any change in allele frequencies in the gene pool of a population (or a species) is called **microevolution**.

Microevolution is always occurring in natural populations because, as you will see in the next section, processes that drive it are always operating. The remaining sections of this chapter explore microevolutionary processes—mutation, natural selection, genetic drift, and gene flow—and their effects. Remember, even though we can recognize patterns of evolution, none of them are purposeful. Evolution simply fills the nooks and crannies of opportunity.

allele frequency Abundance of a particular allele among members of a population.
gene pool All the alleles of all the genes in a population; a pool of genetic resources.
lethal mutation Mutation that drastically alters phenotype; causes death.
microevolution Change in allele frequencies in a population or species.
neutral mutation A mutation that has no effect on survival or reproduction.
population A group of organisms of the same species that live in a specific location and breed with one another more often than they breed with members of other populations.

Take-Home Message

What is microevolution?

» Individuals of a natural population share morphological, physiological, and behavioral traits characteristic of the species.

» Different alleles are the basis of differences in the details of a population's shared traits.

» All alleles of all individuals in a population make up the population's gene pool.

» Change in allele frequencies (microevolution) is always occurring in the gene pools of natural populations.

17.3 Genetic Equilibrium

- Natural populations are always evolving.
- Researchers trace evolution within a population by tracking deviations from a baseline of genetic equilibrium.
- Links to Alleles 12.2, Mendelian inheritance 13.3, Incomplete dominance 13.5, Genetic screening 14.7

Early in the twentieth century, Godfrey Hardy (a mathematician) and Wilhelm Weinberg (a physician) independently applied the rules of probability to population genetics. They realized that, under certain ideal conditions, the frequency of an allele in a sexually reproducing population's gene pool should remain stable from one generation to the next. The stability,

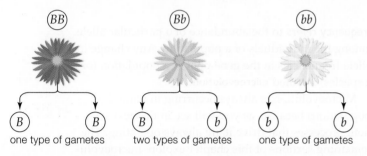

A In this two allele system, *B* specifies dark blue flowers; *b*, white. Plants that are homozygous (*B* or *b*) make one kind of gamete. Heterozygous plants (*Bb*) have light blue flowers and make two kinds of gametes (*B* and *b*).

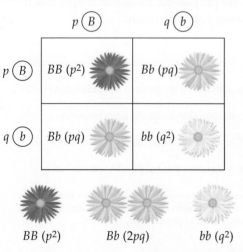

B Say *p* is the proportion of *B* alleles in the gene pool, and *q* is the proportion of *b* alleles. This Punnett square shows that in each generation, the predicted proportion of offspring that will inherit two *B* alleles is *p* × *p*, or *p²*. Likewise, the proportion that will inherit both alleles is 2*pq*, and the proportion that will inherit two *b* alleles is *q²*.

Figure 17.3 Animated Calculating Hardy–Weinberg frequencies. In this example, two alleles of a gene show incomplete dominance over flower color.
Figure It Out: If one-quarter of this population has dark blue flowers and one-quarter has white flowers, what proportion of the next generation will have light blue flowers (assuming genetic equilibrium)?

Answer: Half of the gametes have a B allele; the other half have a b allele. Both p and q = 0.5, so 2pq = 50 percent.

which is called **genetic equilibrium**, can only occur if all of the following five conditions are met:

1. Mutations do not occur.
2. The population is infinitely large.
3. The population is isolated from all other populations of the species (no gene flow).
4. Mating is random.
5. All individuals survive and produce the same number of offspring.

As you can imagine, all five of these conditions never occur in nature, so natural populations are never in genetic equilibrium. Thus, allele frequencies for any gene in a shared gene pool always tend to change.

Applying the Hardy–Weinberg Law

The concept of genetic equilibrium under ideal conditions is called the Hardy–Weinberg law. To see how it works, consider a hypothetical gene that codes for a blue pigment in daisies. A plant homozygous for one allele (*BB*) has dark blue flowers. A plant homozygous for the other allele (*bb*) has white flowers. The *B* allele is incompletely dominant, so a heterozygous plant (*Bb*) has medium-blue flowers (**Figure 17.3A**).

Start with the concept that allele frequencies always add up to one. For a gene with two alleles, the following equation is true:

$$p + q = 1.0$$

where *p* is the frequency of one allele in the population, and *q* is the frequency of the other. Paired alleles assort into different gametes during meiosis (Section 13.3), and those gametes meet up in predictable proportions at fertilization (**Figure 17.3B**). In our example, the predicted fraction of offspring that inherit two *B* alleles (*BB*) is *p* × *p*, or *p²*; the fraction that inherit two *b* alleles (*bb*) is *q²*; and the fraction that inherit one *B* allele and one *b* allele (*Bb*) is 2*pq*. Note that the frequencies of the three genotypes, whatever they may be, add up to 1.0:

$$p^2 + 2pq + q^2 = 1.0$$

Suppose our hypothetical population consists of 1,000 plants: 490 homozygous (*BB*), 420 heterozygous (*Bb*), and 90 homozygous (*bb*). Say each plant makes two gametes. All 980 gametes made by the *BB* individuals will have the *B* allele, as will half of the gametes made by the 420 *Bb* individuals. Thus, the frequency of the *B* allele among the pool of gametes is:

$$p\,(B) = \frac{980 + 420}{2,000 \text{ alleles}} = \frac{1,400}{2,000} = 0.7$$

All 180 gametes made by the *bb* individuals will have the *b* allele, as will half of the gametes made by the 420 *Bb* individuals. Thus, the frequency of the *b* allele among the pool of gametes is:

$$q \, (b) \; = \; \frac{180 \; + \; 420}{2{,}000 \; \text{alleles}} \; = \; \frac{600}{2{,}000} \; = \; 0.3$$

Using our equation, the proportion of individuals in the next generation is predicted to be:

$$
\begin{array}{llll}
BB & (p^2) = & (0.7)^2 & = 0.49 \\
Bb & (2pq) = & 2\,(0.7 \times 0.3) & = 0.42 \\
bb & (q^2) = & (0.3)^2 & = 0.09
\end{array}
$$

These proportions are the same as the ones in the parent population, and in fact as long as the five ideal conditions are occurring, traits specified by the alleles should show up in the same proportions from one generation to the next. If they do not, one or more of the five assumptions is not being met.

Real World Situations

Though genetic equilibrium is a theoretical state, it is often used as a benchmark. As an example, researchers used it to determine the carrier frequency of an allele that causes hereditary hemochromatosis (HH), the most common genetic disorder among people of Irish ancestry. Affected individuals absorb too much iron from food. Symptoms of this autosomal recessive disorder include liver problems, fatigue, and arthritis. The allele's frequency (q) was found to be 0.14. If $q = 0.14$, then p, the frequency of the normal allele, is 0.86. Thus, the carrier frequency, $2pq$, was calculated at 0.24.

Another example: A mutation in the *BRCA2* gene has been linked to breast cancer in adults. A deviation from the predicted phenotype frequencies suggested that this mutation also has effects even before birth. Researchers looked at the allele's frequency among newborn girls. They found fewer homozygotes than expected, based on the number of heterozygotes. Thus, it seems that in homozygous form the mutation impairs the survival of female embryos.

genetic equilibrium Theoretical state in which a population is not evolving.

17.4 Patterns of Natural Selection

- Natural selection occurs in different patterns depending on the organisms involved and their environment.
- Link to Natural selection theory and fitness 16.4

The remainder of this chapter explores the mechanisms and effects of processes that drive evolution, including natural selection. Remember from Section 16.4 that natural selection is a process in which environmental pressures result in the differential survival and reproduction of individuals of a population. It influences the frequency of alleles in a population by operating on phenotypes with a heritable, genetic basis.

We observe different patterns of natural selection, depending on the selection pressures and the organisms involved. Sometimes, individuals with a trait at one extreme of a range of variation are selected against, and those at the other extreme are favored. We call this pattern directional selection. With stabilizing selection, midrange forms are favored, and the extremes are selected against. With disruptive selection, forms at the extremes of the range of variation are favored, and the intermediate forms are selected against. We will discuss these three modes of natural selection, which **Figure 17.4** summarizes, in the following two sections.

Section 17.7 explores sexual selection, a mode of natural selection that operates on a population by influencing mating success. This section also discusses balanced polymorphism, a particular case in which natural selection maintains a relatively high frequency of two or more alleles in a population.

Natural selection and other processes of evolution can alter a population so much that it becomes a new species. We discuss mechanisms of speciation in the final sections of this chapter.

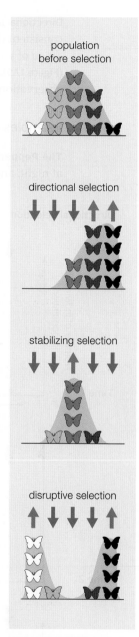

Figure 17.4 Overview of three modes of natural selection.

17.5 Directional Selection

- Changing environmental conditions can result in a directional shift in allele frequencies.
- Link to How experiments work 1.6

Directional selection shifts an allele's frequency in a consistent direction, so phenotypes at one end of a range of variation become more common over time (**Figure 17.5**). The following examples show how field observations provide evidence of directional selection.

Examples of Directional Selection

The Peppered Moth Peppered moths feed and mate at night, and rest motionless on trees during the day.

Directional Selection

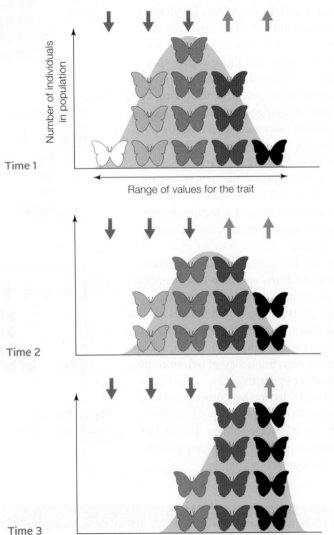

Figure 17.5 Animated Directional selection. The bell-shaped curves indicate continuous variation in a butterfly wing-color trait. *Red* arrows show which forms are being selected against; *green*, forms that are being favored.

Their behavior and coloration help prevent day-flying, moth-eating birds from detecting them. In preindustrial England, most moths were light-colored, and a dominant allele that resulted in darker coloration was rare. At this time, the air was clean, and light-gray lichens grew on the trunks and branches of most trees. Light moths were camouflaged when they rested on the lichens, but dark moths were not (**Figure 17.6A**). By the 1850s, the dark moths had become much more common. Why? The industrial revolution had begun, and smoke emitted by coal-burning factories was beginning to change the environment. Air pollution was killing the lichens. Dark moths were better camouflaged on lichen-free, soot-darkened trees (**Figure 17.6B**).

In the 1950s, H. B. Kettlewell tested the hypothesis that dark-colored moths were at a selective advantage in industrialized areas. He bred dark and light moths in captivity, marked them for easy identification, then released them in several areas. His team recaptured more of the dark moths in the polluted areas and more light ones in the less polluted ones. They also observed predatory birds eating more light-colored moths in soot-darkened forests, and more dark-colored moths in cleaner, lichen-rich forests.

Pollution controls went into effect in 1952. As a result, tree trunks gradually became free of soot, and lichens made a comeback. Kettlewell observed that moth phenotypes shifted too: Wherever pollution decreased, the frequency of dark moths decreased as well. Many other researchers since Kettlewell have confirmed the rise and fall of the dark-colored form of the peppered moth.

A Light-colored peppered moths on a nonsooty tree trunk are hidden from predators; dark ones stand out.

B In places where soot darkens tree trunks, the dark color provides more camouflage than the light color.

Figure 17.6 Animated Directional selection in the peppered moth (*Biston betularia*).

A Mice with light fur are better camouflaged—and more common—in the areas dominated by light-colored granite.

B Mice with dark fur are better camouflaged—and more common—in the areas of dark basalt rock.

Figure 17.7 Directional selection in the rock pocket mouse (*Chaetodipus intermedius*). Predators preferentially eliminate individuals with coat colors that do not match their surroundings in the Sonoran Desert.

Rock Pocket Mice Directional selection also affects the color of rock pocket mice in Arizona's Sonoran Desert. Rock pocket mice are small mammals that spend the day sleeping in underground burrows, emerging at night to forage for seeds. The environment is dominated by light brown granite. There are also patches of dark basalt, the remains of ancient lava flows. Most of the mice in populations that inhabit the dark rock have dark gray coats. Most of the mice in populations that inhabit the light brown rock have light brown coats. The difference arises because mice that match the rock color in each habitat are camouflaged from their natural predators (**Figure 17.7**). Night-flying owls more easily see mice that do not match the rocks, and they preferentially eliminate easily seen mice from each population. Thus, in both habitats, selective predation has resulted in a directional shift in the frequency of alleles that affect coat color.

Antibiotic-Resistant Bacteria Human attempts to control the environment can result in directional selection, as is the case with the warfarin-resistant rats. The use of antibiotics is another example. Prior to the 1940s, scarlet fever, tuberculosis, and pneumonia caused one-fourth of the annual deaths in the United States. Since the 1940s, we have been relying on antibiotics such as penicillin to fight these and other dangerous bacterial diseases. We also use them in other, less dire circumstances. Antibiotics are used preventively in humans, and they are part of the daily rations of millions of cattle, pigs, chickens, fish, and other animals raised on factory farms.

Bacteria evolve at a much accelerated rate compared with humans, in part because they reproduce very quickly. For example, the common intestinal bacteria *E. coli* can divide every 17 minutes. Each new generation is an opportunity for mutation, so the gene pool of a bacterial population varies greatly. Thus, in any population of bacteria, some cells are likely to carry alleles that allow them to survive an antibiotic treatment. When the survivors reproduce, the frequency of antibiotic-resistance alleles increases in the population. A typical two-week course of antibiotics can potentially exert selection pressure on over a thousand generations of bacteria, and antibiotic-resistant strains may be the outcome.

Antibiotic-resistant bacteria have plagued hospitals for many years, and now they are becoming similarly common in schools. Even as researchers scramble to find new antibiotics, this trend is bad news for the millions of people each year who contract a dangerous bacterial disease such as cholera or tuberculosis.

directional selection Mode of natural selection in which phenotypes at one end of a range of variation are favored.

Take-Home Message

What is the effect of directional selection?

» Directional selection causes allele frequencies underlying a range of variation to shift in a consistent direction.

17.6 Stabilizing and Disruptive Selection

- Stabilizing selection is a form of natural selection that maintains an intermediate phenotype.
- Disruptive selection is a form of natural selection that favors extreme forms of a trait.

Natural selection can bring about a directional shift in a population's range of phenotypes. Depending on the environment and the organisms involved, the process may also favor a midrange form of a trait, or it may eliminate the midrange form and favor extremes.

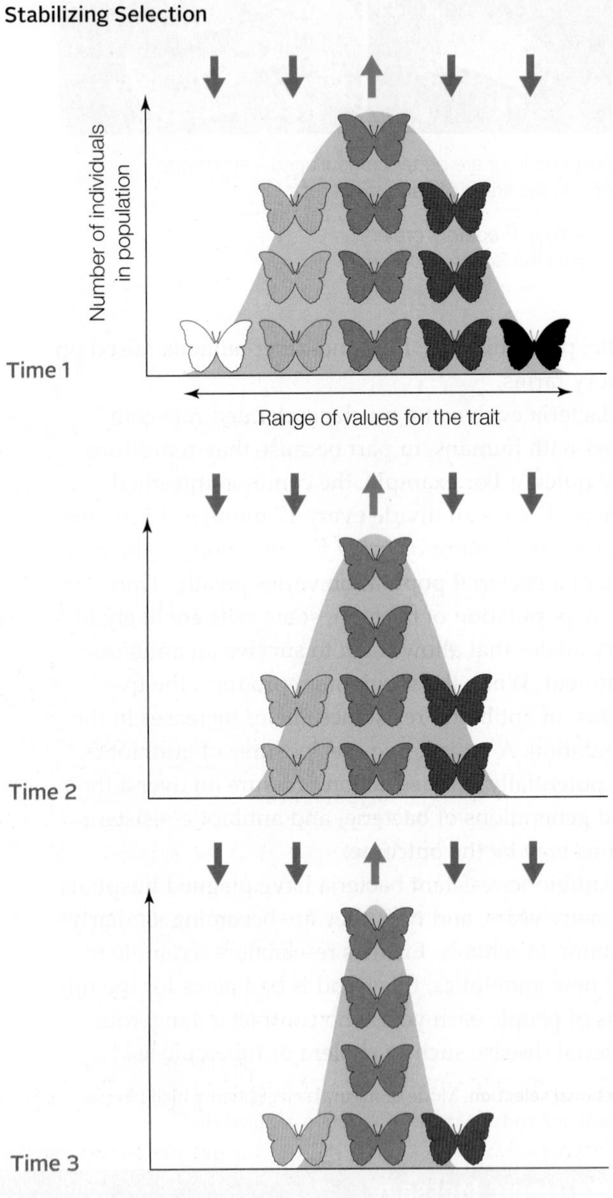

Stabilizing Selection

With **stabilizing selection**, an intermediate form of a trait is favored, and extreme forms are not. This mode of natural selection tends to preserve the midrange phenotypes in a population (**Figure 17.8**). For example, the body weight of sociable weavers (*Philetairus socius*) is subject to stabilizing selection (**Figure 17.9**). Weaver birds build large communal nests in areas of the African savanna. Between 1993 and 2000, Rita Covas and her colleagues investigated selection pressures that operate on sociable weaver body size. They captured, tagged, weighed, and released birds living in communal nests before and after the breeding seasons. Covas's research indicated that optimal body weight in sociable weavers is a trade-off between the risks of starvation and predation. Foraging is not easy in the sparse habitat of an African savanna, and leaner birds do not store enough fat to avoid starvation. A meager food supply selects against birds with low body weight. Fatter birds are more attractive to predators, and not as agile when escaping. Predators select against birds of high body weight. Thus, birds of intermediate weight have the selective advantage, and they make up the bulk of sociable weaver populations.

Stabilizing Selection

Time 1

Number of individuals in population

Range of values for the trait

Time 2

Time 3

Figure 17.8 Animated Stabilizing selection eliminates extreme forms of a trait, and maintains the predominance of an intermediate phenotype in a population. *Red* arrows indicate which forms are being selected against; *green*, forms that are being favored. Compare the data set from a field experiment in **Figure 17.9**.

Figure 17.9 Stabilizing selection in sociable weavers. Graph shows the number of birds (out of 977) that survived a breeding season. **Figure It Out: What is the optimal weight of a sociable weaver?** Answer: About 29 grams

Disruptive Selection

Conditions that favor forms of a trait at both ends of a range of variation drive **disruptive selection**. With this mode of natural selection, intermediate forms are selected against (**Figure 17.10**). Consider the black-bellied seedcracker (*Pyrenestes ostrinus*), a colorful finch species native to Cameroon, Africa. In these birds, there is a genetic basis for bill size. The bill of a typical black-bellied seedcracker, male or female, is either 12 milli-meters wide, or wider than 15 millimeters (**Figure 17.11**).

Disruptive Selection

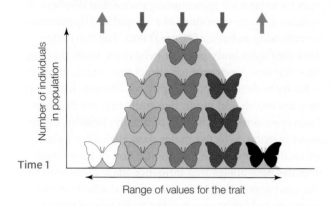

Time 1

Range of values for the trait

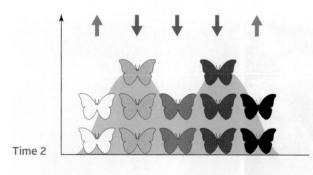

Time 2

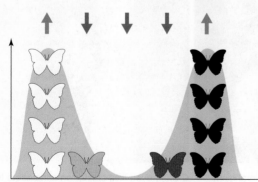

Time 3

Figure 17.10 Animated Disruptive selection eliminates midrange forms of a trait, and maintains extreme forms. *Red* arrows indicate which forms are being selected against; *green*, forms that are being favored.

Birds that have a bill between 12 and 15 millimeters wide are uncommon. Seedcrackers with the large and small bill forms inhabit the same geographic range, and they breed randomly with respect to bill size. It is as if every human adult were 4 feet or 6 feet tall, with no one of intermediate height.

The dimorphism in seedcracker bill size is main-tained by environmental factors that affect feeding performance. The finches feed mainly on the seeds of two types of sedge, a grasslike plant. One sedge pro-duces hard seeds; the other, soft seeds. Small-billed birds are better at opening the soft seeds, but large-billed birds are better at cracking the hard ones. All seeds are abundant during Cameroon's semiannual wet seasons, and all seedcrackers feed on both types of seeds. However, sedge seeds become scarce during the region's dry seasons. As competition for food intensi-fies, each bird focuses on eating the seeds that it opens most efficiently: Small-billed birds feed mainly on soft seeds, and large-billed birds feed mainly on hard seeds. Birds with intermediate-sized bills cannot open either type of seed as efficiently as the other birds, so they are less likely to survive the dry seasons.

Figure 17.11 Animated Disruptive selection in African seedcracker populations. In these birds, a distinct dimorphism in bill size is a result of competition for scarce food during dry seasons. These conditions favor birds with bills that are (**A**) 12 millimeters wide or (**B**) 15 to 20 millimeters wide. Birds with bills of intermediate size are selected against.

disruptive selection Mode of natural selection that favors forms of a trait at the extremes of a range of variation; intermediate forms are selected against.
stabilizing selection Mode of natural selection in which intermedi-ate forms of a trait are favored over extremes.

Take-Home Message

What types of natural selection favor intermediate or extreme forms of traits?

» With stabilizing selection, an intermediate phenotype is favored, and extreme forms are selected against.

» With disruptive selection, an intermediate form of a trait is selected against, and extreme phenotypes are favored.

17.7 Fostering Diversity

- Individuals may be selective agents for their own species.
- Any mode of natural selection may maintain two or more alleles in a population.
- Links to Sickle-cell anemia 9.6, *Drosophila* 10.3, Codominance 13.5

Selection pressures that operate on natural populations are often not as clear-cut as the examples in the previous sections might suggest. An allele may be adaptive in one circumstance but harmful in another, as the story about warfarin resistance in rats illustrates. Even individuals of the same species can play a role.

Nonrandom Mating

Not all evolution is driven by interactions of a species with its environment. Competition for mates is also a selective pressure. Consider how individuals of many sexually reproducing species have a distinct male or female phenotype—a **sexual dimorphism**. Individuals of one sex (often males) tend to be more colorful, larger, or more aggressive than individuals of the other sex. These traits seem puzzling because they take energy and time away from an individual's survival activities. Some are probably maladaptive because they attract predators. Why do they persist?

The answer is **sexual selection**, in which the genetic winners outreproduce others of a population because they are better at securing mates. With this mode of natural selection, the most adaptive forms of a trait help individuals defeat same-sex rivals for mates, or are the ones most attractive to the opposite sex.

For example, the females of some species cluster in defensible groups when they are sexually receptive, and males compete for sole access to the groups. Competition for ready-made harems favors combative males (**Figure 17.12A**). As another example, males or females that are choosy about mates act as selective agents on their own species. The females of some species shop for a mate among males that display species-specific cues such as a specialized appearance or courtship behavior (**Figure 17.12B**). The cues often include flashy body parts or behaviors, traits that can be a physical hindrance (or attract predators). However, a flashy male's survival despite his obvious handicap implies health and vigor, two traits that are likely to improve a female's chances of bearing healthy, vigorous offspring. Selected males pass alleles for their attractive traits to the next generation of males, and females pass alleles that influence mate preference to the next generation of females. Sexual selection can give rise to highly exaggerated traits (**Figure 17.12C**).

A Male elephant seals fight for sexual access to a cluster of females. Males of this species typically compete for access to the clusters of females.

Figure 17.12 Sexual selection in action.

B A male bird of paradise engaged in a flashy courtship display has caught the eye (and, perhaps, the sexual interest) of a female. Female birds of paradise are choosy; a male mates with any female that accepts him.

C Stalk-eyed flies cluster on aerial roots to mate. Females prefer males with the longest eyestalks, a trait that provides no obvious survival advantage other than sexual attractiveness. This male's very long eyestalks (*top*) have captured the interest of the three females below him.

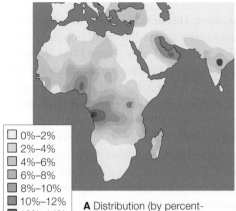

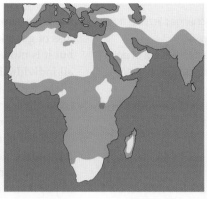

- ☐ 0%–2%
- ☐ 2%–4%
- ☐ 4%–6%
- ☐ 6%–8%
- ☐ 8%–10%
- ☐ 10%–12%
- ☐ 12%–14%
- ☐ >14%

A Distribution (by percentage) of people who carry the sickle-cell allele.

B Distribution of malaria cases (*orange*) reported in Africa, Asia, and the Middle East in the 1920s, before the start of programs to control mosquitoes, which transmit the parasitic protist that causes the disease. Notice the correlation with the distribution of the sickle-cell allele shown in **A**. The photo shows a physician searching for mosquito larvae in Southeast Asia.

Figure 17.13 Malaria and sickle-cell anemia.

Balanced Polymorphism

Any mode of natural selection may result in a **balanced polymorphism**. In this state, multiple alleles are maintained in a population at relatively high frequency. Sexual selection maintains balanced polymorphisms in *Drosophila* fruit flies: Female flies selectively mate with males bearing minority phenotypes in a population. For example, rare white-eyed males are preferred until they become more common than red-eyed males, at which point the red-eyed flies are again preferred. This is an example of frequency-dependent selection, in which the adaptive value of a trait depends on its frequency in a population.

Balanced polymorphism can also be an outcome of a particular environment that favors heterozygotes (individuals with nonidentical alleles). Consider the gene that encodes the beta globin chain of hemoglobin. Hb^A is the normal allele; the codominant Hb^S allele carries a mutation that causes sickle-cell anemia (Section 9.6). Individuals homozygous for the Hb^S allele often die in their teens or early twenties.

Despite being so harmful, the Hb^S allele persists at very high frequency among human populations in tropical and subtropical regions of Asia, Africa, and the Middle East. Why? Populations with the highest frequency of the Hb^S allele also have the highest incidence of malaria (**Figure 17.13**). Mosquitoes transmit the parasitic protist that causes malaria, *Plasmodium*, to human hosts. *Plasmodium* multiplies in the liver and then in red blood cells. The cells rupture and release new parasites during recurring bouts of severe illness.

People who make both normal and sickle hemoglobin are more likely to survive malaria than people who make only normal hemoglobin. In Hb^A/Hb^S heterozygotes, *Plasmodium*-infected red blood cells sometimes sickle. The abnormal shape brings the cells to the attention of the immune system, which destroys them along with the parasites they harbor. By contrast, *Plasmodium*-infected red blood cells of Hb^A/Hb^A homozygotes do not sickle, so the parasite may remain hidden from the immune system.

In areas where malaria is common, the persistence of the Hb^S allele is a matter of relative evils. Malaria and sickle-cell anemia are both potentially deadly. Hb^A/Hb^S heterozygotes are more likely to survive malaria than Hb^A/Hb^A homozygotes. Heterozygotes are not completely healthy, but they do make enough normal hemoglobin to survive. With or without malaria, heterozygotes are more likely to live long enough to reproduce than Hb^S/Hb^S homozygotes. The result is that nearly one-third of people from the most malaria-ridden regions of the world are heterozygous for the Hb^S allele.

balanced polymorphism Maintenance of two or more alleles for a trait at high frequency in a population as a result of natural selection against homozygotes.
sexual dimorphism Distinct male and female phenotypes.
sexual selection Mode of natural selection in which some individuals outreproduce others of a population because they are better at securing mates.

Take-Home Message

How does natural selection maintain diversity?

» With sexual selection, a trait is adaptive if it gives an individual an advantage in securing mates. Sexual selection that reinforces phenotypical differences between males and females sometimes results in exaggerated traits.

» Balanced polymorphism can be an outcome of frequency-dependent selection, or of environmental pressures that favor heterozygotes.

17.8 Genetic Drift and Gene Flow

■ Especially in small populations, random changes in allele frequencies can lead to a loss of genetic diversity.

■ Individuals, along with their alleles, move into and out of populations. This flow of alleles counters genetic change that tends to occur within a population.

■ Links to Probability and sampling error 1.8, Locus 13.2, Ellis–van Creveld syndrome 14.2, Transgenic plants 15.7

Genetic Drift

Genetic drift is random change in an allele's frequency over time, brought about by chance alone. We explain genetic drift in terms of probability—the chance that some event will occur. Sample size is important in probability. For instance, each time you flip a coin, there is a 50 percent chance it will land heads up. With 10 flips, the proportion of times heads actually land up may be very far from 50 percent. With 1,000 flips, that proportion is more likely to be near 50 percent.

The same rule holds for populations: the larger the population, the smaller the impact of random changes in allele frequencies. Imagine two populations, one with 10 individuals, the other with 100. If allele *X* occurs in both populations at a 10 percent frequency, then only one person carries the allele in the small population. If that person dies without reproducing, allele *X* will be lost from that population. However, ten people in the large population carry the allele. All ten would have to die without reproducing for the allele to be lost. Thus, the chance that the small population will lose allele *X* is greater than that for the large population. This effect is a general one: The loss of genetic diversity is possible in all populations, but it is more likely in small ones (**Figure 17.14**). When all individuals of a population are homozygous for an allele, we say that the allele is **fixed**. The frequency of a fixed allele will not change unless mutation or another process introduces a new allele into the population.

Loss of genetic diversity can be dramatic when a few individuals rebuild a population, such as occurs after a **bottleneck**. A bottleneck is a drastic reduction in population size brought about by severe selection pressure. For example, excessive hunting had reduced the population of northern elephant seals (shown in **Figure 17.12A**) to a mere twenty individuals by the late 1890s. The population has recovered since hunting restrictions were implemented, but every individual is homozygous for every allele analyzed to date. Genetic drift after the bottleneck has fixed all of the alleles in this population.

A loss of genetic diversity can also occur when a small group of individuals establishes a new population. If the founding group is not representative of the original population in terms of allele frequencies, then the new population will not be representative of it either, an outcome called the **founder effect**. If the founding group is very small, the new population's genetic diversity may be quite reduced. Imagine that a seabird lands in a population of plants on a mainland. In this population's gene pool, half of the alleles governing flower color specify white flowers; the other half specify yellow flowers. A few seeds stick to the

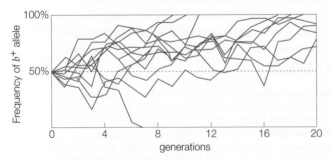

A The size of these populations of beetles was maintained at 10 breeding individuals. Allele b^+ was lost in one population (one graph line ends at 0).

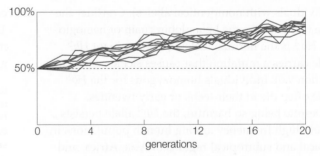

B The size of these populations of beetles was maintained at 100 individuals. Compare the genetic drift in **A**.

Figure 17.14 Animated Genetic drift in flour beetles (*Tribolium castaneum*, shown at *right* on a flake of cereal). Randomly selected beetles heterozygous for alleles b^+ and b were maintained in populations of (**A**) 10 individuals or (**B**) 100 individuals for 20 generations.

Graph lines in (**B**) are smoother than in (**A**), indicating that drift was greatest in the sets of 10 beetles and least in the sets of 100. Notice that the average frequency of allele b^+ rose at the same rate in both groups, an indication that natural selection was at work too: Allele b^+ was weakly favored. **Figure It Out:** In how many populations did allele b^+ become fixed? Answer: Six

Figure 17.15 An Amish child with Ellis–van Creveld syndrome. The syndrome is characterized by dwarfism, polydactyly, and heart defects, among other symptoms. The recessive allele that causes it is common in the Old Order Amish of Lancaster County, an outcome of the founder effect and moderate inbreeding.

bird's feathers. The bird flies to a remote island and drops the seeds, which later sprout and form a new population on the island. If most of the seeds happened to be homozygous for the yellow flower allele, then the frequency of that allele in the new population will be much greater than 50 percent.

Founding populations are often necessarily inbred. **Inbreeding** is nonrandom breeding or mating between close relatives of a population, which share more alleles than nonrelatives do. Inbred populations tend to have unusually high numbers of individuals homozygous for harmful recessive alleles. This is why most societies discourage or forbid incest, or mating between parents and children or between siblings.

The Old Order Amish in Lancaster County, Pennsylvania, offer an example of the effects of inbreeding within human populations. Amish people marry only within their community. Intermarriage with other groups is not permitted, and no "outsiders" are allowed to join the community. As a result, Amish populations are moderately inbred, and many of their individuals are homozygous for harmful alleles. The Lancaster population has an unusually high frequency of a recessive allele that causes Ellis–van Creveld syndrome (**Figure 17.15**). This allele has been traced to a

man and his wife, two of a group of 400 Amish who immigrated to the United States in the mid-1700s. As a result of the founder effect and inbreeding since then, about 1 of 8 people in the Lancaster population is now heterozygous for the allele, and 1 in 200 is homozygous for it.

Gene Flow

Individuals tend to mate or breed most frequently with other members of their own population. However, not all populations of a species are completely isolated from one another, and nearby populations may occasionally interbreed. Also, individuals sometimes leave one population and join another. **Gene flow**, the movement of alleles between populations, occurs in both cases. Gene flow stabilizes allele frequencies, so it counters the effects of mutation, natural selection, and genetic drift.

Gene flow is typical among populations of animals, which tend to be more mobile, but it also occurs in plant populations. Consider the acorns that jays disperse when they gather nuts for the winter (*left*). Every fall, the birds visit acorn-bearing oak trees repeatedly, then bury acorns in the soil of home territories that may be as much as a mile away. The jays transfer acorns (and the alleles carried by the acorns) among populations of oak trees that would otherwise be genetically isolated.

In plants, gene flow also occurs when pollen is transferred from one individual to another, often over great distances. Many opponents of genetic engineering cite gene flow from transgenic organisms into wild populations via pollen transfer. Herbicide-resistance genes and the *Bt* gene (Section 15.7) are now common in weeds and unmodified crop plants; the long-term effects of this gene flow are currently unknown.

bottleneck Drastic reduction in population size as a result of severe selection pressure.
fixed Refers to an allele for which all members of a population are homozygous.
founder effect After a small group of individuals found a new population, allele frequencies in the new population differ from those in the original population.
gene flow The movement of alleles into and out of a population.
genetic drift Change in allele frequencies in a population due to chance alone.
inbreeding Mating among close relatives.

Take-Home Message

How does a population's genetic diversity become reduced?

» Genetic drift, or random change in allele frequencies, can reduce a population's genetic diversity. Its effect is greatest in small populations, such as one that endures a bottleneck.

» Gene flow is the physical movement of alleles into and out of a population. It tends to counter the evolutionary effects of mutation, natural selection, and genetic drift.

17.9 Reproductive Isolation

■ Speciation differs in its details, but reproductive isolating mechanisms are always part of the process.
■ Links to Species 1.5, Zygote 12.2, Meiosis 12.3

Mutation, natural selection, and genetic drift operate on all natural populations, and they do so independently in populations that are not interbreeding. When gene flow does not keep populations alike, different genetic changes accumulate in each one. Sooner or later, the populations become so different that we call them different species. The evolutionary process in which new species arise is called **speciation**.

Evolution is a dynamic, extravagant, messy, and ongoing process that can be challenging for people who like their categories neat. Speciation offers a perfect example, because it rarely occurs at a precise moment in time. Individuals often continue to interbreed even as populations are diverging, and populations that have already diverged may come together and interbreed again.

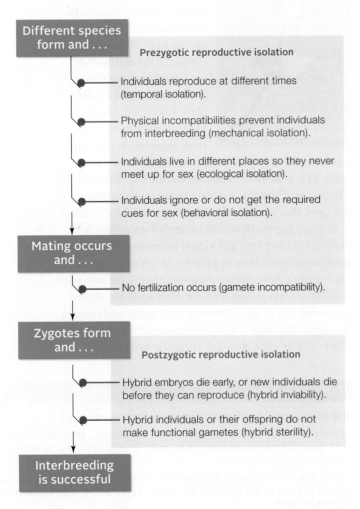

Different species form and . . .

Prezygotic reproductive isolation

— Individuals reproduce at different times (temporal isolation).

— Physical incompatibilities prevent individuals from interbreeding (mechanical isolation).

— Individuals live in different places so they never meet up for sex (ecological isolation).

— Individuals ignore or do not get the required cues for sex (behavioral isolation).

Mating occurs and . . .

— No fertilization occurs (gamete incompatibility).

Zygotes form and . . .

Postzygotic reproductive isolation

— Hybrid embryos die early, or new individuals die before they can reproduce (hybrid inviability).

— Hybrid individuals or their offspring do not make functional gametes (hybrid sterility).

Interbreeding is successful

Figure 17.16 Animated How reproductive isolation prevents interbreeding.

Every time speciation happens, it happens in a unique way, which means that each species is a product of its own unique evolutionary history. However, we can identify some recurring patterns. For example, reproductive isolation is always part of speciation. **Reproductive isolation** refers to the end of gene flow between populations. It is part of the process by which sexually reproducing species attain and maintain their separate identities. By preventing successful interbreeding, reproductive isolation reinforces differences between diverging populations. We say the isolation is prezygotic if pollination or mating cannot occur, or if zygotes cannot form. If hybrids form but are unfit or infertile, the isolation is postzygotic (**Figure 17.16**).

Mechanisms of Reproductive Isolation

Temporal Isolation Some populations cannot interbreed because the timing of their reproduction differs. The periodical cicada (*inset*) offers an example. Cicadas feed on roots as they mature underground, then emerge to reproduce. Three species of cicada reproduce every 17 years. Each has a sibling species with nearly identical form and behavior, except that the siblings emerge on a 13-year cycle instead of a 17-year cycle. Sibling species have the potential to interbreed, but they can only get together once every 221 years!

Mechanical Isolation In some cases, the size or shape of an individual's reproductive parts prevent it from mating with members of another population. For example, black sage (*Salvia mellifera*) and white sage (*Salvia apiana*) grow in the same areas, but hybrids rarely form because the flowers of the two species have become specialized for different pollinators (**Figure 17.17**).

Ecological Isolation Populations adapted to different microenvironments in the same region may be ecologically isolated. For example, two species of manzanita (a plant) native to the Sierra Nevada mountain range rarely hybridize. One species that is better adapted for conserving water inhabits dry, rocky hillsides high in the foothills. The other lives on lower slopes where water stress is not as intense. The physical separation makes cross-pollination unlikely.

Behavioral Isolation In animals, behavioral differences can stop gene flow between related species. For instance, males and females of some species engage in courtship displays before sex. A female recognizes

anthers

stigma

A Black sage is pollinated mainly by honeybees and other small insects.

Figure 17.17
Mechanical isolation in sage.

B The flowers of black sage are too delicate to support larger insects. Big insects access the nectar of small sage flowers only by piercing from the outside, as this carpenter bee is doing. When they do so, they avoid touching the flower's reproductive parts.

C The reproductive parts (anthers and stigma) of white sage flowers are too far away from the petals to be brushed by honeybees, so honeybees cannot pollinate this species. White sage is pollinated mainly by larger bees and hawkmoths, which brush the flower's stigma and anthers as they pry apart the petals to access nectar.

vocalizations and movements of a male of her species as an overture to sex, but females of different species usually do not (**Figure 17.18**).

Gamete Incompatibility Even if gametes of different species do meet up, they often have molecular incompatibilities that prevent a zygote from forming. As you will see in Chapter 29, molecular signals that guide pollen germination in flowering plants are species-specific. Gamete incompatibility may be the primary speciation route of animals that fertilize their eggs by releasing free-swimming sperm in water.

Figure 17.18 Behavioral isolation. A male peacock spider (*Maratus volans*) approaches a female, signaling his intent to mate with her by raising and waving colorful flaps, and gesturing his legs in time with abdominal vibrations. If his species-specific courtship display fails to impress her, she will kill him.

Hybrid Inviability As populations begin to diverge, so do their genes. Even chromosomes of species that diverged recently may have major differences. Thus, a hybrid zygote may end up with extra or missing genes, or genes with incompatible products. These outcomes often disrupt embryonic development. If the hybrid survives, its fitness may be reduced. For example, ligers and tigons (offspring of lions and tigers) have more health problems and a shorter life expectancy than individuals of either parent species.

Hybrid Sterility Some interspecies crosses produce robust but sterile offspring. For example, mating a female horse (64 chromosomes) with a male donkey (62 chromosomes) produces a mule. The mule's 63 chromosomes cannot pair up evenly during meiosis, so this animal makes few viable gametes.

If hybrids are fertile, their offspring usually have lower and lower fitness with each successive generation. A mismatch between nuclear and mitochondrial DNA may be the cause (mitochondrial DNA is inherited from the mother only).

reproductive isolation The end of gene flow between populations.
speciation Evolutionary process in which new species arise.

Take-Home Message

How do species attain and maintain separate identities?

» Speciation is an evolutionary process by which new species form. It varies in its details and duration.

» Reproductive isolation, which occurs by one of several mechanisms, is always a part of speciation.

17.10 Allopatric Speciation

- In allopatric speciation, a physical barrier arises and ends gene flow between populations.
- Links to Galápagos archipelago 16.4, Plate tectonics 16.7

Genetic changes that lead to a new species can begin with physical separation between populations. By **allopatric speciation**, a physical barrier arises and separates two populations, ending gene flow between them (*allo*– means different; *patria*, fatherland). Then, reproductive isolating mechanisms evolve so even if the populations meet up again later, their individuals could not interbreed.

Populations are separated by distance, and gene flow between them is usually intermittent. Whether a geographic barrier can block that gene flow depends on how the species travels (such as by swimming, walking, or flying), and how it reproduces (for example, by internal fertilization or by pollen dispersal).

A geographic barrier can arise in an instant, or over an eon. The Great Wall of China is an example of a barrier that arose abruptly. As it was being built, the wall cut off gene flow among nearby populations of insect-pollinated plants. DNA sequencing comparisons show that trees, shrubs, and herbs on either side of the wall are diverging genetically.

Geographic isolation usually occurs much more slowly. For example, it took millions of years of tectonic plate movements (Section 16.7) to bring the two continents of North and South America close enough to collide. The land bridge where the two continents now connect is called the Isthmus of Panama. When this isthmus formed about 4 million years ago, it cut off the flow of water—and gene flow among populations of aquatic organisms—as it separated one large ocean into what are now the Pacific and the Atlantic oceans (**Figure 17.19**).

Speciation in Archipelagos

New species rarely form on island chains such as the Florida Keys that are in close proximity to a mainland. Being close to a mainland means gene flow is essentially unimpeded between island and mainland populations. By contrast, allopatric speciation is common on archipelagos such as the Hawaiian and Galápagos Islands. These island chains are so geographically isolated that, for most species, gene flow does not occur between island and mainland populations.

Archipelagos are born of hot spots on the ocean floor. Because they are the tops of volcanoes, we can assume that their fiery surfaces are initially barren and inhospitable to life. Later, winds or currents can carry a few individuals of a mainland species to an archipelago. If the individuals reproduce, their descendants may establish a population. The vast expanse of ocean that isolates the island from the mainland is a geo-

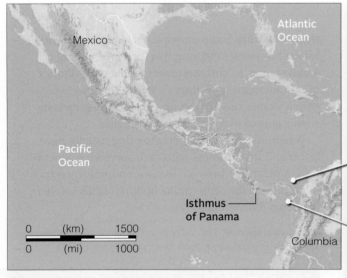

Figure 17.19 Allopatric speciation in snapping shrimp. The Isthmus of Panama (*above*) cut off gene flow among populations of these aquatic shrimp when it formed 4 million years ago. Today, individuals from opposite sides of the isthmus are so similar that they might interbreed, but they are behaviorally isolated: Instead of mating when they are brought together, they snap their claws at one another aggressively. The photos on the *right* show two of the many closely related species that live on opposite sides of the isthmus.

Akepa
(*Loxops coccineus*)

Insects, spiders, nectar; high mountain rain forest

Akekee
(*Loxops caeruleirostris*)

Insects, spiders, nectar; high mountain rain forest

Nihoa finch
(*Telespiza ultima*)

Insects, buds, seeds, flowers, seabird eggs; rocky or shrubby slopes

Palila
(*Loxioides bailleui*)

Mamane seeds, buds, flowers, berries, insects; high mountain dry forests

Maui parrotbill
(*Pseudonestor xanthophrys*)

Insect larvae, pupae; mountain forests, dense underbrush

Apapane
(*Himatione sanguinea*)

Nectar, caterpillars and other insects, spiders; high mountain forests

Poouli (*Melamprosops phaeosoma*)

Tree snails, insects in forest understory

Maui Alauahio
(*Paroreomyza montana*)

Bark or leaf insects, some nectar; high mountain rain forest

Kauai Amakihi (*Hemignathus kauaiensis*)

Bark-picker; insects, spiders, nectar; high mountain rain forest

Akiapolaau
(*Hemignathus munroi*)

Probes, digs insects from big trees; high mountain rain forest

Akohekohe
(*Palmeria dolei*)

Mostly nectar from flowering trees, some insects, pollen; high mountain rain forest

Iiwi
(*Vestiaria coccinea*)

Mostly nectar, some insects; high mountain rain forest

Figure 17.20 Animated Allopatric speciation on an archipelago.

Above, a few of the known species of Hawaiian honeycreepers. Specialized bills and behaviors adapt each species to its particular island niche.

Archipelagos such as the Hawaiian Islands (*right*) are separated from mainland continents by thousands of miles of open ocean—a geographic barrier that prevents gene flow between island colonizers and mainland populations. Further divergences occur as colonizers spread to the other islands in the chain. DNA sequence comparisons suggest that the ancestor of all Hawaiian honeycreepers resembled the housefinch (*Carpodacus*) at *left*.

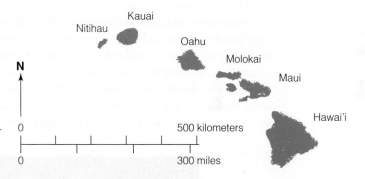

graphic barrier to gene flow. Thus, over generations, the island population diverges from the mainland species. Individuals of the diverging population may in turn colonize other islands in the archipelago. Habitats and selection pressures that differ within and between the islands can foster even more divergences from the ancestral species. Even as the island populations become different species, some individuals may return to an island colonized by their ancestors.

Consider the Hawaiian archipelago, which includes 19 islands and more than 100 atolls that stretch 1,500 miles in the Pacific Ocean. Habitats on the landmasses of this archipelago range from lava beds, rain forests, and grasslands to dry woodlands and snow-capped peaks. One mainland finch species that arrived on the archipelago about 3.5 million years ago found a buf-

fet of fruits, seeds, nectars, tasty insects, and the near absence of competitors and predators. Their descendants thrived. **Figure 17.20** hints at the variation among Hawaiian honeycreepers, descendants of the early colonizers. They and thousands of other species are unique to the Hawaiian archipelago.

allopatric speciation Speciation pattern in which a physical barrier ends gene flow between populations.

Take-Home Message

What happens after a physical barrier arises and prevents populations from interbreeding?

» A physical barrier that intervenes between populations or subpopulations of a species prevents gene flow among them. As gene flow ends, genetic divergences give rise to new species. This process is allopatric speciation.

17.11 Other Speciation Models

■ Populations sometimes speciate even without a physical barrier that bars gene flow between them.
■ Links to Polyploidy 14.6, Fitness 16.4

Sympatric Speciation

In **sympatric speciation**, populations inhabiting the same geographic region speciate in the absence of a physical barrier between them (*sym–* means together).

Sympatric speciation can occur in a single generation when the chromosome number multiplies. Polyploidy typically arises when an abnormal nuclear division during meiosis or mitosis doubles the chromosome number. For example, if the nucleus of a somatic cell in a flowering plant fails to divide during mitosis, the resulting polyploid cell may proliferate and give rise to shoots and flowers. If the flowers can self-fertilize, a new polyploid species may result. Common bread wheat originated after related species hybridized, and then the chromosome number of the hybrid offspring doubled (**Figure 17.21**). Today, about 95 percent of ferns and 70 percent of flowering plant species are polyploid, as well as a few conifers, insects and other arthropods, mollusks, fishes, amphibians, and reptiles.

Sympatric speciation can also occur with no change in chromosome number. The mechanically isolated sage plants you learned about in Section 17.9 speciated with no physical barrier to gene flow. As another example, more than 500 species of cichlid, a freshwater fish, arose in the shallow waters of Lake Victoria. This large freshwater lake sits isolated from river inflow on an elevated plain in Africa's Great Rift Valley. Since

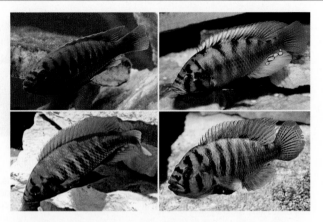

Figure 17.22 Red fish, blue fish: Males of four closely related species of cichlid native to Lake Victoria, Africa.

Hundreds of cichlids speciated in sympatry in this lake. Mutations in genes that affect females' perception of the color of ambient light in deeper or shallower regions of the lake also affect their choice of mates. Female cichlids prefer to mate with brightly colored males of their own species.

Figure It Out: Which form of natural selection is driving sympatric speciation in these cichlids?   Answer: Sexual selection

Lake Victoria formed about 400,000 years ago, it has dried up three times. DNA sequence comparisons indicate that almost all of the cichlid species in this lake arose since the last dry spell, which was 12,400 years ago. How could hundreds of species arise so quickly? In this case, the answer begins with differences in the color of ambient light and water clarity in different parts of the lake. Water absorbs blue light, so the deeper it is, the less blue light penetrates it. The light in shallower, clear water is mainly blue; the light that penetrates deeper, muddier water is mainly red.

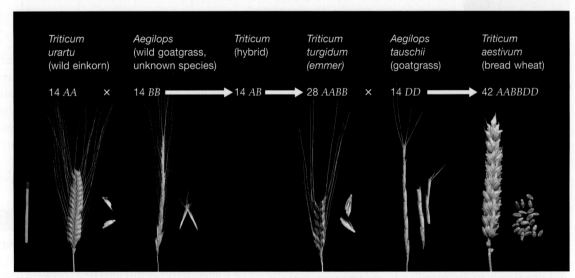

Figure 17.21 Animated Sympatric speciation in wheat.

The wheat genome, which consists of seven chromosomes, occurs in slightly different forms called *A*, *B*, *C*, *D*, and so on.

Many wheat species are polyploid, carrying more than two copies of the genome. Chromosome number is indicated for each species. For example, modern bread wheat (*Triticum aestivum*) is hexaploid, with six copies of the wheat genome: two each of genomes *A*, *B*, and *D* (or 42 *AABBDD*).

| *Triticum urartu* (wild einkorn) | *Aegilops* (wild goatgrass, unknown species) | *Triticum* (hybrid) | *Triticum turgidum* (emmer) | *Aegilops tauschii* (goatgrass) | *Triticum aestivum* (bread wheat) |

14 *AA* × 14 *BB* → 14 *AB* → 28 *AABB* × 14 *DD* → 42 *AABBDD*

A About 11,000 years ago, a diploid wheat (einkorn) hybridized with a diploid species of wild goatgrass.

B Tetraploid (4*n*) emmer arose when the chromosome number of the resulting hybrid doubled.

C Common bread wheat is the result of a hybridization between emmer and a diploid goatgrass.

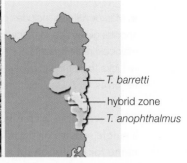

A Giant velvet walking worm, *Tasmanipatus barretti*

B Blind velvet walking worm, *T. anophthalmus*

C The habitats of the worms overlap in a hybrid zone on the island of Tasmania.

T. barretti
hybrid zone
T. anophthalmus

Figure 17.23 Example of parapatric speciation: velvet walking worms in Tasmania.

Victoria cichlids vary in color and in patterning (**Figure 17.22**). Outside of captivity, female cichlids rarely mate with males of other species. Given a choice, they prefer brightly colored males of their own species. Their preference has a molecular basis in genes that encode light-sensitive pigments of the eye. The pigments made by species that live mainly in shallower, clear water are more sensitive to blue light. The males of these species are also the bluest. The pigments made by species that live mainly in deeper, murkier water are more sensitive to red light. Males of these species are redder. In other words, the colors that a female cichlid sees best are the same colors displayed by males of her species. Thus, mutations in genes that affect color perception are likely to affect the choice of mates and of habitats. Such mutations are probably the way sympatric speciation occurs in these fish.

Sympatric speciation has also occurred in greenish warblers of central Asia (*Phylloscopus trochiloides*). A

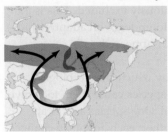

chain of populations of this bird encircles the Tibetan plateau (*left*). Adjacent populations of greenish warblers interbreed easily, except for two populations in northern Siberia. Individuals of these two populations overlap in range, but they do not interbreed because they do not recognize one another's songs—they have become behaviorally isolated. Small genetic differences between the other populations have added up to major differences between the two populations at the ends of the chain. Greenish warbler populations comprising the chain are collectively called a ring species. Ring species present one of those paradoxes for people who like neat categories: Gene flow occurs continuously all around the chain, but the two populations at the ends of the chain are different species. Where should we draw the line that divides those two species?

Parapatric Speciation

With **parapatric speciation**, adjacent populations speciate despite being in contact across a common border. Divergences spurred by local selection pressures are reinforced because hybrids that form in the contact zone are less fit than individuals on either side of it.

Consider velvet walking worms, which resemble caterpillars but may be more related to spiders. These worms are predatory; they shoot streams of glue from their head at insects. Once entangled in the sticky glue, the insects are easy prey for the worms. Two rare species of velvet walking worm are native to the island of Tasmania (**Figure 17.23**). The giant velvet walking worm (*Tasmanipatus barretti*) and the blind velvet walking worm (*Tasmanipatus anophthalmus*) can interbreed, but they only do so in a tiny area where their habitats overlap. Hybrid offspring are sterile, which may be the main reason the two species are maintaining separate identities in the absence of a physical barrier between their adjacent populations.

parapatric speciation Speciation model in which adjacent populations speciate despite being in contact along a common border.
sympatric speciation Pattern in which speciation occurs in the absence of a physical barrier to gene flow.

Take-Home Message

Can speciation occur without a physical barrier to gene flow?

» By a sympatric speciation model, new species arise from a population even in the absence of a physical barrier.

» By a parapatric speciation model, populations maintaining contact along a common border evolve into distinct species.

17.12 Macroevolution

■ Macroevolution includes patterns of change such as one species giving rise to many, the origin of major groups, and major extinction events.

■ Links to Homeotic genes 10.3, Mass extinctions 16.1, Plate tectonics 16.7, Geologic time scale 16.8

Microevolution is change in allele frequencies within a single species or population. **Macroevolution** is our name for evolution on a larger scale: grand evolutionary trends such as flowering plants evolving from seed plants, the dinosaurs disappearing in a mass extinction, hundreds of Hawaiian honeycreepers evolving from a housefinch, and so on.

Patterns of Macroevolution

Stasis With the simplest macroevolutionary pattern, **stasis**, lineages persist for millions of years with little or no change. Consider coelacanths, an order of ancient lobe-finned fish that had been assumed extinct for at least 70 million years until a fisherman caught one in 1938. Modern coelacanths are very similar to fossil specimens hundreds of millions of years old (**Figure 17.24**).

Exaptation Major evolutionary novelties often stem from the adaptation of an existing structure for a completely different purpose. This process is called preadaptation or **exaptation**. Some traits serve a very different purpose in modern species than they did when they first evolved. For example, the feathers that allow modern birds to fly are derived from feathers that first evolved in some dinosaurs. Those dinosaurs could not have used their feathers for flight, but they probably did use them for insulation. Thus, we say that flight feathers in birds are an exaptation of insulating feathers in dinosaurs.

Mass Extinctions By current estimates, more than 99 percent of all species that ever lived are now **extinct**, or irrevocably lost from Earth. In addition to continuing small-scale extinctions, the fossil record indicates that there have been more than twenty mass extinctions, which are simultaneous losses of many lineages. These include five catastrophic events in which the majority of species on Earth disappeared (Section 16.8).

Adaptive Radiation In an evolutionary pattern called **adaptive radiation**, a lineage rapidly diversifies into several new species. Adaptive radiation can occur after individuals colonize a new environment that has a variety of different habitats with few or no competitors. The adaptation of populations to different regions of the new environment produces new species. The Hawaiian honeycreepers that you read about in Section 17.10 arose this way.

Adaptive radiation may occur after a key innovation evolves. A **key innovation** is a new trait that allows its bearer to exploit a habitat more efficiently or in a novel way. The evolution of lungs offers an example, because lungs were a key innovation that opened the way for an adaptive radiation of vertebrates on land.

Adaptive radiation can also occur after geologic or climatic events eliminate some species from a habitat. The surviving species then can exploit resources from which they had been previously excluded. This is the way mammals were able to undergo an adaptive radiation after the dinosaurs disappeared.

Coevolution The process by which close ecological interactions between two species cause them to evolve jointly is called **coevolution**. One species acts as an agent of selection on the other; each adapts to changes in the other. Over evolutionary time, the two species may become so interdependent that they can no longer survive without one another.

Relationships between coevolved species can be quite intricate. Consider the large blue butterfly (*Maculinea arion*), a parasite of ants (**Figure 17.25**). After hatching, the larvae (caterpillars) feed on wild thyme flowers and then drop to the ground. An ant that finds a caterpillar strokes it, whereupon the caterpillar exudes honey. The ant eats the honey and continues to

Figure 17.24 An example of stasis. *Top*, 320-million-year-old coelacanth fossil from Montana. *Bottom*, a live coelacanth (*Latimeria chalumnae*) found near Sulawesi in 1998. The coelacanth lineage has changed very little over evolutionary time.

Rise of the Super Rats (revisited)

The warfarin-resistance allele of *VKORC1* is clearly adaptive in rats exposed to warfarin. Rats that carry it require a lot of extra vitamin K, but being vitamin K deficient is not so bad when compared with being dead from rat poison. However, in the absence of warfarin, rats that carry the allele are at a serious disad-

vantage, because they cannot easily obtain enough vitamin K from their diet to sustain normal blood clotting and bone formation. Thus, the allele's frequency declines quickly in the absence of warfarin. Periodic exposure to the poison maintains the allele in rat populations—an example of how selection pressure can maintain a balanced polymorphism.

How would you vote? One standard animal husbandry practice includes continuously supplementing the feed of healthy livestock with the same antibiotics prescribed for people. Should this practice stop?

---

stroke the caterpillar, which secretes more honey. This interaction continues for hours, until the caterpillar suddenly hunches itself up into a shape that appears (to an ant) very much like an ant larva. The deceived ant then picks up the caterpillar and carries it back to the ant nest, where, in most cases, other ants kill it—except, however, if the ants are of the species *Myrmica sabuleti*. Secretions of the caterpillar fool these ants into treating it just like a larva of their own. For the next 10 months, the caterpillar lives in the nest and grows to gigantic proportions by feeding on ant larvae. After it metamorphoses into a butterfly, the insect emerges from the ground to lay its eggs on wild thyme near another *M. sabuleti* nest, and the cycle starts anew. This relationship between ant and butterfly is typical of coevolved relationships in that it is extremely specific. Any increase in the ants' ability to identify a caterpillar in their nest selects for caterpillars that better deceive the ants, which in turn select for ants that can better identify the caterpillars. Each species exerts directional selection on the other.

Evolutionary Theory

Biologists do not doubt that macroevolution occurs, but many disagree about how it occurs. However we choose to categorize evolutionary processes, the very same genetic change may be at the root of all evolution—fast or slow, large-scale or small-scale. Dramatic jumps in morphology, if they are not artifacts of gaps in the fossil record, may be the result of mutations in homeotic or other regulatory genes. Macroevolution may include more processes than microevolution, or it may not. It may be an accumulation of many microevolutionary events, or it may be an entirely different process. Evolutionary biologists may disagree about these and other hypotheses, but all of them are trying to explain the same thing: how all species are related by descent from common ancestors.

A To a *Myrmica sabuleti* ant, a honey-exuding, hunched-up *Maculinea arion* caterpillar appears to be an ant larva. This deceived ant is preparing to carry the caterpillar back to its nest, where the caterpillar will eat ant larvae for the next 10 months until it pupates.

B A *Maculinea arion* butterfly emerges from the pupa and lays eggs on wild thyme flowers near another nest of *Myrmica sabuleti* ants. Larvae that emerge from the eggs will survive only if an ant colony adopts them.

Figure 17.25 Coevolved species: *Myrmica sabuleti* and *Maculinea arion*.

adaptive radiation A burst of genetic divergences from a lineage gives rise to many new species.

coevolution The joint evolution of two closely interacting species; each species is a selective agent for traits of the other.

exaptation Adaptation of an existing structure for a completely different purpose.

extinct Refers to a species that has been permanently lost.

key innovation An evolutionary adaptation that gives its bearer the opportunity to exploit a particular environment more efficiently or in a new way.

macroevolution Large scale evolutionary patterns and trends.

stasis Evolutionary pattern in which a lineage persists with little or no change over evolutionary time.

Take-Home Message

What is macroevolution?

» Macroevolution comprises large-scale patterns of evolutionary change such as adaptive radiation, the origin of major groups, and mass extinction.

---

18 Information About Species

LEARNING ROADMAP

Where you have been This chapter adds the concept of evolution (Sections 16.2 and 16.3) to taxonomy (1.5). Before starting, you should review DNA sequences (8.4) and sequencing (15.4); the genetic code (9.4); master genes (10.3, 10.4); genomics (15.5), neutral mutations (17.2), genetic equilibrium (17.3); gene flow (17.8); and speciation (17.9).

Where you are now

Phylogeny
Evolutionary biologists can reconstruct the evolutionary history of a group of organisms by identifying shared, heritable traits that evolved in a common ancestor.

Comparing Body Form
Similar body parts in different lineages may indicate descent from a shared ancestor, or they may have evolved independently in response to similar environmental pressures.

```
RDVQFGWLIRNIHANG
RDVNYGWLIRYMHANG
RDVHYGWIIRYMHANG
RDVNYGWLIRNLHANG
RDVNYGWIIRYLHANG
RDVEGGWLLRYMHANG
TDVMNGWMVRSIHANG
```

Comparing Biochemistry
Neutral changes tend to accumulate at a fairly constant rate in DNA. Molecular comparisons help us discover and confirm relationships among species and lineages.

Comparing Development
Patterns of development have a basis in master genes conserved over evolutionary time. Lineages with recent common ancestry often develop in similar ways.

Applications of Phylogeny
Understanding a group's evolutionary history can help us protect endangered species. Applied to agents of infectious disease, it can also reveal large-scale patterns of transmission.

Where you are going You will revisit cladograms and phylogeny throughout Unit IV as you learn about life's diversity. Classification of birds and reptiles is explained in Chapter 25. Chapter 44 returns to population biology, the study of which is important for understanding human impacts on the biosphere (Chapter 48). Our efforts to mitigate that impact include conservation efforts aimed at maintaining biodiversity (Chapter 45) in ecosystems (Chapter 46). Animal development returns in Chapter 42.

18.1 Bye Bye Birdie

Kauai, the first of the big islands of the Hawaiian archipelago, rose above the surface of the sea more than 5 million years ago. A million years later, a few finches reached it after traveling 4,000 kilometers (2,500 miles) across the open ocean. The birds' arrival had been preceded by plenty of insects and plants, but no predators. Their descendants thrived, expanding into habitats along the coasts, through dry lowland forests, and into highland rain forests of the island.

Between 1.8 million and 400,000 years ago, volcanic eruptions created the rest of the archipelago. Descendants of the first finches flew to the new islands, each of which had different foods and nesting sites. Over many generations, the birds evolved as the Hawaiian honeycreepers. Section 17.10 introduced you to some of the unique forms and behaviors that allowed honeycreepers to exploit special opportunities presented by their island habitats.

The first Polynesians arrived on the Hawaiian Islands around 1000 A.D., and Europeans followed in 1778. Hawaii's rich ecosystem was hospitable to all newcomers, including the settlers' dogs, cats, pigs, cows, goats, deer, and sheep. Escaped livestock began to eat and trample rain forest plants that had provided the honeycreepers with food and shelter. Entire forests were cleared to grow imported crops, and plants that escaped cultivation began to crowd out native plants. Mosquitoes accidentally introduced in 1826 spread diseases from imported birds such as chickens to native bird species. Stowaway rats and snakes ate their way through populations of native birds and their eggs. Mongooses deliberately imported to eat the rats and snakes preferred to eat birds and eggs.

The very isolation that had spurred adaptive radiations also made the honeycreepers vulnerable to extinction. The birds had no built-in defenses against predators or diseases of the mainland. Specializations such as extravagantly elongated beaks became hindrances when the birds' habitats suddenly changed or disappeared. Thus, at least 43 honeycreeper species that had thrived on the islands before humans arrived were extinct by 1778. Conservation efforts began in the 1960s, but 26 more species have since disappeared. Today, 35 of the remaining 68 species are endangered (**Figure 18.1**). They are still being pressured by established populations of invasive, non-native species of plants and animals. Rising global temperatures are also allowing mosquitoes to invade high-altitude habitats that had previously been too cold for the insects. Honeycreeper species remaining in these habitats are now succumbing to mosquito-borne diseases.

A The palila (*Loxioides bailleui*) has an adaptation that allows it to feed mainly on the seeds of the mamane plant. The seeds are toxic to most other birds. The one remaining palila population is declining because mamane plants are being trampled by cows and gnawed to death by goats and sheep. Only about 1,200 palila remained in 2010.

B The unusual lower bill of the akekee (*Loxops caeruleirostris*) points to one side, allowing this bird to pry open buds that harbor insects. Avian malaria carried by mosquitoes to higher altitudes is decimating the last population of this species. Between 2000 and 2007, the number of akekee plummeted from 7,839 birds to 3,536.

C This male poouli (*Melamprosops phaeosoma*)—rare, old, and missing an eye—died in 2004 from avian malaria. There were two other poouli alive at the time, but neither has been seen since then.

Figure 18.1 Three honeycreeper species: going, going, and gone.

18.2 Phylogeny

■ Evolutionary history can be reconstructed by studying shared, heritable traits.
■ Links to Taxonomy 1.5, Comparative Morphology 16.2, Evolution 16.3

Our increasing understanding of evolution is prompting a major, ongoing overhaul of the way biologists view life's diversity. Classifying that tremendous diversity into a series of taxonomic ranks is a useful endeavor, in the same way that it is useful to organize a telephone book or contact list in alphabetical order. Today, however, reconstructing evolutionary relationships among organisms has become at least as important as classifying them. **Phylogeny**, the evolutionary history of a species or a group of them, is a kind of genealogy that follows a lineage's evolutionary relationships through time.

Humans were not around to witness the evolution of most species, but we can use evidence to understand events in the past (Section 16.1). Each species bears traces of its own unique evolutionary history in its characters. A **character** is a measurable, heritable trait such as the number of segments in a backbone,

the nucleotide sequence of ribosomal RNA, or any other physical, behavioral, or biochemical feature that can be measured or quantified (**Table 18.1**).

Traditional classification schemes group organisms based on shared characters: birds have feathers, cacti have spines, and so on. Such schemes do not necessarily reflect phylogeny. For example, species that appear very similar are not necessarily closely related (Section 16.2). By contrast, evolutionary biology tries to fit each species into a bigger picture of evolution, because every living thing is related if you just go back far enough in time.

Instead of grouping organisms by shared characters, evolutionary biologists pinpoint what makes the organisms share the characters in the first place: a common ancestor. They determine common ancestry by looking for derived traits. A **derived trait** is a character present in a group under consideration, but not in the group's ancestors. A group whose members share one or more defining derived traits is called a **clade**. By definition, a clade is a **monophyletic group**—one that consists of an ancestor (in which a derived trait evolved) together with all of its descendants.

It is the recent nature of a derived trait that defines a clade. Consider alligators—animals that seem to have more characters in common with lizards than with birds. Thus, we might classify alligators and lizards together based on their appearance. The similarities do indicate shared ancestry, but it is a more distant relationship than alligators have with birds. Derived traits—a gizzard and a four-chambered heart—evolved in the lineage that gave rise to alligators and birds, but not in the one that gave rise to lizards.

Many clades are equivalent to taxonomic rankings—flowering plants, for example, are both a clade and a phylum—but some are not. For example, the

Table 18.1	Examples of Characters		
	Bird	**Bat**	**Dolphin**
Warm Blood	Y	Y	Y
Hair	N	Y	Y
Milk	N	Y	Y
Teeth	N	Y	Y
Wings	Y	Y	N
Feathers	Y	N	N

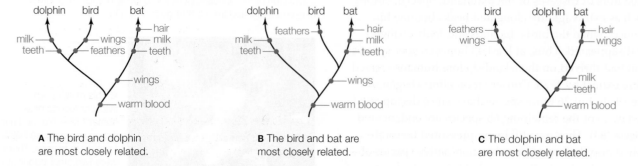

A The bird and dolphin are most closely related.

B The bird and bat are most closely related.

C The dolphin and bat are most closely related.

Figure 18.2 A simple example of parsimony analysis, using the data in **Table 18.1**. There are three possible evolutionary relationships among a bird, bat, and dolphin. The scenario that is most likely to be correct is the one in which the derived traits (in *red*) would have arisen the fewest number of times. **Figure It Out: Which scenario (A, B, or C) is the one most likely to be correct?**

Answer: C. In this scenario, the derived traits evolved 7 times (vs. 9 in A, and 8 in B).

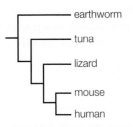

earthworm
tuna
lizard
mouse
human

A Evolutionary connections among clades are represented as lines on a cladogram. Sister groups emerge from a node, which represents a common ancestor.

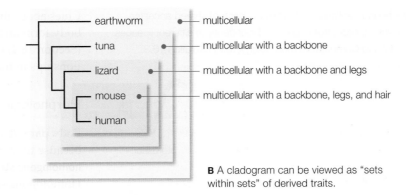

earthworm — multicellular
tuna — multicellular with a backbone
lizard — multicellular with a backbone and legs
mouse — multicellular with a backbone, legs, and hair
human

B A cladogram can be viewed as "sets within sets" of derived traits.

Figure 18.3 Animated An example of a cladogram.

traditional Linnaean class Reptilia ("reptiles") includes crocodiles, tuataras, snakes, lizards, turtles, and tortoises. While it is convenient to classify these animals together, they would not constitute a clade unless birds are also included, as you will see in Chapter 25.

The remaining sections of this chapter explore some of the character comparisons that evolutionary biologists use to group organisms into clades. Remember that evolutionary history does not change because of events in the present: A species' ancestry remains the same no matter how it evolves. However, we can make mistakes when we try to retrace the events in that history. A monophyletic group is defined on the basis of derived traits, and, as with traditional taxonomy, we can be misled by incomplete information. Thus, a clade is necessarily a hypothesis; which organisms it comprises may change when new discoveries are made. As with all hypotheses, the more data that support a grouping, the less likely it is to require revision.

Cladistics

In the big picture of evolution, all clades are interconnected; an evolutionary biologist's job is to figure out where the connections are. Making hypotheses about evolutionary relationships among clades is called **cladistics**. One way of doing this involves the logical rule of simplicity: When there are several possible ways that a group of clades can be connected, the simplest evolutionary pathway is probably the correct one. By comparing all of the possible connections among the clades, we can identify the simplest: the one in which the defining derived traits evolved the fewest number of times (**Figure 18.2**). The process of finding the simplest pathway is called parsimony analysis.

The result of a cladistic analysis is a **cladogram**, a type of **evolutionary tree** that diagrams a summary of our best data-supported hypotheses about how a

group of clades evolved (**Figure 18.3**). We use cladograms to visualize evolutionary trends and patterns. Data from an outgroup (a species not closely related to any member of the group under study) may be included in order to "root" the tree. Each line in a cladogram represents a lineage, which may branch into two lineages at a node. The node represents a

common ancestor of two lineages. Every branch of a cladogram is a clade; the two lineages that emerge from a node on a cladogram are **sister groups**.

character Quantifiable, heritable characteristic or trait.
clade A group whose members share one or more defining derived traits.
cladistics Method of making hypotheses about evolutionary relationships among clades.
cladogram Evolutionary tree diagram that shows evolutionary relationships among clades.
derived trait A novel trait present in a clade but not in the clade's ancestors.
evolutionary tree Diagram showing evolutionary relationships.
monophyletic group An ancestor in which a derived trait evolved, together with all of its descendants.
phylogeny Evolutionary history of a species or group of species.
sister groups The two lineages that emerge from a node on a cladogram.

Take-Home Message

How do evolutionary biologists study life's diversity?

» Evolutionary biologists study phylogeny in order to understand how all species are connected by shared ancestry.

» A clade is a monophyletic group whose members share one or more derived traits. Cladistics is a method of making hypotheses about evolutionary relationships among clades.

» Cladograms and other evolutionary tree diagrams are hypotheses based on our best understanding of the evolutionary history of a group of organisms.

18.3 Comparing Form and Function

■ Physical similarities are often evidence of shared ancestry, but sometimes a trait evolves independently in different lineages.
■ Links to Comparative morphology 16.2, Evolution 16.3

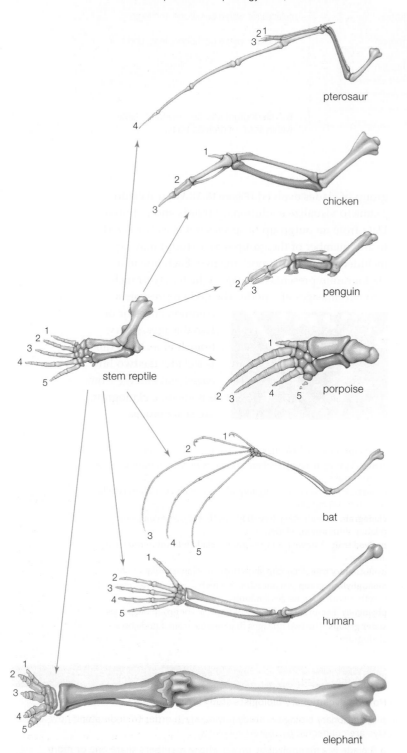

Figure 18.4 Morphological divergence among vertebrate forelimbs, starting with the bones of a stem reptile. The number and position of many skeletal elements were preserved when these diverse forms evolved; notice the bones of the forearms. Certain bones were lost over time in some of the lineages (compare the digits numbered 1 through 5). Drawings are not to scale.

Clues about the history of a lineage are encoded in body form and function, so comparative morphology (Section 16.2) can be used to unravel evolutionary relationships in many cases.

Morphological Divergence

Body parts that appear similar in separate lineages because they evolved in a common ancestor are called **homologous structures** (*hom–* means "the same"). Homologous structures may be used for different purposes in different groups, but the very same genes direct their development. A body part that outwardly appears very different in separate lineages may be homologous in underlying form. For example, even though vertebrate forelimbs vary in size, shape, and function from one group to the next, they clearly are alike in the structure and positioning of bony elements, and in their internal patterns of nerves, blood vessels, and muscles.

Populations that are not interbreeding tend to diverge genetically, and in time these genetic divergences give rise to changes in body form. Change from the body form of a common ancestor is an evolutionary pattern called **morphological divergence**.

Consider the limb bones of modern vertebrate animals. Fossil evidence suggests that many vertebrates are descended from a family of ancient "stem reptiles" that crouched low to the ground on five-toed limbs. This ancestral group's descendants diversified over millions of years, and eventually gave rise to modern reptiles, birds, and mammals. A few lineages that had become adapted to walking on land even returned to life in the seas. During this time, five-toed limbs became adapted for many different purposes (**Figure 18.4**). They became modified for flight in extinct reptiles called pterosaurs and in bats and most birds. In penguins and porpoises, the limbs are now flippers useful for swimming. In humans, five-toed forelimbs became arms and hands with four fingers and an opposable thumb. Among elephants, the limbs are now strong and pillarlike, capable of supporting a great deal of weight. Limbs degenerated to nubs in pythons and boa constrictors, and to nothing at all in other snakes.

Morphological Convergence

Body parts that appear similar in different species are not always homologous; they sometimes evolve independently in lineages subject to the same environmental pressures. The independent evolution of similar

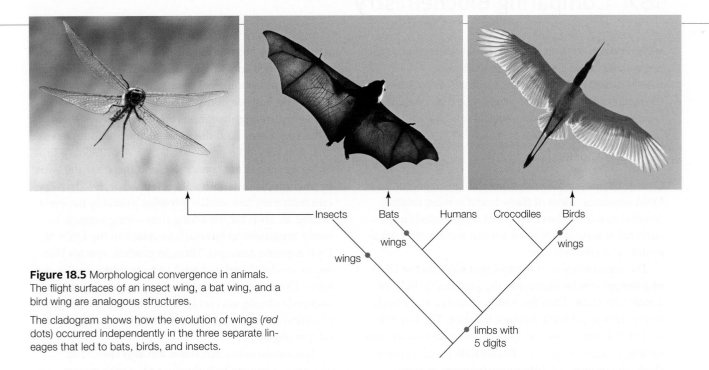

Insects Bats Humans Crocodiles Birds

wings

wings

wings

limbs with
5 digits

Figure 18.5 Morphological convergence in animals. The flight surfaces of an insect wing, a bat wing, and a bird wing are analogous structures.

The cladogram shows how the evolution of wings (*red* dots) occurred independently in the three separate lineages that led to bats, birds, and insects.

body parts in different lineages is called **morphological convergence**. Structures that are similar as a result of morphological convergence are called **analogous structures**. Analogous structures look alike but did not evolve in a shared ancestor; they evolved independently after the lineages diverged.

For example, bird, bat, and insect wings all perform the same function, which is flight. However, several clues tell us that the wing surfaces are not homologous. The wings are adapted to the same physical constraints that govern flight, but each is adapted in a different way. In the case of birds and bats, the limbs themselves are homologous, but the adaptations that make those limbs useful for flight differ. The surface of a bat wing is a thin, membranous extension of the animal's skin. By contrast, the surface of a bird wing is a sweep of feathers, which are specialized structures derived from skin. Insect wings differ even more. An insect wing forms as a saclike extension of the body wall. Except at forked veins, the sac flattens and fuses into a thin membrane. The sturdy, chitin-reinforced veins structurally support the wing. Unique adaptations for flight are evidence that wing surfaces of birds, bats, and insects are analogous structures that evolved after the ancestors of these modern groups diverged (**Figure 18.5**).

analogous structures Similar body structures that evolved separately in different lineages.
homologous structures Body parts or structures that are similar in different lineages because they evolved in a common ancestor.
morphological convergence Evolutionary pattern in which similar body parts evolve separately in different lineages.
morphological divergence Evolutionary pattern in which a body part of an ancestor changes in its descendants.

As another example of morphological convergence, the similar external structures of American cacti and African euphorbias (see **Figure 16.3**) are adaptations to similarly harsh desert environments where rain is scarce. Distinctive accordion-like pleats allow the plant body to swell with water when rain does come. Water stored in the plants' tissues allows them to survive long dry periods. As the stored water is used, the plant body shrinks, and the folded pleats provide it with some shade in an environment that typically has none. Despite these similarities, however, a closer look reveals many differences that indicate the two types of plants are not closely related. For example, cactus spines have a simple fibrous structure; they are modified leaves that arise from dimples on the plant's surface. Euphorbia spines project smoothly from the plant surface, and they are not modified leaves: In many species the spines are actually dried flower stalks (*right*).

Take-Home Message

What clues about phylogeny can comparative morphology reveal?

» In morphological divergence, a body part inherited from a common ancestor becomes modified differently in different lines of descent. Such parts are called homologous structures.

» In morphological convergence, body parts that appear alike evolved independently in different lineages, not in a common ancestor. Such parts are called analogous structures.

18.4 Comparing Biochemistry

■ The kind and number of biochemical similarities among species are clues about evolutionary relationships.

■ Links to Nucleotide sequence 8.4, The genetic code 9.4, DNA sequencing 15.4, DNA fingerprinting and genomics 15.5, Neutral mutations 17.2, Genetic equilibrium 17.3

Molecular Clocks

Over time, inevitable mutations change a genome's DNA sequence. Most of these mutations are neutral. Neutral mutations have no effect on an individual's survival or reproduction, so we can assume they accumulate at a constant rate.

The accumulation of neutral mutations in the DNA of a lineage can be likened to the predictable ticks of a **molecular clock**. Turn the hands of such a clock back, so the ticks wind back through the past. The last tick will be the time when the lineage embarked on its own unique evolutionary road. To calibrate the molecular clock, the number of differences between genomes can be correlated with the timing of morphological changes seen in the fossil record.

Mutations alter the DNA of a linage independently of all other lineages. The more recently two lineages diverged, the less time there has been for unique mutations to accumulate in the DNA of each one. That is why the genomes of closely related species tend to be more similar than those of distantly related ones—a general rule that can be used to estimate relative times of divergence.

DNA and Protein Sequence Comparisons

Comparing biochemical characters such as the nucleotide sequence of a gene or the amino acid sequence of a protein can provide evidence of an evolutionary relationship. Such comparisons are often used in conjunction with morphological data.

As you will see in the next section, some essential genes are highly conserved, which means their DNA sequences (and the proteins they encode) have changed very little or not at all over evolutionary time. Other genes are not conserved, and these are the basis of phenotypic differences that define species. Two species with very few similar proteins probably have not shared an ancestor for a long time—long enough for many mutations to have accumulated in the DNA of their separate lineages. Thus, in general, species that are more closely related tend to have more similar proteins. Evolutionary biologists can compare a protein's sequence among several species, and use the number of amino acid differences as a measure of relative relatedness (**Figure 18.6**).

The amino acids that differ are also clues. For example, a leucine to isoleucine change (a conservative amino acid substitution) may not affect the function of a protein very much, because both amino acids are nonpolar, and both are about the same size. However, the substitution of a lysine (which is basic) for an aspartic acid (which is acidic) may dramatically change the character of a protein. Such nonconservative substitutions—as well as deletions and insertions—often affect phenotype. Most mutations that affect phenotype are selected against, but occasionally one proves adaptive. Thus, the longer it has been since two lineages diverged, the more nonconservative amino acid substitutions we are likely to see when comparing their proteins.

Among species that diverged relatively recently, many proteins have identical amino acid sequences.

```
   honeycreepers (10)...CRDVQFGWLIRNLHANGASFFFICIYLHIGRGIYYGSYLNK--ETWNIGVILLLTLMATAFVGYVLPWGQMSFWG...
        song sparrow...CRDVQFGWLIRNLHANGASFFFICIYLHIGRGIYYGSYLNK--ETWNVGIILLLALMATAFVGYVLPWGQMSFWG...
   Gough Island finch...CRDVQFGWLIRNIHANGASFFFICIYLHIGRGLYYGSYLYK--ETWNVGVILLLTLMATAFVGYVLPWGQMSFWG...
           deer mouse...CRDVNYGWLIRYMHANGASMFFICLFLHVGRGMYYGSYTFT--ETWNIGVLLFAVMATAFMGYVLPWGQMSFWG...
     Asiatic black bear...CRDVHYGWIIRYMHANGASMFFICLFMHVGRGLYYGSYLLS--ETWNIGIILLFTVMATAFMGYVLPWGQMSFWG...
       bogue (a fish)...CRDVNYGWLIRNLHANGASFFFICIYLHIGRGLYYGSYLYK--ETWNIGVVLLLLVMGTAFVGYVLPWGQMSFWG...
                human...TRDVNYGWIIRYLHANGASMFFICLFHIGRGLYYGSFLYS--ETWNIGIILLLATMATAFMGYVLPWGQMSFWG...
   thale cress (a plant)...MRDVEGGWLLRYMHANGASMFLIVVYLHIFRGLYHASYSSPREFVWCLGVVIFLLMIVTAFIGYVLPWGQMSFWG...
        baboon louse...ETDVMNGWMVRSIHANGASWFFIMLYSHIFRGLWVSSFTQP--LVWLSGVIILFLSMATAFLGYVLPWGQMSFWG...
        baker's yeast...MRDVHNGYILRYLHANGASFFFMVMFMHMAKGLYYGSYRSPRVTLWNVGVIIFTLTIATAFLGYCCVYGQMSHWG...
```

Figure 18.6 Alignment of part of the amino acid sequence of mitochondrial cytochrome *b* from twenty species. This protein is a crucial component of mitochondrial electron transfer chains.

The honeycreeper sequence is identical in ten species of honeycreeper; amino acids that differ in the other species are shown in *red*. Gaps in alignment are indicated by dashes. The abbreviations for the amino acids are spelled out in Appendix I.

Figure It Out: Based on this comparison, which species is the most closely related to honeycreepers?

Answer: The song sparrow.

```
...GTAGCCCATATATGCCGCGACGTACAATTCGGCTGACTAATCCGCAACCT...Vestiaria coccinea (iiwi)
...GTAGCCCATGTATGCCGCGACGTACAATTCGGCTGACTAATCCGCAACCT...Himatione sanguinea (apapane)
...GTAGCCCACATATGCCGCGACGTACAATTCGGCTGACTAATCCGCAACCT...Palmeria dolei (akohekohe)
...GTTGCTCACATATGCCGTGACGTACAATTCGGCTGACTAATCCGCAACCT...Oreomystis mana (Hawaii creeper)
...GTAGCCCATATATGCCGCGACGTACAATTCGGCTGACTAATCCGCAACCT...Hemignathus virens (amakihi)
...GTAGCCCACATATGCCGCGACGTACAATTCGGCTGACTAATCCGCAACCT...Hemignathus munroi (akiapolaau)
...GTAGCCCACATATGCCGTGACGTACAGTTCGGCTGACTAATCCGCAACCT...Pseudonestor xanthophrys (Maui parrotbill)
...GTAGCCCACATATGCCGCGACGTACAATTCGGCTGACTAATCCGAAACCT...Paroreomyza montana (Maui alauahio)
...GTAGCCCACATATGCCGCGACGTACAATTCGGCTGACTAATCCGTAATCT...Oreomystis bairdi (akikiki)
...GTAGCCCACATATGCCGCGACGTACAATTCGGCTGACTAATCCGCAACCT...Loxops coccineus (Hawaii akepa)
```

A Part of the mitochondrial cytochrome *b* sequence compared across 10 species of honeycreeper. Differences from the consensus are shown in *red*. Even though the amino acid sequence of cytochrome *b* is identical in all of these species, neutral mutations have accumulated in their separate lineages. The mutations did not change the amino acid sequence of the resulting protein.

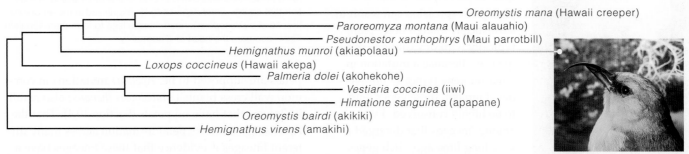

B A sequence comparison can be used to generate evolutionary trees. This one reflects a comparison of 790 nucleotides of the mitochondrial cytochrome *b* DNA sequence from ten honeycreeper species. The length of the branches reflects the number of character changes (here, nucleotide differences), which in turn implies the relative length of time of divergence between the species. The arrangement of branches on such trees may differ depending on the data used to generate them.

Figure 18.7 Example of a DNA sequence comparison.

Nucleotide sequence differences may be instructive in such cases. Even if the amino acid sequence of a protein is identical among species, the nucleotide sequence of the gene that encodes the protein may differ because of redundancy in the genetic code (**Figure 18.7**). For example, a nucleotide substitution that changes one codon from AAA to AAG in a protein-coding region would probably not affect the protein product, because both codons specify lysine. This base substitution is an example of a neutral mutation.

The DNA from nuclei, mitochondria, and chloroplasts can be used in nucleotide comparisons. Mitochondrial DNA can also be used to compare different individuals of the same sexually reproducing animal species. Mitochondria are inherited intact from a single parent, usually the mother, and they contain their own DNA. Thus, any differences in mitochondrial DNA sequences between maternally related individuals are due to mutations, not genetic recombination during fertilization.

It is useful to remember that coincidental homologies are statistically more likely to occur between DNA sequences than amino acid sequences, because there are only four nucleotides in DNA (versus twenty amino acids in proteins). Thus, comparing DNA requires a lot more data than comparing proteins. However, DNA sequencing has become so fast that there is a lot of data available. New genome sequences are compiled continually into online databases accessible by anyone. Comparative genomics studies with such data have shown us (for example) that about 30 percent of the 6,609 genes of yeast cells have counterparts in the human genome. So do 50 percent of the 30,971 genes in *Drosophila* fruit flies, and 40 percent of the 19,023 genes in roundworms.

molecular clock Using molecular change to estimate how long ago two lineages diverged.

Take-Home Message

How does biochemistry reflect evolutionary history?

» Mutations change the nucleotide sequence of a lineage's DNA over time.

» Lineages that diverged long ago generally have more differences between their DNA and proteins than do lineages that diverged more recently.

18.5 Comparing Patterns of Development

■ Similar patterns of embryonic development are an outcome of highly conserved master genes.

■ Links to Master genes in development 10.3, Floral identity gene mutations 10.4, Evolution by gene duplications 14.5

The development of an embryo into the body of a plant or animal is orchestrated by layer after layer of master gene expression. The failure of any single master gene to participate in this symphony of expression can result in a drastically altered body plan, typically with devastating consequences. Because a mutation in a master gene typically unravels development, these genes tend to be highly conserved. Even among lineages that diverged a very long time ago, such genes often retain similar sequences and functions.

Similar Forms in Plants

Homeotic genes encode transcription factors that help sculpt details of the body's form during embryonic development (Section 10.3). A mutation in one homeotic gene can disrupt details of the body's form. For example, any mutation that inactivates a floral identity gene, *Apetala1*, in wild cabbage plants (*Brassica oleracea*) results in mutated flowers. Such flowers form with pollen producing structures (stamens) where

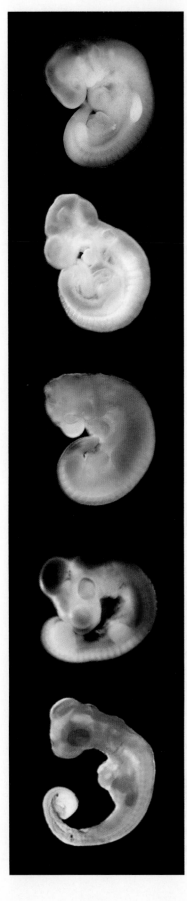

Figure 18.8 Visual comparison of vertebrate embryos. All vertebrates go through an embryonic stage in which they have four limb buds, a tail, and divisions called somites along their back. Embryos *top* to *bottom*: human, mouse, bat, chicken, alligator.

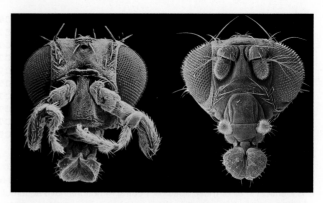

Figure 18.9 Expression of the *antennapedia* gene in the embryonic tissues of the insect thorax causes legs to form. Normally, the gene is never expressed in cells of any other tissue. A mutation that causes *antennapedia* to be expressed in the embryonic tissues of a *Drosophila*'s head causes legs to form there too (*left*). *Right*, compare the head of a normal fly.

petals are supposed to be. *Apetala1* mutations in common wall cress plants (*Arabidopsis thaliana*) also result in flowers that have no petals (Section 10.4). That the *Apetala1* gene affects petal formation across many different lineages is evidence that these lineages have a common ancestor.

Developmental Comparisons in Animals

The embryos of many vertebrate species develop in similar ways. Their tissues form the same way, as embryonic cells divide, differentiate, and interact. For example, all vertebrates go through a stage in which they have four limb buds, a tail, and a series of somites—divisions of the body that give rise to a backbone (**Figure 18.8**). Given that the very same genes direct development in all of the vertebrate lineages, how do the adult forms end up so different? Part of the answer is that there are differences in the onset, rate, or completion of early steps in development. These differences are brought about by variations in the expression patterns of master genes that govern development. The variation has arisen at least in part as a result of gene duplications followed by mutation, the same way that multiple globin genes evolved in primates (Section 14.5).

Hox Genes *Hox* genes are homeotic genes of animals. The pattern of expression of these master genes determines the identity of particular zones along the body axis. *Hox* genes occur in clusters, one after the next on a chromosome, in the order in which they are expressed in a developing embryo. Clustered *Hox* genes are similar, and probably arose by gene duplications. Entire clusters have been duplicated in some lineages. For example, one of the ten *Hox* genes in insects and other arthropods, *antennapedia*, determines the identity of the thorax (the body part with legs). Legs develop wherever *antennapedia* is expressed in an embryo (**Figure 18.9**). Vertebrate animals have multiple sets of the same ten *Hox* genes that occur in arthropods. One vertebrate version of *antennapedia*, the *Hoxc6* gene, determines the identity of the back (as opposed to the neck or tail). Expression of this gene causes ribs to develop on a vertebra (**Figure 18.10**). Vertebrae of the neck and tail normally develop with no *Hoxc6* expression, and no ribs.

Hox genes also regulate limb formation. Body appendages as diverse as crab legs, beetle legs, sea star arms, butterfly wings, fish fins, and mouse feet start out as clusters of cells that bud from the surface of the embryo. The buds form wherever a homeotic gene called *Dlx* is expressed. *Dlx* encodes a transcription factor that signals clusters of embryonic cells to "stick out from the body" and give rise to an appendage. *Hox* genes suppress *Dlx* expression in all parts of an embryo that will not have appendages.

Persistent Juvenile Features A chimpanzee skull and a human skull appear quite similar an early stage. As development continues, both skulls change shape as different parts grow at different rates (**Figure 18.11**). However, the human skull undergoes less pronounced differential growth than the chimpanzee skull does. As a result, a human adult has a rounder braincase, a flatter face, and a less protruding jaw compared with an adult chimpanzee. In its proportions, a human adult skull is more like the skull of an infant chimpanzee than the skull of an adult chimpanzee. The similarity suggests that human evolution involved changes that slowed the rate of development, causing traits that were previously typical of juvenile stages to persist into adulthood.

Juvenile features also persist in other adult animals, notably salamanders called axolotls (*inset*). The larvae of most species of salamander live in water and use external gills to breathe. Lungs that replace the gills as development continues allow the adult to breathe air and live on land. By contrast, axolotls never give up their aquatic lifestyle; external gills and other larval traits persist into adulthood. The closest relatives of axolotls are tiger salamanders, the larvae of which resemble small axolotls.

Figure 18.10 An example of comparative embryology. Expression of the *Hoxc6* gene is indicated by *purple* stain in two vertebrate embryos, chick (*left*) and garter snake (*right*). Expression of this gene causes a vertebra to develop ribs as part of the back. Chickens have 7 vertebrae in their back and 14 to 17 vertebrae in their neck; snakes have upwards of 450 back vertebrae and essentially no neck.

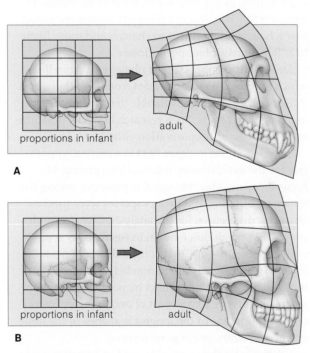

Figure 18.11 **Animated** Morphological differences between two primates. These skulls are depicted as paintings on a rubber sheet divided into a grid. Stretching the sheets deforms the grid. Differences in how they are stretched are analogous to different growth patterns. Shown here, proportional changes during skull development in (**A**) the chimpanzee and (**B**) the human. Chimpanzee skulls change more than human skulls, so the relative proportions in bones of adult and infant humans are more similar than those of adult and infant chimpanzees.

Take-Home Message

How are similarities in development indicative of shared ancestry?

» Similarities in patterns of development are the result of master genes that have been conserved over evolutionary time.

» Some differences between closely related species are a result of changes in the rate or onset of development.

18.6 Applications of Phylogeny Research

■ We use information about phylogeny to understand how to preserve the species that exist today.

■ Links to Gene flow 17.8, Reproductive isolation 17.9

Studies of phylogeny reveal how species relate to one another and to species that are now extinct. In doing so, they inform our understanding of how shared ancestry interconnects all species—including our own. Such studies also have a variety of practical applications, in fields from conservation to medicine.

Conservation Biology

The story of the Hawaiian honeycreepers is a dramatic illustration of how evolution works. It also shows how finding ancestral connections can help species that are still living. The group's reservoir of genetic diversity is dwindling along with the number of its members (**Figure 18.12**). The lowered diversity means the group as a whole is less resilient to change, and more likely to suffer catastrophic species losses.

Deciphering honeycreeper phylogeny can tell us which species are most different from the others, and those are the ones most valuable in terms of preserving genetic diversity. Such research helps us prioritize our resources and conservation efforts.

The cladogram in **Figure 18.13** is a current hypothesis for the evolutionary relationships among 41 honeycreepers. Morphological differences among the skeletons of living and extinct species were used as character differences for the analysis. The cladogram shows that the poouli (shown in **Figure 18.1C**) is an outlying species—one that is most different from the others. Unfortunately, that knowledge came too late; the species is probably extinct by now. Its extinction means the loss of a large part of evolutionary history of the group: One of the longest branches of the honeycreeper family tree is gone forever.

Cladistics analyses are also used to correlate past evolutionary divergences with behavior and dispersal patterns of existing populations. Such studies are useful in conservation efforts. For example, a decline in antelope populations in African savannas is at least partly due to competition with domestic cattle. A cladistic analysis of mitochondrial DNA sequences suggested that current populations of blue wildebeest are genetically less similar than they should be, based on other antelope groups of similar age. Combined with behavioral and geographic data, the analysis helped conservation biologists realize that a patchy distribution of preferred food plants is preventing gene flow among blue wildebeest populations. The absence of

Figure 18.12 The genetic diversity of Hawaiian honeycreepers is dwindling along with their continued extinctions. Deciphering their evolutionary connections helps us in our efforts to preserve the remaining species.

gene flow can lead to a catastrophic loss of genetic diversity in populations under pressure. Restoring appropriate grasses in intervening, unoccupied areas of savanna would allow isolated wildebeest populations to reconnect.

Medical Applications

Researchers study the evolution of viruses and other infectious agents by grouping their biochemical characters into clades. Viruses can mutate every time they infect a host, so their genetic material changes over time. Consider the H5N1 strain of influenza virus, which infects birds and other animals. Humans infected with H5N1 have a very high mortality rate, but to date, human-to-human transmission has been rare. However, the virus replicates in pigs without causing symptoms. Pigs transmit the virus to other pigs—and apparently to humans too. A phylogenetic analysis of H5N1 isolated from pigs showed that the virus "jumped" from birds to pigs at least three times since 2005, and that one of the isolates had acquired the potential to be transmitted among humans. An increased understanding of how this virus evolves may help us find a way to prevent it from spreading.

Take-Home Message

How is studying phylogeny useful?

» Phylogeny research is yielding an ever more specific and accurate picture of how all life is related by shared ancestry.

» Among other applications, phylogeny research can help us preserve species in danger of becoming extinct, and to understand the spread of infectious diseases.

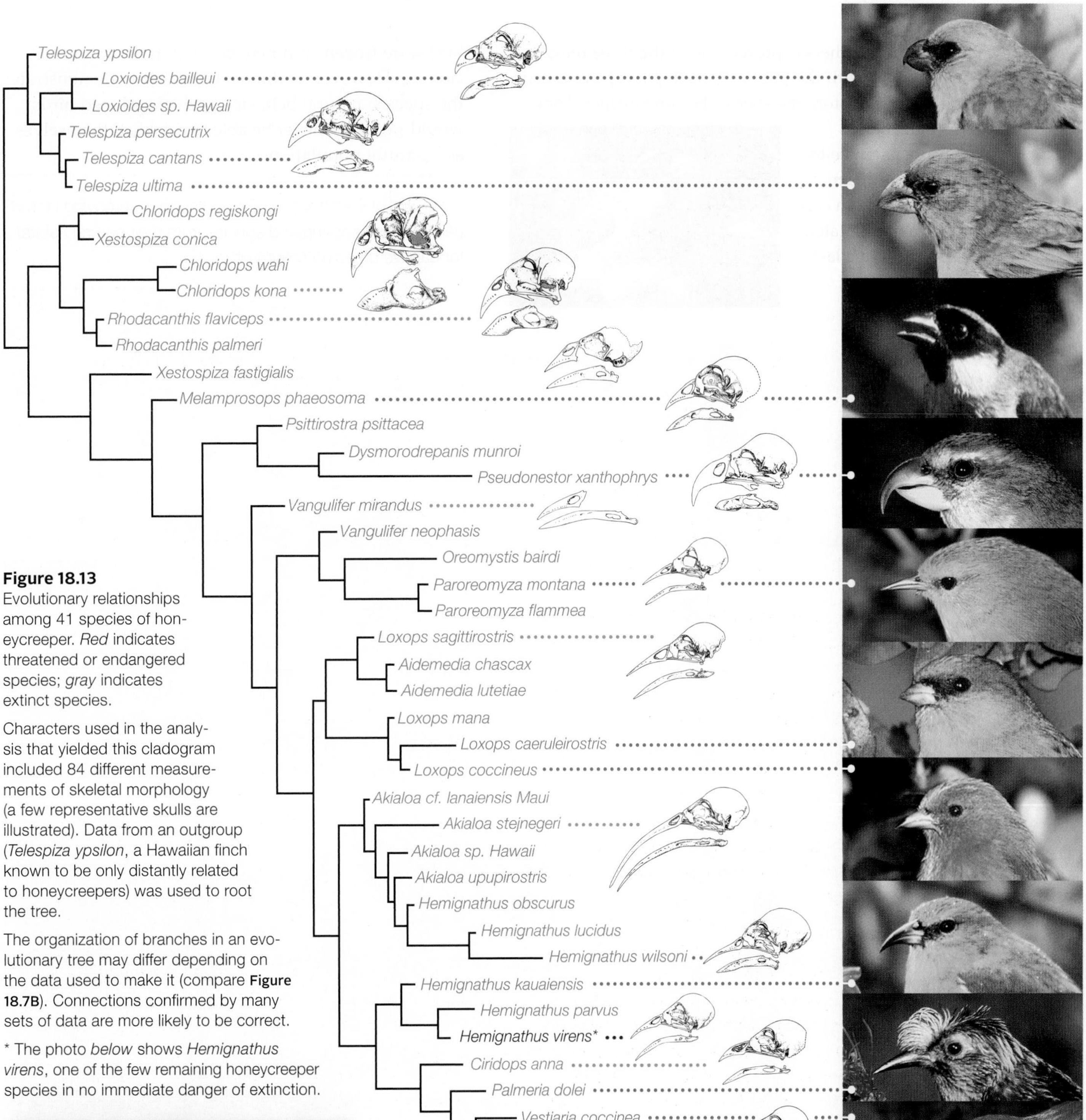
Telespiza ypsilon
Loxioides bailleui
Loxioides sp. Hawaii
Telespiza persecutrix
Telespiza cantans
Telespiza ultima
Chloridops regiskongi
Xestospiza conica
Chloridops wahi
Chloridops kona
Rhodacanthis flaviceps
Rhodacanthis palmeri
Xestospiza fastigialis
Melamprosops phaeosoma
Psittirostra psittacea
Dysmorodrepanis munroi
Pseudonestor xanthophrys
Vangulifer mirandus
Vangulifer neophasis
Oreomystis bairdi
Paroreomyza montana
Paroreomyza flammea
Loxops sagittirostris
Aidemedia chascax
Aidemedia lutetiae
Loxops mana
Loxops caeruleirostris
Loxops coccineus
Akialoa cf. lanaiensis Maui
Akialoa stejnegeri
Akialoa sp. Hawaii
Akialoa upupirostris
Hemignathus obscurus
Hemignathus lucidus
Hemignathus wilsoni
Hemignathus kauaiensis
Hemignathus parvus
Hemignathus virens*
Ciridops anna
Palmeria dolei
Vestiaria coccinea
Himatione sanguinea

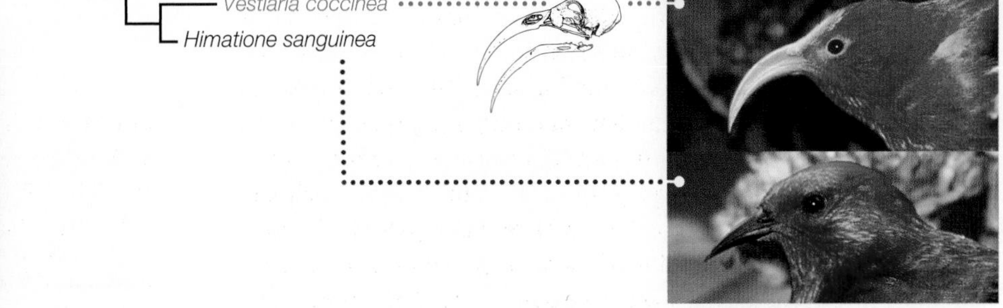

Figure 18.13
Evolutionary relationships among 41 species of honeycreeper. *Red* indicates threatened or endangered species; *gray* indicates extinct species.

Characters used in the analysis that yielded this cladogram included 84 different measurements of skeletal morphology (a few representative skulls are illustrated). Data from an outgroup (*Telespiza ypsilon*, a Hawaiian finch known to be only distantly related to honeycreepers) was used to root the tree.

The organization of branches in an evolutionary tree may differ depending on the data used to make it (compare **Figure 18.7B**). Connections confirmed by many sets of data are more likely to be correct.

* The photo *below* shows *Hemignathus virens*, one of the few remaining honeycreeper species in no immediate danger of extinction.

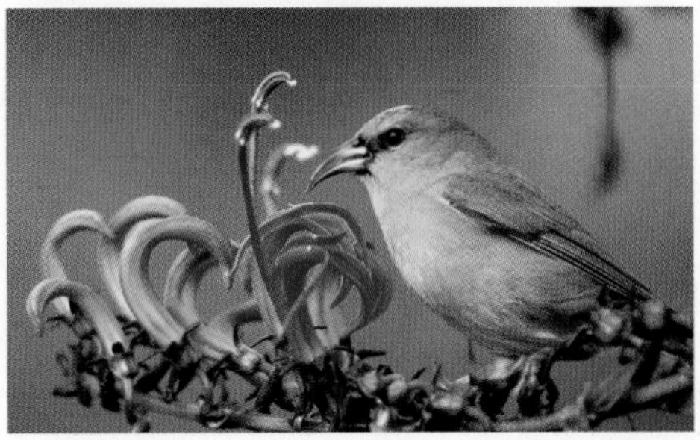

Bye Bye Birdie (revisited)

In 2004, researchers captured one of the three remaining pooulis, with the intent of starting a captive breeding program before the species became extinct. They were unable to capture a female to mate with this male before it died in captivity a month later. Cells from this last  bird were frozen, and may be used in the future for cloning. However, with no parents left to demonstrate the species' natural behavior to chicks, cloned birds would probably never be able to establish themselves as a natural population.

How would you vote? Do you support removing individuals of highly endangered species from their natural habitat for captive breeding programs?

19 Early Evolution

LEARNING ROADMAP

Where you have been This chapter explains how the molecular subunits of life introduced in Section 3.3 could have originated and assembled to form the first cells. It considers the evolution of eukaryotic traits (Sections 4.5–4.12) and draws on your understanding of fossils and the geological time scale (16.5 and 16.8), aerobic respiration (7.2), and photosynthesis (6.4).

Where you are now

The Early Earth
The early Earth was a seemingly inhospitable place. It was continually bombarded by asteroids, and there was no free oxygen or protective ozone layer.

Building Blocks of Life
Under some natural conditions, inorganic molecules bond together to form the amino acids and other small organic subunits that serve as the building blocks of all life.

From Molecules to Protocells
Naturally formed aggregations of organic polymers enclosed in a layer of lipid may have been the forerunners of cells. RNA may have been the first genetic material.

Early Cellular Life
Fossil traces of cells date back more than 3 billion years. An early divergence created two lineages, one leading to modern bacteria, the other to archaea and eukaryotes.

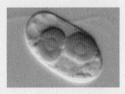

Evolution of Eukaryotic Traits
The internal membranes typical of eukaryotes probably evolved by infolding of the plasma membrane. Mitochondria and chloroplasts descended from bacterial cells.

Where you are going We continue our discussion of the features of bacterial and archaeal cells in Sections 20.6–20.8. You will learn more much about the features of simple eukaryotes when we discuss the protists in Chapter 21. We return to the topic of the ozone layer in Section 48.6 as we discuss the damage done by air pollutants.

19.1 Looking for Life

We live in a vast universe that we have only begun to explore. So far, we know of only one planet that has life—Earth. In addition, biochemical, genetic, and metabolic similarities among Earth's species imply that all evolved from a common ancestor that lived billions of years ago. What properties of the ancient Earth allowed life to arise, survive, and diversify? Could similar processes occur on other planets? These are some of the questions posed by **astrobiology**, the study of life's origins and distribution in the universe.

Astrobiologists study Earth's extreme habitats to determine the range of conditions that living things can tolerate, and they have found species that withstand extraordinary levels of temperature, pH, salinity, and pressure. One team found bacteria living 30 centimeters (1 foot) below the soil surface of Chile's Atacama Desert, the driest place on Earth (**Figure 19.1**). Other researchers drilled 3 kilometers (almost 2 miles) beneath the soil surface in Virginia, and found bacteria thriving at high pressure and temperature. Still others discovered a bacterial community beneath Antarctica's western ice sheet. In short, life seems to thrive nearly anywhere there is a source of energy and carbon.

Knowledge gained from studies of life on Earth informs the search for life elsewhere. On Earth, all metabolic reactions involve interactions among molecules in aqueous solution (dissolved in water). Scientists assume that the same physical and chemical laws operate throughout the universe; thus they consider liquid water an essential requirement for life. As a result, they were excited when a robotic lander discovered water frozen in the soil of Mars, our closest planetary neighbor. Geological formations on Mars also suggest that billions of years ago water flowed across the planet's surface and pooled in giant lakes.

If there is any life on Mars today, it is likely to be underground. Unlike Earth, Mars does not have an **ozone layer**: an atmospheric layer with a high concentration of ozone (O_3). Earth's ozone layer serves as a natural sunscreen, preventing most ultraviolet (UV) radiation emitted by the sun from reaching the planet's surface. Because Mars has no ozone layer, its surface receives intense UV radiation. As explained in Section 8.6, UV radiation can damage DNA. Thus, the UV radiation that reaches Mars probably sterilizes the upper layer of the Martian soil.

Suppose scientists do find evidence that microbial life exists or existed on Mars, or on some other planet. Why would it matter? Science can never prove how life arose, but it can test hypotheses about what could have happened. Discovery of extraterrestrial microbes would lend credence to the idea that nonhuman intelligent life exists somewhere in our universe. The more places microbial life exists, the more likely it is that complex, intelligent life evolved on other planets in the same manner that it did on Earth.

This chapter is your introduction to a slice through time. We begin with Earth's formation and move on to life's chemical origins and the evolution of traits present in modern eukaryotes. The picture we paint here sets the stage for the next unit, which takes you along lines of descent to the present range of biodiversity.

astrobiology The scientific study of life's origin and distribution in the universe.
ozone layer High atmospheric layer rich in ozone; prevents most ultraviolet radiation in sunlight from reaching Earth's surface.

Figure 19.1 The Mars-like landscape of Chile's Atacama Desert. Scientist Jay Quade, visible in the distance at the *right*, was a member of a team that found bacteria living beneath this arid desert soil.

19.2 The Early Earth

- Knowledge of modern chemistry and physics is the basis for scientific hypotheses about early events in Earth's history.
- Links to Elements 2.2, Organic compounds 3.2

Origin of the Universe and the Earth

No one was around to witness the birth of the universe. However, studies of the modern universe allow astronomers and physicists to propose and test ideas about how the universe originated.

The widely accepted **big bang theory** states that the universe began in a single instant, about 13 to 15 billion years ago. In that instant, all existing matter and energy suddenly appeared and exploded outward from a single point. Simple elements such as hydrogen and helium formed within minutes. Then, over millions of years, gravity drew the gases together and they condensed to form giant stars.

Explosions of these early stars scattered the heavier elements from which today's galaxies formed. About 5 billion years ago, a cloud of dust and rocks (asteroids) orbited the star we now call our sun (**Figure 19.2**). The asteroids collided and merged into bigger asteroids. The heavier these pre-planetary objects became, the more gravitational pull they exerted, and the more material they gathered. By about 4.6 billion years ago, this gradual buildup of materials had formed Earth and the other planets of our solar system.

big bang theory Well-supported hypothesis that the universe originated by a nearly instant distribution of matter through space.

Figure 19.2 Artist's depiction of our sun surrounded by a cloud of dust, debris, and gases. Earth and other planets formed from the material in this cloud.

Figure 19.3 An artist's depiction of early Earth.

Conditions on the Early Earth

Planet formation did not clear away all of the debris orbiting the sun, so the early Earth received a constant hail of meteorites. Molten rock and gases spewed continually from volcanoes. Gases released by volcanoes and meteorite impacts became the main components of Earth's early atmosphere.

What was Earth's early atmosphere like? Studies of volcanic eruptions, meteorites, ancient rocks, and other planets suggest that it contained water vapor, carbon dioxide, and gaseous hydrogen and nitrogen. We know that there was little or no oxygen gas (O_2), because the oldest existing rocks show no evidence of iron oxidation (rusting). If oxygen had been present in Earth's early atmosphere, it would have caused rust formation. More important, the presence of O_2 would have interfered with assembly of the organic compounds necessary for life. Oxygen gas would have reacted with and destroyed the compounds as fast as they formed.

In addition to anaerobic conditions, the first steps toward life required water. The two main sources of Earth's water were steam released by volcanic eruptions and other geological processes, and ice that formed in space and was delivered to Earth by comets. As Earth's surface cooled, water pooled in its first seas (**Figure 19.3**). It was in these seas that life first began.

Take-Home Message

What were conditions like on the early Earth?

» Earth's early atmosphere had little or no oxygen.

» Meteorites pummeled the planet's surface, and volcanic activity was more common than it is today.

19.3 Formation of Organic Monomers

- All living things are made from the same organic subunits: amino acids, fatty acids, nucleotides, and simple sugars.
- Link to Organic compounds 3.2

Organic Molecules From Inorganic Precursors

Until the early 1900s, chemists thought that organic molecules possessed a special "vital force" and that only living organisms could make them. Then, in 1925, a chemist synthesized urea, an organic molecule abundant in urine. Later, another chemist synthesized alanine, an amino acid. These reactions proved that nonliving processes could yield organic molecules.

The Source of Life's First Building Blocks

The organic monomers of the molecules of life can form by nonbiological processes. We consider three ways that this might have occurred on the early Earth.

#1 Lightning-Fueled Atmospheric Reactions In 1953, Stanley Miller and Harold Urey tested the hypothesis that lightning could have powered synthesis reactions in Earth's early atmosphere. To simulate this process, they filled a reaction chamber with methane, ammonia, and hydrogen gas, and zapped it with sparks from electrodes (**Figure 19.4**). Within a week, a variety of organic molecules had formed, including

Figure 19.5 A hydrothermal vent on the seafloor. Mineral-rich water heated by geothermal energy streams out, into cold ocean water. As the water cools, dissolved minerals come out of solution and form a chimney-like structure around the vent.

amino acids that are common in living things. The result was hailed as a breakthrough, the discovery of the first steps down the road to life. However, scientists continually revisit ideas in light of new findings. We now know that the mixture of gases used in the Miller–Urey simulation probably did not represent Earth's early atmosphere. More accurate simulations with different mixes of gases produce a far lower yield of amino acids.

#2 Reactions at Hydrothermal Vents By another hypothesis, synthesis of life's building blocks occurred at deep-sea hydrothermal vents. A **hydrothermal vent** is like an underwater geyser, a place where mineral-rich water heated by geothermal energy streams out through a rocky opening in the seafloor (**Figure 19.5**). Günter Wächtershäuser and Claudia Huber synthesized amino acids in a simulated vent environment by combining hot water with carbon monoxide (CO) and potassium cyanide (KCN) in the presence of metal ions.

#3 Delivery From Space Modern-day meteorites that fall to Earth sometimes contain amino acids, sugars, and nucleotide bases, suggesting that life's building blocks may have had an extraterrestrial origin. They may have formed in interstellar clouds and been carried to Earth on meteorites. Keep in mind that during Earth's early years, meteorites fell to Earth thousands of times more frequently than they do today.

hydrothermal vent Underwater opening from which mineral-rich water heated by geothermal energy streams out.

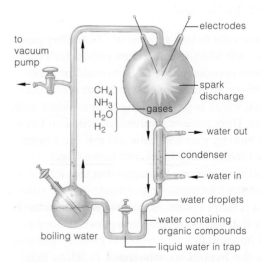

Figure 19.4 Animated Diagram of an apparatus used to test whether organic compounds could have formed spontaneously on the early Earth. Water vapor, hydrogen gas (H_2), methane (CH_4), and ammonia (NH_3) simulated Earth's early atmosphere. Sparks from an electrode simulated lightning.

Figure It Out: What was the source of the nitrogen for the amino acids formed in this apparatus? Answer: Ammonia

Labels in figure: to vacuum pump; electrodes; spark discharge; CH_4 NH_3 H_2O H_2; gases; water out; condenser; water in; water droplets; water containing organic compounds; boiling water; liquid water in trap

Take-Home Message

What was the source of the small organic molecules required to build the first life?

» Small organic molecules that serve as the building blocks for living things can be formed by nonliving mechanisms. For example, amino acids form in reaction chambers that simulate conditions on the early Earth. They are also present in some meteorites.

19.4 From Polymers to Protocells

■ We will never know for sure how the first cells came to be, but we can investigate the possible steps on the road to life.
■ Links to Nucleic acids 3.8, Cell structure 4.2, Cofactors 5.6, Membranes 5.7, Transcript modifications 9.3, Translation 9.5

Properties of Cells

In addition to sharing the same molecular components, all cells have a plasma membrane with a lipid bilayer. They have a genome of DNA that enzymes transcribe into RNA, and ribosomes that translate RNA into proteins. All cells replicate, and pass on copies of their genetic material to their descendants. The many similarities in structure, metabolism, and replication processes among all life provide evidence of descent from a common cellular ancestor.

Time has erased all traces of the earliest cells, but scientists can still investigate this first chapter in life's history. They use their knowledge of chemistry to design experiments that test whether a particular hypothesis about how life began is plausible. Such studies support the hypothesis that cells arose as a result of a stepwise process that began with inorganic materials (**Figure 19.6**). Each step on this hypothetical road to life can be explained by familiar chemical and physical mechanisms that still occur today.

Origin of Metabolism

Modern cells take up organic monomers, concentrate them, and assemble them into organic polymers. Before there were cells, a nonbiological process that concentrated organic subunits would have increased the chance of polymer formation.

By one hypothesis, this process occurred on clay-rich tidal flats. Clay particles have a slight negative charge, so positively charged molecules in seawater stick to them. At low tide, evaporation would have concentrated the subunits even more, and energy from the sun might have caused them to bond as polymers. Amino acids do form short chains under simulated tidal flat conditions.

The **iron–sulfur world hypothesis**, proposed by Günter Wächtershäuser, holds that the first metabolic reactions began on the surface of rocks around hydrothermal vents. These rocks contain iron sulfide and are porous, with many tiny chambers about the size of cells. Metabolism may have begun when iron sulfide in the rocks donated electrons to dissolved carbon monoxide, setting in motion reactions that formed larger organic compounds. Researchers who tested this hypothesis by simulating vent conditions found

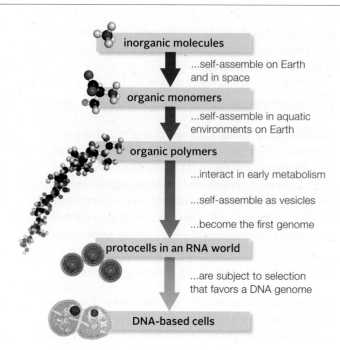

inorganic molecules
...self-assemble on Earth and in space

organic monomers
...self-assemble in aquatic environments on Earth

organic polymers
...interact in early metabolism
...self-assemble as vesicles
...become the first genome

protocells in an RNA world
...are subject to selection that favors a DNA genome

DNA-based cells

Figure 19.6 Proposed sequence for the evolution of cells. Scientists carry out experiments and simulations that test the feasibility of individual steps.

that organic compounds such as pyruvate do form and accumulate in the chambers. In addition, all modern organisms require iron-sulfide cofactors to carry out essential reactions. The universal requirement for these cofactors may be a legacy of life's rocky beginnings.

Origin of the Genome

All modern cells have a genome of DNA. They pass copies of their DNA to descendant cells, which use instructions encoded in the DNA to build proteins. Some of these proteins are enzymes that synthesize new DNA, which is passed along to descendant cells, and so on. Thus, protein synthesis depends on DNA, which is built by proteins. How did this cycle begin?

In the 1960s, Francis Crick and Leslie Orgel addressed this dilemma by suggesting that RNA may have been the first molecule to encode genetic information, a concept known as the **RNA world hypothesis**. Evidence that RNA can both store genetic information and function like an enzyme in protein synthesis, supports this hypothesis. **Ribozymes**, or RNAs that function as enzymes, are common in living cells. For example, the rRNA in ribosomes speeds formation of peptide bonds during protein synthesis. Other ribozymes cut noncoding bits (introns) out of newly formed RNAs. Researchers have also produced self-replicating ribozymes that assemble free nucleotides.

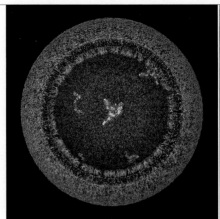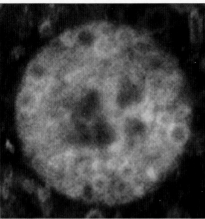

A Illustration of a laboratory-produced protocell with a bilayer membrane of fatty acids and strands of RNA inside.

B Laboratory-formed protocell consisting of RNA-coated clay (*red*) surrounded by fatty acids and alcohols.

C Field-testing a hypothesis about protocell formation. David Deamer pours a mix of small organic molecules and phosphates into a hot acidic pool in Russia.

Figure 19.7 Protocells. Scientists test hypotheses about protocell formation through laboratory simulations and field experiments.

If the earliest self-replicating genetic systems were RNA-based, then why do all organisms now have a genome of DNA? The structure of DNA may hold the answer. Compared to a double-stranded DNA molecule, single-stranded RNA breaks more easily and mutates more often. Thus, a switch from RNA to DNA would make larger, more stable genomes possible.

Origin of the Plasma Membrane

Self-replicating molecules and products of other early synthetic reactions would have floated away from one another unless something enclosed them. In modern cells, a plasma membrane serves this function. If the first reactions took place in tiny rock chambers, the rock would have acted as a boundary. Over time, lipids produced by reactions inside a chamber could have accumulated and lined the chamber wall forming a protocell. A **protocell** is a membrane-enclosed collection of interacting molecules that can takes up material and replicate. Protocells are hypothesized to be the ancestors of cellular life.

Experiments by Jack Szostak and others have shown that protocells can form even without rock chambers. **Figure 19.7A** illustrates one type of protocell that Szostak investigates. **Figure 19.7B** is a photo of a protocell that formed in his laboratory. A membrane of lipid bilayer encloses strands of RNA. The protocell "grows" by adding fatty acids to its membrane and nucleotides to its RNA. Mechanical force causes protocell division.

David Deamer studies protocell formation both in the laboratory and in the field. In the lab, he has shown that the small organic molecules carried to Earth on meteorites can react with minerals and seawater to form vesicles with a bilayer membrane. However, Deamer has yet to locate a natural environment that facilitates the same process. In one experiment, he added a mix of organic subunits to the acidic waters of a clay-rich volcanic pool in Russia (**Figure 19.7C**). The organic subunits bound tightly to the clay, but no vesicle-like structures formed. Deamer concluded that hot acidic waters of volcanic springs do not provide the right conditions for protocell formation. He continues to carry out experiments to determine what naturally occurring conditions would favor this process.

Take-Home Message

What have experiments and simulations revealed about the steps that led to the first cells?

» All living cells carry out metabolic reactions, are enclosed within a plasma membrane, and can replicate themselves.

» Metabolic reactions may have begun when molecules became concentrated on clay particles or in tiny rock chambers near hydrothermal vents.

» RNA can serve as an enzyme, as well as a genome. An RNA world may have preceded evolution of DNA-based genomes.

» Vesicle-like structures with outer membranes form spontaneously when certain organic molecules are mixed with water.

---

iron–sulfur world hypothesis Hypothesis that the metabolic reactions that led to the first cells took place on the porous surface of iron-sulfide-rich rocks at hydrothermal vents.
protocell Membranous sac that contains interacting organic molecules; hypothesized to have formed prior to the earliest life forms.
ribozyme RNA that functions as an enzyme.
RNA world hypothesis Hypothesis that RNA served as the genetic information of early life.

19.5 Life's Early Evolution

■ Fossils and molecular comparisons among modern organisms inform us about the early history of life.

■ Links to Prokaryotic and eukaryotic cells 4.4, 4.5; Photosynthesis 6.4; Fossils 16.5–16.6; Molecular clocks 18.4

Origin of Bacteria and Archaea

How old is life on Earth? Different methods provide different answers. Studies that use accumulated mutations as a molecular clock (Section 18.4) suggest that the last universal common ancestor of all organisms lived about 4.3 billion years ago. The carbon isotope ratios in ancient rocks suggest organisms of some sort were fixing carbon by 3.8 billion years ago. Australian rocks that date back 3.5 billion years hold filamentous structures some scientists call fossil cells (**Figure 19.8A**). However, other scientists contend that these filaments are simply crystal-filled cracks in the rock.

Early fossil cells are similar in size and structure to modern archaea and bacteria. There was little oxygen in the air or water before 2.3 billion years ago, so the first cells must have been anaerobic. They probably used dissolved carbon dioxide as a carbon source and obtained energy by oxidizing inorganic material.

Based on gene differences between existing bacteria and archaea, researchers estimate that the last common ancestor of these groups lived about 3.5 billion years ago. Shortly after they diverged, a new mode of nutrition—photosynthesis—evolved in one bacterial lineage. By one hypothesis, light-capturing pigments used in photosynthesis initially functioned like a sunscreen that protected cells in well-lit waters from the damaging effects of ultraviolet (UV) radiation. Pigments serve this purpose in some modern bacteria. Alternatively, photosynthetic pigments may have first evolved in bacteria living near hydrothermal vents. Some modern bacteria have photosynthetic pigments that capture geothermal energy emitted by such vents, rather than light energy.

By 2.7 billion years ago, the oxygen-producing noncyclic pathway of photosynthesis had evolved in one bacterial lineage: the cyanobacteria (**Figure 19.8B**). Cyanobacteria are a relatively recent branch on the bacterial family tree, so noncyclic photosynthesis presumably arose through mutations that modified the existing cyclic pathway. The two pathways use many of the same enzymes.

When the Proterozoic era began 2.5 billion years ago, cyanobacteria and other photosynthetic bacteria were growing as dense mats in the seas. The mats trapped minerals and sediments. Over many years, continual cell growth and deposition of minerals formed large dome-shaped, layered structures called **stromatolites** (**Figure 19.8C**). Such structures still form in some shallow seas today. The abundance of stromatolites soared during the Proterozoic era. Cyanobacterial populations increased, and so did their waste product: oxygen gas. Oxygen started to accumulate in Earth's waters and air. Sound familiar? Here we pick up the story that we began in Chapter 6.

A Possible cells from 3.5 billion years ago.

B Two types of cyanobacteria from 850 million years ago.

C Artist's depiction of stromatolites. The surface of each stromatolite consists of a mat of living photosynthetic bacteria. Beneath it are layers of earlier generations of bacteria that trapped sediment and precipitated minerals. The *inset photo* shows the layered structure of a fossil stromatolite.

Figure 19.8 Evidence of early cells.

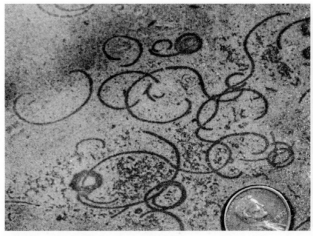

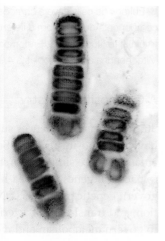

A *Grypania spiralis*; 2.1 billion years ago

B *Tawuia*; 1.6 billion years ago

C *Bangiomorpha pubescens*; 1.2 billion years ago

Figure 19.9 Fossils of some early eukaryotes. All are thought to be algae.

An oxygen-rich atmosphere had three important consequences for life: <u>First</u>, oxygen interferes with the self-assembly of complex organic compounds, so life could no longer arise from nonliving materials.

<u>Second</u>, the presence of oxygen put organisms that tolerated or thrived under aerobic conditions at an advantage. Species that could not adapt to higher oxygen levels became extinct, or were restricted to the remaining low-oxygen environments such as deep ocean sediments. Aerobic respiration evolved and became widespread.

<u>Third</u>, as oxygen enriched the atmosphere, some oxygen molecules broke apart, then recombined as ozone. The ozone accumulated in the upper atmosphere and began to reduce the amount of solar ultraviolet (UV) radiation that reached Earth's surface. UV radiation does not penetrate deep into water, so the lack of an ozone layer did not prevent the evolution of life in the seas, but without the protective effect of the ozone layer, life could not have moved onto land.

The Rise of Eukaryotes

The third domain of life arose when the ancestors of archaea and eukaryotes diverged. The earliest evidence of eukaryotes is lipids in 2.7-billion-year-old rocks. The lipids are **biomarkers** for eukaryotes. A biomarker is a compound made only by one particular type of cell; it is like a molecular signature.

Fossilized coils and blobs so large that they can be seen with the naked eye may also be evidence of early eukaryotes (**Figure 19.9A,B**). The fossil in **Figure 19.9C** is certainly a eukaryote. It is a red alga that lived about

1.2 billion years ago. This alga also has the distinction of being the oldest species known to reproduce sexually, a trait unique to eukaryotes. The alga grew as hairlike strands, with cells at one end forming a holdfast that held it in place. Cells at the strand's opposite end produced sexual spores by meiosis.

Evolution of sexual reproduction and multicellularity were milestones in the history of life. Sex gave some eukaryotic organisms a new way to exchange genes. Multicellularity coupled with cellular differentiation opened the way to evolution of larger bodies that have specialized parts adapted to specific functions.

Trace fossils and biomarkers indicate that sponge-like animals probably evolved by 870 million years ago. By 570 million years ago, animals with more complex bodies shared the oceans with bacteria, archaea, protists, and fungi.

biomarker Molecule produced only by a specific type of cell; its presence indicates the presence of that cell.
stromatolite Dome-shaped structure composed of layers of bacterial cells, their secretions, and sediments.

Take-Home Message

What was early life like and how did it change Earth?

» Life arose by 3–4 billion years ago; it was probably anaerobic and did not have a nucleus.

» An early divergence separated ancestors of modern bacteria from the lineage that would lead to archaea and eukaryotic cells.

» The first photosynthetic cells were bacteria that used the cyclic pathway. Later, the oxygen-producing, noncyclic pathway evolved in cyanobacteria.

» Oxygen accumulation in air and seas halted spontaneous formation of the molecules of life, formed a protective ozone layer, and favored organisms that carried out the highly efficient pathway of aerobic respiration.

19.6 How Did Eukaryotic Traits Evolve?

- Eukaryotic cells have a composite ancestry, with different components derived from different lineages.
- Links to Nucleus 4.6, Endomembrane system 4.7, Mitochondria and chloroplasts 4.9

A Composite Ancestry

Studies of gene sequences suggest that eukaryotes have a composite ancestry. Some nuclear genes in eukaryotes are similar to archaeal genes and others to bacterial genes. Archaea-like nuclear genes tend to govern genetic processes (DNA replication, transcription, and translation). Bacteria-like nuclear genes tend to govern metabolism and membrane formation.

Origin of Internal Membranes

In all eukaryotes, the DNA resides in a nucleus. The outer layer of the nucleus, the nuclear envelope, consists of a double layer of membrane with protein-lined pores that control the flow of material into and out of the nucleus. By contrast, the DNA of archaea and bacteria typically lies unenclosed in the cytoplasm.

The nucleus and endomembrane system probably evolved from infoldings of the plasma membrane (**Figure 19.10A**). Such infoldings can be selectively advantageous because they increase the surface area available for membrane-associated reactions. A few modern bacteria do have internal membrane-enclosed compartments. For example, the marine bacterium *Nitrosococcus oceani* has a system of highly folded internal membranes (**Figure 19.10B**). Enzymes embedded in the membranes allow the cell to meet its energy needs by breaking down ammonia.

An infolded membrane that cordons off a cell's genetic material can help protect the genome from physical or biological threats. Consider *Gemmata obscuriglobus*, one of the few bacteria with a membrane around its DNA (**Figure 19.10C**). Compared to typical bacteria, *G. obscuriglobus* can withstand much higher levels of mutation-causing radiation. Researchers attribute this cell's radiation resistance to the tight packing of its DNA within the membrane-enclosed compartment. Enclosing the genetic material within a membrane could also help protect it from viruses that inject their genetic material into bacteria, or from interference caused by bits of DNA absorbed from the environment. A nuclear membrane would also prevent harmful interactions between genes derived from archeal and bacterial sources.

Evolution of Mitochondria and Chloroplasts

All known eukaryotes have mitochondria or similar organelles thought to be derived from mitochondria. Many have chloroplasts. Both organelles resemble bacteria in their size and structure. Like bacteria, these organelles have their own genome arranged as a circle of DNA. The organelles also behave somewhat independently, duplicating their DNA and dividing at a different time than the cell that holds them. These observations are the basis for the **endosymbiont hypothesis**. This hypothesis holds that mitochondria and chloroplasts evolved as a result of endosymbiosis, a relationship in which one type of cell (the symbiont) lives and replicates inside another cell (the host). The host in an endosymbiotic relationship passes some symbionts along to its descendants when it divides.

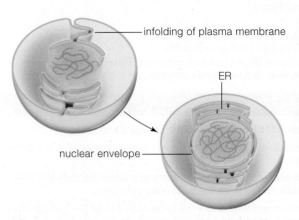
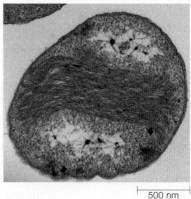
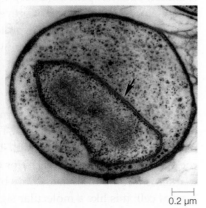

A Infoldings of the plasma membrane could have evolved into the eukaryotic nuclear envelope and endomembrane system.

B Bacterium (*Nitrosococcus oceani*) with highly folded internal membranes visible across its midline.

500 nm

C Bacterium (*Gemmata obscuriglobus*) that has DNA enclosed by a two-layer membrane (indicated by the arrow).

0.2 μm

Figure 19.10 Evolution of internal membranes.

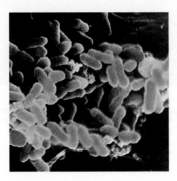

Figure 19.11 *Rickettsia prowazekii*, thought to be a close relative of the bacterial ancestors of mitochondria. It causes the disease typhus in humans. Like mitochondria, *R. prowazekii* divides only inside the cytoplasm of a eukaryotic cell. It cannot carry out glycolysis, but does carry out the Krebs cycle and electron transport phosphorylation.

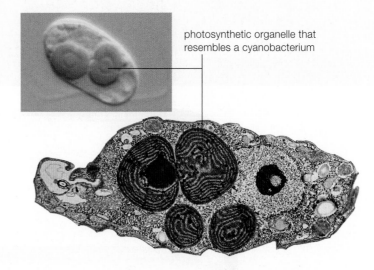

photosynthetic organelle that resembles a cyanobacterium

Figure 19.12 *Cyanophora paradoxa*, a glaucophyte protist with large photosynthetic organelles that make the bacterial protein peptidoglycan.

Genetic similarities between mitochondria and modern aerobic bacteria called rickettsias (**Figure 19.11**) indicate that the two groups share a common ancestor. Presumably, a rickettsia-like cell infected an archaeon or an early eukaryote. The host began to use ATP produced by its aerobic symbiont while the symbiont began to rely on the host for raw materials. Over time, genes that occurred in both the host and symbiont were free to mutate. If a gene lost its function in one partner, a gene from the other could take up the slack. Eventually, the host and symbiont both became incapable of living independently.

Similarly, chloroplasts are structurally and genetically similar to a group of modern oxygen-producing photosynthetic bacteria called cyanobacteria. These similarities cause biologists to infer that chloroplasts evolved from an ancient relative of these cells.

The endosymbiont hypothesis provides a possible explanation for the observed composite genome of eukaryotes. Suppose an archaeal cell engulfed or otherwise partnered with a bacterial cell. Over time, bacterial genes related to metabolism and membrane formation migrated into the nucleus and the corresponding archaeal genes were lost. The result would be a composite genome of the type we observe today.

Evidence of Endosymbiosis in Living Cells

A chance discovery made by microbiologist Kwang Jeon supports the hypothesis that bacteria can evolve into organelles. In 1966, Jeon was studying *Amoeba proteus*, a species of single-celled protist. By accident, one of his cultures became infected by a rod-shaped bacterium. Some infected amoebas died right away. Others kept growing, but only slowly. Intrigued, Jeon maintained those infected cultures to see what would happen. Five years later, the descendant amoebas were

endosymbiont hypothesis Hypothesis that mitochondria and chloroplasts evolved from bacteria.

host to many bacterial cells, yet they seemed healthy. In fact, when treated with bacteria-killing drugs that usually do not harm amoebas, they died.

Experiments confirmed that the amoebas had come to require the bacteria for some life-sustaining function. The amoebas had lost the ability to make an essential enzyme. They now depended on their bacterial endosymbionts to make that enzyme for them.

Glaucophytes provide another example of an evolved dependency. The interior of these single-celled protists is taken up largely by green photosynthetic organelles that resemble cyanobacteria (**Figure 19.12**). The organelle has a layer of peptidoglycan, a material made by some bacteria, but no eukaryotes. The photosynthetic organelles of glaucophytes, like chloroplasts, have evolved a dependence on their host. They cannot survive on their own.

However they arose, early eukaryotic cells had a nucleus, endomembrane system, mitochondria, and—in certain lineages—chloroplasts. These cells were the first protists. Over time, their many descendants came to include the modern protist lineages, as well as the plants, fungi, and animals. The next section provides a time frame for these pivotal evolutionary events.

Take-Home Message

How might the nucleus and other eukaryotic organelles have evolved?

» A nucleus and other organelles are defining features of eukaryotic cells.

» The nucleus and ER may have arisen through modification of infoldings of the plasma membrane.

» Mitochondria and chloroplasts most likely descended from bacteria.

19.7 Time Line for Life's Origin and Evolution

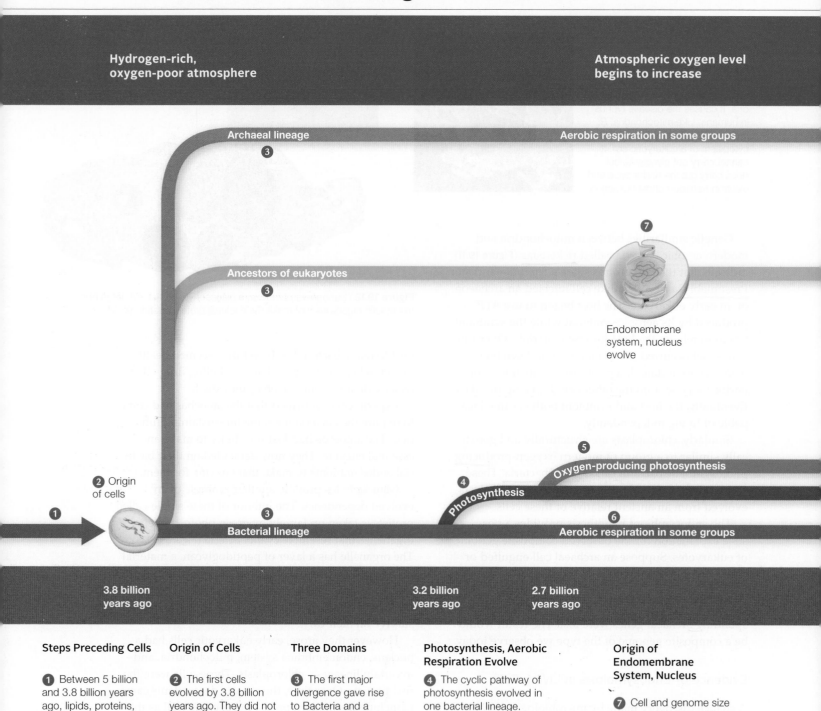

Hydrogen-rich, oxygen-poor atmosphere

Atmospheric oxygen level begins to increase

Archaeal lineage ③

Aerobic respiration in some groups

⑦

Ancestors of eukaryotes ③

Endomembrane system, nucleus evolve

⑤

Oxygen-producing photosynthesis

② Origin of cells

④ Photosynthesis

① ③

Bacterial lineage ⑥ Aerobic respiration in some groups

3.8 billion years ago

3.2 billion years ago

2.7 billion years ago

Steps Preceding Cells

❶ Between 5 billion and 3.8 billion years ago, lipids, proteins, nucleic acids, and complex carbohydrates formed from the simple organic compounds present on early Earth.

Origin of Cells

❷ The first cells evolved by 3.8 billion years ago. They did not have a nucleus or other organelles. Oxygen was scarce; the first cells made ATP by anaerobic pathways.

Three Domains

❸ The first major divergence gave rise to Bacteria and a common ancestor of Archaea and Eukarya. Not long after that, the Archaea and Eukarya diverged.

Photosynthesis, Aerobic Respiration Evolve

❹ The cyclic pathway of photosynthesis evolved in one bacterial lineage.

❺ Noncyclic photosynthesis evolved in a branch from this lineage (cyanobacteria) and oxygen began to accumulate.

❻ Aerobic respiration evolved independently in some bacterial and archaeal groups.

Origin of Endomembrane System, Nucleus

❼ Cell and genome size continued to expand in ancestors of what would become the eukaryotic cells. The endomembrane system, including the nuclear envelope, arose through the modification of cell membranes between 3 and 2 billion years ago.

Figure 19.13 Animated Milestones in the history of life, based on the most widely accepted hypotheses. As you read the next unit on life's past and present diversity, refer to this visual overview. It can serve as a simple reminder of the evolutionary connections among all groups of organisms. Time line not to scale.

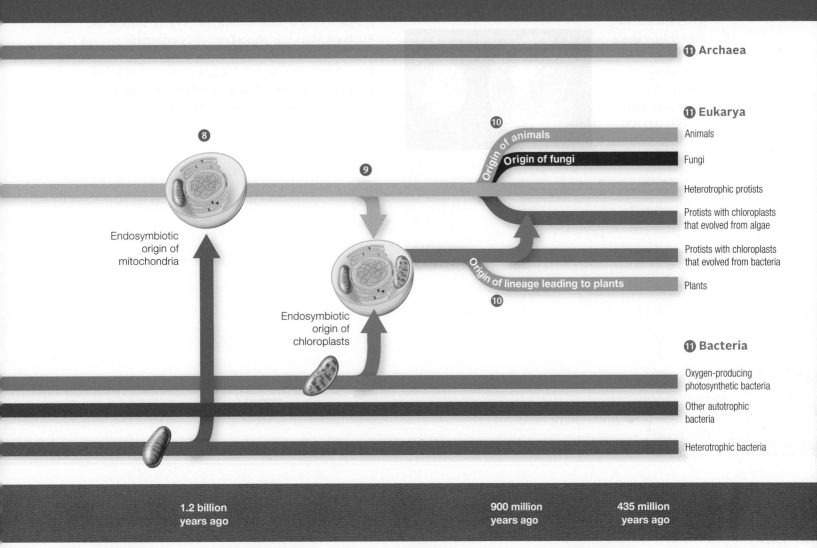

Atmospheric oxygen reaches current levels; ozone layer gradually forms

11 Archaea

11 Eukarya

8

9

10

Origin of animals

Origin of fungi

Animals

Fungi

Heterotrophic protists

Protists with chloroplasts that evolved from algae

Endosymbiotic origin of mitochondria

Endosymbiotic origin of chloroplasts

Origin of lineage leading to plants

Protists with chloroplasts that evolved from bacteria

Plants

10

11 Bacteria

Oxygen-producing photosynthetic bacteria

Other autotrophic bacteria

Heterotrophic bacteria

| 1.2 billion years ago | 900 million years ago | 435 million years ago |

Endosymbiotic Origin of Mitochondria

8 Before about 1.2 billion years ago, an aerobic bacterium entered an anaerobic eukaryotic cell. Over generations, the two species established a symbiotic relationship. Descendants of the bacterial cell became mitochondria.

Endosymbiotic Origin of Chloroplasts

9 By 1.5 billion years ago, a cyanobacterium entered a protist. Over generations, bacterial descendants evolved into chloroplasts. Later, some photosynthetic protists would evolve into chloroplasts inside other protist hosts.

Plants, Fungi, and Animals Evolve

10 By 900 million years ago, representatives of all major lineages—including fungi, animals, and the algae that would give rise to plants—had evolved in the seas.

Lineages That Have Endured to the Present

11 Today, organisms live in nearly all regions of Earth's waters, crust, and atmosphere. They are related by descent and share certain traits. However, each lineage encountered different selective pressures, and unique traits evolved in each one.

Figure It Out: From which lineage are mitochondria descended, bacteria or archaea?

Answer: Bacteria

Looking For Life (revisited)

When it comes to sustaining life, Earth is just the right size. If the planet were much smaller, it would not exert enough gravitational pull to keep atmospheric gases from drift-ing off into space. The photo at the *right* shows the relative sizes of Earth and Mars. As you can see,

Mars is only about half the size of Earth. As a result, it has a much thinner atmosphere. What atmosphere there is consists mainly of carbon dioxide, some nitro-gen, and only a trace amount of oxygen. Thus, if life exists on Mars it is almost certainly anaerobic.

How would you vote? The best way to find out if there is life on Mars is to take soil samples and bring them back to Earth for a thorough analysis. Should we bring samples of soil to Earth for study?

From the Green River Formation near Lincoln, Wyoming, the stunning fossilized remains of a bird trapped in time. During the Eocene, some 50 million years ago, sediments that had been gradually deposited in layers at the bottom of a large inland lake became its tomb. In this same formation, fossilized remains of sycamore, cattails, palms, and other plants suggest that the climate was warm and moist when the bird lived. Fossils from places all around the world yield clues to life's early history.

LEARNING ROADMAP

Where you have been Section 1.4 gave you an early glimpse of the bacteria and archaea. Section 4.4 began our discussion of their structure and Section 19.5 put these groups into an evolutionary time frame. Section 17.5 focused on antibiotic resistance in bacteria. This chapter also discusses bacteriophage, a type of virus that will be familiar from Section 8.3.

Where you are now

Viral Structure and Function
A virus is a noncellular infectious particle that must infect a living cell to replicate. All organisms are susceptible to viral infection.

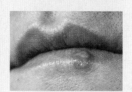

Viruses and Human Health
Viruses can be pathogens, meaning they cause disease. Most viral diseases are mild and pass quickly, but some persist; a few are fatal.

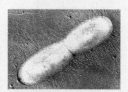

Two Lineages of Simple Cells
Bacteria and archaea are two distinct lineages of asexually reproducing, structurally simple cells that do not have a nucleus. They are extremely diverse and abundant.

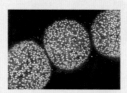

Bacteria
Bacteria play essential ecological roles. They put oxygen into the air, make nitrogen available to plants, and act as decomposers. A minority cause disease.

Archaea
Archaea live in some astonishingly hostile environments such as hot springs and pools of brine. They also live alongside bacteria in soil and in the animal gut.

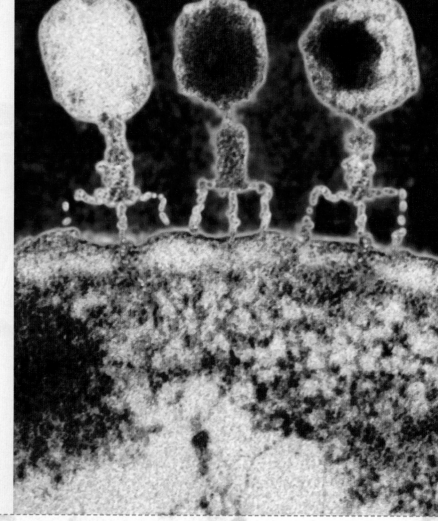

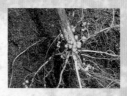

Where you are going You will learn more about the effects of viral and bacterial pathogens in chapters that discuss physiology. For example, Section 37.11 looks at the immune effects of AIDS, and Section 41.10 discusses sexually transmitted diseases. You will also learn about the beneficial effects of the bacteria in your gut (39.1), and the bacteria that partner with plants (28.3). Bacteria play an integral role in food webs (46.3) and biogeochemical cycles (46.5).

20.1 Evolution of a Disease

Billions of years before there were fungi, plants, or animals, Earth's seas were home to two groups of microscopic organisms: <u>bacteria</u> and <u>archaea</u>. These <u>single-celled</u> organisms <u>do not</u> have a <u>nucleus</u> or other typical eukaryotic organelles. Viruses are simpler still, with no chromosomes or metabolic machinery. By many definitions, viruses are not even alive. Despite their simplicity, viruses can evolve because like living organisms they have a genome that can mutate.

In recent years, scientists have learned quite a bit about the origin and evolution of **HIV (human immunodeficiency virus)**. This virus causes the emerging disease AIDS (acquired immune deficiency syndrome). An **emerging disease** is a disease that is relatively new to humans or has newly expanded its range.

HIV was first isolated in the early 1980s. Since then, gene sequence comparisons have revealed that the most common strain (HIV-1) evolved from the strain of simian immunodeficiency virus (SIV) that infects chimpanzees in west central Africa. A recent study investigated the health effects of SIV in a wild chimpanzee population that has been studied by primatologist Jane Goodall and others for many years. Researchers used DNA analysis of chimpanzee feces to identify individual animals and determine whether they were infected by SIV (**Figure 20.1**). This information was combined with observational data from the field. The researchers found that, in this population, SIV reduces fitness. SIV-infected chimpanzees die earlier than unaffected animals and leave fewer offspring.

How did SIV get from chimpanzees into people? Some African populations eat nonhuman primates and it is likely that a person became infected while butchering an infected chimpanzee for use as food. Butchery is a bloody process, and SIV-infected blood could have gotten into a butcher's body through a cut. Presumably the virus survived and mutated inside its unusual host. Over time it became HIV.

So far, the earliest known evidence of HIV infection comes from two tissue samples stored at a hospital in west central Africa. One is a blood sample taken from a man in 1959. The other is a woman's lymph node that was removed in 1960. The viral gene sequences from the two samples differ a bit, which implies that HIV had already been around and mutating by the time these two people became infected. Given the known mutation rate for HIV, researchers estimate that HIV first infected humans in the early 1900s.

emerging disease Disease that is relatively new to a species, or has recently expanded its range.
HIV (human immunodeficiency virus) Retrovirus that causes AIDS.

Figure 20.1 Analyzing wild chimpanzee feces for SIV. Rebecca Rudicell is part of a team that is studying the effects of this virus, from which HIV evolved.

Gene sequence comparisons have also allowed researchers to trace the movement of the virus out of Africa. One recent study concluded that HIV-1 was carried from Africa to Haiti in about 1966. The virus diversified in Haiti and acquired distinctive mutations not seen in Africa. In about 1969, HIV-1 with Haiti-specific mutations was introduced to the United States. It may have arrived in an infected individual or in infected blood. Once there, it spread quietly until AIDS was identified as a threat in 1981.

Today, more than 20 million people worldwide have died from AIDS. About 30 million are currently infected with HIV. The virus infects and replicates inside white blood cells that are essential to immune responses. Eventually, the infected white blood cells die, destroying the body's ability to defend itself. As a result, disease-causing organisms run rampant, causing symptoms of AIDS and health problems that can be fatal. Section 37.11 discusses in detail how AIDS affects the immune system.

Knowing about HIV's ancestry may help us develop new weapons against the virus. For example, although SIV does harm chimpanzees, the effects are not as devastating as untreated HIV in humans. Determining how the chimpanzee immune system fights against SIV may provide insights that we can put to use in our own fight against AIDS.

20.2 Viruses and Viroids

■ A virus consists of nucleic acid and protein. It is smaller than any cell and has no metabolic machinery of its own.

■ Links to Discovery of DNA function 8.3, Endocytosis 5.10

In the late 1800s, biologists studying stunted tobacco plants discovered a new kind of disease-causing agent, or **pathogen**. It was so small that it passed through screens that filtered out bacteria, and it could not be seen with a light microscope. The scientists called this unseen infectious entity a virus, a term that means "poison" in Latin.

Today, we define a **virus** as a noncellular infectious particle that can replicate only in a living cell. A virus is an obligate intracellular parasite. It does not have ribosomes or other metabolic machinery and it cannot make ATP. To replicate, the virus must insert its genetic material into a cell of a specific type of organism. We call that organism its host.

The fact that viruses can only replicate in cells suggests that they evolved from cells. They may be derived from bits of DNA or RNA that escaped. Alternatively, viruses may be remnants of a time before cells. This would explain why many viral genes have no counterpart in cells.

Viral Structure

Viruses are typically so small (about 25 to 300 nanometers) that they can only be seen with an electron microscope. A free viral particle, or virion, always includes a viral genome enclosed within a protein shell, or capsid. The viral genome may be RNA or DNA, and it may be single-stranded or double-stranded.

The capsid consists of many protein subunits that bond together in a repeating pattern, producing a

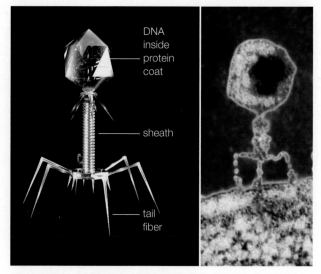

Figure 20.3 Animated Model (*left*) and electron micrograph (*right*) of a bacteriophage, a bacterial virus with a complex structure.

helical or many-sided (polyhedral) shape. The capsid protects the viral genetic material and facilitates its delivery into a host cell. In all viruses, some components of the viral coat bind to proteins at the surface of a host cell. The capsid may also enclose one or more viral enzymes.

Many plant viruses have a helical structure. The tobacco mosaic virus is an example. Its coat proteins bond together in a tight helix around its genetic material, a single strand of RNA (**Figure 20.2**). Viruses typically enter a plant through a wound made by an insect, pruning, or another mechanical injury. They move throughout the plant body and even enter seeds.

Bacteriophages, viruses that infect bacteria, have a complex structure (**Figure 20.3**). Their headlike capsid encloses the viral DNA. Other protein components of the virus allow it to pierce a bacterial cell wall and inject DNA into the cell. You learned earlier how Hershey and Chase used a type of bacteriophage called lambda to identify DNA as the genetic material of all organisms (Section 8.3).

Polyhedral viruses have a many-sided protein coat. Adenoviruses are an example. These animal viruses have a 20-sided capsid with a distinctive protein spike at each corner (**Figure 20.4A**). Adenoviruses frequently cause common colds. They are "naked," or nonenveloped viruses; their capsid is their outermost layer.

In most animal viruses, the capsid is enclosed within an "envelope," a layer of membrane derived from the host cell in which the virus assembled. For example, herpesvirus is an enveloped DNA virus.

Figure 20.2 Animated Tobacco mosaic virus, a virus that infects tobacco (*above*) and related plants. The helical arrangement of the capsid subunits (*right*) gives the virus a rodlike structure. The genome is single-stranded RNA.

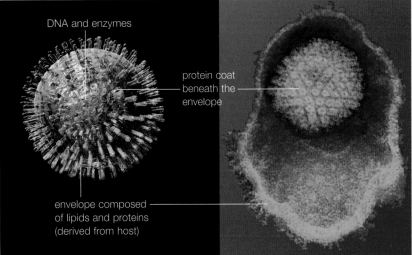

DNA and enzymes

protein coat
beneath the
envelope

envelope composed
of lipids and proteins
(derived from host)

A Model of an adenovirus, a polyhedral virus. Protein subunits form a 20-sided polyhedron around double-stranded DNA.

B Model (*left*) and electron micrograph (*right*) of a herpesvirus, an enveloped virus. The envelope is derived from the nuclear membrane of the cell in which the virus assembled. In the micrograph, the envelope is peeled back to reveal the protein coat beneath.

Figure 20.4 Animated Two animal viruses.

It has a 20-sided capsid surrounded by an envelope made of bits of a host cell's nuclear membrane (**Figure 20.4B**). More frequently, an enveloped virus derives its envelope from the host's plasma membrane.

Ecological Role of Viruses

Everywhere there is life, there are viruses. Viruses infect and replicate in all organisms, no matter how simple or complex. A viral infection often decreases a host's ability to survive and reproduce, so viruses affect ecological interactions throughout the biosphere.

Some viruses assist humans through their effects on other species. For example, we benefit when baculoviruses infect and kill caterpillars that feed on crop species or when bacteriophages kill bacteria that could cause food poisoning.

On the other hand, viruses can have devastating economic effects when they infect livestock or agriculturally important plants. In recent years, outbreaks of influenza among pigs and chickens have led to the slaughter of hundreds of thousands of animals.

Viruses also harm us directly by impairing our health. We discuss viral diseases of humans in Section 20.4.

Viroids

Viroids are small RNAs that cause disease in many commercially valuable plants, including potatoes, tomatoes, citrus, apples, coconuts, avocados, and chrysanthemums. They were discovered in the 1970s by the plant pathologist Theodor Diener. He named the tiny new pathogen he had isolated a viroid, because it seemed like a stripped-down version of a virus.

All viroids are circular, single-stranded RNAs. They are remarkably small, with fewer than 400 nucleotides. By comparison, even the smallest viral genome has thousands of nucleotides. Unlike the genetic material of a virus, viroid RNA does not encode proteins. The viroid is replicated in a plant cell nucleus by the plant's RNA polymerase.

---

bacteriophage Virus that infects bacteria.
pathogen Disease-causing agent.
viroid Small, noncoding, infectious RNA.
virus Noncellular, infectious particle of protein and nucleic acid; replicates only in a host cell.

Take-Home Message

What are viruses and viroids?

» Viruses are noncellular infectious particles made of protein and nucleic acid.

» A virus is an obligate intracellular parasite. It has no metabolic machinery of its own and can multiply only inside living cells.

» Viroids are infectious noncoding RNAs that cause some plant diseases.

20.3 Viral Replication

- A viral infection is like a cellular hijacking; viral genes take over a host cell's machinery and direct it to synthesize viral components that can self-assemble as new viral particles.
- Links to Transcription 9.3, Translation 9.5

Overview of Viral Replication

The details of viral replication processes vary, but all involve the steps outlined in **Table 20.1**. A virus cannot propel itself toward a host, so infection begins with a chance encounter. During attachment, viral proteins bind to receptor proteins on the surface of a host cell. The virus's genetic material enters the host cell and takes over its genetic machinery. Under the influence of the virus, the cell puts aside its normal tasks and turns to replicating and expressing the viral genome. When viral proteins and nucleic acid come into contact, they self-assemble as new virions. The virus either buds from the host cell or escapes when the host cell lyses (breaks open).

Bacteriophage Replication

Bacteriophages replicate in bacteria by two pathways. Both begin when a bacteriophage attaches to a bacterial cell and injects its DNA (**Figure 20.5**). In the **lytic pathway**, viral genes are expressed immediately ❶. The infected host first produces viral components that self-assemble as virus particles. Then, a viral-encoded enzyme breaks down the cell wall causing lysis of the cell—the cell disintegrates and dies.

In the **lysogenic pathway**, viral DNA becomes integrated into the host cell's genome and viral genes are not expressed, so the cell remains healthy ❷. When the cell reproduces, viral DNA is copied and passed on along with the host's genome. Like miniature time bombs, the viral DNA inside these descendant cells awaits a signal to enter the lytic pathway.

Some bacteriophages can only replicate by the lytic pathway. They kill their host cell quickly and are not passed from one bacterial generation to the next. Others embark upon either the lytic or lysogenic pathway, depending on conditions in the host cell.

Table 20.1 Steps in Most Viral Replication Cycles

1. **Attachment**   Proteins on viral particle chemically recognize and lock onto specific receptors at the host cell surface.

2. **Penetration**   Either the viral particle or its genetic material crosses the plasma membrane of a host cell and enters the cytoplasm.

3. **Replication and synthesis**   Viral DNA or RNA directs host to make viral nucleic acids and viral proteins.

4. **Assembly**   Viral components self-assemble as new viral particles.

5. **Release**   The new viral particles are released from the cell.

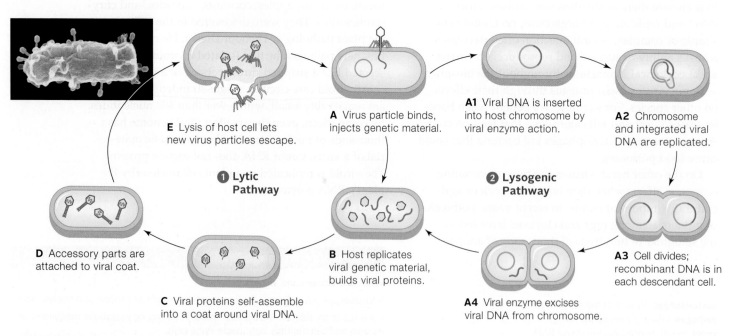

E Lysis of host cell lets new virus particles escape.

A Virus particle binds, injects genetic material.

A1 Viral DNA is inserted into host chromosome by viral enzyme action.

A2 Chromosome and integrated viral DNA are replicated.

❶ **Lytic Pathway**

❷ **Lysogenic Pathway**

D Accessory parts are attached to viral coat.

B Host replicates viral genetic material, builds viral proteins.

A3 Cell divides; recombinant DNA is in each descendant cell.

C Viral proteins self-assemble into a coat around viral DNA.

A4 Viral enzyme excises viral DNA from chromosome.

Figure 20.5 Animated Pathways in the replication cycle of a bacteriophage.
Figure It Out: What is the blue circle in A?   Answer: Bacterial chromosome

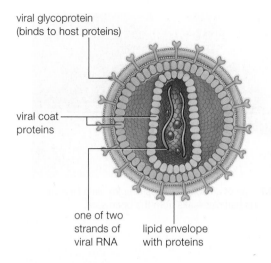

viral glycoprotein
(binds to host proteins)

viral coat proteins

one of two strands of viral RNA

lipid envelope with proteins

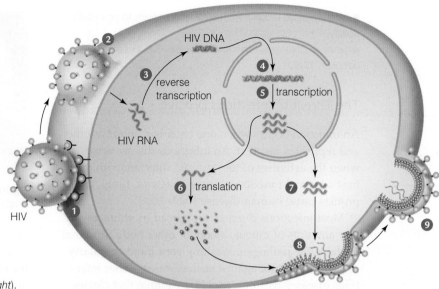

HIV DNA

② reverse transcription

HIV RNA

④ ⑤ transcription

⑥ translation

⑦

⑧

⑨

HIV

Figure 20.6 Animated Replication of HIV, a retrovirus (*right*). The structure of the HIV virion is shown *above*.

Replication of HIV

HIV is an enveloped RNA virus that replicates inside human white blood cells (**Figure 20.6**). It attaches to a cell via a glycoprotein that extends through its envelope ❶. The glycoprotein binds two different proteins on the host cell. After attachment, the viral envelope fuses with the blood cell's plasma membrane, releasing viral enzymes and RNA into the cell ❷.

One of the viral enzymes is **reverse transcriptase**, a polymerase that uses viral RNA as a template to synthesize double-stranded DNA ❸. This DNA enters the nucleus together with another viral enzyme, integrase. Integrase inserts the DNA into one of the host's chromosomes ❹. Once integrated, the viral DNA is replicated and transcribed along with the host genome ❺. Some of the resulting viral RNA is translated into viral proteins ❻ and some becomes the genetic material of new HIV virions ❼. The virions self-assemble at the plasma membrane ❽. As the virus buds from the host cell, some of the host's plasma membrane becomes the viral envelope ❾. Each new virion can then infect another white blood cell. New HIV-infected cells are also produced when an infected cell replicates.

❶ Viral protein binds to proteins at the surface of a white blood cell.

❷ Viral RNA and enzymes enter the cell.

❸ Viral reverse transcriptase uses viral RNA to make double-stranded viral DNA.

❹ Viral DNA enters the nucleus and becomes integrated into the host genome.

❺ Transcription produces viral RNA.

❻ Some viral RNA is translated to produce viral proteins.

❼ Other viral RNA forms the new viral genome.

❽ Viral proteins and viral RNA self-assemble at the host plasma membrane.

❾ New virus buds from the host cell, with an envelope of host plasma membrane.

Drugs designed to fight HIV take aim at steps in viral replication. Some interfere with the way HIV binds to a host cell. Others impair the viral reverse transcriptase. Integrase inhibitors prevent viral DNA from integrating into a human chromosome. Protease inhibitors prevent the processing of newly translated polypeptides into mature viral proteins.

These antiviral drugs lower the number of HIV particles, so a person stays healthier. Less HIV in body fluids also means reduced risk of passing the virus to others. However, no drug eliminates the virus, all have unpleasant side effects, and all must be taken for life.

Take-Home Message

How do viruses replicate?

» A virus binds to a specific type of host cell, and viral genetic material enters the cell. Viral genes direct the production of viral components that then self-assemble as new viral particles.

lysogenic pathway Bacteriophage replication mechanism in which viral DNA becomes integrated into the host's chromosome and is passed to the host's descendants.
lytic pathway Bacteriophage replication mechanism in which a virus replicates in its host and kills it quickly.
reverse transcriptase A viral enzyme that uses RNA as a template to make a strand of cDNA.

20.4 Viruses as Human Pathogens

■ Viral diseases range from merely inconvenient to potentially deadly.

■ Links to DNA replication and repair 8.5–8.6, Directional selection 17.5

The Threat of Infectious Disease

An infection occurs when one organism enters another and replicates inside it. An infectious disease arises when the activities of the "guests" interfere with the host's normal functions. Viruses, bacteria, fungi, and protists cause human disease (**Table 20.2**).

Most infectious diseases are spread by contact with tiny amounts of mucus, blood, or other body fluid that contains the pathogen. Washing your hands regularly is the best defense against such diseases. Other infectious diseases require a **vector**, an animal that carries the pathogen from host to host. Biting insects and ticks are the most important common disease vectors.

Common Viral Diseases

Most viral diseases cause mild symptoms and trouble us only briefly. For example, some adenoviruses infect the membranes of our upper respiratory system and cause common colds. Others colonize the lining of our gut and cause a brief bout of vomiting and diarrhea.

A minority of viral diseases can be more persistent. Various types of herpesviruses cause cold sores, genital herpes, mononucleosis, or chicken pox. Typically

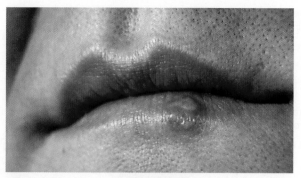

Figure 20.7 Sign of an active herpes simplex virus I infection. Fluid rich in viral particles leaks from the open sore.

the initial infection causes symptoms that subside quickly. However, the virus remains in the body in a latent state, and can reawaken later on. Many people have been infected by herpes simplex virus 1 (HSV-1), which can remain latent in nerve cells for years. When activated, the virus replicates and causes painful "cold sores" on the edge of the lips (**Figure 20.7**). Similarly, HIV can persist in a latent state inside white blood cells that are not actively dividing.

Measles, mumps, rubella (German measles), and chicken pox are childhood diseases that, until recently, were common worldwide. Today, most children in developed countries are protected against these illnesses because they have been vaccinated. Administering a vaccine primes the body to fight off a specific pathogen, a process explained in detail in Section 37.12).

New Flus: Viral Mutation and Reassortment

Like living organisms, viruses have genomes that can be altered by mutation. RNA viruses such as HIV and influenza virus mutate especially quickly. The viral reverse transcriptase makes frequent replication errors. These errors remain uncorrected because the host's proofreading and repair mechanisms evolved to fix errors of transcription and do not operate during reverse transcription.

To keep up with ongoing mutations in influenza viruses, scientists create a new flu shot every year. The flu shot is a vaccine designed to thwart the newly mutated influenza viruses that scientists predict are most likely to pose a threat during the upcoming flu season. Unfortunately, determining which flu strains will be circulating is not an exact science. Even after a flu shot, a person is susceptible to a virus that differs from the virus types targeted by the vaccine.

Table 20.2 Major Causes of Death From Infectious Disease		
Disease	**Type of Pathogen**	**Deaths per year worldwide**
Acute respiratory infections	Viruses, bacteria	4 million
AIDS	Virus (HIV)	2.7 million
Diarrheas	Viruses, bacteria, protists	1.8 million
Tuberculosis	Bacteria	1.6 million
Malaria	Protists	1.3 million
Measles	Virus	164,000
Whooping cough	Bacteria	294,000
Tetanus	Bacteria	204,000
Meningitis	Viruses	173,000
Syphilis	Bacteria	157,000

Influenza subtypes are named after the structure of two proteins at the viral surface (**Figure 20.8**). The glycoprotein hemagglutinin (H) allows the virus to bind to a host cell. The enzyme neuraminidase (N) helps new viral particles to exit from that cell.

In April of 2009, a new version of the H1N1 subtype of influenza appeared unexpectedly. Although the media referred to this virus as "swine flu," it had a composite genome, with genes from a human flu virus, bird flu virus, and two different swine flu viruses. Such novel genomes arise as a result of **viral reassortment**, the swapping of genes between viruses that infect a host at the same time (**Figure 20.9**).

The 2009 H1N1 virus was discovered when it caused an epidemic in Mexico. An **epidemic** is an outbreak of disease in a limited region. Within months, there was a **pandemic**, an outbreak of disease that simultaneously affects people throughout the world. Health officials became concerned because unlike a typical seasonal flu that kills mainly the elderly, the new flu seemed to be causing deaths largely among young, healthy people. Fortunately, initial fears of a high mortality rate proved largely unfounded. Governments quickly released reserves of antiviral drugs to treat those infected. The drugs interfere with neuraminidase function, and so impair the virus's ability to infect cells. A vaccine was created to prevent new infections. The World Health Organization declared the pandemic over in August 2010.

Another strain of influenza, influenza H5N1, is a bird flu that occasionally infects people who have direct contact with birds. When the virus does infect people, the death rate is disturbingly high. From 2003 to 2009, the World Health Organization received reports of 417 human cases of influenza H5N1, mainly in Asia. Of these, 257 (about 60 percent) were fatal. Fortunately, person-to-person transmission of the H5N1 virus is exceedingly rare.

Health officials continue to carefully monitor H5N1 and H1N1 influenza. Either virus could mutate, and their coexistence raises the possibility of a potentially disastrous gene exchange. If H1N1 picked up genes from avian H5N1, the result could be a flu virus that is both easily transmissible and deadly.

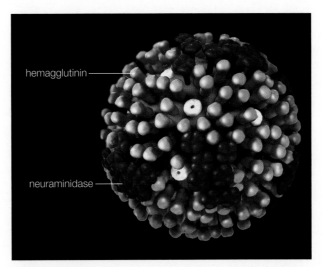

Figure 20.8 Influenza virus. Subtypes such as H1N1 or H5N2 are defined by the structure of viral proteins—hemagglutinin (H) and neuraminidase (N)—that extend through the outer envelope.

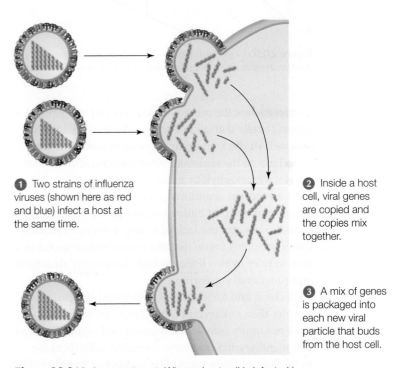

❶ Two strains of influenza viruses (shown here as red and blue) infect a host at the same time.

❷ Inside a host cell, viral genes are copied and the copies mix together.

❸ A mix of genes is packaged into each new viral particle that buds from the host cell.

Figure 20.9 Viral reassortment. When a host cell is infected by two viruses of the same type, copies of viral genes reassort to form new combinations.

epidemic Disease outbreak that occurs in a limited region.
pandemic Disease outbreak with cases worldwide.
vector Of a disease, an animal that transmits a pathogen from one host to the next.
viral reassortment Two related viruses infect the same individual simultaneously and swap genes.

20.5 "Prokaryotes"—Enduring, Abundant, and Diverse

■ The most widespread and abundant forms of life belong to two lineages of single-celled organisms that do not have a nucleus.

■ Links to Prokaryotes 4.4, Aerobic respiration 7.2, Classification systems 18.2, Early life 19.5

Two Lineages of "Prokaryotes"

Biologists have historically divided all life into two groups. Cells without a nucleus were **prokaryotes** and those with a nucleus were eukaryotes (**Figure 20.10A**). More recently we learned that the "prokaryotes" actually constitute two distinct lineages, now referred to as the domains Bacteria and Archaea. Eukaryotes are the third domain (**Figure 20.10B**).

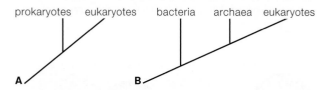

prokaryotes eukaryotes   bacteria   archaea eukaryotes

A **B**

Figure 20.10 Comparison of (**A**) two-domain and (**B**) three-domain trees of life. The three-domain model is now in wide use.

Bacteria are the more well-known and widespread group of cells that do not have a nucleus. **Archaea** are less well studied, and many live in extreme habitats.

In light of the realization that bacteria and archaea are not a monophyletic group, some microbiologists have advocated abandoning the term "prokaryote." They point out that biological groups are defined by shared traits, not the lack of a trait, such as a nucleus. Other scientists argue that the term remains useful as a way to refer to two lineages that share many structural and functional similarities.

Bacteria and archaea are smaller and structurally simpler than eukaryotes. This structural simplicity does not imply inferiority. Bacteria and archaea existed before eukaryotes and have coexisted with them for more than a billion years. The number of bacterial cells currently living on Earth has been estimated at five million trillion trillion. From an evolutionary perspective, bacteria and archaea are highly successful.

Identifying Species and Investigating Diversity

In eukaryotes, a species is defined on the basis of the ability of its members to mate and produce fertile offspring. This definition of a species does not apply to organisms such as bacteria and archaea that typically

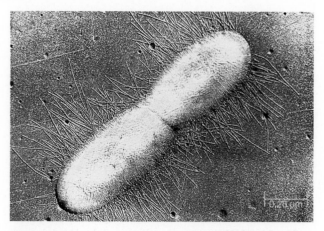

Figure 20.11 Asexual reproduction in *Escherichia coli*, one of the many species of bacteria that lives in the human gut. Meiosis and sexual reproduction do not occur in bacteria or archaea.

reproduce only asexually (**Figure 20.11**). In these groups, a species is defined as a group of individuals that share an ancestor and have a high degree of similarity in many independently inherited traits.

Historically, classification of bacteria was based on numerical taxonomy. By this process, an unidentified cell is compared against a known group on the basis of cell shape, cell wall properties, and metabolism. The more traits the cell shares with the known group, the closer is their inferred relatedness. This approach works best for cells that can be grown in the laboratory, stained, and viewed with a microscope. However, many bacteria and archaea do not grow in the lab.

The relatively new field of **metagenomics** is devoted to assessing microbial diversity by analysis of DNA in samples collected directly from an environment. Metagenomic studies often reveal a remarkable degree of species diversity. For example, air samples collected in two cities in Texas contained about 1,800 different kinds of bacteria.

The Human Microbiome Project is an ongoing metagenomic study of the microorganisms that live in or on the human body. Already, a survey of bacteria that live on the skin of the inner elbow turned up nearly 200 species. Another study found that more than a thousand species of bacteria and a few archaeal species can live in the human gut.

Metabolic Diversity

Autotroph or Heterotroph? Metabolic diversity gives prokaryotes the collective ability to live just about anywhere there is a source of energy and carbon.

Organisms harvest energy and nutrients from the environment in four different ways. All of these nutritional modes occur among bacteria, archaea, or both (**Figure 20.12**). In addition, some bacteria and archaea can switch from one metabolic mode to another.

As you learned in Section 6.1 autotrophs build their own food using carbon dioxide (CO_2) as their carbon source. There are two subgroups of autotrophs: those that obtain energy from light, and those that obtain energy from chemicals.

Photoautotrophs are photosynthetic. They use the energy of light to assemble organic compounds from CO_2 and water. Many bacteria are photoautotrophs, as are plants and photosynthetic protists. As Section 19.6 explained, eukaryotes have chloroplasts that evolved from cyanobacteria, a type of photosynthetic bacteria.

Chemoautotrophs obtain energy by oxidizing (removing electrons from) inorganic molecules such as hydrogen sulfide or methane. They use energy released by this process to build food from CO_2. Chemoautotrophic bacteria and archaea are the main producers in dark environments such as the sea floor. No eukaryotes are known to be chemoautotrophs.

Heterotrophs cannot use inorganic sources of carbon. Instead, they obtain carbon by taking up organic molecules from their environment. **Photoheterotrophs** harvest energy from light, and carbon from alcohols, fatty acids, or other small organic molecules. Heliobacteria that live in the soils of rice paddies are an example.

Chemoheterotrophs obtain both energy and carbon by breaking down carbohydrates, lipids, and proteins. Most bacteria and some archaea are chemoheterotrophs, as are animals, fungi, and nonphotosynthetic protists. All pathogenic bacteria are chemoheterotrophs that extract the organic compounds they need to live from their host. Other bacterial chemoheterotrophs serve as decomposers. They convert organic molecules into inorganic ones. By their activity, decomposers make nutrients that were tied up in organic wastes and remains accessible to plants.

Carbon Source	Energy Source: Light	Energy Source: Chemicals
CO_2	**Photoautotrophs** some bacteria, photosynthetic protists, plants	**Chemoautotrophs** some bacteria, most archaea
Organic molecules	**Photoheterotrophs** some bacteria, some archaea	**Chemoheterotrophs** most bacteria, some archaea, fungi, animals, nonphotosynthetic protists Humans

Figure 20.12 Modes of nutrition in bacteria and archaea.
Figure It Out: Which group of organisms can build their own food from CO_2 in the dark?
Answer: Chemoautotrophs

Aerobe or Anaerobe? With rare exceptions, eukaryotic organisms are aerobes; they rely on aerobic respiration (Section 7.2) and thus require oxygen. By contrast many bacteria and most archaea are anaerobes. Some can tolerate an oxygen-free environment. Others are obligate anaerobes, meaning oxygen either slows their growth or kills them outright. Anaerobes are harmed by oxygen because oxidation reactions damage their biological molecules and, unlike aerobic cells, they do have enzymes that can repair that damage. We find obligate anaerobes in aquatic sediments and the animal gut. They also can infect deep wounds.

Take-Home Message

Why do biologists consider prokaryotes successful?

» Both bacteria and archaea have survived for billions of years and continue to coexist beside the eukaryotes.

» Bacteria and archaea are Earth's most abundant organisms. We are only beginning to appreciate their enormous species diversity.

» Collectively, prokaryotes can live in a wider range of habitats than eukaryotes because they are more diverse in terms of their metabolism. Prokaryotes utilize all four modes of nutrition and may be aerobic or anaerobic.

archaea Most recently discovered and less well-known lineage of single-celled organisms without a nucleus.
bacteria Most diverse and well-known lineage of single-celled organisms without a nucleus.
chemoautotroph Organism that uses carbon dioxide as its carbon source and obtains energy by oxidizing inorganic molecules.
chemoheterotroph Organism that obtains both energy and carbon by breaking down organic compounds.
metagenomics Study of microbial diversity that relies on analysis of DNA samples collected directly from the environment.
photoautotroph Organism that obtains carbon from carbon dioxide and energy from light.
photoheterotroph Organism that obtains its carbon from organic compounds and its energy from light.
prokaryote Member of one of two single-celled lineages (bacteria and archaea) that do not have a nucleus; a bacterium or archaeon.

20.6 Structure and Function of "Prokaryotes"

- Bacteria and archaea are small and structurally simple, but they are well adapted to their environments.
- Links to Discovery of DNA function 8.3, Molecular toolkit 15.2

Cell Size and Structure

Bacteria and archaea are similar in many ways (**Table 20.3**). The typical bacterial or archaeal cell cannot be seen without a light microscope. Thousands can fit on the head of a pin (**Figure 20.13A**). A bacteria is the size of a eukaryotic mitochondrion and, as you know, these organelles are likely descended from bacteria.

Three cell shapes are common: rods, spheres, and spirals. Rod-shaped cells are described as bacilli (singular, bacillus), spherical cells as cocci (singular, coccus), and spiral cells as spirilla (singular, spirillum).

Nearly all bacteria and archaea secrete a semirigid, porous cell wall around their plasma membrane (**Figure 20.13B**). Bacterial cells walls include peptidoglycan, a molecule not found in archaea. Bacteria may also have a slime layer or capsule around the cell wall. Slime helps a cell stick to surfaces. A capsule is tougher and

1 A bacterium has one circular chromosome that attaches to the inside of the plasma membrane.

2 The cell duplicates its chromosome, attaches the copy beside the original, and adds membrane and wall material between them.

3 When the cell has almost doubled in size, new membrane and wall are deposited across its midsection.

4 Two genetically identical cells result.

Figure 20.14 Animated Binary fission, the reproductive mode of bacteria and archaea.

Table 20.3 Traits Bacteria and Archaea Share

1. No nuclear envelope; chromosome in nucleoid

2. Generally a single chromosome (a circular DNA molecule); many species also contain plasmids

3. Cell wall (in most species)

4. Ribosomes distributed in the cytoplasm

5. Asexual reproduction by binary fission.

6. Capacity for gene exchange among existing cells via conjugation, transduction, and transformation.

---

helps some bacterial pathogens evade the immune defenses of their vertebrate hosts.

Bacteria and archaea typically have a single chromosome. This circle of double-stranded DNA attaches to the plasma membrane and resides in a cytoplasmic region called the **nucleoid**. There is no nuclear envelope as in eukaryotes, although at least one bacterial species has a membrane around its nucleoid. Membranes of some bacteria fold inward, but no bacteria or archaea have an endoplasmic reticulum or Golgi apparatus.

Many bacteria and archaea have flagella. The flagella do not bend side to side as in eukaryotes, but rather rotate like a propeller, and they do not contain microtubules. Hairlike filaments called **pili** (singular, pilus) may also extend from the cell surface. Some cells use pili to stick to surfaces. Others glide along by

40 μm

A Bacteria on the head of a pin.

Figure 20.13 Animated Typical bacterial size and structure.

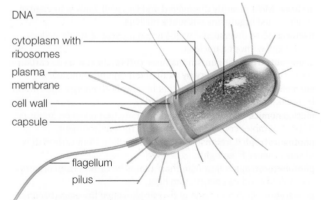

DNA
cytoplasm with ribosomes
plasma membrane
cell wall
capsule
flagellum
pilus

B Rod-shaped bacterium (a bacillus).

using their pili as grappling hooks. A pilus extends out to a surface, sticks to it, then shortens, drawing the cell forward. Another type of retractable pilus draws cells together for gene exchanges, as described below.

Reproduction and Gene Transfers

Bacteria and archaea have staggering reproductive potential. Some can divide every twenty minutes. Both groups reproduce by **binary fission** (Figure 20.14). During this process, a cell replicates its single chromosome and this DNA replica attaches to the plasma membrane adjacent to the parent molecule. The addition of more membrane moves the two DNA molecules apart. Eventually, the membrane and cell wall extend across the cell's midsection and divide the parent cell into two genetically identical descendants.

In addition to inheriting DNA "vertically" from a parent, bacteria and archaea engage in **horizontal gene transfers**, by which an individual acquires genes from another individual. The gene donor can be a cell of the same species or a different species.

Conjugation involves transfer of genes on a plasmid (Figure 20.15). A **plasmid** is a small circle of DNA separate from the chromosome (Section 15.2). During conjugation, a special sex pilus draws two cells together. Then, one cell puts a copy of a plasmid into the other. Both bacteria and archaea have plasmids and can engage in conjugation. Members of the two groups sometimes swap genes in this way.

With **transduction**, bacteriophages move genes between cells. The virus picks up a bit of DNA from one host cell, then transfers the DNA to its next host.

Bacteria and archaea also take up DNA from the environment, a process called **transformation**. For example, Frederick Griffith changed *Streptococcus* bacteria from harmless to deadly by transformation

binary fission Cell reproduction process of bacteria and archaea.
conjugation Mechanism of horizontal gene transfer in which one bacterial or archaeal cell passes a plasmid to another.
horizontal gene transfer Transfer of genetic material among existing individuals.
nucleoid Region of cytoplasm where the DNA is concentrated in a bacterial or archaeal cell.
pilus Protein filament that projects from the surface of some bacterial and archaeal cells.
plasmid Of many bacteria and archaea, a small ring of nonchromosomal DNA replicated independently of the chromosome.
transduction In bacteria and archaea, a mechanism of horizontal gene transfer in which a bacteriophage carries DNA from one cell to another.
transformation In bacteria and archaea, a type of horizontal gene transfer in which DNA is taken up from the environment.

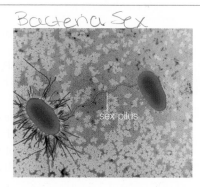

Bacteria Sex

① Conjugation in *E. coli* begins when a cell with a specific type of plasmid extends a sex pilus to another *E. coli* cell that lacks this plasmid. The pilus attaches the cells to one another. When it shortens, the cells are drawn together.

sex pilus

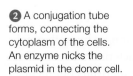
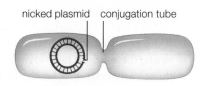

nicked plasmid   conjugation tube

② A conjugation tube forms, connecting the cytoplasm of the cells. An enzyme nicks the plasmid in the donor cell.

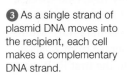
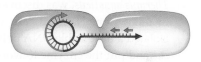

③ As a single strand of plasmid DNA moves into the recipient, each cell makes a complementary DNA strand.

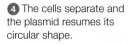

④ The cells separate and the plasmid resumes its circular shape.

Figure 20.15 Animated Conjugation, a mechanism of gene transfer. For clarity, the plasmid's size has been greatly exaggerated and the chromosome is not shown.

(Section 8.3). Griffith mixed the harmless cells with dead cells of a harmful strain. The heat had damaged the membranes of the harmful cells, thus killing them and releasing their DNA. That DNA was picked up by the harmless bacteria and put to use.

The ability of bacteria to acquire new genes has important implications for public health. Suppose a gene for antibiotic resistance arises by mutation in a bacterial cell. This gene not only can be passed on to that cell's descendants, but can also be transferred to other existing cells. Gene transfers increase the rate at which a gene spreads through a population of bacteria, thus enhancing the population's ability to respond to any selective pressure.

Take-Home Message

What structural and functional features do bacteria and archaea share?

» In both lineages, cells are typically walled and a single chromosome is not enclosed in a nucleus.

» Cells reproduce by binary fission and swap genes by conjugation and other mechanisms of horizontal gene transfer.

20.7 Bacterial Diversity

- Most bacteria play important ecological roles. A small minority are human pathogens.
- Links to Photosynthesis 6.4, PCR 15.3, Hydrothermal vents 19.3, Evolution of chloroplasts and mitochondria 19.6

Bacteria that cause human disease often get the spotlight, but most bacteria are either harmless or beneficial. Here we consider a few of the major lineages to give you an idea of bacterial diversity.

The Heat Lovers

If life emerged in thermal pools or near hydrothermal vents, the modern heat-loving bacteria may resemble those early cells. Biochemical comparisons put them near the base of the bacterial family tree. One species, *Thermus aquaticus*, was discovered in a volcanic spring in Yellowstone National Park. Biochemist Kary Mullis isolated a heat-stable DNA polymerase from *T. aquaticus* and put the enzyme to work in the first PCR reactions. He won a Nobel Prize for inventing this process, which is widely used in biotechnology (Section 15.3).

Oxygen-Producing Cyanobacteria

Photosynthesis evolved in many bacterial lineages, but only **cyanobacteria** (**Figure 20.16A**) utilize the noncyclic pathway and release free oxygen. If, as evidence suggests, chloroplasts evolved from ancient cyanobacteria, we have cyanobacteria and their chloroplast descen-

dants to thank for nearly all of the oxygen in Earth's atmosphere (Section 19.6).

When cyanobacteria incorporate the carbon from carbon dioxide into an organic compound, we say that they fix carbon. Some cyanobacteria also carry out **nitrogen fixation**: They incorporate nitrogen from the air into ammonia (NH_3).

Nitrogen fixation is an important ecological service provided only by bacteria. Photosynthetic eukaryotes need nitrogen, but they cannot use the gaseous form ($N\equiv N$) because they do not have an enzyme that can break the molecule's triple bond. They can, however, take up ammonia released by nitrogen-fixing bacteria.

Highly Diverse Proteobacteria

Proteobacteria are the most diverse bacterial group. Some are photoautotrophs that carry out photosynthesis but do not release oxygen. Others are chemoautotrophs. One of these, *Thiomargarita namibiensis,* is the largest bacterium known and can be seen without a microscope (**Figure 20.16B**). It gets energy by stripping electrons from sulfur that it stores in a giant vacuole.

Rhizobium, a chemoheterotroph, lives in roots of legumes such as peas. It gets sugars from its host and in return aids the plant by fixing nitrogen.

Myxobacteria, or slime bacteria, are tiny hunters that live in soil. They glide about as a swarm and feed on other bacteria. When food dwindles, hundreds of thousands of cells form a multicelled fruiting body

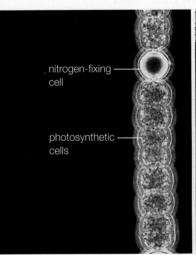
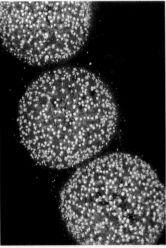
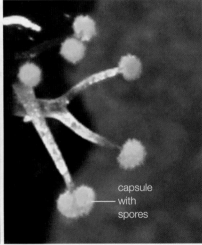
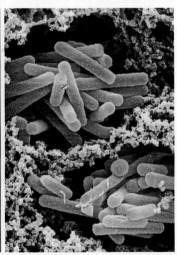

A *Anabaena*, a type of aquatic cyanobacterium. It carries out oxygen-producing photosynthesis and fixes nitrogen.

B *Thiomargarita namibiensis*, the biggest bacterium known. It lives in sea sediments and takes up and stores sulfates (white dots).

C *Chondromyces crocatus*, a myxobacterium, hunts other soil bacteria. When food runs out, thousands of cells form a fruiting body with spores at its tip.

D *Lactobacillus* ferments sugars and produces lactate. The cells shown here were used to turn milk into yogurt. Other lactobacilli are important as decomposers.

Figure 20.16 A sampling of bacterial diversity. Most bacteria do not cause disease.

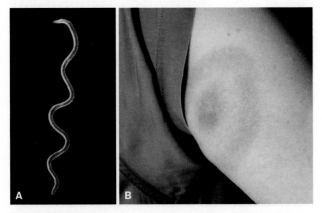

Figure 20.17 Example of a bacterial pathogen. (**A**) Spirochete that causes Lyme disease. (**B**) Bull's-eye rash at the site of a tick bite is often the first sign of infection.

with spores (hard-walled resting structures) at its tips (**Figure 20.16C**).

Escherichia coli is a chemoheterotroph that lives in the mammalian gut. It is part of the normal flora, a collection of microorganisms that typically live in and on a body. Other proteobacteria that enter the body cause disease (**Table 20.4**). One pathogenic group, the rickettsias, is thought to be the closest relatives of mitochondria. Rickettsias live as intracellular parasites.

Thick-Walled, Gram-Positive Bacteria

Gram-positive bacteria are a lineage of thick-walled cells that stain purple when prepared for microscopy by Gram staining. Most are chemoheterotrophs. *Lactobacillus* species ferment sugars and produce lactate (**Figure 20.16D**). The related *Streptococcus* species cause strep throat and impetigo. Actinomycetes grow as threadlike filaments in soil. One genus, *Streptomyces*, is the source of the antibiotic streptomycin. *Clostridium* and *Bacillus* are soil bacteria that form endospores when conditions are unfavorable. An **endospore** contains the cell's genome and a bit of cytoplasm in a protective coat. It can withstand drying, boiling, and radiation. When endospores enter a human body and germinate, the resulting infection by toxin-secreting bacteria can cause fatal anthrax, tetanus, or botulism.

chlamydias Bacteria that are intracellular parasites of vertebrates.
cyanobacteria Oxygen-producing photosynthetic bacteria.
endospore Resistant resting stage of some soil bacteria.
Gram-positive bacteria Lineage of thick-walled bacteria that are colored purple by Gram staining.
nitrogen fixation Incorporation of nitrogen gas into ammonia.
proteobacteria Most diverse bacterial lineage; includes species that carry out photosynthesis, fix nitrogen. Some cause disease.
spirochetes Lineage of bacteria shaped like a stretched-out spring.

Table 20.4 Examples of Disease-Causing Bacteria

Group/Species	Disease	Description
Proteobacteria		
Bordetella pertussis	Whooping cough	Childhood respiratory infection
Neisseria gonorrhoeae	Gonorrhea	Sexually transmitted disease
Rickettsia rickettsii	Rocky Mountain spotted fever	Fever accompanied by rash, spread by ticks
Vibrio cholerae	Cholera	Diarrheal illness
Gram-Positive Bacteria		
Clostridium species	Tetanus, botulism	Toxin released by bacteria causes paralysis
Streptococcus aureus	Impetigo, boils	Blisters, sores on skin
Streptococcus pyogenes	Strep throat	Sore throat, fever, damage to heart valves if not treated
Spirochetes		
Borrelia burgdorferi	Lyme disease	Rash, flulike symptoms spread by ticks
Treponema pallidum	Syphilis	Sexually transmitted disease
Mycobacteria		
Mycobacterium tuberculosis	Tuberculosis	Respiratory disease

Other Groups That Include Human Pathogens

Spirochetes resemble a stretched-out spring (**Figure 20.17A**). One species transmitted by ticks causes Lyme disease (**Figure 20.17B**). Another causes the sexually transmitted disease syphilis.

Chlamydias are small intracellular parasites of vertebrates. One species, *Chlamydia trachomatis*, is known in the United States mainly as a cause of sexually transmitted disease. In developing countries, it is a major cause of blindness.

Mycobacteria are rod-shaped cells with a waxy coat. One species causes tuberculosis, a respiratory disease that kills more than a million people each year.

Take-Home Message

What ecological roles do bacteria play?

» Bacteria benefit us by releasing oxygen, fixing nitrogen, and otherwise participating in nutrient cycles.

» A small minority of the bacterial chemoheterotrophs cause disease.

20.8 Archaea

■ Archaea, the more recently discovered prokaryotic lineage, are found in some very inhospitable places.

■ Links to Pigments 6.2, Chemiosmosis 6.5

Discovery of the Third Domain

The distinctive features of archaea first came to light in the 1970s. Carl Woese was comparing the ribosomal RNAs among what he thought were bacterial species to find out how they related to one another. He discovered that some species fell into a distinct group. Their rRNA gene sequences positioned them between all other bacteria and the eukaryotes. On the basis of this evidence, Woese proposed the three-domain classification system (Section 20.5).

As years went by, evidence in support of Woese's conclusions mounted. Archaea differ from bacteria in the composition of their cell wall and plasma membrane. Like eukaryotes, archaea organize their DNA around histone proteins, which bacteria do not have. The first sequencing of an archaeal genome provided the definitive evidence that archaea and bacteria are distinct lineages—most of this archaeon's genes have no counterpart in bacteria.

Woese has compared the discovery of archaea to the discovery of a new continent, which he and others are now exploring.

Here, There, Everywhere

0.5 μm

Many archaea thrive in seemingly hostile habitats. The first archaeon to have its genome sequenced, *Methanococcus jannaschii* (*left*), was discovered near a hydrothermal vent on the seafloor. It is an **extreme thermophile**, an organism that grows only at a very high temperature. Some archaea that live near hydrothermal vents can grow even at 110°C (230°F). Heat-loving archaea also live in volcanically heated geysers and hot springs (**Figure 20.18A**).

Other archaea are among the **extreme halophiles**, organisms that live in highly salty water. Salt-loving archaea live in the Dead Sea, the Great Salt Lake, and many smaller brine-filled lakes (**Figure 20.18B**). One extreme halophile, *Halobacterium*, has gas-filled vesicles that keep it afloat in well-lit surface waters. Its plasma membrane contains a unique protein, a purple pigment called bacteriorhodopsin. When excited by light, this protein pumps protons (H⁺) out of the cell,

A Thermally heated waters. Pigmented archaea color the rocks in waters of this hot spring in Nevada.

B Highly salty waters. Pigmented extreme halophiles color the brine in this California lake.

C The gut of many animals. Cows belch frequently to release the methane produced by archaea in their stomach.

Figure 20.18 Examples of archaeal habitats.

Evolution of a Disease (revisited)

Infectious diseases that are immediately fatal are rare. This is fortunate and not surprising. Evolutionarily speaking, the pathogens that leave the most descendants win. Think of a person infected with HIV as a factory that makes and distributes virus (**Figure 20.19**). Killing the host would shut down this facility.

Of course, hosts also evolve in response to disease. A disease with a high mortality rate acts as a selective agent, favoring individuals capable of resisting infection or surviving in spite of it. For example, about 10 percent of people of European ancestry have a mutation that lessens the likelihood of infection by HIV. The mutation is absent in American Indian, east Asian, and African populations.

People with this protective mutation currently enjoy a selective advantage as a result of the AIDS epidemic. However, the protective mutation did not become prevalent in Europeans as a result of AIDS. Studies of ancient remains tell us it has been in the northern European gene pool for thousands of years. Like all mutations, it arose randomly. It probably increased to its current frequency in Europe because it provided protection against one of the many historical epidemics that occurred there. Its current selective advantage is simply a matter of luck.

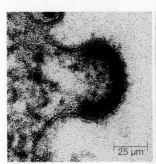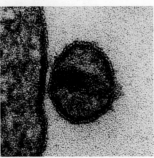

| 25 μm |

Figure 20.19 Micrographs of a new HIV particle budding from an infected white blood cell.

How would you vote? Antiviral drugs help keep people with HIV healthy and lessen the likelihood of viral transmission. However, an estimated 25 percent of HIV-infected Americans do not know they are infected. Annual, voluntary HIV tests with drug treatment for those infected could help curtail the AIDS pandemic. Do you favor an expanded, voluntary testing program?

---

against their gradient. The H$^+$ flows back into the cell through ATP synthases, thus driving the formation of ATP. A similar process drives ATP formation during photosynthesis (Section 6.5). However, *Halobacterium* does not use energy stored in ATP energy to fix carbon dioxide as photosynthetic organisms do. It is a photoheterotroph that obtains carbon by taking up small organic molecules from its environment.

Many archaea, including some extreme thermophiles and extreme halophiles, are **methanogens**, or methane makers. These chemoautotrophs form ATP by pulling electrons from hydrogen gas or acetate. Methane (CH$_4$) gas forms as a product of these reactions. Methanogenic archaea abound in sewage, marsh sediments, and the animal gut (**Figure 20.18C**). All are strictly anaerobic, meaning they cannot live in the presence of oxygen.

By their metabolic activity, methanogens produce 2 billion tons of methane annually. The release of this carbon-containing gas into the air has a major impact on the global carbon cycle.

As biologists continue to explore archaeal diversity, they are finding that archaea live alongside bacteria nearly everywhere. They are more abundant than bacteria in deep, dark ocean waters.

So far, scientists have not discovered any archaea that pose a major threat to human health. However, the presence of archaea may have some ill effects. For example, methanogenic archaea live in the human gut, and their abundance may affect our weight. By taking up hydrogen, methanogens make the gut more hospitable for bacteria that break down complex carbohydrates. As a result, the more methanogens in your gut, the more calories you can extract from food. Some studies have found a correlation between an abundance of gut methanogens and obesity.

extreme halophile Organism adapted to life in a highly salty environment.
extreme thermophile Organism adapted to life in a very high-temperature environment.
methanogen Organism that produces methane gas as a metabolic by-product.

Take-Home Message

What are archaea?

» Archaea are single cells without a nucleus that are closer to eukaryotes than to bacteria. Many live in very hot or very salty habitats, but there are archaea nearly everywhere. Unlike bacteria, they are not a major cause of human disease.

LEARNING ROADMAP

Where you have been This chapter is devoted to the protists, a eukaryotic group introduced in Section 1.4. Section 4.5 described traits of eukaryotic cells and Section 19.6 explored eukaryotic origins. The chapter's discussion of relationships among protist groups will draw on your understanding of cladistics, a topic presented in detail in Section 18.2.

Where you are now

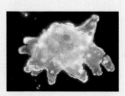

Multiple Unrelated Lineages
Diverse lineages of eukaryotic organisms that are not fungi, plants, or animals are referred to as protists. Protists may be single cells or multicelled, autotrophs or heterotrophs.

Single-Celled Heterotrophs
Single-celled groups include the flagellated protozoans, shelled cells called foraminifera and radiolarians, and the alveolates (ciliates, dinoflagellates, and apicomplexans).

Brown Algae and Relatives
Brown algae are multicelled seaweeds. They are not close relatives of red or green algae, but are relatives of single-celled diatoms and colorless water molds.

Relatives of Land Plants
Red algae and green algae are photosynthetic single cells and multicelled seaweeds. One lineage of multicelled green algae includes the closest living relatives of land plants.

Relatives of Fungi and Animals
Animals and fungi share a common ancestor with the protist group that includes amoebas and slime molds. Choanoflagellates are the sister group to animals.

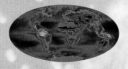

Where you are going We explore the connection between green algae and land plants in more detail when we discuss plant evolution (Sections 22.2 and 22.3). Similarly, the choanoflagellates figure in our discussion of animal origins (Section 24.3). You will be reminded of the photosynthetic protists when we consider primary productivity (Section 46.4) and the carbon cycle (Section 46.7), and of the flagellated protozoans when we look at sexually transmitted diseases (Section 41.10).

21.1 Malaria—From Tutankhamun to Today

Protists are eukaryotes that are not fungi, plants, or animals. As a group, they are sometimes described as "simple" eukaryotes. As you will see, protists are structurally less complex than plants or animals; however, like these organisms, they are well adapted to their environment.

The vast majority of protists are free-living cells or multicelled organisms that play essential roles in ecosystems. However, a few are important agents of human disease. Most notably, the single-celled protist *Plasmodium* causes malaria, a disease that kills more than a million people each year.

Plasmodium is transmitted from person to person by mosquitoes. The parasite invades and multiplies inside human liver cells and oxygen-carrying red blood cells. Infection destroys red blood cells (**Figure 21.1**), resulting in fatigue and weakness. *Plasmodium*-infected blood cells that get into the brain can cause blindness, seizures, coma, and death.

Like AIDS, malaria has an evolutionary connection to a primate disease. The *Plasmodium* species that causes most cases of human malaria is descended from a parasite that infects gorillas in western Africa. Unlike AIDS, malaria has been infecting human populations for millennia. Researchers recently detected bits of *Plasmodium* DNA in the 3,500-year-old mummified remains of the Egyptian pharaoh Tutankhamun (also known as "King Tut").

As Section 17.7 explained, mortality from malaria has driven up the frequency of the hemoglobin allele associated with sickle-cell anemia (HbS) in some human populations. Having one HbS allele reduces the risk of dying from malaria.

Natural selection also acts on *Plasmodium*. In recent years, the protist has become resistant to several widely used antimalarial drugs. Over a longer time scale, *Plasmodium* has evolved an astonishing capacity to affect the behavior of its hosts. Compared to a mosquito free of *Plasmodium*, an infected mosquito is more likely to feed several times a night, and thus more likely to bite several people. The parasite also alters the odor of infected humans, making them especially appetizing to hungry mosquitoes. By manipulating both its insect and human hosts, the protist maximizes the chances that its offspring will reach a new host.

Malaria was common in the southern United States until the 1940s, when crews drained swamps and ponds where mosquitoes breed, and sprayed the insecticide DDT inside millions of homes. Today, nearly all cases of malaria in the United States occur in people who contracted the disease elsewhere.

Malaria remains a threat in many tropical regions. It is endemic (constantly present) in parts of Mexico, Central America, and South America, as well as Asia and in the Pacific Islands. However, the disease takes its greatest toll in Africa. One African child dies of this disease every 30 seconds.

protist General term for member of one of the eukaryotic lineages that is not a fungus, animal, or plant.

Figure 21.1 *Plasmodium*, a protist pathogen that causes malaria. The art at the *left* depicts an infected red blood cell rupturing and releasing new *Plasmodium* cells (*blue*). The oldest evidence of malaria in humans comes from the mummified remains of Tutankhamun that were found in an elaborate tomb (*right*).

21.2 The Many Protist Lineages

■ Protists include many lineages of mostly single-celled eukaryotes, some only distantly related to one another.
■ Links to Classification 18.2, Endosymbiosis 19.6

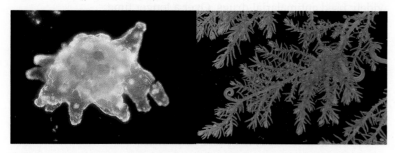

A Amoeba, a single-celled heterotroph.   **B** Red alga, a multicelled autotroph.

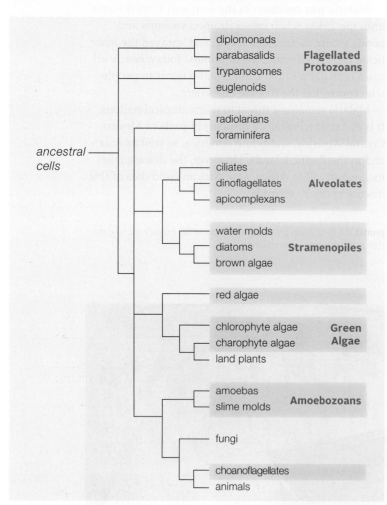

C One proposed eukaryotic family tree with traditional protist groups indicated by *tan boxes*. Notice that the protists are not united as a single lineage.

Figure 21.2 Animated Protist diversity and phylogeny. Protists can be single cells, colonial, or multicellular. Some are autotrophs and others heterotrophs.
Figure It Out: Are land plants more closely related to the red algae or the brown algae?

Answer: Red algae

Classification and Phylogeny

Protists are eukaryotes that are not fungi, plants, or animals. No single trait is unique to this collection of organisms; they are not a clade, or monophyletic group (Section 18.2). Many protist groups are only distantly related to one another. In fact, some are more closely related to plants, fungi, or animals than to other protists. **Figure 21.2** shows two widely dissimilar protists, and also how the various lineages are thought to fit into the eukaryotic family tree. *Tan* boxes denote protist lineages discussed in this book. There are many additional lineages, but this sampling of major groups will suffice to demonstrate protist diversity, ecological importance, and effects on human health.

Cells, Colonies, and Multicellular Organisms

Most protist lineages include only single-celled species (**Table 21.1**). However, colonial protists exist, and multicellularity evolved independently in several lineages. A protist colony consists of cells that live together, but show little integration. Each cell retains all the traits required to survive and reproduce on its own. By contrast, a multicellular organism has a division of labor, with some cells specialized for reproduction, and others specialized for various survival-related tasks. Some of the multicelled protists such as the kelps have large bodies with many types of differentiated cells.

Autotrophs and Heterotrophs

Many heterotrophic protists live in water or soil, where they feed on decaying organic matter or prey on smaller organisms such as bacteria. Other heterotrophic protists live inside larger organisms, including humans. In some cases, protist endosymbionts do no harm or even benefit their host. For example, flagellated protozoans inside the termite gut give these insects their ability to digest wood. Other protists are parasites, and some such as *Plasmodium* parasitize humans and cause disease.

Chloroplasts evolved in autotrophic protists by two slightly different mechanisms. Section 19.6 explained how cyanobacteria could have been engulfed by a heterotrophic cell and later evolved into chloroplasts. We call this process primary endosymbiosis. As far as we know, chloroplasts evolved by primary endosymbiosis only once, in the common ancestor of red algae and green algae. However, chloroplasts evolved by secondary endosymbiosis multiple times, in several protist lineages that engulfed red or green algae.

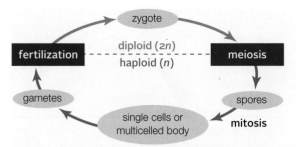

A Haploid-dominant cycle; the zygote is the only diploid cell.

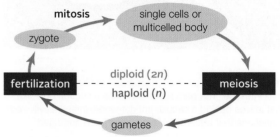

B Diploid dominant cycle; gametes are the only haploid cells.

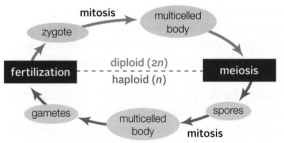

C Alternation of generations; multicelled haploid and diploid stages.

Figure 21.3 Three types of life cycle in sexually reproducing eukaryotes. Collectively, protists utilize all three.

Diverse Life Cycles

Most protists reproduce asexually as long as conditions favor growth, but can switch to sexual reproduction under adverse conditions. There are three types of sexual life cycle in eukaryotes (**Figure 21.3**). Most protists have a cycle dominated by the haploid form (**Figure 21.3A**). Only the zygote is diploid, and haploid spores reproduce asexually. Protist spores are often flagellated cells.

Other protists (and all animals) have a cycle in which diploid forms dominate. Only the gametes are haploid, and diploid cells divide by mitosis (**Figure 21.3B**). Some multicellular algae (and all plants) have an **alternation of generations**, a cycle in which both haploid and diploid bodies form (**Figure 21.3C**).

alternation of generations Of land plants and some algae, a life cycle in which both haploid and diploid multicelled bodies form.

Table 21.1 Characteristics of Some Protist Groups

Protist Group	Organization	Nutritional Mode
Flagellated Protozoans		
Diplomonads	Single cell	Heterotrophs; free-living or parasites of animals
Parabasalids	Single cell	Heterotrophs; free-living or parasites of animals
Kinetoplastids	Single cell	Heterotrophs; mostly parasites; some free-living
Euglenoids	Single cell	Most free-living heterotrophs; some autotrophs, mixotrophs
Radiolarians	Single cell	Free-living heterotrophs
Foraminifera	Single cell	Free-living heterotrophs
Alveolates		
Ciliates	Single cell	Heterotrophs; most free-living, some parasites of animals
Dinoflagellates	Single cell	Autotrophs, mixotrophs, free-living or parasitic heterotrophs
Apicomplexans (sporozoans)	Single cell	Heterotrophs; all parasites of animals
Stramenopiles		
Water molds	Single cell or multicelled	Heterotrophs; free-living or parasites
Diatoms	Single cell	Mostly autotrophs, a few heterotrophs or mixotrophs
Brown algae	Multicelled	Autotrophs
Red Algae	Most multicelled	Autotrophs
Green Algae	Single cell, colonial, or multicelled	Autotrophs
Amoebozoans		
Amoebas	Single cell	Heterotrophs; most free-living, a few parasites of animals
Slime molds	Single-celled and aggregated stages	Heterotrophs
Choanoflagellates	Single cell or colony	Heterotrophs

Take-Home Message

What are protists?

» Protists do not have a specific defining trait; they are a collection of many eukaryotic lineages rather than a clade.

» Most protists are single-celled, but there are multicelled and colonial species. Life cycles vary as does nutrition; some are autotrophs, others are heterotrophs.

21.3 Flagellated Protozoans

- Flagellated protozoans are single-celled species that swim in lakes, seas, and the body fluids of animals.
- Link to Effects of tonicity 5.8

Single-celled heterotrophic protists are commonly referred to as protozoans. **Flagellated protozoans** are single, unwalled cells with flagella. All are entirely or mostly heterotrophic. The life cycle is dominated by haploid cells that reproduce by mitosis. A **pellicle**, a thin layer of elastic proteins just under the plasma membrane, protects the cell and allows it to retain a particular shape.

Anaerobic Flagellates

Diplomonads and parabasalids have multiple flagella and are among the few protists adapted to oxygen-poor waters. Instead of typical mitochondria, they have **hydrogenosomes**, organelles that produce ATP by an anaerobic pathway and release hydrogen gas (H_2) as a by-product. Hydrogenosomes evolve from mitochondria and adapt organisms to anaerobic aquatic habitats. Free-living diplomonads and parabasalids thrive deep in seas and lakes. Others live in animals, where they may be harmful, helpful, or have no effect.

The diplomonad *Giardia lamblia* can infect the human gut, where it causes the disease giardiasis. The protist attaches to the intestinal lining and takes up nutrients (**Figure 21.4**). Symptoms of giardiasis can include cramps, nausea, and severe diarrhea that sometimes persist for weeks. Infected people and animals excrete cysts of *G. lamblia* in their feces. A protistan cyst is a dormant cell with a thick protective wall.

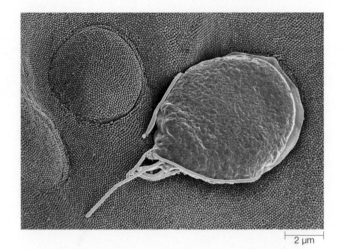

2 µm

Figure 21.4 Colorized electron micrograph of *Giardia lamblia* in the small intestine. The parasite attaches via a suction cup–like disk that leaves behind a circular imprint where intestinal villi (*red*), tiny projections that absorb nutrients, have been damaged.

Ingesting even a few *G. lamblia* cysts in contaminated water can cause an infection.

The parabasalid *Trichomonas vaginalis* infects human reproductive tracts and causes trichomoniasis (**Figure 21.5**). *T. vaginalis* does not make cysts, so it does not survive very long outside the human body. Fortunately for the parasite, sexual intercourse puts it directly into hosts. In the United States, about 6 million people are infected. In women, symptoms include vaginal soreness, itching, and a yellowish discharge. Infected males typically show no symptoms. Untreated infections damage the urinary tract, cause infertility, and increase risk of HIV infection. A dose of an antiprotozoal drug provides a quick cure.

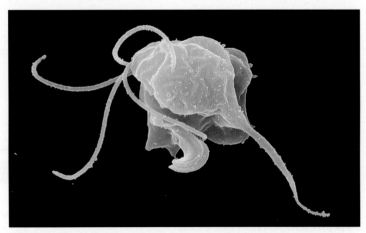

Figure 21.5 Colorized electron micrograph of *Trichomonas vaginalis*, the parabasalid that causes the sexually transmitted disease trichomoniasis. When vaginal secretions from an infected woman are examined under a light microscope, the parasite can be identified by its twitching movements.

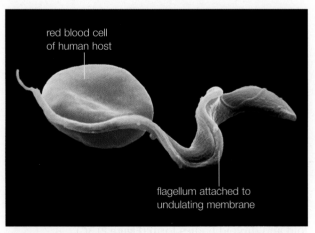

red blood cell of human host

flagellum attached to undulating membrane

Figure 21.6 Colorized electron micrograph of the kinetoplastid *Trypanosoma brucei*. A bite of an infected tsetse fly delivers this parasite into the blood. It lives in the fluid portion of the blood (the plasma) and absorbs nutrients across its body wall.

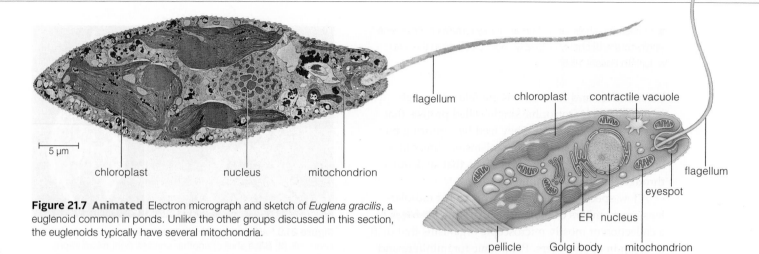

Figure 21.7 Animated Electron micrograph and sketch of *Euglena gracilis*, a euglenoid common in ponds. Unlike the other groups discussed in this section, the euglenoids typically have several mitochondria.

Labels (left image): chloroplast, nucleus, mitochondrion, 5 μm

Labels (right image): flagellum, chloroplast, contractile vacuole, flagellum, eyespot, pellicle, Golgi body, ER, nucleus, mitochondrion

Trypanosomes and Other Kinetoplastids

Kinetoplastids are flagellated protozoans that have a single large mitochondrion. Inside the mitochondrion, near the base of the flagellum, is a clump of DNA. The clump is the kinetoplast for which the group is named.

Some kinetoplastids prey on bacteria in fresh water and seas, but **trypanosomes**, the largest subgroup, are parasitic; they live in plants and animals. Trypanosomes are long, tapered cells with an undulating membrane (**Figure 21.6**). A single flagellum that attaches to this membrane causes its characteristic wavelike motion.

Biting insects act as vectors for many trypanosomes that parasitize humans. Again, a vector is an insect or other animal that carries a pathogen between hosts.

Tsetse flies spread *Trypanosoma brucei*, which causes African trypanosomiasis, commonly known as African sleeping sickness. Infected people are drowsy during daytime and often cannot sleep at night. If untreated, the infection is fatal. Tsetse flies that spread *T. brucei* occur only in sub-Saharan Africa.

Bloodsucking bugs transmit *Trypanosoma cruzi*, the cause of Chagas disease. Untreated infection can harm the heart and digestive organs. Chagas disease is now prevalent in parts of Central and South America, and occurs at low frequency in the southern United States. Large influxes of immigrants from South and Central America into the United States raised concerns about *T. cruzi* contamination of the blood supply. In 2006, blood banks began to test donated blood for *T. cruzi*.

Euglenoids

Euglenoids are flagellated protists closely related to kinetoplastids. Most live in fresh water and none are human pathogens. The majority are tiny predators, a few are parasites of larger organisms, and about a third have chloroplasts (**Figure 21.7**). The structure of euglenoid chloroplasts and the pigments inside them indicate that these organelles evolved from a green alga by secondary endosymbiosis.

Photosynthetic euglenoids can detect light with an eyespot near the base of their long flagellum. They typically revert to heterotrophic nutrition if light levels decline, or conditions for photosynthesis become otherwise unfavorable.

A euglenoid is hypertonic relative to fresh water. As in other freshwater protists, one or more **contractile vacuoles** counter the tendency of water to diffuse into the cell. Excess water collects in contractile vacuoles, which contract and expel it to the outside.

---

contractile vacuole In freshwater protists, an organelle that collects and expels excess water.

euglenoid Flagellated protozoan with multiple mitochondria; may be heterotrophic or have chloroplasts descended from algae.

flagellated protozoan Protist belonging to an entirely or mostly heterotrophic lineage with no cell wall and one or more flagella.

hydrogenosome Organelle that produces ATP and hydrogen gas by an anaerobic pathway; evolved from mitochondria.

pellicle Layer of proteins that gives shape to many unwalled, single-celled protists.

trypanosome Parasitic flagellate with a single mitochondrion and a membrane-encased flagellum.

Take-Home Message

What are flagellated protozoans?

» Flagellated protozoans are single-celled protists with one or more flagella. They are typically heterotrophic and reproduce asexually by mitosis.

» Diplomonads and parabasalids have adapted to life in oxygen-poor waters. Some members of these groups commonly infect humans and cause disease.

» Trypanosomes also include human pathogens that are transmitted by insects. Their relatives, the euglenoids, do not infect humans. Most prey on bacteria, but some have chloroplasts that evolved from green algae.

21.4 Foraminifera and Radiolarians

■ Limestone, chalk, and chert are rocky remains of countless single cells with chalky or glassy shells that lived in the sea.

■ Link to Fossils 16.5

Foraminifera and radiolarians are related lineages of primarily heterotrophic single-celled protists that secrete a sieve-like shell. They feed by capturing bits of organic debris and tiny organisms on threadlike or branching cytoplasmic extensions that stick out through the shell's openings.

A few species of foraminifera and most radiolarians are members of the marine plankton. **Plankton** is a collection of mostly microscopic organisms that drift or swim in open waters. Planktonic foraminifera and radiolarians often have smaller photosynthetic protists such as diatoms or algae living inside them. Sugars provided by these symbionts help meet the host's nutritional needs.

Chalky-Shelled Foraminifera

Foraminifera secrete a calcium carbonate ($CaCO_3$) shell (**Figure 21.8A**). Including its shell, an individual foraminifer can be as large as a grain of sand. Most of the 4,000 or so modern species live on the seafloor, where they probe the water and sediments for prey.

Foraminifera have been living and dying in the world's oceans for hundreds of millions of years, so remains of countless cells have fallen to the sea floor. Over time, geologic processes transform the accumulated calcium carbonate–rich shells of foraminifera and

200 μm

Figure 21.8 Foraminifera. (**A**) Live planktonic foraminiferan with alga cells (*yellow* dots) living on its cytoplasmic extensions. (**B**) Chalk cliffs 90 meters (300 feet) high along the shore in Dover, England, were once seafloor. Chalk consists of remains of foraminifera and other organisms with calcium carbonate–rich shells.

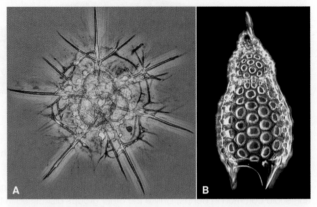

Figure 21.9 Radiolarians. (**A**) Phase contrast micrograph of a living cell. (**B**) Silica shell of another species (light micrograph).

other marine organisms into deposits of limestone and chalk (**Figure 21.8B**). Because foraminiferal shells have changed over time, examining fossil foraminifera can help scientists correlate the age of rock formations in different parts of the world.

Glassy-Shelled Radiolarians

Radiolarians secrete a glassy silica (SiO_2) shell (**Figure 21.9**). The cell has two cytoplasmic layers. The inner layer holds all the typical eukaryotic organelles. The outer cytoplasmic zone has numerous gas-filled vacuoles that keep the cell afloat. Radiolarians are most abundant in nutrient-rich tropical waters where microscopic algae are abundant.

Like foraminifera, radiolarians have lived and died in great numbers for millions of years, and their remains have accumulated on the seafloor. Over time, geologic processes transformed silica-rich deposits of radiolarian shells, called radiolarian ooze, into a rock called chert.

foraminifera Heterotrophic single-celled protists with a porous calcium carbonate shell and long cytoplasmic extensions.
plankton Community of tiny drifting or swimming organisms.
radiolarians Heterotrophic single-celled protists with a porous silica shell and long cytoplasmic extensions.

Take-Home Message

What are foraminifera and radiolarians?

» Foraminifera and radiolarians are related lineages of heterotrophic, single cells that live mainly in seawater.

» Foraminifera make a shell of calcium carbonate and most live on the seafloor.

» Radiolarians have a glassy silica shell; most are planktonic.

21.5 The Ciliates

■ Ciliated cells hunt bacteria, other protists, and one another in freshwater habitats and the oceans.

■ Link to Meiosis 12.3

Three groups of protists—ciliates, dinoflagellates, and apicomplexans—are **alveolates**, members of a lineage characterized by a layer of sacs of unknown function under the plasma membrane. "Alveolus" means sac.

Ciliates, or ciliated protozoans, are highly diverse heterotrophs, with about 8,000 species. They occur just about anywhere there is water, and most are predators (**Figure 21.10**). About a third live in the bodies of animals. Only one ciliate is a known human pathogen: *Balantidium coli* causes nausea and diarrhea.

Paramecium is a freshwater ciliate (**Figure 21.11**). Cilia cover its entire surface and function in feeding and locomotion. Cilia sweep water laden with bacteria, algae, and other food particles into an oral groove at the cell surface, and then to the gullet. Enzyme-filled vesicles digest food in the gullet. Like other ciliates, *Paramecium* has organelles called trichocysts beneath the pellicle that gives it its shape (**Figure 21.11C**). A trichocyst holds a protein thread that can be expelled to help the cell capture prey or fend off a predator.

alveolate Member of a protist lineage having small sacs beneath the plasma membrane; dinoflagellate, ciliate, or apicomplexan.
ciliate Single-celled, heterotrophic protist with many cilia.

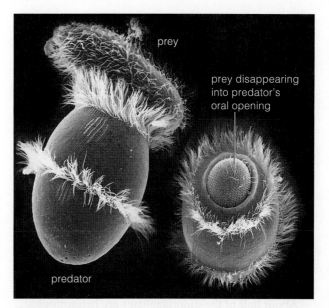

Figure 21.10 *Didinium*, a barrel-shaped ciliate with tufts of cilia, catching (*left*) and engulfing (*right*) *Paramecium*, another ciliate.

Ciliates reproduce asexually and some reproduce sexually. A ciliate has a macronucleus that controls daily function, and one or more small micronuclei. During sexual reproduction, two cells get together and their micronuclei undergo meiosis. The cells swap haploid micronuclei, then use their new combination of micronuclei to form a macronucleus with a mixed genome. After the cells separate, each divides and passes a macronucleus with a mix of genes to its descendant cells.

Take-Home Message

What are ciliates?

» Ciliates are heterotrophic single cells that move about with the help of cilia. Most are free-living predators, but some live inside animals.

» Ciliates, along with dinoflagellates and apicomplexans, are alveolates. This group is characterized by having tiny sacs beneath their plasma membrane.

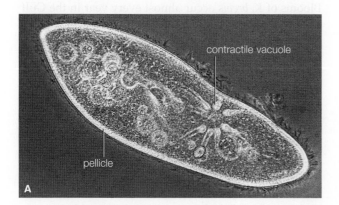

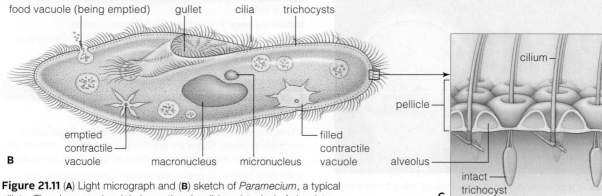

Figure 21.11 (**A**) Light micrograph and (**B**) sketch of *Paramecium*, a typical ciliate. The close-up view (**C**) shows the alveoli (sacs) typical of alveolates.

21.6 Dinoflagellates

■ Dinoflagellates are single-celled heterotrophs and autotrophs. Most whirl about in the seas. Some live inside corals.

The name **dinoflagellate** means "whirling flagellate." These single-celled protists typically have two flagella; one extends out from the base of the cell, and the other wraps around the cell's middle like a belt (**Figure 21.12A**). The combined action of these two flagella causes the cell to rotate as it moves forward. Dinoflagellates are alveolates, and most deposit cellulose in the alveoli beneath their plasma membrane. The cellulose accumulates as thick but porous plates.

Planktonic dinoflagellates live in seas and in places where salt water and fresh water mix. Cells are haploid and usually reproduce asexually. Adverse conditions stimulate sexual reproduction; two cells fuse and form a cyst that later undergoes meiosis.

About half of the dinoflagellates are heterotrophs. Most of these are predators, but some are parasites of fish and aquatic invertebrates.

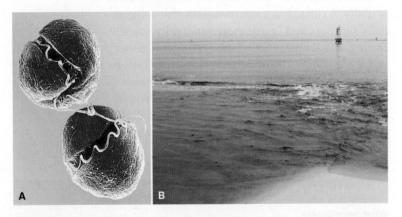

Figure 21.12 Harmful algal blooms. (**A**) Colorized micrograph of *Karenia brevis*, a dinoflagellate that makes a nerve toxin. (**B**) A population explosion, or algal bloom, of *K. brevis* along Florida's Gulf Coast.

Figure 21.13 Tropical waters light up when *Noctiluca scintillans*, a bioluminescent dinoflagellate, is disturbed by something moving through the water.

Photosynthetic dinoflagellates have chloroplasts descended from red algae. A few photosynthetic dinoflagellate species are themselves endosymbionts in cells of reef-building corals. A coral is an invertebrate animal, and its relationship with the protists is mutually beneficial. Dinoflagellates supply their host coral with sugar and oxygen for aerobic respiration. The coral provides the protists with nutrients, shelter, and the carbon dioxide necessary for photosynthesis.

Free-living photosynthetic dinoflagellates or other protists sometimes undergo great increases in population size, a phenomenon known as an **algal bloom**. In habitats enriched with nutrients, as from agricultural runoff, a liter of water may hold millions of cells. Blooms of some species cause "red tides," so named because the abundance of cells tints the water red (**Figure 21.12B**).

Algal blooms can sicken humans and kill aquatic organisms. Aerobic bacteria that feed on algal remains can use up the oxygen in the water, so that aquatic animals suffocate. Some dinoflagellate toxins also kill directly. *Karenia brevis* produces a toxin that binds to transport proteins in the plasma membrane of nerve cells. Eat shellfish contaminated with this toxin and you might end up dizzy and nauseated from neurotoxic shellfish poisoning. Symptoms usually develop hours after the meal and persist for a few days. Blooms of *K. brevis* occur almost every year in the Gulf of Mexico, but the severity and effects vary.

Some dinoflagellates exhibit **bioluminescence**; they convert ATP energy to light energy (**Figure 21.13**). A bioluminescent dinoflagellate emits a flash of light when something disturbs it. The light may protect the protist by startling a predator that was about to eat it. By another hypothesis, the flash of light acts like a car alarm. It attracts the attention of other organisms, including predators that pursue the would-be dinoflagellate eaters.

algal bloom Population explosion of photosynthetic cells in an aquatic habitat.
bioluminescence Production of light by an organism.
dinoflagellate Single-celled, aquatic protist with cellulose plates and two flagella; may be heterotrophic or photosynthetic.

Take-Home Message

What are dinoflagellates?

» Dinoflagellates are a group of mostly marine single-celled alveolate protists. Some are predators or parasites. Others are photosynthetic members of plankton or symbionts in corals.

» An algal bloom—a population explosion of protists—can harm aquatic organisms and endanger human health.

21.7 Cell-Dwelling Apicomplexans

■ As noted in Section 21.1, malaria is a major cause of human death. Here we take a closer look at the protist that causes malaria and at another pathogenic relative.

Apicomplexans are parasitic alveolates that spend part of their life cycle living inside cells of their hosts. Their name refers to a complex of microtubules at their apical (top) end that they use to enter a host cell. They are also sometimes called sporozoans.

Malaria is the most studied apicomplexan disease. **Figure 21.14** shows the life cycle of *Plasmodium*, the agent of malaria. A female *Anopheles* mosquito transmits a motile infective cell (a sporozoite) to a vertebrate such as a human ❶. The sporozoite travels to liver cells, where it reproduces asexually ❷. Some of the resulting cells (merozoites) enter red blood cells and liver cells, where they divide asexually ❸. Others develop into immature gametes, or gametocytes ❹. A mosquito that sucks blood from an infected person takes up gametocytes that later mature as gametes in its gut ❺.

When two gametes fuse, the resulting zygote develops into a new sporozoite ❻. Sporozoites migrate to the insect's salivary glands, where they await transfer to a new vertebrate host.

Fever and weakness usually start a week or two after a bite, when infected liver cells rupture and release merozoites and cellular debris into the bloodstream. After the initial episode, symptoms may sub-side. However, continued infection harms the body and eventually kills the host. Malaria kills more than a million people a year.

Another apicomplexan disease, toxoplasmosis, is caused by *Toxoplasma gondii*. Many infected people have no apparent symptoms. However, an infection can be fatal in immune-suppressed people, such as those with AIDS. A *T. gondii* infection during pregnancy can cause neurological birth defects in offspring.

Most people become infected by eating cysts in undercooked meat. *T. gondii* infects cattle, sheep, pigs, and poultry. Cats that prey on rodents and birds are another source of infection. Keeping a cat indoors and feeding it only commercially prepared food minimizes risk that a cat will be infected. Pregnant women and immune-suppressed people should avoid contact with cat feces, which can contain infective *T. gondii* cysts.

apicomplexan Single-celled alveolate protist that lives as a parasite inside animal cells; some cause malaria or toxoplasmosis.

Take-Home Message

What diseases are caused by apicomplexans?

» Apicomplexans cause the deadly disease malaria, as well as toxoplasmosis.

» Mosquitoes transmit malaria. People get toxoplasmosis when they ingest cysts (resting cells) of the parasite.

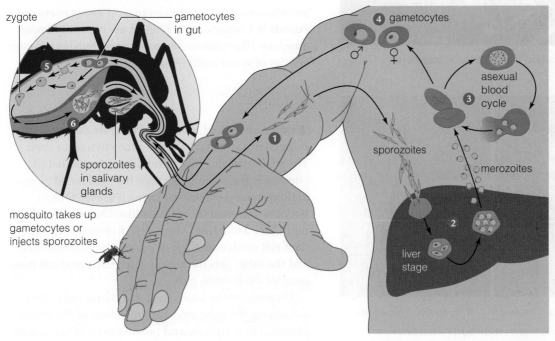

zygote

gametocytes in gut

sporozoites in salivary glands

mosquito takes up gametocytes or injects sporozoites

❹ gametocytes

asexual blood cycle

sporozoites

merozoites

liver stage

Figure 21.14 Animated Life cycle of a *Plasmodium* species that causes malaria. Only the zygote is diploid.

❶ Infected mosquito bites a human. Haploid sporozoites enter blood, which carries them to the liver.

❷ Sporozoites reproduce asexually in liver cells, then mature into merozoites. Merozoites leave the liver and infect red blood cells.

❸ Merozoites reproduce asexually in some red blood cells.

❹ In other red blood cells, merozoites differentiate into gametocytes.

❺ A female mosquito bites and sucks blood from the infected person. Gametocytes in blood enter her gut and mature into gametes, which fuse to form diploid zygotes.

❻ Meiosis of zygotes produces cells that develop into sporozoites. The sporozoites migrate to the mosquito's salivary glands.

21.8 The Stramenopiles

■ Colorless filamentous heterotrophs, photosynthetic single cells, and huge seaweeds belong to the stramenopile lineage.

■ Link to Pigments 6.2

We turn now to another major protist lineage known as the **stramenopiles**. The group name means "straw-haired," and it refers to an anterior flagellum with short hairlike extensions that forms during the life cycle of some members of the lineage.

Most stramenopiles are autotrophs, with chloroplasts that contain a brown accessory pigment called fucoxanthin, along with chlorophylls *a* and *c*. Stramenophile chloroplasts are thought to have evolved from a red alga by secondary endosymbiosis.

Water Molds

The 500 species of **water molds**, or oomycetes, are heterotrophs that form a mesh of nutrient-absorbing filaments. They were once classified with the fungi, because both groups share a similar growth habit. However, water mold filaments consist of diploid cells that have cell walls made of cellulose. By contrast, fungal filaments are haploid and have cell walls of chitin.

Figure 21.15 The water mold *Saprolegnia* growing on an infected fish.

A Dead and dying oaks near Big Sur, California.   **B** Trunk of an infected oak.

Figure 21.16 Effects of the pathogenic water mold *Phytophthora ramorum*.

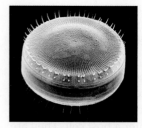

A Scanning electron micrograph of a diatom shell.

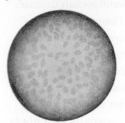

B Light micrograph of a live diatom, with chloroplasts visible through the shell.

C Diatomaceous earth (shells of ancient diatoms) packaged for use as an insecticide.

Figure 21.17 Diatoms.

Most water molds decompose organic matter in aquatic habitats, but some are aquatic parasites. For example, *Saprolegnia* often infects fish in aquariums, fish farms, and hatcheries (**Figure 21.15**). Other water molds live in damp places on land, or in plant tissues.

Phytophthora means "plant destroyer," and this genus of water molds lives up to its name. In the mid-1800s, one species caused an outbreak of disease that destroyed Ireland's potato crop and resulted in widespread famine. Today, *Phytophthora* species cause an estimated 5 billion dollars in crop losses every year. Forests in California and Oregon are currently under attack by *Phytophthora ramorum*, a previously unknown species of water mold (**Figure 21.16**).

Diatoms

Diatoms are diploid, photosynthetic, single-celled or colonial protists that secrete a protective silica shell. The shell consists of two overlapping parts that fit one inside the other, like a shoe box and its lid. Some diatoms have a cylindrical shape (**Figures 21.17A,B**), but others are square or needlelike. Diatoms usually reproduce asexually. During cell division, one descendant cell receives the upper part of the parental shell and the other gets the bottom part. Each new cell then secretes the missing part.

Diatoms live in lakes, seas, and damp soils. They are among the most abundant members of the phytoplankton in temperate and polar waters. When marine

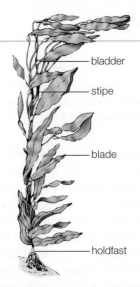

A Giant kelp (*above* and *right*). A holdfast anchors it in place. The stemlike stipe has leaflike blades. Gas-filled bladders make stipes and blades buoyant.

bladder
stipe
blade
holdfast

Gets minerals from the water (No roots)

Figure 21.18 Brown algae.

C *Above, Fucus vesiculosus.* This brown alga, commonly known as bladderwrack or seawrack, lives along rocky North Atlantic coasts. It is harvested for use as an iodine-rich nutritional supplement.

B "Sea palm" (*Postelsia*) a small kelp adapted to the rocky surf zone of the Pacific Northwest. Once overharvested for food, its collection is now tightly regulated.

diatoms die, their shells accumulate on the seafloor. People mine "diatomaceous earth" from places where accumulated deposits of diatom shells have been lifted onto land by geological processes. The silica-rich material is used in filters and cleaners, and as an insecticide that is nontoxic to vertebrates (**Figure 21.17C**). Diatomaceous earth kills crawling insects by nicking their outer covering, causing them to dry out and die.

Brown Algae

Brown algae are multicelled protists that live mainly in cool, coastal waters. Although some brown algae appear plantlike, this similarity in form is not evidence of shared ancestry. Brown algae evolved from a different single-celled ancestor than the lineage that includes the red algae, green algae, and land plants.

About 1,500 species of brown algae range in size from microscopic filaments to giant kelps that stand 30 meters (100 feet) tall. Giant kelps form forestlike stands in coastal waters of the Pacific Northwest (**Figure 21.18A**). Like trees in a forest, giant kelps shelter

a wide variety of other organisms. The kelp life cycle is an alternation of generations in which a diploid, spore-bearing body is the large, long-lived stage. Microscopic gamete-forming bodies form during the shorter, haploid stage of the life cycle.

Brown algae have several commercial uses. Some species are harvested for use as food or as nutritional supplements (**Figure 21.18B,C**). Polysaccharides called algins are extracted from kelps and used to thicken foods, beverages, cosmetics, and other products.

brown alga Multicelled marine protist with a brown accessory pigment (fucoxanthin) in its chloroplasts.
diatom Single-celled photosynthetic protist with a brown accessory pigment (fucoxanthin) and a two-part silica shell.
stramenopiles Protist lineage that includes the photosynthetic diatoms and brown algae, as well as the heterotrophic water molds. Some members of the group have a hairy flagellum.
water mold Heterotrophic protist that grows as nutrient-absorbing filaments.

Take-Home Message

What are stramenopiles?

» Stramenopiles include single-celled diatoms and multicelled brown algae, both important producers.

» Heterotrophic water molds also belong to this group.

21.9 Red Algae

- Red accessory pigments allow the mostly multicellular red algae to survive at greater depths than other algae.
- Link to Pigments 6.2

Nearly all of the more than 4,000 species of **red algae** live in clear, warm seas. Red algae are red because in addition to chlorophylls *a* and *c*, their chloroplasts contain accessory pigments (phycobilins) that absorb blue-green and green light. Light of these wavelengths penetrates deeper into water than other wavelengths, so red algae thrive at greater depths than other algae.

Nearly all red algae are multicelled. Most grow as thin sheets or in a branching pattern (**Figure 21.19**). One subgroup, the coralline algae, has tissues hardened by deposits of calcium carbonate. Their rigid structure contributes to formation of reefs.

Red algae have many commercial uses. Agar is a polysaccharide extracted from cell walls of some red algae. It keeps some baked goods and cosmetics moist, helps jellies set, and is used to make capsules that hold medicines. Carrageenan, another polysaccharide of red algae, is added to soy milk and dairy foods. *Porphyra* is cultivated worldwide. You may be familiar with it in its pressed, dried form as "nori" used to wrap sushi.

Figure 21.19 Branching and sheetlike red algae growing 75 meters (225 ft) beneath the sea surface in the Gulf of Mexico.

red alga Photosynthetic protist; typically multicelled, with chloroplasts containing red phycobilins.

Take-Home Message

What are red algae?

» Red algae are mostly multicelled marine algae that live in clear, warm waters.

» Pigments called phycobilins give red algae their red color and allow them to live at greater depths than other algae.

21.10 Green Algae

- Green algae are the protists most similar to land plants. They include two lineages of single-celled and multicelled species.
- Links to Engelmann's (6.3) and Calvin's (2.2) experiments, Plasmodesmata 4.11, Cell plate formation 11.4

"Green algae" is the informal name for about 7,000 photosynthetic species that range in size from microscopic cells to multicelled filamentous or branching forms more than a meter long.

All green algae resemble land plants in that they store sugars as starch and deposit cellulose fibers in their cell wall. Also like land plants, they have chloroplasts that evolved by primary endosymbiosis from cyanobacteria. These chloroplasts have a double membrane and contain chlorophylls *a* and *b*.

There are two lineages of green algae, the chlorophytes and the charophytes.

Chlorophyte Algae

Chlorophyte algae are the largest lineage of green algae. They include both freshwater and marine species. Some single-celled chlorophytes live in the soil or cling to surfaces, including ice. Others partner with a fungus and form a composite organism known as a lichen. However, most single-celled chlorophytes live in fresh water. Melvin Calvin used one of them, *Chlorella*, to study the light-independent reactions of photosynthesis (Section 2.2). Today *Chlorella* is grown commercially and sold as a health food.

Chlamydomonas is a single-celled species common in ponds. Haploid, flagellated cells (called zoospores) reproduce asexually, as long as nutrients and light are plentiful (**Figure 21.20 ❶**). When conditions do not favor growth, gametes develop and fuse ❷. Fertilization produces a diploid zygote ❸ that develops into a zygospore ❹, a dormant cell with a thick, protective wall. When favorable conditions return, the zygote undergoes meiosis and germinates, releasing the next generation of haploid, flagellated zoospores ❺.

Volvox is a colonial freshwater species. Hundreds to thousands of flagellated cells that resemble those of *Chlamydomonas* are joined together by thin cytoplasmic strands to form a whirling, spherical colony (**Figure 21.21A**). Daughter colonies form inside the parental sphere, which eventually ruptures and releases them.

Other freshwater chlorophytes form long filaments. Theodor Engelmann used one of these, *Cladophora*, in his studies of photosynthesis (Section 6.3).

Some chlorophytes are common "seaweeds." Wispy sheets of *Ulva* cling to coastal rocks worldwide. In some species, the sheets grow longer than your arm,

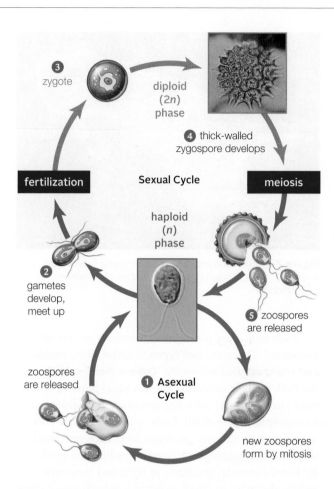

Figure 21.20 Life cycle of *Chlamydomonas*, a single-celled green alga. Cells reproduce asexually until conditions become unfavorable for growth. **Figure It Out: Which of the three types of life cycle does this species have?**

Answer: Haploid dominant

Diagram labels:
③ zygote
diploid (2n) phase
④ thick-walled zygospore develops
fertilization
Sexual Cycle
meiosis
haploid (n) phase
② gametes develop, meet up
⑤ zoospores are released
zoospores are released
① **Asexual Cycle**
new zoospores form by mitosis

A *Volvox* colonies, with many flagellated cells joined by thin strands of cytoplasm. New colonies are visible as bright green spheres inside each parent colony.

B Long, thin sheets of sea lettuce (*Ulva*) are common along coasts where there is little wave action.

C *Codium fragile*, a branching marine alga from Asia that has been accidentally introduced worldwide.

D *Chara*, a stonewort. Gametes, and then zygotes, form in protective jackets of cells on the "branches."

Figure 21.21 Diversity of green algae. Chlorophytes (**A,B,C**) and a charophyte (**D**).

---

but are less than 40 microns thick (**Figure 21.21B**). *Ulva* is commonly known as sea lettuce and is harvested for use in salads and soups.

Other marine chlorophytes have a branching form. *Codium fragile*, commonly known as "dead man's fingers," originated in Asia and was spread worldwide by humans (**Figure 21.21C**). It is considered an invasive species that competes with and harms native species.

Charophyte Algae

Charophyte algae are a mostly freshwater group of green algae. Some are live as single-cells, others form strands of cells, and others have a complex body.

Gene comparisons indicate that charales, or stoneworts, are the charophyte algae most closely related to land plants. Like plants, they have plasmodesmata

(Section 4.11), and divide their cytoplasm by cell plate formation (Section 11.4). *Chara*, native to Florida, is an example (**Figure 21.21D**). Its haploid gametophyte grows as branching filaments in lakes and ponds.

charophyte algae Green algal group that includes the lineage most closely related to land plants.
chlorophyte algae Most diverse lineage of green algae.

Take-Home Message

What are green algae?

» Green algae are photosynthetic single-celled or multicelled protists. Like land plants, they have cellulose in their cell walls, store sugars as starch, and have chloroplasts descended from cyanobacteria.

» Most green algae are chlorophytes. The smaller group of lineages known as charophyte algae are the closest relatives of the land plants.

21.11 Amoebozoans and Choanoflagellates

- Amoebozoans and choanoflagellates are heterotrophic cells with close links to fungi and animals.
- Link to Cell motility 4.10

Amoebozoans

Amoebozoans are shape-shifting cells. Most do not have a cell wall, shell, or pellicle; nearly all undergo dynamic changes in shape. A compact blob of a cell can extend lobes of cytoplasm called pseudopods (Section 4.10) to move about and to capture food.

Amoebas live as single cells. **Figure 21.22** shows *Amoeba proteus*. Like most amoebas, it is a predator in freshwater habitats. Other amoebas live inside animals, and some cause human disease. Each year, about 50 million people suffer from amebic dysentery after drinking water contaminated by *Entamoeba histolytica* cysts. Inadequate sterilization of contact lenses or swimming with lenses can result in an eye infection by *Acanthamoeba*, an amoeba common in soil, standing water, and even tap water.

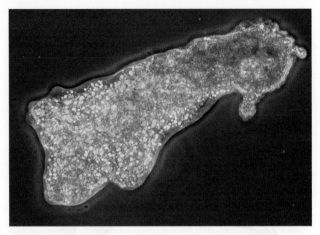

Figure 21.22 *Amoeba proteus*. The cell has no fixed shape. It feeds or shifts position by extending lobes of cytoplasm (pseudopods).

Slime molds are sometimes described as "social amoebas." There are two types, cellular slime molds and plasmodial slime molds. **Cellular slime molds** spend the bulk of their existence as individual haploid amoeboid (amoeba-like) cells. *Dictyostelium discoideum* is an example (**Figure 21.23**). Each cell eats bacteria and reproduces by mitosis ❶. When food runs out, thousands of cells aggregate to form a multicelled mass ❷. Environmental gradients in light and moisture induce the mass to crawl along as a cohesive unit often referred to as a "slug" ❸. When the slug reaches a

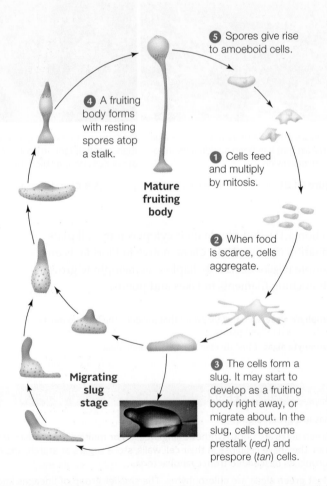

5 Spores give rise to amoeboid cells.

4 A fruiting body forms with resting spores atop a stalk.

Mature fruiting body

1 Cells feed and multiply by mitosis.

2 When food is scarce, cells aggregate.

3 The cells form a slug. It may start to develop as a fruiting body right away, or migrate about. In the slug, cells become prestalk (*red*) and prespore (*tan*) cells.

Migrating slug stage

Figure 21.23 Animated Life cycle of *Dictyostelium discoideum*, a cellular slime mold. During aggregation, the cells secrete and respond to cyclic AMP.

A Mulitinucleated mass (a plasmodium) streaming across a log.

B Fruiting bodies, with resting spores.

Figure 21.24 Plasmodial slime mold (*Physarum*).

Malaria: From Tutankhamun to Today (revisited)

Biting people is a risky business. Each time a mosquito bites, it runs the risk of being detected and swatted. To facilitate speedy feedings and quick getaways, mosquitos inject their host with apyrase, an enzyme that impairs blood clotting. An injection of apyrase before a meal helps ensure that blood will flow smoothly up the insect's proboscis, rather than turning chunky.

When *Plasmodium* has replicated inside a mosquito's gut, it moves to the insect's salivary glands, where its presence decreases the expression of the gene that encodes apyrase. A mosquito with infectious *Plasmodium* in its saliva has to bite more often because blood clots as it feeds. This is bad for the mosquito, but good for *Plasmodium*. A mosquito that is forced to eat many small meals, rather than a single large one, may deliver sporozoites to more new hosts.

Other apicomplexan parasites also alter host behavior to their own benefit. Consider the parasite that causes toxoplasmosis, *Toxoplasma gondii*. It reproduces asexually in rodents, but must enter a cat to complete its life cycle. The parasite's movement from rat to cat is facilitated by changes in the rat's behavior. Unlike healthy rats, those infected by *T. gondii* do not avoid cat-scented areas but actually show a preference for such areas. Exactly how the parasite alters the rodent's behavior is not clear, but we do know that *T. gondii* infects the amygdala, a part of the brain that governs fear and anxiety.

How would you vote? About 40 percent of the world population lives in areas where malaria remains endemic. The fight against malaria continues on many fronts, including the renewed spraying of DDT inside homes. Do the benefits of using DDT to fight malaria in developing countries outweigh the health and ecological costs that caused the United States to ban its use?

---

suitable spot, its component cells differentiate to form a fruiting body. Some cells become a stalk, and others become spores atop it ❹. When a spore germinates, it releases a cell that starts the life cycle anew ❺.

Plasmodial slime molds spend most of their life cycle as a multinucleated mass called a plasmodium. The plasmodium forms when a diploid cell divides its nucleus repeatedly by mitosis, but does not undergo cytoplasmic division. The resulting mass can be as big as a dinner plate. It streams out along the forest floor feeding on microbes and organic matter (**Figure 21.24A**). When food supplies dwindle, a plasmodium develops into spore-bearing fruiting bodies (**Figure 21.24B**).

Choanoflagellates

Aquatic, heterotrophic protists called **choanoflagellates** are thought to be the closest living protist relatives of animals. Their name means "collared flagellate." A choanoflagellate cell (**Figure 21.25**) has a flagellum surrounded by a "collar" of microvilli (tiny threadlike projections that extend from the cell surface). By moving its flagellum, the cell sets up a current that allows its collar to strain food from the water. As you will learn, sponges have cells with a similar structure. We discuss the similarities between choanoflagellates and animals in detail in Section 24.3.

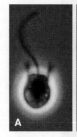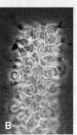

Figure 21.25 (A) Free-living choanoflagellate. A collar of microvilli rings its flagellum. **(B)** Colony of choanoflagellates.

amoeba Single-celled, unwalled protist that extends pseudopods to move and to capture prey.
amoebozoans Lineage of heterotrophic, unwalled protists that live in soils and water; include amoebas and slime molds.
cellular slime mold Soil-dwelling protist that feeds as solitary cells but congregates under adverse conditions to form a cohesive unit that develops into a fruiting body.
choanoflagellates Heterotrophic protists thought to be the sister group of animals; collared cells strain food from water.
plasmodial slime mold Soil-dwelling protist that feeds as a multinucleated mass. Develops into a fruiting body under adverse conditions.

Take-Home Message

What are amoebozoans and choanoflagellates?

» Amoebozoans are a recently recognized lineage that includes amoebas and slime molds. All are unwalled heterotrophs that feed and move by extending cytoplasmic extensions (pseudopods).

» Choanoflagellates are flagellated, heterotrophic cells that are considered to be close relatives of the animals. They resemble sponge cells.

LEARNING ROADMAP

Where you have been This chapter picks up the story of the evolution of plants from green algae (Section 21.10). We return to the topic of plant life cycles, first introduced in Section 12.5. Understanding methods of classification (18.2) will help you get a handle on how plants are now grouped.

Where you are now

Milestones in Plant Evolution
Plants evolved from a green alga. Structural and developmental adaptations allowed plants to colonize land, and to move into increasingly drier land habitats.

Nonvascular Plants
In hornworts, liverworts, and mosses, a gamete-producing body dominates the life cycle, and sperm reach the eggs by swimming through droplets or films of water.

Seedless Vascular Plants
In ferns and related plants, a large spore-producing body with vascular tissues dominates the life cycle. As with bryophytes, sperm swim through water to reach eggs.

Gymnosperms
Gymnosperms are the older of the two existing seed plant lineages. They make pollen and thus do not require water for fertilization. Conifers are the most diverse gymnosperms.

Angiosperms
Angiosperms, the most diverse plant lineage, produce flowers and disperse their seeds inside fruits. Coevolution with insect pollinators contributed to their success.

Where you are going We delve more deeply into the structure and function of vascular plants, especially angiosperms, in Chapters 27–29. The mutually beneficial relationship of plants and their pollinators is discussed in Section 45.3. Mosses and other bryophytes reappear when we consider ecological succession (Section 45.8). We discuss how plants shape communities when we cover biomes (Sections 47.4 to 47.10) and return to human impacts on forests and on plant diversity in Sections 48.3 and 48.4.

The world's great forests influence life in profound ways. Trees are metabolic wizards that produce organic compounds by absorbing energy from the sun, carbon dioxide from the air, and water and dissolved minerals from the soil. By the noncyclic pathway of photosynthesis, they split water molecules and release oxygen. Their oxygen output and carbon uptake sustain the atmosphere. Think of it—every single atom of carbon in a redwood tree that stands a hundred meters high and weighs thousands of tons was taken up from the air.

At the same time, forests act like giant sponges, absorbing water that falls as rain, then releasing it slowly over time. By wicking up water and holding soil in place, forests help prevent erosion, flooding, and sedimentation that can disrupt rivers, lakes, and reservoirs. Remove trees, and the exposed soil washes away or loses nutrients.

All countries rely on their forests for fuel and lumber. **Figure 22.1A** shows a heavily logged forest in the Canadian province of British Columbia. The developing nations of Asia, Africa, and Latin America have fast-growing populations and high demands for food, fuel, and lumber. To meet their own needs and produce materials for sale in the global market, these nations turn to their main resource: local tropical forests. These forests are a biological treasure; they have held 50 percent or more of all land-dwelling species for at least 10,000 years.

Deforestation of tropical forests affects evaporation rates, runoff, and regional patterns of rainfall. In tropical forests, most of the water vapor in the air is released from trees. In heavily logged regions, annual rainfall declines. Rain that does fall swiftly drains away from the exposed, nutrient-poor soil. As the region gets hotter and drier, the fertility and moisture content of the soil decrease.

Like many environmental issues, large-scale deforestation can seem like an overwhelming problem. Wangari Maathai, a Kenyan biologist, suggested a simple solution: Plant new trees (**Figure 22.1B**). Maathai founded the Green Belt Movement, an organization that began by giving poor women in Kenya the resources they needed to plant and care for trees. Maathai reasoned that a woman who plants a tree seedling, cares for it, and watches it grow will value and protect that tree. As many women plant trees, they will observe the improvement in their environment and understand that they have the capacity to bring about positive change.

Maathai began her tree-planting campaign in 1977. She created nurseries that provided seedlings of trees native to Africa. By 2007, members of her organization had planted 40 million trees throughout Africa. These trees provide shade, reduce erosion, and have begun to restore the diverse forests that had been disappearing. Just as important, Maathai's practical solution to a seemingly overwhelming problem has inspired others to do what they can to help stem the tide of environmental degradation.

A Heavily logged forest in British Columbia, Canada.

B Nobel Prize winner Wangari Maathai has inspired and facilitated the planting of billions of trees in reforestation projects worldwide.

Figure 22.1 Cutting and planting trees.

22.2 Plant Ancestry and Diversity

■ Land plants evolved from a green alga. An adaptive radiation that began 500 million years ago produced diverse groups.
■ Links to Ozone layer 19.1, Eukaryote life cycles 21.2, Charophyte algae 21.10

From Algal Ancestors to Embryophytes

Plants are multicelled, typically photosynthetic eukaryotes adapted to life on land. They are close relatives of red algae and green algae, and the charophyte algae are considered their sister group:

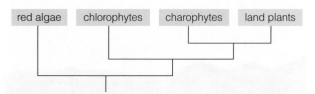

Like all green algae, plants have cell walls made of cellulose, chloroplasts with chlorophylls *a* and *b*, and they store sugars as starch. Shared traits that unite the plants and the charophyte algae include the mechanism by which a new cell wall forms after cell division and the arrangement of flagella on sperm (among those that have flagellated sperm).

A developmental trait defines the clade of land plants. They are called **embryophytes** (meaning embryo-bearing plants), because their embryos form within a chamber of parental tissues and receive nourishment from the parent during early development. In some multi-celled charophytes, such as *Chara*, a zygote forms in a chamber on the parental body, but the zygote is released into the environment and the embryo develops without parental assistance.

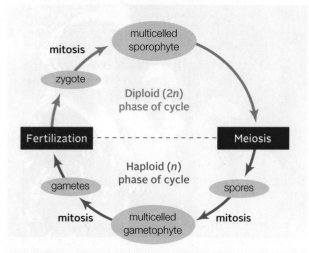

Figure 22.2 Generalized life cycle for land plants. A life cycle in which a muticelled haploid generation alternates with a multi-celled diploid generation is called an alternation of generations.

Table 22.1   Diversity of Modern Land Plants

Bryophytes	
Liverworts	9,000 species
Mosses	15,000 species
Hornworts	100 species
Seedless Vascular Plants	
Lycophytes	1,100 species
Whisk ferns	7 species
Horsetails	25 species
Ferns	12,000 species
Gymnosperms	
Cycads	130 species
Ginkgos	1 species
Conifers	600 species
Gnetophytes	70 species
Angiosperms (Flowering Plants)	
Basal groups (e.g., magnoliids)	9,200 species
Monocots	80,000 species
Eudicots	>180,000 species

An Adaptive Radiation on Land

The first plants evolved about 500 million years ago. By that time, enormous numbers of photosynthetic cells had come and gone, and oxygen-producing species had altered the composition of the atmosphere. High above Earth, the sun's energy had converted some oxygen into a dense ozone layer, which screened out ultraviolet radiation. Before the protective ozone layer had formed, high doses of incoming ultraviolet radiation would have killed any organisms that ventured onto land.

Spores that date to 475 million years ago are the earliest fossil evidence of land plants. A plant spore is a haploid cell that has a thick wall. Like some algae, plants have a life cycle in which a diploid generation alternates with a haploid one (**Figure 22.2**). During the diploid generation, a multicellular **sporophyte** produces spores by meiosis. When these spores germinate (become active), they undergo mitosis and grow into the next generation, a multicelled haploid **gametophyte** that produces gametes by mitosis. When male and female gametes meet up, fertilization occurs and a diploid zygote forms. The zygote undergoes mitosis and develops into a new sporophyte.

In their structure, the oldest fossil spores resemble the spores of modern liverworts. Liverworts, mosses, and hornworts are three early plant lineages informally referred to as **bryophytes** (**Figure 22.3** and **Table 22.1**). In all bryophytes, the gametophyte is larger and longer-lived than the sporophyte.

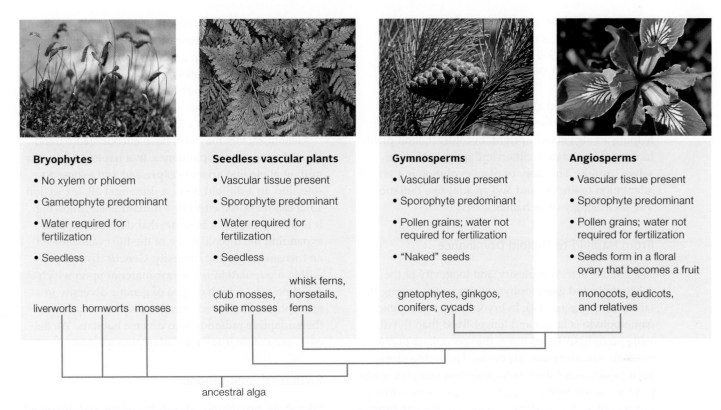

Bryophytes
- No xylem or phloem
- Gametophyte predominant
- Water required for fertilization
- Seedless

liverworts hornworts mosses

Seedless vascular plants
- Vascular tissue present
- Sporophyte predominant
- Water required for fertilization
- Seedless

club mosses,   whisk ferns,
spike mosses   horsetails,
                ferns

Gymnosperms
- Vascular tissue present
- Sporophyte predominant
- Pollen grains; water not required for fertilization
- "Naked" seeds

gnetophytes, ginkgos,
conifers, cycads

Angiosperms
- Vascular tissue present
- Sporophyte predominant
- Pollen grains; water not required for fertilization
- Seeds form in a floral ovary that becomes a fruit

monocots, eudicots,
and relatives

ancestral alga

Figure 22.3 Traits of the major plant groups and relationships among them. Note that bryophytes are a collection of lineages, rather than a clade.

The first vascular plants had evolved by about 430 million years ago. In **vascular plants**, the sporophyte is larger and longer-lived than the gametophyte, and it has specialized internal pipelines that transport water and sugars. Because bryophytes do not have these specialized tissues, they are sometimes described as nonvascular plants.

Seed plants evolved about 385 million years ago, about the same time the first insects evolved. **Seed plants** are vascular plants that hold onto their spores and disperse by releasing seeds. Seed plants are also the only plants that produce pollen. As the next sec-

tion explains, producing seeds and pollen allowed seed plants to expand into dry habitats that seedless plants could not tolerate.

Two lineages of seed plants survived to the present: gymnoperms and angiosperms. The gymnosperms are the more ancient of the two groups. Modern conifers such as pines, firs, and spruces are gymnosperms.

The flowering plants, or angiosperms, arose from a gymnosperm ancestor in the late Jurassic or early Cretaceous. In less than 40 million years, they would replace conifers and their relatives in most habitats.

The modifications in structure, function, and reproductive mode that occurred as different plant groups evolved are the focus of the next section.

bryophyte Member of an early-evolving plant lineage with a gametophyte-dominant life cycle; a moss, liverwort, or hornwort.
embryophytes Land plants; clade of multicelled, photosynthetic species that protect and nourish the embryo on the parental body.
gametophyte A haploid, multicelled body that produces gametes in the life cycle of land plants and some algae.
plants Lineage of multicelled, typically photosynthetic eukaryotes adapted to life on land.
seed plant Plant that produces seeds and pollen; an angiosperm or gymnosperm.
sporophyte Diploid, spore-producing body in the life cycle of land plants and some algae.
vascular plant Plant having specialized tissues (xylem and phloem) that transport water and sugar within the plant body.

Take-Home Message

What traits define plants and how do the various plant lineages differ?

» Land plants are embryophytes; their embryo forms in a chamber on the parental body and is nourished by the parent.

» Three early-evolving plant lineages are known collectively as bryophytes. These nonvascular plants do not produce seeds.

» Vascular plants have internal pipelines that transport water and sugars.

» The first seed plants were gymnosperms. Angiosperms (flowering plants) are the most diverse plant lineage, and the most recently evolved.

22.3 Evolutionary Trends Among Plants

■ Over time, the spore-producing bodies of plants became larger, more complex, and better adapted to dry habitats.

■ Links to Cell walls and cuticle 4.11, Life cycles 12.5, Dominant alleles 13.2, Green algae 21.10

A multicelled green alga absorbs the water it needs across its body surface. Water also buoys algal parts, helping an alga stand upright. In contrast, land plants face the threat of desiccation and must hold themselves upright. The story of plant evolution is a tale of adaptation to life on land and an adaptive radiation into increasingly drier habitats.

From Haploid to Diploid Dominance

The relative size, complexity, and longevity of the sporophyte and gametophyte stages varies among the land plants (**Figure 22.4**). In bryophytes, the haploid gametophyte is larger and longer-lived than the diploid sporophyte. By contrast, the sporophyte dominates all vascular plant life cycles. Flowering plants such as oaks have the largest and most complex sporophytes. An oak tree is a sporophyte that stands many meters tall. Oak gametophytes form inside oak flowers and consist of only a few cells.

What drove the trend toward gametophyte reduction and sporophyte enlargement? Early on, structural

differences between spores and gametes were probably a factor. Plants that release their spores into the environment encase them in a waterproof, decay-resistant wall. By contrast, gametes are unwalled cells. Under dry conditions, spores are more likely to survive than gametes, so increased spore production provides a greater advantage than increased gamete production.

Genetic factors may also have encouraged a trend toward sporophyte dominance. In a haploid body, any mutant allele that arises is expressed and exposed to selection. In a diploid body, a dominant allele (Section 13.2) can mask the effect of a mutant allele, allowing it to persist even if it is somewhat deleterious. Thus, expanding the diploid stage of the life cycle allowed an increase in genetic diversity. Genetic diversity within a population is the raw material upon which selection acts. A high degree of genetic diversity in sporophyte-dominated lineages may have facilitated their adaptive radiation into diverse habitats. We discuss some of their relevant adaptations below.

Structural Adaptations

Like algae, bryophytes absorb the water and dissolved minerals they require across their body surface. They also lose water by evaporation across this surface. In some bryophytes and all vascular plants, the sporophyte secretes a waxy covering, or **cuticle**, that reduces evaporative water loss (**Figure 22.5A**). Closable pores called **stomata** (singular, stoma) extend across the cuticle (**Figure 22.5B**). Stomata open to allow gas exchange for photosynthesis, or close to conserve water.

Early plants had threadlike structures to hold them in place, but these structures did not deliver water and dissolved minerals to the rest of the plant body, as roots do. Moving substances from roots to other body regions requires **vascular tissues**, a system of internal pipelines (**Figure 22.6**). **Xylem** is a vascular tissue that distributes water and mineral ions. **Phloem** is a vascular tissue that distributes sugars made in photosynthetic cells. Most modern plants have xylem and phloem, and thus are vascular plants (tracheophytes).

Vascular tissues not only distribute material, they also provide structural support. An organic compound called **lignin** (Section 4.11) stiffens the walls of xylem cells. Evolution of lignin-stiffened tissue allowed vascular plants to stand taller than bryophytes and to branch, giving them an advantage in spore dispersal.

Most vascular plants also have leaves. Leaves are flattened, aboveground organs that increase the surface area available for intercepting sunlight and for gas exchange. They contain veins of vascular tissue.

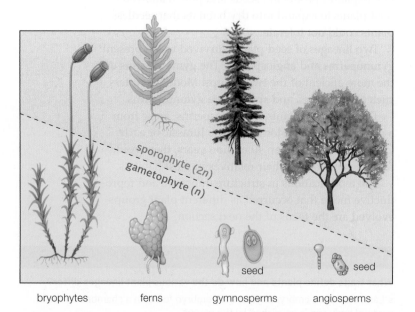

sporophyte (2n)
gametophyte (n)

seed   seed

bryophytes   ferns   gymnosperms   angiosperms

Figure 22.4 Evolutionary trend in plant life cycles. Mosses and other bryophytes put the most energy into making gametophytes. Later evolving groups invested increasingly in making sporophytes.

Figure It Out: A pine tree is a gymnosperm. Which is the larger, more prominent phase in its life cycle, the sporophyte or gametophyte?

Answer: Sporophyte

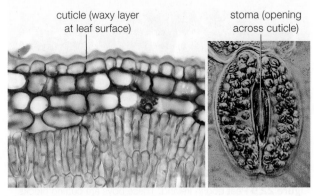

cuticle (waxy layer at leaf surface)

stoma (opening across cuticle)

A Light micrograph showing waxy cuticle (stained *pink*) at a leaf surface. **B** One stoma.

Figure 22.5 Water-conserving adaptations. A secreted cuticle reduces evaporation. Stomata are openings across the cuticle.

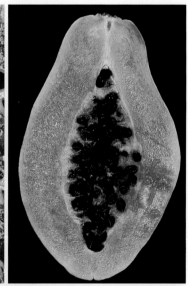

xylem

phloem

A Colorized micrograph of xylem (hollow tubes) in cross-section. **B** Light micrograph of vascular tissue of a squash stem with lignin stained *red*.

Figure 22.6 Vascular tissues. Xylem have walls stiffened with lignin and carry water. Phloem carries dissolved sugars produced by photosynthesis.

Pollen and Seeds

Novel reproductive traits gave one lineage of vascular plants a competitive edge. All bryophytes, and some vascular plants such as the ferns, release sperm that must swim through the environment to eggs. Only seed-bearing vascular plants release pollen grains. A **pollen grain** is a walled, immature gametophyte that will give rise to the male gametes. After pollen grains are released, they travel to female gametophytes on the wind or on the bodies of animals, most often insects. An ability to produce pollen gave seed plants an ability to reproduce even in dry environments.

Bryophytes and seedless vascular plants release spores (**Figure 22.7A**), but seed plants protect spores within their tissues and release seeds. A **seed** consists of an embryo sporophyte and some nutritive tissue enclosed within a waterproof seed coat. Many seeds have features that facilitate their dispersal away from the parent plant. Angiosperms, or flowering plants, disperse the seeds inside a fruit (**Figure 22.7B**). The great majority of modern plants are angiosperms.

A Ferns disperse by releasing spores that form on the underside of leaves.

B Flowering plants such as papayas hold on to their spores and disperse by releasing seeds enclosed in fruit.

Figure 22.7 Mechanisms of dispersal.

cuticle Secreted covering at a body surface.
lignin Material that stiffens cell walls of vascular plants.
phloem In vascular plants, tissue that distributes photosynthetically produced sugars through the plant body.
pollen grain Male gametophyte of a seed plant.
seed Embryo sporophyte of a seed plant packaged with nutritive tissue inside a protective coat.
stomata Closable gaps defined by guard cells on plant surfaces; when open, they allow water vapor and gases to diffuse across the epidermis.
vascular tissue In vascular plants, tissue (xylem and phloem) that distributes water and nutrients through the plant body.
xylem In vascular plants, tissue that distributes water and dissolved minerals through the plant body.

Take-Home Message

What adaptations contributed to plant diversification?

» Plant life cycles shifted from a gametophyte-dominated cycle in bryophytes to a sporophyte-dominated cycle in vascular plants.

» Life on land favored water-conserving features such as a cuticle. In vascular plants, a system of vascular tissue—xylem and phloem—distributes material through the leaves, stems, and roots of sporophytes.

» Bryophytes and seedless vascular plants release spores. Only seed plants release embryos inside protective seeds. Only in the flowering plants do seeds form inside floral tissue that later develops into a fruit.

22.4 Bryophytes

- Three land plant lineages—liverworts, hornworts, and mosses—have a gametophyte-dominated life cycle.
- Link to Nitrogen fixation 20.7

Three lineages—liverworts, hornworts, and mosses—are informally referred to as bryophytes, and also as nonvascular plants. Some mosses have internal pipelines, but none of the bryophytes have true vascular tissue reinforced by lignin, so few are more than 20 centimeters (8 inches) tall.

Mosses

Mosses are the most diverse and familiar group of bryophytes. We will use the life cycle of the moss *Polytrichum* as our example of a bryophyte life cycle (Figure 22.8). Like all bryophytes, this moss has a gametophyte-dominated life cycle.

A moss gametophyte has leaflike green parts that grow from a central stalk ❶. Threadlike **rhizoids** hold the gametophyte in place. The moss sporophyte is a **sporangium** (spore-producing structure) on a stalk ❷.

Figure 22.9 Peat bog in Ireland. These boys are cutting blocks of peat and stacking them to dry for use as fuel.

The sporophyte is not photosynthetic and so depends on the gametophyte to which it is attached for nourishment. Meiosis of cells inside a sporangium yields haploid spores ❸. After dispersal by the wind, a spore germinates and grows into a gametophyte. Multicellular **gametangia** (gamete-producing structures) develop in or on the gametophyte. The moss we are using as our example has separate sexes, with

Figure 22.8 Animated Life cycle of a moss (*Polytrichum*).

❶ The leafy green part of a moss is the haploid gametophyte.

❷ The diploid sporophyte has a stalk and a capsule (sporangium). It is not photosynthetic.

❸ Haploid spores form by meiosis in the capsule, are released, and drift with the winds.

❹ Spores germinate and develop into male or female gametophytes with gametangia that produce eggs or sperm by mitosis.

❺ Sperm swim to eggs.

❻ Fertilization produces a zygote.

❼ The zygote grows and develops into a new sporophyte while remaining attached to and nourished by the female gametophyte.

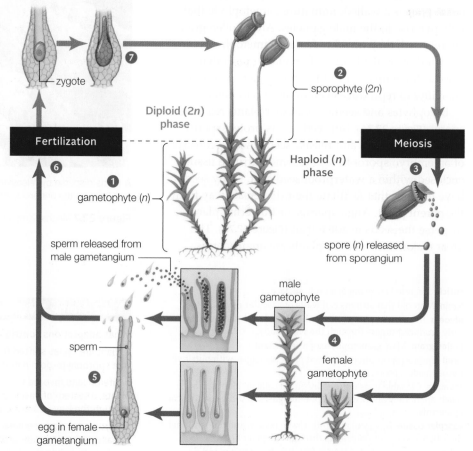

zygote

Diploid (2n) phase

Fertilization ❻

sporophyte (2n) ❷

Meiosis

Haploid (n) phase

❸

gametophyte (n) ❶

sperm released from male gametangium

male gametophyte

spore (n) released from sporangium

sperm

female gametophyte

egg in female gametangium ❺

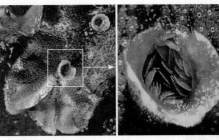

A Thallus (leaflike part) with gemmae cup

B Asexually produced gemmae in cup

C Sperm-producing, umbrella-shaped male gametangia.

spore capsule of sporophyte

stalk of female gametangium

D Female gametangia with attached, dependent sporophytes.

Figure 22.10 Asexual (**A,B**) and sexual (**C,D**) reproduction in the liverwort *Marchantia*. In this genus, sexes are separate.

each gametophyte producing eggs or sperm ❹. Other bryophytes are bisexual. In either case, rain triggers the release of flagellated sperm that swim through a film of water to eggs ❺. Some moss gametangia facilitate sperm movement by attracting mites and crawling insects that unknowingly pick up sperm and move them to adjacent plants. Fertilization inside the egg chamber produces a zygote ❻ that develops into a new sporophyte ❼. Mosses also reproduce asexually by fragmentation when a bit of gametophyte breaks off and develops into a new plant.

The 350 or so species of peat moss (*Sphagnum*) are the most economically important bryophytes. Peat mosses are the dominant plants in **peat bogs** that cover hundreds of millions of acres in high-latitude regions of Europe, Asia, and North America. Many peat bogs have persisted for thousands of years, and layer upon layer of plant remains have become compressed as a carbon-rich material called peat. Blocks of peat are cut, dried, and burned as fuel, especially in Ireland (**Figure 22.9**). Freshly harvested peat moss is also an important commercial product. The moss is dried and added to planting mixes to help soil retain moisture.

Liverworts

Liverworts commonly grow in moist places, often alongside mosses. They may be the most ancient land plant lineage. The oldest known spores of land plants resemble liverwort spores, and genetic comparisons put liverworts near the base of the plant family tree.

Liverwort gametophytes are ribbonlike or leafy. *Marchantia*, a ribbonlike liverwort, is a common pest in greenhouses. It reproduces asexually by producing disks of cells (gemmae) in cups on the gametophyte body (**Figure 22.10A,B**). During sexual reproduction, a male plant produces sperm atop a stalked gametangium (**Figure 22.10C**). Sperm swim to eggs that

have formed in the gametangium of a female plant. Fertilization produces a sporophyte that remains attached to the female gametophyte (**Figure 22.9D**).

Hornworts

A pointy, hornlike sporophyte that can be several centimeters tall gives the hornworts their common name (*right*). The base of the sporophyte is embedded in gametophyte tissues, and spores form in an upright sporangium, or capsule. When spores mature, the tip of the capsule splits, releasing them. The sporophyte grows continually from its base, so it can make and release spores over an extended period.

sporophyte

gametophyte

The sporophyte has chloroplasts and, in some cases, can survive even after the death of the gametophyte. These traits and genetic similarities suggest that hornworts may be the sister group to vascular plants.

gametangium Gamete-producing organ of a plant.
peat bog High-latitude community dominated by *Sphagnum* moss.
rhizoid Threadlike structure that anchors a bryophyte.
sporangium Of plants and fungi, a structure in which haploid spores form by meiosis.

Take-Home Message

What are bryophytes?

» Bryophyte is the common name for three lineages of plants: mosses, liverworts, and hornworts. All are low-growing with no lignin-reinforced vascular tissues. All have flagellated sperm that require a film of water to swim to eggs, and all disperse by releasing spores.

» Bryophytes are unique among land plants in having a life cycle in which the gametophyte is the dominant generation. The sporophyte remains attached to the gametophyte even when mature.

22.5 Seedless Vascular Plants

■ A sporophyte with lignified vascular tissue is the dominant phase in the life cycle of the seedless vascular plants.

■ Link to Gamete formation and fertilization 12.5

Some mosses have internal pipelines that transport fluid within their body. However, only vascular plants have lignin-strengthened vascular tissue with xylem and phloem. This innovation allowed the evolution of larger, branching sporophytes, which are the predominant generation in all vascular plants.

Seedless vascular plants are the oldest vascular plant lineages. Like bryophytes, they have flagellated sperm that swim to eggs, and they disperse by releasing spores directly into the environment.

Two lineages of seedless vascular plants survived to the present. Lycophytes include plants that are commonly known as club mosses and spike mosses, but are not mosses. Monilophytes include whisk ferns, horsetails, and ferns.

Lycophytes and monilophytes diverged from a common ancestor before leaves and roots had evolved, and each developed these features in a different way. For example, lycophytes form spores along the sides of branches. Their leaves have one unbranched vein and probably evolved from a lateral sporangium. In contrast, monilophytes have spores at branch tips. Their leaves have branching veins, so they probably evolved from a branching network of stems.

Club Mosses

Figure 22.11 Club moss (*Lycopodium*) about 20 centimeters (8 inches) tall. Spores form in the cone-shaped strobili at the tip.

Most of the 1,200 modern lycophytes are club mosses. Club mosses of the genus *Lycopodium* grow on the floor of temperate forests (**Figure 22.11**). The sporophytes resemble miniature pine trees and are sometimes called ground pines. The plant has a horizontal stem, or **rhizome**, that runs along the ground. Roots and upright stems with tiny leaves grow from the rhizome. When a plant is several years old, it produces a strobilus seasonally. A **strobilus** is a cone-shaped, spore-producing structure composed of modified leaves.

Lycopodium is gathered from the wild for use in wreaths and bouquets, and an extract of the plant is marketed as an herbal medicine. The spores, which are covered with a waxy coating that ignites easily, are sold as "flash powder" for creating special effects.

Whisk Ferns and Horsetails

Figure 22.12 Whisk fern with sporangia at tips of short lateral branches.

Whisk ferns (*Psilotum*) are native to the southeastern United States. Their sporophytes have underground rhizomes and photosynthetic stems that appear leafless (**Figure 22.12**). Spores form in fused sporangia at the tips of short branches. Florists often use branches of whisk ferns in mixed bouquets.

The 25 *Equisetum* species are known as horsetails or rushes. Their sporophytes have rhizomes and hollow stems with tiny nonphotosynthetic leaves at the joints. Photosynthesis occurs in stems and leaflike branches (**Figure 22.13A**). Deposits of silica in the stem support the plant and give stems a sandpapery texture that helps fend off herbivores. Before scouring powders and pads were widely available, people used stems of *Equisetum* species as pot scrubbers, thus the common name "scouring rush."

Depending on the species, strobili form either at tips of photosynthetic stems or on specialized reproductive stems without chlorophyll (**Figure 22.13B**).

Ferns

With 12,000 or so species, ferns are the most diverse seedless vascular plants. Most live in the tropics. **Figure 22.14** shows the life cycle of a common North

A Photosynthetic stems. The long, needlelike structures are not leaves, but branches.

B Nonphotosynthetic stem with a strobilus at its tip.

Figure 22.13 Horsetails (*Equisetum*).

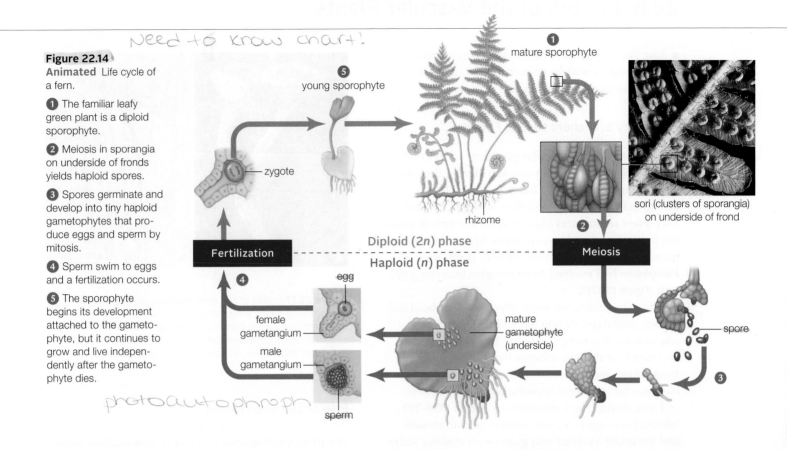

Need to know chart!

Figure 22.14

Animated Life cycle of a fern.

❶ The familiar leafy green plant is a diploid sporophyte.

❷ Meiosis in sporangia on underside of fronds yields haploid spores.

❸ Spores germinate and develop into tiny haploid gametophytes that produce eggs and sperm by mitosis.

❹ Sperm swim to eggs and a fertilization occurs.

❺ The sporophyte begins its development attached to the gametophyte, but it continues to grow and live independently after the gametophyte dies.

photoautophroph

❺ young sporophyte

zygote

❶ mature sporophyte

sori (clusters of sporangia) on underside of frond

rhizome

Diploid (2n) phase
Haploid (n) phase

Fertilization

Meiosis

❷

egg

female gametangium

male gametangium

sperm

❹

mature gametophyte (underside)

spore

❸

American fern. The sporophytes have fronds (leaves) and roots that grow from rhizomes ❶. Spores form in **sori** (singular, sorus), clusters of sporangia on the lower surface of the fronds ❷. After germination, a spore develops into a gametophyte that is typically bisexual and just a few millimeters across ❸. (All seedless nonvascular plants have tiny gametophytes.) Fertilization occurs when flagellated sperm swim to eggs ❹. Fern sperm typically have multiple flagella and some have as many as fifty. A sporophyte begins life attached to the gametophyte ❺. However, it quickly overgrows its parent and survives on its own.

Fern sporophytes vary in their structure and size. The floating fern *Azolla* has fronds only 1 millimeter long (**Figure 22.15A**). Chambers in its fronds shelter nitrogen-fixing cyanobacteria. Southeast Asian farmers grow this fern in rice fields as a natural alternative to chemical fertilizers. Many ferns are **epiphytes**, plants that attach to and grow on another plant but do not withdraw nutrients from it (**Figure 22.15B**). The largest modern-day nonvascular plants are tree ferns (**Figure 22.15C**). Some stand 20 meters (65 feet) high. The "trunk" of a tree fern is a modified rhizome.

Ferns are sold as potted plants, and the trunks of some tree ferns are shredded to produce a potting medium used to grow orchids. Fiddleheads, the young fronds of some ferns, are harvested as food.

A The floating fern *Azolla* fits on a finger.

B Bird's nest fern, one of the many epiphytes.

C Forest of tall tree ferns in Australia.

Figure 22.15 A sampling of fern diversity.

epiphyte Plant that grows on another plant but does not harm it.
rhizome Stem that grows horizontally along or under the ground.
sorus Cluster of spore-producing capsules on a fern leaf.
strobilus (strobili) Of some nonflowering plants such as horsetails and cycads, a cluster of spore-producing structures.

Take-Home Message

What are seedless vascular plants?

» Club mosses and relatives belong to one (lycophytes) seedless vascular lineage. Ferns, horsetails and whisk ferns belong to the other (monilophytes).

» A sporophyte with vascular tissues (xylem and phloem) dominates their life cycle, and they disperse by releasing spores.

22.6 History of the Vascular Plants

■ Seed plants dominate modern forests, but before they evolved, forests of seedless nonvascular plants stood tall.

■ Link to Geologic time scale 16.8

From Tiny Branchers to Coal Forests

The oldest fossils of vascular plants are spores that date to about 450 million years ago, during the late Ordovician period (**Figure 22.16**). *Cooksonia* (**Figure 22.17A**) may be one of the earliest lineages. It stood only a few centimeters high and had a simple branching pattern, and no leaves or roots. Spores formed at branch tips. By the Devonian period, plants such as *Psilophyton* had evolved more complex branching pattern (**Figure 22.17B**).

The oldest forest we know about existed about 385 million years ago, during the middle Devonian, at a site in what is now upstate New York. Fossil stumps and fronds at this site indicate that the plants in this forest stood about 8 meters (26 feet) high and resembled tree ferns in their appearance.

Later, during the Carboniferous period (359–299 million years ago), ancient relatives of club mosses and horsetails evolved into giants with massive stems (**Figure 22.18**). Some stood 40 meters (more than 130 feet) high. After forests of these plants formed, climates changed, and the sea level rose and fell many times. When the waters receded, the forests flourished. After the sea moved back in, submerged trees became buried in sediments that protected them from decomposition. Layers of sediments accumulated one on top of the other. Their weight squeezed the water out of the saturated, undecayed remains, and the compaction generated heat. Over time, pressure and heat transformed the compacted organic remains into **coal**.

It took millions of years of photosynthesis, burial, and compaction to form coal. When you hear about annual production of coal or other fossil fuels, keep in mind that we do not really "produce" these materials, we only extract them. This is why fossil fuels are said to be nonrenewable sources of energy.

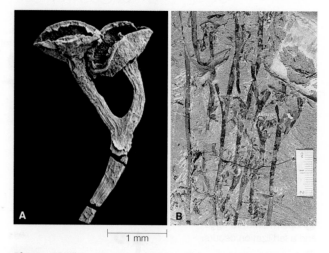

Figure 22.17 Fossils of two early seedless vascular plants. (**A**) *Cooksonia* stems always divided into two equal branches. This plant stood a few centimeters tall. (**B**) *Psilophyton* shows a more complex growth pattern. It branched unequally with a main stem and smaller branches to the side.

Rise of the Seed Plants

The first gymnosperms evolved from a seedless ancestor late in the Devonian. Cycads and ginkgos were early gymnosperms. Conifers evolved a bit later. Angiosperms, or flowering plants, descended from a gymnosperm ancestor by about 120 million years ago, during the reign of the dinosaurs (**Figure 22.19**).

Reproductive traits of seed-bearing plants gave them a selective advantage over earlier lineages. Gametophytes of seedless vascular plants develop in the environment. By contrast, gametophytes of seed plants develop within the protection of a sporophyte body (**Figure 22.20**). Sporangia called **pollen sacs** produce microspores. **Microspores** develop into sperm-producing gametophytes (pollen grains). Sporangia called **ovules** produce megaspores. Egg-producing gametophytes develop from **megaspores**. A sporophyte releases pollen grains, but holds onto its eggs. Wind or animals can deliver pollen from one seed plant to

Origin of first land plants (bryophytes) by 475 mya.	Origin of seedless vascular plants.	Bryophytes and seedless vascular plants diversify. Seed plants arise by 385 mya.	Tree-sized lycophytes and horsetails live in swamp forests. First conifers arise late in Carboniferous.	Ginkgos, cycads appear. Most horsetails and lycophytes disappear by the end of the Permian.	Adaptive radiations of ferns, cycads, conifers; by start of Cretaceous, conifers are dominant trees.	Flowering plants appear in the early Cretaceous, undergo adaptive radiation, and become dominant.
Ordovician	Silurian	Devonian	Carboniferous	Permian	Triassic	Jurassic	Cretaceous	Tertiary
488	443	416	359	299	251	200	146	66

Millions of years ago (mya)

Figure 22.16 A timeline for major events in the evolution of plants.

Figure 22.18 Painting of a Carboniferous coal forest. An understory of ferns is shaded by tree-sized relatives of modern horsetails and club mosses.

Figure 22.19 Early angiosperms such as magnolias (*foreground*) evolved while dinosaurs walked the Earth.

the ovule of another, a process known as **pollination**. Because the sperm of seed plants do not need to swim through a film of water to reach eggs, these plants can reproduce even during dry times.

After pollination and fertilization, an ovule develops into a seed. Releasing seeds puts seed plants at an advantage over spore-bearing plants. A seed contains a multicelled embryo sporophyte and stored food that the embryo can draw on during early development. By contrast, seedless plants release single-celled spores without stored food.

Structural traits also gave seed plants an advantage over seedless lineages. Some seed plants undergo secondary growth (growth in diameter) and produce wood. Wood is lignin-stiffened tissue that strengthens and protects older stems and roots. The giant nonvascular plants that lived in Carboniferous forests did undergo secondary growth. However, because their trunks were softer and more flexible than those of woody seed plants, they could not grow as tall.

coal Fossil fuel formed over millions of years by compaction and heating of plant remains.
megaspore Haploid spore formed in ovule of seed plants; develops into an egg-producing gametophyte.
microspore Haploid spore formed in pollen sacs of seed plants; develops into a sperm-producing gametophyte (a pollen grain).
ovule Of seed plants, sporangium that produces megaspores that develop into egg-producing gametophytes.
pollen sac Of seed plants, sporangium where microspores form and develop into pollen grains.
pollination Arrival of a pollen grain on the egg-bearing part of a seed plant.

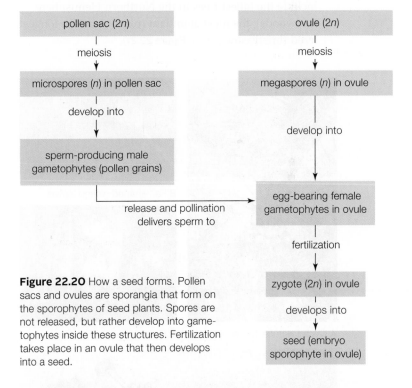

Figure 22.20 How a seed forms. Pollen sacs and ovules are sporangia that form on the sporophytes of seed plants. Spores are not released, but rather develop into gametophytes inside these structures. Fertilization takes place in an ovule that then develops into a seed.

Take-Home Message

What are the major events in the evolutionary history of vascular plants?

» Seedless vascular plants arose and became widespread during the Carboniferous. Coal is the remains of some of these ancient plants.

» The first seed plants evolved in the late Devonian. Because they produced pollen and seeds, they were able to live in drier places than other plants. An ability to produce wood allowed some seed plants to grow very tall.

22.7 Gymnosperms

- Gymnosperms are seed-bearing plants that produce their seeds on the surface of modified leaves.
- Gymnosperms do not make flowers or fruit.

Gymnosperms are vascular seed plants that produce seeds on the surface of ovules. Their seeds are said to be "naked," because unlike those of angiosperms they are not inside a fruit. (*Gymnos* means naked; *sperma* is seed.) However, many gymnosperms enclose their seeds in a fleshy or papery covering.

Conifers

The 600 or so species of conifers are trees and shrubs with woody cones. Conifers typically have needlelike or scalelike leaves with a thick cuticle. They tend to be more resistant to drought and cold than flowering plants and are the main plants in high-latitude forests of the Northern Hemisphere. Most conifers shed some leaves steadily but remain evergreen. Conifers include the tallest trees in the Northern Hemisphere (redwoods), the most abundant (pines), and the longest lived (bristlecone pines, **Figure 22.21**).

Figure 22.21 Ancient bristlecone pine in the Sierra Nevada Mountains. The oldest tree of this species is 4,700 years old.

We mulch our gardens with fir bark, use oils from cedar in cleaning products, and eat the seeds, or "pine nuts," of some pines. Pines also provide lumber for building homes and furniture. Some pines make a sticky resin that deters insects from boring into them. We use this resin to make turpentine, a paint solvent.

Lesser-Known Lineages

Cycads and ginkgos evolved before conifers and were most diverse in dinosaur times. They are the only modern seed plants that have flagellated sperm. Their sperm emerge from pollen grains, then swim in fluid produced by the plant's ovule.

Cycads have palmlike or fernlike leaves and live mainly in the dry tropics and subtropics (**Figure 22.22A**). "Sago palms" commonly used in landscaping and as houseplants are actually a type of cycad.

The only living ginkgo species is *Ginkgo biloba*, the maidenhair tree. It is one of the few gymnosperms that is deciduous (drops all its leaves at once seasonally). *G. biloba* is native to China, but its lovely fan-shaped leaves (**Figure 22.22B**) and resistance to pests and air pollution make it a popular street tree in urban areas.

Gnetophytes include tropical trees, desert shrubs, and leathery vines. *Ephedra*, a twiggy shrub (**Figure 22.22C**), produces ephedrine, a chemical that is sold as an herbal stimulant and weight loss aid. Africa's Namib desert is home to *Welwitschia* (**Figure 22.22D**). It has a taproot, woody stem, and two long, straplike leaves. These leaves split lengthwise repeatedly, giving the plant a shaggy appearance. *Welwitschia* is long-lived. Some plants are more than a thousand years old.

B Fan-shaped ginkgo leaves.

A Cycad with fleshy seeds.

C Pollen cones of *Ephedra*.

D *Welwitschia* with seed cones and two long, wide leaves.

Figure 22.22 Gymnosperm diversity.

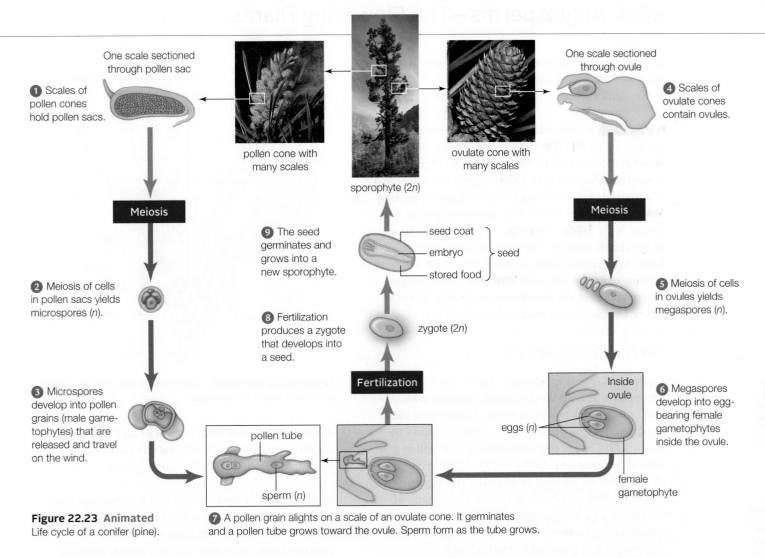

1 Scales of pollen cones hold pollen sacs.

One scale sectioned through pollen sac

pollen cone with many scales

sporophyte (2n)

ovulate cone with many scales

One scale sectioned through ovule

4 Scales of ovulate cones contain ovules.

Meiosis

2 Meiosis of cells in pollen sacs yields microspores (n).

9 The seed germinates and grows into a new sporophyte.

seed coat
embryo } seed
stored food

Meiosis

5 Meiosis of cells in ovules yields megaspores (n).

8 Fertilization produces a zygote that develops into a seed.

zygote (2n)

3 Microspores develop into pollen grains (male gametophytes) that are released and travel on the wind.

Fertilization

Inside ovule

eggs (n)

6 Megaspores develop into egg-bearing female gametophytes inside the ovule.

pollen tube

sperm (n)

female gametophyte

Figure 22.23 **Animated**
Life cycle of a conifer (pine).

7 A pollen grain alights on a scale of an ovulate cone. It germinates and a pollen tube grows toward the ovule. Sperm form as the tube grows.

A Representative Life Cycle

A pine tree is a sporophyte, and its life cycle is typical of conifers (**Figure 22.23**). It produces its spores on specialized strobili that we call cones. The tree makes two types of cones: small, soft pollen cones; and large, woody, ovulate cones. Both have scales arranged around a central axis. Each scale of a pollen cone contains pollen sacs **1**. Microspores that form by meiosis in these sacs **2** develop into pollen grains **3**. Each scale of an ovulate cone holds a pair of ovules **4**. Pollen cones release their tiny pollen grains to drift

with the winds. Meanwhile, meiosis of diploid cells in ovulate cones produces megaspores **5** that develop into egg-producing female gametophytes **6**.

An unpollinated ovule exudes a sugary "pollination droplet" that captures windborne pollen. After pollination, the pollen grain germinates and a pollen tube begins to grow **7**.

The pollen tube grows remarkably slowly. After about a year, the tube reaches and enters the ovule. The sperm cell delivered via the tube fuses with the egg nucleus, forming a zygote **8**. The zygote develops into an embryo sporophyte, which, with the ovule tissues, becomes a seed **9**. The seed germinates and, over many years, develops into a mature sporophyte.

conifer Gymnosperm with nonmotile sperm and woody cones; for example, a pine.
cycad Tropical or subtropical gymnosperm with flagellated sperm, palmlike leaves, and fleshy seeds.
ginkgo Deciduous gymnosperm with flagellated sperm, fan-shaped leaves, and fleshy seeds.
gnetophyte Shrubby or vinelike gymnosperm, with nonmotile sperm; for example, *Ephedra*.
gymnosperm Seed plant that does not make flowers or fruits; for example, a conifer.

22.8 Angiosperms—The Flowering Plants

■ Angiosperms are the most diverse plant lineage and the only plants that make flowers and fruits.

■ Links to Coevolution 17.12, Classification 18.2

Angiosperms are vascular seed plants that make flowers and fruits. A **flower** is a specialized reproductive shoot that consists of modified leaves arranged in three concentric whorls (**Figure 22.24**). Sepals form the outermost whorl; petals, the middle whorl; pollen-bearing **stamens**, the inner whorl. The innermost part of the flower is the **carpel**. The **ovary**, a chamber at the base of the carpel, contains one or more ovules where eggs form. After fertilization, an ovule matures into a seed, and the ovary becomes the **fruit**. The name "angiosperm" refers to the development of seeds inside an ovary. (*Angio–* means enclosed chamber; *sperma*, seed.)

The Angiosperm Life Cycle

Figure 22.25 shows a generalized life cycle for a flowering plant. The upper portion of a stamen is an anther that holds two pollen sacs. Inside those sacs are diploid cells ❶ that give rise to microspores by meiosis ❷. The microspores develop into pollen grains (immature male gametophytes) ❸. A flowering plant's ovules form on the wall of an ovary at the base of a carpel ❹. Meiosis of cells in ovules yields haploid megaspores ❺. A megaspore develops into a female gametophyte that includes a haploid egg, a cell with two nuclei, and a few additional cells ❻.

    Pollination occurs when a pollen grain arrives on a receptive stigma, the uppermost part of the carpel ❼. The pollen grain germinates, and a pollen tube grows through the style (the structure that elevates the stigma) to the ovary at the base of the carpel. Two nonflagellated sperm form inside the tube as it grows.

    Double fertilization occurs when a pollen tube delivers the two sperm into the ovule ❽. One sperm fertilizes the egg to create a zygote. The other sperm fuses with the cell that has two nuclei, forming a triploid (3*n*) cell. After double fertilization, the ovule matures into a seed ❾. The zygote develops into an embryo sporophyte and the triploid cell develops into **endosperm**, a nutritious tissue that will serve as a source of food for the developing embryo. Endosperm also serves as a source of food for humans. We value grains for their starch and protein-rich endosperm.

Factors Contributing to Angiosperm Success

Flowering plants began a spectacular adaptive radiation in the Mesozoic as other plant groups were in

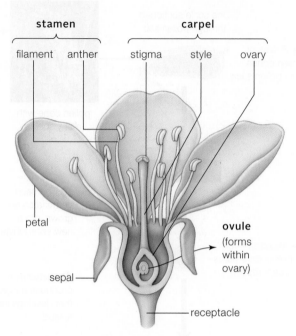

Figure 22.24 Animated Floral components. Anthers produce pollen and ovules produce eggs.

decline. Today, nearly 90 percent of all plant species are flowering plants. Several factors gave angiosperms a selective advantage over gymnosperms.

Accelerated Life Cycle Compared to gymnosperms, most angiosperms have a shorter life cycle. A dandelion or grass can grow from a seed, mature, and produce seeds of its own within a month or so. In contrast, gymnosperms tend to take years to mature. Producing and dispersing seeds quickly helps angiosperms expand their range faster than gymnosperms.

Animal-Pollinated Flowers The evolution of flowers gave angiosperms an edge by facilitating insect-assisted pollination. After seed plants evolved, some insects began feeding on their protein-rich pollen. The plants lost a bit of pollen, but benefited when the insects inadvertently transferred pollen from one plant to another of the same species. An animal that facilitates pollination by transferring pollen between plants is a **pollinator**. Most pollinators are insects, but birds, bats, and other animals can also fulfill this role.

    Over time, many flowering plants coevolved with their pollinators. **Coevolution** refers to the joint evolution of two or more species as a result of their close ecological interactions (Section 17.12). Producing sugary nectar encouraged more pollinator visits, which improved pollination rates and enhanced seed production. Conspicuous petals and distinctive scents that attracted pollinator attention also provided a selective

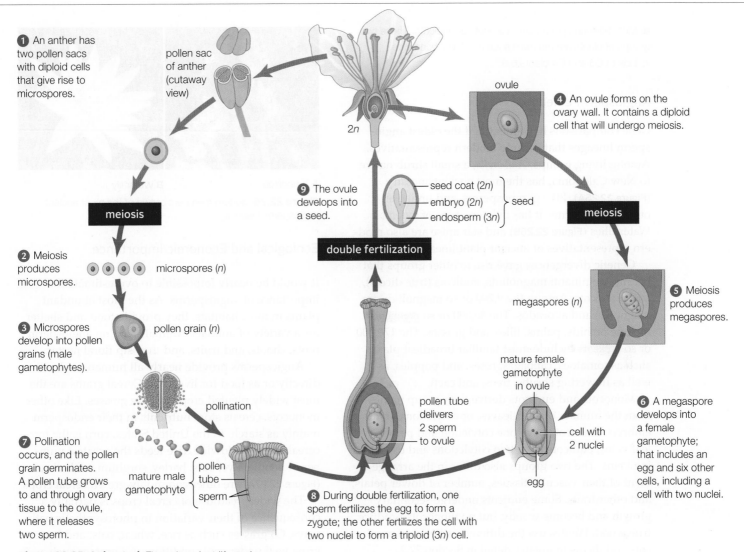

1 An anther has two pollen sacs with diploid cells that give rise to microspores.

pollen sac of anther (cutaway view)

meiosis

2 Meiosis produces microspores.

microspores (*n*)

3 Microspores develop into pollen grains (male gametophytes).

pollen grain (*n*)

pollination

7 Pollination occurs, and the pollen grain germinates. A pollen tube grows to and through ovary tissue to the ovule, where it releases two sperm.

mature male gametophyte { pollen tube — sperm }

2*n*

9 The ovule develops into a seed.

seed coat (2*n*)
embryo (2*n*) } seed
endosperm (3*n*)

double fertilization

ovule

4 An ovule forms on the ovary wall. It contains a diploid cell that will undergo meiosis.

meiosis

megaspores (*n*)

5 Meiosis produces megaspores.

mature female gametophyte in ovule

pollen tube delivers 2 sperm to ovule

cell with 2 nuclei

egg

6 A megaspore develops into a female gametophyte; that includes an egg and six other cells, including a cell with two nuclei.

8 During double fertilization, one sperm fertilizes the egg to form a zygote; the other fertilizes the cell with two nuclei to form a triploid (3*n*) cell.

Figure 22.25 Animated Flowering plant life cycle.

advantage. Mutations that altered floral shapes in ways that maximized the likelihood of pollen transfer onto a pollinator or from a pollinator to a receptive stigma were also favored. At the same time, selection

favored pollinators that were able to seek out nectar-rich flowers and access the nectar reward.

Enhanced Seed Dispersal Fruit production also contributed to angiosperm success. Fleshy, sugary fruits attract animals that carry the fruits (and their enclosed seeds) away from a parent plant. Other fruits have shapes that help them catch the wind or stick to animal fur. As a group, gymnosperms have fewer structural adaptations for seed dispersal.

angiosperms Most diverse seed plant lineage. Only group that makes flowers and fruits.

carpel Of flowering plants, a reproductive structure that produces female gametophytes; consists of a stigma, a style, and an ovary.

coevolution The joint evolution of two closely interacting species; each species is a selective agent for traits of the other.

double fertilization In flowering plants only, one sperm fertilizes an egg to produce a zygote and another fertilizes a diploid cell to create a triploid cell that will develop into endosperm.

endosperm Triploid (3*n*) nutritive tissue in an angiosperm seed.

flower Specialized reproductive shoot of a flowering plant.

fruit Mature ovary of a flowering plant, often with accessory parts; encloses a seed or seeds.

ovary In flowering plants, the enlarged base of a carpel, inside which one or more ovules form and eggs are fertilized.

pollinator Animal that moves pollen, thus facilitating pollination.

stamen Pollen-producing organ of flowering plant.

Take-Home Message

What are the features of angiosperms?

» Angiosperms produce seeds inside the ovaries of flowers. After pollination, an ovary becomes a fruit.

» Angiosperms are the most successful plants. Short life cycles, coevolution with insect pollinators, and diverse fruit structures enhanced their success.

22.9 Angiosperm Diversity and Importance

■ Most flowering plants are monocots or eudicots. These two groups of plants are the main source of food worldwide.
■ Link to C3 and C4 plants 6.8

Angiosperm Lineages

Gene comparisons have identified the oldest angiosperm lineages that include modern representatives. Among living groups, *Amborella*, a small shrub native to New Caledonia, has the most ancient ancestry (**Figure 22.26A**). Like gymnosperms, and unlike all other angiosperms, it has only one type of xylem. Water lilies (**Figure 22.26B**) and star anise are also modern representatives of ancient plant lineages.

Genetic divergences gave rise to other groups that became dominant: magnoliids, eudicots (true dicots), and monocots. Among the 9,200 or so magnoliids are magnolias and avocados. The 80,000 or so **monocots** include orchids, palms, lilies, and grasses. The 170,000 or so **eudicots** include most familiar broadleaf plants such as tomatoes, cabbages, roses, and poppies, as well as flowering shrubs, trees, and cacti.

Monocots and eudicots derive their group names from the number of seed leaves, or cotyledons, in the embryo. Monocots have one cotyledon and parallel leaf veins; eudicots have two cotyledons and branching veins. The two groups also differ in the arrangement of their vascular tissues, number of flower petals, and other traits. Some eudicots undergo secondary growth and become woody, but no monocots produce true wood. We discuss the differences between monocots and dicots in greater detail in Section 27.2.

A *Amborella* **B** Water lily

Figure 22.26 Modern members of two of the most ancient angiosperm lineages.

Ecological and Economic Importance

It would be nearly impossible to overestimate the importance of angiosperms. As the most abundant plants in most habitats, they provide food and shelter for a variety of animals. Animals feed on angiosperm roots, shoots, and fruits, and they sip floral nectar.

Angiosperms provide nearly all human food, either directly or as food for livestock. Cereal grains are the most widely planted crops. All are grasses. Like other monocots, cereals store nutrients in their endosperm mainly as starch. In the United States, corn is the top cereal crop. Worldwide, rice feeds the greatest number of people. Rye, oats, barley, sorghum, and wheat (**Figure 22.27A**) are other commonly grown cereal crops.

The widespread use of cereal crops is partially a consequence of their variation in photosynthetic pathways. C3 grasses such as rice, wheat, oats, and barley grow well under cool conditions. C4 grasses such as

A Harvesting wheat, a monocot.

B A plant geneticist examines quinoa, a eudicot with highly nutritious seeds.

C A field of cotton (a eudicot) ready for harvest.

Figure 22.27 Angiosperms as crops.

Speaking for the Trees (revisited)

A tree is about 20 percent carbon by weight, so enormous amounts of carbon are stored in the living trees of Earth's forests. Additional carbon is tied up in the leaf litter and decaying trees on the forest floor. The United Nations Food and Agriculture Organization estimates that forest plants and soils hold about one and a half times as much carbon as the air. Trees take up most carbon when they are actively growing. Therefore, an acre of tropical forest (where trees grow continually) stores more carbon than an acre of forest in a region with a limited growing season.

The great carbon-storing capacity of tropical forests means that threats to these forests have global implications. When a tropical forest is converted to cropland, the rate of carbon dioxide storage per acre goes down. In addition, if the forest is cleared by burning, the car-

bon that was stored in wood and leaf litter enters the atmosphere. Rising levels of carbon dioxide in Earth's atmosphere contribute to global climate change, by a mechanism we explain in detail in Section 46.8.

How would you vote? International carbon offsets can help discourage tropical deforestation. For example, a company in the United States could agree to offset its greenhouse gas production by funding a tropical reforestation project. Such agreements minimize a company's cost of reducing greenhouse gases, but some opponents of offsets would rather see companies make changes at home. Are international carbon offsets a good idea?

---

corn and sorghum thrive in hot, dry climates. Thus, people throughout the world can grow a cereal grain that is well suited to their climate.

Soybeans, lentils, peas, and peanuts are among the legumes, the second most important source of human food. Legumes are eudicots, and their endosperm stores nutrients as proteins and oils, as well as carbohydrates. Legumes can be paired with grains to provide all the amino acids a human body needs to build proteins. All legumes are C3 plants, so they grow poorly in hot, dry regions.

Another eudicot called quinoa (proounced keen-wah) has been called the most nutritious plant (**Figure 22.27B**). Its seeds are high in protein and it provides all eight amino acids that humans require from their diet. Quinoa has been a staple of diets in Central and South America for thousands of years.

Humans enliven their diet with a variety of plant parts. We dine on leaves of lettuce and spinach, stems of asparagus and celery, immature floral shoots of broccoli and cauliflower, and fleshy fruits of tomatoes and blueberries. Stamens of crocus flowers provide the spice saffron, and cinnamon is the grated bark of a tropical tree. Maple syrup is fluid tapped from a tree's xylem and boiled down to a syrupy consistency.

Fibers from angiosperms are the source of fabrics such as linen, ramie, hemp, burlap, and cotton (**Figure 22.27C**). The fruit of a cotton plant, the cotton boll, is nearly pure cellulose. Fibers from the leaves of agave are used to make sisal rugs.

Oils extracted from the seeds of eudicots such as rapeseed and hemp are used in detergents, skin care products, and as industrial lubricants and fuel.

Eudicot woods are referred to as "hardwoods," as opposed to gymnosperm "softwoods." Furniture and flooring are often made from oak or other hardwoods. Hardwoods are also preferred as firewood.

Some flowering plants make secondary metabolites that humans use as medicines or as mood-altering drugs. A **secondary metabolite** is a compound with no known metabolic role in the organism that makes it. Many plant secondary metabolites are evolved defenses that deter grazing animals. Aspirin is derived from a compound discovered in willows, and digitalis used to slow hearts is derived from foxglove plants. Caffeine in coffee beans and nicotine in tobacco leaves are widely used as stimulants. Illegally grown marijuana is one of the United States' most valuable cash crops. Worldwide, cultivation of opium poppies (the source of heroin) and coca (the source of cocaine) have wide-reaching health, economic, and political effects.

eudicots Most diverse lineage of angiosperms; members have two seed leaves, branching leaf veins.
monocots Highly diverse angiosperm lineage; includes plants such as grasses that have one seed leaf and parallel veins.
secondary metabolite Molecule that is produced by an organism but does not play any known role in its metabolism. Some serve as defense against predation.

Take-Home Message

What are the major angiosperm lineages and how are they important?

» Most flowering plants are monocots or eudicots.

» The grains that serve as staples of the human diet are monocots. Many other crop plants are eudicots. We use the secondary metabolites of some flowering plants as medicines or mood-altering drugs.

LEARNING ROADMAP

Where you have been Fungi are eukaryotes, so their cells have the features you learned about in Section 4.5. Fungi have a cell wall of chitin (Section 3.4), and some are fermenters that assist in the production of bread and alcoholic beverages (7.6). We mentioned lichens in our discussion of cyanobacteria (21.10) and will expand on the information here.

Where you are now

Traits and Classification
Fungi are heterotrophs that secrete enzymes onto organic matter, then absorb the released nutrients. They reproduce sexually and asexually by producing spores.

The Oldest Lineages
Chytrids are an ancient lineage of flagellated fungi. The zygote fungi include diverse molds. The related microsporidia live as parasites inside animal cells.

Sac Fungi and Club Fungi
Many sac fungi and club fungi make complex spore-bearing structures such as mushrooms. Meiosis in cells on these structures produces spores.

Living Together
Many fungi live on, in, or with other species. Some live inside plant roots and benefit their host plant. Others form lichens by living with algae or cyanobacteria.

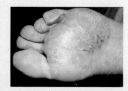

Fungal Pathogens and Toxins
A minority of fungi are parasites, and some of these species cause disease in humans. Fungi also make toxins that can be deadly when eaten.

Where you are going This chapter considers fungus–plant interaction from the fungal point of view. In Section 28.3 we focus on how these interactions affect plant nutrition. Fungal pathogens of humans come up once again when we discuss symptoms of AIDS (Section 37.11). The role of fungi as decomposers is covered in our discussion of nutrient cycling and food webs (Sections 46.2 and 46.3).

23.1 High-Flying Fungi

Fungi are not known for their mobility. You probably don't think of mushrooms and their relatives as world travelers, but some do get around. Fungi produce microscopic spores that can lodge in crevices on tiny dust particles. When winds lift these particles aloft, spores go along for the ride. Dustborne fungal spores can disperse long distances riding winds that swirl high above Earth's surface.

Dust storms in North African deserts sometimes lift fungus-laden particles more than 4.5 kilometers (3 miles) above the desert floor. Winds carry this dust out over the Atlantic Ocean, and sometimes completely across it. Most fungal spores that hitchhike on African dust are harmless, but fungi that cause plant disease occasionally make the journey. For example, winds of a 1978 cyclone introduced sugar cane rust (a fungal disease) from Cameroon to the Dominican Republic. Similarly, winds probably carried coffee rust fungus from Angola to Brazil in 1980.

Today, an African outbreak of an old fungal foe has agricultural officials around the world on edge. The fungus that arouses their concern, *Puccinia graminis*, causes wheat stem rust disease. Like other rust fungi, *P. graminis* is an obligate plant parasite, meaning it can grow and reproduce only in living plant tissue. An infection begins when a spore lands on the leaf of a wheat plant. The spore germinates, and a fungal filament enters the plant through a stoma. As fungal filaments grow through the plant's tissues, they suck up photosynthetic sugars that the plant would normally use to meet its own needs. As a result, an affected plant is stunted and produces little or no wheat. About a week after infection, ten of thousands of rust-colored spores appear at the surface of the infected plant's stem (**Figure 23.1A,B**). Each spore can disperse and infect a new plant.

Wheat stem rust disease routinely decimated crops worldwide until the 1960s, when strains resistant to *P. graminis* became available. The fungus-resistant strains were the product of a plant breeding program headed by Norman Borlaug. The importance of Borlaug's work was highlighted in 1970, when he received the Nobel Peace Prize for his role in preventing food shortages that can contribute to global instability.

Worldwide use of rust-resistant wheats provided a respite from outbreaks of wheat stem rust for decades. Then, in 1999, a new strain of wheat stem rust (called Ug99) was discovered in Uganda, a country in eastern Africa. Ug99 had mutations that allowed it to infect most wheat varieties that were previously considered resistant to wheat stem rust.

By 2009, windblown spores of Ug99 had dispersed to Kenya, Ethiopia, and Sudan, crossed the Red Sea to Yemen, and from there crossed the Persian Gulf to Iran. India, the world's second largest wheat producer, is expected to be affected soon. Most likely, winds will eventually distribute Ug99 throughout the world. Fungicides can help minimize the damage, but are too expensive for farmers in many developing nations.

With one of the world's most important crops at risk, plant scientists set to work (**Figure 23.1C**). They hope to develop new varieties of wheat with genes that will provide resistance against Ug99. Scientists have already bred some resistant wheats. The challenge is multiplying enough seed and deploying it to farmers who currently grow susceptible varieties. Wheat is not of commercial interest to private seed companies, because farmers save their own seed and resow it.

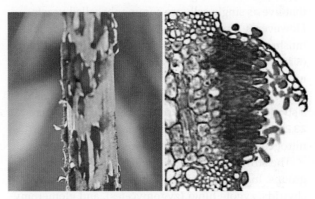

A Wheat stem with rust-colored fungal sporangia on its surface. **B** Micrograph of sporangium releasing spores.

C Plant breeders Peter Njau, Godwin Macharia, and Davinder Singh (*left to right*) are part of an international team of scientists racing to develop and test wheat varieties that resist the Ug99 strain of wheat stem rust.

Figure 23.1 Wheat stem rust, a fungal threat to global food supplies.

23.2 Fungal Traits and Classification

- Fungi are heterotrophs that obtain nutrition by extracellular digestion, and disperse by producing spores.
- Links to Nutrient cycle 1.3, Carbohydrates 3.4

Structure and Function

Like plants, **fungi** have walled cells, spend their lives fixed in place, and produce haploid spores by meiosis. But fungi are more closely related to animals than to plants. Like animals, fungi are heterotrophs that store excess sugars as glycogen. Fungi feed by secreting digestive enzymes onto organic material and absorbing the resulting breakdown products, a mode of nutrition we call extracellular digestion and absorption. Most fungi are **saprobes**, meaning they grow and feed on the remains of other organisms. A lesser number live on or in organisms that are still alive. Some, such as wheat stem rust, are parasites. In other cases, the fungus benefits its host or has no effect.

Fungi may be single-celled or multicelled. Those that live as single cells are commonly called yeasts. However, a typical fungus is multicelled. Molds and mushrooms are familiar examples of multicelled fungi (**Figure 23.2A,B**). They grow as a **mycelium** (plural, mycelia), a network of microscopic interwoven filaments. Each filament, or **hypha** (plural, hyphae), is a strand of walled cells arranged end to end (**Figure 23.2C**). Fungal cell walls consist largely of chitin, a nitrogen-containing polysaccharide (Section 3.4).

The structure of fungal hyphae varies among groups. In the oldest fungal lineages, which include chytrids, zygote fungi (zygomycetes), and glomeromycetes, cells of a hypha do not have cross-walls between them (**Table 23.1**). Each hypha is a long tube full of

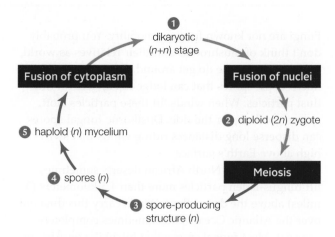

Figure 23.3 Sexual phase of the fungal life cycle.

cytoplasm and nuclei. Hyphae divided into compartments by cross-walls, or septae (singular, septa), evolved in the common ancestor of sac fungi and club fungi. These are the two most diverse lineages, and septate hyphae contributed to their success. The presence of cross-walls made hyphae sturdier, allowing the evolution of larger, more elaborate spore-producing bodies. Septate hyphae also are more resistant to desiccation than nonseptate hyphae.

Life Cycles

Fungi reproduce both asexually and sexually by making spores. Asexually produced spores form by mitosis in sporangia (spore-forming chambers) at the tips of specialized haploid hyphae.

Sexual reproduction begins with cytoplasmic fusion of haploid hyphae to produce a dikaryotic stage (**Figure 23.3 ❶**). **Dikaryotic** means having two genetically distinct types of nuclei (*n+n*) in each cell. Fusion of these two haploid nuclei produces a diploid zygote ❷ that undergoes meiosis, restoring the haploid state. A spore-bearing structure forms ❸ and releases haploid

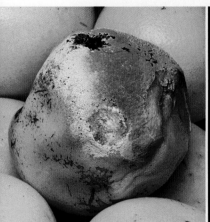  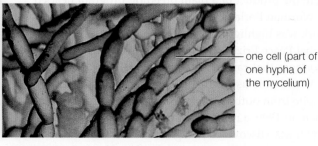

one cell (part of one hypha of the mycelium)

A Green mold (*Penicillium digitatum*) growing on a grapefruit.

B Scarlet hood mushroom (*Hygrophorus*) growing on the forest floor in Virginia.

C Molds and mushrooms grow as a mycelium, a mass of threadlike hyphae. Each hypha is a strand of cells attached one to the other.

Figure 23.2 Multicelled fungi.

Table 23.1 Major Groups of Fungi

Groups with aseptate hyphae composed of haploid cells:

Chytrids

1,000 species. Make spores asexually and sexually. Spores and gametes flagellated. Live in seawater, fresh water, moist soil, and in or on other organisms.

Zygote Fungi (Zygomycetes)

1,100 species. Make spores asexually and sexually. Live in soil, and inside or on other organisms. Some species are human pathogens.

Glomeromycetes

150 species. Not known to reproduce sexually. All live inside plant roots without harming the plant.

Groups with septate hyphae; each hypha may be haploid or dikaryotic:

Sac Fungi (Ascomycetes)

More than 32,000 species. Make spores asexually and sexually. Live in soil, and inside or on other organisms. Some are human pathogens. Some partner with photosynthetic cells and form lichens.

Club Fungi (Basidiomycetes)

More than 26,000 species. Make spores asexually and sexually. Include species with the largest and most complex spore-bearing structures. Live in soil, and inside or on other organisms.

---

spores ❹. The spores germinate and grow into a new haploid mycelium ❺.

In zygote fungi, fusion of the nuclei occurs immediately after cytoplasmic fusion. In sac fungi and club fungi, cytoplasmic fusion is separated from nuclear fusion by an interval of mitosis that produces dikaryotic hyphae. Nuclei fuse and spores form at the tips of dikaryotic cells that are part of a multicelled fruiting body (spore-producing body).

dikaryotic Having two genetically different nuclei in a cell (*n+n*).
fungus Eukaryotic heterotroph with cell walls of chitin; obtains nutrients by extracellular digestion and absorption.
hypha Component of a fungal mycelium; a filament made up of cells arranged end to end.
mycelium Mass of threadlike filaments (hyphae) that make up the body of a multicelled fungus.
saprobe Organisms that feeds on organic wastes and remains.

Take-Home Message

What are characteristics of fungi?

» Fungi are heterotrophs that absorb nutrients from their environment. Some live as single cells; others live as a multicelled mycelium. They disperse by producing spores.

23.3 Flagellated Fungi

■ Chytrids are the only modern fungi with a life cycle that includes flagellated cells.
■ Link to Flagella 4.10

Chytrids include several ancient fungal lineages. They are the only fungi that produce flagellated spores (zoospores). Like animal sperm, chytrids have a single flagellum at the rear of the cell that pushes the cell forward. Chytrids survive in a variety of habitats. Most are soil-dwelling decomposers, and are particularly abundant in some high-elevation soils. Other chytrids live in the gut of mammalian grazers such as sheep and cattle. They assist their hosts by breaking down cellulose. Still others are parasites.

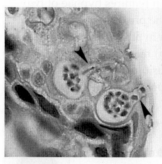 

A Cross-section of infected frog skin; flask-shaped sporangia are indicated by arrows.

B Harlequin frog, one of the species that has been decimated by chytridiomycosis.

Figure 23.4 Frog-killing chytrid (*Batrachochytrium dendrobatidis*).

Many amphibian populations are currently threatened by a parasitic chytrid (*Batrachochytrium dendrobatidis*) that causes chytridiomycosis (**Figure 23.4**). A **mycosis** is a disease caused by a fungus. *B. dendrobatidis* was first discovered in the late 1990s in declining frog populations in Australia and Central America. Since then, the fungus has been linked to deaths of wild frogs across the globe. Each outbreak of *B. dendrobatididis* has far-reaching ecological effects because frogs help control populations of insects and also serve as food for many other animals.

chytrid Fungus with flagellated spores.
mycosis Disease caused by a fungus.

Take-Home Message

What are chytrids?

» Chytrid are fungi that produce flagellated spores. Most are decomposers in soil, but some live in the animal gut, and some are parasites. A parasitic species that kills amphibians is a matter of concern.

23.4 Zygote Fungi and Related Groups

■ Zygote fungi form a branching haploid mycelium on organic material, and inside living plants and animals.

■ Link to Cell division mechanisms 11.2

Zygote Fungi

The 1,100 or so species of **zygote fungi** (zygomycetes) live in damp places. Many are molds, meaning they typically grow over organic matter as a mass of asexually reproducing hyphae. As noted in the Section 23.2, the hyphae of zygote fungi consist of haploid cells without septae between them.

Black bread mold (*Rhizopus stolonifer*) has a typical life cycle (**Figure 23.5**). As long as food is plentiful, a haploid mycelium grows and produces spores asexually ❶. Sexual reproduction requires hyphae of two different mating strains. The proximity of a potential sexual partner causes both hyphae to develop specialized branches (gametangia) that grow toward one another ❷. Cytoplasmic fusion of the gametangia produces a young zygospore that is dikaryotic (it

Figure 23.6 Spore-bearing structures of *Pilobolus*. The name means "hat-thrower." The dark "hats" are sporangia (spore sacs).

contains nuclei from both parents) ❸. As the zygospore matures, the nuclei pair up and fuse. A mature zygospore contains diploid cells and is surrounded by a thick, protective wall ❹. Meiosis occurs as the zygospore germinates. One or more hyphae emerge, each bearing a sporangium with haploid spores ❺.

Many *Rhizopus* species spoil foods, but some are used to produce tempeh, a high-protein fermented soy product with a chewy, meatlike consistency.

A relative of *Rhizopus* has an interesting mechanism of spore dispersal. Grazing horses, cattle, elk, and similar animals ingest *Pilobolus* spores along with grass. The spores pass through the animal's gut and are deposited in feces. They germinate and grow into a mycelium that produces spore-bearing hyphae (**Figure 23.6**). At the tip of each hypha, a tiny dark sporangium perches atop a large fluid-filled vesicle. When fluid pressure builds up in the vesicle, it bursts, hurling the sporangium as far as 2 meters (about 6.5 feet) away.

Pilobolus does not directly harm grazers, but it does facilitate the spread of lungworm, a nematode parasite that infects them. Lungworm larvae excreted by infected grazers often end up in feces alongside *Pilobolus*. The larvae take advantage of the fungus's dispersal ability by climbing to the top of a hypha and hitching a ride inside a sporangium.

Microsporidia—Intracellular Parasites

Microsporidia (single, microsporidium) include about 1,300 named species of single-celled intracellular parasites. Most infect animals, with fish and insects being the most common hosts. Microsporidia were long considered protists, but gene comparisons reveal that they are most likely relatives of the zygote fungi.

A microsporidium spore has a tough coat that allows it to survive adverse conditions for years.

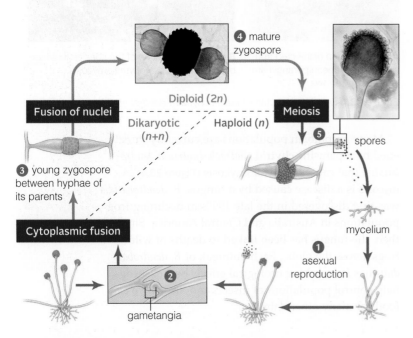

Figure 23.5 Animated Life cycle of *Rhizopus stolonifer*, a black bread mold.

❶ A haploid mycelium grows in size and reproduces asexually by mitotic production of spores in sporangia atop specialized hyphae.

❷ When hyphae of different mating strains (+ and −) come into close proximity, they produce branches (gametangia) that grow toward one another.

❸ Cytoplasmic fusion of the gametangia produces a dikaryotic zygospore, a structure that contains haploid nuclei from both parents.

❹ Fusion of nuclei produces a mature zygospore containing diploid cells.

❺ The zygospore undergoes meiosis, germinates, and produces an aerial hypha with haploid spores at its tip.

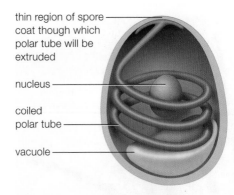

thin region of spore coat though which polar tube will be extruded

nucleus

coiled polar tube

vacuole

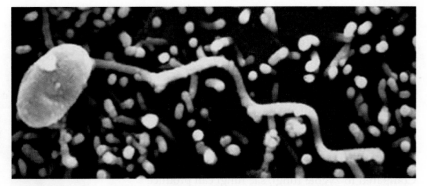

A Diagram of a microsporidian spore.

B Microsporidium with its polar tube extruded (scanning electron micrograph).

Figure 23.7 A microsporidium. Microsporidia are single-celled intracellular parasites of animals.

Beneath the coat, a long tube lies coiled in the cytoplasm (**Figure 23.7**). During infection, the tube uncoils and perforates a host cell. The microsporidium's nucleus and cytoplasm then flow through the tube into the host.

At least fourteen species of microsporidia infect humans, but they generally cause symptoms only in those with a weakened immune system. The intestine is the most frequently affected site. An infection by microsporidia breaks down the intestinal lining and causes diarrhea, abdominal cramping, and nausea.

One microsporidium (*Nosema ceranae*) has recently been implicated in honeybee colony collapse disorder (CCD). Bees in affected colonies decline suddenly in number, and colonies often die off completely. After examining many affected and healthy colonies, researchers concluded that colony collapse disorder can arise as a result of simultaneous infection by *N. ceranae* and a specific virus. Neither pathogen alone is enough to kill a colony, but in concert, they are deadly.

Because honeybees pollinate many crop plants, an ongoing epidemic of CCD has been an economic disaster. Understanding the role of *N. ceranae* will help prevent and control new outbreaks. Diseases caused by microsporidia can be treated with antifungal drugs.

Glomeromycetes—Partners of Plants

Glomeromycetes were previously placed with zygote fungi, but are now considered a separate group. All 150 or so species take part in plant root–fungus partnerships called a **mycorrhiza** (plural, mycorrhizae). A glomeromycete hypha grows into a root and branches inside the wall of a root cell (**Figure 23.8**). Having a fungal roommate does not harm a root cell; the fungus shares nutrients from the soil with its host. We return to fungus–plant associations in Section 23.7.

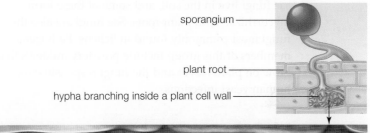

sporangium

plant root

hypha branching inside a plant cell wall

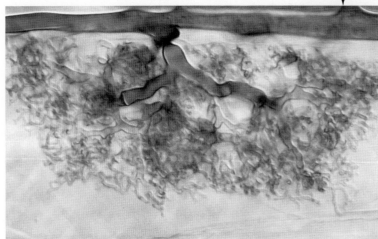

Figure 23.8 Glomeromycete hypha branching inside the cell of a plant root.

glomeromycete Fungus with hyphae that grow inside the wall of a plant root cell.
microsporidium Single-celled spore-forming fungus that is an intracellular animal parasite.
mycorrhiza Mutually beneficial partnership between a fungus and a plant root.
zygote fungus Fungus that forms a zygospore during sexual reproduction.

Take-Home Message

What are zygote fungi and their relatives?

» Zygote fungi form a thick-walled diploid spore when they reproduce sexually. Some spoil food or cause disease.

» Microsporidia are single-celled parasites that invade an animal cell by way of a polar tube.

» Glomeromycetes are mycorrhizal fungi; they partner with plants.

23.5 Sac Fungi—Ascomycetes

- Sac fungi are the most diverse fungal group. There are single-celled and multicelled forms.
- Link to Fermentation 7.6

Sac fungi, or ascomycetes, are the most diverse fungal lineage, with more than 32,000 named species. Some are yeasts (single-celled), but most are multicelled. Hyphae of multicelled species have cross-walls at regular intervals. The walls are porous so material can still flow from one cell of a hypha to another. Compared to zygote fungi, sac fungi can produce larger and more elaborate multicelled structures, and they are better able to survive in dry places.

Sac fungi have a variety of ecological roles. Most sac fungi live in the soil, and some of these form mycorrhizae with plant roots. Sac fungi are also the fungi most commonly found in lichens. Pathogenic members of this group include powdery mildews that grow on plant leaves and the fungi responsible for human yeast infections.

Life Cycle

Sac fungi commonly reproduce by asexual mechanisms. Yeasts often reproduce asexually by **budding**, a process in which mitosis is followed by unequal cytoplasmic division. During budding, a descendent cell with a small amount of cytoplasm buds from its parent (**Figure 23.9A**). Multicelled sac fungi grow as a haploid mycelium that produces spores by mitosis at the tips of specialized hyphae (**Figure 23.9B**). Mitotically produced spores of a sac fungus are called conidiospores or conidia. Conidia is Greek for dust.

haploid spore in ascus

Figure 23.10 Ascocarp and ascospores of the scarlet cup fungus. In this species, meiosis followed by one round of mitosis gives rise to eight haploid spores in each ascus.

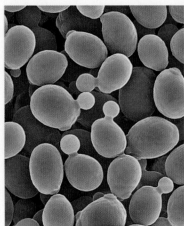
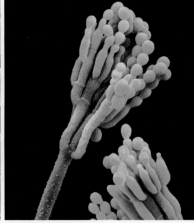

A New cells of *Candida albicans* bud from a parental cell.

B Conidia (asexually produced spores) of *Eupenicillium*.

Figure 23.9 Asexual reproduction in sac fungi.

Sac fungi can also reproduce sexually. As in zygote fungi, sexual reproduction of a multicelled species begins when two haploid hyphae meet and their cells undergo cytoplasmic fusion. In sac fungi, the resulting dikaryotic cell divides by mitosis to produce dikaryotic hyphae. These hyphae intertwine with haploid hyphae to form a spore-producing structure called an ascocarp. Spores form by meiosis inside a saclike cell called an ascus (plural, asci) that develops on the ascocarp (**Figures 23.10** and **23.11**). Meiosis of the ascus produces four spores that, in some groups, divide by mitosis to yield eight.

Human Uses of Sac Fungi

Of all fungal groups, sac fungi have the greatest number of commercial uses. Some are eaten or used to produce foods, and still others produce medicines.

A Morels. Asci line pits on the upper part of the fruiting body.

B Truffles. These ascocarps form underground; asci are inside.

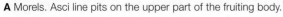
Figure 23.11 Two types of edible ascocarp.

Morels (**Figure 23.11A**) and truffles (**Figure 23.11B**) are two examples of edible ascocarps. Truffles form underground. When spores mature, the fungus gives off an odor like that emitted by an amorous male pig. Female pigs that catch a whiff of this scent disperse truffle spores as they root through the soil in search of a seemingly subterranean suitor. Dogs can also be trained to snuffle out truffles. Searching for truffles can be lucrative. In 2006, a single 1.5-kilogram (about 3-pound) Italian truffle sold for $160,000.

Fermentation reactions of sac fungi help us produce a variety of foods and beverages. A packet of baker's yeast holds spores of *Saccharomyces cerevisiae*. When the spores germinate in bread dough, their metabolic reactions produce carbon dioxide that causes the dough to rise. A different strain of *S. cerevisiae* helps produce beer and wine. One *Aspergillus* species ferments soybeans and wheat in the production of soy sauce and another makes citric acid used as a preservative and to flavor soft drinks. *Penicillium roqueforti* adds tangy blue veins to cheeses such as Roquefort.

Sac fungi have also provided us with some useful medications. Most famously, the initial source of the antibiotic penicillin was the soil fungus *Penicillium chrysogenum*. Another antibiotic, cephalosporin, was first isolated from *Cephalosporium*. Statins from *Aspergillus* help lower cholesterol levels, and cyclosporin from *Trichoderma* helps prevent the rejection of transplanted organs.

Sac fungi that infect plant or animal pests can be used as natural herbicides or pesticides. *Arthrobotrys* provides an example. This predatory sac fungus makes special hyphae with loops that tighten and trap roundworms (**Figure 23.12**). After feeding on a worm, the fungus makes asexual spores.

part of one hypha that forms a nooselike ring   roundworm

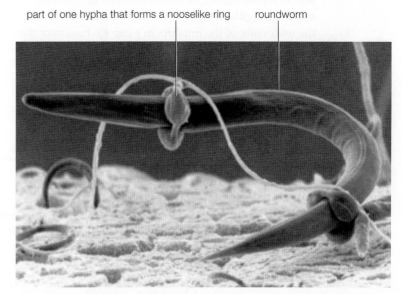

Figure 23.12 A predatory fungus (*Arthrobotrys*) that captures and feeds on roundworms. Rings that form on the hyphae constrict and entrap the worms, then hyphae grow into the captive and digest it.

budding Mechanism of asexual reproduction by which a small cell forms on a parent, then is released.
sac fungi Most diverse group of fungi; sexual reproduction produces spores inside a saclike structure (an ascus).

Take-Home Message

What are sac fungi?

» Sac fungi are the most diverse group of fungi. Some are yeasts, but a haploid mycelium dominates the life cycle of most. Sac fungi that reproduce sexually typically form spores inside an ascus. Yeasts reproduce asexually by budding, and multicelled species by formation of conidia.

» We use sac fungi as sources of food and beverages, as drugs, and as control agents for pest organisms.

23.6 Club Fungi—Basidiomycetes

- Club fungi make the largest and most elaborate fruiting bodies; some familiar mushrooms are examples.
- Link to Plant adaptations to life on land 22.3

Life Cycle

Club fungi seldom reproduce asexually. They produce a spore-bearing structure (a basidiocarp) composed of dikaryotic hyphae. Spores form by meiosis inside club-shaped cells at the tips of these hyphae.

The button mushrooms common in markets are basidiocarps of a club fungus. Haploid hyphae of this fungus grow underground. When hyphae of two different mating strains meet, they fuse, forming a dikaryotic mycelium (**Figure 23.13 ❶**). The mycelium grows through the soil and, when conditions favor sexual reproduction, forms mushrooms ❷. Thin tissue sheets (gills) fringed with club-shaped cells line the underside of the mushroom's cap ❸. Fusion of the nuclei in these dikaryotic cells forms a diploid zygote ❹. The zygote undergoes meiosis, forming four haploid spores ❺. These spores are dispersed by wind, germinate, and start a new cycle ❻.

Ecology and Diversity

There are more than 26,000 named species of club fungi. Many play an essential role as decomposers in forests. Club fungi are the only group capable of breaking down the lignin that makes up the bulk of wood. As a result, these fungi are also important pathogens in forests. The shelf fungus shown in **Figure 23.14A** is feeding on its woody host. Hyphae that extend into the tree's heartwood weaken it, and will eventually kill it.

The honey mushroom (*Armillaria ostoyae*) is another forest species. It helps break down stumps and logs, but it also attacks and kills living trees. In one Oregon forest, the mycelium of a single honey mushroom extends through more than 2,000 acres of soil. This individual fungus has been growing for an estimated 2,400 years.

Jelly fungi (**Figure 23.14B**) also live in forests. These fungi form a gelatinous fruiting body on logs and tree trunks, but they do not feed on wood. Rather, they are parasites that tap into the mycelia of other fungi that feed on wood.

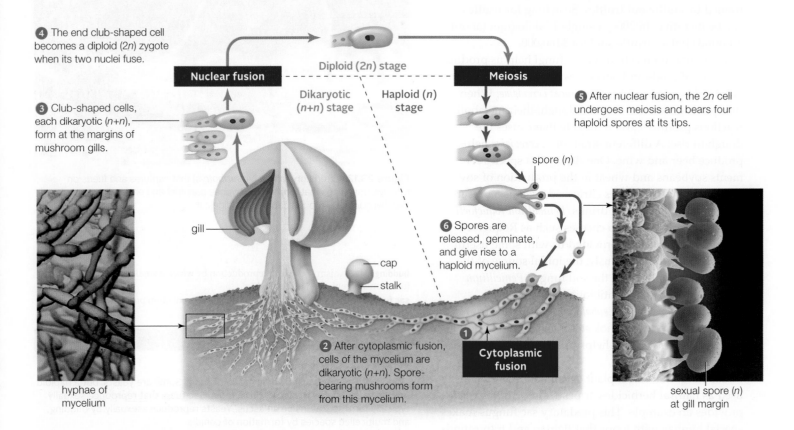

❹ The end club-shaped cell becomes a diploid (2n) zygote when its two nuclei fuse.

Nuclear fusion

Diploid (2n) stage

Meiosis

Dikaryotic (n+n) stage

Haploid (n) stage

❺ After nuclear fusion, the 2n cell undergoes meiosis and bears four haploid spores at its tips.

❸ Club-shaped cells, each dikaryotic (n+n), form at the margins of mushroom gills.

spore (n)

gill

❻ Spores are released, germinate, and give rise to a haploid mycelium.

cap

stalk

❶

❷ After cytoplasmic fusion, cells of the mycelium are dikaryotic (n+n). Spore-bearing mushrooms form from this mycelium.

Cytoplasmic fusion

hyphae of mycelium

sexual spore (n) at gill margin

Figure 23.13 Animated Life cycle of a club fungus having two mating strains of hyphae.

Figure It Out: What are the blue and red dots in this figure? Answer: Genetically different nuclei

A Sulfur shelf fungus   **B** Jelly fungus   **C** Chanterelles   **D** Giant puffball

Figure 23.14 Club fungus fruiting bodies (basidiocarps).

Chanterelles (**Figure 23.14C**) are edible forest mushrooms with a vaselike shape. They partner with tree roots as mycorrhizae and are common in California and the Pacific Northwest.

Puffballs produce some of the largest fruiting bodies (**Figure 23.14D**). They can be as large as a meter across. When young, they have a white, edible flesh. When mature, their covering turns dark and splits, allowing spores to escape.

Stinkhorns elicit help from insects in dispersing their spores. These fungi have a phallic shape and an aroma like rotting meat or feces (**Figure 23.14E**). Flies and beetles attracted by the scent alight on the tip of the fungus, which is covered with spore-laden slime. The insects inadvertently pick up spores and distribute them to new locations.

Though the stinkhorn is extremely unappetizing to humans, it is not poisonous. Many other mushrooms are. Toxins help these species fend off predators. Each year thousands of people become ill after eating poisonous mushrooms that they mistook for an edible species. For instance, untrained foragers sometimes mistake a newly emerged *Amanita phalloides*—the death cap mushroom—for a puffball. Only later does the distinctive cap form (**Figure 23.14F**). Eating an *Amanita* species can cause nausea and abdominal cramping, followed by liver and kidney failure, and even death.

E Stinkhorn   **F** Death cap mushroom

Other toxins made by club fungi alter mood and cause hallucinations. The drug LSD is a compound that was initially isolated from a club fungus. Mushrooms that make psilocybin are eaten for their mind-altering effects. Both psilocybin-containing mushrooms and LSD are illegal in the United States.

club fungi Fungi that have septate hyphae and produce spores by meiosis in club-shaped cells.

Take-Home Message

What are club fungi?

» Club fungi are a group of mostly multicelled fungi. Multicelled species spend most of their life cycle as a dikaryotic mycelium and have the largest, most complex spore-bearing structures of all fungi.

» Club fungi are the only fungi that can break down lignin, and they are important in forests both as decomposers and as pathogens of trees.

23.7 Fungal Partners and Pathogens

■ Fungi directly affect other species as mutualistic partners and as pathogens.

■ Links to Cyanobacteria 20.7, Green algae 21.10

Fungal Partnerships

Lichens A lichen is a composite organism consisting of a sac fungus and cyanobacteria or green algae. The fungus makes up most of a lichen's mass (**Figure 23.15**). Photosynthetic cells that provide the fungus with sugars are held in place by hyphae. If these cells are cyanobacteria, they can also provide fixed nitrogen.

A lichen may be a **mutualism**, an interaction that benefits both partners. However, in some cases the fungus may be parasitically exploiting captive photosynthetic cells that would do better on their own.

Lichens disperse by fragmentation or by releasing packages that contain cells of both partners. In addition, the fungus can release spores. The fungus that

A Leaflike lichen   **B** Pendant lichen

C Encrusting lichen on granite

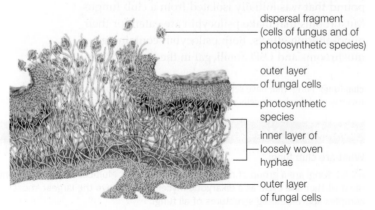

dispersal fragment (cells of fungus and of photosynthetic species)

outer layer of fungal cells

photosynthetic species

inner layer of loosely woven hyphae

outer layer of fungal cells

D A leaflike lichen, shown in cross-section (colorized SEM).

Figure 23.15 Lichen structure and internal organization.

germinates from a spore can form a new lichen only if it alights near an appropriate photosynthetic cell. This is not as unlikely as it may seem; the required algae and bacteria are common as free-living cells.

Lichens colonize places too hostile for most organisms, such as newly exposed bedrock. They break down rock by releasing acids and by holding water that freezes and thaws. When soil conditions improve, plants move in and take root. Long ago, lichens may have preceded plants onto land.

Mycorrhizae Many vascular plants partner with one of the glomeromycete fungi (described in Section 23.4). **Figure 23.8** shows how the hyphae of these fungi branch inside root cells. By contrast, hyphae of mycorrhizal sac fungi and club fungi form a mesh around roots (**Figure 23.16A**). Most forest mushrooms, including truffles and chanterelles, are fruiting bodies of mycorrhizal sac or club fungi.

Hyphae of all mycorrhizal fungi functionally increase the absorptive surface area of their partner. Fungal hyphae are thinner than the smallest roots and can easily grow between soil particles. The fungus takes up nutrients and shares them with the plant, and the plant gives up sugars to the fungus. It is a beneficial trade; many plants do poorly without mycorrhizae (**Figure 23.16B**).

Fungus-Farming Insects Some tropical ants, termites, and beetles raise a crop of fungi. Leaf cutter ants snip bits of leaf and bring it back to their colony, where a club fungus decomposes and feeds on it. Termites and ambrosia beetles tunnel into wood and grow specific fungi in the walls of their tunnels. In all cases, the insects cannot survive without the food produced by the fungi.

Fungal Pathogens

Plant Pathogens Most fungi decompose dead plant material, but some feed on living plants. As noted earlier, airborne spores spread wheat stem rust, a type of club fungus. Powdery mildew, a whitish powder that appears on leaves, is the spores of a sac fungus that feeds on leaf tissues. Fungi cause apple scab, peach leaf curl, and many other diseases of crop plants.

One notorious sac fungus, *Claviceps purpurea*, parasitizes rye and other cereal grains (**Figure 23.17A**). It produces toxic alkaloids that, when eaten, cause a type of food poisoning called ergotism. Symptoms include hallucinations, hysteria, and convulsions. Ergotism may have played a role in witch hunts that occurred in

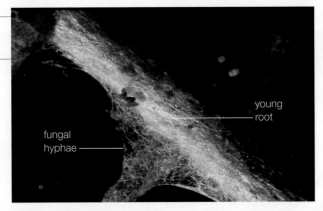

A Mycorrhiza formed by a fungus and root of a hemlock tree.

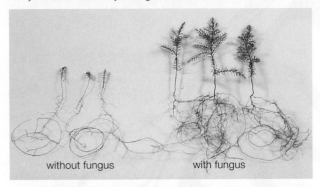

without fungus   with fungus

B Six-month-old juniper seedlings grown with or without mycorrhizal fungi in sterilized, phosphorus-poor soil.

Figure 23.16 Mycorrhiza. See also **Figure 23.8**.

Salem, Massachusetts. Descriptions of the behavior of the supposedly bewitched women resemble symptoms of ergotism.

A fungus that causes chestnut blight was accidentally introduced to the United States from Asia in the early 1900s, and greatly altered North America's eastern forests. Before the fungus arrived, the American chestnut was an important timber species, and its nuts nourished forest animals. Chestnut blight has since destroyed all but a few stands of mature trees.

Human Pathogens Human fungal infections most frequently involve body surfaces. Typically a fungus feeds on the outer layers of skin, secreting enzymes that dissolve keratin, the main skin protein. Infected areas become raised, red, and itchy. For example, several species of fungus infect skin between the toes and on the sole of the foot, causing "athlete's foot" (**Figure 23.17B**). Fungi also cause skin infections misleadingly known as "ringworm." No worm is involved. The ring-shaped lesion is caused by the growth of hyphae outward from the initial infection. Some single-celled fungi normally live in the vagina in low numbers. When they undergo a population explosion, the

lichen Composite organism consisting of a fungus and a single-celled alga or a cyanobacterium.
mutualism Mutually beneficial relationship between two species.

High-Flying Fungi (revisited)

David Schmale of Virginia Tech (*below*) uses remote-controlled model aircraft to sample fungal spores. Using this technique, he has collected spores of more than a dozen species of the sac fungus *Fusarium* from the air. Some members of this genus are plant pathogens, and a few can also infect people.

How would you vote? Different strains of *Fusarium* affect different plants. One strain kills opium poppies that are used to produce opium and heroin; another kills marijuana. Is spraying *Fusarium* spores a good way to reduce supplies of illegally grown drugs?

result is fungal vaginitis (a vaginal yeast infection). Symptoms include itching or burning sensations, a thick, odorless, whitish vaginal discharge, and pain during intercourse.

Fungi seldom cause systemwide disease in otherwise healthy people. However, infections can become life-threatening in people whose immune system is impaired by AIDS, chemotherapy, or drugs that must be taken after an organ transplant.

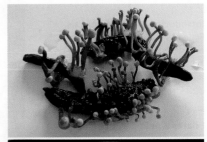

A Grains of rye with ascocarps of *Claviceps purpurea*. Eating grain infected by this species causes a type of food poisoning called ergotism.

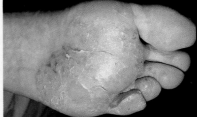

B Athlete's foot, a fungal infection of skin between the toes and on the sole of the foot. Fungi can also infect the skin around the toenails.

Figure 23.17 Signs of infection by pathogenic fungi.

Take-Home Message

What types of close relationships do fungi have with other species?

» Fungi form mutually beneficial partnerships with plants, cyanobacteria, and green algae. Some ants, termites, and beetles farm fungus and rely on them for food.

» Some fungi are pathogens that invade the tissues of plants and animals.

LEARNING ROADMAP

Where you have been Our discussion of animals explains their adaptive traits (Section 16.4) and the role of exaptation (17.12). We draw on earlier discussions of homeotic genes (10.3), comparative genomics (15.5), and patterns of development (18.5).

Where you are now

Animal Traits and Evolution
Animals are multicelled heterotrophs that move about. Early animals were small and structurally simple. More complex multilayered body plans and specialized parts evolved later.

Structurally Simple Groups
Sponges have no body symmetry or tissues. They filter food from the water. The radially symmetrical cnidarians such as jellies have two tissue layers and unique stinging cells.

Introducing Bilateral Animals
Flatworms are bilateral and have organs but do not have a body cavity. Annelid worms and the mollusks have a lined body cavity, and both develop from the same type of larva.

Bilateral Animals That Molt
Roundworms and arthropods molt as they grow. Arthropods, the most diverse animal group, include aquatic crustaceans as well as land-dwelling insects and spiders.

On the Road to Vertebrates
Echinoderms are on the same branch of the animal family tree as vertebrates. They are invertebrates with bilateral ancestors, but adults have a decidedly radial body plan.

Where you are going In Section 25.2, you will learn about invertebrate chordates, the closest relatives of vertebrates. We return to insects as pollinators in Section 29.1. Going forward, we look at invertebrate nervous systems (32.2), vision (33.5) and hormonal control of arthropod molting (34.13). We study the solute-regulating organs of invertebrates in Section 40.2, as well as their skeletal systems (35.2), circulation (36.2), respiration (38.3), and development (42.2). Social insects are the focus of Section 43.9.

24.1 Old Genes, New Drugs

East of Australia, small, reef-fringed islands dot the vast expanse of the South Pacific Ocean. Shelled animals abound in the warm water near their shores. Among them are more than 500 kinds of mollusks called cone snails (*Conus*). The snails have intricately patterned shells highly valued by collectors.

Cone snails also interest biologists. The snails are predators and, like some snakes, they produce venom that helps them subdue their prey. Studies of venoms made by fish-eating cone snails have led to the development of new drugs.

For example, *C. magus* captures a fish by harpooning it, pumping it full of venom, then reeling it in. The snail's venom contains a peptide that interferes with vertebrate pain perception. When injected into a fish, this peptide keeps the fish from writhing and damaging the snail's harpoon. An injectable pain-killing drug called ziconotide (sold under the brand name Prialt) is a synthetic version of the *C. magus* peptide.

C. geographus feeds in a different matter. The snail uses its wide mouth like a net to engulf a fish before injecting it with venom (**Figure 24.1**). A synthetic version of one peptide from this snail's venom is currently being tested as a treatment for epilepsy.

While studying *C. geographus*, University of Utah researchers found that a gene involved in venom synthesis has ancient roots. The gene encodes an enzyme called gamma-glutamyl carboxylase (GGC). Fruit flies and humans have a gene with a similar sequence, so the gene must have originated before snails, insects, and vertebrates diverged from a common ancestor an estimated 500 million years ago. Over time the gene mutated independently in each lineage and its product took on different functions. GGC allows cone snails to make venom and functions in the clotting of human blood. We have yet to figure out what the enzyme does in fruit flies.

This example supports an organizing principle in the study of life. Look back through time, and you discover that all organisms are related. At each branch point in the animal family tree, mutations gave rise to changes in biochemistry, body plans, or behavior. The mutations were the source of unique traits that help define each lineage.

Cone snails are among the **invertebrates**, animals that do not have a backbone. The next chapter continues the story, focusing on vertebrates and their closest invertebrate relatives. Even with two chapters, we can scarcely do justice to so many animal species. We can only consider the scope of diversity and the major evolutionary trends. Keep in mind that although invertebrates are structurally simpler than vertebrates, they are not evolutionarily inferior. The diversity and longevity of the invertebrate lineages attests to how well they are adapted to their environment. About 95 percent of animals are invertebrates.

invertebrate Animal that does not have a backbone.

Figure 24.1 A cone snail, *Conus geographus*, engulfing a fish with its funnel-shaped mouth. When surrounded by the snail's mouth, a fish becomes passive, suggesting that the snail releases some sort of sedative into the water. The snail then harpoons the fish and injects a painkilling, paralyzing venom into it, before swallowing it whole.

The small tube above the mouth is the snail's siphon, through which it takes in water for respiration. The siphon also has sensory receptors that allow the snail to detect the scent of its prey.

24.2 Animal Traits and Body Plans

- All animals are multicelled heterotrophs.
- Animal body plans vary.
- Links to Tissues and organs 1.2, Surface-to-volume ratio 4.2

What Is an Animal?

Animals are multicelled heterotrophs that are motile (move from place to place) during part or all of the life cycle. Their body cells do not have a wall and are typically diploid. Gametes are the only haploid stage in an animal life cycle. **Figure 24.2** shows the relationships among the major animal phyla covered in this book.

Variation in Animal Body Plans

Organization The most ancient animal lineages such as the sponges are organized as aggregations of cells. Later, in the common ancestor of most animals, cells became organized as tissues. A tissue consists of one or more cell types organized in a way that allows them to collectively perform a specific function.

Tissue formation begins in an embryo. Embryos of some animal lineages have two tissue layers: an outer ectoderm and an inner endoderm. Jellyfish and their relatives have this type of organization. A middle embryonic layer, called mesoderm, evolved in later lineages. Evolution of mesoderm allowed an increase in the structural complexity of bodies. Most animal groups have many organs derived from mesoderm.

Body Symmetry The simplest animals such as sponges are asymmetrical, meaning you cannot divide their body into halves that are mirror images.

Some animals such as jellies and sea anemones have **radial symmetry**; their body parts are repeated around a central axis, like the spokes of a wheel. An animal with a radially symmetrical body plan typically has distinctive upper (dorsal) and lower (ventral) surfaces, but no obvious front or back. These animals typically live fixed in place or drift through the water, so food and threats can arrive from any direction.

Most animals have **bilateral symmetry**; many parts are paired, with one on each side of the body. Bilateral animals have distinctive anterior (front) and posterior (back) ends, as well as dorsal and ventral surfaces. Bilateral animals have undergone **cephalization**; nerve cells and sensory cells have become concentrated at their anterior end. In some lineages, this concentration of cells evolved into a brain.

Gut and Body Cavity Most animals have a **gut**: a digestive sac or tube that opens at the body surface. Flatworms and cnidarians have a saclike gut, also called an incomplete digestive system. Their food enters and wastes leave through the same body opening. Most bilateral animals have a tubular gut, also known as a complete digestive system. With this system, food enters through a mouth at one end of the tube and wastes leave via an anus at the other end.

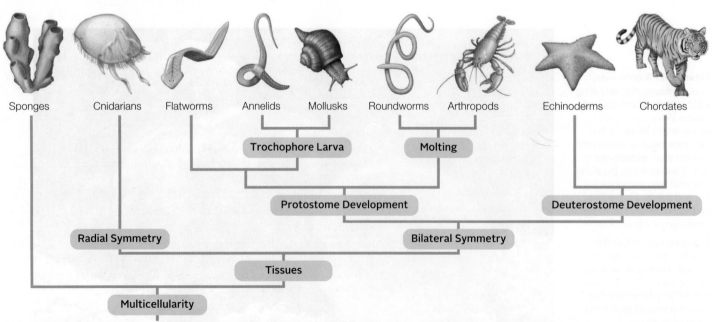

Sponges — Cnidarians — Flatworms — Annelids — Mollusks — Roundworms — Arthropods — Echinoderms — Chordates

Trochophore Larva — Molting — Protostome Development — Deuterostome Development — Radial Symmetry — Bilateral Symmetry — Tissues — Multicellularity

Figure 24.2 An evolutionary tree for major animal phyla. Like all such trees, it is a hypothesis and open to revision in light of new data.

Figure It Out: What invertebrate group is most closely related to chordates?

Answer: Echinoderms

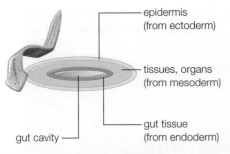

A Acoelomate body plan of a flatworm.

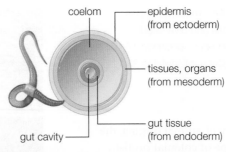

B Coelomate body plan of an earthworm.

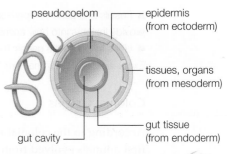

C Pseudocoelomate body plan of a roundworm.

Figure 24.3 Animated Variations in body plans among animals with three tissue layers. Graphics are schematics of a cross-section through the body. Relative width of tissue layers is not shown to scale.

Having a tubular gut has some advantages. Food and waste do not mix. In addition, regions of the tube can became specialized for taking in food, digesting it, absorbing nutrients, or compacting the wastes. Unlike a saclike gut, a tubular gut can carry out all of these digestive tasks simultaneously.

A mass of tissues and organs surrounds a flatworm's gut (**Figure 24.3A**). However, most animals have a "tube within a tube" body plan, in which a fluid-filled body cavity surrounds the gut. Typically this cavity is lined with tissue derived from mesoderm, and is called a **coelom**. Earthworms have this type of body plan (**Figure 24.3B**). A few invertebrates have a **pseudocoelom**, with an incomplete lining. Roundworms are an example (**Figure 24.3C**). Animals having a pseudocoelom were once classified together. However, gene comparisons indicate that this trait evolved independently in several lineages.

Having a fluid-filled coelom or pseudocoelom provides three advantages over an acoelomate body plan. First, nutrients and oxygen can diffuse through the coelomic fluid to cells deep inside the body. Second, muscles can redistribute the fluid to alter the body shape and aid locomotion. Third, organs within a fluid-filled cavity are not hemmed in by a mass of tissue, so they can grow larger and move more freely.

The two major lineages of bilateral animals differ in how their digestive system and coelom form. In the **protostomes**, the first opening that appears on the embryo becomes the mouth, and the second becomes the anus. In **deuterostomes**, the first opening develops into the anus and the second becomes the mouth.

Circulation In small animals, gases and nutrients diffuse through a body. However, diffusion alone cannot move substances fast enough through the body of a big animal to keep it alive. In most animals, a circulatory system speeds distribution of materials. In a closed circulatory system, a heart or hearts propel blood through a continuous system of vessels. Materials diffuse out of vessels and into cells. In an open circulatory system, blood leaves vessels and flows among tissues before returning to the heart. A closed circulatory system allows for faster blood flow and distribution of materials than an open one.

Segmentation Segmentation is common in bilateral animals, meaning similar units are repeated along the length of the body. Segmentation opened the way to evolutionary innovations in body form. When many segments have structures that carry out the same function, some segments can become modified without compromising the body's ability to function.

animals Multicelled heterotrophs with unwalled cells; are motile during part or all of the life cycle.
bilateral symmetry Having paired structures so the right and left halves are mirror images.
cephalization A concentration of nerve and sensory cells at the head end.
coelom Of many animals, a body cavity that surrounds the gut and is lined with tissue derived from mesoderm.
deuterostomes Lineage of bilateral animals in which the second opening on the embryo surface develops into a mouth.
gut A sac or tube in which food is digested.
protostomes Lineage of bilateral animals in which the first opening on the embryo surface develops into a mouth.
pseudocoelom Incompletely lined body cavity of some animals.
radial symmetry Having parts arranged around a central axis, like spokes around a wheel.

Take-Home Message

What are animals?

» Animals are multicelled heterotrophs with bodies made of unwalled, typically diploid cells. Animals are motile for part or all of their life.

» Early animals had an asymmetrical body plan. Later, radially symmetrical and then bilaterally symmetrical bodies evolved.

» Most animals with bilateral symmetry have a fluid-filled coelom.

» The two lineages of bilaterally symmetrical animals, protostomes and deuterostomes, differ in the details of their development, but both develop from an embryo that has three tissue layers.

24.3 Animal Origins and Adaptive Radiation

■ Fossils and gene comparisons among modern species provide insights into how animals arose and diversified.
■ Links to Homeotic genes 10.3, Allopatric speciation 17.10, Exaptation 17.12, Biomarkers 19.5, Choanoflagellates 21.11

Colonial Origins

According to the colonial theory of animal origins, the first animals evolved from a type of colonial protist. At first, all cells in the colony were similar. Each could reproduce and carry out all other essential tasks. Later, mutations resulted in cells that were better at some tasks but did not carry out others. Perhaps these cells captured food more efficiently, but did not reproduce. Colonies that had such interdependent cells and a division of labor were at a selective advantage, and new specialized cell types evolved. Eventually this process produced the first animal.

Modern choanoflagellates, the protists most closely related to modern animals (Section 21.11), give us an idea of what the protist ancestral to animals may have been like. Choanoflagellates resemble sponge cells in their structure and, like sponges, they filter food from the water. Choanoflagellates also have proteins similar to animal adhesion proteins and signaling proteins. Adhesion proteins may help colonial choanoflagellates stick together as a colony. Proteins like those used in animal signaling pathways may help the protists detect molecules associated with food or pathogens. As Section 17.12 explained, by the process of exaptation, traits that evolve in one context are altered over time and become adaptive in a somewhat different or entirely different context in a descendant group.

Early Signs of Animal Life

The earliest evidence of animal life was discovered in rocks found during an oil-drilling project in Oman. The sedimentary rocks, which were laid down as seafloor more than 635 million years ago, contain a biomarker characteristic of sponges. As Section 19.5 explained, a biomarker is a distinctive molecule produced only by a particular type of cell or lineage.

Fossilized bodies show that by 570 million years ago, a diverse collection of multicelled organisms were living in the seas (**Figure 24.4A**). These organisms are collectively referred to as the Ediacarans because their fossils were first discovered in Australia's Ediacara Hills. Most of the Ediacarans have no living descendants, but some are thought to be early representatives of sponges, arthropods (**Figure 24.4B**), or other invertebrate groups.

Figure 24.4 Fossil animals. (**A**) *Spriggina*, an Ediacaran, lived about 570 million years ago. It was about 3 centimeters (1 inch) long. By one hypothesis, it was a soft-bodied ancestor of arthropods, such as trilobites (**B**) that arose during the Cambrian.

The Cambrian Explosion

Animals underwent a dramatic adaptive radiation during the Cambrian (542–488 million years ago). By the end of this period, all major animal lineages were present in the seas. What caused this Cambrian explosion in diversity? Rising oxygen levels in the seas and a warming global climate may have created new habitats favorable to animal life. Also, the supercontinent Gondwana underwent a dramatic rotation. Movement of landmasses altered climates and isolated populations, thus increasing opportunities for allopatric speciation. Biological factors also encouraged speciation. After the first predators arose, prey defenses such as protective hard parts were favored. Evolution of homeotic genes may have sped things along. Mutations in homeotic genes can have dramatic effects on body plans. Some mutations may have produced adaptive traits that better allowed animals to more easily escape predators or to survive in novel habitats.

Take-Home Message

What do we know about animal origins and diversification?

» Animals probably evolved from a colonial protist that resembled modern choanoflagellates.

» Biomarkers characteristic of sponges have been found in rocks that date back more than 635 million years.

» Some representatives of modern animal lineages are found among the Ediacarans. A great adaptive radiation of animals took place during the Cambrian period.

24.4 Sponges

■ Sponges have no tissues or organs. They filter food from water that they draw through pores in their body.

■ Links to Phagocytosis 5.10, Choanoflagellates 21.11

Sponges (phylum Porifera) are aquatic animals that do not have tissues or organs. Most of the more than 5,000 species live in tropical seas. Some are as small as a fingertip, whereas others stand meters tall. An asymmetrical vaselike or columnar shape is most common, but some grow as a thin crust.

An adult sponge is a **sessile animal** (it lives attached to a surface) and has a body riddled with pores (**Figure 24.5**). The outer surface consists of flat cells. Collar cells with a flagellum and a collar of tiny projections (microvilli) line the inner surface. A jellylike matrix lies between the inner and outer layers. As noted earlier, sponge collar cells resemble choanoflagellate cells.

Most sponges feed by filtering food from the water. Movement of the flagella on collar cells draws water through pores in the body wall and into a central cavity. The collar cells trap bits of food such as bacteria in their microvilli, then engulf them by phagocytosis. Digestion is intracellular. Amoeboid cells in the jellylike matrix receive nutrients from collar cells and distribute them through the body.

Amoeboid cells also secrete fibrous proteins and glassy silica spikes. These materials provide structural support and discourage predation. Some sponges that are especially rich in the rubbery protein spongin are harvested for use as bath sponges (*above left*).

A typical sponge is a **hermaphrodite**: an animal that produces both eggs and sperm. It releases sperm into the water (**Figure 24.6A**). Depending on the species, eggs are either released or are held in the parental body until after fertilization. Fertilization produces a zygote that develops into a ciliated larva (**Figure 24.6B**). A **larva** (plural, larvae) is a free-living, sexually immature form that has a different body structure than an adult. Sponge larvae swim about briefly, before settling and developing into adults.

Many sponges also reproduce asexually. Small buds or fragments break away and grow into new sponges. Some freshwater sponges that live in temporary pools

hermaphrodite Animal that produces both eggs and sperm, either simultaneously or at different times in its life.
larva In some animal life cycles, an immature form that has a different body structure than an adult.
sessile animal Animal that lives fixed in place on some surface.
sponge Aquatic invertebrate that has no body symmetry, tissues, or organs; feeds by filtering food from the water.

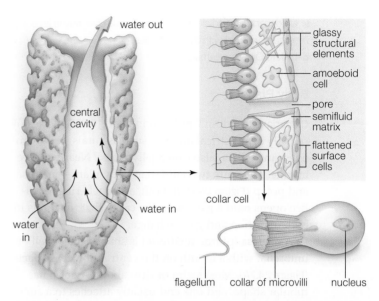

Figure 24.5 Animated Adult sponge, with a porous, asymmetrical body.

B Ciliated sponge larva (colorized scanning electron micrograph).

A Sponge releasing sperm.

Figure 24.6 Sexual reproduction in sponges. Sponges are hermaphrodites.

disperse by producing gemmules, clumps of resting cells inside a hard, protective coat. When a pool dries out, wind disperses gemmules. Those that land in a favorable habitat grow into new sponges.

Take-Home Message

What are sponges?

» Sponges are mostly marine animals that have an asymmetrical body without any tissues or organs.

» Ciliated larvae swim briefly, but adults are sessile filter feeders that strain bits of food from water flowing through their pores.

24.5 Cnidarians—Predators With Stinging Cells

■ Cnidarians are radial aquatic animals with two tissue layers. They capture food with their tentacles.

■ Link to Dinoflagellates 21.6

Body Plans

Cnidarians (phylum Cnidaria) include 10,000 species of radially symmetrical animals such as corals, sea anemones, and jellies (also called jellyfishes). Nearly all are marine. There are two cnidarian body plans—medusa and polyp (**Figure 24.7**). In both, a tentacle-ringed orifice opens onto a **gastrovascular cavity** that functions in both digestion and gas exchange.

A **medusa** (plural, medusae) is shaped like a bell or umbrella, with a mouth on the ventral (lower) surface (**Figure 24.7A**). Most swim or drift about. A **polyp** has tubular shape, and one end usually attaches to a surface (**Figure 24.7B**).

Cnidarians have an embryo with two tissue layers. Their outer epidermis develops from ectoderm, and the inner gastrodermis from endoderm. Mesoglea, a jellylike secreted matrix, fills the space between these tissue layers. Medusae tend to have a lot of mesoglea; polyps usually have less.

Cnidarians are predators that use their tentacles to sting and capture prey. The name Cnidaria is from *cnidos*, the Greek word for nettle, a kind of stinging plant. The tentacles have cells called **cnidocytes** with specialized organelles (nematocysts) that act like a jack-in-the-box (**Figure 24.8**). When something brushes against the cell, a coiled thread pops out and entangles prey or sticks a venomous barb into it. The captured prey is pushed through the mouth, into the saclike gastrovascular cavity. Gland cells of the gastrodermis secrete enzymes that digest the prey, then digestive remains are expelled through the mouth.

Cnidarians are brainless, but interconnecting nerve cells extend through their tissues, forming a **nerve net**. Body parts move when nerve cells signal contractile cells. In a manner analogous to squeezing a water-

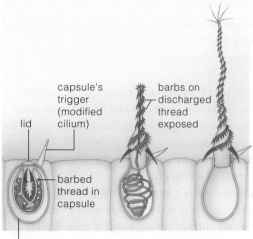

nematocyst (capsule at free surface of epidermal cell)

Figure 24.8 **Animated** Example of cnidocyte action. Mechanical stimulation causes the thread coiled up inside the capsule to spring out and penetrate prey.

filled balloon, the contractions redistribute mesoglea or water trapped in the gastrovascular cavity. A fluid-filled cavity or cellular mass on which contractile cells exert force is a **hydrostatic skeleton**.

Diversity and Life Cycles

There are four cnidarian classes: hydrozoans, anthozoans, cubozoans, and scyphozoans. *Obelia* is a small marine hydrozoan with a life cycle that includes polyp, medusa, and larval stages (**Figure 24.9**). A cnidarian larva, called a planula, is ciliated and bilateral.

Hydra, another hydrozoan, is a freshwater predatory polyp about 20 millimeters (3/4 inch) high (**Figure 24.10A**). There is no medusa stage, and reproduction usually occurs asexually by budding.

Anthozoans such as corals and sea anemones also do not have a medusa stage (**Figure 24.10B**). Gametes form on polyps. Coral reefs are colonies of polyps that enclose themselves in a framework of secreted calcium carbonate. Photosynthetic dinoflagellates (Section 21.6) live in the polyp's tissues and provide it with sugars.

The cubozoans and scyphozoans have a medusa-dominated life cycle. Cubozoans, commonly called box jellies, are active swimmers with structurally complex eyes, complete with a lens. Box jellies that live in the Indo-Pacific ocean produce a powerful venom, and their sting can be deadly to humans (**Figure 24.10C**).

Scyphozoans include most jellies that commonly wash up on beaches. Some are harvested and dried for use as food, especially in Asia. The Portuguese man-of-war (*Physalia*) is a large colonial scyphozoan that consists of many individuals (**Figure 24.10D**).

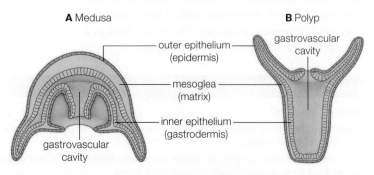

A Medusa **B** Polyp

outer epithelium (epidermis)

gastrovascular cavity

mesoglea (matrix)

inner epithelium (gastrodermis)

gastrovascular cavity

Figure 24.7 **Animated** The two cnidarian body plans.

Figure 24.9 Animated Life cycle of *Obelia*, a hydrozoan that has alternating medusa (*left*) and polyp generations.

1 A medusa makes and releases eggs or sperm. Gametes combine and a zygote forms.

2 The zygote develops into a ciliated bilateral larva.

3 The larva settles and develops into a polyp.

4 The polyp grows and reproduces asexually, eventually producing a branching colony.

5 Some branches of the colony are specialized for capturing and eating prey.

6 Other branches produce and release medusae that begin the sexual phase of the life cycle again.

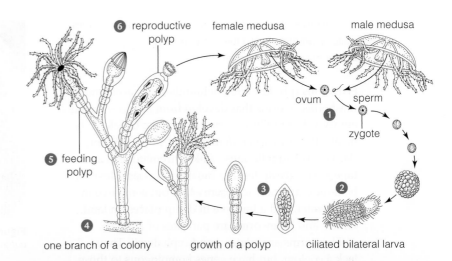

one branch of a colony   growth of a polyp   ciliated bilateral larva

A Hydrozoan. *Hydra*, a fresh-water species capturing and digesting a water flea.

B Anthozoans. A sea anemone (*top*) and polyps of a reef-building coral (*bottom*).

C Cubozoan. The box jelly *Chironex* makes a toxin that can kill a person.

D Portuguese man-of-war (*Physalia*), a colony of scyphozoans. The purplish-blue, air-filled float is a modified polyp.

Figure 24.10 Representatives of the four cnidarian classes.

cnidarian Radially symmetrical invertebrate that has tentacles with stinging cells (cnidocytes).
cnidocyte Stinging cell unique to cnidarians.
gastrovascular cavity Saclike cavity that functions in digestion and gas exchange in cnidarians and flatworms.
hydrostatic skeleton Of soft-bodied invertebrates, a fluid-filled chamber that muscles exert force against, redistributing the fluid.
medusa A bell-like cnidarian body fringed with tentacles.
nerve net A mesh of nerve cells with no central control organ.
polyp A pillarlike cnidarian body topped with tentacles.

Take-Home Message

What are cnidarians?

» Cnidarians are radial, mostly marine predators with two tissue layers. There are two body plans: medusa and polyp. In both, tentacles with cnidocytes surround a mouth that opens onto a gastrovascular cavity. A nerve net interacts with the hydrostatic skeleton to allow movement.

» Cnidarians include sea anemones, jellies, and corals. Reef-building corals secrete the calcium carbonate that forms reefs.

24.6 Flatworms—Simple Organ Systems

■ With this section, we begin our exploration of animals with bilateral symmetry and a three-layer embryo.
■ Link to Levels of organization 1.2

Flatworms (phylum Platyhelminthes) are acoelomate, flat-bodied worms that develop from a three-layered embryo. The phylum name comes from Greek; *platy*– means flat, and *helminth* means worm. Free-living flatworms (turbellarians), flukes (trematodes), and tapeworms (cestodes) are the main classes. Most turbellarians are marine (**Figure 24.11**), but some live in fresh water, and a few live in damp places on land. Flukes and tapeworms are parasites of animals.

Flatworms are bilateral and cephalized. They lack a coelom, but have genes homologous to those that regulate coelom development in other animals. Tapeworms have distinct segments; turbellarians, though not externally segmented, do have internally repeated organs. Most likely, flatworms descended from a segmented, coelomate ancestor, and these traits were lost in some or all lineages as they evolved.

Structure of a Free-Living Flatworm

Planarians, a type of free-living flatworm, commonly live in ponds. They glide along, propelled by the movement of cilia that cover their body surface. A planarian uses a muscular tube called a **pharynx** on its ventral surface to suck food into a highly branched gastrovascular cavity (**Figure 24.12A**). Nutrients and oxygen diffuse from the fine branches to all body cells.

Figure 24.11 A marine flatworm. Brightly colored flatworms are common inhabitants of coral reef ecosystems.

The planarian head has chemical receptors and two light-detecting eyespots. These sensory structures send messages to a simple brain that consists of paired ganglia. A **ganglion** is a cluster of nerve cell bodies. Bundles of nerve fibers called **nerve cords** extend from the head and run the length of the body (**Figure 24.12B**).

A planarian has both female and male sex organs (**Figure 24.12C**). However, individuals cannot fertilize their own eggs. They must mate. Planarians also reproduce asexually. The body splits in two near the middle, then each piece regrows the missing parts. Regrowth also occurs if a planarian is cut in two.

The body fluid of a planarian holds more solutes than its freshwater environment, so water tends to move into the body by osmosis. A system of branch-

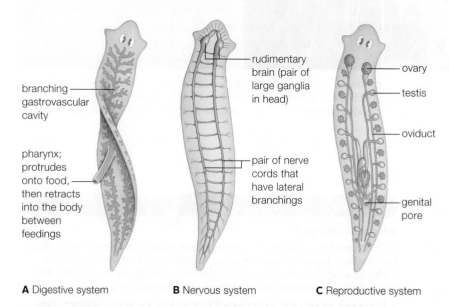

branching gastrovascular cavity

pharynx; protrudes onto food, then retracts into the body between feedings

A Digestive system

rudimentary brain (pair of large ganglia in head)

pair of nerve cords that have lateral branchings

B Nervous system

ovary

testis

oviduct

genital pore

C Reproductive system

pair of highly branched tubules that adjust water and solute levels in body

D Water-regulating system

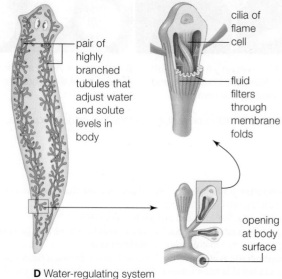

cilia of flame cell

fluid filters through membrane folds

opening at body surface

Figure 24.12 Animated Organ systems of a planarian, one of the flatworms. The body shows internal segmentation (repetition of structures along its length).

Figure It Out: What embryonic tissue layer gives rise to the gut?

Answer: Endoderm. See Figure 24.3A.

Figure 24.13 Animated Life cycle of the beef tapeworm (*Taenia saginata*).

1 A person eats undercooked beef containing a cyst (resting form) of the tapeworm.

2 In the human intestine, the tapeworm uses its barbed scolex to attach to the intestinal wall. It grows by adding new body units (proglottids). Over time, it can become many meters long.

3 Each proglottid produces both eggs and sperm, and can fertilize itself or another. Proglottids that hold fertilized eggs leave the body in feces.

4 Cattle eat grass contaminated with larva-containing proglottids.

5 The larval tapeworm matures, reproduces asexually, and forms a cyst in its host's muscle.

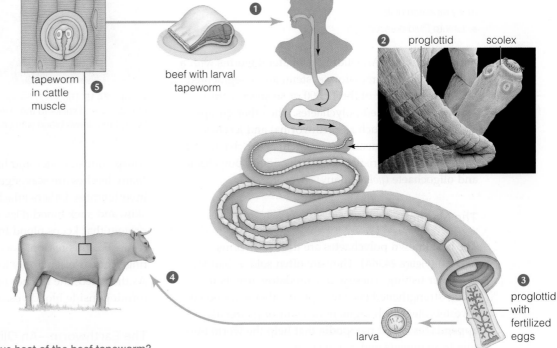

tapeworm in cattle muscle

beef with larval tapeworm

proglottid   scolex

proglottid with fertilized eggs

larva

Figure It Out: What is the definitve host of the beef tapeworm?

Answer: Humans. It reproduces sexually only in humans.

---

ing tubules regulates water and solute levels (**Figure 24.12D**). At the tip of each tubule is a flame cell, whose cilia give it a flickering appearance under the microscope. Action of cilia drives body fluid through tubules that deliver excess water to the body surface.

Flukes and Tapeworms—The Parasites

Flukes and tapeworms are parasitic flatworms. Asexually reproducing forms often live and reproduce in one or more intermediate hosts, then sexual reproduction occurs in the definitive host. For example, aquatic snails are the intermediate host for the blood fluke (*Schistosoma*), and humans are the definitive host. Asexually produced larvae exit a snail and swim about until they contact and burrow into human skin. The flukes take up residence in intestinal veins, resulting in the disease schistosomiasis. Schistosomiasis affects about 200 million people, most of them in Asia.

Like flukes, tapeworms live and reproduce in the vertebrate gut. The tapeworm head has a scolex, a

structure with hooks or suckers that allow the worm to attach to the intestinal wall. Behind the head are body units called proglottids. Unlike planarians and flukes, the tapeworm has no gut. Nutrients from the food its host ingests reach tapeworm cells by diffusing across the worm's body wall.

Figure 24.13 shows the life cycle of a beef tapeworm. Larvae enter the body when a person eats undercooked beef that contains larvae **1**. In the intestine, the tapeworm grows as new proglottids form in the region behind the scolex **2**. Each proglottid produces both eggs and sperm and can either fertilize itself or another proglottid. Older proglottids (farthest from the scolex) contain fertilized eggs. The oldest proglottids break off and exit the body in feces **3**. Fertilized eggs can survive for months before being eaten by cattle, the intermediate host **4**. Inside cattle, the tapeworm forms a cyst in muscle **5**.

flatworm Acoelomate, unsegmented worm; for example, a planarian or tapeworm.
ganglion Cluster of nerve cell bodies.
nerve cord Bundle of nerve fibers that runs the length of the body in many invertebrates.
pharynx Tube connecting the mouth to the digestive tract.

Take-Home Message

What are flatworms?

» Flatworms are bilateral, acoelomate animals with a gastrovascular cavity, a cephalized nervous system, and a system for regulating water and solutes.

» Free-living flatworms include marine species and freshwater planarians.

» Flukes and tapeworms are parasitic flatworms. Both groups include species that infect humans.

24.7 Annelids—Segmented Worms

■ An annelid body is coelomate and segmented; it consists of many repeated units.

■ Link to Fluid pressure 5.8

Annelids (phylum Annelida) are bilateral worms with a coelom, a closed circulatory system, and a segmented body. The majority of the 12,000 or so species are marine worms called polychaetes. The other groups are oligochaetes (such as earthworms) and leeches. Except in leeches, nearly all segments bear chaetae, or chitin-reinforced bristles. Hence the names polychaete and oligochaete (*poly–*, many; *oligo–*, few).

The Marine Polychaetes

The best known polychaetes are the sandworms (*Nereis*) (**Figure 24.14A**). They are often sold as bait for saltwater fishing. These active predators use their chitin-strengthened jaws to capture other soft-bodied invertebrates. Each segment has pair of paddlelike appendages called parapodia that help the worm burrow in sediments and pursue prey.

Other polychaetes have modifications of this basic body plan. Fan worms and feather duster worms live inside a tube of secreted mucus and sand grains. The head protrudes from the tube, and its elaborate tentacles capture food (**Figure 24.14B**). The worms do not crawl much and have only tiny parapodia.

A **trochophore larva**, a larva with a band of cilia above its mouth, forms during polychaete development. Mollusks have the same type of larva, and this developmental similarity is evidence of a close relationship between the annelid and mollusk lineages.

Leeches—Bloodsuckers and Others

Leeches occur in the ocean and damp habitats on land, but are most common in fresh water. Their body lacks

Figure 24.15 The leech *Hirudo medicinalis* feeding on human blood. A leech sticks to the skin with suckers at either end of the body, then draws blood with chitin-hardened jaws.

conspicuous bristles and has a sucker at either end. Many leeches are scavengers or predators of small invertebrates. Others attach to a vertebrate, pierce its skin, and suck blood (**Figure 24.15**). Their saliva has a protein that keeps blood from clotting while the leech feeds. For this reason, doctors who reattach a severed finger or ear often apply leeches to the reattached part. As they feed, the leeches prevent unwanted clots from forming inside blood vessels of the reattached part.

The Earthworm—An Oligochaete

Oligochaetes include marine and freshwater worms, but the land-dwelling earthworms are most familiar. We consider their body in detail as our example of annelid structure (**Figure 24.16**).

An earthworm body is segmented inside and out. The outer layer is a cuticle of secreted proteins. Visible grooves on its surface correspond to internal partitions. A fluid-filled coelom runs the length of the body and is divided into chambers, one per segment.

Gases are exchanged across the body surface, and a closed circulatory system helps distribute oxygen. Five hearts in the anterior of the worm pump the blood.

An earthworm has a complete digestive system. It eats its way through soil, ingesting organic debris and excreting undigested material. Earthworm feces, or "castings," are sold as fertilizer.

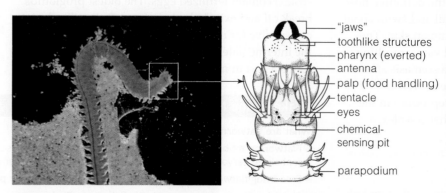

"jaws"
toothlike structures
pharynx (everted)
antenna
palp (food handling)
tentacle
eyes
chemical-sensing pit
parapodium

A The sandworm (*Nereis*) is an active predator that burrows in marine mudflats.

Figure 24.14 Examples of polychaetes.

B A feather duster worm uses tentacles on its head to filter food from the water.

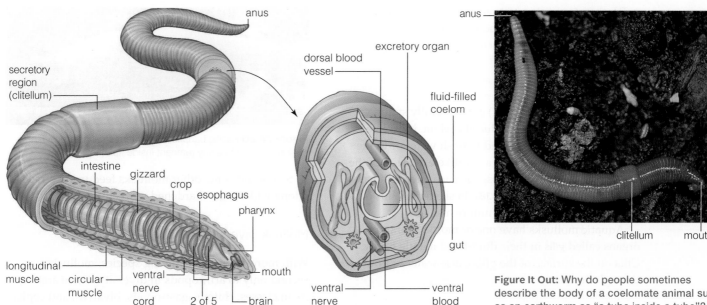

Figure 24.16 **Animated** Earthworm body plan. Each segment contains a coelomic chamber full of organs. A gut, ventral nerve cord, and dorsal and ventral blood vessels run through all coelomic chambers.

Figure It Out: Why do people sometimes describe the body of a coelomate animal such as an earthworm as "a tube inside a tube"?

Answer: One tube, the gut, is inside another tube, the body wall. The fluid-filled coelom lies between the tubes.

The solute composition and volume of coelomic fluid are regulated by **nephridia** (singular, nephridium) that occur in nearly all segments. Each nephridium collects coelomic fluid, adjusts its composition, then expels waste through a pore in the next segment.

An earthworm has a rudimentary "brain," a fused pair of ganglia, that coordinates activities. The brain sends signals via a pair of nerve cords. In response to the nervous commands, muscle contractions put pressure on fluid inside coelomic chambers. This fluid is a hydrostatic skeleton. There are two sets of muscles. Longitudinal muscles parallel the body's long axis and circular ones ring the body. The two types of muscles work in opposition to change segment shape and propel the worm, as illustrated in **Figure 24.17**.

Earthworms are hermaphrodites. A secretory region, the clitellum, produces mucus that glues two worms together while they swap sperm. Later, the clitellum secretes a silky case that encloses the fertilized eggs that a worm deposits in the environment.

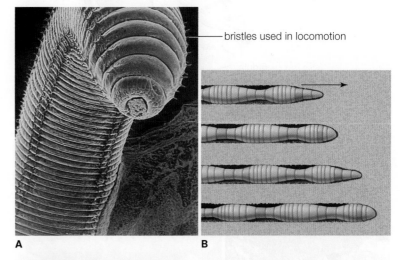

A   B

Figure 24.17 How earthworms move through soil. (**A**) Bristles on sides of the body extend and withdraw as muscle contractions act on coelomic fluid inside each segment. (**B**) Bristles are extended when a segment's diameter is at its widest (when circular muscle is relaxed and longitudinal muscle is contracted). They retract as the segment gets long and thin. A worm's front end is pushed forward, then bristles anchor it and the back of the body is pulled up behind it.

Take-Home Message

What are annelids?

» Annelids are bilateral, coelomate, segmented. They include earthworms, marine polychaete worms, and leeches. All have a tubular gut and a closed circulatory system.

» Larval similarities imply a close relationship between annelids and mollusks.

annelid Segmented worm with a coelom, complete digestive system, and closed circulatory system.
nephridium Of annelids and some other invertebrates, one of many water-regulating units that help control the content and volume of tissue fluid.
trochophore larva Type of ciliated larva that forms during the development of mollusks and some annelids.

24.8 Mollusks—Animals With a Mantle

■ The ability to secrete a protective shell gave mollusks an advantage over other soft-bodied invertebrates.

■ Link to Patterns of development 18.5

General Characteristics

Mollusks (phylum Mollusca) are bilaterally symmetrical invertebrates with a reduced coelom. Most live in seas, but some have adapted to fresh water or to life on land. All have a **mantle**, a skirtlike extension of the upper body wall that encloses a space referred to as the mantle cavity (**Figure 24.18**). In shelled mollusks, the mantle secretes the calcium-rich shell.

Aquatic mollusks have one or more respiratory organs called **gills** in their fluid-filled mantle cavity. Cilia on the surface of the gills cause water to flow through the cavity.

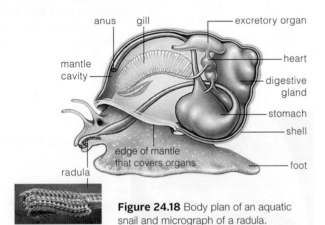

Figure 24.18 Body plan of an aquatic snail and micrograph of a radula.

A Chiton with overlapping plates.

C Bivalve (scallop) with hinged, two-part shell.

B Gastropod with giant "foot."

D Cephalopod (squid), with tentacles.

Figure 24.19 Representatives of the four main mollusk classes.

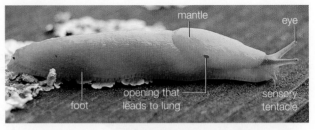

Figure 24.20 Pacific banana slug, a terrestrial gastropod. Its shell is reduced to a tiny remnant inside its mantle.

Snails and some other mollusks feed using a **radula**, a tonguelike organ hardened with chitin.

Mollusk Diversity

With more than 100,000 living species, mollusks are second only to arthropods in diversity. There are four main classes: chitons, gastropods, bivalves, and cephalopods (**Figure 24.19**).

Chitons are probably the most similar to ancestral mollusks. All are marine and have a dorsal shell that consists of eight plates (**Figure 24.19A**). Chitons cling to rocks and scrape up algae with their radula.

With about 60,000 species of snails and slugs, the gastropods are the most diverse mollusks. Their name means "belly foot." Most species glide about on the broad muscular foot that makes up most of the lower body mass (**Figures 24.18** and **24.19B**). A gastropod shell, when present, is one-piece and often coiled.

During development, gastropods undergo a unique rearrangement of body parts called torsion. The body mass twists, putting previously posterior parts, including the anus, up above the head.

Gastropods have a distinct head that usually has eyes and sensory tentacles. In many aquatic species, a part of the mantle forms an inhalant siphon, a tube through which water is drawn into the mantle cavity. The cone snails discussed in Section 24.1 use their siphon to sniff out their prey. Cone snails are predatory, and their radula is modified as a harpoon, but most gastropods are herbivores.

Gastropods include the only terrestrial mollusks (**Figure 24.20**). In land-dwelling snails and slugs, a lung replaces the gill. Glands on the foot continually secrete mucus that protects the animal as it moves across dry, abrasive surfaces. Most mollusks have separate sexes, but land-dwellers tend to be hermaphrodites. Unlike other mollusks, which produce a swimming larva, the embryos of these groups develop directly into adults.

Bivalves include many of the mollusks that end up on our dinner plates, including mussels, oysters, clams, and scallops (**Figure 24.19C**). All bivalves have

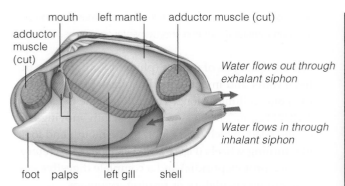

Figure 24.21 Animated Body plan of a clam, a bivalve.

Labels on figure: mouth, left mantle, adductor muscle (cut), adductor muscle (cut), Water flows out through exhalant siphon, Water flows in through inhalant siphon, foot, palps, left gill, shell

A Chambered nautilus, one of six living nautilus species. The shell consists of many gas-filled chambers. The numerous tentacles do not have suckers.

B Octopus. Its eight arms are covered with many individually controlled suckers that allow it to hold onto objects, and to adhere to or walk along surfaces

Figure 24.22 Examples of cephalopods.

a hinged, two-part shell. Powerful adductor muscles hold the valves together (**Figure 24.21**). Contraction of these muscles pulls the two valves shut, enclosing the body and protecting it from predation or drying out. Some scallops can "swim" by repeatedly opening and closing their shell. As the shell shuts, the force of the expelled water causes the scallop to scoot backward. A bivalve has a reduced head, but eyes arrayed around the edge of its mantle alert it to danger.

Bivalves have a large triangular foot commonly used in burrowing. A clam burrows beneath sand and extends its siphons into the water. Like other bivalves, it lacks a radula. It feeds by drawing water into its mantle cavity and trapping bits of food in mucus on its gills. Movement of cilia directs particle-laden mucus to a pair of fleshy extensions called palps that sort out particles and sweep food into the mouth.

Cephalopods include squids (**Figure 24.19D**), nautiluses (**Figure 24.22A**), octopuses (**Figure 24.22B**), and cuttlefish. Cephalopod means "head-footed," and their foot has been modified into arms and/or tentacles that extend from the head. All cephalopods are predators and most have beaklike, biting mouthparts in addition to a radula. They move by jet propulsion. They draw water into the mantle cavity, then force it out through a funnel-shaped siphon.

Five hundred million years ago, large cephalopods with a long conelike shell were the top predators in the seas. Today, nautiluses have a coiled external shell, but other cephalopods have a highly reduced shell or none at all. Jawed fishes evolved 400 million years ago, and competition with this group may

have favored a shift in body form. Cephalopods with the smallest shell could be fastest and most agile. A speedier lifestyle required other changes as well. Of all mollusks, only cephalopods have a closed circulatory system. Competition with fishes also favored improved eyesight. Like vertebrates, cephalopods have eyes with a lens. Because mollusks and vertebrates are not closely related, this type of eye is assumed to have evolved independently in these groups.

Cephalopods include the fastest (squids), biggest (giant squid), and smartest (octopuses) invertebrates. Of all invertebrates, octopuses have the largest brain relative to body size, and the most complex behavior.

Take-Home Message

What are mollusks?

» Mollusks are invertebrates with a bilateral body plan, a reduced coelom, and a mantle that drapes over their internal organs. In most species, the mantle secretes a protective hardened shell.

» Most mollusks are aquatic, but some gastropods have adapted to life on land. In addition to gastropods, mollusks include chitons, bivalves, and cephalopods.

---

gills Folds or body extensions that increase the surface area for respiration.
mantle Of mollusks, extension of the body wall.
mollusk Invertebrate with a reduced coelom and a mantle, including bivalves, gastropods, and cephalopods.
radula Of many mollusks, a tonguelike organ hardened with chitin that is used in feeding.

24.9 Rotifers and Tardigrades—Tiny and Tough

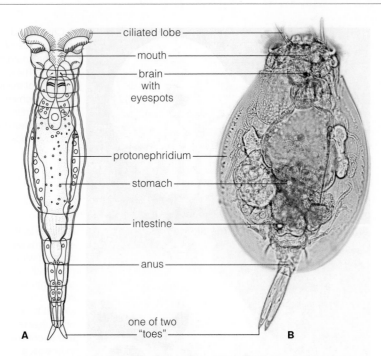

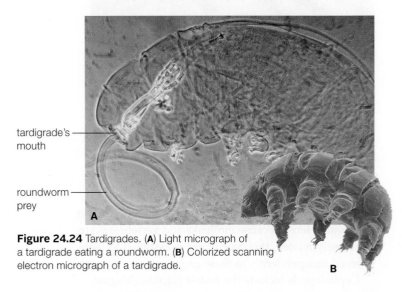

Figure 24.23 (**A**) Body plan of a bdelloid rotifer. (**B**) Micrograph of *Euchlanis*, which secretes a transparent covering around its body.

Figure 24.24 Tardigrades. (**A**) Light micrograph of a tardigrade eating a roundworm. (**B**) Colorized scanning electron micrograph of a tardigrade.

rotifer Tiny bilateral, pseudocoelomate animal with a ciliated head.
tardigrade A tiny coelomate animal with four pairs of legs; in a dormant state, it can survive extremely adverse conditions.

Take-Home Message

What are rotifers and tardigrades?

» Rotifers and tardigrades (water bears) are tiny bilateral animals. Most live in damp habitats or fresh water. An ability to enter a dormant state allows some to survive in an environment that often dries out.

» Rotifers have a pseudocoelom, but genetic comparisons suggest they are closest to annelids and mollusks. Tardigrades have a coelom and molt; they are probably relatives of the roundworms and insects.

■ Rotifers and tardigrades are two phyla of minuscule animals that can withstand extreme environmental conditions.

The 2,150 species of **rotifers** (phylum Rotifera) live in fresh water and in damp land habitats. Most are less than one millimeter long. The group name is Latin for "wheel bearer." It refers to the constantly moving cilia on the head, which direct food to the mouth and look like turning wheels (**Figure 24.23**). There are excretory organs (protonephridia) and a complete digestive system, but no circulatory or respiratory organs.

Digestive and excretory organs are located inside a pseudocoelom. Traditionally, the rotifers and roundworms were grouped together as pseudocoelomates. However, gene comparisons suggest that rotifers are more closely related to annelids and mollusks.

Some rotifers glue themselves to a surface by their toes, but most swim or crawl about. Some species are all female. New individuals develop from unfertilized eggs—a process called parthenogenesis. Other species produce males seasonally or have two sexes.

Tardigrades (phylum Tardigrada) are similarly tiny animals that often live beside rotifers in damp moss and temporary ponds. Commonly called water bears, they waddle about on four pairs of stubby legs (**Figure 24.24**). "Tardigrada" means slow walker.

About 950 tardigrade species have been named. Most suck juices from plants or algae. Some, including the one in **Figure 24.24A**, are predators. They eat roundworms, rotifers, and one another. The digestive system is complete and there are excretory organs, but no circulatory or respiratory organs. The coelom is reduced.

Like roundworms and insects, tardigrades have an external body covering that they molt (shed periodically) as they grow. The molting and gene sequence data suggest tardigrades are related to arthropods and roundworms, but the relationships among these groups are poorly understood.

Tardigrades and rotifers living in habitats that often dry up completely have evolved a remarkable ability. They can survive dry periods by entering a sort of suspended animation. As the habitat dries up, sugar replaces water in their tissues and metabolism slows to a nearly nonexistent pace. In tardigrades, the water content of the body can drop to 1 percent of normal.

Dormant tardigrades can withstand extraordinary heat and cold. They have survived for a few days at −200°C (−328°F) and a few minutes at 151°C (304°F). Also, a tardigrade can remain dormant for years, then revive within a few hours of being placed in water. For all of these reasons, tardigrades are often said to be the toughest animals.

■ Most roundworms act as decomposers in soil and water, but some infect us, our pets, our livestock, or our crops.

Roundworms, or nematodes (phylum Nematoda), are unsegmented worms that have a pseudocoelom and a cuticle-covered, cylindrical body (**Figure 24.25**). The collagen-rich cuticle is periodically molted as the worm grows. Roundworms have a complete digestive system, excretory organs, and a nervous system, but no circulatory or respiratory organs.

Nearly 20,000 named species live in the seas, in fresh water, in damp soil, and inside other animals. Most are free-living decomposers less than a millimeter long. Parasitic species tend to be larger, and some that live inside whales can extend for meters.

Like the fruit fly, the soil roundworm *Caenorhabditis elegans* is frequently used in scientific studies. It is useful as a model for developmental processes because it has the same tissue types as more complex organisms, but it is transparent, has less than 1,000 body cells, and reproduces fast. In addition, its genome is about 1/30 the size of the human genome. Such traits make it easy for scientists to monitor each cell's fate during development. In 2002, the Nobel Prize in Medicine was awarded to scientists whose studies of nematodes revealed how genes control the selective death of cells during normal development.

Several kinds of parasitic roundworms live inside humans. The giant intestinal roundworm, *Ascaris lumbricoides* (**Figure 24.26A**), currently infects more than 1 billion people, mainly in developing tropical nations but also in rural areas of the American Southeast. Most infected people have no symptoms, but a heavy infection can block the intestine.

Lymphatic filariasis is a common tropical disease caused by roundworms. In this case, mosquitoes transmit the worms, which migrate through the body's lymph vessels. Repeated infections injure the valves within these vessels, so lymph pools in lower limbs (**Figure 24.26B**). Elephantiasis, the common name for this disease, refers to fluid-filled, "elephant-like" legs. The swelling is permanent; the lymph vessels remain damaged even if the worms are no longer present.

Pinworms (*Enterobius vermicularis*) are the most common infectious roundworms in the United States. Children are most often affected. The small worms, about the size of a staple, live in the rectum. At night, females crawl out and lay eggs on the skin around the anus. Their movement produces an itching sensation.

Scratching the area to relieve the itch puts eggs onto fingertips and under fingernails. If the eggs get transferred to food or toys and are swallowed, they start a new infection.

Roundworms also affect our livestock, pets, and crop plants. A roundworm that infects pigs can also infect humans who eat undercooked pork, causing trichinosis. Dogs are susceptible to heartworms transmitted by mosquitoes. Cats usually become infected by eating an infected rodent that serves as an intermediate host. Many crop plants are susceptible to nematode parasites. In some cases, the worms suck on plant roots and in others they actually enter the plant. Either way, the result is stunted growth and lower yields.

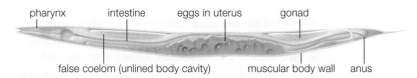

pharynx   intestine   eggs in uterus   gonad

false coelom (unlined body cavity)   muscular body wall   anus

Figure 24.25 Animated Body plan of *Caenorhabditis elegans*, a free-living roundworm used in studies of genetics. Sexes are separate, and this is a female.

A Roundworms (*Ascaris lumbricoides*) passed by an infected child.

B The gross enlargement of this man's leg is a symptom of lymphatic filariasis.

Figure 24.26 Roundworms that parasitize humans.

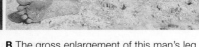

Take-Home Message

What are roundworms?

» Roundworms are unsegmented, pseudocoelomate worms with a secreted cuticle that they molt as they grow.

» Most roundworms are decomposers, but some are parasites. One species is often used by scientists in studies of genetics and development.

---

roundworm Unsegmented pseudocoelomate worm with a cuticle that is molted periodically as the animal grows.

■ Arthropods are the most diverse invertebrate group. A variety of features contribute to their success.

■ Links to Chitin 3.4, Adaptive traits 16.4

Arthropods (phylum Arthropoda) are bilateral, with a reduced coelom. They have a hard, jointed external skeleton, a complete digestive system, an open circulatory system, and respiratory and excretory organs. Sexes are usually separate, although some groups are hermaphrodites.

One lineage, the trilobites, was abundant in seas until the Permian. Modern groups include myriopods such as centipedes, crustaceans, and insects. Here we begin by thinking about the key adaptations that contribute to arthropod success.

Key Arthropod Adaptations

Hardened Exoskeleton Arthropods secrete a cuticle composed of chitin (Section 3.4), proteins, and waxes. It functions as an **exoskeleton**, a hard, external skeleton. The exoskeleton gives the body shape, offers protection from predators, and serves as a framework to which muscles can attach. When some groups invaded land, the exoskeleton also helped them conserve water. Although the arthropod exoskeleton is hard, it does not restrict growth, because—like the roundworms—arthropods molt their cuticle after each growth spurt (**Figure 24.27**). Hormones that regulate molting cause formation of a new cuticle beneath the old one. The new cuticle remains soft until the old one is shed.

Jointed Appendages "Arthropod" means jointed leg. If an arthropod's cuticle were uniformly hard and thick like a plaster cast, it would prevent movement. An arthropod's cuticle thins at regions where two hard body parts meet. When muscles that span one of these joints contract, they cause the cuticle to bend, which in turn causes the parts on either side of the joint to move relative to one another.

Highly Modified Segments In early arthropods, the body segments were distinct, and all appendages were similar. In many of their descendants, the segments became fused into structural units such as a head, a

Figure 24.27 An insect in the process of molting its old exoskeleton. Notice also the jointed legs, antennae, and compound eyes.

thorax (midsection), and an abdomen (hind section). Appendages on some segments became modified for special tasks. For example, among insects, thin extensions of the wall of some segments evolved into wings.

Sensory Specializations Most arthropods have one or more pairs of eyes: organs that sample the visual world. In insects and crustaceans, eyes are compound, with many lenses. With the exception of chelicerates, most arthropods also have paired **antennae** that can detect touch and waterborne or airborne chemicals.

Specialized Developmental Stages The body plan of many arthropods changes during the life cycle. Individuals often undergo **metamorphosis**: Tissues get remodeled as juveniles become adults. Each stage is specialized for a different task. For instance, wingless, plant-eating caterpillars metamorphose into winged butterflies that disperse and find mates. Having such different bodies also prevents adults and juveniles from competing for the same resources.

antenna Of some arthropods, sensory structure on the head that detects touch and odors.
arthropod Invertebrate with jointed legs and a hard exoskeleton that is periodically molted; for example, an insect or crustacean.
exoskeleton Of some invertebrates, hard external parts that muscles attach to and move.
metamorphosis Remodeling of body form during the transition from larva to adult.

Take-Home Message

What are arthropods and what factors contribute to their success?

» Arthropods are the most diverse animal phylum. In addition to modern groups, they include the extinct trilobites.

» A hardened exoskeleton protects the body and prevents water loss on land. The exoskeleton is molted as the animal grows, and it is thin at joints to allow movement. Sensory specializations include antennae and compound eyes.

» In many groups, larvae differ from adults in body form and utilize different resources.

24.12 Chelicerates—Spiders and Their Relatives

■ Chelicerates include spiders, scorpions, and ticks, as well as the marine horseshoe crabs.

■ Link to Disease-causing bacteria 20.7

Chelicerates have a body with two regions, a cephalothorax (fused head and thorax) and an abdomen. Walking legs attach to the cephalothorax. The head has eyes, but no antennae. Paired feeding appendages near the mouth, called chelicerae, give the group its name.

Four species of horseshoe crabs are the only marine chelicerates and members of the oldest surviving arthropod lineage. Horseshoe crabs are bottom feeders that eat clams and worms. A horseshoe-shaped shield covers the cephalothorax and the last segment has evolved into a long spine (**Figure 24.28**).

Arachnids include spiders, scorpions, ticks, and mites (**Figure 24.29**). All have four pairs of walking legs and paired appendages, called pedipalps, between the chelicerae and the first legs. Nearly all live on land.

Scorpions and spiders are venomous predators. Spider chelicerae are fanglike and have venom glands. Of 38,000 spider species, about 30 produce venom harmful to humans. Most spiders benefit us by eating insect pests. Some weave prey-catching webs made of silk produced by paired spinners on their abdomen. Others such as tarantulas (**Figure 24.29A**) are active hunters. Scorpions hunt prey and dispense venom through a stinger on their last segment (**Figure 24.29B**). Their pedipalps have evolved into large claws.

Ticks are parasites of vertebrates (**Figure 24.29C**). They pierce a host's skin with their chelicerae, and then suck blood. Some transmit bacteria that cause Lyme disease or other diseases.

Mites, the smallest arachnids, are usually less than a millimeter long. Dust mites (**Figure 24.29D**) are scavengers that can cause problems for people who are allergic to their feces. Other mites parasitize plants or animals. Some larval mites, commonly called chiggers, attach to human skin at hair follicles and sip our tissue fluids. A mite that burrows beneath the skin causes scabies in humans and mange in dogs.

arachnids Land-dwelling chelicerate arthropods with four pairs of walking legs; spiders, scorpions, mites, and ticks.
chelicerates Arthropod subgroup with specialized feeding structures (chelicerae) and no antennae.

Take-Home Message

What are chelicerates?

» Chelicerates include horseshoe crabs and arachnids. The horseshoe crabs are an ancient marine lineage. Arachnids are land-dwellers with eight walking legs and no antennae.

Figure 24.28 Horseshoe crab. These marine animals are the closest living relatives of the extinct trilobites. A horseshoe crab has five pairs of walking legs, and the final segment on the abdomen has been modified as a long spine that helps it steer as it swims. Horseshoe crabs do not produce venom.

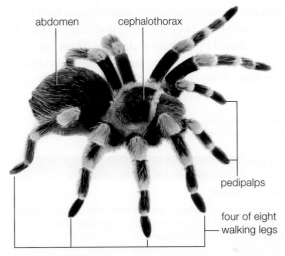

abdomen   cephalothorax

pedipalps

four of eight
walking legs

A Spider (tarantula) with pedipalps that function like arms.

B Scorpion with clawlike pedipalps. The last segment has a venomous stinger.

C Female tick at the tip of a blade of grass waiting for a host to walk by.

D Dust mite, a scavenger (colorized scanning electron micrograph).

Figure 24.29 Arachnids. All have chelicerae, pedipalps, and eight walking legs.

24.13 Myriapods

■ Centipedes and millipedes use their many legs to walk about on land, hunting prey or scavenging.

Myriapod means "many feet," and aptly describes the centipedes and millipedes. Both are nocturnal ground dwellers with an elongated body composed of many similar segments (**Figure 24.30**). The head has a pair of antennae and two simple eyes.

Centipedes have a low-slung, flattened body with a single pair of legs per segment, for a total of 30 to 50. They are fast-moving predators. Their first pair of legs has become modified as fangs that inject paralyzing venom. Most centipedes prey on insects, but some big tropical species eat small vertebrates (**Figure 24.30A**).

Millipedes are slower-moving animals that feed on decaying vegetation. They have a cylindrical body and a cuticle hardened by calcium carbonate. There are two pairs of legs per segment, for a total of a few hundred (**Figure 24.30B**). When threatened, a millipede curls up and secretes a foul-smelling, unpalatable fluid.

myriapod Land-dwelling arthropod with two antennae and an elongated body with many segments; a millipede or centipede.

Take-Home Message

What are myriapods?

» Myriapods are land-dwelling arthropods with two antennae and an abundance of body segments. Centipedes are predators and millipedes are scavengers.

Figure 24.30 (A) A Southeast Asian centipede feeds on its frog prey. The false antennae on the last segment may fool predators into attacking the wrong end. **(B)** A millipede.

24.14 Crustaceans

■ Most marine arthropods are crustaceans. Their amazing diversity and abundance are reflected in their nickname "the insects of the seas."

Crustaceans are a diverse group of mostly marine arthropods that have two pairs of antennae. As in chelicerates, the body has two distinctive regions, a cephalothorax and an abdomen. The crustacean exoskeleton is stiffer than that of other arthropods because it incorporates calcium salts.

You are probably familiar with some of the decapod crustaceans. This group of bottom-feeding scavengers includes the lobsters, crayfish, crabs, and shrimps. Decapods typically have five pairs of walking legs. In some lobsters, crayfish, and crabs, the first pair of legs has become modified into claws (**Figure 24.31**).

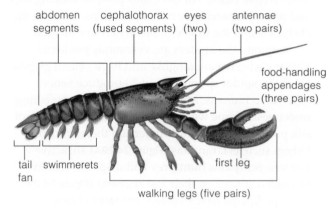

Figure 24.31 Body plan of a lobster (*Homarus americanus*).

An aquatic crustacean begins life as a microscopic, planktonic larva that typically bears little resemblance to the adult animal. The transition from the larval form to the adult form takes place over the course of several molts. For example, the larvae of Dungeness crabs harvested along the Pacific Coast of North America live as plankton for months before taking on the adult form (**Figure 24.32**).

In addition to larvae of decapods, the plankton includes many tiny swimming crustaceans. Krill (euphausiids) are relatives of the decapods and have a shrimplike body up to 6 centimeters long (**Figure 24.33A**). They swim through cool marine waters worldwide. Krill have historically been so abundant that a 30-ton blue whale could sustain itself largely on the krill that it filtered from the water. Recent overharvesting of krill from North Pacific and Antarctic waters has caused steep population declines and may endanger the animals that depend on this resource. Humans harvest krill to feed farm-raised salmon and as a source of omega-3 fatty acids.

Figure 24.32 Animated Development of the Dungeness crab (*Cancer magister*). The crabs live for 8 to 10 years.

❶ After a female crab has mated, she fertilizes eggs with stored sperm and then holds them under her abdomen for 2–3 months until they hatch.

❷ Within an hour of hatching, an early-stage planktonic larva (called a zoea) develops.

❸ The early-stage larva grows and molts repeatedly, altering its form only slightly.

❹ The fifth molt of the early-stage larva produces a late-stage planktonic larva with an altered body form (a megalops).

❺ The late-stage larva molts into a bottom-dwelling juvenile crab. Growth and repeated molting will produce a sexually mature adult in about 2 years.

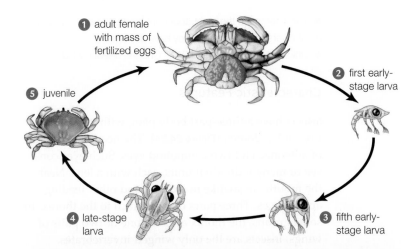

❶ adult female with mass of fertilized eggs
❷ first early-stage larva
❸ fifth early-stage larva
❹ late-stage larva
❺ juvenile

Most copepods are tiny free-living swimmers in seas or lakes (**Figure 24.33B**). They are probably the most numerous of all crustaceans. Some copepods are larger and parasitize fish or marine mammals.

Barnacles are protected by a secreted calcium carbonate shell as well as their exoskeleton. Larval barnacles swim briefly before attaching themselves head down to a surface. A gland in their antennae secretes the glue required to fix them in place. Barnacles attach to rocks, piers, boats, and even whales. They filter food from the water with feathery legs (**Figure 24.33C**). Most barnacles are hermaphrodites and they tend to settle near one another. When the time comes to mate, an individual extend its penis, which is several times the length of its body, out to fertilize its neighbors.

Isopods are a mostly marine group, but you are most likely to be familiar with the species that have adapted to life on land. Sow bugs (*left*) and pill bugs are often found beneath flowerpots or rotting logs. They require a damp habitat where they can feed on organic debris or on soft, young plant parts. Some can be pests of agricultural crops such as soybeans. The species commonly known as pill bugs defend themselves from threats by rolling into a tight ball. As in many other crustaceans, a female retains fertilized eggs on her abdomen during the early stages of their development. However, in terrestrial isopods this affiliation continues after the eggs hatch. The animals live in family groups and recognize members of their group by a shared odor.

crustaceans Mostly marine arthropod group with two pairs of antennae and a calcium-stiffened exoskeleton.

A Antarctic krill (*Euphausia superba*). An individual can be up to 6 centimeters long.

B Female copepod with eggs.

C Barnacles extending feathery legs.

Figure 24.33 Crustaceans that feed on plankton.

Take-Home Message

What are crustaceans?

» Crustaceans are mostly marine arthropods that have two pairs of antennae and an exoskeleton that includes calcium. They include bottom-feeding decapods, planktonic copepods and krill, filter-feeding barnacles, and even some species that have adapted to life on land.

24.15 Insects—Diverse and Abundant

- The insects are the most abundant land arthropods, and the most diverse animal class. They have six legs.
- Links to Malaria 21.1 and 21.7, Chagas disease 21.3

Characteristic Features

Insects have a three-part body plan, with a head, thorax, and abdomen (**Figure 24.34**). The head has one pair of antennae and two compound eyes. Such eyes consist of many individual units, each with a lens. Near the mouth are jawlike mandibles and other feeding appendages. Three pairs of legs attach to the thorax. In some groups, the thorax also has one or two pairs of wings. Insects are the only winged invertebrates.

The group is overwhelmingly terrestrial. A respiratory system consisting of tracheal tubes carries air from openings at the outer surface to deep inside the body. An insect abdomen contains digestive organs, sex organs, and water-conserving excretory organs called **Malpighian tubules**.

Insect developmental patterns vary. The oldest lineages undergo direct development with no change in body form (**Figure 24.35A**). Other insects undergo some type of metamorphosis. With incomplete metamorphosis, the body form changes gradually over multiple molts (**Figure 24.35B**). With complete metamorphosis,

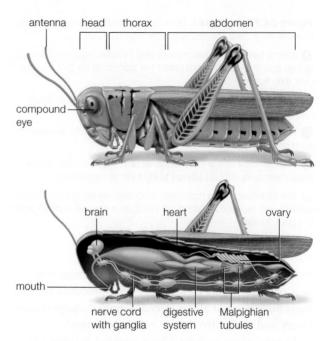

Figure 24.34 Body plan of a female grasshopper.

a larva grows in size without changing form, then becomes an immobile pupa. During pupation, tissues are remodeled to the adult form (**Figure 24.35C**).

The four most diverse insect orders have wings and undergo complete metamorphosis. There are about 150,000 species of flies (Diptera), and at least as many beetles (Coleoptera). The order Hymenoptera includes 130,000 species of wasps, ants, and bees. Moths and butterflies belong to the order Lepidoptera, a group of about 120,000 species. As a comparison, consider that there are about 4,500 species of mammals.

Insect Origins

Until recently, insects were thought to be close relatives of myriapods. Both groups have a single pair of antennae and unbranched legs. Then, new gene comparisons made scientists rethink the connections. The currently favored hypothesis holds that the insects are most closely related to crustaceans. Specifically, insects are thought to be descended from freshwater crustaceans. If this hypothesis is correct, then insects are the crustaceans of the land.

The Importance of Insects

With more than a million species, insects are the most diverse arthropod group, and they are breathtakingly abundant. By one estimate, ants alone make up about 10 percent of the world's terrestrial animal biomass

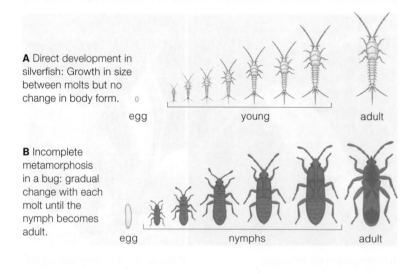

A Direct development in silverfish: Growth in size between molts but no change in body form.

egg · young · adult

B Incomplete metamorphosis in a bug: gradual change with each molt until the nymph becomes adult.

egg · nymphs · adult

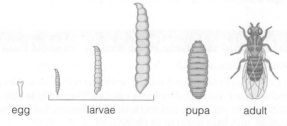

C Complete metamorphosis in a fly: larvae grow, then molt into a pupa, which is remodeled into the adult form.

egg · larvae · pupa · adult

Figure 24.35 The types of insect development.

(the total weight of all living land animals). Given their diversity and numbers, it would be difficult to overestimate the importance of insects.

Ecological Services Most flowering plants are pollinated by members of one of the four most diverse insect orders (**Figure 24.36A**). Other insect orders contain few or no pollinators. By one hypothesis, such interactions between insect groups and flowering plants contributed to an increased rate of speciation in both groups.

Insects are also important as food for wildlife. Most songbirds nourish their nestlings on a diet consisting largely of insects. Migratory songbirds often travel long distances to nest and raise young in areas where insect abundance is seasonally high. Aquatic larvae of insects such as dragonflies, mayflies, and mosquitoes serve as food for trout and other freshwater fish. Amphibians and reptiles feed mainly on insects. Even humans eat insects. In many cultures, they are considered a tasty source of protein.

Insects dispose of wastes and remains. Flies and beetles are quick to discover an animal corpse or a pile of feces (**Figure 24.36B**). They lay their eggs in or on this organic material, and the larvae that hatch devour it. By their actions, these insects keep organic wastes and remains from accumulating, and help distribute nutrients through the ecosystem.

Competitors for Crops Insects are our main competitors for food and other plant products. It is estimated that about a quarter to a third of all crops grown in the United States are lost to insects. Also, in an age of global trade and travel, we have more than just home-grown pests to worry about. Consider the Mediterranean fruit fly (**Figure 24.37A**). The Med fly, as it is known, lays eggs in citrus and other fruits, as well as many vegetables. Damage done to plants and fruits by larvae of the Med fly can cut crop yield in half. Med flies are not native to the United States, and there is an ongoing inspection program for imported produce, but some Med flies still slip in. So far, eradication efforts have been successful, but they have cost hundreds of millions of dollars.

Parasites and Vectors for Disease Some insects are human parasites, and some of these parasites spread disease. The mosquito is probably the deadliest animal. As you know, some mosquito species transmit

insect Six-legged arthropod with two antennae and two compound eyes. Member of the most diverse class of animals.
Malpighian tubule Water-conserving excretory organ of insects.

A Bee transporting pollen from one plant to another.   **B** A dung beetle collecting feces that will nourish its offspring.

Figure 24.36 Helpful insects.

A Med fly, a pest of citrus.   **B** Bedbug, a parasite of humans.

Figure 24.37 Harmful insects.

malaria, which causes millions of deaths each year (Section 21.1). Mosquitoes also serve as vectors for other diseases.

Bites of other insects also spread pathogens. Biting flies transmit African sleeping sickness; biting bugs spread Chagas disease (Section 21.3). Fleas that bite rats and then bite humans can transmit deadly bubonic plague. Body lice transmit typhus.

As far as we know, bedbugs (**Figure 24.37B**) do not transmit disease, although their bites can itch. In addition, a bedbug infestation causes psychological stress and can have a severe negative economic impact on a commercial establishment.

Take-Home Message

What are insects?

» Insects are six-legged arthropods adapted to life on land. They probably evolved from a crustacean ancestor. Tracheal tubes allow insects to breathe air, and Malpighian tubules help them conserve water while eliminating wastes. In some groups, wings evolved as extensions from the body wall.

» Insects are both numerous and diverse. They are food for other animals, pollinate plants, and scavenge wastes and remains. They also compete with us for crops, parasitize us, and infect us with diseases.

24.16 The Spiny-Skinned Echinoderms

■ Echinoderms begin life as bilateral larvae and develop into spiny-skinned, radial adults. All are marine.

■ Link to Patterns of development 18.5

The Protostome–Deuterostome Split

In Section 24.2 we introduced the two major lineages of animals, protostomes and deuterostomes. Here we begin our survey of deuterostome lineages. Echinoderms are the largest group of invertebrate deuterostomes. We will discuss other invertebrate deuterostomes and the vertebrates (also deuterostomes) in the next chapter.

Echinoderm Characteristics and Body Plan

Echinoderms (phylum Echinodermata) include about 6,000 marine invertebrates. Their name means "spiny-skinned" and refers to interlocking spines and plates of calcium carbonate embedded in their skin. Adults are coelomate animals and most have a radial body, with five parts (or multiples of five) around a central axis. The larvae are bilateral, which suggests that the ancestor of echinoderms was a bilateral animal.

Sea stars (also called starfish) are the most familiar echinoderms, and we will use them to illustrate the echinoderm body plan (**Figure 24.38**). Sea stars do not have a brain, but do have a nerve net. Eyespots at the tips of arms detect light and movement.

A typical sea star is an active predator that moves about on tiny, fluid-filled tube feet. Tube feet are part of a **water–vascular system** unique to echinoderms. The system includes a central ring and fluid-filled canals that extend into each arm. Side canals deliver coelomic fluid into muscular ampullae, tiny bulbs that func-tion like the bulb on a medicine dropper. Contraction of an ampulla forces fluid into the attached tube foot, extending the foot. A sea star glides along smoothly as coordinated contraction and relaxation of the ampullae redistribute fluid among hundreds of tube feet.

Sea stars typically feed on bivalve mollusks. A sea star's mouth is on its lower surface. To feed, it slides its stomach out through the mouth and into a bivalve's shell. The stomach secretes acid and enzymes that kill the mollusk and begin the process of digestion. Partially digested food is taken into the stomach, then digestion is completed with the aid of digestive glands in the arms.

Gas exchange occurs by diffusion across the tube feet and tiny skin projections at the body surface. There are no specialized solute-regulating organs.

Sexes are separate. The arms hold sexual organs that release eggs or sperm into the water. Fertilization produces an embryo that develops into a ciliated, bilateral larva. The larva swims briefly, then undergoes metamorphosis into the adult form.

Sea stars and other echinoderms have a remarkable ability to regenerate lost body parts. If a sea star is cut into pieces, any portion with some of the central disk can regrow the missing body parts.

Echinoderm Diversity

Brittle stars are the most diverse and abundant echinoderms (**Figure 24.39A**). They are less familiar than sea stars because they generally live in deeper water. They have a central disk and highly flexible arms that move about in a snakelike way. Most brittle stars are scavengers on the seafloor.

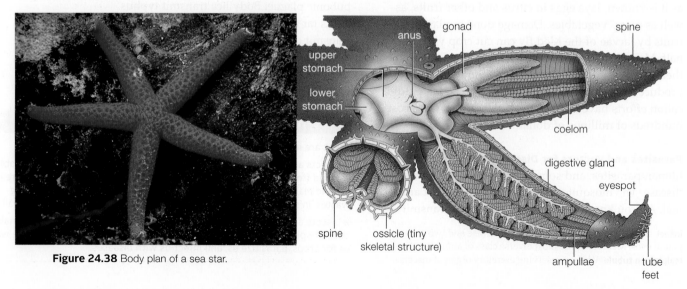

Figure 24.38 Body plan of a sea star.

Old Genes, New Drugs (revisited)

The successful development, production, and adoption of ziconotide, the synthetic cone snail venom used as a painkiller, has inspired scientists to investigate other compounds made by invertebrates.

Nonpredatory invertebrates can also be a source of new medicines. Many invertebrates make chemicals that have antibacterial or antiprotozoal activity. To look for new drugs with these properties, researchers extract compounds from invertebrates, then test the ability of the compounds to kill pathogens cultured in the laboratory.

Researchers also test the effect of each compound on cultured human cells. An ideal candidate drug is one that kills pathogens, but does no harm to human cells. For example, a recent study found that marine sponges make compounds that kill disease-causing protists such as trypanosomes and apicomplexans.

Some of these compounds also have relatively little effect on human cells growing in culture.

A new appreciation of the potential medicinal value of compounds produced by marine invertebrates has many researchers worried about the declining state of our oceans. Pollution, destructive harvesting methods, overharvesting, and climate change are likely to drive many potentially valuable species to extinction before we have time to discover the beneficial compounds they make.

How would you vote? Bottom trawling, a type of fishing in which giant nets are dragged along the seafloor, keeps seafood prices low because it is less labor-intensive than other harvesting methods. It also devastates seafloor populations of invertebrates. Does inexpensive seafood matter more than invertebrate diversity?

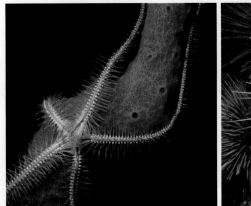
A Brittle star

B Sea urchin

C Sea cucumber

Figure 24.39 Echinoderm diversity

In sea urchins, calcium carbonate plates form a stiff, rounded cover from which spines protrude (**Figure 24.39B**). The spines provide protection and are used in movement. Some urchins graze on algae. Others act as scavengers or prey on invertebrates.

In sea cucumbers, hardened parts have been reduced to microscopic plates embedded in a soft body. Most species have a wormlike body and, like earthworms, they eat their way through sediments and digest any organic material (**Figure 24.39C**). Some deep sea species can swim with the aid of capelike body extensions.

echinoderms Invertebrates with hardened plates and spines embedded in the skin or body, and a water–vascular system.
water–vascular system Of echinoderms, a system of fluid-filled tubes and tube feet that function in locomotion.

Lacking spines or sharp plates, sea cucumbers have alternative defenses. When threatened, they expel a sticky mass of specialized threads or internal organs out through their anus. If this maneuver successfully distracts the predator, the sea cucumber escapes, and its missing parts grow back.

Sea urchin roe (eggs) and sea cucumbers are popular foods in Asia, and overharvesting for this market threatens some species in both groups.

Take-Home Message

What are echinoderms?

» Echinoderms are deuterostome invertebrates that have a radial body as adults. They are brainless and have a unique water–vascular system that functions in locomotion.

LEARNING ROADMAP

Where you have been This chapter introduces the remaining lineages of deuterostomes, a group defined in Section 24.2. We refer to discussions of fossils (16.5), plate tectonics (16.7), the geologic time scale (16.8), and adaptive radiation (17.12). Understanding cladistics (18.2) and how comparative studies help determine relationships (18.3–18.5) is essential.

Where you are now

Characteristics of Chordates
Distinctive embryonic traits characterize the chordates, a group that includes a small number of invertebrates, as well as all animals with a backbone (vertebrates).

The Fishes
Fishes were the first vertebrates and remain the most diverse group. Early fishes were jawless. Evolution of jaws and paired fins opened the way to a great adaptive radiation.

Transition From Water to Land
One group of fishes gave rise to aquatic tetrapods (four-legged walkers). Amphibians typically live on land and breathe air, but return to the water to breed.

The Reptiles
Reptiles have a scaly body and are amniotes: vertebrates with eggs that develop on land. Birds are reptiles with feathers and an ability to regulate their body temperature.

The Mammals
Mammals are amniotes that produce milk and have hair. Some mammals lay eggs, but young of the most diverse group are nourished by their mother as they develop inside her.

Where you are going The next chapter continues the story of mammals, specifically the order Primates. We compare vertebrate brains in Section 32.9, endoskeletons in Section 35.3, and circulatory systems in Section 36.2. Fish gills and bird lungs are covered in Section 38.4. Section 39.2 compares the digestive systems of various vertebrates, and Section 40.2 looks at urinary systems. Sections 40.8 and 40.9 delve into how endotherms and exotherms regulate body temperature, and Section 42.9 explains the function of the mammalian placenta.

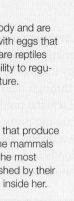

25.1 Transitions Written in Stone

In Darwin's time, one obstacle to acceptance of his theory of evolution by natural selection was the apparent absence of transitional fossils. Skeptics wondered, if new groups evolve from existing ones, then where are the fossils that represent these transitions? In fact, one of these "missing links" was unearthed by workers at a limestone quarry in Germany just one year after Darwin's *On the Origin of Species* was published.

That fossil, about the size of a large crow, looked like a small dinosaur. It had a long, bony tail, three clawed fingers on each forelimb, and a heavy jaw with short, spiky teeth, but it also had feathers (**Figure 25.1A**). The fossil species was named *Archaeopteryx* (ancient winged one). Thus far, eight fossilized members of this species have been unearthed. Radiometric dating indicates that they lived about 150 million years ago.

Archaeopteryx is the most widely known transitional fossil in the bird lineage, but there are others. A fossil from China, *Confuciusornis sanctus* (meaning sacred Confucius bird), had a beak and a short tail with long tail feathers like that of a modern bird (**Figure 25.1B**). Unlike wings of nearly all modern birds, the wings of *C. sanctus* retain long claws at the tips of their digits.

Another fossil takes us back even farther in time and perhaps provides a clue as to which group of dinosaurs gave rise to the birds. In 1994, a farmer in China discovered a fossil of a small dinosaur-like animal with short forelimbs and a long tail (**Figure 25.1C**). Unlike most dinosaurs, this one had areas covered with tiny filaments that resemble downy feathers of modern birds. Researchers named the farmer's fuzzy find *Sinosauropteryx prima*, meaning first Chinese feathered dragon. Given the animal's shape and lack of long feathers, *S. prima* was certainly flightless. If the fuzzy filaments are feathers, they probably served as insulation, as downy feathers do in modern groups.

Scientists interested in the evolutionary transitions that led to modern animal diversity look to fossils for physical evidence of these transitions, and use radiometric dating to determine when transitional species lived. The structure, biochemistry, and genetic traits of living organisms also provide information about the branchings that gave rise to modern animal groups.

Interpretations of fossil data and gene comparisons can vary, and biologists often disagree about the relationships among lineages, both modern and extinct. These disagreements do not call into question Darwin's contention that all animal lineages arose by descent with modification from a common ancestor. Rather, arguments concern the details of where, when, and how the various branchings took place. By using new methods to compare genes, analyzing new fossil finds, or reanalyzing old ones, scientists seek the evidence necessary to bolster currently held hypotheses, revise them, or discard them for new ones.

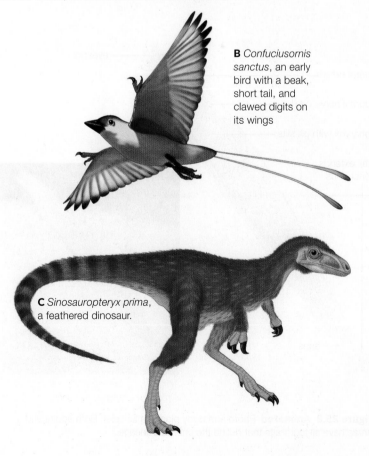

B *Confuciusornis sanctus*, an early bird with a beak, short tail, and clawed digits on its wings

C *Sinosauropteryx prima*, a feathered dinosaur.

A *Archaeopteryx*, an early bird with a long bony tail, teeth, and clawed digits on its wings.

Figure 25.1 Feathered fossil species. The *Archaeopteryx* fossil was discovered in Germany. The reconstructions of *Confuciusornis* and *Sinosauropteryx* are based on fossils unearthed in China.

25.2 Chordate Traits and Evolutionary Trends

■ Chordates, distinguished by their embryonic traits, include two lineages of marine invertebrates, as well as the vertebrates.

■ Link to Animal classification 24.2

Chordate Characteristics

The previous chapter ended with the echinoderms, a deuterostome lineage. The other major deuterostome lineage is the **chordates** (phylum Chordata), a group of bilaterally symmetrical, coelomate animals, with a complete digestive system and a closed circulatory system. Four traits are unique to chordate embryos and define the lineage:

1. A **notochord**, a rod of stiff but flexible connective tissue, extends the length of the body and supports it.
2. A hollow nerve cord parallels the notochord and runs along the dorsal (upper) surface.
3. Gill slits (narrow openings) extend across the wall of the pharynx (the throat region).
4. A muscular tail extends beyond the anus.

Depending on the group, some, none, or all of the defining chordate traits persist in the adult.

Most of the 50,000 or so chordates are **vertebrates**, animals with a backbone. However, the group also includes two lineages of marine invertebrates.

Invertebrate Chordates

Lancelets (subphylum Cephalochordata) are invertebrates with a fish-shaped body (**Figure 25.2**). An adult is about 5 centimeters (2 inches) long. Its nerve cord extends into the head, where a group of nerve cells serves as a simple brain. An eyespot at the end of the nerve cord detects light, but there are no paired sensory organs like those of fishes. Lancelets lie buried up to their mouth in sandy sediments. Cilia lining the pharynx move water into the pharynx and out through gill slits. The gill slits filter food particles out of the water and also function in gas exchange.

In **tunicates** (subphylum Urochordata), only larvae have all the typical chordate traits (**Figure 25.3A**). After metamorphosis, they become barrel-shaped adults with a secreted "tunic" (**Figure 25.3B**). Adults feed by sucking water in through a tube, capturing bits of food on their pharynx, then expelling the filtered water through another tube. Some species are sessile as adults, whereas others drift or swim.

Overview of Chordate Evolution

Until recently, lancelets were considered the closest invertebrate relatives of vertebrates. An adult lancelet

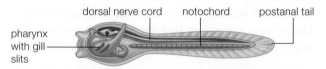

A Free-swimming tunicate larva with all the defining chordate traits.

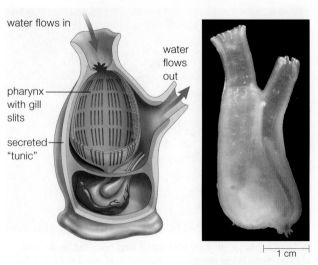

B Adult tunicate. The only defining chordate trait it retains is the pharynx with gill slits. The species shown is sessile as an adult.

Figure 25.3 Animated Larval and adult tunicates.

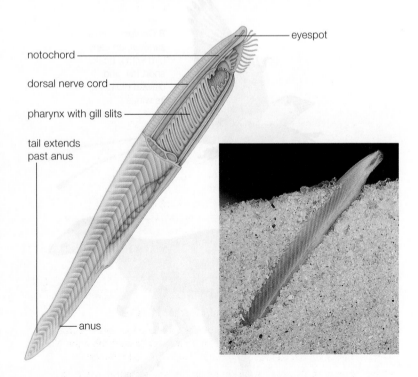

Figure 25.2 Animated Photo and body plan of a lancelet. Both adults and larvae have all four traits that define the chordate lineage.

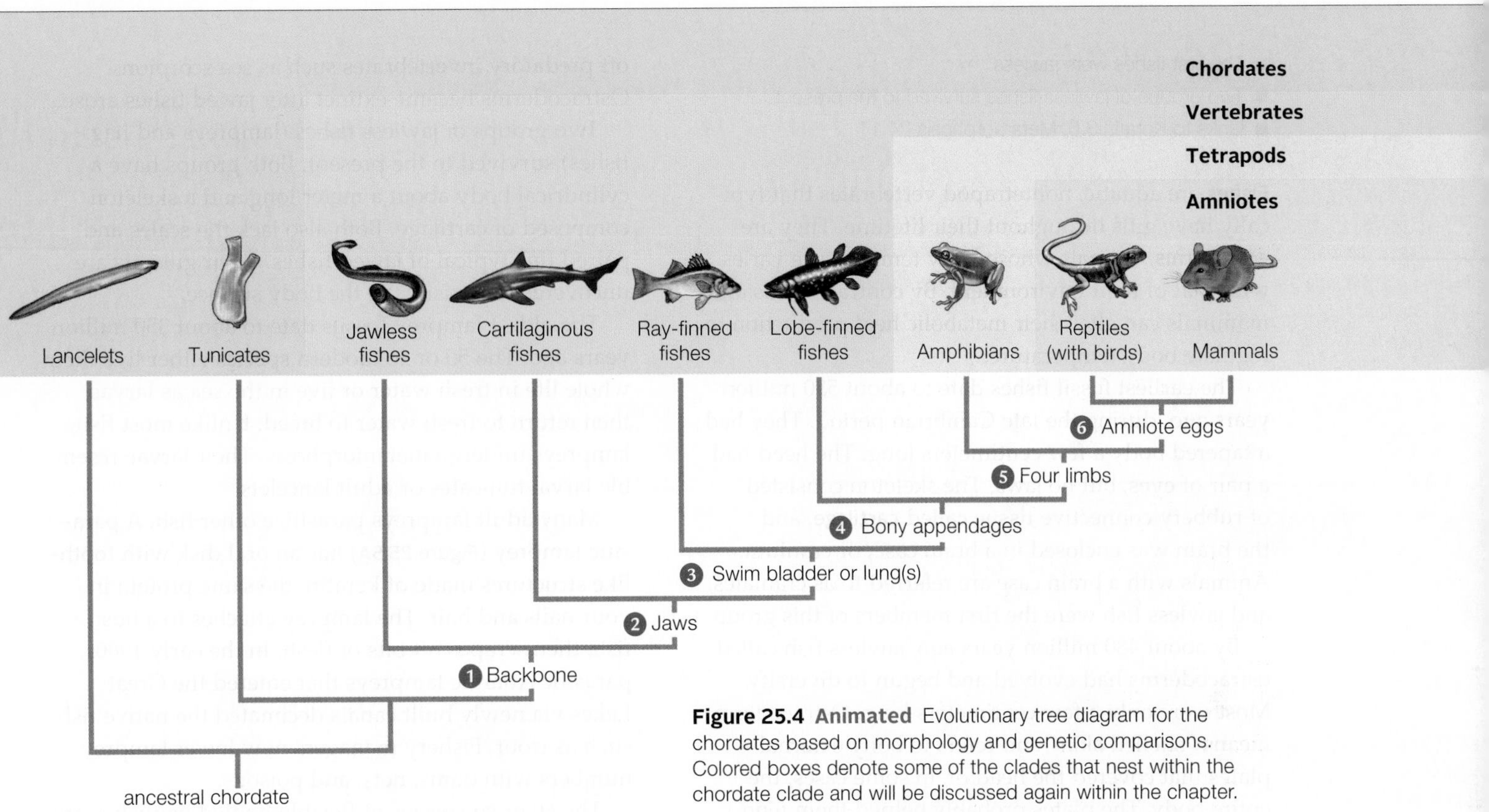

Lancelets   Tunicates   Jawless fishes   Cartilaginous fishes   Ray-finned fishes   Lobe-finned fishes   Amphibians   Reptiles (with birds)   Mammals

6 Amniote eggs

5 Four limbs

4 Bony appendages

3 Swim bladder or lung(s)

2 Jaws

1 Backbone

ancestral chordate

Figure 25.4 Animated Evolutionary tree diagram for the chordates based on morphology and genetic comparisons. Colored boxes denote some of the clades that nest within the chordate clade and will be discussed again within the chapter.

looks more like a fish than an adult tunicate does, but morphological traits can be deceiving. Studies of developmental processes and gene sequences reveal that tunicates are the invertebrate lineage most closely related to the vertebrates (**Figure 25.4**).

Most chordates have a backbone and thus are vertebrates **1**. The backbone and other skeletal elements are components of the vertebrate **endoskeleton**, or internal skeleton. A vertebrate endoskeleton consists of living cells, so it grows with the animal and does not have to be molted.

The first vertebrates were fishes that sucked up or scraped up food. Later, hinged skeletal elements called jaws evolved **2**. Jaws allowed their bearers to exploit new strategies for feeding. The vast majority of fishes and other modern vertebrates have jaws.

Later evolutionary modifications allowed animals to move from water onto land. In one group of fishes, two small outpouchings on the side of the gut wall evolved into lungs: moist, internal sacs that enhance gas exchange with the air **3**. Fins with bony supports inside them evolved in a subgroup of these fishes **4**. The bony fins would later evolve into limbs of the first four-legged walkers, or **tetrapods** **5**.

Early tetrapods spent time on land, but laid their eggs in water. Eggs that enclosed an embryo within a

series of waterproof membranes evolved later. These specialized eggs allowed animals known as **amniotes** to disperse widely on land **6**.

amniote Vertebrate with a unique type of waterproof egg that allows embryos to develop on land.
chordate Animal with an embryo that has a notochord, dorsal nerve cord, pharyngeal gill slits, and a tail that extends beyond the anus. For example, a lancelet or a vertebrate.
endoskeleton Internal skeleton made up of hardened components such as bones.
lancelet Invertebrate chordate that has a fishlike shape and retains all the defining chordate traits into adulthood.
notochord Stiff rod of connective tissue that runs the length of the body in chordate larvae or embryos.
tetrapod Vertebrate with four legs, or a descendant thereof.
tunicate Invertebrate chordate in which the only distinguishing chordate trait retained by adults is a pharynx with gill slits.
vertebrate Animal with a backbone.

Take-Home Message

What traits define the major subgroups of chordates?

» All chordate embryos have a notochord, a dorsal tubular nerve cord, a pharynx with gill slits in its wall, and a tail that extends past the anus. There are two groups of invertebrate chordates: lancelets and tunicates.

» Most chordates also have a backbone and so are vertebrates. Limbs evolved in one lineage that later colonized the land. Amniotes, a tetrapod subgroup with specialized eggs, are the predominant vertebrates on land.

25.3 Jawless Fishes

- The first fishes were jawless.
- Two groups of jawless fishes survived to the present.
- Links to Keratin 3.6, Metamorphosis 24.11

Fishes are aquatic, nontetrapod vertebrates that typically have gills throughout their lifetime. They are **ectotherms**, animals whose body temperature varies with that of their environment. By contrast, birds and mammals can alter their metabolic heat production to regulate body temperature.

The earliest fossil fishes date to about 530 million years ago, during the late Cambrian period. They had a tapered body a few centimeters long. The head had a pair of eyes, but no jaws. The skeleton consisted of rubbery connective tissue called **cartilage**, and the brain was enclosed in a brain case, or cranium. Animals with a brain case are referred to as craniates, and jawless fish were the first members of this group.

By about 480 million years ago, jawless fish called ostracoderms had evolved and begun to diversify. Most were only a few centimeters long. Ostracoderm means "shelled skin" and refers to bony external plates that covered the head or, in some cases, the entire body. The plates probably helped them fend off predatory invertebrates such as sea scorpions. Ostracoderms became extinct after jawed fishes arose.

Two groups of jawless fishes (lampreys and hagfishes) survived to the present. Both groups have a cylindrical body about a meter long and a skeleton composed of cartilage. Both also lack the scales and paired fins typical of jawed fishes. Their gill slits are uncovered and visible at the body surface.

The oldest lamprey fossils date to about 350 million years ago. The 50 or so modern species either live their whole life in fresh water or live in the sea as larvae, then return to fresh water to breed. Unlike most fish, lampreys undergo metamorphosis. Their larvae resemble larval tunicates or adult lancelets.

Many adult lampreys parasitize other fish. A parasitic lamprey (**Figure 25.5A**) has an oral disk with toothlike structures made of keratin, the same protein in your nails and hair. The lamprey attaches to a host fish, then scrapes off bits of flesh. In the early 1900s, parasitic Atlantic lampreys that entered the Great Lakes via newly built canals decimated the native fish such as trout. Fishery managers now lower lamprey numbers with dams, nets, and poisons.

The 60 or so species of flexible-bodied hagfishes are marine bottom-feeders (**Figure 25.5B,C**). Hagfishes have poor eyesight and use sensory tentacles near their mouth to locate worms and carcasses. Their mouth has dental plates covered with sharp barbs of keratin. The barbed plates are used to grab and pull apart food. A frightened hagfish secretes a compound that combines with water to form a gelatinous slime. Slime deters most predators but has not kept humans from harvesting hagfish. Most products labeled as "eelskin" are actually made of hagfish skin.

Relationships among lampreys, hagfishes, and jawed fishes have long inspired debate. The most recent genetic comparisons indicate that hagfishes and lampreys constitute a monophyletic group.

A Lamprey using its oral disk to attach to the glass of an aquarium.

B Hagfish

C Hagfish feeding apparatus.

Figure 25.5 Two lineages of jawless fishes that survived to the present.

cartilage Rubbery connective tissue that is a component of vertebrate skeletons.
ectotherm Animal whose body temperature varies with that of its environment.
fish Gilled aquatic vertebrate that is not a tetrapod.

Take-Home Message

What are jawless fishes?

»Jawless fishes are gilled, aquatic vertebrates with a cartilage skeleton. They do not have jaws or scales.

» Lampreys and hagfishes have hard mouthparts made of keratin. Lampreys undergo metamorphosis and some parasitize other fish as adults. Hagfishes are marine scavengers.

25.4 Evolution of Jawed Fishes

■ The evolution of jaws and fins opened the way to a great diversification of fishes.

■ Link to Adaptive radiation 17.12

The first jawed vertebrates (subphylum Gnathostomata) evolved during the late Silurian period, by about 410 million years ago. Jaws evolved from gill arches, skeletal elements that support a fish's gills (**Figure 25.6**). Fishes with jaws were at an advantage over jawless fishes. Jaws help a fish catch and kill prey, and make it easier to tear large prey into easy-to-swallow chunks.

Jawed fishes were also the first animals with scales and paired fins. **Scales** are hard, flattened structures that grow from and often cover the skin. Fins are flattened appendages used to propel and steer a body while swimming. Most jawed fishes have two pairs of fins: pectoral fins and pelvic fins.

The Devonian period (416–359 million years ago) is called the "Age of Fishes," and placoderms were the most numerous and diverse vertebrates in Devonian seas. About 200 different species of these jawless fishes have been identified from fossils. Placoderm means "tablet skin" and refers to bony armor that covered the animal's head and neck (**Figure 25.7**). A typical placoderm did not have teeth, but sharp bony plates performed the same function. Placoderms grew larger than the jawless fishes that preceded them, and some were enormous. *Dunkleosteus*, which once swam through a shallow sea in what is now Ohio, grew as long as a bus. Placoderms are also the earliest vertebrates for which we have fossil evidence of internal fertilization and development. Scientists recently discovered a fossilized female that apparently died while giving birth. Placoderms became extinct at the end of the Devonian.

Another group of early jawed fish lineages collectively is referred to as acanthodians (spiny fins). Acanthodians arose at about the same time as placoderms, but they were smaller (centimeters long), less diverse, and did not have bony armor. As a result,

scales Flattened structures that grow from and sometimes cover the skin in some groups of vertebrates.

A Cast of a fossil *Dunkleosteus* skull, with a human to illustrate scale.

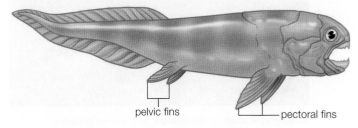

B Placoderms had paired fins that could be moved by muscles.

Figure 25.7 Placoderms, early jawed fishes that are now extinct.

they left fewer fossils than placoderms, so we know less about them. Acanthodians became extinct at the end of the Permian period.

Tiny fossilized scales reveal that small sharks lived during the late Silurian. However, like other jawed fishes, sharks diversified during the Devonian.

Take-Home Message

When did jawed fishes evolve and what traits characterized them?

» Jaws evolved during the Silurian period by the modification of the first pair of gill arches in a jawless ancestor. Jawed fishes were also the first vertebrates with paired fins.

» Placoderms were an early group of jawed fishes that had bony plates on their head and neck. Some grew to great size. The acanthodian lineages were smaller and lacked bony armor.

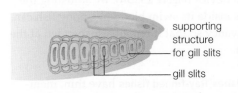
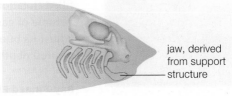
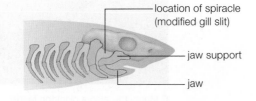

Figure 25.6 Animated Proposed steps in the evolution of jaws.

25.5 Modern Jawed Fishes

- There are two living lineages of jawed fishes, cartilaginous fishes and bony fishes.
- Link to Gene duplications 14.5

Figure 25.8 shows the one hypothesis for relationships among jawed vertebrates, both extinct and living. Cartilaginous fishes may comprise a monophyletic group, but the bony fishes certainly do not.

Cartilaginous Fishes

Cartilaginous fishes are a mostly marine group of jawed fishes with a cartilage skeleton. Their five to seven

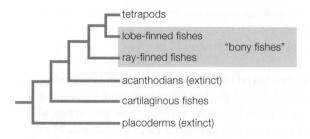

Figure 25.8 Proposed relationships among jawed vertebrates. Note that bony fishes (*blue* box) are not a clade.

A Galápagos shark, a streamlined, fast-moving predator.

B Basking shark, a huge, slow-moving plankton feeder.

C Manta ray, also a plankton feeder.

Figure 25.9 Cartilaginous fishes.

pairs of gill slits are typically uncovered at the body surface. The jaws have teeth that grow in rows and are continually shed and replaced. Sexes are separate. Eggs typically develop in an egg case inside the mother's body. When the young are mature, the egg case ruptures and they are released into the environment. Less commonly, females release egg cases containing developing embryos. In most groups of cartilaginous fishes, a single opening on the ventral surface, called a **cloaca**, functions in reproduction and also serves as the exit for digestive and urinary waste. All living jawless fishes also have a cloaca.

Most of the 850 species of cartilaginous fishes are sharks or rays. The most well-known sharks are speedy predators that chase down and tear apart prey (**Figure 25.9A**). Others are bottom-feeders that suck up invertebrates and act as scavengers. Still others strain plankton from the seawater. The largest living fishes, the whale shark and basking shark, feed in this way (**Figure 25.9B**).

Rays have a flattened body with large pectoral fins. Manta rays glide through seawater and feed by filtering out plankton (**Figure 25.9C**). Stingrays are bottom-feeders. Their barbed tail has a venom gland that serves as a defense against predators.

Bony Fishes

Modern **bony fishes** include members of two lineages: ray-finned fishes and lobe-finned fishes. In both groups, bone replaces cartilage in some or most of the skeleton and gill slits are hidden beneath a gill cover. Most bony fishes have a **swim bladder**: a gas-filled flotation device (**Figure 25.10A**). By adjusting the volume of gas inside its swim bladder, a bony fish can adjust its buoyancy so that it remains suspended at the desired depth.

Ray-Finned Fishes **Ray-finned fishes** have thin, membranous fins with flexible fin supports derived from skin. With about 24,000 living species, they are the most diverse group of modern fishes. Sturgeons are modern representatives of one ancient ray-finned lineage. Humans harvest the eggs of some sturgeons for use as caviar. Gars, predatory fish with an elongated body, are members of another early ray-finned lineage (**Figure 25.10B**).

About 99 percent of all ray-finned fishes belong to the most recently evolved lineage, the teleosts. Teleosts include most fish that we harvest as food, including salmon, sardines, bass, swordfish, trout, tuna, halibut, carp, and cod. Some have a highly modified body plan (**Figure 25.10C,D**).

Comparisons of the teleost genome with that of other ray-finned fishes indicate that the ray-finned lineage underwent whole genome duplication early in its history. Gene duplications can speed evolution by allowing one copy of a gene to mutate and take on a new function while the other copy retains its original role. The diversification of copied genes is thought to have facilitated an adaptive radiation of this group.

Lobe-Finned Fishes Lobe-finned fishes, the fish most closely related to tetrapods, have fleshy fins supported by bones. There are two lineages, the marine coelacanths and the freshwater lungfishes. Coelacanths were abundant from the Devonian through the Cretaceous period, but the group was thought to be extinct until a population was located in 1938. Since then, other populations have been discovered (**Figure 25.11A**). There are at least two species.

Lungfishes (**Figure 25.11B**) have both gills and air sacs, modified outpouchings of the gut wall that func-

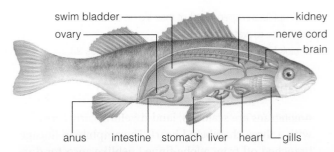

A Body plan of a perch.

B Long-nose gar, a freshwater predator.

C Sea horse. **D** Monkfish, a stealthy marine predator.

Figure 25.10 **Animated** Ray-finned fishes.

tion in respiration. A lungfish fills its air sacs by surfacing and gulping air, then oxygen diffuses from the air in the sacs into the blood.

A Coelacanth. It lives in underwater caves and canyons.

Figure 25.11 Lobe-finned fishes, with fleshy fins supported by bones.

B Australian lungfish.

bony fish Common term for fish with a skeleton containing bone.
cartilaginous fish Jawed fish that has a skeleton of cartilage.
cloaca In some vertebrates, a body opening through which both wastes and gametes exit.
lobe-finned fish Bony fish with fleshy fins supported by bones.
ray-finned fish Bony fish with fin supports derived from skin.
swim bladder Adjustable flotation sac of some bony fish.

Take-Home Message

What are the characteristics of jawed fishes?

» Jawed fishes are cartilaginous fishes and bony fishes. Both groups typically have scales. The ray-finned lineage of bony fishes is the most diverse group of vertebrates. Lobe-finned fishes are the fish closest to the tetrapods.

25.6 Amphibians—First Tetrapods on Land

- Amphibians spend part of their life on land, but most still return to water to breed.
- Link to Homologous structures 18.3, Chytrids 23.3

Adapting to Life on Land

Amphibians are scaleless, land-dwelling carnivores with a three-chambered heart. The amphibian lineage branched off from a lobe-finned fishlike ancestor during the Devonian period. Fossilized species demonstrate how the skeleton of fishes adapted to swimming became modified in the early tetrapods (**Figure 25.12**). The bones inside a lobe-finned fish's pelvic and pectoral fins are homologous with the bones inside amphibian limbs (Section 18.3).

The transition to land was not simply a matter of skeletal changes. Alterations of the inner ear improved detection of airborne sounds. Eyes became protected from drying out by eyelids. Fishes have a two-chamber heart: one chamber receives blood, the other pumps it out. In amphibians, the heart became divided into three chambers: one to receive blood, one to pump blood to the lungs, and one to pump blood to the body. This change increased the rate of blood flow to the body and the efficiency of gas exchange.

What was the advantage of living on land? An ability to survive out of water is useful in seasonally dry places. Individuals on land were also safe from aquatic predators and had access to a new source of food—insects—which also evolved during the Devonian.

Figure 25.13 (**A**) Shasta salamander, with equal-sized forelimbs and hindlimbs. (**B**) A legless caecilian.

Amphibians retain an important tie to the water. Because their eggs dry out easily, most species require standing fresh water or damp soil to breed. Males and females typically congregate at breeding sites, where males squirt sperm onto eggs as a female lays them in the water. The fertilized eggs hatch into larvae that have gills and complete their development to adults with lungs in the water.

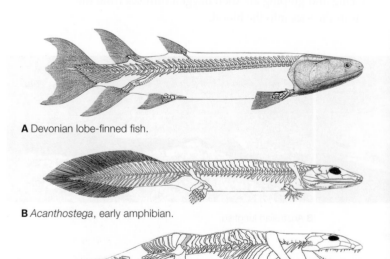

A Devonian lobe-finned fish.

B *Acanthostega*, early amphibian.

C *Ichthyostega*, early amphibian.

Figure 25.12 Animated Skeletons of a Devonian lobe-finned fish and two early amphibians. The painting at right shows what *Acanthostega* (foreground) and *Ichthyostega* (background) may have looked like.

Modern Amphibians

Modern amphibians include salamanders, caecilians, frogs, and toads. All are carnivores as adults, preying mainly on insects and worms.

The 535 species of salamanders and related newts live mainly in North America, Europe, and Asia. In body form, they are the modern group most like early tetrapods. Forelimbs and back limbs are of similar size and there is a long tail (**Figure 25.13A**). As salamanders walk, their body bends from side to side, like the body of a swimming fish. Their ancestors that first ventured onto land probably moved in a similar way.

Larval salamanders look like small versions of adults, except for the presence of gills. Typically, they lose their gills and develop lungs as they mature. However, some salamanders (axolotls) retain gills as adults. Others lose their gills but do not develop lungs; gas exchange occurs across the skin.

Caecilians are close relatives of salamanders that live in the tropics and have adapted to a burrowing way of life. They include about 165 limbless, blind species (**Figure 25.13B**). Most caecilians burrow through soil and use their senses of touch and smell to pursue invertebrate prey.

Frogs and toads belong to the most diverse amphibian lineage, with more than 5,000 species. Long, muscular hindlimbs allow the tailless adults to swim, hop, and make spectacular leaps (**Figure 25.14A**). The much smaller forelimbs help absorb the impact of landings. Toads tend to have shorter hind legs than frogs (**Figure 25.14B**) and are better adapted to dry habitats.

Frogs and toads undergo metamorphosis. A larval frog or toad, commonly called a tadpole or pollywog, has gills and a tail but no limbs (**Figure 25.14C**). All frogs and toads have lungs as adults.

An Ongoing Decline

We are in the midst of an alarming decline in amphibian numbers. Population reductions are best documented in North America and Europe, but declines are happening worldwide. An amphibian's thin, scaleless skin makes it relatively easy for parasites, pathogens, and pollutants to enter the body. Section 23.3 described the effects of an introduced chytrid fungus on frogs. Section 34.1 explains how agricultural chemicals can disrupt amphibian hormone production.

Habitat loss is another threat. In many places, people have filled in low-lying areas that used to collect water from seasonal rains. Such seasonal pools are important breeding sites for many amphibians.

A Long, muscular hindlimbs allow an adult frog to leap.

B American toad. The flattened disk visible behind the eye is the eardrum.

C Tadpole, a gilled larva with a tail.

Figure 25.14 Frogs and toads.

amphibian Tetrapod with a three-chambered heart and scaleless skin; typically develops in water, then lives on land as an air-breathing carnivore.

Take-Home Message

What are amphibians and why are many species in decline?

» Amphibians are carnivorous vertebrates that typically live on land but breed in water. The most diverse group includes the frogs and toads, which undergo metamorphosis from gilled larvae. Salamanders and the closely related caecilians are less diverse lineages.

» An amphibian's scaleless, permeable skin makes it vulnerable to pollutants, and its requirement for water in which to breed makes it sensitive to habitat alteration.

25.7 The Amniotes

■ Amniotes are vertebrates that have adapted to a life lived entirely on land.

■ Links to Asteroid impact 16.1, Cladistics 18.2

About 300 million years ago, during the Carboniferous period, amniotes branched off from an amphibian ancestor. A variety of traits that evolved in this lineage adapted its members to life in dry places. **Amniote eggs** have distinctive internal membranes that keep an embryo moist even away from water (**Figure 25.15**). In addition, amniote skin is rich in keratin, a protein that makes it waterproof. A pair of well-developed kidneys help conserve water, and fertilization usually takes place inside the female's body.

Shortly after amniotes arose, the lineage branched in two. One branch gave rise to the mammals. The other gave rise to a clade that biologists call **reptiles** (**Figure 25.16**). This clade includes turtles, lizards, snakes, crocodilians, and birds.

Dinosaurs are extinct members of a reptile clade that first evolved in the Triassic period. They had unique skeletal features, such as the shape of their pelvis and hips. As Section 25.1 explained, evidence suggests that the first birds evolved from a group of feathered dinosaurs. The painting in **Figure 25.17** shows a Jurassic

Figure 25.15 Snakes hatching from the amniote eggs in which they developed.

scene, with the early bird *Archaeopteryx* and an early mammal. With the exception of the birds, all members of the dinosaur lineage disappeared by the end of the Cretaceous. An asteroid impact is considered the most likely cause of their demise (Section 16.1).

amniote egg Egg with internal membranes that allow the amniote embryo to develop away from water.
dinosaur Reptile lineage abundant in the Jurassic to Cretaceous; now extinct with the exception of birds.
reptile Tetrapod with amniote eggs and scales or feathers.

Take-Home Message

What are the amniotes?

» Amniotes are animals that produce eggs in which the young can develop away from water. They are also adapted to life on land by waterproof skin, highly efficient kidneys, and internal fertilization.

» An early divergence separated the ancestors of mammals from the ancestors of modern reptiles, a group in which biologists include turtles, lizards, snakes, crocodilians, and birds.

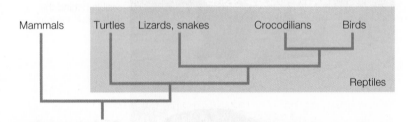

Figure 25.16 Evolutionary tree for the living amniote groups. The *blue* box denotes members of the reptile clade. Note that it includes the birds.

Figure 25.17 Painting of a Jurassic scene. In the middle foreground, the early bird *Archaeopteryx* flies by. *Behind* the birds, a meat-eating dinosaur sizes up a larger plant-eating dinosaur. At the *far right*, an early mammal surveys the scene from its perch on a tree.

Figure It Out: Which of the animals mentioned above do biologists consider amniotes? Which do they group as reptiles?

Answer: All are amniotes. The dinosaurs and birds are reptiles.

25.8 Nonbird Reptiles

■ Reptiles have a scale-covered body. Most have four limbs of approximately equal size, but the snakes are limbless.

General Characteristics

Like fishes, reptiles have scales. However, reptile scales develop from the outer layer of skin (epidermis), whereas fish scales arise from a deeper layer. Male reptiles typically have a penis and fertilize eggs inside a female's body. A cloaca expels waste and functions in reproduction. Females usually lay eggs on land, but females of some lizards and snakes retain eggs in their body and young are born fully developed.

All modern nonbird reptiles are ectotherms. Those that live in temperate regions spend the cold season inactive in a burrow on land, or in the case of some freshwater turtles, in mud at a lake bottom.

Major Groups

Turtles The 300 or so species of turtles have a bony, scale-covered shell that protects their body (**Figure 25.18**). We can see from fossils how turtles have evolved. A 200-million-year-old fossil turtle found in China has a protective plate derived from expanded ribs on its ventral side, but no shell on its back. It also has teeth. Modern turtles do not have teeth. As in birds, a layer of keratin covers their jaws and forms a beak. Some feed on plants and others are predators.

Lizards and Snakes Lizards and snakes constitute the most diverse lineage of modern reptiles. Overlapping scales cover their body. The smallest lizard can fit on a dime (*right*). The largest, the Komodo dragon, grows up to 3 meters (10 feet) long. Most of the 5,000 or so lizards are predators and some produce venom. A Komodo dragon delivers a venomous bite, then trails its prey for hours or days until it collapses.

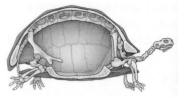

Above, Skeleton of a modern turtle.

Right, Galápagos tortoise. "Tortoise" is the common term for a turtle that lives on land.

Figure 25.18 Animated Turtles.

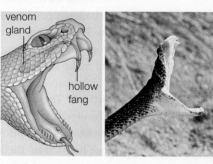

The snakes evolved from short-legged, long-bodied lizards. A few modern snakes retain bony remnants of hindlimbs, but most lack limb bones. All are carnivores and many have flexible jaws that open wide. Fanged snakes subdue prey with venom made in modified salivary glands (**Figure 25.19**).

Figure 25.19 A fanged snake (rattlesnake).

Crocodilians Most reptiles have a three-chambered heart like amphibians do, but crocodiles, alligators, and caimans have a four-chambered heart. Having four chambers keeps oxygen-poor blood from mixing with oxygen-rich blood. All crocodilians spend much of their time in water. They have powerful jaws, a long snout, and sharp peglike teeth (**Figure 25.20**). They clench prey, drag it underwater, tear it apart as they spin around, and gulp the chunks.

Take-Home Message

What are the traits of nonbird reptiles?

» All nonbird reptiles are ectotherms with a cloaca and scales.

» Turtles have a bony shell and a horny beak. Lizards and snakes have overlapping scales on an elongated body, and some produce venom. Crocodilians have a four-chambered heart and a long snout with peglike teeth.

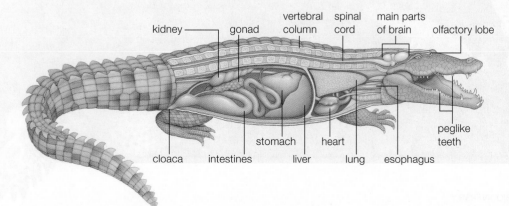

kidney · gonad · vertebral column · spinal cord · main parts of brain · olfactory lobe · cloaca · intestines · stomach · liver · heart · lung · esophagus · peglike teeth

Figure 25.20 Animated Body plan of a crocodile and photo of a caiman.

25.9 Birds—The Feathered Ones

■ In one group of dinosaurs, the scales became modified as feathers. Birds are modern descendants of this group.

■ Links to Aerobic respiration 7.2, Homologous structures 18.3

General Characteristics

Birds, the only living animals that have feathers, first arose during the Jurassic and are thought to be descended from feathered dinosaurs. Feathers are filamentous keratin structures derived from scales. They have a hollow shaft, with thin barbs that in turn have thinner barbules (**Figure 25.21A**).

Many birds have colorful plumage that plays a role in courtship. The color can arise from the structure of the keratin or from deposition of pigments derived from the diet. Dietary carotenoids deposited in feathers make flamingos pink and American cardinals red.

Birds are **endotherms,** which means "heated from within." They maintain their body temperature within a limited range by adjusting their metabolic heat production. A bird's downy feathers serve as insulation that helps slow the loss of metabolic heat in cool environments, and helps prevent heat gain in hot ones.

The bird body has many adaptations for flight (**Figure 25.21**). Like humans, birds stand upright on their hindlimbs, and their wings are homologous to our arms. Each wing is covered with long feathers that extend outward and increase its surface area. The feathers give the wing a shape that helps lift the bird as air passes over it.

Flight muscles connect a large breastbone (sternum), which has a distinctive bony extension, to upper limb bones. Contraction of one set of muscles produces a powerful downstroke that lifts the bird. A less powerful set of muscles contracts to raise the wing.

Most birds are surprisingly lightweight, and this feature helps them become and remain airborne. Air cavities inside a bird's bones keep its body weight low, as does the lack of a bladder (an organ that stores

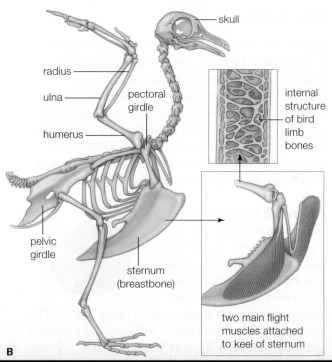

Figure 25.21 Avian adaptations to flight.

(**A**) Structure of a flight feather.

(**B**) A bird's skeleton is made up of lightweight bones with internal air pockets. A wing is a modified forelimb (see **Figure 18.5**). Powerful flight muscles attach to a large breastbone, or sternum.

(**C**) Flight muscles pull wings up and down during flight.

Figure 25.22 Mating house sparrows. A male bird does not have a penis. To inseminate his partner, he must balance on her back and bend his body so his cloaca meets hers.

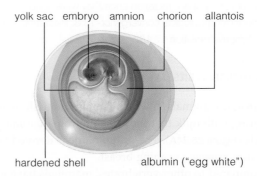

yolk sac   embryo   amnion   chorion   allantois

hardened shell   albumin ("egg white")

Figure 25.23 Animated Bird's egg. The egg has a calcium-hardened shell that encloses the embryo and four characteristic amniote membranes (yolk sac, amnion, chorion, and allantois).

urinary waste in many other vertebrates). Rather than heavy, bony teeth, bird jaws are covered with a lightweight beak made of keratin.

Keeping muscles supplied with ATP that fuels contractions requires plenty of oxygen for aerobic respiration (Section 7.2). A unique system of air sacs keeps air flowing continually through a bird's lungs, maintaining a high oxygen concentration even in active muscles. Birds, like their closest living relatives the crocodilians, have a four-chambered heart that pumps blood in two fully separated circuits.

Flying requires good eyesight and a great deal of coordination. Compared to a lizard of a similar body mass, a bird has much larger eyes and a larger brain.

As in other reptiles, fertilization is internal. However, most male birds do not have a penis. To inseminate a female, a male must press his cloaca against hers, a maneuver poetically described as a cloacal kiss (**Figure 25.22**). Fertilization occurs in the female's body, then she lays an egg with the characteristic amniote membranes (**Figure 25.23**). Nutrients from the yolk and water from the albumin in the egg sustain the developing embryo. Like some turtles and all crocodilians, birds encase their eggs within a rigid shell of calcium carbonate.

Avian Diversity

More than half of the approximately 10,000 living bird species belong to a single order, the perching birds (Passiformes). This group includes familiar backyard birds such as sparrows, crows, jays, starlings,

swallows, finches, robins, warblers, orioles, and cardinals. It also includes the birds of paradise, a tropical group whose males have elaborate plumage that is cited as a product of sexual selection (Section 17.7).

The next most diverse order includes 450 species, the majority of them hummingbirds. Hummingbirds are agile fliers and the only birds capable of flying backward. Adults sustain themselves mainly on floral nectar and are important as pollinators.

Many birds, including some hummingbirds and perching birds, make a seasonal migration; they fly from one region to another in response to a seasonal change, such as a shift in daylength. Migration typically involves spring travel to a breeding site where insects are abundant in the summer, then an autumn flight to an overwintering site. Migration can be arduous. One shorebird monitored by researchers flew from Alaska to New Zealand, a distance of 11,500 kilometers (7,145 miles), without stopping to feed or rest.

At the opposite end of the spectrum are birds that have become flightless, such as the ratite birds and the penguins. The ostrich is an African ratite and the largest living bird. It can weigh up to 150 kilograms (330 pounds). Penguins are aquatic birds that live in coastal waters of the Southern Hemisphere. Penguin wings have become modified as flippers that the birds use to "fly" through water.

bird An animal with feathers.
endotherm Animal that controls its body temperature by varying its production of metabolic heat; for example a bird or mammal.

Take-Home Message

What are birds?

» Birds are the only living animals with feathers. They evolved from dinosaurs. Adaptations for flight include lightweight bones; air sacs that increase the efficiency of respiration; and a four-chambered heart that keeps blood moving rapidly.

25.10 The Rise of Mammals

■ Mammals scurried about while dinosaurs dominated, then underwent an adaptive radiation once they were gone.

■ Links to Morphological convergence 18.3, Plate tectonics 16.7, Adaptive radiation 17.12

Mammalian Traits

Mammals are animals in which females nourish their offspring with milk that they secrete from mammary glands (**Figure 25.24A**). The group name is derived from the Latin *mamma*, meaning breast.

Compared to other vertebrates, mammals have a larger brain for their body size. They also have distinctive middle ear bones and teeth. Only mammals have four different kinds of teeth (**Figure 25.24B**). In other vertebrates, an individual's teeth vary in size, but they are all the same shape. Mammals have incisors that can be used to gnaw, canines that tear and rip flesh, and premolars and molars that grind and crush hard foods. Not all mammals have all four tooth types, but most have some combination. Having a variety of

tooth shapes gives mammals an ability to eat a wider variety of foods than other vertebrates.

Mammals have hair or fur that, like feathers, is made of keratin. Mammals are endotherms, and a coat of fur or head of hair helps them maintain their body temperature. Mammals are the only animals that sweat, although not all mammals do so.

Like birds and crocodilians, mammals have a four-chambered heart that pumps their blood through two fully separate circuits. Gas exchange occurs in a pair of well-developed lungs.

Mammalian Evolution

The lineage from which mammals would evolve (the synapsids) branched off from other reptiles during the Permian period. The earliest mammal fossils date to about 220 million years ago during the late Triassic, before large dinosaurs evolved. Early mammals were tiny shrewlike insectivores.

Two modern mammalian lineages, the **monotremes** (egg-laying mammals) and **marsupials** (pouched mammals), originated in the Jurassic. **Placental mammals** evolved a bit later, during the Cretaceous. Placental embryos grow faster than other mammal embryos and are born more fully formed. Thus, young are less vulnerable to predation.

Continental movements affected mammal dispersal. Monotremes and marsupials evolved while Pangea was intact, and they dispersed across this supercontinent (**Figure 25.25A**). Placental mammals evolved after Pangea had begun to break up (**Figure 25.25B**). As a result, monotremes and marsupial mammals on landmasses that broke early from Pangea lived for millions of years in the absence of placental mammals. For example, Australia split off from Pangea early on and so does not have native placental mammals.

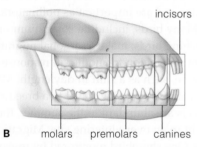

B molars premolars canines

Figure 25.24 Animated Distinctly mammalian traits. (**A**) A human baby, already with a mop of hair, being nourished by milk secreted from the mammary gland in a breast. (**B**) Four types of teeth and a single lower jawbone.

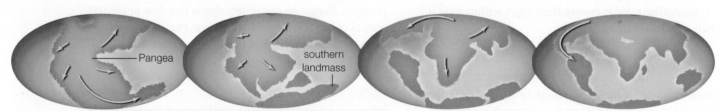

A About 150 million years ago, during the Jurassic, the first monotremes and marsupials evolved and migrated through the supercontinent Pangea.

B Between 130 and 85 million years ago, during the Cretaceous, placental mammals arose and began to spread. Monotremes and marsupials that lived on the southern landmass evolved in isolation from placental mammals.

C Starting about 65 million years ago, mammals expanded in range and diversity. Marsupials and early placental mammals displaced monotremes in South America.

D About 5 million years ago, in the Pliocene, advanced placental mammals invaded South America. They drove most marsupials and the early placental species to extinction.

Figure 25.25 Effects of continental drift on the evolution and distribution of mammalian lineages.

Figure 25.26 Paleocene mammals in a sequoia forest in what is now Wyoming. With the exception of the marsupial on the tree branch, all are members of extinct mammalian lineages.

Figure 25.27 *Indricotherium*, the "giraffe rhinoceros." At 15 tons and 5.5 meters (18 feet) high at the shoulder, it is the largest land mammal we know about. It lived in Asia during the Oligocene and is a relative of the rhinoceros.

Figure 25.28 Example of morphological convergence. (**A**) Australia's spiny anteater, one of only three modern species of monotremes. (**B**) Africa's aardvark and (**C**) South America's giant anteater. All have the ant-snuffling snouts.

A Egg-laying mammal.

B Pouched mammal.

C Placental mammal.

Australia remains a separate continent, and most monotremes and marsupials live there. Other land-masses reunited after millions of years of tectonic movement. When placental mammals entered regions where they were previously unknown, populations of monotremes and marsupials declined. Placental mammals outcompeted their nonplacental counterparts and drove them to local extinction (**Figure 25.25C,D**).

After dinosaurs died out at the end of the Cretaceous, mammals underwent their great adaptive radiation. **Figures 25.26** and **25.27** provide examples of some of the resulting diversity.

Members of different mammal lineages evolved similar body form as they adapted to similar habitats on different continents. For example, Australia's spiny anteater, South America's giant anteater, and Africa's aardvark all hunt ants using their long snout (**Figure 25.28**). This is an example of morphological convergence (Section 18.3).

mammal Animal with hair or fur; females secrete milk from mammary glands.
marsupial Mammal in which young are born at an early stage and complete development in a pouch on the mother's surface.
monotreme Egg-laying mammal.
placental mammal Mammal in which a mother and her embryo exchange materials by means of an organ called the placenta.

Take-Home Message

What are mammals and what is their evolutionary history?

» Mammals are animals that nourish young with milk and have hair or fur. Their four kinds of teeth allow them to eat many different kinds of foods.

» Mammals originated during the late Triassic, then underwent an adaptive radiation after dinosaurs died out at the end of the Cretaceous.

» Continental movements influenced mammal distribution.

25.11 Modern Mammalian Diversity

■ Mammals successfully established themselves on every continent and in the seas.

■ Link to Whale fossils 16.6

Egg-Laying Monotremes

Three species of monotremes still exist. Two are spiny anteaters, and one of these is shown in **Figure 25.27A**. The third species is the platypus (**Figure 25.29A**). A platypus has a beaverlike tail, a ducklike bill, and webbed feet. Platypuses burrow into riverbanks using claws exposed when they retract the web on their feet. Both males and females have spurs on their hind feet. A male's spurs have venom, making them the only venomous mammals.

All female monotremes lay and incubate eggs that have a leathery shell like that of lizards. Offspring hatch in a relatively undeveloped state—tiny, hairless, and blind. Young cling to the mother or are held in a skin fold on her belly. Milk oozes from openings on the mother's skin; monotremes do not have nipples.

Pouched Marsupials

Most of the 240 modern species of marsupials live in Australia and on nearby islands. Groups include kangaroos (**Figure 25.29B**), koalas, and the Tasmanian devil. The opossum (**Figure 25.29C**) is the only marsupial native to North America.

Marsupials develop briefly in their mother's body, nourished by egg yolk and by nutrients that diffuse from maternal tissues. The young are born at an early developmental stage, while they are still tiny. They crawl along their mother's body to a permanent pouch

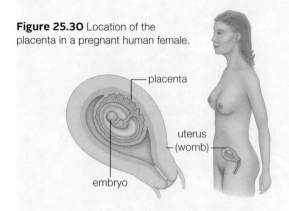

Figure 25.30 Location of the placenta in a pregnant human female.

placenta

uterus (womb)

embryo

on her ventral surface. They attach to a nipple in the pouch, suckle, and grow.

Placental Mammals

Compared to marsupials, placental mammals develop to a far more advanced stage inside their mother's body. An organ called the **placenta** allows materials to pass between maternal and embryonic bloodstreams (**Figure 25.30**). The placenta transfers nutrients more efficiently than diffusion does, allowing the embryo to grow faster. After birth, young suckle milk from nipples on the mother's ventral surface.

With more than 4,000 species, placental mammals are the dominant mammals in most land habitats, and the only mammals that live in the seas. As with other groups, genetic comparisons often produce results that conflict with a traditional classification system that was based largely on morphology. **Figure 25.31** shows examples of major, currently accepted orders.

About 40 percent of all placental mammals are currently classified as rodents (order Rodentia). They

A Platypus, an Australian monotreme.

B Kangaroo, one of many Australian marsupials.

C Opossum, the only marsupial native to North America.

Figure 25.29 Representative nonplacental female mammals with their offspring.

Transitions Written in Stone (revisited)

Vertebrate fossils are big business for rock shops, auction houses, and web sites. Most such fossils are not particularly important to scientists, but some are. For example, one of the few *Archaeopteryx* fossils in existence is privately held. It shows details of the bird's feet that are not visible in other fossils. Some scientists argue that private ownership of such fossils thwarts research and endangers an irreplaceable legacy.

How would you vote? One-of-a-kind vertebrate fossils are in private collections. Is buying and selling scientifically important fossils unethical?

---

include rats, mice, hamsters, squirrels, beavers, porcupines, and guinea pigs. All have teeth specialized for gnawing. Their incisors grow continually, so they are not worn down to nubs by wear.

About 1,000 species of bats constitute the second most diverse order (Chiroptera). Bats are the only mammals that fly. A bat's wing consists of membranes of skin stretched between bones.

Moles and shrews make up the next most diverse order (Soricomorpha). Most moles are adapted to a burrowing way of life, with strong forelimbs for tunneling, reduced eyes, and no external ear flaps. They eat earthworms and insects. Shrews look like mice with spiky teeth, but are not close relatives of rodents. Moles and shrews were traditionally grouped with other small insect eaters such as hedgehogs in an order (Insectivora) that was dismantled when gene comparisons showed its members were not all closely related.

Carnivora means meat eater, and members of this order have enlarged canine teeth suited to tearing at flesh. They include cats, dogs, wolves, bears, foxes, and weasels, as well as marine pinnipeds (fin-footed animals) such as seals, sea lions, and walruses.

Whales and dolphins (Cetacea) have also adapted to life in the sea. Section 16.6 described the fossils that document the whale's transition from land-dwellers to marine mammals. As far as we know, the blue whale, which can be 30 meters (100 feet) long, is the largest mammal that ever lived.

Cetaceans are close relatives of even-toed mammals (Artiodactyla). Most large mammalian grazers such as cattle, deer, and goats are in this order, as are pigs and hippos. Horses, zebras, and rhinos are grazers too, but belong to the odd-toed order (Perissodactyla). Both orders of grazing mammals digest plant material with the assistance of microbes in their gut.

Humans are members of the order Primates. We cover the unique adaptations of this group, and the history of our lineage, in detail in the next chapter.

placenta Of placental mammals, organ that forms during pregnancy and allows diffusion of substances between the maternal and embryonic bloodstreams.

Rodentia (rats, mice, squirrels, porcupines)

Chiroptera (bats)

Soricomorpha (moles and shrews)

Carnivora (dogs, cats, bears, weasels, seals, and walruses)

Cetacea (dolphins, whales)

Artiodactyla (even-toed mammals: deer, cattle, pigs, hippos)

Perissodactyla (odd-toed mammals: horses, zebras, rhinos)

Primates (lemurs, monkeys, apes, humans)

Figure 25.31 Representatives of major orders of placental mammals.

Take-Home Message

What are living mammals like?

» Most mammals living today are placental mammals, with rodents and bats being the most diverse orders. Most mammals live on land, but some have adapted to life in the seas.

LEARNING ROADMAP

Where you have been This chapter focuses on primates, an order of placental mammals (Sections 25.10 and 25.11). It is based largely on information derived from fossils (16.5) and from molecular comparisons (18.4). We discuss primate adaptive traits (16.4) and consider examples of both morphological convergence and divergence (18.3).

Where you are now

The Order Primates
Primates are a mammalian lineage with grasping hands, flexible shoulder joints, and relatively large brains. Like most primates, humans have a movable upper lip.

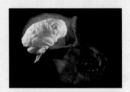

Apes
Our closest living relatives are African apes. Like them we are tailless, but we have a bigger brain, more maneuverable hands, and a body adapted to upright walking.

Early Human Ancestors
In searching for potential human ancestors, researchers look for fossil evidence of upright walking. Several possible ancestors, including australopiths, are known from Africa.

Early Humans
The earliest humans, *Homo habilis*, walked upright but were still small brained. *Homo erectus* had a bigger brain, made more complex tools, and dispersed out of Africa.

Recent Human Species
Modern humans trace their roots to Africa, but their genome bears signs of mating with related populations such as Neanderthals. Until recently, there were multiple *Homo* species.

Where you are going We return often to human evolution in our discussions of anatomy and physiology. For example, we return to the adaptive significance of skin color (Section 31.8), consider the skeletal problems associated with our bipedalism (35.3), see how human populations have adapted to high-altitude environments (38.8), and discuss the evolutionary constraints that affect our ability to control our weight (39.12). We return to nonhuman primates in our discussion of the costs and benefits of social behavior (43.8).

26.1 A Bit of a Neanderthal

Paleoanthropology, the scientific study of prehistoric humans and their relatives, was born in the mid-1800s, when scientists discovered some humanlike fossils in Germany's Neander Valley. They postulated that the remains belonged to an extinct human relative that they called *Homo neanderthalensis*. At the time the fossils were discovered, their age was unknown. We now know that they date to about 40,000 years ago.

Since the discovery of those first Neanderthal bones, many other fossils with similar features have been unearthed, including a few largely complete skeletons. These fossil finds have confirmed that Neanderthals were a distinct population rather than, as once suggested, modern humans deformed by an infectious disease or genetic syndrome. Scientists now consider Neanderthals our closest extinct relatives.

Compared to modern humans, Neanderthals had a shorter, stockier build, with thicker bones and bulkier muscles. A reconstruction of a Neanderthal male, based on material from multiple fossils, stands about 164 centimeters (5'4") tall (**Figure 26.1**). Neanderthals had a braincase that was longer and lower than that of modern humans, but their brain was as big as ours or bigger. Their face had pronounced brow ridges and a large nose with widely spaced nostrils.

We know from fossils that Neanderthals lived in the Middle East, Europe, and in central Asia as far east as Siberia. The last known population lived in seaside caves in Gibraltar until perhaps as recently as 28,000 years ago.

In some regions, Neanderthals existed side by side with our own species (*Homo sapiens*) for thousands of years. People have long speculated about whether the two species mated. Scientists are less interested in the logistics of such matings than in whether they produced fertile offspring whose descendants survive to this day. In other words, they want to know whether Neanderthals contributed in any substantive way to the modern human gene pool.

Advances in our ability to extract, amplify, and sequence DNA now allow us to make the comparisons that can answer this question. In 2010, an international team of scientists headed by Svante Pääbo sequenced DNA extracted from Neanderthal fossils discovered in Croatia, Russia, Spain, and Germany. The sequence data they obtained represents much of the Neanderthal genome.

Pääbo and his team compared Neanderthal genomes with homologous portions of genomes from five modern human populations and from chimpanzees. The results revealed that many of us do, in fact, have

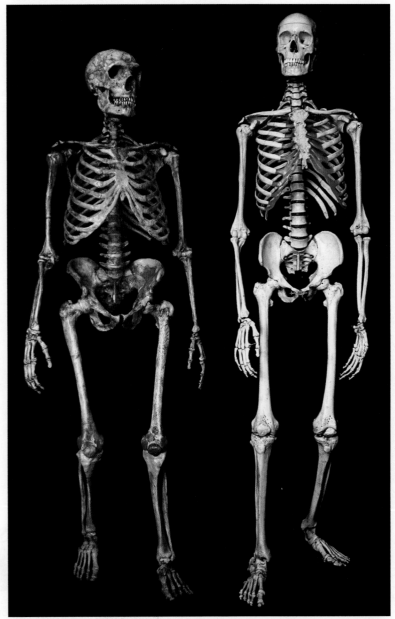

Homo neanderthalensis   *Homo sapiens*

Figure 26.1 Skeletal anatomy of Neanderthal male compared with a modern human male. The Neanderthal skeleton is a reconstruction based on fossils of multiple males. Each color denotes a different fossil.

a Neanderthal in our family tree. Genomes of people from France, China, and Papua New Guinea share with the Neanderthal genome certain mutations that are not present in two modern populations from Africa or in chimpanzees. This suggests that the human–Neanderthal matings that left their mark on our gene pool took place in the Middle East after *H. sapiens* began venturing out of Africa, but before they dispersed to Europe, Asia, and elsewhere.

26.2 Primates: Our Order

- We share our five-digit grasping hands, forward-facing eyes, and a movable upper lip with other primate species.
- Link to Mammals 25.10 and 25.11

Primate Characteristics

Primates are an order of placental mammals that includes humans, apes, monkeys, and close relatives. **Figure 26.2** shows relationships among living groups.

Primates first evolved in warm forests, and many traits characteristic of the group adapt them to life among the branches. Primate shoulders have an extensive range of motion that facilitates climbing. Unlike most mammals, a primate can extend its arms out to its sides, reach above its head, and rotate its forearm at the elbow. With the exception of humans, all living primates have both hands and feet capable of grasping. Mammals often have claws, but the tips of primate fingers and toes typically have touch-sensitive pads protected by flattened nails.

Most mammals have widely spaced eyes set toward the side of the skull, but primate eyes tend to be at the front of the head. As a result, both eyes view the same area, each from a slightly different vantage point. The brain integrates the differing signals it receives from the two eyes to produce a three-dimensional image. A primate's excellent depth perception adapts it to a life spent leaping or swinging from limb to limb.

Compared to other mammals, primates have a large brain for their body size. The regions of the brain devoted to vision and to information processing are expanded, and the area devoted to smell is reduced.

Primates have a varied diet, and their teeth reflect this lack of specialization. They also retain all four types of mammalian teeth (Section 25.10).

Most primates spend their life in a social group that includes adults of both sexes. Females usually give birth to only one or two young at a time and provide care for an extended period after birth.

Origins and Early Branchings

Primates arose before the demise of the dinosaurs, but exactly when remains a matter of debate. A recent study that drew on both fossils and molecular comparisons among living species concluded that the common ancestor of all modern primates lived 85 million years ago. The same study places an early branching about 50 million years ago. This branching created two suborders: the wet-nosed primates (Strepsirrhini) and the dry-nosed primates (Haplorhini).

Modern representatives of the wet-nosed suborder include lemurs of Madagascar, lorises of India and Asia, and galagos of Africa. Wet-nosed primates have a typical mammalian nose. Like a dog, they have nostrils set in an area of continually moist skin, and their

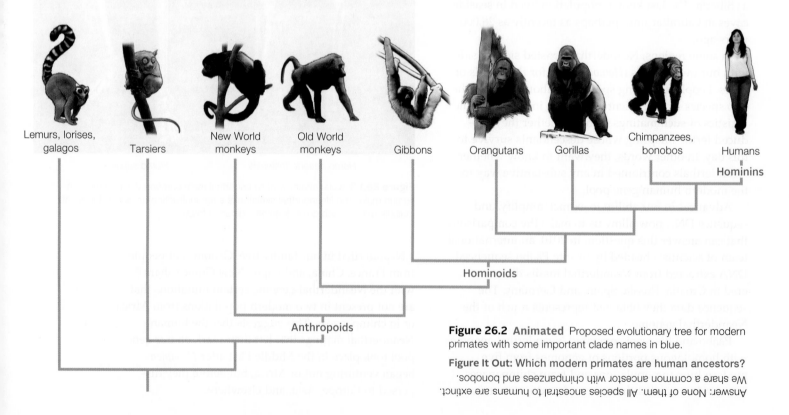

Figure 26.2 Animated Proposed evolutionary tree for modern primates with some important clade names in blue.

Figure It Out: Which modern primates are human ancestors?

Answer: None of them. All species ancestral to humans are extinct. We share a common ancestor with chimpanzees and bonobos.

cleft upper lip attaches tightly to the underlying gum (**Figure 26.3A**). A wet nose aids in detecting scents.

Humans and most other modern primates belong to the dry-nosed suborder. Their nostrils are set in dry, hairy skin. The upper lip is not cleft, and its attachment to the underlying gum is greatly reduced. A movable upper lip allows a wider range of facial expressions and vocalizations.

Tarsiers, the oldest surviving dry-nosed lineage (**Figure 26.3B**), are small, nocturnal insect eaters that live on South Asian islands. Tarsiers were traditionally grouped with lemurs as prosimians (*pro–*, before, *simian*, monkey) because both lineages have claws on some digits. However, DNA comparisons show tarsiers are more closely related to anthropoids than to lemurs.

The **anthropoid** lineage includes monkeys, apes, and humans; "anthropoid" means humanlike. Nearly all anthropoids are diurnal (active during the day) and have good eyesight, including color vision. Most feed primarily on plant material. The group is thought to have originated either in Africa or Asia.

Modern New World monkeys climb through the forests of Central and South America in search of fruits. They have a flat face and a nose with widely separated nostrils that open to the side. A long tail helps them maintain balance. In some species, the tail is prehensile, meaning it grasps things (**Figure 26.3C**).

New World monkeys arrived in South America from West Africa about 35 million years ago. Ancestors of South American rodents such as guinea pigs arrived from Africa at about the same time. Both groups of animals probably crossed the Atlantic on rafts of vegetation. This journey may not have been as daunting as it seems, because the distance between the continents was narrower than it is today. Also, up until 40 million years ago, sea level fluctuations periodically exposed islands that would have been convenient way stations. Crossing the ocean may have taken many generations, with animals colonizing one island after another.

Old World monkeys live in Africa, the Middle East, and Asia. They tend to be larger than New World monkeys and have a longer nose with closely set, downward facing nostrils. Some, such as vervets, are tree-climbing forest dwellers. Others, such as baboons (**Figure 26.3D**), spend most of their time on the ground in grasslands and deserts. Not all Old World monkeys have a tail, but in those that do the tail is typically short and never prehensile.

Apes and humans are tailless primates grouped as **hominoids**. Hominoids are not descended from any existing group of monkeys, but they share a common ancestor with the Old World monkeys.

A Lemur with a wet nose, cleft upper lip.

B Tarsier with dry nose, uncleft upper lip.

C Squirrel monkey (New World monkey) with flat face, prehensile tail.

D Baboon (Old World monkey) with long nose, short tail.

Figure 26.3 Examples of nonanthropoid primates.

anthropoids Primate lineage that includes monkeys, apes, and humans.
hominoids Tailless primate lineage that includes apes and humans.
primates Mammalian order that includes lemurs, tarsiers, monkeys, apes, and humans.

Take-Home Message

What are primates?

» Primates include lemurs, tarsiers, monkeys, apes, and humans. Most primates are tree-dwellers. Traits such as a flexible shoulder joint and grasping hands with nails are adaptations to a climbing lifestyle.

» Compared to other mammals, primates have a relatively large brain with more area devoted to vision and less to smell.

» There are two main subgroups. Lemurs and their relatives belong to the subgroup with a wet nose and a fixed upper lip. Tarsiers, monkeys, apes, and humans belong to the subgroup with a dry nose and a movable upper lip.

26.3 The Apes

■ A tailless body and an upright posture evolved in the common ancestor of humans and other apes.
■ Links to Sexual selection 17.7, Adaptive radiation 17.12, Morphological divergence 18.3

Hominoid Origins and Divergences

Hominoids are tailless primates with an upright posture and relatively large brains. The lineage leading to modern hominoids may have branched off from a common ancestor with Old World monkeys as early as 30 million years ago. By 20 million years ago, during the Miocene, a variety of species now assigned to the genus *Proconsul* lived in the then lush forests of northeast Africa. Like modern apes, *Proconsul* species had flattened molars and were probably tailless. Some were as large as a chimpanzee. However, *Proconsul* retained a monkey-sized brain and a body suited to walking on all fours along limbs.

As the Miocene continued, Africa became hotter and drier. Seasonally dry woodlands and savannas replaced the forests where *Proconsul* had thrived, and new species of apes evolved. Some of these species dispersed to Europe and Asia. The relationship of Europe's Miocene apes to modern species remains unclear. By one hypothesis, modern African apes descend from an ape lineage that radiated in Europe before some members returned to Africa. Alternatively, modern African apes may have evolved in Africa, and European apes may have no modern descendants.

Modern Apes

About 15 species of small apes called gibbons inhabit Southeast Asian forests. They use their elongated arms and permanently curved fingers to swing from limb to limb (**Figure 26.4A**). Gibbons live in family groups with similarly sized males and females. Depending on the species, adults weigh 7 to 14 kg (15 to 30 lb).

Gibbons are sometimes referred to as "lesser apes," in comparison with the larger apes, or "great apes." Unlike gibbons, all great apes are sexually dimorphic, meaning males and females differ in their body form. In this case, males are larger than females.

The forest-dwelling orangutan (*Pongo pygmaeus*) of Sumatra and Borneo is the only surviving Asian great ape. Like gibbons, orangutans are tree dwellers, but they live solitary lives and climb slowly using all four limbs (**Figure 26.4B**). Males are twice as big as females and can weigh 80–90 kg (176–198 lb).

All African great apes (gorillas, chimpanzees, and bonobos) are native to central Africa, live in social groups, and spend most of their time on the ground. When walking, they lean forward and support their weight on their knuckles (**Figure 26.4C** and **Figure 26.5**). Gorillas (*Gorilla gorilla*), the largest living primates, live in forests and feed mainly on leaves.

Chimpanzees (*Pan troglodytes*) (**Figure 26.4D**) and the bonobos, or pygmy chimpanzees (*Pan paniscus*), are our closest living relatives. The chimpanzee/bonobo lineage and the lineage leading to humans diverged

A Gibbon   **B** Orangutan   **C** Gorilla   **D** Chimpanzee

Figure 26.4 Modern ape diversity. A lesser ape (**A**) and three of the great apes (**B–D**).

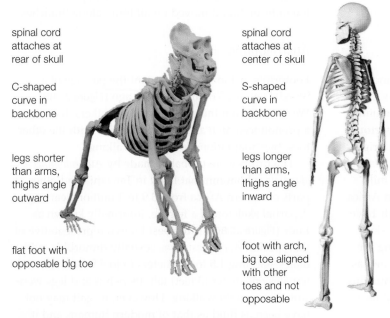

spinal cord attaches at rear of skull

spinal cord attaches at center of skull

C-shaped curve in backbone

S-shaped curve in backbone

legs shorter than arms, thighs angle outward

legs longer than arms, thighs angle inward

flat foot with opposable big toe

foot with arch, big toe aligned with other toes and not opposable

Figure 26.5 Animated Some skeletal differences between a knuckle-walking gorilla (*left*) and a bipedal human (*right*).

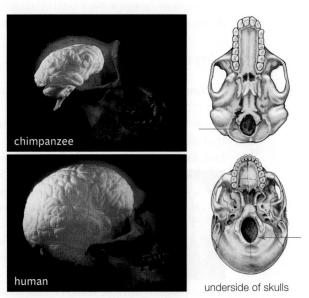

chimpanzee

human

underside of skulls

Figure 26.6 Chimpanzee and human skulls. Humans have a flatter face, a smaller jaw, and a proportionately larger brain. The hole for a human spinal cord (*red* arrow) is centered at the skull's base, rather than toward the rear, as in chimpanzees.

an estimated 8 or so million years ago. Chimpanzees and bonobos live in large social groups. Both eat fruit, but also catch insects and cooperatively hunt small mammals, including monkeys. The two groups differ in their social behavior, with chimpanzees engaging in more intraspecific aggression and bonobos spending more time in nonreproductive sexual acts.

A Human–Ape Comparison

Modern humans are described as naked apes because we have little body hair. A relatively hairless body is a recent adaptation. Other differences between apes and our lineage emerged early in our history.

The first apes walked on all fours like a monkey, with their back parallel to the ground or a branch. A more upright stance evolved in later apes. However, only members of the lineage leading to humans are **bipedal**, or adapted to habitually walking upright. Evolution of upright walking involved many skeletal changes (**Figure 26.5**). For example, the human backbone has an S-shaped curve that keeps the head centered over the feet, and human feet have a pronounced arch and a non-opposable big toe. An ape's spinal cord enters the skull's base near the rear, whereas a human spinal cord enters near the center.

bipedal Adapted to habitually walking upright.

Humans have very flexible hands. A human thumb is longer, stronger, and more maneuverable than that of a chimpanzee. Many monkeys and apes grasp objects in a power grip, but only humans routinely use a precision grip, pinching together the tips of the thumb and forefinger for fine manipulation of objects:

power grip

precision grip

The lineage leading to humans underwent a great expansion of the brain. A human brain is three times the size of a chimpanzee brain (**Figure 26.6**). The human brain also grows for a longer period after birth, and our offspring receive extended parental care.

Take-Home Message

Which human traits are typical of hominoids and which are novel?

» Hominoids share a common ancestor with Old World monkeys, but they are generally larger than monkeys, lack a tail, and have an upright stance. Modern hominoids include Asian apes, African apes, and humans.

» Upright walking (bipedalism) evolved in the lineage leading to humans. This lineage also shows a trend toward a more maneuverable thumb, a larger and more complex brain, and a longer period of postnatal development.

26.4 Rise of the Hominins

■ Humans are upright walkers with big brains, but bipedalism first evolved in our relatively small-brained ancestors.

Early Proposed Hominins

Hominins are the lineage of humans and all extinct species more closely related to humans than to any other group (**Figure 26.7**). Bipedalism is a defining hominin trait, so researchers studying the origins of this group look for fossil evidence that a species walked upright.

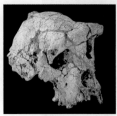 One such species, *Sahelanthropus tchadensis*, lived about 7 million years ago in western Africa (Chad). Only a few fossils have been found, but one is a skull (*left*) in which the opening for the spinal cord is positioned as it is in modern upright walkers. Like modern humans and unlike chimpanzees, *S. tchadensis* had a flat face and small canine teeth.

Another proposed early hominin, *Orrorin tugenensis*, lived about 6 million years ago in East Africa. Two sturdy fossilized femurs (thighbones) are cited as evidence that this species stood upright.

Two species of chimpanzee-sized *Ardipithecus* that lived in East Africa are also considered possible hominins. *A. kadabba* is known from a few fossils that date to 5.8 to 5.2 million years ago. *A. ramidus* left a wealth of fossil remains, including a largely complete skeleton of a female informally known as Ardi (**Figure 26.8A**). *A. ramidus* had smaller teeth than a chimpanzee, and its pelvis is more humanlike than apelike. It probably walked upright on the ground. However, its elongated arms, curved fingers, and an outwardly splayed big toe indicate that it moved on all fours along branches.

Australopiths

Footprints in Tanzania document the passage of a bipedal species 3.6 million years ago (**Figure 26.8B**). We can tell from the prints that the walkers' feet had a pronounced arch and a big toe in line with the other toes, two adaptations to upright walking.

The prints were probably made by *Australopithecus afarensis*, a hominin that lived in Tanzania and other parts of eastern Africa from 3.9 to 3 million years ago. A partial skeleton of a female, informally known as Lucy (**Figure 26.8C**), is the best known representative of this species. *A. afarensis* was sexually dimorphic, with males standing 1.5 to 1.8 meters (5 to 5.5 feet) tall, and females one meter (3 feet) tall. Its pelvis and legs were suited to upright walking. However, its gait may not have been as fluid as that of modern humans and it retained the long arms and curved fingers of a climber.

A. afarensis was an **australopith**, an informal group comprising two genera of hominins that lived in Africa from about 4 million to 1.2 million years ago. *Australopithecus* species were small boned. Their fossil history reveals trends toward smaller teeth and improvements in the ability to walk upright, but little increase in brain size. Most researchers think the first humans descended from an *Australopithecus* species, with *A. afarensis* and *A. africanus* (**Figure 26.9A**) among the most likely ancestors.

Paranthropus descended from an *Australopithecus* ancestor, but this genus had more powerful jaws and

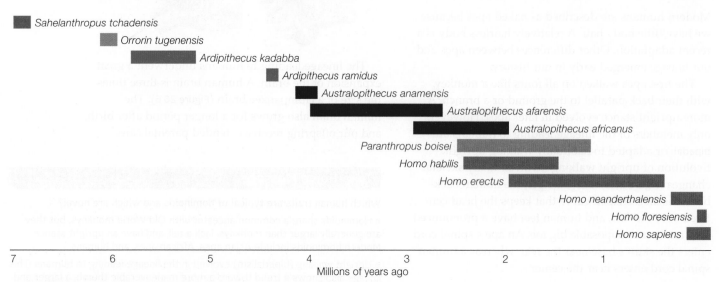

■ Sahelanthropus tchadensis
■ Orrorin tugenensis
■ Ardipithecus kadabba
■ Ardipithecus ramidus
■ Australopithecus anamensis
■ Australopithecus afarensis
■ Australopithecus africanus
■ Paranthropus boisei
■ Homo habilis
■ Homo erectus
■ Homo neanderthalensis
■ Homo floresiensis
■ Homo sapiens

7 6 5 4 3 2 1
Millions of years ago

Figure 26.7 Estimated dates for the origin and extinction of a sampling of hominins and proposed hominins.

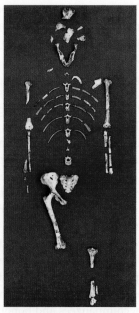

A *Ardipithecus ramidus*. The painting (*right*) is based on a 4.4 million-year-old Ethiopean fossil known as Ardi (*left*). *A. ramidus* may have walked upright on the ground but also climbed on all fours along branches.

B Footprints made 3.6 million years ago by bipedal hominins walking across a bed of volcanic ash. The walkers had a shorter stride than modern humans.

C *Australopithecus afarensis*. This individual, known as Lucy, lived 3.5 million years ago. She belonged to a species of early australopiths that may be ancestral to humans.

Figure 26.8 Evidence of early bipedalism.

A *Australopithecus africanus* 3.2–2.3 million years ago

B *Paranthropus boisei* 2.3–1.2 million years ago

Figure 26.9 Animated Two australopith skulls.

stronger molars with thick enamel (the tooth's hard outer coating). *Paranthropus* skulls have a pronounced crest that served as the attachment point for large jaw muscles (**Figure 26.9B**). Strong teeth and heavily muscled jaws allowed *Paranthropus* to eat hard and fibrous foods when other favored foods were not available. *Paranthropus* species are considered close relatives of early humans, but not possible ancestors.

Factors Favoring Bipedalism

What advantage could have driven an evolutionary shift from walking on all fours to bipedalism? A trend toward a drier and hotter climate was probably a factor. Hominins originated at a time when Africa's lush rain forests were giving way to woodlands interspersed with grassy plains. In this altered habitat, an ability to move efficiently across open ground between trees would have been favored, and upright walking can be energy efficient. Human walkers use less energy than chimpanzees who move on all fours.

Bipedalism has additional advantages. It keeps the body cooler. A bipedal animal gains less heat from the ground than a four-legged walker and also intercepts less warming sunlight. An upright stance also makes it easier to scan the horizon for predators, and hands freed from use in locomotion can be used to gather food and to carry it from place to place.

australopith Informal name for two genera of chimpanzee-sized hominins that lived in Africa between 4 million and 1.2 million years ago.
hominins Modern humans and their closest extinct relatives.

Take-Home Message

What are hominins?

» Hominins include modern humans and extinct members of their lineage. Possible hominin fossils date back as far as 7 million years ago. Early hominins had small brains. Australopiths are a diverse group of African hominins that probably include human ancestors.

26.5 Early Humans

- The genus *Homo* arose in Africa, and early humans dispersed from there to Europe and Asia.
- Link to Reproductive isolation and speciation 17.9

Classifying Fossils—Lumpers and Splitters

When it comes to classifying fossil species, scientists fall into two general camps. The "lumpers" look for similarities among fossils and tend to assign fossils to relatively few species. Lumpers support this approach by pointing out that members of a single species can vary widely in their traits. For example, modern humans share many traits in common, but vary greatly in their height. By contrast, "splitters" focus on differences between fossils and often name new species on the basis of subtle differences. In support of their approach, splitters point out that many modern species differ only slightly in their morphology. For example, chimpanzees and bonobos appear very similar, although bonobos are a bit smaller.

From a scientific standpoint, both approaches are equally valid. Both lumpers and splitters formulate hypotheses about how a particular fossil relates to other known fossils. They then seek evidence that can confirm or refute their hypothesis. Additional fossil discoveries often provide this evidence.

In this section, we discuss the earliest fossil evidence of our own genus *Homo*. Lumpers place these fossils in two species, whereas splitters divide them into four or more.

A *Homo habilis*
2.3–1.4 million years ago

B *Homo erectus*
1.8 million–27,000 years ago

Figure 26.10 Two early *Homo* skulls.

Homo habilis

Humans are members of the genus *Homo*. The most ancient named member of this species is *Homo habilis* (**Figure 26.10A**). Fossils of *H. habilis* date back as far as 2.3 million years before the present. *H. habilis* was named based on a fossil individual discovered in Kenya in 1964. At that time, tool production was considered a diagnostic trait of the genus *Homo*. The name *Homo habilis* means "handy man," and is a reference to the stone tools found near the fossil.

The classification of *H. habilis* continues to inspire debate. *H. habilis* fossils resemble australopiths in body proportions and brain size, and some scientists have argued that they should be classified as a species of *Australopithecus*. Other scientists argue for the species' inclusion in the genus *Homo*, pointing out humanlike hands and other traits. Anatomical variation among the fossils has given rise to a different dispute. Some "splitters" contend that the fossils currently classified as *H. habilis* actually include remains from two different species. They place larger-brained, longer-faced individuals in the species *Homo rudolfensis*.

Homo habilis/rudolfensis persisted in eastern Africa until at least 1.4 million years ago. Thus it overlapped in time and range both with australopiths and with at least one other species of *Homo* (**Figure 26.11**).

Homo erectus

Homo erectus appears in the fossil record beginning about 1.8 million years ago. Its name means "upright man," and like modern humans, *H. erectus* stood on legs that were longer than its arms and it had a relatively big brain (**Figure 26.10B**). The most complete *H. erectus* fossil known is a skeleton of a juvenile male discovered in Kenya (**Figure 26.12A**). Although this individual—commonly referred to as Turkana

Figure 26.11 Painting depicting a band of tool-using *Homo habilis* in an East African woodland. Two australopiths are shown in the distance at the *left*.

boy—apparently died at age nine, his brain was already twice the size of a chimpanzee's.

Early reconstructions of *H. erectus* based on Turkana boy showed a tall, lanky individual. Some scientists postulated that this body form is evidence that *H. erectus* was adapted to long-distance running. More recent fossil finds suggest that Turkana boy's proportions are not typical for his species. For example, analysis of a 1.2-million-year-old, largely intact female pelvis revealed that this *H. erectus* individual had short legs and wide hips. Her anatomy would have made her an unimpressive runner, but her expansive pelvic opening probably facilitated the birth of offspring with large heads.

Whether walking or running, *H. erectus* covered a lot of ground. By 1.75 million years ago, a population of *Homo* had left Africa to become established in what is now Dmanisi, Georgia. Most researchers think the Dmanisi hominins descended from *H. erectus*, although some consider *H. habilis* a likely ancestor. *H. erectus* certainly colonized Indonesia by 1.6 million years ago, and China by 1.15 million years ago.

As with *Homo habilis*, some scientists split *Homo erectus* fossils into two species. The splitters reserve the name *H. erectus* for fossils in Asia, and refer to the similar African fossils as *H. ergaster* (working man).

Early Culture

Culture is a set of learned behaviors that are passed from one individual to another, and from one generation to the next. The earliest hominin cultural trait for which we have fossil evidence is toolmaking. Distinctive scrape marks on bones provide indirect evidence that hominins used sharp stone edges to scrape meat from bones as early as 3.4 million years ago. Given the age of the bones, these early tool users were most likely australopiths.

The first stone tools were sharp flakes chipped from larger rocks. Toolmaking improved about 1.4 million years ago, when African *H. erectus* began to use a succession of carefully placed strikes to sculpt tools in a variety of shapes (**Figure 26.12B**).

Marks on fossilized bones of prey animals indicate that *H. erectus* used stone tools to cut up scavenged animal carcasses, to scrape meat from bones, and to extract marrow. Some researchers have suggested that increased meat consumption may have been a necessary prerequisite to the evolution of a larger brain. Meat is a more concentrated source of energy and protein than plant foods, and thus can better fuel development and maintenance of a large brain.

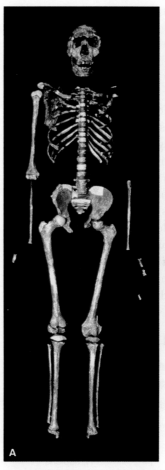

Figure 26.12 *H. erectus.*

(**A**) A 1.53-million-year-old fossil skeleton of a juvenile male from Nariokotome, Kenya.

(**B**) Stone tools from Tanzania that were most likely shaped by *H. erectus*.

Chimpanzees use gestures and calls to communicate, and early humans probably did the same. However, it is unlikely that either *H. habilis* or *H. erectus* had a spoken language. Evolution of a capacity for speech involved changes in the brain, remodeling of the larynx (voice box), and an improved ability to control air flow through the vocal tract. Thus far, scientists have not found fossil evidence of these necessary anatomical modifications in either early species of *Homo*.

culture Learned behaviors transmitted between individuals and down through generations.
human Living or extinct member of the genus *Homo*.

Take-Home Message

What traits characterized early humans?

» *Homo habilis* is the most ancient human species known. It lived in Africa and was similar in many respects to australopiths, but its hands and arms more closely resemble those of modern humans.

» *Homo erectus* is known from Africa, Europe, and Asia. It had a significantly larger brain than earlier hominins. Females had a pelvis that facilitated birth of offspring with large heads.

» Early humans made stone tools and used them to cut up animal carcasses. However, stone tool use may have begun among the australopiths.

26.6 Recent Human Lineages

■ Our status as the only living members of the genus *Homo* came about relatively recently.
■ Links to Gene flow 17.8, Genomics 18.4

Origin and Dispersal of *Homo Sapiens*

A 195,000-year-old fossil found in Ethiopia is the earliest evidence of our species, *Homo sapiens*. Compared to *H. erectus*, we have a higher, rounder skull, a larger brain, and a flatter face with smaller jawbones and teeth. We also have a chin, a protruding area of thickened bone in the middle of our lower jawbone.

Two competing models describe how *H. sapiens* evolved from *H. erectus*. Both attempt to explain the distribution of *H. erectus* and *H. sapiens* fossils, as well as regional genetic differences among modern humans, but they differ in their account of the rate of change and where these changes took place.

A Multiregional model. *H. sapiens* slowly evolves from *H. erectus* in many regions.

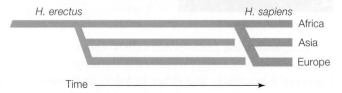

B Replacement model. *H. sapiens* rapidly evolves from one *H. erectus* population in Africa, then disperses and replaces *H. erectus* populations in all regions.

Figure 26.13 Two models for the origin of *H. sapiens*. Arrows represent ongoing gene flow among populations.

According to the **multiregional model**, populations of *H. erectus* in Africa and other regions evolved into populations of *H. sapiens* gradually, over more than a million years. Gene flow among populations maintained the species through the transition to fully modern humans (**Figure 26.13A**). By this model, the variation among modern Africans, Asians, and Europeans began to accumulate soon after their ancestors branched from an ancestral *H. erectus* population.

The **replacement model** holds that *H. sapiens* arose from a single *H. erectus* population in sub-Saharan Africa within the past 200,000 years. Later, bands of *H. sapiens* entered regions occupied by *H. erectus* populations, and replaced them (**Figure 26.13B**). This model emphasizes the similarities among all living humans.

Genomic studies of global populations support the replacement model. Comparisons of mitochondrial genes, and of Y chromosomes, place modern Africans closest to the root of the family tree. The most recent common ancestor of all humans now alive lived in Africa approximately 60,000 years ago. However, though the current human population can trace its lineage to *H. sapiens* in Africa, there is genetic evidence of successful matings between modern humans and Neanderthals. Perhaps some other regional differences in our genome arose as a result of gene exchange with other localized hominin populations.

Our species' range expanded as successive generations traveled along the coasts of Africa, then Eurasia and Australia. Fossils and genetic evidence allow scientists to trace human dispersal routes (**Figure 26.14**). About 15,000 years ago, one small band of humans crossed a now submerged land bridge that connected Siberia with North America.

With each step of their journey, humans faced and overcame extraordinary hardships. During this time, they devised cultural means to survive in inhospitable

Figure 26.14 Some dispersal routes taken by small bands of *Homo sapiens*. This map shows ice sheets and deserts that existed about 60,000 years ago. Some 14,000-year-old fossil feces from a cave in Oregon is reportedly the oldest evidence of humans in North America.

A Bit of a Neanderthal (revisited)

The common conception of Neanderthals as brutish cavemen with poor posture arose from an early reconstruction based on a fossil of an individual deformed by arthritis. Preconceived notions of early humans as "primitive" no doubt contributed to the persistence of this mistaken impression. More recent reconstructions show Neanderthals were shorter than modern humans, but stood upright. Unlike modern humans, they typically lacked a protruding chin (**Figure 26.15**).

Fossils of Neanderthal individuals who survived despite disabilities such as the loss of a limb testify to a compassionate social structure. Some simple burials suggest possible symbolic thought. Several lines of evidence suggest Neanderthals were able to speak.

Neanderthals lived in parts of Europe where winters are cold, and a short stocky body minimized the surface area available for heat loss. Modern Arctic peoples have a similar body shape. Maintaining body temperature in a cool environment burns a lot of energy and requires a lot of calories. Neanderthals gathered shellfish and hunted for meat, using their powerful arms to stab prey with stone-tipped spears.

The lineages leading to Neanderthals and to modern humans diverged from a common ancestor about 500,000 years ago. However, as noted earlier, Neanderthals and early modern humans met and interbred about 60,000 years ago, before humans had dispersed widely from Africa. This meeting probably occurred in the Middle East. There is no genetic evi-

Figure 26.15 Svante Pääbo with a reconstructed Neanderthal skull. Pääbo headed the team that sequenced the Neanderthal genome. Note Pääbo's characteristically *H. sapiens* chin.

dence of later interbreeding when modern humans expanded into Europe 40,000 years ago. The last traces of a Neanderthal population are from caves in Gibraltar and date to 28,000 years ago. Neanderthals may have been outcompeted by the newcomers or killed by diseases they brought with them. Climate changes and volcanic eruptions may also have contributed to their demise by altering the abundance of the large game upon which they depended.

How would you vote? Chimps are irreplaceable models for human responses to diseases and medical procedures. Europe has banned use of chimpanzees in medical research. Should the United States do so too?

---

environments. An unrivaled capacity for modifying the habitat and for language still serves us well, for cultural evolution is ongoing. Hunters and gatherers persist in a few parts of the world, but others have moved from "stone-age" technology to the age of "high tech." Continued coexistence of such diverse groups is a tribute to the behavioral plasticity of the human species.

Premodern *Homo* Species

Our family tree keeps getting bushier. Recent findings suggest that until recently modern humans shared the planet not only with Neanderthals, but with at least one additional hominin relative.

In 2003, scientists discovered some 18,000-year-old hominin fossils on the Indonesian island of Flores. The fossilized individuals were small, only a meter tall, with a heavy brow and a little brain. Scientists who found the fossils assigned them to a new species, *Homo floresiensis*. Other scientists argue that the fossils could be remains of *H. erectus* or *H. sapiens* with a genetic or

nutritional disorder. Still others noted that the fossils resemble some australopiths and suggested that they were descendants of australopiths who migrated out of Africa. Further study of existing fossils, a search for more fossils, and possibly DNA analysis will help test the competing hypotheses.

multiregional model Model that postulates *H. sapiens* populations in different regions evolved from *H. erectus* in those regions.
replacement model Model for origin of *H. sapiens*; humans evolved in Africa, then migrated to different regions and replaced the other hominins that lived there.

Take-Home Message

How are the most recently evolved hominins related?

» Both modern humans and Neanderthals are descendants of *H. erectus*.

» Modern humans originated in Africa, then dispersed into other regions. Early in this dispersal, humans entered the Middle East and mated with the Neanderthals they encountered there.

» A recently discovered Indonesian hominin has been named *Homo floresiensis*, but its ancestry and relationship to other species remain a matter of investigation.

V HOW PLANTS WORK

The sacred lotus, *Nelumbo nucifera*, busily doing what its ancestors did for well over 100 million years—flowering spectacularly during the reproductive phase of its life cycle.

LEARNING ROADMAP

Where you have been This chapter builds on what you learned in Sections 22.3 and 22.9, which introduced angiosperm evolution, structure, and growth. It examines the anatomy of flowering plants in terms of life's organization (1.2), and revisits carbohydrates (3.4), plant cell specializations (4.7, 4.11, 6.8), membrane permeability (5.8), and differentiation (10.2).

Where you are now

Overview of Plant Tissues
Seed-bearing vascular plants have a shoot system; most also have a root system. Both consist of ground, vascular, and dermal tissues. Plants lengthen or thicken at meristems.

Primary Shoots
Monocots and eudicots differ in the organization of tissues. Leaf specializations for sunlight interception, water conservation, and gas exchange support photosynthesis.

Primary Roots
Roots anchor a plant and provide a large surface area for absorbing water and minerals from soil. They conduct water and ions to above-ground plant parts.

Secondary Growth
In many plants, secondary growth thickens branches and roots during successive growing seasons. Tree rings can be used to study past environmental conditions.

Modified Stems
Certain types of stem specializations are adaptations for storing water or nutrients, and for reproduction.

Where you are going In the next chapter, you will see how plant structure contributes to function, including how roots take up water (Section 28.3); how water and solutes move through a plant (28.4, 28.6), and how plants conserve water (28.5). Asexual and sexual reproduction in plants return in Chapter 29. Plants form the basis of almost all food webs (46.3), so they are a critical factor in communities (45.8) and the carbon cycle (46.7). We return to greenhouse gases and climate change in Section 46.8.

27.1 Sequestering Carbon

Carbon in the atmosphere occurs mainly as carbon dioxide gas (CO_2). As you learned in Chapter 6, humans release a lot of carbon dioxide by burning fossil fuels and other plant-derived materials. As a result of these activities, the amount of CO_2 in the atmosphere is increasing exponentially, with unintended and potentially catastrophic effects on Earth's climate.

Efforts are now under way to reduce the amount of carbon dioxide in the atmosphere. These efforts include carbon offsets, which are financial instruments designed to reduce global emissions of carbon dioxide and other carbon-based greenhouse gases. Companies and individuals buy carbon offsets to "offset" activities that release these greenhouse gases. Governments and other large entities buy them to comply with mandated caps on carbon emissions. The funds are then used to support projects aimed at reducing current emissions of the gases, or activities that remove them from the atmosphere. Any process by which carbon dioxide is removed from the atmosphere is called carbon sequestration. Carbon can also be sequestered by capturing carbon dioxide from activities that would otherwise release it into the atmosphere.

Some carbon offsets support forestry activities such as increasing plant density in forests, and replanting deforested areas (Figure 27.1A). Such activities typically cost less than other methods of removing carbon from the atmosphere, in part because they take advantage of the natural ability of plants to sequester carbon. Plants are specialized to absorb carbon dioxide from the air and lock it in their tissues via photosynthesis. Remember from Section 6.1 that photosynthesis harnesses the energy of sunlight to build sugars from carbon dioxide and water. In an actively growing plant, photosynthetically produced sugars are continually remodeled into other carbon-containing compounds that end up in plant parts such as tree trunks, stems, foliage, and roots. For example, carbon is a major part of cellulose (Section 3.4) and lignin (Section 4.11) that make up wood. These sturdy materials reinforce specialized cells and structures that allow a plant to grow tall, and taller plants win the competition for sunlight (Figure 27.1B).

A tree can live for centuries. After it dies, its tissues decompose, and the carbon in them is re-released to the atmosphere. However, plant matter decomposes more slowly than other organic materials, because lignin, cellulose, and other molecules that waterproof and reinforce plant parts are relatively stable. In some forests, 80 percent of the total carbon is in the soil, as part of plant matter in various stages of decay. Currently, forests worldwide absorb about 3 billion tons of CO_2 from the atmosphere every year, about one-third of the amount released by human activities. Cumulatively, they hold more than twice as much carbon as the atmosphere.

A Reforestation effort in Nicaragua funded by carbon offsets.

B Carbon that gets incorporated into durable plant tissues such as wood can stay out of the atmosphere for centuries.

Figure 27.1 Sequestering carbon in trees.

Table 27.1 Comparing Eudicots and Monocots

Eudicots	Monocots

In seeds, two cotyledons (seed leaves of embryo)	In seeds, one cotyledon (seed leaf of embryo)

Flower parts in fours or fives (or multiples of four or five)	Flower parts in threes (or multiples of three)

Leaf veins usually forming a netlike array	Leaf veins usually running parallel with one another
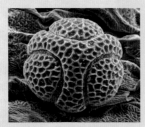	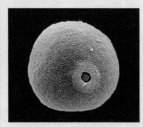
Pollen grains with three pores or furrows	Pollen grains with one pore or furrow
Vascular bundles organized in a ring in ground tissue of stem	Vascular bundles throughout ground tissue of stem

- Most plants consist of roots, stems, and shoots.
- All plant parts consist mainly of ground tissue with vascular tissue threading through it. Dermal tissue covers exposed surfaces of the plant.
- Monocots and eudicots differ in tissue organization.
- Links to Life's levels of organization 1.2, Differentiation 10.2, Plant evolution 22.3, Angiosperms 22.9

With more than 260,000 species (and counting), flowering plants dominate the plant kingdom. Magnoliids, eudicots (true dicots), and monocots (Section 22.9) are the major angiosperm groups. In this chapter, we focus mainly on eudicots and monocots. Eudicots include flowering shrubs and trees, vines, and many non-woody plants such as tomatoes and dandelions. Lilies, orchids, grasses, and palms are examples of monocots.

Monocots and eudicots have the same types of tissues organized in the same systems, but the two lineages differ in many aspects of their tissue organization and hence in their structure (**Table 27.1**). The names of the groups refer to one such difference, the number of seed leaves, or **cotyledons**, in their embryos. The plant embryo in a monocot seed has a single cotyledon; the embryo in a eudicot seed has two.

Overview of Plant Tissues

Like cells of most other multicelled organisms, those in plants are organized as tissues, organs, and organ systems (Section 1.2). The plant body has two organ systems: roots and shoots (**Figure 27.2**). Most shoots are aboveground and most roots are belowground, but as you will see there are many exceptions. A root system consists of a plant's roots, which are structures specialized to absorb water and dissolved minerals as they grow down and outward through the soil. Roots often serve to anchor the plant as well. They are as essential as a plant's shoots, and often at least as extensive. A shoot system comprises stems, leaves, and reproductive organs such as flowers. Stems provide a structural framework for the plant's growth. Leaves are specialized for intercepting sunlight and for photosynthetic

apical meristem Meristem in the tip of a shoot or root.
cotyledon Seed leaf; part of a flowering plant embryo.
dermal tissue Tissue that covers and protects the plant body.
ground tissue Tissue that makes up the bulk of the plant body.
meristem Zone of undifferentiated plant cells; cells of meristem tissue can divide rapidly.
primary growth Plant growth from apical meristems in root and shoot tips.
secondary growth Thickening of older stems and roots.
vascular tissue Tissue that distributes water and nutrients through a plant body.

production of sugars, and flowers are shoots specialized for sexual reproduction.

Roots, stems, leaves, flowers—all plant organs consist of ground, vascular, and dermal tissues (**Figure 27.3**). **Ground tissues** constitute the bulk of the plant, and include the tissues specialized for photosynthesis and for storage. **Vascular tissues** consist of pipelines that thread through ground tissue and distribute water and nutrients to all parts of the plant body. **Dermal tissue** covers and protects the plant's exposed surfaces.

Introducing Meristems

All plant tissues arise from the activity of **meristems**, each a region of undifferentiated cells that can divide rapidly. Cells just under a meristem differentiate and give rise to shoots, stems, and roots as they elongate and mature. New plant parts lengthen by activity at **apical meristems** in shoot tips and root tips. The lengthening of young shoots and roots is called **primary growth**. Some plants also undergo **secondary growth**, which means that their stems and roots thicken over time. The thickening originates at a cylindrical layer of meristem that runs lengthwise through roots and shoots of woody eudicots.

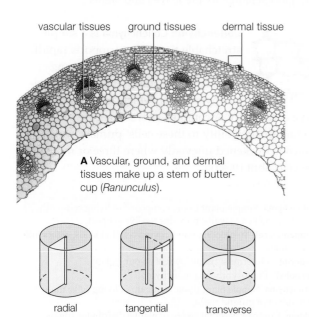

A Vascular, ground, and dermal tissues make up a stem of buttercup (*Ranunculus*).

radial   tangential   transverse

B To simplify interpretation of micrographs, plant parts are typically cut along standard planes. A longitudinal cut along a radius gives a radial section. A cut at right angles to the radius gives a tangential section. A cut perpendicular to the long axis gives a transverse (cross) section.

Figure 27.3 Plant tissues compose plant organs.
Figure It Out: Along which plane was the section in A cut?

Answer: Transverse

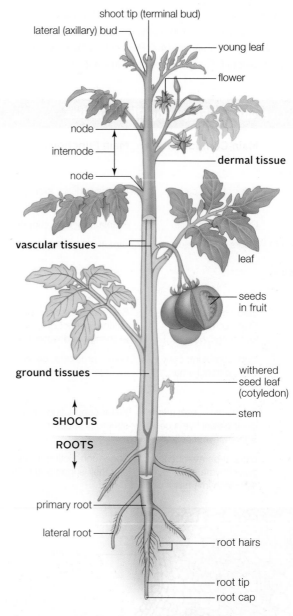

Figure 27.2 Animated Body plan of a tomato plant. Vascular tissues (*purple*) conduct water and solutes. They thread through ground tissues that make up most of the plant body. Dermal tissue covers its surfaces.

Take-Home Message

What is the basic structure of a flowering plant?

» Plants typically have aboveground shoots and stems, and belowground roots, all of which consist of ground, vascular, and dermal tissues. Plant tissues arise from divisions of meristem cells.

» The lengthening of plant parts is called primary growth. Secondary growth is the thickening of older roots and shoots over time.

» Ground tissues make up most of a plant. Vascular tissues that thread through ground tissue distribute water and solutes. Dermal tissue covers and protects plant surfaces.

» Eudicots and monocots have the same types of tissues, but differ somewhat in their pattern of tissue organization.

27.3 Simple and Complex Tissues

■ Different plant tissues form just behind shoot and root tips, and on older stem and root parts.

■ Links to Central vacuole 4.7, Plant cell surface specializations 4.11, Stomata 6.8, Lignin in plant evolution 22.3

Table 27.2	Overview of Flowering Plant Tissues	
Tissue Type	**Main Components**	**Main Functions**
Simple Tissues		
Parenchyma	Parenchyma cells	Photosynthesis, storage, secretion, tissue repair
Collenchyma	Collenchyma cells	Pliable structural support
Sclerenchyma	Fibers or sclereids	Structural support
Complex Tissues		
Dermal		
Epidermis	Epidermal cells, their secretions and outgrowths	Secretion of cuticle; protection; control of gas exchange and water loss
Periderm	Cork cambium; cork cells; parenchyma	Forms protective cover on older stems, roots
Vascular		
Xylem	Tracheids, vessel members; parenchyma cells; sclerenchyma cells	Water-conducting tubes; structural support
Phloem	Sieve-tube members, parenchyma cells; sclerenchyma cells	Sugar-conducting tubes and their supporting cells

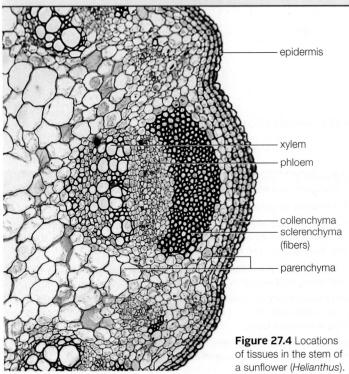

Figure 27.4 Locations of tissues in the stem of a sunflower (*Helianthus*).

— epidermis

— xylem
— phloem

— collenchyma
— sclerenchyma (fibers)

— parenchyma

In plants, cells are organized into simple and complex tissues. Simple tissues consist primarily of one type of cell; complex tissues have two or more cell types working together in a common function (**Table 27.1**). Some of these tissues are visible in the micrograph shown in **Figure 27.4**.

Types of Plant Tissues

Parenchyma Parenchyma is a simple tissue that makes up most of the soft parts inside roots, stems, leaves, and flowers. The shape of parenchyma cells varies with their function, but all have a thin, flexible cell wall (*above*). These cells are alive in mature tissue and can continue to divide, so they play an important role in wound repair.

Parenchyma has various roles depending on where it is located. In leaves and nonwoody stems, for example, chloroplast-containing parenchyma (*left*) is photosynthetic; parenchyma cells in stems and roots store starch, proteins, water, and oils in their large central vacuoles. Parenchyma also has specialized functions such as structural support, nectar secretion, and gas exchange inside leaves and stems.

Collenchyma Collenchyma is a simple, stretchable tissue that supports rapidly growing plant parts such as young stems and leaf stalks. It consists of collenchyma cells, which are elongated and alive at maturity. A complex polysaccharide called pectin imparts flexibility to these cells' primary wall, which is thickened unevenly where three or more of the cells abut (*inset*).

collenchyma Simple plant tissue composed of living cells with unevenly thickened walls; provides flexible support.
companion cell In phloem, parenchyma cell that loads sugars into sieve tubes.
epidermis Outermost tissue layer of a young plant.
mesophyll Photosynthetic parenchyma.
parenchyma Simple plant tissue made up of living cells; main component of ground tissue.
phloem Complex vascular tissue of plants; distributes sugars through its sieve tubes.
sclerenchyma Simple plant tissue that is dead at maturity; its lignin-reinforced cell walls structurally support plant parts.
sieve-tube member Type of cell that composes conducting tubes of phloem.
tracheid Cell that forms water-conducting tubes in xylem.
vessel member Cell that forms water-conducting tubes in xylem.
xylem Complex vascular tissue of plants; its tracheids and vessel members distribute water and mineral ions.

Sclerenchyma Variably shaped cells of **sclerenchyma** are dead at maturity, but the thick, lignin-containing secondary cell walls that remain after the cells die help this simple tissue resist compression. Fibers (*above*) and sclereids are typical sclerenchyma cells. Fibers are long, tapered cells; they occur in bundles that structurally support vascular tissues in

stems and leaves. Fibers flex and twist, but resist stretching. We use fibers of some plants in cloth, rope, paper, and other commercial products. Stubby and sometimes branched sclereids (*left*)

strengthen hard seed coats such as peach pits, and they make pear flesh gritty.

Dermal Tissues The first dermal tissue to form on a plant is **epidermis**, which is usually a single layer of cells on the plant's outer surface (*inset*). Epidermal cells secrete substances such as cutin, a

polymer of fatty acids, on their outward-facing cell walls. The waxy deposits form a waterproof cuticle that helps the plant conserve water and repel pathogens. The epidermis of leaves and young stems includes specialized cells and, often, hairs and other epidermal cell outgrowths. Pairs of specialized epidermal cells form stomata (small gaps across the epidermis) when the cells swell with water (Section 6.8). Plants control the diffusion of water vapor, oxygen, and carbon dioxide across epidermis by opening and closing stomata. In older stems and roots, a complex dermal tissue called periderm replaces epidermis (we return to this topic in Section 27.7).

Vascular Tissues Xylem and phloem are complex vascular tissues (*inset*). Each is composed of elongated conducting tubes often surrounded by sclerenchyma fibers and parenchyma. **Xylem**, which conducts

water and mineral ions, consists of two types of cells, **tracheids** and **vessel members**, that are dead in mature tissue (**Figure 27.5A,B**). The walls of these cells are stiffened and waterproofed with lignin. These walls interconnect to form tubes, and they lend structural support to the plant. Water can move laterally between the tubes as well as upward through them.

Phloem conducts sugars and other organic solutes. Its main cells, **sieve-tube members**, are alive in mature tissue. They connect end to end at sieve plates, forming sieve tubes that transport sugars from photosynthetic cells to all parts of the plant (**Figure 27.5C**).

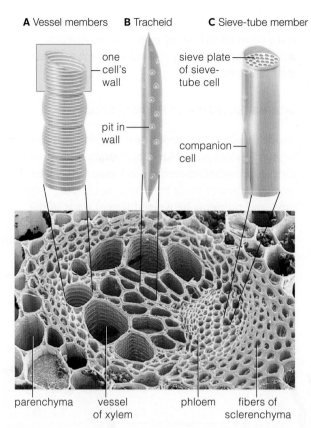

A Vessel members   **B** Tracheid   **C** Sieve-tube member

one cell's wall

pit in wall

sieve plate of sieve-tube cell

companion cell

parenchyma   vessel of xylem   phloem   fibers of sclerenchyma

Figure 27.5 Vascular tissues. Parenchyma cells and fibers of sclerenchyma are also visible in the micrograph. **Figure It Out:** What are the green structures inside the parenchyma cells?   Answer: Chloroplasts

Phloem's **companion cells** are parenchyma cells that actively transport sugars into the sieve tubes.

Ground Tissue Ground tissue, which is defined as everything other than dermal and vascular tissue, accounts for the bulk of a plant. It consists mostly of parenchyma, but can also include other simple tissues. Parenchyma cells that contain chloroplasts compose **mesophyll**, the only photosynthetic ground tissue. The *inset* shows starch-containing amyloplasts in the ground tissue of a potato.

Take-Home Message

What are the main types of plant tissues?

» Dermal tissue covers and protects plant surfaces. Epidermis includes epidermal cells and their secretions and outgrowths.

» Ground tissues make up most of a plant. Cells of parenchyma have diverse roles, including photosynthesis. Collenchyma and sclerenchyma support and strengthen plant parts.

» Vascular tissues distribute water and solutes through ground tissue. In xylem, water and ions flow through tubes of dead tracheid and vessel member cells. In phloem, sieve tubes that consist of living cells distribute sugars.

27.4 Primary Shoot Structure

■ The organization of ground, vascular, and dermal tissues inside stems differs between monocots and eudicots.

■ Primary growth in a stem arises by divisions of apical meristem cells in shoot tips.

Internal Structure of Primary Shoots

Inside the shoots of most flowering plants, the cells of primary xylem and phloem are bundled together as long, multistranded cords called **vascular bundles**. The main function of these bundles is to conduct water, ions, and nutrients between different parts of the plant. Lignin-reinforced walls of tracheids and sclerenchyma fibers in the bundles also play an important role in supporting upright stems.

Vascular bundles extend through the ground tissue of all stems and leaves, but the arrangement of the bundles in these plant parts differs between monocots and eudicots. The vascular bundles of monocot stems are distributed throughout the ground tissue (**Figure 27.6A**). By contrast, the vascular bundles of most eudicot stems form a cylinder that runs parallel with the long axis of the shoot (**Figure 27.6B**). This cylinder divides the ground tissue into regions called cortex and pith. Cortex comprises all ground tissue between the cylinder and epidermis; pith is all ground tissue inside the cylinder.

Primary Growth

During a plant's primary growth, its stems mainly lengthen (as opposed to thickening), and it produces soft organs such as leaves. This growth begins at a shoot's tip, which is called a **terminal bud** (**Figure 27.7**).

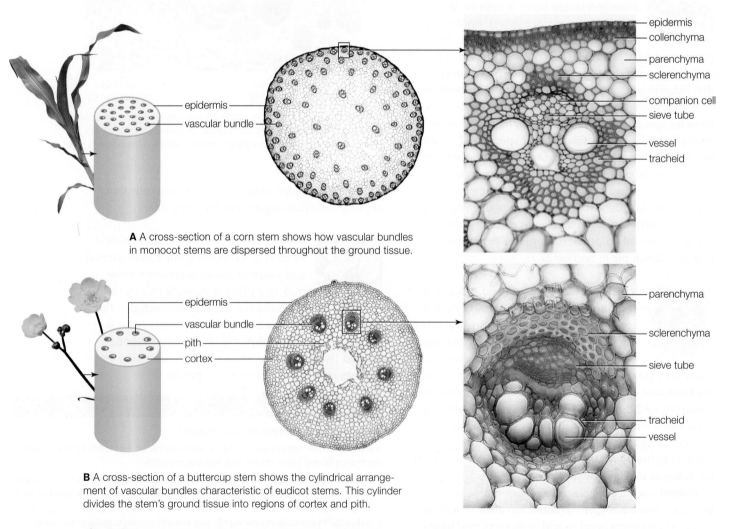

A A cross-section of a corn stem shows how vascular bundles in monocot stems are dispersed throughout the ground tissue.

B A cross-section of a buttercup stem shows the cylindrical arrangement of vascular bundles characteristic of eudicot stems. This cylinder divides the stem's ground tissue into regions of cortex and pith.

Figure 27.6 Comparing the structure of a primary shoot from (**A**) a monocot and (**B**) a eudicot. Cells appear different colors in the sections because they have been stained with different colored dyes that bind to certain carbohydrates and lignin.

A mass of apical meristem lies just below the surface of the bud ❶. Cells at the center of the mass are completely undifferentiated, and they divide continually during the growing season. The divisions push some of the cells toward the periphery of the mass. As a cell is displaced away from the center of the apical meristem, it begins to differentiate. Depending on its location in relationship to the bud's surface and to other cells, the cell will become a component of one of three primary meristem tissues. Cells that reach the bud's outer surface form protoderm ❷; those in its interior form procambium ❸ and ground meristem ❹. Further divisions and differentiation of cells in these primary meristems give rise to mature tissues: protoderm, to dermal tissue; procambium, to vascular tissue; ground meristem, to ground tissue (Figure 27.8).

The shoot grows longer mainly because primary meristem cells elongate as they differentiate and take on their specialized functions. The mass of apical meristem remains on top of the lengthening shoot, continually renewing itself and also producing cells that divide and differentiate below it.

Masses of tissue that bulge out near the sides of the apical meristem develop into leaves. The ones nearest the bud tip may develop into tiny, protective modified leaves called bud scales. As the stem lengthens, more leaves form and mature in orderly tiers, one after the next. A region of stem where leaves form is called a **node**; the region between two successive nodes is an internode. A **lateral bud** (also called an axillary bud) ❺ is a shoot that forms in a leaf axil, the place where a leaf attaches to a stem. Depending on hormonal signals, divisions of apical meristem cells in a lateral bud can give rise to either a branch, a leaf, or a flower.

lateral bud Axillary bud. A bud that forms in a leaf axil.
node A region of stem where leaves form.
terminal bud Shoot tip; a bud at the end of a stem. Main zone of primary growth in a shoot.
vascular bundle Multistranded cord of primary xylem and phloem in the stem or leaf of a plant.

Take-Home Message

How are plant tissues organized inside stems?

» The arrangement of vascular bundles, which are multi-stranded cords of vascular tissue, differs between eudicot and monocot stems.

» Primary growth—lengthening—in shoots arises at apical meristem inside terminal and axillary buds.

» Cell divisions in shoot apical meristem give rise to primary meristem tissues. Further differentiation and enlargement of cells in primary meristem tissues give rise to ground, vascular, and dermal tissues.

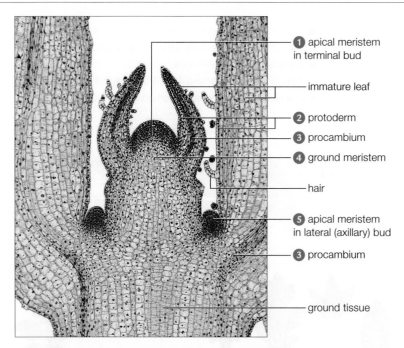

Figure 27.7 A longitudinal cut through the center of a shoot top of *Coleus*, a eudicot. Cell nuclei are stained *red* (nuclei take up most of the volume of the smaller cells in the darker regions).

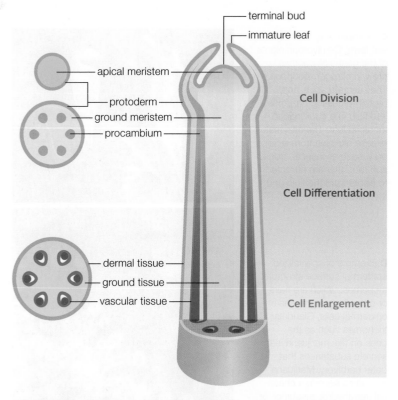

Figure 27.8 Primary growth in a shoot. New tissues arise by divisions of undifferentiated cells in shoot apical meristem. Some of the dividing cells are pushed to the periphery of the apical meristem, where they begin to differentiate as protoderm, procambium, and ground meristem. Cells in these primary meristem tissues divide and fully differentiate into dermal, vascular, and ground tissues, respectively. Regions of mitosis, differentiation, and cell enlargement are indicated.

27.5 A Closer Look at Leaves

■ Most leaves are metabolic factories where photosynthetic cells churn out sugars. They vary in size, shape, surface specializations, and internal structure.

■ Links to Plasmodesmata 4.11, Plant adaptations for photosynthesis 6.8, Water conservation adaptations 22.3

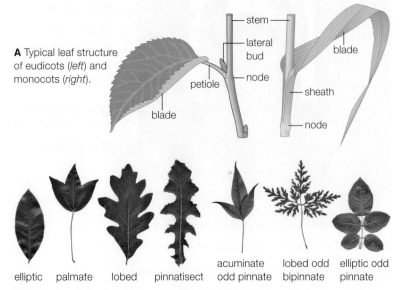

A Typical leaf structure of eudicots (*left*) and monocots (*right*).

stem
lateral bud
node
petiole
blade
blade
sheath
node

elliptic   palmate   lobed   pinnatisect   acuminate odd pinnate   lobed odd bipinnate   elliptic odd pinnate

B Examples of simple and compound leaves. All are from eudicots.

C Example of specialized leaf form. Carnivorous plants of the genus *Nepenthes* grow in nitrogen-poor soil. They secrete acids and protein-digesting enzymes into fluid in a cup-shaped modified leaf. The enzymes release nitrogen from small prey, such as insects, frogs, and rats, that are attracted to odors from the fluid and then drown in it.

D Example of specialized epidermal surface structure: trichomes, which are hairs or other fine outgrowths of epidermal cells. Glandular trichomes such as the ones on this marijuana leaf secrete substances that deter herbivory. Marijuana trichomes secrete a chemical (tetrahydrocannabinol, or THC), that has a psychoactive effect in humans.

Figure 27.9 Leaf structure.

Similarities and Differences

A leaf of duckweed is 1 millimeter (0.04 inch) across; leaves of one palm (*Raphia regalis*) can be 25 meters (82 feet) long. Leaf shapes and orientations on a stem differ: They can be shaped like cups, needles, blades, spikes, tubes, and feathers. They differ in color, odor, and edibility; many form toxins.

Most leaves are thin, with a high surface-to-volume ratio; some reorient themselves during the day in order to stay perpendicular to the sun's rays. Typically, adjacent leaves project from a stem in a pattern that allows sunlight to reach them all. However, the leaves of many plants native to arid regions stay parallel to the sun's rays, reducing heat absorption and thus conserving water. Thick or needlelike leaves of some plants also conserve water.

A typical leaf has a flat blade and, in eudicots, a petiole (or stalk) that attaches it to the stem (**Figure 27.9A**). The leaves of grasses and most other monocots are flat blades, the base of which forms a sheath around the stem. Simple leaves are undivided, although many are lobed; compound leaves have blades divided into leaflets (**Figure 27.9B**).

Fine Structure

The structure of most leaves is adapted to intercept the sun's light and to enhance gas exchange. Some leaves have a highly modified structure that allows them to fulfill an additional specialized function (**Figure 27.9C**).

Epidermis Epidermis covers every leaf surface exposed to the air. This surface tissue may be smooth, sticky, or slimy, with hairs, scales, spikes, hooks, and other specializations (**Figure 27.9D**). A translucent, waxy secreted cuticle slows water loss from the sheetlike array of epidermal cells (**Figure 27.10 ❶**).

Typically, most of a leaf's stomata occur on its lower surface ❷. The guard cells on either side of each stoma are the only photosynthetic cells in leaf epidermis. As you will see in Section 28.5, shape changes of the guard cells close the stomata to prevent water loss, or open the stomata to allow gases to cross the epidermis. Carbon dioxide needed for photosynthesis enters the leaf through stomata, then diffuses through air in the spaces between mesophyll cells. Oxygen released by photosynthesis diffuses in the opposite direction.

Mesophyll The bulk of a leaf consists of mesophyll, which is photosynthetic parenchyma. Plasmodesmata connect the cytoplasm of adjacent cells. Substances can

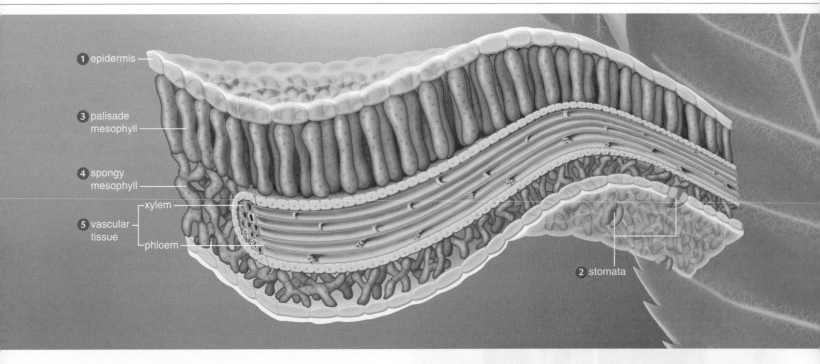

① epidermis
③ palisade mesophyll
④ spongy mesophyll
xylem
⑤ vascular tissue
phloem
② stomata

Figure 27.10 Animated Anatomy of a eudicot leaf.

❶ The upper leaf surface is epidermis with a secreted layer of cuticle.

❷ The lower leaf surface is also covered with cuticle. Gas exchanges between air inside and outside of the leaf occur at stomata positioned mainly in the lower leaf epidermis.

❸ The bulk of the leaf is mesophyll, a type of photosynthetic parenchyma. In many leaves, mesophyll occurs in two distinct forms: elongated palisade mesophyll attached to the upper epidermis, with ❹ spongy mesophyll below it.

❺ Vascular bundles of xylem (*blue*) and phloem (*pink*) form the leaf's veins.

flow rapidly across the walls of adjoining cells through these cell junctions (Section 4.11). Mesophyll in eudicot leaves, which are typically oriented perpendicular to the sun, is arranged in two layers. Palisade mesophyll is attached to the upper epidermis ❸. The elongated and aligned parenchyma cells of this tissue have more chloroplasts than cells of the spongy mesophyll layer below. Cells of spongy mesophyll are more rounded than those of palisade mesophyll, and have very large air spaces between them ❹. Blades of grass and other monocot leaves that grow vertically can intercept light from all directions. The mesophyll in such leaves is typically not divided into distinct layers.

Veins—The Leaf's Vascular Bundles Leaf **veins** are vascular bundles strengthened with fibers ❺. Inside the bundles, continuous strands of xylem rapidly

vein In a leaf, a vascular bundle strengthened with fibers.

 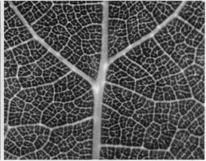

A An *Agapanthus* leaf, with the parallel orientation of veins typical of monocots. Veins of monocot leaves are of similar length.

B A grape leaf, with the netlike array of veins typical of eudicots. A stiffened midrib runs from the petiole to the leaf tip. Smaller veins branch from it.

Figure 27.11 Typical vein patterns in eudicot and monocot leaves. Like umbrella ribs, stiffened veins help maintain leaf shape.

transport water and dissolved ions to mesophyll. Continuous strands of phloem rapidly transport the products of photosynthesis (sugars) away from mesophyll. In most monocots, all veins are similar in length and run parallel with the leaf's long axis (**Figure 27.11A**). In most eudicots, large veins branch into a network of minor veins embedded in mesophyll (**Figure 27.11B**).

Take-Home Message

How does a leaf's structure contribute to its function?

» Leaves are structurally adapted to intercept sunlight and distribute water and nutrients. Components include mesophyll (photosynthetic cells), veins (bundles of vascular tissue), and cuticle-secreting epidermis.

27.6 Primary Root Structure

- Roots anchor a plant, and also provide it with a large surface area for absorbing water and dissolved mineral ions from soil.
- Link to Membrane permeability 5.8

A plant's roots function to anchor the plant, take up water and mineral ions, and transport these substances to aboveground parts of the plant. The root system of a typical plant is at least as extensive as its shoot system. Roots of mesquite (*Prosopis*), a drought-tolerant shrub, can reach more than 50 meters (164 feet) underground. Cacti tend to have shallow roots; some can radiate 15 meters (49 feet) away from the plant. Laid out as a sheet, the root system of a single young rye plant would cover about 600 square meters, or about 6,500 square feet! You will see in Chapter 28 how a large surface area and other specializations adapt roots for absorbing water and mineral ions from the soil.

The first structure to emerge from a seed is a root. In most eudicots, this primary root thickens as it grows longer, and it usually develops many lateral branchings, or secondary roots. The primary root together with its lateral branchings constitutes a **taproot system** (Figure 27.12A). A taproot is generally larger in diameter than any of its branchings. By contrast, the primary root that emerges from a monocot seed is quickly replaced with a mat of adventitious roots, which sprout from the base of the stem. All of the roots in the resulting **fibrous root system** are about equal in diameter—there is no dominant, central root as in a taproot system (Figure 27.12B). Roots in a fibrous system may or may not be branched. The taproot system of some eudicots eventually develops into a fibrous root system as the taproot is replaced by adventitious roots.

Internal Structure

The epidermis of primary roots is the plant's absorptive interface with soil. Many of the specialized cells of this dermal tissue send out fine extensions called **root hairs**, which collectively increase the surface area available for taking up soil water along with dissolved oxygen and mineral ions.

Water and dissolved substances enter a root by crossing epidermis and seeping into parenchyma cells that make up the root's cortex. The water moves from cell to cell of the cortex until it reaches a central column of conductive tissue called the **vascular cylinder**, or **stele**. In a typical monocot, the vascular cylinder divides the root's ground tissue into cortex and pith

A A taproot system consists of a large main root together with its lateral branchings.

B A fibrous root system has similar-sized adventitious roots that branch from the stem.

Figure 27.12 Comparing root systems. *Left*, the taproot system of salsify (*Tragopogon porrifolius*); *right*, the fibrous root system of corn (*Zea mays*).

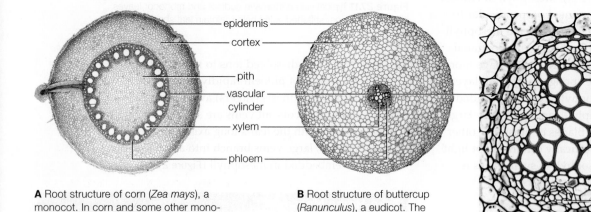

epidermis
cortex
pith
vascular cylinder
xylem
phloem

root cortex
endodermis
pericycle
xylem
phloem

A Root structure of corn (*Zea mays*), a monocot. In corn and some other monocots, the vascular cylinder divides the ground tissue into cortex and pith. This micrograph shows a lateral root forming.

B Root structure of buttercup (*Ranunculus*), a eudicot. The micrograph on the *right* shows the detailed tissue structure of the vascular cylinder in this plant.

Figure 27.13 Structure of roots. Here we compare a primary root from (**A**) a monocot and (**B**) a eudicot.

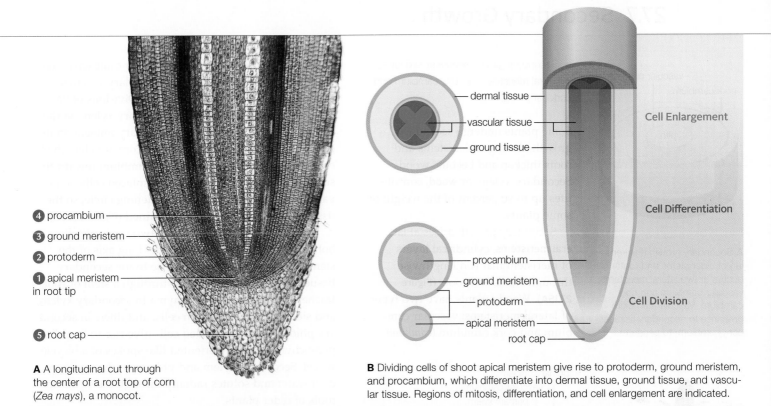

4 procambium
3 ground meristem
2 protoderm
1 apical meristem in root tip
5 root cap

dermal tissue
vascular tissue
ground tissue

Cell Enlargement

procambium
ground meristem
protoderm
apical meristem
root cap

Cell Differentiation

Cell Division

A A longitudinal cut through the center of a root top of corn (*Zea mays*), a monocot.

B Dividing cells of shoot apical meristem give rise to protoderm, ground meristem, and procambium, which differentiate into dermal tissue, ground tissue, and vascular tissue. Regions of mitosis, differentiation, and cell enlargement are indicated.

Figure 27.14 Animated Primary growth in a root. New tissues arise by divisions of undifferentiated cells in root apical meristem. Some of the dividing cells are pushed to the periphery of the apical meristem, where they begin to differentiate as protoderm, procambium, and ground meristem. Cells in these primary meristem tissues divide and fully differentiate into dermal, vascular, and ground tissues, respectively.

(**Figure 27.13A**). By contrast, most of the ground tissue in typical eudicot roots is outside the vascular cylinder. In these plants, the root vascular cylinder is mainly xylem and phloem (**Figure 27.13B**).

The outer boundary of the vascular cylinder is **endodermis**, a layer of cells that separates the root cortex from the conductive tissues inside the vascular cylinder. You will see in Section 28.3 how these cells help control the flow of water and dissolved substances from the soil into the shoot system.

Primary Growth

As in stems, most of the primary growth in roots originates at apical meristems (**Figure 27.14**). Cell division in a mass of apical meristem in a root tip **1** gives rise to primary meristem tissues: protoderm **2**, ground meristem **3**, and procambium **4**. Cells in these three tissues differentiate and enlarge to form dermal, ground, and vascular tissues, respectively. Some protoderm cells give rise to a root cap **5**, a dome-shaped mass of cells that protects the soft, young root as it grows through soil. Cells of the root cap are shed continually as the root grows through soil.

Just inside the endodermis is a layer of cells that form **pericycle**. Pericycle cells retain the capacity to

divide in a direction perpendicular to the longitudinal axis of the root. Lateral roots form from divisions of these cells (a new lateral root branching from pericycle is visible in the micrograph in **Figure 27.12A**).

endodermis In plant roots, a layer of cells just outside the pericycle; separates vascular cylinder from cortex.
fibrous root system Root system composed of an extensive mass of similar-sized roots; typical of monocots.
pericycle Layer of cells just inside root endodermis that can give rise to lateral roots.
root hairs Hairlike, absorptive extensions of a young cell of root epidermis.
stele Vascular cylinder of a root.
taproot system In eudicots, a primary root and all of its lateral branchings.
vascular cylinder A root's central column of vascular tissue; composed of primary xylem and phloem sheathed by pericycle and endodermis.

Take-Home Message

What is the basic structure and function of a plant's root?

» Taproot systems consist of a primary root and lateral branchings. Fibrous root systems consist of adventitious and lateral roots that replace the primary root.

» Inside each root is a vascular cylinder that contains long strands of xylem and phloem.

» Roots provide a plant with a large surface area for absorbing water and dissolved minerals from soil.

27.7 Secondary Growth

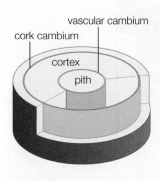

A Secondary growth (thickening of older stems and roots) occurs at two lateral meristems, vascular cambium and cork cambium. Vascular cambium gives rise to secondary tissues; cork cambium, to periderm.

■ Secondary growth occurs at two types of lateral meristem, vascular cambium and cork cambium.

Some plants undergo secondary growth, during which stems and roots thicken and become woody. Secondary xylem, or **wood**, contributes up to 90 percent of the weight of some plants.

Secondary growth arises at **lateral meristems**, cylindrical layers of meristem that run lengthwise through stems and roots (**Figure 27.15A**). Woody plants have two types of lateral meristems: vascular cambium and cork cambium (*cambium*

is the Latin word for change). **Vascular cambium** is the lateral meristem that produces secondary vascular tissue inside older stems and roots. Divisions of vascular cambium cells produce secondary xylem on the cylinder's inner surface, and secondary phloem on its outer surface (**Figure 27.15B**). As the core of xylem thickens, it also displaces the vascular cambium toward the surface of the stem or root. The displaced cells of the vascular cambium divide in a widening circle, so the tissue's cylindrical form is maintained (**Figure 27.15C**).

Vascular cambium consists of two types of cells; both divide perpendicularly to the long axis of the stem. Long, narrow cells give rise to the secondary tissues that extend lengthwise through a stem or root: tracheids, fibers, and parenchyma in secondary xylem; and sieve tubes, companion cells, and fibers in secondary phloem. Small, rounded cells give rise to "rays" of parenchyma, radially oriented like spokes of a bicycle wheel. Secondary xylem and phloem in the rays conduct water and solutes radially through the stems and roots of older plants.

Thin-walled, living parenchyma cells and sieve tubes of secondary phloem lie in a narrow zone outside the vascular cambium. Bands of thick-walled reinforcing fibers are often interspersed through this secondary phloem. The only living sieve tubes lie within a centimeter or so of the vascular cambium.

B In spring, primary growth resumes at terminal and lateral buds. Secondary growth occurs at vascular cambium. Divisions of meristem cells in the vascular cambium expand the inner core of xylem, displacing the vascular cambium (*orange*) toward the surface of the stem or root.

C Pattern of division and differentiation of vascular cambium cells (*orange*). With each division, one descendant cell differentiates and the other remains undifferentiated.

Cells that differentiate on the pith (inner) side of the vascular cambium become part of xylem (X_1, X_2, X_3).

Cells that differentiate on the cortex (outer) side of the vascular cambium become part of phloem (P_1, P_2, P_3).

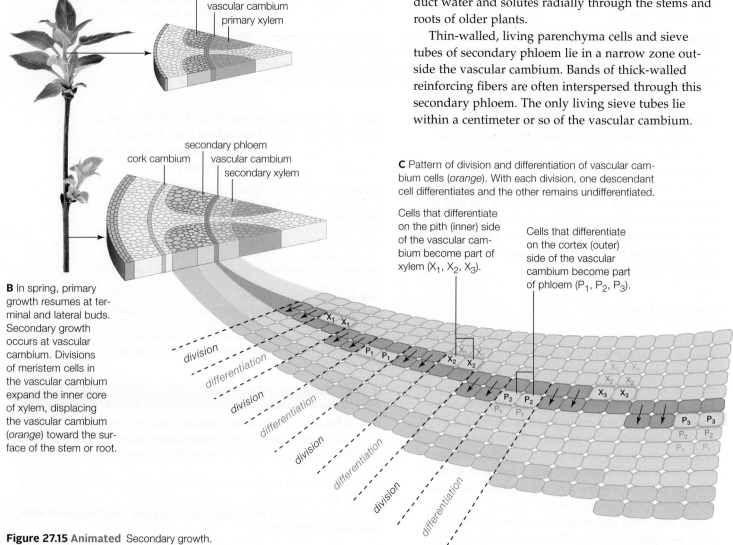

Figure 27.15 Animated Secondary growth.

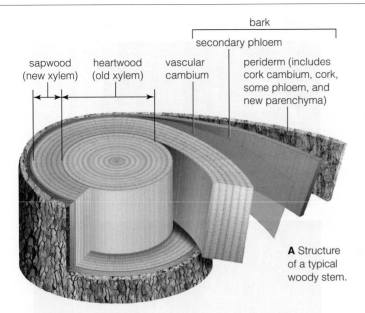

bark
secondary phloem

sapwood (new xylem) | heartwood (old xylem) | vascular cambium | periderm (includes cork cambium, cork, some phloem, and new parenchyma)

A Structure of a typical woody stem.

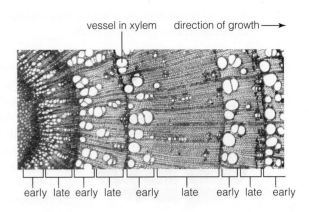

vessel in xylem   direction of growth →

early late early late early late early late early

B Early and late wood in an ash tree. Early wood forms during wet springs. Late wood indicates that a tree did not waste energy making large-diameter xylem cells for water uptake during a dry summer or drought.

Figure 27.16 Structure of a woody stem. The rings visible in the heartwood and sapwood are regions of early and late wood. In most temperate zone trees, one ring forms each year.

As seasons pass, the expanding inner core of xylem continues to put pressure on the stem or root surface. In time, the pressure ruptures the cortex and the outer secondary phloem. Then, another lateral meristem, **cork cambium**, forms and gives rise to periderm. Periderm is a dermal tissue that consists of parenchyma and cork, as well as the cork cambium that produces them. What we call **bark** is secondary phloem and periderm—all living and dead tissue outside the vascular cambium (**Figure 27.16A**). The **cork** component of bark has densely packed rows of dead cells with walls thickened by suberin, a waxy substance. Cork protects, insulates, and waterproofs a stem or root surface. Cork also forms over wounded tissues. When leaves drop from the plant, cork forms at places where the petioles were attached to stems.

Wood's structure and function change as a plant ages. The xylem at its center becomes clogged by tannins, gums, and oils, becoming **heartwood**, a region that provides structural support but does not function in transport. **Sapwood**, the region of functional secondary xylem, lies between heartwood and the vascular cambium. During the spring in temperate zones, dissolved sugars travel from tree roots to buds through sapwood. The sugar-rich fluid is sap. New Englanders collect maple tree sap to make maple syrup.

Vascular cambium is inactive during long periods of cold or drought. During warm weather when water is available, this lateral meristem gives rise to early wood, with large-diameter, thin-walled cells. Late wood, with small-diameter, thick-walled xylem cells, forms during periods of warm, dry weather. A cross-section through an older trunk may reveal alternating bands of early and late wood (**Figure 27.16B**). Each band is a growth ring, or "tree ring." Trees native to regions where growth slows or stops during the winter tend to add one growth ring per year. In tropical regions where weather does not vary seasonally, trees grow at the same rate all year and do not have growth rings.

Oak, hickory, and other eudicot trees that evolved in temperate and tropical zones are hardwoods, with vessels, tracheids, and fibers in xylem. Pines and other conifers are called softwoods because they are less dense than the hardwoods. Their xylem has tracheids and parenchyma rays but no vessels or fibers.

bark Secondary phloem and periderm of woody plants.
cork Component of bark; waterproofs, insulates, and protects the surfaces of woody stems and roots.
cork cambium Lateral meristem that gives rise to periderm.
heartwood Dense, dark accumulation of nonfunctional xylem at the core of older tree stems and roots.
lateral meristem Vascular cambium or cork cambium; cylindrical sheet of meristem that gives rise to plant secondary growth.
sapwood Functional secondary xylem between the vascular cambium and heartwood in an older stem or root.
vascular cambium Lateral meristem that produces secondary xylem and phloem.
wood Accumulated secondary xylem.

Take-Home Message

What is secondary growth?

» Secondary growth thickens the stems and roots of older plants.

» Secondary growth occurs at vascular cambium and cork cambium, which give rise to secondary vascular tissues and periderm, respectively. Wood is mainly accumulated secondary xylem.

27.8 Tree Rings and Old Secrets

■ The relative thicknesses of a tree's rings hold clues to environmental conditions during its lifetime.

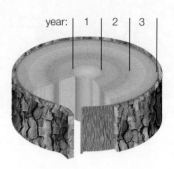

year: 1 2 3

Dendroclimatologists use tree rings to estimate average annual rainfall; to date archaeological ruins; to gather evidence of wildfires, floods, landslides, and glacier movements; and to study the ecology and effects of parasitic insect populations. How? Some tree species, such as redwoods and bristlecone pines, add wood over centuries, one ring per year (*inset*). Count an old tree's rings, and you have an idea of its age. If you know the year in which the tree was cut, you can find out when a particular ring formed by counting the rings backward from the outer edge. Thickness and other features of the rings offer clues about environmental conditions during those years (**Figure 27.17**).

For an example of how we use the information in tree rings, consider the history of early English settlers of the United States. In 1587, 150 settlers arrived at Roanoke Island, which is off the coast of North Carolina. Supply ships that returned in 1589 found the island abandoned; searches of the mainland coast failed to turn up the missing colonists. About twenty years later, a second set of English settlers arrived, this time at Jamestown, Virginia. Although this colony survived, the initial years were difficult. In the summer of 1610 alone, more than 40 percent of the colonists died, many of them from starvation.

Differences in the thicknesses of tree rings from nearby bald cypress trees (*Taxodium distichum*) that had been growing at the time of the Roanoke and Jamestown colonies revealed that both sets of settlers had terrible timing (**Figure 27.17D**). The Roanoke settlers arrived just in time for the worst drought in 800 years, and nearly a decade of severe drought struck Jamestown. We know that the corn crop of the Jamestown colony failed, so similar drought-related crop failures probably occurred at Roanoke. The Jamestown settlers also would have had difficulty finding fresh water, because their colony had been established at the head of an estuary. When the river levels dropped during the drought, the settlers' drinking water would have mixed with ocean water and become salty. Piecing together these bits of evidence gives us an idea of what life must have been like for the early settlers.

direction of growth ⟶

A Pine is a softwood. It grows fast, so it tends to have wider rings than slower-growing species. Note the difference between the appearance of the heartwood and sapwood.

B The rings of this oak tree show dramatic differences in yearly growth patterns over its lifetime. Note the prominent rays.

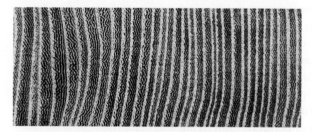

C An elm made this series between 1911 and 1950.

1587–1589   1606–1612

D A section of a bald cypress tree that was living near English colonists when they first settled in North America. Narrower annual rings mark years of severe drought.

Figure 27.17 Tree rings. In most species, each ring corresponds to one year, so the number of rings indicate the age of the tree. Relative thickness of the rings can be used to estimate data such as average annual rainfall long before records of climate were kept.

Take-Home Message

Can tree rings reveal more than just the age of a tree?

» Tree rings reflect conditions in the environment during the time they formed.

27.9 Variations on a Stem

■ Many plants have modified stem structures that function in storage and reproduction.

Specialized stems allow some plants to store nutrients, to reproduce asexually, or both.

Examples of Stem Specializations

Stolons Stolons are stems that branch from the main stem of the plant and grow horizontally along the ground or just under its surface. Stolons often look like roots, but they have nodes (roots do not have nodes). Adventitious roots and leafy shoots that sprout from the nodes develop into new plants. Stolons are commonly called runners because in many plants they "run" along the surface of the soil. The strawberry plant (*Fragaria*) in the photo *above* is reproducing asexually by sending out runners.

Rhizomes Ginger, irises, and some grasses have rhizomes, which are fleshy stems that typically grow under the soil and parallel to its surface. A rhizome is the main stem of the plant, and it also serves as the plant's primary storage tissue. Branches that sprout from nodes grow aboveground for photosynthesis and flowering. The main stems of turmeric plants (*Curcuma longa*), shown in the photo *above*, are underground rhizomes.

Bulbs A bulb is a short section of underground stem encased by overlapping layers of thickened, modified leaves called scales. The scales develop from the base of the bulb, as do roots. The photo (*left*) shows clearly visible scales surrounding the stem at the center of an onion, which is the bulb of an *Allium cepa* plant. Bulb scales contain starch and other substances that a plant holds in reserve when conditions in the environment are unfavorable for growth. When favorable conditions return, the plant uses these stored substances to sustain rapid growth. The dry, paperlike outer scale of many bulbs serves as a protective covering.

Sequestering Carbon (revisited)

Typically, carbon offsets for reforestation are purchased well in advance of their benefit because most trees grow very slowly. It can be many decades before a tree sequesters a substantial amount of carbon in its secondary growth. Critics of carbon offsets point out that carbon sequestration in forests is also impermanent: There is no guarantee that reforested trees will survive to maturity, or that the carbon they have sequestered will not be released again, for example by fire or by logging.

How would you vote? Are carbon offsets helpful, or do they just provide an excuse to continue emitting greenhouse gases?

Corms A corm is a thickened underground stem that stores nutrients. Like a bulb, a corm has a basal plate from which roots grow. Unlike a bulb, a corm is solid rather than layered, and it has nodes from which new plants develop. Taro, also known as arrowroot, is the corm of *Colocasia esculenta* plants (*above*).

Tubers Tubers are thickened portions of underground stolons; they are the plant's primary storage tissue. Tubers are like corms in that they have nodes from which new shoots and roots sprout, but they do not have a basal plate. The photo (*left*) shows how potatoes, which are tubers, grow on stolons of *Solanum tuberosum* plant. Potato "eyes" are the nodes of the tubers.

Cladodes Cacti and other succulents have cladodes, which are flattened, photosynthetic stems that store water. New plants form at the nodes. The cladodes of some plants appear quite leaflike, but most are unmistakably fleshy. The photo (*above*) shows a spiky cladode of a prickly pear plant (*Opuntia*).

Take-Home Message

Are all stems alike?

» Many plants have modified stems that function in storage and reproduction. Stolons, rhizomes, bulbs, corms, tubers, and cladodes are examples.

LEARNING ROADMAP

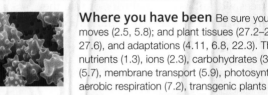

Where you have been Be sure you understand how water moves (2.5, 5.8); and plant tissues (27.2–27.4), structures (27.5, 27.6), and adaptations (4.11, 6.8, 22.3). This chapter revisits nutrients (1.3), ions (2.3), carbohydrates (3.4), receptor proteins (5.7), membrane transport (5.9), photosynthesis (6.4, 6.5, 6.7), aerobic respiration (7.2), transgenic plants (15.7), nitrogen fixation (20.7), and mycorrhiza (23.7).

Where you are now

Plant Nutrients and Soil
Many plant structures are adaptations to limited amounts of water and essential minerals. The amount of water and minerals available for plants depends on soil composition.

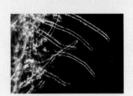

Root Function
Hairs and other structural and functional specializations allow the roots of vascular plants to selectively take up water and minerals from the soil.

Water Movement In Plants
Xylem tubes consist of the interconnected walls of dead cells. The main function of these tubes is to transport water and dissolved minerals from roots to other parts of the plant.

Control of Water Loss
Plant structures such as cuticle and stomata are adaptations that minimize water loss. The opening and closing of stomata balances gas exchange with water conservation.

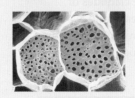

Sugar Movement in Plants
Sieve tubes are stacks of living sieve-tube members. Sugars are loaded into sieve tubes in regions that produce them, and unloaded in regions that use them.

Where you are going Chapter 29 explores flowering plant adaptations for sexual and asexual reproduction. In Chapter 30, plant nutrients are discussed in the context of signaling mechanisms introduced in this chapter. Chapter 32 explains voltage-gated channels. We return to human impacts on the biosphere in Chapter 48. Plant nutrition comes up again in the context of the carbon cycle (Section 46.7), nitrogen cycle (Section 46.9), and phosphorus cycle (46.11).

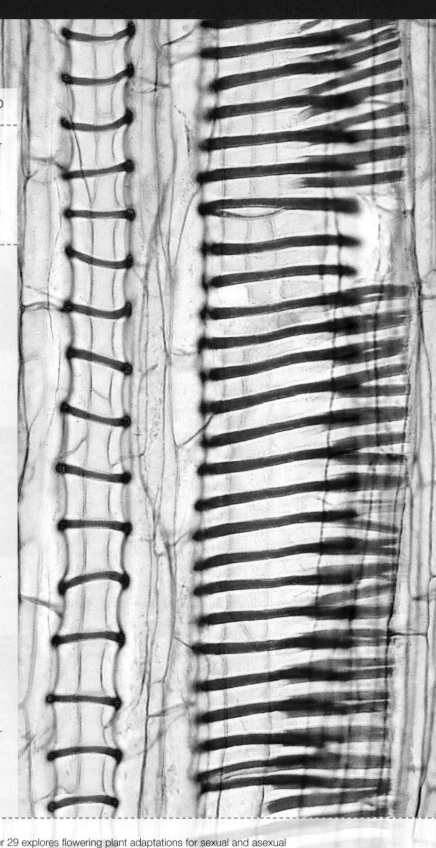

28.1 Leafy Cleanup Crews

From World War I until the 1970s, the United States Army tested and disposed of weapons at J-Field, Aberdeen Proving Ground in Maryland (**Figure 28.1**). Obsolete chemical weapons and explosives were burned in open pits, together with plastics, solvents, and other wastes. Toxic metals, such as lead, arsenic, and mercury heavily contaminated the soil and groundwater. So did highly toxic organic compounds, including trichloroethylene (TCE). TCE damages the nervous system, lungs, and liver, and exposure to large amounts can be fatal. Today, the toxic groundwater is seeping toward nearby marshes and the Chesapeake Bay.

To protect the bay and clean up the soil, the Army and the Environmental Protection Agency turned to phytoremediation: the use of plants to take up and concentrate or degrade environmental contaminants. They planted poplar trees (*Populus trichocarpa*) that cleanse groundwater by removing TCE and other organic pollutants from it.

Like other vascular plants, poplar trees absorb soil water through their roots. Along with the water come nutrients and chemical contaminants, including TCE. Although TCE is toxic to animals, it does not harm the trees. The poplars break down some of the toxin, but they release most of it into the atmosphere. Airborne TCE is the lesser of two evils: It breaks down much more quickly in air than it does in groundwater.

In other types of phytoremediation, groundwater contaminants accumulate in the tissues of the plants, which are then harvested for safer disposal elsewhere.

Analyses at J-Field since the poplars were planted show that the phytoremediation strategy is working splendidly. By 2001, the trees had removed 60 pounds of TCE from the soil and groundwater. In addition, they had helped foster communities of microorganisms that were also breaking down toxins. Researchers estimate that the trees will have removed the majority of contaminants from J-Field by 2030.

Phytoremediation is cheaper and easier on the environment than mechanical methods of cleaning up toxic waste, which include excavation and storage of contaminated soil, and pumping out contaminated groundwater. In general, the best plants for phytoremediation take up many contaminants, grow fast, and grow big. Not very many species with those traits tolerate toxic substances, but genetically engineered ones may increase our number of choices. Alpine pennycress (*Thlaspi caerulescens*) absorbs zinc, cadmium, and other potentially toxic minerals dissolved in soil water. Unlike typical cells, the cells of pennycress plants

store zinc and cadmium inside a central vacuole. Isolated inside these organelles, the toxic elements are kept safely away from the rest of the cells' activities. Pennycress is a small, creeping plant, so its usefulness for phytoremediation is limited. Researchers are working to transfer a gene that confers its toxin-storing ability to larger, faster-growing plants.

Phytoremediation takes advantage of the ability of plants to take up water and solutes (including toxic waste) from the soil. Plants have this capacity not for our benefit, but rather to meet their own metabolic needs. Water and solutes needed by all living cells must move from soil, into roots, and then to other parts of a plant. Sugars that fuel metabolism must also move from where they are produced to where they are used. The structures that allow this movement, and the processes that drive it, are the subject of this chapter.

Figure 28.1 Toxic waste cleanup. *Top*, J-Field during its use as a testing site for weapons. *Bottom*, J-Field today. Poplar trees are helping to remove toxic waste left behind in the soil.

28.2 Plant Nutrients and Availability in Soil

- Plants require elemental nutrients in soil, water, and air.
- Soil type affects the growth of plants.
- Links to Nutrients 1.3, Ions 2.3

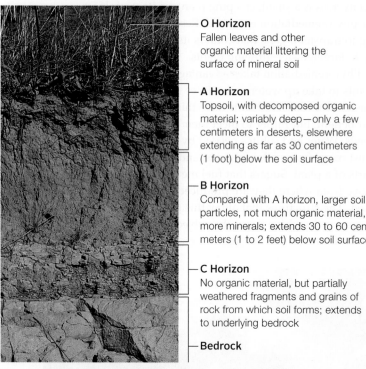

O Horizon
Fallen leaves and other organic material littering the surface of mineral soil

A Horizon
Topsoil, with decomposed organic material; variably deep—only a few centimeters in deserts, elsewhere extending as far as 30 centimeters (1 foot) below the soil surface

B Horizon
Compared with A horizon, larger soil particles, not much organic material, more minerals; extends 30 to 60 centimeters (1 to 2 feet) below soil surface

C Horizon
No organic material, but partially weathered fragments and grains of rock from which soil forms; extends to underlying bedrock

Bedrock

Figure 28.2 From a habitat in Africa, an example of soil horizons.

Plant growth requires the sixteen elements listed in **Table 28.1**. Nine are macronutrients, which means they are required in amounts above 0.5 percent of the plant's dry weight (after all the water has been removed from it). Seven other elements are micronutrients, which make up traces of the plant body. Carbon, oxygen, and hydrogen atoms are abundantly available in carbon dioxide and water. Plants obtain the other nutrients when their roots take up minerals dissolved in soil water.

Properties of Soil

Soil consists of mineral particles mixed with variable amounts of decomposing organic material, or **humus**. The mineral particles form by the weathering of rocks. Humus forms from dead organisms and organic litter: fallen leaves, feces, and so on. Water and air occupy spaces between the mineral particles and organic bits.

Soils differ in the proportions of mineral particles. The particles, which are primarily sand, silt, and clay, differ in size and chemical properties. The biggest sand grains are visible with the naked eye, about one millimeter in diameter. Silt particles are hundreds or thousands of times smaller than sand grains, so they are too small to see. Clay particles are even smaller. A clay particle consists of thin, stacked layers of nega-

Table 28.1 Plant Nutrients and Their Main Functions

Macronutrient	Main Functions	Micronutrient	Main Functions
Carbon Hydrogen Oxygen	Raw materials for photosynthesis. Main components of biological molecules (nucleic acids, proteins, fats, carbohydrates), cofactors, and organic signaling molecules such as hormones	Chlorine	Photosystem II cofactor; signaling; solute and electrochemical balances
Nitrogen	Important component of proteins (including enzymes), nucleic acids, cofactors, photosynthetic pigments	Iron	Component of chlorophyll and many enzymes; particularly important in electron transfers
Potassium	Involved in enzyme activation, pH, opening stomata; also contributes to water–solute balance,* which in turn influences osmosis and turgor	Boron	Cross-links polysaccharides in cell walls; also links cell wall polysaccharides to surface of plasma membrane
Calcium	Stabilizes structure of cell membranes and cell walls; Important in cell signaling, mitosis, stomata function	Manganese	Component of photosystem II that allows it to catalyze photolysis during light reactions of photosynthesis; also functional component of many coenzymes
Magnesium	Functional component of chlorophyll, many enzymes and cofactors; stabilizes structure of cell membranes, cell walls	Zinc	Cofactor for many enzymes; component of many proteins, including auxin receptors and transcription factors
Phosphorus	Important component of phospholipids and nucleic acids (particularly ATP and other coenzymes)	Copper	Cofactor in many proteins and enzymes involved in electron transfer chains (including plastocyanin of light-dependent reactions), lignin synthesis, detoxification of free radicals
Sulfur	Component of proteins, some cofactors	Molybdenum	Cofactor required for nitrogen fixation and for abscisic acid synthesis

* All mineral elements contribute to water–solute balances.

tively charged crystals with sheets of water alternating between the layers. Because clay is negatively charged, it attracts and reversibly binds positively charged mineral ions dissolved in the soil water. Thus, clay retains dissolved nutrients that would otherwise trickle past roots too quickly to be absorbed.

Although sand and silt do not bind mineral ions as well as clay, they are also necessary for plant growth. Water and air occupy spaces between the particles and organic bits in soil. Without enough sand and silt to intervene between tiny particles of clay, soil packs so tightly that it excludes air. Without air, root cells cannot secure enough oxygen for aerobic respiration.

Soils with the best oxygen and water penetration are **loams**, which have roughly equal proportions of sand, silt, and clay. Most plants grow best in loams that contain between 10 and 20 percent humus. Humus affects plant growth because it releases nutrients, and its negatively charged organic acids can trap positively charged mineral ions in soil water. Soil with less than 10 percent humus may not have enough nutrients to support plant growth. Humus also swells and shrinks as it absorbs and releases water, and these changes in size aerate soil by opening spaces for air to penetrate. Soil with more than 90 percent humus, such as that in swamps and bogs, stays so saturated with water that air (and the oxygen in it) is excluded. Very few kinds of plants can grow in these soils.

How Soils Develop Soils develop over thousands of years. They exist in different stages of development in different regions. Most form in layers, or horizons, that are distinct in color and other properties (**Figure 28.2**). Identifying the layers helps us compare soils in different places. For instance, the A horizon, which is **topsoil**, contains the greatest amount of organic matter, so the roots of most plants grow most densely in this layer. Grasslands typically have a deep layer of topsoil; tropical forests do not.

Minerals, salts, and other molecules dissolve in water as it filters through soil. **Leaching** is the process by which water removes soil nutrients and carries them away. Leaching is fastest in sandy soils, which do not bind nutrients as well as clay soils. During

humus Decaying organic matter in soil.
leaching Process by which water moving through soil removes nutrients from it.
loam Soil with roughly equal amounts of sand, silt, and clay.
soil Mixture of various mineral particles and humus.
soil erosion Loss of soil under the force of wind and water.
topsoil Uppermost soil layer; contains the most nutrients for plant growth.

Figure 28.3 Runaway erosion in Providence Canyon, Georgia, a result of poor farming practices combined with soft soil. Settlers who arrived in the area around 1800 plowed the land straight up and down the hills. The furrows made excellent conduits for rainwater, which proceeded to carve out deep crevices that made even better rainwater conduits. The area became useless for farming by 1850. It now consists of about 445 hectares (1,100 acres) of deep canyons that continue to expand at the rate of about 2 meters (6 feet) per year.

heavy rains, more leaching occurs in forests than in grasslands. Why? Grass plants can absorb water more quickly than trees.

Soil erosion is a loss of soil under the force of wind and water. Strong winds, fast-moving water, sparse vegetation, and poor farming practices cause the greatest losses (**Figure 28.3**). For example, each year, about 25 billion metric tons of topsoil erode from croplands in the midwestern United States. The topsoil enters the Mississippi River, which then dumps it into the Gulf of Mexico. Nutrient losses because of this erosion affect not only plants that grow in the region, but also the other organisms that depend on the plants for survival.

Take-Home Message

Where do plants get nutrients they need?

» Plants require nine macronutrients and seven micronutrients, all elements that are available from water, air, and soil.

» Soil consists mainly of mineral particles: sand, silt, and clay. Clay reversibly binds positively charged mineral ions dissolved in soil water.

» Soil contains humus, a reservoir of organic material rich in organic acids. Humus reversibly binds positively charged mineral ions dissolved in soil water.

» Most plants grow best in loams (soils with equal proportions of sand, silt, and clay) that contain 10 to 20 percent humus.

» Leaching and erosion remove nutrients from soil.

28.3 How Do Roots Absorb Water and Nutrients?

■ Root specializations such as hairs, mycorrhizae, and nodules help plants absorb water and nutrients.

■ Links to Plant cell walls and plasmodesmata 4.11, Osmosis 5.8, Membrane crossing mechanisms 5.9, Nitrogen fixation 20.7, Fungal symbionts 23.7, Root structure 27.6

Root Specializations

In actively growing plants, new roots grow into different patches of soil. The roots are not "exploring" the soil; rather, their growth is greatest in areas where water and nutrient concentrations best match the requirements of the particular plant. Root specializations help plants take up water and minerals from soil.

Root Hairs Root hairs are thin extensions of root epidermal cells that enormously increase the surface area available for absorbing water and nutrients (Section 27.6). The hairs are fragile structures no more than a few millimeters long (**Figure 28.4A**). They do not develop into new roots, and last only a few days.

Mycorrhizae As Section 23.7 explains, a **mycorrhiza** (plural, mycorrhizae) is a mutually beneficial interaction between a young root and a fungus. Hyphae, which are filaments of the fungus, form a velvety cloak around roots, or penetrate the root cells. Collectively, the hyphae have a large surface area, so they absorb mineral ions from a larger volume of soil than roots alone (**Figure 28.4B**). The fungus absorbs

some sugars and nitrogen-rich compounds from root cells. The root cells get some scarce minerals that the fungus is better able to absorb.

Root Nodules Some types of soil bacteria form mutually beneficial relationships with clover, peas, and other legumes. Like all other plants, legumes require nitrogen for growth. Nitrogen gas ($N\equiv N$, or N_2) is abundant in air, but plants do not have an enzyme that can break it apart. The bacteria do. The bacterial enzyme uses ATP to convert nitrogen gas to ammonia (NH_3), a process called **nitrogen fixation**. Other soil bacteria convert ammonia to nitrate (NO_3^-), a form of nitrogen that plants readily absorb. You will read more about nitrogen fixation in Section 46.9.

Nitrogen-fixing bacteria infect roots and thus become symbionts in localized swellings called **root nodules** (**Figure 28.4C**). The bacteria, which are anaerobic, share their fixed nitrogen with the plant (**Figure 28.4D**). In return, the plant provides the bacteria with an oxygen-free environment, and shares its photosynthetically produced sugars with them.

Uptake of Soil Water

Water moves from soil, through a root's epidermis and cortex, to the vascular cylinder (**Figure 28.5**). Osmosis drives this movement (Section 5.8). After water and mineral ions enter the vascular cylinder, xylem distributes them to the rest of the plant.

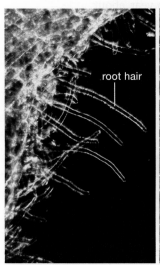

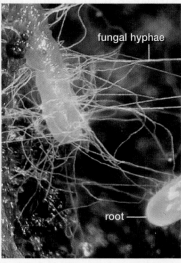

A The hairs on this root of a white clover plant are about 0.2 mm long.

B Fungal hyphae (*white* hairs) extending from the tip of these roots (*tan*) greatly enhance their surface area.

C Anaerobic bacteria in root nodules on this soybean plant fix nitrogen from the air. The nitrogen is shared with the plant.

D Soybean plants in nitrogen-poor soil show how root nodules affect growth. Only the darker green plants on the *right* were infected with nitrogen-fixing *Rhizobium* bacteria.

Figure 28.4 Animated Examples of root specializations that help plants take up nutrients.

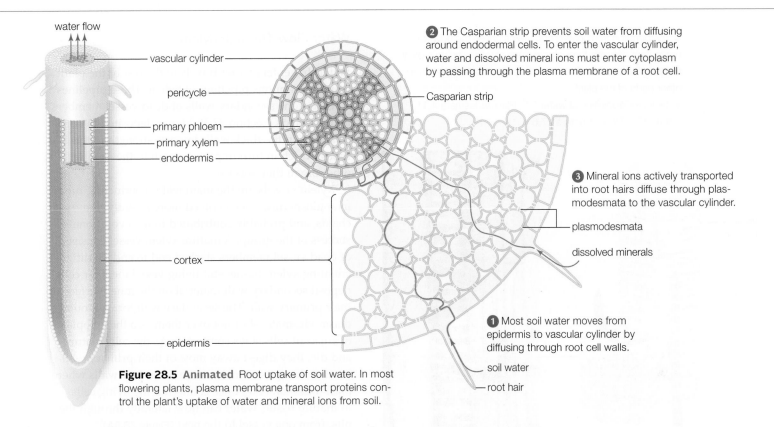

2 The Casparian strip prevents soil water from diffusing around endodermal cells. To enter the vascular cylinder, water and dissolved mineral ions must enter cytoplasm by passing through the plasma membrane of a root cell.

water flow

vascular cylinder

pericycle

primary phloem
primary xylem
endodermis

Casparian strip

3 Mineral ions actively transported into root hairs diffuse through plasmodesmata to the vascular cylinder.

plasmodesmata

dissolved minerals

cortex

1 Most soil water moves from epidermis to vascular cylinder by diffusing through root cell walls.

epidermis

soil water

root hair

Figure 28.5 Animated Root uptake of soil water. In most flowering plants, plasma membrane transport proteins control the plant's uptake of water and mineral ions from soil.

Most of the soil water that enters a root moves through cell walls **1**. Plant cell walls are permeable to water and ions, and they are shared between adjacent cells (Section 4.11). In a root, the cell walls form a pathway from epidermis to vascular cylinder. Thus, soil water can diffuse from epidermis to vascular cylinder without ever entering a cell.

Although soil water can *reach* the vascular cylinder by diffusing through cell walls, it cannot *enter* the cylinder the same way. Why not? A vascular cylinder is separated from root cortex by its endodermis. This tissue consists of a single layer of tightly packed endodermal cells surrounding the pericycle. The endodermal cells secrete a waxy substance into their walls wherever they abut. The substance forms a **Casparian strip**, a waterproof band between the plasma membranes of endodermal cells **2**. A Casparian strip prevents water from diffusing through endodermal cell walls. Thus, soil water can enter a vascular cylinder only by passing through the cytoplasm of an endodermal cell. Water can enter an endodermal cell (or any other cell in the root) by diffusing across its plasma membrane. Once inside root cell cytoplasm, the water diffuses from cell to cell through plasmodesmata until it enters xylem inside the cylinder.

Water can diffuse across a lipid bilayer, but ions cannot (Section 5.8). Minerals dissolved in soil water can enter a cell's cytoplasm only through transporters in its plasma membrane. Thus, transport proteins in root cell membranes control the types and amounts of ions that move from soil water into the body of the plant. This mechanism offers protection from some substances that may damage the plant. In addition, most minerals important for plant nutrition are more concentrated in the root than in soil, so they will not spontaneously diffuse into the root. These mineral ions must be actively transported into root cells. The transport occurs mainly at the membrane of root hairs. Once in cell cytoplasm, mineral ions diffuse from cell to cell through plasmodesmata until they enter xylem in the vascular cylinder **3**.

Casparian strip Waxy, waterproof band that seals abutting cell walls of root endodermal cells.
mycorrhiza Fungus–plant root partnership.
nitrogen fixation Conversion of nitrogen gas to ammonia.
root nodules Swellings of some plant roots that contain nitrogen-fixing bacteria.

Take-Home Message

How do roots take up water and nutrients?

» Root hairs, mycorrhizae, and root nodules greatly enhance a root's ability to take up water and nutrients.

» Transport proteins in root cell plasma membranes control the uptake of water and mineral ions into the vascular cylinder.

28.4 Water Movement Inside Plants

- Evaporation from leaves and stems drives the upward movement of water through pipelines of xylem inside a plant.
- Water's cohesion allows it to be pulled from roots into all other parts of the plant.
- Links to Properties of water 2.5, Plasmodesmata 4.11, Xylem 27.3, Root structure 27.6

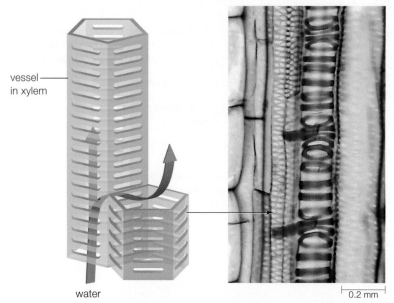

A Pits in the side walls of adjoining vessel members match up, so water can flow between adjacent tubes as well as upward through them. The micrograph shows a longitudinal section of corn (*Zea mays*). The rows of pits are contact faces where vessel elements meet side-to-side.

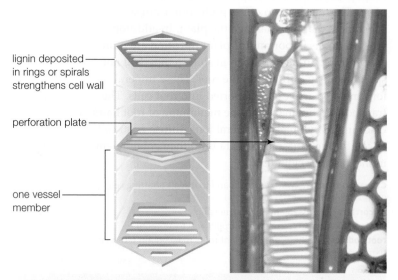

B The thick secondary walls of dead xylem cells connect to make long, water-conducting tubes. Lignin deposited in rings or spirals strengthens the walls. Perforation plates impart strength to a junction where vessel members meet, while also allowing water to flow through the tube. The micrograph shows a perforation plate in a section of magnolia (*Magnolia grandifolia*). Pitted walls of tracheids are also visible.

Figure 28.6 Xylem vessels. Perforated and pitted secondary walls of dead vessel member cells form these water-conducting tubes, which vary in size and in shape.

Water Flow Through Xylem

Water that enters a root travels to the rest of the plant inside continuous pipelines of xylem. These pipelines consist of the secondary walls of dead vessel members and tracheids (Section 27.3). Lignin deposited in rings or spirals in the thick walls imparts strength to the tubes, which in turn impart strength to the stem or other organ they service.

Xylem vessels are the main water-conducting tubes in angiosperms. They evolved more recently than tracheids, and probably contributed to the evolutionary success of the group. A mature xylem vessel consists of dead vessel members stacked end to end. In differentiating xylem tissue, still-living vessel member cells deposit secondary wall material on the inner surface of their primary wall. The secondary wall forms around plasmodesmata—but not over them—so the cytoplasm of adjacent cells stays connected. As the cells mature and die, they digest away most of their primary wall. Small holes (pits) remain in the secondary walls where plasmodesmata had once connected the living cells. In mature tissue, water can flow laterally through the pits, from one vessel to the next (**Figure 28.6A**).

Dying vessel member cells also digest large holes in their end walls. The resulting punctured walls, which are called perforation plates, separate stacked vessel members in mature xylem. Perforation plates allow water to move freely between vessel members in a column. Ladderlike bars of wall material remain in some perforation plates (**Figure 28.6B**). The bars break up air bubbles in water flowing through larger tubes.

Tracheids are generally shorter and thinner than vessels. Their ends are pointy and closed, with no perforation plate. Water moves between tracheids through pits in their walls. Being very narrow, tracheids are more resistant to vertical compression than vessels, so in addition to conducting water they can have a substantial role in structural support.

Pits in xylem tubes are often bordered by accumulated secondary wall, which contains pectins. Pectins are gels: They shrink when dry, and swell when wet. When mineral-rich water flows through a bordered pit, the border swells and eventually plugs the hole. Thus, water tends to flow toward the thirstiest regions of the plant, where bordered pits are dry and open.

The Cohesion–Tension Theory

How does water in xylem move all the way from roots to leaves that may be more than 100 meters (330 feet) above the soil? Tracheids and vessel members are dead at maturity, so these cells cannot be expending any energy to pump water upward against gravity. The movement of water in vascular plants is driven by two features of water: evaporation and cohesion. By the **cohesion–tension theory**, water in xylem is pulled upward by air's drying power, which creates a continuous negative pressure called tension. The tension extends all the way from leaves at the tips of shoots to roots that may be hundreds of feet below (**Figure 28.7**).

Most of the water a plant takes up is lost by evaporation, typically from stomata on the plant's leaves and stems ❶. The evaporation of water from aboveground plant parts, particularly at stomata, is called **transpiration**. Transpiration's effect on water inside a plant is a bit like what happens when you suck a drink through a straw. Transpiration puts negative pressure (pulls) on continuous columns of water in xylem. A column of water resists breaking into droplets as it moves through a narrow conduit such as a straw or a xylem tube. Why? Water molecules are connected by hydrogen bonds, so a pull on one tugs all of them. Thus, the negative pressure created by transpiration (tension) pulls on the entire column of water that fills a xylem tube ❷. Because of water's cohesion, the tension extends from leaves that may be hundreds of feet in the air, down through stems, into young roots where water is being absorbed from soil ❸.

The movement of water through vascular plants is driven mainly by transpiration, but evaporation is only one of many other processes in plants that involve the loss of water molecules. Metabolic pathways that use water also contribute to the negative pressure that results in water movement. For example, water molecules delivered by xylem are used as electron donors in the light reactions of photosynthesis. The reactions split the water molecules into oxygen and hydrogen ions (Section 6.4).

cohesion–tension theory Explanation of how water moves through vascular plants: transpiration creates a tension that pulls a cohesive column of water through xylem, from roots to shoots.
transpiration Evaporation of water from aboveground plant parts.

Take-Home Message

What makes water move inside plants?

» Water moves inside continuous columns of xylem that thread from roots to shoots.

» Transpiration, the evaporation of water from aboveground plant parts, puts columns of water in xylem into a continuous state of tension from shoots to roots.

» Tension pulls water upward through the plant. The collective strength of many hydrogen bonds (cohesion) keeps the water from breaking into droplets as it rises.

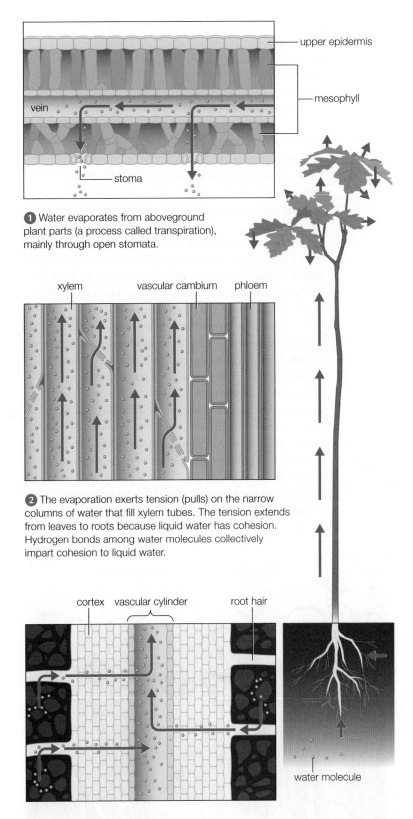

❶ Water evaporates from aboveground plant parts (a process called transpiration), mainly through open stomata.

❷ The evaporation exerts tension (pulls) on the narrow columns of water that fill xylem tubes. The tension extends from leaves to roots because liquid water has cohesion. Hydrogen bonds among water molecules collectively impart cohesion to liquid water.

❸ As long as evaporation continues, the tension it creates drives the uptake of more water molecules from soil.

Figure 28.7 Animated Cohesion–tension theory of water transport in vascular plants.

28.5 Water-Conserving Adaptations of Stems and Leaves

- Water-conserving structures and processes are key to the survival of all land plants.
- Links to Plant cuticle 4.11, Receptor proteins 5.7, Gradients and osmosis 5.8, Gated transporters 5.9, Photosynthesis 6.4, Stomata 6.8, Aerobic respiration 7.2, Land plant adaptations 22.3, Leaf structure 27.5

In land plants, at least 90 percent of the water transported from roots to leaves is lost by evaporation. Only about 2 percent is used in metabolism, but that amount must be maintained or photosynthesis, growth, membrane functions, and other processes will shut down. A plant that is running low on water cannot move around to seek out more, as most animals can. A cuticle and stomata help the plant conserve the water it already holds in its tissues (**Figure 28.8**). Both of these structures restrict the amount of water vapor that diffuses out of the plant's surfaces.

However, the cuticle and stomata also restrict gas exchanges between the plant and the air. Gas exchange is important because the concentrations of carbon dioxide and oxygen in the plant's internal air spaces affect the rate of critical metabolic pathways such as photosynthesis and aerobic respiration. If a plant became impermeable to gases, it could not take in carbon dioxide to run photosynthesis. Neither could it sustain aerobic respiration for very long, because this pathway uses oxygen. Thus, a plant must balance its needs to conserve water with its requirements to exchange gases with the air.

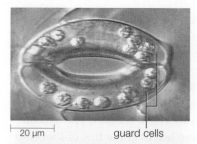

A This stoma is open. When guard cells swell with water, they bend so a gap (the stoma) opens up between them. The gap allows the plant to exchange gases with air.

20 µm   guard cells

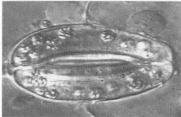

B This stoma is closed. When the guard cells lose water, they collapse against each other so the gap between them closes. A closed stoma limits water loss. It also limits gas exchange.

Figure 28.9 Stomata in action. Whether a stoma is open or closed depends on how much water is plumping up guard cells. The round structures inside the cells are chloroplasts. Guard cells are the only type of plant epidermal cell with these organelles.

A cuticle consists of epidermal cell secretions: a mixture of waxes, pectin, and cellulose fibers embedded in cutin, an insoluble lipid polymer. This waterproof layer is transparent—it absorbs no visible light—so it does not reduce the plant's capacity for photosynthesis. However, a cuticle strongly absorbs light of less than 400 nm, so it does offer protection from UV light. It also protects the plant's surfaces from bacterial and fungal pathogens.

A pair of specialized epidermal cells defines each stoma. When these two **guard cells** swell with water, they bend slightly so a gap forms between them (**Figure 28.9A**). The gap is the stoma. When the cells lose water, they collapse against one another, so the gap closes (**Figure 28.9B**). Even under conditions of high humidity, the interior of a leaf or stem has more water vapor than the environment. Thus, water vapor diffuses out of a stoma whenever it is open. Indentations or other structures that reduce air movement next to a stoma decrease the rate of evaporation from it. The stoma shown in **Figure 28.8** has a ledge over it for this reason.

Stomata open or close based on an integration of cues that include humidity, the level of carbon dioxide inside the leaf, light intensity, and hormonal signals from other parts of the plant. These cues trigger osmotic pressure changes that open or close the guard cells. As an example, when soil water becomes scarce, root cells release abscisic acid (ABA), a plant hormone. ABA travels through the plant's vascular system to leaves and stems, where it binds to receptors on guard cell plasma membranes. ABA binding causes a plant cell to release a burst of nitric oxide (NO). This gas-

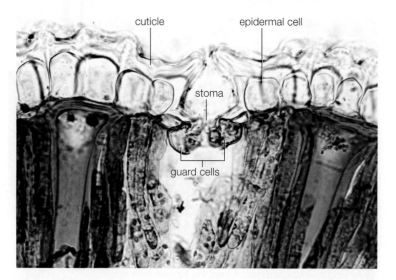

cuticle   epidermal cell

stoma

guard cells

Figure 28.8 Water-conserving structures in a transverse section of pincushion leaf (*Hakea laurina*). In this species, leaf guard cells are recessed beneath a ledge of cuticle. The ledge creates a small cup that traps moist air above the stoma.
Figure It Out: What are the long pink cells?   Answer: Palisade mesophyll

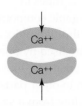
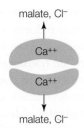
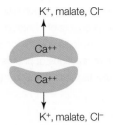
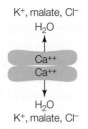

1 ABA binding to its receptor on a guard cell membrane causes the release of nitric oxide (NO). The NO activates calcium ion channels in the membrane, so these ions enter cytoplasm.

2 The influx of calcium ions activates transport proteins that pump negatively charged malate and chloride ions out of the cells. Thus, the overall charge of cytoplasm increases, and the overall charge of extracellular fluid decreases.

3 The resulting voltage change across the guard cell plasma membranes opens gated transport proteins that allow potassium ions to exit the cells.

4 Water follows the solutes by osmosis. The guard cells lose turgor and collapse against one another, so the stoma closes.

Figure 28.10 An osmotic pressure decrease triggered by abscisic acid (ABA) causes stomata to close.

eous signaling molecule activates membrane transport proteins that allow calcium ions to enter guard cell cytoplasm (**Figure 28.10 1**). Cells use calcium ions for signaling purposes, so the concentration of these ions in cytoplasm is tightly regulated. In the presence of ABA, an increase in cytoplasmic calcium ion concentration triggers transport proteins to pump negatively charged malate (an organic acid) and chloride (Cl⁻) out of the cell **2**. As a result, the overall charge of guard cell cytoplasm increases relative to that of the extracellular environment. A difference in charge is called voltage; when the voltage across guard cell plasma membranes changes, gated transport proteins in the membranes open. The opened gates allow potassium ions (K^+) to follow the charge gradient and exit the cells **3**. Water follows the solutes by osmosis, and the stomata close as the guard cells lose turgor and collapse against one another **4**.

A voltage change across guard cell membranes can also trigger the opening of stomata. At sunrise, for example, the stomata of C3 and C4 plants open in response to light absorbed by receptor proteins called phototropins. Light-activated phototropins initiate events that phosphorylate ATP synthases in guard cell membranes. Remember from Section 6.5 that hydrogen ion flow through these transport proteins drives the formation of ATP during the light reactions of photosynthesis. In this case, phosphorylation drives the reverse reaction, in which ATP synthases actively pump hydrogen ions (H^+) out of guard cell cytoplasm (**Figure 28.11A**). The outflow of positively charged hydrogen ions decreases the overall charge of the cytoplasm, and increases the overall charge of the extracel-

lular environment. The resulting voltage change across guard cell plasma membranes causes gated transport proteins in the membranes to open. Potassium ions follow the charge gradient and flow into the cell through the opened gates (**Figure 28.11B**). Water follows by osmosis. Turgor in the guard cells increases as they plump up with water, and the gap between them opens (**Figure 28.11C**). Carbon dioxide in air diffuses into the plant's tissues, and photosynthesis begins.

Chapter 30 returns to plant hormones and other mechanisms of signaling in plants. You will also read more about voltage-gated channels in Chapter 32, in the context of the animal nervous system.

A Light-triggered phosphorylation of ATP synthases causes these transport proteins to pump hydrogen ions (H^+) out of guard cells.

B The resulting change in voltage across the membranes opens gated transport proteins that allow potassium ions to enter the cell.

C Water follows the solutes by osmosis. Turgor increases inside the guard cells, and the stoma opens.

Figure 28.11 Light triggers an osmotic pressure increase that opens stomata.

Take-Home Message

How do land plants conserve water?

» A waxy cuticle covers all epidermal surfaces of the plant exposed to air. It restricts water loss from plant surfaces.

» Plants conserve water by closing their many stomata. Opened stomata allow gas exchanges necessary for photosynthesis and aerobic respiration.

» A stoma stays open when the guard cells that define it are plump with water. It closes when the cells lose water and collapse against each other.

guard cell One of a pair of cells that define a stoma across the epidermis of a leaf or stem.

28.6 Movement of Organic Compounds in Plants

■ Xylem distributes water and minerals through plants, and phloem distributes the organic products of photosynthesis.
■ Links to Carbohydrates 3.4, Osmosis and turgor 5.8, Active transport 5.9, Light-independent reactions of photosynthesis 6.7, Stomata 6.8, Plant vascular tissues 27.3, Plant tissue development 27.4

Movement of Sugars Through Phloem

Phloem is a vascular tissue that consists of fibers, sieve tubes, and companion cells (Section 27.3). Unlike a conducting tube of xylem, a sieve tube in phloem consists of living cells—a stack of sieve-tube members. During differentiation of phloem, plasmodesmata connecting the sieve-tube members in a stack enlarge greatly, sometimes up to 100-fold. The resulting perforated end walls are called **sieve plates**, after their appearance (**Figure 28.12**). A sieve-tube member cell also loses most of its contents during differentiation, including the nucleus, Golgi bodies, cytoskeleton, and most of its ribosomes. How can it stay alive? Each sieve-tube member has an associated companion cell that arises by division of the same parent cell in procambium (Section 27.4). The companion cell retains its nucleus and other components. Many plasmodesmata connect the cytoplasm of the two cells, so the companion cell can provide all metabolic functions necessary to sustain its paired sieve-tube member—for decades, in some cases.

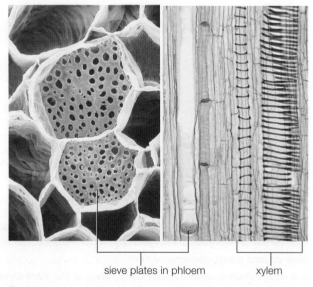

sieve plates in phloem   xylem

Figure 28.12 Sieve plates. The scanning electron micrograph on the *left* shows sieve plates on the ends of two side-by-side sieve-tube members. Sieve plates in two columns of phloem are visible in the light micrograph of squash (*Cucurbita*) on the *right*. Rings of lignin are visible as *red* stripes in xylem vessel members. Older vessels stretch longitudinally as cells in newer tissue lengthen.

Sugar-rich fluid flows through sieve tubes in mature phloem. The movement of organic compounds (mainly sucrose and sugar alcohols) through phloem is called **translocation**. Some sugars are used by photosynthetic cells that make them. The remainder move into sieve tubes, which conduct them to other parts of the plant. Inside sieve tubes, sugars flow from a **source** (a region of the plant where they are being produced) to a **sink** (a region where they are being broken down for energy, remodeled into other compounds, or stored for later use). Photosynthetic tissues are typical source regions; developing shoots and fruits are typical sink regions. However, source and sink regions can change. For example, sugars move from shoots to roots during autumn in preparation for winter dormancy in some plants. As dormancy ends in spring, sugars move in the opposite direction, from roots to budding shoots.

Pressure Flow Theory

Why do dissolved sugars flow from source to sink inside sieve tubes? A pressure gradient between source and sink regions drives the movement. According to the **pressure flow theory**, pressure pushes sugar-rich fluid inside a sieve tube from a source to a sink (**Figure 28.13**). The pressure gradient is set up at source regions, where mesophyll cells produce sugars that move into companion cells, then into sieve tubes ❶.

Depending on the species and the source region, the movement of sugars from cells in source regions into sieve tubes may occur via active transport proteins in the cell membranes or by diffusion through plasmodesmata connecting the cells. Either way, sugar loading at a source region increases the solute concentration of sieve-tube cytoplasm so that it becomes hypertonic with respect to surrounding cells. Water then follows its gradient and moves from the surrounding cells into sieve tubes by osmosis ❷, which increases the volume of fluid inside the tube. The primary cell walls of sieve-tube members cannot expand very much, so the influx of water raises the internal pressure (turgor) of the sieve tube.

Pressure inside a sieve tube can be very high—up to five times higher than that of an automobile tire (**Figure 28.14**). The high fluid pressure pushes sugar-laden cytoplasm inside a sieve tube toward sink regions, where the internal pressure is lower ❸. The pressure in a sieve tube decreases at sink regions because sugars are moving from the sieve tube to sink cells. Water follows, again by osmosis ❹, so the fluid volume inside sieve tube members decreases in these regions. The movement of sugars from sieve tubes to

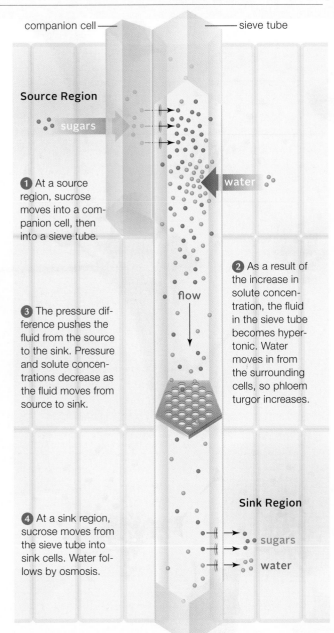

Source Region

sugars

1 At a source region, sucrose moves into a companion cell, then into a sieve tube.

companion cell ——— sieve tube

water

3 The pressure difference pushes the fluid from the source to the sink. Pressure and solute concentrations decrease as the fluid moves from source to sink.

flow

2 As a result of the increase in solute concentration, the fluid in the sieve tube becomes hypertonic. Water moves in from the surrounding cells, so phloem turgor increases.

4 At a sink region, sucrose moves from the sieve tube into sink cells. Water follows by osmosis.

Sink Region

sugars

water

Figure 28.13 Translocation of organic compounds in phloem from source (shaded *pink*) to sink (shaded *yellow*). Water molecules are represented by *blue* balls; sucrose, by *red* balls. The movement of sucrose into and out of phloem occurs by diffusion through plasmodesmata, or by active transport across the cell membranes.

cells in sink regions may occur by diffusion through plasmodesmata or by active transport across the cell membranes, depending on the species and the sink.

pressure flow theory Explanation of how organic compounds move through vascular plants: fluid flow through sieve tubes is driven by a difference in turgor between source and sink regions.
sieve plate Perforated end wall separating stacked sieve-tube members.
sink Region of a plant where sugars are being used or stored.
source Region of a plant where sugars are being produced.
translocation Movement of organic compounds through phloem.

Leafy Cleanup Crews (revisited)

With elemental pollutants such as lead or mercury, the best phytoremediation strategies use plants that take up toxins and store them in aboveground tissues. The toxin-laden plant parts can then be harvested for safe disposal. Researchers have genetically modified some plants to enhance their absorptive and storage capacity. In the case of organic toxins such as TCE, the best phytoremediation strategies use plants with biochemical pathways that break down the compounds to less toxic molecules. Researchers are beefing up these pathways by transferring genes from bacteria or animals into plants, and by enhancing expression of genes that encode molecular participants in the plants' own detoxification pathways.

How would you vote? Plants can be genetically engineered to take up toxins more effectively. Do you support the use of such plants to help clean up toxic waste sites?

Figure 28.14 Honeydew exuding from an aphid after the insect's mouthparts penetrated a sieve tube. High pressure in phloem forced this droplet of sugar-rich fluid out through the terminal opening of the aphid's gut.

Take-Home Message

How do organic molecules move through vascular plants?

» Sieve tubes of phloem consist of sieve-tube members positioned end to end and separated by porous end walls (sieve plates). Each sieve-tube member is connected by many plasmodesmata to a companion cell.

» At source regions, sugars moves from photosynthetic cells into companion cells, then into sieve tubes. Sugars exit sieve tubes at sink regions.

» Concentration and pressure gradients move sugar-laden fluid inside sieve tubes from sources to sinks.

LEARNING ROADMAP

Where you have been This chapter revisits plant tissues (Sections 4.11, 27.2–27.4, 27.9), life cycles (12.5, 22.7), and evolution (17.9, 17.11, 17.12, 22.2, 22.6, 22.8); as well as molecules of life (3.3, 3.4), membrane proteins (5.7), photosynthesis (6.4), respiration (7.5), cloning (8.7), gene control (10.2, 10.4), reproduction (12.2), meiosis (12.3), inheritance (13.2), aneuploidy (14.6), radiometric dating (16.6), and phylogeny (18.2).

Where you are now

Structure and Function of Flowers
Flowers are shoots specialized for reproduction. Modified leaves form their parts. Non-reproductive parts of flowers are specialized to attract and reward pollinators.

Gametes and Fertilization
Male and female gametophytes develop inside the reproductive parts of flowers. Pollination is followed by double fertilization. As in animals, signals are key to sex.

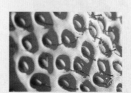

Seeds and Fruit
After fertilization, ovules mature into seeds. As seeds develop, tissues of the ovary and often other parts of the flower mature into fruits, which function in seed dispersal.

Growth and Development
Plant development includes germination, root and shoot development, flowering, fruit formation, and dormancy. All have a genetic basis; all are triggered by environmental cues.

Asexual Reproduction in Plants
Many species of plants reproduce by vegetative reproduction. Humans take advantage of this natural tendency by propagating plants asexually for agriculture and research.

Where you are going Chapter 30 details how plant development (including germination and cyclic patterns of growth) is mediated by signaling molecules and triggered by environmental cues. Chapter 44 discusses limits on population growth. Competitive interactions among species in a community are explained in Chapter 45, and Chapter 47 returns to forests and other biomes. Human impacts on the biosphere are further explored in Chapter 48.

29.1 Plight of the Honeybee

In the fall of 2006, commercial beekeepers worldwide began to notice something was amiss in their honeybee hives. All of the adult worker bees were missing from the hives, but there were no bee corpses to be found. The bees had suddenly (and quite unusually) abandoned their queen and many living larvae. The few young workers that remained were reluctant to feed on abundant pollen and honey stored in their own hive, or in other abandoned hives.

A third of the colonies did not survive the following winter. By spring, the phenomenon had a name: colony collapse disorder (CCD). Since then, honeybee populations have continued to decline. As of 2011, thirty percent of bee colonies in the United States and twenty percent of those in Europe have been affected.

CCD is a problem that affects more than just bees and honey lovers. Nearly all of our food crops are the fruits of angiosperms (Section 22.8). Angiosperms, like other seed-bearing plants, produce pollen (Section 22.6). Many angiosperm species rely on honeybees and other **pollinators** to carry pollen from one plant to another (**Figure 29.1A**). Fruits do not form from unpollinated flowers, and in some species they only form when flowers receive pollen from a different plant of the same species. Even species with flowers that can self-pollinate tend to make bigger fruits and more of them when they are cross-pollinated (**Figure 29.1B**).

Many types of insects pollinate plants, but honeybees are especially efficient pollinators of a wide variety of plant species. They are also the only ones that tolerate living in man-made hives that are loaded onto trucks and transported to wherever crops require pollination. Loss of their portable pollination service is a huge threat to our agricultural economy. Currently, about half of the world's leading crops—about $212 billion worth—depend on pollination, mainly by bees.

CCD is currently in the spotlight because it directly affects our food supply. However, other insect pollinator populations are also dwindling. Habitat loss may be the main factor, but pesticides that harm honeybees also harm other invertebrate pollinators.

Flowering plants rose to dominance in part because they coevolved with animal pollinators. Most flowers are specialized to attract and be pollinated by a specific type (or even a specific species) of pollinator. Those adaptations become handicaps if coevolved pollinator populations decline. Wild animal species that depend on the plants for fruits and seeds will also be affected. Recognizing the prevalence and importance of these interactions is our first step toward finding workable ways to protect them.

pollinator An organism that moves pollen from one plant to another.

A Honeybees are efficient pollinators of a variety of flowers, including those of berry plants (*Rubus*).

B Raspberry flowers can pollinate themselves, but the fruit that forms from a self-fertilized flower is of lower quality than that of a cross-pollinated flower.

The two raspberries on the *left* formed from self-pollinated flowers. The one on the *right* formed from an insect-pollinated flower.

Figure 29.1 Importance of insect pollinators.

29.2 Reproductive Structures

■ Specialized reproductive shoots called flowers consist of whorls of modified leaves.

■ Links to ABC model 10.4, Gametes 12.2, Plant life cycles 12.5 and 22.2, Pollination 22.6, Angiosperms 22.8, Monocots and eudicots 27.2, Lateral buds 27.4

The life cycle of flowering plants (**Figure 29.2**) is dominated by the sporophyte, a diploid spore-producing plant body that grows by mitotic cell divisions of a fertilized egg (Sections 12.5 and 22.2). **Flowers** are the specialized reproductive shoots of angiosperm sporophytes. Spores that form by meiosis inside flowers develop into haploid gametophytes. Gametophytes produce gametes (Section 12.2).

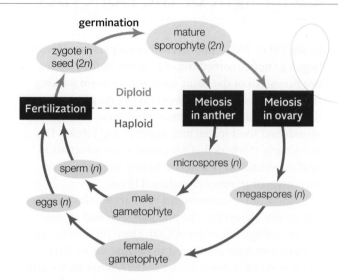

Figure 29.2 Animated Life cycle of a typical flowering plant.

General Anatomy of a Flower

A flower forms when a lateral bud along the stem of a sporophyte develops into a short, modified branch called a receptacle (**Figure 29.3 ❶**). The petals and other parts of a typical flower are modified leaves that form in four spirals or rings (whorls) at the end of the floral shoot. The outermost whorl develops into a **calyx**, which is a ring of leaflike **sepals ❷**. The sepals of most flowers are photosynthetic and inconspicuous; they serve to protect the flower's reproductive parts. Just inside the calyx, petals form in a whorl called the **corolla** (from the Latin *corona*, or crown). **Petals** are usually the largest and most brightly colored parts of a flower ❸. They function mainly to attract pollinators.

A whorl of stamens forms inside the ring of petals. **Stamens** are the structures that produce the plant's male gametophytes. In most flowers, stamens consist of a thin filament with an anther at the tip ❹. Inside a typical **anther** are four elongated pouches called pollen sacs. Meiosis of diploid cells in each sac produces haploid, walled spores. The spores differentiate into pollen grains, which are immature male gametophytes.

The innermost whorl of modified leaves are folded and fused into **carpels**, structures that produce the plant's female gametophytes ❺. Carpels are sometimes called pistils. Many flowers have one carpel; others have several carpels, or several groups of carpels, that may be fused. The upper region of a carpel is a sticky or hairy **stigma** (plural, stigmata) that is specialized to receive pollen grains. Typically, the stigma sits on top of a slender stalk called a style. The lower, swollen region of a carpel is the **ovary**, which contains one or more ovules. An **ovule** is a tiny bulge of tissue in which the haploid female gametophyte forms.

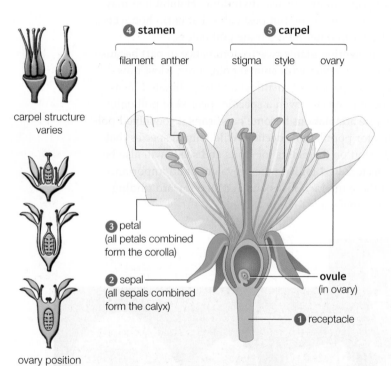

carpel structure varies

ovary position varies

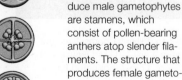

ovule position varies within ovaries

Figure 29.3 Animated Anatomy of a typical flower: cherry (*Prunus avium*).

The structures that produce male gametophytes are stamens, which consist of pollen-bearing anthers atop slender filaments. The structure that produces female gametophytes is the carpel, which consists of ovary, stigma, and style.

A typical *Prunus avium* flower (*above*) has five petals, several stamens, and one carpel. Within the carpel, one ovule forms inside one ovary. Other species have different patterns (*left*).

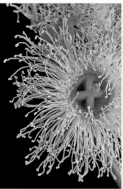

A The irregular blossom of lady's slipper (*Paphiopedilum*).

B Elongated inflorescence of hyacinth (*Hyacinthus orientalis*).

C A daisy (*Gerbera jamesonii*) is a composite of many individual flowers.

D Incomplete flowers of eucalyptus (*Eucalyptus robusta*) have no petals.

E Imperfect flowers: stamen-less female blossoms (*left*) of *Begonia* form on the same plants as carpel-less male blossoms (*right*).

Figure 29.4 Examples of structural variation in flowers.

At fertilization, a diploid zygote forms when male and female gametes meet inside an ovary. The ovule then matures into a seed. The life cycle of the plant is completed when the seed germinates, and a new sporophyte forms and matures.

Diversity of Flower Structure

Mutations in some master genes give rise to dramatic variations in flower structure (Section 10.4). We see many such variations in the range of diversity of flowering plants. For example, the number of floral parts differs between monocots and eudicots (Section 27.2). Flowers may form as solitary blossoms (**Figure 29.4A**) or in clusters called inflorescences (**Figure 29.4B**). A composite flower is an inflorescence of many flowers ("florets") grouped as a single head (**Figure 29.4C**). Cherry blossoms and other regular flowers are symmetric around their center axis: If the flower were cut like a pie, the pieces would be roughly identical. By contrast, irregular flowers are not radially symmetric (**Figure 29.4A**).

A cherry blossom has all four sets of modified leaves (sepals, petals, stamens, and carpels), so it is called a "complete" flower. An "incomplete" flower lacks one or more of these structures (**Figure 29.4D**). Cherry blossoms are also "perfect" flowers, because they have both stamens and carpels. Perfect flowers can be cross-pollinated (fertilized by pollen from other plants), or they can self-pollinate. Self-pollination can be adaptive in situations where plants are widely spaced, such as in newly colonized areas. However, offspring of self-pollinated flowers or plants are generally less vigorous than those of cross-pollinated plants. Accordingly, floral adaptations of many plant species encourage or even require cross-pollination. For example, a perfect flower may release pollen only after its stigma is no longer receptive to being fertilized. Imperfect flowers that lack stamens or carpels (**Figure 29.4E**) cannot self-fertilize. In some species, stamen-bearing and carpel-bearing flowers form on different plants, so fertilization occurs only by cross-pollination.

anther Part of the stamen that produces pollen.
calyx A flower's outer, protective whorl of sepals.
carpel Floral reproductive organ; consists of an ovary, stigma, and often a style.
corolla A flower's whorl of petals; forms within sepals and encloses reproductive organs.
flower Specialized reproductive shoot of a flowering plant.
ovary In flowering plants, the enlarged base of a carpel, inside which one or more ovules form.
ovule Of a seed-bearing plant, a structure inside an ovary in which a haploid female gametophyte forms; after fertilization, matures into a seed.
petal Unit of a flower's corolla; often adapted to attract animal pollinators.
sepal Unit of a flower's calyx; typically photosynthetic.
stamen Floral reproductive organ that consists of an anther and, often, a filament.
stigma Upper part of a carpel; adapted to receive pollen grains.

Take-Home Message

What are flowers?

» Flowers are short reproductive branches of sporophytes. The different parts of a flower (sepals, petals, stamens, and carpels) are modified leaves. Flowers vary in structure.

» The parts of flowers that produce male gametophytes are stamens, which typically consist of a filament with an anther at the tip. Pollen forms inside the anthers.

» The parts of flowers that produce female gametophytes are carpels, which typically consist of stigma, style, and ovary. The haploid female gametophyte forms in an ovule inside the ovary.

29.3 Flowers and Their Pollinators

■ Most flowering plants have coevolved animal pollinators that help them reproduce sexually.

■ Links to Reproductive isolation 17.9, Coevolution 17.12, Coevolution of flowers and pollinators 22.8

Sexual reproduction in angiosperms involves the transfer of pollen, typically from one plant to another. Unlike most animals, plants cannot move about to find a mate, so pollen transfer depends on environmental agents (Section 22.8). The diversity of flower form in part reflects the dependence of angiosperms on pollination vectors. A **pollination vector** is an agent that delivers pollen from an anther to a compatible stigma. Grasses and many other plants are pollinated by wind, which is entirely nonspecific in where it dumps pollen. Such plants often release pollen grains by the billions, insurance in numbers that some of their pollen will reach a receptive stigma.

Other plants enlist the help of animal pollinators (**Figure 29.5**). An animal that is attracted to a particular flower often picks up pollen on a visit, then inadvertently transfers it to another flower on a later visit. The more specific the attraction, the more efficient the transfer of pollen among plants of the same species. Given the selective advantage for flower traits that attract specific pollinators, it is not surprising that about 90 percent of flowering plants have coevolved animal pollinators.

A flower's shape, pattern, color, and fragrance are adaptations to specific pollination vectors. For example, flowers of wind-pollinated plants tend to

A Dandelion
(*Taraxacum officinale*)

B Evening primrose
(*Oenothera biennis*)

Figure 29.6 Bee-attracting patterns. The *top* row shows flowers as we see them. The *bottom* row shows the same flowers photographed with a special filter that allows us to see reflected UV light, here represented by the color *red*. The bull's-eye patterns guide bees directly to the flower's reproductive parts.

be small, nonfragrant, and green, with large stigmata and insignificant petals or sepals. Animal-pollinated flowers tend to have showy petals and many are fragrant (**Table 29.1**). For example, the petals of flowers pollinated by bees are usually bright white, yellow, or blue, with pigments that reflect ultraviolet light. The UV-reflecting pigments are often distributed in bull's-eye patterns that bees recognize as visual guides (**Figure 29.6**). We see these patterns only with special camera filters, because human eyes do not have receptors that respond to UV light.

A Insects. Blueberry bees (*Osmia ribifloris*) are efficient pollinators of a variety of plants, including this barberry (*Berberis*).

B Birds. Pollen accumulates on the feathers of a little wattlebird (*Anthochaera chrysoptera*) sipping nectar from a torch lily (*Kniphofia uvaria*).

C Mammals. The large, white flowers of giant saguaro cactus (*Carnegiea gigantea*) produce sweet nectar. The flowers are visited by bats at night, and by birds and insects during the day.

Figure 29.5 Examples of animal pollinators.

Table 29.1 Common Traits of Flowers Pollinated by Specific Animal Vectors

Floral	Animal Pollination Vector						
	Bats	**Bees**	**Beetles**	**Birds**	**Butterflies**	**Flies**	**Moths**
Color:	Dull white, green, purple	Bright white, yellow, blue, UV	Dull white or green	Scarlet, orange, red, white	Bright, such as red, purple	Pale, dull, dark brown or purple	Pale/dull red, pink, purple, white
Odor:	Strong, musty, emitted at night	Fresh, mild, pleasant	None to strong	None	Faint, fresh	Putrid	Strong, sweet, emitted at night
Nectar:	Abundant, hidden	Usually	Sometimes, not hidden	Ample, deeply hidden	Ample, deeply hidden	Usually absent	Ample, deeply hidden
Pollen:	Ample	Limited, often sticky, scented	Ample	Modest	Limited	Modest	Limited
Shape:	Regular, bowl-shaped, closed during the day	Shallow with landing pad; tubular	Large, bowl-shaped	Large funnel-shaped cups, strong perch	Narrow tube with spur; wide landing pad	Shallow, funnel-shaped or trap-like and complex	Regular; tube-shaped with no lip
Examples:	Banana, agave	Larkspur, violet	Magnolia, dogwood	Fuschia, hibiscus	Phlox	Skunk cabbage, philodendron	Tobacco, lily, some cactuses

Bees and other insect pollinators have an excellent sense of smell, and can follow airborne chemicals to a flower that is emitting them, as can bats (**Figure 29.5C**). Not all flowers smell sweet; odors like dung or rotting flesh beckon beetles and flies.

An animal pollinator visits a flower in search of a reward, which may be **nectar** (a sweet fluid exuded by flowers), nutritious pollen, or even sex (**Figure 29.7**). Nectar is the only food for most adult butterflies, and it is the food of choice for hummingbirds. Honeybees convert nectar to honey, which helps feed the bees and their larvae through the winter. Pollen is an even richer food, with more vitamins and minerals than nectar. Beetles eat pollen directly; bees mix it with honey to make bee bread that they feed to their larvae. Pollen is the only reward for a pollinator of roses, poppies, and some other flowers that do not produce nectar. Some plants produce a tasty but infertile pollen in addition to normal, fertile pollen.

Some flowers have specializations that prevent pollination by anything other than a specific animal pollinator species. For example, nectar or pollen at the bottom of a long floral tube may be accessible only to an insect with a matching feeding device. In some plants, stamens adapted to brush against a pollinator's body or lob pollen onto it will function only when triggered by that pollinator (**Figure 17.17**). Such relationships are to both species' mutual advantage: A flower that captivates the attention of an animal has a pollinator that spends its time seeking out (and pollinating) only those flowers. Pollinators also benefit when they receive an exclusive supply of the flower's reward.

A Female burnet moths (*Zygaena filipendulae*) perch on purple blossoms—preferably pincushion flowers (*Knautia arvensis*)—when they are ready to mate. The visual combination attracts male moths.

B A zebra orchid (*Caladenia cairnsiana*) mimics the scent of a female wasp. Male wasps follow the scent to the flower, then try to copulate with and lift the dark red mass of tissue on the lip. The wasp's movements trigger the lip to tilt upward, which brushes the wasp's back against the flower's stigma and pollen.

Figure 29.7 Intimate connections.

nectar Sweet fluid exuded by some flowers; attracts pollinators.
pollination vector Any agent that moves pollen grains from one plant to another; e.g., wind, bees.

Take-Home Message

What is the purpose of the nonreproductive traits of flowers?

» Most flowering plants coevolved with animal pollinators. The shape, pattern, color, and fragrance of a flower attract the plant's coevolved pollinator.

» Pollinators are often rewarded for visiting a flower, for example, by obtaining pollen or nectar.

29.4 A New Generation Begins

■ In flowering plants, fertilization has two outcomes: It results in a zygote, and it is the start of endosperm.

■ Links to Evolution of seed-bearing plants 22.2, Life cycle of flowering plants 22.8

Figure 29.8 zooms in on the reproduction part of a flowering plant life cycle. The production of female gametes begins when a mass of tissue—the ovule—starts growing on the inner wall of an ovary ❶. One cell in the middle of the mass undergoes meiosis and cytoplasmic division, forming four haploid **megaspores** ❷. Three of the four megaspores typically disintegrate. The remaining megaspore undergoes three rounds of mitosis without cytoplasmic division, the outcome being a single cell with eight haploid nuclei ❸. The cytoplasm of this cell divides unevenly, forming a seven-celled embryo sac that constitutes the female gametophyte ❹. The gametophyte is enclosed and protected by layers of cells, or integuments, that developed from the outer layers of the ovule. One of the cells in the gametophyte, the endosperm mother cell, has two nuclei ($n + n$). Another cell is the egg.

The production of male gametes begins as masses of diploid, spore-producing cells form by mitosis inside anthers. Walls typically develop around the masses, so four pollen sacs form ❺. Each cell in the pollen sacs undergoes meiosis and cytoplasmic division to form four haploid **microspores** ❻. Mitosis and differentiation of a microspore produces a pollen grain. A pollen grain consists of two cells, one inside the cytoplasm of the other, enclosed by a durable coat ❼. The coat protects the cells inside on their journey to meet an egg. After the pollen grains form, they enter **dormancy**, a period of suspended metabolism, before being released from the anther when the pollen sacs split open ❽.

Interactions with a receptive stigma stimulate the pollen grain to germinate. **Germination** is the resumption of metabolic activity after a period of dormancy. Upon germination, one of the two cells in the pollen grain develops into a tubular outgrowth called a pollen tube. The other cell undergoes mitosis and cytoplasmic division, producing two sperm cells (the male gametes) within the pollen tube. A pollen tube together with its contents of male gametes constitutes the mature male gametophyte ❾. The pollen tube grows from its tip down through the style and ovary toward the ovule, carrying with it the two sperm cells. Chemical signals secreted by the female gametophyte guide the tube's growth to the embryo sac within the ovule. Many pollen tubes may grow down into a carpel, but usually only one penetrates an embryo sac. The sperm cells are then released into the sac ❿.

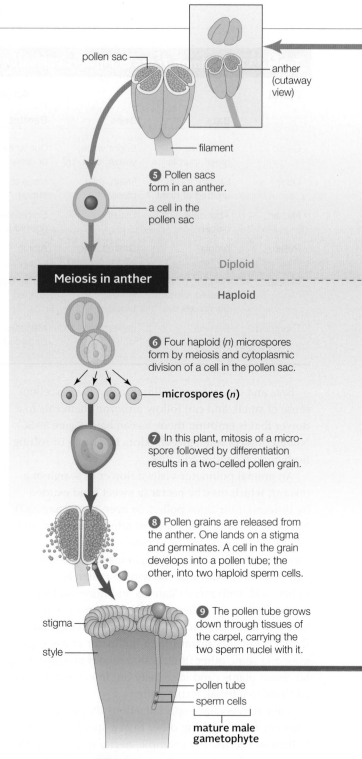

pollen sac

anther (cutaway view)

filament

❺ Pollen sacs form in an anther.

a cell in the pollen sac

Diploid

Meiosis in anther

Haploid

❻ Four haploid (*n*) microspores form by meiosis and cytoplasmic division of a cell in the pollen sac.

microspores (*n*)

❼ In this plant, mitosis of a microspore followed by differentiation results in a two-celled pollen grain.

❽ Pollen grains are released from the anther. One lands on a stigma and germinates. A cell in the grain develops into a pollen tube; the other, into two haploid sperm cells.

stigma

style

❾ The pollen tube grows down through tissues of the carpel, carrying the two sperm nuclei with it.

pollen tube

sperm cells

mature male gametophyte

Flowering plants undergo **double fertilization**, in which one of the sperm cells from the pollen tube fuses with (fertilizes) the egg and forms a diploid zygote; the other fuses with the endosperm mother cell, forming a triploid (3*n*) cell. This cell gives rise to triploid **endosperm**, a nutritious tissue that forms only in seeds of flowering plants. When a seed sprouts, endosperm sustains rapid growth of the sporophyte seedling until true leaves form and begin photosynthesis.

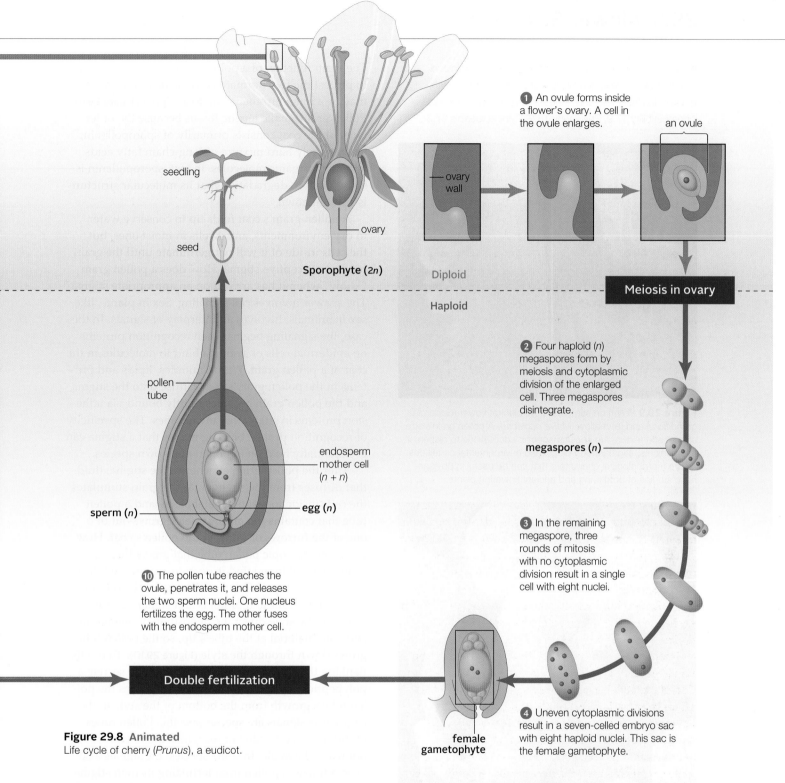

1 An ovule forms inside a flower's ovary. A cell in the ovule enlarges.

an ovule

ovary wall

Diploid

Haploid

Meiosis in ovary

2 Four haploid (*n*) megaspores form by meiosis and cytoplasmic division of the enlarged cell. Three megaspores disintegrate.

megaspores (*n*)

3 In the remaining megaspore, three rounds of mitosis with no cytoplasmic division result in a single cell with eight nuclei.

seedling

seed

Sporophyte (2*n*)

ovary

pollen tube

endosperm mother cell (*n* + *n*)

sperm (*n*)

egg (*n*)

10 The pollen tube reaches the ovule, penetrates it, and releases the two sperm nuclei. One nucleus fertilizes the egg. The other fuses with the endosperm mother cell.

Double fertilization

female gametophyte

4 Uneven cytoplasmic divisions result in a seven-celled embryo sac with eight haploid nuclei. This sac is the female gametophyte.

Figure 29.8 Animated
Life cycle of cherry (*Prunus*), a eudicot.

dormancy Period of temporarily suspended metabolism.
double fertilization Mode of fertilization in flowering plants in which one sperm cell fuses with the egg, and a second sperm cell fuses with the endosperm mother cell.
endosperm Nutritive tissue in the seeds of flowering plants.
germination The resumption of metabolic activity after a period of dormancy.
megaspore Haploid spore that forms in ovule of seed plants; gives rise to an egg-producing gametophyte.
microspore Walled haploid spore of seed plants; gives rise to a sperm-producing gametophyte.

Take-Home Message

How does sexual reproduction occur in flowering plants?

» A pollen grain that germinates on a receptive stigma develops into a pollen tube and two male gametes. As the tube grows into the carpel and enters an ovule, it carries the gametes along with it.

» Double fertilization occurs when one of the male gametes fuses with the egg, the other with the endosperm mother cell.

29.5 Flower Sex

- Species-specific interactions between pollen grain and stigma govern pollen germination and pollen tube growth.
- Links to Receptors and adhesion proteins 5.7, Gamete incompatibility 17.9, Phylogeny 18.2, Plant epidermis 27.3

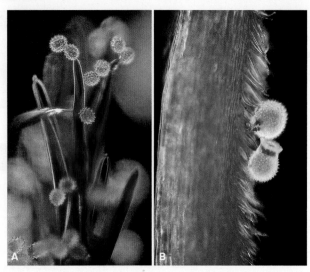

Figure 29.9 Pollen on stigmata of (**A**) blanket flower (*Gaillardia grandiflora*) and (**B**) mallow (*Malva neglecta*). A pollen grain's size, shape, and texture are species-specific adaptations to dispersal mechanisms, stigma morphology, and environmental conditions. All are morphological characters that can be useful in phylogenetic studies of both living and ancient flowering plants.

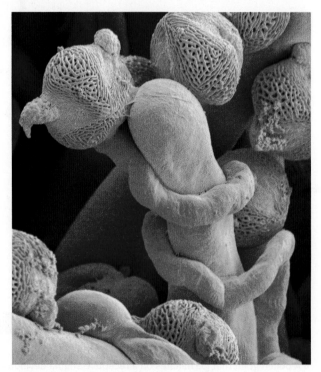

Figure 29.10 Adhesion proteins and signaling molecules guide a pollen tube's growth through carpel tissues to the egg. This electron micrograph shows pollen tubes growing from pollen grains (*orange*) that germinated on stigmata (*yellow*) of prairie gentian (*Gentiana*).

The main function of a pollen grain's durable coat is to protect the two dormant cells inside of it on what may be a long, turbulent ride to a stigma (**Figure 29.9**). Pollen grains make terrific fossils because the outer layer of the coat consists primarily of sporopollenin, an extremely hard mixture of long-chain fatty acids and other organic molecules. In fact, sporopollenin is so resistant to degradation that its molecular structure is still unknown.

A pollen grain's coat folds up to conserve water in dry environments, and swells in moist ones, but the cells inside of it will not germinate until the grain reaches a receptive stigma. How does a pollen grain "know" when it has arrived on an appropriate stigma? The answer involves cell signaling. Sex in plants, like sex in animals, involves an interplay of signals. In this case, the signaling begins when recognition proteins on epidermal cells of a stigma bind to molecules in the coat of a pollen grain. Within minutes, lipids and proteins in the pollen grain's coat diffuse onto the stigma, and the pollen grain becomes tightly bound via adhesion proteins in stigma cell membranes. The specificity of recognition protein binding means that a stigma can preferentially hold on to pollen of its own species.

After the pollen grain attaches to the stigma, fluid that diffuses from the stigma into the grain stimulates the cells inside to resume metabolism, and a pollen tube that contains the male gametes grows out of one of the furrows or pores in the pollen's coat. How does a microscopic pollen tube that grows through centimeters of tissue find its way to a single cell deep inside of the carpel? Cell signaling is involved in this process too. Adhesion proteins and signaling molecules in the style direct the deposition of membrane and wall material at the tube's tip, so the pollen tube grows down through the style (**Figure 29.10**). Two cells flanking the egg in the female gametophyte secrete a polypeptide (aptly named LURE) that guides the pollen tube's growth from the bottom of the style to the egg. These signals are species-specific: Pollen tubes of different species do not recognize them, and will not reach the ovule. In some species, the signals also keep a flower's pollen from fertilizing its own stigma. In these species, only pollen from another flower (or another plant) can give rise to a pollen tube that recognizes the female gametophyte's chemical guidance.

Take-Home Message

What constitutes sex in flowering plants?

» Species-specific molecular signals stimulate pollen germination and guide pollen tube growth to the egg.

29.6 Seed Formation

■ After fertilization, mitotic cell divisions transform a zygote into an embryo sporophyte encased in a seed.

In flowering plants, double fertilization produces a zygote and a triploid (3n) cell. Both begin mitotic cell divisions; the zygote develops into an embryo sporophyte, and the triploid cell develops into endosperm (**Figure 29.11A–C**). When the embryo approaches maturity, the integuments of the ovule separate from the ovary wall and become layers of the protective seed coat. The embryo sporophyte, its reserves of food, and the seed coat have now become a mature ovule, a self-contained package called a **seed** (**Figure 29.11D**). The seed may enter a period of dormancy until it receives signals that conditions in the environment are appropriate for germination.

Seeds as Food

As an embryo is developing, the parent plant transfers nutrients to the ovule. These nutrients accumulate in endosperm mainly as starch with some lipids and proteins. Eudicot embryos transfer nutrients in endosperm to their two cotyledons, so most of the nutrients in a mature eudicot seed are stored in the enlarged cotyledons. By contrast, most of the nutrient reserves in a mature monocot seed remain in endosperm.

The nutrients in endosperm and cotyledons nourish seedling sporophytes. They also nourish humans and other animals. We cultivate cereals—grasses such as rice, wheat, rye, oats, and barley—for their nutritious seeds. The embryo (the germ) contains most of the seed's protein and vitamins, and the seed coat (the bran) contains most of the minerals and fiber. Milling removes bran and germ, leaving only the starch-packed endosperm.

Maize, or corn, is the most widely grown cereal crop. Popcorn pops because the moist endosperm steams when heated; pressure builds inside the seed until it bursts. The cotyledons of bean and pea seeds are valued for their starch and protein; those of coffee and cacao, for their stimulants.

seed Embryo sporophyte of a seed plant packaged with nutritive tissue inside a protective coat.

Take-Home Message

What is a seed?

» After fertilization, the zygote develops into an embryo, the endosperm becomes enriched with nutrients, and the ovule's integuments develop into a seed coat.

» A seed is a mature ovule. It contains an embryo sporophyte.

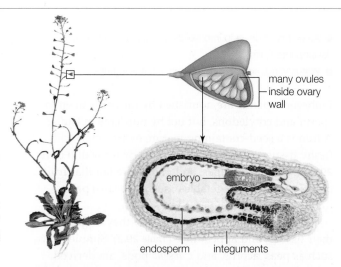

A After fertilization, a *Capsella* flower's ovary develops into a fruit. An embryo surrounded by integuments forms inside each of the ovary's many ovules.

labels: many ovules inside ovary wall; embryo; endosperm; integuments

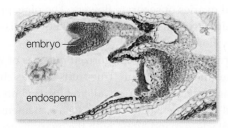

B The embryo is heart-shaped when its two cotyledons start forming. Endosperm tissue expands as the parent plant transfers nutrients into it.

labels: embryo; endosperm

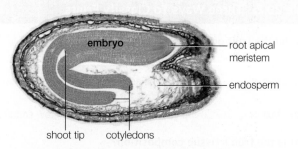

C In eudicots like *Capsella*, nutrients are transferred from endosperm into two cotyledons as the embryo matures. The developing embryo becomes shaped like a torpedo when the enlarging cotyledons bend.

labels: embryo; root apical meristem; endosperm; shoot tip; cotyledons

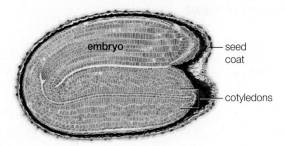

D A layered seed coat that formed from the layers of integuments surrounds the mature embryo and its two enlarged cotyledons.

labels: embryo; seed coat; cotyledons

Figure 29.11 Animated Embryonic development of shepherd's purse (*Capsella*), a eudicot.

29.7 Fruits

- As embryos develop into seeds, tissues around them develop into fruits.
- Water, wind, and animals disperse seeds in fruits.

Embryonic plants are nourished by nutrients in endosperm and cotyledons, but not by nutrients in fruits. A **fruit** is a seed-containing mature ovary, often with fleshy tissues that develop from the ovary wall (**Figure 29.12A**). Apples, oranges, and grapes are familiar fruits, but so are many "vegetables" such as beans, peas, tomatoes, grains, eggplant, and squash.

Botanists categorize fruits by how they originate, their tissues, and appearance (**Table 29.2**). Simple fruits, such as peas, acorns, and *Capsella* pods, are derived from one ovary. Strawberries and other aggregate fruits form from separate ovaries of one flower; they mature as a cluster of fruits. Multiple fruits form from fused ovaries of separate flowers. The pineapple is a multiple fruit that forms from fused ovary tissues of many flowers.

Fruits also may be categorized in terms of which tissues they incorporate. True fruits such as cherries consist only of the ovary wall and its contents. Other floral parts, such as the petals, sepals, stamens, or receptacle, expand along with the ovary in accessory

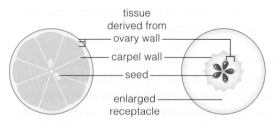

A The tissues of an orange (*Citrus*) develop from the ovary wall.

B The flesh of an apple (*Malus*) is an enlarged receptacle.

Figure 29.12 Parts of a fruit develop from parts of a flower.
Figure It Out: How many carpels were there in the flower that gave rise to the orange in A?
Answer: Eight

fruits. Most of the flesh of an apple, an accessory fruit, is an enlarged receptacle (**Figure 29.12B**).

To categorize a fruit based on appearance, the first step is to describe it as dry or juicy (fleshy). Dry fruits are dehiscent or indehiscent. If dehiscent, the fruit wall splits along definite seams to release the seeds inside. California poppy fruits and pea pods are examples. A dry fruit is indehiscent if the wall does not split open, so the seeds are dispersed inside intact fruits. Acorns and grains (such as corn) are dry indehiscent fruits, as are the fruits of sunflowers, maples, and strawberries. Strawberries are not berries and their fruits are not juicy. A strawberry's red flesh is an accessory to the dry indehiscent fruits on its surface (**Figure 29.13A**). Cherries, almonds, and olives are fleshy fruits, as are individual fruits of blackberries and other *Rubus* species (**Figure 29.13B**). Grapes, tomatoes, and citrus fruits are berries—fleshy fruits produced from one ovary—as are pumpkins, watermelons, and cucumbers. Apples and pears are pomes, in which fleshy tissues derived from the receptacle enclose a core derived from the ovary wall.

Table 29.2 Three Ways To Classify Fruits

How did the fruit originate?

Simple fruit	One flower, single or fused carpels
Aggregate fruit	One flower, several unfused carpels; becomes cluster of several fruits
Multiple fruit	Individually pollinated flowers grow and fuse

What is the fruit's tissue composition?

True fruit	Only ovarian wall and its contents
Accessory fruit	Ovary and other floral parts, such as receptacle

Is the fruit dry or fleshy?

Dry	
Dehiscent	Dry fruit wall splits on seam to release seeds
Indehiscent	Seeds dispersed inside intact, dry fruit wall
Fleshy	
Drupe	Fleshy fruit around hard pit surrounding seed
Berry	Fleshy fruit, often many seeds, no pit
	Pepo: Hard rind on ovary wall
	Hesperidium: Leathery rind on ovary wall
Pome	Fleshy accessory tissues, seeds in core tissue

A A strawberry (*Fragaria*) is not a berry. The flower's carpels turn inside out as the fruits form. The red, juicy flesh is an expanded receptacle; the hard "seeds" on the surface are individual dry fruits.

B Blackberries and fruits of other *Rubus* species are not berries, either. Each is an aggregate of many small drupes.

Figure 29.13 Aggregate fruits.

The function of a fruit is to protect and disperse seeds. Dispersal increases reproductive success by minimizing competition for resources among parent and offspring. Just as flower structure is adapted to certain pollination vectors, so are fruits adapted to certain dispersal vectors: environmental factors such as water or wind, or mobile organisms such as birds or insects.

Adaptations to a specific dispersal mechanism may be reflected in a fruit's form. Fruits dispersed by the wind tend to be lightweight with breeze-catching specializations (**Figure 29.14A**). For example, part of a maple (*Acer*) fruit is a dry, winglike outgrowth of the ovary wall. The fruit breaks in two when it drops from the tree. As the halves drop to the ground, wind catches the wings and spins the attached seeds away. Tufted fruits such as those of thistle, cattail, dandelion, and milkweed may be blown as far as 10 kilometers (6 miles) from the parent plant.

Fruits dispersed by water have water-repellent outer layers. The fruits of sedges (*Carex*) native to American marshlands have seeds encased in a bladderlike envelope that floats (**Figure 29.14B**). Buoyant fruits of the coconut palm (*Cocos nucifera*) have thick, tough husks that can float for thousands of miles in seawater.

The fruits of cocklebur, bur clover, and many other plants have hooks or spines that stick to the feathers, feet, fur, or clothing of more mobile species (**Figure 29.14C**). The dry, podlike fruit of plants such as California poppy (*Eschscholzia californica*) propel their seeds through the air when they pop open explosively (**Figure 29.14D**).

Colorful, fleshy, fragrant fruits attract insects, birds, and mammals that disperse seeds (**Figure 29.14E**). The animal may eat the fruit and discard the seeds, or eat the seeds along with the fruit. Abrasion of the seed coat by teeth or by digestive enzymes in an animal's gut can facilitate germination after the seed departs in feces.

fruit Mature ovary of a flowering plant, often with accessory parts; encloses a seed or seeds.

Take-Home Message

What is a fruit?

» A mature ovary, with or without accessory tissues that develop from other parts of a flower, is a fruit.

» We can categorize a fruit in terms of how it originated, its composition, and whether it is dry or fleshy.

» A fruit's function is seed dispersal. Fruit specializations are adaptations to particular dispersal vectors.

A Fruits adapted to wind dispersal. *Left*, dry outgrowths of the ovary wall of a maple (*Acer*) fruit form "wings" that catch the wind and spin the seeds away from the parent tree. *Right*, wind that lifts the hairy modified sepals of a dandelion (*Taraxacum*) fruit may carry the attached seed miles away from the parent plant.

B Air-filled bladders that encase the seeds of some sedges (*Carex*) allow the fruits to float in their marshy habitats.

C Curved spines make cocklebur (*Xanthium*) fruits stick to the fur of animals (and clothing of humans) that brush past it.

D When dry, the long fruits (pods) of the California poppy (*Eschscholzia californica*) split open suddenly to propel seeds away from the parent plant.

E Red, fleshy fruits of hawthorn plants (*Crataegus*) are an important food for cedar waxwings (*Bombycilla cedrorum*), which disperse the seeds in their feces.

Figure 29.14 Examples of adaptations that aid fruit dispersal.

29.8 Early Development

■ Species-specific patterns of early plant development are triggered by environmental cues.

■ Links to Hydrolysis 3.3, Carbohydrates 3.4, Photosynthesis 6.4, Aerobic respiration 7.5, Gene expression control 10.2, Meristems and cotyledons 27.2

An embryonic plant complete with shoot and root apical meristems forms as part of a seed (Figure 29.15). The embryonic shoot (**plumule**) is separated from the

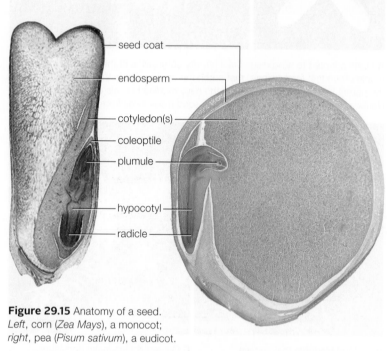

- seed coat
- endosperm
- cotyledon(s)
- coleoptile
- plumule
- hypocotyl
- radicle

Figure 29.15 Anatomy of a seed. *Left*, corn (*Zea Mays*), a monocot; *right*, pea (*Pisum sativum*), a eudicot.

As dormancy ends, cell divisions resume mainly at apical meristems of the plumule (the embryonic shoot) and radicle (the embryonic root). Plumule and radicle are separated by hypocotyl, a section of embryonic stem. In monocot grasses such as corn, the plumule is protected by a sheathlike coleoptile.

embryonic root (**radicle**) by a section of stem called **hypocotyl**. As the seed matures, the embryo typically dries out and enters a period of dormancy. The embryo may remain dormant in its protective seed coat for many years before it resumes metabolic activity and germinates.

Breaking Dormancy

Seed dormancy is a climate-specific adaptation that allows germination to occur when conditions in the environment are most likely to support the growth of a seedling. For example, seeds of many annual plants native to cold winter regions are dispersed in autumn. If the seeds germinated immediately, the tender seedlings would not survive the coming winter. Instead, the seeds remain dormant until spring, when milder temperatures and longer daylength favor the growth of seedling sporophytes. By contrast, the weather in regions near the equator does not vary by season. Seeds of most plants native to such regions do not enter dormancy; they germinate as soon as they become mature.

The triggers for germination, other than the presence of water, differ by species. For example, some seed coats are so dense that they must be abraded or broken (by being chewed, for example) before water can even enter the seed. The seeds of many cool-climate plants require exposure to freezing temperatures; those of some species of lettuce must be exposed to bright light. California poppy and many other wildflower species are

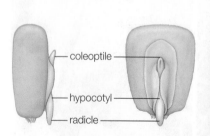

- coleoptile
- hypocotyl
- radicle

A As a corn grain (seed) germinates, its radicle and coleoptile emerge from the seed coat.

Figure 29.16 Animated
Early growth of corn (*Zea mays*), a typical monocot.

B The radicle develops into the primary root. The coleoptile grows upward and opens a channel through the soil to the surface, where it stops growing.

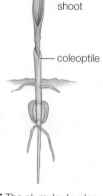

- primary shoot
- coleoptile

C The plumule develops into a primary shoot that emerges from the coleoptile and begins photosynthesis.

Figure 29.17 Animated Early growth of the common bean (*Phaseolus vulgaris*), a typical eudicot.

❶ The seed coat splits and the radicle emerges.

❷ As the hypocotyl emerges from the seed, it bends in the shape of a hook.

❸ The stem lengthens, and the bent hypocotyl drags the two cotyledons upward toward the surface of the soil.

❹ Exposure to sunlight causes the hypocotyl to straighten.

❺ Primary leaves emerge from between the cotyledons as the stem straightens.

❻ The cotyledons wither as the seedling's leaves begin to produce food by photosynthesis. Eventually, they fall off the stem.

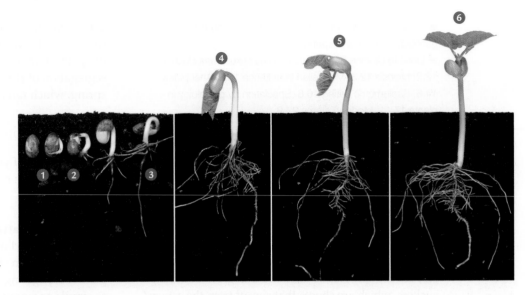

native to regions that experience periodic wildfires. In some of these species, seed germination is inhibited by light and enhanced by smoke; in others, germination does not occur unless the seeds have been previously burned. Germination requirements are evolutionary adaptations to life in a particular environment. All maximize a seedling's chance of survival.

The process of germination begins with water seeping into a seed. The water provokes a series of events that activate hydrolysis enzymes (Section 3.3) in endosperm. These enzymes break down stored starches into sugars. As the seed's tissues swell with water, its coat ruptures, and oxygen diffuses into internal tissues. Meristem cells in the embryo use the sugars and the oxygen for aerobic respiration as they begin to divide rapidly, and the embryonic plant begins to grow. Germination ends when the first part of the embryo—the embryonic root, or radicle—breaks out of the seed coat.

After Germination

The pattern of early growth that occurs after germination varies. For example, the primary (first) shoot of monocot grasses is protected by a rigid **coleoptile** that grows straight upward through soil. The embryo's single cotyledon, which typically stays beneath the soil, functions mainly to transfer endosperm breakdown products to the embryo (**Figure 29.16**). In eudicot seedlings, the primary leaves are not protected by a coleoptile (**Figure 29.17**). In a typical eudicot pattern of development, the radicle emerges from the seed ❶, followed by the hypocotyl. The hypocotyl bends into a hook shape as the stem lengthens ❷. The bent

hypocotyl pulls the cotyledons upward through the soil until it reaches the surface ❸. There, exposure to light causes the hypocotyl to straighten ❹. Primary leaves that emerge from between the cotyledons begin photosynthesis ❺. The cotyledons typically undergo a period of photosynthesis before shriveling ❻; eventually, they fall off the lengthening stem, and leaves produce all of the plant's food.

Germination is the first step in plant development. As a sporophyte grows and matures, its tissues and organs develop in characteristic patterns. Leaves form in predictable shapes and sizes, stems and roots lengthen and thicken, flowering occurs at a certain time of year, and so on. These patterns are outcomes of gene expression, and they also have an environmental component. As you will see in Chapter 30, gene expression in plants is regulated by hormones and other signaling molecules, just as it is in animals.

coleoptile Rigid sheath that protects a growing embryonic shoot of monocots.
hypocotyl Portion of an embryonic stem between cotyledon(s) and embryonic root (radicle).
plumule Embryonic shoot within a seed.
radicle Embryonic root within a seed.

Take-Home Message

What happens during early plant development?

» Mature seeds often undergo a period of dormancy.

» Species-specific environmental cues trigger seed germination.

» The embryonic shoot and root that emerge from a germinating seed tap endosperm (in monocots) or cotyledons (in eudicots) until primary leaves begin producing all of the plant's food.

29.9 Asexual Reproduction of Flowering Plants

- Asexual reproduction permits plants to rapidly produce genetically identical offspring.
- Links to Cloning 8.7, Asexual versus sexual reproduction 12.2, Meiosis 12.3, Mendelian inheritance 13.2, Aneuploidy 14.6, Radiometric dating 16.6, Speciation by polyploidy in plants 17.11, Modified stems 27.9

Most flowering plants can reproduce asexually by **vegetative reproduction**, in which new roots and shoots grow from extensions or pieces of a parent plant. Each new plant is a clone, a genetic replica of its parent. You already know that new roots and shoots can sprout from nodes on modified stems (Section 27.9). This is one example of vegetative reproduction. As another example, forests of quaking aspen (*Populus tremuloides*) are actually stands of clones that grew from root suckers, which are shoots that sprout from the aspens' shallow, cordlike lateral roots. Suckers sprout after above-ground parts of the aspens are damaged or removed. One stand in Utah consists of about 47,000 shoots and stretches for 107 acres (**Figure 29.18**).

Such clones are as close as any organism gets to being immortal. The oldest known plant is a clone: the one and only population of King's holly (*Lomatia tasmanica*), which has several hundred stems along 1.2 kilometers (0.7 miles) of a river gully in Tasmania. Radiometric dating of the plant's fossilized leaf litter show that the clone is at least 43,600 years old—predating the last ice age!

The ancient *Lomatia* is triploid. In general, plants tolerate polyploidy better than animals do, but triploid plants are sterile and can only reproduce asexually.

Why? During meiosis, an odd number of chromosome sets cannot be divided equally between the two spindle poles. If meiosis does not fail entirely, the unequal segregation of chromosomes results in aneuploid offspring, which rarely survive.

Agricultural Applications

For thousands of years, we humans have been taking advantage of the natural capacity of plants to reproduce asexually.

Cuttings and Grafting Almost all houseplants, woody ornamentals, and orchard trees are clones that have been grown from stem fragments (cuttings) of a parent plant. Propagating some plants from cuttings may be as simple as jamming a broken stem into the soil. This method uses the plant's natural ability to form roots and new shoots from stem nodes. Other plants must be grafted. Grafting means inducing a cutting to fuse with the tissues of another plant. Often, the stem of a desired plant is spliced onto the roots of a hardier one.

Propagating a plant from cuttings ensures that offspring will have the same desirable traits as the parent plant. For example, domestic apple trees (*Malus*) are typically grafted because they do not breed true for fruit color, flavor, size, or texture (**Figure 29.19**). Even trees grown from seeds of the same plant can produce fruits that vary dramatically. The genus is native to central Asia, where apple trees grow wild in forests.

In the early 1800s, the eccentric humanitarian John Chapman (known as Johnny Appleseed) planted mil-

Figure 29.18 Quaking aspen (*Populus tremuloides*). A single plant gave rise to this stand of shoots by asexual reproduction. Such clones are connected by underground lateral roots, so water can travel from roots near a lake or river to those in drier soil some distance away.

A Apple trees do not breed true for fruit characters. These fruits of 21 wild apple trees vary in flavor and texture.

B Wild apples still grow in some regions of their native Kazakhstan, but most of the wild forests have been replanted with commercial groves of domesticated European varieties. The remaining wild forests are becoming fragmented by expanding human developments.

C Gennaro Fazio (*left*) and Phil Forsline (*right*) are working to increase the genetic diversity of U.S. apples by cross-breeding them with disease-resistant wild varieties.

Figure 29.19 Apples (*Malus*).

lions of apple seeds in the midwestern United States. He sold the trees to homesteading settlers, who would plant orchards and make hard cider from the apples. About one of every hundred trees produced fruits that could be eaten out of hand. Its lucky owner would graft the tree and patent it. An estimated 16,000 varieties of domesticated apples were produced during this era, but few of them remain: 15 varieties now account for 90 percent of apples in U.S. grocery stores today. These clones are still propagated by grafting.

Grafting is also used to increase the hardiness of a desirable plant. In 1862, the plant louse *Phylloxera* was accidentally introduced into France via imported American grapevines. European grapevines had little resistance to this tiny insect, which attacks and kills the root systems of the vines. By 1900, *Phylloxera* had destroyed two-thirds of the vineyards in Europe, and devastated the wine-making industry. Today, French vintners routinely graft their prized grapevines onto the roots of *Phylloxera*-resistant American vines.

Tissue Culture An entire plant may be cloned from a single cell with **tissue culture propagation**, in which a body cell is coaxed to divide and form an embryo (Section 8.7). The method can yield millions of genetically identical plants from a single specimen. The technique is being used in research intended to improve food crops. It is also used to propagate rare or hybrid ornamental plants such as orchids.

Seedless Fruits In some plants such as figs, blackberries, and dandelions, fruits may form even in the absence of fertilization. In other species, fruit may continue to form after ovules or embryos abort. Seedless grapes and navel oranges are the result of mutations that result in arrested seed development. These plants are sterile, so they are propagated by grafting.

Seedless bananas are triploid (3*n*); they are robust but sterile. Grafting bananas and other monocots is notoriously difficult (if not impossible), so seedless bananas are propagated by cutting off and replanting adventitious shoots that sprout from their corms.

Despite the ubiquity of polyploid plants in nature (Section 17.11), they rarely arise spontaneously. Plant breeders often use the microtubule poison colchicine to artificially increase the frequency of polyploidy. Tetraploid (4*n*) offspring of colchicine-treated plants are then backcrossed with diploid parent plants. The resulting triploid offspring are sterile: They make seedless fruit after pollination (but not fertilization) by a diploid plant, or on their own. Seedless watermelons are produced this way.

tissue culture propagation Laboratory method in which body cells are induced to divide and form an embryo.
vegetative reproduction Growth of new roots and shoots from extensions or fragments of a parent plant; form of asexual reproduction in plants.

Take-Home Message

How do plants reproduce asexually?

» Many plants propagate asexually when new shoots grow from a parent plant or pieces of it. Offspring of such vegetative reproduction are clones.

» Humans propagate plants asexually for agricultural or research purposes by grafting, tissue culture, or other methods.

29.10 Senescence

■ Mature plants of many species drop leaves and other parts in cyclic patterns that reflect seasonal changes in environmental conditions.

■ Links to Hydrocarbon 3.3, Plant cell walls and extracellular matrix 4.11, Deciduous plants 22.7

The process by which leaves or other plant parts are shed is called **abscission**. Abscission may be induced by environmental stress such as infection, injury, water or nutrient deficiency, or extremes of temperature. It also occurs during the normal life cycle of many flowering plants, as part of senescence. **Senescence** means aging, and the term can be applied to cells, individuals, or entire communities. Cycles of growth, abscission, and dormancy are a response to environmental conditions that vary seasonally. Plants that periodically drop their leaves before a period of dormancy are typically native to regions that are too dry or too cold for optimal growth during part of the year.

As you will see in Section 30.9, abscission and dormancy are genetically programmed responses to seasonal changes in the length of night relative to the length of day. The timing of these processes is an adaptation to recurring cycles of weather conditions typical in the plants' native regions. For example, deciduous trees of the northeastern United States begin to lose their leaves in autumn, around September (**Figure 29.20A**). The trees remain dormant during the months of harsh winter weather that would otherwise damage leaves and buds. Growth resumes in spring (April), when milder conditions return.

A Ripening black cherries (*Prunus serotina*) release ethylene that stimulates abscission of the fruits and nearby leaves.

B Remains of an abscission zone between a leaf petiole and twig of horse chestnut (*Aesculus hippocastanum*).

Figure 29.21 Abscission as part of normal life cycles. Abscission occurs in zones.

On the opposite side of the world, many tree species native to tropical monsoon forests of south Asia lose their leaves during the summer dry season, which is between November and May (**Figure 29.20B**). Although the region receives a lot of rain annually, almost none of it falls during this period of the year. Dormancy offers the trees a way to survive the extended drought-like conditions. New growth reappears at the beginning of June, just in time to be supported by the ample water of monsoon rains.

A Leaves of sugar maples (*Acer saccharum*) native to the northeastern U.S. and Canada change color and drop from the trees in September. The trees remain dormant throughout winter, when prolonged cold would otherwise damage tender leaves.

B Teak trees (*Tectona grandis*) native to south Asia lose their leaves and become dormant during the region's summer dry season, which is November through May. New growth that appears in early June is supported by monsoon rains.

Figure 29.20 The timing of dormancy adapts plant species to seasonal cycles of weather in their native regions.

How Abscission Occurs

Leaky pipes provided the first hint of the mechanism of abscission. In the early 1800s, botanists in Europe and the U.S. started reporting that mature, beautiful trees lining city streets and parks were losing leaves at the wrong time of year along with large patches of bark. They quickly discovered that the trees were being affected by coal gas emanating from leaks in newly installed underground pipes. At the time, the gas was a common fuel for interior lighting fixtures and streetlamps. The offending component of coal gas turned out to be ethylene, a small hydrocarbon that was later identified as a plant hormone. The leaky pipes were giving street trees a massive overdose of this hormone.

Small amounts of ethylene produced by a plant cause its fruits and leaves to mature and drop. Let's use a black cherry tree (*Prunus serotina*) native to the eastern United States as an example. In this species, most root and shoot growth occurs between spring and early summer, from April to July. By midsummer, the tree is producing fruits and seeds. In August, the growing season is coming to a close, and the tree is routing nutrients to stems and roots for storage during the forthcoming period of dormancy. The ripe fruits (**Figure 29.21A**) and seeds release ethylene that diffuses into nearby twigs, petioles, and fruit stalks. The ethylene is a signal for cells in these zones to produce enzymes that digest their own walls. The cells bulge as their walls soften, and separate from one another as the extracellular matrix that cements them together dissolves. Tissue in the abscission zone weakens, and the structure above it drops. A scar often remains where the structure had been attached (**Figure 29.21B**).

Ethylene also participates in abscission during plant stress responses. For example, the hormone is released in response to bacterial invasion of a plant's internal tissues. Chapter 30 returns to the topic of how plant hormones mediate this and other processes.

abscission Process by which plant parts are shed. Occurs in response to stress, and, in some species, prior to seasonal dormancy during senescence.
senescence Aging of cells, individuals, or communities.

Take-Home Message

Why do plants drop their leaves and other parts?

» Abscission can be a response to stress. In many species it is also part of normal cycles of dormancy during senescence.

» The timing of seasonal abscission and dormancy is an evolutionary adaptation to recurring cycles of environmental conditions in a particular region.

Plight of the Honeybee (revisited)

Researchers still do not know what causes colony collapse disorder. Many scientists now think that CCD may not be a specific disease, but rather a characteristic of bee colonies suffering from a combination of environmental stresses that include parasites and diseases. Picorna-like viruses, including Israeli acute paralysis virus, are carried by mites that parasitize bees. Parasitic mites and Israeli acute paralysis virus have been detected in many affected hives. Almost all affected colonies are infected with a virus called invertebrate iridescent virus, and *Nosema*, a microsporidium (Section 23.4).

Synthetic pesticides called neonicotinoids may be the primary cause of CCD. Neonicotinoids mimic the neurotoxic (nerve-killing) effect of nicotine, a natural insecticide made by plants such as tobacco (*Nicotiana*). Scientists at the U.S. Department of Agriculture's Bee Research Laboratory showed that neonicotinoids increase bees' vulnerability to *Nosema* infection even in amounts too tiny to detect in the insects themselves.

Neonicotinoids are now the most widely used insecticides in the United States, and they are increasingly used in Europe and other countries—about 2.5 million acres of croplands in total. The chemicals are systemic, which means they are taken up by all tissues of a plant treated with them, including the nectar and pollen that honeybees collect. Even dewdrops that form on neonicotinoid-treated plants apparently contain enough of the insecticide to affect bees.

How would you vote? Systemic pesticides are easy to apply and effective for long periods. They also get into plant nectar and pollen eaten by honeybees. To protect bees and other pollinators, should the use of neonicotinoid pesticides be banned?

LEARNING ROADMAP

Where you have been This chapter revisits *cis* and *trans* molecules (Section 3.5); reactions (5.5, 5.6); cells (4.9, 4.10); membranes (5.7–5.10); pigments and light energy (6.2); photosynthesis (6.5, 6.7); FAD (7.2); gene expression and control (9.2, 10.2, 10.4, 10.6); genetic engineering (10.3, 15.7); and plant structure (27.4, 27.6, 28.4–28.6), development (10.3, 29.7, 29.8), pathogens (23.5), and symbionts (23.7, 28.3).

Where you are now

Plant Hormones
Most development in plants occurs in the adult body. All development, as well as metabolism and other activities, is orchestrated by molecules called hormones.

Mechanisms of Hormone Action
Plant hormones influence cell division, differentiation, expansion, and metabolic activity. They act together, in opposition or synergistically, and often redundantly.

Responses to Environmental Cues
Plants adapt to their environment by altering growth in response to environmental cues, including gravity, contact, and light.

Cycles of Growth
Cyclic, daily patterns of metabolic activity are driven by transcription factor feedback loops. Seasonal changes are integrated responses to daily cycles and night length.

Stress Responses
Plants respond to stress caused by living and nonliving factors in the environment. Hormones are involved in immediate and long-term responses to pathogens.

Where you are going Hormones also govern animal physiology, a topic explored in Chapter 34. Section 31.9 covers negative feedback; Section 32.4, positive feedback; both are in the context of animal homeostasis. Plant immunity shares concepts and molecules with vertebrate immunity, the topic of Chapter 37. Circadian rhythms (Section 34.11) occur in animals as well as plants. The cardiovascular system, its function, and illnesses associated with it are covered in Chapter 36.

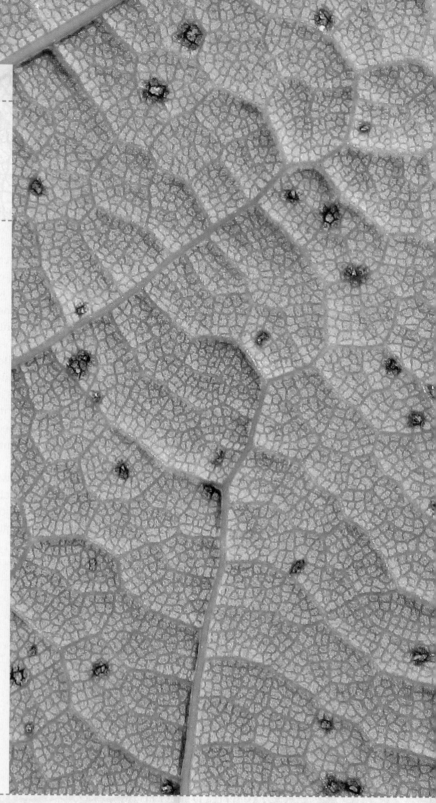

30.1 Prescription: Chocolate

For several centuries, the small islands of the San Blas Archipelago just off the coast of Panama have been inhabited by the Kuna, a tribe of indigenous South Americans. The Kuna have been untroubled by the malaria, yellow fever, and dengue fever that afflict their mainland cousins, primarily because no disease-carrying mosquitoes can survive on their windswept islands. The Kuna also have almost no incidence of the hypertension—high blood pressure and its associated cardiovascular problems—that plagues about one-quarter of people in mainland populations.

Scientists studying the Kuna hypothesized that their resistance to hypertension had a genetic basis, so in 1990, they began to search for an allele unique to the Kuna that affects blood pressure. At the time, the human genome was being sequenced (Section 15.4), and genomic comparisons were becoming a routine avenue of research. However, a decade later, researchers still had not found the allele, and began to look for other correlations. They discovered that genetics really had nothing to do with the low incidence of hypertension in the Kuna. All else being equal, the most striking difference between island-dwelling populations and their mainland relatives is that the islanders consume an unusually large amount of cocoa. The Kuna tend to drink cocoa instead of water, and they live in a hot tropical climate so they drink a lot of it—a minimum of five cups per day.

Cocoa is made from cacao beans, which are the seeds of the *Theobroma cacao* tree (**Figure 30.1**). The seeds and young leaves of this tree have a particularly high content of flavonoids. Flavonoids are second-

epicatechin

ary metabolites: compounds that are not required for the immediate survival of the organism that makes them (Section 22.9). A particular flavonoid called epicatechin is probably responsible for the unusual absence of heart disease among the Kuna.

Cocoa is rich in epicatechin, as are other forms of minimally processed dark chocolate. Epicatechin influences a wide range of biological activities in the human body. This compound has a demonstrably protective effect against oxidative tissue damage that typically occurs after a stroke or a heart attack: A dose of epicatechin can reduce the heart muscle damage that a heart attack causes by 52 percent; and reduce the brain damage that a stroke causes by 32 percent. As an added benefit, epicatechin enhances memory and kills cancer cells.

Plants do not make this seemingly miraculous compound for our benefit. Neither do they make caffeine, resveratrol, curcumin (in coffee, red wine, and turmeric, respectively), or any other medically relevant chemical, for us. Epicatechin serves a role in plant immunity. Pigments, scents, and other chemicals that attract pollinators are also secondary metabolites, as are compounds that deter herbivores and pathogens, attract symbionts, or inhibit the growth of competing individuals. These chemicals are all part of the means by which plants interact with their environments.

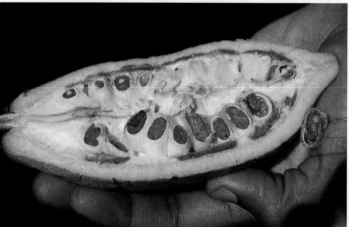

Figure 30.1 The cacao tree, *Theobroma cacao*. *Left*, the reddish pods are cacao fruits. Cocoa and other forms of chocolate are made from the dried, ground seeds of this plant (*right*). The seeds are particularly rich in flavonoids, especially epicatechin.

30.2 Introduction to Plant Hormones

- Plant development and function depend on cell-to-cell communication.
- Plant cells communicate via chemical signals.
- Links to Feedback 5.5, Receptor proteins 5.7, Gene expression controls 10.2, Cell communication in development 10.3, Epigenetics 10.6, Xylem function 28.4, Phloem function 28.6

In typical animals, most development occurs before adulthood. By contrast, in typical plants, most development occurs in the adult body (**Figure 30.2**). Plants, unlike animals, cannot move about to avoid unfavorable conditions. Instead, each plant adapts to a stationary lifestyle by adjusting its development in response to environmental cues, including the availability of water and nutrients, length of night, temperature, the presence of pathogens or herbivores, and gravity.

Development in plants depends on extensive coordination among individual cells, just as it does in animals. A plant is an organism, not just a collection of cells, and as such it develops and functions as a unit. Cells in different tissues or even in different parts of a plant coordinate their activities by communicating with one another. As an example, a leaf being chewed by a caterpillar can signal other parts of the plant to produce appropriate caterpillar-deterring chemicals.

Figure 30.2 Plant versus animal development. A tree grows and adds new branches throughout its adult life, whereas in humans and other animals, most development occurs prior to adulthood.

Chemical Signaling

Much of the cell-to-cell communication within plants involves hormones. A **plant hormone** is an extracellular signaling molecule that exerts its effect at very low concentrations. Hormones direct or affect the development of all plant parts; the direction and rate of growth; immunity and other defensive responses; circadian rhythms; the timing and duration of flowering; fruit and seed formation; aging; and initiating and breaking dormancy, to give some examples. Recent research indicates that plant hormones are never unique controllers of a single process. Instead, their functions overlap in a network of signaling that is sometimes redundant and often antagonistic.

Typically, a hormone released by cells in a localized area of the body alters the activity of cells in a different part of the body. A hormonal signal is detected by receptor proteins (Section 5.7), with each type of receptor binding to a particular hormone.

A cell's response to a hormonal signal depends on the cell and the receptor, and it often varies with the concentration of the hormone. Binding of a plant hormone to its receptor initiates a response in the cell. Typically, the response involves modification of nuclear or mitochondrial DNA that causes a change in gene expression. In some cases, enzyme activity, solute concentration, or another cell function is affected with no change in underlying gene expression patterns.

Different hormones can have synergistic or opposing effects, so a cell's response often depends on the integration of hormonal signals. Plant hormones interact with one another mainly at the transcriptional level. For example, some hormones dampen or enhance the expression of others (Section 10.2). Many hormones inhibit their own expression, a mechanism called negative feedback that is explored in more detail in Chapter 31. Positive feedback loops in which a hormone promotes its own transcription are part of daily cycles of activity, as you will see in Section 30.9.

Hormones operate in both plants and animals, but their structure, origin, and action differ between the two groups. For example, animals have organs specialized for hormone secretion; plants do not. All plant cells have the ability to make and release hormones—the basis of their developmental plasticity—but at any given time, each hormone is produced only in certain regions of the plant body. Animal hormones are defined by their ability to elicit an effect in a dis-

plant hormone Extracellular signaling molecule of plants that exerts its effect at very low concentration.

tant tissue. Similarly, a plant hormone may be released in a region of the body that is very far from cells in a targeted tissue, such as when cells in an actively growing root release cytokinins that keep shoot tips in a concurrent mode of active growth. However, some plant hormones can have very localized effects, such as when ethylene released from a cell in a ripening fruit affects the releasing cell as well as other cells in the same tissue.

Not all molecules that act as hormones in plants act as hormones in animals, and vice versa. For example, plants make phytoestrogens that are structurally similar to estrogens, which are female sex hormones in mammals. The two types of compounds have similar hormonal effects in mammals, but phytoestrogens have nothing to do with sexual reproduction in plants. Rather, they function as part of plant defense mechanisms against fungal attack. The difference does not necessarily stem from differences in the structure of the hormones, but rather in the way the receptors for these compounds work in cells that bear them.

Unlike animals, plants do not have a circulatory system that moves hormones through the body. The movement of most plant hormones from source to target occurs by passive diffusion through local tissues, and by fluid transport through xylem or phloem to more distant areas of the plant body. A few plant hormones move from cell to cell via active transport proteins embedded in plasma membranes.

Table 30.1 gives a preview of some major plant hormones and a few of their known effects. Those effects often vary depending on the tissue. For example, brassinosteroids affect expression of genes governing anther and pollen development in cells of a developing flower. In leaves, the same hormones are part of different signaling pathways that cause cells to die in response to pathogen infection. Recent breakthroughs in plant hormone research have revealed many of the details of such signaling pathways, which we explore in the following sections.

Take-Home Message

What regulates growth and development in plants?

» Plant hormones are signaling molecules that coordinate activities among cells in different parts of the plant body.

» Cells that bear receptors for a hormone—and thus can respond to it—may be in the same tissue as the hormone-releasing cell, or in another region of the plant body.

» Plant hormones are involved in all aspects of growth, development, and function in plants. They often work together, with synergistic or opposing effects on cells.

Table 30.1 Examples of Plant Hormone Effects	
Hormone	**Effect**
Abscisic acid (ABA)	Closes stomata in times of stress
	Inhibits shoot growth
	Inhibits seed germination
	Involved in abiotic stress responses
	Involved in chloroplast movement
Auxin	Coordinates effects of other plant hormones during development of shoots and roots
	Role in apical dominance
	Stimulates cell expansion
	Role in abscission
	Mediates tropisms
	Stimulates meristem cell division
Brassinosteroid	Regulates development of anthers and pollen
	Role in tissue defense
	Role in fruit ripening
Cytokinin	Stimulates differentiation of root meristem cells
	Stimulates division of shoot apical meristem cells
	Inhibits lateral root initiation
Ethylene	Involved in breaking dormancy
	Stimulates flower organ development
	Stimulates fruit ripening
	Stimulates abscission
	Involved in stress responses
Gibberellin	Stimulates cell division, elongation in stems
	Mobilizes food reserves in germinating seeds
	Stimulates flowering
Jasmonic acid	Induces expression of genes used in defense
	Involved in thigmotropic response
Nitric oxide	Abiotic and biotic stress defense responses
	Enhances germination
	Involved in thigmotropic response
Salicylic acid	Activates systemic acquired resistance
Strigolactone	Opposes auxin's effects
	Role in apical dominance

30.3 Auxin: The Master Growth Hormone

■ Auxin coordinates the effects of other plant hormones.
■ Links to Turgor 5.8, Active transport 5.9, Membrane cycling 5.10, Movement through phloem 28.6, Coleoptile 29.8

auxin

There are a few naturally occurring auxins, but the one that occurs most frequently in plants is IAA (indole-3-acetic acid), a small molecule derived from the amino acid tryptophan. In most cases, the term **auxin** refers to IAA. Auxin was first discovered for its ability to promote growth in plants (**Figure 30.3**), and its name is derived from the Greek word for growth. However, it has multiple effects. Auxin plays a critical role in all aspects of plant development, starting with the first division of the zygote. It is involved in polarity and tissue patterning in the embryo, formation of plant parts (primary leaves, shoot tips, stems, and roots), differentiation of vascular tissues, formation of lateral roots (and adventitious roots in some species), and, as you will see in the later sections, shaping the plant body in response to environmental stimuli. Auxin also influences the levels of other plant hormones.

Researchers are still working out the mechanisms of many of auxin's effects, but they do understand one way in which it affects cell elongation. As a plant grows, auxin causes young cells to expand by increasing the activity of transport proteins that pump hydrogen ions from cytoplasm into the cell wall. The resulting increase in acidity softens the wall. Turgor, the pressure exerted by fluid inside the softened wall, stretches the cell irreversibly.

Synthetic auxins have multiple uses. Application of auxin to a cut stem causes roots to develop, so this hormone is often sold as a rooting compound. Auxin can also be used as an herbicide. Applied to leaves, it causes uncontrolled cell division in the plant's api-cal meristems. The resulting growth is unsustainable: Stems curl as they become too long to be supported properly, leaves wither as the plant's resources are diverted to inappropriate growth, and the plant dies. Agent Orange, a defoliant used extensively during the Vietnam War, consisted of a combination of two synthetic auxins. One of those auxins, called 2,4-D, is now widely used as an herbicide on lawns and in corn-fields, where it kills eudicots but not monocots.

Polar Transport

Although present in almost all plant tissues, auxin is unevenly distributed through them. The hormone is made mainly in shoot apical meristems and in young leaves. However, auxin is used in all parts of the plant, so it must be transported from where it is made to where it is needed.

Auxin from shoots is loaded into phloem, travels to roots, and is unloaded into root cells. A different pathway dominates its transport over shorter distances. Auxin diffuses into cells across the plasma membrane, and it is also actively transported through plasma membrane proteins called influx carriers. Once auxin has entered cytoplasm, it can only leave through active transport proteins called efflux carriers.

Unlike most other membrane proteins, efflux carriers are not distributed evenly around a cell's plasma membrane. Efflux carriers in adjacent cells "point" in the same direction, so they direct the flow of auxin (**Figure 30.4A**). In an actively lengthening shoot, for example, efflux carriers are positioned on the side of the cell closest to the base of the stem. Thus, auxin flows from the shoot's apical meristem toward the base of the stem (**Figure 30.4B**). Inside young root

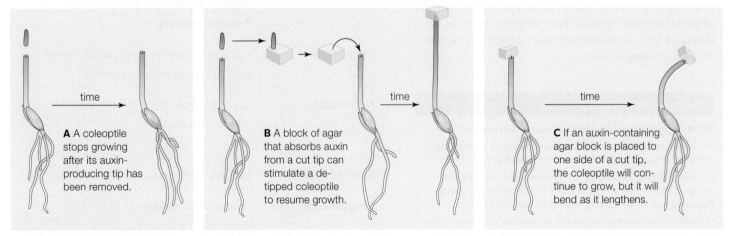

A A coleoptile stops growing after its auxin-producing tip has been removed.

B A block of agar that absorbs auxin from a cut tip can stimulate a de-tipped coleoptile to resume growth.

C If an auxin-containing agar block is placed to one side of a cut tip, the coleoptile will continue to grow, but it will bend as it lengthens.

Figure 30.3 Animated Experiments showing that a coleoptile lengthens in response to auxin produced in its tip.

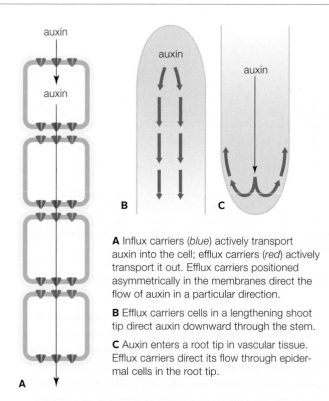

auxin

auxin

auxin

auxin

B

C

A

A Influx carriers (*blue*) actively transport auxin into the cell; efflux carriers (*red*) actively transport it out. Efflux carriers positioned asymmetrically in the membranes direct the flow of auxin in a particular direction.

B Efflux carriers cells in a lengthening shoot tip direct auxin downward through the stem.

C Auxin enters a root tip in vascular tissue. Efflux carriers direct its flow through epidermal cells in the root tip.

Figure 30.4 Polar distribution of auxin is driven by asymmetrical positioning of plasma membrane active transport proteins.

tips, the efflux carriers direct the flow of auxin in the opposite direction, from the root's tip toward the root–shoot interface (**Figure 30.4C**). In both cases, the polar transport of auxin establishes the long axis during the organ's development.

This mechanism of directing auxin flow is unique among plant hormones, and it is important because it establishes auxin concentration gradients across tissues, organs, and the entire plant. The gradients coordinate the actions of other hormones, many of which are expressed in localized patterns that vary even from one cell layer to the next. For example, auxin produced in a shoot's tip supports growth in the shoot's stem and leaves. In the stems, the auxin induces expression of genes whose products promote growth, including enzymes that synthesize gibberellin. In the leaves, the auxin stimulates transcription of an enzyme that breaks down cytokinins, which inhibit cell division. Auxin has a different effect in roots, where it interacts with ethylene to inhibit lengthening of the root and to stimulate root hair development.

apical dominance Growth-inhibiting effect on lateral (axillary) buds, maintained by auxin gradients in growing shoot tips.
auxin Plant hormone with a central role in coordinating responses to other hormones in all stages of growth and development. Indole-3-acetic acid is the most common auxin.

Figure 30.5 A shoot can continue to grow despite losing its tip. The loss of a shoot's tip ends its main supply of auxin—a signal that breaks dormancy in lateral buds.

Efflux carriers are continually recycled by plasma membrane cycling (Section 5.10), so their placement can change over time. This active process allows auxin-directed plant development to be flexible and responsive. Consider how efflux carriers help balance the growth of a plant's apical and lateral buds. When a shoot is lengthening at the tip, its lateral buds are usually dormant, an effect called **apical dominance**. In the shoot's tip, auxin produced by apical meristem is traveling through efflux carriers, and the movement causes the stem to lengthen. In the dormant lateral buds, auxin produced by apical meristem is not traveling, because cells in these buds have few efflux carriers in their membranes.

If a shoot's tip breaks off, its lateral buds begin to grow, an effect exploited by gardeners who pinch off shoot tips to make a plant bushier (**Figure 30.5**). When a shoot loses its tip, it also loses its source of auxin. The decrease in auxin stops the stem from lengthening. It also slows the production of another hormone, strigolactone, in roots. Strigolactone is transported from roots to all of the plant's shoots, where it interrupts the recycling of auxin efflux carriers so fewer of them get inserted into plasma membranes. Thus, strigolactone has a dampening effect on auxin transport—and as a result, it inhibits shoot lengthening. Cells that have a lot of efflux carriers (such as those in lengthening shoot tips) can still transport auxin even when the strigolactone level is high; cells with few efflux carriers (such as those in dormant lateral buds) cannot. When the strigolactone level declines, cells in dormant lateral buds begin to acquire efflux carriers in their membranes, auxin begins to be transported through them, and the buds start to lengthen.

strigolactone

Take-Home Message

What are the main effects of auxin in plants?

» A polar distribution system sets up auxin concentration gradients across a plant's tissues and organs in response to internal and external conditions.

» Auxin gradients coordinate the activities of other plant hormones during growth and development at all stages of the plant life cycle.

30.4 Cytokinin

- Cytokinin stimulates cell divisions in shoot apical meristem, and cell differentiation in root apical meristem.
- A homeostatic balance of cytokinin and auxin is necessary for proper control of meristem activity.
- Link to Pericycle 27.6

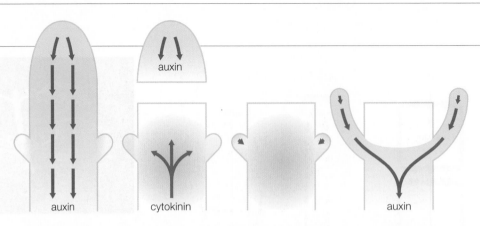

A Auxin flowing through a shoot keeps the level of cytokinin low in the stem.

B Removing the tip ends auxin flow in the stem. As the auxin level declines, the cytokinin level rises.

C The cytokinin stimulates cell division in apical meristem of lateral buds. The cells begin to produce auxin.

D Auxin gradients form and direct the development of the growing lateral buds.

Figure 30.6 Interaction of auxin and cytokinin in the release of apical dominance.

cytokinin

A **cytokinin** is one of a group of plant hormones derived from the nucleotide adenine. Among other functions, a cytokinin affects meristem cell division and differentiation, inhibits the development of lateral roots, and stimulates development of lateral buds. Some cytokinin is synthesized locally in stems, but most is produced in roots and transported to shoots in xylem. Cytokinin binding to its receptor on a cell surface sets in motion a series of phosphorylations that ultimately activate a set of transcription factors in the nucleus. The overlapping effects of the many genes governed by these transcription factors are still being determined.

Cytokinin and auxin work together, often antagonistically, and they influence one another's expression. The relationship between the two hormones is a homeostatic mechanism that dynamically regulates their relative concentrations. This balance is important because it controls organ initiation, embryo formation, and meristem function.

Consider how cytokinin opposes auxin's effect on lateral root formation. Lateral roots grow from pericycle cells (Section 27.6). Only a few pericycle cells give rise to lateral roots, however. This is because the level of cytokinin is normally high in roots compared with auxin. (Most cytokinin is produced in roots; most auxin is produced in shoots.) Both hormones influence transcription of a gene whose product prevents auxin efflux carriers from being inserted into root cell membranes. Auxin inhibits transcription of the gene; cytokinin enhances it. Thus, where the ratio of auxin to cytokinin is high, the gene is not transcribed, and efflux carriers get inserted into root cell membranes. The carriers set up an auxin gradient, which is required to pattern the development of a lateral root. Where the ratio of auxin to cytokinin is low, the gene is transcribed, auxin gradients do not form, and lateral roots do not develop from pericycle.

The two hormones have different effects in roots and shoots. For example, in root apical meristem,

auxin and cytokinin work in opposition to maintain the balance of differentiating and undifferentiated cells. In this context, auxin supports division of undifferentiated meristem cells, and cytokinin signals the cells to differentiate. By contrast, in shoot apical meristem, the two hormones work synergistically. Auxin and cytokinin act on the same set of receptors in shoot apical meristem cells to support cell division and to prevent differentiation.

Perhaps the best-known function of cytokinin is that it stimulates lateral bud growth. Apical dominance varies by species, and it is controlled by environmental cues as much as developmental programs. However, in general, lateral buds tend to remain dormant unless a shoot gets decapitated. Dormancy is maintained by the presence of auxin in the stem. The auxin regulates cytokinin production and transport, and it increases expression of an enzyme that breaks down cytokinin. Once the shoot tip has been removed, the auxin level decreases in the stem, so the cytokinin level rises. The cytokinin moves into lateral buds and stimulates cell divisions there. In each bud, auxin produced by the now-active cells in the apical meristem forms a gradient that directs development (**Figure 30.6**).

cytokinin Plant hormone that promotes cell division in shoot apical meristem and cell differentiation in root apical meristem. Often interacts antagonistically with auxin.

Take-Home Message

What are the main effects of cytokinin in plants?

» Cytokinin stimulates cell divisions in shoot apical meristem, and cell differentiation in root apical meristem.

» Cyokinin and auxin act together and often antagonistically. The cytokinin–auxin balance controls cell division and differentiation in shoot and root apical meristem.

30.5 Gibberellin

- Gibberellins cause stems to elongate and seeds to germinate.
- Links to Repressors 10.2, Shoot structure 27.4

In 1926, researcher E. Kurosawa was studying what Japanese call *bakane*, the "foolish seedling" effect. The stems of rice seedlings infected with a fungus, *Gibberella fujikuroi*, grew twice the length of uninfected seedlings. The abnormally elongated stems were weak and spindly, and eventually they toppled. Kurosawa discovered that he could induce the lengthening experimentally by applying extracts of the fungus to seedlings. Many years later, other researchers purified the substance from fungal extracts that brought about stem lengthening in plants (**Figure 30.7**). They named it gibberellin, after the fungus.

gibberellin

A **gibberellin**, as we now know, is a plant hormone that promotes growth in all flowering plants, among other functions. It causes a stem to lengthen between the nodes by inducing cell division and elongation. Along with auxin, gibberellin influences expansion along the longitudinal axis, so it greatly influences the size of plant organs, and of the plant itself. The short stature of Mendel's dwarf pea plants (Section 13.4) is the result of a mutation that reduces the rate of gibberellin synthesis.

Dwarf phenotype in many other plants is caused by a defective gibberellin receptor. Gibberellin is also involved in slowing the aging of leaves and fruits, breaking dormancy in seeds, germination of seeds, and, in some plants, flowering.

Gibberellin made by cells in young leaves and root tips is transported through phloem to the rest of the plant. Seeds also make this hormone. Seedless grapes tend to be smaller than seeded varieties because their undeveloped seeds do not produce normal amounts of gibberellin. Farmers spray their seedless grape plants with synthetic gibberellin, which increases the size of the resulting fruit.

Figure 30.7
Stem-lengthening effect of gibberellins. The three tall cabbage plants were treated with gibberellins. The two short plants in front of the ladder were not treated.

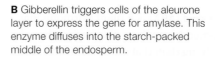

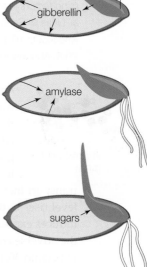

A Absorbed water causes cells of a barley embryo to release gibberellin, which diffuses through the seed into the aleurone layer of the endosperm.

B Gibberellin triggers cells of the aleurone layer to express the gene for amylase. This enzyme diffuses into the starch-packed middle of the endosperm.

C The amylase hydrolyzes starch into sugar monomers, which diffuse into the embryo and are used in aerobic respiration. Energy released by the reactions of aerobic respiration fuels meristem cell divisions in the embryo.

Figure 30.8 Gibberellin and germination, illustrated in a seed of barley (*Hordeum vulgare*).

Gibberellin works by inhibiting inhibitors, thus removing the brakes on some cellular processes. Binding to a gibberellin receptor in endodermal cells sets in motion the destruction of transcription repressors (Section 10.2) in the nucleus. The genes that these proteins repress are still being studied, but their products have overlapping functions involving cell proliferation and expansion, fertility, and germination.

For example, during germination of a barley seed, absorbed water causes cells of the embryo to release gibberellin (**Figure 30.8**). The hormone diffuses into the aleurone, a protein-rich layer of cells surrounding the endosperm. There, gibberellin causes the destruction of a repressor that inhibits transcription of the gene for amylase, an enzyme that hydrolyzes starch into sugar monomers (Section 3.4). The amylase is synthesized and released into the endosperm's starchy interior, where it proceeds to break down stored starch molecules into sugars. The embryo takes up the sugars and uses them for aerobic respiration, which fuels rapid cell divisions at the embryo's meristems.

gibberellin Plant hormone that induces stem elongation and helps seeds break dormancy, among other effects.

Take-Home Message

What are the main effects of gibberellin in plants?

» Gibberellin stimulates cell division and elongation in stems, which causes stems to lengthen between nodes.

» Gibberellin affects the expression of genes for nutrient utilization during seed germination.

30.6 Abscisic Acid

■ Abscisic acid mediates germination, inhibits growth, and it is part of protective responses to stress caused by living and non-living factors in the environment.

■ Links to Redox reactions 5.5, Transcription factors 10.2, Knockouts 10.3, Stomata function 28.5, Seed germination 29.8, Abscission 29.10

abscisic acid (ABA)

Abscisic acid (ABA) is a plant hormone that was named because its discoverers thought it mainly mediated abscission. However, ABA was later discovered to have a much greater role in plant stress responses than in abscission. ABA synthesis, release, and transport increases in response to stress, so that its level rises in the plant's tissues. In turn, the hormone induces expression of genes that help the plant survive the adverse conditions (we return to stress responses in Section 30.10). ABA also has an important role in embryo maturation, stomata closure (Section 28.5), seed and pollen germination, and fruit ripening; and like cytokinin it suppresses lateral root formation.

ABA synthesis begins in chloroplasts, so its concentration is highest in leaves and other photosynthetic parts. Like auxin, ABA diffuses into plant cells, but it can exit a cell only by moving through active transport proteins. Its movement is not polar as is auxin's. Once in vascular tissue, ABA can move through xylem or phloem to all parts of the plant.

ABA receptors occur on the plasma membrane, in cytoplasm, and in the nucleus. ABA binding to its receptors in the nucleus results in a cascade of phosphorylations that activate hundreds of transcription factors. These transcription factors govern the expression of thousands of genes—around 10 percent of the total number in plants and a much larger proportion than any other plant hormone. In general, genes whose expression is repressed by ABA are involved in growth, such as those that encode components of ribosomes, chloroplasts, and cell walls; those whose expression is enhanced by ABA are involved in metabolism, stress responses, and embryonic development. Among the latter are genes that encode and activate a plasma membrane protein called NADPH oxidase.

hydrogen peroxide
H—O=O—H

nitric oxide
N≡O

Figure 30.9 Two small molecules with big effects in living systems: hydrogen peroxide and nitric oxide.

Figure 30.10 ABA prevents seed germination. Part of the ABA signaling pathway in this *Arabidopsis* plant has been knocked out. Its seeds germinated before they had a chance to disperse.

This enzyme transfers electrons from NADPH inside the cell to oxygen molecules outside the cell—a redox reaction that produces oxygen radicals and hydrogen peroxide (**Figure 30.9**). These reactive molecules very quickly activate another plasma membrane enzyme, nitrate reductase, that in turn produces a burst of nitric oxide, a gas. One effect of this burst is the production of an enzyme that breaks down ABA—a feedback loop that dynamically controls ABA signaling.

ABA that accumulates in a seed as it forms, matures, and dries out prevents the seed from germinating too early (**Figure 30.10**). It exerts this effect mainly by inhibiting expression of genes involved in cell wall loosening and expansion—both critical processes for growth of an embryonic plant. ABA also inhibits expression of genes involved in gibberellin synthesis. Thus, a seed cannot germinate until its ABA level declines. Exactly how this decline is triggered is unknown, but as the ABA level falls, cells of the embryo start expressing genes that result in cell wall softening and enlargement. Transcriptional control over gibberellin synthesis genes is also lifted. Hydrogen peroxide (from NADPH oxidase) enhances expression of these genes, so gibberellin is produced, and germination begins as enlarging cells of the embryo start using carbohydrates in endosperm.

abscisic acid (ABA) Plant hormone that inhibits growth and germination. Involved in stomata function and stress responses.

Take-Home Message

What are the main effects of abscisic acid in plants?

» Abscisic acid inhibits germination and growth.

» ABA also stimulates metabolism, stress responses, embryonic development, and stomata closure.

30.7 Ethylene

■ Ethylene is a gas involved in fine-tuning growth responses. It also triggers intermittent processes, including fruit ripening.
■ Links to Plastids 4.9, Membrane permeability 5.8, Genetically engineered plants 15.7, Fruit function 29.7, Abscission 29.10

ethylene

The plant hormone **ethylene** is part of regulatory pathways that govern a wide range of metabolic and developmental processes in plants, including germination, growth, abscission, ripening, and stress responses. Ethylene is a gas that is soluble in water, and it freely crosses lipid bilayers. Cells in all parts of a plant can produce this hormone from methionine and ATP. Each of the two enzymes involved in ethylene synthesis occurs in multiple versions encoded by slightly different genes. Expression of some of these genes is inhibited by ethylene (a negative feedback loop); expression of the others is enhanced by ethylene (a positive feedback loop). Negative feedback loops produce a basal level of ethylene that helps fine-tune ongoing metabolic and developmental processes such as growth and cell expansion. For example, in roots, ethylene produced in a negative feedback loop enhances the transcription of genes involved in producing auxin and its efflux carriers. Remember, auxin inhibits lengthening in roots.

Positive feedback loops produce large amounts of ethylene required for intermittent processes such as germination, fruit ripening, plant defense responses, and abscission. For example, a high level of ethylene governs fruit production in a wide variety of flowering plants (**Figure 30.11**). In one positive feedback loop, ethylene synthesis in flower petals increases until it provokes abscission, so the petals drop.

Ripening of fleshy fruits such as strawberries occurs after a peak of cellular respiration followed by a burst of ethylene produced in another positive feedback

loop. The high level of ethylene stimulates genes whose products make the fruit ripen. Chloroplasts are converted to chromoplasts (Section 4.9) as their (green) chlorophylls break down and carotenoids such as (red) lycopene and (orange and yellow) beta-carotenes accumulate. In a process similar to abscission, cell walls and middle lamellae break down. Starch and organic acids are converted to sugars, and aromatic molecules are produced. The color change, softening, and increased palatability all tempt animals that can disperse seeds.

Because ethylene is a gas, it can diffuse from one fruit to initiate ripening in another. Synthetic ethylene is widely used to artificially ripen fruit. Hard, unripe fruit can be transported long distances with less damage than soft, ripe fruit. Upon arrival at its final destination, the fruit is exposed to ethylene gas, which jump-starts the positive feedback ripening loop.

One of the ripening genes whose expression is induced by ethylene encodes an enzyme that breaks down pectin in plant cell walls. In 1994, the FDA approved "Flavr Savr" tomatoes, which had been genetically engineered to underproduce this enzyme. The tomatoes could be ripened artificially with ethylene, but the delayed softening doubled their shelf life.

ethylene Gaseous plant hormone involved in regulating growth and cell expansion. Participates in germination, abscission, ripening, and stress responses.

Take-Home Message

What are the main functions of ethylene in plants?

» Ethylene produced in negative feedback loops participates in ongoing metabolic and developmental processes.

» Ethylene produced in positive feedback loops is involved in intermittent processes such as abscission, fruit ripening, and defense responses.

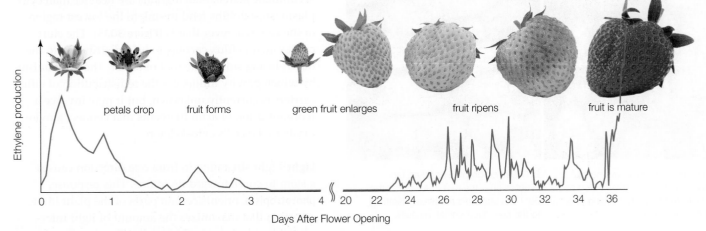

petals drop   fruit forms   green fruit enlarges   fruit ripens   fruit is mature

Ethylene production

Days After Flower Opening

Figure 30.11 Ethylene production during strawberry formation and ripening. Oscillations are normal daily cycles (see Section 30.9).

30.8 Tropisms

- Plants alter growth in response to environmental stimuli. Hormones are typically part of this effect.
- Links to Plastids 4.9, Cytoskeleton 4.11, Pigments 6.2, Light-dependent photosynthesis 6.5, Stomata function 28.5

Plants respond to environmental stimuli by adjusting their growth. In this way, each plant optimizes its opportunities for photosynthesis, absorption of water and nutrients, reproduction, and other factors that affect its fitness. An adjustment of growth in response to environmental stimuli is called a **tropism**. Tropisms are typically mediated by hormones, as the following examples illustrate.

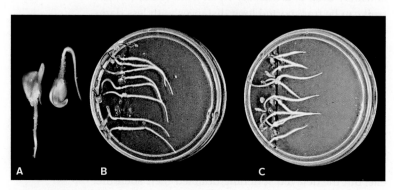

Figure 30.12 Gravitropism.

A Regardless of how a corn seed is oriented in soil, the seedling's primary root grows down, and its primary shoot grows up.

B These seedlings were rotated 90° counterclockwise after they germinated. They adjusted to the change by redistributing auxin, so their direction of growth shifted.

C In the presence of auxin transport inhibitors, seedlings do not adjust their direction of growth after a 90° counterclockwise rotation. Mutations in genes that encode auxin transport proteins have the same effect.

statoliths

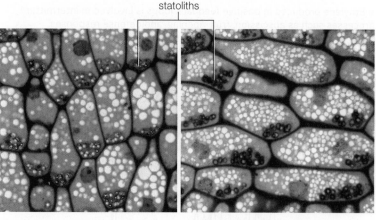

A This micrograph shows heavy, starch-packed statoliths settled on the bottom of gravity-sensing cells in a corn root cap.

B This micrograph was taken ten minutes after the root in **A** was rotated 90°. The statoliths are already settling to the new "bottom" of the cells.

Figure 30.13 Gravity, statoliths, and auxin. **Figure It Out:** In which direction was this root rotated?

Answer: Counterclockwise

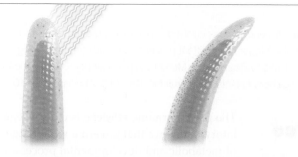

A Sunlight strikes only one side of a coleoptile.

B Auxin flow is directed toward the shaded side, so cells on that side lengthen more.

Figure 30.14 Phototropism. Auxin-mediated differences in cell elongation between two sides of a coleoptile induce bending toward light.

Examples of Environmental Triggers

Gravity Even if a seedling is turned upside down just after germination, its primary root and shoot will curve so the root grows downward and the shoot grows upward (**Figure 30.12**). A growth response to gravity is called **gravitropism**. A root or shoot "bends" because of differences in auxin concentration that occur after asymmetrical auxin transport by efflux carriers. In shoots, auxin enhances cell elongation, so auxin that accumulates on one side of a shoot causes the shoot to bend away from that side. Auxin has the opposite effect in roots: auxin that accumulates on one side of a growing root causes the root to bend toward that side.

Gravity-sensing mechanisms of many organisms are based on organelles called **statoliths**. Plant statoliths are amyloplasts (Section 4.9) stuffed with dense grains of starch. They occur in root cap cells, and also in specialized cells at the periphery of vascular tissues in the stem. Chloroplasts may also function as gravity-sensing organelles in shoots, because they specifically accumulate starch. Starch grains are heavier than cytoplasm, so statoliths tend to sink to the lowest region of the cell, wherever that is (**Figure 30.13**). The shift causes auxin efflux carriers to be redistributed to the down-facing side of the root or shoot. The mechanism by which gravity influences the redistribution of efflux carriers is currently unknown, but it may involve a reorganization of actin filaments that connect amyloplasts to the cell's cytoskeleton.

Light Light streaming in from one direction causes a stem to curve toward its source. This response, **phototropism**, orients certain parts of the plant in the direction that maximizes the amount of light intercepted by its photosynthetic cells. Phototropism in shoots occurs in response to blue light absorbed by

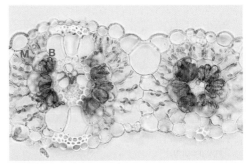 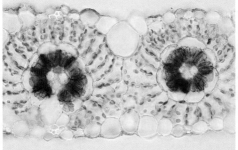 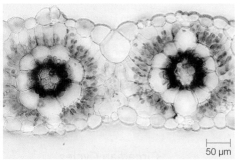

A Before light exposure.

B After 2 hours of light exposure.

C After 16 hours of light exposure.

50 µm

Figure 30.15 Movement of chloroplasts in response to light, as seen in leaf sections of millet (*Eleusine coracana*), a C4 plant. Mesophyll (M), bundle sheath cells (B), and vascular bundles (V).

nonphotosynthetic pigments called phototropins, which also function in stomatal opening in response to sunlight (Section 28.5). In a shoot tip or coleoptile, light-energized phototropins repress transcription of a gene whose product causes auxin efflux carriers to be distributed evenly around the cell membrane. Thus, a cell exposed to a directional light source ends up with efflux carriers positioned mainly on its shady side. The polarization directs auxin flow toward the shaded side of the structure, causing cells on that side to elongate

 more than cells on the illuminated side (**Figure 30.14**). The difference causes the entire structure to bend toward the light as it lengthens (*inset*).

On the interior of a cell, a different mechanism causes chloroplasts to move in a tropic response to light (**Figure 30.15**). Chloroplasts are dragged from one position to another on actin filament tracks of the cytoskeleton. They move away from high-intensity light, an adaptation that minimizes damage from excess electrons accumulating in electron transfer chains of the light reactions (Section 6.5). Chloroplasts move toward low-intensity light, maximizing their exposure to the light for photosynthesis. Blue light drives the movement in both cases.

Leaves or flowers of some plants change position in response to the changing angle of the sun throughout the day, a phototropic response called **heliotropism** (from Greek *helios*, sun). The mechanism that drives heliotropism is not understood, but it may be similar to phototropism because it has been observed to coincide with differential elongation of cells in stems, and it also occurs in response to blue light.

Contact A plant's contact with an object may cause a  change in the direction of its growth, a response called **thigmotropism** (*thigma* means touch in Greek). We see thigmotropism when a vine's tendril touches a wire, for example (*inset*). The mechanism that gives rise to the response is not well understood, but an immediate increase in cytoplasmic concentration of calcium that accompanies contact is likely to play a role in these responses. Plants (and animals) have plasma membrane transport proteins that flood cytoplasm with calcium ions upon mechanical perturbation of the membrane. Several gene products that can sense calcium ions are involved, as well as jasmonic acid and nitric oxide.

Twining tendrils result from unequal growth rates of cells on opposite sides of a shoot. A similar mechanism causes roots to grow away from contact, so they "feel" their way around rocks and other impassable objects in the soil. Mechanical stress, such as by wind exposure, inhibits lengthening of stems and increases their girth.

gravitropism Plant growth in a direction influenced by gravity.
heliotropism Plant parts change position in response to the sun's changing angle through the day.
phototropism Plant growth in a direction influenced by light.
statolith Organelle involved in sensing gravity.
thigmotropism Plant growth in a direction influenced by contact with a solid object.
tropism Adjustment in growth of a plant or its parts in response to an environmental stimulus.

Take-Home Message

How do plants respond to environmental cues?

» Via hormones, plants adjust the direction and rate of growth in response to gravity, light, contact, mechanical stress, and other environmental stimuli.

30.9 Sensing Recurring Environmental Changes

- Shifts in biological activity that recur in 24-hour cycles are mediated by cyclic shifts in gene expression.
- Seasonal shifts in night length trigger seasonal shifts in development in many plants.
- Links to *Cis* and *trans* 3.5, Light energy 6.2, Photosynthesis 6.5 and 6.7, FAD 7.2, Gene expression 9.2 and 10.2, Histone methylation 10.2, Master genes in flowering 10.4

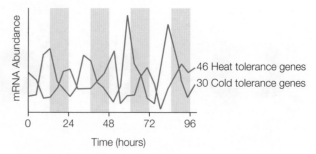

Figure 30.17 Circadian cycles of gene expression for some of the *Arabidopsis* genes involved in temperature stress responses. Plants were kept in constant light; *gray* bars show nighttime hours.

Averaged expression of cold tolerance genes (*blue*) peaks in late afternoon, and that of warm tolerance genes (*red*) peaks in morning. **Figure It Out:** At what time of day is expression of cold tolerance genes lowest?　　　　Answer: Early morning

Circadian Cycles

A **circadian rhythm** is a cycle of biological activity that starts anew every twenty-four hours or so. Circadian means "about a day." For example, a bean plant holds its leaves horizontally during the day but folds them close to its stem at night. A plant exposed to constant light or constant darkness for a few days will continue to move its leaves in and out of the "sleep" position at the time of sunrise and sunset (**Figure 30.16**). Similar mechanisms cause flowers of some plants to open only at certain times of day. For example, the flowers of many bat-pollinated plants unfurl, secrete nectar, and release fragrance only at night. Periodically closing flowers protects the delicate reproductive parts when the likelihood of pollination is typically lowest.

Circadian rhythms are driven by interconnected feedback loops involving transcription factors that directly or indirectly regulate their own expression. These loops give rise to cyclic and predictable oscillations in the levels of more than 30 percent of a plant cell's mRNAs. Transcriptional control in circadian loops often arises by the methylation and demethylation of histones. One recently discovered protein that methylates histones functions in circadian rhythms in both humans and plants.

Cyclic shifts in gene expression underlie cyclic shifts in metabolism, for example between daytime starch-building reactions and nighttime starch-consuming reactions. Components of rubisco, photosystem II, ATP synthase, and other proteins used in photosynthesis (as well as transcription factors that promote their expression) are among many gene products produced during the day and degraded at night.

By contrast, molecules that are needed at night are typically broken down during the day (**Figure 30.17**).

Six types of photoreceptors are known to provide light input into these internal circadian "clocks," but phytochromes and cryptochromes are the best characterized. **Phytochromes** are blue-green pigments that absorb red light at 660 nanometers, upon which their structure changes from a *cis* (inactive) form to a *trans* (active) form. The *trans* form absorbs far-red light at 730 nanometers (which predominates in shade), upon which the pigment changes back to its inactive form. Cryptochromes absorb blue-green and UV light. A nucleotide cofactor (FAD) at the center of a cryptochrome absorbs photons, then gains a hydrogen ion and an electron (becoming FADH). At this point, the cryptochrome is activated and, interestingly, magnetic.

Seasonal Changes

Except at the equator, the length of daylight varies with the season. Days are longer in summer than in winter, and the difference increases with latitude. Plants respond to these seasonal changes in light availability with seasonally appropriate behaviors such as entering or breaking dormancy. **Photoperiodism** refers

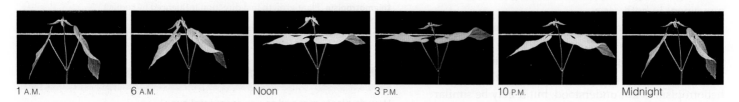

| 1 A.M. | 6 A.M. | Noon | 3 P.M. | 10 P.M. | Midnight |

Figure 30.16 Animated Rhythmic leaf movements by a young bean plant (*Phaseolus*). Physiologist Frank Salisbury kept this plant in darkness for twenty-four hours. Despite the lack of light cues, the leaves kept on folding and unfolding at sunrise (6 A.M.) and sunset (6 P.M.).

to an organism's response to changes in the length of day relative to night.

Flowering is a photoperiodic response in many plants. Such plants are termed long-day or short-day depending on which season they flower, even though it was later discovered that the main trigger for flowering is the length of night, not the length of day (**Figure 30.18**). Long-day plants, such as irises, oats, and clover, flower only when the hours of darkness fall below a critical value, typically in summer. Chrysanthemums, strawberries, and other short-day plants flower only when the hours of darkness are greater than some critical value. Sunflowers, tomatoes, roses, and other day-neutral plants flower independently of photoperiod.

The timing of flowering in photoperiodic plants is governed by circadian cycles of regulatory gene expression, and by phytochrome and cryptochrome light detection. Inputs from both converge on a gene called *CO*, which encodes a transcription factor. All plants have a version of this gene. In long-day plants, the transcription factor induces expression of the *FT* gene (*flowering locus T*) in companion cells. This gene's protein product travels in sieve tubes from leaves to shoot apical meristem, where it activates expression of floral identity genes (Section 10.4).

Long-day plants flower in long-day seasons because the expression of their *CO* gene and the activity of its product peak 8 to 10 hours after dawn—late afternoon during long-day seasons, but after dusk in short-day seasons. After nightfall, the CO protein is broken down. Thus, during short-day seasons, the protein never accumulates to a high enough level to promote flowering in long-day plants.

Cabbage, spinach, and other edible long-day plants grown for their leaves are said to "bolt" when they switch from producing leaves (desirable) to producing flowers (undesirable). Farmers minimize bolting in these plants by growing them in short-day seasons.

Short-day plants have the same *CO* gene, but its product inhibits *FT* gene expression in these plants. Gibberellin stimulates flowering in a separate pathway.

Vernalization

Length of night is not the only cue for flowering. Some plants flower only after exposure to prolonged cold winter temperature, a process called **vernalization** (from Latin *vernalis*, "to make springlike").

Researchers suspect that plants perceive temperature via their plasma membrane, which varies in lipid composition and calcium ion permeability depend-

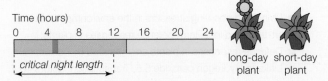

A A flash of red light interrupting a long night causes plants to respond as if the night were short. Long-day plants flower; short-day plants do not.

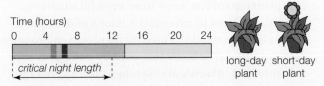

B A flash of far-red light cancels the effect of a red light flash. Short-day plants flower; long-day plants do not.

Figure 30.18 Animated Experiments showing that long- or short-day plants flower in response to night length. Each horizontal bar represents 24 hours. *Blue* bars indicate night length; *yellow* bars, day length. **Figure It Out:** Which type of photoreceptor detected the light flashes in these experiments?

Answer: A phytochrome

ing on temperature. Regardless of the mechanism of detection, a "cold" signal influences gene expression. For example, in plants that undergo vernalization, the *FT* gene is usually silenced by a repressor. Production of the repressor ceases only after the plant is exposed to a long period of cold temperature. In addition, the expression of a gene called *VRN1* increases quantitatively with the duration of cold exposure. The *VRN1* gene encodes a transcription factor that promotes rapid flowering when warm temperatures return and days start lengthening in spring. In plants that do not undergo vernalization, the *VRN1* gene is expressed regardless of exposure to cold.

circadian rhythm A biological activity that is repeated about every 24 hours.
photoperiodism Biological response to seasonal changes in the relative lengths of day and night.
phytochrome A light-sensitive pigment that helps set plant circadian rhythms based on length of night.
vernalization Stimulation of flowering in spring by prolonged exposure to low temperature in winter.

Take-Home Message

How do plants sense and respond to recurring environmental change?

» Plants respond to recurring cues from the environment with recurring cycles of activity such as rhythmic leaf movements.

» Photoreceptors that detect daylight provide input into circadian cycles.

» The main environmental cue for flowering is the length of night relative to the length of day, which varies by the season in most places.

» In some species, prolonged exposure to low temperature stimulates flowering in spring.

30.10 Responses to Stress

- Living and nonliving stressors in the environment provoke short-term and long-term defense responses in plants.
- Defense responses in plants are mediated by hormones.
- Links to Hydrogen peroxide 5.6, Transcription factors 10.2, Sac fungi 23.5, Plant symbionts 23.7 and 28.3, Stomata function 28.5

A plant cannot run away from stressful situations: It either adapts to adverse conditions or dies. Plant stressors are abiotic (caused by nonliving environmental conditions) or biotic (imposed by pathogens and herbivores). Abscisic acid synthesis is triggered by temperature extremes, lack of water, and other abiotic stressors. For example, ABA is part of a response that causes a plant's stomata to close when water is scarce, thus preventing water loss by transpiration. Stomata close after transport proteins move calcium ions into guard cells (Section 28.5). A nitric oxide burst produced in response to ABA activates these proteins.

Nitric oxide also participates in biotic stress responses. For example, cell surface receptors recognize molecules specific to microbial pathogens, including flagellin, a protein component of bacterial flagella. Flagellin binding to one of these receptors triggers a burst of ethylene synthesis. When ethylene is not present, ethylene receptors block transcription of a set of genes involved in defense. When ethylene is present, it binds to its receptors and locks them in an inactive form that marks them for destruction. Thus, ethylene synthesis lifts transcriptional repression of the defense genes. The cell starts producing more receptors that detect bacterial flagella—a positive feedback loop that sensitizes the plant to the presence of more bacteria.

Figure 30.19 Evidence of a hypersensitive response in a leaf. Brown spots are dead tissue where germinating fungi penetrated the leaf's epidermis, triggering a release of hydrogen peroxide and nitric oxide that caused plant cell suicide.

Any additional bacteria that bind to the accumulated flagellin receptors trigger an ABA-mediated nitric oxide burst that immediately closes stomata. Closing stomata defends the plant against bacterial invasion because bacteria cannot penetrate plant epidermis: They can enter plant tissues only through wounds or open stomata.

Beneficial bacteria and fungi avoid triggering a plant's defense responses by engaging in a complex cross-talk during establishment of a symbiotic relationship. For example, *Rhizobium* bacteria in soil are attracted to flavonoids released from root cells. The flavonoids induce expression of bacterial genes involved in forming nitrogen-fixing nodules on the roots—one of which is a substance secreted into the soil that is recognized by the plant as a signal to form a nodule.

A Saliva of a tobacco budworm (*Heliothis virescens*) chewing on a leaf of a tobacco plant (*Nicotiana*) triggers the plant to emit a combination of 11 volatile secondary metabolites.

B Red-tailed wasps (*Cardiochiles nigriceps*) are attracted to the unique chemical signature emitted by the plant. They follow the trail of aromatic chemicals back to the source.

C A wasp that finds a budworm attacks it and deposits an egg inside of it. When the egg hatches, a larva emerges and begins to eat the budworm, which eventually dies.

Figure 30.20 Interspecific plant defenses. Plant cells release a particular combination of volatile secondary metabolites in response to being chewed by a particular insect species. The chemical signature attracts wasps that parasitize the insect.

Prescription: Chocolate (revisited)

Epicatechin has now been shown to have a beneficial effect on a variety of illnesses, including hypertension and other cardiovascular disorders, blood clotting, endothelial cell function, Parkinson's disease, obesity, and even acne. Clinical trials involving chocolate consumption as an intervention in these disorders are now under way. Epicatechin probably works its magic in humans because it activates an enzyme, nitric oxide synthase, that produces nitric oxide in blood vessel walls. Nitric oxide relaxes muscles in the vessel walls, causing the vessels to widen. In mammals, nitric oxide is part of hormonal systems that regulate blood vessel width, and it is necessary for proper function of endothelial cells that make up the vessels. Dysfunction of endothelial cells in blood vessel walls is linked to cardiovascular disease, a primary cause of illness and death in the United States and in many other countries. Thus, increasing nitric oxide production in blood vessel cells is a major area of medical research.

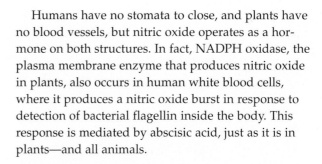

Humans have no stomata to close, and plants have no blood vessels, but nitric oxide operates as a hormone on both structures. In fact, NADPH oxidase, the plasma membrane enzyme that produces nitric oxide in plants, also occurs in human white blood cells, where it produces a nitric oxide burst in response to detection of bacterial flagellin inside the body. This response is mediated by abscisic acid, just as it is in plants—and all animals.

How would you vote? The secondary metabolites of many plants affect the human body, but few of these chemicals are as clearly beneficial—or as extensively tested—as epicatechin. Teas, herbal remedies, and other plant-based dietary supplements are classified by the U.S. government as foods, so they can be (and usually are) sold without being tested either for safety or efficacy. Recent legislation gave the FDA authority to remove harmful supplements from grocery shelves, but the determination of "harmful" in most cases has occurred after some number of consumers have been sickened or killed. Would you take an herbal remedy?

A pathogen that penetrates a plant's epidermis triggers a large surge of hydrogen peroxide and nitric oxide that causes cells in the infected region to commit suicide. This "hypersensitive" response can prevent a pathogen from spreading to other parts of the plant, because it often kills the pathogen along with the infected tissue (**Figure 30.19**).

However, a hypersensitive response is ineffective against pathogens that gain nutrients by killing cells outright. Some sac fungi, for example, use toxins to kill their plant hosts, then absorb nutrients released from their decomposing tissues. Plants have a whole-body, long-term defense response, **systemic acquired resistance**, that increases resistance to attack by a wide range of pathogens as well as tolerance to abiotic stresses. Systemic acquired resistance begins when an infected tissue releases an as yet unknown signal that travels to other parts of the plant, where it triggers

salicylic acid

cells to produce salicylic acid, a molecule similar to aspirin. This hormone increases transcription of hundreds of genes that produce chemicals involved in pathogen resistance, including the flavonoids you learned about in Section 30.1. For example, epicatechin prevents hardening of "pegs" that develop on germinating spores of many fungi. Fungi use these pegs to punch through tough epidermal tissue. Flavonoids related to epicatechin inhibit fatty acid synthesis in bacteria. Resveratrol in

grapes and other plants has antifungal activity. Other gene products include lignin, which increases the strength of plant cell walls; compounds that break down structural components of fungal cell walls; antioxidants; and so on: The chemicals differ by species, but all confer general hardiness to the plant.

jasmonic acid

Wounding of a leaf, such as occurs when an insect chews on it, triggers the production of ABA, hydrogen peroxide, ethylene, and another plant hormone called jasmonic acid. Jasmonic acid in turn increases transcription of genes whose products include volatile chemicals that the plant releases into the air. These secondary metabolites are detected by wasps that parasitize insect herbivores (**Figure 30.20**). Volatile chemicals released in response to herbivory are also detected by neighboring plants, which respond by increasing their own production of ethylene and jasmonic acid.

systemic acquired resistance In plants, inducible whole-body resistance to a wide range of pathogens and abiotic stressors.

Take-Home Message

How do plants respond to stress?

» Abscisic acid is involved in responses to nonliving environmental stresses.

» Detection of plant pathogens can trigger stomatal closure or cell death.

» Systemic acquired resistance triggered by pathogen attacks increases a plant's ability to withstand biotic and abiotic stresses.

VI HOW ANIMALS WORK

How many and what kinds of body parts does it take to function as a lizard in a tropical forest? Make a list of what comes to mind as you start reading Unit VI, then see how resplendent the list can become at the unit's end.

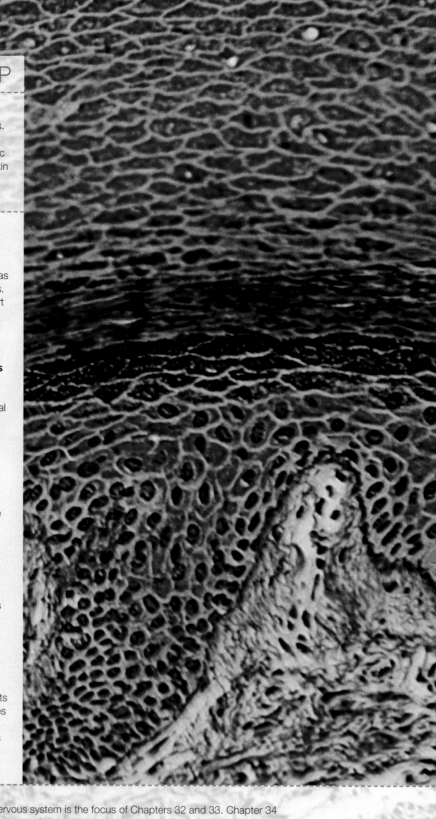

LEARNING ROADMAP

Where you have been With this chapter, we begin our survey of the tissues and organ systems (Section 1.2) in animals. The chapter expands on the nature of animal body plans (24.2) and trends in vertebrate evolution (25.2). We also revisit the topic of cancer (8.6, 11.6) and look again at the evolution of human skin color (14.1).

Where you are now

Animal Organization
Most animals have cells organized as tissues, organs, and organ systems. The components function in concert to maintain conditions in the body's internal environment.

Epithelial and Connective Tissues
Epithelial tissue covers the body's surface and lines its internal tubes. Connective tissue underlies epithelial tissue and supports and connects body parts.

Muscle and Nervous Tissue
Muscle tissue consists of cells that contract in response to signals from nervous tissue. Nervous tissue receives and integrates information from inside and outside the body.

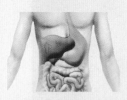

Organ Systems
Vertebrates have a coelom, and many organs reside in body cavities derived from it. Interactions among organ systems sustain life.

Example of an Organ System
Skin is an organ system that protects the body, conserves water, produces vitamin D, and helps maintain body temperature. Temperature control is an example of negative feedback.

Where you are going The nervous system is the focus of Chapters 32 and 33. Chapter 34 describes the function of endocrine glands. Chapter 35 explains how muscles contract and interact with the skeleton. Chapters 36 and 37 consider the transport and immune functions of blood. Chapters 38, 39, and 40 describe how you take in essential substances and eliminate metabolic wastes. Finally, Chapters 41 and 42 describe organs involved in reproduction and how a body develops.

31.1 Stem Cells—It's All About Potential

All of the cells of your body "stem" from stem cells (Figure 31.1). **Stem cells** are self-renewing cells that can either divide and produce more stem cells ❶, or differentiate into specialized cells that characterize specific body parts ❷.

Stem cells vary in their potential to form a new individual or new tissues. A fertilized human egg and the cells produced by its first four divisions are "totipotent," meaning they can develop into a new individual if placed in a womb. Later divisions produce embryonic cells that are "pluripotent." A pluripotent cell does not have the ability to develop into a new individual, but can give rise to into any of the cell types in a body.

Most stem cells in an adult are "unipotent," meaning they yield only one specific type of cells. Some adult stem cells produce new skin cells; others produce new blood cells. However, adults have few stem cells that can make new muscle cells or nerve cells. Thus, heart muscle lost to a heart attack, leg muscles destroyed by muscular dystrophy, or nerves severed in an injured spinal cord are not replaced.

Embryonic stem cells hold great potential as a treatment to repair tissues that are normally not regenerated in the adult body. In the United States, the first clinical trials of such treatments began in 2010. These trials involve stem cells initially harvested from human embryos, and then grown in the laboratory. One trial is testing whether the stem cells can repair nervous tissue in the spinal cord of people with recent injuries. In another, stem cells are being used to regenerate eye tissue in age-related macular degeneration, a common condition that causes blindness. One day, embryonic stem cell treatments might also help people with other nerve and muscle disorders such as heart disease, muscular dystrophy, multiple sclerosis, and Parkinson's disease.

The use of embryonic stem cells remains controversial despite their potential as a universal toolkit for repairing damaged tissues. Many people oppose the use of human embryos for any purpose. Unipotent adult stem cells may offer an alternative to embryonic cells if researchers can find a way to make them dedifferentiate—to turn back their developmental clock so the cells again become pluripotent. Harvesting and culturing the few pluripotent stem cells in adults also offers promise.

Researchers are also investigating methods of controlling cell differentiation. To be of use in clinical treatments, pluripotent stem cells from any source must be induced to differentiate into a desired cell type. By analogy, pluripotent cells are like college freshmen who need to be directed toward a specific major. Investigations into mechanisms of directing pluripotent cell differentiation have the additional benefit of informing us about how normal development gives rise to the many specialized cells that constitute the human body.

stem cell Cell capable of replication or of differentiation into some or all cell types.

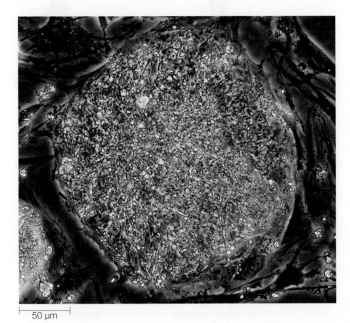

50 μm

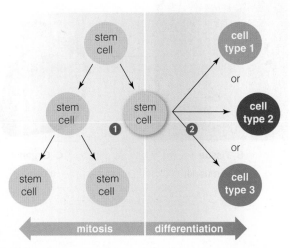

Figure 31.1 Stem cells. Each stem cell can divide to form new stem cells or differentiate to form specialized cell types.

The photo at the *left* shows a colony of human embryonic stem cells growing in a laboratory at the University of Pittsburgh.

31.2 Organization of Animal Bodies

■ Most animal bodies have cells organized as tissues, organs, and organ systems.
■ Physical constraints and evolutionary history influence the structure and function of body parts.
■ Links to Life's levels of organization 1.2, Homeostasis 1.3, Cell junctions 4.11, Diffusion 5.8, Adapting to life on land 25.6

Levels of Organization

In all animals, development produces a body with several to many types of cells (**Figure 31.2A**). An adult human has about 200 different kinds of cells. In most animals, cells of different types are organized in tissues often anchored by extracellular matrix (**Figure 31.2B**). Cell junctions of the types described in Section 4.11 typically connect the cells of a tissue. They hold cells in place and allow them to cooperate in a specific task or tasks.

Four types of tissue occur in all vertebrate bodies:

1. Epithelial tissue covers body surfaces and lines the internal cavities such as the gut.
2. Connective tissue holds body parts together and provides structural support.
3. Muscle tissue moves the body or its parts.
4. Nervous tissue detects stimuli and relays signals.

Different types of cells characterize different tissues. For example, muscle tissue includes contractile cells not found in nervous tissue or epithelial tissue. Typically, animal tissues are organized into organs.

An organ is a structural unit of two or more tissues organized in a specific way and capable of carrying out specific tasks. For example, a human heart is an organ that includes all four tissue types (**Figure 31.2C**). The heart's wall is made up mostly of cardiac muscle tissue. A sheath of connective tissue covers the muscle, and internal chambers are lined with epithelial tissue. The heart receives signals via nervous tissue.

In organ systems, two or more organs and other components interact physically, chemically, or both in a common task. For example, in the vertebrate circulatory system, the force generated by a beating heart (an organ) moves blood (a tissue) through blood vessels (organs), thereby transporting gases and solutes to and from all body cells (**Figure 31.2D**). Multiple organ systems sustain the organism (**Figure 31.2E**).

The Internal Environment

By weight, an animal body is mainly fluid: a water-based solution of salts, proteins, and other solutes. The bulk of this body fluid is intracellular, which means it is inside cells. The remainder is extracellular. **Extracellular fluid** is the environment in which body cells live. It bathes cells and provides them with the substances they require to stay alive. It also functions as a dumping ground for cellular waste. In vertebrates, extracellular fluid consists mainly of **interstitial fluid** (the fluid in spaces between cells) and plasma, the fluid portion of the blood (**Figure 31.3**).

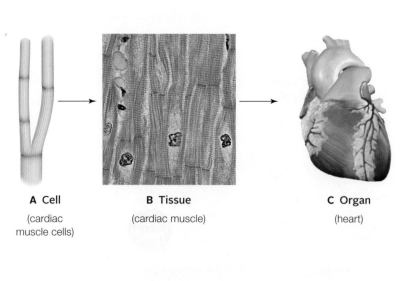
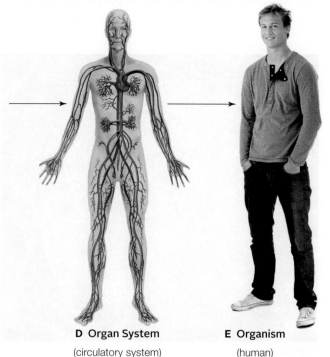

A Cell
(cardiac muscle cells)

B Tissue
(cardiac muscle)

C Organ
(heart)

D Organ System
(circulatory system)

E Organism
(human)

Figure 31.2 Levels of organization in a vertebrate (human) body.

Cells survive only if solute concentrations and temperature of the fluid surrounding them remain within a narrow range. Maintaining conditions of the cell's environment within this range is an important aspect of homeostasis (Section 1.3).

Evolution of Animal Structure

An animal's structural traits (its anatomy) evolve in concert with its functional traits (its physiology). Both types of traits are genetically determined and vary among individuals. In each generation, genes for those traits that best help individuals survive and reproduce in their environment are preferentially passed on. Over many generations, anatomical and structural traits become optimized in ways that reflect their function in a specific environment.

Physical constraints affect evolution of body structure. For example, dissolved substances travel through extracellular fluid by diffusion. Diffusion alone could not sustain a large or thick body because gases, nutrients, and wastes would not move quickly enough through the body to keep up with cellular metabolism. Thus, mechanisms that speed the distribution of materials evolved along with increases in body size. In vertebrates, a circulatory system serves this purpose. The system includes a network of extensively branched blood vessels that extends through the body. Every living cell is close enough to a blood vessel to exchange substances with it by diffusion (**Figure 31.4**).

As another example, vertebrates faced new physical challenges as they left their aquatic habitat for the land (Section 25.6). Gases can only enter or leave an animal's body by diffusing across a moist surface. In an aquatic organism, the surrounding water both delivers oxygen and moistens the respiratory surface. By contrast, a land animal must extract oxygen from air, which can dry a respiratory surface. Evolution of lungs allowed vertebrates to maintain a moist respira-

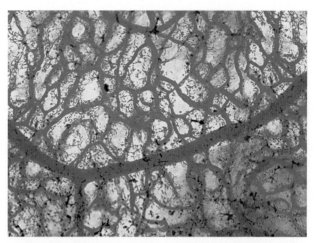

Figure 31.4 Branching blood vessels. The vessels deliver oxygen to within close proximity of all cells in a human body.

tory surface inside their body. Cells inside the lung secrete the fluid that keeps this surface moist.

Lungs are not modified fish gills. Rather, lungs evolved from outpouchings of the gut in fishes ancestral to land vertebrates. As this example illustrates, evolution by natural selection often modifies existing tissues or organs. There is evidence of evolutionary compromise in the anatomy and physiology of many animals. For example, as a legacy of the lungs' ancestral connection to the gut, the human throat connects to both the digestive tract and respiratory tract. As a result of this dual connection, food sometimes goes where air should, and a person chokes. It would be safer if food and air entered the body through separate passageways. However, because evolution modifies existing structures, it often does not produce the most optimal body plan.

extracellular fluid Of a multicelled organism, body fluid that is not inside cells; serves as the body's internal environment.
interstitial fluid Of a multicelled organism, body fluid in spaces between cells.

Take-Home Message

How are animal bodies organized?

» In most animals, cells are organized as tissues. Each tissue consists of cells of a specific type that cooperate in carrying out a particular task. Tissues are organized into organs, which in turn are components of organ systems.

» The animal body consists largely of fluid. The bulk of this fluid is in cells. The fluid outside cells (extracellular fluid) is the body's internal environment. Maintaining the solute concentration and temperature of this fluid is an important facet of homeostasis.

» Many anatomical traits evolved as solutions to physical challenges. However, these solutions are sometimes imperfect because evolution modifies existing structures, rather than building a body plan from the ground up.

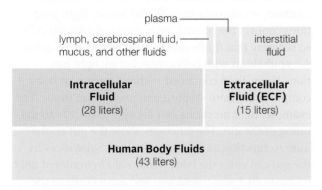

Figure 31.3 Distribution of fluids in a human body.

31.3 Epithelial Tissue

■ Epithelial tissue covers the body's external surfaces and lines internal tubes and cavities.

■ Links to Fibrous proteins 3.6, Cilia 4.10, Cell junctions 4.11, DNA replication errors 8.6

General Characteristics

Epithelial tissue, or an epithelium (plural, epithelia), is a sheetlike layer of cells with little extracellular matrix between them. One surface of the epithelium, referred to as its apical surface, faces the outside world or the interior of a body cavity or tube. The opposite surface, referred to as the basal surface, secretes a noncellular **basement membrane** that attaches the epithelium to an underlying tissue. Blood vessels do not run through an epithelium, so nutrients reach cells by diffusing from vessels in an adjacent tissue.

Most of what you see when you look in a mirror—your skin, hair, and nails—is epithelial tissue or structures derived from it. Hair, fur, nails, hooves, beaks, and feathers all form when specialized epithelial cells produce large amounts of the protein keratin. The visible part of a hoof, hair, or feather consists of the remains of such cells.

Variations in Structure and Function

Epithelial cells may be arranged as a single layer or multiple layers. A simple epithelium is one cell thick, whereas a stratified epithelium includes multiple layers of cells.

Cells of an epithelium are typically described by their shape. Cells in squamous epithelium are flattened or scalelike. (*Squama* is the Latin word for scale.) Cells of cuboidal epithelium are short cylinders that look like cubes when viewed in cross-section. Cells in columnar epithelium are taller than they are wide. **Figure 31.5** shows the three types of simple epithelium and describes their functions.

Simple squamous epithelium facilitates the exchange of materials. It is the thinnest type of epithelium, and gases and nutrients diffuse across it easily. This type of epithelium lines blood vessels and the inner surface of the lungs. By contrast, stratified squamous epithelium has a protective function. It makes up the outermost layer of human skin.

Cells of cuboidal and columnar epithelium function in movement, absorption, or secretion of substances. Those that move substances along the surface of an epithelium have cilia at their apical surface. For example, ciliated epithelial cells in the oviducts propel an egg from an ovary toward the uterus (the womb).

In some epithelia, cells have fingerlike extensions called **microvilli** at their free surface. Microvilli are typically shorter than cilia, do not move, and have an internal framework of actin filaments rather than microtubules. Microvilli increase the surface area across which substances can be detected by, absorbed into, or secreted from a cell.

Three types of intercellular junctions connect cells in animal tissues (Section 4.11). One type, the tight junction, occurs only in epithelial tissue. Tight junctions connect the plasma membranes of adjacent cells so securely that fluids cannot seep between the cells. An epithelium with cells connected by tight junctions keeps fluid contained within a particular body compartment from seeping into underlying tissue. For example, tight junctions join the epithelial cells in the lining of the gut. The junctions allow the gut epithelium to function as a selective barrier. Substances in the gut can enter the body's internal environment only by controlled movement into and across cells of the gut epithelium.

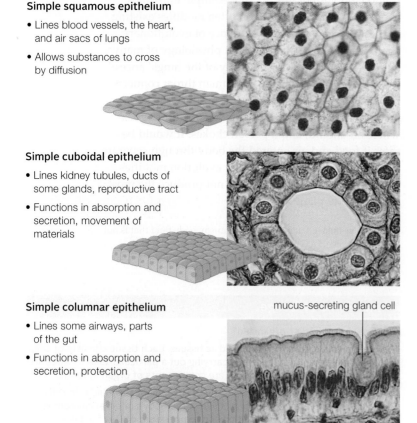

Simple squamous epithelium

• Lines blood vessels, the heart, and air sacs of lungs

• Allows substances to cross by diffusion

Simple cuboidal epithelium

• Lines kidney tubules, ducts of some glands, reproductive tract

• Functions in absorption and secretion, movement of materials

Simple columnar epithelium

• Lines some airways, parts of the gut

• Functions in absorption and secretion, protection

mucus-secreting gland cell

Figure 31.5 Micrographs and drawings of three types of simple epithelia, with examples of their functions and locations.

Figure 31.6 Examples of an exocrine and an endocrine gland.

Figure It Out: Which type of gland is ductless?

Answer: Endocrine glands are ductless and secrete hormones into the blood.

Exocrine gland

parotid gland (secretes saliva)

parotid duct (delivers saliva to mouth)

Endocrine gland

cell that secretes hormone

capillary

thyroid gland (secretes hormones into blood)

Epithelial tissues subject to mechanical stress such as skin epithelium have many adhering junctions. These junctions function like buttons that hold a shirt closed. They connect the plasma membranes of cells at distinct points but do not form a seal between them.

Epithelial Cell Secretions

Specialized epithelial cells called **gland cells** secrete substances that function outside the cell. In most animals, some gland cells cluster as multicelled glands that release substances onto the skin, or into a body cavity or fluid. There are two main types of glands (**Figure 31.6**). **Exocrine glands** have ducts or tubes that deliver their secretions onto an internal or external surface. Exocrine secretions include mucus, saliva, tears, digestive enzymes, earwax, and breast milk. **Endocrine glands** do not have ducts. They release signaling molecules called hormones into a body fluid. Most commonly, hormones enter small blood vessels (capillaries). We discuss the function of endocrine glands in detail in Chapter 34.

basement membrane Secreted material that attaches epithelium to an underlying tissue.
endocrine gland Ductless gland that secretes hormones into a body fluid.
epithelial tissue Sheetlike animal tissue that covers outer body surfaces and lines internal tubes and cavities.
exocrine gland Gland that secretes milk, sweat, saliva, or some other substance through a duct.
gland cell Secretory epithelial cell.
microvilli Thin projections from the plasma membrane of some epithelial cells; increase the cell's surface area.

Carcinomas—Epithelial Cell Cancers

Adult animals make few new muscle cells or nerve cells, but they constantly renew their epithelial cells. For example, each day you lose skin cells and grow new ones to replace them. An adult sheds about 0.7 kilogram (1.5 pounds) of skin each year. Similarly, the lining of your intestine is replaced every four to six days. All those cell divisions provide lots of opportunities for DNA replication errors that can lead to cancer. As a result, epithelium is the animal tissue most likely to become cancerous.

An epithelial cell cancer is called a carcinoma. About 95 percent of skin cancers are carcinomas. Breast cancers are usually carcinomas of epithelial cells that line the milk ducts or of the breast's glandular epithelium. Similarly, most lung cancers arise in cells of the lung's epithelial lining.

Take-Home Message

What are the functions of epithelial tissue?

» Epithelia are sheetlike tissues that line the body's surface and its cavities, ducts, and tubes. They function in protection, absorption, and secretion. Some epithelia have cilia or microvilli at their surface.

» Glands are secretory organs derived from epithelium. Exocrine glands secrete material through a duct onto a body surface or into a body cavity. Endocrine glands secrete hormones into the blood.

» Specialized epithelial cells that produce large amounts of the protein keratin are the source of hair, nails, hooves, and feathers.

» Epithelial tissues undergo continual turnover and are the most frequent site for cancers.

31.4 Connective Tissues

■ Connective tissues connect body parts and provide structural and functional support to other body tissues.
■ Links to Hemoglobin 3.2, Lipids 3.5, Extracellular matrix 4.11, Alternative energy sources in food 7.7

Connective tissues consist of cells in an abundant extracellular matrix. Soft connective tissues hold other tissues in place or connect them to one another. Cartilage, bone tissue, adipose tissue, and blood are specialized connective tissues. We provide a brief overview of the specialized connective tissues here, and describe their function in more detail in the chapters that follow.

Soft Connective Tissues

Fibroblasts, the most common cells in soft connective tissues, secrete a matrix of complex carbohydrates and long fibers of the proteins collagen and elastin. The most abundant soft connective tissue, loose connective tissue, has fibroblasts and fibers dispersed widely in its matrix (**Figure 31.7A**). Loose connective tissue keeps internal organs in place and underlies epithelia. It is densely supplied with blood vessels.

Dense, irregular connective tissue makes up deep skin layers, supports intestinal muscles, and forms capsules around organs that do not stretch, such as

kidneys. Its matrix has fibroblasts and collagen fibers oriented every which way, as in **Figure 31.7B**.

By contrast, dense, regular connective tissue has fibroblasts in orderly rows between parallel, tightly packed bundles of fibers (**Figure 31.7C**). This organization helps prevent tears when the tissue is subject to mechanical stress. Dense, regular connective tissue is the main tissue in tendons and ligaments. Tendons connect skeletal muscle to bones and do not stretch. Ligaments attach one bone to another and are elastic. Like a rubber band, they can be stretched out, then spring back to their original shape. Tendons and ligaments are not well supplied with blood. If they are torn, they are very slow to heal.

Specialized Connective Tissues

All vertebrate skeletons include **cartilage**, which is mainly a rubbery matrix composed of collagen fibers and glycoproteins. Cartilage cells (chondrocytes) secrete the material that surrounds them (**Figure 31.7D**). When you were an embryo, the first skeleton that formed consisted of cartilage. As development continued, bone replaced most of it. Cartilage still supports your nose, throat, and outer ears. It covers the ends of bones at joints and acts as a shock absorber between vertebrae. Blood vessels do not extend through

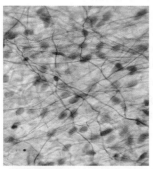

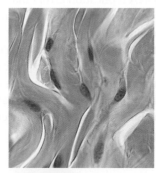

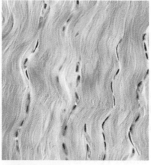

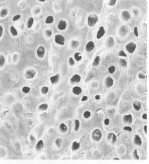

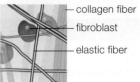
— collagen fiber
— fibroblast
— elastic fiber

— collagen fibers

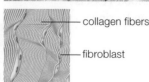
— collagen fibers
— fibroblast

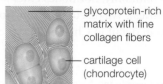
— glycoprotein-rich matrix with fine collagen fibers
— cartilage cell (chondrocyte)

A Loose connective tissue
• Underlies most epithelia
• Provides elastic support and serves as a fluid reservoir

B Dense, irregular connective tissue
• In deep skin layers, around intestine, and in kidney capsule
• Binds parts together, provides support and protection

C Dense, regular connective tissue
• In tendons connecting muscle to bone and ligaments that attach bone to bone
• Provides stretchable attachment between body parts

D Cartilage
• Internal framework of nose, ears, airways; covers the ends of bones
• Supports soft tissues, cushions bone ends at joints, provides a low-friction surface for joint movements

Figure 31.7 Animated Connective tissue structure and function.

cartilage, and little or no cell division occurs in adult cartilage, so torn cartilage does not heal. In addition, production of matrix declines with age.

Adipose tissue stores fats. It consists of cells (adipocytes) with little matrix between them, and it is richly supplied with blood vessels. Most adipose tissue in a human adult is white adipose tissue. Its cells bulge with so much stored fat that their nucleus is typically pushed to one side and flattened (Figure 31.7E). In addition to its role as the body's main energy reservoir, white adipose tissue acts as insulation and cushions body parts. A less abundant tissue, called brown adipose tissue, specializes in producing heat. In human adults, brown adipose tissue is concentrated in the neck and upper chest. Cells of brown adipose tissue store less fat than cells of white adipose tissue and have many specialized mitochondria. Compared to typical mitochondria, those of brown adipose tissue produce less ATP and release more energy as heat.

Bone tissue consists of living cells (osteocytes) in a matrix hardened by calcium and phosphorus (Figure 31.7F). Blood vessels run through channels in the tissue. Bone tissue is the main component of bones, which are organs that interact with skeletal muscles to move a body. Bones also support and protect internal organs. Blood cells form in the spongy interior of some bones.

Blood is considered a connective tissue because its cells and platelets descend from stem cells in bone (Figure 31.7G). Red blood cells filled with hemoglobin transport oxygen (Section 3.2). White blood cells defend the body against pathogens. Platelets are cell fragments that function in clot formation. Cells and platelets drift in plasma, a fluid extracellular matrix consisting mostly of water and dissolved proteins.

adipose tissue Connective tissue that specializes in fat storage.
blood Circulatory fluid; in vertebrates it is a fluid connective tissue consisting of plasma, red blood cells, white blood cells, and platelets.
bone tissue Connective tissue consisting of cells surrounded by a mineral-hardened matrix of their own secretions.
cartilage Connective tissue consisting of cells surrounded by a rubbery matrix of their own secretions.
connective tissue Animal tissue with an extensive extracellular matrix; structurally and functionally supports other tissues.

Take-Home Message

What are connective tissues?

» Various soft connective tissues underlie epithelia, form capsules around organs, and connect muscle to bones or bones to one another.

» A vertebrate skeleton consists of two connective tissues: rubbery cartilage and mineral-hardened bone. Blood is a connective tissue because blood cells form in bone. The cells are carried by plasma, the fluid portion of the blood.

» Adipose tissue is a specialized connective tissue that stores fat.

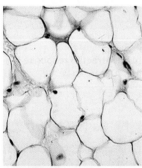

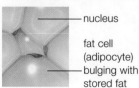

nucleus

fat cell (adipocyte) bulging with stored fat

E Adipose tissue

• Underlies skin and occurs around heart and kidneys

• Serves in energy storage, provides insulation, cushions and protects some body parts

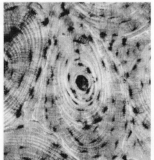

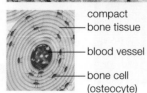

compact bone tissue

blood vessel

bone cell (osteocyte)

F Bone tissue

• Makes up the bulk of most vertebrate skeletons

• Provides rigid support, attachment site for muscles, protects internal organs, stores minerals, produces blood cells

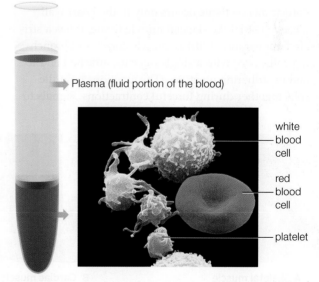

Plasma (fluid portion of the blood)

white blood cell

red blood cell

platelet

G Blood

• Flows through blood vessels, heart

• Distributes essential gases, nutrients to cells; removes wastes from them

31.5 Muscle Tissues

- Muscle moves bodies or propels materials through them.
- Links to Glycogen 3.4, Cell junctions 4.11

Cells of muscle tissues contract (shorten) in response to signals from nervous tissue. ATP provides the energy that fuels muscle contractions. Muscle tissue occurs in most animals, but we focus here on the kinds found in vertebrates. We discuss the mechanism of muscle contraction in detail in Section 35.7.

Skeletal Muscle Tissue

Skeletal muscle tissue interacts with bones to move body parts. It consists of parallel arrays of long, cylindrical cells called muscle fibers, which have a striated, or striped, appearance (**Figure 31.8A**). Muscle fibers are multinucleated and form by cell fusion during embryonic development. Skeletal muscle contracts reflexively, as when you pull your hand away after touching a hot object. More often, its contraction is deliberate, as when you reach for something. Thus, skeletal muscle is commonly described as "voluntary" muscle.

Along with the liver, skeletal muscle is a major site for glycogen storage. Metabolic activity in skeletal muscles is the major source of body heat.

Cardiac Muscle Tissue

Cardiac muscle tissue occurs only in the heart wall (**Figure 31.8B**). Like skeletal muscle tissue, it has a striated appearance. Cardiac muscle consists of branching cells, each with a single nucleus, attached end to end by adhering junctions. The junctions hold the cells together during forceful contractions. Signals to contract pass swiftly from cell to cell at gap junctions that connect the cells along their length. The rapid flow of signals ensures that all cells in cardiac muscle tissue contract as a unit.

Compared to other muscle tissues, cardiac muscle has far more mitochondria. They provide the continually beating heart with a dependable supply of ATP.

Cardiac muscle and smooth muscle tissue are said to be "involuntary" because people cannot deliberately make these tissues contract.

Smooth Muscle Tissue

Many tubular organs, such as the stomach, uterus, and bladder, have **smooth muscle tissue** in their wall. Smooth muscle cells are unbranched, with tapered ends and a single nucleus at their center (**Figure 31.8C**). Smooth muscle tissue is not striated. It contracts more slowly than skeletal muscle, but its contractions can be sustained longer. Contraction of smooth muscle propels material through the gut, reduces the diameter of blood vessels and airways, and closes sphincters (a sphincter is a ring of muscle in a tubular organ).

cardiac muscle tissue Muscle of the heart wall.
skeletal muscle tissue Muscle that pulls on bones and moves body parts; under voluntary control.
smooth muscle tissue Muscle that lines blood vessels and forms the wall of hollow organs.

Take-Home Message

What is muscle tissue?

» Muscle tissue consists of cells that contract in response to nervous signals. Contraction requires ATP.

A Skeletal muscle

- Long, multinucleated, cylindrical cells with conspicuous striping (striations)
- Pulls on bones to bring about movement, maintain posture
- Reflex activated, but also under voluntary control

B Cardiac muscle

- Striated, branching cells (each with a single nucleus) attached end to end
- Found only in the heart wall
- Contraction is not under voluntary control

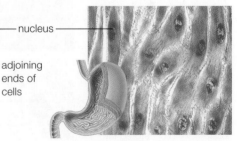

C Smooth muscle

- Cells with a single nucleus, tapered ends, and no striations
- Found in the walls of arteries, the digestive tract, the reproductive tract, the bladder, and other organs
- Contraction is not under voluntary control

Figure 31.8 Animated Three types of muscle tissue.

31.6 Nervous Tissue

■ Nervous tissue detects changes in the internal or external environment, integrates information, and controls the activity of muscle and glands.

Nervous tissue makes up the communication lines of a body. **Neurons** are the signaling cells in nervous tissue. Each neuron has a cell body, a region that contains the nucleus and other organelles. Long cytoplasmic extensions that project from the cell body allow the cell to receive and send electrochemical signals (Figure 31.9).

When a neuron receives sufficient stimulation, an electrical signal travels along its plasma membrane to the ends of specialized cytoplasmic extensions. The electrical signal causes release of chemical signaling molecules from these endings. These molecules diffuse across a small gap to an adjacent neuron, muscle fiber, or gland cell, and alter that cell's behavior.

Your nervous system has more than 100 billion neurons. There are three types. Sensory neurons are excited by specific stimuli, such as light or pressure. Interneurons receive and integrate sensory information. They store information and coordinate responses to stimuli. In vertebrates, interneurons occur mainly in the brain and spinal cord. Motor neurons relay commands from the brain and spinal cord to glands and muscle cells (Figure 31.10).

Neuroglial cells, also called neuroglia, keep neurons positioned where they should be, and provide them with nutrients. Neuroglial cells also wrap around the signal-sending cytoplasmic extensions of most motor neurons. They act as insulation and speed the rate at which signals travel.

nervous tissue Animal tissue composed of neurons and supporting cells; detects stimuli and controls responses to them.
neuroglial cell Cell that supports and assists neurons.
neuron One of the cells that make up communication lines of a nervous system; transmits electrical signals along its plasma membrane and communicates with other cells through chemical messages.

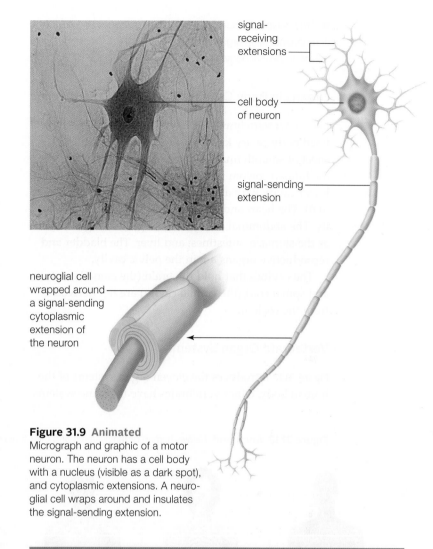

signal-receiving extensions

cell body of neuron

signal-sending extension

neuroglial cell wrapped around a signal-sending cytoplasmic extension of the neuron

Figure 31.9 Animated
Micrograph and graphic of a motor neuron. The neuron has a cell body with a nucleus (visible as a dark spot), and cytoplasmic extensions. A neuroglial cell wraps around and insulates the signal-sending extension.

Take-Home Message

What is nervous tissue?

» Nervous tissue consists of neurons and the cells that support them. Different kinds of neurons detect specific stimuli, integrate information, and issue or relay commands to other tissues.

» The supporting cells in nervous tissue are referred to as neuroglial cells, or neuroglia.

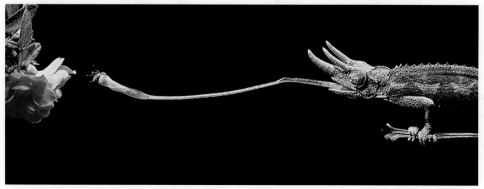

Figure 31.10 Example of a coordinated interaction between skeletal muscle tissue and nervous tissue.

Sensory neurons in the lizard's eyes relay information about the position of a fly to interneurons in the lizard's brain.

Signals from interneurons in the lizard's brain flow to motor neurons, which in turn send stimulatory signals to the muscle fibers of the lizard's long, coiled-up tongue. The tongue uncoils swiftly and precisely to reach the very spot where the fly is perched.

31.7 Organ Systems

■ Organs typically include all four types of tissues and are components of an organ system.

■ Link to Animal body plans 24.2

Organs in Body Cavities

Like other vertebrates, humans are bilateral and have a lined body cavity known as a coelom (Section 24.2). A sheet of smooth muscle called the diaphragm divides the human coelom into an upper thoracic cavity and a lower cavity with abdominal and pelvic regions (**Figure 31.11**). The heart and lungs reside in the thoracic cavity. The abdominal cavity holds digestive organs such as the stomach, intestines, and liver. The bladder and reproductive organs are in the pelvic cavity.

The cavities that hold the brain (the cranial cavity) and spinal cord (the cranial cavity) are not derived from the coelom.

Vertebrate Organ Systems

Figure 31.12 introduces the eleven organ systems of the human body. Other vertebrates have the same systems.

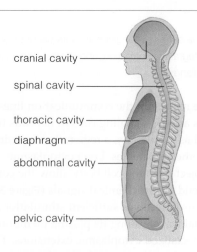

cranial cavity

spinal cavity

thoracic cavity

diaphragm

abdominal cavity

pelvic cavity

Figure 31.11 Animated Main body cavities that hold human organs. **Figure It Out:** Which organs lie in body cavities that are not part of the coelom?

Answer: The spinal cord and brain

Organ systems work cooperatively to carry out specific tasks. For example, organ systems interact to provide cells with essential raw materials and remove wastes (**Figure 31.13**). Food and water enter the body by way of the digestive system, which includes all

Figure 31.12 Animated *Below*, human organ systems and their functions.

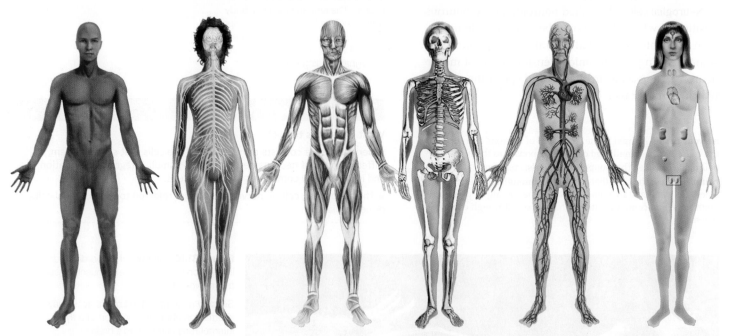

Integumentary System	**Nervous System**	**Muscular System**	**Skeletal System**	**Circulatory System**	**Endocrine System**
Protects body from injury, dehydration, and pathogens; controls its temperature; excretes certain wastes; receives some external stimuli.	Detects external and internal stimuli; controls and coordinates the responses to stimuli; integrates all organ system activities.	Moves body and its internal parts; maintains posture; generates heat by increases in metabolic activity.	Supports and protects body parts; provides muscle attachment sites; produces red blood cells; stores calcium, phosphorus.	Rapidly transports many materials to and from interstitial fluid and cells; helps stabilize internal pH and temperature.	Hormonally controls body functioning; with nervous system integrates short- and long-term activities. (Male testes added.)

components of the tubular gut such as the stomach and intestine, as well as organs that aid digestion such as the pancreas and gallbladder. The digestive system also eliminates undigested wastes.

The respiratory system, which includes lungs and airways that lead to them, takes in oxygen. The heart and blood vessels of the circulatory system deliver nutrients and oxygen to cells, and remove waste carbon dioxide and solutes from them. The circulatory system delivers carbon dioxide to the respiratory system for expulsion in exhalations. The circulatory system also moves excess water, salts, and soluble wastes to the urinary system. Organs of the urinary system include kidneys that filter wastes from the blood. Urine produced by the kidneys is stored in a bladder until it can be eliminated from the body.

Figure 31.13 does not show the nervous, endocrine, muscular, and skeletal systems, but these too help vertebrates obtain essential substances and eliminate wastes. For example, the nervous system detects changes in internal levels of water, solutes, and nutrients. Signals from the nervous and endocrine systems to the kidneys encourage conservation or elimination of water. They also stimulate the muscle contractions that allow you to eat or drink.

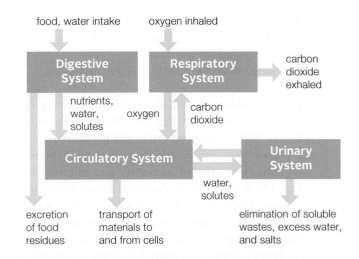

Figure 31.13 Some of the ways that organ systems interact to keep the body supplied with essential substances and eliminate unwanted wastes. Other organ systems that are not shown also take part in these tasks.

Take-Home Message

What are organs and organ systems?

» Organs consist of multiple tissues and are themselves components of organ systems. Cooperative action of organ systems sustains the body.

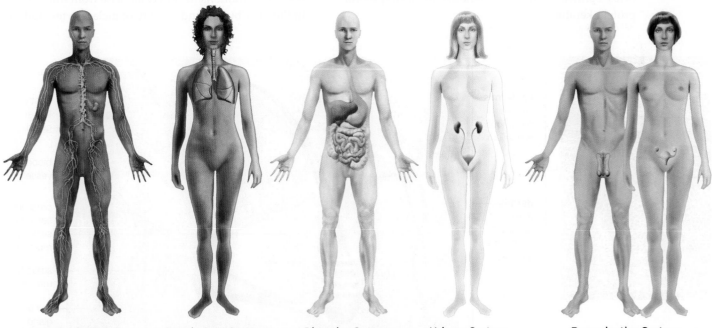

Lymphatic System

Collects and returns some tissue fluid to the bloodstream; defends the body against infection and tissue damage.

Respiratory System

Rapidly delivers oxygen to the tissue fluid that bathes all living cells; removes carbon dioxide wastes of cells; helps regulate pH.

Digestive System

Ingests food and water; mechanically, chemically breaks down food and absorbs small molecules into internal environment; eliminates food residues.

Urinary System

Maintains the volume and composition of internal environment; excretes excess fluid and bloodborne wastes.

Reproductive System

Female: Produces eggs; provides a protected, nutritive environment for the development of new individuals. Male: Produces and transfers sperm to the female. Hormones of both systems also influence other organ systems.

31.8 Human Integumentary System

- In vertebrates, the integumentary system consists of skin, structures derived from skin, and an underlying layer of connective and adipose tissue.
- Links to Human skin color 14.1, *Homo erectus* 26.5

Of all vertebrate organs, the outer body covering called skin has the largest surface area. Skin consists of two layers, a thin upper epidermis and the dermis beneath it (**Figure 31.14**). The dermis connects to the hypodermis, an underlying layer of connective and adipose tissue. The depth of the hypodermis varies among body regions. The hypodermis beneath the skin of eyelids is thin, with few adipose cells. By contrast, the hypodermis of the buttocks is thickened by many adipose cells.

Vertebrate skin has many functions. It contains sensory receptors that keep the brain informed of external conditions. It serves as a barrier to keep out pathogens and it helps control internal temperature. In land vertebrates, skin also helps conserve water. In humans, reactions that produce vitamin D occur in the skin.

Structure of Human Skin

Epidermis is a stratified squamous epithelium with an abundance of adhering junctions and no extracellular matrix. Human epidermis consists mainly of keratinocytes, epithelial cells that synthesize the waterproofing protein keratin.

Figure 31.15 Vitiligo. Lee Thomas, an African American television reporter, has vitiligo. The death of melanocytes has turned his hands white and produced white blotches on his face and arms.

Mitotic cell divisions in deep epidermal layers continually produce new keratinocytes that displace older cells upward toward the skin's surface. As cells move upward, they become flattened, lose their nucleus, and die. Dead keratinocytes at the skin surface form an abrasion-resistant layer that helps prevent water loss.

Melanocytes, another type of epidermal cell, make pigments called melanins and donate them to keratinocytes. Variations in skin color arise from differences in the distribution and activity of melanocytes, and in

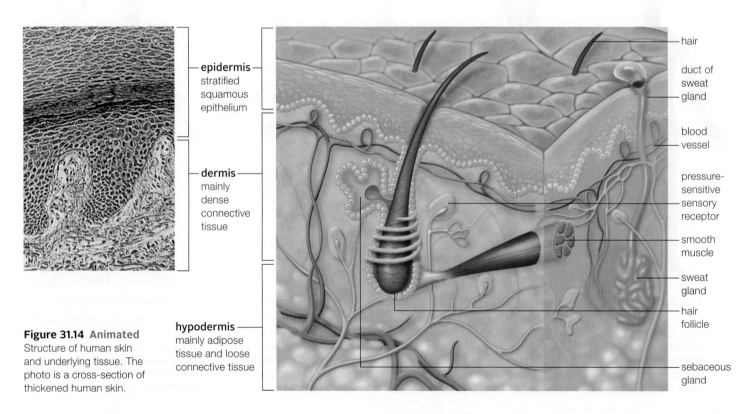

epidermis stratified squamous epithelium

dermis mainly dense connective tissue

hypodermis mainly adipose tissue and loose connective tissue

hair

duct of sweat gland

blood vessel

pressure-sensitive sensory receptor

smooth muscle

sweat gland

hair follicle

sebaceous gland

Figure 31.14 Animated Structure of human skin and underlying tissue. The photo is a cross-section of thickened human skin.

the type of melanin they produce (Section 14.1). One melanin is brown to black. Another is red to yellow. The effect of melanocytes can be seen with vitiligo, a skin disorder in which the destruction of these cells results in light patches of skin (**Figure 31.15**).

Melanin functions as a sunscreen, absorbing ultraviolet (UV) radiation that could damage DNA and other biological molecules. When skin is exposed to sunlight, melanocytes produce more of the brownish-black melanin, resulting in a protective "tan."

Dermis consists primarily of dense connective tissue with stretchy elastin fibers and supportive collagen fibers. Blood vessels, lymph vessels, and sensory receptors weave through the dermis. Dermis is much thicker than epidermis, and more resistant to tearing. Leather is animal dermis that has been treated with chemicals to preserve it.

Sweat glands, sebaceous glands, and hair follicles are pockets of epidermal cells that migrated into the dermis during early development. Sweat is mostly water. As it evaporates, it cools the skin surface and helps keep the body from overheating. Sebaceous glands produce sebum, an oily mix of triglycerides, fatty acids, and other lipids. Sebum helps keep skin and hair soft. It also has antimicrobial properties.

The part of a hair that you see is the keratin-rich remains of dead cells that originated in the follicle at the hair's base. Hair cells divide every 24 to 72 hours, making them among the fastest-dividing cells in the body. As cells at the base of a hair follicle replicate, they push cells above them upward, lengthening the hair. Smooth muscle attaches to each hair, and when this muscle reflexively contracts in response to cold or fright, the hair is pulled upright.

Evolution of Human Skin

Compared to other primates, humans have far more sweat glands and shorter, finer body hairs (**Figure 31.16**). According to one hypothesis, an increase in sweat glands and a loss of body hair occurred in concert with the evolution of bipedalism. During brisk walking or running, the metabolic activity in skeletal muscles produces heat that raises the body temperature. Presumably, when our bipedal ancestors began to run under the hot African sun, individuals with finer hair and more sweat glands were at an advantage because they were less likely to overheat.

dermis Deep layer of skin that consists of connective tissue with nerves and blood vessels running through it.
epidermis Outermost tissue layer; in animals, the epithelial layer of skin.

Figure 31.16 Primate skin. Humans have less body hair and more sweat glands than other primates such as chimpanzees.

When young, our closest primate relatives, chimpanzees and bonobos, have pink skin and a covering of long, black body hair. Our early ancestors probably had similarly pink skin and dark hair. Thus, loss of body hair that facilitated cooling would have created a new selective challenge—an increased exposure to potentially damaging sunlight.

The dark skin now observed in all African populations is considered an adaptation to this challenge. With this in mind, researchers reasoned that determining when dark skin evolved would provide an estimate of when humans lost their body hair. To determine when skin first darkened, the researchers looked at sequence variations in the human *MC1R* gene, which governs melanin deposition. The sequence comparisons indicated that dark skin color evolved by as early as 1.2 million years ago, presumably in concert with hair loss. This date supports the hypothesis that bipedalism selected for hair loss; the loss occurred during the time of *Homo erectus*, the first primate for which we have evidence of long-distance bipedal travel.

Later, as Section 14.1 explained, some populations of humans dispersed from Africa to higher latitudes where sunlight was less intense and their skin color reverted to a more chimpanzeelike pinkness.

Take-Home Message

What are the functions of the integumentary system?

» The integumentary system consists of skin, derivatives of skin such as hair, and underlying connective tissue.

» Skin has sensory receptors that inform the brain about the environment. It also serves as a barrier against pathogens, produces vitamin D, and functions in temperature regulation.

31.9 Negative Feedback in Homeostasis

■ A negative feedback system involving multiple organ systems allows the body to maintain its internal temperature.

■ Link to Metabolic heat 7.5

In vertebrates, homeostasis involves interactions among sensory receptors, the brain, and muscles and glands. A **sensory receptor** is a cell or cell component that detects a specific stimulus. Sensory receptors involved in homeostasis function like internal watchmen that monitor the body for changes. Information from sensory receptors throughout the body flows to the brain. The brain evaluates incoming information, then signals effectors—muscles and glands—to take the necessary actions to keep the body functioning.

Homeostasis often involves **negative feedback**, a process in which a change causes a response that reverses the change. An air conditioner with a thermostat is a familiar nonbiological example of a negative feedback system. A person sets the air conditioner to a desired temperature. When the temperature rises above this preset point, a sensor in the air conditioner detects the change and turns the unit on. When the temperature declines to the desired level, the thermostat detects this change and turns off the air conditioner.

A negative feedback mechanism also keeps your internal temperature near 37°C (98.6°F). Consider what happens when you exercise on a hot day (**Figure 31.17**). Muscle activity generates heat, and your internal temperature rises. Sensory receptors in the skin detect

negative feedback A change causes a response that reverses the change; important mechanism of homeostasis.
sensory receptor Cell or cell component that detects a specific type of stimulus.

the increase and signal the brain, which sends signals that bring about a response. Blood flow shifts, so more blood from the body's hot interior flows to the skin. The shift maximizes the amount of heat given off to the surrounding air. At the same time, sweat glands increase their output. Evaporation of sweat helps cool the body surface. Breathing quickens and deepens, speeding the transfer of heat from the blood in your lungs to the air. Hormonal changes make you feel more sluggish. As your activity level slows and your rate of heat loss increases, your temperature falls.

Sensory receptors also notify the brain when body temperature declines. The brain responds by sending signals that divert blood flow away from the skin and tighten smooth muscles attached to hairs so hairs stand up. With prolonged cold, the brain commands skeletal muscles to contract ten to twenty times a second. This shivering increases heat production by muscles. When you warm up, blood returns to your skin, you stop shivering, and your goose bumps subside.

Through the process of negative feedback, the body can prevent large variations in external temperature from causing similarly large changes inside the body. Negative feedback smooths out variations in body temperature, ensuring that cells can function properly.

Take-Home Message

What is the role of negative feedback in homeostasis?

»Negative feedback prevents dramatic changes in internal conditions. Sensory receptors detect changes and send signals to the brain, which sends signals to muscles and glands. The signals cause a response that reverses the initial change.

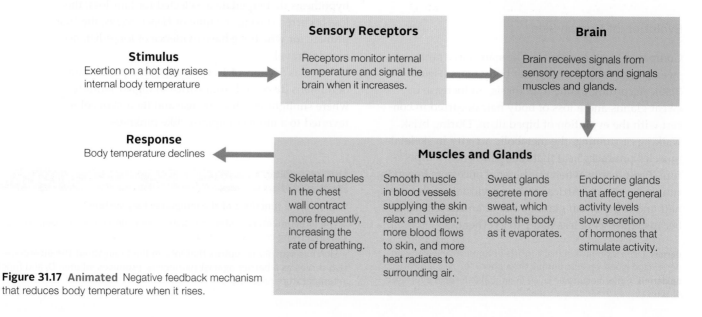

Figure 31.17 **Animated** Negative feedback mechanism that reduces body temperature when it rises.

Stem Cells—It's All About Potential (revisited)

Skin cells are relatively easy to grow in culture and are already in wide use for medical treatments. One currently available product is produced using infant foreskins that were removed during routine circumcisions. The foreskin (a tissue that covers the tip of the penis) provides a rich source of keratinocytes and fibroblasts. These cells are grown in culture with other biological materials (**Figure 31.18**). The resulting product is used to close chronic wounds, help burns heal, and cover sores on patients with epidermolysis bullosa, a genetic disorder that causes skin to slough off.

Researchers also have much more ambitious hopes for cultured fibroblasts. They are working on ways to make these cells behave like embryonic stem cells. Nonembryonic cells that have been altered so they have the properties of embryonic stem cells are referred to as induced pluripotent stem cells (IPSCs).

The first IPSCs were produced using a virus to insert genes into fibroblasts. When the inserted genes were expressed, the resulting proteins caused the cells to dedifferentiate. However, IPSCs produced in this manner are considered unsuitable for clinical use because their genome has been permanently altered by the gene insertion. The insertion could cause the cells to behave in unexpected ways and perhaps even become cancerous.

Researchers needed a way to bring about dedifferentiation without permanently altering the genome. In late 2010, Derrick Rossi of the Harvard Stem Cell Institute reported that he and his team had devised and successfully tested such a method. They had converted fibroblasts to IPSCs by introducing synthetic modified RNAs (rather than genes). The fibroblasts took up the RNAs by endocytosis, then translated them into the proteins that caused dedifferentiation.

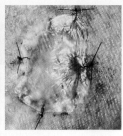

B When placed over a wound, the cells produce growth factors and other proteins that aid healing.

A Apligraf, a living cellular construct with a two-layered structure. The top layer is keratinocytes, and the lower layer is fibroblasts.

Figure 31.18 Cultured skin cells.

Rossi next used the same method to introduce RNAs that caused the former fibroblasts to begin to differentiate as muscle cells.

Rossi's method of creating IPSCs removes one potential barrier to clinical use of these cells, but it remains too early to know if IPSCs are functionally equivalent to embryonic stem cells. For now, most stem cell researchers advocate keeping all options open by continuing to study both embryonic stem cells and IPSCs.

How would you vote? An estimated 500,000 pre-implantation embryos are now stored in fertility clinics in the United States. Many will never be implanted in their mother. They are a potential source of stem cells, or a potential child for a woman who is willing to carry the embryo to term. Should parents of stored embryos be able to donate them for use in embryonic stem cell research?

LEARNING ROADMAP

Where you have been The nervous signals discussed in this chapter involve receptor proteins (5.7) and transport mechanisms (5.9, 5.10). They depend on ion gradients, which are a type of potential energy (5.2). We revisit health applications such as cancer (11.5) and alcohol abuse (5.1) and see examples of PET scans, a technique explained in Section 2.2.

Where you are now

Nervous Systems
In cnidarians, excitable neurons interconnect as a nerve net. Most animals have a bilateral system with a brain or ganglia at the anterior end and one or more nerve cords.

How Neurons Work
A shift in the electric charge across a neuron membrane conveys information along a neuron. Neurons communicate with one another and with other cells via chemical signals.

Vertebrate Nervous System
The brain and spinal cord constitute the central nervous system. The peripheral nervous system includes nerves that connect the brain and spinal cord to the rest of the body.

About the Brain
A human brain includes evolutionarily ancient tissues that govern housekeeping tasks. The more recently expanded cerebral cortex governs analytical thought and language.

The Neuroglial Support System
A variety of cells collectively known as neuroglia assist neurons. The neuroglia greatly outnumber neurons in the brain and play a role in some neurological disorders.

Where you are going We continue our survey of the nervous system with the next chapter, which focuses on the function of sensory neurons and the integration of sensory input in the brain. We discuss the interactions between the nervous and endocrine systems in Chapter 34. We also consider nervous control of skeletal muscle contraction (Chapter 35), heart rate and blood vessel dilation (Chapter 36), breathing (Chapter 38), urination (Chapter 40) and sexual arousal (Chapter 41).

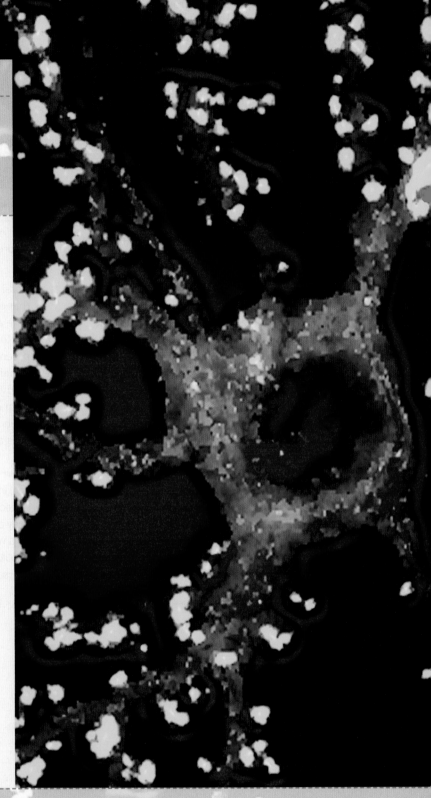

32.1 In Pursuit of Ecstasy

Ecstasy, an illegal drug, can make you feel socially accepted, less anxious, and more aware of your surroundings and of sensory stimuli. It also can leave you dying in a hospital, foaming at the mouth and bleeding from all body openings as your temperature skyrockets. It can send your family and friends spiraling into horror and disbelief as they watch you stop breathing. Lorna Spinks ended life that way when she was nineteen years old. Her anguished parents released the photographs in **Figure 32.1** because they wanted others to know what their daughter did not: Ecstasy can kill.

Ecstasy is a psychoactive drug, meaning it alters psychological functions. Its active ingredient, MDMA (3,4-methylenedioxymethamphetamine), is similar in structure to methamphetamine:

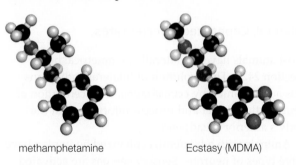

methamphetamine   Ecstasy (MDMA)

Both drugs interfere with the function of neurons (the signaling cells of the nervous system) by disrupting their ability to send, receive, respond to, and clear away chemical signals.

MDMA binds to active transporters in the membranes of brain neurons. The binding results in unnaturally high concentrations of a chemical signaling molecule that influences mood and memory. The overabundance of this molecule is the source of Ecstasy's psychoactive effects: increased energy, empathy, and euphoria, and decreased anxiety.

However, MDMA's effects extend beyond the brain. Under the influence of MDMA, the body behaves as it would in an emergency. The heart races, blood pressure increases, jaw muscles clench, and the pupils widen. Slowed saliva secretion makes the mouth feel dry. Drinking water to alleviate this feeling can cause physiological problems because MDMA also inhibits urine formation and urination. When you drink a large amount of fluid without getting rid of the excess, the extracellular fluid becomes too dilute—a condition that can cause convulsions. On the other hand, drinking too little and being physically active while using MDMA raises the risk of a life-threatening rise in body temperature. Lorna Spinks died after taking two

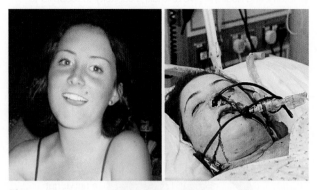

Figure 32.1 Photos of Lorna Spinks alive (*left*), and minutes after her death (*right*). She died after taking two Ecstasy tablets.

Ecstasy tablets because her body temperature soared and caused her organ systems to shut down.

Fortunately, life-threatening complications related to Ecstasy use are rare. When they do occur, rapid medical treatment is essential. If you suspect someone is having a bad reaction to Ecstasy or any other psychoactive drug, get medical help fast and be honest about the cause of the problem. Immediate, informed medical action may save a life.

Studies of animals have shown that repeated doses of MDMA alter and even kill neurons in the brain. In humans, brain abnormalities are correlated with repeated use of this drug. However, it was unclear whether MDMA causes the abnormalities, or existing abnormalities predispose people to use the drug. Establishing a direct cause-and-effect relationship is difficult because researchers cannot ask subjects to take an illegal and possibly dangerous drug.

Researchers in Amsterdam got around this obstacle by carrying out a prospective study. A prospective study follows a group of individuals over time and looks at how a difference in some factor affects them. In this study, the factor under study was MDMA use. The researchers began by scanning the brains of about 200 people who had never used Ecstasy, but who had considered trying the drug. About two years later, the researchers scanned the brains of their subjects once again. By this time, some people had begun using Ecstasy; they served as the experimental group. Others had still not used the drug; they served as a control group. Brain scans of the control group showed no significant changes from the previous scan. However, scans of those who had become Ecstasy users showed a significantly altered pattern of blood flow in the brain. Their scans also suggested that, as in animal studies, the neurons targeted by MDMA had been damaged by exposure to the drug.

32.2 Evolution of Nervous Systems

■ Interacting neurons give animals a capacity to respond to stimuli in the environment and inside their body.
■ Links to Trends in animal evolution 24.2, Chordate traits 25.2, Nervous tissue 31.6

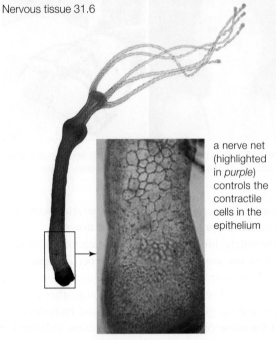

a nerve net (highlighted in *purple*) controls the contractile cells in the epithelium

A Cnidarians such as *Hydra* have a nerve net, but no ganglia or brain to receive and integrate information from multiple neurons.

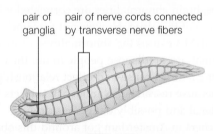

pair of ganglia   pair of nerve cords connected by transverse nerve fibers

B A flatworm such as a planarian has a ladderlike nervous system with two nerve cords and a pair of ganglia in the head.

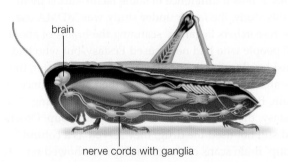

brain

nerve cords with ganglia

C Insects such as grasshoppers have paired ventral nerve cords with a ganglion in each segment. The nerve cords connect to a simple brain.

Figure 32.2 Animated Examples of invertebrate nervous systems. **Figure It Out:** Which of these organisms have a cephalized nervous system? Answer: The planarian and grasshopper.

Of all multicelled organisms, animals respond fastest to external stimuli. The cells called neurons are the key to these rapid responses.

Nerve Nets

Sea anemones, jellies, and other cnidarians have no centralized, controlling organ that functions like a brain. Instead, they have a mesh of interconnected neurons collectively called a **nerve net** (Figure 32.2A). Information can flow in any direction among cells of a nerve net. By causing cells in the body wall to contract, the nerve net can alter the size of the animal's mouth, change its body's shape, or move its tentacles. Echinoderms also have a nerve net with no central controlling organ. A nerve ring surrounds their mouth and branches extend into each arm.

Bilateral, Cephalized Invertebrates

Most animals have a bilaterally symmetrical body (Section 24.2). The evolution of bilateral body plans was accompanied by **cephalization**, a concentration of neurons that detect and process information at the body's anterior (head) end.

Animals with a bilateral, cephalized body plan have three types of neurons. **Sensory neurons** are activated by specific stimuli such as light, touch, or heat. They signal interneurons or motor neurons. **Interneurons** integrate signals from sensory neurons and other interneurons, and send signals of their own to motor neurons. **Motor neurons** control muscles and glands.

Planarians and the other flatworms are the simplest animals with a bilateral, cephalized nervous system. A planarian's head end has a pair of ganglia (Figure 32.2B). A **ganglion** (plural, ganglia) is a cluster of neuron cell bodies. In some cases it functions as an integrating center. A planarian's ganglia receive sig-

central nervous system Brain and spinal cord.
cephalization Concentration of sensory structures and nerve cells at the anterior end of a bilateral body.
ganglion Cluster of nerve cell bodies.
interneuron Neuron that receives signals from and sends signals to other neurons.
motor neuron Neuron that receives signals from another neuron and sends signals to a muscle or gland.
nerve Neuron fibers bundled inside a sheath of connective tissue.
nerve cord Bundle of nerve fibers running the length of a body.
nerve net Of cnidarians, a mesh of interacting neurons with no central control organ.
peripheral nervous system Nerves that extend through the body and carry signals to and from the central nervous system.
sensory neuron Neuron that responds to a specific internal or external stimulus and signals another neuron.

nals from eyespots and chemical-detecting cells on its head. The ganglia also connect to a pair of **nerve cords**, bundles of nerve fibers that run the length of the body. Nerve fibers that cross the body connect the cords.

Insects and other arthropods have paired ventral nerve cords that connect to a simple brain (**Figure 32.2C**). A pair of ganglia in each body segment provides local control over muscles in that segment.

The Vertebrate Nervous System

A single, dorsal nerve cord is one of the defining traits of chordate embryos (Section 25.2). In vertebrates, the anterior region of this nerve cord evolved into a brain. The vertebrate nervous system has two functional divisions (**Figure 32.3**). Most interneurons are in the **central nervous system**—the brain and spinal cord. Nerves that extend through the rest of the body constitute the **peripheral nervous system**. A **nerve** consists of nerve fibers bundled inside a sheath of connective tissue. Peripheral nerves are divided into two functional categories. Autonomic nerves monitor and regulate the body's internal state. They control smooth muscle, cardiac muscle, and glands. Somatic nerves monitor the body's position and conditions in the external environment, and they control skeletal muscle.

The peripheral nervous system consists of 12 cranial nerves that connect to the brain and 31 spinal nerves that connect to the spinal cord (**Figure 32.4**). Most cranial nerves, and all spinal nerves, include fibers of both sensory and motor neurons. For instance, a sciatic nerve runs from the sacral region of the spinal cord, through the buttock, and down the leg. When someone touches your thigh, the sciatic nerve carries signals from receptors in the skin to the spinal cord. When you want to move your leg, your sciatic nerve carries commands for the movement from the spinal cord to skeletal muscles in the leg.

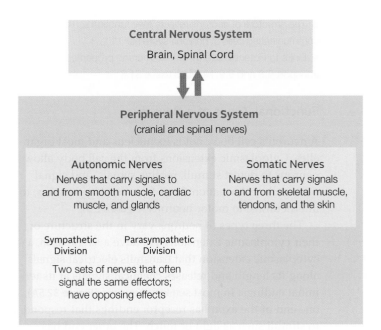

Figure 32.3 Functional divisions of vertebrate nervous systems. The spinal cord and brain are its central portion. The peripheral nervous system includes spinal nerves, cranial nerves, and their branches, which extend through the rest of the body. Peripheral nerves carry signals to and from the central nervous system.

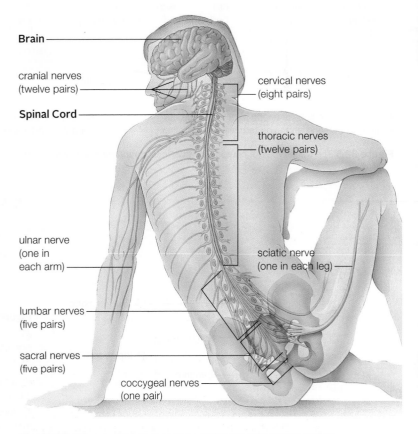

Figure 32.4 Some of the major nerves of the human nervous system.

Take-Home Message

What are the features of animal nervous systems?

» Cnidarians and echinoderms have a simple nervous system, a nerve net with no central integrating organ.

» Bilateral animals have three types of neurons: sensory neurons, interneurons, and motor neurons.

» Flatworms have paired ganglia that serve as an integrating center. Other invertebrates have more complex brains.

» Bilateral invertebrates usually have a pair of ventral nerve cords. In contrast, the chordates have a dorsal nerve cord.

» The vertebrate nervous system includes a well-developed brain, a spinal cord, and peripheral nerves.

32.3 Neurons—The Great Communicators

■ The structure of neurons reflects their function as the communicating cells of nervous systems.
■ Links to Potential energy 5.2, Membrane properties and transport 5.8 and 5.9, Nervous tissue 31.6

Functional Zones of a Neuron

A neuron's cell body holds its nucleus and most organelles. Cytoplasmic extensions from the cell body allow neurons to detect stimuli, receive signals, and signal other cells. Information flows from sensory neurons, to interneurons, to motor neurons (**Figure 32.5**).

The three types of neurons vary in the structure of their cytoplasmic extensions. All have a single **axon**, a cytoplasmic extension that transmits electrical signals along its length and releases chemical signals at its terminal endings. In most sensory neurons (**Figure 32.5A**), one end of the axon has receptor endings that respond to stimuli such as light or touch. The other end has axon terminals, and the cell body lies in between.

Interneurons and motor neurons typically receive signals from other neurons (**Figure 32.5B,C**). They have **dendrites**, branched cytoplasmic extensions that

receive chemical signals and respond by converting them to electrical ones.

Figure 32.6 shows the functional zones of a motor neuron. The dendrites and cell body are an input zone where chemical signals are received and converted to electrical signals ❶. The region of the axon nearest the cell body is a trigger zone ❷. If a sufficiently large electrical signal reaches this zone, electrical signals will flow along the conducting zone of the axon ❸ to the axon terminals. Axon terminals are the output zone: They release signaling molecules that influence the activity of other cells ❹.

Properties of the Neuron Plasma Membrane

The ability of neurons to communicate with one another and with other cells arises from properties of the neuron plasma membrane. A neuron's plasma membrane, like that of other cells, is composed mainly of a lipid bilayer that is impermeable to ions and large molecules. Transport proteins control the movement of ions across the membrane. Also, like other cells, neurons have electrical and concentration gradients across

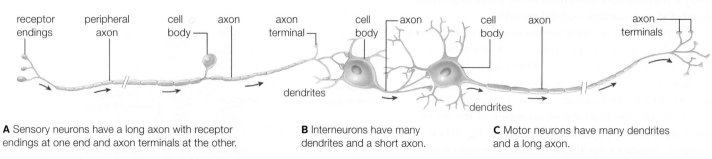

A Sensory neurons have a long axon with receptor endings at one end and axon terminals at the other.

B Interneurons have many dendrites and a short axon.

C Motor neurons have many dendrites and a long axon.

Figure 32.5 The three types of neurons. Arrows indicate the direction of information flow.

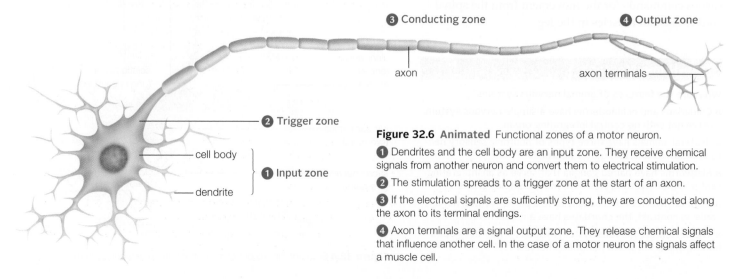

Figure 32.6 Animated Functional zones of a motor neuron.

❶ Dendrites and the cell body are an input zone. They receive chemical signals from another neuron and convert them to electrical stimulation.

❷ The stimulation spreads to a trigger zone at the start of an axon.

❸ If the electrical signals are sufficiently strong, they are conducted along the axon to its terminal endings.

❹ Axon terminals are a signal output zone. They release chemical signals that influence another cell. In the case of a motor neuron the signals affect a muscle cell.

their plasma membrane. Their cytoplasm has more negatively charged components than the interstitial fluid outside the cell.

Negatively charged proteins in the cytoplasm contribute to the gradient across the membrane. Being large and charged, these molecules cannot diffuse across the lipid bilayer of the cell membrane. The distributions of positively charged potassium ions (K$^+$) and positively charged sodium ions (Na$^+$) also play a role in the electrical gradient.

Transport proteins control the movement of sodium and potassium ions into and out of the neuron. Like other cells, neurons have sodium–potassium cotransporters in their plasma membrane (Section 5.9). These active transport proteins use one ATP to pump two potassium ions into the cell and three sodium ions out (**Figure 32.7A**). By moving more positive charges out of the cell than into it, the cotransporter contributes to the charge gradient across the neuron membrane. It also creates concentration gradients for sodium and potassium across the membrane.

Nearly all sodium pumped out of the neuron stays out, as long as the neuron remains at rest. In contrast, some potassium ions follow their concentration gradient by diffusing out of the cell through passive transporters (**Figure 32.7B**). Diffusion of potassium outward adds to the number of positively charged ions in the interstitial fluid, thus heightening the charge difference across the membrane.

In summary, the cytoplasm of a resting neuron has negatively charged proteins that the interstitial fluid lacks. It also has fewer sodium ions (Na$^+$) and more potassium ions (K$^+$). The graphic below illustrates the relative concentrations of the relevant ions in a neuron at rest, with the green ball representing negatively charged proteins:

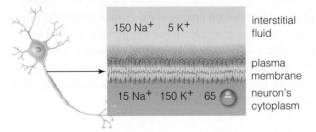

As in a battery, the separated charges constitute potential energy. The voltage difference across a cell membrane is called its **membrane potential** and is measured in thousandths of a volt, or millivolts (mV). An unstimulated neuron has a **resting membrane potential** of about –70 mV. The negative sign indicates that the cytoplasmic region just beneath the neuron membrane

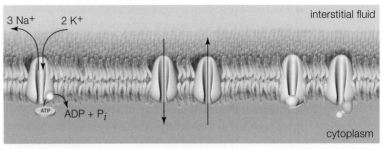

A Sodium–potassium cotransporters actively transport three Na$^+$ out of a neuron for every two K$^+$ they pump in.

B Passive transporters allow K$^+$ ions to move across the plasma membrane, down their concentration gradient.

C Voltage-gated channels for Na$^+$ or K$^+$ are closed in a neuron at rest (*left*), but open when it is excited (*right*).

Figure 32.7 Animated Three types of transport proteins in a neuron membrane. Voltage-gated channels are only in the membrane of the axon.

has more negative ions than the interstitial fluid surrounding the cell.

The transport proteins that contribute to resting membrane potential are found throughout the neuron plasma membrane. However, the axon also has special passive transport proteins called voltage-gated channels (**Figure 32.7C**). These proteins have gates that open at a particular membrane potential (voltage). Ions diffuse through channels in the proteins' interior only when the gates are open. Potassium channels let potassium ions diffuse across the membrane; sodium channels let sodium ions move across. In a resting neuron, gates of both types of channels are closed. How a neuron gets excited and the role of voltage-gated channels in the origin of electrical signals and their movement along the axon are the focus of the next section.

axon Of a neuron, a cytoplasmic extension that transmits electrical signals along its length and releases chemical signals at its endings.
dendrite Of a motor neuron or interneuron, a cytoplasmic extension that receives chemical signals sent by other neurons and converts them to electrical signals.
membrane potential Voltage difference across a plasma membrane.
resting membrane potential Voltage difference across the plasma membrane of an excitable cell that is not receiving stimulation.

Take-Home Message

How does a neuron's structure affect its function?

» Sensory neurons have an axon with one end that responds to a specific stimulus and another that signals other cells.

» Interneurons and motor neurons have many signal-receiving dendrites and one signal-sending axon.

» Transport proteins in the neuron plasma membrane set up electrical and concentration gradients across the membrane of a resting neuron.

» A neuron's axon has special voltage-gated channel proteins that function in the transmission of electrical signals along the axon.

32.4 The Action Potential

■ Movement of sodium and potassium ions through gated channels causes a brief reversal of the membrane potential.

■ Links to Diffusion 5.8, Transport mechanisms 5.9

All cells have a membrane potential, but only neurons and muscle cells are "excitable," meaning that when stimulated, they will undergo an **action potential**—a brief reversal in the electric gradient across the plasma membrane. If you plot the change in membrane potential at any point along an axon over time, the plot of an action potential will look like the graph in **Figure 32.8**. During an action potential, membrane potential shoots up from its resting value (–70 mV) into positive territory. A positive membrane potential means the interior of the cell has more positively charged ions than interstitial fluid. The action potential peaks at +30 mV, then declines once again to resting potential. The whole process takes only a few milliseconds.

Graded Potentials and Reaching Threshold

Changes in membrane potential result from movement of ions across the neuron's plasma membrane. In sensory neurons, a specific stimulus such as touch or light results in ion movement. Tap your wrist, and the pressure stimulates receptor endings of sensory neurons in your skin. The pressure deforms the plasma membrane at the input zone of these neurons and some sodium ions slip across. The ion flow causes a local, graded potential—a slight shift in the voltage difference across the neuron's membrane. "Graded" means the size of this voltage shift varies. In an interneuron or motor neuron, local, graded potentials arise in response to the arrival of signals from another neuron.

When only a few sodium ions cross the neuron membrane, they cause a small, local shift in membrane potential at the neuron's input zone where they entered. As these ions diffuse into the large volume of cytoplasm, the membrane potential of the input zone returns quickly to its resting state. By analogy, think about what would happen if you put a drop of ink into the water at the edge of a swimming pool. Dissipation of the ink would quickly make it invisible.

When a graded potential is large enough, the greater number of ions diffusing from the input zone into the trigger zone will cause a significant shift in the trigger zone's membrane potential. The plasma membrane in the trigger zone has many voltage-gated sodium channels and potassium channels that remain closed when the neuron is at rest ❶. When arrival of enough positively charged ions shifts the voltage difference across the membrane to a particular level,

called the **threshold potential**, the voltage-gated sodium channels open and an action potential gets under way.

An All-or-Nothing Spike

Opening of voltage-gated sodium channels allows Na⁺ to diffuse down its electrical and concentration gradients, from the interstitial fluid into the neuron ❷. Diffusion of sodium into the neuron has a positive feedback effect. With **positive feedback**, a response intensifies the conditions that caused it to occur. In this case, gated sodium channels open in an accelerating way after threshold is reached. As Na⁺ diffuses in, membrane potential increases, causing more gated sodium channels to open:

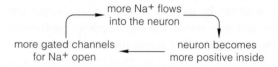

An action potential is an all-or-nothing event, meaning once threshold level is reached, an action potential always occurs. Once gated sodium channels begin opening, the size of the stimulus that brought the neuron to threshold becomes unimportant. Sodium entering the neuron from outside, rather than diffusion of ions from the input zone, drives the feedback cycle.

Every action potential peaks at +30 mV; no more, no less. The potential cannot rise higher because, above a certain voltage, gates on voltage-gated Na⁺ channels shut. About the same time, the voltage-gated potassium channels open ❸. Opening these channels allows K⁺ to diffuse down its concentration and electrical gradients out of the neuron. Outward diffusion of K⁺ causes membrane potential to decline.

Declining membrane potential triggers the closing of gated potassium channels ❹. However, by the time gates close, so many K⁺ have left that the membrane potential is a bit below its resting value in a small area. The rapid diffusion of K⁺ from the adjacent cytoplasm restores the K⁺ concentration, and membrane potential returns to its resting value.

Propagation of an Action Potential

Each action potential is self-propagating and moves from the trigger zone toward the axon terminals. When sodium channels in one part of an axon open, some Na⁺ diffuses into adjoining regions, lowering their membrane potential to threshold level. An action potential cannot move backward, because after

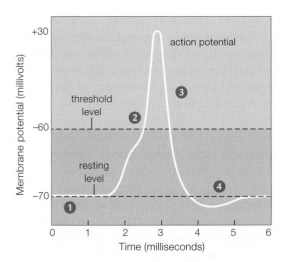

Figure 32.8 **Animated** Action potential. *Above,* a plot of a neuron's membrane potential over time, showing the action potential as a spike, a brief reversal of membrane potential.

Right, illustration of the events at one region of an axon membrane. Numbers on the graph correlate with numbered graphics.

a sodium channel's gate closes, it cannot open for a brief period. However, sodium gates in regions farther along the axon can and do swing open as these regions reach threshold. As gates of sodium channels open in one region after the next, the action potential moves steadily toward the axon terminals. The signal moves without weakening because, as previously noted, all action potentials are the same size.

action potential Brief reversal of the voltage difference across the plasma membrane of a neuron or muscle cell.
positive feedback A response intensifies the conditions that caused its occurrence.
threshold potential Neuron membrane potential at which gated sodium channels open, causing an action potential to occur.

Take-Home Message

What happens during an action potential?

» An action potential begins in the neuron's trigger zone. A strong stimulus decreases the voltage difference across the membrane. This causes gated sodium channels to open, and the voltage difference reverses.

» The action potential travels along an axon as consecutive patches of membrane undergo reversals in membrane potential.

» At each patch of membrane, an action potential ends as sodium channels close and potassium channels open. Potassium ions flow out of the neuron and restore the voltage difference across the membrane.

» Action potentials can move in one direction, toward axon terminals, because gated sodium channels are briefly inactivated after they close.

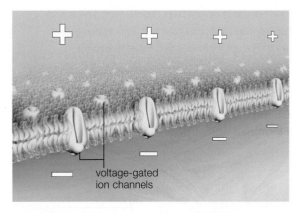

❶ Neuron at rest. Gated channels for Na+ or K+ are closed.

High Na+ in interstitial fluid; high K+ inside neuron.

Interstitial fluid has more positively charged ions than neuron cytoplasm does, making the membrane potential negative.

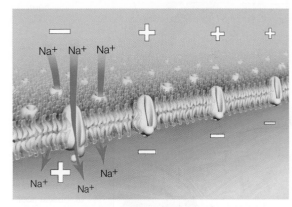

❷ Threshold is reached and Na+ channels open.

Na+ diffuses down its concentration gradient into the neuron.

The Na+ influx increases membrane potential, making the neuron cytoplasm even more positive than the interstitial fluid.

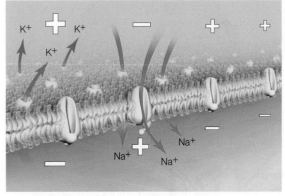

❸ High membrane potential makes Na+ channels close and K+ channels open.

K+ diffuses down its concentration gradient out of the neuron, making neuron cytoplasm once again more negative than the interstitial fluid.

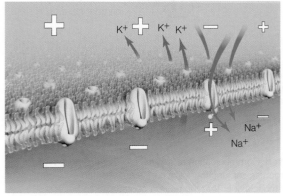

❹ K+ channels close.

Diffusion of Na+ to regions farther along the axon propagates the action potential.

Once closed, Na+ channels are briefly inactivated, preventing action potentials from moving "backward."

32.5 How Neurons Send Messages to Other Cells

■ Action potentials do not pass directly from a neuron to another cell; chemicals carry the signals between cells.
■ Links to Receptor proteins 5.7, Exocytosis 5.10

Chemical Synapses

An action potential travels along a neuron's axon to axon terminals at its tips. Action potentials cannot pass

1 Action potentials flow along the axon of a motor neuron to a neuromuscular junction, where an axon terminal forms a synapse with a muscle fiber.

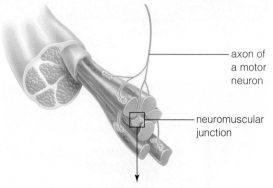

axon of a motor neuron

neuromuscular junction

2 The axon terminal stores neurotransmitter (*green*) inside synaptic vesicles.

3 The arrival of an action potential at an axon terminal causes calcium ions (Ca++) to enter the neuron.

4 The influx of Ca++ causes exocytosis of synaptic vesicles, so neurotransmitter is released into the synaptic cleft.

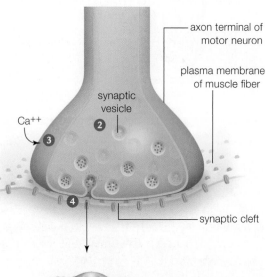

axon terminal of motor neuron

plasma membrane of muscle fiber

synaptic vesicle

Ca++

synaptic cleft

5 The plasma membrane of the post-synaptic cell has receptors that bind neurotransmitter.

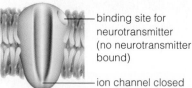

binding site for neurotransmitter (no neurotransmitter bound)

ion channel closed

6 Binding of neurotransmitter opens a channel through the receptor, allowing ions to flow into the postsynaptic cell.

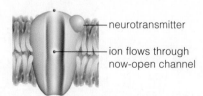

neurotransmitter

ion flows through now-open channel

Figure 32.9 Animated Communication at a synapse. This example depicts what occurs at a neuromuscular junction—a synapse between an axon terminal of a motor neuron and a muscle fiber that the neuron controls.

directly from a neuron to another cell. Instead, chemicals relay the message. The region where an axon terminal signals another cell is called a **synapse**. The signal-sending neuron at a synapse is referred to as the presynaptic cell. A fluid-filled synaptic cleft separates its axon terminal from the signal-receiving zone of the postsynaptic cell. **Figure 32.9** shows a synapse between a motor neuron and a skeletal muscle fiber. This type of synapse is called a **neuromuscular junction**.

Action potentials arrive at a neuromuscular junction by traveling along the axon of a motor neuron to axon terminals **1**. Vesicles inside an axon terminal contain **neurotransmitter**, a type of signaling molecule **2**. Neurotransmitters relay messages between presynaptic and postsynaptic cells. A motor neuron releases acetylcholine (ACh). Other neurotransmitters are released by different neurons.

The plasma membrane of an axon terminal has gated channels for calcium ions (Ca++). When the neuron is resting, these gates are closed, and calcium pumps actively transport calcium out of the cell. Thus, the cytoplasm of a resting neuron has fewer calcium ions than interstitial fluid. When an action potential arrives at an axon terminal, the calcium channels open their gates. Calcium ions can then follow their gradient and diffuse from interstitial fluid into the axon terminal **3**. The concentration of calcium ions in the neuron's cytoplasm increases. This increase stimulates exocytosis, so neurotransmitter-filled vesicles release their contents into the synaptic cleft **4**.

The plasma membrane of a postsynaptic cell has receptors that bind neurotransmitter **5**. At a neuromuscular junction, the receptors are transport proteins that change shape when they bind to ACh. The change opens a channel through the protein. Sodium ions flow through the opening into the muscle cell **6**. Like a neuron, a muscle cell is excitable, meaning it can undergo an action potential. An increase in a muscle cell's cytoplasmic sodium ion concentration drives the membrane potential toward threshold. Once threshold is reached, action potentials stimulate muscle contraction by a process described in detail in Section 35.8.

Cleaning the Cleft

After neurotransmitter molecules do their work, they must be removed from synaptic clefts to make way for new signals. Active transporters pump some of the neurotransmitter back into presynaptic cells or into nearby neuroglial cells. In addition, postsynaptic cells have enzymes in their membrane that break down neurotransmitter. For example, the membranes of

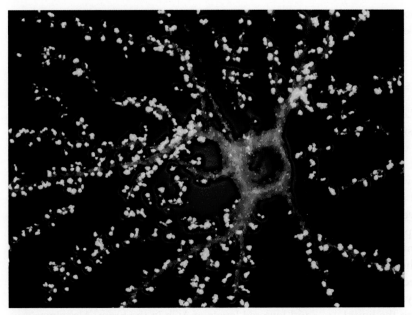

Figure 32.10 Synaptic density. An interneuron stained with a yellow fluorescent dye that indicates the many locations where other neurons synapse on this cell.

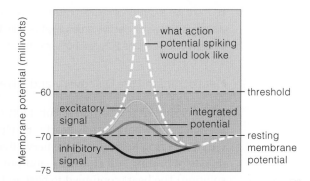

Figure 32.11 Synaptic integration. Excitatory and inhibitory signals arrive at a postsynaptic neuron's input zone at the same time. The colored lines show a postsynaptic cell's response to an excitatory signal (*yellow*), to an inhibitory signal (*purple*) and to both at once (*red*). In this example, summation of the two signals did not lead to an action potential (*white*). **Figure It Out:** Which colored line would you expect to see if a neurotransmitter caused an influx of potassium ions into the postsynaptic cell?

Answer: The purple line. Influx of potassium drives the neuron away from threshold.

muscle cells at a neuromuscular junction contain acetylcholinesterase, an enzyme that breaks down ACh.

Nerve gases such as sarin exert their deadly effects by binding to acetylcholinesterase and preventing it from breaking down ACh. The resulting accumulation of ACh in synaptic clefts causes skeletal muscle paralysis, confusion, headaches, and, when the dosage is high enough, death.

Synaptic Integration

Typically, a postsynaptic cell receives messages from many neurons at the same time (**Figure 32.10**). Certain interneurons in the brain are on the receiving end of synapses with 10,000 neurons! Depending on the type of neurotransmitter released at the synapse and the type of receptor on the postsynaptic cell, an incoming signal may be excitatory or inhibitory. A signal that causes the opening of sodium gates has an excitatory effect because it pushes membrane potential closer to threshold. A signal that opens potassium gates has an inhibitory effect because it nudges the potential away from threshold.

neuromuscular junction Synapse between a neuron and a muscle.
neurotransmitter Chemical signal released by axon terminals of a neuron.
synapse Region where a neuron's axon terminals transmit chemical signals to another cell.
synaptic integration The summation of excitatory and inhibitory signals by a postsynaptic cell.

Through **synaptic integration**, a neuron sums all inhibitory and excitatory signals at its input zone. Incoming synaptic signals can amplify, dampen, or cancel one another's effects. **Figure 32.11** illustrates the integration of two simultaneously arriving signals, one excitatory and the other inhibitory.

Competing signals cause the membrane potential at the postsynaptic cell's input zone to rise and fall. When excitatory signals outweigh inhibitory ones, ions diffuse from the input zone into the trigger zone and drive the postsynaptic cell to threshold. Gated sodium channels swing open, and an action potential occurs.

Neurons also integrate signals that arrive in quick succession from a single presynaptic cell. An ongoing stimulus can trigger a series of action potentials in a presynaptic cell, which will bombard a postsynaptic cell with waves of neurotransmitter.

Take-Home Message

How does information pass between cells at a synapse?

» Action potentials travel to a neuron's output zone. There they stimulate release of neurotransmitters—chemical signals that affect another cell.

» Neurotransmitters are signaling molecules secreted into a synaptic cleft from a neuron's output zone. They may have excitatory or inhibitory effects on a postsynaptic cell.

» Synaptic integration is the summation of all excitatory and inhibitory signals arriving at a postsynaptic cell's input zone at the same time.

» For a synapse to function properly, neurotransmitter must be cleared from the synaptic cleft after the chemical signal has served its purpose.

32.6 A Smorgasbord of Signals

■ There are a variety of neurotransmitters. Neurological disorders and psychoactive drugs interfere with their action.
■ Links to PET scans 2.2, Alcohol's effects 5.1

Neurotransmitter Discovery and Diversity

In the early 1920s, Austrian scientist Otto Loewi was investigating what controls heart rate. He surgically removed a frog heart—with the nerve that controls its rate of beating still attached—and put it in saline solution. The heart continued to beat and, when Loewi stimulated the nerve, the heartbeat slowed a bit.

Loewi suspected stimulation of the nerve caused release of a chemical signal. To test this hypothesis, he put two frog hearts into a saline-filled chamber and stimulated the nerve connected to one of them. Both hearts started to beat more slowly. As expected, the nerve had released a chemical that not only affected the attached heart, but also diffused through the liquid and slowed the beating of the second heart.

Loewi had discovered one effect of the neurotransmitter ACh. ACh acts on skeletal muscle, smooth muscle, the heart, many glands, and the brain. Each type of tissue has a different kind of ACh receptor, so this neurotransmitter can elicit different responses in different cells. Binding of ACh to receptors on skeletal muscle excites the cell and encourages contraction, whereas binding of ACh to cardiac muscle has an inhibitory effect and slows contractions.

The body produces many kinds of neurotransmitters (**Table 32.1**). Norepinephrine and epinephrine (commonly known as adrenaline) prepare the body to respond to stress or excitement. Dopamine influences reward-based learning and acts in fine motor control. Serotonin influences mood and memory. Glutamate is the main excitatory signal in the central nervous system. GABA (gamma aminobutyric acid) has a general inhibitory effect on release of other neurotransmitters.

Some neurons also release **neuromodulators**, molecules that influence the effects of neurotransmitters. One neuromodulator, substance P, enhances pain perception. Enkephalins and endorphins are natural painkillers that are secreted in response to strenuous activity or injury. Endorphins also are released when people laugh, reach orgasm, or get a comforting hug or a relaxing massage.

Disrupted Signaling

Neurological Disorders Many disorders of the nervous system involve disruption of signaling at synapses. A few examples will illustrate how our understanding of neurotransmission can help us develop treatments for these disorders.

Alzheimer's disease, the leading cause of dementia (loss of ability to think), involves damage to neurons and lowered levels of ACh in the brain. Alzheimer's disease begins with forgetfulness. As the disease progresses, a person becomes increasingly confused, cannot communicate, and eventually is incapable of living independently. Drugs that inhibit acetylcholinesterase (the enzyme that normally breaks down ACh) can slow the mental decline in some people.

Damage to dopamine-secreting neurons in the part of the brain that governs motor control results in Parkinson's disease (**Figure 32.12**). Tremors are an early symptom. Later, the sense of balance becomes impaired, and voluntary movement, including speech, becomes difficult. Because the symptoms arise from a dopamine shortage, patients can be treated with a drug (levodopa) that the body converts to dopamine.

Multiple factors contribute to the development of Parkinson's disease. A head injury that causes a loss of consciousness increases the risk, as does exposure to pesticides in drinking water. There are also some rare inherited forms of the disease.

A dopamine level that is lower than normal also plays a role in attention deficit hyperactivity disorder (ADHD). Affected people can have trouble concentrating and controlling impulses. Drugs used to treat ADHD increase dopamine availability in the brain. For example, Ritalin (methylphenidate) acts by preventing the reuptake of dopamine at a synapse.

Interactions among several neurotransmitters, including serotonin, dopamine, and norepinephrine, affect mood. Low levels of one or more of these neurotransmitters can result in depression—a persistent feeling of sadness and inability to experience pleasure. The most widely prescribed antidepressants, including Prozac (fluoxetine) and Paxil (paroxetine), increase the

Table 32.1 Major Neurotransmitters and Their Effects

Neurotransmitter	Examples of Effects
Acetylcholine (ACh)	Induces skeletal muscle contraction, slows cardiac muscle contraction rate, affects mood and memory
Epinephrine and norepinephrine	Speed heart rate; dilate the pupils and airways to lungs; slow gut contractions; increase anxiety
Dopamine	Dampens excitatory effects of other neurotransmitters; has roles in memory, learning, fine motor control
Serotonin	Elevates mood; role in memory
GABA	Inhibits release of other neurotransmitters

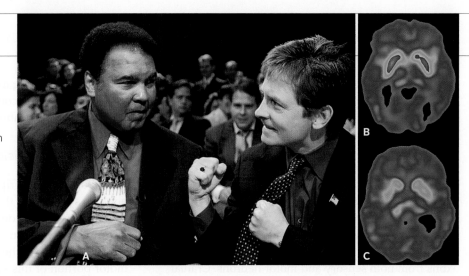

Figure 32.12 Battling Parkinson's disease. (**A**) This neurological disorder affects former heavyweight champion Muhammad Ali, actor Michael J. Fox, and about half a million other people in the United States.

(**B**) A normal PET scan and (**C**) PET scan of a person with Parkinson's disease. *Red* and *yellow* indicate a high level of metabolic activity in neurons that secrete dopamine. Section 2.2 explains how PET scans work.

level of serotonin by blocking proteins that pump it back into axon terminals.

Depression has a genetic component, and families predisposed to depression are also prone to anxiety disorders. People with anxiety disorders are paralyzed with worry or stricken with panic under what most of us view as ordinary circumstances. Anxiety disorders often are treated with antidepressants that prevent serotonin reuptake. Tranquilizers such as Xanax (alprazolam) and Valium (diazepam) alleviate anxiety by enhancing the inhibitory effect of GABA.

Psychoactive Drugs People take psychoactive drugs, both legal and illegal, to alleviate pain, relieve stress, or feel pleasure. Many such drugs are habit-forming. Users often develop tolerance; they have to take increasingly larger or more frequent doses of the drug to obtain the desired effect.

All major addictive drugs stimulate the release of dopamine. When an animal engages in behavior that enhances survival or reproduction, it experiences a pleasurable surge of dopamine. This response helps animals learn to repeat beneficial behaviors. Drugs that cause dopamine release tap into this ancient learning pathway. Drug users inadvertently teach themselves that the drug is essential to their well-being.

Stimulants make users feel alert and energized, but also interfere with fine motor control. Amphetamines, including the MDMA in Ecstasy, act by increasing serotonin, norepinephrine, and dopamine in the brain. In doing so, they inhibit the body's ability to make these neurotransmitters.

Caffeine is a widely used stimulant that blocks receptors for adenosine, a signaling molecule that suppresses brain cell activity and causes drowsiness. Nicotine blocks brain receptors for ACh. Cocaine interferes with the uptake of dopamine, serotonin, and norepinephrine from synaptic clefts and, like nicotine, is highly addictive.

Depressants such as alcohol (ethyl alcohol) and barbiturates slow motor responses by inhibiting ACh release. Alcohol also stimulates the release of endorphins and GABA, causing users to experience a brief euphoria followed by depression.

Narcotic analgesics, such as morphine, codeine, heroin, fentanyl, and oxycodone, mimic a body's natural painkillers. They provide a rush of euphoria and are highly addictive. Ketamine and PCP (phencyclidine) belong to a different class of analgesics. They give users an out-of-body experience and numb the extremities by slowing the clearing of synapses.

Hallucinogens distort sensory perception and bring on a dreamlike state. LSD (lysergic acid diethylamide) binds to serotonin receptors. LSD is not addictive, but users can be injured because they do not perceive and respond to hazards. Flashbacks, or brief distortions of perceptions, may occur years after the last intake of LSD. Two related drugs, mescaline and psilocybin, have weaker effects.

Marijuana consists of parts of *Cannabis* plants. Smoking a lot of marijuana can cause hallucinations. More often, users become relaxed, uncoordinated, and inattentive. The active ingredient, THC (delta-9-tetrahydrocannabinol), alters levels of dopamine, serotonin, norepinephrine, and GABA.

neuromodulator Any signaling molecule that reduces or magnifies the influence of a neurotransmitter on target cells.

Take-Home Message

How do neurological disorders and drugs affect the flow of information in the nervous system?

» Neurological disorders lower the amount of a neurotransmitter or the balance among neurotransmitters.

» Psychoactive drugs act by stimulating release, inhibiting breakdown, or mimicking the action of natural neurotransmitters. Many are addicting, and using them can alter the body's ability to produce neurotransmitter.

32.7 The Peripheral Nervous System

■ Peripheral nerves run through your body and carry information to and from the central nervous system.

Axons Bundled as Nerves

In humans, the peripheral nervous system includes 31 pairs of spinal nerves that connect to the spinal cord and 12 pairs of cranial nerves that connect directly to the brain. Each peripheral nerve consists of axons of many neurons bundled together inside a connective tissue sheath (**Figure 32.13A**). All spinal nerves include axons from both sensory and motor neurons. Cranial nerves may include axons of motor neurons, axons of sensory neurons, or axons of both sensory and motor neurons. Interneurons, remember, are not part of the peripheral nervous system.

The neuroglial cells called Schwann cells wrap like jelly rolls around the axons of most peripheral nerves (**Figure 32.13B**). The Schwann cells collectively form an insulating **myelin sheath** that makes action potentials flow faster. Ions cannot cross a sheathed neural membrane. As a result, ion disturbances associated with an action potential spread through an axon's cytoplasm until they reach a node, a small gap between Schwann cells. At each node, the membrane contains numerous gated sodium channels. When these gates open, the voltage difference reverses abruptly. By jumping from node to node in long axons, a signal can move as fast as 120 meters per second. In unmyelinated axons, the maximum speed is about 10 meters per second.

Functional Subdivisions

We subdivide the peripheral system into the somatic nervous system and the autonomic nervous system.

The sensory portion of the **somatic nervous system** conducts information about external conditions from sensory neurons to the central nervous system, and the motor portion of this system relays commands from the brain and spinal cord to the skeletal muscles. The motor portion is the only part of the nervous system that is under voluntary control. The **autonomic nervous system** is concerned with signals to and from internal organs and to glands.

There are two categories of nerves in the autonomic system: sympathetic and parasympathetic. Both service most organs and work antagonistically, meaning the signals from one type oppose signals from the other (**Figure 32.14**). **Sympathetic neurons** are most active in times of stress, excitement, and danger. Their axon terminals release norepinephrine. **Parasympathetic neurons** are most active in times of relaxation. Release of ACh by their axon terminals promotes housekeeping tasks, such as digestion and urine formation.

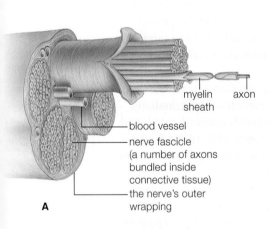

A

myelin sheath

axon

blood vessel

nerve fascicle (a number of axons bundled inside connective tissue)

the nerve's outer wrapping

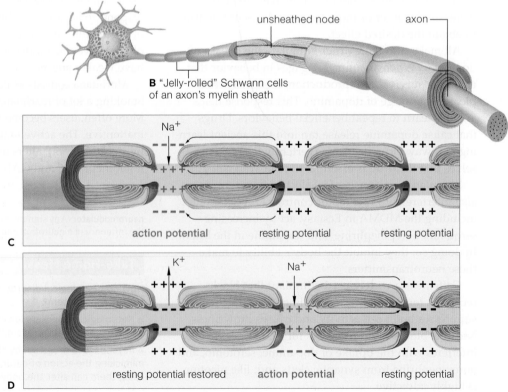

unsheathed node

axon

B "Jelly-rolled" Schwann cells of an axon's myelin sheath

Na^+

++++   action potential   ++++   resting potential   ++++   resting potential

C

K^+ Na^+

++++   resting potential restored   action potential   ++++   resting potential

D

Figure 32.13 Animated (**A**) Structure of one type of nerve. (**B–D**) In axons with a myelin sheath, ions flow across the neural membrane at nodes, or small gaps between the cells that make up the sheath. Many gated channels for sodium ions are exposed to extracellular fluid at the nodes. When excitation caused by an action potential reaches a node, the gates open and sodium rushes in, starting a new action potential. Excitation spreads rapidly to the next node, where it triggers a new action potential, and so on down the axon to the output zone.

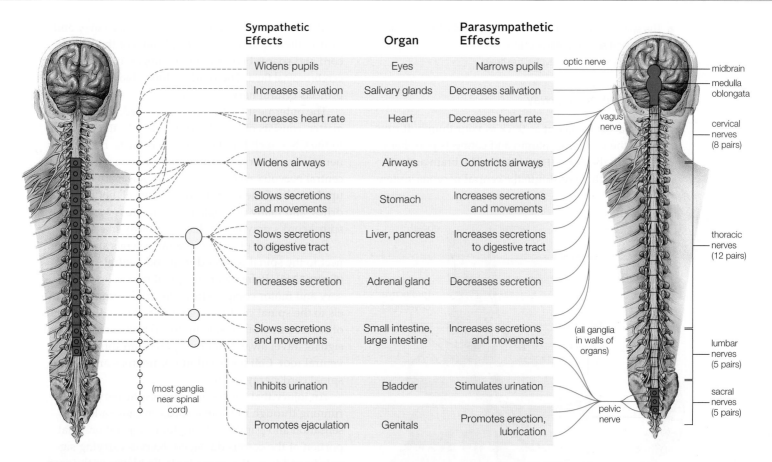

Sympathetic Effects	Organ	Parasympathetic Effects
Widens pupils	Eyes	Narrows pupils
Increases salivation	Salivary glands	Decreases salivation
Increases heart rate	Heart	Decreases heart rate
Widens airways	Airways	Constricts airways
Slows secretions and movements	Stomach	Increases secretions and movements
Slows secretions to digestive tract	Liver, pancreas	Increases secretions to digestive tract
Increases secretion	Adrenal gland	Decreases secretion
Slows secretions and movements	Small intestine, large intestine	Increases secretions and movements
Inhibits urination	Bladder	Stimulates urination
Promotes ejaculation	Genitals	Promotes erection, lubrication

(most ganglia near spinal cord)

optic nerve
vagus nerve
(all ganglia in walls of organs)
pelvic nerve

midbrain
medulla oblongata
cervical nerves (8 pairs)
thoracic nerves (12 pairs)
lumbar nerves (5 pairs)
sacral nerves (5 pairs)

Figure 32.14 Animated Autonomic nerves and their effects. Autonomic signals travel to organs by a two-neuron path. The first neuron has its cell body in the brain or spinal region (indicated in *red* in the two body sketches). This neuron synapses on a second neuron at a ganglion. Sympathetic ganglia are close to the spinal cord. Parasympathetic ganglia are in or near the organs they affect. Axons of the second neuron in the pathway synapse with the organ. **Figure It Out: What effect does stimulation of the vagus nerve (a parasympathetic nerve) have on the heart?**

Answer: It decreases heart rate.

When you are startled or frightened, parasympathetic neurons become less active and sympathetic ones become more active. The unopposed, sympathetic signals raise your heart rate and blood pressure, make you sweat more and breathe faster, and induce adrenal glands to secrete epinephrine. The signals put you in a state of arousal, priming you to fight or make a fast getaway. Hence the term **fight–flight response**.

Opposing sympathetic and parasympathetic signals govern most organs. For instance, both act on smooth muscle cells in the gut wall. As sympathetic neurons release norepinephrine at synapses with these cells, parasympathetic neurons release ACh at other synapses with the same cells. One signal tells the gut to slow its contractions; the other calls for increased activity. Synaptic integration determines the outcome of the conflicting commands.

autonomic nervous system Set of nerves that relay signals to and from internal organs and to glands.
fight–flight response Response to danger or excitement. Activity of parasympathetic neurons declines, sympathetic signals increase, and adrenal glands secrete epinephrine.
myelin sheath Lipid-rich wrappings around axons of some neurons; speeds propagation of action potentials.
parasympathetic neurons Neurons of the autonomic system that encourage "housekeeping" tasks in the body.
somatic nervous system Set of nerves that control skeletal muscle and relay signals from joints and skin.
sympathetic neurons Neurons of the autonomic system that prepare the body for danger or excitement.

Take-Home Message

What is the peripheral nervous system?

» The peripheral nervous system consists of nerves that extend through the body and relay signals to and from the central nervous system.

» Neurons of the somatic part of the peripheral system control skeletal muscle and convey information about the external environment to the central nervous system.

» The autonomic system carries information to and from smooth muscle and cardiac muscle, and to glands. Signals from its two divisions—sympathetic and parasympathetic—have opposing effects.

32.8 The Spinal Cord

- The spinal cord serves as an information highway to and from the brain, and also as a reflex center.
- Spinal reflexes do not involve the brain.

An Information Highway

Your **spinal cord** is about as thick as your thumb. It runs through the vertebral column and connects peripheral nerves with the brain (**Figure 32.15**). The brain and spinal cord together constitute the central nervous system (CNS). Three membranes, called **meninges**, cover and protect these organs. The central canal of the spinal cord and spaces between the meninges are filled with **cerebrospinal fluid**. The fluid cushions blows and thus protects central nervous tissue.

The outermost part of the spinal cord, the **white matter**, consists of bundles of myelin-sheathed axons. In the CNS, such bundles are called tracts, rather than nerves. Tracts carry information from one part of the central nervous system to another. **Gray matter** makes up the bulk of the CNS. It consists of cell bodies, dendrites, and many neuroglial cells. In cross-section, the spinal cord's gray matter has a butterfly-like shape.

Spinal nerves of the peripheral nervous system connect to the spinal cord at dorsal and ventral "roots" (**Figure 32.16**). Remember, all spinal nerves have sensory and motor components. Sensory information travels to the spinal cord through a dorsal root. Cell bodies of sensory neurons reside in dorsal root ganglia. Motor signals travel away from the spinal cord through a ventral root. Cell bodies of motor neurons are in the spinal cord's gray matter.

An injury that severs or damages the nerve fibers running through the spinal cord can cause a loss of sensation and paralysis. Symptoms depend on what portion of the cord is damaged. Nerves carrying signals to and from the upper body lie higher in the cord than nerves that govern the lower body. An injury to the cord in the lower back often paralyzes the legs. An injury to higher cord regions can paralyze all limbs, as well as muscles used in breathing.

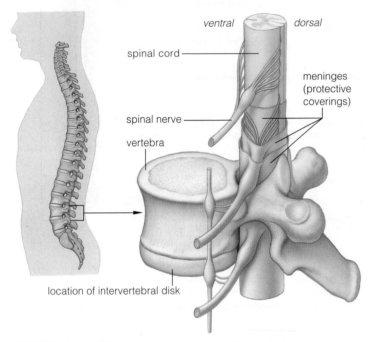

Figure 32.15 Animated Location and organization of the spinal cord.

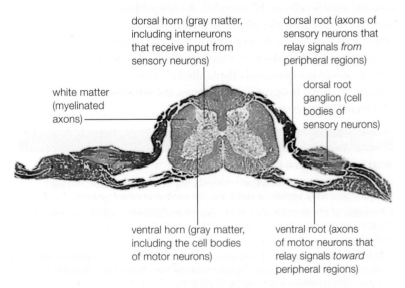

Figure 32.16 Cross-section of the spinal cord.

Reflex Pathways

Reflexes are the simplest and most ancient paths of information flow. A **reflex** is an automatic response to a stimulus, a movement or other action that does not require thought. Basic reflexes do not require any learning. With such reflexes, sensory signals flow to the spinal cord or the part of the brain just above it (the brain stem), which then commands a response by way of motor neurons.

The stretch reflex is a spinal reflex that causes a muscle to contract after some force stretches it (**Figure 32.17**). For example, suppose you hold a bowl as someone drops fruit into it ❶. Adding fruit increases the weight of the bowl, causing the biceps muscle in your upper arm to passively lengthen. Lengthening of the biceps excites endings of sensory receptors called muscle spindles that wrap around a muscle. As a result of this excitation, action potentials flow along the axon of the sensory neuron toward the spinal cord ❷.

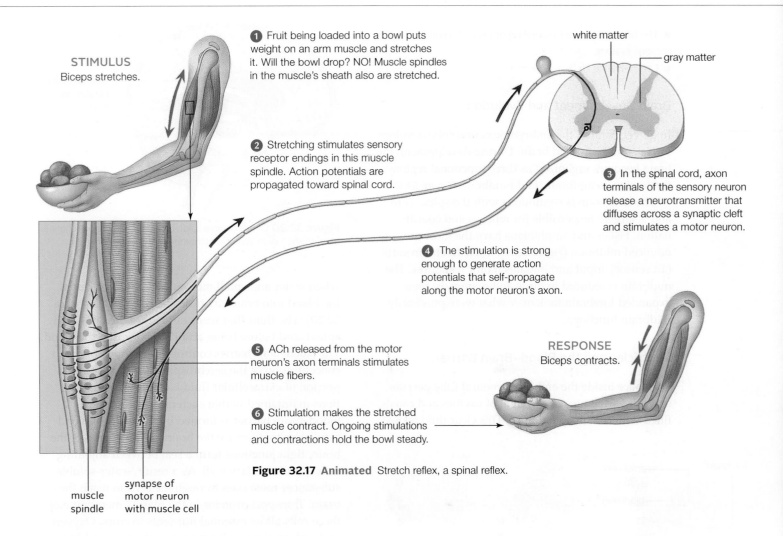

① Fruit being loaded into a bowl puts weight on an arm muscle and stretches it. Will the bowl drop? NO! Muscle spindles in the muscle's sheath also are stretched.

STIMULUS
Biceps stretches.

② Stretching stimulates sensory receptor endings in this muscle spindle. Action potentials are propagated toward spinal cord.

white matter

gray matter

③ In the spinal cord, axon terminals of the sensory neuron release a neurotransmitter that diffuses across a synaptic cleft and stimulates a motor neuron.

④ The stimulation is strong enough to generate action potentials that self-propagate along the motor neuron's axon.

RESPONSE
Biceps contracts.

⑤ ACh released from the motor neuron's axon terminals stimulates muscle fibers.

⑥ Stimulation makes the stretched muscle contract. Ongoing stimulations and contractions hold the bowl steady.

muscle spindle

synapse of motor neuron with muscle cell

Figure 32.17 Animated Stretch reflex, a spinal reflex.

Inside the spinal cord, axons of the sensory neuron synapse on dendrites of one of the motor neurons that controls the now stretched muscle **③**. Signals from the sensory neurons trigger action potentials that travel along the axon of the motor neuron **④**. The axon terminals release ACh at the neuromuscular junction **⑤**. In response to this signal, the biceps contracts and steadies the arm against the additional weight **⑥**.

Another spinal reflex, the withdrawal reflex, helps prevent burns. Touch a hot surface and action potential flows to the spinal cord. Unlike the stretch reflex, the withdrawal response involves a spinal interneuron. A heat-detecting sensory neuron alerts the spinal interneuron, which relays the signal to motor neurons. Before you know it, your biceps contracts, pulling your hand away from the potentially damaging heat.

cerebrospinal fluid Fluid that surrounds the brain and spinal cord and fills cavities within the brain.

gray matter Brain and spinal cord tissue consisting of cell bodies, dendrites, and neuroglial cells.

meninges The three membranes that enclose the brain and spinal cord.

reflex Automatic response that occurs without conscious thought or learning.

spinal cord Portion of central nervous system that connects peripheral nerves with the brain.

white matter Tissue of brain and spinal cord that consists of myelinated axons.

Take-Home Message

What are the functions of the spinal cord?

» Tracts of the spinal cord relay information between peripheral nerves and the brain. The axons involved in these pathways make up the bulk of the cord's white matter. Cell bodies, dendrites, and neuroglia make up gray matter.

» The spinal cord also has a role in some simple reflexes, automatic responses that occur without conscious thought or learning. Signals from sensory neurons enter the cord through the dorsal root of spinal nerves. Commands for responses go out along the ventral root of these nerves.

32.9 The Vertebrate Brain

■ The brain is the main integrating organ in the vertebrate nervous system.
■ Link to Fishes 25.3

Brain Development and Evolution

In all vertebrates, the embryonic neural tube develops into a spinal cord and brain. During development, the brain becomes organized as three functional regions: the forebrain, midbrain, and hindbrain (**Figure 32.18**).

The hindbrain is continuous with the spinal cord and is largely responsible for reflexes and coordination. Fishes and amphibians have the most pronounced midbrain (**Figure 32.19**). Their forebrain sorts out sensory input and initiates motor responses. The midbrain is reduced in birds and mammals; their expanded forebrain took over what were previously midbrain functions.

Ventricles and the Blood–Brain Barrier

The space inside the embryonic neural tube persists in adult vertebrates as a system of cavities and canals filled with cerebrospinal fluid. This clear fluid forms

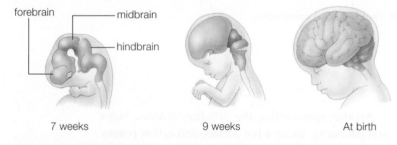

Figure 32.18 Human brain development. At 7 weeks, there is a hollow neural tube with regions that will develop into the forebrain, midbrain, and hindbrain.

7 weeks   9 weeks   At birth

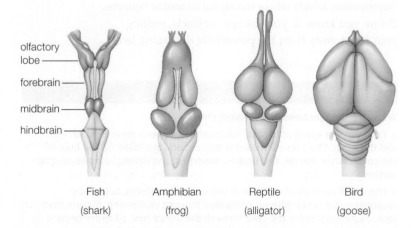

Fish (shark)   Amphibian (frog)   Reptile (alligator)   Bird (goose)

Figure 32.19 Vertebrate brains, dorsal views. Sketches are not to scale.

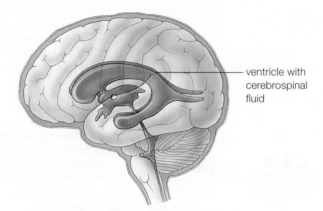

ventricle with cerebrospinal fluid

Figure 32.20 Cerebrospinal fluid. This clear fluid, shown here in *blue*, forms inside ventricles (chambers) within the brain.

when water and small molecules are filtered out of the blood into brain cavities called ventricles (**Figure 32.20**). The fluid that seeps out bathes the brain and spinal cord before being absorbed back into the blood. The **blood–brain barrier** controls the composition and concentration of the cerebrospinal fluid. No other portion of extracellular fluid has solute concentrations maintained within such narrow limits. The blood–brain barrier is formed by the walls of blood capillaries that service the brain. In most parts of the brain, tight junctions form a seal between adjoining cells of the capillary wall. As a result, water-soluble substances must pass through the cells to reach the brain. Transport proteins in the plasma membrane of these cells allow essential nutrients to cross. Oxygen and carbon dioxide diffuse across the barrier, but most waste products such as urea cannot breach it.

The blood–brain barrier cannot protect against all toxins. Nicotine, alcohol, and caffeine exert their effects after they cross the barrier. Mercury also crosses it. Inflammation, a blow to the head, or exposure to certain chemicals harms the barrier and makes it leaky.

The Human Brain

An average human brain weighs 1,240 grams, or 3 pounds. It contains about 100 billion interneurons, and neuroglia make up more than half of its volume.

The **medulla oblongata** is the part of the hindbrain just above the spinal cord (**Figure 32.21**). It governs the strength of heartbeats and the rate of breathing. It also controls reflexes such as swallowing, coughing, vomiting, and sneezing. The adjacent pons also affects breathing. **Pons** means "bridge," and refers to tracts that extend through the pons to the midbrain.

The **cerebellum** lies at the back of the brain and is about the size of a plum. It is densely packed with

Figure 32.21 The human brain.

(**A**) View of a whole brain from above. The meninges have been removed.

(**B**) The right half of a brain that has been sectioned along the central fissure.

(**C**) Major components of the brain and their functions.

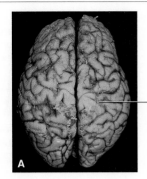

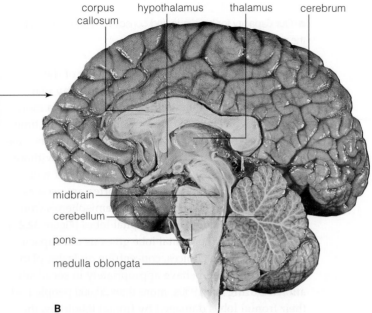

corpus callosum   hypothalamus   thalamus   cerebrum

midbrain
cerebellum
pons
medulla oblongata

B

neurons, having more than all other brain regions combined. The cerebellum controls posture and coordinates voluntary movements. Among other effects, excessive alcohol consumption disrupts coordination by impairing cerebellum function. Police officers evaluate the extent of impairment by asking a person suspected of being drunk to walk a straight line.

In humans, the midbrain is the smallest of the three brain regions. It plays an important role in reward-based learning. The pons, medulla, and midbrain are collectively referred to as the **brain stem**.

The forebrain consists mainly of the **cerebrum**, the largest structure in a human brain. A fissure divides the cerebrum into right and left hemispheres. The two hemispheres communicate via a thick band of nerve tracts called the corpus callosum. Each cerebral hemisphere controls and receives signals from the opposite side of the body. An outer layer of gray matter called the cerebral cortex is responsible for our unique capacities for language and abstract thought.

Most sensory signals destined for the cerebrum pass through the adjacent **thalamus**, which sorts them and sends them to the proper region of the cerebral cortex.

The **hypothalamus** ("under the thalamus") is the center for homeostatic control of the internal environment. It receives signals about the state of the body and regulates thirst, appetite, sex drive, and body temperature. It also interacts with the adjacent pituitary gland as a central control center for the endocrine system.

blood–brain barrier Protective barrier that prevents unwanted substances from entering cerebrospinal fluid.
brain stem The most evolutionarily ancient region of the vertebrate brain; includes the pons, medulla, and midbrain
cerebellum Hindbrain region responsible for coordinating voluntary movements.
cerebrum Forebrain region that controls higher functions.
hypothalamus Forebrain region that controls processes related to homeostasis; control center for endocrine functions.
medulla oblongata Hindbrain region that controls breathing rhythm and reflexes such as coughing and vomiting.
pons Hindbrain region between medulla oblongata and midbrain; helps control breathing.
thalamus Forebrain region that relays signals to the cerebrum.

Forebrain	
Cerebrum	Localizes, processes sensory inputs; initiates, controls skeletal muscle activity; governs memory, emotions, abstract thought
Thalamus	Relays sensory signals to and from cerebral cortex; has a role in memory
Hypothalamus	With pituitary gland, functions in homeostatic control. Adjusts volume, composition, temperature of internal environment; governs behaviors that ensure homeostasis (e.g., thirst, hunger)

Midbrain	Relays sensory input to the forebrain

Hindbrain	
Pons	Bridges cerebrum and cerebellum; also connects spinal cord with forebrain. With the medulla oblongata, controls rate and depth of respiration
Cerebellum	Coordinates motor activity for moving limbs and maintaining posture, and for spatial orientation
Medulla oblongata	Relays signals between spinal cord and pons; functions in reflexes that affect heart rate, blood vessel diameter, and respiratory rate. Also involved in vomiting, coughing, other reflexive functions

C

Take-Home Message

How does the vertebrate brain develop, and what are its functional regions?

» The vertebrate brain develops from a hollow neural tube, the interior of which persists in adults as a system of cavities and canals filled with cerebrospinal fluid.

» Tissue of the embryonic neural tube develops into the hindbrain, forebrain, and midbrain. The hindbrain controls reflexes and coordination. The unique capacities of humans arise in regions of their enlarged forebrain.

32.10 A Closer Look at the Human Cerebrum

■ Our capacity for language and conscious thought arises from the activity of the cerebral cortex.

The **cerebral cortex**, the outermost portion of the cerebrum, is a 2-millimeter-thick layer of gray matter. Compared to other animals, humans have a more highly folded cerebral cortex. Over evolutionary time, the increased folding allowed humans to fit an increasing amount of cortical tissue into their heads, without increasing skull size so much that it interfered with childbirth. Prominent folds in the cortex are used as landmarks to define each cerebral hemisphere's frontal, parietal, temporal, and occipital lobes (**Figure 32.22**).

The cortex of the frontal lobe gives us our capacity to make reasoned choices, concentrate on tasks, plan for the future, and behave appropriately in social situations. During the 1950s, more than 20,000 people had their frontal lobes damaged by frontal lobotomy, the surgical destruction of frontal lobe tissue. One famous advocate of the procedure carried out it out by inserting an ice pick through the bone at the back of the eye and wiggling it. Lobotomy was used to treat mental illness, personality disorders, and even severe headaches. The treatment made patients calmer, but also blunted their emotions and impaired their ability to plan, concentrate, and behave appropriately.

The two hemispheres differ somewhat in their function. In the 90 percent of people that are right-handed, the left hemisphere plays the major role in movement

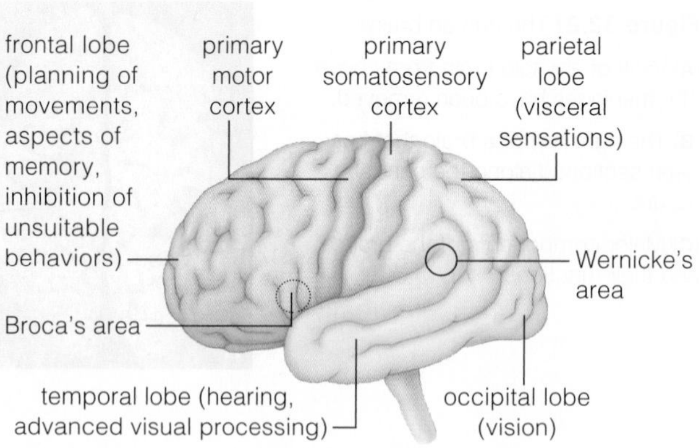

Figure 32.22 Lobes of the brain, with primary receiving and integrating centers of the human cerebral cortex.

frontal lobe (planning of movements, aspects of memory, inhibition of unsuitable behaviors) — primary motor cortex — primary somatosensory cortex — parietal lobe (visceral sensations) — Wernicke's area — Broca's area — temporal lobe (hearing, advanced visual processing) — occipital lobe (vision)

and in language. Broca's area, an association area that translates thoughts into speech, is usually in the left frontal lobe. Damage to Broca's area often prevents a person from speaking, but does not affect understanding of language.

Despite their differences, both hemispheres are flexible. If a stroke or injury damages one side of the brain, the other hemisphere can take on new tasks. People can even function with a single hemisphere.

A **primary motor cortex** at the rear of the frontal lobes controls skeletal muscles. Neurons of this region are laid out like a map of the body, with body parts that make fine movements such as the fingers and thumbs, commanding the greatest area (**Figure 32.23**).

The **primary somatosensory cortex** at the front of the parietal lobe receives sensory input from the skin and joints. Like the motor cortex, it is organized as a map that corresponds to body parts, with the most sensitive parts mapping onto a disproportionately large area.

Perceptions of sound and odor arise in sensory areas of the temporal lobes. Wernicke's area, which is typically in the left temporal lobe, allows comprehension of spoken and written language.

A primary visual cortex at the rear of each occipital lobe receives signals from the eyes.

cerebral cortex Outer gray matter layer of the cerebrum.
primary motor cortex Region of brain's frontal lobes that governs voluntary movements.
primary somatosensory cortex Region of the brain's parietal lobes that receives sensory information from skin and joints.

Take-Home Message

What are the functions of the cerebral cortex?

» The cerebral cortex controls voluntary activity, sensory perception, abstract thought, and language and speech.

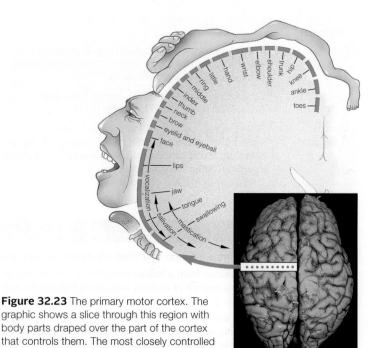

Figure 32.23 The primary motor cortex. The graphic shows a slice through this region with body parts draped over the part of the cortex that controls them. The most closely controlled body parts appear disproportionately large.

32.11 Emotion and Memory

■ A collection of structures in the midbrain plays an important role in emotion and memory.

The Limbic System

The **limbic system** encircles the upper brain stem. It governs emotions, assists in memory, and correlates organ activities with self-gratifying behavior such as eating and sex. That is why the limbic system is known as our emotional-visceral brain, as opposed to the rational cerebral cortex, which can often override the limbic system's "gut reactions."

The limbic system includes the hypothalamus, hippocampus, amygdala, and cingulate gyrus (**Figure 32.24**). The hypothalamus is the major control center for homeostatic responses. It also correlates emotions with visceral activities. The hippocampus functions in the storage and retrieval of memories, including memories of earlier threats. The almond-shaped amygdala helps interpret social cues, and contributes to the sense of self. It is most active during episodes of fear and anxiety, and often it is overactive in people afflicted with panic disorders. The cingulate gyrus has a role in attention and in emotion. It is often smaller and less active than normal in people with schizophrenia.

Evolutionarily, the limbic system is related to the olfactory lobes. Olfactory input causes signals to flow to the hippocampus, amygdala, and hypothalamus as well as to the olfactory cortex. That is one reason why specific odors can call up emotionally significant memories. Information about taste also travels to the limbic system and can similarly trigger emotional responses.

Making Memories

The cerebral cortex receives information continually, but only a fraction of it becomes memories. Memory forms in stages. Short-term memory lasts seconds to hours. This type of memory holds a few bits of

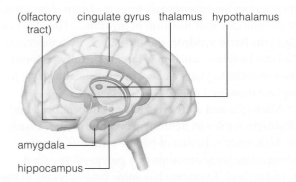

(olfactory tract)   cingulate gyrus   thalamus   hypothalamus

amygdala

hippocampus

Figure 32.24 Limbic system components.

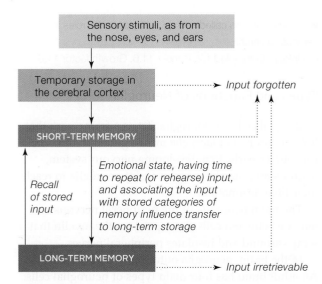

Figure 32.25 Stages in memory processing.

information: a set of numbers, words of a sentence, and so forth. In long-term memory, larger chunks of information get stored more or less permanently (**Figure 32.25**).

Different types of memories are stored and brought to mind by different mechanisms. Repetition of motor tasks creates highly persistent skill memories. Once you learn to ride a bicycle, drive a car, dribble a basketball, or play the piano, you are unlikely to lose that skill. Making skill memories involves neurons in the cerebellum, which controls motor activity.

Declarative memory stores facts and impressions. It helps you remember how a lemon smells and that a quarter is worth more than a dime. Input to declarative memory begins with signals from the sensory cortex to the amygdala. The amygdala screens the input, and sends some signals along to the hippocampus. Declarative memory also involves the temporal lobe, which is why damage to this lobe can cause amnesia.

Emotions influence memory retention. For instance, epinephrine released during times of stress helps convert short-term memories into long-term memories. Thus, traumatic events are often remembered more vividly than more pleasant ones.

limbic system Group of brain structures that govern emotion.

Take-Home Message

What is the cerebral cortex?

» The cerebral cortex, the outer layer of gray matter, has areas that receive and integrate sensory information. It also controls conscious thought and actions.

» The cerebral cortex interacts with the limbic system, a set of brain structures that collectively affect emotions and contribute to memory.

32.12 Neuroglia—The Neurons' Support Staff

■ A variety of cells called neuroglia play essential roles in nervous systems.

■ Links to Cell cycle 11.2, Cancer 11.6, Growth factor 11.6

Types and Functions of Neuroglia

Neuroglial cells, or neuroglia, act as a framework that holds neurons in place; *glia* means glue in Latin. Their role begins early. In a developing nervous system, neurons migrate along highways of neuroglia to reach their final destination.

The main neuroglia of the peripheral nervous system are Schwann cells. They produce the myelin that wraps around and insulates peripheral nerves.

In the brain, neuroglia outnumber neurons 10 to 1. An adult brain has four main types of neuroglial cells: oligodendrocytes, microglia, astrocytes, and ependymal cells. Oligodendrocytes are the functional equivalent of Schwann cells; they make myelin sheaths that insulate axons in the central nervous system.

Microglia are, as the name implies, the smallest neuroglial cells. They continually survey the brain. If brain tissue is injured or infected, microglia become active, motile cells that engulf dead or dying cells and debris. They also produce chemical signals that alert the immune system to threats.

Star-shaped astrocytes are the most abundant cells in the brain (**Figure 32.26**). Scientists have only recently begun to understand their diverse and essential roles. Astrocytes wrap around blood vessels that supply the brain and stimulate formation of the blood–brain barrier. They play a role at synapses, where they take up neurotransmitters released by neurons. They also assist in immune defense, release lactate that fuels the activity of neurons, and synthesize nerve growth factor. Neurons are stopped in G1 of the cell cycle (Section 11.2) and cannot divide, but nerve growth factor encourages a neuron to form new synapses.

Ependymal cells line a brain's fluid-filled cavities (ventricles) and the spinal cord's central canal. The action of cilia on ependymal cells keeps cerebrospinal fluid flowing in a consistent direction.

Neuroglia in Disease

Multiple sclerosis (MS) is an autoimmune disorder that arises when white blood cells mistakenly attack and destroy the myelin sheaths of oligodendrocytes. As myelin becomes replaced by scar tissue, the speed of action potential propagation along the affected axons declines. MS typically appears in young adults, with women affected twice as often as men. The cause of MS is unknown, although genetics certainly plays a role. In the general population, the risk of developing MS is about 1 in 1,000, but a person whose identical twin has MS has a risk of 1 in 4. The fact that one twin can be affected while another is not indicates that environmental factors also influence who develops the disease. MS is incurable and it can cause progressively worsening weakness and fatigue, impaired balance, and vision problems.

Guillain-Barré syndrome is the peripheral nervous system equivalent of MS. It occurs when the immune system attacks and breaks down the myelin of peripheral nerves. Typically, Guillain-Barré syndrome arises after a viral or bacterial infection. Tingling and weakness in the legs are early symptoms, and paralysis sets in within a week or two. The syndrome can be life-threatening because signals from the brain stem cannot reach muscles that function in breathing. The effects of Guillain-Barré syndrome are usually only temporary. Normal nervous function is largely or fully restored over months to years as new myelin sheaths grow to replace damaged ones.

Microglia and astrocytes have both beneficial and destructive roles in neurodegenerative diseases such as Alzheimer's. Brains of people with this disease accumulate large amounts of a polypeptide called beta-amyloid. Everyone has some beta-amyloid in the plasma membranes of their brain cells, but neurons

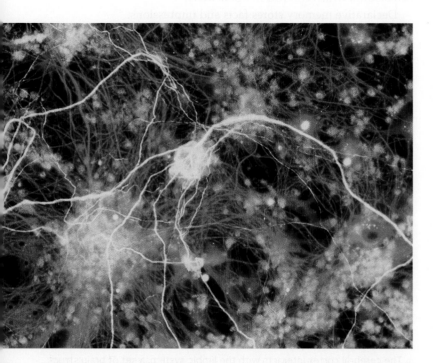

Figure 32.26 Astrocytes (*orange*) and a neuron (*yellow*) in brain tissue. The cells in this fluorescence micrograph were made visible by use of a light-emitting tracer.

In Pursuit of Ecstasy (revisited)

Now that you know a bit more about how the brain functions, take a moment to reconsider effects of MDMA, the active ingredient in Ecstasy. MDMA mainly targets neurons that release the neurotransmitter serotonin. It binds to transporters that pump serotonin from the synaptic cleft back into axon terminals. Binding of MDMA prevents normal reuptake of serotonin and makes the transporter leaky. Both effects increase the amount of serotonin that remains in the synaptic cleft.

Excess serotonin acts on postsynaptic cells and produces the pleasurable sensations that users of Ecstasy seek. However, in the aftermath of the drug use, serotonin-releasing neurons are left with a lowered amount of this essential neurotransmitter. Low serotonin levels are associated with depression, and transient depression is a well-documented aftereffect of Ecstasy use. Depletion of serotonin in cells of the hippocampus may be responsible for MDMA's detrimental effect on memory.

MDMA also impairs reuptake of dopamine and norepinephrine. In animals, it damages the blood–brain barrier that keeps toxins out of the brain. In one study

of rats, the barrier was impaired for as long as 10 weeks after administration of MDMA.

How would you vote? MDMA's ability to make people feel socially comfortable has caused some to advocate for its use in psychotherapy. Others think there are too many negative side effects for MDMA to be useful. Should studies of the potential benefits of MDMA use be carried out?

---

of people with Alzheimer's have an excess amount. Similar to what occurs in prion diseases (Section 3.7), the beta-amyloid polypeptide changes shape and aggregates as large sticky plaques in the brain. Microglia and astrocytes take up and break down beta-amyloid. However, their exposure to beta-amyloid causes them to release chemicals that initiate inflammation and summon white blood cells. Both outcomes can damage and kill neurons. Investigators are trying to find ways to treat Alzheimer's by encouraging neuroglial cells to clear away beta-amyloid. At the same time, they are also trying to find ways to minimize the damaging effects of these defenders.

Brain Tumors

Neurons do not divide and are not replenished, so they do not give rise to tumors. However, uncontrolled division of cells that give rise to neuroglial cells sometimes causes a tumor called a glioma. Tumors can also arise from epithelial cells in the meninges or endocrine glands of the brain, such as the pituitary. In addition, brain tumors can result from arrival of cancerous cells from elsewhere in the body.

Most tumors that originate in the brain are not cancer. However, even a benign (noncancerous) tumor can pose a serious threat. Benign tumors do not metas-

tasize (Section 11.6), but growth of a tumor within the confined space of the skull can put pressure on surrounding nervous tissue and damage neurons.

Some people are concerned that the radio waves emitted by cell phones could cause brain tumors. Epidemiological studies have not shown any increased brain tumor incidence related to cell phone use. However, cell phones are a relatively recent invention, their use has increased since they first became available, and brain tumors can take years to develop. In addition, a recent study found that the radiation emitted by a cell phone increases the metabolic activity of brain cells near the phone. Some health officials now recommend the use of a headset to keep the radiation-emitting part of the phone from coming in close proximity to the brain.

Take-Home Message

What are the functions of neuroglia and how do they affect health?

» Neuroglial cells make up the bulk of the brain. They provide a framework for neurons, insulate neuron axons, assist neurons metabolically, and protect the brain from injury and disease.

» Because neuroglia have essential roles in assisting neurons, diseases that impair neuroglia impair the function of the nervous system.

» Unlike neurons, most types of neuroglia continue to divide. Thus, neuroglia can be a source of brain tumors.

LEARNING ROADMAP

Where you have been This chapter explains how the cone snail venom discussed in Section 24.1 prevents pain. It draws on your knowledge of action potentials (32.4), neuromodulators (32.6), and the human brain (32.9–32.11). We also refer back to morphological convergence (18.3) and other phenomena of vertebrate evolution (25.6).

Where you are now

How Sensory Pathways Work
Information from sensory receptors becomes encoded in the number and frequency of action potentials sent to the brain along particular nerve pathways.

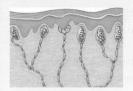

Somatic and Visceral Senses
Somatic sensations are easily localized to skin or joints. Visceral sensations arise in soft organs and are less easily pinpointed.

Chemical Senses
Sensations of smell and taste arise when chemoreceptors bathed in fluid bind molecules of specific substances in the fluid.

Vision
Vision begins with the activation of photoreceptors. Vertebrates have an eye that operates like a film camera. Their sensory pathway starts at the retina and ends in the visual cortex.

Hearing and Balance
Mechanoreceptors in the ear function in hearing and balance. Sound waves excite receptors in the ear's cochlea, and body movements excite receptors in the vestibular apparatus.

Where you are going We return to the function of sensory receptors in later chapters. Section 40.5 describes the role of osmoreceptors in thirst, Section 36.9 discusses the role of pressure receptors in maintaining blood pressure, and Section 40.9 discusses the role of thermoreceptors in thermoregulation. The evolution of visual, auditory, and chemical communication signals are detailed in Section 43.6.

33.1 A Whale of a Dilemma

Imagine yourself in the sensory world of a whale, gliding along 200 meters (650 feet) beneath the ocean surface. Almost no sunlight penetrates to this depth, so vision is extremely limited. Many fishes navigate with the help of a lateral line, a system that detects differences in water pressure. Fishes also use dissolved chemicals as navigational cues. However, a whale has no lateral line and a very poor sense of smell.

Whales navigate by using sounds—acoustical cues. Sound waves move five times faster in water than in air, so water is an ideal medium for transmitting sound. Unlike humans, whales do not have a pair of ear flaps that collect sound waves. Some whales do not even have a canal leading to the ear components inside their head. Others have ear canals, but the canals are packed with wax. A whale hears when its jaws pick up vibrations traveling through water. The vibrations are transmitted from the jaws, through a layer of fat, to a pair of pressure-sensitive middle ears.

Whales rely on sound to communicate and locate food, as well as to find their way around underwater. Killer whales and some other species of toothed whales use echolocation. The whale emits high-pitched sounds and then listens as the echoes bounce off objects, including prey. Its ears are especially sensitive to high-frequency sounds. Baleen whales, including the humpback whale, make very low-pitched sounds that can travel across an entire ocean basin. Thus, their ears are especially sensitive to low-frequency sounds.

Whales attuned to low-frequency sounds are especially threatened by human-created noise in this range. Naval sonar systems that blast pulses of low-frequency sounds to detect submarines have caused some whale strandings (**Figure 33.1**). Autopsies of whales that beached themselves during naval testing of such sonar systems revealed that the whales had blood in their ears and their acoustic fat. This suggests that the intense sounds emitted by the sonar caused the whales to alter their diving pattern in a manner that damaged their internal tissues.

Naval sonar, though dramatic in its effects, is scarce compared with commercial shipping and undersea mining operations that also produce noise pollution. Both emit sounds that can frighten or disorient whales and interfere with their ability to communicate.

As you will learn in this chapter, animals differ in the type and number of sensory receptors that sample the environment, and differ in their perception of it. These sensory systems evolved over countless generations in an environment free of human-generated noise, light, and other sensory distractions. Finding ways to meet human needs without needlessly disrupting the sensory worlds of Earth's other species is an ongoing challenge.

Figure 33.1 A few children drawn to one of the whales that stranded itself during military testing of a new sonar system. Of sixteen stranded whales, six died on the beach. Volunteers pushed the others out to sea. Their fate is unknown.

33.2 Overview of Sensory Pathways

■ An animal's sensory receptors determine what features of the environment it can detect and respond to.

■ Links to Neurons 32.3, Action potentials 32.4

The sensory portion of a vertebrate nervous system consists of sensory neurons that detect stimuli, nerves that carry information about the stimulus to the brain, and brain regions that process such information. A **stimulus** (plural, stimuli) is a form of energy that excites receptor endings of a sensory neuron. When sufficiently powerful, this excitation causes an action potential to occur. The action potential travels along a peripheral nerve to the central nervous system.

A Mechanoreceptors inside a bat's inner ear allow the animal to detect high-pitched sound waves (ultrasound).

B Thermoreceptors in pits above and below a python's mouth allow it to detect body heat, or infrared energy, of its prey.

Figure 33.2 Examples of sensory receptors.

Figure 33.3 A marsh marigold looks yellow to humans (**A**), but photographing it with UV-sensitive film reveals a dark area around the reproductive parts (**B**). This pattern is caused by UV-absorbing pigment that is visible to insect pollinators.

Sensory Receptor Diversity

Animals live in diverse habitats, and the types of stimuli that sensory neurons detect vary among animal groups. We classify sensory neurons based on the stimuli to which they respond.

Mechanoreceptors are sensory neurons that respond to mechanical energy. Some alter their rate of action potentials with shifts in a body's position or acceleration. Other mechanoreceptors fire off action potentials in response to touch or to stretching of a body part. The muscle spindles involved in the human stretch reflex (Section 32.8) are a type of mechanoreceptor.

Still other mechanoreceptors respond to vibrations caused by pressure waves. Hearing involves this type of receptor because sound is a type of pressure wave. Auditory receptors vary in the frequency that they detect. Whales detect ultra-low frequencies that humans cannot hear. Bats emit and respond to sounds too high for humans to perceive (**Figure 33.2A**).

Chemoreceptors detect specific solutes dissolved in a fluid. Nearly all animals have chemoreceptors that help them locate chemical nutrients and avoid ingesting poisons. Chemoreceptors function in the human senses of taste and smell.

Pain receptors, also called nociceptors, detect tissue damage. They have a protective function and are often involved in reflexes that minimize further harm.

Thermoreceptors respond to a specific temperature or fire in response to a temperature change. Pythons and some other snakes have thermoreceptors concentrated in pits on their head (**Figure 33.2B**). The receptors help a snake detect the body heat of prey.

Photoreceptors detect light energy. Humans detect only visible light, but insects and some other animals, including rodents, also respond to ultraviolet light. Insect-pollinated flowers often have UV-absorbing pigments arranged in patterns that are invisible to us, but catch the attention of insects (**Figure 33.3**).

Some types of sensory receptors inform the brain about the state of the internal environment. **Osmoreceptors** monitor the solute concentration in a body fluid such as plasma or cerebrospinal fluid. Other receptors monitor the pressure exerted by blood against the walls of blood vessels.

From Sensing to Sensation

In animals that have a brain, processing of sensory signals gives rise to **sensation**: awareness of a stimulus.

Sensory receptors in skin, in skeletal muscle, or near joints give rise to somatic sensations such as those of touch, warmth, or muscle pain. Visceral sensations, such as the feeling that your bladder or stomach is full, arise from receptors in the walls of internal organs. Sensory receptors restricted to specific sensory organs, such as eyes or ears, function in special senses—vision, smell, balance, hearing, and taste.

When a sensory receptor is stimulated sufficiently, it produces an action potential. Action potentials, remember, are always the same size (Section 32.4). The brain gathers additional information about stimuli in three ways, by noting (1) which nerve pathways are active, (2) the frequency of action potentials traveling on each axon in the pathway, and (3) the number of axons recruited by the stimulus.

An animal's brain interprets action potentials on the basis of where they originate. This is why you may "see stars" if you press on your eyes in a dark room. The pressure causes photoreceptors in the eye to undergo action potentials and send signals along one of two optic nerves to the brain. The brain interprets all signals from an optic nerve as "light."

A strong stimulus causes a receptor to generate action potentials more often and longer than a weak signal does. The same receptors are stimulated by a whisper and a whoop. Your brain interprets the dif-

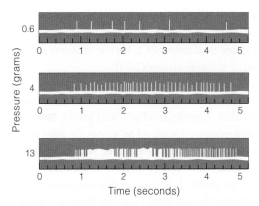

Figure 33.4 Animated Action potential frequency in a pressure-sensitive mechanoreceptor in the skin. A rod was pressed against skin with varying amounts of pressure. Vertical bars above each thick horizontal line represent action potentials. The stronger the pressure, the more action potentials (*white* bars) occur per second.

ference by variations in the frequency of the incoming signals (**Figure 33.4**). In addition, a strong stimulus recruits more sensory receptors, compared with a weak stimulus. A gentle tap on the arm activates fewer receptors than a slap.

Stimulus duration also affects how the stimulus is interpreted. In **sensory adaptation**, sensory neurons stop generating action potentials (or make fewer of them) despite continued stimulation. Walk into a house where an apple pie is in the oven and you will notice the sweet scent of baking apples immediately. Then, within a few minutes, the scent seems to lessen. The odor does not actually change in intensity, but chemoreceptors in your nose adapt to it.

Sensory Perception

Sensory perception arises when the brain assigns meaning to sensory signals. Consider what happens when you watch a person walking away from you. As the distance between you and the person increases, the image of the person on your eye becomes smaller and smaller. You perceive this change in sensation as evidence of increasing distance between you and the person, rather than a sign that the person is shrinking.

chemoreceptor Sensory receptor that responds to a chemical.
mechanoreceptor Sensory receptor that responds to pressure, position, or acceleration.
osmoreceptor Sensory receptor that detects shifts in the solute concentration of a body fluid.
pain receptor Sensory receptor that responds to tissue damage.
photoreceptor Sensory receptor that responds to light.
sensation Detection of a stimulus.
sensory adaptation Slowing or cessation of a sensory receptor response to an ongoing stimulus.
sensory perception The meaning a brain derives from a sensation.
stimulus A form of energy that is detected by a sensory receptor.
thermoreceptor Temperature-sensitive sensory receptor.

Take-Home Message

How do animals detect and process sensory stimuli?

» Sensory neurons undergo action potentials in response to specific stimuli. Different kinds of sensory receptors respond to different types of stimuli.

» Action potentials are all the same size, but which axons are responding, how many are responding, and the frequency of action potentials provide the brain with information about stimulus location and strength.

33.3 Somatic and Visceral Sensations

■ Signals from receptors in the skin, joints, muscles, and internal organs flow through the spinal cord to the brain.
■ Links to Cone snail venom 24.1, Neuromodulators 32.6, Spinal cord 32.8, Cerebral cortex 32.10

Sensory neurons responsible for somatic sensations are located in skin, muscle, tendons, and joints. **Somatic sensations** are easily localized to a specific part of the body. In contrast, **visceral sensations**, which arise from neurons in the walls of soft internal organs, are often difficult to pinpoint. It is easy to determine exactly where someone is touching you, but less easy to say exactly where you feel a stomachache or nausea.

The Somatosensory Cortex

Signals from the sensory neurons involved in somatic sensation travel along axons to the spinal cord, then along tracts in the spinal cord to the brain. The signals end up in the somatosensory cortex, a part of the cerebral cortex. Like the motor cortex (Section 32.10), the somatosensory cortex has neurons arrayed like a map of the body (**Figure 33.5**). Body parts that are disproportionately large in the "body" mapped onto this brain correspond to body regions with the most sensory receptors, such as the fingertips, face, and lips. Body parts that have relatively fewer sensory neurons, such as thighs, are disproportionately small.

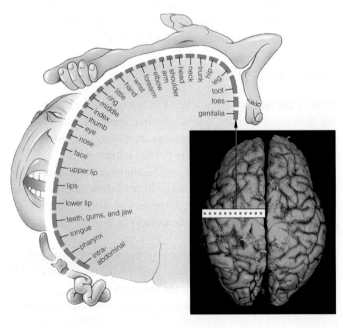

Figure 33.5 A map showing where the different body regions are represented in the human primary somatosensory cortex. This brain region is a narrow strip of the cerebral cortex that runs from the top of the head to just above each ear. Compare **Figure 32.23**.

Receptors Near the Body Surface

As an example of the types of receptors that signal the somatosensory cortex, consider those in the human skin (**Figure 33.6**). Receptor endings of sensory neurons may be either surrounded by some sort of capsule or free (unenclosed) in the tissue. Some of the free nerve endings coil around the roots of hairs in the dermis and can detect even the slightest pressure on the hair. Other free nerve endings detect temperature changes or tissue damage. Free nerve endings also occur in skeletal muscles, tendons, joints, and walls of internal organs. Here, they give rise to sensations that range from itching, to a dull ache, to sharp pain.

Concentric layers of connective tissue wrap around the receptor endings of many sensory receptors. These encapsulated receptors are named for the scientists who first described them. Meissner's corpuscles and Pacinian corpuscles detect touch and pressure in hairless skin regions such as fingertips, palms, and the soles of the feet. Small Meissner's corpuscles in the upper dermis detect light touches. Pacinian corpuscles respond to more pressure. They are larger, deeper in the dermis, and also near joints and in the wall of some organs. Ruffini endings adapt more slowly than Meissner's and Pacinian corpuscles. If you hold a stone in your hand, Ruffini endings inform your brain that the stone is still there even after other receptors have adapted and stopped responding. Ruffini endings also fire when temperature exceeds 45°C (113°F). The bulb of Krause, another encapsulated receptor, responds to touch and cold.

Muscle Sense

Remember those stretch receptors in muscle spindle fibers (Section 32.8)? The more a muscle stretches, the more frequently stretch receptors fire. In concert with receptors in tendons and near movable joints, they inform the brain about positions of the body's limbs.

Pain

Pain is the perception of a tissue injury. Somatic pain begins with signals from pain receptors in skin, skeletal muscles, joints, and tendons. Visceral pain is associated with organs inside body cavities. It occurs as a response to a smooth muscle spasm, inadequate blood flow to an organ, overstretching of a hollow organ such as the stomach, and other abnormal conditions.

Injured or distressed body cells release local signaling molecules such as histamine and prostaglandins.

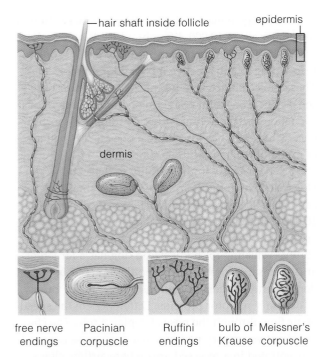

Figure 33.6 **Animated** Sensory receptors in human skin.

free nerve endings   Pacinian corpuscle   Ruffini endings   bulb of Krause   Meissner's corpuscle

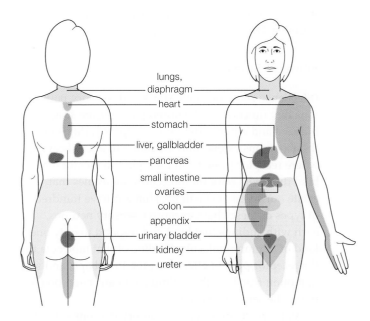

Figure 33.7 Sites of referred pain. Colored regions indicate the area that the brain interprets as affected when specific internal organs are actually distressed.

These molecules stimulate neighboring pain receptors, increasing the likelihood that the receptors will send signals along their axons to the spinal cord. In the spinal cord, the axons synapse with interneurons that relay signals about pain to the somatosensory cortex.

Neuromodulators affect signal transmission at the synapse between a pain-detecting sensory neuron and a spinal interneuron. For example, endorphins interfere with the ability to perceive pain. By contrast, substance P enhances pain perception, making spinal interneurons more likely to send signals about pain on to the sensory cortex. People with fibromyalgia, a disorder characterized by chronic pain in muscles and joints throughout the body, tend to have an elevated level of substance P.

Pain-relieving drugs (analgesics) interfere with pain perception. Aspirin reduces pain by slowing the production of prostaglandins. Synthetic opiates such as morphine mimic the activity of endorphins. The drug ziconotide, a chemical first discovered in the venom of a cone snail (Section 24.1), blocks calcium channels in the axon terminals of pain receptor neurons when injected into the spinal cord. Blocking these channels prevents the neurons from releasing neurotransmitter. Thus, ziconotide inhibits transmission of pain signals to interneurons that would otherwise relay the signals to the brain.

The brain sometimes mistakenly interprets signals from receptors in internal organs as if they were from receptors in the skin or joints. The result is referred pain. The classic example is a pain that radiates from chest across the shoulder and down the left arm during a heart attack (Figure 33.7). The arm is not affected, so why does it hurt? Referred pain occurs because each part of the spinal cord receives sensory input from both skin and internal organs. Skin encounters more painful stimuli than internal organs do, so pain signals from skin flow more frequently along the neural pathway to the brain. The brain tends to attribute signals that arrive along a particular pathway to their most common source, even if they originate elsewhere.

pain Perception of tissue injury.
somatic sensations Sensations such as touch and pain that arise when sensory neurons in skin, muscle, or joints are activated.
visceral sensations Sensations that arise when sensory neurons associated with organs inside body cavities are activated.

Take-Home Message

How do somatic and visceral sensations arise?

» Somatic sensations are signals from sensory receptors in skin, skeletal muscle, and joints. They travel along sensory neuron axons to the spinal cord, then to the somatosensory cortex.

» Visceral sensations originate with the stimulation of sensory neurons in the walls of organs. These signals are relayed to the spinal cord, and then to the brain.

» Pain is the sensation associated with tissue damage. Because pain signals originate most often with somatic sources, the brain sometimes misinterprets visceral pain as if it were caused by a problem in the skin or a joint.

33.4 Chemical Senses

- Both smell and taste begin with chemoreceptors.
- Links to Microvilli 31.3, Limbic system 32.11

Sense of Smell

Olfaction, a sense of smell, starts with sensory neurons that function as chemoreceptors (**Figure 33.8**). In humans and other vertebrates, receptor endings of these neurons extend into the lining of a nasal cavity, where receptor proteins in their plasma membrane bind dissolved odorant molecules (molecules that excite our sense of smell) ❶. Humans have hundreds of types of these chemoreceptive sensory neurons, each with endings that bind and respond to only one kind of odorant molecule. Binding of that molecule to the receptor protein alters the protein's shape, and sets in motion reactions that culminate in an action potential in the sensory neuron.

Axons of chemoreceptive sensory neurons extend through the base of the skull and into the brain's olfactory bulb ❷, where they synapse on interneurons. Each interneuron receives signals from many sensory neurons that all detect the same odorant molecule ❸. In response to excitatory signals from these cells, the interneurons send signals to other brain regions such as the limbic system and the cerebral cortex ❹.

An average person can discriminate between and remember about 10,000 different odors. When you recognize an odor, such as the scent of a rose, you are responding to a distinctive mix of odorant molecules. These molecules excite a unique subset of your nose's sensory neurons, thus triggering a unique pattern of excitation in the cerebral cortex. Through past experience, your brain has learned to associate this pattern of excitatory signals with its source.

Vertebrates vary in their ability to detect odors. Dogs are known for their acute sense of smell. They can detect specific odorant molecules at concentrations 10,000 to 100,000 times lower than humans can. A variety of traits contribute to the dog's superior olfactory ability. Adjusting for body size, a dog has a far larger area of nasal epithelium. This epithelium has both a denser array of sensory neurons and a greater variety of olfactory receptor proteins. Differences in the speed of air flow through the nose also play a role. The chemoreceptive sensory neurons in a dog's nose are most concentrated in a recessed area where airflow slows, so receptor proteins in their membranes have time to bind more odorant molecules. The human nasal cavity, like that of other primates, has no equivalent recessed region. Air flows past our receptor proteins faster, so fewer odorant molecules bind.

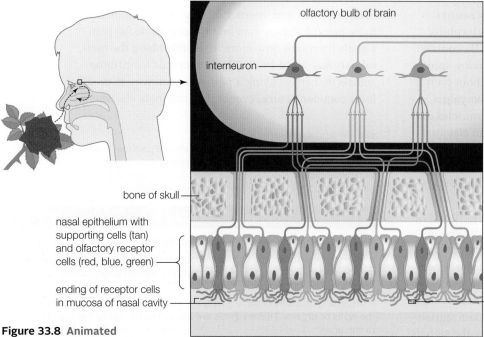

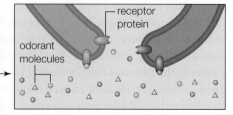

❹ Interneurons in the olfactory lobe relay signals from sensory neurons to other regions of the brain, including the limbic system and cerebral cortex.

❸ In the olfactory bulb, axons of sensory neurons synapse on interneurons. Each interneuron receives signals only from cells with the same type of receptor.

❷ An odorant molecule binds to a receptor protein on a sensory neuron, initiating an action potential that travels along the neuron's axon to the olfactory bulb in the brain.

Figure 33.8 Animated
Human sense of smell. Odorant molecules excite a sensory neuron when they bind to chemoreceptors in its plasma membrane.

❶ Inhaled odorant molecules bind to receptor proteins on chemoreceptive sensory neurons in the nasal cavity. A receptor protein binds only one type of odorant molecule, and each cell has only one type of receptor. This art shows three types of receptor cells (coded *red*, *green*, and *blue*), but there are hundreds.

Sense of Taste

Like olfactory receptors, **taste receptors** are chemoreceptors that detect chemicals dissolved in fluid, but they differ in their structure and location. Taste receptors help animals locate food and avoid poisons. An octopus "tastes" substances with receptors in suckers on its tentacles. A fly "tastes" using receptors in its antennae and feet.

Humans have taste receptors in taste buds on the lining of their mouth and the upper surface of the tongue. Taste buds are located in specialized epithelial structures, or papillae, that are visible as raised bumps (Figure 33.9). Each taste bud contains taste receptor cells and neurons. Receptor-covered microvilli of the taste receptor cells extend out through a pore, and thus come in contact with food molecules in saliva.

The sensation of taste begins with the binding of a molecule to protein receptors on the microvilli of a taste receptor cell. Binding of the molecule causes excitement of the taste receptor cell. This cell in turn excites a sensory neuron within the taste bud, which relays an action potential along its axon to the brain.

You perceive many tastes, but all arise from stimulation of just five classes of sensory receptors. The five primary tastes are *sweet* (elicited by glucose and the other simple sugars), *sour* (acids), *salty* (sodium chloride or other salts), *bitter* (plant toxins, including alkaloids), and *umami* (elicited by amino acids such as glutamate). The food additive MSG (monosodium glutamate) can enhance flavor by stimulating the taste receptors that contribute to the sensation of umami.

Each taste receptor cell is most sensitive to one of the five primary tastes, but taste buds in all regions of the tongue include all five types of cells. Thus, contrary to popular belief, the ability to detect specific tastes does not map to specific regions of the tongue.

Variation in the ability to taste certain bitter compounds arises from genetic differences in the structure of taste receptor proteins. Consider a receptor protein that binds to a synthetic molecule called PTC. This protein is encoded by a gene with two alleles. A dominant allele specifies a version of the receptor that triggers an action potential when it binds PTC. People with this allele perceive PTC as terribly to somewhat

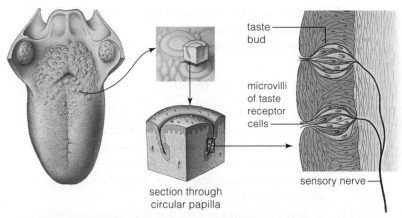

Figure 33.9 Animated Taste receptors in the human tongue. Taste buds are clusters of receptor cells and supporting cells inside special epithelial papillae. One type, a circular papilla, is shown in the section here. The tongue has about 5,000 taste buds, each enclosing as many as 150 taste receptor cells.

bitter. The receptor encoded by the recessive allele does not trigger an action potential when it binds to PTC. People homozygous for this allele find PTC tasteless. PTC is not found in nature, but similar molecules occur in broccoli, cabbage, and related plants. People who can taste PTC are more likely to find these vegetables unpalatable.

Pheromones—Chemical Messages

Many animals release and detect chemicals that function in communication. A **pheromone** is a type of signaling molecule that is secreted by one individual and affects the behavior of other members of its species. For example, female silk moths secrete a sex pheromone and male silk moths have antennae with receptors that help them detect a pheromone-secreting female more than a kilometer upwind.

Reptiles and most mammals have a **vomeronasal organ**, a collection of sensory neurons in the nasal cavity that binds and responds to pheromones. Humans and our closest primate relatives have a reduced version of this organ. Whether humans make and respond to pheromones remains a matter of debate. We discuss the role of pheromones in more detail in Chapter 43.

Take-Home Message

What are the features of the chemical senses?

» Smell and taste both involve stimulation of chemoreceptors by the binding of specific molecules.

» Humans have five types of taste receptors and hundreds of types of olfactory receptors.

» Reptiles and most mammals secrete and detect pheromones, chemicals that function in intraspecific communication.

---

olfaction The sense of smell.
pheromone Chemical that serves as a communication signal among members of an animal species.
taste receptors Chemoreceptors involved in the sense of taste.
vomeronasal organ Pheromone-detecting organ of vertebrates.

33.5 Diversity of Visual Systems

- Many organisms are sensitive to light, but only those with a camera eye see an image as you do.
- Links to Morphological convergence 18.3, Primates 26.2

Requirements for Vision

Vision is detection of light in a way that provides a mental image of objects in the environment. It requires eyes and a brain with the capacity to interpret visual stimuli. Image perception arises when the brain integrates signals regarding shapes, brightness, positions, and movement of visual stimuli.

Eyes are sensory organs that hold photoreceptors. Pigment molecules inside the photoreceptors absorb light energy. That energy is converted to the excitation energy in action potentials that are sent to the brain.

Certain invertebrates, including earthworms, do not have eyes, but they do have photoreceptors dispersed under the epidermis or clustered in parts of it.

Scallops have as many as 60 eyes arrayed along the border of their mantle (**Figure 33.10**). Each eye has

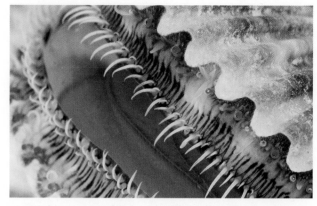

Figure 33.10 A scallop with many blue eyes along the edge of its mantle. The eyes are sensitive to movement and changes in light.

a **lens**, a transparent body that bends light rays from any point in the visual field so that rays converge on photoreceptors. Despite its many eyes, a scallop cannot form a mental image of its surroundings because it does not have a brain to integrate visual information.

Insects have **compound eyes** consisting of many separate units called ommatidia, each with its own lens (**Figure 33.11**). The brain constructs an image based on the intensity of light detected by the different units. Compound eyes do not provide the clearest vision, but they are highly sensitive to movement.

Cephalopod mollusks such as squids and octopuses have the most complex eyes of any invertebrate (**Figure 33.12**). Their **camera eyes** have an adjustable opening that allows light to enter a dark chamber. Each eye's single lens focuses incoming light onto a **retina**, a tissue densely packed with photoreceptors. The retina of a camera eye is analogous to the light-sensitive film used in a traditional film camera. Signals from the photoreceptors in each eye travel along one of the two optic tracts to the brain. Compared to compound eyes, camera eyes can produce a more sharply defined and detailed image.

Vertebrates also have camera eyes, and because they are distant relatives of cephalopod mollusks, camera eyes are presumed to have evolved independently in the two lineages. This is an example of morphological convergence (Section 18.3).

Many animals have eyes on either side of their head, an arrangement that maximizes the area they can see. Predators, including owls, tend to have two forward-facing eyes (**Figure 33.13**). Primates also have eyes at the front of their head. Having eyes that face forward enables depth perception, because the same visual field can be surveyed simultaneously from two slightly different positions. The brain compares the

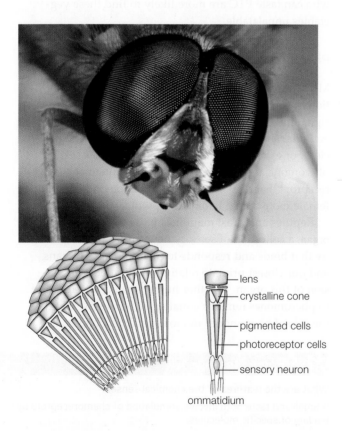

lens
crystalline cone
pigmented cells
photoreceptor cells
sensory neuron
ommatidium

Figure 33.11 The compound eye of a deerfly, with many densely packed, identical units called ommatidia. Each ommatidium has a lens that focuses light on photoreceptor cells. Although the mosaic image produced by such an eye is fuzzy, the eye is very good at detecting movement.

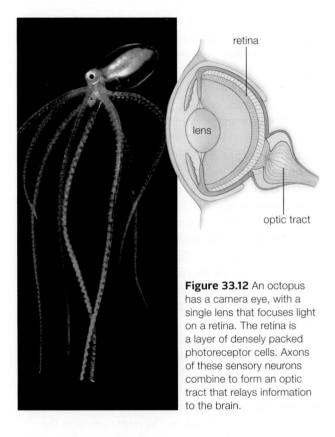

Figure 33.12 An octopus has a camera eye, with a single lens that focuses light on a retina. The retina is a layer of densely packed photoreceptor cells. Axons of these sensory neurons combine to form an optic tract that relays information to the brain.

Figure 33.13 The eyes of owls face forward, and photoreceptors are concentrated near the top of the inner eyeball. Owls mainly look down for prey. When on the ground, they must turn their heads almost upside down to see something above their head.

Figure 33.14 Eyeshine of a black-footed ferret. A reflective layer in this nocturnal predator's eyes enhances its night vision.

overlapping information it receives to determine how far apart objects in the field are.

Animals that are active primarily at night often have large eyes relative to their body size. A large eye captures more available light than a smaller one. Light-reflecting layers also evolved in the eyes of many animals that tend to be active under low-light conditions. When light shines on the eyes of these animals, the reflection from this layer makes their eyes appear to glow, a phenomonen known as "eyeshine" (**Figure 33.14**). Sharks, crocodiles, carnivores, dolphins, deer, and rodents are among the many animals with reflective eyes. The location and structure of the reflective material varies among groups, indicating that this adaptation arose independently in the different lineages. Humans and other dry-nosed primates (Section 26.2) do not have a reflective layer in their eyes.

camera eye Eye with an adjustable opening and a single lens that focuses light on a retina.
compound eye Eye with many units, each having its own lens.
eye Sensory organ that incorporates a dense array of photoreceptors.
lens Disk-shaped structure that bends light rays so they fall on an eye's photoreceptors.
retina Eye layer that contains photoreceptors.
vision Perception of visual stimuli based on light focused on a retina and image formation in the brain.

Take-Home Message

How do animal visual systems differ?

» Some animals such as earthworms have photoreceptors that detect light, but do not form any sort of image.

» Other animals, including insects, have compound eyes. A compound eye has many individual units, each with its own lens. It produces a mosaic image that is fuzzy, but highly sensitive to movement.

» A camera eye with an adjustable opening and a lens that focuses light on a photoreceptor-rich retina provides a richly detailed image. Camera eyes evolved independently in cephalopod mollusks and vertebrates.

33.6 A Closer Look at the Human Eye

■ Protective structures surround the human eye, which is multilayered, with a light-bending cornea, a focusing lens, and a photoreceptor-rich retina.

Anatomy of the Eye

Each human eyeball sits inside a protective, cuplike, bony cavity called the orbit. Skeletal muscles that run from the rear of the eye to the bones of the orbit move the eyeball up and down or side to side.

Eyelids, eyelashes, and tears all help protect the eye's delicate tissues. Periodic blinking is a reflex that spreads a film of tears over the eyeball's exposed surface. Tears are secreted by exocrine glands in the eyelids and consist of water, lipids, salts, and proteins. Among the proteins are enzymes that break down bacterial cell walls and thus help prevent eye infections.

The **conjunctiva**, a protective mucous membrane, lines the inner surface of the eyelids and folds back to cover most of the eye's outer surface. Conjunctivitis, commonly called pinkeye, is an inflammation of this membrane and is most often caused by a viral or bacterial infection.

The eyeball is spherical, and has a three-layered structure (**Figure 33.15**). The front portion of the eye is covered by a **cornea** composed of transparent crystalline proteins. A dense, white, fibrous **sclera** covers the rest of the eye's outer surface.

The eye's middle layer includes the choroid, iris, and ciliary body. The blood vessel–rich **choroid** is darkened by the brownish pigment melanin. This dark layer prevents light reflection within the eyeball.

Attached to the choroid, and suspended behind the cornea, is a muscular, doughnut-shaped **iris**. It too has melanin. Whether your eyes are blue, brown, or green depends on the amount of melanin in your iris.

Light enters the eye's interior through the **pupil**, an opening at the center of the iris. Muscles of the iris reflexively adjust pupil diameter in response to light conditions. Bright light causes the iris muscle encircling the pupil to contract, so the pupil contracts (shrinks). In low light, the spoke-like radial muscle contracts and the pupil dilates (widens). The pupil also widens in response to sympathetic stimulation, which is why drugs that mimic this stimulation cause dilated pupils and a sensitivity to light.

A ciliary body consisting of muscle and secretory cells attaches to the choroid. It holds the lens in place just behind the pupil. The stretchable, transparent lens is about 1 centimeter (1/2 inch) in diameter and bulges outward on both sides.

The eye has two internal chambers. The ciliary body produces a fluid called aqueous humor that fills the anterior chamber and bathes the iris and lens. A jellylike vitreous body fills the larger chamber behind the lens. The innermost layer of the eye, the retina, is at the back of this chamber. The retina contains the light-detecting photoreceptors. We discuss photoreceptor function in the next section.

The cornea and lens both bend incoming light so that rays converge on the retina at the back of the eye. The image formed on the retina is upside down and the mirror image of the real world (**Figure 33.16**). The brain makes the necessary adjustments so you perceive the correct orientation when you view an object.

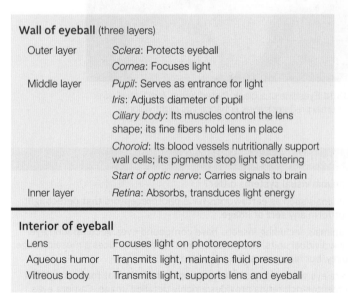

Wall of eyeball (three layers)

Outer layer	*Sclera*: Protects eyeball	
	Cornea: Focuses light	
Middle layer	*Pupil*: Serves as entrance for light	
	Iris: Adjusts diameter of pupil	
	Ciliary body: Its muscles control the lens shape; its fine fibers hold lens in place	
	Choroid: Its blood vessels nutritionally support wall cells; its pigments stop light scattering	
	Start of optic nerve: Carries signals to brain	
Inner layer	*Retina*: Absorbs, transduces light energy	

Interior of eyeball

Lens	Focuses light on photoreceptors
Aqueous humor	Transmits light, maintains fluid pressure
Vitreous body	Transmits light, supports lens and eyeball

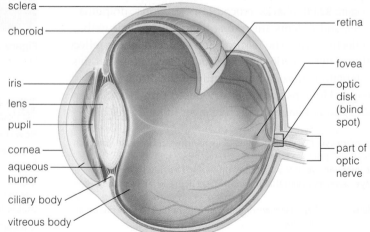

Figure 33.15 Animated Components and structure of the human eye.

Figure 33.16 Pattern of retinal stimulation in the human eye. The curved, transparent cornea changes the trajectory of light rays that enter the eye. As a result, light rays that fall on the retina produce a pattern that is upside down and inverted left to right.

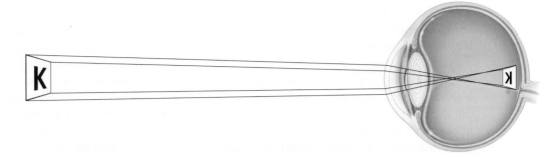

Focusing Mechanisms

With **visual accommodation**, the shape or position of a lens adjusts so that incoming light rays fall on the retina, not in front of it or behind it. Without these adjustments, only objects at a fixed distance would stimulate retinal photoreceptors in a focused pattern. Objects closer or farther away would appear fuzzy.

In fishes and amphibians, the lens of an eye can be shifted forward or back, but its shape does not change. Extending or decreasing the distance between the lens and retina keeps light focused on the retina.

In mammals and other amniotes, a ring-shaped **ciliary muscle** (part of the ciliary body) adjusts the shape of the lens. This muscle encircles the lens and attaches to it by short fibers. When the ciliary muscle relaxes, these fibers are taut, and the lens is under tension and flattened (**Figure 33.17A**). When the ciliary muscle contracts, fibers attached to the lens slacken, allowing the lens to become more round (**Figure 33.17B**).

The curvature of the lens determines the extent to which it bends light rays, and thus where the light falls in the eye. A flat lens focuses light from a distant object onto the retina; the lens must be rounder to focus light from nearby objects. When you read a book, ciliary muscle contracts and fibers that connect this muscle to the lens slacken. The decreased tension on the lens allows it to round up enough to focus light from the page onto your retina. Gaze into the distance

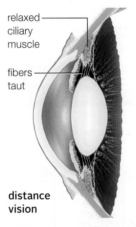

distance vision

A Relaxed ciliary muscle pulls fibers taut; the lens is stretched into a flatter shape that focuses light from a distant object on the retina.

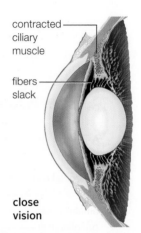

close vision

B Contracted ciliary muscle allows fibers to slacken; the lens rounds up and focuses light from a close object on the retina.

Figure 33.17 How the eye varies its focus. The lens is encircled by ciliary muscle. Elastic fibers attach the muscle to the lens. The shape of the lens is adjusted by contracting or relaxing the ciliary muscle, which increases or decreases the tension on the fibers, and thus changes the shape of the lens. **Figure It Out:** The thicker a lens, the more it bends light. Does the lens bend light more with distance vision or close vision? Answer: Close vision

and ciliary muscle around the lens relaxes, allowing the lens to flatten. Continual viewing of a close object, such as a computer screen or book, keeps ciliary muscle contracted. To reduce eyestrain, take breaks and focus on more distant objects.

choroid A blood vessel–rich layer of the middle eye. It is darkened by the brownish pigment melanin to prevent light scattering.
ciliary muscle A ring-shaped muscle of the eye that encircles the lens and attaches to it by short fibers.
conjunctiva Mucous membrane that lines the inner surface of the eyelids and folds back to cover the eye's sclera.
cornea Clear, protective covering at the front of the vertebrate eye.
iris Circular muscle that adjusts the shape of the pupil to regulate how much light enters the eye.
pupil Adjustable opening that allows light into a camera eye.
sclera Fibrous white layer that covers most of the eyeball.
visual accommodation Process of adjusting lens shape or position so that light from an object falls on the retina.

Take-Home Message

How is the structure of the human eye related to its function?

» The eye consists of delicate tissues that are surrounded by a bony orbit and constantly bathed in infection-fighting tears.

» The cornea at the front of the eye bends light rays, which then enter the eye's interior through the pupil. The diameter of the pupil changes depending on the amount of available light.

» Behind the pupil, the lens focuses light on the retina, the eye's innermost photoreceptor-containing layer. Muscle contractions can alter the shape of the lens to focus light from near or distant objects.

33.7 Light Reception and Visual Processing

■ Processing of visual information begins in the retina and continues along the pathway to the brain.
■ Link to Pigments and properties of light 6.2

As explained in the previous section, the cornea and lens bend light so that it falls on the retina, where it can excite photoreceptors. The retina has two types of photoreceptors. Each has stacks of membranous disks that contain pigment (**Figure 33.18A**). Visual pigments

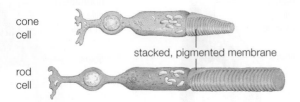

cone cell

stacked, pigmented membrane

rod cell

A The two types of photoreceptors in the retina

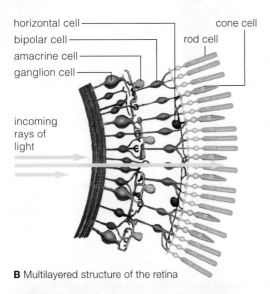

horizontal cell
bipolar cell
amacrine cell
ganglion cell

cone cell
rod cell

incoming rays of light

B Multilayered structure of the retina

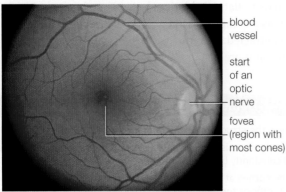

blood vessel

start of an optic nerve

fovea (region with most cones)

C Magnified view of the retina as seen through the pupil

Figure 33.18 Animated Structure of the retina. The two types of photoreceptors, rods and cones, lie at the very rear of the retina, beneath layers of signal-processing neurons.

(called opsins) are derived from vitamin A, which is why a deficiency in this vitamin can impair vision.

Rod cells, the most abundant photoreceptors, detect dim light and respond to changes in light intensity across the visual field. In the human eye, rods tend to be concentrated at the edges of the retina. All rods have the same pigment (rhodopsin), which is most excited by exposure to green-blue light.

Cone cells provide acute daytime vision and allow us to detect colors. There are three types of cone cell, each with a slightly different form of the cone pigment photopsin. One cone pigment absorbs mainly red light, another absorbs mainly blue, and a third absorbs green. Normal human color vision requires all three kinds of cones. The **fovea**, a pit in the central region of the retina, has the greatest density of cones. With normal vision, most light rays are focused on the fovea.

Signal integration and processing begin in the retina, which has a multilayered structure (**Figure 33.18B**). When a pigment in a rod or cone cell absorbs light, signals flow from that cell to neurons in the layer above. These neurons begin the process of visual processing. They respond to specific patterns of photoreceptor activation by sending signals to ganglion cells. Bundled axons of ganglion cells constitute the optic nerve (**Figure 33.18C**). The region of the retina through which the optic nerve exits the eye lacks photoreceptors. It cannot respond to light and thus is a "blind spot." We all have a blind spot in each eye, but do not notice it because the information missed by one eye is provided to the brain by the other.

Signals from the right visual field of each eye travel along an optic nerve to the brain's left hemisphere. Signals from the left visual field travel to the right hemisphere. Each optic nerve ends in a brain region (the lateral geniculate nucleus) that processes signals. From here, signals are conveyed to the visual cortex, where the final integration produces visual sensations.

cone cell Photoreceptor that provides sharp vision and allows detection of color.
fovea Retinal region where cone cells are most concentrated.
rod cell Photoreceptor that is active in dim light; provides coarse perception of image and detects motion.

Take-Home Message

How do we detect and process visual information?

» When stimulated by light, rods and cones send signals to neurons in the layer of retina above them. These neurons process signals, then send messages to the brain.

» The visual cortex integrates signals from the eyes and gives rise to visual sensations.

33.8 Visual Disorders

- A variety of disorders can impair vision.
- Links to X-linked traits 14.4, Embryonic stem cells 31.1

Vision is impaired when light is not focused properly, photoreceptors do not respond as they should, or some aspect of visual processing breaks down.

Color Blindness

Sometimes one or more types of cones fail to develop, or function improperly. The outcome is color blindness. In red–green color blindness, a person has trouble distinguishing reds from greens. This type of color blindness is an X-linked recessive trait. As is the case for other X-linked traits, it shows up more often in males. In the United States, 7 percent of males and 0.4 percent of females are affected.

Lack of Focus

About 150 million Americans have disorders in which the eye does not properly focus light. Astigmatism results from an unevenly curved cornea, which cannot properly focus incoming light on the lens. Nearsightedness occurs when the distance from the front to the back of the eye is longer than normal or when ciliary muscles react too strongly. With either disorder, images of distant objects get focused in front of the retina instead of on it (**Figure 33.19A**).

In farsightedness, the distance from front to back of the eye is unusually short or ciliary muscles are too weak. Either way, light rays from nearby objects get focused behind the retina, rather than where they should be (**Figure 33.19B**).

Glasses, contact lenses, or surgery can correct most focusing problems. About 1.5 million Americans undergo laser surgery (LASIK) annually.

Age-Related Disorders

As a person ages, the lens loses its flexibility, so most people over age forty have a somewhat impaired ability to focus on near objects.

Changes in the structure of proteins in the lens can result in a cataract, a clouding of the lens. Excessive exposure to ultraviolet radiation, smoking, use of steroids, and some diseases such as diabetes promote cataract formation. Typically, both eyes are affected. At first, a cataract scatters light and blurs vision (**Figure 33.20A**). Eventually, the lens may become opaque, causing blindness. Cataract surgery restores normal vision by replacing a clouded lens with a plastic one.

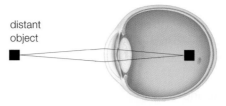
distant object

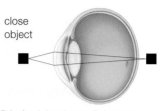
close object

A In nearsightedness, light rays from distant objects converge in front of the retina.

B In farsightedness, light rays from close objects have not yet converged when they arrive at the retina.

Figure 33.19 Focusing problems.

A With cataracts

B With macular degeneration

Figure 33.20 Photos simulating vision with two common visual disorders.

Age-related macular degeneration (AMD) is the leading cause of age-related blindness in the United States. The macula, the part of the retina around and including the fovea, is essential to clear vision. Destruction of photoreceptors in the macula clouds the center of the visual field more than the periphery (**Figure 33.20B**). Some mutations increase the risk of AMD, as do smoking, obesity, and high blood pressure. A vegetable-rich diet may help protect against it. Damage caused by AMD cannot be reversed, but drug injections and laser therapy can slow its progression.

Glaucoma results when too much aqueous humor builds up inside the eyeball. The increased fluid pressure damages blood vessels and ganglion cells. It can also interfere with peripheral vision and visual processing. Although symptoms of glaucoma typically appear in old age, conditions that give rise to the disorder arise earlier. Screening for glaucoma allows doctors to detect and respond to the increased fluid pressure before the damage becomes severe.

Take-Home Message

What causes common visual disorders?

» Missing or malfunctioning cone cells result in color blindness.

» A misshapen eyeball can cause nearsightedness or farsightedness. A lens that has become inflexible with age also causes farsightedness.

» Other age-related vision disorders include cataracts (clouding of the lens), macular degeneration (loss of photoreceptors), and glaucoma (excessive aqueous fluid).

33.9 Sense of Hearing

■ Your ears collect, amplify, and sort out sound waves, which are pressure waves traveling through the air.
■ Link to Evolution of land vertebrates 25.6

Properties of Sound

Hearing is the perception of sound, which is a form of mechanical energy. A sound arises when a vibrating object causes pressure variations in air, water, or some other medium. We can represent the pressure variations as waves. The amplitude of a sound—the magnitude of the pressure waves—determines its intensity or loudness. The frequency of a sound—the number of waves per second—determines pitch (**Figure 33.21**). The more waves per second, the higher the frequency.

The Vertebrate Ear

Water readily transfers vibrations to body tissues, so fishes do not require elaborate ears to detect sounds. When vertebrates left water for land, their capacity to collect and amplify vibrations evolved in response to a new environmental challenge: The transfer of sound waves to body tissues is less efficient in air than in water. Certain features of mammalian ears are evolutionary adaptations that increase the efficiency of this transfer (**Figure 33.22**).

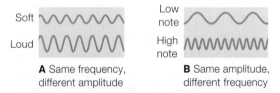

A Same frequency, different amplitude

B Same amplitude, different frequency

Figure 33.21 Wavelike properties of sound.

Unlike amphibians and reptiles, most mammals have an **outer ear** that funnels sound inward ❶. A skin-covered flap of cartilage, called the pinna, projects from the side of the head. The pinna collects sound waves and directs them into the auditory canal, an air-filled passage that connects to the middle ear.

The **middle ear** amplifies and transmits sound waves to the inner ear. An **eardrum**, or tympanic membrane, first evolved in amphibians, in which it is visible as a round area on each side of the head. Sound waves cause an eardrum to vibrate. Behind the eardrum is an air-filled cavity that holds three small bones known as the hammer, anvil, and stirrup (after their shape) ❷. The bones transmit the force of sound waves from the eardrum onto the surface of the oval window, an elastic membrane that is the boundary between the middle and inner ear.

The **inner ear** functions both in hearing and in balance. We discuss structures involved in the sense of balance in the next section. Here we focus on the role of the sound-detecting **cochlea**, a pea-sized, fluid-filled structure that resembles a coiled snail shell (the Greek *koklias* means snail).

Membranes divide the interior of the cochlea into three fluid-filled ducts ❸. When sound waves make the three tiny bones of the middle ear vibrate, the stirrup pushes against the oval window. With each pulse of the sound wave, the oval window bows inward. This movement creates a pressure wave in the fluid inside the cochlea. As such waves travel through the cochlear fluid, they cause the lower wall of the cochlear duct, called the basilar membrane, to vibrate up and down. The **organ of Corti**, an acoustical organ with arrays of hair cells sits on top of this membrane ❹. Each hair cell is a mechanoreceptor with a tuft of

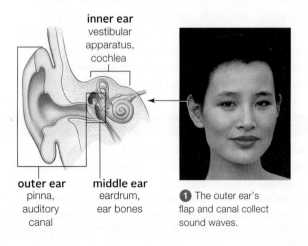

outer ear
pinna, auditory canal

middle ear
eardrum, ear bones

❶ The outer ear's flap and canal collect sound waves.

inner ear
vestibular apparatus, cochlea

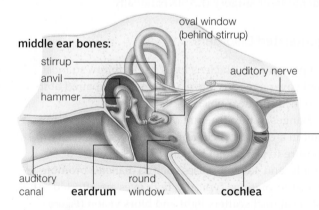

middle ear bones:
stirrup
anvil
hammer

oval window (behind stirrup)

auditory nerve

auditory canal   **eardrum**   round window   **cochlea**

❷ The eardrum and middle ear bones amplify sound.

Figure 33.22 Animated Anatomy of the ear and how we hear.

modified cilia at one end. The cilia project into a tectorial membrane that drapes over them. When the basilar membrane moves, it pushes the cilia against the tectorial membrane and bends them ❺. The bending causes the hair cells to undergo action potentials that travel along an auditory nerve to the brain.

The number of hair cells that fire and the frequency of their signals inform the brain about the volume of a sound. The louder the sound, the more action potentials flow along the auditory nerve to the brain.

cochlea Coiled, fluid-filled structure in the inner ear that holds the sound-detecting organ of Corti.
eardrum The membrane that vibrates in response to pressure waves (sounds), thus transmitting vibrations to the bones of the middle ear.
hearing Perception of sound.
inner ear Fluid-filled vestibular apparatus and cochlea.
middle ear Eardrum and the tiny bones that transfer sound to the inner ear.
organ of Corti An acoustical organ in the cochlea that transduces mechanical energy of pressure waves into action potentials.
outer ear External ear and the air-filled auditory canal.

The brain can determine the pitch of a sound by assessing which part of the basilar membrane is vibrating most. The basilar membrane is not uniform along its length. It is stiff and narrow near the oval window, and broader and more flexible deeper into the coil. High-pitched sounds make the stiff, narrow, closer-in part of the basilar membrane vibrate most. Low-pitched sounds cause vibrations mainly in the wide flexible part close to the membrane's tip. More vibrations make more hair cells in that region fire.

Take-Home Message

How do humans hear?

» Ears that collect and amplify sound waves evolved in some vertebrates that live on land.

» Like most mammals, humans have an outer ear that collects sound waves and directs them to the middle ear. In the middle ear, vibrations of the eardrum are amplified via movement of small bones. These bones set up pressure waves in fluid inside the inner ear's cochlea.

» The organ of Corti inside the cochlea has hair cells that convert pressure waves into action potentials that travel along the auditory nerve to the brain.

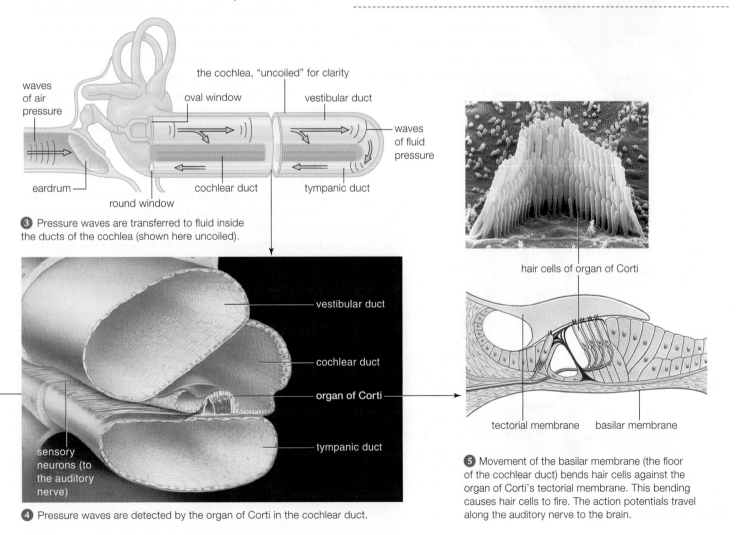

❸ Pressure waves are transferred to fluid inside the ducts of the cochlea (shown here uncoiled).

❹ Pressure waves are detected by the organ of Corti in the cochlear duct.

❺ Movement of the basilar membrane (the floor of the cochlear duct) bends hair cells against the organ of Corti's tectorial membrane. This bending causes hair cells to fire. The action potentials travel along the auditory nerve to the brain.

33.10 Organs of Equilibrium

- Organs inside your inner ear are essential to maintaining posture and a sense of balance.
- Somatic sensory receptors also contribute to balance.

Organs of equilibrium monitor the body's position and motion (**Figure 33.23**). Each vertebrate ear includes these organs inside a fluid-filled sensory structure called the **vestibular apparatus**. The organs are located in three semicircular canals and in two sacs, the saccule and utricle (**Figure 33.23B**).

Like the organ of Corti, organs of the vestibular apparatus have hair cells. Fluid pressure inside the canals and sacs makes the cilia bend. The mechanical energy of this bending deforms the hair cell plasma membrane just enough to let ions slip across and stimulate an action potential. A vestibular nerve carries the sensory input to the brain.

The three semicircular canals are oriented at right angles to one another, so rotation of the head in any combination of directions—front/back, up/down, or left/right—moves the fluid inside them. An organ of equilibrium rests on the bulging base of each canal. The cilia of its hair cells are embedded in a jellylike mass (**Figure 33.23C**). When fluid moves in the canal, it pushes the mass and generates pressure required for initiating action potentials.

The brain receives signals from semicircular canals on both sides of the head. By comparing the number and frequency of action potentials coming from each side of the head, the brain senses dynamic equilibrium: the angular movement and rotation of the head. Among other things, your sense of dynamic equilibrium allows you to keep your eyes locked on an object even when you swivel your head or nod.

Organs in the saccule and utricle act in the sense of static equilibrium. These organs help the brain keep track of the head's position and how fast it is moving in a straight line. They also help keep the head upright and maintain posture. Inside the saccule and utricle, a jellylike layer weighted with calcite overlies the mechanoreceptors (hair cells). When you tilt your head, or start or stop moving, the weighted mass shifts, bending hair cells and altering their rate of action potentials.

The brain also takes into account information from the eyes, and from receptors in the skin, muscles, and joints. Integration of the signals provides awareness of the body's position and motion in space.

A stroke, an inner ear infection, or loose particles in the semicircular canals can cause vertigo, a sensation that the world is moving or spinning around. Vertigo also arises from conflicting sensory inputs, as when you stand at a height and look down. The vestibular apparatus reports that you are motionless, but your eyes report that your body is floating in space.

Mismatched signals can cause motion sickness. On a curvy road, passengers in a car experience changes in acceleration and direction that indicate "motion" to their vestibular apparatus. At the same time, signals from their eyes about objects inside the car tell their brain that the body is at rest. Driving can minimize motion sickness because the driver focuses on sights outside the car such as scenery rushing past, so the visual signals are consistent with vestibular signals.

organs of equilibrium Sensory organs that respond to body position and motion.

vestibular apparatus System of fluid-filled sacs and canals in the inner ear; contains the organs of equilibrium.

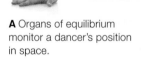

A Organs of equilibrium monitor a dancer's position in space.

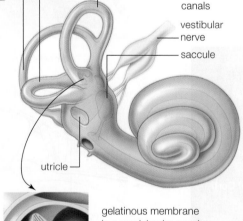

B Vestibular apparatus inside a human ear. Organs inside its fluid-filled sacs and canals contribute to the sense of balance.

semicircular canals

vestibular nerve

saccule

utricle

C Components of one organ in a semicircular canal. Shifts in the position of the head bend hair cells and alter their frequency of action potentials.

gelatinous membrane in a semicircular canal

hair cells with their cilia embedded in membrane

sensory neurons

Figure 33.23 Animated Organs of equilibrium.

Take-Home Message

What gives us our sense of balance?

» Mechanoreceptors in the fluid-filled vestibular apparatus of the inner ear detect the body's position in space, and when we start or stop moving.

A Whale of a Dilemma (revisited)

There's no doubt that human activities generate lots of noise, altering the sensory world through which both humans and animals move. The intensity of a sound is measured in decibels, with an increase of 10 on this scale indicating a tenfold increase in loudness. A normal conversation is about 60 decibels, a food blender operating at high speed is about 90 decibels, and a chain saw is about 100 decibels. Music at a rock concert is about 120 decibels. So is the sound heard through the earbuds of an iPod or similar device cranked up to its maximum volume.

Noise louder than 90 decibels damages hair cells in the cochlea. Humans have about 32,000 such cells at birth, and the number declines with age. Exposure to loud noise accelerates the loss of hair cells and the resulting loss of hearing. Even if noise is not loud enough to deafen you, it can cause harm. Chronic noise impairs concentration and interferes with sleep patterns. It raises anxiety and increases the risk of high blood pressure and other cardiovascular problems.

Marine biologist Susan Parks has been studying how the noise produced by shipping affects whales (Figure 33.24). By placing recorders on whales in a busy shipping channel in the Atlantic, Parks found that an endangered species of whale (the right whale) altered the pitch and the loudness of their calls in response to the level of ambient noise. The whales "shouted" to make themselves heard over the din of ships, although the ships were more than 60 miles away. Making loud calls takes more energy than making quieter ones. Thus, noise pollution is placing an extra burden on this already endangered species.

Figure 33.24 Marine biologist Dr. Susan Parks with the recorder she used to investigate the effect of shipping noise on whale calls. Suction cups hold the device on the whale's body.

Like whales, birds rely heavily on auditory signals. During courtship, man-made noise can interfere with the ability to find and secure a mate. Canadian researchers studied how noisy compressors used to extract oil and gas affect songbirds that share their habitat with the machinery. The study showed that the noise-exposed birds have 15 percent fewer offspring than those in a quiet forest habitat.

How would you vote? Engines of ships and other machines can be quieted, but sound reducing measures have a cost. Should we set maximal noise levels for engines that run outdoors on land or in water?

Endocrine System

LEARNING ROADMAP

Where you have been This chapter focuses on endocrine glands, which were introduced in Section 31.3. Knowing the properties of steroids (3.5), proteins (3.6), and the plasma membrane (5.7) will help you understand hormone action. Among other topics, you will learn how hormones affect human glucose metabolism (7.7) and arthropod molting (24.11).

Where you are now

Vertebrate Endocrine System
Nearly all vertebrates have an endocrine system that includes the same hormone-producing structures. The endocrine system interacts closely with the nervous system.

Hormone Action
A hormone travels in the blood and binds receptors on target cells. Receptor activation converts the hormonal signal to a form that elicits a response in the target cell.

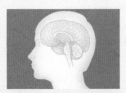

A Master Regulatory Center
In vertebrates, the hypothalamus and pituitary gland deep in the forebrain are connected structurally and functionally. Together, they coordinate activities of many glands.

Other Controlling Factors
Some glands alter their secretion of hormones in response to internal changes such as a shift in blood glucose level, or to external factors such as changes in daylength.

Invertebrate Hormones
Hormones control molting and other events in invertebrate life cycles. Vertebrate hormones and receptors for them first evolved in ancestral lineages of invertebrates.

Where you are going Later chapters discuss the effect of hormones on vertebrate muscle mass (Section 35.1), bone turnover (35.4), adjusting to high altitude (38.8), appetite (Section 39.5) and urine formation (40.5). Sex hormones have a central role in our discussion of gamete formation and reproduction (Chapter 41), and we look at hormonal effects on animal behavior in Section 43.2. We return to the subject of hormone disruptors in our discussion of pollutants in Chapter 48.

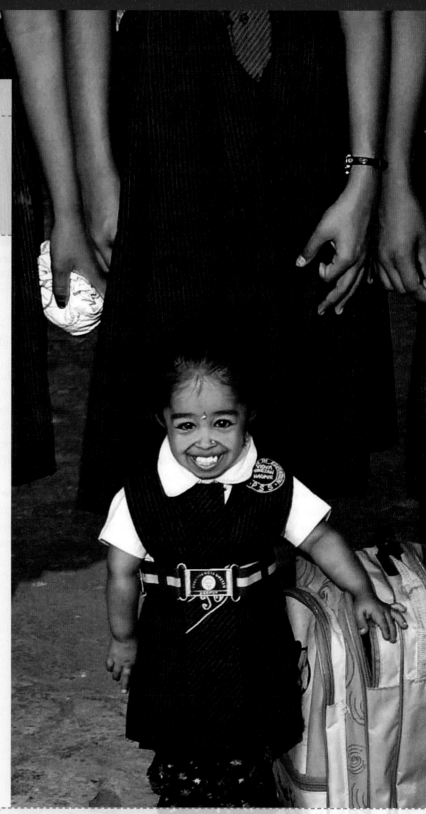

34.1 Hormones in the Balance

We live in a world awash in synthetic chemicals. We drink from plastic bottles, wear clothing made of synthetic fabrics, slather ourselves with synthetic skin products, and eat food treated with synthetic pesticides. Huge numbers of man-made compounds are used to make computers and other electronic gadgets. Synthetic chemicals enter our bodies when we ingest them, inhale them, or absorb them across our skin. How do they affect us? What do we know about the safety of these substances?

We have learned by sad experience that some synthetic chemicals threaten animal and human health. For example, years after we started using them, we realized that DDT (a pesticide) and PCBs (used in electronic products, caulking, and solvents) are endocrine disruptors. **Endocrine disruptors** are molecules that interfere with the action of hormones, signaling molecules secreted by endocrine glands. DDT was banned in 1972 and PCBs in 1979. However, because both chemicals were in wide use for years and are highly stable, they still persist in the environment.

Atrazine, a synthetic herbicide that remains in wide use, is another endocrine disruptor. In the United States, farmers apply about 36,000 metric tons (almost 80 million pounds) every year, mostly to kill weeds in cornfields. Atrazine benefits farmers by reducing the need to till the soil, a labor-intensive process that encourages soil erosion. However, atrazine does not stay put in cornfields. It runs off into streams and ponds and seeps into the ground, contaminating sources of human drinking water.

Tyrone Hayes, a University of California biologist, has studied the hormonal effects of atrazine contamination since 1997 (**Figure 34.1**). His data show that atrazine chemically castrates and femininizes male frogs. He found that laboratory-reared African clawed male frogs exposed to atrazine as tadpoles had low levels of the sex hormone testosterone as adults. Some "males" even had ovaries and produced eggs. Hayes also demonstrated an effect of atrazine in the wild. When he collected leopard frogs from ponds and ditches across the Midwest, he found abnormal sex organs in male frogs from every atrazine-contaminated pond.

Atrazine also affects other aquatic animals. For example, it has feminizing effects on zebrafish. Exposure of fish embryos to atrazine at a level comparable to that in runoff from atrazine-treated fields dramatically increases the percentage that become female.

Atrazine's effect on sexual development could help push some fish and frogs to extinction. If this seems a remote concern, consider this: All vertebrates have similar hormone-secreting glands and endocrine systems. Thus, chemicals that adversely affect one vertebrate are likely to affect others in a similar manner.

endocrine disruptor Chemical that interferes with hormone action.

Figure 34.1 Benefits and costs of herbicide applications. *Left*, atrazine can keep cornfields nearly weed-free; no need for the constant tilling that causes soil erosion. Tyrone Hayes (*right*) discovered that the chemical scrambles amphibian hormonal signals.

34.2 The Vertebrate Endocrine System

■ Animal cells communicate with one another by way of a variety of short-range and long-range chemical signals.
■ Links to Gap junctions 4.11, Glands 31.3, Synapses 32.5, Hypothalamus 32.9, Pain 33.3

Mechanisms of Intercellular Signaling

Cells of an animal body constantly signal one another. Gap junctions allow chemical signals to move directly from the cytoplasm of one cell to that of an adjacent cell. Other cell–cell communication involves signaling molecules that are secreted into interstitial fluid (the fluid between cells). These molecules exert effects only when they bind to a receptor on or inside another cell. A cell with receptors that bind and respond to a specific signaling molecule is a "target" of that molecule.

Some secreted signaling molecules diffuse a short distance through interstitial fluid and bind to nearby cells. For example, most neurons secrete neurotransmitters into the synaptic cleft that separates them from their target—a postsynaptic cell. Only neurons release neurotransmitters, but many cells produce and secrete **local signaling molecules**. These molecules reach nearby targets by diffusion, so they act over a limited distance. Prostaglandins are an example. When released by injured cells, they activate nearby pain receptors.

Animal hormones are longer-range communication molecules. After being secreted into interstitial fluid, hormones enter the blood and circulate throughout the body. Compared to neurotransmitters or local signaling molecules, hormones last longer, travel farther, and exert their effects on a greater number of cells.

Discovery of Hormones

Hormones were first discovered in the early 1900s by physiologists William Bayliss and Ernest Starling. They were studying how the secretion of pancreatic juices is regulated. Bayliss and Starling knew that food mixes with acid in the stomach and that when this acidic mix reaches the small intestine, it stimulates the pancreas to secrete the buffer bicarbonate. However, they did not know how the message that triggered bicarbonate secretion reached the pancreas.

To find out how the small intestine communicated with the pancreas, the scientists did an experiment. They surgically altered a laboratory animal, cutting nerves that carry signals to and from its small intestine. Even with these nerves cut, the small intestine responded to the presence of acid by secreting bicarbonate. This indicated that the signal calling for pancreatic secretion did not travel along nerves.

Starling and Bayliss hypothesized that the small intestine produces a signal that travels in the blood. To test this idea, they exposed small intestinal cells to acid, then made an extract of those cells. Injecting this extract into the bloodstream of another animal caused its pancreas to secrete bicarbonate. The researchers concluded that exposure to acid causes the small intestine to release a chemical signal into the blood. This bloodborne substance encourages the pancreas to secrete bicarbonate into the gut.

The signaling substance is now called secretin. Identifying its mode of action supported a hypothesis that dated back centuries: Blood carries internal secretions that influence the activities of the body's organs.

Starling coined the term "hormone" for glandular secretions such as secretin (the Greek *hormon* means to set in motion). Later researchers identified additional hormones and their sources. Endocrine glands and other structures that secrete hormones are collectively referred to as an animal's **endocrine system**. **Figure 34.2** shows the main glands of a human endocrine system. In addition to the major glands, cells of many internal organs such as the small intestine and heart produce and release hormones.

Conversely, some of the major endocrine glands also have functions unrelated to hormone secretion. For example, the pancreas secretes hormones into the blood, but also secretes digestive enzymes into the small intestine. The gonads (ovaries and testes) produce sex hormones and also make gametes.

The discovery of hormones and the subsequent research on endocrine function have had important medical implications. They allowed treatment of endocrine disorders, such as diabetes. They also allowed the invention of hormonal methods of contraception, a topic we discuss in detail in Chapter 41.

Neuroendocrine Interactions

Both neurons and endocrine cells develop from an embryo's ectodermal layer. Portions of the endocrine system and nervous system are so closely linked that scientists sometimes refer to them collectively as the neuroendocrine system. Both endocrine glands and neurons receive signals from the hypothalamus, a command center in the forebrain (Section 32.9). Most organs respond to both hormones and signals from the nervous system.

Hormones influence brain development, both before and after birth. Hormones can also affect nervous processes such as sleep/wake cycles, emotion, mood, and memory. Conversely, the nervous system

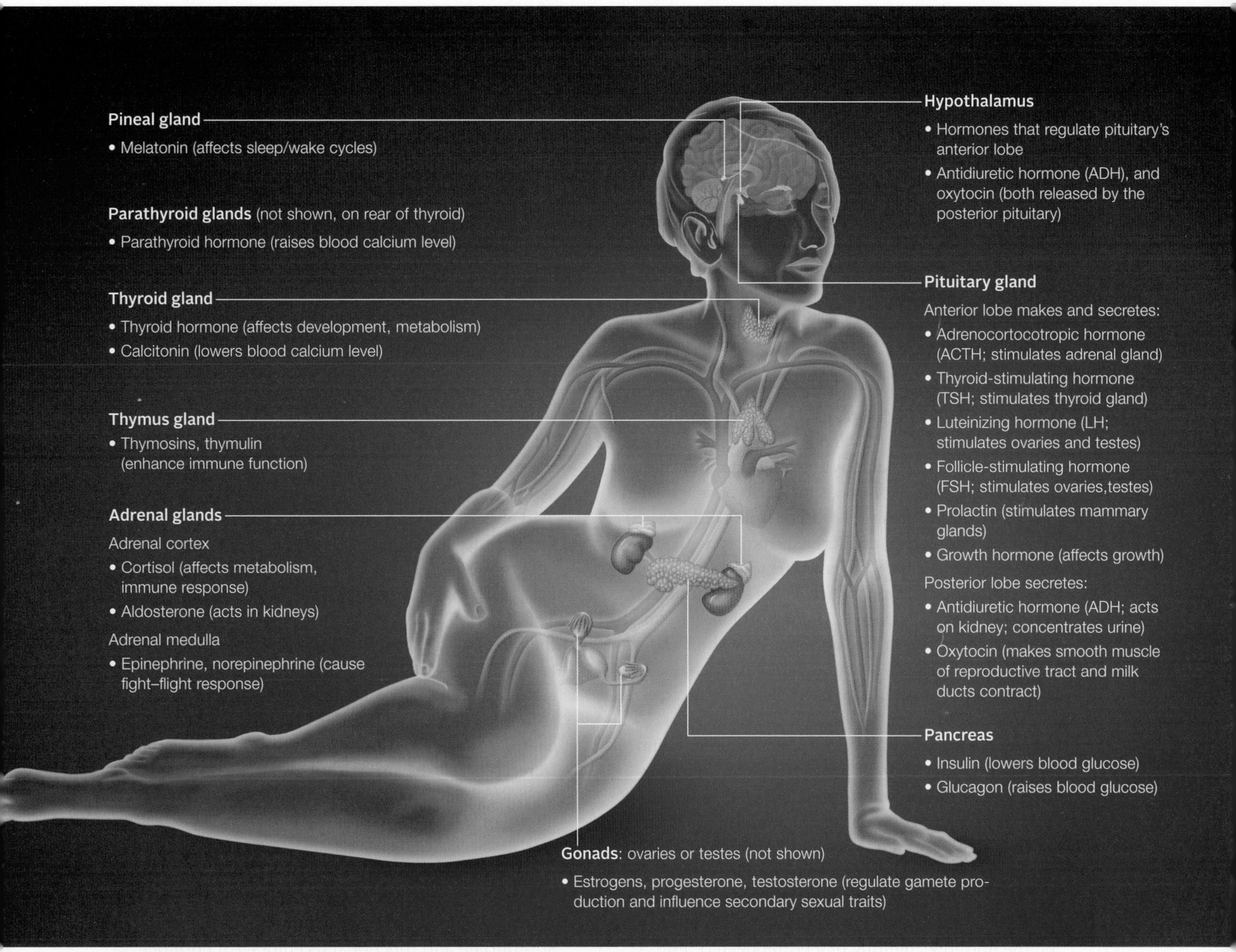

Figure 34.2 Animated Main components of the human endocrine system and the effects of their secretions. Hormone-secreting cells are also present in the glandular epithelia of the stomach, small intestine, liver, heart, kidneys, adipose tissue, skin, placenta, and other organs.

Pineal gland
• Melatonin (affects sleep/wake cycles)

Parathyroid glands (not shown, on rear of thyroid)
• Parathyroid hormone (raises blood calcium level)

Thyroid gland
• Thyroid hormone (affects development, metabolism)
• Calcitonin (lowers blood calcium level)

Thymus gland
• Thymosins, thymulin (enhance immune function)

Adrenal glands
Adrenal cortex
• Cortisol (affects metabolism, immune response)
• Aldosterone (acts in kidneys)
Adrenal medulla
• Epinephrine, norepinephrine (cause fight–flight response)

Hypothalamus
• Hormones that regulate pituitary's anterior lobe
• Antidiuretic hormone (ADH), and oxytocin (both released by the posterior pituitary)

Pituitary gland
Anterior lobe makes and secretes:
• Adrenocortocotropic hormone (ACTH; stimulates adrenal gland)
• Thyroid-stimulating hormone (TSH; stimulates thyroid gland)
• Luteinizing hormone (LH; stimulates ovaries and testes)
• Follicle-stimulating hormone (FSH; stimulates ovaries,testes)
• Prolactin (stimulates mammary glands)
• Growth hormone (affects growth)
Posterior lobe secretes:
• Antidiuretic hormone (ADH; acts on kidney; concentrates urine)
• Oxytocin (makes smooth muscle of reproductive tract and milk ducts contract)

Pancreas
• Insulin (lowers blood glucose)
• Glucagon (raises blood glucose)

Gonads: ovaries or testes (not shown)
• Estrogens, progesterone, testosterone (regulate gamete production and influence secondary sexual traits)

regulates hormone secretion. For example, in a stressful situation, sympathetic nervous stimulation causes increased secretion of some hormones and decreased secretion of others.

animal hormone Intercellular signaling molecule secreted by an endocrine gland or cell; travels in the blood to target cells.
endocrine system Hormone-producing glands and secretory cells of a vertebrate body.
local signaling molecule Chemical signal, such as a prostaglandin, that is secreted by one cell, diffuses through interstitial fluid, and affects nearby cells in an animal body.

Take-Home Message

How do cells of an animal body communicate with one another?

» In all animals, cells release molecules that influence other cells. Each type of signal acts on all target cells that have receptors for it. Hormones are intercellular signaling molecules that travel in the bloodstream.

» Most vertebrates have the same types of hormones produced by similar structures. Collectively, hormone-secreting glands and cells make up an endocrine system.

» Integrated interactions between the nervous system and nearly all endocrine glands coordinate many different functions for the body as a whole.

34.3 The Nature of Hormone Action

■ For a hormone to have an effect, it must bind to receptors on or inside a target cell.

■ Links to Steroids 3.5, Proteins 3.6, Cell membranes 5.7, Promoters 9.3, Sex determination 10.4,

From Signal Reception to Response

Hormone action is a three-step process. A hormone activates a target cell receptor, the signal is transduced (changed into a form that affects target cell behavior), and the cell makes a response:

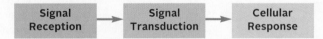

The response summoned up by a hormone declines over time, as the body breaks down and eliminates hormone molecules. Most commonly the hormone is taken up from the blood by the liver and broken down by liver enzymes. Hormones may also be inactivated in the blood, inside target cells, or in the kidneys.

Animal hormones are derived from either cholesterol or amino acids. Cholesterol is the starting material for steroid hormones such as the sex hormones testosterone and estrogen. Amine hormones are modified amino acids. Peptide hormones are short chains of amino acids; protein hormones are longer chains. **Table 34.1** lists a few examples of each.

Intracellular Receptors Intracellular receptors bind hormones that can enter the cell. Steroid hormones are made from cholesterol and, like other lipids, they easily diffuse across a plasma membrane. Once inside a cell, steroid hormones form a hormone–receptor complex by binding to a receptor in the cytoplasm or nucleus. Most often, this hormone–receptor complex binds to a specific promoter (Section 9.3). Depending on the hormone, binding of a hormone–receptor

complex to a promoter increases or decreases the rate of transcription of an adjacent gene or set of genes. **Figure 34.3A** is a simple illustration of this mechanism of steroid hormone action.

Receptors at the Plasma Membrane Most amine hormones, and all peptide or protein hormones, are too big and polar to diffuse across a membrane. They bind to receptors that span a target cell's plasma membrane. Often, this binding activates an enzyme that converts ATP to a nucleotide called cAMP (cyclic adenosine monophosphate). The cAMP functions as a **second messenger**: a molecule that forms inside a cell in response to an external signal and causes some sort of shift in that cell's activity.

For example, when there is too little glucose in the blood, cells in the pancreas secrete a peptide hormone (glucagon). When this hormone binds to receptors in the plasma membrane of target cells, it causes formation of cAMP inside them (**Figure 34.3B**). The cAMP activates an enzyme that activates a different enzyme, setting into motion a cascade of reactions. The last enzyme activated catalyzes breakdown of glycogen into glucose and thus raises the blood glucose level.

Some cells have receptors for steroid hormones at their plasma membrane. Binding of a steroid hormone to these receptors does not influence gene expression. Instead, it triggers a rapid response by way of a second messenger or by affecting the membrane. For example, the steroid hormone aldosterone can bind to receptors at the surface of kidney cells, as well as receptors inside the cell. Aldosterone alters the permeability of kidney cells to certain ions.

Receptor Function and Diversity

A cell can only respond to a hormone for which it has appropriate and functional receptors. All hormone receptors are proteins, and gene mutations can make them less efficient or even nonfunctional. In this case, even though the hormone is present in normal amounts, it will have a lesser or no effect.

For example, typical male genitals will not form in an XY embryo without testosterone, which is a steroid hormone (Section 10.4). XY individuals who have androgen insensitivity syndrome secrete testosterone, but they carry a mutation that affects their testosterone receptors. Without functional receptors, it is as if testosterone is not present. As a result, testes form but do not descend into the scrotum, and the genitals appear female. Such individuals are often raised as females, as discussed in more detail in Chapter 41.

Table 34.1	Categories and Examples of Hormones
Steroids	Testosterone and other androgens, estrogens, progesterone, aldosterone, cortisol
Amines	Melatonin, epinephrine, thyroid hormone
Peptides	Glucagon, oxytocin, antidiuretic hormone, calcitonin, parathyroid hormone
Proteins	Growth hormone, insulin, prolactin, follicle-stimulating hormone, luteinizing hormone

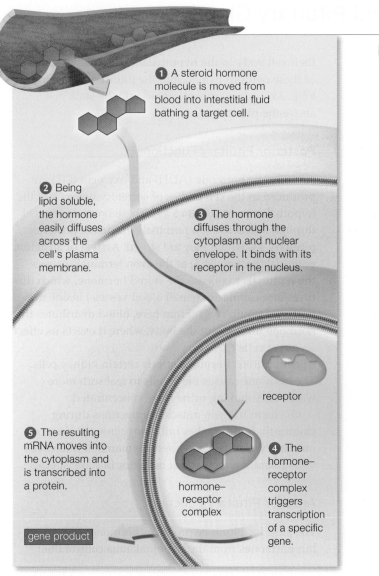

2 Being lipid soluble, the hormone easily diffuses across the cell's plasma membrane.

1 A steroid hormone molecule is moved from blood into interstitial fluid bathing a target cell.

3 The hormone diffuses through the cytoplasm and nuclear envelope. It binds with its receptor in the nucleus.

receptor

5 The resulting mRNA moves into the cytoplasm and is transcribed into a protein.

hormone–receptor complex

4 The hormone–receptor complex triggers transcription of a specific gene.

gene product

A Example of steroid hormone action inside a target cell.

1 A peptide hormone molecule, glucagon, diffuses from blood into interstitial fluid bathing the plasma membrane of a liver cell.

unoccupied glucagon receptor at target cell's plasma membrane

cyclic AMP + P_i

ATP

2 Glucagon binds with a receptor. Binding activates an enzyme that catalyzes the formation of cyclic AMP from ATP inside the cell.

3 Cyclic AMP activates another enzyme in the cell.

4 The enzyme activated by cyclic AMP activates another enzyme, which in turn activates another kind that catalyzes the breakdown of glycogen to its glucose monomers.

5 The enzyme activated by cyclic AMP also inhibits glycogen synthesis.

B Example of peptide hormone action inside a target cell.

Figure 34.3 Mechanisms of hormone action. **Figure It Out:** Which example shows formation of a second messenger and what substance serves as the second messenger?

Answer: Cyclic AMP serves as a second messenger in the peptide hormone example.

Variations in receptor structure also affect responses to hormones. Different tissues have receptor proteins that respond in different ways to binding the same hormone. For example, ADH (antidiuretic hormone) acts on kidney cells and affects urine formation. ADH is sometimes referred to as vasopressin, because it also binds to receptors in the wall of blood vessels and causes the vessels to narrow. In many mammals, ADH helps maintain blood pressure. ADH also binds to brain cells and affects sexual and social behavior, as we will discuss in Section 43.2. This enormous diversity of responses to a single hormone is an outcome of variations in the structure of ADH receptors. In each kind of cell, a different kind of receptor summons up a different cellular response.

second messenger Molecule that forms inside a cell when a hormone or other signaling molecule binds to a receptor at the cell surface; its formation sets in motion reactions that alter some activity of the cell.

Take-Home Message

How do hormones exert their effects on target cells?

» Hormones exert their effects by binding to protein receptors, either inside a cell or at the plasma membrane.

» Steroid hormones often enter a cell and act by altering the expression of specific genes.

» Peptide and protein hormones usually bind to a receptor at the plasma membrane. They trigger formation of a second messenger, a molecule that relays a signal into the cell.

» Variations in receptor structure affect how a cell responds to a hormone.

34.4 The Hypothalamus and Pituitary Gland

■ The hypothalamus and pituitary gland deep inside the brain interact as a central command center.

■ Links to Exocrine glands 31.3, Feedback controls 31.9, Action potentials 32.4, Human brain 32.9

The **hypothalamus** is the main center for control of the internal environment. It lies deep inside the forebrain and connects, structurally and functionally, with the **pituitary gland** (**Figure 34.4**). In humans, the pituitary gland is no bigger than a pea. Its posterior lobe secretes hormones made in the hypothalamus. Its anterior lobe makes its own hormones. **Table 34.2** summarizes the hormones released from the pituitary gland.

The hypothalamus signals the pituitary by way of neurosecretory neurons, specialized neurons that release hormones into the blood. These neurons have their cell body in the hypothalamus. Axons of some of these neurons extend into the pituitary's posterior lobe. Axons from others extend into the stalk just above the pituitary.

Posterior Pituitary Function

Antidiuretic hormone (ADH) and oxytocin (OT) are produced in the cell bodies of secretory neurons of the hypothalamus (**Figure 34.5 ❶**). These hormones travel through axons to axon terminals inside the posterior pituitary ❸, where they are stored. Arrival of an action potential (Section 34.3) at the axon terminals causes the terminals to release the stored hormone, which diffuses into capillaries (small blood vessels) inside the posterior pituitary ❹. From here, blood distributes the hormone throughout the body, where it exerts its effect on target cells ❹.

Antidiuretic hormone affects certain kidney cells. The hormone causes these cells to reabsorb more water, thus making urine more concentrated.

Oxytocin triggers muscle contractions during childbirth. It also makes milk move into the ducts of mammary glands when a female mammal nurses her young, and it affects social behavior in some species.

Anterior Pituitary Function

The anterior pituitary produces hormones of its own, but hormones from the hypothalamus control their

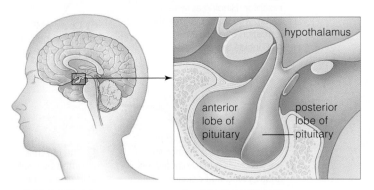

Figure 34.4 Location of the hypothalamus and pituitary gland. The two lobes of the pituitary (anterior and posterior) release different hormones.

Pituitary Lobe	Secretions	Abbreviation	Main Targets	Primary Actions
Posterior Nervous tissue (extension of hypothalamus)	Antidiuretic hormone (vasopressin)	ADH	Kidneys	Induces water conservation as required to maintain extracellular fluid volume and solute concentrations
	Oxytocin	OT	Mammary glands	Induces milk movement into secretory ducts
			Uterus	Induces uterine contractions during childbirth
Anterior Glandular tissue, mostly	Adrenocorticotropic hormone	ACTH	Adrenal glands	Stimulates release of cortisol, an adrenal steroid hormone
	Thyroid-stimulating hormone	TSH	Thyroid gland	Stimulates release of thyroid hormones
	Follicle-stimulating hormone	FSH	Ovaries, testes	In females, stimulates estrogen secretion, egg maturation; in males, helps stimulate sperm formation
	Luteinizing hormone	LH	Ovaries, testes	In females, stimulates progesterone secretion, ovulation, corpus luteum formation; in males, stimulates testosterone secretion, sperm release
	Prolactin	PRL	Mammary glands	Stimulates and sustains milk production
	Growth hormone (somatotropin)	GH	Most cells	Promotes growth in young; induces protein synthesis, cell division; roles in glucose, protein metabolism in adults

Table 34.2 Primary Actions of Hormones Released From the Human Pituitary Gland

secretion. Most hypothalamic hormones that act on the anterior pituitary are **releasing hormones**, which encourage secretion of hormones by target cells. The hypothalamus also produces **inhibiting hormones**, which reduce secretion of hormones by target cells. Hypothalamic releasing and inhibiting hormones are secreted into the stalk that connects the hypothalamus to the pituitary (**Figure 34.6 ❶**). They diffuse into blood and are carried to the anterior lobe of the pituitary ❸. Here, they diffuse out of capillaries and bind to target cells ❹. When stimulated by a releasing hormone, a target cell secretes an anterior pituitary hormone into the blood ❹. An inhibiting hormone has the opposite effect on its target.

Some anterior pituitary hormones target cells inside other glands:

Adrenocorticotropic hormone (ACTH) stimulates the release of hormones by adrenal glands.

Thyroid-stimulating hormone (TSH) regulates the secretion of thyroid hormone by the thyroid gland.

Follicle-stimulating hormone (FSH) and luteinizing hormone (LH) affect sex hormone secretion and gamete production by a male's testes or a female's ovaries.

Prolactin (PRL) targets the mammary glands, which are exocrine glands (Section 31.3). It stimulates and sustains milk production after childbirth.

Growth hormone (GH) has targets in most tissues. It promotes the growth of bone and soft tissues in the young. It also influences metabolism in adults.

Feedback Controls of Hormone Secretion

The hypothalamus and pituitary are involved in many negative feedback control mechanisms. As explained in Section 31.9, with negative feedback control, a stimulus elicits a response that decreases the stimulus. The hypothalamus monitors the concentration of many hormones. When the concentration of one of these hormone declines (the stimulus), the hypothalamus secretes a releasing hormone (the response). The releasing hormone acts on the pituitary or on another gland and results in a rise in that hormone (the decrease of the stimulus). The sections that follow include many examples of negative feedback control.

hypothalamus Forebrain region that controls processes related to homeostasis; control center for endocrine functions.
inhibiting hormone Hormone that is released by one gland and discourages secretion of a hormone by another gland.
pituitary gland Pea-sized endocrine gland in the forebrain that interacts closely with the adjacent hypothalamus.
releasing hormone Hormone that is released by one gland and encourages secretion of a hormone by another gland.

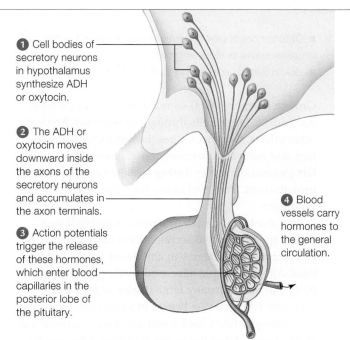

❶ Cell bodies of secretory neurons in hypothalamus synthesize ADH or oxytocin.

❷ The ADH or oxytocin moves downward inside the axons of the secretory neurons and accumulates in the axon terminals.

❸ Action potentials trigger the release of these hormones, which enter blood capillaries in the posterior lobe of the pituitary.

❹ Blood vessels carry hormones to the general circulation.

Figure 34.5 Posterior pituitary function. The posterior pituitary secretes hormones synthesized by cells in the hypothalamus.

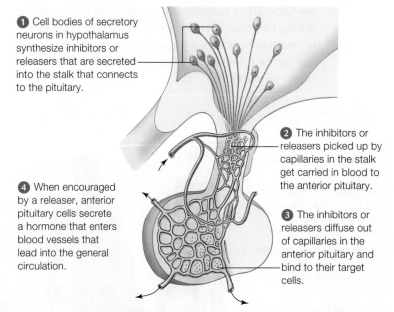

❶ Cell bodies of secretory neurons in hypothalamus synthesize inhibitors or releasers that are secreted into the stalk that connects to the pituitary.

❷ The inhibitors or releasers picked up by capillaries in the stalk get carried in blood to the anterior pituitary.

❸ The inhibitors or releasers diffuse out of capillaries in the anterior pituitary and bind to their target cells.

❹ When encouraged by a releaser, anterior pituitary cells secrete a hormone that enters blood vessels that lead into the general circulation.

Figure 34.6 Animated Anterior pituitary function. The anterior pituitary makes hormones and secretes them in response to hormones from the hypothalamus.

Take-Home Message

How do the hypothalamus and pituitary gland interact?

» Some secretory neurons of the hypothalamus make hormones (ADH, OT) that move through axons into the posterior pituitary, which releases them.

» Other hypothalamic neurons produce releasers and inhibitors that are carried by the blood into the anterior pituitary. These hormones regulate the secretion of anterior pituitary hormones (ACTH, TSH, LH, FSH, PRL, and GH).

34.5 Growth Hormone Function and Disorders

■ Disturbances of growth hormone production or function can cause excessive or reduced growth.

■ Link to Genetic engineering 15.6

Growth hormone (GH) secreted by the anterior pituitary affects target cells throughout the body. Among other effects, GH encourages the production of cartilage and bone and increases muscle mass. Normally, GH production surges during teenage years, causing a growth spurt. The level of the hormone then declines with age.

In some people, a benign (noncancerous) tumor causes the pituitary to secrete excess GH. Excessive GH secretion that begins in childhood results in gigantism. Affected people have a normally proportioned body, but are unusually large (**Figure 34.7A**). When excessive GH secretion begins in adulthood, the result is acromegaly (the Greek word *acro* means extremities; *megas*, large). Adult bones can no longer lengthen, so they become thicker instead. Hands, feet, and facial bones are most often visibly affected (**Figure 34.7B**).

Pituitary dwarfism occurs when the body produces too little GH or receptors do not respond to it properly during childhood. Affected individuals are short but normally proportioned (**Figure 34.7C**). Pituitary dwarfism can be inherited, or it can result from a pituitary tumor or injury.

Human growth hormone can now be made through genetic engineering (Section 15.6). Injections of recombinant human growth hormone (rhGH) increase the growth rate of children who have a naturally low GH level. However, such treatment is expensive ($15,000 to $20,000 a year) and controversial. Some people object to treating short stature as a defect to be cured.

Injections of rhGH are also used to treat adults who have a low GH level because of pituitary or hypothalamic tumors or injury. Injections restore a normal level of GH and help affected individuals maintain a healthy bone and muscle mass. Injections of rhGH have also been touted as a way to slow normal aging or boost athletic performance. However, such uses are not approved by regulatory agencies, have not been shown effective in clinical trials, and can have negative side effects, including increased risk of high blood pressure and diabetes.

Take-Home Message

What are the effects of too much or too little growth hormone?

» Excessive growth hormone causes faster-than-normal bone growth. When the excess occurs during childhood, the result is gigantism. In adults, the result is acromegaly.

» A deficiency of GH during childhood can cause dwarfism.

Figure 34.7 Effects of disrupted growth hormone function.

(**A**) Excessive GH production made Bao Xishun one of the world's tallest men, at 2.36 meters (7 feet 9 inches) tall.

(**B**) A woman before and after she became affected by acromegaly. Notice how her chin elongated.

(**C**) Insufficient growth hormone. At age 16, Jyoti Amge of Nagpur, India, stands not much taller than a classmate's backpack.

age 16

age 52

34.6 Sources and Effects of Other Vertebrate Hormones

■ A cell in a vertebrate body is a target for a diverse array of hormones from endocrine glands and secretory cells.

The next few sections of this chapter describe effects of the main vertebrate hormones released by endocrine glands other than the pituitary. **Table 34.3** provides an overview of this information.

In addition to major endocrine glands, vertebrates have hormone-secreting cells in many internal organs. As noted earlier, cells of the small intestine release secretin, a hormone that acts on the pancreas. Other gut hormones affect appetite and digestion. In addition, adipose (fat) tissue makes leptin, a hormone that acts in the brain and suppresses appetite.

When the oxygen level in blood falls, kidneys secrete erythropoietin, a hormone that stimulates maturation and production of oxygen-transporting red blood cells. Even the heart makes a hormone, atrial natriuretic peptide, which stimulates the kidneys to excrete water and salt.

As you learn about the effects of specific hormones, keep in mind that cells in most tissues have receptors for more than one hormone. The response called up by one hormone may oppose or reinforce that of another. For example, every skeletal muscle fiber has receptors for glucagon, insulin, cortisol, epinephrine, estrogen, testosterone, growth hormone, somatostatin, and thyroid hormone, as well as others. Thus, blood levels of all of these hormones affect a muscle. The muscle integrates the signals and responds accordingly.

Take-Home Message

What are the sources and effects of vertebrate hormones?

» In addition to the pituitary gland and hypothalamus, endocrine glands and endocrine cells secrete hormones. The gut, kidneys, and heart are among the organs that are not considered glands, but do include cells that secrete hormones.

» Most cells have receptors for multiple hormones, and the effect of one hormone can be enhanced or opposed by that of another.

Table 34.3   Sources and Actions of Vertebrate Hormones Discussed in Sections 34.7 to 34.12

Source	Examples of Secretion(s)	Main Target(s)	Primary Actions
Thyroid	Thyroid hormone	Most cells	Regulates metabolism; has roles in growth, development
	Calcitonin	Bone	Lowers calcium level in blood
Parathyroids	Parathyroid hormone	Bone, kidney	Elevates calcium level in blood
Pancreatic islets	Insulin	Liver, muscle, adipose tissue	Promotes cell uptake of glucose; thus lowers glucose level in blood
	Glucagon	Liver	Promotes glycogen breakdown; raises glucose level in blood
	Somatostatin	Insulin-secreting cells	Inhibits secretion of insulin, glucagon, and some gut hormones
Adrenal cortex	Glucocorticoids (including cortisol)	Most cells	Promote breakdown of glycogen, fats, and proteins as energy sources; thus help raise blood level of glucose
	Mineralocorticoids (including aldosterone)	Kidney	Promote sodium reabsorption (sodium conservation); help control the body's salt–water balance
Adrenal medulla	Epinephrine (adrenaline)	Liver, muscle, adipose tissue	Raises blood level of sugar, fatty acids; increases heart rate and force of contraction
	Norepinephrine	Smooth muscle of blood vessels	Promotes constriction or dilation of certain blood vessels; thus affects distribution of blood volume to different body regions
Gonads			
Testes (in males)	Androgens (including testosterone)	General	Required in sperm formation; development of genitals; maintenance of sexual traits; growth, development
Ovaries (in females)	Estrogens	General	Required for egg maturation and release; preparation of uterine lining for pregnancy and its maintenance in pregnancy; genital development; maintenance of sexual traits; growth, development
	Progesterone	Uterus, breasts	Prepares, maintains uterine lining for pregnancy; stimulates development of breast tissues
Pineal gland	Melatonin	Brain	Influences daily biorhythms, seasonal sexual activity
Thymus	Thymosins, thymulin	T lymphocytes	Poorly understood regulatory effect on T lymphocytes

34.7 Thyroid and Parathyroid Glands

■ The thyroid regulates metabolic rate, and the adjacent parathyroids regulate calcium levels.

■ Link to Feedback mechanisms 31.9

Feedback Control of Thyroid Function

The human **thyroid gland** lies at the base of the neck, attached to the trachea, or windpipe (**Figure 34.8**). It secretes two iodine-containing molecules (triiodothyronine and thyroxine) that we refer to collectively as thyroid hormone. Thyroid hormone increases metabolic activity of cells throughout the body. The thyroid gland also secretes calcitonin, a hormone that causes deposition of calcium in bones of growing children. Normal human adults produce little calcitonin.

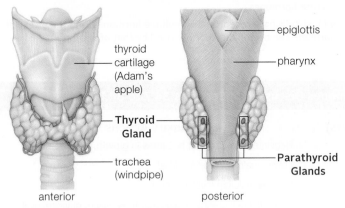

anterior   posterior

Figure 34.8 Location of human thyroid and parathyroid glands.

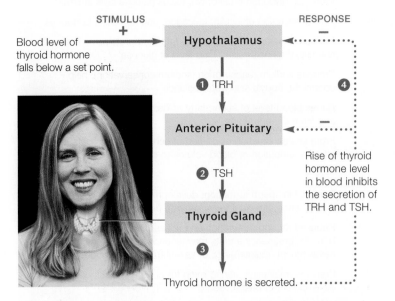

Figure 34.9 Negative feedback loop to the hypothalamus and the pituitary's anterior lobe that governs thyroid hormone secretion. **Figure It Out: What effect does a high thyroid hormone level have on the hypothalamus?**

Answer: It inhibits secretion of TRH.

The anterior pituitary gland and hypothalamus regulate thyroid hormone secretion by a negative feedback loop (**Figure 34.9**). A low blood concentration of thyroid hormone causes the hypothalamus to secrete thyroid-releasing hormone (TRH) ❶. This releasing hormone causes the anterior pituitary to secrete thyroid-stimulating hormone (TSH). TSH in turn stimulates the secretion of thyroid hormone ❸. When the blood level of thyroid hormone rises, secretion of TRH and TSH declines ❹.

Thyroid Disorders

Thyroid hormone increases the metabolic rate, so a deficiency in this hormone (called hypothyroidism) causes fatigue, depression, increased sensitivity to cold temperature, and weight gain. Affected people often have thyroid enlargement, or goiter (**Figure 34.10**).

Thyroid hormone synthesis requires iodine, so adequate iodine intake is essential to normal health and development. A woman's iodine requirement increases during pregnancy. Inadequate maternal iodine intake raises the risk of miscarriage and of infant mortality. It also results in cretinism, a syndrome of mental retardation, stunted growth, and deaf-mutism.

In the United States, use of iodized salt has greatly reduced the incidence of dietary hypothyroidism, so inadequate levels of thyroid hormone most often result from an immune disorder. In some people, the body's white blood cells mistakenly attack the thyroid and destroy its hormone-producing tissue. The result is a type of hypothyroidism known as Hashimoto's disease. Thyroid tumors can also impair thyroid hormone production. Nondietary hypothyroidism is treated by the administration of synthetic thyroid hormone.

An excess of thyroid hormone also has negative health consequences. In Graves' disease, proteins (antibodies) made by white blood cells mimic the action of thyroid-stimulating hormone. As a result, the thyroid produces an excess of thyroid hormone. Affected people have goiter, as well as anxiety, insomnia, heat intolerance, weight loss, and tremors. Protruding eyes are another symptom. Drugs, surgery, or radiation can be used to reduce thyroid hormone output.

Thyroid Disruptors

Some pollutants can impair thyroid function. A sodium–iodide cotransporter normally pumps iodide ions into thyroid cells. However, the transporter binds a chemical pollutant called perchlorate thirty times more strongly than it binds iodide ions. Thus, when

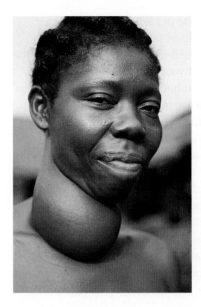

Figure 34.10 Goiter caused by a dietary iodine deficiency. The thyroid also enlarges in Graves' disease, an immune disorder that causes overstimulation of the thyroid.

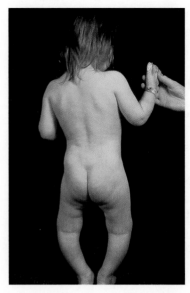

Figure 34.11 Child with rickets. Bowed legs result when parathyroid hormone causes existing bone to break down faster than new bone is deposited.

perchlorate is present, fewer iodide ions enter thyroid cells, and thyroid hormone production declines. Perchlorate seeps into the ground from facilities that manufacture, test, or dispose of military rockets, fireworks, or other explosives. When a high level of perchlorate is detected in a well or other source of drinking water, the source must be closed to ensure public safety. Most municipal water treatment plants cannot remove perchlorate from water.

Aquatic vertebrates do not have the option of avoiding perchlorate-contaminated waters. In frogs, a surge in thyroid hormone triggers metamorphosis from a tadpole (the larval form) to an adult. If a tadpole's thyroid tissue is removed, it will keep on growing, but will never undergo metamorphosis. As you might predict, perchlorate-polluted water also can delay or prevent frog metamorphosis.

The Parathyroid Glands

Four **parathyroid glands**, each about the size of a grain of rice, are located on the thyroid's posterior surface (**Figure 34.8**). The glands release parathyroid hormone (PTH) in response to a decline in the level of calcium in blood. Calcium ions have roles in neuron signaling, blood clotting, muscle contraction, and other essential physiological processes.

PTH targets bone cells and kidney cells. In bones, it induces specialized cells to secrete bone-digesting enzymes. Calcium and other minerals released from the bone enter the blood. In the kidneys, PTH stimulates tubule cells to reabsorb more calcium. It also stimulates secretion of enzymes that activate vitamin

D, transforming it to calcitriol. Calcitriol is a steroid hormone that encourages cells in the intestinal lining to absorb more calcium from food.

Inadequate vitamin D is the most common cause of the nutritional disorder known as rickets. Without adequate vitamin D, a child does not absorb much calcium from food, so formation of new bone slows. At the same time, the lower-than-normal calcium concentration in the blood triggers PTH secretion. When the concentration of PTH rises, the child's body responds by breaking down existing bones. Bowed legs and deformities in pelvic bones are common symptoms of rickets (**Figure 34.11**).

Tumors and other conditions that cause excessive PTH secretion also weaken bone. In addition, they increase risk of kidney stones, because calcium released from bone ends up in the kidney. Disorders that reduce PTH output also lower blood calcium. The resulting seizures and unrelenting muscle contractions can be deadly.

parathyroid glands Four small endocrine glands whose hormone product increases the level of calcium in blood.
thyroid gland Endocrine gland at the base of the neck; produces thyroid hormone, which increases metabolism.

Take-Home Message

What are the functions of the thyroid and parathyroid glands?

» The thyroid gland has roles in regulation of metabolism and in development. Iodine is required to make thyroid hormone.

» The parathyroid glands are the main regulators of blood calcium level.

34.8 Pancreatic Hormones

■ Two pancreatic hormones with opposing effects work together to regulate the level of sugar in the blood.

■ Links to Glycogen 3.4, Alternative energy sources in food 7.7

Regulation of Blood Sugar

The **pancreas** lies in the abdominal cavity, behind the stomach (**Figure 34.12**). Its exocrine cells secrete digestive enzymes into the small intestine. Its endocrine cells group in clusters called pancreatic islets. Each islet contains three types of cells.

Beta cells, the most abundant cells in the pancreatic islets, secrete **insulin**—the only hormone that causes target cells to take up and store glucose. After a meal, the blood level of glucose rises, which stimulates beta cells to release insulin. The main targets are liver, fat, and skeletal muscle cells. Insulin especially stimulates muscle and fat cells to take up glucose. In all target cells, it activates enzymes that function in protein and fat synthesis, and it inhibits the enzymes that catalyze protein and fat breakdown. As a result of its actions, insulin lowers the level of glucose in the blood (**Figure 34.12 ❶–❺**).

Alpha cells secrete the peptide hormone glucagon when the glucose level falls below a set point. **Glucagon** is a peptide hormone that binds to cells in the liver and causes the activation of enzymes that break glycogen into glucose subunits. By its action, glucagon raises the level of glucose in blood (**Figure 34.12 ❻–❿**). The molecular mechanism of glucagon action was illustrated earlier in **Figure 34.3B**.

Delta cells secrete somatostatin. This hormone helps control digestion and nutrient absorption. Also, it can inhibit the secretion of insulin and glucagon.

Some shifts in blood glucose level are typical of all animals that show a discontinuous eating pattern. By working in opposition, glucagon and insulin from the pancreas maintain that level within a range that keeps cells functioning properly.

Diabetes

Diabetes mellitus is a common metabolic disorder. Its name can be loosely translated as "passing honey-sweet water." Diabetics produce sweet urine because their liver, fat, and muscle cells do not take up and

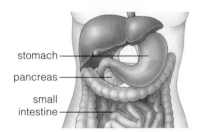

stomach
pancreas
small intestine

Figure 34.12 Animated *Above*, the location of the pancreas. *Right*, how cells that secrete insulin and glucagon work antagonistically to adjust the level of glucose in the blood.

❶ After a meal, glucose enters blood faster than cells can take it up, so blood glucose increases.

❷ The increase stops pancreatic cells from secreting glucagon and ❸ stimulates other cells to secrete insulin.

❹ In response to insulin, adipose and muscle cells take up and store glucose; cells in the liver and muscle make more glycogen.

❺ As a result, the blood level of glucose declines to its normal level.

❻ Between meals, blood glucose declines as cells take it up and use it for metabolism.

❼ The decrease encourages glucagon secretion and ❽ slows insulin secretion.

❾ In the liver, glucagon causes cells to break glycogen down into glucose, which enters the blood.

❿ As a result, blood glucose increases to the normal level.

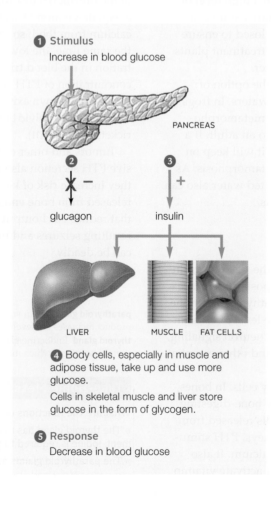

❶ **Stimulus**
Increase in blood glucose

PANCREAS

❷ glucagon   ❸ insulin

LIVER   MUSCLE   FAT CELLS

❹ Body cells, especially in muscle and adipose tissue, take up and use more glucose.

Cells in skeletal muscle and liver store glucose in the form of glycogen.

❺ **Response**
Decrease in blood glucose

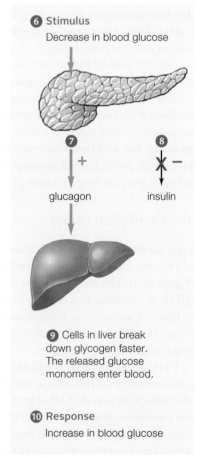

❻ **Stimulus**
Decrease in blood glucose

❼ glucagon   ❽ insulin

❾ Cells in liver break down glycogen faster. The released glucose monomers enter blood.

❿ **Response**
Increase in blood glucose

Table 34.4	Some Complications of Diabetes
Eyes	Changes in lens shape and vision; damage to blood vessels in retina; blindness
Skin	Increased susceptibility to bacterial and fungal infections; patches of discoloration; thickening of skin on the back of hands
Digestive system	Gum disease; delayed stomach emptying that causes heartburn, nausea, vomiting
Kidneys	Increased risk of kidney disease and failure
Circulatory system	Increased risk of heart attack, stroke, high blood pressure, and atherosclerosis
Hands and feet	Impaired sensations of pain; formation of calluses, foot ulcers; tissue death that may require amputation

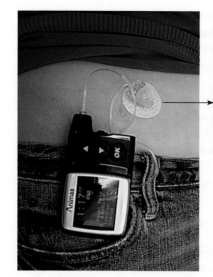

Figure 34.13 An insulin pump. The device is programmed to deliver insulin through a hollow tube that projects through the skin and into the body.

Use of an insulin pump helps smooth out fluctuations in blood sugar, thus lowering the risk of complications that can arise from excessively low or high blood sugar.

store glucose as they should. The resulting high blood sugar, or hyperglycemia, disrupts normal metabolism. When cells do not take up glucose, they have to break down proteins and fats, and breakdown of these substances yields harmful waste products. At the same time, high blood sugar causes some cells to overdose on glucose and produce other harmful substances. Accumulation of harmful molecules causes the complications associated with diabetes (**Table 34.4**).

Type 1 Diabetes There are two main types of diabetes mellitus. Type 1 develops after an autoimmune response. White blood cells mistakenly identify insulin-secreting beta cells as foreign and destroy them. Environmental factors add to a genetic predisposition to the disorder. Symptoms usually appear during childhood and adolescence; thus this metabolic disorder is known as juvenile-onset diabetes. All affected individuals require injections of insulin, and must monitor their blood sugar concentration carefully. New devices called insulin pumps dispense a continuous supply of insulin (**Figure 34.13**).

Type 1 diabetes accounts for only 5 to 10 percent of all reported cases, but it is the most dangerous in the short term. In the absence of a steady supply of glucose, the body of an affected person uses fats and proteins as energy sources. Two outcomes are weight loss and ketone accumulation in the blood and urine.

glucagon Pancreatic hormone that causes cells to break down glycogen and release glucose.
insulin Pancreatic hormone that causes cells to take up glucose and store it as glycogen.
pancreas Organ that secretes digestive enzymes into the small intestine and hormones into the blood.

Ketones are normal acidic products of fat breakdown, but too many can alter the acidity and solute levels of body fluids. This condition, called ketosis, can interfere with normal brain function, and extreme cases may lead to coma or death.

Type 2 Diabetes With type 2 diabetes, the more common form of the disorder, insulin levels are normal or even high. However, target cells do not respond to the hormone, and blood sugar levels remain elevated. Symptoms typically start to develop in middle age, when insulin production declines.

Diet, exercise, and oral medications can control most cases of type 2 diabetes. Even so, if glucose levels are not lowered, pancreatic beta cells receive continual stimulation. Eventually they may falter, and so will insulin production. When that happens, a type 2 diabetic may require insulin injections.

Worldwide, rates of type 2 diabetes are soaring. By one estimate, more than 150 million people are now affected. Western diets and sedentary lifestyles are contributing factors. The prevention of diabetes and its complications is acknowledged as one of the most pressing public health priorities around the world.

Take-Home Message

How do pancreatic hormones maintain the level of glucose in the blood?

» Insulin helps cells take up and store more glucose; it lowers the blood level of glucose.

» Glucagon triggers breakdown of glycogen; it raises blood glucose.

» Diabetes is a metabolic disorder in which the body does not make insulin or the body does not respond to it. As a result, cells do not take up sugar as they should, causing complications throughout the body.

- An adrenal gland has two functional zones: Its outer cortex secretes steroid hormones. Its inner medulla releases molecules that function as neurotransmitters.
- Links to Sympathetic nerves 32.7, Memory 32.11

There are two **adrenal glands**, one above each kidney. (In Latin *ad–* means near, and *renal* refers to the kidney.) Each adrenal gland is the size of a big grape. Its outer layer is the **adrenal cortex** and its inner portion is the **adrenal medulla**. The two regions are controlled by different mechanisms, and secrete different substances.

The Adrenal Cortex

The adrenal cortex releases steroid hormones. One of these, aldosterone, controls sodium and water reabsorption by kidneys. Section 40.5 explains this process in detail. The adrenal cortex also produces and secretes small amounts of sex hormones, which we discuss in Section 34.10. Here we focus on **cortisol**, a hormone that affects metabolism and immune responses.

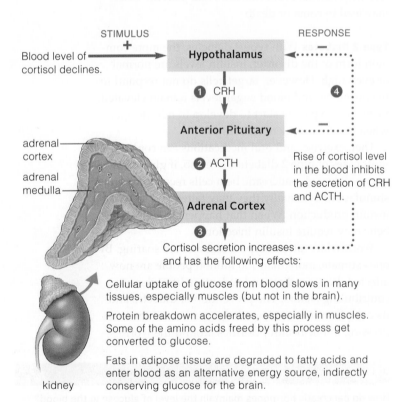

Figure 34.14 Structure of the human adrenal gland. An adrenal gland rests on top of each kidney. The diagram shows a negative feedback loop that governs cortisol secretion.

Figure It Out: What effect would a decrease in ACTH have on the rate of fat breakdown in adipose tissue?

Answer: A decrease in ACTH would cause a decrease in cortisol secretion and less fat breakdown.

A negative feedback loop regulates the cortisol level in blood (**Figure 34.14**). When the cortisol level decreases, the hypothalamus increases its secretion of CRH (corticotropin-releasing hormone) **1**. CRH stimulates the anterior pituitary to secrete ACTH (adrenocorticotropic hormone) **2**, which in turn causes the adrenal cortex to release cortisol **3**. The blood level of cortisol rises. The rise is detected by the hypothalamus, and it stops secreting CRH, the anterior pituitary stops secreting ACTH, and cortisol secretion slows **4**.

Cortisol helps ensure that adequate glucose is available to the brain by inducing liver cells to break down their store of glycogen, and suppressing uptake of glucose by most cells. Cortisol also induces adipose cells to degrade fats, and skeletal muscles to degrade proteins. The breakdown products of these reactions—fatty acids and amino acids—function as energy sources (Section 7.7).

With injury, illness, or anxiety, the nervous system overrides the feedback loop, allowing the amount of cortisol in the blood to soar. In the short term, this response helps get enough glucose to the brain when food supplies are likely to be low. A heightened cortisol level also suppresses inflammatory responses, thus lessening inflammation-related pain.

The Adrenal Medulla

The adrenal medulla contains specialized neurons of the sympathetic division (Section 32.7). Like other sympathetic neurons, those in the adrenal medulla release norepinephrine and epinephrine. However, in this case, the norepinephrine and epinephrine enter the blood and function as hormones, rather than acting as neurotransmitters at a synapse. Epinephrine and norepinephrine released into the blood have the same effect on a target organ as direct stimulation by a sympathetic nerve.

Remember that sympathetic stimulation plays a role in the fight–flight response. Epinephrine and norepinephrine dilate the pupils, increase breathing rate, and make the heart beat faster. They prepare the body to deal with an exciting or dangerous situation.

Stress, Elevated Cortisol, and Health

When an animal is frightened or under physical stress, the nervous system triggers increased secretion of cortisol, epinephrine, and norepinephrine. As these hormones find their targets, they divert resources from longer-term tasks and help the body deal with the immediate threat. This stress response is highly adaptive for short periods of time, as when an animal is fleeing from a predator.

What happens when stress is ongoing? For more than twenty years, neurobiologist Robert Sapolsky and his Kenyan colleagues have been studying how olive baboons (*Papio anubis*) interact and how a baboon's social position influences its hormone levels and health. Olive baboons live in large troops with a clearly defined dominance hierarchy. Individuals on top of the hierarchy get first access to food, grooming, and sexual partners. Those at the bottom of the hierarchy must continually relinquish resources to a higher-ranking baboon or face attack (**Figure 34.15**). Not surprisingly, the low-ranking baboons tend to have chronically high cortisol levels.

Physiological responses to chronic stress interfere with growth, the immune system, sexual function, and cardiovascular function. Chronically high cortisol levels also harm cells in the hippocampus, a brain region central to memory and learning (Section 32.11).

We also see the impact of long-term elevated cortisol levels in humans affected by Cushing's syndrome, or hypercortisolism. This rare metabolic disorder can be triggered by an adrenal gland tumor, oversecretion of ACTH by the anterior pituitary, or chronic use of the drug cortisone. Doctors often prescribe cortisone to relieve chronic pain, inflammation, or other health problems. The body converts it to cortisol.

Figure 34.15 A dominant baboon (*right*) raising the stress level—and cortisol level—of a less dominant member of its troop.

Symptoms of hypercortisolism include a puffy, rounded "moon face" (**Figure 34.16**) and a fat torso. Blood pressure and blood glucose rise. White blood cell counts decline, so affected people are more prone to infections. Thin skin, decreased bone density, and muscle loss are common. Patients with the highest cortisol level also have the greatest reduction in the size of the hippocampus, and the most impaired memory.

Can status-related social stress affect human health? People who are low in a socioeconomic hierarchy tend to have more health problems than those who are better off. These differences persist even after researchers factor out obvious causes, such as variations in diet and access to health care. By one hypothesis, a heightened cortisol level caused by low social status may be one of the links between poverty and poor health.

Cortisol Deficiency

Tuberculosis and other diseases can damage adrenal glands, and slow or halt cortisol secretion. The result is Addison's disease, or hypocortisolism. In developed countries, this disorder more often arises after autoimmune attacks on the adrenal glands. President John F. Kennedy had this form of the disorder. Symptoms include fatigue, depression, weight loss, and darkening of the skin. If cortisol levels get too low, blood sugar and blood pressure fall to life-threatening levels. Addison's disease is treated with synthetic cortisone.

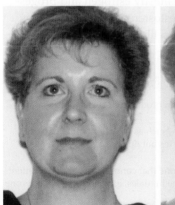

Figure 34.16 Cushing syndrome. *Left*, woman with elevated cortisol levels as a result of an adrenal gland tumor. She has the characteristic puffy moon face. *Right*, the same woman after removal of the tumor lowered her cortisol to normal levels.

adrenal cortex Outer portion of adrenal gland; secretes aldosterone and cortisol.
adrenal gland Endocrine gland located atop the kidney; secretes aldosterone, cortisol, epinephrine, and norepinephrine.
adrenal medulla Inner portion of adrenal gland; secretes epinephrine and norepinephrine.
cortisol Adrenal cortex hormone that influences metabolism and immunity; secretions rise with stress.

Take-Home Message

What are the functions of the hormones secreted by the adrenal glands?

» The adrenal cortex's main secretions are aldosterone, which affects urine concentration, and cortisol, which affects metabolism and stress responses.

» The adrenal medulla releases epinephrine and norepinephrine, which prepare the body for excitement or danger.

» Cortisol secretion is governed by a feedback loop to the hypothalamus and pituitary, but stress breaks that loop and allows the level of cortisol to rise.

» Long-term cortisol elevation harms health. Insufficient cortisol can be fatal.

34.10 The Gonads

- Differences in sex hormone production are the basis for the differences in body form between males and females.
- Sex hormones also regulate gamete production.
- Link to Sexual differentiation 10.4

Gonads are primary reproductive organs, meaning they produce gametes (eggs or sperm). Vertebrate gonads also produce **sex hormones**, steroid hormones that control sexual development and reproduction. **Figure 34.17** shows the location of the human gonads. A male's gonads—his testes (singular, testis)—secrete mainly testosterone. A female's gonads—her ovaries—secrete mainly estrogens and progesterone.

The hypothalamus and anterior pituitary control sex hormone secretion (**Figure 34.18**). In both sexes, the hypothalamus produces gonadotropin-releasing hormone (GnRH). This releasing hormone causes the anterior pituitary to secrete follicle-stimulating hormone (FSH) and luteinizing hormone (LH), which target the gonads and cause them to produce and secrete sex hormones. FSH and LH are sometimes referred to as "gonadotropins." The suffix -*tropin* refers to a substance that has a stimulating effect. In this case, the effect is on the gonads.

As Section 10.4 explained, the presence of an *SRY* gene causes an early human embryo to produce testosterone and develop a male reproductive tract. Embryos that do not make testosterone or do not respond to it develop a female reproductive tract. During childhood, levels of sex hormones are low and, with the exception of their reproductive organs, males and females have similar bodies.

Sex hormone production increases during **puberty**, the period of development when reproductive organs mature and secondary sexual traits appear. **Secondary**

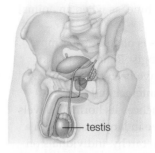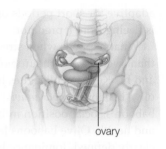
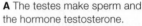

A The testes make sperm and the hormone testosterone.

B The ovaries make eggs and the hormones estrogen and progesterone.

Figure 34.17 Human gonads.

sexual traits are traits that differ between the sexes, but do not have a direct role in reproduction. When a girl undergoes puberty, her ovaries increase their estrogen production, causing development of rounded breasts and hips, and other female secondary sexual traits. Estrogens and progesterone also regulate egg formation and ready the uterus (womb) for pregnancy.

In males, a rise in testosterone output at puberty triggers the onset of sperm production and the development of secondary sexual traits. In humans, these traits include facial hair and a deep voice.

Testes secrete mostly testosterone, but they also make a little bit of estrogen and progesterone. The estrogen is necessary for sperm formation. Similarly, a female's ovaries make mostly estrogen and progesterone, but also a little testosterone. The presence of testosterone contributes to libido—the desire for sex.

gonads Primary reproductive organs (ovaries or testes); produce gametes and sex hormones.
puberty Period when human reproductive organs mature and begin to function.
secondary sexual traits Traits, such as the distribution of body fat, that differ between the sexes but do not have a direct role in reproduction.
sex hormone A steroid hormone that controls sexual development and reproduction; testosterone in males, estrogens or progesterone in females.

Take-Home Message

What are sex hormones?

» Sex hormones are steroid hormones that influence the development of sexual traits and reproduction. Production of sex hormones increases at puberty.

» In both sexes, gonads secrete sex hormones in response to anterior pituitary hormones (FSH and LH), which are in turn released in response to a hypothalamic releasing hormone.

» A male's testes secrete mainly testosterone, and a female's ovaries secrete mainly estrogens and progesterone.

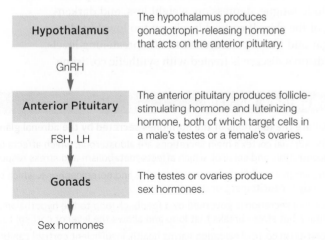

Hypothalamus	The hypothalamus produces gonadotropin-releasing hormone that acts on the anterior pituitary.
↓ GnRH	
Anterior Pituitary	The anterior pituitary produces follicle-stimulating hormone and luteinizing hormone, both of which target cells in a male's testes or a female's ovaries.
↓ FSH, LH	
Gonads	The testes or ovaries produce sex hormones.
↓	
Sex hormones	

Figure 34.18 Central control of sex hormone secretion.

34.11 The Pineal Gland

- At night, darkness allows the pineal gland to secrete a hormone that encourages sleep.
- Link to Human brain 32.9

The **pineal gland** is a pea-sized, pine cone–shaped gland that secretes the hormone **melatonin**—but only during conditions of low light or darkness. Bright light suppresses pineal gland activity. In nonmammalian vertebrates, the pineal gland is close to the surface of the skin and includes photoreceptors that detect light directly. In mammals, including humans, the pineal gland resides deep inside the brain (**Figure 34.19**). It receives information about light indirectly, by way of nervous signals from other regions of the brain.

Circadian Rhythms

The amount of light varies with the time of day, so melatonin secretion follows a **circadian rhythm**, meaning it varies cyclically over an approximately 24-hour interval. Melatonin secretion in turn influences other circadian rhythms. At night, melatonin secretion causes a decline in body temperature and we become sleepy. Just after sunrise, melatonin secretion decreases, body temperature rises, and we awaken. Melatonin's sleep-inducing effects stem from its effects on target cells in the hypothalamus.

Cycles of melatonin secretion that affect sleep/wake rhythms are set by exposure to light, so travelers who fly across many time zones are advised to spend some time in the sun to minimize jet lag. The exposure to light can help them reset their internal clock. Taking supplemental melatonin can also help prevent jet lag and alleviate some types of insomnia.

Routinely working night shifts or having poor sleep habits disrupts melatonin secretion and raises the risk of cancer. Female night-shift workers have lower melatonin levels and a higher risk of breast cancer than their day-shift counterparts, and a woman's risk of breast cancer decreases with the length of her average night's sleep. In men, night-shift work increases the risk of prostate cancer. How could melatonin protect against cancer? Melatonin directly inhibits division of cancer cells in animals. It also suppresses production of sex hormones, and high levels of those hormones encourage the growth of some cancers.

Seasonal Effects

Some people who live at latitudes where daylength fluctuates with the season have seasonal affective disorder (SAD), commonly called the "winter blues."

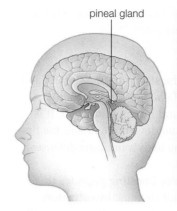

Figure 34.19 Location of the human pineal gland. The gland produces melatonin when the retina is not being stimulated by light. Typically, secretion peaks at about 2 A.M., while we are asleep.

They become depressed in winter, when there is less daylight. SAD arises when a person's rhythm of melatonin secretion gets out of sync with clock time. Exposure to bright light in the morning, a dose of melatonin at an appropriate time, or a combination of the two can diminish or eliminate symptoms of SAD.

Melatonin regulates the onset of seasonal behavior in many animals. For example, it affects the time of year when male birds that live in temperate zones begin to sing. Singing and other courtship behaviors occur only when the male's testosterone level is high. During the winter, long nights cause a rise in the bird's melatonin level. The melatonin acts on the hypothalamus, causing the secretion of an inhibiting hormone that discourages release of FSH and LH by the pituitary. Low levels of these hormones keep the level of testosterone low, and prevent the bird from singing. In spring, daylength increases and the melatonin concentration in the bird's blood declines. With the inhibitory effects of melatonin lifted, the bird's testes secrete more testosterone and the bird begins to sing.

circadian rhythm A physiological change that repeats itself on an approximately 24-hour cycle.
melatonin Hormone secreted by the pineal gland under conditions of darkness; affects sleep/wake cycles and protects against cancer.
pineal gland Endocrine gland in brain that secretes melatonin.

Take-Home Message

What is the role of melatonin secreted by the pineal gland?

» Melatonin secreted by the pineal gland during periods of darkness encourages sleepiness and helps set the body's internal clock.

» Activities that disrupt nighttime melatonin secretion raise the risk of cancer.

» In many animals, seasonal differences in melatonin secretion give rise to seasonal differences in behavior.

34.12 The Thymus

■ Peptides secreted by the thymus early in life are essential to establishing a healthy immune system.
■ Link to Cofactor 5.6

The **thymus** lies beneath the sternum (breastbone) and secretes a complex mix of small polypeptides that enhance wound healing and immune function. It is most active during prenatal development and the first few years of life, when the body's immune defenses are becoming established.

Thymosins are immunity-boosting peptides that were first isolated from calf thymus. Investigators initially thought that thymosins were thymus-specific hormones. However, later investigations revealed that other tissues also make thymosins. Some types of synthetic thymosins show promise as treatment for hepatitis and other viral diseases, as well as for cancer.

Thymulin is a peptide hormone secreted only by epithelial cells of the thymus. It encourages maturation of white blood cells called T cells (short for thymus-derived cells), and enhances T cell function. You will learn about the central role that T cells play in immunity in Chapter 37. Thymulin requires zinc as a cofactor (Section 5.6); thus a diet deficient in zinc can impair immune function.

The size of the thymus peaks at puberty. From puberty until about age 40, the thymus decreases in size, and more and more of its secretory cells are replaced by adipose tissue. By age 40, a normal thymus consists largely of fat, but it will continue to produce a low level of thymulin steadily for the remainder of the individual's life.

The reduced activity of the thymus in one's later years does not have a dramatically adverse effect because a healthy adult retains many T cells that formed at a younger age. However, low thymus activity may contribute to the increased incidence of cancer and infectious disease that accompanies aging. Obesity accelerates the replacement of functional thymus cells by fat, speeding the decline in thymus secretions. Thus, the heightened risk of cancer among the obese may arise in part from impaired thymus output.

thymus Organ beneath the sternum; produces hormones that enhance immune function and is the site of T cell maturation.

Take-Home Message

What is the role of the thymus?

» The thymus produces peptides that enhance immune function and wound healing.

» The thymus is most active in childhood. During adulthood, its secretory tissue is largely replaced by fat.

34.13 Invertebrate Hormones

■ Animal hormones have a long evolutionary history.
■ Links to Introns 9.3, Transcription factor 10.2, Gene duplication 14.5, Cnidarians 24.5, Annelids 24.7, Arthropods 24.11

Evolution of Hormone Diversity

We can trace the evolutionary roots of some vertebrate hormones and hormone receptors back to homologous molecules in invertebrates. For example, receptors for the hormones FSH, LH, and TSH all have a similar structure. The genes that encode these receptors have a similar sequence and all have introns (noncoding DNA) in the same places. The slightly different forms of receptor gene most likely evolved when an ancestral gene was duplicated, then copies of that gene mutated over time (Section 14.5). The ancestral receptor gene apparently evolved very long ago, in the common ancestor of cnidarians and vertebrates. Cnidarians such as sea anemones have a receptor protein gene with a sequence similar to the gene that encodes the vertebrate FSH receptor.

Proteins that serve as vertebrate estrogen receptors also have a long history. Like vertebrates, annelids have an estrogen receptor that binds estrogen and functions as a transcription factor. This suggests that annelids, like vertebrates, are adversely affected by chemical pollutants that disrupt estrogen function.

Hormones and Molting

Other hormones are unique to invertebrates. For example, arthropods, which include crabs and insects, have a hardened external cuticle that they periodically shed as they grow (Section 24.11). Shedding of the old cuticle is called molting. A soft new cuticle forms beneath the old one before the animal molts. Although details vary among arthropod groups, molting is generally under the control of **ecdysone**, a steroid hormone.

The arthropod molting gland produces and stores ecdysone, then releases it for distribution throughout the body when conditions favor molting. Hormone-secreting neurons inside the brain control ecdysone's release. The neurons respond to internal signals and environmental cues, including light and temperature.

Figure 34.20 is an example of the control steps in crabs and other crustaceans. In response to seasonal cues, secretion of a molt-inhibiting hormone declines and ecdysone secretion rises. Ecdysone causes changes in the animal's structure and physiology. The existing cuticle separates from the epidermis and the muscles.

ecdysone Insect hormone with roles in metamorphosis, molting.

Hormones in the Balance (revisited)

Exposing human cells growing in culture to the herbicide atrazine causes their concentration of aromatase to rise. Aromatase is an enzyme that converts testosterone to estrogen, so in the presence of atrazine more testosterone gets converted to estrogen. If atrazine has the same effect in frogs, this would explain the herbicide's feminizing effect on male sex organs.

Other commonly used industrial chemicals called phthalates (pronounced THAL-aytes) can also interfere with sex hormone function. Phthalates are used in soft plastics, cleaning products, and personal care products. In 1999, the U.S. Consumer Product Safety Commission asked manufacturers to stop using phthalates in rattles and toys designed for use by teething infants. The next year, Osvalado Rosario reported that premature breast development in young

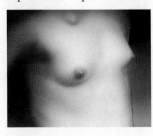 Puerto Rican girls was correlated with high levels of phthalates in their blood. The photo (*left*) shows an affected girl whose breasts began to develop before she reached the age of two.

Phthalates exert their effect by acting like estrogens: They stimulate the development of female secondary sexual traits such as enlarged breasts.

In males, phthalates suppress testosterone secretion. In adult males, a high pthalate level is associated with lowered sperm quality. There are also developmental effects. A woman's chances of giving birth to a son with somewhat feminized genitals, smaller genitals, or an undescended testicle rises with the level of pthalates in her blood.

In 2008, the United States Congress banned the use of phthalates in toys for children under age twelve. However, other soft plastic items still contain phthalates, as do many artificially scented personal care products such as shampoos and lotions.

Pediatrician Sheela Sathyanarayan found that the more scented powders, shampoos, and lotions a mother used on her infant, the higher the level of phthalates in the child's urine. Her findings led the American Academy of Pediatrics to recommend that parents choose unscented products for use on infants and young children. Similarly, many obstetricians recommend that pregnant women use personal care products that are labeled as organic or free of phthalates. Phthalates do not have to be listed as ingredients if they are a component of a fragrance.

How would you vote? Should atrazine use be continued while its health and environmental effects are studied in more detail?

Inner layers of the old cuticle break down. At the same time, cells of the epidermis secrete the new cuticle.

The steps in molting differ a bit in insects, which do not have a molt-inhibiting hormone. Rather, stimulation of the insect brain sets in motion a cascade of signals that trigger the production of molt-inducing ecdysone. Chemicals that mimic ecdysone or interfere with its function are sometimes used as insecticides. When such insecticides run off from fields and get into water, they can affect ecdysone-related responses in noninsect arthropods, such as crayfish or crabs.

Take-Home Message

What types of hormones do invertebrates produce?

» Some invertebrate hormones are homologous to those in vertebrates. Cnidarians and annelids have receptors that resemble those that bind vertebrate hormones.

» Invertebrates also make hormones with no vertebrate counterpart. Hormones that control molting in arthropods are an example.

Absence of suitable stimuli	Presence of suitable stimuli
↓	↓
X organ releases molt-inhibiting hormone (MIH)	Signals from brain inhibit release of MIH
↓	↓
MIH prevents Y organ from making ecdysone	Y organ makes and releases ecdysone
↓	↓
A No molting	**B** Molting

Figure 34.20 Control of molting in crabs. Two hormone-secreting organs play a role, the X organ in the eye stalk and the Y organ at the base of the antennae.

(**A**) In the absence of environmental cues for molting, hormone from the X organ prevents molting. (**B**) When stimulated by proper environmental cues, the brain sends nervous signals that inhibit X organ activity. With the X organ suppressed, the Y organ releases the ecdysone that stimulates molting.

The photo shows a newly molted blue crab with its old shell. The new shell remains soft for about 12 hours, making it a "soft-shelled crab." During this time, the crab is highly vulnerable to predators, including human seafood lovers.

LEARNING ROADMAP

Where you have been This chapter builds on your knowledge of connective (31.4) and muscle (31.5) tissues. You will revisit muscular dystrophy (14.4), and bacterial toxins (20.7) that affect muscles. You hear again about active transport (5.9) and motor proteins (4.10). Nervous control of muscle (32.5) and the effects of hormones on bone (34.7) are also discussed.

Where you are now

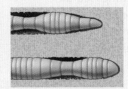

Invertebrate Skeletons
Contractile force exerted against a skeleton moves animal bodies. In many invertebrates, muscles exert force against an external skeleton or a fluid in a confined space.

Vertebrate Skeletons
All vertebrates have an internal skeleton. Bones interact with muscles to move the body. They also protect organs and make blood cells. A joint is a place where two bones meet.

The Muscle–Bone Partnership
Skeletal muscles pull on bones; they cannot push. Muscles often work in opposition, with the action of one reversing that of the other. Tendons connect muscles to bones.

Skeletal Muscle Contraction
Muscle contraction is the result of ATP-driven interactions among protein filaments in a muscle cell. The filaments do not shorten, but rather slide past one another.

Muscle Control and Metabolism
Signals from motor neurons trigger muscle contraction. Muscles make the ATP they require by a variety of metabolic pathways. Exercise improves muscle function.

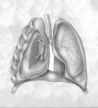

Where you are going Our discussion of circulation includes information about blood cells that form in bone (Section 36.4) and the role skeletal muscles play in moving blood through veins (36.11). You will learn more about vaccines, such as those for polio and tetanus in Section 37.12. The skeletal muscle actions that bring about breathing are the topic of Section 38.5, and Section 40.9 explains how muscle activity helps maintain body temperature.

35.1 Muscles and Myostatin

A mature skeletal muscle cell cannot divide, so new muscle cells do not arise in normal, healthy adults. Thus, a skeletal muscle gets bulkier mainly by enlarging existing cells. Inside each muscle cell, protein filaments are continually built and broken down. Exercise tilts this process in favor of synthesis, so muscle cells get bigger and the muscle gets stronger. Your body is like a machine that improves with use. The more you use your muscles, the more powerful they become.

Hormones affect muscle mass. One effect of the sex hormone testosterone is to encourage muscle cells to build more proteins. Men make much more testosterone than women, which is why men tend to be more muscular. Human growth hormone also stimulates synthesis of muscle proteins. Synthetic versions of natural muscle-building, or "anabolic," hormones can also bulk up muscles, but most sport organizations consider the use of these drugs to be cheating and penalize athletes who use them.

Some people have a natural genetic advantage when it comes to muscle strength. Liam, a young Michigan boy, is one of them (**Figure 35.1A**). His body does not respond normally to myostatin, a protein that slows synthesis of muscle proteins. Without the inhibiting effect of myostatin on protein production, an individual has larger-than-normal muscles and unusual strength. For example, by his first birthday, Liam could do chin-ups. By age 3, he could lift 5-pound (2.3-kilogram) dumbbells and had 40 percent more muscle than an average child his age.

A study of whippets, a type of dog bred for racing, provides another example of myostatin's effect. Whippets homozygous for a mutation that prevents them from making myostatin are heavily muscled (**Figure 35.2B**). These dogs, which are called bully whippets, are not usually raced or bred because they do not match the breeders' ideal of how a whippet should look. However, dogs heterozygous for the mutant allele do race. Compared to normal whippets, these dogs make less myostatin, are more muscular, and are more likely to win races.

Drug companies hope to develop drugs that inhibit myostatin production or activity for use in treatment of muscle-wasting disorders such as muscular dystrophy. Will some athletes use myostatin-inhibiting drugs to push their bodies beyond natural limits? No doubt they will. Some are already buying nutritional supplements that purport to bind to myostatin and reduce its activity. In clinical tests, these supplements had no effect on strength, but hope for an extra edge keeps moving them off the shelf.

A Liam, a Michigan boy whose muscles do not respond to the inhibitory effect of myostatin. As a result he is unusually strong for his age.

B A bully whippet (*left*) does not make myostatin. Such heavily muscled dogs are often euthanized by breeders because they do not conform to the standard for the breed. Bully whippets are homozygous for a two-base deletion mutation in the myostatin gene. The mutation changes an mRNA codon for cysteine to a stop codon (Section 9.4).

A whippet that is homozygous for the wild-type myostatin gene (*right*) has a comparatively lightly muscled body.

Figure 35.1 Disrupted myostatin function. Myostatin is a regulatory protein that discourages production of muscle proteins. Individuals that have bulky muscles as a result of a myostatin deficiency are said to have myostatin-related muscle hypertrophy.

35.2 Invertebrate Skeletons

■ Contractile cells act against skeletons, which can be internal or external.

■ Links to Cnidarians 24.5, Annelids 24.7, Mollusks 24.8, Arthropods 24.11, Echinoderms 24.16

When you think of a skeleton, you probably picture an internal scaffolding of bones, but this is just one type of skeleton. Many invertebrates have a skeleton consisting of a fluid in an enclosed space or of external hardened parts. Animal bodies move when contractile cells exert force against the skeleton.

Hydrostatic Skeletons

Cnidarians and annelids are among the animals with a **hydrostatic skeleton**: a fluid-filled closed chamber or chambers that muscles act against. For example, a sea anemone's body is inflated by water that flows in through its mouth and fills its gastrovascular cavity (**Figure 35.2**). Beating of cilia causes the inward flow of water. Contraction of a ring of muscle around the mouth traps the water inside the body. Contractions of other muscles can redistribute the water and alter body shape. By analogy, think about how squeezing or pulling on a water-filled balloon changes its shape.

An anemone has circular muscles that ring its body and longitudinal ones that run from its top to bottom. Contracting the circular muscles while relaxing the longitudinal ones makes an anemone tall and thin. Contracting the longitudinal muscles while relaxing the circular ones makes it short and wide. The animal can also open its mouth, contract both sets of muscles, and draw in its tentacles. Doing so forces most fluid out of the gastrovascular cavity, so the body shrinks into a protective resting position (**Figure 35.2B**).

In earthworms, a coelom divided into many fluid-filled segments is the hydrostatic skeleton (Section 24.7). Longitudinal and circular muscles put pressure on coelomic fluid in each segment, causing it to become alternately long and narrow or short and wide, so that the worm moves through the soil (**Figure 35.3**).

Exoskeletons

An **exoskeleton** is a stiff body covering to which muscles attach. Bivalve mollusks (Section 24.8) have a hinged two-part shell. A strong muscle attached to the two halves can pull them together, shutting the shell. Some scallops can swim by opening and closing their shell. Each time the shell is pulled shut, water is forced out, and the scallop scoots backwards a bit.

mouth

gastrovascular cavity; the mouth can close and trap fluid inside this cavity

A Water is drawn into the gastrovascular cavity through the mouth. When the cavity is filled and mouth is closed, muscles can act on the trapped fluid and alter body shape. Circular muscles ring the body; longitudinal ones run the length of the body.

Figure 35.2 Animated Hydrostatic skeleton of a sea anemone.

B One anemone inflated with water (*left*) and another that has expelled water from the gastrovascular cavity and pulled in its tentacles (*right*).

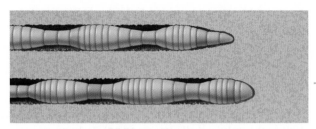

Figure 35.3 How an earthworm moves through the soil. Muscles act on coelomic fluid in individual body segments, causing the segments to change shape. A segment narrows when circular muscle ringing it contracts and longitudinal muscle running its length relaxes. The segment widens when circular muscle relaxes and longitudinal muscle contracts.

Crabs, spiders, insects, and other arthropods have a hinged exoskeleton with attachment sites for sets of muscles that pull on the hardened parts. For example, a fly's wings flap when muscles attached to its thorax alternately contract and relax (**Figure 35.4**).

Redistribution of body fluid also has a role in some arthropod movements. In spiders, muscles attached to the exoskeleton contract and pull the legs inward, but there are no opposing muscles to pull legs out again. Instead, a large muscle of the thorax contracts, which causes blood to surge into the hind legs (**Figure 35.5**). Similarly, redistribution of fluid extends the tubular mouthparts of a butterfly, allowing it to sip nectar.

Endoskeletons

An **endoskeleton** is an internal framework of hardened elements to which the muscles attach. Echinoderms and vertebrates have an endoskeleton. The skeleton of echinoderms such as sea stars (**Figure 35.6**) and sea urchins consists of calcium-carbonate plates embedded in the body wall.

endoskeleton Internal skeleton made up of hardened components such as bones.
exoskeleton Of some invertebrates, hard external parts that muscles attach to and move.
hydrostatic skeleton Of soft-bodied invertebrates, a fluid-filled chamber that muscles exert force against, redistributing the fluid.

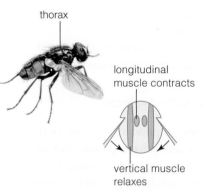

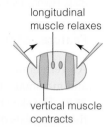

thorax

longitudinal muscle contracts

vertical muscle relaxes

longitudinal muscle relaxes

vertical muscle contracts

A Wings pivot down as the relaxation of vertical muscle and the contraction of longitudinal muscle pull in sides of thorax.

B Wings pivot up when the contraction of vertical muscle and relaxation of longitudinal muscle flatten the thorax.

Figure 35.4 Animated Fly wing movement. Wings attach to the thorax at pivot points. When muscles inside the thorax contract and relax, the thorax changes shape and the wings pivot up and down at their attachment point.

Figure 35.5 Side view of a jumping spider making a leap. When a large muscle in the thorax contracts, volume of the thoracic cavity decreases, forcing blood into the hind legs. The resulting surge of high fluid pressure extends the legs. Some jumping spiders can leap a distance 25 times the length of their body.

element of endoskeleton

Figure 35.6 A sea star. The sketch shows a cross-section through one arm. Hard plates embedded in the body wall form an endoskeleton.

35.3 The Vertebrate Endoskeleton

- All vertebrates have an endoskeleton. In most groups, the endoskeleton consists primarily of bones.
- Links to Vertebrate evolution 25.2, Transition to life on land 25.6, Human bipedalism 26.3, Connective tissues 31.4

Features of the Vertebrate Skeleton

All vertebrates (the fishes, amphibians, reptiles, birds, and mammals) have an endoskeleton (**Figures 35.7** and **35.8**). The skeleton of sharks and other cartilaginous fishes consists of cartilage, a rubbery connective tissue. Other vertebrate skeletons include some cartilage, but most consist mainly of bone tissue (Section 31.4).

The term "vertebrate" refers to the **vertebral column**, or backbone, a feature common to all members of this group. The backbone supports the body, serves as an attachment point for muscles, and protects the spinal cord that runs through a canal inside it. Bony segments called **vertebrae** (singular, vertebra) make up the backbone. **Intervertebral disks** of cartilage between vertebrae act as shock absorbers and flex points.

The vertebral column, along with the bones of the head and rib cage, constitute the **axial skeleton**. The **appendicular skeleton** consists of the pectoral (shoulder) girdle, the pelvic (hip) girdle, and limbs (or bony fins) attached to them.

You learned earlier how vertebrate skeletons have evolved over time. For example, jaws are derived from the gill supports of ancient jawless fishes (Section 25.4). As another example, bones in the limbs of land vertebrates are homologous to those inside the fins of lobe-finned fishes (Section 25.6).

The Human Skeleton

For a closer look at vertebrate skeletal features, think about a human skeleton. The human skull's flattened cranial bones fit together to form the braincase that surrounds and protects the brain (**Figure 35.8A**). The brain and spinal cord connect through an opening called the foramen magnum. In upright walkers such as humans, this opening lies at the base of the skull (Section 26.3). Facial bones include cheekbones and other bones around the eyes, the bone that forms the bridge of the nose, and the bones of the jaw.

Both males and females have twelve pairs of ribs (**Figure 35.8B**). Ribs and the breastbone, or sternum, form a protective cage around the heart and lungs.

The vertebral column extends from the base of the skull to the pelvic girdle (**Figure 35.8C**). In humans, natural selection favored an ability to walk upright and led to modification of the backbone (Section 26.3). Viewed from the side, our backbone has an S shape that keeps our head and torso centered over our feet. By contrast, an ape's backbone has a C-shape.

Maintaining an upright posture requires that vertebrae and intervertebral disks stack one on top of the other, rather than being parallel to the ground, as in four-legged walkers. The stacking puts additional pressure on disks and, as people age, their disks often slip out of place or rupture, causing back pain.

The scapula (shoulder blade) and clavicle (collarbone) are bones of the human pectoral girdle (**Figure 35.8D**). The thin clavicle transfers force from the arms to the axial skeleton. When a person falls on an outstretched arm, the excessive force transferred to the clavicle frequently causes it to fracture or break.

The upper arm has a single bone called the humerus (**Figure 35.8E**). The forearm has two bones, the radius and ulna. Carpals are bones of the wrist, metacarpals are bones of the palm, and phalanges (singular, phalanx) are finger bones.

The pelvic girdle consists of two hip bones that connect to the sacrum to form a basin-shaped ring (**Figure 35.8F**). It protects organs inside the pelvic cavity and supports the weight of the upper body when you stand upright.

Bones of the leg (**Figure 35.8G**) include the femur (thighbone), patella (kneecap), and the tibia and fibula (bones of the lower leg). The fibula serves as a point of muscle attachment, but does not bear weight. Tarsals are ankle bones, and metatarsals are bones of the sole of the foot. Like the bones of the fingers, those of the toes are called phalanges.

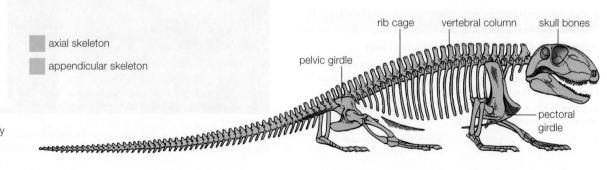

axial skeleton

appendicular skeleton

pelvic girdle   rib cage   vertebral column   skull bones   pectoral girdle

Figure 35.7 Skeletal elements typical of early reptiles.

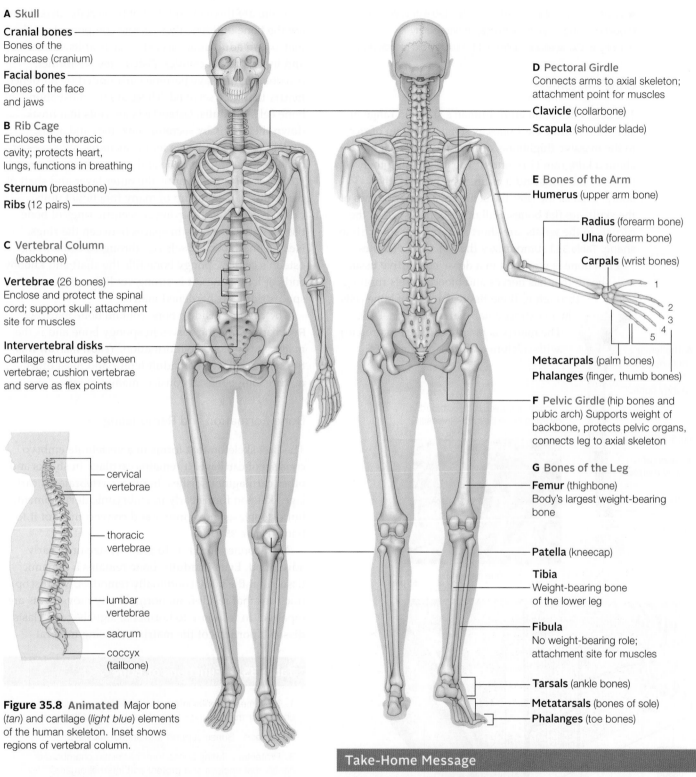

A Skull

Cranial bones
Bones of the
braincase (cranium)

Facial bones
Bones of the face
and jaws

B Rib Cage
Encloses the thoracic
cavity; protects heart,
lungs, functions in breathing

Sternum (breastbone)
Ribs (12 pairs)

C Vertebral Column
(backbone)

Vertebrae (26 bones)
Enclose and protect the spinal
cord; support skull; attachment
site for muscles

Intervertebral disks
Cartilage structures between
vertebrae; cushion vertebrae
and serve as flex points

cervical
vertebrae

thoracic
vertebrae

lumbar
vertebrae

sacrum

coccyx
(tailbone)

D Pectoral Girdle
Connects arms to axial skeleton;
attachment point for muscles

Clavicle (collarbone)
Scapula (shoulder blade)

E Bones of the Arm
Humerus (upper arm bone)

Radius (forearm bone)
Ulna (forearm bone)
Carpals (wrist bones)

1
2
3
4
5

Metacarpals (palm bones)
Phalanges (finger, thumb bones)

F Pelvic Girdle (hip bones and
pubic arch) Supports weight of
backbone, protects pelvic organs,
connects leg to axial skeleton

G Bones of the Leg
Femur (thighbone)
Body's largest weight-bearing
bone

Patella (kneecap)

Tibia
Weight-bearing bone
of the lower leg

Fibula
No weight-bearing role;
attachment site for muscles

Tarsals (ankle bones)
Metatarsals (bones of sole)
Phalanges (toe bones)

Figure 35.8 Animated Major bone
(*tan*) and cartilage (*light blue*) elements
of the human skeleton. Inset shows
regions of vertebral column.

appendicular skeleton Of vertebrates, limb or fin bones and bones
of the pelvic and pectoral girdles.
axial skeleton Of vertebrates, bones of the head, trunk, and tail.
intervertebral disk Cartilage disk between two vertebrae.
vertebrae Bones of the backbone, or vertebral column.
vertebral column Backbone.

Take-Home Message

What type of skeleton is present in humans and other vertebrates?

» The endoskeleton of vertebrates usually consists mainly of bone. Its axial
portion includes the skull, vertebral column, and ribs. Its appendicular part
includes a pectoral girdle, a pelvic girdle, and the limbs.

» Some features of the human skeleton such as an S-shaped backbone are
adaptations to upright posture and walking.

35.4 Bone Structure and Function

■ Bones consist of living cells in a secreted extracellular matrix. Proper diet and exercise help keep them healthy.

■ Links to Extracellular matrix 4.11, Parathyroid glands 34.7

Bone Anatomy

The 206 bones of an adult human's skeleton range in size from middle ear bones as small as a grain of rice to the massive thighbone, or femur, which weighs about a kilogram (2 pounds). The femur and other bones of the arms and legs are long bones. Other bones, such as the ribs, the sternum, and most bones of the skull, are flat bones. Still other bones, such as the carpals in the wrists, are short and roughly squarish in shape. **Table 35.1** summarizes the functions of bones.

Each bone is wrapped in a dense connective tissue sheath that has nerves and blood vessels running through it. Bone tissue consists of bone cells in an extracellular matrix (Section 4.11). The matrix is mainly collagen (a protein) with calcium and phosphorus salts.

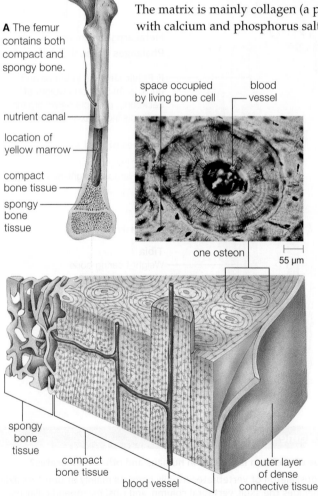

A The femur contains both compact and spongy bone.

nutrient canal

location of yellow marrow

compact bone tissue

spongy bone tissue

space occupied by living bone cell

blood vessel

one osteon

55 µm

spongy bone tissue

compact bone tissue

blood vessel

outer layer of dense connective tissue

B Cross-section through a femur showing osteons. These cylindrical structures consist of concentric layers of compact bone.

Figure 35.9 **Animated** Structure of a human adult femur, or thighbone.

There are three main types of bone cells. **Osteoblasts** are the bone builders; they secrete components of the matrix. An adult bone has osteoblasts at its surface and in its internal cavities. **Osteocytes** are former osteoblasts that have become surrounded by hardened matrix that they secreted. These are the most abundant bone cells in adults. **Osteoclasts** are cells that break down the matrix by secreting enzymes and acids.

A long bone such as a femur includes two types of bone tissue, compact bone and spongy bone (**Figure 35.9**). Compact bone forms the outer layer and shaft of the femur. It is made up of many functional units called osteons, each having concentric rings of bone tissue, with bone cells in spaces between the rings. Nerves and blood vessels run through a canal in the osteon's center. Spongy bone fills the shaft and knobby ends of long bones. It is strong yet lightweight; open spaces riddle its hardened matrix.

The cavities inside a bone contain bone marrow. **Red marrow** fills the spaces in spongy bone and is the major site of blood cell formation. **Yellow marrow** fills the central cavity of an adult femur and most other mature long bones. It consists mainly of fat.

Bone Formation and Remodeling

The first skeleton that forms in a vertebrate embryo consists of cartilage. It remains cartilage in sharks and other cartilaginous fishes. In other vertebrates, a cartilage skeleton forms early in embryonic development, but osteoblasts move into it and convert most of it to bone (**Figure 35.10**).

Many bones continue to grow in size until early adulthood. Even in adults, bone remains a dynamic tissue that the body continually remodels. Microscopic fractures that result from normal body movements are repaired. In response to hormonal signals, osteoclasts dissolve portions of the matrix, releasing mineral

Table 35.1 Functions of Bone
1. Movement. Bones interact with skeletal muscle and change or maintain positions of the body and its parts.
2. Support. Bones support and anchor muscles.
3. Protection. Many bones form hardened chambers or canals that enclose and protect soft internal organs.
4. Mineral storage. Bones are a reservoir for calcium and phosphorus ions. Deposits and withdrawals of these ions help maintain their concentrations in body fluids.
5. Blood cell formation. Only certain bones contain the tissue where blood cells form.

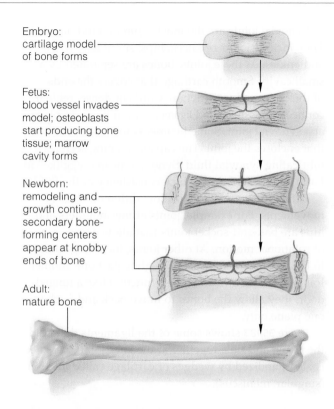

Embryo:
cartilage model
of bone forms

Fetus:
blood vessel invades
model; osteoblasts
start producing bone
tissue; marrow
cavity forms

Newborn:
remodeling and
growth continue;
secondary bone-
forming centers
appear at knobby
ends of bone

Adult:
mature bone

Figure 35.10 Long bone formation. The bone-forming cells are active first in the shaft region, then at the knobby ends. In time, cartilage is left only at the ends.

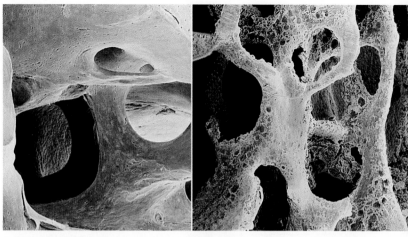

A Normal bone　　　　　　　　　**B** Bone weakened by osteoporosis

Figure 35.11 Osteoporosis. The name of this disorder means "porous bones."

ions into the blood. Osteoblasts secrete new matrix to replace that broken down by osteoclasts.

Bones and teeth contain most of the body's calcium. Hormones regulate calcium concentration in blood by affecting calcium uptake from the gut and calcium release from bone. When the blood calcium level is too high, the thyroid gland secretes calcitonin. This hormone slows the release of calcium into blood by inhibiting osteoclast action. When blood has too little calcium, the parathyroid glands release parathyroid hormone, or PTH (Section 34.7). This hormone stimulates osteoclast activity. It also decreases calcium loss in urine and helps activate vitamin D. The vitamin stimulates cells in the gut lining to absorb calcium.

Other hormones also affect bone turnover. The sex hormones estrogen and testosterone encourage bone deposition. Cortisol, the stress hormone, slows it.

Until an individual is about twenty-four years old, osteoblasts secrete more matrix than osteoclasts break down, so bone mass increases. Bones become denser and stronger. Later in life, bone mass gradually declines as osteoblasts become less active.

Osteoporosis is a disorder in which bone loss outpaces bone formation. As a result, the bones become weaker and more likely to break (**Figure 35.11**). Osteoporosis is most common in postmenopausal women because they no longer produce the sex hormones that encourage bone deposition. However, about 20 percent of osteoporosis cases occur in men.

To reduce your risk of osteoporosis, ensure that your diet provides adequate levels of vitamin D and calcium. Calcium is best obtained from food, rather than supplements, which may increase risk of heart attack. Avoid smoking and excessive alcohol intake, which slow bone deposition. Get regular exercise to encourage bone renewal and avoid an excessive intake of cola soft drinks. Several studies have shown that women who drink more than two such soft drinks a day have a slightly lower than normal bone density.

osteoblast Bone-forming cell; it secretes a collagen-rich matrix that gets mineralized.
osteoclast Bone-digesting cell; it secretes enzymes that digest bone's matrix.
osteocyte A mature bone cell; former osteoblast that has become surrounded by its own secretions.
red marrow Bone marrow that makes blood cells.
yellow marrow Bone marrow that is mostly fat; fills cavity in most long bones.

Take-Home Message

What are the structural and functional features of bones?

» Bones have a variety of shapes and sizes.

» A sheath of connective tissues encloses the bone, and the bone's inner cavity contains marrow. Red marrow produces blood cells.

» All bones consist of bone cells in a secreted extracellular matrix. A bone is continually remodeled; osteoclasts break down the matrix of old bone and osteoblasts lay down new bone. Hormones regulate this process.

35.5 Skeletal Joints—Where Bones Meet

- Bones interact with one another at joints. The range of motion varies with the type of joint.
- Link to Dense connective tissue 31.4

Types and Functions of Joints

A **joint** is an area of contact or near contact between bones. There are three types of joints: fibrous joints, cartilaginous joints, and synovial joints (**Figure 35.12A**).

At fibrous joints, bones are held securely in place by dense, fibrous connective tissue. Fibrous joints hold teeth in their sockets in the jaw.

Pads or disks of cartilage connect bones at cartilaginous joints. The flexible connection allows just a bit of movement. Cartilaginous joints connect vertebrae to one another and connect some ribs to the sternum.

Synovial joints are the most common kind of joints. They include joints of knees, hips, shoulders, wrists, and ankles. At these joints, bones are separated by a small cavity. Smooth cartilage that covers the ends of the bones reduces friction. Cords of dense, regular connective tissue called **ligaments** hold bones in place at a synovial joint. Some ligaments form a capsule that encloses the joint. The capsule's lining secretes a lubricating synovial fluid. Synovial means "egglike" in Latin, and describes the thick consistency of the fluid.

Different synovial joints allow different kinds of movements. For example, joints at the shoulders and hips are ball-and-socket joints that allow a wide range of rotational motion. At other joints, including some in the wrists and ankles, bones glide past one another. Joints at the elbows and knees function like a hinged door; they allow the bones to move back and forth in one plane only.

Figure 35.12B shows some of the ligaments that hold the fibula and tibia together at the knee joint. The knee is also stabilized by wedges of cartilage called menisci (singular, meniscus).

Joint Health

Common Joint Injuries A sprained ankle is the most common joint injury. It occurs when one or more of the ligaments that hold bones together at the ankle joint overstretches or tears. A sprained ankle is usually treated immediately with rest, application of ice, compression with an elastic bandage, and elevation of the affected area. After the ankle heals, exercises may help

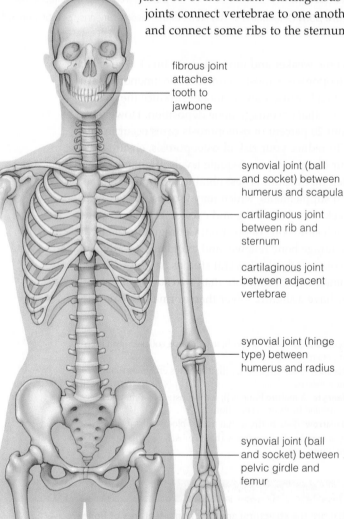

fibrous joint attaches tooth to jawbone

synovial joint (ball and socket) between humerus and scapula

cartilaginous joint between rib and sternum

cartilaginous joint between adjacent vertebrae

synovial joint (hinge type) between humerus and radius

synovial joint (ball and socket) between pelvic girdle and femur

A Examples of the three types of joints.

Figure 35.12 Joint anatomy.

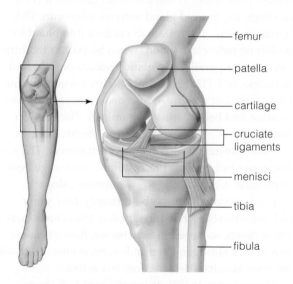

femur

patella

cartilage

cruciate ligaments

menisci

tibia

fibula

B Structure of the knee, a hinge-type synovial joint. Ligaments hold bones in place. Wedges of cartilage called menisci provide additional stability.

strengthen muscles that stabilize the joint and prevent future sprains.

A tear of the cruciate ligaments in the knee joint may require surgery. Cruciate means cross, and these short ligaments cross one another in the center of the joint. They are visible in **Figure 35.12B**. The cruciate ligaments stabilize the knee, and when they are torn completely, bones may shift so the knee gives out when a person tries to stand. A blow to the lower leg, as often occurs in football, can injure a cruciate ligament, but so can a fall or misstep. Female athletes are at a higher risk for cruciate ligament tears than men who play the equivalent sport. For example, female soccer players tear these ligaments four times as often as male soccer players do.

Another common knee injury is a torn meniscus. A meniscus is a C-shaped wedge of cartilage that reduces friction between the bones, cushions them, and helps keep them in place. Each knee has two menisci. A minor tear at the edge of the meniscus may heal on its own, but cartilage repairs itself only very slowly. If a chunk of meniscus cartilage gets torn off, it can drift about in the synovial fluid of the joint and end up jammed into a spot where it interferes with normal function.

A dislocation means bones of a joint are out of place. It is usually highly painful and requires immediate treatment. The bones must be put into proper position and immobilized for a time to allow healing.

Arthritis and Bursitis Arthritis means inflammation of a joint. As you will learn in Chapter 37, inflammation is the body's normal response to injury. However, with arthritis, inflammation—and the associated pain and swelling—becomes chronic.

The most common type of arthritis is osteoarthritis. It usually appears in old age, after cartilage wears down at a frequently used joint. It affects different joints in different people. For example, women who habitually wear high-heeled shoes increase their risk of osteoarthritis of the knees (**Figure 35.13**). Such shoes put added pressure on the cartilage that cushions the knee joint, increasing the chances that it will wear down and fail in later life.

Rheumatoid arthritis is an autoimmune disorder in which the immune system mistakenly attacks the fluid-secreting lining of synovial joints throughout the body. It can occur at any age. Women are two to three times more likely than men to be affected.

Arthritis can be treated with drugs that relieve pain and minimize inflammation. Joints affected by osteoarthritis can also be replaced with artificial, or prosthetic,

Figure 35.13 High heels now, aching knees later. A study by researchers at Tufts University showed that shoes with heels 2.7 inches high increased pressure on the knee joint by 20 to 25 percent over barefoot walking. Wide heels put more pressure on knees than narrow ones, perhaps because women walked more confidently in them.

joints. Knee and hip replacements are now common and allow a person to resume normal activities.

With bursitis, a bursa becomes inflamed. A **bursa** is a fluid-filled sac that reduces friction between parts at a joint. For example, a bursa reduces friction between the patella (kneecap) and femur at a knee joint. Repeating a movement that puts pressure on a particular bursa can cause inflammation. For example, swinging a tennis racket or golf club can lead to inflammation of a bursa in the shoulder or elbow. Continually leaning on an elbow, kneeling to work, or even sitting or standing a certain way can also cause bursitis.

bursa Fluid-filled sac that functions as a cushion between components of some joints.
joint Region where bones meet.
ligament Strap of dense connective tissue that holds bones together at a joint.

Take-Home Message

What are joints?

» Joints are areas where bones meet and interact.

» In the most common type, synovial joints, the bones are separated by a small fluid-filled space and are held together by ligaments of fibrous connective tissue.

35.6 Skeletal–Muscular Systems

- Only skeletal muscles attach to and pull on bones.
- Links to Muscle tissue 31.5, Reflexes 32.8

Skeletal muscles allow us to dance, to smile, and to speak. They are sometimes referred to as voluntary muscles because we control their action at will. However, skeletal muscles also take part in reflex actions such as the stretch reflex (Section 32.8).

A sheath of dense connective tissue encloses each skeletal muscle and extends beyond it to form a cord-like or straplike **tendon**. Most often, tendons attach each end of the muscle to a bone. **Figure 35.14** shows the muscles of the upper arm: the biceps and the triceps. Two tendons attach the upper part of the biceps to the scapula (shoulder blade). At the opposite end of the muscle, a tendon attaches the biceps to the radius in the forearm.

Muscles can only exert force by pulling; they cannot push. Often two muscles work in opposition, meaning the action of one resists or reverses the action of another. For example, the triceps in the upper arm opposes the biceps. When you pull your arm toward your shoulder, the triceps muscle relaxes as the biceps contracts. Contraction of the triceps coupled with relaxation of the biceps reverses this movement, extending the arm.

You can feel the biceps contract if you extend your arm outward, place your other hand over the biceps, then bend the arm at the elbow to bring your hand to your shoulder. Although the biceps shortens only about a centimeter, the hand moves through a much greater distance.

The elbow and many other joints function like a lever, a mechanism in which a rigid structure pivots about a fixed point. The bones are the rigid structures and the joints the fixed points. Use of a lever allows a small force, such as that exerted by a contracting biceps, to overcome a larger one, such as the gravitational force acting on the lower portion of the arm.

The human body has close to 700 skeletal muscles, some near the surface, others deep inside the body wall. **Figure 35.15** shows a sample of some of the largest muscles and describes their functions. Collectively, skeletal muscles account for about 40 percent of the body weight of a young man of average fitness.

Most skeletal muscles attach to and move bones, but some have other functions. Some skeletal muscles attach to and pull on facial skin to cause changes in expression. Others attach to and move the eyeball, or open and close eyelids. The tongue is skeletal muscle, and sphincters of skeletal muscle provide voluntary control of defecation and urination. Skeletal muscles function in respiration and help keep blood circulating through the body.

Bear in mind that only *skeletal* muscle interacts with bone. Smooth muscle is mainly a component of soft internal organs, such as the stomach. Cardiac muscle forms only in the heart wall. Later chapters consider the structure and function of smooth muscle and cardiac muscle.

In addition to its role in movement, skeletal muscle activity generates most of the heat required to maintain your body temperature. You feel hot during vigorous exercise because of the additional heat generated by the metabolic activity of working skeletal muscles. If you become very cold, reflexive contraction of skeletal muscles (shivering), can help warm you.

tendon Strap of dense connective tissue that connects a skeletal muscle to bone.

Take-Home Message

How do muscles and tendons interact with bones?

» Tendons of dense connective tissue attach skeletal muscles to bones.

» Small muscle movements can bring about large movements of bones.

» Muscles can only pull on a bone; they cannot push. At many joints, movement is controlled by a pair of muscles that act in opposition.

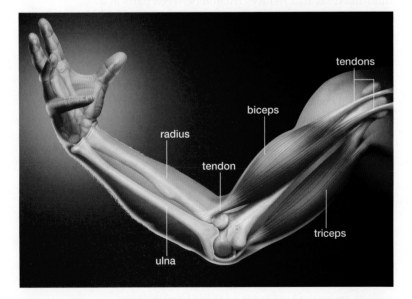

Figure 35.14 Animated Opposing muscles of the upper arm. When the biceps contracts and the triceps relaxes, the forearm is pulled toward the upper arm. When the triceps contracts and the biceps relaxes, the arm straightens at the elbow.
Figure It Out: What bone does the biceps pull on? Answer: The radius

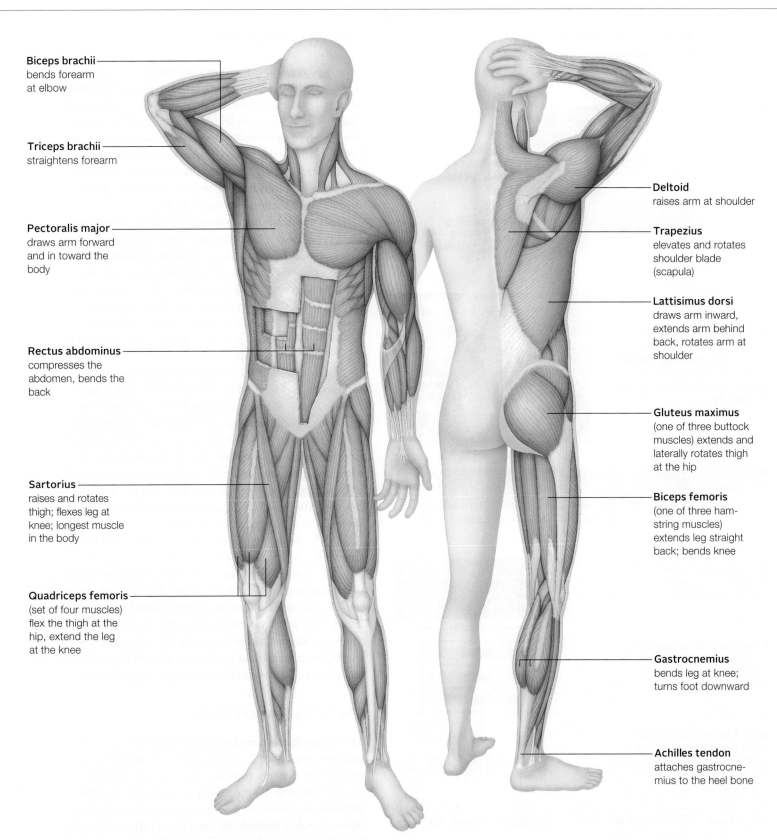

Biceps brachii — bends forearm at elbow

Triceps brachii — straightens forearm

Pectoralis major — draws arm forward and in toward the body

Rectus abdominus — compresses the abdomen, bends the back

Sartorius — raises and rotates thigh; flexes leg at knee; longest muscle in the body

Quadriceps femoris — (set of four muscles) flex the thigh at the hip, extend the leg at the knee

Deltoid — raises arm at shoulder

Trapezius — elevates and rotates shoulder blade (scapula)

Lattisimus dorsi — draws arm inward, extends arm behind back, rotates arm at shoulder

Gluteus maximus — (one of three buttock muscles) extends and laterally rotates thigh at the hip

Biceps femoris — (one of three hamstring muscles) extends leg straight back; bends knee

Gastrocnemius — bends leg at knee; turns foot downward

Achilles tendon — attaches gastrocnemius to the heel bone

Figure 35.15 Major muscles of the human musculoskeletal system. These are the skeletal muscles that gym enthusiasts are familiar with.

Tendons are shown in *light blue*. For example, the Achilles tendon, the largest tendon in the body, attaches muscles in the calf to the heel bone.

35.7 How Does Skeletal Muscle Contract?

- ATP-fueled movements of protein filaments inside a muscle fiber result in muscle contraction.
- Links to Motor protein 4.10, Muscle tissue 31.5

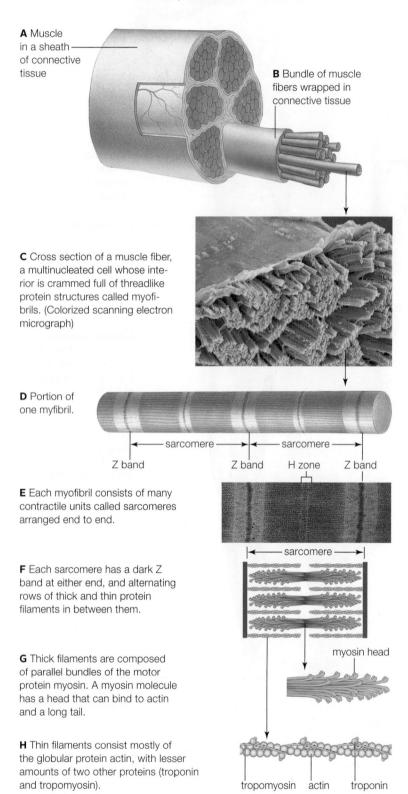

A Muscle in a sheath of connective tissue

B Bundle of muscle fibers wrapped in connective tissue

C Cross section of a muscle fiber, a multinucleated cell whose interior is crammed full of threadlike protein structures called myofibrils. (Colorized scanning electron micrograph)

D Portion of one myfibril.

sarcomere   sarcomere

Z band   Z band   H zone   Z band

E Each myofibril consists of many contractile units called sarcomeres arranged end to end.

sarcomere

F Each sarcomere has a dark Z band at either end, and alternating rows of thick and thin protein filaments in between them.

myosin head

G Thick filaments are composed of parallel bundles of the motor protein myosin. A myosin molecule has a head that can bind to actin and a long tail.

H Thin filaments consist mostly of the globular protein actin, with lesser amounts of two other proteins (troponin and tropomyosin).

tropomyosin   actin   troponin

Figure 35.16 **Animated** Structure of a skeletal muscle.

Structure of Skeletal Muscle

A muscle is enclosed within a sheath of connective tissue, and additional connective tissue encloses bundles of muscle fibers within it (**Figure 35.16A,B**). Each **muscle fiber** is a roughly cylindrical contractile cell that runs the length of the muscle. A skeletal muscle fiber forms before birth by the fusion of many embryonic cells, so it has many nuclei. The nuclei reside at the perimeter of a fiber, and hundreds to thousands of threadlike **myofibrils** fill its interior (**Figure 35.16C**).

Each myofibril has a repeating pattern of dark and light crossbands (**Figure 35.16D**). These bands define the units of muscle contraction called **sarcomeres**. A mesh of cytoskeletal elements called Z bands anchors one sarcomere to another (**Figure 35.16E**). The Z stands for *zwischen*, which is the German word for "between." Each sarcomere is the region between two Z bands, where thick and thin protein filaments are arranged in a regular pattern (**Figure 35.16F**).

A sarcomere's thick filaments are centered in the sarcomere and flanked by thin filaments. Thick filaments consist of myosin (**Figure 35.16G**), a motor protein with a clublike head and a long tail. The myosin head has enzymatic activity; it can bind ATP and break it into ADP and phosphate. The head can also bind actin, the globular protein that is the main component of thin filaments (**Figure 35.16H**). One end of each thin filament attaches to a Z band and the other extends inward toward the sarcomere center.

Muscle fibers, myofibrils, thin filaments, and thick filaments all run parallel with a muscle's long axis. As a result, all sarcomeres in all fibers of a skeletal muscle work together and pull in the same direction. Skeletal muscle and cardiac muscle appear striated because Z lines and other sarcomere components in all their fibers are aligned. Smooth muscle fibers have sarcomeres, but because the sarcomeres are not aligned, smooth muscle does not have a striped appearance.

The Sliding-Filament Model

The **sliding-filament model** explains how interactions between thick and thin filaments bring about muscle contraction. During muscle contraction, the length of thick and thin filaments does not change, and the thick filaments do not change position. Only the thin filaments slide inward. As the Z bands attached to the thin filaments are drawn closer together, the sarcomere shortens (**Figure 35.17**).

Let's take a closer look at the molecular basis for sarcomere contraction. When a sarcomere is relaxed,

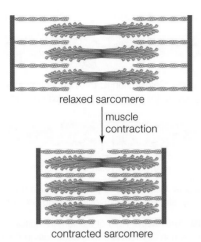

relaxed sarcomere

↓ muscle contraction

contracted sarcomere

Figure 35.17 Animated The sliding-filament model.

Above, during muscle contraction, thin filaments slide inward past thick filaments, reducing the width of the sarcomere.

Right, molecular mechanism of contraction. For clarity, we show only two myosin heads. A head binds repeatedly to an actin filament and slides it toward the center of the sarcomere.

myosin cannot access its binding sites on actin because other components of the thin filament are in the way. It can, however, bind ATP ❶. A myosin molecule with bound ATP is in a low-energy state. Hydrolyzing ATP to ADP and phosphate (P$_i$) energizes the myosin head in a manner analogous to stretching a spring ❷. When a signal from the nervous system excites the muscle (a process we consider in Section 35.8), proteins of the thin filament shift, and myosin heads can bind to actin. This binding creates a cross-bridge between the thin and thick filaments and causes release of the bound P$_i$ ❸. Release of P$_i$ from the myosin head triggers the power stroke ❹. Like a stretched spring returning to its original shape, a myosin head snaps back toward the sarcomere center. As it does, it pulls the attached thin filament along and releases the bound ADP. As thin filaments slide past thick ones toward the sarcomere center, the sarcomere shortens.

A new molecule of ATP can now bind to the myosin head ❺. When it does, the cross-bridge between the thin and thick filaments breaks. The breaking of one cross-bridge does not allow a thin filament to slip backward because other cross-bridges hold it in place. During a contraction, each myosin head repeatedly binds, moves, and releases an adjacent thin filament.

muscle fiber Contractile cell that runs the length of a muscle.
myofibrils Of a muscle, threadlike protein structures consisting of contractile units (sarcomeres) arranged end to end.
sarcomere Contractile unit of muscle.
sliding-filament model Explanation of how interactions among actin and myosin filaments bring about muscle contraction.

❶ A relaxed sarcomere, in which myosin-binding sites on actin are blocked. Myosin heads have bound ATP and are in their low-energy state.

❷ The myosin heads hydrolyze ATP to ADP and P$_i$. The heads are now in a high-energy state, ready to interact with actin, but the myosin-binding sites on actin are blocked.

❸ When a nervous signal excites the muscle, myosin-binding sites open up. Myosin binds to actin, releasing bound phosphate and forming a cross-bridge between thin and thick filaments.

❹ Release of phosphate triggers a power stroke. Myosin heads contract and pull the bound thin filaments inward.

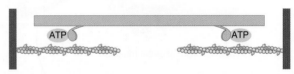

❺ When ATP binds to myosin, myosin releases its grip on actin and returns to its relaxed state, ready to act again.

Take-Home Message

How does a muscle's structure affect its function?

» Sarcomeres are the basic units of contraction in skeletal muscle. Sarcomeres are lined up end to end in myofibrils that run parallel with muscle fibers. These fibers, in turn, run parallel with the whole muscle.

» The parallel orientation of skeletal muscle components focuses a muscle's contractile force in a particular direction.

» Energy-driven interactions between myosin and actin filaments cause the sarcomeres of a muscle cell to shorten and bring about muscle contraction.

» During muscle contraction, the length of actin and myosin filaments does not change, and the myosin filaments do not change position. Sarcomeres shorten because myosin filaments pull neighboring actin filaments inward toward the center of the sarcomere.

35.8 Nervous Control of Muscle Contraction

■ Like neurons, muscle cells are excitable. Action potentials in muscle trigger calcium release that allows contraction.

■ Links to Active transport 5.9, Neuromuscular junctions 32.5

Nervous Control of Contraction

A neuromuscular junction is a synapse between a motor neuron and a muscle fiber (Section 32.5 and **Figure 35.18 ❶, ❷**). For a skeletal muscle to contract, an action potential must travel to a neuromuscular junction and cause the release of acetylcholine (ACh) from a motor neuron's axon terminals. Like a neuron, a muscle fiber is excitable, and the binding of ACh to receptors at its plasma membrane causes an action potential. The action potential travels along the muscle plasma membrane, then down membranous tubules called T tubules. The T tubules convey the action potential to the **sarcoplasmic reticulum**, a type of

❶ A signal travels along the axon of a motor neuron, from the spinal cord to a skeletal muscle.

❷ The signal is transferred from the motor neuron to the muscle at neuromuscular junctions. Here, ACh released by the neuron's axon terminals diffuses into the muscle fiber and causes action potentials.

❸ Action potentials propagate along a muscle fiber's plasma membrane down to T tubules, then to the sarcoplasmic reticulum, which releases calcium ions. The ions promote interactions of myosin and actin that result in contraction.

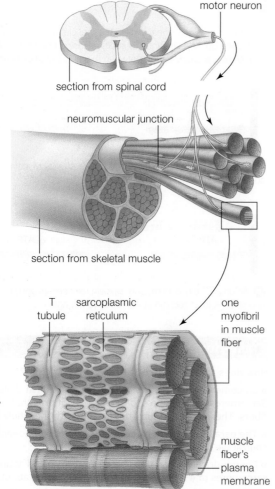

section from spinal cord

motor neuron

neuromuscular junction

section from skeletal muscle

T tubule — sarcoplasmic reticulum

one myofibril in muscle fiber

muscle fiber's plasma membrane

Figure 35.18 Animated Pathway by which the nervous system controls skeletal muscle contraction. A muscle fiber's plasma membrane encloses many individual myofibrils. Tubelike extensions of the membrane (T tubules) connect with part of the calcium-storing sarcoplasmic reticulum that wraps around the myofibrils.

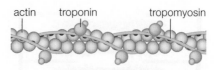

actin   troponin   tropomyosin

A Resting muscle. Calcium ion (Ca^{++}) concentration is low and tropomyosin covers the myosin-binding sites on actin.

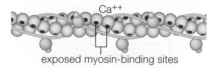

Ca^{++}

exposed myosin-binding sites

B Excited muscle. Ca^{++} binds to troponin, which shifts and moves tropomyosin, exposing myosin-binding sites on actin.

Figure 35.19 Animated Role of calcium in muscle contraction. Configuration of thin filament proteins depends on the Ca^{++} level.

smooth endoplasmic reticulum that wraps around the myofibrils and stores calcium ions **❸**.

Arrival of action potentials opens voltage-gated channels in the sarcoplasmic reticulum, allowing calcium ions to flow out, down their concentration gradient. This raises the calcium concentration around the actin and myosin filaments, allowing them to interact, and muscle contraction occurs.

Calcium ions affect the configuration of troponin and tropomyosin, two thin filament proteins that regulate binding of myosin to actin filaments. **Figure 35.19A** shows a single thin filament in a muscle fiber at rest. Under these circumstances, there is relatively little calcium in the fluid around the thin filament. Tropomyosin, a fibrous protein, wraps around actin and covers myosin-binding sites. Troponin, a globular protein attached to the tropomyosin, has a site that can bind calcium ions.

When an action potential causes release of calcium from the sarcoplasmic reticulum, some of the calcium binds to troponin (**Figure 35.19B**). As a result, troponin changes shape and pulls tropomyosin away from the myosin-binding sites on actin. With the binding sites cleared, actin can bind myosin, and the sliding action described in the previous section takes place. After contraction, calcium ions are pumped back into the sarcoplasmic reticulum.

Motor Units and Muscle Tension

A motor neuron has many axon endings that synapse on different fibers in a muscle. A motor neuron and all of the muscle fibers it synapses with constitute one **motor unit**. Stimulate a motor neuron, and all the muscle fibers on which it synapses will contract. The nervous system cannot make only some of the fibers in a motor unit contract.

A An 1809 painting showing a wounded soldier as he lay dying of tetanus in a military hospital. A bacterial toxin locked his muscles in contraction.

Figure 35.20 Results of disrupted control of skeletal muscles.

B President Franklin Delano Roosevelt had muscles weakened by polio.

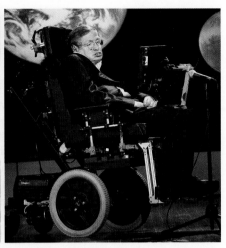

C Physicist Stephen Hawking has ALS. Although paralyzed, he continues to lecture using a speech synthesizer.

The mechanical force generated by a contracting muscle—the **muscle tension**—depends on the number of muscle fibers contracting. Some tasks require more muscle tension than others, so the number of muscle fibers controlled by a single motor neuron varies. In motor units that bring about small, fine movements such as those that control eye muscles, one motor neuron synapses with only 5 or so muscle fibers. By contrast, the biceps of the arm has about 700 muscle fibers per motor unit. Having many fibers contract at once increases the force a motor unit can generate.

Disrupted Control of Skeletal Muscle

Some bacteria make toxins that disrupt the flow of signals from nerves to muscles. *Clostridium botulinum* makes botulinum toxin, a chemical that prevents motor neurons from releasing ACh. Ingesting botulinum-tainted food causes a temporary paralysis that can be fatal if skeletal muscles involved in breathing are involved. Controlled doses of botulinum toxin are used in cosmetic procedures. "Botox" is injected into specific facial muscles to prevent actions that cause wrinkles.

Spores of a related bacterium, *Clostridium tetani*, sometimes colonize wounds. *C. tetani* makes a toxin that acts in the spinal cord by preventing release of a neurotransmitter (GABA) that inhibits motor neurons. As a result, nothing dampens signals to contract, and symptoms of the disease known as tetanus appear. Muscles stiffen and cannot be released from contraction. The fists and jaw stay clenched; hence the common name for the disease, lockjaw. The backbone may become locked in an abnormal arch (**Figure 35.20A**). Death occurs when respiratory and cardiac muscles become locked in contraction. Vaccines have eliminated the disease in the United States, but the annual global death toll is over 200,000.

Polio, a viral disease, also impairs motor neuron function. It most often affects children and can be fatal. People who survive a poliovirus infection may become paralyzed or develop a weakened voluntary muscle response (**Figure 35.20B**). Polio was common worldwide until a vaccine became available in the 1950s. No new cases have arisen in the United States since 1979, but sporadic outbreaks continue in developing countries. In addition, survivors of polio are at risk for postpolio syndrome, a disorder characterized by muscle fatigue and progressive muscle weakness.

Amyotrophic lateral sclerosis (ALS) also kills motor neurons. It is sometimes called Lou Gehrig's disease, after a famous baseball player whose career was cut short by the disease in the late 1930s. ALS usually causes death by respiratory failure within three to five years of diagnosis, but physicist Stephen Hawking has survived more than 40 years since his initial diagnosis (**Figure 35.20C**). The causes of ALS are not known.

motor unit One motor neuron and the muscle fibers it controls.
muscle tension Force exerted by a contracting muscle.
sarcoplasmic reticulum Specialized endoplasmic reticulum in muscle cells; stores and releases calcium ions.

Take-Home Message

How do nervous signals cause muscle contraction?

» A skeletal muscle contracts in response to a signal from a motor neuron. Release of ACh at a neuromuscular junction causes an action potential in the muscle cell.

» An action potential results in release of calcium ions, which affect proteins attached to actin. Resulting changes in the shape and location of these proteins open the myosin-binding sites on actin, allowing cross-bridge formation.

35.9 Muscle Metabolism

■ Muscle contraction requires ATP, which can be supplied by a variety of energy-releasing pathways.

■ Links to Energy-releasing pathways 7.2, Fermentation 7.6, X-linked inheritance 14.4

Energy-Releasing Pathways

Muscle contraction requires ATP, but muscle has a limited amount of ATP. It has a larger store of creatine phosphate, which can transfer a phosphate to ADP and form ATP (**Figure 35.21** ❶). Such phosphate transfers can fuel muscle contraction until other pathways increase ATP output.

Some athletes take creatine supplements to increase the amount of creatine phosphate available to muscle. Research suggests that creatine supplements can enhance performance of tasks that require a quick burst of energy. However, they have no effect on endurance, and the side effects of using such supplements are not fully known.

Most of the ATP used by a muscle during prolonged, moderate activity is produced by aerobic respiration ❷. Glucose derived from stored glycogen fuels five to ten minutes of activity, then the muscle fibers rely on glucose and fatty acids delivered by the blood. Fatty acids are the main fuel for activities that last more than half an hour.

Lactate fermentation is the third source of energy ❸. Some pyruvate is converted to lactate by the fermentation pathway even in resting muscle, but lactate fermentation steps up during exercise. This pathway produces less ATP than aerobic respiration, but has the advantage of operating even when the oxygen level in a muscle is low, as during strenuous exercise.

During such exercise, lactate accumulation raises the acidity in muscles, exciting adjacent pain receptors and causing an uncomfortable burning sensation. After the exercise ends, accumulated lactate enters mitochondria, where it is quickly converted to pyruvate and used in aerobic respiration. The muscle fatigue and soreness that persists for a day or two after strenuous exercise arises from other causes.

Types of Muscle Fibers

As you learned in Section 7.6, we divide muscle fibers into two categories based on how they produce ATP. Red fibers have an abundance of mitochondria and produce ATP mainly by aerobic respiration. They are colored bright red by **myoglobin**, a protein that, like hemoglobin, reversibly binds oxygen. During periods of muscle activity, myoglobin releases bound oxygen,

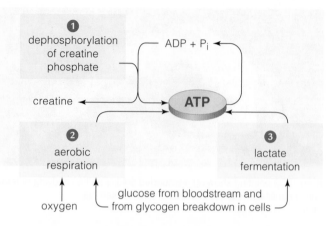

Figure 35.21 Animated Three metabolic pathways that muscles use to obtain the ATP that fuels their contraction.

allowing aerobic respiration to continue even if blood flow is insufficient to meet the muscle's oxygen need. By contrast, white fibers have no myoglobin, have few mitochondria, and make ATP mainly by lactate fermentation. Red fibers fatigue less easily than white fibers, so red fibers tend to predominate in muscles that carry out sustained activities.

Muscle fibers can also be subdivided into fast fibers or slow fibers based on the ATPase activity of their myosin. Myosin of fast fibers hydrolyzes ATP more efficiently than that of slow fibers. Thus, fast fibers can contract more quickly. All white fibers are fast fibers; they react fast and fatigue easily. The muscles that move your eye are mostly white fibers. Red fibers can be either fast or slow. Fast red fibers predominate in the triceps muscle, which must often react quickly. Muscles that play a role in maintaining an upright posture, such as some muscles of the back, consist mainly of slow red fibers.

The mix of fiber types in each skeletal muscle varies between individuals and has a genetic basis. Successful sprinters tend to have a higher-than-average percentage of fast, white fibers in their leg muscles. Marathoners tend to have more slow, red fibers than average.

Effects of Exercise

When unrelenting stimulation keeps a skeletal muscle excited, muscle fatigue follows; the muscle's capacity to generate force declines despite ongoing stimulation. After a few minutes of rest, the fatigued muscle will contract again in response to stimulation.

Aerobic exercise—low intensity, but long duration—makes skeletal muscles more resistant to fatigue.

Muscles and Myostatin (revisited)

Muscular dystrophies are a class of genetic disorders in which skeletal muscles progressively weaken. With Duchenne muscular dystrophy, symptoms begin to appear in childhood. A mutation of a gene on the X chromosome causes Duchenne muscular dystrophy. This gene encodes dystrophin, a plasma membrane protein of muscle fibers. The mutant form of dystrophin allows foreign material to enter a muscle fiber, causing the fiber to break down.

Muscular dystrophy arises in about 1 in 3,500 males. Like other X-linked disorders (Section 14.4), it rarely causes symptoms in females, who nearly always have a normal version of the gene on their other X chromosome. Affected boys usually begin to show signs of weakness by the time they are three years old, and require a wheelchair in their teens. Most die in their twenties of respiratory failure after skeletal muscles involved in breathing become affected.

Treatments that strengthen muscles can help offset the muscle loss that results from muscular dystrophy. Drugs that inhibit myostatin production or impair

myostatin activity may be useful in this regard. One way to learn what sort of effects such drugs might have is to study mice in which the myostatin gene has been knocked out. As one example, the larger mouse in the photo at the *right* is a knockout mutant. Its myostatin gene was disabled by genetic engineering, so it is bigger and more muscular than the unaltered mouse beside it. The bad news is that myostatin-knockout mice have unusually small, stiff, easily torn tendons. Thus, increased tendon injuries are expected to be a side effect of myostatin-inhibiting drugs and conditions.

How would you vote? Supplements that claim to block myostatin are already for sale. Should makers of these products be required to show that they both safe and effective?

Exercise increases the blood supply to a muscle by encouraging the growth of new capillaries. It also increases the number of mitochondria and the amount of myoglobin in red muscle fibers.

Regular episodes of brief, intense exercise such as weight lifting result in increased synthesis of actin and myosin. The increase helps a muscle exert more tension but does not improve endurance.

Exercise also affects health by its effect on a muscle enzyme called lipoprotein lipase (LPL). LPL allows the muscle to take up fatty acids and triglycerides (Section 3.5) from the blood. A low LPL level is associated with

a higher-than-normal blood lipid concentration and a heightened risk of cardiovascular disease.

Muscle production of LPL varies with a person's activity level. Long-distance runners, whose muscles frequently metabolize fatty acids, have high LPL levels. Conversely, people who frequently sit for prolonged periods have a lowered LPL level (**Figure 35.22**). This negative effect persists even if the sitter also gets regular exercise. Apparently, the sustained inactivity of leg muscles during time seated causes these muscles to turn down their transcription of LPL mRNA. To prevent the health problems associated with low LPL, avoid sitting for hours at a time. When you do sit, take periodic short breaks to stand up or walk. Any activity that requires your leg muscles to support your weight encourages the sustained production of LPL.

myoglobin Muscle protein that reversibly binds and stores oxygen.

Figure 35.22 Take a break from sitting. Prolonged sitting alters muscle fiber metabolism and may endanger your health.

Take-Home Message

What factors affect muscle metabolism?

» Muscle contraction requires ATP. When excited, muscle first uses stored ATP, then transfers phosphate from creatine phosphate to ADP to form ATP. With prolonged exercise, aerobic respiration and lactate fermentation provide the ATP.

» Exercise increases blood flow to muscles, the number of mitochondria, production of actin and myosin, and the muscle's ability to take up lipids from the blood for use as an energy source.

LEARNING ROADMAP

Where you have been This chapter expands on the discussion of circulatory systems in Section 24.2. You have already been introduced to blood (31.4) and cardiac muscle (31.5). You will draw on your knowledge of hemoglobin (3.2), diffusion and osmosis (5.8), endocytosis (5.10), autonomic nerves (32.7), sickle-cell anemia (9.6, 17.7), and malaria (21.7).

Where you are now

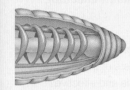

Overview of Circulatory Systems
In an open system, fluid leaves vessels and flows among tissues. In a closed system, blood remains inside vessels and substances are exchanged across vessel walls.

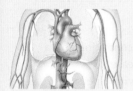

Human Cardiovascular System
A human heart is a muscular pump with four chambers. It pumps blood through two separate circuits: one to the lungs and back, and the other extending through the body.

Blood and Blood Vessels
Blood has oxygen-carrying red cells, white cells that fight pathogens, and platelets that aid clotting, all suspended in plasma. Exchanges with body cells take place at capillaries.

Circulatory Disorders
Circulatory function is impaired by abnormal numbers of blood cells, defective blood cells, narrowing of blood vessels, and alteration of the normal heart rhythm.

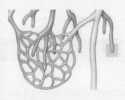

Lymphatic Connections
Fluid that filters out of blood capillaries returns to blood by way of the lymphatic vascular system. Lymphatic organs help defend against pathogens.

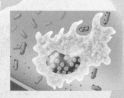

Where you are going White blood cells play an integral role in immunity, the subject of Chapter 37. You will learn more about function of red blood cells in gas exchange in Section 38.7. The role of the blood and lymphatic system in the distribution of nutrients is the topic of Section 39.7, and filtration of the blood by kidneys is covered in Chapter 40. You'll learn about blood loss during menstruation in Section 41.5 and how blood inflates the penis during an erection in Section 41.8. Section 42.9 explains how substances are exchanged between fetal and maternal bloodstreams.

36.1 A Shocking Save

The heart is the body's most durable muscle. It begins to beat during the first month of human development, and keeps on going for a lifetime. Each heartbeat is set in motion by an electrical signal generated by a natural pacemaker in the heart wall. In some people, this pacemaker malfunctions, causing what is called sudden cardiac arrest. Electrical signaling becomes disrupted, the heart stops beating, and blood flow halts. In the United States, sudden cardiac arrest occurs in more than 300,000 people per year. An inborn heart defect causes most cardiac arrests in people under age 36. In older people, heart disease usually causes the heart to stop functioning.

The chance of surviving sudden cardiac arrest rises by 50 percent when cardiopulmonary resuscitation (CPR) is started within four to six minutes of the arrest. With CPR, a person alternates mouth-to-mouth respiration with chest compressions that keep the victim's blood moving. However, CPR cannot restart the heart. That requires a defibrillator, a device that delivers an electric shock to the chest and resets the natural pacemaker. You have probably seen this procedure depicted in hospital dramas.

Matt Nader (**Figure 36.1A**) learned about the importance of CPR and defibrillation when he experienced sudden cardiac arrest during a high school football game. He came off the field after a play, sat on the bench to talk to his coach, and felt a burning pain in his chest. His vision blurred, and then he passed out. Nader's parents, who are physicians, were watching the game and rushed from their seats. They quickly determined that Matt did not have a pulse and, as they examined him, he stopped breathing.

Matt's parents begin CPR on their son. At the same time, someone ran to get the school's automated external defibrillator (AED), a device about the size of a laptop computer (**Figure 36.1B**). The AED provides simple voice commands about how to attach electrodes to a person in distress. It then checks for a heartbeat and, if required, shocks the heart.

The AED restarted Nader's heart, saving his life. Cardiologists determined that his sudden cardiac arrest had been caused by a genetic heart defect. To protect him, they implanted a small defibrillator in his body. It will provide a life-saving shock if his heart stops again.

After his recovery, Matt Nader went to testify before the Texas Legislature about his experience and to advocate for wider availability of AEDs in schools. Thanks in part to his efforts, Texas has passed a law requiring that all high schools have an AED available at athletic events and practices.

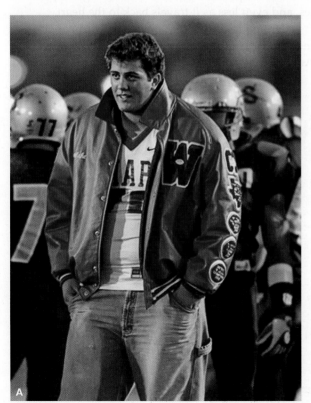

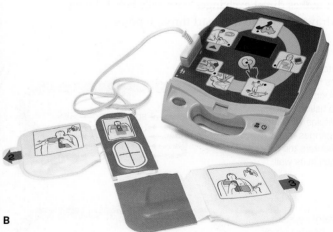

Figure 36.1 Surviving sudden cardiac arrest. (**A**) Matt Nader, a talented high school football player, discovered he had a heart defect when his heart stopped during a game. CPR and quick defibrillation saved his life.

(**B**) One type of automated external defibrillator. Such devices are designed to be simple enough to be used by a trained layperson. AEDs are increasingly available in public places, but they only make a difference if someone uses them.

36.2 The Nature of Blood Circulation

■ Most invertebrates and all vertebrates have a circulatory system that speeds the distribution of materials through the body.
■ Links to Diffusion 5.8, Morphological convergence 18.3

All animals must keep their cells supplied with nutrients and oxygen, and all must get rid of cellular wastes. Some invertebrates, including cnidarians and flatworms (Sections 24.5 and 24.6), rely on diffusion alone to accomplish these tasks. In such animals, nutrients and gases reach cells by diffusing across a body surface and then diffusing through the interstitial fluid (the fluid between cells). Diffusion only works over short distances to move materials quickly, so animals that rely on diffusion to distribute materials have a body plan in which all cells lie close to a body surface.

Open and Closed Circulatory Systems

Evolution of circulatory systems made possible more complex animal body plans. A **circulatory system** is an organ system that speeds the distribution of materials within an animal body. It includes one or more **hearts** (muscular pumps) that propel fluid through vessels extending through the body.

Different types of circulatory systems evolved in different animal lineages. Arthropods and most mollusks have an **open circulatory system**. In such systems, a heart or hearts pump fluid called **hemolymph** into open-ended vessels (**Figure 36.2A**). Hemolymph leaves the vessels and mixes with the interstitial fluid, where it makes direct exchanges with cells before being drawn back into the heart through pores.

By contrast, annelids, cephalopod mollusks, and all vertebrates have a **closed circulatory system** in which a heart or hearts pump fluid through a continuous series of vessels (**Figure 36.2B**). Fluid pumped through such a system is called **blood**. A closed circulatory system distributes substances faster than an open one. It is "closed" because blood does not leave blood vessels to bathe tissues. Instead, most transfers between blood and the cells of other tissues take place by diffusion across the walls of the smallest-diameter blood vessels.

Evolution of Vertebrate Circulation

All vertebrates have a closed circulatory system, with one heart. However, the structure of the heart and the circuits through which blood flows vary among vertebrate groups. In most fishes, the heart has two main chambers, and blood flows in a single circuit (**Figure 36.3A**). One chamber of the heart, an atrium (plural, atria), receives blood. From there, blood enters a ventricle, a chamber that pumps blood out of the heart. Pressure exerted by ventricular contractions drives blood through a series of vessels, into capillaries inside each gill, through capillaries in body tissues and organs, and back to the heart. The pressure imparted to the blood by the ventricle's contraction is dissipated as blood travels through capillaries, so the blood is not under much pressure when it leaves the gill capillaries, and even less as it travels back to the heart.

Adapting to life on land involved coordinated modifications of respiratory and circulatory systems. Amphibians and most reptiles have a three-chambered

A Open circulatory system. A grasshopper's heart pumps hemolymph through a large vessel and out into tissue spaces. Hemolymph mingles with interstitial fluid, exchanges materials, and then reenters the heart through openings in its wall.

B Closed circulatory system. An earthworm's hearts pumps blood through vessels that extend through the body. Exchanges between blood and the tissues take place across the wall of the smallest vessels.

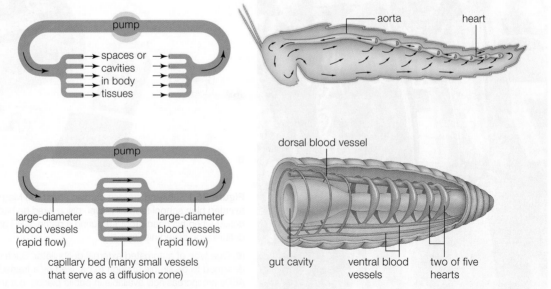

pump
→ spaces or
→ cavities
→ in body
→ tissues

pump
large-diameter blood vessels (rapid flow)
large-diameter blood vessels (rapid flow)
capillary bed (many small vessels that serve as a diffusion zone)

aorta   heart

dorsal blood vessel
gut cavity   ventral blood vessels   two of five hearts

Figure 36.2 Animated Comparison of open and closed circulatory systems.

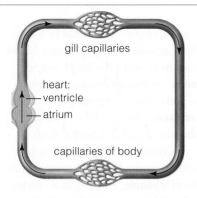

A The fish heart has one atrium and one ventricle. The force of the ventricle's contraction propels blood through the single circuit.

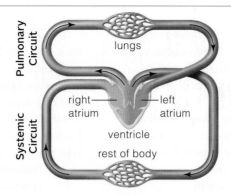

B In amphibians, lizards, snakes, and turtles, the heart has three chambers: two atria and one ventricle. Blood flows in two partially separated circuits. Oxygenated blood and oxygen-poor blood mix a bit in the ventricle.

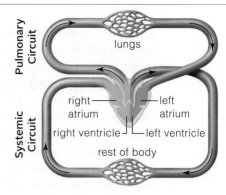

C In crocodilians, birds, and mammals, the heart has four chambers: two atria and two ventricles. Oxygenated blood and oxygen-poor blood do not mix.

Figure 36.3 Animated Variation in vertebrate circulatory systems.

heart, with two atria emptying into one ventricle (Figure 36.3B). A three-chambered heart speeds the rate of flow by moving blood through two partially separated circuits. The force of one contraction propels blood through the **pulmonary circuit** to the lungs, and then back to the heart. (The Latin *pulmo* means lung.) A second contraction sends oxygenated blood through the **systemic circuit**. This circuit extends through capillaries in body tissues and returns to the heart.

In birds and mammals, the single ventricle has been divided into two. Their four-chambered heart has two atria and two ventricles (**Figure 36.3C**). With two fully separate circuits, only oxygen-rich blood flows to tissues. As an additional advantage, blood pressure can be regulated independently in each circuit. Strong contraction of the heart's left ventricle moves blood quickly through the long systemic circuit. At the same time, the right ventricle can contract more gently, protecting the delicate lung tissue that would be blown apart by higher pressure.

The four-chambered heart of mammals and birds is an example of morphological convergence. Birds and mammals do not share an ancestor with such a heart. Rather, this trait evolved independently in the two groups. The enhanced blood flow associated with a four-chambered heart supports the high metabolism of these endothermic (heated from within) animals. Endotherms have higher energy needs than comparably sized ectotherms because they must produce heat to maintain their body temperature. Rapid blood flow in an endotherm's body delivers the large amount of oxygen required to keep heat-generating aerobic reactions going nonstop.

Crocodiles and alligators are ectotherms, but they have a four-chambered heart. This trait, together with features of their respiratory system and the anatomy of fossil crocodilians, suggests that these sluggish, sit-and-wait predators likely evolved from a highly active, endothermic ancestor.

blood Circulatory fluid; in vertebrates it is a fluid connective tissue consisting of plasma and cells that form inside bones.
circulatory system Organ system consisting of a heart or hearts and blood-filled vessels that distribute substances through a body.
closed circulatory system Circulatory system in which blood flows through a continuous system of vessels.
heart Muscular organ that pumps blood through a body.
hemolymph Fluid pumped through an open circulatory system.
open circulatory system Circulatory system in which hemolymph leaves vessels and flows through spaces in body tissues.
pulmonary circuit Circuit through which blood flows from the heart to the lungs and back.
systemic circuit Circuit through which blood flows from the heart to the body tissues and back.

Take-Home Message

How do animals distribute substances to cells throughout the body?

» Most animals have a circulatory system that speeds the distribution of substances through the body.

» Some invertebrates have an open circulatory system, other invertebrates and all vertebrates have a closed circulatory system, in which blood always remains enclosed within the heart or blood vessels.

» Fish have a one-circuit circulatory system. All other vertebrates have a short pulmonary circuit that carries blood to and from the lungs, and a longer systemic circuit that moves blood to and from the body's other tissues.

» A four-chambered heart evolved independently in birds and mammals. Such a heart allows strong contraction of one ventricle to speed blood through the systemic circuit, while a weaker contraction of the other ventricle protects lung tissue.

36.3 Human Cardiovascular System

■ The term "cardiovascular" comes from the Greek *kardia* (for heart) and Latin *vasculum* (vessel). In the human cardiovascular system, the heart pumps blood in two circuits.

■ Links to Glycogen storage 3.4, Alcohol metabolism 5.1

Like other mammals, humans have a four-chambered heart that pumps blood through two circuits. Each circuit includes a network of blood vessels that carries blood from the heart to small vessels where exchanges occur and then back to the heart. **Figure 36.4** shows the location and function of major blood vessels.

In each circuit, the heart pumps blood out of a ventricle and into branching arteries. Wide-diameter blood vessels that carry blood away from the heart and to organs are called **arteries**. Within an organ, arteries branch into smaller vessels called **arterioles**. Arterioles in turn branch into **capillaries**, the smallest vessels. Exchanges between the blood and interstitial fluid take place as blood flows through a capillary bed (a network of capillaries in a tissue). Several capillaries join to form a **venule**, a vessel that carries blood to a vein. **Veins** are large-diameter vessels that return blood to the heart.

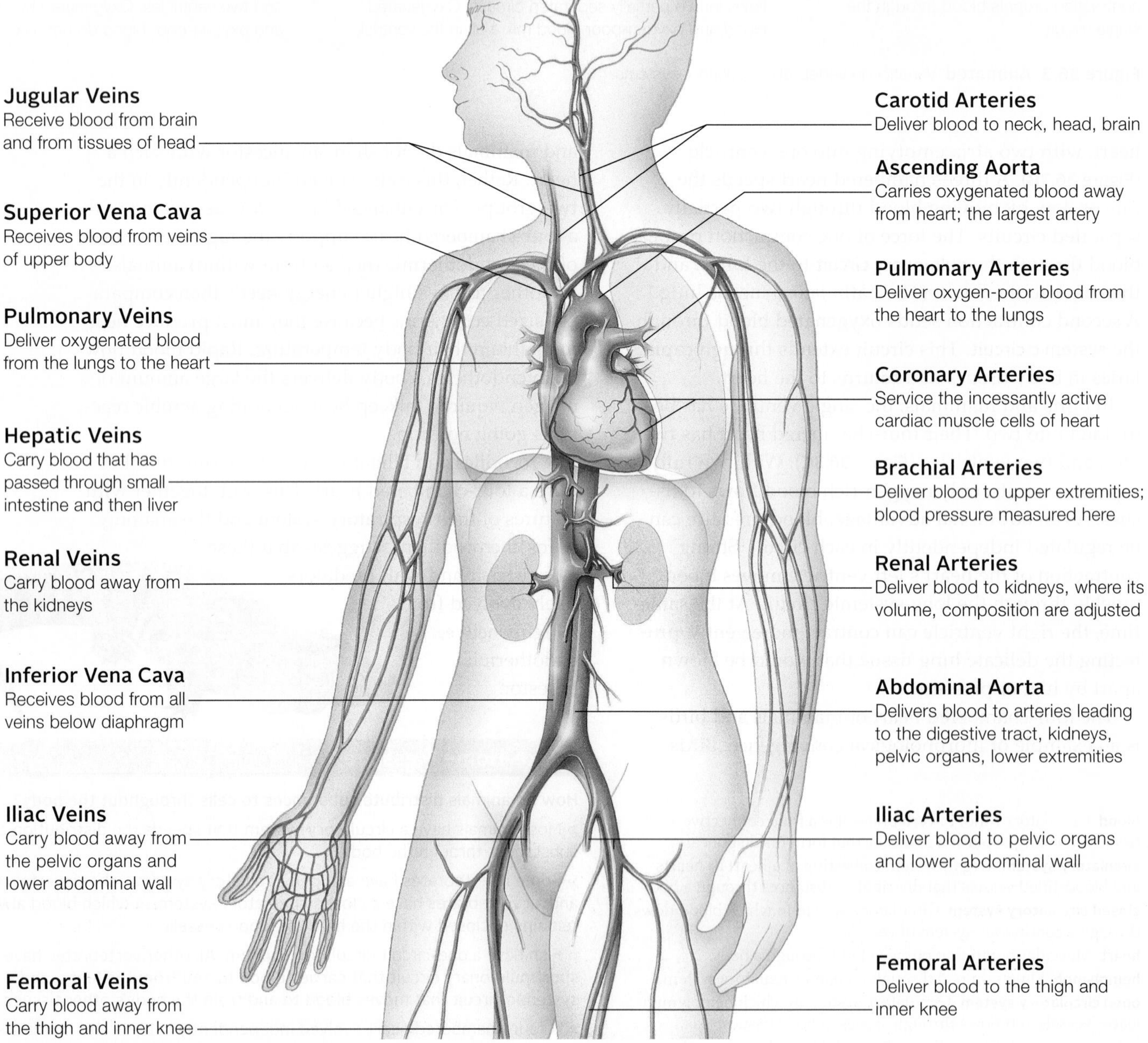

Jugular Veins
Receive blood from brain and from tissues of head

Superior Vena Cava
Receives blood from veins of upper body

Pulmonary Veins
Deliver oxygenated blood from the lungs to the heart

Hepatic Veins
Carry blood that has passed through small intestine and then liver

Renal Veins
Carry blood away from the kidneys

Inferior Vena Cava
Receives blood from all veins below diaphragm

Iliac Veins
Carry blood away from the pelvic organs and lower abdominal wall

Femoral Veins
Carry blood away from the thigh and inner knee

Carotid Arteries
Deliver blood to neck, head, brain

Ascending Aorta
Carries oxygenated blood away from heart; the largest artery

Pulmonary Arteries
Deliver oxygen-poor blood from the heart to the lungs

Coronary Arteries
Service the incessantly active cardiac muscle cells of heart

Brachial Arteries
Deliver blood to upper extremities; blood pressure measured here

Renal Arteries
Deliver blood to kidneys, where its volume, composition are adjusted

Abdominal Aorta
Delivers blood to arteries leading to the digestive tract, kidneys, pelvic organs, lower extremities

Iliac Arteries
Deliver blood to pelvic organs and lower abdominal wall

Femoral Arteries
Deliver blood to the thigh and inner knee

Figure 36.4 Animated Major blood vessels of the human cardiovascular system. Blood vessels carrying oxygenated blood are color-coded *red* and those carrying oxygen-poor blood are color-coded *blue*.

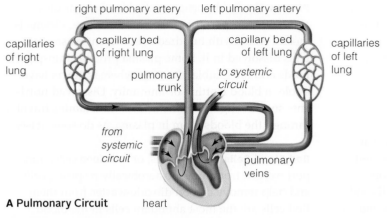

right pulmonary artery   left pulmonary artery

capillaries of right lung

capillary bed of right lung

capillary bed of left lung

capillaries of left lung

pulmonary trunk

to systemic circuit

from systemic circuit

pulmonary veins

A Pulmonary Circuit   heart

Figure 36.5 Animated The two circuits of the human cardiovascular system.

The Pulmonary Circuit

The pulmonary circuit is the shorter of the two circuits. It carries blood to and from the lungs (**Figure 36.5A**). Oxygen-poor blood is pumped out of the heart's right ventricle into pulmonary arteries. One pulmonary artery delivers blood to each lung. As blood flows through pulmonary capillaries, it picks up oxygen and gives up carbon dioxide. Oxygen-rich blood then returns to the heart by way of the pulmonary veins, which empty into the left atrium.

The Systemic Circuit

Oxygenated blood pumped out of the heart travels through the longer systemic circuit (**Figure 36.5B**). The heart's left ventricle pumps blood into the body's largest artery, the **aorta**. Arteries that branch from the aorta carry blood to various body parts. For example, the renal artery delivers blood to the kidneys, and the coronary arteries supply heart cells. Each artery branches into arterioles and then capillaries. Blood gives up oxygen and picks up carbon dioxide as it flows through the capillaries. The oxygen-poor blood that leaves the capillaries flows through venules and veins to the heart's right atrium.

Most blood moving through the systemic circuit flows through only one capillary bed. However, after blood passes through the capillaries in the small intestine, it flows through a vein (the hepatic portal vein) to a capillary bed in the liver. This two-capillary journey allows the blood to pick up glucose and other substances absorbed from the gut, and deliver them to the liver. The liver stores some of the absorbed glucose as glycogen. It also breaks down some absorbed toxins, including alcohol.

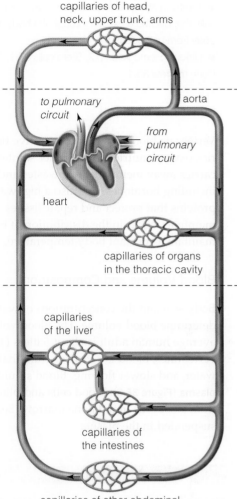

capillaries of head, neck, upper trunk, arms

to pulmonary circuit   aorta

from pulmonary circuit

heart

capillaries of organs in the thoracic cavity

capillaries of the liver

capillaries of the intestines

capillaries of other abdominal organs, lower trunk, legs

B Systemic Circuit

aorta Large artery that receives blood pumped out of the heart's left ventricle.
arteriole Vessel that carries blood from an artery to a capillary.
artery Large-diameter blood vessel that carries blood away from the heart.
capillaries Small blood vessels; exchanges with interstitial fluid take place across their walls.
vein Large-diameter vessel that returns blood to the heart.
venule Small-diameter vessel that carries blood from capillaries to a vein.

Take-Home Message

What are the two circuits of the human cardiovascular system?

» The pulmonary circuit carries oxygen-poor blood from the heart through the pulmonary arteries to arterioles and then capillaries in the lungs. Pulmonary veins return oxygenated blood to the heart.

» The systemic circuit carries oxygenated blood from the heart out the aorta, through branching arteries and to capillaries throughout the body. It returns oxygen-poor blood to the heart by way of venules and veins.

» Most blood traveling through the systemic circuit passes through one capillary bed, but blood that flows through capillaries in the intestines also flows through capillaries in the liver.

36.4 Components and Functions of Blood

■ Tumbling along in the plasma of a vertebrate bloodstream are cells that distribute oxygen through the body and defend the body from pathogens.

■ Links to Hemoglobin 3.2, Stem cells 31.1, Thymus 34.12, Bone marrow 35.4

Functions of Blood

Vertebrate blood is a fluid connective tissue that carries oxygen, nutrients, and other solutes to cells, and carries away their metabolic wastes and secretions, including hormones. It is also a highway for cells and proteins that protect and repair tissues. In birds and mammals, shifts in the distribution of blood flow help maintain a constant body temperature.

Blood Volume and Composition

Body size and the concentrations of water and solutes determine blood volume. The blood volume of an average human adult is about 5 liters (10 pints). In vertebrates, blood is a viscous fluid that is thicker than water, and slower flowing. Blood's fluid portion is **plasma** (**Figure 36.6**). Blood cells and platelets that arise from stem cells in red bone marrow (Section 35.4) are suspended in the plasma.

Plasma The fluid portion of the blood constitutes about 50 to 60 percent of the blood volume. Plasma is mostly water with hundreds of different plasma proteins dissolved in it. Some plasma proteins transport lipids and fat-soluble vitamins, whereas others have a role in blood clotting or immunity. Dissolved nutrients such as sugars, amino acids, and vitamins travel through the bloodstream in plasma, as do some gases.

Red Blood Cells Erythrocytes, or **red blood cells**, transport oxygen from lungs to aerobically respiring cells and help remove carbon dioxide wastes from them. Red cells are the most abundant cells in the blood.

In all mammals, red blood cells lose their nucleus, mitochondria, and other organelles as they mature. A mature red blood cell is a flexible disk with a depression at its center. The flattened shape allows it to slip easily through narrow blood vessels, and facilitates gas exchange. **Hemoglobin**, an iron-containing protein that reversibly binds oxygen, fills the cell's interior.

In addition to hemoglobin, a mature red blood cell contains enough glucose and enzymes to last about 120 days. In a healthy person, ongoing replacements keep red blood cell numbers at a fairly stable level. A **cell count** is a measure of the quantity of cells of one type in 1 microliter (1/1,000,000 liter) of blood. Men

Components	Amounts	Main Functions
Plasma Portion (50–60% of total blood volume)		
1. Water	91–92% of total plasma volume	Solvent
2. Plasma proteins (albumins, globulins, fibrinogen, etc.)	7–8%	Defense, clotting, lipid transport, extracellular fluid volume controls
3. Ions, sugars, lipids, amino acids, hormones, vitamins, dissolved gases, etc.	1–2%	Nutrition, defense, respiration, extracellular fluid volume controls, cell communication, etc.
Cellular Portion (40–50% of total blood volume; numbers per microliter)		
1. Red blood cells	4,600,000–5,400,000	Oxygen, carbon dioxide transport to and from lungs
2. White blood cells:		
Neutrophils	3,000–6,750	Fast-acting phagocytosis
Lymphocytes	1,000–2,700	Immune responses
Monocytes (macrophages)	150–720	Phagocytosis
Eosinophils	100–380	Killing parasitic worms
Basophils	25–90	Anti-inflammatory secretions
3. Platelets	250,000–300,000	Roles in blood clotting

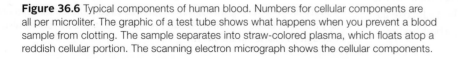

red blood cell   white blood cell   platelet

Figure 36.6 Typical components of human blood. Numbers for cellular components are all per microliter. The graphic of a test tube shows what happens when you prevent a blood sample from clotting. The sample separates into straw-colored plasma, which floats atop a reddish cellular portion. The scanning electron micrograph shows the cellular components.

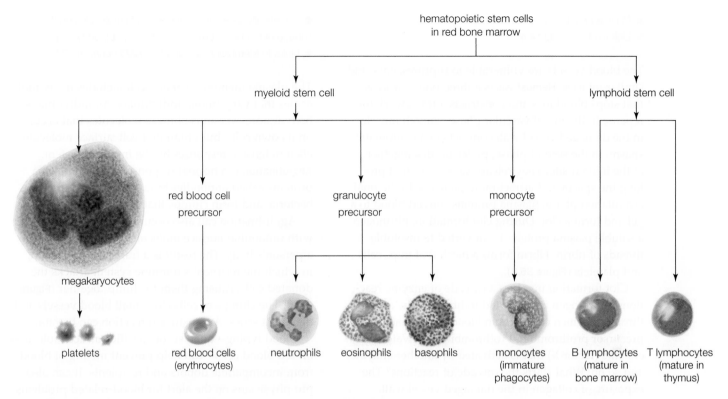

Figure 36.7 Main cellular components of mammalian blood and how they originate.

typically have more red cells than women of reproductive age, who lose blood when they menstruate.

White Blood Cells Leukocytes, or **white blood cells**, carry out ongoing housekeeping tasks and function in defense. The cells differ in their size, nuclear shape, and staining traits (**Figure 36.7**), as well as function. We discuss the role of white blood cells in detail in the next chapter, but provide a brief preview here.

One type of precursor cell in bone marrow gives rise to neutrophils, basophils, and eosinophils. These three types of white blood cells are also called granulocytes because they have cytoplasmic vesicles (granules) filled with pathogen-killing substances. Neutrophils, the most abundant white blood cells, are phagocytes that engulf bacteria and debris. Eosinophils attack larger parasites, such as worms, and have a role in allergies. Basophils secrete chemicals that have a role in inflammation.

cell count Number of cells of one type in 1 microliter of blood.
hemoglobin Iron-containing protein that reversibly binds oxygen.
plasma Fluid portion of blood.
platelet Cell fragment that helps blood clot.
red blood cell Hemoglobin-filled blood cell that transports oxygen.
white blood cell Vertebrate blood cell with a role in housekeeping tasks and defense; a leukocyte.

Monocytes circulate in the blood for a few days, then move into the tissues, where they develop into phagocytic cells called macrophages and dendritic cells. These cells interact with lymphocytes to bring about immune responses described in detail in the next chapter. There are two types of lymphocytes, B cells and T cells. B cells mature in bone, whereas T cells mature in the thymus (Section 34.12).

Platelets Megakaryocytes are ten to fifteen times bigger than other blood cells that form in bone marrow. They break up into membrane-wrapped fragments of cytoplasm called **platelets**. After a platelet forms, it will last five to nine days. When activated, it releases substances needed for blood clotting, as described in detail in the next section.

Take-Home Message

What are the components of human blood and what are their functions?

» Blood consists mainly of plasma, a protein-rich fluid that carries wastes, gases, and nutrients.

» Blood cells and platelets form in bone marrow and are transported in plasma. Red blood cells contain hemoglobin that carries oxygen from lungs to tissues. White cells help defend the body from pathogens. Platelets are cell fragments that have a role in clotting.

36.5 Hemostasis

■ Plasma proteins and platelets interact in clotting.
■ Link to Hemophilia 14.4

The blood vessels are vulnerable to ruptures, cuts, and similar injuries. **Hemostasis** is a three-phase process that stops blood loss and constructs a framework for repairs. In the initial vascular phase, smooth muscle in the damaged vessel wall contracts in an automatic spasm. In the second phase, platelets stick together at the injured site. They release substances that prolong the spasm and attract more platelets. In the final coagulation phase, plasma proteins convert blood to a gel and form a clot. During clot formation, fibrinogen, a soluble plasma protein, is converted to insoluble threads of fibrin. Fibrin forms a mesh that traps cells and platelets (**Figure 36.8**).

Clot formation involves a cascade of enzyme reactions. Fibrinogen is converted to fibrin by the enzyme thrombin, which circulates in blood as the inactive precursor prothrombin. Prothrombin is activated by an enzyme (factor X) that is activated by another enzyme, and so on. What starts the cascade of reactions? The exposure of collagen in the damaged vessel wall.

If a mutation affects any one of the enzymes that acts in the cascade of clotting reactions, the blood may not clot properly. Such mutations cause the genetic disorder hemophilia (Section 14.4).

hemostasis Process by which blood clots in response to injury.

Stimulus
A blood vessel is damaged.

Phase 1 response
A vascular spasm constricts the vessel.

Phase 2 response
Platelets stick together, plugging the site.

Phase 3 response
Clot formation starts:
1. Enzyme cascade results in activation of Factor X.
2. Factor X converts prothrombin in plasma to thrombin.
3. Thrombin converts fibrinogen, a plasma protein, to fibrin threads.
4. Fibrin forms a net that entangles cells and platelets, forming a clot.

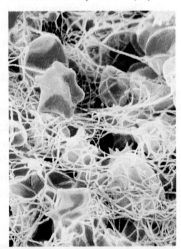

Figure 36.8 Animated The three-phase process of hemostasis. The micrograph shows the result of the final clotting phase—blood cells and platelets in a fibrin net.

Take-Home Message

How does the body respond to blood vessel damage and halt bleeding?

» The vessel constricts, platelets accumulate, and cascading enzyme reactions involving protein components of plasma cause clot formation.

36.6 Blood Typing

■ Genetically determined differences in molecules on the surface of red blood cells are the basis of blood typing.
■ Links to Membrane proteins 5.7, ABO genetics 13.5

The plasma membrane of any cell includes many molecules that vary among individuals. An individual's body ignores versions of these molecules that occur on its own cells, but unfamiliar cell surface molecules elicit defensive responses by the immune system. **Agglutination** is a normal response in which plasma proteins called antibodies bind foreign cells, such as bacteria, and form clumps that attract phagocytes.

Agglutination can also occur when red blood cells with unfamiliar surface molecules are transfused into a person's body. The result is a transfusion reaction, in which the recipient's immune system attacks the donated cells, causing them to clump together (**Figure 36.9**). The clumps of cells clog small blood vessels and damage tissues. A transfusion reaction can be fatal.

Blood typing—analysis of specific surface molecules on red blood cells—can help prevent mixing of blood from incompatible donors and recipients. It can also put physicians on the alert for blood-related problems that can arise during some pregnancies.

ABO Blood Typing

ABO blood typing analyzes variations in one type of glycolipid on the surface of red blood cells. Section 13.5 describes the genetics of these variations. People who have one form of the molecule have type A blood. Those with a different form have type B blood. People with both forms of the molecule have type AB blood. Those who have neither form are type O. See below.

ABO Type	Glycolipid(s) on Red Cells	Antibodies Present
A	A	Anti-B
B	B	Anti-A
AB	Both A and B	None
O	Neither A nor B	Anti-A, Anti-B

If you are blood type O, your immune system treats both type A and type B cells as foreign. You can accept blood only from people who are type O (**Figure 36.9**). However, you can donate blood to anyone. If you are blood type A, your body will recognize type B cells as foreign. If you are type B, your blood will react against type A cells. If you are blood type AB, your immune system treats both type A and type B as "self," so you can receive blood from anyone.

Rh Blood Typing

Rh blood typing is based on the presence or absence of the Rh protein (first identified in blood of *Rh*esus monkeys). If you are type Rh⁺, your blood cells bear this protein. If you are type Rh⁻, they do not.

Normally, Rh⁻ individuals do not have antibodies against the Rh protein. However, they will produce such antibodies if they are exposed to Rh⁺ blood. This can happen during some pregnancies. If an Rh⁺ man impregnates an Rh⁻ woman, the resulting fetus may be Rh⁺. The first time that an Rh⁻ woman carries an Rh⁺ fetus, she will not have antibodies against the Rh protein (**Figure 36.10A**). However, fetal red blood cells may get into her blood during childbirth, causing her to form antibodies that target the Rh protein. If the woman becomes pregnant again, these antibodies may cross the placenta and enter the fetal bloodstream. If the fetus is Rh⁺, the antibodies will attack its red blood cells, with potentially fatal results (**Figure 36.10B**). To prevent this outcome, an Rh⁻ mother who has given birth to an Rh⁺ child is injected with antibodies that destroy any fetal red cells in her body. This treatment prevents her from making antibodies against these cells, and thus averts problems in her next pregnancy.

ABO blood types do not cause a similar condition because maternal antibodies for A and B molecules do not cross the placenta and attack the fetal cells.

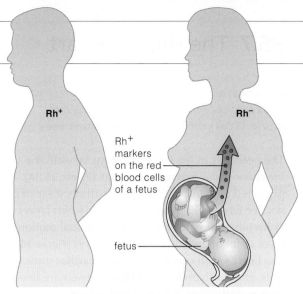

A An Rh⁺ man and an Rh⁻ woman carrying his Rh⁺ child. This is her first Rh⁺ pregnancy, so she has no anti-Rh antibodies. During birth, some of the child's Rh⁺ cells get into her blood.

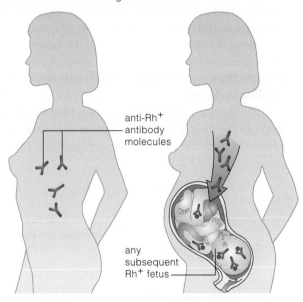

B The foreign marker stimulates antibody formation. If this woman gets pregnant again and if her second fetus (or any other) carries the Rh protein, her anti-Rh antibodies may attack and kill the fetal red blood cells.

Figure 36.10 Animated How Rh differences can complicate pregnancy.

ABO blood typing Method of identifying certain glycoproteins (A or B) on red blood cells; the absence of either type is designated O.
agglutination Clumping of foreign cells after antibodies bind them.
Rh blood typing Method of determining whether Rh, a type of surface recognition protein, is present on an individual's red blood cells. If present, the cell is Rh⁺; if absent, the cell is Rh⁻.

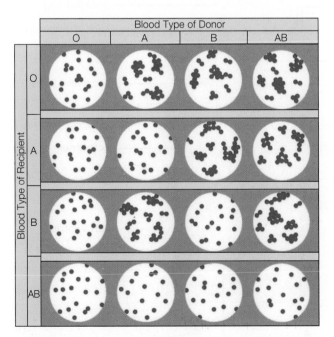

Figure 36.9 Results of mixing blood of the same or differing ABO blood types. Clustered cells indicate an agglutination reaction. **Figure It Out: How many incompatible combinations are shown?**
Answer: Seven.

Take-Home Message

What is a blood type?

» Blood type refers to the kind of surface molecules on red blood cells. Genes determine which form of these molecules an individual has.

» When blood of incompatible types mixes, the immune system attacks the unfamiliar molecules, with results that can be fatal.

36.7 The Human Heart

■ The heart is a durable, muscular pump that contracts in response to its own spontaneous action potentials.

■ Links to Cell junctions 4.11, Sliding-filament model 35.7

The heart lies in the thoracic cavity, beneath the breastbone and between the lungs (**Figure 36.11A**). It is protected and anchored by pericardium, a sac of connective tissue. Fluid between the sac's two layers provides lubrication for the heart's continual motions. A layer of fat offers additional protection (**Figure 36.11B**). The heart's wall consists mostly of cardiac muscle cells, and its chambers and blood vessels are lined with endothelium, a type of epithelium.

Each side of the human heart contains two chambers: an **atrium** that receives blood from veins, and a **ventricle** that pumps blood into arteries (**Figure 36.11C**). Pressure-sensitive valves function like one-way doors to control the flow of blood through the heart. High fluid pressure forces a valve open. When fluid pressure declines, a valve shuts and prevents blood from moving backward.

Two big veins deliver oxygen-poor blood from the body to the right atrium. The **superior vena cava** delivers blood from the upper regions of the body. The **inferior vena cava** delivers blood from lower regions. Blood from the right atrium flows through the right atrioventricular (AV) valve into the right ventricle. The right ventricle pumps it through the pulmonary valve and into the pulmonary trunk, a vessel that branches into two pulmonary arteries. Each **pulmonary artery** carries blood to a lung.

After passing through the lung, the oxygenated blood returns to the left atrium via **pulmonary veins**. The blood then flows through the left atrioventricular (AV) valve into the left ventricle. The left ventricle pumps the blood through the aortic valve into the aorta. From there it flows to tissues of the body.

The Cardiac Cycle

The events that occur from the onset of one heartbeat to another are collectively called the **cardiac cycle** (**Figure 36.12**). During this cycle, the heart's chambers alternate through **diastole** (relaxation) and **systole** (contraction). First, the relaxed atria expand with blood ❶. Fluid pressure forces AV valves to open and blood to flow into the relaxed ventricles, which expand as the atria contract ❷. Once filled, the ventricles contract.

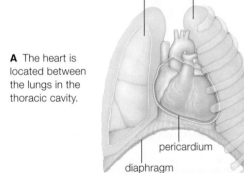

A The heart is located between the lungs in the thoracic cavity.

right lung   left lung

pericardium

diaphragm

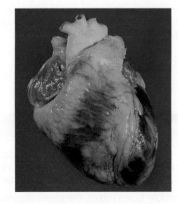

B Heart's outer surface. A bit of fat (light yellow) at the heart's surface is normal.

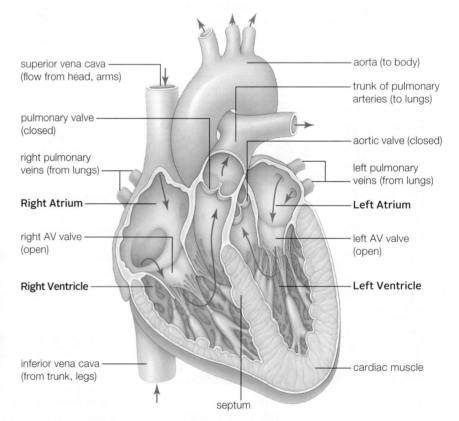

superior vena cava (flow from head, arms)

aorta (to body)

trunk of pulmonary arteries (to lungs)

pulmonary valve (closed)

aortic valve (closed)

right pulmonary veins (from lungs)

left pulmonary veins (from lungs)

Right Atrium

Left Atrium

right AV valve (open)

left AV valve (open)

Right Ventricle

Left Ventricle

inferior vena cava (from trunk, legs)

cardiac muscle

septum

C Cutaway view, showing the heart's internal organization. Arrows indicate the path taken by oxygenated (*red*) and oxygen-poor (*blue*) blood.

Figure 36.11 Animated The human heart.

1 Relaxed atria fill. Fluid pressure opens AV valves and blood flows into the relaxed ventricles.

2 Contracting atria squeeze more blood into the still-relaxed ventricles.

4 As blood flows into the arteries, pressure in the ventricles declines and the aortic and pulmonary valves close.

3 Ventricles start contracting, and rising pressure pushes AV valves shut. A further rise in pressure opens aortic and pulmonary valves.

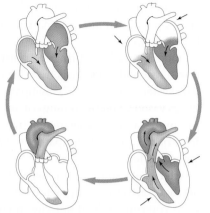

Figure 36.12 Animated Cardiac cycle.

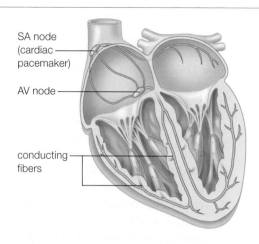

SA node (cardiac pacemaker)

AV node

conducting fibers

Figure 36.13 Animated Cardiac conduction system.

Contraction raises the fluid pressure inside the ventricles and forces the aortic and pulmonary valves to open. Blood flows through these valves and out of the ventricles **3**. Now emptied, the ventricles relax while the atria fill again **4**.

Blood circulation is driven entirely by contracting ventricles; atrial contraction only helps fill the ventricles with blood. The structure of the cardiac chambers reflects their different functions. Atria need only generate enough force to squeeze blood into the ventricles, so they have relatively thin walls. Ventricle walls are more thickly muscled because their contraction has to generate enough pressure to propel blood through an entire cardiovascular circuit. The left ventricle, which pumps blood throughout the long systemic circuit, has thicker walls than the right ventricle, which pumps blood only to the lungs and back.

During the cardiac cycle a "lub-dup" sound can be heard through the chest wall. Each "lub" is the heart's AV valves closing. Each "dup" is the heart's aortic and pulmonary valves closing.

Setting the Pace for Contraction

Like skeletal muscle, cardiac muscle has orderly arrays of sarcomeres that contract by a sliding-filament mechanism. Unlike skeletal muscle, cardiac muscle does not require a signal from the central nervous system to contract. The **sinoatrial (SA) node**, a clump of specialized cells in the wall of the right atrium (**Figure 36.13**), serves as the cardiac pacemaker, generating an action potential about seventy times a minute. Signals from the central nervous system only adjust this rate.

Gap junctions between adjacent muscle cells allow action potentials generated by the SA node to spread across the atria. The action potentials cause the atria to contract. As they do, special noncontractile muscle fibers convey the excitation to the **atrioventricular (AV) node**. This clump of cells is the only electric bridge to the ventricles; action potentials can only reach the ventricles by way of the AV node. The time it takes for a signal to cross this bridge allows blood from the atria to fill the ventricles before the ventricles contract. From the AV node, the signal travels along conducting fibers in the septum between the heart's left and right halves. The fibers extend to the heart's lowest point and up the ventricle walls. The excitatory signal spreads via gap junctions, and the ventricles contract from the bottom up, with a wringing motion.

atrioventricular (AV) node Clump of cells that serves as the electrical bridge between the atria and ventricles.
atrium Heart chamber that receives blood from veins.
cardiac cycle Sequence of contraction and relaxation of heart chambers that occurs with each heartbeat.
diastole Relaxation phase of the cardiac cycle.
inferior vena cava Vein that delivers blood from the lower body to the heart.
pulmonary artery Vessel carrying blood from the heart to a lung.
pulmonary vein Vessel carrying blood from a lung to the heart.
sinoatrial (SA) node Cardiac pacemaker; cluster of specialized cells whose spontaneous rhythmic signals trigger contractions.
superior vena cava Vein that delivers blood from the upper body to the heart.
systole Contractile phase of the cardiac cycle.
ventricle Heart chamber that pumps blood into arteries.

Take-Home Message

How does the structure of the human heart relate to its function?

» The four-chambered heart is a muscular pump partitioned into two halves, each with an atrium and a ventricle. Forceful contraction of the ventricles provides the driving force for blood circulation.

» The SA node is the cardiac pacemaker. Its spontaneous, rhythmic signals make cardiac muscle cells of the heart wall contract in a coordinated fashion.

36.8 Blood Vessel Structure and Function

■ Contracting ventricles put pressure on the blood, forcing it through a series of vessels that vary in their structure.
■ Links to Smooth muscle 31.5, Autonomic nervous system 32.7

Blood vessels vary in their structure and function (**Figure 36.14**). Blood pumped out of ventricles enters arteries, large vessels with thick walls that help keep blood flowing, even between ventricular contractions. An artery wall contains smooth muscle reinforced with elastic connective tissue (**Figure 36.14A**). When a ventricle contracts, it forces blood into the arteries, causing their elastic walls to bulge. When the ventricle relaxes, the artery walls spring back. As the artery walls rebound inward, they push the blood inside them a bit farther away from the heart.

The bulging of an artery with each ventricular contraction is referred to as the **pulse**. You can feel a person's pulse if you place your finger on a pulse point, a place where an artery runs near the body surface. To feel the pulse in your own radial artery, put your fingers on your inner wrist near the base of your thumb.

Blood flows from arteries into arterioles. All blood pumped out of the right half of your heart always travels to your lungs. By contrast, in the systemic circuit, the distribution of blood to particular body parts can be adjusted by altering the diameter of arterioles.

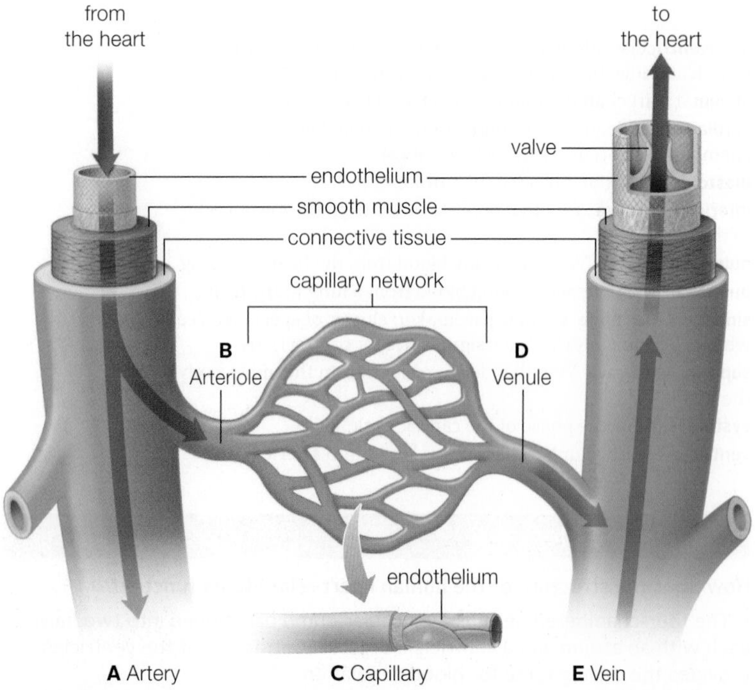

Widening the arterioles that supply an area increases blood flow to that area and narrowing those vessels decreases it.

Smooth muscle that rings each arteriole (**Figure 36.14B**) responds to signals from the autonomic nervous system (Section 32.7). Sympathetic stimulation causes **vasodilation** (widening) of arterioles in the extremities and **vasoconstriction** (narrowing) of arterioles of the gut. Parasympathetic stimulation has the opposite effect.

Arterioles also widen or contract in response to metabolic activity in adjacent tissue. For example, when you run, skeletal muscles in your legs release carbon dioxide and lactic acid (a product of fermentation reactions) into the blood. The presence of these substances causes dilation of the arterioles that deliver blood to the leg muscles.

A capillary is a cylinder of endothelial cells, one cell thick, wrapped in basement membrane (**Figure 36.14C**). Its thin wall and narrow diameter, barely wider than a red blood cell, facilitate exchanges between the blood and the interstitial fluid. Oxygen-carrying red blood cells are forced right up against the capillary wall. We discuss capillary exchange in detail in Section 36.10.

Blood from several capillaries flows into each venule (**Figure 36.14D**). Several thin-walled venules empty into each vein, a large-diameter, low-resistance transport tube that conveys blood to the heart. Many veins, especially those in the legs, have flaplike valves that help prevent backflow (**Figure 36.14E**). These valves automatically snap shut if blood inside the vein reverses its direction.

Figure 36.14 Structural comparison of human blood vessels and the direction of blood flow through them.

pulse Brief stretching of artery walls that occurs when ventricles contract.
vasoconstriction Narrowing of a blood vessel when smooth muscle that rings it contracts.
vasodilation Widening of a blood vessel when smooth muscle that rings it relaxes.

Take-Home Message

How do blood vessels differ in their structure and function?

» Arteries are thick-walled, large-diameter vessels. Stretching and recoil of arteries helps keep blood moving.

» Smooth muscle in the wall of arterioles allows adjustments to blood flow in the systemic circuit.

» Capillaries are narrow tubes of epithelial cells. They are the site of exchanges with interstitial fluid.

» Veins have valves that prevent backflow of blood.

36.9 Blood Pressure

■ Ventricular contractions are the source of blood pressure, which declines throughout a cardiovascular circuit.
■ Link to Medulla oblongata 32.10, Sensory receptors 33.2

Blood pressure is pressure exerted by blood against the wall of the vessel that encloses it. The right ventricle contracts less forcefully than the left ventricle, so blood entering the pulmonary circuit is under less pressure than blood entering the systemic circuit. In both circuits, blood pressure is highest in arteries, and declines over the course of the circuit (**Figure 36.15**).

Blood pressure is usually measured in the brachial artery of the upper arm (**Figure 36.16**). Two pressures are recorded. **Systolic pressure**, the highest pressure of a cardiac cycle, occurs as contracting ventricles force blood into the arteries. **Diastolic pressure**, the lowest blood pressure of a cardiac cycle, occurs when ventricles are relaxed. Blood pressure is measured in millimeters of mercury (mm Hg), a standard unit for measuring pressure, and is written as systolic value/ diastolic value. Normal blood pressure is about 120/80 mm Hg, or "120 over 80."

Blood pressure depends on the total blood volume, how much blood ventricles pump out (cardiac output), and the degree of arteriole dilation. Sensory receptors in the aorta and in carotid arteries signal the medulla oblongata in the hindbrain when blood pressure rises or falls. In a reflexive response, this brain region calls for appropriate changes in cardiac output and arteriole diameter. Vasodilation of arterioles lowers blood pressure; vasoconstriction raises it. This reflex response adjusts blood pressure over the short term. Over the longer term, kidneys adjust blood pressure by regulating the amount of fluid lost in urine and thus determining the total blood volume.

Inability to regulate blood pressure can result in hypertension, in which resting blood pressure remains above 140/90. Chronic high blood pressure makes the heart and kidneys work harder, increasing risk of heart disease or kidney failure.

blood pressure Pressure exerted by blood against a vessel wall.
diastolic pressure Blood pressure when ventricles are relaxed.
systolic pressure Blood pressure when ventricles are contracting.

Take-Home Message

How is blood pressure recorded and regulated?

» Blood pressure is the fluid pressure exerted against a vessel wall. It is recorded as systolic/diastolic pressure.

» Adjustments to arteriole diameter, cardiac output, and blood volume regulate blood pressure.

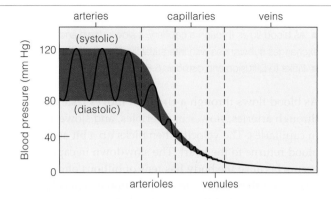

Figure 36.15 Plot of fluid pressure changes as a volume of blood flows through the systemic circuit. Systolic pressure occurs when ventricles contract, diastolic when ventricles are relaxed.

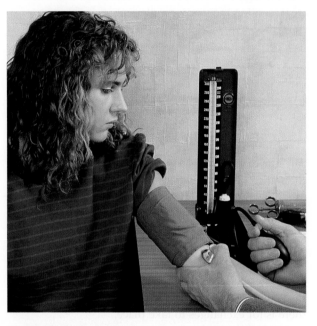

Figure 36.16 Measuring blood pressure. A hollow inflatable cuff attached to a pressure gauge is wrapped around the upper arm. A stethoscope is placed over the brachial artery, just below the cuff.

The cuff is inflated with air to a pressure above the highest pressure of the cardiac cycle, when ventricles contract. Above this pressure, you will not hear sounds through the stethoscope, because no blood is flowing through the vessel.

Air in the cuff is slowly released until the stethoscope picks up soft tapping sounds. Blood flowing into the artery under the pressure of the contracting ventricles—the systolic pressure—causes the sounds. When these sounds start, a gauge typically reads about 120 mm Hg. That amount of pressure will force mercury (Hg) to move up 120 millimeters in a glass column of a standardized diameter.

More air is released from the cuff. Eventually the sounds stop. Blood is now flowing continuously, even when the ventricles are the most relaxed. The pressure when the sounds stop is the lowest during a cardiac cycle, the diastolic pressure, which is usually about 80 mm Hg.

Right, compact monitors are now available that automatically record the systolic/diastolic blood pressure.

36.10 Mechanisms of Capillary Exchange

■ As blood flows through a capillary, it slows down and exchanges substances with interstitial fluid.
■ Links to Diffusion and osmosis 5.8, Exocytosis 5.10

As blood flows through a circuit, it moves fastest through arteries, slower in arterioles, and slowest in capillaries. The velocity then picks up a bit as the blood returns to the heart. The slowdown in capillaries occurs because the body has tens of billions of capillaries, and their collective cross-sectional area is far greater than that of the arterioles that deliver blood to them, or the veins that carry blood away. By analogy, think about what happens if a narrow river (representing few larger vessels) delivers water to a wide lake (representing the many capillaries):

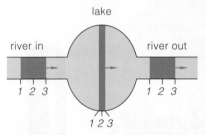

The rate of the flow is constant into and out of the lake: the same volume of water moves from points 1 to 3 in a given interval. However, the velocity of the flow decreases in the lake region. When the volume spreads across a larger cross-sectional area, it flows forward a shorter distance during the specified interval.

Slow flow through capillaries enhances the rate of exchanges between the blood and interstitial fluid. The more time blood spends in a capillary, the more time there is for exchanges to take place.

To move between the blood and interstitial fluid, a substance must cross a capillary wall. Oxygen, carbon dioxide, and small lipid-soluble molecules diffuse across endothelial cells of a capillary. Some larger molecules enter endothelial cells by endocytosis, diffuse through the cell, then escape by exocytosis into the interstitial fluid.

Substances also enter the interstitial fluid when a bit of fluid is forced out of capillaries through spaces between cells of the capillary wall. Blood pressure is highest at the arterial end of a capillary bed, and it is here that pressure forces fluid out between cells (**Figure 36.17A**). The fluid that exits has high levels of oxygen, ions, and nutrients. As blood continues toward the venous end of the capillary, blood pressure falls. Now osmotic pressure is the predominant force. It causes water to move from the interstitial fluid into the hypertonic, protein-rich plasma (**Figure 36.17B**).

Normally, there is a small net outward flow of fluid from capillaries. The lymphatic system (described in Section 36.13) returns the escaped fluid to the blood. If high blood pressure forces excess fluid out of capillaries, or something prevents fluid return, interstitial fluid pools in tissues. The tissue swelling that results is called edema.

Take-Home Message

How does blood exchange materials with interstitial fluid?

» Small molecules cross cells of a capillary by diffusion and larger ones move across by exocytosis.

» Fluid rich in oxygen and nutrients also leaks out between cells of the capillary wall.

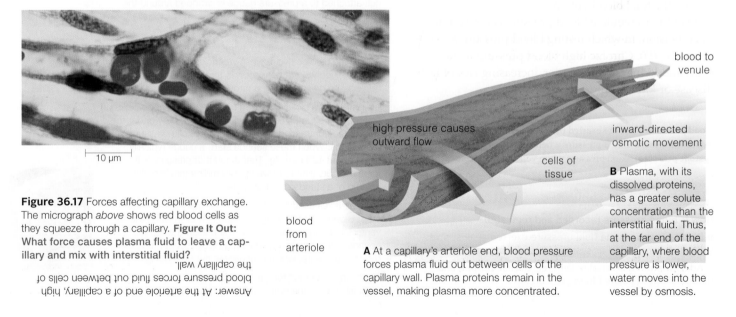

Figure 36.17 Forces affecting capillary exchange. The micrograph *above* shows red blood cells as they squeeze through a capillary. **Figure It Out: What force causes plasma fluid to leave a capillary and mix with interstitial fluid?**

Answer: At the arteriole end of a capillary, high blood pressure forces fluid out between cells of the capillary wall.

A At a capillary's arteriole end, blood pressure forces plasma fluid out between cells of the capillary wall. Plasma proteins remain in the vessel, making plasma more concentrated.

B Plasma, with its dissolved proteins, has a greater solute concentration than the interstitial fluid. Thus, at the far end of the capillary, where blood pressure is lower, water moves into the vessel by osmosis.

36.11 Venous Function

- Veins are the body's largest blood reservoir.
- Skeletal muscle activity helps move blood at low pressure through veins and back to the heart.
- Link to Smooth muscle 31.5

Veins carry blood through the final stretch of a circuit and return it to the heart. By the time blood reaches veins, most of the pressure imparted by ventricular contractions has dissipated. Of all blood vessels, veins have the lowest blood pressure. The vein wall can bulge quite a bit under pressure, much more so than an arterial wall. Thus, veins act as reservoirs for great volumes of blood. When you rest, they hold about 60 percent of the total blood volume.

Several mechanisms help blood at low pressure move through the veins and back toward the heart. First, veins have flaplike valves that help prevent backflow. These valves automatically shut when blood in the vein starts to reverse direction. For example, the valves in the large veins of the leg prevent blood from moving downward in response to gravity when you stand (**Figure 36.18**). In addition, smooth muscle inside a vein's wall contracts in response to signals from the nervous system. The contraction causes the vein to stiffen so it cannot hold as much blood, and the pressure inside it rises, forcing blood toward the heart.

Skeletal muscles used in limb movements also help move blood through veins. When these muscles contract, they bulge and press on veins, squeezing blood toward the heart (**Figure 36.19**). In the same way, exercise-induced deep breathing raises pressure inside veins. As the lungs and thoracic cavity expand during inhalation, adjacent organs are forced against veins. Pressure from this contact forces blood in a vein forward through a valve.

Sometimes one or more valves in a vein become damaged, as by high blood pressure, so blood pools in a vein. When valve damage occurs in a leg, varicose veins may result. With this condition, bulging veins are visible at the skin surface. When valve damage occurs in veins of the rectum or anus, hemorrhoids can be the outcome.

When blood pools in veins because of valve damage or prolonged inactivity, a clot may form. A clot that forms in a blood vessel and stays there is called a thrombus. A clot that breaks loose and travels through vessels to a new location is an embolus. Both types of clot pose a heath risk because they can slow or halt blood flow. For example, an embolus that blocks blood flow in the brain can cause a stroke, in which brain cells die. An embolus in the lung can also be life-threatening.

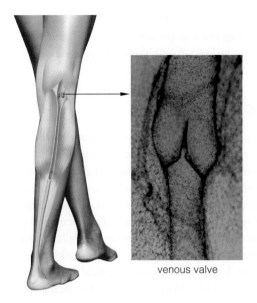

Figure 36.18 Valves in veins prevent the backflow of blood.

blood flow to heart

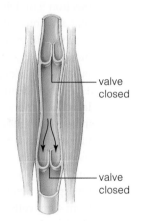

valve open

valve closed

valve closed

valve closed

When skeletal muscles contract, they bulge and press on neighboring veins. This puts pressure on the blood in the vein, forcing it forward through the pressure-sensitive valves.

When skeletal muscles relax, the pressure in neighboring veins declines and pressure-sensitive valves shut, preventing blood from moving backward.

Figure 36.19 How skeletal muscle activity encourages blood flow through veins.

Take-Home Message

What are the functions of veins?

» Veins are the body's main blood reservoir. The amount of blood in the veins changes depending on activity level.

» Blood pressure in veins is low. One-way valves, activity of skeletal muscle, and respiratory muscle action all help move the blood toward the heart.

36.12 Blood and Cardiovascular Disorders

■ Altered blood cell quantity or quality can impair health, as can conditions that interfere with normal blood flow.
■ Links to LDL and HDL 3.6, Thalassemia and sickle-cell anemia 9.6, Malaria 21.7, Diabetes 34.8

Altered Blood Cell Count

In anemias, red blood cells are few or somehow impaired, causing oxygen delivery and metabolism to falter. Shortness of breath, fatigue, and chills are common symptoms. Anemia has many causes. It can arise as a result of blood loss from a wound or an infection by a pathogen that kills red blood cells. For example, the protist that causes malaria enters red cells, divides inside them, and then causes the cell to break apart. A diet with too little iron can cause anemia by preventing the synthesis of iron-containing heme. Sickle-cell anemia arises from a mutation that causes hemoglobin to change shape at a low oxygen concentration (Section 3.6). Thalassemias occur when mutations disrupt or halt synthesis of a globin chain of hemoglobin (Section 9.6). Disrupted globin synthesis results in cells that are unusually small and misshapen.

Polycythemia is an excess of red blood cells. It increases oxygen delivery, but makes blood more viscous and elevates blood pressure.

White cell numbers can also be disrupted. Infectious mononucleosis is a common cause of increased monocyte production. The Epstein–Barr virus infects B lymphocytes and the body produces monocytes in response. Symptoms typically last weeks and include a sore throat, fatigue, muscle aches, and low-grade fever. Leukemias are cancers that originate from stem cells in bone. They cause overproduction of abnormally formed white blood cells that do not function properly. Lymphomas are cancers that originate from B or T lymphocytes. Division of the cancerous lymphocytes can produce tumors in lymph nodes and other parts of the lymphatic system.

Cardiovascular Disease

In the United States, cardiovascular disorders kill about a million people every year. Tobacco smoking tops the list of risk factors for cardiovascular disorders. Other factors include a family history of such disorders, hypertension, a high cholesterol level, diabetes mellitus, and obesity. Regular exercise helps lower the risk of these disorders even when the exercise is not particularly strenuous. Gender is another factor; until about age fifty, males are at greater risk than females.

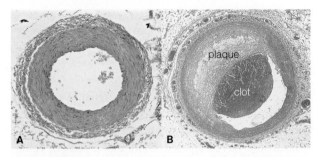

Figure 36.20 Sections from (**A**) a normal artery and (**B**) an artery with its interior narrowed by an atherosclerotic plaque. A clot has adhered to the plaque, further narrowing the vessel.

Atherosclerosis Cardiovascular disease often involves atherosclerosis, in which a buildup of lipids in the arterial wall narrows the lumen, or space inside the vessel. Cholesterol plays a role in this "hardening of the arteries." The human body requires cholesterol to make cell membranes, myelin sheaths, and steroid hormones (Section 3.5). The liver makes enough cholesterol to meet the body's needs, but most of us absorb still more from the food in our gut. Genetics affects how different people's bodies deal with the resulting excess of dietary cholesterol.

Most cholesterol in blood is bound to protein carriers to form complexes known as low-density lipoproteins, or LDLs (Section 3.6). Most cells in the body, including those lining blood vessels, take up LDLs. A lesser amount of cholesterol is bound up in high-density lipoproteins, or HDLs. Cells in the liver take up HDLs and use them to produce bile, a substance the liver secretes into the small intestine. Cholesterol incorporated into bile leaves the body in feces.

Having a high LDL level or a low HDL level raises the risk of atherosclerosis. The first sign of trouble is a mass of lipids that builds up in an artery's endothelium (**Figure 36.20**). Fibrous connective tissue forms over the mass. The resulting atherosclerotic plaque bulges into the vessel's interior, narrowing its diameter and slowing blood flow. A plaque is hard and can rupture an artery wall, thereby triggering clot formation.

Atherosclerosis raises the risk that a blood vessel in the brain, heart, or other organ will become clogged. Interrupted blood flow to brain tissue causes a stroke, and interrupted blood flow to the heart causes a heart attack. In both cases, if the blockage is not removed fast, irreplaceable heart or brain cells die. Drugs that dissolve clots can restore blood flow and minimize cell death, but only if they are administered within an hour of the onset of an attack. Thus, anyone suspected of having a stroke or heart attack should receive prompt medical attention.

Figure 36.21 Two ways of treating blocked coronary arteries, the main cause of heart attacks in older adults.

A Coronary bypass surgery. Veins from another part of the body are used to divert blood past the blockages. This illustration shows a "double bypass," in which veins are placed to divert blood around two blocked coronary arteries.

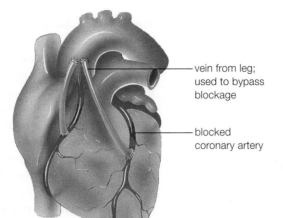

vein from leg; used to bypass blockage

blocked coronary artery

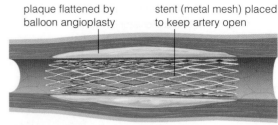

plaque flattened by balloon angioplasty

stent (metal mesh) placed to keep artery open

B Balloon angioplasty and the placement of a stent. After a balloonlike device is inflated in an artery to open it and flatten the plaque, a tube of metal (the stent) is inserted and left in place to keep the artery open.

Clogged coronary arteries can be treated with a bypass or angioplasty. With coronary bypass surgery, doctors open a person's chest and use a blood vessel from elsewhere in the body (usually a leg vein) to divert blood around the clogged coronary artery (**Figure 36.21A**). In laser angioplasty, laser beams vaporize plaques. In balloon angioplasty, doctors inflate a small balloon in a blocked artery to flatten the plaques. A wire mesh tube called a stent is then inserted to keep the vessel open (**Figure 36.21B**). Angioplasty is also used to open partially blocked carotid arteries. These arteries in the neck supply blood to the brain.

Hypertension Hypertension refers to chronically high blood pressure (above 140/90). Often the cause is unknown. Heredity is a factor, and African Americans have an elevated risk. Diet also plays a role; in some people high salt intake causes water retention that raises blood pressure. Hypertension is sometimes described as a silent killer, because people often are unaware they have it. Hypertension makes the heart work harder than normal, which can cause it to enlarge and to function less efficiently. High blood pressure also increases risk of atherosclerosis.

Rhythms and Arrhythmias Electrocardiograms, or ECGs, record the electrical activity of a beating heart (**Figure 36.22**). They can also reveal arrhythmias, which are abnormal heart rhythms caused by malfunction of the SA node.

Bradycardia is a below-average resting cardiac rate. If the resulting slow flow impairs health, implanting an artificial pacemaker can speed the heart rate.

Tachycardia is a faster than normal heart rate. Many people experience palpitations, which are occasional episodes of tachycardia. Palpitations can be brought on by stress, drugs such as caffeine, an overactive thyroid, or an underlying heart problem.

Atrial fibrillation is an arrhythmia in which the atria do not contract normally, but instead quiver. This slows blood flow and increases the risk of clot formation. Often, people who have atrial fibrillation receive anticlotting medication to lower their risk of stroke.

Ventricular contraction is the driving force for blood circulation, so ventricular fibrillation is the most dangerous arrhythmia. Ventricles quiver, and pumping falters or stops, causing loss of consciousness and—if a normal rhythm is not restored—death. A defibrillator can often restore the heart's normal rhythm by resetting the SA node.

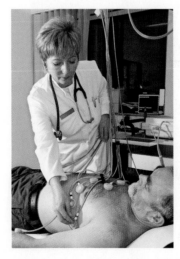

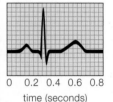

one normal heartbeat

0   0.2 0.4 0.6 0.8
time (seconds)

Figure 36.22 Animated Electrocardiograms. An ECG (*above*) is a graph showing electrical changes that can be detected by electrodes attached to the skin (*left*). The changes arise as a result of the activity of the SA node and the transmission of action potentials through the heart.

Take-Home Message

What factors impair circulatory function?

» Altered blood cell number or quality can alter blood's ability to carry out its functions.

» Atherosclerosis and hypertension raise the risk of heart attack and stroke.

» Problems with the cardiac pacemaker cause arrhythmias.

36.13 Interactions With the Lymphatic System

- Vessels and organs of the lymphatic system interact closely with the circulatory system.
- Link to Thymus gland 34.12

The **lymph vascular system** consists of vessels that collect water and solutes from the interstitial fluid, then deliver them to the circulatory system. It includes lymph capillaries and vessels (**Figure 36.23**). Fluid that moves through these vessels is called **lymph**.

The lymph vascular system serves three functions. First, its vessels are drainage channels for water and plasma proteins that have leaked out of capillaries and must be returned to the circulatory system. Second, it delivers fats absorbed from food in the small intestine to the blood. Third, it transports cellular debris, pathogens, and foreign cells to lymph nodes, where white blood cells assess and respond to them.

Lymph capillaries lie in close proximity to blood capillaries throughout the body (**Figure 36.23B**). A lymph capillary begins as a finger-shaped ending in a tissue. Fluid that leaks out of blood capillaries can enter the ending of a lymph capillary. Gaps between the cells in such endings open sporadically as a result of normal body movements.

Lymph capillaries merge into larger-diameter lymph vessels. Two mechanisms move lymph through these vessels. First, slow wavelike contractions of smooth muscle in the walls of large lymph vessels propel lymph forward. Second, as with veins, the bulging of adjacent skeletal muscles helps move fluid along.

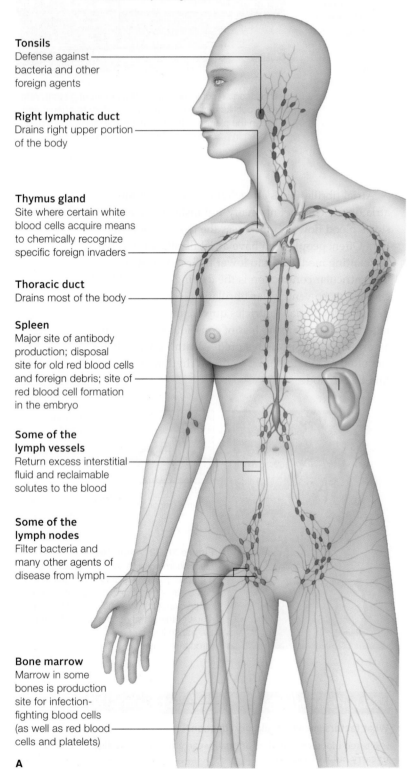

Tonsils
Defense against bacteria and other foreign agents

Right lymphatic duct
Drains right upper portion of the body

Thymus gland
Site where certain white blood cells acquire means to chemically recognize specific foreign invaders

Thoracic duct
Drains most of the body

Spleen
Major site of antibody production; disposal site for old red blood cells and foreign debris; site of red blood cell formation in the embryo

Some of the lymph vessels
Return excess interstitial fluid and reclaimable solutes to the blood

Some of the lymph nodes
Filter bacteria and many other agents of disease from lymph

Bone marrow
Marrow in some bones is production site for infection-fighting blood cells (as well as red blood cells and platelets)

A

Figure 36.23 Animated Components of the lymphatic system.

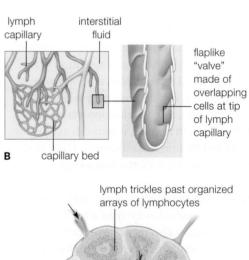

lymph capillary   interstitial fluid

flaplike "valve" made of overlapping cells at tip of lymph capillary

B   capillary bed

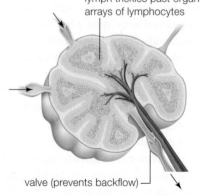

lymph trickles past organized arrays of lymphocytes

valve (prevents backflow)

C Lymph node

A Shocking Save (revisited)

Most cardiac arrests do not occur in a hospital, so the presence of a bystander willing to carry out CPR or to use an AED often means the difference between life and death. Sadly, although most cardiac arrests are witnessed, only about 15 percent of victims get CPR before trained personnel arrive.

Most people have an understandable reluctance to engage in mouth-to-mouth contact with a stranger. This can be a problem with traditional CPR, which calls for a rescuer to alternate between exhaling into a victim's mouth to inflate the lungs and providing chest compressions. A new procedure called CCR (cardiocerebral resuscitation) relies on chest compressions alone to move air into and out of the lungs. Proponents of CCR argue that as long as a person's airway is clear, chest compressions will move enough air into and out of a victim's lungs to oxygenate the blood flowing to the person's heart and brain. Early studies indicate that CCR may be as good as or even better than traditional CPR at saving most people who experience a sudden cardiac arrest.

Sign in an airport indicating where an AED is stored.

Another promising development is the rising availability of AEDs. In 2010, approximately 200,000 AEDs were purchased for installation in public places. The use of such devices is already saving lives. A recent study that looked at the outcome of nearly 14,000 out-of-hospital cardiac arrests concluded that, compared to no treatment, bystander use of an AED increased the likelihood of surviving to discharge from the hospital by about 75 percent. That's the good news. The bad news is that most people still do not know what an AED is, where they are available, or how simple it is to use one. The photo at the *upper right* shows the symbol for an AED in public place.

How would you vote? Knowing how to do CPR and make use of an AED can help save lives. Should high schools include instruction in these skills as a part of their standard curriculum?

Like veins, lymph vessels have one-way valves that prevent backflow.

The largest lymph vessels converge on collecting ducts that empty into veins in the lower neck. Each day these ducts deliver 3 liters of fluid to the blood.

Lymphoid Organs and Tissues

Lymphoid organs and tissues have roles in the body's defense responses to injury and invasion by pathogens. They include the lymph nodes, spleen, and thymus, as well as the tonsils, and some patches of tissue in the wall of the small intestine and appendix.

Lymph nodes are located at intervals along lymph vessels (**Figure 36.23C**). Lymph filters through at least one node before it enters the blood. Large numbers of lymphocytes that formed in bone marrow migrate to the nodes, where they mature. When these cells identify pathogens in the lymph, they provoke an immune response, as described in detail in the next chapter.

Tonsils are two patches of lymphoid tissue at the back of the throat. Adenoids are similar tissue clumps at the rear of the nasal cavity. Tonsils and adenoids help the body respond fast to inhaled pathogens.

The **spleen** is the largest lymphoid organ, about the size of a fist in an average adult. During prenatal development, it functions as a site of red blood cell formation. After birth, the spleen filters pathogens, worn-out red blood cells, and platelets from the many blood vessels that branch through it. The spleen has white blood cells that engulf and digest pathogens and altered body cells. It also has antibody-producing B cells. People can survive removal of their spleen, but they become more vulnerable to infections.

In the thymus gland, T lymphocytes differentiate and become capable of recognizing and responding to particular pathogens. The thymus gland also makes the hormones that influence these actions. It is central to immunity, the focus of the next chapter.

lymph Fluid in the lymph vascular system.
lymph node Small mass of lymphatic tissue through which lymph filters; contains many lymphocytes (B and T cells).
lymph vascular system System of vessels that takes up interstitial fluid and carries it (as lymph) to the blood.
spleen Large lymphoid organ that filters blood.

Take-Home Message

What are the functions of the lymphatic system?

» The lymph vascular system consists of tubes that collect and deliver excess water and solutes from interstitial fluid to blood. It also carries absorbed fats to the blood, and delivers disease agents to lymph nodes.

» The system's lymphoid organs, including lymph nodes, have specific roles in body defenses.

LEARNING ROADMAP

Where you have been This chapter revisits bacteria and viruses (Sections 4.4, 20.1–20.4, 20.7) as pathogens. Cell structure and function (4.7, 4.11, 5.8, 5.10, 34.2), proteins (3.6, 5.7), cDNA (15.2), genetics (9.3), the internal environment (31.2), epithelia and skin (31.3, 31.8), and the circulatory and lymphatic systems (36.4, 36.6, 36.10, 36.13) come up again in the context of diseases (15.10, 32.12, 36.12).

Where you are now

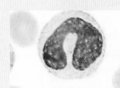

Immune Defenses
Vertebrates have three lines of immune defenses: surface barriers, innate immunity, and adaptive immunity. Leukocytes and signaling molecules function in immune responses.

Surface Barriers
External surfaces of the body come into constant contact with microbial pathogens. Physical, mechanical, and chemical barriers prevent most microbes from entering body tissues.

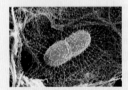

Innate Immunity
Innate immune responses involve a set of general, immediate defenses. Phagocytic white blood cells, plasma proteins, inflammation, and fever quickly rid the body of most invaders.

Adaptive Immunity
In an adaptive immune response, white blood cells interact to destroy specific pathogens or altered cells. Antibodies and other antigen receptors are central to these responses.

Immunity in Our Lives
Vaccines are important in worldwide health programs. Allergies, immune deficiencies, and autoimmune disorders are the result of infections or faulty immune mechanisms.

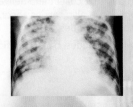

Where you are going Inflammatory diseases of the respiratory system are covered in Section 38.9. Section 39.1 discusses the immunity-enhancing effects of some gut bacteria, and Section 39.10 covers celiac disease, a type of autoimmune disorder. Section 40.9 takes another look at fever. AIDS and other sexually transmitted diseases return in Section 41.10. Cell suicide, or apoptosis, is detailed in Section 42.4, and Section 42.12 describes the passage of antibodies from mother to child. Section 48.7 explains why global climate change may exacerbate pollen allergies.

37.1 Frankie's Last Wish

In October of 2000, Frankie McCullough had known for a few months that something was not quite right. She had not had an annual checkup in many years; after all, she was only 31 and had been healthy her whole life. It never occurred to her to doubt her own invincibility until the moment she saw the doctor's face change as he examined her cervix.

The cervix is the lowest part of the uterus, or womb. Cervical cells can become cancerous, but the process is usually slow. The cells pass through several precancerous stages that are detectable by routine Pap tests. Precancerous and even early-stage cancerous cells can be removed from the cervix before they spread to other parts of the body. However, plenty of women like Frankie do not take advantage of regular exams. Those who end up at the gynecologist's office with pain or bleeding may be experiencing symptoms of advanced cervical cancer, the treatment of which offers only a 9 percent chance of survival. About 3,600 women die of cervical cancer each year in the United States. Many more than that die in places where routine gynecological testing is not common.

What causes cancer? At least in the case of cervical cancer, we know the answer to that question: Healthy cervical cells are transformed into cancerous ones by infection with human papillomavirus (HPV). HPV is a DNA virus that infects skin and mucous membranes. There are about 100 different types of HPV; a few cause warts on the hands or feet, or in the mouth.

About 30 others that infect the genital area sometimes cause genital warts, but usually there are no symptoms of infection. Genital HPV is spread very easily by sexual contact. At least 80 percent of women have been infected with HPV by the age of 50.

A genital HPV infection usually goes away on its own, but not always. A persistent infection with one of about 10 strains is the main risk factor for cervical cancer (**Figure 37.1**). Types 16 and 18 are particularly dangerous: One of the two is found in more than 70 percent of all cervical cancers.

In 2006, the U.S. Food and Drug Administration (FDA) approved Gardasil, a vaccine against four types of genital HPV, including types 16 and 18. The vaccine prevents cervical cancer caused by these HPV strains. It is most effective in girls who have not yet become sexually active, because they are least likely to have become infected with HPV.

The HPV vaccine came too late for Frankie McCullough. Despite radiation treatments and chemotherapy, her cervical cancer spread quickly. She died in 2001, leaving a wish for other people: awareness. "If there is one thing I could tell a young woman to convince them to have a yearly exam, it would be not to assume that your youth will protect you. Cancer does not discriminate; it will attack at random, and early detection is the answer." Almost all women newly diagnosed with invasive cervical cancer have not had a Pap test in five years. Many have never had one.

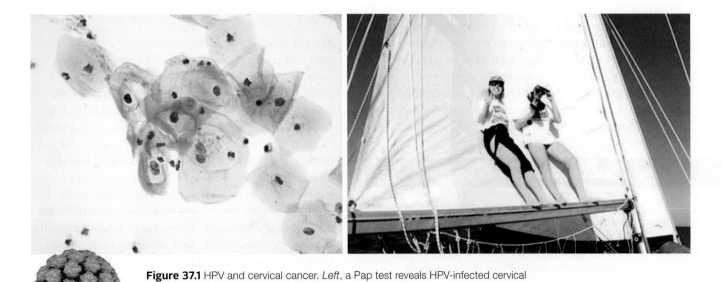

Figure 37.1 HPV and cervical cancer. *Left*, a Pap test reveals HPV-infected cervical cells among normal ones. Infected cells have enlarged, often multiple nuclei surrounded by a clear area. These changes sometimes lead to cervical cancer, which is treatable if detected early enough. The *orange* ball is a model of an HPV16 virus. *Right*, Frankie McCullough (waving) died of cervical cancer in 2001.

37.2 Integrated Responses to Threats

- In vertebrates, the innate and adaptive immune systems work together to combat infection and injury.
- Innate immune mechanisms are fast, general responses to tissue damage and invading microorganisms.
- Adaptive immune mechanisms recognize and respond to potentially billions of specific pathogens.
- Links to Membrane trafficking 5.10, Coevolution of pathogens and hosts 20.1, Neuromodulators 32.6, Local signaling molecules 33.3, Intercellular signaling 34.2, Blood cells 36.4

Humans continually cross paths with a tremendous array of viruses, bacteria, fungi, parasitic worms, and other pathogens, but you need not lose sleep over this. Humans coevolved with these pathogens, so you have defenses that protect your body from them. The evolution of **immunity**, an organism's capacity to resist and combat infection, began well before multicelled eukaryotes evolved from free-living cells. Mutations in the genes for membrane proteins introduced new molecular patterns that were unique in cells of a given type. As multicellularity evolved, so did mechanisms of identifying the patterns as self, or belonging to one's own body.

By about 1 billion years ago, nonself recognition had also evolved. Cells of all modern multicelled eukaryotes bear a set of receptors that collectively can recognize around 1,000 different nonself cues, which are called pathogen-associated molecular patterns (PAMPs). As their name suggests, PAMPs occur mainly on or in pathogens. They include some components of bacterial cell walls, flagellum and pilus proteins, double-stranded RNA unique to some viruses, and so on. A PAMP is an example of **antigen**, which is any molecule or particle recognized by the body as nonself. Binding of a cell's receptors to antigen triggers a set of immediate, general defense responses. In mammals, for example, binding triggers activation of complement. **Complement** is a set of proteins that circulate in inactive form throughout the body. Activated complement can destroy invading cells or mark them

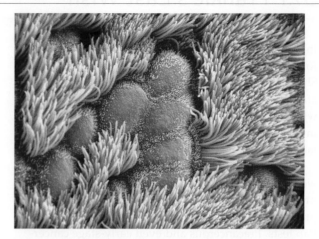

Figure 37.2 One physical barrier to infection: Mucus and the mechanical action of cilia keep pathogens from getting a foothold in the airways to the lungs. Bacteria and other particles get stuck in mucus secreted by goblet cells (*gold*). Cilia (*pink*) on other cells sweep the mucus toward the throat for disposal.

for uptake by phagocytic cells (those that can engulf material by phagocytosis).

Pattern receptors and the responses they initiate are part of **innate immunity**, a set of fast, general defenses against infection. All multicelled organisms start out life with these defenses, which normally do not change within the individual's lifetime.

Vertebrates have another set of defenses carried out by interacting cells, tissues, and proteins. This **adaptive immunity** tailors immune defenses to a vast array of specific pathogens that an individual may encounter during its lifetime. **Table 37.1** compares adaptive and innate immunity.

Three Lines of Defense

The mechanisms of adaptive immunity evolved within the context of innate immunity. The two systems were once thought to operate independently of each other, but we now know they function together. We describe both systems together in terms of three lines of defense. The first line includes the physical, chemical, and mechanical barriers that usually keep pathogens on the outside of the body (**Figure 37.2**). Innate immunity, the second line of defense, begins after tissue is damaged, or after antigen is detected inside the body. Its general response mechanisms quickly rid the body of many invaders. Activation of innate immunity triggers the third line of defense, adaptive immunity. Leukocytes (white blood cells) divide to form huge populations that target a specific antigen and destroy anything bearing it. Some of the white blood cells persist after infection ends. If the same antigen returns, these memory cells mount a secondary response.

Table 37.1 Innate and Adaptive Immunity Compared

	Innate	Adaptive
Response time	Immediate	About a week
Antigen detection	Fixed set of receptors for pathogen-associated molecular patterns (PAMPs)	Antigen receptors produced by gene recombinations
Specificity	About 1,000 PAMPs	Billions of antigens
Persistence	None	Long-term

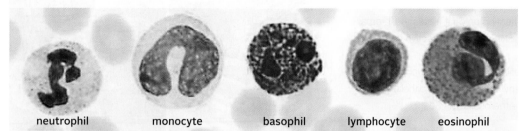

Figure 37.3 Animated Lineup of leukocytes (white blood cells). Staining shows structural details such as lobed nuclei and cytoplasmic granules that contain enzymes, toxins, and signaling molecules.

neutrophil   monocyte   basophil   lymphocyte   eosinophil

The Defenders

White blood cells participate in all immune responses. Many kinds circulate through the body in blood and lymph; a few are shown in **Figure 37.3**. Others populate the lymph nodes, spleen, and other tissues. Some white blood cells are phagocytic, and all are secretory. The secretions include **cytokines**, which are polypeptides and proteins used by cells of the immune system to communicate with one another. Intercellular communication allows white blood cells to coordinate their activities during immune responses. Molecules called interleukins, interferons, and tumor necrosis factors are examples of vertebrate cytokines.

Different types of white blood cells are specialized for specific tasks. **Neutrophils** are the most abundant of the circulating phagocytic cells. Phagocytic **macrophages** that patrol tissue fluids develop from monocytes, which patrol the blood. **Dendritic cells** are phagocytes that alert the adaptive immune system to the presence of antigen in solid tissues.

Some white blood cells have granules, which are secretory vesicles that contain cytokines, destructive enzymes, toxins such as hydrogen peroxide, and local signaling molecules. In response to a trigger such as antigen binding, these cells degranulate (release the contents of their granules) by the process of exocytosis. Neutrophils have granules, as do **eosinophils** that target parasites too big for phagocytosis. **Basophils** and **mast cells** degranulate in response to injury or antigen. Unlike most other leukocytes, mast cells do not wander; they stay anchored in tissues. Mast cells are often closely associated with nerves, and they also degranulate in response to somatostatin and other neuromodulators active in the endocrine and nervous systems.

Lymphocytes are a special category of white blood cells that are central to adaptive immunity. **B cells** (B lymphocytes) and **T cells** (T lymphocytes) have the collective capacity to recognize billions of specific antigens. There are several kinds of T cells, including **cytotoxic T cells** that can kill infected or cancerous body cells (**Figure 37.4**). A different type of lymphocyte called an **NK cell** (natural killer cell) kills cancerous body cells undetectable by cytotoxic T cells.

adaptive immunity In vertebrates, set of immune defenses that can be tailored to specific pathogens encountered by an organism during its lifetime.
antigen A molecule or particle that the immune system recognizes as nonself. Triggers an immune response.
B cell B lymphocyte. Lymphocyte that can make antibodies.
basophil Circulating white blood cell with a role in inflammation.
complement A set of proteins that circulate in inactive form in blood, and when activated play a role in immune responses.
cytokines Signaling molecules secreted by vertebrate white blood cells.
cytotoxic T cell Lymphocyte that kills infected or cancerous cells.
dendritic cell Phagocytic white blood cell that patrols solid tissues and alerts the immune system to the presence of antigen.
eosinophil White blood cell that targets multicelled parasites.
immunity The body's ability to resist and fight infections.
innate immunity Set of inborn, general defenses against infection.
macrophage Phagocytic white blood cell that patrols tissue fluids.
mast cell White blood cell that is anchored in many tissues; factor in inflammation.
NK cell Natural killer cell. Lymphocyte that can kill cancer cells undetectable by cytotoxic T cells.
neutrophil Circulating phagocytic white blood cell.
T cell T lymphocyte. Lymphocyte central to adaptive immunity; some kinds target infected or cancerous body cells.

Figure 37.4
Cytotoxic T cell killing a cancer cell.

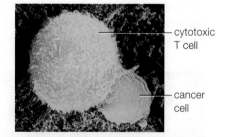

cytotoxic T cell

cancer cell

Take-Home Message

What is immunity?

» The innate immune system is a set of general defenses against a fixed number of antigens.

» Vertebrate adaptive immunity is a system of defenses that can specifically target billions of different antigens.

» White blood cells are central to both systems; signaling molecules such as cytokines integrate their activities.

37.3 Surface Barriers

- A pathogen can cause infection only if it enters the internal environment by penetrating skin or other protective barriers at the body's surfaces.
- Links to Bacteria and biofilms 4.4, Tight junctions 4.11, Fermentation 7.6, Internal environment 31.2, Epithelium 31.3, Hair follicles and skin 31.8, Atherosclerosis 36.12

Your skin is in constant contact with the external environment, so it picks up many microorganisms. It normally teems with about 200 different kinds of yeast, protozoa, and bacteria. They tend to flourish in warm, moist areas, such as between the toes. Huge populations inhabit cavities and tubes that open out on the body's surface, including the eyes, nose, mouth, and anal and genital openings.

Microorganisms that typically live on human surfaces, including the interior tubes and cavities of the digestive and respiratory tracts, are called **normal flora** (**Figure 37.5**). Our surfaces provide them with a stable environment and nutrients. In return, their populations deter more dangerous species from colonizing (and penetrating) body surfaces. Normal flora in the digestive tract help us digest food and make essential nutrients such as vitamins K and B_{12}.

Normal flora are helpful only on body surfaces; they can cause or worsen many conditions when they invade tissues. Consider a major constituent of normal flora, *Propionibacterium acnes* (**Figure 37.5B**), a bacterium that feeds on sebum. Sebum is a greasy mixture of fats, waxes, and glycerides that lubricates hair and skin. Glands in the skin secrete sebum into hair follicles. During puberty, higher levels of steroid hormones trigger an increase in sebum production. Excess sebum combines with dead, shed skin cells and blocks the openings of hair follicles. *P. acnes* can survive on the surface of the skin, but far prefers anaerobic habitats such as the interior of blocked hair follicles. There,

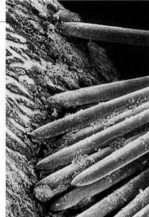
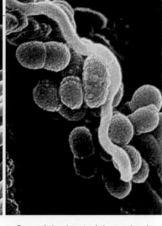

A Micrograph of toothbrush bristles scrubbing plaque on a tooth surface.

B One of the bacterial species in plaque, *Streptococcus mutans*, is a major contributor to tooth decay and periodontitis.

Figure 37.6 Dental plaque.

they multiply to tremendous numbers. Secretions of the flourishing *P. acnes* populations leak into internal tissues of the follicles and initiate inflammation. The resulting pustules are called acne.

A few of the 400 or so species of normal flora in the mouth cause dental **plaque**, a thick biofilm of various bacteria and occasional archaea, their extracellular products, and saliva glycoproteins. Plaque sticks tenaciously to teeth (**Figure 37.6**). Some of the bacteria in plaque are fermenters. The lactic acid they produce dissolves minerals that make up the tooth, resulting in holes called cavities.

In young, healthy people, tight junctions normally seal gum epithelium to teeth. The tight seal prevents oral microorganisms from entering gum tissue. As we age, the connective tissue beneath the epithelium thins, so the seal between gums and teeth weakens. Deep pockets form, and a nasty collection of anaerobic bacteria and archaea tends to accumulate in them. The microorganisms secrete destructive enzymes and acids

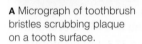

A *Staphylococcus epidermidis*, a common colonizer of human skin.

B *Propionibacterium acnes*, the bacterial cause of acne.

C *Staphylococcus aureus* cells (yellow) stuck in mucus secreted by human nasal epithelial cells.

Figure 37.5 Examples of normal flora.

Table 37.2	Examples of Surface Barriers
Physical	Epithelia that line tubes and cavities such as the gut and eye sockets; intact skin; established populations of normal flora
Mechanical	Mucus; broomlike action of cilia; flushing action of tears, saliva, urination, diarrhea
Chemical	Secretions (sebum, other waxy coatings); low pH of urine, gastric juices, urinary and vaginal tracts; lysozyme

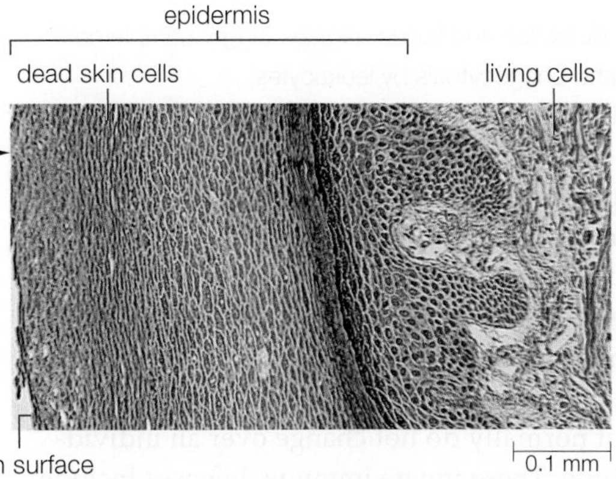

Figure 37.7
One surface barrier to infection: the epidermis of human skin. Its thick, waterproof layer of dead cells keeps normal skin flora from penetrating internal tissues.

that cause inflammation of the surrounding gum tissues, a condition called periodontitis. *Porphyromonas gingivalis* is one of those anaerobic species. As with all species of oral bacteria associated with periodontitis, *P. gingivalis* is also found in atherosclerotic plaque. Periodontal wounds are an open door to the circulatory system and its arteries. Atherosclerosis is now known to be a disease of inflammation. What role oral microorganisms play in atherosclerosis is not yet clear, but one thing is certain: they contribute to the inflammation that fuels coronary artery disease.

Other serious illnesses associated with normal flora include pneumonia; ulcers; colitis; whooping cough; meningitis; abscesses of the lung and brain; and cancers of the colon, stomach, and intestine. The bacterial agent of tetanus, *Clostridium tetani*, passes through our intestines so often that it is considered a normal inhabitant. The bacteria responsible for diphtheria, *Corynebacterium diphtheriae*, was normal skin flora before widespread use of the vaccine eradicated the disease. *Staphylococcus aureus*, a resident of human skin and linings of the mouth, nose, throat, and intestines (**Figure 37.5C**), is also a leading cause of human bacterial disease. Antibiotic-resistant strains of *S. aureus* are now widespread. A particularly dangerous kind, MRSA (methicillin-resistant *S. aureus*), is resistant to a wide range of antibiotics. MRSA is now a permanent resident of most hospitals around the world.

Barriers to Infection

In contrast to body surfaces, blood and tissue fluids of healthy people are typically sterile (free of microorganisms). Surface barriers (**Table 37.2**) usually prevent normal flora from entering the body's internal environment. Epidermis, the tough outer layer of vertebrate skin, is one example of a surface barrier (**Figure 37.7**). Microorganisms flourish on skin's waterproof, oily surface, but they rarely penetrate the thick epidermis.

The thinner epithelial tissues that line the body's interior tubes and cavities also have surface barriers. Sticky mucus secreted by cells of these linings traps microorganisms (**Figure 37.5C**). The mucus contains **lysozyme**, an enzyme that kills bacteria. In the sinuses and respiratory tract, the coordinated beating of cilia sweeps trapped microorganisms away before they have a chance to breach the delicate walls of these structures.

Microorganisms that normally inhabit the mouth resist lysozyme in saliva. Those swallowed are typically killed by gastric fluid, a potent brew of protein-digesting enzymes and acid in the stomach. Those that survive to reach the small intestine are usually killed by bile salts. The hardy ones that reach the large intestine must compete with about 500 resident species that are specialized to live there and have already established large populations. Any that displace normal flora are typically flushed out by diarrhea.

Lactic acid produced by *Lactobacillus* helps keep the vaginal pH outside the range of tolerance of most fungi and other bacteria. Urination's flushing action usually stops pathogens from colonizing the urinary tract.

lysozyme Antibacterial enzyme that occurs in body secretions such as mucus.
normal flora Microorganisms that typically live on human surfaces, including the interior tubes and cavities of the digestive and respiratory tracts.
plaque On teeth, a thick biofilm composed of bacteria, their extracellular products, and saliva proteins.

Take-Home Message

What prevents ever-present microorganisms from entering the body's internal environment?

» Surface barriers keep microorganisms that contact or inhabit vertebrate surfaces from invading the internal environment.

» Skin's tough epidermis is one barrier. Mucus, lysozyme, and often the sweeping action of cilia protect the soft linings of internal tubes and cavities.

» Resident populations of normal flora deter more dangerous microorganisms from colonizing the body's internal and external surfaces.

■ Antigen detection and tissue damage trigger complement activation and phagocytosis by leukocytes.

■ Link to Lysis 20.3

What happens if a pathogen slips by surface defenses and enters the body's internal environment? All animals are normally born with a set of fast-acting, general immune defenses that can keep an invading pathogen from establishing a population in the body's internal environment. All are general defense mechanisms that normally do not change over an individual's lifetime. These innate immune defenses include complement and phagocytic white blood cells.

Complement

Vertebrates make about 30 different kinds of complement proteins, most of which circulate in inactive form through blood and interstitial fluid. One type of complement protein is shown at *left*. Complement was named because the proteins were originally identified by their ability to "complement" the action of antibodies in adaptive immune responses (we return to these responses in Sections 37.6 and 37.8).

One complement protein can recognize and bind to antibodies clustered on the surface of a cell. The binding initiates a series of reactions that ultimately results in the enzymatic cleavage of another complement protein called C3. C3 is extremely abundant, so its cleavage products accumulate very quickly. Some of the fragments attach directly to cell membranes, and they coat an invading pathogen as well as any host cell in the vicinity. C3 fragments become enzymatic when they are bound to a membrane, and these activate other complement proteins, which activate other complement proteins, and so on (**Figure 37.8A**). The cascading reactions quickly produce huge concentrations of activated complement at the membrane's surface.

Some of the activated complement proteins bind to the surface of pathogens, forming a coating that enhances uptake by phagocytic leukocytes. Others assemble into complexes that insert themselves into the lipid bilayer (**Figure 37.8B,C**). The complexes form large transmembrane channels that make the membrane leaky and trigger cell lysis.

Activated C3 fragments also diffuse into surrounding tissues, forming a concentration gradient around the site of infection. The gradient recruits phagocytic leuokocytes to the site of infection.

Years after complement was named, researchers discovered that some complement proteins can target extracellular pathogens independently of antibody-mediated responses. For example, one complement protein functions as a receptor for pathogen-associated molecular patterns. This protein has six binding sites for simple carbohydrates found on the surface

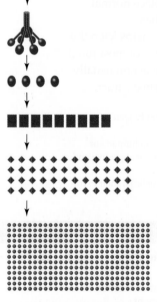

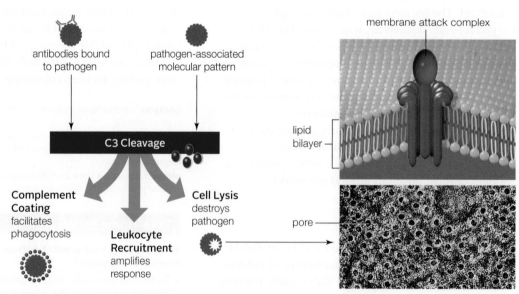

A Activation of one complement protein initiates cascading reactions that activate many other complement proteins.

B Complement cascades activated by antibodies bound to a cell or by PAMPs trigger C3 cleavage. C3 fragments coat pathogens, recruit leukocytes, and activate other complement proteins that assemble as membrane attack complexes.

C Membrane attack complexes insert themselves into lipid bilayers, forming pores in the plasma membrane that bring about lysis of the cell.

Figure 37.8 Animated Complement.

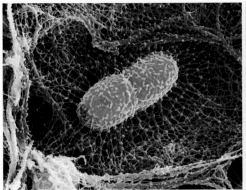

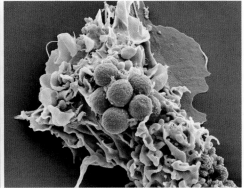

A These *Klebsiella* bacteria in lung tissue have been ensnared by a neutrophil net.

B Macrophage caught in the process of engulfing tuberculosis bacteria.

C A dendritic cell preparing to engulf spores of a fungus, *Aspergillus fumigatus*.

Figure 37.9 Phagocytic leukocytes.

of a variety of pathogens, and it can bind directly to *Candida albicans* (a yeast); HIV and influenza virus; *Salmonella* and *Streptococcus* bacteria; and *Leishmania*, a protozoan parasite. When all six of the protein's binding sites have become bound to carbohydrates on the surface of one of these pathogens, the protein activates an enzyme that cleaves C3, thus initiating the complement cascade. A complement cascade is also triggered by cytoplasmic and mitochondrial proteins that leak out of damaged body cells.

Normal body cells continuously produce proteins that inactivate complement, preventing a complement activation cascade from spreading too far into healthy tissue. Microorganisms do not make these inhibitory proteins, so they are singled out for destruction.

Phagocytic Leukocytes

Neutrophils, macrophages, and dendritic cells are mobile phagocytes. All can follow a chemical trail, a type of movement called **chemotaxis**. For example, these cells bear receptors for C3 fragments, and they follow gradients of this activated complement back to an affected tissue, where they engulf cells and particles coated with complement.

Neutrophils Neutrophils are the most abundant white blood cell, but they are also short-lived. Their turnover rate is phenomenal: The bone marrow of an adult human normally produces about 50 billion new neutrophils every day. This abundance is one reason that neutrophils are among the first responders to injury or infection, collecting within minutes at a site of tissue damage. After a neutrophil engulfs a microorganism, it releases the contents of its granules both into the endocytic vesicle and to the exterior of the cell. Enzymes and toxins released into extracellular fluid destroy all cells in the vicinity, sick and healthy alike. Recently, researchers discovered that neutrophils literally explode in response to a certain combination of signaling molecules and complement, in the process ejecting their nuclear DNA and associated proteins along with the contents of their granules. The mixture solidifies into an extracellular matrix that traps pathogens in the vicinity of the secreted antimicrobial compounds (**Figure 37.9A**). The net is very effective at killing invasive bacteria.

Macrophages and Dendritic Cells Macrophages stationed mainly in interstitial fluid engulf and digest essentially everything except undamaged body cells (**Figure 37.9B**). However, macrophages are more than just scavengers. They also secrete interleukins and other cytokines that alert the immune system to the presence of invading pathogens. A large number of different receptors mediate the diverse responses of macrophages.

Dendritic cells (**Figure 37.9C**) patrol tissues that contact the external environment, such as the lining of respiratory airways. Phagocytosis by dendritic cells plays a critical role in protecting the lungs from pathogens and other harmful particles. However, the main function of dendritic cells is to present antigen to T cells (more on how this works in Section 37.7).

chemotaxis Cellular movement toward or away from a chemical stimulus.

Take-Home Message

What happens after antigen is detected inside the body?

» Antigen (as well as tissue damage) triggers complement activation.

» Activated complement recruits phagocytic leukocytes, coats cells in an affected area, and destroys cells directly.

» Phagocytic leukocytes release antimicrobial compounds and cytokines upon engulfing an antigenic particle.

37.5 Inflammation and Fever

- Complement activation, antigen detection, or tissue damage can trigger inflammation and fever.
- Links to Osmosis 5.8, Neuromodulators 32.6, Local signaling molecules 33.3, Blood 36.4, Capillary function 36.10

Complement activation and cytokines released by phagocytic leukocytes typically trigger inflammation and fever, which are hallmark processes of an innate immune response.

Figure 37.10 Animated Example of inflammation as a response to bacterial infection.

❶ Pattern receptors on mast cells in the tissue recognize and bind to bacterial antigen. The mast cells release signaling molecules (*blue* dots) that cause arterioles to widen. The resulting increase in blood flow reddens and warms the tissue.

❷ The signaling molecules also increase capillary permeability, which allows phagocytes to squeeze through the vessel walls into the tissue. Plasma proteins leak out of the capillaries, and the tissue swells with fluid.

❸ Bacterial antigens activate complement (*purple* dots). Activated complement binds to the bacteria.

❹ Phagocytes in the tissue recognize and engulf the complement-coated bacteria.

Inflammation

Tissue damage or infection trigger **inflammation**, a fast, local response that simultaneously destroys affected tissues and jump-starts the healing process (**Figure 37.10**). Inflammation begins when basophils, mast cells, or neutrophils degranulate, releasing the contents of their granules into an affected tissue. Degranulation can occur in response to a number of stimuli, including pattern receptor binding to antigen, for example on the surface of a bacterium ❶. Fragments of C3 complement trigger degranulation when they bind to receptors on leukocyte plasma membranes. Mast cells (**Figure 37.11A**) also degranulate in response to neuro-modulators released at a site of tissue damage.

Among the substances released by degranulating leukocytes are prostaglandins and histamines. These local signaling molecules have two effects. First, they cause nearby arterioles to widen. As a result, blood flow to the area increases. The increased flow speeds the arrival of more phagocytes, which are attracted to cytokines. The phagocytes migrate out of the blood vessel and into interstitial fluid by moving between

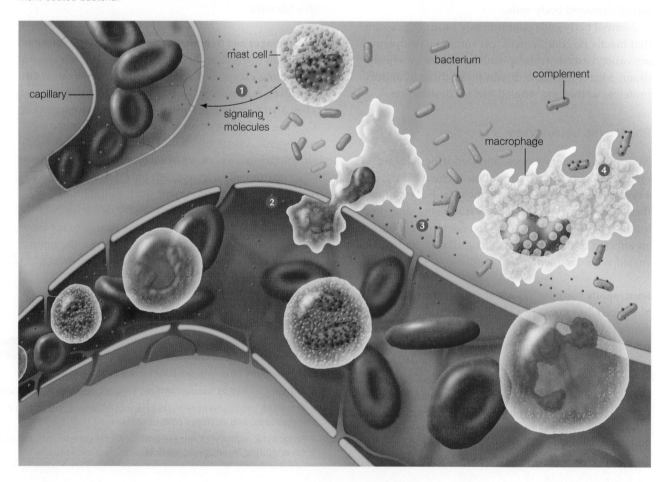

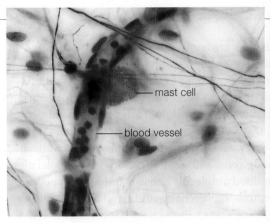

A Mast cells anchored in tissues near blood vessels, nerves, and mucous membranes that border external surfaces degranulate in response to antigen, activated complement, or to signals from the nervous system.

B Lymphocytes (*red*) migrating through membranes and cytoplasm of endothelial cells (*green*). The ability to move directly through other cells allows lymphocytes to exit blood vessels quickly.

C Flushed cheeks are often an outward indication of fever. A raised internal temperature enhances immune defenses and inhibits growth of many pathogens.

Figure 37.11 Some mechanisms of innate immunity.

endothelial cells that make up blood vessel walls ❷ as well as directly through them (**Figure 37.11B**). By the time this occurs, any invading cells have become coated with activated complement ❸, which makes them easy targets for the phagocytes ❹.

Symptoms of inflammation include redness and warmth that are outward indications of the area's increased blood flow. Swelling and pain occur because prostaglandins and histamines widen the spaces between cells in capillary walls, so they make capillaries in an affected tissue leakier to plasma proteins. Plasma proteins that escape from the capillaries make interstitial fluid hypertonic with respect to blood. Water follows by osmosis, and the tissue swells with excess fluid. The swelling causes pain when it puts pressure on nerves.

Inflammation continues as long as its triggers do. Once these stimuli subside, for example after invading bacteria have been cleared from an infected tissue, macrophages begin to produce compounds that suppress inflammation and promote tissue repair. If the stimulus persists, inflammation becomes chronic. Chronic inflammation is not a normal condition. It does not benefit the body, but causes or contributes to many diseases, including asthma, Crohn's disease, rheumatoid arthritis, atherosclerosis, diabetes, and cancer.

Fever

Fever is a temporary rise in body temperature above the normal 37°C (98.6°F) that often occurs in response to infection or serious injury. Some cytokines stimulate brain cells to make and release prostaglandins, which act on the hypothalamus to raise the body's internal temperature set point. As long as the temperature of the body is below the new set point, the hypothalamus sends out signals that cause blood vessels in the skin to constrict, which reduces heat loss from the skin. The signals also trigger an increase in the rate of heartbeat and respiration, as well as reflexive movements called shivering, or "chills," that increase the metabolic heat output of muscles. These responses all raise the body's internal temperature. If the internal temperature rises too much, sweating and flushing (**Figure 37.11C**) quickly lower core temperature to maintain the new set point.

Fever enhances immune defenses by increasing the rate of enzyme activity, thus speeding up metabolism, tissue repair, and formation and activity of phagocytes. In addition, many pathogens multiply more slowly at the higher temperature, so white blood cells can get a head start in the proliferation race against them.

A fever is a sign that the body is fighting something, so it should never be ignored. However, a fever of 40.6°C (105°F) or less does not necessarily require treatment in an otherwise healthy adult. Body temperature usually will not rise above that value, but if it does, immediate hospitalization is recommended. Brain damage or death can occur if core temperature reaches 42°C (107.6°F).

fever An internally induced rise in core body temperature above the normal set point as a response to infection or injury.
inflammation A local response to tissue damage or infection; characterized by redness, warmth, swelling, and pain.

Take-Home Message

How do inflammation and fever function in innate immunity?

» Inflammation occurs when granular white blood cells release local signaling molecules that increase blood flow and attract phagocytes to a site of tissue damage or infection.

» Fever enhances immune defenses while slowing pathogen growth.

37.6 Antigen Receptors

■ Antigen receptors give lymphocytes the potential to recognize billions of different antigens.

■ Links to Protein structure 3.6, Membrane proteins 5.7, Alternative splicing 9.3, Exocrine glands 31.3

If innate immune mechanisms do not quickly rid the body of an invading pathogen, an infection may become established in internal tissues. By that time, long-lasting mechanisms of adaptive immunity have already begun to target the invaders specifically.

Innate immune responses are triggered by leukocytes that detect antigen via antigen receptors. Plasma membrane proteins that recognize PAMPs are one type of antigen receptor. Your T cells bear another type: special antigen receptors called **T cell receptors**, or TCRs. Part of a TCR recognizes antigen as nonself. Another part recognizes certain proteins in the plasma membrane of body cells as self. In humans, these self-proteins are called **MHC markers** (after the genes that encode them), but all vertebrates have them.

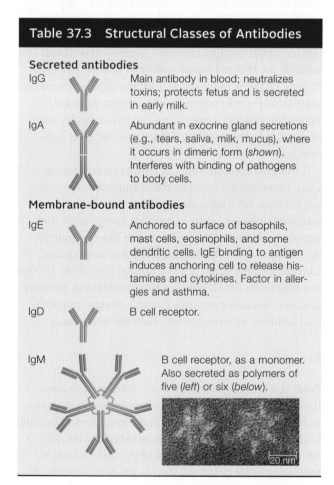

MHC marker

Antibodies are another type of antigen receptor. **Antibodies** are Y-shaped proteins made only by B cells. Each antibody can bind to an antigen. Many antibodies circulate in blood and enter interstitial fluid during inflammation, but they do not kill pathogens directly. Instead, they activate complement and facilitate phagocytosis. They can also prevent pathogens from attaching to body cells, and neutralize some toxic molecules.

An antibody molecule consists of four polypeptides: two identical "light" chains and two identical "heavy" chains (**Figure 37.12A**). Each chain has a variable and a constant region. When the chains fold up together as an intact antibody, the variable regions form two antigen-binding sites that have a specific distribution of bumps, grooves, and charge. These binding sites are the antigen receptor part of an antibody: They bind only to antigen with a complementary distribution of bumps, grooves, and charge (**Figure 37.12B**).

In addition to antigen-binding sites, each antibody also has a constant region that determines its structural identity, or class. There are five antibody classes—IgG, IgA, IgE, IgM, and IgD (Ig stands for

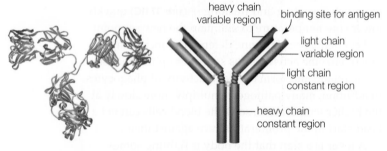

heavy chain variable region
binding site for antigen
light chain variable region
light chain constant region
heavy chain constant region

A An antibody molecule consists of four polypeptide chains, each with a variable and a constant region, joined in a Y-shaped configuration. The variable regions fold up as antigen-binding sites.

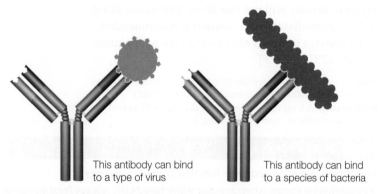

This antibody can bind to a type of virus

This antibody can bind to a species of bacteria

B The antigen-binding sites of each antibody are unique. They only bind to an antigen that has complementary bumps, grooves, and charge distribution.

Figure 37.12 Antibody structure.

Table 37.3	Structural Classes of Antibodies	
Secreted antibodies		
IgG		Main antibody in blood; neutralizes toxins; protects fetus and is secreted in early milk.
IgA		Abundant in exocrine gland secretions (e.g., tears, saliva, milk, mucus), where it occurs in dimeric form (*shown*). Interferes with binding of pathogens to body cells.
Membrane-bound antibodies		
IgE		Anchored to surface of basophils, mast cells, eosinophils, and some dendritic cells. IgE binding to antigen induces anchoring cell to release histamines and cytokines. Factor in allergies and asthma.
IgD		B cell receptor.
IgM		B cell receptor, as a monomer. Also secreted as polymers of five (*left*) or six (*below*).

20 nm

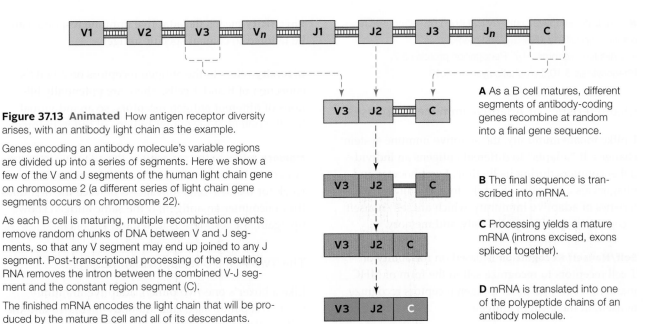

Figure 37.13 Animated How antigen receptor diversity arises, with an antibody light chain as the example.

Genes encoding an antibody molecule's variable regions are divided up into a series of segments. Here we show a few of the V and J segments of the human light chain gene on chromosome 2 (a different series of light chain gene segments occurs on chromosome 22).

As each B cell is maturing, multiple recombination events remove random chunks of DNA between V and J segments, so that any V segment may end up joined to any J segment. Post-transcriptional processing of the resulting RNA removes the intron between the combined V-J segment and the constant region segment (C).

The finished mRNA encodes the light chain that will be produced by the mature B cell and all of its descendants.

A As a B cell matures, different segments of antibody-coding genes recombine at random into a final gene sequence.

B The final sequence is transcribed into mRNA.

C Processing yields a mature mRNA (introns excised, exons spliced together).

D mRNA is translated into one of the polypeptide chains of an antibody molecule.

immunoglobulin, another name for antibody). The different classes serve different functions (**Table 37.3**).

Most of the antibodies circulating in the bloodstream and tissue fluids are IgG, which binds pathogens, neutralizes toxins, and activates complement. IgG is the only antibody that can cross the placenta to protect a fetus before its own immune system is active.

IgA is the main antibody in mucus and other exocrine gland secretions (Section 31.3). IgA is secreted as a dimer (two antibodies bound together), which makes the molecule stable enough to patrol harsh environments such as the interior of the digestive tract. There, IgA encounters pathogens before they contact body cells. Bound to antigen, IgA interacts with mast cells, basophils, macrophages, and NK cells to initiate inflammation.

IgE made and secreted by B cells gets incorporated into the plasma membrane of mast cells, basophils, and some types of dendritic cells. Binding of antigen to membrane-bound IgE triggers the anchoring cell to release the contents of its granules.

B cell receptors are IgM or IgD antibodies bound to B cell membranes. IgM is also secreted as polymers of five or six antibodies. These polymers are particularly efficient at binding antigen and activating complement.

Antigen Receptor Diversity

Humans can make billions of unique antigen receptors. This diversity arises because the genes that encode the receptors do not occur in a continuous

stretch on one chromosome; instead, they occur in several segments on different chromosomes, and there are several different versions of each segment (**Figure 37.13**). The segments are spliced together during B and T cell differentiation, but which version of each segment gets spliced into the antigen receptor gene of a particular cell is random. As a B or T cell differentiates, it ends up with one out of about 2.5 billion different combinations of gene segments.

Before a new B cell leaves bone marrow, it is already making antigen receptors. The constant region of each receptor is embedded in the lipid bilayer of the cell's plasma membrane, and the two arms project into the extracellular environment. In time the B cell bristles with more than 100,000 antigen receptors. T cells also form in bone marrow, but they mature only after they take a tour in the thymus gland (Section 34.12). There, they encounter hormones that stimulate them to make T cell receptors.

antibody Y-shaped antigen receptor protein made only by B cells.
B cell receptor Membrane-bound antibody on a B cell.
MHC markers Self-proteins on the surface of human body cells.
T cell receptor (TCR) Antigen receptor on the surface of a T cell.

Take-Home Message

What are antigen receptors?

» The adaptive immune system has the potential to recognize about 2.5 billion different antigens via receptors on B cells and T cells.

» Antibodies are secreted or membrane-bound antigen receptors. They are made only by B cells.

37.7 Overview of Adaptive Immunity

■ Vertebrate adaptive immunity is defined by self/nonself recognition, specificity, diversity, and memory.

■ Links to Lysosomes 4.7, Recognition proteins 5.7, Phagocytosis 5.10, Lymphatic system 36.13

Characteristics of Adaptive Immune Responses

Unlike innate immunity, the adaptive immune system changes: It "adapts" to different antigens an individual encounters during its lifetime. Lymphocytes and phagocytes interact to effect the four defining characteristics of adaptive immunity, which are self/nonself recognition, specificity, diversity, and memory.

Self/Nonself Recognition is based on the ability of T cell receptors to recognize self in the form of MHC markers. TCRs and other antigen receptors recognize nonself, in the form of antigen.

Figure 37.14 Antigen processing: What happens when a B cell, macrophage, or dendritic cell engulfs an antigenic particle—in this case, a bacterium.

❶ A phagocytic cell engulfs a bacterium.

❷ An endocytic vesicle forms around the bacterium.

❸ The vesicle fuses with a lysosome, which contains enzymes and MHC molecules.

❹ Lysosomal enzymes digest the bacterium to molecular bits.

❺ Inside the vesicle, bits of bacterium bind to MHC molecules.

❻ The vesicle fuses with the cell membrane by exocytosis. When it does, the bacterial antigen–MHC complex becomes part of the plasma membrane.

❼ The antigen–MHC complex displayed on the surface of the phagocyte is a signal that provokes an adaptive immune response targeting the antigen.

Figure It Out: From what organelles do the lysosomes bud?

Answer: From Golgi bodies

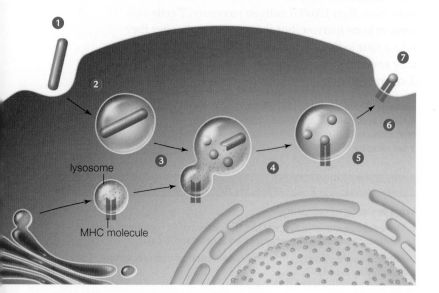

lysosome

MHC molecule

Specificity means the adaptive immune response can be tailored to combat specific antigens.

Diversity refers to the antigen receptors on a body's collection of B and T cells. There are potentially billions of different antigen receptors, so an individual has the potential to counter billions of different threats.

Memory refers to the capacity of the adaptive immune system to "remember" an antigen. It takes about a week for B and T cells to respond in force the first time they encounter an antigen. If the same antigen shows up again, the response is faster and stronger.

The Two Arms of Adaptive Immunity

Like a boxer's one-two punch, adaptive immunity has two separate arms: the antibody-mediated and the cell-mediated immune responses. These two responses work together to eliminate diverse threats.

Why two arms? Not all threats present themselves in the same way. For example, bacteria, fungi, or toxins can circulate in blood or interstitial fluid. These threats are intercepted quickly by B cells and other phagocytes that interact in an **antibody-mediated immune response**. In this response, B cells produce antibodies targeting a specific invader. However, an antibody-mediated immune response is not the most effective way of countering other threats. For example, viruses, bacteria, fungi, and protists that reproduce inside body cells are vulnerable to an antibody-mediated response only when they slip out of one cell to infect others. Such intracellular pathogens are targeted primarily by the **cell-mediated immune response**. In this response, cytotoxic T cells and NK cells detect and destroy infected body cells, or those that have been altered by cancer.

Antigen Processing

Recognition of a specific antigen by T cell receptors is the first step of both antibody-mediated and cell-mediated immune responses. However, T cell receptors can only recognize and bind to antigen that is presented by an antigen-presenting cell. Macrophages, B cells, and dendritic cells do the presenting (**Figure 37.14**). First, one of these cells engulfs something bearing antigen ❶. A vesicle that contains the antigen-bearing particle forms in the cell's cytoplasm ❷ and fuses with a lysosome ❸. Lysosomal enzymes then digest the particle into molecular bits ❹. The lysosomes also contain MHC markers that bind to some of the antigen bits ❺. The resulting antigen–MHC complexes become displayed at the cell's surface when the vesicles fuse with (and become part of) the plasma membrane ❻. The display of MHC markers paired with antigen fragments serves as a call to arms ❼.

Intercepting and Clearing Out Antigen

After engulfing an antigen-bearing particle, a dendritic cell or macrophage migrates to a lymph node, where it presents antigen to T cells. Every day, about 25 billion T cells filter through each node. As you will see shortly, T cells that recognize and bind to antigen presented by a phagocyte initiate an adaptive response.

Antigen-bearing particles in interstitial fluid flow through lymph vessels to a lymph node, where they meet up with resident B cells, dendritic cells, and macrophages. These phagocytes engulf, process, and present antigen to T cells passing through the node. During an infection, the lymph nodes swell because T cells accumulate inside them. When you are ill, you may notice your swollen lymph nodes as tender lumps under the jaw or elsewhere in your body.

A new B or T cell is "naive," which means that no antigen has bound to its receptors yet. In a typical adaptive immune response, a naive T cell recognizes and binds to an antigen–MHC complex displayed on the surface of an antigen-presenting white blood cell. The T cell starts secreting cytokines, which signal all other B or T cells with the same antigen receptor to divide again and again. Huge populations of B and T cells form after a few days; all of the cells recognize the same antigen. Most are **effector cells**, differentiated lymphocytes that act at once. Some are **memory cells**, long-lived B and T cells reserved for future encounters with the antigen. Memory cells can persist for decades after the initial infection ends. If the same antigen enters the body at a later time, these memory cells will initiate a secondary response (**Figure 37.15**). In a secondary immune response, larger populations of effector cells form much more quickly than they did in the primary response.

The tide of battle turns when the effector cells have destroyed most of the antigen-bearing agents.

antibody-mediated immune response Immune response in which antibodies are produced in response to an antigen.
cell-mediated immune response Immune response involving cytotoxic T cells and NK cells that destroy infected or cancerous body cells.
effector cell Antigen-sensitized B cell or T cell that forms in an immune response and acts immediately.
memory cell Long-lived, antigen-sensitized B or T cell that can act in a secondary immune response.

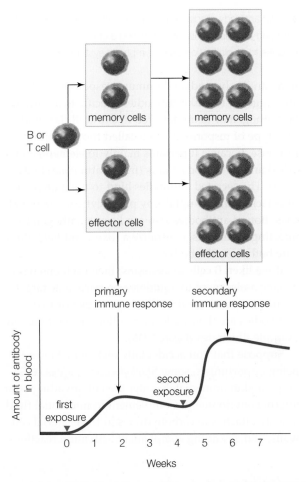

Figure 37.15 Animated Primary and secondary immune responses.

A first exposure to an antigen causes a primary immune response in which effector cells fight the infection.

Memory cells also form in a primary response but are set aside, sometimes for decades. If the antigen returns at a later time, the memory cells initiate a faster, stronger secondary response.

With less antigen present, fewer immune fighters are recruited. Antibody–antigen complexes form large clumps that can be quickly cleared from the blood by the liver and spleen. Complement proteins assist in the cleanup. Immune responses subside after the antigen-bearing particles have been cleared from the body.

Take-Home Message

What is the adaptive immune system?

» Phagocytes and lymphocytes interact to bring about vertebrate adaptive immunity, which has four defining characteristics: self/nonself recognition, specificity, diversity, and memory.

» The two arms of adaptive immunity work together. Antibody-mediated responses target antigen in blood or interstitial fluid; cell-mediated responses target altered body cells.

37.8 The Antibody-Mediated Immune Response

■ In an antibody-mediated immune response, effector B cells form and produce antibodies targeting a specific antigen.
■ Links to Agglutination 36.6, Lymphatic system 36.13

In an antibody-mediated immune response, B cells are triggered to make antibodies specific to a particular pathogen or toxin detected in extracellular fluid. This type of response is often called the humoral response, because it pertains mainly to elements in the blood and other body fluids (from Latin *umor*, body fluid). The secreted antibodies bind to the pathogen, thus facilitating its uptake by phagocytic white blood cells. Bound antibodies also can keep a pathogen from infecting other cells, neutralize a toxin, and help eliminate both from the body.

If we liken B cells to assassins, then each one has a genetic assignment to liquidate one particular target: an antigen-bearing extracellular pathogen or toxin. Antibodies are their molecular bullets, as the following example illustrates (**Figure 37.16**).

Suppose that you accidentally nick your finger. Being opportunists, some *Staphylococcus aureus* cells on your skin immediately enter the cut, invading your internal environment. Complement in interstitial fluid quickly attaches to carbohydrates in the bacterial cell walls, and cascading complement activation reactions begin. Within an hour, complement-coated bacteria tumbling along in lymph vessels reach a lymph node in your elbow. There, they filter past an army of naive B cells.

One of the naive B cells residing in that lymph node makes antigen receptors that recognize a polysaccharide in *S. aureus* cell walls ❶. Via those receptors, the B cell binds to the polysaccharide on one of the bacteria. The complement coating stimulates the B cell to engulf the bacterium. The B cell is now activated.

Meanwhile, more *S. aureus* cells have been secreting metabolic products into interstitial fluid around your cut. The secretions are attracting phagocytic leukocytes. One of these phagocytes, a dendritic cell, engulfs several bacteria, then migrates to the lymph node in your elbow. By the time it gets there, it has digested the bacteria and is displaying their fragments as antigens bound to MHC markers on its surface ❷.

Every hour, about 500 different naive T cells travel through the lymph node, inspecting resident dendritic cells displaying antigen (**Figure 37.17**). Within a couple of hours, one of your T cells has recognized and bound to the *S. aureus* antigen on the dendritic cell ❸. This T cell is called a helper T cell because it helps other lymphocytes produce antibodies and kill pathogens. The helper T cell and the dendritic cell interact for about

Figure 37.16 Animated Example of an antibody-mediated immune response.

❶ The B cell receptors on a naive B cell bind to an antigen on the surface of a bacterium. The bacterium's complement coating triggers the B cell to engulf it. Fragments of the bacterium bound to MHC markers become displayed at the surface of the B cell.

❷ A dendritic cell engulfs the same kind of bacterium that the B cell encountered. Fragments of the bacterium bound to MHC markers become displayed at the surface of the dendritic cell.

❸ The antigen–MHC complexes on the dendritic cell are recognized by TCRs on a naive helper T cell. The two cells interact, and then the T cell begins to divide. Its descendants differentiate into effector helper T cells and memory helper T cells.

❹ TCRs on one of the effector helper T cells recognize and bind to the antigen–MHC complexes on the B cell. Binding makes the T cell secrete cytokines.

❺ The cytokines induce the B cell to divide again and again. Its many descendants differentiate into effector B cells and memory B cells.

❻ The effector B cells begin making and secreting huge numbers of antibodies, all of which recognize the same antigen as the original B cell receptor. The new antibodies circulate throughout the body and bind to the bacteria.

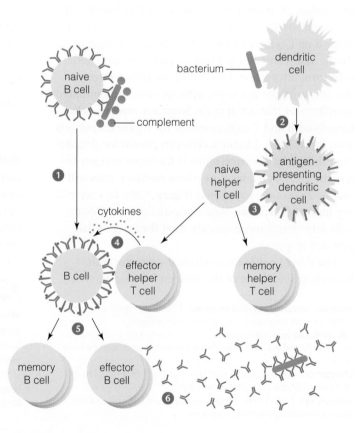

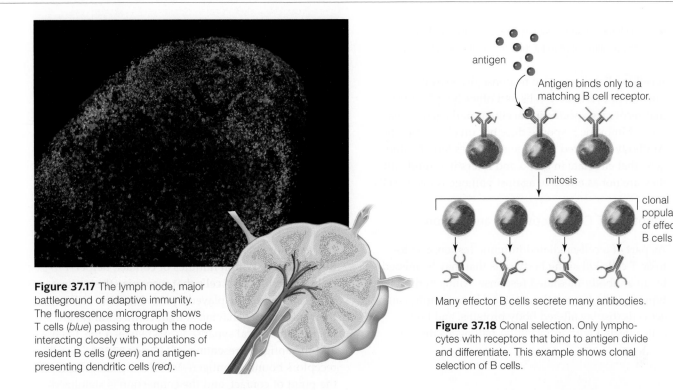

Figure 37.17 The lymph node, major battleground of adaptive immunity. The fluorescence micrograph shows T cells (*blue*) passing through the node interacting closely with populations of resident B cells (*green*) and antigen-presenting dendritic cells (*red*).

antigen

Antigen binds only to a matching B cell receptor.

mitosis

clonal population of effector B cells

Many effector B cells secrete many antibodies.

Figure 37.18 Clonal selection. Only lymphocytes with receptors that bind to antigen divide and differentiate. This example shows clonal selection of B cells.

24 hours. When the two cells disengage, the helper T cell returns to the circulatory system and begins to divide. A gigantic population of identical helper T cells forms. These clones differentiate into effector and memory cells, each of which has receptors that recognize the same *S. aureus* antigen. By the theory of clonal selection, the T cell was "selected" because its receptors bind to the *S. aureus* antigen. T cells with receptors that do not bind the antigen do not divide to form huge clonal populations (**Figure 37.18**).

Let's go back to that B cell in the lymph node. By now, it has digested the bacterium, and it is displaying bits of *S. aureus* bound to MHC molecules on its plasma membrane. The new helper T cells recognize the antigen–MHC complexes displayed by the B cell. One of these helper T cells binds to the B cell. Like long-lost friends, the two cells stay together for a while and communicate ❹. One of the messages that is communicated consists of cytokines secreted by the helper T cell. The cytokines stimulate the B cell to begin mitosis after the two cells disengage. The B cell divides again and again to form a huge population of genetically identical cells, all with receptors that can bind to the same *S. aureus* antigen ❺.

The B cell clones differentiate into effector and memory cells. The effector B cells start working immediately. Instead of making membrane-bound B cell receptors, they start making secreted antibodies ❻. The new antibodies recognize the same *S. aureus* anti-

gen as the original B cell receptor. Antibodies now circulate throughout the body and attach themselves to any *S. aureus* cells. An antibody coating prevents the bacteria from attaching to body cells and brings them to the attention of phagocytic cells for quick disposal. Antibodies also glue the foreign cells together into clumps, a process called agglutination (Section 36.6). The clumps are quickly removed from the circulatory system by the spleen.

The new antibodies also act in innate immune responses. Complement protein that binds to the constant region of two adjacent antibodies (such as two IgG molecules attached to the same bacterium) initiates a complement cascade. The resulting activated complement attracts phagocytic leukocytes, initiates inflammation, coats all cells in the vicinity of an affected tissue, and destroys cells directly.

Take-Home Message

What happens during an antibody-mediated immune response?

» T cells, B cells, and antigen-presenting cells carry out an antibody-mediated immune response.

» Effector B cells that form during an antibody-mediated immune response make and secrete antibodies that recognize and bind antigen-bearing particles in blood or tissue fluids. Antibody binding can neutralize a pathogen or toxin and facilitate its elimination from the body.

» Memory cells also form, and these are reserved for a potential future encounter with the antigen.

37.9 The Cell-Mediated Immune Response

■ In a cell-mediated immune response, cytotoxic T cells and NK cells are stimulated to kill infected or altered body cells.

A cell-mediated immune response involves the production of cytotoxic T cells and other lymphocytes that recognize specific intracellular pathogens. This type of immune response does not involve antibodies: Antibody-mediated immune responses target pathogens that circulate in blood and interstitial fluid, but they are not as effective against pathogens inside cells.

Cytotoxic T Cells: Activation and Action

As part of a cell-mediated immune response, cytotoxic T cells kill ailing body cells that may be missed by an antibody-mediated response. Ailing body cells typically display certain antigens. For example, cancer cells display altered body proteins, and body cells infected with intracellular pathogens display

polypeptides of the infecting agent. Both types of cell are detected and killed by cytotoxic T cells. Cytotoxic T cells also recognize foreign body cells (these T cells are responsible for rejection of transplanted organs).

A typical cell-mediated response starts in interstitial fluid during inflammation, when a dendritic cell recognizes, engulfs, and digests a sick body cell or the remains of one (Figure 37.19). The dendritic cell begins to display antigen that was part of the sick cell, and migrates to the spleen or a lymph node. There, the dendritic cell presents its antigen–MHC complexes to huge populations of naive helper T cells and naive cytotoxic T cells ❶. Some of the naive cells have T cell receptors that recognize the complexes on the dendritic cell. Within minutes of contact, helper T cells and cytotoxic T cells that recognize the antigen–MHC complexes displayed by the dendritic cell stop their migratory behavior. A tight connection called an immunological synapse forms between the T cell and the antigen-presenting cell (Figure 37.20A). T cell receptors bound to antigen–MHC complexes cluster at the point of contact, and the connection is stabilized by adhesion proteins that form in a ring around them (Figure 37.20B). The interaction activates the T cells.

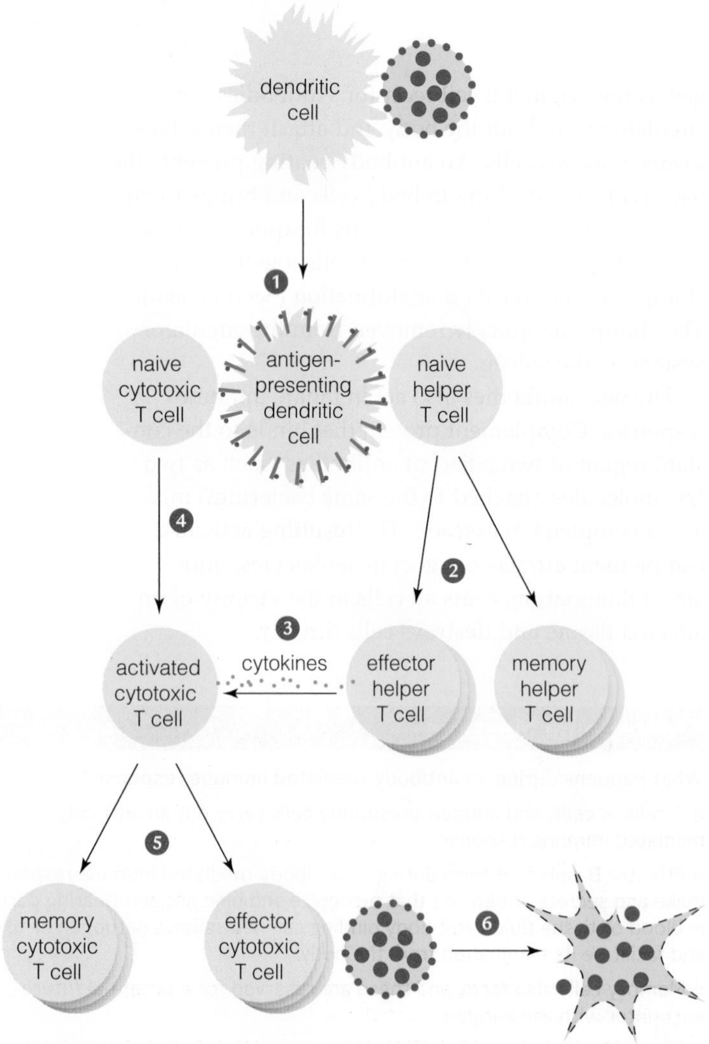

Figure 37.19 Animated An example of a cell-mediated immune response targeting virus-infected cells.

❶ A dendritic cell engulfs a virus-infected cell. Digested fragments of the virus bind to MHC markers, and the complexes become displayed at the dendritic cell's surface. The dendritic cell, now an antigen-presenting cell, migrates to a lymph node.

Receptors on a naive cytotoxic T cell bind to the antigen–MHC complexes on the surface of the dendritic cell. The interaction activates the cytotoxic T cell.

Receptors on a naive helper T cell bind to antigen–MHC complexes on the dendritic cell. The interaction activates the helper T cell, which then begins to divide.

❷ A large population of descendant cells forms. Each cell has T cell receptors that recognize the same antigen. The cells differentiate into effector and memory cells.

❸ The effector helper T cells begin to secrete cytokines.

❹ The activated cytotoxic T cell recognizes these cytokines as a signal to divide.

❺ A large population of descendant cells forms. Each cell bears T cell receptors that recognize the same antigen. The cells differentiate into effector and memory cytotoxic T cells.

❻ The new effector cytotoxic T cells circulate throughout the body. They recognize and kill any body cell that displays the viral antigen–MHC complexes on its surface.

Figure It Out: What do the large red spots represent?

Answer: Viruses

Activated helper T cells begin to divide, and their descendants differentiate into effector and memory helper T cells ❷. The new effector cells can form immunological synapses with antigen-presenting macrophages. The interaction causes the macrophages to increase production of lysosomal enzymes and antimicrobial toxins, thus enhancing the cell's ability to kill pathogens. The macrophage also increases production of inflammatory cytokines, so it is better able to recruit additional phagocytic lymphocytes.

Effector helper T cells also secrete cytokines ❸. Cytotoxic T cells that have been activated by interacting with an antigen-presenting cell ❹ recognize the cytokines as a signal to divide and differentiate. Tremendous populations of effector and memory cytotoxic T cells form ❺. All recognize and bind the same antigen—the one displayed by that first ailing cell.

If B cells are like assassins, then cytotoxic T cells are specialists in cell-to-cell combat. The effector cytotoxic T cells start working immediately. They circulate throughout blood and interstitial fluid, and bind to any other body cell displaying the original antigen together with MHC markers ❻. After a cytotoxic T cell is bound to an ailing cell, the T cell releases protein-digesting enzymes and small molecules called perforins. The perforins assemble into complexes that, like membrane attack complexes, insert themselves into a plasma membrane as a transmembrane channel. The enzymes released by the cytotoxic T cell enter the ailing body cell and induce it to commit suicide.

As occurs in an antibody-mediated response, memory cells form in a primary cell-mediated response. These long-lasting cells do not act immediately. If the antigen returns at a later time, the memory cells will mount a faster, stronger secondary response.

The Role of Natural Killer (NK) Cells

In order to kill a body cell, cytotoxic T cells must recognize the MHC molecules on the surface of the cell. However, some infections or cancer can alter a body cell so that it is missing part or all of its MHC markers. NK cells are crucial for fighting such cells. Unlike cytotoxic T cells, NK cells can kill body cells that lack MHC markers. Cytokines secreted by helper T cells also stimulate NK cell division. The resulting populations of effector NK cells attack body cells tagged by antibodies for destruction. They also recognize certain proteins displayed by body cells that are under stress. Stressed body cells with normal MHC markers are not killed; only those with altered or missing MHC markers are destroyed.

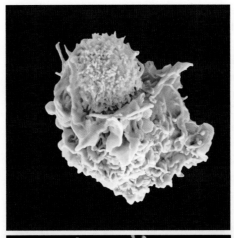

A An antigen-presenting dendritic cell (*blue*) interacting with a T cell (*yellow*). The interaction activates the T cell.

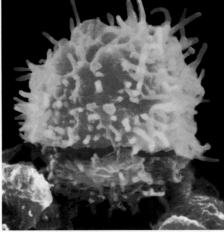

B Composite micrograph showing adhesion proteins (*red* ring) securing the bound TCRs (*green* spot) at the point of contact between the cells.

C At the cell surfaces, TCRs (*yellow*) on the T cell are bound to antigen (*red*) presented in context of MHC markers (*blue*) on the dendritic cell.

Figure 37.20 Zooming in on an immunological synapse: the point of contact between a T lymphocyte and an antigen-presenting cell.

Take-Home Message

What happens during a cell-mediated immune response?

» T cells, NK cells, and antigen-presenting cells carry out a cell-mediated immune response.

» Effector cytotoxic T cells and NK cells that form during a cell-mediated immune response kill infected body cells or those that have been altered by cancer.

» Memory cells also form, and these are reserved for a potential future encounter with the antigen.

37.10 When Immunity Goes Wrong

- An allergy is an immune response to something that is ordinarily harmless to most people.
- Autoimmune disorders occur when an immune response is misdirected against a person's own healthy body cells.
- In immunodeficiency, the immune response is insufficient to protect a person from disease.
- Links to SCID 15.10, Bird flu 20.4, Multiple sclerosis 32.12, Graves' disease 34.7

Despite built-in quality controls and the redundancies of immune system functions, immunity does not always work as well as it should. Its complexity is part of the problem, because there are simply more opportunities for failure to occur in systems with many components. Even small failures in immune function have effects on health.

Allergies

In millions of people, exposure to harmless substances stimulates an immune response. Any substance that is ordinarily harmless yet provokes such responses is an **allergen**. Sensitivity to an allergen is called an **allergy**. Drugs, foods, pollen, dust mites, fungal spores, poison ivy, and venom from bees, wasps, and other insects are among the most common allergens.

Some people are genetically predisposed to allergies. Infections, emotional stress, and changes in air temperature can trigger reactions. A first exposure to an allergen stimulates B cells to make and secrete IgE, which becomes anchored to mast cells and basophils. With later exposures, antigen binds to the IgE. Binding triggers the anchoring cell to release the contents of its granules, and inflammation is initiated. Contact with an allergen may cause the skin to redden, swell, and itch (**Figure 37.21A**). If allergen is detected by mast cells in the lining of the respiratory tract, a copious amount of mucus is secreted and the airways constrict; sneezing, stuffed-up sinuses, and a drippy nose result (**Figure 37.21B**). Antihistamines relieve allergy symptoms by acting on histamine receptors to dampen the effects of histamines. Other drugs can inhibit mast cell degranulation, thus preventing histamine release.

Overly Vigorous Responses

Immune defenses that kill pathogens can also kill body cells, so immune responses are a delicate balance between eliminating pathogen and damaging host tissues. Thus, multiple mechanisms that limit immune responses are always in play. For example, a small amount of the C3 complement is cleaved into activated

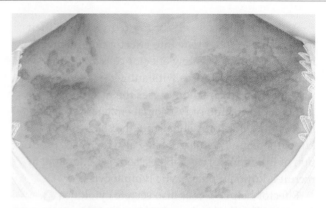

A Itchy welts on the skin called hives are caused by allergies to foods or medicines, or by direct contact with an allergen such as poison ivy.

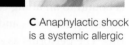

B Inflammation of mucous membranes in respiratory airways due to a pollen allergy causes symptoms of hay fever.

C Anaphylactic shock is a systemic allergic response.

Figure 37.21 Allergies.

fragments at all times. Without the fail-safe mechanism of inhibitory proteins that deactivate the fragments, complement cascades would occur constantly, with disastrous results to body tissues.

Acute illnesses arise when mechanisms that limit immune responses fail. For example, exposure to an allergen sometimes causes a severe, whole-body allergic reaction called anaphylactic shock. Huge amounts of cytokines and histamines released in all parts of the body provoke an immediate, systemic reaction. Too much fluid leaks from blood into tissues, so blood pressure drops too much (a reaction called shock) and tissues swell dramatically. The swelling tissue constricts airways and may block them. Anaphylactic shock is rare but life-threatening and requires immediate treatment (**Figure 37.21C**). It may occur at any time, upon exposure to even a tiny amount of allergen. Risks include any prior allergic reaction.

Severe episodes of asthma or septic shock occur when too many neutrophils degranulate at once. In a "cytokine storm," too many leukocytes release cytokines at the same time. The overdose of cytokines causes immediate, widespread inflammation that rapidly results in organ failure, with potentially fatal results. Cytokine storm triggered by infection with H5N1 influenza virus (Section 20.4) is a reason that this strain of bird flu has an unusually high mortality rate. Why these hyperactive immune responses occur is not well understood.

Autoimmune Disorders

People usually do not make antibodies to molecules that occur on their own, healthy body cells, in part because the thymus has a built-in quality control mechanism that weeds out defective T cell receptors. Thymus cells snip small polypeptides from a variety of body proteins and attach them to MHC markers. Maturing T cells that bind too strongly to one of these peptide–MHC complexes have TCRs that recognize a self protein; those that do not bind at least weakly to the complexes do not recognize MHC markers. Both types of cells die. If this mechanism fails, mature lymphocytes that do not discriminate between self and nonself may be produced. Such lymphocytes can mount an **autoimmune response**, which is an immune response that targets one's own tissues. Autoimmunity is beneficial when a cell-mediated response targets cancer cells, but in most cases it is not (**Table 37.4**).

As an example, Graves' disease is caused by antibodies that target body cells. In this disease, self-reactive antibodies bind to stimulatory receptors on the thyroid gland. The binding causes the gland to release excess thyroid hormone, which quickens the body's overall metabolic rate. Antibodies are not part of the feedback loops that normally regulate thyroid hormone production. So, antibody binding continues unchecked, the thyroid continues to release too much hormone, and the metabolic rate spins out of control. Symptoms of Graves' disease include uncontrollable weight loss; rapid, irregular heartbeat; sleeplessness; pronounced mood swings; and bulging eyes.

A neurological disorder, multiple sclerosis, occurs when self-reactive T cells attack the myelin sheaths of axons in the central nervous system (Section 32.12). Symptoms range from weakness and loss of balance to paralysis and blindness. Specific alleles of MHC genes increase susceptibility, but a bacterial or viral infection may trigger the disorder.

Immunodeficiency

Impaired immune function is dangerous and sometimes lethal. Immune deficiencies render individuals vulnerable to infections by opportunistic agents

Table 37.4 Examples of Autoantibodies Associated With Autoimmune Disorders

Disorder	Autoantibody Target	Affected Area
Crohn's disease	Neutrophil granule proteins	Gastrointestinal tract
Dermatomyositis	tRNA synthesis enzyme	Muscles, skin
Diabetes mellitus type 1	Islet proteins or insulin	Pancreas
Goodpasture's syndrome	Type IV collagen	Kidney, lung
Graves' disease	TSH receptor	Thyroid
Guillain-Barré syndrome	Lipids of ganglia	Peripheral nervous system
Hashimoto's disease	TH synthesis proteins	Thyroid
Idiopathic thrombo-cytopenic purpura	Platelet glycoproteins	Blood (platelets)
Lupus erythematosus	DNA, nuclear proteins	Connective tissue
Multiple sclerosis	Myelin proteins	Central nervous system
Myasthenia gravis	Acetylcholine receptors	Neuromuscular junctions
Pemphigus vulgaris	Cadherin	Skin
Pernicious anemia	Parietal cell glycoprotein	Stomach epithelium
Polymyositis	tRNA synthesis enzyme	Muscles
Primary biliary cirrhosis	Nuclear pore proteins, mitochondria	Liver
Rheumatoid arthritis	Constant region of IgG	Joints
Scleroderma	Topoisomerase	Arteriole endothelium
Ulcerative colitis	Enzyme in neutrophil granules	Large intestine
Wegener's granulomatosis	Enzyme in neutrophil granules	Blood vessels

that are typically harmless to those in good health. Primary immune deficiencies, which are present at birth, are the outcome of mutations. Severe combined immunodeficiencies (SCIDs) are examples. Secondary immune deficiency is the loss of immune function after exposure to an outside agent, such as a virus. AIDS (acquired immune deficiency syndrome, described in the next section) is the most common secondary immune deficiency.

allergen A normally harmless substance that provokes an immune response in some people.
allergy Sensitivity to an allergen.
autoimmune response Immune response that inappropriately targets one's own tissues.

Take-Home Message

What happens when the immune system does not function as it should?

» Normally harmless substances may induce an immune response in some people. Sensitivity to such allergens is called an allergy.

» Misdirected or compromised immunity, which sometimes occurs as a result of mutation or environmental factors, can have severe or lethal outcomes.

■ AIDS is an outcome of interactions between the HIV virus and the human immune system.

■ Links to cDNA 15.2, AIDS 20.1, Viruses 20.2, HIV replication 20.3, Pandemic 20.4

Acquired immune deficiency syndrome, or **AIDS**, is a collection of disorders that occur as a result of infection with HIV, the human immunodeficiency virus. This virus cripples the immune system, so it makes the body very susceptible to infections by other pathogens and to rare forms of cancer. Worldwide, approximately 33.3 million individuals are currently infected with HIV (**Figure 37.22A** and **Table 37.5**). AIDS is now considered to be a pandemic by the World Health Organization (WHO).

There is no way to rid the body of HIV, no cure for those already infected. At first, an infected person appears to be in good health, perhaps fighting "a bout of flu." But symptoms eventually emerge that foreshadow AIDS: fever, many enlarged lymph nodes, chronic fatigue and weight loss, and drenching night sweats. Then, infections caused by normally harmless microorganisms strike. Yeast infections of the mouth, esophagus, and vagina often occur, as well as a form of pneumonia caused by the fungus *Pneumocystis jiroveci*. Gastrointestinal inflammation due to infection by a yeast or virus causes diarrhea. Colored lesions that erupt are evidence of Kaposi's sarcoma, a type of cancer that is common among AIDS patients but rare

among the general population (**Figure 37.22B**). Other cancers are also common, as are infections by cancer-causing viruses such as Epstein-Barr virus. These medical problems are unusual in people with healthy immune systems.

HIV Revisited

HIV is a retrovirus with a lipid envelope. Remember, this type of envelope consists of a small piece of plasma membrane acquired by a virus as it buds from a cell (Section 20.3). Proteins jut from the envelope, span it, and line its inner surface. Just beneath the envelope, more viral proteins enclose two RNA strands and reverse transcriptase enzymes. When a retrovirus particle infects a cell, the reverse transcriptase copies the viral RNA into DNA, which becomes integrated into the host cell's DNA. The host then begins to produce viral proteins.

A Titanic Cellular Struggle HIV mainly infects macrophages, dendritic cells, and helper T cells (**Figure 37.22C**). When virus particles enter the body, dendritic cells engulf them. The dendritic cells then migrate to lymph nodes, where they present processed HIV antigen to naive T cells. An army of HIV-neutralizing IgG antibodies and HIV-specific cytotoxic T cells forms.

We have just described a typical adaptive immune response. It rids the body of most—but not all—of the

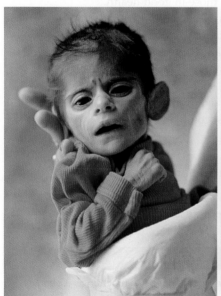

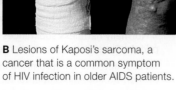

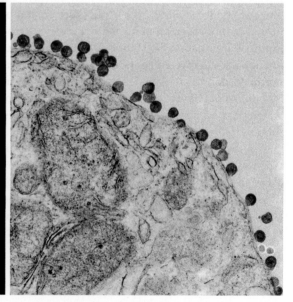

A This baby contracted AIDS from his mother's breast milk.

B Lesions of Kaposi's sarcoma, a cancer that is a common symptom of HIV infection in older AIDS patients.

C HIV (*red*) budding from the surface of a T cell. The virus infects the very cells that would otherwise protect the body from infection.

Figure 37.22 Three views of AIDS.

virus. In this first response, HIV infects a few helper T cells in a few lymph nodes. For years or even decades, the IgG antibodies keep the level of HIV in the blood low, and the cytotoxic T cells kill HIV-infected cells.

Patients are contagious during this stage, although they might show no symptoms of AIDS. HIV persists in a few of their helper T cells, in a few lymph nodes. Eventually, the level of virus-neutralizing IgG in the blood plummets, and the production of T cells slows. Why IgG decreases is still a major topic of research, but its effect is certain: The adaptive immune system becomes less and less effective at fighting the virus. The number of virus particles rises; up to 1 billion viruses are built each day. Up to 2 billion helper T cells become infected. Half of the virus particles are destroyed and half of the helper T cells are replaced every two days. Lymph nodes begin to swell with infected T cells.

Eventually, the battle tilts as the body makes fewer replacement helper T cells and its capacity for adaptive immunity is destroyed. Other types of viruses may make more particles per day, but the immune system eventually demolishes them. HIV demolishes the immune system. Secondary infections and tumors kill the patient.

Transmission HIV is not transmitted by casual contact; most infections are the result of having unprotected sex with an infected partner. The virus occurs in semen and vaginal secretions, and it can enter a sexual partner through epithelial linings of the penis, vagina, rectum, and mouth. The risk of transmission increases by the type of sexual act; for example, anal sex carries 50 times the risk of oral sex. Infected mothers can transmit HIV to a child during pregnancy, labor, delivery, or breast-feeding. HIV also travels in tiny amounts of infected blood in the syringes shared by intravenous drug abusers, or by hospital patients in less developed countries. Many people have become infected via blood transfusions, but this transmission route is becoming rarer as most blood is now tested prior to use for transfusions.

Testing Most AIDS tests check blood, saliva, or urine for antibodies that bind to HIV antigens. These antibodies are detectable in 99 percent of infected people within three months of exposure to the virus. One test can detect viral RNA at about eleven days after exposure. Currently, the only reliable tests are performed in clinical laboratories; home test kits may result in false negatives, which may cause an infected person to unknowingly transmit the virus.

Table 37.5	HIV in Populations Worldwide	
Region	Number Infected	% Adults Infected
Sub-Saharan Africa	22,500,000	5.0
Caribbean Islands	240,000	1.0
Central Asia/East Europe	1,400,000	0.8
Latin America	1,400,000	0.5
North America	1,500,000	0.5
South/Southeast Asia	4,100,000	0.3
Australia/New Zealand	57,000	0.3
Western/Central Europe	820,000	0.2
Middle East/North Africa	460,000	0.2
East Asia	770,000	0.1
Approx. worldwide total	33,300,000	

Source: Joint United Nations Programme HIV/AIDS, 2010 report

Treatments Drugs cannot cure AIDS, but they can slow its progress. Of the twenty or so FDA-approved AIDS drugs, most target processes unique to retroviral replication. For example, RNA nucleotide analogs such as AZT interrupt HIV replication when they substitute for normal nucleotides in the viral RNA-to-DNA synthesis process (Sections 15.2 and 20.3). Other drugs such as protease inhibitors affect different parts of the viral replication cycle.

A three-drug "cocktail" of one protease inhibitor plus two reverse transcriptase inhibitors is currently the most successful AIDS therapy, and has changed the typical course of the disease from a short-term death sentence to a long-term, often manageable illness.

Prevention Preventive use of a drug that contains two reverse transcriptase inhibitors has recently been shown to greatly reduce the HIV infection rate in high risk populations. However, education is still our best option for halting the global spread of AIDS. In most circumstances, HIV infection is the consequence of a choice: either to have unprotected sex, or to use a shared needle for intravenous drugs. Programs that teach people how to avoid these unsafe behaviors are having an effect on the spread of the virus, but overall, our global battle against AIDS is not being won.

AIDS Acquired immune deficiency syndrome. A secondary immune deficiency that develops as the result of infection by the HIV virus.

Take-Home Message

What is AIDS?

» AIDS is a secondary immune deficiency caused by HIV infection. HIV infects lymphocytes and so cripples the human immune system.

37.12 Vaccines

■ Vaccines are designed to elicit immunity to a disease.

Immunization refers to processes designed to induce immunity. In active immunization, a preparation that contains antigen—a **vaccine**—is administered orally or injected. The first immunization elicits a primary immune response, just as an infection would. A second immunization, or booster, elicits a secondary immune response for enhanced immunity. In passive immunization, a person receives antibodies purified from the blood of another individual. The treatment offers immediate benefit for someone who has been exposed to a potentially lethal agent, such as tetanus, rabies, Ebola virus, or a venom or toxin. Because the antibodies were not made by the recipient's lymphocytes, effector and memory cells do not form, so benefits last only as long as the injected antibodies do.

The first vaccine was developed in the late 1700s, a result of desperate attempts to survive smallpox epidemics that swept repeatedly through cities all over the world. Smallpox is a severe disease that kills up to one-third of the people it infects (**Figure 37.23**). Before 1880, no one knew what caused infectious diseases or how to protect anyone from getting them, but there were clues. In the case of smallpox, survivors seldom contracted the disease a second time. They were said to be immune—protected from infection.

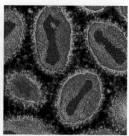

Figure 37.23 Young survivor and the cause of her disease, smallpox viruses. Worldwide use of the vaccine eradicated naturally occurring cases of smallpox; vaccinations for it ended in 1972.

At the time, the idea of acquiring immunity to smallpox was extremely appealing. People had been risking their lives on it for over two thousand years by poking into their skin bits of smallpox scabs or threads soaked in pus from smallpox sores. Some survived the crude practices and became immune to smallpox, but many others did not.

By 1774, it was common knowledge that dairymaids usually did not get smallpox after they had contracted cowpox, a mild disease that affects humans as well as cattle. An English farmer, Benjamin Jesty, made this prediction: If people accidentally infected with cowpox become immune to smallpox, then people deliberately infected with it should become immune too. In a desperate attempt to protect his family from a smallpox epidemic raging in Europe at the time, Jesty collected pus from a cowpox sore on a cow's udder, and poked it into the arm of his pregnant wife and two small children. All survived the smallpox epidemic, though they were from that time on subject to derision and rock peltings by neighbors who were convinced they would turn into cows.

In 1796, Edward Jenner, an English physician, injected liquid from a cowpox sore into the arm of a healthy boy. Six weeks later, Jenner injected the boy with liquid from a smallpox sore. Luckily, the boy did not get smallpox. Jenner's experiment showed directly that the agent of cowpox elicits immunity to smallpox.

Jenner named his procedure "vaccination," after the Latin word for cowpox (*vaccinia*). Though it was still controversial, the use of Jenner's vaccine spread quickly through Europe, then to the rest of the world. The last known case of naturally occurring smallpox was in 1977, in Somalia. The vaccine had eradicated the disease.

Table 37.6	Recommended Immunization Schedule for Children
Vaccine	**Age of Vaccination**
Hepatitis B	Birth
Hepatitis B boosters	1–2 months and 6–18 months
Rotavirus	2, 4, and 6 months
DTP: diphtheria, tetanus, and pertussis (whooping cough)	2, 4, and 6 months
DTP boosters	15–18 months, 4–6 years, and 11–12 years
HiB (*Haemophilus influenzae*)	2, 4, and 6 months
HiB booster	12–15 months
Pneumococcal	2, 4, and 6 months
Pneumococcal booster	12–15 months
Inactivated poliovirus	2 and 4 months
Inactivated poliovirus boosters	6–18 months and 4–6 years
Influenza	Yearly, 6 months and older
MMR (measles, mumps, rubella)	12–15 months
MMR booster	4–6 years
Varicella (chicken pox)	12–15 months
Varicella booster	4–6 years
Hepatitis A series (2 doses)	12–23 months
HPV series (3 doses)	11–12 years
Meningococcal	11–12 years

Centers for Disease Control and Prevention (CDC), 2011

We now know that the cowpox virus is an effective vaccine for smallpox because the antibodies it elicits also recognize smallpox virus antigens. Our knowledge of how the immune system works has allowed us to develop many other vaccines that save millions of lives yearly. These vaccines are an important part of worldwide public health programs (**Table 37.6**).

Progress on an HIV Vaccine

Researchers are using several strategies to develop an HIV vaccine. IgG antibody produced during a normal adaptive response exerts selective pressure on the virus, which has a very high mutation rate because it replicates so fast. The challenge is to develop a vaccine that efficiently recognizes and neutralizes the large number of variants of the HIV virus. At this writing, organizations around the world are testing 25 different HIV vaccines. Most of them consist of isolated HIV proteins or polypeptides, and many deliver the antigens in viral vectors. Live, weakened HIV virus is an effective vaccine in chimpanzees, but the risk of HIV infection from the vaccines themselves far outweighs their potential benefits in humans. Other types of HIV vaccines are notoriously ineffective.

One promising strategy involves reverse engineering HIV antibodies isolated from people with AIDS. Some of these antibodies can neutralize one or two strains of virus, but most are utterly ineffective. Researchers recently isolated three antibodies with a much greater ability to neutralize HIV than other antibodies studied to date; one of them is effective against almost all known strains of HIV. All three antibodies came from a man that has lived with AIDS for more than twenty years. His antibodies are being studied in painstaking detail in order to discover the exact parts of the virus they recognize. Those parts of the virus are being used to create new vaccines. Genes encoding the three antibodies are also being inserted into viral vectors for use in gene therapy (Section 15.10), the idea being that the vector will deliver the genes into body cells, which will then start producing antibodies.

immunization Any procedure designed to promote immunity to a specific disease.
vaccine A preparation introduced into the body in order to elicit immunity to a specific antigen.

Take-Home Message

How do vaccines work?

» Immunity to many diseases can be elicited by administering antigen-bearing vaccines, a process called immunization.

Frankie's Last Wish (revisited)

The Gardasil vaccine consists of viral capsid proteins that self-assemble into virus-like particles (VLPs). The proteins are produced by a genetically engineered yeast, *Saccharomyces cerevisiae*. This yeast carries genes for one surface protein from each of four strains of HPV, so the VLPs have no viral DNA. Thus, the VLPs are not infectious, but the antigenic proteins they consist of elicit an immune response at least as strong as infection with HPV virus.

How would you vote? Vaccines have overwhelmingly reduced suffering and deaths from many infectious diseases, but public confidence is a necessary part of their success. When enough individuals refuse vaccination, outbreaks of preventable and sometimes fatal diseases occur. Should vaccinations be mandatory?

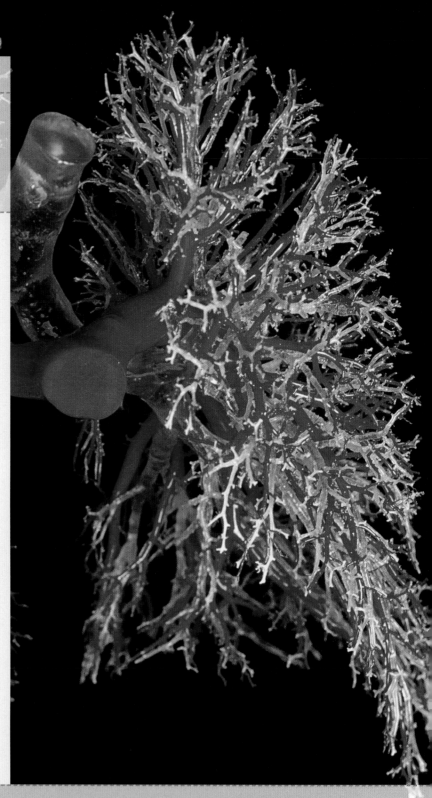

LEARNING ROADMAP

Where you have been Understanding diffusion (Section 5.8) and aerobic respiration (7.2) will help you comprehend gas exchange. You will revisit the role of red blood cells (36.4) and hemoglobin (3.2) and how evolutionary changes that accompanied the move of vertebrates onto land (25.6) allow respiration in specific environments.

Where you are now

Principles of Gas Exchange
Respiration moves oxygen from air or water in the environment to all metabolically active tissues and moves carbon dioxide from those tissues to the outside.

Gas Exchange in Invertebrates
Aquatic invertebrates exchange gases across their body surface or gills. Systems that deliver air to a respiratory surface inside the body adapt other invertebrates to land.

Gas Exchange in Vertebrates
Most vertebrates have either gills or paired lungs. In human lungs, gas exchange occurs in air sacs at the ends of branching airways. Muscle contractions draw air into lungs.

Respiratory Adaptations
Some animals and humans have adapted to life at high altitudes where there is less oxygen. Other animals can go without breathing for hours during deep dives.

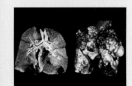

Respiratory Disease and Disorders
Obstructed airways, a faulty respiratory pacemaker, or infectious disease can impair gas exchange. Smoking greatly increases risk of respiratory disease and lung cancer.

Where you are going The iron needed for hemoglobin is an example of an essential dietary mineral, a topic we discuss in Section 39.11. The respiratory system plays a role in acid-base balance (Section 40.6) and in the regulation of body temperature (40.9). Section 42.9 explains how a pregnant woman exchanges gases with her developing child.

It was supposed to be a celebration of life. One evening in December 2010, a group of five young men rented a hotel room for a birthday party. The guest of honor, who was turning nineteen, parked in the garage beneath the room. His vehicle had been having battery problems, so he decided to leave it running for a little while. (Running the engine charges the battery.) The next morning, the men were discovered dead by the maid who came to clean their room.

The cause of death was carbon monoxide poisoning, the most common type of inhalation poisoning deaths both in the United States and worldwide. Carbon monoxide (CO), a colorless, odorless gas, is released when organic material burns (**Figure 38.1**). Being lighter than air, it rises from its point of release. The young men died when CO in exhaust from the engine idling below them rose into their room.

The incidence of accidental CO poisonings typically increases during the winter, when people close up their homes and start up their wood stoves and oil-, gas-, or kerosene-powered heaters. The incidence of CO poisoning also spikes after a power outage, when people turn to gasoline-powered generators.

Inhaled carbon monoxide exerts its adverse effect by interfering with gas exchange. It binds to hemoglobin to form carboxyhemoglobin (COHb). Hemoglobin has a very high affinity for CO, binding it more than two hundred times more tightly than oxygen. In addition to blocking oxygen-binding sites, binding of CO makes hemoglobin hold more tightly to any oxygen that it does bind. Thus, when CO is present, the blood holds less oxygen and delivers little of that to tissues.

The CO-related decline in the rate of oxygen delivery has its most dramatic effects in the brain and heart, the organs with the highest oxygen needs. At a low level, carbon monoxide causes headache, fatigue, irritability, mild nausea, and a shortness of breath. The symptoms are often mistakenly attributed to an infectious disease such as a flu. However, unlike a flu, CO poisoning does not cause a fever. A person with no underlying heart or lung problems will begin to experience symptoms of CO poisoning when the level of COHb in the blood reaches about 15 percent. As this level increases, a person becomes dizzy, light-headed, and confused. Chest pain often occurs because CO binds cardiac myoglobin (the oxygen-storing protein of muscle) even more strongly than it binds hemoglobin. The pain is a signal that heart cells are starved of oxygen. Continued exposure to CO results in seizures and loss of consciousness, followed by coma, cardiac arrest, and death.

Figure 38.1 Two common sources of carbon monoxide exposure—automobile exhaust and cigarette smoke. Carbon monoxide is an invisible odorless gas that binds to hemoglobin and impairs normal gas exchange.

A nonsmoker who lives in an area with little air pollution typically has a blood COHb level of 1 to 2 percent. People exposed to air pollution from heavy traffic have heightened blood levels of CO, as do smokers. Smoking cigarettes increases blood COHb by about 5 percent for each pack smoked per day.

Over the short term, a blood COHb level as low as 4–6 percent can decrease a healthy, young person's capacity to exercise. Over the longer term, an increased COHb level, such as that seen in smokers, alters the heart's structure. The heart must work overtime to make up for CO's suffocating effects on respiratory function.

38.2 The Nature of Respiration

- All animals must supply their cells with oxygen and rid their body of waste carbon dioxide.
- A variety of anatomical, behavioral, and physiological traits allow animals to extract oxygen from air or water around them.
- Links to Hemoglobin 3.2 and 36.4, Diffusion 5.8, Aerobic respiration 7.2, Algal bloom 21.6, Myoglobin 35.9

Gas Exchanges

In Chapter 7 you learned about aerobic respiration, an energy-releasing pathway that requires oxygen (O_2) and produces carbon dioxide (CO_2) as summarized in the equation below:

$$C_6H_{12}O_6 \ + \ O_2 \longrightarrow CO_2 \ + \ H_2O$$

glucose   oxygen   carbon dioxide   water

This chapter focuses on **respiration**, the physiological processes that collectively supply body cells with oxygen from the environment, and deliver waste carbon dioxide to the environment. Respiration depends upon the tendency of gaseous oxygen (O_2) and carbon dioxide (CO_2) to diffuse down their concentration gradients between the external and internal environments.

Gases enter and leave the animal body by diffusing across a thin, moist layer of cells called the **respiratory surface** (**Figure 38.2A**). A typical respiratory surface is one or two cell layers thick. The respiratory surface is thin because gases diffuse quickly only over very short distances. It is kept moist because gases cannot enter a cell unless they first dissolve in fluid. Once dissolved, the gas can diffuse across the cell membrane.

A second exchange of gases occurs internally, at the plasma membrane of body cells (**Figure 38.2B**). Oxygen diffuses from the interstitial fluid into a cell, and carbon dioxide diffuses in the opposite direction. In invertebrates without a circulatory system, oxygen that crosses the respiratory surface reaches body cells by diffusion. In many invertebrates and all vertebrates, a circulatory system enhances movement of gases between the respiratory surface and body cells.

Factors Affecting Diffusion Rates

Several factors affect how much gas diffuses across an animal's respiratory surface in any given interval.

Surface Area The greater the area of a respiratory surface, the more molecules can cross the surface at once. The area of the respiratory surface is often surprisingly large relative to the animal's body size. Branchings and foldings can help an extensive respiratory surface fit into a small volume.

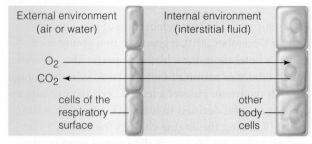

A Cells of the respiratory surface exchange gases with both the external and internal environment.

B Other body cells exchange gases with the internal environment.

Figure 38.2 Two sites of gas exchange during respiration. In some invertebrates, gases simply diffuse between the two sites. However, most animals have a circulatory system. Their blood transports gases rapidly between the two exchange sites.

Ventilation The concentrations of gases on either side of the respiratory surface also affect the rate of gas exchange. The steeper the concentration gradient across this surface, the faster diffusion proceeds. As a result, many animals have mechanisms that keep oxygen-rich air or water flowing over their respiratory surface. For example, your inhalations and exhalations move air rich in carbon dioxide away from the respiratory surface in your lungs and replace it with oxygen-rich air. Many animals that live in water also have mechanisms that keep water flowing past their respiratory surface.

Respiratory Proteins **Respiratory proteins** contain one or more metal ions that reversibly bind oxygen atoms. Oxygen atoms bind to these proteins when the concentration of oxygen is high, and are released when the concentration of oxygen declines. By reversibly binding oxygen, respiratory proteins help maintain a steep concentration gradient for oxygen between cells and the blood. The gradient is steepened because oxygen that is bound to a molecule in solution does not contribute to the concentration of O_2 in that solution.

The iron-containing respiratory protein hemoglobin fills vertebrate red blood cells (Sections 3.2 and 36.4). Annelids, mollusks, and crustaceans do not have red blood cells, but hemoglobin circulates in their blood or hemolymph. The respiratory proteins hemerythrin (which contains iron) and hemocyanin (which contains copper) also aid oxygen transport in some invertebrates. Muscles of vertebrates and some invertebrates contain myoglobin, a heme-containing respiratory protein. By binding oxygen that enters a muscle cell, myoglobin steepens the concentration gradient for oxygen across the cell membrane, and speeds diffusion.

A Air-breathing animals have a consistent oxygen supply.

Figure 38.3 Two animals with different oxygen supplies.

B The oxygen available to aquatic animals such as fish fluctuates with temperature and speed of water flow, and other factors.

Respiratory Medium—Air or Water?

Animals that breathe air have a highly reliable source of oxygen (**Figure 38.3**). Everywhere on Earth, the atmosphere consists of 21 percent oxygen. There is less atmosphere at high altitudes than lower ones, but at any given altitude, the proportion of gases is constant. By contrast, the availability of O_2 in aquatic environments can change dramatically over time and vary over short distances.

Any animal can tolerate only a limited range of environmental conditions. For aquatic animals, dissolved oxygen content of water (DO) is one of the most important factors affecting survival. More oxygen dissolves in cooler, fast-flowing water than in warmer, still water. When water temperature rises too high or moving water becomes stagnant, aquatic species that have high oxygen needs can suffocate.

Pollution can cause DO to decline. A lake enriched with runoff that contains manure or sewage offers a nutrition boost to aerobic bacteria living on the lake bottom. The bacteria are decomposers. As their populations soar, they use oxygen, so the amount available to other species plummets. The same thing can happen after an algal bloom—a population explosion of

protists such as dinoflagellates (Section 21.6). The protists multiply rapidly, then die. Their decomposition depletes the water of oxygen.

In freshwater lakes and streams, aquatic larvae of mayflies and stoneflies are the first invertebrates to disappear when oxygen levels fall. The active predatory lifestyle of these insect larvae requires considerable oxygen. A decline in the number of invertebrates has cascading effects on fishes that feed on them. Some fish are also more directly affected. Trout and salmon are especially intolerant of low oxygen. Carp (including koi and goldfish) are among the fish most tolerant of low oxygen levels; they survive even in warm ponds and tiny goldfish bowls. When the oxygen level falls below 4 parts per million, no fishes can survive. In waters with the lowest oxygen concentration, annelids called sludge worms (*Tubifex*) often are the only animals. They are colored red by large amounts of hemoglobin that help them take up and distribute what little oxygen there is.

Take-Home Message

What is respiration and what factors influence it?

» Respiration supplies cells with oxygen for aerobic respiration and removes carbon dioxide wastes.

» Gases are exchanged by diffusion across a respiratory surface: a thin, moist membrane.

» The area of a respiratory surface and the concentration gradients across it influence the rate of exchange.

» In aquatic animals, the amount of oxygen dissolved in the water can be a limiting factor.

----

respiration Physiological process by which an animal body supplies cells with oxygen and disposes of their waste carbon dioxide.

respiratory protein A protein that reversibly binds oxygen when the oxygen concentration is high and releases it when oxygen concentration is low. Hemoglobin is an example.

respiratory surface Moist surface across which gases are exchanged between animal cells and the external environment.

38.3 Invertebrate Respiration

- Some invertebrates exchange gases across their entire body surface, but most have specialized organs of gas exchange.
- Link to Animal body plans 24.2

A A jellyfish (*top*) and a marine flatworm (*bottom*) have no respiratory organs. All cells in these animals lie close to the body surface and gas exchange takes place by diffusion across that surface.

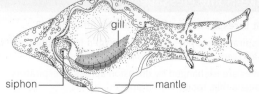

B Gill of the marine sea slug Aplysia, a mollusk. Having a gill increases the surface area for gas exchange. Blood vessels that run through the gill carry gases to and from body tissues.

Figure 38.4 Respiratory surfaces of aquatic invertebrates.

Respiration in Damp or Aquatic Habitats

The oldest invertebrate groups, such as cnidarians and flatworms, have neither respiratory nor circulatory organs (**Figure 38.4A**). These animals live in aquatic or continually damp land environments where their gas exchange needs can be met entirely by **integumentary exchange**: diffusion of gases across their outer body surface. Animals that lack both respiratory and circulatory organs are either small and flat or, when larger, have cells arranged in thin layers. In annelids such as earthworms and sludge worms, a closed circulatory system distributes gases that diffuse across the integument. Integumentary exchange also supplements respiratory organs in many gilled invertebrates.

Gills evolved independently in several invertebrate groups, as well as in vertebrates. **Gills** are filamentous or platelike respiratory organs that increase the surface area available for gas exchange. As hemolymph or blood passes through gills, it exchanges gases with the water surrounding the gill.

Gills may be internal or external. Aquatic mollusks typically have a gill inside their mantle cavity (Section 24.8), but many sea slugs have gills on their body surface (**Figure 38.4B**). Many aquatic arthropods such as lobsters and crabs have feathery gills beneath their exoskeleton, where the delicate tissues are protected from damage. The gills are modified branches of limbs.

Respiratory Adaptations to Life on Land

Snails and slugs that spend some time on land have a lung instead of, or in addition to, their gill. A **lung** is a saclike internal respiratory organ. Inside it, branching tubes deliver air to a respiratory surface supplied by many blood vessels. In snails and slugs, a pore at the side of the body can be opened to allow air into the lung, or shut to conserve water (**Figure 38.5**).

The most successful air-breathing land invertebrates are insects and arachnids, such as spiders. They have a hard integument that helps conserve water but also blocks gas exchange. Insects and some spiders have a **tracheal system** that consists of repeatedly branching, air-filled tubes reinforced with chitin.

Tracheal tubes start at spiracles—small openings across the integument (**Figure 38.6**). There is usually a pair of spiracles per segment: one on each side of the body. They can be opened or closed to regulate the amount of oxygen that enters the body. Substances that clog spiracles are used as insecticides. For example, horticultural oils sprayed on fruit trees kill scale bugs, aphids, and mites by clogging their spiracles.

Figure 38.5 A land snail (*Helix aspersa*) with the opening that leads to its lung visible at the left. Compare to **Figure 24.20**.

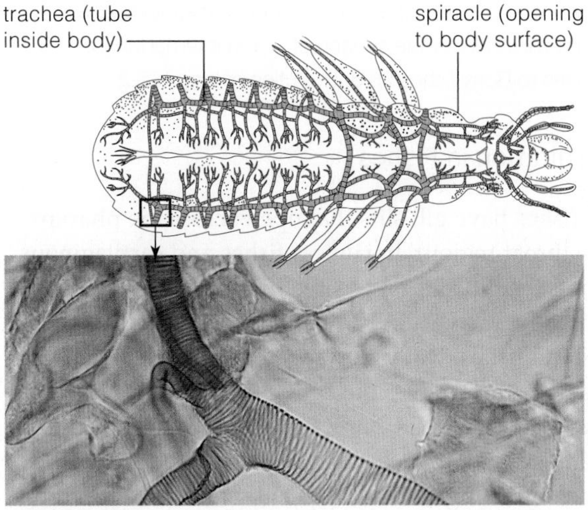

trachea (tube inside body)

spiracle (opening to body surface)

Figure 38.6 Insect tracheal system. Chitin rings reinforce branching, air-filled tubes in such respiratory systems.

At the tips of the finest tracheal branches is a bit of fluid in which gases dissolve. The tips of insect tracheal tubes lie adjacent to body cells, and oxygen and carbon dioxide diffuse between these tubes and the tissues. Because tracheal tubes end right next to cells, insects have no need for a respiratory protein such as hemoglobin to carry gases.

Some insects "breathe" by forcing air into and out of tracheal tubes. For example, when a grasshopper's abdominal muscles contract, organs press on the pliable tracheal tubes and force air out of them. When these muscles relax, pressure on tracheal tubes decreases, the tubes widen, and air rushes in.

Spiders and scorpions often have one or more book lungs in addition to or instead of tracheal tubes. In a book lung, air and hemolymph exchange gases across thin sheets of tissue (**Figure 38.7A**). The air enters the body through a spiracle.

Hemocyanin in a spider's hemolymph picks up oxygen and turns blue-green as it passes through a book lung. It gives up oxygen and becomes colorless in body tissues. Hemocyanin also occurs in the circulatory fluid of many aquatic arthropods (**Figure 38.7B**) and most mollusks.

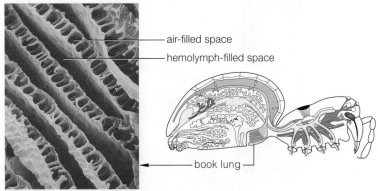

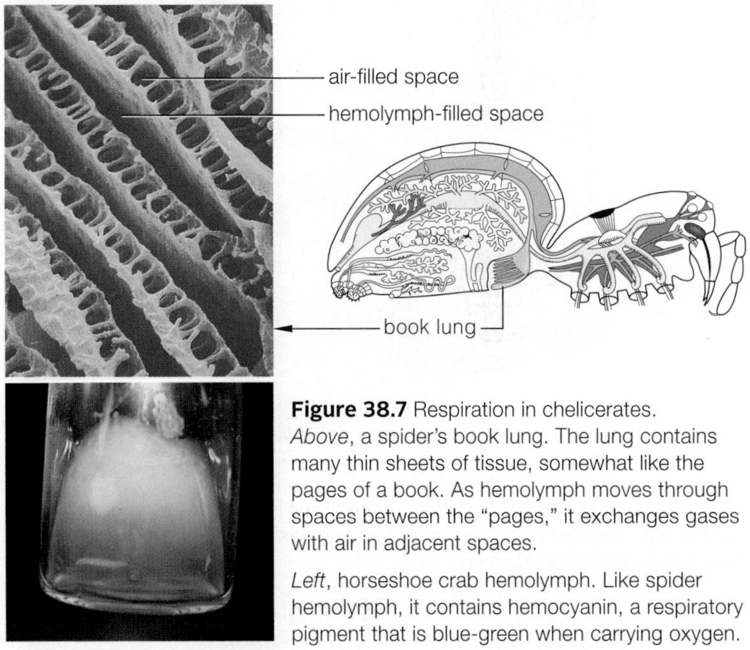

air-filled space

hemolymph-filled space

book lung

Figure 38.7 Respiration in chelicerates. *Above*, a spider's book lung. The lung contains many thin sheets of tissue, somewhat like the pages of a book. As hemolymph moves through spaces between the "pages," it exchanges gases with air in adjacent spaces.

Left, horseshoe crab hemolymph. Like spider hemolymph, it contains hemocyanin, a respiratory pigment that is blue-green when carrying oxygen.

gill Filamentous or branching respiratory organ of some aquatic animals; may be internal or external.
integumentary exchange In some animals, gas exchange across thin, moistened skin or some other external body surface.
lung Internal gas-exchange organ in some air-breathing animals.
tracheal system Of insects and some other land arthropods, tubes that convey gases between the body surface and internal tissues.

Take-Home Message

How do invertebrates exchanges gases with their environment?

» Some invertebrates do not have respiratory organs; all gas exchange takes place across the body wall. Such integumentary exchange also supplements the action of gills in many invertebrates.

» Gills are filamentous organs that increase the surface area for gas exchange in aquatic habitats.

» Some land snails have a lung in their mantle cavity. Land arthropods have tracheal tubes or book lungs, respiratory organs that bring air deep inside their body.

38.4 Vertebrate Respiration

■ Depending on the species, vertebrates exchange gases across gills, skin, or the surface of paired internal lungs.
■ Links to Bony fishes 25.5, Evolution of lungs 25.2

Respiration in Fishes

All fishes have gill slits that open across the pharynx (the throat region). In jawless fishes and cartilaginous fishes, the gill slits are visible from the outside, but bony fishes have a gill cover (Section 25.5).

In all fishes, water flows into the mouth, enters the pharynx, then moves out of the body through the gill slits (**Figure 38.8A**). A bony fish sucks water inward by opening its mouth, closing the cover over each gill, and contracting muscles that enlarge the oral cavity. Water is forced out over the gills when the fish closes its mouth, opens its gill covers, and contracts muscles that reduce the size of the oral cavity.

If you could remove the gill covers of a bony fish, you would see that the gills themselves consist of bony gill arches, each with many gill filaments attached (**Figure 38.8B**). Inside each gill filament are many capillary beds where gases in water are exchanged with gases in blood.

Water flowing over gills and blood flowing through gill capillaries move in opposite directions (**Figure 38.8C**). The result is a **countercurrent exchange**, in which two fluids exchange substances while flowing in opposite directions. For the entire length of the capillary, the water next to the capillary holds more oxygen than the blood flowing inside it (**Figure 38.8D**). As a result, oxygen continually diffuses from the water into the blood. The blood becomes increasingly oxygenated as it passes through the capillary.

Evolution of Paired Lungs

Most tetrapods have paired lungs. The first vertebrate lungs evolved from outpouchings of the gut wall in some bony fishes. Lungs may have helped these fishes survive short trips between ponds. They became increasingly important as aquatic tetrapods began moving onto land (Section 25.6).

Amphibian larvae typically have external gills. Most often, as the animal develops, these gills disappear and are replaced by paired lungs. Amphibians also exchange some gases across their thin-skinned body surface. In all amphibians, most carbon dioxide that forms during aerobic respiration leaves the body across the skin.

All adult frogs have paired lungs. They do not use chest muscles to draw air into their lungs, as you do. Instead, frogs suck in air through their nostrils by lowering the floor of the mouth (**Figure 38.9A**). Then they

countercurrent exchange Exchange of substances between two fluids moving in opposite directions.

A Bony fish with its gill cover removed. Water flows in through the mouth, over the gills, then out through gill slits. Each gill has bony gill arches with many thin gill filaments attached.

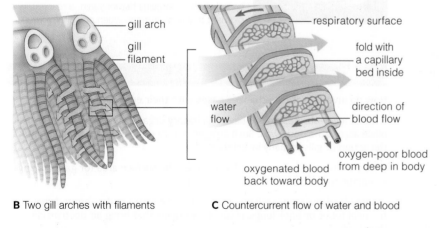

B Two gill arches with filaments

C Countercurrent flow of water and blood

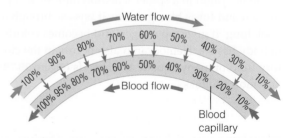

D Oxygen flow from water into a capillary. Percentages indicate the degree of oxygenation of water (*blue*) and blood (*red*). All along the capillary, oxygen flows down its concentration gradient from water into blood.

Figure 38.8 Animated Structure and function of the gills of a bony fish.

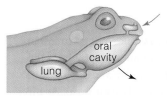

A The frog lowers the floor of its mouth, pulling air into the oral cavity through its nostrils.

B Closing the nostrils and elevating the floor of the mouth pushes air into lungs.

Figure 38.9 Animated How a frog fills its lungs. *Black* arrows show body wall movements. *Blue* arrows show air movement. Frogs push air into their lungs, rather than sucking it in as you do.

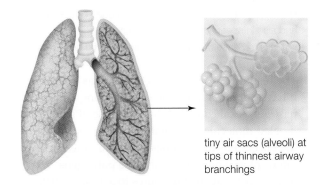

tiny air sacs (alveoli) at tips of thinnest airway branchings

Figure 38.10 Mammalian lungs. Expansion of the chest cavity draws air into branching tubes that end at tiny air sacs, where gas exchange occurs.

close their nostrils and lift the floor of the mouth and throat. Compression of the oral cavity forces air into the lungs (**Figure 38.9B**).

Reptiles (including birds) and mammals are amniotes with a waterproof skin (Section 25.7). Their only respiratory surface is the lining of two well-developed lungs. These lungs inflate when muscles increase the size of the thoracic cavity. As the cavity expands, pressure in the lungs declines and air is sucked inward.

In mammals, the inhaled air flows through increasingly smaller airways until it reaches tiny sacs (**Figure 38.10**). Gas exchange occurs across the walls of these sacs. During exhalation, air retraces its path, flowing out the same way it came in. The lungs do not deflate completely, so a bit of stale air remains behind even after exhalation.

In birds, there are no air sacs, no "dead ends," and no stale air inside a lung. Birds have small, inelastic lungs that do not expand and contract when the bird breathes. Instead, large air sacs attached to the lungs inflate and deflate (**Figure 38.11A**). It takes two breaths to move air through this system. Oxygen-rich air flows through tiny tubes in the lung during inhalations and exhalations. The lining of these tubes is the respiratory surface (**Figure 38.11B**). Continual movement of air over this surface increases the efficiency of gas exchange.

We turn next to the human respiratory system. Its operating principles apply to most vertebrates.

Take-Home Message

How does the structure of vertebrate respiratory organs affect their function?

» Fishes exchange gases with water flowing over their gills. Countercurrent flow of water and blood aids gas exchange.

» Amphibians exchange gases across their skin and push air from their mouth into their lungs. Amniotes suck air into their lungs by expanding the size of their thoracic cavity.

» Air flows into and out of mammalian lungs, but it flows continually through bird lungs.

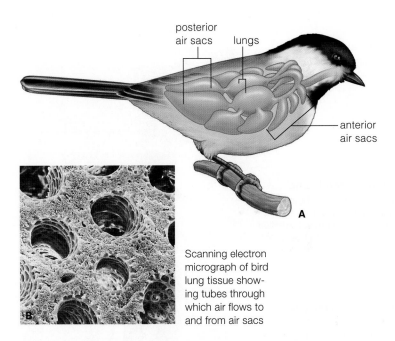

posterior air sacs   lungs

anterior air sacs

Scanning electron micrograph of bird lung tissue showing tubes through which air flows to and from air sacs

Figure 38.11 Animated Respiratory system of a bird.

(**A**) Large air sacs attach to two small, inelastic lungs. Air flows in through many air tubes inside the lung, and into posterior air sacs.

(**B**) The lining of the tiniest of the air tubes, sometimes called air capillaries, is the respiratory surface.

It takes two respiratory cycles to move air through the lungs and air sacs of a bird's respiratory system:

Inhalation 1— Muscles expand chest cavity, drawing air in through nostrils. Most of the air flowing in through the trachea goes to lungs and some goes to posterior air sacs.

Exhalation 1— Anterior air sacs empty. Air from the posterior air sacs moves into lungs.

Inhalation 2 — Air in lungs moves to anterior air sacs and is replaced by newly inhaled air.

Exhalation 2 — Air in anterior air sacs moves out of the body and air from posterior sacs flows into the lungs.

38.5 Human Respiratory System

- The human respiratory system functions in gas exchange, and also in speech, smell, and homeostasis.
- Links to Sense of smell 33.4, Barriers at the body surface 37.3

The respiratory system functions in gas exchange, but it has additional roles. We speak or sing as air moves past our vocal cords. We have a sense of smell because inhaled molecules stimulate olfactory receptors in the nose. Cells in nasal passages and other airways intercept and neutralize airborne pathogens. The respiratory system contributes to acid–base balance by expelling waste CO_2 that can acidify blood. Controls over breathing also help maintain body temperature; water evaporating from airways has a cooling effect.

From Airways to Alveoli

The Respiratory Passageways Take a deep breath. Now look at **Figure 38.12** to get an idea of where the inhaled air went. If you are healthy and sitting quietly, air probably entered through your nose. As air moves through your nostrils, tiny hairs filter out large particles. Mucus secreted by cells of the nasal lining captures most fine particles and airborne chemicals. The action of ciliated cells in the nasal lining helps to remove any inhaled contaminants.

Air from the nostrils enters the nasal cavity, where it is warmed and moistened. It flows next into the **pharynx**, or throat. It continues to the **larynx**, a short

Oral Cavity (Mouth)
Supplemental airway when breathing is labored

Pleural Membrane
Double-layer membrane that separates lungs from other organs; the narrow, fluid-filled space between its two layers has roles in breathing

Intercostal Muscles
At rib cage, skeletal muscles with roles in breathing. There are two sets of intercostal muscles (external and internal)

Diaphragm
Muscle sheet between the chest cavity and abdominal cavity with roles in breathing

A

Nasal Cavity
Chamber in which air is moistened, warmed, and filtered, and in which sounds resonate

Pharynx (Throat)
Airway connecting nasal cavity and mouth with larynx; enhances sounds; also connects with esophagus

Epiglottis
Closes off larynx during swallowing

Larynx (Voice Box)
Airway where sound is produced; closed off during swallowing

Trachea (Windpipe)
Airway connecting larynx with two bronchi that lead into the lungs

Lung (One of a Pair)
Lobed, elastic organ of breathing; site of gas exchange between internal environment and outside air

Bronchial Tree
Increasingly branched airways starting with two bronchi and ending at air sacs (alveoli) of lung tissue

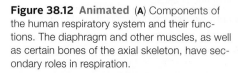
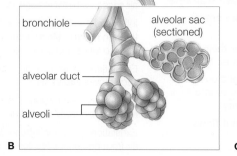
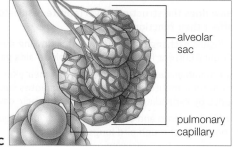

Figure 38.12 Animated (A) Components of the human respiratory system and their functions. The diaphragm and other muscles, as well as certain bones of the axial skeleton, have secondary roles in respiration.

(B,C) Location of alveoli relative to the bronchioles and the lung (pulmonary) capillaries.

bronchiole
alveolar sac (sectioned)
alveolar duct
alveoli

alveolar sac
pulmonary capillary

B

C

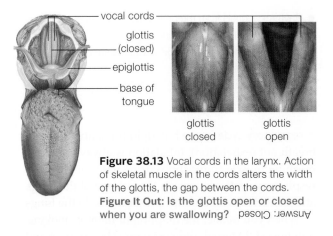

Figure 38.13 Vocal cords in the larynx. Action of skeletal muscle in the cords alters the width of the glottis, the gap between the cords.
Figure It Out: Is the glottis open or closed when you are swallowing? Answer: Closed

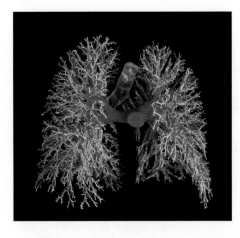

Figure 38.14 Branching airways and blood vessels. This resin cast of the human lungs shows airways of the lung in white, and pulmonary blood vessels in red.

airway commonly known as the voice box because it contains a pair of vocal cords (**Figure 38.13**). A vocal cord is skeletal muscle with a cover of mucus-secreting epithelium. Contraction of the vocal cords narrows the size of the **glottis**, the gap between them.

When the glottis is wide open, air flows through it silently. When muscle contraction narrows the glottis, flow of air outward through the tighter gap makes vocal cords vibrate and produces sounds. The tension on the cords and the position of the larynx determine the sound's pitch. To get a feel for how this works, place one finger on your "Adam's apple," the laryngeal cartilage that sticks out most at the front of your neck. Hum a low note, then a high one. You will feel the vibration of your vocal cords and how laryngeal muscles shift the position of your larynx.

At the entrance to the larynx is an **epiglottis**. When this tissue flap points up, air moves into the **trachea**, or windpipe. When you swallow, the epiglottis flops over, points down, and covers the larynx entrance, so food and fluids enter the esophagus. The esophagus connects the pharynx to the stomach.

The trachea branches into two airways, one to each lung. Each airway is a **bronchus** (plural, bronchi). Its epithelial lining has many ciliated and mucus-secreting cells that fend off respiratory tract infections. Bacteria and airborne particles stick to the mucus. Cilia sweep the mucus toward the throat, where it can be swallowed or expelled by coughing.

The Paired Lungs The rib cage encloses and protects the lungs. A two-layer-thick pleural membrane covers a lung's outer surface and lines the inner thoracic cavity wall. Air inside a lung moves through finer and finer branchings of a "bronchial tree." The branches are called **bronchioles**. At the tips of the finest bronchioles are respiratory **alveoli** (singular, alveolus), little

air sacs where gases are exchanged (**Figure 38.12B,C**). Branching blood vessels are interspersed among the branching airways (**Figure 38.14**). Air in alveoli exchanges gases with blood in adjacent pulmonary capillaries.

Muscles and Respiration The **diaphragm**, a broad dome-shaped smooth muscle beneath the lungs, partitions the coelom into a thoracic cavity and an abdominal cavity. It is the only smooth muscle that can be voluntarily controlled. You can make your diaphragm contract by deliberately inhaling. The diaphragm and **intercostal muscles**, which are skeletal muscles between the ribs, interact to change the volume of the thoracic cavity during breathing.

alveolus Tiny sac; in the mammalian lung, site of gas exchange.
bronchiole Airway that leads from a bronchus to the alveoli.
bronchus Airway connecting the trachea to a lung.
diaphragm Smooth muscle between the thoracic and abdominal cavities; contracts during inhalation.
epiglottis Tissue flap that covers airway during swallowing to prevent food from entering airways.
glottis Opening formed when vocal cords in the larynx relax.
intercostal muscles The skeletal muscles between the ribs; help change the volume of the thoracic cavity during breathing.
larynx Short airway containing the vocal cords (voice box).
pharynx Tube connecting mouth and digestive tract; in vertebrates it is the throat.
trachea Airway to the lungs; windpipe.

Take-Home Message

What roles do the components of the human respiratory system play?

» In addition to gas exchange, the human respiratory system acts in the sense of smell, voice production, body defenses, acid–base balance, and temperature regulation.

» Air enters through the nose or mouth. It flows through the pharynx (throat) and larynx (voice box) to a trachea that branches into two bronchi, one to each lung. Inside each lung, additional branching airways deliver air to alveoli, where gases are exchanged with pulmonary capillaries.

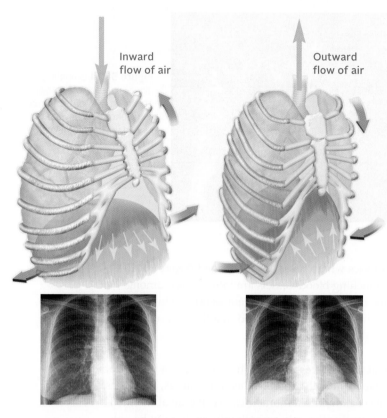

Inward flow of air

Outward flow of air

A Inhalation. Diaphragm contracts, moves down. External intercostal muscles contract, lift rib cage upward and outward. Lung volume expands.

B Exhalation. Diaphragm, external intercostal muscles return to resting positions. Rib cage moves down. Lungs recoil passively.

Figure 38.15 Animated Changes in the size of the thoracic cavity during the respiratory cycle. The x-ray images reveal how inhalation and expiration change the lung volume. **Figure It Out: What effect does contraction of the diaphragm have on the volume of the thoracic cavity?** Answer: It increases the volume.

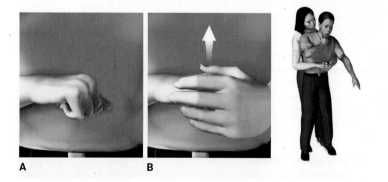

A   **B**

Figure 38.16 Animated How to perform the Heimlich maneuver on an adult who is choking.

1. Determine that the person is actually choking. A person who has an object lodged in his or her trachea cannot cough or speak.

2. Stand behind the person and place one fist below his or her rib cage, just above the navel, with your thumb facing inward as in (**A**).

3. Cover the fist with your other hand as shown in (**B**) and thrust inward and upward. Repeat until the object is expelled.

■ Rhythmic signals from the brain cause muscle contractions that cause air to flow into the lungs.
■ Links to Autonomic signals 32.7, Brain stem 32.9, Chemoreceptors 33.2

The Respiratory Cycle

A **respiratory cycle** is one breath in (inhalation) and one breath out (exhalation). Inhalation is always active, meaning muscle contractions drive it. During the respiratory cycle, changes in the volume of the lungs and thoracic cavity alter air pressure inside the lungs.

When you inhale, the diaphragm flattens, moving downward. External intercostal muscles contract and lift the rib cage up and outward (**Figure 38.15A**). As the thoracic cavity expands, so do the lungs. Pressure in the alveoli falls below atmospheric pressure, and air flows down the pressure gradient, into the airways.

Exhalation is usually passive. When muscles that caused inhalation relax, the lungs passively recoil and lung volume decreases. This compresses alveolar sacs, raising air pressure inside them. Air moves down the pressure gradient, out of the lungs (**Figure 38.15B**).

Exhalation becomes active only when you are exercising vigorously or when you consciously attempt to expel more air. During active exhalation, internal intercostal muscles contract, pulling the thoracic wall inward and downward. At the same time, muscles of the abdominal wall contract, causing intra-abdominal pressure to increase and exerting an upward-directed force on the diaphragm. The volume of the thoracic cavity decreases more than normal, and a bit more air is forced out.

In the **Heimlich maneuver**, a rescuer manually raises the intra-abdominal pressure of a choking person to dislodge an object stuck in the trachea (**Figure 38.16**). By making upward thrusts into the choker's abdomen, a rescuer raises the intra-abdominal pressure, forcing the choker's diaphragm upward. The pressure exerted by the air forced out of the lungs by this maneuver can dislodge the object, allowing the victim to resume normal breathing. The Heimlich maneuver should be performed only if a person cannot speak or cough.

Heimlich maneuver Procedure designed to rescue a choking person; a rescuer presses on a person's abdomen to force air out of the lungs and dislodge an object in the trachea.
respiratory cycle One inhalation and one exhalation.
tidal volume The volume of air that flows into and out of the lungs during a normal inhalation and exhalation.
vital capacity Amount of air moved in and out of lungs with forced inhalation and exhalation.

Respiratory Volumes

The maximum volume of air that the lungs can hold, total lung volume, averages 5.7 liters in men and 4.2 liters in women. Usually lungs are less than half full. **Vital capacity**, the maximum volume that can move in and out in one cycle, is one measure of lung health. **Tidal volume**—the volume that moves in and out in a normal respiratory cycle—is about 0.5 liter (**Figure 38.17**). Your lungs never fully deflate, so the air inside them always is a mix of freshly inhaled air and "stale air" left behind during the previous exhalation. Even so, there is plenty of oxygen for exchange.

Control of Breathing

Neurons in the medulla oblongata of the brain stem act as the pacemaker for inhalation, initiating an action potential 10–20 times per minute. Nerves deliver the action potentials to the diaphragm and intercostal muscles, they contract, and you inhale. Between signals, the muscles relax and you exhale.

Breathing patterns change with activity level. For example, active muscle cells produce CO_2 that enters blood, where it combines with water to forms carbonic acid. Chemoreceptors in the walls of carotid arteries and the aorta detect the rise in acidity and signal the brain (**Figure 38.18**). You breathe faster and deeper, so more carbon dioxide is expelled.

Chemoreceptors in the artery walls also signal the medulla oblongata when the O_2 concentration in the blood falls to a life-threatening level. This control mechanism usually comes into play only in people

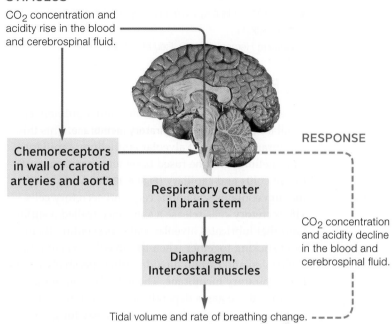

STIMULUS

CO_2 concentration and acidity rise in the blood and cerebrospinal fluid.

Chemoreceptors in wall of carotid arteries and aorta

RESPONSE

Respiratory center in brain stem

CO_2 concentration and acidity decline in the blood and cerebrospinal fluid.

Diaphragm, Intercostal muscles

Tidal volume and rate of breathing change.

Figure 38.18 Respiratory response to increased activity levels. An increase in activity raises the CO_2 output. It also makes the blood and cerebrospinal fluid more acidic. Chemoreceptors in blood vessels and the medulla sense the changes and signal the brain's respiratory center, also in the brain stem.

In response, the respiratory center signals the diaphragm and intercostal muscles. The signals call for alterations in the rate and depth of breathing. Excess CO_2 is expelled, which causes the level of this gas and acidity to decline. Chemoreceptors sense the decline and signal the respiratory center, so breathing is adjusted accordingly.

with severe lung diseases and at very high altitudes, where there is little oxygen in the air.

Reflexes such as swallowing or coughing can briefly halt breathing. Commands from sympathetic nerves make you breathe faster if you are frightened (Section 32.7). Breathing patterns can also be deliberately altered, as when you hold your breath, or break normal breathing rhythm to talk or sing.

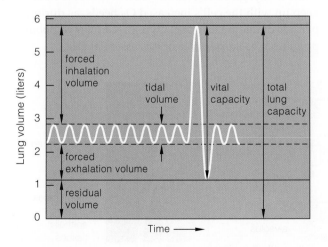

Figure 38.17 Animated Respiratory volumes. In quiet breathing, the tidal volume of air entering and leaving the lungs is only 0.5 liter. Lungs never deflate completely. Even with a forced exhalation, a residual volume of air remains in them.

Take-Home Message

What happens when we breathe?

» Inhalation is always an active process. Contraction of the diaphragm and external intercostal muscles increase the volume of the thoracic cavity. This reduces air pressure in alveoli below atmospheric pressure, so air moves inward.

» Exhalation is usually passive. As muscles relax, the thoracic cavity shrinks back down, air pressure in alveoli rises above atmospheric pressure, and air moves out.

» Only some of the air in the lungs is replaced with each breath. The lungs are never fully emptied of air.

» The brain controls the rate and depth of breathing.

38.7 Gas Exchange and Transport

■ Gases are exchanged by diffusion in alveoli.
■ Red blood cells play a role in transport of both oxygen and carbon dioxide.
■ Links to Hemoglobin 3.2, Red blood cells 36.4

The Respiratory Membrane

Gases diffuse between an alveolus and a pulmonary capillary at the lung's **respiratory membrane**. This thin membrane consists of alveolar epithelium, capillary endothelium, and the fused basement membranes of the alveolus and capillary (**Figure 38.19**). Alveolar epithelium contains squamous cells and secretory cells. The secretory cells release a substance (called a surfactant) that lubricates alveolar walls, preventing them from sticking together when the alveolus has to inflate.

Oxygen and carbon dioxide diffuse passively across the respiratory membrane. The net direction of movement for these gases depends on concentration gradients across the membrane, or as we say for gases, partial pressure gradients. The partial pressure of a gas is its contribution to the pressure exerted by a mix of gases. It is measured in millimeters of mercury (mm Hg). Just as a solute tends to diffuse in response to its concentration gradient, so does a gas. If the partial pressure of a gas differs between two regions, the gas will diffuse from the region of higher partial pressure to the region of lower partial pressure.

Oxygen Transport

Inhaled air that reaches alveoli has a higher partial pressure of O_2 than does blood in pulmonary capillaries. As a result, O_2 tends to diffuse from the air into the blood of these capillaries. After O_2 enters the blood, most diffuses into red blood cells, where it binds to hemoglobin. An iron-containing heme group

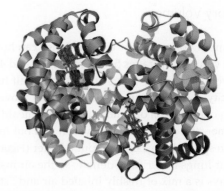

Figure 38.20 Structure of hemoglobin, the oxygen-transporting protein of red blood cells. It consists of four globin chains, each associated with an iron-containing heme group, color-coded *red*.

associates with each of the four globin subunits in hemoglobin (**Figure 38.20**). When O_2 is bound to one or more of hemoglobin's heme groups, we refer to the molecule as **oxyhemoglobin**.

Heme binds oxygen only weakly, and releases it where the partial pressure of O_2 is lower than that in the alveoli. This occurs in the body tissues serviced by systemic capillaries. In **Figure 38.21** compare the *red* boxes for "in alveoli" and "start of system capillaries." Metabolically active tissues also have other traits that encourage release of oxygen from heme: high temperature, low pH, and high CO_2 partial pressure.

The presence of myoglobin, also an iron-containing respiratory protein, helps cardiac muscle and some skeletal muscles take up oxygen. Structurally, myoglobin resembles a globin chain, but its heme binds oxygen more tightly. The O_2 that hemoglobin gives up near a cardiac muscle cell diffuses into the cell and binds to myoglobin inside it. When blood flow cannot keep up with the cell's increased O_2 needs, as during periods of intense exercise, O_2 released by myoglobin provides a backup source of this essential gas.

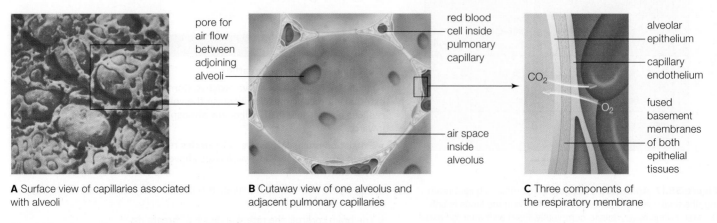

pore for air flow between adjoining alveoli

red blood cell inside pulmonary capillary

air space inside alveolus

CO_2

O_2

alveolar epithelium

capillary endothelium

fused basement membranes of both epithelial tissues

A Surface view of capillaries associated with alveoli

B Cutaway view of one alveolus and adjacent pulmonary capillaries

C Three components of the respiratory membrane

Figure 38.19 Zooming in on the respiratory membrane in human lungs.

Carbon Dioxide Transport

Carbon dioxide diffuses into the blood from any tissue where its partial pressure is higher than it is in blood. This is the case in the body tissues serviced by systemic capillaries, as the boxes color-coded *blue* in **Figure 38.21** show.

Carbon dioxide is transported to the lungs in three forms. About 10 percent remains dissolved in plasma. Another 30 percent reversibly binds with hemoglobin and forms carbaminohemoglobin ($HbCO_2$). However, most CO_2 that diffuses into the plasma—60 percent—is transported as bicarbonate (HCO_3^-). **Figure 38.22** shows this mechanism of transport. Carbon dioxide diffuses out of a body cell and into a systemic capillary, where it enters a red blood cell. Red blood cells have an enzyme, **carbonic anhydrase**, that speeds the hydration and dehydration reactions by which CO_2

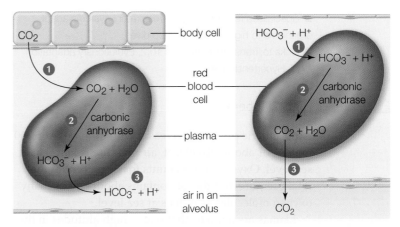

A At a systemic capillary bed:

❶ CO_2 diffuses from a body cell into plasma, then into a red blood cell.

❷ Carbonic anhydrase catalyses formation of bicarbonate (HCO_3^-).

❸ Bicarbonate diffuses into plasma.

B At a pulmonary capillary bed:

❶ Bicarbonate diffuses from plasma into a red blood cell.

❷ Carbonic anhydrase catalyses formation of water and CO_2.

❸ CO_2 diffuses across respiratory membrane into air in an alveolus.

Figure 38.22 Main mechanism of carbon dioxide transport and exchange. A lesser amount of CO_2 travels to the lungs bound to hemoglobin.

is incorporated into, and released from, bicarbonate. In systemic capillaries, where CO_2 partial pressure is high, carbonic anhydrase hydrates CO_2, forming carbonic acid (H_2CO_3), which dissociates into bicarbonate and H^+ (**Figure 38.22A**). Carbonic anhydrase can catalyze the formation of a million molecules of bicarbonate a second. This reaction also occurs spontaneously in the plasma, but at a much slower rate.

In the pulmonary capillaries, where partial pressure of CO_2 is relatively low, carbonic anhydrase catalyzes the reverse reaction, forming water and CO_2 from bicarbonate (**Figure 38.22B**). The CO_2 diffuses across the respiratory membrane and into the air in an alveolus. It then leaves the body in exhalations.

carbonic anhydrase Enzyme in red blood cells that speeds the reversible conversion of carbonic acid into bicarbonate and H^+.
oxyhemoglobin Hemoglobin with oxygen bound to it.
respiratory membrane Membrane consisting of alveolar epithelium, capillary endothelium, and their fused basement membranes; site of gas exchange in lungs.

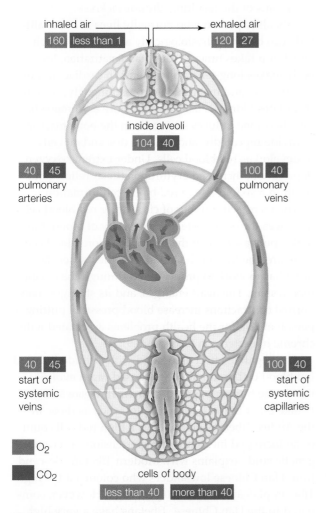

inhaled air → exhaled air
| 160 | less than 1 | | 120 | 27 |

inside alveoli
| 104 | 40 |

| 40 | 45 |
pulmonary arteries

| 100 | 40 |
pulmonary veins

| 40 | 45 |
start of systemic veins

| 100 | 40 |
start of systemic capillaries

■ O_2

■ CO_2

cells of body
| less than 40 | more than 40 |

Figure 38.21 Animated Partial pressures (in mm Hg) for oxygen (*red* boxes) and carbon dioxide (*blue* boxes) in the atmosphere, blood, and tissues.

Take-Home Message

How are gases transported in blood?

» Most oxygen in blood is bound to hemoglobin, which binds oxygen in alveoli where oxygen partial pressure is high, and releases it in tissues where oxygen partial pressure is lower.

» Most carbon dioxide is transported in blood in the form of bicarbonate, nearly all of which forms by enzyme action inside red blood cells.

38.8 Respiration in Extreme Environments

■ Specialized features of some respiratory systems adapt organisms to high altitude or deep dives.

■ Links to Transcription factors 10.2, Evolutionary adaptation 17.1, Hypertension 36.9

High Climbers

Atmospheric pressure decreases with altitude. At 5,500 meters, or about 18,000 feet, air pressure is half that at sea level. Oxygen still accounts for 21 percent of the total pressure, but the air contains about half as many oxygen molecules as it does at sea level.

Llamas are animals that live at high altitudes in the Andes (**Figure 38.23**). Their hemoglobin helps them survive in the "thin air," with its lower oxygen level. Compared to the hemoglobin of humans and most other mammals, llama hemoglobin binds oxygen more efficiently. Also, the lungs and the heart of a llama are unusually large relative to the animal's body size.

Most people live at lower altitudes where there is plenty of oxygen. When they ascend too fast to high altitudes, the delivery of oxygen to cells plummets. Hypoxia, or cellular oxygen deficiency, is the result.

In an acute compensatory response to hypoxia, the brain signals the heart and respiratory muscles to work harder. People breathe faster and more deeply than usual (they hyperventilate). As a result, CO_2 is exhaled faster than it forms, and ion balances in the

Figure 38.24 Tibetan mother and child. Directional selection has adapted Tibetans to life in a high altitude by favoring an allele that prevents excessive red cell production in response to low oxygen.

cerebrospinal fluid get skewed. Shortness of breath, a pounding heart, dizziness, nausea, and vomiting are symptoms of the resulting altitude sickness.

A healthy person who normally lives at a low altitude can become physiologically adjusted to a high one, but it takes time. Through **acclimatization**, the body makes long-term adjustments in cardiac output, and the rate and magnitude of breathing. Hypoxia also stimulates kidney cells to secrete more **erythropoietin**. This hormone induces stem cells in the bone marrow to divide repeatedly, and induces descendant cells to develop as red blood cells. Under extreme oxygen deprivation, increased erythropoietin secretion can result in a sixfold rise in red blood cell formation.

The increased number of circulating red blood cells improves the oxygen-delivery capacity of blood, but it also puts a strain on the heart. The more blood cells there are, the thicker the blood, and the harder the heart has to work to propel blood through the circulatory system. The heart enlarges, and its stronger-than-normal contractions increase blood pressure, putting a person at risk for the health problems associated with chronic hypertension (Section 36.9).

The negative effect associated with altitude-induced red blood cell production has caused rapid evolution in the human population of Tibet (**Figure 38.24**). Unlike other high-altitude peoples, such as those in the Andes, Tibetans do not have a high red cell count or an increased incidence of hypertension. A recent genetic study explains why. Modern Tibetans descend from Han Chinese lowlanders who colonized the high Tibetan plateau about 3,000 years ago. However, compared to the Han Chinese, Tibetans have a very high frequency of a mutant allele at the *ESA1* locus. Eighty-seven percent of Tibetans have the mutant allele,

Figure 38.23 Saturation curve for hemoglobin of humans, llamas, and other mammals. **Figure It Out:** At what partial pressure of oxygen do half the heme groups in human blood have oxygen bound? Answer: 30 mm Hg

compared to six percent of Han Chinese. The *ESA1* gene encodes a transcription factor involved in the signaling pathway by which low oxygen induces red cell production. The mutant allele more common in Tibetans makes the body less likely to produce excessive amounts of red blood cells in response to low oxygen, and thus less likely to suffer from excessive red cell production at high altitude.

Deep-Sea Divers

Water pressure increases with depth. Human divers using tanks of compressed air risk nitrogen narcosis, sometimes called "raptures of the deep." The deeper a diver goes, the more gaseous nitrogen (N_2) dissolves in interstitial fluid. In neurons, this dissolved nitrogen can disrupt signaling, causing a diver to feel euphoric and drowsy. The deeper divers descend, the more weakened and clumsy they become.

Returning to the surface from a deep dive also has risks. As a diver ascends, pressure falls and N_2 moves from interstitial fluid into blood and is exhaled. If a diver rises too fast, N_2 bubbles form inside the body. The resulting decompression sickness, also known as "the bends," usually begins with joint pain. Bubbles of N_2 can slow the flow of blood to organs. If bubbles form in the brain or lungs, the result can be fatal.

Humans who train to dive without oxygen tanks can remain submerged for about three minutes. So far, the human free diving record is about 200 meters (700 feet). Compare that with the depth records for species listed in **Figure 38.25**.

Leatherback sea turtles are among the deep divers. They leave water only to lay eggs (**Figure 38.25A**). The rest of their time is spent in the ocean, where they dive in search of jellyfishes. As a turtle or other air-breathing animal dives deeper and deeper, the weight of more and more water presses down onto the body. Lungs filled with air would collapse inward, so most diving animals move air out of their lungs and into cartilage-reinforced airways before they dive too deep.

Diving deep requires spending long intervals without access to air. The longest dive recorded for a leatherback turtle lasted a little more than an hour. Sperm whales can stay submerged for two hours.

How does a diving animal, whose lungs are emptied of air and who has no access to the surface, supply its cells with the oxygen they need to survive?

acclimatization A body adjusts to a new environment; e.g., after moving from sea level to a high-altitude habitat.
erythropoietin Hormone secreted by kidneys; induces stem cells in bone marrow to give rise to red blood cells.

Species	Maximum Depth
Sperm whale	2,200 meters
Leatherback turtle	1,200 meters
Southern elephant seal	1,620 meters
Weddell seal	741 meters
Bottlenose dolphin	>600 meters
Emperor penguin	565 meters

Figure 38.25 Atlantic leatherback sea turtles returning to the sea after laying eggs. The leathery shell is an adaptation to diving; it bends rather than breaks under extreme pressure. The chart *above* lists a few animal diving records.

First, before the animal dives, it breathes deeply. A sperm whale blows out about 80–90 percent of the air in its lungs with each exhalation; you exhale only about 15 percent. Deep breaths keep oxygen pressure in alveoli high, so more oxygen diffuses into the blood.

Second, diving animals can store great amounts of oxygen inside their blood and muscles. They tend to have a large blood volume relative to their body size, a high red blood cell count, and considerable amounts of myoglobin in their muscles. A skeletal muscle of a bottlenose dolphin has about 3.5 times the amount of myoglobin that a comparable skeletal muscle in a dog has. A muscle in a sperm whale has 7 times as much as the dog muscle.

Third, more oxygen gets distributed to the heart, brain, and other organs that require an uninterrupted supply of ATP for a deep dive. The blood volume and dissolved gases are stored and distributed efficiently with the assistance of valves and plexuses—meshes of blood vessels in local tissues. Metabolic rate and heart rate also decrease. So do oxygen uptake and carbon dioxide formation.

Finally, whenever possible, a diving animal makes the most of its oxygen stores by sinking and gliding instead of actively swimming. It conserves energy by avoiding unnecessary movements.

Take-Home Message

What are some adaptations that aid respiration in extreme environments?

» A big heart and lungs, and hemoglobin with an altered oxygen affinity, adapt some animals to life at high altitudes where oxygen is scarce.

» A high red blood cell count, large amount of myoglobin, and other traits allow some marine animals to hold their breath during long, deep dives.

38.9 Respiratory Diseases and Disorders

■ Genetic disorders, infectious disease, and lifestyle choices can increase the risk of respiratory problems.

■ Links to Tuberculosis 20.7, Marijuana's effects 32.6

Interrupted Breathing A tumor or damage to the brain stem's medulla oblongata can affect respiratory controls and cause apnea. In this disorder, breathing repeatedly stops and restarts spontaneously, especially during sleep. Sleep apnea also occurs when the tongue, tonsils, or other soft tissue obstructs the upper airways. Breathing may stop for up to several seconds many times each night, causing daytime fatigue. The risk of heart attacks and strokes rises with sleep apnea, because blood pressure soars when breathing stops. Obstructive sleep apnea can be reduced by changes in sleeping position or by wearing a mask that delivers pressurized air. Severe cases require surgical removal of tissue that blocks airways.

Sudden infant death syndrome (SIDS) occurs when an infant does not awaken from an episode of apnea. A defect in the medulla oblongata is associated with SIDS. Autopsies reveal that infants who died of SIDS tend to have fewer receptors for the neurotransmitter serotonin than infants who died of other causes. Having fewer of these receptors may impair the medulla's response to potentially deadly respiratory stress. There are also environmental risk factors. Maternal smoking during pregnancy heightens the risk. Infants who sleep on their back are at lower risk than stomach sleepers.

Tuberculosis and Pneumonia Worldwide, about one in three people is currently infected by bacteria that can cause tuberculosis. Most infected people have no symptoms, but about 10 percent of those infected develop "active TB." They cough up bloody

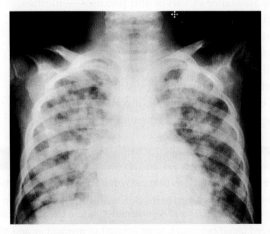

Figure 38.26 X-ray showing pneumonia. Fluid and blood cells fill the lungs. Compare x-rays of clear, healthy lungs in **Figure 38.15**.

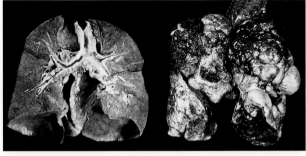

lungs of a nonsmoker   lungs of a smoker with emphysema

Figure 38.27 Smoking's effect on lungs.

mucus, have chest pain, and find breathing difficult. If untreated, active TB can be fatal. Antibiotics can cure most infections, if taken diligently for at least six months. Unfortunately, multi-drug-resistant strains of *Myobacterium tuberculosis* are increasing in frequency.

Pneumonia is a general term for lung inflammation caused by an infection. Bacteria, viruses, and fungi can cause pneumonia. Typical symptoms include a cough, an aching chest, shortness of breath, and fever. An x-ray reveals lungs filled with fluid and white blood cells instead of air (**Figure 38.26**). Treatment and outcome depend on the type of pathogen.

Bronchitis, Asthma, and Emphysema Your bronchi are lined by a ciliated, mucus-producing epithelium that helps protect you from respiratory infections. An inflammation of this epithelium is called bronchitis. Inflamed epithelial cells secrete extra mucus that triggers the coughing reflex. Bacteria can colonize the mucus, leading to more inflammation, more mucus, and more coughing. Bronchitis often arises after an upper respiratory infection. Repeated irritation of the bronchi by cigarette smoke is one of the most frequent causes of chronic bronchitis.

With asthma, an inhaled allergen or irritant triggers inflammation and constriction of the airways, conditions that make breathing difficult. A tendency to have asthma is inherited, but avoiding potential irritants such as cigarette smoke and air pollutants can reduce the frequency of asthma attacks. An acute asthma attack is treated with inhaled drugs that cause dilation of smooth muscle around the airways.

With emphysema, tissue-destroying bacterial enzymes digest the thin, elastic alveolar wall. As these walls disappear, the area of the respiratory surface declines. Over time, the lungs become distended and inelastic, leaving the person constantly feeling short of breath. Some people inherit a genetic predisposition

Carbon Monoxide—A Stealthy Poison (revisited)

Although carbon monoxide binds tightly to hemoglobin, it does not bind irreversibly. If a person with carbon monoxide poisoning is moved into normal atmosphere, about half of the CO bound to hemoglobin in his or her blood will be replaced by oxygen within four to five hours. The speed at which oxygen replaces carbon monoxide at binding sites can be accelerated by giving a person 100 percent oxygen.

To minimize your carbon monoxide exposure, do not smoke and avoid being in a closed area with people who do. Have wood-burning stove and fireplaces, and fossil fuel–burning heaters checked annually. Remember that any fossil fuel–powered engine releases CO, and should not be allowed to run indoors. If you can smell exhaust from a car, boat, heater, stove, or other device, you are also inhaling CO.

How would you vote? Carbon monoxide detectors are devices that sound an alarm in response to an elevated level of CO. It is recommended that homes have a CO detector outside each bedroom. Federal law requires that hotels have smoke detectors, but does not mandate installation of CO detectors. Should CO detectors be required in hotel rooms?

to emphysema. They do not have a functioning gene for an enzyme that inhibits bacterial attacks on alveoli. However, tobacco smoking is by far the main risk factor for emphysema (**Figure 38.27**).

Smoking's Impact Globally, cigarette smoking kills 4 million people each year. By 2030, the number may rise to 10 million, with about 70 percent of the deaths occurring in developing countries. In the United States, the direct medical costs of treating smoke-induced disorders drains $22 billion a year from the economy. As G. H. Brundtland—the former director of the World Health Organization—points out, tobacco is the only legal consumer product that kills half of its regular users. If you are a smoker, you may wish to reflect on the information in **Figure 38.28**.

Smoking marijuana (*Cannabis*) also poses respiratory risks. Compared to nonsmokers, long-term marijuana smokers have an increased incidence of respiratory problems. They tend to show lung damage earlier than cigarette smokers. However, marijuana has not been shown to raise the risk of lung cancer.

Take-Home Message

What causes common respiratory problems?

» Apnea, or interrupted breathing, is caused by tissue obstructing airways or a defective respiratory pacemaker.

» Tuberculosis, a bacterial disease, can be fatal, although most people have no symptoms. Pneumonia can be caused by many different pathogens.

» In asthma and bronchitis, airways become inflamed and constricted. In emphysema, alveolar sacs become distended and inelastic.

Risks Associated With Smoking	Reduction in Risks by Quitting
Shortened life expectancy Nonsmokers live about 8.3 years longer than those who smoke two packs a day from their midtwenties on.	Cumulative risk reduction; after 10–15 years, the life expectancy of ex-smokers approaches that of nonsmokers.
Chronic bronchitis, emphysema Smokers have 4–25 times higher risk of dying from these diseases than do nonsmokers.	Greater chance of improving lung function and slowing down rate of deterioration.
Cancer of lungs Cigarette smoking is the major cause.	After 10–15 years, risk approaches that of nonsmokers.
Cancer of mouth 3–10 times greater risk among smokers.	After 10–15 years, risk is reduced to that of nonsmokers.
Cancer of larynx 2.9–17.7 times more frequent among smokers.	After 10 years, risk is reduced to that of nonsmokers.
Cancer of esophagus 2–9 times greater risk of dying from this.	Risk proportional to amount smoked; quitting should reduce it.
Cancer of pancreas 2–5 times greater risk of dying from this.	Risk proportional to amount smoked; quitting should reduce it.
Cancer of bladder 7–10 times greater risk for smokers.	Risk decreases gradually over 7 years to that of nonsmokers.
Cardiovascular disease Cigarette smoking a major contributing factor in heart attacks, strokes, and atherosclerosis.	Risk for heart attack declines rapidly, for stroke declines more gradually, and for atherosclerosis it levels off.
Impact on offspring Women who smoke during pregnancy increase their risk of miscarriage or of an underweight infant.	When smoking stops before fourth month of pregnancy, risk of stillbirth and lower birth weight eliminated.

Figure 38.28 Risks incurred by smokers and benefits of quitting.

Nutrition and Digestion

LEARNING ROADMAP

Where you have been This chapter explains how your body digests organic polymers (Section 3.3) and obtains vitamins and minerals required to make coenzymes (5.6), hemoglobin (38.7), and some hormones (34.7). You will see how low pH (2.6) and enzymes (5.4) break down food, and how active transport (5.9, 5.10) moves products of digestion across cell membranes.

Where you are now

Overview of Digestive Systems
Some animal digestive systems are saclike, but most are a tube with two openings. Variations in traits of the digestive tract adapt different animals to different diets.

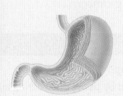

Digestion
Human digestion starts in the mouth, continues in the stomach, and is completed in the small intestine. Secretions of the salivary glands, liver, and pancreas aid digestion.

Absorption of Nutrients
The small intestine is the main site of nutrient and water absorption. The large intestine absorbs some water and concentrates wastes.

Organic Metabolism and Nutrition
Nutrients absorbed from the gut are raw materials used in synthesis of organic polymers. A healthy diet must include essential vitamins and minerals.

Maintaining a Healthy Weight
Weight depends on calories taken in and calories burned. Taking in too many or too few calories endangers health. Genes can predispose people to obesity or eating disorders.

Where you are going Fluid that enters the human gut is absorbed into the blood and any excess is eliminated by the kidneys (Section 40.3). An animal's dietary habits affect its ecological role. For example, dietary requirements are part of an animal's niche (45.4) and determine its position in a food web (46.3). The coevolution of plants and herbivores, and of prey and predators, is discussed in Section 45.5.

39.1 Your Microbial "Organ"

Your body is home to 10 trillion or more individual microorganisms, and the vast majority of them reside in your gut. The species composition of this microbial community varies among individuals and among human populations. Researchers have only recently begun to investigate the extent of these variations and their effects on our health, but some interesting connections have already been uncovered.

For example, we now know that one species of bacteria can cause peptic ulcers. An ulcer is a chronically open, inflamed sore at the body surface or in a body lining. Peptic ulcers arise in the stomach or in the part of the small intestine adjacent to the stomach. During the mid-1980s, two Australian physicians discovered that most people with peptic ulcers are infected by a spiral-shaped bacterial species called *Helicobacter pylori* (**Figure 39.1**). Prior to this discovery, people thought that peptic ulcers were caused by a stressful lifestyle or a diet rich in acidic or spicy foods.

Infection by *H. pylori* is extremely common. Worldwide, more than half of the human population harbor this species. The bacteria live at the surface of the stomach, in mucus secreted by cells of the stomach lining. Most infected people have no symptoms, but 5 to 20 percent of them develop peptic ulcers. Infection by *H. pylori* also increases the risk of stomach cancer sixfold. In 1994, the World Health Organization classified *H. pylori* as the first known bacterial carcinogen.

Treatment with antibiotics can eliminate *H. pylori* from the gut, allow an ulcer to heal, and reduce the risk of gastric cancer. However, clearing *H. pylori* from the gut can also have unintended consequences. Compared to people infected by *H. pylori*, the uninfected are more likely to have gastroesophageal reflux disease (GERD). The main symptom of GERD is a burning pain in the upper abdomen, commonly referred to as heartburn. The pain occurs when acid from the stomach splashes back into the esophagus, the tubular organ between the throat and the stomach. Repeated exposure to acid can damage the tissue of the esophagus and raise the risk of esophageal cancer.

Some gut bacteria also play a role in irritable bowel syndrome (IBS). This noninfectious disorder afflicts about one in five Americans. Affected individuals have frequent abdominal cramps and often feel bloated. Their bowel habits are disrupted, with constipation, diarrhea, or an alternation between these conditions. No single microbial species is thought to cause IBS. Rather, the disorder is associated with a shift in the diversity and proportions of intestinal bacteria. The genera of over- and under-represented bacteria varies

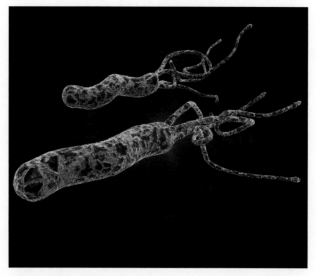

Figure 39.1 *Helicobacter pylori* (artist's representation). The presence of these bacteria in the stomach raises the risk for peptic ulcers and stomach cancer, but reduces the likelihood of heartburn and cancer of the esophagus.

with the type of IBS. However, several studies have found that people suffering from IBS have reduced numbers of *Bifidobacterium* compared with those unaffected by the syndrome.

Additional evidence that a decline in *Bifidobacterium* contributes to IBS comes from studies assessing the effectiveness of a probiotic as a treatment for this disorder. A **probiotic** is a food or supplement that contains living organisms whose presence in the body is thought to enhance health. Clinical studies have demonstrated that IBS patients who receive a probiotic dose of the gut bacteria *Bifidobacterium* have reduced symptoms compared to control patients.

There is also mounting evidence that our gut microbes exert influence far beyond the digestive tract. For example, the presence of *H. pylori* in the gut is associated with a lowered risk of immune-related disorders such asthma and allergies. Similarly, having a sizable gut population of *Bifidobacterium* improves immune function.

Given the impact of gut microbes on health, some have described this microbial population as analogous to an organ. Like our body's other organs, this microbial organ consists of cells whose coordinated function and interactions contribute to our health, and whose disruption can result in illness.

probiotic Food or supplement containing living microorganisms whose presence in the body benefits health.

39.2 The Nature of Digestive Systems

■ Animals are heterotrophs that typically digest food inside their body, but outside of their cells.
■ Links to Beak size and natural selection 17.6, Animal body plans 24.2, Mammalian teeth 25.10

An animal's digestive system is a body cavity or a tube that mechanically and chemically breaks food first into small particles, and then into molecules that can be absorbed into the internal environment. The digestive system also expels unabsorbed residues.

Incomplete and Complete Systems

Flatworms and cnidarians have an **incomplete digestive system**, a saclike gut with one opening. In flatworms, this opening is at the tip of a muscular pharynx (**Figure 39.2A**). Food that enters through this opening is digested, its nutrients are absorbed, then wastes are expelled through the same opening. Such two-way traffic does not allow for regional specialization.

Most invertebrates and all vertebrates possess a **complete digestive system**: a tubular gut with two openings. Food enters at one end; wastes leave through the other. Specialized regions along the tube's length process food, absorb nutrients, and concentrate wastes.

The tubular portion of a frog's complete digestive system (**Figure 39.2B**) includes a mouth, pharynx (throat), esophagus, stomach, small intestine, and large intestine. Digestive wastes exit through a cloaca (as do urinary wastes and gametes). The liver, gallbladder, and pancreas assist digestion by secreting substances into the small intestine.

A complete digestive system carries out five tasks:

1. *Mechanical processing and motility.* Movements that break up, mix, and directionally propel food material.
2. *Secretion.* Release of substances, especially digestive enzymes, into the lumen (the space inside the tube).
3. *Digestion.* Breakdown of food into particles, then to nutrient molecules small enough to be absorbed.
4. *Absorption.* Uptake of digested nutrients and water across the gut wall, into extracellular fluid.
5. *Elimination.* Expulsion of undigested or unabsorbed solid residues.

Diet-Related Structural Adaptations

Natural selection shapes traits of an animal digestive system and so adapts the animal to a specific diet.

Beaks and Bites Birds do not have teeth; a layer of the protein keratin covers their jawbones and forms a beak. The size and shape of a bird's beak determine what kinds of food it can process. Section 17.6 discussed the effect of beak size in African finches. As another example, the pigeon shown in **Figure 39.2C** has a relatively small beak that it uses to pick up seeds from the ground.

Mammals have different types of teeth, depending on their diet. Carnivores tend to have long, sharp canine teeth for killing prey, whereas herbivores typically lack canine teeth and have prominent incisors. The shape of individual teeth also reflects diet. Humans and antelopes both have molars that they use

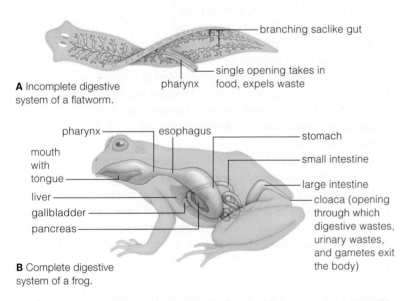

A Incomplete digestive system of a flatworm.

- branching saclike gut
- single opening takes in food, expels waste
- pharynx

- pharynx
- esophagus
- mouth with tongue
- liver
- gallbladder
- pancreas
- stomach
- small intestine
- large intestine
- cloaca (opening through which digestive wastes, urinary wastes, and gametes exit the body)

B Complete digestive system of a frog.

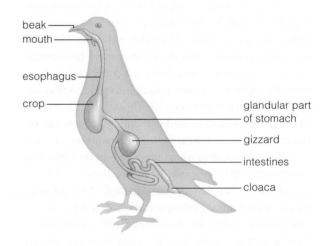

- beak
- mouth
- esophagus
- crop
- glandular part of stomach
- gizzard
- intestines
- cloaca

C Complete digestive system of a bird. The crop is an expandable organ for storing food. The muscular gizzard grinds up food.

Figure 39.2 Animated Examples of animal digestive systems.

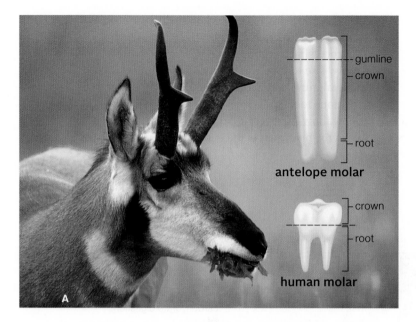

antelope molar

human molar

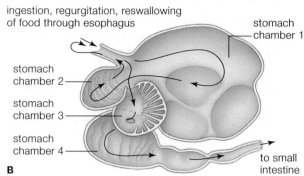

ingestion, regurgitation, reswallowing of food through esophagus

stomach chamber 1

stomach chamber 2

stomach chamber 3

stomach chamber 4

to small intestine

B

Figure 39.3 Animated Dietary adaptations of an antelope, a ruminant. (**A**) Antelope molars have tall crowns because sand mixed with their food wears down their teeth.

(**B**) An antelope's stomach has four chambers. Fermentation reactions carried out by bacteria, archaea, and protozoans in the first two chambers degrade cellulose. The fermented food is repeatedly regurgitated and rechewed as "cud" before continuing on to the next two chambers.

to grind food, but the relative size of the molar crown differs dramatically (**Figure 39.3A**). The large crown of the antelope molar is an adaptation to a diet of plant material that is often mixed with soil particles. Having a tall crown prevents the antelope from wearing its molars down to nubs when it eats.

Gut Specialization Like other seed eaters, a pigeon has a large crop, a saclike food-storing region above the stomach. The bird quickly fills its crop with seeds, then digests them later, in safer places. Birds grind up food inside a gizzard: a stomach chamber lined with hardened plates of keratin. Compared to hawks and other meat-eating birds, seed eaters have bigger gizzards relative to their body size.

Cattle, goats, sheep, and antelopes are ruminants, hoofed grazers with multiple stomach chambers (**Figure 39.3B**). Their digestive system adapts them to a diet of cellulose-rich plant foods. Like other animals, ruminants do not make enzymes that can break down cellulose. However, microbes that live in a ruminant's first two stomach chambers do make these enzymes. Solids accumulate in the second chamber, forming a "cud" that is regurgitated—moved back into the mouth for a second round of chewing. Nutrient-rich fluid moves from the second chamber to the third and fourth chambers, and finally to the intestine.

complete digestive system Tubelike digestive system; food enters through one opening and wastes leave through another.
incomplete digestive system Saclike digestive system; food enters and leaves through the same opening.

Meat takes less time to digest than plant material, so carnivores typically have a shorter gut than grazers. Their stomach can often stretch enormously. A highly expandable stomach allows them to get food into their body fast, so competitors cannot access it (**Figure 39.4**).

Figure 39.4 Python engulfing its prey. Pythons may eat only once or twice a year and some species can swallow a prey animal more than twice their weight.

Take-Home Message

What is a digestive system and how does its structure reflect its function?

» Digestive systems mechanically and chemically degrade food into small molecules that can be absorbed, along with water, into the internal environment. These systems also expel the undigested residues from the body.

» Incomplete digestive systems are a saclike cavity with one opening. Complete digestive systems are a tube with two openings and regional specializations in between.

» Structural variations in bills, teeth, and regions of the gut are adaptations that allow an animal to exploit a particular type or types of foods.

39.3 Overview of the Human Digestive System

■ Humans have a tubular gut. Accessory organs along its length secrete enzymes and other substances that aid in the breakdown of food and absorption of nutrients.

■ Links to Epithelium 31.3, Respiratory tract 38.5

Humans have a complete digestive system (**Figure 39.5**). Mucosa, a mucus-secreting epithelium, lines the tube. Accessory organs such as salivary glands, and the pancreas, liver, and gallbladder produce substances that are secreted into it.

Processing of food begins inside the mouth, or oral cavity. The tongue is a bundle of membrane-covered skeletal muscle attached to the floor of the mouth. It positions food so it can be chewed and swallowed.

Swallowing forces food into the pharynx. A human pharynx, or throat, is the entrance to the digestive and

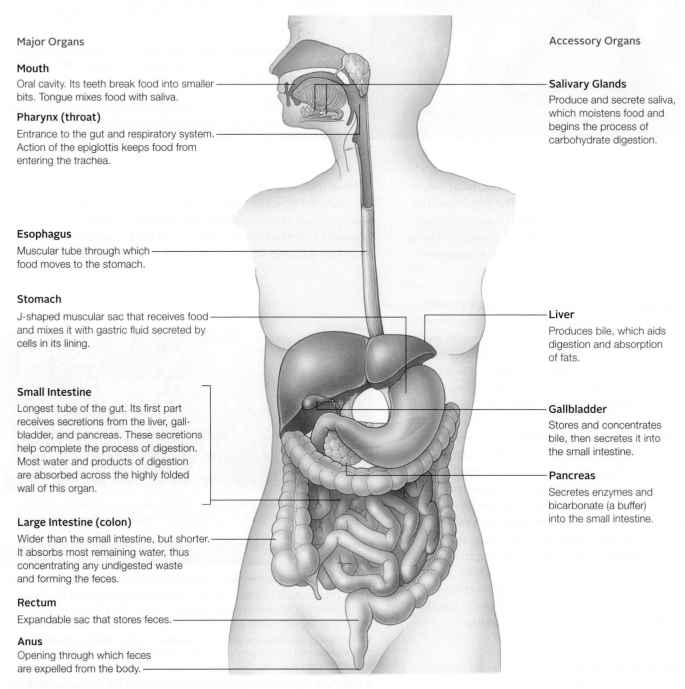

Major Organs

Mouth
Oral cavity. Its teeth break food into smaller bits. Tongue mixes food with saliva.

Pharynx (throat)
Entrance to the gut and respiratory system. Action of the epiglottis keeps food from entering the trachea.

Esophagus
Muscular tube through which food moves to the stomach.

Stomach
J-shaped muscular sac that receives food and mixes it with gastric fluid secreted by cells in its lining.

Small Intestine
Longest tube of the gut. Its first part receives secretions from the liver, gallbladder, and pancreas. These secretions help complete the process of digestion. Most water and products of digestion are absorbed across the highly folded wall of this organ.

Large Intestine (colon)
Wider than the small intestine, but shorter. It absorbs most remaining water, thus concentrating any undigested waste and forming the feces.

Rectum
Expandable sac that stores feces.

Anus
Opening through which feces are expelled from the body.

Accessory Organs

Salivary Glands
Produce and secrete saliva, which moistens food and begins the process of carbohydrate digestion.

Liver
Produces bile, which aids digestion and absorption of fats.

Gallbladder
Stores and concentrates bile, then secretes it into the small intestine.

Pancreas
Secretes enzymes and bicarbonate (a buffer) into the small intestine.

Figure 39.5 Animated Overview of the components of the human digestive system, together with a brief description of their primary functions in digestion.

respiratory tracts (Section 38.5). The presence of food at the back of the throat triggers a swallowing reflex. When you swallow, the flaplike epiglottis flops down and the vocal cords constrict, so the route between the pharynx and larynx is blocked. This reflex keeps food from getting stuck in an airway and choking you.

A muscular tube called the **esophagus** connects the pharynx to the stomach. The esophagus undergoes **peristalsis**, rhythmic muscle contractions that propel material through a tubular digestive organ. A sphincter connects the esophagus to the stomach. Like other **sphincters**, this ring of smooth muscle can relax and open to allow substances through or contract to close off the passageway.

A human gut, or **gastrointestinal tract**, starts at the stomach and extends through the intestines to the tube's terminal opening. The **stomach** is a stretchable sac that stores food, secretes acid and enzymes, and mixes them all together.

The stomach leads to the **small intestine**, the region of the gut where most carbohydrates, lipids, and proteins are digested and where most nutrients and water are absorbed. Secretions from the liver and pancreas assist the small intestine in these tasks.

The **large intestine** absorbs most remaining water and ions, thus compacting wastes. Wastes are briefly stored in a stretchable tube, the **rectum**, before being expelled from the gut's terminal opening, or **anus**.

anus Opening through which waste is expelled from a complete digestive system.
esophagus Muscular tube between the throat and stomach.
gastrointestinal tract The gut. Starts at the stomach and extends through the intestines to the tube's terminal opening.
large intestine or **colon** Wide tubular organ that receives digestive waste from the small intestine and concentrates it as feces.
peristalsis Wavelike smooth muscle contractions that propel food through the digestive tract.
rectum Region where feces are stored prior to excretion.
small intestine Narrowest portion of the digestive tract; site of most digestion and absorption.
sphincter Ring of muscle that controls passage through a tubular organ or body opening.
stomach Muscular organ that secretes gastric fluid, mixes food with it, and controls the flow of food to the small intestine.

Take-Home Message

What type of digestive system do humans have?

» Humans have a complete digestive system with a tubular, mucosa-lined gut.

» Accessory organs positioned adjacent to the gut secrete substances into its interior. These substances aid in digestion of food or absorption of nutrients.

39.4 Digestion in the Mouth

■ Mechanical digestion, the smashing of food into smaller pieces, begins in the mouth. So does chemical digestion, the enzymatic breakdown of food into molecular subunits.

Mechanical digestion begins when teeth rip and crush food. Adults have thirty-two teeth of four types (**Figure 39.6A**). Each tooth consists mostly of bonelike dentin (**Figure 39.6B**). Dentin-secreting cells reside in a central pulp cavity and are serviced by nerves and blood vessels that extend through the tooth's root. Enamel, the hardest material in the body, covers the tooth's exposed crown.

Chemical digestion begins when food mixes with saliva secreted by **salivary glands**, exocrine glands that open into the mouth. Saliva contains enzymes, bicarbonate, and mucins. The enzyme salivary amylase begins the breakdown of starch. Bicarbonate, a buffer, keeps the mouth from becoming too acidic. Mucins are proteins that combine with water and form mucus. The mucus causes chewed-up bits of food to stick together in easy-to-swallow clumps.

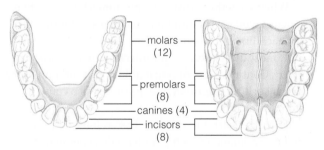

molars (12)
premolars (8)
canines (4)
incisors (8)

A Adult teeth. Lower jaw (*left*) and upper jaw (*right*).

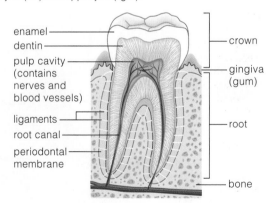

B Cross-section of a human molar. The crown extends above the gum, and the root is embedded in the bone of the jaw at a fibrous joint.

enamel
dentin
pulp cavity (contains nerves and blood vessels)
ligaments
root canal
periodontal membrane

crown
gingiva (gum)
root
bone

Figure 39.6 Structure and function of human teeth.

salivary gland Exocrine gland that secretes saliva into the mouth.

Take-Home Message

What happens to food in the mouth?

» Teeth mechanically break food into smaller particles. Enzymes in saliva begin the chemical digestion of carbohydrates.

39.5 Food Storage and Digestion in the Stomach

■ The stomach stores food and continues the process of digestion that began in the mouth.

■ Links to pH 2.6, Peptide bond 3.6, Enzymes 5.4, Smooth muscle 31.5, Autonomic nerves 32.7

The stomach is a muscular, stretchable sac with a sphincter at either end (**Figure 39.7**). It has three functions. First, it stores food and controls the rate of passage to the small intestine. Second, it mechanically breaks down food. Third, it secretes substances that aid in chemical digestion.

When the stomach is empty, its inner surface is highly folded. As it fills with food, these folds smooth out, increasing the stomach's capacity. In an average adult, the stomach can expand enough to hold about 1 liter of fluid (a little bit more than a quart).

A glandular epithelium, or mucosa, lines the stomach's inner wall. Cells of this lining secrete about 2 liters of **gastric fluid** each day. Gastric fluid includes mucus, hydrochloric acid, and pepsinogen, an inactive form of the protein-digesting enzyme pepsin. Acid lowers the pH to about 2.

When something disrupts the stomach's protective mucus layer, gastric fluid and enzymes can erode the stomach lining, causing an ulcer. As noted in Section 39.1, most ulcers occur after infection by *Helicobacter pylori*. Continual use of nonsteroidal anti-inflammatory drugs such as ibuprofen or aspirin can also cause a stomach ulcer.

Like the heart, the stomach has an internal pacemaker. Spontaneous action potentials generated in the upper portion of the stomach cause the smooth muscle in the stomach wall to contract rhythmically about three times a minute. The contractions mix gastric fluid with food to form a semiliquid mass called **chyme**. They also propel a bit of chyme out through the pyloric sphincter and into the first segment of the small intestine.

Chemical digestion of proteins begins in the stomach. The acidity of chyme denatures proteins (makes them unfold) and exposes their peptide bonds. High acidity also converts the protein pepsinogen into pepsin. Pepsin cuts proteins up into smaller polypeptides.

The stomach steps up or slows down its secretion of acid depending on when and what you eat. The arrival of food in the stomach, especially protein, triggers endocrine cells in the stomach lining to secrete the hormone gastrin into the blood. Gastrin acts on acid-secreting cells of the stomach lining, causing them to increase their output. When the stomach is empty, gastrin secretion and acid secretion decline. This prevents excess acidity from damaging the stomach wall.

Conversely, an empty stomach increases its secretion of another hormone called ghrelin. Ghrelin is a peptide hormone that stimulates the appetite. When people attempt to reduce their weight by eating less, their ghrelin output rises. This is one reason it is difficult to stick to a diet. It may also explain the success of gastric bypass surgery. Such surgery reduces the size of the stomach and the length of the small intestine, so less food can be taken in and fewer nutrients are absorbed. The surgery also removes ghrelin-secreting cells, reducing appetite.

As the stomach fills with food, its wall stretches, causing signals from sensory neurons to travel along the vagus nerve to the brain. Appetite declines as a result. Unfortunately, many of us continue to eat long past the point when we feel hungry.

chyme Mix of food and gastric fluid.
gastric fluid Fluid secreted by the stomach lining; contains digestive enzymes, acid, and mucus.

Take-Home Message

What are the functions of the stomach?

» The stomach receives food from the esophagus and stretches to store it.

» Contractions of stomach muscles break up food and mix it with acidic gastric fluid. They also move the resulting mixture (the chyme) into the small intestine.

» Chemical digestion of proteins begins in the stomach.

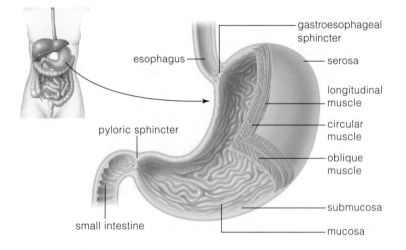

gastroesophageal sphincter

esophagus

serosa

longitudinal muscle

circular muscle

pyloric sphincter

oblique muscle

submucosa

small intestine

mucosa

Figure 39.7 Location and structure of the stomach. The outermost layer, the serosa, is connective tissue covered by epithelium. Beneath the serosa, three layers of smooth muscle differ in their orientation and direction of contraction. Their coordinated action mixes stomach contents with gastric fluid secreted by the mucosa that lines the stomach's interior. The folds shown on the inner surface become smoothed out when the stomach fills with food.

39.6 Structure of the Small Intestine

■ The small intestine has a highly folded lining with many projections that make its surface area enormous.
■ Link to Microvilli 31.3

Chyme forced out of the stomach through the pyloric sphincter enters the duodenum, the initial portion of the small intestine. The small intestine is "small" only in terms of its diameter—about 2.5 cm (1 inch). It is the longest segment of the gut. Uncoiled, it would extend for 5 to 7 meters (16 to 23 feet).

Most digestion and absorption take place at the surface of the small intestine, which is highly folded (**Figure 39.8A**). Unlike the folds of the empty stomach, those of the small intestine are permanent.

The surface of each intestinal fold is covered by **villi** (singular, villus), hairlike multicelled projections about 1 millimeter long (**Figure 39.8B,C**). The millions of villi that project from the intestinal lining give it a furry or velvety appearance. Blood vessels and lymph vessels run through the interior of each villus.

Most of the epithelial cells at the surface of a villus have even tinier projections called **microvilli** (singular, microvillus). The 1,700 or so microvilli at the surface of a cell make its outer edge look like a brush. Thus, these cells are sometimes called **brush border cells**

brush border cell In the lining of the small intestine, an epithelial cell with microvilli at its surface.
microvilli Thin projections from the plasma membrane of some epithelial cells; increase the cell's surface area.
villi Multicelled projections at the surface of each fold in the small intestine.

(**Figure 39.8D,E**). Brush border cells function in both digestion and absorption. Digestive enzymes at the surface of a microvillus break down sugars, protein fragments, and nucleotides. Also at the microvillus surface are many transport proteins that facilitate the movement of nutrients into the microvillus.

Collectively, the many folds and projections of the small intestinal lining increase its surface area by hundreds of times. As a result, the surface area of the small intestine is comparable to that of a tennis court. Like the stomach wall, the wall of the small intestine has layers of smooth muscle. The combined action of these muscles mixes the chyme, propels it forward, and forces it up against the wall of the small intestine, thus enhancing digestion and absorption (**Figure 39.9**).

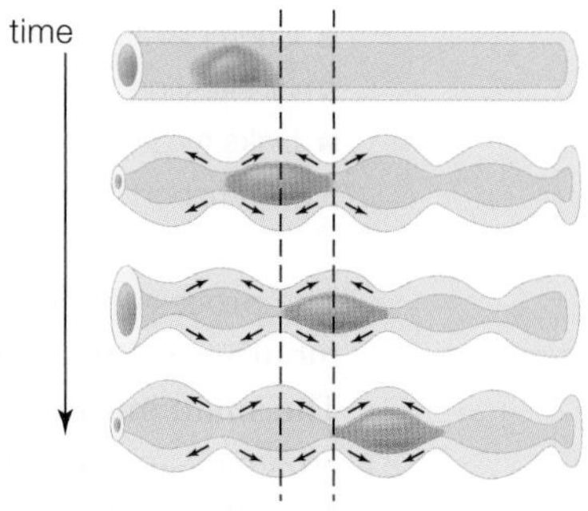

Figure 39.9 Muscle action in the small intestine. Contractions of rings of muscle in the wall of the small intestine causes chyme to slosh back and forth as it progresses through this organ.

Take-Home Message

How does the structure of the small intestine affect its function?

» The surface of the small intestine is highly folded and each fold has many projections (villi). Brush border cells at the surface of a villus have tiny projections (microvilli) at their surface.

» The many folds and projections greatly increase the surface area for the two functions of the small intestine—digestion and absorption.

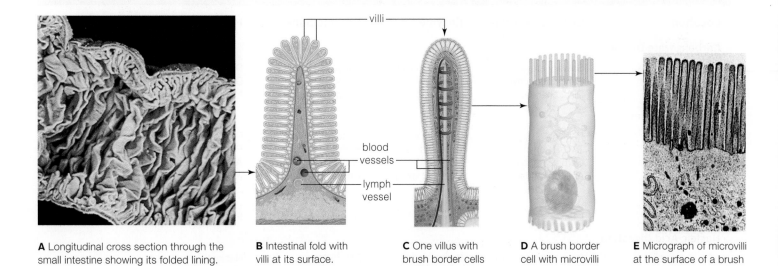

A Longitudinal cross section through the small intestine showing its folded lining.

B Intestinal fold with villi at its surface.

C One villus with brush border cells at its surface.

D A brush border cell with microvilli at its free surface.

E Micrograph of microvilli at the surface of a brush border cell.

Figure 39.8 Structure of the small intestine.
Figure It Out: Are microvilli multicelled or smaller than a cell?

Answer: Microvilli are smaller than a cell. Villi are multicellular.

39.7 Digestion and Absorption in the Small Intestine

■ Chemical and mechanical digestion are completed in the small intestine, and most nutrients are absorbed here.

■ Links to Organic monomers 3.2–3.6, Enzymes 5.4, Osmosis 5.8, Transport proteins 5.9, Secretin 34.6

The process of chemical digestion that began in the mouth is completed in the small intestine (**Table 39.1**). The small intestine receives chyme from the stomach, enzymes and bicarbonate from the pancreas, and bile from the gallbladder. Pancreatic enzymes work in concert with enzymes at the surface of brush border cells to complete the breakdown of large organic compounds into absorbable subunits. The pancreas also secretes bicarbonate into the small intestine, raising the pH of chyme enough for digestive enzymes to function. Pancreatic bicarbonate secretion is controlled by secretin, a hormone that cells of the small intestine release when they are exposed to acid (Section 34.6).

Carbohydrate Digestion and Absorption

In the mouth, salivary amylase broke polysaccharides into disaccharides (two-unit sugars). Pancreatic amylase carries out the same reaction in the small intestine. Enzymes embedded in the plasma membrane of the brush border cells complete the process, splitting disaccharides into monosaccharides (**Figure 39.10 ❶**). Monosaccharides are actively transported into a brush border cell, then into interstitial fluid inside a villus ❷. From here they enter the blood.

Protein Digestion and Absorption

Protein digestion began in the stomach, where pepsin broke proteins into polypeptides. The arrival of polypeptides or fats in the small intestine stimulates it to release the hormone cholecystokinin (CCK) into the blood. CCK causes the pancreas to secrete proteases such as trypsin and chymotrypsin into the small intestine. These enzymes break polypeptides into peptide fragments, then enzymes at the surface of the brush border cell break these fragments into amino acids ❸. Like monosaccharides, amino acids are actively transported into brush border cells, then out into the interstitial fluid. From here they enter the blood ❹.

Fat Digestion and Absorption

Nearly all fat digestion occurs in the small intestine. Here **bile** from the liver increases the effectiveness of lipases secreted by the pancreas. Bile consists mostly of cholesterol and bilirubin, a yellow-orange pigment that forms when the liver breaks down hemoglobin. Bile produced in the liver moves through a duct to the **gallbladder**, an organ that stores it. In addition to encouraging the pancreas to secrete enzymes, the hormone CCK causes smooth muscle of the gallbladder wall to contract, forcing bile out through a duct into the small intestine. Sometimes components of bile accumulate as hard pellets called gallstones. Most gallstones do no harm, but some can be dangerous when

Table 39.1 Summary of Chemical Digestion

Location	Enzymes Present	Enzyme Source	Enzyme Substrate	Main Breakdown Products
Carbohydrate Digestion				
Mouth, stomach	Salivary amylase	Salivary glands	Polysaccharides	Disaccharides
Small intestine	Pancreatic amylase	Pancreas	Polysaccharides	Disaccharides
	Disaccharidases	Intestinal lining	Disaccharides	Monosaccharides* (such as glucose)
Protein Digestion				
Stomach	Pepsins	Stomach lining	Proteins	Protein fragments
Small intestine	Trypsin, chymotrypsin	Pancreas	Proteins	Protein fragments
	Carboxypeptidase	Pancreas	Protein fragments	Amino acids*
	Aminopeptidase	Intestinal lining	Amino acids*	
Lipid Digestion				
Small intestine	Lipase	Pancreas	Triglycerides	Free fatty acids, monoglycerides*
Nucleic Acid Digestion				
Small intestine	Pancreatic nucleases	Pancreas	DNA, RNA	Nucleotides
	Intestinal nucleases	Intestinal lining	Nucleotides	Nucleotide bases, monosaccharides*

* Breakdown products that can be absorbed into the internal environment.

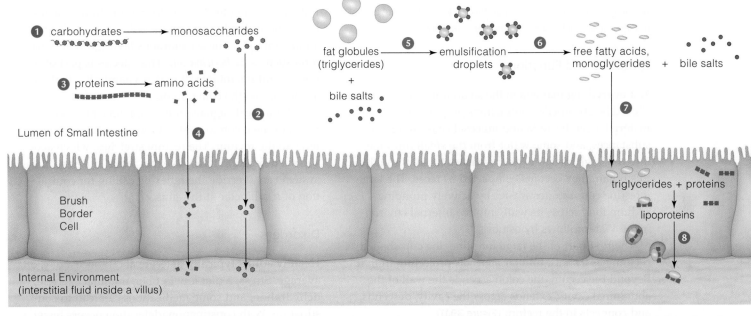

① **carbohydrates** ⟶ **monosaccharides**

③ **proteins** ⟶ **amino acids**

fat globules
(triglycerides)
+
bile salts

⑤ ⟶ **emulsification**
droplets

⑥ ⟶ **free fatty acids,**
monoglycerides + **bile salts**

Lumen of Small Intestine

④

②

⑦

Brush
Border
Cell

triglycerides + proteins

lipoproteins

⑧

Internal Environment
(interstitial fluid inside a villus)

① Enzymes break polysaccharides down to simple sugars, or monosaccharides.

② Monosaccharides are actively transported into brush border cells, then out into interstitial fluid.

③ Proteins are broken into polypeptides, then amino acids.

④ Amino acids are actively transported into brush border cells, then out into interstitial fluid.

⑤ Movements of the intestinal wall break up fat globules into small droplets. Bile salts coat the droplets, so that globules cannot form again.

⑥ Pancreatic enzymes digest the droplets to fatty acids and monoglycerides.

⑦ Monoglycerides and fatty acids diffuse across the plasma membrane's lipid bilayer, into brush border cells.

⑧ In a brush border cell, the products of fat digestion form triglycerides, which associate with proteins. The resulting lipoproteins are then expelled by exocytosis into the interstitial fluid inside the villus.

Figure 39.10 Animated Summary of digestion and absorption in the small intestine.

they block or become lodged in a duct. In this case, the gallbladder or gallstones can be removed surgically.

Bile aids fat digestion by causing **emulsification**, the dispersion of droplets of fat in a fluid. Triglycerides do not dissolve in water and tend to clump together. Movements of the small intestine break big globs of fat into smaller droplets, then bile salts coat the droplets so they remain separate ⑤. Compared to a few big globules, the many smaller droplets present a much greater surface area to the lipases that break triglycerides into fatty acids and monoglycerides ⑥.

Being lipid soluble, fatty acids and monoglycerides produced by fat digestion enter a villus by diffusing across the lipid bilayer of brush border cells ⑦. Inside these cells, triglycerides form and become coated with proteins. The resulting lipoproteins are moved by exocytosis into interstitial fluid inside a villus ⑧. From the interstitial fluid, triglycerides enter lymph vessels that eventually deliver their contents to the blood.

bile Mix of salts, pigments, and cholesterol produced in the liver, then stored and concentrated in the gallbladder; emulsifies fats when secreted into the small intestine.
emulsification Suspension of fat droplets in a fluid.
gallbladder Organ that stores and concentrates bile.

Fluid Absorption

Each day, eating and drinking puts 1 to 2 liters of fluid into your small intestine. Secretions from your stomach, accessory glands, and the intestinal lining add another 6 to 7 liters. About 80 percent of the water that enters the small intestine is absorbed there. Transport of salts, sugars, and amino acids across brush border cells creates an osmotic gradient. Water moves down that gradient from chyme into the interstitial fluid.

Take-Home Message

What are the roles of the small intestine?

» Chemical digestion is completed in the small intestine. Enzymes from the pancreas and enzymes embedded in the membrane of brush border cells break large molecules into smaller, absorbable subunits.

» Small subunits (monosaccharides, amino acids, fatty acids, and monoglycerides) enter the internal environment when they are absorbed into the interstitial fluid in a villus.

» Most fluid that enters the gut is also absorbed across the wall of the small intestine.

39.8 The Large Intestine

■ The large intestine is wider than the small intestine, but also much shorter—only about 1.5 meters (5 feet) long.

Structure and Function

Not everything that enters the small intestine can be or should be absorbed. Contractions propel indigestible material, dead bacteria and mucosal cells, inorganic substances, and some water from the small intestine into the large intestine. As the wastes travel through this organ, they become compacted as **feces**. The large intestine concentrates wastes by actively pumping sodium ions across its wall, into the internal environment. Water follows by osmosis.

The first part of the large intestine, the cup-shaped cecum, has a short, tubular **appendix** projecting from it. Beyond the cecum, the colon ascends the wall of the abdominal cavity, extends across the cavity, descends, and connects to the rectum (**Figure 39.11**).

Contraction of the smooth muscle in the colon wall mixes the colon's contents and propels this material along. Compared with other gut regions, wastes move more slowly through the colon, which also has a moderate pH. These conditions favor growth of bacteria such as *Escherichia coli*. This species is part of our normal gut microbiota. It makes vitamin B_{12} that we absorb across the colon lining.

After a meal, signals from autonomic nerves cause much of the colon to contract forcefully and propel feces to the rectum. The rectum stretches, which activates a defecation reflex to expel feces. The nervous system can override the reflex by calling for contraction of a sphincter at the anus.

Disorders

Healthy adults typically defecate about once a day, on average. Stress, a diet low in fiber, minimal exercise, dehydration, and some medications can lead to constipation. With constipation, defecation occurs fewer than three times a week, is difficult, and yields small, hardened, dry feces. Occasional constipation usually goes away on its own. A chronic problem should be discussed with a doctor. Infection by a viral, bacterial, or protozoan pathogen can cause an episode of diarrhea, the frequent passing of watery feces.

Appendicitis—an inflamed appendix— requires prompt treatment. It often occurs after a bit of feces lodges in the appendix and infection sets in. Removing an inflamed appendix prevents it from bursting and releasing bacteria into the abdominal cavity. Such ruptures can cause a life-threatening infection.

Some people are genetically predisposed to develop colon polyps, small growths on the colon wall (**Figure 39.11B**). Most polyps are benign, but some can become cancerous. If detected in time, colon cancer is highly curable. Blood in feces and dramatic changes in bowel habits may be symptoms of colon cancer and should be reported to a doctor. Also, anyone over the age of 50 should have a periodic colonoscopy, a procedure in which clinicians use a camera to examine the interior of the colon for polyps or cancer.

appendix Worm-shaped projection from the first part of the large intestine.
feces Unabsorbed food material and cellular waste that is expelled from the digestive tract.

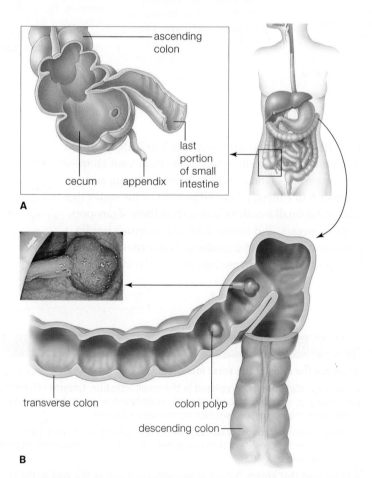

Figure 39.11 The colon. (**A**) Location of cecum and appendix.
(**B**) Sketch and photo of polyps in the transverse colon.

- ascending colon
- last portion of small intestine
- cecum
- appendix
- transverse colon
- colon polyp
- descending colon

A

B

Take-Home Message

What is the function of the colon?

» The colon completes the process of absorption, then concentrates, stores, and eliminates wastes.

39.9 Metabolism of Absorbed Organic Compounds

- Most absorbed organic compounds are broken down for energy, stored, or used to build larger organic compounds.
- Links to Glycogen 3.4, Alcohol metabolism 5.1, Systemic circulation 36.3

Figure 39.12A shows the main routes by which organic molecules from food are shuffled and reshuffled in the body. Living cells constantly recycle some carbohydrates, lipids, and proteins by breaking them apart. They use the resulting breakdown products as energy sources and building blocks. The nervous and endocrine systems regulate this turnover.

The **liver** is a large organ that functions in digestion, metabolism, and homeostasis (**Figure 39.12B**). All blood from the capillaries in the small intestine enters the hepatic portal vein, which delivers it to the liver. The blood flows through capillaries in the liver before returning to the heart (Section 36.3).

The liver helps protect the body against dangerous substances that were ingested or formed as a result of digestion. For example, Chapter 5 explained the role of the liver in detoxifying alcohol, and how alcohol abuse can damage this essential organ. As another example, ammonia (NH_3) is a toxic product of amino acid breakdown. The liver converts ammonia to urea,

liver Large organ that stores glucose as glycogen and releases it as needed. Also produces bile and detoxifies some harmful substances such as alcohol.

a much less toxic compound. Urea is carried by the blood to kidneys and is excreted in the urine.

Most of the body's fat-soluble vitamins, including vitamins A and D, are stored in the liver. The liver also stores glucose. After a meal, liver and muscle cells take up glucose and convert it to glycogen (Section 3.4). Excess carbohydrates and proteins are also converted to fats, which are stored mainly in adipose tissue.

In between meals, the brain takes up much of the glucose circulating in the blood. The brain cannot use fats or proteins as an energy source. Other body cells dip into their stores of glycogen and fat. Adipose cells degrade fats to glycerol and fatty acids, which enter blood. Liver cells break down glycogen and release glucose, which also enters blood. Body cells take up the released fatty acids and glucose and use them to fuel ATP production.

Take-Home Message

What happens to compounds absorbed from the gut?

» Blood carries the absorbed compounds to the liver, which detoxifies dangerous substances and stores vitamins and glucose. The glucose is stored as glycogen.

» Adipose tissue takes up absorbed carbohydrates and proteins and converts them to fats.

» In between meals, the liver breaks down stored glycogen, and releases its glucose subunits into the blood. This ensures that the brain, which can only use carbohydrates as fuel, always has an adequate supply of energy.

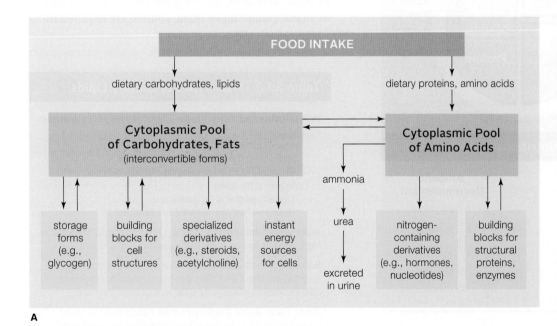

A

B

Liver Functions

Forms bile (assists fat digestion), rids body of excess cholesterol and blood's respiratory pigments

Controls amino acid levels in the blood; converts potentially toxic ammonia to urea

Controls glucose level in blood; major reservoir for glycogen

Removes hormones that served their functions from blood

Removes ingested toxins, such as alcohol, from blood

Breaks down worn-out and dead red blood cells, and stores iron

Stores some vitamins

Figure 39.12 (**A**) Summary of major pathways of organic metabolism. Cells continually synthesize and break down carbohydrates, fats, and proteins. Most urea forms in the liver, an organ that is at the crossroads of organic metabolism (**B**).

39.10 Human Nutritional Requirements

■ You are what you eat—diet profoundly affects your body's structure and function. So what should you eat?

■ Links to Carbohydrates 3.4; Lipids 3.1, 3.5; Proteins 3.6; Lactose intolerance 10.5

USDA Dietary Recommendations

The United States Department of Agriculture (USDA) reviews research on human nutrition and issues a set of dietary guidelines. The guidelines summarize nutrition-related discoveries and explain how dietary choices affect the risk of chronic health problems. At the USDA web site (www.ChooseMyPlate.gov), an interactive program based on these guidelines generates recommendations specific for a person's age, sex, height, weight, and activity level (**Figure 39.13**). Current guidelines recommend eating less refined grains, saturated fats, *trans* fatty acids, sugar or caloric sweeteners, and salt. They also recommend eating more vegetables and fruits with a high potassium and fiber content, more whole grains, and more fat-free or low-fat milk products. The lactose-intolerant (Section 10.5) are advised to choose alternative high-calcium foods.

Energy-Rich Carbohydrates

Fresh fruits, whole grains, and vegetables—especially legumes such as peas and beans—provide abundant complex carbohydrates (Section 3.4). The body breaks the starch in these foods into glucose, your primary source of energy. These foods also provide essential vitamins and fiber. Eating foods high in soluble fiber helps lower one's cholesterol level and may reduce the risk of heart disease. A diet high in insoluble fiber helps prevent constipation.

Foods rich in processed carbohydrates such as white flour, refined sugar, and corn syrup are sometimes said to be full of "empty calories." This is a way of saying that these foods provide little in the way of vitamins or fiber.

You may have noticed breads and other grain-based foods labeled as "gluten-free." Gluten is a protein found in wheat and many other grains. An estimated 1 percent of the population has celiac disease, a genetic disorder in which gluten causes an autoimmune reaction that harms the small intestine's villi. The disease is treated by eliminating gluten from the diet.

Good Fat, Bad Fat

Your body uses lipids to build cell membranes, as energy stores, and as a reservoir for fat-soluble vitamins. Linoleic acid and alpha-linolenic acid are **essential fatty acids**, meaning the human body needs them but cannot make them, so they are required in

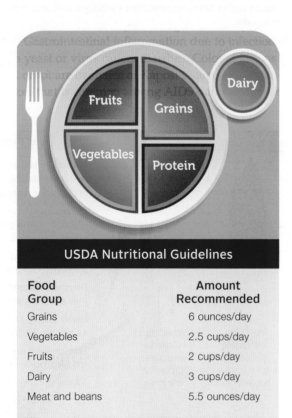

USDA Nutritional Guidelines

Food Group	Amount Recommended
Grains	6 ounces/day
Vegetables	2.5 cups/day
Fruits	2 cups/day
Dairy	3 cups/day
Meat and beans	5.5 ounces/day

Figure 39.13 Example of nutritional guidelines from the United States Department of Agriculture (USDA). These recommendations are for a 20-year-old female, 5 foot 5 inches tall, who weighs 130 pounds and exercises less than 30 minutes a day.

Table 39.2   Main Types of Dietary Lipids

Polyunsaturated Fatty Acids: Liquid at room temperature; essential for health.
    Omega-3 fatty acids
        Alpha-linolenic acid and its derivatives
        Sources: Nut oils, vegetable oils, oily fish
    Omega-6 fatty acids
        Linoleic acid and its derivatives
        Sources: Nut oils, vegetable oils, meat

Monounsaturated Fatty Acids: Liquid at room temperature. Main dietary source is olive oil. Beneficial in moderation.
Saturated Fatty Acids: Solid at room temperature. Main sources are meat and dairy products, palm and coconut oils. Excessive intake may raise risk of heart disease.
***Trans* Fatty Acids (Hydrogenated Fats):** Solid at room temperature. Manufactured from vegetable oils and used in many processed foods. Excessive intake may raise risk of heart disease.

the diet. Both are polyunsaturated fats; their long carbon tails include two or more double bonds (**Table 39.2**). Unsaturated fats are liquid at room temperature (Section 3.5).

We divide the polyunsaturated fatty acids into two categories: omega-3 fatty acids and omega-6 fatty acids. Omega-3 fatty acids, the main fat in oily fish such as sardines, seem to have special health benefits. Studies suggest that a diet high in omega-3 fatty acids can reduce the risk of cardiovascular disease, lessen the inflammation associated with rheumatoid arthritis, and help diabetics control their blood glucose.

Oleic acid, the main fat in olive oil, may also have health benefits. It is monounsaturated, which means its carbon tails have only one double bond. A diet in which olive oil is substituted for saturated fats helps prevent heart disease.

Meats and full-fat dairy products contain large amounts of saturated fats. Overindulging in these foods increases risk of cardiovascular disease and some cancers.

Trans fatty acids, or *trans* fats, are manufactured from vegetable oils. However, they have a molecular structure that makes them even worse for the heart than saturated fats (Section 3.1). All food labels are now required to show the amounts of *trans* fats, saturated fats, and cholesterol per serving (**Figure 39.14**).

Body-Building Proteins

Amino acids are building blocks of proteins (Section 3.6). Your cells can make some amino acids but you must obtain eight **essential amino acids** from food. Essential amino acids are methionine (or cysteine, its metabolic equivalent), isoleucine, leucine, lysine, phenylalanine, threonine, tryptophan, and valine.

Most proteins in meat are "complete," meaning their amino acid ratios match a human's nutritional needs. By contrast, most plant proteins are "incomplete," meaning they lack one or more amino acids essential for the human diet.

The American Dietetic Association states that, with careful planning, a vegetarian diet can provide all essential nutrients for people in any stage of life. To obtain all the required amino acids from plant sources alone, one must combine foods so that the amino acids missing from one component are present in some others. As an example, rice and beans together provide all necessary amino acids, but rice alone or beans alone do not. You do not have to eat the two complementary foods at the same meal, but both should be consumed within a 24-hour period.

Nutrition Facts
Serving Size 1 cup (228g)
Servings Per Container 2

Amount Per Serving

Calories 250	Calories from Fat 110

	% Daily Value*
Total Fat 12g	18%
Saturated Fat 3g	15%
Trans Fat 1.5g	
Cholesterol 30mg	10%
Sodium 470mg	20%
Total Carbohydrate 31g	10%
Dietary Fiber 0g	0%
Sugars 5g	
Protein 5g	

Vitamin A	4%
Vitamin C	2%
Calcium	20%
Iron	4%

* Percent Daily Values are based on a 2,000 calorie diet. Your Daily Values may be higher or lower depending on your calorie needs:

	Calories:	2,000	2,500
Total Fat	Less than	65g	80g
Sat Fat	Less than	20g	25g
Cholesterol	Less than	300mg	300mg
Sodium	Less than	2,400mg	2,400mg
Total Carbohydrate		300g	375g
Dietary Fiber		25g	30g

Figure 39.14 How to read a food label. Information on a food label can be used to ensure that you get the nutrients you need without exceeding recommended limits on less healthy substances such as salt and *trans* fats.

Figure It Out: What amount of the fat in a serving of this macaroni and cheese product comes from the least healthy forms of fat (saturated fat and *trans* fat)? How much is from unhealthy sources?

Answer: Of the total fat content in a serving (12 grams), 3 g are saturated fat and 1.5g are *trans* fat. Thus 4.5g, or 1/3 the fat, is from unhealthy sources.

essential amino acid Amino acid that the body cannot make and must obtain from food.

essential fatty acid Fatty acid that the body cannot make and must obtain from the diet.

Take-Home Message

What are the main types of nutrients that humans require and what is the healthiest way to obtain them?

» A healthy diet provides energy and all necessary building blocks for assembling essential body components.

» Nutritional guidelines are periodically revised in light of new research. Current guidelines call for most calories to come from complex carbohydrates, rather than simple sugars. They also favor fat and protein sources that are low in saturated and *trans* fats.

» A person can obtain all the required nutrients from a vegetarian diet, but doing so requires combining plant foods so amino acids lacking in one are present in the other.

39.11 Vitamins, Minerals, and Phytochemicals

■ In addition to major nutrients, the body requires certain organic and inorganic substances to function properly.

■ Links to Electron transfer chains 5.5, Coenzymes 5.6, Macular degeneration 33.8, Thyroid hormones 34.7, Hemoglobin 38.7

Vitamins are organic substances that are essential in very small amounts; no other substance can carry out their metabolic functions. At a minimum, human cells require the thirteen vitamins listed in **Table 39.3**. Each

has specific roles. For instance, the B vitamin niacin is modified to make NAD, a coenzyme (Section 5.6).

Minerals are naturally occurring inorganic substances. Some have essential metabolic functions and must be obtained from food (**Table 39.4**). As an example, all of your cells use iron as a component of electron transfer chains (Section 5.5). Red blood cells require iron to make oxygen-transporting hemoglobin, and zinc is a cofactor for their carbonic anhydrase

Table 39.3 Major Vitamins: Sources, Functions, and Effects of Deficiencies or Excesses*

Vitamin	Common Sources	Main Functions	Effects of Chronic Deficiency	Effects of Extreme Excess
Fat-Soluble Vitamins				
A	Its precursor comes from beta-carotene in yellow fruits, yellow or green leafy vegetables; also in fortified milk, egg yolk, fish, liver	Used in synthesis of visual pigments, bone, teeth; maintains epithelia	Dry, scaly skin; lowered resistance to infections; night blindness; permanent blindness	Malformed fetuses; hair loss; changes in skin; liver and bone damage; bone pain
D	Inactive form made in skin, activated in liver, kidneys; in fatty fish, egg yolk, fortified milk products	Promotes bone growth and mineralization; enhances calcium absorption	Bone deformities (rickets) in children; bone softening in adults	Retarded growth; kidney damage; calcium deposits in soft tissues
E	Whole grains, dark green vegetables, vegetable oils	Counters effects of free radicals; helps maintain cell membranes; blocks breakdown of vitamins A and C in gut	Lysis of red blood cells; nerve damage	Muscle weakness; fatigue; headaches; nausea
K	Gut bacteria make most of it; also in green leafy vegetables, cabbage	Blood clotting; ATP formation via electron transport	Abnormal blood clotting; severe bleeding (hemorrhaging)	Anemia; liver damage and jaundice
Water-Soluble Vitamins				
B$_1$ (thiamin)	Whole grains, green leafy vegetables, legumes, lean meats, eggs	Connective tissue formation; folate utilization; coenzyme action	Water retention in tissues; tingling sensations; heart changes; poor coordination	None reported from food; possible shock reaction from repeated injections
B$_2$ (riboflavin)	Whole grains, poultry, fish, egg white, milk	Coenzyme action (FAD)	Skin lesions	None reported
B$_3$ (niacin)	Green leafy vegetables, potatoes, peanuts, poultry, fish, pork, beef	Coenzyme action (NAD$^+$)	Contributes to pellagra (damage to skin, gut, nervous system, etc.)	Skin flushing; possible liver damage
B$_6$	Spinach, tomatoes, potatoes, meats	Coenzyme in amino acid metabolism	Skin, muscle, and nerve damage; anemia	Impaired coordination; numbness in feet
Pantothenic acid	In many foods (meats, yeast, egg yolk especially)	Coenzyme in glucose metabolism, fatty acid and steroid synthesis	Fatigue; tingling in hands; headaches; nausea	None reported; may cause diarrhea occasionally
Folate (folic acid)	Dark green vegetables, whole grains, yeast, lean meats; enterobacteria produce some folate	Coenzyme in nucleic acid and amino acid metabolism	A type of anemia; inflamed tongue; diarrhea; impaired growth; mental disorders	Masks vitamin B$_{12}$ deficiency
B$_{12}$	Poultry, fish, red meat, dairy foods (not butter)	Coenzyme in nucleic acid metabolism	A type of anemia; impaired nerve function	None reported
Biotin	Legumes, egg yolk; colon bacteria produce some	Coenzyme in fat, glycogen formation and in amino acid metabolism	Scaly skin (dermatitis); sore tongue; depression; anemia	None reported
C (ascorbic acid)	Fruits and vegetables, especially citrus, berries, cantaloupe, cabbage, broccoli, green pepper	Collagen synthesis; possibly inhibits effects of free radicals; structural role in bone, cartilage, and teeth; used in carbohydrate metabolism	Scurvy; poor wound healing; impaired immunity	Diarrhea, other digestive upsets; may alter results of some diagnostic tests

* Guidelines for appropriate daily intakes are being worked out by the Food and Drug Administration.

(Section 38.7). Iodine is essential for development of a healthy nervous system and synthesis of thyroid hormone (Section 34.7).

Most people can get all the vitamins and minerals they need from a well-balanced diet. Some studies have even found an increased death rate among people who take supplemental antioxidants (beta-carotene, vitamin A, and vitamin E). If you take an over-the-counter multivitamin, be sure the quantities it provides are not excessive.

mineral Naturally occurring inorganic substance; some such as iron are required in small amounts for normal metabolism.
phytochemical Plant molecules that are not an essential part of the human diet, but may reduce risk of certain disorders; e.g., lutein.
vitamin Organic substance required in small amounts for normal metabolism.

In addition to vitamins and minerals, a healthy diet should include a variety of **phytochemicals**, also known as phytonutrients. These organic molecules are found in plant foods and they reduce the risk of certain disorders. For example, leafy green vegetables contain the plant pigments lutein and zeaxanthin. A diet low in these phytochemicals increases the risk of macular degeneration and blindness (Section 33.8).

Take-Home Message

What roles do vitamins, minerals, and phytonutrients play?

» Vitamins are organic molecules with an essential role in metabolism.

» Minerals are inorganic substances with an essential role.

» Phytochemicals are plant molecules that are not essential but may reduce the risk of certain disorders.

Table 39.4   Major Minerals: Sources, Functions, and Effects of Deficiencies or Excesses*

Mineral	Common Sources	Main Functions	Effects of Chronic Deficiency	Effects of Extreme Excess
Calcium	Dairy products, dark green vegetables, dried legumes	Bone, tooth formation; blood clotting; neural and muscle action	Stunted growth; fragile bones; nerve impairment; muscle spasms	Impaired absorption of other minerals; kidney stones in susceptible people
Chloride	Table salt (usually too much in diet)	HCl formation in stomach; contributes to body's acid–base balance; neural action	Muscle cramps; impaired growth; poor appetite	Contributes to high blood pressure in certain people
Copper	Nuts, legumes, seafood, drinking water	Used in synthesis of melanin, hemoglobin, and some transport chain components	Anemia; changes in bone and blood vessels	Nausea; liver damage
Fluorine	Fluoridated water, tea, seafood	Bone, tooth maintenance	Tooth decay	Digestive upsets; mottled teeth and deformed skeleton in chronic cases
Iodine	Marine fish, shellfish, iodized salt, dairy products	Thyroid hormone formation	Enlarged thyroid (goiter) with metabolic disorders	Toxic goiter
Iron	Whole grains, green leafy vegetables, legumes, nuts, eggs, lean meat, molasses, dried fruit, shellfish	Formation of hemoglobin and cytochrome (transport chain component)	Iron-deficiency anemia; impaired immune function	Liver damage; shock; heart failure
Magnesium	Whole grains, legumes, nuts, dairy products	Coenzyme role in ATP–ADP cycle; roles in muscle, nerve function	Weak, sore muscles; impaired neural function	Impaired neural function
Phosphorus	Whole grains, poultry, red meat	Component of bone, teeth, nucleic acids, ATP, phospholipids	Muscular weakness; loss of minerals from bone	Impaired absorption of minerals into bone
Potassium	Diet alone provides ample amounts	Muscle and neural function; roles in protein synthesis and body's acid–base balance	Muscular weakness	Muscular weakness; paralysis; heart failure
Sodium	Table salt; diet provides ample to excessive amounts	Key role in body's salt–water balance; roles in muscle and neural function	Muscle cramps	High blood pressure in susceptible people
Sulfur	Proteins in diet	Component of body proteins	None reported	None likely
Zinc	Whole grains, legumes, nuts, meats, seafood	Component of digestive enzymes; roles in normal growth, wound healing, sperm formation, and taste and smell	Impaired growth; scaly skin; impaired immune function	Nausea, vomiting, diarrhea; impaired immune function and anemia

* Guidelines for appropriate daily intakes are being worked out by the Food and Drug Administration.

39.12 Maintaining a Healthy Weight

■ Maintaining a healthy weight requires balancing energy inputs with energy expenditures.

■ Links to Triglycerides 3.5, Adipose tissue 31.4, Insulin and diabetes 34.8, Inflammation 37.5

What Is a Healthy Weight?

The body mass index (BMI) is a measurement designed to assess increased health risk associated with weight gains. You can calculate your body mass index with this formula:

$$BMI = \frac{weight\ (pounds) \times 703}{height\ (inches)^2}$$

Generally, individuals with a BMI of 25 to 29.9 are said to be overweight. A score of 30 or more indicates obesity: an overabundance of fat in adipose tissue that may lead to severe health problems. The distribution of body fat also helps predict the risks. Fat deposits just above the belt, as in a "beer belly," are associated with an increased likelihood of heart problems.

To maintain a healthy weight, you must balance your caloric intake and energy output. Energy stored in food is expressed as kilocalories, or Calories (with a capital C). One kilocalorie equals 1,000 calories, which are units of heat energy.

Here is a way to calculate roughly how many kilocalories you should take in daily to maintain a preferred weight. First, multiply the weight (in pounds) by 10 if you are not active physically, by 15 if you are moderately active, and by 20 if you are highly active. Next, subtract one of the following amounts from the multiplication result:

Age:	Subtract:
25–34	0
35–44	100
45–54	200
55–64	300
Over 65	400

For example, if you are 25 years old, are highly active, and weigh 120 pounds, you will require 120 × 20 = 2,400 kilocalories daily to maintain weight. If you want to gain weight you will require more; to lose, you will require less.

Why Is Obesity Unhealthy?

Being obese has a negative effect on health. Among other things, it increases the risk of type 2 diabetes, high blood pressure, heart disease, breast and colon cancer, arthritis, and gallbladder problems.

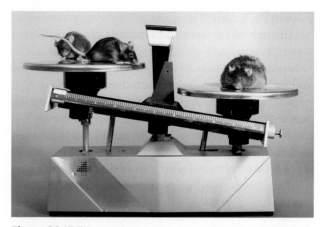

Figure 39.15 Effect of the hormone leptin. Two normal mice (*left*) weigh less than one mutant mouse (*right*) that cannot synthesize leptin. Leptin is made by fat cells and it suppresses appetite.

Why does excess weight have ill effects? As Section 7.7 explained, triglycerides in fat cells are the body's main form of energy storage. Fat cells of people who are at a healthy weight hold a moderate amount of triglycerides and function normally. In obese people, an excess of these molecules distends fat cells and impairs their function. Like cells damaged in other ways, the overstuffed fat cells respond by sending out signals that encourage an inflammatory response (Section 37.5). The resulting chronic inflammation harms organs throughout the body and increases the risk of cancer. Overstuffed fat cells also increase secretion of signals that interfere with the action of insulin. Remember that this hormone encourages cells to take up sugar from the blood (Section 34.8). When insulin becomes ineffective, the result is type 2 diabetes.

Genetics of Obesity

Numerous studies have explored the role that genetics plays in obesity. As one example, Claude Bouchard studied the effect of overeating in twelve pairs of male twins. All were lean young men in their early twenties. For 100 days they did not exercise, and they adhered to a diet that provided 6,000 more kilocalories a week than usual. All the men gained weight, but some gained three times as much as others. Members of each set of twins tended to gain a similar amount, indicating that genes affect the response to overfeeding.

Geneticists have pinpointed some genes that affect weight. The first to be discovered, the *ob* gene, encodes leptin, a hormone made by adipose cells. Leptin acts in the brain and suppresses appetite. Mice that do not have a functional *ob* gene overeat and become fat (**Figure 39.15**). Similarly, people who cannot make leptin

Your Microbial "Organ" (revisited)

For a probiotic such as yogurt to be effective it must include beneficial bacteria or spores capable of surviving a trip through the highly acidic stomach, adhering to the colon wall, and growing amidst competing bacteria. Researchers are studying which bacteria meet these requirements and thus are best suited for use in probiotic foods and pills.

An alternative medical procedure delivers bacteria known to do well in the colon exactly where they are needed. It's called a fecal transplant. A doctor takes feces from a healthy donor and uses an endoscope (a long tube with a camera used in colonoscopy) to insert it deep in the colon of a recipient. The recipient is a person with a chronic debilitating bowel disease, whose colon has been cleansed of bacteria prior to treatment. The goal is to establish a healthy colon microbiota and thus restore the recipient to health. Many patients who have undergone the procedure say it dramatically improved their life, but some doctors worry that the procedure can spread pathogens.

How would you vote? The first controlled clinical trials of fecal transplants are only now getting under way. Is it ethical to carry out such an unusual procedure before a controlled study shows it is safe and effective?

are severely obese. When leptin-deficient mice or humans are injected with leptin, they eat less and slim down as a result.

Human leptin deficiency is extremely rare, but variations in another obesity-related gene, *fto*, are common. About 16 percent of people of European ancestry are homozygous for an *fto* allele that predisposes them to obesity. Compared to people with two low-risk alleles, those homozygous for the high-risk allele are almost twice as likely to be obese. The function of the protein encoded by the *fto* gene is unknown. We do, however, know that the gene is expressed most strongly in the brain.

Genetics can explain why one person is more likely than another to be overweight, but it cannot explain a national trend toward weight gain. Since 1980, the proportion of obese adults in the United States has doubled, and the proportion of obese children has tripled. Our gene pool has not changed, but our habits have. We eat more and move less.

Eating Too Little

Eating too little can be as dangerous as eating too much, or more so. With anorexia nervosa, a person who has access to food routinely eats too little to maintain a weight within 15 percent of normal. Although anorexia nervosa means "nervous loss of appetite," most affected people are obsessed with food and continually hungry. They typically see themselves as fat, even when they are dangerously thin.

Young women are disproportionately affected, and societal pressures to be thin certainly play a role in encouraging weight loss. However, how a person responds to such pressures is determined in part by genetics. Scientists have pinpointed several genes that increase the risk of anorexia.

Anorexia damages organ systems throughout the body. Starved of essential calcium, the body breaks down bone. Inadequate iron intake causes anemia. Heart muscle weakens and heart rhythms can become disrupted. Most anorexia-related deaths occur as a result of sudden cardiac arrest.

Bulimia nervosa, another eating disorder, is also most common among young women. Bulimics tend to be near normal weight, but perceive themselves as heavy. They "binge and purge," eating far too much, then inducing vomiting. Induced vomiting bathes their esophagus and teeth in gastric fluid, making teeth pitted and brittle, and increasing the risk of esophageal cancer. Loss of gastric fluid to vomiting depletes the body of hydrogen ions, making body fluids more basic. The shift in pH can cause apnea, disrupt normal heart rhythm, and lead to convulsions.

Both anorexia and bulimia are treated with counseling and medical attention to the damage done by the disorder. Many patients respond favorably to treatment and are able to maintain a normal weight.

Take-Home Message

How does weight affect health?

» A person who balances caloric intake with energy expenditures will maintain current weight.

» Obesity raises the risk of heart disease, type 2 diabetes, some cancers, and other disorders. These problems may arise because overstuffed adipose cells summon up inflammatory responses in organs throughout the body.

» Anorexia is an eating disorder that results in a lower than normal body weight. It harms organs throughout the body, especially the bones and the heart. Anorexia-induced changes in heart rhythms can be fatal.

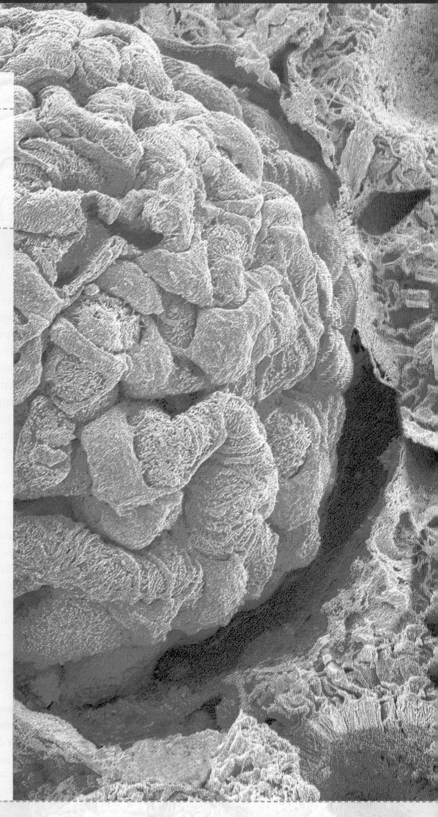

LEARNING ROADMAP

Where you have been This chapter's discussion of fluid balance touches on pH (2.6), osmosis (5.8), aerobic respiration (7.2), protein metabolism (39.9), and body fluids (31.2). Fluid balance involves osmoreceptors (33.2), the hypothalamus (32.9), the pituitary gland (34.4), and adrenal glands (34.9). The discussion of body temperature refers to properties of water (2.5), forms of energy (5.2), and negative feedback controls (31.9).

Where you are now

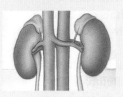

Maintaining the Extracellular Fluid
Animals have organs that maintain the composition and volume of the extracellular fluid within a narrow range. Variations in these organs adapt animals to different habitats.

The Human Urinary System
Humans have two blood-filtering, urine-forming kidneys. Ureters convey urine to a bladder, from which it flows out of the body through the urethra.

What Kidneys Do
Fluid and small solutes from blood enter kidney tubules. Most water and solutes are reabsorbed. Fluid not reabsorbed becomes urine. Hormones adjust urine composition.

Adjusting Body Temperature
Heat lost to the environment, heat gained from the environment, and heat from metabolic activity determine body temperature. Endotherms give off more heat than ectotherms.

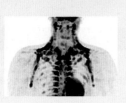

Human Body Temperature
The hypothalamus regulates human body temperature. Sweating and moving more blood to the skin help cool a body. Shivering and moving blood to the core keep it warm.

Where you are going The male urethra does double duty, conveying both urine and sperm to the body surface. We discuss its reproductive function in the next chapter. That chapter also describes the hormone that is detected by urine tests for pregnancy.

40.1 Truth in a Test Tube

Light or dark? Clear or cloudy? A lot or a little? Asking about and examining urine is an ancient art (Figure 40.1). About 3,000 years ago in India, the pioneering healer Susruta reported that some patients formed an excess of sweet-tasting urine that attracted insects. In time, the disorder was named diabetes mellitus, which loosely translates as "passing honey-sweet water." Doctors still diagnose it by testing the sugar level in urine, although they have replaced the taste test with chemical analysis.

Today, physicians routinely check the pH and solute concentrations of urine to monitor their patients' health. Acidic urine suggests metabolic problems. Alkaline urine can indicate an infection. Damaged kidneys will produce urine high in proteins. An abundance of some salts can result from dehydration or trouble with the hormones that control kidney function. Special urine tests detect chemicals produced by cancers of the kidney, bladder, and prostate gland.

Do-it-yourself urine tests are available to test hormone levels. If a woman is hoping to become pregnant, she can use one test to keep track of the amount of luteinizing hormone, or LH, in her urine. About midway through a menstrual cycle, LH triggers ovulation, the release of an egg from an ovary. Another over-the-counter urine test can reveal whether she has become pregnant. Still other tests help older women check for declining hormone levels in urine, a sign that they are entering menopause.

Not everyone is in a hurry to have their urine tested. Olympic athletes can be stripped of their medals when mandatory urine tests reveal they use prohibited drugs. Major League Baseball players agreed to urine tests only after repeated allegations that certain star players took prohibited steroids. The National Collegiate Athletic Association (NCAA) tests urine samples from about 3,300 student athletes per year for any performance-enhancing substances as well as for "street drugs."

If you use marijuana, cocaine, Ecstasy, or other kinds of psychoactive drugs, urine tells the tale. After the active ingredient of marijuana enters blood, the liver converts it to another compound. As kidneys filter blood, they add the compound to newly forming urine. It can take as long as ten days for all molecules of the compound to become fully metabolized and removed from the body. Until that happens, urine tests can detect it.

It is a tribute to the urinary system that urine is such a remarkable indicator of health, hormonal status, and drug use. Each day, a pair of fist-sized kidneys filter all of the blood in an adult human body, and they do so more than forty times. When all goes well, kidneys rid the body of excess water and excess or harmful solutes, including a variety of metabolites, toxins, hormones, and drugs.

So far in this unit, you have considered several organ systems that work to keep cells supplied with oxygen, nutrients, water, and other substances. Turn now to the kinds that maintain the composition, volume, and even the temperature of the internal environment.

Figure 40.1 A seventeenth-century physician examining a urine specimen. Urine's consistency, color, odor, and—at least in the past—taste supply information about a person's health. Urine forms inside kidneys, and it provides clues to abnormal changes in the volume and composition of blood and interstitial fluid.

40.2 Regulating Fluid Volume and Composition

■ All animals constantly acquire and lose water and solutes, yet they must keep the volume and composition of their body fluids stable.

■ Links to Osmosis 5.8, Aerobic respiration 7.2, Coelom 24.2, Extracellular fluid 31.2, Fate of absorbed compounds 39.9

By weight, an animal consists mostly of water, with dissolved salts and other solutes. Fluid outside cells—the extracellular fluid—is the cells' environment. In vertebrates, interstitial fluid (fluid between cells) and plasma (the fluid portion of the blood) constitute the bulk of the extracellular fluid (Section 31.2).

To maintain the solute composition and volume of the extracellular fluid within the range that cells can tolerate, water and solute gains must equal water and solute losses. An animal loses water and solutes in excretions, exhalations, and secretions. It gains water by eating and drinking. In aquatic animals, water also moves into or out of the body by osmosis across the body surface (Section 5.8).

Metabolic wastes, particularly carbon dioxide and ammonia, affect the composition of the extracellular fluid. Aerobic respiration produces water and carbon dioxide. Carbon dioxide diffuses out across the body surface or leaves with the help of respiratory organs.

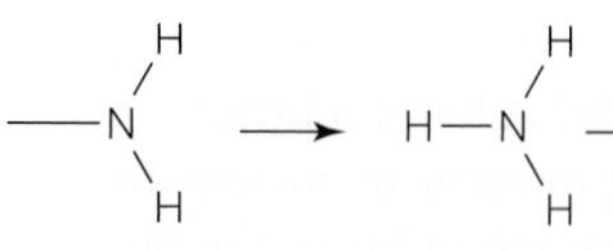

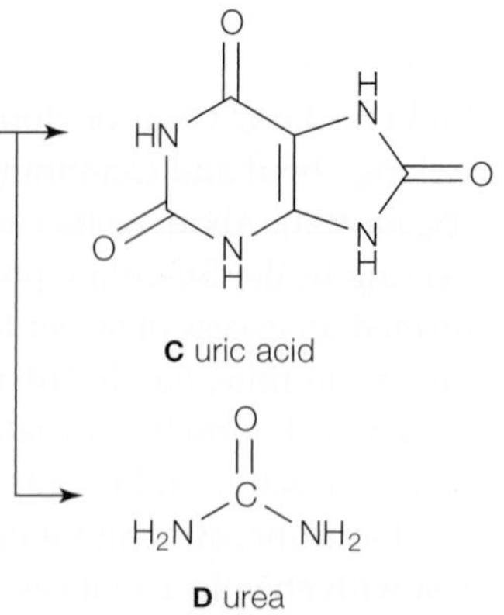

Figure 40.2 Nitrogenous waste products. Amino groups (**A**) cleaved from proteins become toxic ammonia (**B**).

Some animals excrete ammonia directly; others first convert it to uric acid (**C**), or urea (**D**).

A amino group **B** ammonia **C** uric acid **D** urea

Protein breakdown produces **ammonia** (**Figure 40.2**). In most animals, special organs rid the body of ammonia, other unwanted solutes, and any excess water.

Fluid Regulation in Invertebrates

Marine invertebrates usually have body fluids with the same solute concentration as seawater. As a result, osmosis produces no net movement of water into or out of the body.

In planarian flatworms and other freshwater animals, body fluids have a higher solute concentration than the surrounding water, so water enters by osmosis. Planarians have a pair of branching, tubular, excretory organs (protonephridia) that run the length of the body (**Figure 40.3**). Along the tubes are bulbs tipped by ciliated flame cells. Cilia movement draws interstitial fluid into the tubes, propels it along, and forces it out of the body through pores at the body surface.

In most animals, a circulatory system interacts with organs that excrete unwanted solutes. Consider the

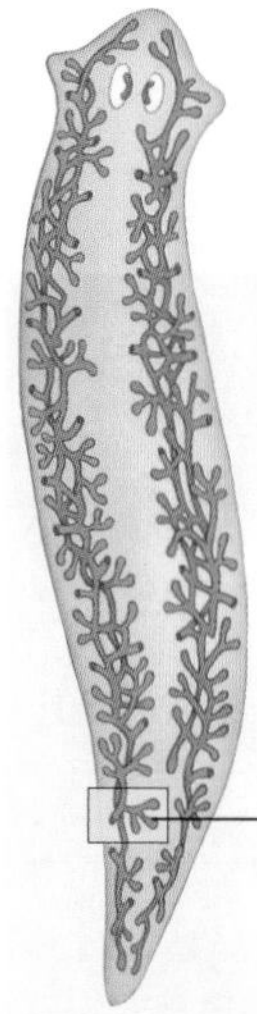

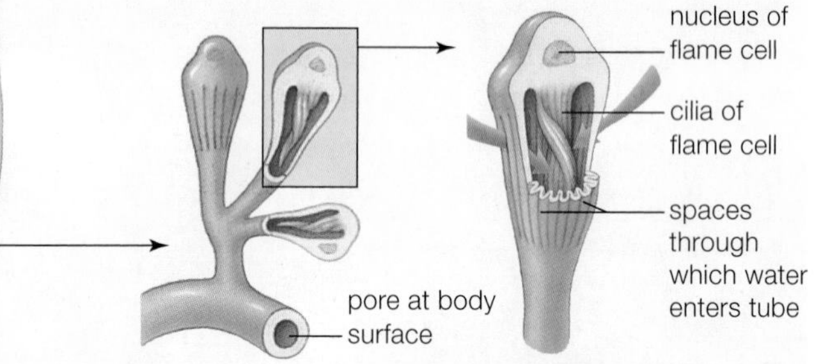

nucleus of flame cell

cilia of flame cell

spaces through which water enters tube

pore at body surface

Figure 40.3 Planarian fluid regulation. Fluid enters bulbs at the tips of branching protonephridia. Each bulb has a flame cell and a tubule cell. Beating of flame cell cilia draws fluid in between the cells. The fluid exits the body through a pore.

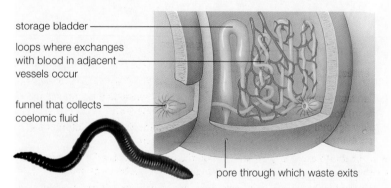

storage bladder

loops where exchanges with blood in adjacent vessels occur

funnel that collects coelomic fluid

pore through which waste exits

Figure 40.4 Earthworm fluid regulation. Coelomic fluid enters a nephridium (*green*). As fluid travels through it, essential solutes leave this tube and enter adjacent blood vessels (*red*). Ammonia-rich waste exits the body through a pore.

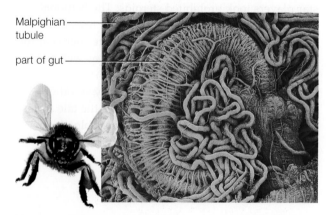

Malpighian tubule

part of gut

Figure 40.5 Insect fluid regulation. A honeybee's Malpighian tubules (*gold*) are outpouchings of the gut (*pink*). Uric acid and other waste solutes move from the hemolymph into the tubule interior. The tubules deliver the wastes to the gut for elimination through the anus.

earthworm, a segmented annelid with a fluid-filled body cavity (a coelom) and a closed circulatory system. Most body segments have a pair of tubular excretory organs called nephridia that collect coelomic fluid (**Figure 40.4**). As fluid flows through a nephridium, essential solutes and some water exit the nephridium and enter adjacent blood vessels. Ammonia remains in the tube and exits the body through a pore.

Land-dwelling arthropods such as insects do not excrete ammonia. Instead, enzymes in their blood convert ammonia to **uric acid** (**Figure 40.2C**). Uric acid and other waste solutes are actively transported into excretory organs called Malpighian tubules, which connect to and empty into the gut (**Figure 40.5**). Ammonia can only be excreted when dissolved in water, but uric acid is excreted as crystals mixed with just a tiny bit of water to produce a thick paste.

Fluid Regulation in Vertebrates

Body fluids of bony fishes are less salty than seawater, but saltier than fresh water. A marine bony fish loses water by osmosis across its body surfaces. It replaces this lost water by gulping seawater, then pumping salt out through its gills (**Figure 40.6A**). Like other vertebrates, bony fishes have a pair of **kidneys**, organs that filter the blood and produce urine. **Urine** is water and soluble wastes. Marine bony fishes produce a small amount of urine. In contrast, a freshwater bony fish produces a large volume of urine because water enters its body by osmosis. Solutes lost in urine are offset by solutes absorbed from the gut, and by sodium ions pumped in across the gills (**Figure 40.6B**).

Waterproof skin and highly efficient kidneys adapt amniotes to life on land. Birds and other reptiles convert ammonia to uric acid, whereas mammals convert most of it to **urea** (**Figure 40.2D**). It takes twenty times more water to excrete 1 gram of urea than to excrete 1 gram of uric acid. Thus, a typical mammal requires more water than a bird or reptile of similar size.

Variations in kidney structure adapt mammals to different habitats. Mammals with limited or no access to fresh water tend to have large kidneys for their size and they lose little water in their urine (**Figure 40.7**).

ammonia Nitrogen-containing compound that is a waste product of amino acid and nucleic acid breakdown.
kidney Organ of the vertebrate urinary system that filters blood, adjusts its composition, and forms urine.
urea Main nitrogen-containing compound in urine of mammals.
uric acid Main nitrogen-containing compound in the urine of insects, as well as birds and other reptiles.
urine Mix of water and soluble wastes formed and excreted by the vertebrate urinary system.

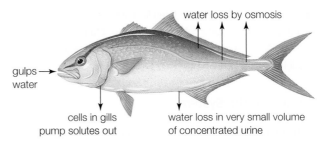

A Marine bony fish with body fluids less salty than the surrounding water; the fish is hypotonic relative to its environment.

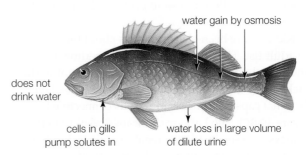

B Freshwater bony fish with body fluids saltier than the surrounding water; the fish is hypertonic relative to its environment.

Figure 40.6 Fluid–solute balance in freshwater and marine bony fishes.

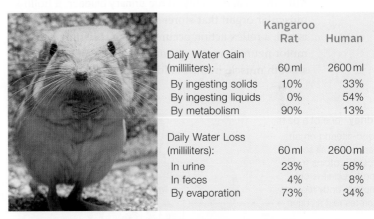

	Kangaroo Rat	Human
Daily Water Gain (milliliters):	60 ml	2600 ml
By ingesting solids	10%	33%
By ingesting liquids	0%	54%
By metabolism	90%	13%
Daily Water Loss (milliliters):	60 ml	2600 ml
In urine	23%	58%
In feces	4%	8%
By evaporation	73%	34%

Figure 40.7 Comparison of water gains and losses for a human and a desert kangaroo rat. In both species, fluid gains must balance fluid losses.

Take-Home Message

How do animals maintain the volume and composition of their body fluid?

» All animals must rid the body of waste carbon dioxide and ammonia; many convert ammonia to urea or uric acid before excreting it.

» Most animals have excretory organs that interact with a circulatory system to remove wastes from the blood and excrete them.

» Invertebrate excretory organs include the ammonia-excreting nephridia of earthworms and the uric acid–excreting Malpighian tubules of insects.

» All vertebrates have two kidneys. The volume of urine and the type of nitrogen-containing wastes excreted (ammonia, urea, or uric acid) vary among groups.

40.3 The Human Urinary System

- Kidneys filter water and solutes from the blood, adjust the volume and composition of this filtrate, and return most of it to the blood. Fluid not returned to the blood becomes urine.
- Links to Epithelium 31.3, Body cavities 31.7

Organs of the System

A human urinary system includes two kidneys, two ureters, one urinary bladder, and one urethra (**Figure 40.8A**). The kidneys filter blood and form urine. The other organs collect and store urine, and convey it to the body surface.

Kidneys are bean-shaped organs about the size of an adult fist. They lie just beneath the peritoneum that lines the abdominal cavity, to the left and right of the backbone (**Figure 40.8B**). The outermost kidney layer, the renal capsule, consists of fibrous connective tissue (**Figure 40.8C**). The Latin *renal* means "relating to the kidneys." Tissue inside the renal capsule is divided into two zones: the outer renal cortex and the inner renal medulla. A renal artery transports blood to each kidney and a renal vein carries blood away from it.

Inside a kidney, urine collects in a central cavity called the renal pelvis. A tubular **ureter** conveys the fluid from each kidney to the **urinary bladder**, a hollow, muscular organ that stores urine. When the bladder is full, a reflex action occurs. Stretch receptors signal motor neurons in the spinal cord. These neurons cause smooth muscle in the bladder wall to contract. At the same time, sphincters encircling the **urethra**, the tube that delivers urine to the body surface, relax. As a result, urine flows out of the body. After age two or three, the brain can override the spinal reflex and prevent urine from flowing through the urethra at inconvenient moments.

In males, the urethra runs the length of the penis. Urine and semen flow through it, but a sphincter cuts off urine flow during erections. In females, the urethra opens onto the body surface between the vagina and the clitoris. A female's urethra is a relatively short tube (about 4 centimeters, or 1.5 inches long), so pathogens move more easily through it to the urinary bladder. That is one reason why women get bladder infections more often than men do.

Nephron Structure

In the next section, we discuss the three processes that rid the body of excess water and solutes in the form of urine. These functions are best understood in light of kidney structure.

A kidney has more than 1 million **nephrons**—microscopically small tubes of cuboidal epithelium associated with capillaries. These kidney tubules are just one cell thick, so substances diffuse easily across them. A kidney tubule begins in the renal cortex, where its wall balloons out and folds back to form a cup-shaped **Bowman's capsule** (**Figure 40.9A,B**). Beyond the cap-

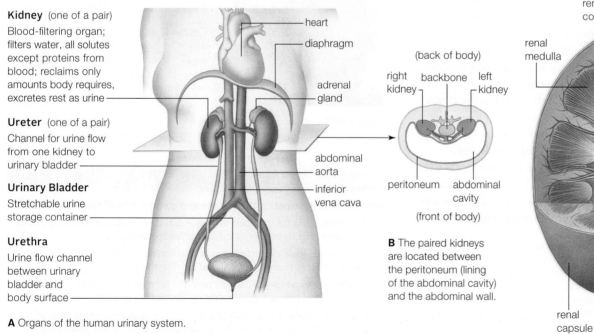

Kidney (one of a pair)
Blood-filtering organ; filters water, all solutes except proteins from blood; reclaims only amounts body requires, excretes rest as urine

Ureter (one of a pair)
Channel for urine flow from one kidney to urinary bladder

Urinary Bladder
Stretchable urine storage container

Urethra
Urine flow channel between urinary bladder and body surface

heart
diaphragm
adrenal gland
abdominal aorta
inferior vena cava

A Organs of the human urinary system.

Figure 40.8 Animated Human urinary system.

(back of body)
right kidney   backbone   left kidney
peritoneum   abdominal cavity
(front of body)

B The paired kidneys are located between the peritoneum (lining of the abdominal cavity) and the abdominal wall.

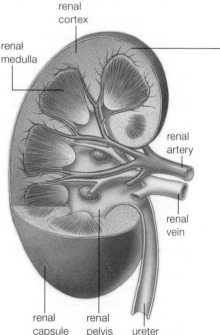

renal cortex
renal medulla
renal artery
renal vein
renal capsule   renal pelvis   ureter

C Structure of a human kidney.

sule, it twists a bit and straightens out as a **proximal tubule** (the part in closest proximity to the beginning of the nephron). After extending down into the renal medulla, the nephron makes a hairpin turn called the **loop of Henle**. The tubule reenters the cortex and twists again, as the **distal tubule** (the part most distant from the start of the nephron). The distal tubules of up to eight nephrons drain into a **collecting tubule**. Many collecting tubules extend through the kidney medulla and open into the renal pelvis. Like the cells lining the small intestine, cells of kidney tubules have microvilli. These tiny extensions increase the surface area for absorption of substances.

Blood vessels absorb substances that leave kidney tubules. Inside each kidney, a renal artery branches into smaller, afferent arterioles. Each arteriole in turn branches into a **glomerulus** (plural, glomeruli), a cluster of capillaries in Bowman's capsule (**Figure 40.9C**). Glomerular capillaries have gaps between the cells in their walls. These gaps make the capillaries about a hundred times more permeable than a typical capillary. As blood flows through the glomerulus, blood pressure forces some fluid out through the gaps in the capillary wall and into Bowman's capsule.

An efferent arteriole carries blood away from the glomerus and branches into **peritubular capillaries** that thread lacily around the nephron (*peri–*, around). These capillaries are the site for exchanges between the filtered fluid that flows through kidney tubules

and the blood. From the peritubular capillaries, blood continues into venules that carry it to the renal vein.

Bowman's capsule Cup-shaped portion of the nephron that encloses the glomerulus and receives filtrate from it.
collecting tubule Kidney tubule that receives filtrate from several nephrons and delivers it to the renal pelvis.
distal tubule Portion of kidney tubule that delivers filtrate to a collecting tubule.
glomerulus Ball of capillaries enclosed by Bowman's capsule.
loop of Henle U-shaped portion of a kidney tubule; it extends deep into the renal medulla.
nephron Kidney tubule and glomerular capillaries; filters blood and forms urine.
peritubular capillaries Capillaries that surround and exchange substances with a kidney tubule.
proximal tubule Portion of kidney tubule that receives filtrate from Bowman's capsule.
ureter Tube that carries urine from a kidney to the bladder.
urethra Tube through which urine from the bladder flows out of the body.
urinary bladder Hollow, muscular organ that stores urine.

Take-Home Message

What are the components of the human urinary system and how do they function?

» The human urinary system has two kidneys, two ureters, a urinary bladder, and a urethra. Kidneys filter the blood and form urine. Urine flows out of the kidney through ureters, and into a hollow, muscular bladder. When the bladder contracts, urine flows out of the body through the urethra.

» The functional unit of the kidneys is the nephron, a microscopic tubule that interacts with two systems of capillaries to filter blood and form urine.

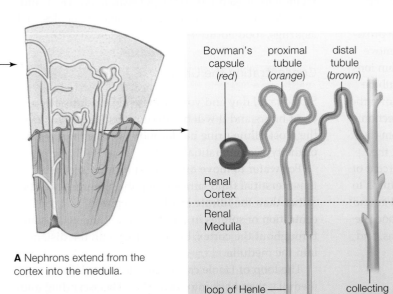

A Nephrons extend from the cortex into the medulla.

Figure 40.9 Animated
Orientation and structure of a nephron, the functional unit of the kidney.

B Tubular portion of one nephron, cutaway view. The tubule starts at Bowman's capsule.

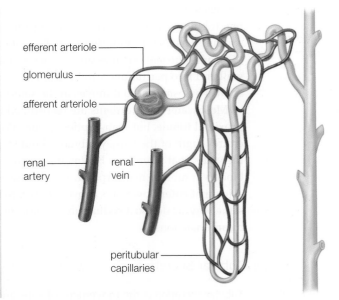

C Blood vessels associated with the nephron. The glomerulus is a ball of capillaries that have unusually leaky walls.

40.4 How Urine Forms

- Urine consists of water and solutes that were filtered from blood and not returned to it, along with solutes secreted from the blood into the nephron's tubular regions.
- Link to Blood pressure 36.9

Urine formation begins when blood pressure drives water and small solutes out of the blood and into a nephron. Variations in permeability along the nephron's tubular parts determine whether components of the filtrate return to blood or leave the body in urine.

Glomerular Filtration

Blood pressure generated by a beating heart drives **glomerular filtration**, the first step in urine formation (**Figure 40.10** and **Figure 40.11** ❶). About 20 percent of the plasma that flows into a glomerulus is forced out through gaps in the capillary walls into Bowman's capsule. The other 80 percent continues to the efferent arteriole. Collectively, the glomerular capillary walls and the inner wall of Bowman's capsule serve as a filter for the blood. Plasma proteins, blood cells, and platelets cannot pass through this filter. They remain in the blood and leave the glomerulus via the efferent arteriole. Filtrate, which is protein-free plasma, drains from Bowman's capsule into the proximal tubule.

Tubular Reabsorption

Only a small fraction of the filtrate actually ends up in urine. **Tubular reabsorption** returns most water and solutes to the blood. Reabsorption begins in the proximal tubule ❷, where active transport proteins move sodium ions (Na^+), chloride ions (Cl^-), potassium ions (K^+), and nutrients such as glucose across the tubule wall and into peritubular capillaries. Water follows the solutes by osmosis, so it moves in the same direction.

Most water and nutrients are reabsorbed from the proximal tubule, but reabsorption occurs along the entire length of the kidney tubule. About 99 percent of the water that filters into Bowman's capsule returns to blood via tubular reabsorption. All glucose and amino acids that enter Bowman's capsule return to blood the same way, as do most sodium ions, chloride ions, and bicarbonate ions.

Tubular Secretion

Tubular secretion is the movement of substances from the blood in peritubular capillaries into the filtrate ❸. Membrane proteins in the walls of peritubular capillaries actively transport substances into the interstitial fluid, from which they cross the epithelium of the kidney tubule and enter the filtrate. Secreted substances include hydrogen ions (H^+), potassium ions (K^+), and breakdown products of foreign organic molecules such as drugs, food additives, and pesticides.

Concentrating the Urine

Sip soda all day and your urine will be dilute; sleep eight hours and it will be concentrated. However, even the most dilute urine has far more solutes than plasma or the typical interstitial fluid.

For water to move out of a nephron by osmosis, the interstitial fluid surrounding the nephron must be saltier than the filtrate inside the nephron. The concentration of solutes in the interstitial fluid is constant throughout the cortex, but increases with the distance into the medulla.

The loop of Henle carries filtrate down into the medulla, then back into the cortex. The ascending and descending arms of the loop differ in their permeabilities to water and sodium. The descending arm of the loop is permeable to water but not to sodium. The loop's ascending arm is impermeable to water and actively pumps salt into the interstitial fluid.

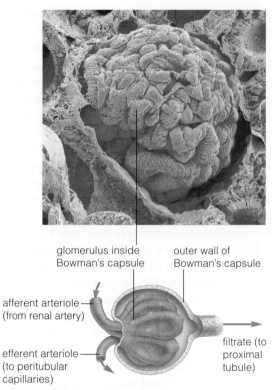

glomerulus inside Bowman's capsule

outer wall of Bowman's capsule

afferent arteriole (from renal artery)

efferent arteriole (to peritubular capillaries)

filtrate (to proximal tubule)

Figure 40.10 Glomerular filtration. Pressure exerted on blood by the beating heart forces protein-free plasma out of the glomerular capillaries and into Bowman's capsule.

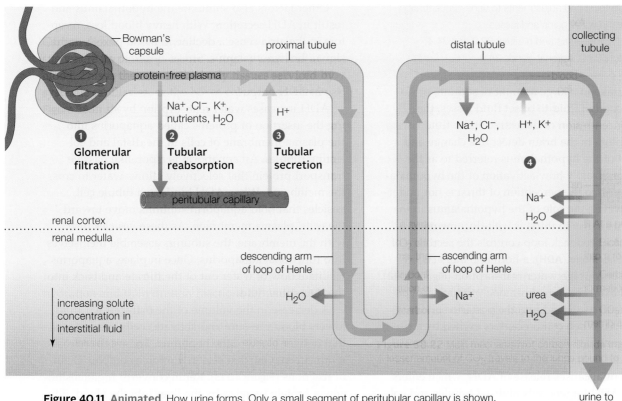

Figure 40.11 Animated How urine forms. Only a small segment of peritubular capillary is shown.
Figure It Out: What process moves H+ from peritubular capillaries into the distal tubule?

Answer: Tubular secretion

1 Glomerular filtration
Protein-free plasma forced out of glomerular capillaries by blood pressure enters Bowman's capsule.

2 Tubular reabsorption
Essential ions, nutrients, water, and some urea in the filtrate return to the blood. *Green* arrows indicate reabsorption.

3 Tubular secretion
Wastes and excess ions are moved from the blood into the filtrate for elimination in urine. *Blue* arrows indicate secretion.

4 Hormones that alter permeability of distal and collecting tubules adjust urine concentration.

A high solute concentration in interstitial fluid draws water out of filtrate as it flows through the descending loop of Henle. Then, ions are actively transported out of the filtrate as it flows through the ascending loop. The ions that leave the ascending loop contribute to the high solute concentration in the interstitial fluid.

Filtrate entering the distal tubule is less concentrated than normal body fluid. The distal tubule delivers this filtrate to the collecting tubule. Like the descending loop of Henle, this tubule extends down into the medulla. In the deepest part of the medulla, urea pumped out of the collecting tubule contributes to the high solute concentration of the interstitial fluid. As urine descends through the collecting tubule, the continually increasing solute concentration of the interstitial fluid around the tubule draws water outward by osmosis.

glomerular filtration First step in urine formation: protein-free plasma forced out of glomerular capillaries by blood pressure enters Bowman's capsule.
tubular reabsorption Substances move from the filtrate inside a kidney tubule into the peritubular capillaries.
tubular secretion Substances move out of peritubular capillaries and into the filtrate in kidney tubules.

The body can adjust how much water is reabsorbed at distal tubules and collecting tubules **4**. When water must be conserved, distal tubules and collecting tubules become more permeable to water, so less leaves in urine. When the body needs to rid itself of excess water, the distal tubule and collecting tubules become less permeable to water and the urine remains dilute. Hormones determine the permeability of the tubules, as described in the next section.

Take-Home Message

How is urine formed and concentrated?

» During glomerular filtration, pressure generated by the beating heart drives water and solutes out of glomerular capillaries and into kidney tubules.

» In tubular reabsorption, water, some ions, glucose, and other solutes move out of the filtrate and return to the blood in peritubular capillaries.

» In tubular secretion, active transport proteins move solutes such as H+ and K+ from peritubular capillaries into the nephron for excretion.

» Differential permeability of the two arms of the loop of Henle sets up a concentration gradient in the interstitial fluid that draws water out of the collecting tubule.

» Urine concentration depends on how much water flows out of the distal tubule and the collecting tubule. Hormones affect the concentration of solutes in urine by their effects on the permeability of these tubules.

40.5 Regulating Thirst and Urine Concentration

- Changes in thirst and adjustments to urine concentration offset solute and water gains and losses.
- Links to Hypothalamus and pituitary 32.9 and 34.4, Osmoreceptors 33.2, Adrenal glands 34.9

When you take in a large amount of sodium, or do not drink enough fluid to offset fluid losses, the sodium concentration of the extracellular fluid rises. Osmoreceptors in the brain detect this change and alert a part of the hypothalamus referred to as the thirst center. Exactly how activation of the hypothalamus gives rise to the perception of thirst is not understood. However, the role of the hypothalamus in the hormonal response to thirst is well-documented.

A negative feedback loop controls the secretion of **antidiuretic hormone (ADH)**, a hormone that acts on kidneys to encourage water reabsorption (**Figure 40.12**). Recall that axons of some hormone-producing hypothalamic neurons extend into the posterior pituitary (Section 34.4). When osmoreceptors excited by a rise in sodium signal the hypothalamus, an action potential travels to the terminals of these axons. Arrival of the action potential causes release of ADH, which enters the blood. ADH targets cells of distal tubules and collecting tubules in the kidney, making them more permeable to water. When ADH is present, more water is reabsorbed and less ends up in urine. Reabsorption of extra water dilutes the extracellular fluid, bringing its sodium concentration back to the optimal level.

Other factors also stimulate the hypothalamus and result in ADH secretion. With heavy blood loss, receptors in the atria sense a decline in blood pressure and signal the hypothalamus. Stress, heavy exercise, or vomiting also cause internal changes that can trigger a rise in ADH output.

ADH increases water reabsorption by stimulating the insertion of proteins called aquaporins into the plasma membrane of cells in the distal and collecting tubules. An aquaporin is a porelike passive transport protein that selectively allows water to cross the membrane. When ADH binds to a tubule cell, vesicles that hold aquaporin subunits move toward the cells' plasma membrane. As these vesicles fuse with the membrane, the subunits assemble themselves into functional aquaporins. Once in place, aquaporins facilitate flow of water out of the filtrate and back into the interstitial fluid.

A decrease in the volume of the extracellular fluid lowers blood pressure and activates cells in arterioles that deliver blood to the nephrons. These cells release renin, an enzyme that sets in motion a complex chain of reactions (**Figure 40.13**). Renin converts angiotensinogen, a protein secreted by the liver into the blood, into angiotensin I. Another enzyme converts angiotensin I to angiotensin II, which acts in the adrenal glands atop the kidneys. The adrenal cortex responds to angiotensin II by secreting the hormone **aldosterone** into the blood. Aldosterone acts on the collecting tubules. It

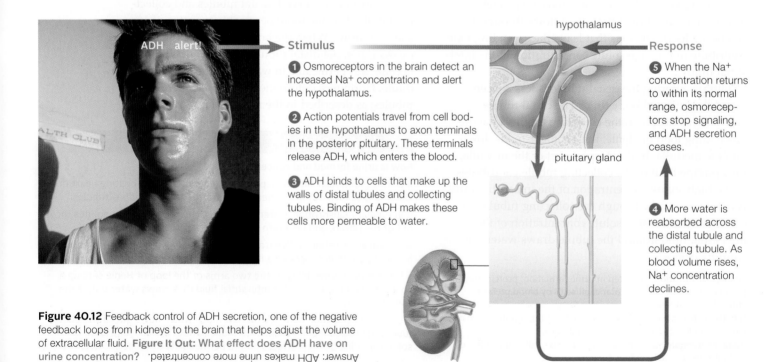

hypothalamus

Stimulus

❶ Osmoreceptors in the brain detect an increased Na⁺ concentration and alert the hypothalamus.

❷ Action potentials travel from cell bodies in the hypothalamus to axon terminals in the posterior pituitary. These terminals release ADH, which enters the blood.

❸ ADH binds to cells that make up the walls of distal tubules and collecting tubules. Binding of ADH makes these cells more permeable to water.

pituitary gland

Response

❺ When the Na⁺ concentration returns to within its normal range, osmoreceptors stop signaling, and ADH secretion ceases.

❹ More water is reabsorbed across the distal tubule and collecting tubule. As blood volume rises, Na⁺ concentration declines.

Figure 40.12 Feedback control of ADH secretion, one of the negative feedback loops from kidneys to the brain that helps adjust the volume of extracellular fluid. **Figure It Out:** What effect does ADH have on urine concentration?   Answer: ADH makes urine more concentrated.

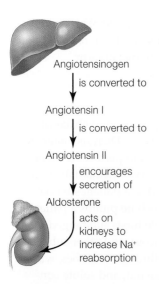

Angiotensinogen

is converted to

Angiotensin I

is converted to

Angiotensin II

encourages
secretion of

Aldosterone

acts on
kidneys to
increase Na⁺
reabsorption

❶ Angiotensinogen made by the liver circulates in the blood. It is converted to angiotensin I by renin, an enzyme relased by kidney arterioles when the blood pressure declines.

❷ Another enzyme converts angiotensin I to angiotensin II.

❸ Among its actions, angiotensin II encourages aldosterone secretion by the adrenal cortex.

❹ Aldosterone acts on kidneys to increase Na⁺ reabsorption; water follows by osmosis.

Figure 40.13 The renin–angiotensin–aldosterone system. Angiotensin II also encourages secretion of ADH and causes an increase in thirst, thus elevating blood volume and pressure.

increases action of sodium–potassium pumps so that more sodium is reabsorbed. Water follows the sodium by osmosis, so less is lost in urine. Retention of additional water raises blood pressure.

Thus, both ADH and aldosterone cause urine to become more concentrated, although they do so by different mechanisms.

Atrial natriuretic peptide (ANP) is a hormone that makes urine more dilute. Muscle cells in the heart's atria release ANP when high blood volume causes the atrial walls to stretch. ANP directly inhibits secretion of aldosterone by acting on the adrenal cortex. It also acts indirectly by inhibiting renin release. In addition, ANP increases the glomerular filtration rate, so more fluid enters kidney tubules.

aldosterone Adrenal hormone that makes kidney tubules more permeable to sodium; encourages sodium reabsorption, thus increasing water reabsorption and concentrating the urine.
antidiuretic hormone (ADH) Hormone released in the posterior pituitary; makes kidney tubules more permeable to water; encourages water reabsorption, thus concentrating the urine.

Take-Home Message

How do hormones affect urine concentration?

» Antidiuretic hormone released by the pituitary causes an increase in water reabsorption. It concentrates the urine.

» Aldosterone released by the adrenal cortex increases salt reabsorption, and water follows. It concentrates the urine.

» Atrial natriuretic peptide released by the heart makes urine more dilute by discouraging secretion of aldosterone and increasing the rate of glomerular filtration.

40.6 Acid–Base Balance

■ The kidneys help maintain the pH of body fluids. They are the only organs that can selectively rid the body of hydrogen ions.
■ Link to pH and buffer systems 2.6

Metabolic reactions such as protein breakdown and lactate fermentation add hydrogen ions (H^+) to the extracellular fluid. Despite these additions, a healthy body can maintain its H^+ concentration within a tight range, a state known as acid–base balance. Buffer systems and adjustments to the activity of respiratory and urinary systems are essential to this balance.

A buffer system involves substances that reversibly bind and release H^+ or OH^- (Section 2.6). In the body, buffer systems minimize pH changes when acidic or basic molecules enter or leave the extracellular fluid.

The pH of human extracellular fluid usually stays between 7.35 and 7.45. In the absence of a buffer, adding acids to this fluid would make its pH decrease. In the presence of bicarbonate, some hydrogen ions combine with bicarbonate to form carbonic acid, which dissociates into carbon dioxide (CO_2) and water:

$$H^+ + \underset{\text{bicarbonate}}{HCO_3^-} \rightleftharpoons \underset{\text{carbonic acid}}{H_2CO_3} \rightleftharpoons CO_2 + H_2O$$

Thus bonded, the hydrogen does not contribute to the pH of the extracellular fluid.

The kidneys contribute to acid–base balance by adjusting the tubular reabsorption of bicarbonate and the tubular secretion of H^+. A decline in pH causes an increase in bicarbonate reabsorption. The bicarbonate moves into peritubular capillaries, where it buffers excess acid. At the same time, tubular secretion of H^+ increases. H^+ leaves the blood and enters the filtrate, where it combines with phosphate or ammonia. The resulting compounds are excreted in the urine.

When the kidney's secretion of H^+ falters, or excess H^+ is formed by metabolic reactions, the pH of body fluids can fall below 7.1, a condition called acidosis. Acidosis can also arise when not enough bicarbonate is reabsorbed.

Take-Home Message

What mechanisms maintain the pH of the extracellular fluid?

» Metabolic reactions produce H^+ that enters extracellular fluid, acidifying it.

» Bicarbonate in the blood serves as a buffer, bonding with H^+ to form CO_2 and water. Thus bicarbonate helps maintain the pH of the extracellular fluid.

» The kidney helps to maintain the pH of the extracellular fluid by adjusting its reabsorption of bicarbonate and secretion of H^+. The H^+ secreted into the filtrate combines with other ions and is excreted in urine.

40.7 When Kidneys Fail

■ Kidney failure can be treated with dialysis, but only a kidney transplant can fully restore function.
■ Link to Protein metabolism 39.9

Causes of Kidney Failure

The vast majority of kidney problems arise as complications of diabetes mellitus or high blood pressure. These disorders damage small blood vessels, including capillaries that interact with nephrons.

High-protein diets force the kidneys to work overtime eliminating excess urea. Such diets also increase the risk for kidney stones. These hardened deposits form when uric acid, calcium, and other wastes settle out of urine and collect in the renal pelvis. Most kidney stones are washed away in urine, but sometimes one lodges in a ureter or the urethra and causes severe pain. A stone that blocks urine flow raises risk of infections and kidney damage.

Kidney function is measured in terms of the rate of filtration through glomerular capillaries. Kidney failure occurs when the filtration rate falls by half. Failure of both kidneys can be fatal because wastes build up

in the blood and interstitial fluid. The pH rises and changes in the concentrations of other ions, most notably Na^+ and K^+, interfere with metabolism.

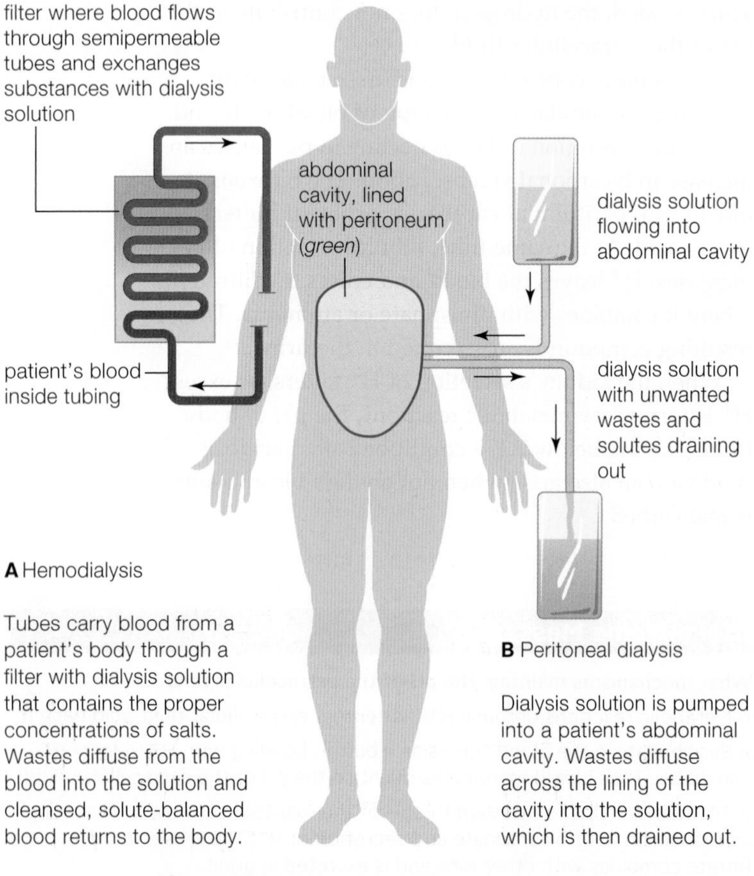

Treating Kidney Failure

Kidney dialysis can restore proper solute balances in a person who has kidney failure. "Dialysis" refers to exchanges of solutes across a semipermeable membrane between two solutions. With hemodialysis, a dialysis machine is connected to a patient's blood vessel (**Figure 40.14A**). The machine pumps a patient's blood through semipermeable tubes submerged in a warm solution of salts, glucose, and other substances. As the blood flows through the tubes, wastes dissolved in the blood diffuse out, and solute concentrations return to normal levels. Cleansed, solute-balanced blood is returned to the patient's body. Typically a person has hemodialysis three times a week at an outpatient dialysis center. By contrast, peritoneal dialysis can be done at home. Each night, dialysis solution is pumped into a patient's abdominal cavity (**Figure 40.14B**). Wastes diffuse across the lining of the cavity (the peritoneum) into the fluid, which is drained out the following morning. With this procedure, the body lining serves as the dialysis membrane.

Kidney dialysis can keep a person alive through an episode of temporary kidney failure. When kidney damage is permanent, dialysis must be continued for the rest of a person's life, or until a donor kidney becomes available for transplant surgery.

Each year in the United States, about 12,000 people are recipients of kidney transplants. More than 40,000 others remain on a waiting list because there is a shortage of donated kidneys. The National Kidney Foundation estimates that every day, 17 people die of kidney failure while waiting for a transplant. Most kidneys for transplants come from deceased donors, but the number of living donors is increasing. One kidney is adequate to maintain good health, so the risks to a living donor are mainly related to the surgery—unless a donor's remaining kidney fails.

filter where blood flows through semipermeable tubes and exchanges substances with dialysis solution

patient's blood inside tubing

abdominal cavity, lined with peritoneum (*green*)

dialysis solution flowing into abdominal cavity

dialysis solution with unwanted wastes and solutes draining out

A Hemodialysis

Tubes carry blood from a patient's body through a filter with dialysis solution that contains the proper concentrations of salts. Wastes diffuse from the blood into the solution and cleansed, solute-balanced blood returns to the body.

B Peritoneal dialysis

Dialysis solution is pumped into a patient's abdominal cavity. Wastes diffuse across the lining of the cavity into the solution, which is then drained out.

Figure 40.14 Two types of kidney dialysis.

Take-Home Message

What causes kidney failure, and how does it affect health?

» Kidney failure most often occurs as a complication of diabetes or high blood pressure.

» Untreated kidney failure is fatal. Dialysis can keep a person with kidney failure alive, but it must be continued until a person dies or receives a kidney transplant.

40.8 Heat Gains and Losses

■ Maintaining the body's core temperature is another aspect of homeostasis. Some animals expend more energy than others to keep their body warm.

■ Links to Properties of water 2.5, Forms of energy 5.2

How the Core Temperature Can Change

Metabolic reactions release heat, so the heat generated by metabolism affects the temperature of an animal's body. Animals also gain heat from, and lose heat to, their surroundings. An animal's internal temperature is stable only when the metabolic heat produced and the heat gained from the environment balance any heat losses to the environment:

$$\begin{array}{ccccccc} \text{change in} & = & \text{heat} & + & \text{heat} & - & \text{heat} \\ \text{body heat} & & \text{produced} & & \text{gained} & & \text{lost} \end{array}$$

Heat is gained or lost at body surfaces by radiation, conduction, convection, and evaporation.

Thermal radiation is emission of heat from a warm object into the space around it. Just as the sun radiates heat energy into space, an animal radiates metabolically produced heat. At rest, a human adult gives off as much heat as a 100-watt incandescent lightbulb.

In **conduction**, heat is transferred within an object or among objects that contact one another. An animal loses heat when it contacts a cooler object, and gains heat when it contacts a warmer one.

In **convection**, heat is transferred by the movement of heated air or water away from the source of heat. As air or water heats up, it rises and moves away from the object, such as a body, that warmed it.

In **evaporation**, a liquid becomes a gas. Energy that powers this process typically comes from the liquid itself in the form of heat, so the remaining liquid cools (Section 2.5). When that liquid is water at a body surface, this cooling helps decrease body temperature.

Endotherms, Ectotherms, and Heterotherms

Fishes, amphibians, and reptiles are **ectotherms**, which means "heated from the outside." The body temperature of ectotherms fluctuates with the temperature of the external environment. That is why the chameleon in **Figure 40.15** is about the same temperature as its surroundings. Ectotherms typically have a low metabolic rate, and they lack insulating fur, hair, or feathers. They regulate internal temperature by altering their position, rather than their metabolism. A lizard, for example, warms up by basking in a sunny spot.

Most birds and mammals are **endotherms**, which means "heated from within." Compared to ectotherms,

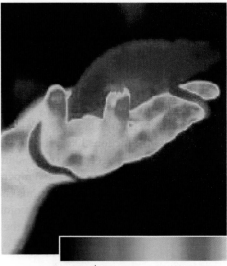

Figure 40.15 Endotherm and ectotherm. Photo of a chameleon (an ectothermic lizard) being held by a human (an endotherm). It was taken using heat-sensitive film.

The lizard blends into the background because it is about the same temperature as its surroundings.

The human is warmer than its surroundings because he or she is producing a large amount of metabolic heat.

cooler---------------->warmer

endotherms have relatively high metabolic rates. For example, a mouse uses thirty times more energy than a lizard of the same body weight. The ability to produce a large amount of metabolic heat helps endotherms remain active in a wider range of temperatures than ectotherms. Fur, hair, or feathers insulate endotherms and minimize heat transfers.

Some birds and mammals are **heterotherms**, which mean they keep their core temperature constant some of the time, but let it fluctuate at other times. For example, hummingbirds have a very high metabolic rate when foraging for nectar during the day. At night, metabolic activity decreases so much that the bird's body may become almost as cool as the surroundings.

conduction Of heat: the transfer of heat between two objects in contact with one another.
convection Transfer of heat by moving molecules of air or water.
ectotherm Animal that controls its internal temperature by altering its behavior; for example, a fish or a lizard.
endotherm Animal controls its internal temperature by adjusting its metabolism; for example, a bird or mammal.
evaporation Transition of a liquid to a gas.
heterotherm Animal that sometimes maintains its temperature by producing metabolic heat, and at other times allows its temperature to fluctuate with the environment.
thermal radiation Emission of heat from an object.

Take-Home Message

How do animals regulate their body temperature?

» Animals can gain heat from the environment, or lose heat to it. They can also generate heat by metabolic reactions.

» Fishes, amphibians, and reptiles are ectotherms that warm themselves mostly by heat gained from the environment.

» Most birds and mammals are endotherms that maintain body temperature with their own metabolic heat.

40.9 Responses to Heat or Cold

- A variety of mechanisms adapt animals to survive in habitats where the temperature fluctuates.
- Links to PET scan 2.2, Adipose tissue 31.4, Feedback control of temperature 31.9, Fever 37.5

The hypothalamus in the brain coordinates organs throughout the body to maintain the body's core temperature via a negative feedback loop (Section 31.9). It functions like a thermostat, receiving input from thermoreceptors in the skin and from others deep in the body. When the temperature deviates from a set point, the hypothalamus calls for responses that return temperature to the set point. When the temperature returns to the set point, the hypothalamus senses the change and stops calling for a response.

Responses to Heat Stress

High external temperature and metabolic heat production by skeletal muscle can raise core temperature. When you become too hot, signals from the hypothalamus cause peripheral vasodilation: Delivery of the blood to the skin increases. As a result, more metabolic heat is given up to the surroundings (Table 40.1).

Another response to heat stress, evaporative heat loss, occurs at moist respiratory surfaces and across skin. Mammals are the only animals that can sweat, although not all mammals do so. Sweat glands are exocrine glands that release water and solutes through pores at the skin's surface. An average-sized adult human has 2.5 million or more sweat glands.

For every liter of sweat produced, about 600 kilocalories of heat energy leave the body by way of evaporative heat loss. Sweat dripping from skin dissipates little heat. Sweat cools you only if it evaporates from your skin. On humid days, the evaporation rate slows, so sweating is less effective at cooling the body.

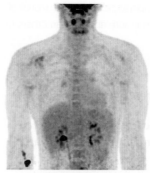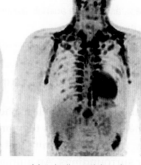

Metabolic activity at 22°C (72°F)

Metabolic activity after two hours at 16°C (61°F)

Figure 40.16 Nonshivering heat production. The PET/CT scans show the metabolic activity of one individual at two temperatures. *Dark* regions indicate areas of highest metabolic activity. In this person, brown adipose tissue in the neck and clavicle region increased metabolic activity in response to cold.

Panting also lowers body temperature by evaporative cooling. It is the major mechanism of cooling in dogs, which have sweat glands mainly on their feet.

Hyperthermia, a dangerous rise in body temperature, occurs when increased peripheral blood flow and evaporative heat loss cannot offset heat gains. Loss of water and salts to sweating change the concentration of body fluids. Blood flow to the gut and liver decreases. Starved of nutrients and oxygen, these organs release toxins that impair normal functions.

To stay safe outside on a hot day, drink plenty of water and avoid excessive exercise. If you must exert yourself, take breaks to monitor how you feel. Wear light-colored clothing that allows evaporation. Avoid direct sunlight, wear a hat, and use sunscreen. Sunburn impairs the skin's ability to transfer heat to the air.

What about a fever? A fever is a response to infection, not an illness itself (Section 37.5). Macrophages activated by tissue damage or pathogens release signaling molecules that stimulate the brain to secrete prostaglandins. These local signaling molecules cause the hypothalamus to allow the core temperature to rise a bit above the normal set point. The rise makes the body less hospitable for pathogens and facilitates immune responses.

Responses to Cold Stress

Selectively distributing blood flow, fluffing up hair or fur, and shivering are typical mammalian responses to cold. Thermoreceptors in skin signal the hypothalamus when things get chilly. The hypothalamus then causes smooth muscle in arterioles that deliver blood to the skin to contract, so less metabolic heat reaches

Table 40.1	Heat and Cold Stress Compared	
Stimulus	**Main Responses**	**Outcome**
Heat stress	Widening of blood vessels in skin; behavioral adjustments; in some species, sweating, panting	Dissipation of heat from body
	Decreased muscle action	Heat production decreases
Cold stress	Narrowing of blood vessels in skin; behavioral adjustments (e.g., minimizing surface parts exposed)	Conservation of body heat
	Increased muscle action; shivering; nonshivering heat production	Heat production increases

Truth in a Test Tube (revisited)

Solutes and nutrients that the body requires are reabsorbed from the filtrate that enters kidney tubules. Water-soluble drugs and metabolites of drugs also travel in the plasma. When filtered into kidney tubules, these compounds are not reabsorbed, so they end up in the urine. How quickly the kidneys clear a substance from the blood depends in part on the efficiency of the kidneys, which varies among individuals. A healthy 35-year-old eliminates drugs from the body about twice as fast as a healthy 85-year-old.

How would you vote? Urine tests detect metabolites of drugs such as marijuana days or even weeks after psychoactive effects of the drug have abated. Should employers be allowed to require potential employees to pass a urine test as a condition of employment?

---

the body surface. When your fingers or toes become chilled, all but 1 percent of the blood that would usually flow to the skin is diverted to other regions of the body. At the same time, muscle contractions make hair (or fur) "stand up." This response creates a layer of still air next to skin and helps reduce heat loss by convection and radiation. Minimizing exposed body surfaces can also prevent heat loss, as when polar bear cubs curl up and cuddle against their mother.

With prolonged exposure to cold, the hypothalamus commands skeletal muscles to contract ten to twenty times each second. Although this **shivering response** increases metabolic heat production, it has a high energy cost.

Brown adipose tissue helps many mammals warm themselves in a cold environment. As noted in Section 31.4, this tissue has mitochondria that release energy as heat, rather than storing it in ATP. The result is **nonshivering heat production**. In human infants, brown adipose tissue accounts for about 5 percent of body weight. As people age, their amount of brown adipose tissue declines, but most adults retain some around their neck and upper chest (**Figure 40.16**).

Thyroid hormone helps to regulate heat production. Exposure to cold increases the secretion of this hormone, which binds to cells of brown adipose tissue and encourages nonshivering heat production. People who make too little thyroid hormone cannot increase their nonshivering heat production, so they often feel cold. They also tend to be overweight, because nonshivering heat production burns a lot of calories.

Hypothermia occurs when normal mechanisms fail to keep core temperature from declining, and the resulting drop in core temperature disrupts normal function. In humans, a decline in body temperature to 95°F (35°C) impairs brain activity. "Stumbles, mumbles, and fumbles" are symptoms of early hypothermia. Severe hypothermia causes loss of consciousness, disrupts heart rhythm, and can be fatal.

Seasonal Dormancy

Some animals enter a period of dormancy during the cold season, a response called **hibernation**. During hibernation, the animal is largely inactive, and its respiratory rate, heart rate, and body temperature decline dramatically. Depending on the species, the animal may reawaken to feed periodically on stored food, or subsist entirely on fat and glycogen accumulated during the prior season.

Mammalian hibernators include some bats, bears, and rodents, such as the groundhog (*Marmota monax*). Many reptiles and amphibians also hibernate during winter. Painted turtles (*Chrysemys picta*) wait out the cold in mud at the bottom of frozen ponds. They go without both food and oxygen, relying on aerobic respiration to fuel their low metabolic needs.

Desert reptiles and amphibians may become dormant during the hot, dry summer, a response called **estivation**. These animals spend this season inactive, inside a burrow deep beneath the soil surface. A desert tortoise, which hibernates in winters and estivates in summers, spends most of its life inactive.

estivation An animal becomes dormant during a hot, dry season.
hibernation An animal becomes dormant during a cold season.
nonshivering heat production Brown adipose tissue releases energy as heat, rather than storing it in ATP.
shivering response In response to cold, rhythmic muscle contractions generate metabolic heat.

Take-Home Message

How do animals adapt to heat and cold?

» Temperature shifts are detected by thermoreceptors that send signals to the hypothalamus, which serves as the body's thermostat.

» Increasing blood flow to the skin, sweating, and panting dissipate heat.

» Shivering and nonshivering heat production warm mammals; fluffing up fur or feathers slows heat loss.

» Some animals become dormant during a cold or hot season.

LEARNING ROADMAP

Where you have been This chapter expands on mechanisms of reproduction (Sections 11.2, 12.1) and gamete formation (12.5). It touches on human sex determination (10.4), Down syndrome (14.6), hypothalamus and pituitary function (34.4), sex hormones (34.10), and autonomic nerves (32.7). We conclude with a discussion of diseases such as HIV/AIDS (20.3).

Where you are now

Modes of Animal Reproduction
Some animals reproduce asexually, but most reproduce sexually. Mechanisms of sexual reproduction vary. Living on land favored fertilization of eggs inside the female body.

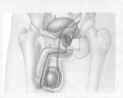

Male Reproductive Function
A human male has testes that make sperm and secrete the sex hormone testosterone. Sperm mixes with secretions from other glands, then leaves the body via the urethra.

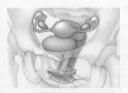

Female Reproductive Function
A human female has ovaries that produce eggs and hormones. Ducts carry eggs toward the uterus, where offspring develop. The vagina receives sperm and is the birth canal.

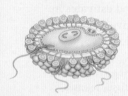

Intercourse and Fertilization
Sexual intercourse involves nervous and hormonal signals. It can lead to pregnancy, which humans use a variety of methods to prevent.

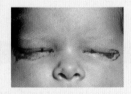

Sexually Transmitted Diseases
Pathogens are passed between partners by sexual interactions and may be transmitted to offspring during birth. Effects of resulting diseases range from discomfort to death.

Where you are going The next chapter continues the story of sexual reproduction with a description of the mechanisms by which sexually reproducing organisms develop. Chapter 43 also touches on animal sexual behavior, courtship, and reproductive success. The evolution of life history traits, such as the number of offspring per reproductive event, is the focus of Section 44.5.

41.1 Intersex Conditions

Santhi Soundarajan was born in a rural area in India in 1981. She overcame poverty and malnutrition to become a competitive runner, and in 2006 represented her country in the Pan Asian Games (**Figure 41.1**). She won a silver medal, but her triumph was short-lived. A few days after the close of the games, the Olympic Council of Asia announced that Soundarajan had been stripped of her medal. Although she had been raised as a female, she has a Y chromosome.

The International Olympics Committee (IOC) began a program of gender testing in 1968. At first, they required that athletes "prove" their femaleness by undergoing a physical exam. In the early 1970s, the committee switched to a less intrusive method. Experts examined a few of an athlete's cells under a microscope for evidence of two X chromosomes. In 1992 the committee upgraded its methods again, this time to a test that detects the *SRY* gene. *SRY* is the gene on the Y chromosome that normally causes development of testes in a human XY embryo (Section 10.4).

The Olympic testing program never uncovered any men deliberately pretending to be women. It did detect athletes who had been brought up as females and thought of themselves as women, but had a Y chromosome. In the 1996 Summer Olympics, 8 of 3,387 women athletes tested positive for an *SRY* gene. Further tests revealed that each of them had some sort of genetic abnormality. The abnormalities prevented testosterone from exerting muscle-building effects, so the women were not thought to have any unfair advantage and were allowed to compete.

Chromosomal sex (XX or XY) determines which gonads (ovaries or testes) form. Hormones secreted by the gonads then shape the genitals and other phenotypic aspects of sex. Generally, when a child is born, a quick look at their genitals (external sex organs) reveals their sex. Males have a penis; females, a vagina.

However, in some people, a mutation results in an **intersex condition**, in which an individual has atypical reproductive anatomy. For example, with complete androgen insensitivity syndrome (AIS), an individual has female genitals, but also has testes inside her abdomen. She is a genetic male who either does not make a normal amount of the male sex hormone testosterone or does not have functional hormone receptors for it. She lacks ovaries and a uterus, does not produce female hormones, and will not begin to menstruate at puberty. She may, however, develop breasts because some testosterone is converted to estrogen. Individuals with complete AIS are brought

Figure 41.1 Indian athlete Santhi Soundarajan rests on the track after the 800-meter race for which she won a silver medal at the Pan Asian Games in 2006. She lost the medal after testing indicated that she has a Y chromosome.

up as women and identify themselves as female. After diagnosis, their testes are usually removed to prevent the possibility of testicular cancer. Undescended testes in intersex individuals are especially prone to cancer.

A variety of other genetic conditions, including partial AIS, result in development of ambiguous genitals, which do not have the appearance typical of either sex. For example, an XY male may have a tiny penis or an XX individual may have a very large clitoris (the organ homologous to the penis). Such anatomy does not affect health. However, it can be disconcerting to parents who worry whether their child will be accepted by peers. As a result, about 2 in 1,000 children undergo plastic surgery to normalize the appearance of their genitals. Such surgery is not without complications; in some cases it can impair later sexual function.

In recent years, some intersex advocates and members of the medical community have suggested that it is unethical to surgically modify children's genitals when they are too young to understand and consent to the procedure. Those with this view advocate recognizing that the appearance of human genitals varies, like that of our eyes, hair, and skin. They suggest abandoning the historical concept of ideal body types, and instead acknowledging the enormous phenotypic variation present in the human population.

intersex condition Genetic abnormality in which an individual has atypical reproductive anatomy.

41.2 Modes of Animal Reproduction

- Sexual reproduction dominates the life cycle of most animals, including many that can also reproduce asexually.
- Links to Sexual and asexual reproduction 11.2, 12.1

Asexual Reproduction

With **asexual reproduction**, a single individual produces offspring. All offspring are genetic replicas (clones) of the parent and identical to one another. Asexual reproduction can be advantageous in a stable environment where the gene combination that makes a parent successful is likely to do the same for its offspring.

Invertebrates reproduce by a variety of asexual mechanisms. In some, a new individual grows on the body of its parent, a process called budding. For example, new hydras bud from existing ones (**Figure 41.2A**). Many corals reproduce by fragmentation, in which a piece of the parent breaks off and develops into a new animal. Some flatworms reproduce by transverse fission: The worm divides in two, leaving one piece headless and one tailless. Each piece then grows the missing body parts.

In some animals, offspring can develop from unfertilized eggs, a process called parthenogenesis. For example, a female aphid (a plant-sucking insect) can give birth to several smaller clones of herself every day. Some fishes, amphibians, lizards, and birds can also produce offspring from unfertilized eggs. No mammals reproduce asexually by natural means.

Sexual Reproduction

With **sexual reproduction**, two parents produce haploid gametes (eggs and sperm) that combine at fertilization. Typically, each resulting offspring has a unique combination of paternal and maternal genes.

On average, only half of a sexually reproducing parent's genes end up in each offspring. Producing gametes takes energy and resources, and many animals spend time and energy finding and courting a mate. What benefits offset these costs? Most animals live where resources and threats change over time. In such environments, producing offspring that differ from both parents and from one another can be advantageous. By reproducing sexually, a parent increases the likelihood that some of its offspring will have a combination of alleles that suits them to the newly changed environment.

Some animals reproduce asexually under favorable conditions, but reproduce sexually when the environment is likely to change. For example, aphids switch to sexual reproduction when their offspring will probably disperse. This occurs with crowding and in autumn.

Gamete Formation and Fertilization

Some sexually reproducing animals make both eggs and sperm; they are **hermaphrodites**. Tapeworms and some roundworms are simultaneous hermaphrodites.

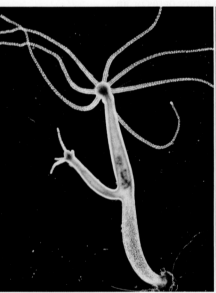

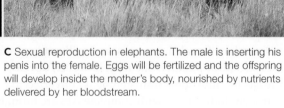

A Asexual reproduction in a hydra. A new individual (*left*) is budding from its parent.

B Sexual reproduction in the barred hamlet, a simultaneous hermaphrodite. Each fish lays eggs and also fertilizes its partner's eggs.

C Sexual reproduction in elephants. The male is inserting his penis into the female. Eggs will be fertilized and the offspring will develop inside the mother's body, nourished by nutrients delivered by her bloodstream.

Figure 41.2 Examples of animal reproduction.

They produce eggs and sperm at the same time, and can fertilize themselves. Earthworms, land snails, and slugs are simultaneous hermaphrodites too, but they require a partner. So do marine fish called hamlets (**Figure 41.2B**). Mating hamlets take turns donating and receiving sperm. Some other fishes are sequential hermaphrodites. They switch from one sex to another during the course of a lifetime. More typically, vertebrates have separate sexes that remain fixed for life; each individual is either male or female.

Most aquatic invertebrates, fishes, and amphibians release gametes into the water, where they combine during **external fertilization**. With **internal fertilization**, sperm fertilize an egg inside the female's body (**Figure 41.2C**). Internal fertilization evolved in most land animals, including insects and the amniotes.

After internal fertilization, a female may lay eggs or retain them inside her body for some portion of their development. Birds eggs develop outside the mother's body, as do eggs of most insects (**Figure 41.3A**). In many sharks, snakes, and lizards, embryos develop while enclosed by an egg sac in the mother's body. The eggs hatch inside the mother shortly before she gives birth (**Figure 41.3B**).

Nourishing Developing Young

A developing animal requires nutrients. Most animals are nourished by yolk, a thick fluid rich in proteins and lipids deposited in the egg during its formation. You have no doubt seen the yolk of a chicken egg. The amount of yolk in a bird egg increases with the time that the egg takes to hatch. Kiwi birds have the longest incubation period of any bird—11 weeks. Their eggs are about two-thirds yolk by volume.

Placental mammals, including humans, have almost yolkless eggs. Instead, a **placenta**, an organ that forms during pregnancy, allows exchange of substances between a mother's blood and that of her developing offspring (**Figure 41.3C**). Nutrients that enter the maternal bloodstream diffuse across the placenta and support the offspring as it develops.

asexual reproduction Reproductive mode by which offspring arise from a single parent only.
external fertilization Sperm and eggs are released into the external environment and meet there.
hermaphrodite Animal that produces both eggs and sperm, either simultaneously or at different times in its life.
internal fertilization A female retains eggs in her body and sperm fertilize them there.
placenta Organ that allows diffusion between the bloodstream of a developing placental mammal and its mother.
sexual reproduction Reproductive mode by which offspring arise from two parents and inherit genes from both.

A Bug depositing fertilized eggs on a pine needle. Yolk in the eggs provides nutrients that will sustain development of the young.

B Lemon shark giving birth to young that recently hatched in her body. The young were nourished by egg yolk, not by their mother.

C An elk examines her newborn calf. The placenta, the organ that allowed nutrients from her blood to diffuse into the calf's blood, is visible at the *left*. It is expelled at birth.

Figure 41.3 How developing offspring are nourished.

Take-Home Message

How do animals reproduce?

» Some animals produce copies of themselves by asexual reproduction. Most reproduce sexually, producing offspring with unique combinations of parental traits. Some switch between asexual and sexual reproduction.

» A sexually reproducing animal may produce eggs and sperm at the same time, produce both types of gametes at different times, or always produce only one type of gamete.

» External fertilization is typical in aquatic animals. Land-dwelling animals typically fertilize eggs inside the female's body.

» Yolk deposited during egg formation sustains the development of most sexually reproducing animals. In placental mammals, nutrients diffuse from maternal blood into the blood of the developing young.

41.3 Reproductive System of Human Males

■ A human male's reproductive system produces hormones and sperm, which it delivers to a female's reproductive tract.

■ Links to Human sex determination 8.2 and 10.4, Sex hormones and puberty 34.10

The Male Gonads

Gonads are gamete-forming organs. A human male's gonads, a pair of **testes** (singular, testis), produce sperm (**Figure 41.4**). Testes also synthesize and secrete the sex hormone **testosterone**. In addition to gonads, a male's reproductive system includes a series of ducts and some accessory glands.

Two testes form on the wall of an XY embryo's abdominal cavity. Before birth, testes descend into the scrotum, a pouch of loose skin suspended below the pelvic girdle. Inside this pouch, smooth muscle encloses the testes. Contraction and relaxation of this muscle in response to threats or temperature adjust the position of the testes. When a man feels cold or afraid, reflexive muscle contractions draw his testes closer to his body. When he feels warm, relaxation of muscle in the scrotum allows his testes to hang lower, so the sperm-making cells do not overheat. These cells function best a bit below normal body temperature.

A male enters puberty—the stage of development when reproductive organs mature—sometime between the ages of 11 and 16 years. His testes enlarge and begin to produce sperm. Secretion of testosterone increases and leads to development of secondary sexual traits: thickened vocal cords that deepen the voice; increased growth of hair on the face, chest, armpits, and pubic region; and an altered distribution of fat and muscle.

Reproductive Ducts and Accessory Glands

Sperm form by meiosis in the testes, a process we discuss in the next section. Here we consider the path sperm travel to the body surface. The journey begins when sperm enter the **epididymis** (plural, epididymides), a coiled duct perched on a testis. The Greek *epi*– means upon and *didymos* means twins. In this context, the "twins" refers to the two testes.

The last region of the epididymis is continuous with the first portion of a vas deferens (plural, vasa deferentia). In Latin, *vas* means vessel, and *deferens*, to carry away. Each **vas deferens** is a duct that carries sperm away from an epididymis, and to a short ejaculatory duct. Ejaculatory ducts deliver sperm to the urethra, the duct that extends through a male's penis to open at the body surface.

The **penis** is the male organ of intercourse. It has a rounded head (the glans) at the end of a narrower shaft. Nerve endings in the glans make it highly sensitive to touch.

Normally, when a man is not sexually excited, a retractable tube of skin called the foreskin covers the glans of his penis. In many cultures, males undergo circumcision, an elective surgical procedure that removes the foreskin. Circumcision reduces the risk of contracting sexually transmitted diseases and of penile cancer, but it is painful, and rare complications require follow-up surgery or impair sexual function.

Three elongated cylinders of spongy tissue fill the interior of the penis. When a male is sexually excited, blood flows into the spongy tissue faster than it flows out. Fluid pressure rises and the normally limp penis becomes stiff and erect.

Sperm stored in the epididymides and first part of the vasa deferentia continue their journey toward the body surface only when a male reaches the peak of sexual excitement and ejaculates. During ejaculation, smooth muscle in the walls of the epididymides and vasa deferentia undergoes rhythmic contractions that propel sperm and accessory gland secretions out of the body as a thick, white fluid called semen.

Semen is a complex mix of sperm, proteins, nutrients, ions, and signaling molecules. Sperm constitute less than 5 percent of semen's volume; the bulk of it consists of secretions from accessory glands. Seminal vesicles, exocrine glands near the base of the bladder, secrete fructose-rich fluid into the vasa deferentia. Sperm use the fructose (a sugar) as their energy source. The prostate gland, which encircles the urethra, is the other major contributor to semen volume. Its secretions help raise the pH of the female reproductive tract, making it more hospitable to sperm. Both seminal vesicles and the prostate gland also secrete prostaglandins, which are local signaling molecules.

The two pea-sized bulbourethral glands secrete a lubricating mucus into the urethra. This mucus helps clear the urethra of urine prior to ejaculation.

Disorders of the Male Reproductive Tract

A young, healthy man has a prostate gland about the size of a walnut. With inflammation or age, this gland can enlarge. Because the urethra runs through the prostate gland, prostate enlargement can narrow this duct and cause difficulty urinating. Medication, laser treatments, and surgery are used to relieve symptoms.

Prostate cancer is a leading cause of cancer death for men, surpassed only by lung cancer. Doctors diag-

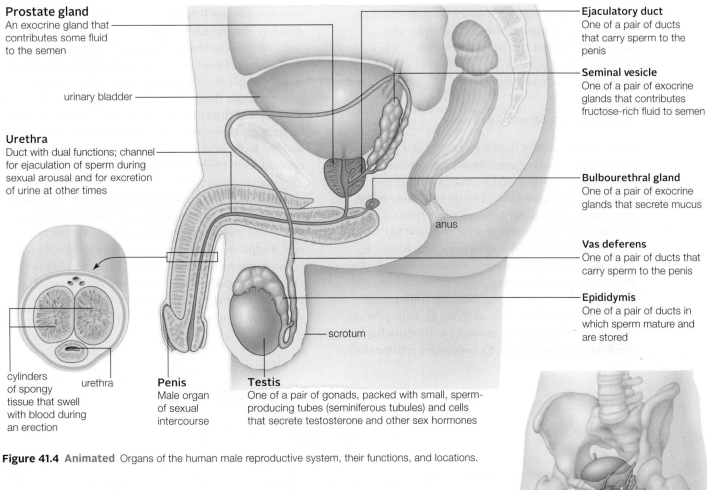

Prostate gland
An exocrine gland that contributes some fluid to the semen

urinary bladder

Urethra
Duct with dual functions; channel for ejaculation of sperm during sexual arousal and for excretion of urine at other times

cylinders of spongy tissue that swell with blood during an erection

urethra

Penis
Male organ of sexual intercourse

Testis
One of a pair of gonads, packed with small, sperm-producing tubes (seminiferous tubules) and cells that secrete testosterone and other sex hormones

scrotum

Ejaculatory duct
One of a pair of ducts that carry sperm to the penis

Seminal vesicle
One of a pair of exocrine glands that contributes fructose-rich fluid to semen

Bulbourethral gland
One of a pair of exocrine glands that secrete mucus

anus

Vas deferens
One of a pair of ducts that carry sperm to the penis

Epididymis
One of a pair of ducts in which sperm mature and are stored

Figure 41.4 Animated Organs of the human male reproductive system, their functions, and locations.

nose prostate cancer with blood tests and by physical examination. The blood test detects antigens made only by prostate cancers. The physical exam involves inserting a finger in the rectum to feel the prostate for any bumps or size change that could indicate the presence of cancer.

Most prostate cancers grow relatively slowly and those in older men often do not require treatment. However, some prostate cancers grow quickly and spread to other organs. Risk factors for prostate cancer include advancing age, a diet rich in animal fats, smoking, and lack of exercise. Genes also play a role.

If a man has an affected father or brother, his own risk of prostate cancer doubles.

Testicular cancer is relatively rare, with 7,000 cases a year in the United States. Even so, it is the most common cancer among men aged 15 to 34. Once a month, after a warm shower or bath, men should examine their testes to detect lumps, enlargement, or hardening.

epididymides A pair of ducts in which sperm formed in testes mature; each empties into a vas deferens.
gonads Organs that produce gametes.
penis Male organ of intercourse; also functions in urination.
semen Sperm mixed with secretions from seminal vesicles and the prostate gland.
testes In animals, gonads that produce sperm.
testosterone Main hormone produced by testes; causes sperm production and development of male secondary sexual traits.
vas deferens One of a pair of long ducts that convey mature sperm toward the body surface.

Take-Home Message

What are the functions of the male reproductive organs?

» A pair of testes, the primary reproductive organs in human males, produce sperm. They also make and secrete the sex hormone testosterone.

» Sperm and secretions from accessory glands form semen. During sexual arousal, semen is propelled through a series of ducts and leaves the body through an opening in the penis.

» One accessory gland, the prostate, frequently becomes enlarged with age. It is also a common site for male cancers.

41.4 Sperm Formation

- In his reproductive years, a male continually produces new germ cells, which undergo meiosis to produce sperm.
- Sperm formation is controlled by hormones.
- Links to Tight junctions 4.11, Growth factors 11.6, Gamete formation 12.5, Sex hormones 34.10

Germ Cells to Sperm Cells

Each testis is smaller than a golf ball. Yet it contains coiled sperm-making **seminiferous tubules** that would extend for 125 meters (longer than a football field) if stretched out (**Figure 41.5A**). Leydig cells (also called interstitial cells) cluster between these tubules and secrete the hormone testosterone (**Figure 41.5B**).

Male germ cells called spermatogonia (singular, spermatogonium) line the inner wall of a tubule (**Figure 41.5C**). In a sexually mature male, these diploid cells undergo mitosis, producing cells that differentiate into diploid primary spermatocytes ❶. Each primary spermatocyte undergoes meiosis I to produce haploid secondary spermatocytes ❷. Secondary spermatocytes undergo meiosis II to produce haploid spermatids that develop into mature sperm ❸.

In addition to the spermatogonia and their descendents, seminiferous tubules contain Sertoli cells. These large, elongated cells are sometimes called "nurse cells" because they surround and support the developing sperm. Adjacent Sertoli cells attach to one another by tight junctions, forming a barrier between blood and the testis interior. Thus, Sertoli cells control which substances in the blood can cross their membrane and affect developing sperm.

A sperm is a haploid cell with very little cytoplasm (**Figure 41.6**). It has a flagellum, or "tail," that allows it to swim toward an egg. Mitochondria in the adjacent midpiece supply the energy required for flagellar movement. A sperm's "head" is packed full of DNA and tipped by an enzyme-containing cap. The enzymes can help a sperm penetrate an oocyte by partly digesting away its outer layer.

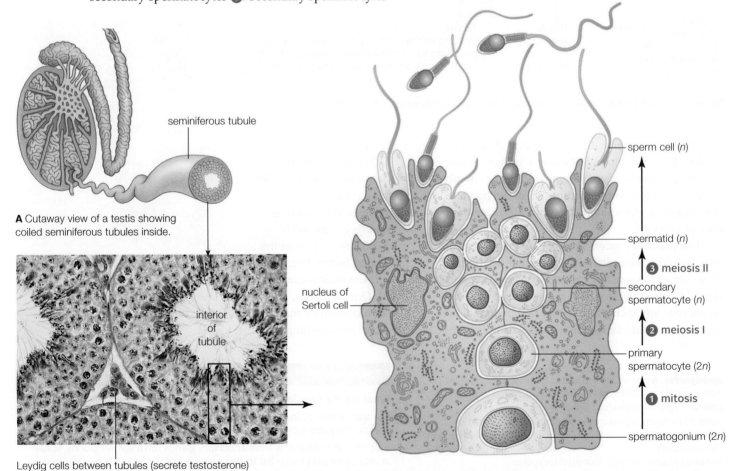

seminiferous tubule

A Cutaway view of a testis showing coiled seminiferous tubules inside.

interior of tubule

nucleus of Sertoli cell

Leydig cells between tubules (secrete testosterone)

B Light micrograph of seminiferous tubules, cross-section.

sperm cell (*n*)

spermatid (*n*)

❸ meiosis II

secondary spermatocyte (*n*)

❷ meiosis I

primary spermatocyte (2*n*)

❶ mitosis

spermatogonium (2*n*)

C Spermatogenesis in seminiferous tubule. Diploid germ cells (spermatogonia) undergo mitosis to renew themselves and to form primary spermatocytes. Primary spermatocytes undergo meiosis to form spermatids that mature into sperm. Sertoli cells (*gold*) support developing sperm.

Figure 41.5 Animated Where and how sperm form.

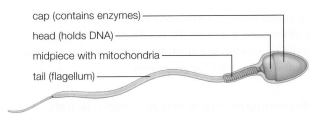

Figure 41.6 Structure of a mature sperm, a male gamete.

Newly formed sperm cannot swim. They are moved out of the testis and into the epididymis by smooth muscle contractions. Cilia of the epididymis lining push the immature sperm further along this duct. The sperm mature and become mobile as they move through the epididymis.

Sperm formation takes about 65 days, from start to finish. An adult male makes sperm continually, so on any given day he has millions of them in various stages of development. A sperm stored in the epididymis remains viable for about a month. After that it breaks down and its components are reabsorbed.

Hormonal Control of Sperm Formation

The male sex hormone testosterone is essential for sperm production. A negative feedback system regulates the hormone's release (**Figure 41.7**). A decline in testosterone concentration prompts the hypothalamus to release **gonadotropin-releasing hormone (GnRH)** ❶. GnRH stimulates cells of the anterior pituitary to secrete luteinizing hormone and follicle-stimulating hormone ❷. These hormones have a role in both male and female reproductive function. In males, **luteinizing hormone (LH)** released by the anterior pituitary binds to Leydig cells that lie in between the seminiferous tubules, stimulating them to secrete testosterone ❸. **Follicle-stimulating hormone (FSH)** targets Sertoli cells. FSH, in combination with testosterone, prompts these cells to produce growth factors and other molecular signals ❹. These substances bathe neighboring male germ cells and encourage sperm development and maturation ❺.

follicle-stimulating hormone (FSH) Pituitary hormone that acts on gonads; stimulates secretion of growth factors by Sertoli cells in males and ovarian follicle maturation in females.
gonadotropin-releasing hormone (GnRH) A hypothalamic hormone that induces the pituitary to release hormones (LH and FSH) that act on the gonads.
luteinizing hormone (LH) An anterior pituitary hormone that acts on the gonads; stimulates testosterone production in males and ovulation in females.
seminiferous tubules Coiled tubules inside a testis that contain male germ cells and produce sperm.

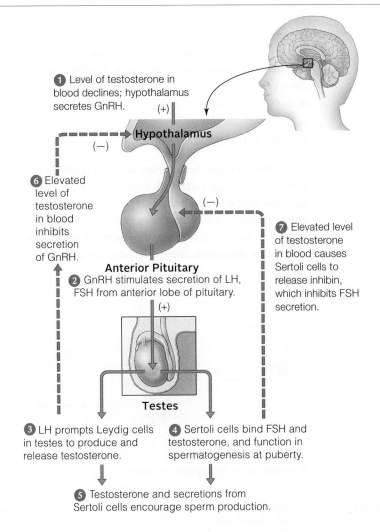

❶ Level of testosterone in blood declines; hypothalamus secretes GnRH.

Hypothalamus

❻ Elevated level of testosterone in blood inhibits secretion of GnRH.

❼ Elevated level of testosterone in blood causes Sertoli cells to release inhibin, which inhibits FSH secretion.

Anterior Pituitary
❷ GnRH stimulates secretion of LH, FSH from anterior lobe of pituitary.

Testes

❸ LH prompts Leydig cells in testes to produce and release testosterone.

❹ Sertoli cells bind FSH and testosterone, and function in spermatogenesis at puberty.

❺ Testosterone and secretions from Sertoli cells encourage sperm production.

Figure 41.7 Animated Hormonal pathways that influence sperm formation. Negative feedback loops control hormonal secretion by the hypothalamus, the anterior lobe of the pituitary gland, and the testes.

When the concentration of testosterone in the blood rises, the hypothalamus detects this rise and slows its secretion of GnRH ❻. The decrease in GnRH then lowers the output of LH and FSH by the testes. In addition, a high sperm count encourages the Sertoli cells to release the hormone inhibin ❼. Inhibin acts on the anterior pituitary, and inhibits FSH secretion.

Take-Home Message

How do sperm form and what role do hormones play?

» Meiosis in germ cells in seminiferous tubules of the testes produces sperm—the haploid male gametes.

» Hormonal control of the process begins with GnRH from the hypothalamus. GnRH causes secretion of the hormones FSH and LH by the pituitary gland.

» FSH and LH act on the testes, where they stimulate release of testosterone and other factors needed for formation and development of sperm.

41.5 Reproductive System of Human Females

■ The reproductive system of human females functions in the production of gametes and sex hormones.

■ The system receives sperm, and has a chamber in which developing offspring are protected and nourished until birth.

■ Sex hormones 34.10

Female Gonads

A female's gonads, her **ovaries**, lie deep inside her pelvic cavity (**Figure 41.8A,B**). Each ovary is about the size and shape of an almond. It produces and releases **oocytes**, which are immature eggs. Ovaries also secrete estrogens and progesterone, the main sex hormones of females. **Estrogens** cause development of female

secondary sexual traits such as enlarged breasts at puberty and maintain the health of cells lining the female reproductive tract. **Progesterone** prepares the reproductive tract for pregnancy.

Reproductive Ducts and Accessory Glands

Adjacent to each ovary is the entrance to an **oviduct**, or fallopian tube. Fingerlike projections surround the opening of an oviduct, and they draw an oocyte inward. Cilia in the oviduct lining then propel the oocyte along the length of the tube. Fertilization usually occurs in an oviduct. Each oviduct opens into the **uterus**, a hollow, pear-shaped organ above the urinary

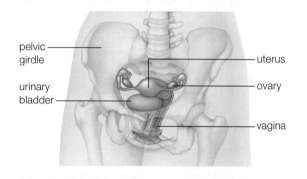

A Location of female reproductive organs.

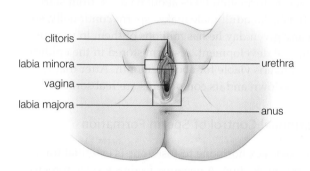

C External sex organs, collectively referred to as the vulva.

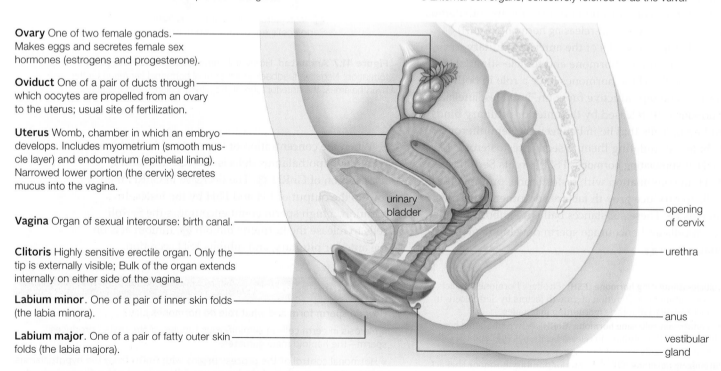

Ovary One of two female gonads. Makes eggs and secretes female sex hormones (estrogens and progesterone).

Oviduct One of a pair of ducts through which oocytes are propelled from an ovary to the uterus; usual site of fertilization.

Uterus Womb, chamber in which an embryo develops. Includes myometrium (smooth muscle layer) and endometrium (epithelial lining). Narrowed lower portion (the cervix) secretes mucus into the vagina.

Vagina Organ of sexual intercourse: birth canal.

Clitoris Highly sensitive erectile organ. Only the tip is externally visible; Bulk of the organ extends internally on either side of the vagina.

Labium minor. One of a pair of inner skin folds (the labia minora).

Labium major. One of a pair of fatty outer skin folds (the labia majora).

B Human female reproductive organs, shown in longitudinal section.

Figure 41.8 Animated Components of the human female reproductive system and their functions.

bladder. A thick layer of smooth muscle, called the myometrium, makes up most of the uterine wall. The uterine lining, or **endometrium**, consists of glandular epithelium, connective tissues, and blood vessels. When a woman becomes pregnant, the embryo attaches to the endometrium and completes its development inside the uterus. The lowest portion of the uterus, a narrowed region called the **cervix**, opens into the vagina.

The **vagina** is a muscular tube that extends to the body's surface and functions as the female organ of intercourse and as the birth canal. Typically, when a girl is born, a membrane called the hymen partially covers the external entrance to her vagina. The hymen thins as a girl approaches puberty. It is stretched open during the first episode of intercourse, if it has not previously been broken by other physical activity.

Two pairs of liplike skin folds, referred to as labia, enclose the surface openings of the vagina and urethra (**Figure 41.8C**). *Labia* is Latin for lips. Adipose tissue fills the thick outer labia majora, which are derived from the same embryonic tissue as the male scrotum. The thin inner folds are the labia minora. The labia minora contain mucus-secreting vestibular glands that keep the entrance to the vagina moist. These glands are the equivalent of the male's bulbourethral glands.

The tip of an erectile organ called the clitoris lies near the anterior junction of the labia minora. The clitoris and penis develop from the same embryonic tissue. Both have an abundance of highly sensitive touch receptors, and both swell with blood and become erect during sexual arousal. Like the penis, the clitoris consists of a head and a shaft. The clitoral head is the visible portion and the shaft is internal.

The Menstrual Cycle

During her reproductive years, a female undergoes cyclic changes in hormone levels and fertility,

cervix Narrow part of uterus that connects to the vagina.
endometrium Lining of uterus.
estrogen Hormone secreted by ovaries; causes development of female sexual traits and maintains the reproductive tract.
menopause Permanent cessation of menstrual cycles.
menstrual cycle Approximately 28-day cycle in which the uterus lining thickens and then, if pregnancy does not occur, is shed.
menstruation Flow of shed uterine tissue out of the vagina.
oocyte Immature egg.
ovary Organ in which eggs form.
oviduct Duct between an ovary and the uterus.
progesterone Hormone secreted by ovaries; prepares the uterus for pregnancy.
uterus Muscular chamber where offspring develop; womb.
vagina Female organ of intercourse and birth canal.

collectively called the **menstrual cycle**. Cycles usually begin at about age twelve. Section 41.6 describes the menstrual cycle in detail, but here is an overview: Every twenty-eight days or so, an oocyte matures and is released from an ovary, and the endometrium thickens in preparation for pregnancy. If the egg is not fertilized, **menstruation** occurs—blood and bits of shed uterine lining flow out through the vagina. This menstrual flow indicates the start of a new cycle.

A woman goes through such cycles until she reaches her late forties or early fifties, when her sex hormone output dwindles. The decline in hormone secretions correlates with the onset of **menopause**, the twilight of a female's fertility.

Disorders of the Female Reproductive Tract

About one-third of women over age thirty have benign uterine tumors called fibroids. Most uterine fibroids cause no symptoms, but some cause pain during menstruation and unusually long and heavy menstrual periods. A woman who needs to change pads or tampons after an hour or two should see a doctor. Surgical removal of fibroids halts excessive bleeding and pain.

Vaginal bleeding between menstrual periods and bleeding after menopause also call for medical attention. Such bleeding can be a sign of uterine cancer, the most common cancer of the female reproductive tract. In the United States, there are about 44,000 new cases of uterine cancer per year and 8,000 deaths. Ovarian cancer is less common (22,000 new cases a year) but causes more deaths (14,000/year) because it often goes undetected until it has enlarged or spread. Cervical cancer (12,000 new cases/year) can be detected early by a Pap smear—a procedure in which cervical cells are examined for changes indicating cancer. Cervical cancer can be prevented by a vaccine against the virus that causes it (Section 37.1). Even so, in 2010 there were 4,200 deaths from cervical cancer in the United States. We return to cervical cancer in Section 41.10, when we discuss sexually transmitted disease.

Take-Home Message

What are the functions of the female reproductive organs?

» A pair of ovaries produce oocytes (eggs). They also make and secrete the hormones estrogen and progesterone.

» About once a month, an oocyte is released from an ovary and enters an oviduct. If fertilization occurs, the resulting embryo develops in the uterus. If fertilization does not occur, the lining of the uterus is shed during menstruation and a new oocyte matures.

» The vagina serves as the female organ of intercourse and as the birth canal.

41.6 Preparations for Pregnancy

- A fertile woman undergoes hormonal changes and releases eggs in an approximately monthly cycle.
- Links to Gamete formation 12.5, Sex hormones 34.10

The Ovarian Cycle

At birth, a girl has about 2 million **primary oocytes**, immature eggs that have entered meiosis but stopped in prophase I. Beginning with her first menstrual cycle, the oocytes mature, typically one at a time, in approximately a 28-day cycle (**Table 41.1**). **Figure 41.9** depicts the events of this cycle in an ovary.

A primary oocyte and the cells that surround it constitute an **ovarian follicle** (**Figure 41.9 ❶**). In the first part of the ovarian cycle—the follicular phase—cells around the oocyte divide repeatedly, while the oocyte enlarges and secretes proteins ❷. As the follicle matures, a fluid-filled cavity opens in the cell layers around the oocyte. Often, more than one follicle begins to develop during the follicular phase, but typically only one goes on to become fully mature.

Follicle maturation requires about 14 days and is under hormonal control. As the follicular phase begins, the hypothalamus secretes GnRH. As in males, this hormone stimulates cells in the anterior pituitary to increase their secretion of FSH and LH (**Figure 41.10A**). Rising levels of FSH and LH in the blood encourage follicle maturation and stimulate follicle cells to secrete estrogens (**Figure 41.10B,C**).

The pituitary detects the rising level of estrogens in blood and responds with an outpouring of LH. The surge of LH causes the primary oocyte to complete meiosis I and undergo cytoplasmic division ❸. One of the resulting haploid cells, the **secondary oocyte**, gets most of the cytoplasm. It continues to metaphase II, then stops and awaits fertilization. The other haploid cell is the first polar body, a cell that will eventually degenerate.

The LH surge also causes the follicle to begin to swell, and eventually to burst ❹. When it does, the secondary oocyte and some of the surrounding follicle cells are released and enter an oviduct. Thus, the midcycle surge of LH is the trigger for **ovulation**, the release of a secondary oocyte from an ovary.

Ovulation is followed by the luteal phase of the ovarian cycle. During this interval, the ruptured follicle becomes a yellowish glandular structure known as the **corpus luteum** ❺. In Latin, *corpus* means body, and *luteum* means yellow.

The corpus luteum secretes a large amount of the sex hormone progesterone, and a lesser amount of estrogens. The high progesterone level feeds back to the hypothalamus and pituitary and reduces secretion of LH and FSH, so a new follicle does not develop.

If pregnancy does not occur, the corpus luteum lasts no more than 12 days. In the final days of the luteal phase, a decline in LH causes it to break down ❻. Once it does, a new follicular phase begins.

The Uterine Cycle

Menstruation coincides with the beginning of the follicular phase in the ovary (**Figure 41.10D**). It usually

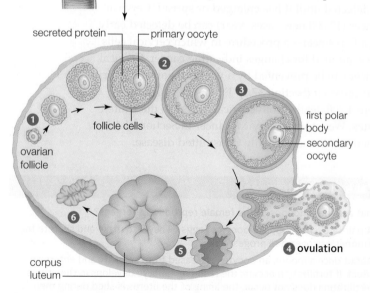

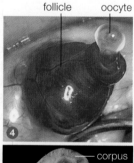

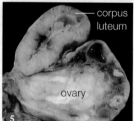

❶ A follicle begins to mature; the primary oocyte enlarges and secretes proteins, and follicle cells around it divide.

❷ Fluid-filled cavity begins to form in the follicle's cell layer.

❸ Primary oocyte completes meiosis I and undergoes unequal cytoplasmic division to form a secondary oocyte and the first polar body.

❹ Ovulation. Follicle wall ruptures, releasing the secondary oocyte and surrounding cells.

❺ Follicle cells left behind after ovulation develop into a corpus luteum that secretes hormones.

❻ If pregnancy does not occur, the corpus luteum degenerates.

Figure 41.9 Animated Ovarian cycle. Graphics in the diagram show the *sequence* of events, not movements. Photos show human ovaries. The mature follicle is about the size of a pea and projects from the ovary surface.

Table 41.1	Events of an Ovarian/Menstrual Cycle Lasting Twenty-Eight Days	
Phase	**Events**	**Day of Cycle**
Follicular phase	Menstruation; endometrium breaks down	1–5
	Follicle matures in ovary; endometrium rebuilds	6–13
Ovulation	Oocyte released from ovary	14
Luteal phase	Corpus luteum forms, secretes progesterone; the endometrium thickens and develops	15–28

lasts for 1 to 5 days. As the follicular phase continues, estrogens secreted by a maturing follicle encourage the uterine lining to repair itself and to begin thickening.

Immediately after ovulation, a follicle's production of estrogen declines until the corpus luteum forms (**Figure 41.10B,C**). During the luteal phase of the cycle, estrogen and progesterone secreted by the corpus luteum cause the uterine lining to thicken and encourage blood vessels to grow through it (**Figure 41.10C,D**). The uterus is now ready for pregnancy.

If pregnancy occurs, the corpus luteum persists and secretes hormones until the placenta forms and assumes this task. If it does not, the corpus luteum degenerates. With no corpus luteum to secrete them, progesterone and estrogen levels plummet. Blood vessels supplying the endometrium wither and the endometrium begins to break down. As this tissue is shed, a new follicular phase gets under way.

corpus luteum Hormone-secreting structure that forms from the cells of a mature follicle that are left behind after ovulation.
ovarian follicle In animals, immature egg and surrounding cells.
ovulation Release of a secondary oocyte from an ovary.
primary oocyte Immature egg that has not completed prophase I.
secondary oocyte A haploid cell produced by the first meiotic division of a primary oocyte; released at ovulation.

Take-Home Message

What cyclic changes occur in the ovary and uterus?

» Every 28 days or so, FSH and LH stimulate maturation of an ovarian follicle.

» A midcycle surge of LH triggers ovulation—the release of a secondary oocyte into an oviduct.

» Estrogen secreted by a maturing follicle causes the endometrium to thicken. After ovulation, progesterone from the corpus luteum encourages endometrial growth.

» If pregnancy does not occur, the corpus luteum breaks down, progesterone levels drop, the endometrial lining is shed, and the cycle begins again.

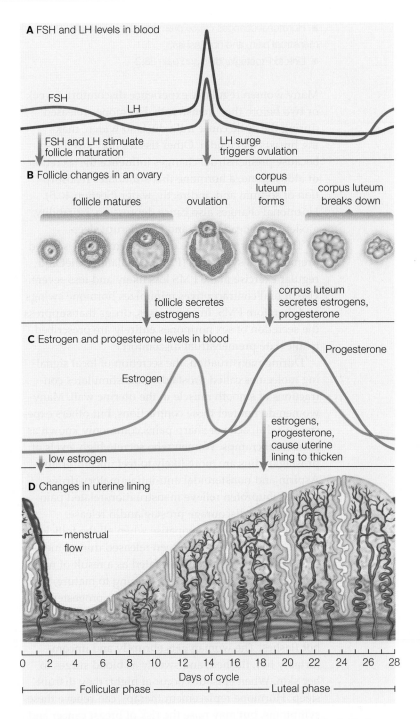

Figure 41.10 Animated Changes in the ovary and uterus correlated with changing hormone levels. We start with the onset of menstrual flow on day one of a twenty-eight-day menstrual cycle.

(**A,B**) Prompted by GnRH from the hypothalamus, the anterior pituitary secretes FSH and LH, which stimulate a follicle to grow and an oocyte to mature in an ovary. A midcycle surge of LH triggers ovulation and the formation of a corpus luteum. A decline in FSH after ovulation stops more follicles from maturing.

(**C,D**) Early on, estrogen from a maturing follicle stimulates repair and rebuilding of the endometrium. After ovulation, the corpus luteum secretes some estrogen and more progesterone that primes the uterus for pregnancy. If pregnancy occurs, the corpus luteum will persist, and its secretions will stimulate the maintenance of the uterine lining.

41.7 PMS to Hot Flashes

■ Hormonal changes cause premenstrual symptoms, menstrual pain, and hot flashes.
■ Link to Prostaglandins and pain 33.3

Many women regularly experience discomfort a week or two before they menstruate. Hormones released during the cycle cause milk ducts to widen, making breasts feel tender. Other tissues may swell also, because premenstrual changes influence the secretion of aldosterone, a hormone that stimulates reabsorption of sodium and, indirectly, water (Section 40.5). Hormonal changes also cause depression, irritability, anxiety, and headaches, and can disrupt sleep.

The regular recurrence of these symptoms is called premenstrual syndrome (PMS). A balanced diet and regular exercise make PMS less likely and less severe. Use of oral contraceptives minimizes hormone swings and therefore PMS. In some cases, drugs that suppress the secretion of sex hormones entirely are prescribed to alleviate premenstrual discomfort.

During menstruation, the secretion of local signaling molecules called prostaglandins stimulates contractions of smooth muscle in the uterine wall. Many women do not feel these contractions, but others experience a dull ache or sharp pains commonly known as menstrual cramps. Women who secrete high levels of prostaglandins are more likely to feel uncomfortable. Aspirin and nonsteroidal anti-inflammatory drugs such as ibuprofen relieve menstruation-related pain because they discourage prostaglandin release.

A woman enters menopause when all the follicles in her ovaries either have been released during menstrual cycles, or have disintegrated as a result of normal aging. With no follicles remaining to mature, the woman no longer produces estrogen or progesterone, so menstrual cycles cease. Hormonal changes around the time of menopause cause many women to have hot flashes. The woman gets abruptly and uncomfortably hot, flushed, and sweaty as blood surges to her skin. When episodes occur at night, they disrupt sleep. Hormone replacement therapy can relieve these symptoms, but may raise the risk of breast cancer and stroke, especially if continued for many years.

Take-Home Message

How do hormonal changes affect non-reproductive health?

» Water retention, breast tenderness, and mood changes associated with premenstrual syndrome arise as a result of changes in hormone level.

» Prostaglandin secretions that cause uterine contractions during menstruation can also cause cramping.

» Hormonal changes during menopause cause hot flashes.

41.8 When Gametes Meet

■ Intercourse involves nervous and hormonal signals and culminates with the delivery of sperm into the vagina.
■ Links to Endorphins 32.6, Autonomic signals 32.7, Limbic system 32.11, Free nerve endings 33.3, Oxytocin 34.4

Sexual Intercourse

For males, intercourse requires an erection. The penis consists mostly of long cylinders of spongy tissue. When a male is not sexually aroused, his penis is limp, because the blood vessels that transport blood to its spongy tissue are constricted. When he becomes aroused, these vessels dilate and veins that carry blood out of the penis constrict. Inward blood flow exceeds outward flow, and the increasing fluid pressure enlarges and stiffens the penis so it can be inserted into a female's vagina.

A man's ability to get and sustain an erection peaks during his late teens. As he ages, he may have episodes of erectile dysfunction. With this disorder, the penis does not stiffen enough for intercourse. Men who have circulatory problems are most often affected. Smoking increases the risk of circulatory problems, so smokers are more likely to have erectile dysfunction. Viagra and similar drugs prescribed for erectile dysfunction help blood enter the penis by relaxing the smooth muscle inside of it. The drugs have some side effects. For example, Viagra affects cone cells in the eyes, causing a temporary shift in color perception.

When a woman becomes sexually excited, blood flow increases to the vaginal wall, labia, and clitoris. Glands in the cervix secret mucus, and vestibular glands on the labia produce a lubricating fluid. The vagina itself does not have any glandular tissue; it is moistened by mucus from the cervix and by plasma fluid that seeps out between the epithelial cells of the vaginal lining.

During intercourse, increased sympathetic stimulation raises the heart rate and breathing rate in both partners. The posterior pituitary steps up its secretion of oxytocin. This hormone acts in the brain, inhibiting signals from the amygdala, a portion of the limbic system involved in fear and anxiety (Section 32.11).

Continued mechanical stimulation of the penis or clitoris excites free nerve endings, which send signals to the spinal cord. With sufficient stimulation, orgasm occurs. During orgasm, a surge of oxytocin causes rhythmic contractions of smooth muscle in the reproductive tract. Endorphins flood the brain and evoke feelings of pleasure. In males, orgasm is usually accompanied by ejaculation, in which the contraction of smooth muscles force the semen out of the penis.

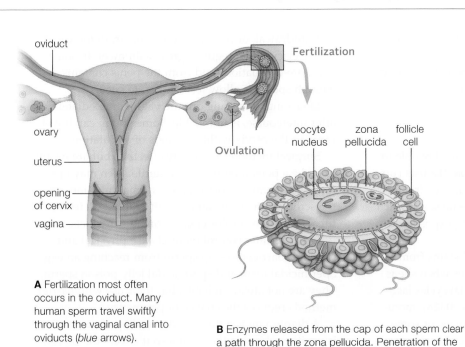

A Fertilization most often occurs in the oviduct. Many human sperm travel swiftly through the vaginal canal into oviducts (*blue* arrows).

Inside an oviduct, the sperm surround a secondary oocyte that was released by ovulation.

B Enzymes released from the cap of each sperm clear a path through the zona pellucida. Penetration of the secondary oocyte by a sperm causes the oocyte to release substances that harden the zona pellucida and prevent other sperm from binding.

C The oocyte nucleus completes meiosis II, forming a nucleus with a haploid maternal genome. The sperm's tail and other organelles degenerate. Its DNA is enclosed by a membrane, forming a haploid nucleus with paternal genes.

Later, the two nuclear membranes will break down and paternal and maternal chromosomes will become arranged on a bipolar spindle in preparation for the first mitotic division.

Figure 41.11 Animated Events in human fertilization. The light micrograph shows a fertilized human oocyte.

Figure It Out: In the micrograph, what are the small cells on the right, just beneath the zona pellucida?   Answer: The polar bodies

Fertilization

Ejaculation puts up to 300 million sperm into the vagina. The sperm swim to the oviducts, usually reaching them within about 30 minutes of ejaculation. Fertilization usually occurs in the upper part of an oviduct (**Figure 41.11A**).

A secondary oocyte released at ovulation has a wrapping of follicle cells over a zona pellucida, a layer of glycoproteins. The sperm makes its way between follicle cells to the protein layer. The binding of receptors in the plasma membrane of a sperm's head to species-specific egg proteins triggers the release of protein-digesting enzymes from the cap on the sperm's head. The collective effect of enzyme release from many sperm clears a passage through the zona pellucida to the oocyte's plasma membrane. Receptors in this membrane bind a sperm's plasma membrane, and the two membranes fuse. The sperm enters the oocyte. Usually only one sperm enters because entry triggers changes to the zona pellucida that prevent other sperm from binding (**Figure 41.11B**).

It only takes one sperm to fertilize an egg, but it takes the binding of many to release enough enzyme to clear a way to the egg plasma membrane. This is why a man who has healthy sperm but a low sperm count can be functionally infertile.

Remember that the secondary oocyte had entered meiosis II, then stopped at metaphase II. Binding of the sperm membrane to the oocyte membrane causes completion of meiosis II. The unequal cytoplasmic division that follows produces a single mature egg— an ovum (plural, ova)—and a polar body.

Inside the ovum cytoplasm, the sperm's tail and organelles degenerate, leaving only the nucleus (**Figure 41.11C**). Chromosomes in the haploid egg and sperm nuclei become the genetic material of the new zygote.

Take-Home Message

What happens during intercourse and fertilization?

» Sexual arousal involves nervous and hormonal signals.

» Ejaculation releases millions of sperm into the vagina. Sperm travel through the uterus toward the oviducts, the site where fertilization most often occurs.

» Penetration of a secondary oocyte by a single sperm causes the oocyte to complete meiosis II, and prevents additional sperm from penetrating it.

» Sperm organelles disintegrate. The DNA of the sperm, along with that of the oocyte, become the genetic material of the zygote.

41.9 Preventing or Seeking Pregnancy

■ Most birth control methods rely on preventing ovulation or blocking fertilization.

■ Infertility most often arises from blocked reproductive ducts or hormonal disorders that prevent ovulation.

Birth Control Options

Emotional and economic factors often lead people to seek ways to control their fertility. **Table 41.2** lists common contraceptive options and compares their effectiveness. Most effective is abstinence—no sex—which has the added advantage of preventing exposure to sexually transmissible pathogens.

With rhythm methods, a woman abstains from sex during her fertile period. She calculates when she is fertile by recording how long menstrual cycles last, checking her temperature daily (**Figure 41.12A**), monitoring the thickness of her cervical mucus, or some combination of these methods. However, cycles vary, so miscalculations are frequent and sperm deposited in the vagina up to three days before ovulation can survive long enough to fertilize an egg.

Withdrawal, or removing the penis from the vagina before ejaculation, requires great willpower. In some men, pre-ejaculation fluid from the penis contains enough sperm to allow fertilization to occur.

Douching, or rinsing out the vagina immediately after intercourse, is unreliable. Some sperm can travel through the cervix within seconds of ejaculation.

Surgical methods are highly effective, but are meant to make a person permanently sterile. Men may opt for a vasectomy. A doctor makes a small incision into the scrotum, then cuts and ties off each vas deferens. A tubal ligation blocks or cuts a woman's oviducts.

Other fertility control methods use physical and chemical barriers to stop sperm from reaching an egg. Spermicidal foam and spermicidal jelly poison sperm. They are not always reliable, but their use with barrier methods reduces the chance of pregnancy.

A diaphragm is a flexible, dome-shaped device that is positioned inside the vagina so it covers the cervix. A diaphragm is relatively effective if it is first fitted by a doctor and used correctly with a spermicide. A cervical cap is a similar but smaller device.

Condoms are thin sheaths worn over the penis or used to line the vagina during intercourse (**Figure 41.12B**). Reliable brands can be 95 percent effective when used correctly with a spermicide. Condoms made of latex also offer some protection against sexually transmitted diseases (STDs). However, even the best ones can tear or leak, reducing effectiveness.

An intrauterine device, or IUD, is inserted into the uterus by a physician. Some IUDs make cervical mucus thicken so sperm cannot swim through it. Others shed copper that interferes with implantation.

Oral contraceptives—birth control pills—are the most common fertility control method in developed countries. "The Pill" is a mixture of synthetic estrogens and progesterone-like hormones that prevents both maturation of oocytes and ovulation (**Figure 41.12C**). When taken diligently, oral contraception is highly effective. It also reduces menstrual cramps and lowers the risk of cancer of the ovaries and uterus. Side effects include nausea, headaches, weight gain, and increased risk of cancer of the breast, cervix, and liver. A birth control patch is a small, flat adhesive patch applied to skin. The patch delivers the same mixture of hormones as an oral contraceptive, and it blocks ovulation the same way.

Hormone injections or implants prevent ovulation. Injections act for several months, whereas an implant such as Implanon lasts for three years (**Figure 41.12C**). Both methods are quite effective, but may cause sporadic, heavy bleeding.

Table 41.2 Common Methods of Contraception

Method	Mechanism of Action	Pregnancy Rate*
Abstinence	Avoid intercourse entirely	0% per year
Rhythm method	Avoid intercourse when female is fertile	25% per year
Withdrawal	End intercourse before male ejaculates	27% per year
Vasectomy	Cut or close off male's vasa deferentia	>1% per year
Tubal ligation	Cut or close off female's oviducts	>1% per year
Condom	Enclose penis, block sperm entry to vagina	15% per year
Diaphragm, cervical cap	Cover cervix, block sperm entry to uterus	16% per year
Spermicides	Kill sperm	29% per year
Intrauterine device	Prevent sperm entry to uterus or prevent implantation	>1% per year
Oral contraceptives	Prevent ovulation	>1% per year
Hormone patches, implants, or injections	Prevent ovulation	>1% per year
Emergency contraception pill	Prevent ovulation	15–25% per use**

* Percentage of users who get pregnant despite consistent, correct use
** Not meant for regular use

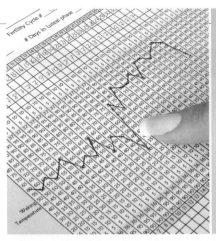

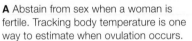

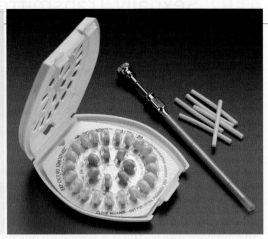

A Abstain from sex when a woman is fertile. Tracking body temperature is one way to estimate when ovulation occurs.

B Use a barrier to prevent sperm from meeting an egg. Male condom (*left*); female condom (*right*).

C Use synthetic female hormones to prevent ovulation. Oral contraceptives (*left*) are taken daily. Implantable contraceptives (*right*) are injected under the skin.

Figure 41.12 Some methods of preventing pregnancy.

If a woman has unprotected sex or a condom breaks, she can turn to emergency contraception. Some so-called "morning-after pills" are available without a prescription to women over age 17. The most widely used emergency contraceptive, Plan B, delivers a large dose of synthetic progesterone that prevents ovulation and interferes with fertilization. It is most effective when taken immediately after intercourse, but has some effect up to three days later. Morning-after pills are not meant for regular use. They cause nausea, vomiting, abdominal pain, headache, and dizziness.

Infertility

About 10 percent of couples in the United States are infertile, which means they do not have a successful pregnancy despite a year of trying. Infertility increases with age. In about two-thirds of cases the problem lies with the female partner. She may have a hormonal disorder that prevents follicles from maturing normally. For example, with polycystic ovarian syndrome, follicles begin to grow, but ovulation does not occur; instead the follicle turns into a cyst.

Some sexually transmitted diseases can scar the oviducts and prevent sperm from reaching eggs. Endometriosis also interferes with oviduct function. With this disorder, endometrial tissue grows in the regions of the pelvis outside the uterus. Like normal endometrial tissue, the misplaced tissue proliferates and bleeds in response to changing hormone levels. The result is pain and scarring that can distort the oviducts or scar them. Endometriois can also make intercourse painful, thus lessening the frequency of intercourse and opportunities for conception.

If fertilization does occur, scarring of oviducts can lead to an embryo implanting in the oviduct, causing a tubal pregnancy that cannot develop to term and threatens the life of the mother (**Figure 41.13**). Fibroids, endometriosis, or other uterine problems can interfere with the ability of the embryo to implant.

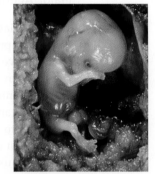

Figure 41.13 Embryo that implanted in an oviduct.

Male infertility arises when a man makes few or abnormal sperm, or sperm are not ejaculated properly. Hormonal disorders that lower testosterone can prevent sperm maturation. Some genetic disorders can interfere with sperm motility. In males, as in females, sexually transmitted disease can scar ducts of the reproductive system, causing blockages that keep sperm from leaving the body. In some men, the valve between the bladder and the urethra does not function properly, so sperm end up in the bladder rather than being ejaculated.

A variety of methods now allow infertile couples to have a family. We discuss this topic more in the following chapter.

Take-Home Message

What can prevent pregnancy?

» Couples can avoid sex entirely or abstain when the woman is fertile.

» Hormones delivered by pills, patches, or injections can prevent a woman from ovulating.

» Temporary barriers such as condoms or a diaphragm keep sperm and egg apart, as do surgical methods such as a vasectomy or tubal ligation.

» Infertility can arise from hormonal disorders, blockages of reproductive ducts, and uterine problems that prevent implantation.

41.10 Sexually Transmitted Diseases

■ Sex acts transfer body fluids in which some human pathogens travel from one host to another.

■ Links to HIV 20.1 and 37.11, HPV 37.1, Bacterial pathogens 20.7, Trichomonads 21.3

Consequences of Infection

Each year, pathogens that cause sexually transmitted diseases, or STDs, infect many millions of Americans (**Table 41.3**). Women are more easily infected than men, and have more complications. For example, pelvic inflammatory disease (PID), a secondary outcome of bacterial STDs, scars the female reproductive tract and can cause infertility, chronic pain, and tubal pregnancies. In addition to threatening a woman's health, an STD can also harm her offspring.

Major Agents of Sexually Transmitted Disease

Trichomoniasis is a common STD caused by the flagellated protozoan *Trichomonas vaginalis*. Infected women have a yellowish discharge and a sore, itchy vagina. Men usually have no symptoms. In both sexes, untreated infection can cause infertility. An antiprotozoal drug can cure the infection. Both partners should be treated to prevent reinfection.

The most common bacterial STD is chlamydia, which is caused by *Chlamydia trachomatis*. Chlamydias are small bacteria that live as intracelluar parasites, relying on their host cell for ATP. In women, an infection of the reproductive tract by *C. trachomatis* most often goes undetected. Some women and most men experience painful urination; most infected men

have a clear or yellow discharge from the penis. Left untreated, a *Chlamydia* infection can scar the reproductive tract and lead to infertility in both sexes. An infection can be passed from a mother to her child as it passes through the birth canal, causing pneumonia and conjunctivitis in the newborn (**Figure 41.14A**). Chlamydia can be cured with antibiotics.

Gonorrhea and syphilis are also bacterial STDs. Less than one week after a male is infected by *Neisseria gonorrhoeae*, yellow pus oozes from his penis and urination becomes frequent and painful. Women are likely to have no early symptoms. In both sexes an ongoing infection can damage reproductive ducts and cause sterility. Gonorrhea is treated with antibiotics, but strains resistant to commonly used antibiotics such as penicillin are increasing in frequency. A persistent gonorrhea infection can spread throughout the body.

Table 41.3	STDs in the United States	
STD	**New Cases/Year**	**Pathogen**
Trichomoniasis	7,400,000	Protozoan
HPV infection	6,000,000	Virus
Chlamydia	1,200,000	Bacteria
Genital herpes	1,000,000	Virus
Gonorrhea	300,000	Bacteria
HIV infection	40,000	Virus
Syphilis	14,000	Bacteria

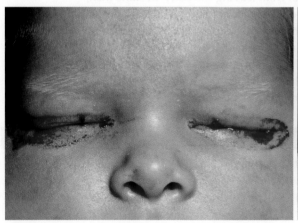

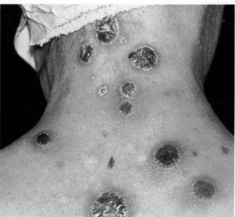

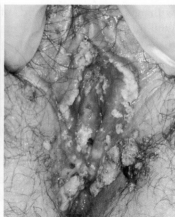

A Passage of pathogens to offspring. This infant's eyes are infected by *Chlamydia* acquired from its mother during birth.

B Spread of a pathogen throughout the body. The chancres (open sores) are a sign of syphilis.

C Warty genitals and increased risk of cancer. The raised white warts on this woman's labia are a sign of HPV.

Figure 41.14 A few downsides of unsafe sex.

Intersex Conditions (revisited)

Until recently, many XY individuals born with a micropenis, a penis less than 2.5 centimeters in length, were reassigned to the female gender. They were given female hormones and underwent multiple surgeries to create a vagina and genitals with a female appearance. These treatments were presumably carried out under the assumption that the pain and risk of complication entailed by such surgery were outweighed by the hardship a man with a small penis

would face. However, a survey of adult individuals diagnosed with micropenis as infants found that those who had undergone surgery to feminize their genitals and were brought up a females were less likely to be satisfied with their sexual function than those who remained males.

How would you vote? Should parents of a child with an intersex condition postpone any non-health-related surgery until the child can consent to it?

Syphilis is caused by *Treponema pallidum*, a spiral-shaped bacterium. During sex with an infected partner, these bacteria get onto the genitals or into the cervix, vagina, or oral cavity. They slip into the body through tiny cuts. One to eight weeks later, many *T. pallidum* cells are twisting about inside a flattened, painless ulcer (a chancre) at the site of infection. If the infection continues untreated, bacteria replicate throughout the body, and skin chancres appear (**Figure 41.14B**) as the liver, bones, and eventually the brain become damaged. Like gonorrhea, syphilis can be treated with antibiotics. As with gonorrhea, the number of antibiotic-resistant cases is on the rise.

Infection by human papillomaviruses (HPV) is widespread in the United States. Many HPV strains cause bumplike growths called warts, including those common on hands or feet. Some strains of the virus are sexually transmitted and they become established wherever sexual behavior introduces them (**Figure 41.14C**). In addition to warts, some strains of HPV cause cancer. HPV infection is the cause of nearly all cervical cancers in women and most anal cancers in homosexual men. In addition, a recent study indicates that introduction of HPV into the mouth by way of oral sex raises the risk of cancers of the mouth and throat. A recently approved vaccine can prevent HPV infection in both males and females, if given before viral exposure.

About 45 million Americans have genital herpes caused by herpes simplex virus 2. An initial infection commonly causes small sores at the site where the virus entered the body. The sores heal, but the infection can be reactivated, causing tingling or itching, which may or may not be accompanied by visible sores. Antiviral drugs cannot cure the infection, but they promote healing of sores and lessen the likelihood of viral reactivation.

Transmission of the herpesvirus from mother to child during birth is rare—the mother must have an

active infection at the time of delivery—but it can have serious consequences. If untreated, an infant infected in this manner has a 40 percent chance of death. Those who survive remain infected, and have a heightened risk of health problems including blindness and impaired brain development. Women who become infected for the first time while pregnant are especially likely to have an active infection at delivery.

Use of latex condoms can lower the risk of transmission of both HPV and genital herpes. However, if the condom does not entirely cover the infected area, transmission of the virus can still occur.

Infection by HIV (or human immunodeficiency virus) can cause AIDS (acquired immune deficiency syndrome). Sections 20.1 and 20.3 described the HIV virus and Section 37.11 explained how its effects on the immune system lead to AIDS. With regard to sexual transmission, oral sex is least likely to transmit an infection. Unprotected anal sex is 5 times more dangerous than unprotected vaginal sex and 50 times more dangerous than oral sex. To reduce the risk of HIV transmission during vaginal or anal sex, doctors recommend use of a latex condom and a lubricant. The lubricant helps prevent small abrasions that could allow the virus to enter the body.

If you think you may have been exposed to an STD, see a doctor as soon as possible. Early diagnosis and treatment minimize the negative health effects.

Take-Home Message

What are causes and effects of common STDs and how are they treated?

» Viruses, bacteria, and protozoans cause common STDs.

» Symptoms range from merely discomforting to life-threatening.

» STDs can cause infertility, and infected women can pass them to their offspring.

» Bacterial and protozoal pathogens can be killed by drugs. Viral infections cannot be cured with drugs, but antiviral drugs reduce the effects of herpes and HIV infection. An HPV vaccine can prevent infection with this virus.

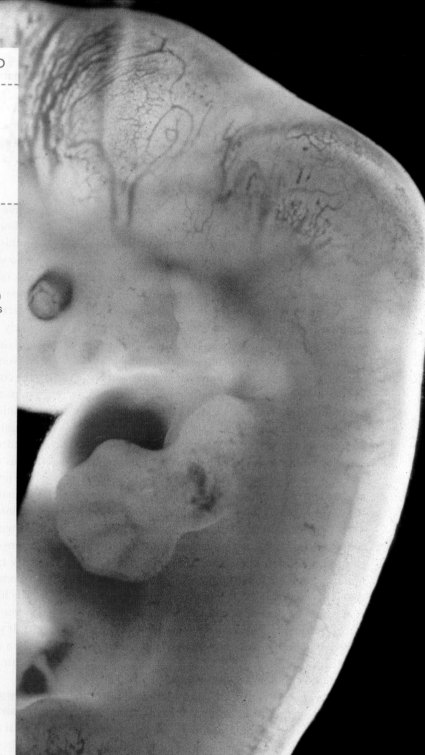

EARNING ROADMAP

Where you are now

Principles of Animal Embryology
All animals develop by similar processes. Localization of materials in eggs sets the stage for differentiation based on cell location and responses to intercellular signals.

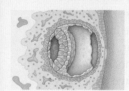

Human Development Begins
A pregnancy starts with fertilization and implantation of a blastocyst in the uterus. Next, a three-layered embryo forms, then development of organs gets under way.

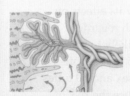

Function of the Placenta
The placenta allows substances to diffuse between the bloodstreams of a mother and her developing child. It also produces hormones that help sustain the pregnancy.

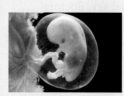

Later Human Development
A fetus appears distinctly human. Harmful substances that get into a mother's blood can cross the placenta and cause birth defects in the developing embryo or fetus.

Birth and Lactation
Positive feedback plays a role in the process of labor, or childbirth. After birth, the newborn is nourished by milk secreted by its mother's mammary glands.

Where you are going In Chapter 43 you will learn that oxytocin, which plays a role in labor and lactation, can also affect social behavior. In Chapter 44 we discuss the reproductive potential of populations and the factors that can influence human population growth.

42.1 Mind-Boggling Births

Some couples that cannot conceive naturally or women who are using sperm from a nonpartner donor turn to *in vitro* fertilization (IVF). This assisted reproductive technology combines an egg with sperm outside the body. Prior to IVF a woman is given hormones to encourage maturation of multiple eggs and to prevent natural ovulation. The mature eggs are removed from her ovaries via a hollow needle inserted into the ovary. The eggs are combined with a partner's or donor's sperm for fertilization. After fertilization, each zygote undergoes mitotic divisions, forming a ball of cells (a blastocyst) that can be placed in a woman's womb to develop to term.

Louise Brown, the first child conceived by IVF, was born in 1978. At the time, much of the public and many scientists were appalled by the idea of what newspapers referred to as "test tube babies." Scientists worried that this new, unnatural procedure would produce children with psychological and genetic defects. Ethicists warned about the societal implications of manipulating human embryos.

Despite the initial reservations, IVF has become widely accepted and practiced. Worldwide, the technique has produced more than 3 million children. The first test tube babies are now adults and have begun having children of their own. Louise Brown is among

them. In 2007, at age 28, she conceived naturally and gave birth to a healthy son (**Figure 42.1A**).

One negative effect of IVF has been an increase in the number of multiple births. Embryos conceived by IVF frequently fail to take hold and develop, so doctors often place several in a woman's womb in the hope that one will survive. If all these embryos do develop, the woman faces the risks associated with a multiple pregnancy. Compared to a woman carrying a single child, she is more likely to undergo miscarriage, stillbirth, or premature birth, and the children that she carries are at an increased risk of low birth weight and birth defects.

Professional organizations have urged fertility doctors to limit the number of embryos that they place in a woman's uterus, but this advice sometimes goes unheeded. In 2009, Nadya Suleman's doctor put twelve embryos conceived by IVF into her uterus. Suleman subsequently gave birth to octuplets (**Figure 42.1B**). The babies were nine weeks premature and all were underweight. Suleman already had six other children, also conceived by IVF, so she is now the mother of fourteen.

in vitro **fertilization** Assisted reproductive technology in which eggs and sperm are united outside the body.

A Louise Brown, the first child conceived by IVF, with her husband and her son, who was conceived naturally.

B Nadya Suleman with her octuplets conceived by IVF.

Figure 42.1 Results of natural and assisted fertilization.

42.2 Stages of Reproduction and Development

■ Animals as different as sea stars and sea otters pass through the same stages in their developmental journey from a single, fertilized egg to a multicelled adult.

■ Links to Gamete formation 12.5, Comparative development 18.5, Animal body plans and germ layers 24.2, Fertilization 41.8

In all sexually reproducing animals, a new individual begins life as a zygote, the diploid cell that forms at fertilization. Development from a zygote to an adult typically proceeds through a series of stages (**Figure 42.2**). An example of these stages in one vertebrate, the leopard frog, appears in **Figure 42.3**.

Sperm penetrates an egg, the egg and sperm nuclei fuse, and a zygote forms.

Mitotic cell divisions yield a ball of cells, a blastula. Each cell gets a different bit of the egg cytoplasm.

Cell rearrangements and migrations form a gastrula, an early embryo that has primary tissue layers.

Organs form as the result of tissue interactions that cause cells to move, change shape, and commit suicide.

Organs grow in size, take on mature form, and gradually assume specialized functions.

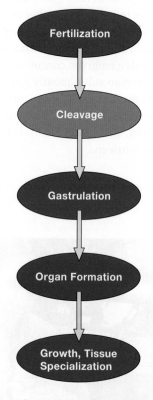

Figure 42.2 Stages of development in vertebrates. Fertilization was described in Section 41.8.

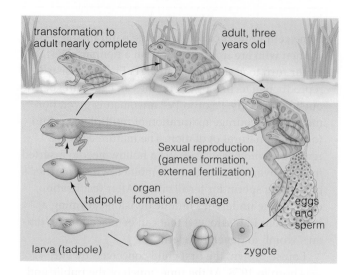

Figure 42.3 Animated *Above*, overview of reproduction and development in the leopard frog. *Opposite page*, a closer look at what occurs during each stage.

The process of reproduction begins when a female frog releases eggs into the water and a male releases sperm onto the eggs. External fertilization produces the zygote. New cells arise when **cleavage** carves up a zygote by repeated mitotic cell divisions ❶. During cleavage, the number of cells increases; however, the zygote's original volume remains unchanged. As a result, cells become more numerous but smaller. The cells that form during cleavage are called blastomeres. They typically become arranged as a **blastula**: a ball of cells that enclose a cavity (blastocoel) filled with their secretions ❷. Tight junctions hold the cells of the blastula together.

During **gastrulation**, cells move about and organize themselves as the layers of the **gastrula** ❸. In most animals and all vertebrates, a gastrula consists of three primary tissue layers, or **germ layers**. The three germ layers give rise to the same types of tissues and organs in all vertebrates (**Table 42.1**). This developmental similarity is evidence of a shared ancestry (Section 18.5).

Ectoderm, the outer germ layer, forms first. It gives rise to nervous tissue and to the outer layer of skin or

Table 42.1	Derivatives of Vertebrate Germ Layers
Ectoderm (outer layer)	Outer layer (epidermis) of skin; nervous tissue
Mesoderm (middle layer)	Connective tissue of skin; skeletal, cardiac, smooth muscle; bone; cartilage; blood vessels; urinary system; gut organs; peritoneum (coelom lining); reproductive tract
Endoderm (inner layer)	Lining of gut and respiratory tract, and organs derived from these linings

blastula Hollow ball of cells that forms as a result of cleavage.
cleavage Mitotic division of an animal cell.
ectoderm Outermost tissue layer of an animal embryo.
endoderm Innermost tissue layer of an animal embryo.
gastrula Three-layered developmental stage formed by gastrulation in an animal.
gastrulation Animal developmental process by which cell movements produce a three-layered gastrula.
germ layer One of three primary layers in an early embryo.
mesoderm Middle tissue layer of a three-layered animal embryo.

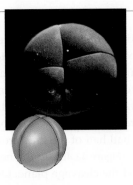

gray crescent

blastula

blastocoel

1 Here we show the first three divisions of cleavage, a process that carves up a zygote's cytoplasm. In this species, cleavage results in a blastula, a ball of cells with a fluid-filled cavity.

2 Cleavage is over when the blastula forms.

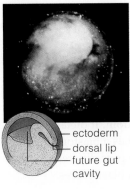
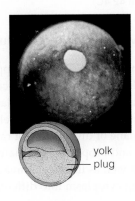
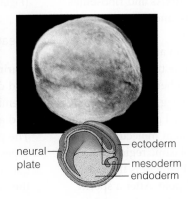

ectoderm
dorsal lip
future gut cavity

yolk plug

neural plate

ectoderm
mesoderm
endoderm

neural tube

notochord
gut cavity

3 The blastula becomes a three-layered gastrula—a process called gastrulation. At the dorsal lip (a fold of ectoderm above the first opening that appears in the blastula), cells migrate inward and start rearranging themselves.

4 Organs begin to form as a primitive gut cavity opens up. A neural tube, then a notochord and other organs, form from the primary tissue layers.

Tadpole, a swimming larva with segmented muscles and a notochord extending into a tail.

Limbs grow and the tail is absorbed during metamorphosis to the adult form.

Sexually mature, four-legged adult leopard frog.

5 The frog's body form changes as it grows and its tissues specialize. The embryo becomes a tadpole, which metamorphoses into an adult.

other body covering. **Endoderm**, the inner layer, is the start of the respiratory tract and gut linings. A third layer called **mesoderm** forms between ectoderm and endoderm. This "middle" layer is the source of muscles, connective tissues, and the circulatory system.

Organ formation begins after gastrulation. The neural tube and notochord characteristic of all chordate embryos form early **4**. Many organs incorporate tissues derived from more than one germ layer. For example, the stomach's epithelial lining is derived from endoderm, and the smooth muscle that makes up the stomach wall develops from mesoderm.

In frogs, as in some other animals, a larva (in this case a tadpole) undergoes metamorphosis, a drastic remodeling of tissues into the adult form **5**.

Take-Home Message

How does an adult vertebrate develop from a single-celled zygote?

» A zygote undergoes cleavage, which increases the number of cells. Cleavage ends with formation of a blastula.

» Rearrangement of blastula cells forms a three-layered gastrula.

» After gastrulation, organs such as the nerve cord begin forming.

» Continued growth and tissue specialization produce the adult body.

42.3 From Zygote to Gastrula

■ Localization of yolk and other material in the egg cytoplasm and lineage-specific cleavage patterns affect early development.
■ Links to Cell cortex 4.10, mRNA transport 10.2, Cytoplasmic division 11.4, Protostomes and deuterostomes 24.2

Components of Eggs and Sperm

A sperm consists of paternal DNA and a bit of equipment that helps it swim to and penetrate an egg. The egg has far more cytoplasm. It also has yolk proteins that will nourish the embryo, mRNAs for proteins essential to early development, tRNAs and ribosomes to translate the maternal mRNAs, and the proteins required to build mitotic spindles.

Cytoplasmic localization occurs in all oocytes: Many cytoplasmic components are not distributed evenly throughout egg cytoplasm, but rather become localized in one region. In a yolk-rich egg, the vegetal pole has most of the yolk and the animal pole has little. In some amphibian eggs, dark pigment molecules accumulate in the cell cortex, a cytoplasmic region just beneath the plasma membrane. Pigment is the most concentrated close to the animal pole. After a sperm penetrates the egg at fertilization, the cortex rotates.

Rotation reveals a gray crescent, a region of the cell cortex that is lightly pigmented (**Figure 42.4A**).

Early in the 1900s, experiments by Hans Spemann showed that substances essential to development are localized in the gray crescent. In one experiment, he separated the first two blastomeres formed at cleavage. Each had half of the gray crescent and developed normally (**Figure 42.4B**). In the next experiment, Spemann altered the cleavage plane. One blastomere got all the gray crescent, and it developed normally. The other lacked gray crescent and got stuck in the blastula stage (**Figure 42.4C**).

Cleavage—The Start of Multicellularity

During cleavage, a furrow appears on the cell surface and defines the plane of the cut. Beneath the plasma membrane, a ring of microfilaments starts contracting, and eventually divides the cell in two (Section 11.4). The plane of division is not random. It dictates what types and proportions of materials a blastomere will get, as well as its size. Cleavage puts different parts of the egg cytoplasm into different blastomeres.

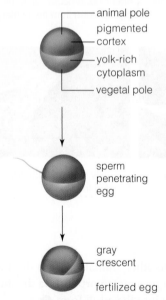

A Many amphibian eggs have a dark pigment concentrated in cytoplasm near the animal pole. At fertilization, the cytoplasm shifts, and exposes a gray crescent-shaped region just opposite the sperm's entry point. The first cleavage normally distributes half of the gray crescent to each descendant cell.

Figure 42.4 Animated Experimental evidence of cytoplasmic localization in an amphibian oocyte.
Figure It Out: Is the gray crescent essential to amphibian development? Answer: Yes

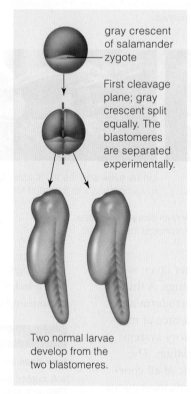

Two normal larvae develop from the two blastomeres.

B In one experiment, the first two cells formed by normal cleavage were physically separated from each other. Each cell developed into a normal larva.

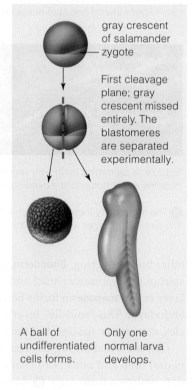

A ball of undifferentiated cells forms.

Only one normal larva develops.

C In another experiment, a zygote was manipulated so one descendant cell received all the gray crescent. This cell developed normally. The other gave rise to an undifferentiated ball of cells.

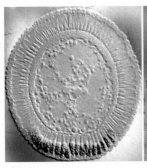

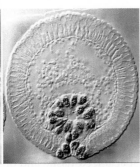

Figure 42.5 Gastrulation in a fruit fly (*Drosophila*). In fruit flies, cleavage is restricted to the outermost region of cytoplasm; the interior is filled with yolk. The series of photographs, all cross-sections, shows sixteen cells (stained gold) migrating inward. The opening the cells move in through will become the fly's mouth. Descendants of the stained cells will form mesoderm. Movements shown in these photos occur during a period of less than 20 minutes.

Each species has a characteristic cleavage pattern. Remember the branching of the coelomate lineage into the protostomes and deuterostomes (Section 24.2)? These two groups differ in certain details of cleavage, such as the angle of divisions relative to an egg's polar axis. The amount of yolk also influences the pattern of division. Insects, frogs, fishes, and birds have yolk-rich eggs. In such eggs, a large volume of yolk slows or blocks some of the cuts. The result is fewer divisions of the yolky part of the egg than the less yolky part. By comparison, the cuts slice right through the nearly yolkless eggs of sea stars and mammals.

Gastrulation

A hundred to thousands of cells may form at cleavage, depending on the species. Starting with gastrulation, cells migrate about and rearrange themselves. **Figure 42.5** shows fruit fly gastrulation. **Figure 42.3** shows gastrulation in a frog embryo.

Hilde Mangold, one of Spemann's students, discovered how gastrulation is regulated in vertebrates. Mangold knew that during gastrulation, cells of a salamander blastula move inward through an open-

cytoplasmic localization Accumulation of different materials in different regions of the egg cytoplasm.

ing on its surface. Cells in the dorsal (upper) lip of the opening are descended from a zygote's gray crescent, and Mangold suspected that these cells made signals that caused gastrulation. To test her hypothesis, she transplanted dorsal lip material from one embryo to another (**Figure 42.6A**). As expected, cells migrated inward at the transplant site, as well as at the usual location (**Figure 416B**). A salamander larva with two joined sets of body parts developed (**Figure 42.6C**). Apparently, the transplanted cells signaled their new neighbors to develop in a novel way. This experiment also explains the results shown in **Figure 42.4C**. Without any gray crescent cytoplasm, an embryo does not have dorsal lip cells to signal the start of gastrulation, so development stops short.

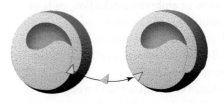

C The embryo develops into a "double" larva, with two heads, two tails, and two bodies joined at the belly.

A Dorsal lip excised from donor embryo, grafted to novel site in another embryo.

B Graft induces a second site of inward migration.

Figure 42.6 Animated Experimental evidence that signals from dorsal lip cells initiate amphibian gastrulation. A dorsal lip region of a salamander embryo was transplanted to a different site in another embryo. A second set of body parts started to form.

42.4 How Specialized Tissues and Organs Form

■ After gastrulation, cells become specialized as their movement and interaction begin to shape tissues and organs.

■ Links to Cell movement 4.10, Cell differentiation and transcription factors 10.2, Neural tube and notochord 25.2

Cell Differentiation

All cells in an embryo are descended from the same zygote, so all have the same genes. How then, do the specialized tissue and organs form? From gastrulation onward, selective gene expression occurs: Different cell lineages express different subsets of genes. Selective gene expression is the key to **differentiation**, the process by which cell lineages become specialized in composition, structure, and function (Section 10.2).

An adult human body has about 200 differentiated cell types. As your eye developed, cells of one lineage turned on genes for crystallin, a transparent protein. These differentiated cells formed the lens of your eye. No other cells in your body make crystallin.

Keep in mind that a differentiated cell still retains the entire genome. That is why it is possible to clone an adult animal—to create a genetic copy—from one of its differentiated cells (Section 8.7).

Responses to Morphogens

Intercellular signals encourage selective gene expression. For example, **morphogens** are signal molecules that can act over a long distance and influence cells in a concentration-dependent manner. They are produced in one area of an embryo and diffuse outward, forming a concentration gradient. Cells nearest the source of a morphogen are exposed to a high morphogen concentration and express one set of genes. Cells more distant from the morphogen source are exposed to a lower concentration and turn on different genes.

The bicoid protein of fruit flies is an example of a morphogen. *Bicoid* is a **maternal effect gene**, meaning it is expressed during egg production and its product influences development. During egg production, this gene is transcribed and bicoid mRNA accumulates at one end of the egg—an example of cytoplasmic localization. After fertilization, the mRNA is translated into bicoid protein, which becomes distributed in a gradient along the length of the cell. This gradient determines the front-to-back axis of the zygote. Where the bicoid protein concentration is highest, the zygote expresses genes that regulate development of anterior body structures. Where it is lowest, the zygote expresses genes that regulate development of posterior body parts.

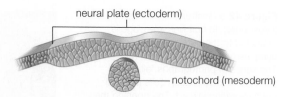

neural plate (ectoderm)

notochord (mesoderm)

1 Chemical signals produced by notochord mesoderm induce the ectoderm above it to thicken and form a neural plate.

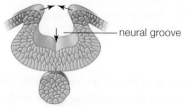

neural groove

2 Changes in cell shape cause edges of the neural plate to fold in toward the plate center, forming a neural groove.

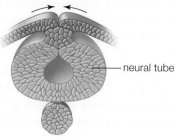

neural tube

3 Further folding causes the edges of the neural plate to meet, forming the neural tube.

Figure 42.7 Animated Neural tube formation in a vertebrate embryo.

The bicoid protein is a transcription factor (Section 10.2) that stimulates the expression of other master genes, which activate other master genes, and so on. This cascade of master gene expression ultimately results in the formation of specific body parts in local regions of the developing embryo (Section 10.3).

Embryonic Induction

Intercellular communication also operates at close range. By the process of **embryonic induction**, cells of one embryonic tissue alter the behavior of cells in an adjacent tissue. For example, the cells of a salamander gastrula's dorsal lip induce adjacent cells to migrate inward and become mesoderm.

After gastrulation, vertebrate organ formation begins with the neural tube, which is the embryonic precursor of the nervous system. Neural tube development is induced by signals from the notochord, which formed earlier from mesoderm (**Figure 42.7**). Development begins when ectodermal cells overlying the notochord elongate, forming a thick neural plate

1. The cell elongation results from the assembly of microtubules inside them. Next, cells at the edges of the neural plate become more wedge-shaped as actin microfilaments at one end constrict. The change in cell shape causes the edges of the plate to fold inward, forming a neural groove **2**. Eventually the edges of the former neural plate meet at the midline, forming the neural tube **3**.

Cell Migrations

Cell migrations are an essential part of development, For example, ectodermal cells that form at the apex of the neural tube (a region called the neural crest), migrate outward to positions throughout the body. The descendants of neural crest cells include the neurons and glial cells of the peripheral nervous system, and the melanocytes in skin.

Cells travel by inching along in an amoeba-like fashion (**Figure 42.8**). Assembly of actin microfilaments at one edge of the cell causes that portion of the cell to protrude forward. Adhesion proteins in the plasma membrane then anchor the advancing portion of the cell to a protein in the membrane of another cell or in the extracelluar matrix. Once the front of the cell is thus anchored, adhesion proteins in the trailing portion of the cell release their grip, and the rear of the cell is drawn forward. How does the cell know where

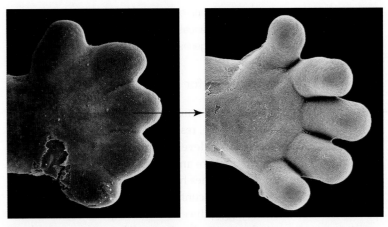

Figure 42.9 Apoptosis in the formation of a human hand. *Left*, forty-eight days after fertilization, tissue webs connect embryonic digits. *Right*, three days later, after apoptosis by cells in those webs, the digits have become fully separated.

to go? It may move in response to a concentration gradient of some chemical signal or it may follow a "trail" of molecules that its adhesion proteins recognize. It stops migrating when it reaches a region where its adhesion proteins hold it tightly in place.

Apoptosis

Programmed cell death, or **apoptosis**, helps sculpt body parts. During apoptosis, cells that are no longer needed self-destruct in response to chemical signals. Depending on the context, an internal or external signal sets in motion a chain of reactions that result in the activation of self-destructive enzymes. Some of these enzymes chop up structural proteins such as cytoskeleton proteins and the histones that organize DNA. Others snip apart nucleic acids. As a result of these activities, the cell dies.

Apoptosis eliminates the webbing between digits of a developing human hand (**Figure 42.9**) and makes the tail of a tadpole disappear. As you will learn later in this chapter, it also eliminates the tail that develops in the early human embryo.

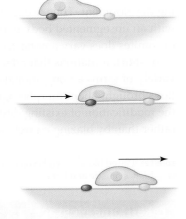

Figure 42.8 How cells migrate. A cell extends its advancing edge and attaches to the underlying substrate (a cell or the extracellular matrix) by way of adhesion proteins. It then breaks the connection between its trailing edge and the substrate, and contracts, pulling its posterior portion forward.

apoptosis Mechanism of programmed cell death.
differentiation Process by which cells become specialized.
embryonic induction Embryonic cells produce signals that alter the behavior of neighboring cells.
maternal effect gene Gene expressed during egg production; its product influences animal development.
morphogen Chemical encoded by a master gene; diffuses out from its source and affects development.

Take-Home Message

What processes produce specialized cells, tissues, and organs?

» All cells in an embryo have the same genes, but they express different subsets of the genome. Selective gene expression is the basis of cell differentiation. It results in cell lineages with characteristic structures and functions.

» Cytoplasmic localization results in concentration gradients of signaling proteins called morphogens. Morphogens activate sets of master genes, the products of which cause embryonic cells to form tissues and organs in specific places.

» Migration, shape changes, and death of cells shape developing organs.

42.5 An Evolutionary View of Development

■ Similarities in developmental pathways among animals are evidence of common ancestry.
■ Links to Homeotic genes 10.3 and 18.5, Somites 18.5

A General Model for Animal Development

Through studies of animals such as roundworms, fruit flies, fish, and mice, researchers have come up with a general model for development. The key point of the model is this: Where and when particular genes are expressed determines how an animal body develops.

First, molecules confined to different areas of an unfertilized egg induce localized expression of master genes in the zygote. Products of these master genes diffuse outward, so concentration gradients for these products form along the head-to-tail and dorsal-to-ventral axes of the developing embryo.

Second, depending on where they fall within these concentration gradients, cells in the embryo activate or suppress other master genes. The products of these genes become distributed in gradients, which affect other genes, and so on.

Third, this positional information affects expression of homeotic genes. As Sections 10.3 and 18.5 explained, all animals have similar homeotic genes. For example, a mouse's *eyeless* gene initiates development of its eyes. Introduce the mouse version of this gene into a fruit fly, and eyes will form in tissues where the introduced gene is expressed.

A Adult zebrafish, an animal used in studies of development.

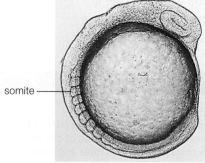 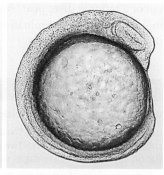

somite —

B Normal zebrafish embryo with somites, bumps of mesoderm that give rise to bone and muscle.

C Embryo with mutation. It cannot form somites and will die in early development.

Figure 42.10 Lethal effect of a mutation in a zebrafish gene (*fused somites*) that functions in early development.

Developmental Constraints and Modifications

The developmental model just described helps explain why we only see certain types of animal body plans. Body plans are influenced by physical constraints. A large body size cannot evolve in an animal without circulatory and respiratory mechanisms that service body cells far from the body surface.

An existing body framework imposes architectural constraints. For example, the ancestors of all modern land vertebrates had a body plan with four limbs. The evolution of wings in birds and bats occurred through modification of existing forelimbs, not by sprouting new limbs. Although it might be advantageous to have both wings and arms, no living or fossil vertebrate with both has been discovered.

Finally, there are phylogenetic constraints on body plans. These constraints are imposed by interactions among genes that regulate development in a lineage. Once master genes evolved, their interactions determined the basic body form. Mutations that dramatically alter these interactions are usually lethal.

Consider the paired bones and skeletal muscles arrayed along a vertebrate's head-to-tail axis. This pattern arises early in development, when the mesoderm on either side of the embryo's neural tube becomes divided into blocks of cells called **somites** (Figure 42.10). The somites will later develop into bones and skeletal muscles. A complex pathway involving many genes governs somite formation. Any mutation that disrupts this pathway so that somites do not form is lethal during development. Thus, we do not find vertebrates with an unsegmented body plan, although the number of somites does vary among species.

In short, mutations that affect development led to a variety of forms among animal lineages. These mutations brought about morphological changes through the modification of existing developmental pathways, rather than by blazing entirely new genetic trails.

somites Paired blocks of embryonic mesoderm that give rise to a vertebrate's muscle and bone.

Take-Home Message

Why are developmental processes and body plans similar among animal groups?

» In all animals, cytoplasmic localization affects expression of sets of master genes shared by most animal groups. The products of these genes cause embryonic cells to form tissues and organs at certain locations.

» Once a developmental pathway evolves, drastic changes to genes that govern this pathway are generally lethal.

42.6 Overview of Human Development

- Like all animals, humans begin life as a single cell and go through a series of developmental stages.
- Links to Placental mammals 25.10, Human fertilization 41.8

Chapter 41 introduced the structure and function of human reproductive organs, and explained how an egg and sperm meet at fertilization to form a zygote. The remaining sections of this chapter will continue this story, with an in-depth look at human development. In this section, we provide an overview of the process and define the stages that we will discuss. Prenatal (before birth) and postnatal (after birth) stages are listed in **Table 42.2**.

It takes about five trillion mitotic divisions to go from the single cell of a zygote to the ten trillion or so cells of an adult human. The process gets under way during a pregnancy that typically lasts an average of thirty-eight weeks from the time of fertilization.

The first cleavage occurs about 12 to 24 hours after fertilization. It takes about one week for a blastula (called a blastocyst in mammals) to form. In humans and other placental mammals, a blastocyst embeds itself in its mother's uterus. As the offspring develops, nutrients diffusing from the maternal bloodstream across the placenta sustain it (Section 25.10).

All major organs, including the sex organs, form during the embryonic period, which ends after eight weeks. The bones of the developing skeleton are laid down as cartilage models, which are then invaded by bone cells that convert the cartilage to bone.

At the end of the embryonic period, the developing individual is referred to as a fetus. In the fetal period, from the start of the ninth week until birth, organs grow and become functional.

We divide the prenatal period into three trimesters. The first trimester includes months one through three; the second trimester, months four through six; the third trimester, months seven through nine.

Births before 37 weeks are considered premature. A fetus born earlier than 22 weeks rarely survives because its lungs are not yet fully mature. About half of births that occur before 26 weeks result in some sort of long-term disability.

After birth, the human body continues to grow and its body parts continue to change in proportion. **Figure 42.11** shows body proportion changes during development. Postnatal growth is most rapid between 13 and 19 years. Sexual maturation occurs at puberty, and bones stop growing shortly thereafter. The brain is the last organ to become fully mature: Portions of it continue to develop until the individual is about 19 to 22 years old.

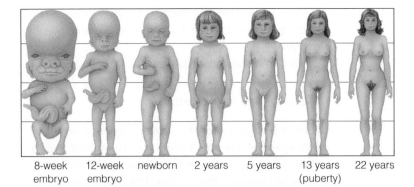

8-week embryo | 12-week embryo | newborn | 2 years | 5 years | 13 years (puberty) | 22 years

Figure 42.11 Observable, proportional changes in prenatal and postnatal periods of human development. Changes in overall physical appearance are slow but noticeable until the teens.

Table 42.2 Stages of Human Development

Prenatal period

Zygote	Single cell resulting from fusion of sperm nucleus and egg nucleus at fertilization.
Blastocyst (blastula)	Ball of cells with surface layer, fluid-filled cavity, and inner cell mass.
Embryo	All developmental stages from two weeks after fertilization until end of eighth week.
Fetus	All developmental stages from ninth week to birth (about 41 weeks after fertilization).

Postnatal period

Newborn	Individual during the first two weeks after birth.
Infant	Individual from two weeks to fifteen months.
Child	Individual from infancy to about ten or twelve years.
Pubescent	Individual at puberty, when secondary sexual traits develop. For girls, occurs between 10 and 15 years; for boys, between 11 and 16 years.
Adolescent	Individual from puberty until about 3 or 4 years later; physical, mental, emotional maturation.
Adult	Early adulthood (between 18 and 25 years); bone formation and growth finished. Changes proceed slowly after this.
Old age	Aging processes result in expected tissue deterioration.

Take-Home Message

How does human development proceed?

» Humans are placental mammals, so offspring develop in the mother's uterus.

» By the end of the second week, the blastocyst is embedded in the uterus.

» By the end of the eighth week, the embryo has all typical human organs.

» Most of a pregnancy is taken up with the fetal period, during which organs grow and begin to function.

■ After a human blastocyst forms, it burrows into the wall of its mother's uterus and a system of membranes forms outside the embryo.

■ Links to Amniotes 25.7, Placenta 41.2, Corpus luteum 41.6

Cleavage and Implantation

For fertilization to occur, a human egg must meet up with sperm in the upper portion of an oviduct (**Figure 42.12**). Cleavage begins within a day, as the zygote is propelled through the oviduct. The zygote divides by mitosis to form two cells, which become four, which become eight, and so on (**Figure 42.13A**). Sometimes a cluster of four or eight cells splits in two and each cluster develops independently, resulting in identical twins. More typically, all of the cells adhere tightly to one another as the divisions continue.

A blastocyst develops about five days after fertilization (**Figure 42.13B**). The blastocyst consists of an outer layer of cells, a cavity (called a blastocoel) filled with their fluid secretions, and an inner cell mass. Of 200 to 250 cells in the human blastocyst, about thirty are part of the inner cell mass that gives rise to the embryo. Other cells give rise to the membranes that surround the developing embryo.

By six days after fertilization, the blastocyst is usually in the uterus. It now expands by cell divisions and uptake of fluid. The expansion ruptures the zona pel-

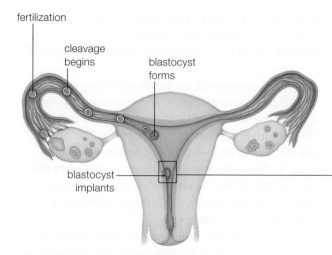

Figure 42.12 Locations of early developmental events.

lucida around it, allowing the blastocyst to slip out of this layer and implant. Implantation begins when the blastocyst attaches to the endometrium and burrows into it (**Figure 42.14**). During implantation, the inner cell mass develops into two flattened layers of cells that are collectively called an embryonic disk (**Figure 42.13C**).

Extraembryonic Membranes

Typical amniote membranes start forming outside the embryo during implantation. A fluid-filled amniotic

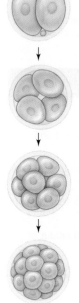

A Days 1-4 Cleavage. Mitotic divisions divide the zygote cytoplasm among an increasing number of cells.

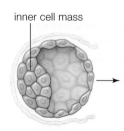

B Days 5-7 Blastocyst forms and expands then sheds its protein coat.

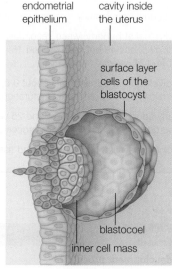

C Days 8-9 Implantation starts. The blastocyst attaches to the uterine lining (the endometrium) and begins to sink into it.

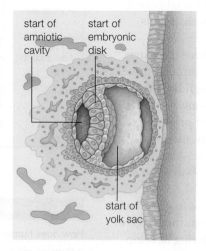

D Days 10-11 The inner cell mass now has two layers. One near the blastocoel will become the yolk sac. The other, the embryonic disk, will give rise to the embryo.

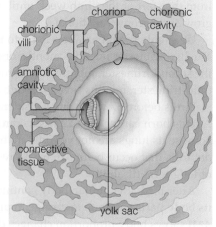

actual size

actual size

E Day 14 Projections from the chorion (chorionic villi) begin to grow into blood-filled spaces in the endometrium. The amniotic cavity has filled with fluid.

Figure 42.13 Animated Cleavage and implantation.

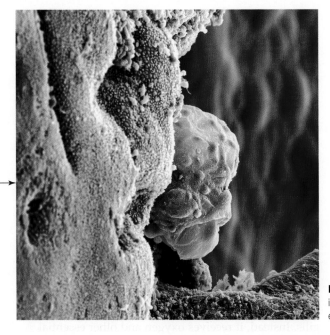

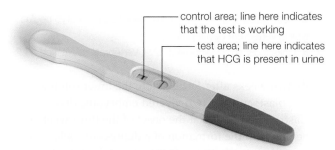

control area; line here indicates that the test is working

test area; line here indicates that HCG is present in urine

Figure 42.15 Urine test for pregnancy. A dipstick containing an absorbent material is dipped in a woman's urine. If the hormone HCG is present in the urine, it reacts with antibodies and dye in the stick, and a colored line appears in the test area. If HCG is not present, a colored line appears only in the control area.

Figure 42.14 Implantation. A human blastocyst (*gold*) burrowing into the endometrium (*red*). The photo is a colorized scanning electron micrograph.

cavity opens up between the embryonic disk and part of the blastocyst surface (**Figure 42.13D**). Many cells migrate around the wall of the cavity and form the **amnion**, a membrane that will enclose the embryo. Fluid in the cavity will function as a buoyant cradle in which an embryo can grow, move freely, and be protected from abrupt temperature changes and any potentially jarring impacts.

As the amnion forms, other cells migrate around the inner wall of the blastocyst and form a lining that becomes the yolk sac. In reptiles and birds, this sac holds yolk. In humans, cells of the yolk sac give rise to the embryo's blood cells and to germ cells.

Before a blastocyst is fully implanted, spaces that open in maternal tissues become filled with blood seeping in from ruptured capillaries. In the blastocyst, a new cavity opens up around the amnion and yolk sac. The lining of this cavity becomes the **chorion**, a membrane folded into many fingerlike projections that extend into blood-filled maternal tissues (**Figure 42.13E**). It will become part of the **placenta**, an organ that func-

tions in exchanges of materials between a mother's bloodstream and that of her developing child.

After the blastocyst is implanted, an outpouching of the yolk sac will become the fourth extraembryonic membrane—the **allantois**. The allantois stores waste in many mammals, but in humans its only function is contributing blood vessels to the umbilical cord.

Early Hormone Production

Once implanted, a blastula releases **human chorionic gonadotropin** (**HCG**). This hormone causes the corpus luteum to keep secreting progesterone and estrogens. These hormones prevent menstruation and maintain the uterine lining. After about three months, the placenta takes over the secretion of HCG.

HCG can be detected in a mother's urine as early as the third week of pregnancy. At-home pregnancy tests include a treated "dipstick" that changes color when exposed to urine that contains HCG (**Figure 42.15**).

allantois Extraembryonic membrane that, in mammals, becomes part of the umbilical cord.
amnion Extraembryonic membrane that encloses an amniote embryo and the amniotic fluid.
chorion Outermost extraembryonic membrane of amniotes; major component of the placenta in placental mammals.
human chorionic gonadotropin (**HCG**) Hormone first secreted by the blastocyst, and later by the placenta; helps maintain the uterine lining during pregnancy.
placenta Organ that forms during pregnancy and allows diffusion of substances between the maternal and embryonic bloodstreams.

Take-Home Message

What occurs during the first two weeks of human development?

» Cleavage produces a blastocyst, which slips out of the zona pellucida and implants itself in the endometrium, the lining of the uterus.

» During implantation, projections from the blastocyst extend into maternal tissues. Connections that will support the developing embryo begin to form.

» The inner cell mass of the blastocyst will become the embryo. Other portions of the blastocyst give rise to four external membranes. The outermost of these is the amnion, which encloses and protects the embryo inside a fluid-filled cavity.

42.8 Emergence of the Vertebrate Body Plan

■ Gastrulation occurs in the third week as the embryo begins typical vertebrate development.

■ Links to Brain development 32.9, Vertebrate circulation 36.2

By two weeks after fertilization, the inner cell mass of a blastocyst is a two-layered embryonic disk. Gastrulation occurs at the onset of the third week. It begins with the formation of a depression called the primitive streak (**Figure 42.16A**). The primitive streak's location indicates the body's head-to-tail axis. During gastrulation, cells migrate across the surface of the embryonic disk and inward through the primitive streak. The three germ layers formed by gastrulation will be the forerunners of all tissues.

After gastrulation, two neural folds appear on the surface of the embryonic disk (**Figure 42.16B**). As described in Section 42.4, these folds overlie the developing notochord and will eventually merge into a neural tube that develops into the spinal cord and brain. If the neural tube does not close as it should, the spinal cord will protrude from the vertebral column at birth. This birth defect is called spina bifida.

By the end of the third week, somites begin to appear on either side of the neural tube (**Figure 42.16C**). These paired segments of mesoderm will develop into the bones, skeletal muscles of the head and trunk, and overlying dermis of the skin.

The heart begins to beat a bit more than three weeks after fertilization. At first it is a linear tube of contractile cells that resembles a fish heart, with a single atrium and a single ventricle (Section 36.2). The sound of its beating is still too faint to hear, even with a stethoscope. As development continues, this tubular heart bends back on itself, forming a heart with three chambers like that of an amphibian, and finally a heart with four chambers. The embryo's first red blood cells are produced by its yolk sac. By the fourth week, the liver will take over the task of blood cell production. The lungs are not functional until after birth, so very little blood enters the pulmonary circuit (Section 36.2) during prenatal development. Instead, most blood flows from the right atrium into the left atrium. Normally, the opening in the atrial septum that allows atrium-to-atrium flow closes at birth or soon after.

Bands of tissue called pharyngeal arches form at the onset of the fourth week (**Figure 42.16D**). These will later contribute to the pharynx, larynx, and the face, neck, mouth, and nose. In fishes, pharyngeal arches develop into gills, but a human embryo never has gills. Instead, it receives oxygen and other essential material by way of the placenta, an organ whose structure and function we discuss in the next section.

Take-Home Message

What happens during weeks three and four of a pregnancy?

» A primitive streak forms on the embryonic disk, and gastrulation produces an embryo with three germ layers.

» The neural tube forms. Somites develop on either side of it.

» The heart forms and begins to beat, pumping blood cells made first in the yolk sac, then later by the liver.

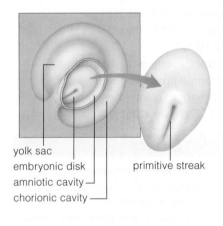

yolk sac
embryonic disk
amniotic cavity
chorionic cavity

primitive streak

A Day 14 Gastrulation begins. A depression called the primitive streak appears on the embryonic disk.

paired neural folds

neural groove (below, notochord is forming)

B Day 16 Neural groove appears as the neural tube begins forming.

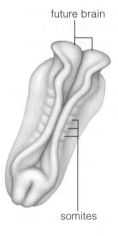

future brain

somites

C Days 18–23 Bumps of mesoderm (somites) form. They will develop into the dermis and most of the axial skeleton.

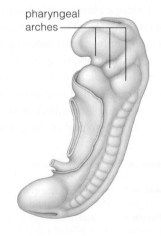

pharyngeal arches

D Days 24–25 Pharyngeal arches develop. They will contribute to the face, neck, larynx, and pharynx.

Figure 42.16 Animated Gastrulation and the onset of organ formation.

42.9 Structure and Function of the Placenta

■ Like other placental mammals, a human mother supplies her developing offspring with nutrients and oxygen by exchanges across the placenta.

■ Link to Corpus luteum 41.6

All exchange of materials between an embryo or fetus and its mother takes place by way of the placenta. This pancake-shaped organ consists of uterine lining, extra-embryonic membranes, and embryonic blood vessels (Figure 42.17). At full term, a placenta covers about a quarter of the uterus's inner surface.

The placenta begins forming early in pregnancy. By the third week, maternal blood has begun to pool in spaces in the endometrial tissue. Chorionic villi—tiny fingerlike projections from the chorion—extend into the pools of maternal blood. Embryonic blood vessels extend through the umbilical cord to the placenta, and into the villi. Embryonic vessels in the villi are surrounded by maternal blood, but maternal and embryonic blood never mixes. Instead, substances move between the maternal and embryonic bloodstreams by diffusing across the walls of the embryonic vessels in the chorionic villi. Oxygen and nutrients diffuse from maternal blood into embryonic blood. Wastes diffuse in the opposite direction, and are disposed of by the mother's body.

The placenta also has a hormonal role. From the third month on, it produces large amounts of HCG, progesterone, and estrogens. These hormones encourage the ongoing maintenance of the uterine lining. Once the placenta takes on this task, the corpus luteum degenerates.

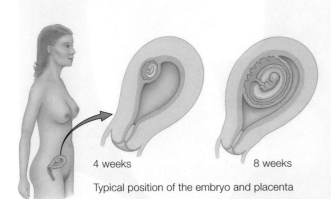

4 weeks   8 weeks

Typical position of the embryo and placenta

Take-Home Message

What is the function of the placenta?

» Vessels of the embryo's circulatory system extend through the umbilical cord to the placenta, where they run through pools of maternal blood.

» Maternal and embryonic blood do not mix; substances diffuse between the maternal and embryonic bloodstreams by crossing vessel walls.

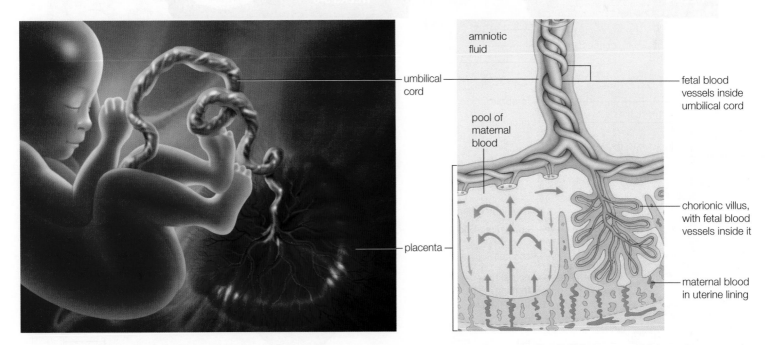

Artist's depiction of the view inside the uterus, showing a fetus connected by an umbilical cord to the pancake-shaped placenta.

Figure 42.17 Life support system of a developing human.

The placenta consists of maternal and fetal tissue. Fetal blood flowing in vessels of chorionic villi exchanges substances by diffusion with maternal blood around the villi. The bloodstreams do not mix.

42.10 Emergence of Distinctly Human Features

■ A human embryo's tail and pharyngeal arches label it as a chordate. The features disappear during fetal development.

■ Link to Sex organ formation 10.4

When the fourth week ends, the embryo is 500 times the size of a zygote, but still less than 1 centimeter long. Growth slows as details of organs begin to fill in. Limbs form; paddles are sculpted into fingers and toes. The umbilical cord and the circulatory system

develop. Growth of the head now surpasses that of all other regions (**Figure 42.18**). Reproductive organs begin forming, as described in Section 10.4. At the end of the eighth week, all organ systems have formed, apoptosis has eliminated the tail, and we define the individual as a human fetus.

In the second trimester, reflexive movements begin as developing nerves and muscles connect. Legs kick, arms wave about, and fingers grasp. The fetus

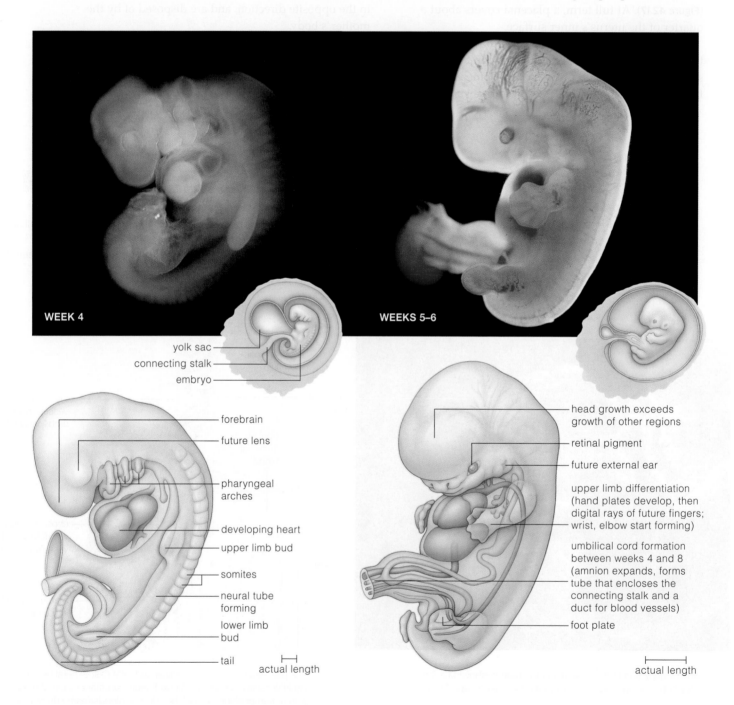

WEEK 4

yolk sac
connecting stalk
embryo

forebrain

future lens

pharyngeal arches

developing heart

upper limb bud

somites

neural tube forming

lower limb bud

tail

actual length

WEEKS 5–6

head growth exceeds growth of other regions

retinal pigment

future external ear

upper limb differentiation (hand plates develop, then digital rays of future fingers; wrist, elbow start forming)

umbilical cord formation between weeks 4 and 8 (amnion expands, forms tube that encloses the connecting stalk and a duct for blood vessels)

foot plate

actual length

Figure 42.18 Human embryo at successive stages of development (not to scale).

frowns, squints, puckers its lips, sucks, and hiccups. When a fetus is five months old, its heartbeat can be heard clearly through a stethoscope positioned on the mother's abdomen. The mother can sense movements of fetal arms and legs.

By now, soft fetal hair (lanugo) covers the skin; most will be shed before birth. A thick, cheesy coating (vernix) protects the skin from abrasion. In the sixth month, eyelids and eyelashes form. Eyes open during the seventh month, the start of the final trimester. By this time, all portions of the brain have formed and have begun to function.

Take-Home Message

What occurs during the late embryonic and the fetal periods?

» The embryo takes on its human appearance by week eight but remains tiny. In the fetal period, organs begin functioning and size increases dramatically.

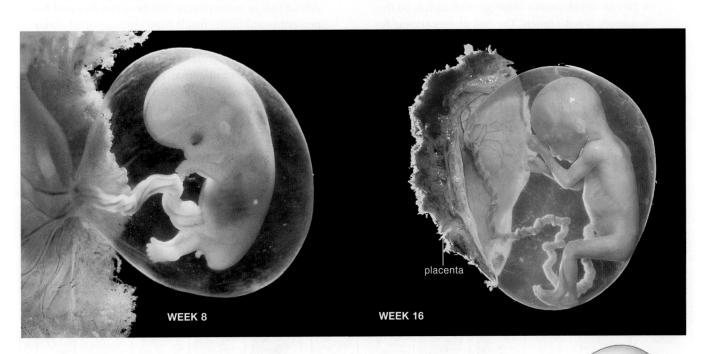

WEEK 8   **WEEK 16**

placenta

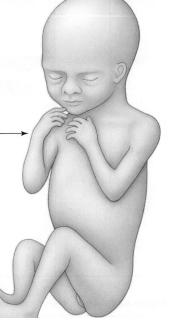

final week of embryonic period; embryo looks distinctly human compared to other vertebrate embryos

upper and lower limbs well formed; fingers and then toes have separated

primordial tissues of all internal, external structures now developed

tail has become stubby

actual length

| Length: | 16 centimeters (6.4 inches) |
| Weight: | 200 grams (7 ounces) |

WEEK 29
| Length: | 27.5 centimeters (11 inches) |
| Weight: | 1,300 grams (46 ounces) |

WEEK 38 (full term)
| Length: | 50 centimeters (20 inches) |
| Weight: | 3,400 grams (7.5 pounds) |

During fetal period, length measurement extends from crown to heel (for embryos, it is the longest measurable dimension, as from crown to rump).

42.11 Miscarriages, Stillbirths, and Birth Defects

■ Many pregnancies do not continue to full term and some that do result in the birth of a child with a birth defect.

■ Links to Apicomplexans 21.7, Thyroid hormone 34.7

What Can Go Wrong

A spontaneous abortion, commonly called a miscarriage, is the death of an embryo or fetus before 20 weeks. Miscarriages that occur during the first month or two of development often go undetected, so their frequency is not known. The risk of miscarriage for pregnancies that have been detected is 12–15 percent. The vast majority of losses occur in the first trimester. On average, a woman who has an apparently healthy embryo on ultrasound at 8 to 11 weeks has a 1.5 percent chance of miscarriage.

Death of a fetus after 20 weeks is called a stillbirth. Most stillbirths take place before labor begins, but some result from complications during delivery.

Disordered development that is not fatal can result in a birth defect. According to the Centers for Disease Control, about 3 percent of children born in the United States have some sort of birth defect. Malformed hearts, neural tube defects, and cleft lip or palate are common examples.

Risk Factors

About half of miscarriages and between five and ten percent of stillbirths result from chromosomal abnormalities (Section 14.6). These abnormalities, and thus the risk of an unsuccessful pregnancy, increase with both paternal and maternal age. The risk of birth defects also rises with parental age.

Other maternal characteristics can also influence the success of a pregnancy. Very thin women and women who report a high level of stress are more likely to miscarry than those who are a bit heavier and happier.

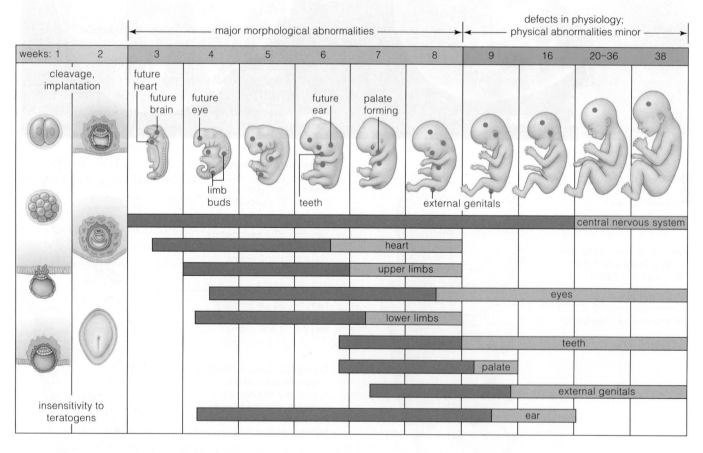

Figure 42.19 Animated Teratogen sensitivity. Teratogens are drugs, infectious agents, and environmental factors that cause birth defects. *Dark blue* signifies the highly sensitive period for an organ or body part; *light blue* signifies periods of less severe sensitivity. For example, the upper limbs are most sensitive to damage during weeks 4 through 6, and somewhat sensitive during weeks 7 and 8.

Figure It Out: Is teratogen exposure in the 16th week more likely to affect the heart or the genitals?   Answer: Genitals

On the other hand, obesity is a risk factor for late still-birth. Having previously given birth to more than five children also raises the risk that a pregnancy will not continue to term.

Some birth defects have an inherited basis, but others result from an environmental factor such as poor nutrition or exposure to a **teratogen**. A teratogen is a toxin or infectious agent that interferes with development. **Figure 42.19** shows the periods when specific organs are the most vulnerable to damage by exposure to teratogens.

Dietary deficiencies affect many developing organs and malnutrition raises the risk of miscarriage. Specific dietary deficiencies are also associated with birth defects. For example, if a mother does not get enough iodine, her newborn may be affected by cretinism, a disorder that affects brain function and motor skills (Section 34.7). A maternal deficiency in folate (folic acid) is a risk factor for neural tube defects.

Certain viral diseases can cause birth defects in the early weeks after fertilization. Rubella, or German measles, is one of them. A woman may avoid the risk of passing on the virus to her child by being vaccinated before she becomes pregnant. Toxoplasmosis caused by an apicomplexan protist is another danger. If the parasite infects an embryo or fetus, it can cause developmental problems, a miscarriage, or stillbirth.

Drinking alcohol raises the risk of miscarriage and of fetal alcohol syndrome, or FAS. Affected individuals have both physical impairments and mental problems (**Figure 42.20**). Most doctors now advise women who are pregnant to avoid alcohol. Women attempting to become pregnant should do the same to avoid inadvertently drinking before their pregnancy is known.

Doctors also advise pregnant women not to smoke, to avoid caffeine, and to stay out of hot tubs. Maternal smoking or heavy exposure to secondhand smoke increases the risk of miscarriage and adversely affects fetal growth and development. High caffeine intake (more than 200 milligrams daily) can raise the risk of miscarriage, as can spending time in a hot tub.

Some prescription medications cause birth defects. Isotretinoin (Accutane) is widely prescribed as a treatment for severe acne. It is similar in its structure to a chemical (retinoic acid) that acts as a morphogen. If taken early in a pregnancy, isotretinoin can cause heart problems or facial and cranial deformities in the embryo. Use of the antidepressant paroxetine (Paxil) or related drugs during early pregnancy increases the likelihood of heart malformations. Taking them later in pregnancy increases risk that an infant will have fatal heart and lung disorders.

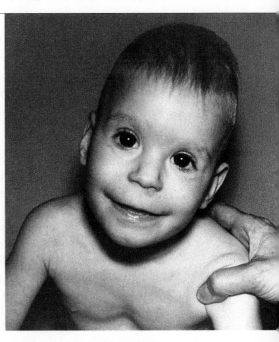

Figure 42.20 A child with fetal alcohol syndrome—FAS. Obvious symptoms are low and prominently positioned ears, improperly formed cheekbones, and an abnormally wide, smooth upper lip. Growth-related complications, heart problems, and nervous system abnormalities are also common.

Morning Sickness

About two-thirds of pregnant women begin to have episodes of nausea with or without vomiting around the sixth week of pregnancy. Although commonly known as morning sickness, the symptoms can occur at any time of day. They typically end by the twelfth week. Although unpleasant, morning sickness generally does not cause health problems and there is some evidence that it may have an adaptive function. Morning sickness most often occurs during the period when an embryo's organs are developing and are most vulnerable to teratogens. Women who have morning sickness are less likely to miscarry than women who are not affected, and women who vomit were more likely to carry a child to term than those who only felt nauseous. The foods women who have morning sickness report they are most likely to avoid—fish, poultry, meat, and eggs—are the ones most likely to be tainted by dangerous microorganisms.

teratogen A toxin or infectious agent that interferes with embryo development and causes birth defects.

Take-Home Message

What causes stillbirths, miscarriages, and birth defects?

» Incorrect chromosome number or other genetic defects can result in the death of an embryo or cause birth defects. Such defects become more likely as parents age.

» Exposure to pathogens and toxins during prenatal development can also be fatal or cause birth defects.

42.12 Birth and Lactation

■ As in other placental mammals, human newborns are at a relatively late state of development and are nourished with nutritious milk secreted from the mother's mammary glands. Shifts in the levels of hormones help control these processes.

■ Links to Positive feedback 32.4, Pituitary hormones 34.4

Labor and Delivery

A mother's body changes as her fetus nears full term, at about 41 weeks after fertilization. Until the last few weeks, her firm cervix helped prevent the fetus from slipping out of her uterus prematurely. Now cervical connective tissue becomes thinner, softer, and more flexible. These changes will allow the cervix to stretch enough to permit the fetus to pass out of the body.

The birth process is known as **labor**. Typically, the amnion ruptures right before birth, so amniotic fluid drains out from the vagina. The cervix dilates. Strong contractions propel the fetus through it, then out through the vagina (**Figure 42.21**).

A positive feedback mechanism operates during labor. When the fetus nears full term, it typically shifts position so that its head puts pressure on the mother's

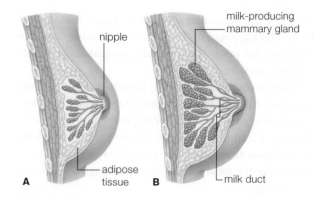

nipple — milk-producing mammary gland

adipose tissue — milk duct

A **B**

Figure 42.22 Cutaway views of (**A**) the breast of a woman who is not pregnant and (**B**) the breast of a lactating woman.

cervix. Receptors inside the cervix sense pressure and signal the hypothalamus, which signals the posterior lobe of the pituitary to secrete **oxytocin**. In a positive feedback loop, oxytocin binds to smooth muscle of the uterus, causing uterine contractions that push the fetus against the cervix. The added pressure triggers more oxytocin secretion, which causes more contractions and more cervical stretching. Forceful uterine contractions continue until the fetus is propelled through the cervix and vagina, and out of the mother's body. Synthetic oxytocin may be administered to induce labor or to increase the strength of contractions.

Strong muscle contractions also detach and expel the placenta from the uterus as the "afterbirth." The umbilical cord that connects the newborn to this mass of expelled tissue is clamped, cut short, and tied. The short stump of cord left in place withers and falls off. The navel marks the former attachment site.

Surgical Delivery

In the United States, more than 30 percent of births involve a cesarean section. During this surgical procedure, a doctor makes an incision in the mother's abdominal wall, then cuts open her uterus to remove the fetus.

A variety of conditions can necessitate a surgical delivery. A placenta that grows in the lower portion of the uterus and covers the cervix can make it impossible for the fetus to exit safely through the vagina. A fetus that is positioned with its feet, rather than its head, at the cervix may also have to be extracted surgically. If the placenta dislodges too early in labor or the umbilical cord becomes kinked, blood flow to the fetus is compromised and surgery may be required to ensure its safety. A cesarean section can also be used to prevent transmission of herpes or another sexually transmitted disease to an infant (Section 41.10).

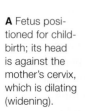
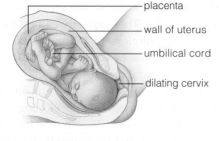

placenta
wall of uterus
umbilical cord
dilating cervix

A Fetus positioned for childbirth; its head is against the mother's cervix, which is dilating (widening).

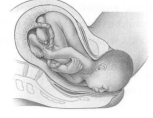

B Muscle contractions stimulated by oxytocin force the fetus out through the vagina.

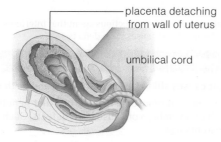

placenta detaching from wall of uterus

umbilical cord

C The placenta detaches from the wall of the uterus and is expelled.

Figure 42.21 Expulsion of fetus and afterbirth during normal delivery. The afterbirth consists of the placenta, tissue fluid, and blood.

Mind-Boggling Births (revisited)

Research into IVF opened the way to a variety of assisted reproductive technologies. If a man makes sperm, but does not ejaculate them or does not ejaculate a large enough number to allow fertilization by normal means, intracytoplasmic sperm injection can put his sperm inside his partner's egg (**Figure 42.23**). A woman who does not make viable eggs but who wishes to carry a child can be implanted with a blastocyst derived from an egg donated by another woman. A woman who has viable eggs but cannot, or does not want to, carry them herself can have her egg fertilized by IVF. The resulting blastocyst can then be implanted in a surrogate mother.

With all IVF procedures, prospective parents can screen blastocysts before they are implanted to avoid genetic defects, or to choose desirable traits such as a particular sex (Section 14.7).

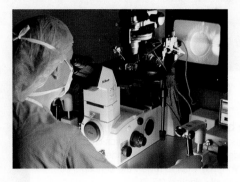

Figure 42.23 *In vitro* fertilization by intracytoplasmic sperm injection. A fertility specialist uses a micromanipulator to insert a single human sperm into an egg. The video screen shows the view through the microscope.

How would you vote? Both organ donation and egg donation entail health risks. Egg donors undergo hormone treatments, and egg extraction requires general anesthesia. Should we prohibit payments to women for donation of eggs, just as we prohibit selling organs?

Nourishing the Newborn

When a woman is not pregnant, her breasts are mostly adipose tissue. Her milk ducts and mammary glands are small and inactive (**Figure 42.22**). During pregnancy, these structures enlarge in preparation for **lactation**, or milk production. **Prolactin**, a hormone secreted by the mother's anterior pituitary, triggers growth of mammary glands during pregnancy.

After birth, a decline in progesterone and estrogens causes milk production to go into high gear. The stimulus of a newborn's suckling stimulates the release of both prolactin and oxytocin. The prolactin now encourages production of milk proteins. The oxytocin stimulates muscles around the milk glands to contract, forcing milk into the milk ducts.

Human milk includes sugar (lactose), easily digested fats and proteins, essential vitamins and minerals, and enzymes such as lipases and amylases that assist in digestion. It has proteins that encourage the growth of beneficial gut bacteria (Section 39.1) and lysozyme (Section 37.3) that kills harmful bacteria. In addition, milk contains maternal antibodies that coat the lining of the newborn's throat and gut lining, lessening the risk of dangerous infections. The infant also has maternal antibodies that entered its bloodstream by crossing the placenta before birth.

Nursing women should keep in mind that drugs and pathogens that enter their body can end up in their milk and affect their child. For example, exposure to nicotine in milk interferes with an infant's ability to sleep, as does exposure to alcohol. Alcohol in milk may also interfere with a child's early motor development. Some viral pathogens, including HIV, can be transmitted via breast milk.

Breast-feeding benefits a mother as well as her child. The oxytocin released in response to suckling causes uterine contractions that help restore the uterus to its pre-pregnancy state. The hormonal effects of breast-feeding also slow the return of menstrual cycles, thus acting as a natural contraceptive. Producing milk takes a lot of energy, so breastfeeding helps a woman lose excess weight that she may have gained during her pregnancy. In addition, breast-feeding lowers a woman's risk of cancers of the reproductive tract and protects against inherited forms of breast cancer that affect premenopausal women.

labor Expulsion of a placental mammal from its mother's uterus by muscle contractions.
lactation Milk production by a female mammal.
oxytocin Pituitary hormone with roles in labor and lactation.
prolactin Hormone that induces mammary gland enlargement during pregnancy and milk production during lactation.

Take-Home Message

What occurs during childbirth and lactation?

» During birth, the hormone oxytocin stimulates muscle contractions that force a fetus out of its mother's body.

» Prolactin promotes mammary gland enlargement and milk production. Oxytocin released in response to suckling causes secretion of milk.

» Both mother and child benefit from breast-feeding. In addition to nourishing the newborn, breast-feeding protects it from infection. The mother benefits from weight loss, contraceptive effects, and decreased cancer risk.

LEARNING ROADMAP

Where you have been This chapter revisits environmental effects on phenotype (Section 13.6) and epigenetics (10.6). It draws on your knowledge of sensory and endocrine systems (Sections 33.4, 34.4). You may wish to review the concepts of adaptation (16.4) and sexual selection (17.7).

Where you are now

Foundations for Behavior
Behavioral variations within or among species often have a genetic basis. Behavior can also be modified by learning.

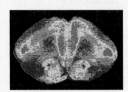

Environmental Influences
An animal's behavioral phenotype may depend on its environment. Animals make simple movements, and in some cases migrate, in response to environmental cues.

Animal Communication
Animals communicate with members of their species by chemical, acoustical, visual, and tactile signals. A signaling system evolves only if it benefits a signal sender and receiver.

Mating and Parental Care
Males and females are subject to different selective pressures with regard to choosing mates. Parental care can increase reproductive success, but it has energetic costs.

Social Behavior
Life in social groups has reproductive benefits and costs. Self-sacrificing behavior has evolved among a few kinds of animals that live in large family groups.

Where you are going In the next chapter you will learn about the effects of behavior on characteristics of populations. Chapter 45 discusses the behavioral interactions between species, such as between predators and their prey. Chapter 48 details some of the ways in which human activities disrupt animal behavior and threaten species with extinction.

43.1 Alarming Bee Behavior

Many animals sting or bite when threatened. For example, honeybees sting skunks, bears, and other animals that attack their hive. Honeybees live in large family groups, gather nectar and pollen from flowers, and store the nectar as honey. A hive's stored food, along with the protein-rich bodies of developing bees, makes it a tempting target for hungry animals. Stinging is an adaptive response, an evolved defense against the threat posed by potential hive raiders.

Humans tend to label an animal whose defensive response is easily triggered by humans as "aggressive." Africanized honeybees, known in popular media as "killer bees," are a prime example. These bees are a hybrid between the European honeybees favored by beekeepers and a bee subspecies native to Africa (**Figure 43.1**). The European bees raised for commercial use have been selectively bred to have a high threat threshold, but Africanized bees retain their natural defensiveness. Both strains of honeybee can sting only once and both make the same kind of venom, but the Africanized bees sting with less provocation, respond to threats in greater numbers, and pursue intruders with more persistence.

Africanized honeybees arose in Brazil in the 1950s. Bee breeders had imported African bees in the hope of breeding an improved pollinator for this region's tropical orchards. Some of the African imports escaped and mated with European honeybees that had already become established there. Descendants of the resulting hybrids expanded northward and reached Texas in 1990. By 2011, Africanized bees had been detected in New Mexico, Nevada, Utah, southern California, Oklahoma, Louisiana, Alabama, Georgia, and Florida.

Honeybees normally live inside cavities in trees or the ground, but they can also nest in spaces in buildings. Vibrations from machinery can trigger a defensive response. In 2010, a Georgia man died after being stung more than 100 times. He had inadvertently disturbed an Africanized bee colony while clearing a lot with a bulldozer.

Such fatalities are extremely rare. However, even a single sting can be life-threatening to someone allergic to honeybee venom. Bee stings are also highly painful, and people who receive a large number of stings may require hospitalization.

What makes Africanized bees so testy? A greater response to alarm pheromone contributes to their fierce reputation. A **pheromone** is a chemical signal that is released into the environment and influences the behavior of other members of the same species. When a honeybee guarding the entrance to a hive detects a

Figure 43.1 Two Africanized honeybees stand guard at their hive entrance. If a threat appears, they will release an alarm pheromone that stimulates hivemates to join an attack.

threat, she releases an alarm pheromone. Bees inside the hive detect this chemical signal and rush out to join the guards in driving off the intruder.

Researchers have tested the response of Africanized honeybees and European honeybees to alarm pheromone by positioning a dark cloth near the entrance of hives and releasing an artificial pheromone. Africanized bees flew out of a hive and attacked the cloth faster than European bees. They also plunged six to eight times as many stingers into the cloth.

The two strains of honeybees also show other behavioral differences. Africanized bees are less picky about where they establish a colony and are more likely to abandon their hive after a disturbance. Of greater concern to beekeepers, the Africanized bees are less interested in storing large amounts of honey.

Such differences among honeybees lead us into the world of animal behavior—the coordinated responses that animals make to stimuli. Scientists study both the proximate and ultimate causes of behavior. Proximate causes are the genetic, developmental, and physiological mechanisms that make a behavior possible. A behavior's ultimate causes are its adaptive significance and evolutionary history. A behavior most often evolves in a population because it increases an animal's own fitness. However, some behavior such as stinging by honeybees has evolved as a result of its benefits to an animal's close relatives.

pheromone Chemical that is emitted by one individual and affects the behavior of another member of the same species.

43.2 Behavioral Genetics

■ Much variation in behavior within or among species results from inherited differences. In a few instances, scientists have even pinpointed genes responsible for the variation.

■ Links to Frequency-dependent selection 17.7, Oxytocin 34.4

Animals differ in which stimuli they perceive and how they respond to them. These differences are shaped by genes that affect the nervous and endocrine systems.

Genetic Variation Within a Species

One way to investigate the genetic basis of behavior is to examine behavioral and genetic differences among members of a single species. Studies of the difference in defensive behavior between Africanized honeybees and European honeybees fall into this category.

Stevan Arnold's study of snake feeding behavior is another example. Garter snakes in the coastal forests of the Pacific Northwest preferentially feed on banana slugs (**Figure 43.2**). Inland, where there are no banana slugs, fishes and tadpoles are favored foods. Arnold found that these food preferences are inborn. When offered a banana slug as their first meal, young coastal snakes eat it, but young inland snakes ignore it.

Arnold hypothesized that inland snakes lack the genetically determined ability to associate the scent of slugs with food. He predicted that if coastal garter snakes were crossed with inland snakes, the resulting offspring would make an intermediate response to slugs. Results from experimental crosses confirmed

Figure 43.2 Coastal garter snake dining on a banana slug. The snake is genetically predisposed to recognize slugs as prey. By contrast, garter snakes from inland regions where there are no banana slugs ignore slugs when experimenters offer them.

this prediction. We do not know which gene or genes underlie this difference in behavior.

We know more about the genetic basis of differences in foraging behavior among fruit fly larvae. The fly's wingless, wormlike larvae feed on yeast that grows on decaying fruit. In wild populations, about 70 percent of fruit fly larvae are "rovers," which means they tend to move around a lot as they feed. Rovers often leave one patch of food to seek another (**Figure 43.3A**). The remaining 30 percent of larvae are "sitters"; they tend to move little once they find food (**Figure 43.3B**). When food is absent, rovers and sitters move the same amount, so both are equally energetic.

The proximate cause of the difference in larval behaviors is a difference in alleles of a gene called *foraging*. Flies with a dominant allele of this gene have the rover phenotype. Sitters are homozygous for the recessive allele. The *foraging* gene encodes an enzyme involved in learning about olfactory cues. Rovers make more of the enzyme than sitters.

The ultimate, or evolutionary, cause of the behavioral variation in larval foraging behavior is natural selection that arises from competition for food. With limited food, a rover does best when surrounded by sitters, and vice versa. Presumably, when there are lots of larvae of one type, all compete for food in the same way. Thus, frequency-dependent selection (Section 17.7) maintains both alleles of the *foraging* gene.

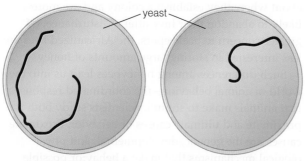

A Rovers (genotype *FF* or *Ff*) move often as they feed. When a rover's movements on a petri dish filled with yeast are traced for 5 minutes, the trail is relatively long.

B Sitters (genotype *ff*) move little as they feed. When a sitter's movements on a petri dish filled with yeast are traced for 5 minutes, the trail is relatively short.

Figure 43.3 Genetic polymorphism for foraging behavior in fruit fly larvae. When a larva is placed in the center of a yeast-filled plate, its genotype at the *foraging* locus influences whether it moves a little or a lot while it feeds. *Black* lines show a representative larva's path.

Genetic Variation Among Species

Comparing behavior of related species can also clarify the basis of a behavior. For example, studies of rodents

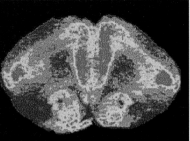

B PET scan of a monogamous prairie vole's brain with many receptors for the hormone oxytocin (*red*).

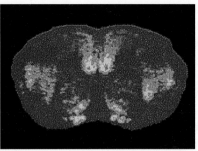

C PET scan of a promiscuous prairie vole's brain with few hormone receptors for oxytocin.

Figure 43.4 Studying the genetic roots of mating and bonding behavior. Voles (*Microtus*) are small rodents in which closely related species vary in their mating and bonding behavior, and in the number and distribution of receptors for the hormone oxytocin.

called voles (**Figure 43.4**) revealed that inherited differences in the number and distribution of certain hormone receptors influence mating and bonding behavior. Most voles are promiscuous. Prairie voles, however, form lifelong, largely monogamous social relationships. For females of this species, loyalty to a mate requires oxytocin, a hormone released during sexual intercourse, labor, and lactation. Responses to oxytocin affect bonding in many mammals. Inject the female member of an established prairie vole pair with a chemical that impairs oxytocin function, and she will no longer associate mainly with her partner.

Females of promiscuous vole species may behave as they do because they are less influenced by oxytocin than prairie voles. When researchers compared the brains of promiscuous and monogamous species, they found a striking difference in the number and distribution of receptors for this hormone (**Figure 43.4B,C**). Monogamous prairie voles have many more oxytocin receptors in parts of the brain associated with social learning than promiscuous voles do.

In male voles, variations in the distribution of receptors for another hormone (arginine vasopressin, or AVP) influence differences in bonding tendency among species. Compared to males of promiscuous vole species, males of monogamous species have more AVP receptors. Further evidence for the role of AVP in male bonding behavior comes from an experiment. Scientists isolated the prairie vole gene for the AVP receptor and used a virus to transfer copies of this gene into the forebrain of male mice. Mice are naturally promiscuous. However, the mice genetically modified by the prairie vole gene gave up their playboy ways. They now preferred a female with whom they had already mated to an unfamiliar female. These results confirmed the role of AVP in fostering monogamy among male rodents.

Human Behavior Genetics

Nearly all human behavioral traits have a polygenic basis and are influenced by the environment. Keep this in mind when you see headlines touting discovery of a gene "for" thrill-seeking behavior, alcoholism, or some other human behavior. Often, studies touted by dramatic headlines show that a particular allele is associated with a small, although statistically significant, increase in the tendency to perform a certain behavior.

Insights from animal behavior studies sometimes help researchers understand human behavioral disorders. For example, evidence of oxytocin's role in animal bonding suggests that impaired oxytocin production or reception may contribute to autism. A person with this disorder has trouble making social and emotional attachments. Scientists are collecting data about oxytocin receptor genes of autistic children and their unaffected siblings. The aim of the genetic study is to determine whether particular alleles of the oxytocin receptor gene raise the likelihood of autism. Oxytocin is also being tested as a treatment for autism. Results of an initial study of an oxytocin nasal spray were encouraging. Researchers found that use of the nasal spray increased the ability of autistic individuals to recognize faces and to interact cooperatively.

Take-Home Message

How do genes influence behavior?

» Behavioral differences within a species can arise from allele differences.

» Differences between closely related species can also have a genetic basis.

» Most human behaviors are complex, polygenic traits. Studies of genetic differences can shed light on predispositions to particular behaviors.

43.3 Instinct and Learning

- Some behaviors are inborn, which means they can be performed without any prior experience.
- Most behaviors can be modified as a result of experience.

Instinctive Behavior

All animals are born with the capacity for **instinctive behavior**—an innate response to a specific and usually simple stimulus. The life cycle of the cuckoo provides several examples of instinct at work. This European bird is a brood parasite. Females lay eggs in nests of other bird species. A newly hatched cuckoo is blind, but contact with an egg laid by its foster parent stimulates an instinctive response. That hatchling maneuvers the egg onto its back, then shoves it out of the nest (**Figure 43.5A**). This behavior removes any potential competition for the foster parent's attention.

A cuckoo's egg-dumping response is a **fixed action pattern**: a series of instinctive, unvarying movements triggered by a specific stimulus, called a sign stimulus. Once the movements have begun, they continue to completion without the need for further cues. A fixed action pattern has survival advantages when it ensures a fixed response to an important stimulus. However, an unthinking response to simple stimuli is not always adaptive. For example, the cuckoo's foster parents are not equipped to note the color and size of their offspring. A simple stimulus—a chick's gaping mouth—induces the fixed action pattern of parental feeding behavior (**Figure 43.5B**).

Time-Sensitive Learning

Learned behavior is behavior that is altered by experience. Some instinctive behavior can be modified with

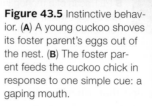

Figure 43.5 Instinctive behavior. (**A**) A young cuckoo shoves its foster parent's eggs out of the nest. (**B**) The foster parent feeds the cuckoo chick in response to one simple cue: a gaping mouth.

learning. A garter snake's initial strikes at prey are instinctive, but the snake learns to avoid dangerous or unpalatable prey. Learning may occur throughout an animal's life, or it may be restricted to a critical period.

Imprinting is a form of instinctive learning that can occur only during a short, genetically determined time period. For example, baby geese learn to follow the large object that bends over them in response to their first peep (**Figure 43.6**). With rare exceptions, this object is their mother. When mature, the geese will seek out a sexual partner that is similar to the imprinted object.

A genetic capacity to learn, combined with actual experiences in the environment, shapes most forms of behavior. For example, a male sparrow has an inborn capacity to recognize his species' song when he hears older males singing it. The young male uses these overheard songs as a guide to fill in details of his own song. Males reared alone sing a simplified version of their species' song. So do males exposed only to the songs of other species.

The sparrow can only learn his species-specific song during a limited period early in life. To learn to sing normally, he must hear a male "tutor" of his own species during his first 50 or so days of life. Hearing a same-species tutor later will not improve his singing.

Most birds must also practice their song to perfect it. In one experiment, researchers temporarily paralyzed throat muscles of zebra finches who were beginning to sing. After being temporarily unable to practice, these birds never mastered their song. In contrast, temporary throat muscle paralysis in very young birds or adults did not impair later song production. Thus, in zebra finches, there is a critical period for song practice, as well as for song learning.

Figure 43.6 Nobel laureate Konrad Lorenz with geese that imprinted on him. The smaller photograph shows results of a more typical imprinting episode.

Conditioned Responses

Nearly all animals are lifelong learners. Most learn to associate certain stimuli with rewards and others with negative consequences.

With **classical conditioning**, an animal's involuntary response to a stimulus becomes associated with another stimulus that is presented at the same time. In the most famous example, Ivan Pavlov rang a bell whenever he fed a dog. Eventually, the dog's reflexive response to food—increased salivation—was elicited by the sound of the bell alone.

With **operant conditioning**, an animal modifies its voluntary behavior in response to consequences of that behavior. This type of learning was first described for conditions in the lab. For example, a rat that presses a lever in a laboratory cage and is rewarded with a food pellet becomes more likely to press the lever again. A rat that receives a shock when it enters a particular area of a cage will quickly learn to avoid that area.

Other Types of Learned Behavior

With **habituation**, an animal learns by experience not to respond to a stimulus that has neither positive nor negative effects. For example, pigeons in cities learn not to flee from the large numbers of people who walk past them.

Many animals learn about the landmarks in their environment and form a sort of mental map. This map may be put to use when the animal needs to return home. For example, a fiddler crab foraging up to 10 meters (30 feet) away from its burrow is able to scurry straight home when it perceives a threat.

Animals also learn the details of their social landscape; they learn to recognize mates, offspring, or competitors by appearance, calls, odor, or some combination of cues. For example, two male lobsters will fight when they meet for the first time (**Figure 43.7**). After the fight, they will recognize one another by scent and behave accordingly, with the loser actively avoiding the winner.

With **observational learning**, an animal imitates the behavior of another individual. For example, Ludwig Huber and Bernhard Voelkl allowed marmoset monkeys to watch another marmoset demonstrate how to open a plastic container and retrieve a treat hidden inside. Marmosets who had seen the demonstrator open the container with its hands imitated this behavior, using their hands in the same way. In contrast, those who had watched a demonstrator open the box with its teeth attempted to do the same (**Figure 43.8**).

Figure 43.7 Social learning. Two male lobsters battle at their first meeting. Later, the loser will remember the odor of the winner and avoid him.

Figure 43.8 Observational learning. A marmoset opens a container using its teeth. After watching one individual perform this maneuver, other marmosets used the same technique.

classical conditioning An animal's involuntary response to a stimulus becomes associated with another stimulus that is presented at the same time.
fixed action pattern Series of instinctive movements elicited by a simple stimulus and carried to completion once begun.
habituation Learning not to respond to a repeated stimulus.
imprinting Learning that can occur only during a specific interval in an animal's life.
instinctive behavior An innate response to a simple stimulus.
learned behavior Behavior that is modified by experience.
observational learning One animal acquires a new behavior by observing and imitating behavior of another.
operant conditioning A type of learning in which an animal's voluntary behavior is modified by the consequences of that behavior.

Take-Home Message

How do instinct and learning shape behavior?

» Instinctive behavior can initially be performed without any prior experience, as when a simple cue triggers a fixed action pattern. Even instinctive behavior may be modified by experience.

» Certain types of learning can occur only at particular times in the life cycle.

» Learning affects both voluntary and involuntary behaviors.

43.4 Environmental Effects on Behavioral Traits

- Environment can influence behavioral phenotypes.
- Link to Environmental and phenotype 13.6, Epigenetics 10.6

Behavioral Plasticity

Phenotypic plasticity refers to the ability of an individual with a specific genotype to express different phenotypes in different environments. Many animals show **behavioral plasticity**; their behavioral traits are altered by environmental factors. An ability to learn is one mechanism of behavioral plasticity, but beneficial behavioral changes also arise in other ways.

For example, parasitism induces a change in the feeding preferences of moth larvae commonly called woolly bear caterpillars (**Figure 43.9**). These caterpillars are leaf eaters. When healthy, they tend to avoid leaves that have a high concentration of bitter, nonnutritive chemicals called alkaloids. However, caterpillars parasitized by fly larvae prefer to eat alkaloid-rich foods.

The caterpillar's ability to switch its aversion to alkaloids on or off depending on the presence or absence of parasites is adaptive. Alkaloids are toxic to caterpillars, but they are even more toxic to caterpillar parasites. Thus, they can function like a medication for parasitized caterpillars, albeit one with negative side effects. Because ingested alkaloids harm a caterpillar's parasites more than the caterpillar, a diet high in alkaloids enhances a parasitized caterpillar's likelihood of surviving to adulthood.

As another example, the social environment affects male behavior in some species of cichlid fishes. During the course of its life, a male cichlid can switch back and forth between two social roles. When in a dominant role, a male cichlid courts and breeds with females, and it confronts other males. Dominance is indicated by bright coloration and a dark eye bar

Figure 43.10 Dominant male cichlids with distinctive eye bars face off. Subordinate males flee from such confrontations.

(**Figure 43.10**). When in a subordinate role, a cichlid flees dominant males, does not court or breed, and has a drab coloration. Whether a male is dominant or subordinate depends on a variety of factors, most importantly whether other dominant males are present. Remove all dominant males, and a subordinate male's brain turns up expression of a regulatory gene. The resulting cascade of changes in gene expression transforms the male's behavioral and physical phenotype from subordinate to dominant within 20 minutes.

Epigenetic Effects

Epigenetic mechanisms cause heritable changes in behavior without altering DNA sequence (Section 10.6). For example, a female rat who received little licking and grooming from her mother early in life has a low tendency to lick and groom her own pups, and her daughters show the same behavior. Researchers have found the lack of tactile stimulation early in a rat's life causes methylation of its DNA. The resulting changes in gene expression suppress oxytocin's effects. The methylated DNA is passed on from generation to generation, disrupting oxytocin's influence and negatively affecting female parental behavior.

behavioral plasticity The behavioral phenotype produced by a genotype depends on conditions in the animal's environment.

Take-Home Message

How does the environment affect behavior?

» With behavioral plasticity, environmental factors influence an individual's behavior during its lifetime.

» The environment can also cause heritable changes in behavior through epigenetic modifications of DNA.

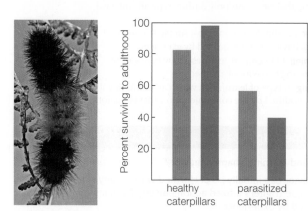

Figure 43.9 Effect of dietary alkaloids on parasitized and healthy woolly bear caterpillars (*above left*). In healthy caterpillars, eating alkaloids decreases survival. In parasitized caterpillars, the alkaloid enhances survival by harming parasites.

43.5 Movements and Navigation

■ All animals move, and some can navigate over extraordinarily long distances.

■ Link to Bird migration 25.9

Taxis and Kinesis

All animals are motile (can move from place to place) during some part of their life cycle. Even simple animals such as planarian flatworms instinctively move toward some stimuli and away from others. An innate directional response is called a **taxis** (plural taxes). Planarians are negatively phototactic (move away from light) and positively geotactic (move toward the pull of gravity). A stimulus may also cause an animal to increase or decrease its movements without regard for direction, a response called **kinesis**. For example, planarians are photokinetic, meaning they move more in light than in the dark. The planarian's innate responses to stimuli direct the worm toward a favorable habitat and help it to remain there.

Migration

Most animals move about daily to find food and avoid predators. Some also migrate. During a **migration**, an animal interrupts its daily pattern of activity to travel in a persistent manner toward a new habitat.

Migration most often involves seasonal movement to and from a breeding site. For example, birds that nest in the Arctic during the summer typically migrate to a temperate or tropical region to spend the winter. Over the course of a lifetime, a migratory bird may migrate many times, traveling each year to breed in the region where it hatched. Marine turtles and some whales also make repeated long-distance journeys to breed at the site where their own life began.

For other animals, the migration to a breeding grounds is a one-way trip. An Atlantic eel spends most of its life in a European or North American river. Then, one fall, it migrates hundreds to thousands of kilometers to the Sargasso Sea, the region of the Atlantic where it hatched 10 to 30 years before. Here, it meets other eels, breeds, and dies. Currents distribute eel larvae throughout the North Atlantic. After the larvae develop into young eels, they swim to a coast and then make their way up a river.

Researchers have only begun to decipher how migrating animals find their way. Some animals have an innate magnetic compass, meaning they use latitudinal variations in Earth's magnetic field to determine direction. In the spring, a caged European robin will hop repeatedly toward the north, even if it has no

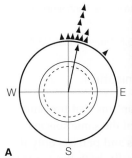

Figure 43.11 Vision-based magnetic compass in European robins. (**A**) Triangles indicate predominant direction of movements made by 12 robins tested indoors in the spring. Arrow indicates average result. (**B**) A robin with a frosted lens in front of its right eye and a clear one over its left eye shows no directional preference in its movements.

view of the outdoors (**Figure 43.11A**). Experiments have shown that the bird actually "sees" directional differences in Earth's magnetic field with its right eye. Impair vision in this eye alone and the bird's compass sense disappears (**Figure 43.11B**). Clouding vision in the bird's left eye has no effect.

Animals can also navigate by the sun and stars. Because the position of celestial bodies shifts over time, use of a sun- or star-based compass requires an innate sense of time. A bird that wants to head north must fly to the left of the sun in the morning, and to the right of the sun in the evening.

To find a specific site, an animal must know not only compass directions, but the location of its goal relative to its current location. It does no good to know where north is if you do not also know whether you are north or south of your destination. Localized variations in the Earth's magnetic field provide this information to some species. An animal may also gauge its progress relative to visual or olfactory landmarks. For example, odor cues help guide eels as they travel from the mouth of the river where they spent their adulthood toward the open ocean.

kinesis Innate response in which an animal speeds up or slows its movement in reaction to a stimulus.
migration An animal ceases an ongoing pattern of daily activity to move in a persistent manner toward a new habitat.
taxis Innate response in which an animal moves toward or away from a stimulus.

Take-Home Message

What factors influence animal movements?

» Innate responses to specific stimuli can influence the rate or direction of animal movements.

» Some animals migrate between habitats. Navigating to a specific site requires an ability to determine compass direction and a mental map.

43.6 Communication Signals

- A variety of evolved signals facilitate interactions among animals of the same species.
- Links to Comparative genomics 15.5, Pheromones 33.4

Evolution of Animal Communication

Communication signals are mechanisms that transmit information between members of the same species. A communication signal evolves and persists only if the exchange of information benefits both the sender and receiver. If signaling is disadvantageous for either party, then natural selection will tend to favor individuals that do not send or respond to it.

Figure 43.12 Prairie dog barking a warning. The bark provides information about whether the threat is an eagle or coyote. Other prairie dogs that hear the alarm call behave accordingly, diving into burrows when the call warns of a flying predator, or standing erect to observe the movements of a ground predator such a coyote.

Figure 43.13 Courtship display in albatrosses. The display involves a series of movements, some accompanied by calls.

Figure 43.14 Threat display of a male collared lizard. This display allows two rival males to assess one another's strengths without engaging in a potentially damaging fight. The display also is called into service to deter potential predators.

Types of Signals

Chemical, acoustical, visual, and tactile cues can transmit information from one member of a species to another. Chemical and acoustical signals typically are effective over longer distances than visual or tactile ones. They also have the added advantage of being effective even in the dark.

As previously noted, pheromones are chemical signals. The honeybee alarm pheromone is an example. So are sex pheromones emitted by females of many species to attract males. Insects detect pheromones with their antennae. In vertebrates, pheromones are detected by a vomeronasal organ on the floor of the nasal cavity. The extent of chemical communication varies among mammals. Their number of functional pheromone receptor genes may provide one clue to the range of this variation. Mice, which have been shown to rely heavily on chemical communication, have 187 such genes, whereas dogs have 8, and humans have 4 or fewer.

Acoustical signals help many male vertebrates, including songbirds, whales, frogs, and some fish, to attract prospective mates. Similarly, male crickets chirp by rubbing their legs together. Male cicadas use organs on their abdomen to make clicking sounds that can exceed 90 decibels.

Birds and mammals may give alarm calls that inform others of potential threats (**Figure 43.12**). Some calls convey more than simply "Danger!" A prairie dog emits one type of bark when it detects an eagle and another when it sees a coyote.

Evolution shapes the properties of acoustical signals depending on whether or not the sender benefits by revealing its position. Sounds that lure mates are typically easily localized, whereas alarm calls are not.

Visual communication is most widespread in animals that have good eyesight and are active during

When bee moves straight up comb, recruits fly straight toward the sun.

When bee moves straight down comb, recruits fly to source directly away from the sun.

When bee moves to right of vertical, recruits fly at 90° angle to right of the sun.

Figure 43.15 Animated Honeybee waggle dance, a tactile display. The orientation of the "waggle run" conveys information about the direction of a food source.

the day. Bird courtship often involves both visual and acoustic signals. For example, courting albatrosses strike coordinated poses while making distinctive calls (**Figure 43.13**). The display assures a prospective mate that the displayer is a member of the correct species and is in good health.

Threat displays are also advertisements of good health, but they serve a different purpose (**Figure 43.14**). Males of many species compete for access to females or to defend territory. When two potential rivals meet, they often engage in a display that demonstrates their strength and how well armed they are. Most often, the males are not evenly matched, and the weaker individual retreats. Both males benefit by avoiding a fight that could lead to injury.

With tactile displays, information is transmitted by touch. For example, after discovering food, a foraging honeybee worker returns to the hive and dances in the dark, surrounded by a crowd of other workers. If the food is more than 100 meters from the hive, she performs a waggle dance, moving in a figure eight (**Figure 43.15**). The orientation of an abdomen-waggling dancer in the straight run of this dance informs other bees about the direction of the food. The speed at which she dances informs them about its distance: the faster her movements, the closer the food.

The same signal may function in more than one context. Dogs and wolves solicit play behavior with a play bow (**Figure 43.16**). A play bow informs an animal's prospective playmate that any seemingly aggressive signals that follow, such as growling, should be interpreted as play behavior.

Eavesdroppers and Counterfeiters

Predators can benefit by intercepting signals sent by their prey. For example, frog-eating bats locate male tungara frogs by listening for their mating calls. This puts the males in a quandary. Female tungara frogs prefer complex calls, but complex calls are easier for

Figure 43.16 A wolf's play bow tells another wolf that its next behavior is meant as play, not aggression.

bats to localize. Thus, when bats are near, male frogs call less, and with less flair. The subdued signal is a trade-off between competing pressures to attract a mate and to avoid being eaten.

Counterfeit signalers can also pose a threat. Fireflies are nocturnal beetles that attract mates by producing flashes of light in a characteristic pattern. When a predatory female sees the flash from a male of the prey species, she flashes back as if she were a female of his own species. If she lures him close enough, she captures and eats him.

communication signal Chemical, acoustical, visual, or tactile cue that is produced by one member of a species and detected and responded to by other members of the same species.

Take-Home Message

What are the benefits and costs of communication signals?

» A communication signal transfers information from one individual to another individual of the same species. Such signals benefit both the signaler and the receiver.

» Signals have a potential cost. Some individuals of a different species benefit by intercepting signals or by mimicking them.

43.7 Mates, Offspring, and Reproductive Success

■ Mating behavior and parental behavior have been shaped by natural selection to maximize reproductive success.

■ Link to Sexual selection and sexual dimorphism 17.7

Mating Systems

Animal mating systems have traditionally been characterized as promiscuous, polygamous, or monogamous. With promiscuity, members of both sexes mate with multiple partners; with polygamy, one sex has multiple partners; and with monogamy, a male and female mate only with one another. In recent years, studies that integrate paternity analysis with behavioral observations have shown that species do not always fit cleanly into these categories. Members of a species often vary in their behavior, and formation of social pair bonds does not preclude extrapair matings.

For example, a paternity study of the prairie voles discussed in Section 43.2 found that 7 percent of "pair bonded" females produced litters of mixed male parentage; and 15 percent of "pair bonded" males sired offspring with females other than their partner. Thus, prairie voles are now described as socially monogamous, but genetically promiscuous. Members of a socially monogamous pair preferentially spend time together, but one or both may also mate with others.

Even social monogamy is rare in most animal groups, with the exception of birds. An estimated 90 percent of birds are socially monogamous. Among this group, the vast majority species that have been examined for paternity patterns are also genetically promiscuous. This is not surprising as having more than one mate benefits both male and female animals by increasing the genetic diversity among their offspring.

Multiple matings can also provide another benefit, an increased number of offspring. This benefit usually applies most strongly to males. Sperm are energetically inexpensive to produce, so the main limit on a male's reproductive success is usually access to mates. By contrast, a female's is mainly limited by her capacity to produce large yolk-rich eggs or, in mammals, to carry developing young.

When one sex is the limiting factor for the other's reproduction, we expect sexual selection to occur. Males often compete to be chosen by females or fight to access them. In many species, females choose males on the basis of their ability to provide necessary resources. For example, a female hangingfly will only mate with a male while feeding on prey that he has provided as a "nuptial gift"(**Figure 43.17A**). Fiddler crab females choose males on the basis of the real estate they control (**Figure 43.17B**). In the breeding season, a male stands at the entrance to his burrow, waving his one giant claw. A female attracted by a male's display inspects his burrow. Only when a burrow has the right location and dimensions does she mate with its owner and lay eggs. Burrow location and size are important because they affect the development of the crab larvae.

A Male hangingfly dangling a moth as a nuptial gift for a potential mate.

B Male fiddler crabs (*top*) wave their one enlarged claw to a attract a female (*bottom*).

C Male sage grouse gather on a communal display ground called a lek, where they dance, puff out their neck, and make booming calls.

Figure 43.17 How to impress a choosy female.

Sage grouse converge at a **lek**, a communal display ground that functions as a dance floor. Males stamp their feet and emit booming calls by puffing and deflating their large neck pouches (**Figure 43.17C**). The most popular males mate with many different females; other males do not mate at all.

In species in which females cluster around a necessary resource, a male may hold a **territory**, a region from which he excludes other males. The territory holder mates with all the females in his territory. Lions, elk, elephant seals, and bison (**Figure 43.18**) have this sort of mating behavior.

Parental Care

Parental care requires time and energy that might otherwise be invested in reproducing again. It evolves only if the benefit to a parent in terms of increased offspring survival is greater than the cost incurred in lost reproductive opportunity.

Mammalian young are born in a relatively helpless state, so parental care is essential. Typically, the female is the sole caregiver (**Figure 43.19A**). A male can leave immediately after mating and reproduce again while a female nurtures his developing offspring.

Males seldom serve as sole caregivers. The midwife toad is an interesting exception. A male holds strings of fertilized eggs around his legs until the eggs hatch (**Figure 43.19B**). He does not lose mating opportunities because he can mate while carrying eggs. Males often carry eggs from more than one female.

Cooperative care of young by both parents occurs most frequently in birds (**Figure 43.19C**). Chicks of birds that cooperate in care of their young tend to hatch while in a relatively helpless state. Chicks of sage grouse and other birds in which females alone provide care tend to hatch when more fully developed.

lek Area where males animals perform courtship displays.
territory Region that an animal or group of animals defends.

Take-Home Message

What factors affect mating systems and parental care?

» The positive effects of genetic diversity among offspring make monogamy rare.

» In many species, a few males monopolize mating opportunities either by enticing females to choose them or by defending a territory with many females.

» Parental care evolves only when the benefit of parental care outweighs the costs in lost opportunity for more offspring.

Figure 43.18 Male bison locked in combat during the breeding season. A few males will mate with many females. Some males will not mate at all.

A Female grizzlies care for their cub for as long as two years. The male takes no part in its upbringing.

B A male midwife toad carries developing eggs.

C A pair of Caspian terns cooperate in caring for their chick.

Figure 43.19 Caring for offspring.

43.8 Living in Groups

■ Animals of many species group together during part or all of their lifetime. Such grouping has both benefits and costs.

■ Link to Primate social behavior 26.2

In many species, individuals live together for part or all of their lifetime. Such groupings evolve only if benefits of close proximity to others outweigh the costs.

Defense Against Predators

In some groups, cooperative responses to predators reduce the net risk to all. Birds, monkeys, meerkats, prairie dogs, and many other animals make alarm calls when they spot a predator. The call alerts other members of the prey species. Once alerted, individuals of the group may hide from the predator or join forces to fend it off.

Presenting a united front can deter predators. Sawfly caterpillars feed in a group and will rear up and vomit partly digested eucalyptus leaves when a hungry bird approaches (**Figure 43.20A**). Birds confronted with such an unappealing cluster eat fewer caterpillars than birds offered caterpillars one at a time. Similarly, when threatened by wolves, musk oxen stand back to back, deterring attack by an imposing display of sharp horns (**Figure 43.20B**).

Whenever animals cluster, some individuals shield others from predators. A **selfish herd** is a temporary group that arises when individuals attempt to hide behind one another. Selfish-herd behavior occurs in nesting bluegill sunfishes. A male sunfish builds a nest by scooping out a depression in the mud on the bottom of a lake. Males compete with one another for nesting sites near the center of a group, with large males taking the innermost locations. Eggs laid here are least likely to be eaten by snails or other egg predators. Smaller males are relegated to nests at the periphery of the breeding colony and bear the brunt of the egg predation.

Improved Feeding Opportunities

Many mammals, including wolves, lions, wild dogs, and chimpanzees, live in social groups and cooperate in hunts (**Figure 43.21**). However, improved hunting success does not seem to be a major advantage for these groups. Field studies of lions and wolves have shown that the weight of prey captured per hunter tends to decline as the number of hunters increases. Individuals probably benefit nutritionally from coordinated efforts to locate prey and to fend off scavengers. They also help to care for one another's young, and jointly protect a territory.

Group living also allows transmission of cultural traits, or behaviors learned by imitation. For example, chimpanzees learn to make and use simple tools by stripping leaves from branches. They use thick sticks to make holes in a termite mound, then insert long, flexible "fishing sticks" into the holes (**Figure 43.22**). The long stick agitates the termites, which attack and cling to it. Chimps withdraw the stick and lick off termites, as a high-protein snack. Different groups of chimpanzees use slightly different tool-shaping and termite-fishing methods. Youngsters of each group learn by imitating the adults.

A Cluster of sawfly caterpillars regurgitating fluid (*yellow*) that predators find unappealing.

B When threatened, musk oxen adults (*Ovibos moschatus*) form a ring of horns, often around their young.

Figure 43.20 Group defenses.

Figure 43.21
Members of a
wolf pack hunt
cooperatively.

Dominance Hierarchies

Many animals that live in permanent groups form
a **dominance hierarchy**. In this type of social system,
dominant animals obtain a greater share of resources
and breeding opportunities than subordinate ones.
Typically, dominance is established by physical con-
frontation. In most wolf packs, one dominant male
breeds with one dominant female. The other members
of the pack are nonbreeding brothers and sisters, aunts
and uncles. All hunt and carry food back to individuals that guard the young in their den.

Why would a subordinate give up resources and
often breeding privileges? Challenging a strong individual can be dangerous, as is living on one's own.
Subordinates get their chance to reproduce by outliv-
ing a dominant peer.

Regarding the Costs

Relatively few animals spend the bulk of their time
in social groups. Why? In most habitats, the costs of
social living outweigh the benefits. Individuals that
live in denser groups compete more for resources.
For example, penguins and many other seabirds
form dense breeding colonies in which competition
for space and food is intense (**Figure 43.23**). Given the
opportunity, a pair of breeding herring gulls will can-
nibalize eggs and even the chicks of their neighbors.

Large social groups also attract more predators, and
individuals that live in dense groups are at a higher
risk of parasites and contagious diseases that jump
from host to host.

Figure 43.22 Chimpanzees (*Pan troglodytes*) using sticks as tools for
extracting tasty termites from a nest. This behavior is learned by imitation.

Figure 43.23 A crowded penguin breeding colony.

Take-Home Message

What are the benefits and costs of living in a social group?

» Living in a social group can provide benefits, as through improved
defenses, shared care of offspring, and greater access to food.

» Costs of group living include increased competition and increased vulnerability to infections.

dominance hierarchy Social system in which resources and mating
opportunities are unequally distributed within a group.
selfish herd Group that forms when individuals hide behind
others to minimize their individual risk of predation.

43.9 Why Sacrifice Yourself?

■ Extreme cases of sterility and self-sacrifice have evolved in only a few groups of insects and one group of mammals. How are genes of the nonreproducers passed on?

A **eusocial animal** lives in a multigenerational family group with a reproductive division of labor. Permanently sterile workers care cooperatively for the offspring of just a few breeding individuals.

Social Insects

Eusocial insects include some bees, and all ants and termites. In some species, the sterile workers are highly specialized in form and function (**Figure 43.24**).

A queen honeybee is the only fertile female in her hive (**Figure 43.25A**). She spends her days laying eggs and secretes a pheromone that makes other females in the hive sterile. The queen's sterile female daughters serve as workers; they feed larvae, clean and maintain the hive, and construct honeycomb from waxy secretions. Workers also gather the nectar and pollen that feed the colony. They guard the hive and will sacrifice themselves to repel any intruders.

New queens and males (drones) are produced seasonally. Drones are stingless and subsist on food gathered by their worker sisters. Each day, drones fly out in search of a mate. The occasional lucky one will meet a virgin queen on her one mating flight. The queen mates with many males during her flight, then uses their stored sperm for years. A drone dies after mating.

Like honeybees, termites live in enormous family groups with a queen who specializes in egg production (**Figure 43.25B**). Unlike the honeybee hive, a termite mound holds sterile males and females. A king supplies the female with sperm. Winged reproductive termites of both sexes develop seasonally.

Social Mole-Rats

Sterility and extreme self-sacrifice are uncommon in vertebrates. The only eusocial mammals are two species of African mole-rat. The best studied is *Heterocephalus glaber*, the naked mole-rat. Clans of this nearly hairless rodent live in burrows in dry parts of East Africa. A reproducing female dominates the clan and mates with one to three males (**Figure 43.25C**). Nonbreeding members live to protect and care for the "queen" and "king" (or kings) and their offspring. Sterile diggers excavate tunnels and chambers. When a digger finds an edible root, it hauls a bit back to the main chamber and chirps. Its chirps recruit others, which help carry food back to the chamber. In this way, the queen, her mates, and her young offspring get fed. Other sterile helpers guard the colony. When a predator appears, they chase and attack it at great risk to themselves.

Evolution of Altruism

A sterile worker in a social insect colony or a naked mole-rat clan shows **altruistic behavior**, which is behavior that enhances another individual's reproductive success at the altruist's expense. How did this behavior evolve? According to William Hamilton's **theory of inclusive fitness**, genes associated with altruism are selected if they lead to behavior that promotes the reproductive success of an altruist's closest relatives.

It is easy to see the genetic advantage of caring for one's own offspring. They have copies of some of your genes. However, recall that in a sexually reproducing diploid species, each offspring typically inherits half its genes from its mother and half from its father. Thus each individual shares only 50 percent of its genes

A An Australian honeypot ant worker. This sterile female is a living storage container.

B Army ant soldier (*Eciton*), also a sterile female, shows her formidable mandibles.

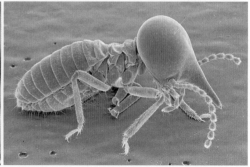

C Eyeless soldier termite (*Nasutitermes*). It shoots gluelike secretions from its nozzle-shaped head. Termite soldiers include both males and females.

Figure 43.24 Specialized ways of serving and defending the colony.

Alarming Bee Behavior (revisited)

When a honeybee worker stings a vertebrate, she sacrifices her life in defense of her family. Only females sting. The worker bee's stinger is a modified version of the queen bee's egg-laying apparatus, or ovipositor. The bee cannot extract her barbed stinger from vertebrate skin. Instead, she tears herself free, leaving her stinger and the attached venom sac behind. The resulting injury to her abdomen results in her death.

Even after the bee has broken free, muscles of the venom sac continue to contract spontaneously, pumping venom into the skin. The honeybee's stinging behavior evolved as a defense that helps the bees fend off vertebrate predators. The immediate burning pain that arises from a sting results from the effect of

mellitin, the most abundant component in bee venom. It is a small peptide that excites the pain receptors (Section 33.2) in a vertebrate's skin. Mellitin molecules insert into the plasma membrane of pain receptors, forming a pore through which ions escape the cell. The ion flow causes action potentials that travel to the brain, giving rise to the sensation of pain.

If you are stung by a honeybee, scraping the stinger from your skin immediately will minimize the amount of venom delivered. If pursued by multiple bees, protect your face with your hands and run to a car or indoor area, where the bees cannot follow you.

How would you vote? The threat Africanized bees pose to public health is real, but minor. Should funding research on the behavior and genetics of these bees be a priority?

with each of its parents. It also shares 50 percent of its genes with any of its siblings (brothers and sisters). By sacrificing its own reproductive success to help rear its siblings, a sterile worker promotes copies of its own "self-sacrifice" genes in these close relatives.

In honeybee, ant, and termite colonies, sterile workers assist fertile relatives with whom they share genes. Honeybees and ants may have an added incentive to make sacrifices. They belong to an order, Hymenoptera, that has a haplodiploid sex determination system: diploid females develop from fertilized eggs and haploid males from unfertilized ones. As a result, female hymenopterans share 75 percent of their genes with their sisters. This may help explain why eusociality has arisen many times in this order.

Inbreeding increases the genetic similarity among relatives and may play a role in mole-rat sociality. A clan is highly inbred as a result of many generations of sibling, mother–son, and father–daughter matings. Researchers are now searching for other factors that select for eusocial behavior in mole-rats. According to one hypothesis, arid habitats and patchy food sources favor mole-rat genes that contribute to cooperation in digging burrows, searching for food, and fending off competitors for resources.

altruistic behavior Behavior that benefits others at the expense of the individual.
eusocial animal Animal that lives in a multigenerational family group with a reproductive division of labor.
theory of inclusive fitness Genes associated with altruism can be advantageous if the expense of this behavior to the altruist is outweighed by the reproductive success of relatives.

A Queen honeybee with her sterile daughters.

B A queen termite dwarfs her offspring and mate. Ovaries fill her enormous abdomen.

C Naked mole-rat queen.

Figure 43.25 Three queens, fertile females in species that have a reproductive division of labor.

Take-Home Message

How can altruistic behavior be selectively advantageous?

» Altruistic behavior may be favored when individuals pass on genes indirectly, by helping relatives survive and reproduce.

» Mechanisms that increase relatedness among siblings may encourage the evolution of eusocial behavior.

VII PRINCIPLES OF ECOLOGY

Lioness and her cub at sunset on the African savanna. What are the consequences of their interactions with each other, with other kinds of organisms, and with their environment? By the end of this last unit, you might find worlds within worlds in such photographs.

LEARNING ROADMAP

Where you have been This chapter considers the structure and growth of populations (Section 1.2). We refer back to sampling error (1.8), directional selection (17.5), gene flow (17.8), evolution of modern humans (26.6), smallpox vaccination (37.12), and animal social behavior (43.8).

Where you are now

The Vital Statistics
Populations can be described in terms of their size, density, distribution, and number of individuals in different age categories.

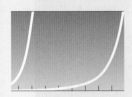

Exponential Rates of Growth
A population's size and reproductive base influence its rate of growth. When the population is increasing at a rate proportional to its size, it is undergoing exponential growth.

Limits on Increases in Number
Density-dependent limiting factors result in logistic growth, in which the rate of growth slows as the population nears its carrying capacity.

Life Histories
Resource scarcity, disease, and predation can restrict population growth. These limiting factors differ among species and shape their life history patterns.

The Human Population
Human populations sidestepped limits to growth by way of global expansion into new habitats, cultural interventions, and innovative technology.

Where you are going We continue our focus on populations in Chapter 45, with a look at the effect of species interactions on population size. Chapter 48 delves into the way in which expansion of the human population is affecting populations of other species.

44.1 A Honking Mess

Canada geese (*Branta canadensis*) were hunted to near extinction in the late 1800s. In the early 1900s, federal laws and international treaties were put in place to protect them and other migratory birds. In recent decades, the number of geese in the United States has soared. For example, Michigan had about 9,000 birds in 1970 and today has more than 300,000.

Canada geese are plant-eating birds, and they often congregate on the grassy lawns of golf courses and parks (**Figure 44.1A**). The geese are considered pests because they damage lawns and produce slimy, green feces that soil shoes, stain clothing, and cloud ponds and lakes. Goose feces may also contain bacteria that can sicken people if ingested.

A more serious problem is the hazard Canada geese pose to air traffic. In January of 2009, both engines of a US Airways flight failed shortly after the plane took off from New York's LaGuardia airport. Fortunately, the pilot was able to land the plane in the nearby Hudson River, where boats safely unloaded all 155 people aboard (**Figure 44.1B**). After the crash, investigators determined that the engine failures occurred after Canada geese were sucked into both engines.

Controlling Canada goose numbers is a challenge, because several different populations spend time in the United States. A **population** is a group of organisms of the same species that occupy a particular area and breed with one another more than they breed with members of other populations. In the past, nearly all Canada geese seen in the United States were migratory. The geese nested in northern Canada, flew to the United States to spend the winter, then returned to Canada. The common name of the species reflects this tie to Canada.

Most Canada geese still migrate, but some populations have lost this trait. Canada geese breed where they were raised, and nonmigratory geese are generally descendants of birds deliberately introduced to a park or hunting preserve. During the winter, migratory birds often mingle with nonmigratory ones. For example, a bird that breeds in Canada and flies to Virginia for the winter may find itself beside geese that have never left Virginia.

Life is more difficult for migratory geese than for nonmigratory ones. Flying hundreds of miles to and from a northern breeding area takes lots of energy and is dangerous. Compared to a migratory bird, one that stays put can devote more energy to producing young.

If the nonmigrant lives in a suburban or urban area, it also benefits from an unnatural abundance of food (grass) and an equally unnatural lack of predators. Not surprisingly, the biggest increases in Canada geese have been among nonmigratory birds that live where humans are plentiful.

In 2006, increasing complaints about Canada geese led the U.S. Fish and Wildlife Service to encourage wildlife managers to look for ways to reduce nonmigratory Canada goose populations, without unduly harming migratory birds. To do so, these biologists are studying which traits characterize each goose population, as well as how populations interact with one another, with other species, and with their physical environment. This sort of information is the focus of the field of population ecology.

A Park in Oakland, California, overrun by Canada geese.

B US Airways Flight 1549 floats in New York's Hudson River after collisions with geese incapacitated both of its engines.

Figure 44.1 Goose troubles.

population Group of organisms of the same species that live in the same area and interbreed.

44.2 Population Demographics

- Ecological factors affect the size, density, distribution, and age structure of a population.
- Links to Sampling error 1.8, Asexual reproduction in plants 29.9, Social groups 43.8

Studying populations requires use of **demographics**—data that describe a population's traits.

Population Size

Population size is the number of individuals in a population. It is often impractical to count all individuals, so biologists frequently use sampling techniques to estimate population size.

Plot sampling estimates the total number of individuals in an area on the basis of direct counts in a small part of the area. For example, ecologists might investigate the number of grass plants in a grassland or clams in a mudflat by measuring the number of individuals in several 1 meter by 1 meter square plots. To determine the total population size, scientists multiply the average number of individuals in the sample plots by the number of plots in the area where the population lives. Size estimates from plot sampling are most accurate for organisms that are not very mobile and live in an area where conditions are uniform.

Scientists use **mark–recapture sampling** to estimate the population size of mobile animals. They capture animals, mark them (**Figure 44.2**), then release them. After allowing a sufficient time to pass for the marked individuals to meld back into the population, the scientists capture animals again. The proportion of marked animals in the second sample is taken to be representative of the proportion marked in the whole population. For example, suppose scientists capture, mark, and release 100 deer in an area. Later, the scientists return and again capture 100 deer. They find 50 of these deer were previously marked, implying that marked deer constitute half of the population. They then infer that the total population is 200 individuals.

Information about the traits of individuals in a sample plot or capture group can be used to infer properties of the population as a whole. For example, if half the recaptured deer are of reproductive age, half of the population is assumed to share this trait. This sort of extrapolation is based on the assumption that the individuals in the sample are representative of the general population in terms of the traits under investigation.

Population Density and Distribution

Population density is the number of individuals per unit area or volume. Examples of population density include the number of dandelions per square meter of lawn, or the number of amoebas per milliliter of pond water. **Population distribution** describes the location of individuals relative to one another. Members of a population may be clumped together, be an equal distance apart, or be distributed randomly.

Clumped Distribution The members of most population are distributed in clumps, which means they are closer to one another than would be predicted by chance alone. A patchy distribution of resources encourages clumping. For example, Canada geese tend to congregate in places with suitable food such as grass and a nearby body of water. A cool, damp, north-facing slope may be covered with ferns, whereas an adjacent drier south-facing slope has none.

Figure 44.2 Florida Key deer marked for a population study.

age structure Of a population, the number of individuals in each of several age categories.
demographics Statistics that describe a population.
mark–recapture sampling Method of estimating population size of mobile animals by marking individuals, releasing them, then checking the proportion of marks among individuals recaptured at a later time.
plot sampling Method of estimating population size of organisms that do not move much by making counts in small plots, and extrapolating from this to the number in the larger area.
population density Number of individuals per unit area.
population distribution Describes whether individuals are clumped, uniformly dispersed, or randomly dispersed in an area.
population size Total number of individuals in a population.
reproductive base Of a population, all individuals who are of reproductive age or younger.

A Clumped distribution of schooling squirrelfish

B Near-uniform distribution of nesting seabirds

C Random distribution of dandelions

Figure 44.3 Population distribution patterns.

Limited dispersal ability increases the likelihood of a clumped distribution. Many plant seeds sprout beneath a parent plant. Asexual reproduction is another source of clusters, as when strawberry plants arise from runners or sea anemones divide in two. Finally, as Section 43.8 explained, many animals benefit by grouping together, as when geese travel in flocks and fish swim in schools (**Figure 44.3A**).

Near-Uniform Distribution Competition for resources can produce a near-uniform distribution, with individuals more evenly spaced than would be expected by chance. Creosote bushes in deserts of the American Southwest grow in this pattern. Competition for water among the root systems keeps the plants from growing in close proximity. Similarly, seabirds in breeding colonies often show a near-uniform distribution. Each bird aggressively repels others that get within reach of its beak as it sits atop its nest (**Figure 44.3B**).

Random Distribution Members of a population are distributed randomly when resources are uniformly available, and proximity to others neither benefits nor harms individuals. For example, when wind-dispersed dandelion seeds land on the uniform environment of a suburban lawn, dandelion plants grow in a random pattern (**Figure 44.3C**). Wolf spider burrows are also randomly distributed relative to one another. When seeking a burrow site, the spiders neither avoid one another nor seek one another out.

Age Structure

The **age structure** of a population refers to the number of individuals in various age categories. Individuals are often categorized as pre-reproductive, reproductive, or post-reproductive. Members of the pre-reproductive category have a capacity to produce offspring when mature. Along with reproductive individuals, they are a population's **reproductive base**.

Effects of Scale and Timing

The scale of the area sampled and the timing of a study can influence the observed demographics. For example, seabirds are spaced almost uniformly at a nesting site, but the nesting sites are clumped along a shoreline. The birds crowd together during the breeding season, but disperse when breeding is over.

Wildlife managers use demographic information to decide how to manage populations. For example, in considering how to manage Canada geese, wildlife managers began by evaluating the size, density, and distribution of nonmigratory populations. Based on this information, the U.S. Fish and Wildlife Service decided to allow destruction of some eggs and nests, and increased hunting opportunities at times when migratory Canada geese are least likely to be present.

Take-Home Message

What characteristics do we use to describe a population and what factors affect these characteristics?

» Each population has characteristic demographics, such as its size, density, distribution pattern, and age structure.

» Characteristics of the population as a whole are often inferred on the basis of a study that determined the traits of a smaller subsample.

» Environmental conditions and interactions among individuals influence a population's demographics, which often change over time.

44.3 Population Size and Exponential Growth

- Populations fluctuate in size as individuals come and go.
- With exponential growth, increases in population size are proportional to the current population size.

Gains and Losses in Population Size

Populations continually fluctuate in size. The number of individuals is increased by births and by **immigration**, the arrival of new residents that previously belonged to another population. The number is decreased by deaths and by **emigration**, the departure of individuals who join another population.

In many animals, young of one or both sexes leave the area where they were born and breed elsewhere. For example, young freshwater turtles typically emigrate from their parental population and become immigrants at another pond some distance away.

By contrast, seabirds typically breed where they were born. However, some individuals may emigrate and end up at breeding sites more than a thousand kilometers away. The tendency of individuals to emigrate to a new breeding site is usually related to resource availability and crowding. As resources decline and crowding increases, the likelihood of emigration rises.

From Zero to Exponential Growth

If we set aside the effects of immigration and emigration, we can define **zero population growth** as an interval during which the number of births is balanced by an equal number of deaths. As a result, population size remains unchanged, with no net increase or decrease in the number of individuals.

We can measure births and deaths in terms of rates per individual, or per capita. *Capita* means head, as in a head count. Subtract a population's per capita death rate (d) from its per capita birth rate (b) and you have the **per capita growth rate**, or r:

$$r = b - d$$

r	b	d
(per capita growth rate)	(per capita birth rate)	(per capita death rate)

Imagine 2,000 mice living in the same cornfield. If 1,000 mice are born each month, then the birth rate is 0.5 births per mouse per month (1,000 births/2,000 mice). If 200 mice die one way or another each month, then the death rate is 200/2,000 or 0.1 deaths per mouse per month. Thus r is 0.5 − 0.1 or 0.4 per mouse per month.

As long as r remains constant and greater than zero, **exponential growth** will occur: Population size will increase by the exact same proportion of its total in every successive time interval. We can calculate population growth (G) for each interval based on the per capita growth rate and the number of individuals in the population (N):

$$G = r \times N$$

G	r	N
(population growth per unit time)	(per capita growth rate)	(number of individuals)

Figure 44.4 Animated Exponential growth in hypothetical population of mice with a per capita rate of growth (r) of 0.4 per mouse per month and a population size of 2,000.

	Starting Population Size		Net Monthly Increase	New Population Size
$G = r \times$	2,000	=	800	2,800
$r \times$	2,800	=	1,120	3,920
$r \times$	3,920	=	1,568	5,488
$r \times$	5,488	=	2,195	7,683
$r \times$	7,683	=	3,073	10,756
$r \times$	10,756	=	4,302	15,058
$r \times$	15,058	=	6,023	21,081
$r \times$	21,081	=	8,432	29,513
$r \times$	29,513	=	11,805	41,318
$r \times$	41,318	=	16,527	57,845
$r \times$	57,845	=	23,138	80,983
$r \times$	80,983	=	32,393	113,376
$r \times$	113,376	=	45,350	158,726
$r \times$	158,726	=	63,490	222,216
$r \times$	222,216	=	88,887	311,103
$r \times$	311,103	=	124,441	435,544
$r \times$	435,544	=	174,218	609,762
$r \times$	609,762	=	243,905	853,667
$r \times$	853,667	=	341,467	1,195,134

A

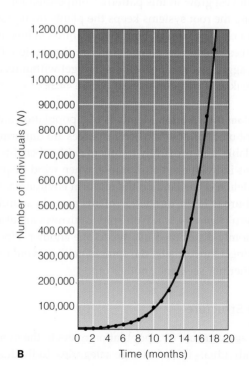

B

Figure 44.5 Effect of deaths on the rate of increase for two hypothetical populations of bacteria. Plot the population growth for bacterial cells that reproduce every half hour with no deaths occurring and you get growth curve 1. Next, plot the population growth of bacterial cells that divide every half hour, with 25 percent dying between divisions, and you get growth curve 2. Deaths slow the rate of increase, but as long as the birth rate exceeds the death rate, exponential growth will continue.

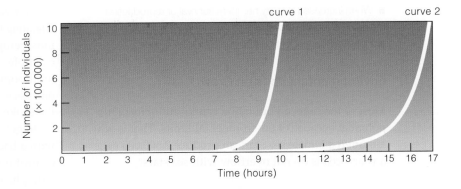

For our mice, *r* is 0.4 mice per month and the initial population size (*N*) is 2,000. Thus, the number of mice added in the first month is 0.4 × 2,000 = 800. This brings the total population of mice to 2,800 (**Figure 44.4A**). The number of mice added in the next month is determined by this larger population size. This time, the population grows by 0.4 × 2,800 = 1,120 individuals, which brings the population size to 3,920. At this growth rate, the number of mice would rise from 2,000 to more than 1 million in under two years! Plot the increases against time and you end up with a J-shaped curve, as shown in **Figure 44.4B**. A J-shaped curve is evidence of exponential growth.

With exponential growth, the number of new individuals added increases each generation, although the per capita growth rate stays the same. Exponential population growth is analogous to the compounding of interest on a bank account. The annual interest *rate* stays fixed, yet every year the *amount* of interest paid increases. Why? The annual interest paid into the account adds to the size of the balance, and the next interest payment will be based on that balance.

In exponentially growing populations, *r* is like the interest rate. Although *r* remains constant, population growth accelerates as the population size increases. When 6,000 individuals reproduce, population growth is three times higher than it was when there were only 2,000 reproducers. With exponential growth, each generation is larger than the prior one.

As another example, think of a single bacterium in a culture flask. After thirty minutes, the cell divides

in two. Those two cells divide, and so on every thirty minutes. If no cells die between divisions, then the population size will double in every interval—from 1 to 2, then 4, 8, 16, 32, and so on. The time it takes for a population to double in size is its doubling time. After 9–1/2 hours, or nineteen doublings, there are more than 500,000 cells. After 10 hours (twenty doublings), there are more than a million. Curve 1 in **Figure 44.5** is a plot of this increase.

Suppose 25 percent of the descendant cells die every thirty minutes. It now requires about seventeen hours, not ten, for that population to reach 1 million. Deaths slow the rate of increase but do not stop exponential growth (curve 2 in **Figure 44.5**). Exponential growth will continue as long as birth rates exceed death rates—as long as *r* is greater than zero.

Biotic Potential

The growth rate for a population under ideal conditions is its **biotic potential**. This is a theoretical value that would hold if shelter, food, and other essential resources were unlimited and there were no predators or pathogens. Microorganisms such as microbes have the highest biotic potentials, whereas large-bodied mammals have some of the lowest. Populations seldom reach their biotic potential because of the effects of limiting factors, a topic we discuss in detail in the next section.

---

biotic potential Maximum possible population growth rate under optimal conditions.
emigration Movement of individuals out of a population.
exponential growth A population grows by a fixed percentage in successive time intervals; the size of each increase is determined by the current population size.
immigration Movement of individuals into a population.
per capita growth rate For some interval, the added number of individuals divided by the initial population size.
zero population growth Interval in which births equal deaths.

Take-Home Message

What determines the size of a population and its growth rate?

» The size of a population depends on its rates of births, deaths, immigration, and emigration.

» Subtract the per capita death rate from the per capita birth rate to get *r*, the per capita growth rate of a population. As long as *r* is constant and greater than zero, a population will grow exponentially. With exponential growth, the number of individuals increases at an ever accelerating rate.

» The biotic potential of a species is its maximum possible population growth rate under optimal conditions.

44.4 Limits on Population Growth

- When increased density hampers survival or reproduction, the result is logistic growth.
- Density-independent factors can also slow growth.

Environmental Limits on Growth

Most of the time, a population cannot fulfill its biotic potential because of environmental limits. That is why, although a female sea star can produce 2,500,000 eggs a year, oceans do not overflow with sea stars.

A Wood ducks build nests only inside hollows of specific dimensions. With the clearing of old-growth forests, the access to natural cavities of the correct size and position is now a limiting factor on wood duck population size.

B Artificial nesting boxes are being placed in preserves to help ensure the health of wood duck populations.

Figure 44.6 Example of a limiting factor.

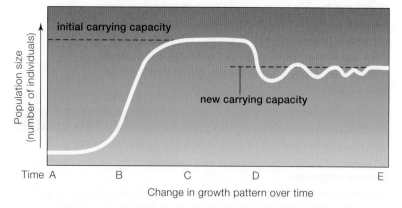

Figure 44.7 Idealized S-shaped curve characteristic of logistic growth. After a rapid growth phase (time B to C), growth slows and the curve flattens as carrying capacity is reached (time C to D).

In the real world, growth curves vary more, as when a change in the environment lowers carrying capacity (time D to E).

An essential resource that is in short supply acts as a **limiting factor** for population growth. Food, mineral ions, refuge from predators, and safe nesting sites are examples (**Figure 44.6**). Many factors can potentially limit population growth. Even so, in any environment, one essential factor will kick in first, and it will act as the brake on population growth.

To get a sense of the limits on growth, start again with a bacterial cell in a culture flask, where you can control the variables. First, enrich the culture medium with glucose and other nutrients required for bacterial growth. Next, let many generations of cells reproduce.

Initially, growth will be exponential. Then it slows, and population size remains relatively stable. After a brief stable period, population size plummets until all the bacterial cells are dead. What happened? The larger population required more nutrients. In time, nutrient levels declined, and cell division halted. Even after cells stopped dividing, existing cells continued to take up and use nutrients. Eventually, the nutrient supply was completely exhausted, and the last cells died out.

Even if you kept freshening the nutrient supply, the population would still eventually collapse. Like other organisms, bacteria generate metabolic wastes. Over time, accumulation of this waste would pollute the habitat and halt growth. Adding nutrients simply substitutes one limiting factor (a clean habitat) for another (lack of nutrients). All natural populations run up against limits eventually.

Carrying Capacity and Logistic Growth

Carrying capacity refers to the maximum number of individuals of a population that a given environment can sustain indefinitely. Ultimately, it means that the *sustainable* supply of resources determines population size. We can use the pattern of **logistic growth**, shown in **Figure 44.7**, to reinforce this point. By this pattern, a small population starts growing slowly in size, then it grows rapidly, then its size levels off as the carrying capacity is reached.

Logistic growth plots out as an S-shaped curve, as shown in **Figure 44.7** (time A to C). In equation form:

population growth per unit time	=	maximum per capita population growth rate	×	number of individuals	×	proportion of resources not yet used

Two Categories of Limiting Factors

Factors that affect population growth fall into two categories: density-dependent and density-independent.

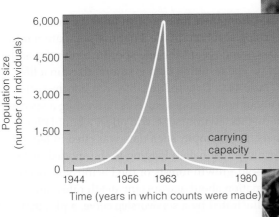

Figure 44.8 Overshoot and crash. A reindeer herd introduced to a small island in 1944 increased in size exponentially, then crashed when a low food supply was coupled with an especially cold and snowy winter in 1963–64.

Density-dependent factors decrease birth rates or increase death rates, and they come into play or worsen with crowding. Competition among members of a population for limited resources leads to density-dependent effects, as does infectious disease. Pathogens and parasites spread most easily when hosts are crowded together. The logistic growth pattern arises from the effects of density-dependent factors on population size.

Density-independent factors also decrease births or increase deaths, but crowding does not influence the likelihood of their occurrence, or the magnitude of their effects. Fires, snowstorms, earthquakes, and other natural disasters affect crowded and uncrowded populations alike. For example, in December 2004, a powerful tsunami (a giant wave caused by an earthquake) hit Indonesia and killed about 250,000 people. The degree of crowding did not make the tsunami any more or less likely to happen, or to strike any particular island.

Density-dependent and density-independent factors can interact to determine the fate of a population. As an example, consider what happened after the 1944 introduction of 29 reindeer to St. Matthew Island, an uninhabited island off the coast of Alaska (**Figure 44.8**). When biologist David Klein first visited the island in 1957, he found 1,350 well-fed reindeer munching on lichens. When he returned in 1963, he counted 6,000 reindeer. The population had soared far above the island's carrying capacity. Lichens had become sparser and the average body size of the reindeer had decreased. When Klein returned again in 1966, bleached-out reindeer bones littered the island and only 42 reindeer were alive. Only one was a male; it had abnormal antlers, which made it unlikely to reproduce. There were no fawns. Klein figured out that thousands of reindeer had starved to death during the winter of 1963–1964. The winter had been unusually harsh, with low temperatures, high winds, and 140 inches of snow. The reindeer had already been in poor condition because of the increased competition for food, so most starved when deep snow covered their food source. By the 1980s, there were no reindeer on the island at all.

carrying capacity Maximum number of individuals of a species that an environment can sustain.
density-dependent factor Factor that limits population growth and has a greater effect in dense populations than less dense ones.
density-independent factor Factor that limits population growth and arises regardless of population density.
limiting factor A necessary resource, the depletion of which halts population growth.
logistic growth Density-dependent limiting factors cause population growth to slow as population size increases.

Take-Home Message

How do environmental factors affect population growth?

» Carrying capacity is the maximum number of individuals of a population that can be sustained indefinitely by the resources in a given environment.

» With logistic growth, population growth is fastest during times of low density, then it slows as the population approaches carrying capacity.

» The effects of density-dependent factors such as disease cause a logistic growth pattern. Density-independent factors such as natural disasters also affect population size.

44.5 Life History Patterns

- Natural selection influences traits such as life span, age at maturity, and the number of offspring per breeding event.
- Link to Directional selection 17.5

Population growth rates are affected by life history traits such as the age at which reproduction begins, how long individuals remain reproductive, number of reproductive events, and the number of offspring produced during each reproductive event.

Life Tables and Survivorship Curves

One way to describe life history traits is to focus on a **cohort**—a group of individuals born during the same interval—from their time of birth until the last one dies. Ecologists often divide a natural population into age classes and record the age-specific birth rates and mortality. The resulting data is summarized in a life table (**Table 44.1**). Data in life tables can inform decisions about how changes, such as harvesting a species or altering its environment, will affect a population's numbers. For example, such data was cited in federal court rulings that halted logging in the spotted owl's habitat—old-growth forests of the Pacific Northwest.

Information about age-specific death rates can also be illustrated by a **survivorship curve**, a plot that shows how many members of a cohort remain alive over time. Ecologists have described three generalized types of curves. A type I curve is convex, indicating survivorship is high until late in life (**Figure 44.9A**). Humans and other large mammals that produce one or two young and care for them show this pattern. A diagonal type II curve indicates that the death rate of the population does not vary much with age (**Figure 44.9B**). In lizards, small mammals, and large birds, old individuals are about as likely to die of disease or predation as young ones. A type III curve is concave, indicating that the death rate for a population peaks early in life (**Figure 44.9C**). Marine animals that release eggs into water have this type of curve, as do plants that release enormous numbers of tiny seeds.

Table 44.1 Life Table for an Annual Plant Cohort*

Age Interval (days)	Survivorship (number surviving at start of interval)	Number Dying During Interval	Death Rate (number dying/ number surviving)	"Birth" Rate During Interval (number of seeds from each plant)
0–63	996	328	0.329	0
63–124	668	373	0.558	0
124–184	295	105	0.356	0
184–215	190	14	0.074	0
215–264	176	4	0.023	0
264–278	172	5	0.029	0
278–292	167	8	0.048	0
292–306	159	5	0.031	0.33
306–320	154	7	0.045	3.13
320–334	147	42	0.286	5.42
334–348	105	83	0.790	9.26
348–362	22	22	1.000	4.31
362–	0	0	0	0
		996		

* *Phlox drummondii*; data from W. J. Leverich and D. A. Levin, 1979.

Life History Strategies

Life history traits are inherited, so they are shaped by natural selection. All organisms take up resources from their environment and use them to grow, to maintain their body, and to reproduce. Resources are never

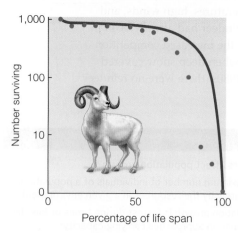

A Type I curve. Mortality is highest very late in life. Data for Dall sheep (*Ovis dalli*).

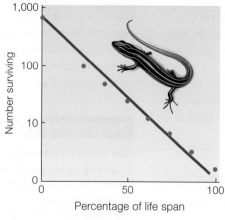

B Type II curve. Mortality does not vary with age. Data for five-lined skink (*Eumeces fasciatus*).

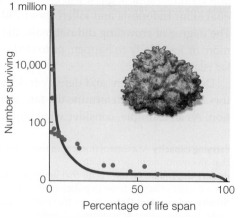

C Type III curve. Mortality is highest early in life. Data for a desert shrub (*Cleome droserifolia*).

Figure 44.9 Animated Survivorship curves. *Blue lines* are theoretical curves. *Red dots* are data from field studies.

r-Selected Species	K-Selected Species
shorter development	longer development
early reproduction	later reproduction
fewer breeding episodes, many young per episode	more breeding episodes, few young per episode
less parental investment per young	more parental investment per young
higher mortality rate, shorter life span	lower mortality rate, longer life span

Figure 44.10 Comparison of traits associated with *r*-selection and *K*-selection. Among mammals, rodents tend to have *r*-selected traits whereas large mammals such as elephants tend to have *K*-selected traits.

unlimited, so organisms constantly face trade-offs. Natural selection favors individuals who maximize their reproductive success, but resources devoted to growth and maintenance are not available for reproduction. How an organism allocates its resources between growth, maintenance, and reproduction over the course of its lifetime is its **life history pattern**.

Life history strategies vary continuously among species, but ecologists have described two theoretical extremes at either end of this continuum. An **r-selected species** is adapted to conditions that change rapidly and unpredictably. Density-independent factors often kill its members, so their populations never reach the carrying capacity of their environment, As a result, competition for resources is minimal, and individuals who maximize their rate of reproduction have the selective advantage. (The "*r*" in *r*-selection refers to the per capita growth rate.) Members of *r*-selected species tend to be small bodied, with a short generation time and life span. They produce multiple offspring at once, maximizing offspring quantity rather than quality.

By contrast, a **K-selected species** is adapted to life in a stable environment. Stable conditions allow population size to approach the carrying capacity of the environment, so density-dependent factors limit individual reproductive success. (The "*K*" in *K*-selection refers to carrying capacity.) *K*-selection favors individuals who maximize their reproductive success by producing high-quality offspring capable of outcompeting others for scarce resources. Compared to *r*-selected species, they tend to develop more slowly and have a larger body and a longer life span. They produce few offspring at a time, maximizing offspring quality rather than quantity, and they usually reproduce repeatedly.

The opposing results of *r*- and *K*- selection can be observed among members of many lineages. Among flowering plants, weedy annuals that colonize vacant sites have traits associated with *r*-selection, whereas flowering trees such as oaks have traits associated with *K*-selection. Among mammals, mice and other rodents have more *r*-selected traits, whereas elephants have more *K*-selected ones (**Figure 44.10**).

Some species have mixes of traits that cannot be explained by *r*-selection or *K*-selection alone. For example, bamboo and century plants (a type of agave) live long and grow to large size, but they reproduce only once, releasing many tiny seeds and then dying. Atlantic eels and Pacific salmon are unusual among vertebrates in also having a one-shot reproductive strategy. Such a strategy is thought to arise when opportunities for reproduction are unlikely to be repeated. In the bamboo and century plant, climate conditions that favor reproduction occur only rarely. In the eels and salmon, physiological changes related to migration between fresh water and salt water make a repeat journey impossible.

cohort Group of individuals born during the same time interval.
K-selected species Species adapted to a stable environment, where population size is often near carrying capacity.
life history pattern A set of traits related to growth, survival, and reproduction such as age-specific mortality, life span, age at first reproduction, and number of breeding events.
r-selected species Species adapted to an environment that changes rapidly and unpredictably, so population size is often far below carrying capacity.
survivorship curve Graph showing the decline in numbers of a cohort over time.

Take-Home Message

How do researchers study and describe life history patterns?

» Tracking a cohort (a group of individuals) from their birth until the last one dies reveals patterns of reproduction and mortality.

» Survivorship curves reveal differences in age-specific survival among species or among populations of the same species.

» Species that maximize offspring quantity are said to be *r*-selected, whereas those who maximize offspring quality are *K*-selected.

44.6 Evidence of Evolving Life History Patterns

■ Predation can serve as a selection pressure that shapes life history patterns.

■ Links to Directional selection 17.5, Gene flow 17.8

A Long-Term Study of Guppies

A long-term study by the evolutionary biologists John Endler and David Reznick illustrates the effect of predation on life history traits. Endler and Reznick studied populations of guppies (*Poecilia reticulata*), small fishes native to shallow freshwater streams (**Figure 44.11**) in the mountains of Trinidad.

For their study sites, the scientists focused on streams with many small waterfalls. These waterfalls serve as natural barriers that prevent guppies in one part of a stream from moving easily to another. As a result, each stream holds several populations of guppies that have very little gene flow between them (Section 17.8).

The waterfalls also keep guppy predators from moving from one part of the stream to another. The main guppy predators are killifishes and cichlids. The two differ in size and prey preferences. The killifish (**Figure 44.11A**) is relatively small and preys mostly on immature guppies. It ignores the larger adults. The cichlids are bigger fish (**Figure 44.11B**). They tend to pursue mature guppies and ignore small ones.

Some parts of the streams hold one type of predator but not the other. Thus, different guppy populations face different predation pressures. Reznick and Endler discovered that guppies in regions with cichlids grow faster and are smaller at maturity than guppies in regions with killifish. Also, guppies hunted by cichlids reproduce earlier, have more offspring at a time, and breed more frequently (**Figure 44.12**).

Were these differences in life history traits genetic, or did some environmental variation cause them? To find out, the biologists collected guppies from both cichlid- and killifish-dominated streams. They reared the groups in separate aquariums under identical predator-free conditions. Two generations later, the groups continued to show the differences observed in

A *Right*, guppy that shared a stream with killifishes (*below*).

B *Right*, guppy that shared a stream with cichlids (*below*).

C Biologist David Reznick at the study site, a freshwater stream in Trinidad.

Figure 44.11 Animated A field study of guppy life history evolution.

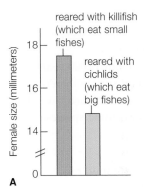
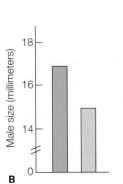
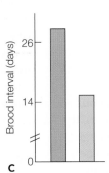
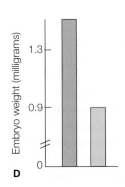

reared with killifish (which eat small fishes)

reared with cichlids (which eat big fishes)

A   B   C   D

Figure 44.12 Experimental evidence of natural selection among guppy populations subject to different predation pressures.

Compared to the guppies raised with killifish (*green* bars), guppies raised with cichlids (*tan* bars) differed in body size and in the length of time between broods.

natural populations. The researchers concluded that differences between guppies preyed on by different predators have a genetic basis.

Reznick and Endler hypothesized that the predators act as selective agents that influence guppy life history patterns. They made a prediction: If life history traits evolve in response to predation, then these traits will change when a population is exposed to a new predator that favors different prey traits.

To test their prediction, they found a stream region above a waterfall that had killifish but no guppies or cichlids. They brought in some guppies from a region below the waterfall where there were cichlids but no killifish. At the experimental site, the guppies that had previously lived only with cichlids were now exposed to killifish. The control site was the downstream region below the waterfall, where relatives of the transplanted guppies still coexisted with cichlids.

Reznick and Endler revisited the stream over the course of eleven years and thirty-six generations of guppies. They monitored traits of guppies above and below the waterfall. The recorded data showed that guppies at the upstream experimental site were evolving. Exposure to a previously unfamiliar predator caused changes in the guppies' rate of growth, age at first reproduction, and other life history traits. By contrast, guppies at the control site showed no such changes. Reznick and Endler concluded that life history traits in guppies can evolve rapidly in response to the selective pressure exerted by predation.

Overfishing of Atlantic Cod

The evolution of life history traits in response to predation pressure is not merely of theoretical interest. It has economic importance. Just as guppies evolved in response to predators, a population of Atlantic codfish (*Gadus morhua*) evolved in response to human fishing

pressure. From the mid-1980s to early 1990s, fishing pressure on the North Atlantic population of codfish increased. As it did, the age of sexual maturity shifted. The proportion of fast-maturing fish that reproduced while young and small increased. Such individuals were at an advantage because both commercial fisherman and sports fishermen preferentially caught and kept larger fish (*above*).

In 1992, declining numbers of cod caused the Canadian government to ban cod fishing in some areas. That ban, and later restrictions, came too late to stop the Atlantic cod population from crashing. The population still has not recovered.

Looking back, it is clear that life history changes were an early sign that the North Atlantic cod population was in trouble. Had biologists recognized what was happening, they might have been able to save the fishery and protect the livelihood of more than 35,000 fishers and associated workers. Ongoing monitoring of the life history data for other economically important fishes may help prevent similar disastrous crashes in the future.

Take-Home Message

What effect does predation have on life history traits?

» When predators prefer large prey, individual prey who reproduce when still small and young are at a selective advantage. When predators focus on small prey, fast-growing individuals have the selective advantage.

44.7 Human Population Growth

■ The size of the human population is at its highest level ever and is expected to continue to increase.

■ Links to Nitrogen fixation 20.7, Evolution of modern humans 26.6, Vaccines 37.12

For most of its history, the human population grew very slowly. The growth rate began to increase about 10,000 years ago, and during the past two centuries, it soared (**Figure 44.13**). Three trends promoted the large increases. First, humans were able to migrate into new habitats and expand into new climate zones. Second, humans developed new technologies that increased the carrying capacity of existing habitats. Third, humans sidestepped some limiting factors that typically restrain population growth.

Early Innovations and Expansions

Modern humans evolved in Africa by about 200,000 years ago. By 43,000 years ago, their descendants were established in much of the world (Section 26.6). Few species can expand into such a broad range of habitats, but our large brains allowed us to master a variety of skills. Humans learned how to start fires, build shelters, make clothing, manufacture tools, and cooperate in hunts. With the advent of language, knowledge of such skills did not die with the individual. Compared to most species, humans have a greater capacity to disperse fast over long distances and to become established in a variety of environments.

The invention of agriculture about 11,000 years ago provided a more dependable food supply than traditional hunting and gathering. A pivotal factor was the domestication of wild grasses, including species ancestral to modern wheat and rice.

From the Industrial Revolution Onward

The rate of human population growth surged in the 1800s as innovations increased the stability of food supplies and decreased the impact of disease for some human populations.

In the middle of the eighteenth century, people learned to harness the energy of fossil fuels to operate machinery. This innovation opened the way to high-yielding mechanized agriculture and to improved food distribution systems. Food production was further enhanced in the early 1900s, when German chemists discovered a way to convert gaseous nitrogen to ammonia. Previously this process had been carried out primarily by nitrogen-fixing bacteria. Use of fertilizers manufactured by this process dramatically increased crop yields. So did the synthetic pesticides that came into widespread use in the mid-1900s.

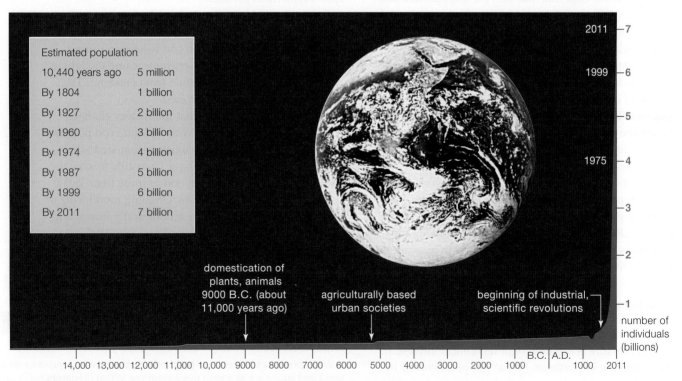

Estimated population	
10,440 years ago	5 million
By 1804	1 billion
By 1927	2 billion
By 1960	3 billion
By 1974	4 billion
By 1987	5 billion
By 1999	6 billion
By 2011	7 billion

domestication of plants, animals 9000 B.C. (about 11,000 years ago)

agriculturally based urban societies

beginning of industrial, scientific revolutions

number of individuals (billions)

14,000 13,000 12,000 11,000 10,000 9000 8000 7000 6000 5000 4000 3000 2000 1000 | B.C. | A.D. | 1000 2011

Figure 44.13 Growth curve (*red*) for the world human population. To check the current world and U.S. population estimates, visit the U.S. Census web site at www.census.gov/main/www/popclock.html.

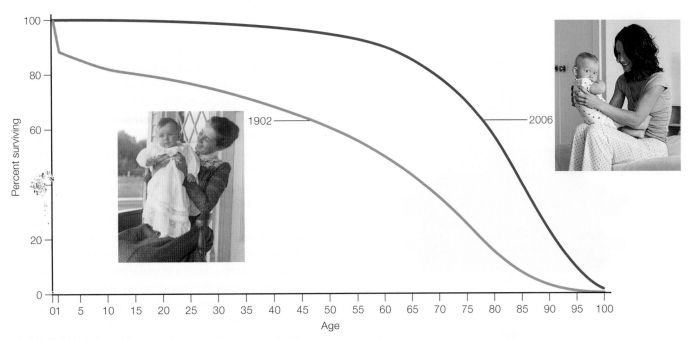

Figure 44.14 Survivorship curves for people in the United States in the early 20th and early 21st centuries. Since 1900, infant mortality has declined by more than 90 percent, and mortality during childbirth has declined 99 percent. (CDC/NCHS, National Vital Statistics System and state death registration data)

Disease has historically dampened human population growth. For example, during the mid-1300s, one-third of Europe's population was lost to a pandemic known as the Black Death. Beginning in the mid-1800s, an increased understanding of the link between microorganisms and illness led to improvements in food safety, sanitation, and medicine. People began to pasteurize foods and drinks, heating them to reduce the numbers of harmful bacteria. They also began to protect their drinking water. The first modern sewer system was constructed in London, England, in the late 1800s. By diverting wastewater downstream of the city's water source, the system lowered the incidence of waterborne diseases such as cholera and typhoid fever. Beginning in the early 1900s, chlorination and other methods of sterilizing drinking water contributed to a further decline in these diseases in the most industrialized nations.

Advances in sanitation also lowered the death rate associated with medical treatment. In the mid-1800s, Ignaz Semmelweis, a German physician, began urging doctors to wash their hands between patients. His advice was largely ignored until after his death, when Louis Pasteur popularized the idea that unseen organisms cause disease. Acceptance of this concept also revolutionized surgery, which had previously been carried out with little regard for cleanliness.

Vaccines helped prevent disease and antibiotics helped treat them. As Section 37.12 explained,

Edward Jenner demonstrated the effectiveness of a vaccine against smallpox in the late 1700s. In 1850, Massachusetts became the first state in the United States to mandate childhood vaccination against this disease. Antibiotics are a more recent development. Large-scale production of penicillin, the first antibiotic to be widely used, did not begin until the 1940s.

Figure 44.14 shows the result of improving nutrition, sanitation, and medicine in the United States: a shift in the survivorship curve. Infant mortality has plummeted and far more people survive to old age.

A worldwide decline in death rates without an equivalent drop in birth rates is responsible for the ongoing explosion in human population size. It took more than 100,000 years for the human population to reach 1 billion in number. Since then, the rate of increase has risen steadily. The population is currently about 7 billion and the United Nations estimates that it will reach 9 billion by 2050.

Take-Home Message

What factors contributed to the increase in the human population size?

» Innovations such as use of fire and clothing allowed early humans to expand into what would otherwise have been inhospitable habitats.

» The invention of agriculture created a stable food supply. More recently, innovations such as chemical fertilizers helped to boost crop yields.

» Improved understanding of the causes of disease led to better sanitation and new medical treatments that have increased the average life span.

44.8 Anticipated Growth and Consumption

■ Human population size—and resource consumption—is expected to continue to rise for many years to come.

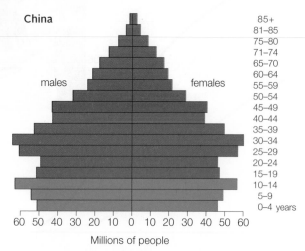

China

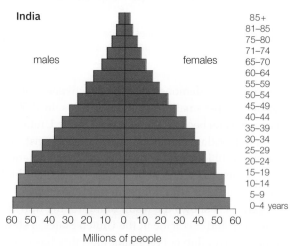

India

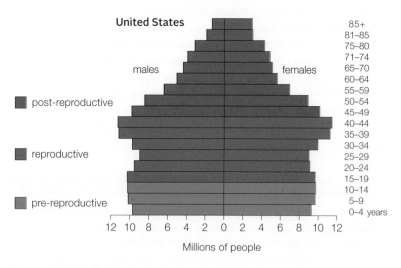

United States

- post-reproductive
- reproductive
- pre-reproductive

Figure 44.15 Animated Age structure for three countries. *Green* bars represent pre-reproductive individuals. The *left* side of each chart indicates males; the *right* side, females. **Figure It Out:** Which country has the largest number of men in the 45 to 49 age group?

Answer: China

Fertility and Age Structure

The **total fertility rate** of a human population is the average number of children born to a woman during her reproductive years. In 1950, the worldwide total fertility rate averaged 6.5. By 2010, it had declined to 2.6. It remains above the **replacement fertility rate**—the average number of children a woman must bear to replace herself with one daughter who reaches reproductive age. At present, the replacement rate is 2.1 for developed countries and as high as 2.5 in some developing countries. (It is higher in developing countries because more daughters die before reaching the age of reproduction.) A population grows as long as its total fertility rate exceeds the replacement rate.

Age structure affects a population's growth rate. **Figure 44.15** shows the age structure for the world's most populous countries. China and India have more than one billion people apiece. Next in line is the United States, with more than 310 million. These three countries differ in age structure, with India having the greatest proportion of young people. The broader the base of an age structure diagram, the greater the anticipated population growth.

Worldwide, about 1.9 billion people are about to enter their reproductive years. Even if every couple from this time forward has no more than two children, population growth will not slow for sixty years.

A Demographic Transition

Demographic factors vary among countries, with the most highly developed countries having the lowest fertility rates and infant mortality, and the highest life expectancy. The **demographic transition model** describes how changes in population growth often unfold in four stages of economic development (**Figure 44.16**). The model predicts that living conditions are harshest during the preindustrial stage, before technological and medical advances become widespread. Birth and death rates are both high, so the growth rate is low ❶.

Next, in the transitional stage, industrialization begins. Food production and health care improve. The death rate drops fast, but the birth rate declines more slowly ❷. As a result, the population growth rate increases rapidly. India is in this stage.

During the industrial stage, when industrialization is in full swing, the birth rate declines. People move from the country to cities, where couples tend to want smaller families. The birth rate moves closer to the death rate, and the population grows less rapidly ❸. Mexico is currently in this stage.

Table 44.2 Ecological Footprints*	
Country	**Hectares per Capita**
United States	8.0
Canada	7.0
France	5.0
United Kingdom	4.9
Japan	4.7
Russian Federation	4.4
Mexico	3.0
Brazil	2.9
China	2.2
India	0.9
World Average	2.7

* 2010 data from www.footprintnetwork.org

In the postindustrial stage, a population's growth rate becomes negative. The birth rate falls below the death rate, and population size slowly decreases **4**.

The demographic transition model is based on the social changes that occurred when western Europe and North America industrialized. Whether it can accurately predict changes in modern developing countries remains to be seen. These countries receive aid from

demographic transition model Model describing changes in birth and death rates that occur as a region becomes industrialized.
ecological footprint Area of Earth's surface required to sustainably support a particular level of development and consumption.
replacement fertility rate Fertility rate at which each woman has, on average, one daughter who survives to reproductive age.
total fertility rate Average number of children the women of a population bear over the course of a lifetime.

the fully industrialized nations, but also must compete with them in a global market.

Resource Consumption

Per capita resource consumption rises with economic and industrial development. Ecological footprint analysis is one method of comparing resource use. An **ecological footprint** is the amount of Earth's surface required to support a particular level of development and consumption in a sustainable fashion. **Table 44.2** shows the per capita global footprint data for 2010. The average person in the United States has an ecological footprint nearly three times that of an average world citizen, and about nine times that of an average person living in India.

Billions of people in less developed nations hope to someday enjoy the same type of lifestyle as the average American. Ecological footprint analysis tells us that, with current technology, Earth may not have enough resources to make those dreams come true. For everyone now alive to live like an average American would require four times the sustainable resources available on Earth.

Take-Home Message

What factors will affect future changes in the human population?

» Fertility rates have been declining but remain above replacement level worldwide. Many young people are about to begin reproducing.

» Industrialization of less developed countries is predicted to decrease their birth rates over time, as it did for currently industrialized nations.

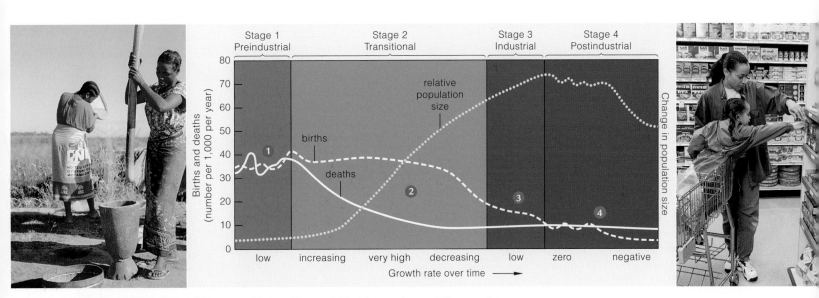

Figure 44.16 Animated Demographic transition model for changes in population growth rates and sizes, correlated with long-term changes in economy.

A Honking Mess (revisited)

In 2009, investigators from the Federal Aviation Agency were asked to find out why both engines of a passenger plane failed, forcing the plane to land in the Hudson River. The pilot had reported a bird strike, and the investigators found bits of feather, bone, and muscle in the plane's wing flaps and engines (*right*). Samples of this tissue were sent to the Smithsonian Institute, which analyzed the DNA. Unique sequences in the DNA identified the tissue in both engines as Canada goose. One engine had female goose DNA. The other had male and

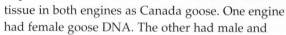

female DNA. Researchers were even able to tell which population of geese these unlucky birds belonged to. The mix of hydrogen isotopes varies with latitude, so the isotope mix in a feather provides information about where that feather developed. The birds were migratory; their feathers had developed in Canada, not in New York.

How would you vote? One way to decrease the number of nonmigratory Canada geese is by hunting them when migratory birds are not present. Do you support relaxing hunting restrictions in regions where Canada geese have become pests?

LEARNING ROADMAP

Where you have been In this chapter, you will see how natural selection (Section 16.4) and coevolution (17.12) shape communities. You will revisit plant–pollinator interactions (22.8), the role of fungi as partners (23.7), and root nodules (28.3). Knowledge of biogeography (16.2) will help you understand how communities in different regions differ.

Where you are now

Community Characteristics
A community consists of all species in a habitat. A habitat's history, its biological and physical characteristics, and the interactions among species affect community structure.

Types of Species Interactions
Commensalism, mutualism, competition, predation, and parasitism are types of interspecific interactions. They influence the population size of participating species.

Successional Changes
Community structure changes over time. When a new community forms, early-arriving species often alter the habitat in a way that facilitates their own replacement.

The Role of Disturbances
Introduction or removal of a species can affect other species within a community. Some species have adapted to a recurring disturbance such as fire.

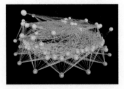

Biogeographic Patterns
Communities in tropical regions hold the greatest number of species. Models based on studies of islands can be used to predict how many species an area will support.

Where you are going The next chapter looks at the flow of energy and resources within communities and between communities and the nonliving portion of their environment. We return to global patterns of species richness as we survey biomes in Chapter 47. The effects of human disturbance on communities are the focus of Chapter 48.

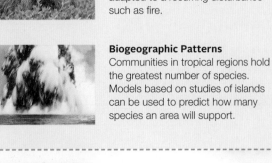

45.1 Fighting Foreign Fire Ants

Like most ants, red imported fire ants (*Solenopsis invicta*) nest in the ground (**Figure 45.1**). Accidentally step on one of these nests, and you will quickly realize your mistake. Like honeybees, fire ants defend their nest by stinging. Venom injected by the stinger causes a burning sensation, and results in formation of reddened bumps that are slow to heal.

S. invicta is native to Brazil and was accidentally introduced to the southeastern United States in the 1930s, when ants arrived as stowaways on a cargo ship. Since then, the ants have gradually expanded their range. In 1997, they were introduced from the Southeast to California.

More recently, *S. invicta* has been found in the Caribbean, Australia, New Zealand, and several countries in Asia. Genetic comparisons of ant populations have demonstrated that the ants involved in these introductions originated in the southeastern United States, rather than Brazil.

Increased dispersal rates among pest species are an unanticipated side effect of increased global trade and improvements in shipping. Speedier ships make quicker trips, increasing the likelihood that pests hidden in cargo holds will survive a journey.

The spread of *S. invicta* concerns ecologists because arrival of this species typically has a negative impact on a region's native species. For example, where *S. invicta* becomes prevalent, native ant species decline. In Texas, the *S. invicta*–induced decline of native ant species threatens the Texas horned lizard. Ants are a staple of the lizard's diet, but it cannot eat the *S. invicta* that have largely replaced its normal prey.

S. invicta has also been implicated in population declines of some bird species, including bobwhite quails and vireos (a songbird). The ants attack and feed on birds' eggs and on nestlings. Ground-nesting birds are at special risk.

Arrival of *S. invicta* can also harm plants. Species whose seeds are dispersed by ants may decline when *S. invicta* replaces the natives. *S. invicta* can also interfere with pollination by displacing or preying on native pollinators such as ground-nesting bees.

Species interactions such as competition between ant species or predation of ants on bird nestlings are one focus of community ecology. A **community** is all the species that live in a region. As you will see, species interactions and disturbances can shift community structure (the types of species and their relative abundances) in small and large ways, some predictable, and others unexpected.

community All species that live in a particular region.

Figure 45.1 Red imported fire ants. The photo at the *right* shows mounds made by colonies of fire ants in a Texas pasture.

45.2 Which Factors Shape Community Structure?

- Community structure refers to the number and relative abundances of species in a habitat.
- Link to Coevolution 17.12

Each species lives in a typical type of place, which we call its **habitat**, and all species in a particular habitat constitute a community. Communities often nest one inside another. For example, a community of microbial organisms lives in the termite gut. A termite is part of a larger community of organisms living on a fallen log. The log-dwellers are part of a larger forest community.

Even communities that are similar in scale differ in their species diversity. There are two components to species diversity. The first, species richness, refers to the number of species. The second is species evenness, or the relative abundance of each species. For example, a pond that has five fish species in nearly equal numbers has a higher species diversity than a pond with one abundant fish species and four rare ones.

Community structure is dynamic, which means that the array of species and their relative abundances in a particular community tend to change over time. Communities change over a long time span as they form and then age. They also change suddenly as a result of natural or human-induced disturbances.

The structure of a community is affected by abiotic (nonbiological) factors such as rainfall, sunlight intensity, and temperature. It varies along gradients in latitude, elevation, and—for aquatic habitats—depth.

Species interactions also influence the types of species in a community and their relative abundances. In some cases, the effect is indirect. For example, when songbirds eat caterpillars, the birds directly reduce the abundance of caterpillars, but they also indirectly benefit the trees that the caterpillars feed on.

We define direct interactions by their effects on both participants (**Table 45.1**). For example, in **commensalism** one species benefits and the other is neither benefited nor harmed. Should evidence of either benefit or harm come to light, the relationship is reclassified. The rela-

Figure 45.2 Commensalism. Epiphytic orchids grow on a tree trunk. The tree provides the orchids with an elevated perch from which they can capture sunlight, and it is neither helped nor harmed by their presence.

tionship between an epiphytic orchid and the plant on which it grows is a commensalism (**Figure 45.2**). Having an elevated perch benefits the orchid, and the tree is unaffected.

Species interactions may be fleeting or a long-term relationship. **Symbiosis** means "living together," and biologists use the term to refer to a relationship in which two species have a prolonged close association that benefits at least one of them. Commensalism, mutualism, and parasitism can be symbiotic relationships, as when commensal, mutualistic, or parasitic microorganisms live inside an animal's gut.

Regardless of whether one species helps or harms another, two species that interact closely for generations can coevolve. As Section 17.12 explained, coevolution is an evolutionary process in which each species acts as a selective agent that shifts the range of variation in the other.

commensalism Species interaction that benefits one species and neither helps nor harms the other.
habitat Type of environment in which a species typically lives.
symbiosis One species lives in or on another in a commensal, mutualistic, or parasitic relationship.

Take-Home Message

What factors affect the types and abundances of species in a community?

» The types and abundances of species in a community are affected by physical factors such as climate and by species interactions.

» A species can be benefited, harmed, or unaffected by its interaction with another species.

Table 45.1	Direct Two-Species Interactions	
Type of Interaction	**Effect on Species 1**	**Effect on Species 2**
Commensalism	Beneficial	None
Mutualism	Beneficial	Beneficial
Interspecific competition	Harmful	Harmful
Predation, herbivory, parasitism, parasitoidism	Beneficial	Harmful

45.3 Mutualism

- A mutualistic interaction benefits both partners.
- Links to Dinoflagellates 21.6, Pollination 22.8, Lichens and mycorrhizae 23.7, Plant mutualisms 29.3, Seed dispersal 29.7

A **mutualism** is an interspecific interaction that benefits both participants. Flowering plants and their pollinators are a familiar example. In some cases, two species coevolve a mutual dependence. For example, each species of yucca plant is pollinated by one species of yucca moth, whose larvae develop only on that plant (**Figure 45.3**). More often, mutualistic relationships are less exclusive. Most flowering plants have more than one pollinator, and most pollinators service more than one species of plant.

Photosynthetic organisms often supply sugars to their nonphotosynthetic partners, as when plants lure pollinators with nectar. In addition, many plants make sugary fruits that attract seed-dispersing animals. Plants also provide sugars to mycorrhizal fungi and nitrogen-fixing bacteria. These symbiotic partners return the favor by supplying their host plant with other essential nutrients. Similarly, photosynthetic dinoflagellates provide sugars to reef-building corals, and photosynthetic bacteria or algae in a lichen feed their fungal partner.

Animals often share ingested nutrients with mutualistic microorganisms that live in their gut. For example, *Escherichia coli* bacteria in your colon share your food, but they also provide you with vitamin K.

Mutualisms can also involve protection. For example, an anemonefish and a sea anemone fend off one another's predators (**Figure 45.4**). As another example, ants protect bull acacia trees from leaf-eating insects. The tree houses the ant in special hollow thorns and provides them with sugar-rich and protein-rich foods.

From an evolutionary standpoint, mutualism is best described as reciprocal exploitation. Each individual increases its fitness by extracting a resource, such as protection or food, from its partner. Resources withdrawn by one partner have a cost to the other, so individuals who cheat—who take benefits without reciprocating—would seem to be at a selective advantage. Of course, if cheaters evolve, selection will favor an ability to detect and avoid them, thus negating the potential benefit of cheating. A recent study demonstrated advantages and disadvantages of cheating in one mutualism. Researchers bred a petunia that "cheated" by making very little nectar. Producing less nectar allowed the plant to make more seeds—but only when pollinated by hand. This advantage disappeared under natural conditions because insect pollinators avoid low-nectar plants in favor of normal ones.

Figure 45.3 An obligate mutualism. Each species of yucca plant (*left*) has a relationship with one species of yucca moth (*right*). After a female moth mates, she collects pollen from a yucca flower and places it on the stigma of another flower, then lays her eggs in that flower's ovary. Moth larvae develop in the fruit that develops from the floral ovary. When mature, they gnaw their way out and disperse. Seeds that larvae did not eat give rise to new yucca plants.

Figure 45.4 Mutual protection. The stinging tentacles of this sea anemone (*Heteractis magnifica*) protect its partner, a pink anemonefish (*Amphiprion perideraion*) from fish-eating predators. In return, the anemonefish chases away fish that eat sea anemone tentacles. The anemonefish secretes a special mucus that prevents the anemone from stinging it.

mutualism Species interaction that benefits both species.

Take-Home Message

What are the effects of participating in a mutualism?

» A mutualism benefits both participants.

» In some cases, two species form an exclusive partnership. In others, a species provides benefits to, and receives benefits from, multiple species.

» Participating in a mutualism has both benefits and costs. Selection favors individuals who maximize their benefits while minimizing their costs.

45.4 Competitive Interactions

- Individuals of different species often compete for access to limiting resources.
- Links to Natural selection 16.4, Limiting factors 44.4

Interspecific competition, or competition between species, is not usually as intense as competition within a species. The requirements of two species might be similar, but they will generally differ more than the requirements of two members of the same species.

Each species requires specific resources and environmental conditions that we refer to collectively as its **ecological niche**. Both physical (abiotic) and biological (biotic) factors define the niche. Aspects of an animal's niche include the temperature range it can tolerate, the species it eats, and the places it can breed. A description of a flowering plant's niche would include its soil, water, light, and pollinator requirements. The more similar the niches of two species are, the more intensely the species will compete.

Competition takes two forms. With interference competition, one species actively prevents another from accessing some resource. As an example, one species of scavenger will often chase another away from a carcass (**Figure 45.5**). As another example, some plants use chemical weapons against potential competition. Aromatic chemicals that ooze from tissues of sagebrush plants, black walnut trees, and eucalyptus trees seep into the soil around these plants. The chemicals prevent other plants from germinating or growing.

In exploitative competition, each species reduces the amount of a resource available to the other by using that resource. For example, deer and blue jays both eat acorns in oak forests. The more acorns birds eat, the fewer there are for deer and vice versa.

Effects of Competition

Species compete most intensely when the supply of a shared resource is the main limiting factor for both (Section 44.4). In the early 1930s, G.F. Gause carried out a series of experiments that led him to describe **competitive exclusion**: Whenever two species require

Figure 45.5 Interspecific competition among scavengers. After facing off over a carcass (*top*), an eagle attacked a fox with its talons (*bottom*). The fox then retreated, leaving the eagle to exploit the carcass.

the same limited resource to survive or reproduce, the better competitor will drive the less competitive species to extinction in that habitat. Gause studied interactions between two species of ciliated protists (*Paramecium*) that compete for bacterial prey. He cultured the species separately and together (**Figure 45.6**). Within weeks, population growth of one species outpaced the other, which went extinct.

When competing species continue to coexist, the presence of each reduces the carrying capacity for the other. For example, the reproductive success of a flowering plant is decreased by competition from other

Figure 45.6 Animated
Competitive exclusion. Growth curves for two *Paramecium* species when grown separately and together. **Figure It Out:** Which species is the better competitor? Answer: *P. aurelia*

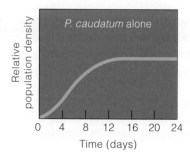
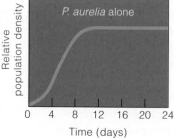
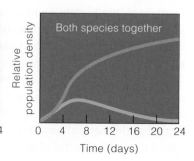

species that flower at the same time and rely on the same pollinators (**Figure 45.7**).

Even strikingly different organisms can influence one another through competition. Wolf spiders and carnivorous plants called sundews compete for insects in Florida swamps. Sundews grown in the presence of wolf spiders make fewer flowers than sundews in spider-free enclosures. Presumably, competition for insect prey reduced the energy the plants could devote to flowering.

Resource Partitioning

Resource partitioning is an evolutionary process by which species become adapted to use a shared limiting resource in a way that minimizes competition. For example, eight species of woodpecker coexist in Oregon forests. All feed on insects and nest in hollow trees, but the details of their foraging behavior and nesting preferences vary. Differences in nesting time also help reduce competitive interactions.

Resource partitioning arises as a result of directional selection on species who share a habitat and compete for a limiting resource. In each species, those individuals who differ most from the competing species with regard to resource use have the least competition and thus leave the most offspring. Over generations, directional selection leads to **character displacement**: The range of variation for one or more traits is shifted in a direction that lessens the intensity of competition for a limiting resource.

Character displacement may explain the variation in body size observed among populations of two species of salamander in the genus *Plethodon* (**Figure 45.8**). In most parts of their ranges, the two species do not overlap and their body sizes are quite similar (*orange* bars). Where the two species do coexist and compete, their difference in body length becomes more pronounced (*purple* bars). One species, *P. cinereus*, gets smaller, and the other *P. hoffmani*, gets larger. The increased difference in body size may be related to a partitioning of food resources. In regions where the species overlap, *P. cinereus* tends to specialize on smaller prey and *P. hoffmani* on larger insect prey.

character displacement Outcome of competition between two species; similar traits that result in competition become dissimilar.
competitive exclusion Process whereby two species compete for a limiting resource, and one drives the other to local extinction.
ecological niche The resources and environmental conditions that a species requires.
interspecific competition Competition between two species.
resource partitioning Species adapt to access different portions of a limited resource; allows species with similar needs to coexist.

Mimulus   *Lobelia*

Figure 45.7 Competing for pollinators. *Mimulus* and *Lobelia* grow together in damp meadows. To test for competition, researchers grew *Mimulus* plants either alone or with *Lobelia*. In mixed plots, pollinator visits to *Lobelia* plants frequently intervened between visits to *Mimulus*. As a result, *Mimulus* in mixed plots produced 37 percent fewer seeds than those grown alone.

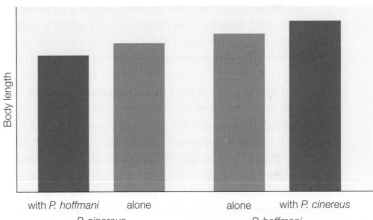

Body length

with *P. hoffmani*   alone   alone   with *P. cinereus*
P. cinereus   P. hoffmani

Figure 45.8 Animated Possible evidence of character displacement in salamanders (*Plethodon*). Where *P. cinereus* (shown at *right*) and *P. hoffmani* coexist, their average body lengths (*purple* bars) differ more than they do in habitats where each species lives alone (*orange* bars).

Take-Home Message

What happens when species compete for resources?

» In some interactions, one species actively blocks another's access to a resource. In other interactions, one species is simply better than another at exploiting a shared resource.

» When two species compete, selection favors individuals whose needs are least like those of the competing species.

45.5 Predator–Prey Interactions

■ The relative abundances of predator and prey populations of a community shift over time in response to species interactions and changing environmental conditions.

■ Link to Coevolution 17.12

Models for Predator–Prey Interactions

Predation is an interspecific interaction in which one species (the predator) captures, kills, and eats another species (the prey). Predation immediately removes a prey individual from the population.

Different types of predators have different functional responses, meaning they differ in how changes in prey density affect the rate at which they kill prey. **Figure 45.9A** shows the three types of functional responses to increases in prey density.

In a type I response, the proportion of prey killed is constant, so the number killed in any given interval depends solely on prey density. Web-spinning spiders and other passive predators tend to show this type of response. As the number of flies in an area increases, more and more become caught in each spider's web. Filter-feeding predators also show a type I response.

In a type II response, the number of prey killed depends on the capacity of predators to capture, eat, and digest prey. As prey density increases, the rate of kills rises steeply at first because there are more prey to catch. Eventually, the rate of increase slows, because each predator is exposed to more prey than it can handle at one time. At this point, rate of kills is limited not by prey density, but by handling time—the amount of time the predator takes to subdue, eat, and digest prey.

Figure 45.9B shows data from a field study of wolf predation on caribou. The wolves had a type II functional response to caribou density. The rate of kills declined at high density because a wolf that just killed a caribou will not hunt again until it has eaten and digested the first prey.

In a type III response, the number of kills increases slowly until prey density exceeds a certain level, then it rises rapidly before leveling off. Three situations can result in a type III response. First, some predators switch among prey, concentrating on the prey species that is most abundant. Second, some predators need to learn how to best capture each prey species; they get more lessons and become more efficient hunters when a particular type of prey is abundant. Third, some prey can escape predation by hiding. Only after prey density rises and some individual prey have no place to hide does the rate of kills increase.

Understanding a predator's functional response helps ecologists predict long-term effects of predation on a prey population. For example, a functional response in which the kill rate rises with prey density can help stabilize prey numbers.

Cyclic Changes in Abundance

A few predator and prey species undergo seemingly coupled cyclic changes in their abundance, with each increase or decline in prey numbers followed by a similar change in predator numbers. The oscillation in populations of Canadian lynx and the snowshoe hare is one example (**Figure 45.10A**). Both populations peak about every ten years, with each peak in hare numbers followed by a peak in lynx numbers.

Such cycles are predicted by a model in which prey abundance regulates predator numbers, and predator abundance regulates prey numbers. According to this model, as prey numbers increase, predators find it increasingly easy to locate and catch prey. As a result, the predator's survival and reproduction increases.

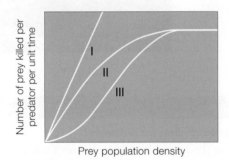

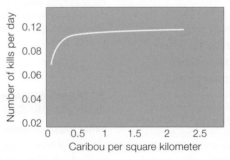

A Three theoretical types of functional response curves.

B Example of a type II response from one winter month in Alaska, during which B. W. Dale and his coworkers observed wolf packs (*Canis lupus*) feeding on caribou (*Rangifer tarandus*).

Figure 45.9 Animated Functional responses of predators to changes in prey density.

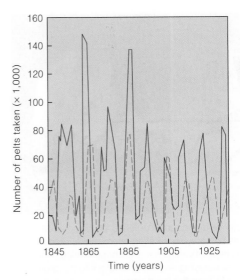

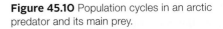

A Abundance of Canadian lynx (*dashed* line) and snowshoe hares (*solid* line), based on the numbers of pelts sold by trappers to Hudson's Bay Company during a ninety-year period.

Figure 45.10 Population cycles in an arctic predator and its main prey.

B Canadian lynx pursuing a snowshoe hare.

After predator numbers increase significantly, prey numbers begin to decline as a result of mortality and decreased reproduction. Predator numbers continue to increase for a while, because there are still plenty of prey. Eventually, prey numbers decline to a point where predator reproduction is affected and the predator population declines. After it does, prey numbers begin to rise, and the cycle begins again.

To determine whether this model accurately explains the cause of fluctuations in lynx and hare populations, Charles Krebs and his coworkers tracked hare densities for ten years in the Yukon River Valley of Alaska. They recorded the numbers of wild hares in control plots and in a variety of experimental plots. One type of experimental plot had fencing that protected hares from mammalian predators such as lynx. Another received fertilizer that increased plant growth. Another provided hares with supplemental food. Yet another both excluded mammalian predators and provided supplemental food.

Excluding mammalian predators, providing supplemental food, and combining these treatments all significantly increased the average density of hares over that in control plots. However, none of these treatments prevented population cycling. In all plots, hare numbers rose and then declined.

predation One species captures, kills, and eats another.

Krebs concluded that the one predator–one prey model did not accurately describe what drives the hare and lynx population cycles. The actual situation is far more complex. For one thing, snowshoe hares have multiple predators. They are prey for owls and hawks, as well as lynx and other mammals. Food availability also affects hare density. Thus, the observed population oscillations may arise as a result of a three-level interaction between plants, the hares that eat them, and the predators that eat hares.

Ecologists continue to investigate the influence of predation on the density of prey populations. Creating new models and testing the ability of these models to describe what we see in nature is an essential part of this process. Understanding predator–prey dynamics is important because such an understanding allows scientists to make valid decisions about how to best manage threatened predator and prey species.

Take-Home Message

How do predator and prey populations change over time?

» Predator populations show three general patterns of response to changes in prey density.

» Population levels of prey may show recurring oscillations.

» The numbers in predator and prey populations vary in complex ways that reflect the multiple levels of interaction in a community.

45.6 An Evolutionary Arms Race

■ Predators select for better prey defenses, and prey select for more efficient predators.

■ Links to Ricin 9.1, Coevolution 17.12, Secondary metabolites 22.9

Coevolution of Predators and Prey

Predator and prey exert selection pressure on one another. Suppose a mutation gives members of a prey species a more effective defense. Over generations, directional selection will cause this mutation to spread through the prey population. If some members of a predator population have a trait that makes them better at thwarting the improved defense, these predators and their descendants will be at an advantage. Thus, predators exert selection pressure that favors improved prey defense, which in turn exerts selection pressure on predators, and so the process continues over many generations.

You have already learned about some defensive adaptations. Many prey species have hard or sharp parts that make them difficult to eat. Think of a snail's shell or a sea urchin's spines. Cnidarians such as sea anemones and corals have stinging cells on their tentacles. Other prey contain chemicals that taste bad or sicken predators. Most defensive toxins in animals are derived from the plants that they eat. For example, a monarch butterfly caterpillar takes up chemicals from the milkweed plant that it feeds on. A bird that later eats the butterfly will be sickened by these chemicals.

Well-defended prey often have **warning coloration**, a conspicuous color pattern that predators learn to avoid. Monarch butterflies are bright orange, and many stinging wasps and bees share a pattern of black and yellow stripes (**Figure 45.11A**). The similar appearance of bees and wasps is an example of **mimicry**, an evolutionary pattern in which one species comes to resemble another. The tendency of well-defended spe-

Figure 45.12 Defense and counter defense. (**A**) *Eleodes* beetles defend themselves by spraying irritating chemicals at predators. (**B**) This defense is ineffective against grasshopper mice, who plunge the chemical-spraying end of the beetle into the ground and devour the insect from the head down.

cies to have similar coloration is called Müllerian mimicry, after the German naturalist who first described the phenomenon. Müller recognized that well-defended species who share the same type of predator would benefit by having a similar appearance. The more often a predator is stung by a black and yellow striped insect, the less likely it is to attack similar looking insects in the future.

In Batesian mimicry, also named for the scientist who first described it, a species mimics the appearance of a species that has a defense the mimic lacks. For example, some stingless insects resemble stinging bees or wasps (**Figure 45.11B–D**). These imposters benefit when predators avoid them after a painful encounter with the better-defended species that they mimic.

Stinging is an example of a defensive behavior that can repel a potential predator. Section 1.7 described how eyespots and a hissing sound protect some butterflies from predatory birds. Similarly, a lizard's tail

A Stinging wasp **B** One of its edible mimics **C** Another edible mimic **D** And another edible mimic

Figure 45.11 Examples of mimicry. Edible insect species often resemble toxic or unpalatable species that are not at all closely related. (**A**) A yellow jacket can deliver a painful sting. Nonstinging wasps (**B**), beetles (**C**), and flies (**D**) benefit by having a similar appearance.

may detach from the body and wiggle a bit as a distraction, allowing the rest of the lizard to escape.

Skunks squirt a foul-smelling, irritating repellent, as do some darkling beetles (**Figure 45.12**). Both skunks and beetles announce their intention to spray by assuming a distinctive posture, with their posterior end pointed toward the threat. Some beetles that cannot spray mimic this posture.

Camouflage is a body shape, color pattern, or behavior that allows an individual to blend into its surroundings and avoid detection. Prey benefit when camouflage hides them from predators (**Figure 45.13A**), and predators benefit when it hides them from prey (**Figure 45.13B,C**).

Other predator adaptations include sharp teeth and claws that can pierce protective hard parts. Speedy prey select for faster predators. For example, the cheetah, the fastest land animal, can run 114 kilometers per hour (70 mph). Its preferred prey, Thomson's gazelles, run 80 kilometers per hour (50 mph).

Coevolution of Herbivores and Plants

With **herbivory**, an animal feeds on plants. The number and type of plants in a community can influence the number and type of herbivores present.

Two types of defenses have evolved in response to herbivory. Some plants have adapted to withstand and recover quickly from the loss of their parts. For example, prairie grasses are seldom killed by native grazers such as bison. The grasses have a fast growth rate and store enough resources in their roots to replace the shoots lost to grazers.

Other plants have traits that deter herbivory. Physical deterrents include spines, thorns, and tough leaves that are difficult to chew. Many plants produce secondary metabolites (Section 22.9) that taste bad to herbivores or sicken them. Ricin, the toxin made by castor bean plants (Section 9.1), makes all eukaryotic herbivores ill. Caffeine in coffee beans and nicotine in tobacco leaves are defenses against insects.

Capsaicin, the compound that makes some peppers "hot," is an evolved defense against seed-eating

B Frilly pink body parts of a flower mantis hide it from insect prey attracted to the real flowers.

A Reedlike defensive posture of a bittern. The bird even sways in the wind like a reed.

C Fleshy protrusions give a predatory scorpionfish the appearance of an algae-covered rock. When algae-eating fish come close for a nibble, they end up as prey.

Figure 45.13 Camouflage in prey and predators.

mammals. Rodents avoid pepper fruits, leaving them to be eaten by birds. The pepper benefits by deterring rodent seed eaters because rodents chew up and kill seeds, whereas birds excrete the seeds alive and intact.

When a particular plant is abundant, selection favors herbivores that have an ability to overcome its defenses. For example, eucalyptus plants are abundant in the koala's habitat, but most herbivores avoid them because their leaves are tough and contain toxic compounds. Koalas have specialized teeth and digestive organs that allow them to access this resource.

camouflage Body coloration, patterning, form, or behavior that helps predators or prey blend with the surroundings and possibly escape detection.
herbivory An animal feeds on plant parts.
mimicry A species evolves traits that make it similar in appearance to another species.
warning coloration In many well-defended or unpalatable species, bright colors, patterns, and other signals that predators learn to recognize and avoid.

> ### Take-Home Message
>
> **How do predation and herbivory influence community structure?**
>
> » In any community, predators and prey coevolve, as do plants and the herbivores that feed on them.
>
> » Defensive adaptations in plants and prey can limit the ability of predators or herbivores to exploit some species in their community.

45.7 Parasites and Parasitoids

■ Predators have only a brief interaction with prey, but parasites live on or in their hosts.

■ Links to Sickle-cell anemia and malaria 17.7, Roundworms 24.10, Ticks 24.12, Lampreys 25.3, Cuckoos 43.3

Parasitism

With **parasitism**, one species (the parasite) benefits by feeding on another (the host), without immediately killing it. Endoparasites such as parasitic roundworms live and feed inside their host (**Figure 45.14A**). An ectoparasite such as a tick feeds while attached to a host's external surface (**Figure 45.14B**).

Parasitism has evolved in members of a diverse variety of groups. Bacterial, fungal, protistan, and invertebrate parasites feed on vertebrates. Lampreys (Section 25.3) attach to and feed on other fish. There are even a few parasitic plants that withdraw nutrients from other plants (**Figure 45.15**).

Parasites usually do not kill their host immediately. In terms of evolutionary fitness, killing a host too fast is bad for the parasite. Ideally, a host will live long enough to give a parasite time to produce offspring.

A Endoparasitic roundworms in the intestine of a host pig.

B Ectoparasitic ticks attached to and sucking blood from a finch.

Figure 45.14 Parasites inside and out.

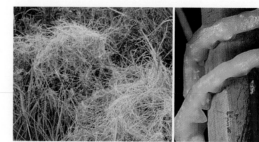

Figure 45.15 Dodder (*Cuscuta*), also known as strangleweed or devil's hair. This parasitic flowering plant has almost no chlorophyll. Leafless stems twine around a host plant during growth, as shown in the close-up at *right*. Modified roots penetrate the host's vascular tissues and absorb water and nutrients from them.

The presence of parasites can weaken a host so it is more vulnerable to predation or less attractive to potential mates. Parasites also affect host populations in other ways. Some cause their host to become sterile. Others shift the sex ratio among host offspring.

Adaptations to a parasitic lifestyle include traits that allow the parasite to locate hosts and to feed undetected. For example, ticks that feed on mammals or birds move toward a source of heat and carbon dioxide, which may be a potential host. A chemical in tick saliva acts as a local anesthetic, preventing the host from noticing the feeding tick. Endoparasites often have adaptations that help them evade a host's immune defenses.

Among hosts, traits that minimize the negative effects of parasites confer a selective advantage. For example, the allele that causes sickle-cell anemia persists at high levels in some human populations because having one copy of the allele increases the odds of surviving malaria (Section 17.7). Grooming and preening behavior are adaptations that minimize the impact of ectoparasites. Some animals produce chemicals that interfere with parasite activity. For example, crested auklets (a type of seabird) produce a citrus-scented secretion that repels ticks.

Strangers in the Nest

With **brood parasitism**, one egg-laying species benefits by having another raise its offspring. The European cuckoos described in Section 43.3 are brood parasites, as are North American cowbirds (**Figure 45.16**). Not having to invest in parental care allows a female cowbird to produce a large number of eggs, in some cases as many as thirty in a single reproductive season.

The presence of brood parasites decreases the reproductive rate of the host species and favors host indi-

Figure 45.16 A cowbird with its foster parent. A female cowbird minimizes her cost of parental care by laying her eggs in the nests of other bird species.

Figure 45.17 Biological control agent: a commercially raised parasitoid wasp about to deposit a fertilized egg in an aphid. The wasp larva will devour the aphid from the inside.

viduals that can detect and eject foreign young. Some brood parasites counter this host defense by producing eggs that closely resemble those of the host species. In cuckoos, different subpopulations have different host preferences and egg coloration. Females of each subpopulation lay eggs that closely resemble those of their preferred host.

Parasitoids

Parasitoids are insects that lay their eggs in other insects. Their larvae develop in the host's body, feed on its tissues, and eventually kill it. As many as 15 percent of all insects may be parasitoids.

Parasitoids reduce the size of a host population in two ways. First, as the parasitoid larvae grow inside their host, they withdraw nutrients and prevent it from reproducing. Second, the presence of these larvae eventually leads to the death of the host.

Biological Pest Controls

Parasites and parasitoids are commercially raised and released in target areas as a form of biological pest control. Biological control involves the use of a pest's natural enemies. It is promoted as an environmentally friendly alternative to chemical pesticides, which kill a wide variety of insects and often have harmful effects

on other animals, including humans. By contrast, species used in biological pest control are selected specifically because they target a limited number of host or prey species. The parasitoid wasp in **Figure 45.17** is an example of a biological control agent. It lays its eggs only in aphids.

For a species to be an effective biological control agent, it must be adapted to take advantage of a specific host species and to survive in that species' habitat. An ideal biological control agent is good at finding hosts, has a population growth rate comparable to that of its host, and produces offspring that are capable of long-range dispersal.

Although biological control of pests has many advantages, introducing a species into a community as a biological control does entail some risks. The introduced biological control agents sometimes go after nontargeted species in addition to, or instead of, those they were expected to control. For example, parasitoids were introduced to Hawaii to control stink bugs that feed on some Hawaiian crops. Instead, the parasitoids decimated the population of koa bugs, Hawaii's largest native bug. Introduced parasitoids have also been implicated in ongoing declines of many native Hawaiian butterfly and moth populations.

Take-Home Message

What are the effects of parasites, brood parasites, and parasitoids?

» Parasites reduce the reproductive rate of host individuals by withdrawing nutrients from them.

» Brood parasites reduce the reproductive rate of hosts by tricking them into caring for young that are not their own.

» Parasitoids reduce the number of host organisms by preventing reproduction and eventually killing the host.

brood parasitism One egg-laying species benefits by having another raise its offspring.
parasitism Relationship in which one species withdraws nutrients from another species, without immediately killing it.
parasitoid An insect that lays eggs in another insect, and whose young devour their host from the inside.

45.8 Ecological Succession

■ Which species are present in a community depends on physical factors such as climate, biotic factors such as which species arrived earlier, and the frequency of disturbances.

■ Links to Mosses 22.4, Lichens 23.7, Nitrogen-fixing bacteria 28.3

Successional Change

The array of species in a community can change over time. Species often alter the habitat in ways that allow other species to come in and replace them. We call this type of change ecological succession.

The process of succession starts with the arrival of **pioneer species**, which are opportunistic colonizers of new or newly vacated habitats. Pioneer species grow and mature fast, and produce many offspring capable of dispersing. Later, other species replace the pioneers. Then the replacements are replaced, and so on.

Primary succession is a process that begins when pioneer species colonize a barren habitat with no soil, such as a new volcanic island or land exposed by the retreat of a glacier (**Figure 45.18**). The earliest pioneers to colonize a new habitat are often mosses and lichens (Sections 22.4 and 23.7), which are small, have a brief life cycle, and can tolerate intense sunlight, extreme temperature changes, and little or no soil. Some hardy annual flowering plants with wind-dispersed seeds are also typical pioneers.

Pioneers help build and improve the soil. In doing so, they often set the stage for their own replacement. For example, many pioneer species partner with nitrogen-fixing bacteria, so they can grow in nitrogen-poor habitats. Seeds of later arrivals find shelter in mats of low-growing pioneer vegetation. Organic wastes and remains that accumulate add volume and nutrients to soil, thereby helping other species to become established. Later successional species often shade and eventually displace earlier ones.

In **secondary succession**, a disturbed area within a community recovers. This process occurs in abandoned agricultural fields and burned forests. Because improved soil is present from the start, secondary succession usually occurs faster than primary succession.

Factors That Influence Succession

The concept of ecological succession was first developed in the late 1800s, and it was hypothesized to be a predictable and directional process. Physical factors such as climate, altitude, and soil type were considered to be the main determinants of which species appeared in what order during succession. In this view, succession culminates in a "climax community," an array of species that persists over time and will be reconstituted in the event of a disturbance.

Ecologists now realize that the species composition of a community changes frequently and unpredictably. Communities do not journey along a well-worn path to a predetermined climax state.

Random events can determine the order in which species arrive in a habitat and thus affect the course of succession. The arrival of a particular species may

A Retreat of a glacier exposes rock and gravel.

B Lichens, mosses, and some annual flowering plants are early pioneers.

C Deciduous trees such as alder, willow, and cottonwood form dense thickets.

D Conifers such as hemlock and spruce move in and overgrow other species.

Figure 45.18 Animated One pathway of primary succession in Alaska's Glacier Bay region.

A The 1980 Mount Saint Helens eruption obliterated the community at the base of this Cascade volcano.

B The first pioneer species arrived less than a decade after the eruption.

Figure 45.19 A natural laboratory for studies of succession.

make it easier or more difficult for others that follow to become established.

Ecologists had an opportunity to investigate these factors after the 1980 eruption of Mount Saint Helens leveled about 600 square kilometers (235 square miles) of forest in Washington State (**Figure 45.19**). Ecologists recorded the natural pattern of colonization and carried out experiments in plots inside the blast zone. The results showed that the presence of some pioneers helped other, later-arriving plants become established. Other pioneers kept the same late arrivals out.

Disturbances also influence a community's species composition. According to the **intermediate disturbance hypothesis**, species richness is greatest in communities where disturbances are moderate in intensity and frequency (**Figure 45.20**). In such habitats, there is enough time for new colonists to arrive and become

established but not enough for competitive exclusion to cause extinctions.

In short, the modern view of succession holds that three types of factors affect communities: (1) physical factors such as climate, (2) chance events such as the order of species arrival, and (3) the frequency and extent of disturbances. The sequence of species arrivals and the frequency and extent of disturbances vary in unpredictable ways between communities. As a result, it is difficult to predict exactly how the composition of any particular community will change in the future.

intermediate disturbance hypothesis Species richness is greatest in communities where disturbances are moderate in their intensity or frequency.
pioneer species Species that can colonize a new habitat.
primary succession A new community colonizes an area where there is no soil.
secondary succession A new community develops in a disturbed site where the soil that supported a previous community remains.

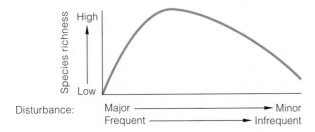

Figure 45.20 Graph showing the pattern predicted by the intermediate disturbance hypothesis. Species richness is highest when disturbances are moderate in intensity and frequency.

Take-Home Message

What is ecological succession?

» Succession is a process in which one array of species replaces another over time. It can occur in a barren habitat (primary succession) or a region where a community previously existed (secondary succession).

» Physical factors affect succession, but so do species interactions and disturbances. As a result, the course that succession will take in any particular community is difficult to predict.

45.9 Species Introduction, Loss, and Other Disturbances

- The loss or addition of even one species may destabilize the number and abundances of species in a community.
- Some species adapted to being disturbed are at a competitive disadvantage if disturbances do not occur.

The Role of Keystone Species

A **keystone species** has a disproportionately large effect on a community relative to its abundance. Robert Paine was the first to describe the effect of a keystone species after his experiments on the rocky shores of California's coast. Species in the rocky intertidal zone withstand pounding surf by clinging to rocks. A rock to cling to is a limiting factor. Paine set up control plots with the sea star *Pisaster ochraceus* and its main prey—chitons, limpets, barnacles, and mussels. Then he removed all sea stars from his experimental plots. Sea stars prey mainly on mussels. With sea stars gone from experimental plots, mussels took over, crowding out seven other species of invertebrates and reducing the diversity of algae living in the plots (**Figure 45.21**). Paine concluded that sea stars are a keystone species. They normally keep the number of prey species in the intertidal zone high by preventing competitive exclusion by mussels.

Keystone species need not be predators. For example, large, herbivorous rodents called beavers can be a keystone species. A beaver cuts down trees by gnawing through their trunk, then uses felled trees to build a dam. Construction of a beaver dam creates a deep

Figure 45.22 Adapted to disturbance. Some woody shrubs, such as this toyon, resprout from their roots after a fire. In the absence of occasional fire, toyons are outcompeted and displaced by species that grow faster but are less fire resistant.

pool where a shallow stream would otherwise exist. By altering the physical conditions in a section of the stream, the beaver affects the types of fish and aquatic invertebrates that can live there.

Adapting to Disturbance

In communities repeatedly subjected to a particular type of disturbance, individuals that withstand or benefit from that disturbance have a selective advantage. For example, some plants in areas subject to periodic fires produce seeds that germinate only after a fire has cleared away potential competitors. Other plants have an ability to resprout quickly after a fire (**Figure 45.22**).

Because different species respond differently to fire, the frequency of fire affects competitive interactions. For example, when naturally occurring fires are suppressed, species adapted to periodic burning lose their competitive edge. They can be overgrown by species that devote all of their energy to growing shoots, rather than storing some in underground parts in readiness for a spurt of regrowth.

Some species are especially sensitive to disturbances to the environment. These **indicator species** are the first to do poorly when conditions change, so they can provide an early warning of environmental degradation. For example, a decline in a trout population can be an early sign of problems in a stream, because trout are highly sensitive to pollutants and cannot tolerate low oxygen levels.

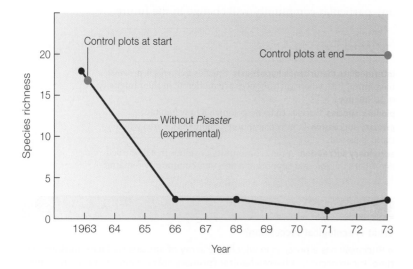

Figure 45.21 Evidence that the sea star *Pisaster* is a keystone species. Removal of *Pisaster* from experimental plots resulted in a decline in species richness (*red dots and line*). Control plots from which *Pisaster* was not removed showed no comparable decline in species richness (*green dots*).

A Kudzu native to Asia is overgrowing trees across the southeastern United States.

B Gypsy moths native to Europe and Asia feed on oaks through much of the United States.

C Nutrias native to South America are abundant in freshwater marshes of the Gulf States. Their population is also increasing in the Pacific Northwest.

Figure 45.23 Three exotic species that are altering natural communities in the United States. To learn more about invasive species in the United States, visit the National Invasive Species Information Center online at www.invasivespeciesinfo.gov.

Species Introductions

The arrival of a new species in a community can also cause dramatic changes. When you hear someone speaking enthusiastically about exotic species, you can safely bet the speaker is not an ecologist. An **exotic species** is a resident of an established community that dispersed from its home range and became established elsewhere. Unlike most imports, which never take hold outside the home range, an exotic species becomes a permanent member of its new community.

More than 4,500 exotic species have become established in the United States. An estimated 25 percent of Florida's plant and animal species are exotics. In Hawaii, 45 percent are exotic. Some species were brought in for use as food crops, to brighten gardens, or to provide textiles. Other species, including red imported fire ants, arrived as stowaways along with cargo from distant regions.

One of the most notorious exotic species is a vine called kudzu (*Pueraria lobata*). Native to Asia, it was introduced to the American Southeast as a food for grazers and to control erosion. However, it quickly

became an invasive weed. Kudzu overgrows trees, telephone poles, houses, and almost everything else in its path (**Figure 45.23A**).

Gypsy moths (*Lymantria dispar*) are an exotic species native to Europe and Asia. They entered the northeastern United States in the mid-1700s and now range into the Southeast, Midwest, and Canada. Gypsy moth caterpillars (**Figure 45.23B**) preferentially feed on oaks. Loss of leaves to gypsy moths can weaken trees, making them more susceptible to parasites and disease.

Large semiaquatic rodents called nutrias (*Myocastor coypus*) were imported from South America to be grown for their fur. They were released into the wild in many states. Along the Gulf of Mexico, nutrias now thrive in freshwater marshes (**Figure 45.23C**), and their voracious appetite threatens the native vegetation. In addition, their burrowing contributes to marsh erosion and damages levees, increasing the risk of flooding.

Take-Home Message

What types of changes alter community structure?

» Removing a keystone species can alter the diversity of species in a community.

» A change in the typical frequency of a disturbance can favor some species over others.

» Introducing an exotic species can threaten native species.

---

exotic species A species that evolved in one community and later became established in a different one.

indicator species A species that is especially sensitive to disturbance and can be monitored to assess the health of a habitat.

keystone species A species that has a disproportionately large effect on community structure.

45.10 Biogeographic Patterns in Community Structure

■ The richness and relative abundances of species differ from one habitat or region of the world to another.

■ Link to Biogeography 16.2

Latitudinal Patterns

Biogeography is the scientific study of how species are distributed in the natural world (Section 16.2). Perhaps the most striking pattern of species richness corresponds with distance from the equator. For most major plants and animal groups, the number of species is greatest in the tropics and declines from the equator to the poles (**Figure 45.24**). Consider just a few factors that help bring about and maintain this pattern.

First, tropical latitudes intercept more intense sunlight and receive more rainfall, and their growing season is longer. As one outcome, resource availability tends to be greatest and most reliable in the tropics. Thus, the tropics support a degree of specialized interrelationships not possible where species are active for shorter periods.

Second, tropical communities have been established for a long time. Some temperate communities did not start forming until the end of the last ice age. The longer a community has been established, the more time there has been for speciation within it.

Third, species richness may be self-reinforcing. The number of species of trees in tropical forests is much greater than in comparable forests at higher latitudes.

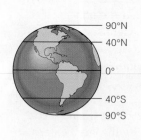
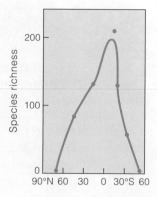

Figure 45.24 Ant species richness by latitude.

Where more plant species compete and coexist, more species of herbivores also coexist, partly because no single herbivore species can overcome all the defenses of all plants. In addition, more predators and parasites can evolve in response to more kinds of prey and hosts. The same principles apply to tropical reefs.

Island Patterns

In the mid-1960s, volcanic eruptions formed a new island 33 kilometers (21 miles) from the coast of Iceland. The island was named Surtsey (**Figure 45.25**). Bacteria and fungi were early colonists. The first vascular plant became established on the island in 1965. Mosses appeared two years later and thrived. The first lichens were found five years after that. The rate of arrivals of new vascular plants picked up after a seagull colony became established in 1986.

The number of species on Surtsey will not continue increasing forever. How many species will there be when the number levels off? The **equilibrium model of island biogeography** addresses this question. According to this model, the number of species living on any island reflects a balance between immigration rates for new species and extinction rates for established ones. The distance between an island and a mainland source of colonists affects immigration rates. An island's size affects both immigration rates and extinction rates.

Consider first the **distance effect**: Islands far from a source of colonists receive fewer immigrants than those closer to a source. Most species cannot disperse very far, so they will not turn up far from a mainland.

Species richness is also shaped by the **area effect**: Big islands tend to support more species than small ones for several reasons. First, more colonists will happen upon a larger island simply by virtue of its size. Second, larger islands are more likely to offer a variety of habitats, such as high and low elevations. This

Figure 45.25 Surtsey, a volcanic island, during the time of its formation. The graph *below* shows the number of vascular plant species found in yearly surveys. Seagulls first began nesting on the island in 1986.

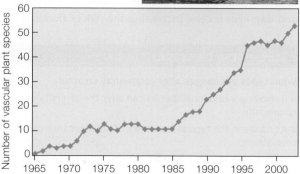

Fighting Foreign Fire Ants (revisited)

Invicta means "invincible" in Latin, and *S. invicta* lives up to its name. So far, red imported fire ants have overcome all control efforts. But scientists have a new weapon. They have imported phorid flies from Brazil for use as a biological control agent (**Figure 45.27**).

Phorid flies are parasitoids and *S. invicta* is their host. A female phorid fly lays an egg in the ant's body. The egg hatches into a larva that grows in and feeds on the ant's soft tissues. Eventually the larva enters the ant's head, causes the head to fall off, then undergoes metamorphosis to an adult fly inside it.

Phorid flies are not expected to kill off all *S. invicta* in affected areas. Rather, the hope is that the flies will reduce the density of invading colonies. Ecologists are also exploring other options, such as introducing imported pathogenic fungi or protists that infect *S. invicta* but not native ants.

How would you vote? Currently, only a small fraction of the crates imported into the United States are inspected to see if they contain exotic insects or other pests. Is paying more to inspect all imported goods an investment that is worth the additional cost?

Figure 45.27 Off with her head! A phorid fly (**A**) lays an egg in a fire ant. The fly larva eventually enters the ant's head and causes it to fall off (**B**).

---

variety makes it more probable that a new arrival will find a suitable habitat. Finally, big islands can support larger populations of species than small islands. The larger a population, the less likely it is to become locally extinct as the result of some random event.

Figure 45.26 illustrates how interactions between the distance effect and the area effect can influence the equilibrium number of species on islands.

Robert H. MacArthur and Edward O. Wilson first developed the equilibrium model of island biogeography in the late 1960s. Since then the model has been modified and its use has been expanded to help scientists think about habitat islands—natural settings surrounded by a "sea" of habitat that has been disturbed by humans. Many parks and wildlife preserves fit this description. Island-based models can help estimate the size of an area that must be set aside as a protected reserve to ensure survival of a species or a community.

area effect Larger islands have more species than small ones.
distance effect Islands close to a mainland have more species than those farther away.
equilibrium model of island biogeography Model that predicts the number of species on an island based on the island's area and distance from the mainland.

Take-Home Message

What are some biogeographic patterns in species richness?

» Generally, species richness is highest in the tropics and lowest at the poles. Tropical habitats have more predictable resources and tropical communities have often been evolving for longer than temperate ones.

» When a new island forms, species richness rises over time and then levels off. The size of an island and its distance from a colonizing source influence the equilibrium number of species it supports.

Figure 45.26 Island biodiversity patterns.

Distance effect: Species richness on islands of a given size declines as distance from a source of colonists rises. *Green* circles are values for islands less than 300 kilometers from the colonizing source. *Orange* triangles are values for islands more than 300 kilometers (190 miles) from a source of colonists.

Area effect: Among islands the same distance from a source of colonists, larger islands tend to support more species than smaller ones.

Figure It Out: Which is likely to have more species, a 100-km² island more than 300 km from a colonizing source or a 500-km² island less than 300 km from a colonist source?

Answer: The 500-km² island

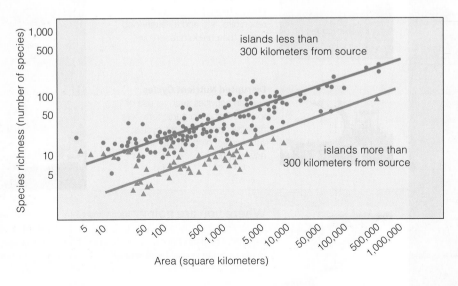

LEARNING ROADMAP

Where you have been This chapter builds on prior discussions of energy flow (Sections 1.3 and 5.2). It returns to the topics of carbon imbalances (6.1), algal blooms (21.6), plant nutrition (28.2), nitrogen fixation (20.7 and 28.3), and mycorrhizal fungi (23.7). Discussions of nutrient cycles will also draw on your knowledge of tectonic plates (16.7).

Where you are now

Organization of Ecosystems
An ecosystem is a community and its physical environment. A one-way flow of energy and a cycling of raw materials among its interacting participants maintain it.

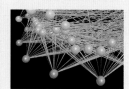

Food Chains and Webs
Food chains are linear sequences of feeding relationships among members of a community. Food chains connect as food webs.

Energy and Materials Flow
Ecosystems differ in the amount of energy captured by producers. Land ecosystems are typically more productive than marine ones.

Biogeological Cycles
Water, carbon, nitrogen, and phosphorus move from environmental reservoirs, into and through food webs, then back to reservoirs.

Disrupted Nutrient Cycles
Human activities such as use of fossil fuel and inorganic fertilizers have disrupted nutrient cycles, with far-reaching effects on the environment.

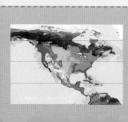

Where you are going The next chapter describes how the differences in climate that influence primary productivity arise. It also describes the types of ecosystems found on land and in aquatic habitats. The final chapter discusses how human activities such as altering nutrient cycles contribute to the ongoing loss of biodiversity.

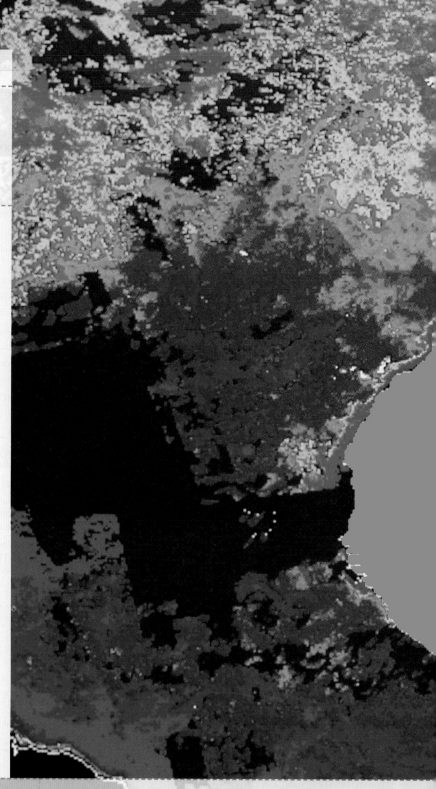

46.1 Too Much of a Good Thing

All organisms require certain elements to build their bodies and carry out metabolic processes. One of those elements, phosphorus, is essential because it is a component of ATP, phospholipids, nucleic acids, and other biological compounds. Plants meet their phosphorus requirement by taking up phosphates dissolved in soil water. Aquatic producers take up phosphates dissolved in the water around them. Animals obtain phosphorus by eating producers or by eating other animals. Thus, phosphorus taken up from the environment passes from one organism to another. When an organism dies and decays, the phosphorus in its body returns to the environment. As a result, phosphorus moves continually from the environment, through organisms, and back to the environment.

In many habitats, a lack of available phosphorus is the main factor limiting producer growth. Thus, fertilizers typically include phosphates. Synthetic phosphate fertilizers are produced by mining rock rich in phosphate and using it to produce phosphoric acid. The acid is then used in fertilizer production.

Phosphate-rich fertilizers boost plant growth, which is why people spread them on lawns and agricultural fields. However, phosphates do not always stay where they were applied. They can run off and pollute rivers or lakes, a consequence depicted in **Figure 46.1**.

Phosphate-containing detergents are another source of phosphate pollution. Compounds such as sodium triphosphate enhance the cleaning power of laundry detergents and dish detergents. Release of wastewater containing these phosphate compounds has the same effect as water polluted by fertilizer.

Addition of nutrients to an aquatic ecosystem is called eutrophication. It can occur slowly by natural processes, or fast as a result of human actions. Because phosphorus is the main limiting factor for algae and cyanobacteria, addition of phosphorus fertilizer to their habitat removes this limitation and allows a population explosion. The result is an algal bloom (Section 21.6) that clouds water and threatens other aquatic species (**Figure 46.2**).

We began using phosphate-rich products when we had no idea of their effects beyond greener lawns and cleaner homes. Today, we confront the simple truth that the daily actions of millions of individuals can disrupt nutrient cycles that have been in operation since long before humans existed. Our species has a unique capacity to shape the environment to our liking. In doing so, we have become major players in the global flows of energy and nutrients even before we fully comprehend how these systems work.

Figure 46.1 Poster created by the Washington State Ecology Department to remind people that phosphates from lawn fertilizer can run off and pollute water.

nitrogen, carbon added

nitrogen, carbon, phosphorus added

Figure 46.2 Effect of phosphate pollution. The photo shows the results of a field experiment in which different combinations of nutrients were added to two artificially separated portions of a lake. Including phosphorus in the mix (lower region) caused an algal bloom that clouded the water.

eutrophication Nutrient enrichment of an aquatic ecosystem.

46.2 The Nature of Ecosystems

- In an ecosystem, energy and nutrients from the environment flow among a community of species.
- Links to Laws of thermodynamics 5.2, Leaching 28.2

Overview of the Participants

Diverse natural systems abound on Earth's surface. In climate, soil type, array of species, and other features, prairies differ from forests, which differ from tundra and deserts. Reefs differ from the open ocean, which differs from streams and lakes. Yet, despite these differences, all systems are alike in many aspects of their structure and function.

We define an ecosystem as an array of organisms and a physical environment, all interacting through a one-way flow of energy and a cycling of nutrients. It is an open system, because it requires ongoing inputs of energy and nutrients to endure (**Figure 46.3**).

All ecosystems run on energy captured by **primary producers.** These autotrophs, or "self-feeders," obtain energy from a nonliving source—generally sunlight—and use it to build organic compounds from inorganic starting materials. Plants and phytoplankton are the main producers. Chapter 6 explains how they capture energy from the sun and use it in photosynthesis building sugars from carbon dioxide and water.

Consumers are heterotrophs that get energy and carbon by feeding on tissues, wastes, and remains of producers and one another. We can describe consumers by their diets. Herbivores eat plants. Carnivores eat the flesh of animals. Parasites live inside or on a living host and feed on its tissues. Omnivores devour both animal and plant materials. **Detritivores,** such as earthworms and crabs, dine on small particles of organic matter, or detritus. **Decomposers** feed on organic wastes and remains and break them down into inorganic building blocks. The main decomposers are bacteria and fungi.

Energy flows one way—into an ecosystem, through its many living components, then back to the physical environment (Section 5.2). Light energy captured by producers is converted to bond energy in organic molecules, which is then released by metabolic reactions that give off heat. This is a one-way process because heat energy cannot be recycled; producers cannot convert heat into chemical bond energy.

In contrast, many nutrients are cycled within an ecosystem. The cycle begins when producers take up hydrogen, oxygen, and carbon from inorganic sources, such as the air and water. They also take up dissolved nitrogen, phosphorus, and other minerals needed to make organic compounds. Nutrients move from

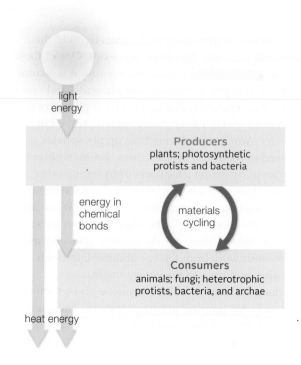

Figure 46.3 Animated Model for ecosystems on land, in which energy flow starts with autotrophs that capture energy from the sun. Energy flows one way, into and out of the ecosystem. Nutrients get cycled among producers and heterotrophs.

producers into the consumers who eat them. After an organism dies, decomposers release nutrients from its tissues into the environment, where they are available to producers once again.

Not all nutrients remain in an ecosystem; typically there are gains and losses. Mineral ions are added to an ecosystem when weathering processes break down rocks, and when winds blow in mineral-rich dust from elsewhere. Leaching and soil erosion remove minerals (Section 28.2). Gains and losses of each mineral tend to balance out over time in most ecosystems.

Trophic Structure of Ecosystems

All organisms of an ecosystem take part in a hierarchy of feeding relationships called **trophic levels** ("troph" means nourishment). When one organism eats another, energy stored in chemical bonds is transferred from the eaten to the eater. All organisms at the same trophic level in an ecosystem are the same number of transfers away from the energy input into that system.

A **food chain** is a sequence of steps by which some energy captured by primary producers is transferred to organisms at successively higher trophic levels. For example, big bluestem grass and other plants are the

Fourth Trophic Level
Third-level consumer

hawk

Third Trophic Level
Second-level consumer

sparrow

Second Trophic Level
Primary consumer

grasshopper

First Trophic Level
Primary producer

big bluestem grass

Figure 46.4 Animated Example of a food chain and corresponding trophic levels in tallgrass prairie, Kansas.

major primary producers in a tallgrass prairie (**Figure 46.4**). They are at this ecosystem's first trophic level. In one food chain, energy flows from bluestem grass to grasshoppers, to sparrows, and finally to hawks. Grasshoppers are primary consumers; they are at the second trophic level. Sparrows that eat grasshoppers are second-level consumers and at the third trophic level. Hawks are third-level consumers, and they are at the fourth trophic level.

At each trophic level, organisms interact with the same sets of predators, prey, or both. Omnivores feed at several levels, so we would partition them among different levels or assign them to a level of their own.

consumer Organism that gets energy and nutrients by feeding on tissues, wastes, or remains of other organisms; a heterotroph.
decomposer Organism that feeds on biological remains and breaks organic material down into its inorganic subunits.
detritivore Consumer that feeds on small bits of organic material.
ecosystem A community interacting with its environment.
food chain Description of who eats whom in one path of energy flow in an ecosystem.
primary producer In an ecosystem, an organism that captures energy from an inorganic source and stores it as biomass; first trophic level.
trophic level Position of an organism in a food chain.

Take-Home Message

What is the trophic structure of an ecosystem?

» An ecosystem includes a community of organisms that interact with their physical environment by a one-way energy flow and a cycling of materials.

» Autotrophs tap into an environmental energy source and make their own organic compounds from inorganic raw materials. They are the ecosystem's primary producers.

» Autotrophs are at the first trophic level of a food chain, a linear sequence of feeding relationships that proceeds through one or more levels of heterotrophs, or consumers.

46.3 The Nature of Food Webs

■ Each food web consists of cross-connecting food chains. Its structure reflects environmental constraints and the inefficiency of energy transfers among trophic levels.

An organism that participates in one food chain usually takes part in others as well. Food chains of an ecosystem cross-connect as a food web. Figure 46.5 shows a small sample of the participants in an arctic food web.

Ecosystems typically support two types of food webs. In grazing food webs, most primary producers are eaten by primary consumers. In detrital food webs, most energy in producers flows directly to detritivores.

Detrital food webs tend to predominate in most land ecosystems. For example, in an arctic ecosystem, primary consumers such as voles, lemmings, and hares eat some living plant parts. However, far more

human (Inuk)

arctic wolf

arctic fox

Higher Trophic Levels

A sampling of carnivores that feed on herbivores and one another

gyrfalcon

snowy owl

ermine

Second Trophic Level

A sampling of primary consumers (herbivores) that eat plants

vole

arctic hare

lemming

mosquito

flea

Parasitic consumers feed at more than one trophic level.

First Trophic Level

Examples of primary producers (plants)

grasses, sedges

purple saxifrage

arctic willow

Detritivores and decomposers (nematodes, annelids, saprobic insects, protists, fungi, bacteria)

Figure 46.5 Animated A very small sampling of organisms in an arctic food web on land.

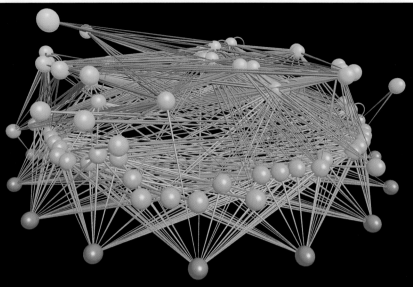

Figure 46.6 Computer model (*right*) for a food web in East River Valley, Colorado (*above*). Balls signify species. Their colors identify trophic levels, with producers (coded *red*) at the bottom and predators (*yellow*) at top. The connecting lines thicken, starting from an eaten species to the eater.

plant matter becomes detritus. Bits of dead plant material sustain detritivores such as roundworms and decomposers that include soil bacteria and fungi.

Grazing food chains tend to predominate in aquatic ecosystems. Zooplankton (heterotrophic protists and tiny animals that drift or swim) consume most of the phytoplankton. Very little phytoplankton ends up on the ocean floor as detritus.

Detrital food chains and grazing food chains interconnect. For example, many animals at higher trophic levels eat both primary consumers and detritivores. Also, after consumers die, their tissues become food for detritivores and decomposers. Decomposers and detritivores also feed on the wastes from consumers.

How Many Transfers?

When ecologists looked at food webs for a variety of ecosystems, they discovered some common patterns. For example, the energy captured by producers usually passes through no more than four or five trophic levels. Even in ecosystems with many species, the number of transfers is limited. Remember that energy transfers are not that efficient (Section 5.2), so energy losses limit the length of a food chain.

Field studies and computer simulations of aquatic and land food ecosystems reveal more patterns. Food chains tend to be shortest in habitats where conditions vary widely over time. Chains tend to be longer in stable habitats, such as the ocean depths where weather has no effect. The most complex webs tend to have a large variety of herbivores, as in grasslands. By comparison, the food webs with fewer connections tend to have more carnivores.

Diagrams of food webs help ecologists predict how ecosystems will respond to change. Neo Martinez and his colleagues constructed the one shown in **Figure 46.6**. By comparing many food webs, they discovered that trophic interactions connect species more closely than ecologists thought. On average, each species in any food web was two links away from all other species. Ninety-five percent of species were within three links of one another, even in large communities with many species. As Martinez concluded in a paper discussing his findings, "Everything is linked to everything else." He cautioned that extinction of any species in a food web may affect many other species.

Take-Home Message

How does energy flow affect food chains and food webs?

» Tissues of living plants and other producers are the basis for grazing food webs. Remains of producers are the basis for detrital food webs.

» Nearly all ecosystems include both grazing food webs and detrital food webs that interconnect as the system's food web.

» The cumulative energy losses from energy transfers between trophic levels limits the length of food chains.

» Even when an ecosystem has many species, trophic interactions link each species with many others.

---

detrital food web Food web in which most energy is transferred directly from producers to detritivores.
food web Set of cross-connecting food chains.
grazing food web Food web in which most energy is transferred from producers to grazers (herbivores).

46.4 Energy Flow

- Primary producers capture energy and take up nutrients, which then move to other trophic levels.
- Links to Lignin 4.11, Endotherms and ectotherms 40.8

Primary Production

The flow of energy through an ecosystem begins with primary production: the rate at which producers capture light energy and convert it into chemical energy. The amount of energy captured by all the producers in an ecosystem is the system's gross primary production. The portion of that energy used for growth and reproduction (rather than for maintenance) is the net primary production of the ecosystem. This is the energy available to the first level consumers.

Any factor that influences the rate of photosynthesis affects primary productivity. As a result, primary production differs among habitats and often varies seasonally (Figure 46.7). Per unit area, the net primary production on land tends to be higher than that in the oceans. However, because oceans cover about 70 percent of Earth's surface, nearly half of the global net primary production comes from marine producers.

Ecological Pyramids

Ecologists often represent the trophic structure of an ecosystem in the form of ecological pyramids. In such diagrams, primary producers collectively form a base for successive tiers of consumers above them.

A biomass pyramid illustrates the dry weight of all organisms at each trophic level. Figure 46.8 shows the biomass pyramid for Silver Springs, an aquatic ecosystem in Florida. Most commonly, primary producers make up most of the biomass in a pyramid, and top carnivores make up very little. If you visited Silver Springs, you would see a lot of aquatic plants but very few gars (the fish that are main top predator in this ecosystem). Similarly, when you walk through a prairie, you see more grams of grass than of hawks.

However, in some aquatic environments, the lowest tier of the biomass pyramid can also be the narrowest. In these environments, the main producers are bacteria and single-celled protists, which reproduce fast, rather than investing in building a big body. Because of their quick turnover, a smaller biomass of phytoplankton can support a greater biomass of zooplankton.

An energy pyramid illustrates how the amount of usable energy diminishes as it is transferred through an ecosystem. Sunlight energy is captured at the base (the primary producers) and declines with successive levels to its tip (the top carnivores). Energy pyramids are always "right-side-up," with their largest tier at the bottom. Such pyramids depict energy flow per unit of water (or land) per unit of time. Figure 46.9 shows the energy pyramid for the Silver Springs ecosystem and the energy flow that this pyramid represents.

A Summary of annual net primary productivity on land and in the oceans.

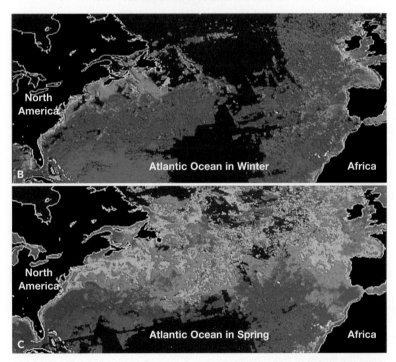

B,C Seasonal changes in net primary productivity of the North Atlantic Ocean.

Figure 46.7 Satellite data showing net primary production. Productivity is coded as *red* (highest) down through *orange*, *yellow*, *green*, *blue*, and *purple* (lowest).

biomass pyramid Diagram that depicts the biomass (dry weight) in each of an ecosystem's trophic levels.
energy pyramid Diagram that depicts the energy that enters each of an ecosystem's trophic levels. Lowest tier of the pyramid, representing primary producers, is always the largest.
primary production The rate at which an ecosystem's producers capture and store energy.

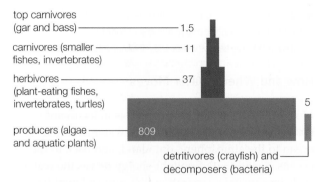

top carnivores (gar and bass) — 1.5

carnivores (smaller fishes, invertebrates) — 11

herbivores (plant-eating fishes, invertebrates, turtles) — 37

producers (algae and aquatic plants) — 809

5

detritivores (crayfish) and decomposers (bacteria)

Figure 46.8 Biomass (in grams per square meter) for Silver Springs, a freshwater aquatic ecosystem in Florida. In this system, primary producers make up the bulk of the biomass.

Ecological Efficiency

Anywhere between 5 and 30 percent of the energy in the tissues of organisms at one trophic level ends up in the tissues of those at the next trophic level. Several factors influence the efficiency of transfers. First, not all energy harvested by consumers is used to build biomass because some is lost as metabolic heat. Second, some components of a body may be unavailable to a consumer. The lignin and cellulose that reinforce bodies of most land plants pass undigested through some consumers, including humans. Similarly, many animals have some biomass tied up in an internal or external skeleton and hair, feathers, or fur—all of which are difficult for carnivores to digest.

Ecological efficiency is usually higher in aquatic ecosystems than on land, in part because the main aquatic producers—photosynthetic protists—do not make difficult-to-digest lignin as most land plants do. In addition, aquatic ecosystems usually have a higher proportion of ectotherms (such as fish). This improves ecological efficiency because ectotherms lose less energy as heat than endotherms do.

Take-Home Message

How does energy flow through ecosystems?

» Primary producers capture energy and convert it into biomass. We measure this process as primary production.

» A biomass pyramid depicts dry weight of organisms at each trophic level in an ecosystem. Its largest tier is usually producers, but the pyramid for some aquatic systems is inverted.

» An energy pyramid depicts the amount of energy that enters each level. Its largest tier is always at the bottom (producers).

» The efficiency of energy transfers tends to be greatest in aquatic systems, where primary producers usually lack lignin and consumers tend to be ectotherms.

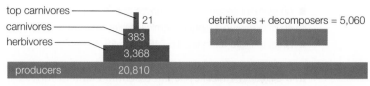

top carnivores — 21
carnivores — 383
herbivores — 3,368
producers — 20,810

detritivores + decomposers = 5,060

A Energy pyramid for the Silver Springs ecosystem. The width of each tier in the pyramid represents the amount of energy that enters each trophic level annually, as shown in detail below.

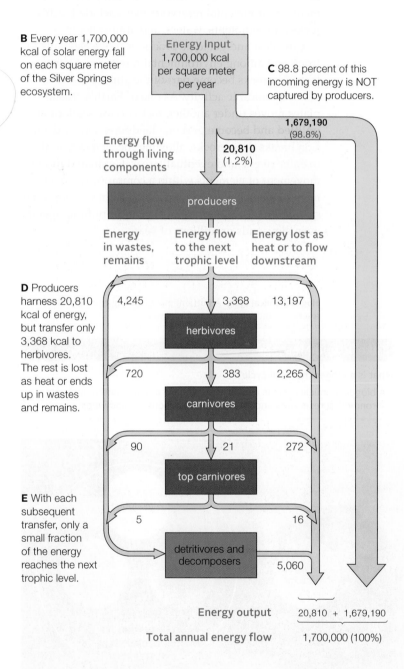

B Every year 1,700,000 kcal of solar energy fall on each square meter of the Silver Springs ecosystem.

Energy Input
1,700,000 kcal per square meter per year

C 98.8 percent of this incoming energy is NOT captured by producers.

1,679,190 (98.8%)

Energy flow through living components

20,810 (1.2%)

producers

Energy in wastes, remains | Energy flow to the next trophic level | Energy lost as heat or to flow downstream

D Producers harness 20,810 kcal of energy, but transfer only 3,368 kcal to herbivores. The rest is lost as heat or ends up in wastes and remains.

4,245 — herbivores — 3,368 — 13,197

720 — carnivores — 383 — 2,265

90 — top carnivores — 21 — 272

E With each subsequent transfer, only a small fraction of the energy reaches the next trophic level.

5 — detritivores and decomposers — 16

5,060

Energy output   20,810 + 1,679,190

Total annual energy flow   1,700,000 (100%)

Figure 46.9 Animated Annual energy flow in Silver Springs measured in kilocalories (kcal) per square meter per year. **Figure It Out:** What percent of the energy carnivores received directly from herbivores was later passed on to top carnivores?

Answer: 21/383 × 100 = 5.5 percent

46.5 Biogeochemical Cycles

- An element essential to life moves between a community and its environment in a biogeochemical cycle.
- Links to Plate tectonics 16.7, Erosion 28.2

In a **biogeochemical cycle,** an essential element moves from one or more environmental reservoirs, through the living component of an ecosystem, and then back to the reservoirs (**Figure 46.10**). Depending on the element, environmental reservoirs may include Earth's rocks and sediments, waters, and atmosphere.

Chemical and geologic processes move elements to, from, and among environmental reservoirs. Elements locked in rocks become part of the atmosphere as a result of volcanic activity. As one of Earth's crustal plates moves under another, rocks on the seafloor are uplifted and become part of a landmass. On land, erosion breaks down rocks, allowing the elements in them to enter rivers, and eventually seas. Compared to the movement of elements within a community, movement of elements among nonbiological reservoirs is far slower. Processes such as erosion and uplifting operate over thousands or millions of years.

biogeochemical cycle A nutrient moves among environmental reservoirs and into and out of food webs.

Take-Home Message

What is a biogeochemical cycle?

» A biogeochemical cycle is the slow movement of a nutrient among its environmental reservoirs and into, through, and out of food webs.

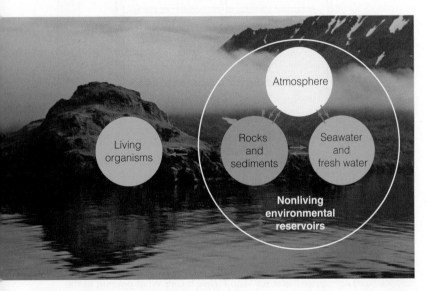

Figure 46.10 Generalized biogeochemical cycle. For all nutrients, the cumulative amount in all environmental reservoirs far exceeds the amount in living organisms.

46.6 The Water Cycle

- Water makes up the bulk of all organisms and serves as a transport medium for many soluble nutrients.
- Links to Properties of water 2.5, Transpiration 28.4

How and Where Water Moves

Most of Earth's water (97 percent) is in its oceans (**Table 46.1**). The **water cycle** moves water from the ocean to the atmosphere, onto land, and back to the oceans (**Figure 46.11**). Sunlight energy drives the water cycle by causing evaporation, the conversion of liquid water to water vapor. Water vapor that enters the cool upper layers of the atmosphere condenses into droplets, forming clouds. When droplets get large and heavy enough, they fall as precipitation—as rain, snow, or hail.

Oceans cover about 70 percent of Earth's surface, so most rainfall returns water directly to the oceans. On land, we define a **watershed** as an area in which all precipitation drains into a specific waterway. A watershed may be as small as a valley that feeds a stream, or as large as the 5.88 million square kilometers of the Amazon River Basin. The Mississippi River Basin watershed includes 41 percent of the continental United States.

Most precipitation that enters a watershed seeps into the ground. Some of this water remains between soil particles as **soil water**. Plant roots can tap into this water source. Soils differ in water holding capacity, with clay-rich soils holding the most water and sandy soils the least. Water that drains through soil layers often collects in **aquifers**. These natural underground reservoirs consist of porous rock layers that hold water. **Groundwater** is water in soil and aquifers. Water that falls on impermeable rock or on saturated soil becomes **runoff.** It flows over the ground into streams. The flow of groundwater and surface water slowly returns water to oceans.

The movement of water causes the movement of other nutrients. Carbon, nitrogen, and phosphorus all have soluble forms that can be moved from place to place by flowing water. As water trickles through soil, it carries nutrient particles from topsoil into deeper soil layers. As a stream flows over limestone, water slowly dissolves the rock and carries carbonates back

aquifer Porous rock layer that holds some groundwater.
groundwater Soil water and water in aquifers.
runoff Water that flows over soil into streams.
soil water Water between soil particles.
water cycle Movement of water among Earth's atmosphere, oceans, and the freshwater reservoirs on land.
watershed Land area that drains into a particular stream or river.

Reservoir	Volume (10^3 km^3)
Ocean	1,370,000
Polar ice, glaciers	29,000
Groundwater	4,000
Lakes, rivers	230
Atmosphere (water vapor)	14

Table 46.1 Water Reservoirs

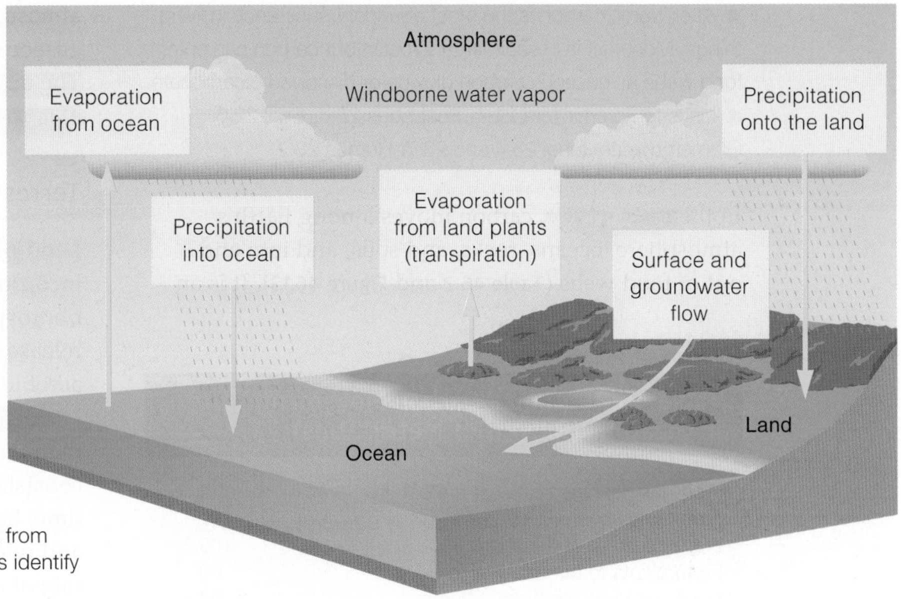

Atmosphere
Evaporation from ocean
Windborne water vapor
Precipitation onto the land
Precipitation into ocean
Evaporation from land plants (transpiration)
Surface and groundwater flow
Land
Ocean

Figure 46.11 Animated The water cycle. Water moves from the ocean to the atmosphere, land, and back. The arrows identify processes that move water.

to the seas where the limestone formed. On a less natural note, runoff from heavily fertilized lawns and agricultural fields carries dissolved phosphates and nitrates into streams and lakes.

Limited Fresh Water

The vast majority of Earth's water is too salty to drink or irrigate crops. If all Earth's water filled a bathtub, the amount of fresh water that could be used each year without decreasing the overall supply would fill a teaspoon. In addition, most fresh water is frozen as ice.

Groundwater supplies drinking water to about half of the United States population. Overdrafts from aquifers are common; water is drawn from aquifers faster than natural processes replenish it. When too much fresh water is withdrawn from a coastal aquifer, saltwater moves in and replaces it. Figure 46.12 shows areas of aquifer depletion and saltwater intrusion. Overdrafts have reduced the volume of water in the Ogallala aquifer by about half. This aquifer stretches from South Dakota to Texas and supplies irrigation water for about 20 percent of the nation's crops. For the past thirty years, withdrawals have exceeded replenishment by a factor of ten.

In the United States, about 80 percent of the water withdrawn from the environment for human use ends up irrigating agricultural fields. Thus, improving the efficiency of irrigation methods would go a long way toward ensuring that future generations have an adequate and safe water supply.

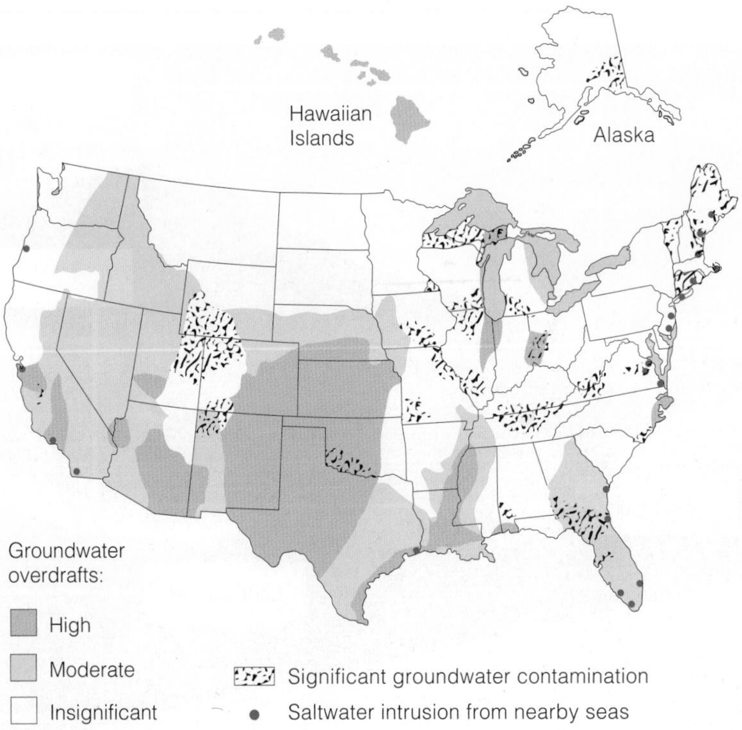

Hawaiian Islands
Alaska
Groundwater overdrafts:
High
Moderate
Insignificant
Significant groundwater contamination
Saltwater intrusion from nearby seas

Figure 46.12 Groundwater troubles in the United States.

Take-Home Message

What is the water cycle and how do human activities affect it?

» Water moves slowly from the world ocean—the main reservoir—through the atmosphere, onto land, then back to the ocean.

» Fresh water makes up only a tiny portion of the global water supply. Excessive water withdrawals threaten many sources of drinking water.

46.7 The Carbon Cycle

■ After water, carbon is the most abundant substance in living things. Most of it is in sedimentary rocks, but carbon can enter food webs as gaseous carbon dioxide or dissolved bicarbonate.
■ Links to Foraminifera 21.4, Peat bogs 22.4, Coal 22.6, Glomeromycete fungi 23.4 and 23.7, Humus 28.2

In the carbon cycle, carbon moves among Earth's atmosphere, oceans, rocks, and soils, and into and out of food webs (Table 46.2 and Figure 46.13). It is an

atmospheric cycle, a biogeochemical cycle in which a gaseous form of the element plays a significant role. The atmosphere holds about 760 gigatons (billion tons) of carbon, mainly in the form of carbon dioxide (CO_2).

Terrestrial Carbon Cycle

Land plants take up CO_2 from the atmosphere and incorporate it into their tissues when they carry out photosynthesis. Plants and other land organisms release CO_2 into the atmosphere by the process of aerobic respiration.

Soil contains at least 1600 gigatons of carbon, more than twice as much as the atmosphere. Soil carbon consists of humus and living soil organisms. Over time, bacteria and fungi in the soil decompose humus and release carbon dioxide into the atmosphere. The rate of decomposition increases with temperature. In the tropics, decomposition proceeds rapidly, so most carbon is stored in living plants, rather than in soil

Table 46.2 Annual Carbon Movement in Gigatons (Billions of Metric Tons)	
From atmosphere to plants (carbon fixation)	120
From atmosphere to ocean	107
From ocean to atmosphere	105
From plants to atmosphere	60
From soil to atmosphere	60
From land to ocean in runoff	0.4
Burial in ocean sediments	0.1

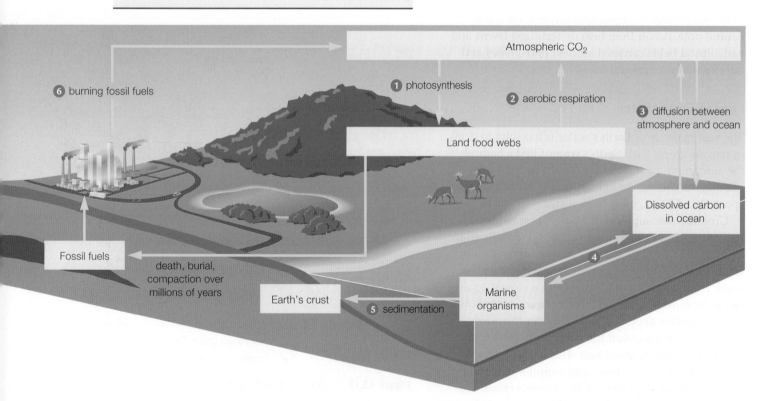

Figure 46.13 Animated The carbon cycle. Most carbon is in Earth's rocks, where it is largely unavailable to living organisms.

❶ Carbon enters land food webs when plants take up carbon dioxide from the air for use in photosynthesis.

❷ Carbon returns to the atmosphere as carbon dioxide when plants and other land organisms carry out aerobic respiration.

❸ Carbon diffuses between the atmosphere and the ocean. Bicarbonate forms when carbon dioxide dissolves in seawater.

❹ Marine producers take up bicarbonate for use in photosynthesis, and marine organisms release carbon dioxide from aerobic respiration.

❺ Many marine organisms incorporate carbon into their shells. After they die, these shells become part of the sediments. Over time, the sediments become carbon-rich rocks such as limestone and chalk in Earth's crust.

❻ Burning of fossil fuels derived from the ancient remains of plants puts additional carbon dioxide into the atmosphere.

	Tropical forest	Temperate forest	Temperate grassland	Cropland
Plants:	16,500 g/m²	8,000 g/m²	720 g/m²	200 g/m²
Soil:	8,300 g/m²	12,000 g/m²	23,600 g/m²	7,900 g/m²

Figure 46.14 Estimates of carbon stored in the plants and soils of different ecosystems.

(Figure 46.14). By contrast, in temperate zone forests and grasslands, soil holds more carbon than living plants. Soil is most carbon-rich in the arctic, where low temperature hampers decomposition, and in peat-bogs, where acidic, anaerobic conditions do the same (Section 22.4).

Conversion of a forest or grassland to cropland reduces the amount of carbon both above and below ground. Tilling soil (mechanically mixing it) increases the rate at which carbon enters the air because it destroys the hyphae of glomeromycete fungi, a type of mycorrhizal fungi (Section 23.7). These hyphae secrete a gluelike glycoprotein (glomalin) that normally holds onto organic material and slows its decomposition.

Marine Carbon Cycle

Seawater holds about 40,000 gigatons of carbon, about fifty times as much as the atmosphere. The main form of carbon in seawater is bicarbonate (HCO_3^-), an ion that forms when CO_2 dissolves in water. Marine producers take up bicarbonate and convert it to CO_2 for use in photosynthesis. Marine organisms release CO_2 produced by aerobic respiration into the water. Some marine organisms such as foraminifera, shelled mollusks, and reef-building corals, also store carbon in calcium carbonate hardened parts.

Marine sediments and sedimentary rocks are Earth's largest carbon reservoir, with more than 65 gigatons. Limestone and other sedimentary rocks form when the calcium carbonate–rich shells of marine

atmospheric cycle Biogeochemical cycle in which a gaseous form of an element plays a significant role.
carbon cycle Movement of carbon, mainly between the oceans, atmosphere, and living organisms.

organisms become compacted over millions of years. Such rocks become part of land ecosystems when movements of tectonic plates uplift portions of the seafloor. Carbon in rocks is not available to producers, so this vast store of carbon has little effect on ecosystems. The geological part of the cycle is completed as erosion breaks down rocks, and rivers carry dissolved carbon to the sea.

Carbon in Fossil Fuels

Fossil fuels such as coal, oil, and natural gas hold an estimated 5,000 gigatons of carbon. Deposits of fossil fuels formed over hundreds of millions of years from carbon-rich remains. High pressure and temperature transformed the remains of ancient land plants to coal (Section 22.6). A similar process transformed the remains of plankton to deposits of oil and natural gas. Until recently, the carbon in fossil fuels, like the carbon in rocks, had little impact on ecosystems. Currently, our burning of this fuel adds billions of tons of CO_2 to the atmosphere every year.

Take-Home Message

How does carbon cycle between its main reservoirs?

» The largest reservoir is sedimentary rock. Carbon moves into and out of this reservoir over very long time spans and is not available to organisms.

» Seawater is the largest reservoir of biologically available carbon. Marine producers take up bicarbonate and convert it to CO_2 for use in photosynthesis.

» On land, large amounts of carbon are stored in soil, especially in arctic regions and in peat bogs.

» The atmosphere holds less carbon than rocks, seawater, or soil. It serves as the source of CO_2 for land producers. Burning of fossil fuels adds carbon to this reservoir.

46.8 Greenhouse Gases and Climate Change

■ Concentrations of gases in Earth's atmosphere help determine the temperature near Earth's surface.

■ Links to Carbon imbalances 6.1, Foraminifera 21.4

The Greenhouse Effect

Carbon dioxide is a greenhouse gas, an atmospheric gas whose ability to absorb and reradiate heat energy helps keeps Earth warm enough to sustain life. The mechanism by which this occurs is referred to as the greenhouse effect (Figure 46.15). When radiant energy from the sun reaches Earth's atmosphere, some energy is reflected back into space ❶. However, more energy passes through the atmosphere and warms Earth's surface ❷. When the warmed surface radiates heat energy, greenhouse gases absorb some of that heat, then emit a portion of it back toward Earth ❸. If greenhouse gases did not exist, heat energy emitted by Earth's surface would escape into space, leaving the planet cold and lifeless.

Changing Carbon Dioxide Concentrations

In 1959, researchers began to measure the atmospheric concentration of CO_2 at an observatory near the top of Mauna Loa, Hawaii's highest volcano. The remote site 3,500 meters (11,000 feet) above sea level was chosen because it is almost free of local airborne contamination and is representative of atmospheric conditions for the Northern Hemisphere. For the first time, researchers saw the effects of carbon dioxide fluctuations for the entire hemisphere. The troughs and peaks seen in the line in Figure 46.16A are annual lows and highs in atmospheric CO_2. The level of CO_2 declines in summer, when the most CO_2 is taken up

for photosynthesis. It rises in winter, when photosynthesis declines but aerobic respiration and fermentation continue.

The researchers also noticed a trend: The average annual level of CO_2 is increasing. As additional sites around the world began to monitor atmospheric CO_2, they too detected an ongoing rise.

To put the current increase in perspective, scientists consider historical changes in atmospheric CO_2. Glacial ice provides one window into the past. Such ice forms when snow falls, then is compressed by new snowfalls above it. In some regions, layers of ice have been laid down one atop the next for hundreds of thousands of years. The result is sheets of ice more than a kilometer thick. To determine what conditions were like in the past, scientists use a hollow drill to remove a long ice core. Air bubbles trapped at different depths within the ice provide snapshots of atmospheric conditions at the time that ice formed. So far, analysis of ice cores has provided a history of atmospheric changes that extends back 800,000 years.

Fossil foraminiferan shells provide information about CO_2 levels in the even more distant past. Foraminifera take up carbon and other elements from seawater and incorporate them into their shells. When atmospheric CO_2 is high, more of it dissolves in the ocean's surface waters, and this increase affects the composition of foraminiferan shells. By studying fossilized shells, scientists can trace changes in atmospheric CO_2 over many millions of years.

Data from glacial ice, foraminiferan fossils, and other sources consistently show that atmospheric CO_2 has risen and fallen many times. These data also show that the current CO_2 concentration is the highest in at least 15 million years.

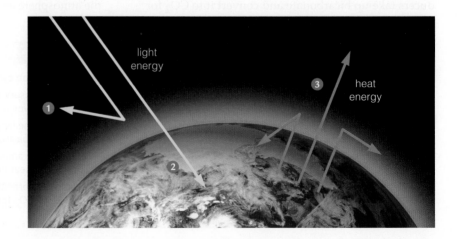

Figure 46.15 Greenhouse effect.

❶ Earth's atmosphere reflects some sunlight energy back into space.

❷ More light energy reaches and warms Earth's surface.

❸ Earth's warmed surface emits heat energy. Some of this energy escapes through the atmosphere into space. The rest is absorbed and then emitted in all directions by greenhouse gases. Some emitted heat warms Earth's surface and lower atmosphere.

Figure It Out: Do greenhouse gases reflect heat energy toward the Earth?

Answer: No. The gases absorb heat energy, then reemit it in all directions.

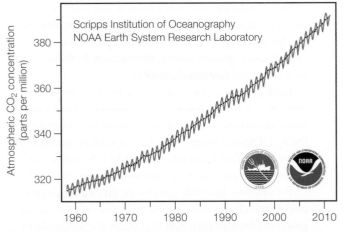

A Atmospheric CO_2 concentration measured at Mauna Loa Observatory. The *red* line shows seasonal highs and lows. *Black* is yearly averages.

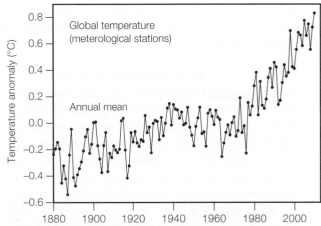

B Global annual mean air temperature based on measurements from meteorological stations. The temperature anomaly is the deviation from the average temperature in the period 1951–1980.

Figure 46.16 Directly measured changes in atmospheric carbon dioxide and global temperature. Source: NASA Goddard Institute for Space Studies.

Changing Climate

Given the greenhouse effect, we would predict that increases in the atmospheric concentration of carbon dioxide and other greenhouse gases would raise the temperature of Earth's surface. In fact, we are in the midst of global climate change, a long-term alteration of Earth's climate. Global warming, an increase in Earth's average surface temperature, is one aspect of this change (**Figure 46.16B**).

Earth's climate has varied greatly over its long history. During ice ages, much of the planet was covered by glaciers. Other periods were warmer than the present, and tropical plants and coral reefs thrived at what are now cool latitudes. Scientists can correlate some historical large-scale temperature changes with shifts in Earth's orbit, which varies in a regular fashion over 100,000 years, and Earth's tilt, which varies over 40,000 years. Changes in solar output and volcanic eruptions also affect Earth's temperature. However, most scientists see no correlation between these factors and the current temperature rise.

In 2007, the Intergovernmental Panel on Climate Change reviewed the results of many scientific studies related to climate change. The panel included hundreds of scientists from all over the world. After reviewing the data, the panel concluded that it is very

likely that a human-induced increase in atmospheric greenhouse gases is responsible for the current warming trend.

Temperature of the land and seas affects evaporation, winds, and currents, so many weather patterns will change as temperature rises. For example, warmer temperatures are correlated with extremes in rainfall patterns: periods of drought interrupted by unusually heavy rains. Rising temperature also elevates sea level. These and other effects of global climate change are the focus of Section 48.7.

Looking forward, atmospheric CO_2 is expected to continue to rise as large nations such as China and India become increasingly industrialized. However, efforts are under way to reduce the damage by increasing the efficiency of processes that require fossil fuels, shifting to alternative energy sources that do not release carbon, such as solar and wind power, and developing innovative ways to store carbon dioxide.

global climate change A long-term change in Earth's climate.
greenhouse gas Atmospheric gas that absorbs heat emitted by Earth's surface and remits it, thus keeping the planet warm.

Take-Home Message

How does carbon dioxide affect climate?

» Carbon dioxide is one of the atmospheric gases that absorb heat and emit it toward Earth's surface, thus keeping the planet warm enough for life.

» The CO_2 level of the atmosphere is currently rising, most likely as a result of human activity, and global mean temperature is rising with it. Changes in temperature affect other climate factors such as patterns of rainfall.

46.9 Nitrogen Cycle

- Gaseous nitrogen makes up about 80 percent of the lower atmosphere, but most organisms cannot use this gaseous form.
- Links to Nitrogen-fixing bacteria 20.7 and 28.3

Nitrogen moves in an atmospheric cycle known as the nitrogen cycle (Figure 46.17). The main reservoir is the atmosphere, which is about 80 percent nitrogen. Triple covalent bonds hold two atoms of nitrogen gas together (N_2, or $N \equiv N$). All organisms need nitrogen to build DNA, RNA and proteins. However, most producers cannot use gaseous nitrogen because they cannot break these bonds. Only certain bacteria can carry out nitrogen fixation. They break the bonds in N_2 and use the nitrogen atoms to form ammonia, which dissolves to form ammonium (NH_4^+) ❶. Nitrogen-fixing cyanobacteria live in aquatic habitats, soil, and as components of lichens. Another group of nitrogen-fixing bacteria forms nodules on the roots of legumes.

ammonification Breakdown of nitrogen-containing organic material resulting in the release of ammonia and ammonium ions.
denitrification Conversion of nitrates or nitrites to gaseous forms of nitrogen.
nitrification Conversion of ammonium to nitrates.
nitrogen cycle Movement of nitrogen among the atmosphere, soil, and water, and into and out of food webs.
nitrogen fixation Incorporation of nitrogen from nitrogen gas into ammonia.

Plants take up ammonium from soil water ❷ and use it in metabolic reactions. Consumers obtain nitrogen by eating plants or one another. Bacterial and fungal decomposers return ammonium to the soil by a process called ammonification ❸.

Nitrification is a two-step process that converts ammonium to nitrates ❹. First, ammonia-oxidizing bacteria and archaea convert ammonium to nitrite (NO_2^-). Other bacteria then take up nitrite and use it in reactions that form nitrates (NO_3^-). Nitrates, like ammonium, are taken up and used by producers ❺.

Ecosystems lose nitrogen by denitrification, a process in which denitrifying bacteria convert nitrate to gaseous forms of nitrogen that enter the atmosphere ❻. Denitrification requires anaerobic conditions, so it typically takes place in deep soil or aquatic sediments.

Take-Home Message

How does nitrogen cycle in ecosystems?

» The atmosphere is the main reservoir for nitrogen, but only nitrogen-fixing bacteria can access it.

» Plants take up ammonium and nitrates from the soil. Ammonium is released by nitrogen-fixing bacteria and by fungal and bacterial decomposers. Bacteria and archaea produce nitrates.

» Nitrogen returns to the atmosphere when denitrifying bacteria convert soluble forms of nitrogen to nitrogen gas.

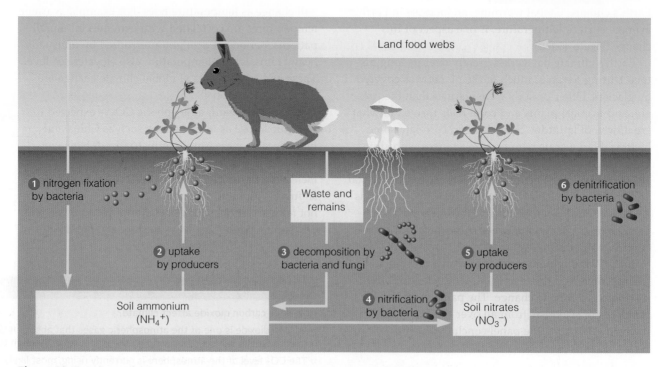

Figure 46.17 Animated Nitrogen cycle on land. Nitrogen becomes available to plants through the activities of nitrogen-fixing bacteria. Other bacterial species cycle nitrogen to plants. They break down organic wastes to ammonium and nitrates.

46.10 Disruption of the Nitrogen Cycle

■ Burning fossil fuels and use of industrially produced nitrogen fertilizers have altered the nitrogen cycle.
■ Links to Ozone layer 19.1, Thyroid gland 34.7

In the early 1900s, German scientists discovered a method of fixing atmospheric nitrogen and producing ammonium on an industrial scale. This process allowed the manufacture of synthetic nitrogen fertilizers that have boosted crop yields and help to feed the rapidly increasing human population. However, use of fertilizers, along with other human activities, has added large amounts of nitrogen-containing compounds to our air and water. Here we consider two of these compounds and their effects.

Figure 46.18 Farmer with a tank of ammonia fertilizer. Ammonia is sprayed into the soil to supply nitrogen and boost crop yield. Bacteria in soil and water can convert ammonia to nitrous oxide that escapes into the air.

Nitrous Oxide—A Double Threat

Air bubbles in ice cores reveal that the atmospheric concentration of nitrous oxide (N_2O) remained about 270 parts per billion (ppb) for at least a thousand years before the industrial revolution. It is now about 325 ppb and rising. The ongoing increase is attributed mainly to burning of fossil fuels, use of synthetic nitrogen fertilizers, and industrial livestock production. Burning fossil fuel releases N_2O directly into the air. Chemical fertilizers and manure from livestock increase atmospheric N_2O by encouraging the growth of bacteria that release this gas (**Figure 46.18**).

An increased concentration of atmospheric N_2O is a matter of concern for two reasons. First, N_2O is a greenhouse gas, and a highly persistent and effective one. It can remain in the atmosphere for more than 100 years, and it has a warming potential 300 times that of CO_2. Second, N_2O contributes to destruction of the ozone layer. As Section 19.1 explained, ozone high in the atmosphere protects life at Earth's surface from the damaging effects of ultraviolet radiation. We discuss ozone destruction in more detail in Section 48.6.

Nitrate Pollution

Nitrate (NO_3^-) is a common water pollutant and it often contaminates sources of drinking water. Nitrates from fertilizers and animal manure run off into streams and lakes or leach (travel down through the soil) into groundwater. In communities that lack a public sewer system, wastewater from homes also contributes to nitrate pollution. In such communities, wastewater from each building flows into an underground septic system. Ideally, bacteria in the system remove most nitrates before the water seeps into the surrounding soil. However, septic system malfunctions

and leaks, which often go undetected, can release nitrate-rich wastewater directly into the ground. By contrast, a sewer system collects wastewater from all buildings in a community and treats it at a central facility before discharging it.

Nitrate has negative effects on human health. Ingested nitrate inhibits iodine uptake by the thyroid gland and may increase the risk of thyroid cancer. A recent study of women in Iowa correlated a high level of nitrate in drinking water with an elevated incidence of thyroid cancer. Nitrate-contaminated drinking water has also been correlated with an increased risk of other cancers, as well as respiratory infections, diabetes, and other disorders.

The Environmental Protection Agency (EPA) mandates that public drinking water contain less than 10 ppm nitrate and requires periodic testing to ensure this standard is met. When water from a well or other public source exceeds this standard, the water must be treated to remove nitrates. Water from private wells is unregulated and thus more likely to have a high nitrate concentration. Owners of private wells are advised to have the nitrate concentration of their water checked regularly, especially in agricultural regions where nitrate pollution from fertilizer is most likely.

Take-Home Message

How have human activities disrupted the nitrogen cycle?

» Burning fossil fuels releases nitrous oxide (N_2O) into the air. Nitrous oxide is a greenhouse gas and it contributes to destruction of the ozone layer.

» Use of synthetic fertilizer encourages the production of nitrous oxide by bacteria. It is also a source of nitrates, which pollute drinking water.

» Wastewater that escapes from septic systems is another source of nitrate pollution.

46.11 The Phosphorus Cycle

- Phosphorus does not occur in gaseous form in habitats that can support life.
- Like nitrogen, phosphorus is taken up by plants only in ionized form, and it, too, is often a limiting factor on plant growth.

Atomic phosphorus is highly reactive, so phosphorus does not occur naturally in its elemental form. Most of Earth's phosphorus is bonded to oxygen as phosphate (PO_4^{3-}), an ion that occurs in rocks and sediments. In the phosphorus cycle, phosphorus passes quickly through food webs as it moves from land to ocean sediments, then slowly back to land (Figure 46.19). Unlike carbon and nitrogen, little phosphorus exists in a gaseous form and its major reservoir is sedimentary rock, so the phosphorus cycle is called a sedimentary cycle.

Weathering and erosion move phosphates from rocks into soil, lakes, and rivers ❶. Leaching and run-off carry dissolved phosphates to the ocean ❷. Here, most phosphorus comes out of solution and settles as deposits along continental margins ❸. Slow movements of Earth's crust can uplift these deposits onto land ❹, where weathering releases phosphates from

phosphorus cycle Movement of phosphorus among Earth's rocks and waters, and into and out of food webs.
sedimentary cycle Biochemical cycle in which the atmosphere plays little role and rocks are the major reservoir.

rocks once again. Phosphate-rich rocks are mined for use in the industrial production of fertilizer.

All organisms require phosphorus. It is a component of all nucleic acids and phospholipids. The biological portion of the phosphorus cycle begins when producers take up phosphate. Land plants take up dissolved phosphate from the soil water ❺. Land animals get phosphates by eating the plants or one another. Phosphorus returns to the soil in the wastes and remains of organisms ❻. Phosphate-rich droppings from seabird or bat colonies are collected and used as a natural fertilizer. Manure from livestock is also rich in phosphorus.

In the seas, phosphorus enters food webs when producers take up phosphate dissolved in seawater ❼. As on land, wastes and remains continually replenish the phosphorus supply ❽.

Take-Home Message

How does phosphorus cycle in ecosystems?

» Rocks are the main phosphorus reservoir. Weathering puts phosphates into water. Producers take up dissolved phosphates.

» Phosphate-rich wastes are a natural fertilizer, and phosphate from rocks can be used to produce fertilizer on an industrial scale.

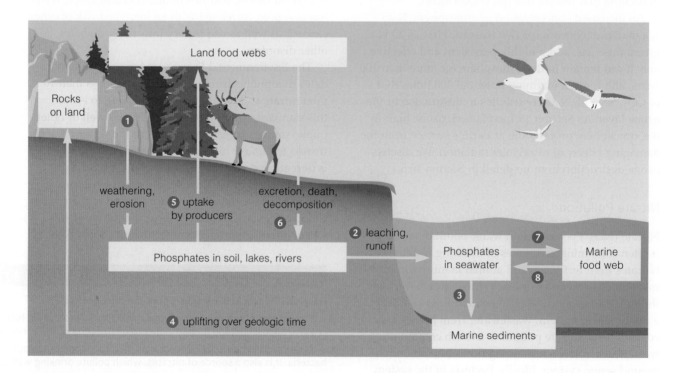

Figure 46.19 Animated The phosphorus cycle. In this sedimentary cycle, phosphorus moves mainly in the form of phosphate ions (PO_4^{3-}).

Too Much of a Good Thing (revisited)

Farmers use industrially produced fertilizers because such products are a relatively inexpensive way to enhance crop yield. However, nitrates and phosphates from these fertilizers can pollute waters, both locally and in distant regions. Fertilizer applied to fields in the Midwest not only pollutes local drinking water, it also flows into streams that feed the Mississippi River. The river then delivers its heavy load of phosphates and nitrates to the Gulf of Mexico.

Most excess phosphates and nitrogen entering the Mississippi come from agriculture, but suburbanites and city dwellers also contribute. In some communities, the sewer system receives both sewage and water from storm drains. Heavy rains can overwhelm such a system, causing an overflow that delivers raw sewage into streams. Rain-related overflows can be prevented by a system that keeps sewage and water from storm drains separate. However, such systems typically do not treat water from storm drains before discharging it. Thus, when fertilizer from lawns enters such a system, it is delivered untreated into the river.

Each summer, the excessive nutrient inputs from the Mississippi into the Gulf of Mexico result in formation of a "dead zone," an area of deep water where the oxygen level is too low to support most marine organisms. In 2010, the dead zone extended across 20,000 square kilometers (7,722 square miles).

The dead zone forms after eutrophication causes an algal bloom (Section 21.6). The high nutrient level encourages a population explosion of marine algae. When the algae die, their remains drift down to the seafloor. Decomposition of these remains by bacteria then depletes the water of oxygen, making it impossible for most animals to survive. Fish can swim away from the low-oxygen zone, but less mobile animals suffocate. There are also indirect effects, as when an increase in jellyfish leads to a decrease in the fish larvae on which the jellies prey. Jellyfish are among the few animals that thrive in anoxic waters.

How would you vote? Opponents of banning the sale of phosphate-rich detergents and fertilizers to homeowners note that most nutrient pollution comes from agriculture and municipal sewage systems. Do you support bans on high-nutrient consumer products as a way to help reduce eutrophication of our waters?

LEARNING ROADMAP

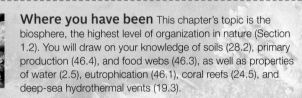

Where you have been This chapter's topic is the biosphere, the highest level of organization in nature (Section 1.2). You will draw on your knowledge of soils (28.2), primary production (46.4), and food webs (46.3), as well as properties of water (2.5), eutrophication (46.1), coral reefs (24.5), and deep-sea hydrothermal vents (19.3).

Where you are now

Air Circulation Patterns
Air circulation patterns start with regional differences in energy inputs from the sun and are affected by Earth's rotation and the distribution of land and seas.

Effects of Ocean and Landforms
Ocean currents distribute heat and moisture and thus influence regional climates. Landforms such as mountains can affect climate by altering air flow patterns.

Land Biomes
Biomes are discontinuous regions characterized mainly by the dominant vegetation. Sunlight intensity, moisture, soil, and evolutionary history vary among biomes.

Aquatic Ecosystems
Water covers most of Earth's surface. Light availability, temperature, and dissolved gases influence primary productivity of aquatic ecosystems.

El Niño Southern Oscillation
Interactions among the atmosphere and ocean can cause cyclical changes in climate that can have effects on human well-being.

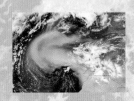

Where you are going In the next chapter, we discuss the many ways in which human activities have disrupted the biosphere. Among other topics, we consider erosion of grasslands, expansion of deserts, the effects of acid rain on temperate forests and lakes, loss of tropical forests, and pollution of the ocean by plastics.

47.1 Warming Water and Wild Weather

Professional surfer Ken Bradshaw has ridden a lot of waves, but one stands out in his memory. In January of 1998, he found himself off the coast of Hawaii riding the biggest wave he had ever seen (**Figure 47.1**). It towered more than 12 meters (39 feet) high.

The giant wave was one manifestation of an El Niño, a recurring climate event in which equatorial waters of the eastern and central Pacific Ocean warm above their average temperature. The term El Niño means "baby boy" and was first used by Peruvian fishermen to describe weather changes and a shortage of fishes that sometimes occurred around Christmas. Scientists now know that during an El Niño, marine currents interact with the atmosphere in ways that influence weather patterns worldwide.

An El Niño affects marine food webs along eastern Pacific coasts. As unusually warm water flows toward these coasts, it displaces currents that would otherwise bring up nutrients from the deep. Without these nutrients, marine primary producers decline in numbers. The dwindling producer populations and warming water cause decreases in numbers of small, cold-water fishes, as well as the larger consumers that rely on them. During the 1997–1998 El Niño, about half of the sea lions on the Galápagos Islands starved to death.

California's population of northern fur seals also suffered a sharp decline.

Rainfall patterns shift during an El Niño. During the winter of 1997–1998, torrential rains caused flooding and landslides along eastern Pacific coasts, while Australia and Indonesia suffered from drought-driven crop failures and raging wildfires. An El Niño typically brings cooler, wetter weather to the American Gulf states, and reduces the likelihood of hurricanes.

An El Niño typically persists for 6 to 18 months. It may be followed by an interval in which the temperature of the eastern Pacific remains near its average, or by a La Niña. During a La Niña, eastern Pacific waters become cooler than average. As a result, the west coast of the United States gets less rainfall and the likelihood of hurricanes in the Atlantic increases.

With this introduction to El Niño/La Niña, we invite you to consider the factors that influence the properties of the biosphere. The biosphere includes all places where we find life on Earth.

biosphere All regions of Earth where organisms live.
El Niño Periodic warming of equatorial Pacific waters and the associated shifts in global weather patterns.
La Niña Periodic cooling of equatorial Pacific waters and the associated shifts in global weather patterns.

Figure 47.1 A powerful El Niño caused this enormous wave in the Pacific. It also affected fish populations, causing sea lion pups (photo at *right*) and seals to starve.

47.2 Global Air Circulation Patterns

■ How much solar energy reaches Earth's surface varies from place to place and with the season.

Climate refers to average weather conditions, such as cloud cover, temperature, humidity, and wind speed, over time. Regional climates differ because many factors that influence winds and ocean currents, such as intensity of sunlight, the distribution of land masses and seas, and elevation, vary from place to place.

Seasonal Effects

Each year, Earth rotates around the sun in an elliptical path (**Figure 47.2**). Seasonal changes in daylength

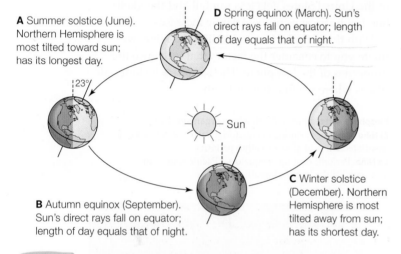

A Summer solstice (June). Northern Hemisphere is most tilted toward sun; has its longest day.

23°

Sun

D Spring equinox (March). Sun's direct rays fall on equator; length of day equals that of night.

C Winter solstice (December). Northern Hemisphere is most tilted away from sun; has its shortest day.

B Autumn equinox (September). Sun's direct rays fall on equator; length of day equals that of night.

Figure 47.2 Earth's tilt and yearly rotation around the sun cause seasonal effects. The 23° tilt of Earth's axis causes the Northern Hemisphere to receive more intense sunlight and have longer days in summer than in winter.

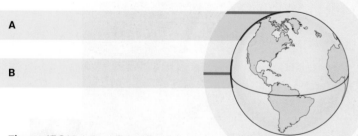

A

B

Figure 47.3 Variation in intensity of solar radiation with latitude. For simplicity, we depict two equal parcels of incoming radiation on an equinox, a day when incoming rays are perpendicular to Earth's axis.

Rays that fall on high latitudes (**A**) pass through more atmosphere (*blue*) than those that fall near the equator (**B**). Compare the length of the *green* lines. Atmosphere is not to scale.

Also, energy in the rays that fall at the high latitude is dispersed over a greater area than energy that falls on the equator. Compare the length of the *red* lines.

and temperature arise because Earth's axis is not perpendicular to the plane of this ellipse, but instead is tilted about 23 degrees. In June, when the Northern Hemisphere is angled toward the sun, this hemisphere receives more intense sunlight and has longer days than the Southern Hemisphere (**Figure 47.2A**). In December, the opposite is true (**Figure 47.2C**). Twice a year—on spring and autumn equinoxes—Earth's axis is perpendicular to incoming sunlight. On these days, every place on Earth has 12 hours of daylight and 12 hours of darkness (**Figure 47.2B,D**).

In each hemisphere, the extent of seasonal change in daylength increases with latitude. At 25° north or south of the equator, the longest daylength is a bit less than 14 hours. By contrast, 60° north or south of the equator, the longest daylength is nearly 19 hours.

Air Circulation and Rainfall

On any given day, equatorial regions receive more sunlight energy than higher latitudes for two reasons. First, fine particles of dust, water vapor, and greenhouse gases absorb some solar radiation or reflect it back into space. Sunlight traveling to high latitudes passes through more atmosphere to reach Earth's surface than light traveling to the equator, so less energy reaches the ground (**Figure 47.3A**). Second, energy in an incoming parcel of sunlight is dispersed over a smaller surface area at the equator than at the higher latitudes (**Figure 47.3B**). As a result, Earth's surface warms more at the equator than at the poles.

Knowing about two properties of air can help you understand how regional differences in surface warming give rise to global air circulation and rainfall patterns. First, as air warms, it becomes less dense and rises. Hot air balloonists take advantage of this effect when they ascend by heating the air inside their balloon. Second, warm air can hold more water than cooler air. This is why you can "see your breath" in cold weather. When you exhale, warm air with moisture from your lungs cools, causing the water in that air to condense as tiny droplets.

The global air circulation pattern begins at the equator, where intense sunlight warms air and causes evaporation from the ocean. The result is an upward movement of warm, moist air (**Figure 47.4A**). As the air from the equator rises to higher altitudes, it cools and flows north and south, releasing moisture as rain that supports tropical rain forests.

By the time the air has reached 30° north or south of the equator, it has given up most moisture and cooled off, so it sinks back toward Earth's surface (**Figure**

Initial Pattern of Air Circulation

D At the poles, cold air sinks and moves toward lower latitudes.

C Air rises again at 60° north and south, where air flowing poleward meets air coming from the poles.

B As the air flows toward higher latitudes, it cools and loses moisture as rain. At around 30° north and south latitude, the air sinks and flows north and south along Earth's surface.

A Warmed by energy from the sun, air at the equator picks up moisture and rises. It reaches a high altitude, and spreads north and south.

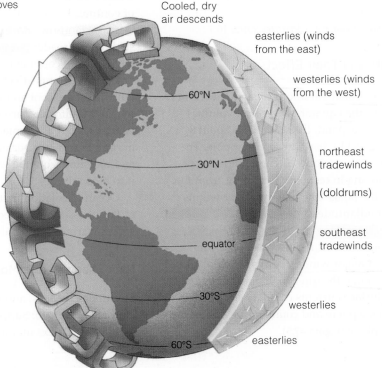

Cooled, dry air descends

easterlies (winds from the east)

westerlies (winds from the west)

60°N

30°N

northeast tradewinds

(doldrums)

equator

southeast tradewinds

30°S

westerlies

60°S

easterlies

Figure 47.4 **Animated** Global air circulation patterns and their effects on climate.

Prevailing Wind Patterns

E Major winds near Earth's surface do not blow directly north and south because of Earth's rotation. Winds deflect to the right of their original direction in the Northern Hemisphere and to the left in the Southern Hemisphere.

Figure It Out: What is the direction of prevailing winds in the central United States?
Answer: From west to east.

47.4B). Many of the world's great deserts, including the Sahara, are about 30° from the equator.

As air continues along Earth's surface toward the poles, it again picks up heat and moisture. At a latitude of about 60°, it rises (**Figure 47.4C**). The resulting rains support temperate zone forests.

Cold, dry air descends near the poles (**Figure 47.4D**). Precipitation is sparse, and polar deserts form.

Surface Wind Patterns

Major wind patterns arise as air in the lower atmosphere moves continually from latitudes where air is sinking toward those where air is rising. Earth's rotation affects the apparent trajectory of these winds. Air masses are not attached to Earth's surface, so the Earth spins beneath them, moving fastest at the equator and most slowly at the poles. Thus, as an air mass moves away from the equator, the speed at which the Earth rotates beneath it continually slows. As a result, major winds trace a curved path relative to the Earth's surface (**Figure 47.4E**). In the Northern Hemisphere, winds curve toward the right; in the Southern Hemisphere,

climate Average weather conditions in a region over a long time period.

they curve toward the left. For example, between 30° north and 60° north, surface air traveling toward the North Pole is deflected right, or toward the east. Winds are named for the direction from which they blow, so the prevailing winds that blow from west to east are westerlies.

Winds blow most consistently between latitudes where air rises. In the regions where two air masses meet and rise, winds are less reliable. For example, air masses from the Northern and Southern hemispheres meet near the equator, causing an area of calm air commonly referred to as the doldrums.

Take-Home Message

How does sunlight affect climate?

» Equatorial regions receive more intense sunlight than higher latitudes.

» Sunlight drives the rise of moisture-laden air at the equator. The air cools as it moves north and south, releasing rains that support tropical forests. Deserts form where cool, dry air descends. Sunlight energy also drives moisture-laden air aloft at 60° north and south latitude. This air gives up moisture as it flows toward the equator or the pole.

» Air flow in the lower atmosphere toward latitudes where air is rising and away from latitudes where it is sinking creates major surface winds. These winds trace a curved path relative to Earth's surface because of Earth's rotation.

47.3 The Ocean, Landforms, and Climates

■ Solar heating and the effects of the wind set the ocean's surface water in motion, resulting in currents that distribute nutrients and affect climate.

■ Links to Properties of water 2.5, Plate tectonics 16.7

Ocean Currents and Their Effects

Latitudinal and seasonal variations in sunlight warm and cool water. At the equator, where vast volumes of water warm and expand, the sea level is about 8 centimeters (3 inches) higher than at either pole. The volume of water in this "slope" is enough to get sea surface water moving in response to gravity, from the equator toward the poles. The moving water warms air above it. At midlatitudes, oceans transfer 10 million billion calories of heat energy per second to the air!

Enormous volumes of water flow as ocean currents. The force of major winds, Earth's rotation, and topography determine the directional movement of these currents. Surface currents circulate clockwise in the Northern Hemisphere and counterclockwise in the Southern Hemisphere (**Figure 47.5**).

Swift, deep, and narrow currents of nutrient-poor water flow away from the equator along the east coast of continents. Along the east coast of North America, warm water flows north, as the Gulf Stream. Slower, shallower, broader currents of cold water parallel the west coast of continents and flow toward the equator.

Ocean currents affect climates. Coasts in the Pacific Northwest are cool and foggy in summer because the cold California current chills the air, so water condenses out as droplets. Boston and Baltimore are muggy in summer because air masses pick up heat and moisture from the warm Gulf Stream, then deliver it to these cities.

Ocean circulation patterns shift over geologic time as landmasses move (Section 16.7). Some worry that global warming could also alter these patterns.

Rain Shadows and Monsoons

Mountains, valleys, and other surface features of the land affect climate. Suppose you track a warm air mass after it picks up moisture off California's coast.

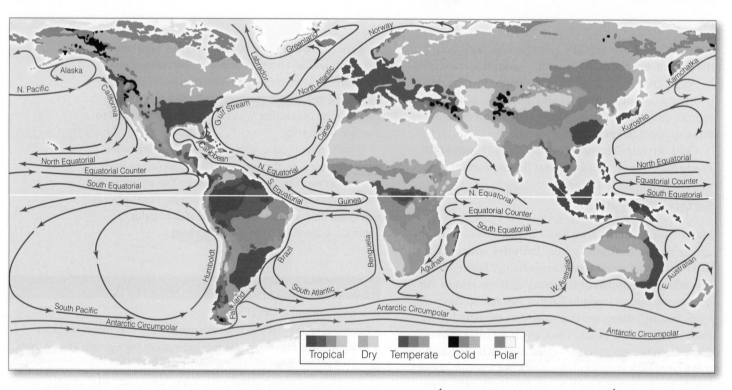

Figure 47.5 Major climate zones correlated with surface currents of the world ocean. Warm surface currents start moving from the equator toward the poles, but prevailing winds, Earth's rotation, gravity, the shape of ocean basins, and landforms influence the direction of flow. Water temperatures, which differ with latitude and depth, contribute to the regional differences in air temperature and rainfall.

A Prevailing winds move moisture inland from the Pacific Ocean.

B Clouds pile up and rain forms on side of mountain range facing prevailing winds.

C Rain shadow on side facing away from the prevailing winds makes arid conditions.

4,000/ 75
3,000/ 85   2,000/ 25
1,800/ 125   1,000/ 25
1,000/ 85
moist habitats
15/ 25

Figure 47.6 Rain shadow effect. On the side of mountains facing away from prevailing winds, rainfall is light. *White* numbers signify elevations, in meters. *Yellow* numbers signify annual precipitation, in centimeters, averaged on both sides of the Sierra Nevada, a mountain range.

It moves inland, as wind from the west, and piles up against the Sierra Nevada. This high mountain range parallels the distant coast. The air cools as it rises in altitude and loses moisture as rain (**Figure 47.6**). The result is a rain shadow—a semiarid or arid region of sparse rainfall on the leeward side of high mountains. "Leeward" is the side facing away from the wind. The Himalayas, Andes, Rockies, and other great mountain ranges cause vast rain shadows.

Differences in the heat capacity of water and land give rise to coastal breezes. In the daytime, water does not warm as fast as the land. Air heated by the warm land rises, and cool offshore air moves in to replace it (**Figure 47.7A**). After sundown, land becomes cooler than the water, so the breezes reverse (**Figure 47.7B**).

Differential heating of water and land also causes monsoons, winds that change direction seasonally. For example, the continental interior of Asia heats up in the summer, so air rises above it. The resulting low pressure draws in moisture from over the warm Indian Ocean to the south, and these north-blowing winds deliver heavy rains. In the winter, the continental interior is cooler than the ocean. As a result, cool, dry winds blow from the north toward southern coasts, causing a seasonal drought.

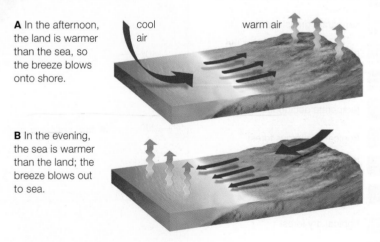

A In the afternoon, the land is warmer than the sea, so the breeze blows onto shore.

cool air   warm air

B In the evening, the sea is warmer than the land; the breeze blows out to sea.

Figure 47.7 Animated Coastal breezes.

monsoon Wind that reverses direction seasonally.
rain shadow Dry region downwind of a coastal mountain range.

Take-Home Message

How do ocean currents arise and how do they affect regional climates?

» Surface ocean currents, which are set in motion by latitudinal differences in solar radiation, are affected by winds and by Earth's rotation.

» Collective effects of air masses, oceans, and landforms determine regional temperature and annual precipitation.

47.4 Biomes

■ Similar communities often evolve in widely separated regions as a result of similar environmental factors.

■ Links to Carbon-fixing pathways 6.8, Morphological convergence 18.3, Soil 28.2, Primary production 46.4

Differences Between Biomes

Biomes are areas of land characterized by their climate and predominant type of vegetation (**Figure 47.8**). A biome is discontinuous; most include areas on different continents. For example, the temperate grassland biome includes North American prairie, South African veld, South American pampa, and Eurasian steppe.

The type of biome characteristic of an area depends mainly on rainfall and temperature. Desert biomes get the least annual rainfall, grasslands and shrublands get more, and forests get the most. Deserts occur where temperatures soar highest and tundra where they drop the lowest.

Soils also influence biome distribution. Soils consist of a mixture of mineral particles and varying amounts of humus. Water and air fill spaces between soil particles. Properties of soils vary depending on the types, proportions, and compaction of particles. Deserts have sandy or gravelly, fast-draining soil with little topsoil. Topsoil is deepest in natural grasslands, where it can be more than one meter thick. This is why grasslands are often converted to agricultural uses.

Climate and soils affect primary production, so primary production varies among biomes (**Figure 47.9**).

Similarities Within a Biome

Unrelated species living in widely separated parts of a biome often have similar body structures that arose by the process of morphological convergence (Section 18.3). For example, cacti with water-storing stems live in North American deserts and euphorbs with

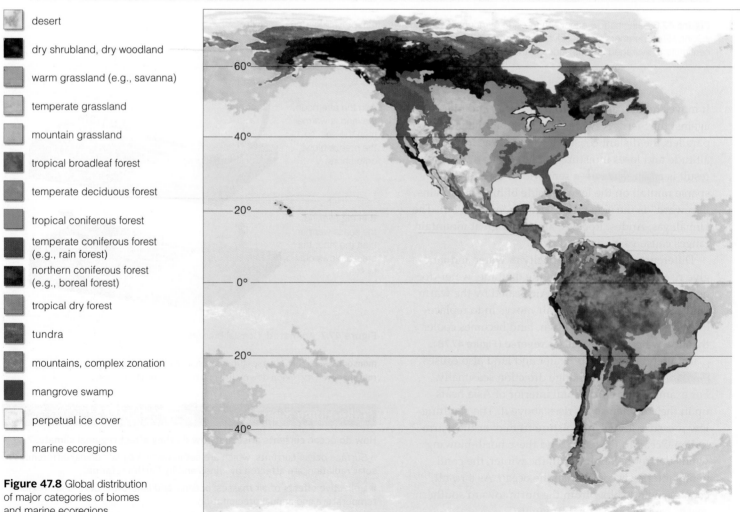

desert

dry shrubland, dry woodland

warm grassland (e.g., savanna)

temperate grassland

mountain grassland

tropical broadleaf forest

temperate deciduous forest

tropical coniferous forest

temperate coniferous forest (e.g., rain forest)

northern coniferous forest (e.g., boreal forest)

tropical dry forest

tundra

mountains, complex zonation

mangrove swamp

perpetual ice cover

marine ecoregions

Figure 47.8 Global distribution of major categories of biomes and marine ecoregions.

water-storing stems live in African deserts. Cacti and euphorbs do not share an ancestor with a water-storing stem. Rather, this feature evolved independently in the two groups as a result of similar selection pressures. Similarly, an ability to carry out C4 photosynthesis evolved independently in grasses growing in warm grasslands on different continents. C4 photosynthesis is more efficient than the more common C3 pathway under hot, dry conditions.

biome Group of regions that may be widely separated but share a characteristic climate and dominant vegetation.

Take-Home Message

What are biomes?

» Biomes are vast expanses of land dominated by distinct kinds of plants that support characteristic communities.

» The global distribution of biomes is a result of topography, climate, and evolutionary history.

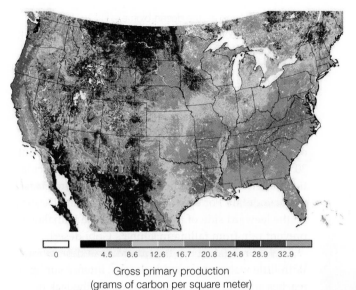

Gross primary production
(grams of carbon per square meter)

Figure 47.9 Remote satellite monitoring of gross primary productivity across the United States. The differences roughly correspond with variations in soil types and moisture.

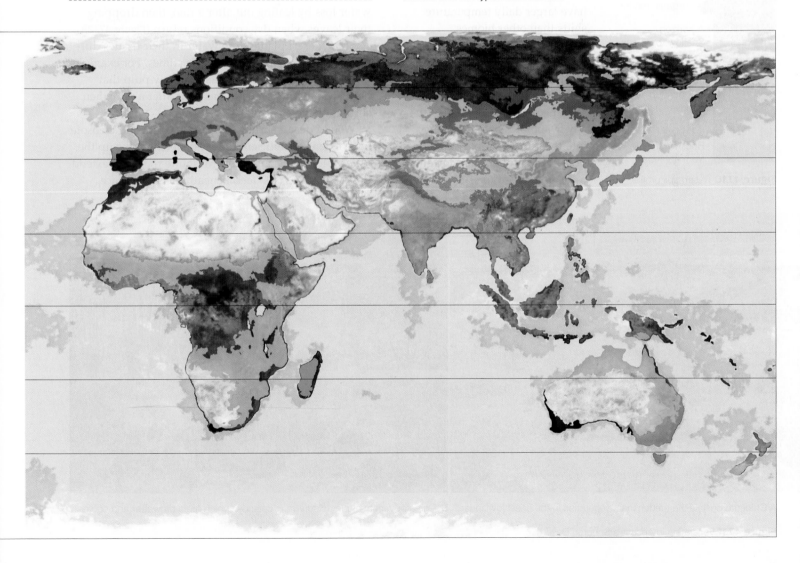

47.5 Deserts

- Low rainfall shapes the desert biome.
- Links to Carbon-fixing pathways 6.8, Atacama Desert 19.1, Cyanobacteria 20.7

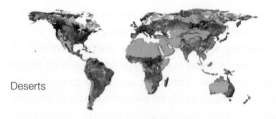

Deserts

Desert Locations and Conditions

Deserts receive an average of less than 10 centimeters (4 inches) of rain per year. They cover about one-fifth of Earth's land surface and many are located at about 30° north and south latitude, where global air circulation patterns cause dry air to sink. Rain shadows also reduce rainfall. For example, Chile's Atacama Desert is on the leeward side of the Andes, and the Himalayas prevent rain from falling in China's Gobi desert.

Lack of rainfall keeps the humidity in deserts low. With little water vapor to block rays, intense sunlight reaches and heats the ground. At night, the lack of insulating water vapor in the air allows the temperature to fall fast. As a result, deserts tend to have larger daily temperature shifts than other biomes.

Desert soils have very little topsoil (Figure 47.10), the layer most important for plant growth. The soils often are somewhat salty, because rain that falls usually evaporates before seeping into the ground. Rapid evaporation allows any salt in rainwater to accumulate at the soil surface.

O horizon:
Pebbles, little organic matter

A horizon:
Shallow, poor soil

B horizon:
Evaporation causes salt buildup; leaching removes nutrients

C horizon:
Rock fragments from uplands

Figure 47.10 Desert soil profile.

Adaptations to Desert Conditions

Despite their harsh conditions, most deserts support some plant life. Diversity is highest in regions where soil moisture is available in more than one season (**Figure 47.11**).

Many desert plants have adaptations that reduce water loss. For example, some have spines or hairs (**Figure 47.12A**). In addition to deterring herbivory, these structures reduce water loss by trapping some water and keeping the humidity around the stomata high. Where rains fall seasonally, some plants reduce their water loss by leafing out after a rain, then dropping their leaves when dry conditions return (**Figure 47.12B**).

Some desert plants store water in their tissues during the wet season, for use in drier times. For example, the stem of a barrel cactus has a spongy pulp that holds water. The cactus stem swells after a rain, then shrinks as the plant uses stored water.

Woody desert shrubs such as mesquite and creosote have extensive, efficient root systems that take up the little water that is available. Mesquite roots can extend up to 60 meters (197 feet) beneath the soil surface.

A Creosote bush is the predominant vegetation in the driest lowlands. **B** A greater variety of plants survive in uplands, which are a bit wetter and cooler.

Figure 47.11 Vegetation in Arizona's Sonoran Desert.

A Barrel cacti are covered by spines that reduce evaporative water loss. The cactus is a CAM plant.

B Ocotillo, a desert shrub, grows leaves on its stems after a rain, then sheds them when conditions become dry again.

Figure 47.12 Perennials adapted to desert conditions.

Alternative carbon-fixing pathways also help desert plants conserve water. Cacti, agaves, and euphorbs are CAM plants. They open their stomata only at night when temperature declines.

Most deserts contain a mix of annuals and perennials (**Figure 47.13**). The annuals are adapted to desert life by a life cycle that allows them to sprout and reproduce in the short time that the soil is moist.

Animals also have adaptations that allow them to conserve water. For example, the highly efficient kidneys of a desert kangaroo rat minimize its water needs (Section 40.2). Most desert animals are not active at the height of the daytime heat (**Figure 47.14**).

The Crust Community

In many deserts, the soil is covered by a desert crust, a community that can include cyanobacteria, lichens, mosses, and fungi. The organisms secrete organic molecules that glue them and the surrounding soil particles together. The crust benefits members of the larger desert community in important ways. Its cyanobacteria fix nitrogen and make ammonia available to plants. The crust also holds soil particles in place. When the fragile connections within the desert crust are broken, the soil can blow away. Negative effects of such disturbance are exacerbated when windblown soil buries healthy crust in an undisturbed area, killing additional crust organisms and allowing more soil to take flight.

desert Biome with little rain and low humidity; plants that have water-storing and water-conserving adaptations predominate.

Figure 47.13 Mojave Desert after the rains. Annual poppies sprout, flower, produce seeds, and die within weeks beneath slow-growing perennial cacti.

A The Sonoran desert tortoise spends much of its life inactive. In hot summer months, it ventures out of its burrow only in cool mornings to feed. During the cold winter, when little food is available, it hibernates.

B Lesser long-nosed bats spend spring and summer in the Sonoran Desert. They avoid the daytime heat by resting in caves.

Figure 47.14 Two Sonoran Desert animals.

Take-Home Message

What are deserts?

» A desert gets very little rain and has low humidity. There is plenty of sunlight, but the lack of water prevents most plants from surviving here.

» The predominant plants in deserts have adaptations that allow them to reduce water lost by transpiration, store water, or access water deep below the soil surface.

» Desert animals often spend the day inactive, sheltering from the heat.

» Desert soils are held in place by a community of organisms that form a desert crust. Disruption of this crust allows wind to strip away soil.

47.6 Grasslands

- Perennial grasses adapted to fire and to grazing are the main plants in grasslands.
- Links to Erosion 28.2, Adaptations to herbivory 45.6

Temperate grasslands and tropical savannas

Grasslands form in the interior of continents between deserts and temperate forests. Their soils are rich, with deep topsoil. Annual rainfall is enough to keep desert from forming, but not enough to support woodlands. Low-growing grasses and other nonwoody plants tolerate strong winds, sparse and infrequent rain, and intervals of drought. Growth tends to be seasonal. Constant trimming by grazers, along with periodic fires, keeps trees and most shrubs from taking hold.

Temperate Grasslands

Temperate grasslands are warm in summer, but cold in winter. Annual rainfall is 25 to 100 centimeters (10–40 inches), with rains throughout the year. Grass roots extend profusely through the thick topsoil and help hold it in place, preventing erosion by the constant winds. North America's grasslands are shortgrass and tallgrass prairies (**Figure 47.15 A,B**).

During the 1930s, much of the shortgrass prairie of the American Great Plains was plowed to grow wheat. Strong winds, a prolonged drought, and unsuitable farming practices turned much of the region into what the newspapers of that time called the Dust Bowl.

Tallgrass prairie has somewhat richer topsoil and slightly more frequent rainfall than shortgrass prairie. Before the arrival of Europeans, it covered about 140 million acres, mostly in Kansas. Nearly all tallgrass prairie has now been converted to cropland. The Tallgrass Prairie National Preserve was created in 1996 to protect the little that remains.

North America's prairies once supported enormous herds of elk, pronghorn antelope, and bison that were prey to wolves. Today, these predators and prey are absent from most of their former range.

Savannas

Savannas are broad belts of grasslands with a few scattered shrubs and trees. Savannas lie between the tropical forests and hot deserts of Africa, India, and Australia. Temperatures are warm year-round. During the rainy season, 90–150 centimeters (35–60 inches) of rain falls. Africa's savannas are famous for their abundant wildlife (**Figure 47.15C**). Herbivores include giraffes, zebras, elephants, a variety of antelopes, and immense herds of wildebeests. Lions and hyenas are carnivores that eat the grazers.

grassland Biome in the interior of continents where grasses and nonwoody plants adapted to grazing and fire predominate.

Take-Home Message

What are grasslands?

» Grasslands are biomes dominated by grasses and other nonwoody plants that can withstand fire and grazing.

A horizon: Alkaline, deep, rich in humus

B horizon: Percolating water enriches layer with calcium carbonates

A Prairie soil profile.

Figure 47.15 Grasslands.

B Bison grazing in a North American prairie.

C Wildebeest grazing on an African savanna.

47.7 Dry Shrublands and Woodlands

- Regions with cool, rainy winters and hot, dry summers support dry shrublands and woodlands.
- Link to Adapting to fire 45.9

Dry shrubland

Dry shrubland is a biome dominated by fire-adapted shrubs. It typically occurs along the western coast of continents, between 30 and 40 degrees north or south latitude. A mild winter brings 25 to 60 centimeters (10–24 inches) of rain, and the summer is hot and dry. California's dry shrublands, called chaparral, are the state's most extensive biological community (**Figure 47.16A**). Dry shrubland also occurs in regions bordering the Mediterranean, as well as Chile, Australia, and South Africa (**Figure 47.16B**).

Plants in dry shrublands tend to have small leathery leaves that help them withstand the summer drought. Many make aromatic oils that help fend off insects, but also make them highly flammable. After a fire, many plants resprout from their roots and new ones germinate from fire-resistant seeds.

dry shrubland Biome dominated by a diverse array of fire-adapted shrubs; occurs in regions with cool, wet winters and a dry summer.
dry woodland Biome dominated by short trees that do not completely shade the ground; occurs in regions with cool, wet winters and a dry summer.

A Oak woodland in California.   **B** Eucalyptus woodland in Australia.

Figure 47.17 Dry woodlands.

Dry shrublands grade into **dry woodlands**, where a bit more winter rain allows trees to grow. In a woodland, trees do not shade as much of the ground as they do in a forest, and they tend to be shorter. Examples of dry woodlands include California's oak woodlands and eucalyptus woodlands of Australia (**Figure 47.17**).

Take-Home Message

What are dry shrublands and woodlands?

» Dry shrublands and woodlands form in areas with a mild, rainy winter and a hot dry summer.

» Dry shrublands are dominated by fire-adapted shrubs, dry woodlands by low-growing trees that allow a lot of sunlight to reach the ground.

A Chaparral in California.   **B** Fynbos, a dry shrubland in the Cape region of South Africa.

Figure 47.16 Dry shrublands. Evergreen shrubs with small, leathery leaves predominate.

47.8 Broadleaf Forests

- Broadleaf (angiosperm) trees are the main plants in temperate and tropical forests.
- Link to Primary production 46.4

Semi-Evergreen and Deciduous Forests

Semi-evergreen forests occur in the humid tropics of Southeast Asia and India. They include broadleaf (angiosperm) trees that retain leaves year-round, and deciduous broadleaf trees. A deciduous plant sheds leaves annually, prior to a season when cold or dry conditions would not favor growth. In semi-evergreen forests, deciduous trees shed their leaves at the start of the dry season.

Where less than 2.5 centimeters (1 inch) of rain falls in the dry season, tropical deciduous forests form. In tropical deciduous forests, most trees shed leaves at the start of the dry season.

Temperate deciduous forests form in the Northern Hemisphere in parts of eastern North America, western and central Europe, and parts of Asia, including Japan. In these regions, 50 to 150 centimeters (about 20–60 inches) of precipitation falls throughout the year. Winters are cool and summers are warm.

Growth of temperate deciduous forests is seasonal. Leaves often turn color before dropping in autumn (Figure 47.18). Winters are cold and trees remain dormant while water is locked in snow and ice. In the spring, when conditions again favor growth, deciduous trees flower and put out new leaves. Also during the spring, leaves that were shed the prior autumn

decay to form a rich humus. Rich soil and a somewhat open canopy that lets sunlight through allow shorter understory plants to flourish.

The temperate deciduous forests of North America are the most species-rich examples of this biome. Different tree species characterize different regions of these forests. For example, Appalachian forests include mainly oaks, whereas beeches and maples dominate Ohio's forests. Animals in North American deciduous forests include grazing deer and seed-eating squirrels and chipmunks, as well as omnivores such as raccoons, opossums, and black bears. Native predators such as wolves and mountain lions have been largely eliminated from the forests.

Tropical Rain Forests

Tropical rain forests of evergreen broadleaf trees form between latitudes 10° north and south in equatorial Africa, the East Indies, Southeast Asia, South America, and Central America. Rain that falls throughout the year sums to an annual total of 130 to 200 centimeters (50 to 80 inches).

Regular rains, combined with an average temperature of 25°C (77°F) and little variation in daylength, allows photosynthesis to continue year-round. Of all land biomes, tropical forests have the greatest primary production. Per unit area, they remove more carbon from the atmosphere than other forests or grasslands.

Tropical rain forest is the most structurally complex and species-rich biome. The forest has a multilayer

Temperate deciduous forest

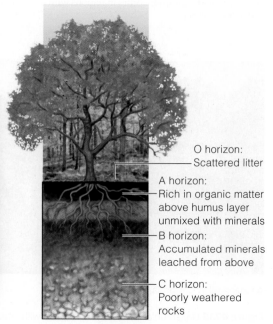

O horizon:
Scattered litter

A horizon:
Rich in organic matter above humus layer unmixed with minerals

B horizon:
Accumulated minerals leached from above

C horizon:
Poorly weathered rocks

Figure 47.18 North American temperate deciduous forest. Forest soil profile at *right*.

O horizon:
Sparse litter

A–E horizons:
Continually leached;
iron, aluminum left
behind impart red
color to acidic soil

B horizon:
Clays with silicates,
other residues of
weathering

Figure 47.19 Tropical rain forest. Forest soil profile at *right*.

Tropical
rain forest

structure (**Figure 47.19**). Its broadleaf trees can stand 30 meters (100 feet) tall. The trees often form a closed canopy that prevents most sunlight from reaching the forest floor. Vines and epiphytes (plants that grow on another plant, but do not withdraw nutrients from it) thrive in the shade beneath the canopy.

Trees of tropical rain forests shed leaves continually, but decomposition and mineral cycling happen so fast in this warm, moist environment that litter does not accumulate. Soils are highly weathered, heavily leached, and very poor nutrient reservoirs.

Deforestation is an ongoing threat to tropical rain forests. Tropical forests are located in developing countries with fast-growing human populations who look to the forest as a source of lumber, fuel, and potential agricultural land. As human populations expand, more and more trees fall to the ax.

Deforestation in any region leaves fewer trees to remove carbon dioxide from the atmosphere. In rain forests, it also causes the extinction of species found nowhere else in the world. Compared to other land biomes, tropical rain forests have the greatest variety and numbers of insects, as well as the most diverse collection of birds and primates. This great species diversity means many species are affected by the loss of any amount of forest. Among the potential losses are species with chemicals that could save human lives. Two chemotherapy drugs, vincristine and vinblastine, were extracted from the rosy periwinkle, a low-growing plant native to Madagascar's rain forests. Today, these drugs help fight leukemia, lymphoma, breast cancer, and testicular cancer. No doubt other similarly valuable species live in the rain forests and will go extinct before we learn how they can help us.

temperate deciduous forest Northern Hemisphere biome in which the main plants are broadleaf trees that lose their leaves in fall and become dormant during cold winters.
tropical rain forest Highly productive and species-rich biome in which year-round rains and warmth support continuous growth of evergreen broadleaf trees.

Take-Home Message

What are broadleaf forests?

» Temperate broadleaf forests grow in the Northern Hemisphere where cold winters prevent year-round growth. Trees lose their leaves in autumn, then remain dormant during the winter.

» Year-round warmth and rains support tropical rain forests, the most productive, structurally complex, and species-rich land biome.

47.9 Coniferous Forests

■ Compared to broadleaf trees, conifers are more tolerant of cold and drought, and can withstand poorer soils. Where these conditions occur, coniferous forests prevail.

■ Link to Conifers 22.7

Conifers (evergreen trees with seed-bearing cones) are the main plants in coniferous forests. Conifer leaves are typically needle-shaped, with a thick cuticle and stomata that are sunk below the leaf surface. These adaptations help conifers conserve water during drought or times when the ground is frozen. As a group, conifers tolerate poorer soils and drier habitats than most broadleaf trees.

The most extensive land biome is the coniferous forest that sweeps across northern Asia, Europe, and North America (**Figure 47.20A**). It is referred to as **boreal forest**, or taiga, which means "swamp forest" in Russian. The conifers are mainly pine, fir, and spruce. Most rain falls in the summer, and little evaporates into the cool summer air. Winters are long, cold, and dry. Moose are dominant grazers in this biome.

boreal forest Extensive high-latitude forest of the Northern Hemisphere; conifers are the predominant vegetation.

Also in the Northern Hemisphere, montane coniferous forests extend southward through the great mountain ranges (**Figure 47.20B**). Spruce and fir dominate at the highest elevations. At lower elevations, the mix becomes firs and pines.

Conifers also dominate temperate lowlands along the Pacific coast from Alaska into northern California. These coniferous forests hold some of the world's tallest trees—Sitka spruce and coast redwoods.

We find other conifer-dominated ecosystems in the eastern United States. About a quarter of New Jersey is pine barrens, a mixed forest of pitch pines and scrub oaks that grow in sandy, acidic soil. Pine forest covers about one-third of the Southeast. Fast-growing loblolly pines dominate these forests and are a major source of lumber and wood pulp. The pines can survive periodic fires that kill most hardwood species. When fires are suppressed, hardwoods outcompete pines.

Take-Home Message

What are coniferous forests?

» Conifers prevail across the Northern Hemisphere's high-latitude forests, at high elevations, and in temperate regions with nutrient-poor soils.

Coniferous forests

B Montane coniferous forest near Mount Rainier, Washington.

A Boreal forest (taiga) in Siberia.

Figure 47.20 Coniferous forests.

47.10 Tundra

■ Low-growing, cold-tolerant plants have only a brief growing season on the tundra.
■ Link to Carbon and global warming 46.8

Arctic tundra

Arctic Tundra

Arctic tundra forms between the polar ice cap and the belts of boreal forests in the Northern Hemisphere. Most is in northern Russia and Canada. Arctic tundra is Earth's youngest biome, having first appeared about 10,000 years ago when glaciers retreated at the end of the last ice age.

Conditions in this biome are harsh; snow blankets the ground for as long as nine months of the year. Annual precipitation is usually less than 25 centimeters (10 inches), but cold temperature keeps the snow that does fall from melting. During a brief summer, plants grow fast under the nearly continuous sunlight (**Figure 47.21**). Lichens and shallow-rooted, low-growing plants are the producers for food webs that include voles, arctic hares, caribou, arctic foxes, wolves, and brown bears. Enormous numbers of migratory birds nest here in the summer.

Only the surface layer of tundra soil thaws during summer. Below that lies permafrost, a frozen layer 500 meters (1,600 feet) thick in places. Permafrost acts as a barrier that prevents drainage, so the soil above it remains perpetually waterlogged. The cool, anaerobic conditions in this soil slow decay, so organic remains can build up. Organic matter in permafrost makes the arctic tundra one of Earth's greatest stores of carbon.

As global temperatures rise, the amount of frozen soil that melts each summer is increasing. With warmer temperatures, much of the snow and ice that would otherwise reflect sunlight is disappearing. As a result, newly exposed dark soil absorbs heat from the sun's rays, which encourages more melting.

Figure 47.21 Arctic tundra in the summer. Permafrost underlies the soil.

Figure 47.22 Alpine tundra. Low-growing, hardy plants at a high altitude in Washington's Cascade range.

Alpine Tundra

Alpine tundra occurs at high altitudes throughout the world (**Figure 47.22**). Even in the summer, some patches of snow persist in shaded areas, but there is no permafrost. The alpine soil is well drained, but thin and nutrient-poor. As a result, primary productivity is low. Grasses and small-leafed, woody shrubs grow in patches where soil has accumulated to a greater depth. These low-growing plants can withstand the strong winds that discourage the growth of trees.

alpine tundra Biome of low-growing, wind-tolerant plants adapted to high-altitude conditions.
arctic tundra Highest-latitude Northern Hemisphere biome, where low, cold-tolerant plants survive with only a brief growing season.
permafrost Continually frozen soil layer that lies beneath arctic tundra and prevents water from draining.

Take-Home Message

What is tundra?

» Arctic tundra prevails at high latitudes, where short, cold summers alternate with long, cold winters.

» Alpine tundra prevails in high, cold mountains across all latitudes.

47.11 Freshwater Ecosystems

■ Freshwater and saltwater provinces cover more of Earth's surface than all land biomes combined. Here we begin our survey of these watery realms.

■ Links to Properties of water 2.5, Aquatic respiration 38.2, Succession 45.8, Eutrophication 46.1, Food webs 46.3, Water cycle 46.6

Lakes

A lake is a body of standing fresh water. If it is sufficiently deep, it can be divided into zones that differ in their physical characteristics and species composition (Figure 47.23). Near shore is the littoral zone, from the Latin *litus* for shore. Here, sunlight penetrates all the way to the lake bottom; aquatic plants and algae that attach to the bottom are the primary producers. The lake's open waters include an upper, well-lit limnetic zone, and—if the lake is deep—a dark profundal zone where light does not penetrate. Primary producers in the limnetic zone can include aquatic plants, green algae, diatoms, and cyanobacteria. These organisms serve as food for rotifers, copepods, and other types of zooplankton. In the profundal zone, where there is not enough light for photosynthesis, consumers feed on organic debris that drifts down from above.

Nutrient Content and Succession A lake undergoes succession; it changes over time (Section 45.8). A newly formed lake is oligotrophic: deep, clear, and nutrient-poor, with low primary productivity (Figure 47.24). Later, as sediments accumulate and plants take root, the lake becomes eutrophic. Eutrophication refers to processes, either natural or artificial, that enrich a body of water with nutrients (Section 46.1).

Seasonal Changes Temperate zone lakes undergo seasonal changes. During winter, a layer of ice forms at the lake surface. Unlike most substances, water is denser as a liquid than as a solid (ice). As water cools, its density increases, until it reaches 4°C (39°F). Below this temperature, additional cooling decreases water's density—which is why ice floats on water (Section 2.5). In an ice-covered lake, water just under the ice is near its freezing point and at its lowest density. The

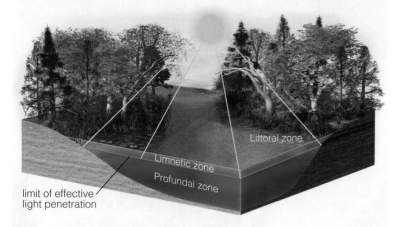

limit of effective light penetration

Littoral zone

Limnetic zone

Profundal zone

Figure 47.23 Animated Lake zonation. A lake's littoral zone extends around the shore to a depth where rooted aquatic plants stop growing. Its limnetic zone is the open waters where light penetrates and photosynthesis occurs. Below that lies the cool, dark profundal zone, where detrital food chains predominate.

Figure 47.24 An oligotrophic lake. Crater Lake in Oregon is a collapsed volcano that filled with snow melt. It began filling about 7,700 years ago. Thus, from a geologic standpoint, it is a young lake.

Oligotrophic Lake	Eutrophic Lake
Deep, steeply banked	Shallow with broad littoral
Large deep-water volume relative to surface-water volume	Small deep-water volume relative to surface-water volume
Highly transparent	Limited transparency
Water blue or green	Water green to yellow- or brownish-green
Low nutrient content	High nutrient content
Oxygen abundant through all levels throughout year	Oxygen depleted in deep water during summer
Not much phytoplankton; green algae and diatoms dominant	Abundant, thick masses of phytoplankton; cyano-bacteria dominant
Aerobic decomposers favored in profundal zone	Anaerobic decomposers in profundal zone
Low biomass in profundal zone	High biomass in profundal zone

densest (4°C) water resides at the bottom of the lake (Figure 47.25A).

In spring, the air warms and ice melts. When the resulting meltwater warms to 4°C, it sinks. Wind blowing across the surface of the lake assists in bringing about a **spring overturn**, during which oxygen-rich water at the surface of the lake moves downward while nutrient-rich water from the lake's depths moves up (Figure 47.25B).

In the summer, a lake's waters form three layers that differ in their temperature and oxygen content (Figure 47.25C). The top layer is warm and oxygen-rich. It overlies the **thermocline,** a thin layer where temperature falls rapidly. Beneath the thermocline is the coolest water. The thermocline acts as a barrier that keeps the upper and lower layers from combining. As a result of the thermocline, oxygen from the lake's surface cannot reach the lake's depths, where decomposition is using up oxygen. At the same time, nutrients from those depths cannot escape into surface waters.

In autumn, the upper layer cools and sinks, and the thermocline disappears. During the **fall overturn**, oxygen-rich water moves down while nutrient-rich water moves up (Figure 47.25D).

Overturns influence primary productivity. After a spring overturn, longer daylength and an abundance of nutrients support the greatest primary productivity. During the summer, vertical mixing ceases. Nutrients do not move up, and photosynthesis slows. By late summer, nutrient shortages limit growth. Fall overturn brings nutrients to the surface and favors a brief burst of photosynthesis. The burst ends as winter brings shorter days and the amount of sunlight declines.

Streams and Rivers

Streams are flowing-water ecosystems that begin as freshwater springs or seeps. As they flow downslope, they grow and merge to form rivers. Rainfall, snowmelt, geography, altitude, and shade cast by plants affect flow volume and temperature.

Properties of a stream or river vary along its length. Streambed composition affects solute concentrations,

fall overturn During the fall, waters of a temperate zone mix. Upper, oxygenated water cools, gets dense, and sinks; nutrient-rich water from the bottom moves up.
lake A body of standing fresh water.
spring overturn In temperate zone lakes, a downward movement of oxygenated surface water and an upward movement of nutrient-rich water in spring.
thermocline Thermal stratification in a large body of water; a cool midlayer stops vertical mixing between warm surface water above it and cold water below it.

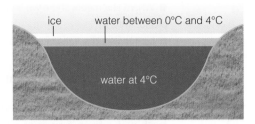

A Winter. Ice covers the thin layer of slightly warmer water just below it. Densest (4°C) water is at bottom. Winds do not affect water under the ice, so there is little circulation.

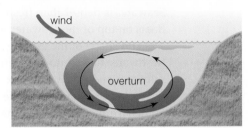

B Spring. Ice thaws. Upper water warms to 4°C and sinks. Winds blowing across water create vertical currents that help overturn water, bringing nutrients up from bottom.

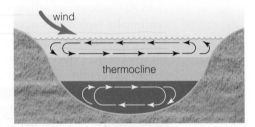

C Summer. Sun warms the upper water, which floats on a thermocline, a layer across which temperature changes abruptly. Upper and lower water do not mix because of this thermal boundary.

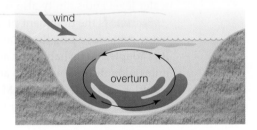

D Fall. Upper water cools and sinks downward, eliminating the thermocline. Vertical currents mix water that was separated during the summer.

Figure 47.25 Seasonal changes in a temperate zone lake.

as when limestone rocks dissolve and add calcium. Water that flows rapidly over rocks mixes with air and holds more oxygen than slower-moving, deeper water. Temperature also affects oxygen content. Cold water holds more oxygen than warm water. As a result, different parts of a stream or river support species with different oxygen needs (Section 38.2).

> Take-Home Message

What factors affect life in freshwater provinces?

» Lakes have gradients in light, dissolved oxygen, and nutrients.

» Primary productivity varies with a lake's age and, in temperate zones, with the season.

» Different conditions along the length of a stream or river create habitat for different organisms.

47.12 Coastal Ecosystems

■ Where sea meets shore we find regions of high primary productivity.

■ Link to Biofilms 4.4, Parenchyma 27.3, Food webs 46.3

Estuaries—Where Fresh and Saltwater Meet

An estuary is a partly enclosed body of water where fresh water from a river or rivers mixes with seawater. Seawater is denser than fresh water, so fresh water floats on top of the seawater where they meet. The size and shape of the estuary, and the rate at which freshwater flows into it, determine how quickly the saltwater and freshwater mix and the effects of tides. Examples of estuaries include San Francisco Bay, Lake Pontchartrain in New Orleans, Boston harbor, and the Chesapeake Bay (**Figure 47.26**).

In all estuaries, an influx of water from upstream continually replenishes nutrients and allows a high level of productivity. Incoming fresh water also carries silt. Where the velocity of water flow slows, the silt falls to the bottom, forming mudflats. Photosynthetic bacteria and protists in biofilms on mud flats often account for a large portion of an estuary's primary production. Plants adapted to withstand changes in water level and salinity also serve as producers.

Spartina is the dominant plant in the salt marshes of many estuaries along the Atlantic coast (**Figure 47.27A**). It is adapted to its habitat by an ability to withstand immersion during high tides and to tolerate salty, waterlogged, anaerobic soil. Modified parenchyma (Section 27.3) with hollow air spaces allows oxygen

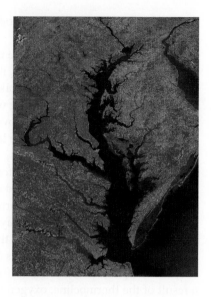

Figure 47.26 An aerial view of the Chesapeake Bay, the largest estuary in the United States. Many rivers empty into the bay, which opens to the Atlantic ocean.

taken in by shoots to diffuse to roots. Salt taken up in water by roots is excreted by specialized glands on the leaves. As a result, the leaves are typically covered with salt crystals. The high salt content of *Spartina* and other estuary plants makes them unpalatable to most herbivores, so detrital food webs predominate.

Estuaries and tidal flats of tropical and subtropical latitudes often support nutrient-rich mangrove wetlands (**Figure 47.27B**). "Mangrove" is the common term for certain salt-tolerant woody plants that live in sheltered areas along tropical coasts. The plants have prop roots that extend out from their trunk and help the plant stay upright in the soft sediments. Specialized

A South Carolina estuary. Cordgrass (*Spartina*) is the dominant plant.

B Florida wetland dominated by red mangroves (*Rhizophora*).

Figure 47.27 Coastal wetlands.

Intertidal zone's upper littoral; submerged only at highest tide of lunar cycle

midlittoral; submerged at each highest regular tide and exposed at lowest tide

lower littoral; exposed only at low tide of lunar cycle

Figure 47.28 Contrasting coasts. (**A,B**) Algae-rich rocky shores where invertebrates abound. (**C**) A sandy shore in Australia shows fewer signs of life. Invertebrates burrow in its sediments.

cells at the surface of some exposed roots allow gas exchange with air.

Estuaries provide important ecological services. Plants in estuaries help slow river flow, thus reducing the risk of flooding. Similarly, they help protect coasts from storm surges. Estuaries' waters serve as nurseries for many species of marine fish and invertebrates. Migratory birds often stop over in estuaries as they travel from one region to another.

Human activities threaten estuary ecosystems. Pollution from farms and cities flows down rivers and into estuary waters. In the tropics, people have traditionally cut mangroves for firewood. A more recent threat is conversion of mangrove wetlands to shrimp farms. The shrimp mainly end up on dinner plates in the United States, Japan, and western Europe.

Rocky and Sandy Coastlines

Rocky and sandy coastlines support ecosystems of the intertidal zone. As with lakes, an ocean's shoreline is described as its littoral zone. The littoral zone can be divided into three vertical regions that differ in their physical characteristics and diversity. The most obvious difference between the regions is the amount of time they spend underwater. The upper littoral zone,

estuary A highly productive ecosystem where nutrient-rich water from a river mixes with seawater.

or splash zone, regularly receives ocean spray but is submerged only during the highest of high tides. It gets the most sun, but holds the fewest species. The midlittoral zone is typically covered by water during an average high tide and dry during a low tide. The lower littoral zone, exposed only during the lowest tide of the lunar cycle, has the most diversity.

You can easily see the zonation along a rocky shore (Figure 47.28A,B). Barnacles live in the midlittoral zone. Algae clinging to rocks are primary producers for the prevailing grazing food web. The primary consumers include a variety of snails.

Zonation is less obvious on sandy shores where detrital food chains start with material washed ashore (Figure 47.28C). Some crustaceans eat detritus in the upper littoral zone. Nearer to the water, other invertebrates feed as they burrow through the sand.

Take-Home Message

What kinds of ecosystems occur along coastlines?

» We find estuaries where rivers empty into seas. The rivers deliver nutrients that foster high productivity.

» Mangrove wetlands are common along shorelines in tropical latitudes.

» Rocky and sandy shores show zonation, with different zones exposed during different phases of the tidal cycle. Diversity is highest in the zone that is submerged most of the time.

47.13 Coral Reefs

- Coral reefs lie just off the coasts of tropical islands and continents.
- Coral reefs are the most diverse marine ecosystems.
- Links to Dinoflagellates 21.6, Corals 24.5, Species introductions 45.9, Global climate change 46.8

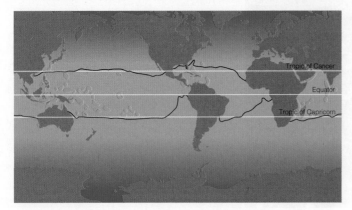

Figure 47.29 Distribution of coral reefs. Reef-building corals live in warm seas, here enclosed in dark lines. The greatest diversity of coral species is in the Indo-Pacifiic region.

Coral reefs are wave-resistant formations that consist primarily of calcium carbonate secreted by generations of coral polyps. Reef-forming corals live mainly in shallow, clear, warm waters between latitudes 25° north and 25° south. About 75 percent of all coral reefs are in the Indian and Pacific oceans (**Figure 47.29**). A healthy reef is home to living corals and a huge number of other species (**Figure 47.30**). Biologists estimate that about a quarter of all marine fish species are associated with coral reefs.

The largest existing reef, Australia's Great Barrier Reef, parallels Queensland for 2,500 kilometers (1,550 miles), and is the largest example of biological architecture. Scientists estimate it began to form about 600,000 years ago. Today it is a string of reefs, some of them 150 kilometers (95 miles) across (**Figure 47.31**). The Great Barrier Reef supports about 500 coral species, 3,000 fish species, 1,000 kinds of mollusks, and 40 kinds of sea snakes.

Photosynthetic dinoflagellates live as mutualistic symbionts inside the tissues of all reef-building corals

Moray eel

Nudibranch (sea slug)

Longnose hawkfish and red sea fan

Banded coral shrimp

Purple tube sponge

Green coral polyp

Figure 47.30 A sample of reef biodiversity.

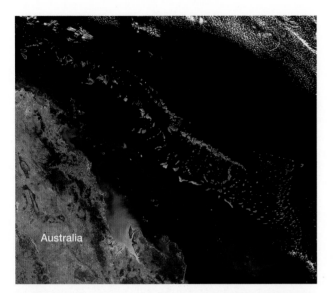

Australia

Figure 47.31 Satellite photo of Australia's Great Barrier Reef.

A Healthy coral reef near Fiji. The coral gets its color from the pigments of symbiotic dinoflagellates that live in its tissues and supply it with sugars.

B "Bleached" reef near Australia. The coral skeletons shown here belong mainly to staghorn coral (*Acropora*), a genus especially likely to undergo coral bleaching.

Figure 47.32 Healthy and damaged coral reefs.

(Section 24.5). The dinoflagellates are protected within the coral's tissues and provide the coral polyp with oxygen and sugars that it depends on. When stressed, coral polyps expel the dinoflagellates. Because the dinoflagellates give a coral its color, expelling these protists turns the coral white, an event called coral bleaching. When a coral is stressed for more than a short time, the dinoflagellate population in a coral's tissues cannot rebound and the coral dies, leaving only its bleached hard parts behind (**Figure 47.32**). The incidence of coral bleaching events has been increasing. Rising sea temperatures and sea level associated with global climate change most likely play a role.

People stress reefs by discharging sewage and other pollutants into coastal waters, by causing erosion that clouds water with sediments, and by destructive fishing practices. Invasive species also threaten reefs. Hawaiian reefs are threatened by exotic algae, including species imported for cultivation during the 1970s.

Assaults on reefs are taking a huge toll. For example, the Indo-Pacific region, the global center for reef diversity, lost about 3,000 square kilometers (1,160 square miles) of living coral reef each year between 1997 and 2003. Reef biodiversity is in danger around the world, from Australia and Southeast Asia to the Hawaiian Islands, the Galápagos Islands, the Gulf of Panama, Kenya, and Florida. The biodiversity on the coral reef offshore from Florida's Key Largo has been reduced by 33 percent since 1970.

coral reef Highly diverse marine ecosystem centered around reefs built by living corals that secrete calcium carbonate.

Take-Home Message

What are coral reefs and how are they threatened?

» Coral reefs form by the action of living corals that lay down a calcium carbonate skeleton. Photosynthetic dinoflagellates in the coral's tissues are necessary for the coral's survival.

» Rising water temperature, pollutants, fishing, and exotic species contribute to loss of reefs.

» Declines in coral reefs will affect the enormous number of fishes and invertebrate species that make their home on or near the reefs.

47.14 The Open Ocean

■ Earth's vast oceans are still largely unexplored. We are only beginning to catalog the diversity they contain.

■ Links to Hydrothermal vents 19.4, Hydrogenosomes 21.3

Pelagic Ecosystems

Like a lake, an ocean shows gradients in light, nutrient availability, temperature, and oxygen concentration. We refer to the open waters of the ocean as the pelagic province (Figure 47.33). These open waters include the neritic zone—the water over continental shelves—together with the more extensive oceanic zone farther offshore. The neritic zone receives nutrients in runoff from land, and is the zone of greatest productivity.

In the ocean's upper, brightly lit waters, photosynthetic microorganisms are the primary producers, and grazing food chains predominate. Depending on the region, some light may penetrate as far as 1,000 meters (3,000 feet) beneath the sea surface. Below that, organisms live in darkness, and organic material that drifts down from above is the basis of detrital food chains.

Our knowledge of the deep pelagic zone is limited because human divers cannot survive at great depths, where the pressure exerted by the weight of the water above them is very high. Thus, exploration of deeper regions requires submersible vessels. Use of such technology has revealed some extraordinary life forms. Figure 47.34 shows two species discovered about 800 meters (2,600 feet) below the surface.

The Seafloor

The benthic province is the ocean bottom—its rocks and sediments. Benthic biodiversity is greatest on the margins of continents, or the continental shelves. The benthic province also includes some largely unexplored concentrations of biodiversity on seamounts and at hydrothermal vents.

Seamounts are undersea mountains that stand 1,000 meters or more tall, but are still below the sea surface (Figure 47.35A). They attract large numbers of fishes and are home to many marine invertebrates (Figure 47.35B). Like islands, seamounts often are home to species that evolved there and are found nowhere else.

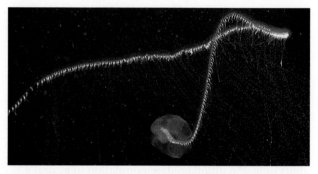

A Giant siphonophore, a colonial cnidarian that can reach 40 meters (130 feet) in length. It emits a bluish bioluminescent glow and snags prey with stinging tentacles.

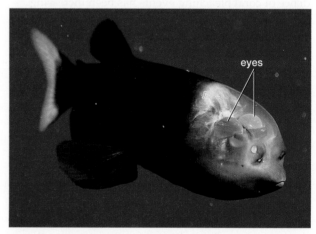

B Barreleye (*Macropinna microstoma*). A transparent dome covers the fish's greenish eyes, which can face upward (as shown) or swivel for a look forward. By one hypothesis, the dome protects the fish's eyes when it steals food from stinging siphonophores.

Figure 47.34 Recently discovered members of the deep pelagic community. They were photographed by remote-controlled submersible vehicles.

air at ocean surface
oceanic zone
neritic zone
sunlit water
"twilight" water
sunless water
0
200
1,000
2,000
4,000
11,000
depth (meters)
Pelagic Province
continental shelf
bathyal zone
abyssal zone
hadal zone
deep-sea trenches
Benthic Province

Figure 47.33 Animated Oceanic zones. Zone dimensions are not drawn to scale.

A Computer model of seamounts on the seafloor off the coast of Alaska. Patton Seamount, at the rear, stands 3.6 kilometers (about 2 miles) tall, with its peak about 240 meters (800 feet) below the sea surface.

B Flytrap anemone discovered on a seamount off the coast of California.

Figure 47.35 Seamounts, undersea islands of diversity. Seamounts formed in the same way as volcanic islands; molten rock erupted through the seafloor.

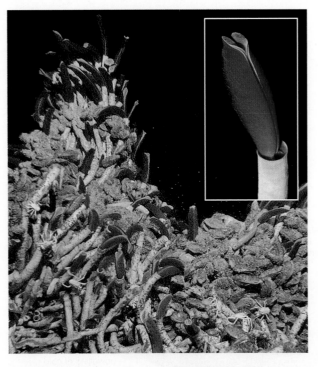

Figure 47.36 Hydrothermal vent community. Tube worms (annelids) and crabs are among the consumers. The producers are chemoautotrophic bacteria and archaea.

There are more than 30,000 seamounts, and scientists have just begun to document species that live on them.

The abundance of life at seamounts makes them attractive to commercial fishing vessels. Fish and other organisms are often harvested by trawling, a fishing technique in which a large net is dragged along the bottom, capturing everything in its path. The process is ecologically devastating; trawled areas are stripped bare of life, and silt stirred up by the giant, weighted nets suffocates filter-feeders in adjacent areas.

Superheated water that contains dissolved minerals spews out from the ocean floor at hydrothermal vents. When this heated, mineral-rich water mixes with cold seawater, the minerals settle out as extensive deposits. Chemoautotrophic bacteria and archaea (Section 20.5) can get energy from these deposits. These microorganisms serve as primary producers for food webs that include invertebrates, such as tube worms and crabs (Figure 47.36). As explained in Section 19.4, one hypothesis holds that life originated on the seafloor in such heated, nutrient-rich places.

benthic province The ocean's sediments and rocks.
hydrothermal vent Rocky, underwater opening where mineral-rich water heated by geothermal energy streams out.
pelagic province The ocean's waters.
seamount An undersea mountain.

Animal life exists even in the deepest sea. A remote-controlled submersible that sampled sediments in the deepest part of the ocean (the Mariana Trench) brought up foraminifera that live 11 kilometers (7 miles) below the surface. Sediment samples from deep in the Mediterranean Sea turned up another surprise, animals that live and reproduce in a habitat devoid of oxygen. The animals, called loriciferans, are less than a millimeter long (Figure 47.37) and are distant relatives of insects and nematodes. They live in sulfur-rich, anoxic sediments. Like some anaerobic protists, they lack mitochondria, and instead have organelles called hydrogenosomes (Section 21.3).

Figure 47.37 A deep-sea loriciferan. This tiny animal can live without oxygen.

Take-Home Message

What factors affect life in ocean provinces?

» Oceans have gradients in light, dissolved oxygen, and nutrients. Nearshore and well-lit zones are the most productive and species-rich.

» On the seafloor, pockets of diversity occur on seamounts and around hydrothermal vents.

» Animal life exists even in the deepest ocean, and some animals have adapted to life in oxygen-free sediments.

47.15 Ocean–Air Interactions

■ Events in the atmosphere and oceans, and on land, interconnect in ways that profoundly affect the world of life.

■ Link to Copepods 24.14

An El Niño Southern Oscillation, or ENSO, is defined by changes in sea surface temperatures and in the air circulation patterns. "Southern oscillation" refers to a seesawing of the atmospheric pressure in the western equatorial Pacific—Earth's greatest reservoir of warm water and warm air. It is the source of heavy rainfall,

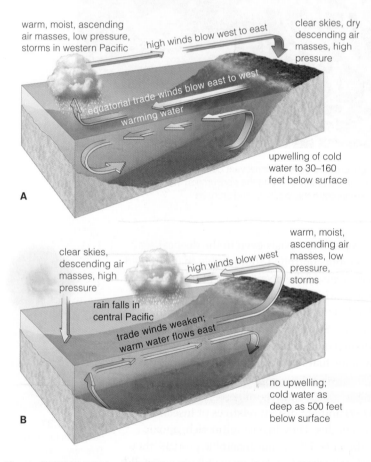

warm, moist, ascending air masses, low pressure, storms in western Pacific

high winds blow west to east

clear skies, dry descending air masses, high pressure

equatorial trade winds blow east to west

warming water

upwelling of cold water to 30–160 feet below surface

A

clear skies, descending air masses, high pressure

high winds blow west

warm, moist, ascending air masses, low pressure, storms

rain falls in central Pacific

trade winds weaken; warm water flows east

no upwelling; cold water as deep as 500 feet below surface

B

Figure 47.38 (**A**) Westward flow of cold, surface water between ENSOs. (**B**) Eastward dislocation of warm water during El Niño.

which releases enough heat energy to drive global air circulation patterns.

Between ENSOs, the warm waters and heavy rains move westward (**Figure 47.38A**). During an ENSO, the prevailing surface winds over the western equatorial Pacific pick up speed and "drag" surface waters east (**Figure 47.38B**). As they do, the westward transport of water slows down. Sea surface temperatures rise, evaporation accelerates, and air pressure falls. These changes affect weather worldwide.

El Niño episodes persist for 6 to 18 months. Often they are followed by a La Niña episode in which the Pacific waters become cooler than usual. Other years, waters are neither warmer nor colder than average.

As noted in Section 47.1, 1997 ushered in the most powerful El Niño event of the century. The average sea surface temperatures in the eastern Pacific rose by 5°C (9°F). This warmer water extended 9,660 kilometers (6,000 miles) west from the coast of Peru.

The 1997–1998 El Niño/La Niña roller-coaster had extraordinary effects on the primary productivity in the equatorial Pacific. With the massive eastward flow of nutrient-poor warm water, photoautotrophs were almost undetectable in satellite photos that measure primary productivity (**Figure 47.39A**).

During the La Niña rebound, cooler, nutrient-rich water moved up to the sea surface and was displaced westward all along the equator. As satellite images revealed, upwelling had sustained an algal bloom that stretched across the equatorial Pacific (**Figure 47.39B**).

During the 1997–1998 El Niño event, 30,000 cases of cholera were reported in Peru alone, compared with only 60 cases from January to August in 1997. People knew that water contaminated by *Vibrio cholerae* causes epidemics of cholera. The disease agent triggers severe diarrhea. Bacteria-contaminated feces enter the water supply and individuals who use the tainted water become infected.

What people did not know was where *V. cholerae* remained between cholera outbreaks. It could not be

Figure 47.39 Satellite data on primary productivity in the equatorial Pacific Ocean. The concentration of chlorophyll in the water was used as the measure.

(**A**) During the 1997–1998 El Niño episode, a massive amount of nutrient-poor water moved to the east, and so photosynthetic activity was negligible.

(**B**) During a subsequent La Niña episode, massive upwelling and westward displacement of nutrient-rich water led to a vast algal bloom that stretched all the way to the coast of Peru.

A Near-absence of phytoplankton in the equatorial Pacific during an El Niño.

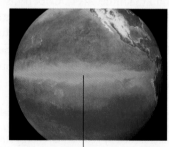

B Huge algal bloom in the equatorial Pacific in the La Niña rebound event.

Warming Water and Wild Weather (revisited)

The United States National Oceanographic and Atmospheric Administration (NOAA) is charged with studying and predicting El Niño events. The agency collects and analyzes sea temperature data from a system of buoys moored in the tropical Pacific Ocean (*below right*). The goal of this research is to determine the way in which the El Niño Southern Oscillation affects global weather patterns and the extent of its effects. Such an understanding could help the scientists to develop a method of predicting when an El

Niño or La Niña event is likely to occur and which regions are at a heightened risk for flooding, drought, hurricanes, or fires as a result. Predicting and planning for such occurrences could help prevent or minimize their harmful effects. The current data on sea surface temperature, as well as information about the monitoring program and its goals, are available on NOAA's website at www.elnino.noaa.gov.

How would you vote? Do you think that funding research into methods of predicting El Niño and La Niña events is a good investment?

found in humans or in water supplies. Even so, the cholera would often break out simultaneously in far-apart places—usually coastal cities where the urban poor draw water from rivers near the sea.

Marine biologist Rita Colwell had been thinking about the fact that humans are not the host between outbreaks. Was there an environmental reservoir for the pathogen? Maybe, but nobody had detected it in water samples subjected to standard culturing.

Then Colwell had a flash of insight: What if no one could find the pathogen because it changes its form and enters a dormant stage between outbreaks?

During one cholera outbreak in Louisiana, Colwell realized that she could use an antibody-based test to detect a protein unique to *V. cholerae*'s surface. Later, tests in Bangladesh revealed bacteria in fifty-one of fifty-two samples of water. Standard culture methods had missed it in all but seven samples.

V. cholerae survives in rivers, estuaries, and seas. As Colwell knew, plankton also thrive in these aquatic environments. She decided to restrict her search for the unknown host to warm waters near Bangladesh, where outbreaks of cholera occur seasonally. It was here that Colwell discovered a dormant *V. cholerae* stage inside copepods, a type of tiny marine crustacean (Section 24.14). Copepods eat phytoplankton, so the abundance of copepods—and of *V. cholerae* cells inside them—increases and decreases with the abundance of phytoplankton.

Colwell suspected that water temperature changes in the Bay of Bengal were tied to cholera outbreaks, so she looked at medical reports for the 1990–1991 and 1997–1998 El Niño episodes. She found the number of reported cholera cases rose four to six weeks after an El Niño event began. El Niño brings warmer water with more nutrients to the Bay of Bengal, encouraging

Figure 47.40 In Bangladesh, Rita Colwell comparing samples of unfiltered and filtered drinking water.

the growth of phytoplankton. This added food then increases the number of cholera-carrying copepods.

Today, Colwell and Anwarul Huq, a Bangladeshi scientist, are investigating salinity and other factors that may relate to outbreaks. Their goal is to design a model for predicting where cholera will occur next. They advised women in Bangladesh to use sari cloth as a filter to remove *V. cholerae* cells from the water (**Figure 47.40**). The copepod hosts are too big to pass through the thin cloths, which can be rinsed in clean water, sun-dried, and used again. This inexpensive, simple method has cut cholera outbreaks by half.

Take-Home Message

What occurs during an El Niño event?

» During an El Niño event, changes in ocean temperatures and winds alter currents, affecting weather, marine food webs, and human health.

Human Impact on the Biosphere

LEARNING ROADMAP

Where you have been In this chapter, we consider how human activities are causing a mass extinction (Section 17.12). In doing so, we reconsider the ecological effects of acid rain (2.6), deforestation (22.1), soil erosion (28.2), aquifer depletion (46.6), and global climate change (46.8).

Where you are now

The Extinction Crisis
Human activities have accelerated the rate of extinctions. Habitat loss, degradation, and fragmentation lead to extinctions, as do species introductions and overharvesting.

Harmful Practices
Plowing grasslands and cutting down forests can have long-term and long-range effects. Loss of plants allows soil erosion, raises soil temperature, and affects rainfall.

Pollutants
Human activities produce pollutants that harm living organisms, destroy the ozone layer, and cause global climate change.

Conserving Biodiversity
All nations have biological wealth that can benefit human populations. Conservation biologists assess biodiversity and investigate the best ways to sustain it.

Reducing Negative Impacts
Individual actions that minimize the use of resources and energy can help reduce the threats to the health of the planet and to biodiversity.

Where you are going This chapter concludes our study of biology, but you will hear about the topics covered here again and again. We face a daunting task—sustaining a growing population without destroying natural processes on which all life depends. We hope that your understanding of these processes and the way that humans affect them will help you decide how best to meet this challenge.

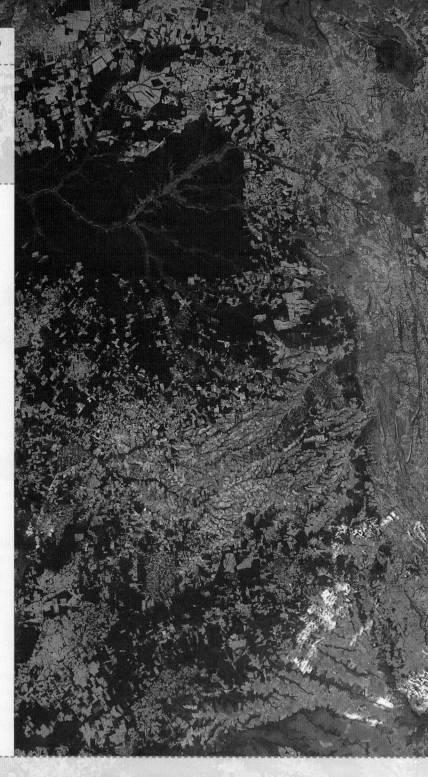

48.1 A Long Reach

We began this book with the story of biologists who ventured into a remote forest in New Guinea, and the many previously unknown species that they encountered. At the far end of the globe, a U.S. submarine surfaced in Arctic waters and discovered polar bears hunting on the ice-covered sea (Figure 48.1). The bears were about 270 miles from the North Pole and 500 miles from the nearest land.

Even such seemingly remote regions are no longer beyond the reach of human explorers—and human influence. You already know that increasing levels of greenhouse gases are raising the temperature of Earth's atmosphere and seas. In the Arctic, the warming is causing sea ice to thin and to break up earlier in the spring. This raises the risk that polar bears hunting far from land will become stranded and unable to return to solid ground before the ice thaws.

Polar bears are top predators and their tissues contain a surprisingly high amount of mercury and organic pesticides. These substances entered the water and air far away, in more populated temperate regions. Winds and ocean currents delivered them to the Arctic. Contaminants also travel north inside the bodies of migratory animals such as seabirds that spend their winters in temperate regions and nest in the Arctic.

In places less remote than the Arctic, effects of human populations have a more direct effect. As we cover more and more of the world with our dwellings, factories, and farms, less appropriate habitat remains for other species. We also put species at risk by competing with them for resources, overharvesting them, and introducing nonnative competitors.

It would be presumptuous to think that we alone have had a profound impact on the world of life. As long ago as the Proterozoic, photosynthetic cells were irrevocably changing the course of evolution by enriching the atmosphere with oxygen. Over life's existence, the evolutionary success of some groups has ensured the decline of others. What is new is the increasing pace of change and the capacity of our own species to recognize and affect its role in this increase.

A century ago, Earth's physical and biological resources seemed inexhaustible. Now we know that many practices put into place when humans were largely ignorant of how natural systems operate take a heavy toll on the biosphere. The rate of species extinctions is on the rise and many types of biomes are threatened. These changes, the methods scientists use to document them, and the ways that we can address them are the focus of this chapter.

Figure 48.1 Three polar bears investigate an American submarine that surfaced in ice-covered Arctic waters.

48.2 The Extinction Crisis

- Extinction is a natural process, but we are accelerating it.
- Links to Geologic time scale 16.8, Mass extinction 16.1, 17.12

Era	Period	
		Major extinction under way
CENOZOIC	QUATERNARY — 1.8 mya — TERTIARY	With high population growth rates and cultural practices (e.g., agriculture, deforestation), humans become major agents of extinction.
	— 65.5 —	*Major extinction event*
MESOZOIC	CRETACEOUS — 145.5 — JURASSIC — 199.6 — TRIASSIC	Slow recovery after Permian extinction, then adaptive radiations of some marine groups and plants and animals on land. Asteroid impact at K–T boundary, 85% of all species disappear from land and seas.
	— 251 —	*Major extinction event*
PALEOZOIC	PERMIAN — 299 — CARBONIFEROUS — 359 —	Pangea forms; land area exceeds ocean surface area for first time. Asteroid impact? Major glaciation, colossal lava outpourings, 90%–95% of all species lost.
		Major extinction event
	DEVONIAN — 416 — SILURIAN — 443 —	More than 70% of marine groups lost. Reef builders, trilobites, jawless fishes, and placoderms severely affected. Meteorite impact, sea level decline, global cooling?
		Major extinction event
	ORDOVICIAN — 488 — CAMBRIAN — 542 —	Second most devastating extinction in seas; nearly 100 families of marine invertebrates lost.
		Major extinction event
	(Precambrian)	Massive glaciation; 79% of all species lost, including most marine microorganisms.

A

Mass Extinction

Extinction, like speciation, is a natural process. Species arise and become extinct on an ongoing basis. Based on several lines of evidence, scientists estimate that 99 percent of all species that have ever lived are now extinct. The rate of extinction picks up dramatically during a **mass extinction**, when a large proportion of Earth's organisms become extinct in a relatively short period of geologic time.

Five great mass extinctions mark the boundaries for the geologic time periods. As you learned in Section 16.1, the mass extinction at the end of the Cretaceous period was most likely caused by an asteroid impact. The greatest mass extinction, in terms of species lost, occurred at the end of the Permian. It took place after increased volcanic activity in what is now Siberia released a large amount of carbon into the atmosphere. It is thought that the resulting warming of Earth's oceans caused the release of methane (natural gas) from frozen deposits deep in the ocean. The ocean became anoxic and methane that entered the air caused explosive conflagrations to occur on land.

Mass extinctions affect all lineages. However, lineages differ in their time of origin, their tendency to branch and give rise to new species, and how long they endure. If we consider the number of species as the measure of success for any lineage, not all lineages are equally successful. **Figure 48.2B** illustrates how the number of species changed over time in some major lineages. Expansion of one lineage sometimes occurs at the same time as contraction of another, as when the gymnosperms declined during the adaptive radiation of angiosperms.

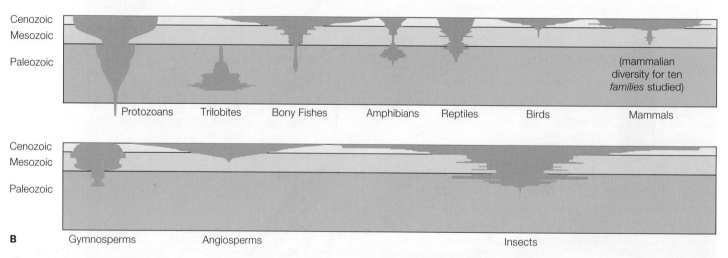

B

Figure 48.2 Animated (A) Dates of the five greatest mass extinctions and recoveries in the past. Compare **Figure 16.15. (B)** Species diversity over time for a sampling of taxa. The width of each *blue* shape represents the number of species in that lineage. Notice the variation among lineages.

A Giant ground sloths. The sloths and other members of North America's megafauna disappeared after humans arrived on the continent.

B The dodo. This flightless bird, larger than a turkey, went extinct in the 1600s, after Europeans discovered the island on which it lived.

Figure 48.3 Drawings of two extinct animals.

The Sixth Great Mass Extinction

We are presently in the midst of a mass extinction. The current extinction rate is estimated to be 100 to 1,000 times above the typical background rate, putting it on a par with the five major extinction events of the past. Unlike previous extinction events, this one is not the inevitable result of a physical catastrophe such as an eruption or asteroid impact. Humans are the driving force behind the current rise in extinctions and our actions will determine the extent of the losses.

Indirect evidence implicates the arrival of humans in the prehistoric decline of many large animals, collectively referred to as megafauna, in Australia and North America. In Australia, the arrival of humans about 40,000 years ago correlates with the onset of extinctions of the continent's largest marsupials, birds, and lizards. In North America, arrival of humans 14,000 years ago was followed by the extinction of large herbivores such as mammoths and mastodons (relatives of elephants), camels, and giant ground sloths (**Figure 48.3A**). Carnivores such as lions and saber-toothed cats also disappeared.

By one hypothesis, humans directly caused declines in herbivorous species such as ground sloths and mammoths by hunting them. These declines then led to the extinction of their predators. Evidence

that humans hunted some megafauna supports this hypothesis. For example, stone spear points and other tools have been found along with mammoth or mastodon remains at many North American sites.

However, not all researchers are convinced that hunting was the most important factor in megafaunal extinctions. The demise of these animals has also been attributed to climate change, a comet impact, or effects of human-introduced pathogens. Continued study of sites dating to the time of these extinction events will help clarify the relative importance of these factors.

More recent extinctions are more certainly attributable to humans. The World Conservation Union has compiled a list of more than 800 documented extinctions that occurred since 1500. As one example, the dodo (**Figure 48.3B**) was a big, flightless bird that lived on the island of Mauritius in the Indian Ocean. Dodos were plentiful in 1600, when Dutch sailors first arrived on the island, but 80 or so years later the birds were extinct. Some were eaten by sailors. However, destruction of nests and habitat by rats, cats, and pigs that accompanied humans probably had a greater effect.

Take-Home Message

How does the current mass extinction differ from previous ones?

» Previous mass extinctions occurred when an event such as an asteroid impact or a series of volcanic eruptions caused a global catastrophe.

» The current extinction crisis is the result of human activity.

mass extinction A large proportion of Earth's organisms become extinct in a relatively short period of geologic time.

48.3 Currently Threatened Species

■ A wide variety of life is currently threatened by human activities.

■ Links to Overfishing of cod 44.6, Exotic species 45.9, Aquifers 46.6, Nitrogen cycle 46.9

An **endangered species** faces extinction in all or part of its range. A **threatened species** is one that is likely to become endangered in the near future. Keep in mind that not all rare species are threatened or endangered. Some species have always been uncommon. A species is considered endangered when one or more of its populations have declined or are declining.

Causes of Species Declines

When European settlers arrived in North America, they found between 3 and 5 billion passenger pigeons. In the 1800s, commercial hunting caused a steep decline in the bird's numbers and the last captive bird died in 1914. We continue to overharvest species. The crash of the Atlantic codfish population (Section

A White abalone   **B** Panda

Figure 48.4 Some threatened and endangered species. To learn about others, visit the International Union for Conservation of Nature and Natural Resources web site at www.iucnredlist.org/.

44.6) is one recent example. The white abalone has suffered a similar fate. This gastropod mollusk is native to kelp forests along the coast of California (**Figure 48.4A**). During the 1970s, overharvesting of this species reduced the population to about 1 percent of its original size. In 2001, the abalone became the first invertebrate listed as endangered by the United States Fish and Wildlife Service. Some white abalone remain in the wild, but population density is too low for effective reproduction. White abalones release their gametes into the water, a strategy that is effective only when many animals live in close proximity. Individuals have been captured for a captive breeding program. If the program succeeds, their offspring can be reintroduced into the wild.

Each species requires a specific habitat, and degradation, fragmentation, or destruction of that habitat also reduces population numbers. An **endemic species**, one that lives only in the region where it evolved, is more likely to go extinct than a species with a more widespread distribution. Also, a species with highly specific needs is more likely to go extinct than one that is a generalist. For example, giant pandas (**Figure 48.4B**) are endemic to China's bamboo forests and feed mainly on bamboo. As bamboo forests have disappeared, so have pandas. Their population, which may once have been as high as 100,000 animals, is now reduced to about 1,600 animals in the wild.

Similarly, logging of forests in the southeastern United States reduced numbers of North America's largest woodpecker (**Figure 48.4C**). The ivory-billed woodpecker was believed to have become extinct in the 1940s until a possible sighting in 2004 led to an extensive hunt for evidence of the bird's survival. This search has produced some blurry photos, snippets of video, and a few recordings of what may or may not be ivorybill calls and knocks. Definitive proof that a population of these birds still exists remains elusive.

Table 48.1 Global List of Threatened Species *

	Described Species	Evaluated for Threats	Found to Be Threatened
Vertebrates			
Mammals	5,416	4,863	1,094
Birds	9,956	9,956	1,217
Reptiles	8,240	1,385	422
Amphibians	6,199	5,915	1,808
Fishes	30,000	3,119	1,201
Invertebrates			
Insects	959,000	1,255	623
Mollusks	81,000	2,212	978
Crustaceans	40,000	553	460
Corals	2,175	13	5
Others	130,200	83	42
Land Plants			
Mosses	15,000	92	79
Ferns and allies	13,025	211	139
Gymnosperms	980	909	321
Angiosperms	258,650	10,771	7,899
Protists			
Green algae	3,715	2	0
Red algae	5,956	58	9
Brown algae	2,849	15	6
Fungi			
Lichens	10,000	2	2
Mushrooms	16,000	1	1

* IUCN–WCU Red List, available online at www.iucnredlist.org

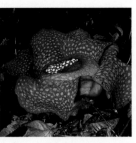

C Ivory-billed woodpecker **D** Texas blind salamander **E** Prairie fringed orchid **F** *Rafflesia*

Habitat degradation threatens Texas blind salamanders (**Figure 48.4D**). The salamander is endemic to the Edwards Aquifer, a series of water-filled, underground limestone formations. Like other aquifers, the salamander's home is threatened by excess withdrawal of water and by groundwater pollution.

Habitat loss and destruction also affect plants. Destruction of North American prairies threatens prairie fringed orchids (**Figure 48.4E**). Destruction of Indonesian rain forest endangers species of *Rafflesia*, plants with world's largest flowers (**Figure 48.4F**).

Deliberate or accidental species introductions are another threat (Section 45.9). Rats that reached islands by stowing away on ships attack and endanger ground-nesting birds. Exotic species also cause problems by outcompeting natives. After European brown trout and eastern brook trout were introduced into California's mountain streams for sport fishing, populations of the native golden trout declined.

The koa bug is the victim of a biological control program gone awry. This Hawaiian native is threatened by an exotic parasitoid fly introduced to attack another bug that is an agricultural pest. Sadly, the fly prefers koa bugs to its intended target.

The decline or loss of one species can endanger others. For example, buffalo once grazed on running buffalo clover in the Midwest. The plants thrived in the open woodlands where buffalo enriched the soil with their droppings and helped to disperse the clover's seeds. Like most endangered species, buffalo clover faces a number of simultaneous threats. In addition to the loss of the buffalo, the clover is threatened by conversion of its habitat to housing developments, competition from introduced plants, and attacks by introduced insects.

endangered species A species that faces extinction in all or a part of its range.
endemic species A species that remains restricted to the area where it evolved.
threatened species Species likely to become endangered in the near future.

The Unknown Losses

In November of 2009, the International Union for Conservation of Nature and Natural Resources (IUCN) reported that of the 48,677 species they had assessed, 36 percent were threatened or endangered. We do not know the level of threat for the vast majority of the approximately 1.8 million named species, or for the countless millions of species yet to be discovered.

Endangered species listings have historically focused on vertebrates. Scientists have only recently begun to consider the threats to invertebrates and to plants. Our impact on protists and fungi is largely unknown, and the IUCN does not address threats to bacteria or archaea (**Table 48.1**).

Microbiologist Tom Curtis is among those making a plea for increased research on microbial ecology and microbial diversity. He argues that we have barely begun to comprehend the vast number of microbial species and to understand their importance. He writes, "I make no apologies for putting microorganisms on a pedestal above all other living things. For if the last blue whale choked to death on the last panda, it would be disastrous but not the end of the world. But if we accidentally poisoned the last two species of ammonia-oxidizers, that would be another matter. It could be happening now and we wouldn't even know . . . " Ammonia-oxidizing bacteria play an essential role in the nitrogen cycle by converting ammonia in wastes and remains to nitrites (Section 46.9).

Take-Home Message

How do human activities endanger existing species?

» Species decline when humans destroy or fragment natural habitat by converting it to human use, or degrade it through pollution or withdrawal of an essential resource.

» Humans also cause declines by overharvesting species and by introducing exotic species that harm native ones.

» Most endangered species are affected by multiple threats.

48.4 Harmful Land Use Practices

■ Human activities have the potential not only to harm individual species, but also to transform entire biomes.
■ Links to Deforestation 22.1, Dustborne pathogens 23.1, Soil erosion 28.2, Transpiration 28.4

Desertification

As human populations increase, greater numbers of people are forced to farm in areas that are ill-suited to agriculture. Others allow livestock to overgraze in grasslands. **Desertification**, a conversion of grassland or woodlands to desertlike conditions, is one result.

Deserts naturally expand and contract over long periods with changes in climate. However, poor agricultural practices that encourage soil erosion can sometimes lead to rapid shifts from grassland or woodland to desert.

For example, during the mid-1930s, large portions of prairie on the southern Great Plains were plowed under to plant crops. This plowing exposed the deep prairie topsoil to the force of the region's constant winds. Coupled with a drought, the result was an economic and ecological disaster. Winds carried more than a billion tons of topsoil aloft as sky-darkening dust clouds turned the region into what came to be known as the Dust Bowl (**Figure 48.5**). Tons of displaced soil fell to earth as far away as New York City and Washington, D.C.

Desertification now threatens vast areas. In Africa, the Sahara desert is expanding south into the Sahel region. Overgrazing in this region strips grasslands of their vegetation and allows winds to erode the soil. Winds carry the soil aloft and westward. Soil particles land as far away as the southern United States and the Caribbean (**Figure 48.6**). As Section 23.1 explained, fungal pathogens can travel on such dust particles.

In China's northwestern regions, overplowing and overgrazing have expanded the Gobi desert so that dust clouds periodically darken skies above Beijing. Winds carry some of the dust across the Pacific to the United States. To hold back the desert, China has now planted billions of trees as a "Green Wall."

Drought encourages desertification, which results in more drought in a positive feedback cycle. Plants cannot thrive in a region where the topsoil has blown away. With less transpiration (Section 28.4), less water enters the atmosphere, so local rainfall decreases.

The best way to prevent desertification is to avoid farming in areas subject to high winds and periodic drought. If these areas must be utilized, methods that do not repeatedly disturb the soil can minimize the risk of desertification.

Deforestation

Deforestation has detrimental effects beyond the immediate destruction of forest organisms. For example, deforestation encourages flooding because water runs off into streams, rather than being taken up by tree roots. Deforestation also raises risk of landslides

Figure 48.5 A giant dust cloud about to descend on a farm in Kansas during the 1930s. A large portion of the southern Great Plains was then known as the Dust Bowl. Drought and poor agricultural practices allowed winds to strip tons of topsoil from the ground and carry it aloft.

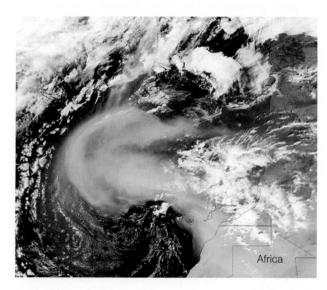

Figure 48.6 Modern-day dust cloud blowing from Africa's Sahara desert out into the Atlantic Ocean. The Sahara's area is increasing as a result of a long-term drought, overgrazing, and stripping of woodlands for firewood.

in hilly areas. Tree roots tend to stabilize the soil. When they are removed, waterlogged soil becomes more likely to slide.

Deforested areas also become nutrient-poor. **Figure 48.7** shows results of an experiment in which scientists deforested a region in New Hampshire and monitored the nutrients in runoff. Deforestation caused a spike in loss of essential soil nutrients such as calcium.

Like desertification, deforestation affects local weather. Temperatures in forests are cooler than in adjacent nonforested areas because trees shade the ground and transpiration causes evaporative cooling. When a forest is cut down, daytime temperatures rise and reduced transpiration results in less rainfall. Once a tropical forest is logged, the resulting nutrient losses and drier, hotter conditions can make it impossible for tree seeds to germinate or for seedlings to survive. Thus, deforestation can be difficult to reverse.

The amount of forested land is currently stable or increasing in North America, Europe, and China, but tropical forests continue to disappear at an alarming rate. In Brazil, increases in the export of soybeans and free-range beef have helped make the country the world's seventh-largest economy. However, this economic expansion has come at the expense of the country's woodlands and forests (**Figure 48.8**). The effect may be long term. Forest loss may alter rainfall patterns, making the region permanently drier.

desertification Conversion of a grassland or woodland to desert.

A plot of forest stripped of vegetation as an experiment.

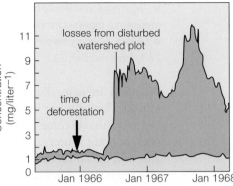

After deforestation, calcium levels in runoff rose sixfold (*dark blue*). An undisturbed control plot in the same forest showed no similar increase during this time (*light blue*).

Figure 48.7 Animated Experimental deforestation of the Hubbard Brook watershed.

Deforestation also contributes to global climate change in two ways. First, trees that are cut to clear land for other uses are often burned, releasing carbon into the atmosphere. Second, conversion of forest to cropland or pasture decreases the rate at which carbon is taken up by plants.

Take-Home Message

What are effects of desertification and deforestation?

» Desertification and deforestation not only destroy habitat, they allow increased erosion and can cause changes in local weather patterns.

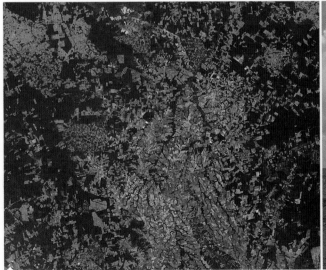

Figure 48.8 Tropical deforestation. Aerial view (*left*) and ground view (*right*) of Mato Grosso, Brazil, a region where tropical woodlands and forest have been cleared at an alarming rate, mainly to create pasture for cattle. Cattle producers have been forced to expand into forests by an increase in the industrial production of soybeans and corn in what was formerly prime cattle-rearing territory.

48.5 Toxic Pollutants

■ Organisms are directly harmed by chemicals and trash we release into the environment.

■ Links to Mercury poisoning 2.1, pH 2.6, Phytoremediation 28.1, Plant nutrition 28.2

Pollutants are natural or man-made substances released into soil, air, or water in greater than natural amounts. Some pollutants come from a few easily identifiable sites, or point sources. A factory that discharges pollutants into the air or water is a point source. Pollutants that come from point sources are the easiest to control: Identify the sources, and you can take action there. Dealing with pollution from nonpoint sources is more challenging. Such pollution stems from widespread release of a pollutant. For example, oil that runs off from roads and driveways is a nonpoint source of water pollution. The diffuse nature of nonpoint pollution makes finding ways to prevent it more difficult.

Acid Rain

Sulfur dioxides and nitrogen oxides are common air pollutants. Coal-burning power plants release most sulfur dioxides. Vehicles and power plants that burn gas and oil emit nitrogen oxides.

In dry weather, airborne sulfur and nitrogen oxides coat dust particles. Dry acid deposition occurs when the coated dust falls to the ground. Wet acid deposition, or **acid rain**, occurs when pollutants combine with water and fall as acidic precipitation. The pH of unpolluted rainwater is 5 or higher (Section 2.6). Acid rain can be ten times more acidic (**Figure 48.9A**).

Acid rain that falls on or drains into waterways, ponds, and lakes affects aquatic organisms. Low pH prevents fish eggs from developing and kills adults. When acid rain falls on forests it burns tree leaves and alters the composition of soils. As acidic water drains through the soil, positively charged hydrogen ions displace positively charged nutrient ions such as calcium, causing nutrient loss. The acidity also causes soil particles to release metals such as aluminum that can harm plants. The combination of poor nutrition and exposure to toxic aluminum weakens trees, making them more susceptible to insects and pathogens, and thus more likely to die. Effects are most pronounced at higher elevations where trees are frequently exposed to clouds of acidic droplets (**Figure 48.9B**).

Biological Accumulation and Magnification

Some pollutants build up inside organisms. By the process of **bioaccumulation**, an organism's tissues store a pollutant taken up from the environment, causing the amount in the body to increase over time. The ability of some plants to bioaccumulate toxic substances makes them useful in phytoremediation of polluted soils (Section 28.1).

In animals, hydrophobic chemical pollutants ingested or absorbed across the skin tend to accumulate in fatty tissues. Because the amount of pollutant in an animal's body increases over time, longer-lived species tend to be more affected by fat-soluble pollutants than shorter-lived ones. Within a species, old individuals tend to have a higher pollutant load than younger ones.

Trophic level also affects the pollutant concentration in an organism's tissues. By the process of **biological**

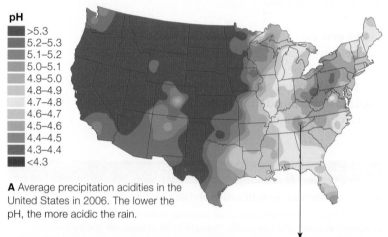

pH
- >5.3
- 5.2–5.3
- 5.1–5.2
- 5.0–5.1
- 4.9–5.0
- 4.8–4.9
- 4.7–4.8
- 4.6–4.7
- 4.5–4.6
- 4.4–4.5
- 4.3–4.4
- <4.3

A Average precipitation acidities in the United States in 2006. The lower the pH, the more acidic the rain.

B Dying trees in Great Smoky Mountains National Park, where acid rain harms leaves and causes loss of nutrients from soil.

Figure 48.9 **Animated** Acid rain. **Figure It Out: Is rain more acidic on the East Coast or the West Coast?**
Answer: The East Coast

magnification, the concentration of a chemical increases as the pollutant moves up a food chain. **Figure 48.10** provides data documenting biological magnification of DDT (an insecticide) in a salt marsh ecosystem during the 1960s. Notice that the concentration of DDT in ospreys, which are fish-eating birds, was 276,000 times higher than that in the water. As a result of bioaccumulation and biological magnification, even very low environmental concentrations of pollutants can have detrimental effects on a species. During the 1960s, ospreys were driven near extinction by widespread use of DDT, because this chemical interferes with their egg shell formation. A ban on DDT use in the United States has allowed the species to recover.

Bioaccumulation and magnification of other pollutants such as methylmercury continue to pose a threat to wildlife and human health (Section 2.1).

The Trouble With Trash

Seven billion people use and discard a lot of stuff. In 2009, the United States generated 243 million tons of garbage. That averages out to about 2 kilograms (4.3 pounds) per person per day.

Plastic and other garbage constantly enters our coastal waters. Foam cups and containers from fast-food outlets, plastic shopping bags, plastic water bottles, and other material discarded as litter ends up in storm drains. From there it is carried to streams and rivers that can convey it to the sea. A seawater sample taken near the mouth of the San Gabriel River in southern California had 128 times as much plastic as

DDT Residues (In parts per million wet weight of organism)	
Osprey	13.8
Green heron	3.57
Atlantic needlefish	2.07
Summer flounder	1.28
Sheepshead minnow	0.94
Hard clam	0.48
Marsh grass	0.33
Flying insects (mostly flies)	0.30
Mud snail	0.26
Shrimps	0.16
Green alga	0.083
Water	0.00005

Figure 48.10 Biological magnification of DDT in an estuary on Long Island, New York, as reported in 1967 by George Woodwell, Charles Wurster, and Peter Isaacson.

Figure 48.11 Recently deceased Laysan albatross chick. Scientists found more than 300 pieces of plastic inside the bird. One had punctured its gut wall, resulting in the bird's death. The chick was fed the plastic by its parents, who gathered the material from the ocean surface, mistaking it for food.

plankton by weight. Waste plastic that gets into oceans harms marine life. For example, seabirds often eat floating bits of plastic and feed them to their chicks, with deadly results (**Figure 48.11**).

acid rain Low-pH rain formed when sulfur dioxide and nitrogen oxides mix with water vapor in the atmosphere.
bioaccumulation An organism accumulates increasing amounts of a chemical pollutant in its tissues over the course of its lifetime.
biological magnification A chemical pollutant becomes increasingly concentrated as it moves up through food chains.
pollutant Substance that is released as result of human activities and harms organisms or disrupts natural processes.

Take-Home Message

What are some ways that pollutants directly harm living organisms?

» Acid rain can make a lake or soil too acidic for life to thrive.

» Even small amounts of fat-soluble chemicals can accumulate in tissues and be magnified from one trophic level to the next.

» Animals can be harmed when they mistake indigestible trash that enters waterways and oceans for food.

48.6 Ozone Depletion and Pollution

■ Ozone is said to be "good up high, but bad nearby." It forms a protective sunscreen in Earth's upper atmosphere, but is a harmful pollutant in air near the ground.

■ Links to UV as mutagen 8.6, Ozone layer 19.1

Depletion of the Ozone Layer

In the upper layers of the atmosphere, between 17 and 27 kilometers (10.5 and 17 miles) above sea level, the ozone (O_3) concentration is so great that scientists refer to this region as the **ozone layer**. The ozone layer benefits living organisms by absorbing most ultraviolet (UV) radiation from incoming sunlight. UV radiation causes mutations (Section 8.6).

In the mid-1970s, scientists noticed that Earth's ozone layer was thinning. Its thickness had always varied seasonally, but now the average level was declining steadily from year to year. By the mid-1980s, the spring ozone thinning over Antarctica was so pronounced that people were referring to the lowest-ozone region as an "ozone hole" (**Figure 48.12A**).

Declining ozone quickly became an international concern. With a thinner ozone layer, people would be exposed to more UV radiation, the main cause of skin cancers. Higher UV levels also harm wildlife, which do not have the option of avoiding sunlight. In addition, exposure to higher than normal UV levels affects plants and other producers, slowing the rate of photosynthesis and the release of oxygen into the air.

Chlorofluorocarbons, or CFCs, are the main ozone destroyers. These odorless gases were once widely used as propellants in aerosol cans, as coolants, and in solvents and plastic foam. In response to the threat posed by ozone thinning, countries worldwide agreed in 1987 to phase out production of CFCs and other ozone-destroying chemicals. As a result of that agreement (the Montreal Protocol), the concentrations of CFCs in the atmosphere are no longer rising dramatically (**Figure 48.12B**). However, because CFCs break down slowly, scientists expect them to significantly impair the ozone layer for several decades. Also, nitrous oxide from fossil fuels breaks down ozone.

Near-Ground Ozone Pollution

Near the ground, where there is naturally little ozone, it is considered a pollutant. Ozone irritates the eyes and respiratory tracts of humans and wildlife. It also interferes with plant growth.

Ground-level ozone forms when nitrogen oxides and volatile organic compounds released by burning or evaporating fossil fuels are exposed to sunlight. Warm temperatures speed the reaction. Thus, ground-level ozone tends to vary daily (being higher in the daytime) and seasonally (being higher in the summer).

To help reduce ozone pollution, avoid putting fossil fuels or their combustion products into the air at times that favor ozone production. On hot, sunny, still days, postpone filling a gas tank or using gas-powered appliances until evening, when there is less sunlight to power conversion of pollutants to ozone.

A Ozone levels in the upper atmosphere in September 2007, the Antarctic spring.

Purple indicates the least ozone, with *blue*, *green*, and *yellow* indicating increasingly higher levels.

Check the current status of the ozone hole at NASA's web site (http://ozonewatch.gsfc.nasa.gov/).

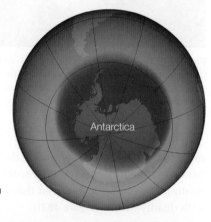

Antarctica

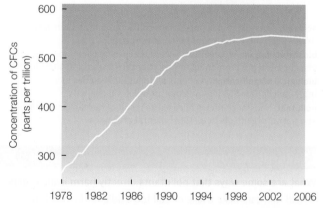

B Concentration of CFCs in the upper atmosphere. These pollutants destroy ozone. A worldwide ban on CFCs has successfully halted the rise in CFC concentration.

Figure 48.12 Animated Ozone and CFCs.

ozone layer High atmospheric layer rich in ozone; prevents most ultraviolet radiation in sunlight from reaching Earth's surface.

Take-Home Message

How do human activities affect ozone levels?

» Certain synthetic chemicals destroy ozone in the upper atmosphere's protective ozone layer. This layer serves as a protective shield against UV radiation.

» Evaporation and burning of fossil fuels increases the amount of ozone in the air near the ground, where ozone is considered a harmful pollutant.

48.7 Effects of Global Climate Change

- Climate change is the most widespread threat to habitats. Among other effects, it melts ice, causing sea level to rise.
- Links to Greenhouse gases 46.8, Coral bleaching 47.13

Ongoing climate change is affecting ecosystems throughout the world. Most notably, average temperatures are increasing. Warming is more pronounced at temperate and polar latitudes than at the equator.

A rising temperature raises sea level. Water expands as it is heated, and heating also melts glaciers (**Figure 48.13**). Together, thermal expansion and the addition of meltwater from glaciers cause sea level to rise. In the past century, the sea level has risen about 20 centimeters (8 inches). As a result, some coastal wetlands are disappearing underwater.

The salinity of the ocean is also changing. During the past 50 years, the upper waters of the Atlantic ocean have become less salty at the poles, where additional meltwater is flowing in. During the same period, subtropical and tropical waters became saltier as warmer ocean temperatures and more intense trade winds encouraged evaporation in these regions. Altered rainfall patterns have also promoted salinity changes; the amount of annual precipitation has risen at high latitudes and fallen at lower ones.

Seawater is becoming more acidic, because when CO_2 dissolves in a solution, the solution's pH declines. Increased acidity can harm marine life by making less calcium carbonate available. Groups ranging from foraminifera, to corals, to bivalve mollusks require calcium carbonate to build their hard parts.

The changing climate is having widespread effects on biological systems. Temperature changes are cues for many temperate zone species. Warmer than normal springs are causing deciduous trees to leaf out earlier, and spring-blooming flowers to flower earlier. Animal migration times and breeding seasons are also shifting. Species arrays in biological communities are changing as warmer temperatures allow some species to shift their range to higher latitudes or elevations. Of course, not all species can move or spread quickly, and warmer temperatures are expected to drive some of these species to extinction. For example, warming of tropical waters is already stressing reef-building corals and increasing the frequency of coral bleaching events.

The warming trend is also expected to affect human health. Deaths from heat stroke are likely to increase, and some infectious diseases will probably become more widespread. One way of predicting how warmer global temperatures will affect the incidence of disease is to look at what happens during an El Niño event, when sea temperature rises. As Section 47.15

A 1941 photo of Muir Glacier in Alaska.

B 2004 photo of the same region.

Figure 48.13 Melting glaciers, one sign of a warming world. Water from melting glaciers contributes to rising sea level.

explained, the incidence of cholera in India rises during an El Niño. Other waterborne diarrheal diseases also increase when waters warm. In addition, elevated temperature and CO_2 concentration are expected to increase production of allergy-inducing plant pollen and fungal spores.

Take-Home Message

What are some biological effects of climate change?

» Warming water, a rising sea level, and changes in the composition of seawater pose threats to aquatic species.

» On land, warming temperatures are altering the timing of flowering and migrations, and the distribution of some species.

» Increased warmth will also encourage the spread of some human pathogens and increase the production of allergy-inducing pollen by plants.

48.8 Conservation Biology

■ Conservation biologists survey and seek ways to protect the world's existing biodiversity.

The Value of Biodiversity

Every nation has several forms of wealth: material wealth, cultural wealth, and biological wealth. Biological wealth is called **biodiversity**. We measure a region's biodiversity at three levels: the genetic diversity within species, species diversity, and ecosystem diversity. Biodiversity is currently declining at all three levels, in all regions.

Conservation biology addresses these declines. The goals of this field of biology are (1) to survey the range of biodiversity, and (2) to find ways to maintain and use biodiversity to benefit human populations. The aim is to conserve biodiversity by encouraging people to value it and use it in ways that do not destroy it.

Why should we protect biodiversity? From a selfish standpoint, doing so is an investment in our future. Healthy ecosystems are essential to our survival. Other organisms produce the oxygen we breathe and the food we eat. They remove waste carbon dioxide from the air and decompose and detoxify wastes. Plants take up rain and hold soil in place, preventing erosion and reducing the risk of flooding. Compounds in wild species can serve as medicines. Wild relatives of crop plants are reservoirs of genetic diversity that plant breeders draw on to protect and improve crops.

Setting Priorities

Protecting biological diversity can pose a challenge. People often oppose environmental protections because they fear that such measures will have adverse economic consequences. However, taking care of the environment can make good economic sense. With a bit of planning, people can both preserve and profit from their biological wealth.

Conservation biologists help us make the difficult choices about which regions should be targeted for protection first. They identify **biodiversity hot spots**, places that have species found nowhere else and are under great threat of destruction. Once identified, hot spots take priority in worldwide conservation efforts.

On a broader scale, conservation biologists define ecoregions, which are land or aquatic regions characterized by climate, geography, and the species found within them. The most widely used ecoregion system was developed by conservation scientists of the World Wildlife Fund. These scientists defined 867 distinctive land ecoregions. **Figure 48.14** shows the locations and conservation status of ecoregions that are considered the top priority for conservation efforts.

The Klamath–Siskiyou forest in southwestern Oregon and northwestern California is one of North America's endangered ecoregions. It is home to many rare conifers. Two endangered birds, the marbled murrelet and the northern spotted owl, nest in old-growth

■ Critical or endangered ecoregion
Vulnerable ecoregion
■ Stable or intact ecoregion
No information available

Figure 48.14 The location and conservation status of the land ecoregions deemed most important by the World Wildlife Fund.

parts of the forest (**Figure 48.15**), and endangered coho salmon breed in streams that run through the forest. Logging threatens all of these species.

By focusing on hot spots and critical ecoregions rather than on individual endangered species, scientists hope to maintain ecosystem processes that naturally sustain biological diversity.

Preservation and Restoration

Worldwide, many ecologically important regions have been protected in ways that benefit local people. The Monteverde Cloud Forest in Costa Rica is one example. During the 1970s, George Powell was studying birds in this forest, which was rapidly being cleared. Powell decided to buy part of the forest as a nature sanctuary. His efforts inspired individuals and conservation groups to donate funds, and much of the forest is now protected as a private nature reserve. The reserve's plants and animals include more than 100 mammal species, 400 bird species, and 120 species of amphibians and reptiles. It is one of the few habitats left for jaguars and ocelots. A tourism industry centered on the reserve provides economic benefits to local people.

Sometimes, an ecosystem is so damaged, or there is so little of it left, that conservation alone is not enough to sustain biodiversity. **Ecological restoration** is work designed to bring about the renewal of a natural ecosystem that has been degraded or destroyed, fully or in part. Restoration work in Louisiana's coastal wetlands is an example. More than 40 percent of the coastal wetlands in the United States are in Louisiana. The marshes are an ecological and economic treasure, but they are troubled. Dams and levees built upstream of the marshes keep back sediments that would normally replenish sediments lost to the sea. Channels cut through the marshes for oil exploration and production have encouraged erosion, and the rising sea level threatens to flood the existing plants. Since the 1940s, Louisiana has lost an area of marshland the size of Rhode Island. Restoration efforts now under way aim to reverse some of those losses (**Figure 48.16**).

Figure 48.15 The Klamath–Siskiyou forest, one of North America's critical ecoregions. Endangered northern spotted owls (*left*) are endemic to this coniferous forest.

Figure 48.16 Ecological restoration in Louisiana's Sabine National Wildlife Refuge. In places where marshland has become open water, sediments are barged in and marsh grasses are planted on them. Squares are new sediment with marsh grass.

biodiversity Of a region, the genetic variation within its species, variety of species, and variety of ecosystems.
conservation biology Field of applied biology that surveys and documents biodiversity, and seeks ways to maintain and use it.
ecological restoration Actively altering an area in an effort to restore or create a functional ecosystem.
biodiversity hot spot Threatened region with great biodiversity that is considered a high priority for conservation efforts.

Take-Home Message

How do we sustain biodiversity?

» Biodiversity is the genetic diversity of individuals of a species, the variety of species, and the variety of ecosystems.

» Conservation biologists identify threatened regions with high biodiversity and prioritize which should be first to receive protection.

» Through ecological restoration, we re-create or renew a biologically diverse ecosystem that has been destroyed or degraded.

48.9 Reducing Negative Impacts

■ The cumulative effects of individual actions will determine the health of our planet.

Ultimately, the health of our planet depends on our ability to live within our limits. The goal is to live sustainably, which means meeting the needs of the present generation without reducing the ability of future generations to meet their own needs.

Promoting sustainability begins with recognizing the environmental consequences of one's lifestyle. People in industrial nations use huge quantities of resources, and the extraction of these resources has negative effects on biodiversity. In the United States, the size of the average family has declined since the 1950s, while the size of the average home has doubled. All of the materials used to build and furnish those larger homes must be extracted from the environment.

For example, an average new home contains about 500 pounds of copper in its wiring and plumbing. Where does that copper come from? Like most other mineral elements used in manufacturing, most copper is mined from the ground (Figure 48.17). Surface mining strips an area of vegetation and soil, creating an ecological dead zone. It puts dust into the air, creates mountains of rocky waste, and can contaminate nearby waterways.

Figure 48.18 Volunteers restoring the Little Salmon River in Idaho so that salmon can migrate upstream to their breeding site.

Globalization makes it difficult to know the source of the raw materials in products you buy. Resource extraction in developing countries is often carried out under regulations that are less strict or less stringently enforced than those in the United States, so the environmental impact is even greater.

Nonrenewable mineral resources are also used in electronic devices such as phones, computers, televi-

Figure 48.17 Resource extraction. Bingham copper mine near Salt Lake City, Utah. This open pit mine is 4 kilometers (2.5 miles) wide and 1,200 meters (0.75 miles) deep. It is the largest man-made excavation on Earth.

A Long Reach (revisited)

The Arctic is not a separate continent like Antarctica, but rather a region that encompasses the northernmost parts of several continents. Eight countries, including the United States, Canada, and Russia, control parts of the Arctic and have rights to its extensive oil, gas, and mineral deposits. Until recently, ice sheets covered the Arctic Ocean, making it difficult for ships to move to and from the Arctic landmass, but those sheets are breaking up as a result of global climate change (**Figure 48.19**). At the same time, ice that covered the Arctic landmass is melting. These changes will make it easier for people to remove minerals and fossil fuels from the Arctic. With the world supply of fuel and minerals dwindling, pressure to exploit Arctic resources is rising. However, conservationists warn that extracting these resources will harm Arctic species such as the polar bear that are already threatened by global climate change.

How would you vote? The Arctic has extensive deposits of minerals and fossil fuel, but extracting these resources might pose a risk to species already threatened by global climate change. Should the United States exploit its share of the Arctic resources or advocate for protection of the Arctic region?

Arctic perennial sea ice in 1979

Arctic perennial sea ice in 2003

Figure 48.19 Declining Arctic perennial ice.

---

sions, and MP3 players. Reducing consumption by fixing existing products is a sustainable resource use, as is recycling. Obtaining nonrenewable materials by recycling reduces the need for extraction of those resources, and it also keep materials out of landfills.

Reducing your energy use is another way to promote sustainability. Fossil fuels such as petroleum, natural gas, and coal supply most of the energy used by developed countries. You already know that burning these nonrenewable fuels contributes to global warming and acid rain. In addition, extracting and transporting these fuels have negative impacts. Oil harms species when it leaks from pipelines, ships, or wells.

Renewable energy sources do not produce greenhouse gases, but they have drawbacks too. For example, dams in rivers of the Pacific Northwest generate renewable hydroelectric power, but they also prevent endangered salmon from returning to streams above the dam to breed. Similarly, wind turbines can harm birds and bats. Panels used to collect solar energy are made using nonrenewable mineral resources, and manufacturing the panels generates pollutants.

To minimize your impact, avoid wasting energy. Look for energy-efficient appliances and lightbulbs. Walking, bicycling, and using public transportation are energy-efficient alternatives to driving. To take a more active role, learn about the threats to ecosystems in your own area. Support efforts to preserve and restore local biodiversity. Many ecological restoration projects are supervised by trained biologists but carried out primarily through the efforts of volunteers (**Figure 48.18**). Keep in mind that unthinking actions of billions of individuals are the greatest threat to biodiversity. Each of us may have little impact on our own, but our collective behavior, for good or for bad, will determine the future of the planet.

Take-Home Message

What can individuals do to reduce their harmful impact on biodiversity?

» Resource extraction and usage have side effects that threaten biodiversity.

» You can save energy and other resources by reducing energy consumption and recycling and reusing materials.

Arctic perennial sea ice in 1979

Arctic perennial sea ice in 2003

Appendix I. The Amino Acids

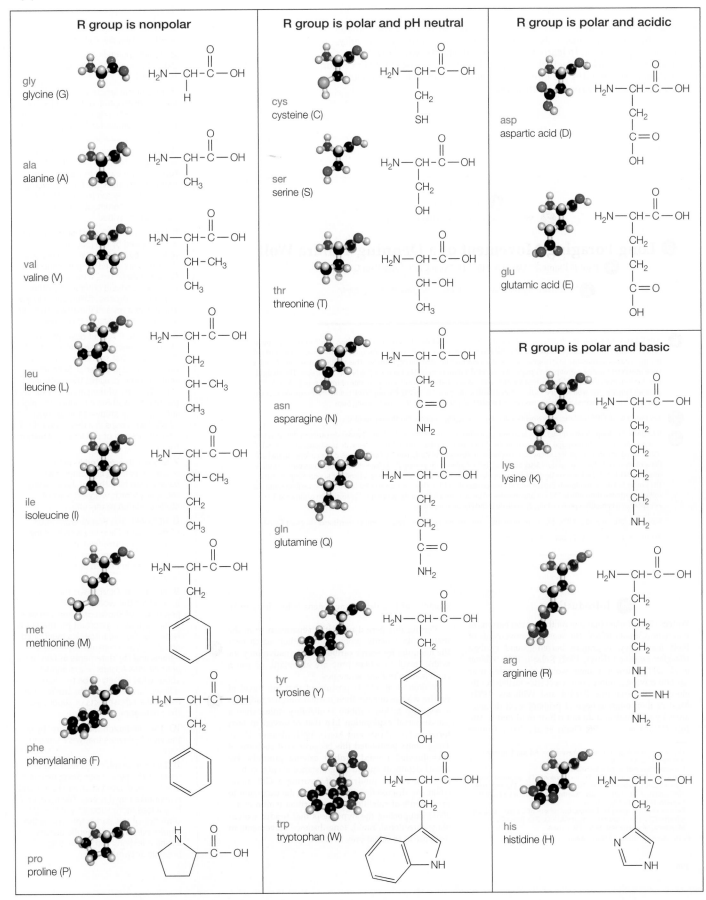

R group is nonpolar

gly
glycine (G)

ala
alanine (A)

val
valine (V)

leu
leucine (L)

ile
isoleucine (I)

met
methionine (M)

phe
phenylalanine (F)

pro
proline (P)

R group is polar and pH neutral

cys
cysteine (C)

ser
serine (S)

thr
threonine (T)

asn
asparagine (N)

gln
glutamine (Q)

tyr
tyrosine (Y)

trp
tryptophan (W)

R group is polar and acidic

asp
aspartic acid (D)

glu
glutamic acid (E)

R group is polar and basic

lys
lysine (K)

arg
arginine (R)

his
histidine (H)

Appendix II. Annotations to A Journal Article

This journal article reports on the movements of a female wolf during the summer of 2002 in northwestern Canada. It also reports on a scientific process of inquiry, observation and interpretation to learn where, how and why the wolf traveled as she did. In some ways, this article reflects the story of "how to do science" told in section 1.5 of this textbook. These notes are intended to help you read and understand how scientists work and how they report on their work.

1 Title of the journal, which reports on science taking place in Arctic regions.

2 Volume number, issue number and date of the journal, and page numbers of the article.

3 Title of the article: a concise but specific description of the subject of study—one episode of long-range travel by a wolf hunting for food on the Arctic tundra.

4 Authors of the article: scientists working at the institutions listed in the footnotes below. Note #2 indicates that P. F. Frame is the corresponding author—the person to contact with questions or comments. His email address is provided.

5 Date on which a draft of the article was received by the journal editor, followed by date one which a revised draft was accepted for publication. Between these dates, the article was reviewed and critiqued by other scientists, a process called peer review. The authors revised the article to make it clearer, according to those reviews.

6 ABSTRACT: A brief description of the study containing all basic elements of this report. First sentence summarizes the *background* material. Second sentence encapsulates the *methods* used. The rest of the paragraph sums up the *results*. Authors introduce the main *subject* of the study—a female wolf (#388) with pups in a den—and refer to later *discussion* of possible explanations for her behavior.

7 Key words are listed to help researchers using computer databases. Searching the databases using these key words will yield a list of studies related to this one.

8 RÉSUMÉ: The French translation of the abstract and key words. Many researchers in this field are French Canadian. Some journals provide such translations in French or in other languages.

9 INTRODUCTION: Gives the background for this wolf study. This paragraph tells of known or suspected wolf behavior that is important for this study. Note that (a) major species mentioned are always accompanied by scientific names, and (b) statements of fact or *postulations* (claims or assumptions about what is likely to be true) are followed by references to studies that established those facts or supported the postulations.

10 This paragraph focuses directly on the wolf behaviors that were studied here.

11 This paragraph starts with a statement of the *hypothesis* being tested, one that originated in other studies and is supported by this one. The hypothesis is restated more succinctly in the last sentence of this paragraph. This is the *inquiry* part of the scientific process—asking questions and suggesting possible answers.

---

(1) ARCTIC

(2) VOL. 57, NO. 2 (JUNE 2004) P. 196–203

(3) Long Foraging Movement of a Denning Tundra Wolf

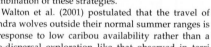

(4) Paul F. Frame,[1,2] David S. Hik,[1] H. Dean Cluff,[3] and Paul C. Paquet[4]

(5) (Received 3 September 2003; accepted in revised form 16 January 2004)

---

(6) ABSTRACT. Wolves (*Canis lupus*) on the Canadian barrens are intimately linked to migrating herds of barren-ground caribou (*Rangifer tarandus*). We deployed a Global Positioning System (GPS) radio collar on an adult female wolf to record her movements in response to changing caribou densities near her den during summer. This wolf and two other females were observed nursing a group of 11 pups. She traveled a minimum of 341 km during a 14-day excursion. The straight-line distance from the den to the farthest location was 103 km, and the overall minimum rate of travel was 3.1 km/h. The distance between the wolf and the radio-collared caribou decreased from 242 km one week before the excursion to 8 km four days into the excursion. We discuss several possible explanations for the long foraging bout.

(7) Key words: wolf, GPS tracking, movements, *Canis lupus*, foraging, caribou, Northwest Territories

(8) RÉSUMÉ. Les loups (*Canis lupus*) dans la toundra canadienne sont étroitement liés aux hardes de caribous des toundras (*Rangifer tarandus*). On a équipé une louve adulte d'un collier émetteur muni d'un système de positionnement mondial (GPS) afin d'enregistrer ses déplacements en réponse au changement de densité du caribou près de sa tanière durant l'été. On a observé cette louve ainsi que deux autres en train d'allaiter un groupe de 11 louveteaux. Elle a parcouru un minimum de 341 km durant une sortie de 14 jours. La distance en ligne droite de la tanière à l'endroit le plus éloigné était de 103 km, et la vitesse minimum durant tout le voyage était de 3,1 km/h. La distance entre la louve et le caribou muni du collier émetteur a diminué de 242 km une semaine avant la sortie à 8 km quatre jours après la sortie. On commente diverses explications possibles pour ce long épisode de recherche de nourriture.

Mots clés: loup, repérage GPS, déplacements, *Canis lupus*, recherche de nourriture, caribou, Territoires du Nord-Ouest

Traduit pour la revue *Arctic* par Nésida Loyer.

---

(9) Introduction

Wolves (*Canis lupus*) that den on the central barrens of mainland Canada follow the seasonal movements of their main prey, migratory barren-ground caribou (*Rangifer tarandus*) (Kuyt, 1962; Kelsall, 1968; Walton et al., 2001). However, most wolves do not den near caribou calving grounds, but select sites farther south, closer to the tree line (Heard and Williams, 1992). Most caribou migrate beyond primary wolf denning areas by mid-June and do not return until mid-to-late July (Heard et al., 1996; Gunn et al., 2001). Conse-

quently, caribou density near dens is low for part of the summer.

During this period of spatial separation from the main caribou herds, wolves must either search near the homesite for scarce caribou or alternative prey (or both), travel to where prey are abundant, or use a combination of these strategies.

Walton et al. (2001) postulated that the travel of tundra wolves outside their normal summer ranges is a response to low caribou availability rather than a pre-dispersal exploration like that observed in territorial wolves (Fritts and Mech, 1981; Messier, 1985). The authors postulated this because most such travel was directed toward caribou calving grounds. We report details of such a long-distance excursion by a breeding female tundra wolf wearing a GPS radio collar. We discuss the relationship of the excursion to movements of satellite-collared caribou (Gunn et al., 2001), supporting the hypothesis that tundra wolves make directional, rapid, long-distance movements in response to seasonal prey availability.

[1] Department of Biological Sciences, University of Alberta, Edmonton, Alberta T6G 2E9, Canada
[2] Corresponding author: pframe@ualberta.ca
[3] Department of Resources, Wildlife, and Economic Development, North Slave Region, Government of the Northwest Territories, P.O. Box 2668, 3803 Bretzlaff Dr., Yellowknife, Northwest Territories X1A 2P9, Canada; Dean_Cluff@gov.nt.ca
[4] Faculty of Environmental Design, University of Calgary, Calgary, Alberta T2N 1N4, Canada; current address: P.O. Box 150, Meacham, Saskatchewan S0K 2V0, Canada

© The Arctic Institute of North America

196

Appendix II

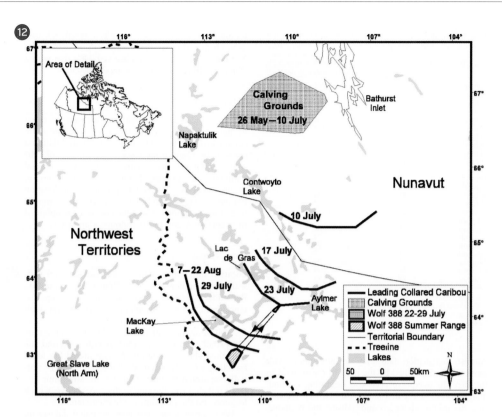

12 This map shows the study area and depicts wolf and caribou locations and movements during one summer. Some of this information is explained below.

13 STUDY AREA: This section sets the stage for the study, locating it precisely with latitude and longitude coordinates and describing the area (illustrated by the map in Figure 1).

14 Here begins the story of how prey (caribou) and predators (wolves) interact on the tundra. Authors describe movements of these nomadic animals throughout the year.

15 We focus on the denning season (summer) and learn how wolves locate their dens and travel according to the movements of caribou herds.

Figure 1. Map showing the movements of satellite radio-collared caribou with respect to female wolf 388's summer range and long foraging movement, in summer 2002.

13 Study Area

Our study took place in the northern boreal forest–low Arctic tundra transition zone (63° 30′ N, 110° 00′ W; Figure 1; Timoney et al., 1992). Permafrost in the area changes from discontinuous to continuous (Harris, 1986). Patches of spruce (*Picea mariana, P. glauca*) occur in the southern portion and give way to open tundra to the northeast. Eskers, kames, and other glacial deposits are scattered throughout the study area. Standing water and exposed bedrock are characteristic of the area.

14 *Details of the Caribou-Wolf System*

The Bathurst caribou herd uses this study area. Most caribou cows have begun migrating by late April, reaching calving grounds by June (Gunn et al., 2001;

Figure 1). Calving peaks by 15 June (Gunn et al., 2001), and calves begin to travel with the herd by one week of age (Kelsall, 1968). The movement patterns of bulls are less known, but bulls frequent areas near calving grounds by mid-June (Heard et al., 1996; Gunn et al., 2001). In summer, Bathurst caribou cows generally travel south from their calving grounds and then, parallel to the tree line, to the northwest. The rut usually takes place at the tree line in October (Gunn et al., 2001). The winter range of the Bathurst herd varies among years, ranging through the taiga and along the tree line from south of Great Bear Lake to southeast of Great Slave Lake. Some caribou spend the winter on the tundra (Gunn et al., 2001; Thorpe et al., 2001).

In winter, wolves that prey on Bathurst caribou do **15** not behave territorially. Instead, they follow the herd throughout its winter range (Walton et al., 2001; Musiani, 2003). However, during denning (May–

Foraging Movement of A Tundra Wolf **197**

Table 1. Daily distances from wolf 388 and the den to the nearest radio-collared caribou during a long excursion in summer 2002.

Date (2002)	Mean distance from caribou to wolf (km)	Daily distance from closest caribou to den
12 July	242	241
13 July	210	209
14 July	200	199
15 July	186	180
16 July	163	162
17 July	151	148
18 July	144	137
19 July[1]	126	124
20 July	103	130
21 July	73	130
22 July	40	110
23 July[2]	9	104
29 July[3]	16	43
30 July	32	43
31 July	28	44
1 August	29	46
2 August[4]	54	52
3 August	53	53
4 August	74	74
5 August	75	75
6 August	74	75
7 August	72	75
8 August	76	75
9 August	79	79

[1] Excursion starts.
[2] Wolf closest to collared caribou.
[3] Previous five days' caribou locations not available.
[4] Excursion ends.

August, parturition late May to mid-June), wolf movements are limited by the need to return food to the den. To maximize access to migrating caribou, many wolves select den sites closer to the tree line than to caribou calving grounds (Heard and Williams, 1992). Because of caribou movement patterns, tundra denning wolves are separated from the main caribou herds by several hundred kilometers at some time during summer (Williams, 1990:19; Figure 1; Table 1).

 Muskoxen do not occur in the study area (Fournier and Gunn, 1998), and there are few moose there (H.D. Cluff, pers. obs.). Therefore, alternative prey for wolves includes waterfowl, other ground-nesting birds, their eggs, rodents, and hares (Kuyt, 1972; Williams, 1990:16; H.D. Cluff and P.F. Frame, unpubl. data). During 56 hours of den observations, we saw no ground squirrels or hares, only birds. It appears that the abundance of alternative prey was relatively low in 2002.

Methods

Wolf Monitoring

We captured female wolf 388 near her den on 22 June 2002, using a helicopter net-gun (Walton et al., 2001). She was fitted with a releasable GPS radio collar (Merrill et al., 1998) programmed to acquire locations at 30-minute intervals. The collar was electronically released (e.g., Mech and Gese, 1992) on 20 August 2002. From 27 June to 3 July 2002, we observed 388's den with a 78 mm spotting scope at a distance of 390 m.

Caribou Monitoring

In spring of 2002, ten female caribou were captured by helicopter net-gun and fitted with satellite radio collars, bringing the total number of collared Bathurst cows to 19. Eight of these spent the summer of 2002 south of Queen Maud Gulf, well east of normal Bathurst caribou range. Therefore, we used 11 caribou for this analysis. The collars provided one location per day during our study, except for five days from 24 to 28 July. Locations of satellite collars were obtained from Service Argos, Inc. (Landover, Maryland).

Data Analysis

Location data were analyzed by ArcView GIS software (Environmental Systems Research Institute Inc., Redlands, California). We calculated the average distance from the nearest collared caribou to the wolf and the den for each day of the study.

Wolf foraging bouts were calculated from the time 388 exited a buffer zone (500 m radius around the den) until she re-entered it. We considered her to be traveling when two consecutive locations were spatially separated by more than 100 m. Minimum distance traveled was the sum of distances between each location and the next during the excursion.

We compared pre- and post-excursion data using Analysis of Variance (ANOVA; Zar, 1999). We first tested for homogeneity of variances with Levene's test (Brown and Forsythe, 1974). No transformations of these data were required.

Results

Wolf Monitoring

Pre-Excursion Period: Wolf 388 was lactating when captured on 22 June. We observed her and two other females nursing a group of 11 pups between 27 June and 3 July. During our observations, the pack consisted of at least four adults (3 females and 1 male) and 11 pups. On 30 June, three pups were moved to a location 310 m from the other eight and cared for by an uncollared female. The male was not seen at the den after the evening of 30 June.

Before the excursion, telemetry indicated 18 foraging bouts. The mean distance traveled during these bouts was 25.29 km (± 4.5 SE, range 3.1–82.5 km). Mean greatest distance from the den on foraging

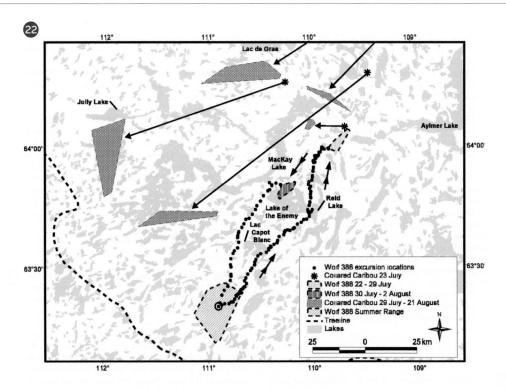

22 The key in the lower right-hand corner of the map shows areas (shaded) within which the wolves and caribou moved, and the dotted trail of 388 during her excursion. From the results depicted on this map, the investigators tried to determine when and where 388 might have encountered caribou and how their locations affected her traveling behavior.

23 The wolf's excursion (her long trip away from the den area) is the focus of this study. These paragraphs present detailed measurements of daily movements during her two-week trip—how far she traveled, how far she was from collared caribou, her time spent traveling and resting, and her rate of speed. Authors use the phrase "minimum distance traveled" to acknowledge they couldn't track every step but were measuring samples of her movements. They knew that she went at least as far as they measured. This shows how scientists try to be exact when reporting results. Results of this study are depicted graphically in the map in Figure 2.

Figure 2. Details of a long foraging movement by female wolf 388 between 19 July and 2 August 2002. Also shown are locations and movements of three satellite radio-collared caribou from 23 July to 21 August 2002. On 23 July, the wolf was 8 km from a collared caribou. The farthest point from the den (103 km distant) was recorded on 27 July. Arrows indicate direction of travel.

bouts was 7.1 km (± 0.9 SE, range 1.7–17.0 km). The average duration of foraging bouts for the period was 20.9 h (± 4.5 SE, range 1–71 h).

The average daily distance between the wolf and the nearest collared caribou decreased from 242 km on 12 July, one week before the excursion period, to 126 km on 19 July, the day the excursion began (Table 1).

23 Excursion Period: On 19 July at 2203, after spending 14 h at the den, 388 began moving to the northeast and did not return for 336 h (14 d; Figure 2). Whether she traveled alone or with other wolves is unknown. During the excursion, 476 (71%) of 672 possible locations were recorded. The wolf crossed the southeast end of Lac Capot Blanc on a small land bridge, where she paused for 4.5 h after traveling for 19.5 h (37.5

km). Following this rest, she traveled for 9 h (26.3 km) onto a peninsula in Reid Lake, where she spent 2 h before backtracking and stopping for 8 h just off the peninsula. Her next period of travel lasted 16.5 h (32.7 km), terminating in a pause of 9.5 h just 3.8 km from a concentration of locations at the far end of her excursion, where we presume she encountered caribou. The mean duration of these three movement periods was 15.7 h (± 2.5 SE), and that of the pauses, 7.3 h (± 1.5). The wolf required 72.5 h (3.0 d) to travel a minimum of 95 km from her den to this area near caribou (Figure 2). She remained there (35.5 km2) for 151.5 h (6.3 d) and then moved south to Lake of the Enemy, where she stayed (31.9 km^2) for 74 h (3.1 d) before returning to her den. Her greatest distance from the den, 103 km, was recorded 174.5 h (7.3 d) after the excursion

Foraging Movement of A Tundra Wolf **199**

24 Post-excursion measurements of 388's movements were made to compare with those of the pre-excursion period. In order to compare, scientists often use *means*, or averages, of a series of measurements—mean distances, mean duration, etc.

25 In the comparison, authors used statistical calculations (F and df) to determine that the differences between pre- and post-excursion measurements were *statistically insignificant*, or close enough to be considered essentially the same or similar.

26 As with wolf 388, the investigators measured the movements of caribou during the study period. The areas within which the caribou moved are shown in Figure 2 by shaded polygons mentioned in the second paragraph of this subsection.

27 This subsection summarizes how distances separating predators and prey varied during the study period.

28 DISCUSSION: This section is the *interpretation* part of the scientific process.

29 This subsection reviews observations from other studies and suggests that this study fits with patterns of those observations.

30 Authors discuss a prevailing *theory* (CBFT) which might explain why a wolf would travel far to meet her own energy needs while taking food caught closer to the den back to her pups. The results of this study seem to fit that pattern.

---

began, at 0433 on 27 July. She was 8 km from a collared caribou on 23 July, four days after the excursion began (Table 1).

The return trip began at 0403 on 2 August, 318 h (13.2 d) after leaving the den. She followed a relatively direct path for 18 h back to the den, a distance of 75 km.

The minimum distance traveled during the excursion was 339 km. The estimated overall minimum travel rate was 3.1 km/h, 2.6 km/h away from the den and 4.2 km/h on the return trip.

24 **Post-Excursion Period:** We saw three pups when recovering the collar on 20 August, but others may have been hiding in vegetation.

Telemetry recorded 13 foraging bouts in the post-excursion period. The mean distance traveled during these bouts was 18.3 km (+ 2.7 SE, range 1.2–47.7 km), and mean greatest distance from the den was 7.1 km (+ 0.7 SE, range 1.1–11.0 km). The mean duration of these post-excursion foraging bouts was 10.9 h (+ 2.4 SE, range 1–33 h).

When 388 reached her den on 2 August, the distance to the nearest collared caribou was 54 km. On 9 August, one week after she returned, the distance was 79 km (Table 1).

Pre- and Post-Excursion Comparison

25 We found no differences in the mean distance of foraging bouts before and after the excursion period (F = 1.5, df = 1, 29, *p* = 0.24). Likewise, the mean greatest distance from the den was similar pre- and post-excursion (F = 0.004, df = 1, 29, *p* = 0.95). However, the mean duration of 388's foraging bouts decreased by 10.0 h after her long excursion (F = 3.1, df = 1, 29, *p* = 0.09).

26 *Caribou Monitoring*

Summer Movements: On 10 July, 5 of 11 collared caribou were dispersed over a distance of 10 km, 140 km south of their calving grounds (Figure 1). On the same day, three caribou were still on the calving grounds, two were between the calving grounds and the leaders, and one was missing. One week later (17 July), the leading radio-collared cows were 100 km farther south (Figure 1). Two were within 5 km of each other in front of the rest, who were more dispersed. All radio-collared cows had left the calving grounds by this time. On 23 July, the leading radio-collared caribou had moved 35 km farther south, and all of them were more widely dispersed. The two cows closest to the leader were 26 km and 33 km away, with 37 km between them. On the next location (29 July), the most southerly caribou were 60 km

farther south. All of the caribou were now in the areas where they remained for the duration of the study (Figure 2).

A Minimum Convex Polygon (Mohr and Stumpf, 1966) around all caribou locations acquired during the study encompassed 85 119 km².

Relative to the Wolf Den: The distance from the **27** nearest collared caribou to the den decreased from 241 km one week before the excursion to 124 km the day it began. The nearest a collared caribou came to the den was 43 km away, on 29 and 30 July. During the study, four collared caribou were located within 100 km of the den. Each of these four was closest to the wolf on at least one day during the period reported.

28 **Discussion**

Prey Abundance

Caribou are the single most important prey of tundra **29** wolves (Clark, 1971; Kuyt, 1972; Stephenson and James, 1982; Williams, 1990). Caribou range over vast areas, and for part of the summer, they are scarce or absent in wolf home ranges (Heard et al., 1996). Both the long distance between radio-collared caribou and the den the week before the excursion and the increased time spent foraging by wolf 388 indicate that caribou availability near the den was low. Observations of the pups' being left alone for up to 18 h, presumably while adults were searching for food, provide additional support for low caribou availability locally. Mean foraging bout duration decreased by 10.0 h after the excursion, when collared caribou were closer to the den, suggesting an increase in caribou availability nearby.

Foraging Excursion

One aspect of central place foraging theory (CPFT) **30** deals with the optimality of returning different-sized food loads from varying distances to dependents at a central place (i.e., the den) (Orians and Pearson, 1979). Carlson (1985) tested CPFT and found that the predator usually consumed prey captured far from the central place, while feeding prey captured nearby to dependants. Wolf 388 spent 7.2 days in one area near caribou before moving to a location 23 km back towards the den, where she spent an additional 3.1 days, likely hunting caribou. She began her return trip from this closer location, traveling directly to the den. While away, she may have made one or more successful kills and spent time meeting her own energetic needs before returning to the den. Alternatively, it may have taken several attempts to make a kill,

200 *P.F. Frame, et al*

which she then fed on before beginning her return trip. We do not know if she returned food to the pups, but such behavior would be supported by CPFT.

 Other workers have reported wolves' making long round trips and referred to them as "extraterritorial" or "pre-dispersal" forays (Fritts and Mech, 1981; Messier, 1985; Ballard et al., 1997; Merrill and Mech, 2000). These movements are most often made by young wolves (1–3 years old), in areas where annual territories are maintained and prey are relatively sedentary (Fritts and Mech, 1981; Messier, 1985). The long excursion of 388 differs in that tundra wolves do not maintain annual territories (Walton et al., 2001), and the main prey migrate over vast areas (Gunn et al., 2001).

Another difference between 388's excursion and those reported earlier is that she is a mature, breeding female. No study of territorial wolves has reported reproductive adults making extraterritorial movements in summer (Fritts and Mech, 1981; Messier, 1985; Ballard et al., 1997; Merrill and Mech, 2001). However, Walton et al. (2001) also report that breeding female tundra wolves made excursions.

Direction of Movement

Possible explanations for the relatively direct route 388 took to the caribou include landscape influence and experience. Considering the timing of 388's trip and the locations of caribou, had the wolf moved northwest, she might have missed the caribou entirely, or the encounter might have been delayed.

A reasonable possibility is that the land directed 388's route. The barrens are crisscrossed with trails worn into the tundra over centuries by hundreds of thousands of caribou and other animals (Kelsall, 1968; Thorpe et al., 2001). At river crossings, lakes, or narrow peninsulas, trails converge and funnel towards and away from caribou calving grounds and summer range. Wolves use trails for travel (Paquet et al., 1996; Mech and Boitani, 2003; P. Frame, pers. observation). Thus, the landscape may direct an animal's movements and lead it to where cues, such as the odor of caribou on the wind or scent marks of other wolves, may lead it to caribou.

Another possibility is that 388 knew where to find caribou in summer. Sexually immature tundra wolves sometimes follow caribou to calving grounds (D. Heard, unpubl. data). Possibly, 388 had made such journeys in previous years and killed caribou. If this were the case, then in times of local prey scarcity she might travel to areas where she had hunted successfully before. Continued monitoring of tundra wolves may answer questions about how their food needs are met in times of low caribou abundance near dens.

Caribou often form large groups while moving south to the tree line (Kelsall, 1968). After a large aggregation of caribou moves through an area, its scent can linger for weeks (Thorpe et al., 2001:104). It is conceivable that 388 detected caribou scent on the wind, which was blowing from the northeast on 19–21 July (Environment Canada, 2003), at the same time her excursion began. Many factors, such as odor strength and wind direction and strength, make systematic study of scent detection in wolves difficult under field conditions (Harrington and Asa, 2003). However, humans are able to smell odors such as forest fires or oil refineries more than 100 km away. The olfactory capabilities of dogs, which are similar to wolves, are thought to be 100 to 1 million times that of humans (Harrington and Asa, 2003). Therefore, it is reasonable to think that under the right wind conditions, the scent of many caribou traveling together could be detected by wolves from great distances, thus triggering a long foraging bout.

Rate of Travel

Mech (1994) reported the rate of travel of Arctic wolves on barren ground was 8.7 km/h during regular travel and 10.0 km/h when returning to the den, a difference of 1.3 km/h. These rates are based on direct observation and exclude periods when wolves moved slowly or not at all. Our calculated travel rates are assumed to include periods of slow movement or no movement. However, the pattern we report is similar to that reported by Mech (1994), in that homeward travel was faster than regular travel by 1.6 km/h. The faster rate on return may be explained by the need to return food to the den. Pup survival can increase with the number of adults in a pack available to deliver food to pups (Harrington et al., 1983). Therefore, an increased rate of travel on homeward trips could improve a wolf's reproductive fitness by getting food to pups more quickly.

Fate of 388's Pups

Wolf 388 was caring for pups during den observations. The pups were estimated to be six weeks old, and were seen ranging as far as 800 m from the den. They received some regurgitated food from two of the females, but were unattended for long periods. The excursion started 16 days after our observations, and it is improbable that the pups could have traveled the distance that 388 moved. If the pups died, this would have removed parental responsibility, allowing the long movement.

Our observations and the locations of radio-collared caribou indicate that prey became scarce in

31 Here our authors note other possible explanations for wolves' excursions presented by other investigators, but this study does not seem to support those ideas.

32 Authors discuss possible reasons for why 388 traveled directly to where caribou were located. They take what they learned from earlier studies and apply it to this case, suggesting that the lay of the land played a role. Note that their description paints a clear picture of the landscape.

33 Authors suggest that 388 may have learned in traveling during previous summers where the caribou were. The last two sentences suggest ideas for future studies.

34 Or maybe 388 followed the scent of the caribou. Authors acknowledge difficulties of proving this, but they suggest another area where future studies might be done.

35 Authors suggest that results of this study support previous studies about how fast wolves travel to and from the den. In the last sentence, they speculate on how these observed patterns would fit into the theory of evolution.

36 Authors also speculate on the fate of 388's pups while she was traveling. This leads to . . .

37 Discussion of cooperative rearing of pups and, in turn, to speculation on how this study and what is known about cooperative rearing might fit into the animal's strategies for survival of the species. Again, the authors approach the broader theory of evolution and how it might explain some of their results.

38 And again, they suggest that this study points to several areas where further study will shed some light.

39 In conclusion, the authors suggest that their study supports the hypothesis being tested here. And they touch on the implications of increased human activity on the tundra predicted by their results.

40 ACKNOWLEDGEMENTS: Authors note the support of institutions, companies and individuals. They thank their reviewers ad list permits under which their research was carried on.

41 REFERENCES: List of all studies cited in the report. This may seem tedious, but is a vitally important part of scientific reporting. It is a record of the sources of information on which this study is based. It provides readers with a wealth of resources for further reading on this topic. Much of it will form the foundation of future scientific studies like this one.

the area of the den as summer progressed. Wolf 388 may have abandoned her pups to seek food for herself. However, she returned to the den after the excursion, where she was seen near pups. In fact, she foraged in a similar pattern before and after the excursion, suggesting that she again was providing for pups after her return to the den.

37 A more likely possibility is that one or both of the other lactating females cared for the pups during 388's absence. The three females at this den were not seen with the pups at the same time. However, two weeks earlier, at a different den, we observed three females cooperatively caring for a group of six pups. At that den, the three lactating females were observed providing food for each other and trading places while nursing pups. Such a situation at the den of 388 could have created conditions that allowed one or more of the lactating females to range far from the den for a period, returning to her parental duties afterwards. However, the pups would have been weaned by eight weeks of age (Packard et al., 1992), so nonlactating adults could also have cared for them, as often happens in wolf packs (Packard et al., 1992; Mech et al., 1999).

Cooperative rearing of multiple litters by a pack could create opportunities for long-distance foraging movements by some reproductive wolves during summer periods of local food scarcity. We have recorded multiple lactating females at one or more tundra wolf dens per year since 1997. This reproductive strategy may be an adaptation to temporally and **38** spatially unpredictable food resources. All of these possibilities require further study, but emphasize both the adaptability of wolves living on the barrens and their dependence on caribou.

Long-range wolf movement in response to caribou **39** availability has been suggested by other researchers (Kuyt, 1972; Walton et al., 2001) and traditional ecological knowledge (Thorpe et al., 2001). Our report demonstrates the rapid and extreme response of wolves to caribou distribution and movements in summer. Increased human activity on the tundra (mining, road building, pipelines, ecotourism) may influence caribou movement patterns and change the interactions between wolves and caribou in the region. Continued monitoring of both species will help us to assess whether the association is being affected adversely by anthropogenic change.

40 ## Acknowledgements

This research was supported by the Department of Resources, Wildlife, and Economic Development, Government of the Northwest Territories; the Department of Biological Sciences at the University of Alberta; the Natural Sciences and Engineering Research Council of Canada; the Department of Indian and Northern Affairs Canada; the Canadian Circumpolar Institute; and DeBeers Canada, Ltd. Lorna Ruechel assisted with den observations. A. Gunn provided caribou location data. We thank Dave Mech for the use of GPS collars. M. Nelson, A. Gunn, and three anonymous reviewers made helpful comments on earlier drafts of the manuscript. This work was done under Wildlife Research Permit – WL002948 issued by the Government of the Northwest Territories, Department of Resources, Wildlife, and Economic Development.

41 ## References

BALLARD, W.B., AYRES, L.A., KRAUSMAN, P.R., REED, D.J., and FANCY, S.G. 1997. Ecology of wolves in relation to a migratory caribou herd in northwest Alaska. Wildlife Monographs 135. 47 p.

BROWN, M.B., and FORSYTHE, A.B. 1974. Robust tests for the equality of variances. Journal of the American Statistical Association 69:364–367.

CARLSON, A. 1985. Central place foraging in the red-backed shrike (Lanius collurio L.): Allocation of prey between forager and sedentary consumer. Animal Behaviour 33:664–666.

CLARK, K.R.F. 1971. Food habits and behavior of the tundra wolf on central Baffin Island. Ph.D. Thesis, University of Toronto, Ontario, Canada.

ENVIRONMENT CANADA. 2003. National climate data information archive. Available online: http://www.climate.weatheroffice.ec.gc.ca/Welcome_e.html

FOURNIER, B., and GUNN, A. 1998. Musk ox numbers and distribution in the NWT, 1997. File Report No. 121. Yellowknife: Department of Resources, Wildlife, and Economic Development, Government of the Northwest Territories. 55 p.

FRITTS, S.H., and MECH, L.D. 1981. Dynamics, movements, and feeding ecology of a newly protected wolf population in northwestern Minnesota. Wildlife Monographs 80. 79 p.

GUNN, A., DRAGON, J., and BOULANGER, J. 2001. Seasonal movements of satellite-collared caribou from the Bathurst herd. Final Report to the West Kitikmeot Slave Study Society, Yellowknife, NWT. 80 p. Available online: http://www.wkss.nt.ca/HTML/08_ProjectsReports/PDF/Seasonal MovementsFinal.pdf

HARRINGTON, F.H., and ASA, C.S. 2003. Wolf communication. In: Mech, L.D., and Boitani, L., eds. Wolves: Behavior, ecology, and conservation. Chicago: University of Chicago Press. 66–103.

HARRINGTON, F.H., MECH, L.D., and FRITTS, S.H. 1983. Pack size and wolf pup survival: Their relationship under varying ecological conditions. Behavioral Ecology and Sociobiology 13:19–26.

HARRIS, S.A. 1986. Permafrost distribution, zonation and stability along the eastern ranges of the cordillera of North America. Arctic 39(1):29–38.

HEARD, D.C., and WILLIAMS, T.M. 1992. Distribution of wolf dens on migratory caribou ranges in the Northwest

Territories, Canada. Canadian Journal of Zoology 70:1504–1510.

HEARD, D.C., WILLIAMS, T.M., and MELTON, D.A. 1996. The relationship between food intake and predation risk in migratory caribou and implication to caribou and wolf population dynamics. Rangifer Special Issue No. 2:37–44.

KELSALL, J.P. 1968. The migratory barren-ground caribou of Canada. Canadian Wildlife Service Monograph Series 3. Ottawa: Queen's Printer. 340 p.

KUYT, E. 1962. Movements of young wolves in the Northwest Territories of Canada. Journal of Mammalogy 43:270–271.

———. 1972. Food habits and ecology of wolves on barren-ground caribou range in the Northwest Territories. Canadian Wildlife Service Report Series 21. Ottawa: Information Canada. 36 p.

MECH, L.D. 1994. Regular and homeward travel speeds of Arctic wolves. Journal of Mammalogy 75:741–742.

MECH, L.D., and BOITANI, L. 2003. Wolf social ecology. In: Mech, L.D., and Boitani, L., eds. Wolves: Behavior, ecology, and conservation. Chicago: University of Chicago Press. 1–34.

MECH, L.D., and GESE, E.M. 1992. Field testing the Wildlink capture collar on wolves. Wildlife Society Bulletin 20:249–256.

MECH, L.D., WOLFE, P., and PACKARD, J.M. 1999. Regurgitative food transfer among wild wolves. Canadian Journal of Zoology 77:1192–1195.

MERRILL, S.B., and MECH, L.D. 2000. Details of extensive movements by Minnesota wolves (*Canis lupus*). American Midland Naturalist 144:428–433.

MERRILL, S.B., ADAMS, L.G., NELSON, M.E., and MECH, L.D. 1998. Testing releasable GPS radiocollars on wolves and white-tailed deer. Wildlife Society Bulletin 26:830–835.

MESSIER, F. 1985. Solitary living and extraterritorial movements of wolves in relation to social status and prey abundance. Canadian Journal of Zoology 63:239–245.

MOHR, C.O., and STUMPF, W.A. 1966. Comparison of methods for calculating areas of animal activity. Journal of Wildlife Management 30:293–304.

MUSIANI, M. 2003. Conservation biology and management of wolves and wolf-human conflicts in western North America. Ph.D. Thesis, University of Calgary, Calgary, Alberta, Canada.

ORIANS, G.H., and PEARSON, N.E. 1979. On the theory of central place foraging. In: Mitchell, R.D., and Stairs, G.F., eds. Analysis of ecological systems. Columbus: Ohio State University Press. 154–177.

PACKARD, J.M., MECH, L.D., and REAM, R.R. 1992. Weaning in an arctic wolf pack: Behavioral mechanisms. Canadian Journal of Zoology 70:1269–1275.

PAQUET, P.C., WIERZCHOWSKI, J., and CALLAGHAN, C. 1996. Summary report on the effects of human activity on gray wolves in the Bow River Valley, Banff National Park, Alberta. In: Green, J., Pacas, C., Bayley, S., and Cornwell, L., eds. A cumulative effects assessment and futures outlook for the Banff Bow Valley. Prepared for the Banff Bow Valley Study. Ottawa: Department of Canadian Heritage.

STEPHENSON, R.O., and JAMES, D. 1982. Wolf movements and food habits in northwest Alaska. In: Harrington, F.H., and Paquet, P.C., eds. Wolves of the world. New Jersey: Noyes Publications. 223–237.

THORPE, N., EYEGETOK, S., HAKONGAK, N., and QITIRMIUT ELDERS. 2001. The Tuktu and Nogak Project: A caribou chronicle. Final Report to the West Kitikmeot/Slave Study Society, Ikaluktuuttiak, NWT. 160 p.

TIMONEY, K.P., LA ROI, G.H., ZOLTAI, S.C., and ROBINSON, A.L. 1992. The high subarctic forest-tundra of northwestern Canada: Position, width, and vegetation gradients in relation to climate. Arctic 45(1):1–9.

WALTON, L.R., CLUFF, H.D., PAQUET, P.C., and RAMSAY, M.A. 2001. Movement patterns of barren-ground wolves in the central Canadian Arctic. Journal of Mammalogy 82:867–876.

WILLIAMS, T.M. 1990. Summer diet and behavior of wolves denning on barren-ground caribou range in the Northwest Territories, Canada. M.Sc. Thesis, University of Alberta, Edmonton, Alberta, Canada.

ZAR, J.H. 1999. Biostatistical analysis. 4th ed. New Jersey: Prentice Hall. 663 p.

Appendix III. Answers to Self-Quizzes and Genetics Problems

Italicized numbers refer to relevant section numbers

CHAPTER 1

1.	a	1.2
2.	c	1.2
3.	c	1.3
4.	Homeostasis	1.3
5.	d	1.3
6.	reproduction	1.3
7.	Inheritance	1.3
8.	b	1.4
9.	a, c, d, e	1.2, 1.3, 1.4
10.	a, b	1.2, 1.4
11.	theory	1.9
12.	b	1.9
13.	b	1.6
14.	b	1.8
15.	c	1.2
	b	1.5
	d	1.9
	e	1.6
	a	1.6
	f	1.8

CHAPTER 2

1.	False (ions are atoms with different numbers of protons and electrons)	2.3
2.	d	2.2
3.	b	2.2
4.	a	2.4
5.	a	2.4
6.	polar covalent	2.4
7.	c, b, a	2.4
8.	c	2.5
9.	H⁺ or OH⁻	2.5
10.	b	2.5
11.	d	2.6
12.	a	2.6
13.	c	2.6
14.	a (hydrogen ion)	2.6
15.	c	2.5
	b	2.2
	d	2.5
	h	2.2
	g	2.5
	f	2.3
	e	2.3
	a	2.3

CHAPTER 3

1.	c	3.2
2.	4	3.2
3.	a	3.3
4.	e	3.4, 3.8
5.	a	3.4
6.	c	3.5
7.	False	3.1, 3.5
8.	b	3.5
9.	e	3.5
10.	d	3.6, 3.8
11.	d	3.7
12.	d	3.8
13.	a	3.8
14.	a (amino acids)	3.6
	b (carbohydrate)	3.4
	c (polypeptide)	3.6
	d (fatty acid)	3.5
15.	c	3.5
	e	3.4
	f	3.5
	d	3.8
	a	3.6
	b	3.8
16.	g	3.6
	a	3.5
	b	3.6
	c	3.5
	d	3.8
	e	3.4
	f	3.8
	i	3.6
	h	3.4
	j	3.5
	k	3.4

CHAPTER 4

1.	c	4.2
2.	c	4.2
3.	b	4.2
4.	False	4.4, 4.7
5.	c	4.4
6.	False (Protists are eukaryotes; by definition, all eukaryotes start life with a nucleus.)	4.5
7.	c	4.2
8.	a	4.7
9.	c, b, d, a	4.7
10.	lipids and proteins	4.7
11.	False (Many cells have walls that surround the plasma membrane.)	4.11
12.	b	4.6, 4.7, 4.9
13.	d	4.10
14.	d	4.11
15.	a	4.11
16.	a	4.10
17.	c	4.10
	d	4.11
	e	4.11
	b	4.11
	a	4.3
18.	g	4.9
	f	4.9
	a	4.4
	e	4.7
	d	4.7
	b	4.7
	c	4.7
	i	4.9
	h	4.10

CHAPTER 5

1.	c	5.2
2.	b	5.2
3.	d	5.2
4.	a	5.3
5.	c	5.3
6.	d	5.4
7.	c	5.3
8.	a	5.4
9.	d	5.5
10.	c	5.5
11.	a	5.6
12.	more/less	5.8
13.	c	5.9
14.	b	5.9
15.	b	5.8
16.	e	5.10
17.	c	5.3
	e	5.10
	f	5.2
	b	5.3
	a	5.6
	g	5.8
	h	5.9
	d	5.9

CHAPTER 6

1.	weed (autotroph; all others heterotrophs)	6.1
2.	a	6.1
3.	b	6.7, 6.1 revisited
4.	a	6.2
5.	a	6.4
6.	d	6.5
7.	b	6.5
8.	b	6.5
9.	c	6.5
10.	c	6.7
11.	b	6.7
12.	e	6.7
13.	f	6.7
14.	f	6.7
	h	6.7
	g	6.5
	d	6.5, 6.6
	e	6.8
	b	6.1, 6.4
	a	6.2
	c	6.1

CHAPTER 7

1.	False	7.2
2.	d	7.2, 7.3
3.	a	7.2, 7.6
4.	c	7.3
5.	b	7.2, 7.4, 7.5
6.	e	7.4
7.	b	7.4
8.	c	7.5
9.	c	7.6
10.	b	7.6
11.	c	7.5
12.	d	7.7
13.	f	7.6
14.	b	7.3
	c	7.6
	a	7.4, 7.6
	d	7.5
15.	b	7.4
	d	7.3
	a	7.3, 7.6
	c	7.2, 7.5
	f	7.5
	e	7.2, 7.4
	g	7.1

CHAPTER 8

1.	b	8.2
2.	centromere	8.2
3.	b	8.2
4.	d	8.2
5.	c	8.4
6.	d	8.4
7.	c	8.5
8.	a	8.5
9.	f	8.5
10.	3′-CCAAAGAAGTTCTCT-5′	8.5
11.	d	8.7
12.	c	8.4
13.	d	8.6
14.	b	8.7
15.	d	8.7
16.	d	8.3
	b	8.1, 8.7
	a	8.4
	f	8.2
	e	8.5
	g	8.5
	c	8.2
	h	8.6

CHAPTER 9

1.	c	9.2
2.	b	9.3
3.	d	9.3
4.	a	9.2
5.	c	9.2
6.	b	9.4
7.	c	9.4
8.	a	9.4
9.	15	9.4
10.	a	9.3
11.	a	9.3
12.	c	9.5
13.	c	9.5
14.	a	9.4, 9.5
15.	e	9.6
16.	c	9.4
	g	9.3
	e	9.5
	a	9.3
	f	9.4
	d	9.3
	b	9.6

CHAPTER 10

1.	d	10.2
2.	d	10.2
3.	d	10.2
4.	b	10.2
5.	b	10.2
6.	g	10.2
7.	c	10.2
8.	d	10.3
9.	d	10.2
10.	c	10.3
11.	b	10.3
12.	b	10.4
13.	c	10.4
14.	b	10.5
15.	True (epigenetics)	10.6
16.	f	10.2
	a	10.5
	b	10.5
	e	10.4
	c	10.2
	d	10.6

CHAPTER 11

1.	d	11.2
2.	b	11.2
3.	d	11.2
4.	e	11.2
5.	f	11.4
6.	c	11.3
7.	d	11.2
8.	a	11.2
9.	c	11.5
10.	b	11.2, 11.3
11.	a	11.2
12.	d	11.6
13.	b	11.6
14.	c	11.4
	f	11.3
	a	11.6
	g	11.4
	b	11.4
	e	11.6
	h	11.5
	d	11.3
15.	d, b, c, a	11.3

CHAPTER 12

1.	b	12.1, 12.2
2.	c	12.1, 12.2
3.	d	12.2, 12.5
4.	b	12.2, 12.3
5.	b	12.4
6.	prophase I	12.3, 12.4
7.	c	12.3
8.	a	12.2
9.	Sister chromatids are still attached (chromosomes are duplicated)	12.3
10.	e	12.4
11.	b	12.3, 12.4, 12.6
12.	b	12.2, 12.3
	c	12.3
	a	12.2
	e	12.5
	d	12.2

CHAPTER 13

1.	b	13.2
2.	a	13.2
3.	b	13.3
4.	c	13.3
5.	b	13.3
6.	a	13.3
7.	b	13.3
8.	d	13.4
9.	c	13.4
10.	c	13.5
11.	b	13.5
12.	d	13.7
13.	False	13.7
14.	b	13.4
	d	13.3
	a	13.2
	c	13.2

CHAPTER 14

1.	b	14.2
2.	b	14.2
3.	a	14.2
4.	b	14.3
5.	False	14.4
6.	d	14.3, 14.4
7.	d	14.4
8.	Y-linked dominant	14.4, 14.5
9.	X mom, Y dad	14.4
10.	d	14.6
11.	b	14.6
12.	d	14.6
13.	True	14.6
14.	c	14.6
15.	c	14.6
	e	14.5
	f	14.6
	b	14.5
	a	14.2
	d	14.6

CHAPTER 15

1.	c	15.2
2.	a	15.2
3.	b	15.2
4.	b	15.3
5.	c	15.3
6.	DNA sequencing	15.4
7.	b	15.4
8.	d	15.4
9.	d	15.5
10.	c	15.6
11.	b	15.8, 15.10
	d	15.2, 15.7
	e	15.10
	f	15.7
12.	c	15.6
13.	b	15.10
14.	a	15.3
	d	15.2
	c	15.2
	e	15.4
	b	15.2
15.	c	15.5
	g	15.7
	d	15.3
	e	15.10
	b	15.1
	a	15.6
	f	15.6

CHAPTER 16

1.	b	16.2
2.	c	16.2
3.	d	16.3, 16.4
4.	b	16.4
5.	b	16.5
6.	e	16.5
7.	d	16.6
8.	Gondwana	16.7
9.	65.5	16.8
10.	a	16.8
11.	a	16.4
	d	16.2, 16.5
	e	16.4

f	16.6
c	16.3
b	16.3
12. a	16.3, 16.5
c	16.7
f	16.7
g	16.5, 16.7
h	16.1

CHAPTER 17

1. a	17.2
2. c	17.8
3. a	17.6
c	17.5
b	17.6
4. populations	17.2
5. d	17.7
6. e	17.7
7. b	17.8
8. a	17.9
9. d	17.7, 17.9
10. c	17.11
11. a	17.10
12. f	17.2, 17.8
13. c	17.8
d	17.7
e	17.2
b	17.8
a	17.12
f	17.12

CHAPTER 18

1. c	18.2
2. c	18.2
3. b	18.2
4. b	18.3
5. d	18.3
6. c	18.4
7. c	18.4
8. b	18.4
9. c	18.2
10. b	18.5
11. e	18.3, 18.4, 18.5
12. b	18.2
g	18.2
d	18.5
c	18.3
e	18.4
f	18.3
a	18.2

CHAPTER 19

1. b	19.2
2. oxygen	19.2
3. b	19.3
4. c	19.4
5. a	19.4
6. ribozyme	19.4
7. a	19.5
8. d	19.5
9. b	19.6
10. d	19.6
11. endosymbiosis	19.6
12. stromatolite	19.5
13. c	19.6
14. b	19.1
15. (1)f	19.2
(2)c	19.4
(3)d	19.5
(4)a	19.5
(5)b	19.7
(6)e	19.7

CHAPTER 20

1. c	20.2
2. b	20.4
3. b	20.3
4. RNA	20.3
5. d	20.5
6. c	20.6
7. c	20.7
8. d	20.5, 20.7
9. b	20.7
10. c	20.7
11. d	20.6
12. b	20.6
13. d	20.4, 20.7
14. pandemic	20.4
15. d	20.8
e	20.6
b	20.2
f	20.6
h	20.8
a	20.4
c	20.6
g	20.6

CHAPTER 21

1. c	21.2
2. b	21.3
3. b	21.4, 21.8
4. c	21.6
5. b	21.8
6. cyanobacteria	21.10
7. c	21.2
8. a	21.5
9. c	21.3, 21.7, 21.11
10. c	21.9
11. a	21.3
12. c	21.11
13. b	21.6
14. e	21.4
c	21.4
a	21.8
b	21.9
d	21.9
15. d	21.3
g	21.7
a	21.6
b	21.8
f	21.8
h	21.9
e	21.10
c	21.11

CHAPTER 22

1. c	22.2
2. a	22.3
3. a	22.3
4. False	22.5
5. a	22.4
6. b	22.4
7. c	22.5
8. a	22.6
9. d	22.4, 22.5
10. b	22.6
11. True	22.6
12. b	22.8
13. d	22.4
c	22.5
a	22.7
b	22.8
14. c	22.6
h	22.3
a	22.2
b	22.2
e	22.8
f	22.8
d	22.5
g	22.5
15. 1 d	22.4
2 e	22.5
3 b	22.7
4 a	22.7
5 c	22.9

CHAPTER 23

1. c	23.2
2. a	23.2
3. a	23.2, 23.4
4. c	23.5
5. d	23.6
6. b	23.2
7. c	23.6
8. d	23.6
9. a	23.3
10. b	23.7
11. a	23.7
12. a	23.4
13. d	23.7
14. b	23.7
15. d	23.2
b	23.2
a	23.3
f	23.5
g	23.6
c	23.7
e	23.4, 23.7

CHAPTER 24

1. True	24.3
2. coelom	24.2
3. a	24.2
4. a	24.4
5. c	24.5
6. d	24.7
7. c	24.6, 24.7
8. c	24.11, Table 24.1
9. two	24.6
10. b	24.8
11. c	24.14
12. d	24.5
13. c	24.7
14. f	24.8
i	24.16
d	24.4
b	24.5
c	24.6
g	24.9
a	24.10
h	24.7
e	24.11

CHAPTER 25

1. notochord, dorsal hollow nerve cord, pharynx with gill slits, tail extending past anus	25.2
2. all	25.2
3. a	25.4
4. c	25.3
5. c	25.6
6. c	25.7
7. f	25.7, 25.8
8. a	25.7
9. c	25.9
10. c	25.9, 25.10
11. b	25.1, 25.9
12. a	25.6
13. j	25.2
i	25.3
h	25.6
f	25.7
c	25.7
g	25.5
d	25.10
a	25.10
e	25.4
b	25.10
14. a	25.3
f	25.4
e	25.6
c	25.7
b	25.1, 25.9
d	25.10
15. d	25.5
a	25.5
f	25.10
e	25.6
c	25.2
b	25.11

CHAPTER 26

1. d	26.2
2. a	26.3
3. c	26.3
4. d	26.2
5. a	26.4
6. c	26.3
7. d	26.2, 26.4
8. c	26.5
9. a	26.6
10. d	26.1
11. b	26.6
12. neanderthals	26.1
13. a	26.1
d	26.4, 26.5
c	26.4
b	26.6
f	26.5
e	26.6
14. f	26.2
b	26.2
c	26.3
e	26.3
d	26.5
a	26.1
g	26.6

CHAPTER 27

1. a	27.2, 27.6
2. d	27.7
3. c	27.3
4. c	27.3
5. a	27.3
6. True	27.3
7. b	27.3
8. b	25.3
9. a	27.3, 27.5
10. b	27.4
11. d	27.6
12. c	27.6
13. b	27.7, 27.8
14. c	27.7
15. b	27.2
d	27.2, 27.7
e	27.3
c	27.3
f	27.6
a	27.7
g	27.4
h	27.2

CHAPTER 28

1. e	28.2
2. b	28.2
3. b	28.3
4. b	28.3
5. e	28.3
6. c	28.4
7. d	28.4
8. b	28.5
9. c	28.5
10. a	28.5
11. d	28.4
12. c	28.6
13. c	28.5
14. e	28.6
15. c	28.5
g	28.2
e	28.6
b	28.3
d	28.4
a	28.4
f	28.6

CHAPTER 29

1. a, b, c	29.1, 29.3
2. b	29.2
3. b	29.2
4. c	29.2
5. b	29.4
6. c	29.4
7. b	29.6
8. a	29.6
9. c	29.6, 29.7
10. c	29.6, 29.8
11. d	29.8
12. c	29.9
13. Berry	29.7
14. Drupe	29.7
15. c	29.4, 29.6
f	29.2
g	29.4
e	29.4
d	29.8
b	29.4
a	29.4

CHAPTER 30

1. d	30.2
2. b	30.3
3. c	30.6, 30.10
4. a	30.3, 30.4, 30.6
5. d	30.7
6. a	30.8
7. d	30.7
8. c	30.9
9. d	30.10
10. b	30.8
d	30.8
a	30.8
c	30.8
f	30.9
e	30.9
11. b	30.7
d	30.4
a	30.3
e	30.5
d	30.6
f	30.6, 30.10
12. c	30.7
e	30.4
b	30.8
a	30.5, 30.9
d	30.6
f	30.10

CHAPTER 31

1. a	31.3
2. a	31.3
3. a	31.3
4. b	31.4
5. d	31.4
6. b	31.4
7. b	31.6
8. c	31.5
9. a	31.4
10. b	31.3
11. c	31.3
12. a	31.3
13. b	31.5
14. a	31.9
15. b	31.3
j	31.3
c	31.4
e	31.8
a	31.6
d	31.5
f	31.5
h	31.4
g	31.8
i	31.2

CHAPTER 32

1. a	32.2
2. c	32.3
3. d	32.4
4. False	32.4
5. a	32.5
6. a	32.5
7. c	32.6
8. c	32.7
9. b	32.7
10. c	32.8
11. a	32.12
12. a	32.12
13. c	32.11
14. a	32.9, 32.10
15. h	32.8
d	32.5
f	32.9
b	32.9
g	32.10
a	32.9
e	32.12
c	32.9

CHAPTER 33

1.	b	33.3
2.	c	33.2
3.	c	33.3
4.	e	33.4
5.	b	33.3
6.	b	33.10
7.	b	33.9
8.	a	33.3
9.	b	33.9
10.	b	33.7
11.	d	33.6
12.	b	33.16
13.	c	33.7
14.	a	33.6
15.	e	33.7
	j	33.9
	b	33.9
	g	33.5
	a	33.6
	h	33.7
	f	33.4
	k	33.3
	i	33.9
	c	33.10
	d	33.4

CHAPTER 34

1.	a	34.2
2.	a	34.3
3.	c, b, a, d	34.4
4.	a	34.4
5.	d	34.5
6.	b	34.7
7.	c	34.7
8.	d	34.8
9.	d	34.8
10.	d	34.9
11.	b	34.11
12.	True	34.6
13.	False	34.4
14.	False	34.13
15.	d	34.9
	f	34.7
	c	34.4
	e	34.8
	a	34.11
	b	34.2

CHAPTER 35

1.	a	35.2
2.	d	35.4
3.	b	35.6
4.	a	35.5
5.	a	35.5
6.	c	35.3
7.	b	35.7
8.	b	35.7
9.	d	35.7
10.	d	35.9
11.	b	35.8
12.	d	35.8
13.	a	35.8
14.	a	35.9
15.	e	35.4
	i	35.7
	g	35.9
	h	35.5
	d	35.7
	c	35.4
	b	35.3
	a	35.8
	f	35.9

CHAPTER 36

1.	d	36.2
2.	d	36.2
3.	b	36.2
4.	a	36.4
5.	b	36.5
6.	d	36.4
7.	b	36.7
8.	b	36.7
9.	a	36.9
10.	c	36.11
11.	left ventricle	36.7
12.	pulmonary artery	36.7
13.	b	36.13
14.	b	36.12
15.	f	36.4
	a	36.13
	e	36.4
	g	36.12
	b	36.7
	c	36.11

CHAPTER 37

1.	e	37.2–37.4
2.	g	37.2, 37.5, 37.7
3.	c	37.2, 37.5, 37.7
4.	d	37.4, 37.5
5.	d	37.2
6.	a, c, d	37.2
7.	b, e, f, g	37.2, 37.7
8.	a	37.7
9.	d	37.6
10.	b	37.7, 37.8
11.	e	37.7, 37.9
12.	b	37.9
13.	c	37.10
14.	b	37.10
15.	b	37.12
16.	e	37.10
	b	37.2
	c	37.9
	d	37.9
	a	37.4, 37.7
	f	37.7

CHAPTER 38

1.	d	38.2
2.	b	38.3
3.	a	38.4
4.	c	38.5
5.	b	38.7
6.	d	38.6
7.	a	38.6
8.	b	38.1
9.	a	38.7
10.	c	38.8
11.	c	38.8
12.	d	38.6
13.	True	38.6
14.	d	38.5
15.	d	38.5
	h	38.5
	f	38.5
	e	38.5
	g	38.5
	c	38.5
	b	38.5
	a	38.5

CHAPTER 39

1.	d	39.2
2.	b	39.5
3.	c	39.6
4.	b	39.7
5.	a	39.7
6.	c	39.8
7.	a	39.5
8.	b	39.7
9.	b	39.11
10.	d	39.9
11.	b	39.6
12.	c	39.10
13.	b	39.12
14.	d	39.5
	e	39.12
	a	39.7
	b	39.7
	c	39.4
	f	39.5
15.	f	39.7
	h	39.4
	c	39.8
	b	39.7
	e	39.6
	a	39.8
	g	39.5
	d	39.7

CHAPTER 40

1.	c	40.2
2.	b	40.2
3.	a	40.3
4.	b	40.3
5.	a	40.4
6.	b	40.4
7.	a	40.5
8.	d	40.5
9.	a	40.6
10.	b	40.5
11.	c	40.3
	a	40.3
	b	40.3
	e	40.5
	d	40.5
12.	d	40.9
13.	b	40.8
14.	b	40.9
15.	mitochondria	40.9

CHAPTER 41

1.	b	41.2
2.	c	41.3
3.	b	41.4
4.	b	41.5
5.	c	41.5
6.	c	41.6
7.	a	41.6
8.	d	41.8
9.	b	41.4
10.	a	41.9
11.	d	41.3
12.	a, c, a, c, a, c, b	41.10
13.	d	41.3
	c	41.3
	h	41.5
	a	41.3
	i	41.5
	e	41.5
	f	41.5
	b	41.3
	g	41.5
14.	a	41.4
	c	41.4
	b	41.5
	d	41.3

CHAPTER 42

1.	b	42.2
2.	b	42.2
3.	d	42.4
4.	c	42.4
	d	42.4
	a	42.2
	b	42.2
	e	42.7
5.	d	42.7
6.	a	42.7
7.	c	42.9
8.	d	42.10
9.	d	42.12
10.	a	42.12
	c	42.12
	b	42.7
11.	c	42.8
12.	d	42.8
13.	a	42.11
14.	d	42.11
15.	4	42.8
	2	42.6
	7	42.10
	3	42.7
	1	42.2
	5	42.4, 42.8
	6	42.8

CHAPTER 43

1.	d	43.2
2.	b	43.2
3.	c	43.5
4.	a	43.6
5.	a	43.1
6.	b	43.7
7.	d	43.8
8.	c	43.8
9.	d	43.9
10.	c	43.9
11.	b	43.9
12.	taxis	43.5
13.	b	43.3
14.	d	43.3
15.	g	43.5
	a	43.3
	b	43.6
	f	43.8
	d	43.3
	c	43.7
	e	43.4

CHAPTER 44

1.	a	44.2
2.	f	44.2
3.	c	44.2
4.	b	44.3
5.	a	44.3
6.	d	44.4
7.	d	44.5
8.	b	44.7
9.	d	44.8
10.	a	44.5
11.	c	44.5
12.	d	44.8
13.	c	44.4
	d	44.3
	a	44.3
	e	44.4
	b	44.4

CHAPTER 45

1.	b	45.2
2.	d	45.2, 45.4
3.	b	45.7
4.	d	45.4
5.	b	45.3
	d	45.7
	c	45.2
	a	45.7
	e	45.4
6.	e	45.4
7.	b	45.8
8.	mimicry	45.6
9.	c	45.8
10.	a	45.10
11.	b	45.4, 45.9
12.	c	45.7
13.	c	45.10
14.	c	45.7
15.	c	45.10
	b	45.8
	d	45.9
	a	45.9
	f	45.9
	e	45.4

CHAPTER 46

1.	a	46.2
2.	b	46.2
3.	c	46.4
4.	d	46.4
5.	b	46.4
6.	c	46.6
7.	b	46.7
8.	d	46.7
9.	d	46.8
10.	a	46.11
11.	c	46.11
12.	d	46.9, 46.11
13.	a	46.9
14.	d	46.7, 46.10
15.	e	46.8
	d	46.9
	b	46.11
	a	46.9
	c	46.7
	f	46.9

CHAPTER 47

1.	b	47.2
2.	d	47.2
3.	d	47.2
4.	a	47.3
5.	c	47.3
6.	b	47.6
7.	d	47.4
8.	c	47.6
9.	a	47.10
10.	less	47.11
11.	d	47.14
12.	c	47.13
13.	c	47.10
14.	b	47.4
15.	d	47.10
	e	47.7
	f	47.5
	c	47.6
	b	47.12
	h	47.9
	i	47.6
	a	47.8
	g	47.14

CHAPTER 48

1.	True	48.2
2.	b	48.2
3.	b	48.3
4.	c	48.5
5.	a	48.5
6.	d	48.5
7.	a	48.6
8.	False	48.5
9.	b	48.7
10.	b	48.6
11.	c	48.8
12.	d	48.8
13.	c	48.8
14.	d	48.9
15.	g	48.8
	a	48.6
	i	48.8
	e	48.5
	d	48.3
	f	48.5
	h	48.7
	b	48.4
	c	48.4

1. a. *AB*
    b. *AB, aB*
    c. *Ab, ab*
    d. *AB, Ab, aB, ab*

2. a. All offspring will be *AaBB*.
    b. 1/4 *AABB*   (25% each genotype)
        1/4 *AABb*
        1/4 *AaBB*
        1/4 *AaBb*
    c. 1/4 *AaBb*   (25% each genotype)
        1/4 *Aabb*
        1/4 *aaBb*
        1/4 *aabb*
    d. 1/16 *AABB* (6.25% of genotype)
        1/8 *AaBB*   (12.5%)
        1/16 *aaBB* (6.25%)
        1/8 *AABb*   (12.5%)
        1/4 *AaBb*   (25%)
        1/8 *aaBb*   (12.5%)
        1/16 *AAbb* (6.25%)
        1/8 *Aabb*   (12.5%)
        1/16 *aabb* (6.25%)

3. a. *ABC*
    b. *ABC, aBC*
    c. *ABC, aBC, ABc, aBc*
    d. *ABC*
        aBC
        AbC
        abC
        ABc
        aBc
        Abc
        abc

4. A mating of two M^L cats yields 1/4 *MM*, 1/2 M^LM, and 1/4 M^LM^L. Because M^LM^L is lethal, the probability that any one kitten among the survivors will be heterozygous is 2/3.

5. a. Both parents are heterozygotes (*Aa*). Their children may be albino (*aa*) or unaffected (*AA* or *Aa*).
    b. All are homozygous recessive (*aa*).
    c. Homozygous recessive (*aa*) father, and heterozygous (*Aa*) mother. The albino child is *aa*, the unaffected children *Aa*.

6. Possible outcomes of an experimental cross between F_1 rose plants heterozygous for height (*Aa*):

3:1 possible ratio of genotypes and phenotypes in F_2 generation

Possible outcomes of a testcross between an F_1 rose plant heterozygous for height and a shrubby rose plant:

Gametes F_1 hybrid:

1:1 possible ratio of genotypes and phenotypes in F_2 generation

7. Yellow is recessive. Because F_1 plants have a green phenotype and must be heterozygous, green must be dominant over the recessive yellow.

8. A mating between a mouse from a true-breeding, white-furred strain and a mouse from a true-breeding, brown-furred strain would provide you with the most direct evidence. Because true-breeding strains of organisms typically are homozygous for a trait being studied, all F_1 offspring from this mating should be heterozygous. Record the phenotype of each F_1 mouse, then let them mate with one another. Assuming only one gene locus is involved, these are possible outcomes for the F_1 offspring:
    a. All F_1 mice are brown, and their F_2 offspring segregate: 3 brown : 1 white. *Conclusion*: Brown is dominant to white.
    b. All F_1 mice are white, and their F_2 offspring segregate: 3 white : 1 brown. *Conclusion*: White is dominant to brown.
    c. All F_1 mice are tan, and the F_2 offspring segregate: 1 brown : 2 tan : 1 white. *Conclusion*: The alleles at this locus show incomplete dominance.

9. The data reveal that these genes do not assort independently because the observed ratio is very far from the 9:3:3:1 ratio expected with independent assortment. Instead, the results can be explained if the genes are located close to each other on the same chromosome, which is called linkage.

10. a. 1/2 red   1/2 pink   white
    b.   red   All pink   white
    c. 1/4 red   1/2 pink   1/4   white
    d.   red   1/2 pink   1/2   white

11. Because both parents are heterozygotes (Hb^AHb^S), the following are the probabilities for each child:
    a. 1/4 Hb^SHb^S
    b. 1/4 Hb^AHb^A
    c. 1/2 Hb^AHb^S

1. The phenotype appeared in every generation shown in the diagram, so this must be a pattern of autosomal dominant inheritance.

2. **a.** Human males (XY) inherit their X chromosome from their mother.

   b. A male can produce two kinds of gametes. Half carry an X chromosome and half carry a Y chromosome. All the gametes that carry the X chromosome carry the same X-linked allele.

   c. A female homozygous for an X-linked allele produces only one kind of gamete.

   d. Fifty percent of the gametes of a female who is heterozygous for an X-linked allele carry one of the two alleles at that locus; the other fifty percent carry its partner allele for that locus.

3. Because Marfan syndrome is a case of autosomal dominant inheritance and because one parent bears the allele, the probability that any child of theirs will inherit the mutant allele is 50 percent.

4. **a.** Nondisjunction might occur during anaphase I or anaphase II of meiosis.

   b. As a result of translocation, chromosome 21 may get attached to the end of chromosome 14. The new individual's chromosome number would still be 46, but its somatic cells would have the translocated chromosome 21 in addition to two normal chromosomes 21.

5. A daughter could develop this muscular dystrophy only if she inherited two X-linked recessive alleles—one from each parent. Males who carry the allele are unlikely to father children because they develop the disorder and die early in life.

6. In the mother, a crossover between the two genes at meiosis generates an X chromosome that carries neither mutant allele.

Appendix IV. Periodic Table of the Elements

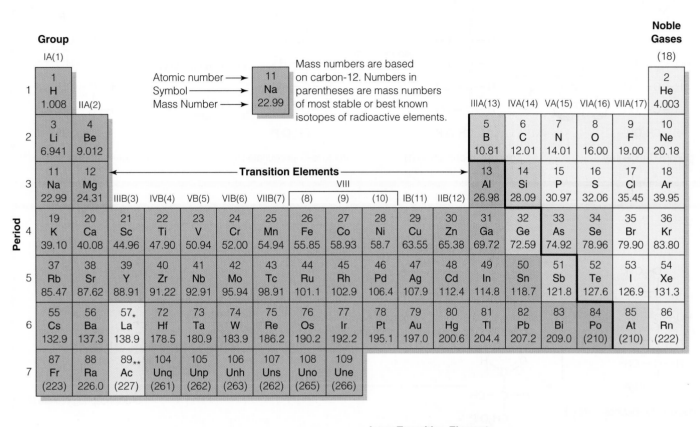

Group

Atomic number ⟶ 11
Symbol ⟶ Na
Mass Number ⟶ 22.99

Mass numbers are based on carbon-12. Numbers in parentheses are mass numbers of most stable or best known isotopes of radioactive elements.

Noble Gases (18)

Transition Elements

Inner Transition Elements

Lanthanide Series 6*	58 Ce 140.1	59 Pr 140.9	60 Nd 144.2	61 Pm (145)	62 Sm 150.4	63 Eu 152.0	64 Gd 157.3	65 Tb 158.9	66 Dy 162.5	67 Ho 164.9	68 Er 167.3	69 Tm 168.9	70 Yb 173.0	71 Lu 175.0
Actinide Series 7**	90 Th 232.0	91 Pa 231.0	92 U 238.0	93 Np 237.0	94 Pu (244)	95 Am (243)	96 Cm (247)	97 Bk (247)	98 Cf (251)	99 Es (252)	100 Fm (257)	101 Md (258)	102 No (259)	103 Lr (260)

Appendix V. A Closer Look at Some Major Metabolic Pathways

Glycolysis

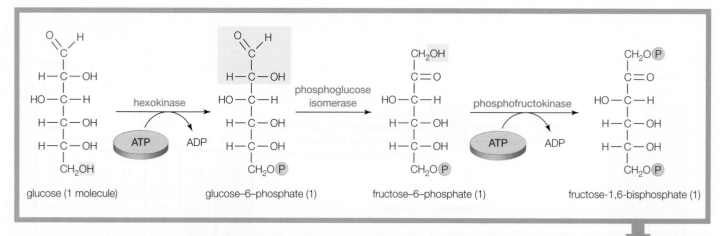

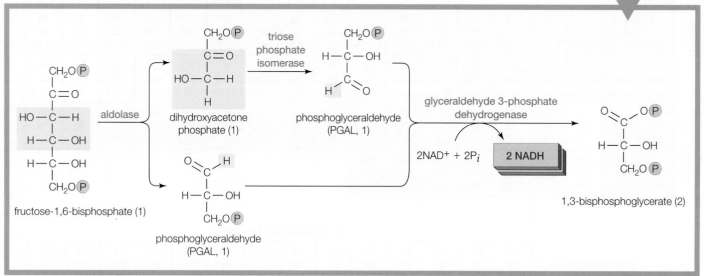

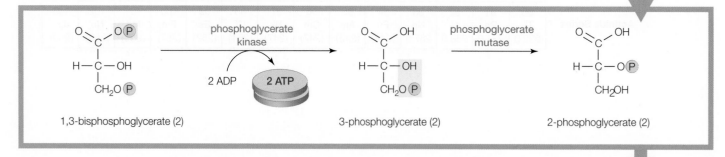

Figure A Glycolysis breaks down one glucose molecule into two 3-carbon pyruvate molecules for a net yield of two ATP. Enzyme names are indicated in *green*; parts of substrate molecules undergoing chemical change are highlighted *blue*.

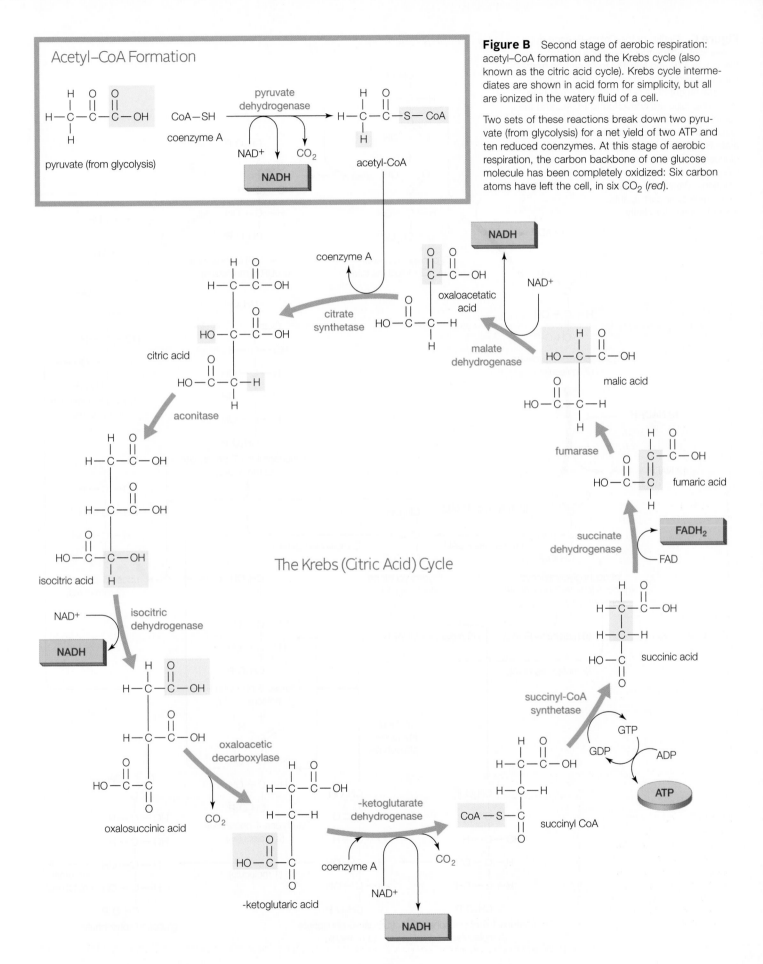

Figure B Second stage of aerobic respiration: acetyl–CoA formation and the Krebs cycle (also known as the citric acid cycle). Krebs cycle intermediates are shown in acid form for simplicity, but all are ionized in the watery fluid of a cell.

Two sets of these reactions break down two pyruvate (from glycolysis) for a net yield of two ATP and ten reduced coenzymes. At this stage of aerobic respiration, the carbon backbone of one glucose molecule has been completely oxidized: Six carbon atoms have left the cell, in six CO_2 (*red*).

Acetyl–CoA Formation

pyruvate (from glycolysis)

pyruvate dehydrogenase

coenzyme A

NAD^+ CO_2

NADH

acetyl-CoA

The Krebs (Citric Acid) Cycle

coenzyme A

citrate synthetase

citric acid

aconitase

isocitric acid

NAD^+

NADH

isocitric dehydrogenase

oxaloacetic decarboxylase

oxalosuccinic acid

CO_2

-ketoglutaric acid

-ketoglutarate dehydrogenase

coenzyme A

CO_2

NAD^+

NADH

oxaloacetatic acid

NAD^+

NADH

malate dehydrogenase

malic acid

fumarase

fumaric acid

succinate dehydrogenase

FADH₂

FAD

succinic acid

succinyl-CoA synthetase

CoA—S—C succinyl CoA

GTP

GDP   ADP

ATP

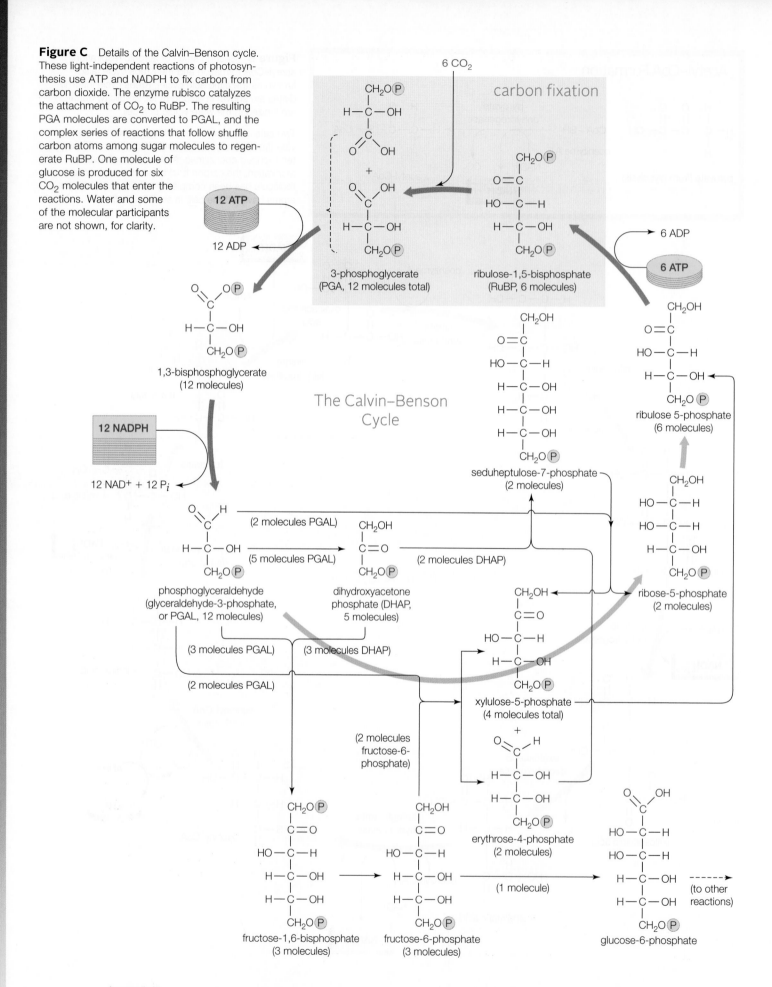

Figure C Details of the Calvin–Benson cycle. These light-independent reactions of photosynthesis use ATP and NADPH to fix carbon from carbon dioxide. The enzyme rubisco catalyzes the attachment of CO_2 to RuBP. The resulting PGA molecules are converted to PGAL, and the complex series of reactions that follow shuffle carbon atoms among sugar molecules to regenerate RuBP. One molecule of glucose is produced for six CO_2 molecules that enter the reactions. Water and some of the molecular participants are not shown, for clarity.

Appendix VI. A Plain English Map of the Human Chromosomes

1
- sweet taste receptors
- Rh blood type
- marijuana receptor
- (anorexia nervosa susceptibility)
- leptin receptor
- TSH β chain
- lamin A (progeria)
- Duffy blood group antigen

2
- LH/choriogonadotropin receptor (micropenis)
- CD8; cytotoxic T cell antigen
- antibody light chain
- lactase
- (cleft palate)
- glucagon

3
- oxytocin receptor
- HIV receptor
- rhodopsin
- (alkaptonuria)
- (sucrose intolerance)
- somatostatin

4
- (achondroplasia)
- (Huntington disease)
- (Ellis-van Creveld syndrome)
- alcohol dehydrogenase (susceptibility to alcoholism)
- red hair color

5
- Cri-du-chat syndrome
- bitter taste receptor
- growth hormone receptor (pituitary dwarfism)
- interleukin-4

6
- (gluten intolerance)
- HLA/MHC
- tumor necrosis factor
- α chains of HCG, FSH, LH, and TSH
- estrogen receptor

7
- cytochrome c
- elastin
- DLX 5/6 homeotic genes
- CFTR (cystic fibrosis)
- leptin (obesity)
- (blue-deficient colorblind)
- TCR β subunit

8
- gonadotropin releasing hormone
- helicase (Werner's syndrome)
- corticotropin releasing hormone

9
- (galactosemia)
- (cerebral palsy)
- (Friedreich ataxia)
- (fructose intolerance)
- ABO blood group

10
- vitamin B-12 receptor
- mannose binding protein
- perforin
- (gluten intolerance)

11
- hemoglobin β chain (sickle cell anemia)
- insulin
- parathyroid hormone
- catalase
- PAX6 (aniridia)
- FSH, β chain
- tyrosinase (albinism)

12
- CD4
- helper T cell antigen
- oncogene KRAS2 (lung cancer, bladder cancer, breast cancer)
- keratins
- lysozyme
- (phenylketonuria)
- aldehyde dehydrogenase (alcohol intolerance)

13
- ribosomal RNA
- BRCA 2 (breast cancer)
- (gastroesophageal reflux)

14
- ribosomal RNA
- presinilin (Alzheimer's)
- TSH receptor
- immunoglobulin heavy chains

15
- ribosomal RNA
- fibrillin 1 (Marfan syndrome)
- (Tay-Sachs disease)

16
- hemoglobin α chain
- DNAse I (lupus)

17
- (Canavan disease)
- p53 tumor antigen
- NF1 (neurofibromatosis)
- serotonin transporter
- BRCA 1 (breast, ovarian cancer)
- Growth hormone

18
- B cell apoptosis regulator (B cell lymphoma)
- myelin basic protein

19
- LDL receptor (coronary artery disease)
- insulin receptor
- brown hair color
- green/blue eye color
- (Warfarin resistance)
- HCG, β chain
- LH, β chain

20
- prion protein (Creutzfeld-Jacob disease)
- oxytocin
- GHRH (acromegaly)

21
- ribosomal RNA
- interferon receptors
- (bipolar disorder, early onset)

22
- ribosomal RNA
- immunoglobulin light chains
- myoglobin

X
- dystrophin (muscular dystrophy)
- (anhidrotic ectodermal dysplasia)
- IL2RG (SCID-X1)
- XIST X chromosome inactivation control
- (hemophilia B)
- (hemophilia A)
- (red-deficient colorblind)
- (green-deficient colorblind)

Y
- sex determining region Y (SRY)
- (no sperm)
- male stature

Haploid set of human chromosomes. The banding patterns characteristic of each type of chromosome appear after staining with a reagent called Giemsa. The locations of some of the 20,065 known genes (as of November, 2005) are indicated. Also shown are locations that, when mutated, cause some of the genetic diseases discussed in the text.

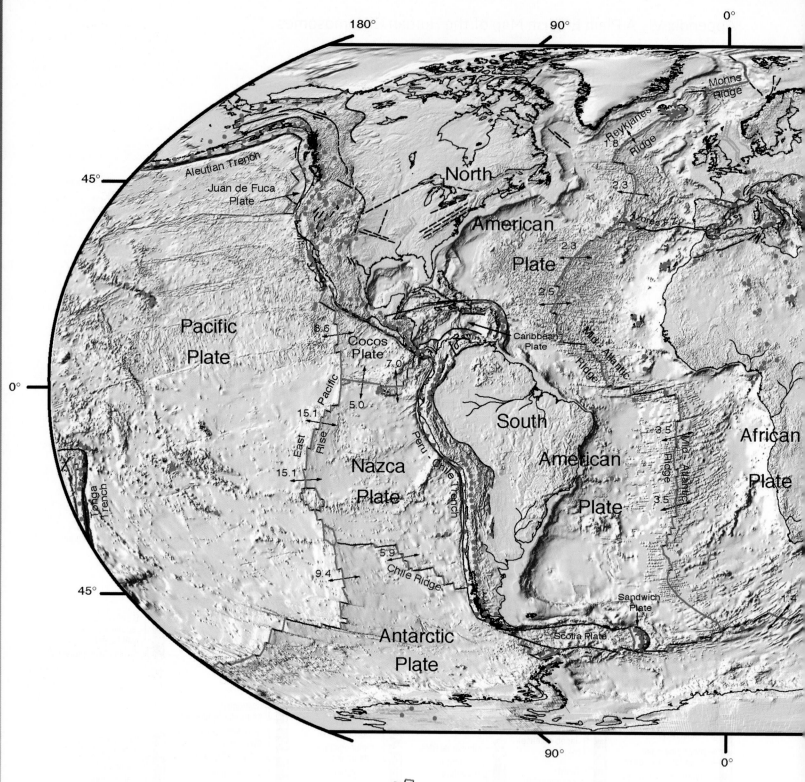

180° 90° 0°

Mohns
Ridge

45°

Reykjanes
Ridge

1.8

Aleutian Trench

Juan de Fuca
Plate

North

2.3

American

Azores F.Z.

Plate

2.3

2.5

Pacific

Caribbean
Plate

Mid-Atlantic

Ridge

Plate

8.6

Cocos
Plate

7.0

African

0°

East
Pacific
Rise

5.0

15.1

South

Plate

3.5

Peru-Chile Trench

American

Nazca

Mid-Atlantic
Ridge

Plate

Plate

15.1

3.5

Tonga
Trench

1.4

5.9

Sandwich
Plate

9.4

Chile Ridge

45°

Scotia Plate

Antarctic

Plate

90° 0°

Appendix VII.
Restless Earth—Life's Changing
Geologic Stage

This NASA map summarizes the tectonic and volcanic
activity of Earth during the past 1 million years. The reconstructions at far right indicate positions of Earth's
major land masses through time.

Actively-spreading ridges and transform faults

Total spreading rate, cm/year

1.4

Major active fault or fault zone; dashed where nature,
location, or activity uncertain

Normal fault or rift; hachures on downthrown side

Reverse fault (overthrust, subduction zones); generalized;
barbs on upthrown side

Volcanic centers active within the last one million years;
generalized. Minor basaltic centers and seamounts omitted.

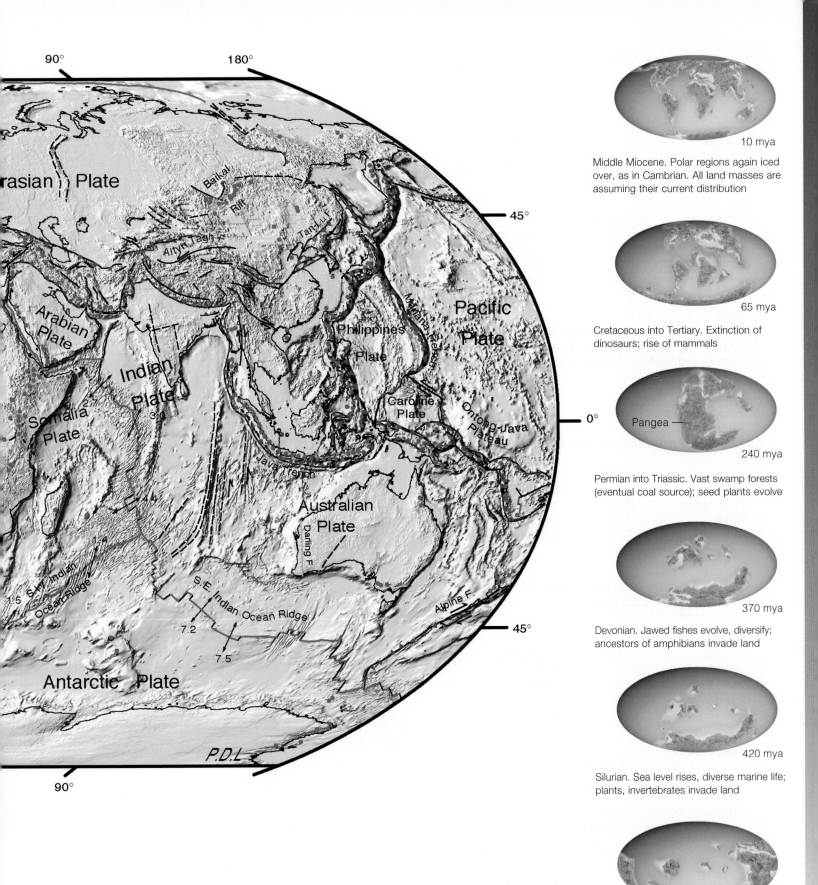

90°   180°

45°

rasian   Plate

Baikal

Rift

Altyn Tagh F.

Tan-Lu F.

Pacific
Plate

Arabian
Plate

Mariana Trench

Indian

Plate

Philippines

2.7

Plate

Somalia

3.0

Plate

Caroline
Plate

Ontong-Java
Plateau

Java   Trench

4.4

Australian
Plate

Darling F.

S.E. Indian Ocean Ridge

1.5 S.W. Indian

Ocean Ridge

7.2

7.5

Alpine F.

45°

Antarctic   Plate

P.D.L

90°

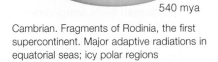
10 mya

Middle Miocene. Polar regions again iced
over, as in Cambrian. All land masses are
assuming their current distribution

65 mya

Cretaceous into Tertiary. Extinction of
dinosaurs; rise of mammals

Pangea

240 mya

Permian into Triassic. Vast swamp forests
(eventual coal source); seed plants evolve

370 mya

Devonian. Jawed fishes evolve, diversify;
ancestors of amphibians invade land

420 mya

Silurian. Sea level rises, diverse marine life;
plants, invertebrates invade land

540 mya

Cambrian. Fragments of Rodinia, the first
supercontinent. Major adaptive radiations in
equatorial seas; icy polar regions

Appendix VII

Appendix VIII. Units of Measure

Length

1 kilometer (km) = 0.62 miles (mi)
1 meter (m) = 39.37 inches (in)
1 centimeter (cm) = 0.39 inches

To convert	multiply by	to obtain
inches	2.25	centimeters
feet	30.48	centimeters
centimeters	0.39	inches
millimeters	0.039	inches

Area

1 square kilometer = 0.386 square miles
1 square meter = 1.196 square yards
1 square centimeter = 0.155 square inches

Volume

1 cubic meter = 35.31 cubic feet
1 liter = 1.06 quarts
1 milliliter = 0.034 fluid ounces = 1/5 teaspoon

To convert	multiply by	to obtain
quarts	0.95	liters
fluid ounces	28.41	milliliters
liters	1.06	quarts
milliliters	0.03	fluid ounces

Weight

1 metric ton (mt) = 2,205 pounds (lb) = 1.1 tons (t)
1 kilogram (kg) = 2.205 pounds (lb)
1 gram (g) = 0.035 ounces (oz)

To convert	multiply by	to obtain
pounds	0.454	kilograms
pounds	454	grams
ounces	28.35	grams
kilograms	2.205	pounds
grams	0.035	ounces

Temperature

Celcius (°C) to Fahrenheit (°F):
$$°F = 1.8 (°C) + 32$$

Fahrenheit (°F) to Celsius:
$$°C = \frac{(°F - 32)}{1.8}$$

	°C	°F
Water boils	100	212
Human body temperature	37	98.6
Water freezes	0	32

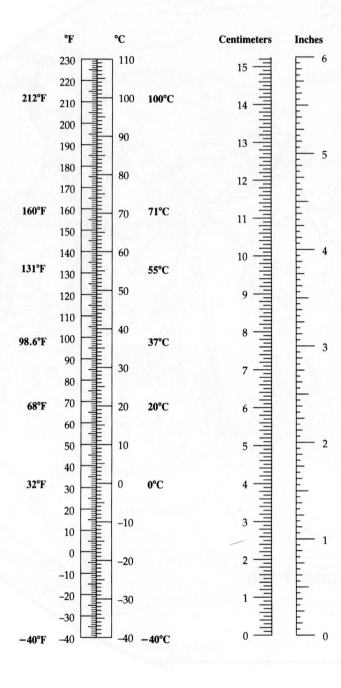

Chapter 1
Summary

Section 1.1 Biology is the systematic study of life. We have encountered only a fraction of the organisms that live on Earth, in part because we have explored only a fraction of its inhabited regions.

Section 1.2 Biologists think about life at different levels of organization. **Emergent properties** appear at successively higher levels. Life emerges at the cellular level. All matter consists of **atoms**, which combine as **molecules**. **Organisms** are individuals that consist of one or more **cells**. Cells of larger multicelled organisms are organized as **tissues**, **organs**, and **organ systems**. A **population** is a group of individuals of a species in a given area; a **community** is all populations of all species in a given area. An **ecosystem** is a community interacting with its environment. The **biosphere** includes all regions of Earth that hold life.

Section 1.3 Life has underlying unity in that all living things have similar characteristics: (1) All organisms require **energy** and **nutrients** to sustain themselves. **Producers** harvest energy from the environment to make their own food by processes such as **photosynthesis**; **consumers** eat other organisms, or their wastes and remains. (2) Organisms keep the conditions in their internal environment within ranges that their cells tolerate—a process called **homeostasis**. (3) **DNA** contains information that guides an organism's form and function, which include **development**, **growth**, and **reproduction**. The passage of DNA from parents to offspring is called **inheritance**.

Section 1.4 The many types of organisms that currently exist on Earth differ greatly in details of body form and function. **Biodiversity** is the sum of differences among living things. **Bacteria** and **archaea** are all single-celled, and their DNA is not contained within a **nucleus**. **Eukaryotes** (**protists**, **plants**, **fungi**, and **animals**) can be single-celled or multicelled. Their DNA is contained within a nucleus.

Section 1.5 Each type of organism has a two-part name. The first part is the **genus** name. When combined with the **specific epithet**, it designates a particular **species**. With **taxonomy**, species are ranked into ever more inclusive **taxa** on the basis of shared **traits**.

Section 1.6 Critical thinking, the self-directed act of judging the quality of information as one learns, is an important part of **science**. Generally, a researcher observes something in nature, uses **inductive reasoning** to form a **hypothesis** (testable explanation) for it, then uses **deductive reasoning** to make a

prediction about what might occur if the hypothesis is correct. Predictions are tested with observations, **experiments**, or both. Experiments typically are performed on an **experimental group** as compared with a **control group**, and sometimes on **models**. Conclusions are drawn from experimental results, or **data**. A hypothesis that is not consistent with data is modified. Making, testing, and evaluating hypotheses are the **scientific method**.

Biological systems are usually influenced by many interacting **variables**. Experiments test how an **independent variable** influences a **dependent variable**.

Section 1.7 Scientific approaches differ, but experiments are typically designed in a consistent way. A researcher changes an independent variable, then observes the effects of the change on a dependent variable. This practice allows the researcher to study a cause-and-effect relationship in a complex natural system.

Section 1.8 A small sample size increases the potential for **sampling error** in experimental results. In such cases, a subset may be tested that is not representative of the whole. Researchers design experiments carefully to minimize sampling error and bias, and they use **probability** rules to check the **statistical significance** of their results. Scientists check and test one another's work, so science is ideally a self-correcting process.

Section 1.9 Science helps us be objective about our observations because it is concerned only with testable ideas about observable aspects of nature. Opinion and belief have value in human culture, but they are not addressed by science. A **scientific theory** is a long-standing hypothesis that is useful for making predictions about other phenomena. It is our best way of describing reality. A **law of nature** describes something that occurs without fail, but our scientific explanation of why it occurs is incomplete.

Self-Quiz

Answers in Appendix III

1. _____ are fundamental building blocks of all matter.
   - a. Atoms
   - b. Molecules
   - c. Cells
   - d. Organisms

2. The smallest unit of life is the _____ .
   - a. atom
   - b. molecule
   - c. cell
   - d. organism

3. All organisms require _____ and _____ to sustain themselves.
   - a. DNA; energy
   - b. cells; raw materials
   - c. nutrients; energy
   - d. DNA; cells

4. _____ is a process that maintains conditions in the internal environment within ranges that cells can tolerate.

5. DNA _____ .
   - a. guides development and functioning
   - b. is the basis of traits
   - c. is transmitted from parents to offspring
   - d. all of the above

Peacock Butterfly Predator Defenses

The photographs on the *right* represent the experimental and control groups used in the peacock butterfly experiment discussed in Section 1.7.

See if you can identify the experimental groups, and match them up with the relevant control group(s).

Hint: Identify which variable is being tested in each group (each variable has a control).

A Wing spots painted out

B Wing spots visible; wings silenced

C Wing spots painted out; wings silenced

D Wings painted but spots visible

E Wings cut but not silenced

F Wings painted, spots visible; wings cut, not silenced

6. A process by which an organism produces offspring is called _____ .

7. _____ is the transmission of DNA to offspring.

8. _____ move around for at least part of their life.
   a. Organisms   c. Consumers
   b. Animals   d. Cells

9. A butterfly is a(n) _____ (choose all that apply).
   a. organism   e. consumer
   b. domain   f. producer
   c. animal   g. hypothesis
   d. eukaryote   h. trait

10. A bacterium is _____ (choose all that apply).
    a. an organism   c. an animal
    b. single-celled   d. a eukaryote

11. A long-standing hypothesis that is used to make predictions about other phenomena is called a _____ .

12. Science addresses only that which is _____ .
    a. alive   c. variable
    b. observable   d. indisputable

13. A control group is _____ .
    a. a set of individuals that have a certain characteristic or receive a certain treatment
    b. the standard against which an experimental group is compared
    c. the experiment that gives conclusive results

14. Fifteen randomly selected students are found to be taller than 6 feet. The researchers concluded that the average height of a student is greater than 6 feet. This is an example of _____ .
    a. experimental error   c. sampling bias
    b. sampling error   d. experimental bias

15. Match the terms with the most suitable description.
    ___ emergent property   a. statement of what a hypothesis leads you to expect
    ___ species   b. type of organism
    ___ scientific theory   c. occurs at a higher organizational level
    ___ hypothesis   d. time-tested hypothesis
    ___ prediction   e. testable explanation
    ___ probability   f. measure of chance

Critical Thinking

1. A person is declared to be dead upon the irreversible cessation of spontaneous body functions: brain activity, or blood circulation and respiration. However, only about 1% of a person's cells have to die in order for all of these things to happen. How can someone be dead when 99% of his or her cells are still alive?

2. Explain the difference between a one-celled organism and a single cell of a multicelled organism.

3. Why would you think twice about ordering from a cafe menu that lists the genus name but not the specific epithet of its offerings? *Hint:* Look up *Homarus americanus, Ursus americanus, Ceanothus americanus, Bufo americanus, Lepus americanus,* and *Nicrophorus americanus.*

4. Once there was a highly intelligent turkey that had nothing to do but reflect on the world's regularities. Morning always started out with the sky turning light, followed by the master's footsteps, which were always followed by the appearance of food. Other things varied, but food always followed footsteps. The sequence of events was so predictable that it eventually became the basis of the turkey's theory about the goodness of the world. One morning, after more than 100 confirmations of the goodness theory, the turkey listened for the master's footsteps, heard them, and had its head chopped off.

   Any scientific theory is modified or discarded upon discovery of contradictory evidence. The absence of absolute certainty has led some people to conclude that "facts are irrelevant—facts change." If that is so, should we stop doing scientific research? Why or why not?

5. In 2005, researcher Woo-suk Hwang reported that he had made immortal stem cells from human patients. His research was hailed as a breakthrough for people affected by degenerative diseases, because stem cells may be used to repair a person's own damaged tissues. Hwang published his results in a peer-reviewed journal. In 2006, the journal retracted his paper after other scientists discovered that Hwang's group had faked their data.

   Does the incident show that results of scientific studies cannot be trusted? Or does it confirm the usefulness of a scientific approach, because other scientists discovered and exposed the fraud?

Chapter 2
Summary

Section 2.1 Mercury in air pollution ends up in the bodies of fish, and in turn, in the bodies of humans. The properties of mercury—and of all substances—arise from the atoms that make them up.

Section 2.2 Atoms consist of **electrons**, which carry a negative **charge**, moving about a **nucleus**. Positively charged **protons** and uncharged **neutrons** reside in the nucleus (**Table 2.2**). The **periodic table** lists **elements** in order of their **atomic number**. **Isotopes** are atoms of an element that differ in the number of neutrons, so they also differ in **mass number**. Researchers can make **tracers** with **radioisotopes**, which spontaneously emit particles and energy by the process of **radioactive decay**.

Section 2.3 Up to two electrons occupy each orbital (volume of space around a nucleus). Which orbital an electron occupies depends on its energy. A **shell model** represents successive energy levels as concentric circles.

Atoms tend to get rid of vacancies. Many do so by gaining or losing electrons, thereby becoming **ions** (charged atoms). **Free radicals** are atoms with unpaired electrons; they tend to be very active chemically.

Section 2.4 A **chemical bond** is an attractive force that unites two atoms as a **molecule**. A **compound** is a molecule that consists of two or more elements. Atoms form different types of bonds depending on their **electronegativity**. An **ionic bond** is a strong association between oppositely charged ions; it arises from the mutual attraction of opposite charges. A molecule that has a separation of charge is said to show **polarity**. Ionic bonds are completely polar. Atoms share a pair of electrons in a **covalent bond**, which is nonpolar if the sharing is equal, and polar if it is not.

Section 2.5 Polar covalent bonds join two hydrogen atoms to one oxygen atom in each water molecule. The polarity causes extensive **hydrogen bonds** to form among water molecules, and this bonding is the basis of unique properties that sustain life: a capacity to act as a **solvent** for **salts** and other polar **solutes**; resistance to **temperature** changes; and **cohesion**.

Hydrophilic substances dissolve easily in water; **hydrophobic** substances do not. **Evaporation** is the transition of a liquid to a gas. A **mixture** is an intermingling of substances.

Section 2.6 A solute's **concentration** refers to the amount of solute in a given volume of fluid. **pH** reflects the number of hydrogen ions (H^+) in a fluid. At neutral pH (7), the amounts of H^+ and OH^- ions are the same.

Table 2.2	Players in the Chemistry of Life
Atoms	Particles that are basic building blocks of all matter; the smallest unit that retains an element's properties.
Proton (p^+)	Positively charged particle of an atom's nucleus.
Electron (e^-)	Negatively charged particle that occupies a defined volume of space (orbital) around an atom's nucleus.
Neutron	Uncharged particle of an atom's nucleus.
Element	Pure substance that consists entirely of atoms with the same, characteristic number of protons.
Isotopes	Atoms of an element that differ in the number of neutrons.
Radioisotope	Unstable isotope that emits particles and energy when its nucleus disintegrates.
Tracer	Molecule that has a detectable substance (such as a radioisotope) attached. Used to track the movement or destination of the molecule in a biological system.
Ion	Atom that carries a charge after it has gained or lost one or more electrons. A single proton without an electron is a hydrogen ion (H^+).
Molecule	Two or more atoms joined in a chemical bond.
Compound	Molecule of two or more different elements in unvarying proportions (for example, water: H_2O).
Mixture	Intermingling of two or more substances in proportions that can vary.
Solute	Substance dissolved in a solvent.
Hydrophilic	Refers to a substance that dissolves easily in water. Such substances consist of polar molecules.
Hydrophobic	Refers to a substance that resists dissolving in water. Such substances consist of nonpolar molecules.
Acid	Compound that releases H^+ when dissolved in water.
Base	Compound that accepts H^+ when dissolved in water.
Salt	Ionic compound that releases ions other than H^+ or OH^- when dissolved in water.
Solvent	Substance that can dissolve other substances.

Acids release hydrogen ions in water; **bases** accept them. A **buffer** keeps a solution within a consistent range of pH. Most cell and body fluids are buffered because most molecules of life work only within a narrow range of pH.

Self-Quiz
Answers in Appendix III

1. Is the following statement true or false? Every atom has an equal number of protons and electrons.

2. In the periodic table, symbols for the elements are arranged according to _____ .
   a. size
   b. charge
   c. mass number
   d. atomic number

Mercury Emissions by Continent By weight, coal does not contain much mercury, but we burn a lot of it. Several industries besides coal-fired power plants contribute substantially to atmospheric mercury pollution. **Figure 2.13** shows mercury emissions by industry from different regions of the world in 2006.

1. About how many tons of mercury were released world-wide in 2006?

2. Which industry tops the list of mercury emitters? Which industry is next on the list?

3. Which region emitted the most mercury from producing cement?

4. About how many tons of mercury were released from gold production in South America?

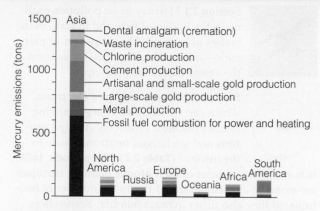

Figure 2.13 Global mercury emissions, 2006. *Source: Global Atmospheric Mercury Assessment: Sources, Emissions and Transport. United Nations Environmental Programme, Chemicals Branch. 2008*

3. A(n) _____ is a molecule into which a radioisotope has been incorporated.
   a. compound   c. salt
   b. tracer   d. acid

4. The measure of an atom's ability to pull electrons away from another atom is called _____ .
   a. electronegativity   c. charge
   b. polarity   d. concentration

5. The mutual attraction of opposite charges holds atoms together as molecules in a(n) _____ bond.
   a. ionic   c. polar covalent
   b. hydrogen   d. nonpolar covalent

6. Atoms share electrons unequally in a(n) _____ bond.

7. Rank the following types of bonds by polarity, with 1 being the least polar, and 3 being the most polar:
   ___ 1   a. ionic
   ___ 2   b. polar covalent
   ___ 3   c. nonpolar covalent

8. A(n) _____ substance repels water.
   a. acidic   c. hydrophobic
   b. basic   d. polar

9. A salt releases ions other than _____ in water.

10. A(n) _____ is dissolved in a solvent.
    a. molecule   c. salt
    b. solute   d. acid

11. Hydrogen ions (H^+) are _____ .
    a. indicated by a pH scale   c. in blood
    b. unbound protons   d. all of the above

12. When dissolved in water, a(n) _____ donates H^+; a(n) _____ accepts H^+.
    a. acid; base   c. buffer; solute
    b. base; acid   d. base; buffer

13. A _____ is a chemical partnership between a weak acid or base and its salt.
    a. covalent bond   c. buffer
    b. hydrogen bond   d. pH

14. What is the name of an atom that has one proton but no neutrons and no electrons?

15. Match the terms with their most suitable description.
    ___ hydrophilic   a. protons > electrons
    ___ atomic number   b. number of protons in nucleus
    ___ hydrogen bond   c. polar; easily dissolves in water
    ___ mass number   d. collectively strong
    ___ temperature   e. protons < electrons
    ___ uncharged   f. protons = electrons
    ___ negative charge   g. measure of molecular motion
    ___ positive charge   h. number of protons and
                             neutrons in atomic nucleus

Critical Thinking

1. Alchemists were medieval scholars and philosophers who were the forerunners of modern-day chemists. Many spent their lives trying to transform lead (atomic number 82) into gold (atomic number 79). Explain why they never did succeed in that endeavor.

2. Draw a shell model of an uncharged nitrogen atom (nitrogen has 7 protons).

3. Polonium is a rare element with 33 radioisotopes. The most common one, ^{210}Po, has 82 protons and 128 neutrons. When ^{210}Po decays, it emits an alpha particle, which is a helium nucleus (2 protons and 2 neutrons). ^{210}Po decay is tricky to detect because alpha particles do not carry very much energy compared to other forms of radiation. They can be stopped by, for example, a sheet of paper or a few inches of air. That is one reason that authorities failed to discover toxic amounts of ^{210}Po in the body of former KGB agent Alexander Litvinenko until after he died suddenly and mysteriously in 2006. What element does an atom of ^{210}Po change into after it emits an alpha particle?

4. Some undiluted acids are more corrosive when they are diluted with water. That is why laboratory workers are told to wipe acid splashes on skin with a towel before washing with copious amounts of water. Explain.

Chapter 3
Summary

Section 3.1 All organisms consist of the same kinds of molecules. Seemingly small differences in the way those molecules are put together can have big effects inside a living organism.

Section 3.2 Under present-day conditions in nature, only living things make complex carbohydrates and lipids, proteins, and nucleic acids. These molecules are **organic**—they consist mainly of carbon and hydrogen atoms.

Section 3.3 Hydrocarbons have only carbon and hydrogen atoms. Carbon chains or rings form the backbone of the molecules of life. **Functional groups** attached to the backbone influence the chemical character, and thus the function of these compounds. **Metabolism** includes all of the processes by which cells acquire and use energy as they make and break the bonds of organic compounds. By metabolic **reactions** such as **condensation**, **enzymes** build **polymers** from **monomers** of simple sugars, fatty acids, amino acids, and nucleotides. Reactions such as **hydrolysis** release the monomers by breaking apart the polymers.

Section 3.4 Enzymes assemble **disaccharides** and **polysaccharides** from simple **carbohydrate** (sugar) monomers. **Cellulose, glycogen,** and **starch** consist of **monosaccharides** bonded in different patterns. Cells use the different kinds for energy, and as structural materials.

Section 3.5 Lipids are fatty, oily, or waxy compounds that cells use for energy and as structural materials. All are nonpolar. **Fats** and some other lipids have **fatty acid** tails; **triglycerides** have three. **Unsaturated fatty acids** have carbon–carbon double bonds; only single bonds link carbons in **saturated fatty acids**. A **lipid bilayer** that consists mainly of **phospholipids** is the structural foundation of all cell membranes. **Waxes** are lipids that are part of water-repellent and lubricating secretions. **Steroids** occur in cell membranes; some are remodeled into other molecules.

Section 3.6 Structurally and functionally, **proteins** are the most diverse molecules of life. The shape of a protein is the source of its function. Protein structure begins as a linear sequence of **amino acids** linked by **peptide bonds** into a **polypeptide** (primary structure). Polypeptides twist into loops, sheets, and coils (secondary structure) that can pack further into functional domains (tertiary structure). Many proteins, including most enzymes, consist of two or more polypeptides (quaternary structure). Fibrous proteins aggregate by the thousands into much larger structures.

Section 3.7 A protein's structure dictates its function, so changes in a protein's structure may also alter its function. Hydrogen bonds and other molecular interactions that are responsible for a protein's shape may be disrupted by shifts in pH or temperature, or exposure to detergent or some salts. If that happens, the protein unravels, or **denatures**, and so loses its function. **Prion** diseases are a consequence of misfolded proteins.

Section 3.8 Nucleotides are small organic molecules consisting of a sugar, a phosphate group, and a nitrogen-containing base. Nucleotides are monomers of **DNA** and **RNA**, which are **nucleic acids**. Some nucleotides have additional functions. For example, **ATP** energizes many kinds of molecules by phosphate-group transfers. DNA encodes heritable information that guides the synthesis of RNA and proteins. RNA molecules interact with DNA to carry out protein synthesis.

Self-Quiz
Answers in Appendix III

1. Organic molecules consist mainly of _____ atoms.
   a. carbon
   b. carbon and oxygen
   c. carbon and hydrogen
   d. carbon and nitrogen

2. Each carbon atom can share pairs of electrons with as many as _____ other atom(s).

3. _____ groups impart polarity to alcohols.
   a. Hydroxyl ($-OH^-$)
   b. Phosphate ($-PO_4$)
   c. Methyl ($-CH_3$)
   d. Sulfhydryl ($-SH$)

4. _____ is a simple sugar (a monosaccharide).
   a. Glucose
   b. Sucrose
   c. Ribose
   d. Starch
   e. both a and c
   f. a, b, and c

5. Which three carbohydrates can be built using only glucose monomers?
   a. Starch, cellulose, and glycogen
   b. Glucose, sucrose, and ribose
   c. Cellulose, steroids, and polysaccharides
   d. Starch, chitin, and DNA
   e. Triglycerides, nucleic acids, and polypeptides

6. Unlike saturated fats, the fatty acid tails of unsaturated fats incorporate one or more _____ .
   a. phosphate groups
   b. glycerols
   c. double bonds
   d. single bonds

7. Is this statement true or false? Unlike saturated fats, all unsaturated fats are beneficial to health because their fatty acid tails kink and do not pack together.

8. Steroids are among the lipids with no _____ .
   a. double bonds
   b. fatty acid tails
   c. hydrogens
   d. carbons

9. Which of the following is a class of molecules that encompasses all of the other molecules listed?
   a. triglycerides
   b. fatty acids
   c. waxes
   d. steroids
   e. lipids
   f. phospholipids

10. _____ are to proteins as _____ are to nucleic acids.
    a. Sugars; lipids
    b. Sugars; proteins
    c. Amino acids; hydrogen bonds
    d. Amino acids; nucleotides

Effects of Dietary Fats on Lipoprotein Levels Cholesterol that is made by the liver or that enters the body from food does not dissolve in blood, so it is carried through the bloodstream by lipoproteins. Low-density lipoprotein (LDL) carries cholesterol to body tissues such as artery walls, where it can form deposits associated with cardiovascular disease. Thus, LDL is often called "bad" cholesterol. High-density lipoprotein (HDL) carries cholesterol away from tissues to the liver for disposal, so HDL is often called "good" cholesterol.

In 1990, Ronald Mensink and Martijn Katan published a study that tested the effects of different dietary fats on blood lipoprotein levels. Their results are shown in **Figure 3.19**.

1. In which group was the level of LDL ("bad" cholesterol) highest?

2. In which group was the level of HDL ("good" cholesterol) lowest?

3. An elevated risk of heart disease has been correlated with increasing LDL-to-HDL ratios. Which group had the highest LDL-to-HDL ratio?

4. Rank the three diets from best to worst according to their potential effect on heart disease.

	Main Dietary Fats			
	cis fatty acids	*trans* fatty acids	saturated fats	optimal level
LDL	103	117	121	<100
HDL	55	48	55	>40
ratio	1.87	2.44	2.2	<2

Figure 3.19 Effect of diet on lipoprotein levels. Researchers placed 59 men and women on a diet in which 10 percent of their daily energy intake consisted of *cis* fatty acids, *trans* fatty acids, or saturated fats.

Blood LDL and HDL levels were measured after three weeks on the diet; averaged results are shown in mg/dL (milligrams per deciliter of blood). All subjects were tested on each of the diets. The ratio of LDL to HDL is also shown.

---

11. A denatured protein has lost its _____ .
    a. hydrogen bonds   c. function
    b. shape   d. all of the above

12. _____ consists of nucleotides.
    a. Sugars   c. DNA
    b. RNA   d. b and c

13. Which of the following is not found in DNA?
    a. amino acids   c. nucleotides
    b. sugars   d. phosphate groups

14. In the following list, identify the carbohydrate, the fatty acid, the amino acid, and the polypeptide:
    a. NH_2—CHR—COOH   c. (methionine)$_{20}$
    b. $C_6H_{12}O_6$ d. $CH_3(CH_2)_{16}COOH$

15. Match the molecules with the best description.
    ___ wax   a. protein primary structure
    ___ starch   b. an energy carrier
    ___ triglyceride   c. water-repellent secretions
    ___ DNA   d. carries heritable information
    ___ polypeptide   e. sugar storage in plants
    ___ ATP   f. richest energy source

16. Match each polymer with the most appropriate set of component monomers.
    ___ protein   a. glycerol, fatty acids, phosphate
    ___ phospholipid   b. amino acids, sugars
    ___ glycoprotein   c. glycerol, fatty acids
    ___ fat   d. nucleotides
    ___ nucleic acid   e. polysaccharide
    ___ carbohydrate   f. sugar, phosphate, base
    ___ nucleotide   g. amino acids
    ___ lipoprotein   h. glucose, fructose
    ___ sucrose   i. lipids, amino acids
    ___ glycogen   j. fatty acids, carbon rings
    ___ wax   k. glucose only

Critical Thinking

1. Lipoproteins are relatively large, spherical clumps of protein and lipid molecules that circulate in the blood of mammals. They are like suitcases that move cholesterol, fatty acid remnants, triglycerides, and phospholipids from one place to another in the body. Given what you know about the insolubility of lipids in water, which of the four kinds of lipids would you predict to be on the outside of a lipoprotein clump, bathed in the fluid portion of blood?

2. In 1976, a team of chemists in the United Kingdom was developing new insecticides by modifying sugars with chlorine (Cl_2), phosgene (Cl_2CO), and other toxic gases. One young member of the team misunderstood his verbal instructions to "test" a new molecule. He thought he had been told to "taste" it. Luckily, the molecule was not toxic, but it was very sweet. It became the food additive sucralose.

Sucralose has three chlorine atoms substituted for three hydroxyl groups of sucrose (table sugar). It binds so strongly to sweet-taste receptors on the tongue that the human brain perceives it as 600 times sweeter than sucrose. Sucralose was originally marketed as an artificial sweetener called Splenda®, but it is now available under several other brand names.

Researchers proved that the body does not recognize sucralose as a carbohydrate by feeding sucralose labeled with ^{14}C to volunteers. Analysis of the radioactive molecules in the volunteers' urine and feces showed that 92.8 percent of the sucralose passed through the body without being altered. Many people are worried that the chlorine atoms impart toxicity to sucralose. How would you respond to that concern?

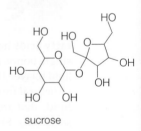

sucrose

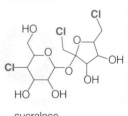

sucralose

Chapter 4
Summary

Section 4.1 Bacteria are found in all parts of the biosphere, including the human body. Huge numbers inhabit our intestines, but most of these are beneficial. A few can cause disease. Contamination of food with disease-causing bacteria can result in food poisoning that is sometimes fatal.

Section 4.2 Cells differ in size, shape, and function, but all start out life with a **plasma membrane, cytoplasm**, and a region of DNA. Most cells have additional components (Table 4.4).

In eukaryotic cells, DNA is contained within a **nucleus**, which is a membrane-enclosed **organelle**. All cell membranes, including the plasma membrane and organelle membranes, are selectively permeable and consist mainly of phospholipids organized as a lipid bilayer. The **surface-to-volume ratio** limits cell size.

By the **cell theory**, all organisms consist of one or more cells; the cell is the smallest unit of life; each new cell arises from another, preexisting cell; and a cell passes hereditary material to its offspring.

Section 4.3 Most cells are far too small to see with the naked eye, so we use microscopes to observe them. Different types of microscopes and techniques reveal different internal and external details of cells.

Section 4.4 Bacteria and archaea, informally grouped as "prokaryotes," are the most diverse forms of life. These single-celled organisms have no nucleus, but they have **nucleoids** and **ribosomes**. Many have a permeable but protective **cell wall** and a sticky capsule, as well as motile structures (**flagella**) and other projections (**pili**). Some have **plasmids** in addition to their single chromosome. Bacteria and other microbial organisms often share living arrangements in **biofilms**.

Section 4.5 Protists, fungi, plants, and animals are eukaryotic. Cells of these organisms start out life with membrane-enclosed organelles, including a nucleus. Organelles compartmentalize tasks and substances that are sensitive or dangerous to the rest of the cell.

Section 4.6 A nucleus protects and controls access to a eukaryotic cell's DNA, which is typically distributed among a characteristic number of **chromosomes**. A **nuclear envelope** surrounds **nucleoplasm**. In the nucleus, ribosome subunits are produced in dense, irregularly shaped **nucleoli**.

Sections 4.7, 4.8 The **endomembrane system** is a system of interacting organelles that includes ER, Golgi bodies, and vesicles. **Endoplasmic reticulum (ER)** is a continuous system of sacs and tubes extending from the nuclear envelope. Ribosome-studded rough ER makes polypeptides; smooth ER assembles lipids and degrades carbohydrates and fatty acids. **Golgi bodies** modify peptides and lipids before sorting them into **vesicles**. Different types of vesicles store, degrade, or transport substances through the cell. Enzymes in **peroxisomes** break down substances such as amino acids, fatty acids, and toxins. **Lysosomes** contain enzymes that break down cellular debris for recycling. Fluid-filled **vacuoles** store and dispose of waste, debris, and toxins. Fluid pressure inside a **central vacuole** keeps plant cells plump, thus keeping plant parts firm.

Section 4.9 Double-membraned **mitochondria** specialize in making ATP by breaking down organic compounds in the oxygen-requiring pathway of aerobic respiration. Different types of **plastids** are specialized for photosynthesis or storage. In eukaryotes, photosynthesis takes place inside **chloroplasts**. Pigment-filled chromoplasts and starch-filled amyloplasts are used for storage, and also serve additional roles.

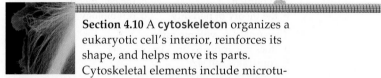

Section 4.10 A **cytoskeleton** organizes a eukaryotic cell's interior, reinforces its shape, and helps move its parts. Cytoskeletal elements include microtubules, microfilaments, and intermediate filaments. Interactions between ATP-driven **motor proteins** and hollow, dynamically assembled **microtubules** bring about cellular movement. A **microfilament** mesh called the **cell cortex** reinforces plasma membranes. Elongating microfilaments bring about movement of **pseudopods**. **Intermediate filaments** lend structural support to cells and tissues.

Centrioles give rise to a special 9+2 array of microtubules inside **cilia** and eukaryotic flagella, then remain beneath these motile structures as **basal bodies**.

Section 4.11 Many cells secrete a complex mixture of fibrous proteins and polysaccharides onto their surfaces. The secretions form an **extracellular matrix (ECM)** that supports cells and tissues, and also functions in cell-to-cell signaling.

Most prokaryotes, protists, fungi, and all plant cells secrete a wall around the plasma membrane. Older plant cells secrete a rigid, **lignin**-containing **secondary wall** inside their pliable **primary wall**. Many eukaryotic cell types also secrete a waxy, protective **cuticle**. **Cell junctions** connect animal cells to one another and to

ECM. **Plasmodesmata** connect the cytoplasm of adjacent plant cells. In animals, **gap junctions** form open channels between adjacent cells; **adhering junctions** anchor cells to one another and to ECM; and **tight junctions** form a waterproof seal between cells in some tissues.

 Section 4.12 We describe the quality of "life" as a set of properties that are collectively unique to living things. Living things consist of cells that engage in self-sustaining biological processes, pass their hereditary material (DNA) to offspring by mechanisms of reproduction, and have the capacity to change over successive generations.

Self-Quiz

Answers in Appendix III

1. Despite the diversity of cell type and function, all cells have these three things in common:
   a. cytoplasm, DNA, and organelles
   b. a plasma membrane, DNA, and proteins
   c. cytoplasm, DNA, and a plasma membrane
   d. carbohydrates, nucleic acids, and proteins

2. Every cell is descended from another cell. This idea is part of _____ .
   a. evolution   c. the cell theory
   b. the theory of heredity   d. cell biology

3. The surface-to-volume ratio _____ .
   a. does not apply to prokaryotic cells
   b. constrains cell size
   c. is part of the cell theory
   d. b and c

4. True or false? Ribosomes are only found in bacteria and archaea.

5. Unlike eukaryotic cells, bacterial cells _____ .
   a. have no plasma membrane   c. have no nucleus
   b. have RNA but not DNA   d. a and c

6. True or false? Some protists start out life with no nucleus.

7. Cell membranes consist mainly of a _____ .
   a. carbohydrate bilayer and proteins
   b. protein bilayer and phospholipids
   c. lipid bilayer and proteins

8. Enzymes contained in _____ break down worn-out organelles, bacteria, and other particles.
   a. lysosomes   c. endoplasmic reticulum
   b. amyloplasts   d. peroxisomes

9. Put the following structures in order according to the pathway of a secreted protein:
   a. plasma membrane   c. endoplasmic reticulum
   b. Golgi bodies   d. post-Golgi vesicles

10. The main function of the endomembrane system is building and modifying _____ and _____ .

11. Is this statement true or false? The plasma membrane is the outermost component of all cells. Explain.

12. Which of the following organelles contains no DNA?
    a. nucleus   c. mitochondrion
    b. Golgi body   d. chloroplast

13. Cytoskeletal elements called _____ form a reinforcing mesh under the nuclear envelope.
    a. intermediate filaments   c. actin filaments
    b. microtubules   d. microfilaments

14. No animal cell has a _____ .
    a. plasma membrane   c. lysosome
    b. flagellum   d. cell wall

15. _____ connect the cytoplasm of plant cells.
    a. Plasmodesmata   c. Tight junctions
    b. Adhering junctions   d. a and b

16. Intermediate filaments are a feature of _____ cells.
    a. eukaryotic   c. animal
    b. all   d. algal

17. Match each term with the best description.
    ___ centriole   a. shows surface details
    ___ ECM   b. feature of secondary walls
    ___ cuticle   c. basal body
    ___ lignin   d. connectivity
    ___ SEM   e. protective covering

18. Match each cell component with its specialization.
    ___ mitochondrion   a. protein synthesis
    ___ chloroplast   b. associates with
    ___ ribosome   ribosomes
    ___ smooth ER   c. fatty acid breakdown
    ___ Golgi body   d. sorts and ships
    ___ rough ER   e. assembles lipids
    ___ peroxisome   f. photosynthesis
    ___ amyloplast   g. ATP production
    ___ flagellum   h. movement
                         i. stores starch

Abnormal Motor Proteins Cause Kartagener Syndrome

An abnormal form of the motor protein dynein causes Kartagener syndrome, a genetic disorder characterized by chronic sinus and lung infections. Biofilms form in the thick mucus that collects in the airways, and the resulting bacterial activities and inflammation damage tissues.

Affected men can produce sperm but are infertile (**Figure 4.25**). Some have become fathers after a doctor injects their sperm cells directly into eggs. Review **Figure 4.25**, then explain how abnormal dynein could cause the observed effects.

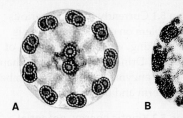

Figure 4.25 Cross-section of the flagellum of a sperm cell from (**A**) a human male affected by Kartagener syndrome and (**B**) an unaffected male.

Critical Thinking

1. In a classic episode of *Star Trek*, a gigantic amoeba engulfs an entire starship. Spock blows the cell to bits before it has a chance to reproduce. Think of at least one problem a biologist would have with this particular scenario.

2. Many plant cells form a secondary wall on the inner surface of their primary wall. Speculate on the reason why the secondary wall does not form on the outer surface of the primary wall.

3. A student is examining different samples with a microscope. She discovers a single-celled organism swimming in a freshwater pond (*right*). What kind of microscope is she using?

4. Which structures can you identify in the organism on the *right*? Is it a prokaryotic or eukaryotic cell? Can you be more specific about the type of cell based on what you know about cell structure? Look ahead to Section 21.3 to check your answers.

Chapter 5
Summary

Section 5.1 Currently the most serious drug problem on college campuses is binge drinking, which is a symptom of alcoholism. Drinking more alcohol than the body's enzymes can detoxify can be lethal in both the short term and the long term.

Section 5.2 Kinetic energy, potential energy, and other forms of **energy** cannot be created or destroyed (**first law of thermodynamics**). Energy can be converted from one form to another and transferred between objects or systems. Energy tends to disperse spontaneously (**second law of thermodynamics**). Some energy disperses with every transfer, usually as heat. **Entropy** is a measure of how much the energy of a system is dispersed.

Living things maintain their organization only as long as they harvest energy from someplace else. Energy flows in one direction through the biosphere, starting mainly from the sun, then into and out of ecosystems. Producers and then consumers use the captured energy to assemble, rearrange, and break down organic molecules that cycle among organisms throughout ecosystems.

Section 5.3 Cells store and retrieve free energy by making and breaking chemical bonds in metabolic reactions, in which **reactants** are converted to **products**. **Endergonic** reactions require a net energy input. **Exergonic** reactions end with a net energy release. **Activation energy** is the minimum energy required to start a reaction.

Section 5.4 Enzymes greatly enhance the rate of reactions without being changed by them, a process called **catalysis**. Enzymes lower a reaction's activation energy by boosting local concentrations of **substrates**, orienting substrates in positions that favor reaction, inducing the fit between a substrate and the enzyme's **active site** (**induced-fit model**), and sometimes excluding water, all of which bring on a substrate's **transition state**. Each type of enzyme works best within a characteristic range of temperature, salt concentration, and pH.

Section 5.5 Cells build, convert, and dispose of substances in enzyme-mediated reaction sequences called **metabolic pathways**. Controls over enzymes allow cells to conserve energy and resources by producing only what they require. **Allosteric** sites are points of control by which a cell adjusts the types and amounts of substances it makes. **Feedback inhibition** is an example of enzyme control. **Redox** (oxidation–reduction) **reactions** in **electron transfer chains** allow cells to harvest energy in manageable increments.

Section 5.6 Most enzymes require **cofactors**, which are metal ions or organic **coenzymes**. Cofactors in some **antioxidants** help them stop reactions with oxygen that produce free radicals. **ATP** functions as an energy carrier between reaction sites in cells. It has three phosphate bonds; when a phosphate group is transferred to another molecule, the energy of the bond is transferred along with it. Phosphate-group transfers (**phosphorylations**) to and from ATP couple reactions that release energy with reactions that require energy. Cells regenerate ATP in the **ATP/ADP cycle**.

Section 5.7 A cell membrane is a mosaic of proteins and lipids (mainly phospholipids) organized as a bilayer. Membranes of bacteria and eukaryotic cells can be described as a **fluid mosaic**; those of archaea are not fluid. Proteins transiently or permanently associated with a membrane carry out most membrane functions. All cell membranes have **transport proteins**. Plasma membranes also incorporate **receptor proteins**, **adhesion proteins**, enzymes, and **recognition proteins**.

Section 5.8 Molecules or ions tend to spread spontaneously (**diffuse**), with the eventual result being a gradual and complete mixing. A **concentration gradient** is a difference in the **concentration** of a substance between adjoining regions of fluid. The steepness of the gradient, temperature, solute size, charge, and pressure influence the diffusion rate.

Osmosis is the diffusion of water across a selectively permeable membrane, from the region with a lower solute concentration (**hypotonic**) toward the region with a higher solute concentration (**hypertonic**). There is no net movement of water between **isotonic** solutions. **Osmotic pressure** is the amount of **turgor** (fluid pressure against a cell membrane or wall) that stops osmosis.

Section 5.9 Gases, water, and small nonpolar molecules can diffuse across a lipid bilayer. Most other molecules, and ions in particular, cross only with the help of transport proteins, which allow a cell or membrane-enclosed organelle to control which substances enter and exit.

The types of transport proteins in a membrane determine which substances can cross it. **Active transport** proteins such as **calcium pumps** use energy, such as a phosphate transfer from ATP, to pump a solute against its concentration gradient. **Passive transport** proteins work without an energy input; a solute's movement is driven by its concentration gradient.

Section 5.10 Substances in bulk and large particles are moved across plasma membranes by processes of exocytosis and endocytosis. With **exocytosis**, a cytoplasmic vesicle fuses with the plasma membrane, and its contents are released to the outside of the cell. The vesicle's membrane lipids and proteins become

part of the plasma membrane. With **endocytosis**, a patch of plasma membrane balloons into the cell, and forms a vesicle that sinks into the cytoplasm. **Pinocytosis** is not as specific as **phagocytosis**, a receptor-mediated pathway by which cells engulf particles or cell debris. Plasma membrane lost by endocytosis is replaced by exocytosis.

Self-Quiz

Answers in Appendix III

1. _____ is life's primary source of energy.
   a. Food   b. Water   c. Sunlight   d. ATP

2. Which of the following statements is not correct?
   a. Energy cannot be created or destroyed.
   b. Energy cannot change from one form to another.
   c. Energy tends to disperse spontaneously.

3. Entropy _____ .
   a. disperses   c. always increases, overall
   b. is a measure of disorder   d. b and c

4. If we liken a chemical reaction to an energy hill, then a(n) _____ reaction is an uphill run.
   a. endergonic   c. catalytic
   b. exergonic   d. both a and c

5. If we liken a chemical reaction to an energy hill, then activation energy is like _____ .
   a. a burst of speed
   b. coasting downhill
   c. a bump at the top of the hill

6. Enzymes _____ .
   a. are proteins, except for a few RNAs
   b. lower the activation energy of a reaction
   c. are changed by the reactions they catalyze
   d. a and b

7. _____ are always changed by participating in a reaction. (Choose all that are correct.)
   a. Enzymes   c. Reactants
   b. Cofactors   d. Coenzymes

8. One environmental factor that influences enzyme function is _____ .
   a. temperature   c. light
   b. wind   d. radioactivity

9. A metabolic pathway _____ .
   a. may build or break down molecules
   b. generates heat
   c. can include an electron transfer chain
   d. all of the above

10. A molecule that donates electrons becomes _____ , and the one that accepts electrons becomes _____ .
    a. reduced; oxidized   c. oxidized; reduced
    b. ionic; electrified   d. electrified; ionic

11. An antioxidant _____ .
    a. prevents other molecules from being oxidized
    b. is necessary in the human diet
    c. balances charge
    d. oxidizes free radicals

12. Ions or molecules tend to diffuse from a region where they are _____ (more/less) concentrated to another where they are _____ (more/less) concentrated.

13. _____ cannot easily diffuse across a lipid bilayer.
    a. Water   c. Ions
    b. Gases   d. all of the above

14. Transporters that require an energy boost help sodium ions across a cell membrane. This is a case of _____ .
    a. passive transport   c. facilitated diffusion
    b. active transport   d. b and c

15. Immerse a human red blood cell in a hypotonic solution, and water _____ .
    a. diffuses into the cell   c. shows no net movement
    b. diffuses out of the cell   d. moves in by endocytosis

16. Vesicles form during _____ .
    a. endocytosis   d. symbiosis
    b. exocytosis   e. a through c
    c. phagocytosis   f. all of the above

17. Match each term with its most suitable description.
    ___ reactant   a. assists enzymes
    ___ phagocytosis   b. forms at reaction's end
    ___ first law of   c. enters a reaction
        thermodynamics   d. requires energy boost
    ___ product   e. one cell engulfs another
    ___ cofactor   f. energy cannot be created
    ___ diffusion   or destroyed
    ___ passive transport g. faster with a gradient
    ___ active transport h. no energy boost required

One Tough Bug The genus *Ferroplasma* consists of a few species of acid-loving archaea. One species, *F. acidarmanus*, was discovered to be the main constituent of slime streamers (a type of biofilm) deep inside an abandoned California copper mine (**Figure 5.32**). *F. acidarmanus* cells living on the surfaces of the streamers use an energy-harvesting pathway that combines oxygen with iron-sulfur compounds in minerals such as pyrite. Oxidizing these minerals dissolves them, so groundwater that seeps into the mine ends up accumulating extremely high concentrations of metal ions such as copper, zinc, cadmium, and arsenic. Another oxidation product of sulfide minerals, sulfuric acid, lowers the pH of the resulting solution to zero.

F. acidarmanus cells are unwalled, yet they are able to maintain their internal pH at a cozy 5 despite living in an environment with a composition essentially the same as hot battery acid. Thus, researchers investigating *Ferroplasma* metabolic enzymes were surprised to discover that most of the cells' enzymes function best at low pH (**Figure 5.33**).

1. What does the dashed line signify?

2. Of the four enzymes profiled in the graph, how many function optimally at a pH lower than 5? How many retain significant function at pH 5?

3. What is the optimal pH for *Ferroplasma* carboxylesterase?

Figure 5.32 A one-meter (3 feet) wide section of a stream deep inside one of the most toxic sites in the United States: Iron Mountain Mine, in California. The water is hot (around 40°C, or 104°F), has a pH of zero, and is heavily laden with arsenic and other toxic metals. The slime streamers growing in it are a biofilm dominated by a species of archaea, *Ferroplasma acidarmanus*.

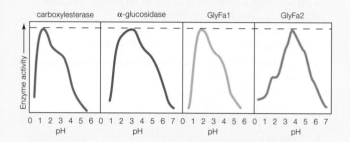

Figure 5.33 pH anomaly of *Ferroplasma* enzymes. The graphs show the pH profiles of four enzymes isolated from *Ferroplasma*. Researchers had expected these enzymes to function best at the cells' cytoplasmic pH (5). *Source:* Golyshina et al., *Environmental Microbiology* (2006) 8(3): 416–425.

Critical Thinking

1. Beginning physics students are often taught the basic concepts of thermodynamics with two phrases: First, you can't win. Second, you can't break even. Explain.

2. How is diffusion like entropy?

3. Water molecules tend to diffuse in response to their own concentration gradient. How can water be more or less concentrated?

4. Dixie Bee wanted to make JELL-O shots for her next party, but felt guilty about encouraging her guests to consume alcohol. She tried to compensate for the toxicity of the alcohol by adding pieces of healthy fresh pineapple to the shots, but when she did, the JELL-O never solidified. What happened? Hint: JELL-O is mainly sugar and a gelatinous mixture of proteins.

5. The enzyme trypsin is sold as a dietary enzyme supplement. Explain what happens to trypsin taken with food.

6. One molecule of catalase can inactivate about 6 million hydrogen peroxide molecules per minute by combining them two at a time. Catalase also inactivates other toxins, including ethanol. Given that its active site specifically binds to hydrogen peroxide, how do you think this enzyme acts on other molecules?

7. Catalase combines two hydrogen peroxide molecules ($H_2O_2 + H_2O_2$) to make two molecules of water. A gas also forms. What is the gas?

8. Hydrogen peroxide bubbles if dribbled on an open cut but does not bubble on unbroken skin. Explain why.

Chapter 6
Summary

Section 6.1 Autotrophs make their own food using energy they get directly from the environment, and carbon from inorganic sources such as CO_2. By metabolic pathways of **photosynthesis**, plants and other autotrophs capture the energy of light and use it to build sugars from water and carbon dioxide. **Heterotrophs** get energy and carbon from molecules that other organisms have already assembled.

Earth's early atmosphere held very little free oxygen, and **chemoautotrophs** were common. When the noncyclic pathway of photosynthesis evolved, oxygen released by **photoautotrophs** permanently changed the atmosphere, and was a selective force that favored evolution of aerobic respiration. Photoautotrophs remove CO_2 from the atmosphere; the metabolic activity of most organisms puts it back. Human activities disrupt this cycle by adding extra CO_2 to the atmosphere. The resulting imbalance is contributing to global warming.

Sections 6.2, 6.3 Visible light is a very small part of the spectrum of electromagnetic energy radiating from the sun. That energy travels in waves, and it is organized as photons. Visible light drives photosynthesis, which begins when photons are absorbed by photosynthetic pigments. **Pigments** are molecules that absorb light of particular **wavelengths** only; photons not captured by a pigment are reflected as its characteristic color. The main photosynthetic pigment, **chlorophyll a**, absorbs violet and red light, so it appears green. Accessory pigments absorb additional wavelengths.

Section 6.4 In **chloroplasts**, the **light-dependent reactions** of photosynthesis occur at a much-folded **thylakoid membrane**. The membrane forms a continuous compartment in the chloroplast's interior (**stroma**) where the **light-independent reactions** occur.

The following diagram summarizes the overall process of photosynthesis (with the noncyclic reactions):

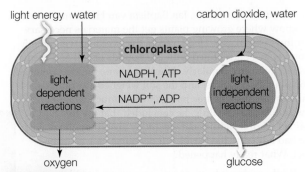

light energy   water

carbon dioxide, water

chloroplast

light-dependent reactions

NADPH, ATP

NADP+, ADP

light-independent reactions

oxygen

glucose

Sections 6.5, 6.6 In the light reactions of photosynthesis, light-harvesting complexes in the thylakoid membrane absorb photons and pass the energy to **photosystems**, which then release electrons.

In the noncyclic pathway, electrons released from photosystem II flow through an electron transfer chain, then to photosystem I. Photon energy causes photosystem I to release electrons, which end up in NADPH. Photosystem II replaces lost electrons by pulling them from water, which then dissociates into H^+ and O_2 (an example of **photolysis**).

In the cyclic pathway, the electrons released from photosystem I enter an electron transfer chain, then cycle back to photosystem I. NADPH does not form.

ATP forms by **electron transfer phosphorylation** in both pathways. Electrons flowing through electron transfer chains cause hydrogen ions to accumulate in the thylakoid compartment. The hydrogen ions follow their gradient back across the membrane through ATP synthases, driving ATP synthesis.

Section 6.7 Carbon fixation occurs in light-independent reactions. Inside the stroma, the enzyme **rubisco** attaches a carbon from CO_2 to RuBP to start the **Calvin–Benson cycle**. This cyclic pathway makes sugars using energy from ATP, carbon and oxygen from CO_2, and hydrogen and electrons from NADPH.

Section 6.8 Environments differ, and so do details of the light-independent reactions. On dry days, plants conserve water by closing their **stomata**. However, when stomata are closed, CO_2 for the light-dependent reactions cannot enter the plant, and O_2 from the light-independent reactions cannot escape it. In **C3 plants**, the resulting high O_2 level in the plant's tissues causes rubisco to attach O_2 instead of CO_2 to RuBP. This pathway, which is called **photorespiration**, reduces the efficiency of sugar production. In **C4 plants**, carbon fixation occurs twice. The first reactions release CO_2 near rubisco, and thus limit photorespiration when stomata are closed. **CAM plants** minimize water loss by opening their stomata and fixing carbon at night.

Self-Quiz

Answers in Appendix III

1. A cat eats a bird, which ate a caterpillar that chewed on a weed. Which organisms are autotrophs? Which ones are heterotrophs?

2. Autotrophs use _____ as an energy source to drive photosynthesis.
   - a. sunlight
   - b. hydrogen ions
   - c. O_2
   - d. CO_2

3. Most of the carbon dioxide used in photosynthesis comes from _____ .
   - a. glucose
   - b. the atmosphere
   - c. rainwater
   - d. photolysis

4. Chlorophyll *a* appears green because it absorbs mainly _____ light.
   - a. violet and red
   - b. green and yellow
   - c. blue and yellow
   - d. blue and violet

Energy Efficiency of Biofuel Production Most corn is grown intensively in vast swaths, which means that farmers who grow it have to use large amounts of fertilizers and pesticides, both of which are typically made from fossil fuels. In 2006, David Tilman and his colleagues published the results of a 10-year study comparing the net energy output of biofuels made from corn, soy, and weeds. For the weed biomass, the researchers grew a mixture of native perennial grasses without irrigation, fertilizer, pesticides, or herbicides, in sandy soil that was so depleted by intensive agriculture that it had been abandoned. Net energy output is the total usable energy in the finished biofuel product minus the total energy required to produce it. Some of their results are shown in Figure 6.16.

1. About how much energy did ethanol produced from one hectare of corn yield? How much energy did it take to grow the corn to make that ethanol?

2. Which of the biofuels tested had the highest ratio of energy output to energy input?

3. Which of the three crops would require the least amount of land to produce a given amount of biofuel energy?

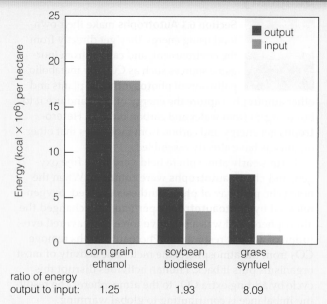

| ratio of energy output to input: | 1.25 | 1.93 | 8.09 |

Figure 6.16 Energy inputs and outputs of biofuels from corn and soy grown on fertile farmland, and grasses grown in infertile soil. One hectare is about 2.5 acres.

---

5. Light-dependent reactions in plants proceed in the _____ .
    a. thylakoid membrane   c. stroma
    b. plasma membrane   d. cytoplasm

6. When a photosystem absorbs light, _____ .
    a. sugar phosphates are produced
    b. electrons are transferred to ATP
    c. RuBP accepts electrons
    d. electrons are ejected from its special pair

7. In the light-dependent reactions, _____ .
    a. carbon dioxide is fixed   d. CO_2 accepts electrons
    b. ATP forms   e. b and c
    c. sugars form   f. a and c

8. What accumulates inside the thylakoid compartment of chloroplasts during the light-dependent reactions?
    a. glucose   c. O_2
    b. hydrogen ions   d. CO_2

9. The atoms in the molecular oxygen released during photosynthesis come from _____ molecules.
    a. glucose   c. water
    b. CO_2 d. O_2

10. Light-independent reactions in plants proceed in the _____ of chloroplasts.
    a. thylakoid membrane   c. stroma
    b. plasma membrane   d. cytoplasm

11. The Calvin–Benson cycle starts when _____ .
    a. light is available
    b. carbon dioxide is attached to RuBP
    c. electrons leave a photosystem II

12. Which of the following substances does *not* participate in the Calvin–Benson cycle?
    a. ATP   d. PGAL
    b. NADPH   e. O_2
    c. RuBP   f. CO_2

13. In the light-independent reactions, _____ .
    a. carbon dioxide is fixed   d. CO_2 accepts electrons
    b. ATP forms   e. b and c
    c. sugars form   f. a and c

14. Match each with its most suitable description.
    ____ PGAL formation   a. absorbs light
    ____ CO_2 fixation   b. converts light to
    ____ photolysis   chemical energy
    ____ ATP forms; NADPH   c. self-feeder
         does not   d. electrons cycle back
    ____ photorespiration   to photosystem I
    ____ photosynthesis   e. problem in C3 plants
    ____ pigment   f. ATP, NADPH
    ____ autotroph   required
                              g. water molecules split
                              h. rubisco function

Critical Thinking

1. About 200 years ago, Jan Baptista van Helmont wanted to know where growing plants get the materials necessary for increases in size. He planted a tree seedling weighing 5 pounds in a barrel filled with 200 pounds of soil and then watered the tree regularly. After five years, the tree weighed 169 pounds, 3 ounces, and the soil weighed 199 pounds, 14 ounces. Because the tree had gained so much weight and the soil had lost so little, he concluded that the tree had gained all of its additional weight by absorbing the water he had added to the barrel, but of course he was incorrect. What really happened?

2. While gazing into an aquarium, you see bubbles coming from an aquatic plant (*right*). What are the bubbles?

3. A C3 plant absorbs a carbon radioisotope (as part of $^{14}CO_2$). In which compound does the labeled carbon appear first? Which compound forms first if a C4 plant absorbs the same radioisotope?

Chapter 7
Summary

Section 7.1 Photosynthesis by early photo-autotrophs changed the composition of Earth's early atmosphere, with profound effects on life's evolution. Organisms that could not tolerate the increased oxygen content persisted only in **anaerobic** habitats. Oxygen-detoxifying pathways evolved, allowing organisms to thrive in **aerobic** conditions. Defects in these pathways, such as occurs in Friedreich's ataxia, can cause serious health problems.

Section 7.2 Most organisms convert chemical energy of carbohydrates to the chemical energy of ATP. Carbohydrate breakdown pathways start in the cyto-plasm with the same set of reactions, **glycolysis**, which converts glucose and other sugars to **pyruvate**. Anaerobic **fermentation** pathways end in cytoplasm and yield two ATP per molecule of glucose. **Aerobic respiration** uses oxygen and yields much more ATP than fermentation. In modern eukaryotes, aerobic respiration is completed inside mitochondria.

Section 7.3 Glycolysis, the first stage of aerobic respiration and fermentation, occurs in the cytoplasm. In the reactions, enzymes use two ATP to convert one molecule of glucose or another six-carbon sugar to two molecules of pyruvate. Two NAD^+ are reduced to NADH. Four ATP also form by **substrate-level phosphorylation**, the direct transfer of a phosphate group from a reaction intermediate to ADP.

The net yield of glycolysis is two pyruvate, two ATP, and two NADH per glucose molecule. The pyruvate may continue in fermentation in the cytoplasm, or it may enter mitochondria and continue in the next steps of aerobic respiration.

Section 7.4 A mitochondrion's inner membrane divides its interior into two fluid-filled spaces: the inner compart-ment, or matrix, and the intermembrane space. The second stage of aerobic respira-tion, acetyl–CoA formation and the **Krebs cycle**, takes place in the matrix. The first steps convert two pyruvate from glycolysis to two acetyl–CoA and two CO_2. The acetyl–CoA enters the Krebs cycle.

It takes two cycles of Krebs reactions to dismantle the two acetyl–CoA. At this stage, all of the carbon atoms in the glucose molecule that entered glycolysis have left the cell in CO_2. During these reactions, electrons and hydro-gen ions are transferred to NAD^+ and FAD, which are thereby reduced to NADH and $FADH_2$. ATP forms by substrate-level phosphorylation.

In total, the breakdown of two pyruvate molecules in the second stage of aerobic respiration yields ten reduced coenzymes and two ATP.

Section 7.5 Aerobic respiration ends in mitochondria. In the third stage of reac-tions, electron transfer phosphorylation, coenzymes that were reduced in the first two stages deliver their cargo of electrons and hydrogen ions to electron transfer chains in the inner mitochondrial membrane. Electrons moving through the chains release energy bit by bit; molecules of the chain use that energy to move H^+ from the matrix to the intermembrane space.

Hydrogen ions that accumulate in the intermembrane space form a gradient across the inner membrane. The ions follow the gradient back to the matrix through ATP synthases. H^+ flow through these transport proteins drives ATP synthesis.

Oxygen combines with electrons and H^+ at the end of the transfer chains, forming water.

Overall, aerobic respiration typically yields thirty-six ATP for each glucose molecule.

Section 7.6 Anaerobic fermentation path-ways begin with glycolysis and finish in the cytoplasm. A molecule other than oxy-gen accepts electrons at the end of these reactions. The end product of **alcoholic fermentation** is ethyl alcohol, or ethanol. The end prod-uct of **lactate fermentation** is lactate. The final steps of fermentation serve to regenerate NAD^+, which is required for glycolysis to continue, but they produce no ATP. Thus, the breakdown of one glucose molecule in either alcoholic or lactate fermentation yields only the two ATP that form in glycolysis reactions.

Skeletal muscle has two types of fibers: red and white. ATP is produced primarily by aerobic respiration in red muscle fibers, so these fibers sustain activities that require endurance. Lactate fermentation in white fibers supports activities that occur in short, intense bursts.

Section 7.7 In humans and other mam-mals, simple sugars from carbohydrate breakdown, glycerol and fatty acids from fat breakdown, and carbon backbones of amino acids from protein breakdown may enter aerobic respiration at various reaction steps.

Self-Quiz
Answers in Appendix III

1. Is the following statement true or false? Unlike animals, which make many ATP by aerobic respiration, plants make all of their ATP by photosynthesis.

2. Glycolysis starts and ends in the _____ .
   a. nucleus   c. plasma membrane
   b. mitochondrion   d. cytoplasm

3. Which of the following metabolic pathways require(s) molecular oxygen (O_2)?
   a. aerobic respiration
   b. lactate fermentation
   c. alcoholic fermentation
   d. all of the above

4. Which molecule does not form during glycolysis?
   a. NADH   b. pyruvate   c. $FADH_2$ d. ATP

5. In eukaryotes, aerobic respiration is completed in the _____ .
   a. nucleus   c. plasma membrane
   b. mitochondrion   d. cytoplasm

6. Which of the following reaction pathways is not part of the second stage of aerobic respiration?
   a. electron transfer   c. Krebs cycle
      phosphorylation   d. glycolysis
   b. acetyl–CoA formation   e. a and d

7. After Krebs reactions run through _____ cycle(s), one glucose molecule has been completely broken down to CO_2.
   a. one   b. two   c. three   d. six

8. In the third stage of aerobic respiration, _____ is the final acceptor of electrons.
   a. water   b. hydrogen   c. oxygen   d. NADH

9. _____ is the final acceptor of electrons in alcoholic fermentation.
   a. Oxygen   c. Acetaldehyde
   b. Pyruvate   d. Sulfate

10. Fermentation pathways make no more ATP beyond the small yield from glycolysis. The remaining reactions serve to regenerate _____ .
   a. FAD   c. glucose
   b. NAD^+ d. oxygen

11. Most of the energy that is released by the full breakdown of glucose to CO_2 and water ends up in _____ .
   a. NADH   c. heat
   b. ATP   d. electrons

12. Your body cells can use _____ as an alternative energy source when glucose is in short supply.
   a. fatty acids   c. amino acids
   b. glycerol   d. all of the above

13. Which of the following is not produced by an animal muscle cell operating under anaerobic conditions?
   a. heat   d. ATP
   b. pyruvate   e. lactate
   c. NAD^+ f. all are produced

14. Match the event with its most suitable description.
   ___ glycolysis   a. ATP, NADH, $FADH_2$,
   ___ fermentation   and CO_2 form
   ___ Krebs cycle   b. glucose to two pyruvate
   ___ electron transfer   c. NAD^+ regenerated, little ATP
      phosphorylation   d. H^+ flows via ATP synthases

15. Match the term with the best description.
   ___ matrix   a. needed for glycolysis
   ___ pyruvate   b. inner space
   ___ NAD^+ c. makes many ATP
   ___ mitochondrion   d. end of glycolysis
   ___ intermembrane   e. reduced coenzyme
      space   f. hydrogen ions
   ___ NADH   accumulate here
   ___ anaerobic   g. no oxygen required

Critical Thinking

1. The higher the altitude, the lower the oxygen level in air. Climbers of very tall mountains risk altitude sickness, which is characterized by shortness of breath, weakness, dizziness, and confusion.

   The early symptoms of cyanide poisoning are the same as those for altitude sickness. Cyanide binds tightly to cytochrome *c* oxidase, a protein complex that is the last component of mitochondrial electron transfer chains. Cytochrome *c* oxidase with bound cyanide can no longer transfer electrons. Explain why cyanide poisoning starts with the same symptoms as altitude sickness.

2. As you learned, membranes impermeable to hydrogen ions are required for electron transfer phosphorylation. Membranes in mitochondria serve this purpose in eukaryotes. Bacteria do not have this organelle, but they do make ATP by electron transfer phosphorylation. How do you think they do it, given that they have no mitochondria?

3. The bar-tailed godwit is a type of shorebird that makes an annual migration from Alaska to New Zealand and back. The birds make each 11,500-kilometer (7,145-mile) trip by flying over the Pacific Ocean in about nine days, depending on weather, wind speed, and direction of travel. One bird was observed to make the entire journey uninterrupted, a feat that is comparable to a human running a nonstop seven-day marathon at 70 kilometers per hour (43.5 miles per hour). Would you expect the flight (breast) muscles of bar-tailed godwits to be light or dark colored? Explain your answer.

4. The bacterium *Escherichia coli* can produce ATP by aerobic respiration and fermentation. Two cultures of *E. coli* are started and held under identical conditions, except that one is kept under anaerobic conditions. At the end of a week, which culture has more bacteria? Why?

Mitochondrial Abnormalities in Tetralogy of Fallot Tetralogy of Fallot (TF) is a genetic disorder characterized by four major malformations of the heart. The circulation of blood is abnormal, so TF patients have too little oxygen in their blood. Inadequate oxygen levels result in damaged mitochondrial membranes, which in turn cause cells to self-destruct.

In 2004, Sarah Kuruvilla and her colleages looked at abnormalities in the mitochondria of heart muscle in TF patients. Some of their results are shown in **Figure 7.13**.

1. In this study, which abnormality was most strongly associated with TF?

2. What percentage of the TF patients had mitochondria that were abnormal in size?

3. Can you make any correlations between blood oxygen content and mitochondrial abnormalities in these patients?

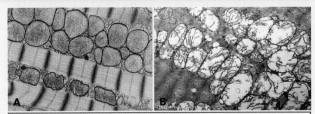

Figure 7.13 Mitochondrial changes in tetralogy of Fallot (TF).

(**A**) Normal heart muscle. Many mitochondria between the fibers provide muscle cells with ATP for contraction. (**B**) Heart muscle from a person with TF shows swollen, broken mitochondria.

(**C**) Mitochondrial abnormalities in TF patients. SPO_2 is oxygen saturation of the blood. A normal value of SPO_2 is 96%. Abnormalities are marked "+".

Patient (age)	SPO_2 (%)	Mitochondrial Abnormalities in TF			
		Number	Shape	Size	Broken
1 (5)	55	+	+	−	−
2 (3)	69	+	+	−	−
3 (22)	72	+	+	−	−
4 (2)	74	+	+	−	−
5 (3)	76	+	+	−	+
6 (2.5)	78	+	+	−	+
7 (1)	79	+	+	−	−
8 (12)	80	+	−	+	−
9 (4)	80	+	+	−	−
10 (8)	83	+	−	+	−
11 (20)	85	+	+	−	−
12 (2.5)	89	+	−	+	−

C

Chapter 8
Summary

 Section 8.1 Making **clones**, or exact genetic copies, of adult animals is now a common practice. The techniques, while improving, are still far from perfect; many attempts are required to produce a clone, and clones that survive often have health problems. The practice continues to raise ethical questions.

 Section 8.2 The DNA of eukaryotes is divided among a characteristic number of **chromosomes** that differ in length and shape. **Histone** proteins organize eukaryotic DNA into **nucleosomes**. When duplicated, a eukaryotic chromosome consists of two **sister chromatids** attached at a **centromere**.

Diploid cells have two of each type of chromosome. **Chromosome number** is the sum of all chromosomes in cells of a given type. A human body cell has twenty-three pairs of chromosomes. Members of a pair of **sex chromosomes** differ among males and females. All others are **autosomes**. Autosomes of a pair have the same length, shape, and centromere location, and they carry the same genes. A **karyotype** can reveal abnormalities in an individual's complement of chromosomes.

 Section 8.3 Eighty years of experiments with bacteria and **bacteriophage** offered solid evidence that deoxyribonucleic acid (DNA), not protein, is the hereditary material of life.

 Section 8.4 A DNA molecule is a polymer of nucleotides, and consists of two strands coiled into a double helix. A DNA nucleotide has a five-carbon sugar (deoxyribose), three phosphate groups, and one of four nitrogen-containing bases after which the nucleotide is named: adenine, thymine, guanine, or cytosine. The bases pair in a consistent way: adenine with thymine (A–T), and guanine with cytosine (G–C). The order of bases along the strands—the **DNA sequence**—varies among species and among individuals.

 Section 8.5 The DNA sequence of an organism's chromosome(s) constitutes genetic information. A cell passes that information to offspring by copying its DNA before it divides. In the process of **DNA replication**, a double-stranded molecule of DNA is copied, and two double-stranded DNA molecules that are identical to the parent are the result. One strand of each molecule is new, and the other is parental; thus the name **semiconservative replication**.

During the replication process, the double helix unwinds and RNA **primers** form on the exposed single strands of DNA. Starting at the primers, **DNA polymerase** uses each strand as a template to assemble new, complementary strands of DNA from free nucleotides.

Synthesis of one strand occurs discontinuously. **DNA ligase** seals any gaps to form a continuous strand.

 Section 8.6 Proofreading by DNA polymerases corrects most base-pairing errors as they occur. Uncorrected replication errors become **mutations**—permanent changes in the nucleotide sequence of a cell's DNA. Cancer arises by mutations.

Environmental agents such as UV light cause DNA damage that can lead to replication errors, so damaged DNA is normally repaired before replication begins.

 Section 8.7 Reproductive cloning technologies produce genetically identical individuals (clones). In **somatic cell nuclear transfer** (SCNT), one cell from an adult is fused with an egg that has had its nucleus removed. The hybrid cell is treated with electric shocks or another stimulus that provokes the cell to divide and begin developing into a new individual. SCNT with human cells, which is called **therapeutic cloning**, produces embryos that are used for stem cell research.

Self-Quiz

Answers in Appendix III

1. Chromosome number _____ .
   a. refers to a particular chromosome pair in a cell
   b. is an identifiable feature of a species
   c. is like a set of books
   d. all of the above

2. Sister chromatids connect at the _____ .

3. A karyotype reveals the _____ of a single cell.
   a. base sequences   c. hereditary information
   b. chromosomes   d. clones

4. The basic unit that structurally organizes a eukaryotic chromosome is the _____ .
   a. higher-order coiling   c. base sequence
   b. double helix   d. nucleosome

5. Which is *not* a nucleotide base in DNA?
   a. adenine   c. uracil   e. cytosine
   b. guanine   d. thymine   f. All are in DNA.

6. What are the base-pairing rules for DNA?
   a. A–G, T–C   c. A–U, C–G
   b. A–C, T–G   d. A–T, G–C

7. Energy that drives DNA synthesis comes from _____ .
   a. ATP   c. DNA nucleotides
   b. DNA polymerase   d. a and c

8. When DNA replication begins, _____ .
   a. the two DNA strands unwind from each other
   b. the two DNA strands condense for base transfers
   c. two DNA molecules bond
   d. old strands move to find new strands

9. DNA replication requires _____ .
   a. template DNA   d. primers
   b. free nucleotides   e. a through c
   c. DNA polymerase   f. all are required

10. Show the complementary strand of DNA that forms on this template DNA fragment during replication:

$$5'—GGTTTCTTCAAGAGA—3'$$

Hershey–Chase Experiments The graph in **Figure 8.16** is reproduced from Hershey and Chase's original 1952 publication that showed DNA is the hereditary material of bacteriophage. The data are from the two experiments described in Section 8.3, in which bacteriophage DNA and protein were labeled with radioactive tracers and allowed to infect bacteria. The virus–bacteria mixtures were whirled in a blender to dislodge the viruses, and the tracers were tracked inside and outside of the bacteria.

1. Before blending, what percentage of ^{35}S was outside the bacteria? What percentage of ^{32}P was outside the bacteria?

2. After 4 minutes in the blender, what percentage of ^{35}S was outside the bacteria? What percentage of ^{32}P was outside the bacteria?

3. How did the researchers know that the radioisotopes in the fluid came from outside of the bacterial cells (extracellular) and not from bacteria that had been broken apart by whirling in the blender?

4. The extracellular concentration of which isotope, ^{35}S or ^{32}P, increased the most with blending? DNA contains much more phosphorus than do proteins; proteins contain much more sulfur than does DNA. Do these results imply that the viruses inject DNA or protein into bacteria? Why or why not?

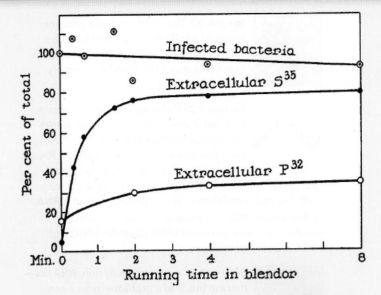

Figure 8.16 Detail of Alfred Hershey and Martha Chase's publication describing their experiments with bacteriophage. "Infected bacteria" refers to the percentage of bacteria that survived the blender. *Source:* "Independent Functions of Viral Protein and Nucleic Acid in Growth of Bacteriophage." *Journal of General Physiology*, 36(1), Sept. 20, 1952.

11. _____ is an example of reproductive cloning.
   a. Somatic cell nuclear transfer (SCNT)
   b. Multiple offspring from the same pregnancy
   c. Artificial embryo splitting
   d. a and c
   e. all of the above

12. The DNA of each species has unique _____ that set it apart from the DNA of all other species.
   a. nucleotides   c. sequences
   b. chromosomes   d. bases

13. Exposure to _____ can cause mutations.
   a. UV light   c. x-rays
   b. cigarette smoke   d. all of the above

14. SCNT with human cells is called _____ .
   a. hCNT   c. iSCNT
   b. therapeutic cloning d. all are correct

15. _____ can be used to produce genetically identical organisms (clones).
   a. SCNT   c. Therapeutic cloning
   b. Embryo splitting   d. all of the above

16. Match the terms appropriately.
   ___ bacteriophage   a. nitrogen-containing base,
   ___ clone   sugar, phosphate group(s)
   ___ nucleotide   b. copy of an organism
   ___ diploid   c. does not determine sex
   ___ DNA ligase   d. only DNA and protein
   ___ DNA polymerase   e. fills in gaps, seals breaks
   ___ autosome   in a DNA strand
   ___ mutation   f. two chromosomes
                            of each type
                         g. adds nucleotides to a
                            growing DNA strand
                         h. can cause cancer

Critical Thinking

1. Matthew Meselson and Franklin Stahl's experiments supported the semiconservative model of replication. These researchers obtained "heavy" DNA by growing *Escherichia coli* with ^{15}N, a radioactive isotope of nitrogen. They also prepared "light" DNA by growing *E. coli* in the presence of ^{14}N, the more common isotope. An available technique helped them identify which of the replicated molecules were heavy, light, or hybrid (one heavy strand and one light).

   Use different colored pencils to draw the heavy and light strands of DNA. Starting with a DNA molecule having two heavy strands, show the formation of the two molecules that result from replication in a ^{14}N-containing medium. Show the four DNA molecules that would form if those two molecules were replicated a second time in the ^{14}N medium. Would the DNA molecules that result from two replications in this medium be heavy, light, or mixed?

2. Woolly mammoths have been extinct for about 10,000 years, but we occasionally find one that has been preserved in Siberian permafrost. Resurrecting these huge elephant-like mammals may be possible by cloning DNA isolated from such frozen remains. Researchers are now studying the DNA of a remarkably intact baby mammoth recently discovered frozen in a Siberian swamp. What are some of the pros and cons, both technical and ethical, of cloning an extinct animal?

3. *Xeroderma pigmentosum* is an inherited disorder characterized by rapid formation of skin sores (*right*) that can develop into cancers. All forms of radiation trigger these symptoms, including fluorescent light, which contains UV light in the range of 320–400 nm. What normal function has been compromised in affected individuals?

Chapter 9
Summary

Section 9.1 The ability to make proteins is critical to all life processes. Molecules such as ricin and other proteins that inactivate ribosomes (RIPs) can be extremely toxic if they enter cells.

Section 9.2 DNA's genetic information is encoded within its base sequence. **Genes** are subunits of that sequence. A cell uses the information in a gene to make an RNA or protein product. The process of **gene expression** involves **transcription** of a DNA sequence to an RNA, and **translation** of the information in an **mRNA**, or **messenger RNA**, to a protein product. Translation requires the participation of **tRNA (transfer RNA)** and **rRNA (ribosomal RNA)**.

Section 9.3 During transcription, **RNA polymerase** binds to a **promoter** near a gene region of a chromosome. The polymerase assembles a strand of RNA by linking RNA nucleotides in the order dictated by the base sequence of the gene. Thus, the new RNA is complementary to the gene from which it was transcribed.

The RNA of eukaryotes is modified before it leaves the nucleus. **Introns** are removed. With **alternative splicing**, some **exons** may be removed also, and the remaining ones spliced in different combinations. A cap and a poly-A tail are also added to a new mRNA.

Section 9.4 mRNA carries DNA's protein-building information. The information consists of a series of **codons**, sets of three nucleotides. Sixty-four codons, most of which specify amino acids, constitute the **genetic code**. Each tRNA has an **anticodon** that can base-pair with a codon, and it binds to the amino acid specified by that codon. rRNA and proteins make up the two subunits of ribosomes.

Section 9.5 Genetic information carried by an mRNA directs the synthesis of a polypeptide during translation. First, an mRNA, an initiator tRNA, and two ribosomal subunits converge. The intact ribosome then catalyzes formation of a peptide bond between successive amino acids, which are delivered by tRNAs in the order specified by the codons in the mRNA. Translation ends when the ribosome encounters a stop codon.

Section 9.6 Insertions, deletions, and **base-pair substitutions** are mutations. The activity of **transposable elements** causes some mutations. A mutation that changes a gene's product may have harmful effects. Sickle-cell anemia, which is caused by a base-pair substitution in the gene for the beta globin chain of hemoglobin, is one example. Beta thalassemia is an outcome of a **frame-shift** mutation in the beta globin gene.

Self-Quiz

Answers in Appendix III

1. A chromosome contains many different gene regions that are transcribed into different _____ .
   - a. proteins
   - b. polypeptides
   - c. RNAs
   - d. a and b

2. A binding site for RNA polymerase is called a _____ .
   - a. gene
   - b. promoter
   - c. codon
   - d. protein

3. Energy that drives transcription is provided mainly by _____ .
   - a. ATP
   - b. RNA nucleotides
   - c. GTP
   - d. all are correct

4. An RNA molecule is typically _____ ; a DNA molecule is typically _____ .
   - a. single-stranded; double-stranded
   - b. double-stranded; single-stranded
   - c. both are single-stranded
   - d. both are double-stranded

5. RNAs form by _____ ; proteins form by _____ .
   - a. replication; translation
   - b. translation; transcription
   - c. transcription; translation
   - d. replication; transcription

6. _____ different codons constitute the genetic code.
   - a. 3
   - b. 64
   - c. 20
   - d. 120

7. Most codons specify a(n) _____ .
   - a. protein
   - b. polypeptide
   - c. amino acid
   - d. mRNA

8. Anticodons pair with _____ .
   - a. mRNA codons
   - b. DNA codons
   - c. RNA anticodons
   - d. amino acids

9. What is the maximum length of a polypeptide encoded by an mRNA that is 45 nucleotides long?

10. _____ are removed from new mRNA transcripts.
   - a. Introns
   - b. Exons
   - c. Telomeres
   - d. Amino acids

RIPs as Cancer Drugs Researchers are taking a page from the structure–function relationship of RIPs in their quest for cancer treatments. The most toxic RIPs, remember, have one domain that interferes with ribosomes, and another that carries them into cells. Melissa Cheung and her colleagues incorporated a peptide that binds to skin cancer cells into the enzymatic part of the *E. coli* enterotoxin RIP. The researchers created a new RIP that specifically kills skin cancer cells, which are notoriously resistant to established therapies. Some of their results are shown in **Figure 9.15**.

1. Which cells had the greatest response to an increase in concentration of the engineered RIP?

2. At what concentration of RIP did all of the different kinds of cells survive?

3. Which cells survived best at 10^{-6} grams per liter RIP?

4. Why are some of the data points linked by straight lines?

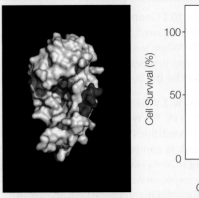
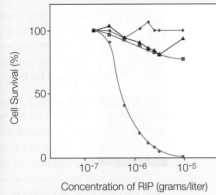

Figure 9.15 Effect of an engineered RIP on cancer cells. The model (*left*) shows the enzyme portion of *E. coli* enterotoxin that has been engineered to carry a small sequence of amino acids (in *blue*) that targets skin cancer cells. (*Red* indicates the active site.) The graph (*right*) shows the effect of this engineered RIP on human cancer cells of the skin (●); breast (◆); liver (▲); and prostate (■). *Source:* Cheung et al. *Molecular Cancer,* 9:28, 2010.

11. Where does transcription take place in a eukaryotic cell?
    a. the nucleus
    b. extracellular fluid
    c. the cytoplasm
    d. b and c are correct

12. Where does translation take place in a typical eukaryotic cell?
    a. the nucleus
    b. extracellular fluid
    c. the cytoplasm
    d. b and c are correct

13. Energy that drives translation is provided mainly by _____ .
    a. ATP
    b. amino acids
    c. GTP
    d. all are correct

14. Each amino acid is specified by a set of _____ bases in an mRNA transcript.
    a. 3   b. 20   c. 64   d. 120

15. _____ are mutations.
    a. Transposable elements
    b. Base-pair substitutions
    c. Insertions
    d. Deletions
    e. b, c, and d are correct
    f. all of the above

16. Match the terms with the best description.
    ___ genetic message   a. coding region
    ___ promoter   b. gets around
    ___ polysome   c. read as base triplets
    ___ exon   d. removed before translation
    ___ genetic code   e. occurs only in groups
    ___ intron   f. complete set of 64 codons
    ___ transposable   g. binding site for RNA
    element   polymerase

Critical Thinking

1. Antisense drugs help us fight some types of cancer and viral diseases. The drugs consist of short mRNA strands that are complementary in base sequence to mRNAs linked to the diseases. Speculate on how antisense drugs work.

2. An anticodon has the sequence GCG. What amino acid does this tRNA carry? What would be the effect of a mutation that changed the C of the anticodon to a G?

3. Each position of a codon can be occupied by one of four nucleotides. What is the minimum number of nucleotides per codon necessary to specify all 20 of the amino acids that are typical of eukaryotic proteins?

4. Using **Figure 9.7**, translate this nucleotide sequence into an amino acid sequence, starting at the first base:

   5'—GGUUUCUUGAAGAGA—3'

5. Translate the sequence of bases in the previous question, starting at the second base.

6. The termination of prokaryotic DNA transcription often depends on the structure of a newly forming RNA. Transcription stops where the mRNA folds back on itself, forming a hairpin-looped structure like the one shown at *right*. How do you think that this structure stops transcription?

Chapter 10

Summary

Section 10.1 Controls over gene expression are required for normal functioning of cells. When gene controls fail, typically as a consequence of some mutations, **cancer** may be the outcome.

Section 10.2 Which genes a cell uses depends on the type of organism, the type of cell, conditions inside and outside the cell, and, in complex multicelled species, the organism's stage of development.

Controls over gene expression help sustain all organisms. They also drive development in multicelled eukaryotes. All cells of an embryo share the same genes. As different cell lineages use different subsets of those genes during development, they become specialized, a process called **differentiation**. Specialized cells form tissues and organs in the adult.

Different molecules and processes govern every step between transcription of a gene and delivery of the gene's product to its final destination. Most controls operate at transcription; **transcription factors** such as **activators** and **repressors** influence transcription by binding to promoters, **enhancers**, or other sequences in chromosomal DNA.

Section 10.3 Local controls over gene expression govern embryonic development of complex, multicelled bodies, a process called **pattern formation**. Various **master genes** are expressed locally in different parts of an embryo as it develops. Their products diffuse through the embryo and affect expression of other master genes, which affect the expression of others, and so on. These cascades of master gene products form a dynamic spatial map of overlapping gradients that spans the entire embryo. Cells differentiate according to their location on the map. Eventually, master gene expression induces local expression of **homeotic genes** that govern development of particular body parts, as revealed by **knockout** experiments in fruit flies (*Drosophila melanogaster*).

Section 10.4 In female mammals, most genes on one of the two X chromosomes are permanently inaccessible. **X chromosome inactivation** balances gene expression between the sexes. Such **dosage compensation** arises because the RNA product of the *XIST* gene shuts down the chromosome that transcribes it. The *SRY* gene determines male sex in humans.

Studies of mutations in *Arabidopsis thaliana* showed that three sets of master genes (*A*, *B*, and *C*) guide cell differentiation in the whorls of a floral shoot so that sepals, petals, stamens, and carpels form.

Section 10.5 Bacterial cells do not have great structural complexity and do not undergo development. Most of their gene controls reversibly adjust transcription rates in response to environmental conditions, especially nutrient availability. **Operons** are examples of bacterial gene controls. The lactose operon governs expression of three genes, the three products of which allow the bacterial cell to metabolize lactose. Two **operators** that flank the promoter are binding sites for a repressor that blocks transcription.

Section 10.6 Epigenetic refers to heritable changes in gene expression that do not involve changes to the DNA sequence. DNA methylations and other epigenetic marks acquired during an individual's lifetime can persist for generations.

Self-Quiz

Answers in Appendix III

1. The expression of a gene may depend on _____ .
   a. the type of organism c. the type of cell
   b. environmental conditions d. all of the above

2. Gene expression in multicelled eukaryotic cells changes in response to _____ .
   a. extracellular conditions c. operons
   b. master gene products d. a and b

3. Binding of _____ to _____ in DNA can increase the rate of transcription of specific genes.
   a. activators; promoters c. repressors; operators
   b. activators; enhancers d. both a and b

4. Proteins that influence gene expression by binding to DNA are called _____ .
   a. promoters   c. operators
   b. transcription factors d. all of the above

5. Polytene chromosomes form in some types of cells that _____ .
   a. have a lot of chromosomes c. are differentiating
   b. are making a lot of protein d. b and c are correct

6. Eukaryotic gene controls govern _____ .
   a. transcription   e. translation
   b. RNA processing   f. protein modification
   c. RNA transport   g. all of the above
   d. mRNA degradation

7. Controls over eukaryotic gene expression guide _____ .
   a. natural selection c. development
   b. nutrient availability d. all of the above

8. The expression of *ABC* genes _____ .
   a. occurs in layers
   b. controls flower formation
   c. causes mutations in flowers
   d. both a and b

9. Cell differentiation _____ .
   a. occurs in all complex multicelled organisms
   b. requires unique genes in different cells
   c. involves selective gene expression
   d. both a and c
   e. all of the above

10. Homeotic gene products _____ .
   a. flank a bacterial operon
   b. map out the overall body plan in embryos
   c. control the formation of specific body parts

Effect of Paternal Grandmother's Food Supply on Infant Mortality Researchers are investigating long-reaching epigenetic effects of starvation, in part because historical data on periods of famine are widely available. Before the industrial revolution, a failed harvest in one autumn typically led to severe food shortages the following winter.

A retrospective study has correlated female infant mortality at certain ages with the abundance of food during the paternal grandmother's childhood. **Figure 10.14** shows some of the results of this study.

1. Compare the mortality of girls whose paternal grandmothers ate well at age 2 with that of those who experienced famine at the same age. Which girl was more likely to die early? How much more likely was she to die?

2. Children have a period of slow growth around age 9. What trend in this data can you see around that age?

3. There was no correlation between early death of a male child and eating habits of his paternal grandmother, but there was a strong correlation with the eating habits of his paternal grand*father*. What does this tell you about the probable location of epigenetic changes that gave rise to this data?

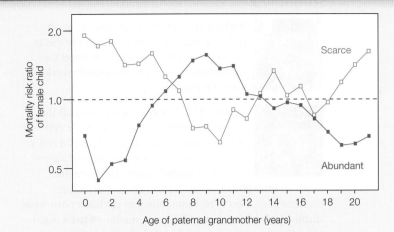

Figure 10.14 Graph showing the relative risk of early death of a female child, correlated with the age at which her paternal grandmother experienced a winter with a food supply that was scarce (*blue*) or abundant (*red*) during childhood. The dotted line represents no difference in risk of mortality. A value above the line means an increased risk; one below the line indicates a reduced risk.

11. A gene that is knocked out is _____ .
   a. deleted
   b. inactivated
   c. expressed
   d. either a or b

12. During X chromosome inactivation, _____ .
   a. female cells shut down
   b. RNA coats chromosomes
   c. pigments form
   d. both a and b

13. A cell with a Barr body is _____ .
   a. a bacterium
   b. a sex cell
   c. from a female mammal
   d. infected by Barr virus

14. A promoter and a set of operators that control access to two or more prokaryotic genes _____ .
   a. occurs in all complex multicelled organisms
   b. is called an operon
   c. involves selective gene expression
   d. regulates gene expression in all organisms

15. True or false? Gene expression patterns are heritable.

16. Match the terms with the most suitable description.
   ___ repressor
   ___ riboswitch
   ___ operator
   ___ Barr body
   ___ differentiation
   ___ methylation
   a. mRNA self-regulation
   b. binding site for repressor
   c. cells become specialized
   d. may cause epigenetic effects
   e. inactivated X chromosome
   f. binds to promoter

Critical Thinking

1. Why are some genes expressed and some not expressed?

2. Children conceived during times of famine tend to be overweight. Suggest a possible reason for this effect.

3. Do the same gene controls operate in bacterial cells and eukaryotic cells? Why or why not?

BRCA Mutations in Women Diagnosed With Breast Cancer				
	BRCA1	BRCA2	No BRCA Mutation	Total
Total number of patients	89	35	318	442
Avg. age at diagnosis	43.9	46.2	50.4	
Preventive mastectomy	6	3	14	23
Preventive oophorectomy	38	7	22	67
Number of deaths	16	1	21	38
Percent died	18.0	2.8	6.9	8.6

4. Investigating a correlation between specific cancer-causing mutations and risk of mortality in humans is challenging, in part because each cancer patient is given the best treatment available at the time. There are no "untreated control" cancer patients, and the idea of what treatments are the best changes quickly as new drugs become available and new discoveries are made.

The table *above* shows some results of a 2007 study in which 442 women who had been diagnosed with breast cancer were checked for *BRCA* mutations, and their treatments and progress were followed over several years. All of the women in the study had at least two affected close relatives, so their risk of developing breast cancer due to an inherited factor was estimated to be greater than that of the general population. Some of the women underwent preventive mastectomy (removal of the noncancerous breast) during their course of treatment. Others had preventive oophorectomy (surgical removal of the ovaries) to prevent the possibility of getting ovarian cancer.

According to this study, is a *BRCA1* or *BRCA2* mutation more dangerous in breast cancer cases? What other data would you have to see in order to make a conclusion about the effectiveness of preventive surgeries?

Chapter 11
Summary

Section 11.1 An immortal line of cancer cells is a legacy of cancer victim Henrietta Lacks. Researchers trying to unravel the mechanisms of cancer continue to work with these cells.

Section 11.2 A **cell cycle** includes all the stages through which a eukaryotic cell passes during its lifetime. Most of a cell's activities, including replication of a diploid cell's **homologous chromosomes**, occur in **interphase**. A cell reproduces by dividing: nucleus first, then cytoplasm. Nuclear division partitions duplicated chromosomes into new nuclei. **Mitosis** maintains the chromosome number. It is the basis of growth, cell replacements, and tissue repair in multicelled species, and **asexual reproduction** in many species.

Section 11.3 Mitosis proceeds in four stages. In **prophase**, the duplicated chromosomes condense. Microtubules form a **spindle**, and the nuclear envelope breaks up. Spindle microtubules drag each chromosome toward the center of the cell. At **metaphase**, all chromosomes are aligned at the spindle's midpoint. During **anaphase**, sister chromatids of each chromosome detach from each other, and spindle microtubules start moving them toward opposite spindle poles. During **telophase**, a complete set of chromosomes reaches each spindle pole. A nuclear envelope forms around each cluster. Two new nuclei, each with the parental chromosome number, are the result.

Section 11.4 In most cases, **cytokinesis** follows nuclear division. A contractile ring pulls the plasma membrane of an animal cell inward, forming a **cleavage furrow** that eventually pinches the cytoplasm in two. In a plant cell, vesicles guided by microtubules to the future plane of division merge as a **cell plate**. The plate expands until it fuses with the parent cell wall, thus becoming a cross-wall that partitions the cytoplasm.

Section 11.5 Telomeres protecting the ends of eukaryotic chromosomes shorten with every cell division. When they become too short, the cell dies. Telomerase can restore the length of a cell's telomeres.

Section 11.6 The products of checkpoint genes are part of the controls governing the cell cycle. When checkpoint mechanisms fail, a cell loses control over its cell cycle, and the cell's descendants form a **neoplasm**. Mutations can turn **proto-oncogenes**, including **growth factor** receptor genes, into **oncogenes**. Such mutations typically disrupt checkpoint gene products or their expression, and can result in neoplasms. Neoplasms may form lumps called **tumors**. Malignant cells break loose from their home tissues and colonize other parts of the body, a process called **metastasis**.

Self-Quiz

Answers in Appendix III

1. Mitosis and cytoplasmic division function in _____ .
   a. asexual reproduction of single-celled eukaryotes
   b. growth and tissue repair in multicelled species
   c. gamete formation in bacteria and archaea
   d. both a and b

2. A duplicated chromosome has _____ chromatid(s).
   a. one   c. three
   b. two   d. four

3. Except for a pairing of sex chromosomes, homologous chromosomes _____ .
   a. carry the same genes   c. are the same length
   b. are the same shape   d. all of the above

4. Most cells spend the majority of their lives in _____ .
   a. prophase   d. telophase
   b. metaphase   e. interphase
   c. anaphase   f. a and c

5. A plant cell divides by the process of _____ .
   a. prophase   d. telophase
   b. metaphase   e. interphase
   c. anaphase   f. cytokinesis

6. The spindle attaches to chromosomes at the _____ .
   a. centriole   c. centromere
   b. contractile ring   d. centrosome

7. Interphase is the part of the cell cycle when _____ .
   a. a cell ceases to function   c. mitosis proceeds
   b. the spindle forms   d. a cell grows and
                                     duplicates its DNA

8. After mitosis, the chromosome number of a descendant cell is _____ the parent cell's.
   a. the same as   c. rearranged compared to
   b. one-half of   d. doubled compared to

HeLa Cells Are a Genetic Mess HeLa cells continue to be an extremely useful tool in cancer research. One early finding was that HeLa cells can vary in chromosome number. The panel of chromosomes on the *right*, originally published in 1989 by Nicholas Popescu and Joseph DiPaolo, shows all of the chromosomes in a single metaphase HeLa cell.

1. What is the chromosome number of this HeLa cell?

2. How many extra chromosomes does this cell have, compared to a normal human body cell?

3. Can you tell that this cell came from a female? How?

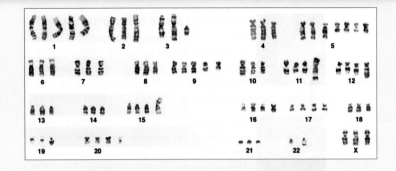

9. Stem cells are immortal as long as they express _____ .
   a. checkpoint genes   c. telomerase
   b. oncogenes   d. growth factors

10. Only _____ is not a stage of mitosis.
    a. prophase b. interphase c. metaphase d. anaphase

11. In intervals of interphase, G stands for _____ .
    a. gap   b. growth   c. Gey   d. gene

12. *BRCA1*, *BRCA2*, and *53BP1* are examples of _____ .
    a. checkpoint genes   c. tumor suppressors
    b. proto-oncogenes   d. all of the above

13. Which one of the following encompasses the other two?
    a. cancer   b. neoplasm   c. tumor

14. Match each term with its best description.
    ___ cell plate   a. lump of cells
    ___ spindle   b. made of microfilaments
    ___ tumor   c. divides plant cells
    ___ cleavage furrow   d. organize(s) the spindle
    ___ contractile ring   e. migrating, metastatic cells
    ___ cancer   f. made of microtubules
    ___ telomere   g. indentation
    ___ centrosomes   h. shortens with age

15. Match each stage with the events listed.
    ___ metaphase   a. sister chromatids move apart
    ___ prophase   b. chromosomes start to condense
    ___ telophase   c. new nuclei form
    ___ anaphase   d. all duplicated chromosomes are
                         aligned at the spindle equator

Critical Thinking

1. When a cell reproduces by mitosis and cytoplasmic division, does its life end?

 2. The eukaryotic cell in the photo on the *left* is in the process of cytoplasmic division. Is this cell from a plant or an animal? How do you know?

3. Exposure to radioisotopes or other sources of radiation can damage DNA. Humans exposed to high levels of radiation face a condition called radiation poisoning. Why do you think that hair loss and damage to the lining of the gut are early symptoms of radiation poisoning? Speculate about why exposure to radiation is used as a therapy to treat some kinds of cancers.

4. Suppose you have a way to measure the amount of DNA in one cell during the cell cycle. You first measure the amount at the G1 phase. At what points in the rest of the cycle will you see a change in the amount of DNA per cell?

Chapter 12
Summary

Section 12.1 Sexual reproduction, which involves the nuclear division mechanism of **meiosis**, mixes up genetic information of two parents. Variation in traits among offspring of sexual reproducers is an evolutionary advantage over genetically identical offspring.

Section 12.2 Offspring of sexual reproducers differ from parents, and often from one another, in details of shared traits. These individuals inherit pairs of chromosomes, one of each pair from the mother and the other from the father. Paired genes on homologous chromosomes in **somatic** cells may vary in DNA sequence, in which case they are called **alleles**. In animals, haploid **gametes** arise from meiosis in **germ cells**. Meiosis shuffles parental DNA, so offspring inherit new combinations of alleles. The fusion of two haploid **gametes** during **fertilization** restores the parental chromosome number in the **zygote**.

Section 12.3 Two divisions, meiosis I and II, halve the parental chromosome number. Chromosomes are duplicated during interphase. In prophase I, chromosomes condense and align tightly with their homologous partners. Each pair of homologues typically undergoes crossing over. Microtubules extending from the spindle poles penetrate the nuclear region and attach to one or the other chromosome of each homologous pair. At metaphase I, all chromosomes are lined up at the spindle equator. During anaphase I, homologous chromosomes separate and move to opposite spindle poles. Two nuclei form during telophase I. The cytoplasm may divide at this point.

Meiosis II occurs in both nuclei that formed in meiosis I. The chromosomes are still duplicated; each still consists of two sister chromatids. The chromosomes condense in prophase II, and align in metaphase II. Sister chromatids of each chromosome are pulled apart from each other in anaphase II, so each becomes an individual chromosome. By the end of telophase II, four haploid nuclei have formed.

Section 12.4 Events in prophase I and metaphase I produce nonparental combinations of alleles. The nonsister chromatids of homologous chromosomes undergo **crossing over** during prophase I: They exchange segments at the same place along their length, so each can end up with a new combination of alleles that was not present in either parental chromosome. The random segregation of maternal and paternal chromosomes into new nuclei also contributes to varia-

tion in traits among offspring. Microtubules can attach the maternal or the paternal chromosome of each pair to one or the other spindle pole. Either chromosome may end up in any new nucleus, and in any gamete. Such chromosome shufflings, along with crossovers during prophase I of meiosis, are the basis of variation in traits among individuals of sexually reproducing species.

Section 12.5 Multicelled diploid bodies are typical in life cycles of plants and animals. A diploid **sporophyte** is a multicelled plant body that makes haploid spores. Spores give rise to haploid **gametophytes**, which are multicelled plant bodies in which haploid gametes form. Germ cells in the reproductive organs of most animals give rise to **sperm** or **eggs**. The fusion of haploid gametes at fertilization results in a diploid zygote.

Section 12.6 Like mitosis, meiosis requires a spindle to move and sort duplicated chromosomes, but meiosis occurs only in cells that are set aside for sexual reproduction. Some parts of meiosis resemble those of mitosis, and may have evolved from them.

Self-Quiz

Answers in Appendix III

1. The main evolutionary advantage of sexual over asexual reproduction is that it produces _____ .
    a. more offspring per individual
    b. more variation among offspring
    c. healthier offspring

2. Meiosis functions in _____ .
    a. asexual reproduction of single-celled eukaryotes
    b. growth and tissue repair in multicelled species
    c. sexual reproduction
    d. both a and b

3. Sexual reproduction in animals requires _____ .
    a. meiosis   c. spore formation
    b. fertilization   d. a and b

4. Meiosis _____ the parental chromosome number.
    a. doubles   c. maintains
    b. halves   d. mixes up

5. Crossing over mixes up _____ .
    a. chromosomes   c. zygotes
    b. alleles   d. gametes

6. Crossing over happens during which phase of meiosis?

7. The stage of meiosis that makes descendant cells haploid is _____ .
    a. prophase I   d. anaphase II
    b. prophase II   e. metaphase I
    c. anaphase I   f. metaphase II

8. Dogs have a diploid chromosome number of 78. How many chromosomes do their gametes have?
    a. 39   c. 156
    b. 78   d. 234

BPA and Abnormal Meiosis In 1998, researchers at Case Western University were studying meiosis in mouse oocytes when they saw an unexpected and dramatic increase of abnormal meiosis events (**Figure 12.13**). The improper segregation of chromosomes during meiosis is one of the main causes of human genetic disorders, which we will discuss in Chapter 14.

The researchers discovered that the spike in meiotic abnormalities began immediately after the mouse facility started washing the animals' plastic cages and water bottles in a new, alkaline detergent. The detergent had damaged the plastic, which began to leach bisphenol A (BPA). BPA is a synthetic chemical that mimics estrogen, the main female sex hormone in animals. BPA is used to manufacture polycarbonate plastic items (including baby bottles and water bottles) and epoxies (including the coating on the inside of metal cans of food).

1. What percentage of mouse oocytes displayed abnormalities of meiosis with no exposure to damaged caging?

2. Which group of mice showed the most meiotic abnormalities in their oocytes?

3. What is abnormal about metaphase I as it is occurring in the oocytes shown in **Figure 12.13B, C, and D**?

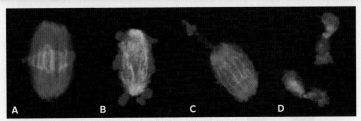

Caging materials	Total number of oocytes	Abnormalities
Control: New cages with glass bottles	271	5 (1.8%)
Damaged cages with glass bottles		
Mild damage	401	35 (8.7%)
Severe damage	149	30 (20.1%)
Damaged bottles	197	53 (26.9%)
Damaged cages with damaged bottles	58	24 (41.4%)

Figure 12.13 Meiotic abnormalities associated with exposure to damaged plastic caging. Fluorescent micrographs show nuclei of single mouse oocytes in metaphase I. (**A**) Normal metaphase; (**B–D**) examples of abnormal metaphase. Chromosomes are stained *red*; spindle fibers, *green*.

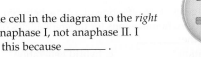

9. The cell in the diagram to the *right* is in anaphase I, not anaphase II. I know this because _____ .

10. _____ contributes to variation in traits among the offspring of sexual reproducers.
    a. Crossing over
    b. Random attachment of chromosomes to spindle poles
    c. Fertilization
    d. both a and b
    e. all are factors

11. Which of the following is one of the most important differences between mitosis and meiosis?
    a. Chromosomes align midway between spindle poles only in meiosis.
    b. Homologous chromosomes pair up only in meiosis.
    c. DNA is replicated only in mitosis.
    d. Sister chromatids separate only in meiosis.

12. Match each term with its description.
    ___ DNA replication
    ___ metaphase I
    ___ allele
    ___ sporophyte
    ___ gamete

    a. different molecular form of a gene
    b. none between meiosis I and meiosis II
    c. all chromosomes are aligned at spindle equator
    d. haploid
    e. does not occur in animals

Critical Thinking

1. Make a simple sketch of meiosis in a cell with a diploid chromosome number of 4. Now try it when the chromosome number is 3.

2. The diploid chromosome number for the body cells of a frog is 26. What would that number be after three generations if meiosis did not occur before gamete formation?

3. Assume you can measure the amount of DNA in the nucleus of a primary oocyte, and then in the nucleus of a primary spermatocyte. Each gives you a mass m. What mass of DNA would you expect to find in the nucleus of each mature gamete (each egg and sperm) that forms after meiosis? What mass of DNA will be (1) in the nucleus of a zygote that forms at fertilization and (2) in that zygote's nucleus after the first DNA duplication?

Chapter 13
Summary

Section 13.1 Cystic fibrosis is usually caused by a particular mutation. The allele persists at high frequency despite its devastating effects. Only those homozygous for the allele have the disorder.

Section 13.2 Each gene occurs at a particular **locus**, or location, on a chromosome. Individuals with identical alleles are **homozygous** for the allele. **Heterozygous** individuals, or **hybrids**, have two non-identical alleles. A **dominant** allele masks the effect of a **recessive** allele on the homologous chromosome. **Genotype** (an individual's particular set of alleles) gives rise to **phenotype**.

Section 13.3 Crossing individuals that breed true for two forms of a trait yields identically heterozygous F_1 offspring. A cross between such offspring is a **monohybrid cross**. The frequency at which the traits appear in the offspring of **testcrosses** can reveal dominance relationships among the alleles associated with those traits. **Punnett squares** are useful in determining the probability of the genotype and phenotype of the offspring of crosses. Mendel's monohybrid cross results led to his **law of segregation** (stated here in modern terms): During meiosis, the genes of each pair on homologous chromosomes separate, so each gamete gets one or the other gene.

Section 13.4 Crossing individuals that breed true for two forms of two traits yields F_1 offspring identically heterozygous for alleles governing those traits. A cross between such offspring is a **dihybrid cross**. Mendel's dihybrid cross results led to his **law of independent assortment** (stated here in modern terms): During meiosis, gene pairs on homologous chromosomes tend to sort into gametes independently of other gene pairs. Crossovers can break up **linkage groups**.

Section 13.5 With **incomplete dominance**, an allele is not fully dominant over its partner on a homologous chromosome, so an intermediate phenotype results. **Codominant** alleles are both fully expressed in heterozygotes; neither is dominant. Codominanace may occur in **multiple allele systems** such as the one underlying ABO blood typing. In **epistasis**, two or more genes often affect the same trait. A **pleiotropic** gene affects two or more traits.

Section 13.6 An individual's phenotype—including appearance as well as behavior—is influenced by environmental factors. Environmental cues alter gene expression by way of cell signaling pathways that ultimately affect gene controls such as chromatin remodeling or RNA interference.

Section 13.7 A trait that is influenced by the products of multiple genes often occurs in a range of small increments of phenotype called **continuous variation**. Continuous variation is indicated by a **bell curve** in the range of values.

Self-Quiz

Answers in Appendix III

1. A heterozygous individual has a _____ for a trait being studied.
   - a. pair of identical alleles
   - b. pair of nonidentical alleles
   - c. haploid condition, in genetic terms

2. An organism's observable traits constitute its _____ .
   - a. phenotype
   - b. variation
   - c. genotype
   - d. pedigree

3. In genetics, F stands for filial, which means _____ .
   - a. friendly
   - b. offspring
   - c. final
   - d. hairlike

4. F_1 offspring of the cross $AA \times aa$ are _____ .
   - a. all AA
   - b. all aa
   - c. all Aa
   - d. 1/2 AA and 1/2 aa

5. The second-generation offspring of a cross between individuals who are homozygous for different alleles of a gene are called the _____ .
   - a. F_1 generation
   - b. F_2 generation
   - c. hybrid generation
   - d. none of the above

6. Refer to question 4. Assuming complete dominance, the F_2 generation will show a phenotypic ratio of _____ .
   - a. 3:1
   - b. 9:1
   - c. 1:2:1
   - d. 9:3:3:1

7. A testcross is a way to determine _____ .
   - a. phenotype
   - b. genotype
   - c. both a and b

8. Assuming complete dominance, crosses between two dihybrid F_1 pea plants, which are offspring from a cross $AABB \times aabb$, result in F_2 phenotype ratios of _____ .
   - a. 1:2:1
   - b. 3:1
   - c. 1:1:1:1
   - d. 9:3:3:1

9. The probability of a crossover occurring between two genes on the same chromosome _____ .
   - a. is unrelated to the distance between them
   - b. decreases with the distance between them
   - c. increases with the distance between them

10. A gene that affects three traits is _____ .
   - a. epistatic
   - b. a multiple allele system
   - c. pleiotropic
   - d. dominant

The Cystic Fibrosis Mutation and Typhoid Fever The Δ*F508* mutation disables the receptor function of the CFTR protein, so it inhibits endocytosis of bacteria into epithelial cells. Endocytosis is an important part of the respiratory tract's immune defenses against common *Pseudomonas* bacteria, which is why *Pseudomonas* infections are a chronic problem in cystic fibrosis patients.

The Δ*F508* mutation also inhibits endocytosis of *Salmonella typhi* into cells of the gastrointestinal tract, where internalization of this bacteria can cause typhoid fever. Typhoid fever is a common worldwide disease. Its symptoms include extreme fever and diarrhea, and the resulting dehydration causes delirium that may last several weeks. If untreated, it kills up to 30 percent of those infected. Around 600,000 people die annually from typhoid fever. Most of them are children.

In 1998, Gerald Pier and his colleagues compared the uptake of *S. typhi* by different types of epithelial cells: those homozygous for the normal allele, and those heterozygous for the Δ*F508* mutation (**Figure 13.17**). (Cells homozygous for the mutation do not take up *S. typhi* bacteria.)

1. Regarding the Ty2 strain of *S. typhi*, about how many more bacteria were able to enter normal cells (those expressing unmutated *CFTR*) than cells expressing the gene with the Δ*F508* deletion?

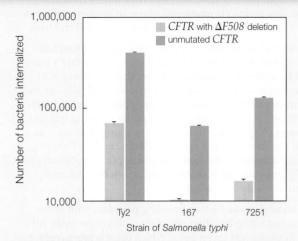

Figure 13.17 In epithelial cells, effect of the *CF* mutation on the uptake of three different strains of *Salmonella typhi* bacteria.

2. Which strain of bacteria entered normal epithelial cells most easily?

3. The Δ*F508* deletion inhibited the entry of all three *S. typhi* strains into epithelial cells. Can you tell which strain was most inhibited?

---

11. _____ alleles are both expressed.
    a. Dominant   c. Pleiotropic
    b. Codominant   d. Hybrid

12. A bell curve indicates _____ in a trait.
    a. epigenetic effects   c. incomplete dominance
    b. pleiotropy   d. continuous variation

13. True or false? All traits are inherited in a predictable way.

14. Match the terms with the best description.
    ___ dihybrid cross   a. *bb*
    ___ monohybrid cross   b. *AaBb* × *AaBb*
    ___ homozygous condition c. *Aa*
    ___ heterozygous condition d. *Aa* × *Aa*

Genetics Problems   *Answers in Appendix III*

1. Assuming that independent assortment occurs during meiosis, what type(s) of gametes will form in individuals with the following genotypes?
    a. *AABB*   b. *AaBB*   c. *Aabb*   d. *AaBb*

2. Refer to problem 1. Determine the frequencies of each genotype among offspring from the following matings:
    a. *AABB* × *aaBB*   c. *AaBb* × *aabb*
    b. *AaBB* × *AABb*   d. *AaBb* × *AaBb*

3. Refer to problem 2. Assume a third gene has alleles *C* and *c*. For each genotype listed, what allele combinations will occur in gametes, assuming independent assortment?
    a. *AABBCC*   c. *AaBBCc*
    b. *AaBBcc*   d. *AaBbCc*

4. Some genes are so vital for development that mutations in them are lethal in homozygous recessives. Even so, heterozygotes can perpetuate recessive, lethal alleles. The allele *Manx* (*M^L*) in cats is an example. Homozygous cats (*M^L M^L*) die before birth. In heterozygotes (*M^L M*), the spine develops abnormally, and the cats end up with no tail (*above*). Two *M^L M* cats mate. What is the probability that any one of their surviving kittens will be heterozygous?

5. Sometimes the gene for tyrosinase mutates so its product is not functional. An individual who is homozygous recessive for such a mutation cannot make melanin. Albinism, the absence of melanin, results. Humans and many other organisms can have this phenotype (*left*). In the following situations, what are the probable genotypes of the father, the mother, and their children?

a. Both parents have normal phenotypes; some of their children are albino and others are unaffected.
b. Both parents are albino and have albino children.
c. The woman is unaffected, the man is albino, and they have one albino child and three unaffected children.

6. Several alleles affect traits of roses, such as plant form and bud shape. Alleles of one gene govern whether a plant will be a climber (dominant) or shrubby (recessive). All F_1 offspring from a cross between a true-breeding climber and a shrubby plant are climbers. If an F_1 plant is crossed with a shrubby plant, about 50 percent of the offspring will be shrubby; 50 percent will be climbers. Using symbols A and a for the dominant and recessive alleles, make a Punnett-square diagram of the expected genotypes and phenotypes in F_1 offspring and in offspring of a cross between an F_1 plant and a shrubby plant.

7. Mendel crossed a true-breeding pea plant with green pods and a true-breeding pea plant with yellow pods. All the F_1 plants had green pods. Which color is recessive?

8. Suppose you identify a new gene in mice. One of its alleles specifies white fur, another specifies brown. You want to see if the two interact in simple or incomplete dominance. What sorts of genetic crosses would give you the answer?

9. In sweet pea plants, an allele for purple flowers (P) is dominant to an allele for red flowers (p). An allele for long pollen grains (L) is dominant to an allele for round pollen grains (l). Bateson and Punnett crossed a plant having purple flowers/long pollen grains with one having white flowers/round pollen grains. All F_1 offspring had purple flowers and long pollen grains. Among the F_2 generation, the researchers observed the following phenotypes:

296 purple flowers/long pollen grains
19 purple flowers/round pollen grains
27 red flowers/long pollen grains
85 red flowers/round pollen grains

What is the best explanation for these results?

10. Red-flowering snapdragons are homozygous for allele R^1. White-flowering snapdragons are homozygous for a different allele (R^2). Heterozygous plants (R^1R^2) bear pink flowers. What phenotypes should appear among first-generation offspring of the crosses listed? What are the expected proportions for each phenotype?
    a. $R^1R^1 \times R^1R^2$ c. $R^1R^2 \times R^1R^2$
    b. $R^1R^1 \times R^2R^2$ d. $R^1R^2 \times R^2R^2$

(Incompletely dominant alleles are usually designated by superscript numerals, as shown, not by uppercase letters for dominance and lowercase letters for recessiveness.)

11. A single allele gives rise to the Hb^S form of hemoglobin. Homozygotes (Hb^SHb^S) develop sickle-cell anemia (Section 9.6). Heterozygotes (Hb^AHb^S) have few symptoms. A couple who are both heterozygous for the Hb^S allele plan to have children. For each of the pregnancies, state the probability that they will have a child who is:
    a. homozygous for the Hb^S allele
    b. homozygous for the normal allele (Hb^A)
    c. heterozygous: Hb^AHb^S

Chapter 14
Summary

Section 14.1 Like most other human traits, skin color has a genetic basis. Minor differences in alleles probably evolved as a balance between vitamin production and protection against harmful UV radiation.

Section 14.2 Geneticists use **pedigrees** to track traits through generations of a family. Such studies can reveal inheritance patterns for alleles that can be predictably associated with specific phenotypes, including genetic abnormalities or disorders. A genetic abnormality is an uncommon version of a heritable trait that does not result in medical problems. A genetic disorder is a heritable condition that sooner or later results in mild or severe medical problems.

Section 14.3 A trait associated with a dominant allele on an autosome occurs in heterozygous individuals, male and female. Such traits appear in every generation of families that carry the allele. A trait associated with a recessive allele on an autosome appears only in homozygous individuals. Such traits occur in both males and females, but they can skip generations.

Section 14.4 An allele is inherited in an X-linked pattern when it occurs on the X chromosome. Most X-linked inheritance disorders are recessive, and these tend to appear in men more often than in women. Heterozygous women have a dominant, normal allele that can mask the effects of the recessive one; men do not. Men can transmit an X-linked allele to their daughters, but not to their sons. Only a woman can pass an X-linked allele to a son.

Section 14.5 Major changes in chromosome structure include **duplications**, **deletions**, **inversions**, and **translocations**. Most major alterations are harmful or lethal in humans. Even so, many major structural changes have accumulated in the chromosomes of all species over evolutionary time.

Section 14.6 Changes in chromosome number are usually an outcome of **nondisjunction**, in which chromosomes fail to separate properly during meiosis. In **aneuploidy**, an individual's cells have too many or too few copies of a chromosome. Except for trisomy 21, which causes Down syndrome, most cases of autosomal aneuploidy are lethal. Some organisms are **polyploid** (they have three or more sets of chromosomes).

Section 14.7 Prospective parents can estimate their risk of transmitting a harmful allele to offspring with genetic screening, in which their pedigrees and genotype are analyzed by a genetic counselor. Prenatal genetic testing can reveal a genetic disorder before birth.

Self-Quiz
Answers in Appendix III

1. Constructing a pedigree is particularly useful when studying inheritance patterns in organisms that _____ .
   a. produce many offspring per generation
   b. produce few offspring per generation
   c. have a very large chromosome number
   d. reproduce sexually
   e. have a fast life cycle

2. Pedigree analysis is necessary when studying human inheritance patterns because _____ .
   a. humans have more than 20,000 genes
   b. of ethical problems with experimenting on humans
   c. inheritance in humans is more complicated than in other organisms
   d. genetic disorders occur in humans
   e. all of the above

3. A recognized set of symptoms that characterize a genetic disorder is a(n) _____ .
   a. syndrome   b. disease   c. abnormality

4. If one parent is heterozygous for a dominant allele on an autosome and the other parent does not carry the allele, any child of theirs has a _____ chance of being heterozygous.
   a. 25 percent   c. 75 percent
   b. 50 percent   d. no chance; it will die

Skin Color Survey of Native Peoples A 2000 study measured average skin color of people native to more than fifty regions, and correlated them to the amount of UV radiation received in those regions. Some of their results are shown in **Figure 14.16**.

1. Which country receives the most UV radiation? The least?

2. The people native to which country have the darkest skin? The lightest?

3. According to these data, how does the skin color of indigenous peoples correlate with the amount of UV radiation incident in their native regions?

Country	Skin Reflectance	UVMED
Australia	19.30	335.55
Kenya	32.40	354.21
India	44.60	219.65
Cambodia	54.00	310.28
Japan	55.42	130.87
Afghanistan	55.70	249.98
China	59.17	204.57
Ireland	65.00	52.92
Germany	66.90	69.29
Netherlands	67.37	62.58

Figure 14.16 Skin color of indigenous peoples and regional incident UV radiation. Skin reflectance measures how much light of 685-nanometer wavelength is reflected from skin; UVMED is the annual average UV radiation received at Earth's surface.

5. Is this statement true or false? A son can inherit an X-linked recessive allele from his father.

6. A trait that is present in a male child but not in either of his parents is characteristic of _____ inheritance.
   a. autosomal dominant
   b. autosomal recessive
   c. X-linked recessive
   d. It is not possible to answer this question without more information.

7. Color blindness is a case of _____ inheritance.
   a. autosomal dominant
   b. autosomal recessive
   c. X-linked dominant
   d. X-linked recessive

8. What do you think the pattern of inheritance of the human *SRY* gene is called?

9. A female child inherits one X chromosome from her mother and one from her father. What sex chromosome does a male child inherit from each of his parents?

10. Nondisjunction may occur during _____ .
    a. mitosis
    b. meiosis
    c. fertilization
    d. both a and b

11. Nondisjunction can result in _____ .
    a. duplications
    b. aneuploidy
    c. crossing over
    d. pleiotropy

12. Nondisjunction can occur during _____ of meiosis.
    a. anaphase I
    b. telophase I
    c. anaphase II
    d. a or c

13. Is this statement true or false? Body cells may inherit three or more of each type of chromosome characteristic of the species, a condition called polyploidy.

14. Klinefelter syndrome (XXY) can be easily diagnosed by _____ .
    a. pedigree analysis
    b. aneuploidy
    c. karyotyping
    d. phenotypic treatment

15. Match the chromosome terms appropriately.
    ___ polyploidy
    ___ deletion
    ___ aneuploidy
    ___ translocation
    ___ syndrome
    ___ nondisjunction during meiosis
    a. symptoms of a genetic disorder
    b. segment of a chromosome moves to a nonhomologous chromosome
    c. extra sets of chromosomes
    d. results in gametes with the wrong chromosome number
    e. a chromosome segment lost
    f. one extra chromosome

Genetics Problems   *Answers in Appendix III*

1. Does the phenotype indicated by the red circles and squares in this pedigree show an inheritance pattern that is autosomal dominant, autosomal recessive, or X-linked?

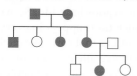

2. Human females are XX and males are XY.
   a. Does a male inherit the X from his mother or father?
   b. With respect to X-linked alleles, how many different types of gametes can a male produce?
   c. If a female is homozygous for an X-linked allele, how many types of gametes can she produce with respect to that allele?
   d. If a female is heterozygous for an X-linked allele, how many types of gametes might she produce with respect to that allele?

3. A mutated allele responsible for Marfan syndrome (Section 13.5) is inherited in an autosomal dominant pattern. What is the chance that any child will inherit it if one parent does not carry the allele and the other is heterozygous for it?

4. Somatic cells of individuals with Down syndrome usually have an extra chromosome 21; they contain forty-seven chromosomes.
   a. At which stages of meiosis I and II could a mistake alter the chromosome number?
   b. A few individuals with Down syndrome have forty-six chromosomes: two normal-appearing chromosomes 21, and a longer-than-normal chromosome 14. Speculate on how this chromosome abnormality may arise.

5. Duchenne muscular dystrophy is associated with a recessive X-linked allele. Unlike color blindness, this disorder nearly always occcurs in males. Suggest why.

6. In the human population, mutation of two genes on the X chromosome causes two types of X-linked hemophilia (A and B). In a few cases, a woman is heterozygous for both mutated alleles (one on each of the X chromosomes). All of her sons should have either hemophilia A or B. However, on very rare occasions, one of these women gives birth to a son who does not have hemophilia, and his one X chromosome does not have either mutated allele. Explain how this boy's X chromosome probably arises.

Chapter 15
Summary

Section 15.1 Personal DNA testing companies identify a person's unique array of **single-nucleotide polymorphisms**. Personal genetic testing may soon revolutionize the way medicine is practiced.

Section 15.2 In **DNA cloning**, **restriction enzymes** cut DNA into pieces, then DNA ligase splices the pieces into plasmids or other **cloning vectors**. The resulting **recombinant DNA** molecules are inserted into host cells such as bacteria. When a host cell divides, it forms huge populations of genetically identical descendant cells (clones), each with a copy of the foreign DNA. The enzyme **reverse transcriptase** is used to transcribe RNA into **cDNA** for cloning.

Section 15.3 A **DNA library** is a collection of cells that host different fragments of DNA, often representing an organism's entire **genome**. Researchers can use **probes** to identify cells that host a specific fragment of DNA. Base pairing between nucleic acids from different sources is called **nucleic acid hybridization**. The **polymerase chain reaction (PCR)** uses **primers** and a heat-resistant DNA polymerase to rapidly increase the number of copies of a targeted section of DNA.

Section 15.4 DNA sequencing reveals the order of bases in a section of DNA. DNA polymerase is used to partially replicate a DNA template. The reaction produces a mixture of DNA fragments of all different lengths. **Electrophoresis** separates the fragments by length into bands. The entire genomes of several organisms have now been sequenced.

Section 15.5 Genomics is providing insights into the function of the human genome. Similarities between genomes of different organisms are evidence of evolutionary relationships, and can be used as a predictive tool in research. **DNA profiling** identifies a person by the unique parts of his or her DNA. An example is the determination of an individual's array of **short tandem repeats** (SNPs). Within the context of a criminal investigation, a DNA profile is called a DNA fingerprint.

Sections 15.6–15.9 Recombinant DNA technology is the basis of **genetic engineering**, the directed modification of an organism's genetic makeup with the intent to modify its phenotype. A gene is modified and reinserted into an individual of the same species, or a gene from one species is inserted into an individual of a different species to make a **transgenic** organism. The result of either process is a **genetically modified organism (GMO)**. Transgenic bacteria and yeast produce medically valuable proteins.

Transgenic crop plants are helping farmers produce food more efficiently. Genetically modified animals produce human proteins, and may one day provide a source of organs and tissues for **xenotransplantation** into humans.

Section 15.10 With **gene therapy**, a gene is transferred into body cells to correct a genetic defect or treat a disease. Potential benefits of genetically modifying humans must be weighed against potential risks. The practice raises ethical issues such as whether **eugenics** is desirable in some circumstances.

Self-Quiz

Answers in Appendix III

1. _____ cut(s) DNA molecules at specific sites.
   - a. DNA polymerase
   - b. DNA probes
   - c. Restriction enzymes
   - d. Reverse transcriptase

2. A _____ is a small circle of bacterial DNA that contains a few genes and is separate from the chromosome.
   - a. plasmid
   - b. chromosome
   - c. nucleus
   - d. double helix

3. Reverse transcriptase assembles a(n) _____ on a(n) _____ template.
   - a. mRNA; DNA
   - b. cDNA; mRNA
   - c. DNA; ribosome
   - d. protein; mRNA

4. For each species, all _____ in the complete set of chromosomes is the _____ .
   - a. genomes; phenotype
   - b. DNA; genome
   - c. mRNA; start of cDNA
   - d. cDNA; start of mRNA

5. A set of cells that host various DNA fragments collectively representing an organism's entire set of genetic information is a _____ .
   - a. genome
   - b. clone
   - c. genomic library
   - d. GMO

6. _____ is a technique to determine the order of nucleotide bases in a fragment of DNA.

7. Fragments of DNA can be separated by electrophoresis according to _____ .
   - a. sequence
   - b. length
   - c. species
   - d. composition

8. PCR can be used _____ .
   - a. to increase the number of specific DNA fragments
   - b. in DNA fingerprinting
   - c. to modify a human genome
   - d. a and b are correct

9. An individual's set of unique _____ can be used as a DNA profile.
   - a. DNA sequences
   - b. short tandem repeats
   - c. SNPs
   - d. all of the above

10. A transgenic organism _____ .
    - a. carries a gene from another species
    - b. has been genetically modified
    - c. both a and b

11. Which of the following can be used to carry foreign DNA into host cells? Choose all correct answers.
    - a. RNA
    - b. viruses
    - c. PCR
    - d. plasmids
    - e. lipid clusters
    - f. blasts of pellets
    - g. xenotransplantation
    - h. sequencing

Enhanced Spatial Learning in Mice with Autism Mutation

Autism is a neurobiological disorder with a range of symptoms that include impaired social interactions and stereotyped patterns of behavior such as hand-flapping or rocking. A relatively high proportion of autistic people—around 10 percent—have an extraordinary skill or talent such as greatly enhanced memory.

Mutations in neuroligin 3, a cell adhesion protein (Section 5.7) that connects brain cells to one another, have been associated with autism. One mutation changes amino acid 451 from arginine to cysteine. Mouse and human neuroligin 3 are very similar. In 2007, Katsuhiko Tabuchi and his colleagues genetically modified mice to carry the same arginine-to-cysteine substitution in their neuroligin 3. The mutation caused an increase in transmission of some types of signals between brain cells. Mice with the mutation had impaired social behavior, and, unexpectedly, enhanced spatial learning ability (**Figure 15.17**).

1. In the first test, how many days did unmodified mice need to learn to find the location of a hidden platform within 10 seconds?

2. Did the modified or the unmodified mice learn the location of the platform faster in the first test?

3. Which mice learned faster the second time around?

4. Which mice had the greatest improvement in memory?

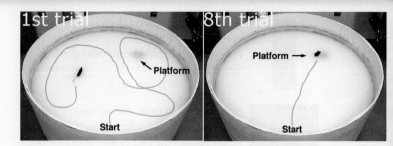

A Mice were tested in a water maze, in which a platform is submerged slightly below the surface of a deep pool of warm water. The platform is not visible to swimming mice. Mice do not particularly enjoy swimming, so they locate a hidden platform as fast as they can. When tested again in the same pool, they use visual cues around the pool's edge to remember the platform's location.

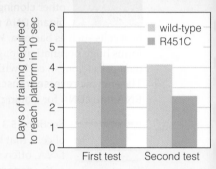

B How quickly mice remember the location of a hidden platform in a water maze is a measure of spatial learning ability.

Figure 15.17 Enhanced spatial learning ability in mice with a mutation in neuroligin 3 (R451C), compared with unmodified (wild-type) mice.

---

12. Transgenic _____ can pass a foreign gene to offspring.
    a. plants
    b. animals
    c. a and b
    d. none are correct

13. _____ can be used to correct a genetic defect.
    a. Cloning vectors
    b. Gene therapy
    c. Cloning
    d. Xenotransplantation
    e. a and b
    f. all of the above

14. Match the method with the appropriate enzyme.
    ___ PCR
    ___ cutting DNA
    ___ cDNA synthesis
    ___ DNA sequencing
    ___ pasting DNA
    a. *Taq* polymerase
    b. DNA ligase
    c. reverse transcriptase
    d. restriction enzyme
    e. DNA polymerase (not *Taq*)

15. Match the terms with the most suitable description.
    ___ DNA profile
    ___ Ti plasmid
    ___ nucleic acid hybridization
    ___ eugenics
    ___ SNP
    ___ transgenic
    ___ GMO
    a. GMO with a foreign gene
    b. an allele contains one
    c. a person's unique collection of short tandem repeats
    d. base pairing of DNA or DNA and RNA from different sources
    e. selecting "desirable" traits
    f. genetically modified
    g. used in some gene transfers

Critical Thinking

1. Restriction enzymes in bacterial cytoplasm cut injected bacteriophage DNA wherever certain sequences occur. Why do you think these enzymes do not chop up the bacterial chromosome, which is exposed to the enzymes in cytoplasm?

2. The *FOXP2* gene encodes a transcription factor associated with vocal learning in mice, bats, birds, and humans. The chimpanzee, gorilla, and rhesus FOXP2 proteins are identical; the human version differs in only 2 of 715 amino acids, a change thought to have contributed to the development of spoken language. In humans, loss-of-function mutations in *FOXP2* result in severe speech and language disorders. In mice, they hamper brain function and impair vocalizations. Mice genetically engineered to carry the human version of *FOXP2* show changes in vocal patterns, and more growth and greater adaptability of neurons involved in memory and learning. Biologists do not anticipate that a similar experiment in chimpanzees would confer the ability to speak, because spoken language is an extremely complex trait. If genetic engineering could produce a talking chimp, how do you think the debate about using animals for research would change?

3. In 1918, an influenza pandemic that originated with avian flu killed 50 million people. Researchers isolated samples of that virus from bodies of infected people preserved in Alaskan permafrost since 1918. From the samples, they sequenced the viral genome, then reconstructed the virus. The reconstructed virus is 39,000 times more infectious than modern influenza strains, and 100 percent lethal in mice.

Understanding how this virus works can help us defend ourselves against other strains that may arise. For example, discovering what makes it is so infectious and deadly would help us design more effective vaccines. Critics of the research are concerned: If the virus escapes the containment facilities (even though it has not done so yet), it might cause another pandemic. Worse, terrorists could use the published DNA sequence and methods to make the virus for horrific purposes. Do you think this research makes us more or less safe?

Chapter 16
Summary

Section 16.1 Events of the ancient past can be explained by the same physical, chemical, and biological processes that operate today. An asteroid impact may have caused a **mass extinction** 65.5 million years ago.

Section 16.2 Expeditions by nineteenth-century **naturalists** yielded increasingly detailed observations of nature. Geology, **biogeography**, and **comparative morphology** of organisms and their **fossils** led to new ways of thinking about the natural world.

Section 16.3 Prevailing belief systems may influence interpretation of the underlying cause of a natural event. The nineteenth-century naturalists proposed **catastrophism** and the **theory of uniformity** in their attempts to reconcile traditional beliefs with physical evidence of **evolution**, or change in a **lineage** over time.

Section 16.4 Humans select desirable traits in animals by selective breeding, or **artificial selection**. Charles Darwin and Alfred Wallace independently came up with a theory of how environments also select traits, stated here in modern terms: A population tends to grow until it exhausts environmental resources. As that happens, competition for those resources intensifies among the population's individuals. Individuals with forms of shared, heritable traits that make them more competitive for the resources tend to produce more offspring. Thus, **adaptive traits** (**adaptations**) that impart greater **fitness** to an individual become more common in a population over generations.

The process in which environmental pressures result in the differential survival and reproduction of individuals of a population is called **natural selection**. It is one of the processes that drives evolution.

Section 16.5 Fossils are typically found in stacked layers of sedimentary rock, with younger fossils in the layers deposited more recently. The fossil record will always be incomplete.

Section 16.6 With **radiometric dating**, the **half-life** of a radioisotope is used to determine the age of rocks and fossils. A sedimentary rock layer's age is estimated by dating volcanic rock above and below it.

Section 16.7 By the **plate tectonics** theory, Earth's crust is cracked into giant plates that carry landmasses to new positions as they move. Earth's landmasses have periodically converged as supercontinents such as **Gondwana** and **Pangea**.

Section 16.8 Transitions in the fossil record are the boundaries of great intervals of the **geologic time scale**, a chronology of Earth's history that correlates geologic and evolutionary events.

Self-Quiz

Answers in Appendix III

1. The number of species on an island depends on the size of the island and its distance from a mainland. This statement would most likely be made by _____ .
   a. an explorer   c. a geologist
   b. a biogeographer   d. a philosopher

2. The bones of a bird's wing are similar to the bones in a bat's wing. This observation is an example of _____ .
   a. uniformity   c. comparative morphology
   b. evolution   d. a lineage

3. Evolution _____ .
   a. is natural selection
   b. is heritable change in a line of descent
   c. can occur by natural selection
   d. b and c are correct

Abundance of Iridium in the K–T Boundary Layer In the late

1970s, geologist Walter Alvarez was investigating the composition of the 1-centimeter-thick layer of clay that marks the Cretaceous–Tertiary (K–T) boundary all over the world. He asked his father, Nobel Prize–winning physicist Luis Alvarez, to help him analyze the elemental composition of the layer. The photo (*inset*) shows Luis and Walter Alvarez with a section of the K–T boundary layer.

The Alvarezes and their colleagues tested samples of the layer taken from formations in Italy and Denmark. The researchers discovered that the K–T boundary layer had a much higher iridium content than the surrounding rock layers. Some of their results are shown in the table in **Figure 16.16**.

Iridium belongs to a group of elements (Appendix IV) that are much more abundant in asteroids and other

Sample Depth	Average Abundance of Iridium (ppb)
+ 2.7 m	< 0.3
+ 1.2 m	< 0.3
+ 0.7 m	0.36
boundary layer	41.6
– 0.5 m	0.25
– 5.4 m	0.30

Figure 16.16 Abundance of iridium in and near the K–T boundary layer in Stevns Klint, Denmark. Many rock samples taken from above, below, and a the boundary layer were tested for iridiu content. Depths are given as meters above or below the boundary layer.

The iridium content of an average Earth rock is 0.4 parts per billion (ppb) of iridium. An average meteorite contains about 550 parts per billion of iridium.

solar system materials than they are in Earth's crust. The Alvarez group concluded that the K–T boundary layer must have originated with extraterrestrial material. They calculated that an asteroid 14 kilometers (8.7 miles) in diameter would contain enough iridium to account for the extra iridium in the K–T boundary layer.

1. What was the iridium content of the K–T boundary layer?

2. How much higher was the iridium content of the boundary layer than the sample taken 0.7 meter above the layer?

4. A trait is adaptive if it _____ .
   a. arises by mutation
   b. contributes to fitness
   c. is passed to offspring
   d. occurs in fossils

5. In which type of rock are you more likely to find a fossil?
   a. basalt, a dark, fine-grained volcanic rock
   b. limestone, composed of sedimented calcium carbonate
   c. slate, a volcanically melted and cooled shale
   d. granite, which forms by crystallization of molten rock below Earth's surface
   e. all are equally likely to hold fossils

6. Which of the following is a fossil?
   a. an insect encased in 10-million-year-old tree sap
   b. a woolly mammoth frozen in Arctic permafrost for the last 50,000 years
   c. mineral-hardened remains of a whalelike animal found in an Egyptian desert
   d. an impression of a plant leaf in a rock
   e. all of the above can be considered fossils

7. If the half-life of a radioisotope is 20,000 years, then a sample in which three-quarters of that radioisotope has decayed is _____ years old.
   a. 15,000   b. 26,667   c. 30,000   d. 40,000

8. Did Pangea or Gondwana form first?

9. The Cretaceous ended _____ million years ago.

10. On the geologic time scale, life originated in the _____ .
    a. Archaean
    b. Proterozoic
    c. Phanerozoic
    d. Cambrian

11. Match the terms with the most suitable description.
    ___ fitness
    ___ fossils
    ___ natural selection
    ___ half-life
    ___ catastrophism
    ___ uniformity
    a. measured by reproductive success
    b. geologic change occurs continuously
    c. geologic change occurs in sudden major events
    d. evidence of life in distant past
    e. survival of the fittest
    f. characteristic of a radioisotope

12. Forces of geologic change include _____ (select all that are correct).
    a. erosion
    b. fossilization
    c. volcanic activity
    d. evolution
    e. catastrophism
    f. tectonic plate movement
    g. wind
    h. asteroid impacts
    i. natural selection
    j. mass extinctions

Critical Thinking

1. Radiometric dating does not measure the age of an individual atom. It is a measure of the age of a quantity of atoms—a statistic. As with any statistical measure, its values may deviate around an average (see sampling error, Section 1.8). Imagine that one sample of rock is dated ten different ways. Nine of the tests yield an age close to 225,000 years. One test yields an age of 3.2 million years. Do the nine consistent results imply that the one that deviates is incorrect, or does the one odd result invalidate the nine consistent ones?

2. If you think of geologic time spans as minutes, life's history might be plotted on a clock such as the one shown on the *right*. According to this clock, the most recent epoch started in the last 0.1 second before noon. Where does that put you?

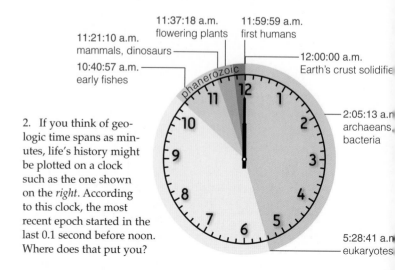

11:37:18 a.m. flowering plants
11:59:59 a.m. first humans
11:21:10 a.m. mammals, dinosaurs
10:40:57 a.m. early fishes
Phanerozoic
12:00:00 a.m. Earth's crust solidifie
2:05:13 a.n archaeans, bacteria
5:28:41 a.n eukaryotes

Chapter 17
Summary

Section 17.1 Populations tend to change along with the selection pressures that are acting on them. Our efforts to control pests such as rats have resulted in directional selection for resistant populations.

Sections 17.2, 17.3 Individuals of a **population** share physical, behavioral, and physiological traits with a heritable basis. All alleles of all their genes form a **gene pool**. New alleles that arise by mutation may be **neutral**, **lethal**, or adaptive.

Microevolution, or changes in the **allele frequencies** of a population, occurs constantly in natural populations by processes of mutation, natural selection, genetic drift, and gene flow. We use deviations from a theoretical **genetic equilibrium** to study how populations evolve.

Sections 17.4–17.6 Natural selection, the process in which environmental pressures result in the differential survival and reproduction of individuals of a population, occurs in recognizable patterns. **Directional selection** shifts the range of variation in traits in one direction. **Stabilizing selection** acts on extreme forms of a trait, so intermediate forms are favored. With **disruptive selection**, intermediate forms of a trait are selected against, and forms at the extremes are favored.

Section 17.7 Sexual dimorphism is an outcome of **sexual selection**, a mode of natural selection in which adaptive traits are those that make their bearers better at securing mates. With **balanced polymorphism**, two or more alleles are maintained at relatively high frequency in a population.

Section 17.8 Random change in allele frequencies—**genetic drift**—can lead to the loss of genetic diversity in a population by causing alleles to become **fixed**. Genetic drift is pronounced in small or **inbreeding** populations. The **founder effect** may occur after an evolutionary **bottleneck**. **Gene flow** counters the effects of mutation, natural selection, and genetic drift.

Section 17.9 The details of **speciation** differ every time it occurs, but **reproductive isolation**, the end of gene flow between populations, is always a part of the process (Table 17.2). The exact moment at which two populations become separate species is often impossible to pinpoint.

Section 17.10 In **allopatric speciation**, a geographic barrier arises and interrupts gene flow between populations. After gene flow ends, genetic divergences that occur independently in the populations can result in separate species.

Section 17.11 Speciation can occur with no barrier to gene flow. In **sympatric speciation**, speciation occurs within the range of the population. Polyploid species of many plants (and a few animals) have originated this way. With **parapatric speciation**, populations in physical contact along a common border speciate.

Section 17.12 Macroevolution refers to large-scale patterns of evolution such as stasis, adaptive radiation, coevolution, and extinction. Major evolutionary novelty typically arises by **exaptation**, in which a lineage uses a structure for a different purpose than its ancestor did. With **stasis**, a lineage changes very little over evolutionary time. A **key innovation** can result in an **adaptive radiation**, or rapid diversification of a lineage into several new species. **Coevolution** occurs when two species act as agents of selection upon one another. A lineage that has become permanently lost from Earth is said to be **extinct**.

Table 17.2 Comparison of Speciation Models

Speciation Model:	Allopatric	Parapatric	Sympatric
Original population(s)			
Initiating event:	physical barrier arises	selection pressures differ	genetic change
Reproductive isolation occurs			
New species arises:	in isolation	in contact along common border	within existing population

Self-Quiz
Answers in Appendix III

1. _____ is the original source of new alleles.
   a. Mutation
   b. Natural selection
   c. Genetic drift
   d. Gene flow
   e. All are original sources of new alleles

2. Evolution can only occur in a population when _____ .
   a. mating is random
   b. there is selection pressure
   c. neither is necessary

3. Match the modes of natural selection with their best descriptions.
   ___ stabilizing a. eliminates extreme forms of a trait
   ___ directional b. eliminates midrange form of a trait
   ___ disruptive   c. shifts allele frequencies in one direction

Resistance to Rodenticides in Wild Rat Populations

Beginning in 1990, rat infestations in northwestern Germany started to intensify despite continuing use of rat poisons. In 2000, Michael H. Kohn and his colleagues tested wild rat populations around Munich. For part of their research, they trapped wild rats in five towns, and tested those rats for resistance to warfarin and the more recently developed poison bromadiolone. The results are shown in **Figure 17.26**.

1. In which of the five towns were most of the rats susceptible to warfarin?

2. Which town had the highest percentage of poison-resistant wild rats?

3. What percentage of rats in Olfen were warfarin resistant?

4. In which town do you think the application of bromadiolone was most intensive?

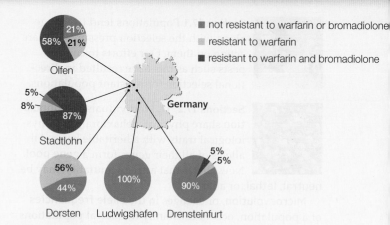

■ not resistant to warfarin or bromadiolone
■ resistant to warfarin
■ resistant to warfarin and bromadiolone

Figure 17.26 Resistance to rat poisons in wild populations of rats in Germany, 2000.

4. Individuals don't evolve, _____ do.

5. Sexual selection, such as occurs when males compete for access to fertile females, frequently influences aspects of body form and can lead to _____ .
   a. male/female differences   c. exaggerated traits
   b. male aggression   d. all of the above

6. The persistence of sickle-cell anemia in a population with a high incidence of malaria is a case of _____ .
   a. bottlenecking   c. natural selection
   b. balanced   d. artificial selection
      polymorphism   e. both b and c

7. _____ tends to keep populations of a species similar to one another.
   a. Genetic drift   c. Mutation
   b. Gene flow   d. Natural selection

8. Which of the following is *not* part of how we define a species?
   a. Its individuals appear different from other species.
   b. It is reproductively isolated from other species.
   c. Its populations can interbreed.
   d. Fertile offspring are produced.

9. Sex in many birds is typically preceded by an elaborate courtship dance. If a male's movements are unrecognized by the female or they do not impress her, she will not mate with him. This is an example of _____ .
   a. reproductive isolation   c. sexual selection
   b. behavioral isolation   d. all of the above

10. The difference between sympatric and parapatric speciation is _____ .
    a. parapatric speciation occurs only in worms
    b. sympatric speciation requires a barrier to gene flow
    c. the extent of habitat overlap between populations
    d. There is no difference between the two modes.

11. A fire devastates all trees in a wide swath of forest. Populations of a species of tree-dwelling frog on either side of the burned area diverge to become separate species. This is an example of _____ .
    a. allopatric speciation   c. sympatric speciation
    b. parapatric speciation   d. adaptive radiation

12. The theory of natural selection does not explain _____ .
    a. genetic drift   d. how mutations arise
    b. the founder effect   e. inheritance
    c. gene flow   f. any of the above

13. Match the evolution concepts.
    ___ gene flow   a. can lead to interdependent species
    ___ sexual   b. changes in a population's allele
        selection   frequencies due to chance alone
    ___ mutation   c. alleles enter or leave a population
    ___ genetic   d. adaptive traits make their bearers
        drift   better at securing mates
    ___ coevolution   e. source of new alleles
    ___ adaptive   f. burst of divergences from one
        radiation   lineage into many

Critical Thinking

1. Rama the cama, a llama–camel hybrid, was born in 1997. The idea was to breed an animal that has the camel's strength and endurance, and the llama's gentle disposition. However, instead of being large, strong, and sweet, Rama is smaller than expected and has a camel's short temper. The breeders plan to mate him with Kamilah, a female cama. Which reproductive isolating mechanism would be at work if it turns out that Kamilah and Rama's offspring are infertile?

2. Two species of antelope, one from Africa, the other from Asia, are put into the same enclosure in a zoo. To the zookeeper's surprise, individuals of the different species begin to mate and produce healthy, hybrid baby antelopes. Explain why a biologist might not view these offspring as evidence that the two species of antelope are in fact one.

3. Some people think that many of our uniquely human traits arose by sexual selection. Over thousands of years, women attracted to charming, witty men perhaps prompted the development of human intellect beyond what was necessary for mere survival. Men attracted to women with juvenile features may have shifted the species as a whole to be less hairy and softer featured than any of our simian relatives. Can you think of a way to test these hypotheses?

Chapter 18

Summary

Section 18.1 The Hawaiian honeycreepers are highly specialized. Their specialization makes them particularly vulnerable to extinction as a result of habitat loss and exotic species introductions that followed human colonization of the Hawaiian Islands.

Section 18.2 Evolutionary biologists reconstruct evolutionary history (**phylogeny**) by comparing physical, behavioral, and biochemical traits, or **characters**, among species. They define a **clade** as a **monophyletic group** that consists of an ancestor in which one or more **derived traits** evolved, together with all of its descendants. **Cladistics** is a method of making hypotheses about the evolutionary history of a group of clades. **Evolutionary tree** diagrams are based on the premise that all organisms are connected by shared ancestry. In a **cladogram**, each line represents a lineage. A lineage branches into two **sister groups** at a node, which represents a shared ancestor.

Section 18.3 Comparative morphology can reveal evolutionary connections among lineages. **Homologous structures** are similar body parts that, by **morphological divergence**, became modified differently in different lineages. Such parts are evidence of a common ancestor. **Analogous structures** are body parts that look alike in different lineages but did not evolve in a common ancestor. By the process of **morphological convergence**, they evolved separately after the lineages diverged.

```
ASFFFICIYLH
ASMFFICLFLH
ASMFFICLFMH
ASFFFICIYLH
ASMFFICLFLH
ASMFLIVVYLH
ASWFFIMLYSH
```
Section 18.4 We can discover and clarify evolutionary relationships through comparisons of nucleic acid and protein sequences, because lineages that diverged recently tend to share more sequences than ones that diverged long ago. Neutral mutations accumulate in DNA at a predictable rate; like the ticks of a **molecular clock**, they help researchers estimate how long ago lineages diverged.

Section 18.5 Master genes that affect development tend to be highly conserved, so similarities in patterns of embryonic development reflect shared ancestry that can be evolutionarily ancient. Mutations in these genes can alter the rate or onset of development.

Section 18.6 Besides offering knowledge about how all species are interrelated, deciphering phylogeny helps us preserve endangered species. Phylogeny research is also used for studying the spread of viruses and other agents of infectious diseases.

Self-Quiz

Answers in Appendix III

1. In cladistics, the only taxon that is always correct as a clade is the _____ .
   a. genus   c. species
   b. family   d. kingdom

2. In evolutionary trees, each branch point represents a(n) _____ .
   a. single lineage   c. point of divergence
   b. extinction   d. adaptive radiation

3. In cladograms, sister groups are _____ .
   a. inbred   c. represented by nodes
   b. the same age   d. members of the same family

4. Through _____ , a body part of an ancestor is modified differently in different lines of descent.
   a. homologous evolution   c. adaptive divergence
   b. morphological divergence   d. morphological convergence

5. Homologous structures among major groups of organisms may differ in _____ .
   a. size   c. function
   b. shape   d. all of the above

6. Neutral mutations are those that do not affect _____ .
   a. amino acid sequence   c. the chances of survival
   b. nucleotide sequence   d. all of the above

7. Mitochondrial DNA sequences are often used in cladistic comparisons of _____ .
   a. different species
   b. individuals of the same species
   c. different taxa

8. Molecular clocks are based on comparisons of the number of _____ mutations between species.
   a. lethal   b. neutral   c. any

9. Cladistics is based on _____ .
   a. reconstructing phylogeny
   b. parsimony analysis of many clades
   c. derived traits

10. By altering steps in the program by which embryos develop, a mutation in a _____ may lead to major differences between adults of related lineages.
    a. derived trait   c. homologous structure
    b. homeotic gene   d. all of the above

11. All of the following data types can be used as evidence of shared ancestry except similarities in _____ .
    a. amino acid sequences   d. embryonic development
    b. DNA sequences   e. form due to convergence
    c. fossil morphologies   f. all are appropriate

12. Match the terms with the most suitable description.
    ___ phylogeny   a. novel trait
    ___ cladogram   b. evolutionary history
    ___ homeotic genes   c. human arm and bird wing
    ___ homologous structures   d. similar across diverse taxa
    ___ molecular clock   e. neutral mutation measure
    ___ analogous structures   f. flight surfaces of insect wing and bird wing
    ___ derived trait   g. sets within sets

Studying Poouli Phylogeny The poouli (*Melamprosops phaeosoma*) was discovered in 1973 by a group of students from the University of Hawaii. Its membership in the honeycreeper clade was—and continues to be—controversial, mainly because its appearance, odor, and behavior are so different from other honeycreepers. It particularly lacks the "old tent" scent characteristic of honeycreepers. A study published in 2001 by Robert Fleischer and his colleagues tried to clarify the relationships of *Melamprosops* to other honeycreepers by comparing bone morphology and DNA sequences. Some of their data, and the resulting cladogram, are shown in **Figure 18.14**.

1. Which two species on the cladogram represent outgroups?

2. According to the cladogram, which species is/are most closely related to the poouli?

3. According to this cladogram, is any honeycreeper species in this study more closely related to the outgroup species than the poouli?

4. Not counting unresolved bases, how many nucleotide differences are there between the poouli and *Paroroeomyza montana* DNA sequences in this example? The *Paroroeomyza montana* and *Loxioides bailleui* sequences?

Figure 18.14 DNA sequence comparison of part of the mitochondrial cytochrome *b* sequence from different honeycreeper species (*below*), and cladogram of honeycreeper phylogeny (*right*) that was constructed using this and other data from the Fleischer study. N indicates an unresolved nucleotide. Differences from the *Melamprosops* sequence are indicated in *red*; gaps in the alignment are dashed.

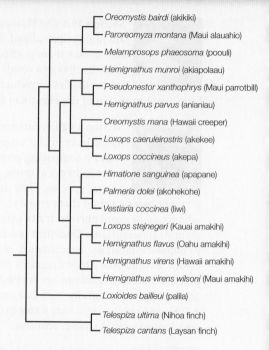

- *Oreomystis bairdi* (akikiki)
- *Paroreomyza montana* (Maui alauahio)
- *Melamprosops phaeosoma* (poouli)
- *Hemignathus munroi* (akiapolaau)
- *Pseudonestor xanthophrys* (Maui parrotbill)
- *Hemignathus parvus* (anianiau)
- *Oreomystis mana* (Hawaii creeper)
- *Loxops caeruleirostris* (akekee)
- *Loxops coccineus* (akepa)
- *Himatione sanguinea* (apapane)
- *Palmeria dolei* (akohekohe)
- *Vestiaria coccinea* (liwi)
- *Loxops stejnegeri* (Kauai amakihi)
- *Hemignathus flavus* (Oahu amakihi)
- *Hemignathus virens* (Hawaii amakihi)
- *Hemignathus virens wilsoni* (Maui amakihi)
- *Loxioides bailleui* (palila)
- *Telespiza ultima* (Nihoa finch)
- *Telespiza cantans* (Laysan finch)

```
Paroroeomyza montana   TCACCAAGACGATCTATTACGCTACACCAGGGAGGATGGCACTCCCACTGGTATATCCACTTGACAG
Loxioides bailleui     TCACTAAGACAGCTTACTACGCCAACTCACGCGAGAGAGCACTCCCGGTGGTATACTCACTTGATAG
Telespiza cantans      TCACCAAGACGGTTCACTATACCACACCAAGTGAGAGAGCACTCNNNNNGGTATACTCACTTTACAG
Hemignathus parvus     TCACCAAGACGACTTATTGTATCAA-CCAAGAGAGANNGCACTCCCAGTGGTAGATTCTCTTGACAG
Hemignathus kauaiensis TCACTAAGACGACTTATTATACCCAACCAAGAGAAAAAGCACCCCCCACGGTAAGTCCTCTTGACCG
Himatione sanguinea    TCGCTTAGACGCCTTATTACGCTAAACCATCACGANNGCACTACTGTTGGTCGATCCTCTTGACAG
Melamprosops phaeosoma TCACTAAGACACCTTCTTATGTTCCATCAAGAGAGANNGCACNNNNNNNGGTATAGCTTCTTGACAG
```

Critical Thinking

1. Species have been traditionally characterized as "primitive" and "advanced." For example, mosses were considered to be primitive, and flowering plants advanced; crocodiles were primitive and mammals were advanced. Why do most biologists of today think it is incorrect to refer to any modern species as primitive?

2. In the late 1800s, a biologist studying animal embryos coined the phrase "ontogeny recapitulates phylogeny," meaning that the physical development of an animal embryo (ontogeny) seemed to retrace the species' evolutionary history (phylogeny). Why would embryonic development retrace evolutionary steps?

3. Synpolydactyly is an abnormality characterized by webbing between partially or completely duplicated fingers or toes (**Figure 18.15**). The same mutations that cause the human phenotype give rise to a similar phenotype in mice. In what family of genes do you think these mutations occur?

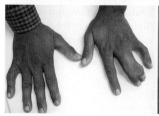
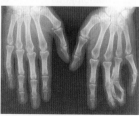

Figure 18.15
Synpolydactyly.

Chapter 19
Summary

Section 19.1 Astrobiology is the study of life's origin and distribution in the universe. The presence of cells in deserts and below Earth's surface suggests life may exist in similar settings on other planets. Mars has some water, but it lacks a protective **ozone layer**, so any Martian life is likely deep in the soil.

Section 19.2 According to the **big bang theory**, the universe formed in an instant 13 to 15 billion years ago (Table 19.1). Earth and other planets formed more than 4 billion years ago. Early in Earth's history, there was little or no oxygen in the air, volcanic eruptions were common, and there was a constant hail of meteorites.

Section 19.3 Laboratory simulations offer indirect evidence that organic compounds could self-assemble in Earth's early atmosphere or in the hot, mineral-rich water around **hydrothermal vents**. Examination of meteorites shows that such compounds might have formed in deep space and reached Earth in meteorites.

Section 19.4 Proteins that speed metabolic reactions might have first formed when amino acids stuck to clay, then bonded under the heat of the sun. Alternatively, the **iron–sulfur world hypothesis** postulates that metabolism began on the surface of rocks near hydrothermal vents.

Membrane-like structures and vesicles form when proteins or lipids are mixed with water. They serve as a model for **protocells**, which may have preceded cells. An RNA world, an interval during which RNA was the genetic material, may have preceded the evolution of DNA-based systems. Discovery of **ribozymes**, RNAs that act as enzymes, lends support to the **RNA world hypothesis**. A subsequent switch from RNA to DNA would have made the genome more stable.

Section 19.5 The first cells evolved when oxygen levels were low, so they probably were anaerobic. An early divergence separated bacteria from the common ancestor of archaea and eukaryotes. Growth of bacterial mats in shallow water formed dome-shaped **stromatolites**. Oxygen released as a by-product of photosynthesis in cyanobacteria changed Earth's atmosphere. The increased oxygen level prevented evolution of new life from nonliving molecules, created a protective ozone layer, and favored cells that carried out aerobic respiration. This ATP-forming metabolic pathway was a key innovation in the evolution of eukaryotic cells. **Biomarkers** and fossils of eukaryotes date back more than 2 billion years.

Sections 19.6, 19.7 Eukaryotes have a composite ancestry. Internal membranes of eukaryotic cells such as a nuclear membrane and endoplasmic reticulum may have evolved by modification of infoldings of the plasma membrane. The **endosymbiont hypothesis** holds that mitochondria evolved from aerobic bacteria and chloroplasts from cyanobacteria.

Protists were the first eukaryotes. Over time, various protist lineages gave rise to plants, fungi, and animals. Evidence from many sources allows us to reconstruct the order of events and make a hypothetical time line for the history of life.

Table 19.1 Major Events in the History of Life

Event	Estimated Time
Universe forms	13 to 15 billion years ago
Our sun forms	5 billion years ago
Earth forms	4.6 billion years ago
First cells	3.2 to 4.3 billion years ago
First eukaryotes	2.8 to 2 billion years ago

Self-Quiz

Answers in Appendix III

1. According to the big bang theory, _____ .
   a. the universe expanded out from a single point
   b. Earth and our sun formed simultaneously
   c. carbon and oxygen were the first elements to form
   d. all of the above

2. An abundance of _____ in the atmosphere would have prevented the spontaneous assembly of organic compounds.

3. Stanley Miller's experiment demonstrated _____ .
   a. the great age of the Earth
   b. that amino acids can assemble spontaneously
   c. that oxygen is necessary for life
   d. that DNA is less stable than RNA

4. The prevalence of iron-sulfide cofactors in living organisms may be evidence that life arose _____ .
   a. in outer space   c. near deep-sea vents
   b. on tidal flats   d. in the upper atmosphere

5. According to one hypothesis, negatively charged clay particles played a role in early _____ .
   a. protein formation   c. photosynthesis
   b. DNA replication   d. oxygen declines

6. An RNA that functions as an enzyme is a(n) _____ .

7. Certain pigments that cells use in photosynthesis may have first helped cells _____ .
   a. capture heat energy at hydrothermal vents
   b. adhere to mineral-rich clays
   c. avoid oxygen
   d. escape potential predators

The Changing Earth Studies of ancient rocks and fossils can reveal changes that have taken place during Earth's existence. **Figure 19.14** shows how asteroid impacts and the composition of the atmosphere are thought to have changed over time. Use this figure and information in the chapter to answer the following questions.

1. Which occurred first, a decline in asteroid impacts, or a rise in the atmospheric level of oxygen (O_2)?

2. How do modern levels of carbon dioxide (CO_2) and O_2 compare to those at the time when the first cells arose?

3. Which is now more abundant, oxygen or carbon dioxide?

4. What do you think accounts for the rise in CO_2 at the far right of the graph?

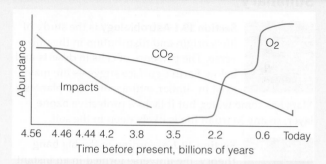

Figure 19.14 How asteroid impacts (*green*), atmospheric carbon dioxide concentration (*pink*), and oxygen concentration (*blue*) changed over geologic time.

8. The evolution of _____ resulted in an increase in the levels of atmospheric oxygen.
   a. DNA-based genomes
   b. endosymbiosis
   c. sexual reproduction
   d. the noncyclic pathway of photosynthesis

9. Mitochondria are thought to be descendants of _____ .
   a. archaea   c. cyanobacteria
   b. aerobic bacteria   d. anaerobic bacteria

10. Infoldings of the plasma membrane into the cytoplasm of some ancestral cells may have evolved into the _____ .
    a. nuclear envelope   c. primary cell wall
    b. ER membranes   d. both a and b

11. By the process of _____ , one cell lives inside another cell and the two become interdependent.

12. A(n) _____ is a dome-shaped structure formed by mats of photosynthetic cells and sediments.

13. The first eukaryotes were _____ .
    a. fungi   c. protists
    b. plants   d. animals

14. Mars is considered a possible candidate for life because it _____ .
    a. has an ozone layer
    b. has water
    c. is about the same size as the Earth
    d. all of the above

15. Chronologically arrange the evolutionary events, with 1 being the earliest and 6 the most recent.
    ___1   a. emergence of the aerobic
    ___2   pathway of photosynthesis
    ___3   b. origin of mitochondria
    ___4   c. origin of protocells
    ___5   d. emergence of the cyclic
    ___6   pathway of photosynthesis
                  e. origin of chloroplasts
                  f. the big bang

Critical Thinking

1. Researchers looking for fossils of the earliest life forms face many hurdles. For example, few sedimentary rocks date back more than 3 billion years. Review what you learned about plate tectonics (Section 16.7). Explain why so few remaining samples of these early rocks remain.

2. Craig Venter has produced what he calls a "synthetic cell." He inserted a laboratory-made copy of the genetic material from bacterial species (*Mycoplasma mycoides*) into a cell of a related species (*M. capricolum*). After this procedure, the synthetic genome took over and the recipient cell acted and replicated like natural *M. mycoides*. By altering the gene sequence inserted into a cell, this technique could be used to create organisms suited to specific tasks. Do you consider an organism created in this manner to be alive? What properties would you assess to make this decision?

Chapter 20
Summary

Section 20.1 AIDS is an **emerging disease** caused by **HIV**. The oldest evidence of this virus comes from central Africa, where chimpanzees are infected by a related virus (SIV). Analysis of viral genes has allowed researchers to trace the spread of HIV.

Section 20.2 A **virus** consists of protein, nucleic acid, and, in some cases, a bit of membrane from a host cell. A virus replicates inside a cell of a specific host type. For example, **bacteriophages** infect bacteria. Some viruses are **pathogens** that cause human disease. Viruses may have evolved before cells, or they may be descended from cells or their components.

Viroids are infectious particles consisting only of a small circle of RNA that does not encode any proteins. They enter plants through a wound and cause disease.

Section 20.3 To replicate, a virus attaches to a host cell and its genetic material enters the cell. Viral genes and enzymes direct host machinery to replicate viral genetic material and make viral proteins. Viral particles self-assemble and are released.

Bacteriophages may multiply by a **lytic pathway**, in which the new viral particles are made fast and released by lysis, or by a **lysogenic pathway**, in which viral DNA becomes part of the host chromosome.

The genetic material of HIV is RNA. The viral enzyme **reverse transcriptase** uses viral RNA as its template to make DNA that the host cell can read.

Section 20.4 Viral diseases may be spread by contact with a viral particle or delivered into the body by a **vector** such as a tick. Most viral diseases such as common colds cause symptoms only briefly. Some viruses persist in the body and reawaken after a latent period. Viral genes mutate and viruses can swap genes in a host individual, a process called **viral reassortment**. An **epidemic** is an outbreak of disease in only a limited region. A **pandemic** is a worldwide outbreak.

Section 20.5 The **prokaryotes** are now known to include two distinct lineages: **bacteria** and **archaea**. Unlike eukaryotes, these lineages do not typically have a nucleus or endomembrane system, and they do not reproduce sexually. **Metagenomics**, the study of DNA in samples drawn directly from the environment, is revealing previously unknown diversity.

Bacteria and archaea are small, abundant, and, as a group, metabolically diverse. Some are aerobic and others cannot tolerate oxygen. **Photoautotrophs** carry out photosynthesis. **Photoheterotrophs** capture light, but they get carbon from organic molecules rather than CO_2. **Chemoautotrophs** build food from CO_2 using energy from inorganic substances. **Chemoheterotrophs** extract both energy and carbon from organic molecules.

Section 20.6 Most bacteria and archaea have cell walls. Many have surface projections such as flagella and **pili**. The chromosome is a circular molecule of DNA that resides in a region of cytoplasm called the **nucleoid**. There may also be one or more **plasmids**, circles of DNA that are separate from the chromosome and carry a few genes. Reproduction occurs by an asexual process called **binary fission**.

Three types of **horizontal gene transfer** move genes between existing cells. **Conjugation** transfers a **plasmid** from one cell to another. Virus-assisted transfer of genes is **transduction**. With **transformation**, cells take up DNA from the environment.

Section 20.7 Bacteria are the most well-known and diverse cells without a nucleus. Many are ecologically important. **Cyanobacteria** produce oxygen as a by-product of photosynthesis. They and other bacteria carry out **nitrogen fixation**, the incorporation of nitrogen from nitrogen gas into ammonia. **Proteobacteria** and **Gram-positive** bacteria are the most highly diverse bacterial lineages. Both include some pathogens. Some Gram-positive bacteria such as those that cause anthrax survive unfavorable conditions as **endospores**. Coiled cells called **spirochetes** and intracellular parasites called **chlamydias** can also cause disease.

Section 20.8 Archaea are the more recently discovered lineage of cells without a nucleus. Comparisons of structure, function, and genetic sequences position them in a separate domain, between eukaryotes and bacteria. Many archaea live in extreme environments. **Halophiles** live in salty waters and **extreme thermophiles** live at very high temperatures. Some archaea are photoheterotrophs with a unique purple protein that captures light energy. Most, including the **methanogens**, are chemoautotrophs. Some archaea live in our bodies, but none are considered pathogens.

Self-Quiz   *Answers in Appendix III*

1. DNA or RNA may be the genetic material of _____ .
   a. a bacterium   b. a viroid   c. a virus   d. an archaeon

2. A viroid consists entirely of _____ .
   a. DNA   b. RNA   c. protein   d. lipids

3. Bacteriophages can kill their host quickly by _____ .
   a. binary fission   c. a lysogenic pathway
   b. a lytic pathway   d. both b and c

4. The genetic material of HIV is _____ .

5. The _____ do not reproduce sexually.
   a. archaea   c. eukaryotes
   b. bacteria   d. both a and b

6. One cell transfers a plasmid to another by _____ .
   a. binary fission   c. conjugation
   b. transformation   d. the lytic pathway

Adapting to Host Defenses Surface proteins called HLAs allow white blood cells to detect HIV particles and fight an infection. In a recent study, scientists tested whether HIV is adapting to this host defense. They did so by looking at the frequency of a specific mutation (I135X) in HIV. This "escape mutation" helps the virus avoid detection by a version of the HLA protein (HLA-B*51) that is common in some regions of the world, but not in others.

Figure 20.20 shows the percentage of HIV-positive people who had HIV with the I135X mutation. Data was collected at medical centers from several parts of the world.

1. What percentage of people with HLA-B*51 in Vancouver had HIV with the escape mutation for this protein?

2. Overall, are people with HLA-B*51 more or less likely than those with other HLAs to have virus with the mutation?

3. Do these results support the hypothesis that the virus is adapting to host defenses?

4. Japan has a high frequency of HLA-B*51; about half the population has it. How might this explain the high frequency of the I135X mutation in Japanese with other HLAs?

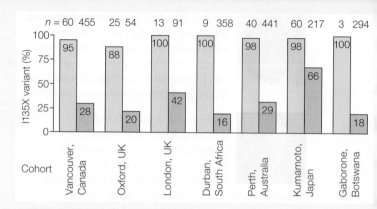

Figure 20.20 Regional variation in the frequency of the I135X escape mutation among HIV-positive people. For each region, *pink* bars represent people whose blood cells have HLA-B*51, and thus cannot detect I135X mutants. *Blue* bars represent people with other versions of the HLA protein. These people have blood cells that can detect and fight HIV even if it has the I135X mutation.

---

7. All _____ are oxygen-releasing photoautotrophs.
   a. spirochetes
   b. chlamydias
   c. cyanobacteria
   d. proteobacteria

8. *E. coli* cells that live in your gut are _____ .
   a. photoautotrophs
   b. photoheterotrophs
   c. chemoautotrophs
   d. chemoheterotrophs

9. _____ are intracellular parasites.
   a. Spirochetes
   b. Chlamydias
   c. Cyanobacteria
   d. Proteobacteria

10. Some Gram-positive bacteria (e.g., *Bacillus anthracis*) survive harsh conditions by forming a(n) _____ .
   a. pilus
   b. heterocyst
   c. endospore
   d. plasmid

11. _____ reproduce by binary fission.
   a. Viruses
   b. Archaea
   c. Bacteria
   d. both b and c

12. A plasmid is a circle of _____ .
   a. RNA   b. DNA   c. either RNA or DNA

13. Which of the following infectious diseases is caused by bacteria?
   a. flu   b. AIDS   c. measles   d. syphilis

14. A worldwide outbreak of a disease is a(n) _____ .

15. Match the terms with their most suitable description.
   ___ methanogen   a. infectious RNA
   ___ nucleoid   b. nonliving infectious particle;
   ___ virus   nucleic acid core, protein coat
   ___ plasmid   c. draws cells together
   ___ extreme   d. releases methane
       halophile   e. region with DNA
   ___ viroid   f. circle of nonchromosomal DNA
   ___ sex pilus   g. rod-shaped cell
   ___ bacillus   h. salt lover

Critical Thinking

1. If a cut or scrape becomes infected, *Staphylococcus aureus* is probably the culprit (*right*). These bacteria often live on the skin and they can cause a problem if they get into a wound. Most "staph" infections can be cured with the antibiotic methicillin. Unfortunately, methicillin-resistant *S. aureus* (MSRA) is on the rise. Antibiotic-resistant staph infections previously occurred mainly in hospitals and nursing homes. Now they are breaking out in schools and health clubs. The bacteria are transmitted by contact with an infected person or something that person has touched, as by sharing towels and razors. The gene conferring methicillin resistance is on a plasmid. Explain why a gene on a plasmid can spread more quickly than a gene that is on the bacterial chromosome.

S. aureus

Wound infected by MRSA

2. The adenoviruses that cause colds do not have a lipid envelope and they tend to remain infectious outside the body for longer than enveloped viruses. "Naked" viruses are also less likely to be rendered harmless by soap and water. Why might possession of an envelope make a virus less hardy?

3. A farmer growing peas can ensure the plants get enough nitrogen by adding a nitrogen fertilizer that contains ammonia, or by applying *Rhizobium* inoculum to seeds or young plants. What are the ecological advantages of encouraging the growth of nitrogen-fixing bacteria rather than using a chemical fertilizer?

4. Scientists on a drilling project in Virginia discovered a new species of bacteria living 3 kilometers (a little less than 2 miles) beneath the soil surface. The temperature here is 75°C (167°F) and the pressure is tremendous. They named the new species *Bacillus infernus*, which means "bacterium from hell." Which of the four possible modes of nutrition could these cells be using?

Chapter 21
Summary

Section 21.1 Protists include many eukaryotic lineages, some only distantly related to one another. *Plasmodium*, the protist that causes malaria, is an example of an important parasitic protist. Malaria has afflicted humans for thousands of years. The parasite alters traits of its human and mosquito hosts in ways that enhance its survival and reproduction.

Section 21.2 Most protists live as single cells, but some are colonial or multicellular. There are autotropic and heterotrophic groups. Life cycles vary: Haploid cells dominate some; diploid cells dominate others; and still others have both haploid and diploid multicelled bodies (an **alternation of generations**).

Section 21.3 Flagellated protozoans are single-celled and mostly or entirely heterotrophic. They have no cell wall, but a **pellicle** gives them a distinct shape. Diplomonads and parabasalids are anaerobic and have **hydrogenosomes** instead of mitochondria. Both include species that infect humans. Like many protists, they usually reproduce asexually by mitosis.

Most **euglenoids** live in fresh water; their **contractile vacuole** rids them of excess water. Some have chloroplasts derived from a green alga.

Trypanosomes are parasites with a single giant mitochondrion and an undulating membrane.

Section 21.4 Foraminifera and **radiolarians** are single-celled heterotrophs with a secreted shell. Foraminifera tend to live on the seafloor, and radiolarians drift as **plankton**. Remains of both groups accumulate in abundance on the seafloor and have been transformed over time into limestone, chalk, and chert.

Sections 21.5–21.7 Alveolates are single cells characterized by tiny sacs (alveoli) beneath the plasma membrane. **Ciliates** are aquatic heterotrophs with many cilia. They reproduce asexually and engage in a process by which they exchange micronuclei.

Dinoflagellates are aquatic heterotrophs and autotrophs with a cellulose covering. Some are capable of **bioluminescence**; they convert ATP energy to light energy. In nutrient-enriched water, dinoflagellates and other photosynthetic protists undergo population explosions known as **algal blooms**.

Apicomplexans such as the organisms that cause malaria and toxoplasmosis are parasites that spend part of their time inside host cells.

Section 21.8 Stramenopiles are named for a flagellum with hairlike filaments. **Water molds** are heterotrophs that grow as a mesh of absorptive filaments. Some are important as plant pathogens.

Diatoms are photosynthetic single cells with a two-part silica shell. Like brown algae, diatoms contain the pigment fucoxanthin. Deposits of ancient diatoms are mined as diatomaceous earth.

All **brown algae** are multicelled. They include tiny strands and giant kelps, which are the largest protists. Algins extracted from brown algae are used as thickeners in foods and other products.

Section 21.9 Most **red algae** are multicelled. They can survive in deeper water than most photoautotrophs because their chloroplasts contain phycobilins. Red algae are the source of carrageenan, agar, and the nori used to wrap sushi.

Section 21.10 Green algae are mostly aquatic single-celled or multicelled autotrophs. Like plants, they have chlorophylls *a* and *b*, and they store excess sugars as starch. Most are **chlorophytes**. **Charophyte algae** include the closest relatives of plants.

Section 21.11 Amoebozoans are heterotrophic cells that do not have a cell wall or pellicle. They move and feed by forming pseudopods. **Amoebas** live in aquatic habitats and the animal gut. There are two lineages of slime molds: **plasmodial slime molds**, which feed as a multinucleated mass, and **cellular slime molds**, which feed as individual amoeba-like cells. When food becomes scarce, cellular slime molds aggregate and move about as a cohesive unit. Both types of slime molds form fruiting bodies that disperse resting spores.

Choanoflagellates are the closest protist relatives of animals. Cells have a flagellum surrounded by microvilli that filter food from the water. They may live on their own or in colonies.

Self-Quiz

Answers in Appendix III

1. Multicelled haploid and diploid bodies form in protists that have a(n) _____ .
   a. haploid-dominant life cycle
   b. diploid-dominant life cycle
   c. alternation of generations

2. Diplomonads and parabasalids often live in anaerobic habitats and their _____ produce ATP aerobically.
   a. trypanosomes   c. mitochondria
   b. hydrogenosomes   d. flagella

3. Radiolarians and diatoms have a shell of _____ .
   a. cellulose   c. calcium carbonate
   b. silica   d. chitin

4. Which of the following might you find in seawater?
   a. an apicomplexan   c. a dinoflagellate
   b. a cellular slime mold d. a euglenoid

5. Diatoms are most closely related to the _____ .
   a. dinoflagellates   c. green algae
   b. water molds   d. red algae

6. Chloroplasts of green algae evolved from _____ .

Summoning Mosquitoes Parasites sometimes alter their host's behavior in a way that increases their chances of transmission to another host.

Dr. Jacob Koella and his associates hypothesized that *Plasmodium* might benefit by making its human host more attractive to hungry mosquitoes when gametocytes are available in the host's blood. Gametocytes taken up by the mosquito will mature into gametes and mate inside its gut.

To test their hypothesis, the researchers recorded the response of mosquitoes to the odor of *Plasmodium*-infected children and uninfected children over the course of 12 trials on 12 separate days. **Figure 21.26** shows their results.

1. On average, which group of children was most attractive to mosquitoes?

2. Which group of children averaged the fewest mosquitoes attracted?

3. What percentage of the total number of mosquitoes were attracted to the most attractive group?

4. Did the data support Dr. Koella's hypothesis?

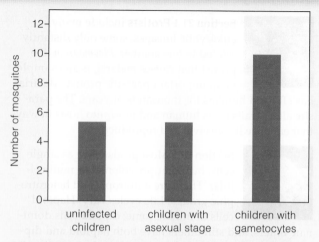

Figure 21. 26 Number of mosquitoes (out of 100) attracted to uninfected children, children harboring the asexual stage of *Plasmodium*, and children with gametocytes in their blood. The bars show the average number of mosquitoes attracted to that category of child over the course of 12 separate trials.

---

7. Green algae are most closely related to _____ .
   a. brown algae   c. red algae
   b. diatoms   d. dinoflagellates

8. To reproduce sexually, two ciliates join and _____ .
   a. exchange nuclei   c. emit light
   b. deter predators   d. infect cells

9. Which species does not cause human disease?
   a. *Toxoplasma gondii*   c. *Dictyostelium discoideum*
   b. *Entamoeba histolytica*   d. *Trichomonas vaginalis*

10. _____ is produced from red algae.
    a. Diatomaceous earth   c. Carrageenan
    b. Chert   d. both b and c

11. The presence of a contractile vacuole is evidence that a protist _____ .
    a. lives in fresh water
    b. reproduces only by asexual means
    c. evolved chloroplasts by secondary endosymbiosis
    d. is a heterotroph

12. The choanoflagellates are considered the sister group to, or closest living relatives of, _____ .
    a. plants   c. animals
    b. fungi   d. dinoflagellates

13. A bioluminescent dinoflagellate _____ .
    a. has a clear silica shell
    b. converts ATP energy to light energy
    c. produces ATP by an anaerobic mechanism
    d. is a multicelled heterotroph

14. Match each term with its most suitable description.
    ___ limestone   a. extracted from kelps
    ___ chert   b. dried red alga
    ___ algin   c. remains of radiolarians
    ___ nori   d. extracted from red algae
    ___ agar   e. remains of foraminifera

15. Match each term with its most suitable description.
    ___ diplomonad   a. protist population explosion
    ___ apicomplexan   b. silica-shelled producer
    ___ algal bloom   c. cells with "collared" flagellum
    ___ diatom   d. no mitochondria, anaerobic
    ___ brown alga   e. closest relative of land plants
    ___ red alga   f. multicelled, with fucoxanthin
    ___ green alga   g. agent of malaria
    ___ choanoflagellate h. deep dweller with phycobilins

Critical Thinking

1. Suppose you vacation in a developing country where sanitation is poor. Having read about parasitic flagellates in water and damp soil, what would you consider safe to drink? What foods might be best to avoid, and which food preparation methods might make them safe to eat?

2. Which groups of protists would you be most likely to find as fossils? Why?

3. Diatoms are the main producers in polar seas. Like other producers, they require dissolved nitrogen and phosphorus to grow. Unlike most producers, their reproduction can be limited by the lack of dissolved silica. Explain why a lack of silica would interfere with diatom reproduction.

4. The parasite *Toxoplasma gondii* is known to alter rodent behavior. This parasite also infects humans, and some researchers suspect that infection may have psychological effects in our species as well. People with the mental illness schizophrenia are more likely than unaffected people to be infected by *T. gondii*. Can you suggest a possible explanation for this finding, other than *T. gondii* increasing a person's risk of becoming schizophrenic?

Chapter 22
Summary

Section 22.1 Forests are reservoirs of plant diversity and play essential ecological roles. However, pressure to log forests is high, especially in tropical regions. The actions of inspired individuals are helping to stem the tide of tropical deforestation and promote awareness of the importance of forests.

Sections 22.2, 22.3 Embryophytes, more commonly known as land **plants** are close relatives of charophyte green algae. Nearly all are photoautotrophs. Life cycles have changed as new lineages evolved. A **gametophyte** dominates the life cycles of the oldest lineages, the **bryophytes**. A **sporophyte** dominates the life cycle of **vascular plants**.

Features that contributed to success on land include **cuticle** and **stomata** that minimize water loss, **xylem** and **phloem** (**vascular tissues**), and **lignin** in cell walls. Evolution of **pollen grains** allowed seed plants to reproduce without standing water. **Seed plants** protect their embryo sporophytes in **seeds**.

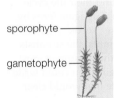

Section 22.4 Three lineages (mosses, liverworts, and hornworts) are collectively referred to as bryophytes. Mosses are the most diverse bryophytes and the main plants in **peat bogs**.

Bryophytes are nonvascular plants (no xylem or phloem). The sporophyte, which produces spores in a **sporangium**, is attached to and nourished by the gametophyte. **Rhizoids** attach a gametophyte to soil or another surface. The gametophyte produces gametes in **gametangia** in or on its surface. Sperm are flagellated and swim to eggs through a film of water.

Section 22.5 Club mosses belong to one lineage of seedless vascular plants; horsetails, whisk ferns, and true ferns belong to another. In both lineages, the sporophyte dominates the life cycle. Roots and aboveground stems grow from **rhizomes**. Spore-bearing structures include the **strobili** of horsetails and the **sori** of ferns. Many ferns live as **epiphytes** attached to another plant. Sperm swim through water to reach eggs.

Section 22.6 The earliest vascular plants were tiny and had a simple branching pattern. By the Carboniferous, swamp forests were dominated by giant lycophytes. The remains of these forests became **coal**.

Seed plants evolved during the Carboniferous. They do not release spores. Their sporophytes have **pollen sacs**, where **microspores** form and develop into pollen grains (male gametophytes). **Megaspores** form inside **ovules** and develop into female gametophytes.

Pollination unites the egg and sperm of a seed plant. A seed is a mature ovule. It includes nutritive tissue and a tough seed coat that protects the embryo sporophyte inside the seed from harsh conditions.

Section 22.7 Gymnosperms include the **conifers** and the lesser known **cycads**, **ginkgos**, and **gnetophytes**. Most conifers are evergreen trees and the group is an important source of lumber. Conifers produce soft, pollen-bearing cones as well as woody, ovulate cones. They are wind pollinated.

Sections 22.8, 22.9 Angiosperms are the most diverse plants. They alone produce **flowers**, which are modified shoots. The **stamens** of a flower produce pollen. An **ovary** at the base of a **carpel** holds one or more ovules. After pollination and **double fertilization**, the flower's ovary becomes a **fruit** that contains one or more seeds. A flowering plant seed includes an embryo sporophyte and **endosperm**, a nutritious tissue.

Factors that contributed to angiosperm success include a short life cycle, **coevolution** with **pollinators**, and a variety of mechanisms for dispersing fruits.

As the dominant plants in most land habitats, flowering plants are ecologically important, as well as essential to human existence. The two major lineages are **monocots** and **eudicots**. The starch-rich endosperm of monocot seeds such as wheat and rice makes them staples of human diets throughout the world. Angiosperms also supply us with vegetables, spices, fiber, furniture, oils, medicines, and mood-altering drugs. Many of the useful plant compounds are **secondary metabolites**, compounds that probably help defend the plant against predation.

Self-Quiz

Answers in Appendix III

1. The first land plants were _____ .
   a. gnetophytes   c. bryophytes
   b. gymnosperms   d. lycophytes

2. Lignin is not found in stems of _____ .
   a. mosses   b. ferns   c. monocots   d. a and b

3. A waxy cuticle helps land plants _____ .
   a. conserve water   c. reproduce
   b. take up carbon dioxide   d. stand upright

4. True or false? Ferns produce seeds inside sori.

5. _____ attach mosses to soil.
   a. Rhizoids   c. Roots
   b. Rhizomes   d. Strobili

6. Bryophytes alone have a relatively large _____ and an attached, dependent _____ .
   a. sporophyte; gametophyte
   b. gametophyte; sporophyte

7. Club mosses, horsetails, and ferns are _____ plants.
   a. multicelled aquatic   c. seedless vascular
   b. nonvascular seed   d. seed-bearing vascular

In the Section 22.4 diagram: sporophyte, gametophyte

A Global View of Deforestation The United Nations Food and Agriculture Organization (UNFAO) recognizes the importance of forests to human populations and keeps track of forest abundance. **Figure 22.28** provides UNFAO data on the amount of forested land in various regions and globally for 1990, 2000, and 2010.

1. How many hectares of forested land were there in North America in 2000?

2. In which region(s) did the amount of forested land increase between 1990 and 2010?

3. Where was the most forest lost between 1990 and 2000? Between 2000 and 2010?

4. In 2002, China embarked on an ambitious campaign to reverse deforestation by planting trees. Do you see evidence that this campaign is successful?

Region	Forested Area (in millions of hectares)		
	1990	2000	2010
Africa	750	709	674
Asia	576	570	592
Europe	989	998	1,005
Oceania	199	198	191
North America	676	677	678
Central America	26	22	19
South America	946	882	864
World total	4,168	4,061	4,033

Figure 22.28 Changes in forested area in selected regions and globally. One hectare is 2.47 acres. The full report on the world's forests is online at www.fao.org/forestry/en/.

8. Coal consists primarily of compressed remains of the _____ that dominated Carboniferous swamp forests.
   a. seedless vascular plants
   b. conifers
   c. flowering plants
   d. hornworts

9. The sperm of _____ swim to eggs.
   a. mosses
   b. ferns
   c. conifers
   d. a and b

10. A seed is a(n) _____ .
    a. female gametophyte
    b. mature ovule
    c. mature pollen tube
    d. immature microspore

11. True or false? Only seed plants produce pollen.

12. Which angiosperm lineage includes the most species?
    a. magnoliids
    b. eudicots
    c. monocots
    d. water lilies

13. Match the terms appropriately.
    ___ bryophyte
    ___ seedless vascular plant
    ___ gymnosperm
    ___ angiosperm
    a. has seeds, but no fruits
    b. has flowers and fruits
    c. has xylem and phloem, but no pollen
    d. gametophyte dominates

14. Match the terms appropriately.
    ___ ovule
    ___ cuticle
    ___ gametophyte
    ___ sporophyte
    ___ fruit
    ___ endosperm
    ___ rhizome
    ___ sorus
    a. gamete-producing body
    b. spore-producing body
    c. where eggs form
    d. underground stem
    e. mature ovary
    f. nutritive tissue in seed
    g. where fern spores form
    h. waxy layer

15. Place these groups in order of their appearance with the oldest lineage first and the most recently evolved last.
    1 ___
    2 ___
    3 ___
    4 ___
    5 ___
    a. conifers
    b. cycads
    c. eudicots
    d. mosses
    e. ferns

Critical Thinking

1. Early botanists admired ferns but found their life cycle perplexing. In the 1700s, they learned to propagate ferns by sowing what appeared to be tiny dustlike "seeds" from the undersides of fronds. Despite many attempts, the scientists could not find the pollen source, which they assumed must stimulate these "seeds" to develop. Imagine you could write to one of these botanists. Compose a note that would clear up their confusion.

2. With the exception of the bryophytes, the dominant stage in land plants is the diploid sporophyte. By one hypothesis, diploid dominance was favored because it allowed a greater level of genetic diversity. Suppose that a recessive mutation arises. It is mildly disadvantageous now, but it will be useful in some future environment. Explain why such a mutation would be more likely to persist in a plant with a sporophyte-dominated life cycle, such as a fern, than in a moss.

3. A pine tree produces far more pollen than an apple or cherry tree produces. Explain why a fruit tree does not need to make as much pollen as a pine to ensure pollination.

4. Compared to vascular plants, bryophytes are more easily harmed by environmental contaminants and acid rain. Thus the diversity of these plants is sometimes used to assess pollution levels. What is it about the structure or function of a moss that would make it more susceptible to damage from air pollution than a horsetail or a cactus?

5. Today, the tallest mosses reach a maximum height of 20 centimeters (8 inches) or so. So far as we know from fossils, there were no giants among their ancestors. Lignin and vascular tissue first evolved in relatives of club moss, and some extinct species stood 40 meters (130 feet) high. Among modern seed plants, *Sequoia* (a gymnosperm) and *Eucalyptus* (an angiosperm) can be more than 100 meters (330 feet) high. Explain how the evolution of vascular tissues and lignin would have allowed a dramatic increase in plant height. How might being tall give one plant species a competitive advantage over another?

Chapter 23
Summary

Section 23.1 Fungi disperse by releasing spores. Winds can distribute spores far from their point of origin. Windblown dispersal of a new strain of wheat stem rust fungus threatens global food supplies.

Section 23.2 Fungi are heterotrophs that secrete digestive enzymes on organic matter and absorb released nutrients. Most feed on organic remains (are **saprobes**), but some live in or on other organisms.

Fungi are more closely related to animals than to plants. They include single-celled yeasts and multicelled species. In the multicelled species, spores germinate and give rise to filaments called **hyphae**. The filaments typically grow as an extensive mesh called a **mycelium**. Depending on the group, a mycelium may be haploid or **dikaryotic** (*n+n*).

Section 23.3 Chytrids are an ancient group of fungi and the only fungi that produce flagellated spores. A chytrid that causes a **mycosis** threatens amphibian populations worldwide.

Section 23.4 The hyphae of **zygote fungi**, which include common molds, are continuous tubes with no cross-walls (septae). A thick-walled, diploid zygospore forms during sexual reproduction. Meiosis of cells inside the zygospore produces haploid spores that germinate and produce a haploid mycelium. Mycelia also produce asexual spores.

Microsporidia are relatives of zygote fungi and live inside animal cells, sometimes causing disease. **Glomeromycetes**, also relatives of zygote fungi, partner with plant roots in a relationship called a **mycorrhiza.**

Section 23.5 Sac fungi are the most diverse group of fungi. They include single-celled yeasts and multicelled species that have hyphae with cross-walls. Sac fungi usually grow as haploid hyphae and produce asexual spores called conidia. Sexual spores are produced in asci. In multicelled species, these saclike structures form on an ascocarp consisting of intertwined haploid and dikaryotic hyphae. Sac fungi are economically important in production of food and medicines.

Section 23.6 The mostly multicelled **club fungi** have hyphae with cross-walls. This group produces the largest and most complex fruiting bodies. Only club fungi can break down the lignin in wood, so they are important decomposers and pathogens in forests.

Typically, a dikaryotic mycelium dominates the life cycle. It grows by mitosis and, in some species, extends through a vast volume of soil. When conditions favor reproduction, a basidiocarp, also made up of dikaryotic hyphae, develops. Haploid spores form by meiosis at the tips of club-shaped cells in the basidiocarp.

A mushroom is a familiar basidiocarp, but club fungus basidiocarps come in a wide variety of shapes and sizes. Some produce toxins that can sicken or kill people.

Section 23.7 Many fungi spend their life closely associated with another species.

A **lichen** is a composite organism that usually consists of a sac fungus and one or more photoautotrophs, such as green algae or cyanobacteria. The fungus, which makes up the bulk of the lichen, obtains nutrients from its photosynthetic partner. Lichens fix nitrogen, and they contribute to the breakdown of rocks to soil.

In a mycorrhiza, hyphae surround or penetrate a root and supplement the plant's surface area for absorbing water and minerals. The fungus shares some absorbed mineral ions with the plant and gets sugars in return,

making the relationship a **mutualism**.

Both plants and animals can be infected by fungi. In humans, most fungal infections occur at a body surface such as the skin, mouth, or vaginal lining. Such infections seldom pose a severe threat unless the immune system is impaired.

Self-Quiz

Answers in Appendix III

1. All fungi _____ .
   a. are multicelled
   b. form flagellated spores
   c. are heterotrophs
   d. all of the above

2. Most fungi obtain nutrients from _____ .
   a. nonliving organic matter
   b. living plants
   c. living animals
   d. photosynthesis

3. In _____ , a hypha has no cross-walls.
   a. zygote fungi
   b. sac fungi
   c. club fungi
   d. all of the above

4. The yeasts whose fermentation reactions produce the carbon dioxide that makes bread rise are a type of _____ .
   a. chytrid
   b. zygote fungus
   c. sac fungus
   d. club fungus

5. In most _____ , an extensive dikaryotic mycelium is the longest-lived phase of the life cycle.
   a. chytrids
   b. zygote fungi
   c. sac fungi
   d. club fungi

6. The mycelium of a multicelled fungus is a mesh of filaments, each called a _____ .
   a. septa
   b. hypha
   c. spore

7. A mushroom is _____ .
   a. the digestive organ of a club fungus
   b. the only part of the fungal body made of hyphae
   c. a reproductive structure that releases sexual spores
   d. the only diploid phase in the club fungus life cycle

8. Spores released from a mushroom's gills are _____ .
   a. diploid
   b. triploid
   c. dikaryotic
   d. haploid

Fighting a Forest Fungus The honey mushroom, *Armillaria ostoyae*, acts as a parasite of living trees because it draws nutrients from them. If the tree dies, the fungus continues to dine on its remains. Hyphae grow out from the roots of infected trees and roots of dead stumps. If these hyphae contact roots of a healthy tree, they can invade and cause a new infection.

Canadian forest pathologists hypothesized that removing stumps after logging could help prevent tree deaths. To test this hypothesis, they carried out an experiment. In half of a forest they removed stumps after logging. In a control area, they left stumps behind. For more than 20 years, they recorded tree deaths and whether *A. ostoyae* caused them. **Figure 23.18** shows the results.

1. Which tree species was most often killed by *A. ostoyae* in control forests? Which was least affected by the fungus?

2. For the most affected species, what percentage of deaths did *A. ostoyae* cause in control and in experimental forests?

3. Looking at the overall results, do the data support the hypothesis? Does stump removal reduce effects of *A. ostoyae*?

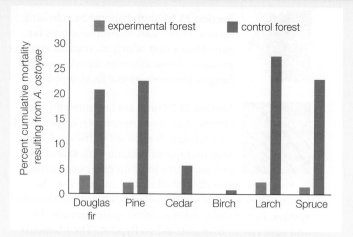

Figure 23.18 Results of a long-term study of how logging practices affect tree deaths caused by the fungus *A. ostoyae*. In the experimental portion of the forest, whole trees—including stumps—were removed (*brown* bars). The control portion of the forest was logged conventionally, with stumps left behind (*blue* bars).

---

9. _____ are fungi that produce flagellated spores.
   a. Chytrids
   b. Sac fungi
   c. Zygote fungi
   d. Club fungi

10. Nitrogen-fixing cyanobacteria often interact with a fungus as a _____ .
   a. mycelium
   b. lichen
   c. mycorrhiza
   d. mycosis

11. _____ are mycorrhizal fungi with hyphae that grow into a root cell and branch inside it.
   a. Glomeromycetes
   b. Chytrids
   c. Zygote fungi
   d. Club fungi

12. The _____ are all intracellular parasites that penetrate a host cell with a polar tube.
   a. microsporidia
   b. mycorrhizae
   c. chytrids
   d. sac fungi

13. Chestnut blight _____ .
   a. altered the species composition of eastern forests
   b. was caused by an introduced fungal pathogen
   c. is a mycosis
   d. all of the above

14. Ergotism _____ .
   a. is a mutually beneficial relationship between species
   b. is a type of food poisoning
   c. affects only those with weak immune systems
   d. affects mainly the surface of the skin

15. Match the terms appropriately.
   ___ hypha
   ___ chitin
   ___ chytrid
   ___ sac fungus
   ___ club fungus
   ___ lichen
   ___ mycorrhiza

   a. produces flagellated spores
   b. component of fungal cell walls
   c. partnership between a fungus and one or more photoautotrophs
   d. filament of a mycelium
   e. fungus–plant partnership
   f. forms sexual spores in an ascus
   g. can break down lignin

Critical Thinking

1. Researchers working in a Brazilian rain forest recently found eight species of bioluminescent mushrooms at a single site. The mushrooms continually emit a faint glow that, although undetectable in daylight, makes them visible at night. Suggest a mechanism by which glowing in the dark could benefit a mushroom. Why do you think so many species with this unusual trait live in the same region?

2. Inhaling spores of *Aspergillus fumigatus* can trigger asthma attacks and may cause severe disease in people whose immune system is impaired. Until recently, scientists did not think this fungus reproduced sexually. New evidence that sexual reproduction does occur suggests that it will be harder to develop drugs to fight infections. Explain why.

3. Health professionals refer to fungal skin diseases as "tineas" and name them according to the region affected. As **Table 23.2** shows, fungi thrive on most body surfaces. Fungal skin diseases are persistent, in part because fungi can penetrate deeper layers of skin than can ointments and creams. There are fewer antifungal drugs than antibacterial ones, and antifungals often have more severe side effects. Reflect on the evolutionary relationships among bacteria, fungi, and humans. Why it is harder to fight fungi than bacteria?

Table 23.2 Fungal Diseases of Skin

Disease	Infected Body Parts
Tinea corporis (ringworm)	Trunk, limbs
Tinea pedis (athlete's foot)	Feet, toes
Tinea capitis	Scalp, eyebrows, eyelashes
Tinea cruris (jock itch)	Groin, perianal area
Tinea barbae	Bearded areas
Tinea unguium	Toenails, fingernails

Chapter 24
Summary

Section 24.1 Invertebrates (with no backbone) are the most diverse animal group. Vertebrates evolved from an invertebrate ancestor. As lineages evolved, genes that were present in ancestral lineages mutated and their products took on new functions.

Sections 24.2, 24.3 Animals are multicelled heterotrophs with unwalled cells. Table 24.1 summarizes the traits of groups covered in this chapter. Some have no body symmetry; others have **radial symmetry**, like a wheel. Most animals have **bilateral symmetry** and **cephalization**, a concentration of nerves and sensory structures at the head end. There are two lineages of bilateral animals, **protostomes** and **deuterostomes**. Both develop from a three-layered embryo. Both also digest food inside a **gut**, which may be saclike or tubular. The gut is usually located inside a fluid-filled cavity (a **coelom** or a **pseudocoelom**).

Animals most likely evolved from a colonial species similar to choanoflagellates, a type of protist. The oldest animal fossils (of Ediacarans) date back about 600 million years. A great adaptive radiation during the Cambrian gave rise to most modern lineages.

Section 24.4 Sponges are asymmetrical and do not have tissues. They filter food from water and are **hermaphrodites**: each makes eggs and sperm. Adults are **sessile animals**, but the ciliated **larvae** swim.

Section 24.5 Cnidarians have two radially symmetrical body forms: **medusa** and **polyp**. Both have tentacles with **cnidocytes** that help them catch prey. Both also have two tissues with a jellylike layer between them that functions as a **hydrostatic skeleton**. A **nerve net** controls movements, and a **gastrovascular cavity** functions in both respiration and digestion.

Section 24.6 Flatworms are bilateral acoelomate worms with organ systems. They include free-living species and parasitic tapeworms and flukes. **Nerve cords** connect to **ganglia** in the head that serve as a control center. A **pharynx** on their lower surface leads to a gastrovascular cavity.

Section 24.7 Annelids are segmented worms and leeches. Circulatory, digestive, solute-regulating, and nervous systems extend through coelomic chambers. **Nephridia** regulate body fluid composition. Like mollusks, annelids have a **trochophore larva**.

Section 24.8 Mollusks are a highly diverse group of animals with a **mantle**, an extension of the body wall. Most have **gills** inside the mantle cavity and feed using a food-scraping **radula**. Subgroups include chitons, gastropods, bivalves, and cephalopods.

Section 24.9 Rotifers and **tardigrades** are tiny animals of damp or aquatic habitats. Both groups can dry out and survive long periods of adverse conditions.

Section 24.10 The **roundworms** (nematodes) have an unsegmented body, a cuticle that is molted, a complete gut, and a false coelom. Some are parasites of humans.

Sections 24.11–24.15 Arthropods, the largest phylum of animals, have a jointed **exoskeleton**. Most have one or more pairs of sensory **antennae**. Development often includes **metamorphosis**, a change in body form between juveniles and adults.

Chelicerates include marine horseshoe crabs and the eight-legged, land-dwelling **arachnids**. **Crustaceans** have an exoskeleton hardened with calcium, and most are marine. **Myriapods** are centipedes and millipedes. **Insects**, the most diverse arthropods, include the only

Animal Phylum	Representative Groups	Living Species	Organization	Symmetry	Digestion	Body Circulation
Porifera	Barrel sponges, encrusting sponges	8,000	Connected cells	None	Intracellular	Diffusion
Cnidaria	Sea anemones, jellyfishes, corals	11,000	2 tissue layers	Radial	Saclike gut	Diffusion
Platyhelminthes	Planarians, tapeworms, flukes	15,000	2 tissue layers	Bilateral	Saclike gut	Diffusion
Annelida	Polychaetes, earthworms, leeches	15,000	3 tissue layers	Bilateral	Complete gut	Closed system
Mollusca	Snails, slugs, clams, octopuses	110,000	3 tissue layers	Bilateral	Complete gut	Open in most, closed in cephalopods
Rotifera	Rotifers	2,150	3 tissue layers	Bilateral	Complete gut	Diffusion
Tardigrada	Water bears	950	3 tissue layers	Bilateral	Complete gut	Diffusion
Nematoda	Pinworms, hookworms	20,000	3 tissue layers	Bilateral	Complete gut	Closed system
Arthropoda	Spiders, crabs, insects	>1,000,000	3 tissue layers	Bilateral	Complete gut	Open system
Echinodermata	Sea stars, sea urchins, sea cucumbers	6,000	3 tissue layers	Larvae bilateral; adults radial	Complete gut	Open system

Table 24.1 Comparison of Invertebrate Groups Surveyed in This Chapter

Sustainable Use of Horseshoe Crabs Eggs of horseshoe crabs are the main food of some migratory shorebirds. People have traditionally used horseshoe crabs for bait, but they now serve an additional purpose. An extract of horseshoe crab blood is used to test the safety of injectable drugs. To keep horseshoe crab populations stable, blood is extracted from captured animals, which are then returned to the wild. Concerns about the survival of animals after bleeding led researchers to do an experiment. They compared survival of animals captured and maintained in a tank with that of animals captured, bled, and kept in a similar tank. **Figure 24.40** shows the results.

1. In which trial did the most control crabs die? In which did the most bled crabs die?

2. Looking at the overall results, how did the mortality of the two groups differ?

3. Based on these results, would you conclude that bleeding harms horseshoe crabs more than capture alone does?

	Control Animals		Bled Animals	
Trial	Number of crabs	Number that died	Number of crabs	Number that died
1	10	0	10	0
2	10	0	10	3
3	30	0	30	0
4	30	0	30	0
5	30	1	30	6
6	30	0	30	0
7	30	0	30	2
8	30	0	30	5
Total	200	1	200	16

Figure 24.40 Mortality of young male horseshoe crabs kept in tanks during the 2 weeks after their capture. Blood was taken from half the animals on the day of their capture. Control animals were handled, but not bled. This procedure was repeated 8 times with different sets of horseshoe crabs.

---

winged invertebrates. Tracheal tubes and **Malpighian tubules** adapt them to life on land. Insects pollinate plants, dispose of wastes, and serve as food, but some eat crops or transmit disease.

Section 24.16 Echinoderms are deuterostomes. Their skin is hardened with bits of calcium carbonate. A **water–vascular system** with tube feet helps most glide about. Adults are radial with some bilateral features, and the larvae are bilateral.

Self-Quiz

Answers in Appendix III

1. True or false? Animal cells do not have walls.

2. A body cavity fully lined with tissue derived from mesoderm is a _____ .

3. Flatworms, annelids, and roundworms are all _____ .
   a. protostomes
   b. deuterostomes

4. A _____ has an asymmetrical body and a ciliated larva.
   a. sponge
   b. roundworm
   c. cnidarian
   d. flatworm

5. Cnidarians alone have _____ .
   a. a hydrostatic skeleton
   b. ciliated larvae
   c. nematocysts
   d. Malpighian tubules

6. Earthworms are most closely related to _____ .
   a. tapeworms
   b. roundworms
   c. flatworms
   d. leeches

7. Earthworm nephridia have the same role as _____ .
   a. gemmules of sponges
   b. mandibles of insects
   c. flame cells of planarians
   d. tube feet of echinoderms

8. Which invertebrate phylum includes the most species?
   a. mollusks
   b. roundworms
   c. arthropods
   d. flatworms

9. How many tissue layers does a jellyfish embryo have?

10. A slug is a terrestrial _____ .
    a. arthropod
    b. gastropod
    c. cephalopod
    d. copepod

11. A barnacle is a shelled _____ .
    a. gastropod
    b. cephalopod
    c. crustacean
    d. copepod

12. _____ include the only winged invertebrates.
    a. Cnidarians
    b. Echinoderms
    c. Choanoflagellates
    d. Arthropods

13. The _____ and _____ have similar larvae.
    a. cnidarians/arthropods
    b. echinoderms/rotifers
    c. annelids/mollusks
    d. flatworms/roundworms

14. Match the organisms with their descriptions.
    ___ mollusks   a. complete gut, pseudocoelom
    ___ echinoderms   b. nematocyst producers
    ___ sponges   c. simplest organ systems
    ___ cnidarians   d. no tissues, filters out food
    ___ flatworms   e. jointed exoskeleton
    ___ tardigrades   f. mantle over body mass
    ___ roundworms   g. survive freezing, drying out
    ___ annelids   h. segmented worms
    ___ arthropods   i. tube feet, spiny skin

Critical Thinking

1. Flatworms, roundworms, and annelids all include some species that parasitize mammals. There are no such parasites among sponges, cnidarians, mollusks, and echinoderms. Propose a plausible explanation for this observation.

2. A massive die-off of lobsters in the Long Island Sound was blamed on pesticides sprayed to control the mosquitoes that carry West Nile virus. Why might a chemical designed to kill insects also harm lobsters?

3. Most animals that are hermaphrodites cannot fertilize their own eggs, but tapeworms can. Explain the advantages and disadvantages of self-fertilization.

Chapter 25
Summary

Section 25.1 Fossils provide evidence of evolutionary transitions between major animal groups. For example, fossils show that feathers, a defining trait of modern birds, evolved in a dinosaur ancestor of birds. Fossils also document how birds lost their long tail and teeth, and evolved a beak.

Section 25.2 Four embryonic features define the **chordates**: a **notochord**, a dorsal hollow nerve cord, a pharynx with gill slits, and a tail that extends past the anus. Depending on the group, these features may or may not persist in adults. Two groups of marine invertebrates, **tunicates** and **lancelets**, are chordates, but most chordates are **vertebrates** and have a backbone as part of their **endoskeleton**.

Early vertebrates that swam in the seas gave rise to **tetrapods** that walked on land. **Amniotes** are a tetrapod subgroup that adapted to a life spent entirely on land.

Section 25.3 Fishes are gilled aquatic vertebrates. They are **ectotherms**, meaning their body temperature fluctuates with the temperature of their environment. The earliest fishes were jawless and had a skeleton of **cartilage**. Lampreys and hagfishes are modern jawless fishes.

Section 25.4 Jaws evolved as a modification of the first set of gill arches in a jawless ancestor. Jawed fishes were the first vertebrates with **scales** and paired fins. Placoderms, a group of armored jawed fishes, dominated seas during the Devonian period but are now extinct.

Section 25.5 Cartilaginous fishes such as sharks have a skeleton that does not include bone. Products of their reproductive, digestive, and urinary systems exit the body by way of a single opening called the **cloaca**.

Bony fishes have some bone in their skeleton and a **swim bladder** that helps them regulate their buoyancy. There are two lineages. The **ray-finned fishes** have thin fins. They include teleosts, the most diverse lineage of modern fishes. **Lobe-finned fishes** have bones in their fins. Coelacanths and lungfishes are modern representatives of this group.

Section 25.6 Tetrapods, or four-legged walkers, evolved from lobe-finned bony fishes. **Amphibians** are scaleless, carnivorous tetrapods that live on land, but need water to reproduce. Their heart has three chambers; adults typically have lungs. Young salamanders look like miniature adults. Frog and toads are the most diverse amphibians. They undergo metamorphosis from a gilled aquatic larva with a tail.

Section 25.7 Amniotes are tetrapods with waterproof skin and highly efficient kidneys. Their **amniote eggs** allow embryos to develop on land. **Reptiles** were the first amniote lineage. They include the now extinct **dinosaurs** and their bird descendants. An early branching from the reptile lineage gave rise to the ancestors of modern mammals.

Section 25.8 All modern, nonbird reptiles are ectotherms with scales. A bony, protective shell evolved in the turtles. Lizards and snakes constitute the most diverse lineage. Most are predators and some are venomous. Crocodilians are predators that spend much of their time in water. Unlike other nonbird reptiles, the crocodilians have a four-chambered heart that prevents oxygen-poor and oxygen-rich blood from mixing.

Section 25.9 Birds are the only living animals with feathers. Like their closest living relatives, the crocodilians, they have a four-chamber heart. Like mammals, they are **endotherms**: They maintain their body temperature by regulating their metabolic heat production. The bird body has been highly modified for flight. Front limbs are modified as wings that are moved by muscles attached to a keeled sternum. A system of air sacs keeps air flowing continually through lungs.

Perching birds are the most diverse lineage. Some bird lineages such as penguins and ostriches have become flightless. Other birds migrate seasonally, typically moving between a breeding area and an area where they spend the winter.

Sections 25.10, 25.11 Mammals nourish young with milk secreted by mammary glands, have fur or hair, and have more than one kind of tooth. Three lineages are egg-laying mammals (**monotremes**), pouched mammals (**marsupials**), and **placental mammals**, the most diverse group. A **placenta** is an organ that facilitates exchange of substances between the embryonic and maternal blood.

Placental mammals develop faster than other mammals and are born at a later stage of development. Thus, placental mammals have become the dominant mammal group in most environments. Rodents, bats, and moles and shrews constitute the three most diverse lineages.

Self-Quiz

Answers in Appendix III

1. List the four distinguishing chordate traits.

2. Which of these traits are retained by an adult lancelet?

3. Vertebrate jaws evolved from _____ .
   a. gill supports   b. ribs   c. scales   d. teeth

4. Lampreys and sharks both have _____ .
   a. jaws   d. a swim bladder
   b. a bony skeleton   e. a four-chambered heart
   c. a cranium   f. lungs

Sounding a Feathery Alarm Australian crested pigeons are flocking birds with unusual flight feathers. As the bird takes off, vibration of these feathers produces a metallic-sounding whistle. The sound is loudest when the birds flee in alarm, and biologists Mae Hingee and Robert Magrath suspected that it serves as a warning signal. To find out, they recorded flight whistles of birds taking off casually and in response to a hawk model. The researchers then played back flight whistles and a control sound (a parrot call adjusted to be as loud as an alarmed flight whistle) and observed whether pigeons flew away (**Figure 25.32**).

1. Did the pigeons react differently to nonalarmed flight whistles than to alarmed flight whistles?

2. How did lowering the amplitude (loudness) of the alarmed flight sound affect the birds' response?

3. What evidence suggests that the birds were not merely fleeing whenever they heard a loud sound?

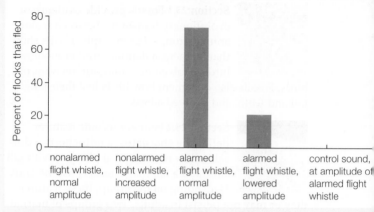

Figure 25.32 Percent of pigeon flocks that fled in response to playback of a sound at a feed Fifteen flocks were each exposed to all five sounds. The control sound was the call of a local parrot adjusted to the same loudness as the alarmed flight whistles.

5. Tetrapods evolved from _____ .
   a. sharks
   b. teleosts
   c. lobe-finned fishes
   d. placoderms

6. Reptiles and birds belong to one major lineage of amniotes, and _____ belong to another.
   a. sharks
   b. frogs and toads
   c. mammals
   d. salamanders

7. Reptiles are adapted to life on land by _____ .
   a. waterproof skin
   b. internal fertilization
   c. efficient kidneys
   d. amniote eggs
   e. both a and c
   f. all of the above

8. The closest living relatives of birds are _____ .
   a. crocodilians   b. mammals   c. snakes   d. lizards

9. Among living animals, only birds have _____ .
   a. a cloaca
   b. a four-chamber heart
   c. feathers
   d. amniote eggs

10. Feathers and hair consist of _____ .
    a. chitin
    b. cartilage
    c. keratin
    d. lignin

11. Unlike *Archaeopteryx*, modern birds have _____ .
    a. a long bony tail
    b. a beak
    c. teeth
    d. feathers

12. _____ eggs are typically released into the water.
    a. Frog
    b. Turtle
    c. Shark
    d. all of the above

13. Match the organisms with the appropriate description.
    ___ lancelets
    ___ lampreys
    ___ amphibians
    ___ lizards
    ___ birds
    ___ sharks
    ___ monotremes
    ___ marsupials
    ___ placoderms
    ___ placental mammals
    a. pouched mammals
    b. most diverse mammal lineage
    c. feathered amniotes
    d. egg-laying mammals
    e. extinct jawed fishes
    f. ectothermic amniotes
    g. cartilaginous fishes
    h. first land tetrapods
    i. living jawless fishes
    j. invertebrate chordates

14. Arrange the groups in order in which they first evolved.
    ___ 1 (earliest)
    ___ 2
    ___ 3
    ___ 4
    ___ 5
    ___ 6 (most recent)
    a. Jawless fishes
    b. Birds
    c. Dinosaurs
    d. Placental mammals
    e. Amphibians
    f. Jawed fishes

15. Match each structure with its description.
    ___ cloaca
    ___ swim bladder
    ___ fur
    ___ pectoral fins
    ___ endoskeleton
    ___ placenta
    a. adjusts buoyancy
    b. nourishes embryo
    c. supports body
    d. multipurpose exit from body
    e. homologous to forelimbs
    f. insulates body

Critical Thinking

1. In 1798, a stuffed platypus specimen was delivered to the British Museum. Reports that it laid eggs added to the confusion. To modern biologists, a platypus is clearly a mammal. It has fur and the females produce milk. Young animals have typical mammalian teeth that are replaced by hardened pads of the "bill" as the animal matures. Why do you think modern biologists can more easily accept that a mammal can have some seemingly reptilian traits?

2. Controlling for body size, would you expect a flightless bird to produce eggs that are larger than, smaller than, or the same size as eggs produced by a bird that flies?

3. Why is it more difficult to determine the sex of a newly hatched canary than a newborn puppy?

4. Many sharks and whales are highly active predators, but the largest shark (the basking shark) and the largest whale (the blue whale) are leisurely swimmers that filter food, mainly plankton, from the water. Why do you think large body size evolved to a greater extent in the filter-feeding members of these lineages than in the predatory members?

Chapter 26
Summary

Section 26.1 Fossils provide information about human relatives that are now extinct, such as Neanderthals. DNA extracted from fossils can also provide information. For example, extracting DNA from Neanderthal fossils and comparing it to human DNA showed that Neanderthals and modern humans mated just after modern humans left Africa.

Section 26.2 Primates are a mammalian order adapted to climbing. They have flexible shoulder joints, grasping hands tipped by nails, and good depth perception. They rely on vision more than smell. There are two main subgroups. Lemurs and their relatives are wet-nosed primates with a typical mammalian nose and a fixed upper lip. Most primates belong to the dry-nosed subgroup and have a movable upper lip.

Monkeys, apes, and humans share a common ancestor and are grouped as **anthropoids**. Apes and humans do not have tails. They are grouped as **hominoids** and are more closely related to Old World monkeys than to New World monkeys.

Section 26.3 Fossils show that *Proconsul*, an early ape, was living in Africa by about 20 million years ago. Fossil apes have also been found in Europe.

Modern apes include the gibbons and orangutans of Asia, and the gorillas, chimpanzees, and bonobos of Africa. Chimpanzees and bonobos are our closest living relatives. All live in social groups and are sexually dimorphic. Our lineage and the chimpanzee/bonobo lineage diverged from a common ancestor an estimated 8 million years ago.

Proconsul walked on all fours, but gorillas, chimpanzees, and bonobos are knuckle walkers. Humans are **bipedal**, meaning they habitually walk and stand upright. Human skeletons have adaptations related to bipedalism. The position of the hole that attaches our spinal cord to our brain is at the base of our skull, we have a foot with an arch and a nonopposable big toe, and our backbone has an S-shaped curve.

Humans also have a much larger brain and more flexible hands than any living ape.

Section 26.4 Hominins include modern humans and their extinct bipedal relatives. The earliest proposed hominins date to 6 or 7 million years ago and are known from fossil fragments found in Africa. *Ardipithecus* lived 5.8 to 5.2 million years ago and left many fossils, including a largely intact female skeleton. It probably walked upright when on the ground, but more often walked on all fours along branches.

Footprints dating to 3.6 million years ago are evidence of the passage of a fully bipedal species. It was most likely one of the **australopiths**, a group of hominins that lived in Africa 4.1 to 1.2 million years ago and left many fossils. Some are considered likely ancestors of humans. Although australopiths were upright walkers, they were no brainier than a chimpanzee. Evolution of bipedalism preceded the evolution of a larger brain.

Bipedalism evolved at a time when the African climate was warming and grasslands were replacing forests. Bipedalism increases the efficiency of movement on the ground, keeps a body cooler than four-legged walking, and facilitates transport of materials.

Section 26.5 Humans are members of the genus *Homo*. The earliest named human species, *Homo habilis*, appeared in Africa by 2.3 million years ago. *H. habilis* had an australopith-like build and a small brain.

Homo erectus evolved by 1.8 million years ago. It had a significantly larger brain and a body proportioned more like that of a modern human. Females were shorter than males, with broad hips capable of delivering infants with large heads. *H. erectus* dispersed out of Africa and left fossils in Europe, China, and Indonesia.

Stone tool production is an aspect of **culture**, a set of learned behaviors passed down from one generation to the next. Evidence of stone tool use dates back to 3.4 million years ago, suggesting it probably began among the australopiths. Hominins used tools to scrape meat from animal bones. By one hypothesis, increased meat consumption provided energy necessary to develop larger bodies and brains. *H. habilis* and *H. erectus* probably did not have a spoken language.

Section 26.6 The earliest fossils of our own species, *Homo sapiens*, are from Africa and date to 195,000 years ago. *H. sapiens* is distinguished by a larger brain, flatter face, smaller jaw, and a chin.

There are two models for the origin of *H. sapiens*, which is thought to be descended from *H. erectus*. By the **multiregional model**, *H. sapiens* evolved from *H. erectus* populations that had already spread into many separate regions. The **replacement model** postulates that *H. sapiens* evolved in Africa, then dispersed and replaced *H. erectus* populations elsewhere. Gene comparisons among modern populations support the hypothesis of an African origin for *H. sapiens*, but evidence that humans who left Africa mated with Neanderthals in the Middle East raises the possibility that regional differences in modern populations may be the result of matings with other hominins.

Newly discovered fossils of short, small-brained hominins that lived 18,000 years ago on an island in Indonesia have been named as a new species, *Homo floresiensis*. However, not all scientists agree that the fossils belong to a new species of *Homo*.

Self-Quiz
Answers in Appendix III

1. New World monkeys _____ .
   a. lack a tail
   b. are bipedal
   c. live only in Africa
   d. are dry-nosed primates
   e. are human ancestors
   f. all of the above

Neanderthal Hair Color The *MC1R* gene regulates pigmentation in humans (Sections 14.1 and 15.1 revisited), so loss-of-function mutations in this gene affect hair and skin color. A person with two mutated alleles for this gene makes more of the reddish melanin than the brownish melanin, resulting in red hair and pale skin. DNA extracted from two Neanderthal fossils contains a mutated *MC1R* allele that has not yet been found in humans. To see how the Neanderthal mutation affects the function of the *MC1R* gene, Carles Lalueza-Fox and her team introduced the allele into cultured monkey cells (**Figure 26.16**).

1. How did *MCR1* activity in monkey cells with the mutant allele differ from that in cells with the normal allele?

2. What does this imply about the mutation's effect on Neanderthal hair color?

3. What purpose do the cells with the gene for green fluorescent protein serve in this experiment?

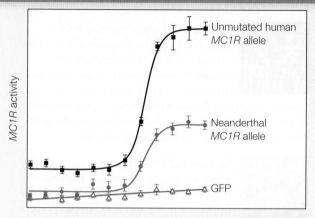

Figure 26.16 *MC1R* activity in monkey cells transgenic for an unmutated *MC1R* gene, the Neanderthal *MC1R* allele, or the gene for green fluorescent protein (GFP). GFP is not related to *MC1R*.

2. The closest relatives of bonobos are _____ .
   a. chimpanzees
   b. humans
   c. tarsiers
   d. Old World monkeys

3. An S-shaped backbone is an adaptation to _____ .
   a. tool use
   b. climbing trees
   c. bipedalism
   d. carnivory

4. Like most mammals, a(n) _____ has a wet nose.
   a. New World monkey
   b. Old World monkey
   c. chimpanzee
   d. lemur

5. The 3.6-million-year-old footprints left by bipedal walkers in Tanzania were probably made by _____ .
   a. australopiths
   b. Neanderthals
   c. *Homo floresiensis*
   d. *Homo erectus*

6. The position where a spinal cord enters the skull provides evidence about whether a fossil species _____ .
   a. was nocturnal
   b. was carnivorous
   c. walked upright
   d. all of the above

7. Australopiths are _____ .
   a. extinct
   b. placental mammals
   c. hominoids
   d. all of the above

8. The oldest *Homo* fossils found outside of Africa are assigned to the species _____ .
   a. *H. sapiens*
   b. *H. habilis*
   c. *H. erectus*
   d. *H. floresiensis*

9. A prominent chin is typical in _____ .
   a. *H. sapiens*
   b. *H. habilis*
   c. *H. erectus*
   d. *H. floresiensis*

10. Compared to modern humans, Neanderthals had a _____ brain and were _____ .
    a. smaller, taller
    b. smaller, shorter
    c. similar-sized, taller
    d. similar-sized, shorter

11. The oldest *Homo sapiens* fossil was found in _____ .
    a. the Middle East
    b. Africa
    c. Indonesia
    d. Europe

12. Fill in the blank. _____ had body proportions like those of modern people who live in the Arctic.

13. Match each group with its description.
    ___ Neanderthals   a. our closest extinct relatives
    ___ australopiths   b. modern humans
    ___ hominins   c. includes all other groups listed
    ___ *Homo sapiens*   d. nonhuman but bipedal
    ___ *Homo habilis*   e. short-statured, from Indonesia
    ___ *Homo floresiensis* f. most ancient human species

14. Place the events in order.
    ___1 (earliest)   a. Neanderthals disappear
    ___2   b. monkeys reach the New World
    ___3   c. early apes colonize Europe
    ___4   d. *Homo erectus* leaves Africa
    ___5   e. divergence of lineages leading to
    ___6   humans and to chimpanzees
    ___7 (most recent)   f. divergence of wet-nosed and
                             dry-nosed primates
                          g. *Homo sapiens* cross land bridge
                             and enter North America

Critical Thinking

1. Male aggression is rare in bonobo society and common in chimpanzee society. Various authors have argued that either one species or the other should be considered a model for "natural" human behavior. Explain why, from the standpoint of relatedness, there is no reason to that think one of these species is a better model for human behavior than the other.

2. Researchers recently extracted mitochondrial DNA from a fossil pinky finger found in a cave in Siberia and compared it to mitochondrial DNA (mitoDNA) from modern humans. The sequence of the finger DNA differed from modern human DNA at an average of 385 sites. Previous studies have shown that Neanderthal mitoDNA differs from modern human mitoDNA at an average of 202 sites, and chimpanzee mitoDNA differs from human mitoDNA at an average of 1,462 sites. Using this information, draw a tree that shows the relationship between chimpanzees, humans, Neanderthals, and the owner of this finger.

Chapter 27
Summary

Section 27.1 Plants naturally remove carbon from the atmosphere and sequester it in their tissues. Carbon locked in wood and other durable plant tissues can stay out of the atmosphere for centuries.

Section 27.2 Most flowering plants have aboveground shoots, including stems, leaves, and flowers. Most also have belowground roots. Shoots and roots are composed of ground, vascular, and dermal tissues. **Ground tissue** stores materials, functions in photosynthesis, and structurally supports a plant. Tubes in **vascular tissue** conduct substances to and from living cells. **Dermal tissue** protects plant surfaces. Monocots and eudicots have the same tissues organized in different ways. For example, they differ in how xylem and phloem are distributed through ground tissue, in the number of petals in flowers, and in the number of **cotyledons**.

All plant tissues originate at **meristems**, which are regions of undifferentiated cells that retain their ability to divide. **Primary growth** (or lengthening) arises from apical meristems. **Secondary growth** (or thickening) arises from lateral meristems.

Section 27.3 Complex tissues consist of two or more types of cell; simple tissues consist of one cell type. The living, thin-walled cells in **parenchyma** have diverse roles in ground tissue. **Mesophyll** is photosynthetic parenchyma. Living cells in **collenchyma** have sturdy, flexible walls that support fast-growing plant parts. Cells in **sclerenchyma** die at maturity, but their lignin-reinforced walls remain and support the plant. **Vessel members** and **tracheids** of **xylem** are dead at maturity; their perforated, interconnected walls conduct water and dissolved minerals. The **sieve-tube members** of phloem remain alive at maturity. These cells interconnect to form sieve tubes that conduct sugars. **Companion cells** load sugars into the sieve tubes. **Epidermis** is a dermal tissue that covers and protects the outer surfaces of young plant parts. Epidermis is replaced by periderm in older roots and shoots.

Section 27.4 New shoots form by the activity of apical meristems in **terminal buds** and also in **lateral buds** that form in **nodes**. **Vascular bundles** extending through stems conduct water, ions, and nutrients between different parts of the plant, and also function in support. In most eudicot stems, vascular bundles form a ring that divides the ground tissue into cortex and pith. In monocot stems, the vascular bundles are distributed throughout ground tissue.

Section 27.5 Leaves are photosynthetic factories that contain mesophyll and stiffened vascular bundles (**veins**) between their upper and lower epidermis. Air spaces around mesophyll cells allow gas exchange for photosynthesis. Water vapor and gases cross the cuticle-covered epidermis at stomata.

Section 27.6 Roots absorb water and mineral ions for the entire plant. Inside each is a **vascular cylinder**, or **stele**, enclosed by **endodermis**. **Root hairs** increase the surface area of roots. Most eudicots have a **taproot system** of a primary root with lateral branchings; many monocots have a **fibrous root system** that consists of similar-sized adventitious roots that branch from the stem and lateral roots. Lateral roots arise from divisions of **pericycle** cells inside the vascular cylinder.

Section 27.7 Woody plants thicken (add secondary growth) by cell divisions in **lateral meristems**. Cell divisions in **vascular cambium** add secondary xylem (**wood**) and secondary phloem. Cell divisions in **cork cambium** give rise to periderm (**cork** is part of this periderm). **Bark** is periderm and secondary phloem of a woody stem. Wood is classified by its location and function, as in **heartwood** or **sapwood**.

Section 27.8 In many trees, one ring forms each season. Tree rings hold information about environmental conditions that prevailed while the rings were forming. For example, the relative thicknesses of the rings reflect the relative availability of water.

Section 27.9 Stem specializations such as rhizomes, corms, tubers, bulbs, cladodes, and stolons are adaptations that function in storage or reproduction in many types of plants.

Self-Quiz
Answers in Appendix III

1. Roots and shoots lengthen by divisions of cells in _____ .
   - a. apical meristems
   - b. lateral meristems
   - c. vascular cambium
   - d. cork cambium

2. In many plant species, older roots and stems thicken by divisions of cells in _____ .
   - a. apical meristems
   - b. cork cambium
   - c. vascular cambium
   - d. both b and c

3. _____ conducts water and minerals through a plant, and _____ conducts sugars.
   - a. Phloem; xylem
   - b. Cambium; phloem
   - c. Xylem; phloem
   - d. Xylem; cambium

4. Mesophyll consists of _____ .
   - a. waxes and cutin
   - b. lignified cell walls
   - c. parenchyma cells
   - d. cork but not bark

5. Which one of the following cell types is alive in mature plant tissue?
   - a. companion cells
   - b. sclerenchyma cells
   - c. tracheids
   - d. vessel members

6. True or false? A dermal tissue called periderm replaces epidermis in woody stems and roots.

Tree Rings and Droughts Douglas fir trees (*Pseudotsuga menziesii*) are exceptionally long-lived, and particularly responsive to rainfall levels. Researcher Henri Grissino-Mayer sampled Douglas firs in El Malpais National Monument, in west central New Mexico. Pockets of vegetation in this site have been surrounded by lava fields for about 3,000 years, so they have escaped wildfires, grazing animals, agricultural activity, and logging. Grissino-Mayer compiled tree ring data from old, living trees, and dead trees and logs to generate a 2,129-year annual precipitation record (**Figure 27.18**).

1. The Mayan civilization began to suffer a massive population loss around 770 A.D. Do these tree ring data reflect a drought condition at this time? If so, was that condition more or less severe than the "dust bowl" drought (1933–1939)?

2. One of the worst population catastrophes ever recorded occurred in Mesoamerica between 1519 and 1600 A.D., when around 22 million people native to the region died. Which period between 137 B.C. and 1992 had the most severe drought? How long did that drought last?

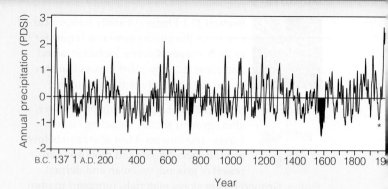

Figure 27.18 A 2,129-year annual precipitation record inferred from compiled tree ring data in El Malpais National Monument, New Mexico. Data was averaged over 10-year intervals; graph correlates with other indicators of rainfall collected in all parts of North America.

PDSI: Palmer Drought Severity Index: 0, normal rainfall; increasing numbers mean increasing excess of rainfall; decreasing numbers mean increasing severity of drought.

* A severe drought contributed to a series of catastrophic dust storms that turned the midwestern United States into a "dust bowl" between 1933 and 1939.

---

7. In phloem, fluid flows through _____ .
   a. collenchyma cells   c. vessels
   b. sieve tubes   d. tracheids

8. Xylem and phloem are _____ tissues.
   a. ground   b. vascular   c. dermal   d. both b and c

9. Most photosynthetic tissue is _____ .
   a. mesophyll   c. collenchyma
   b. chlorophyll   d. sclerenchyma

10. Vascular bundles are arranged in a cylinder in _____ .
   a. monocot stems   c. monocot roots
   b. eudicot stems   d. eudicot roots

11. In a(n) _____ , the primary root is typically the largest.
   a. lateral meristem   c. fibrous root system
   b. adventitious root system d. taproot system

12. The main function of root hairs is to _____ .
   a. conduct water from soil to aboveground shoots
   b. "feel" the soil for water that may be scarce
   c. increase the root's surface area for absorption
   d. anchor the plant in soil

13. Tree rings occur when _____ .
   a. there are droughts during the time the rings form
   b. environmental conditions influence xylem cell size
   c. heartwood alternates with sapwood

14. Bark is mainly _____ .
   a. periderm and cork   c. periderm and phloem
   b. cork and wood   d. cork cambium and
                                  phloem

15. Match the plant parts with the best description.
   ___ apical meristem   a. secondary growth
   ___ lateral meristem   b. source of primary growth
   ___ xylem   c. distribution of sugars
   ___ phloem   d. source of secondary growth
   ___ vascular cylinder e. distribution of water
   ___ wood   f. central column in eudicot roots
   ___ cortex   g. seed leaf
   ___ cotyledon   h. mainly parenchyma

Critical Thinking

1. Is the plant with the yellow flower *above* a eudicot or a monocot? What about the plant with the purple flower?

2. Oscar and Lucinda meet in a tropical rain forest and fall in love, and he carves their initials into the bark of a tiny tree. They never do get together, though. Ten years later, still heartbroken, Oscar searches for the tree. Given what you know about primary and secondary growth, will he find the carved initials higher relative to ground level? If he goes berserk and chops down the tree, what kinds of growth rings will he see?

3. Was the transverse section shown at *right* taken from a stem or a root? Monocot or eudicot?

4. Aboveground plant surfaces are covered with a waxy cuticle. Why do roots lack this protective coating?

5. Why do eudicot trees tend to be wider at the base than at the top?

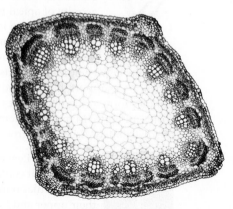

Chapter 28
Summary

Section 28.1 The ability of plants to take up substances from soil and water is the basis for phytoremediation, the removal of toxic substances from a contaminated area with the help of plants.

Section 28.2 Plant growth requires steady sources of elemental nutrients obtainable from carbon dioxide, water, and **soil**. The availability of water and mineral ions in soil depends on its proportions of sand, silt, and clay, and also on its **humus** content. **Loams** consist of roughly equal proportions of sand, silt, and clay. **Leaching** and **soil erosion** deplete nutrients from soil, particularly **topsoils**.

Section 28.3 Root hairs greatly increase a root's surface area for absorption of soil water. Fungi are symbionts with young roots in **mycorrhizae**, which enhance a plant's ability to absorb mineral ions from soil. **Nitrogen fixation** by bacteria in **root nodules** gives a plant extra nitrogen. In both cases, the microorganisms receive some of the plant's sugars.

Roots control the movement of water and dissolved mineral ions into the vascular cylinder. Endodermal cells that form the outer layer of the cylinder deposit a waterproof band, a **Casparian strip**, in their abutting walls. The strip keeps water from diffusing around endodermal cells into the vascular cylinder. Ions can enter xylem inside the cylinder only by moving through the plasma membrane of at least one root cell. Thus, the uptake of ions is controlled by active transport proteins in root cell plasma membranes.

Section 28.4 Water and dissolved mineral ions flow through xylem from roots to shoot tips. Xylem consists of tubes formed from the perforated and lignin-reinforced secondary walls of dead tracheid and vessel member cells. **Transpiration** is the evaporation of water from plant parts into air. Evaporation mainly occurs at stomata.

The **cohesion–tension theory** explains how water moves through plants: Transpiration pulls water upward by creating a continuous negative pressure (tension) inside xylem from leaves to roots. Hydrogen bonds among water molecules keep the columns of fluid continuous inside the narrow vessels.

Section 28.5 A cuticle and stomata balance a plant's loss of water with its needs for gas exchange. A stoma, which is a gap across the cuticle-covered epidermis of leaves and other plant parts, is defined by a pair of **guard cells**. A stoma may be surrounded by an indentation, protrusions, or other specializations that reduce air flow around it.

Closed stomata limit the loss of water, but also prevent the gas exchange required for photosynthesis and aerobic respiration. Environmental signals cause stomata to open or close. The signals trigger guard cells to pump ions into or out of their cytoplasm; water follows the ions (by osmosis). Water moving into guard cells plumps them, which opens the stoma between them. Water diffusing out of the cells causes them to collapse against each other, so the stoma closes.

Section 28.6 Organic compounds (mainly sugars) move through a plant by **translocation** in sieve tubes, which consist of stacked sieve-tube members separated by perforated **sieve plates**.

By the **pressure flow theory**, the movement of sugar-rich fluid through a sieve tube is driven by a pressure gradient between **source** and **sink** regions. The gradients are set up as sugars move into companion cells, then into adjacent sieve tubes at source regions. Water that follows by osmosis increases sieve-tube turgor at these regions. The sugars are unloaded from sieve tubes at sink regions. Water that leaves the tubes by osmosis reduces sieve-tube turgor at these regions.

Self-Quiz
Answers in Appendix III

1. Carbon, hydrogen, and oxygen are _____ for plants.
   a. macronutrients
   b. micronutrients
   c. trace elements
   d. required elements
   e. both a and d

2. Decomposing matter in soil is called _____ .
   a. loam
   b. humus
   c. topsoil
   d. nutrients

3. The nutrition of some plants depends on a mutually beneficial association between a root and a fungus. The association is known as a _____ .
   a. root nodule
   b. mycorrhiza
   c. root hair
   d. root hypha

4. A _____ strip between abutting endodermal cell walls forces water and solutes to move through these cells rather than around them.
   a. cutin   b. Casparian   c. cohesion   d. cellulose

5. A vascular cylinder consists of cells of the _____ .
   a. exodermis
   b. endodermis
   c. root cortex
   d. xylem and phloem
   e. b and d
   f. all of the above

6. Water evaporation from plant parts is called _____ .
   a. translocation
   b. expiration
   c. transpiration
   d. tension

7. Water transport from roots to leaves occurs mainly because of _____ .
   a. pressure flow
   b. differences in source and sink solute concentrations
   c. the pumping force of xylem vessels
   d. transpiration, tension, and cohesion of water
   e. a and b

8. A waxy cuticle is secreted by _____ .
   a. ground tissue
   b. epidermal cells
   c. a stoma
   d. root hairs

9. When guard cells swell, _____ .
   a. transpiration ceases
   b. sugars enter phloem
   c. a stoma opens
   d. root cells die

TCE Uptake by Transgenic Poplar Trees Plants used for phytoremediation take up organic pollutants from the soil or air, then transport the chemicals to plant tissues, where they are stored or broken down. Researchers are now designing transgenic plants with enhanced ability to take up or break down toxins. In 2007, Sharon Doty and her colleagues published the results of their efforts to design plants useful for phytoremediation of soil and air containing organic solvents. The researchers used *Agrobacterium tumefaciens* (Section 15.7) to deliver a mammalian gene into poplar plants. The gene encodes cytochrome P450, a type of heme-containing enzyme involved in the metabolism of a range of organic molecules, including solvents such as TCE. The results of one of the researchers' tests on these transgenic plants are shown in **Figure 28.15**.

1. How many transgenic plants did the researchers test?

2. In which group did the researchers see the slowest rate of TCE uptake? The fastest?

3. On day 6, what was the difference between the TCE content of air around transgenic plants and that around vector control plants?

4. Assuming no other experiments were done, what two explanations are there for the results of this experiment? What other control might the researchers have used?

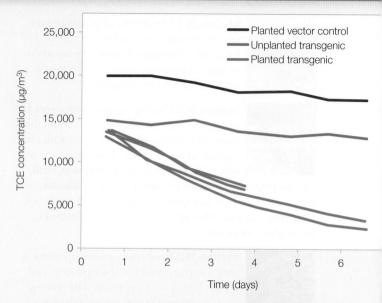

Figure 28.15 Results of tests on transgenic poplar trees. Planted trees were incubated in sealed containers with an initial 15,000 micrograms of TCE (trichloroethylene) per cubic meter of air. Samples of the air in the containers were taken daily and measured for TCE content. Controls included a tree transgenic for a Ti plasmid with no cytochrome P450 in it (vector control), and a bare-root transgenic tree (one that was not planted in soil).

---

10. Stomata open in response to light when _____ .
    a. ions flow into guard cell cytoplasm
    b. ions flow out of guard cell cytoplasm
    c. water evaporates out of guard cells

11. Tracheids are part of _____ .
    a. cortex   c. phloem
    b. mesophyll   d. xylem

12. Sieve tubes are part of _____ .
    a. cortex   c. phloem
    b. mesophyll   d. xylem

13. When soil is dry, guard cells respond to _____ by collapsing against one another, so stomata close.
    a. air temperature   c. abscisic acid
    b. low humidity   d. oxygen

14. Transport of photosynthetically produced sugars from leaves to roots occurs by _____ .
    a. pressure flow
    b. differences in source and sink solute concentrations
    c. the pumping force of xylem vessels
    d. transpiration, tension, and cohesion of water
    e. a and b

15. Match the concepts of plant nutrition and transport.
    ____ stomata
    ____ nutrient
    ____ sink
    ____ root system
    ____ hydrogen bonds
    ____ transpiration
    ____ translocation

    a. evaporation from plant parts
    b. harvests soil water and nutrients
    c. balance water loss with gas exchange
    d. cohesion in water transport
    e. sugars unloaded from sieve tubes
    f. organic compounds distributed through the plant body
    g. required element

Critical Thinking

1. What are the three structures in the micrographs in **Figure 28.16**? From which tissue(s) do they originate?

2. Successful home gardeners, like farmers, make sure that their plants get enough nitrogen from either fertilizer or nitrogen-fixing bacteria. Which biological molecules incorporate nitrogen? Nitrogen deficiency stunts plant growth; leaves yellow and then die. How would nitrogen deficiency cause these symptoms?

3. You just returned home from a three-day vacation. Your severely wilted plants tell you they were not watered before you left. Being aware of the cohesion–tension theory of water transport, explain what happened to them.

4. When you dig up a plant to move it from one spot to another, the plant is more likely to survive if some of the soil around the roots is transferred along with the plant. Formulate a hypothesis that explains this observation.

5. If a plant's stomata are made to stay open at all times, or closed at all times, it will die. Why?

6. Allen is studying the rate of water uptake by tomato plant roots. He notices that several environmental factors, including wind and relative humidity, affect the rate. Explain how they might do so.

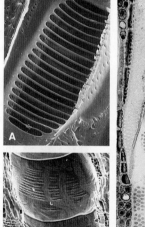
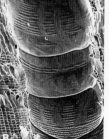

Figure 28.16 Name the mystery structures.

Chapter 29
Summary

 Section 29.1 Colony collapse disorder (CCD) is killing honeybees. Declines in populations of bees and other **pollinators** negatively affect plant populations as well as other animal species that depend on the plants, including humans. Widely used neonicotinoid pesticides that increase bee susceptibility to infection may be the cause of CCD.

 Section 29.2 Flowers consist of whorls of modified leaves at the ends of specialized branches of angiosperm sporophytes. A **calyx** of **sepals** surrounds a **corolla** of **petals**, which in turn surround **stamens** and **carpels**. A carpel consists of a **stigma**, often a style, and an **ovary** in which one or more **ovules** develop. Spores produced by meiosis in ovules develop into female gametophytes; those produced in **anthers** develop into immature male gametophytes (pollen grains). Adaptations of many flowers restrict self-pollination.

 Section 29.3 A flower's shape, pattern, color, and fragrance typically reflect an evolutionary relationship with a particular **pollination vector**, often a coevolved animal. Coevolved pollinators receive **nectar**, pollen, or another reward for visiting a flower.

 Section 29.4 Meiosis of diploid cells inside pollen sacs of anthers produces haploid **microspores**. Each microspore develops into a pollen grain that is released from pollen sacs after a period of **dormancy**. Mitosis and cytoplasmic division of a cell in an ovule produces four **megaspores**, one of which gives rise to the female gametophyte. One of the seven cells of the gametophyte is the egg; another is the diploid endosperm mother cell.

A pollen grain arrives on a receptive stigma at pollination. Upon **germination**, a pollen grain forms a pollen tube that contains two sperm cells. The tube grows through carpel tissues to the egg. In **double fertilization**, one of the sperm cells in the pollen tube fertilizes the egg, forming a zygote; the other fuses with the endosperm mother cell and gives rise to triploid **endosperm**.

 Section 29.5 Species-specific molecular cues on a pollen grain's durable coat are part of an interplay of cell signals that trigger germination on a receptive stigma. Molecules produced by the carpel guide pollen tube growth to the egg.

 Section 29.6 As a zygote develops into an embryo, endosperm collects nutrients from the parent plant, and the ovule's protective layers develop into a seed coat. A **seed** is a mature ovule: an embryo sporophyte and endosperm enclosed within a seed coat. Nutrients in endosperm or cotyledons make seeds a nutritious food source.

 Section 29.7 As an embryo sporophyte develops, the ovary wall and sometimes other tissues mature into a **fruit** that encloses the seeds. Fruit specializations are adaptations to seed dispersal by specific vectors such as wind, water, or animals. A fruit can be categorized by tissue of origin, composition, and whether it is dry or fleshy.

 Section 29.8 Mature seeds often undergo a period of dormancy. Their arrested metabolism does not end until species-specific environmental cues such as light, smoke, or passage through an animal's gut trigger germination. The **radicle** emerges from the seed coat at the end of germination; other patterns of early development vary. For example, only some monocot **plumules** are sheathed by a **coleoptile**; only eudicot **hypocotyls** form a hook that pulls cotyledons up through soil.

 Section 29.9 Many types of flowering plants can reproduce asexually by **vegetative reproduction**. The offspring produced by asexual reproduction are clones of the parent. Many plants are propagated commercially by grafting; the common laboratory technique of **tissue culture propagation** is used to produce some valuable ornamentals.

 Section 29.10 Senescence is aging of cells, individuals, or communities. In many species, senescence includes cycles of **abscission** and dormancy that are adaptations to seasonal changes in environmental conditions. Abscission of plant parts can also occur in response to stress.

Self-Quiz

Answers in Appendix III

1. A pollinator may receive _____ in return for a visit to a flower of a coevolved plant (choose all that apply).
   - a. pollen
   - b. nectar
   - c. pesticides
   - d. fruit

2. In plants, the structures that produce male gametophytes are called _____ ; those that produce female gametophytes are called _____ .
   - a. pollen; flowers
   - b. stamens; carpels
   - c. anthers; stigma
   - d. microspores; megaspores

3. A flower's _____ has one or more ovaries in which eggs develop, fertilization occurs, and seeds mature.
   - a. pollen sac
   - b. carpel
   - c. coleoptile
   - d. sepal

4. The transfer of pollen grains from one flower to another is called _____ .
   - a. fertilization
   - b. abscission
   - c. pollination
   - d. reproduction

5. Meiosis of cells in pollen sacs forms haploid _____ .
   a. megaspores   c. stamens
   b. microspores   d. sporophytes

6. Meiosis in an ovule produces _____ megaspores.
   a. two   c. four
   b. six   d. eight

7. The three main parts of a seed are the _____ .
   a. pollen grain, egg, and seed coat
   b. embryo, endosperm, and seed coat
   c. megaspores, microspores, and ovule
   d. embryo, cotyledons, and nutrients

8. The seed coat forms from the _____ .
   a. integuments   c. endosperm
   b. coleoptile   d. sepals

9. Seeds are mature _____ ; fruits are mature _____ .
   a. ovaries; ovules   c. ovules; ovaries
   b. ovules; stamens   d. stamens; ovaries

10. Cotyledons develop as part of _____ .
    a. carpels   c. embryo sporophytes
    b. accessory fruits   d. petioles

11. In some species, exposure to _____ is a trigger for seed germination.
    a. light   c. smoke
    b. cold   d. all of the above

12. A new plant forms from a stem that broke off of the parent plant. This is an example of _____ .
    a. parthenogenesis   c. vegetative reproduction
    b. grafting   d. nodal growth

13. Wanting to impress friends with her sophisticated knowledge of botany, Dixie Bee prepares a plate of tropical fruits for a party and cuts open a papaya (*Carica papaya*). Soft skin and soft fleshy tissue enclose many seeds in a slimy tissue (*inset*). Knowing her friends will ask her how to categorize this fruit, she panics, runs to her biology book, and opens it to Table 29.2 in Section 29.7. What does she find out?

14. Having succeeded in spectacularly impressing her friends, Dixie Bee prepares a platter of peaches (*inset*) for her next party. How will she categorize this fruit?

15. Match the terms with the most suitable description.

___ ovule
___ receptacle
___ double fertilization
___ anther
___ plumule
___ mature female gametophyte
___ mature male gametophyte

a. pollen tube together with its contents
b. embryo sac of seven cells, one with two nuclei
c. starts out as cell mass in ovary; may become a seed
d. embryonic shoot
e. pollen sacs inside
f. base of floral shoot
g. formation of zygote and first cell of endosperm

Critical Thinking

1. Label the parts of the flower shown *below*.

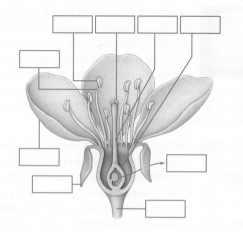

2. Would you expect wind, bees, birds, bats, butterflies, or moths to pollinate a dull-colored flower with no scent? Explain your choice.

3. Is the seedling shown on the *left* a monocot or eudicot? How can you tell?

4. All but one species of large-billed birds native to New Zealand's tropical forests are now extinct. Numbers of the surviving species, the kereru, are declining rapidly due to the habitat loss, poaching, predation, and interspecies competition that wiped out the other native birds. The kereru remains the sole dispersing agent for several native trees that produce big seeds and fruits. One tree, the puriri (*Vitex lucens*), is New Zealand's most valued hardwood. Explain, in terms of natural selection, why we might expect to see no new puriri trees in New Zealand.

Who's the Pollinator? *Massonia depressa* is a low-growing succulent plant native to the desert of South Africa. The dull-colored flowers of this monocot develop at ground level, have tiny petals, emit a yeasty aroma, and produce a thick, jellylike nectar. These features led researchers to suspect that desert rodents such as gerbils pollinate this plant (**Figure 29.22**).

To test their hypothesis, the researchers trapped rodents in areas where *M. depressa* grows and checked them for pollen. They also put some plants in wire cages that excluded mammals, but not insects, to see whether fruits and seeds would form in the absence of rodents. The results are shown in **Figure 29.23**.

1. How many of the 13 captured rodents showed some evidence of pollen from *M. depressa*?

2. Would this evidence alone be sufficient to conclude that rodents are the main pollinators for this plant?

3. How did the average number of seeds produced by caged plants compare with that of control plants?

4. Do these data support the hypothesis that rodents are required for pollination of *M. depressa*? Why or why not?

Figure 29.23 *Right*, results of experiments testing rodent pollination of *M. depressa*. (**A**) Evidence of visits to *M. depressa* by rodents. (**B**) Fruit and seed production of *M. depressa* with and without visits by mammals. Mammals were excluded from plants by wire cages with openings large enough for insects to pass through. 23 plants were tested in each group.

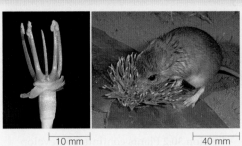

Figure 29.22 The dull, petal-less, ground-level flowers of *Massonia depressa* are accessible to rodents such as this gerbil, who push their heads through the stamens to reach the nectar. Note the pollen on the gerbil's snout.

| 10 mm | 40 mm |

A

Type of rodent	Number caught	# with pollen on snout	# with pollen in feces
Namaqua rock rat	4	3	2
Cape spiny mouse	3	2	2
Hairy-footed gerbil	4	2	4
Cape short-eared gerbil	1	0	1
African pygmy mouse	1	0	0

B

	Mammals allowed access to plants	Mammals excluded from plants
Percent of plants that set fruit	30.4	4.3
Average number of fruits per plant	1.39	0.47
Average number of seeds per plant	20.0	1.95

Chapter 30
Summary

Section 30.1 Secondary metabolites are not required for immediate survival of the organism that produces them. Some plant secondary metabolites attract pollinators or symbionts or function in defense. A few of these compounds, including a number of flavonoids, are beneficial to human health.

Section 30.2 Plants continue to develop throughout their lifetime. **Plant hormones** are signaling molecules that affect cellular activities. They can promote or arrest development by stimulating or inhibiting cell division, differentiation, or enlargement. Some have roles in reproduction or responses to stress. Hormones typically work together, either synergistically or in opposition, to coordinate the activity of cells in all parts of the plant body during development and defense.

Section 30.3 Auxin coordinates the action of other hormones, and it has different effects in shoots and roots. A transport system that is responsive to internal and external con-ditions distributes auxin directionally. This polar distribution of auxin directs development of plant organs, such as when auxin gradients maintain **apical dominance** in a growing shoot tip.

Section 30.4 Cytokinin promotes cell division in shoot apical meristem, and differentiation in root apical meristem. This hormone acts together with auxin, often antagonistically, to balance growth with development in shoot and root tips.

Section 30.5 Gibberellin lengthens stems between nodes by inducing cell division and elongation. It also stimulates production of enzymes that break down endosperm during germination.

Section 30.6 Abscisic acid (ABA) influences the expression of thousands of genes. Its effects in plants include inhibiting seed germination and growth; and promoting metabolism, embryonic development, and responses to stress.

Section 30.7 Ethylene is produced in negative and positive feedback loops. Negative feedback loops are part of regulating ongoing metabolism and development; positive loops provoke special processes such as abscission, ripening, and defense responses.

Section 30.8 In **tropisms**, plants adjust the direction and rate of growth in response to hormone-mediated environmental cues. In **gravitropism**, roots grow down and stems grow up in response to gravity.

Statoliths are part of this response. In **heliotropism** and other **phototropisms**, stems and leaves bend toward or away from light. Blue light is the trigger for such phototropic responses. In some plants, the direction of growth changes in response to contact (**thigmotropism**). Growth may also be affected by mechanical stress.

Section 30.9 Internal timing mechanisms called **circadian rhythms** are driven by feedback loops in which transcription factors influence their own expression in daily cycles. Nonphotosynthetic pigments such as cryptochromes and **phytochromes** provide light input into circadian clocks. **Photoperiodisms** are integrated responses to circadian rhythms and night length. Some plants require prolonged exposure to cold before they can flower, a process called **vernalization**.

Section 30.10 Plants can respond to abiotic and biotic stresses. ABA is part of the stress response that closes stomata when water is scarce. A nitric oxide burst is part of a hypersensitive defense response that closes stomata or kills the cell in response to pathogen detection. Pathogen-triggered **systemic acquired resistance** increases the plant's resilience to biotic and abiotic stress in general.

Self-Quiz

Answers in Appendix III

1. Plant hormones _____ .
   a. often have multiple, overlapping effects
   b. are active in developing plant embryos
   c. are active in adult plants
   d. all of the above

2. _____ is the hormone in most rooting compounds.
   a. Gibberellin   c. Cytokinin
   b. Auxin   d. ABA

3. Which of the following statements is false?
   a. Auxin and gibberellin promote stem elongation.
   b. Cytokinin promotes cell division in shoot tips.
   c. Abscisic acid promotes water loss and dormancy.
   d. Ethylene promotes fruit ripening and abscission.

4. Which combination of hormones is best for lateral root growth?
   a. A high auxin level and low ABA and cytokinin levels
   b. A low auxin level and high ABA and cytokinin levels
   c. A high level of auxin, ABA, and cytokinin

5. Ethylene differs from other hormones in that it _____ .
   a. is a gas   c. operates during ripening
   b. participates in defense d. none of the above

6. Heliotropism is a form of _____ .
   a. phototropism   c. photoperiodism
   b. gravitropism   d. a and c

7. Sunlight resets biological clocks in plants by activating and inactivating _____ .
   a. phototropins   d. chlorophylls
   b. phytochromes   e. b and c
   c. cryptochromes   f. all of the above

Volatile Secondary Metabolites in Plant Stress Responses

In 2007, researchers Casey Delphia, Mark Mescher, and Consuelo De Moraes (pictured at *left*) published a study on the production of different volatile chemicals by tobacco plants (*Nicotiana tabacum*) in response to predation by two types of insects: western flower thrips (*Frankliniella occidentalis*) and tobacco budworms (*Heliothis virescens*). Their results are shown in **Figure 30.21**.

1. Which treatment elicited the greatest production of volatiles?

2. Which volatile chemical was produced in the greatest amount? What was the stimulus?

3. Which one of the chemicals tested is most likely produced by tobacco plants in a nonspecific response to predation?

4. Are any chemicals produced in response to predation by budworms, but not in response to predation by thrips?

Volatile Compound Produced	Treatment					
	C	T	W	WT	HV	HVT
Myrcene	0	0	0	0	17	22
β-Ocimene	0	433	15	121	4,299	5,315
Linalool	0	0	0	0	125	178
Indole	0	0	0	0	74	142
Nicotine	0	0	233	160	390	538
β-Elemene	0	0	0	0	90	102
β-Caryophyllene	0	100	40	124	3,704	6,166
α-Humulene	0	0	0	0	123	209
Sesquiterpene	0	7	0	0	219	268
α-Farnesene	0	15	0	0	293	457
Caryophyllene oxide	0	0	0	0	89	166
Total	0	555	288	405	9,423	13,563

Figure 30.21 Volatile (airborne) compounds produced by tobacco plants (*Nicotiana tabacum*) in response to predation by different insects. Plants were untreated (C), attacked by thrips (T), mechanically wounded (W), mechanically wounded and attacked by thrips (WT), attacked by budworms (HV), or attacked by budworms and thrips (HVT). Values are nanograms/day.

8. In some plants, flowering is a _____ response.
   a. phototropic
   b. gravitropic
   c. photoperiodic
   d. thigmotropic

9. Stomata close in response to _____ .
   a. ABA
   b. nitric oxide
   c. bacteria
   d. all of the above

10. Match the response with its main trigger.
    ___ phototropism   a. mechanical stimulation
    ___ gravitropism   b. blue light
    ___ thigmotropism   c. angle of sunlight
    ___ heliotropism   d. gravity
    ___ vernalization   e. night length
    ___ photoperiodism   f. a long period of cold

11. Match the hormone with the characteristic.
    ___ ethylene   a. efflux carriers set up gradients
    ___ cytokinin   b. produced in positive and
    ___ auxin   negative feedback loops
    ___ gibberellin   c. affects expression of more
    ___ ABA   genes than any other hormone
    ___ nitric oxide d. works antagonistically to auxin
                         in apical dominance
                      e. big in stems
                      f. active in bursts

12. Match the the hormone with the observation.
    ___ ethylene   a. Your cabbage plants bolt (they
    ___ cytokinin   form elongated flowering stalks).
    ___ auxin   b. The potted plant in your room
    ___ gibberellin   is leaning toward the window.
    ___ ABA   c. The last of your apples is getting
    ___ nitric oxide   really mushy.
                      d. The seeds of your roommate's
                         marijuana plant do not germinate
                         no matter what he does to them.
                      e. Lateral buds are sprouting.
                      f. Your lettuce plants develop
                         brown spots on their leaves.

Critical Thinking

1. Professional gardeners often soak seeds in hydrogen peroxide before planting them. Why?

2. Photosynthesis sustains plant growth, and inputs of sunlight sustain photosynthesis. Why, then, do seedlings that germinate in a fully darkened room grow taller than seedlings of the same species that germinate in full sun?

3. Belgian scientists discovered that certain mutations in common wall cress (*Arabidopsis thaliana*) cause excess auxin production. Predict the impact on the plant's phenotype.

4. Cattle in industrial dairy farms are typically given rBST, or recombinant bovine somatotropin. This animal hormone increases a cow's milk production, but there is concern that such hormones may have unforeseen effects on milk-drinking humans. There is no similar concern about plant hormones in the plant foods we eat. Why not?

5. The oat coleoptiles *below* have been modified: either cut or placed in a light-blocking tube. Which ones will still bend toward a directional light source?

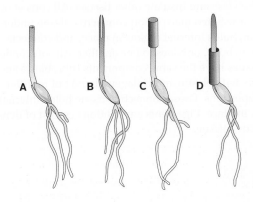

Chapter 31
Summary

 Section 31.1 All cells in an animal body are derived from **stem cells**, cells that can divide or differentiate into a specialized cell type. The first divisions of a fertilized egg yield totipotent cells that can form any tissue or develop into a new individual. Later embryos have pluripotent cells that can still form any tissue. After birth, cells are less versatile. Researchers hope to use embryonic stem cells to produce new cells of types that are not normally replaced in adults. They are also trying to making adult cells behave like embryonic stem cells.

 Section 31.2 Most animals have four types of tissues organized as organs and organ systems. **Extracellular fluid** serves as the body's internal environment. In humans, it consists mainly of **interstitial fluid** and plasma. Animal structure has been influenced both by physical constraints and evolutionary history.

 Section 31.3 Epithelial tissue covers the body surface and lines its internal tubes and cavities. Epithelial cells have little extracellular matrix between them. An epithelium has a free apical surface. Its basal surface secretes a **basement membrane** that attaches it to underlying connective tissue. Tight junctions occur only in epithelial cells. The gut is lined by an epithelium with tight junctions. Gut epithelium functions as a selective barrier by controlling which substances move into the body's internal environment. Some ciliated epithelial cells move materials across their surface. Others have **microvilli** that increase their surface area for absorption or secretion. Hair, fur, and nails are keratin-rich remains of specialized epithelial cells.

Gland cells are epithelial cells whose secretions act outside the cell. Ductless **endocrine glands** secrete hormones into blood. **Exocrine glands** secrete products such as milk or saliva through ducts.

 Section 31.4 Connective tissues "connect" tissues to one another, both functionally and structurally. Different types bind, organize, support, strengthen, protect, and insulate other tissues. All consist of cells in a secreted matrix. Soft connective tissue underlies skin, holds internal organs in place, and connects muscle to bone, or bones to one another. All soft connective tissues have the same components (fibroblasts and a matrix with elastin and collagen fibers) but in different proportions. Loose connective tissue holds internal organs in place. Ligaments and tendons consist of dense connective tissue.

Rubbery **cartilage** and mineral-hardened **bone tissue** are components of the skeleton. Fat stored in **adipose tissue** is the body's main energy reservoir. **Blood** consists of fluid plasma, cells, and platelets. It is considered a connective tissue because blood cells and platelets arise from stem cells in bone.

 Section 31.5 Muscle tissues contract and move a body or its parts. Muscle contraction is a response to signals from the nervous system and is fueled by ATP.

Skeletal muscle consists of long fibers with multiple nuclei and has a striated (striped) appearance. Skeletal muscles, which pull on bones, are under voluntary control. They have large stores of glycogen. The metabolic reactions carried out by skeletal muscle are the main source of body heat.

Cardiac muscle, found only in the heart wall, consists of branching cells and has a striated appearance. Gap junctions allow signals to travel fast between the cells.

Smooth muscle tissue is found in the walls of tubular organs and some blood vessels. Its cells taper at both ends and are not striated.

 Section 31.6 Nervous tissue makes up the communication lines through the body. It consists of **neurons** that send and receive electrochemical signals, and **neuroglial cells** that support them. A neuron has a central cell body and long cytoplasmic extensions that send and receive signals. Sensory neurons detect information, interneurons integrate and assess information about internal and external conditions, and motor neurons command muscles and glands.

 Section 31.7 Vertebrates are bilateral and coelomate and many of their internal organs reside inside a body cavity derived from the coelom. An organ system consists of two or more organs that interact chemically, physically, or both in tasks that help keep individual cells as well as the whole body functioning smoothly. All vertebrates have the same set of organ systems.

 Section 31.8 The integumentary system consists of skin and structures such as hair that are derived from it. It functions in protection, temperature control, detection of shifts in external conditions, vitamin production, and defense against pathogens. The outermost layer of skin, the **epidermis**, is a stratified squamous epithelium consisting mainly of keratinocytes. Melanocytes produce the melanin that gives skin its color and serves as a natural sunblock. The deeper **dermis** consists mainly of dense connective tissue and contains blood vessels, nerves, and muscles. Underlying the dermis is the hypodermis, a layer of connective tissue and adipose cells.

Sweat glands and hair follicles are collections of epidermal cells that descended into the dermis during development. Compared to our closest related primate relatives, we have more sweat glands and finer, shorter body hair. These traits helped our early ancestors in Africa disperse heat generated by walking and running under hot conditions. The reduction in body hair was accompanied by an increase in melanin deposition that protected the skin against sunlight. Later, when some humans moved to regions with less sunlight, their skin color reverted to a lighter state.

Section 31.9 Homeostasis requires **sensory receptors** that detect changes, an integrating center (the brain), and effectors (muscles and glands) that bring about responses. **Negative feedback** often plays a role in homeostasis: A change causes the body to respond in a way that reverses the change.

Self-Quiz

Answers in Appendix III

1. _____ tissues are sheetlike with one free surface.
   - a. Epithelial
   - b. Muscle
   - c. Nervous
   - d. Connective

2. _____ are found only in epithelial tissue.
   - a. Tight junctions
   - b. Adhering junctions
   - c. Gap junctions
   - d. all of the above

3. Glands are specialized _____ tissue.
   - a. epithelial
   - b. muscle
   - c. nervous
   - d. connective

4. A rubbery secreted matrix of glycoproteins and collagen surrounds the cells in _____ .
   - a. bone
   - b. cartilage
   - c. adipose tissue
   - d. blood

5. Blood cells develop from stem cells in _____.
   - a. epidermis
   - b. dermis
   - c. cartilage
   - d. bone

6. Your body's main energy reservoir is _____ .
   - a. glycogen stored in cardiac muscle
   - b. lipids stored in adipose tissue
   - c. starch stored in skeletal muscle
   - d. phosphorus stored in bone

7. Cytoplasmic extensions of _____ send and receive chemical messages.
   - a. neuroglial cells
   - b. neurons
   - c. fibroblasts
   - d. melanocytes

8. _____ muscle pulls on bones and _____ muscle regulates the diameter of blood vessels.
   - a. Skeletal/cardiac
   - b. Smooth/cardiac
   - c. Skeletal/smooth
   - d. Smooth/skeletal

9. Straps of dense, regular connective tissue _____ .
   - a. connect muscles to bones
   - b. produce blood cells
   - b. underlie the skin
   - d. lack fibroblasts

10. Microvilli _____ some epithelial cells.
    - a. provide pigment to
    - b. increase surface area of
    - c. wrap tightly around
    - d. receive signals from

11. Tears are an _____ secretion released by specialized _____ tissue cells.
    - a. endocrine/epithelial
    - b. endocrine/connective
    - c. exocrine/epithelial
    - d. exocrine/connective

12. Cancers most commonly arise in _____ tissue.
    - a. epithelial
    - b. muscle
    - c. nervous
    - d. connective

13. Adhering junctions attach cells of _____ end to end.
    - a. cartilage
    - b. cardiac muscle
    - c. loose connective tissue
    - d. nervous tissue

14. With negative feedback, detection of a change brings about a response that _____ the change.
    - a. reverses
    - b. accelerates
    - c. has no effect on
    - d. mimics

15. Match each term with the most suitable description.
    - ___ exocrine gland
    - ___ endocrine gland
    - ___ fibroblast
    - ___ melanocyte
    - ___ neuron
    - ___ smooth muscle
    - ___ skeletal muscle
    - ___ blood
    - ___ keratinocyte
    - ___ extracelluar fluid
    - a. signaling cell in nervous tissue
    - b. secretion through duct
    - c. collagen-producing cell
    - d. contraction is involuntary
    - e. pigment-producing cell
    - f. main source of metabolic heat
    - g. main cell type in epidermis
    - h. fluid connective tissue
    - i. includes interstitial fluid, lymph
    - j. secretes hormones

Critical Thinking

1. Many people oppose the use of animals for testing the safety of cosmetics. They argue that alternative test methods are available, such as the use of lab-grown tissues in some cases. Given what you learned in this chapter, speculate on the advantages and disadvantages of tests that use lab-grown tissues as opposed to living animals.

2. Radiation and chemotherapy drugs preferentially kill cells that divide frequently, most notably cancer cells. These cancer treatments also cause hair to fall out. Why?

3. Each level of biological organization has emergent properties that arise from the interaction of its component parts. For example, cells have a capacity for inheritance that molecules making up the cell do not. What are some emergent properties of specific types of tissues?

4. The micrograph to the *right* shows cells from the lining of an airway leading to the lungs. The *gold* cells are ciliated and the darker *brown* ones secrete mucus. What type of tissue is this? How can you tell?

Cultured skin for healing wounds Diabetes is a disorder in which the blood sugar level is not properly controlled. Among other effects, this disorder reduces blood flow to the lower legs and feet. As a result, about 3 million diabetes patients have ulcers, or open wounds that do not heal, on their feet. Each year, about 80,000 require amputations.

Several companies provide cultured cell products designed to promote the healing of diabetic foot ulcers. **Figure 31.19** shows the results of a clinical experiment that tested the effect of the cultured skin product shown in **Figure 31.18** versus standard treatment for diabetic foot wounds. Patients were randomly assigned to either the experimental treatment group or the control group and their progress was monitored for 12 weeks.

1. What percentage of wounds had healed at 8 weeks when treated the standard way? When treated with cultured skin?

2. What percentage of wounds had healed at 12 weeks when treated the standard way? When treated with cultured skin?

3. How early was the healing difference between the control and treatment groups obvious?

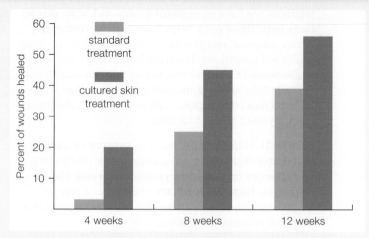

Figure 31.19 Results of a multicenter study of the effects of standard treatment versus use of a cultured cell product for diabetic foot ulcers. Bars show the percentage of foot ulcers that had completely healed.

Chapter 32
Summary

Section 32.1 Ecstasy is a type of amphetamine. Its pleasurable effects stem from its action at synapses where serotonin is released. Use of the drug can harm neurons at these synapses and also cause other detrimental, and occasionally deadly, side effects.

Section 32.2 Neurons are electrically excitable cells that signal other cells by means of chemical messages. **Sensory neurons** detect stimuli. **Interneurons** relay signals between neurons. **Motor neurons** signal muscles and glands.

Cnidarians have a **nerve net**. Most animals have a bilateral nervous system with **cephalization**; they have paired **ganglia** or a brain at the head end. Chordates have a dorsal **nerve cord**. A vertebrate **central nervous system** (CNS) consists of the brain and spinal cord. **Nerves** that run through the body and connect to CNS constitute the **peripheral nervous system**.

Sections 32.3–32.4 A neuron's **dendrites** receive signals and its **axon** transmits signals. Neurons maintain a **resting membrane potential** across their plasma membrane. An **action potential** is a brief reversal of the membrane potential that occurs after membrane potential increases to the **threshold potential**. An action potential occurs when voltage-gated sodium channels open, allowing sodium ions to follow their concentration gradient and diffuse into the neuron. The increasing flow of sodium is an example of **positive feedback**. The influx causes voltage-gated potassium channels to open, so potassium ions follow their concentration gradient and diffuse out of the neuron. All action potentials are the same size and travel in one direction: away from the cell body and toward the axon terminals.

Section 32.5 Neurons send chemical signals to cells at **synapses**. A motor neuron communicates with a muscle fiber at a **neuromuscular junction**. The arrival of an action potential at a presynaptic cell's axon terminal triggers the release of a chemical signaling molecule called a **neurotransmitter**. Neurotransmitter diffuses to receptors on a postsynaptic cell and binds to them. A postsynaptic cell often receives signals from many presynaptic cells; its response is determined by **synaptic integration** of all of these signals.

Section 32.6 Different kinds of neurons produce different neurotransmitters. The structure of receptors for a particular neurotransmitter also varies, so the same neurotransmitter can elicit different effects in different cells. Some neurons produce **neuromodulators** that can alter the effects of neurotransmitters.

With some neurological disorders, there is an abnormal amount of a neurotransmitter. Psychoactive drugs interfere with signaling at synapses. Dopamine plays a role in reward-based learning, and drug addiction taps into this learning pathway.

Section 32.7 Nerves are bundles of axons that carry signals. **Myelin sheaths** enclose most axons and increase the speed of action potential propagation along the length of an axon.

The peripheral nervous system is functionally divided into the **somatic nervous system**, which controls skeletal muscles, and the **autonomic nervous system**, which controls internal organs and glands.

Sympathetic neurons of the autonomic system are most active in times of stress or danger, when they cause a **fight–flight response**. During less stressful times the **parasympathetic neurons** are most active. Organs receive signals from both types of neurons.

Section 32.8 Like the brain, the **spinal cord** consists of **white matter** (with myelinated axons) and **gray matter** (with cell bodies, dendrites, and neuroglia). The spinal cord and brain are enclosed by **meninges** and cushioned by **cerebrospinal fluid**. Spinal reflexes involve peripheral nerves and the spinal cord. A **reflex** is an automatic response to stimulation; it does not require conscious thought or learning.

Sections 32.9–32.11 The neural tube of a vertebrate embryo develops into the spinal cord and brain. Evolutionarily, the **brain stem** is the oldest type of brain tissue. It includes the **pons** and **medulla oblongata**, which control reflexes involved in breathing and other essential tasks. The **cerebellum** acts in motor control. The **thalamus** and **hypothalamus** function in homeostasis. A **blood–brain barrier** protects the brain from many harmful chemicals.

The corpus callosum connects the two halves of the **cerebrum**. The **cerebral cortex**, the most recently evolved region of the brain, governs complex functions. The **primary somatosensory cortex** receives sensory input from the skin, and the **primary motor cortex** controls voluntary movements. The cerebral cortex interacts with the **limbic system** in emotion and memory.

Section 32.12 Neuroglial cells make up the bulk of the brain. They insulate axons of the CNS, produce growth factors that encourage synapse formation, contribute to the blood brain barrier, and carry out other supportive tasks. Uncontrolled divisions of neuroglia can cause brain tumors, and damage to neuroglia causes neurological disorders such as multiple sclerosis.

Self-Quiz
Answers in Appendix III

1. _____ relay messages from the brain and spinal cord to muscles and glands.
    a. Motor neurons   b. Interneurons   c. Sensory neurons

Effect of MDMA on Neural Development Animal studies are often used to assess effects of prenatal exposure to illicit drugs. For example, Jack Lipton used rats to study the behavioral effect of prenatal exposure to MDMA, the active ingredient in Ecstasy. He injected female rats with either MDMA or saline solution when they were 14 to 20 days pregnant. This is the period when their offspring's brains were forming. When those offspring were 21 days old, Lipton tested their ability to adjust to a new environment. He placed each young rat in a new cage and used a photobeam system to record how much each rat moved around before settling down. **Figure 32.27** shows his results.

1. Which rats moved around most (caused the most photobeam breaks) during the first 5 minutes in a new cage, those prenatally exposed to MDMA or the controls?

2. How many photobeam breaks did the MDMA-exposed rats make during their second 5 minutes in the new cage?

3. Which rats moved around most during the last 5 minutes?

4. Does this study support the hypothesis that MDMA affects a developing rat's brain?

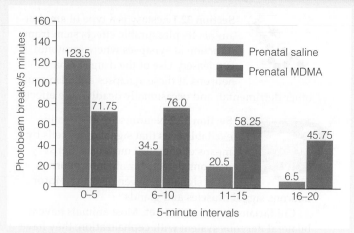

Figure 32.27 Effect of prenatal exposure to MDMA on activity levels of 21-day-old rats placed in a new cage. Movements were detected when the rat interrupted a photobeam. Rats were monitored at 5-minute intervals for a total of 20 minutes. *Blue* bars are average numbers of photobeam breaks for rats whose mothers received saline; *red* bars are for rats whose mothers received MDMA.

2. When a neuron is at rest, _____ .
   a. it is at threshold potential
   b. gated sodium channels are open
   c. sodium–potassium pumps are operating
   d. both a and c

3. Action potentials occur when _____ .
   a. a neuron receives adequate stimulation
   b. more and more sodium gates open
   c. sodium–potassium pumps kick into action
   d. both a and b

4. True or false? Action potentials vary in their size.

5. Neurotransmitters are released by _____ .
   a. axon terminals   c. dendrites
   b. the cell body   d. the myelin sheath

6. What chemical is released by axon terminals of a motor neuron at a neuromuscular junction?
   a. ACh   b. serotonin   c. dopamine   d. epinephrine

7. Which neurotransmitter is important in reward-based learning and drug addiction?
   a. ACh   b. serotonin   c. dopamine   d. epinephrine

8. Skeletal muscles are controlled by _____ .
   a. sympathetic neurons   c. somatic nerves
   b. parasympathetic neurons d. both a and b

9. When you sit quietly on the couch and read, output from _____ neurons prevails.
   a. sympathetic   b. parasympathetic

10. Cell bodies of the sensory neurons that deliver signals to the spinal cord are in the _____ .
    a. brain   b. skin   c. dorsal root ganglia   d. meninges

11. Which of the following are not in the brain?
    a. Schwann cells   b. astrocytes   c. microglia

12. Neurons are arrested in _____ of the cell cycle.
    a. G1   b. anaphase   c. prophase   d. S phase

13. _____ plays a role in emotion and memory.
    a. The brain stem   c. The limbic system
    b. Broca's area   d. The somatosensory cortex

14. Commands to move your right arm start in the _____.
    a. left frontal lobe   c. right temporal lobe
    b. right occipital lobe   d. left parietal lobe

15. Match each item with its description.
    ___ gray matter   a. start of brain, spinal cord
    ___ neurotransmitter   b. connects the hemispheres
    ___ pons   c. protects brain and spinal
    ___ corpus callosum   cord from some toxins
    ___ cerebral cortex   d. type of signaling molecule
    ___ neural tube   e. brain's myelin makers
    ___ oligodendrocytes   f. brain stem structure
    ___ blood–brain   g. most complex integration
        barrier   h. cell bodies and dendrites

Critical Thinking

1. In humans, the axons of some motor neurons extend more than a meter, from the base of the spinal cord to the big toe. What are some of the functional challenges involved in the development and maintenance of such impressive cellular extensions?

2. Some survivors of disastrous events develop post-traumatic stress disorder (PTSD). Symptoms include nightmares about the experience and suddenly feeling as if the event is recurring. Brain-imaging studies of people with PTSD showed that their hippocampus was shrunken and their amygdala unusually active. Given these changes, what other brain functions might be disrupted in PTSD?

3. In human newborns, especially premature ones, the blood–brain barrier is not yet fully developed. Why is this one reason to pay careful attention to the diet of infants?

Chapter 33
Summary

 Section 33.1 An animal's sensory system adapts it to life in a particular environment. Human activities have dramatically altered sensory environments of some species, with harmful effects.

 Section 33.2 The types of sensory receptors that an animal has determine the types of **stimuli** it detects and can respond to. Stimulation of a sensory receptor causes action potentials. **Mechanoreceptors** respond to mechanical energy such as touch. **Pain receptors** respond to tissue damage. **Thermoreceptors** are sensitive to temperature. **Chemoreceptors** fire when a dissolved chemical binds. **Osmoreceptors** sense and respond to water concentration. **Photoreceptors** respond to light.

The brain evaluates action potentials from sensory receptors based on which nerves deliver them, their frequency, and the number of axons firing in any given interval. Continued stimulation of a receptor may lead to a diminished response (**sensory adaptation**). **Sensation** is detection of a stimulus, whereas **sensory perception** involves assigning meaning to a sensation.

 Section 33.3 **Somatic sensations** are easy to pinpoint and arise from receptors in the skin, joints, and skeletal muscles. Signals from these receptors flow to an area of the cerebral cortex where interneurons are organized like maps of individual parts of the body surface. **Visceral sensations** originate from receptors in the walls of soft organs and are less easily pinpointed. **Pain** is the perception of tissue damage. With referred pain, the brain mistakenly attributes signals that come from an internal organ to the skin or muscles.

 Section 33.4 The senses of taste and **olfaction** (smell) involve molecules binding to chemoreceptor proteins. Binding triggers action potentials, and these signals are relayed to the cerebral cortex and limbic system. In vertebrates, **olfactory receptors** line the nasal passages and **taste receptors** are concentrated in taste buds on the tongue and lining of the mouth. There are five classes of taste receptors, but hundreds of types of olfactory receptors.

Pheromones are chemical signals that act as social cues among many animals. A **vomeronasal organ** functions in detection of pheromones in many vertebrates.

 Section 33.5 An **eye** is a sensory organ that contains a dense array of photoreceptors. **Vision** requires eyes and brain centers capable of processing the visual information from the eyes.

Insects have a **compound eye**, with many individual units. Each unit has a **lens**, a structure that bends light so that it falls on photoreceptors. Like squids and octopuses, humans have **camera eyes**, with an adjustable opening that lets in light, and a single lens that focuses the light on a photoreceptor-rich **retina**.

The arrangement and structure of vertebrate eyes vary. Depth perception arises when two forward-facing eyes send the brain information about the same viewed area. Large eyes and a light-reflecting layer in the eye adapt animals to low light conditions.

 Section 33.6 A human eye sits in a bony orbit and is protected by eyelids lined by the **conjunctiva**. This membrane also covers the **sclera**, or white of the eye. The clear, curved **cornea** at the front of the eye bends incoming light. Light enters the eye's interior through the **pupil**, an adjustable opening in the center of the muscular, doughnut-shaped **iris**. Light that enters the eye falls on the retina. The retina sits on a pigmented **choroid** that minimizes reflections inside the eye.

With **visual accommodation**, the **ciliary muscle** adjusts the shape of the lens so that light from a near or distant object falls on the retina's photoreceptors.

 Sections 33.7, 33.8 Humans have two types of photoreceptors. **Rod cells** detect dim light and are important in coarse vision and peripheral vision. **Cone cells** detect bright light and colors, and they provide a sharp image. The greatest concentration of cones is in the portion of the retina called the **fovea**.

The rods and cones interact with other cells in the retina that begin processing visual information before sending it to the brain. Visual signals travel to the cerebral cortex along two optic nerves. There are no photoreceptors in the eye's blind spot, the area where the optic nerve begins.

Common vision disorders result from defective or degenerated photoreceptors, misshapen eyes, a clouded lens, or excess aqueous humor.

 Section 33.9 **Hearing** is the perception of sound, which is a form of mechanical energy. Sound waves are pressure waves. We perceive variations in their amplitude as differences in loudness and variations in wave frequency as differences in pitch.

Human ears have three functional regions. The pinna of the **outer ear** collects sound waves. The **middle ear** contains the **eardrum** and a set of tiny bones that amplify sound waves and transmit them to the inner ear. In the **inner ear**, pressure waves elicit action potentials inside a **cochlea**. This coiled structure with fluid-filled ducts holds the mechanoreceptors responsible for hearing in its **organ of Corti**.

Pressure waves traveling through the fluid inside the cochlea bend hair cells of the organ of Corti. The brain gauges the loudness of a sound by the number of action potentials it elicits. It determines a sound's pitch by which part of the cochlea's coil the signals arrive from.

Section 33.10 Organs of equilibrium detect effects of gravity and acceleration to determine body positions and motions. The **vestibular apparatus** is a system of fluid-filled sacs and canals in the inner ear. The sense of dynamic equilibrium arises when body movements cause shifts in the fluid, which causes cilia of hair cells to bend. Static equilibrium depends on signals from hair cells that lie beneath a weighted, jellylike mass. A shift in head position or a sudden stop or start shifts the mass, bends the hair cells, and alters the rate at which they undergo action potentials.

15. Match each structure with its description.

___ rod cell	a. protects eyeball
___ cochlea	b. transmits vibration to bone
___ eardrum	c. functions in balance
___ lens	d. detects pheromones
___ sclera	e. detects dim light
___ cone cell	f. contains chemoreceptors
___ taste bud	g. focuses rays of light
___ Pacinian corpuscle	h. detects color
___ pinna	i. collects sound waves
___ vestibular	j. sorts out sound waves
apparatus	k. detects touch
___ vomeronasal organ	

Self-Quiz

Answers in Appendix III

1. The pain of heartburn is an example of a _____ .
   a. somatic sensation   c. sensory adaptation
   b. visceral sensation   d. spinal reflex

2. _____ is defined as a decrease in the response to an ongoing stimulus.
   a. Perception   c. Sensory adaptation
   b. Visual accommodation   d. Somatic sensation

3. Which is a somatic sensation?
   a. taste   c. touch   e. a through c
   b. smell   d. hearing   f. all of the above

4. Chemoreceptors play a role in the sense of _____ .
   a. taste   c. touch   e. both a and b
   b. smell   d. hearing   f. all of the above

5. In the _____ , neurons are arranged like maps that correspond to different parts of the body surface.
   a. retina   c. basilar membrane
   b. somatosensory cortex   d. occipital lobe

6. Mechanoreceptors in the _____ send signals to the brain about the body's position relative to gravity.
   a. eye   b. ear   c. tongue   d. nose

7. The middle ear functions in _____ .
   a. detecting shifts in body position
   b. amplifying and transmitting sound waves
   c. sorting sound waves out by frequency
   d. all of the above

8. Substance P _____ .
   a. increases pain-related signals
   b. is a natural painkiller
   c. is the active ingredient in aspirin

9. The organ of Corti contains receptors that signal in response to _____ .
   a. heat   b. sound   c. light   d. pheromones

10. Night vision begins with stimulation of _____ .
    a. hair cells   b. rod cells   c. cone cells   d. neuroglia

11. Visual accommodation involves adjustment to the shape or position of the _____ .
    a. conjunctiva   b. retina   c. orbit   d. lens

12. When you view a close object your lens gets _____ .
    a. flatter   b. rounder   c. darker   d. cloudier

13. Defective _____ cause color blindness.
    a. hair cells   b. rod cells   c. cone cells   d. neuroglia

14. _____ causes the pupil to widen.
    a. Low light   b. Bright light

Critical Thinking

1. Like other nocturnal carnivores, the ferret shown in **Figure 33.14** has light-reflecting material in its choroid. Explain why the presence of reflective material in this layer of the eye maximizes the degree to which light excites photoreceptors. Explain also why having reflective material in this layer causes the perceived image to be somewhat blurry.

2. A compound extracted from the leaves of the shrub *Stevia rebaudiana* shows promise as a natural sugar substitute. The compound is 300 times sweeter than sugar, but it also has a slight bitter aftertaste. Given what you know about taste receptors, explain how a compound can be perceived as both sweet and bitter.

3. The strength of Earth's magnetic field and its angle relative to the surface vary with latitude. Diverse species sense these differences and use them as cues for assessing their location and direction of movement. Behavioral experiments have shown that sea turtles, salamanders, and spiny lobsters use information from Earth's magnetic field during their migrations. Whales and some burrowing rodents also seem to have a magnetic sense. Evidence about humans is contradictory. Suggest an experiment to test whether humans can detect a magnetic field.

4. After a leg injury, pain makes a person avoid putting too much weight on the affected leg. Shielding the injury gives it time to heal. An injured insect shows no such shielding response when its leg is injured. Some have cited the lack of such a response as evidence that insects do not feel pain. On the other hand, insects do produce substances similar to encephalins, our natural painkillers. Is the presence of these compounds in insects sufficient evidence to conclude that they do feel pain?

Occupational Hearing Loss Frequent exposure to loud noise of a particular pitch can cause loss of hair cells in the part of the cochlea's coil that responds to that pitch. People who work with or around noisy machinery are at risk for such frequency-specific hearing loss. Taking precautions such as using ear plugs to reduce sound exposure is important. Noise-induced hearing loss can be prevented, but once it occurs it is irreversible because dead or damaged hair cells are not replaced.

Figure 33.25 shows the threshold decibel levels at which sounds of different frequencies can be detected by an average 25-year-old carpenter, a 50-year-old carpenter, and a 50-year-old who has not been exposed to on-the-job noise. Sound frequencies are given in hertz (cycles per second). The more cycles per second, the higher the pitch.

1. Which sound frequency was most easily detected by all three people?

2. How loud did a 1,000-hertz sound have to be for the 50-year-old carpenter to detect it?

3. Which of the three people had the best hearing in the range of 4,000 to 6,000 hertz? Which had the worst?

4. Based on this data, would you conclude that the hearing decline in the 50-year-old carpenter was caused by age or by job-related noise exposure?

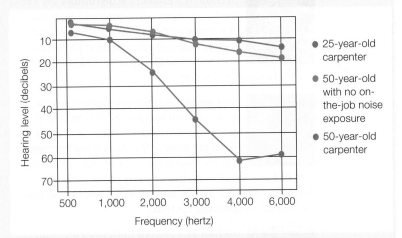

Figure 33.25 Effects of age and occupational noise exposure. The graph shows the threshold hearing capacities (in decibels) for sounds of different frequencies (given in hertz) in a 25-year-old carpenter (*blue*), a 50-year-old carpenter (*red*), and a 50-year-old who did not have any on-the-job noise exposure (*brown*).

Chapter 34
Summary

Section 34.1 Endocrine disruptors are synthetic substances that interfere with the endocrine system. The widely used synthetic chemicals atrazine and phthalates are endocrine disruptors. Both disrupt the body's balance of sex hormones.

Section 34.2 Animal cells communicate with adjacent body cells by way of gap junctions, neurotransmitters, and **local signaling molecules**. An **animal hormone** travels through the blood and can convey signals between cells in distant parts of the body. All hormone-secreting glands and cells in a body constitute the animal's **endocrine system**.

Section 34.3 Steroid hormones are lipid soluble and derived from cholesterol. They can enter cells and bind to receptors inside them. Peptide and protein hormones are derived from amino acids. They bind to receptors in the cell membrane. Binding triggers the formation of a **second messenger**, a molecule that elicits changes inside the cell. For a cell to respond to a hormone, it must have functional receptors for that hormone. Variation in the protein receptors for a hormone allow a hormone to elicit different responses.

Sections 34.4–34.6 The **hypothalamus**, deep in the forebrain, is structurally and functionally linked with the **pituitary gland**. Axons of neurons with cell bodies in the hypothalamus extend into the posterior pituitary, where they release antidiuretic hormone and oxytocin. Antidiuretic hormone concentrates the urine by acting in the kidney. Oxytocin targets smooth muscle in mammary glands and the uterus.

Releasing hormones and **inhibiting hormones** secreted by the hypothalamus control the secretion of hormones made by the anterior lobe of the pituitary. Four anterior pituitary hormones target other glands (the adrenal cortex, the thyroid gland, and gonads). Another anterior pituitary hormone encourages milk production. Growth hormone (GH) secreted by the anterior pituitary acts throughout the body. Pituitary gigantism and dwarfism result from mutations that affect GH secretion or receptors for this hormone.

In addition to the major endocrine glands, there are hormone-secreting cells in tissues and organs throughout the body. Most cells have receptors for, and are influenced by, many different hormones.

Section 34.7 A feedback loop to the anterior pituitary and hypothalamus governs the **thyroid gland** in the base of the neck. Production of thyroid hormone requires iodine. The hormone is required for normal development and influences metabolic rate. A lower-than-normal level of thyroid hormone slows metabolism and causes weight gain. The thyroid also produces calcitonin, a hormone that causes calcium deposition in bone during childhood.

Four **parathyroid glands** make a hormone that acts on bone and kidney cells and raises the blood calcium level.

Section 34.8 The **pancreas**, located inside the abdominal cavity, has both exocrine and endocrine functions. Beta cells secrete the hormone **insulin** when the blood glucose level is high. Insulin stimulates uptake of glucose by muscle and liver cells. When the blood glucose level is low, alpha cells secrete **glucagon**, a hormone that calls for glycogen breakdown and glucose release by the liver. The two hormones work in opposition to keep the blood glucose concentration within an optimal range.

Diabetes occurs when the body does not make insulin or its cells do not respond to it.

Section 34.9 There is an **adrenal gland** atop each kidney. The gland's outer layer, or **adrenal cortex**, secretes aldosterone, which acts on the kidney, and **cortisol**, the stress hormone. Cortisol secretion is governed by a negative feedback loop to the anterior pituitary and hypothalamus. In times of stress, the nervous system overrides feedback controls that normally keep the cortisol level stable. Norepinephrine and epinephrine released by neurons of the **adrenal medulla** influence organs as sympathetic stimulation does; they cause a fight–flight response. Long-term elevation of cortisol level, as a result of either stress or a disorder, is harmful to health. A lack of cortisol can be fatal.

Section 34.10 The **gonads** (ovaries or testes) make gametes and secrete the **sex hormones**. Ovaries secrete mostly estrogens and progesterone. Testes secrete mostly testosterone. Sex hormone output rises at **puberty**, and the heightened concentration of these hormones encourages development of **secondary sexual traits** such as facial hair or rounded breasts.

In both sexes, follicle-stimulating hormone (FSH) and luteinizing hormone (LH) from the anterior pituitary govern sex hormone production. LH and FSH secretion are in turn controlled by gonadotropin-releasing hormone from the hypothalamus.

Sections 34.11, 34.12 Light suppresses secretion of **melatonin** by the **pineal gland** in the brain. Melatonin secretion has a **circadian rhythm** and results in changes in body temperature and in drowsiness. Melatonin also has a protective effect against cancer.

The **thymus** in the chest produces hormones that help some white blood cells (T cells) mature. It is essential to life in children, but largely inactive in adults over age 40.

Section 34.13 Some vertebrate hormone receptors are homologous to those in invertebrates, but **ecdysone**, a steroid hormone that affects molting in arthropods, has no vertebrate counterpart.

Sperm Counts Down on the Farm Contamination of water by agricultural chemicals affects reproductive function in some animals. Are there effects on humans? Epidemiologist Shanna Swan and her colleagues studied sperm from men in four cities in the United States (**Figure 34. 21**). The men were partners of women who had become pregnant and were visiting a prenatal clinic, so all were fertile. Of the four cities, Columbia, Missouri, is located in the county with the most farmlands. New York City in New York is in an area with no agriculture.

1. In which cities did researchers record the highest and lowest sperm counts?

2. In which cities did samples show the highest and lowest sperm motility (ability to move)?

3. Aging, smoking, and sexually transmitted diseases adversely affect sperm. Could differences in any of these variables explain the regional differences in sperm count?

4. Do these data support the hypothesis that living near farmlands can adversely affect male reproductive function?

	Location of clinic			
	Columbia, Missouri	Los Angeles, California	Minneapolis, Minnesota	New York, New York
Average age	30.7	29.8	32.2	36.1
Percent nonsmokers	79.5	70.5	85.8	81.6
Percent with history of STD	11.4	12.9	13.6	15.8
Sperm count (million/ml)	58.7	80.8	98.6	102.9
Percent motile sperm	48.2	54.5	52.1	56.4

Figure 34.21 Characteristics of men in four cities. All men were partners of women who visited prenatal health clinics, and so were presumably fertile. STD stands for sexually transmitted disease.

Self-Quiz

Answers in Appendix III

1. _____ are signaling molecules that travel through the blood and affect distant cells in the same individual.
    - a. Hormones
    - b. Neurotransmitters
    - c. Pheromones
    - d. Local signaling molecules
    - e. both a and b
    - f. a through d

2. A _____ is synthesized from cholesterol and can diffuse across the plasma membrane.
    - a. steroid hormone
    - b. hormone receptor
    - c. peptide hormone
    - d. all of the above

3. Match each pituitary hormone with its target.
    - ___ antidiuretic hormone
    - ___ oxytocin
    - ___ luteinizing hormone
    - ___ growth hormone
    - a. gonads (ovaries, testes)
    - b. mammary glands, uterus
    - c. kidneys
    - d. most body cells

4. Releasing hormones secreted by the hypothalamus cause secretion of hormones by the _____ pituitary lobe.
    - a. anterior
    - b. posterior

5. In adults, too much _____ can cause acromegaly.
    - a. melatonin
    - b. cortisol
    - c. insulin
    - d. growth hormone

6. A diet lacking in iodine can cause _____ .
    - a. rickets
    - b. a goiter
    - c. diabetes
    - d. gigantism

7. Low blood calcium triggers secretion by _____ .
    - a. adrenal glands
    - b. ovaries
    - c. parathyroid glands
    - d. the thyroid gland

8. _____ lowers blood sugar levels; _____ raises it.
    - a. Glucagon; insulin
    - b. Insulin; glucagon

9. The _____ has endocrine and exocrine functions.
    - a. hypothalamus
    - b. parathyroid gland
    - c. pineal gland
    - d. pancreas

10. Secretion of _____ suppresses immune responses.
    - a. melatonin
    - b. antidiuretic hormone
    - c. thyroid hormone
    - d. cortisol

11. Exposure to bright light lowers blood _____ levels.
    - a. glucagon
    - b. melatonin
    - c. thyroid hormone
    - d. parathyroid hormone

12. True or false? Some heart cells and kidney cells secrete hormones.

13. True or false? Only women make follicle-stimulating hormone (FSH).

14. True or false? All hormones secreted by arthropods such as crabs and insects are also secreted by vertebrates.

15. Match the term listed at left with the most suitable description at right.
    - ___ adrenal medulla
    - ___ thyroid gland
    - ___ posterior pituitary gland
    - ___ pancreatic islets
    - ___ pineal gland
    - ___ prostaglandin
    - a. affected by daylength
    - b. a local signaling molecule
    - c. secretes hormones made in the hypothalamus
    - d. source of epinephrine
    - e. secrete insulin, glucagon
    - f. hormones require iodine

Critical Thinking

1. A tumor in the pituitary gland can cause a woman to produce milk even when she is not pregnant. In which lobe of the pituitary would such a tumor be located, and what hormone would it affect?

2. Sex hormone secretion is governed by a negative feedback loop to the hypothalamus and pituitary, similar to that for thyroid hormone or cortisol. Because of this, a veterinarian can use a blood test to find out whether a female dog has been neutered. Dogs who still have their ovaries have a lower blood level of luteinizing hormone (LH) than dogs who have been neutered. Explain why removing a dog's ovaries would result in an elevated level of LH.

3. A diabetic who injects too much insulin can lose consciousness. Explain why injecting excess insulin could impair brain function. Glucagon reverses this effect. Explain how.

Chapter 35
Summary

Section 35.1 Muscle cells do not divide, but they can enlarge by adding proteins. Some hormones encourage this process and myostatin discourages it. Mutations that alter myostatin's effect lead to increased muscle mass and muscle strength.

Section 35.2 Nearly all animals move by applying the force of muscle contraction to their skeletal elements. A **hydrostatic skeleton** is a confined fluid upon which muscle contractions act. An **exoskeleton** consists of hardened parts at the body surface. An **endoskeleton** consists of hardened parts inside the body.

Section 35.3 Like other vertebrates, humans have an endoskeleton. A vertebrate **axial skeleton** includes the skull, **vertebral column** (backbone), and a rib cage. The **appendicular skeleton** is the pelvic girdle, pectoral girdle, and paired limbs. The vertebral column consists of segments called **vertebrae**, with **intervertebral disks** between them. The spinal cord runs through the vertebral column and connects with the brain through a hole in the base of the skull. Placement of this hole, and other features of the human skeleton, are adaptations that facilitate upright walking.

Section 35.4 Bones are organs rich in collagen, calcium, and phosphorus. In addition to having a role in movement, they store minerals and protect organs. Some have **red marrow** that makes blood cells; most have **yellow marrow**. In a human embryo, bones develop from a cartilage model. Even in adults, bones are continually remodeled. **Osteoblasts** are cells that synthesize bone, whereas **osteoclasts** break bone down. **Osteocytes** are former osteoblasts enclosed in a matrix of their secretions.

Section 35.5 A **joint** is an area of close contact between bones. One or more **ligaments** hold bones together at most joints. Bits of cartilage and fluid-filled **bursae** cushion joints.

Section 35.6 Bones move when skeletal muscles pull on them. Skeletal muscles are surrounded by a sheath of connective tissue that extends beyond the muscle as a **tendon**. Tendons attach a muscle to a bone or, in some cases, to skin. A muscle can exert force in only one direction; it can pull but not push. Thus, skeletal muscles with opposing actions often work as a pair.

Section 35.7 The internal organization of a skeletal muscle promotes a strong, directional contraction. Many **myofibrils** fill the interior of a **muscle fiber**. A myofibril consists of **sarcomeres**, units of muscle contraction, lined up along its length. Each sarcomere has parallel arrays of thick and thin filaments.

The **sliding-filament model** describes how ATP-driven sliding of thin filaments past thick filaments shortens the sarcomere and brings about muscle contraction. Thin filaments consist mainly of the globular protein actin. Thick filaments are composed of the motor protein myosin. In an excited muscle, myosin binds to actin and uses energy released by the hydrolysis of ATP to move the thin filament toward the sarcomere center.

Section 35.8 A motor neuron and all the muscle fibers it controls constitute a **motor unit**. Signals from motor neurons result in action potentials in muscle fibers, which in turn cause the **sarcoplasmic reticulum** to release stored calcium ions. The flow of this calcium into the cytoplasm causes two proteins associated with the thin filaments to shift in such a way that actin and myosin heads can interact and bring about a muscle contraction.

Diseases such as polio and toxins such as botulinum toxin can interfere with motor neuron function and cause skeletal muscle weakness and paralysis.

Section 35.9 Muscle fibers produce the ATP they require for contraction by way of three pathways: dephosphorylation of creatine phosphate, aerobic respiration, and lactate fermentation. Muscle fibers differ in which pathway predominates and how fast they hydrolyze the resulting ATP. Red fibers depend mainly on aerobic respiration. They have many mitochondria and contain **myoglobin**, a protein that can bind and release oxygen. White fibers depend mainly on lactate fermentation and react fast.

Exercise can increase circulation to muscles, increase the number of myofibrils in muscle fibers, and enhance the ability of fibers to burn lipids as fuel. Long periods of uninterrupted sitting have a detrimental effect on the metabolic activity of fibers in leg muscles.

Self-Quiz
Answers in Appendix III

1. A hydrostatic skeleton consists of _____ .
   a. a fluid in an enclosed space
   b. hardened plates at the surface of a body
   c. internal hard parts
   d. none of the above

2. Bones are _____ .
   a. mineral reservoirs
   b. skeletal muscle's partners
   c. sites where blood cells form
   d. all of the above

3. Bones move when _____ muscles contract.
   a. cardiac
   b. skeletal
   c. smooth
   d. all of the above

4. A ligament connects _____ .
   a. bones at a joint
   b. a muscle to a bone
   c. a muscle to a tendon
   d. a tendon to bone

Building Better Bones Tiffany, shown in **Figure 35.23**, was born with multiple fractures in her arms and legs. By age six, she had undergone surgery to correct more than 200 bone fractures. Her fragile, easily broken bones are symptoms of osteogenesis imperfecta (OI), a genetic disorder caused by a mutation in a gene for collagen. As bones develop, collagen forms a scaffold for deposition of mineralized bone tissue. The scaffold forms improperly in children with OI. **Figure 35.23** also shows the results of a test of a new drug. Treated children, all less than two years old, were compared to similarly affected children of the same age who were not treated with the drug.

1. An increase in vertebral area during the 12-month period of the study indicates bone growth. How many of the treated children showed such an increase?

2. How many of the untreated children showed an increase in vertebral area?

3. How did the rate of fractures in the two groups compare?

4. Do these results shown support the hypothesis that this drug, which slows bone breakdown, can increase bone growth and reduce fractures in young children with OI?

Treated child	Vertebral area in cm^2 (Initial)	(Final)	Fractures per year
1	14.7	16.7	1
2	15.5	16.9	1
3	6.7	16.5	6
4	7.3	11.8	0
5	13.6	14.6	6
6	9.3	15.6	1
7	15.3	15.9	0
8	9.9	13.0	4
9	10.5	13.4	4
Mean	11.4	14.9	2.6

Control child	Vertebral area in cm^2 (Initial)	(Final)	Fractures per year
1	18.2	13.7	4
2	16.5	12.9	7
3	16.4	11.3	8
4	13.5	7.7	5
5	16.2	16.1	8
6	18.9	17.0	6
Mean	16.6	13.1	6.3

Figure 35.23 Results of a clinical trial of a drug treatment for osteogenesis imperfecta (OI), which affects the child shown at *right*. Nine children with OI received the drug. Six others were untreated controls. Surface area of certain vertebrae was measured before and after treatment. Fractures occurring during the 12 months of the trial were also recorded.

5. Parathyroid hormone stimulates _____ .
   a. osteoclast activity
   b. bone deposition
   c. red blood cell formation
   d. all of the above

6. The _____ attaches to the pelvic girdle.
   a. radius
   b. sternum
   c. femur
   d. tibia

7. The _____ is the basic unit of contraction.
   a. osteoblast
   b. sarcomere
   c. twitch
   d. myosin filament

8. In sarcomeres, phosphate-group transfers from ATP activate _____ .
   a. actin   b. myosin   c. troponin   d. Z bands

9. A sarcomere shortens when _____ .
   a. thick filaments shorten
   b. thin filaments shorten
   c. both thick and thin filaments shorten
   d. none of the above

10. ATP for muscle contraction can be formed by _____ .
    a. aerobic respiration
    b. lactate fermentation
    c. creatine phosphate breakdown
    d. all of the above

11. A virus causes _____ .
    a. osteoporosis
    b. polio
    c. botulism
    d. all of the above

12. A motor unit is _____ .
    a. a muscle and the bone it moves
    b. two muscles that work in opposition
    c. the amount a muscle shortens during contraction
    d. a motor neuron and the muscle fibers it controls

13. _____ from a motor neuron excites a muscle fiber.
    a. ACh   b. GABA   c. calcium   d. phosphate

14. A regimen of aerobic exercise increases the _____ .
    a. blood supply to a muscle
    b. number of fibers in a muscle
    c. myostatin in a muscle
    d. all of the above

15. Match the words with their defining feature.
    ___ osteoblast
    ___ myofibrils
    ___ myoglobin
    ___ joint
    ___ myosin
    ___ red marrow
    ___ metacarpals
    ___ sarcoplasmic reticulum
    ___ creatine phosphate

    a. stores and releases calcium
    b. all in the hands
    c. site of blood cell production
    d. hydrolyzes ATP
    e. bone-forming cell
    f. donates phosphate to ADP
    g. binds and releases oxygen
    h. area of contact between bones
    i. muscle fiber's threadlike parts

Critical Thinking

1. A friend is training for a marathon. Knowing that you are taking biology, she asks if creatine supplements will improve her performance. What do you tell her?

2. After death, a person no longer makes ATP. As a result, calcium stored in the sarcoplasmic reticulum diffuses down its concentration gradient into the muscle cytoplasm. The result is rigor mortis—an unbreakable state of muscle contraction. Explain why the contraction occurs and why it is irreversible.

3. Mike's younger brother had Duchenne muscular dystrophy and died at age 16. Mike, now 26 years old and healthy, wants to start a family. He worries that his sons might be at risk for muscular dystrophy. His wife's family has no history of this disorder. Are Mike's concerns are well founded?

Chapter 36
Summary

Section 36.1 When the heart stops pumping, blood flow halts and brain cells begin to die from lack of oxygen. CPR can keep some oxygenated blood moving to cells, but it cannot restart a heart. Reestablishing the normal rhythm requires a shock from a defibrillator.

Section 36.2 A **circulatory system** moves substances through a body by way of a fluid transport medium called **blood**. Some invertebrates have an **open circulatory system**, in which **hemolymph** pumped out of vessels mixes with interstitial fluid. In other invertebrates and all vertebrates, a **closed circulatory system** confines blood inside a **heart** and blood vessels. All exchanges between blood and interstitial fluid take place across the walls of the smallest vessels.

In fish, blood flows through a single circuit. Evolution of a two-circuit system accompanied the evolution of lungs. The **pulmonary circuit** moves blood to the lungs and back. The longer **systemic circuit** moves blood to other body tissues and back.

Section 36.3 Each circuit of the human cardiovascular system has **arteries** that move blood from the heart to **arterioles**. The largest artery in the body is the **aorta**. Arterioles supply **capillaries**. Capillaries drain into **venules** that feed into veins. **Veins** carry blood back to the heart. Most blood flows through only one capillary system, but blood in intestinal capillaries will later flow through liver capillaries. The liver stores nutrients and neutralizes some bloodborne toxins.

Sections 36.4–36.6 Blood consists of **plasma**, blood cells, and platelets. The number of cells in a given volume of blood is the **cell count**. Plasma is mostly water. **Red blood cells** contain **hemoglobin** that functions in oxygen transport. **White blood cells** help defend the body against damaged cells and pathogens. **Platelets** function in **hemostasis** (clotting).

The body mounts an attack against any cells that bear unfamiliar molecules, causing **agglutination**, or a clumping of cells. **ABO blood typing** helps match the blood of donors and recipients to avoid blood transfusion problems. **Rh blood typing** and the appropriate treatment prevent problems that can arise when maternal and fetal Rh blood types differ.

Section 36.7 The human heart is a double pump partitioned into two halves, each with an **atrium** above a **ventricle**. The **superior vena cava** and **inferior vena cava** are large veins that fill the right atrium. Blood from the right atrium flows into the right ventricle, which pumps blood into **pulmonary arteries** that carry it to the lungs. Blood returns to the heart's left atrium through **pulmonary veins**. The left atrium empties into the left ventricle, which pumps blood into the aorta and through the systemic circuit.

During one **cardiac cycle**, all heart chambers undergo rhythmic relaxation (**diastole**) and contraction (**systole**). Contraction of the ventricles provides the force that propels blood through blood vessels.

The **sinoatrial (SA) node** in the right atrium wall is the cardiac pacemaker. The **atrioventricular (AV) node** serves as an electrical bridge to the ventricles. The delay between atrial contraction in response to SA signaling and ventricular contraction in response to signals arriving via the AV node allows the ventricles to fill fully before they contract.

Sections 36.8–36.11 **Blood pressure** results from the force exerted by ventricular contraction and it declines as blood proceeds through a circuit. It is usually recorded as **systolic pressure** over **diastolic pressure**. Thick-walled arteries smooth out pressure variations. A **pulse** is a brief expansion of an artery caused by ventricular contraction. Arterioles are the main site for regulation of flow through the systemic circuit. **Vasodilation** of arterioles supplies more blood to a region. **Vasoconstriction** decreases the blood supply.

Blood flow slows in capillaries because of their collectively large cross-sectional area. Substances leave a capillary by diffusion, by exocytosis, or in fluid that seeps out between cells. Fluid that seeps out of a capillary at the arterial end is balanced by osmotic uptake of water nearer the vein end.

Veins return blood to the heart. Blood pressure in veins is low, but one-way valves and pressure on veins from skeletal muscles and respiratory movements help move blood forward.

Section 36.12 In a blood disorder, an individual has too many, too few, or abnormal red or white blood cells. Abnormal heart rhythms can slow or halt blood flow. Flow is also impaired when atherosclerosis narrows the interior of a blood vessel.

Section 36.13 Some fluid that leaves capillaries enters the **lymph vascular system**. The fluid, now called **lymph**, is filtered by **lymph nodes**. The **spleen** filters the blood and removes any old red blood cells.

Self-Quiz
Answers in Appendix III

1. All vertebrates have _____ .
   a. an open circulatory system
   b. a two-chambered heart
   c. lungs
   d. none of the above

2. In _____ blood flows through two completely separate circuits.
   a. birds   b. mammals   c. fish   d. both a and b

The Stroke Belt Risk of death by stroke is not distributed evenly across the United States. Epidemiologists refer to a swath of states in the Southeast as the "stroke belt" because of the increased incidence of stroke deaths there. By one hypothesis, the high rate of deaths from stroke in this region results largely from a relative lack of access to immediate medical care. Compared to other parts of the country, more stroke-belt residents live in rural settings with few medical services.

Figure 36.24 compares the rate of stroke deaths in stroke-belt states (Alabama, Arkansas, Georgia, Mississippi, North Carolina, South Carolina, and Tennessee) with that of New York State. It also breaks down death risk in each region by ethnic group and sex.

1. How does the rate of stroke deaths among blacks living in the stroke-belt compare with whites in the same region?

2. How does the rate of stroke deaths among blacks living in New York compare with whites in the same region?

3. Which group has the higher rate of stroke deaths, blacks living in New York, or whites living in the stroke belt?

4. Do these data support the hypothesis that poor access to care causes the high rate of death by stroke in the stroke belt?

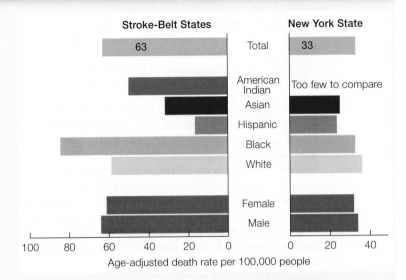

Figure 36.24 Comparison of the age-adjusted rate of deaths by stroke in southeastern "stroke-belt" states and in New York State. *Source*: National Vital Statistics System—Mortality (NVSS-M), NCHS, CDC.

---

3. The _____ circuit carries blood to and from lungs.
   a. systemic
   b. pulmonary

4. The fluid portion of blood is _____ .
   a. plasma
   b. lymph
   c. interstitial fluid
   d. hemolymph

5. Platelets function in _____ .
   a. oxygen transport
   b. hemostasis
   c. thermal regulation
   d. both a and b

6. Most oxygen in blood is transported _____ .
   a. in red blood cells
   b. in white blood cells
   c. bound to hemoglobin
   d. both a and c

7. Blood flows directly from the left atrium to _____ .
   a. the aorta
   b. the left ventricle
   c. the right atrium
   d. the pulmonary arteries

8. Contraction of _____ drives the flow of blood through the aorta and pulmonary arteries.
   a. atria
   b. ventricles

9. Blood pressure is highest in the _____ and lowest in the _____ .
   a. arteries; veins
   b. arterioles; venules
   c. veins; arteries
   d. capillaries; arterioles

10. At rest, the largest volume of blood is in _____ .
    a. arteries
    b. capillaries
    c. veins
    d. arterioles

11. Which of the four heart chambers has the thickest wall?

12. Which artery carries oxygen-poor blood?

13. Lymph nodes filter _____ .
    a. blood
    b. lymph
    c. plasma
    d. all of the above

14. Which is more dangerous?
    a. atrial fibrillation
    b. ventricular fibrillation

15. Match the terms with their definition.
    ___ bone marrow
    ___ thoracic duct
    ___ hemoglobin
    ___ HDL
    ___ SA node
    ___ veins
    ___ aorta

    a. returns lymph to blood
    b. cardiac pacemaker
    c. main blood volume reservoir
    d. largest artery
    e. fills red blood cells
    f. source of all blood cells
    g. delivers cholesterol to the liver for use in bile

Critical Thinking

1. The highly publicized deaths of a few airline travelers led to warnings about economy-class syndrome. The idea is that sitting motionless for long periods on flights allows blood to pool and clots to form in legs. More recent studies suggest that long-distance flights cause problems in about 1 percent of air travelers, and that the risk is the same regardless of whether a person is in a first-class seat or an economy seat. Physicians suggest that air travelers drink plenty of fluids and periodically get up and walk around the cabin. Given what you know about blood flow in the veins, explain why these precautions can lower the risk of clot formation.

2. Mitochondria occupy about 40 percent of the volume of human cardiac muscle but only about 12 percent of the volume of skeletal muscle. Explain this difference.

3. In some people the valve between an atrium and ventricle does not close properly. This condition can be diagnosed by listening carefully to the heart. The listener will hear a whooshing sound called a murmur when the ventricle of the affected chamber contracts. What causes this sound?

Chapter 37
Summary

Section 37.1 Screenings, treatments, and vaccines for diseases such as cervical cancer are a direct outcome of our increasing understanding of the way the human body interacts with pathogens.

Section 37.2 A pathogen that breaches surface barriers triggers **innate immunity**, a set of general defenses that can prevent pathogens from becoming established in the body. **Adaptive immunity**, which can specifically target billions of different **antigens**, follows. **Complement** and signaling molecules such as **cytokines** help coordinate the activities of white blood cells—**dendritic cells**, **macrophages**, **neutrophils**, **basophils**, **mast cells**, **eosinophils**, B and T cells (including **cytotoxic T cells**) and **NK cells**—in **immunity**.

Section 37.3 Vertebrates can fend off pathogens such as those that cause dental **plaque** at body surfaces with physical, mechanical, and chemical barriers (including **lysozyme**). Most **normal flora** do not cause disease unless they penetrate inner tissues.

Section 37.4 An innate immune response can be triggered by activated complement or by phagocytic leukocytes that engulf a pathogen or other antigen-bearing particle. Complement activated by the presence of antigen or tissue damage attracts phagocytes (which follow gradients of complement by **chemotaxis**), coats pathogens and nearby cells, and forms membrane attack complexes that induce cell death.

Section 37.5 Inflammation begins when granular leukocytes in infected or damaged tissue release prostaglandins and histamine. These signaling molecules attract phagocytes and increase blood flow to the affected area. Plasma proteins that leak from capillaries cause swelling and pain.

Prolonged exposure to an inflammation-provoking stimulus can cause chronic inflammation, a condition that is associated with many diseases. **Fever** fights infection by increasing the metabolic rate and by slowing pathogen replication.

Section 37.6 Random splicing of antigen receptor genes gives rise to antigen receptors (**T cell receptors**, and **B cell receptors** and other **antibodies**) that collectively have the ability to recognize billions of specific antigens. T cell receptors that recognize **MHC markers** are the basis of self/nonself discrimination.

Section 37.7 B cells and T cells carry out adaptive immune responses. **Antibody-mediated** and **cell-mediated immune responses** work together to rid the body of a specific pathogen (**Figure 37.24**).

Phagocytic lymphocytes present antigen in adaptive responses. **Effector cells** form and target the antigen-bearing particles in a primary response. **Memory cells** that also form are reserved for a later encounter with an antigen, in which case they trigger a faster, stronger secondary response.

Section 37.8 B cells, assisted by T cells, antigen-presenting cells, and signaling molecules, carry out antibody-mediated immune responses. Antibodies that recognize a specific antigen are produced by B cells during these responses. Antibody binding causes agglutination of foreign cells, and tags antigen-bearing particles for phagocytosis.

Section 37.9 T cells, NK cells, and antigen-presenting cells carry out cell-mediated immune responses. Cytotoxic T cells and NK cells that form in cell-mediated responses kill body cells that have been altered by infection or cancer.

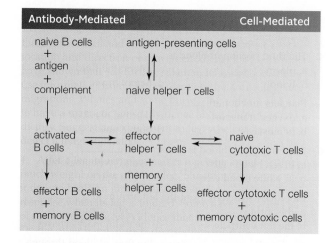

Figure 37.24 Overview of key interactions between antibody-mediated and cell-mediated immune responses.

Section 37.10 Allergens are normally harmless substances that induce immune responses; hypersensitivity to an allergen is called **allergy**. A malfunction in the immune system or its checks and balances can cause dangerous acute illnesses, or chronic and sometimes deadly diseases. Immune deficiency is a reduced capacity to mount an immune response. In an **autoimmune response**, a body's own cells are inappropriately recognized as foreign and attacked.

Section 37.11 AIDS is caused by the human immunodeficiency virus (HIV). An initial adaptive immune response to infection rids the body of most—but not all—of the virus. The virus infects helper T cells, so it eventually cripples the adaptive immune system.

 Section 37.12 Immunization with **vaccines** designed to elicit immunity to specific diseases saves millions of lives every year as part of worldwide health programs.

Self-Quiz
Answers in Appendix III

1. Which of the following is not among the first line of defenses against infection?
  a. resident bacterial populations   d. skin
  b. tears, saliva, gastric fluid   e. complement
  c. urine flow   f. lysozyme

2. Which of the following is not considered to be part of an innate immune response?
  a. phagocytic cells   e. inflammation
  b. fever   f. complement activation
  c. histamines   g. antigen-presenting cells
  d. cytokines   h. none of the above

3. Which of the following is not considered to be part of an adaptive immune response?
  a. phagocytic cells   e. antigen receptors
  b. antigen-presenting cells   f. complement activation
  c. histamines   g. antibodies
  d. cytokines   h. none of the above

4. Activated complement proteins _____ .
  a. form pore complexes   c. attract phagocytes
  b. promote inflammation   d. all of the above

5. _____ trigger immune responses.
  a. Cytokines   d. Antigens
  b. Lysozymes   e. Histamines
  c. Immunoglobulins   f. all of the above

6. Which of the following are defining characteristics of innate immunity? Choose all that apply.
  a. General defense   e. Memory
  b. Diversity   f. Self/nonself discrimination
  c. Fixed   g. Targets specific antigens
  d. Immediate   h. All are characteristic

7. Which of the following are defining characteristics of adaptive immunity? Choose all that apply.
  a. General defense   e. Memory
  b. Diversity   f. Self/nonself discrimination
  c. Fixed   g. Targets specific antigens
  d. Immediate   h. All are characteristic

8. A dendritic cell engulfs a bacterium, then presents bacterial bits on its surface along with a(n) _____ .
  a. MHC marker   c. T cell receptor
  b. antibody   d. antigen

9. Antibodies are _____ .
  a. antigen receptors   c. proteins
  b. made only by B cells   d. all of the above

10. Antibody-mediated responses work against _____ .
  a. intracellular pathogens   d. both a and b
  b. extracellular pathogens   e. both b and c
  c. cancerous cells   f. a, b, and c

11. Cell-mediated responses work against _____ .
  a. intracellular pathogens   d. both a and b
  b. extracellular pathogens   e. both a and c
  c. cancerous cells   f. a, b, and c

12. _____ are targets of cytotoxic T cells.
  a. Extracellular virus particles in blood
  b. Virus-infected body cells or tumor cells
  c. Parasitic flukes in the liver
  d. Bacterial cells in pus
  e. Pollen grains in nasal mucus

13. Allergies occur when the body responds to _____ .
  a. pathogens   c. normally harmless substances
  b. toxins   d. all of the above

14. Which type of antibody is responsible for allergies?
  a. IgA   c. IgM
  b. IgE   d. IgG

15. Vaccines are designed to elicit _____ .
  a. antibodies   c. viruses
  b. immunity   d. a and b

16. Match the immune cell with the best description.
  ___ mast cell   a. antigen presenter
  ___ eosinophil   b. targets parasitic worms
  ___ helper T cell   c. activates cytotoxic T cells
  ___ NK cell   d. kills body cells with no
  ___ dendritic cell   MHC markers
  ___ memory cell   e. factor in allergic reactions
     f. long-lasting immunity

Critical Thinking

1. Before each flu season, you get a flu shot, which consists of a vaccine against several strains of influenza virus. This year, you get "the flu" anyway. What happened? There are at least three explanations.

2. Monoclonal antibodies are produced by immunizing a mouse with a particular antigen, then removing its spleen. Individual B cells producing mouse antibodies specific for the antigen are isolated from the mouse's spleen and fused with cancerous B cells from a myeloma cell line. The resulting hybrid myeloma cells ("hybridoma" cells) are cloned: Individual cells are grown in tissue culture as separate cell lines. Each line produces and secretes antibodies that recognize the antigen to which the mouse was immunized. These monoclonal antibodies can be purified and used for research or other purposes.
    Monoclonal antibodies are sometimes used in passive immunization. They are effective, but only in the immediate term. Antibodies produced by one's own immune system can last up to about six months in the bloodstream, but monoclonals delivered in passive immunization often last for less than a week. Why the difference?

Cervical Cancer Incidence in HPV-Positive Women In 2003, Michelle Khan and her coworkers published their findings on a 10-year study in which they followed cervical cancer incidence and HPV status in 20,514 women. All women who participated in the study were free of cervical cancer when the test began. Pap tests were taken at regular intervals, and the researchers used a DNA probe hybridization test to detect the presence of specific types of HPV in the women's cervical cells.

The results are shown in **Figure 37.25** as a graph of the incidence rate of cervical cancer by HPV type. HPV-positive women are often infected by more than one type, so the data were sorted into groups based on the women's HPV status ranked by type: either positive for HPV16; or negative for HPV16 and positive for HPV18; or negative for HPV16 and 18 and positive for any other cancer-causing HPV; or negative for all cancer-causing HPV.

1. At 110 months into the study, what percentage of women who were not infected with any type of cancer-causing HPV had cervical cancer? What percentage of women who were infected with HPV16 also had cervical cancer?

2. In which group would women infected with both HPV16 and HPV18 fall?

3. Is it possible to estimate from this graph the overall risk of cervical cancer that is associated with infection of cancer-causing HPV of any type?

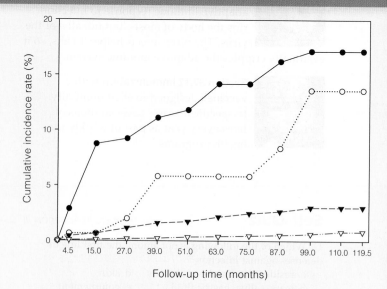

Figure 37.25 Cumulative incidence rate of cervical cancer correlated with HPV status in 20,514 women aged 16 years and older.

The data were grouped as follows:

Closed circles (●): HPV16 positive
Open circles (○): HPV16 negative and HPV18 positive
Closed triangles (▼): All other cancer-causing HPV types combined
Open triangles (▽): no cancer-causing HPV type detected

Chapter 38
Summary

Section 38.1 Carbon monoxide is the most commonly inhaled poison. It binds to hemoglobin and prevents normal gas transport and exchange. Burning of fossil fuels and other organic material, including tobacco, releases carbon monoxide.

Section 38.2 Respiration is a physiological process by which O_2 enters the internal environment and CO_2 leaves it. Both gases diffuse across a **respiratory surface**. Enlarging the respiratory surface, moving air or water past that surface, and having **respiratory proteins** speeds the rate of gas exchange.

Air-breathing animals have a more constant supply of oxygen than aquatic animals. The amount of oxygen in water is an important factor in determining biodiversity.

Section 38.3 Some invertebrates do not have special respiratory organs and rely on **integumentary exchange**, diffusion of gases across the body surface. **Gills** enhance respiration in other aquatic invertebrates. On land, **lungs** and **tracheal systems** assist in gas exchange with the air.

Section 38.4 Water flowing over fish gills exchanges gases with blood flowing in the opposite direction inside gill capillaries. This **countercurrent exchange** is highly efficient. Most amphibians have lungs, and also exchange gases across the skin. Reptiles (including birds) and mammals rely on lungs for gas exchange. In birds, air sacs connected to lungs keep air flowing continually through them.

Section 38.5 In humans, air flows through two nasal cavities and a mouth into the **pharynx**, then the **larynx**, and the **trachea** (windpipe). The larynx contains the vocal cords, movements of which alter the size of the opening (the **glottis**) between them. When you swallow, the position of the **epiglottis** at the entrance to the larynx shifts, keeping food out of the trachea. The trachea branches into two **bronchi** that enter the lungs. These two airways branch into **bronchioles**. At the ends of the finest bronchioles are thin-walled **alveoli**, the site of gas exchange. The **diaphragm** at the base of the thoracic cavity and the **intercostal muscles** between ribs are involved in breathing.

Section 38.6 A **respiratory cycle** is one inhalation and one exhalation. Inhalation is active. As muscle contractions expand the chest cavity, pressure in lungs decreases below atmospheric pressure, and air flows into the lungs. These events are reversed during exhalation, which usually is passive.

If a person is choking, the **Heimlich maneuver** can be used to expel food from the trachea.

Tidal volume is normally far less than **vital capacity**. The lungs never fully deflate. The medulla oblongata in the brain stem adjusts the rate and magnitude of breathing.

Section 38.7 In human lungs, gases diffuse across a thin **respiratory membrane** that separates the air in alveoli from the blood in pulmonary capillaries. As red blood cells pass through these vessels, hemoglobin binds O_2 to form **oxyhemoglobin**. In systemic capillaries, hemoglobin releases O_2, which diffuses into cells.

Also in systemic capillaries, CO_2 diffuses from cells into the blood. Most CO_2 reacts with water inside red blood cells to form bicarbonate. The enzyme **carbonic anhydrase** catalyzes this reaction, which is reversed in the lungs. There, CO_2 forms and is expelled from the body in exhalations.

Section 38.8 The amount of available oxygen declines with altitude. Physiological changes that occur in response to high altitude are called **acclimatization**. They include altered breathing patterns and an increase in **erythropoietin**, a hormone that stimulates red blood cell formation. Over the longer term, genetic changes can adapt a population to life at a high altitude, as among Tibetans.

A variety of adaptive mechanisms allow some turtles and marine mammals to hold their breath for long periods while making deep dives.

Section 38.9 Respiratory disorders include apnea and sudden infant death syndrome (SIDS). Respiratory diseases include tuberculosis, pneumonia, bronchitis, and emphysema. Smoking increases risk of many respiratory problems. It is a leading cause of debilitating disease and deaths.

Self-Quiz
Answers in Appendix III

1. Respiratory proteins such as hemoglobin _____ .
   a. contain metal ions
   b. occur only in vertebrates
   c. increase the efficiency of oxygen transport
   d. both a and c

2. In _____ gas exchange occurs at the body surface and gas is distributed by diffusion alone.
   a. earthworms   c. frogs
   b. flatworms   d. insects

3. Countercurrent flow of water and blood increases the efficiency of gas exchange in _____ .
   a. fishes   c. birds
   b. amphibians   d. all of the above

4. In human lungs, gas exchange occurs at the _____ .
   a. bronchi   c. alveolar sacs
   b. bronchioles   d. pleural sacs

5. What type of metal ion associates with hemoglobin?
   a. zinc   c. copper
   b. iron   d. none of the above

Risks of Radon Radon is a colorless, odorless gas emitted by many rocks and soils. It is formed by the radioactive decay of uranium and is itself radioactive (Section 2.2). There is some radon in the air almost everywhere, but routinely inhaling a lot of it raises the risk of lung cancer. Radon also seems to increase cancer risk far more in smokers than in nonsmokers. **Figure 38.29** is an estimate of how radon in homes affects risk of lung cancer mortality. Note that this data shows only the death risk for radon-induced cancers. Smokers are also at risk from lung cancers that are caused by tobacco.

1. If 1,000 smokers were exposed to a radon level of 1.3 pCi/L over a lifetime (the average indoor radon level), how many would die of a radon-induced lung cancer?

2. How high would the radon level have to be to cause approximately the same number of cancers among 1,000 nonsmokers?

3. The risk of dying in a car crash is about 7 out of 1,000. Is a smoker in a home with an average radon level (1.3 pCi/L) more likely to die in a car crash or of radon-induced cancer?

Radon level (pCi/L)	Risk of cancer death from lifetime radon exposure	
	Never smoked	Current smokers
20	36 out of 1,000	260 out of 1,000
10	18 out of 1,000	150 out of 1,000
8	15 out of 1,000	120 out of 1,000
4	7 out of 1,000	62 out of 1,000
2	4 out of 1,000	32 out of 1,000
1.3	2 out of 1,000	20 out of 1,000
0.4	>1 out of 1,000	6 out of 1,000

Figure 38.29 Estimated risk of lung cancer death as a result of lifetime radon exposure. Radon levels are measured in picocuries per liter (pCi/L). The Environmental Protection Agency considers a radon level above 4 pCi/Liter to be unsafe. For information about testing your home for radon and what to do if the radon level is high, visit the EPA's Radon Information Site at www.epa.gov/radon.

6. When you breathe quietly, inhalation is _____ and exhalation is _____ .
   a. passive; passive
   b. active; active
   c. passive; active
   d. active; passive

7. During inhalation _____ .
   a. the thoracic cavity expands
   b. the diaphragm relaxes
   c. atmospheric pressure declines
   d. both a and c

8. _____ binds to hemoglobin even more strongly than oxygen does.
   a. Carbon dioxide
   b. Carbon monoxide
   c. Oxyhemoglobin
   d. Carbonic anhydrase

9. Most oxygen transported in blood _____ .
   a. is bound to hemoglobin
   b. combines with carbon to form carbon dioxide
   c. is in the form of bicarbonate
   d. is dissolved in the plasma

10. At high altitudes, _____ .
    a. nitrogen bubbles out of the blood
    b. hemoglobin has fewer oxygen-binding sites
    c. there are fewer O_2 molecules than at low altitudes.
    d. both b and c

11. The hormone erythropoietin causes _____ .
    a. increased heart rate
    b. deeper breathing
    c. red blood cell formation
    d. all of the above

12. _____ in arteries sense changes in the acidity of the blood.
    a. Mechanoreceptors
    b. Neurotransmitters
    c. Photoreceptors
    d. Chemoreceptors

13. True or false? Human lungs retain some air even after forced exhalation.

14. The diaphragm is a _____ muscle.
    a. smooth
    b. skeletal
    c. dome-shaped
    d. a and c

15. Match the words with their descriptions.
    ___ trachea   a. muscle of respiration
    ___ pharynx   b. gap between vocal cords
    ___ epiglottis   c. between bronchi and alveoli
    ___ larynx   d. windpipe
    ___ bronchus   e. voice box
    ___ bronchiole   f. keeps food out of airway
    ___ glottis   g. airway leading to lung
    ___ diaphragm   h. throat

Critical Thinking

1. The red blood cell enzyme carbonic anhydrase contains the metal zinc. Humans obtain zinc from their diet, especially from red meat and some seafoods. A zinc deficiency does not reduce the number of red blood cells, but it does impair respiratory function by reducing carbon dioxide output. Explain why a zinc deficiency has this effect.

2. Look again at **Figure 38.21**. Notice that the oxygen and carbon dioxide content of blood in pulmonary veins is the same as at the start of the systemic capillaries. Notice also that systemic veins and pulmonary arteries have equal partial pressures. Explain the reason for the similarities.

3. Respiration supplies cells with oxygen needed for aerobic respiration. Explain oxygen's role in this energy-releasing pathway. Where is it used and what is its function?

4. A developing fetus gets oxygen from its mother's blood. Fetal capillaries run through pools of maternal blood in an organ called the placenta. As fetal blood runs through the capillaries, it exchanges substances with the maternal blood around the capillary. The hemoglobin made by a fetus is different than that made after birth. Fetal hemoglobin binds oxygen more strongly at low oxygen levels than does normal hemoglobin. How would fetal hemoglobin's somewhat higher affinity for oxygen benefit the fetus?

Chapter 39
Summary

Section 39.1 The human gut contains trillions of living microbes. The numbers and types of these organisms affect digestive function and also influence general health by their effects on the immune system. **Probiotics** are foods that contain living bacteria whose presence in our gut benefits our health.

Section 39.2 A digestive system breaks food down into molecules small enough to be absorbed into the internal environment. It also stores and eliminates unabsorbed materials. Some invertebrates have an **incomplete digestive system**: a saclike gut with a single opening. Most animals and all vertebrates have a **complete digestive system**: a tube with two openings and specialized areas between them. Variations in the structure of vertebrate digestive systems are adaptations to particular diets. For example, the multiple stomachs of **ruminants** allow them to digest grasses.

Section 39.3 The human pharynx is the entrance to the digestive and respiratory systems. **Peristalsis** moves food down the **esophagus** and through a **sphincter** into the **stomach**, the start of the **gastrointestinal tract**. From the stomach, material moves to the **small intestine**. The **large intestine** concentrates undigested wastes, which are stored in the **rectum** until expelled through the **anus**.

Sections 39.4, 39.5 Digestion starts in the mouth, where food is broken into bits and mixed with saliva from **salivary glands**. Saliva contains salivary amylase, which begins the process of starch digestion. Protein digestion begins in the stomach, a muscular sac with a glandular lining that secretes **gastric fluid**. This fluid contains acid, enzymes, and mucus. It mixes with food to form **chyme**.

Sections 39.6, 39.7 The small intestine is the longest portion of the gut and has the largest surface area. Its highly folded lining has many **villi** at its surface. Each multicelled villus has a covering of **brush border cells**. These cells have **microvilli** that increase their surface area for digestion and absorption.

Chemical digestion is completed in the small intestine through the action of enzymes from the pancreas, bile from the gallbladder, and enzymes embedded in the plasma membrane of brush border cells.

Carbohydrates are broken into monosaccharides, which are actively transported across brush border cells, and enter the blood. Similarly, proteins are broken down to amino acids, which are actively transported across these cells and enter the blood.

Bile made in the liver and stored and concentrated in the **gallbladder** aids in the **emulsification** of fats.

Monoglycerides and fatty acids diffuse into the brush border cells. Here, they recombine as triglycerides, which get a protein coat and are moved by exocytosis into interstitial fluid. The lipoproteins then enter lymph vessels that deliver them to blood.

The small intestine is also the site of most water absorption. Water moves out of the gut by osmosis.

Section 39.8 More water and ions are absorbed in the large intestine, or colon, which compacts undigested solid wastes as **feces**. Feces are stored in the rectum, a stretchable region just before the anus. The **appendix** is a short extension from the first part of the large intestine.

Section 39.9 The small organic compounds absorbed from the gut are stored, used in biosynthesis or as energy sources, or excreted by other organ systems. Blood that flows through the small intestine travels next to the liver, which eliminates ingested toxins and stores excess glucose as glycogen.

Sections 39.10, 39.11 Food must provide both energy and raw materials, including **essential amino acids** and **essential fatty acids**. It must also include two additional types of compounds needed for metabolism: **vitamins**, which are organic, and **minerals**, which are inorganic. **Phytochemicals** are plant molecules that may improve health or prevent certain disorders.

Section 39.12 Maintaining body weight requires balancing energy intake and energy output. Obesity raises the risk of health problems because stressed adipose cells cause inflammation. Anorexia and bulimia are eating disorders that also threaten health.

Self-Quiz
Answers in Appendix III

1. A digestive system functions in _____ .
   a. secreting enzymes
   b. absorbing compounds
   c. eliminating wastes
   d. all of the above

2. The enzyme pepsin is activated in the _____ .
   a. mouth
   b. stomach
   c. small intestine
   d. large intestine

3. Most nutrients are absorbed in the _____ .
   a. mouth
   b. stomach
   c. small intestine
   d. large intestine

4. Bile has a role in _____ digestion and absorption.
   a. carbohydrate
   b. fat
   c. protein
   d. amino acid

5. Monosaccharides and amino acids absorbed from the gut enter _____ .
   a. blood vessels
   b. lymph vessels
   c. fat droplets
   d. both b and c

6. Bacteria in the _____ make essential vitamins.
   a. stomach
   b. small intestine
   c. large intestine
   d. esophagus

Human adaptation to a starchy diet The human *AMY-1* gene encodes salivary amylase, an enzyme that breaks down starch. The number of copies of this gene varies, and people who have more copies generally make more enzyme. In addition, the average number of *AMY-1* copies differs among cultural groups.

George Perry and his colleagues hypothesized that duplications of the *AMY-1* gene would confer a selective advantage in cultures in which starch is a large part of the diet. To test this hypothesis, the scientists compared the number of copies of the *AMY-1* gene among members of seven cultural groups that differed in their traditional diets. **Figure 39.16** shows their results.

1. Starchy tubers are a mainstay of Hadza hunter–gatherers in Africa, whereas fishing sustains Siberia's Yakut. Almost 60 percent of Yakut had fewer than 5 copies of the *AMY1* gene. What percent of the Hadza had fewer than 5 copies?

2. None of the Mbuti (rain-forest hunter–gathers) had more than 10 copies of *AMY-1*. Did any European Americans?

3. Do these data support the hypothesis that a starchy diet favors duplications of the *AMY-1* gene?

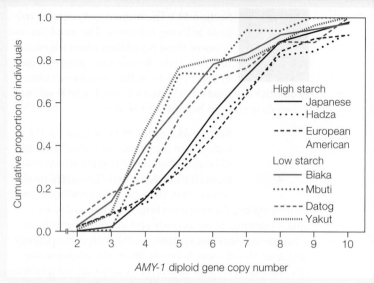

Figure 39.16 Number of copies of the *AMY-1* gene among members of cultures with traditional high-starch or low-starch diets. The Hadza, Biaka, Mbuti, and Datog are tribes in Africa. The Yakut live in Siberia.

---

7. The low pH of the _____ aids in protein digestion.
   a. stomach
   b. small intestine
   c. large intestine
   d. esophagus

8. Most water that enters the gut is absorbed across the lining of the _____ .
   a. stomach
   b. small intestine
   c. large intestine
   d. esophagus

9. _____ are inorganic substances; some have metabolic roles that no other substance can fulfill.
   a. Fats
   b. Minerals
   c. Vitamins
   d. Simple sugars

10. Blood that flows through capillaries in the small intestine flows next through vessels in the _____ .
    a. stomach
    b. heart
    c. pancreas
    d. liver

11. Tiny filaments called _____ increase the surface area of a brush border cell.
    a. villi
    b. microvilli
    c. cilia
    d. flagella

12. _____ is (are) a good source of omega-3 fatty acids that can reduce the risk of heart disease.
    a. Oatmeal
    b. Legumes
    c. Fish
    d. Corn syrup

13. What is the most common cause of anorexia-related death?
    a. cancer
    b. cardiac arrest
    c. pneumonia
    d. stroke

14. Match each substance with its digestive function.
    ___ gastrin   a. stimulate pancreatic secretions
    ___ leptin   b. emulsifies fats
    ___ secretin   c. breaks down polysaccharides
    ___ bile   d. encourages gastric secretions
    ___ salivary amylase e. curbs appetite
    ___ pepsin   f. cleaves peptide bonds

15. Match each organ with its digestive function.
    ___ gallbladder   a. final stop for digestive waste
    ___ salivary gland   b. makes bile
    ___ colon   c. compacts undigested residues
    ___ liver   d. adds enzymes to small intestine
    ___ small   e. absorbs most nutrients
        intestine   f. stores, secretes bile
    ___ rectum   g. secretes gastric fluid
    ___ stomach   h. secretes enzyme that starts
    ___ pancreas   starch digestion

Critical Thinking

1. The reported incidence of anorexia has soared during the past 20 years. Why is it unlikely that a rise in the frequency of alleles that put people at risk for anorexia is responsible for this rise in reported cases?

2. A python can survive by eating a large meal once or twice a year (**Figure 39.4**). When it does eat, microvilli in its small intestine lengthen fourfold and its stomach pH drops from 7 to 1. Explain the benefits of these changes.

3. Starch and sugar have the same number of calories per gram. However, not all vegetables are equally calorie dense. For example, a serving of boiled sweet potato provides about 1.2 calories per gram, while a serving of kale yields only 0.3 calories per gram. What could account for the difference in the calories your body obtains from these two foods?

4. **Figure 5.13** shows how pH affects activity of the enzyme pepsin. Review that figure and detemine the optimal pH for this enzyme. What lowers the pH to this level?

5. Many people who are lactose-intolerant find that eating yogurt does not cause the same cramps and flatulence that occurs when they consume milk. How might the fermentation of milk by yogurt-producing bacteria alter its effect on some lactose intolerant people?

Chapter 40
Summary

Section 40.1 The composition of urine provides information about health and about chemicals taken into the body. Hormones and drugs that enter the blood are filtered out by kidneys and end up in the urine.

Section 40.2 Plasma and interstitial fluid are the main components of the extracellular fluid. Maintaining the volume and composition of the extracellular fluid is an essential aspect of homeostasis. Organisms must balance solute and fluid gains with solute and fluid losses. They also must eliminate metabolic wastes such as the **ammonia** produced by the breakdown of proteins and nucleic acids.

Animals in different habitats face different challenges. Those living in water lose or gain water by osmosis. On land, the main challenge is avoiding dehydration. Insects and reptiles (including birds) conserve water by converting ammonia to **uric acid** crystals, which can be eliminated with very little water. Mammals excrete **urea** dissolved in a lot of water. All vertebrates have a pair of **kidneys** that filter the blood and produce **urine**.

Section 40.3 A human urinary system consists of two kidneys, a pair of **ureters**, a **urinary bladder**, and the **urethra**. Kidney **nephrons** are small, tubular structures that interact with nearby capillaries to form urine. Each nephron starts at **Bowman's capsule** in a kidney's outer region, or renal cortex. The nephron continues as a **proximal tubule**, a **loop of Henle** that descends into and ascends from the renal medulla, and a **distal tubule** that drains into a **collecting tubule**.

Bowman's capsule and capillaries of the **glomerulus** that it cups around serve as a blood-filtering unit. Most filtrate that enters Bowman's capsule is reabsorbed into **peritubular capillaries** around the nephron. The portion of the filtrate not returned to blood is excreted as urine.

Section 40.4 Urine formation begins when **glomerular filtration** delivers protein-free plasma into a kidney tubule. Most water and solutes are returned to the blood by **tubular reabsorption**. Substances that are not reabsorbed, and substances added to the filtrate by **tubular secretion**, end up in the urine. The amount of water reabsorbed across the distal tubule and collecting tubule can be altered, and affects the concentration of the urine.

Section 40.5 Hormones regulate the urine's concentration and composition. A region of the hypothalamus regulates thirst. In response to an increase in the solute concentration, the hypothalamus signals the pituitary gland to release **antidiuretic hormone**, a hormone that increases the reabsorption of water through passive transport proteins called aquaporins. **Aldosterone**, a hormone secreted by the adrenal cortex, increases sodium reabsorption. Water follows the sodium by osmosis. Thus, both antidiuretic hormone and aldosterone concentrate the urine.

Atrial natriuretic peptide, a hormone secreted by the heart in response to high blood volume and pressure, slows secretion of aldosterone and thus makes urine more dilute.

Sections 40.6, 40.7 The urinary system helps regulate **acid–base balance** by eliminating H$^+$ in urine and reabsorbing bicarbonate, which has a role in the body's main **buffer system**.

When kidneys fail, frequent dialysis or a kidney transplant is required to sustain life.

Section 40.8 Animals produce metabolic heat. They also gain or lose heat by **thermal radiation**, **conduction**, and **convection** and lose it by **evaporation**. **Ectotherms** such as reptiles control core temperature mainly by behavior; their temperature varies with that of the external environment. **Endotherms** (most mammals and birds) regulate temperature by controlling production and loss of metabolic heat. **Heterotherms** control core temperature only part of the time.

Section 40.9 The hypothalamus is the main center for temperature control. Widening of blood vessels in the skin, sweating, and panting are responses to heat. Mammals alone can sweat, but not all mammals have this ability. A fever is elevation of body temperature as a defensive response to infection.

Exposure to cold makes blood vessels in the skin constrict, causes hair to stand upright, and elicits a **shivering response**. Long-term exposure to cold can alter metabolism and encourage **nonshivering heat production** by brown adipose tissue. The metabolic activity of brown adipose tissue is regulated by the thyroid gland.

Hibernation is dormancy during a cold season, and **estivation** is dormancy during a hot one.

Self-Quiz

Answers in Appendix III

1. An insect's _____ deliver nitrogen-rich waste to its gut.
   a. nephridia   c. Malpighian tubules
   b. nephrons   d. contractile vacuoles

2. Body fluids of a marine bony fish have a solute concentration that is _____ its surroundings.
   a. higher than   b. lower than   c. equal to

3. Bowman's capsule, the start of the tubular part of a nephron, is located in the _____ .
   a. renal cortex   c. renal pelvis
   b. renal medulla   d. renal artery

4. Fluid that enters Bowman's capsule flows directly into the _____ .
   a. renal artery   c. distal tubule
   b. proximal tubule   d. loop of Henle

Pesticide residues in urine Products labeled as "organic" fill an increasing amount of space on supermarket shelves. Food that carries the USDA's organic label must be produced without pesticides, such as the insecticides malathion and chlorpyrifos, which farmers often use on conventionally grown fruits, vegetables, and many grains.

Does eating organic food significantly affect the level of pesticide residues in the body? Chensheng Lu of Emory University used urine testing to find out (**Figure 40.17**). For fifteen days, the urine of twenty-three children (aged 3 to 11) was monitored for breakdown products of pesticides. During the first five days, children ate their normal diet of conventionally grown foods. The next five days, they ate organic versions of the same foods and drinks. For the final five days, they returned to their conventional diet.

1. During which phase of the experiment did the children's urine contain the lowest level of the malathion metabolite?

2. During which phase of the experiment was the maximum level of the chlorpyrifos metabolite detected?

3. Did switching to an organic diet lower the amount of pesticide residues excreted by the children?

Study Phase	No. of Samples	Malathion Metabolite		Chlorpyrifos Metabolite	
		Mean (µg/liter)	Maximum (µg/liter)	Mean (µg/liter)	Maximum (µg/liter)
1. Conventional	87	2.9	96.5	7.2	31.1
2. Organic	116	0.3	7.4	1.7	17.1
3. Conventional	156	4.4	263.1	5.8	25.3

Figure 40.17 *Above*, levels of metabolites (breakdown products) of malathion and chlorpyrifos detected in the urine of children taking part in a study of the effects of an organic diet. The difference in the mean level of metabolites in the organic and convential phases of the study was statistically significant. *Left*, the USDA organic food label.

4. Even in the conventional phases of this experiment, the highest pesticide metabolite levels detected were far below those known to be harmful. Given these data, would you spend more to buy organic foods?

---

5. Blood pressure forces water and small solutes into Bowman's capsule during _____ .
  a. glomerular filtration   c. tubular secretion
  b. tubular reabsorption   d. both a and c

6. Kidneys return most of the water and small solutes back to blood by way of _____ .
  a. glomerular filtration   c. tubular secretion
  b. tubular reabsorption   d. both a and b

7. ADH binds to receptors on distal tubules and collecting tubules, making them _____ permeable to _____ .
  a. more; water   c. more; sodium
  b. less; water   d. less; sodium

8. Increased sodium reabsorption _____ .
  a. will make urine more concentrated
  b. will make urine more dilute
  c. is stimulated by aldosterone
  d. both a and c

9. Secretion of H^+ into kidney tubules _____ .
  a. makes the extracellular fluid less acidic
  b. makes the urine less acidic
  c. can cause acidosis
  d. both a and c

10. Aquaporins are passive transport proteins that facilitate the reabsorption of _____ by kidney tubules.
  a. sodium   c. aldosterone
  b. water   d. ADH

11. Match each structure with a function.
  ___ ureter   a. start of nephron
  ___ Bowman's capsule   b. delivers urine to body surface
  ___ urethra
  ___ collecting tubule   c. carries urine from kidney to bladder
  ___ pituitary gland   d. secretes ADH
     e. target of aldosterone

12. The main control center for maintaining mammalian body temperature is in the _____ .
  a. anterior pituitary   c. adrenal gland
  b. renal cortex   d. hypothalamus

13. An animal with a low metabolism that maintains its temperature mainly by adjusting its behavior is _____ .
  a. an endotherm   b. an ectotherm

14. How does the human body respond to an increase in core body temperature?
  a. Decreased blood flow to the skin
  b. Increased sweating
  c. Increased activity of brown adipose tissue
  d. Increased output of thyroid hormone

15. Which organelle in brown adipose tissue gives this tissue its capacity for generating heat?

Critical Thinking

1. The kangaroo rat kidney efficiently produces a very small volume of urine (Section 40.2). Compared to a human, its nephrons have a loop of Henle that is proportionally much longer. Explain how a long loop helps the rat conserve water.

2. In cold habitats, ectotherms are few and endotherms often have morphological adaptations to cold. Compared to closely related species that live in warmer areas, cold dwellers tend to have smaller appendages. Also, animals adapted to cool climates tend to be larger than relatives in warmer places. For example, the largest bear is the polar bear and the largest penguin is Antarctica's emperor penguin.

  Think about heat transfers between animals and their habitat, then explain why smaller appendages and larger overall body size are advantageous in very cold climates.

Chapter 41
Summary

Section 41.1 Some mutations that affect sex hormone function can cause **intersex conditions**, in which an individual has atypical reproductive anatomy. Such conditions do not generally affect health.

Section 41.2 Asexual reproduction produces genetically identical copies of the parent. **Sexual reproduction** has higher energetic costs, and a parent has fewer of its genes represented among the offspring. However, sexual reproduction produces variable offspring, which may be advantageous in environments where conditions fluctuate.

Most animals that reproduce sexually have separate sexes, but some are **hermaphrodites** that produce both eggs and sperm. With **external fertilization**, gametes are released into water. Most animals on land have **internal fertilization**; gametes meet in a female's body. Offspring may develop inside or outside the maternal body. **Yolk** helps nourish developing young of most animals. In placental mammals, young are sustained by nutrients delivered across the **placenta**.

Sections 41.3, 41.4 A human male's **gonads** are his **testes**. They produce sperm and **testosterone**. Sperm form in the **seminiferous tubules** of a testis and mature in an **epididymis** that opens into a **vas deferens**. Secretions from the seminal vesicles and prostate gland join with sperm to form **semen**. Semen is expelled from the body through the **penis**.

Gonadotropin-releasing hormone (GnRH) released by the hypothalamus causes the anterior pituitary gland to secrete **luteinizing hormone (LH)** and **follicle-stimulating hormone (FSH)**. These hormones influence gamete formation in both males and females.

Sections 41.5–41.7 The human female's gonads are **ovaries**. Ovarian follicles produce **oocytes** and secrete **estrogens** and **progesterone**. An **oviduct** conveys an oocyte released at **ovulation** to the **uterus**. The **cervix** of the uterus opens into the **vagina**, which serves as the organ of intercourse and the birth canal.

From puberty until **menopause**, a woman has an approximately monthly **menstrual cycle**. During each cycle, FSH causes a **primary oocyte** that formed before birth to mature and a follicle to secrete estrogen. The estrogen causes thickening of the **endometrium**, the lining of the uterus. A surge of LH triggers ovulation of a **secondary oocyte**. After ovulation, the remains of the follicle, now called the **corpus luteum**, secrete progesterone that primes the uterus for pregnancy. When the corpus luteum breaks down, **menstruation** occurs.

In some women, hormonal changes that occur during the menstrual cycle cause symptoms of PMS. Hormonal changes during menopause result in hot flashes and other symptoms.

Section 41.8 Hormones and autonomic nerve impulses govern the physiological changes that occur during arousal and intercourse. Ejaculation puts millions of sperm into the vagina. Usually only one penetrates the secondary oocyte, causing it to complete meiosis II. The nucleus of the resulting ovum and that of the sperm supply the genetic material of the zygote.

Section 41.9 Humans can choose to prevent pregnancy by abstaining from sex entirely; or to reduce the chance of pregnancy by abstaining during a woman's fertile period, surgically severing reproductive ducts, placing a physical or chemical barrier at the entrance to the uterus, or administering synthetic female sex hormones to prevent ovulation. Infertility can arise as a result of hormone disorders that affect sperm or egg production, blocked reproductive ducts, lack of sperm motility, or uterine problems that interfere with embryonic implantation or development.

Section 41.10 Sexual intercourse can pass viral, bacterial, and protozoan pathogens between partners. The consequences of a sexually transmitted disease range from mild discomfort to sterility and systemic disease that can be fatal. Infected mothers can transmit pathogens to their offspring.

Self-Quiz

Answers in Appendix III

1. Sexual reproduction _____ .
   a. requires internal fertilization
   b. produces offspring that vary in their traits
   c. is more efficient than asexual reproduction
   d. puts all of a parent's genes in each offspring

2. Semen contains secretions from the _____ .
   a. adrenal gland   c. prostate gland
   b. pituitary gland   d. all of the above

3. Male germ cells undergo meiosis in the _____ .
   a. urethra   c. prostate gland
   b. seminiferous tubules   d. vasa deferentia

4. The female _____ is derived from the same embryonic tissue as the male penis.
   a. cervix   b. clitoris   c. vagina   d. oviduct

5. The cervix is the entrance to the _____ .
   a. oviducts   b. vagina   c. uterus   d. scrotum

6. During a menstrual cycle, a midcycle surge of _____ triggers ovulation.
   a. estrogens   b. progesterone   c. LH   d. FSH

7. The corpus luteum develops from _____ and secretes hormones that cause the lining of the uterus to thicken.
   a. follicle cells   c. a primary oocyte
   b. polar bodies   d. a secondary oocyte

Preventing an Intersex Condition Adrenal glands normally make a little testosterone, but a mutation in the gene for the enzyme 21-hydroxylase causes them to produce too much of it. A female with 21-hydroxylase deficiency is exposed to high levels of testosterone during development. This hormone can enlarge her clitoris and cause her labia to fuse, giving her genitals a more male appearance.

The drug dexamethasone slows the adrenal glands' testosterone production. **Figure 41.15** shows data from a study in which doctors gave this drug to pregnant women carrying daughters with 21-hydroxylase deficiency. Sixteen of these women had previously given birth to a daughter with 21-hydroxylase deficiency. These daughters (sisters of the treated newborns) serve as a point of comparison.

1. How many daughters produced by dexamethasone-treated pregnancies had normal female genitals?

2. How many phenotypically normal girls had the women's earlier untreated pregnancies produced?

3. How many women who previously had girls with level 4 or 5 masculinization saw an improvement with treatment?

4. Do the data support the hypothesis that giving the drug dexamethasone to a pregnant woman can reduce the effects of her developing daughter's 21-hydroxylase deficiency?

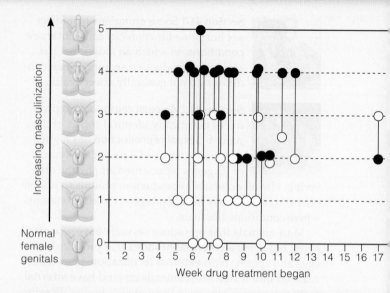

Figure 41.15 Degree of masculinization of 21-hydroxylase deficient females exposed to dexamethasone in the womb (*open* circles), compared to that of older affected sisters who were untreated during development (*dark* circles). Graphics along the side depict appearance of the newborn's genital area.

8. A male has an erection when _____ .
   a. muscles running the length of the penis contract
   b. Leydig cells release a surge of testosterone
   c. the posterior pituitary releases oxytocin
   d. spongy tissue inside the penis fills with blood

9. Sertoli cells inside seminiferous tubules _____ .
   a. secrete testosterone   c. are a type of germ cell
   b. nurture sperm   d. all of the above

10. Birth control pills deliver synthetic _____ .
    a. estrogens and progesterone
    b. LH and FSH
    c. testosterone
    d. oxytocin and prostaglandins

11. Sperm in an epididymis pass next into the _____ .
    a. prostate gland
    b. urethra
    c. seminiferous tubules
    d. vas deferens

12. Match each disease with the type of organism that causes it. The choices can be used more than once.
    ___ chlamydial infection   a. bacteria
    ___ AIDS   b. protist
    ___ syphilis   c. virus
    ___ genital warts
    ___ gonorrhea
    ___ genital herpes
    ___ trichomoniasis

13. Match each structure with its description.
    ___ testis   a. conveys sperm out of body
    ___ epididymis   b. secretes semen components
    ___ labia majora   c. stores sperm
    ___ urethra   d. produces testosterone
    ___ vagina   e. produces estrogens and
    ___ ovary   progesterone

    ___ oviduct   f. usual site of fertilization
    ___ prostate gland   g. lining of uterus
    ___ endometrium   h. fat-padded skin folds
                         i. birth canal

14. Match each hormone with its source.
    ___ FSH and LH   a. pituitary gland
    ___ GnRH   b. ovaries
    ___ estrogens   c. hypothalamus
    ___ testosterone   d. testes

Critical Thinking

1. Drugs that inhibit signals of sympathetic neurons may be prescribed for males who have high blood pressure. How might such drugs interfere with sexual performance?

2. In most groups of birds, males do not have a penis. Both males and females have a single opening, called a cloaca, through which wastes leave the body. The male's sperm also exit through this opening. During mating, a male perches on a female's back and bends his abdomen under, so his cloaca covers hers. This action is referred to as a "cloacal kiss." Some birds even carry out this feat in midair. Flightless birds such as ostriches and kiwis do have a penis. What types of information would help you determine whether the common reptile ancestor of all birds had a penis?

3. Fraternal twins are nonidentical siblings that form when two eggs mature and are released and fertilized at the same time. Explain why an increased level of FSH raises the likelihood of fraternal twins.

4. Some sperm mitochondria do get into an egg during fertilization, but they do not persist. As sperm mature, their mitochondria become tagged with a protein (ubiquitin) that signals the egg to destroy them. What organelle would you expect to be involved in this destruction process?

Chapter 42
Summary

Section 42.1 *In-vitro* fertilization (IVF) is a procedure in which eggs and sperm are combined outside the mother's body. The resulting early embryo is implanted in her uterus to develop. Use of IVF allows infertile couples to have children, but placement of multiple embryos increases the risk of multiple births.

Section 42.2 All sexually reproducing animals have similar stages of development. After fertilization, **cleavage** produces a **blastula**. The blastula undergoes **gastrulation** to produce a **gastrula** with two or three primary tissue layers, or **germ layers**. In vertebrates, three germ layers form: **ectoderm** (outer layer), **mesoderm** (middle layer), and **endoderm** (inner layer.)

Section 42.3 Cytoplasmic localization is a feature of all oocytes. Cleavage distributes different portions of the egg cytoplasm to different cells and forms the blastula. Protostomes and deuterostomes differ in the details of their cleavage pattern, and the amount of yolk in an egg also influences cleavage. Experimental manipulation of amphibian development demonstrated that material localized in one region of the zygote is essential to gastrulation. Cells derived from this region produce signals that initiate gastrulation.

Sections 42.4, 42.5 Differentiation begins after gastrulation: different cell lineages begin to express different genes. Maternal effect genes produce mRNAs that are localized in different parts of a zygote. Some of these mRNAs encode **morphogens** that influence gene expression in a concentration-dependent fashion. Signals from adjacent cells cause **embryonic induction**. Cell migrations and **apoptosis** (programmed cell death) help shape organs.

Studies of many animals have led to a general model of development. In all animals, interactions among master genes constrain development. For example, in all vertebrates, paired blocks of mesoderm called **somites** form and give rise to skeletal muscles and bone.

Section 42.6 Human prenatal development occurs over a period of nine months. Organs take shape during the embryonic period, which concludes at the end of the eighth week. For the remainder of the pregnancy, the fetus grows larger and organs take on their specialized roles. Growth and development continue after birth (in the postnatal period).

Section 42.7 Human fertilization occurs in an oviduct. Cleavage produces a blastocyst that buries itself in the uterine wall. Membranes form outside the blastocyst and support its development. The **amnion** encloses and protects the embryo in a fluid-filled sac. The **chorion** and **allantois** become part of the **placenta**, which allows exchange of substances between maternal and fetal bloodstreams. An implanted blastula makes the hormone **human chorionic gonadotropin**.

Sections 42.8–42.11 Gastrulation occurs after implantation. The first organ to form, the neural tube, later becomes the brain and spinal cord. Somites form on either side of the neural tube.

By the end of the eighth week, the embryo has lost its tail and pharyngeal arches and has a distinctly human appearance. It continues to grow in size and its organs continue to mature during the fetal period.

The placenta consists of extraembryonic membranes and endometrium. It allows embryonic blood to exchange substances with the maternal blood by diffusion. Oxygen, nutrients, and **teratogens** cross the placenta, so a mother's health, nutrition, and lifestyle can affect the growth and development of her future child. Many pregnancies end in miscarriage before they are even detected.

Section 42.12 Hormones typically induce **labor** at about 41 weeks. Positive feedback controls the secretion of **oxytocin**, a hormone that causes contractions that expel the fetus and then the afterbirth. A cesarean section delivers a fetus for whom vaginal delivery would be impossible or dangerous.

Prolactin regulates maturation of the mammary glands and oxytocin causes **lactation**. Breast-feeding has multiple benefits for both mother and child.

Self-Quiz

Answers in Appendix III

1. The end product of cleavage is a _____ .
   - a. zygote
   - b. blastula
   - c. gastrula
   - d. gamete

2. The outermost germ layer in a frog embryo gastrula is the _____ .
   - a. endoderm
   - b. ectoderm
   - c. mesoderm
   - d. dermis

3. A morphogen _____ .
   - a. diffuses through an embryo
   - c. has different effects at different concentrations
   - b. influences gene expression
   - d. all of the above

4. Match each term with the most suitable description.
   - ___ apoptosis
   - ___ embryonic induction
   - ___ cleavage
   - ___ gastrulation
   - ___ implantation
   - a. blastomeres form
   - b. cellular rearrangements form primary tissues
   - c. cells die on cue
   - d. cells influence neighbors
   - e. blastocyst burrows into the uterus

5. In humans, fertilization typically occurs in the _____ .
   - a. vagina   b. uterus   c. cervix   d. oviduct

6. The _____ , a fluid-filled sac, surrounds and protects a human embryo and keeps it from drying out.
   - a. amnion   b. allantois   c. yolk sac   d. chorion

Birth Defects and Multiple Births People considering fertility treatments should be aware that such treatments raise the risk of multiple births, and that multiple pregnancies are associated with an increased risk of some birth defects.

Figure 42.24 shows the results of Yiwei Tang's study of birth defects reported in Florida from 1996 to 2000. Tang compared the incidence of various defects among single and multiple births. She calculated the relative risk for each type of defect based on type of birth, and corrected for other differences that might increase risk such as maternal age, income, race, and medical care during pregnancy. A relative risk of less than 1 means that multiple births pose less risk of that defect occurring. A relative risk greater than 1 means multiples are more likely to have a defect.

1. What was the most common type of birth defect in the single-birth group?

2. Was that type of defect more or less common among the multiple-birth newborns than among single births?

3. Tang found that multiples have more than twice the risk of single newborns for one type of defect. Which type?

4. Does a multiple pregnancy increase the relative risk of chromosomal defects in offspring?

	Prevalence of Defect		Relative Risk
	Multiples	Singles	
Total birth defects	358.50	250.54	1.46
Central nervous system defects	40.75	18.89	2.23
Chromosomal defects	15.51	14.20	0.93
Gastrointestinal defects	28.13	23.44	1.27
Genital/urinary defects	72.85	58.16	1.31
Heart defects	189.71	113.89	1.65
Musculoskeletal defects	20.92	25.87	0.92
Fetal alcohol syndrome	4.33	3.63	1.03
Oral defects	19.84	15.48	1.29

Figure 42.24 Prevalence, per 10,000 live births, of various types of birth defects among multiple and single births. Relative risk for each defect is given after researchers adjusted for the mother's age, race, previous adverse pregnancy experience, education, Medicaid participation during pregnancy, as well as the infant's sex and number of siblings.

7. The placenta consists of _____ .
   a. embryonic tissue
   b. maternal tissue
   c. a combination of maternal and embryonic tissue

8. During the fetal period, _____ .
   a. gastrulation ends
   b. somites form
   c. heartbeats begin
   d. eyes open

9. Human milk contains _____ .
   a. sugars
   b. lysozyme
   c. antibodies
   d. all of the above

10. Match each hormone with its action(s).
   ___ prolactin
   ___ oxytocin
   ___ human chorionic gonadotropin
   a. causes milk production
   b. sustains corpus luteum
   c. causes contraction of uterus, milk ducts

11. Pharyngeal arches of a human embryo will later develop into _____ .
   a. gills
   b. lungs
   c. structures of the head and neck

12. Prior to birth, red blood cells are made by the _____ .
   a. bones
   b. liver
   c. yolk sac
   d. both b and c

13. The vast majority of miscarriages _____ .
   a. occur in the first trimester
   c. are caused by exposure to a pathogen
   b. can be prevented by a cesarean section
   d. threaten the life of the mother

14. Morning sickness _____ .
   a. is an infectious disease
   c. can threaten the life of the mother
   b. typically occurs in the final trimester
   d. may help prevent ingestion of harmful substances

15. Number these events in human development in the correct order.
   ___ gastrulation is completed
   ___ blastocyst forms
   ___ tail is reabsorbed
   ___ implantation occurs
   ___ zygote forms
   ___ neural tube forms
   ___ pharyngeal arches form

Critical Thinking

1. By UNICEF estimates, each year 110,000 people are born with birth defects as a result of prenatal rubella infections. If a woman becomes infected during the first trimester of her pregnancy, her child may be born deaf or blind. Infections later in pregnancy do not increase the risk of these effects. Explain why.

2. The most common ovarian tumors in young women are ovarian teratomas. The name comes from the Greek word *teraton*, which means monster. What makes these tumors "monstrous" is the presence of well-differentiated tissues, most commonly bones, teeth, fat, and hair. Early physicians suggested that teratomas arose as a result of nightmares, witchcraft, or intercourse with the devil. Unlike all other tumors, which arise from somatic cells, teratomas arise from germ cells. Explain why a tumor derived from a germ cell (but not a somatic cell) can give rise to a teratoma.

3. A nursing mother who has an alcoholic drink will secrete alcohol into her milk for two to three hours afterward. Alcohol also decreases oxytocin production. What effect would decreased oxytocin have on breast-feeding?

Chapter 43
Summary

Section 43.1 Stinging is an adaptive behavior in which a worker honeybee sacrifices her life for her relatives' benefit. Her behavior is elicited by a chemical signal called a **pheromone**. Africanized honeybees make a greater and more sustained response to alarm pheromone than European bees.

Section 43.2 Behavior starts with genes that influence the development and activity of the nervous and endocrine systems. Studies of behavioral differences within a species or among closely related species can shed light on the proximate and ultimate causes of a behavior.

Section 43.3 Instinctive behavior occurs without a prior experience. A **fixed action pattern** is an instinctive response to a simple stimulus. **Learned behavior** arises in response to experience. **Imprinting** is time-sensitive learning; with **habituation**, an animal learns to disregard certain stimuli; **classical conditioning** forms an association between two stimuli, and **operant conditioning** forms one between a behavior and its outcome; **observational learning** involves imitation.

Section 43.4 Behavioral traits can be influenced by an animal's environment. **Behavioral plasticity** allows an animal to respond appropriately to environmental pressures that can vary. Epigenetic behavioral effects arise in response to a particular environment and are passed on to offspring.

Section 43.5 Kinesis (random movement) and **taxis** (directional movement) are instinctual responses to stimuli. During **migration**, an animal travels from one habitat to another. Navigating to a specific location requires a mental map and an ability to sense compass directions.

Section 43.6 Communication signals are adaptations that convey information between individuals of the same species. Evolution influences properties of these signals, as when courtship calls are easily localized, but alarm calls are not. Predators sometimes take advantage of the communications of their prey.

Section 43.7 Genetic monogamy, in which a male and female mate only with each other, is extremely rare in animals. Social monogamy is rare in most groups, with the exception of birds. Males maximize their reproductive success by mating with as many females as possible. Females maximize theirs by mating with high-quality mates. Females may choose males on the basis of resources they offer, or may assess mate quality by their courtship performances, as when males perform at a **lek**. When large numbers of females cluster in a defensible area, males compete with one another to control a **territory**.

Parental care has reproductive costs in terms of reduced frequency of reproduction. It is adaptive when benefits to a present set of offspring offset this cost. Most often, females are the sole caregivers of their young.

Section 43.8 Some animals come together as a **selfish herd** in response to the threat of predation. Others live in social groups and benefit by cooperating in predator detection, defense, and rearing young. With a **dominance hierarchy**, resource and mating opportunities are distributed unequally among group members. Individuals that live in large groups also have an increased risk of disease and parasitism, and more intense competition for resources.

Section 43.9 Ants, termites, and some other insects as well as mole-rats are **eusocial animals**: They live in colonies with overlapping generations and have a reproductive division of labor. Most colony members do not reproduce; they assist their relatives instead. The **theory of inclusive fitness** states that **altruistic behavior** can be perpetuated when altruistic individuals help their reproducing relatives. Altruists help perpetuate the genes that lead to their altruism by promoting reproductive success of close relatives who also carry copies of these genes.

Self-Quiz

Answers in Appendix III

1. Genes affect the behavior of individuals by _____ .
   a. influencing the development of nervous systems
   b. affecting the kinds of hormones in individuals
   c. determining which stimuli can be detected
   d. all of the above

2. Stevan Arnold offered slug meat to newborn garter snakes from different populations to test his hypothesis that the snakes' response to slugs _____ .
   a. was shaped by indirect selection
   b. is an instinctive behavior
   c. is based on pheromones
   d. is an epigenetic effect

3. An animal that uses a celestial compass needs _____ .
   a. an ability to sense the Earth's magnetic field
   b. pheromone receptors
   c. an internal clock
   d. an acute sense of hearing

4. The honeybee dance language transmits information about distance to food by way of _____ signals.
   a. tactile   c. acoustical
   b. chemical   d. visual

5. A _____ is a chemical that conveys information between individuals of the same species.
   a. pheromone   c. hormone
   b. neurotransmitter d. all of the above

Spread of Africanized Bees Honeybees disperse by forming new colonies. An old queen leaves the hive along with a group of workers. These bees fly off, find a new nest site, and set up a new hive. Meanwhile, at the old hive, a new queen emerges, mates, and takes over. A new hive can be several kilometers from the old one.

Africanized honeybees form new colonies more often than European ones, a trait that contributes to their spread. Africanized bees also spread by taking over existing hives of European bees. In addition, in areas where European and Africanized hives coexist, European queens are more likely to mate with Africanized males, thus introducing Africanized traits into the colony. **Figure 43.26** shows the counties in the United States in which Africanized honeybees became established from 1990 through 2009.

1. Where in the United States did Africanized bees first become established?

2. In what states did Africanized bees first appear in 2005?

3. Why is it likely that human transport of bees contributed to the spread of Africanized honeybees to Florida?

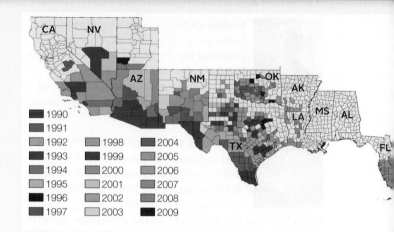

Figure 43.26 The spread of Africanized honeybees in the United States, from 1990 through 2009. The USDA adds a county to this map only when the state officially declares bees in that county Africanized. Bees can be identified as Africanized on the basis of morphological traits analysis of their DNA.

6. In most _____ , males and females cooperate in care of the young.
   a. mammals   c. insects
   b. birds   d. all of the above

7. Generally, living in a social group costs the individual in terms of _____ .
   a. competition for food, other resources
   b. vulnerability to contagious diseases
   c. competition for mates
   d. all of the above

8. Social behavior evolves because _____ .
   a. social animals are more advanced than solitary ones
   b. under some conditions, the costs of social life to individuals are offset by benefits to the species
   c. under some conditions, the benefits of social life to an individual offset the costs to that individual
   d. under most conditions, social life has no costs to an individual

9. Eusocial insects _____ .
   a. live in extended family groups
   b. include termites, honeybees, and ants
   c. have a reproductive division of labor
   d. all of the above

10. Helping other individuals at a reproductive cost to oneself might be adaptive if those helped are _____ .
    a. members of another species
    b. competitors for mates
    c. close relatives
    d. counterfeit signalers

11. A honeybee worker is a _____ .
    a. fertile female   c. fertile male
    b. sterile female   d. sterile male

12. Fill in the blank. A _____ is an instinctive movement toward or away from a stimulus.

13. With _____ , the consequences of a voluntary behavior teach an animal to repeat or avoid that behavior.
    a. classical conditioning   c. imprinting
    b. operant conditioning   d. instinct

14. The gaping mouth of a young cuckoo is a(n) _____ that elicits feeding behavior by a foster parent.
    a. epigenetic trait   c. threat display
    b. altruistic act   d. sign stimulus

15. Match the terms with their most suitable description.
    ___ migration   a. time-limited learning
    ___ imprinting   b. requires staying in touch
    ___ tactile display   c. rare in animals
    ___ selfish herd   d. learning not to respond
    ___ habituation   e. same genes, different behaviors
    ___ genetic   f. each hides behind others
        monogamy   g. move to new habitat
    ___ behavioral
        plasticity

Critical Thinking

1. For billions of years, the only bright objects in the night sky were stars or the moon. Night-flying moths used them to navigate in a straight line. Today, the instinct to fly toward bright objects causes moths to exhaust themselves fluttering around streetlights and banging against brightly lit window-panes. This behavior is not adaptive, so why does it persist?

2. Damaraland mole-rats are fur-covered relatives of naked mole-rats. In their clans, too, nonbreeding individuals of both sexes cooperatively assist one breeding pair. Even so, breeding individuals in wild Damaraland mole-rat colonies usually are unrelated, and few subordinates move up in the hierarchy to breeding status. Researchers suspect that ecological factors, not genetic ones, were the most important selective force in Damaraland mole-rat altruism. Explain why.

Chapter 44
Summary

Section 44.1 A **population** is a group of individuals of the same species that live in the same area and tend to interbreed. Canada geese in the United States include migratory populations and resident ones.

Section 44.2 Demographics are statistics used to describe a population. We can estimate **population size** by using a sampling method such as **plot sampling** or **mark–recapture sampling**. Other demographics include **population density** and **population distribution**. Most populations have a clumped distribution. **Age structure** describes the proportion of individuals in each age category. The size of the **reproductive base** affects population growth.

Section 44.3 Immigration and **emigration** affect population size. The per capita birth rate minus the per capita death rate is a population's **per capita growth rate** (r). **Zero population growth** occurs when birth rate equals death rate.

With **exponential growth**, the population size increases at a fixed rate. In each interval, the number of individuals added is some fixed proportion of the current population size. The time required for a population to double is the doubling time. The maximum possible rate of increase is a species' **biotic potential**.

Section 44.4 Limiting factors constrain the growth of natural populations. **Density-independent factors** are conditions or events that are unaffected by crowding. **Logistic growth** occurs because of constraints imposed by **density-dependent factors**. With logistic growth, a population size rises exponentially, then levels off as it nears **carrying capacity**.

Sections 44.5, 44.6 The time to maturity, number of reproductive events, number of offspring per event, and life span are aspects of a **life history pattern**. A **cohort** is a group of individuals that were born at the same time. Three types of **survivorship curves** are common: a high death rate late in life, a constant rate at all ages, or a high rate early in life. Life histories have a genetic basis and are subject to natural selection. At low population density, ***r*-selection** favors quickly producing as many offspring as possible. At a higher population density, ***K*-selection** favors investing more time and energy in fewer, higher-quality offspring. Most species have a mixture of both *r*-selected and *K*-selected traits.

Sections 44.7, 44.8 The human population is about 7 billion. Expansion into new habitats and the invention of agriculture allowed early increases. Later, medical and technological innovations raised carrying capacity and sidestepped many limiting factors. Today, the global **total fertility rate** is declining, but it remains above the **replacement fertility rate**. The **demographic transition model** predicts economic development may slow population growth. World resource consumption will probably continue to rise because a highly developed nation has a much larger **ecological footprint** than a developing one.

Self-Quiz

Answers in Appendix III

1. Most commonly, individuals of a population show a _____ distribution within their habitat.
   a. clumped   c. nearly uniform
   b. random   d. none of the above

2. The rate at which population size grows or declines depends on the rate of _____ .
   a. births   c. immigration   e. a and b
   b. deaths   d. emigration   f. all of the above

3. Suppose 200 fish are marked and released in a pond. The following week, 200 fish are caught and 100 of them have marks. There are about _____ fish in this pond.
   a. 200   b. 300   c. 400   d. 2,000

4. A population of worms is growing exponentially in a compost heap. Thirty days ago there were 400 worms and now there are 800. How many worms will there be thirty days from now, assuming conditions remain constant?
   a. 1,200   b. 1,600   c. 3,200   d. 6,400

5. For a given species, the maximum rate of increase per individual under ideal conditions is its _____ .
   a. biotic potential   c. environmental resistance
   b. carrying capacity   d. density control

6. _____ is a density-independent factor that influences population growth.
   a. Resource competition   c. Predation
   b. Infectious disease   d. Harsh weather

7. A life history pattern is a set of adaptations that influence the individual's _____ .
   a. longevity   c. age at reproductive maturity
   b. fertility   d. all of the above

8. The human population is now about 7 billion. It reached 6 billion in _____ .
   a. 2007   b. 1999   c. 1802   d. 1350

9. Compared to the less developed countries, the highly developed ones have a higher _____ .
   a. death rate   c. total fertility rate
   b. birth rate   d. resource consumption rate

Iguana Decline In 1987, Martin Wikelski began a long-term study of marine iguanas in the Galápagos Islands. He marked iguanas on two islands—Genovesa and Santa Fe—and collected data on how their body size, survival, and reproductive rates varied over time. He found that because iguanas eat algae and have no predators, deaths usually result from food shortages, disease, or old age. In January 2001, an oil tanker ran aground and leaked a small amount of oil into the waters near Santa Fe. **Figure 44.17** shows the number of marked iguanas that Wikelski and his team counted in their study populations just before the spill and about a year later.

1. Which island had more marked iguanas at the time of the first census?

2. How much did the population size on each island change between the first and second census?

3. Wikelski concluded that changes on Santa Fe were the result of the oil spill, rather than sea temperature or other climate factors common to both islands. How would the census numbers be different from those he observed if an adverse event had affected both islands?

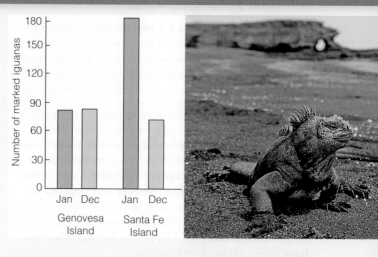

Figure 44.17 Shifting numbers of marked marine iguanas on two Galápagos islands. An oil spill occurred near Santa Fe just after the January 2001 census (*green* bars). A second census was carried out in December 2001 (*tan* bars).

10. Species that live in unpredictable habitats are more likely to show traits that are favored by _____ .
    a. *r*-selection   b. *K*-selection

11. All members of a cohort are the same _____ .
    a. sex   b. size   c. age   d. weight

12. The ecological footprint of a person in the United States is about _____ that of a person in India.
    a. half   b. twice   c. one-ninth   d. nine times

13. Match each term with its most suitable description.
    ___ carrying capacity
    ___ exponential growth
    ___ biotic potential
    ___ limiting factor
    ___ logistic growth

    a. maximum rate of increase per individual under ideal conditions
    b. population growth plots out as an S-shaped curve
    c. maximum number of individuals sustainable by the resources in a given environment
    d. population growth plots out as a J-shaped curve
    e. essential resource that restricts population growth when scarce

Critical Thinking

1. Think back to Section 44.6. When researchers moved guppies from populations preyed on by cichlids to a habitat with killifish, the life histories of the transplanted guppies evolved. They came to resemble those of guppy populations preyed on by killifish. Males became gaudier; some scales formed larger, more colorful spots. How might a decrease in predation pressure on sexually mature fish allow this change?

2. The age structure diagrams for two hypothetical populations are shown at *right*. Describe the growth rate of each population and discuss the current and future social and economic problems that each is likely to face.

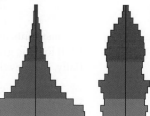

Chapter 45
Summary

Section 45.1 All the species that live in an area constitute a **community**. Introduction of a new species to a community can have negative effects on the species that evolved in that community. Global trade has distributed species such as red fire ants from Brazil to new habitats where they are pests.

Section 45.2 Each species occupies a certain **habitat** characterized by physical features and by the array of other species living in it. Species interactions affect community structure. **Commensalism** is a common interaction in which one species benefits and the other is unaffected. A **symbiosis** is an interaction in which one species lives in or on another.

Section 45.3 In a **mutualism**, both species benefit from an interaction. Some mutualists cannot complete their life cycle without the interaction. Mutualists who maximize their own benefits while limiting the cost of cooperating are at a selective advantage.

Section 45.4 A species's roles and requirements define its unique **ecological niche**. When two species with similar niches share a habitat, **interspecific competition** harms both. When the species are very similar, **competitive exclusion** may occur. Competing species with similar requirements become less similar when directional selection causes **character displacement**. This change in traits allows **resource partitioning**.

Sections 45.5, 45.6 **Predation** benefits a predator at the expense of the prey it captures, kills, and eats. Predators and their prey exert selective pressure on one another. Both predators and prey can benefit from **camouflage**. Some well-defended prey have **warning coloration**. With **mimicry**, well-defended species evolve a similar appearance or vulnerable species come to resemble better defended ones. Plant traits such as thorns or unpalatable secondary metabolites allow plants to escape or survive **herbivory**.

Section 45.7 **Parasitism** involves feeding on a living host. It benefits the parasite at the expense of the host. **Brood parasites** steal parental care from another species. **Parasitoids** are insects that lay eggs in or on a host insect. Parasites and parasitoids are often used in **biological control** of pest species.

Section 45.8 Ecological succession is the sequential replacement of one array of species by another over time. **Primary succession** happens in newly formed, vacant habitats. **Secondary succession** occurs in disturbed ones. The first species to become established are **pioneer species**. The pioneers may help, hinder, or have no effect on later colonists.

The older idea that all communities eventually reach a predictable climax state has been replaced by models that emphasize the role of chance and disturbances. The **intermediate disturbance hypothesis** holds that disturbances of moderate intensity and frequency maximize species diversity.

Section 45.9 **Keystone species** are especially important in maintaining the composition of a community. The removal of a keystone species or introduction of an **exotic species**—one that evolved in a different community—can alter community structure permanently. An **indicator species** is especially sensitive to environmental change, so its presence or absence can provide early information about an environment's degradation. Some species are adapted to a particular type of periodic disturbance such as wildfires. Such species are at a competitive disadvantage if the disturbance stops occurring.

Section 45.10 Species richness, the number of species in a given area, varies with latitude. The **equilibrium model of island biogeography** predicts the number of species that an island will sustain based on the **area effect** and the **distance effect**. Scientists can use this model to predict the number of species that islands of habitat such as parks can sustain.

Self-Quiz
Answers in Appendix III

1. The type of physical environment in which a species typically lives is its _____ .
   - a. niche
   - b. habitat
   - c. community
   - d. population

2. Which cannot be a symbiosis?
   - a. mutualism
   - b. parasitism
   - c. commensalism
   - d. interspecific competition

3. A tick is a(n) _____ .
   - a. parasitoid
   - b. ectoparasite
   - c. endoparasite

4. _____ can lead to resource partitioning.
   - a. Mutualism
   - b. Parasitism
   - c. Commensalism
   - d. Interspecific competition

5. Match the terms with the most suitable descriptions.
   - ___ mutualism
   - ___ parasitism
   - ___ commensalism
   - ___ predation
   - ___ interspecific competition

   - a. one free-living species feeds on another and usually kills it
   - b. two species interact and both benefit by the interaction
   - c. two species interact and one benefits while the other is neither helped nor harmed
   - d. one species feeds on another that it lives on or in
   - e. two species access a resource

Testing Biological Control Ant-decapitating phorid flies are just one of the biological control agents used to battle imported fire ants. Researchers have also enlisted the help of *Thelohania solenopsae*, another natural enemy of the ants. This microsporidian (Section 23.4) is a parasite that infects ants and shrinks the ovaries of the colony's egg-producing female (the queen). As a result, a colony dwindles in numbers and eventually dies out.

Are these biological controls useful against imported fire ants? To find out, USDA scientists treated infested areas with either traditional pesticides or pesticides plus biological controls (both flies and the parasite). The scientists left some plots untreated as controls. **Figure 45.28** shows the results.

1. How did population size in the control plots change during the first four months of the study?

2. How did population size in the two types of treated plots change during this same interval?

3. If this study had ended after the first year, would you conclude that biological controls had a major effect?

4. Would your conclusion differ at the end of the time period shown?

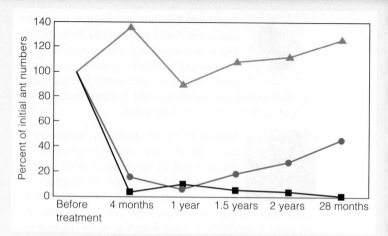

Figure 45.28 A comparison of two methods of controlling red imported fire ants. The graph shows the numbers of red imported fire ants over a 28-month period. *Orange* triangles represent untreated control plots. *Green* circles are plots treated with pesticides alone. *Black* squares are plots treated with pesticide and biological control agents (phorid flies along with a microsporidian parasite).

6. Lizards that eat flies they catch on the ground and birds that catch and eat flies in the air are engaged in _____ competition.
   a. exploitative   d. interspecific
   b. interference   e. both a and d
   c. intraspecific   f. both b and c

7. By a currently favored hypothesis, species richness of a community is greatest between physical disturbances of _____ intensity or frequency.
   a. low   c. high
   b. intermediate   d. variable

8. With _____ , one species evolves to look like another.

9. Growth of a forest in an abandoned corn field is an example of _____ .
   a. primary succession   c. secondary succession
   b. resource partitioning   d. competitive exclusion

10. Species richness is greatest in communities _____ .
   a. near the equator   c. near the poles
   b. in temperate regions   d. that recently formed

11. If you remove a species from a community, the population size of its main _____ is likely to increase.
   a. parasite   b. competitor   c. predator

12. _____ steal parental care.
   a. Mutualists   c. Brood parasites
   b. Commensalists   d. Predators

13. The oldest established land communities are _____ .
   a. in the Arctic   c. in the tropics
   b. in temperate zones   d. on volcanic islands

14. Biological control of pest species _____ .
   a. has no side effects   c. uses natural enemies
   b. involves mutualists   d. requires use of chemicals

15. Match the terms with the most suitable descriptions.
   ___ area effect   a. native species with large effect
   ___ pioneer   b. first species established in a
       species   new habitat
   ___ indicator   c. more species on large islands
       species   than small ones at same distance
   ___ keystone   from the source of colonists
       species   d. species that is especially sensitive
   ___ exotic   to changes in the environment
       species   e. allows competitors to coexist
   ___ resource   f. often outcompete, displace native
       partitioning   species of established community

Critical Thinking

1. With antibiotic resistance rising, researchers are looking for ways to reduce use of these drugs. Some cattle once fed antibiotic-laced food now get probiotic feed that can bolster populations of helpful bacteria in the animal's gut. The idea is that if a large population of beneficial bacteria is in place, then harmful bacteria cannot become established or thrive. Which ecological principle is guiding this practice?

2. Flightless birds that live on islands often have relatives on the mainland that can fly. The island species presumably evolved from fliers that, in the absences of predators, lost their ability to fly. Many island populations of flightless birds are now in decline as a result of rats and other exotic species that have been introduced to their previously isolated island habitats. Despite the current selection pressure in favor of flight, no flightless island bird species has yet regained the ability to fly. Why is this unlikely to happen?

Chapter 46
Summary

Section 46.1 Inorganic substances such as phosphorus serve as nutrients for living organisms. Excessive inputs of nutrients as a result of human activities can alter ecosystem dynamics, as when phosphate from detergents or fertilizers causes **eutrophication**.

Section 46.2 An **ecosystem** consists of an array of organisms along with nonliving components of their environment. There is a one-way flow of energy into and out of an ecosystem, and a cycling of materials among resident species. All ecosystems have inputs and outputs of energy and nutrients.

Sunlight supplies energy to most ecosystems. **Primary producers** convert sunlight energy into chemical bond energy. They also take up the nutrients that they, and all consumers, require. Herbivores, carnivores, omnivores, **decomposers**, and **detritivores** are **consumers**.

Energy moves from organisms at one **trophic level** to organisms at another. Organisms are at the same trophic level if they are an equal number of steps away from the energy input into the ecosystem. A **food chain** shows one path of energy and nutrient flow among organisms. It depicts who eats whom.

Section 46.3 Food chains interconnect as **food webs**. In a **grazing food webs**, most energy captured by producers flows to herbivores. In **detrital food webs**, most energy flows from producers directly to detritivores and decomposers. Both types of food webs interconnect in nearly all ecosystems. The efficiency of energy transfers is low, so most food chains have no more than four or five trophic levels.

Section 46.4 The **primary production** of an ecosystem is the rate at which producers capture and store energy in their tissues. It varies with climate, seasonal changes, and nutrient availability.

Energy pyramids and **biomass pyramids** depict how energy and organic compounds are distributed among the organisms of an ecosystem. All energy pyramids are largest at their base. If producers get eaten as fast as they reproduce, the biomass of consumers can exceed that of producers, so the biomass pyramid is upside down.

Section 46.5 In a **biogeochemical cycle**, water or some nutrient moves from an environmental reservoir, through organisms, then back to the environment. Chemical and geological processes move nutrients between environmental reservoirs.

Section 46.6 In the **water cycle**, evaporation, condensation, and precipitation move water from its main reservoir—oceans—into the atmosphere, onto land, then back to oceans. **Runoff** is water that flows over ground into streams. A **watershed** is an area where all precipitation drains into a specific waterway. Water in **aquifers** and in the soil is **groundwater**. Only a small fraction of Earth's water is available as fresh water, and most of that is frozen.

Section 46.7 The **carbon cycle** moves carbon mainly among seawater, the air, soils, and living organisms in an **atmospheric cycle**. The largest carbon reservoir is sedimentary rocks, but living organisms cannot take up carbon from this reservoir. The largest reservoir of biologically available carbon is the ocean. Aquatic producers take up dissolved bicarbonate and convert it to CO_2. Plants take up CO_2 from air.

When land organisms die, their wastes and remains contribute to the carbon in soil, which exceeds the amount in the atmosphere. The amount of carbon in soil differs among ecosystems, with the carbon content being highest in regions where decomposition is slowest.

Human use of fossil fuels is moving carbon from a reservoir that was previously inaccessible to living organisms into the air and water.

Section 46.8 The **greenhouse effect** refers to the ability of certain gases to trap heat in the lower atmosphere and thus warm Earth's surface. Carbon dioxide is a greenhouse gas and human activities are putting increasingly larger amounts of it into the atmosphere. Direct measurements of the atmosphere, studies of air bubbles in ice cores, and the composition of fossil foraminiferan shells indicate that atmospheric CO_2 is currently at its highest in at least 15 million years. The ongoing rise in carbon dioxide is thought to contribute to global warming, which is one aspect of an ongoing **global climate change**.

Section 46.9 The **nitrogen cycle** is an atmospheric cycle. Air is the main reservoir for N_2, a gaseous form of nitrogen accessible only to nitrogen-fixing bacteria. By the process of **nitrogen fixation**, bacteria take up N_2 and incorporate it into ammonia that producers can take up and use. Ammonia is also formed by the **ammonification** of organic remains by bacteria and fungi. Bacteria also carry out **nitrification**, converting ammonium to nitrite and then nitrate, which plants can also take up. **Denitrification** of nitrate by bacteria converts nitrate in soil or water to a gaseous form.

Section 46.10 The industrial fixation of nitrogen to produce chemical fertilizers, the production of livestock, and the burning of fossil fuels add nitrogen-containing compounds to the air and water.

Nitrous oxide is a greenhouse gas that also contributes to destruction of the protective ozone layer. It is formed by burning fossil fuels and by the activity of bacteria in habitats enriched by nitrogen fertilizer.

Nitrate is a soluble compound that frequently pollutes our sources of drinking water. Ingestion of excess nitrate has been tied to a variety of ailments, including thyroid cancer. Nitrate enters water as a result of fertilizer use and inadequate treatment of sewage.

Section 46.11 The **phosphorus cycle** is a **sedimentary cycle**. Earth's crust is the largest reservoir of this element and it does not occur as a gas in any significant quantity. Producers cannot access the phosphate tied up in rocks. They obtain phosphorus by taking up dissolved phosphates. Lack of phosphate often limits plant growth. Deposits of bat and bird wastes are mined as a natural fertilizer, and phosphate-rich rocks are used to produce fertilizer on an industrial scale.

Self-Quiz

Answers in Appendix III

1. In most ecosystems, the primary producers use energy from _____ to build organic compounds.
   a. sunlight
   b. heat
   c. breakdown of wastes and remains
   d. breakdown of inorganic substances in the habitat

2. Organisms at the lowest trophic level in a tallgrass prairie are all _____ .
   a. two steps away from the original energy input
   b. autotrophs   d. both a and b
   c. heterotrophs   e. both a and c

3. All organisms at the top trophic level _____ .
   a. capture energy from a nonliving source
   b. obtain carbon from a nonliving source
   c. would be at the top of an energy pyramid
   d. all of the above

4. Primary productivity is affected by _____ .
   a. nutrient availability   c. temperature
   b. amount of sunlight   d. all of the above

5. Efficiency of energy transfers in aquatic ecosystems is typically higher than in land ecosystems because _____ .
   a. aquatic food webs include more endotherms
   b. algae do not make lignin
   c. primary production cannot occur in water
   d. all of the above

6. Most of Earth's fresh water is _____ .
   a. in lakes and streams   c. frozen as ice
   b. in aquifers and soil   d. in bodies of organisms

7. Earth's largest carbon reservoir is _____ .
   a. the atmosphere   c. seawater
   b. sediments and rocks   d. living organisms

8. Carbon is released into the atmosphere by _____ .
   a. photosynthesis   c. burning fossil fuels
   b. aerobic respiration   d. b and c

9. Greenhouse gases _____ .
   a. slow the escape of heat energy from Earth into space
   b. are produced by natural and human activities
   c. are at higher levels than they were 100 years ago
   d. all of the above

10. The _____ cycle is a sedimentary cycle.
    a. phosphorus   c. nitrogen
    b. carbon   d. water

11. Earth's largest phosphorus reservoir is _____ .
    a. the atmosphere   c. sedimentary rock
    b. the ocean   d. living organisms

12. Plant growth requires uptake of _____ from the soil.
    a. nitrogen   d. both a and c
    b. carbon   e. all of the above
    c. phosphorus

13. Nitrogen fixation converts _____ to _____ .
    a. nitrogen gas; ammonia   d. ammonia; nitrates
    b. nitrates; nitrites   e. nitrogen gas; nitrogen
    c. ammonia; nitrogen gas   oxides

14. Burning fossil fuel releases _____ into the air.
    a. carbon dioxide   c. phosphates
    b. nitrous oxide   d. a and b

15. Match each term with its most suitable description.
    ___ carbon dioxide   a. contains triple bond
    ___ nitrate   b. mined from sedimentary rock
    ___ phosphate   c. marine carbon source
    ___ nitrogen gas   d. soluble form of nitrogen
    ___ bicarbonate   e. greenhouse gas
    ___ nitrous oxide   f. greenhouse gas and ozone
                             destroyer

Changes in the Air To assess the impact of human activity on the carbon dioxide level in Earth's atmosphere, it helps to take a long view. One useful data set comes from deep core samples of Antarctic ice. The oldest ice core that has been fully analyzed dates back a bit more than 400,000 years. Air bubbles trapped in the ice provide information about the gas content in Earth's atmosphere at the time the ice formed. Combining ice core data with more recent direct measurements of atmospheric carbon dioxide—as in **Figure 46.20**—can help scientists put current changes in the atmospheric carbon dioxide into historical perspective.

1. What was the highest carbon dioxide level between 400,000 B.C. and 0 A.D.?

2. During this period, how many times did carbon dioxide reach a level comparable to that measured in 1980?

3. The industrial revolution occurred around 1800. What was the trend in carbon dioxide level in the 800 years prior to this event? What about in the 175 years after it?

4. Was the rise in the carbon dioxide level between 1800 and 1975 larger or smaller than the rise between 1980 and 2007?

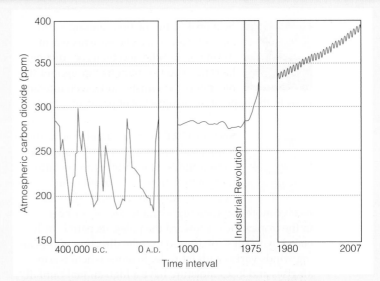

Figure 46.20 Changes in atmospheric carbon dioxide levels (in parts per million). Direct measurements began in 1980. Earlier data are based on ice cores.

Critical Thinking

1. Marguerite plants a vegetable garden in Maine. Eduardo plants a similar garden in Florida. Based on climate alone, which garden would you expect to have a higher annual primary productivity? What other factors could affect primary production in each garden?

2. Where does your drinking water come from? An aquifer or an aboveground reservoir? What area is included within your watershed? Visit the Science in Your Watershed site at http://water.usgs.gov/wsc to find out.

3. Scientists study bubbles trapped in ancient glacial ice to determine how concentrations of nitrogen and carbon dioxide gas have changed over time. However, bubbles in glacial ice cannot provide information about changes in phosphorus. Explain why air samples are not useful for this purpose and propose an alternative method to study how the amount of phosphorus in a region has changed over time.

4. Nitrogen-fixing bacteria live throughout the ocean, from its sunlit upper waters to 200 meters (650 feet) beneath its surface. Recall that nitrogen is a limiting factor in many habitats. What effect would an increase in populations of marine nitrogen-fixers have on carbon uptake and primary productivity in those waters?

5. As noted in Section 46.7, mycorrhizal fungi help prevent carbon in the soil from escaping into the atmosphere. These fungi also benefit a host plant by providing it with a share of the phosphorus and nitrogen that their hyphae take up from soil. Most crop plants are capable of forming a relationship with mycorrhizal fungi, if any are present in the soil. Some people have suggested that inoculating soil with fungal spores could help reduce the use of industrially produced fertilizer. What are some potential advantages of using fungi, rather than chemical fertilizers, to enhance plant growth?

Chapter 47
Summary

Section 47.1 Interactions among Earth's air and waters, as when oceans warm during an **El Niño** and then cool during a **La Niña**, affect conditions throughout the **biosphere**. Scientists study such interactions in the hope of devising ways of predicting them.

Section 47.2 Global air circulation patterns influence **climate** and the distribution of communities. The patterns are set into motion by latitudinal variations in incoming solar radiation. Tropical latitudes receive more sun than higher latitudes. As air heated in the tropics moves toward the poles, its path is influenced by Earth's rotation so that the direction of prevailing winds varies with latitude. Regions where warm air rises and loses moisture have a high annual rainfall; regions where cool air descends are dry.

Section 47.3 Latitudinal and seasonal variations in sunlight warm up seawater and create surface currents that are affected by prevailing winds. The currents distribute heat energy worldwide and influence the weather patterns. Ocean currents, air currents, and landforms interact in shaping global temperature zones, as where the presence of coastal mountains causes a **rain shadow** or where **monsoon** rains fall seasonally.

Section 47.4 Biomes are discontinuous areas characterized by a particular type of vegetation. Differences in climate, elevation, and soil properties affect the distribution of biomes.

Section 47.5 Deserts form around latitudes 30° north and south, where rainfall is sparse and falls seasonally. Desert plants include annuals that grow fast after seasonal rains and perennials that are adapted to withstand a seasonal drought. Desert crust, a community of organisms in the upper soil layer, helps hold soil in place.

Sections 47.6–47.7 Grasslands form in the interior of midlatitude continents. North America's grasslands are prairies. Prairies have a highly fertile soil and most have now been converted to agricultural use. Africa has savannas, which include widely dispersed trees. Both grasslands and savannas support herds of grazing animals.

Southwest coasts of continents have cool, rainy winters and a hot, dry summer. Depending on the amount of rain that falls, they support **dry woodlands** or **dry shrub-** lands such as California's chaparral. Shrubland plants are adapted to periodic fire.

Sections 47.8-47.10 Broadleaf trees are angiosperms. Most trees in **temperate deciduous forests** shed their leaves all at once just before a cold winter prevents growth. Broadleaf trees in **tropical rain forests** can grow year-round, and these enormously productive forests are home to a large number of species.

Conifers withstand cold and drought better than broadleaf trees and dominate Northern Hemisphere high-latitude **boreal forests**.

Farther north in this hemisphere, **arctic tundra** overlies **permafrost**. **Alpine tundra** is a similar biome that occurs at high altitudes.

Section 47.11 Most **lakes**, streams, and other aquatic ecosystems have gradients in the penetration of sunlight, water temperature, and in dissolved gases and nutrients. These characteristics vary over time and affect primary productivity. In temperate zone lakes, a **spring overturn** and a **fall overturn** cause vertical mixing of waters and trigger a burst of productivity. In summer, a **thermocline** prevents upper and lower waters from mixing.

Sections 47.12–47.14 Coastal zones support diverse ecosystems. Among these, the coastal wetlands, **estuaries**, and **coral reefs** are especially productive.

Life persists throughout the ocean. Diversity is highest in sunlit waters at the top of the **pelagic province**, which is the ocean's waters. On the seafloor (in the **benthic province**), diversity is high near deep-sea **hydrothermal vents** and on **seamounts**. Even the deepest ocean sediments support life.

Section 47.15 The El Niño Southern Oscillation triggers changes in rainfall and other weather patterns around the world. These changes can harm human health, as when warming waters increase the likelihood of cholera epidemics.

Self-Quiz

Answers in Appendix III

1. The Northern Hemisphere is most tilted toward the sun in _____ .
   a. spring   b. summer   c. autumn   d. winter

2. Which latitude will have the most hours of daylight on the summer solstice?
   a. 0° (the equator)   c. 45° north
   b. 30° north   d. 60° north

3. Warm air _____ and it holds _____ water than cold air.
   a. sinks; less   c. sinks; more
   b. rises; less   d. rises; more

Sea Temperatures To predict the effect of El Niño or La Niña events in the future, the National Oceanographic and Atmospheric Administration collects information about sea surface temperature (SST) and atmospheric conditions. They compare monthly temperature averages in the eastern equatorial Pacific Ocean to historical data and calculate the difference (the degree of anomaly) to determine if El Niño conditions, La Niña conditions, or neutral conditions are developing. El Niño is a rise in the average SST above 0.5°C. A decline of the same amount is La Niña. **Figure 47.41** shows data for nearly 39 years.

1. When did the greatest positive temperature deviation occur during this time period?

2. What type of event, if any, occurred during the winter of 1982–1983? What about the winter of 2001–2002?

3. During a La Niña event, less rain than normal falls in the American West and Southwest. In the time interval shown, what was the longest interval without a La Niña event?

4. What type of conditions were in effect in the fall of 2007 when California suffered severe wildfires?

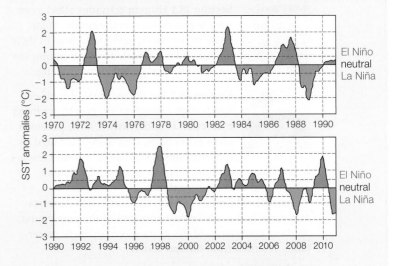

Figure 47.41 Sea surface temperature anomalies (differences from the historical mean) in the eastern equatorial Pacific Ocean. A rise above the dotted *red* line is an El Niño event, a decline below the *blue* line is La Niña.

---

4. A rain shadow is a reduction in rainfall _____ .
   a. on the inland side of a coastal mountain range
   b. during an El Niño event
   c. that results from global warming

5. The Gulf Stream is a current that flows _____ along the _____ coast of the United States.
   a. north to south; east   c. south to north; east
   b. north to south; west   d. south to north; west

6. _____ have a deep layer of humus-rich topsoil.
   a. Deserts   c. Rain forests
   b. Grasslands   d. Seamounts

7. Biome distribution depends on _____ .
   a. climate   c. soils
   b. elevation   d. all of the above

8. Grasslands most often are found _____ .
   a. at 30° north and south   c. in interior of continents
   b. at high altitudes   d. all of the above

9. Permafrost underlies _____ .
   a. arctic tundra   c. boreal forest
   b. alpine tundra   d. all of the above

10. Warm, still water holds _____ oxygen than cold, fast-flowing water.

11. Chemoautotrophic bacteria and archaea are the primary producers for food webs _____ .
    a. in mangrove wetlands   c. on coral reefs
    b. at seamounts   d. at hydrothermal vents

12. Corals rely on symbiotic _____ for sugars.
    a. fungi   c. dinoflagellates
    b. amoebas   d. green algae

13. What biome borders on boreal forest?
    a. savanna   c. tundra
    b. taiga   d. chaparral

14. Unrelated species in geographically separated parts of a biome may resemble one another as a result of _____ .
    a. morphological divergence   c. resource partitioning
    b. morphological convergence   d. coevolution

15. Match the terms with the most suitable description.
    ___ tundra   a. broadleaf forest near equator
    ___ chaparral   b. partly enclosed by land; where
    ___ desert   fresh water and seawater mix
    ___ savanna   c. African grassland with trees
    ___ estuary   d. low-growing plants at
    ___ boreal forest   high latitudes or elevations
    ___ prairie   e. dry shrubland
    ___ tropical rain   f. at latitudes 30° north and south
        forest   g. mineral-rich, superheated
    ___ hydrothermal   water supports communities
        vents   h. conifers dominate
                        i. North American grassland

Critical Thinking

1. London, England, is at the same latitude as Calgary in Canada's province of Alberta. However, the mean January temperature in London is 5.5°C (42°F), whereas in Calgary it is minus 10°C (14°F). Compare the locations of these two cities and suggest a reason for this temperature difference.

2. Increased industrialization in China has environmentalists worried about air quality elsewhere. Are air pollutants emitted in Beijing more likely to end up in eastern Europe or the western United States? Why?

3. The use of off-road recreational vehicles may double in the next twenty years. Enthusiasts would like increased access to government-owned deserts. Some argue that it's the perfect place for off-roaders because "There's nothing there." Explain whether you agree, and why.

Chapter 48
Summary

Section 48.1 Human activities affect even remote places such as the Arctic. Polar bears in the Arctic have pollutants in their bodies and are threatened by thinning sea ice, one effect of global climate change.

Section 48.2 The current rate of species loss is high enough to suggest that a **mass extinction** is under way. Unlike previous mass extinctions, which are attributed to physical causes such as an asteroid impact or a volcanic eruption, this one is caused by human actions and thus is preventable.

Arrival of humans in Australia and North America may have had a role in the extinction of the megafauna on these continents. Many more-recent extinctions certainly resulted from human activity.

Section 48.3 Endangered species currently face a high risk of extinction. **Threatened species** are likely to become endangered in near the future. **Endemic species**, which evolved in one place and are present only in that habitat, are highly vulnerable to extinction. Species with highly specialized needs are also especially vulnerable. Our knowledge of existing species is limited and biased toward vertebrates. We know little about the abundance and diversity of microbial species that carry out essential ecosystem processes.

Human activities cause habitat loss, degradation, and fragmentation that can endanger a species. Humans also directly reduce populations by overharvesting. In most cases, a species becomes endangered because of multiple factors. Sometimes, a decline in one species as a result of human activity leads to decline of another species.

Section 48.4 Desertification and deforestation dramatically alter a habitat and can even affect local climate. The changes caused by deforestation are especially difficult to reverse.

Section 48.5 Burning fossil fuels, especially coal, releases sulfur dioxide and nitrogen oxides into the atmosphere. These **pollutants** mix with water vapor in the air, then fall to earth as **acid rain**. The resulting increase in the acidity of soils and waters can sicken or kill organisms.

With **bioaccumulation**, a pollutant taken up from the environment becomes stored in an organism's tissues, so older animals have a higher pollutant concentration than younger ones. With **biological magnification**, a pollutant increases in concentration as it is passed up a food chain. As a result of bioaccumulation and biological magnification, the tissues of an organism can have a far higher concentration of a pollutant than the environment.

Plastic that washes into or is dumped into oceans is another type of pollutant. It poses a threat to organisms that mistake it for food.

Section 48.6 The **ozone layer** in the upper atmosphere protects against incoming UV radiation. Chemicals called CFCs were banned when they were found to cause thinning of the ozone layer. Near the ground, where the ozone concentration is naturally low, ozone emitted as a result of fossil fuel use is considered a pollutant. It irritates animal respiratory tracts and interferes with photosynthesis by plants.

Section 48.7 Global climate change is raising the sea level, altering the salinity of upper ocean waters, and making the sea more acidic. It is also affecting the distribution and behavior of terrestrial species, allowing some to move into higher elevations or latitudes and altering the timing of migrations and flowering. Global climate change is also expected to have negative effects on human health by increasing heat-related deaths and encouraging outbreaks of some infectious diseases.

Section 48.8 We recognize three levels of **biodiversity**: genetic diversity, species diversity, and ecosystem diversity. All are threatened. The field of **conservation biology** surveys the range of biodiversity, investigates its origins, and identifies ways to maintain and use it in ways that benefit human populations.

Given that resources are limited, biologists attempt to identify **biodiversity hot spots**, regions rich in endemic species and under a high level of threat. Biologists also identify ecoregions, larger regions characterized by their physical characteristics as well as the species in them. The biologists prioritize ecoregions, with the goal of identifying those whose conservation will ensure that a representative sample of all of Earth's current biomes remains intact. When an ecosystem has been totally or partially degraded, **ecological restoration** can help restore biodiversity.

Section 48.9 Individuals can help sustain biodiversity by limiting their energy use and material consumption. Reuse and recycling can help protect species by reducing the use of destructive practices that are required to extract resources.

Self-Quiz
Answers in Appendix III

1. True or false? Most species that evolved have already become extinct.

2. Dodos were driven to extinction _____ .
   a. when humans arrived in North America
   b. by overharvesting and introduced species
   c. as a result of global warming
   d. by volcanic eruptions

3. An _____ species has population levels so low it is at great risk of extinction in the near future.
   a. endemic   c. indicator
   b. endangered   d. exotic

Pervasive PCBs Winds carry chemical contaminants produced and released at temperate latitudes to the Arctic, where the chemicals enter food webs. By the process of biological magnification (Section 48.5), top carnivores in arctic food webs—such as polar bears and people—end up with high doses of these chemicals. For example, indigenous arctic people who eat a lot of local wildlife tend to have unusually high levels of hormone-disrupting polychlorinated biphenyls (PCBs) in their bodies. The Arctic Monitoring and Assessment Programme studies the effects of these chemicals on health and reproduction. **Figure 48.20** shows the effect of PCBs on the sex ratio at birth in indigenous human populations in the Russian Arctic.

1. Which sex is was most common in offspring of women with less than one microgram per milliliter of PCB in serum?

2. At what PCB concentrations were women more likely to have daughters?

3. In some Greenland villages, nearly all recent newborns are female. Would you expect PCB levels in those villages to be above or under 4 micrograms per milliliter?

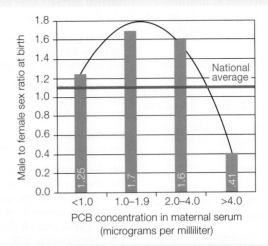

Figure 48.20 Effect of maternal PCB concentration on the sex ratio of newborns in indigenous populations in the Russian Arctic. The red line indicates the average sex ratio for births in Russia—1.06 males per female.

4. Sulfur dioxide released by coal-burning power plants contributes to _____ .
   a. ozone destruction   c. acid rain
   b. sea level rise   d. desertification

5. As a result of _____ , an old animal usually has more pollutants in its body than a young one.
   a. bioaccumulation   b. biological magnification

6. With biological magnification, a _____ will have the highest pollutant load.
   a. producer   c. secondary consumer
   b. primary consumer   d. top-level consumer

7. An increase in the size of the ozone hole would be expected to _____ .
   a. increase skin cancers   c. both a and b
   b. reduce respiratory disorders

8. True or false? All pollutants are synthetic chemicals.

9. Global climate change is causing _____ .
   a. a decrease in sea level   c. acid rain
   b. glacial melting   d. all of the above

10. The Montreal Protocol banned use of _____ , which contribute(s) to ozone depletion.
    a. DDT   c. fossil fuels
    b. CFCs   d. sulfur dioxides

11. A highly threatened region that is home to many unique species is a(n) _____ .
    a. ecoregion   c. hot spot
    b. biome   d. community

12. Biodiversity refers to _____ .
    a. genetic diversity   c. ecosystem diversity
    b. species diversity   d. all of the above

13. Restoring a marsh that has been damaged by human activities is an example of _____ .
    a. biological magnification   c. ecological restoration
    b. bioaccumulation   d. globalization

14. Individuals help sustain biodiversity by _____ .
    a. reducing consumption   c. recycling materials
    b. reusing materials   d. all of the above

15. Match the terms with the most suitable description.
    ___ hot spot
    ___ ozone
    ___ biodiversity
    ___ acid rain
    ___ endemic species
    ___ biological magnification
    ___ global climate change
    ___ deforestation
    ___ desertification

    a. good up high; bad nearby
    b. tree loss alters rainfall pattern and is difficult to reverse
    c. can increase dust storms
    d. evolved in one region and remains there
    e. coal-burning is major cause
    f. results in highest pollution level at top trophic level
    g. has lots of threatened species
    h. cause of rising sea level
    i. genetic, species, and ecosystem diversity

Critical Thinking

1. Many biologists think that global climate change resulting from greenhouse gas emissions is the single greatest threat to existing biodiversity. List some negative effects that climate change could have on native species in your area.

2. Two arctic marine mammals that live in the same waters differ in the level of pollutants in their bodies. Bowhead whales have a lower pollutant load than ringed seals. What are some factors that might explain this difference?

2. In one seaside community in New Jersey, the U.S. Fish and Wildlife Service suggested removing feral cats (domestic cats that live in the wild) in order to protect some endangered wild birds (plovers) that nested on the town's beaches. Many residents were angered by the proposal, arguing that the cats have just as much right to exist as the birds. Explain why you agree or disagree with this reasoning.

Glossary of Biological Terms

ABO blood typing Method of identifying certain glycoproteins (A or B) on red blood cells; the absence of either type is designated O. **590**

abscisic acid Plant hormone that inhibits growth and germination. Important participant in stomata function and plant responses to biotic and abiotic stress. **479**

abscission Process by which plant parts are shed. Occurs in response to stress, and, in some species, prior to seasonal dormancy during senescence. **469**

acclimatization A body adjusts to a new environment; e.g., after moving from sea level to a high-altitude habitat. **641**

acid Substance that releases hydrogen ions in water. **32**

acid rain Low-pH rain formed when sulfur dioxide and nitrogen oxides mix with water vapor in the atmosphere. **820**

action potential Brief reversal of the voltage difference across the plasma membrane of a neuron or muscle cell. **510**

activation energy Minimum amount of energy required to start a reaction. **75**

activator Regulatory protein that increases the rate of transcription when it binds to a promoter or enhancer. **153**

active site Of an enzyme, pocket in which substrates bind and a reaction occurs. **76**

active transport Energy-requiring mechanism in which a transport protein pumps a solute across a cell membrane against its concentration gradient. **86**

adaptation (adaptive trait) A heritable trait that enhances an individual's fitness in a particular environment. **243**

adaptive immunity In vertebrates, set of immune defenses that can be tailored to specific pathogens encountered by an organism during its lifetime. **604**

adaptive radiation Pattern of macroevolution in which a burst of genetic divergences from a lineage gives rise to many new species. **274**

adaptive trait *See* adaptation.

adenine *See* nucleotide.

adhering junction Cell junction composed of adhesion proteins; anchors cells to each other and extracellular matrix. **68**

adhesion protein Membrane protein that helps cells stick together in animal tissues. **83**

adipose tissue Connective tissue that specializes in fat storage. **495**

adrenal cortex Outer portion of adrenal gland; secretes aldosterone and cortisol. **558**

adrenal gland Endocrine gland located atop the kidney; secretes aldosterone, cortisol, epinephrine, and norepinephrine. **558**

adrenal medulla Inner portion of adrenal gland; secretes epinephrine and norepinephrine. **558**

aerobic Involving or occurring in the presence of oxygen. **108**

aerobic respiration Oxygen-requiring metabolic pathway that breaks down carbohydrates to produce ATP. Includes glycolysis, acetyl–CoA formation, the Krebs cycle, and electron transfer phosphorylation. **109**

age structure Of a population, the distribution of individuals among different age categories. **736**

agglutination Clumping of foreign cells after antibodies bind them. **590**

AIDS Acquired immune deficiency syndrome. A collection of diseases that develops after infection with HIV weakens the immune system. **623**

alcoholic fermentation Anaerobic carbohydrate breakdown pathway that produces ATP and ethanol. **116**

aldosterone Adrenal hormone that makes kidney tubules more permeable to sodium; encourages sodium reabsorption, thus increasing water reabsorption and concentrating the urine. **671**

algal bloom Population explosion of photosynthetic cells in an aquatic habitat. **331**

allantois Extraembryonic membrane that, in mammals, becomes part of the umbilical cord. **705**

alleles Forms of a gene with slightly different DNA sequences; may encode slightly different versions of the gene's product. Basis of variation in shared traits among sexual reproducers. **176**

allele frequency Abundance of a particular allele in a population. **257**

allergen A normally harmless substance that provokes an immune response in some people. **621**

allergy Sensitivity to an allergen. **621**

allopatric speciation Speciation pattern in which a physical barrier ends gene flow between populations. **271**

allosteric Describes a region of an enzyme that can bind a regulatory molecule and is not the active site. **79**

alpine tundra Biome dominated by low-growing, wind-tolerant plants adapted to high-altitude conditions. **801**

alternation of generations A life cycle that alternates between two multicellular body forms, a haploid gametophyte and a diploid sporophyte. **325**

alternative splicing RNA processing event in which some exons are removed or joined in alternate combinations. **141**

altruistic behavior Behavior that benefits another at the expense of the individual who carries it out. **728**

alveolate Member of a protist lineage having small sacs beneath the plasma membrane; dinoflagellate, ciliate, or apicomplexan. **329**

alveolus Tiny sac; in the mammalian lung, site of gas exchange. **635**

amino acid Small organic compound consists of a carboxyl group, an amine group, and a characteristic side group (R), all typically bonded to the same carbon atom. Twenty kinds are common subunits of proteins. **44**

ammonia Nitrogen-containing compound produced by amino acid and nucleic acid breakdown. **664**

ammonification Breakdown of nitrogen-containing organic material resulting in the release of ammonia and ammonium ions. **782**

amnion Extraembryonic membrane that encloses an amniote embryo and the amniotic fluid. **705**

amniote Vertebrate with a unique type of waterproof egg that allows embryos to develop on land. **395**

amniote egg Egg with internal membranes that allow an amniote embryo to develop away from water. **402**

amoeba Single-celled, unwalled protist that extends pseudopods to move and to capture prey. **336**

amoebozoans Lineage of heterotrophic, unwalled protists that live in soils and water; include amoebas and slime molds. **336**

amphibian Tetrapod with scaleless skin and a three-chambered heart; typically develops in water, then lives on land as an air-breathing carnivore. **400**

anaerobic Occurring in the absence of oxygen; oxygen-free. **107**

analogous structures Similar body parts or structures that evolved independently in different lineages. **281**

anaphase Stage of mitosis during which sister chromatids separate and move to opposite spindle poles. **166**

aneuploidy A chromosome abnormality in which an individual's cells carry too many or too few copies of a particular chromosome. **212**

angiosperms Most diverse seed plant lineage. Only group that makes flowers and fruits. **352**

animal Multicelled consumer that develops through a series of stages and moves about during part or all of its life cycle. **8, 370**

animal hormone Intercellular signaling molecule secreted by an endocrine gland or cell; travels in the blood to target. **546**

annelid Segmented worm with a coelom, complete digestive system, and closed circulatory system. **378**

antenna Of some arthropods, sensory structure on the head that detects touch and odors. **384**

anther Part of the stamen that produces pollen. **454**

anthropoids Primate lineage that includes monkeys, apes, and humans. **413**

antibody Y-shaped antigen receptor protein made only by B cells during an antibody-mediated immune response. Each binds specifically to the antigen that triggered its production. **613**

antibody-mediated immune response Adaptive immune response in which antibodies are produced that target a specific antigen. **614**

anticodon Set of three nucleotides in a tRNA; base-pairs with mRNA codon. **143**

antidiuretic hormone (ADH) Hormone released by the posterior pituitary; makes kidney tubules more permeable to water; encourages water reabsorption, thus concentrating the urine. **671**

antigen A molecule or particle that the immune system recognizes as nonself; triggers an immune response. **604**

antioxidant Substance that prevents oxidation of other molecules. **80**

anus Opening through which waste exits a complete digestive system. **649**

aorta Largest artery of the systemic system; receives blood from the heart's left ventricle. **587**

apical dominance Growth-inhibiting effect on lateral (axillary) buds, maintained by auxin gradients in growing shoot tips. **475**

apical meristem Mass of meristem just under the surface of shoot and root tips; lengthens these plant parts during primary growth. **427**

apicomplexan Single-celled alveolate protist that lives as a parasite in animal cells; some cause malaria or toxoplasmosis. **331**

apoptosis Mechanism of programmed cell death. **701**

appendicular skeleton Of vertebrates, limb or fin bones and bones of the pelvic and pectoral girdles. **568**

appendix Worm-shaped projection from the first part of the large intestine. **654**

aquifer Porous rock layer that holds some groundwater. **776**

arachnids Land-dwelling chelicerate arthropods with four pairs of walking legs; for example, spiders and ticks. **385**

archaea *See* archaeon.

archaeon Member of archaea, the most recently discovered and less well-known lineage of single-celled organisms without a nucleus. More closely related to eukaryotes than to bacteria. **8, 314**

arctic tundra High-latitude Northern Hemisphere biome, where low, cold-tolerant plants survive with only a brief growing season. **801**

area effect Larger islands have more species than small ones. **766**

arteriole Vessel that carries blood from an artery to a capillary. **586**

artery Large-diameter blood vessel that carries blood away from the heart. **586**

arthropod Invertebrate with jointed legs and a hard exoskeleton that is periodically molted; for example, an insect or crustacean. **384**

artificial selection Process whereby humans choose traits that they favor in a domestic species by selective breeding. **242**

asexual reproduction Reproductive mode by which offspring arise from a single parent only. **164, 678**

astrobiology Scientific study of life's origin and distribution in the universe. **291**

atmospheric cycle Biogeochemical cycle in which a gaseous form of an element plays a significant role. **778**

atom Fundamental particle that is a building block of all matter. Consists of varying numbers of protons, neutrons, and electrons. **4, 24**

atomic number Number of protons in the atomic nucleus; determines the element. **24**

ATP Adenosine triphosphate. Nucleotide that has an adenine base, a ribose sugar, and three phosphate groups. Subunit of RNA; also has an important role as an energy-carrying coenzyme. **47, 81**

ATP/ADP cycle Process by which cells regenerate ATP. ADP forms when ATP loses a phosphate group, then ATP forms again as ADP gains a phosphate group. **81**

atrioventricular (AV) node Clump of cells that serves as the electrical bridge between the atria and ventricles. **593**

atrium Heart chamber that receives blood from veins. **592**

australopith Informal name for two genera of chimpanzee-sized hominins that lived in Africa between 4 million and 1.2 million years ago. **416**

autoimmune response Immune response that inappropriately targets one's own tissues. **621**

autonomic nervous system Set of nerves that relay signals to and from internal organs and to glands. **516**

autosome Any chromosome other than a sex chromosome. **125**

autotroph Organism that makes its own food using carbon from inorganic molecules such as CO_2, and energy from the environment. **93**

auxin Plant hormone with a central role in coordinating responses to other hormones in all stages of growth and development. Indole-3-acetic acid (IAA) is the most common natural auxin. **474**

axial skeleton Of vertebrates, bones of the head, trunk, and tail. **568**

axon Of a neuron, a cytoplasmic extension that transmits electrical signals along its length and releases chemical signals at its endings. **508**

B cell B lymphocyte. Only lymphocyte that makes antibodies; also functions as an antigen-presenting cell in adaptive immune responses. **605**

B cell receptor Membrane-bound IgM or IgD antibody on a B cell; functions in antibody-mediated immune responses. **613**

B lymphocyte *See* B cell.

bacteria *See* bacterium.

bacteriophage Type of virus that infects bacteria. **127, 308**

bacterium Member of bacteria, the most diverse and well-known lineage of single-celled organisms with no nucleus. **8, 314**

balanced polymorphism Maintenance of two or more alleles for a trait at high frequency in a population. **264**

bark In woody plants, all tissue (secondary phloem and periderm) that lies outside the vascular cambium. **437**

basal body Organelle at the base of a cilium or flagellum; derived from a centriole. **65**

base Substance that accepts hydrogen ions in water. **32**

basement membrane Secreted material that attaches epithelium to an underlying tissue. **492**

base-pair substitution Mutation in which a single base pair changes. **146**

basophil Circulating granular leukocyte with a role in inflammation. **605**

behavioral plasticity Behavioral phenotype produced by a genotype varies with an animal's environment. **720**

bell curve Bell-shaped curve; typically results from graphing frequency versus distribution for a trait that varies continuously. **200**

benthic province The ocean's sediments and rocks. **808**

big bang theory Well-supported hypothesis that the universe originated by a nearly instant distribution of matter through space. **292**

bilateral symmetry Having paired structures so the right and left halves are mirror images. **370**

bile Mix of salts, pigments, and cholesterol produced in the liver, then stored and concentrated in the gallbladder; emulsifies fats when secreted into the small intestine. **652**

binary fission Cell reproduction process of bacteria and archaea. **317**

bioaccumulation An organism accumulates increasing amounts of a chemical pollutant in its tissues over the course of its lifetime. **820**

biodiversity Scope of variation among living organisms; includes genetic variation within species, variety of species, and variety of ecosystems. **8, 824**

biodiversity hot spot Threatened region with great biodiversity that is considered a high priority for conservation. **824**

biofilm Community of microorganisms living in a shared mass of secretions. **55**

biogeochemical cycle A nutrient moves among environmental reservoirs and into and out of food webs. **776**

biogeography Study of patterns in the geographic distribution of species and communities. **238**

biological magnification A chemical pollutant becomes increasingly concentrated as it moves up through food chains. **821**

biology The scientific study of life. **3**

bioluminescence Production of light by an organism. **330**

biomarker Molecule produced only by a specific type of cell; its presence indicates the presence of that cell. **297**

biomass pyramid Diagram that depicts the biomass (dry weight) in each of an ecosystem's trophic levels. **774**

biome Group of regions that may be widely separated but share a characteristic climate and dominant vegetation. **792**

biosphere All regions of Earth's waters, crust, and air where organisms live. **5, 787**

biotic potential Maximum possible rate of population growth under optimal conditions. **737**

bipedal Adapted to habitually walking upright. **415**

bird A winged vertebrate with feathers; belongs to the reptile clade. **404**

blastula Hollow ball of cells that forms as a result of cleavage. **696**

blood Circulatory fluid; in vertebrates it is a fluid connective tissue consisting of plasma, red blood cells, white blood cells, and platelets. **495, 584**

blood–brain barrier Protective barrier that prevents unwanted substances from entering cerebrospinal fluid. **520**

blood capillary *See* capillary, blood.

blood pressure Pressure exerted by blood against a vessel wall. **595**

bond *See* chemical bond, covalent bond, hydrogen bond, ionic bond.

bone tissue Connective tissue consisting of cells surrounded by mineral-hardened secretions that form a matrix. **495**

bony fish Common term for fish with a skeleton containing bone. **398**

boreal forest Extensive high-latitude forest of the Northern Hemisphere; conifers are the predominant vegetation. **800**

bottleneck Drastic reduction in population size as a result of severe selection pressure. **266**

Bowman's capsule Cup-shaped portion of the nephron that encloses the glomerulus and receives filtrate from it. **666**

brain stem Most evolutionarily ancient region of the vertebrate brain; includes the pons, medulla, and midbrain. **521**

bronchiole Airway that leads from a bronchus to the alveoli. **635**

bronchus Airway connecting the trachea to a lung. **635**

brood parasitism One egg-laying species benefits by having another raise its offspring. **760**

brown alga Multicelled marine protist with a brown accessory pigment (fucoxanthin) in its chloroplasts. **333**

brush border cell In the lining of the small intestine, an epithelial cell with microvilli at its surface. **651**

bryophyte Member of a plant lineage with a gametophyte-dominant life cycle; a moss, liverwort, or hornwort. **340**

budding In some yeasts, process of asexual reproduction by which a small cell forms on a parent, then is released. **362**

buffer system Set of chemicals that can keep the pH of a solution stable by alternately donating and accepting ions that contribute to pH. **32**

bursa Fluid-filled sac that functions as a cushion between components of some joints. **573**

C3 plant Type of plant that uses only the Calvin–Benson cycle to fix carbon. **102**

C4 plant Type of plant that minimizes photorespiration by fixing carbon twice, in two cell types. **103**

calcium pump Active transport protein that pumps calcium ions across a cell membrane. **87**

Calvin–Benson cycle Second stage of photosynthesis: light-independent reactions that form sugars by fixing carbon from CO_2. Runs in the stroma of chloroplasts on the chemical energy of ATP and the energy of electrons donated by NADPH. **101**

calyx A flower's outer, protective whorl of sepals. **454**

CAM plant Type of C4 plant that conserves water by fixing carbon twice, at different times of day. **103**

camera eye Eye with an adjustable opening and a single lens that focuses light on a retina. **534**

camouflage Body coloration, patterning, form, or behavior that helps predators or prey blend with the surroundings and possibly escape detection. **759**

cancer Disease that occurs when the uncontrolled growth of body cells that can invade other tissues physically and metabolically disrupts normal body functioning. **151**

capillaries Smallest blood vessels; exchanges with interstitial fluid take place across their walls. **586**

carbohydrate Molecule that consists primarily of carbon, hydrogen, and oxygen atoms in a 1:2:1 ratio. **40**

carbon cycle Movement of carbon, mainly between the oceans, atmosphere, and living organisms. **778**

carbon fixation Process by which carbon from an inorganic source such as carbon dioxide gets incorporated into an organic molecule. **101**

carbonic anhydrase Enzyme in red blood cells that speeds the reversible conversion of carbonic acid into bicarbonate and H^+. **639**

cardiac cycle Sequence of contraction and relaxation of heart chambers that occurs with each heartbeat. **592**

cardiac muscle tissue Muscle of the heart wall. **496**

carpel Of flowering plants, a reproductive structure that produces female gametophytes; consists of an ovary, a stigma, and often a style. **352, 454**

carrying capacity Maximum number of individuals of a species that an environment can sustain. **738**

cartilage Connective tissue consisting of cells surrounded by a rubbery matrix of their own secretions. **396, 494**

cartilaginous fish Jawed fish that has a skeleton of cartilage. **398**

Casparian strip Waxy, waterproof band that seals abutting cell walls of root endodermal cells. **445**

catalysis The acceleration of a reaction by a molecule (such as an enzyme) that is unchanged by participating in the reaction. **76**

catastrophism Now-abandoned hypothesis that catastrophic geologic forces unlike those of the present day shaped Earth's surface. **240**

cDNA DNA synthesized from an RNA template by the enzyme reverse transcriptase. **221**

cell Smallest unit of life; at minimum, consists of plasma membrane, cytoplasm, and DNA. **4**

cell cortex Mesh of cytoskeletal elements that reinforces a plasma membrane. **64**

cell count Number of cells of one type in 1 microliter of blood. **588**

cell cycle A series of events from the time a cell forms until its cytoplasm divides; comprises G1, S, G2, and mitosis. **164**

cell junction Structure that connects a cell to another cell or to extracellular matrix. **67**

cell-mediated immune response Adaptive immune response in which effector cytotoxic T cells and NK cells form and destroy infected or cancerous body cells. **614**

cell plate After nuclear division in a plant cell, a disk-shaped structure that forms a cross-wall between the two new nuclei. **168**

cell theory Theory that all organisms consist of one or more cells, which are the basic unit of life; all cells come from division of preexisting cells; and all cells pass hereditary material to offspring. **51**

cellular slime mold Soil-dwelling protist; feeds as solitary cells that congregate under adverse conditions and develop into a spore-bearing fruiting body. **336**

cellulose Polysaccharide that is a major structural material in plants. **40**

cell wall Semirigid but permeable structure that surrounds the plasma membrane of some cells. **54**

central nervous system Brain and spinal cord. **507**

central vacuole Large, fluid-filled vesicle in many plant cells. **61**

centriole Barrel-shaped organelle from which microtubules grow. **65**

centromere Constricted region in a eukaryotic chromosome where sister chromatids are attached. **124**

cephalization Concentration of sensory structures and nerve cells at the anterior (head) end of a bilateral body. **370, 506**

cerebellum Hindbrain region that coordinates voluntary movements. **520**

cerebral cortex Outer gray matter layer of the cerebrum. **522**

cerebrospinal fluid Fluid that surrounds the brain and spinal cord and fills cavities within them. **518**

cerebrum Forebrain region that controls higher functions. **521**

cervix Narrow part of uterus that connects to the vagina. **685**

character Quantifiable, heritable characteristic or trait. **278**

character displacement Outcome of competition between two species; similar traits that result in competition become dissimilar. **755**

charge Electrical property of matter. Particles with opposite charges attract, and those with like charges repel. **24**

charophyte algae Green algal group that includes the lineage most closely related to land plants. **335**

chelicerates Arthropod subgroup with specialized feeding structures (chelicerae) and no antennae. **385**

chemical bond An attractive force that arises between two atoms when their electrons interact; can be ionic or covalent. **28**

chemoautotroph Organism that makes its own food using carbon from inorganic sources such as carbon dioxide, and energy from chemical reactions. **104, 315**

chemoheterotroph Organism that obtains both energy and carbon by breaking down organic compounds. **315**

chemoreceptor Sensory receptor that responds to a chemical. **528**

chemotaxis Cellular movement toward or away from a chemical stimulus. **609**

chlamydias Bacteria that are intracellular parasites of vertebrates. **319**

chlorophyll a Most common photosynthetic pigment in plants, photosynthetic protists, and bacteria. **94**

chlorophyte algae Most diverse lineage of green algae. **334**

chloroplast Organelle of photosynthesis in the cells of plants and many protists. Its two outer membranes enclose the thylakoid membrane and stroma. **62, 97**

choanoflagellates Heterotrophic protists thought to be the sister group of animals; collared cells strain food from water. **337**

chordate Animal with an embryo that has a notochord, dorsal nerve cord, pharyngeal gill slits, and a tail that extends beyond the anus. For example, a lancelet or a vertebrate. **394**

chorion Outermost extraembryonic membrane of amniotes; major component of the placenta in placental mammals. **705**

choroid Blood vessel–rich layer of the middle eye. It is darkened by the brownish pigment melanin to prevent light scattering. **536**

chromatin Collective term for a cell's DNA and its associated proteins. **59**

chromosome A structure that consists of a molecule of double-stranded DNA and associated proteins; carries part or all of a cell's genetic information. **59, 124**

chromosome number The sum of all chromosomes in a cell of a given type. **124**

chyme Mix of food and gastric fluid. **650**

chytrid Fungus with flagellated spores. **359**

ciliary muscle A ring-shaped muscle of the eye that encircles the lens and attaches to it by short fibers. **537**

ciliate Single-celled, heterotrophic protist with many cilia. **329**

cilium Short, movable structure that projects from the plasma membrane of some eukaryotic cells. **64**

circadian rhythm A physiological change that repeats itself on an approximately 24-hour cycle. **482, 561**

circulatory system Organ system consisting of a heart or hearts and blood-filled vessels that distribute substances through a body. **584**

clade A group whose members share one or more defining derived traits; a monophyletic group. **278**

cladistics Method of making hypotheses about evolutionary relationships among clades. **279**

cladogram Diagram that shows evolutionary relationships among clades. **279**

classical conditioning An animal's involuntary response to a stimulus becomes associated with another stimulus that is presented at the same time. **719**

cleavage Mitotic division of an animal cell. **696**

cleavage furrow Indentation where a contractile ring is pinching an animal cell in two during cytoplasmic division. **168**

climate Average weather conditions in a region over a long time period. **788**

cloaca In some vertebrates, a body opening through which both wastes and gametes exit. **398**

clone Genetically identical copy of an organism. **123**

cloning vector A DNA molecule that can accept foreign DNA and get replicated inside a host cell. **221**

closed circulatory system Circulatory system in which blood flows through a continuous system of vessels. **584**

club fungi Fungi that have septate hyphae and produce spores by meiosis in club-shaped cells. **364**

cnidarian Radially symmetrical invertebrate that has tentacles with stinging cells (cnidocytes). **374**

cnidocyte Stinging cell unique to cnidarians. **374**

coal Fossil fuel formed over millions of years by compaction and heating of plant remains. **348**

cochlea In the inner ear, the coiled, fluid-filled structure that holds the sound-detecting organ of Corti. **540**

codominant Refers to two alleles that are both fully expressed in heterozygotes and neither is dominant over the other. **196**

codon In mRNA, a sequence of three nucleotides that codes for an amino acid or stop signal during translation. **142**

coelom Of many animals, a body cavity that surrounds the gut and is lined with tissue derived from mesoderm. **371**

coenzyme An organic cofactor. **80**

coevolution The joint evolution of two closely interacting species; each species is a selective agent for traits of the other. **274, 352**

cofactor A metal ion or an organic molecule that associates with an enzyme and is necessary for its function. **80**

cohesion Property of a substance that arises from the tendency of its molecules to resist separating from one another. **30**

cohesion–tension theory Explanation of how water moves through vascular plants: transpiration creates a tension that pulls a cohesive column of water through xylem, from roots to shoots. **447**

cohort Group of individuals born during the same time interval. **740**

coleoptile Rigid sheath that protects a growing embryonic shoot of monocots. **465**

collecting tubule Kidney tubule that receives filtrate from several nephrons and delivers it to the renal pelvis. **667**

collenchyma Simple plant tissue composed of living cells with unevenly thickened walls; provides flexible support. **428**

colon *See* large intestine.

commensalism Species interaction that benefits one species and neither helps nor harms the other. **752**

communication signal Chemical, acoustical, visual, or tactile cue that is produced by one member of a species and detected and responded to by other members of the same species. **722**

community All species in a particular region. **5, 751**

companion cell In phloem tissue, a parenchyma cell that loads sugars into an associated sieve-tube member and provides it with metabolic support. **429**

comparative morphology Study of body plans and structures among groups of organisms. **239**

competitive exclusion Process whereby two species compete for a limiting resource, and one drives the other to local extinction. **754**

complement A set of proteins that circulate in inactive form in blood, and when activated play a role in immune responses. **604**

complete digestive system Tubelike digestive system; food enters through one opening and wastes leave through another. **646**

compound Molecule that consist of atoms of more than one element. **28**

compound eye Eye with many units, each having its own lens. **534**

concentration The number of molecules or ions per unit volume. **32, 84**

concentration gradient Difference in concentration between adjoining regions of fluid. **84**

condensation Enzymatic reaction in which two molecules become bonded together; water also forms. **38**

conduction Of heat: the transfer of heat between two objects in contact with one another. **673**

cone cell Photoreceptor that provides sharp vision and allows detection of color. **538**

conifer Gymnosperm with nonmotile sperm and woody cones; for example, a pine. **350**

conjugation Mechanism of horizontal gene transfer in which one bacterial or archaeal cell passes a plasmid to another. **317**

conjunctiva Mucous membrane that lines the inner surface of the eyelids and folds back to cover the eye's sclera. **536**

connective tissue Animal tissue with an extensive extracellular matrix; structurally and functionally supports other tissues. **494**

conservation biology Field of applied biology that surveys and documents biodiversity, and seeks ways to maintain and use it. **824**

consumer Organism that gets energy and nutrients by feeding on tissues, wastes, or remains of other organisms; a heterotroph. **6, 770**

continuous variation Range of small differences in a shared trait. **200**

contractile vacuole In freshwater protists, an organelle that collects and expels excess water. **327**

control group In an experiment, a group of individuals who are not exposed to the independent variable being tested. **13**

convection Transfer of heat by moving molecules of air or water. **673**

coral reef Highly diverse marine ecosystem centered around reefs built by living corals that secrete calcium carbonate. **806**

cork In woody plants, tissue that waterproofs, insulates, and protects surfaces of stems and roots. Component of bark. **437**

cork cambium In woody plants, lateral meristem that gives rise to periderm. **437**

cornea Clear, protective covering at the front of the vertebrate eye. **536**

corolla A flower's whorl of petals; forms within sepals and encloses reproductive organs. **454**

corpus luteum Hormone-secreting structure that forms from the cells of a mature follicle that are left behind after ovulation. **686**

cortisol Adrenal cortex hormone that influences metabolism and immunity; secretions rise with stress. **558**

cotyledon Seed leaf; structure that stores nutrients in a flowering plant embryo. Monocots have one; eudicots, two. **426**

countercurrent exchange Exchange of substances between two fluids moving in opposite directions. **632**

covalent bond Chemical bond in which two atoms share a pair of electrons. **29**

critical thinking Judging the quality of information before allowing it to guide one's beliefs and actions. **12**

crossing over Process in which homologous chromosomes exchange corresponding segments during prophase I of meiosis. Gives rise to new combinations of parental alleles among offspring of sexual reproducers. **180**

crustaceans Mostly marine arthropod group with two pairs of antennae and a calcium-stiffened exoskeleton. **386**

culture Learned behaviors transmitted between individuals and down through generations. **419**

cuticle Secreted covering at a body surface. **66, 342**

cyanobacteria Oxygen-producing photosynthetic bacteria. **318**

cycad Tropical or subtropical gymnosperm with flagellated sperm, palmlike leaves, and fleshy seeds. **350**

cytokines Signaling molecules secreted and recognized by vertebrate leukocytes; function to coordinate the activities of leukocytes during immune responses. **605**

cytokinesis Cytoplasmic division. **168**

cytokinin Plant hormone that promotes cell division in shoot apical meristem and cell differentiation in root apical meristem. Often interacts antagonistically with auxin. **476**

cytoplasm Semifluid substance enclosed by a cell's plasma membrane. **50**

cytoplasmic localization Accumulation of different materials in different regions of the egg cytoplasm. **698**

cytosine *See* nucleotide.

cytoskeleton Dynamic framework of protein filaments that support, organize, and move eukaryotic cells and their internal structures. **64**

cytotoxic T cell Phagocytic lymphocyte that targets and kills infected or cancerous body cells. **605**

data Experimental results. **13**

decomposer Organism that feeds on biological remains and breaks organic material into inorganic subunits. **770**

deductive reasoning Logical process of using a general premise to draw a conclusion about a specific case. **12**

deletion Mutation in which one or more base pairs are lost. **146**

demographics Statistics that describe a population. **734**

demographic transition model Model describing changes in birth and death rates that occur as a region becomes industrialized. **746**

denature To unravel the shape of a biological molecule. **46**

dendrite Of a motor neuron or interneuron, a cytoplasmic extension that receives chemical signals sent by other neurons and converts them to electrical signals. **508**

dendritic cell Phagocytic lymphocyte that patrols solid tissues; important antigen-presenting cell in adaptive immunity. **605**

denitrification Conversion of nitrates or nitrites to gaseous forms of nitrogen. **782**

density-dependent factor Factor that limits population growth and has a greater effect in dense populations than less dense ones. **739**

density-independent factor Factor that limits population growth and arises regardless of population density. **739**

deoxyribonucleic acid *See* DNA.

dependent variable In an experiment, a variable that is presumably affected by the independent variable under investigation. **13**

derived trait A novel trait present in a clade but not in any of the clade's ancestors. **278**

dermal tissue Tissue that covers and protects the plant body; e.g., epidermis, periderm. **427**

dermis Deep layer of skin that consists of connective tissue with nerves and blood vessels running through it. **501**

desert Biome with little rain and low humidity; plants that have water-storing and water-conserving adaptations predominate. **794**

desertification Conversion of a grassland or woodland to desert. **818**

detrital food web Food web in which most energy is transferred directly from producers to detritivores. **772**

detritivore Consumer that feeds on small bits of organic material. **770**

deuterostomes Lineage of bilateral animals in which the second opening on the embryo surface develops into a mouth: includes echinoderms and vertebrates. **371**

development Multistep process by which the first cell of a new individual becomes a multicelled adult. **7**

diaphragm Smooth muscle between the thoracic and abdominal cavities; contracts during inhalation. **635**

diastole Relaxation phase of the cardiac cycle. **592**

diastolic pressure Blood pressure when ventricles are relaxed. **595**

diatom Single-celled photosynthetic protist with a brown accessory pigment (fucoxanthin) and a two-part silica shell. **332**

differentiation Process by which cells of a multicelled organism become specialized. **152, 700**

diffusion Spontaneous spreading of molecules or ions in a liquid or gas. **84**

dihybrid cross Cross between two individuals identically heterozygous for two genes; for example $AaBb \times AaBb$. **194**

dikaryotic Having two genetically different nuclei in a cell ($n+n$). **358**

dinoflagellate Single-celled, aquatic protist with cellulose plates and two flagella; may be heterotrophic or photosynthetic. **330**

dinosaurs Reptile lineage abundant in the Jurassic to Cretaceous; now extinct with the exception of birds. **402**

diploid Having two of each type of chromosome characteristic of the species ($2n$). **124**

directional selection Mode of natural selection in which phenotypes at one end of a range of variation are favored. **260**

disaccharide Polymer of two sugar subunits. **40**

disruptive selection Mode of natural selection in which forms of a trait at the extremes of a range of variation are favored; intermediate forms are selected against. **263**

distal tubule Portion of kidney tubule that delivers filtrate to a collecting tubule. **667**

distance effect Islands close to a mainland have more species than those farther away. **766**

DNA Deoxyribonucleic acid. Nucleic acid that consists of two chains of nucleotides (adenine, guanine, thymine, and cytosine) twisted into a double helix. Carries hereditary information. **7, 47**

DNA cloning Set of procedures that uses living cells to make many identical copies of a DNA fragment. **220**

DNA library Collection of cells that host different fragments of foreign DNA, often representing an organism's entire genome. **222**

DNA ligase Enzyme that seals gaps or breaks in double-stranded DNA. **131**

DNA polymerase DNA replication enzyme that uses a DNA template to assemble a complementary strand of DNA from free nucleotides. **130**

DNA profiling Identifying an individual by analyzing the unique parts of his or her DNA. **226**

DNA replication Process by which a cell duplicates its DNA before it divides. **130**

DNA sequence Order of nucleotide bases in a strand of DNA. **129**

DNA sequencing Method of determining the order of nucleotides in a fragment of DNA. **224**

dominance hierarchy Social system in which resources and mating opportunities are unequally distributed within a group. **727**

dominant Refers to an allele that masks the effect of a recessive allele paired with it. **191**

dormancy Period of temporarily suspended metabolism. **458**

dosage compensation Theory that the inactivation of one of the two X chromosomes in the cells of females equalizes X chromosome gene expression between the sexes. **156**

double fertilization In flowering plants only, one sperm fertilizes an egg to produce a zygote and another fertilizes a diploid cell to create a triploid cell that will develop into endosperm. **352, 458**

dry shrubland Biome dominated by a diverse array of fire-adapted shrubs; occurs in regions with cool, wet winters and a dry summer. **797**

dry woodland Biome dominated by short trees that do not completely shade the ground; occurs in regions with cool, wet winters and a dry summer. **797**

duplication Chromosomal structure change in which a section is repeated. **201**

eardrum The membrane that vibrates in response to pressure waves (sounds), thus transmitting vibrations to the bones of the middle ear. **540**

ecdysone Insect hormone with roles in metamorphosis, molting. **562**

echinoderms Invertebrates with hardened plates and spines embedded in the skin or body, and a water–vascular system. **390**

ecological footprint Area of Earth's surface required to sustainably support a particular level of development and consumption. **747**

ecological niche The resources and environmental conditions that a species requires. **754**

ecological restoration Actively altering an area in an effort to restore or create a functional ecosystem. **825**

ecosystem A community interacting with its environment through a one-way flow of energy and cycling of materials. **5, 770**

ectoderm Outermost tissue layer of an animal embryo. **696**

ectotherm Animal that controls its internal temperature by altering its behavior; for example, a fish or a lizard. **396, 673**

effector cell Antigen-sensitized B cell or T cell that forms in an immune response and acts immediately. **615**

egg Mature female gamete, or ovum. **182**

El Niño Periodic warming of equatorial Pacific waters and the associated shifts in global weather patterns. **787**

electron Negatively charged subatomic particle that occupies orbitals around an atomic nucleus. **24**

electron transfer chain Array of enzymes and other molecules that accept and give up electrons in sequence, thus releasing the energy of the electrons in usable increments. **79**

electron transfer phosphorylation Process in which electron flow through electron transfer chains sets up a hydrogen ion gradient that drives ATP formation. Also called chemiosmosis. **98**

electronegativity Measure of the ability of an atom to pull electrons away from other atoms. **28**

electrophoresis Technique that separates DNA fragments by size. **224**

element A pure substance that consists only of atoms with the same number of protons. **24**

embryonic induction Embryonic cells produce signals that alter the behavior of neighboring cells. **700**

embryophytes Land plants; clade of multicelled, photosynthetic species that protect and nourish the embryo on the parental body. **340**

emergent property A characteristic of a system that does not appear in any of the system's component parts. **4**

emerging disease Disease that is relatively new to a species, or has recently expanded its range. **307**

emigration Movement of individuals out of a population. **736**

emulsification Suspension of fat droplets in a fluid. **653**

endangered species A species that faces extinction in all or a part of its range. **816**

endemic species A species that remains restricted to the area where it evolved. **816**

endergonic Describes a reaction that requires a net input of free energy to proceed. **74**

endocrine disruptor Chemical that interferes with hormone action. **545**

endocrine gland Ductless gland that secretes hormones into a body fluid. **493**

endocrine system Hormone-producing glands and secretory cells of a vertebrate body. **546**

endocytosis Process by which a cell takes in a small amount of extracellular fluid by the ballooning inward of its plasma membrane. **88**

endoderm Innermost tissue layer of an animal embryo. **697**

endodermis In the vascular cylinder of plant roots, a layer of cells just outside the pericycle whose abutting walls are sealed by a Casparian strip. Helps control the movement of water and ions into root xylem. **435**

endomembrane system Series of interacting organelles (endoplasmic reticulum, Golgi bodies, vesicles) between nucleus and plasma membrane. **60**

endometrium Lining of uterus. **685**

endoplasmic reticulum (ER) Organelle that comprises a continuous system of sacs and tubes extending from the nuclear envelope. Smooth ER makes lipids and breaks down carbohydrates and fatty acids; rough ER modifies polypeptides made by ribosomes on its surface. **60**

endoskeleton Internal skeleton made up of hardened components such as bones. **395, 567**

endosperm Triploid (3*n*) nutritive tissue in the seeds of flowering plants. **352, 458**

endospore Resistant resting stage of some soil bacteria. **319**

endosymbiont hypothesis Hypothesis that mitochondria and chloroplasts evolved from bacteria. **298**

endotherm Animal that controls its body temperature by varying its production of metabolic heat; for example a bird or mammal. **404, 673**

energy The capacity to do work. 6, 72

energy pyramid Diagram that depicts the energy that enters each of an ecosystem's trophic levels. Lowest tier of the pyramid, representing primary producers, is always the largest. 774

enhancer Binding site in DNA for proteins that enhance the rate of transcription. 152

entropy Measure of how much the energy of a system has become dispersed. 72

enzyme Protein or RNA that speeds up a chemical reaction without being changed by it. 38

eosinophil Granular leukocyte that targets multicelled parasites. 605

epidemic Disease outbreak that occurs in a limited region. 313

epidermis Outermost tissue layer; in animals, the epithelial layer of skin. 429, 500

epididymides A pair of ducts in which sperm formed in testes mature; each empties into a vas deferens. 680

epigenetic Refers to heritable changes in gene expression that are not the result of changes in DNA sequence. 160

epiglottis Tissue flap that covers airway during swallowing to prevent food from entering airways. 635

epiphyte Plant that grows on another plant but does not harm it. 347

epistasis Effect in which a trait is influenced by multiple genes. Also called polygenic inheritance. 197

epithelial tissue Sheetlike animal tissue that covers outer body surfaces and lines internal tubes and cavities. 492

equilibrium model of island biogeography Model that predicts the number of species on an island based on the island's area and distance from the mainland. 766

erythropoietin Hormone secreted by kidneys; induces stem cells in bone marrow to give rise to red blood cells. 640

esophagus Muscular tube between the throat and stomach. 649

essential amino acid Amino acid that the body cannot make and must obtain from food. 657

essential fatty acid Fatty acid that the body cannot make and must obtain from the diet. 657

estivation An animal becomes dormant during a hot, dry season. 675

estrogen Hormone secreted by ovaries; causes development of female sexual

traits and maintains the reproductive tract. 684

estuary A highly productive ecosystem where nutrient-rich water from a river mixes with seawater. 804

ethylene Gaseous plant hormone involved in regulating growth and cell expansion. Participates in germination, abscission, ripening, and stress responses. 479

eudicots Most diverse lineage of angiosperms; members have two seed leaves, branching leaf veins. 354

eugenics Idea of deliberately improving the genetic qualities of the human race. 232

euglenoid Flagellated protozoan with multiple mitochondria; may be heterotrophic or have chloroplasts descended from algae. 327

eukaryote Organism that consists of one or more cells that characteristically have a nucleus. 8

eukaryotic flagella *See* flagellum.

eusocial animal Animal that lives in a multigenerational family group with a reproductive division of labor. 728

eutrophication Nutrient enrichment of an aquatic ecosystem. 769

evaporation Transition of a liquid to a gas. 30, 673

evolution Change in a line of descent. 240

evolutionary tree Diagram showing evolutionary relationships. 279

exaptation Evolutionary adaptation of an existing structure for a completely different purpose. 274

exergonic Describes a reaction that ends with a net release of free energy. 74

exocrine gland Gland that secretes milk, sweat, saliva, or some other substance through a duct. 493

exocytosis Process by which a cell expels a vesicle's contents to extracellular fluid. 88

exon Nucleotide sequence that remains in an RNA after post-transcriptional splicing. 141

exoskeleton Of some invertebrates, hard external parts that muscles attach to and move. 384, 566

exotic species A species that evolved in one community and later became established in a different one. 765

experiment A test designed to support or falsify a prediction. 13

experimental group In an experiment, a group of individuals who are exposed to an independent variable. 13

exponential growth A population grows by a fixed percentage in successive time intervals; so the size of each increase is determined by the current population size. 736

external fertilization Sperm and eggs are released into the external environment and meet there. 679

extinct Refers to a species that has been permanently lost. 274

extracellular fluid Of a multicelled organism, body fluid that is not inside cells; serves as the body's internal environment. 490

extracellular matrix (ECM) Complex mixture of cell secretions that supports cells and tissues; also has roles in cell signaling. 66

extreme halophile Organism adapted to life in a highly salty environment. 320

extreme thermophile Organism adapted to life in a very high-temperature environment. 320

eye Sensory organ that incorporates a dense array of photoreceptors. 534

fall overturn During the fall, waters of a temperate zone mix. Upper, oxygenated water cools, gets dense, and sinks; nutrient-rich water from the bottom moves up. 803

fat Lipid that consists of a glycerol molecule with one, two, or three fatty acid tails. 42

fatty acid Organic compound that consists of a chain of carbon atoms with an acidic carboxyl group at one end. Carbon chain of saturated types has single bonds only; that of unsaturated types has one or more double bonds. 42

feces Unabsorbed food material and cellular waste that is expelled from the digestive tract. 654

feedback inhibition Mechanism in which a change that results from some activity decreases or stops the activity. 78

fermentation Metabolic pathway that breaks down carbohydrates to produce ATP; does not require oxygen. Starts with glycolysis. 108

fertilization Fusion of two gametes to form a zygote. 177

fever An internally induced rise in core body temperature above the normal set point as a response to infection or injury. 611

fibrous root system Root system typical of monocots; composed of an extensive mass of similar-sized roots. 434

fight–flight response Response to danger or excitement. Activity of parasympathetic neurons declines, sympathetic signals increase, and adrenal glands secrete epinephrine. **517**

first law of thermodynamics Energy cannot be created or destroyed. **72**

fish Gilled aquatic vertebrate that is not a tetrapod. **396**

fitness Degree of adaptation to an environment, as measured by an individual's relative genetic contribution to future generations. **242**

fixed Refers to an allele for which all members of a population are homozygous. **266**

fixed action pattern Series of instinctive movements elicited by a simple stimulus and carried to completion once begun. **718**

flagellated protozoan Protist belonging to an entirely or mostly heterotrophic lineage with no cell wall and one or more flagella. **326**

flagellum Long, slender cellular structure used for motility. **55**

flatworm Acoelomate, unsegmented worm; for example, a planarian or tapeworm. **376**

flower Specialized reproductive shoot of a flowering plant. **352, 454**

fluid mosaic Model of a cell membrane as a two-dimensional fluid of mixed composition (of proteins and lipids). **82**

follicle-stimulating hormone (FSH) Pituitary hormone that acts on gonads; stimulates secretion of growth factors by Sertoli cells in males and ovarian follicle maturation in females. **683**

food chain Description of who eats whom in one path of energy flow through an ecosystem. **770**

food web Set of cross-connecting food chains. **772**

foraminifera Heterotrophic single-celled protists with a porous calcium carbonate shell and long cytoplasmic extensions. **328**

fossil Physical evidence of an organism that lived in the ancient past. **239**

founder effect After a small group of individuals found a new population, allele frequencies in the new population differ from those in the original population. **266**

fovea Retinal region where cone cells are most concentrated. **538**

frameshift Any type of mutation that causes the reading frame of mRNA codons to shift. **147**

free radical Atom with an unpaired electron. Most are highly reactive and can damage the molecules of life. **27**

fruit Mature ovary of a flowering plant, often with accessory parts; encloses a seed or seeds. **352, 462**

functional group A group of atoms bonded to a carbon of an organic compound; imparts a specific chemical property to the molecule. **38**

fungus Eukaryotic single-celled or multicelled heterotroph with cell walls of chitin; obtains nutrients by extracellular digestion and absorption. **8, 358**

gallbladder Organ that stores and concentrates bile. **652**

gametangium Gamete-producing organ of a plant. **344**

gamete Mature, haploid reproductive cell; e.g., an egg or a sperm. **176**

gametophyte A haploid, multicelled, gamete-producing body that forms in the life cycle of land plants and some algae. **182, 340**

ganglion Cluster of nerve cell bodies. **376, 506**

gap junction Cell junction that forms a channel across the plasma membranes of adjoining animal cells. **67**

gastric fluid Fluid secreted by the stomach lining; contains digestive enzymes, acid, and mucus. **650**

gastrointestinal tract The gut. Starts at the stomach and extends through the intestines to the tube's terminal opening. **649**

gastrovascular cavity Saclike cavity that functions in digestion and gas exchange in cnidarians and flatworms. **374**

gastrula Three-layered animal developmental stage formed by gastrulation. **696**

gastrulation Animal developmental process that transforms a two-layered blastula into a three-layered gastrula. **696**

gene DNA sequence that encodes an RNA or protein product. **138**

gene expression Process by which the information in a gene becomes converted to an RNA or protein product. **139**

gene flow The movement of alleles into and out of a population. **267**

gene pool All the alleles of all the genes in a population; a pool of genetic resources. **257**

gene therapy Treating a genetic defect or disorder by transferring a normal or modified gene into the affected individual. **232**

genetic code Complete set of sixty-four mRNA codons. **142**

genetic drift Change in allele frequencies in a population due to chance alone. **266**

genetic engineering Process by which deliberate changes are introduced into an individual's genome. **228**

genetic equilibrium Theoretical state in which a population is not evolving. **258**

genetically modified organism (GMO) Organism whose genome has been modified by genetic engineering. **228**

genome An organism's complete set of genetic material. **222**

genomics The study of genomes. **226**

genotype The particular set of alleles carried by an individual. **191**

genus A group of species that share a unique set of traits; also the first part of a species name. **10**

geologic time scale Chronology of Earth's history; correlates geologic and evolutionary events of the ancient past. **250**

germ cell Immature reproductive cell that gives rise to haploid gametes when it divides. **177**

germ layer One of three primary layers in an early embryo (a gastrula). **696**

germination The resumption of metabolic activity after a period of dormancy. **458**

gibberellin Plant hormone that induces stem elongation and helps seeds break dormancy, among other effects. **477**

gill Filamentous or branching respiratory organ of some aquatic animals; may be internal or external. **380, 630**

ginkgo Deciduous gymnosperm with flagellated sperm, fan-shaped leaves, and fleshy seeds. **350**

gland cell Secretory epithelial cell. **493**

global climate change A long-term change in Earth's climate. **781**

glomeromycete Fungus with hyphae that grow inside the wall of a plant root cell. **361**

glomerular filtration First step in urine formation: protein-free plasma forced out of glomerular capillaries by blood pressure enters Bowman's capsule. **688**

glomerulus Ball of capillaries enclosed by Bowman's capsule. **667**

glottis Opening formed when vocal cords in the larynx relax. **635**

glucagon Pancreatic hormone that causes cells to break down glycogen and release glucose. **556**

glycogen Polysaccharide that serves as an energy reservoir in animal cells. **41**

glycolysis Set of reactions in which glucose or another sugar is broken down to two pyruvate for a net yield of two ATP.

First step of aerobic respiration, fermentation; occurs in cytoplasm. **108**

gnetophyte Shrubby or vinelike gymnosperm, with nonmotile sperm; for example, *Ephedra*. **350**

Golgi body Organelle that modifies polypeptides and lipids; also sorts and packages the finished products into vesicles. **61**

gonadotropin-releasing hormone (GnRH) A hypothalamic hormone that induces the pituitary to release hormones (LH and FSH) that act on the gonads. **683**

gonads Primary reproductive organs (ovaries or testes); produce gametes and sex hormones. **560, 680**

Gondwana Supercontinent that existed before Pangea, more than 500 million years ago. **249**

Gram-positive bacteria Lineage of thick-walled bacteria that are colored purple by Gram staining. **319**

grassland Biome in the interior of continents where grasses and nonwoody plants adapted to grazing and fire predominate. **796**

gravitropism Plant growth in a direction influenced by gravity. **480**

gray matter Brain and spinal cord tissue consisting of cell bodies, dendrites, and neuroglial cells. **518**

grazing food web Food web in which most energy is transferred from producers to grazers (herbivores). **772**

greenhouse gas Atmospheric gas that absorbs heat emitted by Earth's surface and remits it, thus keeping the planet warm. **780**

ground tissue Every plant tissue that is not dermal or vascular tissue; makes up the bulk of the plant body and includes most photosynthetic cells. **427**

groundwater Soil water and water in aquifers. **776**

growth In multicelled species, an increase in the number, size, and volume of cells. In single-celled species, an increase in the number of cells. **7**

growth factor Molecule that stimulates mitosis and differentiation. **170**

guanine *See* nucleotide.

guard cell One of a pair of cells that define a stoma across the epidermis of a leaf or stem. **448**

gut A sac or tube in which food is digested. **370**

gymnosperm Seed plant that does not make flowers or fruits; for example, a conifer. **350**

habitat Type of environment in which a species typically lives. **752**

habituation Learning not to respond to a repeated stimulus. **719**

half-life Characteristic time it takes for half of a quantity of a radioisotope to decay. **246**

haploid Having one of each type of chromosome characteristic of the species. **177**

hearing Perception of sound. **540**

heart Muscular organ that pumps blood through a body. **584**

heartwood Accumulated nonfunctional xylem at the core of older tree stems and roots. **437**

Heimlich maneuver Procedure designed to rescue a choking person; a rescuer presses on a person's abdomen to force air out of the lungs and dislodge an object in the trachea. **636**

heliotropism Plant parts change position in response to the sun's changing angle through the day. **481**

hemoglobin Iron-containing protein that reversibly binds oxygen. **588**

hemolymph Fluid pumped through an open circulatory system. **584**

hemostasis Process by which blood clots in response to injury. **590**

herbivory An animal eats plant parts. **759**

hermaphrodite Animal that produces both eggs and sperm, either simultaneously or at different times in its life. **373, 678**

heterotherm Animal that sometimes maintains its temperature by producing metabolic heat, and at other times allows its temperature to fluctuate with the environment. **673**

heterotroph Organism that obtains energy and carbon from organic compounds assembled by other organisms. **93**

heterozygous Having two different alleles of a gene. **191**

hibernation An animal becomes dormant during a cold season. **675**

histone Type of protein that associates with eukaryotic DNA and structurally organizes chromosomes. **124**

HIV (Human immunodeficiency virus) Retrovirus that causes AIDS. **307**

homeostasis Set of processes by which an organism keeps its internal conditions within tolerable ranges. **7**

homeotic gene Type of master gene; its expression controls the formation of specific body parts during development. **154**

hominins Modern humans and their closest extinct relatives. **416**

hominoids Tailless primate lineage that includes apes and humans. **413**

homologous chromosomes Chromosomes with the same length, shape, and set of genes. **165**

homologous structures Body parts or structures that are similar in different lineages because they evolved in a common ancestor. **280**

homozygous Having identical alleles of a gene. **191**

horizontal gene transfer Transfer of genetic material among existing individuals. **317**

hormone *See* animal hormone, plant hormone.

human Living or extinct member of the genus *Homo*. **418**

human chorionic gonadotropin (HCG) Hormone first secreted by the blastocyst, and later by the placenta; helps maintain the uterine lining during pregnancy. **705**

human immunodeficiency virus *See* HIV.

humus Decaying organic matter in soil. **442**

hybrid The offspring of a cross between two individuals that breed true for different forms of a trait; a heterozygous individual. **191**

hydrocarbon Compound that consists only of carbon and hydrogen atoms. **38**

hydrogen bond Attraction between a covalently bonded hydrogen atom and an electronegative atom taking part in a separate covalent bond. **30**

hydrogenosome Organelle that produces ATP and hydrogen gas by an anaerobic pathway; evolved from mitochondria. **326**

hydrolysis Enzymatic reaction with water that causes a molecule to break into smaller subunits. **38**

hydrophilic Describes a substance that dissolves easily in water. **31**

hydrophobic Describes a substance that resists dissolving in water. **31**

hydrostatic skeleton Of soft-bodied invertebrates, a fluid-filled chamber that muscles exert force against, redistributing the fluid. **374, 566**

hydrothermal vent Underwater opening from which mineral-rich water heated by geothermal energy streams out. **293, 809**

hypertonic Describes a fluid that has a high overall solute concentration relative to another fluid. **84**

hypha Component of a fungal mycelium; a filament made up of cells arranged end to end. **358**

hypocotyl Portion of an embryonic stem between cotyledon(s) and embryonic root (radicle). **464**

hypothalamus Forebrain region that controls processes related to homeostasis; control center for endocrine functions. **521, 550**

hypothesis Testable explanation of a natural phenomenon. **12**

hypotonic Describes a fluid that has a low overall solute concentration relative to another fluid. **84**

immigration Movement of individuals into a population. **736**

immunity The body's ability to resist and fight infections. **604**

immunization A procedure designed to promote immunity to a specific disease. E.g., vaccination. **624**

imprinting Learning that can occur only during a specific interval in an animal's life. **718**

inbreeding Mating among close relatives. **267**

incomplete digestive system Saclike digestive system; food enters and leaves through the same opening. **646**

incomplete dominance Effect in which one allele is not fully dominant over another, so the heterozygous phenotype is between the two homozygous phenotypes. **196**

independent variable Variable that is controlled by an experimenter in order to explore its relationship to a dependent variable. **13**

indicator species A species that is especially sensitive to disturbance and can be monitored to assess the health of a habitat. **764**

induced-fit model The concept that substrate binding to an active site of an enzyme improves the fit between the two molecules. **76**

inductive reasoning Arriving at a conclusion based on one's observations. **12**

inferior vena cava Vein that delivers blood from the lower body to the heart. **592**

inflammation A local response to tissue damage or infection; characterized by redness, warmth, swelling, and pain. **610**

inheritance Transmission of DNA to offspring. **7**

inhibiting hormone Hormone that is released by one gland and discourages secretion by another gland. **551**

innate immunity Set of inborn, general defenses against infection that do not normally change over an individual's

lifetime. Comprises surface barriers, complement, inflammation, fever, and phagocytic leukocytes. **604**

inner ear Fluid-filled vestibular apparatus and cochlea. **540**

insect Six-legged arthropod with two antennae and two compound eyes. Member of the most diverse class of animals. **388**

insertion Mutation in which one or more base pairs become inserted into DNA. **146**

instinctive behavior An innate response to a simple stimulus. **718**

insulin Pancreatic hormone that causes cells to take up glucose and store it as glycogen. **556**

integumentary exchange In some animals, gas exchange across thin, moistened skin or some other external body surface. **630**

intercostal muscles The skeletal muscles between the ribs; help change the volume of the thoracic cavity during breathing. **635**

intermediate disturbance hypothesis Species richness is greatest in communities where disturbances are moderate in their intensity or frequency. **763**

intermediate filament Stable cytoskeletal element that structurally supports cells and tissues. **64**

internal fertilization A female retains eggs in her body and sperm fertilize them there. **679**

interneuron Neuron that receives signals from and sends signals to other neurons. **506**

interphase In a eukaryotic cell cycle, the interval between mitotic divisions when a cell enlarges, roughly doubles the number of its cytoplasmic components, and replicates its DNA. **164**

intersex condition Genetic abnormality in which an individual has atypical reproductive anatomy. **677**

interspecific competition Competition between two species. **754**

interstitial fluid Of a multicelled organism, body fluid in between cells. **490**

intervertebral disk Cartilage disk between two vertebrae. **568**

intron Nucleotide sequence that is excised from an RNA during post-transcriptional splicing. Intervenes between exons. **141**

inversion Structural rearrangement of a chromosome in which part of it becomes oriented in the reverse direction. **210**

invertebrate Animal that does not have a backbone. **369**

in vitro **fertilization** Assisted reproductive technology in which eggs and sperm are united outside the body. **695**

ion Atom that carries a charge because it has an unequal number of protons and electrons. **27**

ionic bond Type of chemical bond that consists of a strong mutual attraction between oppositely charged ions. **28**

iris Circular muscle that adjusts the shape of the pupil to regulate how much light enters the eye. **536**

iron–sulfur world hypothesis Hypothesis that the metabolic reactions that led to the first cells took place on the porous surface of iron-sulfide-rich rocks at hydrothermal vents. **294**

isotonic Describes two fluids with identical solute concentrations. **84**

isotopes Forms of an element that differ in the number of neutrons. **25**

joint Region where bones meet. **572**

karyotype Image of an individual's complement of chromosomes arranged by size, length, shape, and centromere location. **125**

key innovation An evolutionary adaptation that gives its bearer the opportunity to exploit a particular environment more efficiently or in a new way. **274**

keystone species A species that has a disproportionately large effect on community structure. **764**

kidney Organ of the vertebrate urinary system that filters blood, adjusts its composition, and forms urine. **665**

kinesis Innate response in which an animal speeds up or slows its movement in reaction to a stimulus. **721**

kinetic energy The energy of motion. **72**

knockout An experiment in which a gene is deliberately inactivated in a living organism. **155**

Krebs cycle Cyclic pathway that, along with acetyl–CoA formation, breaks down pyruvate to carbon dioxide during aerobic respiration. In eukaryotes, occurs in mitochondrial matrix. **112**

***K*-selected species** Species adapted to a stable environment, where population size is often near carrying capacity. **741**

La Niña Periodic cooling of equatorial Pacific waters and the associated shifts in global weather patterns. **787**

labor Expulsion of a placental mammal from its mother's uterus by muscle contractions. **712**

lactate fermentation Anaerobic carbohydrate breakdown pathway that produces ATP and lactate. 116

lactation Milk production by a female mammal. 713

lake A body of standing fresh water. 803

lancelet Invertebrate chordate that has a fishlike shape and retains all the defining chordate traits into adulthood. 395

large intestine or **colon** Wide tubular organ that receives digestive waste from the small intestine and concentrates it as feces. 650

larva In some animal life cycles, an immature form that has a different body structure than an adult. 374

larynx Short airway containing the vocal cords (voice box). 634

lateral bud Axillary bud. A bud that forms in a leaf axil. 431

lateral meristem Cylindrical sheet of meristem in older shoots and roots that thickens these plant parts during secondary growth. E.g., vascular cambium; cork cambium. 436

law of independent assortment During meiosis, members of a pair of genes on homologous chromosomes tend to be distributed into gametes independently of other gene pairs. 194

law of nature Generalization that describes a consistent natural phenomenon for which there is incomplete scientific explanation. 18

law of segregation The members of each pair of genes on homologous chromosomes end up in different gametes during meiosis. 193

leaching Process by which water moving through soil removes nutrients from it. 443

learned behavior Behavior that is modified by experience. 718

lek Area where males animals perform courtship displays. 725

lens Disk-shaped structure that bends light rays so they fall on an eye's photoreceptors. 534

lethal mutation Mutation that drastically alters phenotype; causes death. 257

lichen Composite organism consisting of a fungus and a single-celled alga or a cyanobacterium. 366

life history pattern A set of traits related to growth, survival, and reproduction such as age-specific mortality, life span, age at first reproduction, and number of breeding events. 741

ligament Strap of dense connective tissue that holds bones together at a joint. 572

light-dependent reactions First stage of photosynthesis: A series of reactions that collectively convert light energy to chemical energy of ATP and NADPH. A noncyclic pathway includes electron transfer phosphorylation and produces oxygen; a cyclic pathway does not. 97

light-independent reactions Second stage of photosynthesis: A series of reactions that collectively use ATP and NADPH to assemble sugars from water and CO_2. 97

lignin Organic compound that stiffens cell walls of vascular plants. 66, 342

limbic system Group of brain structures that govern emotion. 523

limiting factor A necessary resource, the depletion of which halts population growth. 738

lineage Line of descent. 240

linkage group All genes on a chromosome. 195

lipid Fatty, oily, or waxy organic compound; e.g., a fat, steroid, or wax. 42

lipid bilayer Double layer of lipids arranged tail-to-tail; forms the structural foundation of all cell membranes. In eukaryotes, consists mainly of phospholipids. 43

liver Large organ that stores glucose as glycogen and releases it as needed. Also produces bile and detoxifies some harmful substances such as alcohol. 655

loam Soil with roughly equal amounts of sand, silt, and clay. 443

lobe-finned fish Bony fish with fleshy fins supported by bones. 399

local signaling molecule Chemical signal, such as a prostaglandin, that is secreted by one cell, diffuses through interstitial fluid, and affects nearby cells in an animal body. 546

locus Location of a gene on a chromosome. 191

logistic growth Density-dependent limiting factors cause population growth to slow as population size increases. 738

loop of Henle U-shaped portion of a kidney tubule; it extends deep into the renal medulla. 667

lung Internal gas-exchange organ in some air-breathing animals. 630

luteinizing hormone (LH) An anterior pituitary hormone that acts on the gonads; stimulates testosterone production in males and ovulation in females. 683

lymph Fluid in the lymph vascular system. 600

lymph node Small mass of lymphatic tissue through which lymph filters; contains many lymphocytes (B and T cells). 601

lymph vascular system System of vessels that takes up interstitial fluid and carries it (as lymph) to the blood. 600

lysogenic pathway Bacteriophage replication mechanism in which viral DNA becomes integrated into a host's chromosome and is passed to the host's descendants. 310

lysosome Enzyme-filled vesicle that functions in intracellular digestion of waste, debris, and ingested particles. 61

lysozyme Antibacterial enzyme that occurs in body secretions such as mucus. 607

lytic pathway Bacteriophage replication mechanism in which a virus replicates in its host and kills it quickly. 310

macroevolution Large-scale evolutionary patterns and trends; e.g., coevolution, exaptation, mass extinction. 274

macrophage Phagocytic leukocyte that patrols mainly interstitial fluid. Engulfs everything except undamaged body cells; presents antigen. 605

Malpighian tubule Water-conserving excretory organ of insects. 388

mammal Animal with hair or fur; females secrete milk from mammary glands. 406

mantle Of mollusks, extension of the body wall. 380

mark–recapture sampling Method of estimating population size of mobile animals by marking individuals, releasing them, then checking the proportion of marks among individuals recaptured later. 734

marsupial Mammal in which young are born at an early stage and complete development in a pouch on the mother's surface. 406

mass extinction Extinction of a large proportion of Earth's organisms in a relatively short period of geologic time. 237, 814

mass number Total number of protons and neutrons in the nucleus of an element's atoms. 24

mast cell Granular leukocyte anchored in solid tissues near blood vessels, nerves, and mucous membranes that border external surfaces; factor in inflammation and allergies. 605

master gene Gene encoding a product that affects the expression of many other genes. 154

maternal effect gene Gene expressed during egg production; its product influences animal development. 700

mechanoreceptor Sensory receptor that responds to pressure, position, or acceleration. 528

medulla oblongata Hindbrain region that controls breathing rhythm and reflexes such as coughing and vomiting. **520**

medusa A bell-like cnidarian body fringed with tentacles. **374**

megaspore Haploid spore formed in ovule of seed plants; develops into an egg-producing gametophyte. **348, 458**

meiosis Nuclear division process that halves the chromosome number. Basis of sexual reproduction. **175**

melatonin Hormone secreted by the pineal gland under conditions of darkness; affects sleep/wake cycles and protects against cancer. **561**

membrane potential Voltage difference across a plasma membrane. **509**

memory cell Long-lived, antigen-sensitized B or T cell that can act in a secondary immune response. **615**

meninges The three membranes that enclose the brain and spinal cord. **518**

menopause Permanent cessation of menstrual cycles. **685**

menstrual cycle Approximately 28-day cycle in which the uterus lining thickens and then, if pregnancy does not occur, is shed. **685**

menstruation Flow of shed uterine tissue out of the vagina. **685**

meristem Zone of undifferentiated plant cells that retain the ability to divide rapidly; gives rise to all plant tissues during primary and secondary growth. **426**

mesoderm Middle tissue layer of a three-layered animal embryo. **696**

mesophyll Photosynthetic parenchyma. **428**

messenger RNA (mRNA) A type of RNA that carries a protein-building message. The mRNA of eukaryotes is modified after transcription. **138**

metabolic pathway Series of enzyme-mediated reactions by which cells build, remodel, or break down an organic molecule. **78**

metabolism All the enzyme-mediated chemical reactions by which cells acquire and use energy as they build and break down organic molecules. **38**

metagenomics Study of microbial diversity that relies on analysis of DNA samples collected directly from the environment. **314**

metamorphosis Remodeling of body form during the transition from larva to adult. **384**

metaphase Stage of mitosis at which the cell's chromosomes are aligned midway between poles of the spindle. **166**

metastasis The process in which cancer cells spread from one part of the body to another. **171**

methanogen Organism that produces methane gas as a metabolic by-product. **321**

MHC markers Self-proteins that mark human body cells and are recognized by T cell receptors. All vertebrates have versions of these proteins. **612**

microevolution Change in allele frequency; driven by mutation, natural selection, genetic drift, and gene flow. **257**

microfilament Reinforcing cytoskeletal element; a fiber of actin subunits. **64**

microspore Walled spore formed in pollen sacs of seed plants; develops into a sperm-producing gametophyte (a pollen grain). **349, 459**

microsporidia Single-celled fungi that live as intracellular animal parasites and form flagellated spores. **360**

microtubule Cytoskeletal element involved in cellular movement; hollow filament of tubulin subunits. **64**

microvilli Thin projections from the plasma membrane that increase the surface area of some epithelial cells. **493, 651**

middle ear Eardrum and the tiny bones that transfer sound to the inner ear. **540**

migration An animal ceases an ongoing pattern of daily activity to move persistently toward a new habitat. **721**

mimicry A species evolves traits that make it similar in appearance to another species. **758**

mineral Naturally occurring inorganic substance; some such as iron are required in small amounts for normal metabolism. **657**

mitochondrion Organelle that produces ATP by aerobic respiration in eukaryotes. **62**

mitosis Nuclear division mechanism that maintains the chromosome number. Basis of body growth and tissue repair in multi-celled eukaryotes; also asexual reproduction in some plants, animals, fungi, and protists. Occurs in stages: prophase, metaphase, anaphase, telophase. **164**

mixture An intermingling of two or more types of molecules. **30**

model Analogous system used for testing hypotheses. **12**

molecular clock Using molecular change to estimate time of divergence. **282**

molecule A group of two or more atoms joined by chemical bonds. **4, 28**

mollusk Invertebrate with a reduced coelom and a mantle, including bivalves, gatropods, and cephalopods. **380**

monocots Highly diverse angiosperm lineage; includes plants such as grasses that have one seed leaf and parallel veins. **354**

monohybrid cross Cross between two individuals identically heterozygous for one gene; for example $Aa \times Aa$. **192**

monomers Molecules that are subunits of polymers. **38**

monophyletic group An ancestor in which a derived trait evolved, together with all of its descendants. **278**

monosaccharide Simple sugar that can be used as a monomer of polysaccharides. **40**

monotreme Egg-laying mammal. **406**

monsoon Wind that reverses direction seasonally. **791**

morphogen Chemical encoded by a master gene; diffuses out from its source and affects development. **700**

morphological convergence Evolutionary pattern in which similar body parts evolve independently in different lineages. Gives rise to analogous structures. **281**

morphological divergence Evolutionary pattern in which a body part of an ancestor changes in its descendants. **281**

motor cortex *See* primary motor cortex.

motor neuron Neuron that receives signals from another neuron and sends signals to a muscle or gland. **506**

motor protein Type of energy-using protein that interacts with cytoskeletal elements to move the cell's parts or the whole cell. **64**

motor unit One motor neuron and the muscle fibers it controls. **578**

multiple allele system Gene for which three or more alleles persist in a population. **196**

multiregional model Model that postulates *H. sapiens* populations in different regions evolved from *H. erectus* in those regions. **420**

muscle fiber Contractile cell that runs the length of a muscle. **576**

muscle tension Force exerted by a contracting muscle. **579**

mutation Permanent change in the nucleotide sequence of DNA. **132**

mutualism Mutually beneficial interaction between two species. **366, 753**

mycelium Mass of threadlike filaments (hyphae) that make up the body of a multicelled fungus. **358**

mycorrhiza Mutually beneficial partnership between a fungus and a plant root. **361, 444**

mycosis Disease caused by a fungus. **359**

myelin sheath Lipid-rich wrappings around axons of some neurons; speeds propagation of action potentials. **516**

myofibrils Of a muscle, threadlike protein structures consisting of contractile units (sarcomeres) arranged end to end. **576**

myoglobin Muscle protein that reversibly binds and stores oxygen. **580**

myriapod Land-dwelling arthropod with two antennae and an elongated body with many segments; a millipede or centipede. **386**

natural killer cell *See* NK cell.

natural selection A process in which environmental pressures result in the differential survival and reproduction of individuals of a population who vary in the details of shared, heritable traits. **243**

naturalist Person who observes life from a scientific perspective. **238**

nectar Sweet fluid exuded by some flowers; attracts pollinators. **457**

negative feedback A change causes a response that reverses the change; important mechanism of homeostasis. **502**

neoplasm An accumulation of cells that have lost control over their cell cycle. **170**

nephridium Of annelids and some other invertebrates, one of many water-regulating units that help control the content and volume of tissue fluid. **379**

nephron Kidney tubule and glomerular capillaries; filters blood and forms urine. **666**

nerve Neuron fibers bundled inside a sheath of connective tissue. **506**

nerve cord Bundle of nerve fibers that runs the length of the body in many invertebrates. **376, 506**

nerve net A mesh of nerve cells with no central control organ. **374, 506**

nervous tissue Animal tissue composed of neurons and supporting neuroglial cells; detects stimuli and controls responses to them. **497**

neuroglial cell Cell that supports and assists neurons. **497**

neuromodulator Any signaling molecule that reduces or magnifies the influence of a neurotransmitter on target cells. **514**

neuromuscular junction Synapse between a neuron and a muscle. **512**

neuron One of the cells that make up communication lines of a nervous system; transmits electrical signals along its plasma membrane and communicates with other cells through chemical messages. **497**

neurotransmitter Chemical signal released by axon terminals of a neuron. **512**

neutral mutation A mutation that has no effect on survival or reproduction. **257**

neutron Uncharged subatomic particle in the atomic nucleus. **24**

neutrophil Abundant circulating phagocytic, granular leukocyte. **605**

niche *See* ecological niche.

nitrification Conversion of ammonium to nitrates. **762**

nitrogen cycle Movement of nitrogen among the atmosphere, soil, and water, and into and out of food webs. **762**

nitrogen fixation Incorporation of nitrogen gas into ammonia. **318, 444, 762**

NK cell Natural killer cell. Lymphocyte that can kill cancer cells undetectable by cytotoxic T cells. **605**

node A region of stem where leaves form. **431**

nondisjunction Failure of sister chromatids or homologous chromosomes to separate during nuclear division. **212**

nonshivering heat production Brown adipose tissue releases energy as heat, rather than storing it in ATP. **675**

normal flora Microorganisms that typically live on human surfaces, including the interior tubes and cavities of the digestive and respiratory tracts. **606**

notochord Stiff rod of connective tissue that runs the length of the body in chordate larvae or embryos. **394**

nuclear envelope Nuclear membrane. A double membrane that constitutes the outer boundary of the nucleus. Pores in the membrane control which substances can cross. **58**

nucleic acid Single-stranded or double-stranded chain of nucleotides joined by sugar–phosphate bonds; e.g., DNA, RNA. **47**

nucleic acid hybridization Base-pairing between DNA or RNA from different sources. **222**

nucleoid Region of cytoplasm where the DNA is concentrated inside a bacterium or archaeon. **54, 316**

nucleolus In a cell nucleus, a dense, irregularly shaped region where ribosomal subunits are assembled. **59**

nucleoplasm Viscous fluid enclosed by the nuclear envelope. **59**

nucleosome A length of DNA wound twice around a spool of histone proteins. Basic unit of structural organization of eukaryotic chromosomes. **124**

nucleotide Organic compound that consists of a five-carbon sugar (e.g., ribose, deoxyribose) bonded to one or more phosphate groups and a nitrogen-containing base (e.g., adenine, guanine, cytosine, thymine, uracil) after which it is named. Various types function as coenzymes, signaling molecules, and monomers of nucleic acids. **47**

nucleus Of a eukaryotic cell, a double-membraned organelle that encloses the cell's DNA. **8, 51** Of an atom; core region occupied by protons and neutrons. **24**

nutrient Substance that an organism needs for growth and survival, but cannot make for itself. **7**

observational learning One animal acquires a new behavior by observing and imitating behavior of another. **719**

olfaction The sense of smell. **532**

oncogene Gene that helps transform a normal cell into a tumor cell. **170**

oocyte Immature egg. **684**

open circulatory system Circulatory system in which hemolymph leaves vessels and flows through spaces in body tissues. **584**

operant conditioning A type of learning in which an animal's voluntary behavior is modified by the consequences of that behavior. **719**

operator Part of an operon; a DNA binding site for a repressor. **158**

operon Group of genes together with a promoter–operator DNA sequence that controls their transcription. **158**

organ In multicelled organisms, a grouping of tissues engaged in a collective task. **5**

organ of Corti An acoustical organ in the cochlea that transduces mechanical energy of pressure waves into action potentials. **540**

organ system In multicelled organisms, set of organs engaged in a collective task. **5**

organelle Structure that carries out a specialized metabolic function inside a cell. E.g., nucleus, mitochondrion. **50**

organic Describes a compound that consists primarily of carbon and hydrogen atoms. **37**

organism An individual that consists of one or more cells. **4**

organs of equilibrium Sensory organs that respond to body position and motion. **542**

osmoreceptor Sensory receptor that detects shifts in the solute concentration of a body fluid. **529**

osmosis The diffusion of water across a selectively permeable membrane in response to a concentration gradient. **84**

osmotic pressure Amount of turgor that prevents osmosis into cytoplasm or other hypertonic fluid. **85**

osteoblast Bone-forming cell; it secretes a collagen-rich matrix that becomes mineralized. **570**

osteoclast Bone-digesting cell; it secretes enzymes that digest bone's matrix. **570**

osteocyte A mature bone cell; former osteoblast that has become surrounded by its own secretions. **570**

outer ear External ear and the air-filled auditory canal. **540**

ovarian follicle In an animal ovary, an immature egg and surrounding cells. **686**

ovary In flowering plants, the enlarged base of a carpel, inside which one or more ovules form and eggs are fertilized. **352, 454** In animals, female reproductive organ in which eggs form. **684**

oviduct Duct between an ovary and the uterus. **684**

ovulation Release of a secondary oocyte from an ovary. **686**

ovule Of a seed-bearing plant, a structure inside an ovary in which a haploid female gametophyte forms; after fertilization, matures into a seed. **348, 454**

oxidation–reduction reaction *See* redox reaction.

oxyhemoglobin Hemoglobin with oxygen bound to it. **638**

oxytocin Pituitary hormone with roles in labor and lactation. **712**

ozone layer High atmospheric layer rich in ozone; prevents most ultraviolet radiation in sunlight from reaching Earth's surface. **291, 822**

pain Perception of tissue injury. **530**

pain receptor Sensory receptor that responds to tissue damage. **528**

pancreas Organ that secretes digestive enzymes into the small intestine and hormones into the blood. **556**

pandemic Disease outbreak with cases worldwide. **313**

Pangea Supercontinent that formed about 250 million years ago; included all major land masses on Earth. **248**

parapatric speciation Speciation model in which adjacent populations speciate despite being in contact along a common

border. Hybrids that form in the contact zone have reduced fitness. **273**

parasitism Relationship in which one species withdraws nutrients from another, without immediately killing it. **760**

parasitoid An insect that lays eggs in another insect, and whose young devour their host from the inside. **761**

parasympathetic neurons Neurons of the autonomic system that encourage "housekeeping" tasks in the body. **516**

parathyroid glands Four small endocrine glands whose hormone product increases the level of calcium in blood. **555**

parenchyma Simple plant tissue made up of living cells; main component of ground tissue. **428**

passive transport Mechanism by which a concentration gradient drives the movement of a solute across a cell membrane through a transport protein. Requires no energy input. **86**

pathogen Disease-causing agent. **308**

pattern formation Process by which a complex body forms from local processes during embryonic development. **154**

PCR *See* polymerase chain reaction.

peat bog High-latitude community dominated by Sphagnum moss. **345**

pedigree Chart of family connections that shows the appearance of a trait through generations. **204**

pelagic province The ocean's waters. **808**

pellicle Layer of proteins that gives shape to many unwalled, single-celled protists. **326**

penis Male organ of intercourse; also functions in urination. **680**

peptide bond A bond between the amine group of one amino acid and the carboxyl group of another. Joins amino acids in proteins. **44**

per capita growth rate For some interval, the added number of individuals divided by the initial population size. **736**

pericycle Layer of cells just inside root endodermis that can give rise to lateral roots. **435**

periodic table Tabular arrangement of the elements by atomic number. **24**

peripheral nervous system Nerves that extend through the body and carry signals to and from the central nervous system. **506**

peristalsis Wavelike smooth muscle contractions that propel food through the digestive tract. **649**

peritubular capillaries Capillaries that surround and exchange substances with a kidney tubule. **667**

permafrost Continually frozen soil layer that lies beneath arctic tundra and prevents water from draining. **801**

peroxisome Enzyme-filled vesicle that breaks down amino acids, fatty acids, and toxic substances. **61**

petal Unit of a flower's corolla; often adapted to attract animal pollinators. **454**

pH A measure of the number of hydrogen ions in a fluid. **32**

phagocytosis "Cell eating"; an endocytic pathway by which a cell engulfs particles such as microbes or cellular debris. **88**

pharynx Tube connecting mouth and digestive tract; in vertebrates it is the throat. **376, 634**

phenotype An individual's observable traits; product of genotype, epigenetics, and environmental influences. **191**

pheromone Chemical that serves as a communication signal among members of an animal species. **533, 715**

phloem Complex vascular tissue of plants; distributes sugars through its sieve tubes. **343, 428**

phospholipid A lipid with a phosphate group in its hydrophilic head, and two nonpolar fatty acid tails; main constituent of eukaryotic cell membranes. **43**

phosphorus cycle Movement of phosphorus among Earth's rocks and waters, and into and out of food webs. **784**

phosphorylation Transfer of a phosphate group from one molecule to another. **80**

photoautotroph Photosynthetic autotroph; obtains carbon from carbon dioxide and energy from light. **104, 315**

photoheterotroph Organism that obtains carbon from organic compounds and energy from light. **315**

photolysis Any process by which light energy breaks down a molecule. **99**

photoperiodism Biological response to seasonal changes in the relative lengths of day and night. **483**

photoreceptor Sensory receptor that responds to light. **529**

photorespiration Reaction in which rubisco attaches oxygen instead of carbon dioxide to ribulose bisphosphate. **103**

photosynthesis Metabolic pathway by which most autotrophs use light energy to make sugars from carbon dioxide and water. *See* light-dependent reactions, light-independent reactions. **6, 93**

photosystem Cluster of pigments and proteins that work as a unit to convert light energy to chemical energy in photosynthesis. Two types (I and II) occur in thylakoid membranes. 98

phototropism Plant growth in a direction influenced by light. 480

phylogeny Evolutionary history of a species or group of species. 279

phytochemical Plant molecules that are not an essential part of the human diet, but may reduce risk of certain disorders; e.g., lutein. 659

phytochrome A light-sensitive pigment that helps set plant circadian rhythms based on length of night. 482

pigment An organic molecule that can absorb light of certain wavelengths. 95

pilus Protein filament that projects from the surface of some bacteria and archaea. 55, 317

pineal gland Melatonin-secreting endocrine gland in the brain. 561

pinocytosis Endocytosis of bulk materials. 89

pioneer species Species that can colonize a new habitat. 762

pituitary gland Pea-sized endocrine gland in the forebrain that interacts closely with the adjacent hypothalamus. 550

placenta Of placental mammals, organ that forms during pregnancy and allows diffusion of substances between the maternal and embryonic bloodstreams. 408, 679, 705

placental mammal Mammal in which a mother and her embryo exchange materials by means of an organ called the placenta. 406

plankton Community of tiny drifting or swimming organisms. 328

plant A multicelled producer, typically photosynthetic and adapted to life on land. 8, 341

plant hormone Extracellular signaling molecule of plants that exerts its effect at very low concentration. 472

plaque On teeth, a thick biofilm composed of bacteria, their extracellular products, and saliva proteins. 606

plasma Fluid portion of blood. 588

plasma membrane A cell's outermost membrane. 50

plasmid Of many bacteria and archaea, a small ring of DNA replicated independently of the chromosome. 54, 317

plasmodesmata Cell junctions that connect the cytoplasm of adjacent plant cells. 67

plasmodial slime mold Soil-dwelling protist that feeds as a multinucleated mass and develops into a spore-forming fruiting body under adverse conditions. 337

plastid Category of double-membraned organelle in plants and algal cells. Different types specialize in storage or photosynthesis; e.g., chloroplast, amyloplast. 62

plate tectonics Theory that Earth's outer layer of rock is cracked into plates, the slow movement of which rafts continents to new locations over geologic time. 249

platelet Cell fragment that helps blood clot. 589

pleiotropic Refers to a gene that influences multiple traits. 197

plot sampling Method of estimating population size of organisms that do not move much by making counts in small plots, and extrapolating from this to the number in the larger area. 734

plumule Embryonic shoot within a seed. 464

polarity Separation of charge into distinct positive and negative regions. 28

pollen grain Male gametophyte of a seed plant. 343

pollen sac Of seed plants, structure where microspores form and develop into pollen grains. 348

pollination Arrival of a pollen grain on the egg-bearing part of a seed plant. 349

pollination vector Any agent that moves pollen grains from one plant to another; e.g., wind, bees. 456

pollinator Organism that moves pollen from one plant to another, thus facilitating pollination. 352, 453

pollutant Substance that is released as result of human activities and harms organisms or disrupts natural processes. 820

polymer Molecule that consists of multiple monomers. 38

polymerase chain reaction (PCR) Method that uses heat-tolerant *Taq* polymerase to rapidly generate many copies of a specific section of DNA. Used in DNA profiling, among other applications. 222

polyp A pillarlike cnidarian body topped with tentacles. 375

polypeptide Chain of amino acids linked by peptide bonds. 44

polyploid Having three or more of each type of chromosome characteristic of the species. 212

polysaccharide Polymer of monosaccharides; e.g., cellulose, starch, or glycogen. 40

pons Hindbrain region between medulla oblongata and midbrain; helps control breathing. 521

population A group of organisms of the same species that live in a specific location and breed with one another more often than they breed with members of other populations. 5, 256, 733

population density Number of individuals per unit area. 734

population distribution Describes whether individuals are clumped, uniformly dispersed, or randomly dispersed in an area. 734

population size Total number of individuals in a population. 734

positive feedback A response intensifies the conditions that caused its occurrence. 510

potential energy Stored energy. 73

predation One species captures, kills, and eats another. 756

prediction Statement, based on a hypothesis, about a condition that should exist if the hypothesis is correct. 12

pressure flow theory Explanation of how organic compounds move through vascular plants: the flow of fluid through sieve tubes is driven by a difference in turgor between source and sink regions. 450

primary growth Plant lengthening that originates at apical meristems in root and shoot tips. 426

primary motor cortex Region of brain's frontal lobes that governs voluntary movements. 522

primary oocyte Immature egg that has not completed prophase I. 686

primary producer In an ecosystem, an organism that captures energy from an inorganic source and stores it as biomass; first trophic level. 770

primary production The rate at which an ecosystem's producers capture and store energy. 774

primary somatosensory cortex Region of the brain's parietal lobes that receives sensory information from skin and joints. 522

primary succession A new community colonizes an area where there is no soil. 762

primary wall The first cell wall of young plant cells. 66

primates Mammalian order that includes lemurs, tarsiers, monkeys, apes, and humans. 412

primer Short, single strand of DNA that base-pairs (hybridizes) with a complementary DNA sequence. 130, 222

prion Infectious protein. 46

probability Chance that a particular outcome of an event will occur; depends on the total number of possible outcomes. 17

probe Short fragment of DNA labeled with a tracer; designed to hybridize with a nucleotide sequence of interest. 222

probiotic Food or dietary supplement containing living microorganisms whose presence in the body benefits health. 645

producer Organism that makes its own food using energy and simple raw materials from the environment; an autotroph. 6

product A molecule that remains at the end of a reaction. 74

progesterone Hormone secreted by ovaries; prepares the uterus for pregnancy. 684

prokaryote A member of one of the two single-celled lineages (bacteria and archaea) that do not have a nucleus. 314

prolactin Hormone that induces mammary gland enlargement during pregnancy and milk production during lactation. 713

promoter A DNA sequence to which RNA polymerase binds. 140

prophase Stage of mitosis during which chromosomes condense and become attached to a newly forming spindle. 166

protein Organic compound that consists of one or more polypeptide chains. 44

proteobacteria Most diverse bacterial lineage; includes species that carry out photosynthesis, fix nitrogen. Some cause disease. 318

protist General term for member of one of the eukaryotic lineages that is not a fungus, animal, or plant. 8, 323

protocell Membranous sac that contains interacting organic molecules; hypothesized to have formed prior to the earliest life forms. 295

proton Positively charged subatomic particle that occurs in the nucleus of all atoms. 24

proto-oncogene Gene that, by mutation, can become an oncogene. 170

protostomes Lineage of bilateral animals in which the first opening on the embryo surface develops into a mouth. 371

proximal tubule Portion of kidney tubule that receives filtrate from Bowman's capsule. 667

pseudocoelom Incompletely lined body cavity of some invertebrates. 371

pseudopod "False foot." A temporary protrusion that helps some eukaryotic cells move and engulf prey. 65

puberty Period when human reproductive organs mature and begin to function. 560

pulmonary artery Vessel carrying blood from the heart to a lung. 592

pulmonary circuit Circuit through which blood flows from the heart to the lungs and back. 585

pulmonary vein Vessel carrying blood from a lung to the heart. 592

pulse Brief stretching of artery walls that occurs when ventricles contract. 594

Punnett square Diagram used to predict the genetic and phenotypic outcome of a cross or mating. 192

pupil Adjustable opening that allows light into a camera eye. 536

pyruvate Three-carbon end product of glycolysis. 108

radial symmetry Having parts arranged around a central axis, like spokes around a wheel. 370

radicle Embryonic root within a seed. 464

radioactive decay Process by which an atom emits energy and/or subatomic particles when its nucleus spontaneously disintegrates. 24

radioisotope Isotope with an unstable nucleus. 24

radiolarians Heterotrophic, single-celled, marine protists with a porous silica shell and long cytoplasmic extensions. 328

radiometric dating Method of estimating the age of a rock or fossil by measuring the content and proportions of a radioisotope and its daughter elements. 246

radula Of many mollusks, a tonguelike organ hardened with chitin that is used in feeding. 380

rain shadow Dry region downwind of a coastal mountain range. 791

ray-finned fish Bony fish with fin supports derived from skin. 398

reactant A molecule that enters a reaction. 74

reaction Process of molecular change. 38

receptor protein Plasma membrane protein that triggers a change in the cell's activities when it binds to a particular substance outside of the cell. 83

recessive Refers to an allele with an effect that is masked by a dominant allele on the homologous chromosome. 191

recognition protein Plasma membrane protein that identifies a cell as belonging to self (one's own tissue or body). 83

recombinant DNA A DNA molecule that contains genetic material from more than one organism. 220

rectum Region where feces are stored prior to excretion. 649

red alga Photosynthetic protist; typically multicelled, with chloroplasts containing red phycobilins. 334

red blood cell Hemoglobin-filled blood cell that transports oxygen. 588

red marrow Bone marrow that makes blood cells. 570

redox reaction Oxidation–reduction reaction, in which one molecule accepts electrons (it becomes reduced) from another molecule (which becomes oxidized). 79

reflex Automatic response that occurs without conscious thought or learning. 518

releasing hormone Hormone that is released by one gland and encourages hormone secretion by another gland. 551

replacement fertility rate Fertility rate at which each woman has, on average, one daughter who survives to reproductive age. 746

replacement model Model for origin of *H. sapiens*; humans evolved in Africa, then migrated to different regions and replaced the other hominins that lived there. 420

repressor Regulatory protein that blocks transcription. 152

reproduction Processes by which individuals produce offspring. 7

reproductive base Of a population, all individuals who are of reproductive age or younger. 735

reproductive cloning Technology that produces genetically identical individuals. 134

reproductive isolation The end of gene flow between populations; always part of speciation. 268

reptile Tetrapod with amniote eggs and scales or feathers. 402

resource partitioning Species evolve ways of utilizing different portions of a limited resource; allows species with similar needs to coexist. 755

respiration Physiological process by which an animal body supplies cells with oxygen and disposes of their waste carbon dioxide. 628

respiratory cycle One inhalation and one exhalation. 636

respiratory membrane Membrane consisting of alveolar epithelium, capillary endothelium, and their fused basement membranes; site of gas exchange in lungs. 638

respiratory protein A protein that reversibly binds oxygen when the oxygen concentration is high and releases it when oxygen concentration is low. Hemoglobin is an example. **628**

respiratory surface Moist surface across which gases are exchanged between animal cells and the external environment. **628**

resting membrane potential Voltage difference across the plasma membrane of an excitable cell that is not receiving stimulation. **509**

restriction enzyme Type of enzyme that cuts specific nucleotide sequences in DNA. **220**

retina Eye layer that contains photoreceptors. **534**

reverse transcriptase An enzyme that synthesizes a strand of cDNA using mRNA as a template. **221, 311**

Rh blood typing Method of determining whether Rh, a type of surface recognition protein, is present on an individual's red blood cells. If present, the cell is Rh$^+$; if absent, the cell is Rh$^-$. **591**

rhizoid Threadlike structure that anchors a bryophyte. **344**

rhizome Stem that grows horizontally along or under the ground. **346**

ribonucleic acid *See* RNA.

ribosomal RNA (rRNA) Type of RNA that, together with structural proteins, composes ribosomal subunits. As part of ribosomes, catalyzes formation of peptide bonds. **138**

ribosome Organelle of protein synthesis. Has two subunits that consist of rRNA and proteins **54**

ribozyme RNA that functions as an enzyme. **294**

RNA Ribonucleic acid. Nucleic acid, a polymer of nucleotides (adenine, guanine, cytosine, and uracil). Various types have roles in gene expression and controls over it. Most are single-stranded; some are enzymatic. **47**

RNA polymerase Enzyme that carries out transcription. **140**

RNA world hypothesis Hypothesis that RNA served as the genetic information of early life. **294**

rod cell Photoreceptor that is active in dim light; provides coarse perception of image and detects motion. **538**

root hairs Hairlike, absorptive extensions of a young cell of root epidermis. **434**

root nodules Swellings of some plant roots that contain symbiotic nitrogen-fixing bacteria. **444**

rotifer Tiny bilateral, pseudocoelomate, aquatic animal with a ciliated head. **382**

roundworm Unsegmented pseudocoelomate worm with a cuticle that is molted periodically as the animal grows. **383**

r-selected species Species adapted to an environment that changes rapidly and unpredictably, so population size is often far below carrying capacity. **741**

rubisco Ribulose bisphosphate carboxylase. Carbon-fixing enzyme of the Calvin–Benson cycle. Operates inefficiently under high oxygen conditions (*see* photorespiration). **101**

runoff Water that flows over soil into streams. **776**

sac fungi Most diverse group of fungi; sexual reproduction produces spores inside a saclike structure (an ascus). **362**

salivary gland Exocrine gland that secretes saliva into the mouth. **649**

salt Ionic compound that releases ions other than H$^+$ and OH$^-$ when it dissolves in water. **31**

sampling error Difference between results derived from testing an entire group of events or individuals, and results derived from testing a subset of the group. **16**

saprobe Organism that feeds on organic wastes and remains. **358**

sapwood Functional secondary xylem between the vascular cambium and heartwood in an older stem or root. **437**

sarcomere Contractile unit of muscle. **576**

sarcoplasmic reticulum Specialized endoplasmic reticulum in muscle cells; stores and releases calcium ions. **578**

saturated fatty acid Fatty acid that contains no carbon–carbon double bonds in its hydrocarbon tail. **42**

scales Flattened structures that grow from and sometimes cover the skin in reptiles and some fish. **397**

science Systematic study of the observable world. **12**

scientific method Making, testing, and evaluating hypotheses. **13**

scientific theory Hypothesis that has not been disproven after many years of rigorous testing. **18**

sclera Fibrous white layer that covers most of the eyeball. **536**

sclerenchyma Simple plant tissue that is dead at maturity; its lignin-reinforced cell walls structurally support plant parts. E.g., fibers, sclereids. **429**

seamount An undersea mountain. **808**

second law of thermodynamics Energy tends to disperse spontaneously (the entropy of a system tends to increase). **72**

second messenger Molecule that forms inside a cell when a hormone or other signaling molecule binds to a receptor at the cell surface; its formation sets in motion reactions that alter some activity of the cell. **548**

secondary growth Thickening of older stems and roots; originates at lateral meristems. **427**

secondary metabolite Molecule that is produced by an organism but does not have any known role in its metabolism. Some serve as defense against predation. **355**

secondary oocyte A haploid cell produced by the first meiotic division of a primary oocyte; released at ovulation. **686**

secondary sexual traits Traits, such as the distribution of body fat, that differ between the sexes but do not have a direct role in reproduction. **560**

secondary succession A new community develops in a disturbed site where the soil that supported a previous community remains. **762**

secondary wall Lignin-reinforced wall that forms inside the primary wall of a plant cell. **66**

sedimentary cycle Biochemical cycle in which the atmosphere plays little role and rocks are the major reservoir. **784**

seed Embryo sporophyte of a seed plant packaged with nutritive tissue inside a protective coat. **343, 461**

seed plant Plant that produces seeds and pollen; angiosperm or gymnosperm. **341**

selfish herd Group that forms when individuals hide behind others to minimize their individual risk of predation. **726**

semen Sperm mixed with secretions from seminal vesicles and the prostate gland. **680**

semiconservative replication Describes the process of DNA replication, which produces two copies of a DNA molecule; one strand of each copy is new, and the other is a strand of the original DNA. **131**

seminiferous tubules Coiled tubules inside a testis that contain male germ cells and produce sperm. **682**

senescence Aging of cells, individuals, or communities. **468**

sensation Detection of a stimulus. **529**

sensory adaptation Slowing or cessation of a sensory receptor response to an ongoing stimulus. **529**

sensory neuron Neuron that responds to a specific internal or external stimulus and signals another neuron. **506**

sensory perception The meaning a brain derives from a sensation. **529**

sensory receptor Cell or cell component that detects a specific type of stimulus. **502**

sepal Unit of a flower's calyx; typically leaflike and photosynthetic. **454**

sequence *See* DNA sequence.

sessile animal Animal that lives fixed in place on some surface. **373**

sex chromosome Member of a pair of chromosomes that differs between males and females. **125**

sex hormone A steroid hormone that controls sexual development and reproduction; testosterone in males, estrogens or progesterone in females. **560**

sexual dimorphism Distinct male and female phenotype. **264**

sexual reproduction Reproductive mode by which offspring arise from two parents and inherit genes from both. **164, 678**

sexual selection Mode of natural selection in which some individuals outreproduce others of a population because they are better at securing mates. **264**

shell model Model that helps us visualize how electrons populate atoms. **26**

shivering response In response to cold, rhythmic muscle contractions generate metabolic heat. **675**

short tandem repeats In chromosomal DNA, sequences of 4 or 5 bases repeated multiple times in a row. **227**

sieve plate Perforated end wall separating stacked sieve-tube members. **450**

sieve-tube member Type of elongated cell in phloem tissue. Alive at maturity, its perforated end walls (sieve plates) connect with other sieve-tube members to form tubes that conduct sugar-rich fluid from sources to sinks. **429**

single-nucleotide polymorphism (SNP) One-base DNA sequence variation carried by a measurable percentage of a population. **219**

sink Region of a plant where sugars are being used or stored. **450**

sinoatrial (SA) node Cardiac pacemaker; cluster of specialized cells whose spontaneous rhythmic signals trigger contractions. **593**

sister chromatid One of two attached DNA molecules of a duplicated eukaryotic chromosome. **124**

sister groups The two lineages that emerge from a node on a cladogram. **279**

skeletal muscle tissue Muscle that pulls on bones and moves body parts; under voluntary control. **496**

sliding-filament model Explanation of how interactions among actin and myosin filaments bring about muscle contraction. **576**

slime mold *See* cellular slime mold, plasmodial slime mold.

small intestine Narrowest portion of the digestive tract; site of most digestion and absorption. **649**

smooth muscle tissue Muscle that lines blood vessels and forms the wall of hollow organs. **496**

soil Mixture of various mineral particles and humus. **442**

soil erosion Loss of soil under the force of wind and water. **443**

soil water Water between soil particles. **776**

solute A dissolved substance. **31**

solution Homogeneous mixture. **31**

solvent Substance that can dissolve other substances. **31**

somatic Relating to the body. **176**

somatic cell nuclear transfer (SCNT) Reproductive cloning method in which genetic material is transferred from an adult somatic cell into an unfertilized, enucleated egg. **134**

somatic nervous system Set of nerves that control skeletal muscle and relay signals from joints and skin. **516**

somatic sensations Sensations such as touch and pain that arise when sensory neurons in skin, muscle, or joints are activated. **530**

somatosensory cortex *See* primary somatosensory cortex.

somites Paired blocks of embryonic mesoderm that give rise to a vertebrate's muscle and bone. **702**

sorus Cluster of spore-producing capsules on a fern leaf. **347**

source Region of a plant where sugars are being produced. **450**

speciation Evolutionary process in which new species arise. **268**

species Unique type of organism; designated by genus name and specific epithet. **10**

specific epithet Second part of a species name; differentiates the species from other members of the genus. **10**

sperm Mature male gamete. **182**

sphincter Ring of muscle that controls passage through a tubular organ or body opening. **649**

spinal cord Portion of central nervous system that connects peripheral nerves with the brain. **518**

spindle Dynamically assembled and disassembled network of microtubules that moves chromosomes during nuclear division. **166**

spirochetes Lineage of bacteria shaped like a stretched-out spring. **319**

spleen Large lymphoid organ that filters blood. **601**

sponge Aquatic invertebrate that has no body symmetry, tissues, or organs; feeds by filtering food from the water. **373**

sporangium Of plants and fungi, a structure in which haploid spores form by meiosis. **344**

sporophyte Diploid, spore-producing body in the life cycle of land plants and some algae. **182, 340**

spring overturn In temperate zone lakes, a downward movement of oxygenated surface water and an upward movement of nutrient-rich water in spring. **803**

stabilizing selection Mode of natural selection in which intermediate forms of a trait are favored over extremes. **262**

stamen Pollen-producing organ of a flowering plant; consists of an anther and, often, a filament. **352, 454**

starch Polysaccharide that serves as an energy reservoir in plant cells. **40**

stasis Evolutionary pattern in which a lineage persists with little or no change over evolutionary time. **274**

statistically significant Refers to a result that is statistically unlikely to have occurred by chance. **17**

statolith Organelle involved in sensing gravity. **480**

stele *See* vascular cylinder. **434**

stem cell Cell capable of replication or of differentiation into some or all cell types. **489**

steroid Type of lipid with four carbon rings and no fatty acid tails. **43**

stigma Upper part of a carpel; adapted to receive pollen grains. **454**

stimulus A form of energy that is detected by a sensory receptor. **528**

stomach Muscular organ that secretes gastric fluid, mixes food with it, and controls the flow of food to the small intestine. **649**

stomata Closable gaps formed by guard cells in plant epidermis; when open, they allow water vapor and gases to diffuse across plant surfaces. Singular, stoma. **102, 342**

stramenopiles Protist lineage that includes the photosynthetic diatoms and brown alga, as well as the heterotrophic water molds. Some members of the group have a hairy flagellum. **332**

strobilus (strobili) Of some nonflowering plants such as horsetails and cycads, a cluster of spore-producing structures. **346**

stroma Semifluid matrix between the thylakoid membrane and the two outer membranes of a chloroplast. The light-independent reactions take place in it. **97**

stromatolite Dome-shaped structure composed of layers of bacterial cells, their secretions, and sediments. **296**

substrate A molecule that is specifically acted upon by an enzyme. **76**

substrate-level phosphorylation A reaction that transfers a phosphate group from a substrate directly to ADP, thus forming ATP. **110**

superior vena cava Vein that delivers blood from the upper body to the heart. **592**

surface-to-volume ratio A relationship in which the volume of an object increases with the cube of the diameter, and the surface area increases with the square. Constrains cell size and shape. **51**

survivorship curve Graph showing the decline in numbers of a cohort over time. **740**

swim bladder Adjustable flotation sac of some bony fish. **398**

symbiosis One species lives in or on another in a commensal, mutualistic, or parasitic relationship. **752**

sympathetic neurons Neurons of the autonomic system that prepare the body for danger or excitement. **516**

sympatric speciation Pattern in which speciation occurs in the absence of a physical barrier to gene flow. **272**

synapse Region where a neuron's axon terminals transmit chemical signals to another cell. **512**

synaptic integration The summation of excitatory and inhibitory signals by a postsynaptic cell. **513**

systemic acquired resistance In plants, inducible whole-body resistance to a wide range of pathogens and abiotic stressors. **485**

systemic circuit Circuit through which blood flows from the heart to the body tissues and back. **585**

systole Contractile phase of the cardiac cycle. **592**

systolic pressure Blood pressure when ventricles are contracting. **595**

T cell T lymphocyte. Lymphocyte central to adaptive immunity. **605**

T cell receptor (TCR) Antigen receptor on the surface of a T cell; basis of self/nonself discrimination. Initiates an adaptive immune response when it recognizes antigen–MHC marker complexes on the surface of an antigen-presenting cell. **612**

T lymphocyte *See* T cell.

taproot system In eudicots, a primary root and all of its lateral branchings. **434**

tardigrade Tiny coelomate animal with four pairs of legs; can survive adverse conditions in a dormant state. **382**

taste receptors Chemoreceptors involved in the sense of taste. **533**

taxis Innate response in which an animal moves toward or away from a stimulus. **721**

taxon Ranked group of organisms that share a unique set of features; each comprises a set of the next lower taxon. E.g., domain, kingdom, phylum, etc. **10**

taxonomy Science of systematically naming and classifying species. **10**

telomere Noncoding, repetitive DNA sequence at the end of a chromosome. **169**

telophase Stage of mitosis during which the two sets of chromosomes arrive at opposite spindle poles and decondense as new nuclei form. **166**

temperate deciduous forest Northern Hemisphere biome in which the main plants are broadleaf trees that lose their leaves in fall and become dormant during cold winters. **798**

temperature Measure of molecular motion, which increases with heat. **30**

tendon Strap of dense connective tissue that connects skeletal muscle to bone. **574**

teratogen A toxin or infectious agent that interferes with embryonic development and causes birth defects. **711**

terminal bud Shoot tip; a bud at the end of a stem. Main zone of primary growth in a shoot. **430**

territory Region that an animal or group of animals defends. **725**

testcross Method of determining genotype by tracking a trait in the offspring of a cross between an individual of unknown genotype and an individual known to be homozygous recessive. **192**

testes In animals, gonads that produce sperm. **680**

testosterone Main hormone produced by testes; causes sperm production and development of male secondary sexual traits. **680**

tetrapod Vertebrate with four legs, or a descendant thereof. **395**

thalamus Forebrain region that relays signals to the cerebrum. **521**

theory, scientific *See* scientific theory.

theory of inclusive fitness Genes associated with altruism can be advantageous if the expense of this behavior to the altruist is outweighed by the resulting increased success of relatives. **728**

theory of uniformity Idea that gradual repetitive processes occurring over long time spans shaped Earth's surface. **241**

therapeutic cloning The use of SCNT to produce human embryos for research purposes. **135**

thermal radiation Emission of heat from an object. **673**

thermocline Thermal stratification in a large body of water; a cool midlayer stops vertical mixing between warm surface water above it and cold water below it. **803**

thermoreceptor Temperature-sensitive sensory receptor. **528**

thigmotropism Plant growth in a direction influenced by contact with a solid object. **481**

threatened species Species likely to become endangered in the near future. **816**

threshold potential Neuron membrane potential at which gated sodium channels open, causing an action potential to occur. **510**

thylakoid membrane A chloroplast's highly folded inner membrane system; forms a continuous compartment in the stroma. The light-dependent reactions occur at the thylakoid membrane. **97**

thymine *See* nucleotide.

thymus Thymus gland. Organ beneath the sternum; produces hormones that enhance immune function and is the site of T cell maturation. **562**

thyroid gland Endocrine gland at the base of the neck; produces thyroid hormone, which increases metabolism. **554**

tidal volume The volume of air that flows into and out of the lungs during a normal inhalation and exhalation. **637**

tight junctions Arrays of fibrous proteins that join epithelial cells and collectively prevent fluids from leaking between them. **67**

tissue In multicelled organisms, specialized cells organized in a pattern that allows them to perform a collective function. 4

tissue culture propagation Laboratory method in which body cells are induced to divide and form an embryo. 467

topsoil Uppermost soil layer; contains the most nutrients for plant growth. 443

total fertility rate Average number of children the women of a population bear over the course of a lifetime. 746

tracer Substance with a detectable component, such as a molecule labeled with a radioisotope. 25

trachea Airway to the lungs; windpipe. 635

tracheal system Of insects and some other land arthropods, tubes that convey gases between the body surface and internal tissues. 630

tracheid Narrow, tapered cell in xylem tissue; dead at maturity, its pitted side walls remain connected with other tracheids to form water-conducting tubes and provide structural support. 429

trait Physical, biochemical, or behavioral characteristic of an organism or species. 10

transcription RNA synthesis; process by which an RNA is assembled from nucleotides using a DNA template. 138

transcription factor Regulatory protein that influences transcription; e.g., an activator or repressor. 152

transduction In bacteria and archaea, a mechanism of horizontal gene transfer in which a bacteriophage carries DNA from one cell to another. 317

transfer RNA (tRNA) A type of RNA that delivers amino acids to a ribosome during translation. 138

transformation In bacteria and archaea, a type of horizontal gene transfer in which DNA is taken up from the environment. 317

transgenic Refers to a genetically modified organism that carries a gene from a different species. 228

transition state Point during a reaction at which substrate bonds reach their breaking point and the reaction will run spontaneously to completion. 76

translation Process by which a polypeptide chain is assembled from amino acids in the order specified by an mRNA. 138

translocation, chromosomal Structural change of a chromosome in which a broken piece gets reattached in the wrong location. 210

translocation Movement of organic compounds through phloem. 450

transpiration Evaporation of water from aboveground plant parts. 447

transport protein Protein that passively or actively assists specific ions or molecules across a cell membrane. 83

transposable element Segment of chromosomal DNA that can spontaneously move to a new location. 147

triglyceride A fat with three fatty acid tails. 42

trochophore larva Type of ciliated larva that forms during the development of mollusks and some annelids. 378

trophic level Position of an organism in a food chain. 770

tropical rain forest Highly productive and species-rich biome in which year-round rains and warmth support continuous growth of evergreen broadleaf trees. 798

tropism Adjustment in growth of a plant or its parts in response to an environmental stimulus. 480

trypanosome Parasitic flagellate with a single mitochondrion and a membrane-encased flagellum. 327

tubular reabsorption Substances move from the filtrate inside a kidney tubule into the peritubular capillaries. 668

tubular secretion Substances move out of peritubular capillaries and into the filtrate in kidney tubules. 668

tumor A neoplasm that forms a lump in the body. 170

tunicate Invertebrate chordate in which the only distinguishing chordate trait retained by adults is a pharynx with gill slits. 394

turgor Pressure that a fluid exerts against a wall, membrane, or other structure that contains it. 85

unsaturated fatty acid Fatty acid that has one or more carbon–carbon double bonds in its hydrocarbon tail. 42

uracil *See* nucleotide.

urea Main nitrogen-containing compound in urine of mammals. 665

ureter Tube that carries urine from a kidney to the bladder. 666

urethra Tube through which urine from the bladder flows out of the body. 666

uric acid Main nitrogen-containing compound excreted by insects, as well as birds and other reptiles. 665

urinary bladder Hollow, muscular organ that stores urine. 666

urine Mix of water and soluble wastes formed and excreted by the vertebrate urinary system. 665

uterus Muscular chamber where offspring develop; womb. 684

vaccine A preparation introduced into the body in order to elicit immunity to a specific antigen. 624

vacuole A membrane-enclosed, saclike organelle filled with fluid; functions to isolate or dispose of waste, debris, or toxins. 61

vagina Female organ of intercourse and birth canal. 685

variable In an experiment, a characteristic or event that differs among individuals or over time. 13

vascular bundle Multistranded cord of primary xylem and phloem in a stem or leaf. The arrangement of vascular bundles in monocots and eudicots differs. 430

vascular cambium Lateral meristem that produces secondary xylem and phloem. 436

vascular cylinder Stele. A root's central column of vascular tissue; composed of primary xylem and phloem sheathed by pericycle and endodermis. 434

vascular plant Plant having specialized tissues (xylem and phloem) that transport water and sugar within the plant body. 341

vascular tissue In vascular plants, tissue (xylem and phloem) that distributes water and nutrients through the plant body. 342, 427

vas deferens One of a pair of long ducts that convey mature sperm toward the body surface. 680

vasoconstriction Narrowing of a blood vessel when smooth muscle that rings it contracts. 594

vasodilation Widening of a blood vessel when smooth muscle that rings it relaxes. 594

vector Of a disease, an animal that transmits a pathogen from one host to the next. 312

vegetative reproduction Form of asexual reproduction in plants; growth of new roots and shoots from extensions or fragments of a parent plant. 466

vein In plant leaves, a vascular bundle strengthened with fibers. In animals, a large-diameter vessel that returns blood to the heart. 433, 586

ventricle Heart chamber that pumps blood into arteries. 592

venule Small-diameter vessel that carries blood from capillaries to a vein. 586

vernalization Stimulation of flowering in spring by prolonged exposure to low temperature in winter. 483

vertebrae Bones of the backbone, or vertebral column. **568**

vertebral column Backbone. **568**

vertebrate Animal with a backbone. **395**

vesicle Small vacuole; different kinds store, transport, or degrade their contents. **60**

vessel member Elongated cell in xylem tissue; dead at maturity, its perforated walls remain connected end-to-end with other vessel members to form water-conducting tubes. **429**

vestibular apparatus System of fluid-filled sacs and canals in the inner ear; contains the organs of equilibrium. **542**

villi Multicelled projections at the surface of each fold in the small intestine. **651**

viral reassortment Two related viruses infect the same individual simultaneously and swap genes. **313**

viroid Small, noncoding, infectious RNA. **309**

virus Noncellular, infectious particle of protein and nucleic acid; replicates only in a host cell. **308**

visceral sensations Sensations that arise when sensory neurons associated with organs inside body cavities are activated. **530**

vision Perception of visual stimuli based on light focused on a retina and image formation in the brain. **534**

visual accommodation Process of adjusting lens shape or position so that light from an object falls on the retina. **537**

vital capacity Amount of air moved in and out of lungs with forced inhalation and exhalation. **636**

vitamin Organic substance required in small amounts for normal metabolism. **658**

vomeronasal organ Pheromone-detecting organ of vertebrates. **533**

warning coloration In many unpalatable or well-defended species, bright colors, patterns, and other signals that predators learn to recognize and avoid. **758**

water cycle Movement of water among Earth's oceans, atmosphere, and the freshwater reservoirs on land. **776**

water mold Heterotrophic protist that grows as nutrient-absorbing filaments. **332**

watershed Land area that drains into a particular stream or river. **776**

water–vascular system Of echinoderms, a system of fluid-filled tubes and tube feet that function in locomotion. **390**

wavelength Distance between the crests of two successive waves. **94**

wax Water-repellent mixture of lipids with long fatty acid tails bonded to long-chain alcohols or carbon rings. **43**

white blood cell Vertebrate blood cell with a role in housekeeping tasks and defense; a leukocyte. **589**

white matter Tissue of brain and spinal cord that consists of myelinated axons. **518**

wood Accumulated secondary xylem. **436**

X chromosome inactivation Shutdown of one of the two X chromosomes in the cells of female mammals. **156**

xenotransplantation Transplantation of an organ from one species into another. **231**

xylem Complex vascular tissue of plants; its tracheids and vessel members distribute water and mineral ions. **342, 429**

yellow marrow Bone marrow consisting mainly of fat; fills cavity in most adult long bones. **570**

zero population growth Interval in which births equal deaths. **736**

zygote Diploid cell formed by fusion of two gametes; the first cell of a new individual. **177**

zygote fungus Fungus that forms a zygospore during sexual reproduction. **361**

Art Credits and Acknowledgments

This Page constitutes an extension of the book copyright page. We have made every effort to trace the ownership of all copyrighted material and secure permission from copyright holders. In the event of any question arising as to the use of any material, we will be pleased to make the necessary corrections in future printings. Thanks are due to the following authors, publishers, and agents for permission to use the material indicated.

TABLE OF CONTENTS **Page iii** top, Bob Jensen Photography; middle, Jeff Vanuga/Corbis; bottom, John Easley. **Page v** top, Bob Jensen Photography; bottom, John Easley. **Page vii** top, NASA; bottom right, © Exactostock/ SuperStock. **Page viii** from left, © R. Calentine/ Visuals Unlimited; © Dylan T. Burnette and Paul Forscher; Hemoglobin models: PDB ID: 1GZX; Paoli, M., Liddington, R., Tame, J., Wilkinson, A., Dodson, G.; Crystal structure of T state hemoglobin with oxygen bound at all four haems. *J.Mol.Bio.*, v256, pp. 775–792, 1996. **Page ix** from left, Professors P. Motta and T Naguro/ Photo Researchers, Inc.; Andrew Syred/ Photo Researchers, Inc. **Page x** from left, © Jose Luis Riechmann; Ed Reschke; Image courtesy of Carl Zeiss Micro-Imaging, Thornwood, NY; Moravian Museum, Brno. **Page xi** from left, Photo courtesy of The Progeria Research Foundation; Photo courtesy of MU Extension and Agricultural Information; © Louie Psihoyos/ Getty Images. **Page xii** from left, J. A. Bishop, L. M. Cook; © Jack Jeffrey Photography; © James C. Gaither, www.flickr.com/people/jim-sf/. **Page xiii** from left, © Janet Iwasa; Eye of Science/ Photo Researchers, Inc.; SciMAT/ Photo Researchers, Inc.; Malaria illustration by Drew Berry, The Walter and Eliza Hall Institute of Medical Research. **Page xiv** from left, Jane Burton/ Bruce Coleman Ltd.; Smithsonian Institution Department of Botany, G.A. Cooper @ USDA-NRCS PLANTS Database; John Hodgin. **Page xv** from left, Dave Fleetham/Tom Stack & Associates; CDC/ Piotr Naskrecki; Courtesy of Gregory B. Skomal; D. & V. Blagden/ ANT Photo Library. **Page xvi** from left, © Frank Vincken, Courtesy of Max Planck Institute for Evolutionary Anthropology; Courtesy of Dr. Thomas L. Rost; M. I. Walker/ Photo Researchers, Inc. **Page xvii** from left, Dr. Keith Wheeler/ Photo Researchers, Inc.; John Alcock, Arizona State University; © Brian Johnston; Michael Clayton, University of Wisconsin. **Page xviii** from left, © Michael Shore Photography; AP Images; photos.com. **Page xix** from left, Mitch Reardon/ Photo Researchers, Inc.; Professor P. Motta/ Department of Anatomy/ La Sapienza, Rome/ Photo Researchers, Inc. **Page xx** from left, National Cancer Institute/ Photo Researchers, Inc.; Biophoto Associates/ Photo Researchers, Inc.; Dr. Richard Kessel and Dr. Randy Kardon/ Tissues & Organs/ Visuals Unlimited; Dr. Oliver Schwartz, Institute Pasteur/ Photo Researchers, Inc. **Page xxi** from left, Martin Dohrn, Royal College of Surgeons/ Photo Researchers, Inc.; Francois Gohier/ Photo Researchers, Inc.; © camilla$$/ Shutterstock.com; Microslide courtesy Mark Nielsen, University of Utah. **Page xxii** from left, © MicroScan/ Phototake; NASA/JPL-Caltech/L. Hermans; © Rodger Klein/ Peter Arnold, Inc.; Courtesy of Elizabeth Sanders, Women's Specialty Center, Jackson, MS. **Page xxiii** from left, © Lennart Nilsson/

Bonnierforlagen AB; From "Magneto-reception of Directional Information in Birds Requires Nondegraded Vision," *Current Biology*, vol. 20 (pp.1–4), © 2010 Elsevier Ltd. All rights reserved, Reprinted by permission; © Jacques Langevin/ Corbis Sygma. **Page xxiv** from left, Bob and Miriam Francis/ Tom Stack & Associates; © W. Perry Conway/ Corbis; Graphic created by FoodWeb3D program written by Rich Williams courtesy of the Webs on the Web project (www.foodwebs.org). **Page xxv** from left, NASA Goddard Space Flight Center Scientific Visualization Studio; © George M. Sutton/ Cornell Lab of Ornithology; © Lee Prince/ Shutterstock.com.

CHAPTER 1 **Page 2** © Sergei Krupnov, www.flickr.com/photos/7969319@N03; **Learning Roadmap**, from top, © Mauritius/ SuperStock; © Umberto Salvagnin, www.flickr.com/photos/kaibara; Dr. Marina Davila Ross, University of Portsmouth; From Meyer, A., Repeating Patterns of Mimicry. PLoS Biology Vol. 4, No. 10, e341 doi:10.1371/journal.pbio.0040341; © Antje Schulte; © Raymond Gehman/ Corbis; NASA. **1.1** Tim Laman/ National Geographic Stock. **1.2** (3-4) © Umberto Salvagnin, www.flickr.com/photos/kaibara; (5) California Poppy, © 2009, Christine M. Welter; (6) Lady Bird Johnson Wildflower Center; (7) Michael Szoenyi/ Photo Researchers, Inc.; (8) Photographers Choice RF/ SuperStock; (9) © Sergei Krupnov, www.flickr.com/photos/7969319@N03; (10) © Mark Koberg Photography; (11) NASA. **1.3** above, © Victoria Pinder, www.flickr.com/photos/vixstarplus. **1.4** Dr. Marina Davila Ross, University of Portsmouth. **1.5** (A) Clockwise from top left, Dr. Richard Frankel; Photo Researchers, Inc.; © Susan Barnes; www.zahnarzt-stuttgart.com; (B) left, ArchiMeDes; right, © Dr. Harald Huber, Dr. Michael Hohn, Prof. Dr. K.O.Stetter, University of Regensburg, Germany; (C) Protists, left, © Lewis Trusty/ Animals Animals; right, clockwise from top left, M I Walker/ Photo Researchers, Inc.; Courtesy of Allen W. H. Bé and David A. Caron; © Emiliania Huxleyi photograph, Vita Pariente, scanning electron micrograph taken on a Jeol T330A instrument at Texas A&M University Electron Microscopy center; Oliver Meckes/ Photo Researchers, Inc.; © Carolina Biological Supply Company; Plants, left © John Lotter Gurling / Tom Stack & Associates; right © Edward S. Ross; Fungi, left, © Robert C. Simpson/ Nature Stock; right, © Edward S. Ross; Animals, from left, © Tom & Pat Leeson, Ardea London Ltd.; Thomas Eisner, Cornell University; © Martin Zimmerman, Science, 1961, 133:73-79, © AAAS.; © Pixtal/ SuperStock. **1.6** from left, Joaquim Gaspar; © kymkemp.com; Nigel Cattlin/Visuals Unlimited, Inc.; Courtesy of Melissa S. Green, www.flickr.com/photos/henkimaa; © Gordana Sarkotic. **1.8** From Meyer, A., Repeating Patterns of Mimicry. PLoS Biology Vol. 4, No. 10, e341 doi:10.1371/journal.pbio.0040341. **1.9** (A) Volker Steger/ Photo Researchers, Inc.; (B) Cape Verde National Institute of Meteorology and Geophysics and the U.S. Geological Survey; (C) © Roger W. Winstead, NC State University; (D) Photo by Scott Bauer, USDA/ARS. **1.10** © Superstock. **1.11** (A) © Matt Rowlings, www.eurobutterflies.com; (B) © Adrian Vallin; (C) © Antje Schulte. **1.12** © Bruce Beehler/ Conservation International. **1.13** © Gary Head. **Table 1.3** © Raymond Gehman/ Corbis. **1.15** Courtesy East Carolina University.

CHAPTER 2 **Page 22**, © Vicki Rosenberg, www.flickr.com/photos/roseofredrock; **Learning Roadmap**, from top, © Victoria Pinder, www.flickr.com/photos/vixstarplus; © Michael S. Yamashita/ Corbis; © Bill Beatty/ Visuals Unlimited; © Herbert Schnekenburger; W. K. Fletcher/ Photo Researchers, Inc. **2.1** left, Michael Grecco/ Picture Group; inset, Theodore Gray/ Visuals Unlimited, Inc; right, © Kim Westerskov, Photographer's Choice/ Getty Images. **2.4** Brookhaven National Laboratory. **Page 27** © Michael S. Yamashita/ Corbis. **2.8** (B) left, Gary Head; center, © Bill Beatty/ Visuals Unlimited. **Page 30** © Herbert Schnekenburger. **2.11** (C) www.flickr.com/photos/roseofredrock. **2.12** © JupiterImages Corporation. **Page 33** left, © Jared C. Benedict, www.flickr.com/photos/redjar; right, W. K. Fletcher/ Photo Researchers, Inc.

CHAPTER 3 **Page 34**, © JupiterImages Corporation; **Learning Roadmap**, from top, © JupiterImages Corporation; © JupiterImages Corporation; From the collection of Jamos Werner and John T. Lis. **3.1** © ThinkStock/ SuperStock. **3.2** (B) © JupiterImages Corporation. **3.4** (A) © National Cancer Institute; (B) Hemoglobin models: PDF ID: 1GZX; Paoli, M., Liddington, R., Tame, J., Wilkinson, A., Dodson, G., Crystal structure of T state hemoglobin with oxygen bound at all four hems. J.Mol.Bio., v256, pp. 775–792, 1996. **3.5** (A) © Exactostock/ SuperStock. **3.7** left, **3.8** (C) © JupiterImages Corporation. **3.13** right, Tim Davis/ Photo Researchers, Inc. **Page 43** left, Kenneth Lorenzen. **Page 45** left, From: *Structure of the Rotor of the V-Type Na+-ATPase from Enterococcus hirae* by Murata, et al. Science 29 April 2005: 654-659. DOI:10.1126/science.1110064. Used with permission; right, This lipoprotein image was made by Amy Shih and John Stone using VMD and is owned by the Theoretical and Computational Biophysics Group, NIH Resource for Macromolecular Modeling and Bioinformatics, at the Beckman Institute, University of Illinois at Urbana-Champaign. Labels added to the original image by book author. **3.17** (A) © Lily Echeverria/Miami Herald; (B) Sherif Zaki, MD PhD, Wun-Ju Shieh, MD PhD; MPH/ CDC. **Page 47** © JupiterImages Corporation.

CHAPTER 4 **Page 48**, © Dylan T. Burnette and Paul Forscher; **Learning Roadmap**, from top, Courtesy of © Johannes Kästner, Universität Stuttgart; R. Calentine/ Visuals Unlimited; Astrid & Hanns-Frieder Michler/ Photo Researchers, Inc.; Don W. Fawcett/ Visuals Unlimited; © ADVANCELL (Advanced In Vitro Cell Technologies; S.L.) www.advancell.com; Michael Clayton/ University of Wisconsin, Department of Botany. **4.1** left, © JupiterImages Corporation; right, Stephanie Schuller/ Photo Researchers, Inc. **4.3** Courtesy of © Johannes Kästner, Universität Stuttgart. **4.6** (A–B, D–E) © Jeremy Pickett-Heaps, School of Botany, University of Melbourne; (C) © Prof. Franco Baldi. **4.7** Whale, © Dorling Kindersley/ the Agency Collection/ Getty Images; Giraffe, © Ingram Publishing, SuperStock; Man, © JupiterImages Corporation; Dog, © Pavel Sazonov/ Shutterstock.com; Guinea Pig, © Sascha Burkard/ Shutterstock.com; Frog, Ant, photos.com; Louse, Edward S. Ross; Eukaryotic cell, Bruce Runnegar, NASA Astrobiology Institute; HPV particle, CDC. **4.9** (A) Rocky Mountain Laboratories, NIAID, NIH; (B) © R. Calentine/ Visuals Unlimited; (C) Cryo-EM image of Haloquadratum wals-byi, isolated from Australia. Courtesy of Zhuo Li (City of Hope, Duarte, California, USA), Mike L. Dyall-Smith (Charles Sturt University, Australia), and Grant J. Jensen (California Institute of Technology, Pasadena, California, USA); (D–E) © K.O. Stetter & R. Rachel, Univ. Regensburg. **4.10** © Dennis Kunkel Microscopy, Inc./ Phototake. **Page 56,** Astrid & Hanns-Frieder Michler/ Photo Researchers, Inc. **4.11** (A) Dr. Gopal Murti/ Photo Researchers, Inc.; (B) M.C. Ledbetter, Brookhaven National Laboratory. **4.13** right, © Kenneth Bart. **4.14** (A) Don W. Fawcett/ Visuals Unlimited; (B) © Martin W. Goldberg, Durham University, UK. **Page 59,** Andrew Syred/ Photo Researchers, Inc. **4.15** from left, © Kenneth Bart; (#2,3) Don W. Fawcett/ Visuals Unlimited; (#4) Micrograph, Gary Grimes. **4.16** Courtesy of © Conner's Way Foundation, www.connersway.com. **Page 63,** © David T. Webb. **4.17** Micrograph, Keith R. Porter. **4.18** (A) © Dr. Ralf Wagner, www.dr-ralf-wagner.de; (B) Dr. Jeremy Burgess/ Photo Researchers, Inc. **4.19** (D) © Dylan T. Burnette and Paul Forscher. **4.21** (A) right, Don W. Fawcett/ Photo Researchers, Inc. **4.23** George S. Ellmore. **4.24** © ADVANCELL (Advanced In Vitro Cell Technologies; S.L.) www.advancell.com. **Page 68,** Above, © R. Llewellyn/ Super-Stock, Inc.

CHAPTER 5 **Page 70,** © Greg Epperson/ Shut-terstock.com; **Learning Roadmap,** from top, © Victoria Pinder, www.flickr.com/photos/vix-starplus; © JupiterImages Corporation; Hemoglobin models: PDB ID: 1GZX; Paoli, M., Liddington, R., Tame, J., Wilkinson, A., Dodson, G.; Crystal structure of T state hemoglobin with oxygen bound at all four haems. *J.Mol.Bio.*, v256, pp. 775–792, 1996; © JupiterImages Corporation; Andrew Lambert Photography/ Photo Researchers, Inc.; Photo Researchers, Inc. **5.1** © BananaStock/ SuperStock. **5.3** © JupiterImages Corporation. **5.5** © Greg Epperson/ Shutterstock.com. **5.8** © Westend61/SuperStock. **5.10** Hemoglobin models: PDB ID: 1GZX; Paoli, M., Liddington, R., Tame, J., Wilkinson, A., Dodson, G.; Crystal structure of T state hemoglobin with oxygen bound at all four haems. *J.Mol.Bio.*, v256, pp. 775–792, 1996. **5.15** Lisa Starr, after David S. Goodsell, RCSB, Protein Data Bank. **5.16** (A) © JupiterImages Corporation. **Page 84** Andrew Lambert Photography/ Photo Researchers, Inc. **5.23** (A–C) M. Sheetz, R. Painter, and S. Singer, Journal of Cell Biology, 70:193 (1976) by permission of The Rockefeller University Press.; (D–E) Claude Nuridsany & Marie Perennou/ Photo Researchers, Inc. **5.28** © R.G.W. Anderson, M.S. Brown and J.L. Goldstein. Cell 10:351 (1977). **5.29** (A) Photo Researchers, Inc. **5.31** (A) Southern Illinois University/ Photo Researchers, Inc.; (B) Martin M. Rotker/ Photo Researchers, Inc.; (C) Stuart Clark/ The Sunday Times/ nisyndication. **Page 91,** Above, © BananaStock/ SuperStock.

CHAPTER 6 **Page 92,** © Photodisc/ Getty Images; **Learning Roadmap,** from top, © JupiterImages Corporation; Lisa Starr; Martin Zimmerman, Science, 1961, 133:73-79, © AAAS. **6.1** left, © Roger W. Winstead, NC State University; right, Photo by Peggy Greb/ USDA. **6.2** left, © Photodisc/ Getty Images. **6.3** above, © Larry West/ FPG /Getty Images. **6.4** (A) Jason Sonneman. **6.5** above, © Dr. Ralf Wagner, www.dr-ralf-wagner.de. **6.11** (A) above, © iStockphoto.com/ farbenrausch; below Masahiro Yamada, Michio Kawasaki, Tatsuo Sugiyama, Hiroshi Miyake,

Mitsutaka Taniguchi; *Differential Positioning of C4 Mesophyll and Bundle Sheath Chloroplasts: Aggregative Movement of C4 Mesophyll Chloroplasts in Response to Environmental Stresses*: Plant and Cell Physiology; (2009) 50(10): 1736-1749; (B) above, © JC Schou/ www.biopix.dk; below, Eri Maai, Shouu Shimada, Masahiro Yamada, Tatsuo Sugiyama, Hiroshi Miyake, Mitsutaka Taniguchi; *The avoidance and aggregative movements of mesophyll chloroplasts in C4 monocots in response to blue light and abscisic acid*; Journal of Experimental Botany, doi:10.1093/jxb/err008, by permission of Oxford University Press. **6.13** © Lisa Starr. **6.14** (E) © Richard Uhlhorn Photography. **6.15** © Brooke Schreier Ganz. **Page 114,** Section 6.1, Photo by Peggy Greb/ USDA; Section 6.5–6.6, © Dr. Ralf Wagner, www.dr-ralf-wagner.de; Section 6.8, Lisa Starr.

CHAPTER 7 **Page 106,** © sportgraphic/ Shut-terstock.com; **Learning Roadmap,** from top, © Dr. Ralf Wagner, www.dr-ralf-wagner.de; © Elena Boshkovska/ Shutterstock.com; © shabaneiro/ Shutterstock.com. **7.1** above, Professors P. Motta and T Naguro/ Photo Researchers, Inc.; below, © Louise Chalcraft-Frank and FARA. **7.7** right, Photo Researchers, Inc. **7.10** above left, © Elena Boshkovska/ Shutterstock.com; above right, © optimarc/ Shutterstock.com; below, Dr. Dennis Kunkel/ Visuals Unlimited. **7.11** (B) Courtesy of © William MacDonald, M.D.; (C) © sportgraphic/ Shutterstock.com. **Page 118** © shabaneiro/ Shutterstock.com.

CHAPTER 8 **Page 122,** Patrick Landmann/ Photo Researchers, Inc.; **Learning Roadmap,** from top, © University of Washington Department of Pathology; A C. Barrington Brown © 1968 J. D. Watson; Courtesy of Cyagra, Inc., www.cyagra.com; Michael Clayton/ University of Wisconsin, Department of Botany. **8.1** © James Symington. **8.2** top, Andrew Syred/ Photo Researchers, Inc.; #3, B. Hamkalo; #5, O. L. Miller, Jr., Steve L. McKnight. **8.3** (A) © University of Washington Department of Pathology. **8.4** Patrick Landmann/ Photo Researchers, Inc. **8.5** (A) below, Eye of Science/ Photo Researchers, Inc. **8.8** A C. Barrington Brown © 1968 J. D. Watson. **8.11** Olga Shovman, Andrew C. Riches, Douglas Adamson, and Peter E. Bryant. *An improved assay for radiation-induced chromatid breaks using a colcemid block and calyculin-induced PCC combination.* Mutagenesis (2008) 23(4): 267-270 first published online March 6, 2008 doi:10.1093/mutage/gen009, by permission of Oxford University Press. **8.14** Courtesy of Cyagra, Inc., www.cyagra.com. **Page 135** top, Ben Glass, courtesy of © BioArts International. **8.15** Courtesy of Cyagra, Inc., www.cyagra.com.

CHAPTER 9 **Page 136,** O. L. Miller; **Learning Roadmap,** from top, From: *Structure of the Rotor of the V-Type Na+-ATPase from Enterococcus hirae* by Murata, et al. Science 29 April 2005: 654-659. DOI:10.1126/science.1110064. Used with permission; Dr. Gopal Murti/ Photo Researchers, Inc.; © Rollin Verlinde/ Vilda. **9.1** right, (A) Vaughan Fleming/ Photo Researchers, Inc.; (B) Steve Hurst/ USDA-NRCS PLANTS Database; (C) Courtesy of Zhang Zhugang; (D) Stephanie Schuller/ Photo Researchers, Inc.; (E) Dr. Kari Lounatmaa/ Photo Researchers, Inc. **9.5** O. L. Miller. **9.12** (B) Kiseleva and Donald Fawcett/ Visuals Unlimited. **9.14** (B) left, Dr. Gopal Murti/ Photo Researchers, Inc.; right, © Ben Rose/ WireImage/ Getty Images. **Page 148** Above, Vaughan Fleming/ Photo Researchers, Inc.

CHAPTER 10 **Page 150,** FlyEx Database, urchin.spbcas.ru/flyex/index.jsp; **Learning Roadmap,** from top, Patrick Landmann/ Photo Researchers, Inc.; From the collection of Jamos Werner and John T. Lis; Courtesy of Dr. Barbara Jennings, UCL Cancer Institute, www.ucl.ac.uk; © Jose Luis Riechmann/ Nederlands Fotomuseum, all rights reserved; © Gary Roberts/ world-widefeatures.com. **10.1** above, Courtesy of Robin Shoulla and Young Survival Coalition; below, from the archives of www.breastpath.com, courtesy of J.B. Askew, Jr., M.D., P.A. Reprinted with permission, copyright 2004 Breastpath.com. **Page 153** From the collection of Jamos Werner and John T. Lis. **10.4** (A–E) © Maria Samsonova and John Reinitz; (F) © Jim Langeland, Jim Williams, Julie Gates, Kathy Vorwerk, Steve Paddock and Sean Carroll, HHMI, University of Wisconsin-Madison. **10.6** (A–B) David Scharf/ Photo Researchers, Inc.; (C) Eye of Science/ Photo Researchers, Inc.; (D) Courtesy of the Aniridia Foundation International, www.aniridia.net; (E) M. Bloch. **Page 155** Courtesy of Dr. Barbara Jennings, UCL Cancer Institute, www.ucl.ac.uk. **10.7** (A) © Dr. William Strauss; (B) Bernard Cohen, MD, Dermatlas; www.dermatlas.org. **10.9** (A) Juergen Berger, Max Planck Institute for Developmental Biology, Tuebingen, Germany; (B) © Jose Luis Riechmann. **10.11** PDB ID: 1CJG; Spronk, A.A.E.M., Bovin, A.M.J.J., Radha, P.K., Melacini, G., Boelens, R., Kaptien, R.: The Solution Structure of Lac Repressor Headpiece 62 Complexed to a Symmetrical Lac Operator Structure (London) 7 pp. 1483, (1999). Also PDB ID: 1LBI; Lewis, M., Chang, G., Horton, N.C., Kercher, M.A., Pace, H.C., Schumacher, M.A., Brenan, R.G., Lu, P.: Crystal structure of the lactose operon repressor and its complexes with DNA and inducer. *Science* 271 pp. 1247 (1966); lactose pdb files form the Hetero-Compound Information Center-Uppsala (HIC-Up). **10.13** © Marius Meijboom/ Nederlands Fotomuseum, all rights reserved. **Page 161** Courtesy of Robin Shoulla and Young Survival Coalition.

CHAPTER 11 **Page 162,** © Carolina Biological Supply Company/ Phototake; **Learning Roadmap,** from top, Andrew Syred/ Photo Researchers, Inc.; Michael Clayton/ University of Wisconsin, Department of Botany; Michael Clayton/ University of Wisconsin, Department of Botany; © ISM/ Photototakeusa; From *Expression of the epidermal growth factor receptor (EGFR) and the phosphorylated EGFR in invasive breast carcinomas.* breast-cancer research.com/content/10/3/R49; © Lennart Nilsson/ Bonnierforlagen AB. **11.1** left, Dr. Paul D. Andrews/ University of Dundee; right, Courtesy of the family of Henrietta Lacks. **11.3** © Carolina Biological Supply Company/ Phototake. **11.5** left (all), Michael Clayton/ University of Wisconsin, Department of Botany; right (all) Ed Reschke. **11.8** © ISM/ Photototakeusa. **11.9** From *Expression of the epidermal growth factor receptor (EGFR) and the phosphorylated EGFR in invasive breast carcinomas.* breast-cancer research.com/content/10/3/R49. **11.10** © Phillip B. Carpenter, Department of Biochemistry and Molecular Biology, University of Texas - Houston Medical School. **11.12** (A) Ken Greer/ Visuals Unlimited; (B) Biophoto Associates/ Photo Researchers, Inc.; (C) James Stevenson/ Photo Researchers, Inc. **Page 172** Dr. Paul D. Andrews/University of Dundee.

CHAPTER 12 **Page 174,** MRC NIMR/Wellcome Images; **Learning Roadmap,** from top, Michael Clayton/ University of Wisconsin, Department of Botany; Image courtesy of Carl Zeiss MicroImaging, Thornwood, NY; With thanks to the John

Minden Pictures/ Getty Images; © Taro Taylor, www.flickr.com/photos/tjt195; © Panoramic Images/ Getty Images. **Page 281** © James C. Gaither, www.flickr.com/people/jim-sf/. **18.7** © Jack Jeffrey Photography. **18.8** from top, © Lennart Nilsson/ Bonnierforlagen AB; Courtesy of Anna Bigas, IDIBELL-Institut de Recerca Oncologica, Spain; From *Embryonic staging system for the short-tailed fruit bat, Carollia perspicillata, a model organism for the mammalian order Chiroptera, based upon timed pregnancies in captive-bred animals.* C.J. Cretekos et al., Developmental Dynamics Volume 233, Issue 3, July 2005, Pages: 721–738. Reprinted with permission of Wiley-Liss, Inc. a subsidiary of John Wiley & Sons, Inc.; Courtesy of Prof. Dr. G. Elisabeth Pollerberg, Institut für Zoologie, Universität Heidelberg, Germany; USGS. **18.9** left, © Visuals Unlimited; right, © Jürgen Berger, Max-Planck-Institut for Developmental Biology, Tübingen. **Page 303** © NaturePL/ SuperStock. **18.10** Courtesy of Ann C. Burke, Wesleyan University. **18.12** John Steiner/ Smithsonian Institution. **18.13** left, © Eric Nishibayashi Photography; right, Third from top, Bill Sparklin/Ashley Dayer; all others, © Jack Jeffrey Photography. **Page 288** above, Bill Sparklin/ Ashley Dayer.

CHAPTER 19 **Page 290**, Chase Studios/ Photo Researchers, Inc.; **Learning Roadmap**, from top, Painting by William K. Hartmann; Courtesy of the University of Washington; From Hanczyc, Fujikawa, and Szostak, *Experimental Models of Primitive Cellular Compartments: Encapsulation, Growth, and Division*; www.sciencemag.org, Science 24 October 2003; 302;529, Figure 2, Page 619. Reprinted with permission of the authors and AAAS.; Chase Studios/ Photo Researchers, Inc.; © University of California Museum of Paleontology; © Courtesy of Isao Inouye, Institute of Biological Sciences, University of Tsukuba; M I Walker/ Photo Researchers, Inc. **19.1** Photo by Julio Betancourt/ U.S. Geological Survey. **19.2** Painting by William K. Hartmann. **19.5** Courtesy of the University of Washington. **19.7** (A) © Janet Iwasa; (B) From Hanczyc, Fujikawa, and Szostak, *Experimental Models of Primitive Cellular Compartments: Encapsulation, Growth, and Division*; www.sciencemag.org, Science 24 October 2003; 302;529, Figure 2, Page 619. Reprinted with permission of the authors and AAAS; (C) Photo by Tony Hoffman, courtesy of David Deamer. **19.8** (A) Stanley M. Awramik; (B) upper, © University of California Museum of Paleontology; (C) Chase Studios/ Photo Researchers, Inc.; inset, © Dr. J Bret Bennington/ Hofstra University. **19.9** (A–B) © Bruce Runnegar, NASA Astrobiology Institute; (C) © N.J. Butterfield, University of Cambridge. **19.10** (B) Photo provided by M.G. Klotz, Department of Biology, University of Louisville, KY, USA (mgkmicro.com); (C) Courtesy of John Fuerst, University of Queensland, originally published in Archives of Microbiology vol 175, p 413–429 (Lindsay MR, Webb RI, Strous M, Jetten MS, Butler MK, Forde RJ, Fuerst JA. *Cell compartmentalisation in planctomycetes: novel types of structural organisation for the bacterial cell*, Arch Microbiol. 2001 Jun;175(6):413–29). **19.11** CNRI/ Photo Researchers, Inc. **19.12** above, © Courtesy of Isao Inouye, Institute of Biological Sciences, University of Tsukuba; below, © Robert Trench, Professor Emeritus, University of British Columbia. **Page 302** above, NASA/JPL.

CHAPTER 20 **Page 306**, Eye of Science/ Photo Researchers, Inc.; **Learning Roadmap**, from top, © Susan Barnes; Dr. Linda Stannard, UCT/ Photo Researchers, Inc.; CDC/ Dr. Hermann; CNRI/ Photo Researchers, Inc.; © Dr. Manfred Schloesser,

Max Planck Institute for Marine Microbiology; © Savannah River Ecology Laboratory; © Wally Eberhart/ Visuals Unlimited. **20.1** © Steve Wood–UAB Creative. **20.2** left, Courtesy D. Shew, Reproduced by permission from Compendium of Tobacco Diseases, 1991, American Phytopathological Society, St. Paul, MN. **20.3** right, Eye of Science/ Photo Researchers, Inc. **20.4** (A) Dr. Richard Feldmann/ National Cancer Institute; (B) left, Russell Knightly/ Photo Researchers, Inc.; right, Dr. Linda Stannard, UCT/ Photo Researchers, Inc. **20.5** Photo Researchers, Inc. **20.7** CDC/ Dr. Hermann. **20.8** CDC/ Dan Higgins. **20.11** CNRI/ Photo Researchers, Inc. **20.13** (A) Tony Brian and David Parker/ Photo Researchers, Inc. **20.15** (1) Dr. Dennis Kunkel / Visuals Unlimited. **20.16** (A) P. W. Johnson and J. MeN. Sieburth, Univ. Rhode Island/ BPS; (B) © Dr. Manfred Schloesser, Max Planck Institute for Marine Microbiology; (C) Courtesy of © Dr. Rolf Müller and Dr. Klaus Gerth; (D) SciMAT/ Photo Researchers, Inc. **20.17** (A) Stem Jems/ Photo Researchers, Inc.; (B) CDC/ James Gathany. **Page 320** Courtesy Jack Jones, *Archives of Microbiology*, Vol. 136, 1983, pp. 254–261. Reprinted by permission of Springer-Verlag. **20.18** (A) © Savannah River Ecology Laboratory; (B) Courtesy of Benjamin Brunner; (C) Dr. John Brackenbury/ Photo Researchers, Inc. **20.19** Micrographs Z. Salahuddin, National Institutes of Health.

CHAPTER 21 **Page 322**, © Kim Taylor/ Bruce Coleman, Inc./ Photoshot; **Learning Roadmap**, from top, © Courtesy of Isao Inouye, Institute of Biological Sciences, University of Tsukuba; Astrid & Hanns-Frieder Michler/ Photo Researchers, Inc.; Dr. Dennis Kunkel/ Visuals Unlimited; © Lewis Trusty/ Animals Animals; Wim van Egmond/ Visuals Unlimited; Courtesy of www.hiddenforest.co.nz; NASA's Earth Observatory. **21.1** left, Malaria illustration by Drew Berry, The Walter and Eliza Hall Institute of Medical Research; right, © Corbis/ SuperStock. **21.2** (A) Astrid & Hanns-Frieder Michler/ Photo Researchers, Inc.; (B) D. P. Wilson/ Eric & David Hosking. **21.4** CDC/ Dr. Stan Erlandsen. **21.5** Dr. Dennis Kunkel/ Visuals Unlimited. **21.6** Oliver Meckes/ Photo Researchers, Inc. **21.7** left, P. L. Walne and J. H. Arnott, Planta, 77:325–354, 1967. **21.8** (A) Courtesy of Allen W. H. Bé and David A. Caron; (B) © Klaus Wagensonner, www.flickr.com/photos/33646230@N03. **21.9** (A) © Franz Neidl; (B) Wim van Egmond/ Visuals Unlimited. **21.10** Gary W. Grimes and Steven L'Hernault. **21.11** (A) Courtesy James Evarts. **21.12** (A) Dr. David Phillips/ Visuals Unlimited; (B) © Peter M. Johnson/ www.flickr.com/photos/pmjohnso. **21.13** left, © Frank Borges Llosa/ www.frankley.com; right, Wim van Egmond/ Visuals Unlimited. **Page 331** Gary Head. **21.15** Heather Angel. **21.16** (A) © Susan Frankel, USDA-FS; (B) Dr. Pavel Svihra. **21.17** (A) Science Museum of Minnesota; (B) Wim van Egmond/ Visuals Unlimited; (C) Image courtesy of Woodstream Corporation. **21.18** (A) right, © Lewis Trusty/ Animals Animals; (B) Steven C. Wilson/ Entheos; (C) Garo/ Photo Researchers, Inc. **21.19** Image courtesy of FGBNMS/UNCW-NURC. **21.20** Courtesy of Professeur Michel Cavalla. **21.21** (A) © Kim Taylor/ Bruce Coleman, Inc./ Photoshot; (B) © Lawson Wood/ Corbis; (C) Montery Bay Aquarium; (D) Wim van Egmond/ Visuals Unlimited. **21.22** M I Walker/ Photo Researchers, Inc. **21.23** Carolina Biological Supply Company. **21.24** (A) Edward S. Ross; (B) Courtesy of www.hiddenforest.co.nz. **Page 337** Photo by James Gathany, Centers for Disease Control. **21.25** David Patterson, courtesy micro*scope/ microscope.mbl.edu.

CHAPTER 22 **Page 338**, Dr. Annkatrin Rose, Appalachian State University; **Learning Roadmap**, from top, Wim van Egmond/ Visuals Unlimited; Courtesy of © Professor T. Mansfield; Jane Burton/ Bruce Coleman Ltd.; David C. Clegg/ Photo Researchers, Inc.; Fletcher and Baylis/ Photo Researchers, Inc.; Smithsonian Institution Department of Botany, G.A. Cooper @ USDA-NRCS PLANTS Database; © Cathlyn Melloan/ Stone/ Getty Images. **22.1** (A) T. Kerasote/ Photo Researchers, Inc.; (B) © William Campbell/ TimePix/ Getty Images. **22.3** © Christine Evers. **22.5** (A) Michael Clayton/ University of Wisconsin, Department of Botany; (B) Courtesy of © Professor T. Mansfield. **22.6** (A) Andrew Syred/ Photo Researchers, Inc.; (B) © M.I. Walker/ Wellcome Images. **22.7** (A) © Christine Evers; (B) Gary Head. **22.8** Jane Burton/ Bruce Coleman Ltd. **22.9** © Fred Bavendam/ Peter Arnold, Inc. **22.10** (A–B) © Wayne P. Armstrong, Professor of Biology and Botany, Palomar College, San Marcos, California; (C) National Park Services, Martin Hutten; (D) Dr. Annkatrin Rose, Appalachian State University. **Page 345** © University of Wisconsin-Madison, Department of Biology, Anthoceros CD. **22.11** © Martin LaBar, www.flickr.com/photos/martinlabar. **22.12** © Gerald D. Carr. **22.13** (A) © William Ferguson; (B) © Christine Evers. **22.14** A. & E. Bomford/ Ardea, London. **22.15** (A) © S. Navie; (B) David C. Clegg/ Photo Researchers, Inc.; (C) © Klein Hubert/ Peter Arnold, Inc. **22.17** (A) © Reprinted with permission from Elsevier; (B) Patricia G. Gensel. **22.19** © Karen Carr Studio/ www.karencarr.com. **22.21** © Dave Cavagnaro/ Peter Arnold, Inc. **22.22** (A) © M. Fagg, Australian National Botanic Gardens; (B) Michael P. Gadomski/ Photo Researchers, Inc.; (C) Courtesy of © Christine Evers; (D) Fletcher and Baylis/ Photo Researchers, Inc. **22.23** top from left, R. J. Erwin/ Photo Researchers, Inc.; Robert Potts, California Academy of Sciences; Robert & Linda Mitchell Photography. **22.26** (A) © Sangtae Kim, Sungshin University; (B) Smithsonian Institution Department of Botany, G.A. Cooper @ USDA-NRCS PLANTS Database. **22.27** (A) Photo USDA; (B) © Dan Fairbanks; (C) Photo USDA. **Page 355** R. Bieregaard/Photo Researchers, Inc.

CHAPTER 23 **Page 356**, Eye of Science/ Photo Researchers, Inc.; **Learning Roadmap**, from top, © Elena Boshkovska/ Shutterstock.com; Micrograph Garry T. Cole, University of Texas, Austin/ BPS; John Hodgin; © Chris Worden; Mark E. Gibson/ Visuals Unlimited; Dr. P. Marazzi/ Photo Researchers, Inc.; Courtesy of Mark Holland, Salisbury University. **23.1** (A) Photo by Yue Jin/ USDA; (B) Courtesy of Charles Good, Ohio State University at Lima; (C) Courtesy of © International Maize and Wheat Improvement Center (CIMMYT). All rights reserved. **23.2** (A) Photo by Scott Bauer/ USDA; (B) Robert C. Simpson/ Nature Stock; (C) Micrograph Garry T. Cole, University of Texas, Austin/ BPS. **23.4** (A) CDC; (B) Courtesy of Ken Nemuras. **23.5** Micrographs Ed Reschke. **23.6** John Hodgin. **23.7** (A) From RUSSELL/WOLFE/ HERTZ/STARR. Biology, 1E. © 2008 Brooks/ Cole, a part of Cengage Learning, Inc. Reproduced by permission. www.cengage.com/permissions; (B) www.dpd.cdc.gov/dpdx. **23.8** above, From RUSSELL/WOLFE/HERTZ/STARR. Biology, 1E. © 2008 Brooks/Cole, a part of Cengage Learning, Inc. Reproduced by permission. www.cengage.com/permissions; below, © Dr. Mark Brundrett, The University of Western Australia. **23.9** (A) Dr. Dennis Kunkel/ Visuals Unlimited; (B) © Dennis Kunkel Microscopy, Inc. **23.10** above, © North Carolina State University, Department of Plant

NA (2006) *Essential Endocrinology & Diabetes*, edn 5. **34.19** right, © iStockphoto.com/ Diane Diederich. **Page 563** Dr. Carlos J. Bourdony. **34.20** © Kevin Fleming/ Corbis.

CHAPTER 35 Page 564, Dr. Dennis Kunkel/ Visuals Unlimited, Inc.; **Learning Roadmap**, from top, Ed Reschke; © prodakszyn/ Shutterstock. com. **35.1** (A) The Muskegon Chronicle; (B) left, @ Stuart Isett; right, © Lakatos Sandor/ Shutterstock.com. **35.2** (B) Linda Pitkin/ Planet Earth Pictures. **35.4** (A) left, Stephen Dalton/ Photo Researchers, Inc. **35.5** D. A. Parry, *Journal of Experimental Biology*, 1959, 36:654. **35.6** Herve Chaumeton/ Agence Nature. **35.9** (A) right, Micrograph Ed Reschke. **35.11** Professor P. Motta/ Department of Anatomy/ La Sapienza, Rome/ Photo Researchers, Inc. **35.13** © JupiterImages Corporation. **35.16** (C) Dr. Dennis Kunkel/Visuals Unlimited; (E) Don Fawcett/ Visuals Unlimited. **35.20** (A) Painting by Sir Charles Bell, 1809, courtesy of Royal College of Surgeons, Edinburgh; (B) Courtesy of the Franklin D. Roosevelt Presidential Library & Museum, Hyde Park, New York; (C) NASA and Professor Stephen Hawking. **Page 581** Se-Jin Lee, Johns Hopkins University School of Medicine. **35.22** © prodakszyn/ Shutterstock.com.

CHAPTER 36 Page 582, National Cancer Institute/ Photo Researchers, Inc.; **Learning Roadmap**, from top, Malaria illustration by Drew Berry, The Walter and Eliza Hall Institute of Medical Research; © Lennart Nilsson/ Bonnierforlagen AB; Biophoto Associates/ Photo Researchers, Inc. **36.1** (A) Courtesy of the family of Matt Nadar; (B) Courtesy of ZOLL Medical Corporation. **Page 585** © Anan Kaewkhammul/ Shutterstock.com. **36.6** right, National Cancer Institute/ Photo Researchers, Inc. **36.8** Professor Pietro M. Motta/ Photo Researchers, Inc. **36.11** (B) C. Yokochi and J. Rohen, *Photographic Anatomy of the Human Body*, 2nd Ed., Igaku-Shoin, Ltd., 1979. **36.14** From RUSSELL/WOLFE/HERTZ/STARR. Biology, 1E. © 2008 Brooks/Cole, a part of Cengage Learning, Inc. Reproduced by permission. www.cengage. com/permissions. **36.16** above, Sheila Terry/ Photo Researchers, Inc.; below, Courtesy of Oregon Scientific, Inc. **36.17** left © Lennart Nilsson/ Bonnierforlagen AB. **36.18** left, Lisa Starr, using © Photodisc/ Getty Images photograph; right, Dr. John D. Cunningham/ Visuals Unlimited. **36.20** (A) Ed Reschke; (B) Biophoto Associates/ Photo Researchers, Inc. **36.22** © Alexander Raths/ Shutterstock.com. **Page 601** © Christine Evers.

CHAPTER 37 Page 602, © Fiona Pragoff/ Wellcome Images; **Learning Roadmap**, from top, Eye of Science/ Photo Researchers, Inc.; © Antonio Zamora, www.scientificpsychic.com; Juergen Berger/ Photo Researchers, Inc.; © 2010, Papayannopoulos et al. Originally published in J. Cell Biol. 191:677-691. doi: 10.1083/jcb.201006052 (Image by Volker Brinkman and Abdul Hakkim); James Hicks, Centers for Disease Control and Prevention; CDC/ Dr. Joel D. Meyer. **37.1** left, Biomedical Imaging Unit, Southampton General Hospital/ Photo Researchers, Inc.; right, In memory of Frankie McCullough; below, CDC. **37.2** Dr. Richard Kessel and Dr. Randy Kardon/ Tissues & Organs/ Visuals Unlimited. **37.3** © Antonio Zamora, www.scientificpsychic.com. **37.4** Dr. A. Liepins/ Photo Researchers, Inc. **37.5** (A) © David Scharf, 1999. All rights reserved; (B) Kwangshin Kim/ Photo Researchers, Inc.; (C) Juergen Berger/ Photo Researchers, Inc. **37.6** www.zahnarzt-

stuttgart.com. **37.7** John D. Cunningham/ Visuals Unlimited. **37.8** above, Courtesy K.B.M. Reid; (C) below, Robert R. Dourmashkin, courtesy of Clinical Research Centre, Harrow, England. **37.9** (A) © 2010, Papayannopoulos et al. Originally published in J. Cell Biol. 191:677-691. doi: 10.1083/ jcb.201006052 (Image by Volker Brinkman and Abdul Hakkim); (B) Photo Researchers, Inc.; (C) Prof. Matthias Gunzer/ Photo Researchers, Inc. **37.11** (A) Alvin Telser/ Visuals Unlimited, Inc.; (B) © Dr. David Becker/ Wellcome Images; (C) © Fiona Pragoff/ Wellcome Images. **Page 612** Table 37.3, © R. Dourmashkin/ Wellcome Images. **37.17** Image courtesy of Dr Fabien Garcon, The Babraham Institute. **37.20** (A) Dr. Oliver Schwartz, Institute Pasteur/ Photo Researchers, Inc; (B) Courtesy of Dr. Michael Dustin, New York University School of Medicine. From Grakoui A, Bromley SK, Sumen C, Davis MM, Shaw AS, Allen PM, Dustin ML. *The immunological synapse: A molecular machine controlling T cell activation.* Science. 1999; 285:221-7. **37.21** (A) Dr. P. Marazzi/ Photo Researchers, Inc.; (B) © Colin Hawkins/ Cultura/ Getty Images; (C) Hayley Witherell. **37.22** (A) © Peter Turnley/ Corbis; (B) © Zeva Oelbaum/ Peter Arnold, Inc.; (C) © R. Dourmashkin/ Wellcome Images. **37.23** left, James Hicks, Centers for Disease Control and Prevention; right, Eye of Science/ Photo Researchers, Inc. **Page 625** above, CDC.

CHAPTER 38 Page 626, Martin Dohrn, Royal College of Surgeons/ Photo Researchers, Inc.; **Learning Roadmap**, from top, © John Lund/ Getty Images; © D. E. Hill; © Brian McDairmant/ ardea.com; O. Auerbach/ Visuals Unlimited. **38.1** above, © iStockphoto.com/ Ralph125; below, © Wallenrock/ Shutterstock.com. **38.3** (A) John Lund/ Getty Images; (B) © iStockphoto.com/ GlobalP. **38.4** (A) above, Douglas Faulkner/ Sally Faulkner Collection; below, Peter Parks/ Oxford Scientific Films; (B) John Glowczwski/ University of Texas Medical Branch. **38.5** Aydin Örstan. **38.6** below, Micrograph Ed Reschke. **38.7** left above, © D. E. Hill; below, Florida Conservation Commission/ Fish and Wildlife Research Institute. **38.11** Micrograph H. R. Duncker, Justus-Liebig University, Giessen, Germany. **38.13** right, Photographs, Courtesy of Kay Elemetrics Corporation. **38.14** Martin Dohrn, Royal College of Surgeons/ Photo Researchers, Inc. **38.15** below (A-B) Charles McRae, MD/ Visuals Unlimited. **38.18** C. Yokochi and J. Rohen, *Photographic Anatomy of the Human Body*, 2nd Ed., Igaku-Shoin, Ltd., 1979. **38.19** (A) R. Kessel/ Visuals Unlimited. **38.22** From RUSSELL/WOLFE/HERTZ/STARR. Biology, 1E. © 2008 Brooks/Cole, a part of Cengage Learning, Inc. Reproduced by permission. www.cengage. com/permissions. **38.23** Francois Gohier/ Photo Researchers, Inc. **38.24** © Brian McDairmant/ ardea.com. **38.25** Christian Zuber/ Bruce Coleman, Ltd. **38.26** CDC/ Dr. Joel D. Meyer. **38.27** O. Auerbach/ Visuals Unlimited. **Page 643** © Wallenrock/ Shutterstock.com.

CHAPTER 39 Page 644, © Gunter Ziesler/ Bruce Coleman, Inc.; **Learning Roadmap**, from top, © Exactostock/ SuperStock; Microslide courtesy Mark Nielsen, University of Utah; USDA; Courtesy of Dr. Jeffrey M. Friedman, Rockefeller University; © D. A. Rintoul. **39.1** © Michael-Taylor/ Shutterstock.com. **39.3** (A) © W. Perry Conway/ Corbis. **39.4** © Gunter Ziesler/ Bruce Coleman, Inc. **39.8** (A) Microslide courtesy Mark Nielsen, University of Utah; (E) D. W. Fawcett/

Photo Researchers, Inc. **39.11** (B) National Cancer Institute. **39.13** USDA. **39.14** left, USDA; right, © Food Collection/ SuperStock. **39.15** Courtesy of Dr. Jeffrey M. Friedman, Rockefeller University. **Page 661** © camilla$$/ Shutterstock.com.

CHAPTER 40 Page 662, © MicroScan/ Phototake; **Learning Roadmap**, from top, C. Yokochi and J. Rohen, *Photographic Anatomy of the Human Body*, 2nd Ed., Igaku-Shoin, Ltd., 1979; Tom McHugh/ Photo Researchers, Inc.; NASA/JPL-Caltech/L. Hermans; From *Cold-Activated Brown Adipose Tissue in Healthy Men*. van Marken Lichtenbelt, Wouter D. et al. Article DOI: 10.1056/ NEJMoa0808718. © 2009 The New England Journal of Medicine. Reprinted with permission; © iStockphoto.com/ Ronnie Comeau. **40.1** © Archivo Iconografico, S.A./ Corbis. **40.4** below, © jerrysa/ Shutterstock.com. **40.5** left, © Stephen Dalton/ Photo Researchers, Inc.; right, Susumu Nishinaga / Photo Researchers, Inc. **40.7** Tom McHugh/ Photo Researchers, Inc. **40.10** above, © MicroScan/ Phototake. **40.12** left, Evan Cerasoli. **40.15** NASA/JPL-Caltech/L. Hermans. **40.16** From *Cold-Activated Brown Adipose Tissue in Healthy Men*. van Marken Lichtenbelt, Wouter D. et al. Article DOI: 10.1056/NEJ-Moa0808718. © 2009 The New England Journal of Medicine. Reprinted with permission. **Page 675** Lawrence Lawry/ Photo Researchers, Inc.

CHAPTER 41 Page 676, Courtesy of Dr. M. Jacques Donnez and Dr. Jean-Christophe.Lousse, Catholic University of Louvain, Belgium. From Fertility and Sterility, *Laparoscopic observation of spontaneous human ovulation*, Volume 90, Issue 3, Pages 833-834, September 2008.; **Learning Roadmap**, from top, Francis Leroy, Biocosmos/ Photo Researchers, Inc.; © Rodger Klein/ Peter Arnold, Inc.; Western Ophthalmic Hospital/ Photo Researchers, Inc.; © Steve Kaufman/ Corbis. **41.1** © AFP/ Getty Images. **41.2** (A) Biophoto Associates/ Photo Researchers, Inc.; (B) © Rodger Klein/ Peter Arnold, Inc.; (C) © Gabriela Staebler Wildlife Photography. **41.3** (A) R. Scott Cameron, Advanced Forest Protection, Inc., Bugwood.org; (B) © Doug Perrine/ seapics.com; (C) NPS Yellowstone/ Becky Wyman. **41.5** (B) Ed Reschke; (C) From RUSSELL/WOLFE/HERTZ/STARR. Biology, 1E. © 2008 Brooks/Cole, a part of Cengage Learning, Inc. Reproduced by permission. www. cengage.com/permissions. **41.9** (4) Courtesy of Dr. M. Jacques Donnez and Dr. Jean-Christophe. Lousse, Catholic University of Louvain, Belgium. From Fertility and Sterility, *Laparoscopic observation of spontaneous human ovulation*, Volume 90, Issue 3 , Pages 833-834, September 2008.; (5) Courtesy of Ed Uthman. **41.11** (C) above, Courtesy of Elizabeth Sanders, Women's Specialty Center, Jackson, MS. **41.12** (A) © iStockphoto. com/ Ever; (B) © D. A. Wagner Productions LLC/ Phototake; (C) © SuperStock/ SuperStock. **41.13** Dr. E. Walker/ Photo Researchers, Inc. **41.14** (A) © Western Ophthalmic Hospital/ Photo Researchers, Inc.; (C) CNRI/ Photo Researchers, Inc.; (C) CDC/ Joe Millar. **Page 693** © Laura Dwight/ Corbis.

CHAPTER 42 Page 694, © Lennart Nilsson/ Bonnierforlagen AB; **Learning Roadmap**, from top, © Maria Samsonova and John Reinitz; Dr. Maria Leptin, Institute of Genetics, University of Koln, Germany; © Lennart Nilsson/ Bonnierforlagen AB; © Don Mason/ Corbis. **42.1** (A) © Craig Hibbert / Mail on Sunday / Solo Syndication; (B)

CHAPTER 48 **Page 812**, NASA courtesy of the MODIS Rapid Response Team at Goddard Space Flight Center; **Learning Roadmap**, from top, © George M. Sutton/ Cornell Lab of Ornithology; NOAA; Frederica Georgia/ Photo Researchers, Inc.; © Lee Prince/ Shutterstock.com; NASA. **48.1** U.S. Navy photo by Chief Yeoman Alphanso Braggs. **48.3** (A) Jaime Chirinos/ Photo Researchers, Inc.; (B) Mansell Collection/ Time, Inc./ Getty Images. **48.4** (A) John Butler, NOAA; (B) Jeffrey Sylvester/ FPG/ Getty Images; (C) © George M. Sutton/ Cornell Lab of Ornithology; (D) Joe Fries, U.S. Fish & Wildlife Service; (E) © Dr. John Hilty; (F) Edward S. Ross. **48.5** NOAA. **48.6** Satellite image courtesy of GeoEye. **48.7** left, USDA Forest Service, Northeastern Research Station. **48.8** left, NASA courtesy of the MODIS Rapid Response Team at Goddard Space Flight Center; right, Courtesy of Guido van der Werf, Vrije Universiteit Amsterdam. **48.9** (B) Frederica Georgia/ Photo Researchers, Inc. **48.10** Gary Head. **48.11** Claire Fackler/ NOAA. **48.12** (A) NASA. **48.13** National Snow and Ice Data Center, W. O. Field. **48.15** David Patte, USFWS; inset, USFWS. **48.16** Diane Borden-Bilot, U.S. Fish and Wildlife Service. **48.17** © Lee Prince/ Shutterstock.com. **48.18** Mountain Visions/ NOAA. **48.19** NASA.

APPENDIX IX **1.1**, Tim Laman/ National Geographic Stock. **1.2**, California Poppy, 2009, Christine M. Welter. **1.3**, Victoria Pinder, www.flickr.com/ photos/vixstarplus. **1.4**, Courtesy of Allen W. H. BeÅL and David A. Caron. **1.5**, From Meyer, A., Repeating Patterns of Mimicry. PLoS Biology Vol. 4, No. 10, e341 doi:10.1371/journal.pbio.0040341. **1.6**, Roger W. Winstead, NC State University. **1.7**, Adrian Vallin. **1.8**, Gary Head. **1.9**, Raymond Gehman/ Corbis. Scientific Paper; Adrian Vallin, Sven Jakobsson, Johan Lind and Christer Wiklund, Proc. R. Soc. B (2005 272, 1203, 1207). Used with permission of The Royal Society and the author. **2.1**, Kim Westerskov, Photographer's Choice/ Getty Images. **2.3**, Michael S. Yamashita/ Corbis. **2.5**, Herbert Schnekenburger. **2.6**, W. K. Fletcher/ Photo Researchers, Inc. **3.1**, ThinkStock/ SuperStock. **3.3**, Exactostock/ Super- Stock. **3.4**, JupiterImages Corporation. **3.5**, Kenneth Lorenzen. **3.6**, From: *Structure of the Rotor of the V-Type Na+-ATPase from Enterococcus hirae* by Murata, et al. Science 29 April 2005: 654-659. DOI:10.1126/science.1110064. Used with permission. **3.7**, Sherif Zaki, MD PhD, Wun-Ju Shieh, MD PhD; MPH/ CDC. **4.1**, JupiterImages Corporation. **4.2**, Courtesy of Johannes KaÅNstner, UniversitaÅNt Stuttgart. **4.3**, photos.com. **4.4**, Dennis Kunkel Microscopy, Inc./ Phototake. **4.5**, Astrid & Hanns- Frieder Michler/ Photo Researchers, Inc. **4.6**, Don W. Fawcett/ Visuals Unlimited. **4.7–8**, Don W. Fawcett/ Visuals Unlimited. **4.9**, Dr. Ralf Wagner, www.dr-ralf-wagner. de. **4.10**, Dylan T. Burnette and Paul Forscher. **4.11**, ADVANCELL (Advanced In Vitro Cell Technologies; S.L.) www.advancell.com. **4.12**, R. Llewellyn/ SuperStock, Inc. **4.25** From "Tissue & Cell", Vol. 27, pp.421– 427, Courtesy of Bjorn Afzelius, Stockholm University. Critical Thinking, P. L. Walne and J. H. Arnott, Planta, 77:325-354, 1967. **5.2**, Greg Epperson/ Shutterstock.com. **5.3**, Westend61/ SuperStock. **5.4**, Hemoglobin models: PDB ID: 1GZX; Paoli, M., Liddington, R., Tame, J., Wilkinson, A., Dodson, G.; Crystal structure of T state hemoglobin with oxygen bound at all four haems. *J.Mol.Bio.*, v256, pp. 775–792, 1996. **5.5**, JupiterImages Corporation. **5.8**, Andrew Lambert Photography/ Photo Researchers, Inc. **5.10**, R.G.W. Anderson, M.S.

Brown and J.L. Goldstein. Cell 10:351 (1977). **5.32** Courtesy of Dr. Katrina J. Edwards. **6.1**, Photo by Peggy Greb/ USDA. **6.5–6.6**, Dr. Ralf Wagner, www.dr-ralf-wagner.de. **6.8**, Lisa Starr. Critical Thinking, E.R. Degginger. **7.1**, Louise Chalcraft-Frank and FARA. **7.6**, Elena Boshkovska/ Shutterstock.com. **7.7**, shabaneiro/ Shutterstock.com. **7.13** (A) Steve Gschmeissner/ Photo Researchers, Inc.; (B) Images Paediatr Cardiol. **8.1**, James Symington. **8.2**, Andrew Syred/ Photo Researchers, Inc. **8.3**, Eye of Science/ Photo Researchers, Inc. **8.4**, A C. Barrington Brown 1968 J. D. Watson. **8.7**, Courtesy of Cyagra, Inc., www.cyagra.com. **8.16**, the Rockefeller Press, *The Journal of General Physiology*, 1952, 36:39–56. Critical Thinking, Shahbaz A. Janjua, MD, Dermatlas; www.dermatlas.org. **9.1**, Steve Hurst/ USDA-NRCS PLANTS Database. **9.3**, O. L. Miller. **9.5**, Kiseleva and Donald Fawcett/ Visuals Unlimited. **9.6**, Dr. Gopal Murti/ Photo Researchers, Inc. **10.1**, From the archives of www.breastpath.com, courtesy of J.B. Askew, Jr., M.D., P.A. Reprinted with permission, copyright 2004 Breastpath.com. **10.2**, From the collection of Jamos Werner and John T. Lis. **10.3**, Maria Samsonova and John Reinitz. **10.4**, Dr. William Strauss. **10.6**, Marius Meijboom/ Nederlands Fotomuseum, all rights reserved. **11.1**, Courtesy of the family of Henrietta Lacks. **11.2**, Carolina Biological Supply Company/ Phototake. **11.3**, Michael Clayton/ University of Wisconsin, Department of Botany. **11.5**, ISM/ Phototake-usa. **11.6**, James Stevenson/ Photo Researchers, Inc. **Data Analysis Activities**, Courtesy of Dr. Thomas Ried, NIH and the American Association for Cancer Research; **Critical Thinking**, micrograph, D. M. Phillips/ Visuals Unlimited. **12.1**, JupiterImages Corporation. **12.2**, Image courtesy of Carl Zeiss MicroImaging, Thornwood, NY. **12.3**, With thanks to the John Innes Foundation Trustees, computer enhanced by Gary Head. **12.5**, David M. Phillips/ Photo Researchers, Inc. **12.6**, MRC NIMR/Wellcome Images. **12.13 (A–D)**, Reprinted from *Current Biology*, Vol 13, (Apr 03), Authors Hunt, Koehler, Susiarjo, Hodges, Ilagan, Voight, Thomas, Thomas and Hassold, Bisphenol A Exposure Causes Meiotic Aneuploidy in the Female Mouse, pp. 546–553, 2003 Cell Press. Published by Elsevier Ltd. With permission from Elsevier. **13.1**, Courtesy of Steve & Ellison Widener and Breathe Hope, breathehope.tamu.edu. **13.2**, Moravian Museum, Brno.**13.5**, John Daniels/ ardea.com. **13.6**, Dr. Christian Laforsch. **13.7**, Medioimages/Photodisc/ Jupiter Images; Genetic Problems #4, Leslie Falteisek/ Clacritter Manx. Genetic Problems #5, Rick Guidotti, Positive Exposure. **14.1**, Gary Roberts/ worldwidefeatures.com. **14.2**, Courtesy of Irving Buchbinder, DPM, DABPS, Community Health Services, Hartford CT. **14.3**, Photo courtesy of The Progeria Research Foundation. **14.6**, Ciarra, photo by Michelle Harmon. **14.7**, Lennart Nilsson/ Bonnierforlagen AB. **15.1**, Courtesy of Illumina, Inc., www.illumina.com. **15.2**, Professor Stanley Cohen/ Photo Researchers, Inc. **15.4**, Patrick Landmann/ Photo Researchers, Inc. **15.5**, The Sanger Institute. Wellcome Images. **15.6–9**, Photo courtesy of MU Extension and Agricultural Information. **15.10**, Corbis/ SuperStock. **15.17**, Courtesy of Dr. S. Thuret, Kings College London. **16.1**, Brad Snowder. **16.2**, Wolfgang Kaehler/ Corbis. **16.3**, Painting by George Richmond. **16.4**, 2004 Arent, commons.wikimedia.org/wiki/File:Glyptodon-1. jpg. **16.5**, Louie Psihoyos/ Getty Images. **16.7**, Ron Blakey and Colorado Plateau Geosystems, Inc. **16.8**, Michael Pancier. Data Analysis Activities, Lawrence Berkeley National

Laboratory.**17.1**, Reuters NewMedia, Inc./ Corbis. **17.2–3**, Alan Solem. **17.4–6**, Courtesy of Hopi Hoekstra, University of California, San Diego. **17.7**, Bruce Beehler. **17.8**, Ashok Khosla, www.seeingbirds.com. **17.9**, Alvin E. Staffan/ Photo Researchers, Inc. **17.10**, Jack Jeffrey Photography. **17.11**, Photo by J. Honegger, courtesy of S. Stamp, E. Merz, www.sortengarten/ ethz.ch. **17.12**, Jeremy Thomas/ Natural Visions. **18.1**, Jack Jeffrey Photography. **18.2**, National Geographic/ SuperStock. **18.3**, Taro Taylor, www.flickr.com/photos/tjt195. **18.5**, Courtesy of Ann C. Burke, Wesleyan University. **18.6**, John Steiner/ Smithsonian Institution. **18.15**, Courtesy of Dr. Sajid Malik, from www.biomedcentral.com/1471-2350/8/78. **19.1**, Photo by Julio Betancourt/ U.S. Geological Survey. **19.3**, Courtesy of the University of Washington. **19.4**, From Hanczyc, Fujikawa, and Szostak, *Experimental Models of Primitive Cellular Compartments: Encapsulation, Growth, and Division*; www.sciencemag.org, Science 24 October 2003; 302;529, Figure 2. Reprinted with permission of the authors and AAAS. **19.5**, Chase Studios/ Photo Researchers, Inc. **19.6–7**, Courtesy of Isao Inouye, Institute of Biological Sciences, University of Tsukuba.**20.1**, Micrographs Z. Salahuddin, National Institutes of Health. **20.2**, Russell Knightly/ Photo Researchers, Inc. **20.4**, CDC/ Dan Higgins. **20.5**, CNRI/ Photo Researchers, Inc. **20.6**, Dr. Dennis Kunkel / Visuals Unlimited. **20.7**, Dr. Manfred Schloesser, Max Planck Institute for Marine Microbiology. **20.8**, Savannah River Ecology Laboratory. **20.20**, From Yuka Kawashima et al., "Adaptation of HIV-1 to human leukocyte antigen class I," *Nature*, vol. 458(7238), 2009 Nature Publishing Group, Reproduced by permission. Critical Thinking #1 above, CDC/ Janice Haney Car; below, CDC/ Bruno Coignard, M.D.; Jeff Hageman, M.H.S. **21.1**, Malaria illustration by Drew Berry, The Walter and Eliza Hall Institute of Medical Research. **21.2**, Astrid & Hanns- Frieder Michler/ Photo Researchers, Inc. **21.3**, Dr. Dennis Kunkel/ Visuals Unlimited. **21.4**, Franz Neidl. **21.5–7**, Dr. David Phillips/ Visuals Unlimited. **21.8**, Garo/ Photo Researchers, Inc. **21.9**, Image courtesy of FGBNMS/UNCW-NURC. **21.10**, Wim van Egmond/ Visuals Unlimited. **21.11**, Courtesy of www.hiddenforest.co.nz. **22.1**, William Campbell/ TimePix/ Getty Images. **22.2–3**, Andrew Syred/ Photo Researchers, Inc. **22.5**, Martin LaBar, www.flickr.com/photos/martinlabar. **22.7**, Dave Cavagnaro/ Peter Arnold, Inc. **22.8–9**, Smithsonian Institution Department of Botany, G.A. Cooper @ USDA-NRCS PLANTS Database. **22.28**, Clinton Webb. **23.1**, Courtesy of Charles Good, Ohio State University at Lima. **23.2**, Micrograph Garry T. Cole, University of Texas, Austin/ BPS. **23.3**, Courtesy of Ken Nemuras. **23.4**, Micrograph Ed Reschke. **23.5**, North Carolina State University, Department of Plant Pathology. **23.6**, Eye of Science/ Photo Researchers, Inc. **23.7**, Mark E. Gibson/ Visuals Unlimited; Dr. P. Marazzi/ Photo Researchers, Inc. **24.1**, K.S. Matz. **24.4**, Marty Snyderman/ Planet Earth Pictures. **24.5**, Jeffrey L. Rotman/ Corbis. **24.7**, J. Solliday/ BPS. **24.8**, B. Borrell Casals/ Frank Lane Picture Agency/ Corbis. **24.10**, CDC/ Henry Bishop. **24.11–15**, Photo by Scott Bauer/ USDA.**24.16**, Herve Chaumeton/ Agence Nature. **25.1**, P. Morris/ Ardea London. **25.2**, Runk & Schoenberger / Grant Heilman, Inc. **25.3**, Heather Angel. **25.4**, Courtesy of John McNamara, www.paleodirect.com. **25.5**, Wernher Krutein/ photovault.com. **25.6**, Stephen Dalton/ Photo Researchers, Inc. **25.7**, Z. Leszczynski/ Animals Animals. **25.8**, S. Blair Hedges, Pennsylvania

Index